THE ELECTRICAL ENGINEERING HANDBOOK

THE ELECTRICAL ENGINEERING HANDBOOK

WAI-KAI CHEN

EDITOR-IN-CHIEF

ELSEVIER
ACADEMIC
PRESS

AMSTERDAM • BOSTON • HEIDELBERG • LONDON
NEW YORK • OXFORD • PARIS • SAN DIEGO
SAN FRANCISCO • SINGAPORE • SYDNEY • TOKYO

Academic Press is an imprint of Elsevier

Elsevier Academic Press
200 Wheeler Road, 6th Floor, Burlington, MA 01803, USA
525 B Street, Suite 1900, San Diego, California 92101-4495, USA
84 Theobald's Road, London WC1X 8RR, UK

This book is printed on acid-free paper. ∞

Library of Congress Cataloging-in-Publication Data
Application submitted

British Library Cataloguing in Publication Data
A catalogue record for this book is available from the British Library

ISBN: 0-12-170960-4

For all information on all Academic Press publications visit our Web site at www.books.elsevier.com

04 05 06 07 08 09 9 8 7 6 5 4 3 2 1

Printed in the Unitec

Contents

Contributors

Rashid Ansari
Department of Electrical and Computer Engineering
University of Illinois at Chicago
Chicago, Illinois, USA

Faisal Bashir
Department of Electrical and Computer Engineering
University of Illinois at Chicago
Chicago, Illinois, USA

Magdy Bayoumi
The Center for Advanced Computer Studies
University of Louisiana at Lafayette
Lafayette, Louisiana, USA

Christopher J. Bett
Raytheon Integrated Defense Systems
Tewksbury, Massachusetts, USA

Anjan Bose
College of Engineering and Architecture
Washington State University
Pullman, Washington, USA

Erik Brockmeyer
IMEC
Leuven, Belgium

Francky Catthoor
IMEC
Leuven, Belgium

Morris Chang
Department of Electrical and Computer Engineering
Iowa State University
Ames, Iowa, USA

Peter Y. K. Cheung
Department of Electrical and Electronic Engineering
Imperial College of Science, Technology, and Medicine
London, UK

Weng Cho Chew
Center for Computational Electromagnetics
Department of Electrical and Computer Engineering
University of Illinois at Urbana-Champaign
Urbana, Illinois, USA

George A. Constantinides
Department of Electrical and Electronic Engineering
Imperial College of Science, Technology, and Medicine
London, UK

Koen Danckaert
IMEC
Leuven, Belgium

Tarek Darwish
The Center for Advanced Computer Studies
University of Louisiana at Lafayette
Lafeyette, Louisiana, USA

Nirod K. Das
Department of Electrical and Computer Engineering
Polytechnic University
Brooklyn, New York, USA

Eduardo A.B. da Silva
Program of Electrical Engineering
Federal University of Rio de Janeiro
Rio de Janeiro, Brazil

William R. Deal
Northrup Grumman Space Technologies
Redondo Beach, California, USA

Franco De Flaviis
Department of Electrical and Computer Engineering
University of California at Irvine
Irvine, California, USA

Bob C. Degeneff
Department of Computer, Electrical, and Systems Engineering
Rensselaer Polytechnic Institute
Troy, New York, USA

John R. Deller, Jr.
Department of Electrical and Computer Engineering
Michigan State University
East Lansing, Michigan, USA

Rodolfo E. Diaz
Department of Electrical Engineering
Ira A. Fulton School of Engineering
Arizona State University
Tempe, Arizona, USA

Paulo S. R. Diniz
Program of Electrical Engineering
Federal University of Rio de Janeiro
Rio de Janeiro, Brazil

Shantanu Dutt
Department of Electrical and Computer Engineering
University of Illinois at Chicago
Chicago, Illinois, USA

Mohamed Elgamel
The Center for Advanced Computer Studies
University of Louisiana at Lafayette
Lafayette, Louisiana, USA

Jay Farrell
Department of Electrical Engineering
University of California
Riverside, California, USA

Eby G. Friedman
Department of Electrical and Computer Engineering
University of Rochester
Rochester, New York, USA

Vijay K. Garg
Department of Electrical and Computer Engineering
University of Illinois at Chicago
Chicago, Illinois, USA

Turan Gönen
College of Engineering and Computer Science
California State University, Sacramento
Sacramento, California, USA

Oscar R. González
Department of Electrical and Computer Engineering
Old Dominion University
Norfolk, Virginia, USA

Ravi S. Gorur
Department of Electrical Engineering
Arizona State University
Tempe, Arizona, USA

Susan C. Hagness
Department of Electrical and Computer Engineering
University of Wisconsin
Madison, Wisconsin, USA

Fran Hanchek
Intel Corporation
Portland, Oregan, USA

John Hansen
Department of Electrical and Computer Engineering
Michigan State University
East Lansing, Michigan, USA

Xudong He
School of Computer Science
Florida International University
Miami, Florida, USA

Bonnie S. Heck
School of Electrical and Computer Engineering
Georgia Institute of Technology
Atlanta, Georgia, USA

Gerald T. Heydt
Department of Electrical Engineering
Arizona State University
Tempe, Arizona, USA

Yu-Hen Hu
Department of Electrical and Computer Engineering
University of Wisconsin-Madison
Madison, Wisconsin, USA

Yih-Fang Huang
Department of Electrical Engineering
University of Notre Dame
Notre Dame, Indiana, USA

Sorin A. Huss
Integrated Circuits and Systems Laboratory
Computer Science Department
Darmstadt University of Technology
Darmstadt, Germany

Tatsuo Itoh
Department of Electrical Engineering
University of California, Los Angeles
Los Angeles, California, USA

David R. Jackson
Department of Electrical and Computer Engineering
University of Houston
Houston, Texas, USA

Gang Jin
Ford Motor Company
Dearborn, Michigan, USA

Jian-Ming Jin
Center for Computational Electromagnetics
Department of Electrical and Computer Engineering
University of Illinois at Urbana-Champaign
Urbana, Illinois, USA

Atul G. Kelkar
Department of Mechanical Engineering
Iowa State University
Ames, Iowa, USA

Mladen Kezunovic
Department of Electrical Engineering
Texas A & M University
College Station, Texas, USA

Shashank Khanvilkar
Department of Electrical and Computer Engineering
University of Illinois at Chicago
Chicago, Illinois, USA

Ashfaq Khokhar
Department of Electrical and Computer Engineering
University of Illinois at Chicago
Chicago, Illinois, USA

Yean-Woei Kiang
Department of Electrical Engineering
National Taiwan University
Taipei, Taiwan

Surin Kittitornkun
King Mongkut's Institute of Technology Ladkrabang
Bangkok, Thailand

Ivan S. Kourtev
Department of Electrical and Computer Engineering
University of Pittsburgh
Pittsburgh, Pennsylvania, USA

Chidamber Kulkani
IMEC
Leuven, Belgium

Sun-Yung Kung
Department of Electrical Engineering
Princeton University
Princeton, New Jersey, USA

Fred C. Lee
Center for Power Electronics Systems
The Bradley Department of Electrical Engineering
Virginia Polytechnic Institute and State University
Blacksburg, Virginia, USA

Hsueh-Jyh Li
Department of Electrical Engineering
National Taiwan University
Taipei, Taiwan

Xiaoqiu Li
Cummins Engine
Columbus, Indiana, USA

Stanley R. Liberty
Academic Affairs
Bradley University
Peoria, Illinois, USA

Yao-Nan Lien
Department of Computer Science
National Chengchi University
Taipei, Taiwan

Derong Liu
Department of Electrical and Computer Engineering
University of Illinois at Chicago
Chicago, Illinois, USA

Wayne Luk
Department of Electrical and Electronic Engineering
Imperial College of Science, Technology, and Medicine
London, UK

Erik A. McShane
Department of Electrical and Computer Engineering
University of Illinois at Chicago
Chicago, Illinois, USA

Gelson V. Mendonça
Department of Electronics
COPPE/EE/Federal University of Rio de Janeiro
Rio de Janeiro, Brazil

Veena Misra
Department of Electrical and Computer Engineering
North Carolina State University
Raleigh, North Carolina, USA

Tadao Murata
Department of Computer Science
University of Illinois at Chicago
Chicago, Illinois, USA

Lode Nachtergaele
IMEC
Leuven, Belgium

David J. Nagel
Department of Electrical and Computer Engineering
The George Washington University
Washington, D.C., USA

Krishna Naishadham
Massachusetts Institute of Technology
Lincoln Laboratory
Lexington, Massachusetts, USA

Ajoy Opal
Department of Electrical and Computer Engineering
University of Waterloo
Waterloo, Ontario, Canada

Raúl Ordóñez
Department of Electrical and Computer Engineering
University of Dayton
Dayton, Ohio, USA

Mehmet C. Öztürk
Department of Electrical and Computer Engineering
North Carolina State University
Raleigh, North Carolina, USA

Kevin M. Passino
Department of Electrical and Computer Engineering
The Ohio State University
Columbus, Ohio, USA

Melinda Piket-May
Department of Electrical and Computer Engineering
University of Colorado
Boulda, Colorado, USA

Yongxi Qian
Department of Electrical Engineering
University of California, Los Angeles
Los Angeles, California, USA

Vesna Radisic
Microsemi Corporation
Los Angeles, California, USA

P.K. Rajan
Department of Electrical and Computer Engineering
Tennessee Technological University
Cookeville, Tennessee, USA

Federico Rota
Department of Electrical and Computer Engineering
University of Illinois at Chicago
Chicago, Illinois, USA
Politecnico di Torina, Italy

Michael Sain
Department of Electrical Engineering
University of Notre Dame
Notre Dame, Indiana, USA

Patrick M. Sain
Raytheon Company
EI Segundo, California, USA

Sheppard Joel Salon
Department of Electrical Power Engineering
Renssalaer Polytechnic Institute
Troy, New York, USA

Rolf Schaumann
Department of Electrical and Computer Engineering
Portland State University
Portland, Oregan, USA

Dan Schonfeld
Department of Electrical and Computer Engineering
University of Illinois at Chicago
Chicago, Illinois, USA

Cheryl B. Schrader
College of Engineering
Boise State University
Boise, Idaho, USA

Michael Schröter
Institute for Electro Technology and Electronics Fundamentals
University of Technology
Dresden, Germany

Arun Sekar
Department of Electrical and Computer Engineering
Tennessee Technological University
Cookeville, Tennessee, USA

Yi Shang
Department of Computer Science
University of Missouri-Columbia
Columbia, Missouri, USA

Krishna Shenai
Department of Electrical and Computer Engineering
University of Illinois at Chicago
Chicago, Illinois, USA

Nabeel Shirazi
Xilinx, Inc.
San Jose, California, USA

Michael Shur
Department of Electrical, Computer, and Systems Engineering
Rensselaer Polytechnic Institute
Troy, New York, USA

Jennie Si
Department of Electrical Engineering
Arizona State University
Tempe, Arizona, USA

Marcio G. Siqueira
Cisco Systems
Sunnyvale, California, USA

John R. Smith
IBM
T. J. Watson Research Center
Hawthorne, New York, USA

Thanos Stouraitis
Department of Electrical and Computer Engineering
University of Patras
Rio, Greece

M.N.S. Swamy
Department of Electrical and Computer Engineering
Concordia University
Montreal, Quebec, Canada

Allen Taflove
Department of Electrical and Computer Engineering
Northwestern University
Chicago, Illinois, USA

Krishnaiyan Thulasiraman
School of Computer Science
University of Oklahoma
Norman, Oklahoma, USA

Kevin Tomsovic
School of Electrical Engineering and Computer Science
Washington State University
Pullman, Washington, USA

Ljiljana Trajković
School of Engineering Science
Simon Fraser University
Vancouver, British Columbia, Canada

Malay Trivedi
Department of Electrical and Computer Engineering
University of Illinois at Chicago
Chicago, Illinois, USA

Franco Trovo
Department of Electrical and Computer Engineering
University of Illinois at Chicago
Chicago, Illinois, USA
Politecnico di Torina, Italy

Ruediger Vahldieck
Laboratory for Electromagnetic Fields and Microwave Electronics
Swiss Federal Institute of Technology
Zurich, Switzerland

Lucia Valbonesi
Department of Electrical and Computer Engineering
University of Illinois at Chicago
Chicago, Illinois, USA

Arnout Vandercappelle
IMEC
Leuven, Belgium

Mani Venkatasubramanian
School of Electrical Engineering and Computer Science
Washington State University
Pullman, Washington, USA

Jiri Vlach
Department of Electrical and Computer Engineering
University of Waterloo
Waterloo, Ontario, Canada

Benjamin W. Wah
Computer and Systems Research Laboratory
University of Illinois at Urbana-Champaign
Urbana, Illinois, USA

Yih-Chen Wang
Lucent Technologies
Naperville, Illinois, USA

Keith W. Whites
Department of Electrical and Computer Engineering
South Dakota School of Mines and Technology
Rapid City, South Dakota, USA

Chang-Hee Won
Department of Electrical Engineering
University of North Dakota
Grand Forks, North Dakota, USA

Ke Wu
Department of Electrical and Computer Engineering
Ecole Polytechnique
Montreal, Quebec, Canada

Hung-Yu David Yang
Department of Electrical and Computer Engineering
University of Illinois at Chicago
Chicago, Illinois, USA

Gary G. Yen
Intelligent Systems and Control Laboratory
School of Electrical and Computer Engineering
Oklahoma State University
Stillwater, Oklahoma, USA

Stephen Yurkovich
Center for Automotive Research
The Ohio State University
Columbus, Ohio, USA

Mona E. Zaghloul
Department of Electrical and Computer Engineering
The George Washington University
Washington, D.C., USA

Xunwei Zhou
Center for Power Electronics Systems
The Bradley Department of Electrical Engineering
Virginia Polytechnic Institute and State University
Blacksburg, Virginia, USA

Lei Zhu
Department of Electrical and Computer Engineering
Ecole Polytechnique
Montreal, Quebec, Canada

Preface

Purpose

The purpose of *The Electrical Engineering Handbook* is to provide a comprehensive reference work covering the broad spectrum of electrical engineering in a single volume. It is written and developed for the practicing electrical engineers in industry, government, and academia. The goal is to provide the most up-to-date information in classical fields of circuits, electronics, electromagnetics, electric power systems, and control systems, while covering the emerging fields of VLSI systems, digital systems, computer engineering, computer-aided design and optimization techniques, signal processing, digital communications, and communication networks. This handbook is not an all-encompassing digest of everything taught within an electrical engineering curriculum. Rather, it is the engineer's first choice in looking for a solution. Therefore, full references to other sources of contributions are provided. The ideal reader is a B.S. level engineer with a need for a one-source reference to keep abreast of new techniques and procedures as well as review standard practices.

Background

The handbook stresses fundamental theory behind professional applications. In order to do so, it is reinforced with frequent examples. Extensive development of theory and details of proofs have been omitted. The reader is assumed to have a certain degree of sophistication and experience. However, brief reviews of theories, principles, and mathematics of some subject areas are given. These reviews have been done concisely with perception. The handbook is not a textbook replacement, but rather a reinforcement and reminder of material learned as a student. Therefore, important advancement and traditional as well as innovative practices are included.

Since the majority of professional electrical engineers graduated before powerful personal computers were widely available, many computational and design methods may be new to them. Therefore, computers and software use are thoroughly covered. Not only does the handbook use traditional references to cite sources for the contributions, but it also contains *relevant* sources of information and tools that would assist the engineer in performing his/her job. This may include sources of software, databases, standards, seminars, conferences, and so forth.

Organization

Over the years, the fundamentals of electrical engineering have evolved to include a wide range of topics and a broad range of practice. To encompass such a wide range of knowledge, the handbook focuses on the key concepts, models, and equations that enable the electrical engineer to analyze, design, and predict the behavior of electrical systems. While design formulas and tables are listed, emphasis is placed on the key concepts and theories underlying the applications.

The information is organized into nine major sections, which encompass the field of electrical engineering. Each section is divided into chapters. In all, there are 79 chapters involving 113 authors, each of which was written by leading experts in the field to enlighten and refresh knowledge of the mature engineer and educate the novice. Each section contains introductory material, leading to the appropriate applications. To help the reader, each article includes two important and useful categories: defining terms and references. *Defining terms* are key definitions and the first occurrence of each term defined is indicated in boldface in the text. The *references* provide a list of useful books and articles for following reading.

Locating Your Topic

Numerous avenues of access to information contained in the handbook are provided. A complete table of contents is presented at the front of the book. In addition, an individual table of contents precedes each of the nine sections. The reader is urged to look over these tables of contents to become familiar with the structure, organization, and content of the book. For example, see Section VII: Signal Processing, then Chapter 7: VLSI Signal Processing, and then Chapter 7.3: Hardware Implementation. This tree-like structure enables the reader to move up the tree to locate information on the topic of interest.

The Electrical Engineering Handbook is designed to provide answers to most inquiries and direct inquirer to further sources and references. We trust that it will meet your need.

Acknowledgments

The compilation of this book would not have been possible without the dedication and efforts of the section editors, the publishers, and most of all the contributing authors. I particularly wish to acknowledge my wife, Shiao-Ling, for her patience and support.

Wai-Kai Chen
Editor-in-Chief

Editor-in-Chief

Wai-Kai Chen, Professor and Head Emeritus of the Department of Electrical Engineering and Computer Science at the University of Illinois at Chicago. He received his B.S. and M.S. in electrical engineering at Ohio University, where he was later recognized as a Distinguished Professor. He earned his Ph.D. in electrical engineering at University of Illinois at Urbana-Champaign.

Professor Chen has extensive experience in education and industry and is very active professionally in the fields of circuits and systems. He has served as visiting professor at Purdue University, University of Hawaii at Manoa, and Chuo University in Tokyo, Japan. He was Editor-in-Chief of the *IEEE Transactions on Circuits and Systems, Series I and II*, President of the IEEE Circuits and Systems Society, and is the Founding Editor and Editor-in-Chief of the *Journal of Circuits, Systems and Computers.* He received the Lester R. Ford Award from the Mathematical Association of America,

Dr. Wai-Kai Chen

the Alexander von Humboldt Award from Germany, the JSPS Fellowship Award from Japan Society for the Promotion of Science, the National Taipei University of Science and Technology Distinguished Alumnus Award, the Ohio University Alumni Medal of Merit for Distinguished Achievement in Engineering Education, the Senior University Scholar Award and the 2000 Faculty Research Award from the University of Illinois at Chicago, and the Distinguished Alumnus Award from the University of Illinois at Urbana/Champaign. He is the recipient of the Golden Jubilee Medal, the Education Award, and the Meritorious Service Award from IEEE Circuits and Systems Society, and the Third Millennium Medal from the IEEE. He has also received more than dozen honorary professorship awards from major institutions in Taiwan and China.

A fellow of the Institute of Electrical and Electronics Engineers (IEEE) and the American Association for the Advancement of Science (AAAS), Professor Chen is widely known in the profession for his *Applied Graph Theory* (North-Holland), *Theory and Design of Broadband Matching Networks* (Pergamon Press), *Active Network and Feedback Amplifier Theory* (McGraw-Hill), *Linear Networks and Systems* (Brooks/Cole), *Passive and Active Filters: Theory and Implements* (John Wiley), *Theory of Nets: Flows in Networks* (Wiley-Interscience), *The Circuits and Filters Handbook* (CRC Press) and *The VLSI Handbook* (CRC Press).

Dr. Wai-Kai Chen

I

CIRCUIT THEORY

Krishnaiyan Thulasiraman
School of Computer Science,
University of Oklahoma,
Norman, Oklahoma, USA

Circuit theory is an important and perhaps the oldest branch of electrical engineering. A circuit is an interconnection of electrical elements. These include passive elements, such as resistances, capacitances, and inductances, as well as active elements and sources (or excitations). Two variables, namely voltage and current variables, are associated with each circuit element. There are two aspects to circuit theory: **analysis and design**. Circuit analysis involves the determination of current and voltage values in different elements of the circuit, given the values of the sources or excitations. On the other hand, circuit design focuses on the design of circuits that exhibit a certain prespecified voltage or current characteristics at one or more parts of the circuit. Circuits can also be broadly classified as **linear or nonlinear circuits**.

This section consists of five chapters that provide a broad introduction to most fundamental principles and techniques in circuit analysis and design:

- Linear Circuit Analysis
- Circuit Analysis: A Graph-Theoretic Foundation
- Computer-Aided Design
- Synthesis of Networks
- Nonlinear Circuits.

1

Linear Circuit Analysis

P.K. Rajan and Arun Sekar
Department of Electrical and Computer Engineering, Tennessee Technological University, Cookeville, Tennessee, USA

1.1 Definitions and Terminology

An **electric charge** is a physical property of electrons and protons in the atoms of matter that gives rise to forces between atoms. The charge is measured in coulomb [C]. The charge of a proton is arbitrarily chosen as positive and has the value of 1.601×10^{-19} C, whereas the charge of an electron is chosen as negative with a value of -1.601×10^{-19} C. Like charges repel while unlike charges attract each other. The electric charges obey the principle of conservation (i.e., charges cannot be created or destroyed).

A **current** is the flow of electric charge that is measured by its flow rate as coulombs per second with the units of ampere [A]. An ampere is defined as the flow of charge at the rate of one coulomb per second (1 A = 1 C/s). In other words, current $i(t)$ through a cross section at time t is given by dq/dt, where

$q(t)$ is the charge that has flown through the cross section up to time t:

$$i(t) = \frac{dq(t)}{dt} \, [\mathrm{A}]. \qquad (1.1)$$

Knowing i, the total charge, Q, transferred during the time from t_1 to t_2 can be calculated as:

$$Q = \int_{t1}^{t2} idt \; [\mathrm{C}]. \qquad (1.2)$$

The voltage or potential difference (V_{AB}) between two points A and B is the amount of energy required to move a unit positive charge from B to A. If this energy is positive, that is work is done by external sources against forces on the charges, then V_{AB} is positive and point A is at a higher potential with respect to B. The voltage is measured using the unit of volt [V]. The voltage between two points is 1 V if 1 J (joule) of work is required to move 1 C of charge. If the voltage, v, between two points is constant, then the work, w, done in moving q coulombs of charge between the two points is given by:

$$w = vq \; [\mathrm{J}]. \qquad (1.3)$$

Power (p) is the rate of doing work or the energy flow rate. When a charge of dq coulombs is moved from point A to point B with a potential difference of v volts, the energy supplied to the charge will be $v\ dq$ joule [J]. If this movement takes place in dt seconds, the power supplied to the charge will be $v\ dq/dt$ watts [W]. Because dq/dt is the charge flow rate defined earlier as current i, the power supplied to the charge can be written as:

$$p = vi \; [\mathrm{W}]. \qquad (1.4)$$

The energy supplied over duration $t1$ to $t2$ is then given by:

$$w = \int_{t1}^{t2} vi\, dt \; [\mathrm{J}]. \qquad (1.5)$$

A **lumped electrical element** is a model of an electrical device with two or more terminals through which current can flow in or out; the flow can pass *only* through the terminals. In a two-terminal element, current flows through the element entering via one terminal and leaving via another terminal. On the other hand, the voltage is present across the element and measured between the two terminals. In a multiterminal element, current flows through one set of terminals and leaves through the remaining set of terminals. The relation between the voltage and current in an element, known as the v–i relation, defines the element's characteristic. A circuit is made up of electrical elements.

Linear elements include a v–i relation, which can be linear if it satisfies the homogeneity property and the superposition principle. The homogeneity property refers to proportionality; that is, if i gives a voltage of v, ki gives a voltage of kv for any arbitrary constant k. The superposition principle implies additivity; that is, if i_1 gives a voltage of v_1 and i_2 gives a voltage of v_2, then $i_1 + i_2$ should give a voltage $v_1 + v_2$. It is easily verified that $v = Ri$ and $v = L\, di/dt$ are linear relations. Elements that possess such linear relations are called linear elements, and a circuit that is made up of linear elements is called a linear circuit.

Sources, also known as **active elements,** are electrical elements that provide power to a circuit. There are two types of sources: (1) independent sources and (2) dependent (or controlled) sources. An independent voltage source provides a specified voltage irrespective of the elements connected to it. In a similar manner, an independent current source provides a specified current irrespective of the elements connected to it. Figure 1.1 shows representations of independent voltage and independent current sources. It may be noted that the value of an independent voltage or an independent current source may be constant in magnitude and direction (called a direct current [dc] source) or may vary as a function of time (called a time-varying source). If the variation is of sinusoidal nature, it is called an alternating current (ac) source.

Values of dependent sources depend on the voltage or current of some other element or elements in the circuit. There are four classes of dependent sources: (1) voltage-controlled voltage source, (2) current-controlled voltage source, (3) voltage-controlled current source and (4) current-controlled current source. The representations of these dependent sources are shown in Table 1.1.

Passive elements consume power. Names, symbols, and the characteristics of some commonly used passive elements are given in Table 1.2. The v–i relation of a linear resistor, $v = Ri$,

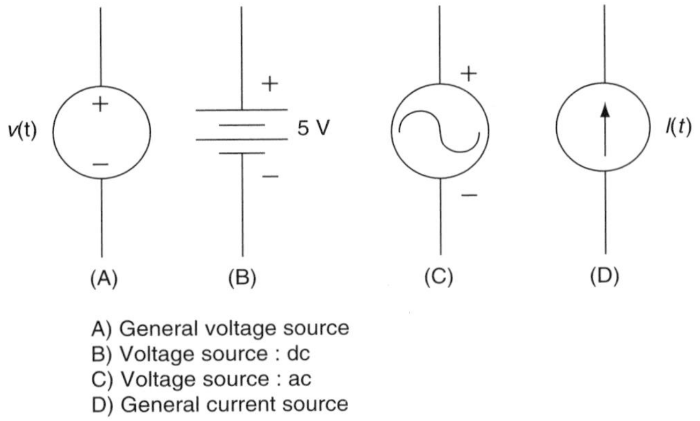

FIGURE 1.1 Independent Voltage and Current Sources

TABLE 1.1 Dependent Sources and Their Representation

Element	Voltage and current relation	Representation
Voltage-controlled voltage source	$v_2 = a\ v_1$ a: Voltage gain	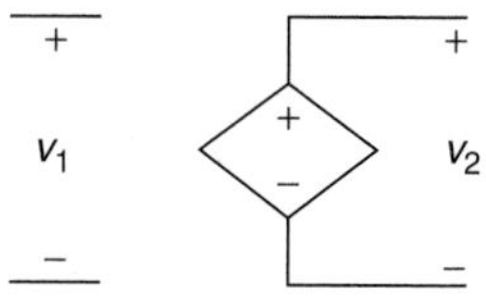
Voltage-controlled current source	$i_2 = g_t v_1$ g_t: Transfer conductance	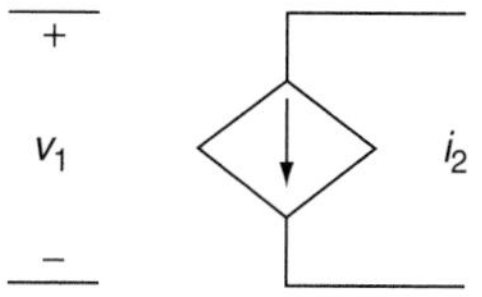
Current-controlled voltage source	$v_2 = r_t i_1$ rt: Transfer resistance	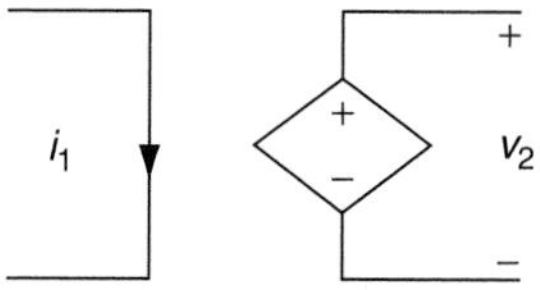
Current-controlled current source	$i_2 = bi_1$ b: Current gain	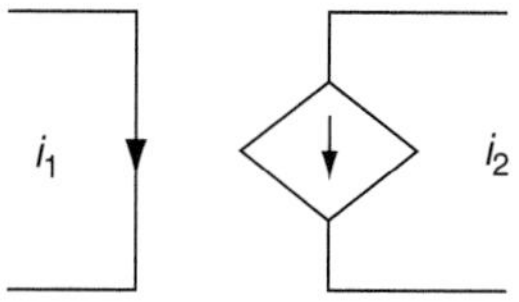

TABLE 1.2 Some Passive Elements and Their Characteristics

Name of the element	Symbol	The v–i relation	Unit
Resistance: R	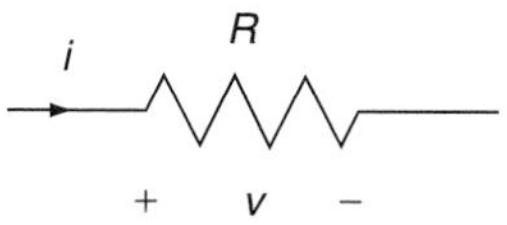	$v = Ri$	ohm [Ω]
Inductance: L	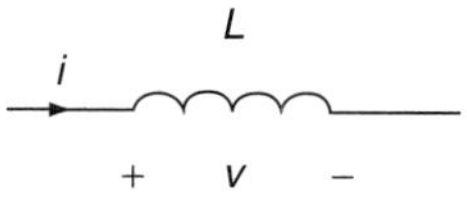	$v = L\ di/dt$	henry [H]
Capacitance: C	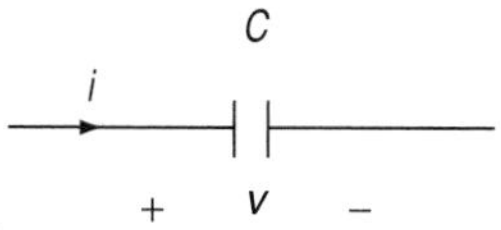	$i = C\ dv/dt$	farad [F]
Mutual Inductance: M	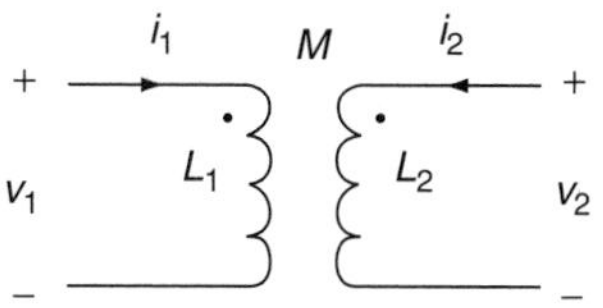	$v_1 = M\ di_2/dt + L_1\ di_1/dt$ $v_2 = M\ di_1/dt + L_2\ di_2/dt$	henry [H]

is known as Ohm's law, and the linear relations of other passive elements are sometimes called generalized Ohm's laws. It may be noted that in a passive element, the polarity of the voltage is such that current flows from positive to negative terminals. This polarity marking is said to follow the passive polarity convention.

A **circuit** is formed by an interconnection of circuit elements at their terminals. A **node** is a junction point where the

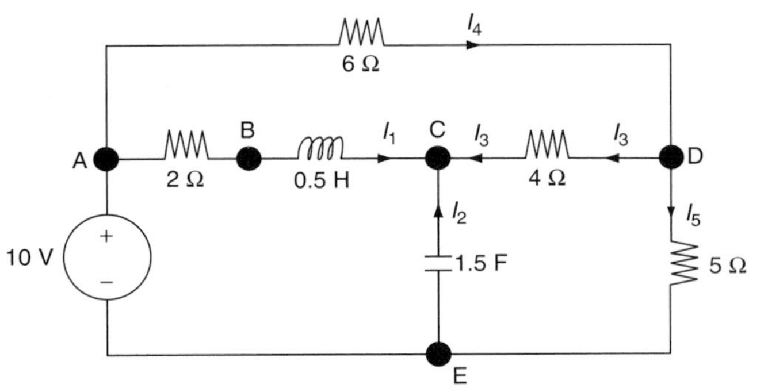

FIGURE 1.2 Example Circuit Diagram

terminals of two or more elements are joined. Figure 1.2 shows A, B, C, D, and E as nodes. A **loop** is a closed path in a circuit such that each node is traversed only once when tracing the loop. In Figure 1.2, ABCEA is a loop, and ABCDEA is also a loop. A **mesh** is a special class of loop that is associated with a window of a circuit drawn in a plane (**planar circuit**). In the same Figure ABCEA is a mesh, whereas ABCDEA is not considered a mesh for the circuit as drawn. A **network** is defined as a circuit that has a set of terminals available for external connections (i.e., accessible from outside of the circuit). A **pair of terminals** of a network to which a source, another network, or a measuring device can be connected is called a **port** of the network. A network containing such a pair of terminals is called a **one-port network**. A network containing two pairs of externally accessible terminals is called a **two-port network**, and multiple pairs of externally accessible terminal pairs are called a **multiport network**.

1.2 Circuit Laws

Two important laws are based on the physical properties of electric charges, and these laws form the foundation of circuit analysis. They are Kirchhoff's current law (KVL) and Kirchhoff's voltage law (KCL). While Kirchhoff's current law is based on the principle of conservation of electric charge, Kirchhoff's voltage law is based on the principle of energy conservation.

1.2.1 Kirchhoff's Current Law

At any instant, the algebraic sum of the currents (i) entering a node in a circuit is equal to zero. In the circuit in Figure 1.2, application of KCL at node C yields the following equation:

$$i_1 + i_2 + i_3 = 0 \tag{1.6}$$

Similarly at node D, KCL yields:

$$i_4 - i_3 - i_5 = 0. \tag{1.7}$$

1.2.2 Kirchhoff's Voltage Law

At any instant, the algebraic sum of the voltages (v) around a loop is equal to zero. In going around a loop, a useful convention is to take the voltage drop (going from positive to negative) as positive and the voltage rise (going from negative to positive) as negative. In Figure 1.2, application of KVL around the loop ABCEA gives the following equation:

$$v_{AB} + v_{BC} + v_{CE} + v_{EA} = 0. \tag{1.8}$$

1.3 Circuit Analysis

Analysis of an electrical circuit involves the determination of voltages and currents in various elements, given the element values and their interconnections. In a linear circuit, the v–i relations of the circuit elements and the equations generated by the application of KCL at the nodes and of KVL for the loops generate a sufficient number of simultaneous linear equations that can be solved for unknown voltages and currents. Various steps involved in the analysis of linear circuits are as follows:

1. For all the elements except the current sources, assign a current variable with arbitrary polarity. For the current sources, current values and polarity are given.
2. For all elements except the voltage sources, assign a voltage variable with polarities based on the passive sign convention. For voltage sources, the voltages and their polarities are known.
3. Write KCL equations at $N - 1$ nodes, where N is the total number of nodes in the circuit.
4. Write expressions for voltage variables of passive elements using their v–i relations.
5. Apply KVL equations for $E - N + 1$ independent loops, where E is the number of elements in the circuit. In the case of planar circuits, which can be drawn on a plane paper without edges crossing over one another, the meshes will form a set of independent loops. For nonplanar circuits, use special methods that employ topological techniques to find independent loops.
6. Solve the $2E$ equations to find the E currents and E voltages.

The following example illustrates the application of the steps in this analysis.

Example 1.1. *For the circuit in Figure 1.3, determine the voltages across the various elements.* Following step 1, assign the currents I_1, I_2, I_3, and I_4 to the elements. Then apply the KCL to the nodes A, B, and C to get $I_4 - I_1 = 0$, $I_1 - I_2 = 0$, and $I_2 - I_3 = 0$. Solving these equations produces $I_1 = I_2 = I_3 = I_4$. Applying the v–i relation characteristics of the nonsource elements, you get $V_{AB} = 2\ I_1$, $V_{BC} = 3\ I_2$, and $V_{CD} = 5\ I_3$. Applying

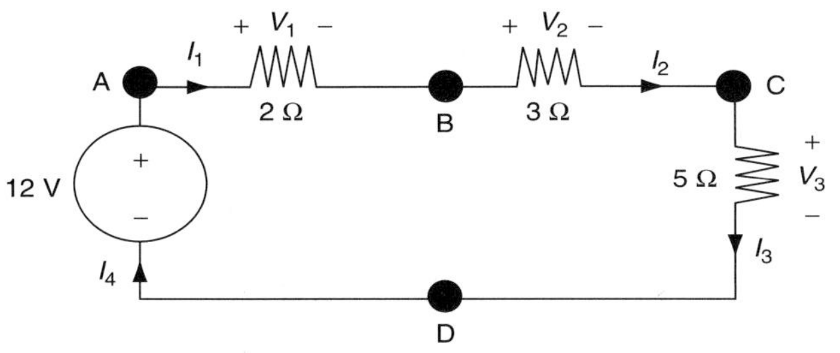

FIGURE 1.3 Circuit for Example 1.1

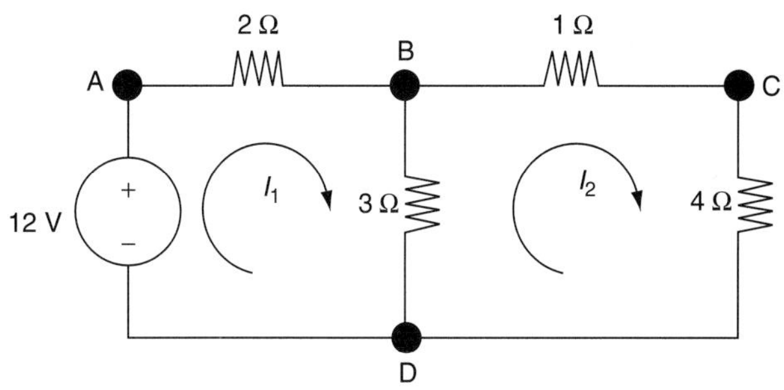

FIGURE 1.4 Circuit for Example 1.2

the KVL to the loop ABCDA, you determine $V_{AB} + V_{BC} + V_{CD} + V_{DA} = 0$. Substituting for the voltages in terms of currents, you get $2\,I_1 + 3\,I_1 + 5\,I_1 - 12 = 0$. Simplifying results in $10\,I_1 = 12$ to make $I_1 = 1.2$ A. The end results are $V_{AB} = 2.4$ V, $V_{BC} = 3.6$ V, and $V_{CD} = 6.0$ V.

In the above circuit analysis method, 2E equations are first set up and then solved simultaneously. For large circuits, this process can become very cumbersome. Techniques exist to reduce the number of unknowns that would be solved simultaneously. Two most commonly used methods are the loop current method and the node voltage method.

1.3.1 Loop Current Method

In this method, one distinct current variable is assigned to each independent loop. The element currents are then calculated in terms of the loop currents. Using the element currents and values, element voltages are calculated. After these calculations, Kirchhoff's voltage law is applied to each of the loops, and the resulting equations are solved for the loop currents. Using the loop currents, element currents and voltages are then determined. Thus, in this method, the number of simultaneous equations to be solved are equal to the number of independent loops. As noted above, it can be shown that this is equal to $E - N + 1$. Example 1.2 illustrates the techniques just discussed. It may be noted that in the case of planar circuits, the meshes can be chosen as the independent loops.

Example 1.2. *In the circuit in Figure 1.4, find the voltage across the 3-Ω resistor.* First, note that there are two independent loops, which are the two meshes in the circuit, and that loop currents I_1 and I_2 are assigned as shown in the diagram. Then calculate the element currents as $I_{AB} = I_1$, $I_{BC} = I_2$, $I_{CD} = I_2$, $I_{BD} = I_1 - I_2$, and $I_{DA} = I_1$. Calculate the element voltages as $V_{AB} = 2\,I_{AB} = 2\,I_1$, $V_{BC} = 1\,I_{BC} = 1\,I_2$, $V_{CD} = 4\,I_2$, and $V_{BD} = 3\,I_{BD} = 3(I_1 - I_2)$. Applying KVL to loops 1 (ABDA) and 2 (BCDB) and substituting the voltages in terms of loop currents results in:

$$5\,I_1 - 3\,I_2 = 12$$
$$-3\,I_1 + 8\,I_2 = 0.$$

Solving the two equations, you get $I_1 = 96/31$ A and $I_2 = 36/31$ A. The voltage across the 3-Ω resistor is $3(I_1 - I_2) = 3(96/31 - 36/31) = 180/31$ A.

Special case 1

When one of the elements in a loop is a current source, the voltage across it cannot be written using the *v–i* relation of the element. In this case, the voltage across the current source should be treated as an unknown variable to be determined. If a current source is present in only one loop and is not common to more than one loop, then the current of the loop in which the current source is present should be equal to the value of the current source and hence is known. To determine the remaining currents, there is no need to write the KVL equation for the current source loop. However, to determine the voltage of the current source, a KVL equation for the current source loop needs to be written. This equation is presented in example 1.3.

Example 1.3. *Analyze the circuit shown in Figure 1.5 to find the voltage across the current sources.* The loop currents are assigned as shown. It is easily seen that $I_3 = -2$. Writing KVL equations for loops 1 and 2, you get:

Loop 1: $2(I_1 - I_2) + 4(I_1 - I_3) - 14 = 0 =>$
$6\,I_1 - 2\,I_2 = 6.$

Loop 2: $I_2 + 3(I_2 - I_3) + 2(I_2 - I_1) = 0 =>$
$-2\,I_1 + 6\,I_2 = -6.$

Solving the two equations simultaneously, you get $I_1 = 3/4$ A and $I_2 = -3/4$ A. To find the V_{CD} across the current source, write the KVL equation for the loop 3 as:

$$4(I_3 - I_1) + 3(I_3 - I_2) + V_{CD} = 0 =>$$
$$V_{CD} = 4\,I_1 + 3\,I_2 - 7\,I_3 = 14.75 \text{ V}.$$

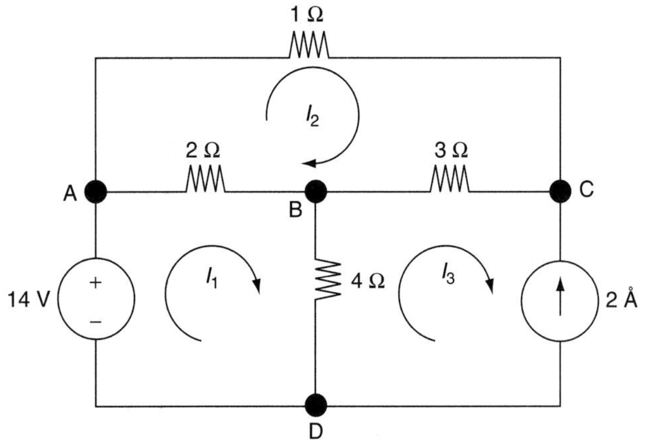

FIGURE 1.5 Circuit for Example 1.3

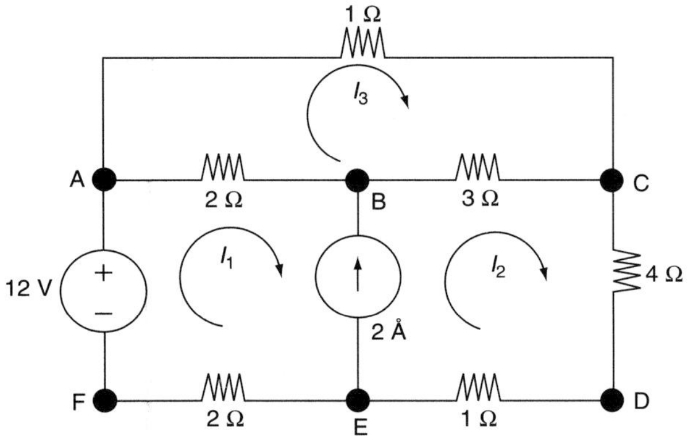

FIGURE 1.6 Circuit for Example 1.4

Special case 2

This case concerns a current source that is common to more than one loop. The solution to this case is illustrated in example 1.4.

EXAMPLE 1.4. *In the circuit shown in Figure 1.6, the 2 A current source is common to loops 1 and 2.* One method of writing KVL equations is to treat V_{BE} as an unknown and write three KVL equations. In addition, you can write the current of the current source as $I_2 - I_1 = 2$, giving a fourth equation. Solving the four equations simultaneously, you determine the values of I_1, I_2, I_3, and V_{BE}. These equations are the following:

Loop 1: $2(I_1 - I_3) + V_{BE} + 2\ I_1 - 12 = 0$
$=> 4\ I_1 - 2\ I_3 + V_{BE} = 12.$
Loop 2: $3(I_2 - I_3) + 4\ I_2 + I_2 - V_{BE} = 0$
$=> 8\ I_2 - 3\ I_3 - V_{BE} = 0.$
Loop 3: $I_3 + 3(I_3 - I_2) + 2(I_3 - I_1) = 0$
$=> -2\ I_1 - 3\ I_2 + 6\ I_3 = 0.$
Current source relation: $-I_1 + I_2 = 2.$

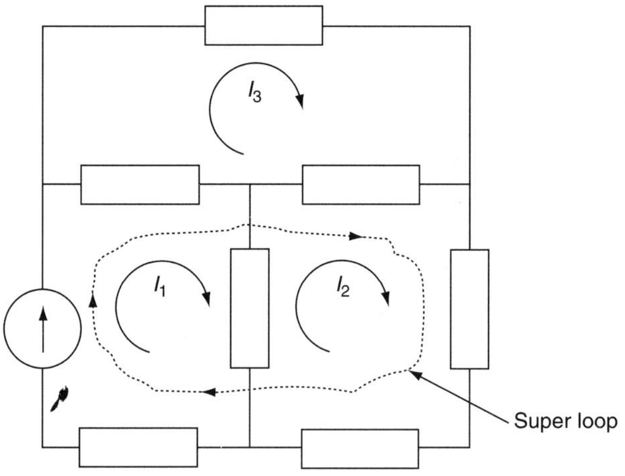

FIGURE 1.7 Circuit in Figure 1.6 with the Super Loop Shown as Dotted Line

Solving the above four equations results in $I_1 = 0.13$ A, $I_2 = 2.13$ A, $I_3 = 1.11$ A, and $V_{BE} = 13.70$ V.

Alternative method for special case 2 (Super loop method): This method eliminates the need to add the voltage variable as an unknown. When a current source is common to loops 1 and 2, then KVL is applied on a new loop called the **super loop**. The super loop is obtained from combining loops 1 and 2 (after deleting the common elements) as shown in Figure 1.7. For the circuit considered in example 1.4, the loop ABCDEFA is the super loop obtained by combining loops 1 and 2. The KVL is applied on this super loop instead of KVL being applied for loop 1 and loop 2 separately. The following is the KVL equation for super loop ABCDEFA:

$$2(I_1 - I_3) + 3(I_2 - I_3) + 4\ I_2 + I_2 + 2\ I_1 - 12 = 0$$
$$=> 4\ I_1 + 8\ I_2 - 5\ I_3 = 12.$$

The KVL equation around loop 3 is written as:

$$-2\ I_1 - 3\ I_2 + 6\ I_3 = 0.$$

The current source can be written as:

$$-I_1 + I_2 = 2.$$

Solving the above three equations simultaneously produces equations $I_1 = 0.13$ A, $I_2 = 2.13$ A, and $I_3 = 1.11$ A.

1.3.2 Node Voltage Method (Nodal Analysis)

In this method, one node is chosen as the reference node whose voltage is assumed as zero, and the voltages of other nodes are expressed with respect to the reference node. For example, in Figure 1.8, the voltage of node G is chosen as the

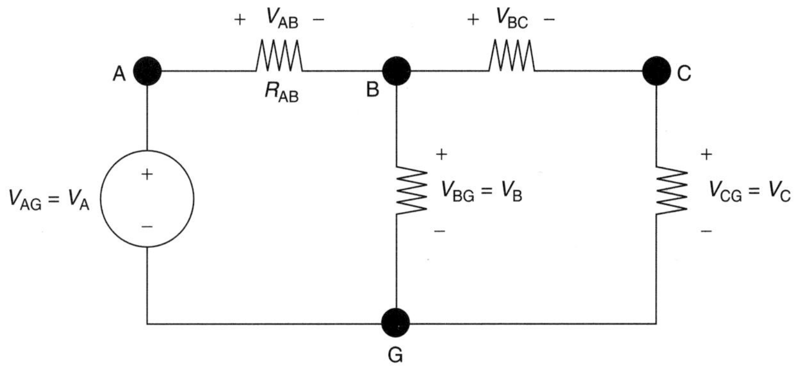

FIGURE 1.8 Circuit with Node Voltages Marked

reference node, and then the voltage of node A is $V_A = V_{AG}$ and that of node B is $V_B = V_{BG}$ and so on. Then, for every element between two nodes, the element voltages may be expressed as the difference between the two node voltages. For example, the voltage of element R_{AB} is $V_{AB} = V_A - V_B$. Similarly $V_{BC} = V_B - V_C$ and so on. Then the current through the element R_{AB} can be determined using the *v–i* characteristic of the element as $I_{AB} = V_{AB}/R_{AB}$. Once the currents of all elements are known in terms of node voltages, KCL is applied for each node except for the reference node, obtaining a total of *N*–1 equations where *N* is the total number of nodes.

Special Case 1

In branches where voltage sources are present, the *v–i* relation cannot be used to find the current. Instead, the current is left as an unknown. Because the voltage of the element is known, another equation can be used to solve the added unknown. When the element is a current source, the current through the element is known. There is no need to use the *v–i* relation. The calculation is illustrated in the following example.

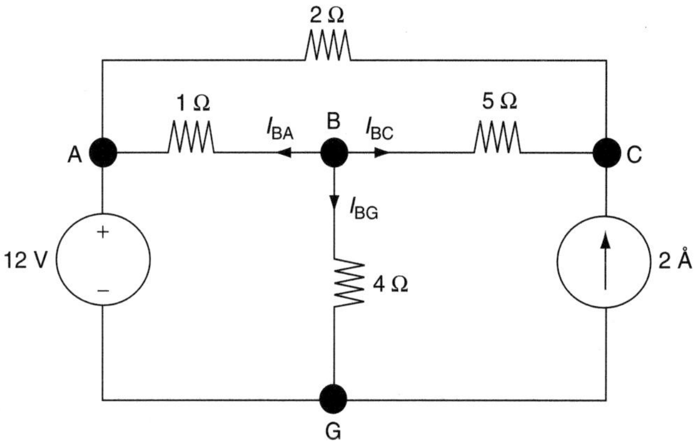

FIGURE 1.9 Circuit for Example 1.5

Example 1.5. In Figure 1.9, solve for the voltages V_A, V_B, and V_C with respect to the reference node G. At node A, $V_A = 12$. At node B, KCL yields:

$$I_{BA} + I_{BG} + I_{BC} = 0 \Rightarrow$$
$$(V_B - V_A)/1 + V_B/4 + (V_B - V_C)/5 = 0 \Rightarrow$$
$$-V_A + (1 + 1/4 + 1/5)V_B - V_C/5 = 0.$$

Similarly at node C, KCL yields:

$$V_A/2 - V_B/5 + (1/5 + 1/2)V_C = 2.$$

Solving the above three equations simultaneously results in $V_A = 12$ V, $V_B = 10.26$ V, and $V_C = 14.36$ V.

Super Node: When a voltage source is present between two nonreference nodes, a super node may be used to avoid introducing an unknown variable for the current through the voltage source. Instead of applying KCL to each of the two nodes of the voltage source element, KCL is applied to an imaginary node consisting of both the nodes together. This imaginary node is called a super node. In Figure 1.10, the super node is shown by a dotted closed shape. KCL on this super node is given by:

$$I_{BA} + I_{BG} + I_{CG} + I_{CA} = 0 \Rightarrow (V_B - V_A)/1$$
$$+ V_B/3 + V_C/4 + (V_C - V_A)/2 = 0.$$

In addition to this equation, the two voltage constraint equations, $V_A = 10$ and $V_B - V_C = 5$, are used to solve for V_B and V_C as $V_B = 9$ V and $V_C = 4$ V.

1.4 Equivalent Circuits

Two linear circuits, say circuit 1 and circuit 2, are said to be equivalent across a specified set of terminals if the voltage–current relations for the two circuits across the specified terminals are identical. Now consider a composite circuit

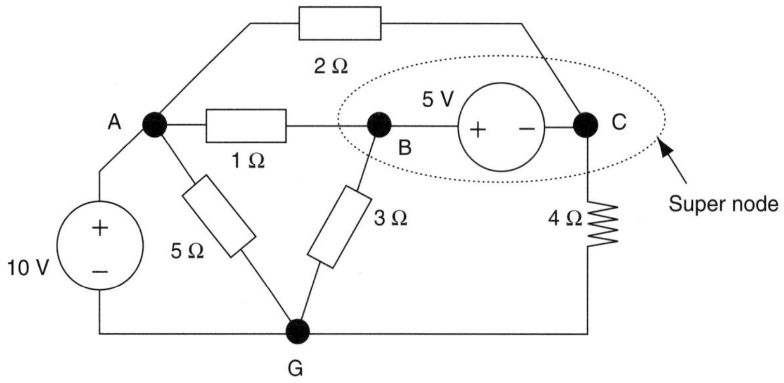

FIGURE 1.10 Circuit with Super Node

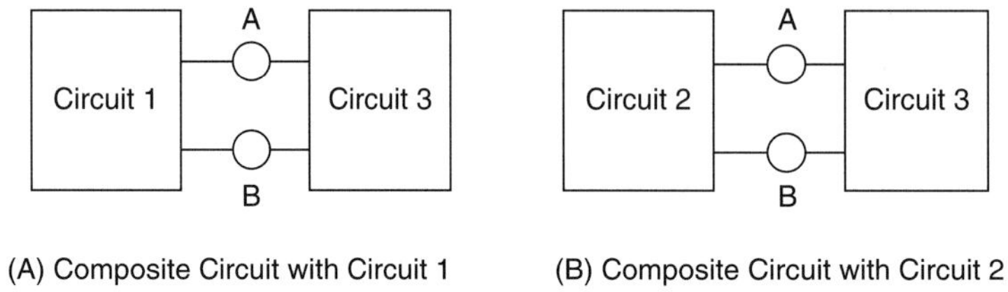

FIGURE 1.11 Equivalent Circuit Application

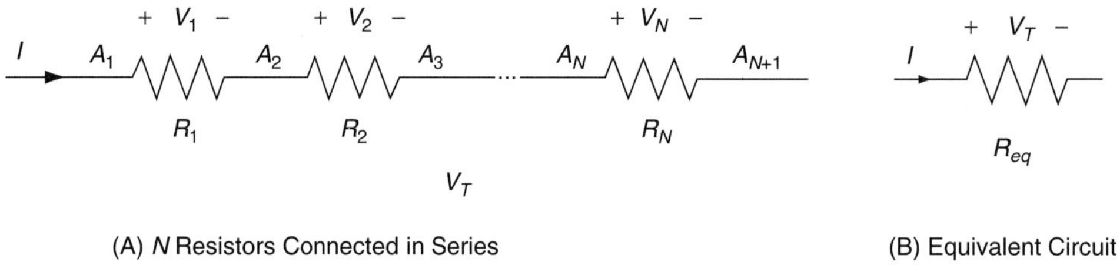

FIGURE 1.12 Resistances Connected in Series

consisting of circuit 1 connected to another circuit, circuit 3, at the specified terminals as shown in Figure 1.11(A). The voltages and currents in circuit 3 are not altered if circuit 2 replaces circuit 1, as shown in Figure 1.11(B). If circuit 2 is simpler than circuit 1, then the analysis of the composite circuit will be simplified. A number of techniques for obtaining two-terminal equivalent circuits are outlined in the following section.

1.4.1 Series Connection

Two two-terminal elements are said to be **connected in series** if the connection is such that the same current flows through both the elements as shown in Figure 1.12. When two resistances R_1 and R_2 are connected in series, they can be replaced by a single element having an equivalent resistance of sum of the two resistances, $R_{eq} = R_1 + R_2$, without affecting the voltages and currents in the rest of the circuit. In a similar manner, if N resistances R_1, R_2, ..., R_N are connected in series, their equivalent resistance will be given by:

$$R_{eq} = R_1 + R_2 + \ldots + R_N. \tag{1.9}$$

Voltage Division: When a voltage V_T is present across N resistors connected in series, the total voltage divides across the resistors proportional to their resistance values. Thus

$$V_1 = V_T \frac{R_1}{R_{eq}}, \ V_2 = V_T \frac{R_2}{R_{eq}}, \ \ldots, \ V_N = V_T \frac{R_N}{R_{eq}}, \tag{1.10}$$

where $R_{eq} = R_1 + R_2 + \ldots + R_N$.

1.4.2 Parallel Connection

Two-terminal elements are said to be **connected in parallel** if the same voltage exists across all the elements and if they have two distinct common nodes as shown in Figure 1.13. In the case of a parallel connection, conductances, which are reciprocals of resistances, sum to give an equivalent conductance of G_{eq}:

$$G_{eq} = G_1 + G_2 + \ldots + G_N, \tag{1.11}$$

or equivalently

$$\frac{1}{R_{eq}} = \frac{1}{R_1} + \frac{1}{R_2} + \ldots + \frac{1}{R_N}. \tag{1.12}$$

Current Division: In parallel connection, the total current I_T of the parallel combination divides proportionally to the conductance of each element. That is, the current in each element is proportional to its conductance and is given by:

$$I_1 = I_T \frac{G_1}{G_{eq}}, \; I_2 = I_T \frac{G_2}{G_{eq}}, \ldots, \; I_N = I_T \frac{G_N}{G_{eq}}, \tag{1.13}$$

where $G_{eq} = G_1 + G_2 + \ldots + G_N$.

1.4.3 Star–Delta (Wye–Delta or T–Pi) Transformation

It can be shown that the star subnetwork connected as shown in Figure 1.14 can be converted into an equivalent delta subnetwork. The element values between the two subnetworks are related as shown in Table 1.3. It should be noted that the star subnetwork has four nodes, whereas the delta network has only three nodes. Hence, the star network can be replaced in a circuit without affecting the voltages and currents in the rest

TABLE 1.3 Relations Between the Element Values in Star and Delta Equivalent Circuits

Star in terms of delta resistances	Delta in terms of star resistances
$R_1 = \frac{R_b R_c}{R_a + R_b + R_c}$	$R_a = \frac{R_1 R_2 + R_2 R_3 + R_3 R_1}{R_1}$
$R_2 = \frac{R_c R_a}{R_a + R_b + R_c}$	$R_b = \frac{R_1 R_2 + R_2 R_3 + R_3 R_1}{R_2}$
$R_3 = \frac{R_a R_b}{R_a + R_b + R_c}$	$R_c = \frac{R_1 R_2 + R_2 R_3 + R_3 R_1}{R_3}$

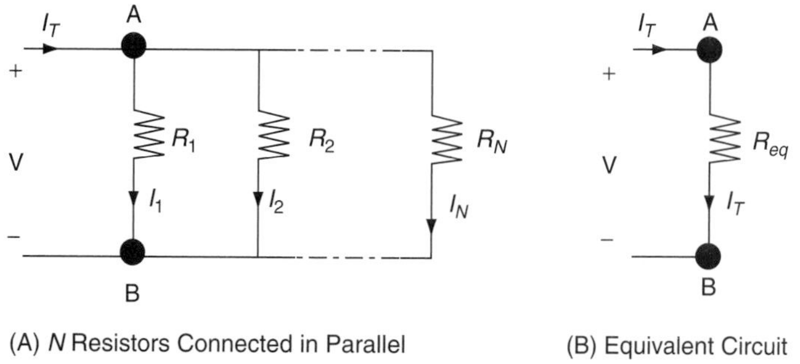

FIGURE 1.13 Resistances Connected in Parallel

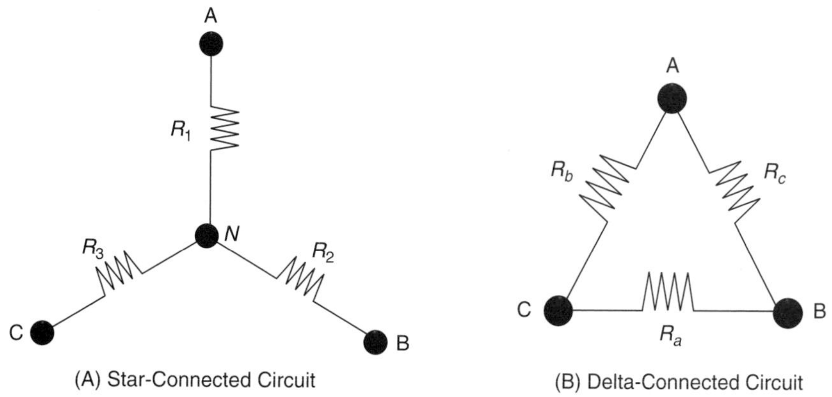

FIGURE 1.14 Star and Delta Equivalent Circuits

of the circuit *only* if the central node in the star subnetwork is not connected to any other circuit node.

1.4.4 Thevenin Equivalent Circuit

A network consisting of linear resistors and dependent and independent sources with a pair of accessible terminals can be represented by an equivalent circuit with a voltage source and a series resistance as shown in Figure 1.15. V_{TH} is equal to the open circuit voltage across the two terminals A and B, and R_{TH} is the resistance measured across nodes A and B (also called **looking-in resistance**) when the independent sources in the network are deactivated. The R_{TH} can also be determined as $R_{TH} = V_{oc}/I_{sc}$, where V_{oc} is the open circuit voltage across terminals A and B and where I_{sc} is the short circuit current that will flow from A to B through an external zero resistance connection (short circuit) if one is made.

1.4.5 Norton Equivalent Circuit

A two-terminal network consisting of linear resistors and independent and dependent sources can be represented by an equivalent circuit with a current source and a parallel resistor as shown in Figure 1.16. In this figure, I_N is equal to the short circuit current across terminals A and B, and R_N is the looking-in resistance measured across A and B after the independent sources are deactivated. It is easy to see the following relation between Thevenin equivalent circuit parameters and the Norton equivalent circuit parameters:

$$R_N = R_{TH} \text{ and } I_N = V_{TH}/R_{TH}. \tag{1.14}$$

1.4.6 Source Transformation

Using a Norton equivalent circuit, a voltage source with a series resistor can be converted into an equivalent current source with a parallel resistor. In a similar manner, using Thevenin theorem, a current source with a parallel resistor can be represented by a voltage source with a series resistor. These transformations are called source transformations. The two sources in Figure 1.17 are equivalent between nodes B and C.

1.5 Network Theorems

A number of theorems that simplify the analysis of linear circuits have been proposed. The following section presents, without proof, two such theorems: the superposition theorem and the maximum power transfer theorem.

1.5.1 Superposition Theorem

For a circuit consisting of linear elements and sources, the response (voltage or current) in any element in the circuit is the algebraic sum of the responses in this element obtained by applying one independent source at a time. When one independent source is applied, all other independent sources are deactivated. It may be noted that a deactivated voltage source behaves as a short circuit, whereas a deactivated current source behaves as an open circuit. It should also be noted that the dependent sources in the circuit are not deactivated. Further, any initial condition in the circuit is treated as an appropriate independent source. That is, an initially charged

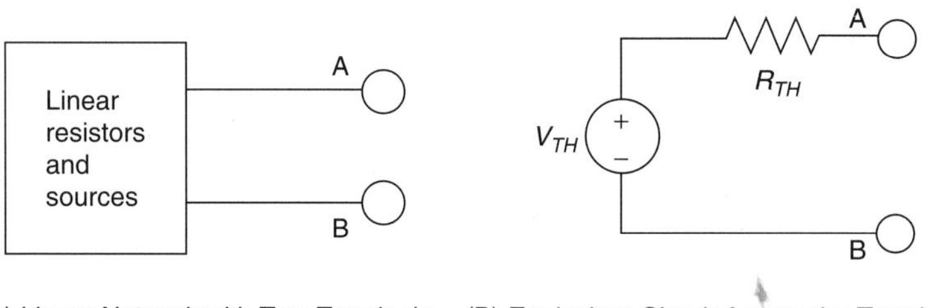

(A) Linear Network with Two Terminals (B) Equivalent Circuit Across the Terminals

FIGURE 1.15 Thevenin Equivalent Circuit

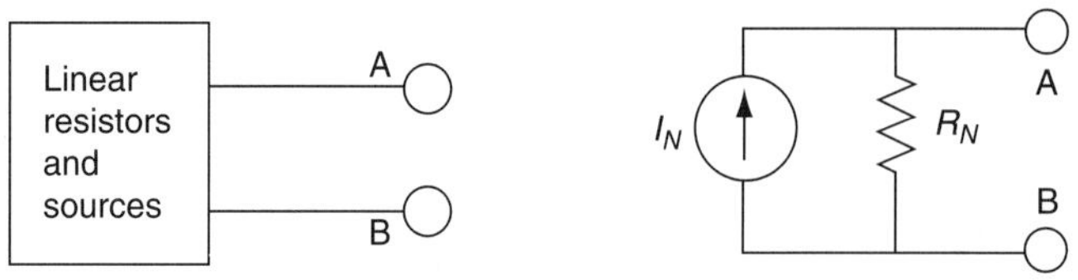

(A) Linear Network with Two Terminals (B) Equivalent Circuit Across AB in (A)

FIGURE 1.16 Norton Equivalent Circuit

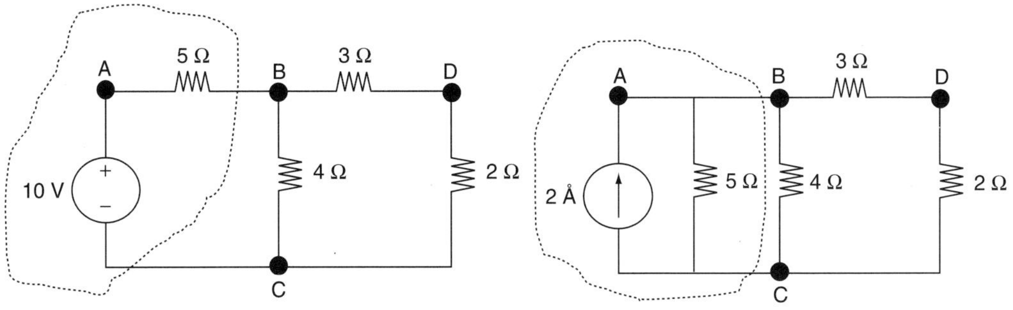

FIGURE 1.17 Example of Source Transformation

capacitor is replaced by an uncharged capacitor in series with an independent voltage source. Similarly, an inductor with an initial current is replaced with an inductor without any initial current in parallel with an independent current source. The following example illustrates the application of superposition in the analysis of linear circuits.

EXAMPLE 1.6. *For the circuit in Figure 1.18(A), determine the voltage across the* 3-Ω *resistor.* The circuit has two independent sources, one voltage source and one current source. Figure 1.18(B) shows the circuit when voltage source is activated and current source is deactivated (replaced by an open circuit). Let V_{31} be the voltage across the 3-Ω resistor in this circuit. Figure 1.18(C) shows the circuit when current source is activated and voltage source is deactivated (replaced by a short circuit). Let V_{32} be the voltage across the 3-Ω resistor in this circuit. Then you determine that the voltage across the 3-Ω resistor in the given complete circuit is $V_3 = V_{31} + V_{32}$.

1.5.2 Maximum Power Transfer Theorem

In the circuit shown in Fig. 1.19, power supplied to the load is maximum when the load resistance is equal to the source resistance.

It may be noted that the application of the maximum power transfer theorem is not restricted to simple circuits only. The theorem can also be applied to complicated circuits as long as the circuit is linear and there is one variable load. In such cases, the complicated circuit across the variable load is replaced by its equivalent Thevenin circuit. The maximum power transfer theorem is then applied to find the load resistance that leads to maximum power in the load.

1.6 Time Domain Analysis

When a circuit contains energy storing elements, namely inductors and capacitors, the analysis of the circuit involves the solution of a differential equation.

1.6.1 First-Order Circuits

A circuit with a single energy-storing element yields a first-order differential equation as shown below for the circuits in Figure 1.20.

Consider the RC circuit in Figure 1.20(A). For $t > 0$, writing KVL around the loop, the result is equation:

$$Ri + v_c(0) + \frac{1}{c}\int_0^t i dt = v_s. \tag{1.15}$$

Differentiating with respect to t yields:

$$R\frac{di}{dt} + \frac{1}{c}i = 0. \tag{1.16}$$

The solution of the above homogeneous differential equation can be obtained as:

$$i(t) = Ke^{-1/RCt}. \tag{1.17}$$

The value of K can be found by using the initial voltage $v_c(0)$ in the capacitor as:

$$K = i(0) = \frac{v_s - v_c(0)}{R}. \tag{1.18}$$

Substituting this value in the expression for $i(t)$ determines the final solution for $i(t)$ as:

$$i(t) = \frac{v_s - v_c(0)}{R} e^{-(1/RC)t}. \tag{1.19}$$

This exponentially decreasing response $i(t)$ is shown in Figure 1.21(A). It has a time constant of $\tau = RC$. The voltage $V_c(t)$ is also shown in Figure 1.21(B)

In a similar manner, the differential equation for $i(t)$ in the RL circuit shown in Figure 1.20(B) can be obtained for $t > 0$ as:

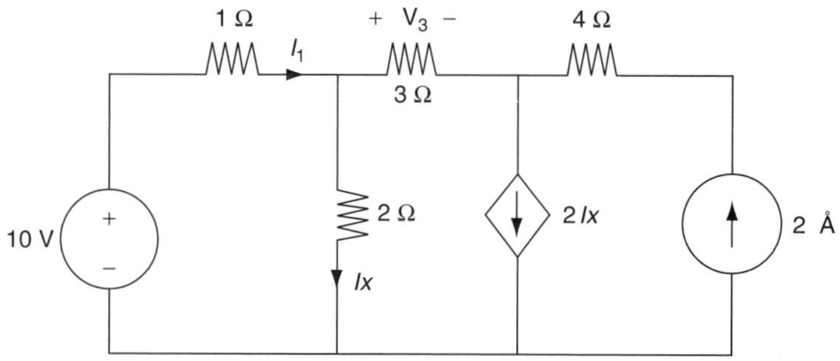

(A) Original Circuit

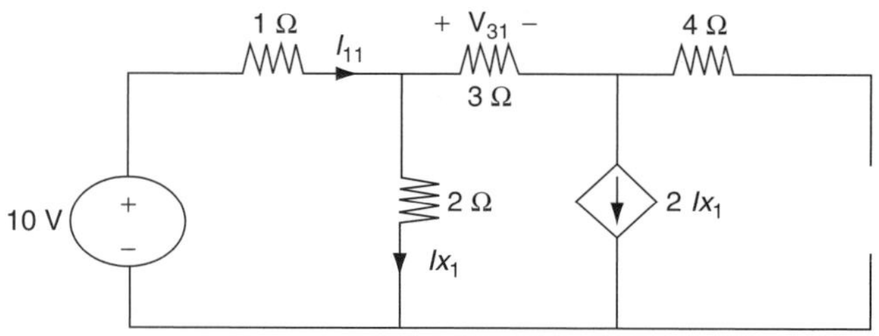

(B) Circuit When Voltage Source Is Activated and Current Source Is Deactivated

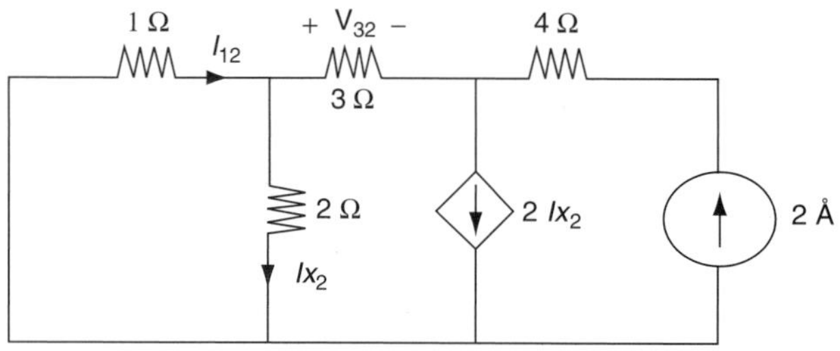

(C) Circuit When Current Source Is Activated and Voltage Source Is Deactivated

FIGURE 1.18 Circuits for Example 1.6

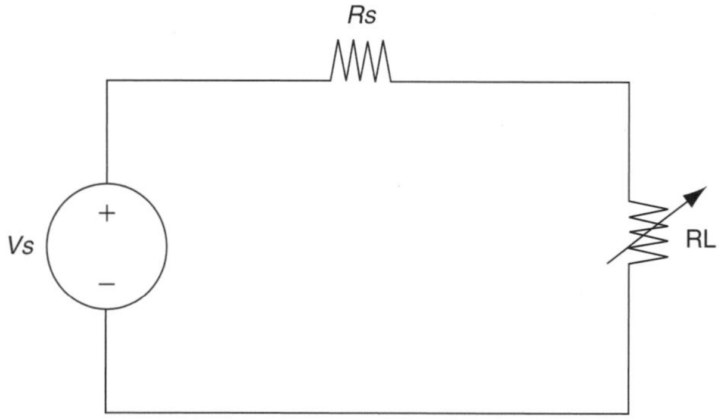

FIGURE 1.19 Circuit with a Variable Load Excited by a Thevenin Source

$$Ri + L\frac{di}{dt} = v_s. \tag{1.20}$$

Because this is a nonhomogeneous differential equation, its solution consists of two parts:

$$i(t) = i_n(t) + i_f(t), \tag{1.21}$$

where $i_n(t)$, the natural response (also called the complementary function) of the circuit, is the solution of the homogeneous differential equation:

$$L\frac{di_n}{dt} + Ri_n = 0. \tag{1.22}$$

The forced response (also called the particular integral) of the circuit, $i_f(t)$, is given by the solution of the nonhomoge-

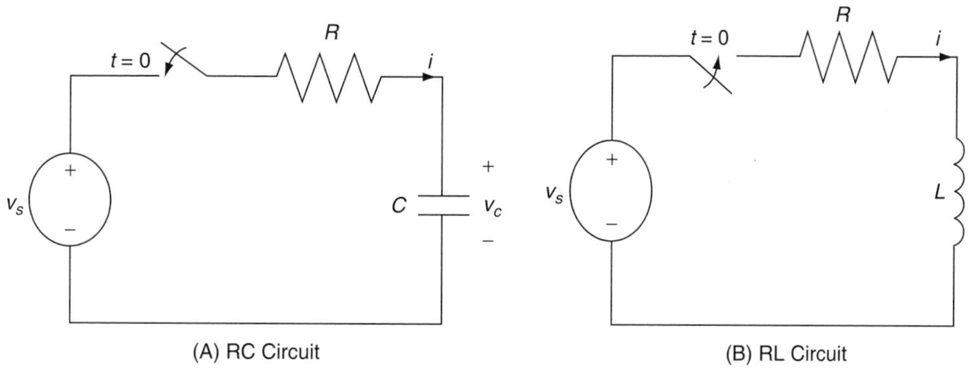

FIGURE 1.20 Circuits with a Single Energy-Storing Element

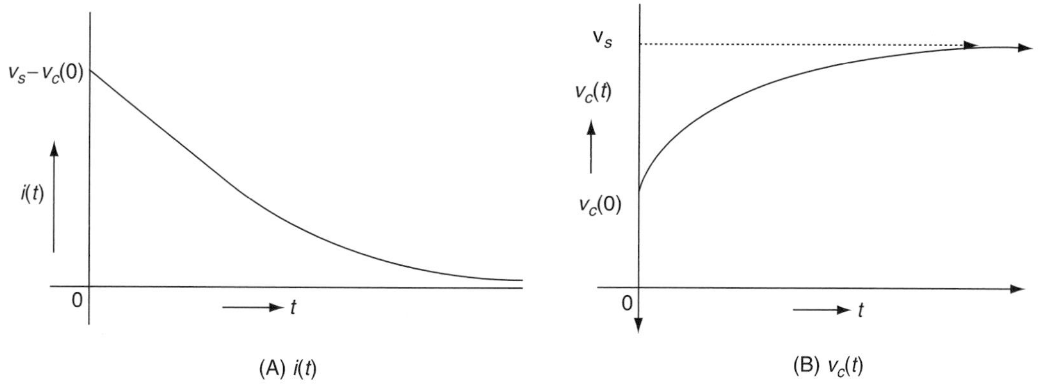

FIGURE 1.21 Response of the Circuit in Figure 1.20(A)

neous differential equation corresponding to the particular forcing function v_s. If v_s is a constant, the forced response in general is also a constant. In this case, the natural and forced responses and the total response are given by:

$$i_n(t) = Ke^{-R/Lt},\ i_f(t) = \frac{v_s}{R},\ \text{and}\ i(t) = Ke^{-(R/L)t} + \frac{v_s}{R}. \tag{1.23}$$

K is found using the initial condition in the inductor $i(0) = I_0$ as $i(0) = K + v_s/R$, and so $K = I_0 - v_s/R$. Substituting for *K* in the total response yields:

$$i(t) = \left(I_0 - \frac{v_s}{R}\right)e^{-(R/L)t} + \frac{v_s}{R}. \tag{1.24}$$

The current waveform, shown in Figure 1.22, has an exponential characteristic with a time constant of *L/R* [s].

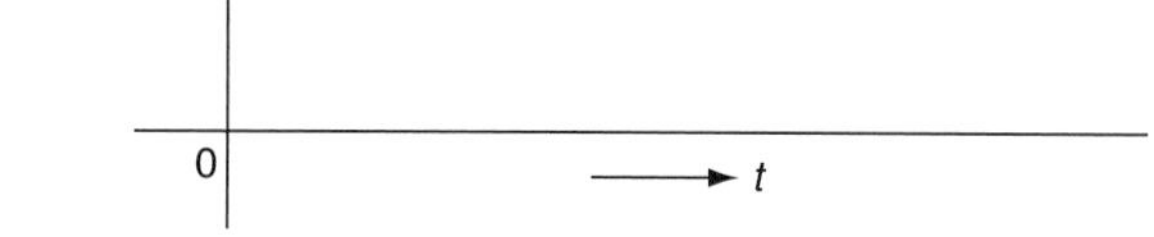

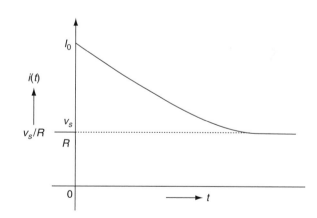

FIGURE 1.22 Response of the Circuit Shown in Figure 1.20(B)

1.6.2 Second-Order Circuits

If the circuit contains two energy-storing elements, *L* and/or *C*, the equation connecting voltage or current in the circuit is a second-order differential equation. Consider, for example, the circuit shown in Figure 1.23.

Writing KCL around the loop and substituting $i = Cdv_c/dt$ results in:

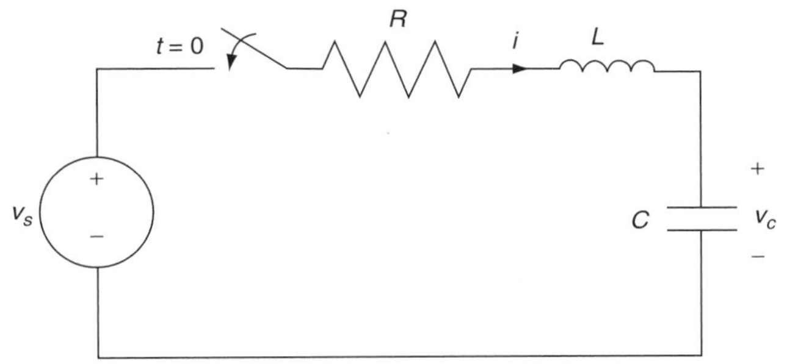

FIGURE 1.23 Circuit with Two Energy-Storing Elements

$$LC\frac{d^2v_c}{dt} + RC\frac{dv_c}{dt} + v_c = v_s. \quad (1.25)$$

This equation can be solved by either using a Laplace transform or a conventional technique. This section illustrates the use of the conventional technique. Assuming a solution of the form $v_n(t) = Ke^{st}$ for the homogeneous equation yields the characteristic equation as:

$$LCs^2 + RCs + 1 = 0. \quad (1.26)$$

Solving for s results in the characteristic roots written as:

$$s_{1,2} = -\frac{R}{2L} \pm \sqrt{\left(\frac{R}{2L}\right)^2 - \frac{1}{LC}}. \quad (1.27)$$

Four cases should be considered:

Case 1: $(R/2L)^2 > (1/LC)$. The result is two real negative roots s_1 and s_2 for which the solution will be an overdamped response of the form:

$$v_n(t) = K_1e^{s_1t} + K_2e^{s_2t}. \quad (1.28)$$

Case 2: $(R/2L)^2 = (1/LC)$. In this case, the result is a double root at $s_0 = -R/2L$. The natural response is a critically damped response of the form:

$$v_n(t) = (K_1t + K_2)e^{s_0t}. \quad (1.29)$$

Case 3: $0 < (R/2L)^2 < (1/LC)$. This case yields a pair of complex conjugate roots as:

$$s_{1,2} = -\frac{R}{2L} \pm j\sqrt{\frac{1}{LC} - \left(\frac{R}{2L}\right)^2} = -\sigma \pm j\omega_d. \quad (1.30)$$

The corresponding natural response is an underdamped oscillatory response of the form:

$$v_n(t) = Ke^{-\sigma t}\cos(\omega_d t + \theta). \quad (1.31)$$

Case 4: $R/2L = 0$. In this case, a pair of imaginary roots are created as:

$$s_{1,2} = \pm j\sqrt{\frac{1}{LC}} = \pm\, j\omega_0. \quad (1.32)$$

The corresponding natural response is an undamped oscillation of the form:

$$v_n(t) = K\cos(\omega_0 t + \theta). \quad (1.33)$$

The forced response can be obtained as $v_f(t) = V_s$. The total solution is obtained as:

$$v_c(t) = v_n(t) + V_s. \quad (1.34)$$

The unknown coefficients K_1 and K_2 in cases 1 and 2 and K and θ in cases 3 and 4 can be calculated using the initial values on the current in the inductor and the voltage across the capacitor.

Typical responses for the four cases when V_s equals zero are shown in Figure 1.24. For circuits containing energy-dissipating elements, namely resistors, the natural response in general will die down to zero as t goes to infinity. The component of the response that goes to zero as time t goes to infinity is called the **transient response.** The forced response depends on the forcing function. When the forcing function is a constant or a sinusoidal function, the forced response will continue to be present even as t goes to infinity. The component of the total response that continues to exist for all time is called **steady state response.** In the next section, computation of steady state responses for sinusoidal forcing functions is considered.

1.6.3 Higher Order Circuits

When a circuit has more than two energy-storing elements, say n, the analysis of the circuit in general results in a differential equation of order n. The solution of such an equation follows steps similar to the second-order case. The characteristic equation will be of degree n and will have n roots. The natural response will have n exponential terms. Also, the forced response will in general have the same shape as the forcing function. The Laplace transform is normally used to solve such higher order circuits.

1.7 Laplace Transform

In the solution of linear time-invariant differential equations, it was noted that a forcing function of the form K_ie^{st} yields an output of the form K_oe^{st} where s is a complex variable. The function e^{st} is a complex sinusoid of exponentially varying amplitude, often called a damped sinusoid. Because linear equations obey the superposition principle, the solution of a linear differential equation to any forcing function can be found by superposing solutions to component-damped sinusoids if the forcing function is expressed as a sum of damped

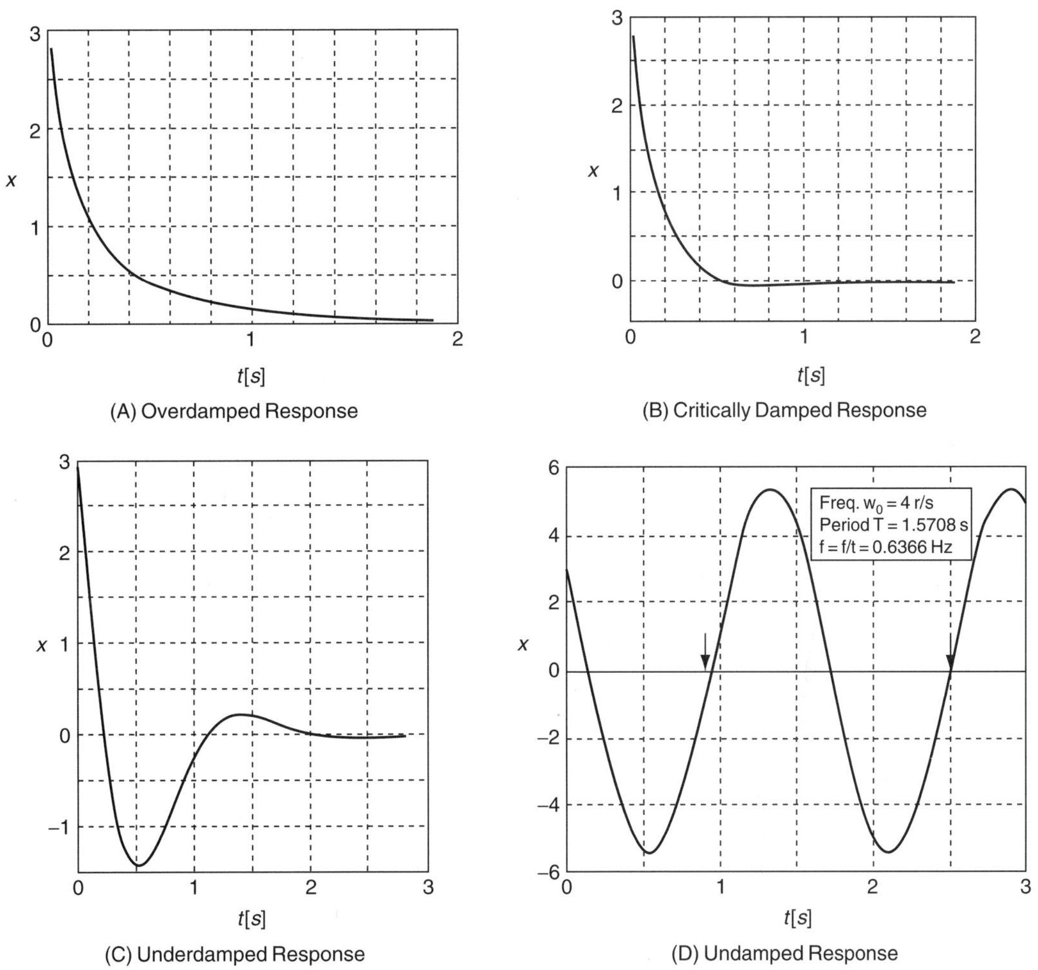

FIGURE 1.24 Typical Second-Order Circuit Responses

sinusoids. With this objective in mind, the Laplace transform is defined. The **Laplace transform** decomposes a given time function into an integral of complex-damped sinusoids.

1.7.1 Definition

The Laplace transform of $f(t)$ is defined as:

$$F(s) = \int_{0}^{\infty} f(t)e^{-st}\,dt. \tag{1.35}$$

The inverse Laplace transform is defined as:

$$f(t) = \frac{1}{2\pi j} \int_{\sigma_0 - j\infty}^{\sigma_0 + j\infty} F(s)e^{st}\,dt. \tag{1.36}$$

$F(s)$ is called the Laplace transform of $f(t)$, and σ_0 is included in the limits to ensure the convergence of the improper integral. The equation 1.36 shows that $f(t)$ is expressed as a sum (integral) of infinitely many exponential functions of complex frequencies (s) with complex amplitudes (phasors) $\{F(s)\}$. The complex amplitude $F(s)$ at any frequency s is given by the integral in equation 1.35. The Laplace transform, defined as the integral extending from zero to infinity, is called a single-sided Laplace transform against the double-sided Laplace transform whose integral extends from $-\infty$ to $+\infty$. As transient response calculations start from some initial time, the single-sided transforms are sufficient in the time domain analysis of linear electric circuits. Hence, this discussion considers only single-sided Laplace transforms.

1.7.2 Laplace Transforms of Common Functions

Consider

$$f(t) = Ae^{-at} \text{ for } 0 \leq t \leq \infty, \tag{1.37}$$

then

$$F(s) = \int_0^\infty Ae^{-at}e^{-st}dt = \left. \frac{Ae^{-(a+s)t}}{-(a+s)} \right|_0^\infty = \frac{Ae^{-\infty} - Ae^{-0}}{-(a+s)} = \frac{A}{s+a}. \quad (1.38)$$

In this equation, it is assumed that Re(s) – Re(a). In the region in the complex s-plane where s satisfies the condition that Res > Rea, the integral converges, and the region is called the **region of convergence of F(s)**. When $a = 0$ and $A = 1$, the above $f(t)$ becomes $u(t)$, the unit step function. Substituting these values in equation 1.38, the Laplace transform of $u(t)$ is obtained as $1/s$. In a similar way, letting $s = -j\omega$, the Laplace transform of $Ae^{j\omega t}$ is obtained as $A/(s - j\omega)$. Expressing $\cos(\omega t) = (e^{j\omega t} + e^{-j\omega t})/2$, we get the Laplace transform of $A\cos(\omega t)$ as $A\,s/(s^2 + \omega^2)$. In a similar way, the Laplace transform of $A\sin(\omega t)$ is obtained as $A\,\omega/(s^2 + w^2)$. Transforms for some commonly occurring functions are given in Table 1.4. This table can be used for finding forward as well as inverse transforms of functions.

As mentioned at the beginning of this section, the Laplace transform can be used to solve linear time-invariant differential equations. This will be illustrated next in example 1.7.

Example 1.7. *Consider the second-order differential equation and use the Laplace transform to find a solution:*

$$\frac{d^2f}{dt^2} + 6\frac{df}{dt} + 8f = 4e^{-t} \quad (1.39)$$

with initial conditions $f(0) = 2$ and $\frac{df}{dt}(0) = 3$.

Taking the Laplace transform of both sides of the above differential equation produces:

TABLE 1.4 Laplace Transforms of Common Functions

$f(t)$, for $t \geq 0$	$F(s)$
A	$\frac{A}{s}$
$Ae^{-\sigma t}$	$\frac{A}{s+\sigma}$
$A\,t$	$\frac{A}{s^2}$
$\sin(\omega t)$	$\frac{\omega}{s^2+\omega^2}$
$\cos(\omega t)$	$\frac{s}{s^2+\omega^2}$
$e^{-\sigma t}\cos(\omega t)$	$\frac{s+\sigma}{(s+\sigma)^2+\omega^2}$
$e^{-\sigma t}\sin(\omega t)$	$\frac{\omega}{(s+\sigma)^2+\omega^2}$

$$s^2F(s) - sf(0) - \frac{df}{dt}(0) + 6(sF(s) - f(0)) + 8F(s) = \frac{4}{(s+1)}. \quad (1.40)$$

Substituting for the initial values, you get:

$$s^2F(s) - 2s - 3 + 6sF(s) - 12 + 8F(s) = \frac{4}{(s+1)}. \quad (1.41)$$

$$(s^2 + 6s + 8)F(s) = \frac{(2s+15)(s+1)+4}{(s+1)} = \frac{(2s^2+17s+19)}{(s+1)}. \quad (1.42)$$

$$F(s) = \frac{(2s^2+17s+19)}{(s^2+6s+8)(s+1)} = \frac{(2s^2+17s+19)}{(s+2)(s+4)(s+1)}. \quad (1.43)$$

Applying partial fraction expansion, you get:

$$F(s) = \frac{7/2}{s+2} + \frac{-17/6}{s+4} + \frac{4/3}{s+1}. \quad (1.44)$$

Taking the inverse Laplace transform using the Table 1.5, you get:

$$f(t) = \frac{-7e^{-2t}}{2} \frac{-17e^{-4t}}{6} + \frac{4e^{-t}}{3} \text{ for } t > 0. \quad (1.45)$$

It may be noted that the total solution is obtained in a single step while taking the initial conditions along the way.

1.7.3 Solution of Electrical Circuits Using the Laplace Transform

There are two ways to apply the Laplace transform for the solution of electrical circuits. In one method, the differential equations for the circuit are first obtained, and then the differential equations are solved using the Laplace transform. In the second method, the circuit elements are converted into s-domain functions and KCL and KVL are applied to the s-domain circuit to obtain the needed current or voltage in the s-domain. The current or voltage in time domain is obtained using the inverse Laplace transform. The second method is simpler and is illustrated here.

Let the Laplace transform of $\{v(t)\} = V(s)$ and Laplace transform of $\{i(t)\} = I(s)$. Then the s-domain voltage current relations of the R, L, and C elements are obtained as follows. Consider a resistor with the v–i relation:

$$v(t) = R\,i(t). \quad (1.46)$$

Taking the Laplace transform on both the sides yields:

$$V(s) = R\,I(s). \quad (1.47)$$

TABLE 1.5 Properties of Laplace Transforms

Operations	$f(t)$	$F(s)$
Addition	$f_1(t) + f_2(t)$	$F_1(s) + F_2(s)$
Scalar multiplication	$Af(t)$	$AF(s)$
Time differentiation	$d/dt\{f(t)\}$	$sF(s) - f(0)$
Time integration	$\int_0^\infty f(t)dt$	$\frac{F(s)}{s}$
Convolution	$\int_0^\infty f_1(t-\tau)f_2(\tau)d\tau$	$F_1(s)F_2(s)$
Frequency shift	$f(t)e^{-at}$	$F(s+a)$
Time shift	$f(t-a)u(t-a)$	$e^{-as}F(s)$
Frequency differentiation	$-tf(t)$	$\frac{dF}{ds}$
Frequency integration	$\frac{f(t)}{t}$	$\int_s^\infty F(s)ds$
Scaling	$f(at),\ a > 0$	$\frac{1}{a}F(\frac{s}{a})$
Initial value	$f(0^+)$	$\lim_{s\to\infty} sF(s)$
Final value	$f(\infty)$	$\lim_{s\to 0} sF(s)$

Note: The $u(t)$ is the unit step function defined by $u(t) = 0$ for $t < 0$ and $u(t) = 1$ for $t > 0$.

Defining the impedance of an element as $V(s)/I(s) = Z(s)$ produces $Z(s) = R$ for a resistance. For an inductance, $v(t) = L\ di/dt$. Taking the Laplace transform of the relation yields $V(s) = sL\,I(s) - Li(0)$, where $i(0)$ represents the initial current in the inductor and where $Z(s) = sL$ is the impedance of the inductance. For a capacitance, $i(t) = c\ dv/dt$ and $I(s) = sc\ V(s) - cv(0)$, where $v(0)$ represents the initial voltage across the capacitance and where $1/sc$ is the impedance of the capacitance.

Equivalent circuits that correspond to the s-domain relations for R, L, and C are shown in Table 1.6 and are suitable for writing KVL equations (initial condition as a voltage source) as well as for writing KCL equations (initial condition as a current source). With these equivalent circuits, a linear circuit can be converted to an s-domain circuit as shown in the example 1.8.

It is important first to show that the KCL and KVL relations can also be converted into s-domain relations. For example, the KCL relation in s-domain is obtained as follows: At any node, KCL states that:

$$i_1(t) + i_2(t) + i_3(t) + \ldots + i_n(t) = 0. \tag{1.48}$$

By applying Laplace transform on both sides, the result is:

$$I_1(s) + I_2(s) + I_3(s) + \ldots + I_n(s) = 0, \tag{1.49}$$

which is the KCL relation for s-domain currents in a node. In a similar manner, the KVL around a loop can be written in s-domain as:

$$V_1(s) + V_2(s) + \ldots + V_n(s) = 0, \tag{1.50}$$

where $V_1(s)$, $V_2(s)$, ..., $V_n(s)$ are the s-domain voltages around the loop. In fact, the various time-domain theorems considered earlier, such as the superposition, Thevenin, and Norton theorems, series and parallel equivalent circuits and voltage and current divisions are also valid in the s-domain. The loop current method and node voltage method can be applied for analysis in s-domain.

EXAMPLE 1.8. *Consider the circuit given in Figure 1.25(A) and convert a linear circuit into an s-domain circuit.* You can obtain the s-domain circuit shown in Figure 1.25(B) by replacing each element by its equivalent s-domain element. As noted previously, the differential relations of the elements on application of the Laplace transform have become algebraic relations. Applying KVL around the loop, you can obtain the following equations:

$$2I(s) + 0.2sI(s) - 0.4 + \frac{6}{s}I(s) + \frac{3}{s} = \frac{10}{s}. \tag{1.51}$$

$$(2 + 0.2s + \frac{6}{s})I(s) = \frac{10}{s} - \frac{3}{s} + 0.4 = \frac{7 + 0.4s}{s}. \tag{1.52}$$

$$I(s) = \frac{2s + 35}{s^2 + 10s + 30} = \frac{2(s+5) + 25}{(s+5)^2 + (\sqrt{5})^2}. \tag{1.53}$$

$$i(t) = e^{-5t}\{2\cos(\sqrt{5}t) + 5\sqrt{5}\sin(\sqrt{5}t)\} \quad \text{for } t > 0. \tag{1.54}$$

From 1.51 to 1.54 equations, you can see that the solution using the Laplace transform determines the natural response and forced response at the same time. In addition, the initial conditions are taken into account when the s-domain circuit is set up. A limitation of the Laplace transform is that it can be used only for linear circuits.

1.7.4 Network Functions

For a one-port network, voltage and current are the two variables associated with the input port, also called the driving port. One can define two driving point functions under zero initial conditions as:

$$\text{Driving point impedance } Z(s) = \frac{V(s)}{I(s)}.$$

$$\text{Driving point admittance } Y(s) = \frac{I(s)}{V(s)}.$$

In the case of two-port networks, one of the ports may be considered as an input port with input signal $X(s)$ and the other considered the output port with output signal $Y(s)$. Then the transfer function is defined as:

TABLE 1.6 *s*-Domain Equivalent Circuits for *R*, *I*, and *C* Elements

Time domain	Laplace domain KVL	Laplace domain KCL
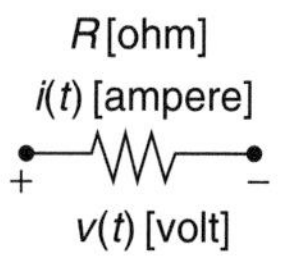 $v(t) = R\,i(t)$	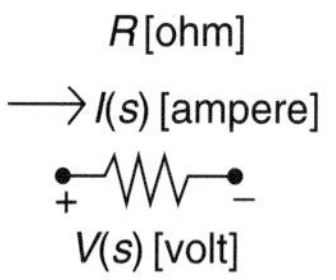 $V(s) = RI(s)$	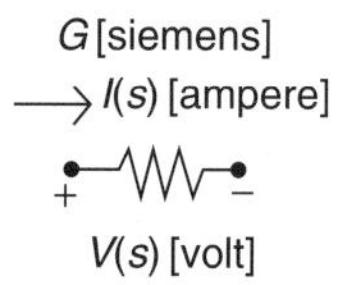 $I(s) = G\,V(s)$
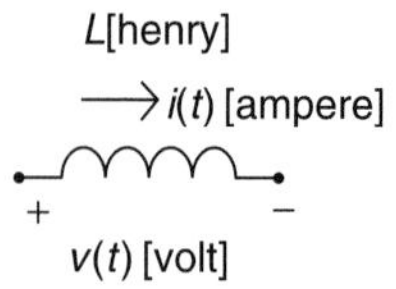 $v(t) = L\,di/dt$	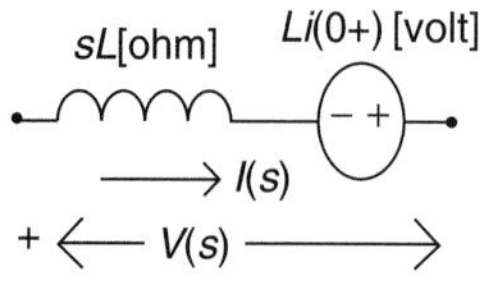 $V(s) = (sL)\,I(s) - Li(0+)$	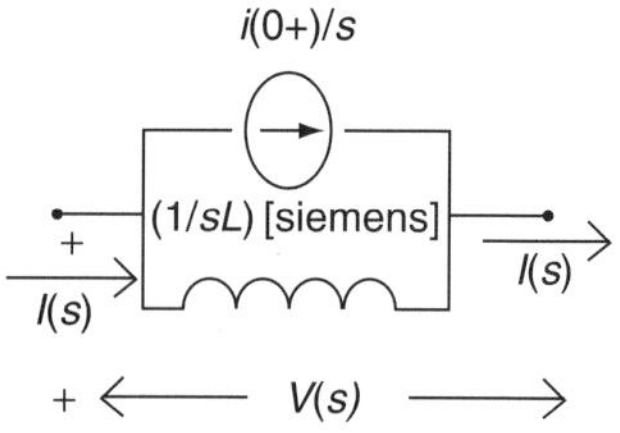 $I(s) = (1/sL)\,V(s) + [i(0+)/s]$
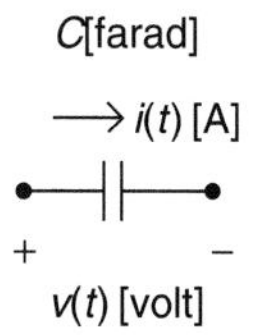 $i(t) = C\,dv/dt$	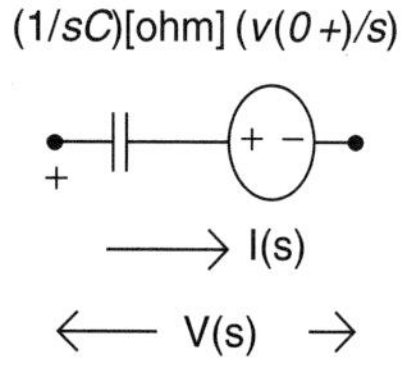 $V(s) = (1/sC)I(s) + (v(0+)/s)$	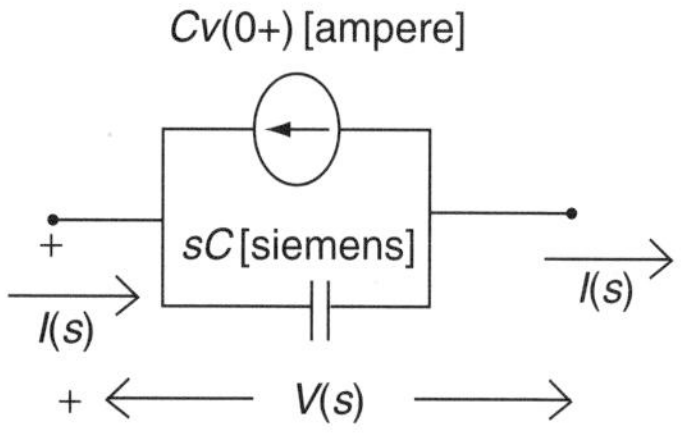 $I(s) = (sC)\,V(s) - Cv(0+)$

Note: [A] represents ampere, and [*V*] represents volt.

$$H(s) = \frac{Y(s)}{X(s)}, \text{ under zero initial conditions.}$$

In an electrical network, both $Y(s)$ and $X(s)$ can be either voltage or current variables. Four transfer functions can be defined as:

Transfer voltage ratio $G_v(s) = \dfrac{V_2(s)}{V_1(s)}$,
under the condition $I_2(s) = 0$.

Transfer current ratio $G_i(s) = \dfrac{I_2(s)}{I_1(s)}$,
under the condition $V_2(s) = 0$.

Transfer impedance $Z_{21} = \dfrac{V_2(s)}{I_1(s)}$,
under the condition $I_2(s) = 0$.

Transfer admittance $Y_{21} = \dfrac{I_2(s)}{V_1(s)}$,
under the condition $V_2(s) = 0$.

1.8 State Variable Analysis

State variable analysis or state space analysis, as it is sometimes called, is a matrix-based approach that is used for analysis of circuits containing time-varying elements as well as nonlinear elements. The state of a circuit or a system is defined as a set of a minimum number of variables associated with the circuit; knowledge of these variables along with the knowledge of the input will enable the prediction of the currents and voltages in all system elements at any future time.

1.8.1 State Variables for Electrical Circuits

As was mentioned earlier, only capacitors and inductors are capable of storing energy in a circuit, and so only the variables associated with them are able to influence the future condition of the circuit. The voltages across the capacitors and the currents through the inductors may serve as state variables. If loops are not solely made up of capacitors and voltage sources, then the voltages across all the capacitors are independent

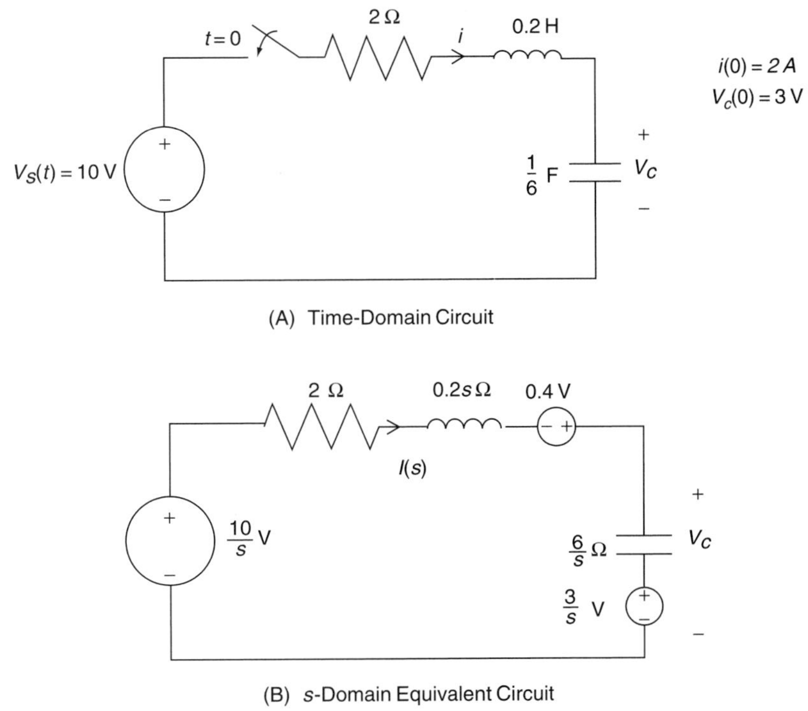

FIGURE 1.25 Circuit for Example 1.8

variables and may be taken as state variables. In a similar way, if there are no sets of inductors and current sources that separate the circuit into two or more parts, then the currents associated with the inductors are independent variables and may be taken as state variables. The following examples assume that all the capacitor voltages and inductor currents are independent variables and will form the set of state variables for the circuit.

1.8.2 Matrix Representation of State Variable Equations

Because matrix techniques are used in state variable analysis, the state variables are commonly expressed as a vector $\boldsymbol{x}$, and the input source variables are expressed as a vector $\boldsymbol{r}$. The output variables are denoted as $\boldsymbol{y}$.

Once the state variables $\boldsymbol{x}$ are chosen, KVL and KCL are used to determine the derivatives $\dot{x}$ of the state variables and the output variables y in terms of the state variables $\boldsymbol{x}$ and source variables $\boldsymbol{r}$. They are expressed as:

$$\begin{aligned} \dot{x} &= Ax + Br. \\ y &= Cx + Dr. \end{aligned} \tag{1.55}$$

The $\dot{x}$ equation is called the **state dynamics equation**, and the y equation is called the **output equation**. A, B, C, and D are appropriately dimensioned coefficient matrices. This set of equations, where the derivative of state variables is expressed as a linear combination of state variables and forcing functions, is said to be in **normal form**.

Example 1.9. *Consider the circuit shown in Figure 1.26 and solve for state variable equations in matrix form.*

Taking v_L and i_C as state variables and applying KVL around loop ABDA, you get:

$$\frac{di_L}{dt} = \frac{1}{L} v_L = \frac{1}{L}(v_a - v_C) = \frac{1}{L} v_a - \frac{1}{L} v_C. \tag{1.56}$$

Similarly, by applying KCL at node B, you get:

$$\frac{dv_C}{dt} = \frac{1}{C}(i_C) = \frac{1}{C}(i_L + i_1). \tag{1.57}$$

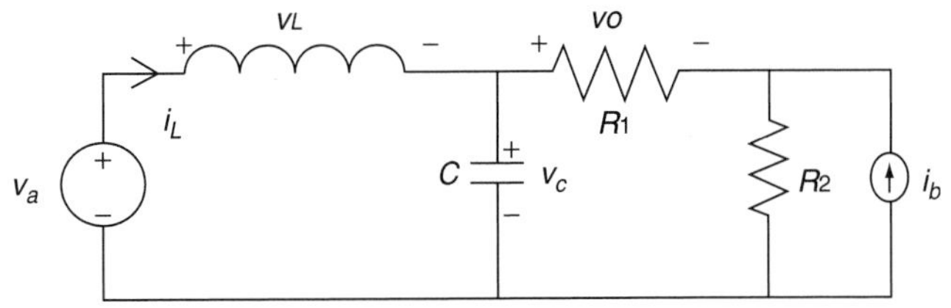

FIGURE 1.26 Circuit for Example 1.9

The current i_1 can be found either by writing node equation at node C or by applying the superposition as:

$$i_1 = \frac{R_2}{R_1 + R_2} i_b - \frac{1}{R_1 + R_2} v_c. \tag{1.58}$$

Substituting for i_1 in equation 1.57, you get:

$$\frac{dv_C}{dt} = \frac{1}{C} i_L + \frac{1}{C(R_1 + R_2)} i_b - \frac{1}{C(R_1 + R_2)} v_c. \tag{1.59}$$

The output v_o can be obtained as $-i_1 R_1$ and can be expressed in terms of state variables and sources by employing equation 1.58 as:

$$v_o = -\frac{R_1 R_2}{R_1 + R_2} i_b + \frac{R_1}{R_1 + R_2} v_c. \tag{1.60}$$

Equations 1.56, 1.59, and 1.60 can be expressed in matrix form as:

$$\begin{bmatrix} \frac{di_L}{dt} \\ \frac{dv_C}{dt} \end{bmatrix} = \begin{bmatrix} 0 & -\frac{1}{L} \\ \frac{1}{C} & \frac{1}{C(R_1 + R_2)} \end{bmatrix} \begin{bmatrix} i_L \\ v_C \end{bmatrix} + \begin{bmatrix} \frac{1}{L} & 0 \\ 0 & \frac{1}{C(R_1 + R_2)} \end{bmatrix} \begin{bmatrix} v_a \\ i_b \end{bmatrix}. \tag{1.61}$$

$$[v_o] = \begin{bmatrix} 0 & \frac{R_1}{R_1 + R_2} \end{bmatrix} \begin{bmatrix} i_L \\ v_C \end{bmatrix} + \begin{bmatrix} 0 & -\frac{R_1 R_2}{R_1 + R_2} \end{bmatrix} \begin{bmatrix} v_a \\ i_b \end{bmatrix}. \tag{1.62}$$

The ordering of the variables in the state variable vector $\boldsymbol{x}$ and the input vector $\boldsymbol{r}$ is arbitrary. Once the ordering is chosen, however, it must remain the same for all the variables in every place of occurrence. In large circuits, topological methods may be employed to systematically select the state variables and write KCL and KVL equations. For want of space, these methods are not described in this discussion. Next, this chapter briefly goes over the method for solving state variable equations.

1.8.3 Solution of State Variable Equations

There are many methods for solving state variable equations: (1) computer-based numerical solution, (2) conventional differential equation time-domain solution, and (3) Laplace transform s-domain solution. This chapter will only present the Laplace transform domain solution.

Consider the state dynamics equation:

$$\dot{x} = Ax + Br. \tag{1.63}$$

Taking the Laplace transform on both sides yields:

$$sX(s) - x(0) = AX(s) + BR(s), \tag{1.64}$$

where $X(s)$ and $R(s)$ are the Laplace transforms of $x(t)$ and $r(t)$, respectively, and $x(0)$ represents the initial conditions.

Rearranging equation 1.64 results in:

$$\begin{aligned} (sI - A)X(s) &= x(0) + BR(s) \\ X(s) &= (sI - A)^{-1}x(0) + (sI - A)^{-1}BR(s). \end{aligned} \tag{1.65}$$

Taking the inverse Laplace transform of $X(s)$ yields:

$$x(t) = \phi(t)x(0) + \phi(t) * Br(t), \tag{1.66}$$

where $\phi(t)$, the inverse Laplace transform of $\{(sI - A)^{-1}\}$, is called the state transition matrix and where * represents the time domain convolution.

Expanding the convolution, $x(t)$ can be written as:

$$x(t) = \phi(t)x(0) + \int_0^t \phi(t - \tau)Br(\tau)d\tau. \tag{1.67}$$

Once $x(t)$ is known, $y(t)$ may be found using the output equation 1.60.

1.9 Alternating Current Steady State Analysis

1.9.1 Sinusoidal Voltages and Currents

Standard forms of writing sinusoidal voltages and currents are:

$$v(t) = V_m \cos(\omega t + \alpha)[\mathrm{V}]. \tag{1.68a}$$

$$i(t) = I_m \cos(\omega t + \beta)[\mathrm{A}]. \tag{1.68b}$$

V_m and I_m are the maximum values of the voltage and current, ω is the frequency of the signal in radians/second, and α and β are called the phase angles of the voltage and current, respectively. V_m, I_m, and ω are positive real values, whereas α and β are real and can be positive or negative. If α is greater than β, the voltage is said to lead the current, or the current to lag the voltage. If α is less than β the voltage is said to lag the current, or the current to lead the voltage. If α equals β, the voltage and current are in phase.

1.9.2 Complex Exponential Function

Define $X_m e^{j(\omega t+\phi)}$ as a complex exponential function. By Euler's theorem,

$$e^{j\theta} = \cos\theta + j\sin\theta.$$
$$X_m e^{j(\omega t+\phi)} = X_m[\cos(\omega t+\phi) + j\sin(\omega t+\phi)].$$
$$x(t) = X_m \cos(\omega t+\phi).$$
$$= \text{Real}\ [X_m e^{j(\omega t+\phi)}] = \text{Real}\ [(X_m e^{j\phi})e^{j\omega t}]. \quad (1.69)$$

The term $(X_m e^{j\phi})$ is called the phasor of the sinusoidal function $x(t)$. For linear RLCM circuits, the forced response is sinusoidal at the input frequency. Since the natural response decays exponentially in time, the forced response is also the steady state response.

1.9.3 Phasors in Alternating Current Circuit Analysis

Consider voltage and current waves of the same frequency:

$$v(t) = V_m \cos(\omega t+\alpha)[\text{V}].$$
$$i(t) = I_m \cos(\omega t+\beta)[\text{A}].$$

Alternative representation is by complex exponentials:

$$v(t) = \text{Real}\ [V_m e^{j\alpha}]e^{j\omega t}[\text{V}].$$
$$i(t) = \text{Real}\ [I_m e^{j\beta}]e^{j\omega t}[\text{A}].$$

Phasor voltage and current are defined as:

$$\boldsymbol{V} = V_m e^{j\alpha}[\text{V}]. \quad (1.70\text{a})$$
$$\boldsymbol{I} = I_m e^{j\beta}[\text{A}]. \quad (1.70\text{b})$$

Since a phasor is a complex number, other representations of a complex number can be used to specify the phasor. These are listed in Table 1.7.

Addition of two voltages or two currents of the same frequency is accomplished by adding the corresponding phasors.

1.9.4 Phasor Diagrams

Since a phasor is a complex number, it can be represented on a complex plane as a *vector* in Cartesian coordinates. The length of the vector is the magnitude of the phasor, and the direction is the phase angle. The projection of the vector on the *x*-axis is the real part of the phasor, and the projection on the *y*-axis is the imaginary part of the phasor in rectangular form as noted in Figure 1.27. The graphical representation is called the phasor diagram and the vector is called the phasor.

TABLE 1.7 Representation of Phasor Voltages and Currents

Phasor	Exponential form	Rectangular form	Polar form
V	$V_m e j\alpha$	$V_m \cos\alpha + jV_m \sin\alpha$	$V_m \angle\alpha[\text{V}]$
I	$I_m e j\beta$	$I_m \cos\beta + jI_m \sin\beta$	$I_m \angle\beta[\text{A}]$

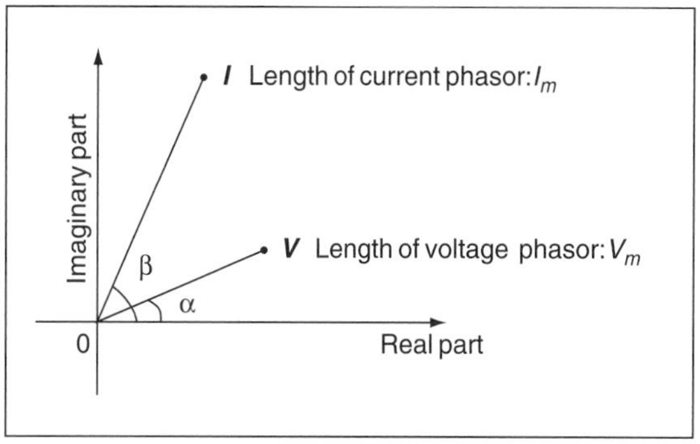

FIGURE 1.27 Phasor Diagram

1.9.5 Phasor Voltage–Current Relationships of Circuit Elements

Voltage $v(t)$ and current $i(t)$ are sinusoidal signals at a frequency of ω rad/s, whereas $\boldsymbol{V}$ and $\boldsymbol{I}$ are phasor voltage and current, respectively. The *v–i* and **V–I** relations for *R*, *L*, and *C* elements are given in Table 1.8.

1.9.6 Impedances and Admittances in Alternating Current Circuits

Impedance $\boldsymbol{Z}$ is defined as the ratio of phasor voltage to phasor current at a pair of terminals in a circuit. The unit of impedance is ohms. **Admittance** $\boldsymbol{Y}$ is defined as the ratio of phasor current to phasor voltage at a pair of terminals of a circuit. The unit of admittance is siemens. $\boldsymbol{Z}$ and $\boldsymbol{Y}$ are complex numbers and reciprocals of each other. Note that phasors are also complex numbers, but phasors represent time-varying sinusoids. Impedance and admittance are time invariant and frequency dependent. Table 1.9 shows the impedances and admittances for *R*, *L*, and *C* elements. It may be noted that the phasor impedance for an element is obtained by substituting $s = j\omega$ in the *s*-domain impedance $Z(s)$ of the corresponding element.

TABLE 1.8 Element Voltage–Current Relationships

Element	Time domain	Frequency domain
Resistance: *R*	$v = Ri$	$\boldsymbol{V} = R\boldsymbol{I}$
Inductance: *L*	$v = Ldi/dt$	$\boldsymbol{V} = j\omega L\boldsymbol{I}$
Capacitance: *L*	$i = Cdv/dt$	$\boldsymbol{I} = j\omega C\boldsymbol{V}$

TABLE 1.9 Impedances and Admittances of Circuit Elements

Element	Resistance: R	Inductor: L	Capacitor: C
Z [ohm]	R	$j X_L$ $X_L = \omega L$ (X_L is the inductive reactance.)	$j X_C$ $X_C = -1/\omega\, C$ (X_C is the capacitive reactance.)
Y [siemens]	$G = 1/R$ (G is the conductance.)	$j B_L$ $B_L = -1/\omega\, L$ (B_L is the inductive susceptance.)	$j B_C$ $B_C = \omega C$ (B_C is the capacitive susceptance.)

1.9.7 Series Impedances and Parallel Admittances

If n impedances are in series, their equivalent impedance is given by $Z_{eq} = (Z_1 + Z_2 + \ldots + Z_n)$. Similarly, the equivalent admittance of n admittances in parallel is given by $Y_{eq} = (Y_1 + Y_2 + \ldots + Y_n)$.

1.9.8 Alternating Current Circuit Analysis

Before the steady state analysis is made, a given time-domain circuit is replaced by its phasor-domain circuit, also called the **phasor circuit**. The phasor circuit of a given time-domain circuit at a specified frequency is obtained as follows:

- Voltages and currents are replaced by their corresponding phasors.
- Circuit elements are replaced by impedances or admittances at the specified frequency given in Table 1.9.

All circuit analysis techniques are now applicable to the phasor circuit.

1.9.9 Steps in the Analysis of Phasor Circuits

- Select mesh or nodal analysis for solving the phasor circuit.
- Mark phasor mesh currents or phasor nodal voltages.
- Use impedances for mesh analysis and admittances for nodal analysis.
- Write KVL around meshes (loops) or KCL at the nodes.
  KVL around a mesh: The algebraic sum of phasor voltage drops around a mesh is zero.
  KCL at a node: The algebraic sum of phasor currents leaving a node is zero.
- Solve the mesh or nodal equations with complex coefficients and obtain the complex phasor mesh currents or nodal voltages. The solution can be obtained by variable elimination or Cramer's rule. Remember that the arithmetic is complex number arithmetic.

1.9.10 Methods of Alternating Current Circuit Analysis

All methods of circuit analysis are applicable to alternating current (ac) phasor circuits. Phasor voltages and currents are the variables. Impedances and admittances describe the element voltage–current relationships. Some of the methods are described in this section.

Method of superposition: Circuits with multiple sources of the same frequency can be solved by using the mesh or nodal analysis method on the phasor circuit at the source frequency. Alternatively, the principle of superposition in linear circuits can be applied. First, solve the phasor circuit for each independent source separately. Then add the response voltages and currents from each source to get the total response. Since the responses are at the same frequency, phasor addition is valid.

Voltage and current source equivalence in ac circuits:
An ac voltage source in series with an impedance can be replaced across the same terminals by an equivalent ac current source of the same frequency in parallel with an admittance, as shown in Figure 1.28. Similarly, an ac current source in parallel with an admittance can be replaced across the same terminals by an equivalent ac voltage source of the same frequency in series with an impedance.

Current and voltage division in ac circuits:
For two impedances in series:

$$V_1 = (\frac{Z_1}{Z_1 + Z_2})V, \quad V_2 = (\frac{Z_2}{Z_1 + Z_2})V$$

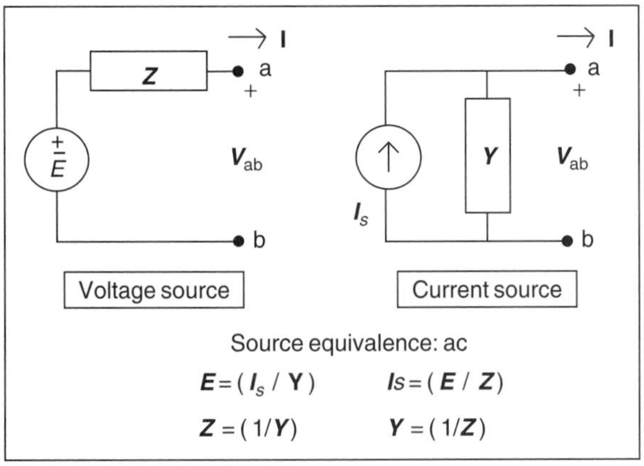

FIGURE 1.28 Source Transformations

For n impedances in series

$$\mathbf{V}_1 = \left(\frac{\mathbf{Z}_1}{\sum_{i=1}^{n} \mathbf{Z}_i}\right)\mathbf{V}, \quad \mathbf{V}_2 = \left(\frac{\mathbf{Z}_2}{\sum_{i=1}^{n} \mathbf{Z}_i}\right)\mathbf{V}, \ldots, \mathbf{V}_n = \left(\frac{\mathbf{Z}_n}{\sum_{i=1}^{n} \mathbf{Z}_i}\right)\mathbf{V}.$$

For two admittances in parallel

$$\mathbf{I}_1 = \left(\frac{\mathbf{Y}_1}{\mathbf{Y}_1 + \mathbf{Y}_2}\right)\mathbf{I}, \quad \mathbf{I}_2 = \left(\frac{\mathbf{Y}_2}{\mathbf{Y}_1 + \mathbf{Y}_2}\right)\mathbf{I}$$

For n admittances in parallel:

$$\mathbf{I}_1 = \left(\frac{\mathbf{Y}_1}{\sum_{i=1}^{n} \mathbf{Y}_i}\right)\mathbf{I}, \quad \mathbf{I}_2 = \left(\frac{\mathbf{Y}_2}{\sum_{i=1}^{n} \mathbf{Y}_i}\right)\mathbf{I}, \ldots, \mathbf{I}_n = \left(\frac{\mathbf{Y}_n}{\sum_{i=1}^{n} \mathbf{Y}_i}\right)\mathbf{I}.$$

Thevenin theorem for ac circuits: A phasor circuit across a pair of terminals is equivalent to an ideal phasor voltage source $\mathbf{V}_{oc}$ in series with an impedance $\mathbf{Z}_{Th}$, where $\mathbf{V}_{oc}$ is the open circuit voltage across the terminals and where $\mathbf{Z}_{Th}$ is the equivalent impedance of the circuit across the specified terminals.

Norton theorem for ac circuits: A phasor circuit across a pair of terminals is equivalent to an ideal phasor current source $\mathbf{I}_{sc}$ in parallel with an admittance $\mathbf{Y}_N$, where $\mathbf{I}_{sc}$ is the short circuit current across the terminals and where $\mathbf{Y}_N$ is the equivalent admittance of the circuit across the terminals.

Thevenin and Norton equivalent circuits of a linear phasor circuit are shown in Figure 1.29.

1.9.11 Frequency Response Characteristics

The voltage gain G_v at a frequency ω of a two-port network is defined as the ratio of the output voltage phasor to the input voltage phasor. In a similar manner, the current gain G_i is defined as the ratio of output current phasor to input current phasor. Because the phasors are complex quantities that depend on frequency, the gains, voltages, and currents are written as $G(j\omega)$, $V(j\omega)$, and $I(j\omega)$. Frequency response is then defined as the variation of the gain as a function of frequency. The above gain functions are also called transfer functions and written as $H(j\omega)$. $H(j\omega)$ is a complex number and can be written in polar form as follows:

$$H(j\omega) = A(\omega)/\phi(\omega) \tag{1.71}$$

where $A(\omega)$ is the gain magnitude function, $|H(j\omega)|$, and $\phi(\omega)$ is the phase function given by the argument of $H(j\omega)$. In addition to the above gain functions, a number of other useful ratios (network functions) among the voltages and currents of two port networks can be defined. These definitions and their nomenclature are given in Figure 1.30.

1.9.12 Bode Diagrams

Bode diagrams are graphical representations of the frequency responses and are used in solving design problems. Magnitude and phase functions are shown on separate graphs using logarithmic frequency scale along the x-axis. Logarithm of the frequency to base 10 is used for the x-axis of a graph. Zero frequency will correspond to negative infinity on the logarithmic scale and will not show on the plots. The x-axis is graduated in $\log_{10}\omega$ and so every decade of frequency (e.g., ... 0.001, 0.01, 0.1, 1, 10, 100, ...) is equally spaced on the x-axis.

The gain magnitude, represented by decibels defined as $20 \log_{10}[A(\omega)]$, is plotted on the y-axis of magnitude plot. Since $A(\omega)$ dB can be both positive and negative, the y-axis has both positive and negative values. Zero dB corresponds to a magnitude function of unity. The y-axis for the phase function uses a linear scale in radians or degrees. Semilog graph paper makes it convenient to sketch Bode plots.

Bode plots are easily sketched by making asymptotic approximations first. The frequency response function $H(j\omega)$ is a rational function, and the numerator and denominator are factorized into first-order terms and second-order terms with complex roots. The factors are then written in the standard form as follows:

$$\text{First-order terms: } (j\omega + \omega_0) \rightarrow \omega_0\left(1 + j\frac{\omega}{\omega_0}\right).$$

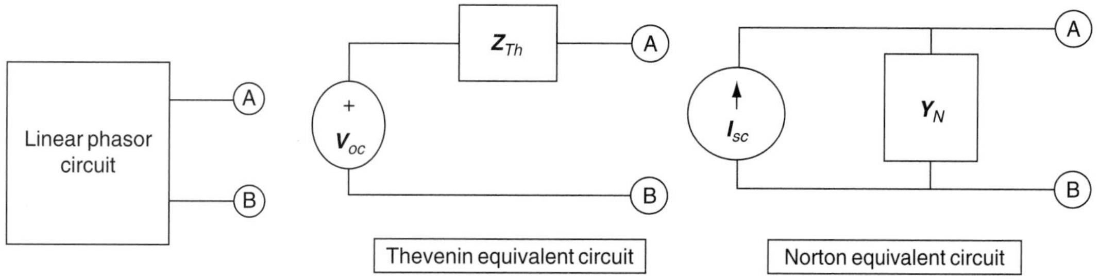

FIGURE 1.29 Thevenin and Norton Equivalent Circuits

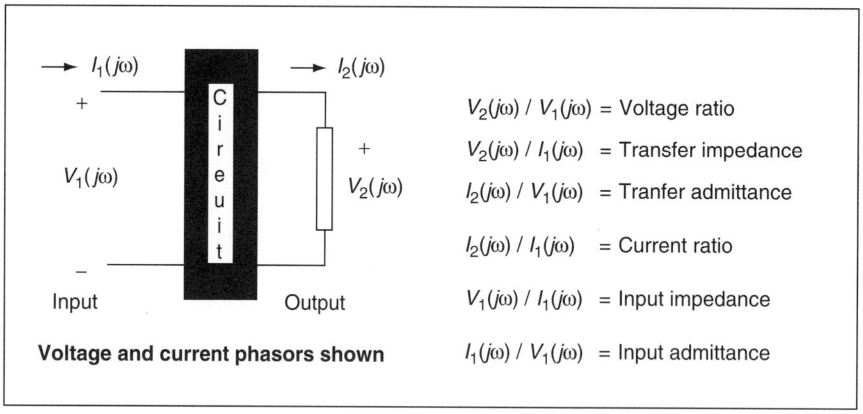

FIGURE 1.30 Frequency Response Functions
Notes: The frequency response ratios are called network functions.
When the input and output are at different terminals, frequency response function is also called a transfer function.

Second-order terms: $[(\omega_0^2 - \omega^2) + j(2\zeta\omega_0\omega)]$

$$\rightarrow \omega_0^2\left[\left(1 - \left(\frac{\omega}{\omega_0}\right)^2\right) + j\left(2\zeta\left(\frac{\omega}{\omega_0}\right)\right)\right].$$

Here ζ is the damping ratio with a value of less than 1. The magnitude and phase Bode diagrams are drawn by adding the individual plots of the first- and second-order terms. Asymptotic plots are first easily sketched by using approximations. For making asymptotic Bode plots, the ratio ω/ω_0 is assumed to be much smaller than one or much larger than one, so the Bode plots become straight line segments called asymptotic approximations. The normalizing frequency ω_0 is called the corner frequency. The asymptotic approximations are corrected at the corner frequencies by calculating the exact value of magnitude and phase functions at the corner frequencies.

The first- and second-order terms can occur in the numerator or denominator of the rational function $H(j\omega)$. Normalized plots for these terms are shown in figures 1.31 and 1.32. Bode diagrams for any function can be made using the normalized plots as building blocks. Figure 1.31 shows the Bode diagrams for a first-order term based on the following equations:

$$\text{(i) } H(j\omega) = \left(1 + j\frac{\omega}{\omega_0}\right). \quad \text{(ii) } H(j\omega) = \left(1 + j\frac{\omega}{\omega_0}\right)^{-1}.$$

The magnitude and phase plots of a denominator second-order term with complex roots are given in Figure 1.32. If the term is in the numerator, the figures are flipped about the x-axis, and the sign of the y-axis calibration is reversed.

1.10 Alternating Current Steady State Power

1.10.1 Power and Energy

Power is the rate at which energy E is transferred, such as in this equation:

$$p(t) = \frac{dE}{dt}$$

Energy is measured in Joules [J]. A unit of power is measured in watts [W]: 1W = 1J/s. When energy is transferred at a constant rate, power is constant. In general, energy transferred over time t is the integral of power.

$$E = \int_0^T p(t)dt$$

1.10.2 Power in Electrical Circuits

Power in a two-terminal circuit at any instant is obtained by multiplying the voltage across the terminals by the current through the terminals:

$$p(t) = v(t)i(t). \tag{1.72}$$

If the voltage is in volts [V] and the current in amperes [A], power is in watts. In direct current (dc) circuits under steady state, the voltage V and current I are constant and power is also constant, given by $P = VI$. In ac circuits under steady state, V

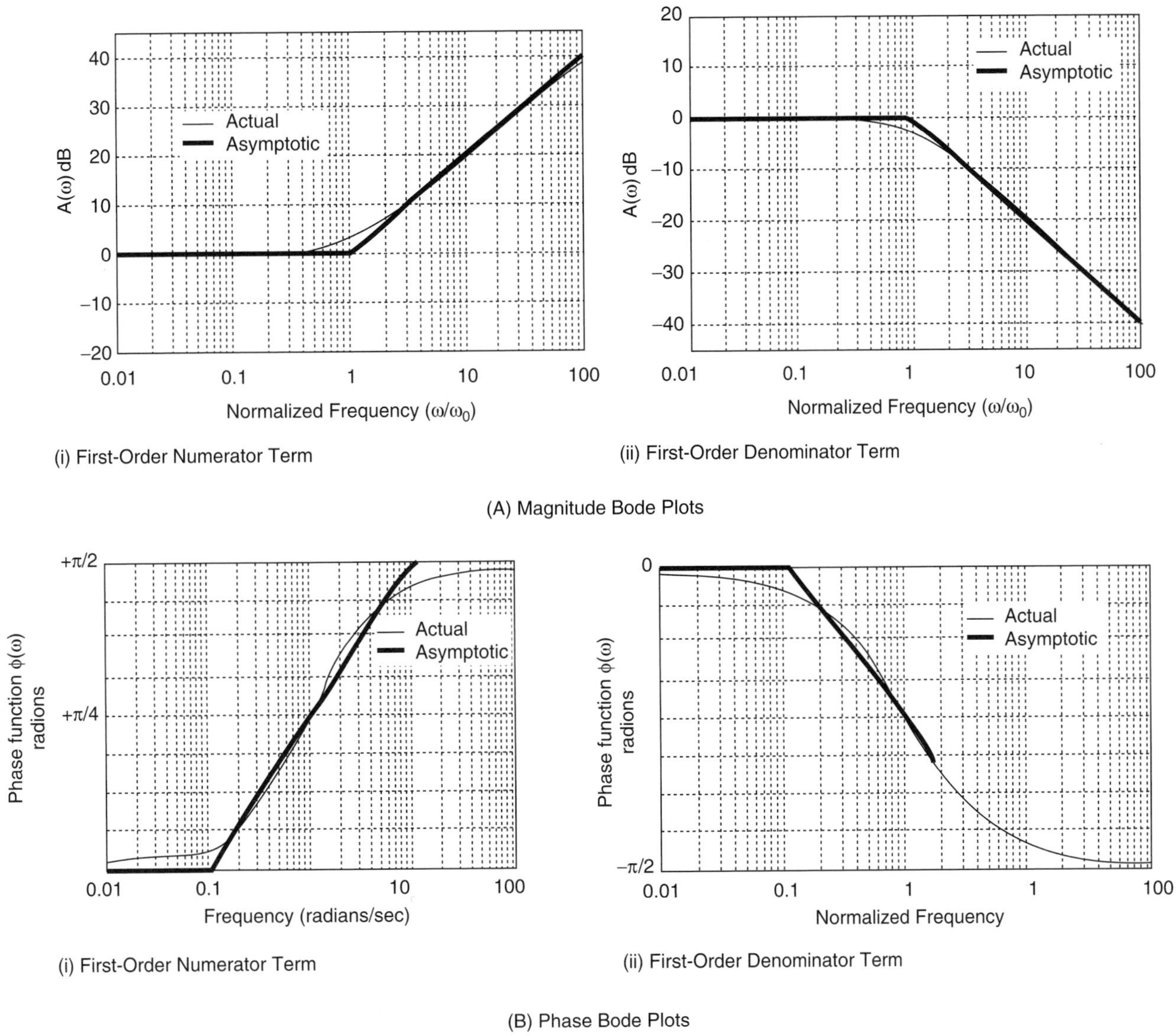

FIGURE 1.31 Bode Diagrams of First-Order Terms

and I are sinusoidal functions of the same frequency. Power in ac circuits is a function of time. For $v(t) = V_m \cos(\omega t + \alpha)$ [V] and $i(t) = I_m \cos(\omega t + \beta)$[A],

$$\begin{aligned} p(t) &= [V_m \cos(\omega t + \alpha)][I_m \cos(\omega t + \beta)] \\ &= 0.5V_m I_m\{\cos(\alpha - \beta) + \cos(2\omega t + \alpha + \beta)\} \end{aligned} \tag{1.73}$$

Average power P is defined as the energy transferred per second and can be calculated by integrating $p(t)$ for 1s. Since the voltage and current are periodic signals of the same frequency, the average power is the energy per cycle multiplied by the frequency f. Energy per cycle is calculated by integrating $p(t)$ over one cycle, that is, one period of the voltage or current wave.

$$\begin{aligned} P &= f\left[\int_0^T p(t)dt\right] = \frac{1}{T}\left[\int_0^T p(t)dt\right] \\ &= \frac{1}{T}\int_0^T 0.5V_m I_m\{\cos(\alpha - \beta) + \cos(2\omega t + \alpha + \beta)\}dt. \end{aligned}$$

Evaluating the integral yields average power:

$$P = 0.5V_m I_m \cos(\alpha - \beta)[\mathrm{W}]. \tag{1.74}$$

The $p(t)$ is referred to as the **instantaneous power**, and P is the average power or **ac power**. When a sinusoidal voltage of peak value of V_m [V] is applied to a 1Ω resistor, the average power

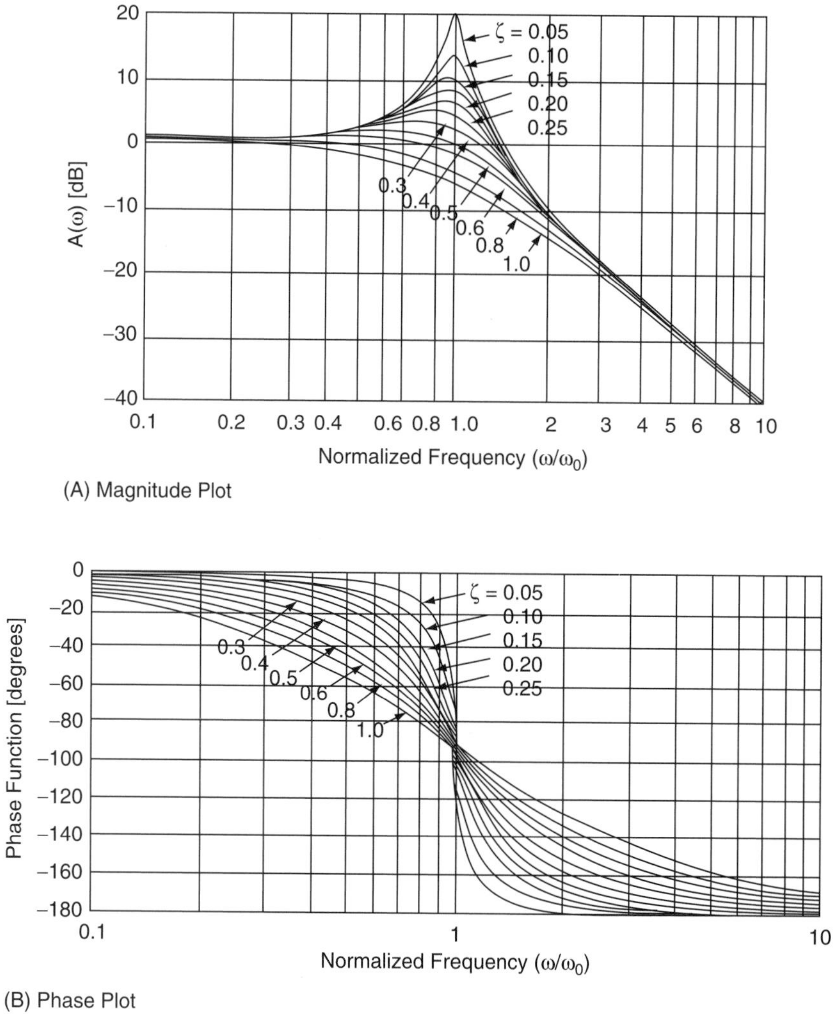

FIGURE 1.32 Bode Diagrams of $H(j\omega) = [(\omega_0^2 - \omega^2) + j(2\zeta\omega_0\omega)]^{-1}$

dissipated in the resistor is 0.5 V_m^2 [W]. Similarly, when a sinusoidal current of I_m [A] passes through a one ohm resistor, the average power dissipated in the resistor is 0.5 I_m^2 [W].

Root mean square (RMS) or effective value of an ac voltage or current is defined as the equivalent dc voltage or current that will dissipate the same amount of power in a $1-\Omega$ resistor.

$$\begin{aligned} V_{RMS} &= \sqrt{0.5}V_m = 0.707V_m. \\ I_{RMS} &= \sqrt{0.5}V_m = 0.707I_m. \end{aligned} \tag{1.75}$$

For determining whether the power is dissipated (consumed) or supplied (delivered), source or load power conventions are used. In a load, the current flowing into the positive terminal of the voltage is taken in calculating the dissipated power. For a source, the current leaving the positive terminal of the voltage is used for evaluating the supplied power.

1.10.3 Power Calculations in AC Circuits

The average power of a resistance R in an ac circuit is obtained as $0.5\ \mathrm{V_m^2/R}$ [W] or $0.5\,\mathrm{R\ I_m^2}$ [W], with V_m the peak voltage across the resistance and I_m the peak current in the resistance. The average power in an inductor or a capacitor in an ac circuit is zero.

Phasor voltages and currents are used in ac calculations. Average power in ac circuits can be expressed in terms of phasor voltage and current. Using the effective values of V and I for the phasors, the following definitions are given:

Using the notation $\boldsymbol{V} = V\angle\alpha$ [V], $\boldsymbol{I} = I\angle\beta$ [A] and conjugate $\boldsymbol{I}$, denoted by $\boldsymbol{I}^* = \boldsymbol{I}\angle-\beta$ [A], you get:

TABLE 1.10 AC Power in Circuit Elements

Element	Voltage–current relationship	Complex power: $S = VI^*$	Average power: P	Reactive power: Q	Power factor
Resistance R [ohm]	$\mathbf{V} = R\mathbf{I}$	$V\,I/0^0$	I^2R	0	Unity
Inductive reactance $X_L = \omega L$ [ohm]	$\mathbf{V} = jX_L\mathbf{I}$	$V\,I/90^0$	0	I^2X_L	Zero-lagging
Capacitive susceptance $B_C = \omega C$ [siemens]	$\mathbf{I} = jB_C\mathbf{V}$	$V\,I/-90^0$	0	$-V^2B_C$	Zero-leading
Impedance $\mathbf{Z} = (R + jX) = Z/\theta$ [ohm]	$\mathbf{V} = \mathbf{Z}\,\mathbf{I}$	$V\,I/\theta = I^2Z$	I^2R	I^2X	$\cos\theta$ $\theta > 0$ Lagging $\theta < 0$ Leading
Admittance $\mathbf{Y} = (G + jB) = Y/\phi$ [siemens]	$\mathbf{I} = \mathbf{Y}\,\mathbf{V}$	$V\,I/-\phi = V^2\mathbf{Y}$	V^2G	V^2B	$\cos\phi$ $\phi > 0$ Leading $\phi < 0$ Lagging

Notes: Bold letters refer to phasors or complex numbers. Units: Voltage in volts, current in amperes, average power in watts, reactive power in volt ampere reactive, and complex power in voltampere

$$\text{Average power } P = V\,I\cos(\alpha - \beta) = \text{Real}\{\mathbf{V}\,\mathbf{I}^*\}[\text{W}] \quad (1.76a)$$

$$\text{Reactive power } Q = V\,I\sin(\alpha - \beta) = \text{Imaginary}\{\mathbf{V}\,\mathbf{I}^*\}[\text{VAR}] \quad (1.76b)$$

$$\text{Apparent power } S = V\,I[\text{VA}] \quad (1.76c)$$

$$\text{Complex power } S = (P + jQ) = \{\mathbf{V}\,\mathbf{I}^*\}[\text{VA}] \quad (1.76d)$$

In the set of equations [W] stands for watts, [VAR] for volt-ampere reactive, and [VA] for voltampere. Average power is also called **active power**, **real power** or simply **power**. Reactive power is also referred to as **imaginary power**. Reactive power is a useful concept in power systems because the system voltage is affected by the reactive power flow. The average power in an inductor or capacitor is zero. By definition, the reactive power taken by an inductor is positive, and the reactive power taken by a capacitor is negative. Complex power representation is useful in calculating the power supplied by the source to a number of loads connected in the system.

Power factor is defined as the ratio of the average power in an ac circuit to the apparent power, which is the product of the voltage and current magnitudes.

$$\text{power factor} = \frac{\text{average power}}{\text{apparent power}} = \frac{P}{S} \quad (1.77)$$

Power factor (PF) has a value between zero and *unity*. The nature of the power factor depends on the relationship between the current and voltage phase angles as:

$(\alpha - \beta) > 0$ PF is lagging.

$(\alpha - \beta) = 0$ PF is unity (UPF).

$(\alpha - \beta) < 0$ PF is leading.

$(\alpha - \beta) = \pi/2$ PF is zero-lagging.

$(\alpha - \beta) = -\pi/2$ PF is zero-leading.

As reactive power $Q = V\,I\,\sin(\alpha - \beta)$,

its sign depends on the nature of PF:
Q is positive for lagging PF.
Q is negative for leading PF.
Q is zero for UPF.

The PF of an inductive load is lagging and that of a capacitive load is leading. A pure inductor has zero-lagging power factor and absorbs positive reactive power. A pure capacitor has zero-leading power factor and absorbs negative reactive power or delivers positive reactive power. Table 1.10 summarizes the expressions for various power quantities in ac circuit elements. Loads are usually specified in terms of P and PF at a rated voltage. Examples are motors and household appliances. Alternatively, the apparent P and PF can be specified.

PF plays an important role in power systems. The product of V and I is called the apparent power. The investment cost of a utility depends on the voltage level and the current carried by the conductors. Higher current needs larger, more expensive conductors. Higher voltage means more insulation costs. The revenue of a utility is generally based on the amount of energy in Kilowatt per hour sold. At low power factors, the revenue is low since power P and energy sold is less. For getting full benefit of investment, the utility would like to sell the highest possible energy, that is, operate at unity power factor all the time. Utilities can have a tariff structure that penalizes the customer for low power factor. Most electromagnetic equipment such as motors have low lagging PF and absorb large reactive powers. By connecting capacitors across the terminals of the motor, part or all of the reactive power absorbed by the motor can be supplied by the capacitors. The reactive power from the supply will be reduced, and the supply PF improved. Cost of the capacitors is balanced against the savings accrued due to PF improvement.

2

Circuit Analysis: A Graph-Theoretic Foundation

Krishnaiyan Thulasiraman
School of Computer Science, University of Oklahoma, Norman, Oklahoma, USA

M.N.S. Swamy
Department of Electrical and Computer Engineering, Concordia University, Montreal, Quebec, Canada

2.1 Introduction

The theory of graphs has played a fundamental role in discovering structural properties of electrical circuits. This should not be surprising because graphs, as the reader shall soon see, are good pictorial representations of circuits and capture all their structural characteristics. This chapter develops most results that form the foundation of graph theoretic study of electrical circuits. A comprehensive treatment of these developments may be found in Swamy and Thulasiraman (1981). All theorems in this chapter are stated without proofs.

The development of graph theory in this chapter is self-contained except for the definitions of standard and elementary results from set theory and matrix theory.

2.2 Basic Concepts and Results

2.2.1 Graphs

A graph $G = (V, E)$ consists of two sets: a finite set $V = (v_1, v_2, \ldots v_n)$ of elements called **vertices** and a finite set $E = (e_1, e_2, \ldots e_n)$ of elements called **edges**. If the edges of G are identified with ordered pairs of vertices, then G is called a **directed** or an **oriented graph**; otherwise, it is called an **undirected** or an **unoriented graph**.

Graphs permit easy pictorial representations. In a pictorial representation, each vertex is represented by a dot, and each edge is represented by a line joining the dots associated with the edge. In directed graphs, an orientation or direction is assigned to each edge. If the edge is associated with the ordered pair (v_i, v_j), then this edge is oriented from v_i to v_j. If an edge e connects vertices v_i and v_j, then it is denoted by $e = (v_i, v_j)$. In a directed graph, (v_i, v_j) refers to an edge directed from v_i to v_j. An undirected graph and a directed graph are shown in Figure 2.1. Unless explicitly stated, the term **graph** may refer to a directed graph or an undirected graph.

The vertices v_i and v_j associated with an edge are called the **end vertices** of the edge. All edges having the same pair of end vertices are called **parallel edges.**

In a directed graph, parallel edges refer to edges connecting the same pair of vertices v_i and v_j the same way from v_i to v_j or from v_j to v_i. For instance, in the graph of Figure 2.1(A), the edges connecting v_1 and v_2 are parallel edges. In the directed graph of Figure 2.1(B), the edges connecting v_3 and v_4 are parallel edges. However, the edges connecting v_1 and v_2 are not parallel edges because they are not oriented in the same way.

If the end vertices of an edge are not distinct, then the edge is called a **self-loop**. The graph of Figure 2.1(A) has one self loop, and the graph of Figure 2.1(B) has two self-loops. An edge is said to be **incident on** its end vertices. In a directed

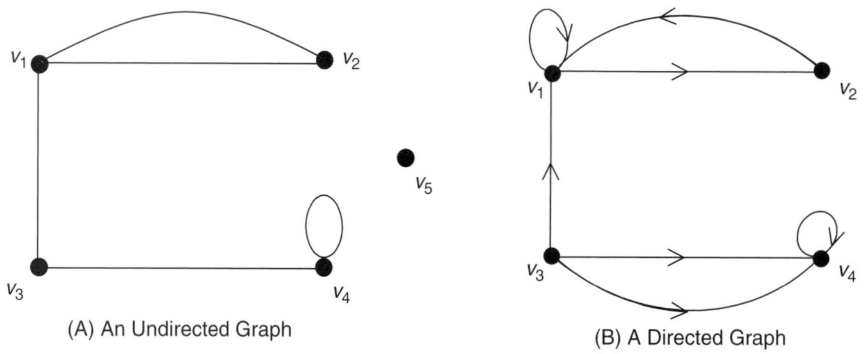

FIGURE 2.1 Graphs: Easy Pictorial Representations

graph, the edge (v_i, v_j) is said to be **incident out** of v_i and is said to be **incident into** v_j. Vertices v_i and v_j are adjacent if an edge connects v_i and v_j.

The number of edges incident on a vertex v_i is called the **degree** of v_i and is denoted by $d(v_i)$. In a directed graph, $d_{in}(v_i)$ refers to the number of edges incident into the vertex v_i, and it is called the **in-degree**. In a directed graph, $d_{out}(v_i)$ refers to the number of edges incident out of the vertex v_i or the **out-degree**. If $d(v_i) = 0$, then v_i is said to be an **isolated vertex**. If $d(v_i) = 1$, then v_i is said to be a **pendant vertex**.

A self-loop at a vertex v_i is counted twice while computing $d(v_i)$. As an example in the graph of Figure 2.1(A), $d(v_1) = 3$, $d(v_4) = 3$, and v_5 is an isolated vertex. In the directed graph of Figure 2.1(B), $d_{in}(v_1) = 3$ and $d_{out}(v_1) = 2$.

Note that in a directed graph, for every vertex v_i,

$$d(v_i) = d_{in}(v_i) + d_{out}(v_i).$$

2.2.2 Basic Theorems

Theorem 2.1: (1) The sum of the degrees of the vertices of a graph G is equal to $2\,m$, where m is the number of edges of G, and (2) in a directed graph with m edges, the sum of the in-degrees and the sum of out-degrees are both equal to m.

The following theorem is known to be the first major result in graph theory.

Theorem 2.2: The number of vertices of odd degree in any graph is even.

Consider a graph $G = (V', E')$. The graph $G' = (V', E')$ is a subgraph of G if $V' \subseteq V$ and $E' \subseteq E$. As an example, a graph G and a subgraph of G are shown in Figure 2.2.

In a graph G, a **path P** connecting vertices v_i and v_j is an alternating sequence of vertices and edges starting at v_i and ending at v_j, with all vertices except v_i and v_j being distinct. In a directed graph, a path P connecting vertices v_i and v_j is called a **directed path** from v_i to v_j if all the edges in P are oriented in the same direction when traversing P from v_i toward v_j.

If a path starts and ends at the same vertex, it is called a **circuit**. In a directed graph, a circuit in which all the edges are oriented in the same direction is called a **directed circuit**. It is often convenient to represent paths and circuits by the sequence of edges representing them.

For example, in the undirected graph of Figure 2.3(A), P: e_1, e_2, e_3, e_4 is a path connecting v_1 and v_5, and C: e_1,

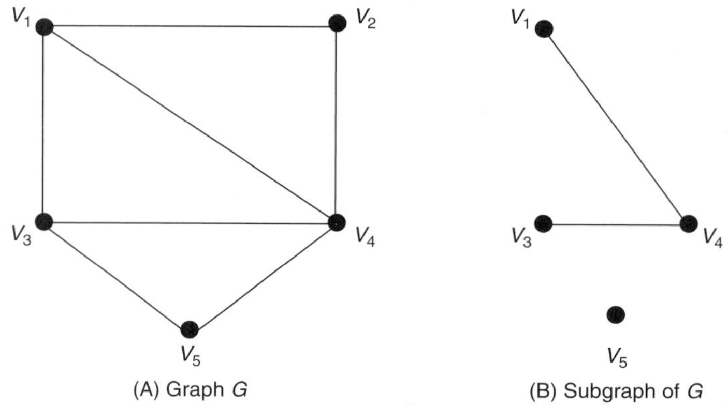

FIGURE 2.2 Graphs and Subgraphs

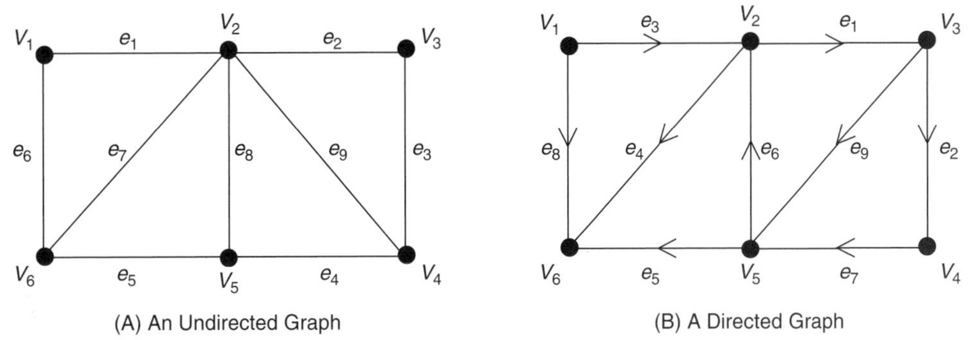

FIGURE 2.3 Graphs Showing Paths

e_2, e_3, e_4, e_5, e_6 is a circuit. In the directed graph of Figure 2.3(B) P: e_1, e_2, e_7, e_5 is a directed path and C: e_1, e_2, e_7, e_6 is a directed circuit. Note that C: e_7, e_5, e_4, e_1, e_2 is a circuit in this directed graph, although it is not a directed circuit. Similarly, P: e_9, e_6, e_3 is a path but not a directed path.

A graph is **connected** if there is a path between every pair of vertices in the graph; otherwise, the graph is not connected. For example, the graph in Figure 2.2(A) is a connected graph, whereas the graph in Figure 2.2(B) is not a connected graph.

A **tree** is a graph that is connected and has no circuits. Consider a connected graph G. A subgraph of G is a **spanning tree** of G if the subgraph is a tree and contains all the vertices of G. A tree and a spanning tree of the graph of Figure 2.4(A) are shown in Figures 2.4(B) and (C), respectively. The edges of a spanning tree T are called the **branches** of T. Given a spanning tree of connected graph G, the **cospanning tree** relative to T is the subgraph of G induced by the edges that are not present in T. For example, the cospanning tree relative to the spanning tree T of Figure 2.4(C) consists of these edges: e_3, e_6, e_7. The edges of a cospanning tree are called **chords**.

It can be easily verified that in a tree, exactly one path connects any two vertices. It should be noted that a tree is minimally connected in the sense that removing any edge from the tree will result in a disconnected graph.

Theorem 2.3: A tree on n vertices has $n - 1$ edges.

If a connected graph G has n vertices and m edges, then the **rank** ρ and **nullity** μ of G are defined as follows:

$$\rho(G) = n - 1.$$
$$\mu(G) = m - n + 1.$$

The concepts of rank and nullity have parallels in other branches of mathematics such as matrix theory.

Clearly, if G is connected, then any spanning tree of G has $\rho = n - 1$ branches and

$$\mu = m - n + 1 \text{ chords.}$$

2.2.3 Cuts, Circuits, and Orthogonality

Introduced here are the notions of a cut and a cutset. This section develops certain results that bring out the dual nature of circuits and cutsets.

Consider a connected graph $G = (V, E)$ with n vertices and m edges. Let V_1 and V_2 be two mutually disjoint nonempty subsets of V such that $V = V_1 \cup V_2$. Thus V_2 is the complement of V_1 in V and vice versa. V_1 and V_2 are also said to form a partition of V. Then the set of all those edges that have one end vertex in V_1 and the other in V_2 is called a **cut** of G and is denoted by $< V_1, V_2 >$. As an example, a graph and a cut $< V_1, V_2 >$ of G are shown in Figure 2.5.

The graph G, which results after removing the edges in a cut, will not be connected. A **cutset** S of a connected graph G is a minimal set of edges of G, such that removal of S disconnects G. Thus a cutset is also a cut. Note that the minimality property of a cutset implies that no proper subset of a cutset is a cutset.

Consider a spanning tree T of a connected graph G. Let b denote a branch of T. Removal of branch b disconnects T into two trees, T_1 and T_2. Let V_1 and V_2 denote the vertex sets of T_1 and T_2, respectively. Note that V_1 and V_2 together contain all the vertices of G. It is possible to verify that the cut $< V_1, V_2 >$ is a cutset of G and is called the **fundamental cutset** of G with respect to branch b of T. Thus, for a given graph G and a spanning tree T of G, we can construct $n - 1$ fundamental cutsets, one for each branch of T. As an example, for the graph shown in Figure 2.5, the fundamental cutsets with respect to the spanning tree $T = [e_1, e_2, e_6, e_8]$ are the following:

Branch e_1: (e_1, e_3, e_4).
Branch e_2: (e_2, e_3, e_4, e_5).
Branch e_6: (e_6, e_4, e_5, e_7).
Branch e_8: (e_8, e_7).

Note that the fundamental cutset with respect to branch b contains b. Furthermore branch b is not present in any other fundamental cutset with respect to T.

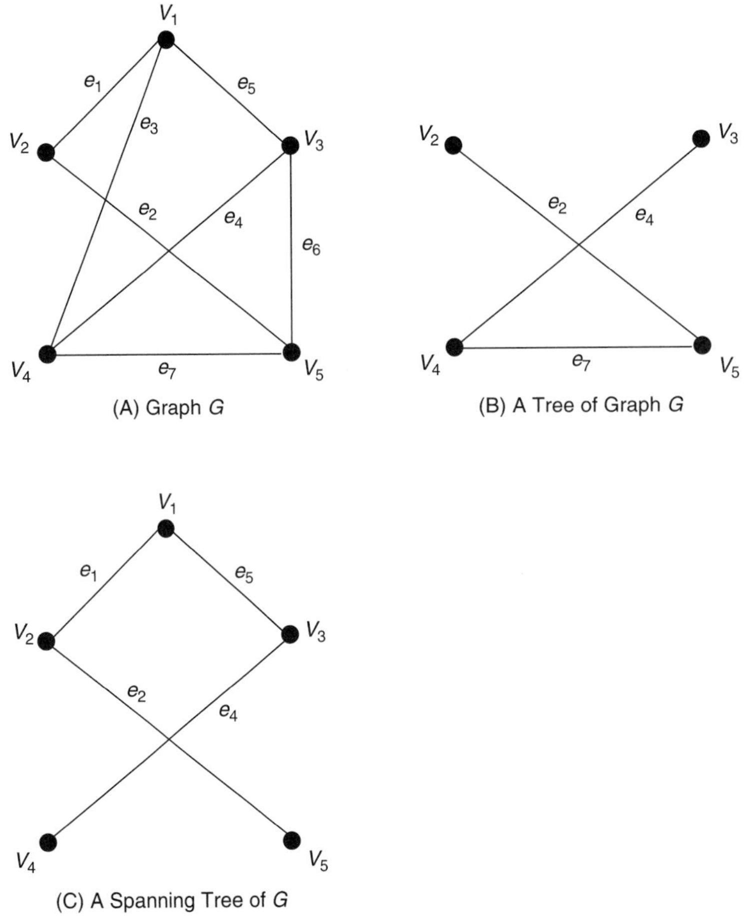

FIGURE 2.4 Examples of Tree Graphs

This section next identifies a special class of circuits of a connected graph G. Again let T be a spanning tree of G. Because exactly one path exists between any two vertices of T, adding a chord c to T produces a unique circuit. This circuit is called the **fundamental circuit** of G with respect to chord c of T. Note again that the fundamental circuit with respect to chord c contains c, and that the chord c is not present in any other fundamental circuit with respect to T. As an example, the set of fundamental circuits with respect to the spanning tree $T = (e_1, e_2, e_6, e_8)$ of the graph shown in Figure 2.5. is the following:

$$\text{Chord } e_3\text{: } (e_3, e_1, e_2).$$
$$\text{Chord } e_4\text{: } (e_4, e_1, e_2, e_6).$$
$$\text{Chord } e_5\text{: } (e_5, e_2, e_6).$$
$$\text{Chord } e_7\text{: } (e_7, e_8, e_6).$$

2.2.4 Incidence, Circuit, and Cut Matrices of a Graph

The incidence, circuit and cut matrices are coefficient matrices of Kirchhoff's voltage and current laws that describe an electrical network. This section defines these matrices and presents some properties of these matrices that are useful in studying electrical networks.

Incidence Matrix

Consider a connected directed graph G with n vertices and m edges and with no self-loops. The **all-vertex incidence matrix** $A_c = [a_{ij}]$ of G has n rows, one for each vertex, and m columns, one for each edge. The element a_{ij} of A_c is defined as follows:

$$a_{ij} = \begin{cases} 1, & \text{if } j\text{th edge is incident out of the } i\text{th vertex.} \\ -1, & \text{if } j\text{th edge is incident into the } i\text{th vertex.} \\ 0, & \text{if the } j\text{th edge is not incident on the } i\text{th vertex.} \end{cases}$$

As an example, a graph and its A_c matrix are shown in Figure 2.6.

From the definition of A_c it should be clear that each column of this matrix has exactly two nonzero entries, one $+1$ and one -1, and therefore, any row of A_c can be obtained from the remaining rows. Thus,

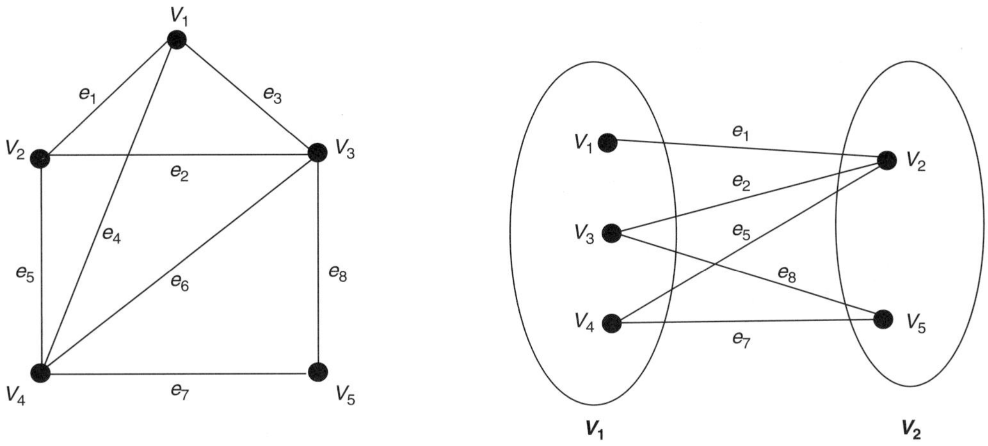

(A) Graph G (B) Cut <V_1, V_2> of G

FIGURE 2.5 A Graph and a Cut

	e_1	e_2	e_3	e_4	e_5	e_6	e_7
v_1	1	0	0	0	0	1	−1
v_2	−1	1	0	0	0	0	0
v_3	0	−1	1	0	1	0	0
v_4	0	0	−1	−1	0	−1	0
v_5	0	0	0	1	−1	0	1

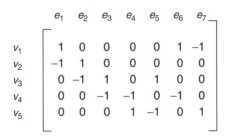

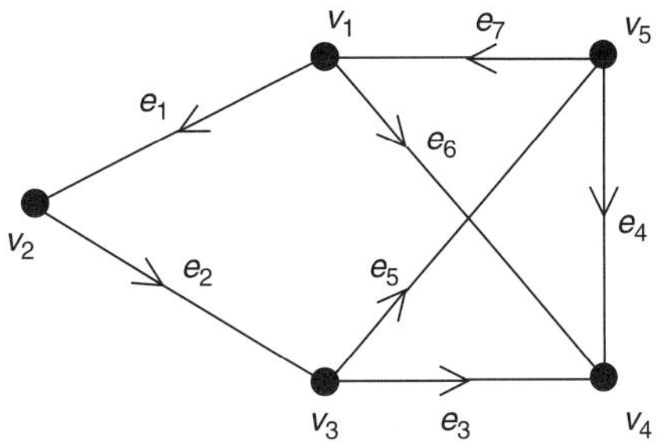

FIGURE 2.6 A Directed Graph

$$\text{rank}(A_c) \leq n - 1.$$

An $(n-1)$ rowed submatrix of A_c is referred to as an **incidence matrix** of G. The vertex that corresponds to the row that is not in A_c is called the reference vertex of A.

Cut Matrix

Consider a cut (V_a, V_b) in a connected directed graph G with n vertices and m edges. Recall that $< V_a, V_b >$ consists of all those edges connecting vertices in V_a to V_b. This cut may be assigned an orientation from V_a to V_b or from V_b to V_a. Suppose the orientation of (V_a, V_b) is from V_a to V_b. Then the orientation of an edge (v_i, v_j) is said to agree with the cut orientation if $v_i \in V_a$, and $v_j \in V_b$.

The **cut matrix** $Q_c = [q_{ij}]$ of G has m columns, one for each edge and has one row for each cut. The element q_{ij} is defined as follows:

$$q_{ij} = \begin{cases} 1 & \text{if the } j\text{th edge is in the } i\text{th cut and its} \\ & \text{orientation agrees with the cut orientation.} \\ -1 & \text{if the } j\text{th edge is in the } i\text{th cut and its} \\ & \text{orientation does not agree with cut orientation.} \\ 0 & \text{if the } j\text{th edge is not in the } i\text{th cut.} \end{cases}$$

Each row of Q_c is called the **cut vector**. The edges incident on a vertex form a cut. Hence, the matrix A_c is a submatrix of Q_c. Next, another important submatrix of Q_c is identified.

Recall that each branch of a spanning tree T of connected graph G defines a fundamental cutset. The submatrix of Q_c corresponding to the $n-1$ fundamental cutsets defined by T is called the **fundamental cutset matrix** Q_f of G with respect to T.

Let $b_1, b_2, \ldots, b_{n-1}$ denote the branches of T. Assume that the orientation of a fundamental cutset is chosen to agree with that of the defining branch. Arrange the rows and the columns of Q_f so that the ith column corresponds to the fundamental cutset defined by b_i. Then the matrix Q_f can be displayed in a convenient form as follows:

$$Q_f = [U|Q_{fc}],$$

where U is the unit matrix of order $n-1$, and its columns correspond to the branches of T. As an example, the fundamental cutset matrix of the graph in Figure 2.6 with respect to the spanning tree $T = (e_1, e_2, e_5, e_6)$ is:

$$Q_f = \begin{bmatrix} e_1 & e_2 & e_5 & e_6 & e_3 & e_4 & e_7 \\ 1 & 0 & 0 & 0 & -1 & -1 & -1 \\ 0 & 1 & 0 & 0 & -1 & -1 & -1 \\ 0 & 0 & 1 & 0 & 0 & -1 & -1 \\ 0 & 0 & 0 & 1 & 1 & 1 & 0 \end{bmatrix}$$

It is clear that the rank of Q_f is $n-1$. Hence,

$$\text{rank}(Q_c) \geq n-1.$$

Circuit Matrix

Consider a circuit C in a connected directed graph G with n vertices and m edges. This circuit can be traversed in one of two directions, clockwise or counterclockwise. The direction chosen for traversing C is called the orientation of C. If an edge $e = (v_i, v_j)$ directed from v_i to v_j is in C, and if v_i appears before v_j in traversing C in the direction specified by the orientation of C, then the orientation agrees with the orientation of e.

The **circuit matrix** $B_c = [b_{ij}]$ of G has m columns, one for each edge, and has one row for each circuit in G. The element b_{ij} is defined as follows:

$$b_{ij} = \begin{cases} 1 & \text{if the } j\text{th edge is in the } i\text{th circuit and its} \\ & \text{orientation agrees with the circuit orientation.} \\ -1 & \text{if the } j\text{th edge is in the } i\text{th circuit and its} \\ & \text{orientation does not agree with circuit orientation.} \\ 0 & \text{if the } j\text{th edge is not in the } i\text{th circuit.} \end{cases}$$

The submatrix of B_c corresponding to the fundamental circuits defined by the chords of a spanning tree T is called **fundamental circuit matrix** B_f of G with respect to the spanning tree T.

Let $c_1, c_2, c_3, \ldots, c_{m-n+1}$ denote the chords of T. Arrange the columns and the rows of B_f so that the ith row corresponds to the fundamental circuit defined by the chord c_i and the ith column corresponds to the chord c_i. Then orient the fundamental circuit to agree with the orientation of the defining chord. The result is writing B_f as:

$$B_f = [U|B_{ft}],$$

where U is the unit matrix of order $m-n+1$, and its columns correspond to the chords of T.

As an example, the fundamental circuit matrix of the graph shown in Figure 2.6 with respect to the tree $T = (e_1, e_2, e_5, e_6)$ is given here.

$$B_f = \begin{matrix} & \begin{matrix} e_3 & e_4 & e_7 & e_1 & e_2 & e_5 & e_6 \end{matrix} \\ \begin{matrix} e_3 \\ e_4 \\ e_7 \end{matrix} & \begin{bmatrix} 1 & 0 & 0 & 1 & 1 & 0 & -1 \\ 0 & 1 & 0 & 1 & 1 & 1 & -1 \\ 0 & 0 & 1 & 1 & 1 & 1 & 0 \end{bmatrix} \end{matrix}$$

It is clear from this example that the rank of B_f is $m-n+1$. Hence,

$$\text{rank}(B_c) \geq m-n+1.$$

The following results constitute the foundation of the graph-theoretic application to electrical circuit analysis.

Theorem 2.4 (orthogonality relationship): (1) A circuit and a cutset in a connected graph have an even number of common edges; and (2) if a circuit and a cutset in a directed graph have $2k$ common edges, then k of these edges have the same relative orientation in the circuit and the cutset, and the remaining k edges have one orientation in the circuit and the opposite orientation in the cutset.

Theorem 2.5: If the columns of the circuit matrix B_c and the columns of the cut matrix Q_c are arranged in the same edge order, then

$$B_c Q_c^t = 0.$$

Theorem 2.6:

$$\text{rank}(B_c) = m-n+1.$$
$$\text{rank}(Q_c) = n-1.$$

Note that it follows from the above theorem that the rank of the circuit matrix is equal to the nullity of the graph, and the rank of the cut matrix is equal to the rank of the graph. This result, in fact, motivated the definitions of the rank and nullity of a graph.

2.3 Graphs and Electrical Networks

An electrical network is an interconnection of electrical network elements, such as resistances, capacitances, inductances, voltage, and current sources. Each network element is associ-

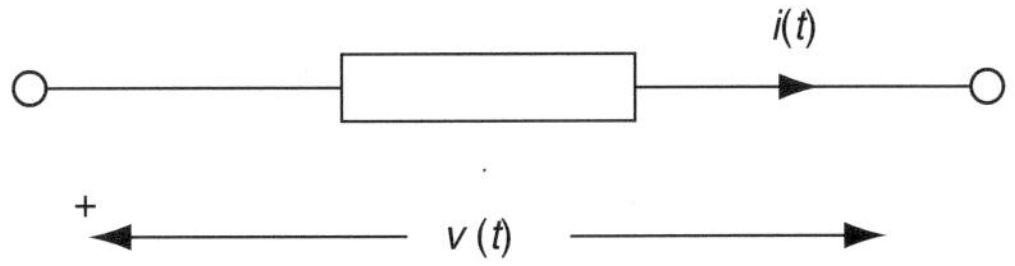

FIGURE 2.7 A Network Element with Reference Convention

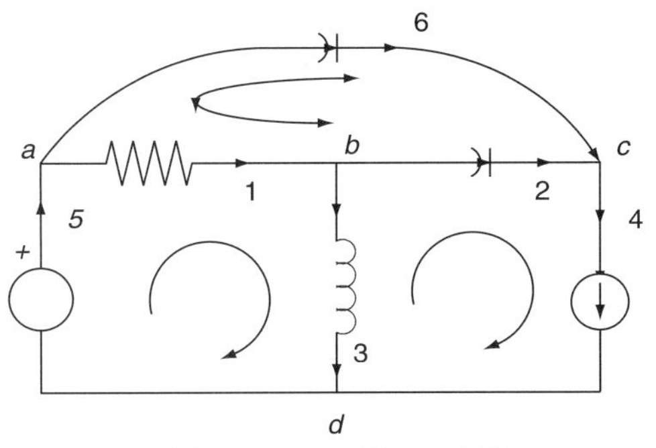
6
a
b
c
1
2
5
4
+
3
d
(A) An Electrical Network N

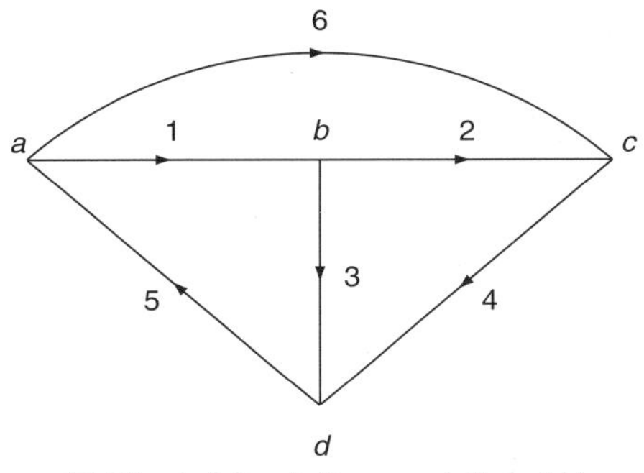
6
a
1
b
2
c
3
5
4
d
(B) Directed Graph Representation of N

FIGURE 2.8 Graph Examples of KVL and KCL Equations

ated with two variables: the voltage variable $v(t)$ and the current variable $i(t)$. Reference directions are also assigned to the network elements as shown in Figure 2.7 so that $i(t)$ is positive whenever the current is in the direction of the arrow and so that $v(t)$ is positive whenever the voltage drop in the network element is in the direction of the arrow. Replacing each element and its associated reference direction by a directed edge results in the directed graph representing the network. For example, a simple electrical network and the corresponding directed graph are shown in Figure 2.8.

The physical relationship between the current and voltage variables of network elements is specified by Ohm's law. For voltage and current sources, the voltage and current variables are required to have specified values. The linear dependence among the voltage variables in the network and the linear dependence among the current variables are governed by Kirchhoff's voltage and current laws:

Kirchhoff's voltage law (KVL): The algebraic sum of the voltages around any circuit is equal to zero.

Kirchhoff's current law (KCL): The algebraic sum of the currents flowing out of a node is equal to zero.

As examples, the KVL equation for the circuit 1, 3, 5 and the KCL equation for vertex b in the graph of Figure 2.8 are the following:

$$\text{Circuit 1, 3, 5.} \quad V_1 + V_3 + V_5 = 0$$
$$\text{Vertex } b: \quad -I_1 + I_2 + I_3 = 0$$

It can be easily seen that KVL and KCL equations for an electrical network N can be conveniently written as:

$$A_c I_e = 0.$$
$$B_c V_e = 0.$$

A_c and B_c are, respectively, the incidence and circuit matrices of the directed graph representing N, I_e, and V_e, respectively, and these are the column vectors of element currents and voltages in N. Because each row in the cut matrix Q_c can be expressed as a linear combination of the rows of the matrix, in the above A_c can be replaced by Q_c. Thus, the result is as follows:

$$\text{KCL: } Q_c I_e = 0.$$
$$\text{KVL: } B_c V_e = 0.$$

Hence, KCL can also be stated as: the algebraic sum of the currents in any cut of N is equal to zero.

If a network N has n vertices and m elements and its graph is connected, then there are only $(n-1)$ linearly independent cuts and only $(m-n+1)$ linearly independent circuits. Thus, in writing KVL and KCL equations, only B_f and Q_f, respectively, need to be used. Thus, we have

$$\text{KCL: } Q_f I_e = 0$$
$$\text{KVL: } B_f V_e = 0$$

Note that KCL and KVL equations depend only on the way network elements are interconnected and not on the nature of the network elements. Thus, several results in electrical network theory are essentially graph-theoretic in nature. Some of those results of interest in electrical network analysis are presented in the remainder of this chapter. Note that for these results, a network N and its directed graph representation are both denoted by N.

2.3.1 Loop and Cutset Transformations

Let T be a spanning tree of an electrical network. Let I_c and V_t be the column vectors of chord currents and branch currents with respect to T.

1. **Loop transformation:**

$$I_e = B_f^t I_c.$$

2. **Cutset transformation:**

$$V_e = Q_f^t V_t.$$

If, in the cutset transformation, Q_f is replaced by the incidence matrix A, then we get the **node transformation** given below:

$$V_e = A^t V_n,$$

where the elements in the vector V_n can be interpreted as the voltages of the nodes with respect to the reference node r. (Note: The matrix A does not contain the row corresponding to the node r).

The above transformations have been extensively employed in developing different methods of network analysis. Two of these methods are described in the following section.

2.4 Loop and Cutset Systems of Equations

As observed earlier, the problem of network analysis is to determine the voltages and currents associated with the elements of an electrical network. These voltages and currents can be determined from Kirchhoff's equations and the element voltage–current (v–i) relations given by Ohm's law. However, these equations involve a large number of variables. As can be seen from the loop and cutset transformations, not all these variables are independent. Furthermore, in place of KCL equations, the loop transformation can be used, which involves only chord currents as variables. Similarly, KVL equations can be replaced by the cutset transformation that involves only branch voltage variables. Taking advantage of these transformations enables the establishment of different sytems of network equations known as the loop and cutset systems.

In deriving the loop system, the loop transformation is used in place of KCL in this case, the loop variables (chord currents) serve as independent variables. In deriving the cutset system, the cutset transformation is used in place of KVL, and the cutset variables (tree branch voltages) serve as the independent variables in this case. It is assumed that the electrical network N is connected and that N consists of only resistances, (R), capacitances (C), inductances (L), including mutual inductances, and independent voltage and current sources. It is also assumed that all initial inductor currents and initial capacitor voltages have been replaced by appropriate sources. Further, the voltage and current variables are all Laplace transforms of the complex frequency variable s. In N, there can be no circuit consisting of only independent voltage sources, for if such a circuit of sources were present, then by KVL, there would be a linear relationship among the corresponding voltages; this would violate the independence of the voltage sources. For the same reason, in N, there can be no cutset consisting of only independent current sources. Hence, there exists in N a spanning tree containing all the voltage sources but not current sources. Such a tree is the starting point for the development of both the loop and cutset systems of equations.

Let T be a spanning tree of the given network such that T contains all the voltage sources but no current sources. Partition the element voltage vector V_e and the element current vector I_e as follows:

$$V_e = \begin{bmatrix} V_1 \\ V_2 \\ V_3 \end{bmatrix} \quad \text{and} \quad I_e = \begin{bmatrix} I_1 \\ I_2 \\ I_3 \end{bmatrix}$$

The subscripts 1, 2, and 3 refer to the vectors corresponding to the current sources, RLC elements, and voltage sources, respectively. Let B_f be the fundamental circuit matrix of N, and let Q_f be the fundamental cutset matrix of N with respect to T. The KVL and the KCL equations can then be written as follows:

$$\text{KVL: } B_f V_e = \begin{bmatrix} U & B_{12} & B_{13} \\ 0 & B_{22} & B_{23} \end{bmatrix} \begin{bmatrix} V_1 \\ V_2 \\ V_3 \end{bmatrix} = 0.$$

$$\text{KCL: } Q_f I_e = \begin{bmatrix} Q_{11} & Q_{12} & 0 \\ Q_{21} & Q_{22} & U \end{bmatrix} \begin{bmatrix} I_1 \\ I_2 \\ I_3 \end{bmatrix} = 0.$$

2.4.1 Loop Method of Network Analysis

Step 1: Solve the following for the vector I_l. Note that I_l is the vector of currents in the nonsource chords of T.

$$Z_l \, I_l = -B_{23} V_3 - B_{22} Z_2 B_{12} I_1, \qquad (2.1)$$

where Z_2 is the impedance matrix of RLC elements and

$$Z_l = B_{22} Z_2 B_{22}.$$

Equation 2.1 is called the **loop system of equations.**

Step 2: Calculate I_2 using:

$$I_2 = B_{12}I_1 + B_{22}I_l. \tag{2.2}$$

Then

$$V_2 = Z_2 I_2. \tag{2.3}$$

Step 3: Determine V_1 and I_3 using the following:

$$V_1 = -B_{12}V_2 - B_{13}V_3. \tag{2.4}$$

$$I_3 = B_{13}I_1 + B_{23}I_l. \tag{2.5}$$

Note that I_1 and V_3 have specified values because they correspond to current and voltage sources, respectively.

2.4.2 Cutset Method of Network Analysis

Step 1: Solve the following for the vector V_b. Note that V_b is the vector of voltages in the nonsource branches of T.

$$Y_b\ V_b = -Q_{11}I_1 - Q_{12}Y_2Q_{22}V_b\,, \tag{2.6}$$

where Y_2 is the admittance matrix of *RLC* elements and

$$Y_b = Q_{12}Y_2Q_{12}$$

Equation 2.6 is called the **cutset system of equations.**
Step 2: Calculate V_2 using:

$$V_2 = Q_{12}V_b + Q_{22}V_3. \tag{2.7}$$

Then

$$I_2 = Y_2Y_2. \tag{2.8}$$

Step 3: Determine V_1 and I_3 using the following:

$$V_1 = Q_{21}V_b + Q_{21}V_3 \tag{2.9}$$

$$I_3 = -Q_{21}I_1 - Q_{22}I_2. \tag{2.10}$$

Note that I_1 and V_3 have specified values because they correspond to current and voltage sources.

This completes the cutset method of network analysis.

The following discussion illustrates the loop and cutset methods of analysis on the network shown in Figure 2.9(A). The graph of the network is shown in Figure 2.9(B). The chosen spanning tree T consists of edges 4, 5 and 6. Note that T contains the voltage source and has no current source. The fundamental circuit and the fundamental cutset matrices with respect to T are given below in the required partioned form:

$$B_f = \begin{array}{c} \begin{matrix} 1 & 2 & 3 & 4 & 5 & 6 \end{matrix} \\ \begin{bmatrix} 1 & 0 & 0 & -1 & -1 & 1 \\ 1 & 1 & 0 & 1 & 0 & -1 \\ 0 & 0 & 1 & 1 & -1 & 0 \end{bmatrix} \end{array}$$

$$Q_f = \begin{array}{c} \begin{matrix} 1 & 2 & 3 & 4 & 5 & 6 \end{matrix} \\ \begin{bmatrix} 1 & -1 & 1 & 1 & 0 & 0 \\ 1 & 0 & 1 & 0 & 1 & 0 \\ -1 & 1 & 0 & 0 & 0 & 1 \end{bmatrix} \end{array}$$

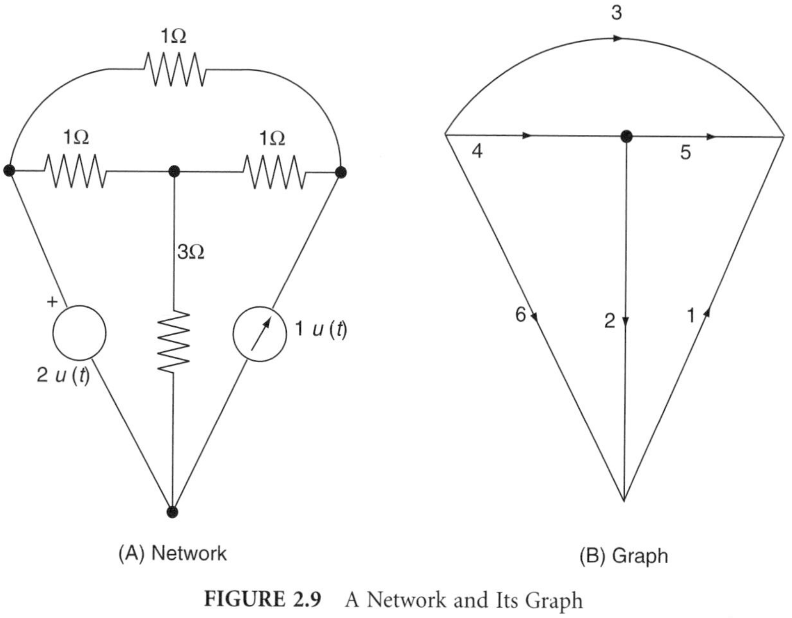

FIGURE 2.9 A Network and Its Graph

From these matrices results the following:

$$B_{12} = [0 \quad 0 \quad -1 \quad -1]$$
$$B_{13} = [1]$$
$$B_{22} = \begin{bmatrix} 1 & 0 & 1 & 0 \\ 0 & 1 & -1 & -1 \end{bmatrix}$$
$$B_{23} = \begin{bmatrix} -1 \\ 0 \end{bmatrix}$$
$$Q_{11} = \begin{bmatrix} -1 \\ 0 \end{bmatrix}$$
$$Q_{12} = \begin{bmatrix} -1 & 1 & 1 & 0 \\ 0 & 1 & 0 & 1 \end{bmatrix}$$
$$Q_{21} = [-1]$$
$$Q_{22} = [1 \quad 0 \quad 0 \quad 0].$$

The following also results:

$$Z_2 = \begin{bmatrix} 3 & 0 & 0 & 0 \\ 0 & 1 & 0 & 0 \\ 0 & 0 & 1 & 0 \\ 0 & 0 & 0 & 1 \end{bmatrix}$$
$$Y_2 = \begin{bmatrix} 1/3 & 0 & 0 & 0 \\ 0 & 1 & 0 & 0 \\ 0 & 0 & 1 & 0 \\ 0 & 0 & 0 & 1 \end{bmatrix}$$
$$v_6 = 2\,\text{V}$$
$$i_1 = 1\text{A}$$

Loop Method

Edges 2 and 3 are nonsource chords. So,

$$I_l = \begin{bmatrix} i_2 \\ i_3 \end{bmatrix}$$

and substituting

$$Z_l = B_{22} Z_2 B_{22}^t = \begin{bmatrix} 4 & -1 \\ -1 & 3 \end{bmatrix}$$

in equation 2.1 yields the following loop system of equations:

$$\begin{bmatrix} 4 & -1 \\ -1 & 3 \end{bmatrix} \begin{bmatrix} i_2 \\ i_3 \end{bmatrix} = \begin{bmatrix} 3 \\ -2 \end{bmatrix}$$

Solving for i_2 and i_3 yields:

$$I_l = \begin{bmatrix} i_2 \\ i_3 \end{bmatrix} = 1/11 \begin{bmatrix} 7 \\ -5 \end{bmatrix}$$

Using equation 2.2 results in:

$$I_2 = \begin{bmatrix} i_2 \\ i_3 \\ i_4 \\ i_5 \end{bmatrix} = 1/11 \begin{bmatrix} 7 \\ -5 \\ 1 \\ -6 \end{bmatrix}$$

Then using $V_2 = Z_2 I_2$ yields:

$$V_2 = \begin{bmatrix} v_2 \\ v_3 \\ v_4 \\ v_5 \end{bmatrix} = 1/11 \begin{bmatrix} 21 \\ -5 \\ 1 \\ -6 \end{bmatrix}$$

Finally, equations 2.4 and 2.5 yield the following:

$$V_1 = [v_1] = -27/11.$$
$$I_3 = [i_6] = 4/11$$

Cutset Method

Edges 4 and 5 are the nonsource branches. So,

$$V_b = \begin{bmatrix} v_4 \\ v_5 \end{bmatrix}$$

and substituting

$$Y_b = \begin{bmatrix} 7/3 & 1 \\ 1 & 2 \end{bmatrix}$$

in equation 2.6 yields the following cutset system of equations:

$$Y_b V_b = \begin{bmatrix} 7/3 & 1 \\ 1 & 2 \end{bmatrix} \begin{bmatrix} v_4 \\ v_5 \end{bmatrix} = \begin{bmatrix} -1/3 \\ -1 \end{bmatrix}$$

Solving for V_b yields:

$$V_b = \begin{bmatrix} v_4 \\ v_5 \end{bmatrix} = 1/11 \begin{bmatrix} 1 \\ -6 \end{bmatrix}$$

From equation 2.7, the result is:

$$V_2 = \begin{bmatrix} v_2 \\ v_3 \\ v_4 \\ v_5 \end{bmatrix} = 1/11 \begin{bmatrix} 21 \\ -5 \\ 1 \\ -6 \end{bmatrix}$$

Then using $I_2 = Y_2V_2$ yields:

$$I_2 = \begin{bmatrix} i_2 \\ i_3 \\ i_4 \\ i_5 \end{bmatrix} = 1/11 \begin{bmatrix} 7 \\ -5 \\ 1 \\ -6 \end{bmatrix}$$

Finally, using equations 2.9 and 2.10, the end results are:

$$V_1 = [v_1] = -27/11$$
$$I_3 = [i_6] = 4/11$$

This completes the illustration of the loop and cutset methods of circuit analysis.

Node Equations

Suppose a network N has no independent voltage sources. A convenient description of N with the node voltages as independent variables can be obtained as follows.

Let A be the incidence matrix of N with vertex v_r as reference. Let A be partitioned as $A = [A_{11}, A_{12}]$, where the columns of A_{11} and A_{12} correspond, respectively, to the RLC elements and current sources. If I_1 and I_2 denote the column vectors of RLC element currents and current source currents, then KCL equations for N become:

$$A_{11}I_1 = -A_{12}I_2.$$

By Ohm's law:

$$I_1 = Y_1V_1,$$

where V_1 is the column vector of voltages of RLC elements and Y_1 is the corresponding admittance matrix. Furthermore, by the node transformation yields:

$$V_1 = A_{11}^tV_n,$$

where V_n is the column vector of node voltages. So the result from the KCL equations is the following:

$$(A_{11}Y_1A_{11}^t)V_n = -A_{12}I_2.$$

The above equations are called **node equations**. The matrix $A_{11}Y_1A_{11}^t$ is called the **node admittance matrix** of N.

2.5 Summary

This chapter has presented an introduction to certain basic results in graph theory and their application in the analysis of electrical circuits. For a more comprehensive treatment of other developments in circuit theory that are primarily of a graph-theoretic nature, refer to works by Swamy and Thulasiraman (1981) and Chen (1972). A publication by Seshu and Reed (1961) is an early work that first discussed many of the fundamental results presented in this chapter. A review by Watandabe and Shinoda (1999) is the most recent reference summarizing the graph theoretic contributions from Japanese circuit theory researchers.

References

Chen, W.K. (1972). *Applied graph theory.* Amsterdam: North-Holland.

Seshu, S., and Reed M.B. (1961). *Linear graphs and electrical networks.* Reading, MA: Addison-Wesley.

Swamy, M.N.S., and Thulasiraman, K. (1981). *Graphs, networks, and algorithms.* New York: Wiley Interscience.

Watanabe, H., and Shinoda, S. (1999). Soul of circuit theory: A review on research activities of graphs and circuits in Japan. *IEEE Transactions on Circuits and Systems 45,* 83–94.

3

Computer-Aided Design

Ajoy Opal
Department of Electrical and Computer Engineering, University of Waterloo, Waterloo, Ontario, Canada

3.1 Introduction

Hand analysis of large circuits, especially if they contain nonlinear elements, can be tedious and prone to error. In such cases, computer-aided design (CAD) is necessary to speed up the analysis procedure and to obtain accurate results. This chapter describes the algorithms used in CAD programs. Such programs are available from commercial companies, such as *Mentor Graphics*, *Cadence*, and *Avant*!

Typically, CAD programs are based on formulating the equations using matrix methods, and the equations are then solved using numerical methods. For ease in programming, methods that are general and straightforward are preferred. This chapter assumes that the reader is familiar with methods for solving a set of simultaneous linear equations.

3.2 Modified Nodal Analysis

The first step in circuit analysis is formulation of equations. This section modifies the nodal analysis method described in Chapter 1 of this book's Section 1 to include different types of elements in the formulation. Recall that in nodal analysis, Kirchhoff's current law (KCL) is used to write equations at each node in terms of the nodal voltages and element values. The equations are written for one node at a time, and the node voltages are unknown variables. In CAD, it is useful to consider one element at a time and to develop the matrix equations on this basis. Assume that the nodes in a circuit are labeled by consecutive integers from 1 to N such that there are the N nodal voltages in the circuit. The reference node is usually labeled zero. This presentation uses the lowercase letters j, k, l, and m to denote nodes in the circuit. Voltages will be denoted by V and branch currents by I. Consider a linear resistor R connected between nodes j and k as in Figure 3.1.

This element will affect the KCL equations at nodes j and k, and the contributions will be:

$$\text{KCL at Node } j\text{:} \ldots + GV_j - GV_k \ldots = 0.$$
$$\text{KCL at Node } k\text{:} \ldots - GV_j + GV_k \ldots = 0.$$

The V_j and V_k are the nodal voltages at the two nodes $V = V_j - V_k$ and $G = 1/R$, and the ellipses $(\ldots)$ denote contributions from other elements connected to that node. Thus, the entries in the nodal admittance matrix due to the resistor are in four locations:

$$\begin{array}{r c} & \begin{array}{cc} \text{Col. } V_j & \text{Col. } V_k \end{array} \\ \begin{array}{r} \text{Row } j\text{:} \\ \text{Row } k\text{:} \end{array} & \begin{bmatrix} G & -G \\ -G & G \end{bmatrix} \end{array}$$

Row and Col. denote the row and the column in the matrix, respectively. If one end of the resistor, for example node k, is connected to the reference or ground node, then the

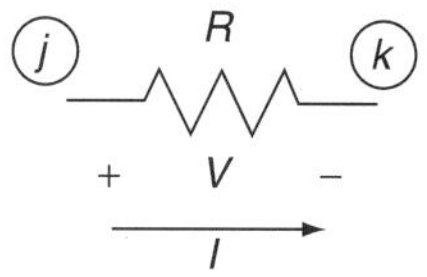

FIGURE 3.1 Linear Resistor

corresponding column and row are not present in the admittance matrix, and there is only one entry in the matrix at location (j, j). The above is called the **circuit stamp** for the element and can be used for all elements that can be written in admittance form (i.e., the current through the element can be written in terms of nodal voltages and the element value).

Next, consider an independent voltage source for which the current through it cannot be written in terms of nodal voltages, such as in Figure 3.2.

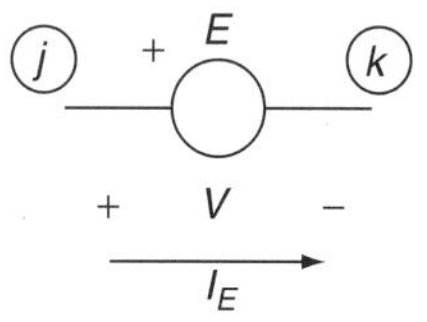

FIGURE 3.2 Independent Voltage Source

In this case, the KCL equations at nodes j and k are listed as:

$$\begin{aligned} &\text{KCL at node } j\text{:} \ldots + I_E \ldots = 0. \\ &\text{KCL at node } k\text{:} \ldots - I_E \ldots = 0. \end{aligned} \tag{3.1}$$

The current I_E is included in the formulation. This increases the number of unknown variables by one. Hence, another equation is needed so that the number of unknowns and the number of equations are the same and can be solved. The additional equation for the voltage source is the following:

$$V_j - V_k = E. \tag{3.2}$$

The circuit stamp for this element includes both equations 3.1 and 3.2 and becomes:

$$\begin{array}{r} \\ \\ \text{Row } j\text{:} \\ \text{Row } k\text{:} \\ \text{Row } I_E\text{:} \end{array} \begin{array}{c} \begin{array}{ccc} \text{Col.} & \text{Col.} & \text{Col.} \\ V_j & V_k & I_E \end{array} \\ \begin{bmatrix} & & 1 \\ & & -1 \\ 1 & -1 & \end{bmatrix} \end{array} \quad \begin{array}{c} \text{RHS} \\ \\ \begin{bmatrix} \\ \\ E \end{bmatrix} \end{array}$$

Row I_E denotes the additional equation 3.2 added to nodal analysis, and Col. I_E is the additional column in matrix due to the current through the voltage source. The vector RHS denotes the right-hand side and contains known values of sources. This method is called **modified nodal analysis (MNA)** because of the modifications made to nodal analysis to allow different types of elements in the formulation. The circuit stamps for other common elements are given in Table 3.1 and can be found in CAD books. In this table, rows labeled with lowercase letters j, k, l, and m represent KCL equations at that node, and V_j, V_k, V_l, and V_m represent nodal voltages at the corresponding nodes. All rows and columns labeled by a current I represent additional rows and columns required for MNA.

As examples, the MNA equations for the linear circuit are in Figure 3.3.

Scanning of the circuit shows that it has three nodal voltages and that it needs three branch currents for MNA formulation. The branch currents are required, one each, for the voltage source, the inductor in impedance form, and the VVT. The total number of variables is six and can be computed before the matrices are formulated. The MNA equations are as follows:

$$\begin{bmatrix} G & -G & 0 & 1 & 0 & 0 \\ -G & G & 0 & 0 & 1 & 0 \\ 0 & 0 & 0 & 0 & 0 & 1 \\ 1 & 0 & 0 & 0 & 0 & 0 \\ 0 & 1 & 0 & 0 & 0 & 0 \\ 0 & -\mu & 1 & 0 & 0 & 0 \end{bmatrix} \begin{bmatrix} V_1 \\ V_2 \\ V_3 \\ I_E \\ I_L \\ I_{VVT} \end{bmatrix} + \begin{bmatrix} 0 & 0 & 0 & 0 & 0 & 0 \\ 0 & C & -C & 0 & 0 & 0 \\ 0 & -C & C & 0 & 0 & 0 \\ 0 & 0 & 0 & 0 & 0 & 0 \\ 0 & 0 & 0 & 0 & -L & 0 \\ 0 & 0 & 0 & 0 & 0 & 0 \end{bmatrix} \frac{d}{dt} \begin{bmatrix} V_1 \\ V_2 \\ V_3 \\ I_E \\ I_L \\ I_{VVT} \end{bmatrix} = \begin{bmatrix} 0 \\ 0 \\ 0 \\ 1 \\ 0 \\ 0 \end{bmatrix} E, \ \boldsymbol{v}(0) = \begin{bmatrix} V_1(0) \\ V_2(0) \\ V_3(0) \\ I_E(0) \\ I_L(0) \\ I_{VVT}(0) \end{bmatrix}$$

The first three rows represent KCL equations at the three nodes. The last three rows represent the characteristic equations from the independent voltage source, inductor in impedance form, and VVT, respectively. The equations for linear time-invariant circuits are in the form:

$$\boldsymbol{Gv}(t) + \boldsymbol{C}\frac{d}{dt}\boldsymbol{v}(t) = \boldsymbol{du}(t) \text{ and } \boldsymbol{v}(0) = \boldsymbol{v}_0, \tag{3.3}$$

where all conductances and constants arising in the formulation are stored in the matrix $\boldsymbol{G}$ and all capacitor and inductor values are stored in the matrix $\boldsymbol{C}$. In addition, the connection of the input source to the circuit is in $\mathbf{d}$, the vector of nodal voltages and some branch currents needed for MNA is $\boldsymbol{v}(t)$, and the initial condition vector is $\boldsymbol{v}_0$.

TABLE 3.1 Circuit Stamps for MNA

Element	Symbol	Matrix entries
Current source	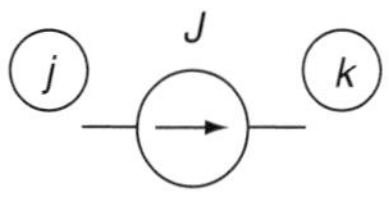	$\begin{array}{r|cc|c} & V_j & V_k & \text{RHS} \\ \hline \text{Row } j: & & & -J \\ \text{Row } k: & & & J \end{array}$
Voltage source	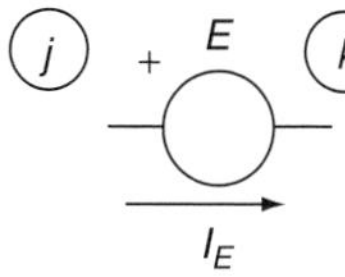	$\begin{array}{r|ccc|c} & V_j & V_k & I_E & \text{RHS} \\ \hline \text{Row } j: & & & 1 & \\ \text{Row } k: & & & -1 & \\ \text{Row } I_E: & 1 & -1 & & E \end{array}$
Admittance $I = Y(V_j - V_k)$	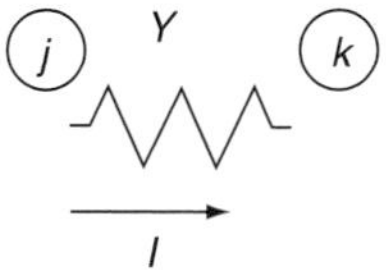	$\begin{array}{r|cc} & V_j & V_k \\ \hline \text{Row } j: & Y & -Y \\ \text{Row } k: & -Y & Y \end{array}$
Impedance $V_j - V_k = ZI_Z$	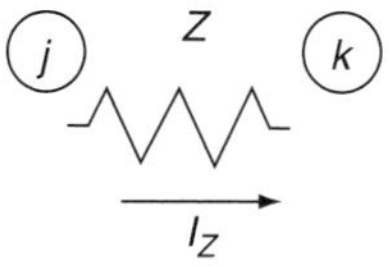	$\begin{array}{r|ccc} & V_j & V_k & I_Z \\ \hline \text{Row } j: & & & 1 \\ \text{Row } k: & & & -1 \\ \text{Row } I_Z: & 1 & -1 & -Z \end{array}$
Voltage-controlled current source (VCT) $J = g(V_j - V_k)$	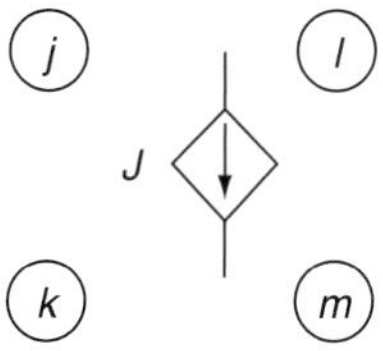	$\begin{array}{r|cccc} & V_j & V_k & V_l & V_m \\ \hline \text{Row } j: & & & & \\ \text{Row } k: & & & & \\ \text{Row } l: & g & -g & & \\ \text{Row } m: & -g & g & & \end{array}$
Voltage-controlled voltage source (VVT) $V_l - V_m = \mu(V_j - V_k)$	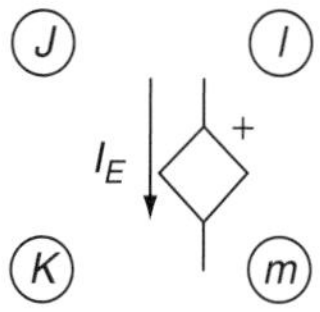	$\begin{array}{r|ccccc} & V_j & V_k & V_l & V_m & I_E \\ \hline \text{Row } j: & & & & & \\ \text{Row } k: & & & & & \\ \text{Row } l: & & & & & 1 \\ \text{Row } m: & & & & & -1 \\ \text{Row } I_E: & -\mu & \mu & 1 & -1 & \end{array}$
Current-controlled current source (CCT) $J = \beta I_{jk}$	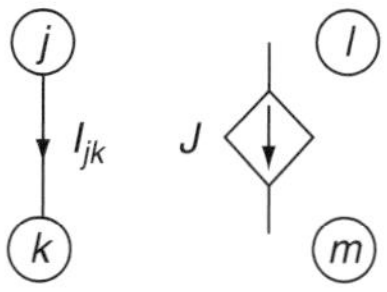	$\begin{array}{r|ccccc} & V_j & V_k & V_l & V_m & I_{jk} \\ \hline \text{Row } j: & & & & & 1 \\ \text{Row } k: & & & & & -1 \\ \text{Row } l: & & & & & \beta \\ \text{Row } m: & & & & & -\beta \\ \text{Row } I_{jk}: & 1 & -1 & & & \end{array}$
Current-controlled voltage source (CVT) $V_l - V_m = rI_{jk}$	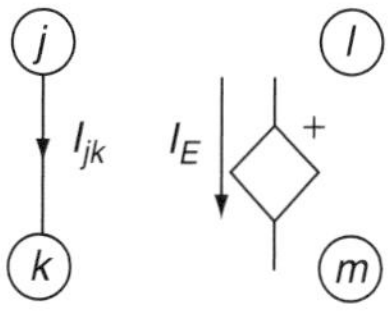	$\begin{array}{r|cccccc} & V_j & V_k & V_l & V_m & I_{jk} & I_E \\ \hline \text{Row } j: & & & & & 1 & \\ \text{Row } k: & & & & & -1 & \\ \text{Row } l: & & & & & & 1 \\ \text{Row } m: & & & & & & -1 \\ \text{Row } I_{jk}: & 1 & -1 & & & & \\ \text{Row } I_E: & & & 1 & -1 & -r & \end{array}$

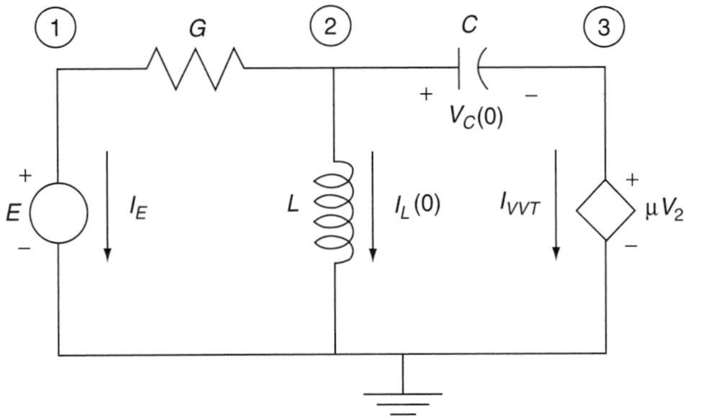

FIGURE 3.3 Linear Circuit Example

For sinusoidal steady state response of a linear time invariant circuit, the input is:

$$u(t) = e^{j\omega t}$$

Here, the initial conditions $\boldsymbol{v}_0$ are ignored, and phasor analysis is used for computing the results. In this case, equation 3.3 becomes:

$$(\boldsymbol{G} + j\omega \boldsymbol{C})\boldsymbol{V}(j\omega) = \boldsymbol{d}, \quad (3.4)$$

where $\boldsymbol{V}(j\omega)$ is the phasor representation of $\boldsymbol{v}(t)$. The matrix equation 3.4 is a set of simultaneous linear equations and are solved numerically for $\boldsymbol{V}(j\omega)$. The solution, in general, is a complex quantity with magnitude $|\boldsymbol{V}(j\omega)|$ and phase angle $\angle \boldsymbol{V}(j\omega)$:

$$\boldsymbol{V}(j\omega) = |\boldsymbol{V}(j\omega)| \angle \boldsymbol{V}(j\omega).$$

The magnitude is also plotted in decibels [dB] and computed as:

$$|\boldsymbol{V}(j\omega)|_{\text{dB}} = 20 \, \log |\boldsymbol{V}(j\omega)|.$$

Using the values $L = 1$ H, $G = 1$ S, $C = 1$ F, $\mu = 0.5$, the magnitude of V_2 is plotted in Figure 3.4.

3.3 Formulation of MNA Equations of Nonlinear Circuits

Formulation of MNA equations of nonlinear circuits follows the same steps as equation formulation for linear circuits. Consider the circuit in Figure 3.5 with a nonlinear resistor defined by $I_R = g(V_R) = 0.001(V_R)^3$ and a nonlinear capacitor defined by $Q_C = q(V_C) = 0.001(V_C)^3$. The current through the capacitor is given by $I_C = (d/dt)Q_C$. Both the nonlinear elements are called voltage-controlled elements because the current through the resistor and charge on the capacitor are nonlinear functions of the voltage across the individual elements. Let the input be a DC source $\mathrm{E}(t) = 1$ V.

The MNA equations for the circuit in Figure 3.5 are the following:

$$\begin{aligned} \frac{V_1 - V_2}{1000} + I_E = 0 = f_1. \\ \frac{V_2 - V_1}{1000} + \frac{(V_2)^3}{1000} + \frac{d}{dt} Q_C(V_2) = 0 = f_2. \\ V_1 - E = 0 = f_3. \end{aligned} \quad (3.5)$$

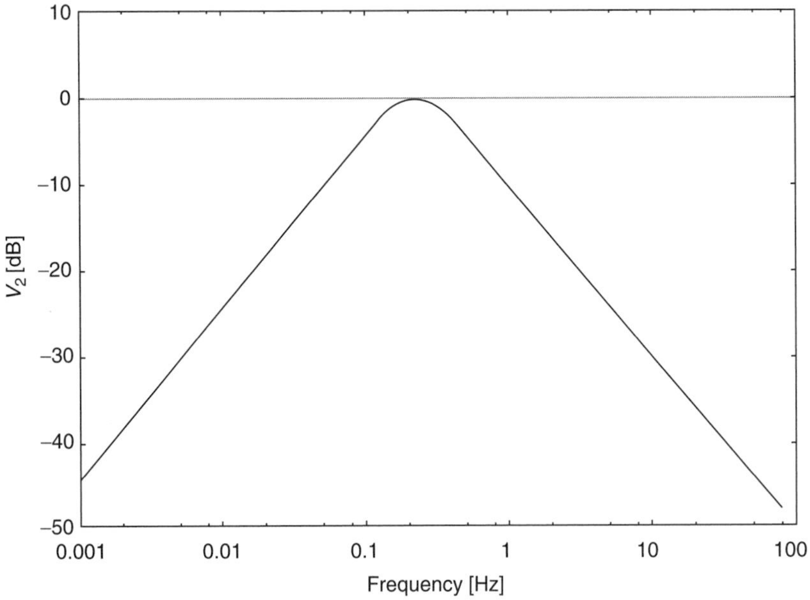

FIGURE 3.4 Frequency Response of a Linear Circuit

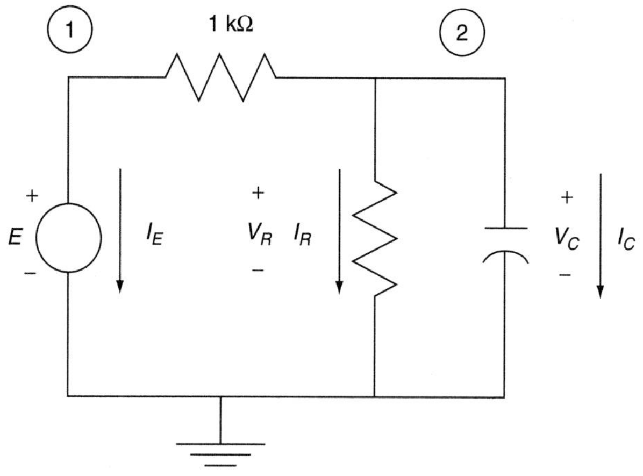

FIGURE 3.5 Nonlinear Circuit

The first two equations are the KCL equations at nodes 1 and 2 respectively, and the last equation is the characteristic equation for the voltage source. In general, all the equations are nonlinear and are collectively written in the form

$$f(\dot{v},\ v,\ t) = \begin{vmatrix} f_1(\dot{v},\ v,\ t) \\ f_2(\dot{v},\ v,\ t) \\ f_3(\dot{v},\ v,\ t) \end{vmatrix} = \mathbf{0}, \quad (3.6)$$

where the unknown vector and initial conditions are the following:

$$v = \begin{bmatrix} V_1 \\ V_2 \\ I_E \end{bmatrix} \text{and } v(0) = v_0.$$

3.4 A Direct Current Solution of Nonlinear Circuits

For a direct current (dc) solution of nonlinear circuits, the derivatives of the network variables and time are set to zero, and equation 3.6 reduces to:

$$f(\mathbf{0},\ v,\ 0) = x(v) = \mathbf{0}.$$

For example, the equations for the circuit in Figure 3.5 are the following:

$$x(v) = \begin{bmatrix} x_1(V_1,\ V_2,\ I_E) \\ x_2(V_1,\ V_2,\ I_E) \\ x_3(V_1,\ V_2,\ I_E) \end{bmatrix} = \begin{bmatrix} \dfrac{V_1 - V_2}{1000} + I_E \\ \dfrac{V_2 - V_1}{1000} + 0.001(V_2)^3 \\ V_1 - E \end{bmatrix} = \begin{bmatrix} 0 \\ 0 \\ 0 \end{bmatrix} = \mathbf{0}. \quad (3.7)$$

The nonlinear equation 3.7 is solved numerically using an iterative method called the Newton–Raphson (NR) method. Let v^0 denote the initial guess and v^i the result of the ith iteration for the solution of equation 3.7. The calculation of the next iteration value v^{i+1} is attempted such that $x(v^{i+1}) \approx \mathbf{0}$. Expanding $x(v^{i+1})$ in a Taylor series around the point v^i gives Equation 3.8.

Equation 3.8 shows that only the first two terms in the series have been used and other higher order terms have been ignored. Equation 3.8 can be rewritten in the form:

$$J^i \Delta v^i = -x(v^i), \quad (3.9)$$

where

$$J^i = \begin{bmatrix} \dfrac{\partial x_1}{\partial V_1} & \dfrac{\partial x_1}{\partial V_2} & \dfrac{\partial x_1}{\partial I_E} \\ \dfrac{\partial x_2}{\partial V_1} & \dfrac{\partial x_2}{\partial V_2} & \dfrac{\partial x_2}{\partial I_E} \\ \dfrac{\partial x_3}{\partial V_1} & \dfrac{\partial x_2}{\partial V_2} & \dfrac{\partial x_3}{\partial I_E} \end{bmatrix} = \begin{bmatrix} \dfrac{1}{1000} & \dfrac{-1}{1000} & 1 \\ \dfrac{-1}{1000} & \dfrac{1+3(V_2^i)^2}{1000} & 0 \\ 1 & 0 & 0 \end{bmatrix}$$

$$\Delta v^i = \begin{bmatrix} V_1^{i+1} - V_1^i \\ V_2^{i+1} - V_2^i \\ I_E^{i+1} - I_E^i \end{bmatrix},$$

and J^i is called the Jacobian of the system, and Δv^i is the change in the solution at the ith iteration. From the solution of equation 3.9, Δv^i is obtained and $v^{i+1} = v^i + \Delta v^i$ is computed. Note that equation 3.9 is a linear system of equations, and the Jacobian can be created using circuit stamps, as done

$$x(v^{i+1}) \approx \begin{bmatrix} x_1(v^i) + \dfrac{\partial x_1}{\partial V_1}(V_1^{i+1} - V_1^i) + \dfrac{\partial x_1}{\partial V_2}(V_2^{i+1} - V_2^i) + \dfrac{\partial x_1}{\partial I_E}(I_E^{i+1} - I_E^i) \\ x_2(v^i) + \dfrac{\partial x_2}{\partial V_1}(V_1^{i+1} - V_1^i) + \dfrac{\partial x_2}{\partial V_2}(V_2^{i+1} - V_2^i) + \dfrac{\partial x_2}{\partial I_E}(I_E^{i+1} - I_E^i) \\ x_3(v^i) + \dfrac{\partial x_3}{\partial V_1}(V_1^{i+1} - V_1^i) + \dfrac{\partial x_3}{\partial V_2}(V_2^{i+1} - V_2^i) + \dfrac{\partial x_3}{\partial I_E}(I_E^{i+1} - I_E^i) \end{bmatrix} \approx \mathbf{0} \quad (3.8)$$

previously for linear circuits. The MNA circuit stamp for a nonlinear voltage-controlled resistor $I_R = g(V_R)$, shown in Figure 3.1, is as follows:

$$\begin{array}{lcc} & \text{Col. } V_j & \text{Col. } V_k \\ \text{Row } j: & \dfrac{\partial g}{\partial V_j} & -\dfrac{\partial g}{\partial V_k} \\ \text{Row } k: & -\dfrac{\partial g}{\partial V_j} & \dfrac{\partial g}{\partial V_k} \end{array} \qquad \text{RHS} \begin{bmatrix} -g(V_R^i) \\ \\ g(V_R^i) \end{bmatrix}$$

The $V_R = V_j - V_k$ and the linear resistor are special cases of the nonlinear resistor. The difference from MNA formulation of linear circuits is that, for nonlinear circuits, there exists an entry in RHS for all elements. For a current-controlled nonlinear resistor in the form $V_R = r(I_R)$, the impedance form of MNA formulation is used, and the current through the resistor I_R is added to the vector of the unknowns. The additional equation is $V_j - V_k - r(I_R) = 0$.

The Newton–Raphson method is summarized in the following steps:

1. Set the iteration count $i = 0$, and estimate the initial guess of $\boldsymbol{v}^0$.
2. Calculate the Jacobian $\boldsymbol{J}^i$ and right-hand side of equation 3.9, which is $-\boldsymbol{x}(\boldsymbol{v}^i)$.
3. Solve equation 3.9 for $\Delta \boldsymbol{v}^i$.
4. Update the solution vector $\boldsymbol{v}^{i+1} = \boldsymbol{v}^i + \Delta \boldsymbol{v}^i$.
5. Calculate the error in the solution as the norm of the vector $\|\boldsymbol{x}(\boldsymbol{v}^{i+1})\|$. A common method is to use the norm defined as $L_2 = \|\boldsymbol{x}(\boldsymbol{v}^{i+1})\|_2 = \sqrt{\Sigma_j (x_j)^2}$. If the error is small then stop, or instead set $i = i + 1$ and go to step 2.

Under fairly general conditions, it can be shown that if the initial guess is close to the solution, then the Newton–Raphson method converges quadratically to the solution. For the circuit in Figure 3.6, if the initial guess $\boldsymbol{v}^0 = [0 \quad 0 \quad 0]^T$ is used, then the iterations for nodal voltage V_2 are given in Table 3.2.

Note that the last column in the table shows that the right-hand side of equation 3.9 approaches zero with each iteration, indicating that the final value is being reached and the iterations can be stopped when the value is sufficiently close to zero. Also note that with each iteration, the change in the solution is becoming smaller, indicating that one is getting closer to the exact solution. Since the circuit is nonlinear, the Jacobian must be computed for each iteration and is usually the most expensive portion of the entire solution method. If a better initial guess is made, such as $\boldsymbol{v}^0 = [1.0 \quad 0.7 \ -0.003]^T$, then the NR method will converge to the final value in fewer iterations.

The algorithm just described is the basic Newton–Raphson method that works well in most, but not all, cases. Modifications to this method as well as to other methods for solving nonlinear equations can be found in references listed at the end of this chapter.

TABLE 3.2 Newton–Raphson Iterations

Iteration i	V_2^{i+1}	ΔV_2^i	$\|\boldsymbol{x}(\boldsymbol{v}^{i+1})\|_2$
0	1.0000	1.0000	0.0316
1	0.7500	−0.2500	0.0131
2	0.6860	−0.0640	0.0030
3	0.6823	−0.0037	0.0002
4	0.6823	−0.0000	0.0000

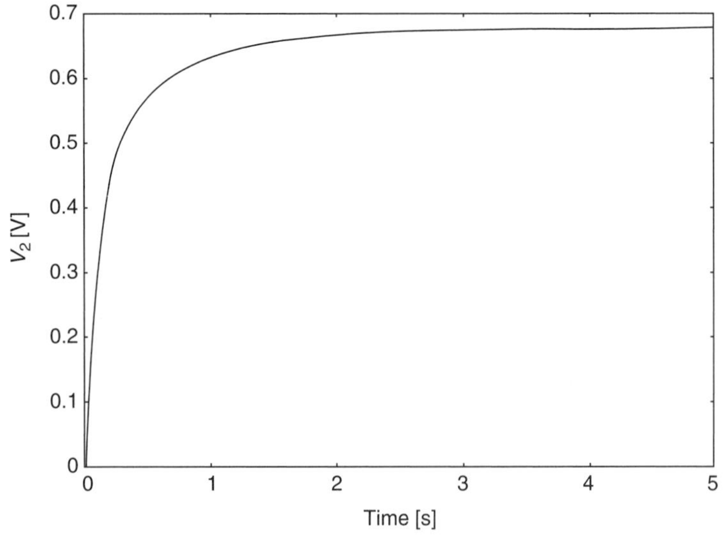

FIGURE 3.6 Time Response of a Nonlinear Circuit

3.5 Transient Analysis of Nonlinear Circuits

Another very useful analysis is the time response of nonlinear circuits. This requires the solution of the differential equation 3.6, and once again numerical methods are used. Usually the solution of the network at time $t = 0$ is known, and the response is calculated after a time step of h at $t + h$. All numerical solution methods replace the time derivative in equation 3.6 with a suitable approximation, and then the resulting equations are solved. Expanding the solution in a Taylor series, the following is obtained:

$$\mathbf{v}(t+h) = \mathbf{v}(t) + h\frac{d}{dt}\mathbf{v}(t) + \ldots \text{ higher order terms.}$$

Assuming that the time step h is small and ignoring the higher order terms as being small, the method called the Euler forward method is used. The method approximates the derivative with:

$$\dot{\mathbf{v}}(t) = \frac{d}{dt}\mathbf{v}(t) \approx \frac{\mathbf{v}(t+h) - \mathbf{v}(t)}{h}. \tag{3.10}$$

Substituting equation 3.10 in equation 3.6 yields:

$$f\left(\frac{\mathbf{v}(t+h) - \mathbf{v}(t)}{h},\ \mathbf{v}(t),\ t\right) = 0 = \mathbf{x}(\mathbf{v}(t+h)),$$

which is a nonlinear equation with the unknown value $\mathbf{v}(t+h)$. It can be written in the form $\mathbf{x}(\mathbf{v}(t+h)) = 0$ and solved using the Newton–Raphson method given previously. Experience has shown that this method works well with small step sizes; however, large step sizes give results that diverge from the true solution of the network. Such methods are called unstable and are generally not used in integration of differential equations. A better solution method is to use the Euler backward method for which the derivative is approximated by a Taylor series expansion around the time point $\mathbf{v}(t+h)$:

$$\mathbf{v}(t) = \mathbf{v}(t+h) + (-h)\frac{d}{dt}\mathbf{v}(t+h) + \ldots \text{ higher order terms.}$$

This equation can be rewritten in the form:

$$\dot{\mathbf{v}}(t+h) = \frac{d}{dt}\mathbf{v}(t+h) \approx \frac{\mathbf{v}(t+h) - \mathbf{v}(t)}{h}. \tag{3.11}$$

Substituting equation 3.11 in equation 3.6 results in:

$$f\left(\frac{\mathbf{v}(t+h) - \mathbf{v}(t)}{h},\ \mathbf{v}(t+h),\ t+h\right) = 0 = \mathbf{x}(\mathbf{v}(t+h)), \tag{3.12}$$

which is a nonlinear equation in the variable $\mathbf{v}(t+h)$ and is solved using the Newton–Raphson method. The last step is called the discretization of the differential equation, and this step changes it into an algebraic equation. For example, substituting equation 3.11 in equation 3.5 results in:

$$\begin{bmatrix} \dfrac{V_1(t+h) - V_2(t+h)}{1000} + I_E(t+h) \\ \dfrac{V_2(t+h) - V_1(t+h)}{1000} + \dfrac{(V_2(t+h))^3}{1000} + \dfrac{Q_C(V_2(t+h)) - Q_C(V_2(t))}{h} \\ V_1(t+h) - E(t+h) \end{bmatrix} = \mathbf{0},$$

where the Jacobian will be:

$$J^i = \begin{bmatrix} \dfrac{1}{1000} & \dfrac{-1}{1000} & 1 \\ \dfrac{-1}{1000} & \dfrac{1 + 3(V_2^i)^2}{1000} + \dfrac{3(V_2^i)^2}{1000h} & 0 \\ 1 & 0 & 0 \end{bmatrix}$$

The explicit dependence on time has been dropped for simplicity. The Jacobian can be created using circuit stamps for the nonlinear elements as done in the case of dc solution of nonlinear circuits. For a nonlinear voltage-controlled capacitor, such as:

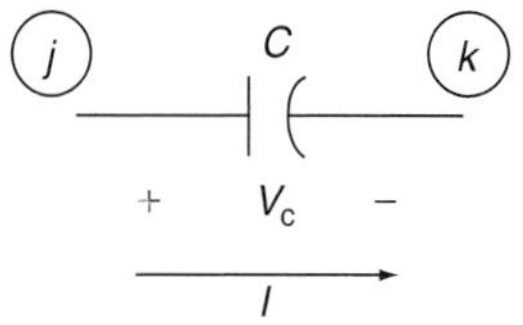

defined by the equation $Q_C = q(V_C) = 0.001(V_C)^3$ where $V_C = V_j - V_k$, the circuit stamp is obtained by writing KCL equations at nodes j and k:

$$\text{KCL at node } j{:} \ldots + \frac{d}{dt}Q_C(t+h) \ldots = 0.$$

$$\text{KCL at node } k{:} \ldots - \frac{d}{dt}Q_C(t+h) \ldots = 0.$$

Using the Euler backward method, the differentials are discretized by:

$$\text{KCL at node } j{:} \ldots + \frac{Q_C(t+h) - Q_C(t)}{h} \ldots = 0.$$

$$\text{KCL at node } k{:} \ldots - \frac{Q_C(t+h) - Q_C(t)}{h} \ldots = 0.$$

The Jacobian and RHS entries become:

$$\begin{array}{ll} & \begin{array}{ccc} \text{Col.} & \text{Col.} & \\ V_j & V_k & RHS \end{array} \\ \begin{array}{l} \text{Row } j: \\ \\ \text{Row } k: \end{array} & \begin{bmatrix} \frac{1}{h}\frac{\partial Q_C}{\partial V_j} & -\frac{1}{h}\frac{\partial Q_C}{\partial V_k} \\ -\frac{1}{h}\frac{\partial Q_C}{\partial V_j} & \frac{1}{h}\frac{\partial Q_C}{\partial V_k} \end{bmatrix} \begin{bmatrix} \frac{-Q_C(V_C^i(t+h)) - Q_C(V_C(t))}{h} \\ \frac{Q_C(V_C^i(t+h)) - Q_C(V_C(t))}{h} \end{bmatrix} \end{array} \tag{3.13}$$

Note that in deriving equation 3.13, the use of the chain rule of differentiation and then discretizing $(d/dt)V_C$ has been avoided:

$$\frac{d}{dt}Q_C = \frac{d}{dV_C}Q_C\frac{d}{dt}V_C.$$

The value of $(d/dt)V_C$ has also been discretized. Discretization applied directly to the charge of the capacitor (and flux in case of inductors) leads to better results that maintain charge (and flux) conservation in the numerical solution of electrical circuits. Additional details can be found in work by Kundert (1995).

For the circuit in Figure 3.5, let the capacitor be initially discharged such that $V_2(0) = 0$ and the correct initial condition vector is $\boldsymbol{v}(0) = [1.0\ 0.0 - 0.001]^T$. For this circuit, it is expected that the capacitor will charge up from 0 V to a finite value with time. All nodal voltages in the circuit will remain less than the source voltage of 1 V. If step $h = 0.1$ s is taken, and the initial guess is $\boldsymbol{v}^0(h) = \boldsymbol{v}(0)$, the Newton–Raphson iteration results will be obtained as shown in Table 3.3.

Once the solution at $t = 0.1$ s is obtained, the values from the circuit conditions at that time are used, another time step is taken, and the response at $t = 2h = 0.2$ s is calculated. This procedure is repeated until the end time of interest is reached for computing the response. Although the circuit has a continuous response, the numerical methods are used to compute the response at discrete instants of time. It is expected that if small time steps are taken, then no interesting features of the response will be missed. Note that at each time step, the Newton–Raphson method is used to solve a set of nonlinear equations. Moreover, since the circuit is nonlinear, the Jacobian has to be recomputed at each iteration and each time step. The complete time domain response of the circuit is given in

TABLE 3.3 Newton–Raphson Iterations for $t = 0$ and $h = 0.1$ s

Iteration i	V_2^{i+1}	ΔV_2^i
0	1.0000	1.0000
1	0.6765	−0.3235
2	0.4851	−0.1914
3	0.4006	−0.0845
4	0.3835	−0.0171
5	0.3828	−0.0007

Figure 3.6. The final value $V_2 = 0.6823$ V is reached at about $t = 8$ s. Note that the final value is the same as that obtained in the dc solution above.

Choosing a large step size $h = 10$ s and calculating the time response will give the results $V_2(10) = 0.6697$ V, $V_2(20) = 0.6816$ V, $V_2(30) = 0.6823$ V,.... Note that even though the result at $t = 10$ s is not accurate, the error does not grow with each time step and the correct value is reached after a few more steps. Such time domain integration methods are called stable and are preferred over the Euler Forward method that may not give correct results for large time steps.

Finally, this section discusses some issues concerning the use of numerical methods for the solution of differential equations and gives some recommendations for obtaining correct results:

1. The discretization equation 3.11 is an approximation, since only the first two terms in the Taylor series expansion are included. As a result of this approximation, the numerically computed results contain an error at each time step, called the local truncation error (LTE). In general, the smaller the time step used, the smaller will be the LTE. The normal expectation is that if the error is reduced at each time step, then the overall error in the numerical results will also be small.
2. This discussion concerning the LTE suggests that if a discretization formula is used that matches more than two terms in the Taylor series expansion, then more accurate results will be obtained. This is also correct, provided the same step size is used for both discretization methods, or alternatively the second method will allow larger step sizes for a given error.
3. This discussion also suggests that it is not necessary to use the same time step when determining the time domain solution of a circuit over a long period of time. It is possible to change the time step during the computation based on error and accuracy requirements.
4. The Newton–Raphson method is used at each time step to solve a set of nonlinear equations. For this method, a good initial guess gives the solution in fewer iterations. Such formulae, called predictor formulae, are available and reduce the overall computation needed for the solution.
5. The issue of stability of the response with large step sizes has already been mentioned. A stable integration method is preferred. Stable methods ensure that the error at each time step does not accumulate over time and reasonably accurate results are obtained.

Additional reasons for these recommendations, as well as other integration methods for time domain solution of nonlinear circuits, can be found in the references at the end of this chapter.

References

Kundert, K. (1995). *The designer's guide to Spice and Spectre.* Boston: Kluwer Academic Publishers.

Ruehli, A. (Ed.). (1986). *Circuit analysis, simulation, and design vol. 3: Advances in CAD for VLSI parts 1 and 2.* Amsterdam: North-Holland.

Vlach, J. and Singhal, K. (1994). *Computer methods for circuit analysis and design.* New York: Van Nostrand Reinhold.

4

Synthesis of Networks

Jiri Vlach
Department of Electrical and Computer Engineering, University of Waterloo, Waterloo, Ontario, Canada

4.1 Introduction

Synthesis of electrical networks is an area of electrical engineering in which one attempts to find the network from given specifications. In most cases, this is applied to filters constructed with various elements.

This chapter begins with a few words about history. Before the second world war, the main communication products were radios working with amplitude modulation. Stations transmitted on various frequencies, and it was necessary to get only the desired one and suppress all the others. This led to the development of various filters. Two men contributed fundamentally to the filter design theory: Darlington, from the United States, and Cauer, from Germany.

As time went by, it was discovered that inductor capacitor (LC) filters were not suitable for many applications, especially in low frequency regions where inductors are large and heavy. An idea occurred: replace the inductors by active networks composed of amplifiers with feedback by means of resistors and/or capacitors. This was the era of active networks, approximately in the 1960s to 1980s of the 20th century.

Technological advances and miniaturization led to integrated circuits, with the attempt to get complete filters on a chip. Here it turned out that using resistors was not convenient, and a new idea emerged of switched capacitor networks. In these networks, the resistors are replaced by rapidly switched capacitors that act approximately as resistors. In such networks, we have only capacitors and transistors work as amplifiers or switches. Capacitors and transistors are suitable for integration.

New developments continue to find their ways. This Chapter limits explanations mostly to LC filters. Switched capacitor networks and later theoretical developments are too complex to cover in this short contribution.

Returning now to filter synthesis, this discussion must first clarify the appropriate theoretical tools. Filters are linear devices that allow the use of linear network theories. Specifications of filters are almost always given as frequency domain responses, which leads to the use of Laplace transform. This chapter assumes that the reader is at least somewhat familiar with the concept of the transformation because the discussion deals with the complex plane, the complex variable s, frequency domain responses, poles, and zeros. A sufficient amount of information is given so that a reader can refresh his or her knowledge.

4.2 Elementary Networks

Filters can be built with passive elements that do not need any power supply to retain their properties and with active elements that work only when electrical power is supplied from a battery or from a power supply.

The passive elements are inductors (L), capacitors (C), and resistors (R). Resistance of the resistor is measured in ohms. Its inverse value is a conductance, $G = 1/R$. For inductors and capacitors, we use the Laplace transform; in such a case, we speak about impedances of these elements: $Z_L = sL$ and $Z_C = 1/sC$. The inverse of the impedance is the admittance, $Y_C = sC$ and $Y_L = 1/sL$. The subscripts used here are for clarification only and are usually not used. Symbols for these elements are in Figure 4.1. Such networks have two terminals, sometimes called a port. Notice the voltage signs and positive directions of the currents. This is an important convention because it is also valid for independent voltage and current sources, shown in Figure 4.2. We use the letter E for the independent voltage source and the letter J for the independent current source to distinguish them from voltages and currents anywhere inside the network.

From the above description of a one-port network, we can refer to the concept of a two-port network, usually drawn so that the input port is on the left and the output port is on the right. We can introduce a general symbol for the two-port as shown in Figure 4.3. Notice that in such a case we speak about the input voltage across the left one-port, V_1, and the output voltage across the right one-port, V_2. The plus sign is on top and minus sign is at the bottom. We also indicate the currents I_1 and I_2 as shown. Notice how the currents flow into the two-port and how they leave it. It is very important to know these directions.

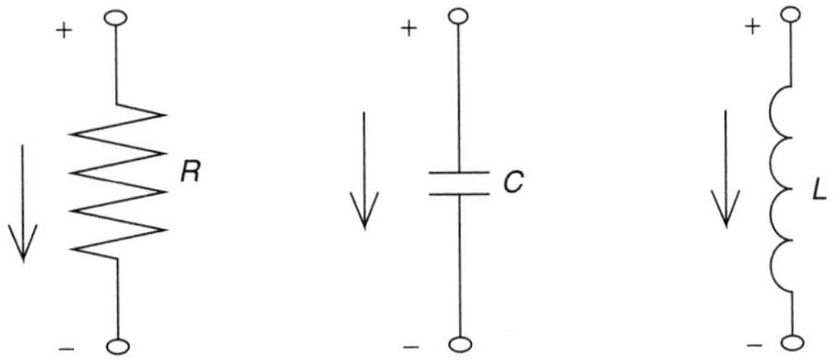

FIGURE 4.1 Symbols for a Resistor, Capacitor, and Inductor

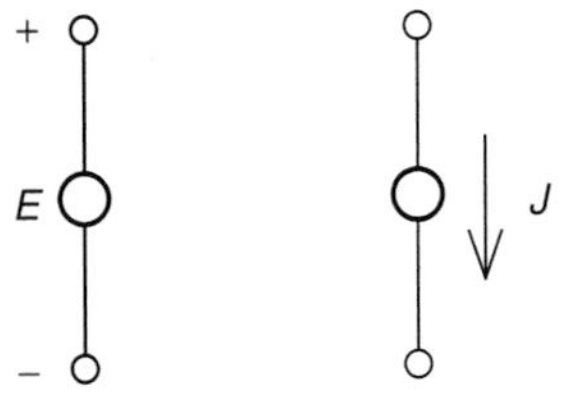

FIGURE 4.2 Symbols for Independent Voltage (E) and Independent Current (J) Source

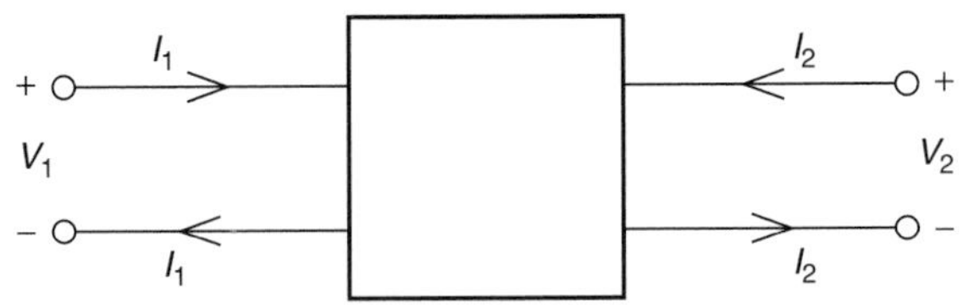

FIGURE 4.3 Symbol for a Two-Port

There exist four elementary two-ports, shown in Figure 4.4. They are the most simplified forms of amplifiers. The voltage-controlled voltage source, *VV*, has its output controlled by the input voltage:

$$V_2 = \mu V_1. \tag{4.1}$$

The μ is a dimensionless constant. Another elementary two-port is a voltage-controlled current source, *VC*, described by the equation:

$$I_2 = gV_1, \tag{4.2}$$

where g is transconductance and has the dimension of a conductance. The third elementary two-port is the current-controlled voltage source, *CV*, defined by:

$$V_2 = rI_1, \tag{4.3}$$

where r represents transresistance and has the dimension of a resistor. The last such two-port is a current-controlled current source, *CC*, defined by:

$$I_2 = \alpha I_1, \tag{4.4}$$

where α is a dimensionless constant.

Any number of elements can be variously connected; if we consider on such a network an input port and an output port, we will have a general two-port. This two-port has four variables, V_1, V_2, I_1, and I_2, as in Figure 4.3. Any two can be selected as independent variables, and the other two will be dependent variables. For instance, we can consider both currents as independent variables, which will make the voltages dependent variables. To couple them in a most general form, we can write the following:

$$\begin{aligned} V_1 &= z_{11}I_1 + z_{12}I_2. \\ V_2 &= z_{21}I_1 + z_{22}I_2. \end{aligned} \tag{4.5}$$

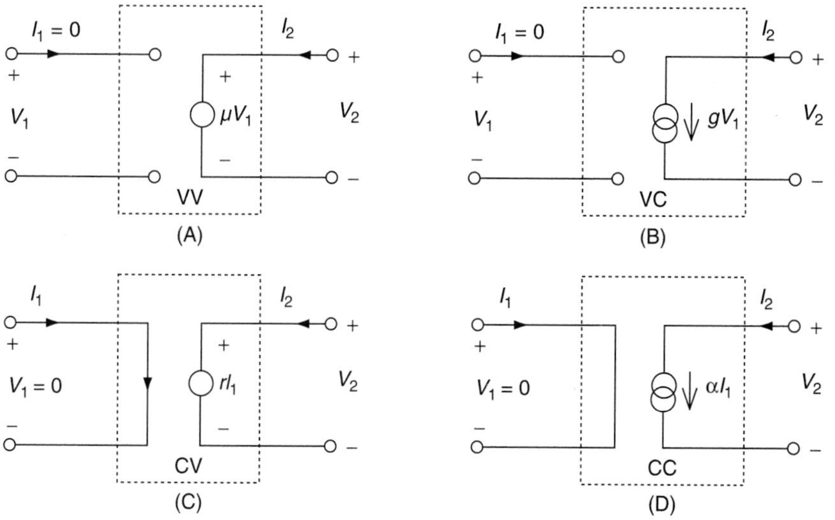

FIGURE 4.4 Simplest Two-Ports

In equation 4.5, the z_{ij} have dimensions of impedances, and we speak about the impedance description of the network. The equations are usually cast into a matrix equation:

$$\begin{bmatrix} V_1 \\ V_2 \end{bmatrix} = \begin{bmatrix} z_{11} & z_{12} \\ z_{11} & z_{22} \end{bmatrix} \begin{bmatrix} I_1 \\ I_2 \end{bmatrix} \tag{4.6}$$

Another way of expressing the dependences is to select the voltages as independent variables and the currents as dependent variables and present them in the form of two equations:

$$\begin{aligned} I_1 &= y_{11}V_1 + y_{12}V_2. \\ I_2 &= y_{21}V_1 + y_{22}V_2. \end{aligned} \tag{4.7}$$

This is an admittance description of a two-port, and in matrix form it is:

$$\begin{bmatrix} I_1 \\ I_2 \end{bmatrix} = \begin{bmatrix} y_{11} & y_{12} \\ y_{21} & y_{22} \end{bmatrix} \begin{bmatrix} V_1 \\ V_2 \end{bmatrix} \tag{4.8}$$

There exist additional possibilities on how to couple the variables. They can be found in any textbook on network theory.

For demonstration, we use the network in Figure 4.5 and derive its z_{ij} parameters. We attach a voltage source V_1 on the left and nothing on the right. This means that $I_2 = 0$, and the voltage V_2 appears in the middle of the network as indicated. We see that $V_1 = (Z_1 + Z_3)I_1$ and $z_{11} = V_1/I_1 = Z_1 + Z_3$. In addition, since the voltage in the middle of the network is V_2, we can write $V_2 = I_1Z_3$ and $z_{21} = V_2/I_1 = Z_3$. In the next step, we place the voltage source on the right and proceed similarly. The result is:

$$\begin{bmatrix} z_{11} & z_{12} \\ z_{21} & z_{22} \end{bmatrix} = \begin{bmatrix} Z_1 + Z_3 & Z_3 \\ Z_3 & Z_2 + Z_3 \end{bmatrix} \tag{4.9}$$

In the next example, we consider the connection of a two-port to a loading resistor, as sketched in Figure 4.6. We wish to find the input impedance of the combination. After we connect them, as indicated by the dashed line, there is the same voltage, V_2, across the second port and across the resistor. The current flowing into the resistor will be $-I_2$ and thus $V_2 = -I_2R$. Inserting into the second equation of the set (4.5), we obtain this equation:

$$I_2 = \frac{-z_{21}}{R + z_{22}} I_1.$$

Replacing I_2 in the first equation by this result, we get:

$$Z_{in} = V_1/I_1 = \frac{z_{11}R + z_{11}z_{22} - z_{12}z_{21}}{R + z_{22}}. \tag{4.10}$$

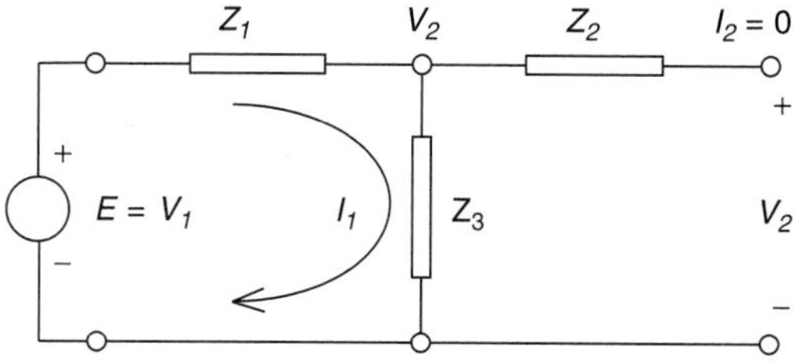

FIGURE 4.5 Finding Z Parameters for the Network

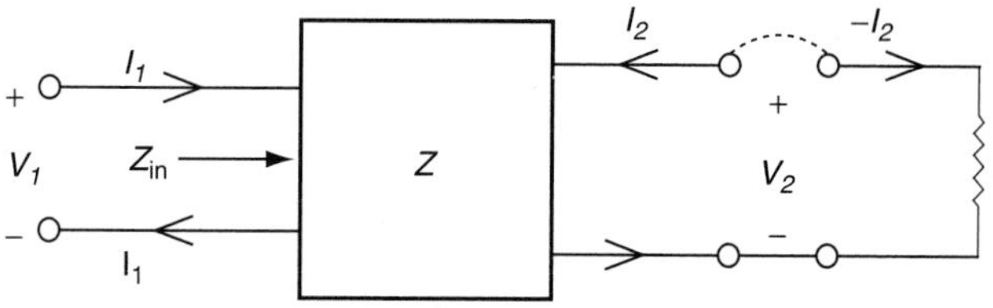

FIGURE 4.6 Obtaining Input Impedance of a Two Port Loaded by a Resistor

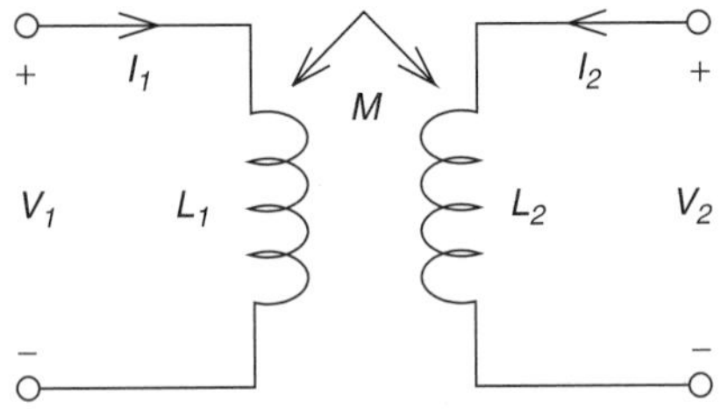

FIGURE 4.7 Symbol for a Technical Transducer

We will conclude this section by introducing two more two-ports, namely the ideal and the technical transformer[5]. The ideal transformer, often used in the synthesis theory, is described by the equations:

$$\begin{aligned} V_2 &= nV_1. \\ I_2 &= -\frac{1}{n}I_1. \end{aligned} \tag{4.11}$$

We note that we cannot put equation 4.11 into any of the two matrix forms that we introduced (impedance or admittance form). A technical transformer is realized by magnetically coupled coils. The device is described by the equations:

$$\begin{aligned} V_1 &= sL_1I_1 + sMI_2. \\ V_2 &= sMI_1 + sL_2I_2 \end{aligned}. \tag{4.12}$$

The L_1 and L_2 are the primary and secondary inductances, and M is the mutual inductance. These variables can be expressed in the impedance form as:

$$\begin{bmatrix} V_1 \\ V_2 \end{bmatrix} = \begin{bmatrix} sL_1 & sM \\ sM & sL_2 \end{bmatrix} \begin{bmatrix} I_1 \\ I_2 \end{bmatrix} \tag{4.13}$$

Figure 4.7 Shows the technical transformer.

4.3 Network Functions

In the second section, we introduced two-port networks with their input and output variables. We now introduce the concept of networks with only one source and one output. As an example, take the network in Figure 4.8. The output voltage is:

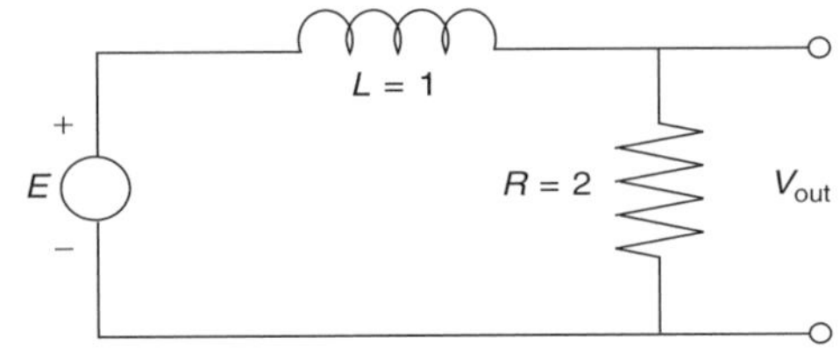

FIGURE 4.8 Introducing a Network Function

$$V_{out} = \frac{2}{s+2}E, \tag{4.14}$$

where E is any independent signal voltage. If we do not consider a specific signal but rather divide the equation by E, we get the voltage transfer function:

$$T_V = \frac{V_{out}}{E} = \frac{2}{s+2}. \tag{4.15}$$

This is one of the possible network functions. Had we chosen a larger network, the ratio would be in the form:

$$T_V = \frac{V_{out}}{E} = \frac{N(s)}{D(s)}, \tag{4.16}$$

where $N(s)$ and $D(s)$ are polynomials in the variable s.

One network may have many network functions. The number depends on where we apply the independent source, which source we select, and where we take the output. To be able to define the network function, we must satisfy two conditions: (1) the network is linear, and (2) there are zero initial conditions on capacitors and inductors.

If these conditions are satisfied, then we define a general network function as the ratio:

$$\text{Network function} = \frac{\text{output}}{\text{input}}. \tag{4.17}$$

Depending on the type of independent input source (voltage or current) and the type of the output (voltage or current), we can define the following network functions:

1. Voltage transfer is $T_V = \dfrac{V_{out}}{E}$.
2. Current transfer is $T_I = \dfrac{I_{out}}{J}$.
3. Transfer impedance is $Z_{TR} = \dfrac{V_{out}}{J}$.
4. Transfer admittance is $Y_{TR} = \dfrac{I_{out}}{E}$.
5. Input impedance is $Z_{in} = \dfrac{V_{in}}{J}$.
6. Input admittance is $Y_{in} = \dfrac{I_{in}}{E}$.

The input impedance is the inverse of the input admittance, but this is not true for the transfer impedances and admittances.

As an example, we take the network in Figure 4.9. Using nodal analysis, we can write the equations:

$$\begin{aligned} V_1(G_1 + G_2 + sC) - G_2V_2 &= J. \\ -G_2V_1 + (G_2 + 1/sL)V_2 &= 0. \end{aligned}$$

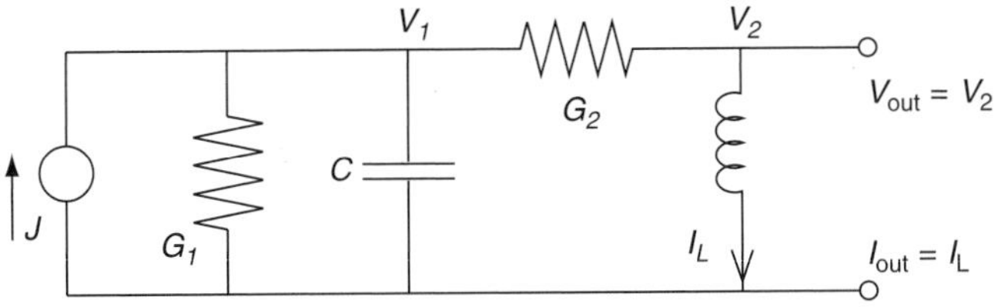

FIGURE 4.9 Finding Two Network Functions of One Network

They can also be written in matrix form:

$$\begin{bmatrix} (G_1 + G_2 + sC) & -G_2 \\ -G_2 & (G_2 + 1/sL) \end{bmatrix} \begin{bmatrix} V_1 \\ V_2 \end{bmatrix} = \begin{bmatrix} J \\ 0 \end{bmatrix}$$

The denominator of the transfer function is the determinant of the system matrix:

$$D = sCG_2 + (G_1G_2 + C/L) + (G_1 + G_2)/sL.$$

Using Cramer's rule, we obtain:

$$V_{out} = \frac{G_2 J}{D}.$$

Since $I_{out} = V_{out}/sL$, we also get

$$I_{out} = (G_2/sL)/D.$$

We remove the complicated fractions by multiplying the numerator and the denominator by sL, with the result:

$$Z_{TR} = \frac{sLG_2}{s^2LCG_2 + s(LG_1G_2 + C) + G_1 + G_2}.$$

$$T_I = \frac{G_2}{s^2LCG_2 + s(LG_1G_2 + C) + G_1 + G_2}.$$

This example demonstrated that the network functions are ratios of polynomials in the variable s. If the elements are given by their numerical values, we can find roots of such polynomials. Roots of the numerator polynomial are called **zeros**, and roots of the denominator are called **poles**. From mathematics, we know that a polynomial has many zeros just as the highest power of the s variable. In the above example, the denominator is a polynomial of second degree, and the network has two poles. Since the numerator in this example is only a constant, the network will not have any (finite) zeros.

In general, the poles (or zeros) can be real or can appear in complex conjugate pairs. We can draw them in the complex plane, and we can use crosses to mark the poles and use small circles to mark the zeros. Figure 4.10 shows the pole-zero plot of a network with one real and two complex conjugate poles and with two purely imaginary zeros.

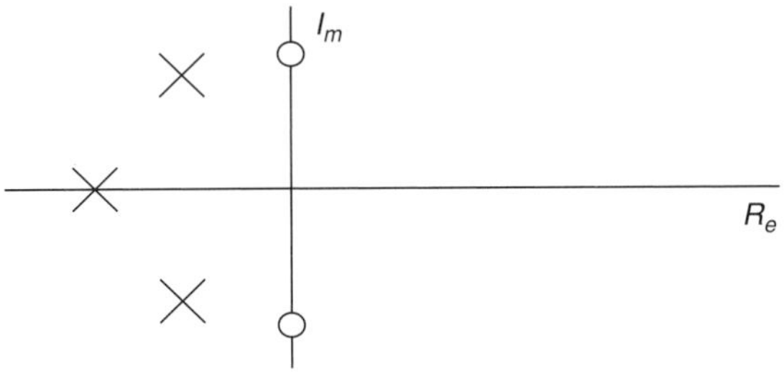

FIGURE 4.10 Pole and Zero Positions of Some Low-Pass Filter

4.4 Frequency Domain Responses

Let us now consider any one of the above network functions and denote it by the letter F. The function will be the ratio of two polynomials in the variables s, $N(s)$, and $D(s)$. If we use a root-finding method, we can also get the poles, p_i, and zeros, z_i. As a result, for a general network function, we can use one of the two following forms:

$$F(s) = \frac{N(s)}{D(s)} = K\frac{\prod(s - z_i)}{\prod(s - p_j)}. \tag{4.18}$$

To obtain various network responses, we substitute

$$s = j\omega. \tag{4.19}$$

$$\omega = 2\pi f = \frac{2\pi}{T}. \tag{4.20}$$

In equations 4.19 and 4.20, f is the frequency of the signal we applied, and $f = 1/T$. Using one frequency and substituting it into equation 4.18, we obtain one complex number with a real and imaginary part:

$$F = x + jy = \mathrm{R}e^{j\phi} \tag{4.21}$$

Here

$$R = \sqrt{x^2 + y^2} \tag{4.22}$$

and

$$\tan\phi = \frac{y}{x}. \tag{4.23}$$

This is sketched in Figure 4.11.

If we evaluate the absolute value or the angles for many frequencies and plot the frequency on the horizontal axis and the resulting values on the vertical axis, we obtain the amplitude and the phase responses of the network.

Calculation of the frequency responses is always done by a computer. We assume that the reader knows how this calculation is done, but we show the responses for the filter in Figure 4.12.

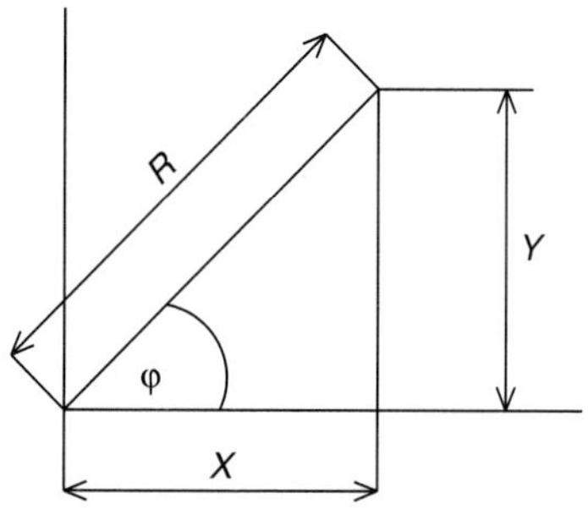

FIGURE 4.11 Finding the Absolute Value and Phase in Complex Plane

We have several possibilities for plotting them: the horizontal axis can be linear or logarithmic, the amplitude responses can be in absolute values, $|F_i| = |F(j\omega_i)|$, or $|F_{i,\mathrm{dB}}| = 20\log|F_i|$ (in decibels, which is a logarithmic scale). The four possible plots of the same responses are in Figures 4.13(A) through 4.13(D). Clearly, selecting a correct scale and plot is very important for proper presentation and understanding.

The phase responses can also be plotted, but we have a couple of problems: the tangent function is periodic, and the plots experience jumps. For this reason, the phase response is rarely plotted, and we prefer to use the group delay, defined by:

$$\tau = -\frac{d\phi}{d\omega}. \tag{4.24}$$

Plot of the group delay for the filter in Figure 4.12 is in Figure 4.14. We see that the amplitude response approximates fairly well a constant in the passband, but the group delay has a large peak. Not every program has a built-in evaluation of the group delay, but there is a remedy: every computer has a polynomial root-finding routine. We can calculate the poles and zeros and use the second formula in equation 4.18. Let there be M zeros

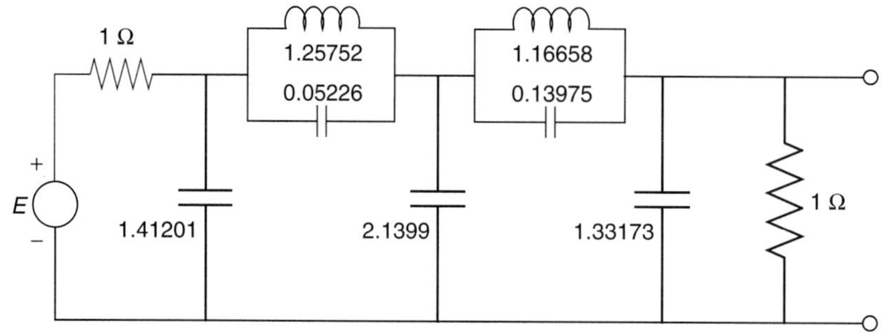

FIGURE 4.12 Example of a Low-Pass Filter with Transfer Zeros

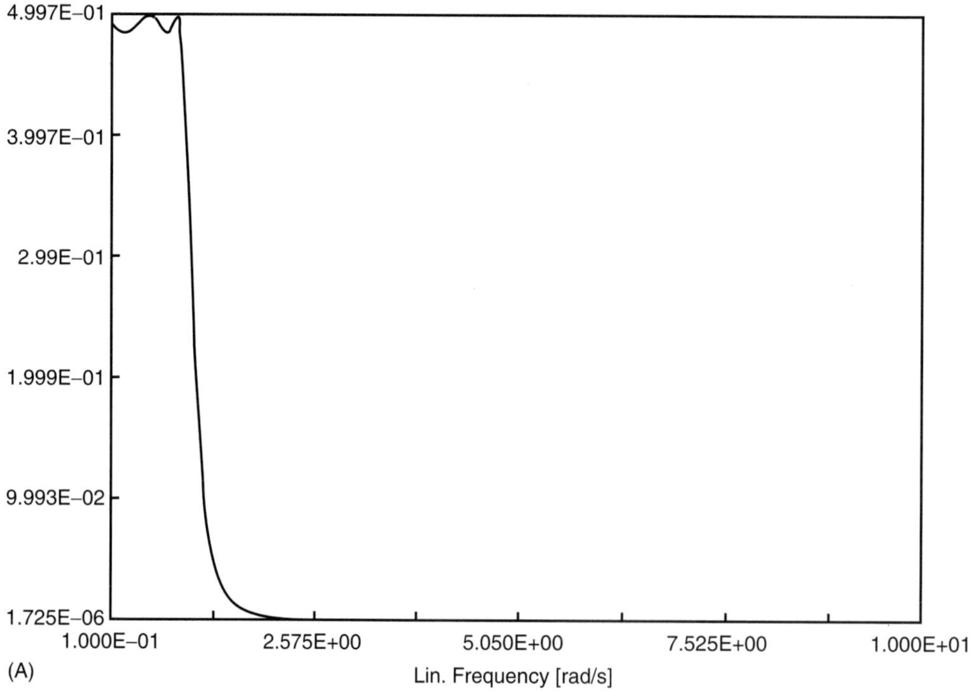

FIGURE 4.13 (A) LC Filter with Transmission Zeros AC Analysis, Magnitude of V_4, Linear Frequency Scale

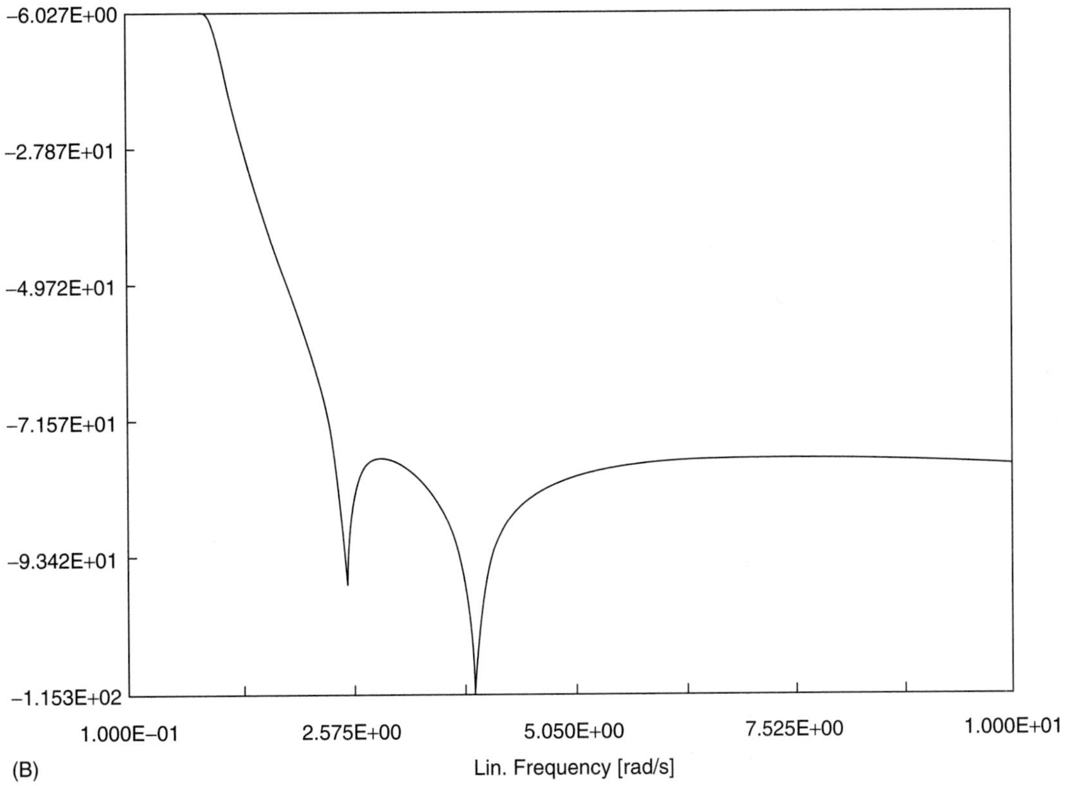

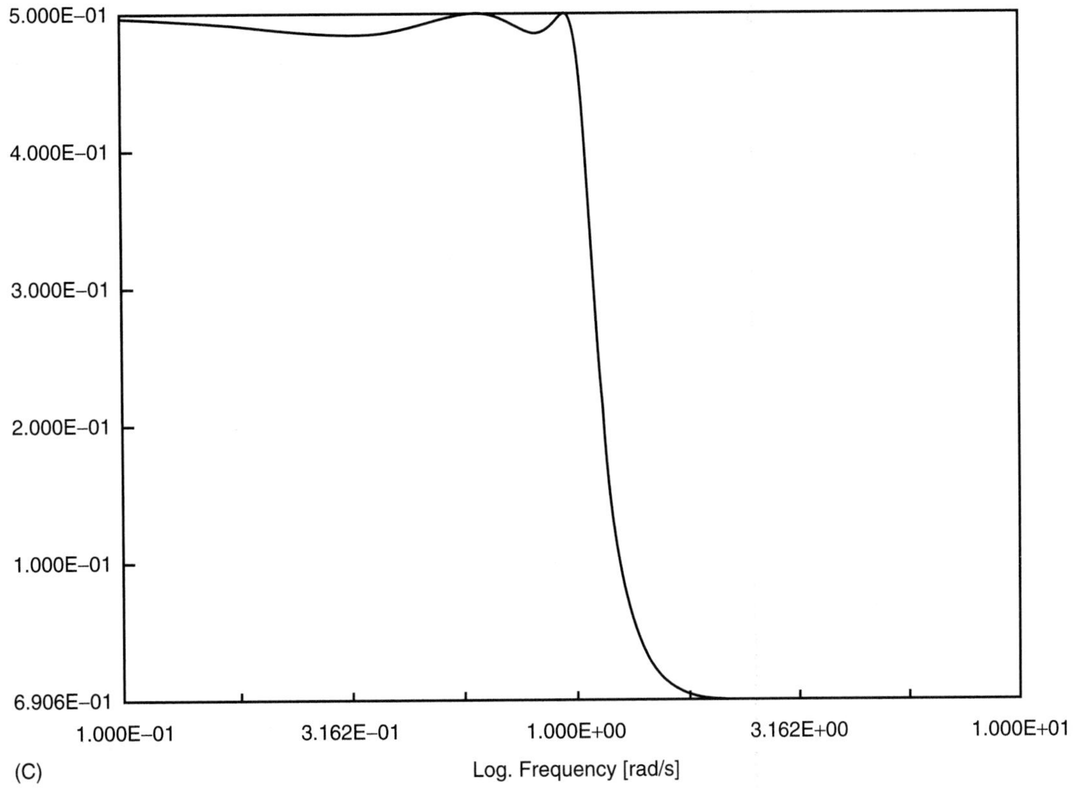

FIGURE 4.13 (cont'd) (B) Same Filter, dB of V_4, Linear Frequency Scale (C) Same Filter, Magnitude of V_4, Logarithmic Frequency Scale

continued

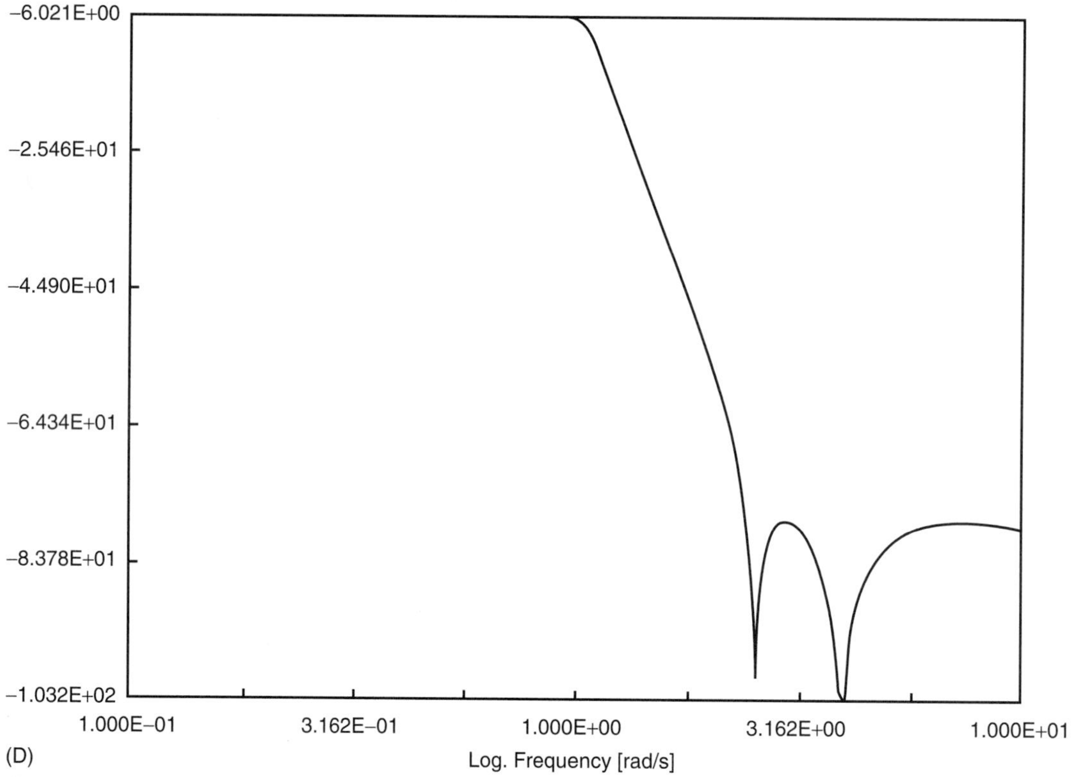

FIGURE 4.13 (cont'd) (D) Same Filter, dB of V_4, k Logarithmic Scale

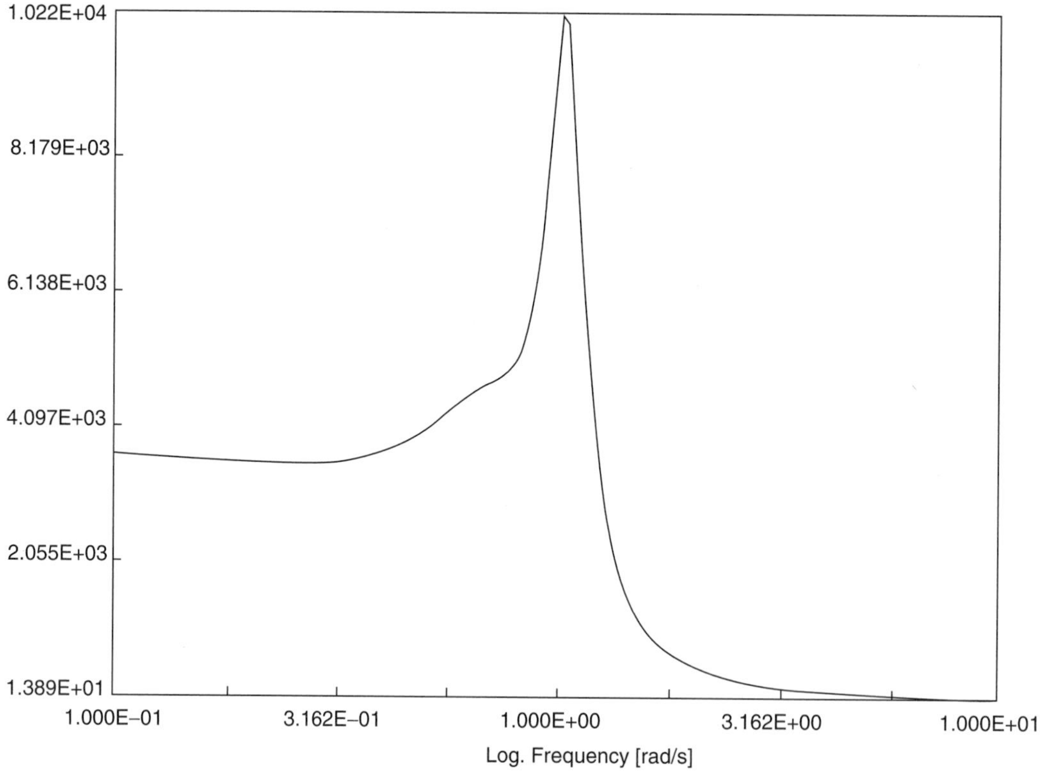

FIGURE 4.14 LC Filter, Group Delay of V_4, in ms, Logarithmic Frequency Scale

$z_i = \alpha_i + \beta_i$ and N poles $p_i = \gamma_i + \delta_i$. Substituting $s = j\omega$ and taking the absolute value, we arrive at the formula:

$$|F| = K \left[\frac{\prod_{i=1}^{M} [\alpha_i^2 + (\omega - \beta_i)^2]}{\prod_{i=1}^{N} [\gamma_i^2 + (\omega - \delta_i)^2]} \right]^{1/2} \quad (4.25)$$

The phase response will be as follows:

$$\Phi = \sum_{i=1}^{M} \arctan \frac{\beta_i - \omega}{\alpha_i} - \sum_{i=1}^{N} \arctan \frac{\delta_i - \omega}{\gamma_i}. \quad (4.26)$$

Differentiating equation 4.26 with respect to ω_o provides the formula for the group delay:

$$\tau(\omega) = \sum_{i=1}^{M} \frac{\alpha_i}{\alpha_i^2 + (\omega - \beta_i)^2} - \sum_{i=1}^{N} \frac{\gamma_i}{\gamma_i^2 + (\omega - \delta)^2}. \quad (4.27)$$

All formulas are easily programmed and plotted.

4.5 Normalization and Scaling

Normalization is one of the most useful concepts in linear networks. It allows us to reduce one frequency and the network impedance to unit values without losing any information. Consider first the scaling of impedances. If we increase (decrease) the impedance of every element of a filter described by its voltage or current transfer function, the input–output relationship should not change. It is thus customary to apply scaling, which reduces one (any) resistor to the value of 1. We can also scale the frequency so that the cutoff frequency of a low-pass filter or the center frequency of a band-pass filter is reduced to 1 rad/s.

To derive the necessary formulas, we will denote the scaled values by the subscript s and leave the values to be used for realization without subscript. Impedance scaling means that the impedance of every network element must be scaled by the same constant, k. This immediately leads to $R_s = R/k$. For frequency normalization, we introduce the formula:

$$\omega_s = \frac{\omega}{\omega_0}, \quad (4.28)$$

where ω_0 is a constant by which we wish to scale the frequency. Now consider the impedance of a scaled inductor by writing:

$$Z_{L,s} = j\omega \frac{L}{k} = j\frac{\omega}{\omega_0}\frac{\omega_0 L}{k} = j\omega_s(\frac{\omega_0 L}{k}).$$

Similarly, for a capacitor we obtain:

$$Z_{C,s} = \frac{1}{j\omega Ck} = \frac{1}{j\frac{\omega}{\omega_0}(\omega_0 Ck)} = \frac{1}{j\omega_S C_S}.$$

The results can be collected in the following formulas:

$$\begin{aligned} R_S &= \frac{R}{k}, \quad G_S = Gk. \\ L_S &= \frac{L\omega_0}{k}. \\ C_S &= C\omega_0 k. \end{aligned} \quad (4.29)$$

If the network has amplifiers, then:

- VV and CC remain unchanged;
- VC transconductance g is multiplied by k; and
- CV transresistance r is divided by k.

The scaling makes it possible to provide numerous tables of filters, all normalized so that one resistor is equal to 1 Ω and some frequency is normalized to $\omega_S = 1$. From such tables, the reader selects a suitable filter and modifies the values to suit the frequency band and impedance level.

4.6 Approximations for Low-Pass Filters

Before we go into the details of the approximations, we first pose a question: what kind of properties should a lowpass filter have to transfer, without distortion, a certain frequency band and suppress completely all other frequencies. The thought comes immediately to mind that all components should be amplified equally. This is true but not sufficient. In addition, the phase response must be a straight line; thus, the group delay must be a constant for all frequencies of interest. The conditions are sketched in Figure 4.15. We will see that this is impossible to achieve and that we have to accept some approximations when comparing with Figures 4.13(A) through 4.13(D) and Figure 4.14.

A lowpass filter, as indicated by the name, passes some frequency band starting from zero frequency and suppresses higher frequencies. One type of low-pass filter, called the polynomial filter, is described by transfer functions having a general form:

$$T(s) = K \frac{1}{\text{polynomial in } s}. \quad (4.30)$$

The polynomial must have all its roots (poles of the filter) in the left half of the complex plane to make the filter stable.

One such filter was suggested by Butterworth (Schaumann, 1990). He wanted to determine what the coefficients were of the polynomial in equation 4.30 to get a maximally flat amplitude low-pass transfer at $\omega = 0$. To make the problem unique, he also requested that the output of the filter at $\omega = 1$ should be 3 dB less than at $\omega = 0$.

We will not go into detail about how the polynomials were and are found; the steps are in any book on filters. Instead, we show the responses in Figure 4.16 for orders $n = 2$ to 10. We have used a logarithmic horizontal scale, and the responses start at $\omega = 0.1$. As we see, the approximation of a constant at low frequencies is reasonable for a high n, but the group delay

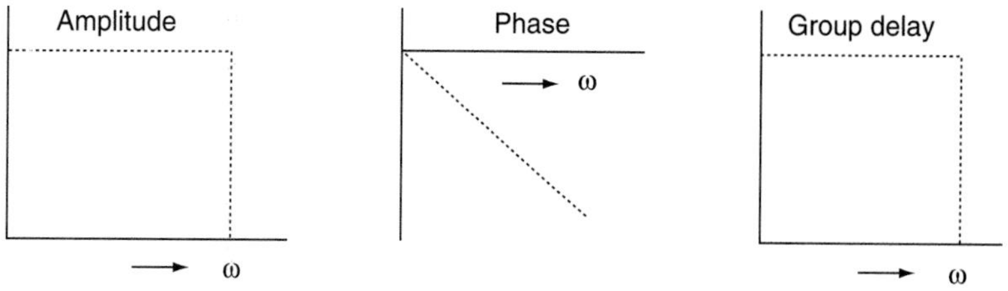

FIGURE 4.15 Ideal Responses for Amplitide, Phase, and Group Delay

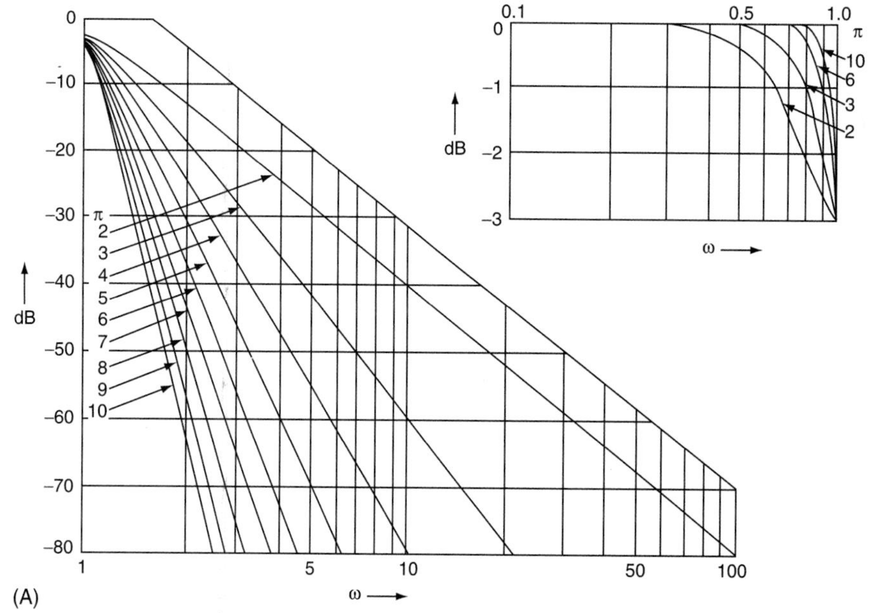

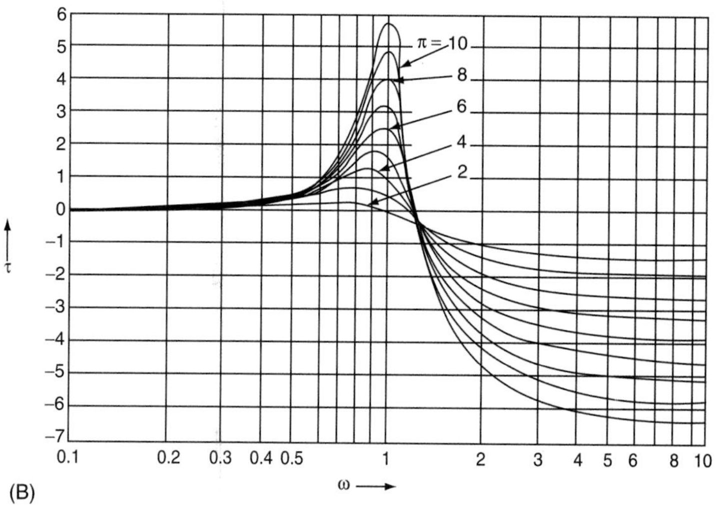

n	τ(0)
2	1.414
3	1.388
4	2.613
5	3.236
6	3.863
7	4.493
8	5.125
9	5.758
10	6.392

FIGURE 4.16 (A) Selectivity Curves of Maximally Flat Filters (B) Group Delay of Maximally Flat Filters

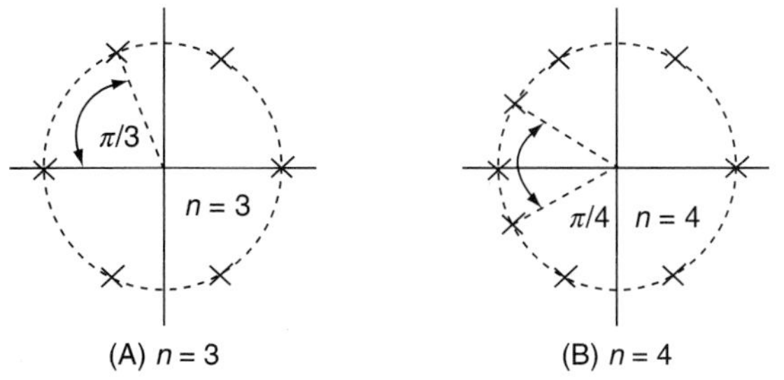

FIGURE 4.17 Poles' Positions for Maximally Flat (Butterworth) Filters, $n = 3$ and $n = 4$

turns out to have a peak at approximately $\omega = 1$, and the peak grows with the order n.

The poles of the normalized Butterworth (or maximally flat) filter lie on a circle with a radius $r = 1$. Figure 4.17 shows the situation for $n = 3$ and $n = 4$. Extension to higher degrees is easy to extrapolate from these two figures: all odd powers will have one real pole, and all even powers will have only complex conjugate poles. The poles of a few filters are in Table 4.1.

Another well-known types of polynomial filters include the Chebychev filters; they use special polynomials discovered by Russian scientist Chebychev. They are also described by the general Formula 4.30, but the requirements are different. In the passband, the amplitude response oscillates between two limits, like in Figure 4.18. The ripple is normally expressed in decibels. At $\omega = 1$, the response always drops by the amount of specified ripple, as in the figure. Compared with the maximally flat filters, Chebychev filters have better attenuation outside the passband and approximate a constant better in the frequency band from 0 to 1. Table 4.1 gives pole positions of Chebychev filters with 0.5 dB ripple.

If the requirements on the suppression of signals outside the passband are more severe, polynomial filters are not sufficient. Much better results are obtained with filters having transfer zeros placed on the imaginary axis. Because a pair of imaginary axis complex conjugate zeros is expressed by terms $(s + j\omega_i)(s - j\omega_i) = s^2 + \omega_i^2$, the transfer function will have the form:

$$T(s) = K\frac{\prod(s^2 + \omega_i^2)}{\text{polynomial in } s}. \tag{4.31}$$

A sketch of the pole-zero plot of some such lowpass filter was shown in Figure 4.10. As always, the roots of the polynomial must be in the left half of the plane. There are many possibilities for selecting the zeros and poles, all beyond the scope of this presentation. Fortunately, there exist numerous tables of filters in literature. Best known are the Cauer-parameter filters with equiripple responses in the passband and with equal minimal suppressions in the stopband. We have analyzed one such filter: see Figure 4.12 and its amplitude responses in Figures 4.13(A) through 4.13(D).

Before closing this section, we provide a general network for low-pass filters described by Formula 4.30. The elements are like those in Figure 4.19. Depending on the order of the filter, the last element before the load resistor will be either an inductor in series or a capacitor in parallel. We can have filters with different values of the resistors and also a design with only one resistor, R_L, but that does not change the general structure of the filter. Table 4.2 gives element values for normalized Butterworth filters with $R_E = R_L = 1$.

TABLE 4.1 Poles of Filters

Order	Butterworth	Chebychev 0.5 dB ripple
2	$-0.7071068 \pm j0.7071068$	$-0.7128122 \pm j1.0040425$
3	-1.0000000	-0.6264565
	$-0.5000000 \pm j0.8660254$	$-0.3132282 \pm j1.0219275$
4	$-0.3826834 \pm j0.9238795$	$-0.1753531 \pm j1.0162529$
	$-0.9238795 \pm j0.3826834$	$-0.4233398 \pm j0.4209457$
5	-1.0000000	-0.3623196
	$-0.3090170 \pm j0.9510565$	$-0.1119629 \pm j1.0115574$
	$-0.8090170 \pm j0.5877852$	$-0.2931227 \pm j0.6251768$
6	$-0.2588190 \pm j0.9659258$	$-0.0776501 \pm j1.0084608$
	$-0.7071068 \pm j0.7071068$	$-0.2121440 \pm j0.7382446$
	$-0.9659258 \pm j0.2588190$	$-0.2897940 \pm j0.2702162$
7	-1.0000000	-0.2561700
	$-0.2225209 \pm j0.9749279$	$-0.0570032 \pm j1.0064085$
	$-0.6234898 \pm j0.7818315$	$-0.1597194 \pm j0.8070770$
	$-0.9009689 \pm j0.4338837$	$-0.2308012 \pm j0.4478939$

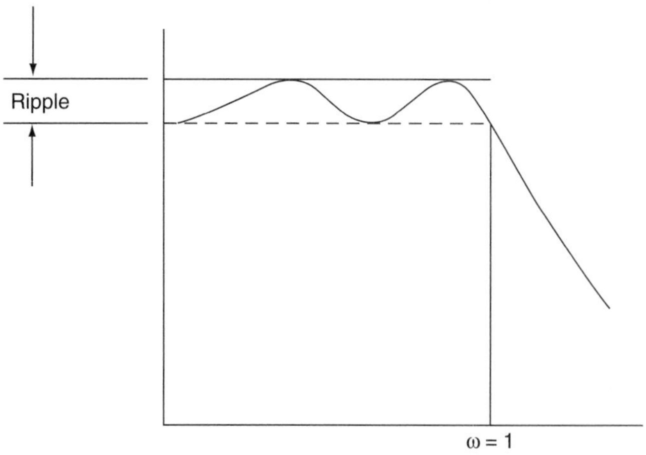

FIGURE 4.18 Amplitude Response of a Chebycher Filter

4.7 Transformations of Inductor Capacitor Low-Pass Filters

We spoke about low-pass filters and normalization previously because many normalized LC low-pass filters can be found in

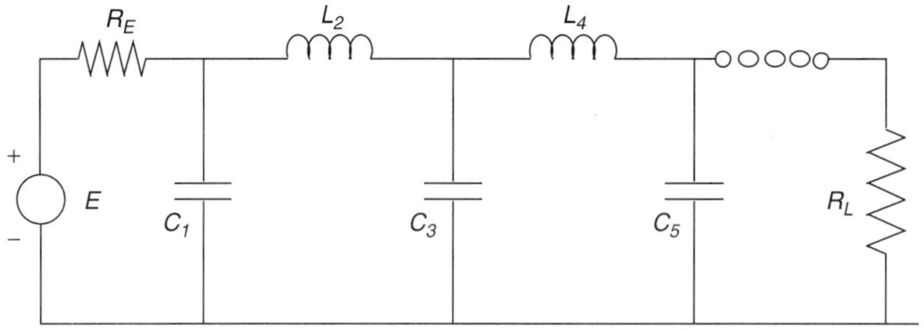

FIGURE 4.19 Realization of Polynomial Low-Pass Filters

TABLE 4.2 Elements Values of Butterworth Filters

Order	C_1	L_2	C_3	L_4	C_5	L_6	C_7
2	1.4142	1.4142					
3	1,0000	2,0000	1.0000				
4	0.7654	1.8478	1.8478	0,7654			
5	0.6180	1.6180	2,0000	1.6180	0.6180		
6	0.5176	1.4142	1.8319	1.8319	1.4142	0.5176	
7	0.4450	1,2470	1.8019	2.0000	1.8019	1.2470	0.4450

the literature, such as in Zverev (1976). One selects a suitable scaled filter and transforms it to the desired frequency and impedance level.

There are, however, additional possibilities for inductor capacitor (LC) low-pass filters. They can be transformed into band-pass, high-pass, or band-stop filters. We introduce these filters now.

4.7.1 Low-Pass into a High-Pass Filter

Consider the transformation:

$$z = \frac{\omega_0}{s}, \tag{4.32}$$

where z describes the complex frequency variable of the original scaled low-pass filter, possibly taken from tables, and s is the variable of the network to be realized. The meaning of ω_0 will become clear later. Multiply both sides of equation 4.32 by L. On the left is the impedance of an inductor. Simple arithmetic operations lead to:

$$zL = \frac{\omega_0 L}{s} = \frac{1}{s/\omega_0 L} = \frac{1}{sC'},$$

where on the right is an impedance of a capacitor with the value $C' = 1/\omega_0 L$. Considering multiplication of equation 4.32 by C, we get:

$$zC = \frac{\omega_0 C}{s} = \frac{1}{s/\omega_0 C} = \frac{1}{sL'}.$$

On the left is the admittance of a capacitor, and on the right is the admittance of an inductor: $L' = 1/\omega_0 C$. The transformation changes each capacitor into an inductor and vice versa; the filter is transformed into a high-pass.

We must still understand the meaning of ω_0. When evaluating the frequency domain responses, we always substitute $s = j\omega$. The frequency of interest is the cutoff frequency of the original low-pass filter, $z = j$. Inserting into equation 4.32, we get $-\omega_0 = \omega$. The sign only means that negative frequencies, existing in mathematics but not in reality, transform into positive frequencies and vice versa. The low-pass cutoff frequency of $\omega_s = 1$ transforms into the highpass cutoff frequency ω_0.

4.7.2 Low-Pass into a Band-Pass Filter

Consider next the transformation:

$$z = \frac{s}{\Delta} + \frac{\omega_0^2}{s\Delta} = \frac{s^2 + \omega_0^2}{s\Delta}. \tag{4.33}$$

In equation 4.33, z belongs to the original low-pass normalized filter. Multiply both sides by L to represent the impedance of an inductor in the z variable:

$$zL = \frac{sL}{\Delta} + \frac{\omega_0^2 L}{s\Delta} = \frac{sL}{\Delta} + \frac{1}{s\Delta/\omega_0^2 L} = sL' + \frac{1}{sC'}.$$

This means that the original inductor impedance was transformed into a series connection of two impedances. One of the elements is an inductor $L'_{ser} = L/\Delta$ and the other is a capacitor $C'_{ser} = \Delta/\omega_0^2 L$. Proceeding similarly for the admittance of a capacitor, we get:

$$zC = \frac{sC}{\Delta} + \frac{\omega_0^2 C}{s\Delta} = s\frac{C}{\Delta} + \frac{1}{s\Delta/\omega_0^2 C} = sC' + \frac{1}{sL'}.$$

The admittance of the capacitor is changed into the sum of two admittances, indicating a parallel connection. One of the elements is a capacitor $C'_{par} = C/\Delta$, and the other is an inductor

$$L'_{par} = \Delta/\omega_0^2 C.$$

To find additional properties of the transformation, we first form the product:

$$L'_{ser}C'_{ser} = L'_{par}C'_{par} = \frac{1}{\omega_0^2}.$$

This equation shows that the resonant frequencies of the parallel and series tuned circuits are the same, ω_0. Next, multiply equation 4.33 by the denominator to get:

$$s^2 - sz\Delta + \omega_0^2 = 0.$$

This is a quadratic equation with two solutions:

$$s_{1,2} = \frac{z\Delta}{2} \pm \sqrt{\frac{z^2\Delta^2}{4} - \omega_0^2}.$$

Now consider special points of the original filter. For $z = 0$, we get

$$s_{1,2}(z = 0) = \pm j\omega_0.$$

Zero frequency of the original filter is transformed into $\pm\omega_0$ point on the imaginary axis. Taking next the cutoff frequency of the low-pass filter, $z = \pm j$, we get four points:

$$s_{1,2,3,4} = \pm\frac{j\Delta}{2} \pm j\sqrt{\omega_0^2 + \frac{\Delta^2}{4}}.$$

Only two of these points will be on the positive imaginary axis:

$$j\omega_1 = j\frac{\Delta}{2} + j\sqrt{\omega_0^2 + \frac{\Delta^2}{4}}.$$

$$j\omega_2 = -j\frac{\Delta}{2} + j\sqrt{\omega_0^2 + \frac{\Delta^2}{4}}.$$

Their difference is the following:

$$\omega_1 - \omega_2 = \Delta. \tag{4.34}$$

Their product is:

$$\omega_1\omega_2 = \omega_0^2. \tag{4.35}$$

Thus, ω_0 is the (geometric) center of the two frequencies, and Δ is the passband of the transformed filter.

4.7.3 Low-Pass into a Band-Stop Filter

Consider a transformation similar to equation 4.33:

$$\frac{1}{z} = \frac{s}{\Delta} + \frac{\omega_0^2}{s\Delta} = \frac{s^2 + \omega_0^2}{s\Delta}. \tag{4.36}$$

It transforms a low-pass into a band-stop. All the above steps remain valid, and only the elements will be transformed differently. Divide both sides by L to get:

$$\frac{1}{zL} = \frac{s}{\Delta L} + \frac{\omega_0^2}{s\Delta L}.$$

On the left is an admittance, and on the right is the sum of two admittances, indicating that the elements will be in parallel: $C'_{par} = 1/\Delta L$ and $L'_{par} = \Delta L/\omega_0^2$. A similar division by C will result in:

$$\frac{1}{zC} = \frac{s}{\Delta C} + \frac{\omega_0^2}{s\Delta C}$$

On the left side is an impedance, and on the right side is the sum of two impedances, indicating that the elements will be in series $L'_{ser} = 1/\Delta C$ and $C'_{ser} = \Delta C/\omega_0^2$. All three transformations are summarized in Figure 4.20 for easy understanding. As an example, we will take the third order maximally flat filter from Table 4.2 and its realization in Figure 4.21(A). We then transform it into a band-pass filter with center frequency $\omega_0 = 1$ rad/s and with the bandwidth $\Delta = 0.1$ rad/s. The transformed band-pass filter is in Figure 4.21(B) and its amplitude response in Figure 4.22. Using equations from section 4.5, the band-pass filter can be transformed to any impedance level and any center frequency with a bandwidth equal to one tenth of the center frequency.

4.8 Realizability of Functions

Synthesis is a process in which we have a given function, and we try to find a network whose properties will be described by that function. In most cases, we try to do so for a voltage transfer function. The first question that comes into mind is whether any function can be realized as a network composed of passive elements: capacitors, inductors, resistors, and transformers. Rather obviously, the answer is no.

To make the problem treatable, we must start with a simple network function and establish conditions that the function must satisfy. The simplest network function is an impedance (or admittance) but since one is the inverse of the other, we can restrict ourselves to only one of them: the impedance. A large amount of work went into establishing the necessary and sufficient conditions for an impedance function composed of

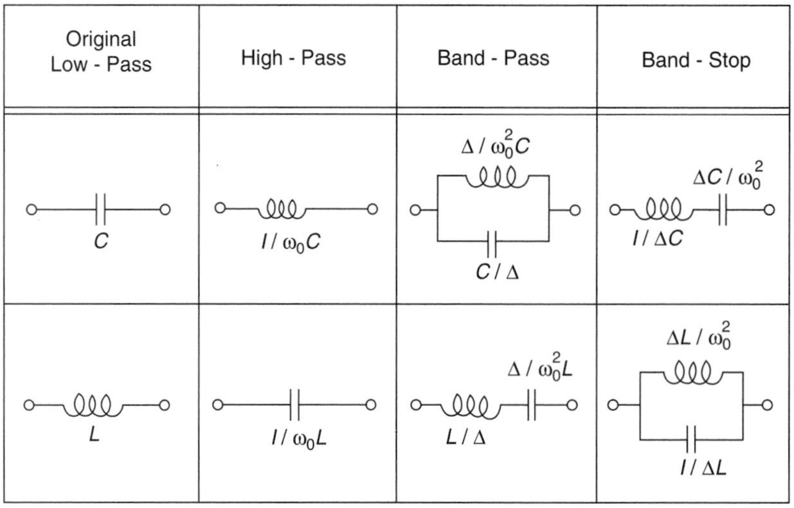

FIGURE 4.20 Transformations of Low-Pass Filters

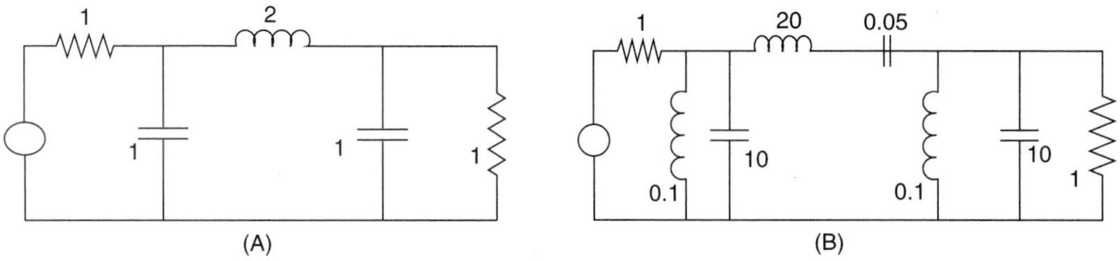

FIGURE 4.21 Transformation of a Butterworth Low-Pass Filter into a Band-Pass

L, C, and R elements, possibly with an ideal transformer. We will explain the conditions as a set of rules without trying to establish the reasons for these rules.

A rational function in the variable s can be realized as an impedance (admittance) if it satisfies the following rules:

1. The degree of the numerator and denominator may differ by at most one.
2. $Z(s)$ may not have any poles or zeros in the right half of the plane.
3. Poles and zeros in the left half of the plane may be multiple.
4. Poles on the imaginary axis must be simple and must have positive residues.
5. The function must satisfy the condition:

$$Re\,Z(s) \geq 0 \text{ if } Re\,s \geq 0. \tag{4.37}$$

The first condition is easy to establish by inspection. The second condition is a standard requirement for stability. Because the impedance and the admittance are the inverse of each other, the same must apply for the zeros. Stability is not destroyed by multiple poles in the left half plane, as stated in point 3. Point 4 would call for partial fraction expansion of the function. Unfortunately, all these steps are only necessary but not sufficient. The only necessary and sufficient condition is point 5, which is difficult to test.

All of this information may seem very discouraging, but, fortunately, synthesis of completely arbitrary impedances is almost never needed. In most cases, we need to realize LC impedances (admittances), and the rules are considerably simplified here. Again, without trying to provide a proof, let us state that any LC impedance can be realized in the form of the circuit in Figure 4.23(A) and any LC admittance in the form of the network in Figure 4.23(B).

Consider the network Figure 4.23(A). The tuned circuits have the impedance:

$$Z_i = \frac{1}{C_i}\frac{s}{s^2 + 1/L_iC_i}. \tag{4.38}$$

We usually define:

$$k_i = \frac{1}{2C_i}. \tag{4.39}$$

$$\omega_i^2 = \frac{1}{L_iC_i}. \tag{4.40}$$

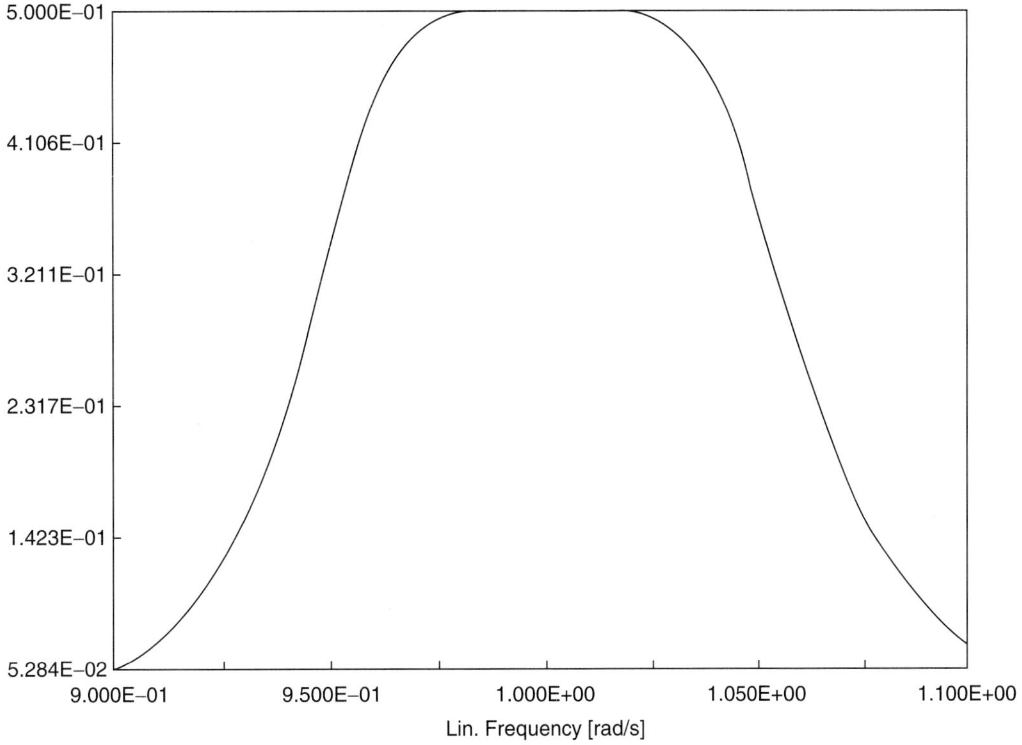

FIGURE 4.22 Amplitude Response of the Band-Pass Filter in Figure 4.21

The variable k_i represents the residues of the poles on the imaginary axis, $p_i = \pm j\omega_i$, and ω_i is the resonant frequency of the circuit. This gives us the possibility of writing a general LC impedance in the form:

$$Z_{LC} = sL + \frac{1}{sC} + \sum \frac{2k_i s}{s^2 + \omega_i^2}. \qquad (4.41)$$

Any of the components in equation 4.41 may be missing. A similar expression could be derived for the network in Figure 4.23(B). We could now use a computer, plot various impedances, and study the results. We will summarize them for you:

1. Z_{LC} or Y_{LC} have only simple poles on the imaginary axis.
2. The residues of the poles are positive.
3. The zeros and poles on the imaginary axis must alternate.
4. Z_{LC} or Y_{LC} are expressed as a ratio of an even and an odd polynomial.
5. A pole or a zero may appear at the origin or at infinity, but the alternating nature of the poles and zeros must be preserved.

These results can be summarized by the simple sketches in Figure 4.24.

Synthesis of filters is usually based on the two-port theory. Using overall expressions for the whole filter, we separate the LC two-port from the loading resistors and apply synthesis to find the elements of the LC two-port. Again, as could be expected, some restrictions apply for the overall network, but the conditions are much more relaxed than for the impedances. A voltage or current transfer function can be realized as an LC two-port with loading resistors if the following statements are true:

1. The degree of the numerator, M, and denominator, N, satisfy $M \leq N$.
2. Zeros are usually on the imaginary axis, but theoretically they can be anywhere.
3. Poles must be in the left half of the plane.

Networks composed of passive elements always have $z_{12} = z_{21}$. In addition, z_{11} and z_{22} will have the same poles as $z_{12} = z_{21}$. Both z_{11} and z_{22} can have additional poles (called private poles), as we will show later. If the network components are only L and C, then z_{11} and z_{22} must satisfy the conditions on LC impedances discussed previously.

We now indicate the simplest method to extract the LC two-port from the network function. Consider the network in Figure 4.25. We wish to find the transfer impedance V_2/I_1. To do so, we need the second equation from the equation set 4.5,

$$V_2 = z_{21}I_1 + z_{22}I_2. \qquad (4.42)$$

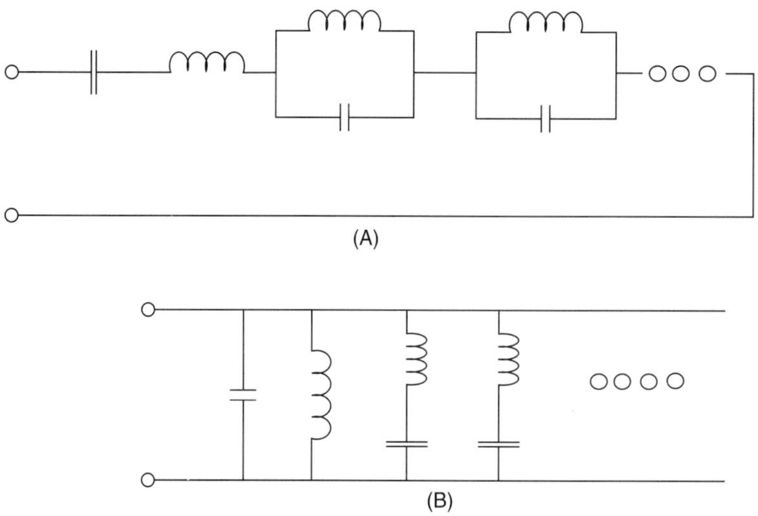

FIGURE 4.23 A General Form of an LC: (A) Impedance and (B) Admittance

After the indicated connection, we have:

$$-I_2 = \frac{V_2}{R}.$$

Inserting into equation 4.42 and eliminating I_2, we obtain:

$$Z_{TR} = \frac{V_2}{I_1} = \frac{z_{21}}{1 + z_{22}/R}. \quad (4.43)$$

As we explained in section 4.5, we can simplify the expression by normalizing the impedance level to $R = 1$. Now consider a function having the form:

$$f = \frac{g}{m + n}, \quad (4.44)$$

where m collects all the even terms of the denominator and n collects all the odd terms. All roots of the denominator must be in the left half of the plane. We can rewrite equation 4.44 in one of the two forms:

$$f = \frac{g/m}{1 + m/n}. \quad (4.45)$$

or

$$f = \frac{g/m}{1 + n/m}. \quad (4.46)$$

The ratios in the denominator, n/m or m/n, satisfy the conditions imposed on z_{22} of an LC network. The problem is reduced to the synthesis of z_{22}, taking into account the properties of z_{21}. We will describe such synthesis in section 4.10.

We have chosen this primitive case for demonstration only. Practical synthesis steps usually require voltage transfer with a loading resistor and with an input resistor. The steps that extract properties of the LC two-port are more complicated but are available in books focusing on synthesis of filters.

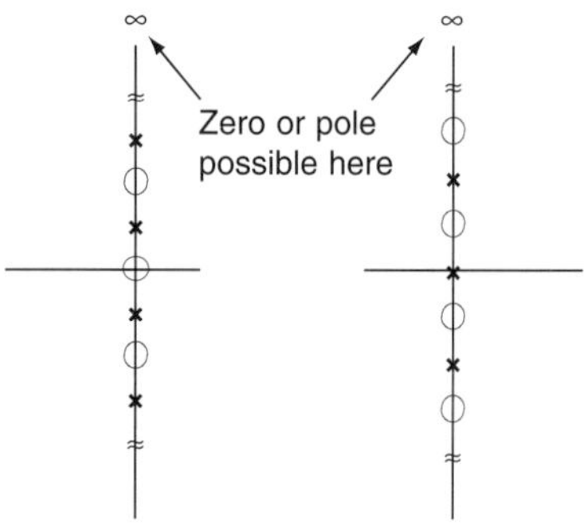

FIGURE 4.24 Possible Positions of LC Network Poles and Zeros

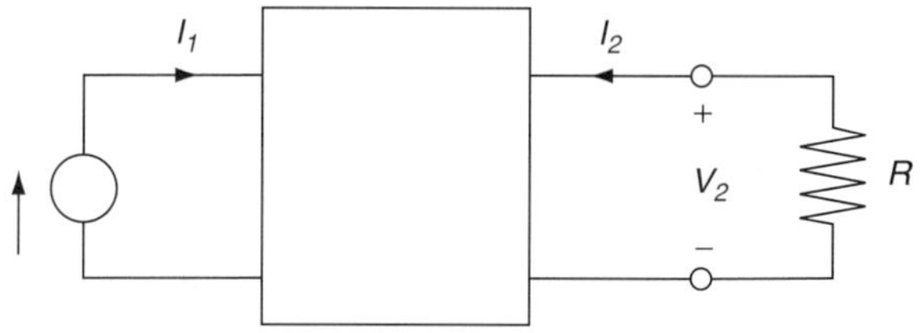

FIGURE 4.25 Deriving Z_{TR} for a Loaded Two-Port

4.9 Synthesis of LC One-Ports

For LC impedances or admittances, we stated the rules that the function is represented by the ratio of an even and odd polynomial and the poles and zeros must interchange, be simple, and have positive residues. To find the partial fractions, we must use a root-finding routine and actually get the zeros and poles first. To calculate the residues of the poles, we use the following formulas:

$$\text{Pole at the origin:} \quad k_0 = sZ(s)|_{s=0}. \tag{4.47}$$

$$\text{Pole at infinity:} \quad k_\infty = \frac{Z(s)}{s}|_{s=\infty}. \tag{4.48}$$

$$\text{Two poles at } \pm j\omega_i\text{: } 2k = \frac{s^2+\omega_i^2}{p} Z(s)|_{s^2=-\omega_i^2}. \tag{4.49}$$

In formulas 4.47 through 4.49, it is first necessary to cancel the additional terms against equal terms in $Z(s)$ and then substitute the indicated values for s. Suppose now that we consider an admittance in the form:

$$Y(s) = k_\infty s + \frac{k_0}{s} + \sum_{i=1}^{n} \frac{2k_i s}{s^2+\omega_i^2}. \tag{4.50}$$

The first term is $C = k_\infty$, and the second is $L = 1/k_0$. The remaining terms can be written as:

$$Y_i(s) = \frac{2k_i s}{s^2+\omega_i^2} = \frac{1}{s/2k_i + \omega_i^2/2k_i s}, \tag{4.51}$$

where

$$L_i = \frac{1}{2k_i}. \tag{4.52}$$

$$C_i = \frac{2k_i}{\omega_i^2}. \tag{4.53}$$

The admittance realized this way was shown in Figure 4.23(B). Should we consider the function in equation 4.50 as an impedance, similar steps would lead to the network in Figure 4.23(A). These networks have the name Foster canonical forms. Another method, originally due to Cauer, gives impedances or admittances in the form of a ladder. As an example, consider the impedance:

$$Z = \frac{s^4+10s^2+9}{s^3+4s} = \frac{(s^2+1)(s^2+9)}{s(s^2+4)}. \tag{4.54}$$

Subtract from equation 4.54 an impedance sL (a pole at infinity). This gives the following:

$$Z_2 = \frac{s^4+10s^2+9}{s^3+4s} - sL = \frac{s^4(1-L)+s^2(10-4L)+9}{s^3+4s}.$$

We could now select L to be anywhere between zero and one, but if we select $L = 1$, the function simplifies to:

$$Z_2 = \frac{6s^2+9}{s^3+4s}.$$

The subtraction means that we have realized the first element in series, $L = 1$, and removed a pole at infinity. The remaining function can now be inverted to:

$$Y_2 = \frac{s^3+4s}{6s^2+9}.$$

This equation has a pole at infinity. We can continue by subtracting from it the admittance of a capacitor:

$$Y_3 = Y_2 - sC = \frac{s^3(1-6C)+s(4-9C)}{6s^2+9}.$$

If we select $C = 1/6$, we remove again a pole at infinity and realize a capacitor in parallel, $C = 1/6$, with the result:

$$Y_3 = \frac{5s/2}{6s^2+9}.$$

Y_3 can be inverted again:

$$Z_3 = \frac{12s}{5} + \frac{18}{5s}.$$

The first element is an inductor $L = 12/5$, the second a capacitor $C = 5/18$, both connected in series. The resulting network is in Figure 4.26.

The above expansion started from the highest powers. There exists another possibility by starting from the lowest power. Rewrite the impedance from 4.54 in the form:

$$Z = \frac{9+10s^2+s^4}{4s+s^3}. \tag{4.55}$$

and subtract $1/sC$, a pole at the origin:

$$Z_2 = Z - \frac{1}{sC} = \frac{(9-4/C)+s^2(10-1/C)+s^4}{4s+s^3}.$$

If we select $C = 4/9$ as a capacitor in series, the first term disappears, and we remove the pole at $s = 0$ to get:

$$Z_2 = \frac{31s/4+s^3}{4+s^2}.$$

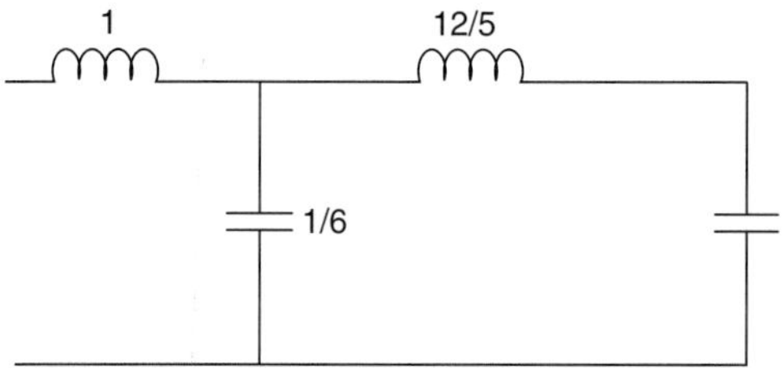

FIGURE 4.26 Synthesis of Equation 4.54 by Extracting Poles at Infinity

The remainder can be inverted and the admittance of an inductor subtracted again:

$$Y_3 = Y_2 - \frac{1}{sL} = \frac{(4 - 31/4L) + s^2(1 - 1/L)}{\frac{31}{4}s + s^3}.$$

The choice $L = 31/16$, connected in parallel, removes the first term and another pole at the origin. The process can be continued, with the resulting network shown in Figure 4.27. It is not necessary to continue one type of the expansion to the end. It is always possible, at any step, to rewrite the remaining function like we did going from equations 4.54 to 4.55 and continue. In addition, we need not remove one of the terms completely. All this indicates that synthesis of LC impedances or admittances is far from a unique procedure.

Let us return to equation 4.54, where we subtracted $L = 1$; we subtract now only $L = 0.5$. The result will be as follows:

$$Z_2 = \frac{0.5s^4 + 8s^2 + 9}{s^3 + 4s} = \frac{0.5(s^2 + 1.22)(s^2 + 14.78)}{s^3 + 4s}. \quad (4.56)$$

We see that the partial removal did not simplify the function but shifted zeros of the original function into new positions. Normally, if no other conditions are imposed, we would always remove the full value because then the resulting number of elements is minimal. Partial removal is used in the synthesis of two-port networks, as is shown later in this chapter.

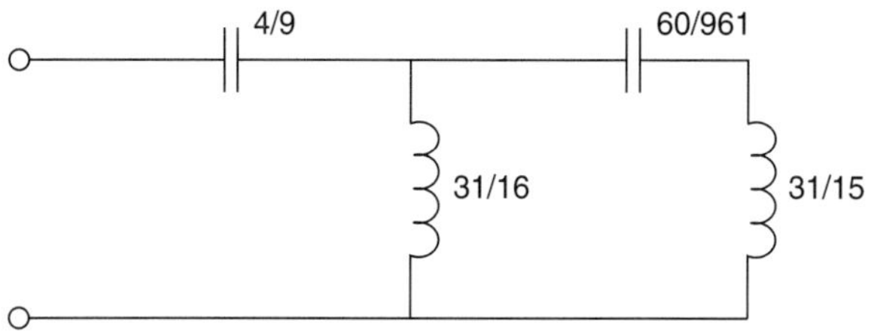

FIGURE 4.27 Synthesis of Equation 4.55 by Extracting Poles at Zero

4.10 Synthesis of LC Two-Port Networks

4.10.1 Transfer Zeros at Infinity

In this section, we indicate synthesis steps of two-ports. It is a fairly complicated procedure, and we will try to explain directly with examples.

An LC two-port is described by its impedance or admittance parameters, equations 4.5 or 4.7. We will use the impedance parameters and assume that the functions have been decomposed into partial fractions. Because for passive networks $z_{12} = z_{21}$, we need (for full description only):

$$z_{11} = \frac{k_0^{(11)}}{s} + \sum \frac{2k_i^{(11)}s}{s^2 + \omega_i^2} + k_\infty^{(11)}s$$

possibly plus some LC impedance.

$$z_{22} = \frac{k_0^{(22)}}{s} + \sum \frac{2k_i^{(22)}s}{s^2 + \omega_i^2} + k_\infty^{(22)}s$$

possibly plus some LC impedance.

$$z_{12}(= z_{21}) = \frac{k_0^{(12)}}{s} + \sum \frac{2k_i^{(12)}s}{s^2 + \omega_i^2} + k_\infty^{(12)}s.$$

The words *possibly in addition LC impedance* indicate that additional terms may exist in z_{11} and/or z_{22}. These are called private impedances and do not influence z_{12}. They would appear as in Figure 4.28(A) for impedance parameters and as in Figure 4.28(B) for admittance parameters. After their removal, all z_{ij} must have the same poles, the residues of z_{11} and z_{22} must be positive, and residues of z_{12} may be positive or negative.

Before we start with the synthesis example, let us state that a removal of an element, which is supposed to influence z_{12} (or y_{12}), must be connected as in Figure 4.29(A) or 4.29(B). Let us now have the following two functions:

$$z_{11} = s + \frac{9}{4s} + \frac{15}{4}\frac{s}{s^2 + 4}.$$
$$z_{12} = \frac{1}{4s} + \frac{-s/4}{s^2 + 4} = \frac{1}{s(s^2 + 4)}. \quad (4.57)$$

The first term in z_{11} is a private impedance (the pole at infinity does not appear in z_{12}) and can be removed as an inductor $L = 1$ in series. The remaining functions are these:

$$z_{11}^* = Z_1 = \frac{9}{4s} + \frac{15}{4}\frac{s}{s^2 + 4} = \frac{6s^2 + 9}{s(s^2 + 4)}.$$

$$z_{12} = \frac{1}{s(s^2 + 4)}.$$

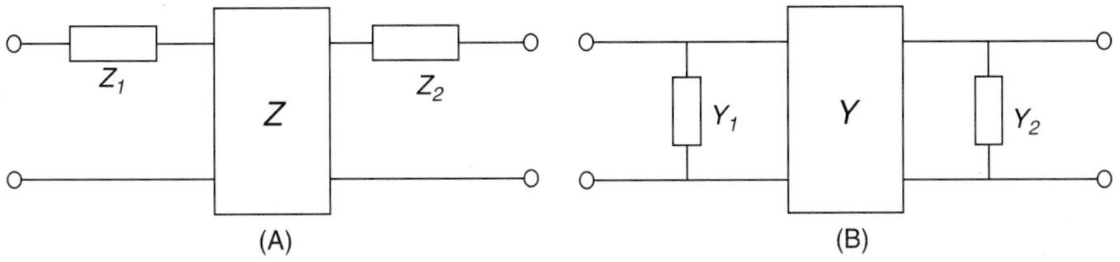

FIGURE 4.28 Removing Private (A) Impedances and (B) Admittances

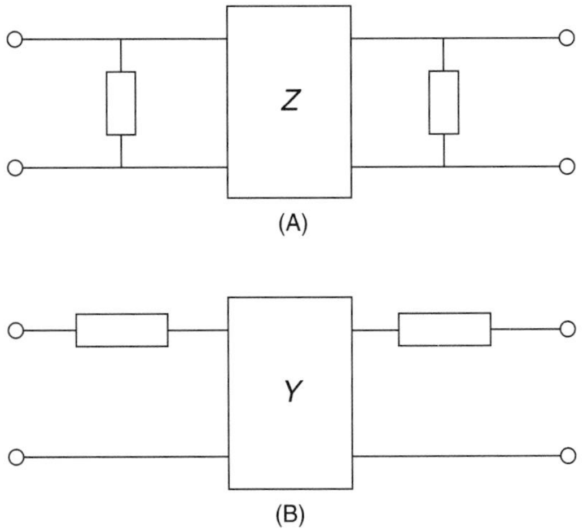

FIGURE 4.29 Removals Affecting Transfer Immittances and Connection of the Last Element for (A) Impedance Description and (B) Admittance Description

The poles of both functions are the same, z_{12} has only a constant in the numerator, and three powers of s are in the denominator. If we insert infinity for any s in the denominator, the function will become zero, and we say that z_{12} has three zeros at infinity. As was shown in the previous section, we cannot remove zeros as elements, but we can remove them as poles of inverted functions. Subtracting the admittance sC from $Y_1 = 1/Z_1$ leads to:

$$Y_2 = \frac{s^3 + 4s}{6s^2 + 9} - sC = \frac{s^3(1 - 6C) + s(4 - 9C)}{6s^2 + 9}$$

Selecting $C = 1/6$ (realized as a capacitor in parallel) reduces the admittance to

$$Y_2 = \frac{5s}{12s^2 + 18}.$$

and takes care of one of the transfer zeros of z_{12}. The remaining function, Y_2, has a zero at infinity. We can invert and remove a pole at infinity from:

$$Z_3 = \frac{1}{Y_2} - sL = \frac{s^3(12 - 5L) + 18}{5s}.$$

We realize $L = 12/5$ as an inductor in series. This has taken care of the second transfer zero of z_{12}. The remaining function is $Z_4 = 18/5\,s$. By inverting it, we have $Y_4 = 5\,s/18$ and remove the last transfer zero as a pole at infinity by connecting a capacitor $C = 5/18$ in parallel. The whole network is in Figure 4.30. Let us now check our result logically. Inductors in series obstruct transfer of high frequencies and capacitors in parallel represent short circuits at high frequencies. Hence, each of the capacitors in parallel and the middle inductor in series will indeed help in suppressing high frequencies and create together three zeros at infinity.

We have used the above more complicated procedure to prepare the reader for synthesis of two-ports with finite transfer zeros. We could have achieved the same by dividing the numerator by the denominator in each of the steps. In fact, if the reader knows expansions into continued fractions, he or she should try it on $1/Z_1$. It is the fastest way for the removal of transfer zeros at infinity.

4.10.2 Transfer Zeros on the Imaginary Axis

As we have demonstrated in equation 4.56, partial removal does not reduce the degree of the function but shifts the zeros to different places. This feature is used in the synthesis. We will indicate the method in an example. Consider:

$$z_{11} = \frac{(s^2+2)(s^2+6)}{s(s^2+3)} = \frac{s^4 + 8s^2 + 12}{s(s^2+3)} = s + \frac{4}{s} + \frac{s}{s^2+3}.$$
$$z_{12} = \frac{(s^2+1)(s^2+6)}{s(s^2+3)} = s + \frac{4}{3s} + \frac{2s}{3(s^2+3)}. \quad (4.58)$$

The transfer impedance has transfer zeros at $s = \pm j1$ and $s = \pm j2$. Both z_{11} and z_{12} have the same poles. There are no private poles in z_{11}; the functions are ratios of even and odd polynomials, zeros, and poles of z_{11} interchange. Thus, all the conditions for realizability are satisfied.

In the first step, we wish to take into consideration the transfer zero at $j1$. We substitute this value into z_{11} to obtain:

$$z_{11}(j1) = \frac{5}{2j}. \quad (4.59)$$

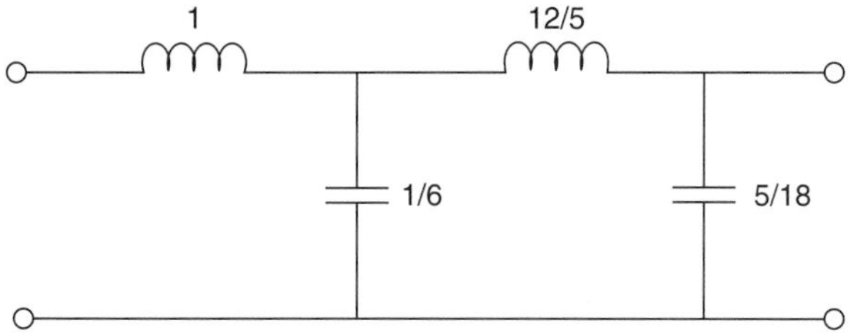

FIGURE 4.30 Synthesis of Equation 4.57

Equation 4.59 behaves as a capacitance; at $\omega_1 = 1$, we have $5/2j = 1/j\omega_1 C = 1/j1C$ from which $C = 2/5$. The capacitance and z_{11} have a pole at the origin, and we subtract:

$$Z_2 = z_{11} - \frac{5}{2s} = \frac{s^4 + 11s^2/2 + 9/2}{s(s^2+3)}. \tag{4.60}$$

The intention is to get the numerator polynomial with zeros at $s = \pm j1$. We can now divide the numerator of equation 4.60 by the term $s^2 + 1$ and obtain the decomposition:

$$Z_2 = \frac{(s^2+1)(s^2+9/2)}{s(s^2+3)}. \tag{4.61}$$

Notice the neat trick we used to shift the zeros by first evaluating equation 4.59. Using the above steps, we have realized the left capacitor in Figure 4.31. Z_2 now has the same number of zeros as the transfer function. We remove them as poles of the inverted function:

$$Y_2 = \frac{s(s^2+3)}{(s^2+1)(s^2+9/2)}. \tag{4.62}$$

To be removed is an admittance of a series-tuned circuit of the form $2k_i s/(s^2 + \omega_i^2)$. This is done by the formula 4.49:

$$2k_1 = \frac{(s^2+1)}{s}\frac{s(s^2+3)}{(s^2+1)(s^2+9/2)}\Big|_{s^2=-1} = \frac{4}{7}. \tag{4.63}$$

The element values of the tuned circuit are:

$$Y_{tc} = \frac{4s/7}{s^2+1} = \frac{1}{7s/4 + 7/4s},$$

with the result $L = 7/4$ and $C = 4/7$. The series tuned circuit, connected in parallel, is on the left of Figure 4.31. In the next step, we remove the expression for the tuned circuit from Y_2:

$$Y_3 = \frac{s(s^2+3)}{(s^2+1)(s^2+9/2)} - \frac{4s/7}{s^2+1} = \frac{3s}{7(s^2+9/2)}. \tag{4.64}$$

Then we return to the impedance:

$$Z_3 = \frac{7}{3}\frac{s^2+9/2}{s}.$$

There is another transfer zero to be removed, $s = j2$. The procedure is repeated by first evaluating $Z_3(j2) = 7/j12$. It behaves as a capacitance equal to $7/j12 = 7/j\omega_2 C_2 = 7/j2C_2$ with the result $C_2 = 6/7$. The impedance corresponding to the capacitance has a pole at the origin, and Z_3 has the same pole, so a partial removal of the pole is possible:

$$Z_4 = Z_3 - \frac{1}{sC_2} = \frac{7(s^2+9/2)}{3s} - \frac{7}{6s} = \frac{7(s^2+4)}{3s}.$$

Because now Z_4 has a zero at the proper place, we can remove it as a pole of a tuned circuit:

$$Y_4 = \frac{3s}{7s^2+28},$$

with the result $L = 7/3$ and $C = 3/28$, see Figure 4.31.

4.11 All-Pass Networks

In filter design, we place the transfer zeros almost always on the imaginary axis because this secures largest suppression of the signal in the stopband. In most cases, a low-pass filter should have approximately constant transfer of low frequencies and rapid suppression of frequencies beyond the specified passband. This may result in almost constant transfer of the desired frequencies, but the group delay response is then far from ideal. We have analyzed one such filter, Figure 4.12, with its amplitude responses in Figure 4.13(A) through Figure 4.13D and with the group delay in Figure 4.14.

In special cases, it may be necessary to compensate for the nonideal group delay response, and this can be done with all-pass networks. They modify only the group delay and pass all frequencies without attenuation.

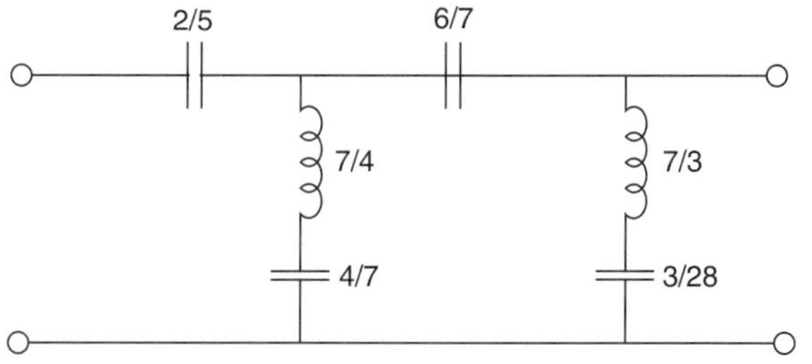

FIGURE 4.31 Synthesis of Equation 4.58

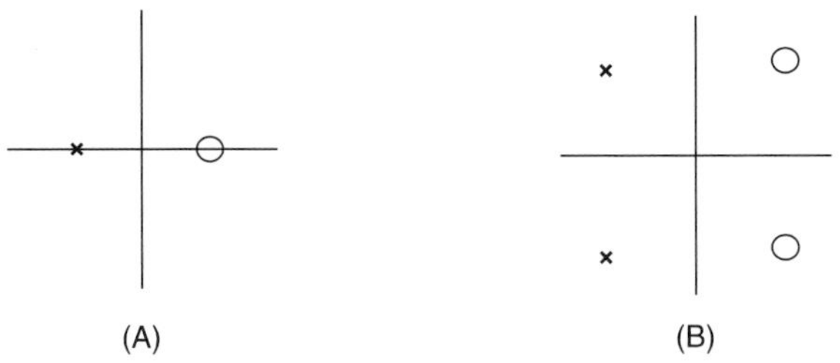

FIGURE 4.32 Possible Pole-Zero Positions for All-Pass Networks

Consider the symmetrical pole and zero positions sketched in Figure 4.32(A) and Figure 4.32(B) and the formulas for amplitude and group delay responses, equations 4.25 and 4.27. Symmetrical positions of both axes make $\alpha_i = -\gamma_i$ and $\beta_i = \delta_i$. When these are inserted into equation 4.25, the numerator cancels against the denominator and the expression in large brackets becomes equal to one. Inserted into equation 4.27, the two terms add and:

$$\tau_{\text{all-pass}} = 2\sum_{i=1}^{M} \frac{\alpha_i}{\alpha_i^2 + (\omega - \beta_i)^2}. \tag{4.65}$$

For further study, we will use the two-port theory and the impedance parameters. Consider the network in Figure 4.33, and remove the dotted resistor. Without it, we have two voltage dividers:

$$V_2 = V_B - V_A = \frac{Z_1 - Z_2}{Z_1 + Z_2} V_1.$$

The current flowing from the source into the two branches will be:

$$I_1 = \frac{2V_1}{Z_1 + Z_2}$$

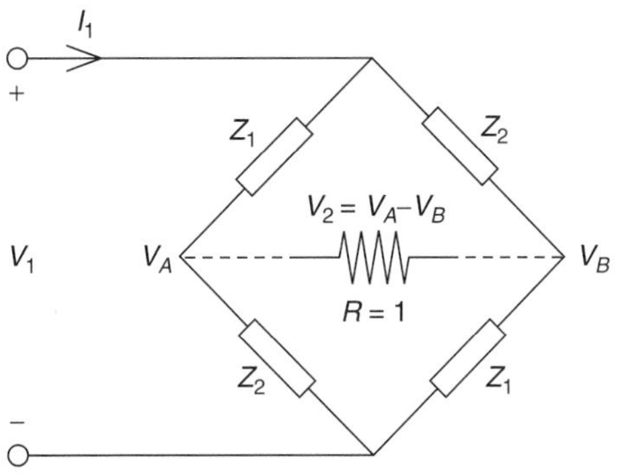

FIGURE 4.33 Deriving Properties of an All-Pass Network

and the entries of the two-port impedance matrix are:

$$Z = \begin{bmatrix} \dfrac{Z_2 + Z_1}{2} & \dfrac{Z_2 - Z_1}{2} \\ \dfrac{Z_2 - Z_1}{2} & \dfrac{Z_2 + Z_1}{2} \end{bmatrix} \tag{4.66}$$

If we connect the dotted resistor, the situation changes:

$$Z_{TR} = \frac{z_{21}}{1 + z_{22}}, \tag{4.67}$$

derived already in equation 4.43, and the input impedance becomes:

$$Z_{in} = \frac{z_{11} + z_{11}z_{22} - z_{12}z_{21}}{1 + z_{22}}, \tag{4.68}$$

obtained in equation 4.10. Let us now impose the condition:

$$Z_1 Z_2 = 1. \tag{4.69}$$

It changes the matrix of equation 4.66 into:

$$Z = \begin{bmatrix} \dfrac{1 + Z_1^2}{2Z_1} & \dfrac{1 - Z_1^2}{2Z_1} \\ \dfrac{1 - Z_1^2}{2Z_1} & \dfrac{1 + Z_1^2}{2Z_1} \end{bmatrix} \tag{4.70}$$

Inserting these entries into equation 4.68, we get, after a few simple algebraic steps, the surprising result:

$$Z_{\text{in}} = 1. \tag{4.71}$$

The input impedance of the combination is equal to the loading resistor, irrespective of what the value of Z_1, as long as we maintain the condition that $Z_2 = 1/Z_1$. Inserting this condition into equation 4.67 results in:

$$Z_{TR} = \frac{1 - Z_1}{1 + Z_1}. \tag{4.72}$$

Let us now take the special case $Z_1 = sL$. The transfer impedance becomes $Z_{TR} = (1 - sL)/(1 + sL)$, and we get the design parameters shown in Figure 4.34. The figure uses an accepted practice to draw only two of the impedances and to indicate the presence of the other two by dashed lines.

Suppose we take for Z_1 a parallel tuned circuit with elements L and C. In such a case,

$$Z_1 = \frac{sL}{s^2LC + 1}. \tag{4.73}$$

Inserting into equation 4.72, we get:

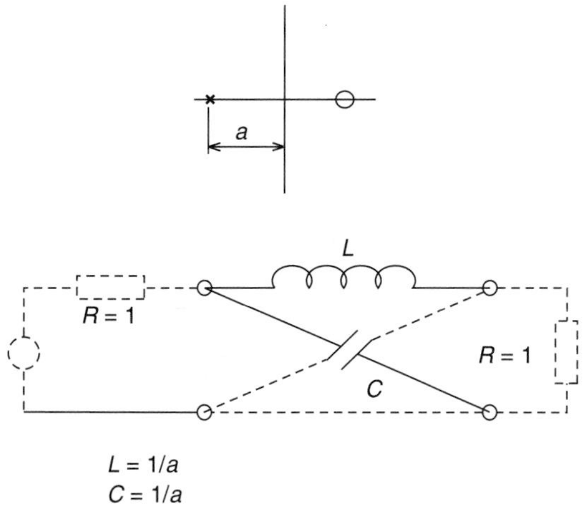

FIGURE 4.34 Position of One All-Pass Pole-Zero and Network Realization

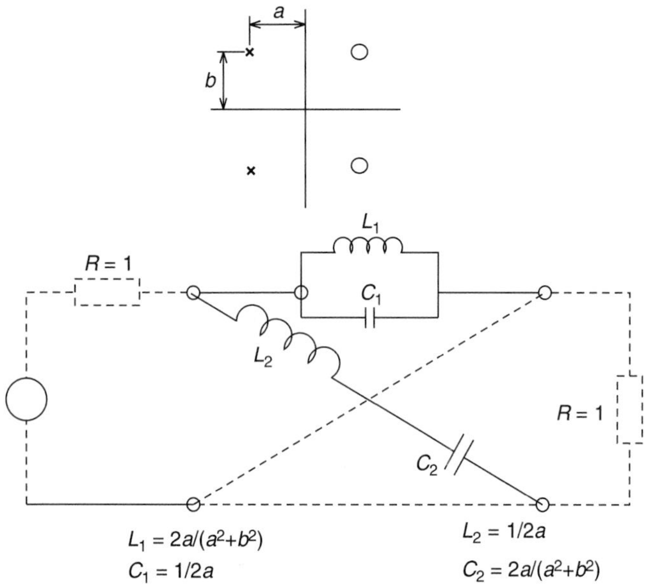

FIGURE 4.35 Positions of Two All-Pass Pole-Zero and Network Realization

$$Z_{TR} = \frac{s^2 - s/C + 1/LC}{s^2 + s/C + 1/LC}. \tag{4.74}$$

Comparing the denominator with

$$(s + a + jb)(s + a - jb) = s^2 + 2as + (a^2 + b^2),$$

we conclude that $C = 1/2a$ and $L = 2a/(a^2 + b^2)$. The network and its design values are in Figure 4.35. Remember that the loading resistor was normalized to unit value, and for other values of R, we must also scale impedances of these elements.

The above networks have one considerable disadvantage: the output is taken between two nodes and not with respect to ground. Connection to filters would have to be done through a transformer. All-pass networks with input and output taken with respect to ground also exist but require a transformer in the all-pass.

4.12 Summary

We have presented simplified steps for the synthesis of LC filters. The intention was to provide enough information for a general understanding and for encouraging the reading of more advanced books. The material should be sufficient for practical situations when tables of filters can be used. We have intentionally skipped synthesis of RC networks. The theory is available but is almost never used for designs, and is in many respects similar to what has been presented. Synthesis of RC networks can be found in specialized books. All references except Mitra (1969), which is an early text concerning active network synthesis, are classic texts about passive network synthesis.

References

Balabanian, N. (1958). *Network synthesis.* Englewood Cliffs, NJ: Prentice Hall.

Guillemin, E.A. (1957). *Synthesis of passive networks.* New York: John Wiley & Sons.

Mitra, S.K. (1969). *Analysis and synthesis of linear active networks.* New York: John Wiley & Sons.

Schaumann, R., Ghausi, M.S., and Laker, K.R. (1990). *Design of analog filters: passive, active RC, and switched capacitor.* Eaglewood Cliffs, NJ: Prentice Hall.

Temes, G.C., and Mitra, S.K. (1973). *Modern filter theory and design.* New York: Wiley-Interscience.

Van Valkenberg, M.E. (1960). *Introduction to modern network synthesis.* New York: John Wiley & Sons.

Weinberg, L. (1962). *Network analysis and synthesis.* New York: McGraw-Hill.

Zverev, A.I. (1976). *Handbook of filter synthesis.* New York: John Wiley & Sons.

5

Nonlinear Circuits

Ljiljana Trajković
School of Engineering Science, Simon Fraser University, Vancouver, British Columbia, Canada

5.1 Introduction

Nonlinearity plays a crucial role in many systems and qualitatively affects their behavior. For example, most circuits emanating from the designs of integrated electronic circuits and systems are nonlinear. It is not an overstatement to say that the majority of interesting electronic circuits and systems are nonlinear.

Circuit nonlinearity is often a desirable design feature. Many electronic circuits are designed to employ the nonlinear behavior of their components. The nonlinearity of the circuits' elements is exploited to provide functionality that could not be achieved with linear circuit elements. For example, nonlinear circuit elements are essential building blocks in many well-known electronic circuits, such as bistable circuits, flip-flops, static shift registers, static RAM cells, latch circuits, oscillators, and Schmitt triggers: they all require nonlinear elements to function properly. The global behavior of these circuits differs from the behavior of amplifiers or logic gates in a fundamental way: they must possess multiple, yet isolated, direct current (dc) operating points (also called equilibrium points). This is possible only if a nonlinear element (such as a transistor) is employed.

A **nonlinear circuit** or a **network** (a circuit with a relatively large number of components) consists of at least one nonlinear element, not counting the voltage and current independent sources. A circuit element is called nonlinear if its constitutive relationship between its voltage (established across) and its current (flowing through) is a nonlinear function or a nonlinear relation. All physical circuits are nonlinear. When analyzing circuits, we often simplify the behavior of circuit elements and model them as linear elements. This simplifies the model of the circuit to a linear system, which is then described by a set of linear equations that are easier to solve. In many cases, this simplification is not possible and the nonlinear elements govern the behavior of the circuit. In such cases, approximation to a linear circuit is possible only over a limited range of circuit variables (voltages and currents), over a limited range of circuit parameters (resistance, capacitance, or inductance), or over a limited range of environmental conditions (temperature, humidity, or aging) that may affect the behavior of the circuit. If global behavior of a circuit is sought, linearization is often not possible and the designer has to deal with the circuit's nonlinear behavior (Chua *et al.*, 1987; Hasler and Neirynck, 1986; Mathis, 1987; Willson, 1975).

As in the case of a **linear circuit**, Kirchhoff's voltage and current laws and Telegen's theorem are valid for nonlinear circuits. They deal only with a circuit's topology (the manner in which the circuit elements are connected together). Kirchhoff's voltage and current laws express linear relationships between a circuit's voltages or currents. The nonlinear relationships between circuit variables stem from the elements' constitutive relationships. As a result, the equations governing the behavior of a nonlinear circuit are nonlinear. In cases where a direct current response of the circuit is sought, the governing equations are nonlinear algebraic equations. In cases where transient behavior of a circuit is examined, the governing equations are nonlinear differential-algebraic equations.

Analysis of nonlinear circuits is more difficult than the analysis of linear circuits. Various tools have been used to understand and capture nonlinear circuit behavior. Some approaches employ quantitative (numerical) techniques and the use of circuit simulators for finding the distribution of a circuit's currents and voltages for a variety of waveforms supplied by sources. Other tools employ qualitative analyses methods, such as those dealing with the theory for establishing the number of dc operating points a circuit may possess or dealing with the analysis of stability of a circuit's operating point.

5.2 Models of Physical Circuit Elements

Circuit elements are models of physical components. Every physical component is nonlinear. The models that capture behavior of these components over a wide range of values of voltages and currents are nonlinear. To some degree of simplification, circuit elements may be modeled as linear elements that allow for an easier analysis and prediction of a circuit's behavior. These linear models are convenient simplifications that are often valid only over a limited range of the element's voltages and currents.

Another simplification that is often used when seeking to calculate the distribution of circuit's voltages and currents, is a piecewise-linear approximation of the nonlinear element's characteristics. The piecewise-linear approximation represents a nonlinear function with a set of nonoverlapping linear segments closely resembling the nonlinear characteristic. For example, this approach is used when calculating circuit voltages and currents for rather simple nonlinear electronic circuits or when applying certain computer-aided software design tools.

Nonlinear circuit elements may be classified based on their constitutive relationships (resistive and dynamical) and based on the number of the element's terminals.

5.2.1 Two-Terminal Nonlinear Resistive Elements

The most common two-terminal nonlinear element is a **nonlinear resistor**. Its constitutive relationship is:

$$f(i,\ v) = 0, \tag{5.1}$$

where f is a nonlinear mapping. In special cases, the mapping can be expressed as:

$$i = g(v), \tag{5.2}$$

when it is called a voltage-controlled resistor or:

$$v = h(i), \tag{5.3}$$

when it is a current-controlled resistor.

Examples of nonlinear characteristics of nonlinear two-terminal resistors are shown in Figure 5.1(A). Nonlinear resistors commonly used in electronic circuits are diodes: exponential, zener, and tunnel (Esaki) diodes. They are built from semiconductor materials and are often used in the design of electronic circuits along with linear elements that provide appropriate biasing for the nonlinear components. A circuit symbol for an exponential diode and its characteristic are given in Figure 5.1(B). The diode model is often simplified and the current flowing through the diode in the **reverse biased region** is assumed to be zero. Further simplification leads to a model of a switch: the diode is either *on* (voltage across the diode terminals is zero) or *off* (current flowing through the diode is zero). These simplified diode models are also nonlinear. Three such characteristics are shown in Figure 5.1(B). The zener diode, shown in Figure 5.1(C), is another commonly used electrical component. Tunnel diodes have more complex nonlinear characteristic, as shown in Figure 5.1(D). This nonlinear element is qualitatively different from exponential diodes. It permits multiple values of voltages across the diode for the same value of current flowing through it. Its characteristic has a region where its derivative (differential resistance) is negative. Silicon-controlled rectifier, used in many switching circuits, is a two-terminal nonlinear element with a similar behavior. It is a current-controlled resistor that permits multiple values of currents flowing through it for the same value of voltage across its terminals. A symbol for the silicon-controlled rectifier and its current-controlled characteristic are shown in Figure 5.1(E).

Two simple nonlinear elements, called **nullator** and **norator**, have been introduced mainly for theoretical analysis of nonlinear circuits. The nullator is a two-terminal element defined by the constitutive relations:

$$v = i = 0. \tag{5.4}$$

The current and voltage of a norator are not subject to any constraints. Combinations of nullators and norators have been used to model multiterminal nonlinear elements, such as transistors and operational amplifiers.

Voltage- and current-independent sources are also considered to belong to the family of nonlinear resistors. In contrast to the described nonlinear resistors, which are all passive for the choice of their characteristics presented here, voltage and current independent sources are active circuit elements.

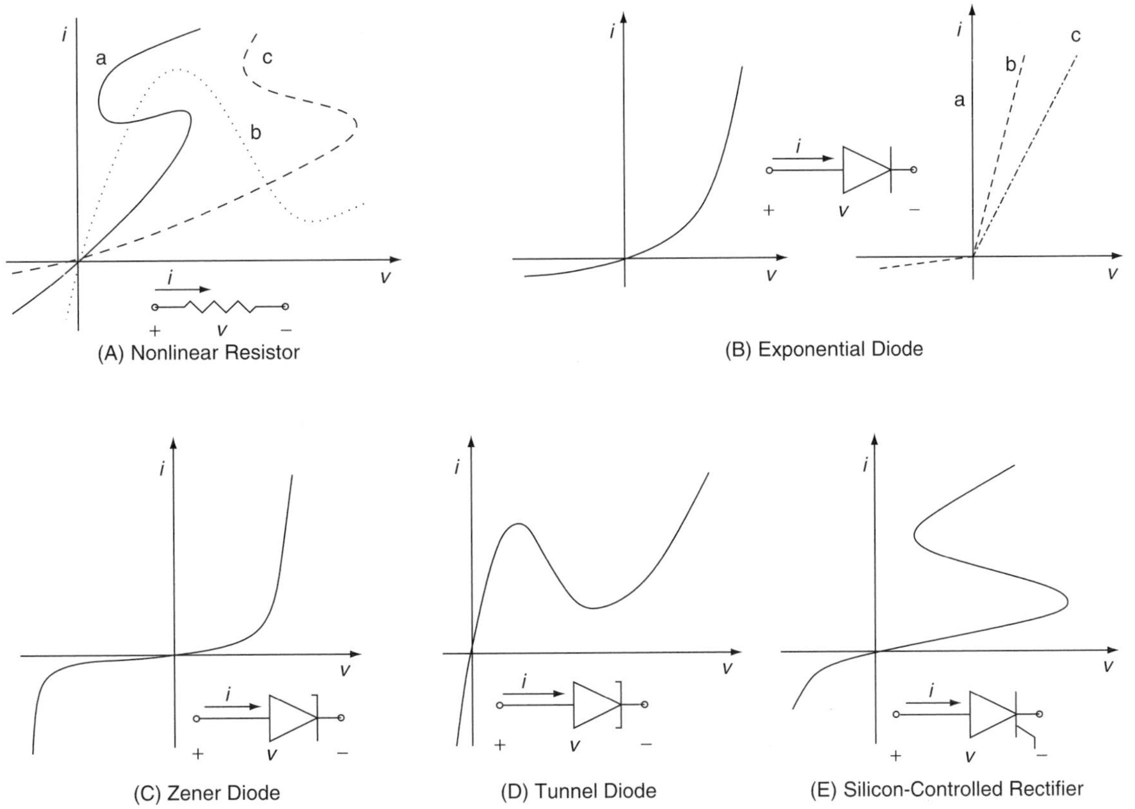

FIGURE 5.1 Two-Terminal Nonlinear Circuit Elements: (A) Nonlinear resistor and examples of three nonlinear constitutive characteristics (a. general, b. voltage-controlled, and c. current-controlled). (B) Exponential diode and its three simplified characteristics. (C) Zener diode and its characteristic with a breakdown region for negative values of the diode voltage. (D) Tunnel diode and its voltage-controlled characteristic. (E) Silicon-controlled rectifier and its current-controlled characteristic.

5.2.2 Two-Terminal Nonlinear Dynamical Elements

Nonlinear resistors have a common characteristic that their constitutive relationships are described by nonlinear algebraic equations. **Dynamical nonlinear elements**, such as nonlinear inductors and capacitors, are described by nonlinear differential equations. The constitutive relation of a nonlinear capacitor is:

$$f(v,\ q) = 0, \tag{5.5}$$

where the charge q and the current i are related by:

$$i = \frac{dq}{dt}. \tag{5.6}$$

The constitutive relation for a nonlinear inductor is:

$$f(\phi,\ i) = 0, \tag{5.7}$$

where the flux ϕ and the voltage v are related by:

$$v = \frac{d\phi}{dt}. \tag{5.8}$$

A generalization of the nonlinear resistor that has memory, called a **memristor**, is a one-port element defined by a constitutive relationship:

$$f(\phi, q) = 0, \tag{5.9}$$

where the flux ϕ and charge q are defined in the usual sense. In special cases, the element may model a nonlinear resistor, capacitor, or inductor.

5.2.3 Three-Terminal Nonlinear Resistive Elements

A three-terminal nonlinear circuit element commonly used in the design of electronic circuits is a bipolar junction transistor (Ebers and Moll, 1954; Getreu, 1976). It is a current-controlled nonlinear element used in the design of various integrated circuits. It is used in circuit designs that

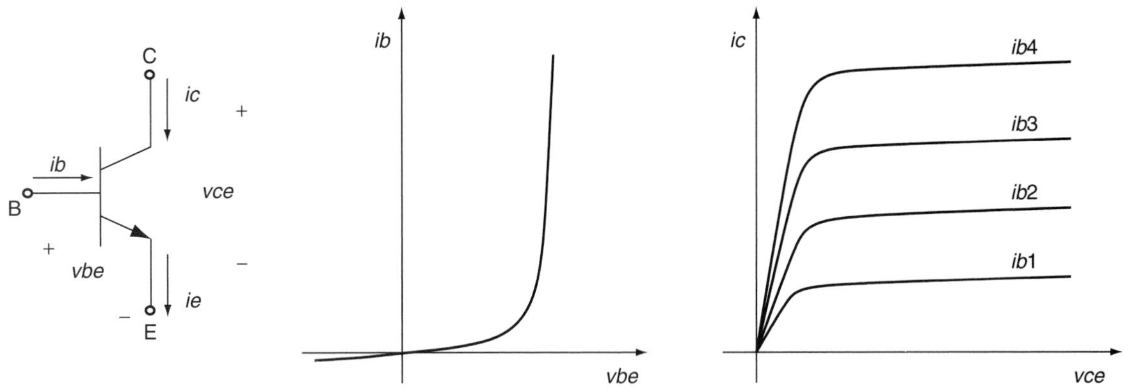

FIGURE 5.2 Three-Terminal Nonlinear Resistive Element: A Circuit Symbol and Characteristics of an *n–p–n* Bipolar Junction Transistor

rely on the linear region of the element's characteristic (logic gates and arithmetic circuits) to designs that rely on the element's nonlinear behavior (flip-flops, static memory cells, and Schmitt triggers). A circuit symbol of the bipolar junction transistor and its simplified characteristics are shown in Figure 5.2. The large-signal behavior of a bipolar junction transistor is usually specified graphically by a pair of graphs with families of curves. A variety of field-effect transistors (the most popular being a metal-oxide semiconductor) are voltage-controlled nonlinear elements used in a similar design fashion as the bipolar junction transistors (Massobrio and Antognetti, 1993).

5.2.4 Multiterminal Nonlinear Resistive Elements

A commonly used multiterminal circuit element is an **operational amplifier** (op-amp). It is usually built from linear resistors and bipolar junction or filed-effect transistors. The behavior of an operational amplifier is often simplified and its circuit is modeled as a linear multiterminal element. Over the limited range of the op-amp input voltage, the relationship between the op-amp input and output voltages is assumed to be linear. The nonlinearity of the op-amp circuit elements is employed in such a way that the behavior of the op-amp can be approximated with a simple linear characteristic. Nevertheless, even in the case of such simplification, the saturation of the op-amp characteristic for larger values of input voltages needs to be taken into account. This makes the op-amp a nonlinear element. The op-amp circuit symbol and its simplified characteristic are shown in Figure 5.3.

5.2.5 Qualitative Properties of Circuit Elements

Two simple albeit fundamental attributes of circuit elements are **passivity** and **no-gain**. They have both proven useful in applying certain modern mathematical techniques for analyzing and solving nonlinear circuit equations. A circuit element is passive if it absorbs power (i.e., if at any operating point, the net power delivered to the element is nonnegative). A nonlinear resistor is passive if for any pair (v, i) of voltage and current, its constitutive relationship satisfies:

$$vi \geq 0. \tag{10}$$

Otherwise, the element is active. Hence, a nonlinear resistor is passive if its characteristic lies in the first and third quadrant. A circuit is passive if all its elements are passive.

A multiterminal element is a **no-gain** element if every connected circuit containing that element, positive resistors, and independent sources possesses the no-gain property. A circuit possesses the no-gain property if, for each solution to the dc network equations, the magnitude of the voltage between any pair of nodes is not greater than the sum of the magnitudes of the voltages appearing across the network's independent sources. In addition, the magnitude of the current flowing through any element of the network must not be greater than the sum of the magnitudes of the currents flowing through the network's independent sources. For example, when considering a multiterminal's large-signal dc behavior, Ebers-Moll modeled bipolar junction transistors, filed-effect transistors, silicon-controlled rectifiers, and many other three-terminal devices are passive and incapable of producing voltage or current gains (Gopinath and Mitra, 1971; Willson, 1975).

Clearly, the no-gain criterion is more restrictive than passive. That is, passivity always follows as a consequence of no-gain, although it is quite possible to cite examples (e.g., the ideal transformer) of passive components capable of providing voltage or current gains.

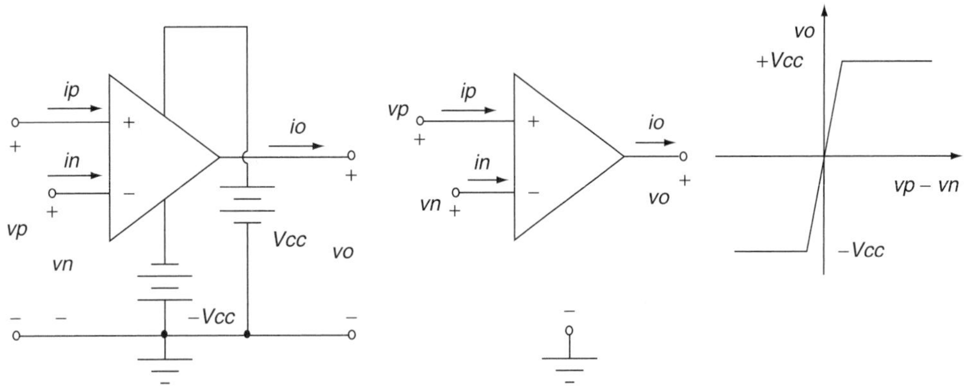

FIGURE 5.3 Multiterminal Nonlinear Circuit Element: An Op-Amp with Five Terminals, its Simplified Circuit Symbol, and an Ideal Characteristic

5.3 Voltages and Currents in Nonlinear Circuits

Circuit voltages and **currents** are solutions to nonlinear differential-algebraic equations that describe circuit behavior. Of particular interest are circuit equilibrium points, which are solutions to associated nonlinear algebraic equations. Hence, circuit equilibrium points are related to dc operating points of a resistive circuit with all capacitors and inductors removed. If capacitor voltages and inductor currents are chosen as state variables, there is a one-to-one correspondence between the circuit's equilibrium points and dc operating points (Chva *et al.*, 1987).

Analyzing nonlinear circuits is often difficult. Only a few simple circuits are adequately described by equations that have a closed form solution. Contrary to linear circuits, which consist of linear elements only (excluding the independent current and voltage sources), nonlinear circuits may possess multiple solutions or may not possess a solution at all (Willson, 1994). A trivial example is a circuit consisting of a current source and an exponential diode, where the value of the dc current supplied by the current source is more negative than the asymptotic value of the current permitted by the diode characteristic when the diode is reverse-biased. It is also possible to construct circuits employing the Ebers-Moll modeled (Ebers and Moll, 1954) bipolar junction transistors whose dc equations may have no solution (Willson, 1995).

5.3.1 Graphical Method for Analysis of Simple Nonlinear Circuits

Voltages and currents in circuits containing only a few nonlinear circuit elements may be found using graphical methods for solving nonlinear equations that describe the behavior of the circuit. A simple nonlinear circuit consisting

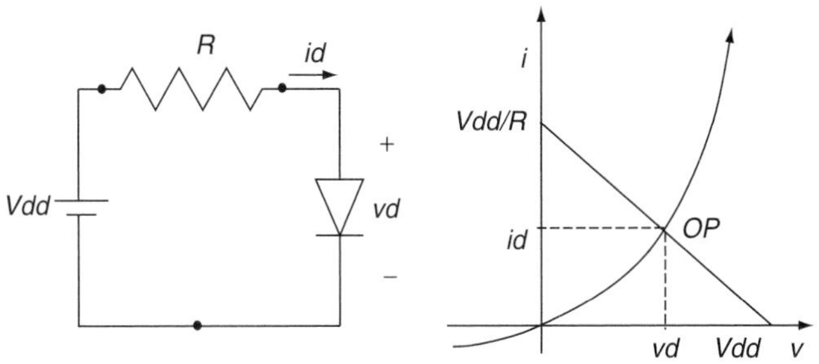

FIGURE 5.4 Simple Nonlinear Circuit and a Graphical Approach for Finding its DC Operating Point (OP): The circuit's load line is obtained by applying Kirchhoff's voltage law. The intersection of diode's exponential characteristic and the load line provides the circuit's dc operating point.

of a constant voltage source, a linear resistor, and an exponential diode is shown in Figure 5.4. Circuit equations can be solved using a graphical method. The solution is the circuit's dc operating point, found as the intersection of the diode characteristics and the "load line." The load line is obtained by applying Kirchhoff's voltage law to the single circuit's loop.

Another simple nonlinear circuit, shown in Figure 5.5, is used to rectify a sinusoidal signal. If the diode is ideal, the sinusoidal signal propagates unaltered and the current flowing through the diode is an ideally rectified sinusoidal signal. The circuit's steady-state response can be found graphically. The diode is *off* below the value $v_s(t) = V_{dd}$. For values $v_s(t) \geq V_{dd}$ the diode is *on*. Hence,

$$i_d(t) = 0 \text{ for } v_s(t) \leq V_{dd}.$$
$$i_d(t) = \frac{v_s(t) - V_{dd}}{R} \text{ for } v_s(t) > V_{dd}. \tag{5.11}$$

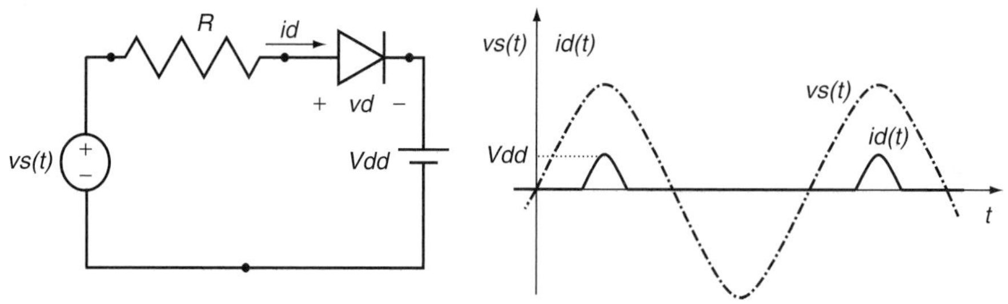

FIGURE 5.5 A Simple Rectifier Circuit with an Ideal Diode: The steady-state solution can be found graphically.

5.3.2 Computer-Aided Tools for Analysis of Nonlinear Circuits

In most cases, a computer program is used to numerically solve nonlinear differential-algebraic circuit equations. The most popular circuits analysis program is SPICE (Massobrio and Antognetti, 1993; Quarles *et al.*, 1994; Vladimirescu, 1994). The original **SPICE code** has been modified and enhanced in numerous electronic design automation tools that employ computer programs to analyze complex integrated circuits containing thousands of nonlinear elements, such as bipolar junction and field-effect transistors.

One of the most important problems when designing a transistor circuit is to find its dc operating point(s) (i.e., voltages and currents in a circuit when all sources are dc sources). The mathematical problem of finding a nonlinear circuit's dc operating points is described by a set of nonlinear algebraic equations constructed by applying Kirchhoff's voltage and current laws and by employing the characteristic of the circuit elements. A common numerical approach for finding these operating points is the Newton–Raphson method and its variants. These methods require a good starting point and sometimes fail. This is known as the dc convergence problem. In the past decade, several numerical techniques based on continuation, parameter-embedding, or homotopy methods have been proposed to successfully solve convergence problems that often appear in circuits possessing more than one dc operating point (Melville *et al.*, 1993; Trajković, 1999; Wolf and Sanders, 1996; Yamamura and Horiuchi, 1990; Yamamura *et al.*, 1999).

5.3.3 Qualitative Properties of Circuit Solutions

Many fundamental issues arise in the case of nonlinear circuits concerning a solution's existence, uniqueness, continuity, boundedness, and stability. Mathematical tools used to address these issues and examine properties of nonlinear circuits range from purely numerical to geometric (Smale, 1972). Briefly addressed here are solutions to a circuit's equations. Described are several fundamental results concerning their properties.

Existence and Uniqueness of Solutions

Circuits with nonlinear elements may have multiple discrete dc operating points (equilibriums). In contrast, circuits consisting of positive linear resistors possess either one dc operating point or, in special cases, a continuous family of dc operating points. Many resistive circuits consisting of independent voltage sources and voltage-controlled resistors, whose *v*–*i* relation characteristics are continuous strictly monotone-increasing functions, have at most one solution (Duffin, 1947; Minty, 1960; Willson, 1975). Many transistor circuits possess the same property based on their topology alone. Other circuits, such as flip-flops and memory cells, possess feedback structures (Nielsen and Willson, 1980). These circuits may possess multiple operating points with an appropriate choice of circuit parameters and biasing of transistors (Trajković and Willson, 1992). For example, a circuit containing two bipolar transistors possesses at most three isolated dc operating points (Lee and Willson, 1983). Estimating the number of dc operating points or even their upper bounds for an arbitrary nonlinear circuit is still an open problem (Lagarias and Trajković, 1999).

Continuity and Boundedness of Solutions

Properties such as continuity and boundedness of solutions are often desired in cases of well-behaved circuits. One would like to expect that "small" changes in the circuit's output will result from the "small" change of the circuit's input and that the bounded circuit's input leads to the bounded circuit's output. They are intimately related to the stability of circuit solutions.

Stability of Solutions

Once a solution to a circuit's currents and voltages is found, a fundamental question arises regarding its stability. A solution may be stable or unstable, and stability is both a local and a global concept (Hasler and Neirynck, 1986). Local stability usually refers to the solution that, with sufficiently close initial conditions, remains close to other solutions as time increases. A stronger concept of stability is asymptotic stability: if initial conditions are close, the solutions converge toward each other as time approaches infinity. Although no rate of convergence is implied in the case of asymptotic stability, a stronger concept of exponential stability implies that the rate of convergence toward a stable solution is bounded from above by an exponential function. Note that the exponential convergence to a solution does not necessarily imply its stability. Contrary to nonlinear circuits, stability (if it exists) of linear time-invariant circuits is always exponential. Linear time-invariant circuits whose natural frequencies have negative real parts have stable solutions. A global asymptotic (complete) stability implies that no matter what the initial conditions are, the solutions converge toward each other. Hence, a completely stable circuit possesses exactly one solution.

A fundamental observation can be made that there exist dc operating points of transistor circuits that are unstable: if the circuit is biased at such an operating point and if the circuit is augmented with *any* configuration of positive-valued shunt capacitors and/or series inductors, the equilibrium point of the resulting dynamic circuit will always be unstable (Green and Willson, 1992, 1994). A simple example is the common latch circuit that, for appropriate parameter values, possesses three dc operating points: two stable and one unstable.

5.4 Open Problems

In the area of analysis of resistive circuits, interesting and still unresolved issues are: Does a given transistor circuit possess exactly one, or more than one dc operating point? How many operating points can it possess? What simple techniques can be used to distinguish between those circuits having a unique operating point and those capable of possessing more than one? What can we say about the stability of an operating point? How can circuit simulators be used to find all the solutions of a given circuit? These and other issues have been and still are a focal point of research in the area of nonlinear circuits. The references cited provide more detailed discussions of the many topics considered in this elementary review of nonlinear circuits.

References

Chua, L.O., Desoer, C.A., and Kuh, E.S. (1987). *Linear and nonlinear circuits.* New York: McGraw-Hill.

Duffin, R.J. (1947). Nonlinear networks IIa. *Bulletin of the American Mathematical Society 53*, 963–971.

Ebers, J.J., and Moll, J.L. (1954). Large-signal behavior of junction transistors. *Proceedings IRE 42*, 1761–1772.

Getreu, I. (1976). *Modeling the bipolar transistor.* Beaverton, OR: Tektronix, 9–23.

Gopinath, B., and Mitra, D. (1971). *When are transistors passive? Bell Systems Technological Journal 50*, 2835–2847.

Green, M., and Willson, A.N. Jr. (1992). How to identify unstable dc operating points. *IEEE Trans. Circuits System-I 39*, 820–832.

Green, M., and Willson, A.N. Jr. (1994). (Almost) half of any circuit's operating points are unstable. *IEEE Transactions on Circuits and Systems-I 41*, 286–293.

Hasler, M., and Neirynck, J. (1986). *Nonlinear circuits.* Norwood, MA: Artech House.

Lagarias, J.C., and Trajković, Lj. (1999). Bounds for the number of dc operating points of transistor circuits. *IEEE Transactions on Circuits and Systems* 46(10), 1216–1221.

Lee, B.G., and Willson, A.N. Jr. (1983). All two-transistor circuits possess at most three dc equilibrium points. *Proceedings of the 26th Midwest Symposium Circuits and Systems*, 504–507.

Massobrio, G. and Antognetti, P. (1993). *Semiconductor device modeling with spice.* (2d Ed.). New York: McGraw-Hill.

Mathis, W. (1987). *Theorie nichtlinearer netzwerke.* Berlin: Springer-Verlag.

Melville, R.C., Trajković, Lj., Fang, S.C., and Watson, L.T. (1993). Artificial parameter homotopy methods for the dc operating point problem. *IEEE Transactions on Circuits and Systems 12*(6), 861–877.

Minty, G.J. (1960). Monotone networks. *Proceedings of the Royal Society (London)*, Series A *257*, 194–212.

Nielsen, R.O., and Willson, A.N., Jr. (1980). A fundamental result concerning the topology of transistor circuits with multiple equilibria. *Proceedings of IEEE 68*, 196–208.

Quarles, T.L., Newton, A.R., Pederson, D.O., and Sangiovanni-Vincentelli, A. (1994). *SPICE 3 version 3F5 user's manual.* Berkeley: Department of EECS, University of California.

Smale, S. (1972). On the mathematical foundations of electrical circuit theory. *Journal of Differential Geometry 7*, 193–210.

Trajković, Lj., and Willson, A.N., Jr. (1992). Theory of dc operating points of transistor networks. *International Journal of Electronics and Communications 46*(4), 228–241.

Trajković, Lj. (1999). Homotopy methods for computing dc operating points. In J.G. Webster (Ed.), *Encyclopedia of electrical and electronics engineering*, New York: John Wiley & Sons.

Vladimirescu, A. (1994). *The SPICE book.* New York: John Wiley & Sons.

Willson, A.N., Jr. (1975). *Nonlinear networks.* New York: IEEE Press.

Willson, A.N., Jr. (1975). The no-gain property for networks containing three-terminal elements. *IEEE Transactions on Circuits and Systems 22*, 678–687.

Wolf, D., and Sanders, S. Multiparameter homotopy methods for finding dc operating points of nonlinear circuits. *IEEE Transactions on Circuits and Systems 43*, 824–838.

Yamamura, K., and Horiuchi, K. (1990). A globally and quadratically convergent algorithm for solving nonlinear resistive networks (1990). *IEEE Transactions of Computer-Aided Design 9*(5), 487–499.

Yamamura, K., Sekiguchi, T., and Inoue, T. (1999). A fixed-point homotopy method for solving modified nodal equations. *IEEE Transactions on Circuits and Systems 46*, 654–665.

II

ELECTRONICS

Krishna Shenai
Department of Electrical and Computer Engineering, University of Illinois at Chicago, Chicago, Illinois, USA

With the emergence of digital information age, there is an ever increasing need to integrate various technologies to efficiently perform advanced communication, computing, and information processing functions. Efficient generation and utilization of electric energy is of paramount importance in an energy and environmentally conscious society. This section deals with recent developments in these areas with relevant background information.

In Chapter 1, an account of voltage regulation modules (VRMs) used in powering microchips is provided with a particular emphasis on circuit switching topologies. Chapter 2 deals with noise in mixed-signal electronic systems with a focus on submicron ultra-large scale integrated (ULSI) system-on-a-chip (SOC) technologies. Noise in semiconductor devices and passive components becomes a key issue in addressing system-level signal integrity concerns. A detailed review and progress on metal-oxide-silicon field-effect transistors (MOSFETs) is discussed in Chapter 3 with emphasis on important performance and reliability degradation parameters. These devices form the basic electronic building blocks in a majority of microsystems.

Filters are used in a variety of signal processing circuits. Chapter 4 provides an account of latest advances in the design and application of active filters in microsystems. Although bipolar junction transistors (BJTs) are not as prominent as MOSFETs, nevertheless they are used in a variety of mixed-signal and poweer circuits. Chapter 5 discusses the state-of-the-art on diodes and BJTs in these circuits. Chapter 6 provides insight into basic semiconductor material physics and how it relates to device performance and reliability. The section concludes with a detailed account of the evolution and current state-of-the-art of power semiconductor devices for discrete as well as integrated applications. Power semiconductor devices are playing an increasingly important role in microsystems, especially for performing the power management function.

1

Investigation of Power Management Issues for Future Generation Microprocessors*

Fred C. Lee and
Xunwei Zhou
*Center for Power Electronics Systems,
The Bradley Department of
Electrical Engineering,
Virginia Polytechnic Institute
and State University,
Blacksburg, Virginia, USA*

1.1 Introduction

Advances in microprocessor technology pose new challenges for supplying power to these devices. The evolution of microprocessors began when the high-performance Pentium processor was driven by a nonstandard power supply of less than 5 V instead of drawing its power from the 5-V plane on the motherboard (Goodfellow and Weiss, 1997).

Low-voltage power management issues are becoming increasingly critical in state-of-the-art computing systems. The current generation of high-speed CMOS processors (e.g., Alpha, Pentium, and Power PC) operate at above 300 MHz with 2.5- to 3.3-V output voltage. Future processors will be designed with even lower logic voltages of 1 to 1.5 V and increases in current demands from 13 A to 50 to 100 A (Zhang *et al.*, 1996). Meanwhile, operating frequencies will increase to above 1 GHz. These demands, in turn, require special power supplies, voltage regulator modules (VRMs), to provide lower voltages with higher current capabilities for microprocessors. Table 1.1 shows the specifications for present and future VRMs.

As the speed of the processors increases, the dynamic loading of the VRMs is also significantly increased. Future microprocessors are expected to exhibit higher current slew rates of up to 5 A/ns. These slew rates represent a severe problem for the large load changes that are encountered when systems transfer from the sleep mode to the active mode and vice versa. In these cases, the parasitic impedance of the power supply connection to the load and the equivalent series resistor (ESR) and equivalent series inductor (ESL) of the capacitors have a dramatic effect on VRM voltage (Zhang *et al.*, 1996). If this impedance is not low enough, the supply voltage may fall out of the required range during the transient period. Moreover, the total voltage tolerance will be much

* This work is supported by Intel, Texas Instruments, National Semiconductors Inc., SGS Thomson, and Delta Electronics Inc.
* This work made use of ERC Shared and Facilities supported by the National Science Foundation under Award Number EEC-9731677.

TABLE 1.1 Specifications for Present and Future VRMs

Voltage/Current	Present	Future
Output voltage	2.1 ~ 3.5 V	1 ~ 1.5 V
Load current	0.3 ~ 13 A	1 ~ 50 A
Output voltage tolerance	± 5%	± 2%
Current slew at decoupling capacitors	1 A/nS*	5 A/ns

* Current slew rate at today's VRM output is 30 MS (1997).

tighter. Currently, the voltage tolerance is 5% (for a 3.3 V VRM output with a voltage deviation of ± 165 mV). In the future, the total voltage tolerance will be 2% (for a 1.1 V VRM output with a voltage deviation requirement of only ± 33 mV). All of these requirements pose serious design challenges and require VRMs to have very fast transient responses. Figure 1.1 shows today's VRMs and the road map of microprocessors' development.

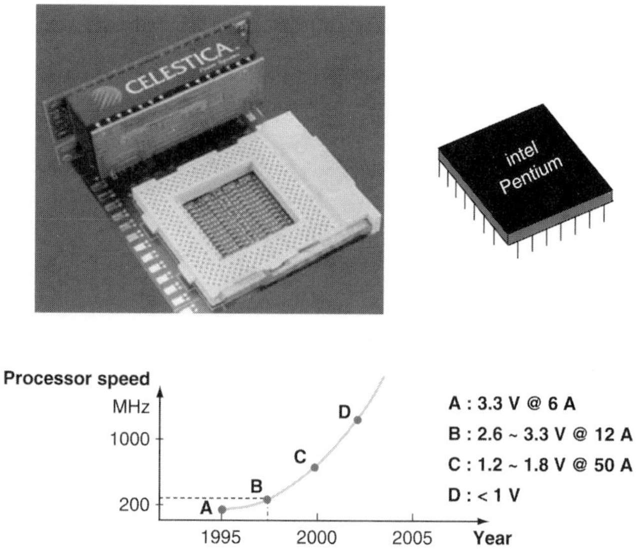

FIGURE 1.1 Today's VRM and Processor Road Map

Today's VRMs are powered up from the 5-V or 12-V outputs of silver boxes that are used for supplying various parts of the system, such as the memory chips, the video cards, and some sub buses. Future VRMs will be required to provide lower voltages and higher currents with tighter voltage regulations. The traditional centralized power system, the silver box, will no longer meet the stringent requirements for VRM voltage regulation because of the distributed impedance associated with a long power bus and the parasitic ringing due to high-frequency operation. On the other hand, with much heavier loads in the future, the bus loss becomes significant. To maintain system stability, a huge silverbox output capacitance is also needed. At the same time, to avoid the interaction between different outputs, a very large VRM input filter capacitance is required. Figure 1.2 shows the trend of computer power system architecture. In the future, a distributed power system (DPS) with a high-voltage bus, 12 V or 48 V, can be the solution for servers' and workstations' power systems. High-performance, high-input-voltage VRMs, however, must be developed.

To meet future requirements, a number of critical issues must be addressed. For example, advanced power devices and control technologies are needed for high-efficiency and high-frequency operations. Today's vertical power device technology cannot provide acceptable levels of conversion efficiency at a multimegahertz level due to its high conduction and switching and gate drive losses. This chapter addresses important issues of advanced VRM topologies for fast transient responses and low ripple voltages as well as advanced packaging technologies for improving power density and thermal management. In addition, this discussion covers the limitations of today's tech-

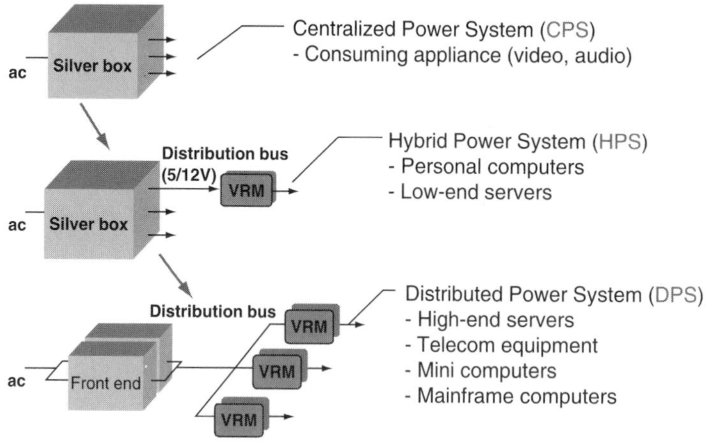

FIGURE 1.2 The Trend of Computer Power System Architecture

nologies, including VRM topologies, power devices, and power systems.

1.2 Limitations of Today's Technologies

1.2.1 Limitations of Present VRM Topologies

Most of today's VRMs use **conventional buck** or **synchronous rectifier buck topologies**. Figure 1.3 shows the conventional buck circuit, which is the most cost-effective approach. Usually, Schottky diodes are used as rectifiers. The top metal-oxide-semiconductor field effect transitor (MOSFET) transfers energy from the input, and the bottom rectifier conducts the inductor current. The control regulates the output voltage by modulating the conduction interval at the top MOSFET. Figure 1.4 shows the synchronous rectifier buck circuit. This topology increases the efficiency by replacing the rectifier with a low Rds(on) MOSFET. The synchronous switch is controlled by the complementary signal of the top switch's gate signal. The synchronous rectifier buck always operates in continuous current mode. Its transient response is faster than that of a conventional buck converter. Conventional VRMs use large output filter inductances, $2 \sim 4\ \mu H$, to reduce ripple.

Figure 1.5 shows the practical VRM load model (processor model). The packaging capacitor is the parasitic capacitor inside the microprocessor package. There are several decoupling capacitors near and around the microprocessors to reduce noise and maintain voltage regulation. Bulk capacitors are VRM output capacitors. In very high-speed and high-current systems, physical capacitors can no longer be simplified as ideal capacitors. The parasitic parameters play very important roles. The capacitor should be considered as an ideal capacitor in series with an ESR and an ESL, as shown in Figure 1.6 (Wong, 1997).

There are interconnection parasitic inductances and resistances between bulk capacitors and decoupling capacitors and between decoupling capacitors and packaging capacitors. Future microprocessor load transitions will have a 5-A/ns slew rate. In this case, all these parasitics have a significant

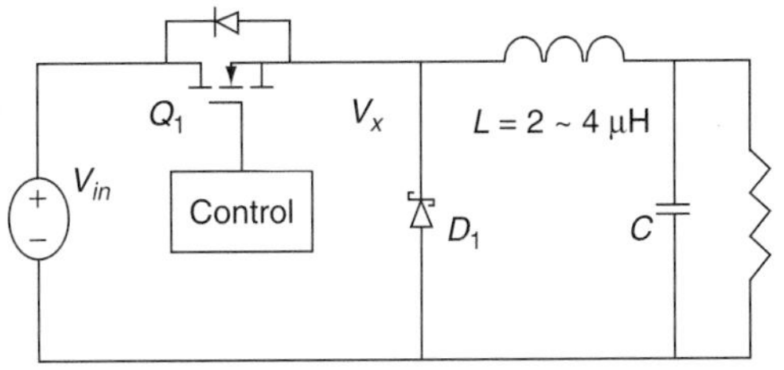

FIGURE 1.3 Conventional Buck Converter

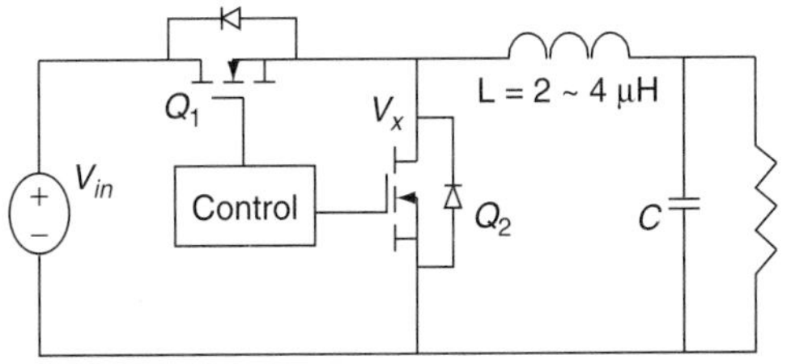

FIGURE 1.4 Synchronous Rectifier Buck Converter

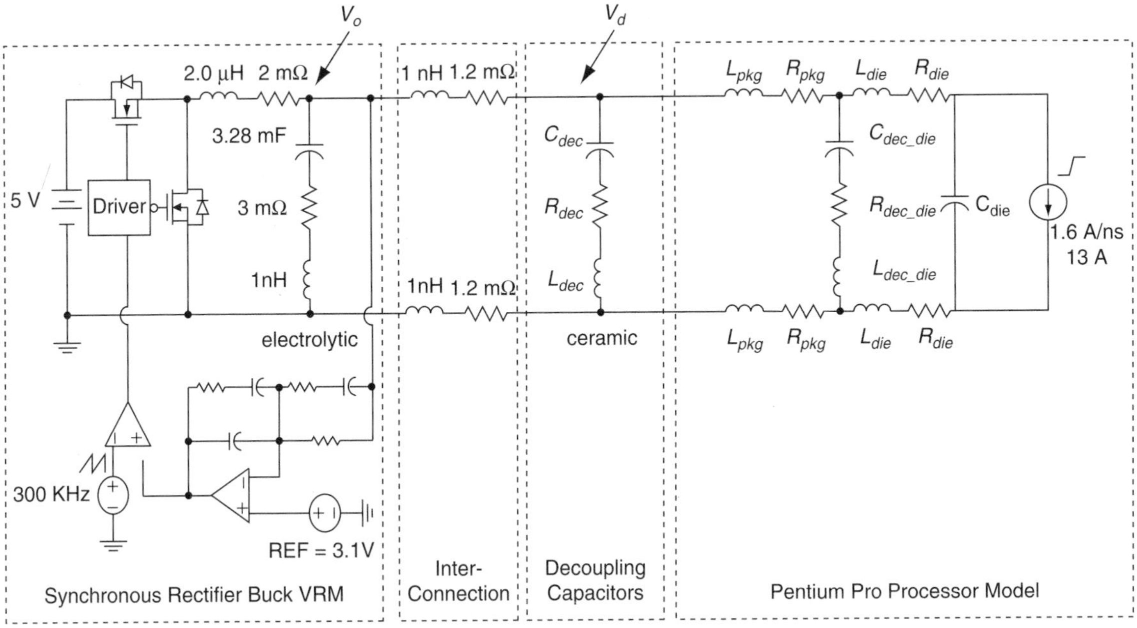

FIGURE 1.5 Practical VRM Load: Pentium Pro Processor Model

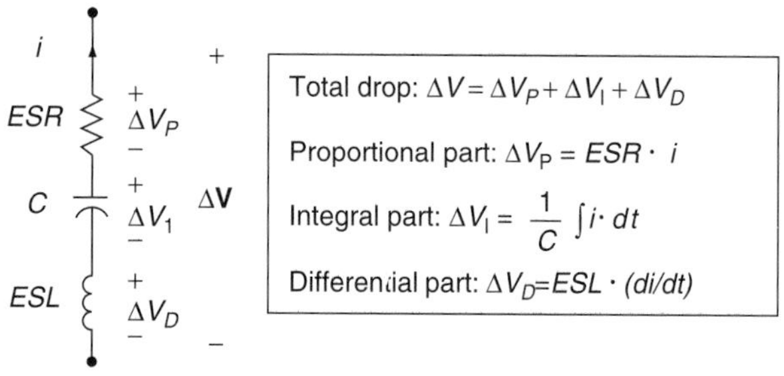

FIGURE 1.6 The Voltage Drop of a Capacitor Divided into Three Parts: Proportional, Integral, and Differential

effect on the VRM transient voltage. Figure 1.7 shows the transient response of a synchronous rectifier VRM. The VRM's input voltage is 5 V, and its load changes from 0.8 A to 30 A. It is obvious that for future microprocessor loads, today's VRM topologies cannot meet the 2% transient requirement.

During the transient, there are three spikes in the voltage drop (Wong, 1997). The first high-frequency spike is dominated by loop 1, shown in Figure 1.8, which combines the parasitic of the packaging capacitors and decoupling capacitors and the interconnection between them. The second spike is controlled by loop 2, which combines the parasitic of the decoupling capacitors and VRM bulk capacitors and the interconnection between them. The third spike is decided by loop 3, which combines the parasitic of the VRM output filter inductor and the VRM output bulk capacitors.

The transient limitation of today's VRM topologies comes from their large output filter inductance. During the transient, this large inductor limits the energy transfer speed so that the capacitors have to store or discharge all the energy from the load. For future microprocessors, because of heavier load currents, higher load transient slew rates, and tighter voltage tolerance requirements, more decoupling capacitors will be required to reduce the second spike, and more VRM output bulk capacitors will be required to reduce the third spike. As a result, to meet future specifications, 23 times the decoupling capacitors will be needed, and 3 times VRM bulk capacitors will be needed (Wong, 1997). The VRM will be very large and expensive. The space of the VRM, however, is limited and the real estate of the motherboard is expensive. The need for a large quantity of capacitors makes VRMs, which use these topologies, impractical for future microprocessors.

1.2.2 Limitations from Power Devices

Another limitation of today's VRMs is efficiency. Figure 1.9 shows the conventional VRM efficiency at a 2-V output. Using IRL3803 as switches, with an on-resistance of 6 mΩ and a 30-V

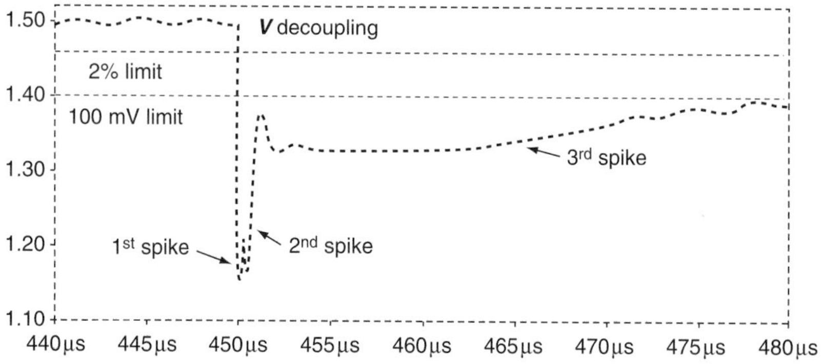

FIGURE 1.7 Transient Response of Conventional VRMs ($V_{in} = 5$ V, load: 0.8 A to 30 A, and $fs = 300$ kHz)

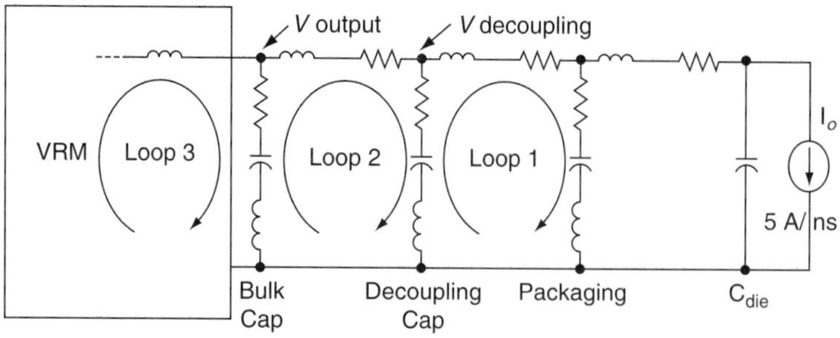

FIGURE 1.8 The Processor Model: Three Resonant Loops with Different Resonant Frequencies

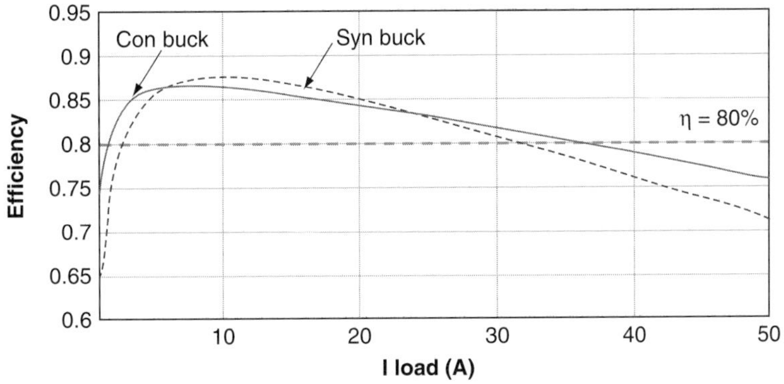

FIGURE 1.9 Conventional VRM Efficiency ($V_{in} = 5\,V$, $V_o = 2\,V$, $fs = 300$ kHz, and switches: IRL3803)

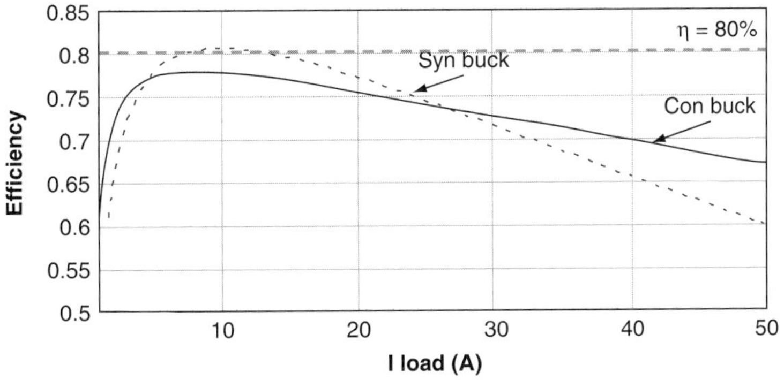

FIGURE 1.10 Conventional VRM Efficiency ($V_{in} = 5\,V$, $V_o = 1.2\,V$, $fs = 300$ kHz, and switches: IRL3803)

voltage rating, the conventional VRM cannot meet the 80% efficiency requirement at heavy load. For lower output voltages, it will be even more difficult to meet the efficiency requirement. Figure 1.10 shows the conventional VRMs' efficiency at a 1.2-V output. Their efficiency is lower than 80% in the whole load range.

This limitation stems from today's power devices' technology. Based on vertical power MOSFET technology, most of today's low-voltage power MOSFETs are available at a rating of 30 V. Roughly, the total power loss of a power device can be divided into three parts: conduction loss, gate drive loss, and switching loss. Figure 1.11 shows the relationship between conduction loss, gate drive loss, and switching loss. For this kind of low-voltage, high-current application, conduction loss contributes a large percentage of the total loss. When only one IRL3803 is used, the MOSFET's conduction loss is 25 times the gate drive loss plus the switching loss. To reduce conduction and total loss, more switches need to be paralleled. This does not necessarily mean, however, that more parallel switches equals lower total loss. When five IRL3803 are paralleled, the total loss is reduced to the minimum. After this point, adding more switches will not improve efficiency. Figure 1.12 shows the number of parallel

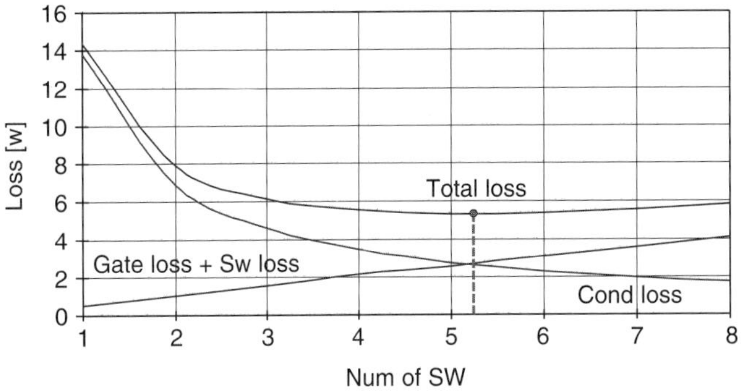

FIGURE 1.11 Switching Loss, Gate Drive Loss, and Conduction Loss of an IRL3803 Versus a Parallel Switch Number ($V_{in} = 5\,V$, $fs = 300$ kHz, and I load $= 50\,A$)

switches needed to meet the 80% efficiency requirement when the input voltage is 5 V. In Figure 1.12 (A), when the output voltage is 2 V, paralleling two switches can enable the device to meet the efficiency requirement at heavy load. However, when the output voltage is 1.2 V, as shown in Figure 1.12 (B), no matter how many switches are in parallel, the VRM cannot meet the 80% efficiency requirement at heavy

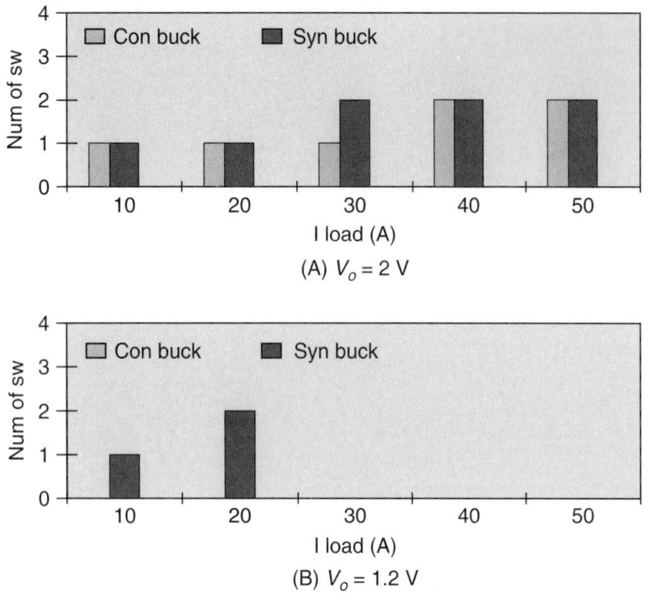

FIGURE 1.12 The Number of Parallel Switches Needed to Meet the Efficiency Requirement
($V_{in} = 5$ V, $fs = 300$ kHz, and efficiency >80%)

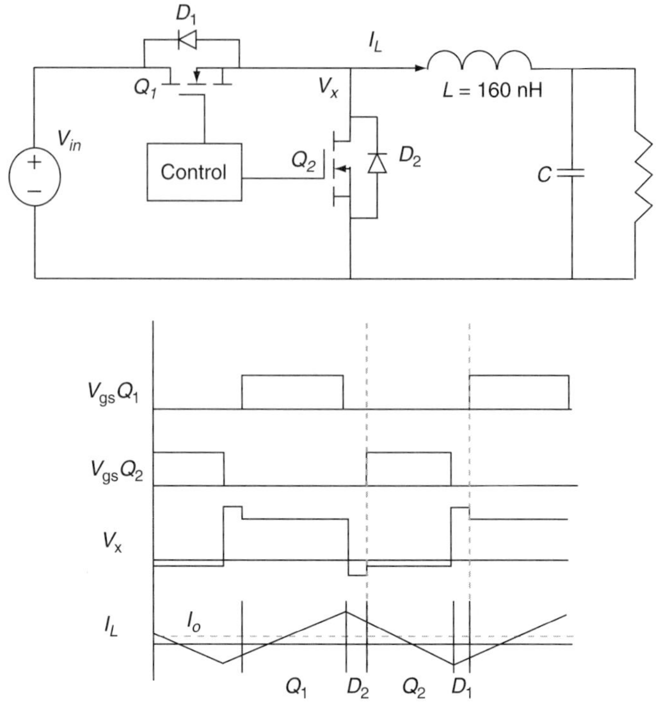

FIGURE 1.13 Quasi-Square-Wave (QSW) VRM Topology

load (>30). This limitation is due to the high figure of merit (FOM) of today's devices. The FOM is equal to Rds(on) times gate charge (Qg). For today's device technology, the lowest FOM value is around 300(mΩ × nC). With such a high FOM value, power devices not only limit the VRM's efficiency but also limit the VRM's ability to operate at higher operating frequencies. Most of today's VRMs operate at a switching frequency lower than 300 kHz. This low switching frequency causes slow transient responses and creates a need for very large energy storage components.

1.3 Advanced VRM Topologies

1.3.1 Low-Input-Voltage VRM Topologies

A Fast VRM Topology: The Quasi-Square-Wave (QSW) VRM

To overcome the transient limitation occurring in conventional VRMs, a smaller output filter inductance is the most desirable option for increasing the energy transfer speed. Figure 1.13 shows the quasi-square-wave (QSW) circuit. The QSW topology keeps the VRM output inductor current touching zero in both sleep and active modes. When Q_1 turns on, the inductor current is charged to positive by the input voltage. After Q_1 turns off and before Q_2 turns on, the inductor current flows through Q_2's body diode. When Q_2 turns on, the inductor current is discharged to negative. After Q_2 turns off and before Q_1 turns on, the inductor current flows through Q_1's body diode. Compared with conventional buck and synchronous buck topologies, the output filter inductance is reduced significantly. At a 13-A load and a 300-kHz switching frequency, the QSW circuit needs only a 160-nH inductor, as compared with the $2 \sim 4\,\mu$H inductor needed in the conventional design. This small inductance makes the VRM transient response much faster. Figure 1.14 shows the transient response of the QSW topology. The third spike of the output voltage becomes insignificant, and the second spike is reduced significantly.

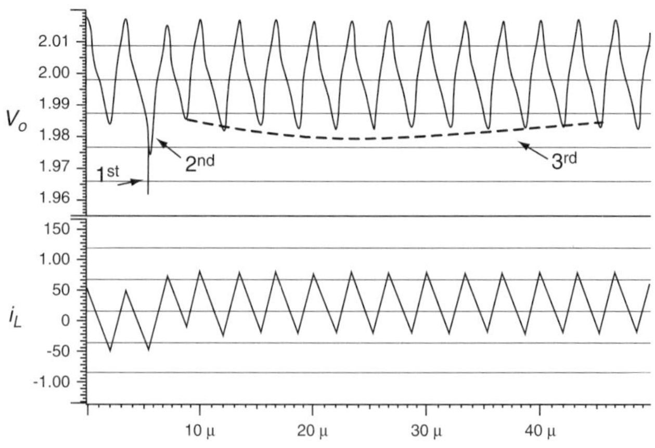

FIGURE 1.14 Transient Response of the QSW

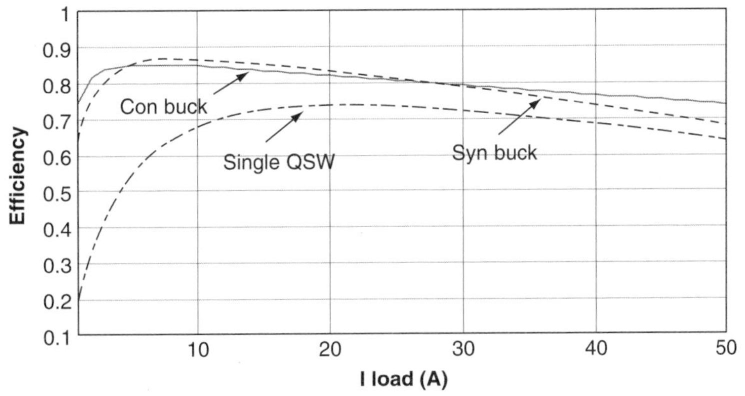

FIGURE 1.15 Efficiency of the QSW Compared with a Conventional VRM ($V_{in} = 5\,\text{V}$, $V_o = 2\,\text{V}$, and $fs = 300$ kHz)

There are two disadvantages to this fast VRM topology. The first one is the large current ripple. A huge VRM output filter capacitance is needed to suppress the steady-state ripple. The smaller inductance results in a faster transient response but requires a larger bulk capacitance. The second disadvantage is its low efficiency. In Figure 1.15, due to the large ripple current, the QSW switch has a larger conduction loss. Its efficiency is lower than that of a conventional VRM.

A Fast VRM with a Small Ripple: The Interleaved QSW VRM

To meet both the steady state and transient requirements, a novel VRM topology, the interleaved QSW, is introduced in Figure 1.16. The interleaved QSW topology naturally cancels the output current ripple and still maintains the fast transient response characteristics of the QSW topology. A smaller capacitance is needed compared to both the single-module QSW VRM and the conventional VRM. The more modules in parallel, the better the ripple canceling effect. Figure 1.17 shows a four-module interleaved QSW VRM. Figure 1.18 shows its transient response. The results show that this technique can meet future transient requirements without a large steady-state voltage ripple. Compared with the single-module QSW topology, the efficiency is improved significantly. Figure 1.19 shows the efficiency comparison results.

Figure 1.20 shows a four-module interleaved QSW VRM prototype picture. In the VRM, an integrated magnetic design is used. Every two inductors use one magnetic core. As a result, in total, two magnetic cores are used for these four channel inductors. Figure 1.21 shows the integrated magnetic structure (Chen 1999). By taking advantage of the interleaving technology, the AC flux of the two inductors is canceled out in the center leg. As a result, the core loss and center leg crossing area are reduced. The planar core structure

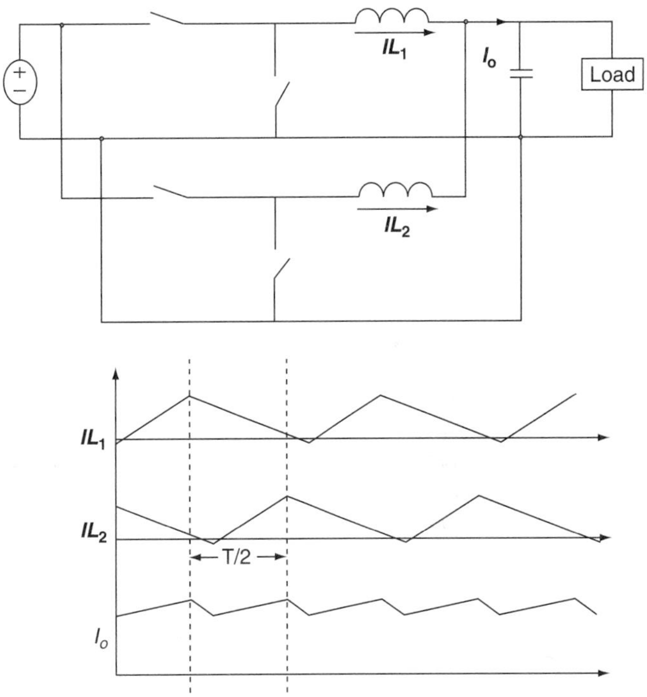

FIGURE 1.16 Current Ripple Cancelling Effect of Interleaved QSW

makes a very low-profile VRM. This kind of low-profile magnetic also has good thermal management. In the magnetic design, the PCB trace is used as the inductor winding. This approach is very cost-effective. In addition, the termination loss is eliminated.

Table 1.2 compares the four-module interleaved QSW VRM design with the conventional VRM. Since the necessary capacitance is reduced, the VRM power density is dramatically increased by six times. Also, since each module handles a lower current, the circuit will be packaged more easily. The test results in Figure 1.22 show the transient response of the

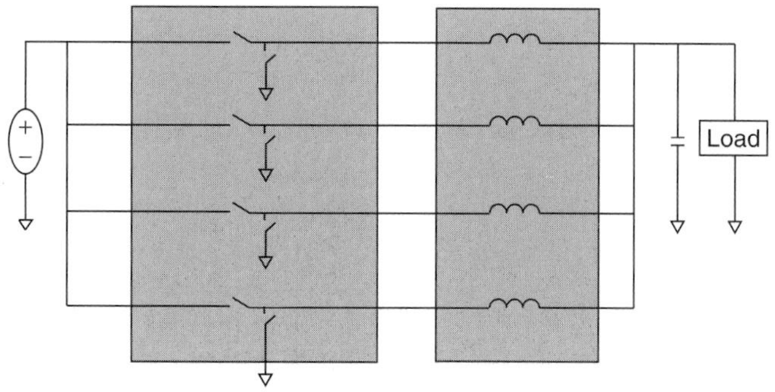

FIGURE 1.17 A Four-Module Interleaved QSW VRM

four-module interleaved QSW VRM compared to today's commercial VRM. The interleaved QSW topology can not only reduce output current ripple, but it can also reduce input current ripple.

1.3.2 High-Input-Voltage VRM Topologies

Most of today's VRMs draw power from the 5-V output of the silver box. This bus voltage is too low for future low-voltage, high-current processors' applications. A distributed power system (DPS) with a high bus voltage can be the solution for future computer systems, such as servers' and workstations' power systems. In a high-voltage bus-distributed system, the bus conduction loss is lower and the distribution bus is easy to

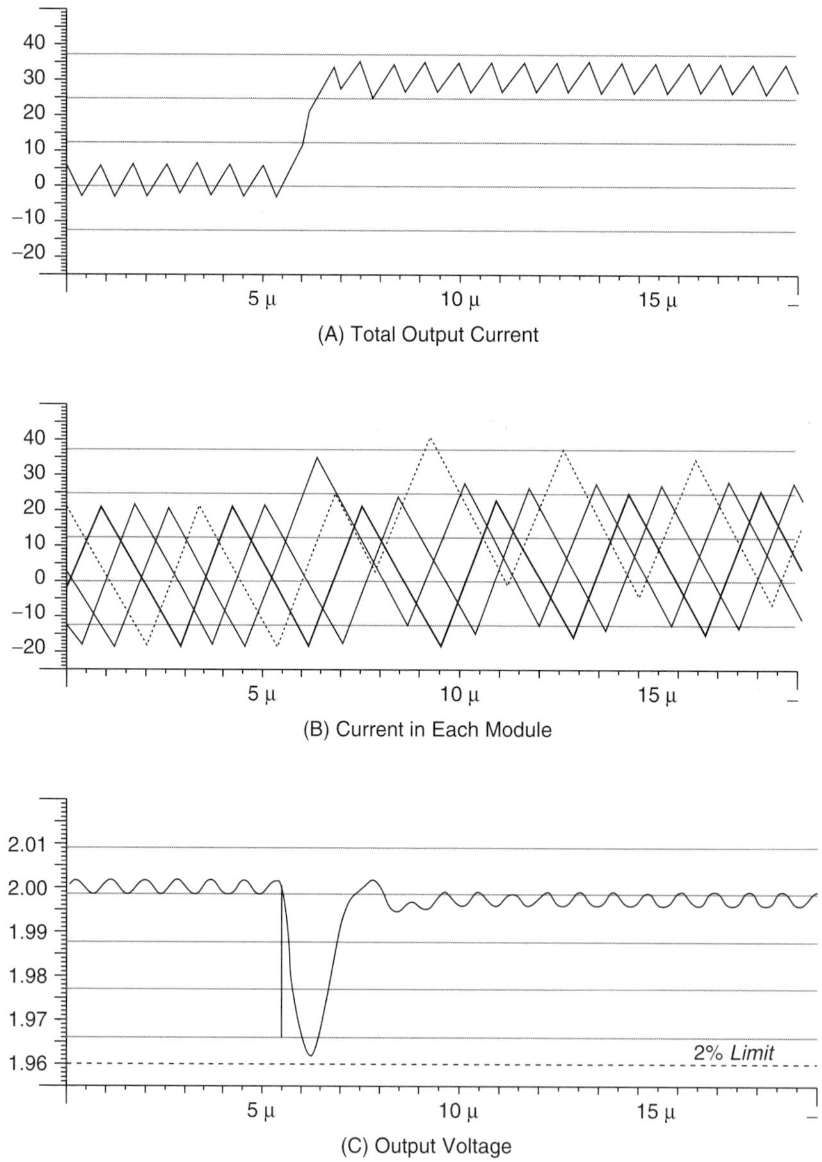

FIGURE 1.18 Transient Response of the Four-Module Interleaved QSW

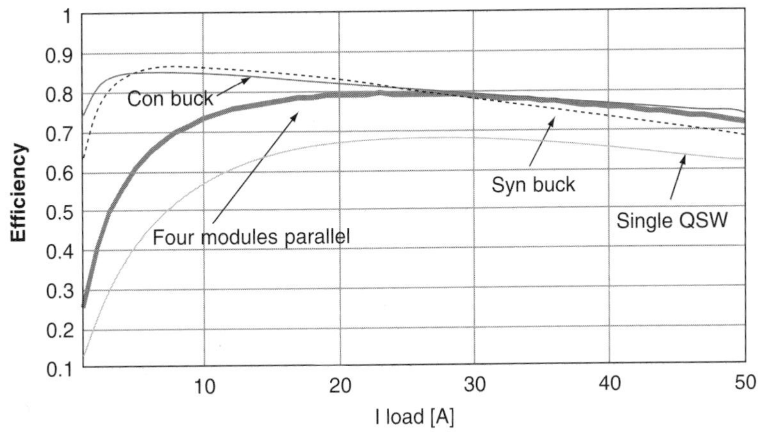

FIGURE 1.19 Efficiency Comparison

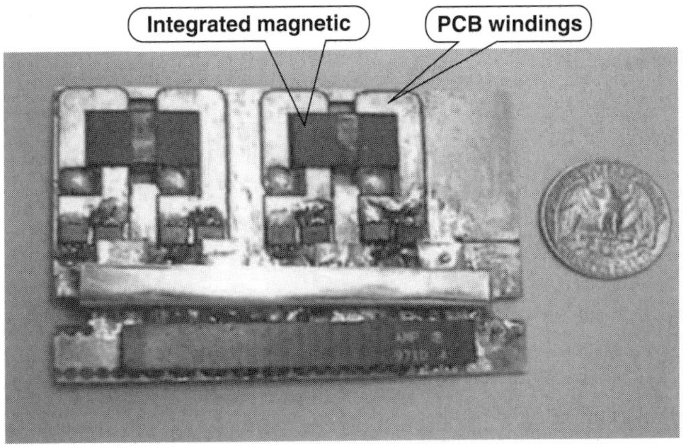

FIGURE 1.20 The Four-Module Interleaved QSW VRM

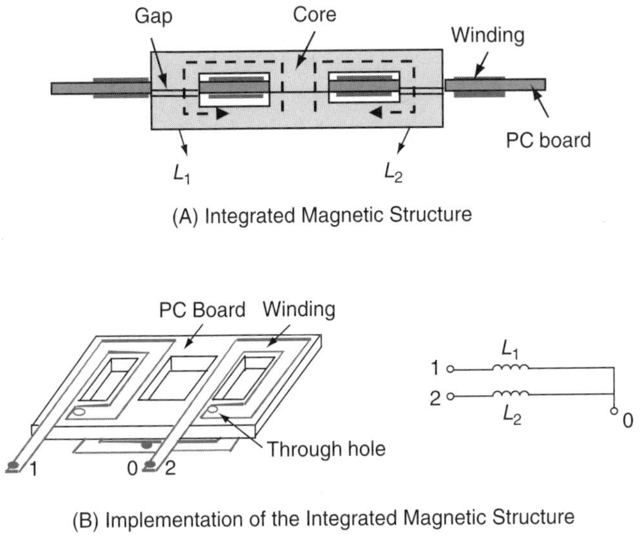

FIGURE 1.21 Integrated Magnetic Structure

TABLE 1.2 Design Comparison of the Interleaved QSW VRM and the Conventional VRM

	Interleaved QSW	Conventional VRM
V_{in}	5	5
Bulk capacitance	1200 μF	8000 μF
Output inductance	320 nH (×4)	3.8 μH
Transient voltage drop:	50 mV	150 mV
V_o @ load	2 V @ 30 A	2 V @ 13 A
Power-stage-power density (W/in^3)	30	3 ~ 5

design. On the other hand, in this kind of system, the transient of the load-end converter will have less effect on the bus voltage and thus less effect on the other load-end converters. As a result, for future applications, the high-input-voltage VRMs' input filter size can be dramatically reduced. Figure 1.23 shows the results. When the bus voltage increases from 5 V to 48 V, the VRM input filter capacitance can be reduced from 10 mF to 10 μF.

For high-voltage bus systems, advanced high-input-voltage VRM topologies must be developed. For example, if the buck converter shown in Figure 1.3 is used in a high-voltage bus system, the VRM's duty cycle is very asymmetrical. When the input voltage is 12 V and the output voltage is 2 V, the VRM duty cycle is only 0.16. Figure 1.24 shows the asymmetrical transient response of a synchronous buck VRM. When the load changes from a light load to a heavy load, since the inductor charging voltage ($V_{in} - V_o$) is high, the VRM's step-down voltage drop is small. When the load changes from a heavy load to a light load, since the inductor discharging voltage V_o, is low, the VRM's step-up voltage drop is large. This asymmetrical transient response makes the output filter overdesigned. Besides, it is difficult to optimize efficiency in an asymmetrical duty cycle converter. Figure 1.25 shows the efficiency comparison for different input

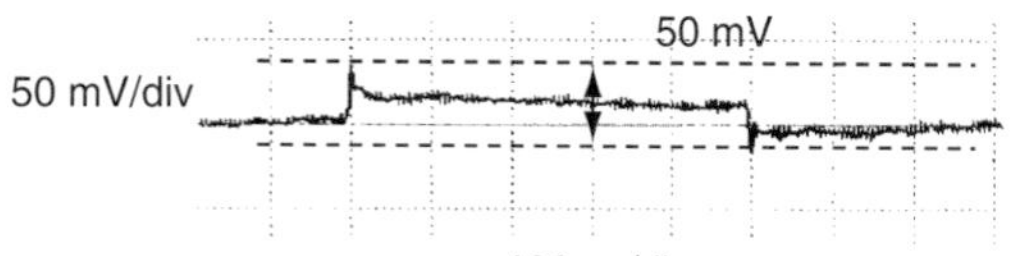

(A) Transient Response of the Four-Module Interleaved QSW

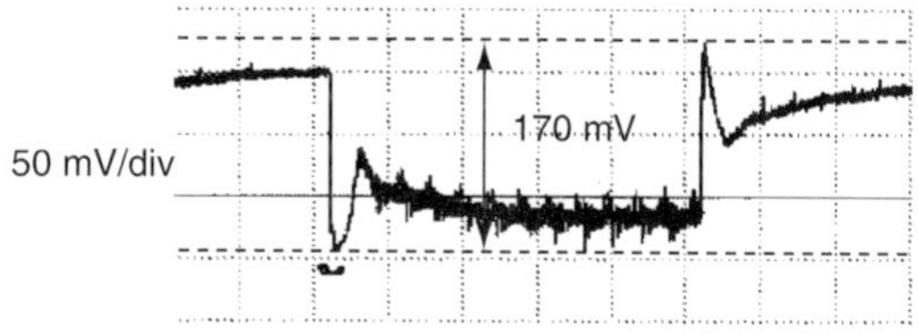

(B) Transient Response of the Conventional VRM

FIGURE 1.22 Transient Response Test Results

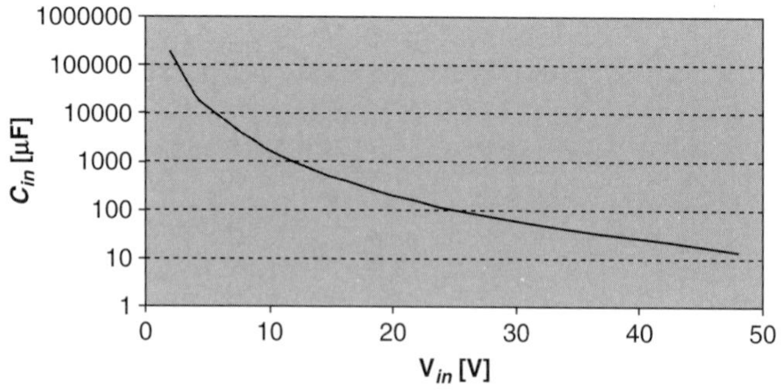

FIGURE 1.23 Input Filter Capacitance Versus Input Voltage

voltages. With a higher input voltage, the VRM has a lower efficiency.

1.3.3 A Nonisolated, High-Input-Voltage VRM Topology: The Center-Tapped Inductor VRM

With advanced inductor designs, the VRM duty cycle can be adjusted. Figure 1.26 shows a changed inductor structure, a center-tapped inductor, that can improve the VRM duty cycle. Figure 1.27 shows an improved VRM topology with the center-tapped inductor structure and a loss-less snubber that can absorb the leakage inductance energy as well as reduce ripple and voltage spikes (Zhou, 1999). With this topology, VRMs do not need high-side gate drives, thus improving VRM efficiency at high-input-voltages. To reduce input and output current ripple, an interleave technique can be used, which is shown in Figure 1.28. Figure 1.29 shows the efficiency of this two-channel interleaved center-tapped inductor VRM efficiency. Its input voltage is 12 V, and its output voltage is 1.2 V. In the test, IRL3103D1 is used as top switch, and two MTP75N03DHs in parallel are used as a synchronous rectifier. This topology can achieve 80% efficiency at 1.2 V at a 60 A output.

1.3.4 An Isolated High-Input-Voltage VRM Topology: The Push-Pull Forward Topology

Another approach to adjusting the VRM duty cycle in high-voltage bus systems is to use transformers. By designing the transformer's turns ratio, VRMs can simply adjust their duty cycles to optimize efficiency and cancel the ripple.

Figure 1.30 shows the push-pull forward VRM topology (Zhou, 1999). In the primary side, switch and transformer windings are alternately connected in a circle. A capacitor is connected between any of two interleaved terminations. The left two terminations are connected to the input and ground, respectively. The two primary windings have the same turns. Figure 1.31 explains the operation of this converter.

Compared with conventional high-input-voltage VRM topologies, the push-pull forward converter has a higher power density, faster transient response, and higher efficiency. For example, compared with the forward flyback shown in Figure 1.32 and asymmetrical half-bridge shown in Figure 1.33, which are fourth-order systems, the push-pull forward converter is a second-order system. Therefore, its control is simpler, and its transient is faster. As a result, the output filter inductance and capacitance needed are significantly reduced. Due to its reduced input current ripple, the converter's input filter can also be reduced.

Compared with half-bridge converters and asymmetrical and symmetrical half-bridge converters, the push-pull forward topology has a larger transformer's turns ratio. For example, if VRM input voltage is 48 V and output voltage is 3.3 V, for both the symmetrical and asymmetrical half-bridges, the transformer's turns ratio is 3 to 1. For the push-pull forward converter, the transformer's turns ratio is 6 to 1. Therefore, the conduction loss of the primary switches in a push-pull forward converter is lower. As a result, the push-pull forward converter has a higher efficiency. At heavy load, the efficiency of the push-pull forward converter is 2 to 3% higher than that of either half-bridge converter.

Figures 1.34(A) and 1.34(B) show the efficiencies of the push-pull forward converter under 12 V and 48 V input voltages, respectively. For a 12-V input, this converter can achieve 81% efficiency at an output of 1.2 V at 60 A. For a 48-V input, this converter can achieve 83.6% efficiency at an output of 1.2 V at 60 A.

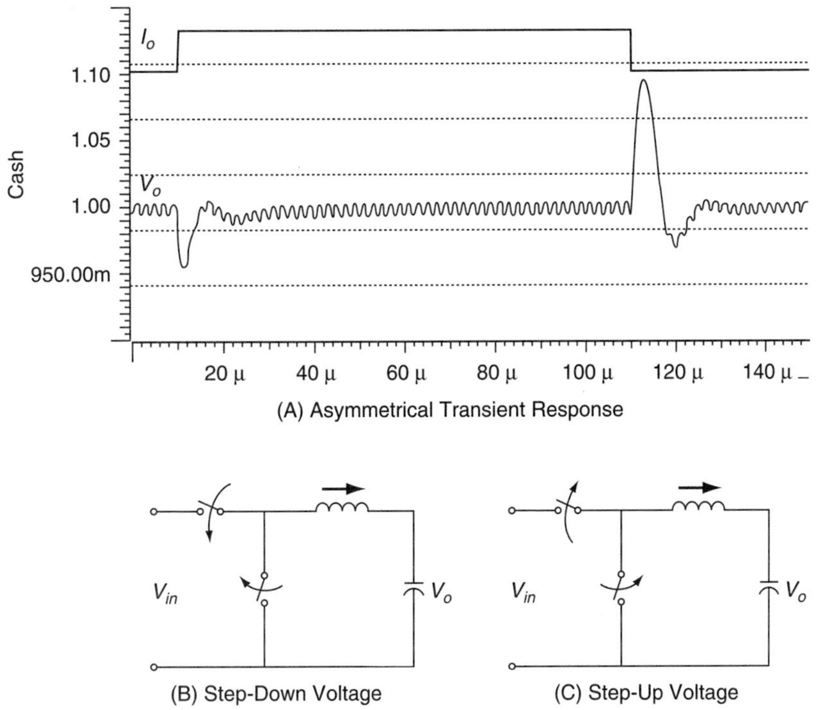

FIGURE 1.24 An Asymmetrical Transient Response from Low-Output-Voltage and High-Input-Voltage

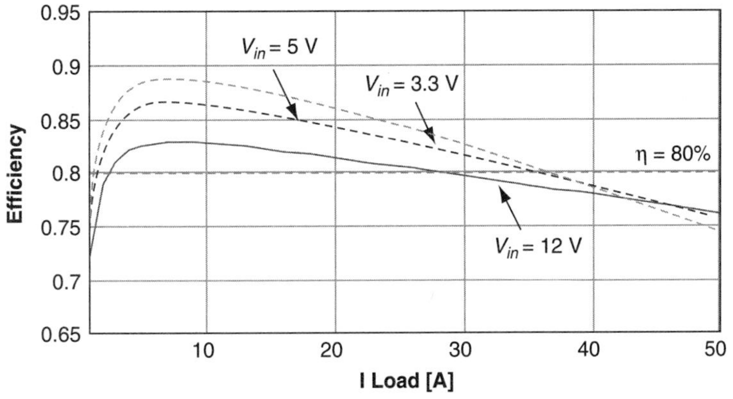

FIGURE 1.25 Efficiency Comparison of Buck Converters ($V_o = 2\,\text{V}$, $fs = 300\,\text{kHz}$, and switches: IRL3803)

1.4 Future VRMs

1.4.1 High-Frequency and High-Power-Density VRMs

To develop low-cost, high-efficiency, low-profile, high-power density, fast-transient-response, board-mount VRM modules for future generation microprocessor loads, high operating frequencies are desirable. Figure 1.35 shows the transient response of the interleaved QSW VRM when it operates at 1 MHz. Obviously, the voltage spike is reduced significantly. Figure 1.36 shows the inductance and capacitance needed in the interleaved QSW VRM when it operates at a high switching frequency. At 10 MHz, the inductance needed is only 9.25 nH and the capacitance needed is only 5.26 μF. With such a small inductance and capacitance, very high-power-density VRMs can be created, and energy storage costs can be dramatically reduced. Because of today's device technology, however, most VRMs' operating frequencies are lower than 300 kHz. Even at this frequency, the VRM cannot meet efficiency requirements. When the frequency is increased, the resulting VRM efficiency levels are shown in Figure 1.37. At 10 MHz, the VRM will only

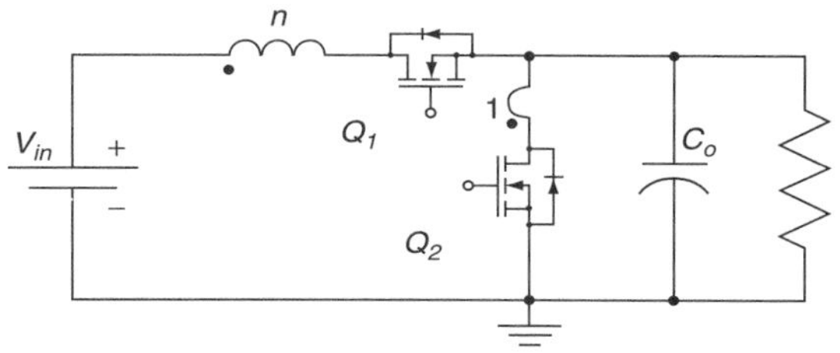

FIGURE 1.26 Center-Tapped Inductor Structure

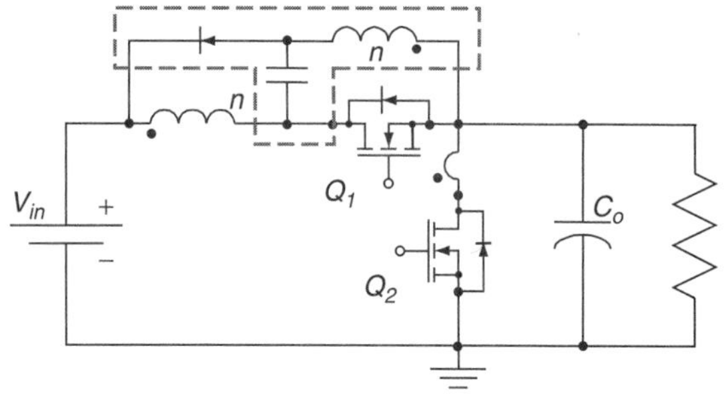

FIGURE 1.27 Improved VRM Topology with a Center-Tapped Inductor and a Lossless Snubber

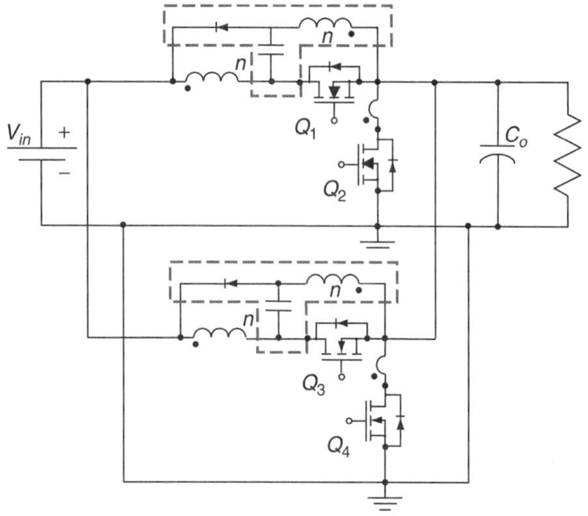

FIGURE 1.28 Interleaved Center-Tapped Inductor VRM

have 40% efficiency. This efficiency makes thermal management and packaging very difficult.

For future microprocessor applications, the power device must have a smaller FOM value $[< 100(\text{m}\Omega \times nC)]$ and a lower miller charge. With improved device technologies, such as the SOI LDDMOS technology (Huang, 1998), future VRM efficiency will be higher than 90% at several megahertz operating frequencies. Figure 1.38 shows the difference between a vertical DMOS and the proposed LDD MOSFET on SOI structure, whose power density is higher than $100\,\text{W/in}^3$. Table 1.3 shows the VRM efficiency comparison based on today's device technology and the improved LDDMOS technology.

Although advanced topologies have very fast transients, and future device techniques can operate at very high frequencies,

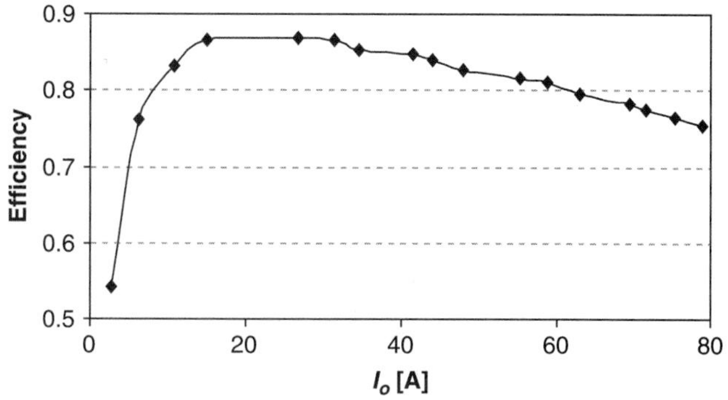

FIGURE 1.29 Efficiency of the Two-Channel Interleaved Center-Tapped Inductor VRM

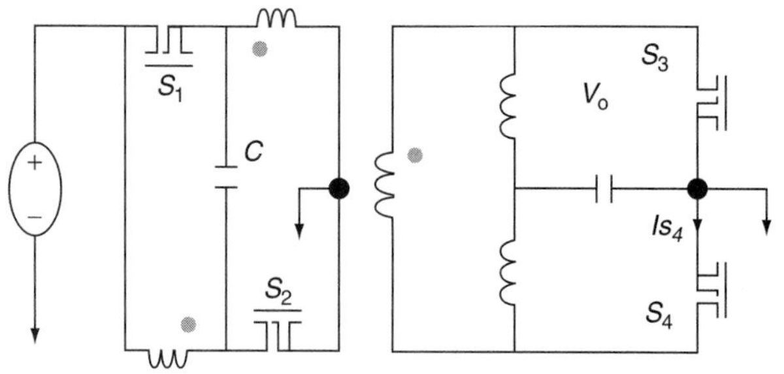

FIGURE 1.30 A Novel Topology: The Push-Pull Forward Converter

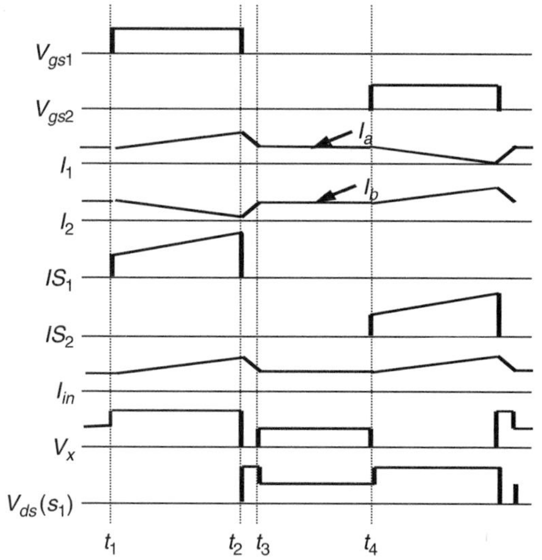

FIGURE 1.31 The Operation of the Push-Pull Forward Converter

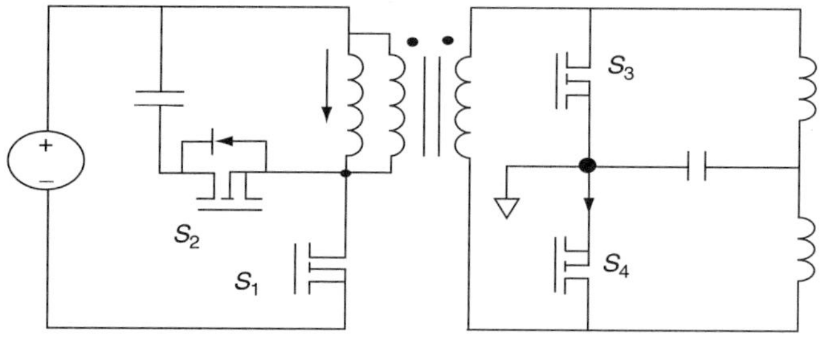

FIGURE 1.32 Flyback Forward Converter

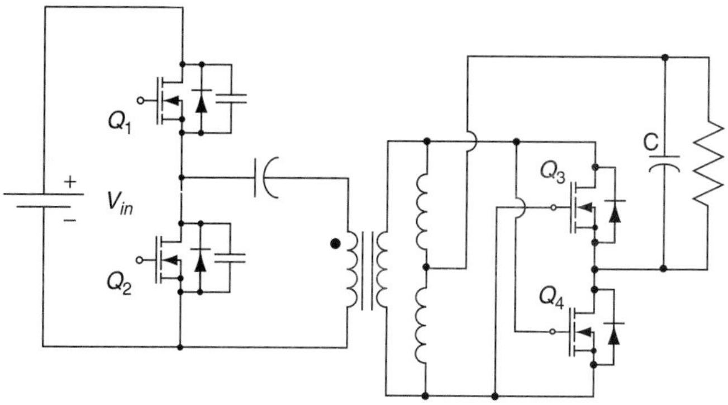

FIGURE 1.33 Asymmetrical Half-Bridge Converter

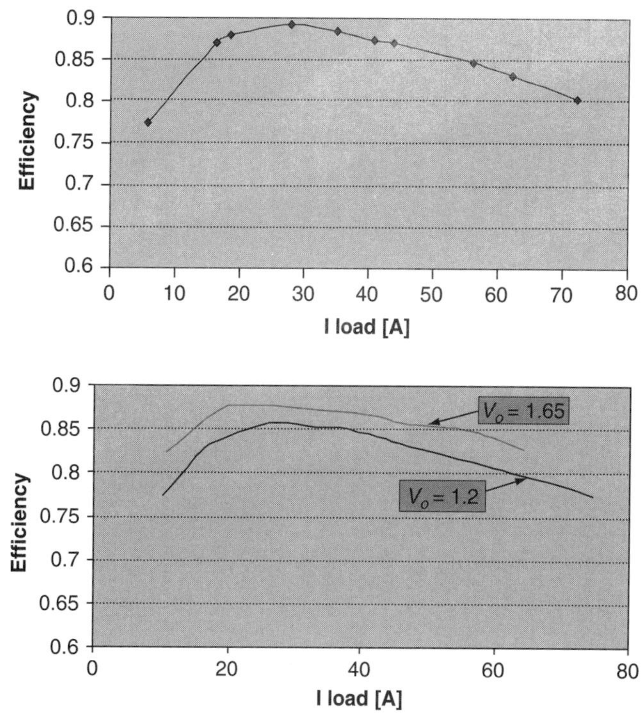

FIGURE 1.34 Push-Pull Forward Efficiency (A) Efficiency of the Push-Pull Forward Converter ($V_{in} = 48$ V, $V_o = 1.2$ V, $fs = 100$ kHz; secondary side uses four MTP75N03DHL, and primary side uses two IRF630). (B) Efficiency of the Push-Pull Forward Converter ($V_{in} = 12$ V, $fs = 100$ kHz; secondary side uses four MTP75N03DHL, and primary side uses two IRL3103D1).

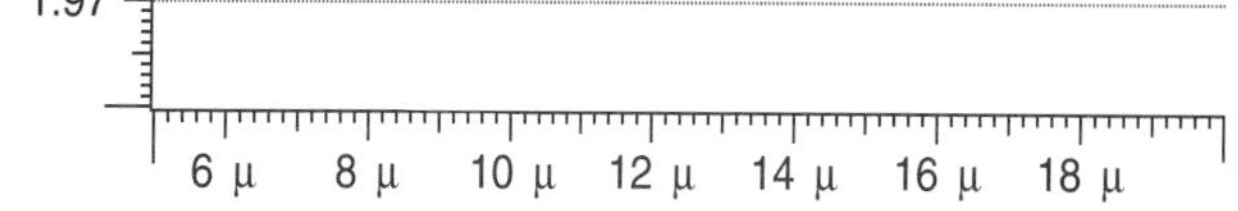

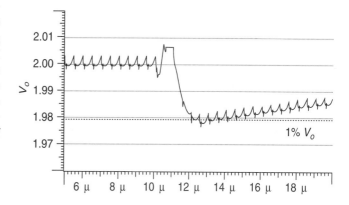

FIGURE 1.35 Transient Response of the Interleaved QSW ($V_{in} = 5$ V, $V_o = 2$ V, and $fs = 1$ MHz)

to minimize the effects of the interconnections an innovative design with a possible integration of the VRM and the processor is still the key to meeting the ever-increasing demand for processor performance and speed.

The integration of the VRM with the processor can take either a hybrid or a monolithic approach. In the hybrid approach, the VRM can be made as a silicon chip with all the control functions. Figure 1.39 shows an integration-packaging example. As shown in Figure 1.39(A) and (B), several VRM chips can be placed in parallel and be mounted close to the microprocessor on the same cartridge. Ceramic capacitors with small ESRs and ESLs can be used as the output capacitors and can be placed on the PCB board next to the processor. By connecting the output of the VRM and the power input of the processor via a path through a magnetic material sheet, the small output inductor can also be created. With this kind of packaging approach, interconnection parasitics can be minimized. For future applications, some other advanced packaging technologies, such as flip-chip technology, can also be used.

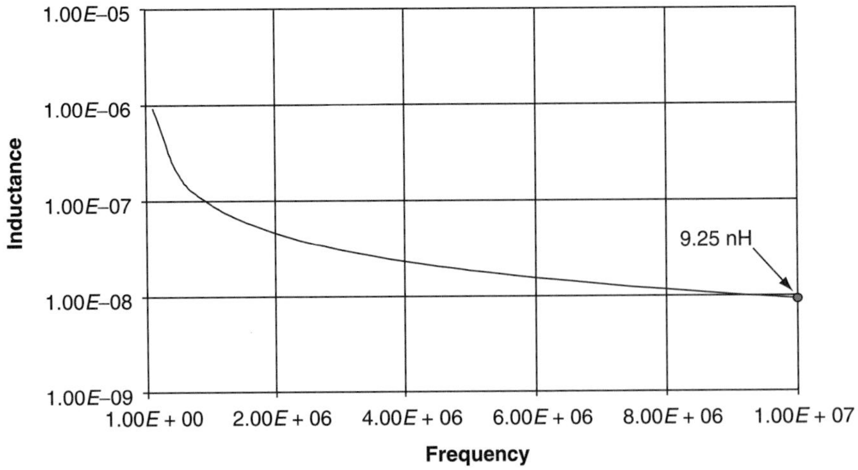

(A) Inductance Needed Versus Frequency

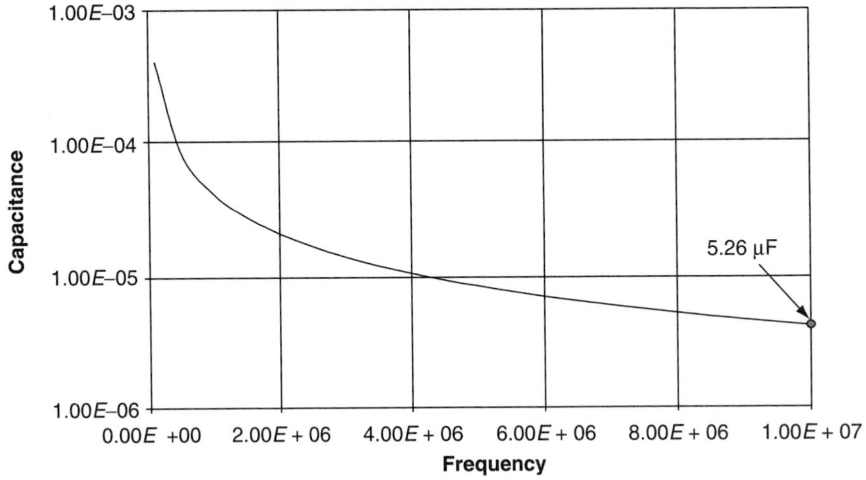

(B) Capacitance Needed Versus Frequency

FIGURE 1.36 Inductance and Capacitance Needed in the Interleaved QSW VRM Topology at a High Operating Frequency

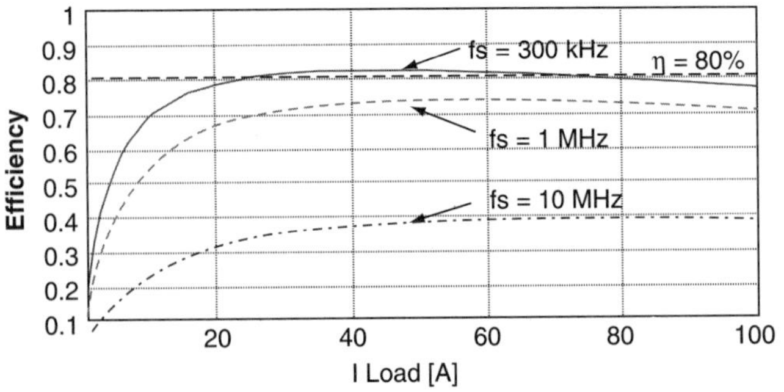

FIGURE 1.37 VRM Efficiency Based on Today's Device Technology ($V_{in} = 5\,V$, $V_o = 2\,V$, and Switches 5 IRL3803 in parallel)

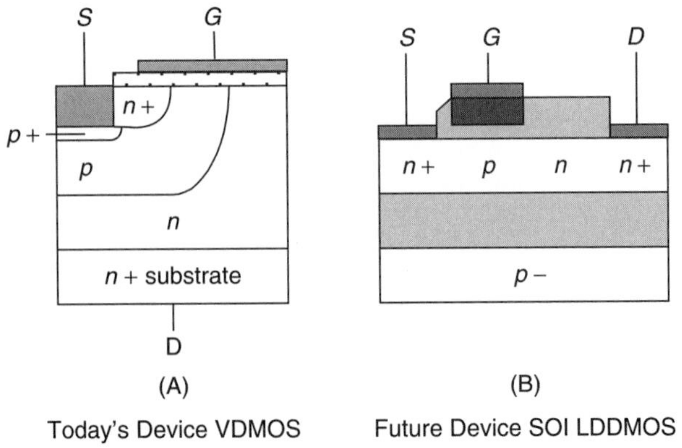

FIGURE 1.38 Future Power Device Technology

TABLE 1.3 VRM Efficiency Comparisons

$V_{in} = 5V$ $V_o = 2V$	BV [V]	FOM [m$\Omega \cdot$ nC]	Optimized efficiency for interleaved QSW VRM		
			300 kHz	1 MHz	10 MHz
LDDMOS	10	77	95%	91%	88%
Today's device	30	473	87%	79%	60%

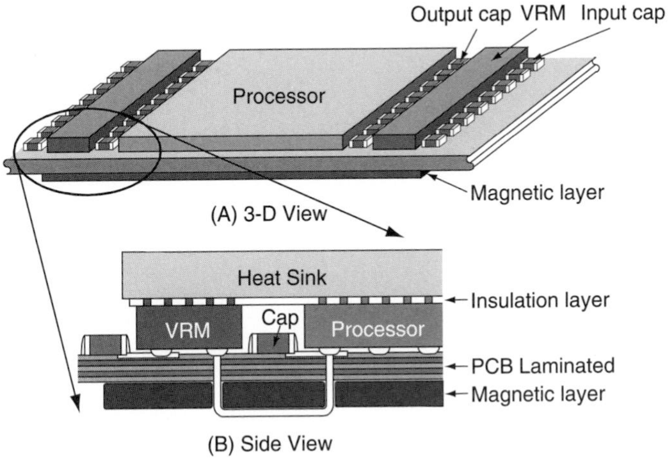

FIGURE 1.39 Hybrid Approach

1.5 Conclusions

For future microprocessor applications, there are many power management issues that need to be addressed, such as those concerning VRM topologies, power device technologies, and packaging technologies. To meet future requirements, VRMs should have high-power densities, high efficiencies, and fast transient performances. To achieve this target, advanced VRM topologies, advanced power devices, and advanced packaging technologies must be developed.

References

Goodfellow, S. and Weiss, D. (1997). Designing power systems around processor specifications. *Electronic Design*, 53–57.

Zhang, M. T., Jovanovic, M. M., and Lee, F. C. (1996). Design consign considerations for low-voltage on-board dc/dc modules for next generations of data processing circuits. *IEEE Transactions on Power Electronics* (*11*) 2, 328–337.

Wong, P., Zhou, X., Chen, J., Wu, H., Lee, F. C., and Chen, D. Y. (1997). *VRM transient study and output filter design for future processors.* VPEC Seminar.

Hoshi, M., Shimoida, Y., Hayami, Y., and Mihara, T. (1995). Low on-resistance LDMOSFET with DSS (a drain window surrounded by source windows) pattern layout. *IEEE Conference ISPSD.*, 63–67.

Chen, W., Lee, F. C., Zhou, X., and Xu, P. (1999). Integrated planar inductor scheme for multimodule interleaved quasi-square-wave dc/dc converter. *IEEE PESC.* Vol. 2, 759–762.

Zhou, X. Ph. D. (July, 1999) dissertation. Low-voltage, high-efficiency, fast-transient voltage regulator module.

Zhou, X., Yang, B., Amoroso, L., Lee F. C., and Wong, P. (1999). A novel high-input-voltage. High efficiency and fast transient voltage regulator module: Push-pull forward converter. *APEC.*

Huang, A. Q., Sun, N. X., Zhang, B., Zhou, X., and Lee, F. C. (1998). Low voltage power devices for future VRM. *ISPSD*, 395–398.

Rozman, A., and Fellhoelter, K. (1995). Circuit considerations for fast, sensitive, low-voltage loads in a distributed power system. *IEEE APEC.* 34–42.

O'Connor, J. (1996). Converter optimization for power low-Voltage, high-performance microprocessors. *IEEE APEC.*, 984–989.

Efand, T., Tasi, C., Erdeljac, J., Mitros, J., and Hutter, L. (1997). A performance comparison between new reduced surface drain RSD LDMOS and RESURF and conventional planar power devices rated at 20 V. *IEEE Conference ISPSD*, 185–188.

Zhou, X., Zhang, X., Liu, J., Wong, P. L., Chen, J., Wu, H. P., Amoroso, L., Lee, F. C., and Chen, D. Y. (1998). Investigation of candidate VRM topologies for future microprocessors. *IEEE APEC.* 145–150.

Watson, R., Chen, W., Hua, G., and Lee, F. C. (1996). Component development for a high-frequency ac distributed power system. *Proceedings of the 1996 Applied Power Electronics Conference*, 657–663.

Von Jouanne, A. (1995). The effect of long motor leads on PWM inverter-fed ac drive system. *IEEE APEC.* 592–597.

Matsuda, H., Hiyoshi, M., and Kawamura, N., (1997). Pressure contact assembly technology of high-power devices. *Proceedings ISPSD* 17–24.

2

Noise in Analog and Digital Systems

Erik A. McShane and
Krishna Shenai
Department of Electrical and Computer Engineering, University of Illinois at Chicago, Chicago, Illinois, USA

2.1 Introduction

Noise from an analog (or **small-signal**) perspective is a random time-varying signal that is generated in all passive and active electronic devices. It can be represented as either a current or a voltage, $i_n(t)$ or $v_n(t)$. Over a fixed time period, t, the average value of noise is zero. Analog noise, therefore, is commonly presented in terms of its **mean-square** value (I^2_{rms} or V^2_{rms}). It is sometimes also described by its **root-mean-square** value (I_{rms} or V_{rms}).

Noise in digital (or **large-signal**) circuits is the perturbation of one nonlinear signal, the **noise victim**, by a second nonlinear signal, the **noise aggressor**. The perturbation is introduced by a parasitic coupling path that is resistive, capacitive, or inductive in nature. Since digital systems are typically characterized by voltage levels, digital noise is presented in terms of voltage and voltage transients, $v_n(t)$ and dv_n/dt.

The impact of noise differs for analog and digital systems and leads to unique equivalent circuit models and analytical techniques. Analog noise is treated by linearized expressions that correspond to small-signal equivalent circuit parameters. Digital noise is analyzed by large-signal expressions that define logic transitions. In this chapter, the characteristics of noise are presented for both systems. Several examples are included for each type of system.

For further reading, several respected books are listed in the Bibliography section of this chapter. These provide a more thorough background on analog and digital systems, network theory, and the introduction of noise to classical systems analysis.

2.2 Analog (Small-Signal) Noise

Noise in analog systems is typically characterized by its impact as a small-signal perturbation. In this form, noise is considered to be an independent alternating current (ac) source (either voltage or current). Table 2.1 lists the nomenclature for discussing analog noise. These measures are derived and defined throughout the chapter.

The relationships as a function of time for noise voltage, square noise voltage, and mean-square noise voltage are illustrated in Figure 2.1; similar waveforms can also be shown for representating noise current. If multiple noise sources are present, they are considered to be independent and uncorrelated. Therefore, the cumulative noise contribution can be expressed in mean-square units by:

$$\overline{i_n^2} = \overline{i_{n1}^2} + \overline{i_{n2}^2}, \tag{2.1}$$

as the sum of the individual mean-square components. In *rms* terms, the total noise is represented by:

$$i_{n,\,rms} = \sqrt{i^2_{n1,\,rms} + i^2_{n2,\,rms}}. \tag{2.2}$$

TABLE 2.1 Nomenclature in Analog (Small-Signal) Noise Analysis

Name	Current-referred symbol	Units	Voltage-referred symbol	Units
Noise signal	$i_n(t)$	A	$v_n(t)$	V
Mean-square noise signal	$\overline{i_n^2(t)}$	A^2_{rms}	$\overline{v_n^2(t)}$	V^2_{rms}
Root-mean-square noise signal	$\sqrt{\overline{i_n^2(t)}}$ or $i_{n,\ rms}$	A_{rms}	$\sqrt{\overline{v_n^2(t)}}$ or $v_{n,\ rms}$	V_{rms}
Noise power spectral density	$S_I(f)$	A^2/Hz	$S_V(f)$	V^2/Hz
Root noise power spectral density	$\sqrt{S_I(f)}$	$A/\sqrt{Hz}$	$\sqrt{S_V(f)}$	$V/\sqrt{Hz}$

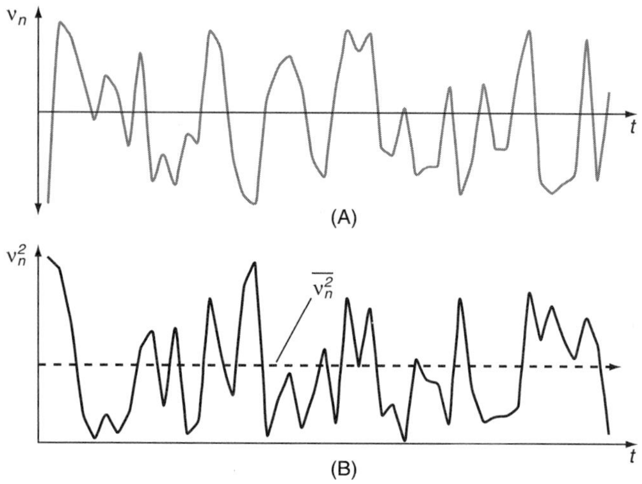

FIGURE 2.1

Some noise also exhibits a frequency dependence. If the noise mean-square value of spectral components in a narrow bandwidth (Δf) is determined, then the ratio of that value to Δf is defined as the **noise power spectral density** (the noise **mean-square per bandwidth**). Mathematically, the definition is expressed in terms of noise current as:

$$S_I(f) = \lim_{\Delta f \to 0} \frac{\overline{i_n^2(t)}}{\Delta f}. \qquad (2.3)$$

The relationship can be reversed to define the mean-square noise current in terms of the noise power spectral density by:

$$\overline{i_n^2} = \int_{f_1}^{f_2} S_I(f)df. \qquad (2.4)$$

Equivalent expressions can be derived for $S_V(f)$, the noise power in terms of noise voltage.

2.2.1 Noise Categories

Analog small-signal noise can be subdivided into two categories: frequency independent (or **white**) noise and frequency dependent (or **pink**) noise. The spectrum of the former is constant to frequencies as high as 10^{14} Hz. In the latter, the noise power is a function of $1/f^n$, where typically $n = 1$ (although some forms of noise obey a power relationship with $n \geq 2$). The total noise power in an element, as shown in Figure 2.2, is the combination of both, and a clear knee frequency is observed. The terms S_{Iw} and S_{Ip} are used to refer to the specific contributions of white and pink noise, respectively, to the total noise power.

Table 2.2 lists the common characteristics of small-signal noise. The expressions are developed more fully in the following subsections.

2.2.2 White Noise

White noise sources are composed of **thermal** (also called **Johnson** or **Nyquist**) noise and **shot** noise. Thermal noise is generated by all resistive elements and is associated with the random motion of carriers in an electronic element. It is a function only of temperature. Shot noise is associated with the discretization of charged particles and is always present in semiconductor *pn* junctions. It is a function only of dc current flow.

The thermal noise of a resistance, *R*, can be modeled as an ideal resistor and an independent current or voltage source, as shown in Figure 2.3. The Thevenin equivalent noise voltage is given by:

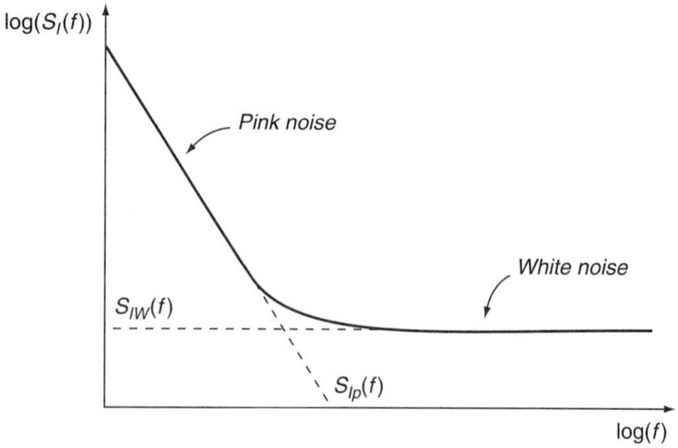

FIGURE 2.2

TABLE 2.2 Characteristics of Analog (Small-Signal) Noise

Name	Type	Dependence	Expression
Thermal (or Johnson or Nyquist)	White	Temperature	$\overline{v_n^2} = 4\hat{k}TR\Delta f$
Shot	White	Current	$\overline{i_n^2} = 2qI\Delta f$
Capacitor bandwidth-limited	White	Temperature, bandwidth	$v_{n,\,\text{rms}} = \sqrt{\frac{\hat{k}T}{C}}$
Inductor bandwidth-limited	White	Temperature, bandwidth	$i_{n,\,\text{rms}} = \sqrt{\frac{\hat{k}T}{L}}$
Flicker (or $1/f$)	Pink	Current	$\overline{v_n^2} = K_R \frac{R_{\text{sq}}^2}{A} V^2 \frac{\Delta f}{f}$
			$\overline{i_n^2} = \frac{K_D I}{A} \frac{\Delta f}{f}$
			$\overline{v_{n,\,\text{eq}}^2} = \frac{K_M}{WLC_{\text{ox}}^2} \frac{\Delta f}{f}$
Popcorn	Pink	Recombination	$\propto \frac{1}{f^2}$

Constants:

$\hat{k} = 1.38 \times 10^{-23}\ \text{V} \cdot \text{C/K}$

$q = 1.6 \times 10^{-19}\ \text{C}$

$K_R \approx 5 \times 10^{-24}\ \text{cm}^2/\Omega^2$

$K_D \approx 10^{-21}\ \text{cm}^2 \cdot \text{A}$

$$K_M \approx \begin{cases} 10^{-33}\text{C}^2/\text{cm}^2 (p\text{JFET}) \\ 10^{-32}\text{C}^2/\text{cm}^2 (\text{PMOSFET}) \\ 4 \times 10^{-31}\text{C}^2/\text{cm}^2 (\text{NMOSFET}) \end{cases}$$

$$\overline{v_n^2} = 4\hat{k}TR\Delta f, \tag{2.5}$$

where $\hat{k}$ is Boltzmann's constant ($1.38 \times 10^{-23}\ \text{V} \cdot \text{C/K}$), T is absolute temperature (in degrees Kelvin), and Δf is the noise bandwidth. The Norton equivalent noise current can also be obtained. It is expressed as:

$$\overline{i_n^2} = 4\hat{k}T\frac{1}{R}\Delta f. \tag{2.6}$$

A MOSFET's channel resistance also produces a thermal noise current according to equation 2.6. Since it is informative to compare the MOSFET noise current with the small-signal input stimulus, the output noise current can be referred back to the source using the amplifier gain expression:

$$i_d = g_m v_{gs}, \tag{2.7}$$

such that the equivalent input noise voltage is as follows:

$$\overline{v_{n,\,\text{eq}}^2} = \frac{4\hat{k}T}{Rg_m^2}\Delta f. \tag{2.8}$$

In the saturation regime, the channel resistance can be approximated by $1/g_m$, and the channel pinch-off reduces the effective noise to about 66% of its nominal value. Therefore, equation 2.8 can be simplified to:

$$\overline{v_{n,\,\text{eq}}^2} = \frac{8\hat{k}T}{3}\frac{1}{g_m}\Delta f. \tag{2.9}$$

Both MOSFET channel thermal noise source configurations are shown in Figure 2.4. It should be noted that the input noise voltage representation is valid only at low frequencies. As the frequency rises, the admittance of gate parasitic capacitances (C_{GD} and C_{GS}) becomes higher; $\overline{v_{n,\,\text{eq}}^2}$ is dropped partly on the gate impedance and partly on the output resistance of the

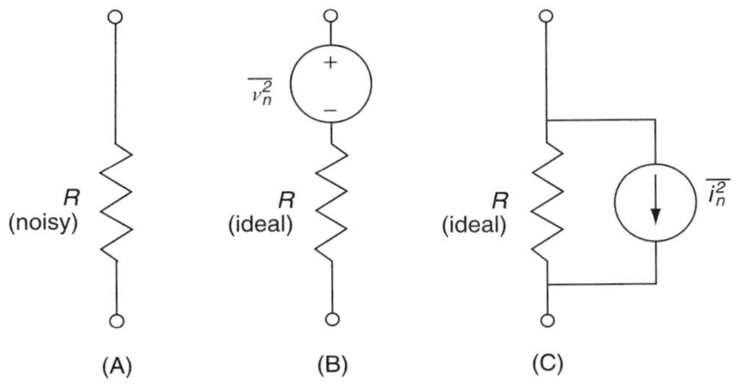

FIGURE 2.3

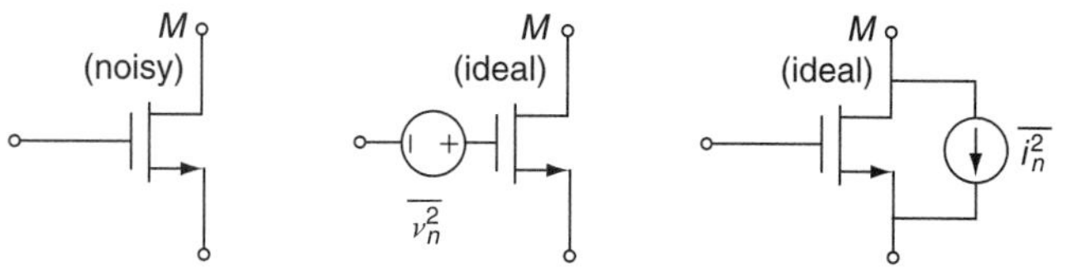

FIGURE 2.4

preceding device. Therefore, the ideal noiseless MOSFET is stimulated by an erroneously low noise voltage, resulting in inaccurate predictions of output noise.

Likewise, the shot noise of a junction diode, D, can be modeled as an ideal diode and an independent current source, as shown in Figure 2.5. The Norton equivalent noise current is given by

$$\overline{i_n^2} = 2qI\Delta f, \tag{2.10}$$

where I is the dc current conducted by the diode and Δf is the noise bandwidth. Noise is generated by the junction in both forward- and reverse-biased conditions since either a forward current or a leakage current is present.

Many analog circuits such as filters, sample-and-hold amplifiers, and switched-capacitor networks employ a resistor or a conductance that is bandwidth-limited by a capacitance or inductance. Just as the reactance of the inductor or capacitor limits the bandwidth of the input signal, it also affects the noise bandwidth. In the case of an RC circuit, the *rms* noise voltage is expressed as:

$$v_{n,\,rms} = \sqrt{\frac{\hat{k}T}{C}}. \tag{2.11}$$

A complementary result can be obtained for RL circuits, with the *rms* noise current expressed as:

$$i_{n,\,rms} = \sqrt{\frac{\hat{k}T}{L}}. \tag{2.12}$$

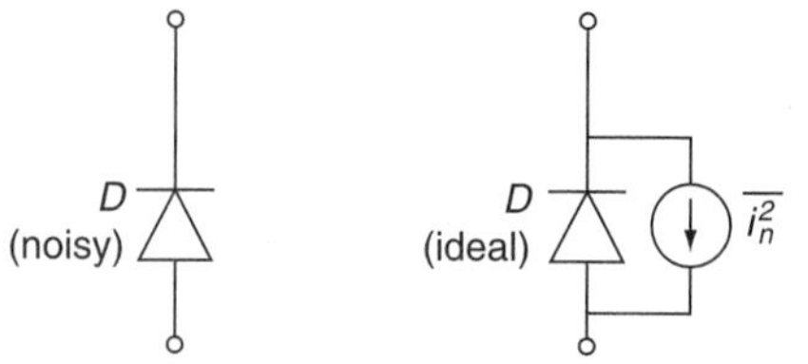

FIGURE 2.5

The expressions of equations 2.11 and 2.12 are interesting in that the resulting noise voltage or current is independent of the resistance or conductance. The total noise can be reduced only by choosing larger values for the reactive elements.

2.2.3 Pink Noise

Pink noise sources consist of **flicker** (also called $1/f$) **noise** and **popcorn noise**. Flicker noise pertains to the conductive properties of an electronic element. Other theories attribute flicker noise to interface traps present at oxide-semiconductor interfaces or to fluctuations in mobility. The noise is a function of the material homogeneity, its volume, the current, and the frequency. Popcorn noise, on the other hand, is proportional to $1/f^2$ and is an indication of poor semiconductor manufacturing. It is associated with distinct recombination processes and appears as a series of low-frequency noise bursts. Because it is uncommon, popcorn noise is not discussed further.

Flicker noise is proportional to current and inversely proportional to volume. Since electronic devices are three-dimensional structures, planar resistors and semiconductor junctions have unique expressions and scaling factors. For example, the flicker noise voltage of a resistor, R, is given as:

$$\overline{v_n^2} = K_R \frac{R_{sq}^2}{A} V^2 \frac{\Delta f}{f}, \tag{2.13}$$

where R_{sq} is the sheet resistance, A is the planar area of the resistor, V is the dc voltage, and K_R is a technology scaling constant ($\approx 5 \times 10^{-24}\ \text{cm}^2/\Omega^2$). The flicker noise current of a junction diode is similarly found as:

$$\overline{i_n^2} = \frac{K_D I}{A} \frac{\Delta f}{f}, \tag{2.14}$$

where A is the cross-sectional area of the diode, I is the dc current, and K_D is a diode scaling constant ($\approx 10^{-21}\ \text{cm}^2 \cdot \text{Å}$).

Finally, the input referred flicker noise voltage in a MOSFET is found by an expression that is derived from equation 2.14. It is given by:

$$\overline{v_{n,\text{eq}}^2} = \frac{K_M}{WLC_{ox}^2} \frac{\Delta f}{f}, \tag{2.15}$$

where W and L are, respectively, the gate width and length, C_{ox} is the oxide capacitance per unit area, and technology constant K_M is found experimentally. Recent empirical values for three types of FET are:

p-type JFET	$\approx 10^{-33}\ \text{C}^2/\text{cm}^2$
PMOSFET	$\approx 10^{-32}\ \text{C}^2/\text{cm}^2$
NMOSFET	$\approx 4 \times 10^{-31}\ \text{C}^2/\text{cm}^2$

2.3 Digital (Large-Signal) Noise

Noise in digital systems is typically characterized by the integrity of the voltages that represent the logic 0 and logic 1 states. Noise can manifest itself either during switching (when it may result in switching times that exceed theoretical results) or in static conditions. Even in CMOS logic, which nominally provides fully rail-to-rail logic levels, dynamic noise events can momentarily perturb a voltage level. Deep submicron technologies, which typically employ supply voltages of under 1.8 V, may be especially susceptible to logic glitches or errors that are induced by random coupled noise.

Table 2.3 lists the nomenclature used for discussing digital noise. These terms are derived and defined throughout the chapter.

2.3.1 Noise Categories

Several categories of noise are present in digital circuits. These include **series resistance**, dynamic charge sharing of hard and soft nodes, and ***I/O* integrity**, which are the types discussed in this chapter. They can appear in any combination, but for simplicity, the analysis below considers each noise source independently. All are strongly dependent on the specifications of the semiconductor manufacturing process, the physical arrangement of gates and interconnections on an integrated circuit, the switching frequency, and the relative activity level of adjacent gates or interconnections.

Table 2.4 lists the fundamental characteristics of large-signal noise. The expressions derived below demonstrate the impact of noise on either the input logic level (V_{IL} or V_{IH}) or the output logic level (V_{OL} or V_{OH}). Through these four parameters, noise indirectly affects the low and high noise margins.

2.3.2 Series Resistance

The effect of **series resistance** is to degrade output voltage levels (push them away from the rails). Series resistance can be present in the supply rail and/or the ground rail. Each series resistance has a different effect on the output characteristics. Ground rail resistance degrades V_{OL} and supply rail resistance degrades V_{OH}.

For example, consider the CMOS inverter shown in Figure 2.6(A) that includes a ground resistance R_{series}. The output voltage, V_O, can be modeled [see Figure 2.6(B)] as a resistor–divider network such that:

$$V_O = \frac{R_{\text{NMOS}} + R_{\text{series}}}{R_{\text{NMOS}} + R_{\text{PMOS}} + R_{\text{series}}} Vdd, \qquad (2.16)$$

where R_{NMOS} and R_{PMOS} are the output resistances of the inverter gates and R_{series} is the ground resistance. If the inverter input is a logic 1, the NMOS is strongly conducting and therefore $R_{\text{NMOS}} \approx 0$. Then equation 2.16 can be simplified to:

$$V_O = \frac{R_{\text{series}}}{R_{\text{PMOS}} + R_{\text{series}}} Vdd. \qquad (2.17a)$$

The output voltage is increased by the presence of the series resistance. A complementary result can be obtained for a supply rail resistance and an input of logic 0, which yields:

TABLE 2.3 Nomenclature in Digital (Large-Signal) Noise Analysis

Name	Symbol	Units
Input low (highest voltage acceptable as a logic ‘0’)	V_{IL}	V
Input high (lowest voltage acceptable as a logic ‘1’)	V_{IH}	V
Output low (nominal voltage produced as a logic ‘0’)	V_{OL}	V
Output high (nominal voltage produced as a logic ‘1’)	V_{OH}	V
Low-level noise margin	NM_L	V
High-level noise margin	NM_H	V

TABLE 2.4 Characteristics of Digital (Large-Signal) Noise

Name	Expression (CMOS)		Expression (NMOS)
V_{OL}	0		$\frac{1}{k}\frac{V_{TD}^2}{2(Vdd - V_{TN})}$
V_{OH}	Vdd		Vdd
V_{IL}	$\frac{3Vdd + 3V_{TP} + 5V_{TN}}{8}$		$V_{TN} - \frac{V_{TD}}{\sqrt{k(1+k)}}$
V_{IH}	$\frac{5Vdd + 5V_{TP} + 3V_{TN}}{8}$		$V_{TN} - \frac{2V_{TD}}{\sqrt{3k}}$
NM_L		$V_{IL} - V_{OL}$	
NM_H		$V_{OH} - V_{IH}$	

$$V_O = \frac{R_{\text{NMOS}}}{R_{\text{NMOS}} + R_{\text{series}}} Vdd. \tag{2.17b}$$

Supply and ground rails are designed to have minimal resistance to alleviate such effects as series resistance. This is accomplished by using low resistivity materials, increasing the cross-sectional area of the conductors, using shorter interconnection lengths, and applying other techniques. Two conditions, however, may still produce large voltage deviations. One condition is high-current *I/O* buffers, and the second condition is simultaneous switching of many parallel gates. The latter situation is illustrated in Figure 2.6(C).

If n gates (inverters in this example) with a common series resistance simultaneously switch the expressions of equations 2.17a and 2.17b, they can be modified as:

$$V_O = \frac{nR_{\text{series}}}{R_{\text{PMOS}} + nR_{\text{series}}} Vdd. \tag{2.18a}$$

and

$$V_O = \frac{R_{\text{NMOS}}}{R_{\text{NMOS}} + nR_{\text{series}}} Vdd. \tag{2.18b}$$

The effect of simultaneous switching becomes more pronounced as device density on integrated circuits rises and die sizes also increase.

2.3.3 Dynamic Charge Sharing

The phenomenon of **dynamic charge sharing** occurs when two functionally unrelated nodes become coupled due to a capacitance or mutual inductance. Coupling appears between any two interconnected lines that are adjacent or that overlap. In general, noise coupling is proportional to the cross-sectional area that is shared by the two lines. Coupling is also proportional to frequency and inversely proportional to the separation. Therefore, as integrated circuits are scaled to smaller dimensions and operate at higher frequencies, the detrimental effects of charge sharing become more severe.

Until recently, coupling was exclusively capacitive in nature. As integrated circuits reach frequencies of 1 GHz and above, however, inductive coupling also becomes significant. Results in this chapter, though, are limited to conventional capacitive coupling.

Consider, as shown in Figure 2.7(A), the two nodes v_n (the **noise aggressor** or source) and v_v (the **noise victim**) that are coupled by a capacitor C_c. Node v_v is also tied to ground through a generic admittance Y. In digital circuits, v_v can be either a **floating node** (such as found in dynamic logic, pass transistor logic, or memory cells) or a **driven node** (such as found in static logic). The admittance Y is therefore equivalent to either a second coupling capacitance C_p (from the floating node to the ground plane) or a resistance r_a (representing the output impedance of the gate driving the node). Both cases are illustrated in Figure 2.7.

A floating node's change in voltage at v_v that is produced by a signal at v_n is expressed as:

$$v_v = \frac{v_n C_c}{C_c + c_p}. \tag{2.19}$$

Because c_p is usually just the input capacitance of a logic gate, it is very small. Any value of C_c can cause significant charge-sharing to occur. Active nodes are much less susceptible to disruption by noise coupling since the equivalent output resistance, r_a, of the driving gate may be large ($> 10\,\text{k}\Omega$). The noise coupled to the victim node is written as:

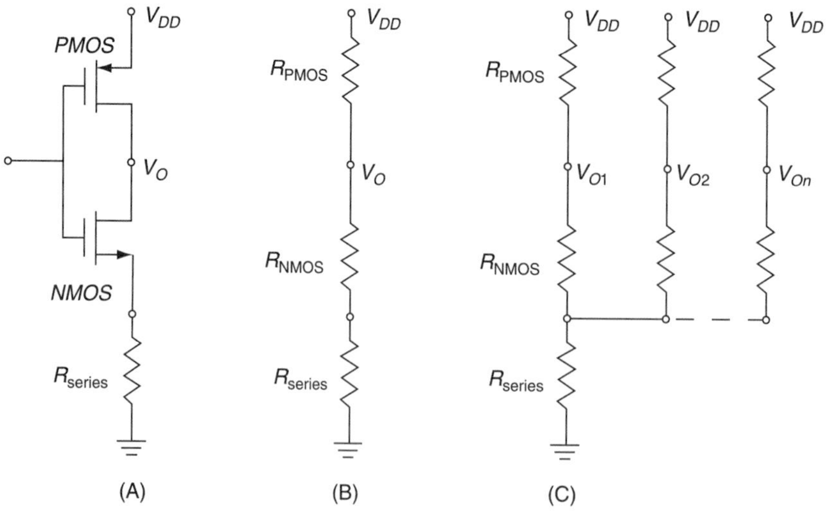

FIGURE 2.6

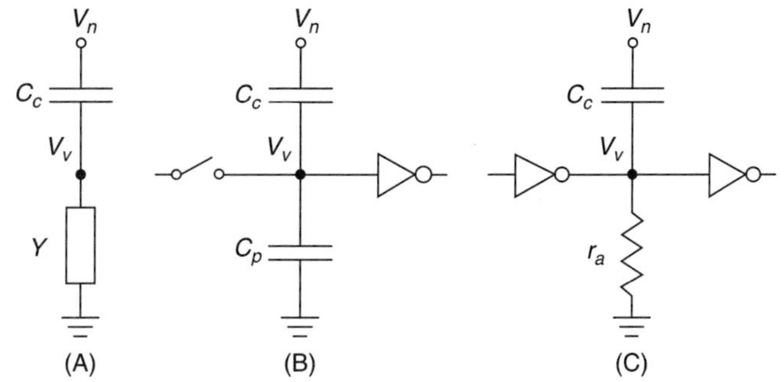

FIGURE 2.7

$$v_v = \frac{v_n r_a C_c s}{1 + r_a C_c s}. \tag{2.20}$$

Because typically $|r_a C_c s| \ll 1$, equation 2.20 can be reduced to:

$$v_v = v_n r_a C_c s. \tag{2.21}$$

Clearly for an actively driven node, a much larger coupling capacitance is necessary to produce a significant perturbation of the victim node.

2.3.4 Noise Margins

The *I/O* noise margins, NM_L and NM_H, refer to the ability of a logic gate to accommodate input noise without producing a faulty logic output. The input noise threshold levels, V_{IL} andV_{IH}, are by convention defined as the input voltages that result in a slope of -1 in the dV_O/dV_I response. This is shown in Figure 2.8. As is clear from Table 2.4, the noise margins of CMOS logic gates are larger than for comparable NMOS technologies. This is evident because CMOS delivers rail-to-rail outputs, whereas the V_{OL} is a circuit constraint in NMOS.

The noise margins of a CMOS gate can be found by first examining the dc transfer curve shown in Figure 2.8. From graphical analysis, the V_{IL} occurs when the PMOS is in its linear regime and the NMOS is in its saturation regime. Since a CMOS gate is complementary in operation, the V_{IH} by symmetry occurs when the PMOS is in its saturation regime and the NMOS is in its linear regime.

Considering first the CMOS V_{IL}, begin by equating the NMOS and PMOS currents:

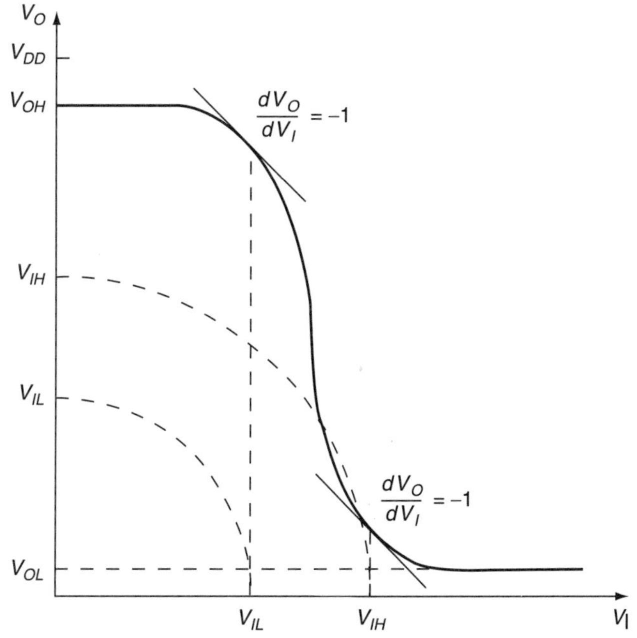

FIGURE 2.8

$$k_n \frac{W_n}{L_n} \frac{(V_I - V_{Tn})^2}{2} = k_p \frac{W_p}{L_p} \left(V_I - Vdd - V_{Tp} - \frac{V_O - Vdd}{2} \right) (V_O - Vdd). \quad (2.22)$$

Assuming that the inverter is designed to have a balanced transfer curve such that:

$$k_n \frac{W_n}{L_n} = k_p \frac{W_p}{L_p}, \quad (2.23)$$

then equation 2.22 reduces to a simpler form such that:

$$V_{IL} = \frac{3Vdd + 3V_{TP} + 5V_{TN}}{8}. \quad (2.24)$$

Considering next the CMOS V_{IH}, equating the NMOS and PMOS currents results in:

$$k_n \frac{W_n}{L_n} (V_I - V_{Tn} - \frac{V_O}{2}) V_O = k_p \frac{W_p}{L_p} \frac{(V_I - Vdd - V_{Tp})^2}{2}. \quad (2.25)$$

Again using the assumption of equation 2.23, this expression can also be reduced and rearranged, yielding the form:

$$V_{IH} = \frac{5Vdd + 5V_{TP} + 3V_{TN}}{8}. \quad (2.26)$$

The noise margins of an NMOS inverter can be found using similar methods. The derivations are not shown here but the steps are identified. Beginning with V_{IH} and examining through graphical techniques the output characteristics, the NMOS inverter is found to be equivalent to the CMOS case; that is, the driver (enhancement mode) is in the linear regime and the load (depletion mode) is in the saturation regime. Assuming that the inverter pull-up: pull-down ratio is *k*, then:

$$V_{IH} = V_{TN} - \frac{2V_{TD}}{\sqrt{3k}}. \quad (2.27)$$

Considering the NMOS V_{IL}, the driver and load bias regimes are exchanged (as in CMOS), and the result is as follows:

$$V_{IL} = V_{TN} - \frac{V_{TD}}{\sqrt{k(1+k)}}. \quad (2.28)$$

Note that the NM_H of NMOS and CMOS inverters are similar since both achieve $V_{OH} \approx Vdd$. Because an NMOS inverter V_{OL} is not zero (100 mV – 500 mV are typical values), however, the NM_L of NMOS is considerably lower than for a CMOS inverter.

Bibliography

Chen, W.K. (2000). *The VLSI handbook.* CRC Press.

Geiger, R.L., Allen, P.E., and Strader, N.R. (1990). *VLSI design techniques for analog and digital circuits.* New York: McGraw-Hill.

Laker, K.R., and Sansen, W.M.C. (1994). *Design of analog integrated circuits and systems.* New York: McGraw-Hill.

Tsividis, Y. (1996). *Mixed analog-digital VLSI devices and technology: An introduction.* New York: McGraw-Hill.

Vyemura, J.P. (1988). *Fundamentals of MOS digital integrated circuits.* Reading, MA: Addison-Wesley.

3

Field Effect Transistors

Veena Misra and
Mehmet C. Öztürk
Department of Electrical and Computer Engineering, North Carolina State University, Raleigh, North Carolina, USA

3.1 Introduction

The concept of modulating the electrical current in a semiconductor by an external electrical field was first proposed by Julius Lilienfeld in the 1930s. Years later, William Shockley, a scientist at Western Electric, led a research program to create a semiconductor device based on this "field effect" concept. To replace bulky vacuum tubes then used in telephone switching. Two scientists in Shockley's team, W. Brattain and J. Bardeen, invented the point contact transistor in 1947. Subsequently, Shockley invented the bipolar-junction transistor (BJT). Shockley later developed the junction field effect transistor (JFET); however, the JFET was not able to challenge the dominance of BJT in many applications. The development of field effect devices continued with moderate progress until the early 1960s, when the metal-oxide-silicon field effect transistor (MOSFET) emerged as a prominent device. This device quickly became popular in semiconductor memories and digital integrated circuits. Today, the MOSFET dominates the integrated circuit technology, and it is responsible for the computer revolution of the 1990s.

The first half of this chapter is dedicated to the basic theory of the MOSFET, beginning with its fundamental building block, the MOS capacitor. The second half of the chapter presents an overview of less common field effect devices used only in specific applications. Their coverage is limited to a qualitative explanation of the operating principles and basic equations. The reader is referred to other books available in literature to obtain detailed information.

3.2 Metal-Oxide-Silicon Capacitor

The metal-oxide-silicon (MOS) capacitor is at the core of the complementary metal-oxide-silicon (CMOS) technology. MOSFETs rely on the extremely high quality of the interface between SiO_2, the standard gate dielectric, and silicon (Si). Before presenting the MOSFET in detail, it is essential to achieve a satisfactory understanding of the MOS capacitor fundamentals.

The simplified schematic of the MOS capacitor is shown in Figure 3.1. The structure is similar to a parallel plate capacitor in which the bottom electrode is replaced by a semiconductor. When Si is used as the substrate material, the top electrode, known as the gate, is typically made of polycrystalline silicon (polysilicon), and the dielectric is thermally grown silicon dioxide. In MOS terminology, the substrate is commonly referred to as the *body.*

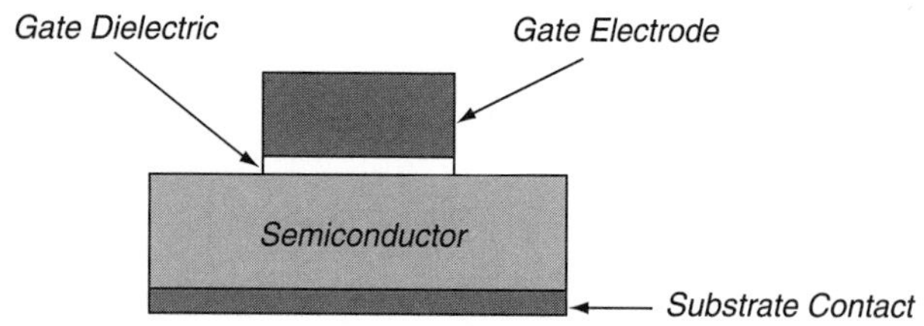

FIGURE 3.1

3.2.1 The Ideal MOS Capacitor

The energy band diagram of an ideal p-type substrate MOS capacitor at zero bias is shown in Figure 3.2. In an ideal MOS capacitor, the metal work function, ϕ_m, is equal to the semiconductor work function, ϕ_s. Therefore, when the Fermi level of the semiconductor, E_{FS}, is aligned with the Fermi level of the gate, E_{Fm}, there is no band bending in any region of the MOS capacitor. Furthermore, the gate dielectric is assumed to be free of any charges, and the semiconductor is uniformly doped.

Figure 3.3 shows the energy band diagram and the charge distribution in an ideal MOS structure for different gate-to-body voltages (V_{GB}).

With a negative gate bias (Figure 3.3(A)), the gate charge, Q_G, is negative. The source of this negative charge is electrons supplied by the voltage source. In an MOS capacitor, charge neutrality is always preserved. This requires:

$$Q_G + Q_C = 0, \tag{3.1}$$

where Q_C is the charge induced in the semiconductor. Therefore, a net positive charge, Q_C, must be generated in the silicon substrate to counterbalance the negative charge on the gate. This is achieved by accumulation of the positively charged holes under the gate. This condition, where the majority carrier concentration is greater near the $Si{-}SiO_2$ interface compared to the bulk, is called **accumulation**.

Under an applied negative gate bias, the Fermi level of the gate is raised with respect to the Fermi level of the substrate by an amount equal to qV_{GB}. The energy bands in the semiconductor bend upward, bringing the valence band closer to the Fermi level that is indicative of a higher hole concentration under the dielectric. It is important to note that the Fermi level in the substrate remains invariant even under an applied bias since no current can flow through the device due to the presence of an insulator.

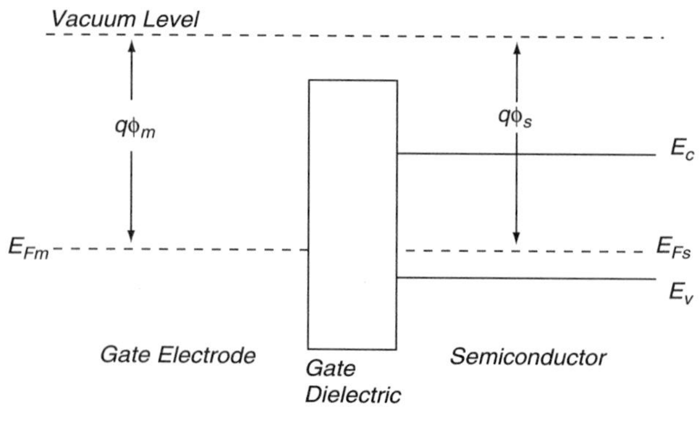

FIGURE 3.2

The applied gate voltage is shared between the gate dielectric and the semiconductor such that:

$$V_G = V_{ox} + V_c, \tag{3.2}$$

where V_{ox} and V_c are the voltages that drop across the oxide and the semiconductor, respectively. The band bending in the oxide is equal to qV_{ox}. The electric field in the oxide can be expressed as:

$$E_{ox} = \frac{V_{ox}}{t_{ox}}, \tag{3.3}$$

where t_{ox} is the oxide thickness. The amount of band bending in the semiconductor is equal to $q\psi_s$, where ψ_s is the surface potential and is negative when the band bending is upward.

Figure 3.3(B) shows the energy band diagram and the charge distribution for a positive gate bias. To counterbalance the positive gate charge, the holes under the gate are pushed away, leaving behind ionized, negatively charged acceptor atoms, which creates a **depletion region**. The charge in the depletion region is exactly equal to the charge on the gate to preserve charge neutrality. With a positive gate bias, the Fermi level of the gate is lowered with respect to the Fermi level of the substrate. The bands bend downward, resulting in a positive surface potential. Under the gate, the valence band moves away from the Fermi level indicative of hole depletion. When the band bending at the surface is such that the intrinsic level coincides with the Fermi level, the surface resembles an intrinsic material. The surface potential required to have this condition is given by:

$$\psi_s = \phi_F = \frac{1}{q}(E_i - E_F), \tag{3.4}$$

where

$$\phi_F = \frac{kT}{q}\ln\frac{N_A}{n_i} \tag{3.5}$$

Under a larger positive gate bias, the positive charge on the gate increases further, and the oxide field begins to collect thermally generated electrons under the gate. With electrons, the intrinsic surface begins to change into an n-type inversion layer. The negative charge in the semiconductor is comprised of ionized acceptor atoms in the depletion region and free electrons in the inversion layer. As noted above, at this point, the electron concentration at the surface is still less than the

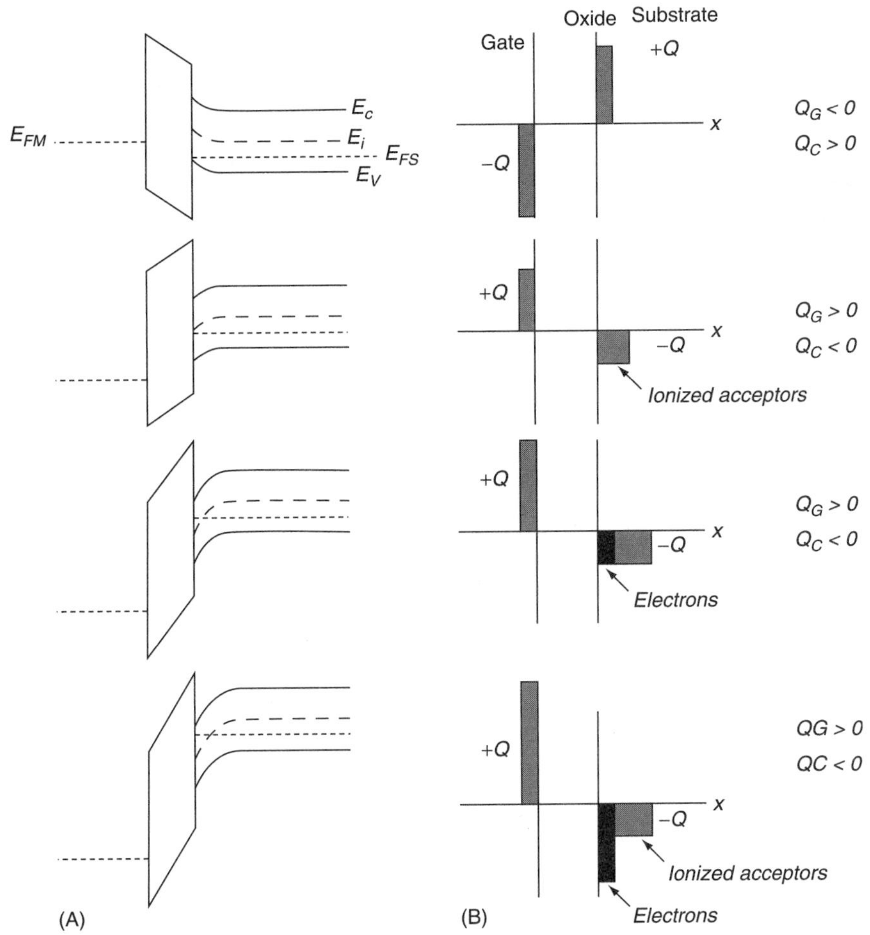

FIGURE 3.3

hole concentration in the neutral bulk. Thus, this condition is referred to as **weak inversion** and $\psi_s = \phi_F$ is defined as the onset of weak inversion and is shown in Figure 3.3(B).

As the gate bias is increased further, the band bending increases. The depletion region becomes wider, and the electron concentration in the inversion layer increases. When the electron concentration is equal to the hole concentration in the bulk, a **strong inversion** layer is said to form. The surface potential required to achieve strong inversion is equal to:[1]

$$\psi_s = 2\phi_F = \frac{2}{q}(E_i - E_F). \tag{3.6}$$

The inversion and depletion charge variation with ψ_s is shown in Figure 3.4. The electron concentration in the inversion layer is an exponential function of the surface potential and is given by:

[1] In some texts, $\psi_s = 2\phi_F$ is taken as the onset of moderate inversion. For strong inversion, the surface potential is required to be $\sim 6\,\mathrm{kT/q}$ above this level.

$$n_{inv} \approx N_A e^{q(\psi_s - 2\phi_F)/kT} \tag{3.7}$$

On the other hand, the charge density in the depletion region is written as:

$$Q_D = -qN_A W_D, \tag{3.8}$$

where W_D is the depletion region width is given by:

$$W_D = \sqrt{\frac{2\varepsilon_s}{qN_A}}\sqrt{\psi_s}. \tag{3.9}$$

Therefore, the charge density in the depletion region is a weak function of the surface potential. Consequently, when the gate bias is further increased beyond the value required to reach strong inversion, the extra positive charge on the gate can be easily compensated by new electrons in the inversion layer. This eliminates the need to uncover additional acceptor atoms in the depletion region. After this point, the depletion region width and, hence, the band bending can increase only

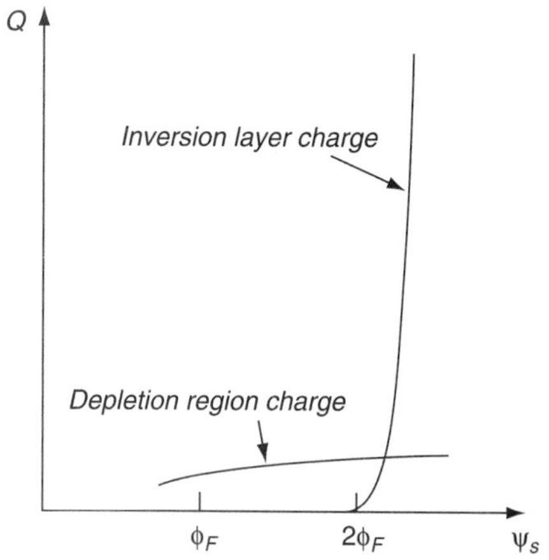

FIGURE 3.4

negligibly. The surface potential is pinned to its maximum value, which is a few kT/q over $2\phi_F$. For simplicity, $2\phi_F$ is generally taken as the maximum value for the surface potential, ψ_s.[2] The maximum depletion layer width, $W_{D,\,max}$, reached at the onset of strong inversion is given by:

$$W_{D,\,max} = \sqrt{\frac{2\varepsilon_s}{qN_A}}\sqrt{2\phi_F}. \tag{3.10}$$

3.2.2 Deviations from the Ideal Capacitor

In practical devices, the work function of the metal, ϕ_m, is not equal to that of the semiconductor, ϕ_s. Consider a MOS capacitor with $\phi_m < \phi_s$ at zero gate bias (gate connected to the substrate). Electrons in the metal reside at energy levels above the conduction band of the semiconductor. As a result, electrons flow from the metal into the semiconductor until a potential that counterbalances the difference in the work functions is built up between the two plates. This induces a negative charge under the gate dielectric accompanied by a downward band bending and, hence, a positive surface potential.

If an external voltage equal to this difference is applied to the gate, the net charge in the semiconductor disappears and the bands return to their flat position. This voltage is defined as the flatband voltage, V_{FB}, and can be expressed as:

$$V_{FB} = q\phi_{ms} = q(\phi_m - \phi_s). \tag{3.11}$$

[2] A more accurate definition for the onset of strong inversion is the surface potential at which the inversion layer charge is equal to the depletion region charge, which is ~ 6 kT/q above the onset of moderate inversion.

Another modification to the above picture comes from charges that reside in the dielectric. These charges may originate from processing, from defects in the bulk, or from charges that exist at the interface between Si and SiO_2. The charges in the oxide or at the interface induce opposite charges in the semiconductor and the gate. This can cause the bands to bend up or down. Therefore, the flat-band voltage will have to be adjusted to take this into account. Assuming that all oxide fixed charges reside at the Si–SiO_2 interface, the flat-band voltage can be expressed as:

$$V_{FB} = \phi_{ms} - \frac{Q_o}{C_{ox}}, \tag{3.12}$$

where Q_o is the oxide charge and C_{ox} is the oxide capacitance.

In addition, defects located at the Si–SiO_2 interface may not be fixed in their charge state and may vary with the surface potential of the substrate. These defects are referred to as **fast interface traps** and have an impact on switching characteristics of MOSFETs as will be discussed later.

Based on the values of ψ_s, different regions of operation can be defined and are shown in Table 3.1.

3.2.3 Small-Signal Capacitance

If the gate voltage is changed by a small positive amount, a small positive charge will flow into the gate. To preserve charge neutrality, an equal amount of charge must flow out of the semiconductor. The relation of the small change in charge due to a small change in voltage is defined as the small-signal capacitance and can be written as:

$$C_{gb} \equiv \frac{dQ_G}{dV_{GB}}. \tag{3.13}$$

The equivalent circuit for the total gate capacitance is shown in Figure 3.5. The total gate capacitance consists of the oxide and semiconductor capacitances, where the semiconductor capacitance is the sum of the depletion, the inversion capacitance, and the interface states capacitance.

3.2.4 Threshold Voltage

The threshold voltage of an MOS capacitor is the gate voltage, V_{GB}, required to create strong inversion (i.e., $\psi_s = 2\phi_F$) under the gate. Figure 3.6 shows the inversion charge as a function of V_{GB}. The straight-line extrapolation of this charge to the *x*-axis

TABLE 3.1 Regions of Operation of MOS Capacitor for *p*-Substrate

Accumulation	$\psi_s < 0$	$V_{GB} < V_{FB}$	$Q_c > 0$
Depletion	$\psi_s > 0$	$V_{GB} > V_{FB}$	$Q_c < 0$
Inversion	$\psi_s > 0$	$V_{GB} >> V_{FB}$	$Q_c < 0$

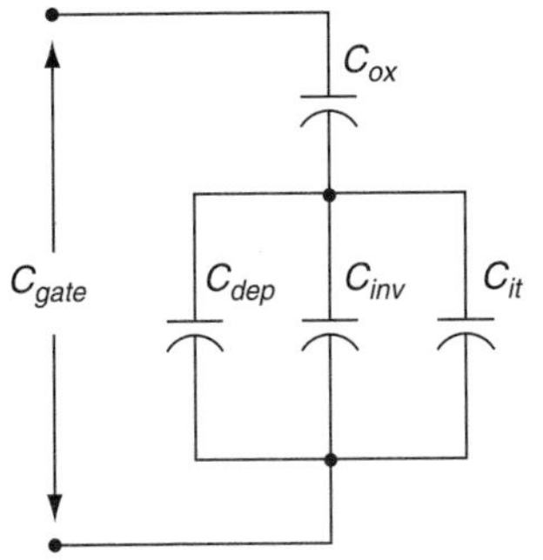

FIGURE 3.5

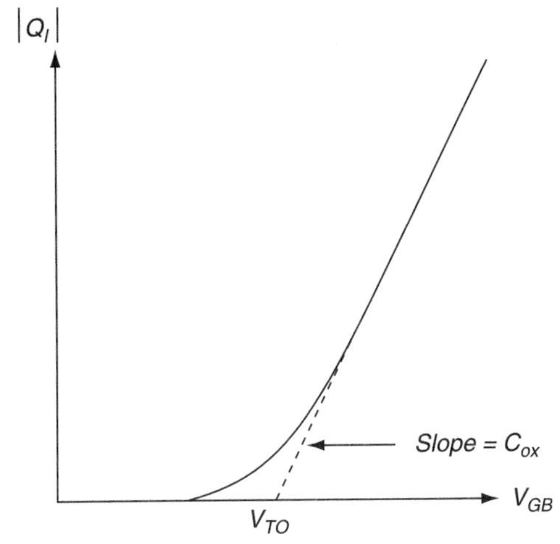

FIGURE 3.6

is called the extrapolated threshold voltage, V_{TO}. When the semiconductor (body) is at ground potential ($V_B = 0$), V_{TO} is given by:

$$V_{TO} = V_{FB} + 2\phi_F + \gamma\sqrt{2\phi_F}, \tag{3.14}$$

where γ is referred to as the body-effect coefficient and can be expressed as:

$$\gamma = \frac{\sqrt{2q\varepsilon_S N_A}}{C_{OX}}. \tag{3.15}$$

With a body bias, the threshold voltage is given by:

$$V_T = V_{FB} + 2\phi_F + \gamma\sqrt{2\phi_F + V_B}. \tag{3.16}$$

As shown in the above equation, the threshold voltage increases when a back bias is applied. A positive bias on the substrate results in a wider depletion region and assists in balancing the gate charge. This causes the electron concentration in the inversion layer to decrease. Thus, a higher gate voltage is needed to achieve the onset of inversion resulting in an increase of the threshold voltage. In addition, the doping concentration and oxide thickness can also have an impact on the threshold voltage dependence on back bias. Lower doping concentrations and thinner oxides result in a weaker dependence of back bias on threshold voltage.

3.3 Metal-Oxide-Silicon Field Effect Transistor

The MOSFET is the building block of the current ultra large-scale integrated circuits (ICs). The growth and complexity of MOS ICs have been continually increasing over the years to enable faster and denser circuits. Shown in Figure 3.7 is the simplified schematic of a MOSFET. The device consists of a MOS gate stack formed between two *pn* junctions called the **source** and **drain** junctions. The region under the gate is often referred to as the channel region. The length and the width of this region are called the **channel length** and **channel width**, respectively. An inversion layer under the gate creates a conductive path (i.e., channel) from source to drain and turns the transistor on. When the channel forms right under the gate dielectric, the MOSFET is referred to as a **surface channel MOSFET**. Buried channel MOSFETs in which the channel forms slightly beneath the surface are also used, but they are becoming less popular with the continuous downscaling of MOSFETs. A fourth contact to the substrate is also formed and referred to as the **body contact**.

The standard gate dielectric used in Si-integrated circuit industry is SiO_2. The gate electrode is heavily doped *n*-type or *p*-type polysilicon. The source and drain regions are formed by ion-implantation.

We shall first consider MOSFETs with uniform doping in the channel region. Later in the chapter, however, we shall learn how nonuniform doping profiles are used to enhance the performance of MOSFETs.

The MOSFET shown in Figure 3.7 is an *n*-channel MOSFET, in which electrons flow from source to drain in the channel

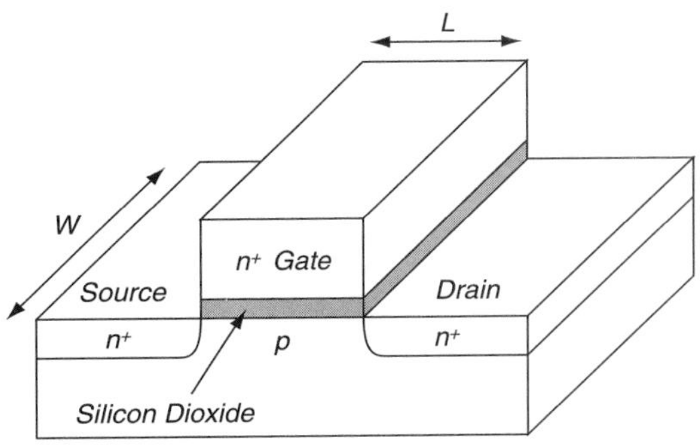

FIGURE 3.7

TABLE 3.2 Dopant Types for Different Regions of *n*-Channel and *p*-Channel MOSFETs

	n-channel MOSFET	*p*-channel MOSFET
Substrate (channel)	*p*	*N*
Gate electrode	*n*+	*p*+
Source and drain	*n*+	*p*+

induced under the gate oxide. Both *n*-channel and *p*-channel MOSFETs are extensively used. In fact, CMOS IC technology relies on the ability to use both devices on the same chip. Table 3.2 shows the dopant types used in each region of the two structures.

3.3.1 Device Operation

Figure 3.8 shows an *n*-channel MOSFET with voltages applied to its four terminals. Typically, $V_S = V_B$ and $V_D > V_S$. Shown in Figure 3.9 is the typical $I_{DS} - V_{DS}$ characteristics of such a device. For simplicity, we assume that the body and the source terminals are tied to the ground (i.e. $V_{SB} = 0$. This yields.

$$\begin{aligned} V_{GS} &= V_{GB} - V_{SB} = V_{GB} = V_G. \\ V_{DS} &= V_{DB} - V_{SB} = V_{DB} = V_D. \end{aligned} \tag{3.17}$$

As shown, at low V_{DS}, the drain current increases almost linearly with V_{DS}, resulting in a series of straight lines with slopes increasing with V_{GS}. At high V_{DS}, the drain current saturates and becomes independent of V_{DS}.

In this section, we present a description of the device operation complete with important equations valid in different regions of operation. We shall refer to Figure 3.10, which

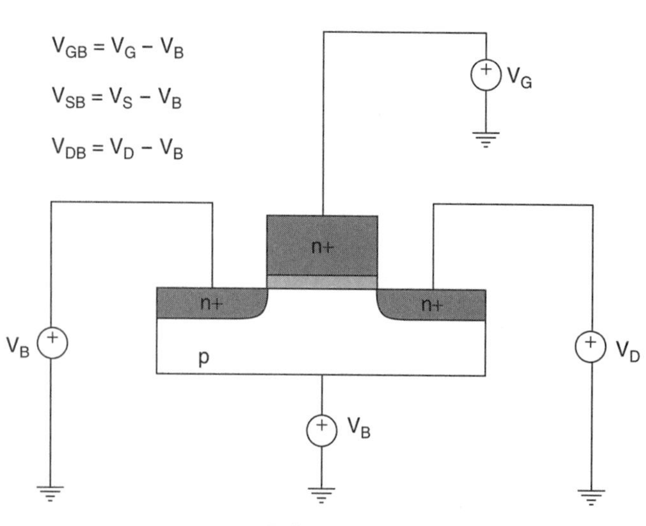

FIGURE 3.8

shows the inversion layer and the depletion regions that form under the inversion layer and around the source and drain junctions. Now we can begin to discuss MOSFET regions of operation in detail.

Linear Region

With a small positive voltage on the drain and no bias on the gate (i.e., $V_{DS} > 0$ and $V_{GS} = 0$), the drain is a reverse-biased *pn* junction. Conduction band electrons in the source region encounter a potential barrier determined by the built-in potential of the source junction. As a result, electrons cannot enter the channel region and, hence, no current flows from the source to the drain. This is referred to as the *off* state. With a small positive bias on the gate, band bending in the channel region ($\psi_s > 0$) brings the conduction band in the channel region closer to the conduction band in the source region, thus reducing the height of the potential barrier to electrons. Electrons can now enter the channel and a current flow from source to drain is established.

In the low-drain-bias regime, the drain current increases almost linearly with drain bias. Indeed, here the channel resembles an ideal resistor obeying Ohm's law. The channel resistance is determined by the electron concentration in the channel, which is a function of the gate bias. Therefore, the channel acts like a voltage-controlled resistor whose resistance is determined by the applied gate bias. As shown in Figure 3.9, as the gate bias is increased, the slope of the *I–V* characteristic gradually increases due to the increasing conductivity of the channel. We obtain different slopes for different gate biases. This region where the channel behaves like a resistor is referred to as the **linear region** of operation. The drain current in the linear regime is given by:

$$I_{D,\,lin} = \frac{W}{L}\mu C'_{ox}\left[(V_{GS} - V_T)V_{DS} - \frac{\alpha}{2}V_{DS}^2\right], \tag{3.18}$$

where V_T is the threshold voltage, C'_{ox} is the gate capacitance per unit area, α is a constant, and μ is the effective channel mobility (which differs from bulk mobility). We shall deal with the concept of effective channel mobility later in this chapter.

Threshold voltage in the above equation is defined as:

$$V_T = V_{FB} + 2\phi_F + \gamma\sqrt{2\phi_F + V_{SB}}. \tag{3.19}$$

For small V_{DS}, the second term in the parenthesis can be ignored, and the expression for drain current reduces to:

$$I_{D,\,lin} = \frac{W}{L}\mu C'_{ox}(V_{GS} - V_T)V_{DS}, \tag{3.20}$$

which is a straight line with a slope equal to the channel conductance:

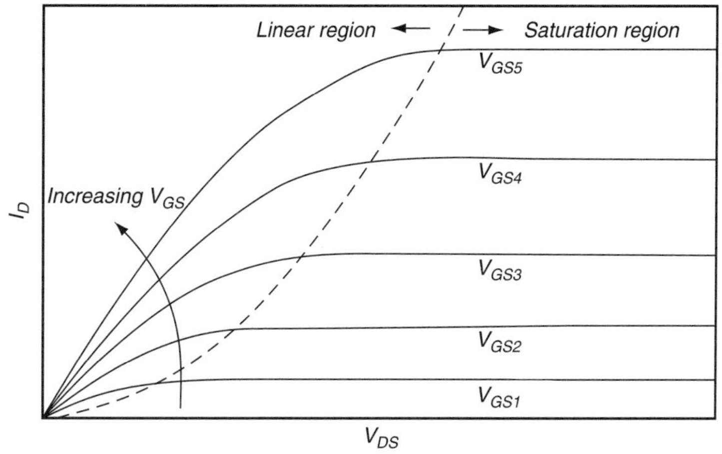

FIGURE 3.9

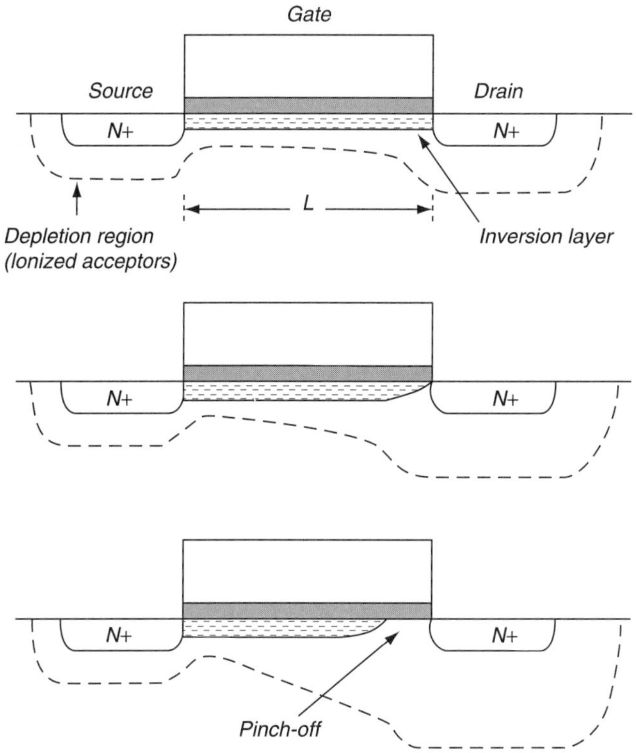

FIGURE 3.10

$$\sigma_c = \frac{W}{L}\mu C'_{ox}(V_{GS} - V_T). \tag{3.21}$$

Saturation Region

For larger drain biases, the drain current saturates and becomes independent of the drain bias. Naturally, this region is referred to as the **saturation region**. The drain current in saturation is derived from the linear region current shown in equation 3.18, which is a parabola with a maximum occurring at $V_{D,sat}$ given by:[3]

$$V_{D,sat} = \frac{V_{GS} - V_T}{\alpha}. \tag{3.22}$$

To obtain the drain current in saturation, this $V_{D,sat}$ value can be substituted in the linear region expression, which gives:

$$I_{D,\ sat} = \frac{W}{L}\mu C'_{ox}\frac{(V_{GS} - V_T)^2}{2\alpha}. \tag{3.23}$$

As V_{DS} increases, the number of electrons in the inversion layer decreases near the drain. This occurs due to two reasons. First, because both the gate and the drain are positively biased, the potential difference across the oxide is smaller near the drain end. Because the positive charge on the gate is determined by the potential drop across the gate oxide, the gate charge is smaller near the drain end. This implies that the amount of negative charge in the semiconductor needed to preserve charge neutrality will also be smaller near the drain. Consequently, the electron concentration in the inversion layer drops. Second, increasing the voltage on the drain increases the depletion width around the reverse-biased drain junction. Since more negative acceptor ions are uncovered, a fewer number of inversion layer electrons are needed to balance the gate charge. This implies that the electron density in the inversion layer near the drain would decrease even if the charge density on the gate was constant. The reduced number of carriers causes a reduction in the channel conductance, which is reflected in the smaller slope of $I_{DS} - V_{DS}$ characteristics as V_{DS} approaches $V_{D,sat}$, and the MOSFET enters the saturation region. Eventually, the inversion layer completely

[3] The V_{DS} at which the linear drain current parabola reaches its maximum can be found by setting $\partial I_{DS}/\partial V_{DS}$ equal to zero.

disappears near the drain. This condition is called **pinch-off**, and the channel conductance becomes zero. As shown in Figure 3.9, $V_{D,sat}$ increases with gate bias. This results because a larger gate bias requires a larger drain bias to reduce the voltage drop across the oxide near the drain end. As given in equation 3.22, $V_{D,sat}$ increases linearly with V_{GS}.

As V_{DS} is increased beyond $V_{D,sat}$, the width of the pinch-off region increases. However, the voltage that drops across the inversion layer remains constant and equal to $V_{D,sat}$. The portion of the drain bias in excess of $V_{D,sat}$ appears across the pinch-off region. In a long channel MOSFET, the width of the pinch-off region is assumed small relative to the length of the channel. Thus, neither the length nor the voltage across the inversion layer change beyond the pinch-off, resulting in a drain current independent of drain bias. Consequently, the drain current saturates. In smaller devices, this assumption falls apart and leads to **channel length modulation**, which is discussed later in the chapter.

From the above discussion, it is also evident that the electron distribution is highest near the source and lowest near the drain. To keep a constant current throughout the channel, the electrons travel slower near the source and speed up near the drain. In fact, in the pinch-off region, the electron density is negligibly small. Therefore, in this region, to maintain the same current level, the electrons have to travel at much higher speeds to transport the same magnitude of charge.

An important figure of merit for MOSFETs is transconductance, g_m, in the saturation regime, which is defined as:

$$g_m = \frac{\partial I_{D,sat}}{\partial V_{GS}} = \frac{W}{L}\frac{\mu C'_{ox}}{\alpha}(V_{GS} - V_T). \tag{3.24}$$

Transconductance is a measure of the responsiveness of the drain current to variations in gate bias.

Subthreshold Region: MOSFET in Weak Inversion

When the surface potential at the source end is sufficient to form an inversion layer but the band bending is less than what is needed to reach strong inversion (i.e., $\phi_F < \Psi_s < 2\phi_F$), the MOSFET is said to operate in weak inversion. This region of operation is commonly called the **subthreshold region** and plays an important role in determining switching characteristics of logic circuits. When an *n*-channel MOSFET is in weak inversion, the drain current is determined by diffusion of electrons from the source to the drain. This is because the drift current is negligibly small due to the low lateral electric field and small electron concentration in weak inversion.

Even though the electron concentration in weak inversion is small, it increases exponentially with gate bias. Consequently, the drain current in weak inversion also rises exponentially, and it can be expressed as:

$$I_{DS} = \frac{W}{L} I' e^{q(V_{GS}-V_T)/nkT}\left(1 - e^{-qV_{DS}/kT}\right), \tag{3.25}$$

where I' and n are constants defined as:

$$I' = \mu \frac{\sqrt{2q\varepsilon_S N_A}}{2\sqrt{2\phi_F + V_{SB}}}\phi^2 \tag{3.26}$$

$$n = 1 + \frac{\gamma}{2\sqrt{2\phi_F + V_{SB}}}. \tag{3.27}$$

When the drain bias is larger than a few kT/q, the dependence on the drain bias can be neglected, and the above equation reduces to:

$$I_{DS} = \frac{W}{L} I' e^{q(V_{GS}-V_T)/nkT}, \tag{3.28}$$

which yields an exponential dependence on gate bias. When $\log I_{DS}$ is plotted against gate bias, we obtain:

$$\log I_{DS} = \log\left(\frac{W}{L}I'\right) + \frac{q}{kT}\frac{V_{GS} - V_T}{n}, \tag{3.29}$$

which is a straight line with a slope:

$$\frac{1}{S} = \frac{1}{n}\frac{q}{kT}. \tag{3.30}$$

The parameter S in the above equation is the MOSFET **subthreshold swing**, which is one of the most critical performance figures of MOSFETs in logic applications. It is highly desirable to have a subthreshold swing as small as possible since this is the parameter that determines the amount of voltage swing necessary to switch a MOSFET from its *off* state to its *on* state. This is especially important for modern MOSFETs with supply voltages approaching 1.0 V.

Figure 3.11 shows typical subthreshold characteristics of a MOSFET. As predicted by the above model, $\log I_{DS}$ increases linearly with gate bias up to V_T. In strong inversion, the subthreshold model is no longer valid, and either equation 20 or 23 must be used depending on the drain bias. When $V_{GS} = V_T$, $\log I_{DS}$ deviates from linearity. In practice, this is a commonly used method to measure the threshold voltage. In terms of device parameters, subthreshold swing can be expressed as:

$$S \cong \frac{kT}{q}\left(1 + \frac{C'_{dep} + C'_{it}}{C'_{ox}}\right)\ln(10), \tag{3.31}$$

where C'_{dep} is depletion region capacitance per unit area of the MOS gate determined by the doping density in the channel region, and C'_{it} is the interface trap capacitance. Lower

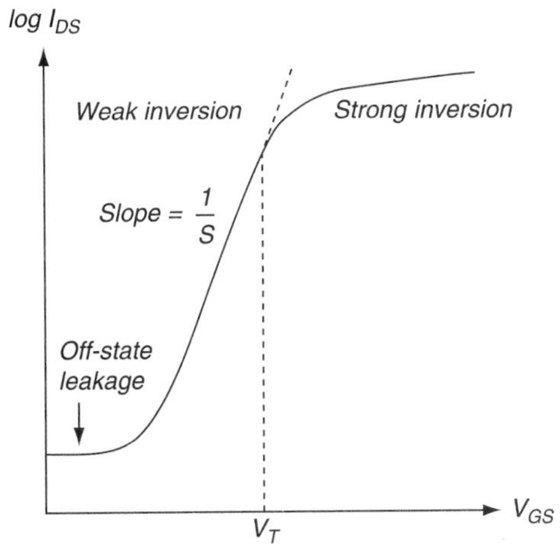

FIGURE 3.11

channel doping densities yield wider depletion region widths and hence smaller values for C'_{dep}. Another critical parameter is the gate oxide thickness, which determines C'_{ox}. To minimize S, the thinnest possible oxide must be used. The most widely used unit for S is mV/decade. Typical values range from 60 to 100 mV/decade.

The lower limit of the subthreshold current is determined by the leakage current that flows from source to drain when the transistor is off. This current, which is referred to as the **off-state leakage**, is determined by the source-to-channel potential barrier as well as the leakage currents of the source and drain junctions. For this reason, it is critical to be able to form low leakage junctions.

3.3.2 Effective Channel Mobility

The mobility term used in the previous equations differs from the bulk mobility due to additional scattering mechanisms associated with the surface, including interface charge scattering, coulombic scattering, and surface roughness scattering. These mechanisms tend to lower the carrier mobility in the channel. These mechanisms also tend to exhibit a dependence on the vertical electric field, E_y. At low fields, interface and coulombic scattering are the dominant mechanisms. At high fields, the surface roughness scattering dominates. Since the vertical field varies along the channel, mobility also varies. The dependence on the vertical field is generally expressed as:

$$\mu = \frac{\mu_o}{1 + \alpha E_y}, \tag{3.32}$$

where μ_o is roughly half of the bulk mobility and α is roughly 0.025 μm/V at room temperature. This mobility equation is not useful in device equations because E_y varies along the channel. To be able to use the standard equations, the field dependence is lumped into a constant termed the **effective channel mobility, μ_{eff}**. A simple yet intuitive model for μ_{eff} is as follows:

$$\mu_{eff} \approx \frac{\mu_o}{1 + \theta(V_{GS} - V_T)}, \tag{3.33}$$

where θ is a constant inversely proportional to the gate oxide thickness, t_{ox}. A typical plot of effective mobility as a function of gate bias is shown in Figure 3.12 (A). The effect of mobility degradation on $I_{DS} - V_{GS}$ characteristics is shown in Figure 3.12(B).

3.3.3 Nonuniform Channels

Typical channels in today's MOSFETs are nonuniform, which consist of a higher doped region placed underneath the Si–SiO_2 interface. This is done primarily to optimize the threshold voltage value while keeping the substrate concentration low. The new threshold voltage can be expressed as:

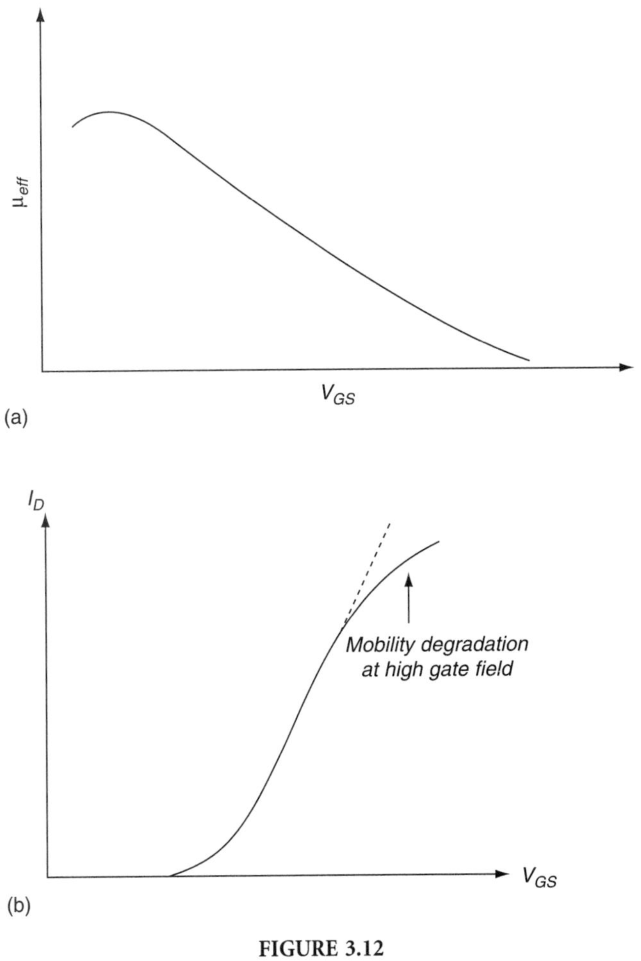

FIGURE 3.12

$$V_T = V'_{FB} + 2\phi_F + \gamma\sqrt{2\phi_F + V_B}, \quad (3.34)$$

where, V'_{FB} is given by:

$$V'_{FB} = V_{FB} + \frac{qM}{C_{ox}}. \quad (3.35)$$

The variable M is the implant dose. This equation assumes that the threshold adjust implant is very shallow and behaves as a sheet of fixed charge located under the gate oxide. In some cases, additional doped layers are placed in the channel to suppress punchthrough, a term that will be discussed later.

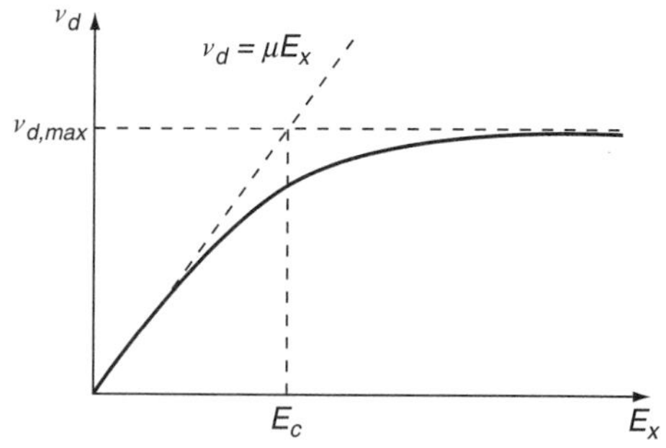

FIGURE 3.13

3.3.4 Short Channel Effects

MOSFETs are continually downscaled for higher packing density, higher device speed, and lower power consumption. The scaling methods are covered later in this chapter in a dedicated subsection. When physical dimensions of MOSFETs are reduced, the equations for drain current have to be modified to account for the so-called **short channel effects**. The three primary short channel effects included in this chapter are the following:

1. *Velocity saturation*: Limits the maximum carrier velocity in the channel to the saturation velocity in Si
2. *Channel length modulation*: Causes the drain current to increase with drain bias in the saturation region
3. *Drain-induced barrier lowering (DIBL)*: Causes the threshold voltage to change from its long channel value with dependence on device geometry as well as drain bias

Velocity Saturation

As discussed earlier, in long channel MOSFETs, the drain current saturates for V_{DS} larger than $V_{D,sat}$. The potential drop across the inversion layer remains at $V_{D,sat}$, and the horizontal electric field, E_x, along the channel is fixed at:

$$E_x = \frac{V_{Dsat}}{L}. \quad (3.36)$$

The electron drift velocity as a function of applied field is shown in Figure 3.13. At low fields, the drift velocity is given by:

$$v_d = \mu E_x, \quad (3.37)$$

where μ is generally referred to as the low-field mobility. At high fields, the velocity saturates due to phonon scattering. In a short channel device, the electric field across the channel can become sufficiently high such that carriers can suffer from velocity saturation. The effect of velocity saturation on MOSFET drain current can be severe. In short channel MOSFETs, it is impossible to overcome this effect. Thus, it is safe to say that all devices used in modern logic circuits suffer from velocity saturation to some extent.

To account for this phenomenon, I_{DS} has to be obtained using velocity dependence on the electric field. A simple yet intuitive model for I_{DS} contains the original long channel drain current expression in the linear regime (equation 18) divided by a factor that accounts for velocity saturation. This formula is written as:

$$I_{DS}(\text{with velocity saturation}) = \frac{I_{DS}(\text{without velocity saturation})}{1 + V_{DS}/LE_c} \quad (3.38)$$

where E_c is the critical field above which velocity saturation occurs (Figure 3.13) and is given by:

$$E_c = \frac{v_{sat}}{\mu}. \quad (3.39)$$

The $V_{D,sat}$ value obtained using the modified current is smaller than the long channel $V_{D,sat}$ value. The drain current in saturation can be obtained from the linear region expression by assuming the drain current saturates for $V_{DS} > V_{D,sat}$:

$$I_{D,sat} \approx W\mu C_{ox}(V_{GS} - V_T)E_C. \quad (3.40)$$

It is important to note that with velocity saturation, the dependence of drain current on V_{GS} is linear instead of square, as it is the case for the long channel MOSFET.

Channel Length Modulation

After pinch-off occurs at the drain end, the length of the inversion layer and, hence, the channel resistance continually decrease as the drain bias is raised above $V_{D,sat}$. In long channel devices, the length of the pinch-off region is negligibly small. Since the voltage drop across the channel is pinned at $V_{D,sat}$, drain current increases only negligibly. However, in short channel devices, this pinch-off region can become a significant fraction of the total channel length. This reduction of channel length manifests itself as a finite slope in the

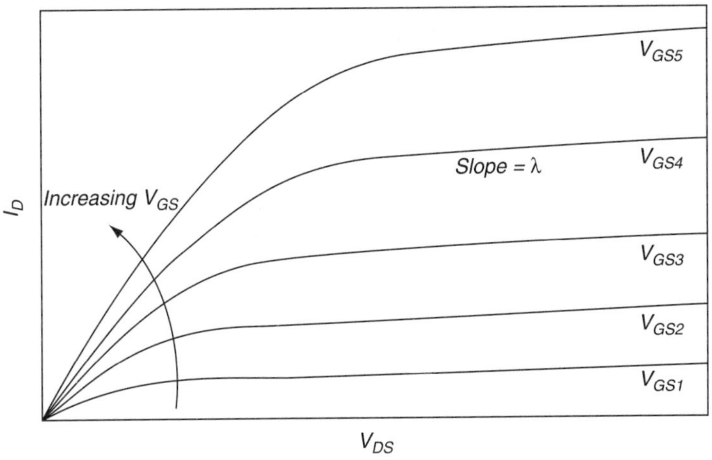

FIGURE 3.14

$I_{DS} - V_{DS}$ characteristics beyond $V_{D,sat}$ as shown in Figure 3.14. A commonly used expression for MOSFET drain current in saturation with channel length modulation is the following:

$$I_{D,sat} = \frac{W}{L}\mu C'_{ox}\frac{(V_{GS} - V_T)^2}{2\alpha}(1 + \lambda V_{DS}), \quad (3.41)$$

where λ is known as the channel length modulation factor expressed as:

$$\lambda = \frac{1}{E_o L}, \quad (3.42)$$

and E_o is the magnitude of the electric field at the pinch-off point. If both velocity saturation and channel length modulation are taken into account, the saturation drain current can be written as:

$$I_{D,sat} \approx W\mu C_{ox}(V_{GS} - V_T)E_C(1 + \lambda V_{DS}) \quad (3.43)$$

Drain-Induced Barrier Lowering

Another major short channel effect deals with the reduction of the threshold voltage as the channel length is reduced. In long channel devices, the influence of source and drain on the channel depletion layer is negligible. However, as channel lengths are reduced, overlapping source and drain depletion regions start having a large effect on the channel depletion region. This causes the depletion region under the inversion layer to increase. The wider depletion region is accompanied by a larger surface potential, which makes the channel more attractive to electrons. Therefore, a smaller amount of charge on the gate is needed to reach the onset of strong inversion, and the threshold voltage decreases. This effect is worsened when there is a larger bias on the drain since the depletion region becomes even wider. This phenomenon is called drain-induced barrier lowering (DIBL). Figure 3.15 shows the variation of surface potential from source to drain for a short channel and a long channel MOSFET. As shown, for the long channel device, the potential is nearly constant throughout the channel and is determined only by the gate bias.[4] The height of the potential barrier for electrons at the source end is given by:

$$\phi_{BL} \approx 2\phi_F. \quad (3.44)$$

For the smaller device, the surface potential is larger due to additional band bending. In addition, the surface potential is

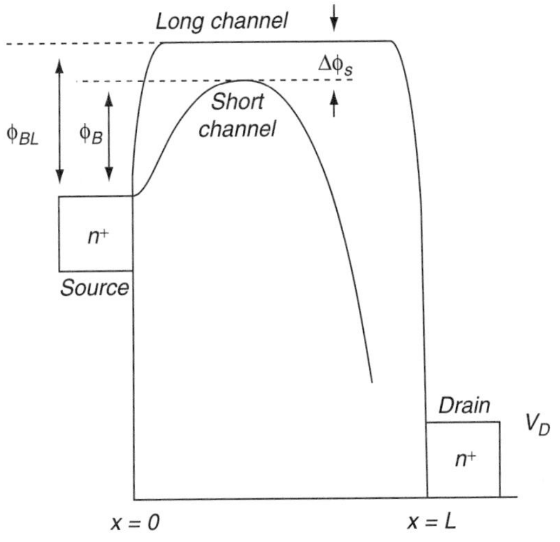

FIGURE 3.15

[4] The channel potential actually increases gradually toward the drain due to small yet finite resistance of the inversion layer. When pinch-off occurs at the drain end, majority of the drain-to-source bias appears across the pinch-off region due to its much larger resistivity.

no longer constant throughout the channel. With a larger surface potential, the barrier for electrons between the source and the channel is reduced and is given by:

$$\phi_B = \phi_{BL} - \Delta\phi_B. \tag{3.45}$$

The change in threshold voltage due to DIBL has been modeled Liu *et al.* (1993) as:

$$\Delta V_T \approx -[3(\phi_{bi} - 2\phi_F) + V_{DS}]e^{-L/\lambda_{\text{DIBL}}}, \tag{3.46}$$

where ϕ_{bi} is the built-in potential of the drain junction and:

$$\lambda_{\text{DIBL}} = \sqrt{\frac{\varepsilon Si^{t}_{ox} W_D}{\varepsilon_{ox}\beta}}. \tag{3.47}$$

In the above equation, W_D is the depletion region width under the inversion layer, and β is a fitting parameter. The model indicates that ΔV_T is a strong function of the λ_{DIBL} term in the exponential and must be minimized. This requires higher doping density in the channel to reduce W_D and down-scaling of the gate oxide thickness, t_{ox}. The model does not include the well-known dependence of ΔV_T on source/drain junction depths although some modifications have been suggested.

The impact of DIBL on V_T can be measured by plotting the subthreshold characteristics for increasing values of V_{DS} as shown in Figure 3.16. The subthreshold slope degrades and the impact of drain bias on ΔV_T increases as the channel length is reduced. This methodology is valid only until the subthreshold slope remains intact. A very large subthreshold swing implies that the device cannot be turned off. This phenomenon is called **punchthrough** and occurs roughly when the source and drain depletion regions meet. Since the depletion region becomes wider at larger drain biases, the onset of punchthrough is reached sooner. Punchthrough can occur either at or below the surface depending on the doping profile in the channel region. Surface punchthrough occurs for uniformly doped substrates and results in the loss of gate control of the channel region, causing the device to fail. Bulk punchthrough occurs for ion-implanted channels that have a higher doping concentration at the surface. In this condition, there still exists a channel that is gate controlled, but the background leakage becomes very high and is strongly dependent on V_{DS}. The common solution is to add a shallow punchthrough implant typically placed immediately below the threshold adjust implant. This raises the doping density under the channel while keeping the bulk doping density as low as possible. The goal is to minimize junction depletion region capacitance, which has an impact on device switching speed.

Similarly, for a fixed drain bias, a larger body bias increases the reverse bias on the drain junction and, thus, has the same effect as increasing the drain bias. Lowering of the threshold voltage due to DIBL is demonstrated in Figure 3.17. Note that the effect of body bias on threshold voltage for large channel MOSFETs is determined by equation 3.15. Here we are interested in lowering of the threshold voltage due to body bias only for short channel lengths.

In addition to the short channel effects described above, the scaling of the channel width can also have a large effect on the threshold voltage, and its nature depends on the isolation technology (Tsividis, 1999). In the case of LOCOS isolation, the threshold voltage increases as the channel width decreases. On the other hand, for shallow trench isolation, the threshold voltage decreases as the channel width decreases. In today's technologies, shallow trench isolation is the predominant isolation technique.

3.3.5 MOSFET Scaling

Scaling of MOSFETs is necessary to achieve (a) low power, (b) high speed, and (c) high packing density. Several scaling methodologies have been proposed and used, which are

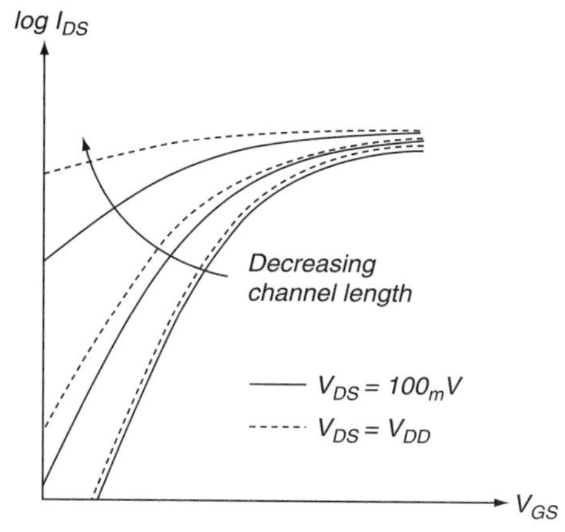

FIGURE 3.16

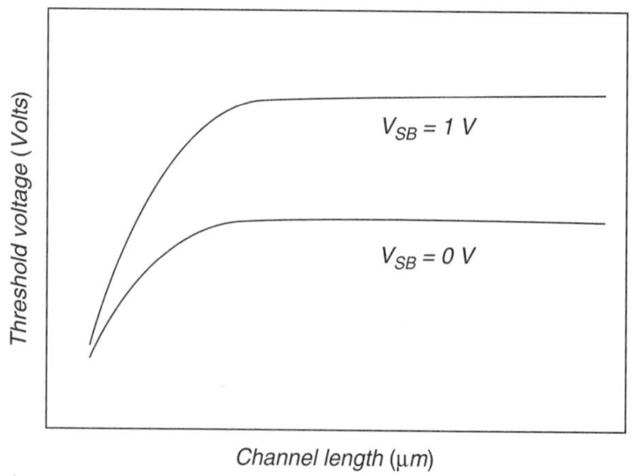

FIGURE 3.17

derivatives of constant field scaling. In constant-field scaling, the supply voltage and the device dimensions (both lateral and vertical) are scaled by the same factor, κ such that the electric field remains unchanged. This results in scaling of the drain current or current drive. At the same time, gate capacitance is also scaled due to reduced device size. This provides a reduction in the gate charging time, which directly translates into higher speed. Furthermore, power dissipation per transistor is also reduced.

Constant field scaling provides a good framework for CMOS scaling without degrading reliability. However, there are several parameters, such as the kT/q and an energy gap, that do not scale with reduced voltages or dimensions and present challenges in device design. Since the junction built-in potential, ϕ_{bi}, and the surface potential, ψ_s, are both determined by the bandgap, they do not scale either. Consequently, depletion region widths do not scale as much as other parameters, which results in worsened short channel effects.

Another parameter that does not scale easily is the threshold voltage. This sets the lower limit for the power-supply voltage since a safe margin between the two parameters is required for reliable device operation. Other parameters that do not scale well include the off-state current and subthreshold slope.

Although **constant field scaling** provides a reasonable guideline, scaling of the voltages by the same factor as the physical dimensions is not always practical. This is due to the inability to scale the subthreshold slope properly as well as standardize voltage levels of prior generations. This is the basis for **constant voltage scaling** and other modified scaling methodologies. One problem with constant voltage scaling is that the oxide field increases since the t_{ox} is also scaled by κ. To reduce this problem, the oxide thickness is reduced by κ' where $\kappa' < \kappa$. Modifications of the constant voltage and constant field scaling have also been tried to avoid high-field problems. For example, in **quasi-constant voltage scaling**, physical dimensions are scaled by κ, whereas the voltages are scaled by a different factor, κ'. Furthermore, since depletion layers do not scale proportionately, the doping concentration, N_A, must be increased by more than what is suggested by constant-field scaling. This is called **generalized scaling** and the doping scaling factor is greater than κ. The scaling rules discussed above are shown in Table 3.3.

It should be noted that there are other effects that can limit the performance of scaled devices. These include (a) reduction of the effective oxide capacitance due to finite thickness of the inversion layer, (b) depletion in the polysilicon gate, (c) quantum effects in the inversion layer that increase V_T, and (d) quantum mechanical tunneling of carriers through the thin oxide, which is a critical issue for $t_{ox} < 50$ A.

3.3.6 Modern MOSFETs

Today's MOSFET has evolved through many modifications to provide solutions for parasitic problems that emerge during scaling. A modern n-channel MOSFET structure is shown in Figure 3.18. The gate dielectric is SiO_2 and is grown by thermal oxidation. As a result of continued scaling, the gate oxide has thinned down to the extent that direct tunneling has become a serious concern. Currently, alternate high-κ dielectrics are being considered as replacements for SiO_2. The goal is to achieve higher capacitance without compromising gate leakage. The standard gate electrode is heavily doped polysilicon defined by anisotropic reactive ion etching. Extension junctions are formed by ion-implantation using the polysilicon gate as an implant mask. Because a separate masking step is not needed to define the junction regions, the approach is named **self-aligned polysilicon gate technology**. At the same time, the junction sheet resistance must be sufficiently low not to increase the device series resistance. Extension junctions must be as shallow as possible to reduce DIBL. When formed by ion-implantation, shallow junctions require low doses and low energies. Unfortunately, low dose results in higher series resistance. Sidewall spacers are formed by deposition of a dielectric (SiO_2 or Si_3N_4) followed by anisotropic reactive ion etching. The spacers must be as thin as possible to minimize the series resistance contribution of the extension junctions. Deep source and drain junctions are formed by ion-implantation following spacer formation. Because they are separated from the channel by the spacers, their contribution to DIBL is negligible. These junctions must be sufficiently deep to allow formation of good quality silicide contacts. Both Ti and Co silicides are used as source/drain contact materials. Silicides are formed by an approach referred to as **self-aligned-silicide** (SALICIDE), which selectively forms the silicide on the junctions as well as polysilicon. Silicide formation consumes the Si substrate in the deep source/drain regions and can lead to excessive junction leakage. To avoid this, the deep source/drain regions are required to be about 50 nm deeper than the

TABLE 3.3 Scaling Rules for Four Different Methodologies

Parameters	Constant field scaling $1 < \kappa' < \kappa$	Constant voltage scaling $1 < \kappa' < \kappa$	Quasi-constant voltage scaling $1 < \kappa' < \kappa$	Generalized scaling $1 < \kappa' < \kappa$
W, L	$1/\kappa$	$1/\kappa$	$1/\kappa$	$1/\kappa$
t_{ox}	$1/\kappa$	$1/\kappa'$	$1/\kappa$	$1/\kappa$
N_A	κ	κ	κ	κ^2/κ'
$Vdd,\ V_T$	$1/\kappa$	1	$1/\kappa'$	$1/\kappa'$

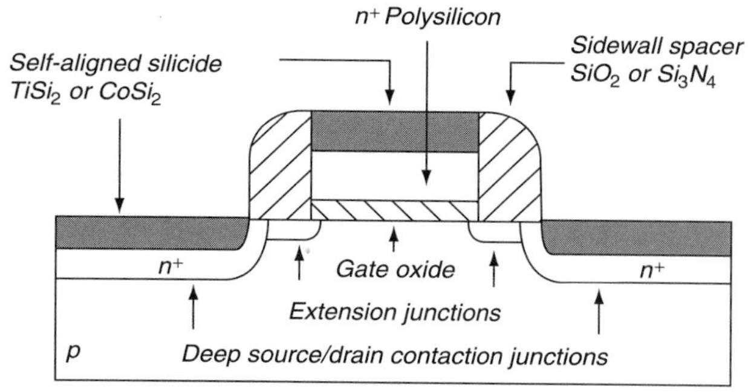

FIGURE 3.18

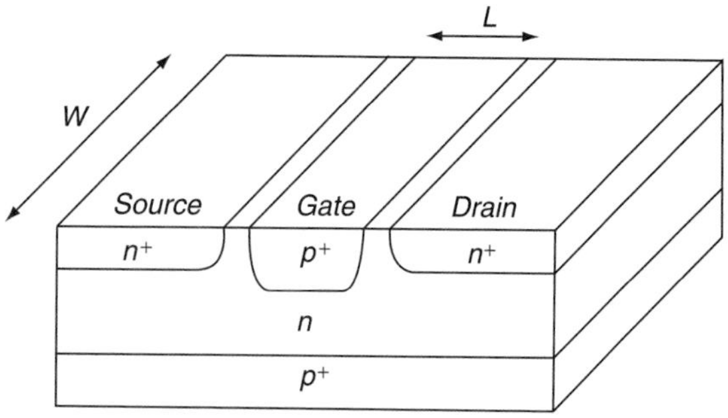

FIGURE 3.19

amount consumed. An important advantage of the SALICIDE process is that the entire junction area up to the sidewall spacer is used for contact formation, which translates into low contact resistance. The channel doping profile typically consists of the threshold adjust and punchthrough stop implants. The gate electrode is very heavily doped to avoid polysilicon depletion.

The MOSFET continues to maintain its dominance in today's digital ICs. To ensure continued benefits from MOSFET scaling, several structural and material changes are required. These include new gate dielectric materials to reach equivalent SiO_2 thicknesses less than 1 nm, new doping technologies that provide ultra-low resistance and ultra-shallow junctions, as well as metal gate electrodes that do not suffer from depletion effects.

3.4 Junction Field Effect Transistor

The junction field effect transistor (JFET) was first proposed by William Shockley in 1952. The first demonstration was made the following year by Dacey and Ross. Due to the large popularity of the bipolar junction transistor (BJT) at the time, major advancements in JFET fabrication did not occur until the 1970s. With the introduction of MOSFET, the use of JFET remained limited to specific applications.

Figure 3.19 shows a simplified schematic of the JFET. The device consists of an n-type channel region sandwiched between two p^+n junctions that serve as the gate. The p^+ regions can be tied together to form a dual-gate JFET. The heavily doped $n+$ source and drain regions serve as low resistivity ohmic contacts to the n-type channel. As such, a conductive path exists between the source and the drain contacts. The device is turned off by depleting the channel off the free carriers.

The $I_{DS} - V_{DS}$ characteristics of a JFET are very similar to those of a MOSFET. Figure 3.20 illustrates the operation of a JFET in different regions. The depletion region widths of the two p^+n junctions are determined by the gate bias and the potential variation along the channel. The widths of the depletion regions at the source end are determined mainly by the built-in potential of the junction. The depletion regions are wider at the drain end due to the large positive bias applied to the drain terminal. Figure 3.20(A) corresponds to the linear region of the JFET. Even though the channel depth is reduced at the drain end, the n-type channel extends from the source to the drain. The channel resembles a resistor, and the current is a linear function of the drain bias. The depth and, hence, the resistance of the channel is modulated by the gate bias. The drain current in the linear region can be expressed as:

$$I_{D,\,lin} = \frac{G_o}{2V_P}(V_G - V_T)V_D, \tag{3.48}$$

where V_P is the pinch-off voltage corresponding to the potential drop across the gate junction when the two depletion

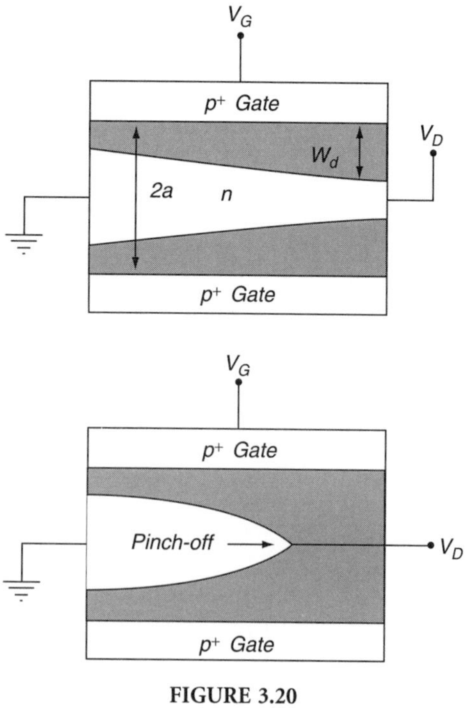

FIGURE 3.20

regions meet at the drain end. The variable G_o is the full channel conductance with a channel depth equal to $2a$. These can be expressed as:

$$G_o = 2aq\mu_n N_D \frac{W}{L} \tag{3.49}$$

and

$$V_P = \frac{qN_D a^2}{2\varepsilon_s}. \tag{3.50}$$

When the two depletion regions meet at the drain end, the JFET enters the saturation region. As shown in Figure 3.20(B), the pinch-off point moves toward the source as the drain bias is raised. The potential at the pinch-off point, however, remains pinned at:

$$V_{D,sat} = V_P - \psi_{bi} + V_G = V_G - V_T. \tag{3.51}$$

The built-in potential ψ_{bi} is a function of the doping concentrations in both sides of the junction. For large devices, the reduction in channel length due to pinch-off can be negligibly small. Hence the channel conductance remains approximately the same with a fixed voltage equal to $V_{D,sat}$ across the conductive channel. The drain current no longer increases with drain bias, and JFET is said to operate in the saturation region. The drain current after pinch-off can be expressed as:

$$I_{D,sat} = \frac{G_o}{4V_P}\left(1 + \frac{V_G}{V_P}\right)^2. \tag{3.52}$$

In JFETs with short channel lengths, the electric field along the channel can be large enough to cause carrier velocity saturation even before pinch-off. This requires a drain bias of:

$$V_D = E_c L, \tag{3.53}$$

where E_c is the critical field for velocity saturation.

3.5 Metal-Semiconductor Field Effect Transistor

A MESFET is very similar to a metal-semiconductor field effect transistor (MOSFET). The main difference is that in a MESFET, the MOS gate is replaced by a metal-semiconductor (Schottky) junction. A self-aligned GaAs MESFET structure is shown in Figure 3.21. The heavily doped source and drain junctions are formed in an n-type epitaxial layer formed on semi-insulating GaAs, which provides low parasitic capacitance. The n-type channel region has a thickness, a, which is typically less than 200 nm. The source

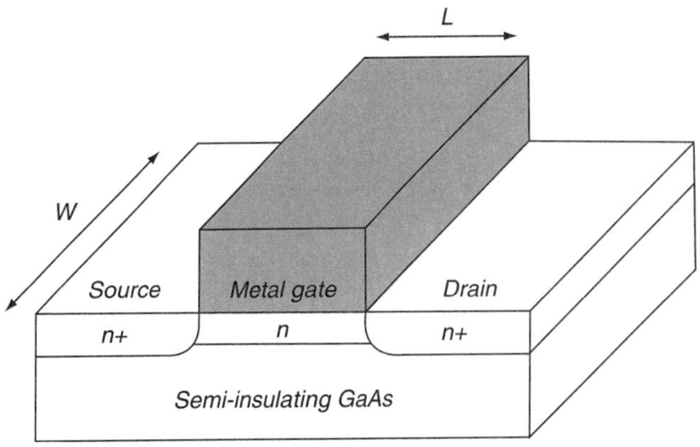

FIGURE 3.21

and drain junctions are formed by ion-implantation followed by annealing. Common source and drain contact materials are AuGe alloys. Popular Schottky gate metals include Al, Ti-Pt-Au-layered structure, Pt, W, and WSi_2. However, in a self-aligned MESFET, since the source and drain implantation and implant annealing must be performed after gate formation, refractory metals that can withstand the annealing temperatures are preferred.

MESFET is better suited to materials that do not have a good dielectric to form a high-quality MOS gate. Today, MESFETs are commonly fabricated on compound semiconductors, predominantly GaAs. Because of high mobility of carriers in GaAs and low capacitance due to the semi-insulating GaAs, MESFETs have the speed advantage over MOSFETs. They are used in microwave applications, which require high-speed devices. Application areas include communication, high-speed computer, and military systems.

Although the MESFET structure resembles the MOSFET, its operation is much closer to that of the JFET. Figure 22 shows the MESFET cross-section in different regions of operation. The MESFET has a conductive path between the source and the drain since the channel is of the same conductivity type as the junctions. Therefore, to turn-off a MESFET, a sufficiently high gate bias must be applied to deplete the channel. The depletion region under the gate modulates the width and the resistance of the conductive channel from source to drain. As such, in a MESFET, the Schottky gate plays the exact same role that the pn junction gate plays in a JFET.

The width of the depletion region is wider at the drain end since the source is grounded and a finite bias is applied to the drain. At small drain biases, the channel is conductive from source to drain. However, the width of the channel is smaller at the drain end due to the wider depletion region. The depletion region width under the gate is given by:

$$W_d(x) = \sqrt{\frac{2\varepsilon_s[\psi_{bi} + \psi(x) - V_G]}{qN_D}}, \tag{3.54}$$

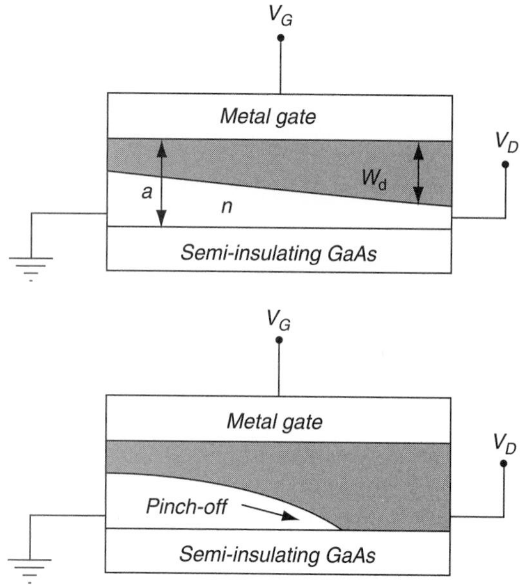

FIGURE 3.22

where ψ_{bi} is the band bending in the semiconductor due to the metal-semiconductor work function difference, and $\psi(x)$ is the channel potential with respect to the source. In this regime, the MESFET channel resembles a voltage, controlled resistor. The drain current varies linearly with the drain bias. The gate bias changes the width of the depletion region to modulate the width and the resistance of the channel. This is the linear region of operation for a MESFET.

For a long channel MESFET ($L \gg a$), the drain current in the linear region is approximately equal to:

$$I_{D,\,lin} \approx \frac{G_i}{2V_P}(V_G - V_T)V_D, \tag{3.55}$$

where

$$G_i = \frac{W}{L} q\mu_n N_D a \tag{3.56}$$

is the channel conductance without depletion under the Schottky gate. The variable V_P in equation 3.55 is called the pinch-off voltage, which is the net potential across the gate junction when the depletion region width under the gate is exactly equal to the channel depth, a. The pinch-off voltage is given by:[5]

$$V_P = \frac{qN_D a^2}{2\varepsilon_s}. \tag{3.57}$$

[5] This comes from the depletion region of equation 3.54 by replacing the potential term with V_P and the depletion region width with the channel depth.

Finally, V_T is the threshold voltage, and it is given by:

$$V_T = \psi_{bi} - V_P$$

In saturation, the drain current is given by:

$$I_{D,\,sat} \approx \frac{G_i}{4V_P}(V_G - V_T)^2, \tag{3.59}$$

yielding a transconductance of:

$$g_{m,\,sat} = G_i\left(1 - \sqrt{\frac{\psi_{bi} - V_G}{V_P}}\right). \tag{3.60}$$

When pinch-off occurs, the voltage at the pinch-off point is pinned at V_P even when the channel length is continually reduced at higher drain biases. In long channel MESFETs, the length of the pinch-off region is negligibly small and can be ignored. This results in saturation of the drain current for voltages beyond V_P. In smaller devices, however, the length and the resistance of the channel decreases as the pinch-off region becomes wider. Similar to channel length modulation in MOSFETs, this results in a gradual increase in drain current with applied drain bias.

In short channel devices, the horizontal field in the channel is high enough to reach the velocity saturation regime, which degrades the drain current as well as the transconductance. Velocity saturation can be reached even before pinch-off. With velocity saturation, the drain current in the saturation region can be expressed as:

$$I_{D,\,sat} = qv_s WN_D a\left(1 - \sqrt{\frac{\psi_{bi} - V_G}{V_P}}\right), \tag{3.61}$$

which gives a transconductance of:

$$g_{m,\,sat} \approx \frac{qv_s WaN_D}{2\sqrt{V_P(\psi_{bi} - V_G)}}. \tag{3.62}$$

3.6 Modulation-Doped Field Effect Transistor

The modulation-doped field effect transistor (MODFET) is also known as the high-electron mobility transistor (HEMT). The device relies on the ability to form a high-quality heterojunction between a wide bandgap material lattice matched to a narrow bandgap material. The preferred material system is AlGaAs–GaAs; however, MODFETs have also been demonstrated using other material systems including $Si{-}Si_xGe_{1-x}$. The device was developed in the

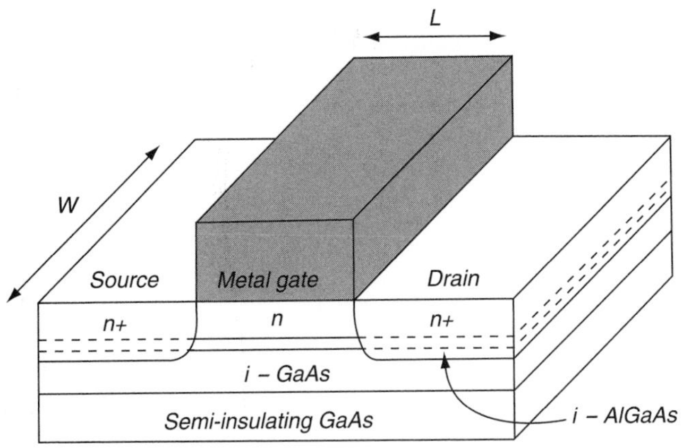

FIGURE 3.23

1970s. Enhanced mobility was first demonstrated by Dingle in 1978.

Figure 3.23 shows the simplified schematic of a self-aligned MODFET. The current conduction takes place in the undoped GaAs layer. The *n*-type AlGaAs layer located under the metal (Schottky barrier) gate is separated from the undoped GaAs by a thin undoped AlGaAs that acts as a buffer layer. A two-dimensional (2-D) electron gas is formed in GaAs immediately under the AlGaAs. High mobility results from the absence of ionized impurity scattering in the undoped layer. The thickness of the undoped AlGaAs buffer layer is critical. The buffer layer must be sufficiently thin to allow electrons to diffuse from the *n*-type AlGaAs into GaAs. At the same time, it must be sufficiently thick to place the 2-D electron gas sufficiently far away from the ionized impurities in the doped AlGaAs layer. The thickness of this layer is typically around 80 A.

The semiconductor layers are typically grown by molecular beam epitaxy (MBE); however, metal-organic chemical vapor deposition (MOCVD) has also shown to be feasible.

The energy band diagram at the onset of threshold is shown in Figure 3.24. The threshold is typically taken as the gate bias at which the conduction band of the GaAs layer coincides with the Fermi level. The 2-D electron gas forms at the heterointerface immediately under the undoped AlGaAs buffer layer and resembles the inversion layer that forms under the gate dielectric of a MOSFET.

The threshold voltage can be expressed as:

$$V_T = \phi_{bn} - V_P - \frac{\Delta E_C}{q}, \tag{3.63}$$

where V_P is the pinch-off voltage for the AlGaAs layer given by:

$$V_P = \frac{q_N D x_d^2}{2\varepsilon_s}. \tag{3.64}$$

The drain current in the linear regime is given by:

$$I_{D,\,lin} \cong \frac{\mu_n C_o W(V_G - V_T)V_D}{L}, \tag{3.65}$$

where

$$C_o = \frac{\varepsilon_S}{x_d + x_{ud} + \Delta d}. \tag{3.66}$$

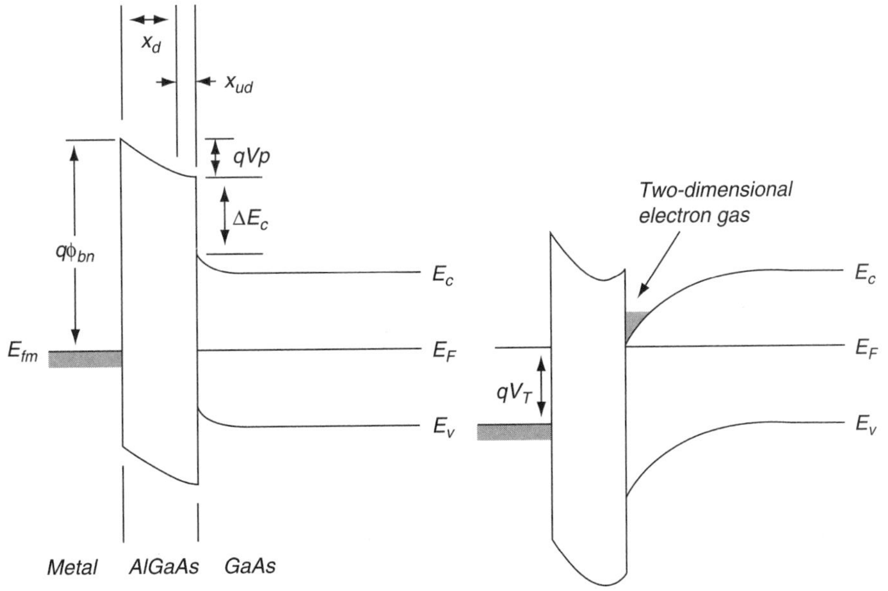

FIGURE 3.24

In the above equations, x_d and x_{ud} are the doped and undoped AlGaAs thicknesses, and Δd is the thickness of the 2-D electron gas that is typically less than 100 A. The threshold voltage can be made positive or negative by adjusting x_d. Hence, both enhancement and depletion mode devices are possible. The transconductance in the linear region is given by:

$$g_{m,\,lin} \equiv \frac{dI_{D,\,lin}}{dV_G} = \frac{\mu_n C_o W V_D}{L}. \tag{3.67}$$

When the drain bias is sufficiently large, electron concentration in the 2-D gas is reduced to zero at the drain end and the drain current saturates with V_D. This occurs at a drain bias of:

$$V_{D,\,sat} = V_G - V_T. \tag{3.68}$$

The drain current in saturation is given by:

$$I_{D,\,sat} = \frac{\mu_n C_o W}{2L}(V_G - V_T)^2, \tag{3.69}$$

yielding a transconductance of:

$$g_{m,\,sat} \equiv \frac{dI_{D,\,sat}}{dV_G} = \frac{\mu_n C_o W(V_G - V_T)}{L}. \tag{3.70}$$

In practical devices, the lateral electric field can be high enough that the carriers suffer from velocity saturation even before the drain current saturates. This is especially an issue for MODFETs due to high mobilities achieved in the 2-D electron gas since high mobility also implies low E_c. With velocity saturation, the drain current saturates at a drain bias of:

$$V_{D,\,sat} \approx E_c L. \tag{3.71}$$

The drain current is then given by:

$$I_{D,\,sat} = WC_o(V_G - V_T)v_s, \tag{3.72}$$

where v_s is the saturated drift velocity. This yields a transconductance of:

$$g_{m,\,sat} \equiv \frac{dI_{D,\,sat}}{dV_G} = WC_o v_s, \tag{3.73}$$

It is interesting to note that the saturation current is independent of channel length, and the transconductance is independent of both L and V_G. The device, however, benefits from a velocity overshoot that provides higher drive current.

References

Arora, N. (1983). *MOSFET models for VLSI circuits simulation: Theory and practice.* New York: Springer-Verlag Wien.

Grove, A.S. (1967) *Physics and technology of semiconductor devices.* New York: John Wiley & Sons.

Ng, K.K. (1995). *Complete guide to semiconductor devices.* New York: McGraw-Hill.

Tsividis, Y.P. (1999). *Operation and modeling of the MOS transistor.* New York: McGraw-Hill.

Z.H. Liu, et al., Threshold voltage model for deep submicrometer MOSFETs. *IEEE Transactions on Electron Devices 40*, 86–95.

4

Active Filters

Rolf Schaumann
Department of Electrical and Computer Engineering, Portland State University, Portland, Oregon, USA

4.1 Introduction

Electrical filters are circuits designed to shape the magnitude and/or phase spectrum of an input signal to generate a desired response at the output. Thus, a frequency range can be defined as the **passband** where the input signal should be transmitted through the filter undistorted and unattenuated (or possibly amplified); **stopbands** require transmission to be blocked or the signal to be at least highly attenuated. In the frequency domain, the transmission characteristic is described by the **transfer function**:

$$T(s) = \frac{V_{out}}{V_{in}} = \frac{N_m(s)}{D_n(s)} = \frac{b_m s^m + b_{m-1} s^{m-1} + \cdots + b_1 s + b_0}{s^n + a_{n-1} s^{n-1} + \cdots + a_1 s + a_0}. \quad (4.1)$$

The nth-order function $T(s)$ is a ratio of two polynomials in frequency $s = j\omega$. In a passband, the transfer function should be approximately a constant, typically unity (i.e., $|T(j\omega)| \approx 1$), so that the attenuation is $\alpha = -20 \log |T(j\omega)| \approx 0$ dB. In the stopband, we need $|T(j\omega)| << 1$ so that the attenuation α is large, say $\alpha = 60$ dB for $|T(j\omega)| = 0.001$. $N_m(s)$ and $D_n(s)$, with $n \geq m$, are chosen so that the attenuation specification is as prescribed (Schaumann and van Valkenburg, 2001). All numerator coefficients b_j are real and, for stable circuits, all denominator coefficients a_i are positive. As a realistic example, consider a low-pass filter. Its passband attenuation must vary by no more than 0.1 dB in $0 \leq f \leq 21$ kHz; the stopband attenuation must be at least 22 dB in $f \geq 26$ kHz, and it is required to have 12.5-dB low-frequency gain. These requirements call for the fifth-order elliptic function (Zverev, 1967):

$$T(s) = \frac{9.04(s^4 + 2,718.9s^2 + 1,591,230.9)}{s^5 + 40.92s^4 + 897.18s^3 + 28,513.8s^2 + 400,103.6s + 3,411,458.7}, \quad (4.2)$$

where the frequency is normalized with respect to 1 krad/s. A sketch of the function is shown in Figure 4.1.

The denominator polynomial of $T(s)$ in equation 4.1 can be factored to display its n roots, the **poles**, which are restricted to the left half of the s plane. Assuming that n is even and keeping conjugate complex terms together, the factored expression is as written here:

$$D_n(s) = s^n + a_{n-1} s^{n-1} + \cdots + a_1 s + a_0 = \prod_{i=1}^{n/2} \left(s^2 + s\frac{\omega_{0i}}{Q_i} + \omega_{0i}^2 \right). \quad (4.3)$$

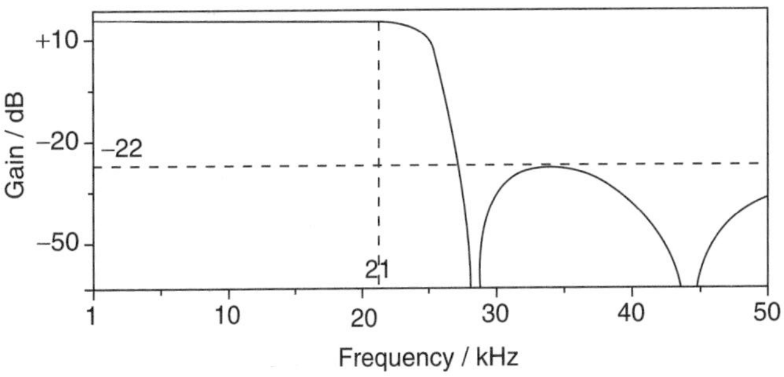

FIGURE 4.1 Sketch of the Elliptic Low-Pass Filter Characteristic of Equation 4.2

The variable ω_{0t} represents the pole frequencies, and Q_i refers to the pole quality factors of the conjugate complex pole pairs:

$$p_i, p_i^* = -\omega_{0i}\left(\frac{1}{2Q_i} \pm j\sqrt{1 - \frac{1}{4Q_i^2}}\right). \tag{4.4}$$

Note that $Q_i > 0$ for stable circuits with poles in the left half of the s plane and that the poles are complex if $Q_i > 1/2$. If n is odd, the product in equation 4.3 contains one first-order term. Similarly, $N_m(s)$ may be factored to give:

$$\begin{aligned} N_m(S) &= b_m s^m + b_{m-1} s^{m-1} + \cdots + b_1 s + b_0 \\ &= \prod_{j=1}^{m/2} (k_{2j} s^2 + k_{1j} s + k_{0j}), \end{aligned} \tag{4.5}$$

where a first-order term (i.e., $k_{2j} = 0$) will appear if m is odd. The roots of $N_m(s)$ are the **transmission zeros** of $T(s)$ and may lie anywhere in the s plane. Thus, the signs of k_{ij} are unrestricted. For example, $k_{1j} = 0$ for transmission zeros on the $j\omega$ axis. Assuming now $m = n$, the following is true:

$$\begin{aligned} T(s) &= \frac{N_m(s)}{D_n(s)} = \frac{\prod_{j=1}^{m/2} (k_{2j} s^2 + k_{1j} s + k_{0j})}{\prod_{i=1}^{n/2} (s^2 + s\omega_{0i}/Q_i + \omega_{0i}^2)} \\ &= \prod_{i=1}^{n/2} \frac{k_{2i} s^2 + k_{1i} s + k_{0i}}{s^2 + s\omega_{0i}/Q_i + \omega_{0i}^2} = \prod_{i=1}^{n/2} T_i(s). \end{aligned} \tag{4.6}$$

In case $m < n$, the numerator of equation 4.6 has $(n - m)/2$ factors equal to unity. The objective is now to design filter circuits that realize this function.

For the fifth-order example function in equation 4.2, factoring results in three sections:

$$\begin{aligned} T(s) &= T_1(s) T_2(s) T_3(s) \\ &= \frac{38.73}{s + 16.8} \frac{0.498(s^2 + 29.2^2)}{s^2 + 19.4s + 20.01^2} \frac{0.526(s^2 + 43.2^2)}{s^2 + 4.72s + 22.52^2}, \end{aligned} \tag{4.7}$$

where the gain constants are determined to equalize[1] the signal level throughout the filter and to realize a 12.5-dB passband gain. Obtained were two second-order low-pass functions, each with a finite transmission zero (at 29.2 kHz and at 43.2 kHz), and one first-order low-pass.

[1] Since the signal level that an op-amp can handle without distortion is finite and the circuits generate noise, dynamic range is limited. To maximize dynamic range, the first- and second-order blocks in equation 4.7 are cascaded in Figure 4.4 in the order of increasing Q values, and the gain constants are chosen such that the signal level throughout the cascade stays constant.

4.2 Realization Methods

Typically, filter applications have sharp transitions between passbands and stopbands and have complex poles close to the $j\omega$ axis; that is, their quality factors, Q_i, are large. The realization of such high-Q poles requires circuit impedances that change their values with frequency very rapidly. This problem has been solved traditionally with **resonance**: the implementation requires inductors, L, and capacitors, C, because RC circuits have poles only on the negative real axis in the s-plane where $Q \le 1/2$. Because inductors are large and bulky, LC circuits cannot easily be miniaturized (except at the highest frequencies) so that other approaches are needed for filters in modern communications and controls systems. A popular and widely used solution that avoids bulky inductors makes use of the fact that complex high-Q pole pairs can be implemented by combining RC circuits with **gain**. Gain in active circuits is most commonly provided by the operational amplifier, the op-amp, or sometimes by the operational transconductance amplifier (OTA). It is very easy to see that complex poles can indeed be obtained from **active RC** circuits. Consider an inverting lossy integrator, as shown in Figure 4.2(A), and a noninverting lossless (ideal) integrator, as shown in Figure 4.2(B). Using conductances, $G = 1/R$, and assuming the op-amps are ideal, the lossy and lossless integrators, respectively, realize the functions:

$$\frac{V_B}{V_1} = -\frac{G}{sC + G_q} = -\frac{1}{s\tau + q}$$

and

$$\frac{V_L}{V_B} = \frac{G}{sC} = \frac{1}{s\tau}. \tag{4.8}$$

The integrator time constant is $\tau = RC$ and the loss term is $q = R/R_q$. If these two integrators are connected in a two-integrator loop shown in Figure 4.3, the function realized is:

$$V_L = \left(-\frac{1}{s\tau + q} \times \frac{1}{s\tau}\right)(KV_{in} + V_L)$$

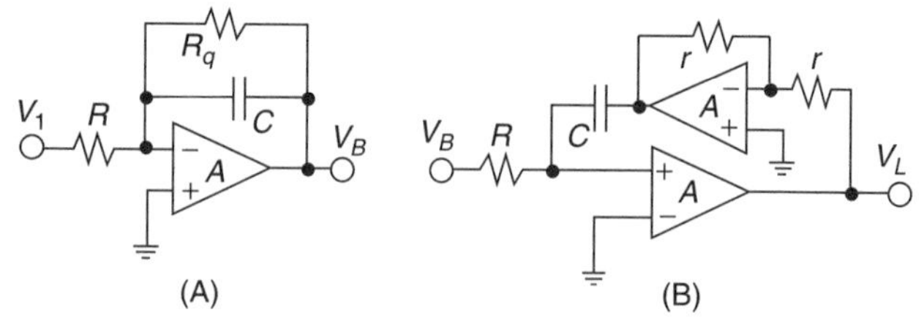

FIGURE 4.2 Integrators (A) Lossy Inverting Integrator (B) Lossless Noninverting Integrator

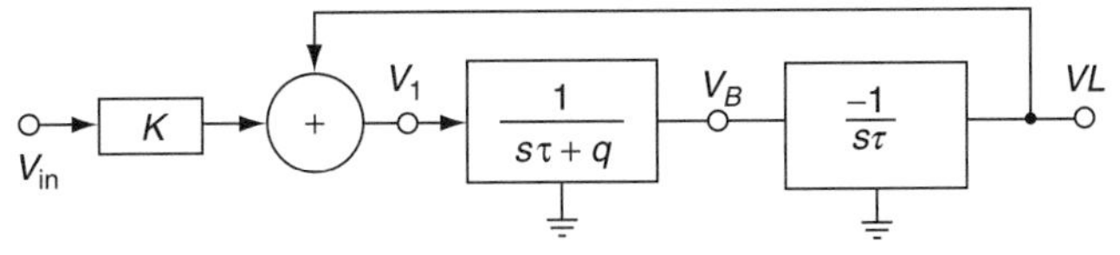

FIGURE 4.3 Two-Integrator Loop Realizing a Second-Order Active Filter (Note that one of the blocks in inverting so that the loop gain is negative for stability).

or

$$\frac{V_L}{V_{in}} = -\frac{K}{s^2\tau^2 + s\tau q + 1}. \tag{4.9a}$$

If we compare the denominator of equation 4.9a with a pole-pair factor in equation 4.3, we notice that the pole frequency is implemented as $\omega_{0i} = 1/\tau = 1/(RC)$, and the pole quality factor is $Q_i = 1/q = R_q/R$. Clearly, we may choose R_q larger than R, so that $Q_i > 1$ and the poles are complex. In addition, a band-pass function is realized at the output V_B because according to Figure 4.3, we have $V_B = s\tau V_L$ to give with equation 4.9a:

$$\frac{V_B}{V_{in}} = -\frac{s\tau K}{s^2\tau^2 + s\tau q + 1}. \tag{4.9b}$$

After this demonstration that inductors are not required for high-Q filters, we need to consider next how to implement practical high-order active filters as described in equation 4.6. We shall discuss the two methods that are widely used in practice: **cascade design** and **ladder simulation**.

When analyzing the circuits in Figure 4.2 to obtain equation 4.8, we did assume **ideal op-amps** with infinite gain A and infinite bandwidth. The designer is cautioned to investigate very carefully the validity of such an idealized model. Filters designed with ideal op-amps will normally not function correctly except at the lowest frequencies and for moderate values of Q. A more appropriate op-amp model that is adequate for most filter applications uses the finite and frequency-dependent gain $A(s)$:

$$A(s) = \frac{\omega_i}{s + \sigma} \approx \frac{\omega}{s}. \tag{4.10}$$

That is, the op-amp is modeled as an integrator. The op-amp's -3-dB frequency, σ, (typically less than 100 Hz), can be neglected for most filter applications as is indicated on the right-hand side of equation 4.10. The ω_t is the op-amp's gain-bandwidth product (greater than 1 MHz). To achieve predictable performance, the operating frequencies of active filters designed with op-amps should normally not exceed about $0.1\omega_t$. The Ackerberg-Mossberg and *GIC* second-order sections discussed below have been shown to be optimally insensitive to the finite value of ω and to behave very well even when designed with real operational amplifiers.

4.2.1 Cascade Design

In cascade design, a transfer function is factored into low-order blocks or sections as indicated in equation 4.6. The low-order blocks are then connected in a chain, or are **cascaded** as in Figure 4.4, such that the individual transfer functions multiply:

$$\begin{aligned}\frac{V_{out}}{V_{in}} &= \frac{V_1}{V_{in}} \times \frac{V_2}{V_1} \times \frac{V_3}{V_2} \times \cdots \times \frac{V_{out}}{V_{n/2-1}} \\ &= T_1 T_2 T_3 \cdots T_{n/2-1} = \prod_{i=1}^{n/2} T_i,\end{aligned} \tag{4.11}$$

as required by equation 4.6. This simple process is valid provided that the individual sections do not interact. Specifically, section T_{i+1} must not load section T_i. In active filters, this is generally not a problem because the output of a filter section is taken from an op-amp output. The output impedance of an op-amp is low, ideally zero, and therefore can drive the next stage in the cascade configuration without loading effects. A major advantage of cascade design is that transfer functions with zeros anywhere in the s-plane can be obtained. Thus, arbitrary transfer functions can be implemented. Ladder simulations have somewhat lower passband sensitivities to component tolerances, but their transmission zeros are restricted to the $j\omega$ axis.

To build a cascade filter, we need to identify suitable second-order (and first-order, if they are present) filter sections that realize the factors in equation 4.6. The first- and second-order functions, respectively, are the following:

$$T_1(s) = \frac{as + b}{s + \sigma} \tag{4.12a}$$

and

$$T_2(s) = \frac{k_2 s^2 + k_1 s + k_0}{s^2 + s\omega_0/Q + \omega_0^2}. \tag{4.12b}$$

In the first-order function, T_1, a and b may be positive, negative, or zero, but σ must be positive. Similarly, in the second-order function, the coefficients k_i can be positive, negative, or zero, depending on where T_2 is to have transmission zeros. For example, T_2 can realize a low-pass ($k_2 = k_1 = 0$), a high-pass ($k_1 = k_0 = 0$), a band-pass ($k_2 = k_0 = 0$), a band-rejection or

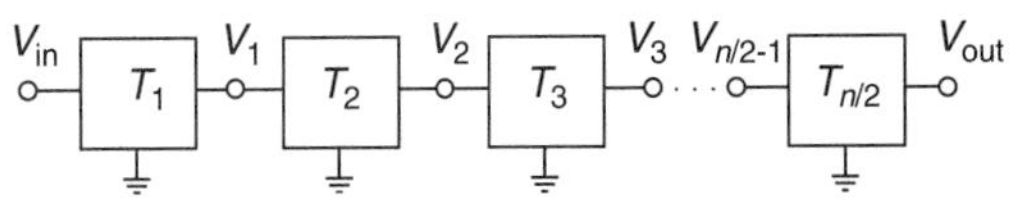

FIGURE 4.4 Realizing a High-Order Filter as a Cascade Connection of Low-Order Sections

"notch" filter ($k_1 = 0$) with transmission zeros at $\pm j\sqrt{k_0/k_2}$, and an all-pass or delay equalizer ($k_2 = 1, k_0 = \omega_0^2, k_1 = -\omega_0/Q$). An example for the choice of coefficients of a first-order low-pass and two second-order notch circuits was presented in equation 4.7.

First-Order Filter Sections

The function T_1 in equation 4.12a is written as:

$$T_1(s) = \frac{V_2}{V_1} = -\frac{sC_1 + G_1}{sC_2 + G_2} = -\frac{(C_1/C_2)s + 1/(C_2R_1)}{s + 1/(C_2R_2)}, \quad (4.13)$$

from which we can identify:

$$a = \frac{C_1}{C_2}, \quad b = \frac{1}{C_2R_1}, \quad \sigma = \frac{1}{C_2R_2} \quad (4.14)$$

Note, however, that the circuit is inverting and causes a (normally unimportant) phase shift of 180°. The example of the first-order low-pass in equation 4.7 is realized by the circuit in Figure 4.5 with $C_1 = 0$, $R_2 = 2.31R_1$, and $C_2R_2 = 1/(16.8\,\text{krad/s})$. The components can be determined if C_2 is chosen (e.g., as 1 nF). We obtain $R_1 = 4.10\,\text{k}\Omega$ and $R_2 = 9.47\,\text{k}\Omega$.

Often, a zero in the right half of the plane is needed (i.e., $b/a < 0$). For this case, the circuit in Figure 4.5(B) can be used. It realizes, with $\sigma = 1/(RC)$:

$$T_1(s) = \frac{V_2}{V_1} = \frac{s - (R_F/R)\sigma}{s + \sigma}. \quad (4.15)$$

For $R_F = R$, the circuit in Figure 4.5(B) realizes a first-order all-pass function that changes the phase but not the magnitude of an input signal.

Second-Order Filter Sections. The literature on active filters contains a large number of second-order sections, the so-called **biquads**, which can be used to realize T_2 in equation 4.12b. Among those, we shall only present three circuits that have proven themselves in practice because of their versatility and their low sensitivities to component tolerances and to finite ω_t-values of the op-amps. The first is the four-amplifier **Ackerberg-Mossberg biquad** that may be called a universal filter because it can implement any kind of second-order transfer function. Then we shall consider the more restrictive two-amplifier **GIC section** that is derived from an **RLC** prototype and has been found to have excellent performance. Finally, we shall present the single-amplifier **Delyiannis-Friend biquad**, which is more sensitive than the previous two circuits but may be used in applications where considerations of cost and power consumption are critical.

The Ackerberg-Mossberg Biquad. The Ackerberg-Mossberg biquad is a direct implementation of the two-integrator loop in Figure 4.3 with the two integrators of Figure 4.2. Adding further a summer that combines all amplifier output voltages weighted by different coefficients, we obtain the circuit in Figure 4.6. Using routine analysis,[2] we can derive the transfer function of the circuit as:

$$\frac{V_{out}}{V_{in}} = -\frac{as^2 + s\omega_0/Q[a - b(kQ)] + \omega_0^2[a - (c - d)k]}{s^2 + s\omega_0/Q + \omega_0^2}. \quad (4.16)$$

The parameters a, b, c, d, and k, and the quality factor Q are given by the resistors shown in Figure 4.6, and the pole frequency equals $\omega_0 = 1/(RC)$. If we compare the function with

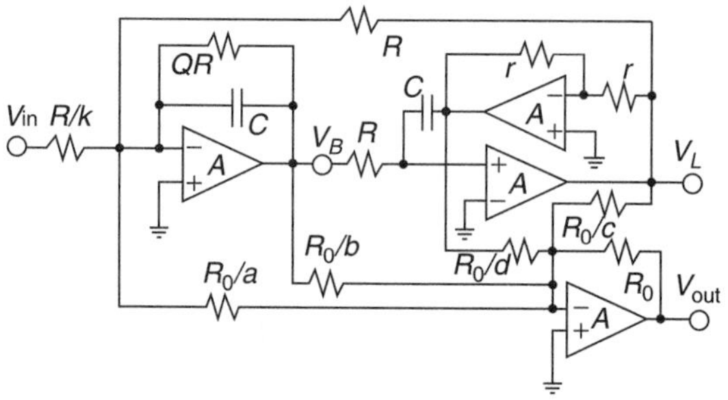

FIGURE 4.6 The Ackerberg-Mossberg Biquad with Output Summer

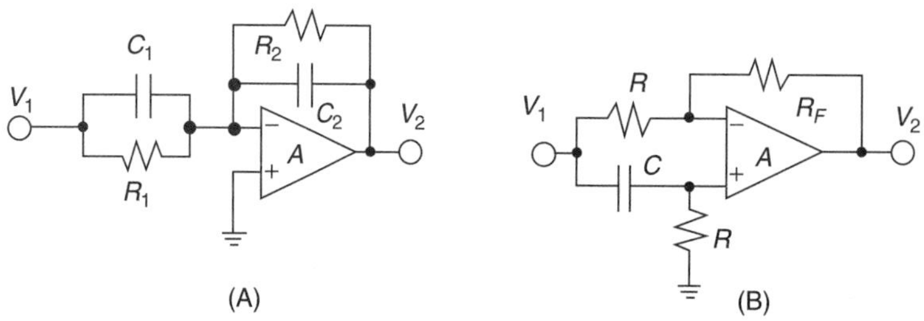

FIGURE 4.5 Active Circuits Realizing the Bilinear Function. (A) Realization for Equation 4.13. (B) Realization for Equation 4.15.

[2] The op-amps are again assumed ideal. This assumption is justified because the combination of the two particular integrators in the Ackerberg-Mossberg circuit causes cancellation of most errors induced by finite ω_t.

T_2 in equation 4.12b, it becomes apparent that an arbitrary set of numerator coefficients can be realized by the Ackerberg-Mossberg circuit. For example, a notch filter with a transmission zero on the $j\omega$ axis requires that $a = bkQ$ by equation 4.16. Notice that the general configuration with the summer is not required for low-pass and band-pass functions: a low-pass function is obtained directly at the output V_L and a band-pass at V_B as we saw in equations 4.9a and 4.9b.

As a design exercise, let us realize T_2 and T_3 in our example function in equation 4.7. We equate formula 4.16 with:

$$T_2(s) = \frac{0.498(s^2 + 29.2^2)}{s^2 + 19.4s + 20.01^2}$$

and

$$T_3(s) = \frac{0.526(s^2 + 43.2^2)}{s^2 + 4.72s + 22.52^2}.$$

With $\omega_{02} = 20.01$ (normalized) and $Q_2 = 20.01/19.4 = 1.03$ for T_2, we find by comparing coefficients that $a = 0.498$, $b = a/(kQ) = 0.498/(1.03 \cdot k)$, and $a - (c - d)k = 1.06$. Choosing for convenience, $k = 1$ and $c = 0$ yields $b = 0.483$ and $d = 0.562$. The choice of $C = 1\,\text{nF}$ leads to $R = 7.95\,\text{k}\Omega$. The value r for the inverter is uncritical, as is R_0; let us pick $r = R_0 = 5\,\text{k}\Omega$; the remaining resistor values are then determined. Similarly, we have for T_3 $\omega_{03} = 22.52$ (normalized) and $Q_3 = 4.77$. We choose again $k = 1$ and $c = 0$ to get $a = 0.526$, $b = a/Q = 0.118$, and $d = 1.41$. The choice of $C = 1\,\text{nF}$ results in $R = 7.07\,\text{k}\Omega$, and $r = R_o = 5\,\text{k}\Omega$ settles the remaining resistor values. If these two second-order sections and the first-order block determined earlier are cascaded according to Figure 4.4, the resulting filter realizes the prescribed function of equation 4.2.

The GIC Biquad A very successful second-order filter is based on the RLC band-pass circuit in Figure 4.7(A) that realizes the following:

$$\frac{V_{out}}{V_{in}} = \frac{G}{G + sC + 1/(sL)} = \frac{sG/C}{s^2 + sG/C + 1/(LC)}. \quad (4.17)$$

Since inductors have to be avoided, circuits that use only capacitors, resistors, and gain were developed whose input impedance looks inductive. The concept is shown in Figure 4.7(B) where the box labeled GIC is used to convert the resistor R_L to the inductive impedance Z_L. The **general impedance converter (GIC)** generates an input impedance as $Z_L = sTR_L$, where T is a time constant. For the inductor L in the RLC circuit, we then can substitute a GIC loaded by a resistor $R_L = L/T$.

One way to develop a GIC is to construct a circuit that satisfies the following two-port equations (refer to Figure 4.7(B)):

$$V_1 = V_2, \quad I_1 = I_2/(sT). \quad (4.18)$$

In that case, we have:

$$Z_L = \frac{V_1}{I_1} = sT\frac{V_2}{I_2} = sTR_L, \quad (4.19)$$

exactly as desired. A circuit that accomplishes this feat is shown in the dashed box in Figure 4.7(C). Routine analysis yields $V_2 = V_1$ and $I_2 = s(C_2R_1R_3/R_4)I_1 = sTI_1$ to give the inductive impedance:

$$Z_L(s) = \frac{V_1}{I_1} = s\left(C_2\frac{R_1R_3}{R_4}\right)R_L = sTR_L. \quad (4.20)$$

Normally one chooses in the GIC band-pass filter identical capacitors to save costs: $C = C_2$. Further, it can be shown that the best choice of resistor values (to make the simulated inductor optimally insensitive to the finite gain-bandwidth product, ω_t, of the op-amps) is the following:

$$R_1 = R_3 = R_4 = R_L = 1/(\omega_0 C_2), \quad (4.21)$$

where ω_0 is a frequency that is critical for the application, such as the center frequency in our band-pass case of Figure 4.7 or the passband corner in a low-pass filter. The simulated inductor is then equal to $L = C_2R_1^2$. Finally, the design is completed by choosing $R = QR_1$ to determine the quality factor.

A further small problem exists because the output voltage in the RLC circuit of Figure 4.7(A) is taken at the capacitor node. In the active circuit, this node (V_1 in Figure 4.7(C)), is not an op-amp output and may not be loaded without disturbing the filter parameters. A solution is readily obtained. The voltage V_{out} in Figure 4.7(C) is evidently proportional to V_1 because $V_2 = V_1$: $V_{out}/V_1 = 1 + R_4/R_L = 2$. Therefore,

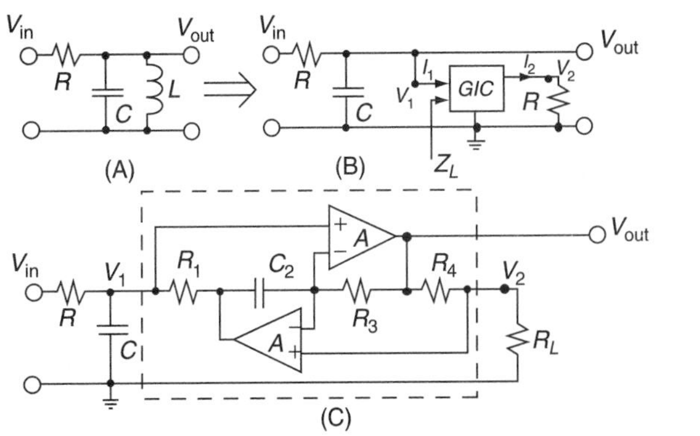

FIGURE 4.7 (A) RLC Prototype Band-Pass Filter (B) Inductor Simulation with a General Impedance Converter (C) The Final GIC Band-Pass Filter with *GIC* Enclosed

we may take the filter output at V_{out} for a preset gain of $V_{out}/V_1 = 2$. The band-pass transfer function realized by the circuit is then

$$\frac{V_{out}}{V_{in}} = \frac{2s/(CR)}{s^2 + s/(CR) + 1/(C^2R_1^2)} = \frac{2s\omega_0/Q}{s^2 + s\omega_0/Q + 1/\omega_0^2}. \quad (4.22)$$

Transfer functions other than a band-pass can also be obtained by generating additional inputs to the GIC band-pass kernel in Figure 4.7(c).

The Single-Amplifier Biquad On occasion, considerations of cost and power consumption necessitate using a single op-amp per second-order section. In that case, the single-amplifier biquad (SAB) of Figure 4.8 may be used. Depending on the choice of components, the circuit can realize a variety of, but not all, different transfer functions. The function realized by this circuit is written as:

$$T(s) = b\frac{s^2 + s\frac{\omega_0}{Q_0}\left[1 + 2Q_0^2\left(1 - \frac{a/b}{1-K}\right)\right] + \omega_0^2}{s^2 + s\frac{\omega_0}{Q_0}\left(1 - 2Q_0^2\frac{K}{1-K}\right) + \omega_0^2}. \quad (4.23)$$

The pole frequency equals $\omega_0 = 1/(C\sqrt{R_1R_2})$. To optimize the performance, $R_2 = 9R_1$ and $Q_0 = 1.5$ are chosen. The realized pole quality factor then becomes:

$$Q = \frac{1.5}{1 - 4.5K/(1-K)}. \quad (4.24)$$

Evidently, Q is very sensitive to the tap position, K, of the resistor R and must be adjusted carefully. This filter is particularly useful for building low-Q gain or delay equalizers to improve the performance of a transmission channel.

4.2.2 Realization of Ladder Simulations

Because of their low passband sensitivity to component tolerances, LC ladders have been found to be among the best performing filters that can be built. Therefore, a tremendous effort has gone into designing active filters that retain the excellent performance of LC ladders without having to build and use inductors. A ladder filter consists of alternating series

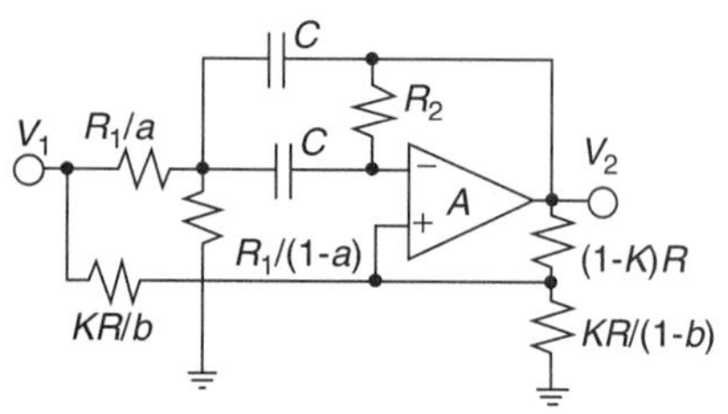

FIGURE 4.8 The Delyiannis-Friend Single-Amplifier Biquad (SAB)

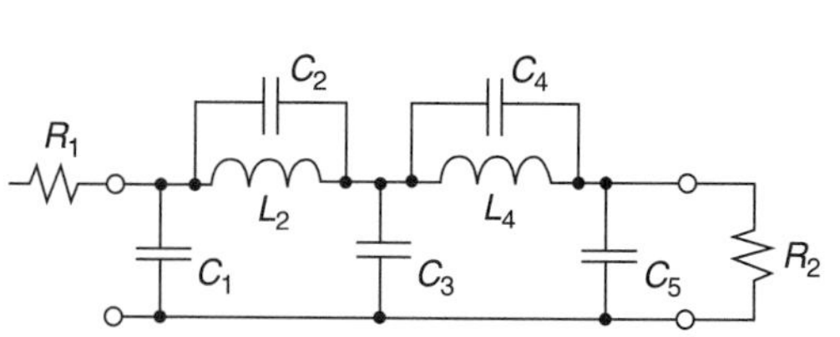

FIGURE 4.9 Typical (Low-Pass) LC Ladder Filter: This circuit is of the form required to realize the transfer function of equation 4.2 with the attenuation curve of Figure 4.1.

and shunt immittance branches that, as the name implies, are lossless and contain only inductors and capacitors. A typical structure of an LC ladder is shown in Figure 4.9. The circuit consists of five ladder arms (C_1, $L_2\|C_2$, C_3, $L_4\|C_4$, and C_5) and source and load resistors R_1 and R_2. It can realize the fifth-order low-pass function of equation 4.2 with the transfer behavior sketched in Fig. 2.6.1. The two transmission zeros at 29.2 kHz and 43.2 kHz are realized by the parallel resonance frequencies of $L_2\|C_2$ and $L_4\|C_4$. The question to be addressed is how to implement such a circuit without the use of inductors.

The problem is attacked in two different ways. The **element replacement method** eliminates the inductors by electronic circuits whose input impedance is inductive over the relevant frequency range. We encountered this method in Figure 4.7, where the need for an inductor was avoided when L was replaced by a GIC terminated in a resistor. The **method of operational simulation** makes use of the fact that inductors and capacitors fundamentally perform the function of integration, 1/s. Specifically, the voltage V across an inductor generates the current $I = V/(sL)$, and the current I through a capacitor is integrated to produce the voltage $V = I/(sC)$. Thus, we should need only to employ electronic integrators, as in Figure 4.2, to arrive at an inductorless simulation of an LC ladder.

The Element-Replacement Method

In a complete analogy to the GIC biquad discussed earlier, we now create the inductive impedance Z_L at the input of a GIC that is loaded by a resistor R [see Figure 4.10(A)]. The GIC circuit used here is shown in Figure 4.10(B). Notice that compared to Figure 4.7(C), we interchanged the capacitor and the resistor in positions 2 and 4. From equation 4.20, the time constant equals $T = C_4R_1R_3/R_2$. This change is immaterial as far as Z_L is concerned, but the selection is optimal for this case. Note that we have attached the label 1:sT to the GIC boxes to help us keep track of their orientation in the following development: the sT side faces the resistive load. It was shown in equation 4.18 that the GIC implements the two-port equations for the element choice in Figure 4.10(B):

$$V_1 = V_2,\ I_2 = C_4\frac{R_1R_3}{R_2} = sTI_1. \quad (4.25)$$

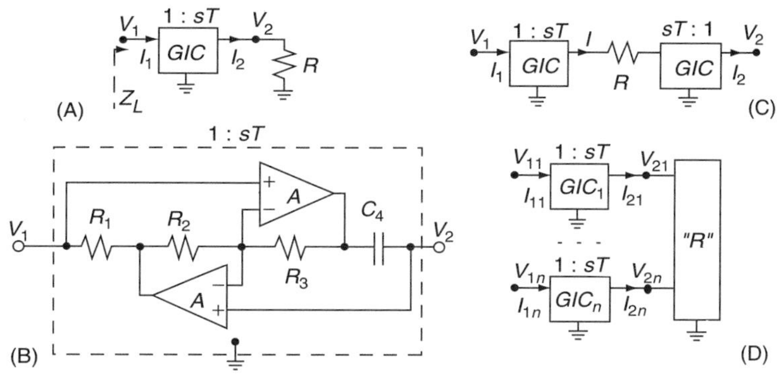

FIGURE 4.10 Circuits for the Element-Replacement Method

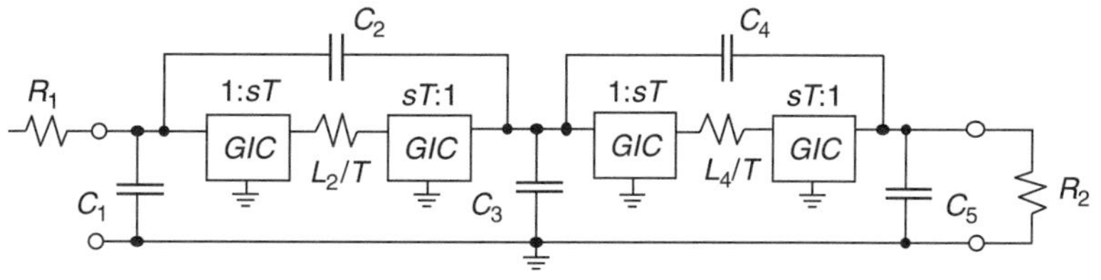

FIGURE 4.11 Inductors in the LC Ladder of Figure 4.9: The inductors simulated by resistors of value $L_i/T, i = 2, 4$, which are embedded between two GICs.

Consequently, the input impedance of the circuit in Figure 4.10(A) is that of a grounded inductor $L = TR$:

$$Z_L = \frac{V_1}{I_1} = sT\frac{V_2}{I_2} = sTR = sL, \tag{4.26}$$

and this scheme can be used to replace any **grounded** inductor in a filter. A **floating** inductor can be implemented by using two GICs as shown in Figure 4.10(C). It is a simple extension of the circuit in Figure 2.10(A), obtained by noting that the resistor is grounded as in Figure 4.10(A) for $V_2 = 0$, according to equation 4.25, input and output voltages of the GIC are the same. The same is true for $V_1 = 0$. Because the voltage $\Delta V = V_1 - V_2$ appears across the series resistor R, we have $I = (V_1 - V_2)/R$. Observing the orientation of the two GIC boxes, we find further $I_2 = I/(sT)$ and:

$$I_1 = I_2 = \frac{V_1 - V_2}{sTR} = \frac{V_1 - V_2}{sL}. \tag{4.27}$$

Evidently, this equation describes a floating inductor of value $L = TR$. This method affords us a simple way to implement floating inductors L by embedding resistors of value L/T between two GICs. Specifically, for the filter in Figure 4.9, we obtain the circuit in Figure 4.11.

We still remark that this method is relatively expensive in requiring two GICs (four op-amps) for each floating inductor and one GIC (two op-amps) for each grounded inductor. A simplification is obtained by the circuit illustrated in Figure 4.10(D). We have shown a resistive network R that has a $1{:}sT$ GIC inserted in each input lead. The network R is described by the equation:

$$\mathbf{V}_{2k} = \mathbf{R}\mathbf{I}_{2k}, \tag{4.28}$$

where $\mathbf{V}_{2k}$ and $\mathbf{I}_{2k}$ are the vectors of the input voltages and currents, respectively, identified in Figure 4.10(D) and $\mathbf{R}$ is the resistance matrix describing the network and linking the two vectors. By equation 4.25, we have from the GICs the relationships $\mathbf{V}_{1k} = \mathbf{V}_{2k}$ and $\mathbf{I}_{2k} = sT\mathbf{I}_{1k}$ so that with equation 4.28, we obtain the result:

$$\mathbf{V}_{1k} = \mathbf{V}_{2k} = \mathbf{R}\mathbf{I}_{2k} = \mathbf{R}sT\mathbf{I}_{1k} = sT\mathbf{R}\mathbf{I}_{1k} \tag{4.29}$$

or

$$\mathbf{V}_{1k} = sT\mathbf{R}\mathbf{I}_{1k} = s\mathbf{L}\mathbf{I}_{1k}. \tag{4.30}$$

This equation implies that a resistive network embedded in GICs appears like an inductive network **of the same topology** with the inductance matrix $\mathbf{L} = T\mathbf{R}$, that is, each inductor L is replaced by a resistor of value L/T. This procedure is the **Gorski-Popiel** method. Its significance is that it permits complete inductive subnetworks to be replaced by resistive subnetworks embedded in GICs, rather than requiring each inductor to be treated separately. Depending on the topology

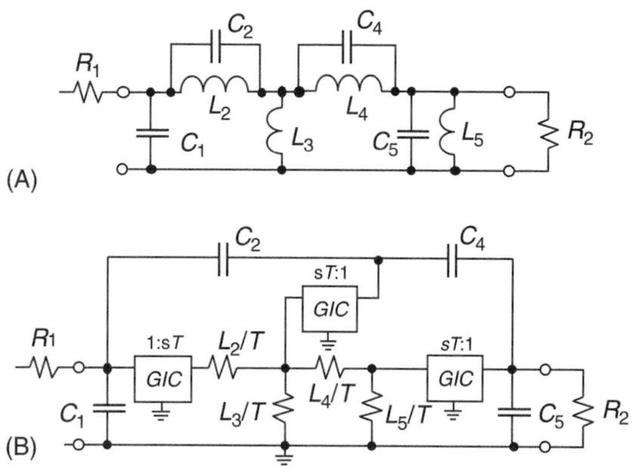

FIGURE 4.12 (A) *LC* Band-Pass Ladder (B) Realization by Gorski-Popiel's Element-Replacement Method

of the *LC* ladder, Gorski-Popiel's method may save a considerable number of GICs, resulting in reduced cost, power consumption, and noise. The simulation of the band-pass ladder in Figure 4.12(A) provides an example for the efficiencies afforded by Gorski-Popiel's procedure. Converting each inductor individually would require six GICs, whereas identifying first the inductive subnetwork consisting of all four inductors leads to only three GICs for the active simulation of the ladder shown in Figure 4.12(B).

Operational Simulation or Signal-Flow Graph Implementation of LC Ladders

As mentioned earlier, in the SFG simulation of an LC ladder filter, each inductor and capacitor are interpreted as signal-flow integrators that can be implemented by an RC/op-amp circuit. The method is completely general in that it permits arbitrary ladder arms to be realized, but in this chapter, we shall illustrate the procedure only on low-pass ladders.

Consider the general section of a ladder shown in Figure 4.13(A). It can be represented by the equations:

$$V_1 = \frac{I_0 - I_2}{Y_1},\quad V_3 = \frac{I_2 - I_4}{Y_3},\quad V_5 = \frac{I_4 - I_6}{Y_5}. \tag{4.31a}$$

$$I_2 = \frac{V_1 - V_3}{Z_2},\quad I_4 = \frac{V_3 - V_5}{Z_4}. \tag{4.31b}$$

For example, in all-pole low-pass filters, all branches Y_i and Z_j are single capacitors and inductors, respectively, so that the five expressions in equations 4.31(A) and 4.31(B) represent integration. Generally, however, Y_i and Z_j may be arbitrary LC immittances. Equations 4.31(A) and 4.31(B) indicate that differences of voltages and currents that need to be formed, which requires more complicated circuitry than simple summing. To achieve operation with only summers, we recast equations 4.31(A) and 4.31(B) as follows by introducing a number of multiplications with (-1):

$$V_1 = \frac{I_0 + (-I_2)}{Y_1},$$
$$(-V_3) = \frac{(-I_2) + I_4}{Y_3},\quad V_5 = \frac{I_4 + (-I_6)}{Y_5}. \tag{4.32a}$$

$$(-I_2) = \frac{V_1 + (-V_3)}{-Z_2},\quad I_4 = \frac{(-V_3) + V_5}{-Z_4}. \tag{4.32b}$$

Figure 4.13(B) shows the resulting block diagram.

To be able to sum only voltages, we next convert the currents in equations 4.32(A) and 4.32(B) into voltages. This step is accomplished by multiplying all currents by a scaling resistor R and, simultaneously, by multiplying the admittances by R to obtain so-called "transmittances" $t_Y = 1/(RY)$ or $t_Z = R/Z$ as is shown in equation 4.33:

$$\begin{aligned}
V_1 &= \frac{RI_0 + (-RI_2)}{RY_1} \Rightarrow v_1 = t_{Y1}[v_{10} + (-v_{12})].\\
(-RI_2) &= \frac{V_1 + (-V_3)}{-Z_2/R} \Rightarrow -v_{12} = -t_{Z2}[v_1 + (-v_3)].\\
(-V_3) &= \frac{(-RI_2) + RI_4}{RY_3} \Rightarrow -v_3 = t_{Y3}[(-v_{12}) + v_{14}].\\
RI_4 &= \frac{(-V_3) + V_5}{-Z_4/R} \Rightarrow v_{14} = -t_{Z4}[(-v_3) + v_5].\\
V_5 &= \frac{RI_4 + (-RI_6)}{RY_5} \Rightarrow v_5 = t_{Y5}[v_{14} + (-v_{16})].
\end{aligned} \tag{4.33}$$

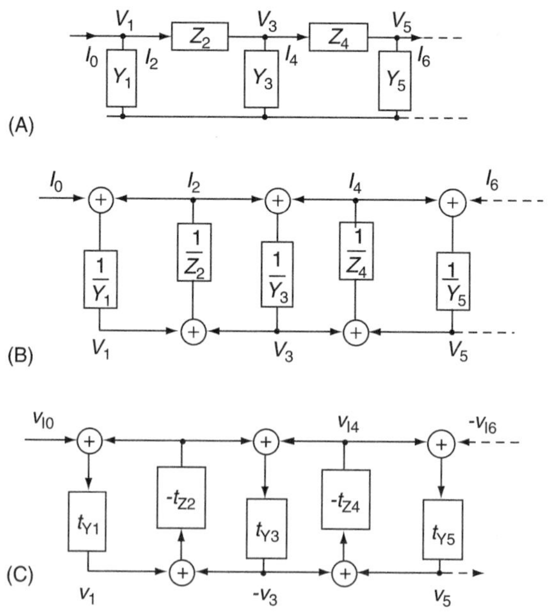

FIGURE 4.13 (A) Section of a General Ladder (B) Signal-Flow Graph Representation for Equation 4.32 (C) Signal-Flow Graph Representation for Equation 4.33

For consistency, we have labeled all voltages, the signals of the SFG implementation, by lowercase symbols and have used the subscripts I, Z, and Y to be reminded of the origin of the dimensionless signals, v, and transmittances, t. Observe that all equations have the same format, $v_k = t_k \times (v_{k-1} + v_{k+1})$, that the signs of all signals are consistent, and (for low-pass ladders) that all t_Z transmittances are inverting and the t_Y transmittances are noninverting integrators: $t_Y = 1/(RY) = 1/(sRC)$, $t_Z = -R/Z = -R/(sL)$. Note also that this process may lead to a normally unimportant sign inversion in the realization. Equations 4.33 are represented by the signal-flow graph in Figure 4.13(C) (Martin and Sedra, 1978). Also observe that inverting and noninverting integrators alternate so that all loops have negative feedback (the loop-gain is negative) and are stable. In low-pass filters, the boxes, together with the summing junctions, form summing integrators that now have to be implemented.

The integrators can be obtained from the circuits n Figure 4.2 by adding a second input. As drawn in Figure 4.13(C), the inverting integrators point upward and the noninverting integrators downward. We used this orientation for the circuits in Figure 4.14. Apart from the sign, both circuits realize the same two-input lossy integrator functions,

$$V_2 = \pm \frac{a_1 V_{11} + a_2 V_{12}}{s\tau + q}, \tag{4.34}$$

where the minus sign is for the inverting integrator in Figure 4.14(A), and the plus sign for the noninverting integrator in Figure 4.14(B). We used a scaling resistor R_a to define $\tau = C_A R_a$, $q = R_a/R_q$, the gain constants $a_1 = R_a/R_1$, and $a_2 = R_a/R_2$ that multiply the two input voltages V_{11} and V_{12}. These gain constants can be adjusted to maximize the dynamic range of the active filter, but the treatment goes beyond the scope of this chapter. We will for simplicity set $a_1 = a_2 = 1$ by selecting $R_a = R_1 = R_2 = R$ to get $\tau = C_A R$ and $q = R/R_q$. Remember still that the LC ladder is lossless. All *internal* integrators are, therefore, lossless (i.e. $q = 0$, $R_q = \infty$). A finite value for R_q is needed only for the first and last ladder arms where R_q serves to realize source and load resistors.

As an example for the signal-flow graph technique, consider the low-pass ladder in Figure 4.15(A). The component values are $R_S = R_L = 1\,\text{k}\Omega$, $C_1 = C_3 = 20\,\text{nF}$, and $L_2 = 30\,\text{mH}$. The filter has a flat passband; the 1-dB passband corner is at 9.6 kHz.

The first step in the realization is a Norton source transform to convert the voltage source into a current source, so that the source resistor is placed in parallel to C_1. This transformation removes one mesh from the circuit and simplifies the design as in Figure 4.15(B). Let us write the describing equations to understand how the components in the active circuit are derived from the ladder. Choosing as the scaling resistor R_S, we obtain

$$V_1 = \frac{R_S(V_{in}/R_S) + (-I_L)R_S}{R_S G_S + sC_1 R_S},$$

$$-I_L R_S = \frac{V_1 + (-V_2)}{sL_2/R_S}, \quad -V_2 = \frac{-I_L R_S}{R_S G_L + sC_3 R_S}.$$

These equations need to be compared with those of the integrators, equation 4.34, to determine the element values. Since $C_1 = C_3$ and $R_S = R_L$, the first and last lossy integrators are the same. The comparison yields $C_1 R_S = C_3 R_S = \tau = C_A R_a$, $q_1 = q_3 = G_S/R_S = 1$, $C_2 R_S = \tau_2 = L_2/R_S$, and $q_2 = 0$. If we select[3] $R_a = R_S = 1\,\text{k}\Omega$, we have $C_1 = C_3 = 20\,\text{nF}$ and $C_2 = L_2/R_S^2 = 30\,\text{nF}$. The value of r is unimportant; let us choose $r = 1\,\text{k}\Omega$ so that all resistors have the same value. The final circuit is shown in Figure 4.15(C). Notice that each loop combines the inverting and noninverting integrators, which we used earlier to construct the Ackerberg-Mossberg

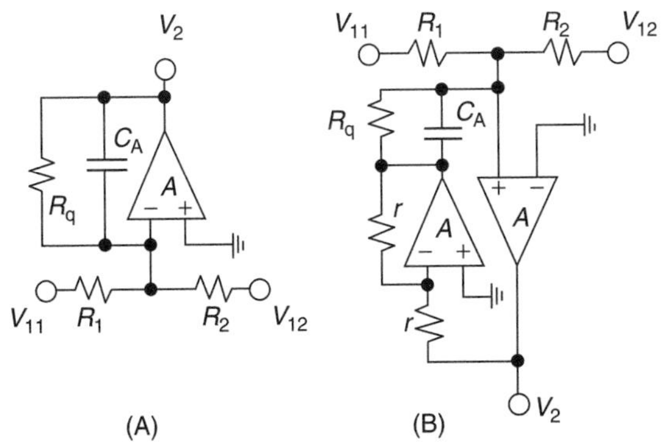

FIGURE 4.14 (A) Inverting Lossy Summing Integrator (B) Noninverting Lossy Summing Integrator

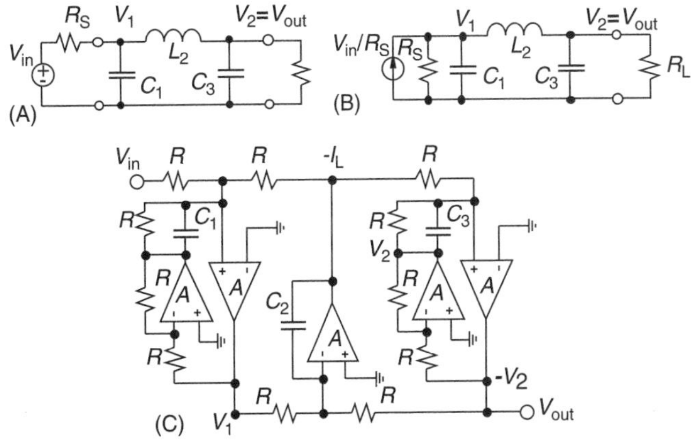

FIGURE 4.15 Realizing a Low-Pass Ladder by the Signal-Flow Graph Technique. (a) The original ladder. (b) Source transformation of the ladder. (c) Final active circuit with $R = 1\,\text{k}\Omega$, $C_1 = C_3 = 20\,\text{nF}$, $C_2 = L_2/R_1^2 = 30\,\text{nF}$.

[3] Alternatively, we could choose different resistor values and arrive at identical capacitors.

biquad (see Figure 4.6 without the summer). As a result, we conclude that the performance of this simulated ladder can be expected to be similarly insensitive to the op-amp's finite gain-bandwidth products. Using 741-type op-amps with $f_t \approx 1.5\,\text{MHz}$, the performance of the active circuit is acceptable to over 300 kHz, and the performance of the two circuits in Figure 4.15(A) and 4.15(C) is experimentally indistinguishable until about 90 kHz when the op-amp effects begin to be noticeable. As a final comment, observe that the output V_{out} in the active realization is inverting. If this phase shift of 180° is not acceptable, the output may be taken at the node labeled V_2.

4.2.3 Transconductance-*C* (OTA-*C*) Filters

The finite value of the op-amp's gain-bandwidth product ω_t causes the frequency range of the active filters discussed so far to be limited to approximately $0.1\,\omega_T$. Using inexpensive op-amps in which f_t is of the order 1.5 to 3 MHz is not sufficient because the frequency range is too low for most communications applications. Although relatively economical wide-band amplifiers are available with f_t values up to about 100 MHz, a different solution is available that is especially attractive for filter circuits compatible with integrated-circuit (IC) technology. For these cases, gain is not obtained from operational voltage amplifiers but from operational transconductance amplifiers, also labeled OTAs or transconductors. These circuits are voltage-to-current converters characterized by their transconductance parameter g_m that relates the output current to the (normally differential) input voltage:

$$I_{\text{out}} = g_{\text{m}}(V_{\text{in}}^{+} - V_{\text{in}}^{-}). \tag{4.35}$$

Simple transconductors with bandwidths in the GHz range have been designed in standard CMOS technology (Szczepanski *et al.*, 1997); in turn, they permit communication filters to be designed with operating frequencies in the hundreds of megahertz. There are several additional reasons for the popularity of the use of transconductors in filter design. Among them are the small OTA circuitry, the particularly easy filter design techniques that normally permit all OTA cells to be identical, and the compatibility with digital (CMOS) integrated-circuit technology: "OTA-C" filters use no resistors.

A small-signal model of the OTA is shown in Figure 4.16(A) including parasitic input and output capacitors (of the order 0.01 to 0.05 pF) and the finite output resistance (of the order of 0.1 to 2 MΩ). We have shown the lower output terminal grounded because for simplicity we shall assume that designs are single-ended for the sake of this discussion. In practice, IC analog circuitry is designed in differential form for better noise immunity, linearity, and dynamic range. Conversion to differential circuitry is straightforward. The customary circuit symbol for the OTA is shown in Figure 4.16(B).

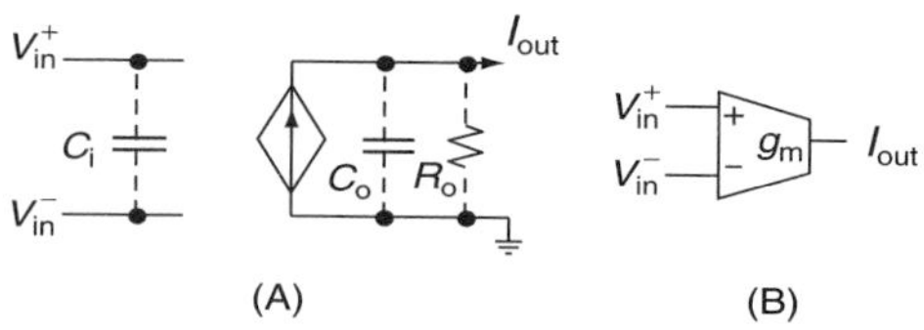

FIGURE 4.16 (A) Small-Signal Model of a Transconductor: $I_{out} = g_m(V_{\text{in}}^{+} - V_{\text{in}}^{-})$ (B) Circuit Symbol

Apart from minor differences having to do largely with IC implementation, active filter design with OTAs is the same as filter design with op-amps. As before, we have available the cascade method with first and second-order sections, or we can use ladder simulations. For the latter, it can be shown that element replacement and the signal flow-graph technique lead to identical OTA-*C* filter structures, so only one of the two methods needs to be addressed.

Cascade Design

The only difference between this cascade design method and the one presented earlier (in Section 4.2.1) is the use of only OTAs and capacitors. In this case, the filter sections do not normally have low output impedance, but the input impedance[4] is as a rule very large, so cascading remains possible. General first- and second-order sections are shown in Figure 4.17. The circuit in Figure 4.17(A) realizes the function:

$$T_1(s) = \frac{V_{\text{out}}}{V_{\text{in}}} = \frac{saC + g_{\text{m}1}}{sC + g_{\text{m}2}}, \tag{4.36}$$

and the second-order section in Figure 4.17(B) implements:

$$T_2(s) = \frac{V_{\text{out}}}{V_{\text{in}}} = \frac{s^2 bC_1C_2 + s(bC_2 g_{\text{m}2} - aC_1 g_{\text{m}3}) + g_{\text{m}1}g_{\text{m}3}}{s^2 C_1C_2 + sC_2 g_{\text{m}2} + g_{\text{m}3}g_{\text{m}4}}. \tag{4.37}$$

Normally, one selects $g_{\text{m}3} = g_{\text{m}4} = g_{\text{m}}$; the pole frequency and pole quality factor are then:

$$\omega_0 = \frac{g_{\text{m}}}{\sqrt{C_1C_2}}, \quad Q = \frac{\omega_0 C_1}{g_{\text{m}2}} = \sqrt{\frac{C_1}{C_2}\frac{g_{\text{m}}}{g_{\text{m}2}}}. \tag{4.38}$$

Since ω_0 is proportional to g_{m}, which in turn depends on a bias current or voltage, the pole frequency can be tuned by varying the circuit's bias. Automatic electronic tuning techniques have been developed for this purpose. In addition, note that *Q*, as a dimensionless parameter, is set by ratios of like components

[4] The input of an OTA-*C* filter is normally a parasitic capacitor ($\approx 0.05\,\text{pF}$) of a CMOS transistor. Thus, even at 100 MHz, the input impedance is still larger than 30 kΩ. Furthermore, as the topology of the first- and second-order circuits in Figure 4.17 indicates, these input capacitors can be absorbed in the circuit capacitors of the previous filter section.

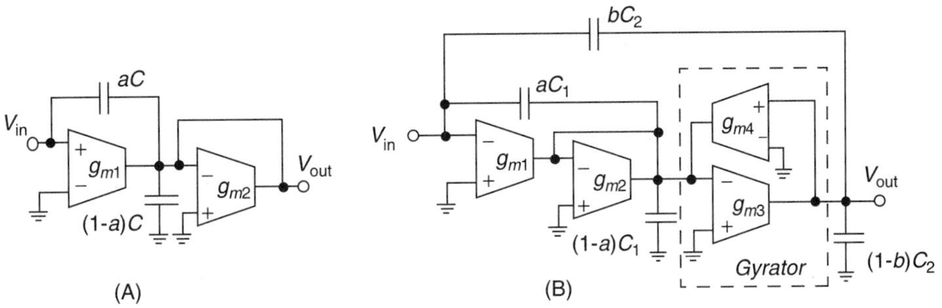

FIGURE 4.17 General OTA-*C* Filter Sections. (A) First-order. (B) Second-order.

and, therefore, is designable quite accurately. Difficulties arise only for large values of Q where parasitic effects can cause large deviations and tuning must be used.

Evidently, a variety of transfer functions can be realized by choice of the components, the coefficients a and b, and the values and signs of the transconductances, g_{mi}. For instance, comparing equation 4.37 with T_2 in equation 4.12 shows that the second-order functions can be realized with arbitrary coefficients. Negative values of g_{mi} are obtained by simply inverting the polarity of their input connections; this must be done with caution, of course, to make certain that the transfer functions' denominator coefficients stay positive.

An interesting observation can be made about the two circuits in Figure 4.17. Notice that all internal nodes of the circuits are connected to a circuit capacitor. This means that the input and output capacitors of all transconductance cells (see Figure 4.16(A)) can be absorbed in a circuit capacitor and do not generate parasitic poles or zeros, which could jeopardize the transfer function. The designer must, however, "predistort" the capacitors by reducing their nominal design values by the sum of all parasitic capacitors appearing at the capacitor nodes. Thus, if a high-frequency design calls for $C = 1.1\,\text{pF}$ and the sum of all relevant parasitics equals 0.2 pF, for example, the circuit should be designed with $C = (1.1 - 0.2)\text{pF} = 0.9\,\text{pF}$. The unavoidable parasitics will restore the effective capacitor to its nominal value of 1.1 pF.

An additional interesting observation can be made related to the structure of the second-order circuit in Figure 4.17(B). We have labeled the circuit in the dashed box a **gyrator**. A gyrator is a circuit whose input impedance is proportional to the reciprocal of the load impedance. Figure 4.18(A) shows the situation for a load capacitor. From this circuit, we derive that the input impedance equals:

$$Z_{in} = \frac{V_1}{I_1} = \frac{1}{g_m^2} sC = s\frac{C}{g_m^2} \Rightarrow sL. \qquad (4.39)$$

That is, it realizes an inductor of value $L = C/g_m^2$. Consequently, we recognize that the second-order OTA-*C* circuit is derived from the passive RLC band-pass stage in Figure 4.7(A) in the same way as the GIC filter in Figure

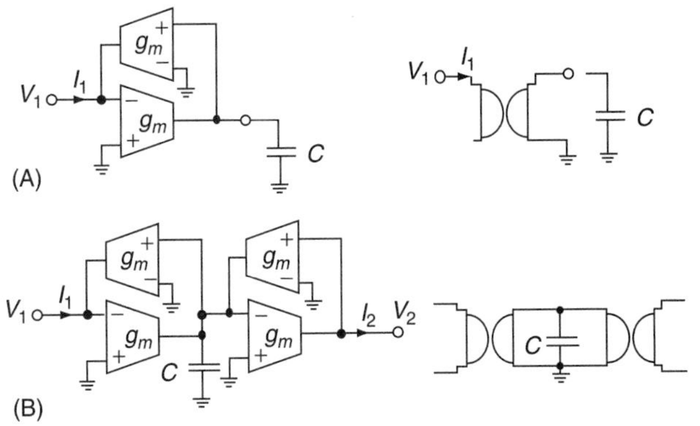

FIGURE 4.18 Capacitor-Loaded Gyrators and their Symbols. (A) Grounded Inductor. (B) Floating Inductor.

4.7(C). Assume[5] for the following discussion that $a = b = 0$ in the circuit in Figure 4.17(B). We observe then that g_{m1} converts the input voltage V_{in} into a current that flows through C_1 and the resistor[6] $1/g_{m2}$. In parallel with the capacitor, C_1, is the inductor $L = C_2/(g_{m3}g_{m4})$. The parallel RLC connection of $1/g_{m2}$, C_1, and $L = C_2/(g_{m3}g_{m4})$ is driven by the current $g_{m1}V_{in}$. The OTA-based grounded inductor can be used in any filter application in the same way as the GIC-based inductor of Figure 4.10(A), except that the OTA-based circuit can be employed at much higher frequencies.

Ladder Simulation

The analogy of the present treatment with the one summarized in Figure 4.10 should let us suspect that floating inductors can be simulated as well by appropriate gyrator connections. Indeed, we only need to connect two gyrators with identical g_m values to a grounded capacitor as shown in Figure 4.18(B) to obtain a floating inductor of value $L = C/g_m^2$. For example,

[5] The fractions a and b of the capacitors C_1 and C_2 are connected to the input to generate a general numerator polynomial from the band-pass core circuit. The method is analogous to the one used in Figure 4.8 for the resistors R_1 and KR.

[6] It is easy to show that a transconductor g_m with its output current fed back to the inverting input terminal acts like a resistor $1/g_m$.

using this approach on the ladder in Figure 4.9 results in the circuit in Figure 4.19, in complete analogy with the GIC circuit in Figure 4.11. Notice that all g_m cells in the simulated ladder are the same, a convenience for matching and design automation. The circuits in Figures 4.11 and 4.19 have identical performance in all respects, except that the useful frequency range of the OTA-based design is much larger. Even when using fast amplifiers with, say, $f_t \approx 100\,\text{MHz}$, the useful operating frequencies of the GIC filter in Figure 4.11 will be less than about 10 MHz, whereas it is not difficult to achieve operation at several 100 MHz with the OTA-based circuit. A signal-flow graph method need not be discussed here because, as we stated, the resulting circuitry is identical to the one obtained with the element replacement method.

The treatment of transconductance-C filters in this chapter has necessarily been brief and sketchy. Many topics, in particular the important issue of automatic tuning, could not be addressed for lack of space. The reader interested in this important modern signal processing topic is referred to the literature in the References section for further details.

References

Martin, K. and Sedra, A.S. (1978). Design of signal-flow-graph (SFG) active filters. *IEEE Transactions on Circuits and Systems 25,* 185–195.

Schaumann, R. (1998). Simulating lossless ladders with transconductance-C circuits. *IEEE Transactions on Circuits and Systems* II *45*(3), 407–410.

Schaumann, R., Ghausi, M.S., and Laker, K.R. (1990). *Design of analog filters: Passive, Active RC, and Switched Capacitor.* Englewood Cliffs, NJ: Prentice-Hall.

Szczepanski, S., Jakusz, J., and Schaumann, R. (1997). A linear fully balanced CMOS OTA for VHF filtering applications. *IEEE Transactions on Circuits and Systems,* II *44,* 174–187.

Schaumann, R., and Valkenburg, M.V. (2001). *Modern active filter design.* New York: Oxford University.

Tsividis, Y., and Voorman, J.A. (Eds.). (1993). *Integrated continuous-time filters: Principles, design and implementations.* IEEE Press.

Zverev, A. (1967). *Handbook of filter synthesis.* New York: John Wiley & Sons.

5

Junction Diodes and Bipolar Junction Transistors

Michael Schröter
Institute for Electro Technology and Electronics Fundamentals, University of Technology, Dresden, Germany

5.1 Junction Diodes

Applications of junction diodes are possible for switches, demodulators, rectifiers, limiters, (variable) capacitors, nonlinear resistors, level shifters, and frequency generation. Special diode types exist that are optimized for the particular application. The cross-section of a discrete diode is shown in Figure 5.1(A); diode structures in integrated circuits also contain a parasitic junction.

5.1.1 Basic Equations

The *pn*-junction and *pn*-diode can be treated theoretically by applying the basic semiconductor equations (Sze, 1981; Selberherr, 1984) consisting of Poisson's equation as well as the continuity equations for electrons and holes; these are complemented by material equations for carrier transport, recombination, and generation. Sufficiently simple analytical equations describing the direct current (dc) and small-signal high-frequency terminal behavior are then obtained by simplifications, such as the **regional approach** (Schilling, 1969). The latter corresponds to a partitioning of the device structure into space-charge regions (SCR) and neutral regions (NR) in which only a subset of the basic equations needs to be solved. V and V_j are the applied dc terminal voltage and the dc voltage across the junction, respectively; a (possible) time dependence is indicated by lowercase letters (i.e., v and v_j).

For the sake of clarity, the regional approach is applied here to the one-dimensional (1-D) structure of a *pn*-junction, assuming an abrupt doping profile as shown in Figure 5.1(B) that also contains the respective dimensions used in the subsequent analysis. The lateral (y-) dimensions of the junction define the diode area A shown in Figure 5.1(A).

Space Charge Region

Contacting a p- and n-doped region leads to a large gradient of the carrier densities at the junction. The result is a diffusion current across the junction and an electric field in the opposite direction, due to the ionized doping atoms left behind, which form the SCR. Therefore, in thermal equilibrium ($V = 0$), the corresponding drift current equals the diffusion current, resulting in a zero net current through the junction. The occurring electric field leads to a built-in voltage:

$$V_D = V_T \ln \frac{N_D N_A}{n_{i0}^2}, \tag{5.1}$$

with V_T as the thermal voltage and n_{i0} as intrinsic carrier density. An applied voltage across the junction (positive in the direction defined in Figure 5.1) either decreases ($V > 0$) or increases ($V < 0$) the internal junction voltage, which is then given by superposition as $V_D - V_j$ (with $V_j = V$ here).

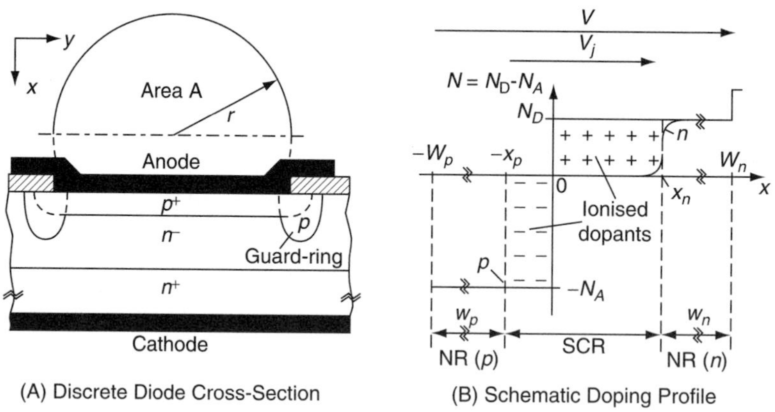

FIGURE 5.1 Junction Diodes. (A) This figure shows a cross-section of a discrete pn-diode with a guard-ring for preventing perimeter breakdown. (B) Illustrated here is a schematic doping profile (assuming an abrupt junction) along a one-dimensional (1-D) cut through the center with definitions of neutral region (NR), space-charge region (SCR), and the respective dimensions.

Important properties of the junction can be calculated from solving Poisson's equation in the SCR (Lindmayer and Wrigley, 1965; Sze, 1981). From this and overall space-charge neutrality, one obtains the width of the SCR

$$w = \sqrt{\frac{2\varepsilon}{q}\left(\frac{N_A + N_D}{N_A N_D}\right)(V_D - V_j)} = w_0\sqrt{1 - \frac{V_j}{V_D}} \qquad (5.2)$$

and the maximum electric field that determines the breakdown characteristics of a diode:

$$E_m = \sqrt{\frac{2q}{\varepsilon}\left(\frac{N_A N_D}{N_A + N_D}\right)(V_D - V_j)}. \qquad (5.3)$$

At a forward bias that is not too high ($V_j < 0.7V_D$), the depletion capacitance can be calculated by considering the SCR as two plates with opposite charges and located the distance w apart:

$$C_j(V) = \frac{\varepsilon}{w}A = A\sqrt{\frac{q\varepsilon}{2}\left(\frac{N_A N_D}{N_A + N_D}\right)\frac{1}{V_D - V_j}} = \frac{C_{j0}}{(1 - V_j/V_D)^z}, \qquad (5.4)$$

with $z = 0.5$ in this case of an abrupt doping profile and with C_{j0} as zero-bias capacitance. For the often occurring case of highly nonsymmetrical profiles (e.g., $N_A \gg N_D$), $w_0 = \sqrt{2\varepsilon V_D/(qN_D)}$, and $C_{j0} = A\sqrt{\varepsilon q N_D/(2V_D)}$. The corresponding depletion charge can be described by:

$$Q_j(V) = \int_0^{V_j} C_j dV = \frac{C_{j0}V_D}{1 - z}\left[\left(1 - \frac{V_j}{V_D}\right)^{1-z} - 1\right]. \qquad (5.5)$$

At high forward bias, the depletion charge eventually becomes completely neutralized and disappears, leading to a drop of $C_j(V_j)$; a typical curve is shown in Figure 5.2(A),

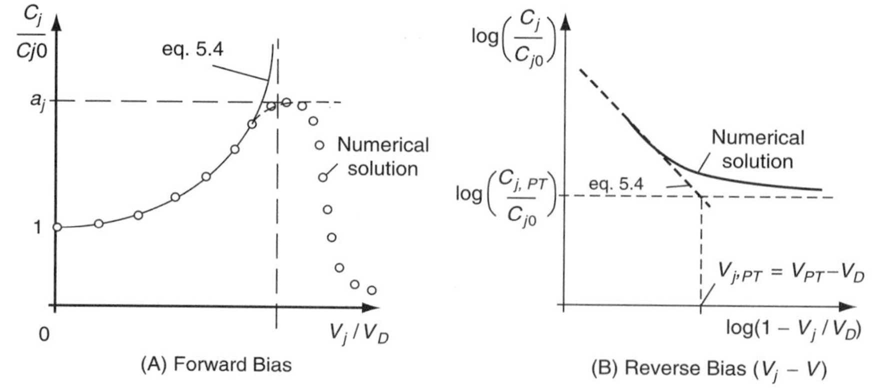

FIGURE 5.2 Normalized Depletion Capacitance Characteristics

containing the numerical solution of Poisson's equation as reference. At high reverse bias, the SCR of the lightly-doped side of the junction extends to the highly doped access region; this **punchthrough effect** results in an (almost) constant depletion capacitance $C_{j\cdot PT} = \varepsilon A/W_n$ (for $N_A \gg N_D$) and is easily visible in a log-plot shown in Figure 5.2(B); $V_{PT} = qN_A W_n^2/(2\varepsilon)$ is the punchthrough voltage.

Current–Voltage Relationship (Direct Current)

Consider, for instance, the NR of the n-side of the junction. With applied bias, $V > 0$ holes are injected into the NR at x_n. Using the law-of-mass and the neutrality condition (for low injection, i.e., $V_j = V$), the injected excess minority (hole) carrier density is $\Delta p_n(x_n) = p_{n0}[\exp(V_j/V_T) - 1]$, with p_{n0} as equilibrium carrier density. Combining continuity and transport equations of holes yields a second-order differential equation for calculating the spatial dependence of Δp_n (Lindmayer and Wrigley, 1965; Sze, 1981). Under the assumption of a finite width $w_n(= W_n - x_n)$ of the NR, the solution is:

$$\Delta p_n = p_{n0}\left[\exp\left(\frac{V_j}{V_T} - 1\right)\right] \left[\sinh\left(\frac{W_n - x}{L_p}\right) / \sinh\left(\frac{w_n}{L_p}\right)\right]. \quad (5.6)$$

Because there is no electric field in the considered case of abrupt doping profiles, the hole current density equals the diffusion component; the current density, injected at x_n, is then given by:

$$J_p(x_n) = -qD_p \left.\frac{dp}{dx}\right|_{x_n} = \frac{qD_p p_{n0}}{L_p}\coth\left(\frac{w_n}{L_p}\right)\left[\exp\left(\frac{V_j}{V_T}\right) - 1\right]. \quad (5.7)$$

A similar expression can be derived for the electron component. Combination yields the well-known diode equation:

$$I_d = I_S\left[\exp\left(\frac{V_j}{mV_T}\right) - 1\right], \quad (5.8)$$

with an ideality factor m (= 1 here) and the saturation current:

$$I_S = Aq\left[\frac{D_p P_{n0}}{L_p}\coth\left(\frac{w_n}{L_p}\right) + \frac{D_n n_{p0}}{L_n}\coth\left(\frac{w_p}{L_n}\right)\right] \quad (5.9)$$

as parameters. For real devices, m is usually slightly larger than one due to recombination loss and nonabrupt doping profiles, for example. A typical characteristic is shown in Figure 5.3 where the various bias regions are indicated. At very low forward bias recombination dominates and results in $m = 2$. At high-current densities and high injection, respectively, deviations from the ideal I–V behavior of equation 5.8 occur because of conductivity modulation, resulting in $m = 2$, as well as because of series resistances (see later sections in this chapter). At certain reverse bias $V < 0$, the current starts to increase again due to avalanche breakdown at lower doping concentrations or tunneling at higher doping concentrations (e.g., Zener diodes) (Sze, 1981).

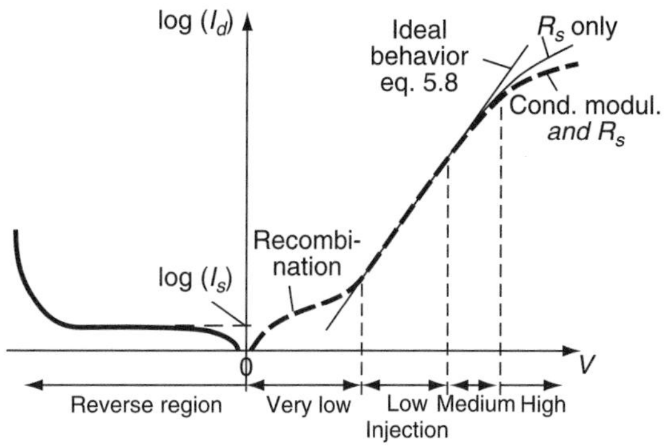

FIGURE 5.3 Typical Current–Voltage Relationship. The V is the terminal voltage applied from anode to cathode.

Diffusion Charge and Capacitance

Minority carriers injected into a NR result in a diffusion charge Q_d and an associated diffusion capacitance of C_d. These are strong functions of bias and can become much larger than Q_j and C_j, respectively, at high forward bias. Combining continuity and transport equation, assuming the small-signal case, with transformation into frequency domain ($\partial/\partial t \rightarrow j\omega$) yields again a second-order differential equation that enables the calculation of the frequency and spatially dependent injected excess carrier densities as well as the corresponding injected small-signal terminal current densities. Adding both the hole and electron current components yields the total small-signal terminal current $I_d(\omega)$. Division by the applied small-signal voltage $V(\omega)$ gives the small-signal admittance $Y_d(\omega) = G_d + j\omega C_d$ of the pn-diode. For the most interesting case of a strongly nonsymmetrical junction (e.g., $N_A \gg N_D$) with a short NR width w_n and frequencies that are not too high (i.e., $j\omega\tau_n \ll 1$), one obtains for the case of low injection (Lindmayer and Wrigley, 1965), assuming $V > 4V_T$, with the dc bias current I_d:

$$G_d = \frac{I_d}{mV_T}$$

and

$$C_d = \frac{w_n^2}{3D_p}G_d. \quad (5.10)$$

The diffusion charge is $Q_d = \tau_d I_d$ if the storage time $\tau_d = w_n^2/(3D_p)$ is current independent.

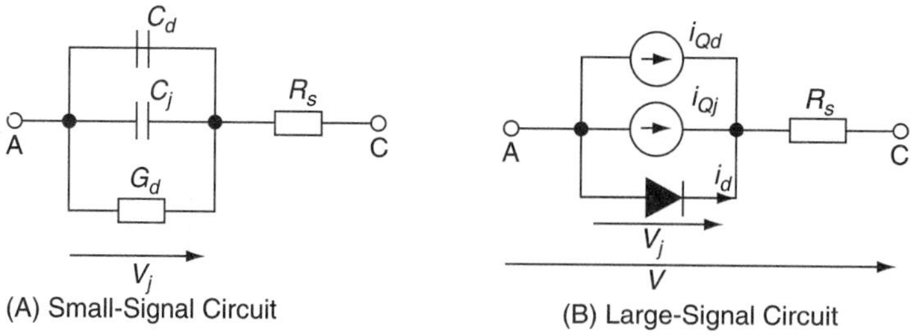

FIGURE 5.4 Equivalent Circuits. (A) This figure shows a small-signal equivalent circuit of a *pn*-diode. (B) Illustrated in this figure is a large-signal equivalent circuit. A and C denote the anode and cathode terminals. $i_{Qd} = dQ_d/dt$ and $i_{Qj} = dQ_j/dt$.

Series Resistances

Both the neutral regions (see Figure 5.1) and the contacts themselves have a finite resistivity, which results in additional parasitic series resistances. Physically, these can be expressed by the sheet resistances of the (access) regions as well as the area-specific contact resistances and the dimensions of the corresponding regions and contacts, respectively.

5.1.2 Equivalent Circuit

A generic small-signal diode-equivalent circuit (EC) is given in Figure 5.4. C_d and the junction conductance G_d have been defined in equation 5.10 and C_j is given by equation 5.4. R_s is the series resistance, which is a lumped representation of all contributing components discussed before. The voltage v_j across the junction is smaller than the applied terminal voltage v due to the voltage drop across R_s. In a realistic diode structure, additional parasitic elements have to be taken into account, such as a sidewall capacitance and—in integrated circuits—a parasitic *pn*-junction that is connected to a different terminal. For tunneling diodes, the effect of a negative differential resistance is usually taken into account in the small-signal EC by an inductance in series to R_s.

5.2 Bipolar Junction Transistor

The generic expression of **bipolar transistor** is used for various types of this device. The most important type is the vertical *npn* bipolar transistor that is available as (a) silicon-based homojunction version (*npn*-BJT) and (b) SiGe or III–V material-based heterojunction version (HBT). Less often employed versions are the lateral *pnp* (LPNP) and vertical *pnp* (VPNP) transistor. In this chapter, emphasis is placed on the *npn*-BJT for presenting a basic theory that can be applied to the other transistor versions with few modifications.

Bipolar transistors are employed wherever speed and drive capability are required. Major application areas include high-speed data links, such as in fiber-optic communications; RF-front-ends, such as mixers and amplifiers in wireless communications; and drivers, bias circuitry, and high-speed logic in general, and in particular in BiCMOS technologies. Both discrete and integrated solutions can be found. An example for a standard integrated BJT implemented in a self-aligning double-polysilicon process is shown in Figure 5.5 (Klose *et al.*, 1993; Yamaguchi *et al.*, 1994). The self-alignment scheme allows submicron emitter window widths b_{E0} with a μm-lithography yielding transit frequencies f_T of 25 GHz and beyond for present production processes. Due to the recent trend to integrate BJTs and silicon-germanium (SiGe) into BiCMOS processes, bipolar transistors benefit from the advanced CMOS lithography tools. The tools enable a significant reduction in parasitics and size and, as a result, the transistors increase in speed ranging from $f_T \approx 30$ GHz (Racanelli *et al.*, 1999) up to ≈ 90 GHz (Freeman *et al.*, 1999), thus possibly even obviating the need for self-alignment schemes. The structure in Figure 5.5 can be divided into the internal transistor T_i (s. blow-up), the peripheral transistor T_p, and the external transistor regions.

Figure 5.6(A) contains the typical doping profile under the emitter for the transistor in Figure 5.5. Two versions are possible: constant collector doping N_C, which yields a high-voltage device, and a selectively implanted collector (SIC), which yields a high-speed device. The transistor speed can be further increased by (a) employing an epitaxial base and (b) adding a graded Ge profile. Both measures significantly reduce the base transit time. A further improvement is achieved by forming a Ge step at the BE junction in Figure 5.6(B) that prevents holes from being back-injected into the emitter and results in a "real" HBT. Here, the BE profile can be inverted without affecting the current gain despite a much higher base doping. The latter allows a significant reduction in base resistance.

The basic operation principle can be derived already for the 1-D structure under the emitter. Classical transistor theory is discussed first; however, it becomes invalid at medium to high current densities because in all practical transistors, $N_C \ll N_B$, so that more general equations are also presented. To allow a generic representation of the typical characteristics and a comparison of technologies, a collector current **density** $J_C = I_C/A_E$ is used in the following considerations, where A_E is the (effective) emitter area that is larger than the emitter window area A_{E0}. The controlling terminal voltages of the 1-D transistor are denoted as $V_{B'E'}$ and $V_{B'C'}$.

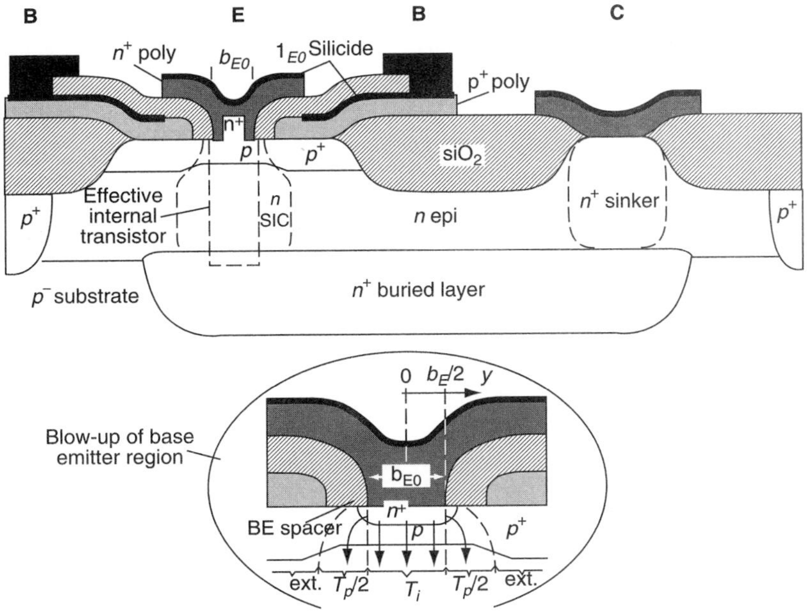

FIGURE 5.5 Schematic Cross-Section of a Self-Aligned Double-Polysilicon Bipolar Transistor Structure. The emitter window area is $A_{E0} = b_{E0|E0}$, and the effective (electrical) internal transistor and its effective emitter width b_E are defined later.

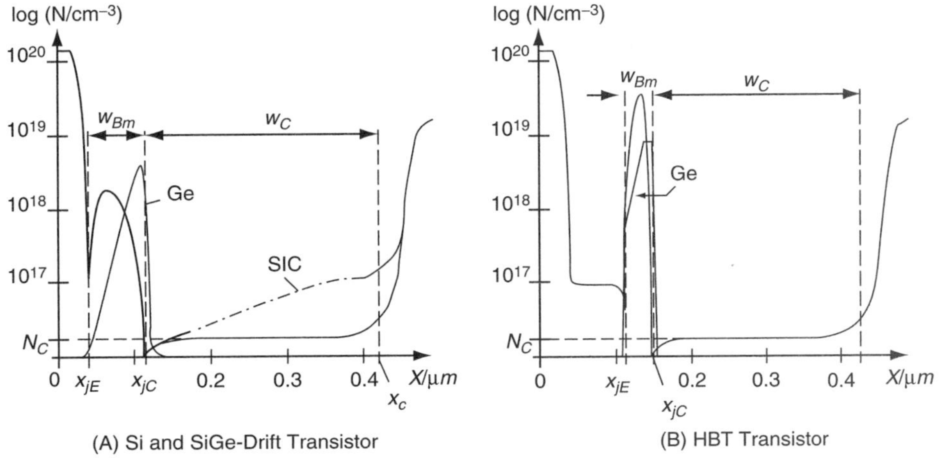

FIGURE 5.6 Typical Doping Profiles Under the Emitter (of the Internal Transistor). (A) This graph shows the standard Si and SiGe-drift transistor (typical Ge content ≈ 7–15%). (B) Shown here is a SiGe HBT (typical Ge content ≈ 20–30%) transistor. The location x = 0 defines the surface of the monosilicon and the E′ terminal; w_{Bm} and w_C are the metallurgical widths of base and collector, respectively; x_c is the C′ terminal. Scales are approximate numbers. The SIC does not exist in the external transistor region. The collector substrate junction is not shown here

5.2.1 Basic Equations

Based on the 1-D structure, the basic BJT action can be explained as follows. For normal forward operation ($V_{B'E'} > 0$ and $V_{B'C'} < 0$), electrons are injected from the emitter across the BE SCR into the neutral base. The carriers traverse the base by a combination of drift and diffusion, the partition of which depends on the electric field in the base, and then enter the BC SCR, where they are pulled toward the collector with high velocity. Since in today's processes, recombination of minorities (electrons in this case) in the base can be neglected, the (1-D) current I_C at the collector ($x = x_c$) equals the current injected into the base and, for useful bias conditions $V_{B'E'}$, even the current entering the BE SCR at the emitter side; this current component is often called forward **transfer current** I_{Tf}. Similarly, holes are injected from the base across the BE SCR into the neutral emitter, constituting the base current I_B; a portion of the associated holes recombine in the emitter (Auger recombination), while the remaining portion arrives at the emitter contact. Under practically useful bias conditions $V_{B'E'}$, recombination in the BE SCR is negligible. The emitter current is then $I_E = I_C + I_B$, and the dc common-emitter current gain is $B = I_C/I_B$, which becomes bias dependent at low and high injection (or J_C).

For the theoretical treatment of the BJT, the basic semiconductor equations can be employed (cf., *pn*-diode). Once again, a regional approach is useful for arriving at reasonably simple analytical equations that describe the electrical behavior. Results of classical transistor theory are presented first to provide a feeling for the basic analytical formulations of various characteristics (Lindmayer and Wrigley, 1965; Pritchard, 1967; Philips, 1962).

Direct Current Behavior of a BJT

It is a fair assumption in present technologies to neglect recombination in the base of a "useful" bipolar transistor. According to equation 5.6, the excess electron density injected into the base of a transistor with a simplified box doping profile $\overline{N}_B$ reads, after series expansion of the sinh:

$$\begin{aligned}\Delta n(x) &= \frac{n_{iB}^2(w_B - x)}{N_B w_B}\left[\exp\left(\frac{V_{B'E'}}{V_T}\right) - 1\right] \\ &= n_{Be}\left(1 - \frac{x}{w_B}\right).\end{aligned} \tag{5.11}$$

The corresponding (quasistatic) electron current flowing through the base to the collector is represented by:

$$i_{Tf} = I_S\left[\exp\left(\frac{v_{B'E'}}{m_f V_T}\right) - 1\right], \tag{5.12}$$

with an ideality coefficient $m_f = 1$ for the considered assumptions. The saturation current is as follows:

$$I_S = \frac{qA_E D_n n_{iB}^2}{N_B w_B}. \tag{5.13}$$

The current depends on base doping and neutral base width. The variation of the latter with the applied voltages $v_{B'C'}$ and $v_{B'E'}$ is known as forward and reverse **Early effect**. In most transistor models, $m_f > 1$ is supposed to account for the reverse Early effect (variation of w_B with $v_{B'E'}$), whereas the $v_{B'C'}$ dependence is approximated analytically, such as $w_B = w_{B0}(1 + v_{B'C'}/V_{Ef})$ with w_{B0} as neutral base width in equilibrium and V_{Ef} as forward **Early voltage**.

At high-injection, a solution for i_{Tf} exists that is similar to the one above, but with $2D_n$ and $m_f = 2$ rather than D_n and $m_f = 1$. Unfortunately, such a discontinuous description of the transfer current is not very useful for compact models and circuit simulation.

In classical theory, similar equations as above are also derived for reverse operation, leading to $i_{Tr} = -I_S(\exp[v_{BCi}/(m_r V_T)] - 1)$, which is superimposed with i_{Tf} to yield i_T. Since in practically useful transistors $N_C \ll \overline{N_B}$, the control voltage v_{BCi} differs from the terminal voltage $v_{B'C'}$ by a voltage drop that results from the strongly nonlinear bias dependent internal epicollector resistance. Similarly, the base charge and transit time are controlled by v_{BCi}, which is very difficult to model accurately. This together with the high-injection problem mentioned before requires an improvement of classical theory, which is described next.

"Modern" bipolar transistor models employ, one way or the other, the integral charge-control relation (ICCR), which was first proposed by Gummel (1970). Under the assumption of a negligible time dependence $\partial n/\partial t$ in the 1-D electron continuity equation, the (electron) transfer current from the emitter contact $x = 0$ to the internal collector contact at x_c can be written in a simple form (Rein *et al.*, 1985):

$$i_T = I_S\frac{\exp(v_{B'E'}/V_T) - \exp(v_{B'C'}/V_T)}{Q_p/Q_{p0}} = i_{Tf} - i_{Tr}, \tag{5.14}$$

with the saturation current:

$$I_S = (qA_E)^2 V_T \frac{\overline{\mu_n n_i^2}}{Q_{p0}}. \tag{5.15}$$

The hole charge at equilibrium ($V_{B'E'} = V_{B'C'} = 0$) is:

$$Q_{p0} = qA_E\int_0^{x_c} p_0\,dx \approx qA_E\int N_B dx = qA_E\overline{N_B}w_{B0}. \tag{5.16}$$

The last two terms containing the base doping concentration N_B and its average value $\overline{N_B}$, respectively, and the zero-bias width w_{B0} of the neutral base show the correspondence to the classical solution of equation 5.13.

The bias-dependent total hole charge in the (1-D) transistor is as follows:

$$Q_p = Q_{p0} + Q_{jEi} + Q_{jCi} + Q_f + Q_r. \tag{5.17}$$

This equation 5.17 can be expressed by several components, namely the base-emitter depletion charge and base-collector depletion charge, respectively:

$$Q_{jEi} = \int_0^{V_{B'E'}} C_{jEi}dV \quad\text{and}\quad Q_{jCi} = \int_0^{V_{B'C'}} C_{jCi}dV, \tag{5.18}$$

as well as the "forward" and "reverse" minority charge:

$$Q_f = \int_0^{V_{B'E'}} C_{dE} dv = \int_0^{I_{Tf}} \tau_f di \quad \text{and}$$
$$Q_r = \int_0^{V_{B'C'}} C_{dC} dv = \int_0^{I_{Tr}} \tau_r di. \tag{5.19}$$

In the above integrands, C_{jEi} and C_{jCi} are the BE and BC depletion capacitance, C_{dE} and C_{dC} are the BE and BC diffusion capacitance, and τ_f and τ_r are the minority storage times, which are often called transit times. The latter are strongly bias-dependent small-signal quantities, which will be discussed later in more detail. From a comparison of equations 5.14 and 5.12, it becomes obvious that there is no single Early voltage (i.e., the Early voltage is bias-dependent and modeled by Q_{jCi}).

At this point, it is important to note that the ICCR permits a self-consistent description of nonideal effects via those small-signal quantities that can (a) be measured quite easily with standard methods and equipment and (b) also determine the dynamic (i.e., frequency and large-signal transient) behavior of the transistor. Moreover, from a circuit simulation and numerical point of view, the ICCR provides the basis of a continuous description of the transfer current via a single-piece equation *if* the charges are modeled continuously differentiable. For Si-based HBTs, certain extensions of equation 5.17 are required, which are discussed in Schröter *et al.* (1993) and Friedrich and Rein (1999).

Figure 5.7(A) shows the schematic dependence of the forward dc transfer current density on $V_{B'E'}$. In a "useful" process, I_T/A_E behaves almost ideally at low current densities and is often described in compact models in the form:

$$i_T \cong I_S \exp\left(\frac{v_{B'E'}}{m_f V_T}\right), \tag{5.20}$$

where m_f is an ideality factor close to 1. For Si bipolar processes, however, this equation is only valid over a fairly small bias range because m_f changes with bias according to the ICCR. At high current densities, the curve bends downward due to so-called high-current effects that are discussed later in more detail.

The base current i_{jBEi} across the BE junction of a modern bipolar transistor can be described by two components:

$$i_{jBEi} = I_{BEiS}\left[\exp\left(\frac{v_{B'E'}}{m_{BEi}V_T}\right) - 1\right] + I_{REiS}\left[\exp\left(\frac{v_{B'E'}}{m_{REi}V_T}\right) - 1\right], \tag{5.21}$$

which are shown in Figure 5.7(A). The first component represents the back injection into the (neutral) emitter, with an ideality factor m_{BEi} that is usually slightly larger than one. This component dominates at low- to high-current densities and remains unaffected by high-current effects for many processes; that is, it keeps increasing almost ideally with $V_{B'E'}$. The second component represents the recombination loss in the BE SCR, with an ideality factor m_{REi} that is close to two. This component dominates for Si BJTs only at very low-current densities. The component i_{jBCi} across the BC junction can be described similarly. The total internal dc base current is then $I_{Bi} = I_{jBEi} + I_{jBCi}$.

The dc current gain B is (obviously) a derived quantity rather than a basic physical parameter of a transistor. Therefore, for modeling purposes, it is easier to describe separately the collector and base current rather than B, as is still often attempted in compact models. For circuit design purposes, only an average value of B is usually required and can be readily provided.

Depletion Capacitances

For reverse and forward biases the capacitances C_{jEi} and C_{jCi} follow the classical voltage dependence given in equation 5.4. For typical capacitance curves as a function of bias, see Figure 5.2. In BJTs, the correct description of C_{jEi} is of most importance for the forward bias region up to the peak value

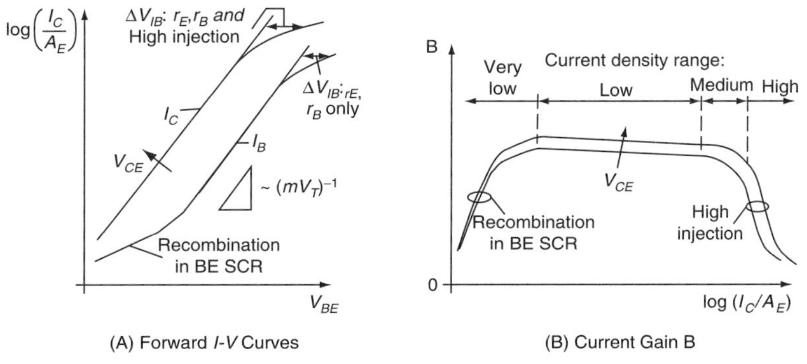

FIGURE 5.7 Sketch of the dc Characteristics and Definition of Various Bias Regions. (A) This schematic diagram shows forward $I - V$ curves with total base current I_B (see later). (B) This diagram shows current gain B. The arrow indicates an increase in V_{CE}. Typical units: I_C/A_E in $mA/\mu m^2$.

($a_j C_{jEi0}$), since C_{jEi} strongly determines the charge storage and dynamic behavior in that region. At higher bias, the diffusion capacitance starts to dominate, and the exact modeling of C_{jEi} becomes academic. C_{jEi} is modeled differently in the various transistor models (Schröter, 1998). For the above mentioned reasons, a voltage-independent approximation after the peak (shown by dashed line in Figure 5.2(A)) is sufficient and numerically efficient as well as consistent with the determination of τ_f from f_T. An accurate description of C_{jCi} is usually most important at reverse and low forward bias of the BC junction. For transistors with a lightly doped collector (around $10^{16}\,\text{cm}^{-3}$), the SCR can already punch through the collector at quite low reverse bias, leading to an almost voltage-independent value of C_{jCi} and, therefore, a deviation from the classical equation (refer to Figure 5.2(B). At high collector current densities, the mobile carriers start to compensate for the ionized doping in the collector. As a result, the electric field distribution in the BC SCR changes, and C_{jCi} becomes dependent on i_T in a complicated way. Since in this bias region the minority charge storage dominates the dynamic behavior, an accurate modeling of the above effect on C_{jCi} is usually of little importance for practical applications.

Minority Charge and Transit Time

Classical theory assumes that minority charge storage only occurs in the neutral base; according to equation 5.11, the injected electron density decreases linearly, leading to the stored minority charge $Q_{nB} = qA_E n_{Be}(w_B/2)$. Derivative with respect to I_{Tf} gives the base transit time of a diffusion transistor (i.e., a transistor with a spatially constant base doping) operated at low current densities:

$$\tau_{nB} = \frac{w_B^2}{2D_{nB}}. \tag{5.22}$$

In advanced transistors, τ_{nB} is about two to three times smaller due to a built-in drift field and, thus, is only a portion of the total transit time τ_{f0}. The result of a more detailed analysis, which additionally takes into account both Early effect and delay time through the BC SCR, yields (Schröter, 1998).

$$\tau_{f0} = \tau_0 + \Delta\tau_{0h}(c-1) + \tau_{Bfvl}\left(\frac{1}{c} - 1\right), \tag{5.23}$$

with $c = C_{jCi0}/C_{jCi}(V_{B'C'})$ and τ_0, $\Delta\tau_{0h}$, and τ_{Bfvl} as time constants. Figure 5.8(A) shows the possible voltage dependence of τ_{f0} for the two different cases mentioned previously.

With increasing collector current density, so called high-current effects (e.g., high-injection in the collector region) occur, leading to additional charge storage (Kirk, 1962; Schröter and Lee, 1999). Figure 5.8(B) contains typical characteristics of the transit time $\tau_f = \tau_{f0}(V_{C'B'}) + \Delta\tau_f(I_C, V_{C'E'})$ as a function of collector current density. The sharp increase of τ_f due to high-current effects can be described by a critical current I_{CK} (Schröter and Walkey, 1996) that also serves as a useful aid for designing high-speed switching circuits (Rein, In press). A rough approximation of the corresponding maximum value of the τ_f increase is $\Delta\tau_{f,\,\max} \approx w_C(w_{Bm} + w_C/2)/(2D_{nC})$. (Kirk, 1962). Beyond I_{CK}, a drop in current gain and transit frequency f_T also occur. The latter can be used to determine $\tau_f(I_C, V_{C'E'})$. (Rein, 1983). The total minority charge for forward also operation is given by $Q_f(I_C, V_{C'E'}) = \int_0^{I_{Tf}} \tau_f\, di$ (= area under the τ_f curve), from which the BE diffusion capacitance can be calculated as $C_{dE} = \partial Q_f/\partial V_{B'E'} = \tau_f g_m$ with the transconductance $g_m = \partial I_T/\partial V_{B'E'}$ at constant $V_{C'E'}$.

5.2.2 Internal Transistor and Base Resistance

The base region under the emitter has a finite resistivity that is characterized by the internal base sheet resistance r_{SBi}. The bias dependence of the latter can be reasonably well approximated by:

$$r_{SBi} = r_{SBi0}\, Q_{p0}/Q_p, \tag{5.24}$$

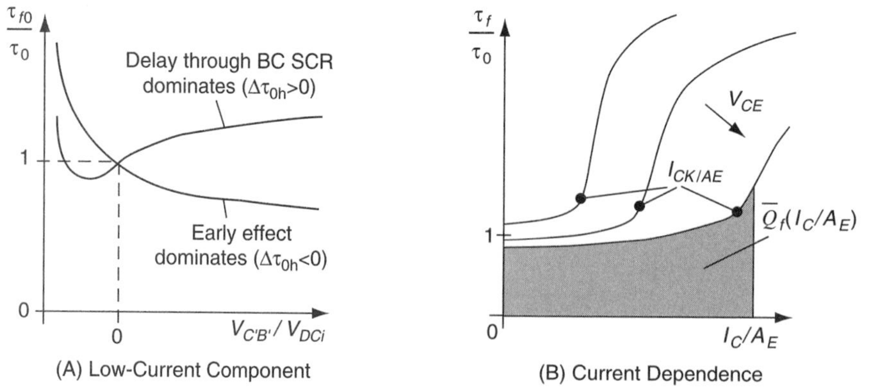

FIGURE 5.8 Bias Dependence of the Forward Transit Time. (A) This graph shows the low-current component for different cases ($V_{C'B'} = V_{CB}$). (B) Notice the current dependence shown here; high-current effects start for $I_C > I_{CK}$. Typical units: τ_f in ps; $\overline{Q}_f$ in fC/μm^2.

with r_{SBi0} as zero-bias value. This automatically includes changes due to the changes of the SCRs and conductivity modulation at high collector current densities (Rein and Schröter, 1991).

The base current entering the internal base region horizontally from the side at $y = \pm b_E/2$ (see Figure 5.5) causes a voltage drop in the direction of the center ($y = 0$) that is called **emitter current crowding**. Because the hole injection into the emitter is distributed laterally, the associated lumped internal base resistance r_{Bi} has to be calculated from a differential equation (Ghosh, 1965; Rein and Schröter, 1989). The formulation for r_{Bi} depends on both the mode of operation (e.g., dc transient and small-signal low or high-frequency) and the emitter geometry (e.g., rectangular, square, and circular). The various cases are discussed below.

Direct Current Operation ($\omega = 0$)

A set of equations describing the geometry dependence and emitter current crowding is given in (Schröter, 1991). The relations are based on analytical solutions for both a long rectangular emitter stripe (used in high-speed applications, for example) and a small square emitter geometry (used in low-power applications, for example):

$$r_{Bi} = r_{SBi}\frac{b_E}{l_E}g_i(b_E, l_E)\psi(\eta), \tag{5.25}$$

where the basic expression for the dc current crowding function and factor, respectively, are taken from the square-emitter case for simplicity reasons:

$$\psi(\eta) = \frac{\ln(l + \eta)}{\eta} \quad \text{and} \quad \eta = \frac{r_{SBi}}{g_\eta(b_E, l_E)}\frac{b_E}{l_E}\frac{I_{Bi}}{V_T}. \tag{5.26}$$

The functions $g_i = 1/12 - (1/12 - 1/28.6)b_E/l_E$ and $g_\eta = 21.8(b_E/l_E)^2 - 13.5b_E/l_E + 20.3$ describe in a smooth form the geometry dependence of the various possible cases. The results are valid for transistors with n_E emitter and $n_B = n_E + 1$ base contacts (then: $I_{Bi} \rightarrow I_{Bi}/n_B$). For example, $g_i = 12$ for a long stripe and $g_i = 28.6$ for a square emitter window. An extension to the case of a single base contact (i.e., $n_E = n_B = 1$) is given in (Schröter, 1992). Note that in advanced high-speed transistors with narrow emitter widths, dc current crowding becomes negligible.

Small-Signal High-Frequency Operation

At sufficiently high frequencies (hf) the ac base current shunts the center region under the emitter; this effect is called **hf emitter current crowding** (Pritchard, 1958). The small-signal resistance at low frequencies is the following:

$$r_{bi} = r_{Bi} + I_{Bi}\frac{dr_{Bi}}{dI_{Bi}}. \tag{5.27}$$

Assuming negligible dc emitter current crowding and $l_E \gg b_E$, the distributed charging of the capacitances can be represented by a distributed RC network. The solution of the corresponding linear differential equation yields an infinite series for the admittance as a function of frequency. Truncating after the first frequency term leads to the impedance:

$$z_{bi} = r_{Bi}(1 - j\omega r_{Bi}C_{rBi}) \approx r_{Bi}(1 + j\omega r_{Bi}C_{rBi})^{-1}. \tag{5.28}$$

The last expression corresponds to a parallel circuit with a lumped representation of hf emitter current crowding that is modeled by a capacitance $C_{rBi} = (C_{jEi} + C_{jCi} + C_{dE} + C_{dC})/5$ and the dc resistance r_{Bi}.

Large-Signal Transient Operation

During (fast) switching, the base resistance can change significantly over time. At the beginning of a switching process, only the portion at the emitter edge is charged or discharged, leading initially to a small value of the effective internal base resistance $z_{Bi}(t)$. With a continuing turn-on process, the region toward the center starts to get charged, too, and z_{Bi} increases with time, until it approaches the dc value at the end of the turn-on process. For the turn-off process, the discharge of the emitter edge region continues, thus making this region high-ohmic compared to the center region that acts as a conducting transistor. Therefore, z_{Bi} not only starts to become larger than r_{Bi} but also keeps increasing during the remainder of the turn-off process. Overall, for switching:

$$z_{Bi}(t)|_{\text{turn-on}} \leq r_{Bi} \leq z_{Bi}(t)|_{\text{turn-off}}. \tag{5.29}$$

In circuit simulators, usually the dc value is employed for r_{Bi}, which has turned out to be a reasonable approximation for differential-type switching circuits (Schröter and Rein, 1995; Rein, In press).

5.2.3 Emitter Periphery Effects

The lateral outdiffusion of the emitter doping forms the emitter periphery junction with an associated depletion capacitance C_{jEp}. In addition, a portion of the base current, I_{jBEp}, coming from the base contact is back-injected into the emitter already across the peripheral junction. Electrons injected across the emitter periphery junction pass through the external base region under the BE spacer before they enter the external BC depletion region (refer to Figure 5.5). The associated carrier transport and charge storage lead to a perimeter transfer current I_{Tp} and a corresponding diffusion capacitance. Therefore, in most cases, the emitter periphery acts like a second transistor almost in parallel to the internal one but with lower performance. To avoid a two-transistor model, one can combine I_{Ti} and I_{Tp} into a single transfer current source, $I_T = I_{Ti} + I_{Tp} = \bar{I}_{Ti}A_{E0} + I'_{Tp}P_{E0}$, by defining an effective electrical emitter width (for $l_{E0} \gg b_{E0}$) and area, respectively (Rein, 1984):

$$b_E = b_{E0} + 2\gamma_C \quad \text{and} \quad A_E = A_{E0}(1 + \gamma_C P_{E0}/A_{E0}), \quad (5.30)$$

with the emitter window perimeter P_{E0}. The ratio $\gamma_C = I'_{Tp}/\bar{I}_{Ti}$ is bias-independent at low-current densities. Employing the effective emitter dimensions in all geometry-dependent equations of the internal transistor defines a procedure for constructing a consistent one-transistor model (Schröter and Walkey, 1996). Merging I_{Ti} and I_{Tp} together with the corresponding minority charges into single elements I_T and C_{dE} requires a modification also of r_{Bi} into r^*_{Bi} to obtain the correct time constants (Koldehoff *et al.*, 1993).

5.2.4 Extrinsic Transistor Regions

The regions outside of the (effective) internal transistor are usually described by lumped elements, which will be discussed briefly (see Figures 5.5 and 5.9). For dc operation, the series resistances of the emitter (r_E) as well as of the external base (r_{Bx}) and collector (r_{Cx}) have to be taken into account besides possible current components across the junctions (such as i_{jBCx}); r_{Cx} results from the buried layer and sinker; advanced bipolar transistors r_{Bx} is dominated by the component under the BE spacer (refer to blow-up in Figure 5.5).

For hf operation, the junction capacitances of the external BC region (C_{jCx}) and the collector–substrate region (C_{jS}) are required. With shrinking device dimensions, bias-independent isolation (partially oxide) capacitances caused by the BE spacer (C_{Eox}) as well as by the B-contacting region over the epicollector (C_{Cox}) are of increasing importance. As frequencies increase, a substrate coupling resistance (r_{su}) and—for lightly doped substrate material—the capacitance (C_{su}) that is associated with the substrate permittivity become important (Lee, 1999; Pfost, 1996). In some processes, a parasitic substrate transistor can turn on; however, only a simple transport model (i_{TS}, i_{jSC}, C_{dS}) is usually required to indicate this undesired action to the designer.

From a processing point of view, the goal is to reduce the dimensions of the parasitic regions as much as possible to minimize their impact on electrical performance.

5.2.5 Other Effects

Depending on the operation, a number of other effects can occur that may have to be taken into account for certain applications. Increasing the speed of bipolar transistors comes partially at the expense of lower breakdown voltages in the BC and BE junctions due to larger doping concentrations and shallower junction depths. Therefore, the corresponding effect of BC avalanche breakdown (i_{AVL} in Figure 5.9), which usually occurs in the internal transistor, often needs to be taken into account (Maes *et al.*, 1990; Rickelt and Rein, 1999; Schröter *et al.*, 1998). For some applications, such as varactors, a BE breakdown current (i_{BEt}), which in high-speed transistors is mostly determined by tunneling across the BE periphery junction, can also become important for circuit design (Schröter *et al.*, 1998).

Temperature variations or self-heating in the device cause most characteristics to change more or less significantly with temperature. The corresponding shift in dc *I–V* characteristics and depletion capacitances is mostly determined by the temperature dependence of the bandgap. For instance, typical values for $\partial V_{BE}/\partial T$ at constant I_C and the temperature coefficient (TC) of built-in voltages are in the range of −1.5 to −2.5 mV/K (for silicon). The TC of the current gain is positive for most Si–BJTs, about zero for SiGe bipolar transistors, and negative for III–V (e.g., AlGaAs) HBTs. Changes in series resistances and transit time are predominantly determined by the temperature dependence of the respective mobilities and the thermal voltage V_T. Self-heating occurs at bias points where a high power $P \approx I_C V_{CE}$ is generated and can be described to first-order by a separate network consisting of thermal resistance, R_{th}, and thermal capacitance, C_{th}.

For many applications, such as low-noise amplifiers and oscillators, noise plays an important role. Fundamental noise mechanisms are thermal noise in series resistances, shot noise in junction and transfer currents, and flicker noise. The latter is a phenomenological model for taking into account (physically usually not well described) low-frequency noise sources at surfaces and interfaces (Markus and Kleinpenning, 1995; Deen *et al.*, 1999; Chen *et al.*, 1998). The high-frequency (hf) noise performance of modern bipolar transistors is mostly determined by the noise in $r_B(= r^*_{Bi} + r_{Bx})$, I_{jBE}, and I_T as well as by the parasitic capacitances.

5.2.6 Equivalent Circuits

Figure 5.9 shows a physics-based large-signal equivalent circuit (EC) of a bipolar transistor, containing the elements that were discussed before (Koldchoff *et al.*, 1993; Schröter and Rein, 1995; Schröter and Lee, 1998.). The BC capacitance $C_{BCx} = C_{jCx} + C_{Cox}$ has been split across r_{Bx} to account for distributed hf effects in the external BC region in as simple as possible EC topology (Π-representation). In practice, the full complexity of the EC is often not required and certain effects (i.e., their associated elements, such as those for the substrate transistor or i_{BEt}) can be turned off. In many cases, the transistors are even operated outside any critical bias regions, such as the high-current and breakdown regions, so that a simplified EC [refer to Figure 5.9(B)] and a set of model equations are satisfactory and enable a better analytical insight into the circuit performance for the designer. The simplified model can be constructed from the physics-based model by merging elements and simplifying equations at the expense of a smaller bias, geometry, and temperature validity range. A scalable modeling approach is described in Schröter *et al.* (1999).

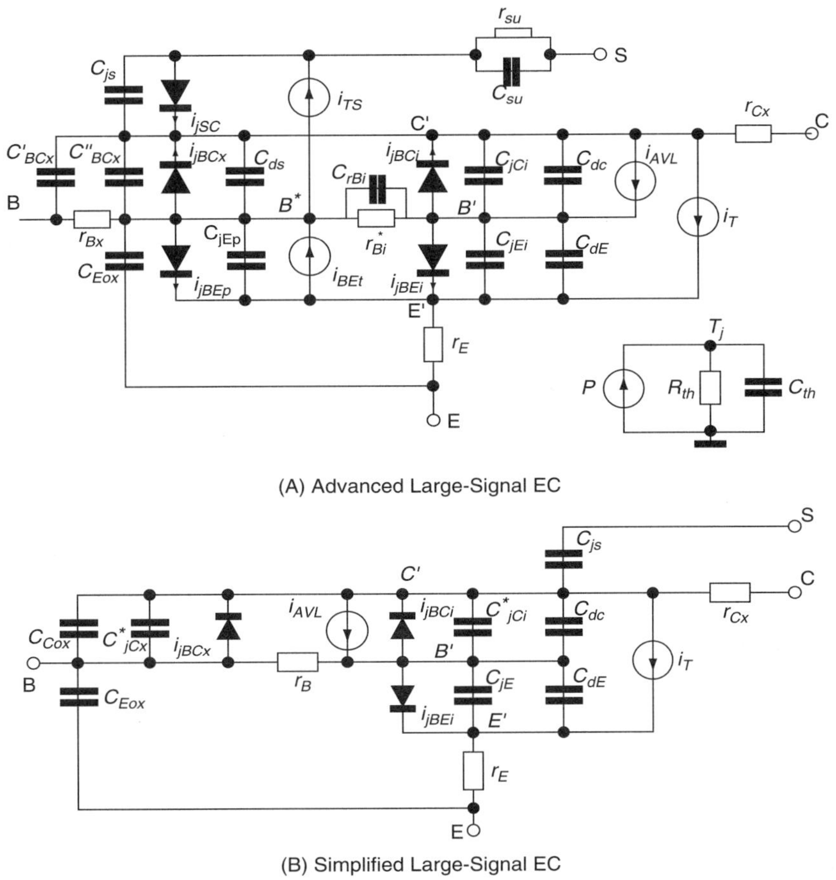

FIGURE 5.9 (A) Shown here is a large-signal physics-based EC of an advanced compact model for circuit design; (B) In comparison, here is a simplified large-signal EC obtained by merging elements and simplifying equations.

5.2.7 Some Typical Characteristics and Figures of Merits

This chapter contains a brief overview and discussion of the typical behavior of some of the most important characteristics of a bipolar transistor. Due to the lack of space, the overview is by no means complete, and the reader is referred to the literature in the Reference list at the end of the chapter. To facilitate in practice an easier understanding for circuit designers and a meaningful comparison of processes, it is recommended to choose the collector current density $J_C = I_C/A_E$ and the voltage V_{CE} as independent variables for defining the bias points of a figure of merit (FoM).

The forward dc ("Gummel") characteristics and the current gain were already shown in Figure 5.7. For most processes, the voltage drop ΔV_{IB} is caused by only r_E and r_B, whereas ΔV_{IC} contains in addition the influence of high-current effects, causing B to drop at higher J_C. Therefore, $I_C(V_{BE})$ should *not* be used for determining the series resistances!

Figure 5.10 shows typical forward output characteristics for constant I_B (with linear spacing ΔI_B). The thin solid line (labeled with $I_{CK}|A_E$) indicates the boundary between the quasisaturation region and the **normal** forward operation. The slope of the curves determines the dc output conductance. At sufficiently high V_{CE}, avalanche breakdown in the BC SCR causes J_C to increase rapidly; this effect is more pronounced in $J_C(V_{CE})$ if I_B rather than V_{BE} is held constant. Often the CE breakdown voltage for an open base, BV_{CEO}, is used to characterize the breakdown performance of a process. At high J_C *and* V_{CE}, self-heating occurs that leads J_C to increase (in Si) or decrease (in GaAs) already before the onset of breakdown.

The most important FoMs characterizing the high-frequency performance of a transistor are discussed next. For those frequencies of f that are not too high, the small-signal current gain in common-emitter (CE) configuration (i.e., E-grounded with the signal applied to either B or C) can be described quite accurately as a function of frequency by:

$$\underline{\beta} = \frac{\beta_0}{1 + j(f/f_\beta)} = \frac{\beta_0}{1 + j(\beta_0 f/f_T)}, \qquad (5.31)$$

with f_β as 3-dB-corner frequency, f_T as (quasistatic) transit frequency, and β_0 as low-frequency current gain. The above

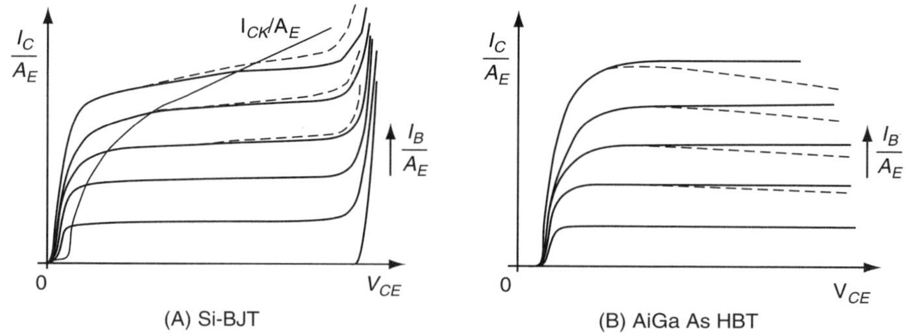

FIGURE 5.10 Sketch of the Output Characteristics. The impact of self-heating is shown by the dashed lines. Si-based transistors possess a much smaller BV_{CEO} value than GaAs based transistors of comparable performance. Typical units: I_C/A_E in mA/μm^2; V_{CE} in V.

equation allows a simple method for determining f_T (Gummel, 1969):

$$f_T = f\, Im^{-1}\left(\frac{1}{\underline{\beta}(f)}\right), \tag{5.32}$$

which—in addition—can be performed at fairly low frequencies (down to $f \approx f_\beta$), thus avoiding sophisticated de-embedding methods (Koolen, *et al.* 1991). The following is a quite accurate analytical expression for f_T:

$$f_T = \frac{1}{2\pi}\frac{g_m}{C_{BE} + C_{BC} + g_m[\tau_f + r_{Cx}C_{BC} + (r_E + r_B/\beta_0)(C_{BC} + C_{Eox})]}, \tag{5.33}$$

where the transconductance g_m is proportional to I_C at low-current densities.

The maximum frequency of oscillation, f_{max}, depends on the power gain chosen for its extraction. According to theory, though, the magnitude of only the unilateral power gain, G_U, is expected to show a 20 dB per decade decrease in frequency in a certain frequency range. In practice, a deviation of G_U (and also of $|\underline{\beta}|$) from this ideal behavior occurs at higher frequencies due to parasitics, such as C_{BC} and nonquasistatic effects (TeWinkel, 1973; Schröter and Rein, 1995). From G_U, the following equation can be derived for CE configuration (Armstrong and French, 1995):

$$f_{max} = \frac{f_T}{1 + C_L/C_{BC}}\sqrt{2\left[\sqrt{1 + \frac{(1 + C_L/C_{BC})^2}{8\pi f_T C_{BC}(r_B + 1/g_m)}} - 1\right]}. \tag{5.34}$$

C_L is the capacitive load at the transistor output and includes C_{jS}. For small ratios C_L/C_{BC}, the simple classical equation $f_{max} = \sqrt{f_T/(8\pi C_{BC} r_B)}$ is obtained. The typical bias dependence of f_T and f_{max} is shown in Figure 5.11(A). Although f_T shows a moderate dependence on emitter geometry, f_{max} is a strong function of the emitter dimensions and is usually evaluated for transistors with $l_{E0} \gg b_{E0}$. An important FoM characterizing the noise behavior of a transistor is the noise factor F that can also be calculated as a function of frequency if the EC topology and element noise sources are given (Vendelin, 1990). F depends on a number of transistor parameters as well as on the generator impedance Z_G and increases at high frequencies as shown in Figure 5.11(B). Toward low frequencies, F increases again due to flicker noise. For Si and SiGe bipolar transistors, typical values of the corner frequency f_{fn} are in the order of 1 to 10 kHz, which is much lower than in MOS transistors, for example. Of particular interest for circuit applications is the minimum noise factor F_{min} at high frequencies, which is obtained by determining the minimum of F with respect to real and imaginary parts of Z_G. At high frequencies, F_{min} can be expressed analytically as a function of dc bias, transistor parameters, and frequency dependent y-parameters (Voinigescu *et al.*, 1997); a reasonable simplification is as follows:

$$F_{min} \cong 1 + \frac{m_f}{\beta_0} + \frac{f}{f_T}\sqrt{\frac{2I_C}{V_T}(r_E + r_B)\left(1 + \frac{f_T^2}{\beta_0 f^2}\right) + \frac{m_f^2 f_T^2}{\beta_0 f^2}}. \tag{5.35}$$

5.2.8 Other Types of Bipolar Transistors

Heterojunction bipolar transistors (HBTs) show excellent high-frequency behavior and are in many respects superior to BJTs. Conventionally, HBTs have been fabricated in III–V materials such as GaAs (Ali and Gupta, 1991); major applications of those HBTs are power amplifiers. More recently, also SiGe HBTs have started entering into applications that used to be dominated by III–V semiconductors; however, so far the majority of those "HBTs" are actually transistors with an increased drift field in the base (Harame *et al.*, 1995) rather than "true" HBTs (Schüppen *et al.*, 1996) as shown in Figure 5.6(B).

Besides vertical *npn* transistors, variants of *pnp* transistors are offered in processes, since they are required for certain circuit applications, such as bias networks or push-pull stages. Only few truly complementary bipolar processes do exist (Yamaguchi *et al.*, 1994); that is, in most cases the *pnp*

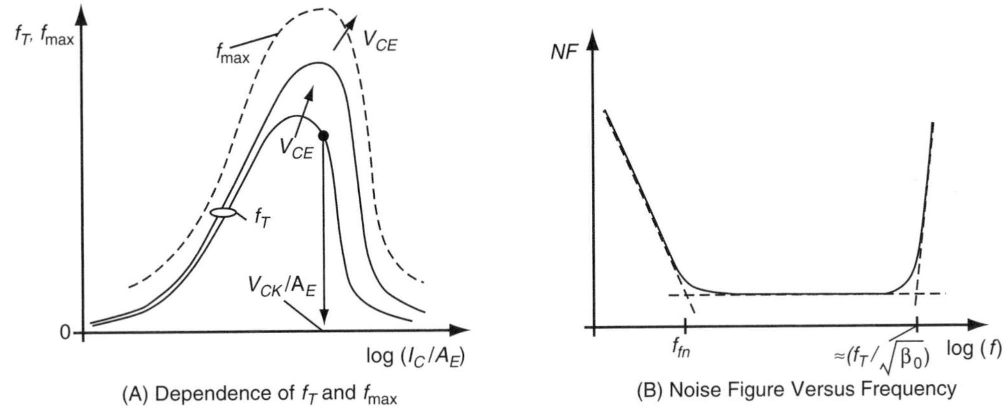

FIGURE 5.11 Typical Characterization of the (Small-Signal) High-Frequency Performance of Transistors. (A) This figure shows bias dependence of transit frequency f_T and maximum oscillation frequency f_{max}. (B) Shown here is a sketch of noise figure [= 10log(F)] versus frequency. Typical units: f, f_T, f_{max} in GHz; *NF* in dB.

transistors have a much lower performance than their *npn* counterparts. Lateral *pnp* transistors usually are quite slow (f_T around 0.2 to 1 GHz) and consume significant space on a chip. Vertical *pnp* transistors can be about a factor of 10 faster than LPNPs but suffer often from a large collector series resistance and associated high-current effects already at quite low collector current densities.

In all cases, the fundamental transistor action is the same, so that the same basic modeling approach can be used. For accurate circuit simulation, however, modifications or extensions of both the EC and some of the element equations are required.

References

Ali, F., and Gupta, A. (Eds.). (1991). HEMTs and HBTs: Devices, fabrication, and circuits. Boston: Artech House.

Armstrong, G., and French, W. (1995). A model for the dependence of maximum oscillation frequency on collector to substrate capacitance in bipolar transistors. *Solid-State Electron. 38,* 1505–1510.

Chen, X.Y., Deen, M.J., Yan, Z.X., and Schröter, M. (1998). Effects of emitter dimensions on low-frequency noise in double-polysilicon BJTs. *Electronics Letters, 34*(2), 219–220.

Deen, M.J., Rumyantsev, S.L., and Schröter, M. (1999). On the origin of $1/f$ noise in polysilicon emitter bipolar transistors. *J. Appl. Phys. 85,* 1192–1195.

Freeman, G., et al. (1999). A 0.18-μm 90-GHz f_T SiGe HBT BiCMOS, ASIC-compatible, copper interconnect technology for RF and microwave applications. IEDM, Washington DC.

Friedrich, M., and Rein, H.M. (1999). Analytical current–voltage relations for compact SiGe HBT models. *IEEE Trans. Electron Dev. 46,* 1384–1401.

Ghosh, H.N. (1965). A distributed model of the junction transistor and its application in the prediction of the emitter-base diode characteristic, base impedance, and pulse response of the device. *IEEE Trans. Electron Dev. 12,* 513–531.

Gummel, H.K. (1970). A charge-control relation for bipolar transistors. *BSTJ 49,* 115–120.

Gummel, H.K. (1969). On the definition of the cutoff frequency f_T. *Proc. IEEE 57,* 2159.

D. Harame et al. (1995). Si/SiGe epitaxial base transistors. *IEEE Trans. Electron Dev. 40,* 455–482.

Kirk, C.T. (1962). A theory of transistor cutoff frequency fall-off at high current densities. *IEEE Trans. Electron Dev. 9,* 914–920.

Klose, H., et al. (1993). B6HF: A. 0.8 micron 25GHz/25ps bipolar technology for mobile radio and ultra fast data link IC products. *Proc BCTM Minn,* 125–128.

Koldehoff, A., Schröter, M., and Rein, H.M., (1993). A compact bipolar transistor model for very high-frequency applications with special regard to narrow stripes and high current densities. *Solid-State Electron. 36,* 1035–1048.

Koolen, M., et al. (1991). An improved de-embedding technique for on-wafer high-frequency characterization. *Proc. BCTM* 188–191.

Lee, T.Y., Fox, R. Green, K., and Vrotsos, T. (1999). Modeling and parameter extraction of BJT substrate resistance. *Proc. BCTM* 101–103.

Lindmayer, J., and Wrigley, C.Y. (1965). *Fundamentals of semiconductor devices.* Princeton, New Jersey: Van Nostrand.

Maes, W., DeMeyer, K., and van Overstraaten, R. (1990). Impact ionization in silicon: A review and update. *Solid-State Electron. 33,* 705–718.

Markus and Kleinpenning, (1995). Low-frequency noise in polysilicon bipolar transistors. *IEEE Trans. Electron Dev. 42,* 720–727.

Pfost, M., Rein, H.M., and Holzwarth, T. (1996). Modeling substrate effects in the design of high-speed Si bipolar ICs. *IEEE J. Solid-State Circuits. 31,* 1493–1502.

Philips, A.B. (1962). *Transistor engineering.* New York: McGraw Hill.

Pritchard, R.L. (1967). *Transistor characteristics.* New York: McGraw Hill.

Pritchard, R.L. (1958). Two-dimensional current flow in junction transistors at high frequencies. *Proc. IEEE 46,* 1152–1160.

Racanelli, M., et al. (1999). BC35—a 0.35-μm, 30-GHz production RF BiCMOS technology. *Proc. BCTM,* 125–128.

Rein, H.M. (1984). A simple method for separation of the internal and external (peripheral) currents of bipolar transistors. *Solid-State Electronics 27,* 625–632.

Rein, H.M. (1983). Proper choice of the measuring frequency for determining f_T of bipolar transistors. *Solid-State Electronics. 26,* 75–82 and 929.

Rein, H.M. (In press). Si and SiGe bipolar ICs for 10–40 Gb/s optical-fiber TDM links. *Int. J. High-Speed Electronics and Systems.*

Rein, H.M. and Schröter, M. (1989). Base spreading resistance of square emitter transistors and its dependence on current crowding. *IEEE Trans. Electron Dev. 36,* 770–773.

Rein, H.M. and Schröter, M. (1991). Experimental determination of the internal base sheet resistance of bipolar transistors under forward-bias conditions. *Solid-State Electron. 34,* 301–308.

Rein, H.M., Stübing, H., and Schröter, M. (1985). Verification of the integral charge-control relation for high-speed bipolar transistors at high current densities. *IEEE Trans. Electron Dev. 32,* 1070–1076.

Rickelt, M., and Rein, H.M. (1999). Impact-ionization induced instabilities in high-speed bipolar transistors and their influence on the maximum usable output voltage. *Proc. BCTM* 54–57.

Schilling, R.B. (1969). A regional approach for computer-aided transistor design. *IEEE Trans. Electron Education. 12,* 152–161.

Schröter, M. (1998). *Bipolar transistor modeling for the design of high-speed integrated circuits.* Monterey/Lausanne: MEAD.

Schröter, M. (1992). Modeling of the low-frequency base resistance of single base contact bipolar transistors. *IEEE Trans. Electron Dev. 39,* 1966–1968.

Schröter, M. (1991). Simulation and modeling of the low-frequency base resistance of bipolar transistors in dependence on current and geometry. *IEEE Trans. Electron Dev. 38,* 538–544.

Schröter, M., et al. (1999). Physics- and process-based bipolar transistor modeling for integrated circuit design. *IEEE Journal of Solid-State Circuits 34,* 1136–1149.

Schröter, M., Friedrich, M., and Rein, H.M. (1992). A generalized integral charge control relation and its application to compact models for silicon based HBTs. *IEEE Trans. Electron Dev. 40,* 2036–2046.

Schröter, M., and Lee, T.Y. (1998). HICUM—A physics-based scaleable compact bipolar transistor model. *Presentation to the Compact Model Council.* www.eigroup.org/cmc

Schröter, M., and Lee, T.Y. (1999). A physics-based minority charge and transit time model for bipolar transistors. *IEEE Trans. Electron Dev. 46,* 288–300.

Schröter, M., and Rein, H.M. (1995). Investigation of very fast and high-current transients in digital bipolar circuits by using a new compact model and a device simulator. *IEEE J. Solid-State Circuits 30,* 551–562.

Schröter, M. and Walkey, D.J. (1996). Physical modeling of lateral scaling in bipolar transistors. *IEEE J. Solid-State Circuits 31,* 1484–1491.

Schröter, M., Yan, Z., Lee, T.Y., and Shi, W. (1998). A compact tunneling current and collector break-down model. *Proc. IEEE Bipolar Circuits and Technology Meeting,* 203–206.

Schüppen, A. et al. (1996). SiGe Technology and components for mobile communication systems. *Proc. BCTM,* 130–133.

Selberherr, S. (1984). *Semiconductor device modeling.* Springer: Wien.

Sze, S. (1981). Physics of semiconductor devices. New York: John Wiley & Sons.

TeWinkel, J. (1973). Extended charge-control model for bipolar transistors. *IEEE Trans. Electron Dev. 20,* 389–394.

Vendelin, G., Pavio, A., and Rohde, U. (1990). *Microwave circuit design.* Toronto: John Wiley & Sons.

Voinigescu, S., et al., A scaleable high-frequency noise model for bipolar transistors and its application to optimal transistor sizing for low-noise amplifier design. *IEEE J. Sol.-St. Circ. 32,* 1430–1439.

Yamaguchi, T., et al. (1994). Process investigations for a 30 GHz f_T submicrometer double poly-Si bipolar technology. *IEEE Trans. Electron Dev. 41,* 321–328.

6

Semiconductors

Michael Shur
Department of Electrical, Computer, and Systems Engineering, Rensselaer Polytechnic Institute, Troy, New York, USA

6.1 History of Semiconductors

In 1821, the German physicist Tomas Seebeck first noticed unusual properties of semiconductor materials, such as lead sulfur (PbS). In 1833, the English physicist Michael Faraday reported that resistance of semiconductors decreased as temperature increased (opposite to what happens in metals, in which resistance increases with temperature).

In 1873, the British engineer Willoughby Smith discovered that the resistivity of selenium, a semiconductor material, is very sensitive to light. In 1875, Werner von Siemens invented a selenium photometer, and, in 1878, Alexander Graham Bell used this device for a wireless telephone communication system. The discovery of a bipolar junction transistor by the American scientists John Bardeen, Walter Houser Brattain, and William Bradford Shockley in 1947 led to a revolution in electronics, and, soon after, semiconductor devices largely replaced electronic tubes.

6.2 Dielectrics, Semiconductors, and Metals

When atoms form solids, their electron energy states split into many close energy levels that form allowed **energy bands.** In each energy band, allowed energy levels are very close to each other, and the electron energy can vary continuously. These bands are separated by **forbidden energy gaps** as shown in Figure 6.1. The position and extent of allowed and forbidden energy gaps determine the properties of solids.

The important property of electrons is determined by the rule that is called the **Pauli exclusion principle.** According to this principle, not more than two electrons with different spins can occupy each energy state. Electrons occupy the lowest energy levels first. In semiconductors and dielectrics, almost all the states in the lowest energy bands are filled by electrons, whereas the energy states in the higher energy bands are, by and large, empty. The lower energy bands with mostly filled energy states are called the **valence bands**. The higher energy bands with mostly empty energy states are called **conduction**

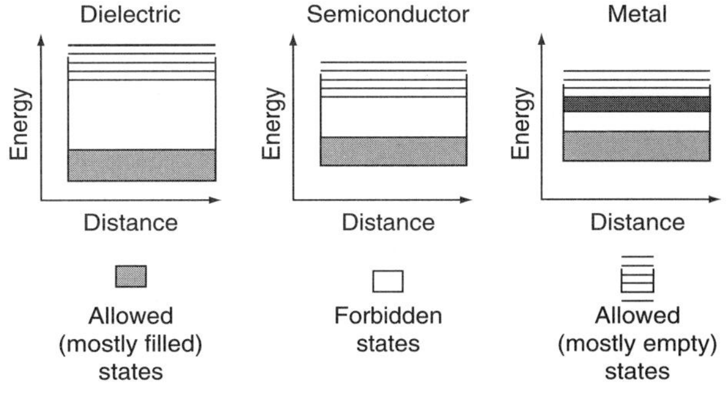

FIGURE 6.1 Band structures of a Dielectric, a Semiconductor, and a Metal. The shaded regions represent energy levels filled with electrons (two per state to satisfy the Pauli exclusion principle).

bands. The difference between the highest valence band and the lowest conduction band is called the **energy band gap** or the **energy gap**. An electron in a valence band needs the energy equal to or higher than the energy gap to experience a transition from the valence to the conduction band.

In a dielectric, the energy gap, E_g, is large so that the valence bands are completely filled and conduction bands are totally devoid of electrons. Typically for a dielectric, E_g is larger then 5 to 6 eV. For semiconductors, energy band gaps vary between 0.1 eV and 3.5 eV. The energy gap of **silicon** (Si), which is the most important semiconductor material, is approximately 1.12 eV at room temperature. The energy gap of silicon dioxide—the most widely used dielectric material in microelectronics—is 9 eV. In a metal, the lowest conduction band is partially filled with many electrons (usually one electron per atom), and the metal's resistance is very small.

6.2.1 Chemical Bonds and Crystal Structure

The most important semiconductor is silicon. Silicon belongs to the fourth column of the Periodic Table. Elements belonging to the fourth column of the Periodic Table, such as Si, Ga, As, Ge, and N, have only *s* and *p* electrons in their valence shells (shells with the largest value of the principal quantum number, *n*, for a given atom). It takes eight valence electrons to fill up all the states in these two valence subshells (two *s* electrons and six *p* electrons). In the Periodic Table, the elements with completely filled *s* and *p* valence subshells correspond to inert gases. When atoms are combined together in a solid, they may share or exchange valence electrons, forming **chemical bonds**. In silicon, germanium, and related compound semiconductors, these bonds are formed in such a way that neighboring atoms share their valence electrons having (on average) completed *s* in the valence shell. In these semiconductors, each atom forms four bonds with four other atoms (four nearest neighbors) and shares two valence electrons with each of them (i. e., an atom shares eight valence electrons with all four nearest neighbors). Since each atom has four nearest neighbors, it is **tetrahedrally coordinated** as shown in Figure 6.2.

If all atoms in a crystal are identical (as in Si or Ge, for example) the electrons shared in a bond must spend, on average, the same time on each atom. This corresponds to a purely homopolar (covalent) bond. In a compound semiconductor, such as GaAs, bonding electrons spend a greater fraction of time on the anions (i. e., negatively charged atoms). This situation corresponds to a partially heteropolar (partially ionic) bond.

The quantitative measure of the type of a chemical bond is ionicity, f_i. The ionicity is zero for a purely covalent bond and unity for a purely heteropolar bond. The ionicity values $0 < f_i < 1$ can be assigned to any binary chemical compound. Elemental semiconductors, such as Si, Ge, or diamond, have zero ionicity. According to the ionicity scale proposed by Phillips in 1973, $f_i = 0.177$ for silicon carbide (SiC) and $f_i = 0.31$ for GaAs.

There are very few elemental semiconductors (e.g., Si, Ge, and C). Many compound materials (such as GaAs, InP, AlGaAs, and GaN), however, also have the tetrahedral bond configuration seen in Figure 6.3 and exhibit semiconducting properties.

Some of these semiconductors, such as GaAs and AlGaAs or GaN and AlGaN, can be grown as films on top of each other to form heterostructures. Hence, very many different material combinations with different properties are available to a semiconductor device designer. As can be seen from Figure 6.2, each silicon atom is located at the center of the tetrahedron formed by four other silicon atoms. In Si, such an arrangement is formed by two interpenetrating face-centered cubic sublattices of silicon atoms, shifted with respect to each other by one fourth of the body diagonal. A diamond crystal (one of the crystalline modifications of carbon) has the same crystal structure, which is traditionally called the **diamond crystal structure** (see Figure 6.4). Another important semiconductor—germanium (Ge)—also has the diamond crystal structure.

Most compound semiconductors have the **zinc blende** crystal structure seen in Figure 6.5 that is very similar to the diamond structure. This structure contains two kinds of atoms, A and B, with each species forming a face-centered cubic lattice. These mutually penetrating face-centered cubic (fcc) lattices of element A and element B are shifted relative to each other by a quarter of the body diagonal of the unit cell cube.

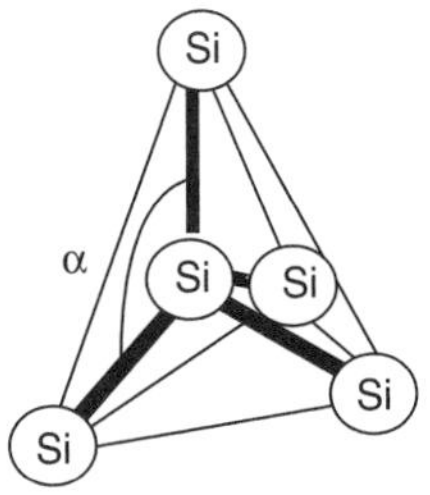

FIGURE 6.2 Tetrahedral Bond Configuration in Si; $\alpha = 108^\circ 29'$

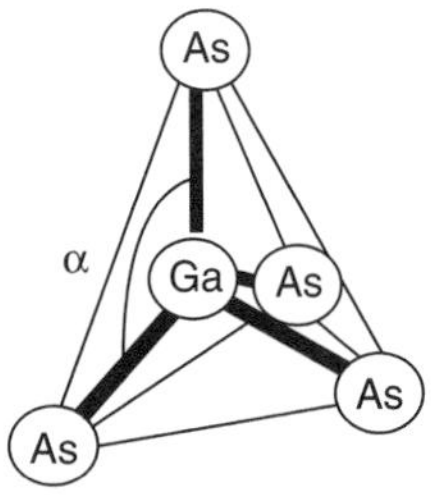

FIGURE 6.3 Tetrahedral Bond Configuration in GaAs; $\alpha = 108^\circ 29'$

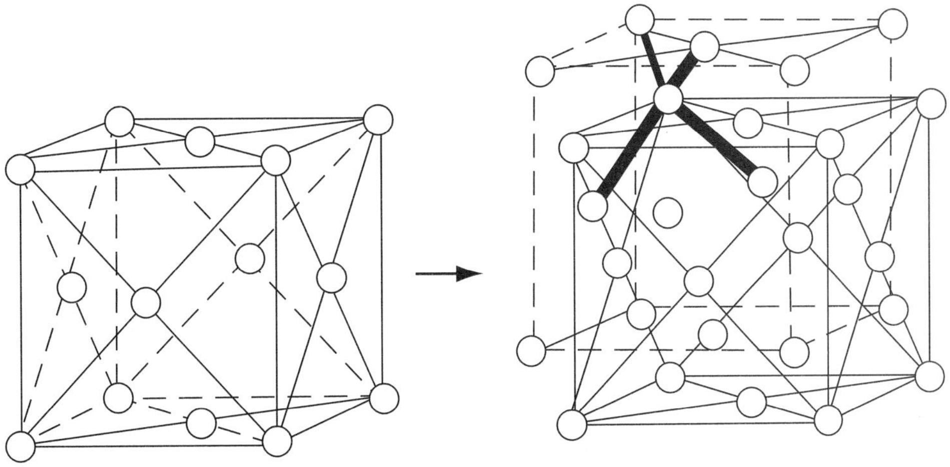

FIGURE 6.4 The Face-Centered Cubic (fcc) and Diamond Structure. This structure is formed by two interpenetrating fcc lattices shifted by one quarter of the body diagonal. Thick lines show the bonds formed between an atom and its four nearest neighbors (tetrahedral bond configuration). Figure adapted From Shur (1996).

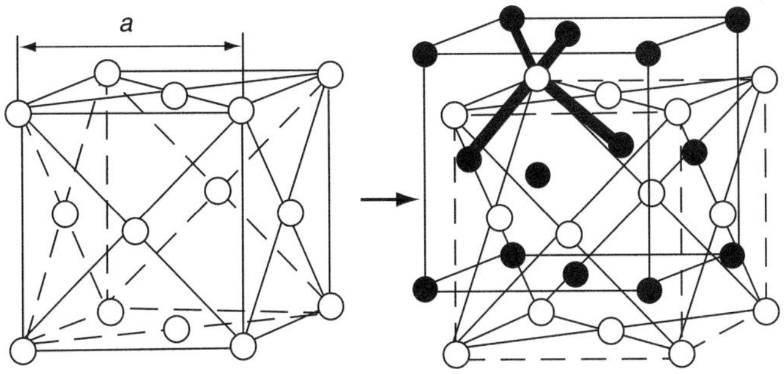

FIGURE 6.5 The Face-Centered Cubic (fcc) and Zinc Blende Structure. This structure is formed by two interpenetrating fcc lattices shifted by one quarter of the body diagonal (similar to diamond structure). Thick lines show the bonds between an atom and its four nearest neighbors (tetrahedral bond configuration). The a is the lattice constant. Figure adapted from Shur (1996).

6.2.2 Electrons and Holes

In the absence of an electric field, electrons in a semiconductor move randomly because of thermal motion. The average velocity of this motion is called thermal velocity. An electric field accelerates electrons and causes a component of the electron velocity to go in the direction of the electric field. This velocity of the directed electronic motion in the electric field is called the **drift velocity**. This drift velocity results in an electric current.

To the first order, the difference between the motion of conduction band electrons and free electrons can be approximately described by introducing an electron **effective mass**, m_n, which is different from the **free electron mass**, $m_e = 9.11 \times 10^{-31}$ kg. For example, in gallium arsenide (GaAs), $m_n \sim 0.067\, m_e$.

A valence band electron can change its energy and velocity if and only if it can be promoted to an empty energy level. The vacant levels in the valence band allow electrons in the valence band to move as shown in Figure 6.6.

The motion of a vacant energy level in the valence band can be visualized as the motion of a fictitious particle (called a **hole**). The absence of a negative electronic charge, $-q$, corresponds to a positive charge $q = -(-q)$. Hence, the hole has a positive charge with a magnitude equal to the electronic charge. To the first order, holes can be treated as particles with a certain effective mass, m_p.

6.2.3 Band Diagrams

Usually, conducting electrons occupy states in the conduction band that are close to the lowest energy in the conduction band, E_c, which is called the **conduction band edge** or the **bottom of**

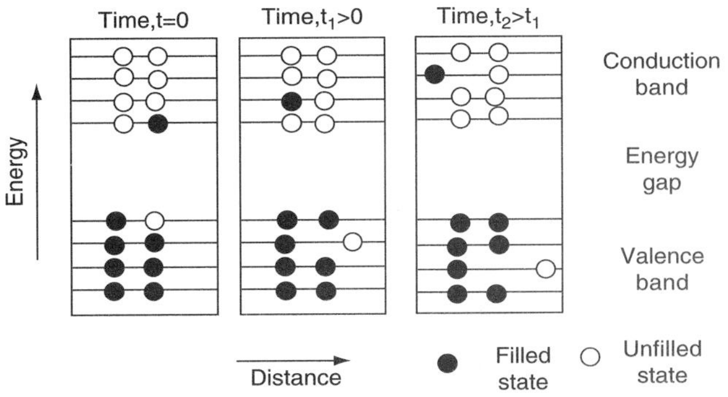

FIGURE 6.6 Motion of Electrons in an Electric Field. Solid circles represent electrons occupying energy levels; open circles represent energy levels available for electrons.

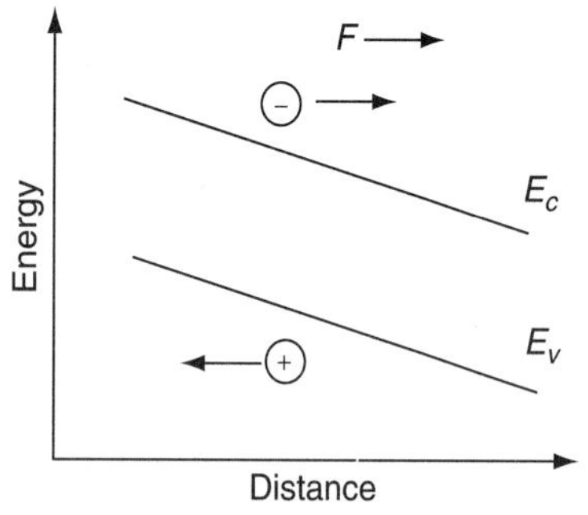

FIGURE 6.7 Band Diagram of a Semiconductor in an Electric Field

the conduction band. Holes occupying states in the valence band are close to the highest energy in the valence band, E_v. E_v is called the **valence band edge** or the **top of the valence band.** The dependencies of E_c and E_v in a semiconductor device on position are called **band diagrams.** Band diagrams are very useful for illustrating properties and understanding behavior of semiconductor materials and devices. As an example, Figure 6.7 shows a band diagram of a semiconductor in an electric field.

In an electric field, F, the bands are tilted, with the slope $-qF$. The arrows in Figure 6.7 represent the directions of forces exerted on the electrons and holes by the electric field. (Since electrons are charged negatively and holes are charged positively, they move in opposite directions.)

In free space, an electron energy is as follows:

$$E = m_e v^2/2, \tag{6.1}$$

where m_e is the free electron mass and v is the electron velocity. According to the basics of quantum mechanics, an electron has both particle-like and wave-like properties and its momentum, $p = m_e v$, can be related to its wave vector, k:

$$p = \hbar k, \tag{6.2}$$

where $\hbar = 1.055 \times 10^{-34}$ js is the Planck constant. Hence, equation 6.1 can be rewritten as:

$$E = \hbar^2 k^2 / 2m_e. \tag{6.3}$$

In semiconductor materials, $E(k)$ dependence is more complex (see Figure 6.8). Near the lowest point of the conduction band, the dependence of the electron energy on the wave vector can still be approximated by a parabolic function similar to that for an electron in free space (see equation 6.3 and Figure 6.8). The curvature of this dependence, however, is usually quite different from that for an electron in free space. Moreover, different crystallographic directions are not equivalent, and this curvature may depend on direction. These features can be accounted for by introducing an inverse effective mass tensor with components defined as follows:

$$\frac{1}{m_{i,j}} = \frac{1}{\hbar^2} \frac{\partial^2 E_n}{\partial k_i \partial k_j}, \tag{6.4}$$

k_i and k_j are the projections of the wave vector **k**. When $E(k)$ depends only on the magnitude of **k**, and not on the k direction, this tensor reduces to a scalar inverse effective mass, $1/m_n$. This is the case for GaAs, where dependence of the electron energy on the wave vector near the lowest point of the conduction band can be approximated by the following function:

$$E_n(k) = E_c + \frac{\hbar^2 k^2}{2m_n}. \tag{6.5}$$

In Si, however, the lowest minimum point of the conduction band is quite anisotropic. In this case, the simplest equation for $E(k)$ is obtained by expressing k in terms of two components: k_l and k_t, which is perpendicular to k_l. These components are called **longitudinal and transverse components** of k, respectively. (This takes into account the crystal symmetry, which makes all possible directions of k_t equivalent.) Now the inverse effective mass tensor reduces to two components: $1/m_l$ and $1/m_t$, where m_l and m_t are called the **longitudinal effective mass** and **transverse effective mass**, respectively. In this case, the equation for $E(k)$ is given by:

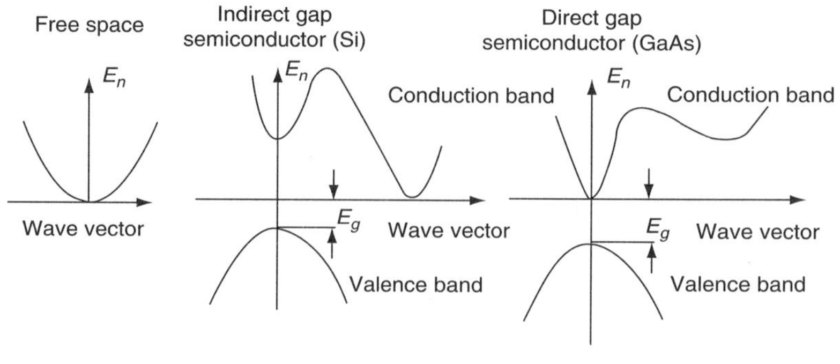

FIGURE 6.8 Qualitative Energy Spectra for Electrons in Free Space. Si and GaAs. Shaded states in valence bands are filled with valence electrons. Figure adapted from Shur (1996).

$$E_n(k) = E_c + \frac{\hbar^2}{2}\left(\frac{k_l^2}{m_l} + \frac{k_t^2}{m_t}\right). \qquad (6.6)$$

For Si, $m_l \sim 0.98\, m_e$ and $m_t \sim 0.189\, m_e$, where m_e is the free electron mass, so that these two effective masses are quite different indeed!

For holes, an equation similar to equation 6.5 is written as:

$$E_p(k) = E_v - \frac{\hbar^2 k^2}{2m_p}. \qquad (6.7)$$

Equation 6.7 can be used for the valence bands in cubic semiconductors. Here E_v is the energy corresponding to the top of the valence band, and m_p is called a **hole** effective mass.

Doping

In a pure semiconductor, the electron concentration in the conduction band and the hole concentration in the valence band are usually very small compared to the number of available energy states. These concentrations can be changed by many orders of magnitude by **doping**, which means adding to a semiconductor impurity atoms that can "donate" electrons to the conduction band (such impurities are called donors) or "accept" electrons from the valence band creating holes (such impurities are called acceptors). Both donors and acceptors are referred to as **dopants.** A donor atom has more electrons available for bonding with neighboring atoms than is required for an atom in the host semiconductor. For example, a silicon atom has four electrons available for bonding and forms four bonds with the four nearest silicon atoms. A phosphorus atom has five electrons available for bonding. Hence, when a phosphorus atom replaces a silicon atom in a silicon crystal, it can bond with the four nearest neighbors and "donate" one extra electron to the conduction band.

An acceptor atom has fewer electrons than are needed for chemical bonds with neighboring atoms of the host semiconductor. For example, a boron atom has only three electrons available for bonding. Hence, when a boron atom replaces a silicon atom in silicon, it "accepts" one missing electron from the valence band, creating a hole.

A semiconductor doped with donors is called an ***n*-type** semiconductor. A semiconductor doped with acceptors is called a ***p*-type** semiconductor. Often, especially at room temperature or elevated temperatures, each donor in an *n*-type semiconductor supplies one electron to the conduction band, and the electron concentration, n, in the conduction band is approximately equal to the donor concentration, N_d. In a similar way, at room temperature or elevated temperatures, each acceptor creates one hole in the valence band, and the hole concentration, p, in the valence band of a *p*-type semiconductor is approximately equal to the acceptor concentration, N_a. If both donors and acceptors are added to a semiconductor, they "compensate" each other, since electrons supplied by donors occupy the vacant levels in the valence band created by the acceptor atoms. In this case, the semiconductor is called **compensated.** In a compensated semiconductor, the largest impurity concentration "wins": if $N_d > N_a$, the compensated semiconductor is *n*-type with the effective donor concentration, $N_{deff} = N_d - N_a$; if $N_d < N_a$, the compensated semiconductor is *p*-type with the effective acceptor concentration, $N_{aeff} = N_a - N_d$.

6.3 Electron and Hole Velocities and Mobilities

Electrons and holes experience a chaotic random thermal motion. The average kinetic energy of thermal motion per one electron is $3k_BT/2$, where T is temperature in degrees Kelvin and k_B is the Boltzmann constant. The electron **thermal velocity,** v_{thn}, is found by equating the electron kinetic energy to $3k_BT/2$:

$$\frac{m_n v_{thn}^2}{2} = \frac{3k_BT}{2}, \qquad (6.8)$$

where m_n is the electron effective mass. From equation 6.8:

$$v_{thn} = \left(\frac{3k_BT}{m_n}\right)^{1/2}. \qquad (6.9)$$

Thermal velocities v_{thn} and v_{thp} represent average magnitudes of electron and hole velocities caused by the thermal motion. Since the directions of this thermal motion are random, the same number of carriers crosses any device cross-section in any direction so that the electric current is equal to zero.

In an electric field, the electrons and holes acquire an additional drift velocity caused by the electric field, which is superimposed on the chaotic thermally induced velocities. In a uniform semiconductor and in a weak electric field, F, the drift velocities, v_n and v_p, of the electrons and holes are proportional to the electric field:

$$v_n = -\mu_n \mathbf{F}. \qquad (6.10)$$

and

$$v_p = \mu_p \mathbf{F}. \qquad (6.11)$$

The μ_n and μ_p are called the electron and hole mobilities, respectively. (The direction of v_p coincides with the direction of the electric field, and the direction of v_n is opposite to the direction of the electric field, since holes are positively charged and electrons are negatively charged.) The total charge of the electrons crossing a unit area of a semiconductor per second is qnv_n, where n is the electron concentration in the conduction band. The charge of the holes crossing a unit area of a semiconductor per second is qpv_p, where p is the hole concentration in the valence band. Hence, using equations 6.1 and 6.2, we obtain the following expressions for the current density, j, and conductivity, σ for a semiconductor:

$$j = \sigma F, \tag{6.12}$$

where

$$\sigma = q\mu_n n + q\mu_p p. \tag{6.13}$$

Often, the hole contribution to the conductivity of an n-type semiconductor can be neglected because the hole concentration is many orders of magnitude smaller than the electron concentration. Likewise, the electron contribution to the conductivity of a p-type semiconductor can often be neglected.

Equation 6.12 is called **Ohm's law**. We can rewrite this equation in a more familiar form:

$$I = V/R, \tag{6.14}$$

where $I = jS$ is the total current, S is the sample cross-section, $V = FL$ is the potential difference across the sample (here a constant electric field is assumed), and L is the sample length. The **sample resistance** is the following:

$$R = \frac{L}{\sigma S}. \tag{6.15}$$

Remember that, in semiconductors, Ohm's law is only valid when an electric field is weak, which is rarely the case in modern semiconductor devices.

To the first order, equation 6.10 can be interpreted by using the second law of motion for an electron in the conduction band:

$$m_n \frac{dv_n}{dt} = -qF - m_n \frac{v_n}{\tau_{np}}. \tag{6.16}$$

The first term in the right-hand side of equation 6.16 represents the force from the electric field. The second term in the right-hand side of the same equation represents the loss of the electron momentum, $m_n v_n$, due to electron collisions with impurities and/or lattice vibrations; τ_{np} is called the momentum relaxation time. From equation 6.16, in a steady state:

$$v_n = -\mu_n F,$$

where

$$\mu_n = \frac{q\tau_{np}}{m_n}. \tag{6.17}$$

The τ_{np} is an average time between random collisions interrupting these free electron flights. The average distance that an electron travels between two collisions is called the **mean free path**. In relatively weak electric fields when the electron drift velocity is much smaller than the thermal velocity, the mean free path is given by:

$$\lambda_n = v_{thn}\tau_{np}. \tag{6.18}$$

The momentum relaxation time, τ_{np}, can be approximately expressed as:

$$\frac{1}{\tau_{np}} = \frac{1}{\tau_{ii}} + \frac{1}{\tau_{ni}} + \frac{1}{\tau_{\text{lattice}}} + \ldots, \tag{6.19}$$

where the terms on the right-hand side represent momentum relaxation times due to different scattering processes such as **ionized impurity scattering** (τ_{ii}), **neutral impurity scattering** (τ_{ni}), and **lattice vibration scattering** (τ_{lattice}).

In a low electric field, τ_{np} and m_n are independent of the electric field, and the electron drift velocity, v_n, is proportional to the electric field. However, in high electric fields when electrons may gain considerable energy from the electric field, τ_{np} and, in certain cases, m_n become strong functions of the electron energy and, hence, of the electric field. In a high electric field, electrons gain energy from the field, and their average energy exceeds the thermal energy (they become "hot").

Hot electrons transfer energy into thermal vibrations of the crystal lattice. Such vibrations can be modeled as harmonic oscillations with a certain frequency, ω_l. The energy levels of a harmonic oscillator are equidistant with the energy difference between the levels equal to $E_l = \hbar\omega_l$. Hence, the scattering process for a hot electron can be represented as follows. The electron accelerates in the electric field until it gains enough energy to excite lattice vibrations:

$$\frac{m_n v_{n\max}^2}{2} = E_n - E_o \approx \hbar\omega_l, \tag{6.20}$$

where $v_{n\max}$ is the maximum electron drift velocity. Then the scattering process occurs, and the electron loses all the excess energy and all the drift velocity. Hence, the electron drift velocity varies between zero and $v_{n\max}$, and average electron drift velocity ($v_n = v_{n\max}/2$) becomes nearly independent of the electric field:

$$v_n \approx \sqrt{\frac{\hbar\omega_l}{2m_n}} = v_{sn}. \tag{6.21}$$

The v_{sn} is called the **electron saturation velocity**.

Figure 6.9 shows the dependencies of the electron drift velocity on the electric field for several semiconductors.

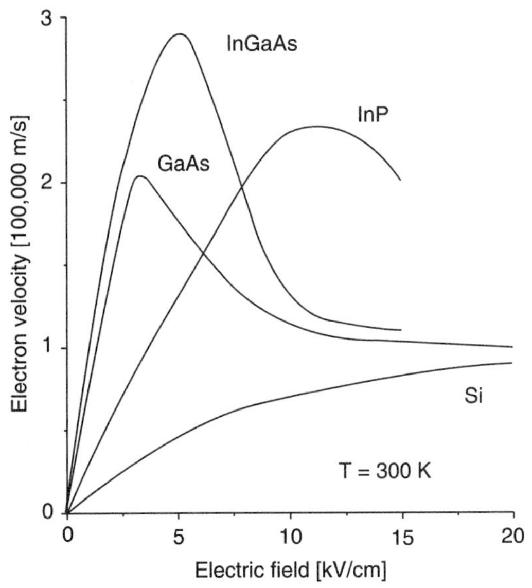

FIGURE 6.9 Velocity Versus Electric Field for Several Semiconductors. Figure adapted from Shur (1996).

As can be seen from Figure 6.9, in Si, the drift velocity saturates in high electric fields, as expected. In many compound semiconductors, such as GaAs, InP, and InGaAs, the electron velocity decreases with an electric field in a certain range of high electric fields. (In this range of the electric fields, the electron differential mobility, $\mu_{dn} = dv_n/dF$, is negative.) In these semiconductors, the central valley of the conduction band (Γ) is the lowest as shown in Figure 6.8. In high electric fields, hot electrons acquire enough energy to transfer from the central valley of the conduction band (where the effective mass and, hence, the density of states is relatively small) into the satellite valleys, where electrons have a higher effective mass and, hence, a larger density of states but a smaller drift velocity.

As can be seen from Figure 6.9, compound semiconductors have a potential for a higher speed of operation than silicon because electrons in these materials may move faster. The negative differential mobility observed in many compound semiconductors can be used for generating microwave oscillations at very high frequencies.

Modern semiconductor devices are often so small that their dimensions are comparable to or even smaller than the electron mean free path. In this case, electrons may acquire higher velocities than those shown in Figure 6.9, since transient effects associated with acceleration of carriers become important. In the limiting case of very short devices, the electron transit time may become so small that most electrons will not experience any collisions during the transit. Such a mode of electron transport is called **ballistic transport**.

Figure 6.10 shows the dependence of the electron and hole mobilities in Si on temperature for different electron concentrations.

6.3.1 Impact Ionization

In high electric fields, electron-hole pairs in a semiconductor are often generated by **impact ionization**. In this process, an electron (or a hole) acquires enough energy from the electric field to break a bond and promote another electron from the valence band into the conduction band. The electron impact ionization generation rate, G_{ni}, is proportional to the electron concentration, n, to the electron velocity, v_n, and to the **impact ionization coefficient**, α_{ni}:

$$G_{ni} = \alpha_{ni} n v_n, \tag{6.22}$$

where α_{ni}, is given by the following empirical expression:

$$\alpha_{ni} = \alpha_{no} \exp\left[-(F_{in}/F)^{m_{in}}\right]. \tag{6.23}$$

In a similar fashion, when the impact ionization is caused by holes, the result is as follows:

$$G_{pi} = \alpha_{pi} p v_p. \tag{6.24}$$

$$\alpha_{pi} = \alpha_{po} \exp\left[-(F_{ip}/F)^{m_{ip}}\right]. \tag{6.25}$$

Here F is an electric field and α_{no}, α_{po}, F_{in}, F_{ip}, m_{in}, and m_{ip} are constants that depend on semiconductor material and, in certain cases, even on the direction of the electric field. The impact ionization coefficients for Si calculated by using

TABLE 6.1 The Parameters of Some Important Semiconductors

	a [Å]	ε_r [rel.]	ρ [g/cm^3]	E_g [eV]	m_n	m_p	μ_n [cm^2/Vs]	μ_p [cm^2/Vs]
Si	5.43	11.8	2.33	1.12	1.08	0.56	1350	480
Ge	5.66	16.0	5.32	0.67	0.55	0.37	3900	1900
GaAs	5.65	13.2	5.31	1.42	0.067	0.48	8500	400
InP	5.87	12.1	4.79	1.35	0.080	–	4000	100

Properties of Important Semiconductors. Note: a: Lattice constant, ε_r: Dielectric constant, ρ: Density, E_g: Energy band gap, m_n: Electron effective mass (density of states effective mass for Si and Ge and central valley effective mass for GaAs and InP), m_p: Hole effective mass (density of states effective mass), μ_n: Electron mobility, μ_p: hole mobility. All are at 300 K. Table adapted from Fjeldy *et al.* (1998).

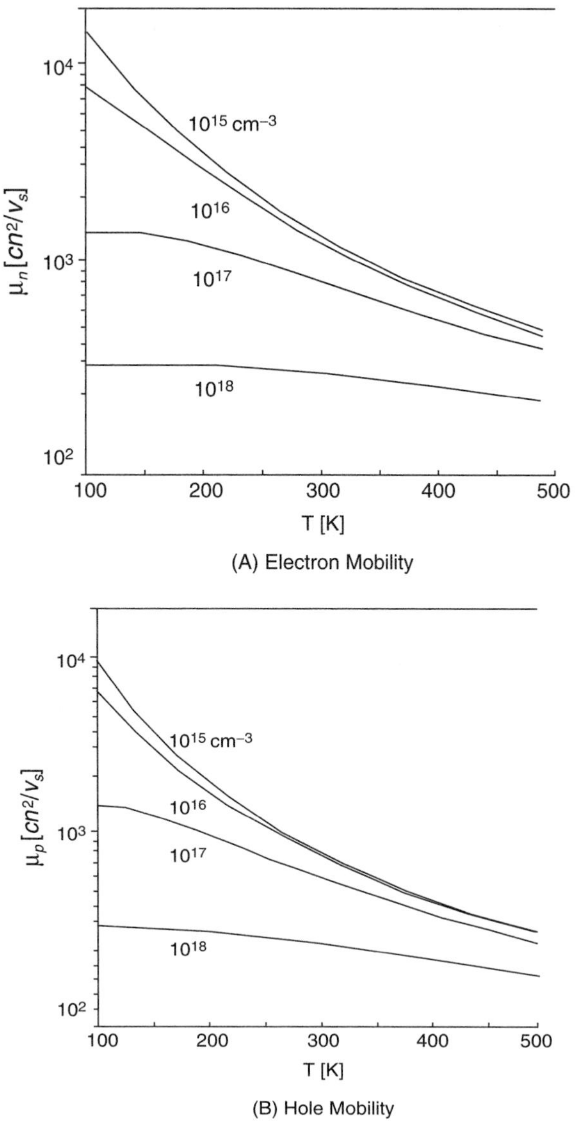

FIGURE 6.10 Electron and Hole Mobilities. These mobilities are shown in *n*-type, (A) and *p*-type (B) silicon, respectively, versus temperature for different impurity concentrations. Figure adapted from Fjeldly *et al.* (1998).

equations 6.3, 6.8, and 6.14 and equations 6.3, 6.8, and 6.16 are shown in Figure 6.11.

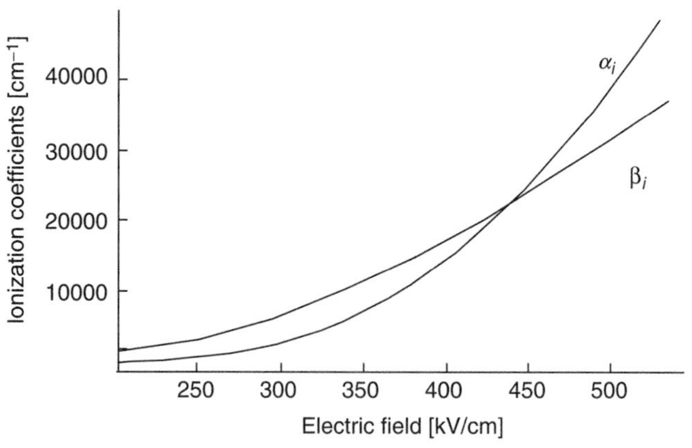

FIGURE 6.11 Impact Ionization Coefficients for Si; $\alpha_0 = 3,\ 318\ \text{cm}^{-1}$, $F_{no} = 1,174\,\text{kV/cm}$, $m_{in} = 1$. Figure adapted from Shur (1996).

6.3.2 Optical Properties

Generation of electron-hole pairs in a semiconductor can be achieved by illuminating a semiconductor sample with light with photon energies larger than the energy gap of the semiconductor. The light is partially reflected from the semiconductor surface and partially absorbed in the semiconductor (see Figure 6.12).

The light intensity P_l (measured in W/m^2) is given by:

$$P_l = P_o \exp(-\alpha x), \tag{6.26}$$

where P_o is the light intensity at the surface. The distance $1/\alpha$ is called the **light penetration depth** (see Figure 6.12(B)). If $1/\alpha \gg L$, where L is the sample dimension, the generation rate of electron-hole pairs is nearly uniform in the sample. Absorption coefficient, α, is a strong function of the wavelength. (Typically, α is measured in the range from 10^2 to $10^6\ \text{cm}^{-1}$.) The number of generated electron-hole pairs is proportional to the number of absorbed photons. Since the energy of each photon is $\hbar\omega$, the generation rate is given by:

$$G = Q_e \frac{\alpha P_l}{\hbar\omega}, \tag{6.27}$$

where Q_e is equal to the average number of electron-hole pairs produced by one photon. Q_e is called the **quantum efficiency.**

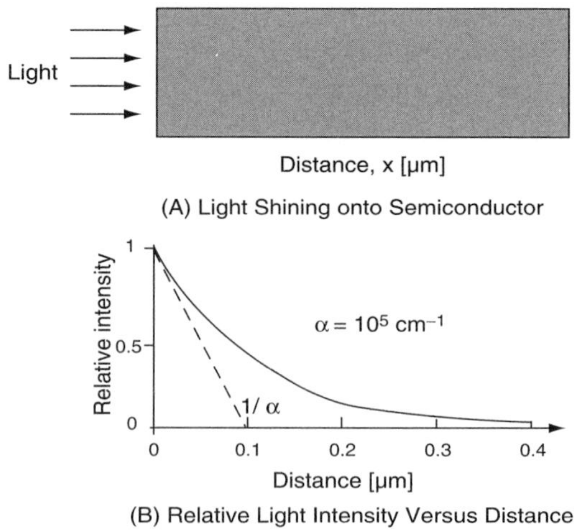

FIGURE 6.12 Light Effects on Semiconductors. Figure adapted from Shur (1996).

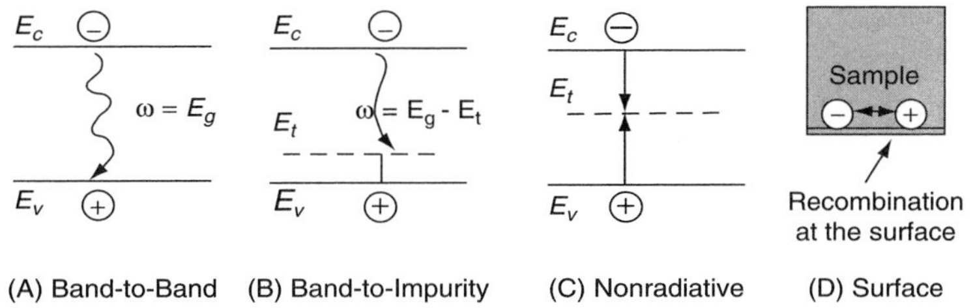

FIGURE 6.13 Recombination Processes According to Electron-Hole Pair Generation. The figures show (A) direct (band-to-band) radiative recombination, (B) radiative band-to-impurity recombination, (C) nonradiative recombination via impurity (trap) levels, and (D) surface recombination. Figure adapted from Shur (1996).

For $\hbar\omega > E_g$, the quantum efficiency Q_e is often fairly close to unity.

Under equilibrium conditions, $G = R$, where R is the recombination rate; that is, the generation of electron-hole pairs is balanced by different recombination processes that include **direct (band-to-band) radiative recombination, radiative band-to-impurity recombination, nonradiative recombination via impurity (trap) levels, Auger recombination**, and **surface recombination.** These processes (except for the Auger recombination discussed below) are schematically shown in Figure 6.13.

In the radiative recombination processes, recombining electron-hole pairs emit light, and the energy released during this recombination process is transformed into the energy of photons. Since during band-to-band recombination, the energy released by each recombining electron-hole pair is either equal to the energy gap or is slightly higher, the frequency of the emitted radiation can be estimated as:

$$\hbar\omega = E_g. \tag{6.28}$$

The radiative band-to-band recombination rate is proportional to the np product:

$$R = A\left(np - n_i^2\right), \tag{6.29}$$

since both electrons and holes are required for the band-to-band recombination (just like the number of marriages in a city should be proportional to both the number of eligible men and to the number of eligible women). In practical light-emitting semiconductor devices, radiative band-to-impurity recombination may be more important than radiative band-to-band recombination. The radiative band-to-impurity recombination rate is given by:

$$R = p/\tau_r, \tag{6.30}$$

where p is the concentration of the electron-hole pairs, $\tau_r = 1/(B_r N_t)$, N_t is the concentration of impurities involved in this radiative recombination process, and B_r is a constant.

In many cases, the dominant recombination mechanism is recombination via traps, especially in indirect semiconductors such as silicon. Shockley and Reed were first to derive the famous (but somewhat long) expression for the recombination rate related to traps:

$$R = \frac{pn - n_i^2}{\tau_{pl}(n + n_l) + \tau_{nl}(p + p_l)}. \tag{6.31}$$

Here, n_i is the intrinsic carrier concentration, and τ_{nl}, τ_{pl} are electron and hole lifetimes given by:

$$\tau_{nl} = \frac{1}{v_{thn}\sigma_n N_t} \text{ and } \tau_{pl} = \frac{1}{v_{thp}\sigma_p N_t}, \tag{6.32}$$

where v_{thn} and v_{thp} are electron and hole thermal velocities, σ_n and σ_p are effective capture cross-sections for electrons and holes, and N_t is the trap concentration. The carrier concentrations corresponding to the position of the Fermi level coinciding with the trap level can be written as follows:

$$n_l = N_c \exp\left(\frac{E_t - E_c}{k_B T}\right) \text{ and } p_l = N_v \exp\left(\frac{E_v - E_t}{k_B T}\right). \tag{6.33}$$

Here N_c and N_v are the effective densities of states in the conduction and valence bands, respectively, and E_t is the energy of the trap level. When electrons are minority carriers ($n = n_p \ll p \approx N_a = p_p$, where N_a is the concentration of shallow ionized acceptors, $p \gg p_l$, $p \gg n_l$), equation 6.31 reduces to:

$$R = \frac{n_p - n_{po}}{\tau_{nl}}, \tag{6.34}$$

where $n_{po} = n_i^2/N_a$. When holes are minority carriers ($p = p_n \ll n \sim N_d = n_n$, where N_d is the concentration of shallow ionized donors, $n \gg n_l$, $n \gg p_l$:

$$R = \frac{p_n - p_{no}}{\tau_{pl}}, \tag{6.35}$$

where $p_{no} = n_i^2/N_d$. The electron lifetime, τ_{nl}, in p-type silicon and the hole lifetime, τ_{pl}, in n-type silicon decrease with increasing doping. At low doping levels, this decrease may be explained by higher trap concentrations in the doped semiconductor. We can crudely estimate electron and hole lifetimes in Si as follows:

$$\tau_{nl}(\mathrm{s}) = \frac{10^{12}}{N_a(\mathrm{cm}^{-3})} \text{ and } \tau_{pl}(\mathrm{s}) = \frac{3 \times 10^{12}}{N_d(\mathrm{cm}^{-3})}. \tag{6.36}$$

At high doping levels, however, the lifetimes decrease faster than the inverse doping concentration. The reason is that a different recombination mechanism, called **Auger recombination**, becomes important. In an Auger process, an electron and a hole recombine without involving trap levels, and the released energy (of the order of the energy gap) is transferred to another carrier (a hole in p-type material and an electron in n-type material). Auger recombination is the inverse of impact ionization, where energetic carriers cause the generation of electron-hole pairs. Since two electrons (in n-type material) or two holes (in p-type material) are involved in the Auger recombination process, the recombination lifetime associated with this process is inversely proportional to the square of the majority carrier concentration; that is:

$$\tau_{nl} = \frac{1}{G_p N_a^2} \text{ and } \tau_{pl} = \frac{1}{G_n N_d^2}, \tag{6.37}$$

for p-type material and for n-type material, respectively, where $G_p = 9.9 \times 10^{-32}\,\mathrm{cm}^6/\mathrm{s}$ and $G_n = 2.28 \times 10^{-32}\,\mathrm{cm}^6/\mathrm{s}$ for silicon.

As stated previously, in many cases the recombination rate can be presented as $R = (p_n - p_{no})/\tau_{pl}$ (in n-type material in which holes are minority carriers). In this case, the diffusion of the minority carriers can be described by the **diffusion equation**:

$$D_p \frac{\partial^2 p_n}{\partial x^2} - \frac{p_n - p_{no}}{\tau_{pl}} = 0. \tag{6.38}$$

The solution of this second-order linear differential equation is given by:

$$\Delta p = A \exp(x/L_p) + B \exp(-x/L_p), \tag{6.39}$$

where $\Delta p = p_n - p_{no}$, A and B are constants to be determined from the boundary conditions, and

$$L_p = \sqrt{D_p \tau_{pl}} \tag{6.40}$$

is called the **hole diffusion length**. A similar diffusion equation applies to electrons in p-type material, where electrons are minority carriers. The **electron diffusion length** is then given by:

$$L_n = \sqrt{D_n \tau_{nl}}. \tag{6.41}$$

6.4 Important Semiconductor Materials

Today, most semiconductor devices are made of silicon. However, submicron devices made of compound semiconductors, such as gallium arsenide or indium phosphide, successfully compete for applications in microwave and ultra-fast digital circuits. Other semiconductors, such as mercury cadmium telluride are utilized in infrared detectors. Silicon carbide, aluminum nitride, and gallium nitride promise to be suitable for power devices operating at elevated temperatures and in harsh environments. Hydrogenated amorphous silicon and related compounds have found important applications in the display industry and in photovoltaics. Ternary compounds, quaternary compounds, and heterostructure materials provide another important dimension to semiconductor technology. Developing new semiconductor materials and using their unique properties in electronic industry is a major challenge that will exist for many years to come.

References

Fjeldly, T., Ytterdal, T., and Shur, M.S. (1998). Introduction to device and circuit modeling for VLSI. New York: John Wiley & Sons.

Levinshtein, M.E., and Simin, G.S. (1992). *Getting to know semiconductors.* Singapore: World Scientific.

Pierret, R.F. (1988). *Semiconductor Fundamentals,* (2d Ed.): *Modular series on solid state devices, Vol. 1.* Reading, MA: Addison-Wesley.

Shur, M.S. (1996). *Introduction to electronic devices.* New York: John Wiley & Sons.

Streetman, B.G. (1995). *Solid state electronics devices,* (4th Ed.). Englewood Cliffs, NJ: Prentice Hall.

Wolfe, C.M., Holonyak, N.J., and Stillman, G.E. (1989). *Physical properties of semiconductors.* Englewood Cliffs, NJ: Prentice Hall.

7

Power Semiconductor Devices

Maylay Trivedi and
Krishna Shenai
Department of Electrical and Computer Engineering, University of Illinois at Chicago, Chicago, Illinois, USA

7.1 Introduction

Power semiconductor devices are used as switches in power electronics applications. Ideal switches arbitrarily block large forward and reverse voltages, with zero current flow in the off-state, arbitrarily conduct large currents with zero voltage drop in the on-state, and have negligible switching time and power loss. Material and design limitations prevent semiconductor devices from operating as ideal switches. It is important to understand the operation of these devices to determine how much the device characteristics can be idealized. Available semiconductor devices could be either controllable or uncontrollable. In an **uncontrollable device**, such as the diode, on-and off-states are controlled by circuit conditions. Devices like BJTs, MOSFETs, and their combinations can be turned on and off by control signals, and hence, are **controllable devices. Thyristors** belong to an intermediate category defined by latch-on being determined by a control signal and turn-off governed by external circuit conditions. This chapter presents a summary of the design equations, terminal characteristics, and the circuit performance of contemporary and emerging power devices.

7.2 Breakdown Voltage

Figure 7.1(A) shows the cross-section of a typical **power diode.** A *p-n* junction is formed when an *n*-type region in a semiconductor crystal abuts a *p*-type region in the same crystal. Rectification properties of the device result from the presence of two types of carriers in a semiconductor: electrons and holes. Power devices are required to block high voltage when the anode is biased negative with respect to the cathode. Application of a reverse bias removes carriers from the junction and creates a depletion region devoid of majority carriers. The width of the depletion layer depends on the applied bias, V_R, and doping of the region, N_D, as:

$$W_D = \sqrt{\frac{2\varepsilon(V_R + V_{bi})}{qN_D}}, \tag{7.1}$$

where V_{bi} is the built-in potential. A high electric field is formed in the depletion layer. This field rises as the reverse voltage is increased until a critical electric field, E_C, is reached, at which point the device breaks down. The critical electric field for Si is roughly $2 \times 10^5\,\text{V/cm}$. A nonpunchthrough diode has drift region doping (N_D) and width ($W_{D,pp}$) parameters such that the critical electric field is reached before the entire region is depleted. In such cases, the breakdown voltage, V_{BD}, and depletion width at breakdown, W_D, for a given doping level are expressed as:

$$V_{BD} = 5.34 \times 10^{13} N_D^{-3/4} \tag{7.2a}$$

and

$$W_{D,pp} = 2.67 \times 10^{10} N_D^{-7/3}. \tag{7.2b}$$

In some power devices, it is preferable to use a punchthrough structure to support the voltage. In this case, the entire drift region is depleted at the time of breakdown. The breakdown voltage of a punchthrough device with a certain doping (N_D) and width $W_D(< W_{D,pp})$ is given as:

$$V_{BD} = E_C W_D - \frac{q N_D W_D^2}{2\varepsilon}. \tag{7.3}$$

The thickness, W_D, of the punchthrough structure is smaller than that of the nonpunchthrough device with the same breakdown voltage. A thinner drift region is preferable for bipolar power devices, such as P-i-N diodes operating under high-level injection conditions.

Planar diodes are formed by implantation and diffusion of impurities through a rectangular diffusion window. A cylindrical junction is formed at the straight edges of the diffusion window, and a spherical junction is formed at each of the corners. The breakdown voltage of practical devices can be significantly lower than the ideal values by the occurrence of high electric fields at the device edges. With respect to the parallel-plane breakdown voltage, $V_{BD,pp}$, the breakdown voltage of cylindrical ($V_{BD,CYL}$) and spherical ($V_{BD,SP}$) junctions with a junction depth x_j is given as the following two equations:

$$\frac{V_{BD,CYL}}{V_{BD,PP}} = \frac{1}{2}\left[\left(\frac{x_j}{W_D}\right)^2 + 2\left(\frac{x_j}{W_D}\right)^{6/7}\right] \ln\left[1 + 2\left(\frac{x_j}{W_D}\right)^{8/7}\right] - \left(\frac{x_j}{W_D}\right)^{6/7}. \tag{7.4a}$$

$$\frac{V_{BD,SP}}{V_{BD,PP}} = \left(\frac{x_j}{W_D}\right)^2 + 2.14\left(\frac{x_j}{W_D}\right)^{6/7} - \left[r\left(\frac{x_j}{W_D}\right)^3 + 3\left(\frac{x_j}{W_D}\right)^{13/7}\right]^{2/3}. \tag{7.4b}$$

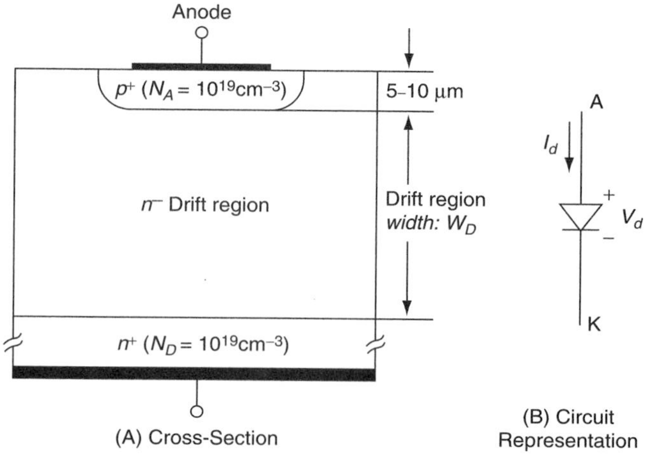

FIGURE 7.1 Typical Power Diode (P-i-N)

EXAMPLE 7.1. The asymmetrically doped p^+n diode shown in Figure 7.1 should be designed to support a reverse voltage of 500 V by forming a p^+ junction 5 μm deep into the n-type material. For avalanche breakdown at the parallel-plane junction, the drift region doping and maximum depletion width at breakdown are obtained using equation 7.3 to yield $5 \times 10^{14}\,\text{cm}^{-3}$ and 36 μm respectively.

Considering that the p^+ junction is 5 μm deep into the n-type drift region, the diode forms a cylindrical junction at the edge, with a radius of 5 μm. This gives a ratio of 0.14 with the ideal parallel plane breakdown depletion width. Then, using equation 7.4a, the finite device will have a cylindrical breakdown voltage that is 40% of the ideal parallel plane breakdown. In this case, the finite device has a breakdown voltage of 200 V.

Special device termination structures have been devised to reduce the electric field strength at the device edge and to raise the breakdown voltage close to the ideal value. These include floating field rings, field plates, bevelled edges, ion-implanted edge terminations, and so forth.

7.3 P-i-N Diode

The **P-i-N diode** has a punchthrough structure as described earlier and has an n^+ region at the end of the drift region opposite from the rectifying junction. The n^+ region provides an ohmic cathode contact and injects electrons into the drift region under high-level injection. When the diode is forward biased, the p^+ anode injects minority holes into the drift region, resulting in current flow. The forward bias current flow is determined by recombination of the minority carriers injected into the drift region. The device conducts in low-level injection regime when the injected minority carrier density is much lower than background doping. The magnitude of this current is written as:

$$J_F = \frac{qD_p p_{on}}{L_p}\left(e^{\frac{qVF}{kT}} - 1\right), \tag{7.5}$$

where V_F is the voltage drop across the rectifying junction.

Away from the junction in the drift region, current flow is dominated by majority carriers under the influence of an electric field. This causes a resistive voltage drop. During on-state current flow, when injected minority carrier density exceeds background doping concentration, the device is said to be conducting in high-level injection. The current flow in high-level injection is given by:

$$J_F = \frac{2qL_a n_i}{\tau_a}\tanh\left(\frac{W_D}{2L_a}\right)\left(e^{\frac{qVF}{kT}} - 1\right), \tag{7.6}$$

where τ_a and L_a are the ambipolar carrier lifetime and diffusion length, respectively.

7.3.1 Switching Characteristics

A power diode requires a finite time to switch from the blocking state (reverse bias) to the on-state (forward bias) and vice versa. This is because of the finite time required for establishment and removal of steady-state carrier distribution in the drift region. During turn-on, the terminal voltage, V_A, first rises to a value V_{FP} that is much higher than the static forward drop before decaying to the steady-state value, V_F, determined by the static characteristics. This process is called forward recovery, and V_{FP} is known as the forward recovery voltage. When the diode is turned off, the current changes at a rate dI/dt determined by the external circuit conditions. The diode current displays a negative peak before decaying to zero. Correspondingly, device voltage rises to the off-state voltage, V_R. The process of negative current flow during device turn-off is referred to as the **reverse recovery process**. The current, I_{rr}, is the reverse recovery current peak, and t_{rr} is the reverse recovery time. These quantities are expressed as:

$$I_{rr} = \sqrt{\frac{2Q_{rr}(di_R/dt)}{S+1}} \tag{7.7a}$$

and

$$t_{rr} = \sqrt{\frac{2Q_{rr}(S+1)}{di_R/dt}}, \tag{7.7b}$$

where S is the "snappiness" factor defined in Figure 7.2. The variable Q_{rr} represents the reverse charge extracted out of the diode and is indicated by the shaded area in Figure 7.2.

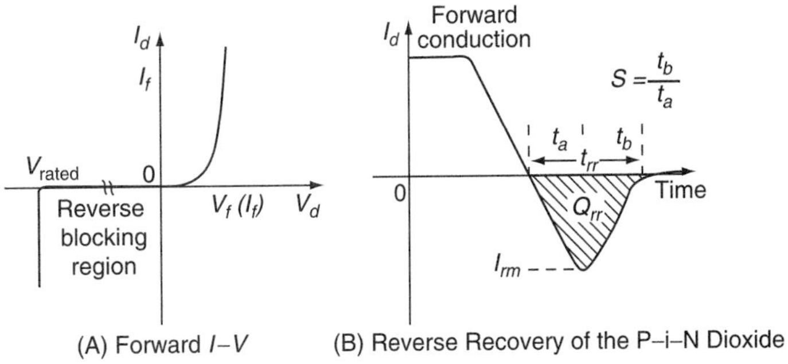

FIGURE 7.2 Reverse and Forward Biases of a Power Node

The extracted charge, Q_{rr}, is a fraction of the total charge $Q_F(=\tau_F I_F)$ stored in the diode during forward bias.

Example 7.2. The diode designed in example 7.1 has a high-level carrier lifetime of 500 ns. At a current density of 100 A/cm^2, assuming conduction in high-level injection, the voltage drop across the diode is roughly given by equation 7.2, where $L_a = \sqrt{(D_a\tau_a)}$. The ambipolar diffusion constant, D_a, for a diode at room temperature is 9 cm^2/V-s. Then, for a lifetime of 500 ns, the ambipolar diffusion length, L_a, is 21 μm, and the forward voltage of the diode is 0.82 V.

This diode is turned off from an on-state current density of 100 A/cm^2 with a turn-off dJ/dt of 100 A/cm^{2}μs. Assuming that all the excess charge is removed from the device, $Q_{rr} \sim Q_F = \tau_F J_F$. Thus, a charge of 50 μC is removed from the device. For a diode with S = 1, the reverse recovery current has a magnitude of 71 A/cm^2 and a duration of 0.7 μs.

7.4 Schottky Diode

A P-i-N diode displays large reverse recovery transients because of charge accumulation in the drift region brought about by bipolar current flow. A majority carrier diode may be formed by making a rectifying junction between a metal and the semiconductor. Such a device, as shown in Figure 7.3, is referred to as the **Schottky diode**. The rectifying characteristics of a Schottky diode are identical to a P-i-N diode, although the fundamental physics of operation are quite different. The forward current of a Schottky diode is expressed as:

$$J_F = A^* T^2 e^{\frac{q\phi_{Bo}}{kT}}\left(e^{\frac{qV_F}{kT}} - 1\right), \tag{7.8}$$

where A^* is the effective Richardson's constant and where ϕ_{Bo} is the metal-semiconductor barrier height defined as the difference between the metal and semiconductor work

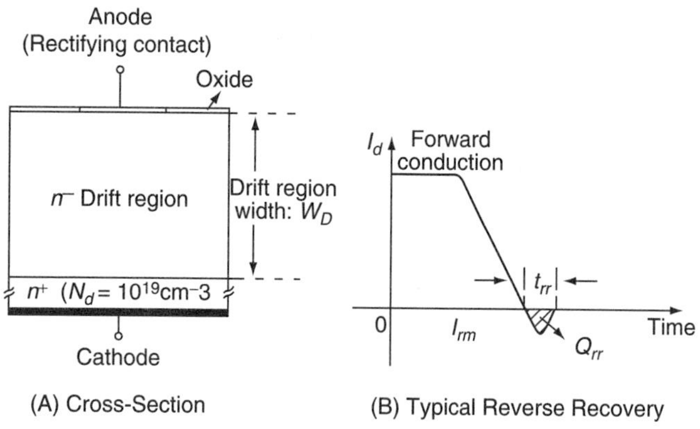

FIGURE 7.3 A Schottky Diode

functions. Current flow is dominated by majority carriers. The Schottky diode has a negligible stored charge in the drift region because of a lack of minority carriers. The switching transients do not include forward and reverse recovery. A finite transient recovery is observed because of the capacitance associated with the depletion region. However, the reverse leakage current is considerably higher than for P-i-N diodes. Furthermore, at higher voltage rating, the Schottky diode has a considerable voltage drop in the drift region, expressed as:

$$V_{dr} = \frac{J_F W_D}{q\mu_n N_D}. \tag{7.9}$$

EXAMPLE 7.3. A Schottky diode is designed with the same drift region parameters indicated in example 7.1. The Schottky contact has a barrier height of 0.8 eV. Conduction in a Schottky diode takes place by majority carriers. Forward voltage drop of the diode is obtained from equations 7.8 and 7.9 as:

$$V_F = \frac{kT}{q}\ln\left(\frac{J_F}{A^* T^2 e^{q\phi Bo/kT}}\right) + R_s J_F = 0.49 + 3.1 = 3.59\,\text{V},$$

where $R_s = \frac{W_D}{q\mu_n N_D}$ is the specific series resistance of the drift region and $A^* = 146\,\text{Å/cm}^2\text{-K}^2$ for Si. A 200-V Schottky diode requires a drift region with a doping of $1.7 \times 10^{15}\,\text{cm}^{-3}$ and thickness of 12.5 μm. For a barrier height identical to the previous case, the forward voltage of this diode is $V_F = 0.49 + 0.32 = 0.81\,\text{V}$.

The voltage drop across the series resistor in the Schottky diode rises rapidly with the breakdown voltage rating. This restricts the operating voltage range of Schottky diodes even as reverse recovery limits high-frequency operation of P-i-N diodes. Hybrid devices such as junction barrier Schottky and merged P-i-N–Schottky diodes have been developed to expand the operating range of power diodes by taking advantage of the properties of Schottky and P-i-N diodes.

7.5 Power Bipolar Transistor

The vertical power bipolar junction transistor (BJT) has a four-layer structure of alternating *p*-type and *n*-type doping as shown in Figure 7.4. The transistor has three terminals labeled **collector, base**, and **emitter.** The need for a large off-state blocking voltage and high on-state current-carrying capability is responsible for the changes in structure over the logic-level counterpart. The transistor is commonly operated in a common emitter mode. The steady-state static *I–V* characteristics of the power BJT are shown in Figure 7.5. The BJT is turned on by injecting current in the base terminal and is a current-controlled device. This causes electrons to flow from the emitter to collector. The maximum collector current for a given base current is determined by the current gain, $I_{C,\,\max} = \beta I_B$:

$$I_C = \frac{A_q D_n N_B}{W_B}\exp\left(\frac{qV_{BE}}{kT}\right). \tag{7.10}$$

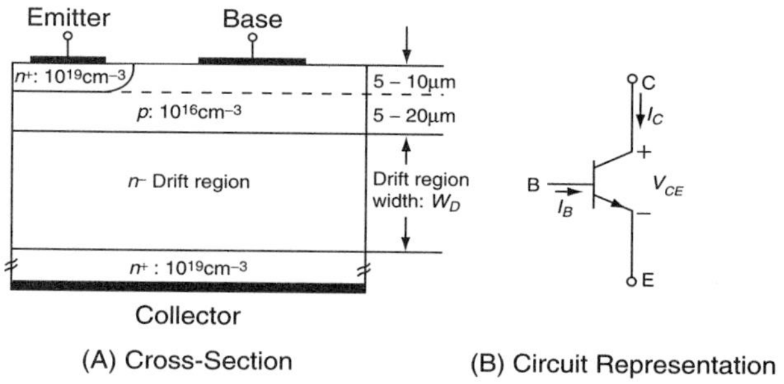

FIGURE 7.4 Power Bipolar Junction Transistor

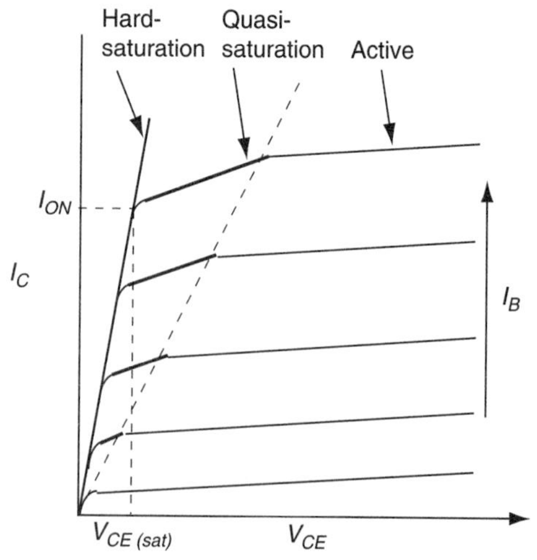

FIGURE 7.5 Output Characteristics of Power BJT

The static output characteristics of the BJT has three distinct regimes of conduction. From its off-state current, the transistor enters a forward active region when the base–emitter (B–E) junction is forward biased by a base current. With the base-collector (B–C) junction reverse biased, a collector current of magnitude $\beta_F I_B$ flows through the device. For the same base current, if the collector voltage is reduced such that the B–C junction is barely forward biased, the device enters the quasi-saturation mode of operation. The boundary for this transition is $V_{CE} = V_{BE,\,on} + I_C R_D$, where R_D is the resistance of the drift region. Forward biasing the B–C junction results in minority carrier injection into the collector drift region. The increase in effective base width reduces the transistor current gain and, hence, the collector current. The device conducts in hard saturation when the entire collector drift region is in high-level injection. Any additional reduction in collector voltage drives the transistor further into hard saturation. The transistor has very little on-state voltage in hard saturation and quasisaturation and is operated in this condition for practical application.

Since BJTs are controlled by current, base current must be supplied continuously to maintain conduction, resulting in power loss. Power BJTs have a low current gain ($\beta \sim 5 - 20$). The gain can be enhanced by connecting the transistors in Darlington configuration, as shown in Figure 7.6. The output of one transistor drives another transistor, resulting in higher overall current gain. The current capability, however, is accompanied by a deterioration in forward voltage and switching speed.

7.5.1 Switching Characteristics

The **clamped inductive load circuit**, a typical switching application of a controllable power switch, is shown in Figure 7.7. The switching waveforms of the power BJT, used as a switch in

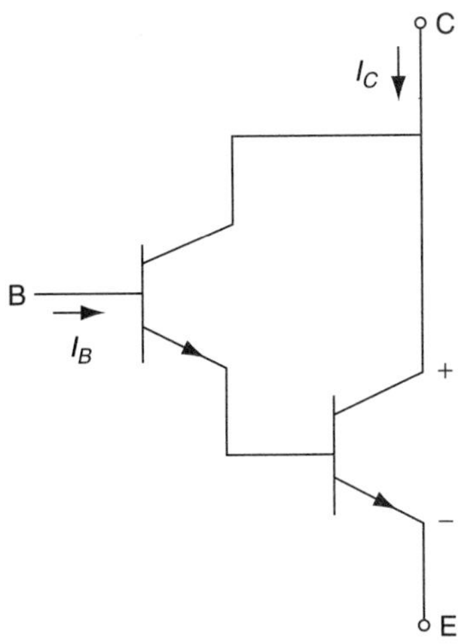

FIGURE 7.6 Darlington Configuration

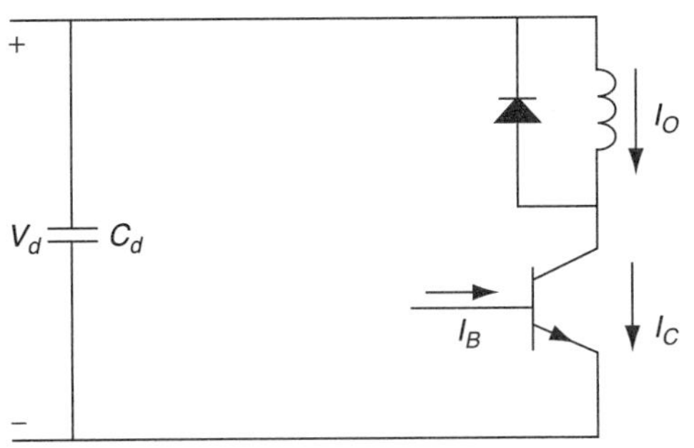

FIGURE 7.7 Clamped Inductive Load Circuit

the circuit, are shown in Figure 7.8. A large inductor represents a typical inductive load that is driven by the circuit. The inductor is approximated as a constant current source I_O. For a given collector current, the base current should be high enough to ensure conduction in hard or quasisaturation.

After an initial delay because of the B–E capacitance, the charge in the base begins to buildup, and the collector current rises to its on-state value with the voltage clamped to V_d by the diode. The voltage beings to fall rapidly once the device attains on-state current. As the charge is injected into the drift region, the gain of the transistor reduces and slows down the rate of fall of voltage. This phase increases the overall device turn-on time. The turn-off process involves removal of all the stored charge in the transistor. The base current is driven negative to remove the charge during turn-off. The negative base current rapidly removes excess charge from the device. The base current reversal is usually achieved gradually because abrupt reversal leads to a long current tail, thus increasing switching loss. When the base current is reversed, the collector current remains constant for a storage time, t_s, until the drift region comes out of high-level injection at one of its ends. After this interval, the device enters quasisaturation, and voltage starts rising slowly. Once the charge distribution reduces to zero at the collector-base junction, the voltage rises rapidly until the diode clamps it to the bus voltage. The collector current then falls off toward zero and device is cut off.

For higher current or voltage-handling capability, devices can be connected in parallel or series. Connecting BJTs in parallel can lead to instabilities because of a negative temperature coefficient of the on-state resistance. During conduction, if the temperature of one device rises, its resistance reduces, causing it to pass more current than other transistors. This, in turn, leads to a further rise in current, eventually resulting in a thermal runaway condition that destroys the device.

7.6 Thyristor

The vertical cross-section and symbol of a thyristor are shown in Figure 7.9. The thyristor is a three-junction *p-n-p-n*

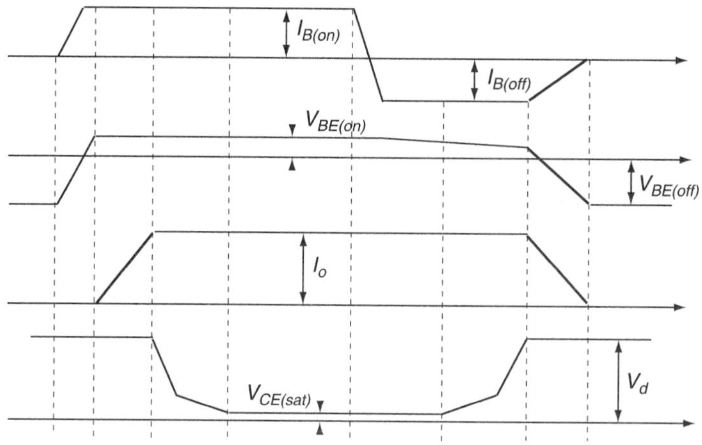

FIGURE 7.8 Switching Waveforms of a Power BJT During Clamped Inductive Turn-Off

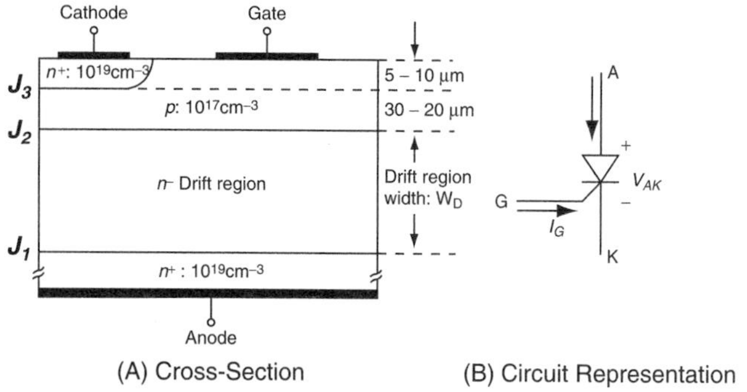

FIGURE 7.9 Diagram of a Thyristor

structure. The thyristor may be viewed as two bipolar transistors connected back to back. The collector current of one transistor provides base current to the other transistor. In the reverse direction, the thyristor operates similarly to a reverse-biased diode. When a forward polarity volage is applied, the junction J_3 (between the gate and drift region) is reverse biased, and the thyristor is in the forward blocking mode. The device can be triggered into latch-up by injecting a hole current from the gate electrode. Considering I_{CO} to be the leakage current of each transistor in cut-off condition, the anode current can be expressed in terms of gate current as:

$$I_A = \frac{\alpha_2 I_G + I_{CO1} + I_{CO2}}{1 - (\alpha_1 + \alpha_2)}, \tag{7.11}$$

where α is the common base current gain of the transistor ($\alpha = I_C/I_E$). From equation 7.11, the anode current becomes arbitrarily large as $(\alpha_1 + \alpha_2)$ approaches unity. As the anode–cathode voltage increases, the depletion region expands and reduces the neutral base width of the n_1 and p_2 regions. This causes a corresponding increase in the α of the two transistors. If a positive gate current of sufficient magnitude is applied to the thyristor, a significant amount of electrons will be injected across the forward-biased junction, J_3, into the base of the $n_1p_2n_2$ transistor. The resulting collector current provides base current to the $p_1n_1p_2$ transistor. The combination of the positive feedback connection of the *npn* and *pnp* BJTs and the current-dependent base transport factors eventually turn the thyristor on by regenerative action. Among the power semiconductor devices known, the thyristor shows the lowest forward voltage drop at large current densities. The large current flow between the anode and cathode maintains both transistors in saturation region, and gate control is lost once the thyristor latches on.

7.6.1 Transient Operation

When gate current is injected in the thyristor, finite time is required by the regenerative action to turn the thyristor on under the influence of the gate current (see Figure 7.10). A time delay is observed before the current starts rising. After the delay time, the anode current rapidly rises toward its on-state value at a rate limited only by external current elements, and the device voltage collapses. The finite time required for the charge plasma to spread through the device causes a tail in on-state voltage. The thyristor cannot be turned off by the gate and turns off naturally when the anode current is forced to change direction. A thyristor exhibits turn-off reverse recovery characteristics just like a diode. Excess charge is removed once the current crosses zero and attains a negative value at a rate determined by external circuit elements. The reverse recovery peak is reached when either junction J_1 or J_3 becomes reverse biased. The reverse recovery current starts decaying, and the anode–cathode voltage rapidly attains its off-state value.

Because of the finite time required for spreading or collecting the charge plasma during turn-on or turn-off stage, the maximum dI/dt and dV/dt that may be imposed across the device are limited in magnitude. Further, device manufacturers

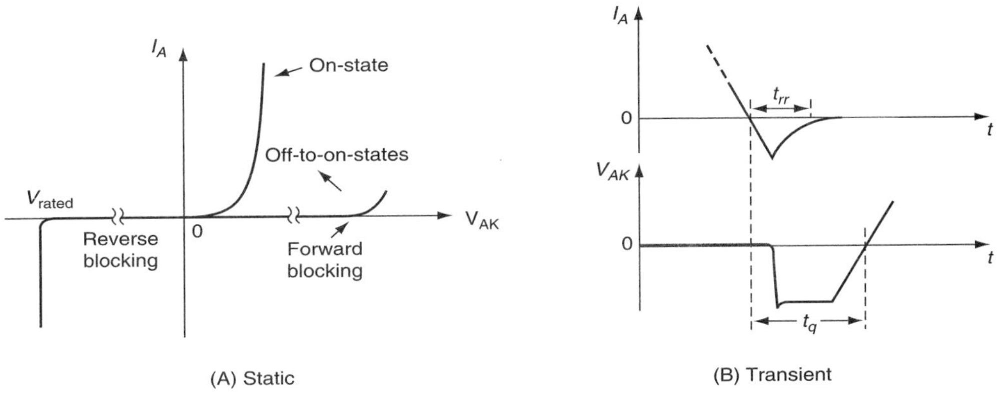

FIGURE 7.10 Typical Characteristics of a Thyristor

specify a circuit-commutated recovery time, t_q, for the thyristor, which represents the minimum time for which the thyristor must remain in its reverse blocking mode before forward voltage is reapplied.

Thyristors come in various forms depending on the application requirements. They can be made to handle large voltage and current with low on-state forward drop. Such thyristors typically find applications in circuits requiring rectification of line frequency voltage and currents. Faster recovery thyristors can be made at the expense of increased on-state voltage drop.

7.7 Gate Turn-Off Thyristor

The gate turn-off thyristor (GTO) adds gate-controlled turn-off capability to the thyristor. In a GTO, modifications are made to ensure that the base current of the transistor $n_1p_2n_2$ is briefly made less than the value needed to maintain saturation ($I_{B2} < I_C/\beta$). This forces the transistor to go into the active region. The regenerative action pulls both transistors out of saturation into the active region and eventually into turn-off mode. A GTO requires complex interdigitated layout of gate and cathode. Anode short structures further suppress regeneration. The turn-off capability of a GTO is gained at the expense of on-state performance. Other variations in structure, such as the reverse-conducting thyristor (RCTs), asymmetrical silicon-controlled-rectifier (ASCR), light-triggered thyristor (LTT), and so forth, are applied in many modern power electronic applications.

7.8 Metal-Oxide-Semiconductor Field Effect Transistor

The power metal-oxide-semiconductor field effect transistor (MOSFET) is perhaps the most thoroughly investigated and optimized power semiconductor device. Figure 7.11 shows the cross-section of a vertical n-channel power MOSFET. The MOSFET is a voltage-controlled device as opposed to a BJT that is a current-controlled device. In an off-state mode, the depletion layer expands in the drift region and supports the blocking voltage. As the depletion layer grows, it pinches off the region between the p-wells and isolates the gate oxide from the high voltage appearing at the drain. The channel region is formed by implantation and diffusion of p-type impurities in the window for the source region. The p-type region also isolates the source and the drain. Application of a gate voltage higher than a threshold value creates an inversion channel under the gate oxide that supports current flow from source to the drain. For an ideal metal-oxide-semiconductor structure, the threshold voltage is expressed as:

$$V_{TH} = \sqrt{\frac{4\varepsilon kTN_D \ln(N_D/n_i)}{\varepsilon_{ox}/t_{ox}}} + \frac{2kT}{q}\ln\left(\frac{N_D}{n_i}\right). \qquad (7.12)$$

The maximum current supported per unit cell for a given gate voltage is limited by the threshold voltage and channel length.

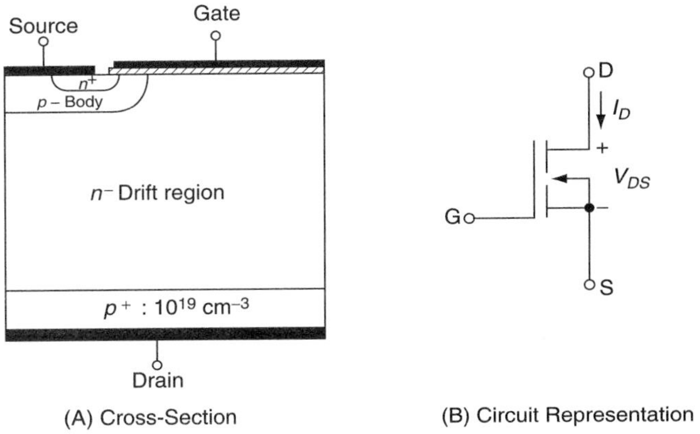

FIGURE 7.11 Power MOSFET

Several unit cells of the kind shown in Figure 7.11 are connected in parallel to achieve the desired current rating. The cells are connected in a variety of geometrical layouts, such as stripes, squares, and hexagonal shapes. The source n^+ region and p-body region are shorted. This creates a parasitic diode connected from the drain to the source of the MOSFET. This integral diode is useful in several power electronics applications, such as half-bridge and full-bridge PWM converters.

The MOSFET is a majority carrier device. In the case of the n-channel MOSFET shown in Figure 7.11, the electrons traveling from source to drain have to overcome the resistance offered by the channel, the accumulation region at the surface, the neck or JFET region between the p-wells, the drift region, and the substrate, as indicated in Figure 7.12(A). The overall resistance of the device is determined by the contributions of the individual components. Thus, the following is true:

$$R_{on} = R_{ch} + R_{acc} + R_{neck} + R_{drift} + R_{sub}. \quad (7.13a)$$

$$\frac{L_{ch}}{2ZC_{ox}\mu_{eh}(V_{GS} - V_{TH})} + \frac{L_G}{8ZC_{ox}\mu_{acc}(V_{GS} - V_{TH})} + \frac{x_D}{q\mu_n ZL_G N_D} + R_{on}$$
$$= \frac{\ln[1 + 2(W_D/L_G)\tan\alpha]}{q\mu_n ZN_D \tan\alpha} + \frac{W_{sub}}{q\mu_n ZL_{cell}N_{sub}}. \quad (7.13b)$$

The resistance of low-voltage MOSFETs is dominated by the resistance of the channel and neck regions. Higher voltage MOSFETs, however, have thicker drift regions with low doping. The resistance is then dominated by the drift region. This limits the application of MOSFETs at higher voltage levels. The I–V characteristics of the power MOSFET are shown in Figure 7.12(B). Increasing the gate voltage raises the saturation current level as given by:

$$I_{D,\ sat} = \mu_{ch} C_{ox} Z (V_{GS} - V_{TH})^2 / L_{ch}. \quad (7.14)$$

Example 7.4. A power MOSFET has a p-base region that is uniformly doped at a concentration of $1 \times 10^{17}\ \text{cm}^{-3}$ with a gate oxide thickness of 1000 Å. Using equation 7.14, the threshold voltage of this device is 5.65 V at room temperature. A fixed charge density at the semiconductor–oxide interface introduces additional charge and shifts the threshold voltage of the device by:

$$\Delta V_{TH} = -\frac{Q_{ox}}{C_{ox}}.$$

Thus, a charge density of $1 \times 10^{11}\ \text{cm}^{-2}$ would cause a shift of -0.47 V in the threshold voltage. Saturation current of the MOSFET is determined by the channel. For a MOSFET with channel length of 2 μm, the channel width required to have a saturation current of 1 A at a gate voltage of 7 V can be obtained with the help of equation 7.14 to be 6.4 cm. The channel on-resistance is 0.68 Ω.

7.8.1 Switching Performance

The MOSFETs require continuous application of gate-source voltage above the threshold voltage to be in an on-state mode. The gate is isolated from the semiconductor by the oxide. Gate current flows only during transition from on-to off-state and vice versa when the gate capacitor charges or discharges. This implies reduced power loss in the control circuit and simple gate control. The switching transients are determined by the time required to charge and discharge various capacitors in the MOSFET structure. The capacitors appear where the gate oxide overlaps the semiconductors and junction capacitance because of the depletion layer, as indicated in Figure 7.12.

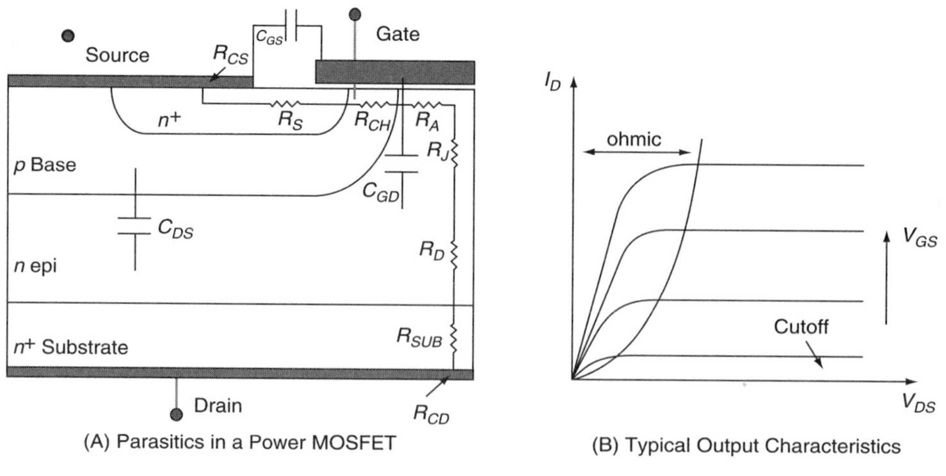

FIGURE 7.12 MOSFET Device

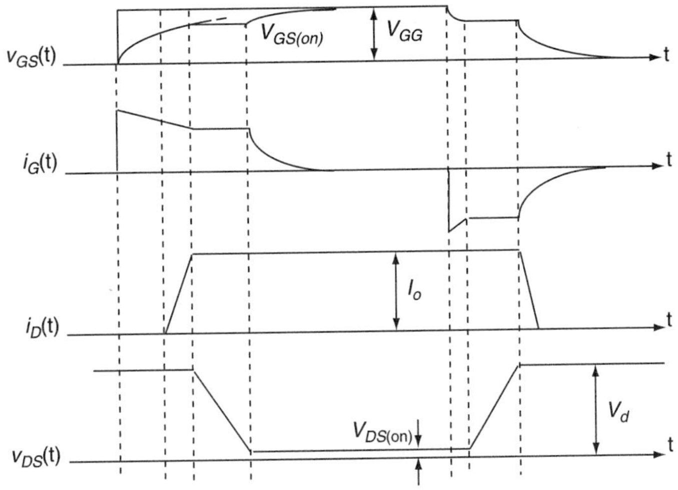

FIGURE 7.13 Switching Waveforms of Power MOSFET

Important switching waveforms of the MOSFET are indicated in Figure 7.13 for an inductive load, similar to the one in Figure 7.7. The gate pulse makes transitions between V_{GP} and V_{GN} through the gate resistor, R_G. Initially, the device is off. When the gate pulse is turned on high, the capacitor C_{GD} and C_{GS} charge exponentially toward V_{GP} through R_G. The MOSFET turns on when the gate-source voltage crosses V_{TH}. The device goes from cut-off to saturation region. As the gate voltage rises, the MOSFET current also increases, as determined by equation 7.14, until it is conducting I_O. The drain voltage remains at V_D as long as the free-wheeling diode is conducting. Once the MOSFET attains the current I_O, the drain current is clamped. Since the device is conducting in saturation, the gate-source voltage is also clamped. The capacitor C_{GD} now starts discharging through the gate resistor. The gate current is given as:

$$I_G = \frac{V_{GP} - V_{GS,\,on}}{R_G} = C_{GD}\frac{dV_{GD}}{dt} = C_{GD}\frac{dV_{CE}}{dt}, \qquad (7.15)$$

where $V_{GS,\,on}$ is the gate-source voltage required to maintain a drain current I_O in saturation. The gate voltage starts rising exponentially again when the drain voltage reduces to on-state value.

The turn-off of the MOSFET involves the inverse sequence of the events that occur during turn-on, and the turn-off waveforms are also shown in Figure 7.16. During the switching transient, the maximum power loss occurs during the cross-over between voltage and current states. This power loss originates from the charging and discharging of capacitors. Since these capacitances do not vary with temperature, the switching power loss in MOSFET is independent of temperature. The on-state resistance and, hence, conduction loss, however, do vary with temperature.

Example 7.5. The MOSFET designed in example 7.4 has a parasitic capacitance of $C_{ISS} = 2.5\,\text{nF}$, $C_{RSS} = 200\,\text{pF}$, and $C_{OSS} = 600\,\text{pF}$. A gate pulse of 10 V is applied through a gate resistor of 10 Ω to the MOSFET to turn it on from an off-state voltage of 100 V to an on-state current of 500 mA. Based on example 7.4, the $V_{GS,\ on}$ of the MOSFET at this current level is about 7 V. During the turn-off phase, the current rapidly rises to its on-state value. The device voltage then reduces from 100 V to the on-state value in the phase of the Miller plateau. Then, according to equation 7.18, the time required for the device voltage to drop is 67 ns. Accordingly, the turn-on energy loss is as follows:

$$E_{on} = 1/2 I_{on} V_{Bus} t_{on} = 1.7\ \mu\text{J}.$$

7.8.2 Limits of Operation

Static breakdown of power MOSFETs occurs when the reverse voltage exceeds the static breakdown. As explained in equation 7.13, the on-resistance of the power MOSFET increases rapidly with the device blocking voltage because of increased drift region resistance. The dependence of resistance on breakdown voltage is nearly cubic ($\sim V_{BD}^{2.5-2.7}$). The on-resistance has a positive temperature coefficient because of a reduction in carrier mobility at higher temperature. This permits paralleling of MOSFETs without risk of instability. The device conducting higher current heats up and is forced to share its current equitably with the MOSFETs that are in parallel with it.

The MOSFET structure has a parasitic *npn* bipolar transistor, as can be identified from Figure 7.11. The base of this transistor is shorted to the source of the MOSFET through a

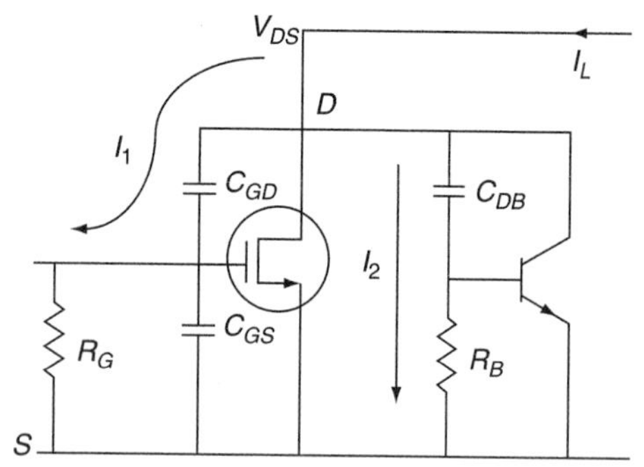

FIGURE 7.14 Circuit Representation of MOSFET with Parasitic Elements

resistor that is the lateral resistance of the *p*-base. The equivalent circuit model of the MOSFET, including the parasitic transistor, is shown in Figure 7.14. Under rapid transients of voltage at the drain, a displacement current is induced through the parasitic capacitors of the MOSFET along two paths, as shown. The current I_1 may rise V_{GS} above V_{TH}, thus causing spurious turn-on of the MOSFET. Likewise, the current I_2 may cause a resistive drop across R_B that is high enough to turn the parasitic bipolar transistor on. This places a limit on the maximum dV/dt across the MOSFET. Reliability of MOSFET is also affected by the power dissipation under transient conditions. This power dissipation leads to a temperature rise in the device. The temperature range of operation and highest current and voltage stress are determined by the extent of device self-heating.

7.8.3 Improved Structures

Apart from channel resistance, on-resistance of the MOSFET has contributions from the neck region between the *p*-wells and spreading resistance in the drift region. Various structures have been designed to reduce the on-resistance toward the ideal limit. Two notable device structures are the trench MOSFET and the lateral MOSFET shown in Figure 7.15. The trench MOSFET ensures a nearly one-dimensional current flow and eliminates the neck region. This structure also achieves a much lower cell size and higher packing density. Two of the major markets where power MOSFETs are finding increasing use are portable electronics and automotive applications. In such applications, the power MOSFET must be integrated with other circuit elements for logic-controlled or analog applications. This requires all terminals to be available at the top surface for integration. The lateral MOSFET, shown in Figure 7.15(B) satisfies this requirement.

7.9 Insulated Gate Bipolar Transistor

Although power BJTs have low on-state resistance, their operation involves circuit complexity and significant losses in the control circuitry because of the current-controlled nature of device operation. On the other hand, MOSFETs involve relatively simple gate drive circuits but have significant losses during forward conduction at higher blocking voltages. Improved overall performance is achieved by merging the advantages of bipolar conduction with MOS-gate control. Several hybrid MOS-gate transistors have been introduced. Among

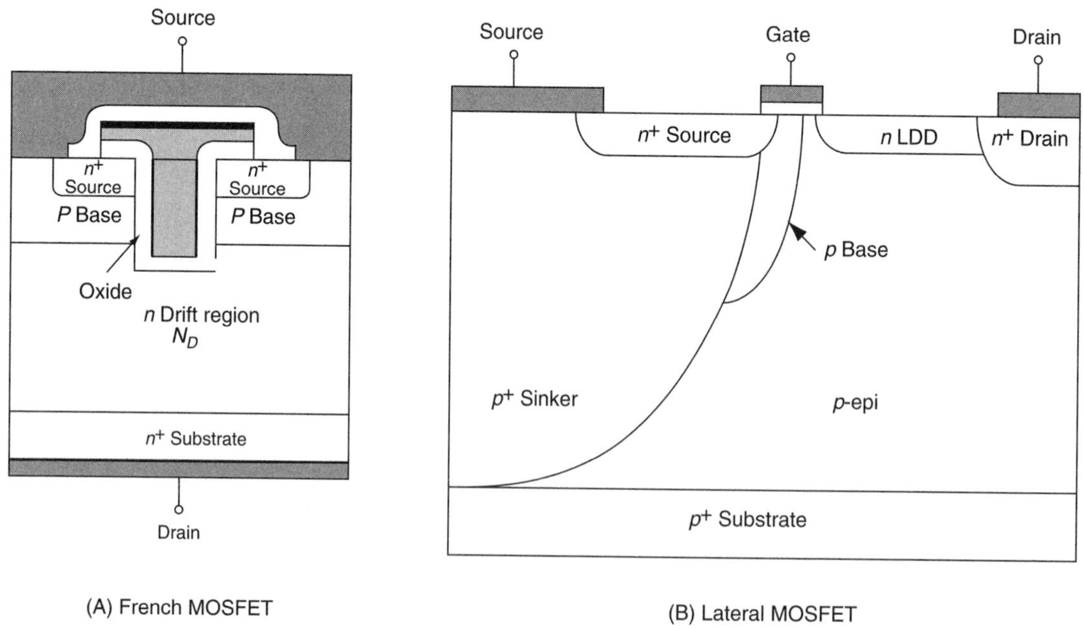

FIGURE 7.15 Cross-Sections of MOSFETs

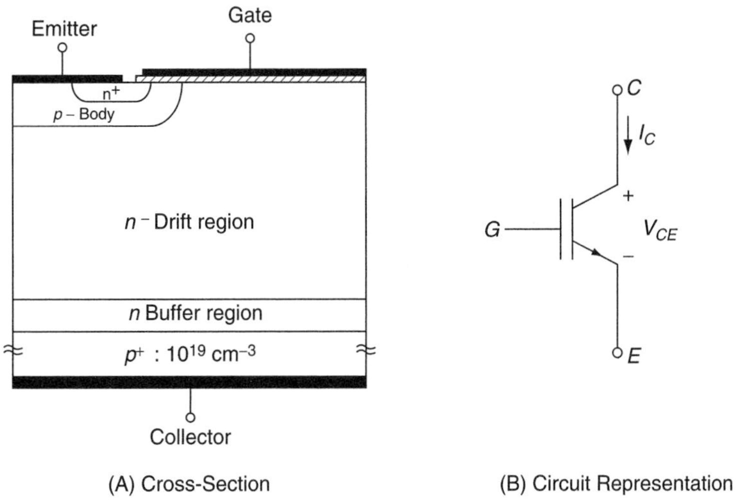

FIGURE 7.16 Insulated Gate Bipolar Transistor

these devices, the insulated gate bipolar transistor (IGBT) is the most popular and widely used device. A vertical cross-section of the IGBT is shown in Figure 7.16(A). The structure is similar to a vertical diffused MOSFET. The main difference is the p^+ layer at the back of the device that forms the collector of the device. This layer forms a *pn* junction, which injects minority carriers into the drift region under normal conduction. The structure at the top of the device is identical to a MOSFET in all respects including processing and cell geometry.

The forward blocking operation of the IGBT is identical to a power MOSFET. The presence of a rectifying junction at the backside also provides the device with reverse voltage blocking capability. A nonpunchthrough (NPT) drift region design has symmetrical forward and reverse blocking capability. The on-state operation of the IGBT can be understood by an equivalent representation of a MOSFET and a power diode connected in series. When a positive gate voltage is applied, an inversion layer is formed under the gate that facilitates the injection of electrons from the emitter into the drift region. Positive voltage at the collector simultaneously induces hole injection into the drift region from the p^+ backside layer. Double injection into the drift region gives the device its excellent on-state properties. Just like a power diode, the drift region conductivity is modulated significantly during conduction, which helps reduce the on-state resistance compared to the power MOSFET. The NPT IGBT has a thick drift region to support the high voltage. The same voltage blocking capability can be achieved by use of a much thinner drift region if an n^+ buffer layer is introduced between the drift region and p^+ substrate. These punchthrough (PT) IGBTs exhibit stronger conductivity modulation by virtue of a thinner drift region.

The on-state characteristics of the IGBT are depicted in Figure 7.17. For a given gate voltage, the saturation current is determined by the MOS channel, as expressed in equation 7.14. The device is operated in the linear region, where the contributions to voltage drop are from the MOS channel, the drift region, and the backside junction drop:

$$V_{CE,\,on} = V_{p-n} + V_{MOS} + V_{drift}. \tag{7.16}$$

Neglecting the voltage drop in the drift region because of conductivity modulation yields:

$$V_{CE,\,on} = \frac{kT}{q}\ln\left(\frac{I_C\tau}{qWL_{cell}Zn_i}\right) + \frac{I_C L_{ch}}{\mu_{ch}C_{ox}Z(V_{GS} - V_{TH})}. \tag{7.17}$$

A more detailed and accurate analysis can be done by treating the IGBT as a power BJT whose base current is provided by a MOSFET. A parasitic thyristor is formed in the

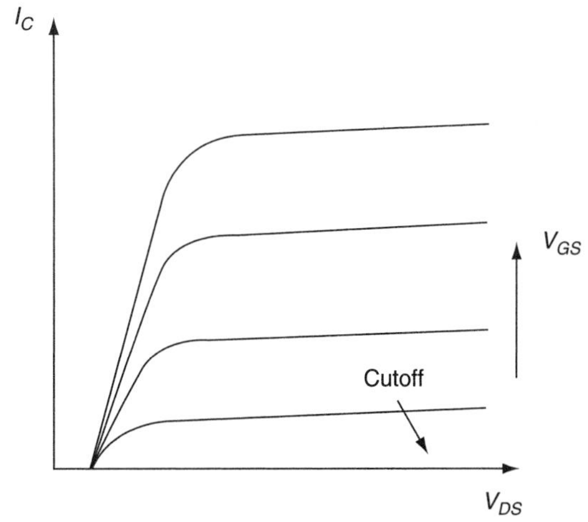

FIGURE 7.17 Output Characteristics of IGBT

IGBT by the junctions between the p^+ substrate, n-type drift, p-base, and n^+ emitter regions. Under normal operation, a fraction of the holes injected into the drift region by the p^+ substrate are collected by the p-base. These holes travel laterally in the p-base before being collected at the emitter contact. The lateral voltage drop may be sufficient to turn on the p-base n^+ emitter junction, eventually latching the parasitic thyristor. Several improvements have been made in the structure of present-day IGBTs to incorporate very high latch-up immunity.

7.9.1 Transient Operation

The switching waveforms of an IGBT in the clamped inductive circuit are shown in Figure 7.18. The turn-on waveforms are similar to those of the MOSFET in Figure 7.13. Two distinct phases are observed while the IGBT turns on. The first is the rapid reduction in voltage as the capacitor C_D discharges, similar to a MOSFET. A finite time is required for high-level injection conditions to set in the drift region. The gate voltage starts rising again only after the transistor comes out of its saturation region into the linear region.

During turn-off, the current is held constant while the collector voltage rises. This is followed by a sharp decline in collector current as the MOS channel turns off. The turn-off transient until this phase is similar to a MOSFET. However, the IGBT has excess charge in the drift region during on-state conduction that must be removed to turn the device off. This high concentration of minority carriers stored in the n-drift region supports current flow even after the MOS channel is cut off. As the minority carrier density decays due to recombination, it leads to a gradual reduction in the collector current and results in a current tail. For an on-state current of I_{ON}, the magnitude of the current tail and turn-off time are roughly expressed as:

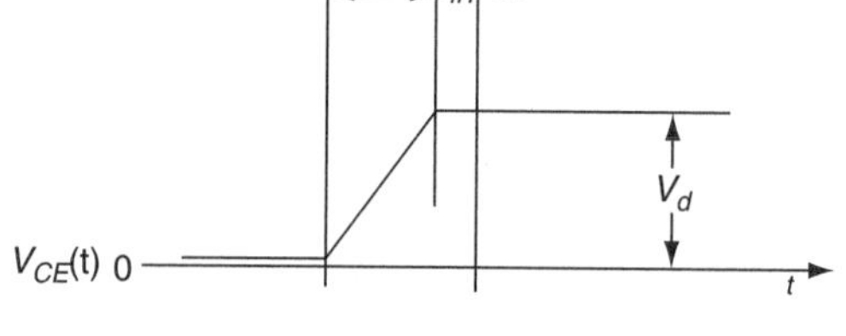

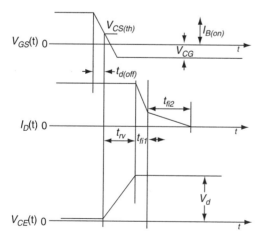

FIGURE 7.18 Typical Switching Waveforms During IGBT Clamped Inductive Load Turn-Off

$$I_C(t) = \alpha_{pnp} I_{on} e^{-t/\tau HL}. \quad (7.18a)$$

$$t_{off} = \tau_{HL} \ln(10\alpha_{pnp}). \quad (7.18b)$$

In these equations, $\alpha_{pnp} = \text{sech}(W_D/L_a)$ is the gain of the bipolar transistor.

The PT IGBT has a thin drift region and better on-state performance than NPT IGBT. High gain of the *pnp* transistor can lead to potential thyristor latch-up. For this reason, PT IGBT is subjected to lifetime control techniques to reduce the gain of the bipolar transistor. On the other hand, NPT IGBT has a wide drift region, and the gain is low even without lifetime control. Low lifetime and thin drift region make the PT IGBT more sensitive to temperature variation than NPT IGBT. Further, being bipolar in nature, IGBTs are vulnerable to thermal runaway in certain conditions, depending on the relative importance of MOSFET and bipolar transistor parameters. Hence, paralleling IGBTs is a challenge. The delay in turn-on and current tail during turn-off are significant sources of power loss in an IGBT. Several structures have been designed for faster removal of charge from the drift region during turn-off. Controlling the carrier lifetime within a tightly controlled band in the drift region has also been used to obtain a better turn-off between conduction and switching characteristics.

Example 7.6. An IGBT is designed with the same parameters as the MOSFET in example 7.4 by replacing the n^+ substrate with a p^+ substrate. A symmetrical structure is designed for both devices with a voltage blocking capability of 500 V. The drift region has a minority carrier lifetime of 1 μs. We can compare the on-state performance of the IGBT and the MOSFET.

From Example 7.3, the voltage drop across the drift region at a current density of $100\,\text{Å/cm}^2$ is 3.1 V. With technological limitations and following the design guidelines, the unit cell dimension of a 500 V MOSFET is about 8 μm. Then, considering that the channel has an on-state resistance of 0.68 Ω (example 7.4), the channel drop is 0.35 V. Thus, the overall MOSFET voltage drop is 3.45 V. The IGBT will have the same channel drop. However, the voltage across the diode in the equivalent representation of the IGBT is given by equation 7.19 to be 0.43 V. Thus, the overall voltage-drop of the IGBT is 0.78 V. This illustrates the superior on-state performance of IGBTs over MOSFETs at higher voltage levels.

7.10 Other MOS-Gate Devices

In the on-state, the IGBT operates as a bipolar transistor driven by a MOSFET. The on-state characteristics can be significantly improved by thyristor-like operation. This is achieved by the MOS-controlled thyristor (MCT) shown in Figure 7.19(A). Application of a negative gate voltage creates a p-channel

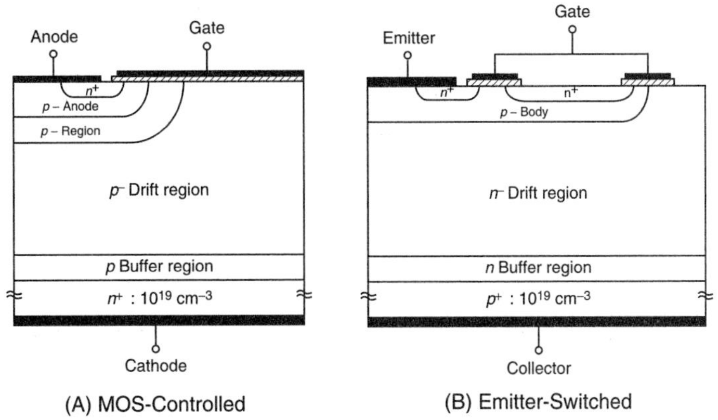

FIGURE 7.19 Cross-Section of Thyristors

under the gate. This channel provides the gate current to the vertical *n-p-n-p* thyristor that latches due to regenerative feedback and conducts a large forward current with a low forward voltage drop. During on-state conduction, the *p*-channel loses control over MCT performance once the thyristor latches on, and the *I–V* characteristics are independent of gate voltage beyond the threshold voltage V_{TH} of the channel. MCT turn-off is accomplished by applying a high voltage of reverse polarity at the gate. This creates an *n*-channel under the gate. If the resistance of the channel is small enough, it will divert the electron current from the thyristor. This effectively raises the holding current of the thyristor, thus forcing it to turn-off.

Although MCTs have significantly improved current-handling capability, they do not match the switching performance of IGBTs. This is due to the increased resistance in the device during turn-on and higher current density. During the turn-off stage, the thyristor may fail to turn off if resistance of the diverting channel is high. In addition the channel has to be formed uniformly and abruptly to effectively cut off the feedback in the thyristor.

The MCT has an uncontrolled on-state performance; hence, its forward-biased safe operating area (FBSOA) cannot be defined. The emitter-switched thyristor (EST) has been proposed to achieve control over the thyristor on-state conduction. The cross-section of an EST is shown in Figure 7.19(B). The EST has an n^+ floating emitter region in addition to the basic IGBT structure. When the inversion channel is formed on application of a gate voltage, device performance is identical to an IGBT. At higher current levels, hole flow in the *p*-base region under the floating region forward biases the p-n^+ junction, causing latch-up of the vertical thyristor. The thyristor current flows through the MOS channel before reaching the emitter terminal. This provides control over the thyristor current as resistance of the channel can be varied by changing the gate voltage. On-state thyristor operation of EST results in a much higher current density than IGBT. The switching mechanism in an EST is identical to an IGBT. Because of a higher level of drift region conductivity modulation, however, large turn-off power results.

Several other MOS-gated structures, such as base resistance-controlled thyristor (BRT), insulated gate-controlled thyristor (IGCT), injection-enhanced transistor (IEGT), MOS-turn-off thyristor (MTO), and so forth have been introduced in the recent past to further optimize the on-state and switching performance at high voltage and current levels.

7.11 Smart Power Technologies

The development of MOS-gate driven power devices has greatly simplified the gate drive circuits. The devices have made it possible to integrate the gate drive circuit into a monolithic chip. With the current technology, it is also possible to add other functions, such as protection against adverse operating conditions and logic circuits to interface with microprocessors. In a typical smart power control chip, the sensing and protection circuits are usually implemented using analog circuits with high-speed bipolar transistors. These circuits must sense adverse temperatures, currents, and voltages. This circuitry helps detect situations like thermal runaway, impact ionization, insufficient gate drive, and so forth. Present day Smart Power chips are manufactured using a junction isolation technology. Efforts are in progress to make lateral structures with high breakdown voltages and thin epitaxial layers. The dielectric isolation technology is being perfected to replace the junction isolation technology to achieve fewer parasitics, compactness, and a higher degree of integration.

7.12 Other Material Technologies

Numerous power circuit applications exist in which the device temperature can rise significantly above room temperature. Further, several applications require devices that can handle large blocking voltage exceeding 5 kV. Inherent material limitations make it impossible to use silicon-based devices beyond 200°C. These limitations have led to the search for new materials that can handle large voltages. Wide band gap materials with high carrier mobilities are ideally suited for such applications. Presently, SiC appears to be the most promising material to replace Si for high-voltage, high-temperature applications. With the commercial availability of high-quality silicon carbide wafers with lightly doped epitaxial layers, the fabrication of power devices has become viable. Power switches and rectifiers fabricated from silicon carbide offer tremendous promise for reduction of power losses in the device. All the devices discussed in this chapter have been demonstrated using SiC technology. However, the predicted superior performance of SiC devices has not been achieved on a large scale because of limitations in wafer quality and processing technology. Much work needs to be done before a manufacturable technology can be commercialized.

Bibliography

Baliga, B.J. (1996). *Power semiconductor devices.* Boston: PWS Publishing.

Grant, A. and Gowar, J. (1989). *Power MOSFETs—Theory and applications.* New York: John Wiley & Sons.

Sze, M. (1981). *Physics of semiconductor devices.* New York: John Wiley & Sons.

Undeland, M.T. and Robbins, W. (1996). *Power electronics—Design, converters, and applications.* New York: John Wiley & Sons.

III

VLSI SYSTEMS

Magdy Bayoumi
The Center for Advanced Computer Studies, University of Louisiana at Lafayette, Lafayette, Louisiana, USA

This Section covers the broad spectrum of VLSI arithmetic, custom memory organization and data transfer, the role of hardware description languages, clock scheduling, low-power design, micro electro mechanical systems, and noise analysis and design. It has been written and developed for practicing electrical engineers in industry, government, and academia. The goal is to provide the most up-to-date information in the field.

Over the years, the fundamentals of the field have evolved to include a wide range of topics and a broad range of practice. To encompass such a wide range of knowledge, the section focuses on the key concepts, models, and equations that enable the design engineer to analyze, design, and predict the behavior of large-scale systems. While design formulas and tables are listed, emphasis is placed on the key concepts and the theories underlying the processes. In order to do so, the material is reinforced with frequent examples and illustrations.

The compilation of this section would not have been possible without the dedication and efforts of the section editor and contributing authors. I wish to thank them all.

Wai-Kai Chen
Editor

1

Logarithmic and Residue Number Systems for VLSI Arithmetic

Thanos Stouraitis
Department of Electrical and Computer Engineering, University of Patras, Greece

1.1 Introduction

Very large-scale integrated circuit (VLSI) arithmetic units are essential for the operations of the data paths and/or the addressing units of microprocessors, digital signal processors (DSPs), as well as data-processing application-specific integrated circuits (ASICs) and programmable integrated circuits. Their optimized realization, in terms of power or energy consumption, area, and/or speed, is important for meeting demanding operational specifications of such devices.

In modern VLSI design flows, the design of standard arithmetic units is available from design libraries. These units employ binary encoding of numbers, such as one's or two's complement, or sign magnitude encoding to perform additions and multiplications. If nonstandard operations are required, or if high performance components are needed, then the design of special arithmetic units is necessary. In this case, the choice of arithmetic system is of utmost importance.

The impact of arithmetic in a digital system is not only limited to the definition of the architecture of arithmetic circuits. Arithmetic affects several levels of the design abstraction because it may reduce the number of operations, the signal activity, and the strength of the operators. The choice of arithmetic may lead to substantial power savings, reduced area, and enhanced speed.

This chapter describes two arithmetic systems that employ nonstandard encoding of numbers. The logarithmic number system (LNS) and the residue number system (RNS) are singled out because they have been shown to offer important advantages in the efficiency of their operation and may be at the same time more power- or energy-efficient, faster, and/or smaller than other systems.

Although a detailed comparison of performance of these systems to their counterparts is not offered here, one must keep in mind that such comparisons are only meaningful when the systems under question cover the same dynamic range and present the same precision of operations. This necessity usually translates in certain data word lengths, which, in their turn, affect the operating characteristics of the systems.

1.2 LNS Basics

Traditionally, LNS has been considered as an alternative to floating-point representation (Koren, 1993; Stouraitis, 1986). The organization of an LNS word is shown in Figure 1.1.

The LNS maps a linear real number X to a triplet as follows:

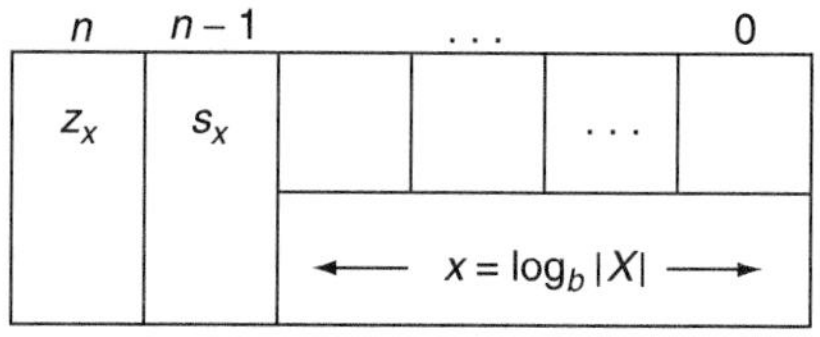

FIGURE 1.1 The Organization of an $(n+1)$-bit LNS Digital Word

$$X \xrightarrow{\text{LNS}} (z_x,\ s_x,\ x = \log_b |X|), \tag{1.1}$$

where s_x is the sign of X, b is the base of the logarithmic representation, and z_x is a single-bit flag, which, when asserted, denotes that X is zero. A zero flag is required because $\log_b X$ is not a finite number for $X = 0$. Similarly, since the logarithm of a negative number is not a real number, the sign information of X is stored in flag s_x. Logarithm $x = \log_b |X|$ is encoded as a binary number, and it may comprise a number of k integer and l fractional bits.

The inverse mapping of a logarithmic triple $(z_x,\ s_x, x)$ to a linear number X is defined by:

$$(z_x,\ s_x,\ x) \xrightarrow{\text{LNS}^{-1}} X{:}\, X = (1 - z_x)(-1)^{s_x} b^x. \tag{1.2}$$

1.2.1 LNS and Linear Representations

Two important issues in a finite word length number system are the **range** of numbers that can be represented and the **precision** of the representation (Koren, 1993).

Let (k, l, b)–LNS denote an LNS of integer and fractional word lengths k and l, respectively, and of base b. These three parameters determine the properties of the LNS and can be computed so that the LNS meets certain specifications. For example, for a (k, l, b)–LNS to be considered as equivalent to an n-bit linear fixed-point system, the following two restrictions may be posed:

1. The two representations should exhibit equal average representational error.
2. The two representations should cover equivalent data ranges.

The average representational error, ε_{ave}, is defined as:

$$\varepsilon_{\text{ave}} = \frac{\sum_{X=X_{\min}}^{X_{\max}} \varepsilon_{\text{rel}}(X)}{X_{\max} - X_{\min} + 1}, \tag{1.3}$$

where $X_{\min}$ and $X_{\max}$ define the range of representable numbers in each system and where $\varepsilon_{\text{rel}}(X)$ is the relative representational error of a number X encoded in a number system. This error is, in general, a function of the value of X and it is defined as:

$$\varepsilon_{\text{rel}}(X) = \frac{|X - \hat{X}|}{X}, \tag{1.4}$$

in which X is the actual value and $\hat{X}$ is the corresponding value representable in the system. Notice that $X \neq \hat{X}$ due to the finite length of the words. Assuming that the logarithm of X is represented as a two's complement number, the relative representational error $\varepsilon_{\text{rel, LNS}}$ for a (k, l, b)–LNS is independent of X and, therefore, is equal to the average representational error. It is given by [refer to Koren (1993) for the case $b = 2$].

$$\varepsilon_{\text{ave, LNS}} = \varepsilon_{\text{rel, LNS}} = b^{2^{-l}} - 1. \tag{1.5}$$

Due to formula 1.3, the average representational error for the n-bit linear fixed-point case is given by:

$$\varepsilon_{\text{ave, FXP}} = \frac{1}{2^n - 1} \sum_{i=1}^{2^n - 1} \frac{1}{i}, \tag{1.6}$$

which, by computing the sum on the right-hand side, can be written as:

$$\varepsilon_{\text{ave, FXP}} = \frac{\psi(2^n) + \gamma}{2^n - 1}, \tag{1.7}$$

where γ is the Euler gamma constant and function ψ is defined through:

$$\psi(x) = \frac{d}{dx} \ln \Gamma(x), \tag{1.8}$$

where $\Gamma(x)$ is the Euler gamma function.

In the following, the maximum number representable in each number system is computed and used to compare the ranges of the representations. Notice that different figures could also have been used for range comparison, such as the ratio $X_{\max}/X_{\min}$ (Stouraitis, 1986). The maximum number representable by an n-bit linear integer is $2^n - 1$; therefore the upper bound of the fixed-point range is given by:

$$X_{\max}^{\text{FXP}} = 2^n - 1. \tag{1.9}$$

The maximum number representable by a (k, l, b)-LNS encoding 1.1 is as follows:

$$X_{\max}^{\text{LNS}} = b^{2^k + 1 - 2^{-l}}. \tag{1.10}$$

Therefore, according to the equivalence restrictions posed above, to make an LNS equivalent to an n-bit linear fixed-point representation, the following inequalities should be simultaneously satisfied:

$$X_{\max}^{\text{LNS}} \geq X_{\max}^{\text{FXP}}. \tag{1.11}$$

$$\varepsilon_{\text{ave, LNS}} \leq \varepsilon_{\text{ave, FXP}}. \tag{1.12}$$

Hence, from equations 1.5 and 1.7 through 1.10 the following equations are obtained:

$$l = \left\lceil -\log_2 \log_b \left(1 + \frac{\psi(2^n) + \gamma}{2^n - 1}\right)\right\rceil. \tag{1.13}$$

$$k = \lceil \log_2(\log_b(2^n - 1) + 2^{-l} - 1)\rceil. \tag{1.14}$$

Values of k and l that correspond to various values of n for various values of b can be seen in Table 1.1, where for each base, there is a third column (explained next).

Although the word lengths k and l computed via equations 1.13 and 1.14 meet the posed equivalence specifications of equations 1.11 and 1.12, LNS is capable of covering a significantly larger range than the equivalent fixed-point representation. Let n_{eq} denote the word length of a fixed-point system that can cover the range offered by an LNS defined through equations 1.13 and 1.14. Equivalently, let n_{eq} be the smallest integer, which satisfies:

$$2^{n_{\text{eq}}} - 1 \geq b^{2^k + 1 - 2^{-l}}. \tag{1.15}$$

From equation 1.15, it follows that:

$$n_{\text{eq}} = \lceil (2^k + 1 - 2^{-l}) \log_2 b \rceil. \tag{1.16}$$

It should be stressed that when $n_{\text{eq}} \geq n$, the precision of the particular fixed-point system is better than that of the LNS derived by equations 1.13 and 1.14. Equation 1.16 reveals that the particular LNS, while meeting the precision of an n-bit linear representation, in fact covers the range provided by an n_{eq}-bit linear system.

Of course, the average (relative) error is not the only way to compare the accuracy of computing systems. Especially true for signal processing systems, one may use the signal-to-noise ratio (SNR), assuming that quantization errors represent noise, to compare the precision of two systems. In that case, by equating the SNRs of the LNS and the fixed-point system that covers the required dynamic range, the integer and fractional word lengths of the LNS may be computed.

1.2.2 LNS Operations

Mapping of equation 1.1 is of practical interest because it can simplify certain arithmetic operations (i.e., it can reduce the implementation complexity, also called **strength**, of several operators). For example, due to the properties of the logarithm function, the multiplication of two linear numbers, $X = b^x$ and $Y = b^y$, is reduced to the addition of their logarithmic images, x and y.

The basic arithmetic operations and their LNS counterparts are summarized in Table 1.2, where, for simplicity and without loss of generality, the zero flag z_x is omitted and it is assumed that $X > Y$. Table 2 reveals that, while the complexity of most operations is reduced, the complexity of LNS addition and LNS subtraction is significant. In particular, for $d = |x - y|$, LNS addition requires the computation of the nonlinear function:

$$s_a(d) = \log_b(1 + b^{-d}), \tag{1.17}$$

and subtraction requires the computation of the nonlinear function:

$$s_s(d) = \log_b(1 - b^{-d}). \tag{1.18}$$

Equations 1.7 and 1.8 substantially limit the data word lengths for which LNS can offer efficient VLSI implementations. The

TABLE 1.1 Correspondence of n, k, l, and n_{eq} for Various Bases b

n	$b = 1.5$			$b = 2$			$b = 2.5$		
	k	l	n_{eq}	k	l	n_{eq}	k	l	n_{eq}
5	3	2	6	3	3	9	2	3	7
6	4	3	10	3	4	9	2	4	7
7	4	4	10	3	5	9	3	5	12
8	4	5	10	3	5	9	3	6	12
9	4	5	10	4	6	17	3	7	12
10	5	6	20	4	7	17	3	7	12
11	5	7	20	4	8	17	3	8	12
12	5	8	20	4	9	17	4	9	23
13	5	9	20	4	10	17	4	10	23
14	5	10	20	4	11	17	4	11	23
15	5	11	20	4	12	17	4	12	23

TABLE 1.2 Basic Linear Arithmetic Operations and Their LNS Counterparts

	Linear operation	Logarithmic operation
Multiply	$W = XY = b^x b^y = b^{x+y}$	$w = x + y,\ s_w = s_x \text{ XOR } s_y$
Divide	$W = \frac{X}{Y} = \frac{b^x}{b^y} = b^{x-y}$	$w = x - y,\ s_w = s_x \text{XOR } s_y$
Root	$W = \sqrt[m]{X} = \sqrt[m]{b^x} = b^{\frac{x}{m}}$	$w = \frac{x}{m},\ m, \text{integer}, s_w = s_x$
Power	$W = X^m = (b^x)^m$	$w = mx,\ m, \text{integer}, s_w = s_x$
Add	$W = X + Y = b^x + b^y = b^x(1 + b^{y-x})$	$w = x + \log_b(1 + b^{y-x}),\ s_w = s_x$
Subtract	$W = X - Y = b^x - b^y = b^x(1 - b^{y-x})$	$w = x + \log_b(1 - b^{y-x}),\ s_w = s_x$

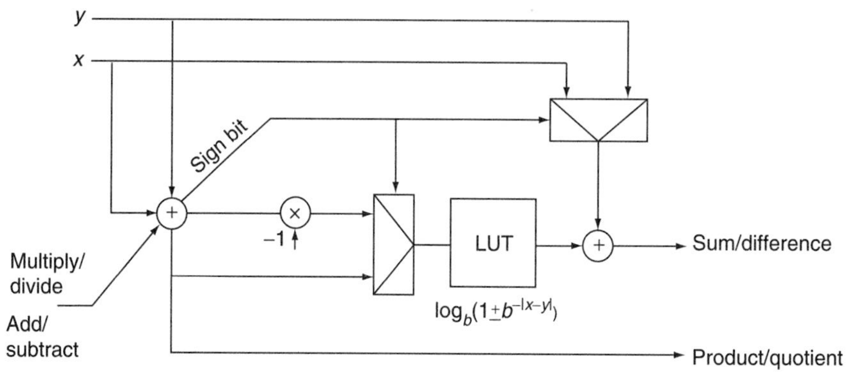

FIGURE 1.2 The Organization of a Basic LNS Processor: the processor comprises an adder, two multiplexers, a sign-inversion unit, a look-up table, and a final adder. It may perform the four operations of addition, subtraction, multiplication, or division.

organization of an LNS processor that can perform the four basic operations of addition, subtraction, multiplication, or division is shown in Figure 1.2. Note that to implement LNS subtraction (i.e., the addition of two quantities of opposite sign) a different memory look-up table (LUT) is required.

The main complexity of an LNS processor is the implementation of the LUTs for storing the values of the functions $s_a(d)$ and $s_s(d)$. A straightforward implementation is only feasible for small word lengths. A different technique can be used for larger word lengths based on the partitioning of an LUT into an assortment of smaller LUTs. The particular partitioning becomes possible due to the nonlinear behavior of the addition and subtraction functions, $\log_b(1 + b^{-d})$ and $\log_b(1 - b^{-d})$, respectively, that are depicted in Figure 1.3 for $b = 2$. By exploiting the different minimal word length required by groups of function samples, the overall size of the LUT is compressed, leading to a LUT organization of Figure 1.4. In addition to the above techniques, reduction of the size of memory can be achieved by proper selection of the base of the logarithms. It turns out that the same bases that yield minimum power consumption for the LNS arithmetic unit by reducing the bit activity, as mentioned in the next section, also result in minimum LUT sizes.

To use the benefits of LNS, a conversion overhead is required in most cases to perform the forward LNS mapping defined by equation 1.1. It is noted that conversions of equations 1.1 and 1.2 are required if an LNS processor receives input or transmits output linear data in digital format. Since all arithmetic operations can be performed in the logarithmic domain, only an initial conversion is imposed; therefore, as the amount of processing implemented in LNS grows, the contribution of the conversion overhead to power dissipation and to area–time complexity becomes negligible because it remains constant.

In stand-alone DSP systems, the adoption of a different solution to the conversion problem is possible. In particular, the LNS forward and inverse mapping overhead can be mitigated by converting the analog data directly into digital logarithms.

LNS Arithmetic Example

Let $X = 2.75$, $Y = 5.65$, *and* $b = 2$. *Perform the operations* $X \cdot Y$, $X + Y$, $\sqrt{X}$ *and* Y^2 *using the LNS.*

Initially, the data are transferred to the logarithmic domain as implied by equation 1.1:

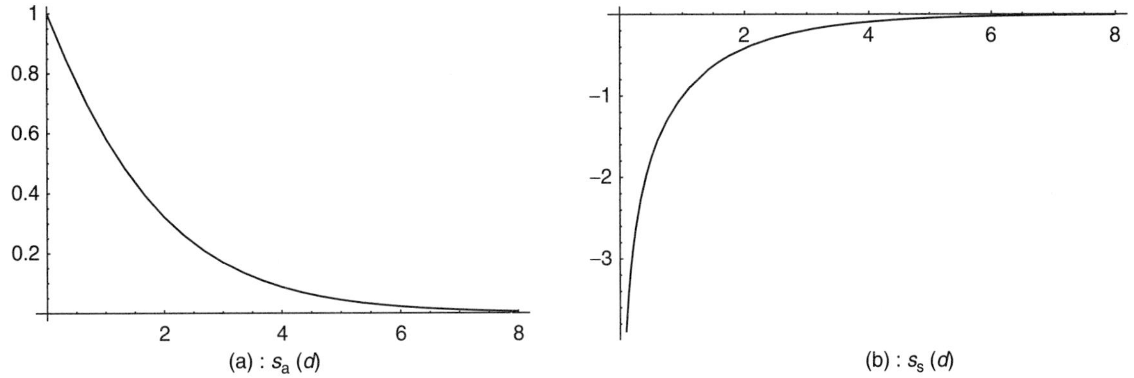

FIGURE 1.3 The Functions $s_a(d)$ and $s_s(d)$: Approximations required for LNS addition and subtraction.

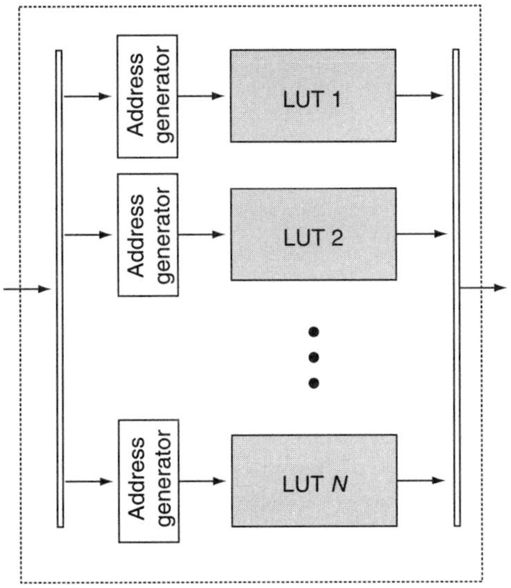

FIGURE 1.4 The Partitioning of the LUT: The partitioning stores the addition and subtraction functions into a set of smaller LUTs, which leads to memory compression.

$$X \xrightarrow{\text{LNS}} (z_x,\ s_x, x = \log_2 |X|) = (0,\ 0, x = \log_2 2.75) = (0,\ 0,\ 1.4594). \tag{1.19}$$

$$Y \xrightarrow{\text{LNS}} (z_y,\ s_y,\ y = \log_2 |Y|) = (0,\ 0,\ y = \log_2 5.65) = (0,\ 0,\ 2.4983). \tag{1.20}$$

Using the LNS images from equations 1.19 and 1.20, the required arithmetic operations are performed as follows: The logarithmic image w of the product $W = X \cdot Y$ is given by:

$$w = x + y = 1.4594 + 2.4983 = 3.9577. \tag{1.21}$$

As both operands are of the same sign (i.e., $s_x = s_y = 0$), the sign of the product is $s_w = 0$. In addition, because $z_x \neq 1$ and $z_y \neq 1$, the result is non-zero (i.e., $z_w = 0$).

To retrieve the actual result W from equation 1.21, inverse conversion of 1.2 is used as follows:

$$W = (1 - z_w)(-1)^{s_w} 2^w = 2^{3.9577} = 15.5377. \tag{1.22}$$

By directly multiplying X by Y, it is found that $W = 15.5375$. The difference of 0.0002 is due to round-off error during the conversion from linear to the LNS domain.

The calculation of the logarithmic image w of $W = \sqrt{X}$ is performed as follows:

$$w = \frac{1}{2}x = \frac{1}{2}1.4594 = 0.7297. \tag{1.23}$$

The actual result is retrieved as follows:

$$W = 2^{0.7297} = 1.6583. \tag{1.24}$$

The calculation of the logarithmic image w of $W = X^2$ can be done as:

$$w = 2 \cdot 1.4594 = 2.9188. \tag{1.25}$$

Again, the actual result is obtained as:

$$W = 2^{2.9188} = 7.5622. \tag{1.26}$$

The operation of logarithmic addition is rather awkward, and its realization is usually based on a memory LUT operation. The logarithmic image w of the sum $W = X + Y$ is as follows:

$$w = \max(x,\ y) + \log_2 (1 + 2^{\min(x, y) - \max(x, y)}) \tag{1.27}$$

$$= 2.4983 + \log_2 (1 + 2^{-1.0389}) \tag{1.28}$$

$$= 3.0704. \tag{1.29}$$

The actual value of the sum $W = X + Y$ is obtained as:

$$W = 2^{3.0704} = 8.4001. \tag{1.30}$$

1.2.3 LNS and Power Dissipation

Power dissipation minimization is sought at all levels of design abstraction, ranging from software and hardware partitioning down to technology-related issues. The average power dissipation in a circuit is computed via the relationship:

$$P_{\text{ave}} = a f_{\text{clk}} C_L V_{\text{dd}}^2, \tag{1.31}$$

where f_{clk} is the clock frequency, C_L is the total switching capacitance, V_{dd} is the supply voltage, and a is the average activity in a clock period.

LNS is applicable for low-power design because it reduces the complexity of certain arithmetic operators and the bit activity.

Power Dissipation and LNS Architecture

LNS exploits properties of the logarithm function to reduce the strength of several arithmetic operations; thus, it leads to complexity savings. By reducing the area complexity of operations, the switching capacitance C_L of equation 1.31 can be reduced. Furthermore, reduction in latency allows for further reduction in supply voltage, which also reduces power dissipation (Chandrakasan and Brodersen, 1995). A study of the impact of the choice of the number system on the QRD-RLS algorithm revealed that LNS offers accuracy comparable to that of floating-point operations but only at a fraction of the switched capacitance per iteration of the algorithm (Sacha and Irwin, 1998). The reduction of average switched capacitance of LNS systems stems from the simplification of basic arithmetic operations, shown in Table 1.2. It can be seen that n-bit multiplication and division are reduced to $(k + l)$-bit addition and subtraction, respectively, while the computation of roots and powers is reduced to division and multiplication by a constant, respectively. For the common cases of square root or square, the operation is reduced to left or right shift respectively. For example, assume that a n-bit carry-save array multiplier, which has a complexity of $n^2 - n$ 1-bit full adders (FAs), is replaced by an n-bit adder, assuming $k + l = n$ has a complexity of n FAs for a ripple-carry implementation (Koren, 1993). Therefore, multiplication complexity is reduced by a factor r_{C_L}, given as:

$$r_{C_L} = \frac{n^2 - n}{n} = n - 1. \tag{1.32}$$

Equation 1.32 reveals that the reduction factor r_{C_L} grows with the word length n.

Addition and subtraction, however, are complicated in LNS because they require an LUT operation for the evaluation of $\log_b(1 \pm b^{y-x})$, although different approaches have been proposed in the literature (Orginos *et al.*, 1995; Paliouras and Stouraitis, 1996). An LUT operation requires a ROM of $n \times 2^n$ bits, a size that can inhibit use of LNS for large values of n. In an attempt to solve this problem, efficient table reduction techniques have been proposed (Taylor *et al.*, 1988). As a result of the above analysis, applications with a computational load dominated by operations of simple LNS implementation can be expected to gain power dissipation reduction due to the LNS impact on architecture complexity.

Since multiplication–additions are important in DSP applications, the power requirements of an LNS and a linear fixed-point adder–multiplier have been compared. It has been reported that approximately a two times reduction in power dissipation is possible for operations with word sizes of 8 to 14 bits (Paliouras and Stouraitis, 2001). Given a sufficient number of consecutive multiplication–additions, the LNS implementation becomes more efficient from the low-power dissipation viewpoint, even when a constant conversion overhead is taken into consideration.

Power Dissipation and LNS Encoding

The encoding of data through logarithms of various bases implies variations in the bit activity (i.e., the a factor of equation 31 and, therefore, the power dissipation) (Paliouras and Stouraitis, 1996, 2001).

Assuming a uniform distribution of linear n-bit input numbers, the distribution of bit assertions of the corresponding LNS words reveals that LNS can be exploited to reduce the average activity. Let $p_{0\to1}(i)$ be the bit assertion probabilities (i.e., the probability of the ith bit transition from 0 to 1). Assuming that data are temporarily independent, it holds that:

$$p_{0\to1}(i) = p_0(i)p_1(i) = (1 - p_1(i))p_1(i), \tag{1.33}$$

where $p_0(i)$ and $p_1(i)$ is the probability of the ith bit being 0 or 1, respectively. Due to the assumption of uniform data distribution, it holds that:

$$p_0(i) = p_1(i) = \frac{1}{2}, \tag{1.34}$$

which, due to equation 1.33, gives:

$$p_{0\to1}(i) = \frac{1}{4}. \tag{1.35}$$

Therefore, all bits in the linear fixed-point representation exhibit an equal $p_{0\to1}(i)$, $i = 0, 1, \ldots, n - 1$.

Activities of the bits in an LNS-encoded word are quantified under similar assumptions. Since there is an one-to-one correspondence of linear fixed-point values to their LNS images defined by equation 1.1, the LNS values follow a probability function identical to the fixed-point case. In fact, the LNS mapping can be considered as a continuous transformation of the discrete random variable *X*, which is a word in the linear representation, to the discrete random variable *x*, an LNS word. Hence, the two discrete random variables follow the same probability function (Peebles, 1987).

The $p_{0\to1}^{\mathrm{LNS}}$ probabilities of bit assertions in LNS words, however, are not constant as $p_{0\to1}(i)$ of equation 1.35; they depend on the significance of the *i*th bit. To evaluate the probabilities $p_{0\to1}^{\mathrm{LNS}}(i)$, the following experiment is performed. For all possible values of *X* in a *n*-bit system, the corresponding $\lfloor \log_b X \rfloor$ values in a (*k*, *l*, *b*)-LNS format are derived, and probabilities $p_1(i)$ for each bit are computed. Then, $p_{0\to1}^{\mathrm{LNS}}(i)$ is computed as in equation 1.33. The actual assertion probabilities for the bits in an LNS word, $p_{0\to1}^{\mathrm{LNS}}(i)$, are depicted in Figure 1.5. It can be seen that $p_{0\to1}(i)$ for the more significant bits is substantially lower than $p_{0\to1}(i)$ for the less significant bits. Moreover, it can be seen that $p_{0\to1}(i)$ depends on *b*. This behavior, which is due to the inherent data compression property of the logarithm function, leads to a reduction of the average activity in the entire word. The average activity savings percentage, S_{ave}, is computed as:

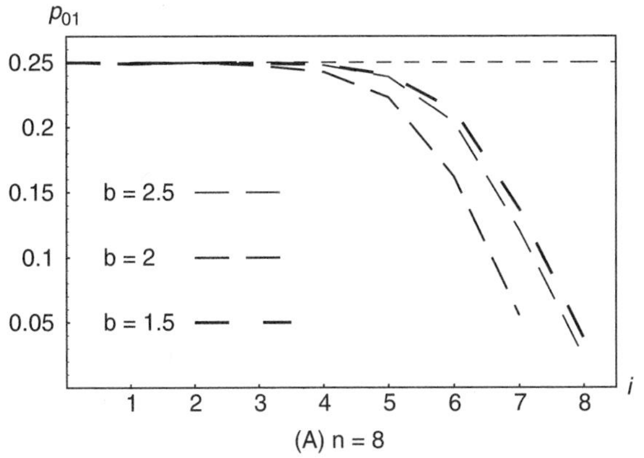

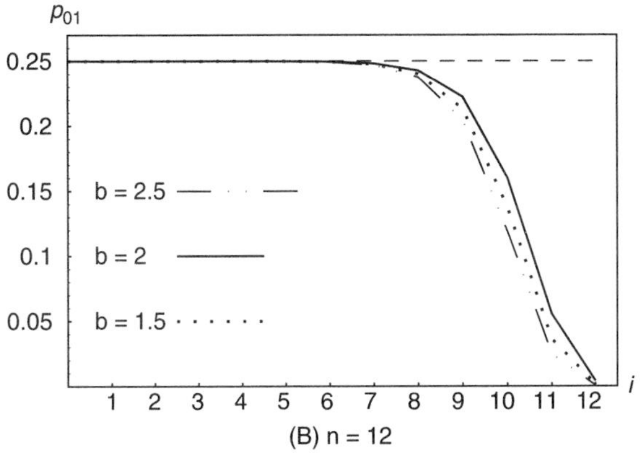

FIGURE 1.5 Activities Against Bit Significance *i* (in an LNS Word for $n = 8$ and $n = 12$) and Various Values of the Base *b*. The horizontal dashed line is the activity of the corresponding *n*-bit fixed-point system

$$S_{\mathrm{ave}} = \left(1 - \frac{\sum_{i=0}^{k+l-1} p_{0\to1}^{\mathrm{LNS}}(i)}{0.25n}\right) 100\%, \qquad (1.36)$$

where $p_{0\to1}^{\mathrm{FXP}}(i) = 1/4$ for $i = 0, 1, \ldots, n-1$; the word lengths *k* and *l* are computed via equations 1.13 and 1.14, and *n* denotes the length of the fixed-point system. The savings percentage S_{ave} is demonstrated in Figure 1.6(A) for various values of *n* and *b*, and the percentage is found to be more than 15% in certain cases.

As implied by the definition of n_{eq} in equation 1.16, however, the linear system that provides an equivalent range to that of a (*k*, *l*, *b*)-LNS, requires n_{eq} bits. If the reduced precision of a (*k*, *l*, *b*)-LNS, compared to an n_{eq}-bit fixed-point system, is acceptable for a particular application, S'_{ave} is used to describe the relative efficiency of LNS, instead of equation 1.36, where:

$$S'_{\mathrm{ave}} = \left(1 - \frac{\sum_{i=0}^{k+l-1} p_{0\to1}^{\mathrm{LNS}}(i)}{0.25n_{\mathrm{eq}}}\right) 100\%. \qquad (1.37)$$

Savings percentage S'_{ave} is demonstrated in Figure 1.6(B) for various values of *n* and *b*. Savings are found to exceed 50% in some cases. Notice that Figure 1.6 reveals that, for a particular word length *n*, the proper selection of logarithm base *b* can significantly affect the average activity. Therefore, the choice of *b* is important in designing a low-power LNS-based system.

Finally, it should be noted that overhead is imposed for linear-to-logarithmic and logarithmic-to-linear conversion. Conversion overhead contributes additional area and time complexity as well as power dissipation. As the number of operations grows, however, the conversion overhead remains constant; therefore, the overhead's contribution to the overall budget becomes negligible.

1.3 The Residue Number System

A different concept than the nonlinear logarithmic transformation is followed by mapping of data to appropriately selected finite fields. This may be achieved through the use of one of the many available versions of the residue number system (RNS) (Szabo and Tanaka, 1967). RNS arithmetic faces difficulties with sign detection, division, and magnitude comparison. These difficulties may outweigh the benefits it presents for addition, subtraction, and multiplication as far as general computing is concerned. Its use in specialized computations, like those for signal processing, offers many advantages. RNS has been used to offer superior fault tolerance capabilities as well as high-speed, small-area, and/or significant power-dissipation savings in the design of signal processing architectures for FIR filters (Freking and Parhi, 1997) and other circuits (Chren, 1998). RNS may even reduce the computational load in complex-number processing (Taylor *et al.*, 1985), thus

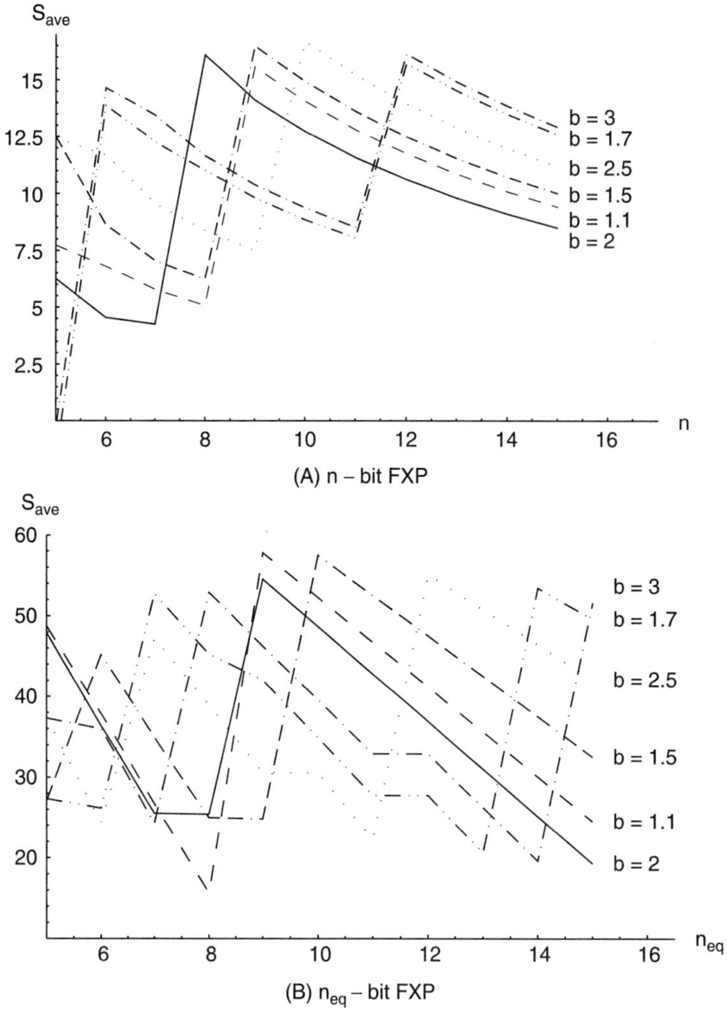

FIGURE 1.6 Percentage of Average Activity Reduction from Use of LNS. The percentage is compared to n-bit and to n_{eq}-bit linear fixed-point system for various bases b of the logarithm. The diagram reveals that the optimal selection of b depends on n, and it can lead to significant power dissipation reduction.

providing speed and power savings at the algorithmic level of the design abstraction.

1.3.1 RNS Basics

The RNS maps a natural number X in the range $[0,\ M-1]$, with $M = \prod_{i=1}^{N} m_i$, to an N-tuple of **residues** x_i:

$$X \xrightarrow{\text{RNS}} \{x_1,\ x_2,\ \ldots,\ x_N\}, \tag{1.38}$$

where $x_i = \langle X \rangle_{m_i}$, $\langle \cdot \rangle_{m_i}$ denotes the mod m_i operation and where m_i is a member of the set of the co-prime integers $B = \{m_1,\ m_2,\ \ldots,\ m_N\}$ called **moduli**. Co-prime integers' greatest common divisor is $gcd(m_i,\ m_j) = 1,\ i \neq j$. The set of RNS moduli is called the **base** of RNS. The modulo operation $\langle X \rangle_m$ returns the integer remainder of the integer division x div m (i.e., an integer k such that $x = m \cdot l + k$) where l is an integer.

RNS is of interest because basic arithmetic operations can be performed in a digit-parallel carry-free manner, such as in:

$$z_i = \langle x_i \circ y_i \rangle_{m_i}, \tag{1.39}$$

where $i = 1,\ 2,\ \ldots,\ N$ and where the symbol $\circ$ stands for addition, subtraction, or multiplication. Every integer in the range $0 \leq X < \prod_{i=1}^{N} m_i$ has a unique RNS representation.

Inverse conversion may be accomplished by means of the Chinese remainder theorem (CRT) or the mixed-radix conversion (Soderstrand *et al.*, 1986). The CRT retrieves an integer from its RNS representation as:

$$X = \left\langle \sum_{i=1}^{N} \overline{m_i} \langle \overline{m_i}^{-1} x_i \rangle_{m_i} \right\rangle_M, \quad (1.40)$$

where $\overline{m_i} = \frac{M}{m_i}$, $M = \Pi_{i=1}^{N} m_i$, and $\overline{m_i}^{-1}$ is the multiplicative inverse of $\overline{m_i}$ modulo m_i (i.e., an integer such that $\langle \overline{m_i} \cdot \overline{m_i}^{-1} \rangle_{m_i} = 1$).

Using an associated mixed radix system, inverse conversion may also be performed by translating the residue representations to a mixed radix representation. By choosing the RNS moduli to be the weights in the mixed radix representation, the inverse mapping is facilitated by associating the mixed radix system with the RNS. Specifically, an integer $0 < X < M$ can be represented by N mixed radix digits $(x'_1, \ldots, x'_N)$ as:

$$X = x'_N(m_{N-1}m_{N-2}\ldots m_1) + \ldots + x'_3(m_2m_1) + x'_2m_1 + x'_1, \quad (1.41)$$

where $0 \leq x'_i < m_i$, $i = 1 \ldots N$, and the x'_i can be generated sequentially from the x_i using only residue arithmetic, such as in:

$$\begin{aligned} x'_1 &= \langle X \rangle_{m_1} = x_1 \\ x'_2 &= \langle m_1^{-1}(X - x'_1) \rangle_{m_2} \\ x'_3 &= \langle m_2^{-1}(m_1^{-1}(X - x'_1)) - x'_2) \rangle_{m_3}, \end{aligned} \quad (1.42)$$

and so on, or as in the following:

$$\begin{aligned} x'_1 &= x_1 \\ x'_2 &= \langle (x_2 - x'_1)m_1^{-1}m_2 \rangle_{m_2} \\ x'_3 &= \langle ((x_3 - x'_1)m_1^{-1}m_3 - x'_2)m_2^{-1}m_3 \rangle_{m_3} \\ &\vdots \\ x'_N &= \langle (\ldots((x_N - x'_1)m_1^{-1}m_N - x'_2)m_2^{-1} \\ &\quad m_N - \ldots - x'_{N-1})m_{N-1}^{-1}m_N \rangle_{m_N}. \end{aligned} \quad (1.43)$$

The digits x'_i can be generated sequentially through residue subtraction and multiplication by the fixed m_i^{-1}. The sequential nature of calculation increases the latency of the residues conversion to binary numbers.

The set of RNS moduli is often chosen so that the implementation of the various RNS operations (e.g., addition, multiplication, and scaling) becomes efficient. A common choice is the set of moduli $\{2^n - 1, 2^n, 2^n + 1\}$, which may also form a subset of the base of RNS.

RNS Arithmetic Example

Consider the base $B = \{3, 5, 7\}$ *and two integers* $X = 10$ *and* $Y = 5$. *The RNS images of* X *and* Y *are as written here:*

$$\begin{aligned} X \xrightarrow{\text{RNS}} \{x_1, x_2, x_3\} &= \{\langle 10 \rangle_3, \langle 10 \rangle_5, \langle 10 \rangle_7\} \\ &= \{1, 0, 3\}. \end{aligned} \quad (1.44)$$

$$Y \xrightarrow{\text{RNS}} \{y_1, y_2, y_3\} = \{\langle 5 \rangle_3, \langle 5 \rangle_5, \langle 5 \rangle_7\} = \{2, 0, 5\}. \quad (1.45)$$

The RNS image of the sum $Z = X + Y$ is obtained as:

$$\begin{aligned} Z \xrightarrow{\text{RNS}} \{z_1, z_2, z_3\} &= \{\langle 1 + 2 \rangle_3, \langle 0 + 0 \rangle_5, \langle 3 + 5 \rangle_7\} \\ &= \{0, 0, 1\}. \end{aligned} \quad (1.46)$$

To retrieve the integer that corresponds to the RNS representation {0, 0, 1} by applying the CRT of equation 1.40, the following quantities are precomputed: $M = 3 \cdot 5 \cdot 7 = 105$, $\overline{m_1} = \frac{105}{3} = 35$, $\overline{m_2} = \frac{105}{5} = 21$, $\overline{m_3} = \frac{105}{7} = 15$, $\overline{m_1}^{-1} = 2$, $\overline{m_2}^{-1} = 1$, and $\overline{m_3}^{-1} = 1$. The value of the sum in integer form is obtained by applying equation 1.40

$$\begin{aligned} Z = X + Y &= \langle 35\langle 2 \cdot 0 \rangle_3 + 21\langle 1 \cdot 0 \rangle_5 + 15\langle 1 \cdot 1 \rangle_7 \rangle_{105} \\ &= \langle 15 \rangle_{105} = 15. \end{aligned} \quad (1.47)$$

To verify the result of equation 1.46, notice that $X + Y = 10 + 5 = 15$ and that:

$$15 \xrightarrow{\text{RNS}} \{\langle 15 \rangle_3, \langle 15 \rangle_5, \langle 15 \rangle_7\} = \{0, 0, 1\} = \{z_1, z_2, z_3\}, \quad (1.48)$$

which is the result obtained in equation 1.46. The same integer may be retrieved by using an associated mixed radix system defined by equation 1.41 as:

$$Z = z'_3 \cdot 15 + z'_2 \cdot 3 + z'_1,$$

with $0 \leq z'_1 < 3$, $0 \leq z'_2 < 5$, $0 \leq z'_3 < 7$ and the following:

$$\begin{aligned} z'_1 &= z_1 = 0 \\ z'_2 &= \langle 3^{-1}(z_2 - z'_1) \rangle_5 = \langle 2 \cdot z_2 \rangle_5 = 0 \end{aligned}$$

and

$$\begin{aligned} z'_3 &= \langle 5^{-1}[3^{-1}(z_3 - z'_1) - z'_2] \rangle_7 = \langle (z_3 - z'_1) - 3 \cdot z'_2 \rangle_7 \\ &= \langle (1 - 0) - 3 \cdot 0 \rangle_7 = 1 \end{aligned} \quad (1.49)$$

so that $Z = 1 \cdot 15 + 0 \cdot 3 + 0 = 15$.

1.3.2 RNS Architectures

The basic architecture of an RNS processor in comparison to a binary counterpart is depicted in Figure 1.7. This figure shows that the word length n of the binary counterpart is partitioned into N subwords, the residues, that can be processed

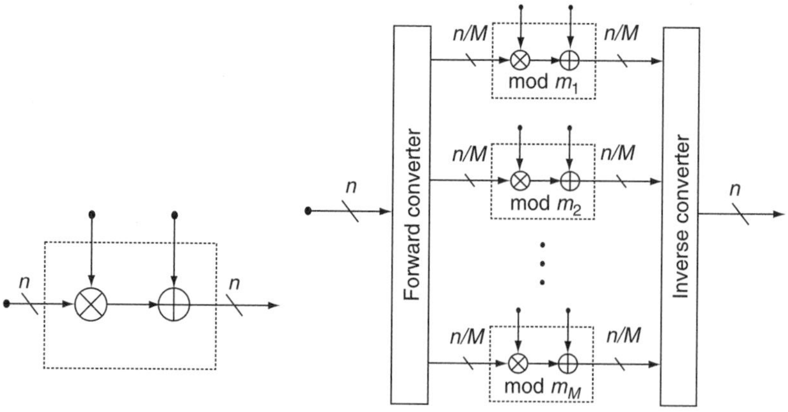

(A) Structure of a Binary Architecture (B) Corresponding RNS Processor

FIGURE 1.7 Basic Architectures

independently and are of word length significantly smaller than n. The architecture in Figure 1.7 assumes, without loss of generality, that the moduli are of equal word length. The ith residue channel performs arithmetic modulo m_i.

Most implementations of arithmetic units for RNS consist of an accumulator and a multiplier and are based on ROMs or PLAs. Bayoumi *et al.* (1983) have analyzed the efficiency of various VLSI implementations of RNS adders. Moreover, implementations of arithmetic units that operate in a finite integer ring $R(m)$ and that are called AU_ms are offered in the literature (Stouraitis, 1993). They are less costly and require less area and lower hardware complexity and power consumption. They are based on continuously decomposing the residue bits that correspond to powers of 2 that are larger than or equal to 2^n, until they are reduced to a set of bits that correspond to a sum of powers of 2 that is less than 2^n, where $n = \lceil \log_2 m \rceil$. This decomposition is implemented by using full adder (FA) arrays. For all moduli, the FA-based AU_ms are shown to execute much faster as well as have much smaller hardware complexity and time-complexity products than ROM-based general multipliers. Since the AU_ms use full adders as their basic units, they lead to modular and regular designs, which are inexpensive and easy to implement in VLSI.

1.3.3 Error Tolerance in RNS Systems

Because there is no interaction among digits (channels) in residue arithmetic, any errors generated at one digit cannot propagate and contaminate other channels during subsequent operations, given that no conversion has occurred from the RNS to a weighted representation.

In addition, because there is no weight associated with the RNS residues (digits), if any digit becomes corrupted, the associated channel may be easily identified and dealt with.

Based on the amount of redundancy that is built in an RNS processor, the faulty channels may be replaced or just isolated, with the rest of the system operating in a "soft failure" mode, being allowed to gracefully degrade into accurate operations of reduced dynamic range. Provided that the remaining dynamic range contains the results, there is no problem with this degradation.

The more redundant an RNS is, the easier it is to identify and correct errors. A redundant RNS (RRNS) uses a number r of moduli in addition to the N standard moduli that are necessary for covering the desired dynamic range. All $N + r$ moduli must be relatively prime. In an RRNS, a number X is presented by a total of N nonredundant residue digits $\{X_2, \ldots, X_N\}$ plus r redundant residue digits $\{X_{N+1} \ldots, X_{N+r}\}$. Of the total number of states, $M_R = \Pi_{i=1}^{N+r} m_i$ is represented by the RRNS. The $M = \Pi_{i=1}^{N} m_i$ first states constitute its "legitimate range," while any number that lies in the range $(M,\ M_R)$, is called "illegitimate."

Any single error moves a legitimate number X into an illegitimate number X'. Once it is verified that the number being tested is illegitimate, its digits are discarded one by one, until a legitimate representation is found. The discarded digit whose omission results in the legitimate representation is the erroneous one. A correct digit can then be produced by extending the base of the reduced RNS that produced the legitimate representation. The above error-locating-and-correcting procedure can be implemented in a variety of ways. Assuming that the mixed radix representations of all the reduced RNS representations can be efficiently generated, the legitimate one can be easily identified by checking the highest order mixed radix digit against zero. If it is zero, the representation is legitimate.

Mixed radix representations associated with the RNS numbers can be used to detect overflows as well as to detect and correct errors in redundant RNS systems. For example, to detect overflows, a redundant modulus m_{N+1} is added to the base and the corresponding highest order mixed radix digit a_{N+1} is found and compared to zero. Assuming that the number being tested for overflow is not large enough to overflow the augmented range of the redundant system, overflow occurs whenever a_{N+1} is not zero.

1.3.4 RNS and Power Dissipation

RNS may reduce power dissipation because it reduces the hardware cost, the switching activity, and the supply voltage (Freking and Parhi, 1997). By employing binary-like RNS filter structures (Ibrahim, 1994), it has been reported that RNS reduces the bit activity up to 38% in (4×4)-bit multipliers. As the critical path in an RNS architecture increases logarithmically with the equivalent binary word length, RNS can tolerate a larger reduction in the supply voltage than the corresponding binary architecture while achieving a particular delay specification. To demonstrate the overall impact of the RNS on the power budget of an FIR filter, Freking and Parhi (1997) report that a filter unit with 16-bit coefficients and 32-bit dynamic range, operating at 50 MHz, dissipates 26.2 mW on average for a two's complement implementation, while the RNS equivalent architecture dissipates 3.8 mW. Hence, power dissipation reduction becomes more significant as the number of filter taps increases, and a 3-fold reduction is possible for filters with more than 100 taps.

Low-power may also be achieved via a different RNS implementation. It has been suggested to one-hot encode the residues in an RNS-based architecture, thus defining **one-hot RNS** (OHR) (Chren, 1998). Instead of encoding a residue value x_i in a conventional positional notation, an $(m-1)$-bit word is employed. In this word, the assertion of the ith bit denotes the residue value x_i. The one-hot approach allows for a further reduction in bit activity and power-delay products using residue arithmetic. OHR is found to require simple circuits for processing. The power reduction is rendered possible since all basic operations (i.e., addition, subtraction, and multiplication) as well as the RNS-specific operations of scaling (i.e., division by constant), modulus conversion, and index computation are performed using transposition of bit lines and barrel shifters. The performance of the obtained residue architectures is demonstrated through the design of a direct digital frequency synthesizer that exhibits a power-delay product reduction of 85% over the conventional approach (Chren, 1998).

RNS Signal Activity for Gaussian Input

The bit activity in an RNS architecture with positionally encoded residues has been experimentally studied for the encoding of 8-bit data using the base {2, 151}, which provides a linear fixed-point dynamic range of approximately 8.24 bits. Assuming data sampled from a Gaussian process, the bit assertion activities of the particular RNS, an 8-bit sign-magnitude, and an 8-bit two's-complement system are measured and compared. The results are depicted in Figure 1.8 for 100 Monte Carlo runs. It is observed that RNS performs better than two's complement representation for anticorrelated data and slightly worse than sign-magnitude and two's complement representations for uncorrelated and correlated sequences.

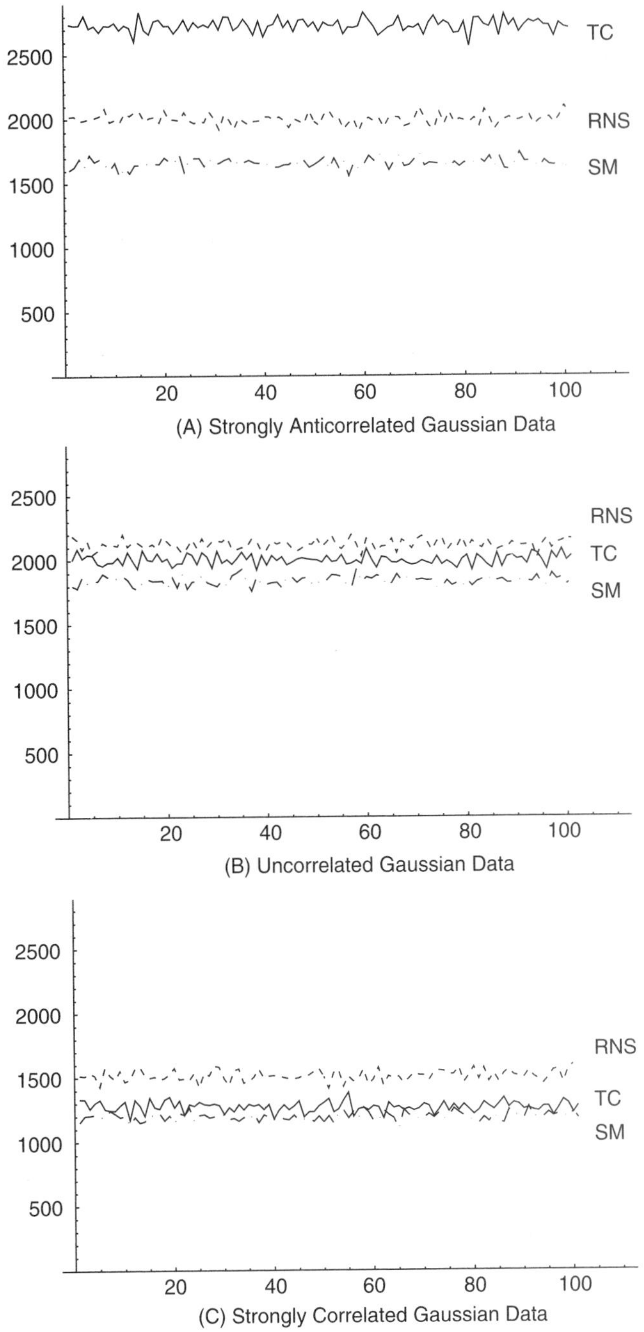

FIGURE 1.8 Number of Low-to-High Transitions. (A) This figure shows strongly anticorrelated ($\rho = -0.99$) Gaussian data for two's complement, RNS, and sign-magnitude number systems for 100 Monte Carlo runs. (B) Shown here are uncorrelated ($\rho = 0$) Gaussian data; (C) This figure illustrates strongly correlated ($\rho = 0.99$) Gaussian data

References

Bayoumi, M.A., Jullien, G.A., and Miller, W.C. (1983). Models of VLSI implementation of residue number system arithmetic modules. *Proceedings of 6th Symposium on Computer Arithmetic*, 412–415.

Chandrakasan, A.P., and Brodersen, R.W. (1995). *Low power digital CMOS design.* Boston: Kluwer Academic Publishers.

Chren, W.A., Jr., (1998). One-hot residue coding for low delay-power product CMOS design. *IEEE Transactions on Circuits and Systems—Part II 45*, 303–313.

Freking, W.L., and Parhi, K.K. (1997). Low-power FIR digital filters using residue arithmetic. *Proceedings of Thirty-first Asilomar Conference on Signals, Systems, and Computers* 739–743.

Ibrahim, M.K. (1994). Novel digital filter implementations using hybrid RNS-binary arithmetic. *Signal Processing 40*, 287–294.

Koren, I. (1993). *Computer arithmetic algorithms.* Englewood Cliffs, NJ: Prentice Hall.

Orginos, I., Paliouras, V., and Stouraitis, T. (1995). A novel algorithm for multioperand logarithmic number system addition and subtraction using polynomial approximation. *Proceedings of International Symposium on Circuits and Systems*, III.1992–III.1995.

Paliouras, V., and Stouraitis, T. (2001). Signal activity and power consumption reduction using the logarithmic number system. *Proceedings of IEEE International Symposium on Circuits and Systems*, II.653–II.656.

Paliouras, V., and Stouraitis, T. (2001). Low-power properties of the logarithmic number system. *Proceedings of the 15th Symposium on Computer Arithmetic (ARITH15)*, 229–236.

Paliouras, V., and Stouraitis, T. (1996). A novel algorithm for accurate logarithmic number system subtraction. *Proceedings of International Symposium on Circuits and Systems. 4*, 268–271.

Peebles, P.Z. Jr. (1987). *Probability, random variables, and random signal principles.* New York: McGraw-Hill.

Soderstrand, M.A., Jenkins, W.K., Jullien, G.A., and Taylor, F.J. (1986). *Residue number arithmetic: Modern applications in digital signal processing.* New York: IEEE Press.

Stouraitis, T., Kim, S.W., and Skavantzos, A. (1993). Full adder-based units for finite integer rings. *IEEE Transactions on Circuits and Systems—Part II 40*, 740–745.

Taylor, F., Gill, R., Joseph, J., and Radke, J. (1988). A 20-bit logarithmic number system processor. *IEEE Transactions on Computers, 37*, 190–199.

Taylor, F.J., Papadourakis, G., Skavantzos, A., and Stouraitis, T. A radix-4 FFT using complex RNS arithmetic. *IEEE Transactions on Computers C-34*, 573–576.

Sacha, J.R., and Irwin, M.J. (1998). The logarithmic number system for strength reduction in adaptive filtering. *Proceedings of International Symposium on Low-Power Electronics and Design*, 256–261.

Stouraitis, T. (1986). *Logarithmic number system: Theory, analysis and design.* Ph.D. diss., University of Florida.

Szabó, N., and Tanaka, R. (1967). *Residue arithmetic and its applications to computer technology.* New York: McGraw-Hill.

2

Custom Memory Organization and Data Transfer: Architectural Issues and Exploration Methods

Francky Catthoor,
Erik Brockmeyer,
Koen Danckaert,
Chidamber Kulkani,
Lode Nachtergaele, and
Arnout Vandecappelle
*IMEC,
Leuven, Belgium*

2.1 Introduction

Because of the eternal push for more complex applications with correspondingly larger and more complicated data types, storage technology is taking center stage in more and more applications (Lawton, 1999), especially in the information technology area, including multimedia processing, network protocols, and telecom terminals. In addition, the access speed, size, and power consumption associated with the available storage devices form a severe bottleneck in these systems, especially in an embedded context. In this chapter, several building blocks for memory storage will be investigated first, with the emphasis on their internal architectural organization. Section 2 presents a general classification of the main memory components for customized organizations, including register files and on-chip SRAM and DRAMs. Next, Section 3. explains off-chip and global hierarchical memory organization issues.

Apart from the storage architecture itself, the way data are mapped to these architecture components is important for a good overall memory management solution. Actually, these issues are gaining in importance in the current age of deep submicron technologies, where technology and circuit solutions are not sufficient on their own to solve the system design bottlenecks. In current practice, designers usually go for the highest speed implementation for most submodules of a complex system, even when real-time constraints apply for the global design. Moreover, the design tools for exploration support (e.g., compilers and system synthesis tools) focus mainly on the performance aspect. The system cost, however can often be significantly reduced by system-level code transformations or trading-off cycles spent in different submodules. Therefore, the last four sections of this chapter are devoted to different aspects of data transfer and storage exploration: code rewriting techniques to improve data reuse and access locality (Section 4), how to meet real-time bandwidth constraints (Section 5), custom memory organization design (Section 6), and data layout reorganization for reduced memory size (Section 7). The main emphasis lies on custom processor contexts. Realistic multimedia and telecom applications are used to demonstrate the impressive effects of such techniques.

2.2 Custom Memory Components

In a custom processing context, many different options are open to the designer for the organization of the data storage and the access to it. Before going into the usually hierarchical organization of this storage (see section 2–3), this section presents the different components that make up the primitives of the hierarchy.

The goal of a storage device is in general to store a number of n-bit data words for a short- or long-term period. Under control of the address unit(s), these data words are transferred to the custom processing units (called processors further on) at the appropriate point in time (cycle), and the results of the operations are then written back in the storage device for future use. Due to the different characteristics of storage and access, different styles of devices have been developed.

2.2.1 General Principles and Storage Classification

A very important difference can be made between memories (for frequent and repetitive use) concerning short-term and long-term storage. Short-term storage devices are in general located very close to the operators and require a very small access time. Consequently, they should be limited to a relatively small capacity (less than 32 words typically) and are taken up usually in feedback loops over the operators (e.g., RegfA-BufA-BusA-EXU-RegfA in Figure 2.1) or at the input of execution units. This is typically called foreground memory. The devices for long-term storage are in general meant for (much) larger capacities (from 64- to 16-M words) and take a separate cycle for read or write access (Figure 2.2). Both categories are described in more detail in the next subsections.

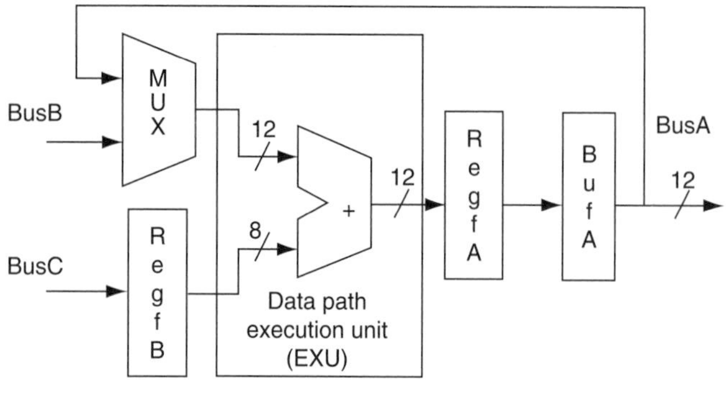

FIGURE 2.1 Register File in Feedback Loop of Custom Data Path

Six other important distinctions between short-term and long-term storage devices can be made using the tree-like "genealogy" of storage devices presented in Figure 2.3:

1. **Read-only and read or write (R/W) access**: Some memories are used only to store constant data, such as read-only memories or ROMs. Good alternatives include programmable logic arrays (PLA) or multilevel logic circuits, especially when the amount of data is relatively small. In most cases, data needs to be overwritable at high speeds, which means that *read* and *write* are treated with the same priority (R/W access),

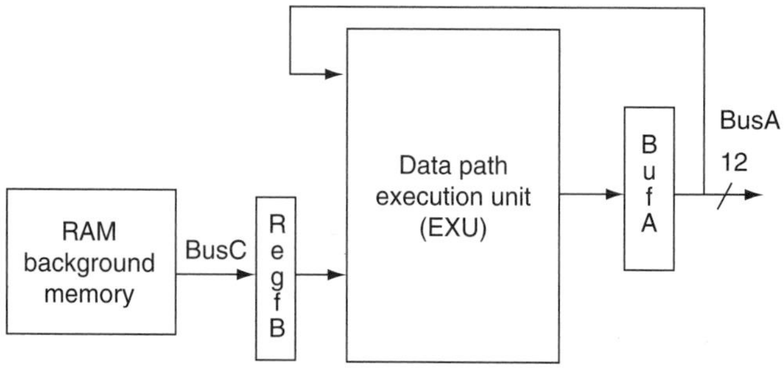

FIGURE 2.2 Large Capacity Background Memory Communicating with Custom Data Path

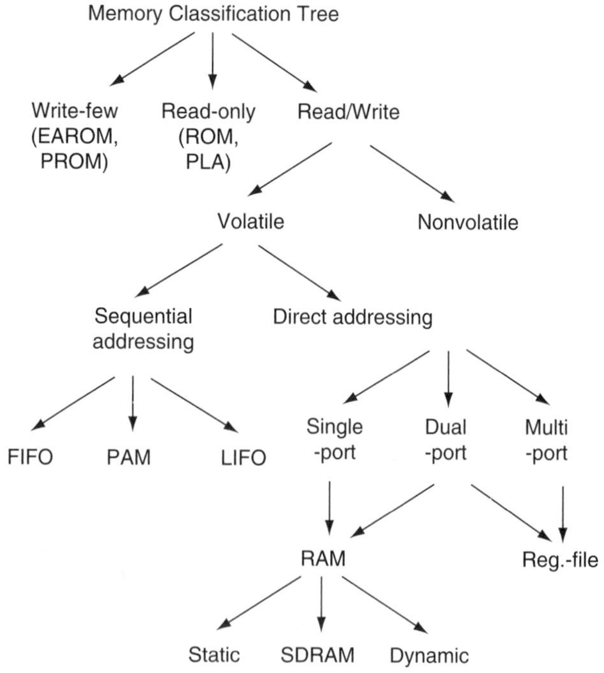

FIGURE 2.3 Storage Classification

such as in random-access memories or RAMs. In some cases, the ROMs can be made electrically alterable (= write-few) with high energies (EAROM) or programmable by means of fuses (PROM), for example. Only the R/W memories will be discussed further on.

2. **Volatile or not**: For R/W memories, the data are usually removed once the power goes down. In some cases, this can be avoided, but these nonvolatile options are expensive and slow. Examples are magnetic media and tapes that are intended for slow access of mass data. This discussion is restricted to the most common case on chip, namely, the volatile one.
3. **Address mechanism**: Some devices require only sequential addressing, such as the first-in first-out (FIFO) queue or first-in last-out (FILO) or stack structures, which put a severe restriction on the order in which the data are read out. Still, this restriction is acceptable for many applications. A more general but still sequential access order is available in a pointer-addressed memory (PAM). In most cases, however, the address sequence should be random (including repetition), which requires a direct addressing scheme. Then, an important requirement is that in this case, the access time should be independent of the address selected.
4. **The number of independent addresses and corresponding gateways (busses) for access**: This parameter can be one (single-port), two (dual-port), or even more (multiport) ports. Any of these ports can be for reading only, writing only, or R/W. Of course, the area occupied will increase considerably with the number of ports.
5. **Internal organization of the memories**: The memory can be meant for capacities that remain small or that can become large. Here, a trade-off is usually involved between speed and area efficiency. The registerfiles in subsection 2.2.2 constitute an example of the fast small capacity organizations that are usually also dual-ported or even multiported. The queues and stacks are meant for medium-sized capacities. The RAMs in subsection 2.2.3 can become extremely large (up to 256 Mb for the state-of-the-art RAM) but are also much slower in random access.
6. **Static or dynamic**: For R/W memories, the data can remain valid as long as Vdd is on (static cell in SRAM), or the data should be refreshed every 1.2 ms (dynamic cell in DRAM). In the dynamic class, high-throughput Synchronous DRAMs (SDRAM) are also available. Circuit level issues are discussed in overview articles concerning SRAMs, such as in Evans (1995), and for DRAMs, such as in Itoh (1997) and Prince (1994).

In the following subsections, the most important R/W-type memories and their characteristics are investigated in more detail.

2.2.2 Register Files and Local Memory Organization

This subsection discusses the register file and local memory organization. An illustrative organization for a dual port register file with two address busses, where the separate read and write addresses are generated from an address calculation unit (ACU), is shown in Figure 2.4. In this case, two data busses (A and B) are used but only in one direction, so the write and read addresses directly control the port access. In general, the number of different address words can be smaller than the number of port(s) when they are shared (e.g., for either read or write), and the busses can be bidirectional. Additional control signals decide whether to write or read and for which port the address applies. The number of address bits per word is $\log_2(N)$. The register file of Figure 2.4 can be used very efficiently in the feedback loop of a data path as already illustrated in Figure 2.1. In general, the file is used only for the storage of temporary variables in the application running on the data path (sometimes also referred to as execution unit). Such register files (regfiles) are also used heavily in most modern general-purpose RISCs and especially for modern multimedia-oriented signal processors that have regfiles up to 128 locations.[1] For multimedia-oriented VLIW processors or recent super-scalar processors, regfiles with a very large access bandwidth, up to 17 ports, are provided (Jolly, 1991). Application-specific instruction-set processors (ASIPs) and custom processors make heavy use of regfiles for the same purpose. It should be noted that although it has the clear advantage of very fast access, the number of data words to be stored should be minimized as much as possible due to the power- and area-intensive structure of such register files (both due to the decoder and the cell overhead). Detailed circuit issues will not be discussed here (for review, see Weste and Eshraghian [1993]).

After this brief discussion of the local foreground memories, we will now proceed with on- and off-chip background memories of the random access type.

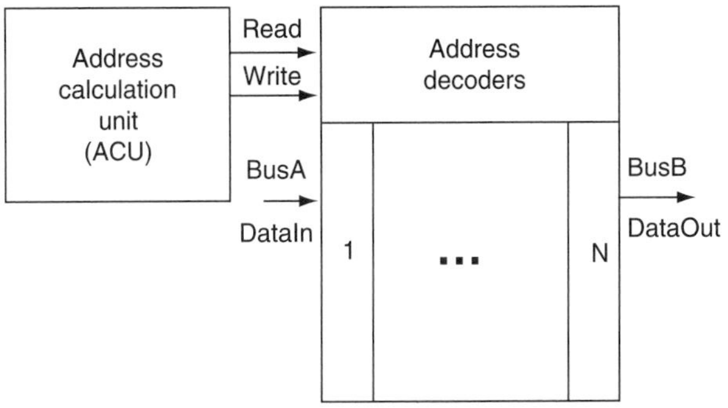

FIGURE 2.4 Regfile with Both R and W Addresses

[1] In this case, it becomes doubtful whether a register file is really a good option due to the very high area and power penalty.

2.2.3 RAM Organization

A large variety of possible types of RAMs for use as main memories has been proposed in the literature; research on RAM technology is still very active, as demonstrated by the results presented at International Solid-State Circuits (ISSCC) and Custom Integrated Circuit (CICC) conferences. Summary articles are available by Comerford and Watson (1992), Prince (1994), Oshima *et al.* (1997), and Lawton (1999). The general organization depends on the number of ports, but usually single-port structures are encountered in the large density memories. These structures will also be a restriction here. Most other distinguishing characteristics are related to the circuit design (and the technological issues) of the RAM cell, the decoder, and the auxiliary R/W devices. In this subsection, only the general principles are discussed.

For a B-bit organized RAM with a capacity of 2^k words, the floor plan in Figure 2.5 is the basis for everything. Notice the presence of read and write amplifiers that are necessary to drive or sense the (very) long bit-lines in the vertical direction. Notice also the presence of write-enable (WE) signal for controlling the R/W option and a chip select (CS) control signal that is mainly used to save power.

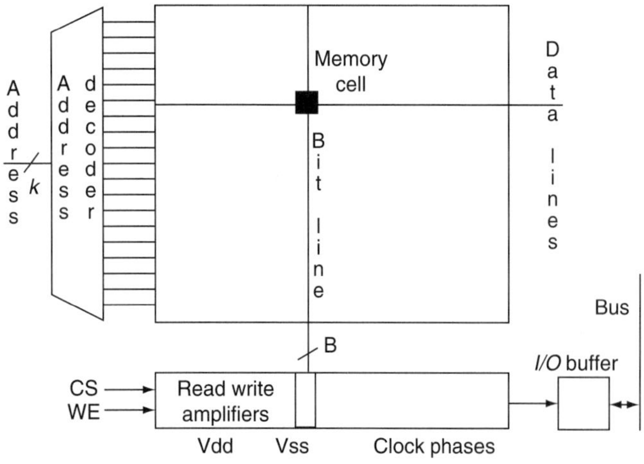

FIGURE 2.5 Basic Floor Plan for B-Bit RAM with 2^k Words

For large-capacity RAMs, the basic floor plan of Figure 2.5 leads to a very slow realization because of bit-lines that are too long. For this purpose, postdecoding is applied. This leads to the use of an *X* and a *Y* decoder (Figure 2.6), the flexibility of the floor plan shape is now used to end up with a near square (dimensions *x* by *y*), which makes use of the chip area in the most optimal way and reduces the access time and power (wire-length related). To achieve this, the following equations can be applied for a *k*-bit total address of $x + y = k$ and $x = y + \log 2B$, leading to $x = (\log 2B + k)/2$ and $y = k - x$ for maximal "squareness."

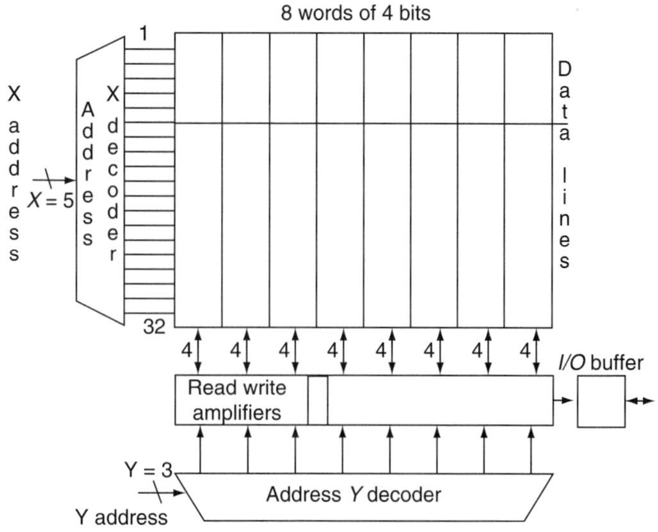

FIGURE 2.6 Example of Postdecoding

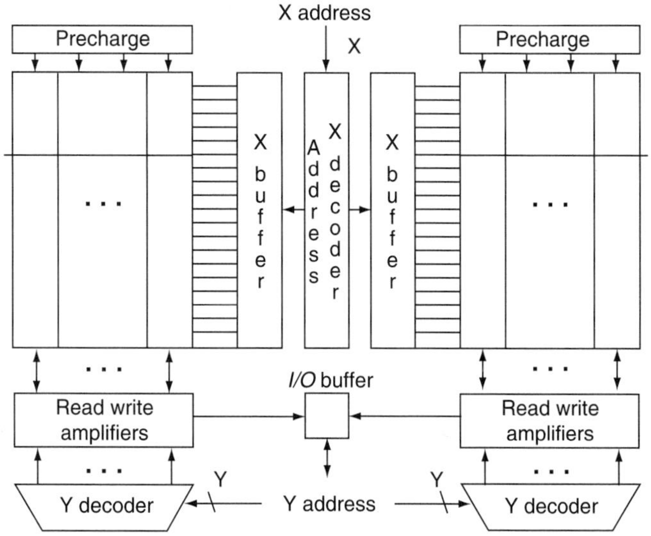

FIGURE 2.7 Partitioning of Memory Matrix Combined with Post-decoding

This breakup of a large memory plane into several subplanes is very important also for low-power memories. In that case, however, care should be taken to enable only the memory plane that contains data needed in a particular address cycle. If possible, the data used in successive cycles should also come from the same plane because activating a new plane takes up a significant amount of extra precharge power. An example floor plan in Figure 2.6 is drawn for $k = 8$ and $B = 4$. The memory matrix can be split up in two or more parts to reduce the length of the word lines also by a factor 2 or more. This results in a typically heavily partitioned floor plan (Itoh, 1997) as shown in a simplified form in Figure 2.7.

It should also be noted that the word length of data stored in the RAM is usually matched to the requirement of the application in the case of an on-chip RAM embedded in an ASIC. For RAM chips, this word organization normally has been

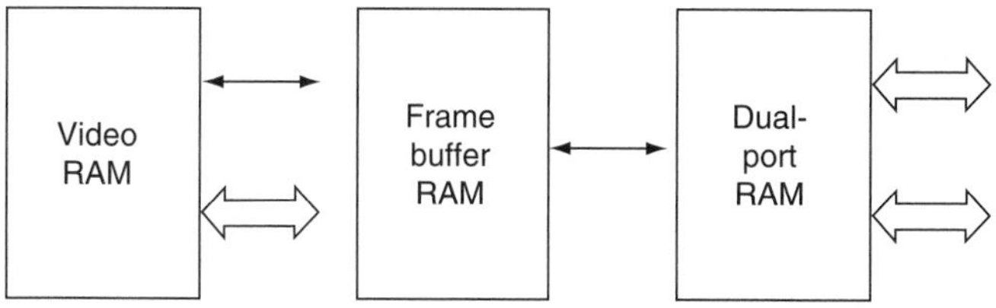

FIGURE 2.8 Specialized Storage Units. Thin arrows are serial ports; wide arrows represent parallel ports.

standardized to a few choices only. Most large RAMs are 1-bit RAMs. However, with the push for more application-specific hardware, 4-bit (nibble), 8-bit (byte), and 16-bit RAMs are now commonly produced. For the on-chip, clearly more options can be made available in a memory generator.

In addition to the mainstream evolution of these SRAMs, DRAMs, and SDRAMs, more customized large capacity high bandwidth memories are proposed that are intended for more specialized purposes. Examples are video RAMs, very wide DRAMs, and SRAMs with more than two ports (see the eight-port SRAM described by Takayanagi *et al.*, 1996, and review Prince, 1994). See also Figure 2.8.

2.3 Off-Chip and Global Hierarchical Memory Organization

Custom processors used in a data-dominated application context are driven from a customized memory organization. In programmable instruction-set processors, this organization is usually constructed according to a rigorous hierarchy that can be considered almost like a bidirectional pipeline from disk[2] over main memory and *L2*/*L1* cache to the multiport register file. Still, the pipeline gets saturated and blocked more and more due to the large latencies that are introduced compared to the CPU clock cycles in current process technologies. This happens especially for the off-chip memories in the pipeline. In a custom processor context, more options are open to increase the on-chip bandwidth to the data, but off-chip similar restrictions apply for the off-chip. These issues are discussed in more detail in Subsection 2.3.1.

Next, a brief literature overview is presented of some interesting evolutions of current and future stand-alone RAM memory circuit organizations in different contexts. In particular, three different "directions" will be highlighted describing the apparent targets of memory vendors. They differ namely in terms of their mostly emphasized cost functions: power, speed, or density. Some of these directions can be combined, but several of them are noncompatible. One conclusion of this literature study is that the main emphasis in almost all directions is on the access structure and not on the way the actual storage takes place in the "cell." This access structure, which includes the "selection" part and the "input/output" part, dominates the overall cost in terms of power, speed, and throughput. When power-sensitive memories are constructed, the area contribution also becomes more balanced between the heavily distributed memory matrices and the access structures.

2.3.1 The External Data Access Bottleneck

Most older DRAMs are of the pipelined page-mode type. Pipelined memory implementations improve the throughput of a memory, but they don't improve the latency. For instance, in an extended data output (EDO) RAM, the address decoding and the actual access to the storage matrix (of the previous memory access) are done in parallel. In that case, the access sequence to the DRAM is also important. For example, the data sheet of an EDO memory chip specifies the sequence 10-2-2-23-2-2-2, where the numbers represent clock cycles. Each curly bracket indicates four bus cycles of 64-bit each—that is one cache line. The first sequence, 10-2-2-2, specifies the timing if the page is first opened and accessed four times. The second sequence, 3-2-2-2, specifies the timing if the page was already open and accessed four additional times—that means no other memory page was opened and accessed in between those times. The last sequence repeats as long as memory is accessed in the same page.

The data sheet in question relates to a memory bus running at 66 MHz. Taking another processing speed, say a 233-MHz processor, the timing becomes 35-7-7-711-7-7-7 in processor clocks (refer to Table 2.1).

Synchronous DRAMs (SDRAM) can have a larger pipeline that provides a huge theoretical bandwidth. Basically, the internal state machine enables to enlarge the pipeline. An SDRAM can sustain this high throughput rate by data access interleaving over several banks. However, the address needs to be known several cycles prior to when the data are actually needed. Otherwise the data path will stall, canceling the advantage of the pipelined memory. As already mentioned, modern stand-alone DRAM chips, which are often of this SDRAM type, also offer low-power solutions, but this comes at a price. Internally they contain banks and a small cache with a (very) wide width connected to the external high-speed bus

TABLE 2.1 Memory Architecture and Timing for a System Using the Pentium II Processor and EDO Memory

Bus	Bus clocks	CPU clocks at 233 MHz	Total CPU clocks (bandwidth)
L1 cache	1-1-1-1	1-1-1-1	4 (1864 MB/s)
L2 cache	5-1-1-1	10-2-2-2	16 (466 MB/s)
EDO memory	10-2-2-2	35-7-7-7	56 (133 MB/s)
	3-2-2-2	11-7-7-7	32 (233 MB/s)
SDRAM	11-1-1-1	39-4-4-4	51 (146 MB/s)
	2-1-1-1	7-4-4-4	19 (392 MB/s)

[2] This is possibly with a disk cache to hide the long latency.

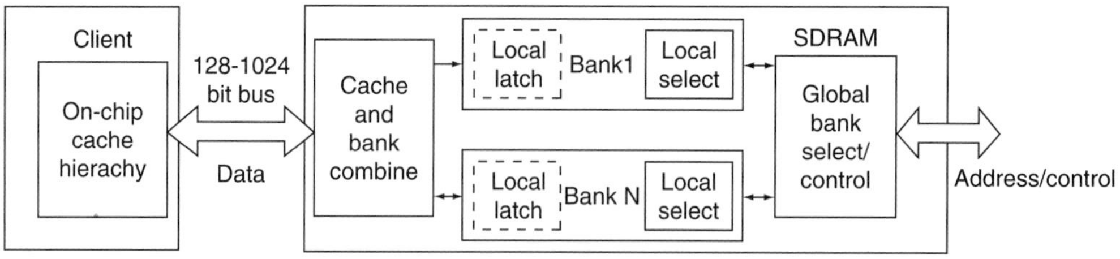

FIGURE 2.9 External Data Access Bottleneck Illustration with SDRAM

(see Figure 2.9) (Kim *et al.*, 1998; Kirihata *et al.*, 1998). So the low-power operation per bit is only feasible when the chips operate in burst mode with entire (parts of) memory column(s) transferred over the external bus. This is not directly compatible with the actual use of the data in the processor data paths, so without a buffer to the processors, most of the data words that are exchanged would be useless (and discarded). Obviously, the effective energy consumption per useful bit becomes very high in that case, and the effective bandwidth is low.

Therefore, a hierarchical and typically much more power-hungry intermediate memory organization is needed to match the central DRAM to the data ordering and bandwidth requirements of the processor data paths. The reduction of the power consumption in fast random-access memories is not as advanced as in DRAMs; moreover, this type of memory is saturating because many circuit and technology level tricks have already been applied also in SRAMs. As a result, fast SRAMs keep on consuming on the order of watts for high-speed operation around 500 MHz. So the memory-related system power bottleneck remains a very critical issue for data-dominated applications.

From the process technology point of view, this problem is not so surprising, especially for submicron technologies. The relative power cost of interconnections is increasing rapidly compared to the transistor (active circuit) related components. Clearly, local data paths, and controllers themselves contribute little to this overall interconnect compared to the major data/instruction busses and the internal connections in the large memories. Hence, if all other parameters remain constant, the energy consumption (and also the delay or area) in the storage and transfer organization will become even more dominant in the future, especially for deep submicron technologies. The remaining basic limitation lies in transporting the data and the control (like addresses and internal signals) over large on-chip distances and in storing them.

One last technological recourse to try to alleviate the energy-delay bottleneck is to embed the memories as much as possible on-chip. This has been the focus of several recent activities, such as in the Mitsubishi announcement of an SIMD processor with a large distributed DRAM in 1996 (Tsuruda *et al.*, 1996), followed by the offering of the "embedded DRAM" technology by several other vendors, and the IRAM initiative of Dave Patterson's group at U.C. Berkeley (Patterson *et al.*, 1997). The results show that the option of embedding logic on a DRAM process leads to a reduced power cost and an increased bandwidth between the central DRAM and the rest of the system. This is indeed true for applications where the increased processing cost is allowed (Wehn *et al.*, 1998). It is a one-time drop, however after which the widening energy-delay gap between the storage and the logic will keep on progressing due to the unavoidable evolution of the relative interconnect contributions (see previous discussion).

So on the longer term, the bottleneck should be broken also by other means. In sections 2.4 to 2.7 it is shown that this breakdown is feasible with quite spectacular effects at the level of the system-design methodology. The price paid will be increased design complexity that can however be offset with appropriate design methodology support tools.

2.3.2 Power Consumption Issues

In many cases, a processor requires one or more large external memories, mostly of the (S)DRAM type, to store long-term data. For data-dominated applications, the total system power cost in the past was for the most part attributed to the presence of these external memories on the board. Because of the heavy push toward lower power solutions to keep the package costs low, and also for mobile applications or due to reliability issues, the power consumption of such external DRAMs has been reduced significantly. This has also been a very active research topic.

The main concepts are a hierarchical organization with:

- Usually more than 32 divisions for RAM sizes above 16 Mb (Itoh *et al.*, 1995).
- Wide memory words to reduce the access speed (Amrutur and Horowitz, 1994).
- Multidivided arrays (both for word line and data line) with up to 1024 divisions in a single matrix (Sugibayashi *et al.*, 1993).
- Low on-chip Vdd up to 0.5 V (Yamagata *et al.*, 1995; Yamauchi *et al.*, 1996), with Vdd/2 precharge (Yamagata *et al.*, 1995).
- Negative word line drive (NWD) (Itoh *et al.*, 1995) or over-VCC grounded data storage (OVGS) (Yamauchi *et al.*, 1996) to reduce the leakage.[3]

[3] But both of these concepts requires several on-chip voltages to be fabricated, some of them negative.

- Level-controllable (Yamagata *et al.*, 1995) or boosted (Morimura and Shibata, 1996) word line for further reducing leakage,
- Multi-input selectively precharged NAND decoder (Morimura and Shibata, 1996),
- Driving level-controllable (Yamagata *et al.*, 1995) or boosted (Morimura and Shibata, 1996) word line instead of purely dynamic decoder of the past,
- Reduced bit/word line swing[4],
- Differential bus drivers,
- Charge recycling in the *I/O* buffers (Morimura and Shibata, 1996)

Because these principles distribute the power consumption from a few "hot spots" to all parts of the architecture (Itoh *et al.*, 1997), the end result is indeed a very optimized design for power where every piece consumes about an equal amount. An explicit power distribution is provided by Seki *et al.* (1993): 47% for everything to do with address selection and buffering (several subparts), 15% in the cells, 15% in the data line and sensing, and 23% in the output buffer.

The end result is a very low-power SRAM component (e.g., 5 mW for 1 M SRAM operating at 10 MHz for a 1-V 0.5-μm CMOS technology [Morimura and Shibata, 1996], and 10 mW for 1-M SRAM operating at 100 MHz for a longer term future 0.5-V 0.35-μm CMOS technology [Yamauchi *et al.*, 1996]. Combined with the advance in process technology, all this has also led to a remarkable reduction of the DRAM-related power: from several watts for the 16–32-Mb generation to about 100 mW for 100-MHz operation in a 256-Mb DRAM.

It is expected, however, that not much more can be gained because the "bag of tricks" now contains only the more complex solutions with a smaller return-on-investment. Note that the combination of all these approaches indicates a very advanced circuit technology, which still outperforms the current state-of-the-art in data path and logic circuits for low-power design (at least in industry). Hence, it can be expected that the relative power in the nonstorage parts can be more drastically reduced for the future (on condition that similar investments are done).

2.3.3 Synchronization and Access Times for High-Speed Off- and On-Chip RAMs

An important aspect of RAMs are the access times, both for read and write. In principle, these should be balanced as much as possible as the worst case determines the maximum clock rate from the point of view of the periphery. It should be noted that the RAM access itself can be either "asynchronous" or "clocked." Most individual RAM chips are of the asynchronous type. The evolution of putting more and more RAM on chip has led to a state defined by stand-alone memories that are nearly only DRAMs. In that category, the most important subclass is becoming the SDRAMs (see previous section). In that case, the bus protocol is fully synchronous, but the internal operation of the DRAM is still partly asynchronous. For the more conventional DRAMs that also have an asynchronous interface, the internal organization involves special flags, which signal the completion of a read or write and thus the "readiness" for a new data and/or address word. In this case, a distinction has to be made between the address access delay t_{AA} from the moment when the address changes and when the actual data are available at the output buffer and the chip (RAM) access delay t_{ACS} from the moment when the chip select (CS) goes up and the data are available. Ideally, $t_{ACS} = t_{AA}$, but in practice t ACS is the largest. So special "tricks" have to be applied to approach this ideal.

For on-chip (embedded) RAMs as used in ASICs, a clocked RAM is typically preferred as it is embedded in the rest of the (usually synchronous) architecture. These are nearly always SRAMs. Sometimes they are also used as stand-alone *L*2 cache, but the trend is to embed that cache level as well on the processor chip. A possible timing (clock) diagram for such a RAM is illustrated in Figure 2.10. The highest speeds can almost be achieved with embedded SRAMs. These memories are used in the so-called *L*1 cache of the modern microprocessors, where hundreds of megahertz are required to keep up with the speed of the CPU core. Representative examples of this category are a 2-ns (333-MHz) 256-kb (semi) set-associative cache (Covino *et al.*, 1996) that requires 8.7 W in a 2.5-V 0.5-μm CMOS technology and a 500-MHz 288-kb directly mapped cache (Furumochi *et al.*, 1996) requiring 1 W in a 2.5-V 0.25-μm CMOS technology.

2.3.4 High-Density Issues

The evolution at the process technology side has been the most prominent in the DRAM, leading to new device structures like the recessed-array stacked capacitor technology (Horiguchi *et al.*, 1995) resulting in a 1-Gb version in 0.16-μm TLM CMOS and the single-electron Coulomb cell (Yano *et al.*, 1996) for even larger sizes. Multilevel logic is also being

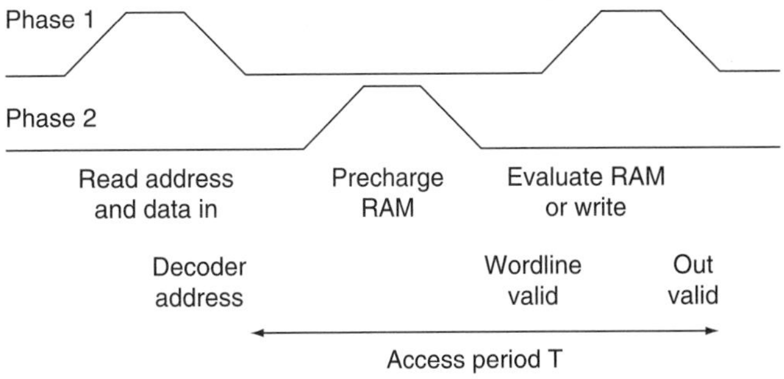

FIGURE 2.10 Timing Diagram for Synchronous Clocked RAM

[4] The reduction equals up to only 10% of the Vdd (Itoh *et al.*, 1995), which means less than 0.1 V for low voltage RAMs.

considered to put even more bits per square micrometer (Okuda and Murotani, 1997). This research direction is still mostly focused on the cell itself. SOI type devices are being investigated (Kuge *et al.*, 1996) to keep the power reasonable, but recognized is that low voltage CMOS does much to achieve this (Horiguchi *et al.*, 1995).

2.4 Code Rewriting Techniques to Improve Data Reuse and Access Locality

An efficiently used memory hierarchy is of primary importance in optimizing data transfer and storage. To exploit such a memory hierarchy, the code to be mapped should expose maximal data reuse possibilities. Code rewriting techniques, consisting of loop and data flow transformations, are essential to achieve this. The primitive loop transformations are also used in modern optimizing and parallelizing compilers to enhance the temporal and spatial locality for cache performance and to expose the inherent parallelism of the algorithm to the outer (for asynchronous parallelism) or inner (for synchronous parallelism) loops (Amarasinghe *et al.*, 1995; Wolf, 1992; Banerjee et al. 1993). Other application areas are communication-free data allocation techniques (Chen and Sheu, 1994) and optimizing communications in general (Gupta *et al.*, 1996).

It is thus no surprise that these code rewriting techniques are also at the heart of our data transfer and storage (DTSE) methodology (Catthoor *et al.*, 1998). As the first step (after the preprocessing and pruning) in the script, they are able to significantly reduce the required amount of storage and transfers. First, in the global data-flow transformation step, redundant accesses to large data types are removed. Moreover, the exploration space for the other steps is increased mainly by breaking/moving the critical data-flow bottlenecks and by exploiting algebraic laws. Next, global loop transformations are applied that significantly improve the overall regularity and access locality of the code. This enables the next step of the script, the data reuse step, to arrive at the desired reduction of storage and transfers. During this step, hierarchical data reuse copies are added to the code, exposing the different levels of reuse that are inherently present (but not directly visible) in the transformed code. A custom memory hierarchy can then be designed on which these copies can be mapped in an optimal way (see sections 2.5 and 2.6).

Crucial in our methodology is that the code transformations have to be applied globally (i.e., with the entire algorithm as scope). This is in contrast with most existing loop transformation research for which the scope is limited to one procedure or even one loop nest. This application can enhance the locality in that loop nest, but it does not solve the global data flow and associated buffer space needed between the loop nests or procedures. To allow these global transformations, a step has to be taken before the application is partitioned over different HW/SW components.

2.4.1 Methodology

Data-Flow and Loop Transformations

In a first step, the code has to be transformed to expose maximal locality and potential data reuse. This is done by globally applying data-flow and loop transformations. Loop transformations are well known and include loop interchange, merging, folding, and skewing. Data-flow transformations are less known and include, for example, algebraic law exploitation, recomputation (in terms of data transfer and storage, which can be beneficial to recompute certain data whenever they are needed instead of storing them in memory), and bottleneck removal (of which an illustration is shown in section 2.4.2). We will now give a simple example of applying loop transformations for data transfer and storage optimization.

The example is a three-level loop nest, in which a horizontal filter (with length L) is applied to a two-dimensional array. The initial description is the following (for simplicity, all filter coefficients are 1):

```
for (j = 0; j <= N-L; ++j)
  for (i = 0; i < N; ++i) {
    b [i] [j] = 0;                      (2.1)
    for (k = 0; k < L; ++k)
      b[i][j] += a[i] [j + k];
  }
```

It is clear that code 2.1 has a bad locality because L new values of array *a*[][] have to be read from memory for each iteration of the *i*-loop. These values are reused only for each iteration of the outer *j*-loop. When this algorithm is implemented directly, it will result in high storage and bandwidth requirements (assuming that N is large): either the *a*[][] values are read from main memory each time they are needed, or a large buffer of $N \times L$ locations is implemented to avoid this.

By interchanging the *i*- and *j*-loop, we can achieve a much better locality. Indeed, for each iteration of *j*, $L - 1$ values of *a*[][] can be reused from the previous iteration, and only one new value has to be read from main memory. Only for each iteration of the outer *i*-loop, L new values of *a*[][] are needed. This will lead to significantly reduced storage and bandwidth requirements. The transformed code is as follows:

```
for (i = 0; i < N; ++i)
  for (j = 0; j <= N-L; ++j) {
    b [i] [j] = 0;                      (2.2)
    for (k = 0; k < L; ++k)
      b [i] [j] += a [i] [j + k];
  }
```

Since exploring the search space for these transformations is tedious and error-prone, we are developing a systematic method for automating the global loop transformation step. Some information is available in (Catthoor *et al.*, 2000). This work, however, will still be in a research stage at the publication time of this book, and currently the step is applied manually.

Adding Hierarchical Data Reuse Copies

When the example transformed code 2.2 is executed on a processor with a small cache, it will perform much better than the initial code. To map it on a custom memory hierarchy, however, we have to know the optimal size of the different levels of this hierarchy. To this end, signal copies (buffers) are added to the code to make the data reuse explicit. For the example, this results in the following code (the initialization of *a_buf*□ has been left out for simplicity):

```
int a_buf [L];
int b_buf;
for (i = 0; i < N; ++i)
  {initialize a_buf}
  for (j = 0; j <= N-L; ++j) {
    b_buf = 0;                                  (2.3)
    a_buf [(j + L-1) % L] = a [i] [j + L-1];
    for (k = 0; k < L; ++k)
      b_buf += a_buf[(j + k) % L];
    b [i] [j] = b_buf;
  }
```

In code 2.3, two data reuse buffers are present:

- *a_buf*□(*L* words) for the *a*□□ signals
- *b_buf* (1 word) for the *b*□□ signals

In the general case, more than one level of data reuse buffers is possible for each signal (see Section 2.4.2 for an example). We have developed a formal methodology (Wuytack *et al.*, 1998) for which all possible buffers are arranged in a tree. For each signal, such a tree is generated, and an optimal alternative is selected.

2.4.2 Illustration on Cavity Detection Application

The cavity detection algorithm, which will be used as an example throughout this section, is a medical image processing application that extracts contours from images to help physicians detect brain tumors. The initial algorithm consists of a number of functions, each of which has an image frame as input and one as output, as shown in Figure 2.11. In the first function, a horizontal and vertical **GaussBlur** step is performed, in which each pixel is replaced by a weighted average of itself and its neighbors. In the second function called **ComputeEdges**(), for each pixel, the difference with all eight neighbors is computed, and the pixel is replaced by the maximum of these differences. In the last function called **DetectRoots**(), the image is first reversed. To this end, the maximum value of the image is computed, and each pixel is replaced by the difference between this maximum value and itself. Next, for each pixel we look at whether a neighbor pixel is larger than itself. If this is the case, the output pixel is false; otherwise it is true. The complete cavity detection algorithm contains some more functions, but these have been left out here for simplicity:

```
void GaussBlur (unsigned char image_in [M] [N],
unsigned char gauss_xy [M] [N]) {
  // Perform horizontal horizontal and vertical
  GaussBlur step on each pixel
  unsigned char gauss_x [M] [N];
  for (y = 0; y < M; ++y)
    for (x = 0; x < N; ++x)
      gauss_x [y] [x] = . . . // Apply
      horizontal GaussBlur
  for (y = 0; y < M; ++y)
    for (x = 0; x < N; ++x)                          (2.4)
      gauss_xy [y] [x] = . . . // Apply
      vertical GaussBlur
}
void ComputeEdges (unsigned char gauss_xy [M] [N],
unsigned char comp_edge [M] [N]) {
  // Replace every pixel with the maximum differ-
  ence with its neighbors
  for (y = 0; y < M; ++y)
    for (x = 0; x < N; ++x)
      comp_edge [y] [x] = . . .
}
void Reverse (unsigned char comp_edge [M] [N], un-
signed char ce_rev [M] [N]) {
  for (y = 0; y < M; ++y) // Search for the maximum
  value that occurs : maxval
    for (x = 0; x < N; ++x)
      maxval = . . .
  for (y = 0; y < M; ++y) // Subtract every pixel
  value from this maximum value
    for (x = 0; x < N; ++x)
      ce_rev [y] [x] = maxval - comp_edge [y] [x];
}
void DetectRoots (unsigned char comp_edge [M] [N],
unsigned char image_out [M] [N]) {
  unsigned char ce_rev [M] [N];
  Reverse (comp_edge, ce_rev); // Reverse image
  for (y = 0; y < M; ++y)
    for (x = 0; x < N; ++x)
      image_out [y] [x] = . . .
      // image_out [y] [x] is true if no neighbors
      are bigger than ce_rev [y] [x]
}
```

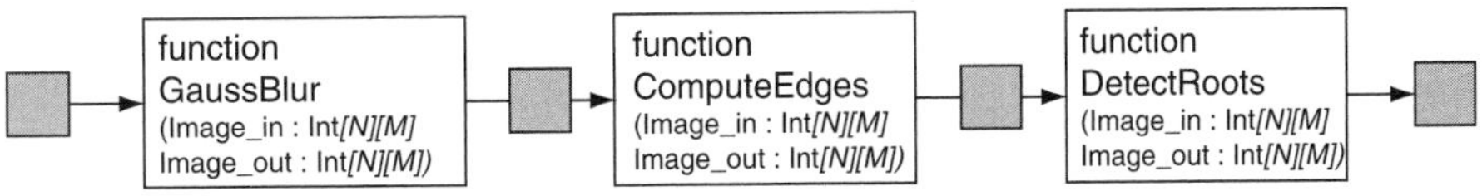

FIGURE 2.11 Initial Cavity Detection Algorithm

```
void main () {
  unsigned char image_in [M] [N], gauss_xy [M] [N],
  comp_edge [M] [N], image_out [M] [N];
  GaussBlur (image_in, gauss_xy);
  ComputeEdges (gauss_xy, comp_edge);
  DetectRoots (comp_edge, image_out);
}
```

Transformed Cavity Detection Algorithm

We will now briefly describe how data-flow and loop transformations can introduce more locality and possibilities for data reuse into the code. These transformations are explained in more detail by Catthoor (2000).

First, a data-flow transformation is applied to remove the computation of the maximum of the whole image in the function Reverse(). Although this computation is negligible in terms of CPU time, it is very important in terms of data transfer and storage because the whole image has to read from memory and be stored again before the next function DetectRoots() can proceed.

Next, a global loop folding and merging transformation is applied to all the *y*-loops in the algorithm. As a result, the functions will not work on a whole image at once anymore but on a line-per-line pipelining base. This is shown in Figure 2.12.

Finally, a similar global loop folding and merging transformation is applied to the *x*-loops. The result is that the algorithm will now work on a fine-grain (pixel-per-pixel) pipelining base, as illustrated in Figure 2.13.

Adding Hierarchical Data Reuse Copies

From the above figures that illustrate the loop transformations, two levels of data reuse can be identified: line buffers and pixel buffers. As explained above, we will insert explicit buffers into the code to exploit the potential data reuse in an optimal way.

Line Buffers

For each of the functions, a buffer of three lines can be implemented, in which the line being processed is stored together with the previous and the next line:

- The horizontal GaussBlur is done on an incoming pixel, and the result is stored in the buffer gauss_x_lines □□.
- Next, the vertical GaussBlur is performed (on one pixel) in this buffer, and the result is stored in gauss_xy_lines □□.
- Then ComputeEdges() can be executed in that buffer, the result of which is stored in comp_edge_lines □□.
- Finally, DetectRoots () is executed in comp_edge_lines □□, and the resulting pixel is stored in the output image.

The resulting code is quite complex after inserting all necessary preambles and postambles or conditions; therefore, we give only the code for the heart of the loop nest here:

```
void cav_detect (unsigned char image_in [M] [N],
                 unsigned char image_out [M] [N])
{
  unsigned char gauss_x_lines [3] [N] ;
  unsigned char gauss_xy_lines [3] [N] ;
  unsigned char comp_edge_lines [3] [N] ;
for (y=0; y<=M − 1 + 3; ++y) {

  for (x=0; x <= N − 1 + 2; + +x) {                    (2.5)
    gauss_x_lines [y % 3] [x] = . . . // Apply hori-
    zontal GaussBlurr
    gauss_xy_lines [(y−1) % 3] [x] = . . . // Apply
    vertical GaussBlur
    comp_edge_lines  [(y−2)  %  3]  [x−1]  = . . . //
    Replace with max difference // with neighbors
    image_out [y−3] [x−2] = . . . // Is true if
    no neighbors are smaller than //
    comp_edge_lines [(y−3) % 3] [x−2] ;
```

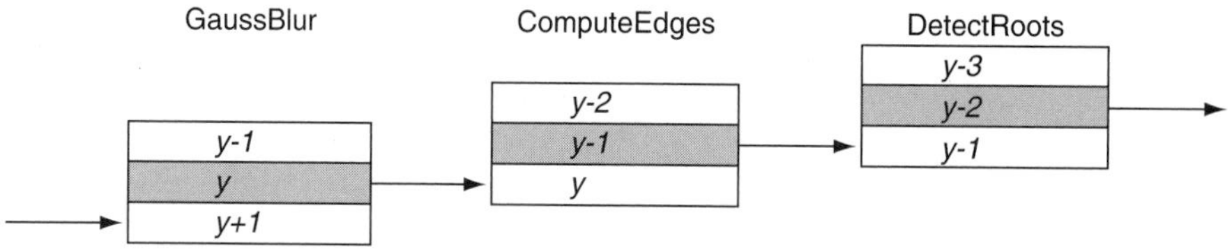

FIGURE 2.12 Cavity Detection Algorithm After *y*-Loop Transformation

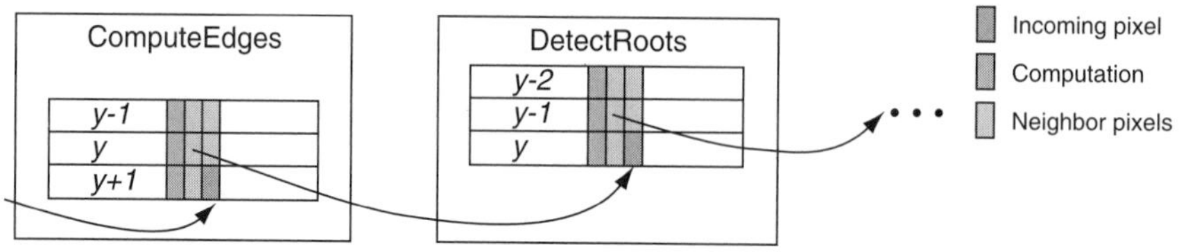

FIGURE 2.13 Cavity Detection Algorithm After *y*- and *x*-Loop Transformation

Pixel Buffers

In Figure 2.13, we can see that a second level of data reuse can be identified; that is, we can see the pixels in the neighborhood of the pixel being processed:

- For the horizontal GaussBlur, a buffer of three pixels in_pixels [] can be implemented, storing the last used values of the incoming image.
- For the vertical Gauss Blur, no such buffer is possible. However, the output of this step is stored into a three by three pixels buffer: gauss_xy_pixels[][].
- This buffer is used in ComputeEdges (), and the result of that step is again stored in a three by three buffer (i.e., comp_edge_pixels[][])
- Finally, DetectRoots () is performed on this buffer, and the result is stored in the output image.

The final code is (again without initializations and preambles and postambles) as follows:

```
void cav_detect (unsigned char image_in[M] [N],
          unsigned char image_out [M] [N])
{
  unsigned char gauss_x_lines [3] [N] ;
  unsigned char gauss_xy_lines [3] [N] ;
  unsigned char comp_edge_lines [3] [N] ;
  unsigned char in_pixels [3] ;
  unsigned char gauss_xy_pixels [3] [3] ;
  unsigned char comp_edge_pixels [3] [3] ;
  for (y=0; y<=M-1+3; ++y) {
    for (x=0; x<=N-1+3; ++x) {                    (2.6)
      in_pixels[x % 3] = image_in[y] [x];
      gauss_x_lines [y % 3] [x-1] =... // Apply
      horizontal Gauss Blur
      gauss_xy_pixels [(y-1) % 3] [(x-1) % 3]
        = gauss_xy_lines [(y-1) % 3] [x-1] =...
                // Apply vertical GaussBlur
      comp_edge_pixels [(y-2) % 3] [(x-2) % 3]
        = comp_edge_lines [(y-2) % 3] [x-2] =...
            // Replace with max difference with
            neighbors
      image_out[y-3] [x-3] =... // Is true if no
      neighbors are smaller than
                // comp_edge_pixels [(y-3)
                % 3] [(x-3) % 3] ;
    }
  }
}
```

Results

When processing the different versions presented above with our ATOMIUM access counting tool, we obtain the results presented in Table 2.2. This shows clearly that the final version is much better suited for mapping onto a custom memory hierarchy. The results in Section 2.6 will further substantiate this.

TABLE 2.2 Results of Loop and Data Reuse Transformations for Cavity Detection Application

	Accesses to		
Version	Images	Line buffers	Pixel buffers
Initial	84,954,834	0	0
Transformed + line buffers	5,229,068	29,346,281	0
Transformed + line + pixel buffers	2,618,880	11,787,272	37,208,041

2.5 How to Meet Real-Time Bandwidth Constraints

In data transfer intensive applications, a costly and difficult issue to solve concerns getting the data to the processor quickly enough. A certain amount of parallelism in the data transfers is usually required to meet the application's real-time constraints. Parallel data transfers, however, can be very costly. Therefore, the trade-off between data transfer bandwidth and data transfer cost should be carefully explored. This section describes the potential trade-offs involved and also introduces a new way to systematically trade-off the data transfer cost with other components (illustrated for total system energy consumption).

In our application domain, an overall target storage cycle budget is typically imposed and corresponds to the overall throughput. In addition, other real-time constraints that restrict the ordering freedom can be present. In data transfer and storage intensive applications, the memory accesses are often the limiting factor to the execution speed, both in custom hardware and instruction-set processors (software). Data processing can easily be sped up through pipelining and other forms of parallelism. Increasing the memory bandwidth, on the other hand, is much more expensive and requires the introduction of different hierarchical layers, typically also involving multiport memories. These memories, however, cause a large penalty in area and energy. Because memory accesses are so important, it is even possible to make an initial system-level performance evaluation based solely on the memory accesses to complex data types (Catthoor *et al.*, 1998). Data processing is then temporarily ignored except for the fact that it introduces dependencies between memory accesses. This section focuses on the trade-off between cycle budget distribution over different system components and gains in the total system energy consumption. Before discussing this topic, a number of other issues need to be introduced.

Defining such a memory system based on a high-level specification is far from trivial when taking real-time constraints into account. High-level tools will have to support the definition of the memory system (see section 2.6). An accurate data-transfer scheduling is needed to derive a suitable memory architecture in the given (timing) constraints. This as such is an impossible task to do by hand, especially when taking sophisticated cost models into account.

The cost models can be based on area or power. Using both the models provides a clear trade-off in the power, area, and performance design space. In this section, we focus on the (cost) consideration and optimization freedom of all the three axes (Subsection 2.5.1). Moreover, the effective use of the trade-off is shown in Subsection 2.5.2. These trade-offs will be illustrated with real examples.

In Subsection 2.5.2, we will also use our tools to create Pareto curves visualizing the useful trade-off space between the cycle budget assigned to a given system submodule and its corresponding energy and/or area consumption (i.e., involving three search space axes). As far as we know, no other systematic (automatable) approach is available in literature to solve this important design problem. These curves can then be used to optimize the trade-offs between different system components at the global system level (see Subsection 2.5.3). These trade-offs are illustrated with real-life demonstrators. Prior to this, in Subsection 2.5.1, we will show how different cycle budgets assigned for the data transfer and storage requirements allow to (significantly) modify the required energy both in fully customizable processors and in predefined instruction-set processor contexts.

2.5.1 The Cost of Data Transfer Bandwidth

High performance applications do need a high bandwidth for their data. This high memory bandwidth requirement will typically involve a high system cost. A clear insight in these costs is necessary to optimize the memory subsystem. The cost will nearly always boil down to energy and area/size. These are the relevant costs for the end user and manufacturer. A high memory bandwidth can be obtained in several ways. Basically, two classes of solutions exist to obtain a high bandwidth: high access speed over a single port and high parallelism. Both come with large costs in energy and area.

Technology-wise, the basic access speed of memory planes increases slowly (less than 10% per year [Przybylski, 1997]). In contrast, the speed of the data path is doubling every 18 months (Moore's law). As a result, the performance gap between memory and data path increases rapidly. Advanced memory implementations are used to keep up with the path speed, but, in many cases, these can be quite energy consuming. Note that the maximal power consumption is not relevant in this context but that the actual energy consumed to execute a given application in a specific time period (cycle budget) is important. In many cases, from the application point of view, the external memory organization is used in an inefficient way. Sometimes data caches can overcome the problems partly at runtime, but these caches have a big area and energy overhead, especially when they are set-associative (Kamble and Ghose, 1997; Ko *et al.*, 1995; Su and Despain, 1995). Future memory organization solutions will have to provide more effective bandwidth at a reasonable actual energy consumption.

An alternative approach is to lower the application demand in terms of peak bandwidth. Lowering the peak bandwidth does not have to affect the overall application performance or functionality. Tool support is then needed to fully exploit this. Note also that the larger the available cycle budget, the more freedom exists to optimize the memory access ordering. Hence, the cycle budget (or performance) can be traded off for energy reduction.

Cost Model for High-Speed Memory Architectures

New memory architectures are developed to provide a higher basic access speed. To have a better understanding, we first briefly explain a model reflecting the virtual operation of a modern SDRAM[5] (see Figure 2.14). We abstract the physical implementation details that we cannot influence (e.g., the bitwidth, the circuits used, the floor-plan, and the memory plane organization). Some of these have already been mentioned in Subsection 2.3.1. The SDRAM consists of several banks that are accessed in an interleaved fashion to sustain a high throughput (pipelining principle). Every bank typically has several planes that can be activated separately. The **virtual plane** is defined as the total storage matrix entity that is activated (and consumes energy) when a particular memory access is performed. The planes contain a page that is copied to a local latch register for fast access.

It is also beneficial to have successive accesses located in a single page. Typically a factor 2 to 4 in energy can be saved this way. In an initial phase of a memory access, the row is selected. This row (= page) is sensed[6] and partly or entirely stored in a local register. Next, the proper column is transferred to the

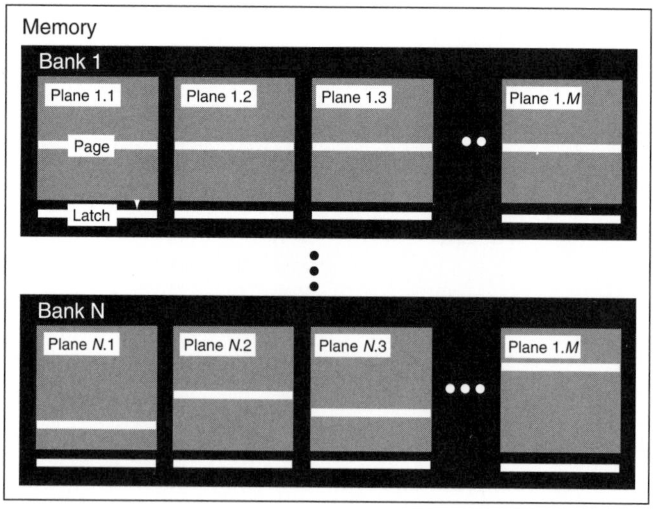

FIGURE 2.14 High-Level Synchronous DRAM Organization

[5] RAMBUS and other alternative architectures work with the same basic concept.

[6] Note that the sensing is an energy-consuming phase of an access.

output buffer. When the next access is in the same page, the row selection and sensing can be skipped. A high page hit rate is obtained when the next three conditions are fulfilled. First, the accesses need to be localized (local in every array). Second, the data layout should match the page mapping. Third, the memory access order in different arrays should access these page members together. Obviously, array access ordering and data layout freedom form the key factors.

Large memories consist of multiple planes that can be powered up separately. Multiple planes are used to avoid bit lines that are too long. By lowering switching between different planes, the energy is saved further. This should again be enabled by optimizing the access ordering and the data placement in the memories.

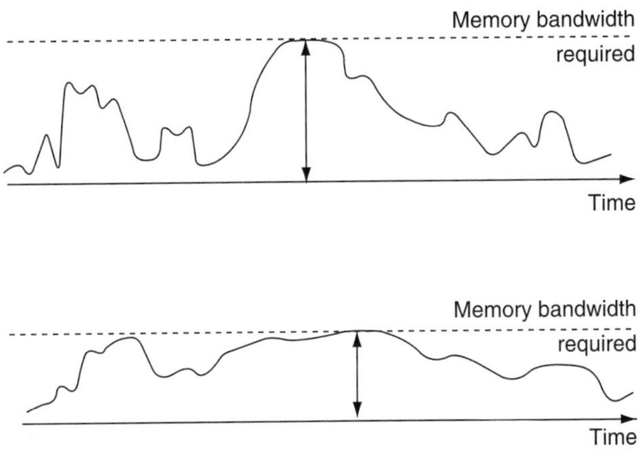

FIGURE 2.15 Balancing the Bandwidth and Lowering the Cost

Costs in Highly Parallel Memory Architectures

The throughput can be enlarged by increasing the number of bits being transferred per access. Different signals can be transferred from different memories. Moreover, multiport memories can provide a higher memory bandwidth. The main problem here is creating the parallelism and efficient use of parallel data.

Increasing the number of memories has a positive influence on both bandwidth and memory energy consumption. But the interconnect complexity increases, causing higher costs. Obviously, there must be a trade-off. The definition of the memory architecture and assignment of signals to these memories has a large impact on the available bandwidth.

Multiported memories increase the bandwidth roughly linearly with the number of ports. A complete double wiring accomplishes the improvement. Unfortunately, this will also heavily increase the area of the memory matrix. Due to this increase, the average wire length increases as well. Therefore, the energy consumption and the cost in general will go up rapidly when using multiport memories. To this end, multiport memories have to be avoided as much as possible in energy-efficient solutions.

By packing different data elements together, a larger bandwidth can be obtained (Vandecappelle *et al.*, 1999). Consequently, the data bus widens and more data are transferred. These elements can belong to the same signal (packing), or they can belong to different signals (merging). In addition to the wider bus, such a coupling of two elements can cause many redundant transfers when only one of the elements is needed. Needless to say, this option should be considered very carefully.

Balance Memory Bandwidth

The previous two subsections focused on how to use the memories efficiently and on what can be done when we change the memory (architecture). In addition, much can be gained by balancing the bandwidth load over the application.

The required bandwidth is as large as the maximum bandwidth needed by the application (see Figure 2.15). When a very high demand is present in a certain part of the code (e.g., a certain loop nest), the entire application suffers. By flattening this peak, a reduced overall bandwidth requirement is obtained (see lower part of Figure 2.15).

The rebalancing of the bandwidth load is again performed by reordering the memory accesses (also across loop scopes). Moreover, the overall cycle distribution for the loops has a large impact on the peak bandwidth. Tool support is indispensable for this difficult task. An accurate memory access ordering is needed for defining the needed memory architecture.

2.5.2 Energy Cost Versus Cycle Budget Trade-Off

The previous subsection presented the most crucial factors influencing the data transfer and storage ("memory") related cost. Cost clearly increases when a higher memory bandwidth is needed. In most cases, designers of real-time systems assume that the time for executing a given (concurrently executed) task is predefined by "some" initial system decision step. Usually, this is only seemingly the case. When several communicating tasks or threads are co-operating concurrently, then trade-offs can be made between the cycle budget and timing constraints assigned to each of them, which makes the overall problem much more complex (see Subsection 2.5.3). Moreover, in data-dominated applications, the memory subsystem is not the only cost for the complete system. In particular, after our DTSE optimization steps, the data path is typically not negligible anymore, especially for instruction-set processor realization (see Subsection 2.5.3).

This subsection explains how to use the available trade-offs between cycle budget for a given task and the related memory cost. This leads to the use of **Pareto curves**. Moreover, several detailed real-life demonstrators are worked out. The data transfer and storage-related cycle budget ("memory cycle budget") is strongly coupled to the memory system cost (both energy and memory size/area are important here, as mentioned earlier). A Pareto curve is a powerful instrument for making the right trade-offs. The Pareto curve only represents the potentially

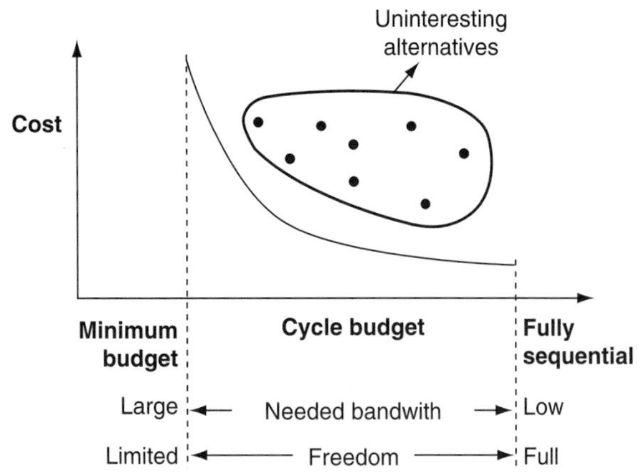

FIGURE 2.16 Pareto Curve for Trading of Memory Cycle Budget Versus Cost

interesting points in a search space with multiple axes and excludes all the solutions that have an equally good or worse solution for all the axes.

The memory cost increases when lowering the cycle budget (see Figure 2.16). When lowering the cycle budget, a more energy consuming and more complex (larger) memory architecture is needed to deliver the required bandwidth. Note that the lower cycle budget is not obtained by reducing the number of memory accesses, as in the case of the algorithmic changes and data-flow transformations (discussed in Section 2.4). These high-level code transformation related optimizations can be performed platform- and cycle-budget independent, prior to the trade-off that is the focus here. Therefore, during the currently discussed step in the system design trajectory, the amount of data transferred to the data-path remains equal over the complete cycle budget range. Nevertheless, some control flow transformations and data reuse optimizations can be beneficial for energy consumption and not for the cycle budget (or vice versa). These type of optimizations still have to be considered in the trade-off.

The interesting range of such a Pareto graph on the cycle budget axis should be defined from the critical path up to the fully sequential memory access path. In the fully sequential case, all the memory accesses can be transferred over a single port (lowest bandwidth). However, the number of memories is fully free, and the memory architecture can be freely defined. Therefore, the energy is the lowest. In the other extreme, the critical path, only a limited memory access ordering is valid. Many memory accesses are performed in parallel. Consequently, the constraints on the memory system are large, and a limited freedom is available. As a result, the energy consumption goes up. Because of a restricted cycle budget, the power itself increases even more of course, but the latter involves a more traditional trade-off that is not our focus here.

This energy-performance trade-off can be made on real-life applications thanks to the use of powerful novel tools that have been developed (IMEC, n.d.). Publications have focused on their use to generate a single near optimal solution in the entire search space. Their use in this new Pareto curve context is illustrated on a binary tree predictive coder (BTPC) used in offset image compression. Another demonstrator based on a digital audio broadcast (DAB) receiver is discussed by Brockmeyer *et al.* (1999) and Brockmeyer *et al.* (2000). Both demonstrators consist of several pages of *real* code containing multiple loops and arrays (in the range of 10 to 20) and a large number of memory access instructions (100 to 200).

BTPC is a lossless or lossy image compression algorithm based on multiresolution (Robinson, 1997). The image is successively split into a high-resolution image and a low-resolution quarter image, where the low-resolution image is split up further. The pixels in the high-resolution image are predicted based on patterns in the neighboring pixels. The remaining error is then expected to achieve high compression ratios with an adaptive Huffman coder.

Figure 2.17 shows the trade-off for the complete cycle budget range. The trade-off is obtained by letting the tool explore the cycle budget starting from the fully sequential mode and then progressing through the most interesting memory organizations (from a cost point of view) to reduce the cycle budget. Differently optimized code specifications are displayed and are derived by applying several system-level DTSE transformations prior to the individual Pareto curve generation. More information about these different versions of the code can be found in Brockmeyer (2000) and Vandecappelle (1999).

The code optimizations are performed in several steps defined by the upper part of our DTSE script (Catthoor, 1998). Platform-independent optimizations (as discussed in Section 2.4) reduce the number of memory accesses and improve the

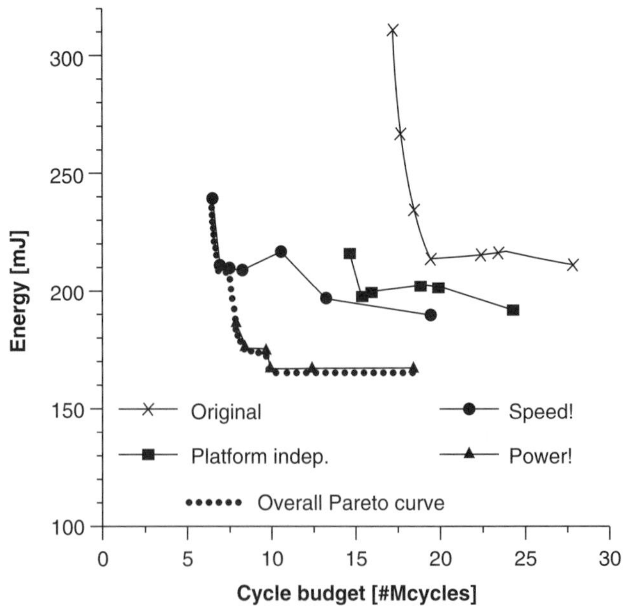

FIGURE 2.17 Pareto Curve for Binary Tree Predictive Coder

data locality (box curve). Next, two different optimization paths are followed: an energy-optimized one and a performance-optimized one. The main difference is the modified use of the available memory hierarchy: using the technique of Diguet *et al.* (1997), another intermediate signal is created and assigned differently for the speed-optimized implementation. As a consequence, an energy-intensive multiport memory is added in the memory hierarchy, delivering more bandwidth and improving the worst-case execution path of the application.

The overall Pareto curve obviously combines both differently optimized "sub-Pareto" results. For the high cycle budget and low performance situations, the low energy implementation is better. Only for a very high performance, the speed optimized implementation should be used. This kind of trade-off was first discussed by Brockmeyer *et al.* (2000). We will now also show how these curves can be effectively used to significantly improve the overall system design of data-dominated applications.

2.5.3 Exploiting the Use of the Pareto Curves

The previous subsections have shown which factors influence the memory cost. The memory cost increases when a higher memory bandwidth is needed (see Subsection 2.5.2). The memory subsystem, however, is not the only cost for the complete system. Data path is not always neglectable. Moreover, communicating tasks or threads make the problem more complex.

In this subsection, we will not go into detail on obtaining the information for the curve itself (see Section 2.4 and review IMEC [n.d.] and Brockmeyer [1999]), but we will show how to use them effectively in a low-power design context. Moreover, detailed real-life examples are worked out.

Trading off Cycles Between Memory Organization and Data Path

In case of a single thread and the memory subsystem being dominant in energy, the trade-off can be based solely on the memory organization Pareto curve. The given cycle budget defines the best memory architecture and its corresponding energy and area cost. Due to the use of latency hiding and a "system-level pipeline" between the computation and data communication, most of the cycles where memory access takes place are also useful for the data path. Still, to reduce the cost of the data path realization, it is important to leave some of the total cycle budget purely available for computation cycles (see further).

We will now illustrate how the assignment of spare cycles to the data path can have a significant impact on its energy consumption. Assigning too little indeed introduces an energy waste in the functional units when implemented in a low-power setup. For instance, several previous papers (see Chandrakasan *et al.* [1996] and its references) have motivated that functional units should operate at several voltages. The number of different Vdds has been restricted to a few specific values only because of practical reasons. Assume now that we either have a single unit that can operate part of the time at a higher Vdd and partly at a lower Vdd, or similarly assume two parallel units with different Vdds over which all the operations can be distributed (see Figure 2.18). This duplication will require some extra area, but on the entire chip this will involve a (very) small overhead, and for energy reasons this can be motivated. The high Vdd unit will clearly allow a higher speed. In the simple example of Figure 2.18, 110 million instructions need to be executed for the entire algorithm. Most instructions can be executed in the low energy functional unit in a very loose data path schedule (in total 1.0 s results in consuming 850 mJ). The same functionality, executed 25% faster, consumes twice as much energy due to the requirement to execute more on the high-speed functional unit (in total 0.75 s results in 1600 mJ). The assignment of cycles to memory accesses and to the data path is important for the overall energy consumption. Obviously, the contributions of both functions are summed. This motivates a trade-off between the memory system and the data path itself.

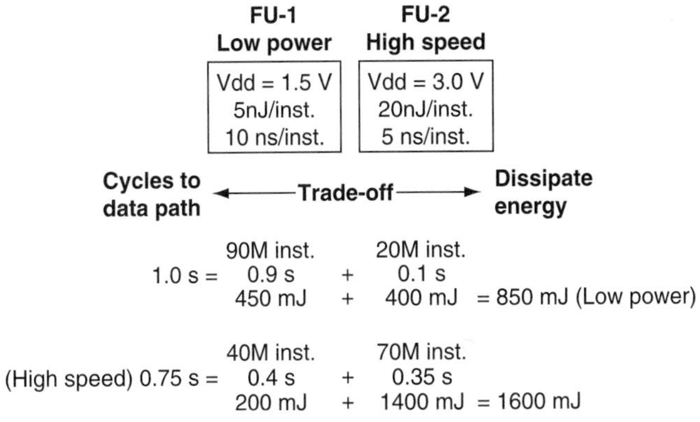

FIGURE 2.18 Data Path Example of a Performance Versus Energy Trade-Off

The reality, however, is still more complex. A certain percentage of the overall time can be spent to memory accesses, and the remaining part to computational issues, taking into account the system pipelining and the (large) overlap between memory accesses and computation (see Figure 2.19). The more time the overall budget is committed to the memory accesses, the more restrictions are imposed on the data path schedule that has less freedom for the ordering and the assignment to the functional units. As a result (see dotted curve), the Pareto curve for the data path usually exhibits a cost overhead that is small and constant for small to medium memory cycle budgets and that only starts to increase sharply when the memory cycle budget comes close to the globally available cycle budget (see Figure 3.19). Due to the fact that Pareto curve, are monotonous in their behavior, the sum of these two curves should exhibit a (unique) minimum.[7] This also advocates for an

[7] This sum usually will occur in the region where the two curves cross, but this is not necessarily the case.

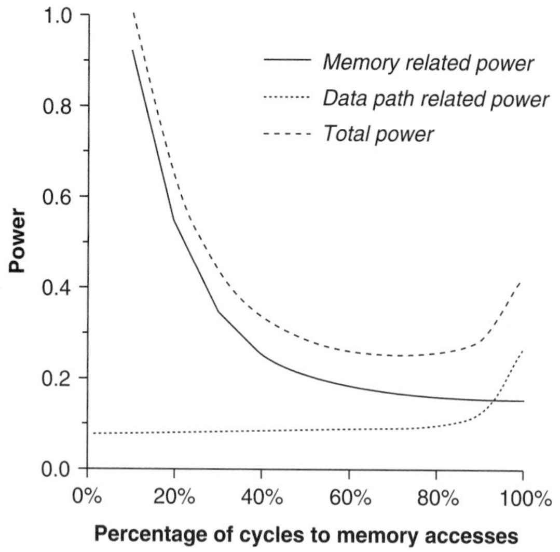

FIGURE 2.19 Trading Off Cycles for Memory Accesses Versus Cycles for Data Path Operations. Due to dependency and ordering, constraints cannot be executed in parallel with any data transfer.

accurate energy versus computation cycle budget modeling for the complex data path functions.

Energy Trade-Offs Between Concurrent Tasks

The performance–energy function can be used even better to trade off cycles assigned to different tasks in a complex system. Our Pareto curves should indeed be generated per concurrently executed task. It is then clearly visible that assigning too few cycles to a single task causes the cost of the entire system to increase. The Pareto curves also show the minimal (realistic) cycle budget per task. The cycle and energy estimates can clearly help the designer to assign concurrent tasks to processors and to distribute the cycles in the processors over the different tasks (see Figure 2.20). Minimizing the overall energy in a heterogeneous multiprocessor is then possible by applying Pareto function minimization on all the energy-cycle functions together. Even when multiple tasks have to be assigned to a single processor under control of an operating system, interesting trade-offs are feasible (a detailed example is given [Brockmeyer *et al.*, 2000]).

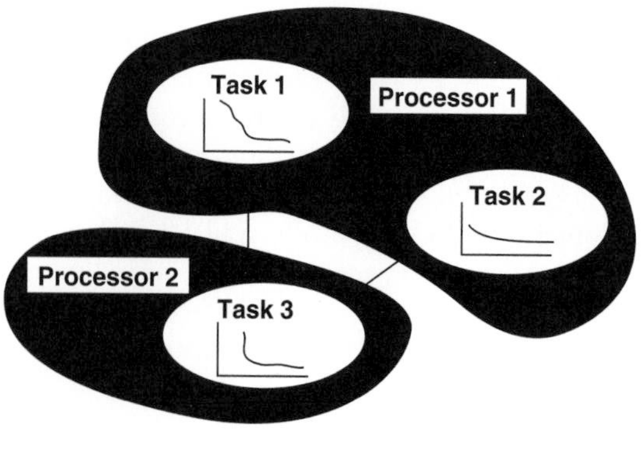

FIGURE 2.20 Trade-Off Cycles: Assigned Tasks

2.6 Custom Memory Organization Design

Many designs of VLSI systems assume a default organization of background memories. In contrast, with a customized memory organization, huge reductions are possible in the cost of the data storage. It is no trivial task, however, to achieve this cost reduction. This section first shows the cost reduction potential on the BTPC demonstrator. Then it explains the difficulties to achieve this reduction. Finally, a methodological approach to solve these problems is proposed.

2.6.1 Impact

In data-transfer intensive applications, a very high peak bandwidth to the background memories is needed. Spreading out the data transfers over time, as discussed in Subsection 2.5.1, helps to alleviate this problem, but the required bandwidth can still be high. The high-speed memory required to meet this bandwidth is a very important energy consumer and takes up much of the chip or board area (implying a higher manufacturing or assembly cost). In addition, purchasing a high-performance memory component is several times more expensive than a slower memory designed with older technology. The memory organization cost can be heavily reduced if it is possible to customize the memory architecture itself. Instead of having a single memory operating at high speed, it is better in terms of energy consumption, for example, to have two or more memories that can be accessed in parallel. Moreover, for embedded instruction-set processors, a partly customized memory architecture is possible: the processor core can be combined with one or more SRAMs or embedded DRAMs on the chip, and different configurations of the off-chip memories can also be explored. Although an instruction-set processor usually has only a single bus to connect with these memories, the clock frequency of the processor and the bus is often much higher than that of the memories, thus, the accesses to different memories can be interleaved over the bus.

An example of the impact of a custom memory organization is shown in Table 2.3 for the BTPC application introduced in

TABLE 2.3 Impact of a Custom Memory Organization for the BTPC Application

Memory architecture	On-chip area	On-chip energy	Off-chip energy
Default memory architecture One off-chip memory, One dual-port on-chip memory	200 mm^2	334 mJ	380 mJ
Custom memory architecture Three off-chip memories, Four single-port on-chip memories	111 mm^2	87 mJ	128 mJ

Subsection 2.5.2. The application's many memory accesses put a heavy load on the memory system. Therefore, the default memory configuration contains a high-speed off-chip DRAM as bulk memory and a dual-port SRAM for the arrays that are small enough to fit there. For this configuration, we have estimated the area of the on-chip SRAM and the energy consumption of both, based on the vendor's data-sheets and how many times each array is accessed. Then, we have also designed a custom memory organization for this application. In contrast to the default memory architecture, the custom memory organization can meet the real-time constraint with a much lower energy consumption and area cost. Slower, smaller, and more energy-efficient memories can be used. These results, however, are only attainable by careful systematic exploration of the alternatives, as the following subsection shows.

2.6.2 Difficulties

Although a custom memory organization has the potential to significantly reduce the system cost, achieving these cost reductions is far from trivial, especially manually. Figure 2.21 illustrates the problem of finding a custom memory architecture for an application. In this case, the application is specified using the C programming language. This specification contains arrays and accesses to them that have to be mapped into a background memory architecture. Designing such a custom memory architecture means deciding how many memories to use and which type. In addition, the memory accesses have to be ordered in time, so that the real-time constraints (the cycle budget) are met. Finally, each array must be assigned to a memory to enable arrays to be accessed in parallel as required to meet the real-time constraints.

For a given application, one can conceive numerous memory organizations: many memories or few memories, arrays stacked together into a memory or spread out over different memories, and memories with multiple ports to access the arrays in parallel, for example. Very few of these many alternative memory organizations, however, are acceptable. Most of them are actually very costly or dissipate a lot of energy. Many of them also fail to meet the real-time constraints: they don't provide the processor with the required data fast enough. Finally, some of them may be infeasible due to, for example, constraints on the size of the available memories.

To tackle the complexity of finding a custom memory organization, a methodological approach is called for. Only few techniques for allocation and assignment of background memories have been proposed as opposed to the abundance of work on the scalar level (Stok, 1994; Pavlin and Knight, 1989; Verhaegh, 1995). As usual, when resources (memories) have to be used under timing constraints, there is a phase coupling problem between the scheduling of the resource and the allocation of it. Some techniques start from a given schedule (Lippens *et al.*, 1993). Others optimize the scheduling itself (Passos and Sha, 1995; Verhaegh, 1995), minimizing the parallelism of memory accesses. Still another technique bases the memory allocation on a bandwidth requirement estimation step (Balasa *et al.*, 1994).

In this discussion, we present an approach in two steps (Catthoor *et al.*, 1998; Slock *et al.*, 1997; Vandecappelle *et al.*, 1999). First, there is the **storage cycle budget distribution** (SCBD) step that does a partial ordering of the application's memory accesses. The ordering is driven by a cost function based on the nature of the conflicts that are created to leave as much freedom as possible for the next step. In the **memory allocation and assignment** (MAA) step, the memory architecture is defined by allocating memories and assigning arrays to them. To connect the two steps by propagating constraints from SCBD to MAA, we have developed the model of an **(extended) conflict graph**.

2.6.3 Storage Cycle Budget Distribution

The first step in custom memory organization design is to make sure the real-time constraints are met at minimal cost. All the memory accesses in the application have to be made within the given real-time constraint. A certain amount of memory bandwidth is needed to meet this constraint. The goal of the SCBD step is to order the memory accesses in the available time such that the resulting memory bandwidth is as cheap as possible.

To meet the real-time constraints, the data storage components of the system (i.e., the memory architecture) must be able to provide the required data to the data processing

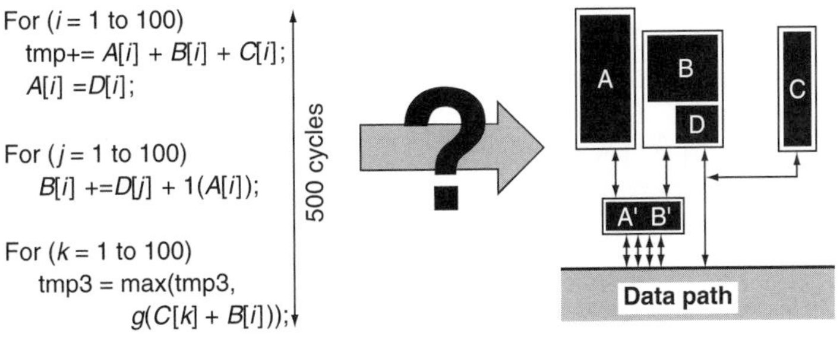

FIGURE 2.21 Problem of Defining a Memory Architecture with Timing Constraints and Minimal Cost

components fast enough. In addition, the data processing components must work fast enough. However, the two can be cleanly separated. Most of the transfers can take place during the processing of other data; there may be a small amount of processing or a small amount of transfers that cannot be overlapped, but between these, there is a broad range of exploration, as indicated in Subsection 2.5.2.

It is important to explore the ordering freedom of the data transfers before scheduling the data processing itself, and not to do it the other way around. First of all, the cost of the data transfers is much higher than the cost of data processing, especially in data-transfer intensive applications. In addition, the constraints coming from first ordering the data transfers do not limit the scheduling freedom of the data processing too much, whereas the reverse is certainly not true.

To keep complexity reasonable, it is necessary at this stage to work at the level of arrays and not to expand them into scalars. In data-transfer intensive applications, the size of the arrays can go up to several millions of words, making scalar approaches completely infeasible. On the other hand, treating the arrays as a unit hardly limits the freedom; indeed, they will in any case be mapped into memories in a uniform way. The only disadvantage is that the analysis needs to be more advanced to be able to deal correctly with arrays (Balasa *et al.*, 1997; Catthoor *et al.*, 1998). This type of an analysis falls outside the scope of this book, however.

The ordering of the application's memory accesses in time determines how much memory bandwidth is needed. This memory bandwidth comes at a cost. Therefore, an estimation model of the memory bandwidth cost is required to steer the ordering of the memory accesses. Scheduling algorithms can then be used to optimize this cost; for example, the ones proposed by us (Wuytack *et al.*, 1996; Brockmeyer *et al.*, 2000; Omnès *et al.*, 2000), directly support the background memory cost in a global way. Manually exploring possible orderings and evaluating the resulting memory bandwidth is infeasible, so tool support is essential for this step.

The model for the memory bandwidth requirement is that of a conflict graph. It is derived from the ordering of the memory accesses, as shown in Figure 2.22. Its nodes represent the application's arrays, while its edges indicate that two arrays are accessed in parallel. It models the constraints on the memory architecture (see Subsection 2.6.4). An example is

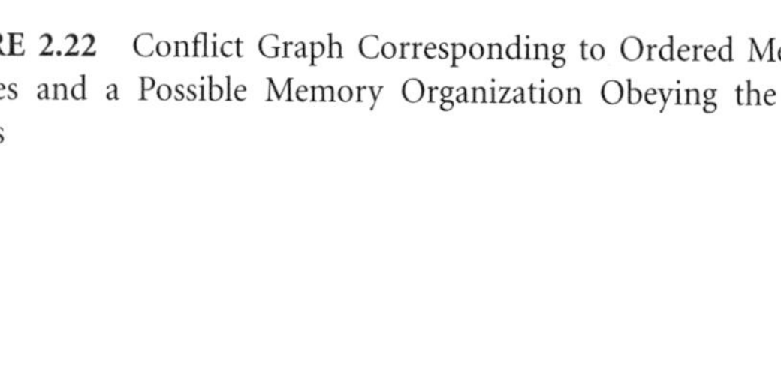

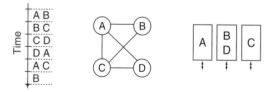

FIGURE 2.22 Conflict Graph Corresponding to Ordered Memory Accesses and a Possible Memory Organization Obeying the Constraints

given in Figure 2.22, where array A cannot be stored in the same single-port memory as any of the other arrays: it is in conflict with them. However, there is no edge between B and D, so these two arrays can be stored in one memory. Note that to support multiport memories, this model must be extended with hyperedges (Catthoor *et al.*, 1998; Wuytack *et al.*, 1996).

From the conflict graph, it is possible to estimate the cost of the memory bandwidth required by the specified ordering of the memory accesses. Three main components contribute to the cost:

1. When two accesses to the same array occur in parallel, it means there is a **self-conflict**, a loop in the conflict graph. Such a self-conflict forces a two-ported memory in the memory architecture. Two-ported and multi-ported memories are much more costly; therefore self-conflicts carry a very heavy weight in the cost function.
2. In the absence of self-conflicts, the conflict graph's **chromatic number** indicates the minimal number of single-port memories required to provide the necessary bandwidth. For example, in Figure 2.22, the chromatic number of the graph is three: A, B, and C need three different colors to color the graph, for example. That means also a minimum of three single-port memories are required, even though three accesses never occur in parallel. More memories make the design (potentially) more costly; therefore, the conflict graph's chromatic number is the second component of the cost.
3. Finally, the more conflicts exist in the conflict graph, the more costly the graph is. Indeed, every conflict takes away some freedom for the assignment of arrays to memories (see also Subsection 2.6.4). In addition, not every conflict carries the same weight. For instance, a conflict between a very small, very heavily accessed array and a very large, lightly accessed array is not so costly because it is a good idea for energy efficiency to split these two up over different memories. A conflict between two arrays of similar size, however, is fairly costly because then the two cannot be stored in the same memory and potentially have to be combined with other arrays that do not match very well.

This abstract cost model makes it possible to evaluate and explore the ordering of memory accesses without going to a detailed memory architecture. Indeed, the latter is in itself a rather complex task, so combining the two is not feasible.

2.6.4 Memory Allocation and Assignment

In a custom memory architecture, the designer can choose how many memories to use, what their dimensions should be, and how many ports they have, for example. This decision, taking into account the constraints derived in the SCBD step, is the focus of the memory allocation and assignment step. The problem can be subdivided into two problems to be solved.

First, memories must be allocated: a number of memories is chosen from the available memory types and the different port configurations, different types are possibly mixed with each other, and some memories may be multiported. The dimensions of the memories are however only determined in the second stage. When arrays are assigned to memories, their sizes can be added up and the maximum bit width can be taken to determine the required size and bit width of the memory. With this decision, the memory organization is fully determined.

Allocating more or less memories has an effect on the chip area and the energy consumption of the memory architecture as shown in Figure 2.23 for single-port memories. With multiport memories, the number of possibilities becomes even much larger, but that will be left out of this discussion. Large memories consume more energy per access than small memories due to the longer word- and bit-lines. Therefore, the energy consumed by a single large memory containing all the data is much greater than the energy consumed when the data is distributed over several smaller memories. The area of the one-memory solution is also often higher when different arrays have different bit widths. For example, when a 6-bit and an 8-bit array are stored in the same memory, two bits are unused for every 6-bit word. By storing the arrays in different memories, one six-bits wide and the other eight bits wide, this overhead can be avoided. Note that the steepness of the curves heavily depends on the particular application and also on the memory library.

The other end of the spectrum is to store all the arrays in different memories. This leads to a relatively high energy consumption because then the external global interconnection lines connecting all these (small) memories with each other and with the data paths become an important contributor. Likewise, the area occupied by the memory system increases, due to the interconnections and the fixed address decoding and other overhead per memory.

Obviously, the interesting memory allocations lie somewhere between the two extremes. The area and the energy function reach a minimum somewhere but at a different point (the area typically at less memories than the energy). In between the two minima lies the useful exploration region to trade off the area with energy consumption.

The cost of the memory organization does not only depend on the allocation of memories, however, but also on the assignment of arrays to the memories (the previous discussion assumes an optimal assignment). When several memories are available, many ways exist to assign the arrays to them. The paragraph on the cost of conflicts in Subsection 2.6.3 demonstrates some of these issues. In addition, the optimal assignment of arrays to memories depends on the memories used. For example, the energy consumption of some memories is very sensitive to their size, while for others it is not. In the former case, it may be advantageous to accept some wasted bits to keep the heavily accessed memories very small, whereas in the latter case, the reverse may be true. To find a (near optimal signal-to-memory assignment, a huge number of possibilities have to be considered. Therefore, a tool to explore the memory organization possibilities based on a good memory library is indispensable for this step.

Both the memory allocation and the signal-to-memory assignment have to take into account the constraints generated by the SCBD tool to meet the real-time requirements. For the memory allocation, this means that a certain minimum number of memories is needed (again, for multiport memories, the situation is more complicated). This minimum is derived from the conflict graph's chromatic number and is also indicated in Figure 2.23. For the signal-to-memory assignment, conflicting arrays cannot be assigned to the same memory. When there is a conflict between two arrays that in the optimal assignment were stored in the same memory, a less efficient assignment must be accepted. Thus, more conflicts make the memory organization potentially more expensive.

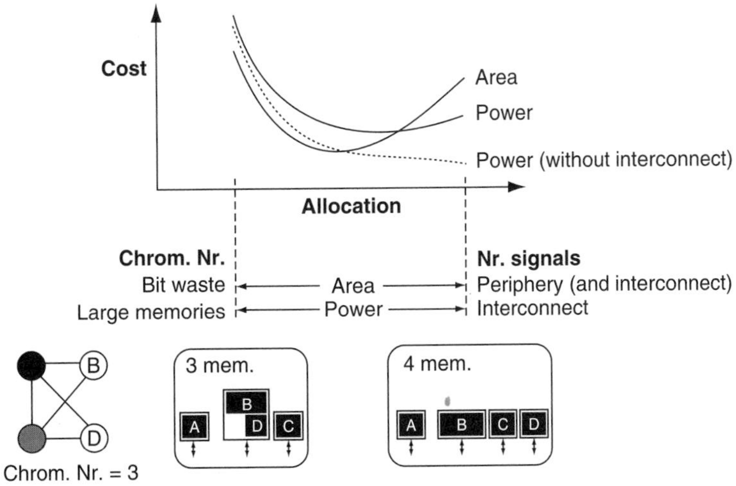

FIGURE 2.23 Exploration Space for the Number of Memories

2.6.5 Results

Manually exploring alternative custom memory organizations is infeasible, as discussed. Tool support is imperative for any application that is of realistic size. We have developed such tools such as the BTPC application and the cavity detector, and applied them to several realistic examples. The combination of the SCBD and MAA steps yields the Pareto curves for the BPTC demonstrator discussed in Section 2.5.

Figure 2.24 shows two results for the cavity detector application. The dashed curve is based on the code obtained after the DTSE optimizations introduced in Section 2.4. This assumes that all loops are executed one after the other.

Parallelizing some of the loops, which currently has to be done manually, allows to go to lower cycle budgets without increasing the (energy) cost. This is shown in the filled curve, where the four most important small loops are parallelized in a merge.

Figure 2.24 also shows the conflict graph and the optimal memory organization for three points on the cavity detector's Pareto curve. With lowering cycle budgets, more and more parallelism is required to meet the constraint. This is reflected by the additional conflicts in the conflict graph. The memory architecture needs more and more ports to provide the required parallelism and thus becomes more and more power-hungry. These results clearly substantiate the importance and the powerful exploration capability of the applied methodology and tools.

2.7 Data Layout Reorganization for Reduced Memory Size

In this section, we present some strategies that are capable of reducing the required memory sizes and the related power consumption of data-dominated multimedia applications. The main objective of these strategies is to reuse memory as much as possible by obtaining an optimal storage order for each of the arrays present in a program by means of data layout transformations that we have called "inplace data mapping." We assume that the arrays are already assigned to memories in the preceding MAA step of Section 2.6.

In the past, techniques for estimating memory reuse have been proposed (Cierniak and Li, 1995; Eisenbeis *et al.*, 1991; Wilde and Rajopadhye, 1996). These techniques however have provided only a memory reuse model and not a (complete) strategy for actually reusing memory space inside an array as well as between different arrays. This is due to the lack of exact mathematical model(s) of memory usage. In contrast, we have presented exact mathematical models for memory usage in examples presented by Catthoor *et al.* (1998), which allows us to perform more aggressive optimizations.

In this section, we first present an example illustration of the data layout reorganization strategy in Subsection 2.7.1. How this technique is actually realized is discussed in Subsection 2.7.2. Experimental results on a voice coder algorithm are given in Subsection 2.7.3. To conclude a summary is provided in Subsection 2.7.4.

2.7.1 Example Illustration

In this subsection, we illustrate how in-place mapping on arrays with partly nonoverlapping lifetimes allows the reduction of the required memory size for an algorithm. We use the autocorrelation part of a linear predictive coding vocoder algorithm introduced in Section 2.7.3.

1. **Initial Algorithm:** The initial algorithm is shown below in Figure 2.25. Note that we have two signals that are responsible for most of the memory accesses and that are dominant in size, namely ac_inter[] and

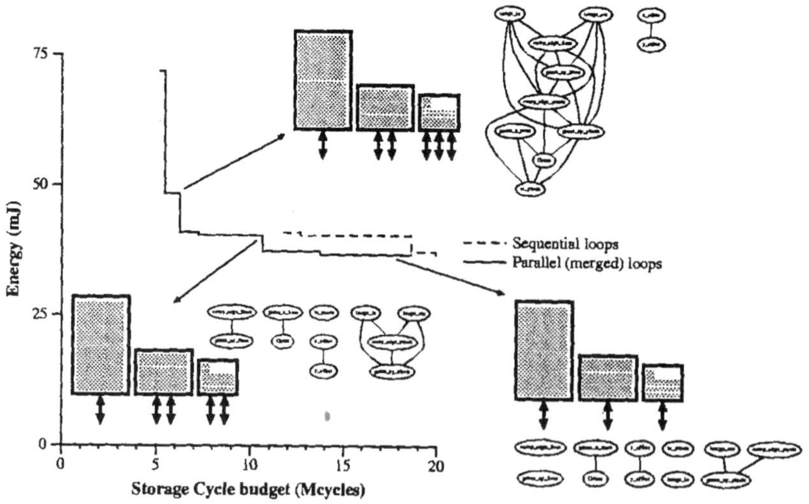

FIGURE 2.24 Pareto curve for Custom Memory Organization of the Cavity Detector (After DTSE Optimizations). This figure also shows the resulting memory organization for some of the Pareto points.

```
for(i6=0; i6 <11; i6 ++) {
  ac_inter [i6][i6] =Hamwind[0]*Hamwind [i6];
  ac_inter [i6] [i6+1] = Hamwind[1] * Hamwind
  [i6+1];
  for (i7 = (i6+2); i7<2400; i7++)
    ac_inter [i6] [i7] =ac_inter [i6] [i7-1] +
                        ac_inter [i6] [i7-2] +
                (Hamwind [i7-i6] * Hamwind [i7]);
  AutoCorr[i6] = ac_inter[i6] [2399] ;
}
for (i8=0; i8<10; i8++) {
  v[0] [i8] = AutoCorr [i8+1];
  u[0] [i8] = AutoCorr [i8];
}
```

FIGURE 2.25 Pseudocode for the Autocorrelation Function

Hamwind, with respective sizes of 26,400 and 2400 integer elements. Observe that the loop nest has non-rectangular access patterns dominated by accesses to the temporary variable ac_inter[]. Thus to reduce power, we need to reduce the number of memory accesses to ac_inter[]. This is possible only by first reducing the size of ac_inter[] and then placing this signal in a local memory.

2. **In-place data mapping:** Note that ac_inter[] has a dependency only on two of its earlier values; thus only three (earlier) integer values need to be stored for computing each of the autocorrelated values. Thus, by performing intrasignal in-place data mapping, as shown in Figure 2.26, we are able to drastically reduce the size of this signal from 26,400 to 33 integer elements. The signal AutoCorr[] is a temporary signal. By reusing the memory space of signal ac_inter[] for storing AutoCorr[], we can further reduce the total memory space required. This is achieved by intersignal in-place mapping of array AutoCorr[] on ac_inter[]. Thus, initially ac_inter[] could not have been accommodated in the on-chip local memory due to the large size of this signal, but now we have removed this

```
for (i6=0; i6<11; i6++) {
  ac_inter[i6] [i6%3] =Hamwind [0] * Hamwind [i6];
  ac_inter [i6] [(i6+1)%3] = Hamwind [1] * Hamwind
  [i6+1];
  for (i7 = (i6+2); i7<2400; i7++)
    ac_inter[i6] [i7%3] =ac_inter[i6] [(i7-1)%3]+
                ac_inter [i6] [(i7-2)%3] +
                (Hamwind [i7-i6] * Hamwind [i7]);
}
for (i8=0; i8<10; i8++) {
  v[0] [i8] =ac_inter [i8+1] [2]; /* 2399 % 3 = 2 */
  u[0] [i8] = ac_inter [i8] [2];
}
```

FIGURE 2.26 Pseudocode for the In-Place Mapped Autocorrelation Function

problem. This results both in reduced memory size and a reduction in the associated power consumption. Note that in Section 2.4, the data reuse step introduces additional buffers to exploit the reuse of data in a signal, whereas in-place mapping step exploits the reuse in as well as between signals to reduce the size of the existing signal.

2.7.2 Data Layout Reorganization Methodology

The data layout reorganization step comprises two phases, namely intrasignal in-place data mapping and intersignal in-place data mapping. These two phases are explained next.

Intrasignal In-Place Mapping

In a first phase, we optimize the internal organization of each array separately, resulting in a partially fixed storage order for each array. The intrasignal in-place mapping phase comprises calculation of the address reference window for-every array as shown in Figure 2.27 (A) and identification of an (optimal) intra-array storage order as shown in Figure 2.27 (D) (Greef *et al.*, 1997; Catthoor *et al.*, 1998). The address reference window represents the maximum distance between two addresses being occupied by the array at the same time.[8] This is calculated using the geometrical domains for each array referred to as **occupied address/time domains** (OATD's) as shown in Figure 2.27 (A) (Greef *et al.*, 1997).

The conventional (nonwindowed) memory allocation for arrays is shown in Figure 2.27 (B). We call this type of allocation **static allocation**. Figure 2.27 (D) shows the intrasignal in-place mapped arrays, also called **static windowed allocation** of memory space.

Thus, we observe that intrasignal in-place mapping helps reduce the memory size occupied by the individual array variable(s). In practice, we have observed that loop transformations such as loop fusion, loop tiling (Ciernak and Li, 1995; Lam *et al.*, 1991), and data reuse decisions (Diguet *et al.*, 1997) (see Section 2.4) help improve the temporal locality of data. Intrasignal in-place data mapping reduces the memory size of the data with good temporal locality. Thus, this technique is complementary to the loop transformations used for enhancing locality of data.

Intersignal In-place Mapping

The next step, intersignal in-place data mapping, optimizes the storage order between different arrays. We also refer to this step as **dynamic allocation**[9] and **dynamic windowed allocation** in Figure 2.27.

This step involves reusing the memory space between different arrays, both for arrays with disjoint and partially

[8] In our example illustration in Subsection 2.7.1, this distance was 3.

[9] Note that this step is performed at compile time, in contrast to traditional dynamic allocation provided by languages like C.

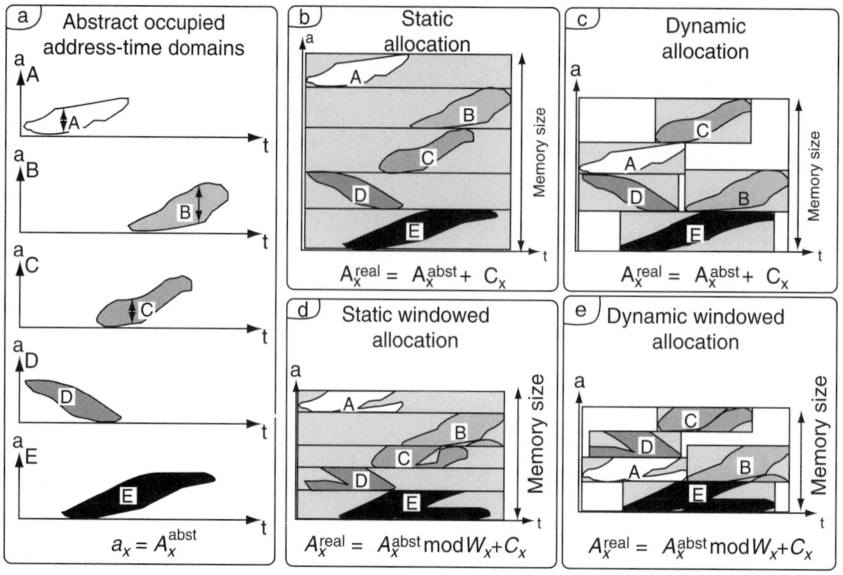

FIGURE 2.27 Intrasignal and Intersignal In-Place Data Mapping

overlapped lifetimes (Greef *et al.*, 1997). The criterion that identifies (or checks) arrays with disjoint lifetimes is called **compatibility**, and the criterion that identifies arrays with partially overlapped life times that can be merged[10] is called **mergeability**. This step is illustrated in Figures 2.27(C) and 2.27(E). In Subsection 2.7.1, the arrays ac_inter□ and AutoCorr□ had partially overlapped lifetimes and were mergeable. This resulted in the reuse of memory space occupied by array ac_inter□.

Figures 2.27(C) and 2.27(E) show that the combination of intra signal and intersignal in-place data mapping allows for an even larger reduction in the required memory size for the arrays as compared to intrasignal in-place mapping alone.

2.7.3 Voice Coder Algorithm and Experimental Results

Most of the communication applications like GSM require digital coding of voice for efficient, secure storage and transmission. We have used a linear predictive coding (LPC) vocoder, also called a voice coder, that provides an analysis–synthesis approach for speech signals (Rabiner and Schafer, 1988). In our algorithm, speech characteristics (energy, transfer function, and pitch) are extracted from a frame of 240 consecutive samples of speech. Consecutive frames share 60 overlapping samples. All information to reproduce the speech signal with an acceptable quality is coded in 54 bits. The general characteristics of the algorithm are presented in Figure 2.28. It is worth noting that the LPC analysis and pitch detection steps use autocorrelation to obtain the best matching predictor values and pitch (using a Hamming window). A part of this routine was used in Subsection 2.7.1 as an example illustration.

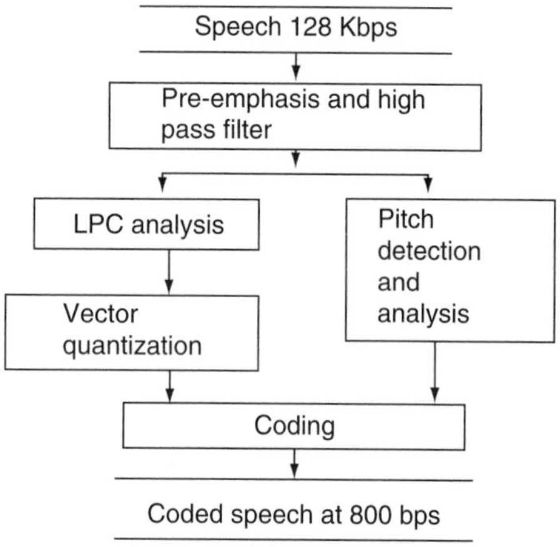

FIGURE 2.28 The Vocoder Algorithm and the Comprising Modules

Experimental Results

We now present the experimental results on the above described vocoder algorithm. Here, the initial algorithm represents the reference initial algorithm, the in-place mapped algorithm represents the initial algorithm with in-place data mapping, and the globally optimized algorithm comprises both in-place data mapping and loop transformations.

[10] Alternatively, memory space can be reused since all the R/W to particular addresses of one array are over just before the start of the second one for those addresses. If this is possible, we consider these arrays to be mergeable.

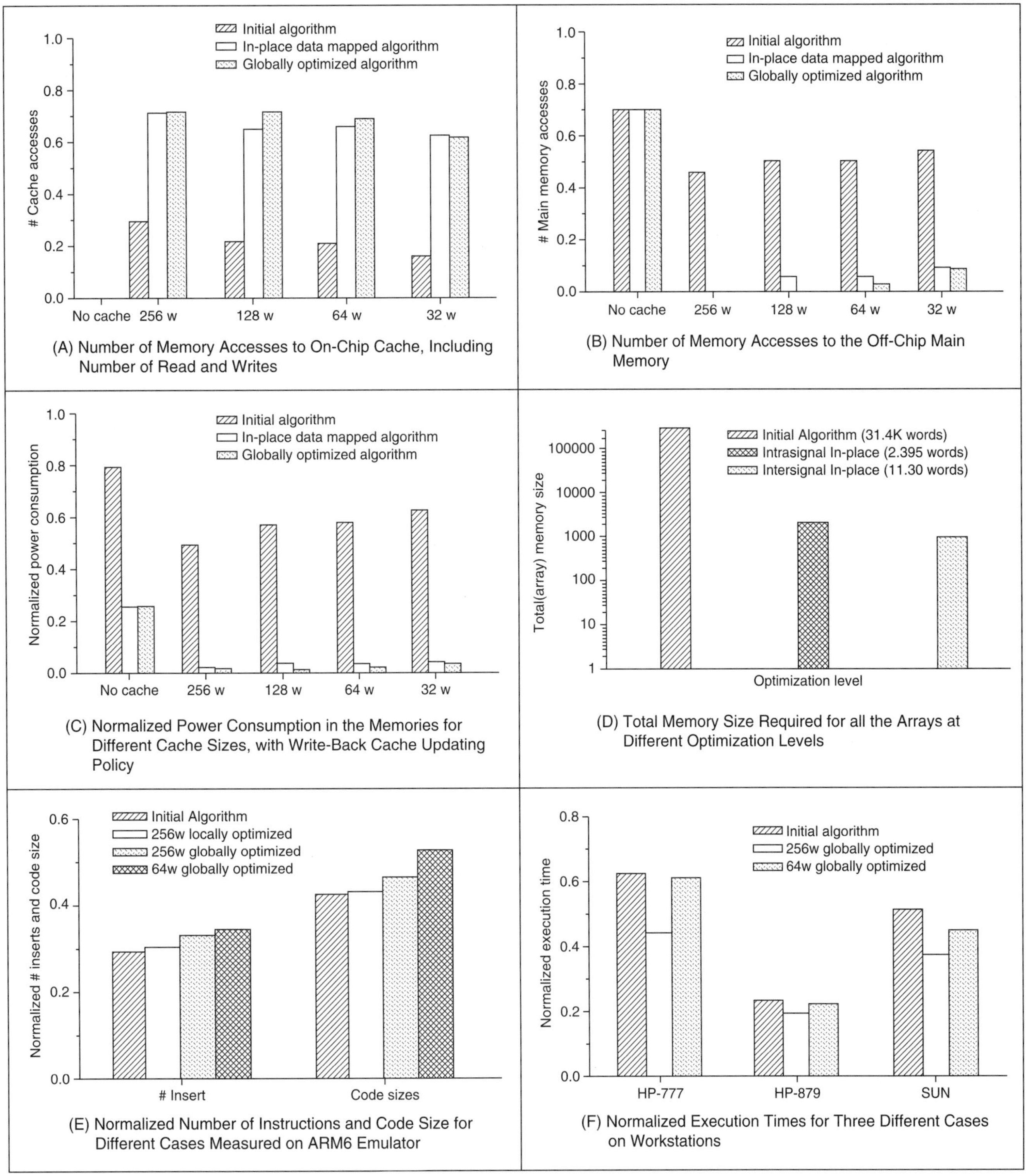

(A) Number of Memory Accesses to On-Chip Cache, Including Number of Read and Writes

(B) Number of Memory Accesses to the Off-Chip Main Memory

(C) Normalized Power Consumption in the Memories for Different Cache Sizes, with Write-Back Cache Updating Policy

(D) Total Memory Size Required for all the Arrays at Different Optimization Levels

(E) Normalized Number of Instructions and Code Size for Different Cases Measured on ARM6 Emulator

(F) Normalized Execution Times for Three Different Cases on Workstations

FIGURE 2.29 Experimental Results for the Vocoder Algorithm

Figure 2.29(A) shows the number of R/W to the on-chip local memory. Note that the number of accesses to the local memory is much larger for the globally optimized case, whereas it is much lower for the initial algorithm. Similarly, the in-place data mapped algorithm has a significantly larger number of accesses to the local memory. In Figure 2.29(B), we observe a similar phenomenon, but now the number of memory accesses is reduced for the optimized cases. We observe that for the initial algorithm, we have more accesses to the main memory than to the local memory. In contrast, for the globally optimized case, we access the main memory only a bit more than once per location. The rest of the accesses are to the local memory. This is due to the increased data locality that is brought about by the loop transformations and reduced working set sizes from data layout reorganization. Figure 2.29(C) gives the total power consumption in the memories for the different local memory sizes and optimization levels. We observe two issues here first, the globally optimized algorithm always performs better, and second the lowest power consumption for the globally optimized case is for a local memory size of 128 words and not for any higher size. This is because as we increase the local memory size, the number of memory accesses to this local memory does not increase proportionately and the number of accesses to the main memory remains the same. Hence, the total power consumption increases due to the increase in size of the local memory. Pure performance optimization does not lead to the best power solution.

Figure 2.29(D) shows the reduction in memory size for the two different steps of in-place data mapping obtained using a prototype tool. We observe that the initial algorithm requires 314Kwords, the intrasignal in-place mapped version needs 2387 words, and the intersignal in-place mapped version (after intrasignal in-place mapping) needs 1130 words. The benefits of these two steps are clearly evident. Figure 2.29(E) presents the comparison of the number of instructions and the code size for different cases. We observe that the overhead in code size and instructions are less than 15% for the most optimized case (64 words globally optimized), which is very acceptable for these application kernels because typically also a larger amount of additional purely control dominated code is present in the program memory. To conclude, we have executed the different algorithms on three different workstations to observe the performance gains. As shown in Figure 2.29(F), we observe an increase in performance by 20% on average, which is obtained apart from the large gains in power consumption.

2.7.4 Summary

The main conclusions from Section 2.7 are the following:

1. Aggressive data layout reorganization based on advanced array lifetime analysis allows designer(s) to reduce the required memory size of algorithms systematically.
2. Experimental results on a voice coder (vocoder) algorithm show that this technique is able to reduce the total memory size significantly and also reduce the associated power consumption. Automation of this technique has been demonstrated in a prototype tool that is being developed into a mature compiler program.

References

Amarasinghe, S., Anderson, J., Lam, M., and Tseng, C. (1995). The SUIF compiler for scalable parallel machines. *Proceedings of the seventh SIAM Conference on Parallel Processing for Scientific Computing.*

Amrutur, B., and Horowitz, M. (1994). Techniques to reduce power in fast wide memories. *Symposium on Low-Power Electronics.*

Balasa, F., Catthoor, F., de Man, H. (1997). Practical solutions for counting scalars and dependences in ATOMIUM—A memory management system for multidimensional signal processing. *IEEE Transactions on computer-Aided Design CAD-16* (2) 133–145.

Balasa, F., Catthoor, F., and de Man, H. (1994). Dataflow-driven memory allocation for multidimensional processing systems. *Proceedings IEEE International Conference on Computer-Aided Design.*

Banerjee, U., Eigenmann, R., Nicolau, A., and Padua, D. (1993). Automatic program parallelisation. *Proceedings of the IEEE 81*(2).

Brockmeyer, E., Wuytack, S., Vandecappelle, A., and Catthoor, F. (2000). Low-power storage for hierarchical graphs. *Proceedings for 3rd ACM/IEEE Design and Test in Europe Conference*

Brockmeyer, E., D'Eer, J., Busa, N., Catthoor, F., Lippens, P., and Huiskens, J. (1999). Code transformations for reduced data transfer and storage in low-power realization of DAB synchro core. *Patmos* 6–8.

Catthoor, F., Danckaert, K., Kulkarni, C., and Omnes, T. (2000). Data transfer and storage architecture issues and exploration in multimedia processors. In Y. H. Yu (Ed.), *Programmable digital signal processors: Architecture, programming, and applications.* New York: Marcel Dekker.

Catthoor, F., Wuytack, S., de Greef, E., Balasa, F., Nachtergaele, L., and Vandecappelle, A. (1998). Custom memory management methodology—Exploration of memory organization for embedded multimedia system design. Boston: Kluwer Academic.

Chandrakasan, A., Gutnik, V., and Xanthopoulos, J. (1996). Data-driven signal processing: An approach for energy-efficient computing. *Proceedings of the IEEE International Symposium on Low Power Electronics and Design* 347–352.

Chen, T. S., and Sheu, J. P. (1994). Communication-free data allocation techniques for parallelizing compilers on multicomputers. *IEEE Transactions on Parallel and Distributed Systems 5*(9) 924–938.

Cierniak, M., and Li, W. (1995). Unifying data and control transformations for distributed shared-memory machines. *Proceedings of the SIGPLAN'95 Conference on Programming Language Design and Implementation* 205–217.

Comerford, R., and Watson, G. (Eds.). (1992). Memory catches up *IEEE Spectrum* 34–57.

Covino, J. *et al.*, (1996). A 2-ns zero-wait state 32-kB semiassociate L1 cache. *Proceedings of the IEEE International Solid-State Circuits Conference.* 154–155.

Danckaert, K., Masselos, K., Catthoor, F., de Man, H., and Goutis, C. (1999). Strategy for power-efficient design of parallel systems. *IEEE Transactions on VLSI Systems 7*(2) 258–265.

de Greef, E., Catthoor, F., and de Man, H. (1997). Memory size reduction through storage order optimization for embedded parallel multimedia applications. *International Parallel Proceedings Symposium (IPPS)* 84–98.

Diguet, J. P., Wuytack, S., Catthoor, F., and de Man, H. (1997). Formalized methodology for data reuse exploration in hierarchical memory mappings. *Proceedings of the IEEE International Symposium on Low Power Electronics and Design* 30–35.

Eisenbeis, C., Jalby, W., Windheiser, P., and Bodin, F. (1991). A strategy for array management in local memory. *Proceedings of the 4th Workshop on Languages and Compilers for Parallel Computing.*

Eto, S. *et al.* (1998). A 1-Gb SDRAM with ground-level precharged bit line and non-boosted 2.1-V wordline. *Proceedings of the IEEE International Solid-State Circuits Conference SC-33,* 1697–1702.

Evans, R., and Franzon, P. (1995). Energy consumption modeling and optimization for SRAMs. *IEEE Journal of Solid-State Circuitry. SC-30*(5), 571–579.

Furumochi, K. *et al.*, (1996). A 500-MHz 288-kb CMOS SRAM macro for on-chip cache. *Proceedings of the IEEE International Solid-State Circuits Conference.* 156–157.

Gupta, M., Schonberg, E., and Srinivasan, H. (1996). A unified framework for optimizing communication in data-parallel programs. *IEEE Transactions on Parallel and Distributed Systems 7*(7) 689–704.

Horiguchi, M. *et al.* (1995). An experimental 220-MHz 1-Gb DRAM. *Proceedings of the IEEE International Solid-State Circuits Conference.* 252–253.

IMEC. (year). Multimedia compilation project. Retrieved from http://www.imec.be/acropolis.

Itoh, K. (1997). Low voltage memory design. *Proceedings of the IEEE International Symposium on Low Power Electronics and Design* (Monterey, CA).

Itoh, K., Nakagome, Y., Kimura, S., and Watanabe, T. (1997). Limitations and challenges of multigigabit DRAM chip design. *IEEE Journal of Solid-State Circuitry SC-32,* 624–634.

Itoh, K., Sasaki, K., and Nakagome, Y. (1995). Trends in low-power RAM circuit technologies. *Proceedings of the IEEE International Solid-State Circuits Conference, 83*(4), 524–543.

Jolly, R. (1991). A 9-ns 1.4 GB/s 17-ported CMOS register file. *IEEE Journal of Solid-State Circuitry, SC-26*(10) 1407–1412.

Kamble, M., and Ghose, K. (1997). Analytical energy dissipation models for low power caches. *Proceedings of the IEEE International Symposium on Low Power Electronics and Design.* 143–148.

Kim, C. *et al.* (1998). A 64-Mbit, 640-MB/s bidirectional data-strobed, double data rate SDRAM with a 40-mW DLL for a 256-MB memory system. *IEEE Journal of Solid-State Circuitry. SC-33,* 1703–1710.

Kirihata, T. *et al.* (1998). A 220 mm2, four- and eight-bank, 256-Mb SDRAM with single-sided stitched WL architecture. *IEEE Journal of Solid-State Circuitry. SC-33,* 1711–1719.

Ko, U., Balsara, P., and Nanda, A. (1995). Energy optimization of multilevel processor cache architectures. *Proceedings of the IEEE International Workshop on Low Power Design* 45–50.

Kuge, S., Morishita, F., Suruda, T., Tomishima, S., Tsukude, M., Yamagata, T., and Arimoto, K. (1996). SOI-DRAM circuit technologies for low-power high-speed multigiga scale memories. *IEEE Journal of Solid-State Circuitry SC-31,* 586–596.

Lam, M., Rothberg, E., and Wolf, M. (1991). The cache performance and optimizations of blocked algorithms. *Proceedings ASPLOS-IV* 63–74.

Lawton, G. (1999). Storage technology takes the center stage. *IEEE Computer Magazine 32*(11) 10–13.

Lippens, P., van Meerbergen, J., Verhaegh, W., and van der Werf, A. (1993). Allocation of multiport memories for hierarchical data streams. *Proceedings IEEE International Conference on Computer-Aided Design.* 728–735.

Morimura, H., and Shibata, N. A. (1996). A I-V I-MB SRAM for portable equipment. *Proceedings of the IEEE International Symposium on Low Power Electronics and Design* 61–66.

Omnes, T. J. F., Franzetti, T., and Catthoor, F. (2000). Interative algorithms for minimizing bandwidth in high throughput telecom and multimedia. In *Proceedings of the 37th ACM/IEEE Design Automation Conference*

Okuda, T., and Murotani, T. (1997). A four-level storage 4-Gb DRAM. *IEEE Journal of Solid-State Circuitry. SC-32*(11), 1743–1747.

Oshima, Y., Sheu, B., and Jen, S. (1997). High-speed memory architectures for multimedia applications. *IEEE Circuits and Devices Magazine 1,* 8–13.

Passos, N., and Sha, E. (1995). Push-up scheduling: Optimal polynomial-time resource constrained scheduling for multidimensional applications. *Proceedings IEEE International Conference Computer-Aided Design* 588–591.

Patterson, D. A., Anderson, T., Cardwell, N., Fromm, R., Keeton, K., Kozyrakis, C., Thomas, R., and Yelick, K. (1997). Intelligent RAM (IRAM): Chips that remember and compute. *Proceedings of the IEEE International Solid-State Circuits Conference* 224–225.

Paulin, P., and Knight, J. (1989). Force-directed scheduling for the behavioral synthesis of ASICs *IEEE Transactions on Computer-Aided Design 8*(6), 661–679.

Prince, B. (1994). Memory in the fast lane. *IEEE Spectrum* 38–41.

Przybylski, S. (1997). New DRAM architectures. *Proceedings of the IEEE International Solid-State Circuits Conference*

Rabiner, L. R., and Schafer, R. W. (1988). Digital signal processing of speech signals. Englewood cliffs, NJ: Prentice Hall.

Robinson, J. (1997). Efficient general-purpose image compression with binary tree predictive coding. *IEEE Transactions on Image Processing 6*(4), 601–608.

Seki, T., Itoh, E., Furukawa, C., Maeno, I., Ozawa, T., Sano, H., and Suzuki, N. (1993). A 6-ns 1-Mb CMOS SRAM with latched sense amplifier. *IEEE Journal of Solid-State Circuits SC-28*(4) 478–483.

Slock, P., Wuytack, S., Catthoor, F., and de Jong, G. (1997). Fast and extensive system-level memory exploration for ATM applications. In *Proceedings 10th ACM/IEEE International Symposium on System Synthesis* 74–81.

Stok, L. (1994). Data path synthesis. *Integration: The VLSI journal 18,* 1–71.

Su, C., and Despain, A. (1995). Cache design trade-offs for power and performance optimization: A case study *Proceedings of the IEEE International Symposium on Low Power Electronics and Design* 63–68.

Sugibayashi, T. *et al.* (1993). A 30-ns 256-Mb DRAM with a multi-divided array structure. *IEEE Journal of Solid-State Circuitry. SC-28*(11) 1092–1096.

Takayanagi, T. *et al.* (1996). 350-MHz time-multiplexed 8-port SRAM and word-size variable multiplier for multimedia DSP. *Proceedings of the IEEE International Solid-State Circuits Conference.* 150–151.

Tsuruda, T., Kobayashi, M., Tsukude, M., Yamagata, T., and Arimoto. (1996). High-speed, high-bandwith design methodologies for on-chip DRAM core multimedia system LSIS. *Proceedings for IEEE International Solid-State Circuits Conference* 265–268.

Vandecappelle, A., Miranda, M., Brockmeyer, E., Catthoor, F., and Verkest, D. (1999). Global multimedia system design exploration using accurate memory organization feedback. *Proceedings for the 36th ACM/IEEE Design Automation Conference*

Verhaegh, W. (1995). *Multidimensional periodic scheduling.* Ph.D. diss., T. U. Eindhoven.

Verhaegh, W., Lippens, P., Aarts, E., Korst, J., van Meerbergen, J., and van der Werf, A. (1995). Improved force-directed scheduling in high-throughput digital signal processing. *IEEE Transactions on Computer Aided Design 14*(8), 945–960.

Wehn, N., and Hein, S. (1998). Embedded DRAM architectural tradeoffs. *Proceedings of First ACM/IEEE Design and Test in Europe Conference* 704–708.

Weste, N. and Eshraghian, K. (1993). Principles of CMOS VLSI design (2d ed.). Reading, MA: Addison-Wesley.

Wilde, D., and Rajopadhye, S. (1996). Memory reuse analysis in the polyhedral model. *Proceedings EuroPar Conference 1128*, 389–397.

Wolf, M. (1992). Improving locality and parallelism in nested loops. (1992) Ph.D. diss, Standford University.

Wuytack, S., Catthoor, F., De Jong, G., Lin, B., and de Man, H. (1996). Flow graph balancing for minimizing the required memory bandwidth. In *Proceedings 9th ACM/IEEE International Symposium on System Synthesis* 127–132.

Wuytack, S., Catthoor, F., de Jong, G., and de Man, H. (1999). Minimizing the required memory bandwidth in VLSI system realizations. *IEEE Transactions on VLSI Systems 7*(4), 433–441.

Wuytack, S., Diguet, J. P., Catthoor, F., and de Man, H. (1998). Formalized methodology for data reuse exploration for low-power hierarchical memory mappings. *IEEE Transactions on VLSI Systems 6*(4) 529–537.

Yamagata, T., Tomishima, S., Tsukude, M., Hashizume, X., and Arimoto, K. (1995). Circuit design techniques for low-voltage operating and/or giga-scale DRAMs. *Proceedings of the IEEE International Solid-State Circuits Conference* 248–249.

Yamauchi, H., Iwata, T., Akamatsu, H., and Matuszawa, A. (1996). A 0.5/100-MHz over-VCC grounded data storage (OVGS) SRAM cell architecture with boosted bit line and offset source overdriving schemes. *Proceedings of the IEEE International Symposium on Low Power Electronics and Design* 49–54.

Yano, K. *et al.* (1996). Single-electron memory integrated circuit for giga-to-tera bit storage. *Proceedings of the IEEE International Solid-State Circuits Conference.* 266–267.

3

The Role of Hardware Description Languages in the Design Process of Multinature Systems

Sorin A. Huss
Integrated Circuits and Systems Laboratory, Computer Science Department, Darmstadt University of Technology, Darmstadt, Germany

3.1 Introduction

In the past, information processing systems migrated from computing centers via desktops and laptops into embedded systems, which are a well-hidden part of not only transportation or production systems but even of devices for entertainment and leisure usage. These embedded systems derive their functionality from a cooperation of components that operate either **time-continuously** or **event-discretely**. In addition, embedded systems are products from different engineering disciplines, such as mechanical, electrical, and computer engineering. Examples of such systems are not only robots or automotive engine control units, for instance, and can be part of a scuba diver's equipment, as will be shown subsequently.

The engineering of such heterogeneous or multinature systems is not to be accomplished by well-known "identify and assign design tasks" approaches any more. A paradigm shift toward a holistic system view is mandatory due to the tight interaction of components that determines the overall performance of the system.

Specification, conceptualization, and assessment of design decisions are of utmost importance during the design process. In addition, these tasks have to be performed in a fast, reliable, and cost-efficient manner. Generation and application of models play an important role in the design process. Different models, abstraction levels, and accuracy requirements are to be considered and have to be handled efficiently regardless of the operation domains of all these models and engineering disciplines involved.

Starting with a real-world problem to be solved, a first conceptual model is derived that addresses the relevant properties of the problem. In general, this model is denoted in a mathematical way and has to be represented such that an execution becomes possible. At this point, the question of description languages gains interest. After some refinement steps at different abstraction and accuracy levels, the model is to be validated in the envisaged context, serving as a virtual prototype of the system to be produced.

This chapter addresses the refinement steps toward such a virtual prototype. It is organized as follows. Section 3.2 details the design process and the levels of abstraction to be covered in engineering multinature, mixed-mode embedded systems. Section 3.3 is a short introduction to the main features of the hardware description language VHDL–AMS (IEEE Standard 1076.1–1999) aimed to represent models for event-discrete and/or time-continuous operating heterogeneous systems. In Section 3.4, system modeling and simulation is demonstrated for an electronic depth gauge to be used by scuba divers. This application highlights the problems of combining multinature, mixed mode components into a virtual prototype. Finally,

Section 3.5 concludes the chapter and recommends related literature for further reading.

3.2 Design Process and Levels of Abstraction

The design process for a technical product may, in general, be roughly subdivided into three main phases: conceptualization, concept refinement, and implementation. Inputs to the process are the design requirements, eventually resulting in the design results. A more detailed view to this "black box" model of the design process, as depicted in Figure 3.1, results in an identification of a process chain consisting of generating and analyzing activities to be performed iteratively in the outlined steps of conceptualization, refinement, and implementation.

The design requirements consist of a description of the envisaged functionality of the new product and sometimes of constraints referring to the final implementation, such as an exploitation of commercially available subsystems. This set of information is commonly known as the technical product specification, and it is still denoted in an informal way (i.e., written in a natural language and augmented by some tables

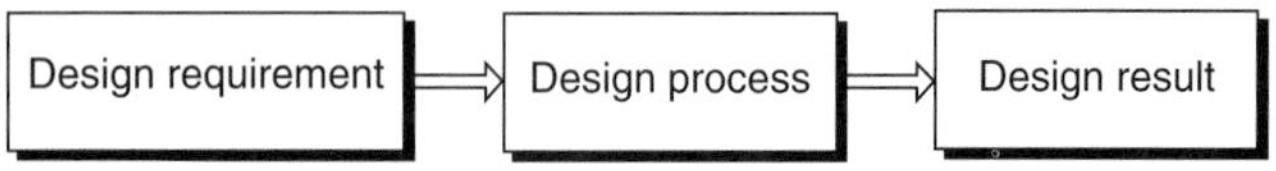

FIGURE 3.1 Black Box Model of a Design Process

and diagrams). Nontechnical specifications such as cost frames and design deadlines are important for the design process as well. They are, therefore, viewed as additional inputs to the design process as outlined in Figure 3.1.

The first and most important activities of the systems engineer during the conceptual phase are formalization of the specification, determination of solution strategies, and partitioning of the overall task into independent subtasks. These tasks are forwarded to design teams specializing in different areas, such as analog circuit design or real-time software engineering. Figure 3.2 depicts the design flow during conceptualization and refinement.

Modeling plays a central role in this design phase. Different model instantiations, abstraction levels, and accuracy requirements have to be dealt with during concept refinement. In addition, multinature systems in general operate time-continuously, but the information processing inherent to most such systems consists of digital hardware and software modules, which are best represented in a time- or even-discrete way. Levels of abstraction are well-defined in the digital domain. They are summarized in Table 3.1.

Views, abstraction levels, modeling concepts, structural primitives, and time models are related to the observable values produced by models of digital systems. These models may be denoted in a hardware description language such as Verilog or VHDL. We will focus on the latter in the following paragraphs for reasons given in the next section.

Abstraction levels for time-continuous systems are not yet that well agreed upon in terms of counterparts for digital circuits and systems. Table 3.2 introduces and discusses four levels of design concepts and observable signals beginning with the highest abstraction level.

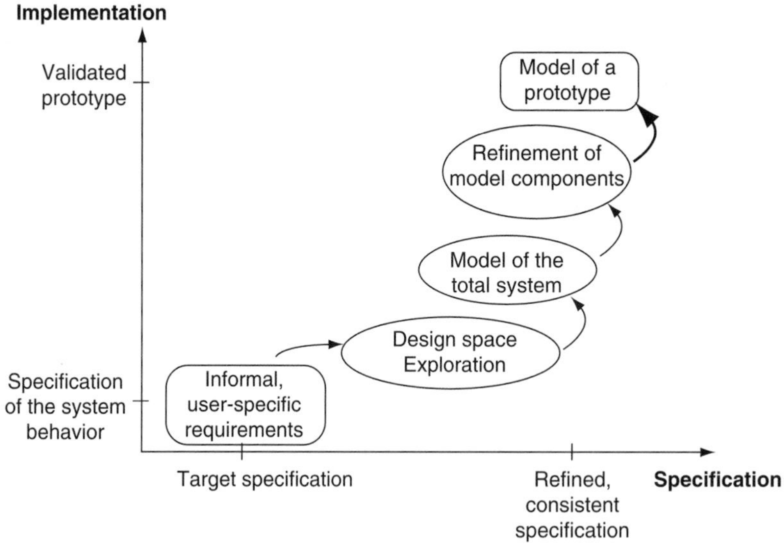

FIGURE 3.2 Design Flow for System Design

TABLE 3.1 Levels of Abstraction in Digital Systems

View	Level	Modeling concept	Structural primitive	Time model	Observable values
	System	Co-operating processors	CPU, memory, busses	Causality	Free definable
Imperative	Algorithmic level (Chip)	Parallel algorithms	Controller, RAM, ROM, UART	Discrete (fine/course granularity)	Interpreted words (free definable)
	Register transfer	Guarded commands	Register, counter, ALU, multiplexer	Discrete (course granularity)	Bit fields (not interpreted, multivalued)
	Gates	(Boolean) logic equations	Gates, flip-flop	Discrete (fine granularity)	Multivalued logic
Reactive	Switch	Discrete equations	Switch, discrete capacitors	Continuous (increase or decrease times)	Discrete (value, strength)
	Circuits	Differential equations	Transistors, *R*, *L*, *C*	Continuous (mathematical exact)	Continuous (voltage, current)

TABLE 3.2 Levels of Abstraction in Analog Systems

Level	Modeling concept	Observable signals
Functional	• Description of signal flow equivalent to data flow in digital circuits • Input/Output relations by means of mathematical functions	• Time- and value-continuous • Conservative laws *not* considered
Behavioral	• Equations that describe the relations between port variables of an entity	• Time- and value-continuous • Conservative laws considered
Macromodel	• Hierarchical composition using ideal functional blocks	• Same properties as at a behavioral level
Circuit	• Hierarchical composition based on discrete basic elements	• Same properties as at a behavioral level

These abstraction levels seem to be strongly related to analog circuit design only—especially levels three and four. However, an introduction of the abstraction levels one and two—functional and behavioral—is well suited for modeling purposes in other engineering domains as demonstrated in the last section of this chapter.

The interrelation between conceptual models according to Figure 3.3 and the executing simulator is highly complex. An appropriate support of model representations is thus mandatory. This is covered by highly expressive modeling languages such as VHDL–AMS, which originated recently from the well-known hardware description language VHDL aimed at denoting executable models of digital systems. Figure 3.4 highlights this relationship.

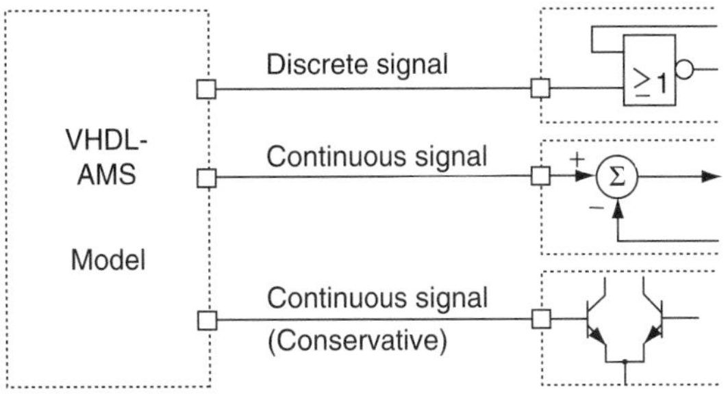

FIGURE 3.4 Combination of Signal Classes

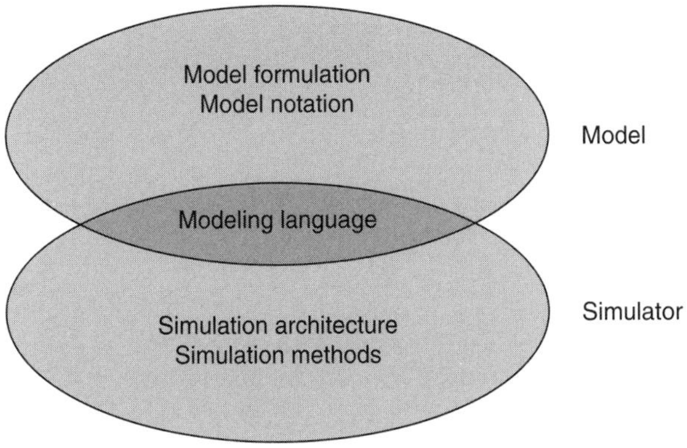

FIGURE 3.3 Interrelationship of Model and Simulator

3.3 Fundamentals of VHDL–AMS

The basic concepts of VHDL form the foundation for the extension of this hardware description language to the modeling domain of analog mixed signal (i.e., AMS), resulting in the new IEEE Standard 1076.1–1999. These concepts of VHDL are summarized here:

- Model composition of communicating, concurrently active "design entities"

- Strict separation of the unique interface description and the functional description(s) of an entity
- Execution of the model description based on the stimulus/answer paradigm (Events are only stimuli considered, resulting in an emphasis on digital circuits and systems.)
- Definition of the simulation process on top of an event-driven simulation cycle:

Simulation Process

```
  while transactions remain to be processed
  increase time until a transaction has to be
  processed
  update the state from the previous transac-
  tions
  determine events caused by updates
  resume processes sensitive to these events
end while
```

The description of a design entity consists of the entity declaration, one or more architectural bodies, and an optional configuration declaration. Syntactical details are as follows:

Entity Syntax

```
EntityDeclaration ::=
  entity Identifier is
    EntityHeader
    EntityDeclarativePart
  [begin
    Entity StatementPart]
  end [entity] [EntitySimpleName];
EntityHeader ::=
  [FormalGenericClause]
  [FormalPortClause]
GenericClause ::=
  generic (GenericList);
PortClause ::=
  port (PortList);
```

Architecture Syntax

```
ArchitectureBody ::=
  architecture Identifier of EntityName is
    ArchitectureDeclarativePart
  begin
    ArchitectureStatementPart
  end [architecture] [ArchitectureSimpleName];
ArchitectureStatement ::= ConcurrentStatement
```

The IEEE Standard 1076 of VHDL has been considerably extended to cope with modeling requirements in the time-continuous domain and especially with mixed-signal, multi-nature systems (i.e., event-discrete and time-continuous behaviors for heterogeneous models). This objective is depicted in Figure 3.4.

One problem arises when combining signal flow and physical models: the semantic of connections must be determined and implemented by appropriate methods as shown in Figure 3.5.

Obviously, the range of VHDL is by far not sufficient to suit all of these requirements. Therefore, extensions to design objects and declarations, to assignments, to attributes, and especially to the simulation cycle were necessary. These extensions to the original VHDL language are summarized in Figure 3.6.

Descriptions of value- and time-continuous variables and the enforcement of conservation in different physical domains, such as Kirchhoff's laws in the electrical domain, are possible when introducing the basic concepts of quantities and terminals.

The object class **quantity** has the following properties: representation of time-dependent functions, partial continuity with a finite number of discontinuities in one time interval, assignment of physical dimension, and existence of derivatives and integrals. Quantities may be defined as free, branch, or source quantities. Figure 3.7 shows their classification. Note that by means of this single basic concept, mixed descriptions in terms of domains, modes, and technologies are achievable.

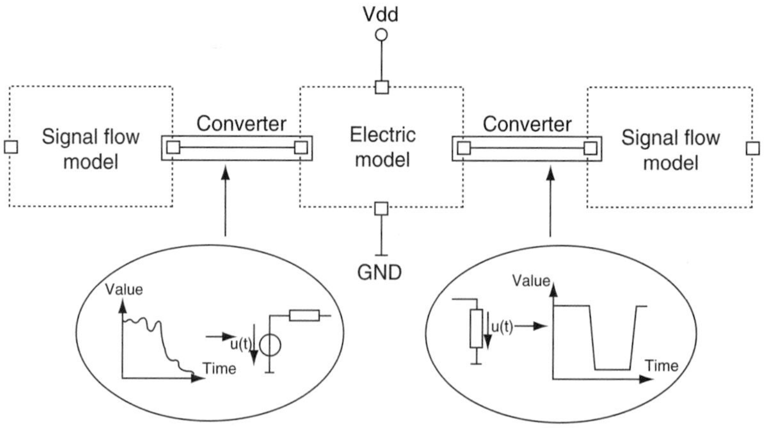

FIGURE 3.5 Connecting Models at Functional and Behavioral Levels

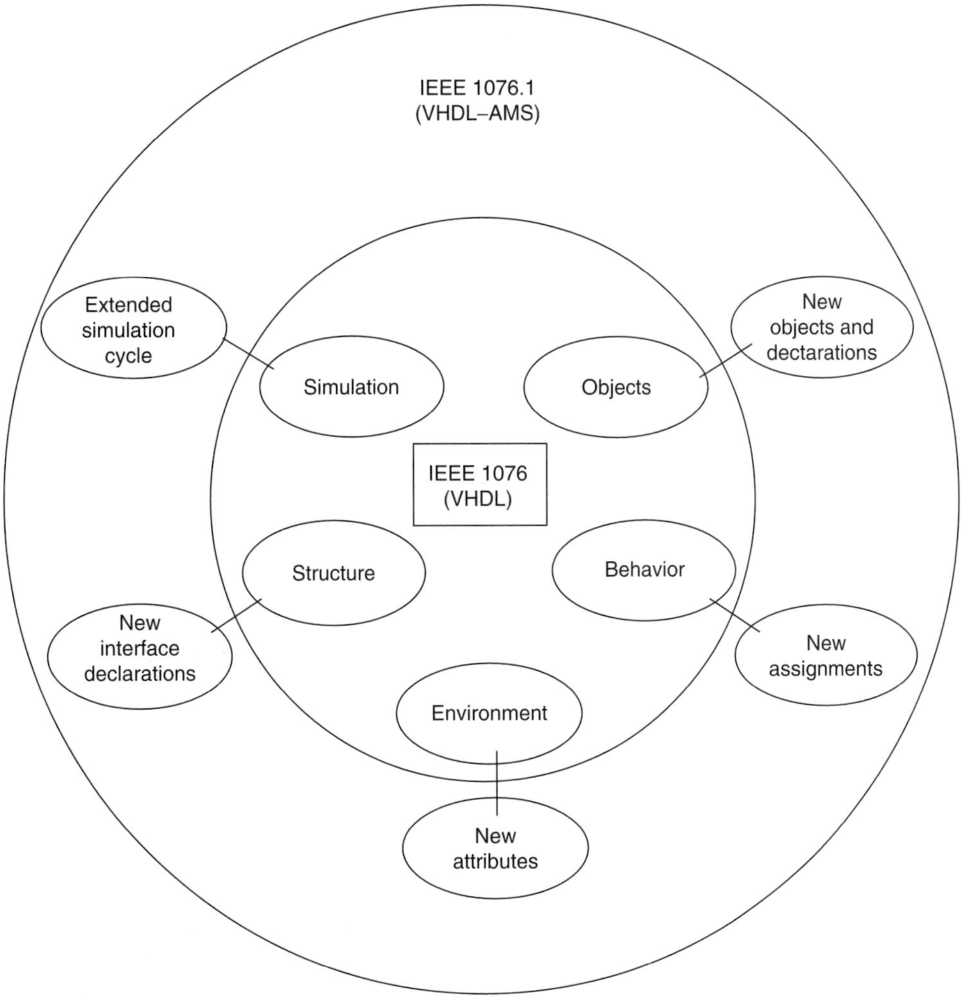

FIGURE 3.6 Extension to IEEE Standard 1076–1993

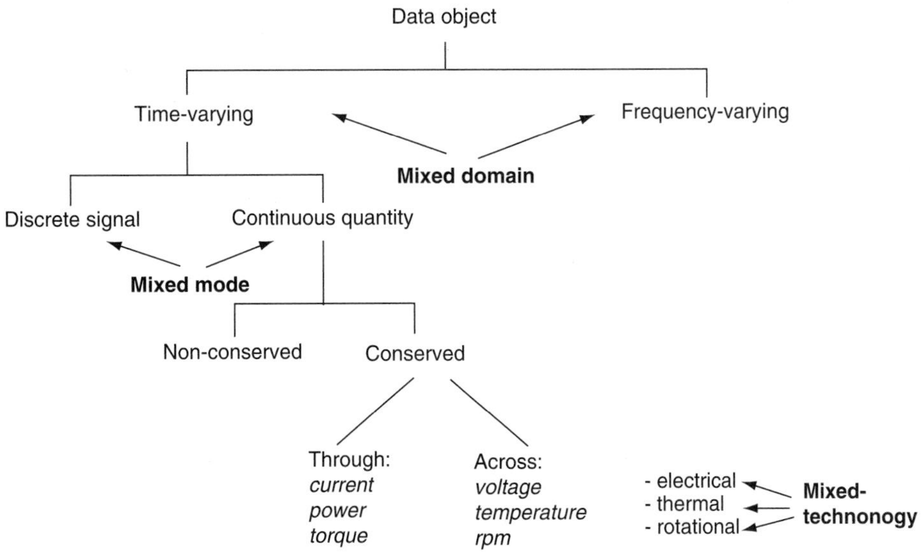

FIGURE 3.7 Classification of Data Objects

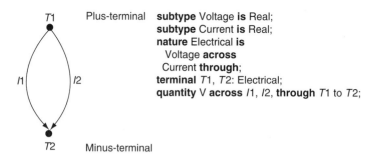

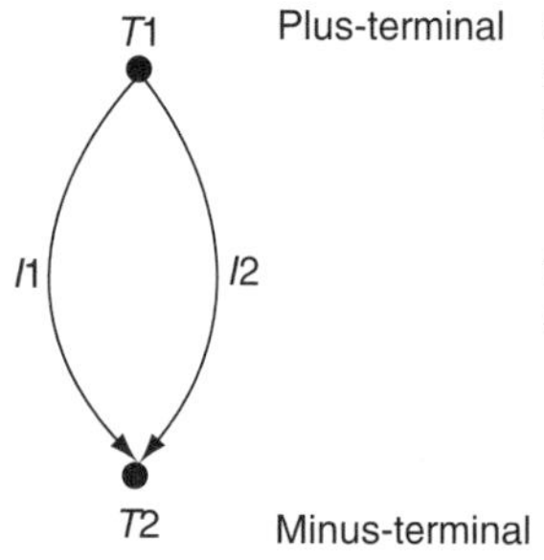


```
subtype Voltage is Real;
subtype Current is Real;
nature Electrical is
  Voltage across
  Current through;
terminal T1, T2: Electrical;
quantity V across I1, I2, through T1 to T2;
```

FIGURE 3.8 Simple Example for Across and Through Quantities

The enforcement of conservative laws is of considerable interest when modeling physical systems, except for the functional level. Enforcement has been tackled in VHDL–AMS by means of a graph-based concept. Branch quantities with reference to terminals and natures are used for this purpose, as detailed in the following:

- Quantities between two **terminals** represent the unknown variables (branch quantities) in a conservative system of equations:
  a. **Across** quantity: Difference of potentials
  b. **Through** quantity: Quantity of flow in a branch
- Terminals are assigned to a physical domain, known as **nature**, which has to be defined appropriately.

Figure 3.8 shows how two branches between two terminals related to the nature electrical are defined.

Conservation laws have to be enforced independently of a user-defined partitioning of the entire model into design entities. This implies that the interface description of entities requires an enhancement. The connection pins of an entity, denoted as **ports** in VHDL terminology, may now carry waveforms of different qualities: digital signals and continuous quantities of some physical domain with or without conservation requirements as given in Table 3.3. The new interface description syntax of VHDL–AMS is given here; note that the compatability for VHDL 1076 is accomplished by a blank PortAttribute, which is interpreted as the **signal** attribute:

```
port (PortAttribute NameList: [Mode] Type)
PortAttribute : : =
  signal | quantity | terminal
```

TABLE 3.3 Values of Mode Subject to PortAttribute

PortAttribute	Mode	Remarks
Signal	**in, out, inout**	
Quantity	**in, out, inout**	
Terminal	—	No mode may be specified because a terminal is always bidirectional

The description of the functionality of a model is stated in the architectural body. In addition to the concurrent statements of VHDL used for digital circuits and systems (e.g., component instances, signal assignments, and process statements), simultaneous statements are available:

```
architectureStatement ::=
  concurrentStatement | simultaneousStatement
simultaneousStatement ::=
  simpleSimStatement | compoundSimStatement
```

Simultaneous statements can be either simple or compound. The latter form may consist of a **procedural**, a **case**, or an **if** statement. The simple statement as defined above unveils two interesting properties. First, it is a noncausal formulation of an equation. This means that the == symbol denotes equality in the mathematical sense and not an assignment such as the <= symbol in digital data flow descriptions. As a consequence, the modeler does not need to formulate explicit equations. The more general implicit differential algebraic equation (DAE) form may be used directly instead. Second, the other interesting property is related to the optional tolerance aspect of the statement. This is intended to specify the numerical accuracy of the equality, which is a very general way to deal with numerical problems. Usually, a modeler is interested only in accuracy aspects but not in selecting solution methods for (e.g., numerical) integration. The implementors of a VHDL–AMS simulation engine can provide the appropriate algorithms, whereas a modeler can specify accuracy levels only by means of tolerance aspects.

Statements in the architectural body of a design entity may consist of, as already detailed, both concurrent and simultaneous statements. This is of special interest to a modeler because it supports a consistent description of design objects operating in mixed-mode without an artificial separation into event-discrete and time-continuous parts. Mixing up event-oriented and time-continuous behavior in many cases results in dealing with model discontinuities. This problem is addressed by the "break" statement used for both initialization and for discontinuity handling. A simple example for a 1-bit D/A converter given in Figure 3.9 illustrates its usage.

```
entity DAC is
  generic (VHigh: Real := 5.0);
  port(signal S: in Bit;
    terminal A: Electrical)
end entity DAC;

architecture Simple of DAC is
  quantity V across I through A;

begin
  if S = '0' use
    V == 0.0;
  else
    V == VHigh
  end use;
  break on S;
end architecture Simple;
```

FIGURE 3.9 Model of a Simple D/A Converter

The mixed-mode property of the entity DAC is easily visible from the interface description. There is both a signal and a terminal definition at ports. Obviously, the model is aimed to the behavioral abstraction level. The branch quantities related to the terminal A linked to the **nature Electrical** are defined in the architectural body. There are two architecture statements in the executable part of the architectural body: a compound simultaneous statement (if…use…else…end use) and the break statement that is sensitive to an event on the signal *S*. The output voltage [V] of the D/A converter is recalculated each time the signal value of *S* changes. Note that no additional equations or definitions are required to enforce conservation laws at port A.

The modeling features of VDHL–AMS have been highlighted so far for electrical circuits only. A simple multinature modeling example will now demonstrate that all required concepts and syntax elements are already present for that purpose. First, we need to extend our set of **natures**. Figure 3.10 summarizes the definitions for two more physical domains. They are given as part of a package identified by the name **NaturePkg**:

A combined electrical and thermal model of a diode demonstrates how the effects of nonself-heating on its electrical characteristics may be modeled for a multinature application in an easy and consistent way. This model is outlined in Figures 3.11 and 3.12, respectively.

Passing of parameter values is accomplished by means of the "generic" mechanism of the entity declaration. There is one terminal port for the nature Thermal and two terminal ports for the nature Electrical. The associated branch quantities are defined in the architectural body. In addition, the two free quantities *Q* and *VT* are defined as well. The first three simultaneous equations denote the electrical behavior of the nonlinear diode. Note the differential equation relating the charge *Q* to the current *IC*. The thermal voltage is now determined by means of the constants Boltzmann and ElemCharge taken

```
package NaturePkg is
  - physical constants
  constant Eps0 :Real := 8.8542e-12;
           - vacuum permitivity [F/m]
  constant Mu0 : Real := 1.256e-6;
           - vacuum permeability [H/m]
  constant Boltzmann : Real := 1.381e-23; - [J/K]
  constant ElemChange : Real := 1.602e-19; - [C]
  constant AmbTemp : Real := 300.0; - [K]

  - electrical systems
  subtype Voltage is Real; - subtypes for voltage
  subtype Current is Real; - and current

  nature Electrical is Voltage across Current
    through Electrical_reference reference;
  alias Ground is Electrical_reference;

  - fluidic systems
  subtype Pressure is Real; - subtypes for pressure
  subtype FlowRate is Real; - and flow rate

  nature Fluid is Pressure across FlowRate through
    Fluid_reference reference;
  alias FluidGround is Fluid_reference;

  - thermal systems
  subtype Temperature is Real; - subtypes for
    temperature
  subtype PowerTh is Real; - and power

  nature Thermal is Temperature across PowerTh
    through Thermal_reference reference;
  alias ThermalGround is Thermal_reference;
end package NaturePkg;
```

FIGURE 3.10 Package Example for Some Natures

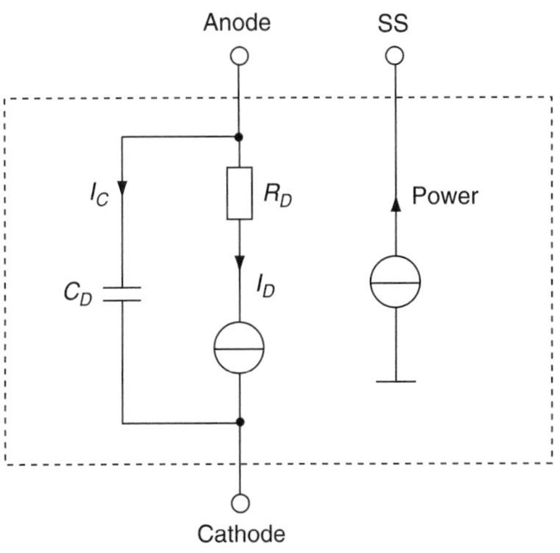

FIGURE 3.11 Structural Model of a Diode

```
entity Diode Th is
  generic(ISO        : Real : = 1.0E-14;
     N,Area        : Real := 1.0;
     Tau,CJO,Phi,RD : Real := 0.0  );
  port(terminal Anode, Cathode : Electrical;
     terminal SS : Thermal  );
  end entity DiodeTh;

architecture BNRealTh of DiodeTh is
  quantity V across ID,IC through Anode to Cathode;
  quantity Temp across Power through ThermalGround
  to SS;
  quantity Q : Charge;
  quantity VT : Voltage;

  begin
   ID == Area*ISO*(exp( (V-RD*ID)/(N*VT))-1.0);
   Q == Tan*ID-2.0*CJO*sqrt(Phi*(Phi-V));
   IC == Q'dot;
   Temp*Boltzmann/ElemCharge == VT;
   V*ID == Power;
end architecture BNRealTh;
```

FIGURE 3.12 Behavioral Multinature Model Code

from NaturePkg as a function of the temperature Temp, which, in turn, is one of the thermal branch quantities. The second one, Power, is calculated from the electrical branch quantities *V* and *ID* in the last equation. Note the noncausal notation of these equations. The diode current *ID* thus depends on the junction temperature via the thermal voltage *VT*.

3.4 Systems Modeling: A Multinature Example

The increasing complexity of information processing products and the pressure of design to market requirements are the main reasons for a change in the design process of these products, resulting in an emphasis of timely design validation of virtual prototypes instead of breadboarding. This reveals possible concept errors long before an implementation of real prototypes. On the other hand, such products are too complex to be modeled at lower (i.e., more accurate) abstraction levels. A need for a sensible mix of models on different abstraction levels thus arises and results in the necessity to represent and to execute models both in different domains and at several abstraction levels. The purpose of such a system model, known as virtual prototype, is to support an assessment of the conceptualization and the refinement according to the design flow outlined in Figure 3.1 in terms of meeting initial target specifications. The process of model definition, refinement, and assessment is demonstrated in the following by means of a depth gauge, a safety critical device for deep water divers.

The demonstrator was built with system simulation of multinature design objects in mind. It is a fairly complex but still easily comprehensive example on how to construct a simulation model in three physical domains featuring both time-continuous and event-discrete behavior. In addition, synthesizable models of digital signal processing units are incorporated at register and algorithmic levels as defined in Table 3.1.

All VHDL–AMS models are operational and were thoroughly tested by means of the ADVanceMS simulator from Mentor Graphics. Unfortunately, the complete set cannot be presented for space restriction reasons. All models of this depth gauge and additional information on other complex multinature modeling projects are available at the author's site. (www.vlsi.informatik.tu-darmstadt.de/vhdlams/)

3.4.1 Partitioning the System

The models are subdivided in two ways. The whole system is partitioned into smaller entities that are modeled as a unit; furthermore, there are several models (i.e., entities, which are each modeled as a unit; furthermore, there are several models (i.e., architectures) for each unit. These architectures differ in terms of abstraction levels and exactness or represent different implementation approaches. The architectures suited for system simulation are presented in the following.

The functional partitioning of the model of a depth gauge denoted Virtual Diving Computer (vdc) is based on

- functional classification
- entity reuse
- complexity
- testability.

The different architectures can be classified using the architecture name. Most of them are self-explanatory.

- The **simple** architecture is reduced to the fundamental features and it is modeled at the functional level. It yields low simulation cost and idealized results.
- The **digital_algorithmic** architecture is a high-level description using a synthesizable subset of VHDL. These models must be passed to a scheduling and resource allocation tool prior to mapping a target library. For the demonstrator, this is done with the Behavioral Compiler from Synopsys.
- The **digital_RTL** architecture is a register-transfer level description using again a synthesizable subset of VHDL. These models can be multiclock or even asynchronous designs and are easily mapped to target libraries. This is tested with the Design Compiler from Synopsys.
- The **beh** architecture is a behavioral model of analog or mixed-signal components. This model reproduces timing and signal waveforms of the entity, including major hazards and glitches.
- The **eldo** architecture is used to incorporate Spice (i.e., macro and circuit) level models.

3.4.2 Models of Functional Units

The Testbench

The entity denoted **testbench** is the top-level entity. It contains the environment and the device under test. The **testbench** is self-contained and used to define the interconnections between the single layers. Moreover, it selects the architectures to be used. The code is given in the following:

```
library IEEE;
use ieee.std_logic_1164.all;
use ieee.std_logic_arith.all;

library disciplines;
use disciplines.electromagnetic_system.all;
use disciplines.Fluidic_system.all;
use disciplines.Thermal_system.all;

use work.all;

entity vdc_testbench is
end vdc_testbench;

architecture simple of vdc_testbench is
  constant T_ADC_Width : natural := 6;
  constant U_T_min : emf := 0.0;
  constant U_T_max : emf := 2.72;
  constant P_ADC_Width : natural := 12;
  constant U_P_min : emf := -0.03;
  constant U_P_max : emf := 0.188;
  constant Depth_Width : natural := 10;

-Derived constants:
  constant DSP_P_Min : integer := integer(1000.0*
  U_P_Min);
  constant DSP_P_Max : integer := integer(1000.0*
  U_P_Max);
  constant DSP_T_Min : integer := integer(1000.0*
  U_T_Min);
  constant DSP_T_Max : integer := integer(1000.0*
  U_T_Max);

-Signals and terminals (internal to testbench):
  terminal Pressure_In : fluidic;
  terminal Temp_In : thermal;
  terminal U_Temp, U_Pressure : electrical;

  terminal a_Depth, a_Error : electrical;
  signal Clk : std_logic := '0';
  signal nClk : std_logic;   -not(Clk)
  signal Rst : std_logic;  -Reset
  signal T_Data : std_logic_vector((T_ADC_Width-
  1) downto 0);
  signal P_Data : std_logic_vector((P_ADC_Width-
  1) downto 0);
  signal D_Data : std_logic_vector((Depth_Width-
  1) downto 0);

begin
Rst <= '1', '0' after 5 ns;  -Initialization
Clk <=not(Clk) after 1 ms;  -Clock Generator
nClk <=not(Clk);  -Inverted Clocksignal

-Structural description:
  ENVIRONMENT : entity vdc_sources(beh_LakeDive)
    port map (Pressure_In, Temp_In);
  TSENSOR : entity vdc_tsens
    port map (Temp_In, U_Temp);
  PSENSOR : entity vdc_psens
    port map (Pressure_In, Temp_In, U_Pressure_P,
    U_Pressure_N);
  T_ADC : entity vdc_ADC
    generic map (U_T_Min, U_T_Max, T_ADC_Width)
    port map (U_Temp, Electrical_Ground, Clk, Rst,
    T_Data);
  P_ADC : entity vdc_ADC
    generic map (U_P_Min, U_P_Max, P_ADC_Width)
    port map (U_Pressure_P, U_Pressure_N, Clk, Rst,
    P_Data);
  DSP  : entity vdc_dsp
    generic map (Depth_Width, DSP_P_Min, DSP_
    P_Max, DSP_T_Min, DSP_T_Max)
    port map (nClk, Rst, P_Data, T_Data, D_Data);
  TESTER : entity vdc_tester
    port map (Rst, D_Data, a_Depth, a_Error, Pres-
    sure_In);
end simple;
```

The Environmental Model

The environment is modeled to produce all the input data needed to test the depth gauge. In consequence, only observable effects for the device under test are reproduced.

The generic **timescale** can be used as a workaround for simulators that do not allow to change the resolution limit for the type **time**. Simulations based on femtoseconds reduce the guaranteed simulation time to 2.14 μs—typical application employments are several hours or even days. This requires range extensions beyond 2^{31} or changing the resolution limit to milliseconds. If both are not feasible, a **timescale** to shrink the simulation's duration may be used.

The port terminals **Pressure_sensor** and **Temp_sensor** are modeled as sources that guarantee pressure and temperature potentials without side effects. These potentials can be gained from the scuba diver's position. To simplify the model, only the resulting potentials are described in the following two architectures. The first is characteristic for lake diving, and the second is aimed at ocean diving.

```
library disciplines;
use disciplines. Fluidic_system.all;
use disciplines. Thermal_system.all;

entity vdc_sources is
  generic(timescale : real := 1.0);
  port(terminal Pressure_sensor : fluidic;
      terminal Temp_sensor : thermal);
end vdc_sources;
```

The model for the lake dive outlined in the following reproduces the strong coherence between diving depth and water temperature. This temperature profile is typical for deep lakes during the summertime: the surface is heated by the sun, whereas wind and day/night changes form a several meter deep warm surface layer. Below this layer, the temperature drops rapidly below 10°C.

```
library disciplines;
use disciplines.Fluidic_system.all;
use disciplines.Thermal_system.all;

library IEEE;
use IEEE.math_real.all;

architecture beh_LakeDive of vdc_sources is
  constant surfacePressure : pressure := 0.0;
  constant metalimnion : pressure : = 1.2;
  constant surfaceTemp : temperature :=21.0;
function LakeTemp(
  constant Tsurface : real;
  constant Psurface : real;
  constant Pmetalimnion : real;
     p      :real)
  return real is

  constant Tc : real := 5.0 / (Pmetalimnion -Psur-
  face);
  constant Tb : real := (Tsurface -10.0) /
  (exp (TC* Psurface)
  -exp (TC* Pmetalimnion));
  constant Ta : real := Tsurface -Tb* exp(TC* Psur-
  face);
  constant Td : real := 4.0;
  constant Tf : real := (Tb* TC* exp (TC* metalim-
  nion)) / (10.0 -Td);
  constant Te : real := (10.0 -Td)/exp (Tf* meta-
  limnion);

  - The 6 constants are calculated based on:
  - Temp(p = Psurface) == surface Temp
  - Temp(p = Pmetalimnion) == 10.0
  - Temp(p- >infty) == 4.0
  - Temp'dot(p = Psurface)*exp(5.0) == Temp'dot(p
             = Pmetalimnion)
  - Temp(metalimnion -delta) == Temp(metalimnion
             + delta)
  - Temp'dot(metalimnion -delta) == Temp'dot
             (metalimnion + delta)

  variable res : real;

begin - Lake Temp
  if (p < Pmetalimnion) then
    res := Ta + Tb* exp(Tc* p);
  else
    res := Td + Te* exp(Tf* p);
  end if;
  return res;
end Lake Temp;

quantity
  p across
  P_flow through
  Pressure_sensor;
quantity
  Temp across
  T_P through
  Temp_sensor;
begin
  if (now < 0.1 * timescale) use          - diving starts at the surface
    p == surfacePressure;
  elsif (now < 3.0 * timescale) use       - descending along the
    p'dot == 0.3 / timescale;               lake's profile
  elsif (now < 16.0 * timescale) use      - descending along the
    p'dot == 0.1 / timescale;             - lake's profile
  elsif (now < 18.0* timescale) use       - ground time: 2 min
    p'dot == 0.0;                         - (cold, dark, very few fish)
  elsif (p > 0.4) use                     - ascending up to 4 m
    p'dot == -0.75 / timescale;
  elsif (p > 0.3) use                     - diving in the epilimnion
    p'dot == -0.005 / timescale;          - (much fish, crayfish and insects)
  elsif (p > surfacePressure) use
    p'dot == -0.04 / timescale;           - ascending to the surface
  else
    p == surfacePressure;                 - diving ends at the surface
  end use;
  Temp == LakeTemp(surface Temp, surfacePressure,
  Metalimnion, p);
end beh_LakeDive;
```

Temperature Sensor

The temperature sensor we used is a platinum resistor made in thin-layer technology on Al_2O_3 substrate. We use a 1-kΩ sensor with a temperature sensitivity (Tk) of $3.85 \cdot 10^3/°C$ (as specified in IEC 751). These sensors are quite popular in automotive applications. This results in low cost and good availability of subcomponents. The sensor is used in a voltage divider together with the constant resistor **Rfix**. The temperature information is represented in the output voltage **Uout**. The multinature property is thus obvious as shown in Figure 3.13.

The **simple** architecture is based on the Sensor Nite PT1000 from Heraeus. Thus, the modeling process is constrained to an exploitation of commercially available subsystems:

```
library IEEE;
use ieee.math_real.all;
library disciplines;
```

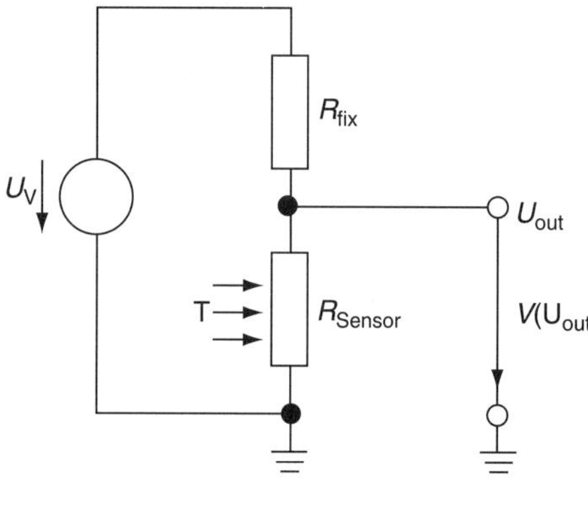

FIGURE 3.13 Temperature Sensor

```
use disciplines.electromagnetic_system.all;
use disciplines.Thermal_system.all;

entity vdc_tsens is
generic (
  R0 : real := 1000.0;
  Rfix : real := 1000.0;
  UV : emf := 5.0;
  Tk : real := 3.85e-3);
port(terminal T_ln : thermal;
    terminal Uout : electrical);
end vdc_tsens;

library IEEE;
use ieee.math_real.all;
library disciplines;
use disciplines.electromagnetic_system.all;
use disciplines.Thermal_system.all;

architecture simple of vdc_tsens is

quantity R1 : real;

quantity
  Temp across
  T_ln;

quantity
  Uout_V across
  Uout_I through
  Uout;

begin
  R1 == R0 * (1.0 + Tk* Temp);
  Uout_V == UV * R1 / (R1 + Rfix);
end simple;
```

Digital Block

The depth is determined in the digital part from data obtained through pressure and temperature measurements and then represented as electrical quantities. In the first intermediate step, the actual temperature is calculated using the temperature data represented in 6-bit. This is followed by the calculation of the actual pressure from the pressure data given in 12-bit. The conversion of the specified pressure to a depth below the water surface is obtained by means of following equation:

$$\text{depth} = \frac{1\ \text{m}}{0.1\ \text{bar}} \cdot (\text{pressure} - 1\ \text{bar}) \tag{3.1}$$

The model outlined in this section implements digital signal transformations aimed to compensate for the nonlinearities of the used sensors only. The depth is calculated with a guaranteed accuracy of 0.1 m.

```
library ieee;
use ieee.std_logic_1164.all;

entity vdc_dsp is
  generic (D_Res : integer := 10;
      P_min : integer := -30; -in mV
      P_max : integer := 188; -in mV
      T_min : integer := 0; -in mV
      T_max : integer := 2720; -in mV
      Tk0 : integer := 260; -(Rfix)/(R0*Tk)
      Tk1 : integer := 5000; -UV in mV
      Tk2 : integer := -260; --1/Tk
      Tp0 : integer := -10; -UOffset in mV
      Tp1 : integer := 10 -UV *k0 in mV
      );
  port(Clk : in std_logic;
    Reset : in std_logic;
    Pdata : in std_logic_Vector;
    Tdata : in std_logic_Vector;
    Depth : out std_logic_Vector((D_Res-1) downto
    0));
end vdc_dsp;
```

The **digital_algorithmic** architecture makes use of the resource sharing and scheduling features of the synthesis products from Synopsys, which results in a relatively small circuit design. The synthesized code consists of one multiplication unit, one division unit, and some glue logic. The code given in the following (algorithmic level) code and is fully synthesizable by the Behavioral Compiler.

```
library ieee;
use ieee.std_logic_arith.all;
use ieee.std_logic_1164.all;

library DW02;
use DW02.DW02_components.all;

architecture digital_algorithmic of vdc_dsp is

begin -behaviorDescription
  process
    variable k1 : integer;
    variable k2 : integer;
    variable Temp, Pressure : integer;
    variable i : integer;
    variable i32_1, i32_2, i32_3 : integer;
    variable b : bit;
    variable s32_1 : signed (31 downto 0);
    variable u32_1 : unsigned (31 downto 0);

begin
  Depth <= (others=> '0');
  main_loop: loop
    wait until clk'event and clk='1';
    if reset='1' then
      exit main_loop;
    end if;
    Temp := conv_integer(unsigned(TData));
    i32_1 := Temp* (T_Max - T_Min);
    i32_2 := T_Min* (2** TData'length);
    Temp := i32_1 + i32_2;

Pressure := conv_integer(unsigned(PData));
    i32_1 := Pressure* (P_Max - P_Min);
    i32_2 := P_Min* (2** PData'length);
    Pressure := i32_1 + i32_2;
```

```
-correction of sensor error:
-28-bit resolution needed
    i32_1 := Tk0* Temp;
    i32_2 := Tk1* (2** TData'length);
    i32_2 := i32_2 - Temp;

    Temp := conv_integer(conv_signed(i32_1,32) /
    conv_signed(i32_2,32));
    Temp := Temp + Tk2;
    Temp := Temp + 10;

-accuracy of Temp is 1 deg C
-Temp is T' = T + 10 deg C (because of T in (-10;50))

-k2 by linear approx.
    s32_1 := conv_signed(3 * Temp,32);
    u32_1 := conv_unsigned(5,32);
    i32_1 := conv_integer(shr(s32_1, u32_1));
    k2 := 253 + i32_1;

-k2 is k2' = k2*256

-correction of sensor error:

-first guess:k1' = 256:

    k1 := 256;

i32_1 := (Pressure - Tp0 * (2 ** PData'length));
    i32_1 := i32_1 * (25 * (2 ** (18 - PData'
    length)));
    i32_2 := Tp1 * k1;
    i32_2 := i32_2 * k2;
    i32_3 := conv_integer(conv_signed(i32_1,32)
    / conv_signed(i32_2,32));

-calculation of k1'using the approximation:
    if (i32_3 < 320) then
      k1 := 256;
    else
      s32_1 := conv_signed(i32_3,32);
      u32_1 := conv_unsigned(14,32);
      i32_1 := conv_integer(shr(s32_1, u32_1));
      k1 :=i32_1 + 251;
    end if;

    i32_1 := (Pressure - Tp0 * (2 ** PData'length));
    i32_1 := i32_1 * (25 * (2 ** (18 - PData'
length)));
    i32_2 := Tp1 * k1;
    i32_2 := i32_2 * k2;
    i32_3 := conv_integer(conv_signed(i32_1,32)
    / conv_signed(i32_2,32));

-i32_3=depth in factors of 10 cm

    Depth <= std_logic_vector(conv_unsigned
    (i32_3, Depth'length));
  end loop main_loop;
 end process;
end digital_algorithmic;
```

This example highlights the features of VHDL in denoting high levels but these features indicate still synthesizable models for the digital domain.

3.4.3 Simulation Results

Aberration Monitoring

The aberration monitor takes the environmental pressure data and the depth output of the digital block to calculate an absolute error and an analog value corresponding to the digital depth information. The results of the monitoring are depicted in the third trace as part of the overall simulation results shown in Figure 3.14.

```
library IEEE;
use ieee.std_logic_1164.all;
library disciplines;
use disciplines.electromagnetic_system. all;
use disciplines.Fluidic_system.all;
entity vdc_tester is
  port (signal Reset : in std_logic;
    signal Depth : in std_logic_vector;
    terminal aDepth, aError : electrical;
    terminal act_Pressure : fluidic);
end vdc_tester;

library IEEE;
use ieee.std_logic_1164.all;
use ieee.std_logic_arith.all;
library disciplines;
use disciplines.electromagnetic_system.all;
use disciplines.Fluidic_system.all;

architecture simple of vdc_tester is
quantity
  act_P_V across
  act_Pressure;
quantity theDepth across Depth_I through aDepth;
quantity theError across Error_I through aError;
quantity myDepth : real;
begin
  if reset = '1' use
    myDepth == 0.0;
    theDepth == 0.0;
    theError == 0.0;
  else
    myDepth == 0.1* emf(conv_integer(unsigned
    (Depth)));
    theDepth == myDepth;
    theError == act_P_V -0.1* myDepth;
  end use;
end simple;
```

The first plot visualizes the results of system simulation for a lake dive. Note the temperature profile of the lake (second trace).

In contrast, the second plot depicts the simulation results for an ocean dive. The system's model of the gauge is as before, but the environment has changed. The gauge model, therefore, acts as a virtual prototype being exercised in different contexts. The accuracy of the model is visible from the third trace in each plot. The absolute error (i.e., the difference between the actual and the measured depth) is in the range of the least significant bit of the discretized value.

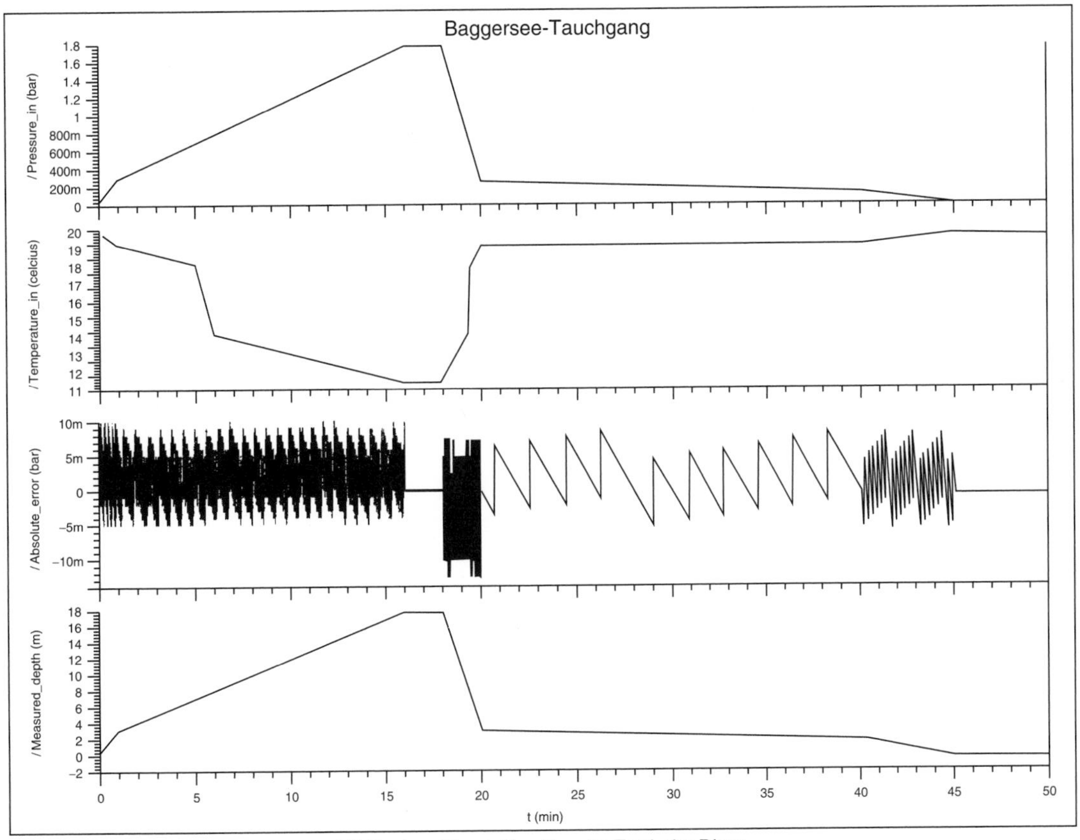

(A) System Simulation Results For Lake Diver

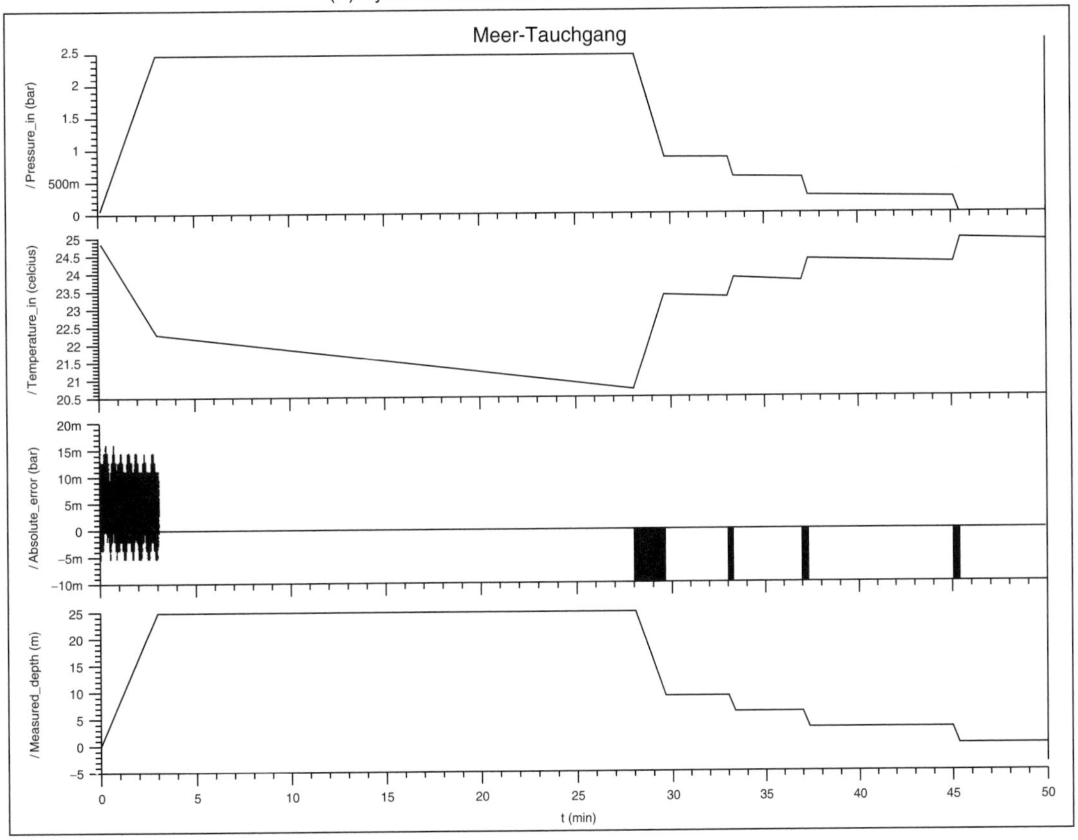

(B) System Simulation Results For Ocean Diver

FIGURE 3.14 Results of System Simulation

3.5 Conclusion and Further Readings

The design of embedded systems is a highly complex task due to the tight interaction of their functional units, which, in general, originate from several engineering domains. Generation and assessment of models play an important role in the whole design process. When dealing with modeling issues, questions about concepts, representation, efficiency, and accuracy have to be addressed.

An introduction to modeling of multinature systems is best given by discussing the role of models and of associated abstraction levels in the design process as presented in the first section. The representation of abstract models should then allow for an execution and assessment of the captured specification. VHDL–AMS has been introduced as a highly expressive description language for the representation of mixed time-continuous and event-discrete operating embedded systems. The advantages for a design engineer, which result from the proposed modeling methodology, have been demonstrated for a practical multinature application example.

Plenty of good literature is available on digital hardware description languages, especially on VHDL, and on synthesis algorithms that exploit such languages as a means for specifying the functionality of digital components and systems. Continuous time modeling, however, and mixed-signal or multinature modeling are not yet covered sufficiently. In Cellier (1991), the reader will find a comprehensive discussion of this domain. An introduction to underlying numerical algorithms for solving differential-algebraic equations is given in (Brenan *et al.*, 1989). Modeling issues with an emphasis on analog circuits are covered in Mantooth and Fiegenbaum (1995) and Bergé *et al.* (1995). The language reference of VHDL–AMS is given by the IEEE Computer Society (1999). The Mentor Graphics (1999) user's manual is a good reference for a commercial simulator. Finally, in Vachoux *et al.* (1997), the modeling of multinature systems using VHDL–AMS is addressed by several authors, thus giving an in-depth insight into this evolving area.

References

Vachoux, J.-M., Bergé, O., Levia, J., and Rouillard. (Eds.). (1997). *Analog and mixed-signal hardware description languages.* London: Kluwer Academic Publishers.

Bergé, J.M., Levia, O., Rouillard, J. (Eds.) (1995). *Modeling in analog design.* London: Kluwer Academic Publishers.

Brenan, K.E., Campell, S.L., Petzold, L.R. (1989). New York: *Numerical solution of initial-value problems in differential-algebraic equations.* Philadelphia: Society for Industrial and Applied Mathematics.

Cellier, F.E. (1991). *Continuous system modeling.* New York: Springer-Verlag.

IEEE Computer Society. (1999). *IEEE standard VHDL analog and mixed-signal extensions*, IEEE Std 1076.1.

Mantooth, H.A., and Fiegenbaum, M. (1995). *Modeling with an analog hardware description language.* London: Kluwer Academic Publishers.

Mentor Graphics Corporation. (1999). *ADVance MS user's manual.*

4

Clock Skew Scheduling for Improved Reliability*

Ivan S. Kourtev
Department of Electrical and Computer Engineering, University of Pittsburgh, Pittsburgh, Pennsylvania, USA

Eby G. Friedman
Department of Electrical and Computer Engineering, University of Rochester, Rochester, New York, USA

4.1 Introduction

Most high performance digital integrated circuits implement data processing algorithms based on the iterative execution of basic operations. Typically, these algorithms are highly parallelized and pipelined by inserting clocked registers at specific locations throughout the circuit. The synchronization strategy for these clocked registers in the vast majority of VLSI/ULSI-based digital systems is a fully synchronous approach. It is common for the computational process in these systems to be spread over hundreds of thousands of functional logic elements and tens of thousands of registers.

For such synchronous digital systems to function properly, the many thousands of switching events require a strict temporal ordering. This strict ordering is enforced by a global synchronization signal known as the **clock signal** that must be delivered to every register at a precise relative time. The delivery function is accomplished by a circuit and interconnect structure commonly known as a **clock distribution network** (Friedman, 1995, 1997).

Multiple factors affect the propagation delay of the data signals through the combinational logic gates and interconnect. Because the clock distribution network is composed of logic gates and interconnection wires, the signals in the clock distribution network are delayed. Moreover, the dependence of the correct operation of a system on the signal delay in the clock distribution network is far greater than on the delay of the logic gates. Recall that by delivering the clock signal to registers at precise times, the clock distribution network essentially quantizes the operational time of a synchronous system into clock periods, thereby permitting the simultaneous execution of operations.

The nature of the on-chip clock signal has become a primary factor limiting circuit performance, causing the clock distribu-

*This research was supported in part by the National Science Foundation under Grant No. MIP-9423886 and Grant No. MIP-9610108, a grant from the New York State Science and Technology Foundation to the Center for Advanced Technology-Electronic Imaging Systems, and by grants from the Xerox Corporation, IBM Corporation, and Intel Corporation.

tion network to become a performance bottleneck in high speed VLSI systems. The primary source of the load for the clock signals has shifted from the logic gates to the interconnect, thereby changing the physical nature of the load from a lumped capacitance (C) to a distributed resistive-capacitive (RC) load (Bakoglu, 1990; Bothra *et al.*, 1993). These interconnect impedances degrade the on-chip signal waveform shapes and increase the path delay. Furthermore, statistical variations of the parameters characterizing the circuit elements along the clock and data signal paths (caused by the imperfect control of the manufacturing process and the environment) introduce ambiguity into the signal timing that cannot be neglected. All of these changes have a profound impact on both the choice of synchronous design methodology and on the overall circuit performance. Among the most important consequences are increased power dissipated by the clock distribution network and increasingly challenging timing constraints that must be satisfied to avoid timing violations (Friedman, 1995; Gaddis and Lotz, 1996; Gronowski *et al.*, 1996; Vasseghi *et al.*, 1996; Bowhill *et al.*, 1995). Therefore, the majority of the approaches used to design a clock distribution network focus on simplifying the performance goals by targeting minimal or zero global clock skew (Neves and Friedman, 1993; Xi and Dai, 1996; Neves and Friedman, 1996), which can be achieved by different routing strategies (Jackson *et al.*, 1990; Tsay, 1993; Chou and Cheng, 1995; Ito *et al.*, 1995), buffered clock tree synthesis, symmetric n-ary trees (Gaddis and Lotz, 1996) (most notably H-trees), or a distributed series of buffers connected as a mesh (Friedman, 1995, 1997; Bowhill *et al.*, 1995).

4.2 Background

A **synchronous digital system** is a network of combinational logic and storage registers whose input and output terminals are interconnected by wires. An example of a synchronous system is shown in Figure 4.1. The sets of registers and logic gates of this specific system are outlined in Figure 4.1 and consist of the four registers, R_1 through R_4, and the four logic gates, G_1 through G_4, respectively. For clarity, the clock distribution network and clock signals to the registers are not shown, and the details of the registers and logic gates are also omitted.

A sequence of connected logic gates (no registers) is called a **signal path**. For example, in Figure 4.1, one signal path begins at the register R_1 and propagates through the logic gates G_1 and G_2 before reaching the register R_3. Other signal paths can also be identified in the system shown in Figure 4.1. Every signal path in a synchronous system is delimited by a pair of registers—one register each for the start and the end of the path. Such a pair of registers is called a **sequentially adjacent pair** and is defined next:

Definition 4.1 Sequentially adjacent pair of registers: *For an arbitrary* ordered *pair of registers* $\langle R_i, R_f \rangle$ *in a synchronous circuit, one of the following two situations can be observed. Either there exists at least one signal path that connects some output of* R_i *to some input of* R_f*, or inputs of* R_f *cannot be reached from outputs of* R_i *through a signal path.*[1] *In the former case—denoted by* $R_i \rightsquigarrow R_f$*—the pair of registers* $\langle R_i, R_f \rangle$ *is called a sequentially adjacent pair of registers, and switching events at the output of* R_i *can possibly affect the input of* R_f *during the same clock period. A sequentially adjacent pair of registers is also referred to as a local data path* (Friedman, 1995).

4.2.1 Permissible Range of Clock Skew

The clock signal is delivered to both the initial, R_i, and final, R_f, registers of any sequentially adjacent pair of registers $R_i \rightsquigarrow R_f$ in a circuit. The delays of the clock signal to these two registers, R_i and R_f, may be different. This difference in clock signal delay is accounted for by the concept of **clock skew** defined next:

Definition 4.2 Clock skew: *In a given digital synchronous circuit, the clock skew* $T_{Skew}(i, j)$ *between the registers* R_i *and* R_j *is defined as the algebraic difference*:

$$T_{Skew}(i, j) = t_{cd}^{i} - t_{cd}^{j}, \tag{4.1}$$

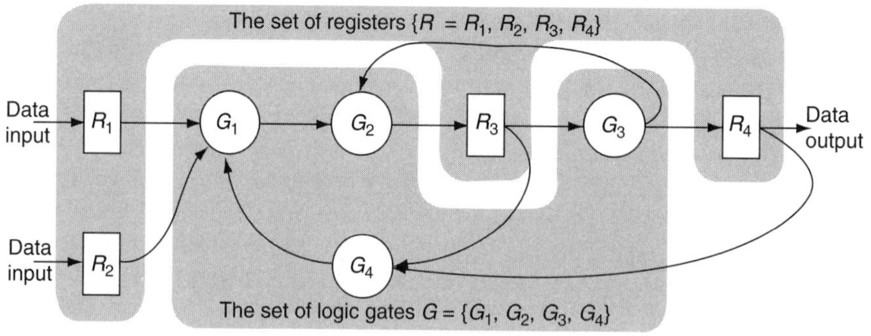

FIGURE 4.1 A Simple Synchronous Digital Circuit with Four Registers and Four Logic Gates

[1] This refers to propagating through a sequence of logic elements only.

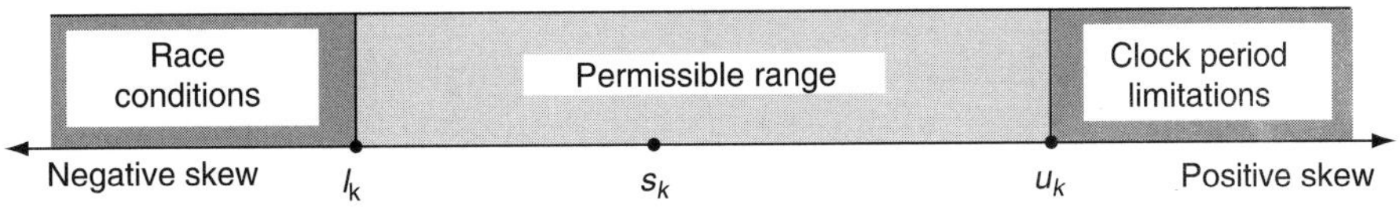

FIGURE 4.2 The Permissible Range of the Clock Skew of a Local Data Path. A timing violation exists if $s_k \notin [l_k,\ u_k]$.

where C_i and C_j are the clock signals driving the registers R_i and R_j, respectively, and t_{cd}^i and t_{cd}^j are the delays of the clock signals C_i, and C_j, respectively.

In definition 4.2, the clock delays, t_{cd}^i and t_{cd}^j, are with respect to an arbitrary—but necessarily the same—reference point. A commonly used reference point is the source of the clock distribution network on the integrated circuit. Note that the clock skew $T_{Skew}(i,\ j)$ as defined in definition 4.2 obeys the antisymmetric property:

$$T_{Skew}(i,\ j) = -T_{Skew}(j,\ i). \tag{4.2}$$

Technically, the clock skew $T_{Skew}(i,\ j)$ can be calculated for any ordered pair of registers $\langle R_i,\ R_j\rangle$. However, the clock skew between a nonsequential pair of registers has little practical value. The clock skew $T_{Skew}(i,\ j)$ between sequentially adjacent pairs of registers $R_i \rightsquigarrow R_j$ (that is, a local data path) is of primary practical importance since $T_{Skew}(i,\ j)$ is part of the timing constraints of these paths (Friedman, 1995, 1990; Afghani and Svensson, 1990; Unger and Tan, 1986; Kourtev, 1999; Kourtev and Friedman, 1996).

For notational convenience, clock skews in a circuit are frequently denoted throughout this work with the small letter s with a single subscript. In such cases, the clock skew s_k corresponds to a uniquely identified local data path k in the circuit, where the local data paths have been numbered 1 through a certain number p. In other words, the skew s_1 corresponds to the local data path one, the skew s_2 corresponds to the local data path two, and so on.

Previous research (Neves and Friedman, 1996a, 1996b) has indicated that tight control over the clock skews rather than the clock delays is necessary for the circuit to operate reliably. The timing constraints of sequentially adjacent pairs of registers are used to determine a **permissible range** of allowable clock skew for each signal path (Friedman, 1995; Neves and Friedman, 1996a, 1996b; Afghani and Svensson, 1990; Kourtev, 1999). The concept of a permissible range for the clock skew s_k of a data path $R_i \rightsquigarrow R_f$ is illustrated in Figure 4.2.

Each signal data path has a unique permissible range associated with it.[2] The permissible range is a continuous interval of valid skews for a specific path. As illustrated in Figure 4.2, every permissible range is delimited by a lower and upper bound of the clock skew. These bounds—denoted by l_k and u_k, respectively—are determined based on the timing parameters and timing constraints of the individual local data paths. Note that the bounds l_k and u_k also depend on the operational clock period for the specific circuit. When $s_k \in [l_k,\ u_k]$—as shown in Figure 4.2—the timing constraints of this specific k-th local data path are satisfied. The clock skew s_k is not permitted to be in either the interval $(-\infty,\ l_k)$ because a race condition will be created or the interval $(u_k,\ +\infty)$ because the minimum clock period will be limited.

Furthermore, note that the reliability of a circuit is related to the probability of a timing violation occurring for any local data path $R_i \rightsquigarrow R_f$. This observation suggests that the reliability of any local data path $R_i \rightsquigarrow R_f$ of a circuit (and therefore of the entire circuit) is increased in two ways:

1. By choosing the clock skew s_k for the kth local data path as far as possible from the borders of the interval $[l_k,\ u_k]$; that is, by (ideally) positioning the clock skew s_k in the middle of the permissible range: $s_k = \frac{1}{2}(l_k + u_k)$
2. By increasing the width $(u_k - l_k)$ of the permissible range of the local data path $R_i \rightsquigarrow R_f$

Even if the clock signals can be delivered to the registers in a given circuit with arbitrary delays, it is generally not possible to have all clock skews in the middle of the permissible range as suggested above. The reason behind this characteristic is that inherent structural limitations of the circuit create linear dependencies among the clock skews in the circuit. These linear dependencies and the effect of these dependencies on a number of circuit optimization techniques are examined in detail in Section 4.3.

4.2.2 Graphical Model of a Synchronous System

Many different fully synchronous digital systems exist. It is virtually impossible to describe the variety of all past, current, or future such systems depending on the circuit manufacturing technology, design style, performance requirements, and multiple other factors. A system **model** of these fully synchronous digital systems is required so that the system properties can be fully understood and analyzed from the perspective of clock skew scheduling and clock tree synthesis while permitting unnecessary details to be abstracted.[3]

[2] Section 4.2 later shows that it is more appropriate to refer to the permissible range of a sequentially adjacent pair of registers. There may be more than one local data path between the same pair of registers, but circuit performance is ultimately determined by the permissible ranges of the clock skew between pairs of registers.

[3] As a matter of fact, the graph model described here is quite universal and can be successfully applied to a variety of circuit analysis and optimization purposes.

In this section, a graphical model used to represent fully synchronous digital systems is introduced. The purpose of this model is twofold. First, the model provides a common abstract framework for the automated analysis of circuits by computers, and second, it permits a significant reduction of the size of the data that needs to be stored in the computer memory when performing analysis and optimization procedures on a circuit. This graph-based model can be arrived at in a natural way by observing what constitutes relevant system information (in terms of the clock skew scheduling problem). For example, it is sufficient to know that a pair of registers $\langle R_i, R_j \rangle$ are sequentially adjacent, whereas knowing the specific functional information characterizing the individual logic gates along the signal paths between R_i and R_j is not necessary.

Consider, for instance, the system shown in Figure 4.1. This system is completely described (for the purpose of clock skew scheduling) by the timing information describing the four registers, four logic gates, ten wires (nets), and the connectivity of these wires to the registers and logic gates. Consider next the abstract representation of this system shown in Figure 4.3. Note that the registers R_1 through R_4 are represented by the vertices of the graph shown in Figure 4.3. However, the logic gates and wires have been replaced in Figure 4.3 by arrows, or arcs, representing the signal paths among the registers. The four logic gates and ten nets in the original system have been reduced to only six local data paths represented by the arcs. For clarity, each arc, or edge, is labeled with the logic gates[4] along the signal path represented by this specific arc.

The type of data structure shown in Figure 4.3 is known as a multigraph (West, 1996) since there may be more than one edge between a pair of vertices in the graph. To simplify data storage and the relevant analysis and optimization procedures, this multigraph is reduced to a simple graph (West, 1996) model by imposing the following restrictions:[5]

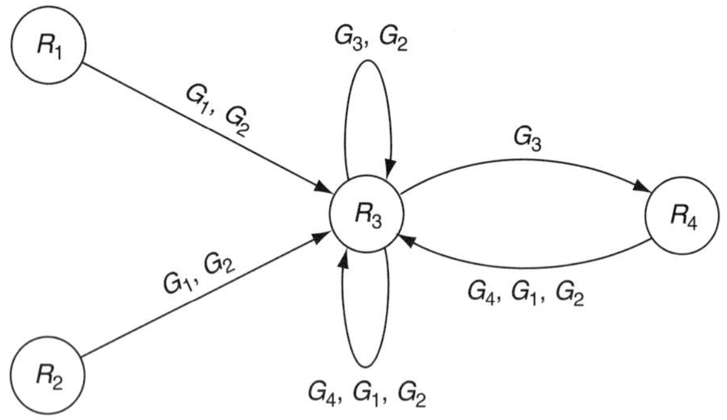

FIGURE 4.3 A Directed Multigraph Representation of the Synchronous System Shown in Figure 4.1. The graph vertices correspond to the registers R_1, R_2, R_3, and R_4, respectively.

[4] The arcs are labeled in the order in which the traveling signals pass through the gates.

[5] The restrictions refer to the model itself and *not* to the ability of the model to represent features of the circuits.

- Either one or zero edges can exist between any two different vertices of the graph;
- There cannot be self-loops or edges that start and end at the same vertex of the graph;
- Additional labelings (or markings) of the edges are introduced to represent the timing constraints of the circuit.

With the above restrictions, a formal definition of the circuit graph model is as follows:

Definition 4.3 Circuit graph: *A fully synchronous digital circuit C is represented as the connected undirected simple graph G_C. The graph G_C is the ordered sixtuple $G_C = \langle V^{(C)}, E^{(C)}, \mathbf{A}^{(C)}, h_l^{(C)}, h_u^{(C)}, h_d^{(C)} \rangle$, where:*

- $V^{(C)} = \{v_1, \ldots v_r\}$ *is the set of vertices of the graph* G_C;
- $E^{(C)} = \{e_1, \ldots e_p\}$ *is the set of edges of the graph* G_C;
- $\mathbf{A}^{(C)} = [a_{ij}^{(C)}]_{r \times r}$ *is the symmetric adjacency matrix of* G_C.

Each vertex from $V^{(C)}$ represents a register of the circuit C. There is exactly one edge in $E^{(C)}$ for every sequentially adjacent pair of registers in C. The mappings $h_l^{(C)}: E^{(C)} \mapsto \mathbb{R}$ and $h_u^{(C)}: E^{(C)} \mapsto \mathbb{R}$ to the set of real numbers $\mathbb{R}$ assign the lower and upper permissible range bounds, $l_k, u_k \in \mathbb{R}$, respectively, for the sequentially adjacent pair of registers indicated by the edge $e_k \in E$. The edge labeling $h_d^{(C)}$ defines a direction of signal propagation for each edge v_x, e_z, v_y.

Note that in a fully synchronous digital circuit, there are no purely combinational signal cycles; that is, it is impossible to reach the input of any logic gate G_k by only starting at the output of G_k and going through a sequence of combinational logic gates (Friedman, 1995; Kourtev, 1999).

Naturally, all registers from the circuit C are preserved when constructing the circuit graph G_C as described in definition 4.3—these registers are enumerated 1 through r, and a vertex v_i is created in the graph for each register R_i. Alternatively, an edge between two vertices is added in the graph if there is one or more local data paths between these two vertices. The self-loops are discarded because the clock skew of these local data paths is always zero and cannot be manipulated in any way.

The graph G_C for any circuit C can be determined by either direct inspection of C, or by first building the circuit multigraph and then modifying the multigraph to satisfy definition 4.3. Consider, for example, the circuit multigraph shown in Figure 4.3—the corresponding circuit graph is illustrated in Figure 4.4. Observe the labelings of the graph edges in Figure 4.4. Each edge is labeled with the corresponding permissible range of the clock skew for the given pair of registers. An arrow is drawn next to each edge to indicate the order of the registers in this specific sequentially adjacent pair—recall that the clock skew as defined in definition 4.2 is an algebraic difference. As shown in the rest of this section, either direction of an edge can be selected as long as the proper choices of lower and upper clock skew bounds are made.

In most practical cases, a unique signal path (a local data path) exists between a given sequentially adjacent pair of regis-

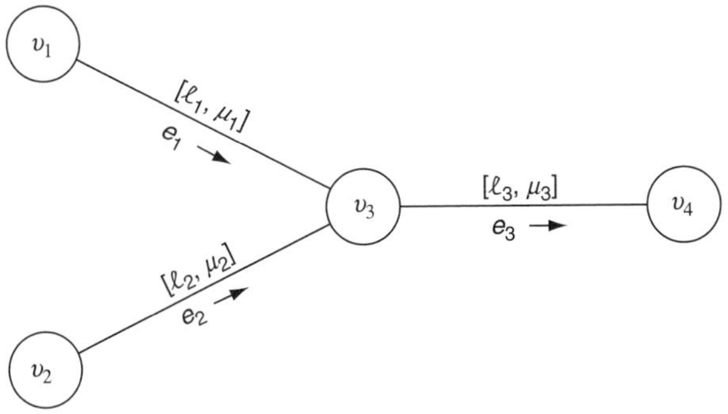

FIGURE 4.4 A Graph Representation of the Synchronous System Shown in Figure 4.1 According to Definition 4.3. The graph vertices v_1, v_2, v_3, and v_4 correspond to the registers R_1, R_2, R_3, and R_4, respectively.

ters $\langle R_i, R_j \rangle$. In these cases, the labeling of the corresponding edge is straightforward. The permissible range bounds l_k and u_k are computed as shown in (Friedman, 1995; Neves and Friedman, 1996a, 1996b; Afghani and Svensson, 1990; Kourtev, 1999), and the direction of the arrow is chosen so as to coincide with the direction of the signal propagation from R_i to R_j. With these choices, the clock skew is computed as $s = t^i_{cd} - t^j_{cd}$. In Figure 4.4 for example, the direction labelings of both e_1 and e_2 can be chosen from v_1 to v_3 and from v_2 to v_3, respectively.

Multiple signal paths between a pair of registers, R_x and R_y, require a more complicated treatment. As specified before, there can be only one edge between the vertices v_x and v_y in the circuit graph. Therefore, a methodology is presented for choosing the correct permissible range bounds and direction labeling for this single edge. This methodology is illustrated in Figure 4.5 and is a two-step process. First, multiple signal paths in the same direction from the register R_x to the register R_y are replaced by a single edge in the circuit graph according to the transformation illustrated in Figure 4.5(A). Next, two-edge cycles between R_x and R_y are replaced by a single edge in the circuit graph according to the transformation illustrated in Figure 4.5(B).

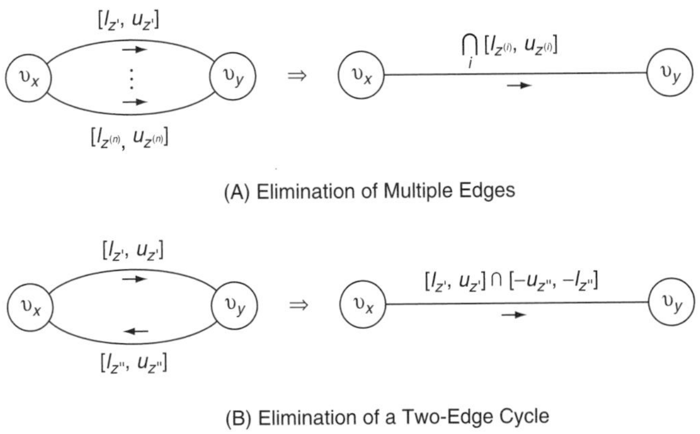

FIGURE 4.5 Transformation Rules for the Circuit Graph

In the former case, Figure 4.5(A), the edge direction labeling is preserved while the permissible range for the new single edge is chosen such that the permissible ranges of the multiple paths from R_x to R_y are simultaneously satisfied. As shown in Figure 4.5(A), the new permissible range $[l_z, u_z]$ is the intersection of the multiple permissible ranges $[l_{z'}, u_{z'}]$ through $[l_{z^{(n)}}, u_{z^{(n)}}]$ between R_x and R_y. In other words, the new lower bound is $l_z = \max_i \{l_{z^{(i)}}\}$ and the new upper bound is $u_z = \min_i \{u_{z^{(i)}}\}$.

In the latter case, Figure 4.5(B), an arbitrary choice for the edge direction can be made—the convention adopted here is to choose the direction toward the vertex with the higher index. For the vertex v_y, the new permissible range has a lower bound $l_z = \min(l_{z'}, -u_{z''})$ and an upper bound $u_z = \max(u_{z'}, -l_{z''})$. It is straightforward to verify that any clock skew $s \in [l_z, u_z]$ satisfies both permissible ranges $[l_{z'}, u_{z'}]$ and $[l_{z''}, u_{z''}]$ as shown in Figure 4.5(B). The process for computing the permissible ranges of a circuit graph (Friedman, 1995; Neves and Friedman, 1996a, 1996b; Afghani and Svensson, 1990; Kourtev, 1999), and the transformations illustrated in Figure 4.5 have linear complexity in the number of signal paths because each signal path is examined only once.

Note that the terms, *circuit* and *graph*, are used throughout the rest of this chapter interchangeably to denote the same fully synchronous digital circuit. In addition, note that for brevity, the superscript C, when referring to the circuit graph G_C of a circuit C, is omitted unless a circuit is explicitly indicated. The terms, *register* and *vertex*, are also used interchangeably as are *edge, local data path, arc,* and *a sequentially adjacent pair of registers.* On a final note, it is assumed that the graph of any circuit considered here is connected. If this is not the case, each of the disjoint connected portions of the graph (circuit) can be individually analyzed.

4.2.3 Clock Scheduling

The process of nonzero clock skew scheduling is discussed in this section. The following substitutions are introduced for notational convenience:

Definition 4.4 *Let C be a fully synchronous digital circuit, and let R_i and R_f be a sequentially adjacent pair of registers (i.e., $R_i \rightsquigarrow R_f$). The long path delay $\hat{D}^{i,f}_{PM}$ of a local data path $R_i \rightsquigarrow R_f$ is defined as:*

$$\hat{D}^{i,f}_{PM} = \begin{cases} (D^{Fi}_{CQM} + D^{i,f}_{PM} + \delta^{F,f}_S + 2\Delta_L), & \text{if } R_i, R_f \text{ are flip-flops.} \\ (D^{Li}_{CQM} + D^{i,f}_{PM} + \delta^{Lf}_S + \Delta_L + \Delta_T), & \text{if } R_i, R_f \text{ are latches.} \end{cases} \quad (4.3)$$

The D^{Fi}_{CQM} and D^{Li}_{CQM} are the maximum clock-to-output delay propagation times of the flip-flop and latch i, respectively; $D^{i,f}_{PM}$ is the maximum logic propagation delay from R_i to R_f, δ^{Ff}_S and δ^{Lf}_S are the setup times of the flip-flop and latch f, respectively; and Δ_L and Δ_T are the maximum deviations of the leading and trailing edges of the clock signal, respectively. Similarly, the short delay $\check{D}^{i,f}_{Pm}$ of a local data path *$R_i \rightsquigarrow R_f$ is defined as:*

$$\hat{D}_{Pm}^{i,f} = \begin{cases} (D_{Pm}^{i,f} + D_{CQ}^{Fi} - \delta_H^{Ff} - 2\Delta_L), \text{ if } R_i \text{ and } R_f \text{ are flip-flops.} \\ (+D_{CQm}^{Li} + D_{Pm}^{i,f} - \delta_H^{Lf} - \Delta_L - \Delta_T), \text{ if } R_i \text{ and } R_f \text{ are latches.} \end{cases} \quad (4.4)$$

The D_{CQ}^{Fi} and D_{CQ}^{Li} are the minimum clock-to-output delay propagation times of the flip-flop and latch i, respectively; $D_{Pm}^{i,f}$ is the minimum logic propagation delay from R_i to R_f; and δ_H^{Ff} and δ_H^{Lf} are the hold times of the flip-flop and latch f, respectively.

For example, using the notations described in definition 4.4, the timing constraints of a local data path $R_i \rightsquigarrow R_f$ with flip-flops are (Friedman, 1995; Kourtev, 1999) as follows:

$$T_{Skew}(i, f) \leq T_{CP} - \hat{D}_{PM}^{i,f}. \quad (4.5)$$

$$-\hat{D}_{Pm}^{i,f} \leq T_{Skew}(i, f). \quad (4.6)$$

For a local data path $R_i \rightsquigarrow R_f$ consisting of the flip-flops R_i and R_f, the setup and hold-time violations are avoided if equations 4.5 and 4.6, respectively, are satisfied.

The clock skew $T_{Skew}(i, f)$ of a local data path $R_i \rightsquigarrow R_f$ can be either positive or negative. Note that negative clock skew may be used to effectively speed up a local data path $R_i \rightsquigarrow R_f$ by allowing an additional $T_{Skew}(i, f)$ amount of time for the signal to propagate from the register R_i to the register R_f. However, excessive negative skew may create a hold-time violation, thereby creating a lower bound on $T_{Skew}(i, f)$ as described by equation 4.6 and illustrated by l in Figure 4.2. A hold-time violation is a clock hazard or a race condition, also known as **double clocking** (Friedman, 1995; Fishburn, 1990). Similarly, positive clock skew effectively decreases the clock period T_{CP} by $T_{Skew}(i, f)$, thereby limiting the maximum clock frequency and imposing an upper bound on the clock skew as illustrated by u in Figure 4.2.[6] In this case, a clocking hazard known as **zero clocking** may be created (Friedman, 1995; Fishburn, 1990).

Examination of the constraints, 4.5 and 4.6, reveals a procedure for preventing clock hazards. Assuming equation 4.5 is not satisfied, a suitably large value of T_{CP} can be chosen to satisfy constraint 4.5 and prevent zero clocking. Also note that unlike equation 4.5, equation 4.6 is independent of the clock period T_{CP} (or the clock frequency). Therefore, T_{CP} cannot be changed to correct a double clocking hazard. Rather, a redesign of the entire clock distribution network may be required (Neves and Friedman, 1996a).

Both double and zero clocking hazards can be eliminated if two simple choices characterizing a fully synchronous digital circuit are made. Specifically, if equal values are chosen for all clock delays, then the clock skew $T_{Skew}(i, f) = 0$ for each local data path $R_i \rightsquigarrow R_f$ becomes zero:

$$\forall \langle R_i, R_f \rangle : t_{cd}^i = t_{cd}^f \Rightarrow T_{Skew}(i, f) = 0. \quad (4.7)$$

Therefore, equations 4.5 and 4.6 become:

$$T_{Skew}(i, f) = t_{cd}^i - t_{cd}^f = 0 \leq T_{CP} - \hat{D}_{PM}^{i,f} \quad (4.8)$$

$$-\hat{D}_{Pm}^{i,f} \leq 0 = T_{Skew}(i, f) = t_{cd}^i - t_{cd}^f. \quad (4.9)$$

Note that equation 4.8 can be satisfied for each local data path $R_i \rightsquigarrow R_f$ in a circuit if a sufficiently large value—larger than the greatest value $\hat{D}_{PM}^{i,f}$ in a circuit—is chosen for T_{CP}. Furthermore, equation 4.9 can be satisfied across an entire circuit if it can be ensured that $\hat{D}_{Pm}^{i,f} \geq 0$ for each local data path $R_i \rightsquigarrow R_f$ in the circuit. The timing constraints of these equations can be satisfied because choosing a sufficiently large clock period T_{CP} is always possible and $\hat{D}_{Pm}^{i,f}$ is positive for a properly designed local data path $R_i \rightsquigarrow R_f$. The application of this zero clock skew methodology of equations 4.7 through 4.9 has been central to the design of fully synchronous digital circuits for decades (Friedman, 1995, 1997; Lee and Kong, 1997). By requiring the clock signal to arrive at each register R_j with approximately the same delay t_{cd}^j, these design methods have become known as **zero clock skew methods.**[7]

As shown by previous research (Friedman, 1995, 1992; Neves and Friedman, 1993, 1996a, 1996b; Xi and Dai, 1996; Kourtev and Friedman, 1999b), both double and zero clocking hazards may be removed from a synchronous digital circuit even when the clock skew is not zero [$T_{Skew}(i, f) \neq 0$ for some or all local data paths $R_i \rightsquigarrow R_f$]. As long as equations 4.5 and 4.6 are satisfied, a synchronous digital system can operate reliably with nonzero clock skews, permitting the system to operate at higher clock frequencies while removing all race conditions.

The vector column of clock delays $T_{CD} = [t_{cd}^1, t_{cd}^2, \ldots]^T$ is called a **clock schedule** (Friedman, 1995; Fishburn, 1990). If T_{CD} is chosen such that equations 4.5 and 4.6 are satisfied for every local data path $R_i \rightsquigarrow R_f$, T_{CD} is called a **consistent clock schedule**. A clock schedule that satisfies equation 4.7 is called a **trivial clock schedule**. Note that a trivial clock schedule T_{CD} implies global zero clock skew since for any i and f, $t_{cd}^i = t_{cd}^f$; thus, $T_{Skew}(i, f) = 0$.

An intuitive example of nonzero clock skew being used to improve the performance and reliability of a fully synchronous digital circuit is shown in Figure 4.6. Two pairs of sequentially adjacent flip-flops, $R_1 \rightsquigarrow R_2$ and $R_2 \rightsquigarrow R_3$, are shown in Figure 4.6, where both zero skew and nonzero skew situations are illustrated in Figures 4.6(A) and 4.6(B), respectively. Note that the local data paths made up of the registers R_1 and R_2 and of R_2 and R_3, respectively, are connected in series (R_2 being common to both $R_1 \rightsquigarrow R_2$ and $R_2 \rightsquigarrow R_3$). In each of the Fig-

[6] Positive clock skew may also be thought of as increasing the path delay. In either case, positive clock skew ($T_{Skew} > 0$) increases the difficulty of satisfying equation 4.5.

[7] Equivalently, it is required that the clock signal arrive at each register at approximately the same time.

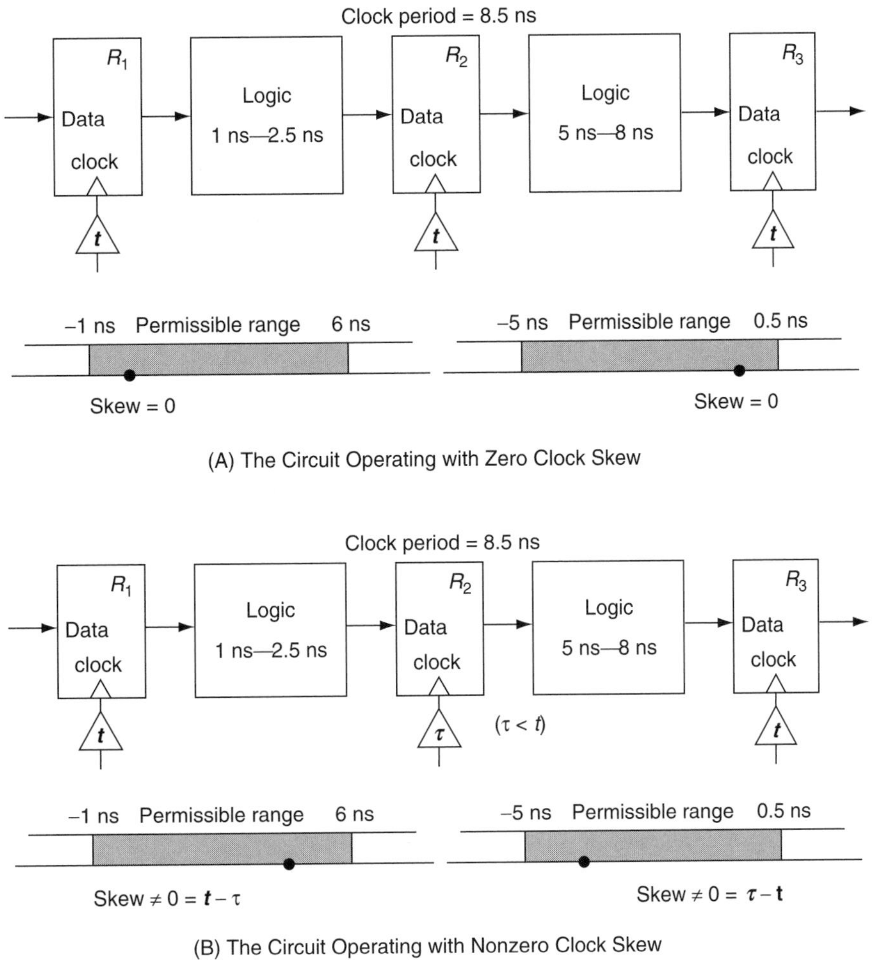

FIGURE 4.6 Application of Nonzero Clock Skew to Improve Circuit Performance (a Lower Clock Period) or Circuit Reliability (Increased Safety Margins in the Permissible Range)

ures 4.6(A) and 4.6(B), the permissible ranges of the clock skew for both local data paths, $R_1 \rightsquigarrow R_2$ and $R_2 \rightsquigarrow R_3$, are lightly shaded under each circuit diagram. As shown in Figure 4.6, the target clock period for this circuit is $T_{CP} = 8.5$ ns.

The zero clock skew points (skew = 0) are indicated in Figure 4.6(A)—zero skew is achieved by delivering the clock signal to each of the registers R_1, R_2 and R_3 with the same delay t (symbolically illustrated by the buffers connected to the clock terminals of the registers). Observe that while the zero clock skew points fall within the respective permissible ranges, these zero clock skew points are dangerously close to the lower and upper bounds of the permissible range for $R_1 \rightsquigarrow R_2$ and $R_2 \rightsquigarrow R_3$, respectively. A situation could be foreseen in which, for example, the local data path $R_2 \rightsquigarrow R_3$ has a longer than expected delay (longer than 8 ns), thereby causing the upper bound of the permissible range for $R_2 \rightsquigarrow R_3$ to decrease below the zero clock skew point. In this scenario, a setup violation will occur on the local data path $R_2 \rightsquigarrow R_3$.

Consider next the same circuit with nonzero clock skew applied to the data paths $R_1 \rightsquigarrow R_2$ and $R_2 \rightsquigarrow R_3$, as shown in Figure 4.6(B). Nonzero skew is achieved by delivering the clock signal to the register R_2 with a delay $\tau < t$, where t is the delay of the clock signal to both R_1 and R_3. By applying this delay of $\tau < t$, positive $(t - \tau > 0)$ and negative $(\tau - t < 0)$ clock skews are applied to $R_1 \rightsquigarrow R_2$ and $R_2 \rightsquigarrow R_3$, respectively. The corresponding clock skew points are illustrated in the respective permissible ranges in Figure 4.6(B). Comparing Figure 4.6(A) to Figure 4.6(B), observe that a timing violation is less likely to occur in the latter case. For the previously described setup timing violation to occur in Figure 4.6(B), the deviations in the delay parameters of $R_2 \rightsquigarrow R_3$ would have to be much greater in the nonzero clock skew case than in the zero clock skew case. If the precise target value of the nonzero clock skew $\tau - t < 0$ is not met during the circuit design process, the safety margin from the skew point to the upper bound of the permissible range would be much greater.

Therefore, there are two identifiable benefits of applying a nonzero clock skew. First, the safety margins of the clock skew (that is the distances between the clock skew point and the bounds of the permissible range) in the permissible ranges of a data path can be improved. The likelihood of correct circuit operation in the presence of process parameter variations and operational conditions is improved with these increased margins. In other words, the circuit reliability is improved. Second, without changing the logical and circuit structure, the performance of the circuit can be increased by permitting a higher maximum clock frequency (or lower minimum clock period).

Friedman (1989) first presented the concept of negative nonzero clock skew as a technique to increase the clock frequency and circuit performance across sequentially adjacent pairs of registers. Soon afterward, Fishburn (1990) first suggested an algorithm for computing a consistent clock schedule that is nontrivial. It is shown in both Friedman's and Fishburn's studies that by exploiting negative and positive clock skew in a local data path $R_i \rightsquigarrow R_f$, a circuit can operate with a clock period T_{CP} smaller than the clock period achievable by a zero clock skew schedule while satisfying the conditions specified by equations 4.5 and 4.6. In fact, Fishburn (1990) determined an optimal clock schedule by applying linear programming techniques to solve for T_{CD} to satisfy these equations while minimizing the objective function $F_{\text{objective}} = T_{CP}$.

The process of determining a consistent clock schedule T_{CD} can be considered as the mathematical problem of minimizing the clock period T_{CP} under the constraints of equations 4.5 and 4.6. However, there are important practical issues to consider before a clock schedule can be properly implemented. A clock distribution network must be synthesized such that the clock signal is delivered to each register with the proper delay to satisfy the clock skew schedule T_{CD}. Furthermore, this clock distribution network must be constructed to minimize the deleterious effects of interconnect impedances and process parameter variations on the implemented clock schedule. Synthesizing the clock distribution network typically consists of determining a topology for the network, together with the circuit design and physical layout of the buffers and interconnect that make up a clock distribution network (Friedman, 1995, 1997).

4.3 Clock Scheduling for Improved Reliability

The problem of determining an optimal clock skew schedule for a fully synchronous VLSI system is considered in this section from the perspective of improving system reliability. An original formulation of the clock skew scheduling problem by the authors is introduced as a constrained quadratic programming (QP) problem (Kourtev and Friedman, 1999a, 1999b).

The operation of a fully synchronous digital system was discussed in Section 4.1. Briefly, for such systems to function properly, a strict temporal ordering of the many thousands of switching events in the circuit is required. This strict ordering is enforced by a global synchronizing clock signal delivered to every register in a circuit by a clock distribution network. Algorithms for determining a nonzero clock skew schedule that satisfy the tighter timing constraints of high-speed VLSI complexity systems can be found in (Friedman, 1995; Neves and Friedman 1996a, 1996b; Afghani and Svensson, 1990; Kourtev, 1999; Fishburn, 1990).

In this section, a different class of clock skew scheduling algorithms is introduced. In these algorithms, the primary objective is to improve circuit reliability by maximizing the tolerance to process parameter variations. Improvements are achieved by first choosing an objective clock skew value for each local data path. A consistent clock schedule is found by applying the proposed optimization algorithm. Unlike a performance optimization approach, the algorithm presented in this section minimizes the *least square* error between the computed and objective clock skew schedules.[8] A secondary objective of the clock skew scheduling algorithm developed in this chapter is to increase the system-wide clock frequency.

This section begins with an alternative formulation of the clock skew scheduling problem as a quadratic programming problem discussed in detail in Section 4.3.1. The mathematical procedures used to determine the clock skew schedule are developed and analyzed in Section 4.4.

4.3.1 Problem Formulation

Existing algorithms for clock skew scheduling are reviewed in this section.[9] The classical linear programming approach for minimizing only the clock period T_{CP} of a circuit is first described in the subsection entitled Clock Scheduling for Maximum Performance. A new quantitative measure to compare different clock schedules is introduced in Further Improvement. This section is concluded by sketching the clock skew scheduling problem as an efficiently solvable problem in Clock Scheduling as a Quadratic Programming Problem.

Recall the short delay $\hat{D}_{Pm}^{i,j}$ and long delay $\hat{D}_{PM}^{i,j}$ of a local data path $R_i \rightsquigarrow R_j$ introduced in definition 4.4. Using the substitutions, equations 4.3 and 4.4, the timing constraints of a local data path $R_i \rightsquigarrow R_f$ are rewritten in equations 4.5 and 4.6. A pair of constraints such as in these latter equations must be satisfied for each local data path in a circuit for this circuit to operate correctly. Furthermore, the local data path timing constraints lead to the concept of a permissible range introduced in

[8] In a performance optimization strategy, the starting point is the set of timing constraints and the objective of the clock.

[9] Scheduling algorithms is to determine a feasible clock schedule and clock distribution network given these constraints (Friedman, 1995; Neves and Friedman 1996a, 1996b; Afghani and Svensson, 1990; Kourtev, 1999; Fishburn, 1990).

Section 4.2.1 and illustrated in Figure 4.2. Formally, the lower and upper bounds of the permissible range of a local data path $R_i \rightsquigarrow R_j$ are the following:

$$l_{i,j} = -\hat{D}_{Pm}^{i,j}. \quad (4.10)$$

$$u_{i,j} = T_{CP} - \hat{D}_{PM}^{i,j}. \quad (4.11)$$

Also defined here for notational convenience are the width $w_{i,j}$ and middle $m_{i,j}$ of the permissible range. Specifically, these are the equations:

$$w_{i,j} = u_{i,j} - l_{i,j} = T_{CP} - \left(\hat{D}_{PM}^{i,j} - \hat{D}_{Pm}^{i,j}\right). \quad (4.12)$$

$$m_{i,j} = \frac{1}{2}(l_{i,j} + u_{i,j}) = \frac{1}{2}\left(T_{CP} - \hat{D}_{PM}^{i,j} - \hat{D}_{Pm}^{i,j}\right). \quad (4.13)$$

Recall from Section 4.2.3 that it is frequently possible to make two simple choices (concerning equation 4.7) characterizing the clock skews and delays in a circuit, such that both zero and double clocking violations are avoided. Specifically, if equal values are chosen for all clock delays and a sufficiently large value—larger than the longest delay $\hat{D}_{PM}^{i,f}$—is chosen for T_{CP}, neither of these two clocking hazards will occur. Formally,

$$\forall \langle R_i, R_f \rangle : t_{cd}^i = t_{cd}^f = \text{Const} \quad (4.14)$$

and

$$R_i \rightsquigarrow R_f \Rightarrow T_{CP} > \hat{D}_{PM}^{i,f}, \quad (4.15)$$

and with equations 4.14 and 4.15, the timing constraints 4.5 and 4.6 for a hazard-free local data path $R_i \rightsquigarrow R_f$ become:

$$\hat{D}_{PM}^{i,f} < T_{CP}. \quad (4.16)$$

$$\hat{D}_{Pm}^{i,f} > 0. \quad (4.17)$$

Next, recall that each clock skew $T_{Skew}(i, f)$ is the difference of the delays of the clock signals t_{cd}^i and t_{cd}^f. These delays are the tangible physical quantities that are implemented by the clock distribution network. The set of all clock delays in a circuit can be denoted as the vector column:

$$\mathbf{t_{cd}} = \begin{bmatrix} t_{cd}^1 \\ t_{cd}^2 \\ \vdots \end{bmatrix},$$

which is called a clock skew schedule or simply a clock schedule (Friedman, 1995; Fishburn, 1990; Kourtev and Friedman, 1997b). If $\mathbf{t_{cd}}$ is chosen such that equations 4.5 and 4.6 are satisfied for every local data path $R_i \rightsquigarrow R_j$, $\mathbf{t_{cd}}$ is called a *feasible* clock schedule. A clock schedule that satisfies equation 4.7 (respectively, equations 4.14 and 4.15) is called a trivial clock schedule. Again, a trivial $\mathbf{t_{cd}}$ implies global zero clock skew since for any i and f, $t_{cd}^i = t_{cd}^f$; thus, $T_{Skew}(i, f) = 0$. Also, observe that if $[t_{cd}^1 \; t_{cd}^2 \; \ldots]^t$ is a feasible clock schedule (trivial or not), $[c + t_{cd}^1 \; c + t_{cd}^2 \; \ldots]^t$ is also a feasible clock schedule where $c \in \mathbb{R}^1$ is any real constant.

An alternative way to refer to a clock skew schedule is to specify the vector of all clock skews in a circuit corresponding to a set of clock delays $\mathbf{t_{cd}}$ as specified above. Denoted by $\mathbf{s}$, the vector column of clock skews is $\mathbf{s} = [s_1 \; s_2 \; \ldots]^t$ where the skews $s_1, s_2, \ldots s_n$ of all local data paths in the circuit are enumerated. Typically, the dimension of $\mathbf{s}$ is different from the dimension of $\mathbf{t_{cd}}$ for the same circuit. If a circuit consists of r registers and p local data paths, for example, then $\mathbf{s} = [s_1 \ldots s_p]^t$ and $\mathbf{t_{cd}} = [t_{cd}^1 \ldots t_{cd}^r]^t$ for this circuit. Therefore, the clock skew schedule refers to either $\mathbf{t_{cd}}$ or $\mathbf{s}$, where the precise reference is usually apparent from the context.

Note that $\mathbf{t_{cd}}$ must be known to determine each clock skew within $\mathbf{s}$. The inverse situation, however, is *not* true; that is, the set of all clock skews in a circuit need not be known to determine the corresponding clock schedule $\mathbf{t_{cd}}$. As is shown in Sections 4.3.1 and 4.4, a small subset of clock skews (compared to the total number of local data paths or clock skews) uniquely determines all the skews in a circuit as well as the different feasible clock schedules $\mathbf{t_{cd}}$. Finally, note that a given feasible clock schedule $\mathbf{s}$ allows for many possible implementations $\mathbf{t_{cd}} = [c + t_{cd}^1 \; c + t_{cd}^2 \; \ldots]^t$ where any specific constant c implies a different $\mathbf{t_{cd}}$ but the same $\mathbf{s}$. Thus, the term clock schedule is used to refer to $\mathbf{t_{cd}}$ where the choice of the real constant $c \in \mathbb{R}^1$ is arbitrary.

Clock Scheduling for Maximum Performance

The linear programming (LP) problem of computing a feasible clock skew schedule while minimizing the clock period T_{CP} of a circuit can be found in (Friedman, 1995; Neves and Friedman 1996a, 1996b; Afghani and Svensson, 1990; Kourtev, 1999; Fishburn, 1990). With T_{CP} as the value of the objective function being minimized, this problem is formally defined as problem LCSS:

Problem LCSS__________ (LP Clock skew scheduling)

$$\min T_{CP}$$

$$\text{subject to: } t_{cd}^i - t_{cd}^j \leq T_{CP} - \hat{D}_{PM}^{i,j} \quad (4.18)$$

$$t_{cd}^i - t_{cd}^j \geq -\hat{D}_{Pm}^{i,j}.$$

To develop additional insight into problem LCSS, consider a circuit C_1 consisting of four registers, R_1, R_2, R_3, and R_4, and five local data paths, $R_1 \rightsquigarrow R_2$, $R_1 \rightsquigarrow R_3$, $R_3 \rightsquigarrow R_2$, $R_3 \rightsquigarrow R_4$, and $R_4 \rightsquigarrow R_2$. Let the long and short delays for this circuit be[10] $\hat{D}_{Pm}^{1,2} = 1$, $\hat{D}_{PM}^{1,2} = 3$, $\hat{D}_{Pm}^{1,3} = 2$, $\hat{D}_{PM}^{1,3} = 4$, $\hat{D}_{Pm}^{3,2} = 5$, $\hat{D}_{PM}^{3,2} = 7$,

[10] The times used in this section are all assumed to be in the same time unit. The actual time unit—e.g., picoseconds, nanoseconds, microseconds, milliseconds, seconds—is irrelevant and is therefore omitted.

TABLE 4.1 Clock Schedule t^1_{cd}: Clock Skews and Permissible Ranges for the Example Circuit C_1 (for the Minimum Clock Period $T_{CP} = 5$)

Local data path	Permissible range	Clock skew
$R_1 \rightsquigarrow R_3$	$[-2, 1]$	$t^1_{cd} - t^3_{cd} = 1 - 0 = 1$
$R_3 \rightsquigarrow R_4$	$[-2.5, 0]$	$t^3_{cd} - t^4_{cd} = 0 - 2.5 = -2.5$
$R_1 \rightsquigarrow R_2$	$[-1, 2]$	$t^1_{cd} - t^2_{cd} = 1 - 2 = -1$
$R_3 \rightsquigarrow R_2$	$[-5, -2]$	$t^3_{cd} - t^2_{cd} = 0 - 2 = -2$
$R_4 \rightsquigarrow R_2$	$[-2, 1]$	$t^4_{cd} - t^2_{cd} = 2.5 - 2 = 0.5$

$\hat{D}^{3,4}_{Pm} = 2.5$, $\hat{D}^{3,4}_{PM} = 5$, $\hat{D}^{4,2}_{Pm} = 2$, and $\hat{D}^{4,2}_{PM} = 4$. Solving problem LCSS (4.18) yields a feasible clock schedule t^1_{cd} for the minimum achievable clock period $T_{CP} = 5$:

$$\min T_{CP} = 5 \rightarrow t^1_{cd} = \begin{bmatrix} t^1_{cd} \\ t^2_{cd} \\ t^3_{cd} \\ t^4_{cd} \end{bmatrix} = \begin{bmatrix} 1 \\ 2 \\ 0 \\ 2.5 \end{bmatrix}.$$

These results are summarized in Table 4.1 along with the actual permissible range for each local data path for the minimum value of the clock period $T_{CP} = 5$ (recall that the permissible range depends on the value of the clock period T_{CP}).

Note that most of the clock skews (specifically, the first four) listed in Table 1 are at one end of the corresponding permissible range. This positioning is due to the inherent feature of linear programming that seeks the objective function extremum at a vertex of the solution space. In practice, however, this situation can be dangerous since correct circuit operation is strongly dependent on the accurate implementation of a large number of clock delays—effectively, the clock skews—across the circuit. It is quite possible that the actual values of some of these clock delays may fluctuate from the target values—due to manufacturing tolerances as well as variations in temperature and supply voltage—thereby causing a catastrophic timing failure of the circuit. Observe that while zero clocking failures can be corrected by operating the circuit at a slower speed (higher clock period T_{CP}), double clocking violations are race conditions that are catastrophic and render the circuit nonfunctional.

Maximizing Safety

Frequently in practice, a target clock period T_{CP} is established for a specific circuit implementation. Making the target clock period smaller may not be a primary design objective. If this is the case, alternative optimization strategies may be sought such that the resulting circuit is more tolerant to inaccuracies in the timing parameters. Two different classes of timing parameters are considered—the local data path delays and the clock delays (respectively, the clock skews). Note first that the clock skew scheduling process depends on accurate knowledge of the short and long path delays ($\hat{D}^{i,j}_{Pm}$ and $\hat{D}^{i,j}_{PM}$) for every local data path $R_i \rightsquigarrow R_j$. Second, provided the path delay information is predictable, correct circuit operation is contingent upon the accurate implementation of the computed clock schedule t_{cd}. Both of these factors must be considered if reliable circuit operation under various operating conditions is to be attained.

One way to achieve the specified goal of higher circuit reliability is to artificially shrink the permissible range of each local data path by an equal amount from either side of the interval and determine a feasible clock skew schedule based on these new timing constraints. This idea has been addressed by Fishburn (1990) as the problem of maximizing the minimum *slack* (over all inequalities of equations 4.5 and 4.6) or the amount by which an inequality exceeds the limit. Formally, the problem can be expressed as the linear programming problem LCSS–SAFE:

Problem LCSS–SAFE __________ (LP Clock skew scheduling for safety)

$$\begin{aligned} &\max M \\ &\text{subject to: } t^i_{cd} - t^j_{cd} + M \le T_{CP} - \hat{D}^{i,j}_{PM} \\ &\qquad t^i_{cd} - t^j_{cd} - M \ge -\hat{D}^{i,j}_{Pm} \\ &\qquad M \ge 0 \end{aligned} \tag{4.19}$$

To gain additional insight into problem LCSS–SAFE, consider again the circuit example used in Table 4.1. Two solutions of problem LCSS–SAFE are listed in Table 4.2 for two different values of the clock period, $T_{CP} = 6.5$ and $T_{CP} = 6$, respectively. These results are shown—denoted by t^2_{cd} and t^3_{cd}, respectively—in columns 2 through 5 and 6 through 9 for $T_{CP} = 6.5$ (clock schedule t^2_{cd}) and $T_{CP} = 6$ (clock schedule t^3_{cd}), respectively. For the specific value of T_{CP}, the permissible range is listed in columns 2 and 6, respectively, and the clock skew solution is listed in columns 3 and 7, respectively.

TABLE 4.2 Solution of Problem LCSS-SAFE for the Example Circuit C_1 for Clock Periods $T_{CP} = 6.5$ and $T_{CP} = 6$, respectively

	$t^2_{cd} \rightarrow T_{CP} = 6.5$, $M = 1$; $t^2_{cd} = [\frac{3}{2}\ \frac{3}{2}\ 0\ \frac{1}{2}]^t$				$t^3_{cd} \rightarrow T_{CP} = 6$, $M = 2/3$; $t^3_{cd} = [\frac{4}{3}\ \frac{5}{3}\ 0\ \frac{1}{3}]^t$			
1	2	3	4	5	6	7	8	9
$R_1 \rightsquigarrow R_3$	$[-2, 2.5]$	1.5	0.25	1.25	$[-2, 2]$	4/3	0	4/3
$R_3 \rightsquigarrow R_4$	$[-2.5, 1.5]$	−0.5	−0.5	0	$[-2.5, 1]$	−1/3	−3/4	5/12
$R_1 \rightsquigarrow R_2$	$[-1, 3.5]$	0	1.25	1.25	$[-1, 3]$	−1/3	1	4/3
$R_3 \rightsquigarrow R_2$	$[-5, -0.5]$	−1.5	−2.75	1.25	$[-5, -1]$	−5/3	−3	4/3
$R_4 \rightsquigarrow R_2$	$[-2, 2.5]$	−1	0.25	1.25	$[-2, 2]$	0	−4/3	4/3

Notes: **1:** Local data path; **2, 6:** Permissible range; **3, 7:** Clock skew solution for this local data path; **4, 8:** Ideal clock skew value for this path (middle of permissible range); **5, 9:** Distance (absolute value) of the clock skew solution from the actual clock skew

Note that there are two additional columns of data for either value of T_{CP} in Table 4.2. First, an "ideal" objective value of the clock skew is specified for each local data path in columns 4 and 8, respectively. This objective value of the clock skew is chosen in this example to be the value corresponding to the middle $m_{i,j}$ (note equation 4.13) of the permissible range of a local data path $R_i \rightsquigarrow R_j$ in a circuit with a clock period T_{CP}. The middle point of the permissible range is equally distant from either end of the permissible range, thereby providing the maximum tolerance to process parameter variations. Second, the absolute value of the distance $|T_{Skew}(i, j) - m_{i,j}|$ between the ideal and actual values of the clock skew for a local data path is listed in columns 5 and 9, respectively. This distance is a measure of the difference between the ideal clock skew and the scheduled clock skew. Note that in the general case, it is virtually impossible to compute a clock schedule t_{cd} such that the clock skew T_{Skew} (i, j) for each local data path $R_i \rightsquigarrow R_j$ is exactly equal to the middle $m_{i,j}$ of the permissible range of this path. The reasons for this characteristic are due to structural limitations of the circuits as highlighted in Section 4.4.

Further Improvement

Problem LCSS–SAFE (see equation 4.19) provides a solution to the clock skew scheduling problem for a case in which circuit reliability is of primary importance and clock period minimization is not the focus of the optimization process. As shown in the previous subsection, a certain degree of safety may be achieved by computing a feasible clock schedule subject to artificially smaller permissible ranges (as defined in equation 4.19). Problem LCSS–SAFE, however, is a brute force approach since it requires that the same absolute margins of safety are observed for each permissible range regardless of the range width. Therefore, this approach does not consider the individual characteristics of a permissible range and does not differentiate among local data paths with wider and narrower permissible ranges.

It is possible to provide an alternative approach to clock skew scheduling that considers all permissible ranges and also provides a natural quantitative measure of the quality of a particular clock schedule. Consider, for instance, a circuit with a target clock period T_{CP}. Furthermore, denote an *objective* clock skew value for a local data path $R_i \rightsquigarrow R_j$ by $g_{i,j}$, where it is required that $l_{i,j} \leq g_{i,j} \leq u_{i,j}$ (recall the lower equation 4.10 and upper equation 4.11 bounds of the permissible range). For most practical circuits, it is unlikely that a feasible clock schedule can be computed that is exactly equal to the objective clock schedule for each local data path. Multiple linear dependencies among clock skews in each circuit exist—those linear dependencies define a solution space such that the clock schedule $s = \left[g_{i_1,j_1} \quad g_{i_2,j_2} \quad \cdots\right]^t$ most likely is *not* within this solution space (unless the circuit is constructed of only nonrecursive feed-forward paths). If t_{cd} is a feasible clock schedule, however, it is possible to evaluate how close a realizable clock schedule is to the objective clock schedule by computing the sum:

$$\varepsilon = \sum_{R_i \rightsquigarrow R_j} \left[T_{Skew}(i, j) - g_{i,j}\right]^2, \qquad (4.20)$$

over all local data paths in the circuit.

Note that ε, as defined in equation 4.20 is the total least squares error of the actual clock skew as compared to the objective clock skew. This error permits any two different clock skew schedules to be compared. Moreover, the clock skew scheduling problem can be considered as a problem of minimizing ε of a clock schedule t_{cd} given the clock period T_{CP} and an "ideal" clock schedule $\left[g_{i_1,j_1} \quad g_{i_2,j_2} \quad \cdots\right]^t$ subject to any specific circuit design criteria. The flexibility permitted by such a formulation is far greater since the ideal schedule $\left[g_{i_1,j_1} \quad g_{i_2,j_2} \quad \cdots\right]^t$ can be any clock schedule that satisfies a specific target circuit.

Consider, for instance, the solution of LCSS–SAFE listed in Table 4.2 for $T_{CP} = 6.5$ and $T_{CP} = 6$. Computing the total error (as defined by equation 4.20) for both solutions gives $\varepsilon_{6.5} = 6.25$ and $\varepsilon_6 = \frac{1049}{,144} = 7.2847$. Next, consider an alternative clock schedule t_{cd}^2 for $T_{CP} = 6.5$ as follows:

$$T_{CP} = 6.5 \rightarrow t_{cd}^{2'} = \begin{bmatrix} t_{cd}^1 \\ t_{cd}^2 \\ t_{cd}^3 \\ t_{cd}^4 \end{bmatrix} = \begin{bmatrix} 43/32 \\ 38/32 \\ 0 \\ 31/32 \end{bmatrix}. \qquad (4.21)$$

It can be verified that with $t_{cd}^{2'}$ as specified, $\varepsilon_{6.5}$ improves to $\frac{675}{128} = 5.2734$ from 6.25 for t_{cd}^2 (columns two equations 4.2 through 4.5 in Table 4.2). Similarly, an alternative clock schedule t_{cd}^3 for the clock period $T_{CP} = 6$ is

$$T_{CP} = 6.5 \rightarrow t_{cd}^{3'} = \begin{bmatrix} t_{cd}^1 \\ t_{cd}^2 \\ t_{cd}^3 \\ t_{cd}^4 \end{bmatrix} = \begin{bmatrix} 35/32 \\ 54/32 \\ 0 \\ 39/32 \end{bmatrix}. \qquad (4.22)$$

Again, using $t_{cd}^{3'}$ leads to an improvement of ε_6 to 6.1484 as compared to 7.2847 for the solution of LCSS–SAFE t_{cd}^3 (see Table 4.2, columns 6 through 9).

Clock Scheduling as a Quadratic Programming Problem

As discussed in the previous three subsections, a common design objective is ensuring reliable system operation under a target clock period. It is possible to redefine the problem of clock skew scheduling for this case. The input data for this redefined problem consists of:

- The clock period of the circuit T_{CP};
- The circuit connectivity and delay information (i.e., all local data paths $R_i \rightsquigarrow R_j$ and the short and long delays $\hat{D}_{Pm}^{i,j}$ and $\hat{D}_{PM}^{i,j}$, respectively); and
- An objective clock schedule $g = \begin{bmatrix} g_{i_1, j_1} \\ g_{i_2, j_2} \\ \vdots \end{bmatrix}$.

Given this information, the optimization goal is to compute a feasible clock schedule s^* (respectively t_{cd}^*) so as to minimize the least square error between the computed clock schedule $\mathbf{s}^*$ and the objective clock schedule g. Recall that the least square error ε_τ (described by equation 4.20) is defined as the sum of the squares of the distances (algebraic differences) between the actual and objective clock skews over all local data paths in the circuit. This problem is described in a formal framework in the following section. Also in the following section, the mathematical algorithm to solve this revised problem is explained in greater detail.

4.4 Derivation of the QP Algorithm

The formulation of clock skew scheduling as a quadratic programming problem is described in detail in this section. First, the graph model introduced in Section 4.2.2 is further analyzed in Section 4.4.1. The linear dependencies among the clock skews and the fundamental set of cycles are introduced and analyzed in Section 4.4.2. Finally, the quadratic programming problem is formulated and solved in Section 4.4.3.

4.4.1 The Circuit Graph

As discussed in Section 4.2.2, a synchronous circuit C can be represented as the simple undirected graph $g_C = \{V^{(C)}, E^{(C)}, A^{(C)}, h_l^{(C)}, h_u^{(C)}, h_d^{(C)}\}$, where $V_C = \{v_1, \ldots, v_r\}$ is the set of vertices of the graph, $E_C = \{e_1, \ldots, e_p\}$ is the set of edges of the graph, and the symmetric $r \times r$ matrix A_C—called the adjacency matrix—contains the graph connectivity (West, 1996). Vertices from g_C correspond to the registers of the circuit C, and the edges reflect the fact that pairs of registers are sequentially adjacent. Note the cardinalities $|V_C| = r$ and $|E_C| = p$—the circuit C has r registers and p local data paths. The adjacency matrix $A_C = [a_{ij}]_{r\times r}$ is a square matrix of order $r \times r$ where both the rows and columns of A correspond to the vertices of g_C. As previously mentioned, for notational convenience s_j denotes the clock skew corresponding to the edge $e_j \in E_C$. Specifically, if the vertices v_{i_1} and v_{i_2} correspond to the sequentially adjacent pair of registers $R_{i_1} \rightsquigarrow R_{i_2}$ connected by the j-th edge e_j, then:

$$s_j \stackrel{\text{def}}{=} T_{Skew}(i_1, i_2).$$

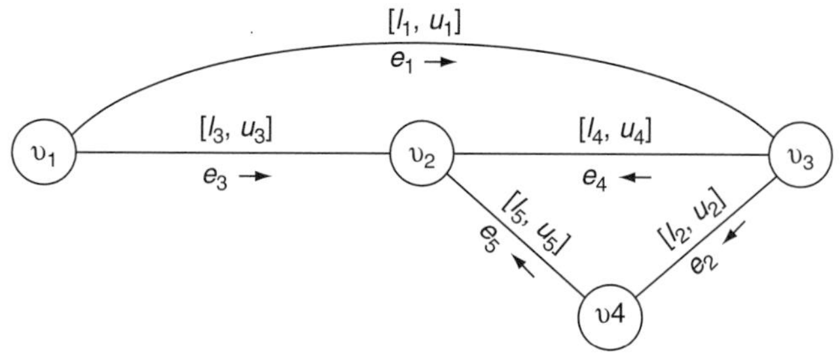

FIGURE 4.7 Circuit Graph of the Simple Example Circuit C_1

To illustrate these concepts, the graph g_{C_1} of the small circuit example C_1 introduced previously is illustrated in Figure 4.7 (note the enumeration and labeling of the edges as specified in definition 4.3). For this example, $r = 4$, $p = 5$, and the adjacency matrix is the following:

$$A_{C_1} = \begin{array}{c} \\ v_1 \\ v_2 \\ v_3 \\ v_4 \end{array} \begin{array}{c} \begin{array}{cccc} v_1 & v_2 & v_3 & v_4 \end{array} \\ \begin{bmatrix} 0 & 1 & 1 & 0 \\ 1 & 0 & 1 & 1 \\ 1 & 1 & 0 & 1 \\ 0 & 1 & 1 & 0 \end{bmatrix} \end{array}.$$

Observe that in general, the elements of A_C are defined as:

$$a_{ij}^{u} = \begin{cases} 1 & \textit{if there is an edge } e_k \textit{ connecting the vertices } v_i \textit{ and } v_j. \\ 0 & \textit{otherwise.} \end{cases} \quad (4.23)$$

In addition, note that the adjacency matrix as defined in equation 4.23 is always symmetric. The edges of g_C have no direction, so each edge between vertices v_i and v_j is shown in both of the rows corresponding to i and j. In addition, all diagonal elements of the adjacency matrix are zeros since self-loop edges are excluded by the required circuit graph properties described in Section 4.2.2. As a final reminder and without any loss of generality, it is assumed that a circuit has a connected graph (West, 1996). In other words, a circuit does not have isolated groups of registers. If a specific circuit has a disconnected graph, then each connected subgraph (subcircuit) can be considered separately.

4.4.2 Linear Dependence of Clock Skews

Consider the circuit graph of C_1 illustrated in Figure 4.7. The clock skews for the local data paths $R_3 \rightsquigarrow R_2$, $R_3 \rightsquigarrow R_4$, and $R_4 \rightsquigarrow R_2$ are $s_4 = T_{Skew}(3, 2) = t_{cd}^3 - t_{cd}^2$, $s_2 = T_{Skew}(3, 4) = t_{cd}^3 - t_{cd}^4$, and $s_5 = T_{Skew}(4, 2) = t_{cd}^4 - t_{cd}^2$, respectively. Note that $s_4 = s_2 + s_5$ (that is the clock skews s_2, s_4, and s_5 are **linearly dependent**). In addition, note that other sets of linearly dependent clock skews can be identified within C_1, such as s_1, s_3, and s_4.

Generally, large circuits contain many feedback and feed-forward signal paths. Thus, many possible linear dependencies among clock skews—such as those described in the previous paragraph—are typically present in such circuits. A natural question arises as to whether there exists a minimal set[11] of linearly independent clock skews that uniquely determines all clock skews in a circuit. (The existence of any such set could lead to substantial improvements in the run time of the clock scheduling algorithms as well as permit significant savings in storage requirements when implementing these algorithms on a digital computer.) It is generally possible to identify multiple minimal sets in any circuit. Consider C_1, for example—it can be verified that $\{s_3, s_4, s_5\}$, $\{s_1, s_3, s_5\}$, and $\{s_1, s_4, s_5\}$ each are sets with the property that (a) the clock skews in the set are linearly independent and (b) C_1 can be expressed as a linear combination of the clock skews that exist in the set.

Let C be a circuit with graph g_C and let $v_{i_0}, e_{j_0}, v_{i_1}, \ldots, e_{j_{z-1}}, v_{i_z} \equiv v_{i_0}$ be an arbitrary sequence of vertices and edges. Formally, the condition for linear dependence of the clock skews, $s_{j_0}, s_{j_1}, \ldots, s_{j_{z-1}}$, is as written here:

$$\left.\begin{array}{c}\prod_{k=0}^{z-1} a_{i_k j_k} \neq 0 \\ (i_z = i_0) \neq i_1 \neq \ldots \neq i_{z-1}\end{array}\right\} \Rightarrow \sum_{k=0}^{z-1} \pm T_{Skew}(i_k, j_k) = 0, \quad (4.24)$$

where the proof of equation 4.24 is trivial by substitution. The product on the left side of this equation requires that there exists an edge between every pair of vertices v_{i_k} and $v_{i_{k+1}} (k = 0, \ldots, z-1)$. The sum can be interpreted[12] as traversing the vertices of the cycle $C = v_{i_0}, e_{j_0}, v_{i_1}, \ldots, e_{j_z - 1}, v_{i_z} \equiv v_{i_0}$ in the order of appearance in C and adding the skews along C with a positive or negative sign, depending on whether the direction labeled on the edge coincides with the direction of traversal.

Typically, multiple cycles can be identified in a circuit graph and an equation—such as equation 4.24—can be written for each of these cycles. Referring to Figure 4.7, three such cycles are as follows:

$$C_1 = v_1, e_1, v_3, e_2, v_4, e_5, v_2, e_3, v_1$$
$$C_2 = v_2, e_4, v_3, e_2, v_4, e_5, v_2$$
$$C_3 = v_1, e_1, v_3, e_4, v_2, e_3, v_1.$$

These cycles can be identified and the corresponding linear dependencies written as:

$$\text{cycle } C_1 \rightarrow s_1 + s_2 - s_3 \quad + s_5 = 0 \quad (4.25)$$
$$\text{cycle } C_2 \rightarrow \quad s_2 \quad - s_4 + s_5 = 0 \quad (4.26)$$
$$\text{cycle } C_3 \rightarrow s_1 \quad - s_3 + s_4 \quad = 0. \quad (4.27)$$

[11] Such that the removal of any element from the set destroys the property.

[12] Note the similarity with Kirchhoff's voltage law (KVL or loop equations) for an electrical network (Chan *et al.*, 1972).

Note that the order of the summations in equations 4.25 through 4.27 has been intentionally modified from the order of cycle traversal to highlight an important characteristic. Specifically, observe that equation 4.25 is the sum of equations 4.26 and 4.27 that is, there exists a linear dependence not only among the skews in the circuit C but also among the cycles (or sets of linearly dependent skews).

Note that any minimal set of linearly independent clock skews must not contain a cycle (as defined by equation 4.24), for if the set contains a cycle, the skews in the set would not be linearly independent. Furthermore, any such set must span all vertices (registers) of the circuit, or it is not possible to express the clock skews of any paths in and out of the vertices not spanned by the set. Given a circuit C with r registers and p local data paths, these conclusions are formally summarized in the following two results from graph theory (West, 1996; Reingold *et al.*, 1997):

1. *Minimal set of linearly independent clock skews*: A minimal set of clock skews can be identified such that (a) the skews in the set are linearly independent and (b) every skew in C is a linear combination of the skews from the set. Such a minimal set is any **spanning tree** of g_C and consists of exactly $r-1$ elements (recall that a spanning tree is a subset of edges such that all vertices are spanned by the edges in the set). These $r-1$ skews (respectively, edges) in the spanning tree are referred to as the **skew basis** while the remaining $p-(r-1) = p-r+1$ skews (edges) of the circuit are referred to as **chords**. Note that there is a unique path between any two vertices such that all edges of the path belong to the spanning tree.
2. *Minimal set of independent cycles*: A minimal set of cycles (where a cycle is as defined by equation 4.24) can be identified such that (a) the cycles are linearly independent and (b) every cycle in C is a linear combination of the cycles from the set. Each choice of a spanning tree of g_C determines a unique minimal set of cycles, where each cycle consists of exactly one chord $\{v_{i_1}, e_j, v_{i_2}\}$ plus the unique path that exists in the spanning tree between the vertices v_{i_1} and v_{i_2}. Since there are $p-(r-1) = p-r+1$ chords, a minimal set of independent cycles consists of $p-r+1$ cycles. The minimal set of independent cycles of a graph is also called a fundamental set of cycles (West, 1996; Chan *et al.*, 1972; Reingold *et al.*, 1997).

To illustrate the aforementioned properties, observe the two different spanning trees of the example in circuit C_1 outlined with the thicker edges in Figure 4.8 (the permissible ranges and direction labelings have been omitted from Figure 4.8 for simplicity). The first tree is shown in Figure 4.8(A) and consists of the edges $\{e_3, e_4, e_5\}$ and the independent cycles C_2 (equation 4.26) and C_3 (equation 4.27). As previously explained, both C_2 and C_3 contain precisely one of the skews

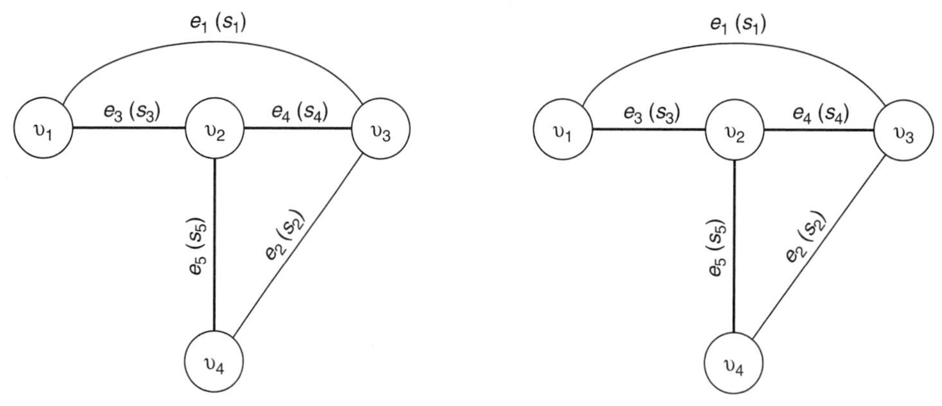

(A) Spanning Tree $\{e_3, e_4, e_5\}$ (B) Spanning Tree $\{e_1, e_3, e_5\}$

FIGURE 4.8 Two Spanning Trees. These trees show the corresponding minimal sets of linearly independent clock skews and linearly independent cycles for the circuit example C_1. Edges from the spanning tree are indicated with thicker lines

not included in the spanning tree—s_2 for C_2 and s_1 for C_3. Similarly, the second spanning tree $\{e_1, e_3, e_5\}$ is illustrated in Figure 4.8(B). The independent cycles for the second tree are C_1 (equation 4.25) and C_3 (equation 4.27)—generated by s_2 and s_4, respectively.

Let a circuit C with r registers and p local data paths be described by a graph g, and let a skew basis (spanning tree) for this circuit (graph) be identified. For the remainder of this work, it is assumed that the skews have been enumerated such that those skews from the skew basis have the highest indices.[13] Introducing the notation s^b for the basis and s^c for the chords, the clock schedule s can be expressed as:

$$s = \begin{bmatrix} s^c \\ s^b \end{bmatrix} = [\underbrace{\overbrace{s_1 \cdots s_{p-r+1}}^{p-r+1}}_{\text{Chords}} \underbrace{\overbrace{s_{p-r+2} \cdots s_p}^{r-1}}_{\text{Basis}}]^t, \tag{4.28}$$

where

$$s^c = \begin{bmatrix} s_1 \\ \vdots \\ s_{p-r+1} \end{bmatrix} \text{ and } s^b = \begin{bmatrix} s_{p-r+2} \\ \vdots \\ s_p \end{bmatrix}. \tag{4.29}$$

Note that the case illustrated in Figure 4.8(A) is precisely the type of enumeration just described by equations 4.28 and 4.29—e_1, e_2 (s_1, s_2) are the chords and e_3, e_4, e_5 (s_3, s_4, s_5) are the basis.

With the notation and enumeration as specified above, let $n_b = r - 1$ be the number of skews (edges) in the basis, and let $n_c = p - r + 1 = p - n_b$ be the number of chords (equal to the number of cycles). The set of linearly independent cycles is $C_1, \ldots, C_{n_c}$ and the clock skew dependencies for these cycles are these:

$$\begin{aligned} \text{cycle } C_1 &= v_{i_0^1}, e_{j_0^1}, v_{i_1^1}, \ldots, e_{j_{z-1}^1}, v_{i_0^1} \rightarrow 0 = \sum_{k=i_0^1}^{i_{z-1}^1} \pm s_{j_k^1}. \\ &\vdots \\ \text{cycle } C_{n_c} &= v_{i_0^{nc}}, e_{j_0^{nc}}, v_{i_1^{nc}}, \ldots, e_{j_{z-1}^{nc}}, v_{i_0^{nc}} \rightarrow 0 = \sum_{k=i_0^{nc}}^{i_{z-1}^{nc}} \pm s_{j_k^{nc}}. \end{aligned} \tag{4.30}$$

Note that the sums in equation 4.30 can be written in matrix form:

$$Bs = 0, \tag{4.31}$$

where $B = [b_{ij}]_{n_c \times p}$ is a matrix of order $n_c \times p$. The matrix B will be called the **circuit connectivity matrix**, and each row of B corresponds to a cycle of the circuit graph and contains elements from the incidence matrix A combined with zeros depending on whether a skew (an edge) belongs to the cycle. Note that since each cycle contains exactly one chord, the cycles can always be permuted such that the cycles appear in the order of the chords i.e. C_1 corresponds to e_1, C_2 corresponds to e_2, and so on). If this correspondence is applied, the matrix B can be represented as:

$$B = [I_{n_c} C_{n_c \times n_b}], \tag{4.32}$$

where the submatrix I_{n_c} is an identity[14] matrix of dimension $n_c \times n_c$, thereby permitting equation 4.31 to be rewritten as:

[13] Such enumeration is always possible since the choice of indices for any enumeration (including this example) is arbitrary.

[14] Recall that an identity matrix I_n is a square $n \times n$ matrix such that the only nonzero elements are on the main diagonal and are all equal to one.

$$Bs = [I \; C]\begin{bmatrix} s^c \\ s^b \end{bmatrix} = s^c + Cs^b = 0. \tag{4.33}$$

Consider, for instance, the choice of spanning tree illustrated in Figure 4.8(A). There are two independent cycles denoted by C_1 (corresponding to C_2 in equation 4.26) and C_2 (corresponding to C_3 in equation 4.27). The matrix relationship of equation 4.31 for this case is the following:

$$s_1 - s_3 + s_4 = 0 \leftarrow \text{cycle } C_1 = v_1, e_1, v_3, e_4, v_2, e_3, v_1.$$
$$+ s_2 - s_4 + s_5 = 0 \leftarrow \text{cycle } C_2 = v_3, e_2, v_4, e_5, v_2, e_4, v_3.$$

The matrices B and C, respectively, are

$$B = [I_2 \; C_{2\times3}] = \begin{bmatrix} 1 & 0 & -1 & 1 & 0 \\ 0 & 1 & 0 & -1 & 1 \end{bmatrix}, \quad C = \begin{bmatrix} -1 & 1 & 0 \\ 0 & -1 & 1 \end{bmatrix}. \tag{4.34}$$

From an algebraic standpoint (Bretscher, 1996), equation 4.31 requires that any clock schedule s must necessarily be in the kernel ker(B) of the linear transformation B: $\mathbb{R}^p \rightarrow \mathbb{R}^{n_c}$ (that is $s \in \ker B$). The inverse situation, however, is not true; that is, an arbitrary element of the kernel is *not* necessarily a feasible clock schedule. Furthermore, note that B is already in reduced row echelon form (Bretscher, 1996) so the rank of B is rank(B) $= n_c$. Thus, the dimension of ker (B) is [33] as follows:

$$\begin{aligned} \dim(\ker(B)) &= \text{columns of } B - \text{rank}(B) \\ &= p - \text{rank}(B) \\ &= p - n_c = n_b. \end{aligned} \tag{4.35}$$

Therefore, equation 4.31 is referred to here as the circuit kernel equation.

This last result expressed by equation 4.35 demonstrates that there are $n_b = r - 1$ linearly independent skews in a circuit. Furthermore, considering that the matrix C is:

$$C = \begin{bmatrix} | & & | \\ c_1 & \cdots & c_{n_b} \\ | & & | \end{bmatrix},$$

one possible basis for ker(B) can be written from inspection:

$$\text{basis for } \ker(B) = \underbrace{\begin{bmatrix} -c_1 \\ 1 \\ 0 \\ \vdots \\ 0 \end{bmatrix} \begin{bmatrix} -c_2 \\ 0 \\ 1 \\ \vdots \\ 0 \end{bmatrix} \cdots \begin{bmatrix} -c_{n_b} \\ 0 \\ 0 \\ \vdots \\ 1 \end{bmatrix}}_{n_b \text{vectors}}. \tag{4.36}$$

Any feasible clock schedule $s \in \ker(B)$ can be expressed as a linear combination of the vectors from the basis of the kernel:

$$s = \begin{bmatrix} s^c \\ s^b \end{bmatrix} = s_1^b \begin{bmatrix} -c_1 \\ 1 \\ 0 \\ \vdots \\ 0 \end{bmatrix} + s_2^b \begin{bmatrix} -c_2 \\ 0 \\ 1 \\ \vdots \\ 0 \end{bmatrix} + \ldots + s_{n_b}^b \begin{bmatrix} -c_{n_b} \\ 0 \\ 0 \\ \vdots \\ 1 \end{bmatrix} = \begin{bmatrix} -Cs^b \\ s^b \end{bmatrix}, \tag{4.37}$$

where the scalars, $s_1^b, s_2^b, \ldots, s_{r'}^b$, in equation 4.37 are the elements of the vector s^b as defined by equation 4.28:

$$s^b = \begin{bmatrix} s_1^b \\ s_2^b \\ \vdots \\ s_{n_b}^b \end{bmatrix} = \begin{bmatrix} s_{n_c+1} \\ s_{n_c+2} \\ \vdots \\ s_p \end{bmatrix}. \tag{4.38}$$

Observe that either knowing or deliberately choosing s^b not only provides sufficient information to determine the corresponding s^c (respectively, the entire s) but also permits computation of the clock delays t_{cd} to implement the desired clock schedule s. Specifically, the dependencies among the clock skews in the branches (the local data paths) and the clock delays to the vertices (the registers) can be described in matrix form as follows:

$$s^b = T_{n_b \times r} t_{cd} \tag{4.39}$$

Note that each skew is the difference of two clock delays so that each row of the matrix T in equation 4.39 contains exactly two nonzero elements. These two nonzero elements are 1 and -1, respectively, depending on which two clock delays determine the clock skew corresponding to this equation (or row in the matrix). Also note that equation 4.39 is a consistent linear system (the rows correspond to linearly independent skews in the circuit) with fewer equations than the r unknown clock delays t_{cd}. Therefore, equation 4.39 has an infinite number of solutions all corresponding to the same clock schedule s.

Finding a solution t_{cd} of equation 4.39 is now a straightforward matter. For example, setting $t_{cd}^r = 0$ and rewriting this equation to account for this substitution:

$$t_{cd}^r = 0 \Rightarrow s^b = T^*_{n_b \times n_b} t_{cd} = T^*_{n_b \times n_b}, \tag{4.40}$$

yields a consistent linear system with the same number of variables as equations where the matrix $T^*_{n_b \times n_b}$ is the matrix $T_{n_b \times r}$ with the rightmost column deleted. The most efficient way to solve the system characterized by equation 4.40 with the highest accuracy is by back substitution (only addition/subtraction operations are necessary). In the software implementation of this algorithm discussed in this work, t_{cd} is computed in an efficient way by traversing the edges of the spanning tree.

This section concludes by illustrating the concepts discussed with a small circuit example C_1; the circuit graph gC_1 is shown in Figure 4.7, and the respective spanning tree is shown in Figure 4.8(A). For this circuit, $r = 4$, the number of local data paths is $p = 5$, and $n_b = 4 - 1 = 3$. The clock schedule is as follows:

$$\mathrm{s} = \begin{bmatrix} \mathrm{s}^c \\ \mathrm{s}^b \end{bmatrix}, \text{ where } \mathrm{s}^c = \begin{bmatrix} s_1 \\ s_2 \end{bmatrix}, \mathrm{s}^b = \begin{bmatrix} s_3 \\ s_4 \\ s_5 \end{bmatrix}. \quad (4.41)$$

The independent cycles are C_2 (from equation 4.26) and C_3 (from equation 4.27) and the matrices B and C are as defined in equation 4.34. A basis for the kernel of B has a dimension $n_b = 3$ and consists of the vectors:

$$\begin{bmatrix} 1 \\ 0 \\ 1 \\ 0 \\ 0 \end{bmatrix}, \begin{bmatrix} -1 \\ 1 \\ 0 \\ 1 \\ 0 \end{bmatrix}, \text{ and } \begin{bmatrix} 0 \\ -1 \\ 0 \\ 0 \\ 1 \end{bmatrix}. \quad (4.42)$$

Any clock schedule is in $\ker(B)$ and can be expressed as a linear combination of the vectors from the kernel basis:

$$s = s_3^b \begin{bmatrix} 1 \\ 0 \\ 1 \\ 0 \\ 0 \end{bmatrix} + s_4^b \begin{bmatrix} -1 \\ 1 \\ 0 \\ 1 \\ 0 \end{bmatrix} + s_5^b \begin{bmatrix} 0 \\ -1 \\ 0 \\ 0 \\ 1 \end{bmatrix}. \quad (4.43)$$

Consider, for instance, the clock skew schedule for $T_{CP} = 6.5$ shown in Table 4.2. Substituting $s_3 = 0$, $s_4 = -1.5$, and $s_5 = -1$ into equation 4.43 yields the clock schedule:

$$s = 0 \begin{bmatrix} 1 \\ 0 \\ 1 \\ 0 \\ 0 \end{bmatrix} - 1.5 \begin{bmatrix} -1 \\ 1 \\ 0 \\ 1 \\ 0 \end{bmatrix} - 1 \begin{bmatrix} 0 \\ -1 \\ 0 \\ 0 \\ 1 \end{bmatrix} = \begin{bmatrix} 1.5 \\ -0.5 \\ 0 \\ -1.5 \\ -1 \end{bmatrix}. \quad (4.44)$$

Finally, the clock delays t_{cd} are derived from the underdetermined linear system (as described by equation 4.39):

$$s^b = \begin{bmatrix} 0 \\ -1.5 \\ -1 \end{bmatrix} = \begin{bmatrix} 1 & -1 & 0 & 0 \\ 0 & -1 & 1 & 0 \\ 0 & -1 & 0 & 1 \end{bmatrix} \begin{bmatrix} t_{cd}^1 \\ t_{cd}^2 \\ t_{cd}^3 \\ t_{cd}^4 \end{bmatrix}, \quad (4.45)$$

where setting $t_{cd}^4 = 0$ yields:

$$s^b = \begin{bmatrix} 0 \\ -1.5 \\ -1 \end{bmatrix} = \begin{bmatrix} 1 & -1 & 0 \\ 0 & -1 & 1 \\ 0 & -1 & 0 \end{bmatrix} \begin{bmatrix} t_{cd}^1 \\ t_{cd}^2 \\ t_{cd}^3 \end{bmatrix} \Rightarrow \begin{matrix} t_{cd}^1 = 1 \\ t_{cd}^2 = 1 \\ t_{cd}^3 = -0.5. \end{matrix} \quad (4.46)$$

Interestingly, the clock schedule $\begin{bmatrix} 1 & 1 & -1/2 & 0 \end{bmatrix}^t$ differs from the solution shown in Table 4.2 by only a constant of $c = -1/2$, namely,

$$\begin{bmatrix} 1 & 1 & -\frac{1}{2} & 0 \end{bmatrix}^t = \begin{bmatrix} c+\frac{3}{2} & c+\frac{3}{2} & c+0 & c+\frac{1}{2} \end{bmatrix}^t. \quad (4.47)$$

4.4.3 Optimization Problem and Solution

Recall the intuitive definition of clock skew scheduling as a quadratic programming (QP) problem first introduced in Section 4.3. In this section, the QP formulation is formalized, and the solution of the problem is explained in detail.

Problem QP-1__________(QP Clock Skew Scheduling)

Let C be a circuit with r registers, p local data paths, and a target clock period T_{CP}, and let the local data paths be enumerated as:

$$p \text{ local data paths} \begin{cases} \text{path}_1 & \rightarrow & R_{i_1} \rightsquigarrow R_{j_1} \\ \vdots & & \\ \text{path}_p & \rightarrow & R_{i_p} \rightsquigarrow R_{j_p}. \end{cases} \quad (4.48)$$

For each local data path $\text{path}_k(R_{i_k} \rightsquigarrow R_{j_k})$ in C, let the lower bound l_{i_k, j_k}, upper bound u_{i_k, j_k}, width w_{i_k, j_k}, and middle m_{i_k, j_k} of the permissible range of this path, respectively, be defined as in equations 4.10, 4.11, 4.12 and 4.13, respectively. For simplicity, these parameters of the permissible range are denoted with a single subscript corresponding to the number of the respective local data path; that is, for the $\text{path}_k \equiv R_{i_k} \rightsquigarrow R_{j_k}$, $l_{i_k, j_k} = l_k$, $u_{i_k, j_k} = u_k$, $w_{i_k, j_k} = w_k$, and $m_{i_k, j_k} = m_k$. Furthermore, let the circuit graph of C be $\mathcal{G}C$, let the skew basis s^b and chords s^c be identified in $\mathcal{G}C$ (according to equation 4.28), and let the corresponding independent set of cycles be described by the matrix $B = [I \quad C]$ as defined in equation 4.32. Let an objective clock schedule be $g = [g_1 \ \dots \ g_p]^t = [m_1 \ \dots \ m_p]^t$, and let $l = [l_1 \ \dots \ l_p]^t$ and $u = [u_1 \ \dots \ u_p]^t$ be the vectors of the lower and upper bounds, respectively, of the permissible ranges. Find a feasible clock schedule s that minimizes the least square error ε between s and g. Formally, it is shown that:

$$\begin{aligned} \min \ \varepsilon &= \sum_{k=1}^{p} (s_k - g_k)^2, \\ \text{subject to: } Bs &= 0 \\ l &\leq s \\ s &\leq u, \end{aligned} \quad (4.49)$$

where the inequalities in equation 4.49 are treated in terms of each component (that is, $l_1 \leq s_1 \leq u_1$, $l_2 \leq s_2 \leq u_2$, and so on).

Problem QP-1 is a constrained QP problem with bounded variables—methods such as **active constraints** exist for solving such problems (Ciarlet and Lions, 1990; Farebrother, 1988; Osborne, 1985; Fletcher, 1987; Björck, 1996). These methods are both analytically and numerically challenging, so a two-phase solution process is suggested here such that a constrained version of problem QP-1 is solved initially. If the result is not feasible, a rapidly converging iterative refinement of the objective g is performed until the feasibility of s is satisfied. This two-phase process is defined formally as:

$$\begin{aligned} &\text{Phase 1} \rightarrow \min \varepsilon = \sum_{k=1}^{p} (s_k - g_k)^2, \\ &\qquad \text{subject to } Bs = 0. \\ &\text{Phase 2} \rightarrow \text{Iterative refinement of } s, \end{aligned} \tag{4.50}$$

where Phase 1 is an equality-constrained quadratic optimization problem expressed as the following problem QP-2:

Problem QP-2 __________ (QP Clock skew scheduling)

$$\begin{aligned} &\min\ \varepsilon = (s - g)^2 = \sum_{k=1}^{p} (s_k - g_k^{\tau})^2 \\ &\text{subject to: } Bs = 0. \end{aligned} \tag{4.51}$$

Problem QP-2 is representative of a broader class of optimization problems where the function that is minimized is a distance in the Euclidean space $\mathbb{R}^n$. One typical problem that arises in a variety of situations, for instance, is the linear least squares problem. The objective of the linear least squares problem is to find $x^* \in \mathbb{R}^n$ such that the Euclidean distance between $Dx^* \in \mathbb{R}^m$ and $b \in \mathbb{R}^m$ is as small as possible. The matrix D is an $m \times n$ matrix, and the system $Dx = b$ is typically inconsistent. The function being minimized in the linear least squares problem is the following:

$$\sum_{i=1}^{m} (d_i^t x - b_i)^2, \text{ where } D^t = \begin{bmatrix} | & & | \\ d_1 & \dots & d_m \\ | & & | \end{bmatrix}.$$

It is well known (Bretscher, 1996; Björck, 1996) that if the kernel of D is $\ker(D) = \{0\}$, then x^* is the solution of the consistent system $D^t Dx = D^t b$.

The quadratic programming problem QP-2 is solved by applying the classical method of **Lagrange multipliers** for constrained optimization (Fletcher, 1987; Björck, 1996; Lawson and Hanson, 1974). To start, note that minimizing the objective function ε in equation 4.51 is equivalent to minimizing the function:

$$\varepsilon^* = s^t s - 2g^t s.$$

For a quick proof of this equivalence, consider expanding the value of ε:

$$\begin{aligned} \varepsilon &= (s - g)^2 \\ &= (s)^2 - 2g^t s + (g)^2 \\ &= s^t s - 2g^t s + g^t g, \end{aligned} \tag{4.52}$$

where the inner product $g^t g$ in equation 4.52 is a numeric constant. Therefore, if a value $s = s^*$ exists that minimizes ε^* in equation 4.52, s^* also minimizes ε. Note that since $\varepsilon^* = \varepsilon - g^t g$, the two minimums are related by:

$$\min(\varepsilon^*) = \min(\varepsilon) - g^t g. \tag{4.53}$$

Thus, problem QP-2 is transformed into the following problem QP-3:

Problem QP-3 __________ (QP Clock skew scheduling)

$$\begin{aligned} &\min\ \varepsilon^* = s^t s - 2g^t s, \\ &\text{subject to: } Bs = 0. \end{aligned} \tag{4.54}$$

To apply the method of Lagrange multipliers to problem QP-3, the vector $\lambda = [\lambda_1 \dots \lambda_{n_c}]^t$ is introduced, where each multiplier λ_i in λ corresponds to the i-th equality constraint from $Bs = 0$. The Lagrangian function $\mathcal{L}(s, \lambda)$ is introduced next:

$$\begin{aligned} L(s, \lambda) &= \varepsilon^* + \lambda^t Bs \\ &= s^t s - 2g^t s + \lambda^t Bs, \end{aligned} \tag{4.55}$$

where the term $\lambda^t Bs$ in equation 4.55 is the sum of all equality constraints for the product of the ith constraint times the multiplier λ_i.

Any extremum of ε^* must be a stationary point of the Lagrangian $\mathcal{L}(s, \lambda)$ (Bretscher, 1996); that is, the first derivatives of $\mathcal{L}(s, \lambda)$ with respect to $s_i\ i \in \{1, \dots, p\}$ and $\lambda_j\ j \in \{1, \dots, n_c\}$, must be zero. Formally, if the differential operator is denoted as ∇, then any stationary point (s^*, λ^*) of $\mathcal{L}(s, \lambda)$ is a solution of the system of equations:

$$\nabla \mathcal{L}(s, \lambda) = 0 \Rightarrow \left| \begin{aligned} &\nabla_s \mathcal{L}(s, \lambda) = 0 \\ &\nabla_\lambda \mathcal{L}(s, \lambda) = 0. \end{aligned} \right. \tag{4.56}$$

In the general case of a QP problem with any type of constraints, systems such as equation 4.56 can be nonlinear and difficult to solve. In the case of linear constraints, however, a solution can be derived in a straightforward manner. To this end, consider the derivatives $\nabla_s \mathcal{L}(s, \lambda)$ and $\nabla_\lambda \mathcal{L}(s, \lambda)$ of the Lagrangian:

$$\begin{aligned} \nabla_s \mathcal{L}(s, \lambda) &= \nabla_s (s^t s - 2g^t s + \lambda^t Bs) \\ &= 2s - 2g + (\lambda^t B)^t \\ &= 2s - 2g + B^t \lambda, \end{aligned} \tag{4.57}$$

$$\text{and } \nabla_\lambda \mathcal{L}(s, \lambda) = \nabla_\lambda (s^t s - 2gs + \lambda^t Bs) = Bs. \tag{4.58}$$

Note that equations 4.57 and 4.58 contain p and n_c equations, respectively (recall that s and λ have p and n_c variables, respectively). Therefore, the solution of equation 4.56 requires finding exactly $p + n_c = 2p - n_b = 2p - r + 1$ variables.

Substituting equations 4.57 and 4.58 back into 4.56 yields the linear system:

$$\left| \begin{array}{r} 2s + B^t\lambda = 2g \\ Bs = 0, \end{array} \right. \tag{4.59}$$

which can be conveniently written in matrix form:

$$\begin{bmatrix} 2I_p & B^t \\ B & 0 \end{bmatrix} \begin{bmatrix} s \\ \lambda \end{bmatrix} = 2 \begin{bmatrix} g \\ 0 \end{bmatrix}. \tag{4.60}$$

Solving equation 4.60 by Gauss-Jordan elimination is straightforward by premultiplying with $1/2B$ the first row of the system described by equation 4.60 and subtracting the result from the second row, thereby yielding:

$$\begin{bmatrix} 2I_p & B^t \\ B & 0 \end{bmatrix} \begin{bmatrix} s \\ \lambda \end{bmatrix} = 2 \begin{bmatrix} g \\ 0 \end{bmatrix} \Rightarrow \begin{bmatrix} 2I_p & B^t \\ 0 & BB^t \end{bmatrix} \begin{bmatrix} s \\ \lambda \end{bmatrix} = 2 \begin{bmatrix} g \\ Bg \end{bmatrix}. \tag{4.61}$$

A natural way to solve the linear system described by equation 4.61 is by back substitution,[15] such that λ is initially computed, followed by the computation of s. The Lagrange multipliers λ are determined from the equation $(BB^t)\lambda = 2Bg$ in the second row of equation 4.61, where the right-hand side $2Bg$ is a nonzero vector; that is, $Bg \neq \{0\}$. The opposite situation, $Bg = \{0\}$, is highly unlikely to occur since $Bg = \{0\}$ means that $g \in \ker(B)$, which in turn means (recall equation 4.35 through 4.38) that the objective clock schedule g is feasible and no optimization needs to be performed.[16]

Therefore, the equation $(BB^t)\lambda = 2Bg$ in equation 4.61 can have either no solutions or exactly one solution depending on whether the matrix BB^t is singular or not. In other words, the nonsingularity of BB^t is a *necessary* and *sufficient* condition for the existence of a unique solution $\left[\hat{s}^t\ \hat{\lambda}^t\right]^t$ of equation 4.60. If the product BB^t is denoted by M, note that the symmetric $n_c \times n_c$ matrix:

$$M = BB^t = [I \quad C] \begin{bmatrix} I \\ C^t \end{bmatrix} = I + CC^t \tag{4.62}$$

is strictly positive-definite and thus nonsingular. Therefore, the system 4.60 is guaranteed to have a unique solution of:

$$\hat{\lambda} = 2M^{-1}Bg \tag{4.63}$$

and

$$\hat{s} = -\frac{1}{2}B^t\lambda + g = -(B^tM^{-1}B)g + g, \tag{4.64}$$

where the matrix M is as introduced in equation 4.62.

To gain further insight into the solution described by equations 4.60 through 4.64, consider substituting equation 4.32 for B into equation 4.60 and representing the vector column g of the objective clock skew schedule as:

$$g = \begin{bmatrix} g^c \\ g^b \end{bmatrix}, \tag{4.65}$$

where g^c and g^b correspond to s^c and s^b; that is, g_1 is the objective value of the clock skew s_1, g_2 is the objective value of the clock skew s_2, and so on. With these substitutions, the system represented by equation 4.60 can be written as:

$$\begin{bmatrix} 2I_{n_c} & 0 & I_{n_c} \\ 0 & 2I_{n_b} & C^t \\ I_{n_c} & C & 0 \end{bmatrix} \begin{bmatrix} s^c \\ s^b \\ \lambda \end{bmatrix} = K \begin{bmatrix} s^c \\ s^b \\ \lambda \end{bmatrix} = 2 \begin{bmatrix} g^c \\ g^b \\ 0 \end{bmatrix}, \tag{4.66}$$

where the coefficient matrix K on the left is symmetric. In equation 4.66, the Gaussian elimination step described by equation 4.61 is equivalent to multiplying by 1/2 the first row of K, premultiplying by $1/2\ C$ the second row of K, and subtracting both of these rows from the third row:

$$\begin{array}{l} \begin{bmatrix} 2I & 0 & I \\ 0 & 2I & C^t \\ I & C & 0 \end{bmatrix} \begin{bmatrix} s^c \\ s^b \\ \lambda \end{bmatrix} = 2 \begin{bmatrix} g^c \\ g^b \\ 0 \end{bmatrix} \Rightarrow \\ \begin{bmatrix} 2I & 0 & I \\ 0 & 2I & C^t \\ 0 & 0 & I + CC^t \end{bmatrix} \begin{bmatrix} s^c \\ s^b \\ \lambda \end{bmatrix} = 2 \begin{bmatrix} g^c \\ g^b \\ g^c + Cg^b \end{bmatrix}. \end{array} \tag{4.67}$$

Observe that the linear system of equation 4.67 is simply a more detailed technique for rendering the linear system described by equation 4.61, where the first row has been expanded into the first two rows of equation 4.67:

$$BB^t = [I \quad C] \begin{bmatrix} I \\ C^t \end{bmatrix} = I + CC^t \tag{4.68}$$

$$Bg = [I \quad C] \begin{bmatrix} g^c \\ g^b \end{bmatrix} = g^c + Cg^b. \tag{4.69}$$

With the matrix M as defined in equation 4.62, the solution of equation 4.67 is the following:

[15] This is true because the coefficient matrix is an upper triangular matrix.

[16] The chances of g being feasible for a large real circuit are infinitesimally small.

$$\hat{\lambda} = 2M^{-1}Bg, \tag{4.70}$$

$$\hat{s}^b = -\frac{1}{2}C^t\hat{\lambda} + g^b, \tag{4.71}$$

$$\hat{s}^c = -\frac{1}{2}\hat{\lambda} + g^c. \tag{4.72}$$

As a final note, observe that the solution described by equations 4.63 and 4.64 is not only a stationary point of the Lagrangian function $\mathcal{L}(s, \lambda)$ (i.e., a potential local minimizer), but the solution is also a *global* minimizer of ε^* in equation 4.54 (Björck, 1996). As a matter of fact, problem QP-3 belongs to a broader class of optimization problems for which the function being minimized is of the form $f(x) = x^t Zx + y^t x$ (note that in the case of problem QP-3, the matrix Z is the positive-definite identity matrix I_p). A proof can be found in (Björck, 1996): if Z is positive-definite, a solution process similar to the process represented by equations 4.55 through 4.64 can be applied to obtain a unique global minimizer of $f(x) = x^t Zx + y^t x$. Björck (1996) provides a most thorough treatment of this subject as well as proofs of the existence and uniqueness of the solution.

4.5 Practical Considerations

A new formulation of clock skew scheduling as a QP problem was introduced in Sections 4.3 and 4.4. Recall that in this formulation, a feasible and consistent clock schedule is found that is close[17] to a previously chosen "ideal" objective clock schedule.

In this section, a computer methodology is presented for the solution of the QP clock scheduling problem introduced in Section 4.3. Different computer implementations are analyzed and compared in detail in Subsection 4.5.1. It is shown that the QP problem can be efficiently solved, and three computer algorithmic procedures for this solution are discussed. These three algorithms are demonstrated to have $O(r^3)$ run time complexity and $O(r^2)$ storage complexity, where r is the number of registers in the circuit. The numerical constants of the leading terms in these complexity expressions are derived as a function of the ratio of the number of local data paths to the number of registers in the circuit, thereby permitting a suitable algorithm to be chosen for a specific circuit. Furthermore, the methodology presented in Section 4.3 is extended here to account for two important details of practical interest. The circuit graph model is revisited in Section 4.5.3, where it is shown that certain clock skews from the basis are unconstrained[18], and this information is integrated into the mathematical framework described in Section 4.3. Section 4.5.4, demonstrates how to efficiently handle the timing constraints of the *I/O* registers of a circuit and reviews the necessary modifications to the mathematical optimization procedure.

4.5.1 Computational Analysis

The solution to problem QP-3 was described in Section 4.4.3 in purely mathematical terms and without consideration of any computational aspects. Naturally, the solution described by equations 4.63 and 4.64 is determined from a program running on a digital computer. In this section, the time and memory requirements of three different computer implementations are analyzed in greater detail. The run time complexity N of these algorithms is considered to be dependent on the number of multiplicative (multiple and divide) floating point operations. Similarly, the memory complexity M is considered to be the largest number of floating point storage units that must be stored in memory at any time during the execution of the specific algorithm.[19]

It is shown here that the run time complexity of all three algorithms described in this section is $O(r^3)$, where r is the number of registers in the circuit. Furthermore, it is shown that the numerical constant of the leading r^3 term in these complexity expressions is a function of the ratio:

$$k = \frac{p}{r}. \tag{4.73}$$

The ratio in 4.73 refers to the number of local data paths p to the number of registers r in a circuit. Similarly, the memory complexity of all three algorithms is $O(r^2)$ where the numerical constant of the r^2 term is a function of k introduced in equation 4.73. This relationship is exploited to determine the most efficient algorithm for a specific circuit.

Note that formally, the Lagrange multipliers λ are *not* required for the solution of problem QP-3 because the objective of the procedures described here is to determine a feasible clock schedule s. Since the existence and uniqueness of a clock schedule $\hat{s}$ satisfying problem QP-3 have been established in Section 4.4.3, this clock schedule can be directly computed by evaluating the rightmost expression in equation 4.64:

$$\hat{s} = -(B^t M^{-1} B)g + g. \tag{4.74}$$

As an alternative, a sequential approach can be adopted such that the Lagrange multipliers $\hat{\lambda}$ are computed first, followed by computing $\hat{s}$ (consisting of $\hat{s}^b$ and $\hat{s}^c$) using $\hat{\lambda}$.

In the former case (a straightforward computation of $\hat{s}$), the complexity of evaluating the expression described by equation 4.74 determines the complexity of the overall solution. In the

[17] The clock schedule is close in a Euclidean sense.

[18] These skews are independent from other skews in the circuit. Nevertheless, these skews must satisfy the permissible range requirement.

[19] Memory transfers between main and secondary storage are, of course, always an option. For the quickest execution, however, all data must reside in the main storage.

latter case (computing $\hat{\lambda}$ first), both $\hat{s}^b$ and $\hat{s}^c$ can be computed quickly because these computations involve only the addition and subtraction operations (recall that all nonzero elements of the matrix C are either 1 or -1). Therefore, in the case of computing $\hat{\lambda}$ and $\hat{s}$ in this order, the complexity of the overall solution of problem QP-3 is dominated by the computation of the Lagrange multipliers $\hat{\lambda}$.

Three computational algorithms for solving problem QP-3 are described in the following three sections. The first two algorithms—called LMCS-1 and LMCS-2, respectively—compute $\hat{\lambda}$ and $\hat{s}$ in this order according to the dependence relationship $\hat{s} = 1/2B^t\hat{\lambda}$ described by equation 4.64. The third algorithm—called CSD—computes the clock schedule $\hat{s}$ directly as described by equation 4.74. The algorithms LMCS-1 and LMCS-2 are described in the following first two subsections. The last two subsections describe algorithm CSD, which is shown to be superior to both of the other algorithms, and offer a comparative summary of the results.

Algorithm LMCS-1

As mentioned previously, this algorithm for solving problem QP-3 consists of eliminating $\hat{\lambda}$ from $M\lambda = 2Bg$ (see equation 4.63) then computing $\hat{s}$ according to equation 4.64. To determine the value of the Lagrange multipliers $\hat{\lambda}$ corresponding to the minimization of ε_{τ}^* in problem QP-3, consider the linear system:

$$M\lambda = (BB^t)\lambda = (I + CC^t)\lambda = 2Bg = 2(g^c + Cg^b), \quad (4.75)$$

which corresponds to the last row of equations 4.61 and 4.67, respectively. As mentioned previously in Section 4.4.3, the symmetric matrix M is always positive-definite[20] and nonsingular, thereby permitting exactly one solution $\hat{\lambda}$ of the linear system described by equation 4.75.

This linear system is a large square system of the type $Ax = b$, where $b \in \mathbb{R}^n$ is a column vector and where the coefficient matrix $A \in \mathbb{R}^{n\times n}$ is dense. Typically, the most effective approach to computing the solution $\hat{x} \in \mathbb{R}^n$ of such systems consists of performing a triangular **decomposition**[21] of the coefficient matrix A followed by the successive solution of two relatively "easy" to solve square linear systems of order $n \times n$. The triangular decomposition of A is of the form $A = LU$, where L and U are a lower triangular and an upper triangular matrix, respectively (Golub and Loan, 1996; Forsythe and Moler, 1967). The solution of $Ax = LUx = b$ is obtained next by first computing the intermediate solution $\hat{y}$ of the system $Ly = b$. Finally, $\hat{x}$ is the solution of the system $Ux = \hat{y}$. Because of the triangularity of the matrices L and U, the vectors $\hat{y}$ and $\hat{x}$ can be computed with relatively little effort. The components of the intermediate solution $\hat{y}$ are obtained by solving the system $Ly = b$—referred to as forward elimination (Golub and Loan, 1996; Forsythe and Moler, 1967)—since the first equation of $Ly = b$ involves only y_1, the second only y_1 and y_2, and so on. Similarly, the components of $\hat{x}$ are obtained from the system $Ux = \hat{y}$ in the reverse order $x_n, x_{n-1}, \ldots, x_1$. The process of solving $Ux = \hat{y}$ for $\hat{x}$ is also called back substitution (Golub and Loan, 1996; Forsythe and Moler, 1967).

Furthermore, the symmetry and positive-definiteness of M can be exploited to obtain a special form of the LU triangular decomposition of M so that the lower and upper triangular matrices in the decomposition are the transpose of each other. This alternative decomposition is known as the **Cholesky** decomposition of M and permits M to be uniquely represented (Golub and Loan, 1996) as the product:

$$BB^t = M = L_1L_1^t, \quad (4.76)$$

where L_1 is a lower triangular matrix. The Cholesky decomposition is computationally more efficient than a general LU decomposition in that the Cholesky decomposition requires about half of the computation time of a general LU decomposition. Finally, the Cholesky decomposition has useful properties related to issues of numerical stability and accuracy. (An in-depth treatment of this subject is discussed in Golub and Loan [1996] and Forsythe and Moler [1967].)

As mentioned previously, the complexity of algorithm LMCS-1 is dominated by the complexity of computing the Lagrange multipliers $\hat{\lambda}$. This computation of $\hat{\lambda}$ consists of a total of:

$$N_1(r, k) = \frac{1}{6}(k-1)^3r^3 + (k-1)^2r^2. \quad (4.77)$$

These Multiplications are distributed among tasks as follows:

Task	← Number multiplications
a. Computing the Cholesky decomposition L_1 of M	$\leftarrow \frac{1}{6}n_c^3 = \frac{1}{6}(k-1)^3r^3$.
b. Forward elimination of ξ from $L_1\xi = 2Bg$	$\leftarrow \frac{1}{2}n_c^2 = \frac{1}{2}(k-1)^2r^2$.
c. Back substitution of $\hat{\lambda}$ from $L_1^t\lambda = \xi$	$\leftarrow \frac{1}{2}n_c^2 = \frac{1}{2}(k-1)^2r^2$.

The maximum memory usage of the algorithm LMCS-1 is:

$$M_1(r, k) = \frac{1}{2}(k-1)^2r^2, \quad (4.78)$$

floating point elements. This memory is used during different tasks in LMCS-1 as follows:

a. Matrix M $\leftarrow \frac{1}{2}(p-r)^2 = \frac{1}{2}(k-1)^2r^2$.

b. Cholesky decomposition L_1 of $M \leftarrow L_1$ overwrites M as is computed.

[20] The positive-definiteness of M follows from $M = BB^t$ where $B = [I \quad C]$ has linearly independent rows. Therefore, the kernel of B^t is $\ker(B^t) = \{0\}$, and the value of the quadratic form $x^tMx = x^tBB^tx$ is positive for any value of $x \neq 0$.

[21] The nonsingularity of A, L, and U is assumed in this discussion.

A numerical example is offered in Appendix A to illustrate how algorithm LMCS-1 is applied to the circuit example C_1 introduced earlier.

Algorithm LMCS-2

The algorithm LMCS-2 described in this section is similar to algorithm LMCS-1 described in the previous subsection: both algorithms follow the same general course of computation. Specifically, algorithm LMCS-2 also first eliminates $\hat{\lambda}$ from $M\lambda = 2Bg$ (see equation 4.63) and next computes $\hat{s}$ according to equation 4.64. To determine the value of the Lagrange multipliers $\hat{\lambda}$, equation 4.63 is solved by finding the matrix inverse M^{-1} and then multiplying the right-hand side (2Bg) by M^{-1}:

$$\hat{\lambda} = M^{-1}(2Bg). \tag{4.79}$$

Note that the matrix inverse $M^{-1} = (I + CC^t)^{-1}$ in equation 4.79 can be expressed using the Sherman–Morrison–Woodbury formula (Golub and Loan, 1996):

$$(D + EF^t)^{-1} = D^{-1} - D^{-1}E(I + F^tD^{-1}E)^{-1}F^tD^{-1}, \tag{4.80}$$

where $D \in \mathbb{R}^{n\times n}$, $E \in \mathbb{R}^{n\times k}$, $F \in \mathbb{R}^{n\times k}$, and both D and $(I + F^tD^{-1}E)$ are nonsingular. When applied to the matrix $M^{-1} = (I + CC^t)^{-1}$, the Sherman–Morrison–Woodbury formula described by equation 4.80 yields:[22]

$$\left.\begin{array}{l} D = I \\ E = F = C \\ N = I + C^tC \end{array}\right\} \Rightarrow \begin{array}{l} M^{-1} = (I + CC^t)^{-1} \\ \quad = I - C(I + C^tC)^{-1}C^t. \\ \quad = I - CN^{-1}C^t \end{array} \tag{4.81}$$

Note that in equation 4.81, not only can the matrix inverse $N^{-1} = (I + C^tC)^{-1}$ be computed more quickly than M^{-1} (the dimension of N is $n_b \times n_b$ versus $n_c \times n_c = (k-1)r \times (k-1)r$ for M), but the computation of this inverse N^{-1} matrix does *not* have to be explicitly performed to evaluate the product $CN^{-1}C^t$ in equation 4.81. Let the Cholesky decomposition of $N = I + C^tC$ be the following:

$$N = L_2L_2^t, \tag{4.82}$$

and substitute equation 4.82 into the product $C(I + C^tC)^{-1}C^t = CN^{-1}C^t$ in equation 4.81, then:

$$\begin{aligned} M^{-1} &= I - CN^{-1}C^t \\ &= I - C(L_2L_2^t)^{-1}C^t \\ &= I - (C(L_2^t)^{-1})(L_2^{-1}C^t) \\ &= I - X^tX, \end{aligned} \tag{4.83}$$

where X is used to denote the product $(L_2^{-1}C^t)$. The matrix X can be computed by forward elimination according to the matrix equation $L_2X = C^t$, while the product $CN^{-1}C^t$ is equal to the product X^tX. Also, observe that the matrix M^{-1} can be computed one row at a time, thereby drastically reducing the storage requirements of the algorithm. The *j*th row of M^{-1} is computed and used to calculate the Lagrange multiplier $\hat{\lambda}_j$ as the inner product of this *j*th row of M^{-1} and the vector $2Bg$. The memory used to store the elements of the *j*th row of M^{-1} is then overwritten with the elements of the $(j+1)$-th row of M^{-1} and so on. The rows of the matrix M^{-1} can be stored in a disk to permit the rows to be retrieved for future execution.

Just as in algorithm LMCS-1, the complexity of algorithm LMCS-2 is dominated by the complexity of computing the Lagrange multipliers $\hat{\lambda}$. This computation of $\hat{\lambda}$ consists of a total of:

$$N_2(r, k) = \left[\frac{1}{6} + \frac{1}{2}(k-1) + \frac{1}{2}(k-1)^2\right]r^3 + (k-1)r^2. \tag{4.84}$$

These multiplications are distributed among the following tasks:

Task	← Number multiplications
a. Computing the Cholesky decomposition L_2 of N	$\leftarrow \frac{1}{6}r^3$.
b. Forward elimination of X from $L_2X = C^t$	$\leftarrow \frac{1}{2}r^2(p-r) = \frac{1}{2}(k-1)r^3$.
c. Evaluate $M^{-1} = I - X^tX$	$\leftarrow \frac{1}{2}r(p-r)^2 = \frac{1}{2}(k-1)^2r^3$.
d. Evaluate $\hat{\lambda} = M^{-1}(2Bg)$	$\leftarrow (p-r)^2 = (k-1)^2r^2$.

The maximum memory usage of algorithm LMCS-2 is as written here:

$$M_2(r, k) = (k - \frac{1}{2})r^2 + (k-1)r, \tag{4.85}$$

floating point elements. This memory usage is distributed among different tasks in LMCS-2 as follows:

a. Matrix N	← Requires $1/2r^2$ storage units
b. Cholesky decomposition L_2 of N	← L_2 overwrites N as is computed
c. Matrix X from $L_2X = C^t$	← L_2 Requires $r(p-r) = (k-1)r^2$ storage units
d. Matrix $M^{-1} = I - X^tX$	← Requires $(p-r) = (k-1)r$ storage units for one row of M

Refer to Appendix B for a numerical example illustrating algorithm LMCS-2 as applied to the circuit example C_1 introduced previously.

Algorithm CSD

Unlike algorithms LMCS-1 and LMCS-2, the clock schedule $\hat{s}$ is computed directly in algorithm CSD (i.e., without first computing the Lagrange multipliers $\hat{\lambda}$). With this strategy, the clock schedule $\hat{s}$ is determined according to equation 4.74:

[22] Note that $I + C^tC$ is positive-definite, thus nonsingular.

$$Z = B^t M^{-1} B \Rightarrow \hat{s} = -Zg + g = (-Z + I)g, \qquad (4.86)$$

where the matrix Z is introduced in equation 4.86 to simplify the notation. To evaluate Z, the expression described by equation 4.81 is substituted for M^{-1} into equation 4.86, and the product $Z = B^t M^{-1} B$ is evaluated using the same technique as in equations 4.82 and 4.83:

$$\begin{aligned} Z = B^t M^{-1} B &= B^t(I - CN^{-1}C^t)B \\ &= B^t B - B^t C N^{-1} C^t B \\ &= B^t B - B^t C (L_2^t)^{-1} L_2^{-1} C^t B \\ &= B^t B - Y^t Y. \end{aligned} \qquad (4.87)$$

The notation

$$Y = L_2^{-1} C^t B \qquad (4.88)$$

is introduced in equation 4.87 for simplicity, whereas such as for the previously described algorithm LMCS-2, the matrix Y can be eliminated according to the equation $L_2 Y = C^t B$.

The clock schedule $\hat{s}$ can be computed if the operations described by equations 4.86 through 4.88 are carried on literally. These expressions, however, can be manipulated to significantly reduce both the run time and memory requirements for algorithm CSD. Initially, note that computing each clock skew s_i requires evaluating the inner product of two dense p-element-long vectors—the ith row of the matrix $(-Z + I)$ and g. The evaluation of this inner product requires p multiplications, where p is the number of local data paths in the circuit. Recall, however, that the values of the clock skews from the basis s^b provide sufficient information to reconstruct all clock skews s in a quick fashion. Specifically, once the skews from the basis s^b are known, the skews s^c in the chords of the circuit may be derived through the operation described by equation 4.33:

$$\begin{bmatrix} I & C \end{bmatrix} \begin{bmatrix} s^c \\ s^b \end{bmatrix} = s^c + Cs^b = 0 \Rightarrow s^c = -Cs^b. \qquad (4.89)$$

Since only the basis s^b is evaluated, only the *last* n_b rows of the matrix $(-Z + I)$ are computed, thereby yielding significant savings of computation time. (Note that computing one row of Z requires the evaluation of p row elements, each row requiring r' multiplications in the product $Y^t Y$.) These concepts are illustrated graphically in Figure 4.9.

The complexity of the evaluation of $(-Z + I) = (-B^t B + Y^t Y + I)$ can be reduced further by examining the computation of Y. Typically, the direct evaluation of Y—by forward elimination from $L_2 Y = C^t B$—requires $1/2pr^2 = 1/2kr^3$ multiplications. This number can be reduced by noting that:

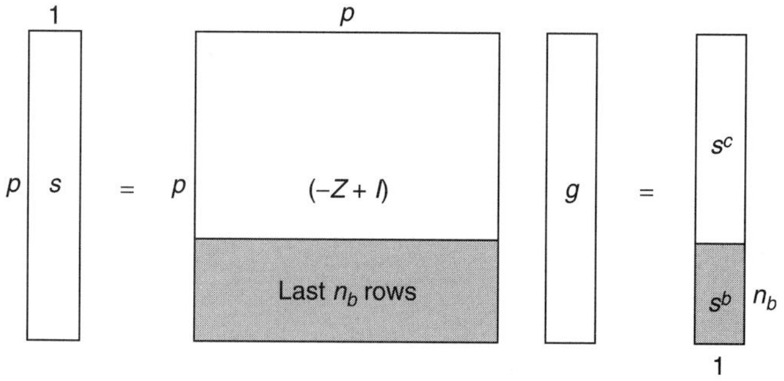

FIGURE 4.9 Computation of the Clock Schedule Basis s^b by Computing Only the Last n_b Rows of the Matrix $-Z + I$

$$\begin{aligned} C^t B &= C^t \begin{bmatrix} I & C \end{bmatrix} = \begin{bmatrix} C^t & CC^t \end{bmatrix} \\ &= \begin{bmatrix} C^t & N - I \end{bmatrix} = \begin{bmatrix} C^t & L_2 L_2^t - I \end{bmatrix} \qquad (4.90) \\ \text{and } Y &= \begin{bmatrix} Y_1 & Y_2 - Y_3 \end{bmatrix}, \qquad (4.91) \end{aligned}$$

where the matrices Y_1, Y_2, and Y_3 can be eliminated from the following dependencies, respectively:

$$\begin{aligned} &L_2 Y_1 = C^t \leftarrow \text{compute } Y_1 && \rightarrow \text{requires } (1/2)(k-1)r^3 \text{ multiplications} && (4.92) \\ &L_2 Y_2 = N \leftarrow \quad Y_2 = L_2^t && \rightarrow \text{already computed} && (4.93) \\ &L_2 Y_3 = I \leftarrow \text{compute } Y_3 && \rightarrow \text{requires } \tfrac{1}{6}r^3 + \tfrac{1}{6}(3r^2 + 2r) \text{ multiplications} && (4.94) \end{aligned}$$

Finally, the following transformations of equations 4.95 through 4.97 are used to evaluate the matrix $(-Z + I)$:

$$\begin{aligned} B^t B &= \begin{bmatrix} I \\ C^t \end{bmatrix} \begin{bmatrix} I & C \end{bmatrix} = \begin{bmatrix} I & C \\ C^t & C^t C \end{bmatrix} = \begin{bmatrix} I & C \\ C^t & N - I \end{bmatrix} \\ &= I + \begin{bmatrix} O & C \\ C^t & N - 2I \end{bmatrix}, \end{aligned} \qquad (4.95)$$

$$\begin{aligned} V = Y^t Y &= \begin{bmatrix} Y_1^t \\ Y_2^t - Y_3^t \end{bmatrix} \begin{bmatrix} Y_1 & Y_2 - Y_3 \end{bmatrix} \\ &= \begin{bmatrix} V_{11} & V_{12} \\ (L_2 L_2^{-1} C^t - L_2^{-t} L_2^{-1} C^t) & (L_2 L_2^t + L_2^{-t} L_2^{-1} - L_2 L_2^{-1} - L_2^{-t} L_2^t) \end{bmatrix} \\ &= \begin{bmatrix} V_{11} & V_{12} \\ C^t - (L_2^{-t} L_2^{-1}) C^t & N - 2I + L_2^{-t} L_2^{-1} \end{bmatrix}, \end{aligned} \qquad (4.96)$$

and

$$\begin{aligned} -Z + I &= -B^t B + Y^t Y + I \\ &= -I - \begin{bmatrix} O & C \\ C^t & N - 2I \end{bmatrix} \\ &\quad + \begin{bmatrix} V_{11} & V_{12} \\ C^t - (L_2^{-t} L_2^{-1}) C^t & N - 2I + L_2^{-t} L_2^{-1} \end{bmatrix} + I \\ &= \begin{bmatrix} \cdots & \cdots \\ -(L_2^{-t} L_2^{-1}) C^t & L_2^{-t} L_2^{-1} \end{bmatrix}. \end{aligned} \qquad (4.97)$$

Note that only the last r' rows of $(-Z+I)$ are shown in equation 4.97 because only these r' rows are required to compute s^b. Moreover, note that the matrix $Y_1 = L_2^{-1}C^t$ does not require evaluation. Only $Y_3 = L_2^{-1}$ must be determined (from $L_2Y_3 = I$) because $L_2^{-t} = (L_2^{-1})^t$.

The computation of the clock schedule $\hat{s}$ in algorithm CSD consists of a total of:

$$N_3(r, k) = \frac{1}{2}r^3 + \frac{1}{3}(3k+4)r^2 + \frac{1}{2}r - \frac{1}{6}. \quad (4.98)$$

These multiplications are distributed among the following tasks:

Task	← Number multiplications
a. Computing the Cholesky decomposition L_2 of N	← $\frac{1}{6}r^3$.
b. Forward elimination of $Y_3 = L_2^{-1}$ from $L_2Y_3 = I$	← $\frac{1}{6}r^3 + \frac{1}{2}r^2 + \frac{1}{3}r$.
c. Evaluate the product $L_2^{-t}L_2^{-1}$	← $\frac{1}{6}r^3 + \frac{1}{6}(5r^2 + r - 1)$.
d. Evaluate s^b	← $rp = kr^2$.

The maximum memory usage of algorithm CSD is as follows:

$$M_3(r, k) = r^2, \quad (4.99)$$

floating point elements. This memory usage is distributed among different tasks in CSD as follows:

a. Matrix N	← requires $1/2r^2$ storage units
b. Cholesky decomposition L_2 of N	← L_2 overwrites N as is computed
c. Matrix $L_2^{-1} = Y_3$	← L_2^{-1} overwrites L_2 as is computed
d. Product $L_2^{-t}L_2^{-1}$	← requires $1/2r^2$ storage units

A numerical example is provided in Appendix C to illustrate algorithm CSD.

4.5.2 Summary of the Proposed Algorithms

This section concludes with a brief synopsis of the run time and memory requirements of the three algorithms for solving problem QP-3 described in the previous subsections. To summarize the results, each of the three algorithms, LMCS-1, LMCS-2, and CSD, requires $O(r^3)$ floating-point multiplicative operations and $O(r^2)$ floating-point storage units. The numerical constant of the leading terms in the polynomial expressions for both the run time and memory complexity is a function of the ratio $k = p/r$, which is the ratio of the number of local data paths to the number of registers in a circuit.

To gain further insight into the proposed algorithms, the numerical constants of the leading terms in the polynomial run time complexity expressions are plotted versus k in Figure 4.10. Similarly, the numerical constants of the leading terms in the polynomial memory complexity expressions are plotted versus k in Figure 4.11. Note that algorithm CSD outperforms both of the other two LMCS algorithms where the superiority of algorithm CSD is particularly evident with respect to the speed of execution. Thus, algorithm CSD is the algorithm of choice for solving problem QP-3 as introduced in Section 5.4.3.

4.5.3 Unconstrained Basis Skews

Consider again the example circuit C_1 introduced in Section 4.3 (the graph of C_1 is shown in Figure 4.7). A modified version of C_1 with one additional edge—the edge e_6—is shown in Figure 4.12. Also shown with thicker edges in Figure 4.12 is a spanning tree for the modified circuit C_1. Note that the basis edge e_6 does *not* belong to any of the fundamental cycles of the circuit depicted in Figure 4.12. In fact, the edge e_6 does not belong to any cycles of the circuit in Figure 4.12 at all. Such basis edges that do not belong to any cycles are called **isolated** while the rest of the basis edges are called **main**. Note that any isolated edge must necessarily by definition be a basis edge.[23]

Theoretically, a circuit with r registers (the vertices in the circuit graph) may have any number n_i of isolated basis edges

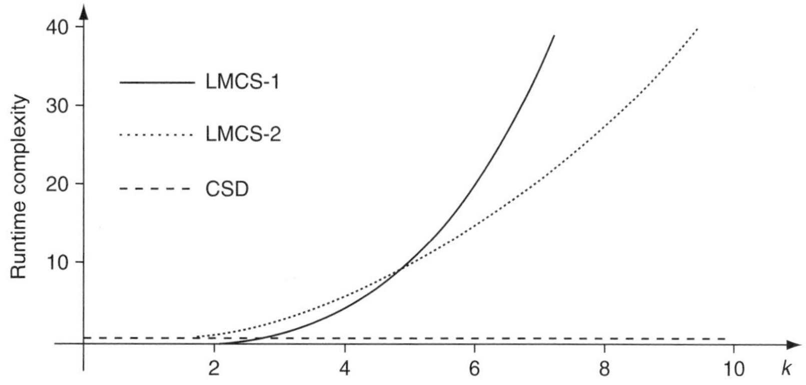

FIGURE 4.10 The Numerical Constants (as Functions of $k = p/r$) of the Term r^3. These are in the run time complexity expressions for the algorithms LMCS-1, LMCS-2, and CSD, respectively.

[23] A chord edge is already a part of a cycle and cannot be isolated.

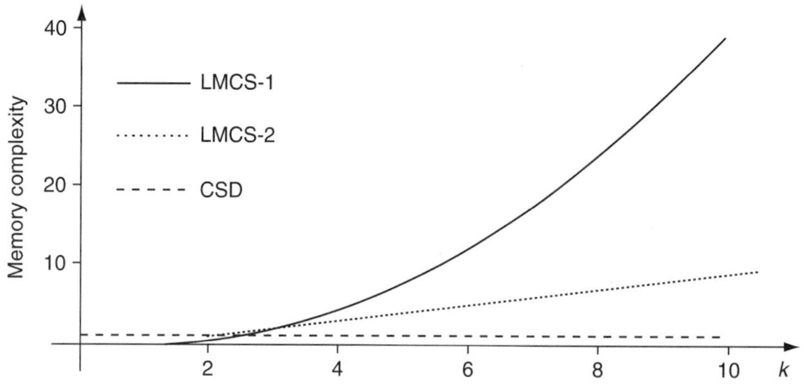

FIGURE 4.11 The Numerical Constants (as Functions of $k = p/r$) of the Term r^2. These are in the memory complexity expressions for the algorithms LMCS-1, LMCS-2, and CSD, respectively.

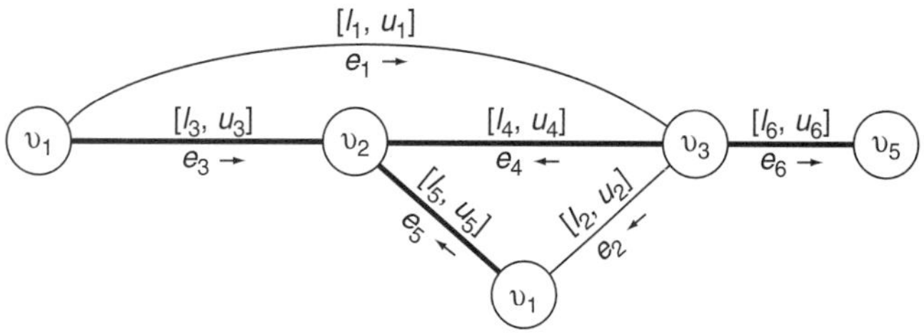

FIGURE 4.12 Modified Example Circuit C_1 to Include an Additional Edge e_6. (C_1 was originally introduced in Section 4.3 and illustrated in Figure 4.7.)

where n_i ranges from zero to $r - 1 = n_b$. A circuit with $n_i = n_b = r - 1$ isolated basis edges does not have any cycles whatsoever—all edges of such circuits are basis edges, and there are no chord edges to complete a cycle. A simple example of such a circuit is a shift register.

Note that since isolated edges do not belong to a cycle, the clock skews on these edges are linearly independent of any other clock skews in the circuit. Intuitively, the clock skew of an isolated edge can be assigned to be any value without contradicting the linear dependencies among the skews in a circuit. Observe, for example, equation 4.31 written for the modified circuit C_1 shown in Figure 4.12:

$$Bs = \begin{bmatrix} 1 & 0 & -1 & 1 & 0 & 0 \\ 0 & 1 & 0 & -1 & 1 & 0 \end{bmatrix} \begin{bmatrix} s_1 \\ s_2 \\ s_3 \\ s_4 \\ s_5 \\ s_6 \end{bmatrix} = 0. \qquad (4.100)$$

All of the elements in the sixth column of B are zeros. Therefore, if s_1 through s_5 are such that equation 4.100 is satisfied, the choice of s_6 does not invalidate this equation.

This fact can be exploited in the mathematical solution of problem QP-1 to decrease the number of variables, thereby decreasing the run time and memory requirements. The only requirement is that the basis skews (the edges) must be enumerated such that the isolated skews are last. In other words, the clock skew vector for equation 4.28 becomes:

$$s = \begin{bmatrix} s^c \\ s^b \\ s^i \end{bmatrix} = \Big[\underbrace{\overbrace{s_1 \ldots s_{n_c}}^{n_c}}_{\text{Chords}} \overbrace{\underbrace{\overbrace{s_{n_c} + 1 \ldots s_p - n_i}^{n_b - n_i}}_{\text{Main basis}} \underbrace{\overbrace{s_{p-n_i} + 1 \ldots s_p}^{n_i}}_{\text{Isolated basis}}}^{\text{Basis with } n_b \text{ elements}} \Big]^t, \qquad (4.101)$$

where s^b stands for the main basis, and the isolated basis is denoted by s^i. With this specific choice of clock skew enumeration, the B matrix in equation 4.31 becomes:

$$B = [\, B_1 \quad 0 \,], \qquad (4.102)$$

where 0 in equation 4.102 is a zero matrix of dimension $n_c \times n_i$.

With this notation, it is straightforward to show that the matrix M in equation 4.62 becomes:

$$M = BB^t = B_1 B_1^t \qquad (4.103)$$

The solution to problem QP-1 equations 4.63 and 4.64 is as follows:

$$\hat{\lambda} = 2M^{-1}Bg = 2M^{-1}[\, B_1 \quad 0 \,] \begin{bmatrix} g^c \\ g^b \\ g^i \end{bmatrix} = 2M^{-1}B_1 \begin{bmatrix} g^c \\ g^b \end{bmatrix}. \qquad (4.104)$$

$$\hat{s} = g - \left(B^t M^{-1} B\right) g = \begin{bmatrix} (I - B_1 M^{-1} B_1^t) \begin{bmatrix} g^c \\ g^b \end{bmatrix} \\ g^i \end{bmatrix} \qquad (4.105)$$

As can be observed in equations 4.104 and 4.105, the following is true:

1. The choice of the objective isolated basis skews g^i has *no* effect on either the Lagrange multipliers of equation 4.104 or the chords and main skew basis from the equation 4.105 solution.
2. The final solution for the clock skews s^i in the isolated basis edges corresponds precisely to the objective skew values g^i for these edges.

Therefore, the isolated basis edges can be completely excluded from consideration when solving problem QP-1. Equations 4.104 and 4.105 demonstrate that the final clock skew values of these edges can be chosen arbitrarily provided these values satisfy the permissible range requirements.

4.5.4 *I/O* Registers and Target Delays

The clock skew scheduling methodology discussed in Sections 4.3 and 4.4 is based on the assumption that complete connectivity and timing information is available for all local data paths in a circuit. This condition may, however, not be realistic. Consider, for example, the input and output registers (also called the *I/O* registers) in a VLSI system. Some *I/O* registers are illustrated in Figure 4.13, where the registers R_1 and R_5 are an input register and an output register, respectively, of the circuit *C*. The register R_3 shown in Figure 4.13 is an internal register because all of the other registers to which R_3 is connected (via local data paths) are inside the circuit *C*.

The timing of the *I/O* registers is less flexible than the timing of the internal registers. Consider, for example, the local data path $R_6 \rightsquigarrow R_1$ shown in Figure 4.13. The register R_6 is outside the circuit *C* that contains the registers R_1 through R_5. It is possible to apply a clock schedule to *S* that specifies a clock delay t_{cd}^1 to the register R_1. However, the timing information for the local data path $R_6 \rightsquigarrow R_1$ is *not* considered when scheduling the clock signal delays to the registers in *C* (including t_{cd}^1). Therefore, a timing violation may occur on the local data path $R_6 \rightsquigarrow R_1$ illustrated in Figure 4.13.

One strategy to overcome this difficulty is to include in the clock scheduling process the timing information of those local data paths that cross the boundaries of the circuit *C*. This approach does not change the nature of the clock scheduling algorithm but rather only the number of timing constraints. Such an optimization scenario, however, is difficult to conceive due to the many instances in which *C* may be used. Therefore, a preferable approach is to set the clock signal delay to the *I/O* registers (such as t_{cd}^1 to R_1) to a specific value with respect to the clock source (shown as the clock pin in Figure 4.13). If this value is specified, all of the necessary timing information is available to avoid any timing violations of the local data paths, such as the path $R_6 \rightsquigarrow R_1$ shown in Figure 4.13. Equivalently, a group of registers (e.g., the *I/O* registers) may be defined, which require that the clock signal be delivered to all of the registers in such a group with the same delay. Application-specific integrated circuits (ASICs) and intellectual property (IP) blocks are good examples of circuits for which the aforementioned strategy may be useful.

Given the difficulty in knowing *a priori* all timing contexts of an integrated circuit, a preferred solution may be to require that all *I/O* registers are clocked at the same time (zero skew). More specifically, all possible explicit clock delay requirements for registers in the circuit fall into one of the following categories:

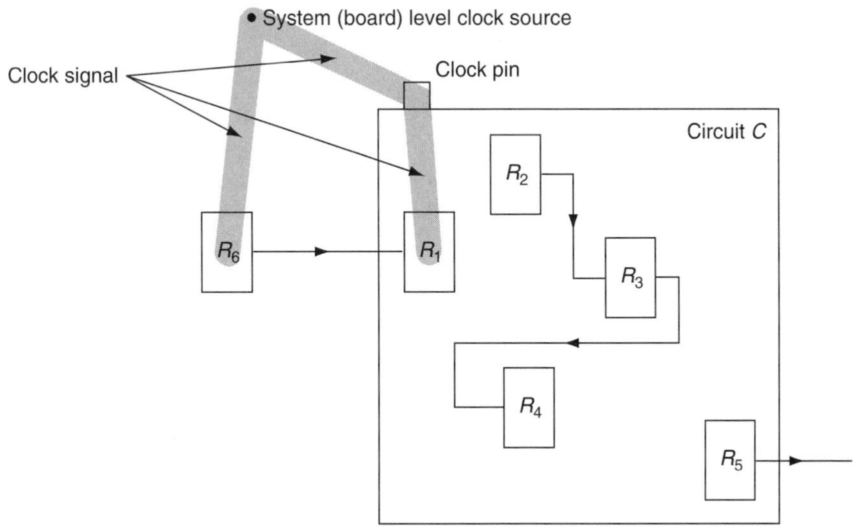

FIGURE 4.13 *I/O* Registers in a VLSI Integrated Circuit. Note that the *I/O* registers form part of the local data paths between the inside of the circuit and the outside of the circuit.

1. **Zero skew island**: The island refers to a group of registers with equal delays.
2. **Target delays**: $t_{cd}^{k_1} = \delta_{k_1}, \ldots, t_{cd}^{k_\alpha} = \delta_{k_\alpha}$, where $k_\alpha \leq r$ and $\delta_{k_1} \ldots \delta_{k_\alpha}$ are explicitly specified clock signal delay constants.
3. **Target skews**: $s_{j1} = \sigma_{j1}, \ldots, s_{j\beta} = \sigma_{j\beta}$, where $j_\beta < n_b$ and $\sigma_{j1} \ldots \sigma_{j\beta}$ are explicitly specified clock skew constants.

Zero skew islands can be satisfied by collapsing the corresponding graph vertices into a single vertex while eliminating all edges among vertices in the island. Note that in this case, it must be verified that zero skew is in the permissible range of each in-island path.[24] Alternatively, the target delays are converted to target skews (category 3 above) for sequentially adjacent pairs or by adding a "fake" edge. Thus, an algorithm to handle only target skews is necessary.

Note first that target values for only $n_f \leq n_b$ skews can be independently specified. As n_f approaches n_b, the freedom to vary all skews decreases, and it may become impossible to determine any feasible s. Given $n_f \leq n_b$, (a) the basis can always be chosen to contain all target skews by using a spanning tree algorithm with edge swapping and (b) the edge enumeration can be accomplished such that the target skews appear last in the basis. The problem is now similar to equation 4.51 except for the change of the circuit kernel equation:

$$C = \begin{bmatrix} C_1 & C_2 \end{bmatrix} \Rightarrow Bs = \begin{bmatrix} I & C_1 & C_2 \end{bmatrix} \begin{bmatrix} \hat{s}^c \\ \hat{s}^b \\ \sigma \end{bmatrix} \Rightarrow \hat{B}\hat{s} + C_2\sigma = 0, \tag{4.106}$$

where $\hat{B} = [I \quad C_1]$, $\hat{s} = [\hat{s}^c/\hat{s}^b]$, $\hat{s}^c = s^c$, and $\hat{s}^b$ is s^c with the last n_f elements removed. The matrix C_2 in equation 4.106 consists of the last n_f columns of C, while the target skew vector σ is an n_f-element vector of target skews whose elements are ordered in the order of the target edges. The linear system of equation 4.60 becomes:

$$\left| \begin{array}{l} 2\hat{s} + \hat{B}^t \hat{m} = 2\hat{g} \\ \hat{B}\hat{s} + C_2\sigma = 0 \end{array} \right. \Rightarrow \begin{bmatrix} 2I & \hat{B}^t \\ \hat{B} & 0 \end{bmatrix} \begin{bmatrix} \hat{s} \\ \hat{m} \end{bmatrix} = \begin{bmatrix} 2\hat{g} \\ -C_2\sigma \end{bmatrix}, \tag{4.107}$$

with solution:

$$\begin{aligned} \hat{m}^* &= 2\hat{M}^{-1}(\hat{B}\hat{g} + C_2\sigma) \\ \hat{s}^* &= (I - \hat{B}^t\hat{M}^{-1}\hat{B})\hat{g} - \hat{B}^t\hat{M}^{-1}C_2\sigma. \end{aligned} \tag{4.108}$$

[24] Normally, this would be the case. However, in an aggressive circuit design with a short clock period, it may so happen that zero skew is designed to be out of the permissible range, most likely creating a setup time violation. In these circuits, negative skew is used to increase the overall system-wide clock frequency, thereby removing the setup violation.

4.6 Experimental Results

A quadratic programming formulation of the clock skew scheduling problem was developed in Section 4.3. This QP problem can be efficiently solved by applying the mathematical procedures developed in Section 4.4. Algorithm CSD as applied to the QP problem described in Section 5.4.3 has been implemented as a C + + program and applied to ISCAS'89 and ISCAS'93 benchmark circuits as well as to industrial circuits (IC1, IC2, and IC3). Results from the application of this computer program are described in this section. Certain characteristics of the implementation are initially described in Section 4.6.1. Graphical illustrations of representative results are shown in Section 4.6.2.

4.6.1 Description of Computer Implementation

The results described in this section are obtained from the execution of a computer implementation of algorithm CSD introduced in Section 4.5. Without unnecessary details, this computer implementation consists of the sequential execution of the following major steps:

Step 1. The circuit timing and connectivity data are read in and compressed and stored in a binary database. The database can be used for fast data access in subsequent algorithmic applications of the same circuit. Furthermore, the data size of the database permits significant space and time savings if the circuit data are exchanged.

Step 2. The circuit data are examined, and the circuit graph is built according to the graph model described in Section 4.2.2. An adjacency lists data structure (Cormen *et al.*, 1989) stored in memory and is used for fast access of the circuit graph data.

Step 3. The circuit graph is transformed according to the transformation rules described in Section 4.2.2 and illustrated in Figure 4.5. Within this step, the permissible range bounds are calculated and directions for the graph edges are determined.

Step 4. The circuit graph is traversed to determine the edges in the skew basis s^b and in the skew chords s^c. This graph traversal is accomplished by using a depth-first search (West, 1996; Reingold *et al.*, 1997; Cormen *et al.*, 1989) algorithm—the classical traversal algorithm of choice for building a spanning tree. Three additional important tasks are accomplished during the traversal step:

1. For circuits with more than one connected disjoint subcircuit, these connected disjoint parts are identified and marked. This step does not incur any computational overhead—it is an inherent feature of the depth-first search graph traversal algorithm to separate a graph into disjoint pieces (if any).

2. The skew basis and chords of each disjoint connected circuit subgraph are identified and enumerated.
3. The circuit connectivity matrix B (actually, only the nonidentity matrix C portion of B) is derived for each disjoint connected circuit subgraph. Recall that C contains only elements from the set $\{-1, 0, 1\}$, thus permitting an efficient bit compression scheme to be used to store C in a small amount of memory.

Step 5. Using C, the matrix N is computed as described by equation 4.81.

Step 6. The Cholesky factorization L_2 of N is calculated as described by equation 4.82. Simple, yet efficient, algorithms for computing the Cholesky factorization have long been known and can be found in multiple sources (Farebrother, 1988; Lawson and Hanson, 1974; Golub and Loan, 1996; Forsythe and Moler, 1967; Cormen *et al.*, 1989). Recall that the matrix N is guaranteed to be positive-definite by construction. Therefore, the real (no complex numbers) Cholesky decomposition is guaranteed to exist.

Step 7. The objective clock skews are chosen at the center of the permissible range for all local data paths. The actual clock skews (a consistent clock schedule) are calculated as described by equation 4.97 and as illustrated in Figure 4.9. At this point, each clock skew is verified against the respective permissible range. If all skews are in the respective permissible range bounds, the algorithm concludes. Otherwise, the objective clock skews are modified, and the calculation is repeated again. Only the calculation described in this step must be repeated since all matrices have now been computed.

Different objective clock schedule modification strategies can be used. The most effective strategy to modify the objective clock schedule—resulting in the fastest convergence toward a feasible schedule—is as follows. All objective clock skews are slightly increased or decreased depending on whether the respective calculated clock skew is larger or smaller than the objective one. Using this strategy, a feasible solution is typically reached within a few iterations.

Step 8. The actual clock delays to the individual registers are calculated by traversing the spanning tree (basis) of the circuit graph. The clock delay of the first register is arbitrarily chosen (zero in this implementation). As the spanning tree is traversed, additional vertices adjacent to the current vertex are visited. The clock delay of the visited vertex is determined trivially because both the clock delay of the current vertex and the clock skew of the edge between the current and visited vertex are known.

The results of the application of the algorithm to these circuits are summarized in Table 4.3. For each circuit, the following data is listed—the circuit name in column **1**; the number of disjoint subgraphs in column **2**; and the number of vertices, edges, chords (cycles), main and isolated basis, and target clock period in nanoseconds in columns **3** through **8**, respectively. The number of iterations to reach a solution is listed in column **9**. The average value of ε in equation 4.51 (i.e., $\sqrt{\varepsilon/p}$) is listed in column **10**. The run time in minutes for the mathematical portion of the program is shown in column **11** for a 170-MHz Sun Ultra 1 workstation.

4.6.2 Graphical Illustrations of Results

The application of the computer implementation described in Section 4.6.1 to many of the circuits listed in Table 4.3 is graphically illustrated. Immediately following are illustrations of two circuits shown in Figures 4.14 and 4.15, respectively. Three histograms for a circuit are shown in each graphical illustration. These histograms are as follows:

(A) The distribution of the zero clock skews in the permissible range for the clock period listed in Table 4.3 is illustrated in Figures 4.14(A) and 4.15(A).

(B) The distribution of the nonzero clock skews in the permissible range after one iteration of problem QP-2—as described in Step 7 in Section 4.6.1—is shown in Figures 4.14(B) and 4.15(B). Note that there are frequent lower bound and upper bound violations of the permissible range. These violations are represented by the dark leftmost and rightmost regions, respectively, where the number of violations is also indicated.

(C) The final distribution of the nonzero clock skews within the permissible range—no timing violations—is illustrated in Figures 4.14(C) and 4.15(C). There is a noticeable improvement because most clock skews are concentrated around the center of the permissible range. The majority of the clock skews are within 10% of the safest clock skew value at the center of the respective permissible range of each local data path.

The computer algorithms used to solve the QP problems are numerically exemplified in this appendix for the circuit example C_1 introduced in Section 4.3. The algorithms LMCS-1, LMCS-2, and CSD—introduced in Section 4.5—are illustrated in Appendices A, B, and C, respectively. Note that in all cases, the same final clock schedule is achieved.

Appendix A: Algorithm LMCS-1

Algorithm LMCS-1, originally introduced in Section 4.5, is illustrated numerically in this part of the appendix. Consider the circuit example C_1 introduced in Section 4.3; the corresponding circuit graph and spanning tree are shown in Figures 4.7 and 4.8(A), respectively. The matrices B, M, and L_1 in the Cholesky decomposition of M are as follows:

TABLE 4.3 Experimental Results of the Application of the QP-based Clock Scheduling Algorithm to Both Benchmark and Industrial Circuits

Circuit	# Subcircuits	r	p	n_c	n_m	n_i	T_{CP} [nanoseconds]	# Iterations	$\sqrt{\frac{\varepsilon}{p}}$	Run time [min]
1	2	3	4	5	6	7	8	9	10	11
s1196	7	18	20	9	8	3	20.8	5	3.19	1
s1238	7	18	20	9	8	3	20.8	5	3.19	1
s13207	49	669	3068	2448	581	39	85.6	20	18.92	5
s1423	2	74	1471	1399	72	0	92.2	20	60.9	3
s1488	1	6	15	10	5	0	32.2	1	0.87	1
s1494	1	6	15	10	5	0	32.8	1	0.88	1
s15850	15	597	14257	13675	546	36	116	10	70.6	21
s15850.1	22	534	10830	10318	478	34	81.2	9	31.44	19
s208.1	1	8	28	21	7	0	12.4	1	1.22	1
s27	1	3	3	1	2	0	6.6	1	0.71	1
s298	1	14	54	41	12	1	13	1	1.16	1
s344	1	15	68	54	14	0	27	4	4.91	1
s349	1	15	68	54	14	0	27	4	4.91	1
s35932	1	1728	4187	2460	1727	0	34.2	20	60.4	27
s382	1	21	113	93	20	0	14.2	6	1.59	2
s38417	11	1636	28082	26457	1443	182	69	20	32.35	31
s38584	2	1452	15545	14095	1400	50	94.2	11	29.1	29
s386	1	6	15	10	5	0	17.8	1	0.82	1
s400	1	21	113	93	20	0	14.2	8	1.6	1
s420.1	1	16	120	105	15	0	16.4	20	1.95	1
s444	1	21	113	93	20	0	16.8	2	1.05	1
s510	1	6	15	10	5	0	16.8	1	0.85	1
s526	1	21	117	97	20	0	13	2	1.26	1
s526n	1	21	117	97	20	0	13	2	1.26	2
s5378	1	179	1147	969	158	20	28.4	20	8.79	3
s641	1	19	81	63	18	0	83.6	5	11.67	1
s713	1	19	81	63	18	0	89.2	6	12.74	1
s820	1	5	10	6	4	0	18.6	1	0.71	1
s832	1	5	10	6	4	0	19	2	0.66	1
s838.1	1	32	496	465	31	0	24.4	3	3.68	3
s9234	3	228	2476	2251	222	3	75.8	20	16.67	4
s9234.1	2	211	2342	2133	205	4	75.8	20	18.6	4
s953	4	29	135	110	25	0	23.2	3	1.93	2
s1269	1	37	251	215	36	0	51.2	20	12.73	2
s1512	1	57	405	349	56	0	39.6	4	4.43	3
s3271	1	116	789	674	107	8	40.4	3	3.64	5
s3330	1	132	514	383	61	70	34.8	4	3.4	5
s3384	25	183	1759	1601	151	7	85.2	5	15.5	7
s4863	1	104	620	517	103	0	81.2	8	39.85	3
s6669	20	239	2138	1919	218	1	128.6	3	20.67	6
s938	1	32	496	465	31	0	24.4	2	3.41	2
s967	4	29	135	110	25	0	20.6	2	1.76	2
s991	1	19	51	33	18	0	96.4	3	8.58	1
IC1	1	500	124750	124251	499	0	8.2	2	1.51	30
IC2	1	59	493	435	58	0	10.3	3	1.82	4
IC3	34	1248	4322	3108	1155	59	5.6	2	1.43	2

$$
\begin{aligned}
B &= \begin{bmatrix} 1 & 0 & -1 & 1 & 0 \\ 0 & 1 & 0 & -1 & 1 \end{bmatrix}, \\
M = BB^t = L_1 L_1^t &= \begin{bmatrix} 3 & -1 \\ -1 & 3 \end{bmatrix}, \\
L_1 &= \frac{1}{\sqrt{3}} \begin{bmatrix} 3 & 0 \\ -1 & 2\sqrt{2} \end{bmatrix}.
\end{aligned} \tag{4.109}
$$

The right-hand side of equation 4.75 is the following:

$$
\begin{aligned}
2Bg &= \begin{bmatrix} 1 & 0 & -1 & 1 & 0 \\ 0 & 1 & 0 & -1 & 1 \end{bmatrix} \left(\frac{1}{4} [1 \; -2 \; 5 \; -11 \; 1]^t \right) \\
&= \begin{bmatrix} -\frac{15}{2} \\ 5 \end{bmatrix},
\end{aligned} \tag{4.110}
$$

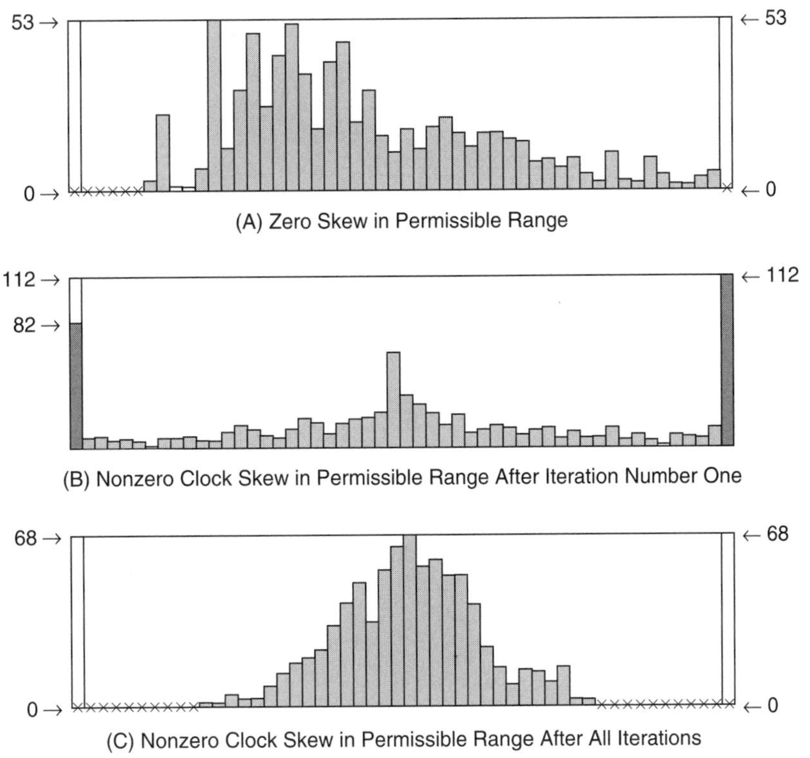

FIGURE 4.14 Circuit s3271 with $r = 116$ Registers and $p = 789$ Local Data Paths. The target clock period is $T_{CP} = 40.4$ ns.

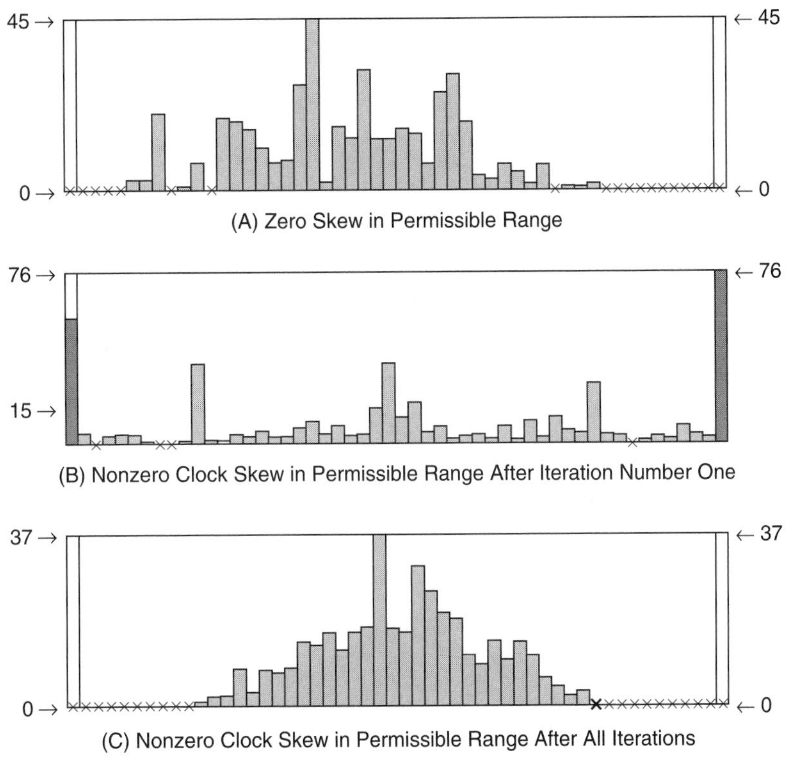

FIGURE 4.15 Circuit s1512 with $r = 57$ Registers and $p = 405$ Local Data Paths. The target clock period is $T_{CP} = 39.6$ ns.

where the vector of the objective clock skews $g = \frac{1}{4}[1 \quad -2 \quad 5 \quad -11 \quad 1]^t$ is as specified in column four of Table 4.2. The lower triangular system:

$$L_1\xi = \frac{1}{\sqrt{3}}\begin{bmatrix} 3 & 0 \\ -1 & 2\sqrt{2} \end{bmatrix}\begin{bmatrix} \xi_1 \\ \xi_2 \end{bmatrix} = 2Bg = \begin{bmatrix} -15/2 \\ 5 \end{bmatrix}, \quad (4.111)$$

is solved first, yielding (note the order of elimination of unknowns):

$$\xi_1 = -\frac{15}{2\sqrt{3}}, \text{ then } \xi_2 = 5 + \frac{\sqrt{3}}{3}\xi_1 = \frac{5\sqrt{3}}{4\sqrt{2}}. \quad (4.112)$$

Finally, the upper triangular system:

$$L_1^t\lambda = \frac{1}{\sqrt{3}}\begin{bmatrix} 3 & -1 \\ 0 & 2\sqrt{2} \end{bmatrix}\begin{bmatrix} \lambda_1 \\ \lambda_2 \end{bmatrix} = \begin{bmatrix} \xi_1 \\ \xi_2 \end{bmatrix} = \frac{5\sqrt{3}}{2}\begin{bmatrix} -1 \\ \sqrt{2}/4 \end{bmatrix}, \quad (4.113)$$

is solved through back substitution (note the order of elimination of λ_1 and λ_2) to find the Lagrange multipliers:

$$\lambda_2 = \frac{15}{16}, \text{ then } \lambda_1 = \frac{1}{\sqrt{3}}\left(-\frac{5\sqrt{3}}{2} + \frac{\sqrt{3}}{3}\lambda_2\right) = -\frac{35}{16}. \quad (4.114)$$

The clock skew schedule s is determined from the first row of equation 4.59:

$$s = \begin{bmatrix} s_1 \\ s_2 \\ s_3 \\ s_4 \\ s_5 \end{bmatrix} = -\frac{1}{2}B^t\lambda + g = \frac{1}{32}\left(\begin{bmatrix} 35 \\ -15 \\ -35 \\ 50 \\ 15 \end{bmatrix} + \begin{bmatrix} 8 \\ -16 \\ 40 \\ -88 \\ 8 \end{bmatrix}\right) = \frac{1}{32}\begin{bmatrix} 43 \\ -31 \\ 5 \\ -38 \\ -7 \end{bmatrix}. \quad (4.115)$$

The clock delays t_{cd} corresponding to this schedule are determined from equations 4.39 and 4.40:

$$s^b = \begin{bmatrix} t_{cd}^1 - t_{cd}^2 \\ t_{cd}^3 - t_{cd}^2 \\ t_{cd}^4 - t_{cd}^2 \end{bmatrix} = \begin{bmatrix} 1 & -1 & 0 & 0 \\ 0 & -1 & 1 & 0 \\ 0 & -1 & 0 & 1 \end{bmatrix}\begin{bmatrix} t_{cd}^1 \\ t_{cd}^2 \\ t_{cd}^3 \\ t_{cd}^4 \end{bmatrix} = \frac{1}{32}\begin{bmatrix} 5 \\ -38 \\ -7 \end{bmatrix} \Rightarrow \begin{bmatrix} t_{cd}^1 \\ t_{cd}^2 \\ t_{cd}^3 \\ t_{cd}^4 \end{bmatrix} = \frac{1}{32}\begin{bmatrix} 43 \\ 38 \\ 0 \\ 31 \end{bmatrix}. \quad (4.116)$$

Note that the solution given by equation 4.116 corresponds precisely to the solution described by equation 4.21. As a matter of fact, the clock schedule in the latter equation, 4.21, is obtained by applying algorithm LMCS-1 to the circuit C_1.

Appendix B: Algorithm LMCS-2

To illustrate algorithm LMCS-2 (introduced in Section 4.5), consider again the circuit example C_1 introduced in Section 4.3; the corresponding circuit graph and spanning tree are shown in Figures 4.7 and 4.8(A), respectively. The matrices $N = I + C^tC$, L_2, and X, respectively, in equation 4.83 are as follows:

$$N = L_2L_2^t = \begin{bmatrix} 2 & -1 & 0 \\ -1 & 3 & -1 \\ 0 & -1 & 2 \end{bmatrix}, \quad L_2 = \frac{1}{\sqrt{2}}\begin{bmatrix} 2 & 0 & 0 \\ -1 & \sqrt{5} & 0 \\ 0 & -\frac{2}{\sqrt{5}} & -\frac{4}{\sqrt{5}} \end{bmatrix}, \quad \text{and } X = \frac{1}{2\sqrt{10}}\begin{bmatrix} -2\sqrt{5} & 0 \\ 2 & -4 \\ 1 & 3 \end{bmatrix}. \quad (4.117)$$

The matrix inverse $M^{-1} = I - CN^{-1}C^t = I - X^tX$ and the Lagrange multipliers $\hat{\lambda}$, respectively, are as written here:

$$M^{-1} = I - X^tX = \frac{1}{8}\begin{bmatrix} 3 & 1 \\ 1 & 3 \end{bmatrix}, \quad \hat{\lambda} = M^{-1}(2Bg) = \begin{bmatrix} -15/2 \\ 5 \end{bmatrix} = \frac{1}{16}\begin{bmatrix} -35 \\ 15 \end{bmatrix}. \quad (4.118)$$

The values of λ_1 and λ_2 given by equation 4.118 are the same as derived in equations 4.114, and the clock schedule is determined in the same way as described by equations 4.115 and 4.116.

Appendix C: Algorithm CSD

Algorithm CSD was introduced in Section 4.5. To illustrate algorithm CSD, consider again the small circuit example C_1 introduced in Section 4.3; the corresponding circuit graph and spanning tree are shown in Figures 4.7 and 4.8(A), respectively. The matrix N and the Cholesky decomposition L_2 of N are as described by equation 4.117. The matrices $Y_3 = L_2^{-1}$, described by equation 4.94, and $L_2^{-t}L_2^{-1}$, are the following, respectively:

$$Y_3 = L_2^{-1} = \frac{1}{2\sqrt{10}}\begin{bmatrix} \frac{2}{\sqrt{5}} & 0 & 0 \\ 2 & 4 & 0 \\ 1 & 2 & 5 \end{bmatrix}, \quad L_2^{-t}L_2^{-1} = \frac{1}{8}\begin{bmatrix} 5 & 2 & 1 \\ 2 & 4 & 2 \\ 1 & 2 & 5 \end{bmatrix}. \quad (4.119)$$

The matrix $(-Z + I)$ is written as:

$$-Z + I = \begin{bmatrix} \ldots & & \ldots & \\ \frac{1}{8}\begin{bmatrix} -3 & -1 \\ 2 & -2 \\ 1 & 3 \end{bmatrix} & & \frac{1}{8}\begin{bmatrix} -5 & -2 & -1 \\ -2 & -4 & -2 \\ -1 & -2 & -5 \end{bmatrix} \end{bmatrix}. \quad (4.120)$$

Therefore, the clock schedule basis s^b is the following:

$$s^b = \frac{1}{8}\begin{bmatrix} 3 & 1 & 5 & 2 & 1 \\ -2 & 2 & 2 & 4 & 2 \\ -1 & -3 & 1 & 2 & 5 \end{bmatrix} \left(\frac{1}{4} \begin{bmatrix} 1 \\ -2 \\ 5 \\ -11 \\ 1 \end{bmatrix} \right) = \frac{1}{32}\begin{bmatrix} 5 \\ -38 \\ -7 \end{bmatrix}. \quad (4.121)$$

References

Afghani, M., and Svensson, C. (1990). A unified clocking scheme for VLSI Systems. *IEEE Journal of Solid State Circuits SC-25*, 225–233.

Bakoglu, H.B. (1990). *Circuits, interconnections, and packaging for VLSI.* Reading, MA: Addison-Wesley.

Bretscher, O. (1996). *Linear algebra with applications.* Englewood Cliffs, New Jersey: Prentice Hall.

Björck, A. (1996). *Numerical methods for least squares problems.* Amsterdam: North-Holland.

Bothra, S., Rogers, B., Kellam, M., and Osburn, C.M. (1993) Analysis of the effects of scaling on interconnect delay in ULSI circuits. *IEEE Transactions on Electron Devices ED-40*, 591–597.

Bowhill, W.J. *et al.* (1995). Circuit implementation of a 300-MHz 64-bit second-generation CMOS Alpha CPU. *Digital Technical Journal* 7(1), 100–118.

Chan, S.P., Chan, S.Y. and Chan, S.G. (1972). *Analysis of linear networks and systems: A matrix-oriented approach with computer applications.* Reading, MA: Addison-Wesley.

Chou, N.C., and Cheng, C.K. (1995). On general zero-skew clock net construction. *IEEE Transactions on Very Large Scale Integration (VLSI) Systems VLSI-3*, 141–146.

Ciarlet, P.G., and Lions, J.L. (Eds.). (1990). *Handbook of numerical analysis, Vol. I.* Amsterdam: North-Holland.

Cormen, T.H., Leiserson, C.E., and Rivest, R.L., (1989). *Introduction to algorithms.* Cambridge, MA: MIT Press.

Farebrother, R.W. (1988). *Linear least square computations.* New York: Marcel Dekker.

Fishburn, J.P. (1990). Clock skew optimization. *IEEE Transactions on Computers C-39*, 945–951.

Fletcher, R. (1987). *Practical methods of optimization.* New York: John Wiley & Sons.

Forsythe, G., and Moler, C.B. (1967). *Computer solution of linear algebraic systems.* Englewood Cliffs, New Jersey: Prentice Hall.

Friedman, E.G. (Ed.). (1997). *High performance clock distribution networks.* Norwell, MA: Kluwer Academic Publishers.

Friedman, E.G. (1995). *Clock distribution networks in VLSI circuits and systems.* IEEE Press.

Friedman, E.G. (1992). The application of localized clock distribution design to improving the performance of retimed sequential circuits. *Proceedings of the IEEE Asia-Pacific Conference on Circuits and Systems*, 12–17.

Friedman, E.G. (1990). *Performance limitations in synchronous digital systems.* Ph.D. diss., University of California, Irvine. *Dissertations Abstracts International 50*(9), 3067–B.

Gaddis, N., and Lotz, J. (1996). A 64-b Quad-Issue CMOS RISC microprocessor. *IEEE Journal of Solid-State Circuits SC–31*, 1697–1702.

Golub, G.H., and Loan, C.F.V. (1996). *Matrix computations.* Baltimore, Maryland: Johns Hopkins University Press.

Gronowski, P.E. *et al.* (1996). A 433–MHz 64-bit Quad-Issue RISC microprocessor. *IEEE Journal of Solid-State Circuits SC–31*, 1687–1696.

Ito, N., Sugiyama, H., and Konno, T. (1995). ChipPRISM: Clock routing and timing analysis for high-performance CMOS VLSI chips. *Fujitsu Scientific and Technical Journal 31*, 180–187.

Jackson, M.A.B., Srinivasan, A., and Kuh, E.S. (1990). Clock Routing for High-Performance ICs. *Proceedings of the ACM/IEEE Design Automation Conference* 573–579.

Kourtev, I.S. (1999). *Enhanced algorithms for nonzero clock skew scheduling.* Ph.D. diss., University of Rochester.

Kourtev, I.S., and Friedman, E.G. (2000). System timings. In W.K. Chen (Ed.), *The VLSI handbook.* CRC Press.

Kourtev, I.S., and Friedman, E.G. (1999a). Clock skew scheduling for improved reliability via quadratic programming. *Proceedings of the IEEE/ACM International Conference on Computer-Aided Design*, 239–243.

Kourtev, I.S., and Friedman, E.G. (1999b). A quadratic programming approach to clock skew scheduling for reduced sensitivity to process parameter variations. *Proceedings of the IEEE International ASIC/SOC Conference*, 210–215.

Kourtev, I.S., and Friedman, E.G. (1997a). Simultaneous clock scheduling and buffered clock tree synthesis. *Proceedings of the IEEE International Symposium on Circuits and Systems*, 1812–1815.

Kourtev, I.S., and Friedman, E.G. (1997b). Topological synthesis of clock trees for VLSI-based DSP systems. *Proceedings of the IEEE Workshop on Signal Processing Systems*, 151–162.

Lawson, C.L., and Hanson, R.J. (1974). *Solving least squares problems.* Englewood Cliffs, New Jersey: Prentice Hall.

Lee, T.C., and Kong, J. (1997). The new line in IC design. *IEEE Spectrum*, 52–58.

Leiserson, C.E., and Saxe, J.B. (1998). A mixed-integer linear programming problem which is efficiently solvable. *Journal of Algorithms 9*, 114–128.

Neves, J.L., and Friedman, E.G. (1996a) Design methodology for synthesizing clock distribution networks exploiting nonzero localized clock skew. *IEEE Transactions on Very Large Scale Integration (VLSI) Systems VLSI-4*, 286–291.

Neves, J.L., and Friedman, E.G. (1996b). Optimal clock skew scheduling tolerant to process variations. *Proceedings of the ACM/IEEE Design Automation Conference*, 623–628.

Neves, J.L., and Friedman, E.G. (1993). Topological design of clock distribution networks based on nonzero clock skew specification. *Proceedings of the IEEE Midwest Symposium on Circuits and Systems*, 468–471.

Osborne, M.R. (1985). *Finite algorithms in optimization and data analysis.* New York: John Wiley & Sons.

Reingold, E.M., Nievergelt, J., and Deo, N. (1977). *Combinatorial algorithms: Theory and practice.* Englewood Cliffs, New Jersey: Prentice-Hall.

Tsay, R.S. (1993). An exact zero-skew clock routing algorithm. *IEEE Transactions on Computer-Aided Design of Integrated Circuits and Systems CAD-12,* 242–249.

Unger, S.H., and Tan, C.J. (1986). Clocking schemes for high-speed digital systems. *IEEE Transactions on Computers C-35,* 880–895.

Vasseghi, N., Yeager, K., Sarto, E., and Seddighnezhad, M. (1996). A 200-Mhz Superscalar RISC microprocessor. *IEEE Journal of Solid-State Circuits SC-31,* 1675–1686.

West, D.B. (1996). *Introduction to graph theory.* Englewood Cliffs, New Jersey: Prentice Hall.

Xi, J.G., and Dai, W.W.M. (1996). Useful-skew clock routing with gate sizing for low-power design. *Proceedings of the ACM/IEEE Design Automation Conference,* 383–388.

5

Trends in Low-Power VLSI Design

Tarek Darwish and
Magdy Bayoumi
The Center for Advanced Computer Studies, University of Louisiana at Lafayette, Lafayette, Louisiana, USA

5.1 Introduction

As advances in lithography and fabrication of the N-type metal oxide superconductor (NMOS) technology became possible in the 1970s, the bipolar digital logic, transistor-transistor logic (TTL) lost the battle in the digital design world for exactly the same reasons that caused older technologies, such as the vacuum tube technology, to retire. Circuits implemented in the NMOS technology outperformed the corresponding TTL circuits in terms of power dissipation. One of the main aspects of power consumption is that it puts an upper limit on the number of gates that can be reliably integrated on a single package for any technology. As technology advanced, chips grew, and it was possible to integrate more functions into one chip. Just as for TTL, newer technology, called CMOS, threatened to replace NMOS in the 1980s because CMOS proved to consume even less power. With further advances in technology and fabrication, the integration densities and the rate at which chips operate have increased drastically, causing power consumption to be of primary concern. In addition, the new requirements set by device portability, reliability, and costs have helped in alleviating the power consumption threat in CMOS circuits. Because the power problem is getting more concerning, very large-scale integrated circuit (VLSI) designers need to develop new efficient techniques to reduce the power dissipation in current and future technologies, a task that is full of challenges but yet exciting to explore.

5.2 Importance of Low-Power CMOS Design

With advances in CMOS technology, the potential packing densities increase as the feature size of the MOS devices shrinks, as shown in Figure 5.1. These increases and decreases validate what Gordon Moore once said in the 1960s: the number of transistors that can be integrated on a single die would grow exponentially with time (Moore, 1965). The example that amazingly proved his visionary prediction is best illustrated by tracking the historical evolution of integrated circuit (IC) design in the company he founded in 1972, Intel, and by using the trends in memory evolution. Such observations are evident in Figure 5.2. Figure 5.2(A) shows the trend in the IC logic complexity evolution for Intel processors in the last two decades, whereas Figure 5.2(B) shows the memory integration density as a function of time.

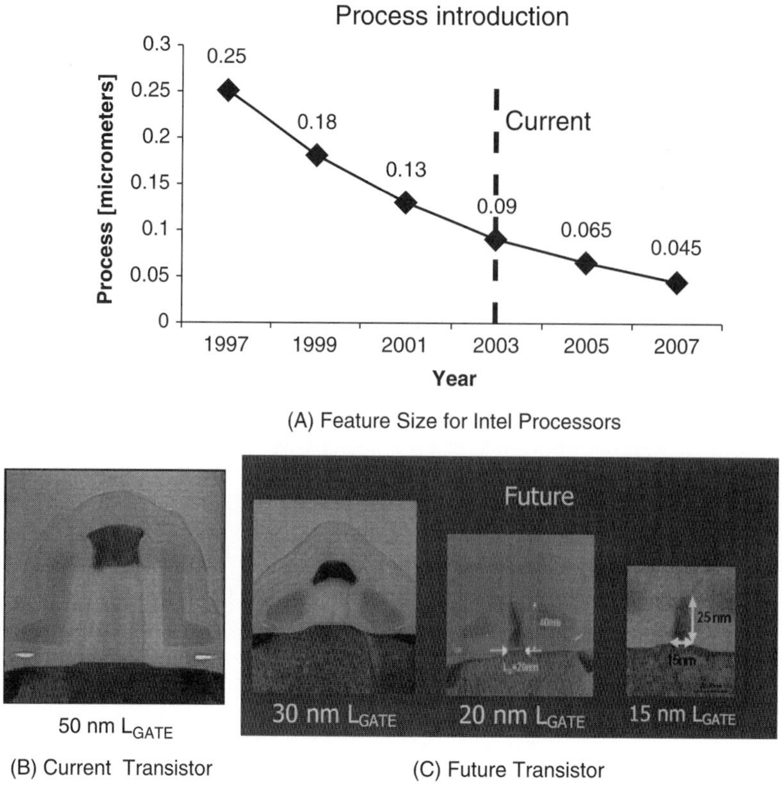

FIGURE 5.1 Packing Density and Feature Size (A) The curve shows the past, current, and future trends for the feature size of the Intel Processes. (B) This image shows a transistor laid out in the 0.09 micrometer or 90 nanometer technology. The gate for that transistor is 50nm. (C) The transistor for future processors is shown here. Courtesy of Intel Corporation.

This increase in the number of transistors per package allowed more functions to be integrated and increased the total logic density of a chip. For example, Figure 5.3(A) indicates that the logic density for the Intel microprocessors doubles every process generation. At the same time, the frequency of operation has dramatically increased due to the device scaling, Figure 5.3(B) shows the introduction of performance boosting techniques, such as pipelining, super-pipelining, and parallel architectures. This increase in operating frequency was driven by the need for fast digital systems realizing powerful personal workstations, sophisticated computer graphics, and multimedia capabilities, such as real-time speech recognition and real-time video (Chandrakasan *et al.*, 1992). When CMOS technology was introduced, it was believed that the power consumption problem was solved. The major concerns of the VLSI designer were mainly speed, cost, and reliability; power consideration was mostly of secondary importance (Pedram, 1995). But as major concerns were being met, the power dissipated in a chip has increased from one generation to another. Figure 5.4 shows the power dissipation of the Intel family of microprocessors.

Figure 5.5 shows some projections for future processors if power continues to dissipate at the same rate. The figure also shows the power density behavior during the last two decades. With the current processors, we have reached the power density of a hot plate. Without limiting power dissipation, we will someday have nuclear reactor-equivalent power dissipated in chips we use at home or in our mobile devices!

Such observations are in contradiction with the scaling theory. VLSI technology scaling has evolved at an amazingly fast pace during the last 30 years; the minimum device size keeps shrinking by a factor $k = 0.7$ per technology generation. The basic scaling theory, known as **constant field scaling** mandates the synergistic scaling of geometric features and silicon doping levels to maintain a constant field across the gate oxide of the MOS transistor (Benini *et al.*, 2001; Panasonic, 2000). According to the constant field scaling theory, power dissipation scales as k^2 and power density (i.e., power dissipated per unit area) remains constant while speed increases as k. Such contradiction between the theory and conclusion from Figure 5.4 and Figure 5.5 has two causes. First, die size has been steadily increasing with technology (Figure 5.6), causing an increase in total average power, as shown in Figure 5.4. Second, supply voltage has scaled much more slowly than device size because supply voltage levels have been standardized and because faster transistor operation can be obtained for higher supply voltage levels.

Hence, this large amount of power dissipation is one of the main motivations for VLSI designers to develop new techniques to reduce the power consumed inside chips; power

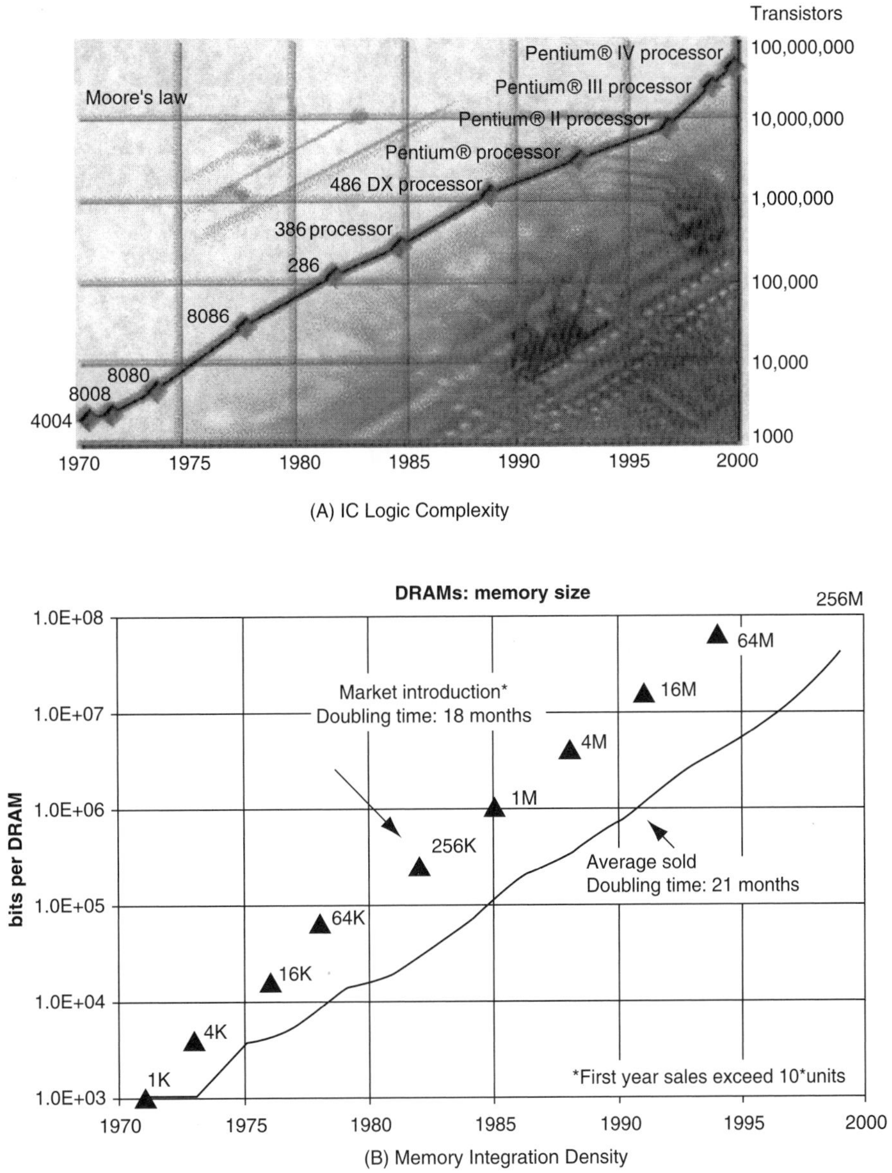

FIGURE 5.2 This Figure Shows the Number of Transistors Per Chip is Increasing. (A) Shows the number of transistors per Intel processor. The horizontal axis shows the time of introducing a chip, and the vertical axis shows the corresponding number of transistors included. Courtesy of Intel Corporation. (B) Shows the number of transistors per DRAM memory chip. Similarly, the horizontal axis shows the time of introducing a chip, and the vertical axis shows the corresponding number of transistors included.

is being given comparable weight to area and speed considerations (Najm, 1994). Several important factors have also contributed to this increased concern about power. One of these primary driving factors has been the remarkable success and growth of the personal computing device (e.g., Pass portable desktops, audio- and video-based multimedia products, etc.) and wireless communications systems (e.g., personal digital assistants, personal communicators, etc.) that demand high-speed computation and complex functionality with low-power consumption (Pedram, 1995). Unfortunately, this rapid development in VLSI has not been reflected in developments in battery technology, and without low-power design techniques, current and future portable devices will suffer from either very short battery life or very heavy battery pack (Pedram, 1995; Najm, 1994; Athas *et al.*, 1994; Landman and Rabaey, 1995; Tsul *et al.*, 1995; Chang and Pedram, 1997; Bajwa *et al.*, 1997).

It is clear that in the absence of low-power design techniques, portable and handheld products would suffer from a very short battery life, and packing and cooling them would be very difficult (Tsui *et al.*, 1995; Ko *et al.*, 1995). These factors lead to an unavoidable increase in the cost of the product, as shown in Figure 5.7. In addition, reliability is strongly affected

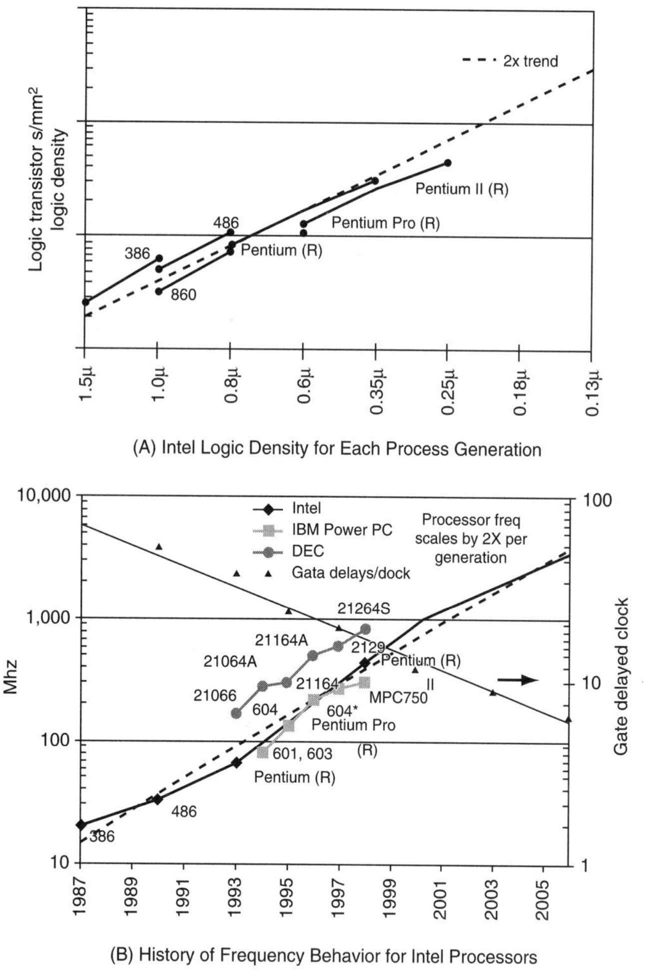

FIGURE 5.3 Logic Density and Performance Boosting. (A) The logic density, another measurement for number of transistors per chip, doubles every new process generation. The trend is clear for Intel microprocessors. The horizontal axis shows the process generation, and the vertical axis shows the logic density for microprocessors manufactured at the corresponding process. Courtesy of Intel Corporation. (B) This figure shows the frequency trend for VLSI processors. Almost same trend can be seen for each generation of processors from companies such as Intel, IBM, and DEC. The left vertical axis shows the frequency, the right axis shows the gate delay which decreases for new processes. The gate delay trend is projected through the straight line curve. Courtesy of Intel Corporation.

by power consumption. Usually, increased power dissipation implies high temperature operation, which, in turn, may induce several failure mechanisms in the system (Landman and Rabaey; 1995; Chang and Pedram, 1997; Bajwa *et al.*, 1997).

5.3 Sources of Power Consumption in CMOS

Two types of power consumption exist for digital CMOS. The first, the **dynamic power component,** may be thought of as useful because it establishes information by charging and discharging signal lines; the second type, consisting of **short-circuit** and **static power components,** is waste and comes from short-circuit and leakage currents that flow directly from the power supply to ground. Figure 5.8 shows the relative behavior of these power components with respect to the input switching activity for an inverter. The following subsections introduce the different power components and give their parametric equations. Much of this material can be found in (Smith, 1997; Rabaey, 1996; Weste and Eshraghian, 1993; Kang and Leblebici, 1999).

5.3.1 Dynamic Power Dissipation

The **dynamic power dissipation** is the power required for the circuit to perform its anticipated tasks. In other words, it is the power needed for charging and discharging all nodes in a CMOS circuit. This power is only consumed when the circuit input signals change. In CMOS circuits, the dyanmic power dominates the total power dissipation. Such characteristic is greatly affected by current processes or the deep sub-micron processes (DSM), for which the ratio of leakage power to dynamic power is increasing. More details about this issue are presented in Subsection 5.3.3. Dynamic power dissipation is illustrated in Figure 5.9 for a simple static CMOS inverter. When the input signal falls, the PMOS transistor switches on while the NMOS transistor switches off, creating a path from the supply voltage to the output capacitance, thus allowing the output load to charge up to the supply voltage. On the other hand, when the input signal rises, the opposite scenario occurs: the NMOS transistor switches on, and the PMOS transistor switches off, creating a direct path from the output load to ground and allowing the output load to discharge.

If C_L represents the total capacitance charged per cycle, then the dynamic power dissipation is as follows:

$$P_{\text{dynamic}} = C_L \times Vdd \times Vdd \times f \times \alpha, \qquad (5.1)$$

where Vdd is the supply voltage level, f is the frequency of operation, and α is the switching activity of the capacitive node C_L on each clock cycle.

Even though the above analysis is for a simple inverter, a similar approach can be followed to evaluate the dynamic power dissipation for a circuit with n nodes; in such a case, the total dynamic power dissipation is evaluated as follows:

$$P_{\text{dynamic (total)}} = Vdd \times Vdd \times f \times \left(\sum_{i=1}^{n} \alpha_i \times C_i \right), \qquad (5.2)$$

where α_i is the switching activity of node i with capacitance C_i. If C_{eff} represents the average switching capacitance per cycle, then the corresponding power dissipation equation represents the average dynamic power dissipation:

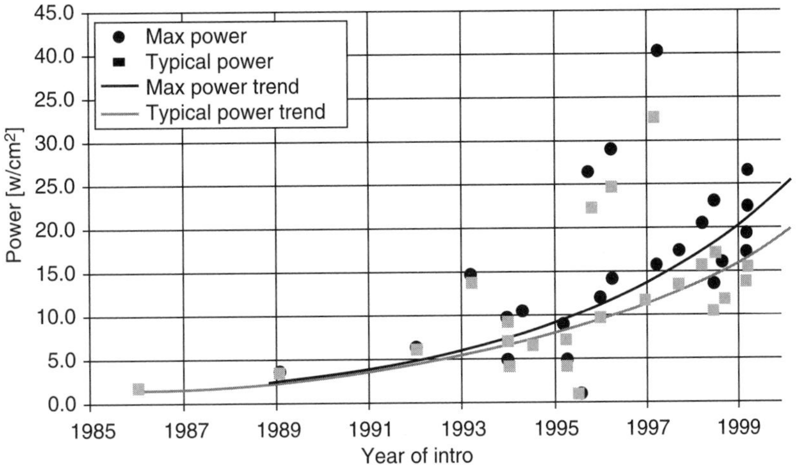

FIGURE 5.4 Power Trends for the Intel Processors Through Time. Courtesy of Intel Corporation.

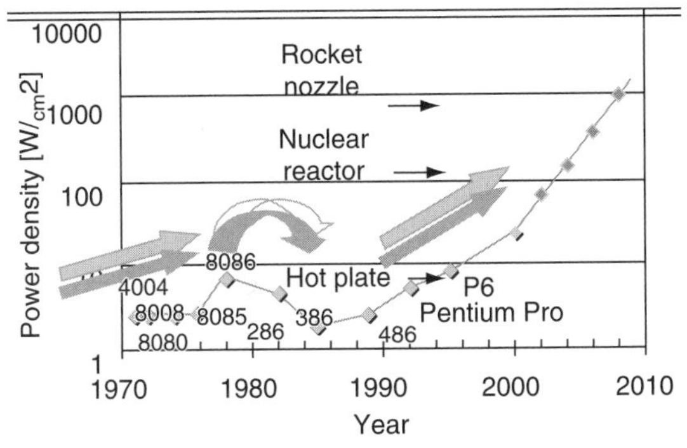

FIGURE 5.5 Power Density Trend for the Intel Processors with Future Projections

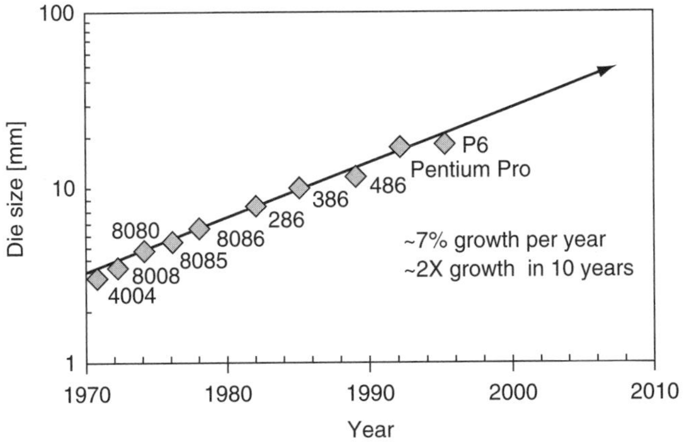

FIGURE 5.6 Die Size Increase Through Time

$$P_{\text{dynamic(average)}} = Vdd \times Vdd \times f \times C_{eff}. \quad (5.3)$$

Even though the device dimensions, and thus the device capacitances, are scaled down with newer technologies, the total integration per chip is increasing, causing the total capacitance per chip to increase. In addition, two of the main requirements of current applications are speed and real-time operation. As such, the frequency of operation is meant to increase drastically especially in the microprocessor area. In addition, supply voltage is not scaled at the same rate as the device scaling for a specific technology. From these facts, it is quite easily understandable that in scaling technology, the relative power is increasing with respect to that of predecessor technologies (refer to Figure 5.4).

As Section 5.4 and Section 5.7 explain, these equations are vital in investigating new techniques targeting lower power dissipation of CMOS chips and are important for addressing the dynamic power dissipation that is the dominant factor in the total power dissipation of CMOS circuits.

5.3.2 Short Circuit Power Dissipation

The dynamic power dissipation equation is derived usually by assuming that the inputs have zero rise and fall times, or, in other words, that the NMOS and PMOS devices are never on simultaneously. But in reality, such assumption is not valid, and input signals have nonzero rise and fall times; hence, a direct current path exists between *Vdd* and *GND* for a short period of time during input switching, in which case the PMOS and the NMOS devices are simultaneously conducting. This power component is consumed without attributing to the circuit behavior; thus, it is considered redundant.

Figure 5.10 shows the short circuit power dissipation for the case of the simple static CMOS inverter. The short circuit power dissipation can be stated as follows:

$$P_{\text{short circuit}} = I_{sc} \cdot Vdd, \quad (5.4)$$

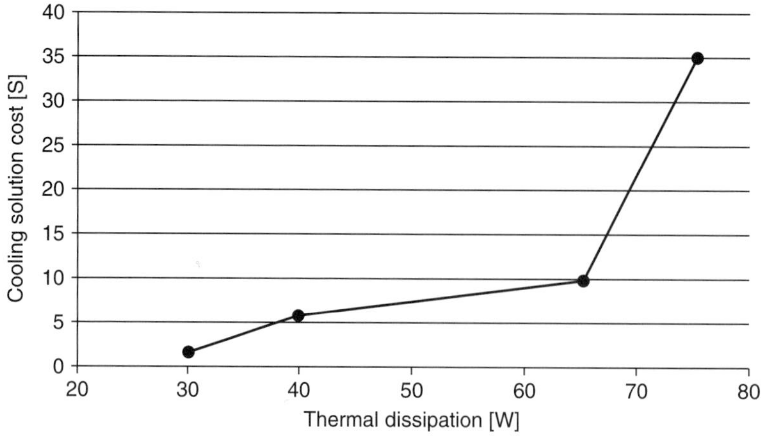

FIGURE 5.7 Chip Cooling Costs as a Function of Power Dissipation

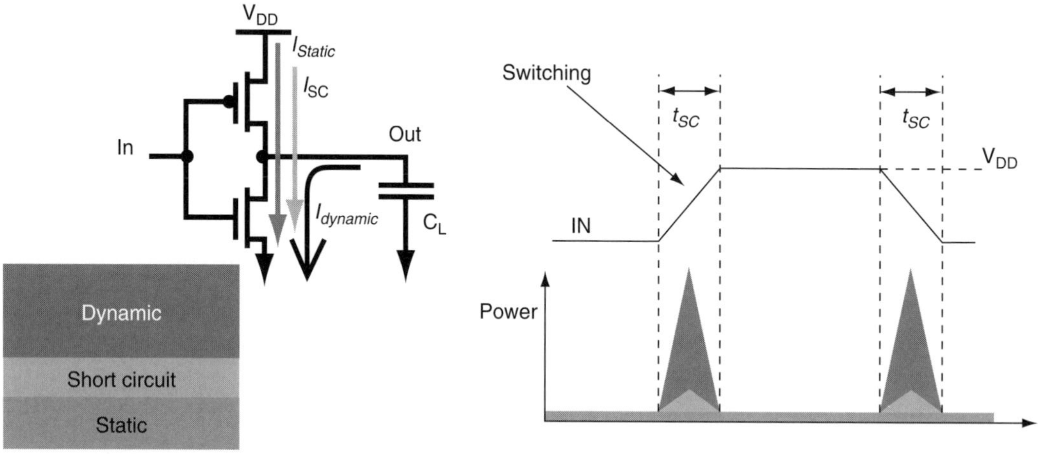

FIGURE 5.8 Power Consumption Components for CMOS Inverter and Their Behavior with Respect to the Input Switching Activity

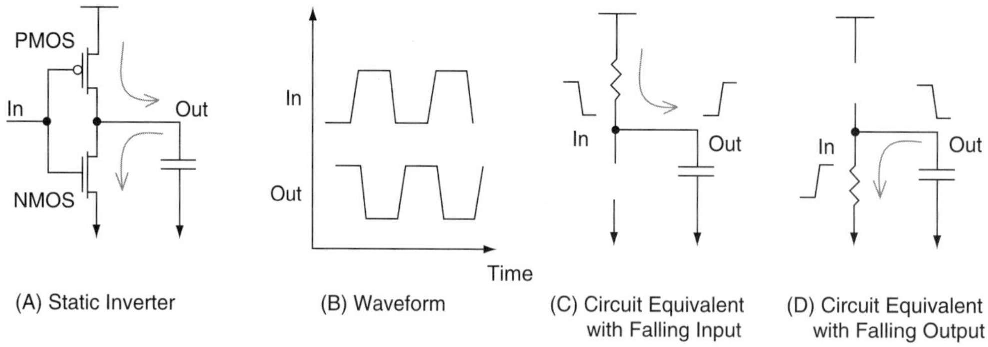

FIGURE 5.9 Dynamic Power Dissipation. (A) This figure shows a static CMOS inverter, the output charges using the PMOS transistor, and discharges through the NMOS transistor. (B) The waveform shows the function of the inverter output (out) with respect to the input (in) (C) Shown here is the circuit equivalent when the input falls and the corresponding output rises. (D) The figure illustrates the circuit equivalent when the input rises and the corresponding output falls.

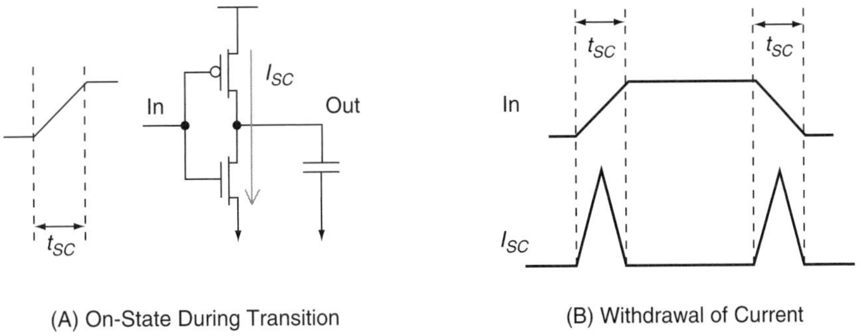

FIGURE 5.10 Short Circuit Power Dissipation. (A) The nonzero input transition time causes both transistors to be on for the duration of the transition, t_{sc}. (B) As a result, a current, i_{sc}, is withdrawn from the supply voltage.

where I_{sc} is the current leaking when both transistors are on during switching. For a system with n gates, the total short circuit power dissipation can be expressed as follows:

$$P_{\text{short circuit (total)}} = Vdd \times \sum_{i=1}^{n} \alpha_i \times I_{sc_i}. \quad (5.5)$$

5.3.3 Static Power Dissipation

Ideally, the static power consumption of static CMOS circuits is assumed to be zero, as the PMOS and the NMOS devices are never on simultaneously in steady-state operation. But in reality, the drain current through the CMOS transistor does not drop to zero once the gate voltage goes below the threshold voltage. There is, unfortunately, always a leakage current, which is primarily determined by the fabrication technology. Historically, the contribution of the static power dissipation to the overall power dissipation is in general very small and can be ignored. However, with the current technologies, for which reductions in the device threshold voltage are employed for achieving better performance, the percentage of the static power dissipation from the total power has increased as shown in Figure 5.11.

A variety of leakage currents are steadily flowing through various parts of the transistor, including a subthreshold current through the channel. Those currents are small, but they are becoming increasingly important in low-power applications. Not only do many circuits need to operate under very low current drains, but subthreshold currents are also getting relatively larger as transistor sizes shrink.

Leakage currents come from a variety of effects in the transistor (Wolfe, 2002; Wei *et al.*, 2000), as in Figure 5.12:

- The weak inversion current (also known as subthreshold current) is carried through the channel when the gate voltage is below threshold, and it flows between the source and drain of a MOS transistor. It increases exponentially with the reduction of the device threshold voltage, making it critical for low-voltage, low-power circuit design.

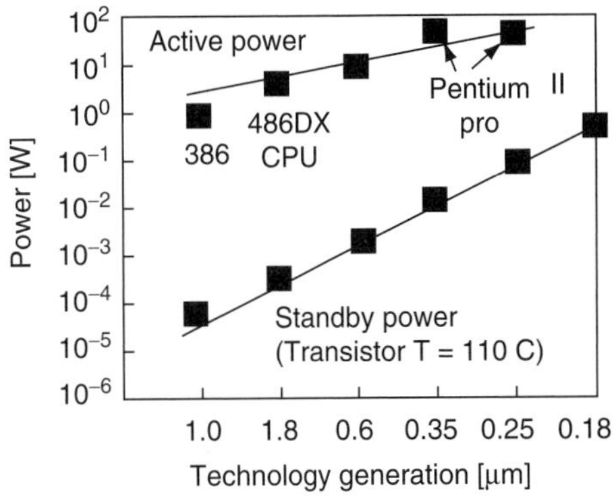

FIGURE 5.11 Static Power Evolution with Respect to Technology

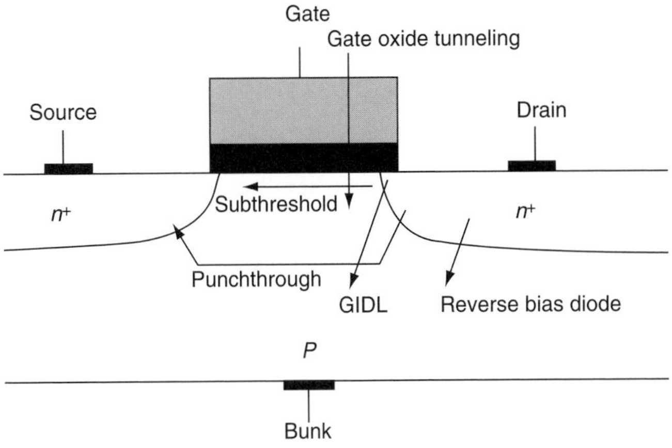

FIGURE 5.12 The Different Sources of Leakage Current in MOS Transistors

- Reverse bias pn junctions in the transistor, such as the one between the drain and its well, carry small reverse bias currents. Carrying these currents results from the minority carrier drift near the edge of the depletion region and the electron hole pair generation in the depletion region.

- Gate-induced drain leakage (GIDL) current exists around the electric field under the gate/drain overlap.
- Punchthrough currents flow when the source and drain depletion regions connect in the channel.
- Gate oxide tunneling currents are caused by the high electric fields in the gate.

Reverse-bias current, GIDL, punchthrough, and gate-oxide tunneling are all caused by high electric fields, and, for low-voltage circuits, they can be neglected. Only the subthreshold voltage is the dominant component of the leakage current for low-voltage, low-power circuits.

If I_{static} is the total current leaking in steady-state, then the static power dissipation component can be written as:

$$P_{static} = I_{static} \times Vdd. \tag{5.6}$$

Thus, the total amount of power consumed by a CMOS circuit can be stated as follows:

$$\begin{aligned} P_{total} &= P_{dynamic} + P_{short\ circuit} + P_{static} \\ &= C_L \times Vdd \times Vdd \times f \times \alpha + I_{sc} \times Vdd + I_{static} \times Vdd \\ &= Vdd(C_L \times Vdd \times f \times \alpha + I_{sc} + I_{static}) \end{aligned} \tag{5.7}$$

To develop techniques to reduce the power dissipation, the factors present in equation 5.7 need to be thoroughly analyzed.

5.4 Power Consumption Considerations

The dependence of the power consumption on many parameters allows the VLSI designer to have a large number of techniques available to reduce the power dissipation. From the equations presented in the previous section, the following parameters should be considered when designing for low power: power-supply voltage, device-threshold voltage, physical node capacitance, and switching frequency.

5.4.1 Supply Voltage Level

Equations 5.3, 5.5, 5.6, and 5.7 show that all power consumption components (static, short circuit, and dynamic) linearly depend on the **supply voltage level**, *Vdd*, except the dynamic power component. The dynamic power component is not only in quadratic relation with *Vdd*, but it is also the dominant component of the total power dissipation. As such, scaling down the supply voltage has a profound effect in reducing the total power dissipation of the whole circuit in hand (Chang and Pedram, 1997; Wei *et al.*, 2000). Indeed, reducing the supply voltage is the key to low-power operation, even after taking into account the modifications to the system architecture, which is required to maintain the computational throughput (Chandraksan *et al.*, 1992). Once applied, scaling down will result in power saving that has a global effect, experienced not only in one subcircuit or block of the chip but throughout the entire design. Because of this, designers are often willing to sacrifice increased physical capacitance or circuit activity for reduced voltage. Unfortunately, a speed penalty is paid for supply voltage reduction, with delays drastically increasing as *Vdd* approaches the threshold voltage V_t of the devices. This tends to limit the useful range of *Vdd* to a minimum of about 2 to 3 V_t (Pedram, 1995; Chandraksan *et al.*, 1992). Effectively, speed optimization techniques are applied and then the supply voltage is scaled which bring the design back to its original timing but with a lower power requirement Benini *et al.*, 2001.

5.4.2 Device Threshold Voltage

When scaling the supply voltage down, one can compensate for the speed loss by altering the **device threshold voltage**. Reducing the V_t (achieved by changing the substrate and channel dopant concentrations) allows the supply voltage to be scaled down without loss in speed. Unfortunately, such scaling leads to an exponential increase in the subthreshold leakage current (Wei *et al.*, 2000). The limit of how low the V_t can go is determined by the requirement to set adequate noise margins and control the increase in subthreshold leakage currents, as shown in Figure 5.13. The optimum V_t must be determined based on the current drives at low supply voltage operation and control of the leakage currents (Pedram, 1995). If the threshold voltage is reduced, then the whole characteristic curve moves toward the left, as Figure 5.8 illustrates. Thus for low threshold voltages, the device cannot be properly switched off (when $V_{gs} = 0\ V$) and there is significant short circuit current. The subthreshold currents can result in significant static power dissipation. Essentially, subthreshold leakage occurs due to carrier diffusion between the source and the drain when the gate-source voltage V_{gs} has exceeded the weak inversion point but is still below the threshold voltage V_t, where drift is dominant.

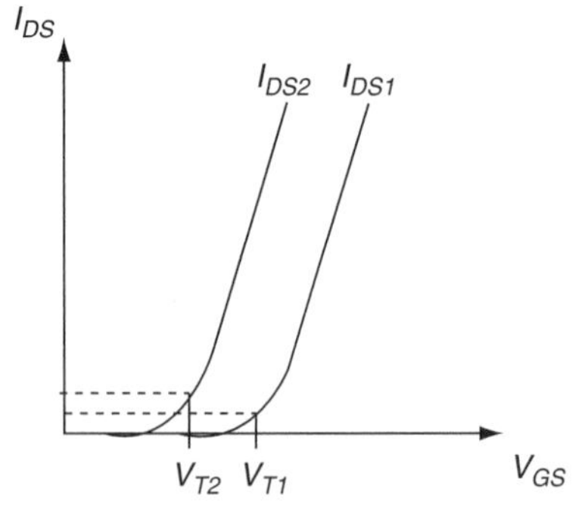

FIGURE 5.13 ($I_{DS} - V_{GS}$) Characteristic Curve for a Typical MOS Transistor. As the threshold voltage V_T decreases, the curve shifts to the left, and the subthreshold current increases.

5.4.3 Physical Capacitance

Dynamic power component is the power expended while charging and discharging the capacitive nodes. This physical capacitance is attributed to the transistor parasitic capacitance and the interconnect capacitance. Reducing these capacitive nodes could result in substantial power reductions across the whole system, as equation 53 shows. So, minimizing capacitances offers another option for minimizing power consumption. Estimating this capacitance at the behavioral or logical levels of abstraction, however, is difficult and imprecise if gate mapping into a cell library is not done (Pedram, 1995).

The capacitance load of a circuit is made of two components: intrinsic and extrinsic. **Intrinsic capacitance** refers to internal parasitic capacitances of the circuit, and they consist of drain and source to substrate and overlap capacitances. The extrinsic capacitances are the parasitic capacitances of the fan-out, driven by the circuit (normally gate and overlap capacitances) and the routing capacitances resulting from the routing between the circuit and its fanout gates (Rabaey, 1996). Accordingly, capacitances can be kept at a minimum by using less logic, smaller devices, and fewer and shorter interconnects.

5.4.4 Switching Frequency

Dynamic power dissipation is only consumed when there is switching activity at some nodes in a CMOS circuit. For example, a chip may contain an enormous amount of capacitive nodes, but if there is no switching in the circuit, then no dynamic power will be consumed (Chandraksan *et al.*, 1992). The **switching frequency**, which is $\alpha \times f$, determines how often the switching occurs. Normally, f increases with technology scaling, but for a given technology, double-edge-triggered flip-flops may be used to reduce the dynamic power dissipation of the clock distribution network, which could in turn result in significant savings in the total power dissipation. The node switching activity per cycle, α has two components: **useful data switching activity** (UDSA) and **redundant spurious switching activity** (RSSA).

The UDSA can be interpreted as the probability that a useful power consuming transition will occur during a single data period. The calculation of this term is difficult because it depends on the input switching activity, logic function implemented, and the spatial and temporal correlations among the circuit inputs. Several attempts to estimate this switching activity can be found in the literature (Najm, 1994).

For certain logic styles, however, RSSA or glitching can be an important source of signal activity. **Glitching** refers to spurious and unwanted transitions that occur before a node settles down to its final steady-state value that occurs due to partially resolved functions. Glitching often arises when paths with unbalanced propagation delays converge at the same point in the circuit. Because glitching can cause a node to make several power-consuming transitions, it should be avoided whenever possible (Pedram, 1995).

In some work, the capacitive loads and switching frequency are combined to give the switched capacitance per cycle (SCPS). Usually, optimization approaches that have a lower impact on performance, but that allow significant power savings, are those targeting the minimization of the SCPS.

5.5 Energy Versus Power

The terms **power consumption** and **energy consumption** are often interchanged. It is important to distinguish between the two especially in the context of mobile applications. Mobile systems run on the limited energy available in a battery. The energy consumed by the system determines the length of the battery life, so the reduction in the energy consumed by executing system's tasks is necessary.

Energy is a measure of the total number of joules [J] dissipated by a circuit, whereas **power** refers to the number of joules dissipated over a certain amount of time. The average power consumed by a circuit is given by:

$$P_{av} = \frac{1}{T}\int_0^T P(t)dt = \frac{V_{\text{Supply}}}{T}\int_0^T i(t)dt = I_{av} \times V_{\text{Supply}}, \quad (5.8)$$

where P is the average power, I is the average current, and V_{supply} is the supply voltage.

Because power is the rate at which energy is consumed, the energy equation can be written as:

$$E = P \times t = V \times I \times t. \quad (5.9)$$

The energy consumed by a circuit to perform a specific task is as follows:

$$E_s = P \times T_S = P \times \frac{N_s}{f} = P \times N_s \times \tau, \quad (5.10)$$

where T_s is the total time needed to finish the task s. N_s is the number of clock cycles taken by the task s, f is the operating frequency, and τ is the clock period.

A battery is simply a storage component that contains a certain amount of power available for a circuit to dissipate. The power that can be drawn from a battery depends on other things besides the charge itself. Figure 5.14 shows battery discharge lines for three different discharge rates with three different values for the average current (Panasonic, 2000).

Figure 5.14 shows that the power drawn from a battery highly depends on the average current. More current means more power. So, average current should be reduced to a minimum. Energy sources, namely batteries, are more efficient

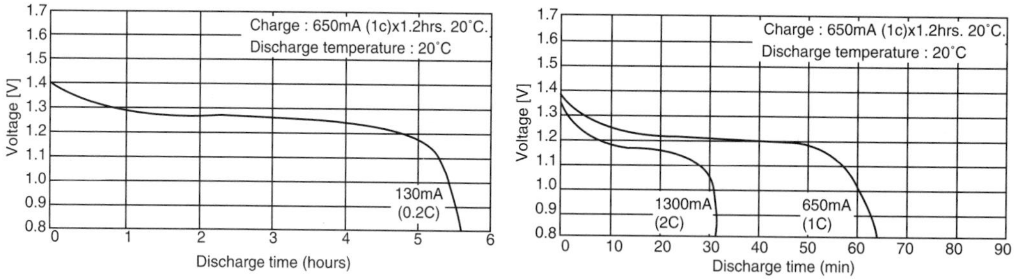

FIGURE 5.14 Indicates that the Battery Life Depends on the Amount of Average Current Discharged, Noting that the Charging Process was Identical in the Three Cases and More Current Means More Power Withdrawn; the figure shows three different average discharge currents values: 130 mA, 650 mA, and 1300 mA. The 130 mA average discharge current gave the longest discharge time of all three and the 1300 mA average discharge current gave the shortest time.

when they are discharged with current bursts instead of a steady flow.

5.6 Optimization Metrics

During the optimization process, power minimization is never the only optimization objective. Performance is always implicitly of ultimate importance in the design of any digital system. Unfortunately, meeting both requirements is very hard if not impossible. Major techniques that target one objective have side effects with respect to the other. For this reason, several metrics for joint power performance have been proposed in the past. In many designs, the power-delay product (PDP, or the energy) is an acceptable metric. Energy minimization rules outdesign choices that heavily compromise performance to reduce power consumption. When performance has priority over power consumption, the energy-delay product (EDP) can be adopted to tightly control performance degradation (Burd and Brodersen, 1996).

Alternatively, a constrained optimization approach can be taken. In this case, performance degradation is acceptable to a given *bound* that is represented by performance or timing constraints. Thus, power minimization requires optimal exploitation of the slack on performance constraints. Besides power versus performance, another key trade-off in VLSI design involves **power versus flexibility** (Benini *et al.*, 2001). Several researchers have observed that application-specific designs are characterized by orders of magnitude more power efficient than general-purpose systems programmed to perform the same computation (Nielsen *et al.*, 1994). On the other hand, flexibility (programmability) is often an indispensable requirement, and designers must strive to achieve maximum power efficiency without compromising flexibility (Benini *et al.*, 2001).

5.7 Techniques for Power Reduction

In general, power minimization targets maximum instantaneous power or average power. As stated in Section 5.5, the latter impacts battery lifetime and heat dissipation system cost, while the former constrains power grid and power supply circuit design (Benini *et al.*, 2001).

Optimizations can be achieved by facing the power problem from different perspectives: design and technology. Enhanced design capabilities mostly impact switching and short-circuit power; technology improvements, on the other hand, contribute to reductions of all three components (Benini *et al.*, 2001).

From a design perspective, circuit designers can choose from a number of options to reduce the power dissipation ranging from the highest levels of abstraction (architecture level) to the lowest levels of abstraction (physical or technology level) (Alidina *et al.*, 1994). In each level of abstraction, there exists a number of techniques that can be employed to effectively reduce the power dissipation of CMOS circuits, all of which can be obtained from the power dissipation equations mentioned in the previous section. As each technique has its own figure of merit, it usually affects one or more aspect of the system being optimized. As such, attempts are always being made to find a compromise that can effectively present a fair and reasonable solution among conflicting objectives like power dissipation and performance, where improving one generally means degrading the other!

Actually, addressing the power problem from the very early stages of design development offers enhanced opportunities to obtain significant reductions of the power budget and to avoid costly redesign steps. Power-conscious design flows must then be adopted; these require, at each level of the design hierarchy, the exploration of different alternatives as well as the availability of power estimation tools that could provide accurate feedback on the quality of each design choice (Benini and de Micheli, 1999). This section studies the various techniques that can be effectively used to reduce the power consumption during the VSLI system design cycle. The stages during the design involve the **system, architectural, logic, circuit**, and **physical levels**.

5.7.1 System Level

At the highest level of abstraction, a system is viewed as an integrated entity that consists of a hardware infrastructure executing software programs. The hardware platform mainly consists of execution units, storage units, and communication and interface networks, whereas the software part consists of application and system software. Addressing the power dissipation at this level has the greatest influence on power dissipation; as much as a 400% power savings could be achieved (Stammerman *et al.*, 2001). The effective techniques at this level include but are not limited to instruction-level optimization, hardware–software codesign, memory design techniques, dynamic power management, and variable-voltage techniques. Most of the material related to system-level optimization techniques can be found in Benini and de Micheli (1999) and (2000).

Instruction-Level Optimization

Techniques involving **instruction-level optimization** target programmable cores such as general-purpose microprocessors, DSP, and microcontrollers. In these techniques, power optimization is achieved by selecting a minimum power instruction mix for executing application software. Tiwari *et al.* (1994) describe a power dissipation reduction of up to 40% that could be achieved by rewriting code. In another work by Woo *et al.* (2001), an instruction encoding technique is used for low-power instruction fetches. The basic approach is that instruction fields of a binary program are re-encoded so that the number of bits switched from the neighboring fields of consecutive instructions is minimized; this technique allowed up to a 62% reduction in switching activities. Similar techniques can be found in Stammerman *et al.* (2001) and Choi and Chatterjee (2001). Another approach that can reduce the power dissipation of a system is devising a low-power system compiler in which high-level code is transformed into machine code characterized by its low-power consumption. The design of such a compiler requires knowledge of the hardware architecture that will execute the machine code.

Hardware–Software Codesign

Hardware–Software Codesign techniques target system-on-chip (SoC) design or embedded core design that involves integration of general-purpose microprocessors, DSP structures, programmable logic (FPGA), ASIC cores, memory block peripherals, and interconnection buses on one chip. Traditionally, a system is divided into hardware and software sections that are designed independently except for some common standards required for compatibility concerns, shown in Figure 5.15(A). With systems growing larger and power consumption becoming of great importance, a new wave came forth to consider the whole system design process and attempt to partition the various tasks of the system between hardware and software from the early stages of the design process to reduce many design problems, as indicated by Figure 5.15(B). These techniques attempt to find an optimal partitioning and assignment of tasks between software running on microprocessors or DSPs and hardware implemented on ASIC or FPGA for a given application. In Henkel (1999), a system-level power optimization approach that deploys hardware–software partitioning based on a fine-grained (instruction/operation-level) power estimation analysis is proposed that achieves up to 94% savings in energy consumption. In this approach, the system tasks are partitioned into software

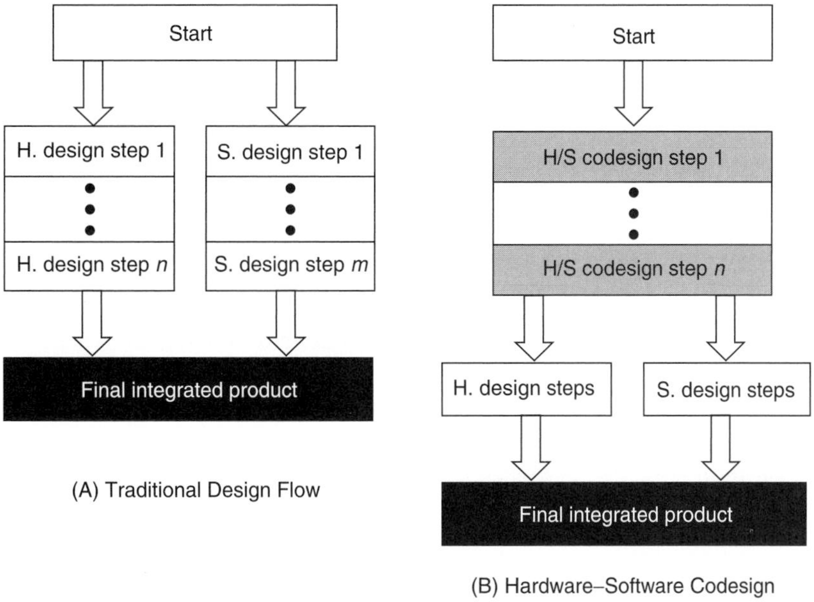

FIGURE 5.15 Traditional and Modern Designs (A) This figure shows the traditional design flow, in which hardware and software sections are designed independently. (B) Illustrated here is a concurrent design flow that considers both hardware and software solutions to create efficient designs.

sections running on a general microprocessor core and into hardware sections implemented in application-specific cores to minimize the total power dissipation.

Memory Design Techniques

Memory design techniques techniques are mainly focused on reducing the power consumed by memories, such as creatively exploiting caching to reduce power consumption (Pedram and Rabaey, 2002). Increasing memory blocks on-chip can reduce the overall power consumption of a processor because the power dissipation of an external memory access is at least an order of magnitude higher than that of an on-chip access (Ko *et al.*, 1998). The discrepancy is due mainly to the relatively large capacitive overhead in driving device *I/O*, board traces, and discrete memory components. For optimizing internal memories, Lee and Tiwari (1996) analyze memory allocation of variables in embedded DSP software to maximize simultaneous data transfers from different memory banks to registers. A memory allocation technique based on simulated annealing is proposed. Experimental results demonstrate that energy reduction of up to 47% can be achieved using this approach. In work by Wen-Tsong *et al.* (1999), a relation between energy consumption and cache design (i.e., cache size, line size, set associativity, and tiling) is established.

Dynamic Power Management

Dynamic power management techniques allow systems or system's blocks to be placed in low-power sleep modes when the systems are inactive. Normally, not all blocks of a system participate in performing different functions, and it is useful to shut down inactive blocks to reduce power consumption. For example, in a general microprocessor, a register-based instruction does not access the data memory and some other modules, so these components can be deactivated to achieve power savings. In this case, more control is involved, but it is true that the waking up of these deactivated blocks is expensive in terms of speed or performance.

Variable-Voltage Techniques

As stated before, the most effective way to reduce power consumption is to lower the supply voltage with **variable voltage techniques**. With the availability of almost "limitless" number of transistors, parallelism can be employed at the chip level, providing significant power savings by reducing the supply voltage level (Chandraksan *et al.*, 1992).

If both the technology and architecture are fixed, the determining factor in power dissipation is timing: at higher voltages, the circuits operate faster but use more energy. Some techniques adjust the circuit during operation to achieve power savings. For example, with the use of a dynamically adjustable power supplies technique (Gutnik and Chandraksan, 1997), power can be reduced substantially without sacrificing performance in fixed-throughput applications by slowing the clock and lowering the supply voltage instead of idling when the computational workload varies, as shown in Figure 5.16. The basic idea is to lower the supply voltage and slow the clock during reduced workload periods instead of working at a fixed speed and idling.

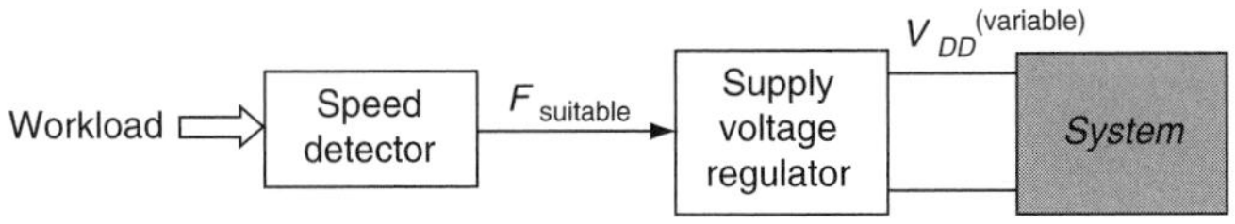

FIGURE 5.16 Variable Voltage Scheme. A speed detector is used to sense the suitable frequency and execute some workload; consequently, the supply voltage regulator will generate a suitable supply voltage level for this specific frequency.

5.7.2 Architectural Level

At the architectural level, a digital system consists of the structural view of the data path and a logical view of the control unit of a circuit. The data path consists of a set of interconnected functional units (arithmetic, logic, memory, and registers) and steering units (multiplexers and busses), while the control unit sends signals to the data path to schedule the appropriate sequence of operations in time.

At this level of abstraction, the VLSI designer has a lot of freedom in employing many techniques to reduce the power dissipation. Normally, decisions at this level have a global effect on the whole system. For example, inactive hardware modules may be automatically turned off to save power; modules may be provided with the optimum supply voltage and interfaced by means of level converters. In the following subsections, different techniques that are applied at this level are presented, namely parallelism and pipelining exploitation, block-disabling techniques, clock gating, and intercommunication and interconnect optimization.

Parallelism and Pipelining Exploitation

One effective technique to reduce energy dissipation is to reduce the supply voltage level by employing **pipelining** and **parallelism** which has an immense effect in reducing the overall power consumption and compensating for the resultant speed loss. This technique is used in general for data path sections. Assuming performance reduces by half when the supply voltage is lowered, the hardware can be duplicated to produce outputs at twice the slower rate to achieve the initial performance before lowering the supply voltage level, as shown in Figure 5.17.

Block-Disabling Technique and Clock Gating

Because not all blocks in an architecture are simultaneously operating, one can disable the blocks (**block-disabling technique**) that are not in use during some particular clock cycles with the objective of limiting the power consumption. In fact, idling or stopping these units causes a decrease in the overall

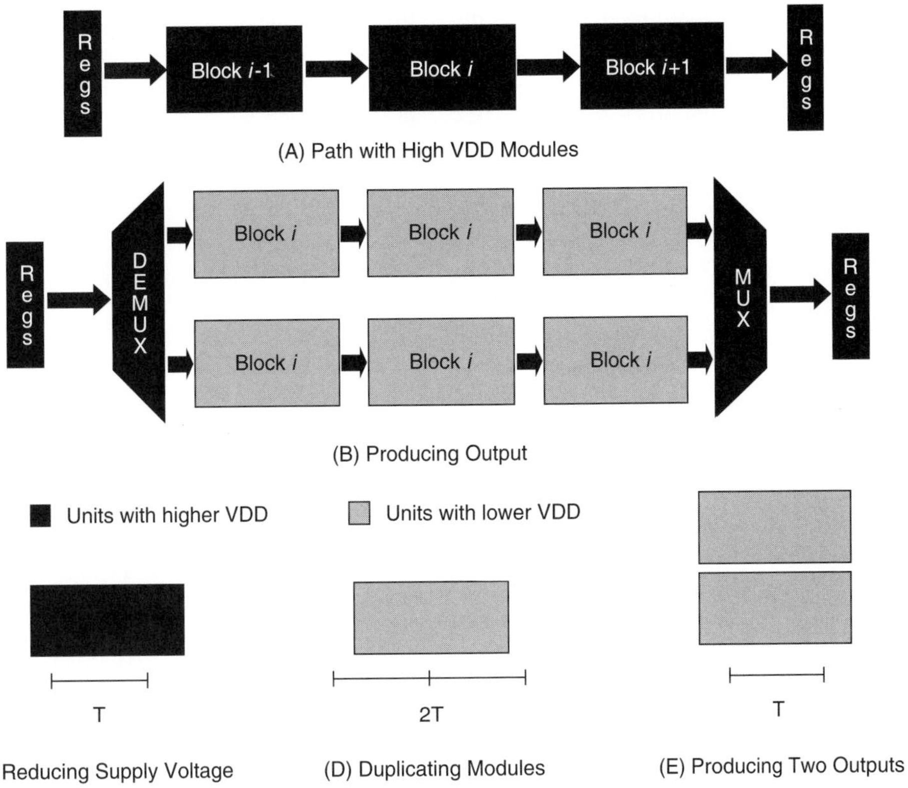

FIGURE 5.17 Exploiting Parallelism. (A) This figure shows a path with modules supplied from a high Vdd; the path length is T (C). Reducing the supply voltage causes the path to produce the output after $2T$ seconds (D). Duplicating the modules with the lower Vdd (B) causes the parallel structure to produce two outputs every $2T$ seconds, which is equivalent to one output every T seconds (E).

switching capacitance of the system, thus reducing the switching component of the power dissipation. Optimization techniques based on this principle belong to the broad class of dynamic power management methods. Clock gating provides a way to selectively stop the clock and thus force the original circuit to make no transition whenever the computation to be carried out by a hardware unit at the next clock cycle is useless.

Intercommunication and Interconnect Optimization

A significant portion of the power in high performance CMOS VLSI circuits is dissipated in busses or global communications (Mehendale *et al.*, 1998; Winzker, 1998). Global communication typically involves the switching of large capacitive loads, such as *I/O* ports and the clock distribution network, that inherently require significant power (Mehendale *et al.*, 1998a; Stan and Burleson, 1997). Some of the techniques used to reduce the power dissipation in communication networks inside a chip include **data encoding techniques** and **low-swing signaling**.

In (Stan and Burleson, 1997), data encoding or compression methods are used to reduce the power dissipation by reducing the switching activities on global busses. Depending on the data type, compression can have a very high impact on switching activity. For example, an address bus tends to have sequential behavior, and a Gray code is optimal in such a case (Mehendale *et al.*, 1998a, 1998b; Su *et al.*, 1994) (as much as a 50% reduction is cited in (Mehendale *et al.*, 1998a). Another similar technique is the asymptotic zero-transition encoding used in Benini *et al.* (1997). Taking this further, there is really no need to actually send a new sequential address for burst transfers on a synchronous bus; only the first address is actually transferred, whereas the next addresses in the burst are locally generated. Another low-power encoding scheme is to add redundancy in a controlled manner such that the correlation in time between successive data values reduces the switching activity. One example of these techniques is the bus-invert encoding (BIE) (Fujiwara and Pradhan, 1990; Stan and Burleson, 1995) used for buses with lengths $K > 2$. The idea is to use one extra bus line, called an invert. The method computes the Hamming distance between the present *encoded* bus value and the next data value. If the distance is greater than $K/2$, then the transmitter *inverts* the next value and signals the inversion with invert $= 1$. Otherwise, the next value is not inverted and invert $= 0$. The value of invert must be transmitted over the bus to recover the correct information at the receiver; hence, the method increases the number of bus lines by one.

The BIE uses redundancy in space (number of bus lines) for reducing the switching activity. One problem with such encoding techniques is that the required number of extra bus lines increases exponentially as the transition activity is reduced. An alternative is to keep the number of bus lines constant and inject redundancy in time by using extra transfer cycles. In the deep-submicron era, interconnect wires (more specifically global wires, such as buses and clock signals) and the associated driver and receiver circuits are responsible for a large fraction of the energy consumption of an integrated circuit (Zhang and Rabaey, 1998). The low-swing signaling technique is used for information transfers on long wires where wires achieve low-power consumption as they assume reduced voltage swings on the transmitted signals. Such technique requires special transmitter and receiver circuits to provide and sense the low swing voltages on signals flowing in these busses; normal circuits cannot detect these low swing circuits correctly. Examples of these interfacing circuits can be found in (Nakagome *et al.*, 1993; Burd, 1998; Colshan and Jaroun, 1994; Hiraki *et al.*, 1995).

5.7.3 Logic Gate Level

Logic synthesis is the process by which a behavioral or RTL design is transformed into a logic gate level net list using a predefined technology library (Devadas *et al.*, 1994). The trivial attempt for low-power design is to target a library in which the components are designed to be low power. However, other techniques are needed since there are opportunities that are not exploited by the use of a low-power library.

Constrained Optimization

As mentioned in Section 5.6, power optimization can take the form of a constrained optimization problem where performance degradation is acceptable to a given bound. Thus, power minimization requires optimal exploitation of the slack on performance constraints. To this end, many CAD tools that implement optimization algorithms were developed to optimize an architecture from a power point of view given some timing constraints with other specifications.

Path Equalization: Lower Vdd

Between two consecutive registers, one can identify a number of paths, the most important of which is the critical path. The critical path is characterized by the fact that it takes the most time, with respect to the other paths, to traverse between these registers. The other paths will be faster, and one can lower the supply voltage for these fast paths until their delay becomes similar to the critical path, such as in Figure 5.18.

Path Equalization: Resizing

For path equalization, another alternative to lowering the supply voltage is to use gate resizing (See Figure 5.18). Sizing here again focuses on the noncritical paths or the fast paths where gates along these fast paths are downsized to reduce their input capacitances. Reducing these capacitances yields savings in power consumption. Resizing does not always imply downsizing. Power can also be reduced by enlarging heavily loaded gates to increase their output slew rates. Fast transitions minimize short-circuit power of the gates in the fan-out of the gate that has been sized up, but the gate's capacitance is increased. In most cases, resizing is a complex optimization problem involving a trade-off between output switching power and internal short circuit power on several gates at the same time with the constraint being set on performance by the critical path timing.

Glitch Avoidance and Local Transformations

Spurious switching, as mentioned before, is due to the existence of unequal paths to outputs. Path equalization techniques help in reducing these glitches, consequently resulting in less power dissipation. Other logic-level power minimization techniques involve local transformations, including refactoring, remapping, phase assignment, and pin swapping. All the material in this section is from Benini *et al.* (2001). All these techniques can be classified as local transformations. They are applied on gate net-lists and focus on nets with large switching capacitance. Most of these techniques replace a gate or a small group of gates around the target net in an effort to reduce capacitance and switching activity. Similarly to resizing, local transformations must carefully balance short circuit and output power consumption.

Logic-level power minimization is relatively well studied and understood. Unfortunately, due to the local nature of most logic-level optimizations, a large number of transformations have to be applied to achieve sizable power savings. This is a time-consuming and uncertain process during which uncertainty is caused by the limited accuracy of power estimation. In many cases, the savings produced by a local move are below the "noise floor" of the power estimation engine. As a consequence, logic-level optimization does not result in massive power reductions. Savings are in the 10 to 20% range, on average.

5.7.4 Circuit Level

Only local optimizations are possible at this level; more specifically, the low-power techniques are applied for simple primitive components that assume some specific input and output characteristics, such as input rise and fall times and output load capacitance. The techniques at this level are **library cell design, transistor sizing**, and **circuit design style**.

Library Cell Design

The low-power **library cell design** lies at the heart of the circuit level techniques to reduce power consumption. From the power viewpoint, the most critical cells in a design library are the timing elements—flip-flops and latches—

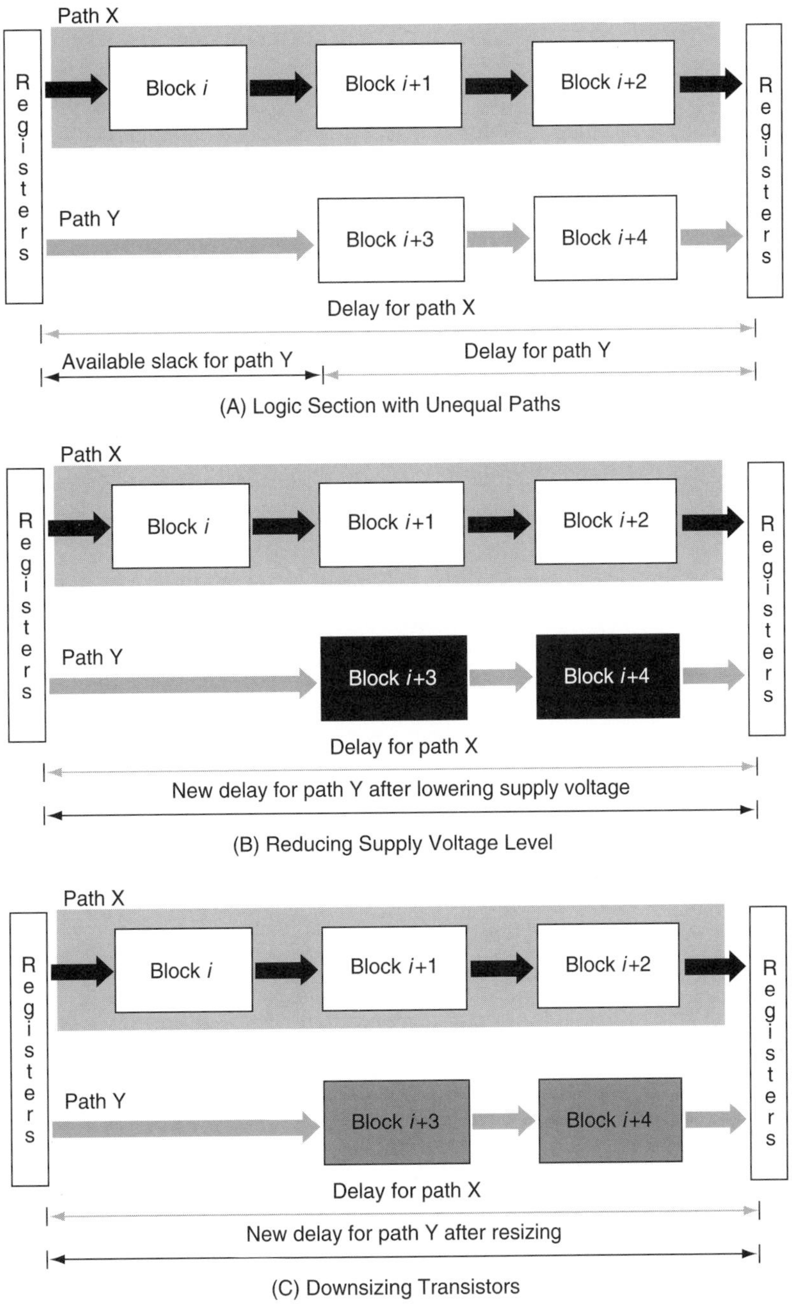

FIGURE 5.18 (A) This Part of the Figure Shows Logic Section with Unequal Paths to the Output. (B) One Can Reduce the Supply Voltage Level of the Units on the Fast Section, in this Case Path Y, to Match the Speed of the Stage Critical Path, Path X. (C) Alternatively, one Can Downsize the Transistors of the Units on Path Y where Speed Is Bound by that of the Critical Path to Reduce the Capacitances. In either case, one can achieve a considerable power reduction on path Y.

because of two reasons. First, these timing elements are extensively used in almost all digital systems where many storage units and pipelining are used. Second, it is well known that a significant fraction of the system power is dissipated in the clock distribution network that drives all timing elements. Consequently, the design of low-power sequential primitives is of great importance where flip-flop design for low power focuses on minimizing clock load and reducing the internal power when the clock is toggled (Pedram and Rabacy, 2002).

Another extensively used element in VLSI designs is the adder cell, which is considered the basic building block for all arithmetic and DSP functions. The literature describes many adder cells with different characteristics such as transistor count, speed, power consumption, and driving capabilities. The choice of a library element not only depends on its power characteristics but depends on the other characteristics of the chosen component and the design specification (such as area and speed). The choice also depends on the context in which such an element is used, which specifies the inputs driving the library element and the output loads of the driven components.

Transistor Sizing

While building a library, it is quite useful to design the components with different sizes for a wide range of gate loads; each component is thus sized to optimally drive a certain load. **Transistor sizing** at the circuit level complements design techniques at the gate level, where logic level sizing can help the synthesis tool in selecting the component with optimum sizing for low-power consumption. Another sizing example is to effectively drive a large load from a source with low driving capability. One could insert a large buffer between the driving circuit and the load to effectively drive the load, but this solution will not solve the problem; it will only change the problem to effectively drive the large buffer. A traditional solution for this problem is to use a chain of successively larger buffers, each driving an appropriate load, as shown in Figure 5.19. Too many buffers result in more stages with large propagation delay and more capacitive nodes with more power consumption. On the other hand, two few buffers will lead to output slopes that are not steep, resulting in larger delays and more short circuit power consumption. One traditional solution to effectively drive the large load can be found in work by Mead and Conway (1980) as well as by Cong and Koh (1994).

Circuit Design Style

Currently, the majority of digital VLSI systems are designed using **static CMOS (SCMOS) circuit style**. The main reasons for adopting SCMOS in designing VLSI systems are robustness (i.e., low sensitivity to noise) and low power with no static power consumption. SCMOS circuits are characterized by almost ideal voltage transfer characteristic (VTC) curves—symmetrical shape, full logic swing, and high noise margins.

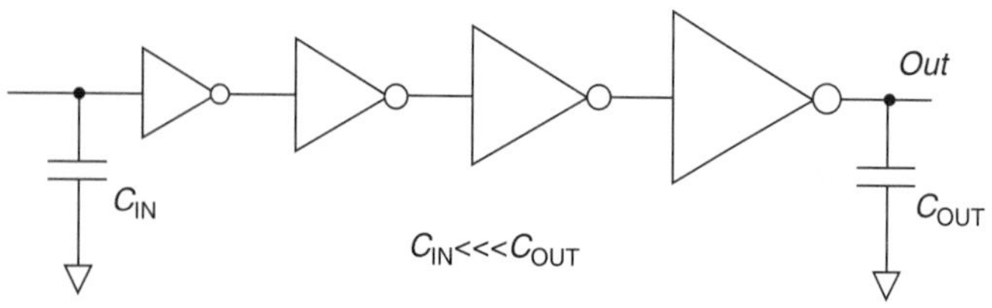

FIGURE 5.19 Multistage Buffer Insertion with Buffer Sizing to Effectively Drive the Large Load

Such characteristics simplify the design process considerably and open the door for design automation. Another major attractor for SCMOS, the main concern of this chapter, is its low-power consumption feature. The static power consumption component is almost absent in the steady-state operation mode, except for some leakage that is mentioned in Section 5.3. The world of the SCMOS, however, is not so perfect; there are some disadvantages associated with SCMOS circuit style, more specifically performance and number of transistors. The SCMOS design family has almost double the number of transistors compared to other families and has inferior performance especially when compared to dynamic CMOS (DCMOS) style. Figure 5.20 compares the general structure for SCMOS and DCMOS styles.

Despite the aforementioned advantages over SCMOS, DCMOS is not widely used because of more serious problems, namely higher power consumption, charge sharing, cascading, and noninverting drawbacks. The higher power consumption results from dynamic circuits requiring two phases of operation: precharging and evaluation. Due to the inherent structure of dynamic circuits, output nodes are precharged every clock cycle, which results in a switching activity equivalent to the system operating frequency. Since the power consumption is proportional to switching activity, more dynamic and short circuit power consumption are associated with DCMOS. Charge sharing is the process by which the output node charge

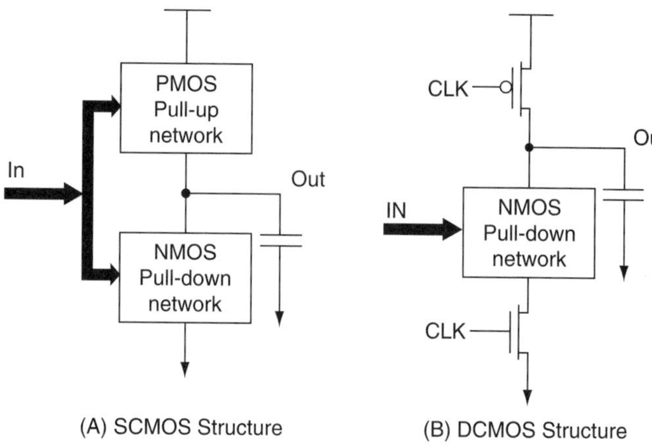

FIGURE 5.20 (A) SCMOS general structure has two complementary networks: PMOS pull-up network for charging the output into Vdd, and NMOS pull-down network to discharge the output to *GND*. The SCMOS style requires $2N$ transistors to implement a function. (B) DCMOS general structure has only an NMOS pull-down structure. The DCMOS style operates in two phases controlled by a clocking signal, *CLK*. When $CLK = 0$, the DCMOS gate is in the precharge state, where the PMOS transistor controlled by *CLK* is on allowing the output to charge to *Vdd*. At the same time, the pull-down network is off since the NMOS transistor controlled also by *CLK* is off ($CLK = 0$). When $CLK = 1$, the evaluate state takes place, allowing the output to be evaluated as a function of the input. DCMOS style requires $N + 2$ transistors to implement a function.

leaks to some internal nodes via paths set by certain changing inputs during the evaluation cycle. This problem results in spurious glitches at the outputs not only to affect output stages driven by the dynamic stage but also to result in additional power dissipation. Many solutions, such as static bleeder and precharging internal nodes, will help alleviate the charge-sharing problem but will result in more power dissipation and more transistors!

Lastly, DCMOS circuits cannot be cascaded; they have to be interleaved by inverting functions for proper operation, a solution provided by Domino logic circuits. In addition, Domino circuits are noninverting in nature and cannot implement all functions. All these problems have aided in ruling out the use of DCMOS and favoring the use of SCMOS style. Another approach to design low-power circuits is to use complementary-pass-transistor logic (CPL) or transmission gate theory. CPL features low-power consumption and a reduced number of transistors. In addition, certain functions are more efficiently implemented using this family, such as XOR gates and MUXs. One major drawback of this family is the weak driving capability with reduced swing outputs, which limits the number of CPL stages that can be cascaded.

5.7.5 Physical Level

Techniques at this level are technology dependent and are produced at the fabrication stage. Ideally speaking, scaling the MOS transistor feature size causes the power to scale down by the same scaling factor and the power density to remain constant. But such observation is too far from reality when using actual implementation products, and it lends itself to two reasons: (1) historically, the supply voltage level scaling does not follow the feature size scaling theory, and (2) more transistors are incorporated per chip with the die size increasing from one generation to the next. For older technologies down to 0.5 micrometer technology, the supply voltage level has been standardized (5 V, then 3.3 V), but for the deep-submicron, DSM, regions when feature size is less than 0.5 micrometers, probably the most widely successful power reduction technique is the power-driven voltage scaling. At this point, it is important to mention that transistor speed does not depend on Vdd alone but also on the gate overdrive, as shown in this equation:

$$V_{\text{Gate Over Drive}} = Vdd - V_T, \tag{5.9}$$

where V_T is the device threshold voltage set more or less by the technology. It is important to mention also that the current flowing across the transistor channel is inversely proportional to V_T so a decrease in V_T will lead to an increase in the current and faster switching, as shown in Figure 5.21. At this level, one can reduce the power consumption by using techniques such as joint supply voltage–threshold voltage reduction as well as multiple threshold and variable threshold techniques (Pedram and Rabaey, 2002).

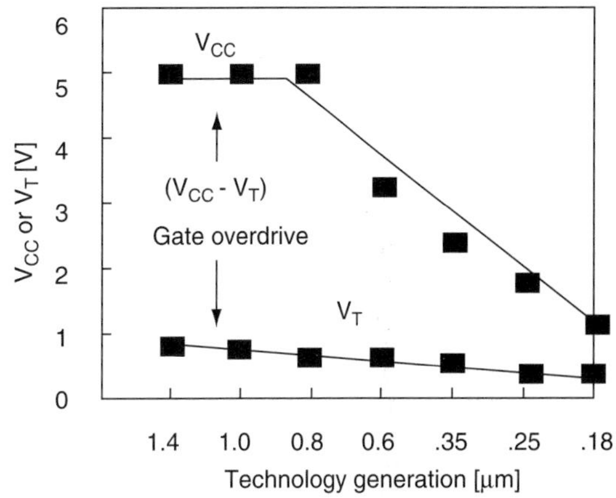

FIGURE 5.21 Supply Voltage Scaling and Threshold Voltage Scaling with Technology Advancement. The gate overdrive is shown.

In the joint supply voltage–threshold voltage reduction, the supply voltage can be reduced to decrease the power consumption and at the same time decrease the threshold voltage to compensate for the speed loss. The decrease in the threshold voltage, however, will cause more static power dissipation due to the increase in the leakage current. One alternative is to use multithreshold devices where different devices with different threshold voltages are used across the design to limit the amount of leakage current. Another alternative is the use of variable threshold devices where the threshold voltage is dynamically controlled through substrate biasing.

Acknowledgments

The authors acknowledge the support of the United States Department of Energy (DoE) EET APP program (DE-97ER12220) and the Louisiana Information Technology Initiative.

References

Alidina, M., Monteiro, J., Devadas, S., Ghosh, A., and Papaefthymiou, M. (1994). Precomputation-based sequential logic optimization for low power. *IEEE Transactions on VLSI Systems 2*(4), 426–436.

Athas, W.C., Svensson, L.L., Koller, J.G., Tzartzanis, N., and Chou, E. Y.C. (1994). Low-power digital systems based on adiabatic-switching principles. *IEEE Transactions on VLSI Systems 2*(4), 398–407.

Bajwa, R., Hiraki, M., Kojima, H., Gorny, D., Nitta, K., Shridhar, A., Seki, K., and Sasaki, K. (1997). Instruction buffering to reduce power in processors for signal processing. *IEEE Transactions on VLSI Systems 5*(4), 417–424.

Benini, L., and de Micheli, G. (1999). System-level power optimization: Techniques and tools. ACM ISLPED, San Diego, California, 288–293.

Benini, L., and de Micheli, G. (2000). System-level power optimization: Techniques and tools. *ACM Transactions on Design Automation of Electronic Systems (TODAES) 5*(2), 115–192.

Benini, L., de Micheli, G., and Macii, E. (2001). Designing low-power circuits: Practical recipes. *IEEE Circuits and Systems Magazine 1*(1), 6–25.

Benini, L., de Micheli, G., Macii, E., Sciuto, D., and Silvano, C. (1997). Asymptotic zero-transition activity encoding for address busses in low-power microprocessor-based systems. *Proceedings of GLS-VLSI*, 77–82.

Burd, T. (1998). *Energy efficient processor system design*. Ph.D. Diss. University of California at Berkeley.

Burd, T., and Brodersen, R. (1996). Processor design for portable systems. *Journal of VLSI Signal Processing Systems 13*(3), 203–221.

Chandrakasan, A.P., Sheng, S., and Brodersen, R.W. (1992). Low-power CMOS digital design. *IEEE Journal of Solid-State Circuits 27*(4), 473–484.

Chang, J.M., and Pedram, M. (1997). Energy minimization using multiple supply voltages. *IEEE Transactions on VLSI Systems 5*(4), 436–443.

Choi, K.W., and Chatterjee, A. (2001). Efficient instruction-level optimization methodology for low-power embedded systems. *International Symposium on Systems Synthesis*, 147–152.

Colshan, R., and Jaroun, B. (1994). A novel reduced swing CMOS bus interface circuit for high-speed low-power VLSI systems. *Proceedings of IEEE International Symposium on Circuits and Systems*, 351–354.

Cong, J., and Koh, C.K. (1994). Simultaneous driver and wire sizing for performance and power optimization. *IEEE Transactions on VLSI Systems 2*(4), 398–407.

Devadas, S., Ghosh, A., and Keutzer, K. (1999). Logic synthesis. McGraw-Hill.

Fujiwara, E., and Pradhan, D. (1990). Error-control coding in computers. *IEEE Computer 23*(7), 63–72.

Gutnik, V., and Chandrakasan, A. (1997). Embedded power supply for low-power DSP. *IEEE Transactions on VLSI Systems 5*(4), 425–435.

Henkel, J. (1999). A low-power hardware/software partitioning approach for core-based embedded systems. DAC, 122–127.

Hiraki, M., Kojima, H., Misawa, H., Akazawa, T., and Hatano, Y. (1995). "Data-dependent logic swing internal bus architecture for ultra low-power LSIs. *IEEE Journal of Solid-State Circuits 30*(4), 397–402.

Kang, S.M., and Leblebici, Y. (1999). *CMOS digital integrated circuits analysis and design*. McGraw-Hill.

Ko, U., Balsara, P., and Lee, W. (1995). Low-power design techniques for high-performance CMOS adders. *IEEE Transactions on VLSI Systems 3*(2), 327–333.

Ko. U., Balsara, P., and Nanda, A. (1998). Energy optimization of multilevel cache architectures for RISC and CISC processors. *IEEE Transactions on VLSI Systems 6*(2), 299–308.

Landman, P.E., and Rabaey, J.M. (1995). Architectural power analysis: The dual-bit type method. *IEEE Transactions on VLSI Systems 3*(2), 173–187.

Lee, M.T.C., and Tiwari, V. (1996). A memory allocation technique for low-energy embedded DSP software. *IEEE Symposium on Low-Power Electronics*, San Diego, California, 24–25.

Mead, C., and Conway, L. (1980). *Introduction to VLSI systems*. Addison-Wesley.

Mehendale, M., Sherlekar, S., and Venkatesh, G. (1998a). Low-power realization of FIR filters on programmable DSPs. *IEEE Transactions on VLSI Systems 6*(4), 546–553.

Mehendale, M., Sherlekar, S., and Venkatesh, G. (1998b). Extensions to programmable DSP architectures for reduced power dissipation. *Proceedings of the 11th International Conference on VLSI Design*, 37–42.

Moore, G.E. (1965). Cramming more components onto integrated circuits. *Electronics 38*(8).

Najm, F.C. (1994). A survey of power estimation techniques in VLSI circuits. *IEEE Transactions on VLSI Systems 2*(4), 446–455.

Nakagome, Y., Itoh, K., Isoda, M., Takeuchi, K., and Aoki, M. (1993). Sub-1-V swing internal bus architecture for future low-power ULSIs. *IEEE Journal of Solid-State Circuits. 28*(4), 414–419.

Nielsen, L., Niessen, C., Sparso, J., and Berkel, K. (1994). Low power operation using self-timed circuits and adaptive scaling of the supply voltage. *IEEE Transactions on VLSI Systems 2*(4), 391–397.

Panasonic. (2000). Panasonic nickel metal hydride batteries technical handbook.

Pedram, M. (1995). Design technologies for low-power VLSI. *Encyclopedia of Computer Science and Technology*, 73–96.

Pedram, M., and Rabaey, J. (2002). *Power aware design methodologies*. Kluwer Academic Publishers.

Rabaey, J. (1996). *Digital integrated circuits: A design perspective*. Prentice Hall.

Shiue, W.T., and Chakrabarti, C. (1999). Memory exploration for low-power embedded systems. *Proceedings of the 36th ACM/IEEE Conference on Design Automation*, 140–145.

Smith, M.J.S. (1997). *Application-specific integrated circuits*. Addison-Wesley.

Stammermann, A., Kurse, L., Nebel, W., Praatsch, A., Schmidt, E., Schulte, M., and Schulz, A. (2001). System-level optimization and design space exploration for low power. ACM ISSS, 142–146.

Stan, M., and Burleson, W. (1995). Bus-invert coding for low-power *I/O. IEEE Transactions on VLSI Systems 3*, 49–58.

Stan, M., and Burleson, W. (1997). Low-power encodings for global communication in CMOS VLSI. *IEEE Transactions on VLSI Systems 5*(4), 444–455.

Su, C., Tsui, C., and Despain, A. (1994). Saving power in the control path of embedded processors. *IEEE Design Test Computer 4*, 24–31.

Tiwari, V., Malik, S., and Wolfe, A. (1994). Power analysis of embedded software: A first step towards software power minimization. *IEEE Transactions on VLSI Systems 2*(4), 437–445.

Tsui, C.Y., Monterio, J., Pedram, M., Devadas, S., Despain, A., and Lin, B. (1995). Power estimation methods for sequential logic circuits. *IEEE Transactions on VLSI Systems 3*(3), 404–416.

Wei, L., Roy, K., and De, V. (2000). Low-voltage, low-power CMOS design techniques for deep submicron ICs. *Thirteen International Conference on VLSI Design*, 24–29.

Weste, N., and Eshraghian, K. (1993). *Principles of CMOS VLSI design: A system perspective*. Addison-Weseley.

Winzker, M. (1998). Low-power arithmetic for the processing of video signals. *IEEE Transactions on VLSI Systems 6*(3), 493–497.

Wolfe, W. (2002). *Modern VLSI design system-on-chip design*. Prentice Hall.

Woo, S., Yoon, J., and Kim, J. (2001). Low-power instruction encoding techniques. *Proceedings of the SoC Design Conference 2001*.

Zhang, H., and Rabaey, J. (1998). Low-swing interconnect interface circuits. *Proceedings of International Symposium on Low-Power Electronics and Design*, 161–166.

6

Production and Utilization of Micro Electro Mechanical Systems

David J. Nagel and
Mona E. Zaghloul
Department of Electrical and Computer Engineering, The George Washington University, Washington, D.C., USA

6.1 Introduction

Many micro electro mechanical systems (MEMS) devices are now on the market. They are available for use as components in diverse systems with many applications. Some MEMS devices are used primarily for their planned employment. Systems of micromirrors for transmitting and displaying information are examples. Other MEMS components are finding opportunistic applications in the hands of creative system engineers. Pressure sensors and accelerometers are prime examples. MEMS are already important devices in several major industries. Both the variety of available mechanical microdevices and their applications will grow rapidly in the foreseeable future.

This chapter first provides an overview of MEMS. The sequential steps in their production are then reviewed, namely design, simulation, fabrication, packaging, and testing. The reliability of MEMS is considered from the viewpoint of their performance in actual applications. The applications of micromachined static MEMS, micromechanical MEMS sensors and actuators, and functional MEMS systems are also surveyed. Examples of MEMS use in the transportation, communication, analytical, and medical industries are provided. The Appendix at the end of this chapter lists more than 30 books on MEMS that have been published in the past decade.

6.2 Overview of MEMS

Micro electro mechanical systems are devices that have static or moveable components with some dimensions on the scale of a micrometer to nanometer. For comparison, a human hair is about 80 micrometers in diameter. MEMS combine microelectronics and micromechanics and sometimes microoptics and micromagnetics. A decade ago, hundreds of MEMS components were prototyped and dozens were commercially available. Many people then viewed the technology as the classical "solution looking for a problem to solve." The situation changed markedly during the 1990s when the market for MEMS took off in a manner reminiscent of the sales growth of integrated circuits in the 1960s. At present, roughly 100 million MEMS components are being sold annually. Patents on MEMS are being granted at the global rate of about one per workday or 200 annually. While MEMS devices will not be used as commonly as integrated circuits, they will be found in a great diversity of products and installations. Just as most people in technological societies own products with integrated circuits and microlasers, pervasive ownership and use of MEMS are clearly in prospect. It is estimated that there are 1.6 MEMS devices per person in the United States (*1*). Micromechanical devices will both improve the performance of existing systems and enable entirely new applications.

Dozens of companies make and sell MEMS. Now that MEMS devices are available in a greater variety and in large numbers, it is possible for the applications engineer to incorporate them in many different products. Most products are the systems for which particular MEMS were designed, but others are targets of opportunity. That is, the MEMS components are simply devices that are available for incorporation into whatever systems can use them.

Figure 6.1 shows the dollar volume for MEMS products as a function of year as compiled and projected by several studies. It is striking that the compilations vary so widely in their absolute value and in their rate of growth. The variance in values is due to differences in what is included in the study: for example, are micromachined ink jet print heads counted as MEMS or not? Some studies also include products enabled by the availability of MEMS devices. However, all the studies indicate that the production of MEMS is already a multibillion dollar industry that doubles every 2 to 4 years. The associated growth rates of 17 to 35% are noteworthy.

The first micromechanical device made by modern manufacturing techniques was demonstrated in the mid-1960s. Significant commercial production of MEMS started in the 1980s, with the appearance of pressure sensors for automotive and medical applications. Then, in the 1990s, microaccelerometers were mass-produced for air bag triggers in cars. Volume production of microfluidic devices also began in that decade. Optical MEMS for displays came to market a few years ago, and micromirrors for switching signals in fiber networks are poised to enter the volume production stage. MEMS switches for control of radio-frequency signals and microwave signals is the next predictable market for MEMS devices. Devices for dense data storage and many new medical applications are also in prospect.

The integration of microelectronics and micromechanics is a historic advance in the technology of small-scale systems and is very challenging for designers and producers of MEMS. We will show examples of monolithic (single substrate) and hybrid (two substrate) MEMS later in this review. We now pause to

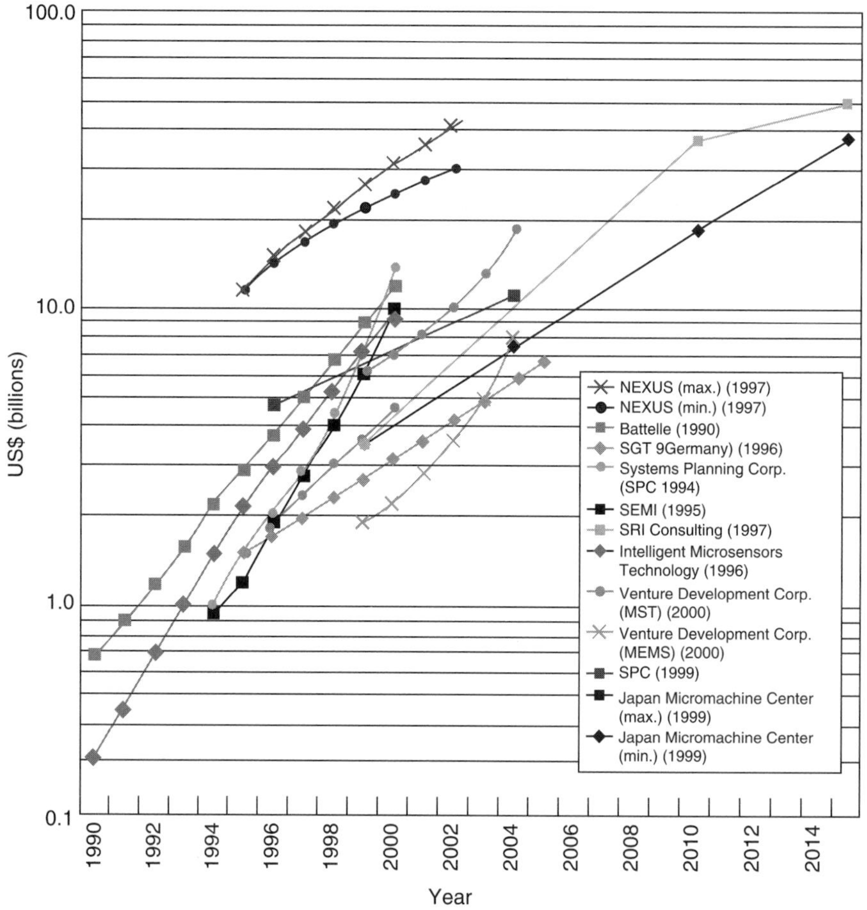

FIGURE 6.1 The Actual and Projected Commercial Values for the Manufacturing of MEMS and Microsystems. Data are from several studies published from 1994 to 2000. Figure adapted from Marshall (2000).

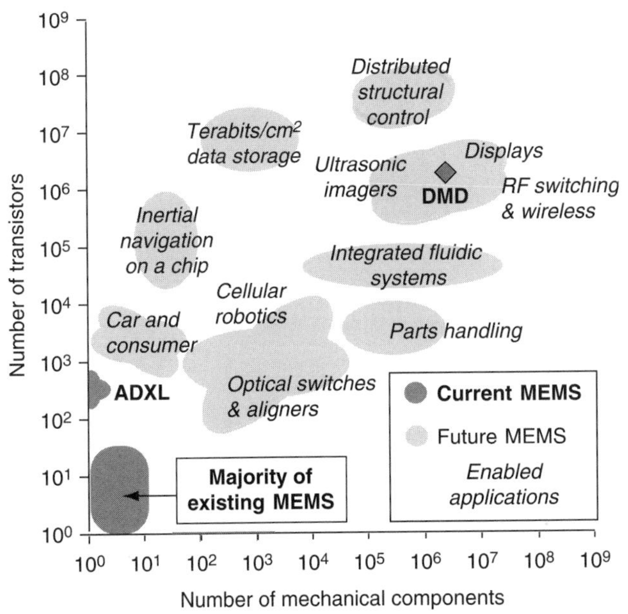

FIGURE 6.2 Plot by Gabriel: The Number of Electronic and Mechanical Components in Current and Future MEMS Devices. Applications enabled by emerging MEMS are shown by the shaded regions. Figure adapted from Gabriel (1995).

appreciate the scope and impact of integrating microelectronics and micromechanics. One way to do this is to consider Figure 6.2. It is arguably the most important graphic in the entire field of MEMS. We note that the microelectronic personal computer and Internet revolutions all occurred using components without moving parts. The addition of micromachined parts to microelectronics opens up a large and very important parameter space to technological development and exploitation. The figure also shows the regions into which fall both current and future applications of MEMS. Most commercial MEMS have only a few electronic and mechanical components. Two commercially important exceptions are the ADXL microaccelerometers from Analog Devices and the Digital Mirror Display (DMD) from Texas Instruments. These are reviewed in the upcoming paragraphs. The variety of potential applications shown in Figure 6.2 is remarkable. The impacts of MEMS will increasingly be felt in many industries and by many consumers.

This chapter surveys some of the current and projected applications of MEMS. Before getting to these applications, we consider in the next section the means by which MEMS are produced. On a macroscopic scale, carpenters do not make engines, and machinists do not produce buildings. That is, the characteristics and performance of large engineered systems depend on the materials and tools used to make them. On a microscopic scale, the uses of MEMS similarly depend on the performance specifications that result from the design and manufacturing phases, especially the behavior of materials and processes at small scales. Section 6.3 surveys the computer and other tools that are available currently for designing, simulating, manufacturing, packaging, and testing MEMS devices. Having some appreciation of these key aspects of MEMS technologies, we then review diverse applications of MEMS structures, sensors, actuators, and systems on a chip. One potential but uncertain use of MEMS is their employment for information storage devices with terabit per square centimeter data densities, which could challenge the dominance of magnetic storage systems.

There are now many books on MEMS and related technologies. Some are quite comprehensive, and others are more detailed. The Bibliography contains a list of the books on or closely related to MEMS. A few Web sites are also good gateways to obtaining information on MEMS.

6.3 From Design to Reliable MEMS Devices

The sequential steps common to making many engineered components also apply to MEMS: **design**, **simulation**, **fabrication**, **packaging**, and **testing**. Figure 6.3 shows these steps. The results of these actions determine both the performance and the price of devices like integrated circuits and MEMS as well as their reliability. Hence, they are considered a necessary prelude to understanding practical applications of MEMS.

6.3.1 Design and Simulation

Computer-aided design (CAD) software is necessary not only for the design of systems and devices that go into MEMS devices but also for the simulation of their expected behavior. Design of MEMS devices is markedly different from the design of integrated circuits. For MEMS devices, the designer should be completely aware of the processing steps and of the materials to be used and their properties. This is in contrast to the design of integrated circuits, in which case the design and the processing steps are separate, and the designer does not have to know the exact details of the process steps. However, for MEMS devices design, the designer has to make many decisions, some determined by processing, and be aware of many constraints to realize the desired device.

The first step in designing an MEMS device is to decide on the materials to be used for a particular device. There are varieties of materials to choose from, and the designer has to choose materials that are compatible and that yield the maximum performance of the device. The designer should have an understanding of the physical properties and knowledge of the physical or chemical techniques required to etch or to deposit these materials. In addition, the designer should know the effects of any of these processing steps on the properties of the materials used and, hence, on the performance of the

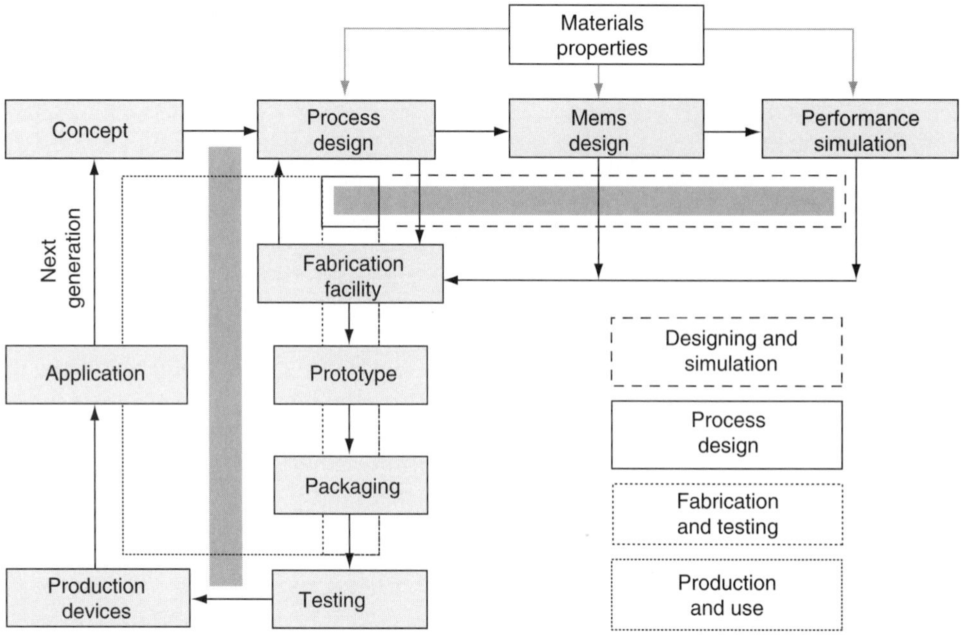

FIGURE 6.3 Flow Diagrams for the Design and Simulation (Horizontal Grey Rectangle) and for the Production, Packaging, Testing, and Use of MEMS Devices (Vertical Grey Rectangle). Loops for designing and simulation, process design, fabrication and testing, and production and use are shown by the solid and dashed lines. Figure adapted from Nagel (1999).

device. After specifying all the materials properties to be used and the processing steps to be followed, the designer then has to decide on the dimensions of the design so that the optimal device dimensions will yield the maximum performance. The dimensions of the device have to be compatible with the process steps and also with the general design rules of the particular technology.

After choosing the materials and dimensions, the designer is ready to start simulating the device to verify that the specified materials and dimensions would yield the required performance. The relationship of the design and simulation phases of MEMS development can be considered by reference to the diagram in Figure 6.3. As noted in that figure, material properties impact both the design and simulation phases. That is, the properties must be known adequately for the particular processes and equipment in the fabrication facility or else the simulated performance will not be accurate.

There are now many software packages and suites available for the design and simulation of MEMS devices. In a very fundamental way, they are more complicated than the software for design of either solely electronic devices or solely mechanical devices. This is due to the close coupling of both electrical and mechanical effects in many MEMS. Consider a microcantilever that is pulled down by electrostatic forces. Its simulation has to take into account both the flow of electrical charge and mechanical elasticity in an iterative and self-consistent fashion. Thermal, optical, magnetic, fluidic, and other mechanisms are also active in some MEMS and have to be handled self-consistently in the simulation phase.

Two basic approaches addressing the need for specialized software for the design and simulation of MEMS have been taken in the past decade. In the first, CAD design tools and available software from electronics were modified to accommodate the requirement for MEMS design. In the second, software from mechanics was applied to MEMS. Some of the software from Tanner Tools VLSI design suite is used for MEMS (e.g., MEMSPRO and the popular mechanical engineering software from ANSYS). New suites of software developed specifically for MEMS have also been marketed. Most of them include electronic, mechanical, and thermal mechanisms, and some have other physical or chemical mechanisms. In addition, some of them also have process simulation tools. Examples are the software that is available from Algor, CFD Research Corporation, Coventor (formerly called Microcosm Technologies), MEMSCaP, and Corning IntelliSense Corporation. We will summarize some of the features of the two software suites, those from MEMSCaP and Corning IntelliSense Corporation, as illustrations of current capabilities.

MEMSCaP is based on CADENCE set of tools, which are the most popular IC design tools (www.memscap.com). MEMSCaP enables the design flow either from the bottom up or the top down and incorporates the MEMS design environment into an existing and well-known IC environment. It permits easy IP or design reuse and the ability to exchange data between multidisciplinary teams. Because the MEMSCaP simulator is based on CADENCE environment, the designer can simulate a MEMS device using the IC schematics or hardware definition language (HDL) approaches. Models can be generated from ANSYS finite element codes or from written analytical equations. Behavior models and scalable symbolic views can be generated. Verilog-A models can also be generated. MEMS Xplorer in MEMSCaP offers a complete environment for MEMS design by combining leading-edge IC design environment (CADENCE) and FEM tools (ANSYS). The generated model can be used to perform optimization simulations inside the environment or to realize a system-level simulation. In addition, emulators are available for isotropic or anisotropic etching and for sacrificial etching.

Intellisuite is another integrated software system that assists designers in optimizing MEMS devices by providing the needed manufacturing databases, allowing them to model the entire device manufacturing sequence, and to simulate the behavior without having to enter a manufacturing facility (http://www.corningintellisense.com). The system can be used to predict the behavior of complex structures to within a 10% accuracy by utilizing fully coupled thermoelectromechanical analysis. Intellisuite incorporates three extensive material, design, and fabrication databases from respected sources. These databases are integrated, and users can easily add custom information to any of them. Intellisuite uses a process sequence builder to emulate the actual process steps that go into building a device in a fabrication facility. For example, the steps in a typical MEMS process (e.g., laying down the silicon substrate, cleaning, deposition, mask layout, implantation, and etching) are incorporated in the software database. Once the user inputs the process sequence or uses a process template, an expert system process checker debugs the process to ensure completeness and continuity of the defined steps. A solid modeler will then create a three-dimensional structure upon combining the fabrication information from the process sequence builder and planar geometry information from the mask layout editor. Surface and volume meshes are generated automatically for the purpose of simulation of the device performance. The simulation allows the specification of voltage, temperature, mechanical stress, thermal stress, and residual stress for a fully coupled analysis. At any process step, the designer can access the material database, MEMaterial, to investigate the effects of process parameters on the structure. For example, the user can investigate the stress of a silicon nitride thin film by varying parameters such as the temperature and pressure of deposition, the radio frequency and power, and gas partial pressures. Once an acceptable stress is achieved, the material properties and fabrication information are automatically used for device simulation.

It is clear from this discussion that CAD tools for MEMS vary widely in the materials parameters and mechanisms that they include, the details of design and simulation of devices, and the fabrication facilities with which they interface. The choice of which software to use for MEMS design and simulation is still challenging.

6.3.2 Materials and Processes

Many options for the fabrication of MEMS devices are now available. This stage, however, is also a daunting part of producing MEMS. The first complexity involves the wide range of materials and associated processes that can be employed.

The fact that the field of MEMS largely grew out of the IC industry has been noted often. There is no doubt that the use of fabrication processes and associated equipment developed initially for semiconductor industry has given the MEMS industry the impetus needed to overcome the massive infrastructure requirements. Less discussed is the additional fact that the field of MEMS has gone far beyond the materials and processes used for IC production. The situation is indicated schematically in Figure 6.4. Less than ten materials, notably silicon and its oxide and nitride, and a similar number of processes, such as lithography and ion implantation, are generally employed to make ICs. The set of materials for IC devices has expanded to include, for example, both low and high dielectric constant materials. However, compared to MEMS a relatively small number of materials and processes are employed now to make IC chips.

Many MEMS can be made with the same set of materials and processes that are used for microelectronics. One of the hallmarks of the emerging MEMS industry, however, is the use of numerous other materials and processes. Substrates other than silicon are being employed for MEMS. Silicon carbide has been demonstrated to be a good basis for many mechanisms that can stand higher temperature service than silicon. In addition there is early and strong interest in the use of synthetic diamond substrates. MEMS are also being made out of plastic, glass, and ceramics. Processes for taking MEMS made out of various materials and putting them on plastic or steel tapes have also been developed.

Diverse materials can be used in MEMS devices. Although aluminum and copper are the metals used in IC devices,

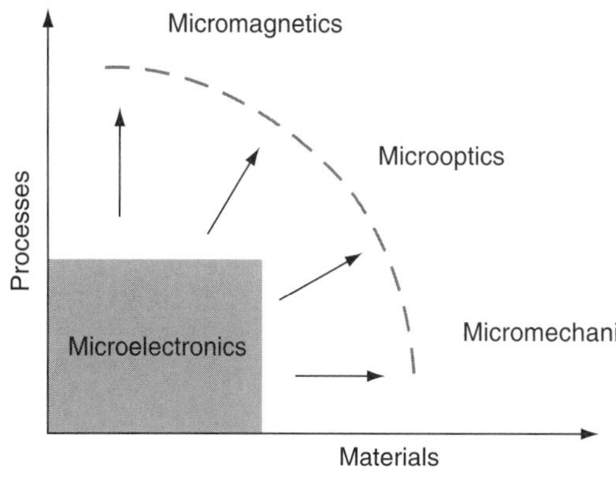

FIGURE 6.4 MEMS Materials and Processes. The number of materials and processes employed to make other microtechnologies greatly exceeds those used to manufacture integrated circuits.

micromachining of many other metals and alloys has been demonstrated. The so-called "shape memory" alloys are used for actuators in several types of MEMS (e.g., microfluidic pumps and valves). Magnetic materials have been incorporated into some MEMS devices. Piezoelectric materials are especially attractive for MEMS because of their electrical–mechanical reciprocity. That is, application of a voltage to a piezoelectric material deforms it, and application of a strain produces a voltage. Zinc oxide and lead zirconium titinate (PZT) are important piezoelectric materials for MEMS. Many other examples of materials employed for MEMS could be given. However, the point is clear: micromechanics are made of many more kinds of materials than microelectronics.

Because materials go hand in hand with the processes for producing and modifying them, the range of processes for making MEMS has also widened far beyond those found in the IC industry. There is something more fundamental at work, however, when it comes to processes for making MEMS. Integrated circuits are monolithic, and, despite having 30 layers in some cases, are made by largely two-dimensional (2-D) thin film processes that yield what some people call 2.5-D structures. By contrast, micromechanical devices must have space between their parts so they can move, and the dimension perpendicular to the substrate is often very fundamentally necessary for their performance. Development of processes to make micrometer-scale parts that can move relative to each other was the breakthrough that enabled MEMS. Such micromachining processes fall into three major categories, which are now reviewed briefly:

A. **Surface micromachining** involves the buildup of micromechanical structures on the surface of a substrate by deposition, patterning, and etching processes. The key step is the etching away of a formerly deposited and patterned sacrificial layer to free the mechanism. This process was first demonstrated almost 40 years ago, when a MOSFET with a cantilever mechanical gate was produced (Nathanson *et al.* 1967). The most common sacrificial material now is silicon dioxide, which is conveniently dissolved from under a moveable part using hydrofluoric acid. Surface micromachining has been used to produce an amazing variety of micromechanical devices, some of which are now in large-scale production. Microaccelerometers and MEMS angle rate sensors are examples.

B. As the name implies, **bulk micromachining** involves etching into the substrate to produce structures of interest. It can be done with either wet or "dry" plasma processes, either of which can attack the substrate in any direction (isotropically) or in preferred directions (anisotropically). Bulk micromachining has two primary variants. The first depends on the remarkable property of some wet chemical etches to attack single crystal silicon as much as 600 times faster along some crystallographic directions compared to others. This anisotropic process is called orientation-dependent etching (ODE). It was known long before the emergence of MEMS technologies and has become a mainstay of the industry. ODE is especially useful for producing thin membranes that serve as the sensitive element in MEM pressure sensors. It is employed for production of these and other commercial MEMS devices. The second approach to bulk micromachining is using plasma-based etching processes that attack the substrate, usually silicon, in preferential directions. Deep reactive ion etching (DRIE) is a plasma process that is used increasingly to make MEMS. It can produce structures that are over ten times as deep as they are wide.

C. Producing structures and mechanisms on a micrometer scale is possible with a **collection of numerous and varied techniques**. Laser-induced etching and deposition of materials, electroetching and electroplating, ultrasonic and electron discharge milling, ink-jetting, molding and embossing are all available to the MEMS designer. Wafer-to-wafer bonding to buildup MEMS devices, or to vacuum package them, is also an important process.

Like integrated circuits, MEMS devices are made using creative combinations of materials and processes noted in this subsection. Some remarkable micromechanisms have been demonstrated, largely in academic fabrication facilities, and commercialized using diverse foundries.

6.3.3 Fabrication Foundries

The next concern after designing and simulating MEMS, that is, deciding on the materials they will contain and the processes needed to make them, is which fabrication facility to employ. The designer can choose from a wide range of facilities. Sometimes a standard IC facility can be used with post-processing of the CMOS to remove material from the bulk substrate and create a suspended structure on top of an etched pit. For example, the micro-hot plate shown in Figure 6.5 was realized in CMOS technology made through MOSIS, with a postprocessing step of bulk micromachining to produce the suspended thin film with a resistive heater. The small mass of the heated element permits temperature changes of over 300 °C in a few milliseconds. Many MEMS structures and devices have been produced by such postprocessing of CMOS chips. Other techniques, which are not compatible with CMOS, were also used, in which case surface micromachining techniques produced mechanical structures on top of the substrate. Micromirrors fabricated using surface micromachining are example of such devices.

Several foundries exist specifically or primarily for the production of MEMS. The fact that the design rules in MEMS are roughly two generations behind those in ICs is significant. This enables MEMS foundries to buy used equipment from the microelectronics industry. Mass production of many MEMS is done using 100- and 150-mm wafers. Several companies and organizations in the United States and abroad offer fabrication services for MEMS that are somewhat analogous to what MOSIS does for ICs. Companies include BFGoodrich

FIGURE 6.5 Optical and Electron Micrographs. These images are of an optical micrograph (top) of elements in a micro-hot plate array and a scanning electron micrograph (bottom) of one of the elements. Images courtesy of Parameswaram *et al.* (1991).

Advanced MicroMachines (Ohio), CMP (France), Cronos Integrated Microsystems (the former Microelectronics Center of North Carolina, known as MCNC), CSEM (Switzerland), Institute of Microelectronics (Singapore), Corning IntelliSense (Massachusetts), ISSYS (Michigan), Kionix (New York), Standard Microsystems (New York), and Surface Technology Systems (United Kingdom).

Most of these foundries have all the facilities in-house to produce complete MEMS devices. The wide variety of materials and processes that can be designed into MEMS means, however, that it is not always possible to find all the needed tools under one roof. Hence, several years ago, the Defense Advanced Research Projects Agency instituted a new type of foundry service called the MEMS-Exchange. This organization contracts with diverse industrial and academic fabrication facilities for a wide range of services, such as deposition of zinc oxide piezoelectric materials. The MEMS designer can draw from any of the facilities and processes. A completed design and funding are sent to the MEMS-Exchange, which handles scheduling, production, billing, and other factors, such as the protection of proprietary designs.

6.3.4 Packaging and Testing

The various systems and materials used for sealed packaging of electronic chips certainly are complex. Packaging of MEMS is usually more difficult because of their mechanical motions or other reasons. During the manufacturing of ICs, some testing can be done at the wafer level as well as after packaging. With most MEMS, performance can only be tested after packaging. Hence, both the packaging and testing of MEMS are significantly more complicated and more expensive than for ICs. Some aspects of both are surveyed in the following paragraphs.

Packaging of MEMS is more challenging than the already quite complex packaging of ICs for a sequence of reasons. Some MEMS can be sealed in prescribed atmospheres in the same kinds of packages as ordinary chips. Microaccelerometers are an example. Many MEMS, however, require vacuum packaging to avoid air damping of vibratory structures or deleterious thermal conductivity. Microresonators for frequency standards and uncooled infrared sensor arrays fall in this category. The MEMS that are used to measure ambient effects must be open in some way to atmospheric conditions. Pressure sensors are an example; they have compliant seals that transmit pressure but not humidity or chemical vapors. However, MEMS designed to measure the latter quantities must be fully open to the atmosphere. Maintenance of performance over their shelf lives and during use is a problem. Some MEMS, such as those mounted on the surface of a structure to control motion, notably vibrations or airflow, are not in separate packages. They have to be mounted on open surfaces, so they offer another set of challenges.

The testing of MEMS devices is intrinsically more complex than the testing of ICs because of the integrated electronic and mechanical characters of MEMS. That is, both the electronic and the micromechanical aspects of a chip with moving parts have to be tested at some point in the prototype and mass production stages. For ICs, testing involves only electronic inputs and outputs although temperature, radiation, and other effects are also of interest. For MEMS, the inputs can also be vibrations and other accelerations or particular conditions of pressure, humidity, and chemical vapor composition, among many other ambient parameters. The production of the required input conditions for the operational testing of diverse MEMS devices is complex and costly as is the measurement of "cross" sensitivities (e.g., the influence of unavoidable temperature and humidity variations on chemical sensors). Only limited testing of MEMS can be done at the wafer level. Testing of fully packaged devices is usually needed. The ability of MEMS to withstand aging of the sensitive components can be problematic to determine in a testing laboratory. The calibration of MEMS sensors, to obtain their quantitative input–output relationships, is another important and nontrivial part of their testing. Setting up the correct environments for MEMS calibration and determining their conditions independent of the MEMS to be calibrated are also challenging. Testing and calibration of MEMS are frequently done in the same experimental arrangements.

In summary, just as the design, simulation, and fabrication of MEMS are more complex than similar functions for microelectronics, the packaging and testing of MEMS are also more difficult than those functions for integrated circuits. The fact that MEMS are newer than ICs and have lower volume further complicates these functions for MEMS. It is widely reported that the packaging of MEMS is often more than half, sometimes 90%, of the cost of the devices. Similar general figures for the testing and calibration of MEMS do not seem to be available.

6.3.5 Reliability

The long-term reliability of MEMS components is essential to their utility for consumer, industrial, and military applications, no matter whether they are used continuously or intermittently. Reliability can be either an enabler for applications or else a "show stopper." Hence, there have been questions and concerns about the reliability of MEMS devices from about 1980 when they first appeared on the market. A great deal of work has been done to determine the ability of MEMS to operate for long times and in diverse environments. This section is a summary of some of the highlights of that work. A few references on MEMS reliability include Miller *et al.* (1998), Stark (1999), Lawton (2000), and the www.sandia.gov/mems/micromachine/biblog_char.html web site.

There are three basic approaches to determining the reliability of MEMS. The first is to make **physics-based estimates** of the likelihood of different potential failures in various conditions. This bottom-up approach to reliability can yield useful understanding of the relevant parameters, but it is plagued with large uncertainties. The second approach involves **laboratory or factory testing** for significant periods of time, as indicated in Figure 6.6. It has the advantage of being experimental, and it permits the calculation of statistics from the measured times or cycles to failure. Further, the ambient conditions can be varied in this type of testing to obtain information on the susceptibility to diverse kinds of failures of particular MEMS that are operated in specific conditions.

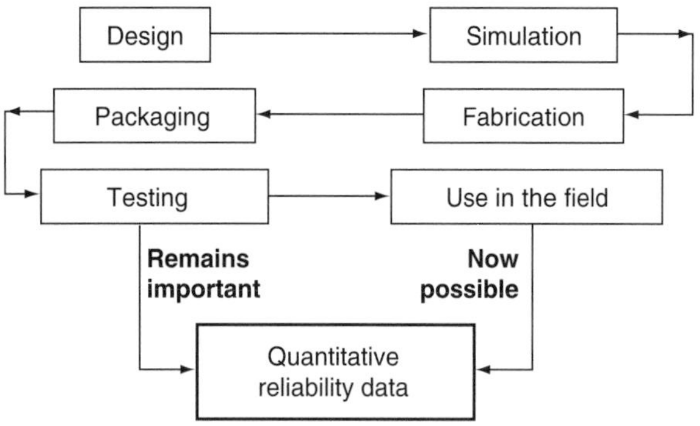

FIGURE 6.6 Types of Testing. Testing in the laboratory or factory and use in the field both produce experimental data on the reliability of MEMS.

This class of reliability testing, however, rarely duplicates the conditions that will actually be encountered by MEMS components in their "real-world" employment. Further, companies that do lifetime and reliability testing on their devices are often reluctant to disclose the data, the methods by which the data were obtained, or the conditions of the tests.

The third and most important and relevant data on reliability of engineering components and systems in general and MEMS in particular is **user field data**. In the mid-1990s, significant data of this kind only existed on micromachined pressure sensors. Manifold air pressure sensors had already proven to be reliable in automobile engines, and disposable blood pressure sensors also operated satisfactorily in their one-time uses. The situation for reliability data on MEMS used by customers has changed quantitatively since the middle of the last decade, as indicated in Figure 6.6. Tens of millions of MEMS devices, most notably air bag triggers in cars and micromirrors in conference room projectors, have proven to be remarkably reliable. It is now possible to make quantitative estimates of the reliability of some major MEMS devices in use by large numbers of customers.

MEMS accelerometers are made or sold by dozens of companies. Of these micro-accelerometers, those from Analog Devices are noteworthy and will be discussed in the subsection on MEMS sensors. Unlike most commercial microaccelerometers, the analog devices accelerometer (ADXL) components have the microelectronics and micromechanics integrated on the same silicon substrate. The company has sold a 100 million of the components, primarily for triggers for air bags, since the mid-1990s. An employee of Analog Devices stated that there have been only a small number of failures involving deployment in an accident or deployment without a crash. Assuming that the annual sales rate is constant and that cars are driven about 300 hours per year, a simple calculation gives reliability better than one failure per 10^{10} hrs more than 1 million years). The rate of failure could also be calculated per turn-on of the car (when the ADXL goes through a self-test routine). Since the deceleration history of cars in actual use is generally unknown, it is not possible to compute failures per various levels of deceleration. The rarity of failures, however, ensures that such numbers would be very small. Further, because the ADXL MEMS components are part of an air bag subsystem, it is not clear whether it was the MEMS device or some other part of the subsystem that failed. It is possible that failure of the MEMS device is rarely the cause of an improper event.

The Digital Mirror Device (DMD) developed and sold by Texas instruments has also been on the market for several years. It is a display device used in projectors for conference rooms and movie theaters. It will be noted in the subsection on actuators. The DMD contains more than one-half million micromirrors in one version and more than one million mirrors in another. There are reports of more than 100,000 of the DMD already being sold. Each of the mirrors has an operating goal of 10^{10} cycles for more than 10 years (essentially running at 30 Hz for that entire period). No specifics on DMD failures are available, but the number is said to be small. Here again, simple estimates yield a remarkable value for the temporal failure rate, namely one failure in 10^{14} cycles, which exceeds 10^9 hours. The salient point is that DMDs that pass the manufacturer's tests have proven to be very reliable in actual use.

Optical MEMS switches for fiber communications systems have been developed and tested by a wide variety of small and large companies. Their testing data, however, is not available, and such switches have not yet found widespread use in actual systems. In a similar fashion, MEMS switches for radio-frequency (RF) and microwave systems have been widely developed and tested. Some company test data are available, even though the RF MEMS switches have yet to find their way into large numbers of commercial products. Raytheon has reported testing capacitive-style stitches to over 10^8 cycles. Motorola tested metal-to-metal RF switches to over 4×10^9 cycles. Such reliability is adequate for some applications of RF MEMS switches but not for the most demanding defense applications. The Defense Advanced Research Projects Agency currently has a development program underway to produce MEMS switches for RF systems that will work for 10^{12} cycles.

The already large and fast-growing database on MEMS reliability, both during pre-use laboratory tests and during field use, indicates that sophisticated MEMS devices can be very reliable. This high reliability derives from the same wafer production technologies that make microelectronics reliable as well as relatively cheap. The general situation on MEMS reliability can be summarized along the lines of the different kinds of MEMS. Those devices that are micromachined but do not have moving parts (like nozzles and microchannels) and those devices that actually move but do not touch (like pressure and acceleration sensors) are highly reliable. Even some MEMS components, notably the DMD, that involve internal contact for many cycles can be very reliable. Only in the most demanding situations, namely actual contact for about a trillion cycles, do MEMS remain to be proven as reliable.

The much-improved state of knowledge about MEMS reliability in recent years does not mean that some concerns about MEMS reliability will not persist. Engineering designers in many large and necessarily conservative applications have to be concerned about the specifics of the devices *and* conditions relevant to their application. The reliability of MEMS devices depends on their packaging. For example, the long-term performance of RF resonators and infrared imagers depends on long-lived vacuum packaging. Shock and vibration resistance of MEMS devices is critical to many prolonged applications, and much more data are needed in this arena. MEMS with built-in test capabilities are sometimes possible, and they are highly desirable for improved system-level reliability. The small size and low cost of MEMS devices make it possible to enhance system-level reliability by the use of redundant devices, which has also been done with microelectronics in satellites for a long time.

6.4 Diversity of MEMS Applications

The MEMS structures and devices that result from the sequence of design, simulation, fabrication, packaging, and testing can be classified into four major groups. The first includes **passive**, or **nonmoving, structures**. The second and third involve **sensors** and **actuators** that have micromechanical components. These are conceptually reciprocal: sensors respond to the world and provide information and actuators use information to influence something in the world. The fourth class includes **systems that integrate both sensors and actuators** to provide some useful function. This classification, like most, is imperfect. For example, some devices that are dominantly sensors have actuators built into them for self-testing. Air bag triggers are an example. However, the taxonomy provides a simple but quite comprehensive framework for considering MEMS devices.

The applications of MEMS are conveniently grouped according to major industries. This approach also raises questions. For example, should the transportation industry, which has very large components, (e.g., the automotive or aerospace industries) be considered as one entity or be subdivided into its units? Similarly, the communications industry embraces both fiber optic and wireless radio-frequency networks. It is simple and useful to group applications of MEMS as indicated in Figure 6.7. We will provide an overview of the types of MEMS in each of the four major categories, including details on particular applications in some of the large industries shown in the top of Figure 6.7.

6.4.1 MEMS: Structures

Immobile structures made by micromachining generally provide a means for guiding the motion of signals or fluids. Optical waveguides that channel photons along the surfaces of chips are a prime example. Waveguides for transmission of electronic signals at microwave frequencies are another. Channels in systems for the analyses of both gaseous and liquid samples are also micromachined structures with no moving parts. Microstructures find applications in the communications, analytical, and medical industries. Static structures integrated with electronics form the basis of infrared imagers that are of use in the transportation industry, for civilian security, and military operations. Because these imagers are sensors, they will be discussed in the next subsection.

Micromachining is already being used commercially to produce passive channels in microfluidics devices, in which control of the motion of the sample and reagent fluids is

	Transportation	Communications	Analytical and medical	Other
MEMS structures	Infrared imagers	Optical and RF signal guides	Microfilters, Microchannels, and micromixers	
MEMS sensors	Pressure, acceleration and angular rate	Power sensors		Many
MEMS actuators	Aerodynamic flow control	Displays, optical switches and RF switches and filters	Micro pumps and valves	
MEMS systems			Point of care analytical systems	Data storage

FIGURE 6.7 Matrix Relating the Four Categories of MEMS to Applications Arenas. Most of the devices and applications that are listed are already commercially significant. Others are under development.

accomplished by application of voltages or externally generated pressures. A few established and about 20 new companies are offering systems, such as the so-called "lab on a chip," for analysis of biomedical materials in research laboratories and for clinical medicine. These companies include Agilent, the formed Hewlett Packard division (which is teamed with Caliper Technologies), and newcomers, such as Aclara, Micronics, and Nanogen. Figure 6.8 shows a microfluidics substrate from Micronics. The approach to DNA, protein, and cellular analyses taken by these companies generally involves the use of small plastic or glass microfluidics systems for mixing of samples and reagents and for separations, in some cases. The microfluidic elements are inserted into desktop machines that provide other needed functions, such as power, control, and display. The microfluidic structures are disposable, so they avoid sample cross contamination. The small size of the key microfluidic elements raises the possibility of having entire handheld systems. This means that medical analyses can be performed immediately at the point of care rather than later at a central laboratory. Only very small samples (e.g., a drop of blood) are needed for microfluidic analyzer systems. Cephied, another microfluidics start-up, is designing a handheld blood analyzer containing micromachined fluidic elements.

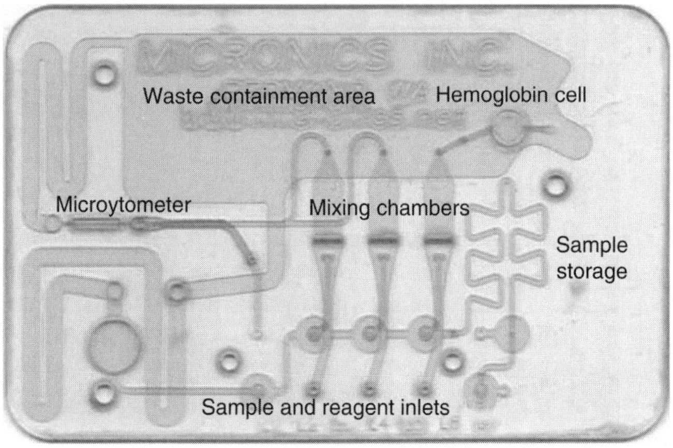

FIGURE 6.8 Photograph of the Plastic Microfluidic Element from Micronics. It is the size of a credit card. Image courtesy of Micronics.

6.4.2 MEMS: Sensors

Most MEMS sensors involve moving elements. However, detectors with no moving parts for sensing optical, infrared, and microwave signals are also made by micromachining. We will review important newly developed infrared sensor arrays before turning to the more familiar and already commercially successful micromechanical MEMS sensors.

The detection of infrared radiation with arrays of small elements enables night vision, which has numerous civil and military applications, especially for security and safety. Warm objects, such as humans and animals, emit infrared radiation. There are two basic approaches to imaging infrared radiation with arrays of small elements. The first involves cooling an array of narrow band-gap semiconductor elements to low temperatures to reduce the thermal noise level and allow measurement of weak thermal radiation. The second is to isolate the infrared absorbing element from surrounding solids that would couple energy from the element. Doing so involves etching away a sacrificial layer under each picture element (pixel), leaving it suspended by fine legs that conduct relatively little heat from the element. The legs are also the current paths used for measuring the resistance of the material on each element that absorbs infrared radiation. The associated heating, actually only a few one-thousandths of a degree, is enough to produce a measurable change in the resistance of the pixel. This MEMS approach has the tremendous advantage of not requiring a cooling system, so it is simpler and cheaper than semiconductor infrared detector arrays although it is not as sensitive to weak signals.

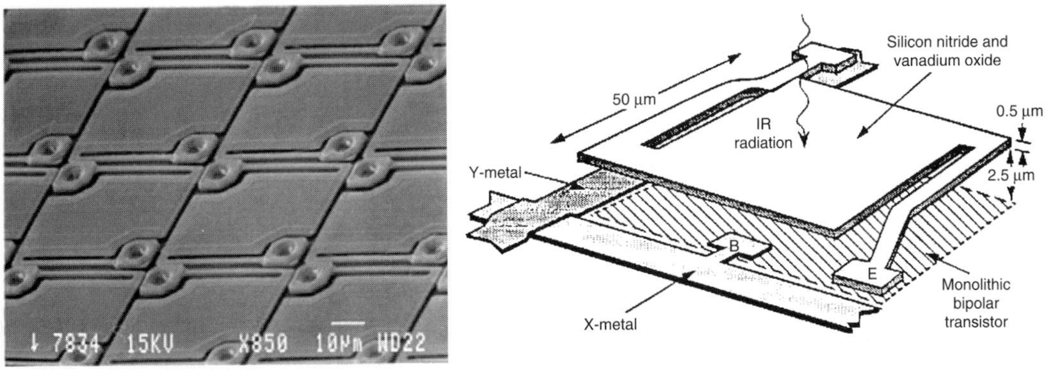

FIGURE 6.9 Honeywell Array and Schematic. This figure shows the scanning electron micrograph and schematic of the micromachined pixels in the Honeywell uncooled infrared imaging array. The sacrificial layer under the pixels was 2.5, micrometers thick. Figure adapted from Wood *et al.* (1993).

Honeywell and Raytheon developed uncooled, micromachined thermal imagers. Figure 6.9 includes both a micrograph of part of the Honeywell array and a schematic showing its construction. In the recent past, micromachined arrays of uncooled infrared detectors were incorporated into some Cadillac vehicles. The systems permit the driver to see the thermal images from people or animals in the road far beyond the range of the headlights.

Micromachined devices with moving parts are at the heart of most of the MEMS sensors already prototyped, many of which are in production. Basically, some condition, be it physical, chemical, or biological, induces motion in the MEMS device that can be sensed by any of several mechanisms. For example, resistive or capacitive sensing of the deflection of a thin, small membrane is used in various MEMS to measure pressure and ultrasound levels. The most important MEMS sensors detect pressure and acceleration; devices for measurement of angular rates are now moving to large-scale production, largely for automotive applications.

FIGURE 6.10 Pressure Sensors. Photograph of various micromachined pressure sensors from Silicon MicroStructures, now part of ORI Systems. Image courtesy of Novasensor.

MEMS pressure sensors cover a wide range of pressures and have very diverse applications. Commercial parts can measure pressures from about 1% of one atmosphere to over one thousand atmospheres. Figure 6.10 shows a variety of devices from one of several manufacturers. Two major areas of application are in the automotive and medical industries. MEMS pressure sensors are employed for measurement of the air pressure in engine manifolds, oil and fuel pressures, and the pressure in tires. Medical uses include the measurement of blood pressure. Medical applications are based on the fact that MEMS pressure sensors can be made cheaply and discarded after use to avoid contamination. Such measurements can be done on catheters inserted into the body because of the small size of the sensors or outside of the body when a needle and tube permit access to the internal blood pressure. Because MEMS pressure sensors are catalog items, they are available for a wide variety of uses, many of them well beyond the initially envisioned applications. They can be found in small weather stations, monitors for heating, ventilation and airconditioning systems, sensors for water and other fluid systems, and recreational engineering projects such as payloads for amateur rockets.

Microaccelerometers, like MEMS pressure sensors, have been on the market for several years, with well over 10 million sold annually. The first market for MEMS accelerometers was the automobile industry for air bag triggers. Initially, single axis devices were offered, and then two axis components became available. The next big application for the two-dimensional micro-accelerometers was in personal entertainment systems to detect the motion of joy sticks. Packaging of microaccelerometers started with transistor cans. Next came surface mount packages, and leadless devices are now available. Figure 6.11 shows exterior of a two-axis accelerometer with a range of plus or minus twice-normal gravity and a micrograph of the chip. The integration of the single moving element and its associated electronics in and on the silicon substrate is note-

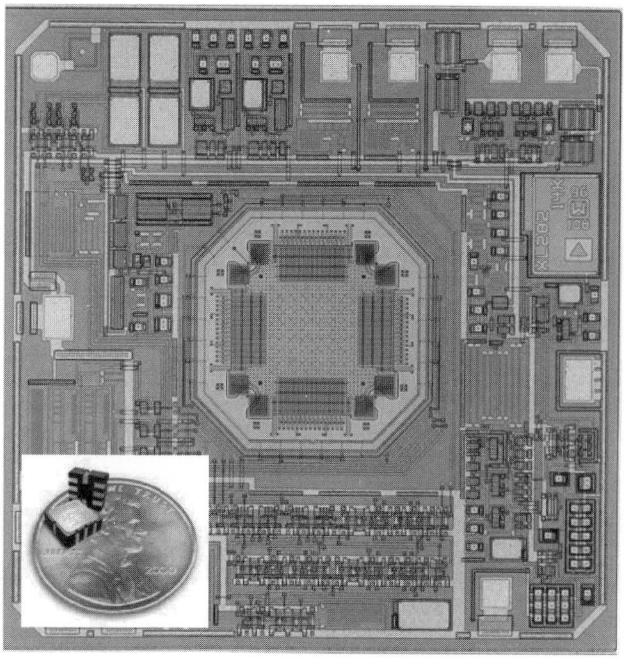

FIGURE 6.11 Two-Axis Accelerometer. These are Photographs of the exterior of a new two-axis microaccelerometer in leadless packages on a penny and a micrograph of the chip from Analog Devices. In this device, the microelectronic and micromechanical components are tightly integrated on the silicon substrate. Image courtesy of Analog Devices.

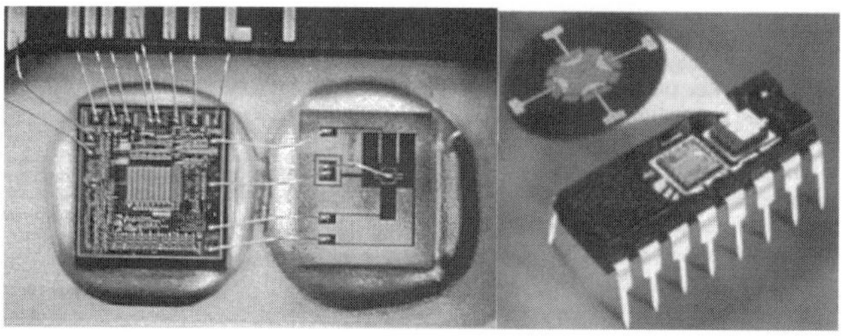

FIGURE 6.12 Commercial microaccelerometers of Ford (Left) and Motorola, (Right). The electronics and mechanics are on separate substrates bridged by wire bonds. Image (Left) courtesy of Spranger and Kemp (1995); image (Right) courtesy of Motorola.

worthy. Figure 6.12 shows two other commercial microaccelerometers in which the manufacturers have used separate substrates for the electronic and mechanical elements. A remarkable feature of the available microaccelerometers is the very low mass, on the order of a microgram, for the moving element. A cube of water 100 μm on an edge weighs 1 μg! Despite this miniscule mass, the sensitive electronics enable the devices to measure a few one-thousandths of normal gravity. Prototype MEMS devices have measured accelerations exceeding 10,000 times normal gravity.

The microaccelerometers are the best example of MEMS components being employed for opportunistic applications. The ability to measure tilt has led to their incorporation into a sensor system for workers who have to lift heavy objects. If the person bends his or her back excessively while lifting, the system sounds a warning. MEMS accelerometers are being used by researchers in sleep studies. A glove equipped with such components on each finger is under development as a training aid for American Sign Language. Microaccelerometers are being used for studies of other motions of the human body, especially in sports. They are also being incorporated into sports equipment, including golf clubs and tennis racquets. Amateur scientists have used them to record accelerations during roller-coaster rides and to make backyard seismometers.

Integrated into the MEMS accelerometers from Analog Devices are capacitor structures that act as electrostatic actuators. When a voltage is applied to them, the moving part of the device deflects, which produces a signal similar to that generated by an input acceleration of a specific value. This feature enables a built-in test capability that allows the device to automatically assay its performance. This is an example of the melding of actuators into MEMS devices that are made for sensing.

6.4.3 MEMS: Actuators

There are many MEMS components whose primary function is to move some small structure to accomplish a useful function. The signals that cause such motion are entirely electronic in the case of electrostatic actuators. In some cases, electrical energy is first turned into thermal or magnetic energy to induce motion on the micrometer scale. Some MEMS actuators depend only on thermal effects. One type is based on the atomic-scale phase transformation of "shape memory" alloys that automatically assume a form given to them at one temperature when they are returned to that temperature. Whatever the physical origin of the small but effective forces in MEMS actuators, interesting and useful changes in positions of a micromechanical structure can result. Some of these will be discussed in the rest of this section.

MEMS and optical signals have a natural synergism. Many small mirrors for redirecting light beams and fine-scale diffraction gratings that can be dynamically reconfigured to analyze optical spectra have been demonstrated and are being manufactured. Micromechanisms with dimensions from a few to a few hundred micrometers are small enough to be moved easily by microactuators and large enough to manipulate small beams of light. In most cases of commercial interest, the actuation is accomplished by application of electrostatic forces to nearby parts in the MEMS device.

The first micromirror device to make it to market is also one of the most remarkable. It is called the digital mirror device (DMD) and made by Texas Instruments for display of images.

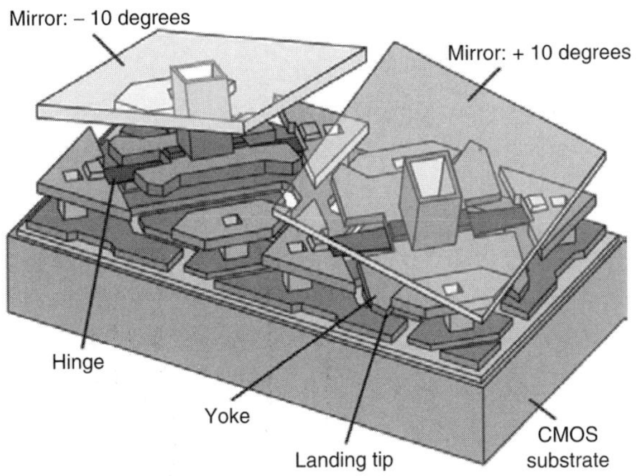

FIGURE 6.13 Drawing of Two Pixels of the Digital Mirror Device Manufactured by Texas Instruments. The torsional hinges are 5 by 1 μm in area and about 100 nms thick. The individual mirrors are 16 μm square. Over 500,000 of them are found in a single device, making this the system with the most moving parts produced in the history of mankind. The inventor, Larry Hornbeck, and the company received television Emmy Awards in 1998 for outstanding achievement in engineering development. Figure provided by Hornbeck (1993).

The operation can be understood with reference to the schematic shown in Figure 6.13. Each picture element of the DMD has three operational levels. The lowest is electronic. It contains the CMOS electronics that control the operation of the device by applying voltages that produce electrostatic forces on the movable elements. The second layer is mechanical. It has the torsional hinges that allow the mirror to be tipped 10 degrees one way or the other. The top layer is optical, consisting of the aluminized mirrors that reflect the light falling on it either into a beam dump or else through a projection system onto the screen. The light originates in a xenon arc lamp and then passes through a rotating red–green–blue filter that cycles at 60 Hz. During each color phase of each cycle, the mirrors tilt to pass light to the screen for no time, some time, or all of the time. This serves to adjust the relative brightness of each color during each cycle.

DMD have been incorporated into what are called display engines and sold to manufacturers of conference room projectors for several years. It is estimated that some tens of thousands of these display engines are sold annually. Texas Instruments has started testing their use in movie theaters. This will enable digital distribution of movies so that handling of celluloid rolls for distribution of movies to theaters will no longer be necessary. In the future, DMD might be employed in homes as a replacement for the large and heavy tubes in current television sets. The bright and high-resolution character of images displayed using DMD technology is attractive. The DMD, however, is in competition with other MEMS display devices and with liquid crystal display technologies.

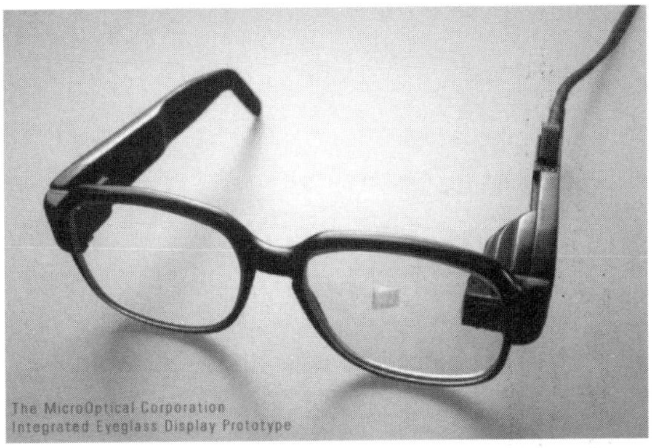

FIGURE 6.14 Prototype Integrated Eyeglass Display from MicroOptical Corporation. The mirror in the middle of one lens permits viewing of a video image produced in the structure on the left ear piece of the glasses. Image courtesy of MicroOptical.

Another very different optical display might be enabled by the small size and high performance of MEMS actuators. Figure 6.14 is a photograph of a prototype display integrated into a pair of eyeglasses. The video signal fed to the left earpiece produces a very small image that can be viewed by looking at the small mirror mounted in the center of the left lens. The wearer can look at the world in a normal manner through most of the lens or view dynamic colored images by looking at the mirror on the one lens. A two-dimensional MEMS scanning micromirror is a candidate technology for production of the image. This approach to a "heads-up display" is substantially lighter than its competitors. Repair personnel could view manuals while working on a system such as an automobile. Voice activation would be used for turning pages and enlarging the images of diagrams.

Micromirrors are also at the heart of optical signal routers for a coming generation of the Internet that will contain "all optical" components. Optical fibers can carry more than one hundred signals using different photon wavelengths ("colors"). Switching this flood of information requires converting the light signals for each color to electronic signals, manipulating them and regenerating light pulses for continued transmission down other fibers. The great advantage of MEMS mirrors for fiber network switches is that switching can be done entirely in the optical domain. There is no need to turn the optical signals into electrical signals for routing and then to produce other light pulses for further transmission. Hence, this is called "all optical" networking (even though the production and detection of the light bundles still involves electronics!). The MEMS approach is faster and requires less hardware. During the year 2000, because of the explosive growth of the Internet, six small MEMS companies were bought by suppliers of hardware for fiber communications systems. Two of them sold for $750 million, one went for $1.25 billion, and another sold for $3.25 billion (with all of these values in stock prices

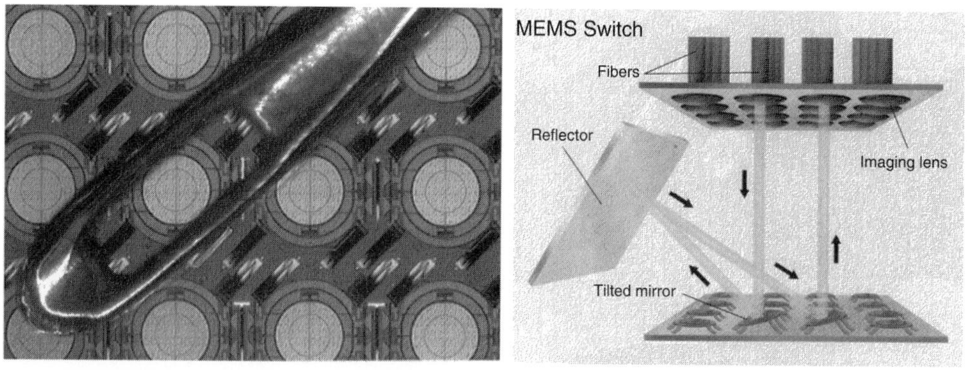

FIGURE 6.15 Optical Micrograph of a Sewing Needle. The needle is atop the MEMS mirrors in the optical switch (left) developed by Lucent. The schematic (right) shows how tilting the mirrors routes signals from one fiber to another, whatever the wavelength of the light. Images courtesy of Bishop *et al.* (2001).

for that time period). Clearly, there was exploding interest in MEMS components for fiber optic networks. Lucent Technologies has been one of the major developers of MEMS mirror switches for optical networks. The company's technology is shown in Figure 6.15. Lucent and other companies making switches to transfer optical signals from one fiber to another are field-testing programs. Major, indeed massive, usage of such switches might occur in the next few years.

The small actuators that result from MEMS technology are also useful for switching microwave-frequency electronic signals on chips. Micromechanisms can be employed to make tunable capacitors and resonators for manipulation of such signals. These technologies for the microwave and nearby frequencies are now under intense development and should come to market within a few years. Such devices are expected to have major impacts on wireless communication systems and short-range radars for automobile collision avoidance, among other products.

A wide variety of MEMS switch designs for high frequencies has been prototyped. In general, they have very low insertion losses (when closed) and excellent isolation (when open). Raytheon has developed a capacitively coupled switch, as shown in Figure 6.16. Rockwell has produced MEMS metal-to-metal switches as well as interdigitated capacitors for use in microwave circuits. The University of Michigan has the highest frequency MEMS resonators. A device based on a beam with both ends free (similar to a xylophone key) has exhibited resonance at 92 MHz. Its vibratory displacement is on the order of a nanometer. Other designs based on the expansion and contraction of disc-shaped structures were developed at the University of Michigan for operation at frequencies approaching 1 GHz. Few microwave MEMS devices are on the market now, but this situation will change markedly in the coming years because of the wireless revolution.

The final application of microactuators that deserves attention is in the field of microfluidics. Micrometer-scale actuators have been used to move pumps and valves in microfluidic systems. The small forces available from most MEMS actuators limit the pressures that can be developed and switched and also limit the flow rates, but the performance is adequate for small-scale analytical systems, notably the "lab on a chip."

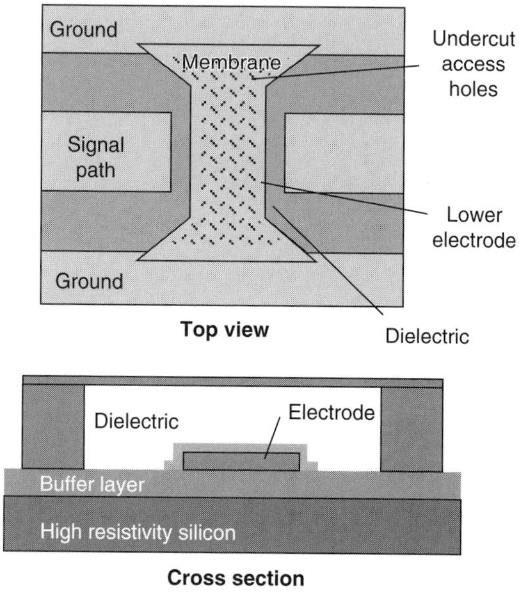

FIGURE 6.16 Micrograph (top) and schematic of the Raytheon MEMS microwave switch. The electrode under the flexible membrane is the actuator. The capacitance of the switch varies from near zero (open) to 3.4 pF (closed). The signal path is about 50 μm wide. Figure adapted from Goldsmith (2000).

6.4.4 Integrated MEMS: Systems

The last letter in MEMS stands for systems, which is appropriate because the MEMS device is quite complex in itself. MEMS

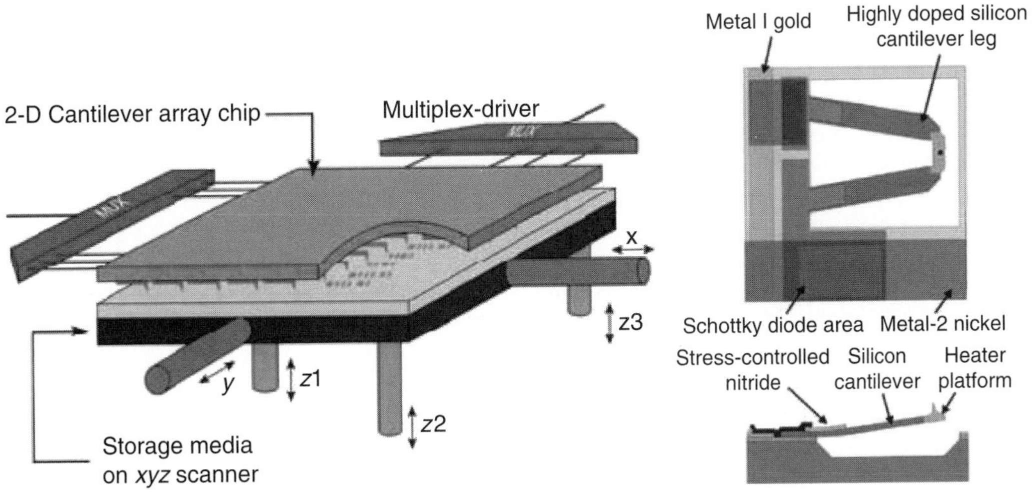

FIGURE 6.17 "Millipede" Data Storage Technology Developed by IBM. Schematic (left) of the electromagnetically actuated stage and recording media under an array of microcantilevers with fine tips for thermally recording and reading nanoscale bits of information. Schematics (right) of the microcantilevers with the heated probe tip for writing bits and later reading them. Figures courtesy of IBM.

devices, however, are only components that are used in larger and more complex systems. That is, individual sensors or actuators can be used as components and incorporated into subsystems or systems to perform some useful function. The accelerometer in the air bag subsystem of an automobile and the DMD in a projection system in a theater are examples. It is also possible to closely couple both MEMS sensors and actuators into miniature systems all on one substrate. These are called "systems on a chip." Microfluidics with all the needed functionality on a substrate, including pumps and valves, as well as channels, mixers, separators and detectors, are under development for compact analyzers. These will be relatively evolutionary advances over current microfluidic chips. High-density data storage systems with both actuation and sensing functions represent a more revolutionary example of integrated microsystems.

In the last decade, it was shown that scanning instruments with very sharp tips could modify surfaces as well as make images of them, both on an atomic scale. The techniques involve the use of scanning tunneling or atomic force microscopes called **proximal probes.** The ability to make a dot with several atoms or molecules of a material (a bit) at a location on a surface raises the possibility of producing integrated MEMS devices that could store terabits of information in a square centimeter. Such a density would exceed the projected capabilities of magnetic storage. Hence, MEMS technology holds the possibility of displacing part of the giant hard disk market. Thousands of fine tips would be employed in a square centimeter. Each tip would be mounted on a micromachined and electrostatically actuated mechanism that can be independently scanned in two dimensions, as sketched in Figure 6.17. Tips would first write bits and later sense their presence or absence, analogous to the function of the magnetic head in a disk drive. The individual bits would have dimensions and spacings of a few molecular diameters. This nascent technology has been called **nanocuneiform** after the production of marks in clay tablets by people in Mesopotamia (at about $10\,\text{bits/cm}^2$). Methods for erasing bits are contemplated, so the MEMS data storage devices would be comparable in behavior to current magnetic media. This approach to data storage is being pursued by IBM and a few other companies. The initial MEMS data storage products would offer densities near $30\,\text{Gbits/cm}^2$. It is interesting that while an entirely new MEMS technology is being developed to displace disk drives, MEMS actuators could be incorporated into the arms in magnetic drives to give finer control of the read/write heads, which decreases the track spacing and increases storage density.

6.5 Summary

The variety of commercially available MEMS and their applications have both increased dramatically in recent years. The production of MEMS is now more than a 20-billion dollar industry worldwide, with about 100 million devices marketed annually. Although this industry grew out of the microelectronics industry, it is more complex in many important ways. Most fundamentally, it requires the integration of both microelectronics and micromechanics. Many MEMS involve several closely coupled mechanisms, some of which behave differently on the micrometer spatial scale than on familiar macroscopic scales. This complicates both the design and simulation of MEMS. Complications also arise from the much wider variety of materials and processes used to make MEMS compared to microelectronics. Because many MEMS have to be open to the atmosphere, their packaging, calibration, and testing are all complex. Questions about the long-term reliability of MEMS are being answered as MEMS devices spend more years in use by consumers and industries.

Despite such engineering challenges, MEMS offer high performance and are small, require low power, and are relatively cheap. They both improve on some existing applications and enable entirely new systems. Some of their applications are targeted from the outset of design, but others are opportunistic; that is, the large number of MEMS components on the market make them available to design engineers for a very wide variety of uses.

It is noteworthy that many large companies have strong positions in the manufacturing and use of MEMS. Several of them were mentioned in this chapter. Being a relatively new field, MEMS technology has also been the basis for the creation of many new companies, especially in the optical networking industry, a significant fraction of which have been bought by larger companies.

The transportation industry has benefited most from the availability of MEMS to date. The impact of MEMS on the analytical and medical industries has already been significant, and major growth in these arenas can be expected. The promises of optical MEMS in the fiber communication industry, and of RF MEMS in the wireless industry, have yet to be realized. Their impact on these communication modes, however, is only a matter of time and timing. Defense industries will similarly see ever increasing utilization of MEMS.

The entire field of MEMS has "turned the corner" in the past decade from gradual growth to rapid exponential growth. MEMS are not likely ever to be so important as integrated circuits. However, they will be a part of the fabric of life in the new century, embedded in many products owned by most people in technological societies.

Appendix: Books on MEMS

A great deal of information on the design, simulation, fabrication, packaging, testing, and applications for MEMS is available. Information can be found in journals, conference proceedings, and on the World Wide Web. In the past several years many books on MEMS have been published. They are listed in this Appendix. It is possible to group them by subject. This is complicated by the fact that some books cover several main subdivisions of MEMS. Hence, we content ourselves to list them chronologically by the year of their appearance.

1995: INTEGRATED OPTICS, MICROSTRUCTURES, & SENSORS. Massood Tabib-Azar (Ed.). Kluwer Academic Publishers. (399 pages).

1996: MICROMACHINES (A NEW ERA IN MECHANICAL ENGINEERING). Iwao Fujimasa. Oxford University Press. (156 pages).

1997: MICROMECHANICS AND MEMS (Classical and Seminal Papers to 1990). William S. Trimmer (Ed.). IEEE Press. (701 pages).

1997: FUNDAMENTALS OF MICROFABRICATION. Marc Madou. CRC Press. (589 pages).

1997: HANDBOOK OF MICROLOTHOGRAPHY, MICROMACHINING, AND MICROFABRICATION, VOL. 2. MICROMACHINING AND MICROFABRICATION. P. Rai-Choudhury (Ed.). SPIE Press. (692 Pages).

1998: MICROMACHINED TRANSDUCERS SOURCEBOOK. Gregory T. A. Kovacs. WCB McGraw Hill. (911 pages).

1998: MICROACTUATORS. Massood Tabib-Azar. Kluwer Academic Publishers. (287 pages).

1998: MODERN INERTIAL TECHNOLOGY. (2d Ed.). Anthony Lawrence. Springer. (278 pages).

1998: METHODOLOGY FOR MODELING AND SIMULATION OF MICROSYSTEMS. Bartlomiej F. Romanowicz. Kluwer Academic Publishers. (136 pages).

1999: SELECTED PAPERS ON OPTICAL MEMS. Victor M. Bright and Brian J. Thompson (Eds.). SPIE Press. Milestone Series, Vol. MS. 153. (644 Pages).

1999: MICROSYSTEM TECHNOLOGY IN CHEMISTRY AND LIFE SCIENCES. A. Manz and H. Becker (Eds.). Springer. (252 pages).

2000: AN INTRODUCTION TO MICRO ELECTRO MECHANICAL SYSTEMS ENGINEERING. Nadim Maluf. Artech House. (265 pages).

2000: MEMS AND MOEMS TECHNOLOGY AND APPLICATIONS. Prosenjit Rai-Choudhury (Ed.). SPIE Press. Vol. PM85. (528 Pages).

2000: ELECTROMECHANICAL SYSTEMS, ELECTRIC MACHINES, AND APPLIED MECHATRONICS. Sergey Edward Lyshevski. CRC Press. (800 pages).

2000: HANDBOOK OF MICRO/NANO TRIBOLOGY (2d Ed.). Bharat Bhushan. CRC Press. (880 pages).

2001: MEMS HANDBOOK. Mohamed Gad-El-Hak (Ed.). CRC Press. (1368 pages).

2001: MICROSYSTEM DESIGN. Stephen D. Senturia. Kluwer Academic Publishers. (689 pages).

2001: MEMS AND MICROSYSTEMS: DESIGN AND MANUFACTURE. Tai-Ran Hsu. McGraw-Hill College Division. (448 pages).

2001: MECHANICAL MICROSENSORS. M. Elwenspoek and R. Wiegerink. Springer. (295 pages).

2001: NANO- AND MICRO ELECTRO MECHANICAL SYSTEMS. Sergey Edward Lyshevski. CRC Press. (338 pages).

2001: MICROFLOWS: FUNDAMENTALS & SIMULATIONS (2d Ed.). George Em Karniadakis and Ali Berskok. Springer Verlag. (360 Pages).

2001: MICROSENSORS, MEMS, AND SMART DEVICES. Julian W. Gardner, Vijay K. Varadan, and Osama O. Awadelkarim. John Wiley & Sons. (528 pages).

2001: MICROSTEREOLITHOGRAPHY AND OTHER FABRICATION TECHNIQUES FOR 3-D MEMS. Vijay K. Varadan, Xiaoning Jiang, and Vasundara V. Varadan. New York: John Wiley & Sons. (260 Pages).

2002: FUNDAMENTALS OF MICROFABRICATION (The Science of Miniaturization) (2d Ed.). Marc Madou. CRC Press. (752 pages).

2002: MEMS AND NEMS: SYSTEMS, DEVICES, AND STRUCTURES. Sergey Edward Lyshevsky. CRC Press. (480 pages).

2002: MICROFLUIDIC TECHNOLOGY & APPLICATIONS. Michael Koch, Alan Evans, and Arthur Brunnschweiler. Research Studies Press. (321 Pages).

2002: FUNDAMENTALS AND APPLICATIONS OF MICROFLUIDICS. Nam-Trung Nguyen and Steven T. Wereley. Artech House. (471 Pages).

2002: MICROELECTROFLUIDIC SYSTEMS MODELING AND SIMULATION. Tianhao Zhang, Krishenedu Chakrabarty, Richard B. Fair, and Sergey Edward Lyshevsky. CRC Press. (288 pages).

2002: MODELING MEMS AND NEMS. John A. Pelesko and David H. Bernstein. CRC Press. (376 Pages).

2002: NANOELECTROMECHANICS IN ENGINEERING AND BIOLOGY. Michael Pycraft Hughes. CRC Press. (400 pages).

2002: OPTICAL MICROSCANNERS & MICROSPECTROMETERS USING THERMAL BIMORPH ACTUATORS. Gerhard Lammel, Sandra Schweizer, and Phillipe Renaud. Klewer. (280 pages).

2003: RF MEMS THEORY, DESIGN, AND TECHNOLOGY. Gabriel M. Rebeiz. Wiley-Interscience. (483 pages).

2003: RF MEMS AND THEIR APPLICATIONS. Vijay K. Varadan, K. J. Vinoy, and K. A. Jose John Wiley & Sons. (394 pages).

References

Analog Device, Inc. http://www.analog.com/

Bishop, D.J., *et al.* (2001). The rise of optical switching. *Scientific American* 88–94.

Corning IntelliSense, Inc. http://www.corningintellisense.com/

Gabriel, K.J. (1995). Engineering microscopic machines. *Scientific american: Technology in the 21st century 273* (3), 150–153.

Goldsmith, C.L. (2000). Private communication.

Hornbeck, L.J. (1993). Current status of the digital micromirror device (DMD) for projection television applications. *International Electron Devices Meeting. Technical Digest.*

Lawton, R.A. (Ed.). (2000). MEMS reliability for critical applications. SPIE *4180.*

Marshall, S. (2000). *Micromachine devices newsletter 5* (12), 12.

Memscap, Inc. http://www.memscap.com

MEMS Industry. (2001). *MEMS industry group annual report 2001.* www.memsindustrygroup.org

Micronics, Inc. http://www.micronics.net/

Micro Optical Corporation. http://www.microopticalcorp.com

Miller, W.M., Tanner, D.M., Miller, S.L., and Peterson, K.A. (1998). MEMS reliability: The challenge and the promise. *Fourth Annual The Reliability Challenge* 4.1–4.7.

Motorola, Inc. http://e-www.motorola.com/sensors/ index.html

Nagel, D.J. (1999). Design of MEMS and microsystems. In B. Courtois *et al.* (Eds.), *Design, test, and microfabrication of MEMS and MOEMS SPIE 3680*, 20–29.

Nathanson, H.C., *et al.* (1967). The resonant gate transistor. *IEEE Transactions on Electron Devices ED-14* (3), 117–133.

Novasensor, Inc. http://www.novasensor.com/

Parameswaram, M. *et al.* (1991). Micromachined thermal radiation emitter from a commercial CMOS process. *IEEE Electron Device Letters 12*, 57–59.

IBM, Inc. http://www.research.ibm.com/resources/news/20020611_millipede.shtml

Sprangler, L., and Kemp, C.J. (1995). ISAAC-integrated silicon automotive accelerometer. *Technical Digest 8th International Conference Solid-State Sensors and Actuators* 585–588.

Stark, B. (Ed.). (1999). MEMS reliability assurance guidelines for space applications. *Jet Propulsion Laboratory Report*, 99–1.

Wood, R.A. *et al.* (1993). High-performance uncooled microbolometer focal planes. *IRIS Detector Conference.*

7

Noise Analysis and Design in Deep Submicron Technology

Mohamed Elgamel
and Magdy Bayoumi
The Center for Advanced Computer Studies, University of Louisiana at Lafayette, Lafayette, Louisiana, USA

7.1 Introduction

Traditionally, area minimization and speed maximization were the only axes with respect to which a design's effectiveness was measured. Low-power, high-throughput, and computationally intensive circuits are critical application domains (Rabaey and Pedram, 1996). In addition to these three design parameters, **area**, **speed**, and **power**, there are two design metrics that have been of great importance to current designs. These metrics are **noise** and **reliability**. The five metrics are shown in Figure 7.1 with an arrow associated with each to show whether the particular metric should be increased or decreased.

The current trend of technology scaling is predicted to continue. Feature sizes will continue to shrink to very deep submicrometer (VDSM) dimensions and clock frequencies to increase. Shrinking feature size implies not only shorter gate lengths but also decreasing interconnect pitch and device threshold voltages (Shepard and Narayanan, 1998). Now a single integrated circuit (IC) can contain an entire system (a system-on-a-chip [SOC]), and the interconnections can have many interleaved signal layers and multiple planes of interconnect. Die sizes remain relatively constant. Reduction in the top and bottom areas of a minimum-width wire means that total wire capacitance is decreasing. Resistance, however, is increasing faster, despite efforts not to scale metal thickness. Practical efforts to control resistor-capacitor (RC) delays through the use of low-resistivity metal (copper), low-dielectric-constant insulators, and wide, thick wiring will require future interconnection analysis to consider inductance and inductive coupling. It is projected that use of a lower resistivity metal (copper) and that replacement of silicon dioxide ($k \sim 4$) with various insulating materials of progressively lower dielectric constant ($k \sim 2 - 3$) will be adopted in the chip fabrication process. Use of copper will reduce degradation of signal propagation delay time due to voltage drop on power lines. Use of low k dielectric will decrease the degradation of signal propagation delay time by reducing capacitive coupling.

Although technology scaling results in lower threshold voltages, the threshold voltage magnitude determines noise immunity in these circuits.

7.1.1 Noise

The term **noise** in digital VLSI systems has come to mean any unwanted deviation in the voltages and currents at various nodes in a circuit. When noise acts against a stable logic level on a circuit node, it can transiently destroy logical information carried by the node. If this ultimately causes an incorrect

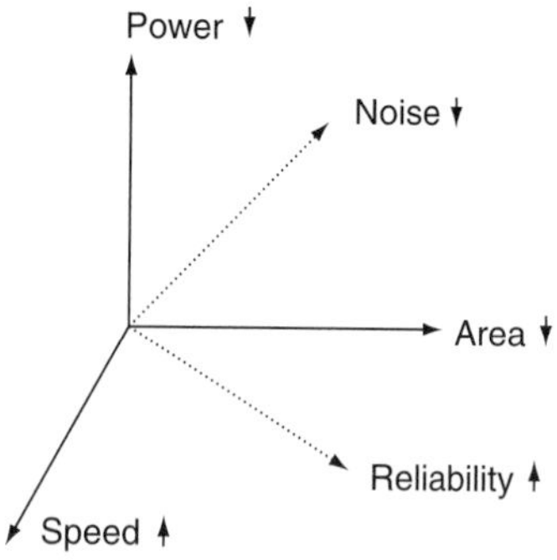

FIGURE 7.1 Design Metrics

machine state stored in a latch, functional failure will result. Even when noise does not cause functional failure, it has an impact on timing, affecting both delay and slew. Noise can also cause the dissipation of extra power due to incorrect switching.

Digital circuits create deterministic noise several orders of magnitude greater than noise from stochastic physical sources. Problems due to these noise sources were first observed in mixed signal applications that plunged highly noise sensitive analog circuits into a noisy digital environment. Although digital circuits create much more noise than analog circuits, digital systems are prevalent because they are inherently immune to noise. Figure 7.2 shows valid voltage ranges for defining the digital 0 and 1.

The inherent noise immunity of digital circuits is due to the presence of high-gain restoring logic gates such as the inverter shown in Figure 7.3, which has a very nonlinear voltage transfer characteristic. As power supply levels have

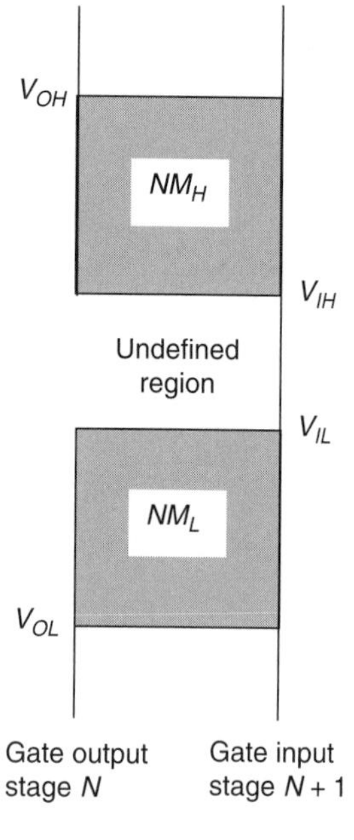

FIGURE 7.2 A Range of Analog Voltages Defining the Digital 0 and 1

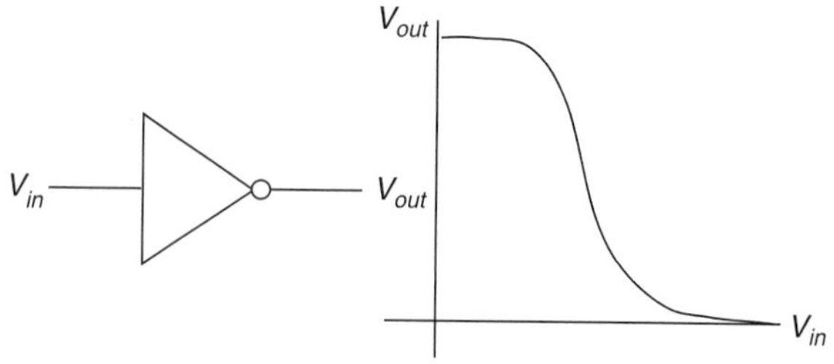

FIGURE 7.3 The Inherent Noise Immunity of Digital Circuits

decreased, however, this advantage has diminished. Thus, the problem of noise has increased in importance such that on-chip noise is the main research area for continuing the growth in integrated circuit density and performance.

With technology scaling, noise is a problem affecting all types of designs from custom microprocessors to standard-cell application specific integrated circuits (ASICs). A noise analysis solution must be capable of analyzing tens of millions of transistors, considering both circuit and interconnect noise, and evaluating the distinct noise tolerances of each node in the circuit. Successful design methodologies incorporate a three-level noise strategy. The first line of defense is a set of noise avoidance rules to guide circuit and interconnect design. These rules should prevent most noise problems without introducing too much area or timing constraints. Next, a detailed static noise analysis of the design should find all possible noise failures. Finally, careful circuit simulation should determine whether the design could tolerate some failures flagged by static noise analysis.

7.1.2 Reliability

One of the most important attributes of any system is its reliability. It is imperative to design reliability into complex chips even more carefully as the chip functionality increases almost without limit. This important issue was addressed by Kang (2000) with emphasis on incorporating reliability from early design phases rather than treating reliability assurance as a back-end manufacturing process. As computer-aided design (CAD) tools have played key roles in developing integrated circuits and systems, new reliability analysis tools need to be developed and used in SOC development. For more than a decade, new CAD capabilities have been developed with reliability focus, in particular to address reliability concerns about hot carrier-induced degradation of circuits, electromigration, electrostatic discharge (ESD), cross talk, leakage currents, and high-power dissipation.

It was also reported in the International Technology Roadmap for Semiconductors (Semtech, 1999) that computer-aided design (CAD) tools would need to incorporate contextual reliability considerations in the design of new products and technologies. It is essential that advances in failure mechanism understanding and modeling, which result from the use of improved test methodologies, be used to provide input data for these new CAD tools. With these data and smart reliability CAD tools, the impact on product reliability of design selections can be evaluated. New CAD tools need to be developed that can calculate degradation in electrical performance of the circuit over time. The inputs used would be the predicted resistance increases in interconnect wires and vias in the circuit based on the following:

- Wire length
- Current densities expected for the currents required by the circuit
- Calculated local operating temperature, which includes the effects of joule heating in the circuit element and elsewhere

These tools will need to become an integral part of the circuit designer's tool set to help predict product reliability before processing begins and to develop solutions that anticipate technology and thereby accelerate their introduction.

7.2 Noise Sources

Serious on-chip electrical problems are being encountered in deep submicron (DSM). These problems include signal distortion along coupled interconnect lines, voltage variations in the power supply distribution, substrate coupling, charge sharing, charge leakage, process variation, thermal noise, and alpha particles. Each one is a major source of on-chip noise in VLSI circuits.

7.2.1 Interconnect Crosscapacitance Noise

Interconnect crosscapacitance noise refers to charge injected in quiet wires (victims) by neighboring switching wires (aggressors) through the capacitance between them (crosscapacitance). The resulting noise has the form of a pulse in which the leading edge is determined by the switching slew of the aggressor, and the trailing edge is determined by the restoring time constant of the victim. This is perceived to be the most significant source of noise in current processes (see Figure 7.4). If the net is a dynamic node, the restoring time is infinite, and the node will never recover. It can lead to setup violations in downstream latches or flip-flops. In addition, when the aggressors are switching in the same direction of the victim, the transport delay and slew decreases, leading to hold-time violations.

7.2.2 Charge Sharing Noise

Charge sharing noise is caused by charge redistribution between a dynamic evaluation node and intermediate nodes in pull-up or pull-down logic stack, as in Figure 7.5. This primarily impacts domino nodes, weakly driven pass gate latches, and dynamic latches. The primary technology variable here is the ratio of junction capacitance to gate and interconnect capacitance. For most circuits, this noise is not dramatically changing with technology scaling.

7.2.3 Charge Leakage Noise

Charge leakage noise is mainly composed of subthreshold conduction in nominally off transistors. This current can either charge/discharge a dynamic node or cause the stable state of a weakly held node to be significantly different from rails. This is mainly a concern for wide domino NOR, PLA, and memory arrays. This current rises exponentially with reduction in the threshold voltage and is becoming very significant in DSM. It is helped greatly by feedback devices.

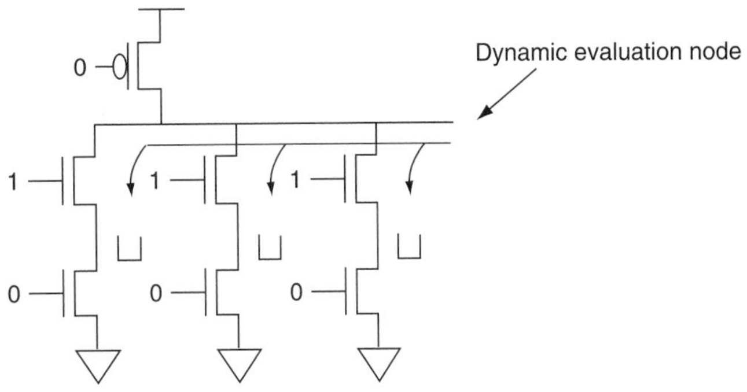

FIGURE 7.5 Charge Sharing Noise

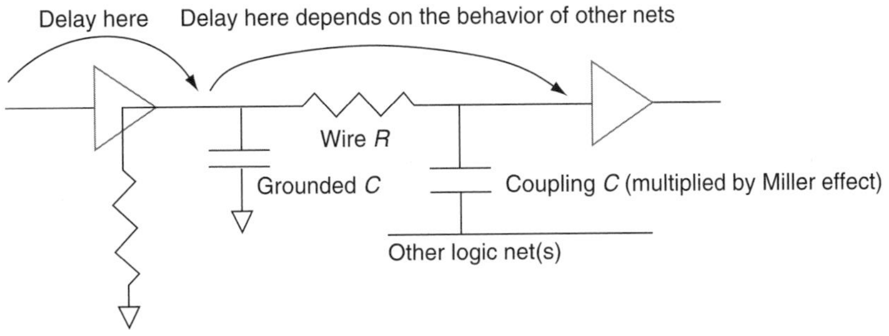

FIGURE 7.4 Various Noise Sources for Digital Circuits

7.2.4 Power Supply Noise

Power supply noise is the difference between the local voltage references of the driver and the receiver. The increased amount of current on power supply lines causes a raise in IR drop on voltage references. This makes the gate have higher sensibility to noise spikes. This low frequency component (IR drop) is managed well by flip-chip C4 packaging, which provides a very low resistance current path. Besides, the higher speed transients allowed by scaled transistor sizes are associated to higher *L di/dt* due to the large inductances of the package and on-chip. Furthermore, the decoupling capacitance of the circuit is decreased due to the reduced sizes of the gates. Power supply noise is a dominant factor in the design of wide domino circuits and in circuits using contention where the alternating current (ac) logic level is shifted with respect to power supply rails. To counter these problems in high-speed designs, several physical design techniques have been proposed: (1) sizing up the P/G lines to accommodate the large current peaks and to minimize the IR and *L di/dt* voltage variations in these lines (Su *et al.*, 2000), (2) increasing the number of P/G pins, and (3) deploying decoupling capacitors in the P/G lines (Bakoglu, 1990; Zhao *et al.*, 2001). In addition, a designer could perform clock skew scheduling to minimize the number of simultaneous switching (Lam *et al.*, 2002). He or she could also use copper in place of aluminum to overcome the increased resistance of scaled interconnect.

7.2.5 Mutual Inductance Noise

Mutual inductance noise occurs when signal switching causes transient current to flow through the loop formed by the signal wire and current return path (Intel, 2001), thereby creating a changing magnetic field (see Figure 7.6). This induces a voltage on a quiet line, which is in or near this loop. These noise sources can be cumulative if there are several signals switching simultaneously in a bus. Mutual inductance is a long-range phenomenon and hence is worse in the presence of wide busses. High-speed switching and synchronous bus structures are making this noise very significant in current technologies. Inductive noise can combine with capacitive noise to cause even worse noise than that shown in Figure 7.6. Because the analysis of inductive effects is highly dependent on layout and is quite complex, the approach is usually to design the problem out through rules rather than analyze arbitrary configurations.

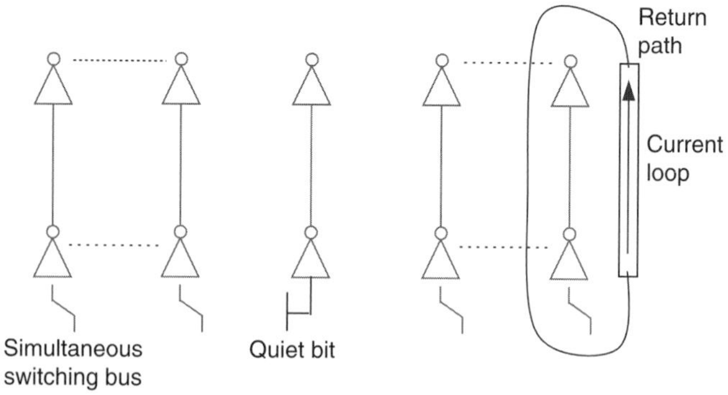

FIGURE 7.6 Mutual Inductance Noise From Simultaneous Switching on a Wide Bus

7.2.6 Thermal Effects

Thermal effects are an inseparable aspect of electrical power distribution and signal transmission through the interconnects due to self-heating (or joule heating) caused by the flow of current. Thermal effects impact interconnect design and reliability in the following ways. First, they limit the maximum allowable RMS current density (since the RMS value of the current density is responsible for heat generation) in the interconnects to limit the temperature increase. Second, interconnect lifetime (reliability), which is limited by electromigration (EM) (transport of mass in metals under an applied current density), has an exponential dependence on the inverse metal temperature. Hence, temperature rise of metal interconnects due to self-heating phenomena can also limit the maximum allowed average current density because EM capability is dependent on the average current density (Liew *et al.*, 1990). Third, thermally induced open-circuit metal failure under short-time high peak currents (including ESD) is also a reliability concern (Banerjee *et al.*, 1997) and can introduce latent EM damage that has important reliability implications (Banerjee *et al.*, 1996). It has been argued that thermal effects will increasingly dominate interconnect design rules that specify maximum current densities for circuit designers (Chatterjee *et al.*, 1995). Recently, Hunter (1997) has followed up on this issue by solving the EM lifetime equation for Al-Cu and the ID heat equation in a self-consistent manner. In this approach, both EM and self-heating can be comprehended simultaneously.

7.2.7 Process Variation

The typical characteristics of a process like the gate oxide process can vary among wafers or even on a single die. The devices and their properties are defined only in a certain margin and, hence, affect the performance of the circuit.

7.3 Noise Reduction Techniques

Noise reduction techniques can be categorized into four main categories: **signal-encoding schemes**, which have been proposed to minimize transition activities on buses; **circuit techniques** to make circuits more immune to noise; **interconnect structures techniques**, which are changing the interconnect topology, wire sizing and spacing, and buffer locations; and

high-level synthesis techniques. Of course, a designer can also overcome noise by using new materials in interconnect, including fiber optics or electromagnetic transmission and three-dimensional interconnects that use multiple levels of active devices. New packaging methodologies can also be used to reduce noise (Sematech, 2001).

The most common techniques for reducing noise in digital circuits include disallowing:

- Pass gates at the ends of long wires
- Long wire runs feeding domino gate inputs
- Single *n*-FET or *p*-FET pass gates because of the V_t voltage drop they cause
- High-beta static circuits feeding low-beta static circuits or vice versa

7.3.1 Signal Encoding Techniques

Increased coupling effect between interconnects in ultra deep submicron technology not only aggravates the power-delay metrics but also deteriorates the signal integrity due to capacitive and inductive cross talk noises. Conventional approaches to interconnect synthesis are aimed at optimal interconnect structures in terms of interconnect topology, wire width and spacing, and buffer location and sizes (Cong, 1999).

Signal encoding schemes have been proposed to minimize transition activities on busses while ignoring crosscoupled capacitances. When statistical properties are unknown a priori, the bus-invert method (Stan and Burleson, 1995) and the on-line adaptive scheme (Benini *et al.*, 1999) can be applied to encode randomly distributed signals. On the other hand, highly correlated access patterns exhibit a spatiotemporal locality that can be exploited for energy reduction (Panda and Dutt, 1996) in Gray code (Mehta *et al.*, 1996; Su *et al.*, 1994), the TO method (Benini *et al.*, 1997), the working-zone encoding (Musoll *et al.*, 1998), the combined bus-invert/TO (Benini *et al.*, 1998), and the coupling-driven method (Kim *et al.*, 2000). Lower bounds for minimum achievable transition activity have been derived for noiseless busses in (Ramprasad *et al.*, 1999) and for noisy busses in Hegde and Shanbhag (1998). In Zhang *et al.*, (1998), a segmentation method was introduced to reduce power consumption. Specification transformation approaches have also been used to reduce the number of memory accesses at the behavioral level (Catthoor *et al.*, 1994). The effectiveness of various encoding schemes was compared at the system level by Fornaciari *et al.* (1999).

Bus-Invert Encoding

The **bus-invert encoding** has been introduced to reduce the bus activity: the encoding is derived from the Hamming distance between the consecutive binary numbers. If the Hamming distance of the two consecutive binary numbers is more than half of the word length, the latter binary number is sent in inverted polarity by asserting an additional signal line that indicates bus inversion (Stan and Burleson, 1995). The number can be used to reduce the weight (the number of ones or zeros) of the binary numbers if the bus-inversion decision is made when the weight is more than half of the bus width. The bus-invert method is as follows:

1. Compute the Hamming distance (the number of bits in which they differ) between the present bus value (also counting the present invert line of Figure 7.7) and the next data value.
2. If the Hamming distance is larger than $n/2$, set invert equal to 1 (and thus make the next bus value equal to the inverted next data value).
3. Otherwise let the invert equal 0 (and make the next bus value equal to the next data value).
4. At the receiver side, the contents of the bus must be conditionally inverted according to the invert line unless the data are not stored encoded as they are (e.g., in a RAM). In any case, the value of the invert must be transmitted over the bus (the method increases the number of bus lines from n to $n+1$).

The **Gray code** has only a 1-bit difference in consecutive numbers for addressing. Due to locality of program execution, Gray code addressing can significantly reduce the number of bit switches. Experimental results show that for typical programs running on an RISC microprocessor, using Gray code addressing reduces the switching activity at the address lines by 30 to 50% compared to using normal binary code addressing.

In **TO** Encoding, the bus transitions are reduced by freezing the address lines when consecutive patterns are found to be sequential. An extra bus line is employed to inform the receiver side regardless if the current pattern is sequential.

The basis of the **working zone encoding** (WZE) technique is as follows:

1. The WZE takes into account the locality of the memory references: applications favor a few working zones of their address space at each instant. In such cases, a reference can be described by an identifier of the working zone and by an offset. This encoding is sent through the bus.
2. The offset can be specified with respect to the base address of the zone or to the previous reference to

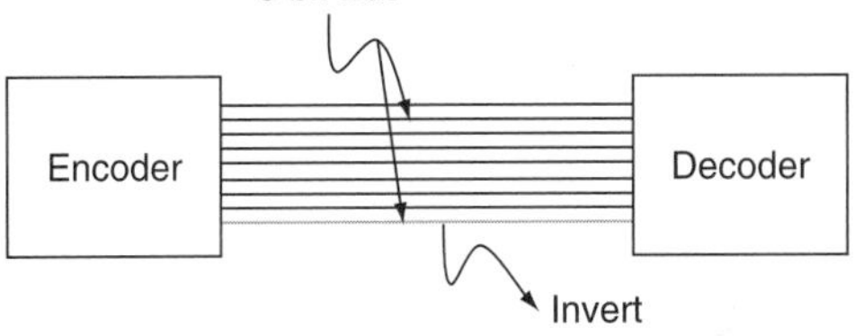

FIGURE 7.7 Invert Signal in Bus-InvertMethod

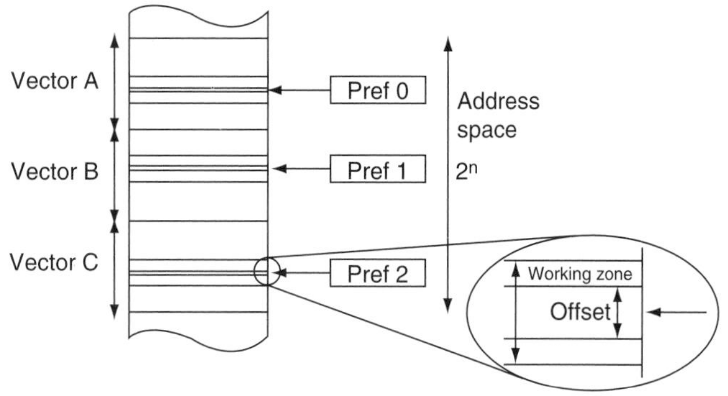

FIGURE 7.8 Address Space for Three Vectors. Adapted from Musoll *et al.* (1998). © 2004 IEEE.

that zone. Because all small offsets should be encoded in a one-hot code, the latter approach is the most convenient. As a simple example, consider an application that works with three vectors (A, B, and C) as shown in Figure 7.8. Memory references are often interleaved among the three vectors and frequently close to the previous reference to the vector. Thus, if both the sender and the receiver had three registers (henceforth named p) holding a pointer to each active working zone, the sender would only need to send:

a. The offset of the current memory reference with respect to the one associated with the current working zone
b. An identifier of the current

To reduce the number of transitions, the offset is encoded in a one-hot code. Because the one-hot code produces two transitions if the previous reference was also in the one-hot code and an average of $n/2$ transitions when the previous reference is arbitrary, using a transition-signaling code reduces the number of transitions (Musoll *et al.*, 1998).

Coupling-Driven Signal Encoding

The key idea in **coupling-driven signal encoding** is that transforming the signal sequences traveling on-chip busses that are closely placed could alleviate coupling effects (Kim *et al.*, 2000). Small blocks of encoding and decoding logic are employed at the transmitter and receiver of on-chip busses as shown in Figure 7.9.

There are four types of possible transitions when dynamic charge distribution is considered over cross coupling capacitances as in Figure 7.10. Two parallel wires are placed with minimum spacing. A type I transition occurs when one of the signals switches while the other stays unchanged such that the cross coupling capacitance is then charged up to $k_I C_x V$, where the coefficient k_1 is introduced as a reference for other types of transition. In a type II transition, one bus switches from low to high while the other switches from high to low. The effective capacitance will be larger than k_1 by a factor of k_2, the value of which is usually $k_2 = 2$. In a type III transition, both signals switch simultaneously, and C_x will not be charged. Because of possible misalignment of the two transitions, however, the amount of power consumption varies according to the dynamic characteristics by a factor of k_3. In a type IV transition, there is no dynamic charge distribution over cross coupling capacitance. Thus, k_4 is set to zero.

There are some assumptions. First, synchronous latches are located at the transmitter side, thus all the transitions take place at the same time on the bus. The simultaneous transitions exclude type III transitions by setting $k_3 = 0$. The achieved results are on the lower end of power savings. Second, statistics on the information source are not given in advance. Hence, this scheme is suitable for data bus encoding, where it is difficult to extract accurate probabilistic information offline. An enumeration method is employed to represent the coupling effect. If a bus line Bi is located between two other lines, a signal transition on B_i can trigger charge shifts on both coupling capacitances connected to B_{i-1} and B_{i+1}, respectively. In other words, two couplings can be initiated at most by a

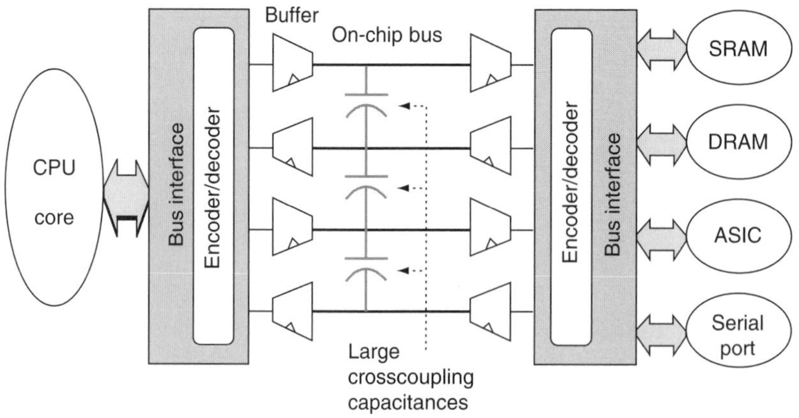

FIGURE 7.9 Tightly Crosscoupled On-Chip Busses in a System-Level Chip Design. Adapted from Kim et al. (2000). © 2004 IEEE.

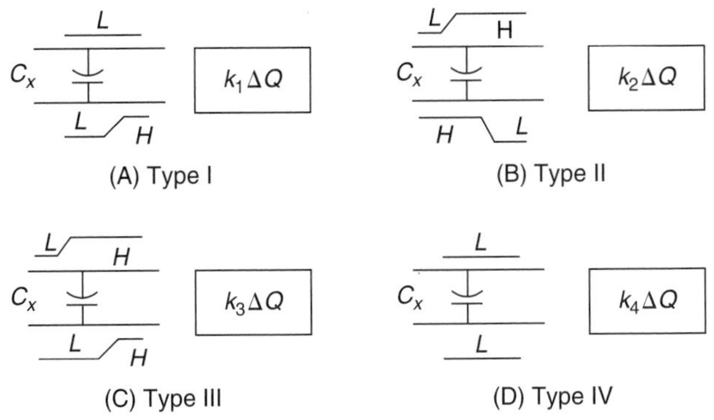

FIGURE 7.10 Transition Types. (A) This circuit shows single line switching; (B) Both lines are switching in the opposite direction; (C) Both lines are switching in the same direction; (D) No switching is occurring here.

signal transition. Thus, 2(*N*-1) bits are sufficient to represent the whole set of couplings in an *N*-bit bus per bus cycle. According to the types of correlated transition between neighboring busses, the coupling encoder generates a codeword as follows: 00 for a type III or IV transition, 01 for a type I transition, and 11 for a type II transition. The reason 11 is assigned to a type II transition is that switchings in different directions require the change in polarity of the charge stored in the coupling capacitance, hence consuming about twice the amount of charge required for a type I transition. The codeword 11, instead of 10, helps to make a decision on data inversion using a majority voter because the majority voter outputs high when at least eight input lines are high out of fifteen inputs. The majority voter can be implemented by using either full-adder circuitry or resistors and a voltage comparator. The control signal inv can be transmitted to the receiver using extra bus lines or extra transfer cycles. One problem of additional bus lines for control is the area overhead that may not be allowed due to physical constraints. In some cases, widening the space between signal bus lines can reduce the coupling effects more effectively than introducing extra control lines because the coupling capacitance is inversely proportional to net space. Temporal redundancy is an alternative and uses extra clock cycles to transfer control signals.

7.3.2 Circuit Techniques

One way to effectively increase noise immunity is to increase the switching threshold voltage V_{th} of the gate. V_{th} is defined as the input voltage at which the output changes state. Increasing the V_{th}, on the other hand, has an adverse effect on the performance, such as on speed and power consumption, which are the prime features of dynamic circuits.

Gated-Vdd Technique

Gated-Vdd is used to decrease the leakage power. The key idea for gated-Vdd (Powell *et al.*, 2000) is to introduce an extra transistor in the supply voltage (Vdd) or the ground (Gnd) path as in. The extra transistor is turned on in the used section and off in the unused section gated. It maintains the performance and advantages of low-power supply and threshold voltages while reducing leakage and leakage energy dissipation. The fundamental reason for the reduction in leakage is the stacking effect of self-reverse-bias series-connected transistors. However, the transistor impacts the switching speed due to a nonzero voltage drop across the gated-Vdd transistor between the supply rails and the "virtual Gnd" for NMOS gated Vdd (Figure 7.11) or the "virtual Vdd" for PMOS gated-Vdd.

Dual Threshold Technique

Dual threshold technique (Kao and Chandrakasan, 2000) is used to reduce leakage power by assigning high threshold voltage to some transistors in noncritical paths and using low threshold transistors in critical paths. To achieve the best leakage power savings under target performance constraints, an algorithm is presented for selecting and assigning an optimal high threshold voltage. Results show that dual threshold technique is good for power reduction.

Dynamic Threshold Technique

Dynamic threshold MOS (DTMOS), as noted by Fariborz *et al.* (1997), is a scheme that allows for a self-adjusting threshold voltage. By tying the gate to the body, the threshold voltage decreases as the gate voltage increases and vice versa. In this manner, a higher zero-bias threshold can be used to reduce the leakage current. In a speed-adaptive threshold-voltage CMOS (SA-Vt CMOS) circuit miyazaki (1998) the substrate bias is controlled so that delay in the circuit stays constant. Distributions of device speeds are squeezed under fast-operation conditions. With a ring oscillator using 0.25-mm CMOS devices as a test circuit, the worst-case operating frequency was improved from 20 MHz to 55 MHz, and the fluctuation of the operating frequency was suppressed from 44 to 15%; the supply-voltage variation was under 0.1 V with a 1.8-V supply voltage.

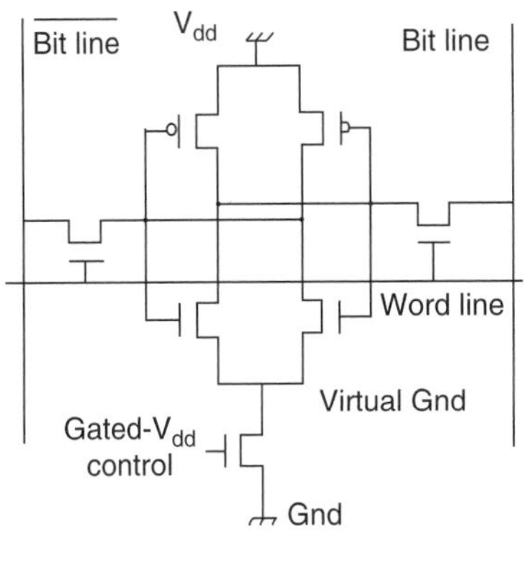

FIGURE 7.11 NMOS Gated-Vdd. Adapted from Powell *et al.* (2000).

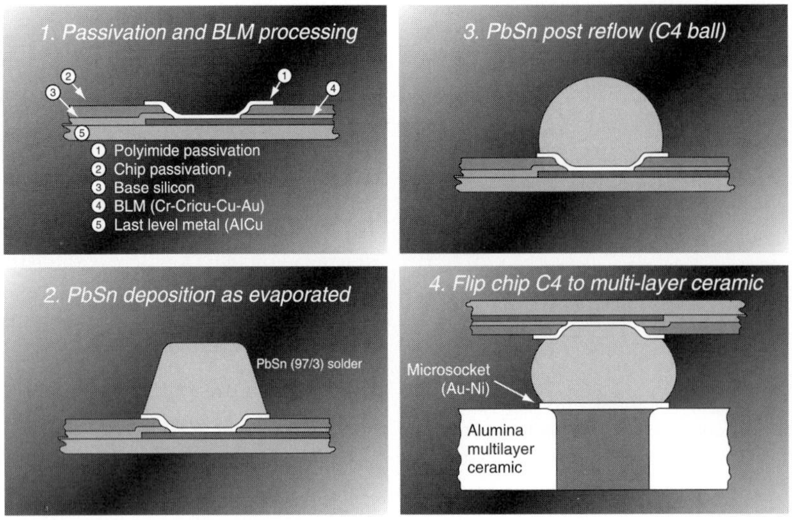

FIGURE 7.12 C4 Flip-Chip

C4 Flip-Chip

The **C4 flip-chip** is used to manage the IR drop. IBM researchers developed C4 technology in the 1960s (see Figure 7.12). The bonding process is characterized by the soldering of silicon devices directly to a substrate (e.g., organic). The chip faces the substrate as opposed to wire bonding, hence the name flip-chip bonding. There are two main salient features of this packaging methodology. One involves solder bumps that are distributed on metal terminals of the chip itself. These solder bumps are typically composed of 97% lead and 3% tin. The substrate has identically placed metal pads on its surface. The second feature is that the chip is turned over and the metal pads are aligned to solder bumps; metal reflow is used to form connectivity between the substrate and chip.

The advantages of C4 technology are numerous and important. They include (comparisons refer to wire bonding):

- Increased *I/O* density (C4 bumps may be placed over the entire area of the chip, called area array, rather than simply the periphery)
- Self-aligning process step (due to surface tension)
- Reduced die size for previously pad-limited designs
- Reduced simultaneous switching noise due to smaller inductance of bumps compared to wire leads
- Better thermal properties as the backside of the wafer is now available for heat sinking
- Much better power distribution capabilities as circuits in the middle of the die can now access Vdd/Gnd directly
- Low cost and high throughput (all connections for one chip are made simultaneously in C4 as opposed to one-by-one in wire bonding)
- Shorter wire lengths and fewer global wires ease wiring requirements

The main drawbacks to C4 regard the use of lead in the solder bumps that lead to the emission of alpha particles, which, in turn can lead to circuit failure in sensitive circuits such as DRAMs. Restricting the placement of solder bumps over these sensitive areas' however, can minimize this effect. Research is ongoing to find alternate materials for solder bumps as well. In addition, the use of C4 packaging allows designers to do many different things in the floor planning and routing stages of a design. Commercial tools for place-and-route and other such designs are predicated on the use of peripheral wire bonding for *I/O* pads. New tools need to be in place for designers to take full advantage of flip-chip's advantages.

Pseudo-CMOS

Pseudo-CMOS, shown in Figure 7.13, provides the logic capability of Domino and the noise robustness of CMOS.

PMOS Pull-Up Technique

The **PMOS pull-up technique** (DiSouza, 1996) (as shown in Figure 7.14) uses a pull-up device to increase the source

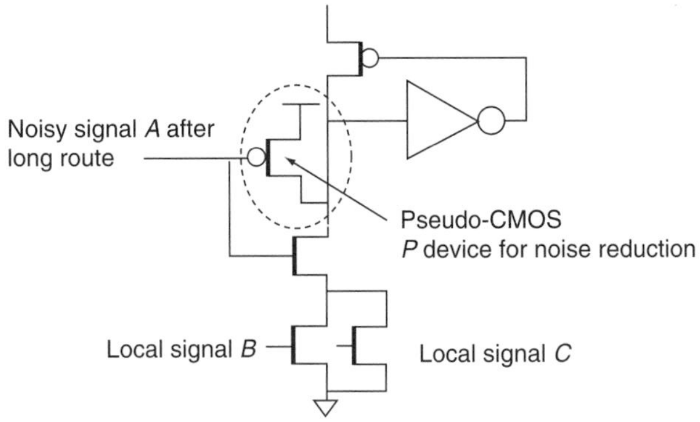

FIGURE 7.13 Pseudo-CMOS

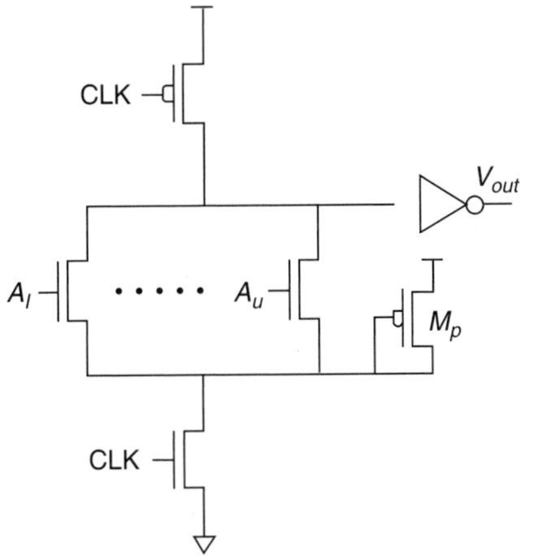

FIGURE 7.14 The PMOS Pull-Up Technique

potential of the NMOS network, thereby increasing the transistor threshold voltage V_t and V_{st} during the evaluate phase. This technique suffers from large-static power dissipation.

CMOS Inverter Technique

The **CMOS inverter technique** (Covino, 1997) uses a PMOS transistor for each input, thereby adjusting V_{st} to equal that of a static circuit. This technique cannot be used for dynamic NOR-type circuits because certain logic combinations will short V_{dd} to Ground (GND).

The Mirror Technique

The **mirror technique** (Wang and Shanbhag, 1999) (see Figure 7.16) uses two identical NMOS evaluation networks and one additional NMOS transistor M_1 to pull up the source node of the upper NMOS network to $V_{dd} - V_t$ during the precharge phase, thereby increasing V_{st}. The mirror technique guarantees zero dc power dissipation, but a speed penalty is incurred if the transistors are not resized.

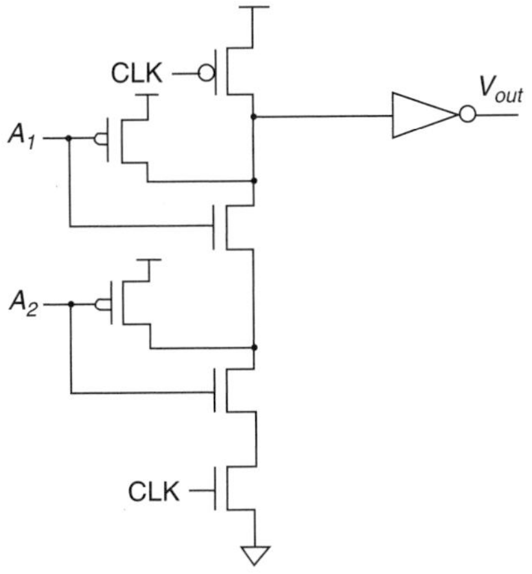

FIGURE 7.15 The CMOS Inverter Technique. Adapted from Covino (1997).

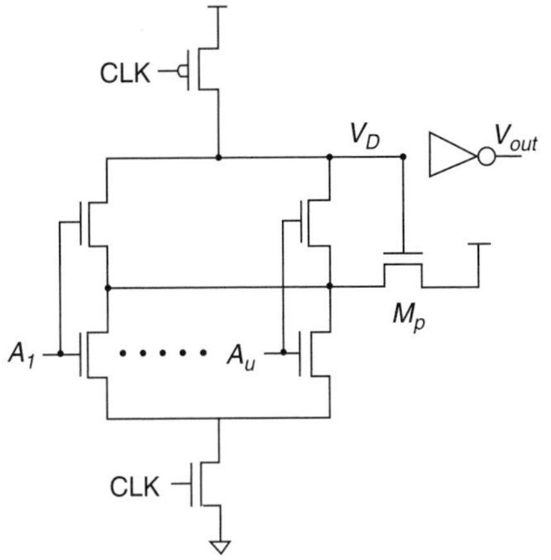

FIGURE 7.16 The Mirror Technique. Adapted from Wang and Shanbhag (1997). © 2004 IEEE.

The Twin Transistor Technique

The **twin transistor technique** (see Figure 7.17) uses an extra transistor for every transistor in the pull-down network to pull up the source potential (Balamurugan and Shanbag, 1999, 2000). The twin transistor technique consumes no dc power.

7.4 Noise Analysis Algorithms

Some amount of noise is unavoidable in digital circuits. The question concerns whether the noise causes function failure.

7.4.1 Small-Signal Unity Gain Failure Criteria

Traditional analysis of noise margins rely on the **small-signal unity gain** failure criteria [4]. As illustrated in Figure 7.18, for a small change in input noise to a circuit biased at an operating point, the resultant change in output noise is measured. If |d

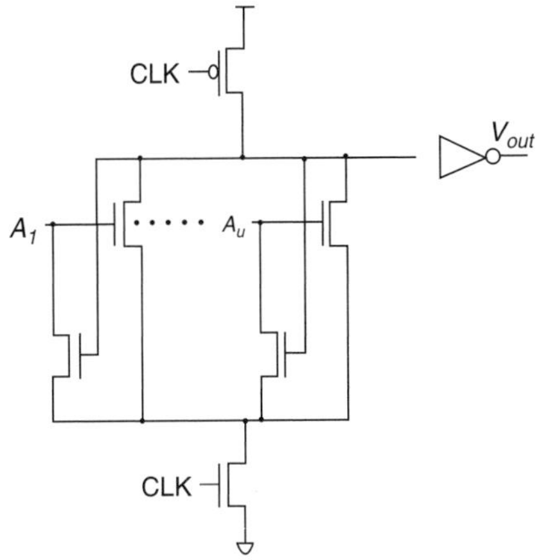

FIGURE 7.17 The Twin Transistor Technique. Adapted from Balamurugan and Shanbhag (1999). © 2004 IEEE.

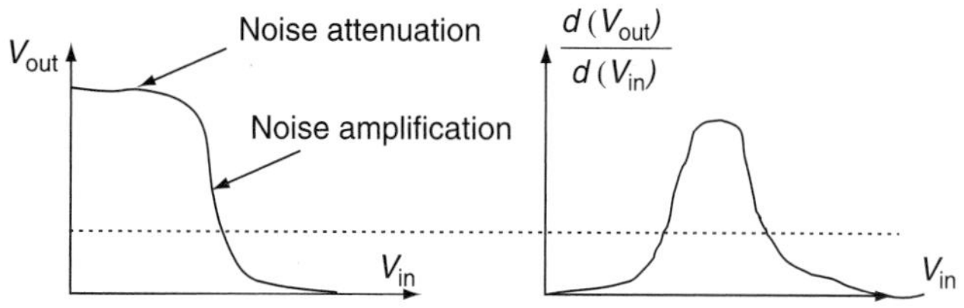

FIGURE 7.18 Direct Current Transfer Function of an Inverter Illustrating Small-Signal Unity Gain

(output)/d (input)| > 1 then the circuit is considered unstable. Unity gain is a good design metric but is neither necessary nor sufficient for noise immunity.

Most aggressively designed paths have some noise-sensitive stages interspersed with quiet stages. Some noise amplification needs to be allowed in the sensitive stage because it should be attenuated in the quiet stage.

7.4.2 Intel Failure Criteria

In this subsection, guidelines and steps followed by Intel in determining a circuit failure are listed.

- Break the circuit into circuit stages.
- Track the noise propagation across these stages by ac circuit simulation.
- Measure if any circuit stage failed because of the following:
  - Injected noise
  - Noise propagated from previous stages
- Noise can propagate across any number of stages, eliminating the need for any unity gain budgeting.
- Combination of noise sources and simultaneous noise on multiple inputs should be considered.
- The peak noise in the event of two simultaneous couplers on a line is larger than the sum of these two events.
- The simultaneous occurrence of different parallel inputs have to be considered.
- New transistor level (symbolic circuit) simulation should be found because SPICE is very slow.
- Solving the differential equation symbolically in a piecewise linear manner can decrease the CPU time needed with good accuracy.

7.4.3 Design Flow

The traditional design flow in Figure 7.19 (Sylvester and Keutzer, 1998) contains a separation between the logic synthesis step and the physical design step. For designs with aggressive performance goals, Sylvester and Keutzer (1998)

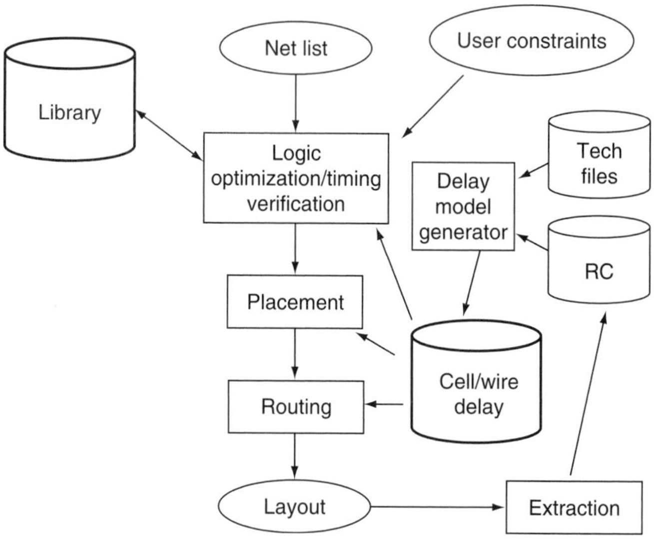

FIGURE 7.20 Today's High-Performance Logical/ Physical Flow. Adapted from Sylvester and Keutzer (1998). © 2004 IEEE.

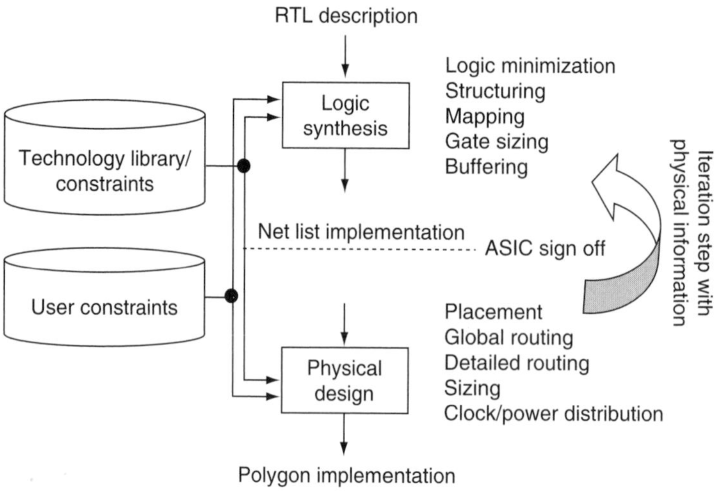

FIGURE 7.19 Traditional ASIC Design Flow

found that several iterations between synthesis and physical design are required to converge to a desired implementation.

As a result, design teams have begun to bring more of the backend design flow in-house, and the handoff to the semiconductor–vendor occurs only at the end. This approach is shown in Figure 7.20 and is known as the customer-owned tooling (COT) approach.

Acknowledgments

The authors acknowledge the support of the U.S. Department of Energy (DoE) EETAPP program (DE97ER 12220) and the Governor's Information Technology Initiative.

References

Bakoglu, H.B. (1990). *Circuits, interconnections, and packaging for VLSI.* Addison-Wesley.

Balamurugan, G., and Shanbhag, N.R. (1999). Energy-efficient dynamic circuit design in the presence of cross talk noise. *The 1999 International Symposium on Low-Power Electronics and Design.*

Balamurugan, G., and Shanbhag, N.R. (2000). A noise-tolerant dynamic circuit design technique. *The 2000 Custom Integrated Circuits Conference.*

Banerjee, K., Amerasekera, A., and Hu, C. (1996). Characterization of VLSI circuit interconnect heating and failure under ESD conditions. *International Reliability Physics Symposium.*

Banerjee, K., Amerasekera, A., Cheung, N., and Hu, C. (1997). High-current failure model for VLSI interconnects under short-pulse stress conditions. *IEEE Electron Device Letters 18*(9).

Benini, L., de Micheli, G., Macii, E., Sciuto, D., and Silvano, C. (1997). Asymptotic zero-transition activity encoding for address busses in low-power microprocessor-based systems. *The Great Lakes Symposium VLSI,* 77–82.

Benini, L., de Micheli, G., Macii, E., Sciuto, D., and Silvano, C. (1998). Address bus encoding techniques for system-level power optimization. *Design, Automation, and Test in Europe* 861–866.

Benini, L., Macii, A., Macii, E., Poncino, M., and Scarsi, R. (1999). Synthesis of low-overhead interfaces for power-efficient communication over wide busses. *ACM/IEEE Design Automation Conference.* 128–133.

Catthoor, F., Franssen, F., Wuytack, S., Nachtergaele, L., and Man, H.D. (1994). Global communication and memory optimizing transformations for low-power signal processing systems. *VLSI Signal Processing VII* 178–187.

Chatterjee, P.K., Hunter, W.R., Amerasekera, A., Aur, S., Duvvury, C., Nicollian, P.E., Yang, L.M., and Yang, P. (1995). Trends for deep submicron VLSI and their implications for reliability. *International Reliability Physics Symposium.*

Cong, J. (1999). An interconnect-centric design flow for nanometer technologies. *International Symposium VLSI Technology, Systems, and Applications.*

Covino, J.J. (1997). Dynamic CMOS circuits with noise immunity. *U.S. Patent 5650733.*

D'Souza, G.P. (1996). Dynamic logic circuit with reduced charge leakage. *U.S. Patent 5483181,*

Fariborz, A., Dennis, S., Stephen, P., Jaffrey, B., Ping, K., and Chen-ming, H. (1997). Dynamic threshold-voltage MOSFET (DTMOS) for ultra-low voltage VLSI. *IEEE Transactions on Electron Devices 44* (3).

Fornaciari, W., Sciuto, D., and Silvano, C. (1999). Power estimation for architectural exploration of HW/SW communication on system-level buses. *International Workshop on Hardware/Software Codesign* 152–156.

Hegde, R., and Shanbhag, N.R. (1998). Energy-efficiency in presence of deep submicron noise. *IEEE/ACM International Conference Computer-Aided Design* 228–234.

Intel (2001). Interconnect and noise immunity design for the Pentium IV processor. http://developer.intel.com/technology/itj/q12001/articles/art5.htm

Hunter, W.R. (1997). Self-consistent solutions for allowed interconnect current density-part I: Implications for technology evolution. *IEEE Transactions on Electron Devices 44* (2).

Kang, S. (2000, November). Design-in-reliability for giga-scale SOC integration. http://www.ece.iit.edu/~awang/seminar/00f/kang.html

Kao, J.T., and Chandrakasan, A.P. (2000). Dual-threshold voltage techniques for low-power digital circuits. *IEEE Journal of Solid-State Circuits 35,* 1009–1018.

Kim, K.M., Baek, K.H., Shanbhag, N.R., Liu, C.L., and Kang, S. (2000) Coupling-driven signal encoding scheme for low-power interface design. *The International Conference on Computer-Aided Design.*

Lam, W.C.D., Cheng-Kok Koh; Tsao, and C.-W.A., (2002). Power-supply noise suppression via clock skew scheduling. *International Symposium on Quality Electronic Design,* 355–360.

Liew, B.K., Cheung, N.W., and Hu, C. (1990). Projecting interconnect electromigration lifetime for arbitrary current waveforms. *IEEE Transactions Electron Devices 37* (5).

Mehta, H., Owens, R.M., and Irwin, M.J., Some issues in gray code addressing. *The Great Lakes Symposium VLSI,* 178–180.

Miyazaki, M., Mizuno, H., and Ishibashi, K. (1998). A delay distribution squeezing scheme with speed-adaptive threshold-voltage CMOS (SA-Vt CMOS) for low-voltage LSIs. *The 1998 International Symposium on Low-Power Electronics and Design.*

Musoll, E., Lang, T., and Cortadella, J. Working-zone encoding for reducing the energy in microprocessor address busses. *IEEE Transactions on VLSI Systems 6*(4), 568–572.

Panda, P.R., and Dutt, N.D. (1996). Reducing address bus transitions for low power memory mapping. *European Design and Test Conference* 63–37.

Powell, M.D., Yang, S.H., Falsafi, B., Roy, K., and Vijayku-mar, T.N. (2000). Gated-Vdd: A circuit technique to reduce leakage in cache memories. *International Symposium on Low-Power Electronics and Design.*

Rabaey, J., and Pedram, M. (1996). *Low-power design methodologies.* Kluwer.

Ramprasad, S., Shanbhag, N.R., and Hajj, I.N. (1999). Information-theoretic bounds on average signal transition activity. *IEEE Transactions on VLSI.*

Sematech. (1999). *International roadmap for semiconductors 1999 edition: interconnect* http://public.itrs.net/files/1999 SIA Roadmap/Int.pdf

Sematech. (2001). *International technology road map for semiconductors, 2001 edition: Interconnect.* http:/public.itrs.net/files/2001/ITRS/Interconnect.pdf

Shepard K.L., and Narayanan, V. (1998). Conquering noise in deep submicron digital design. *IEEE Design Test Computing,* (15) 51–62.

Stan, M.R., and Burleson, W.P. (1995). Bus-invert coding for low-power *I/O. IEEE Transactions on VLSI Systems* 49–58.

Su, C.L., Tsui, C.Y., and Despain, A.M. Saving power in the control path of embedded processors. *IEEE Design and Test of Computers 11* (4), 24–30.

Su, H.H., Gala, K., and Sapatnekar, S. (2000). Fast analysis and optimization of power/ground network. *Proceedings for the International Conference on Computer-Aided Design* 477–480.

Sylvester, D., and Keutzer, K. (1998). Getting to the bottom of deep submicron. *ICCAD* 203–211.

Wang, L., and Shanbhag, N.R. (1999). Noise-tolerant dynamic circuit design. *IEEE International Symposium* on *Circuits and Systems,* 549–552.

Zhang, Y., Ye, W., and Irwin, M.J. (1998). An alternative architecture for on-chip global interconnect: Segmented bus power modeling. *Asilomar Conference on Signals, Systems, and Computers* 1062–1065.

Zhao, S., Roy, K., and Koh, C.K. (2001). Decoupling capacitance allocation for power supply noise suppression. *Proceedings for the 2001 International Symposium on Physical Design* 66–71.

8

Interconnect Noise Analysis and Optimization in Deep Submicron Technology

Mohamed Elgamel
and Magdy Bayoumi
The Center for Advanced Computer Studies, University of Louisiana at Lafayette, Lafayette, Louisiana, USA

8.1 Introduction

The current trend of technology scaling is predicted to continue. Feature sizes will continue to shrink and clock frequencies will continue to increase. Shrinking feature size implies not only shorter gate lengths but also decreasing interconnect pitch and device threshold voltages (Shepard and Narayanan, 1998). The total wire area capacitance will decrease due to the reduction of the areas of a minimum-width wire. Resistance, however, is increasing faster despite efforts not to scale metal thickness. Moreover, the crosscoupling capacitance between neighboring wires will increase due to the decrease in spacing between wires. The expectation for the die sizes is to remain constant in spite of the feature size shrinking. This is due to the integration of more functionality on a single chip. Practical efforts to control RC delays through the use of a low-resistivity metal (copper), low-dielectric-constant insulators, and thick wiring that is wide will require interconnection analysis to consider inductance and inductive coupling. Figures 8.1 and 8.2 show the difference between older silicon technology and the current deep submicron technology.

The function of interconnects or a wiring system is to distribute clock and other signals and to provide power/ground to and among the various circuits/systems functions on a chip. Current leading-edge logic processors have seven to eight levels of high-density interconnect, and current leading-edge memory has three levels (Sematech, 1999). There are three types of wiring to distribute the clock and signal functions (local, intermediate, and global). Local wiring, consisting of very thin lines, connects gates and transistors in an execution unit or a functional block (such as embedded logic, cache memory, or address adder) on the chip. Local wires usually span a few gates and occupy first and sometimes second metal layers. Intermediate wiring provides clock and signal distribution in a functional blocks with typical lengths up to 3 to 4 mm. Intermediate wires are wider and taller than local wires to provide lower resistance signal/clock paths. Global wiring provides clock and signal distribution between the functional blocks and delivers power/ground to all functions on a chip. Figure 8.3 shows the delay of local and global wiring in future generations. Repeaters can be incorporated to mitigate the delay in global wiring but consume power and chip area (Sematech, 1999).

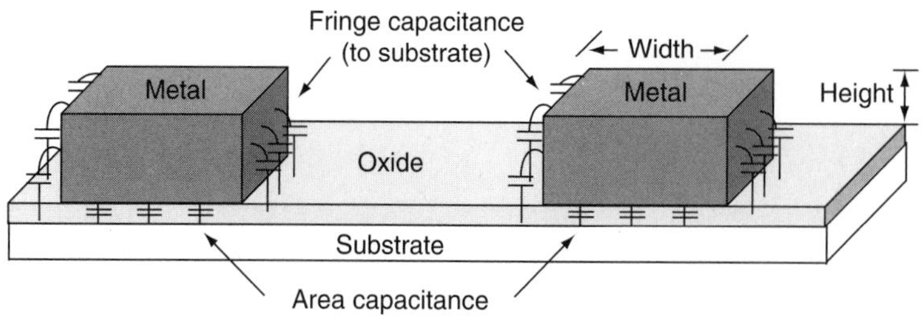

FIGURE 8.1 Older Silicon Technology. Adapted from Synopsys (1999).

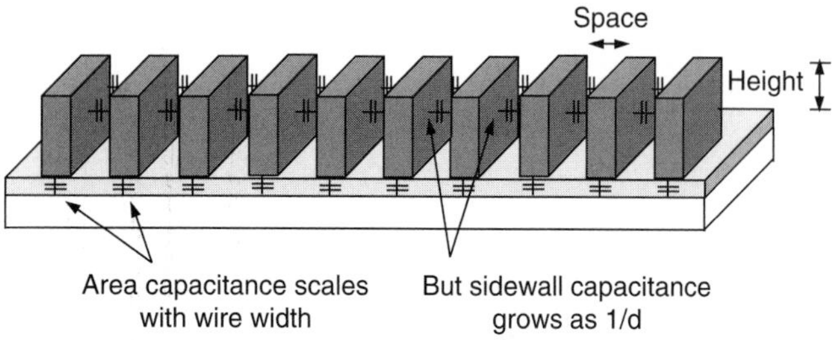

FIGURE 8.2 Deep Submicron Technology. Adapted from Synopsys (1999).

In the long term, new design or technology solutions (such as coplanar wave guides, free space RF, and optical interconnect) will be needed to overcome the performance limitations of traditional interconnect (Sematech, 1999). Inductive effects will also become increasingly important as frequency of operation increases, and additional metal patterns or ground planes may be required for inductive shielding. As supply voltage is scaled or reduced, cross talk has become an issue for all clock and signal wiring levels; the near term solution adopted by the industry is the use of thinner metallization to lower line-to-line capacitance. This approach is more effective for the lower resistivity copper metallization, for which reduced aspect ratios (A/R) can be achieved with less sacrifice in resistance

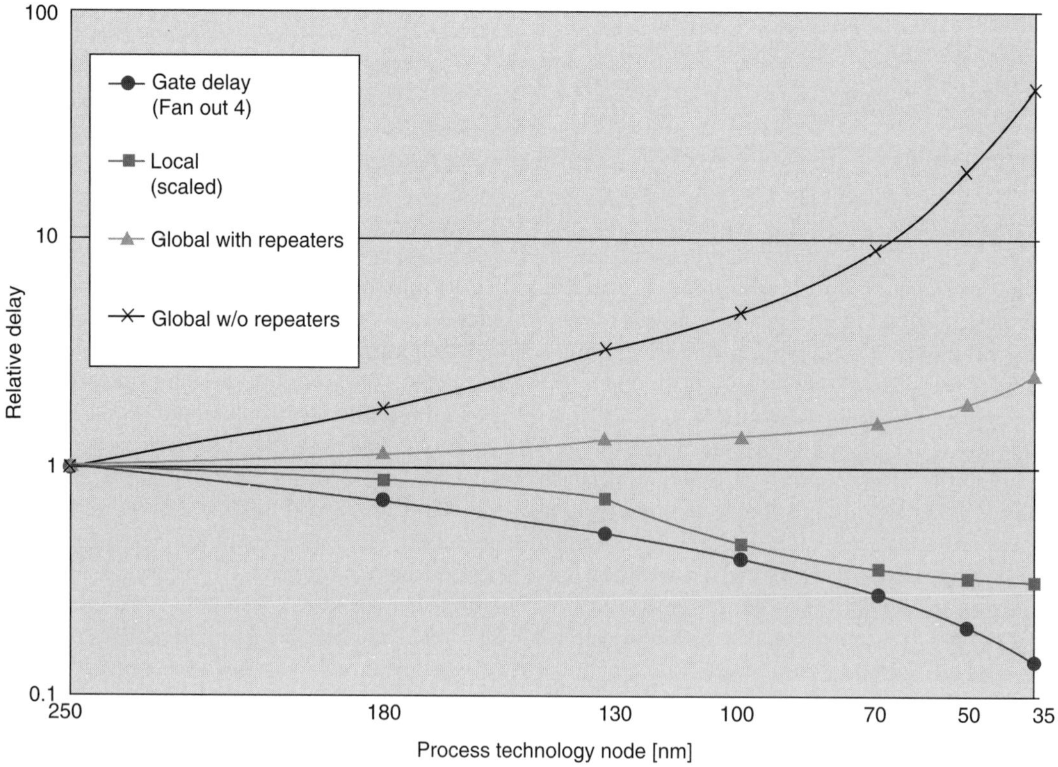

FIGURE 8.3 Delay for Local and Global Wiring Versus Feature Size. Adapted from Sematech (1999).

as compared with aluminum metallization. The 1999 Roadmap [25] reflects this design trend by featuring reduced aspect ratios (as an alternative means of reducing capacitance) and less aggressive scaling of a dielectric constant.

With technology scaling, noise is a problem affecting all types of designs from custom microprocessors to standard-cell application-specific integrated circuit (ASICs). A noise analysis solution must be capable of analyzing tens of millions of transistors, considering both circuit and interconnect noise, and evaluating the distinct noise tolerances of each node in the circuit.

In this chapter, we survey new trends for interconnect synthesis and optimization techniques. We emphasize interconnect delay and noise models, techniques and algorithms to optimize them, and advantages and drawbacks of each. Finally, we focus on which area we can extend our research.

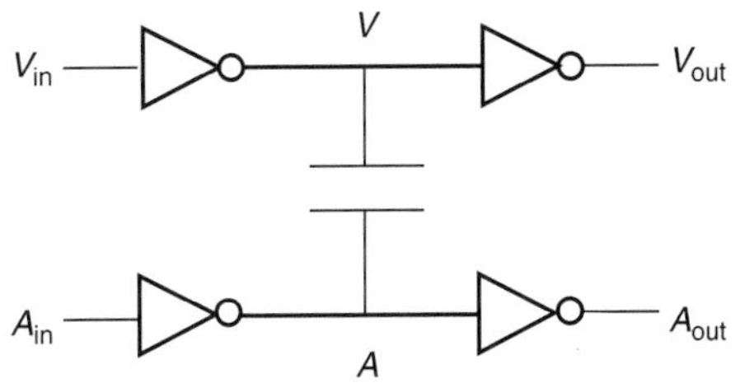

FIGURE 8.4 A Simple Coupling Circuit Structure

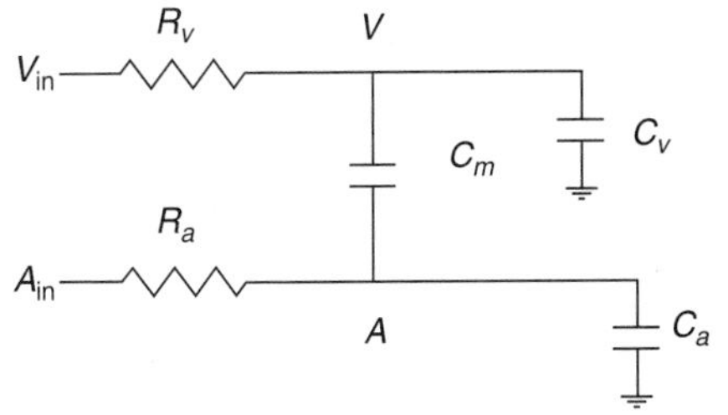

FIGURE 8.5 Capacitive Coupling Model (for the Circuit in Figure 8.4)

8.2 Interconnect Noise Models

Analytical expressions are preferred in analyzing interconnect noise because simulation is always expensive and ineffective to use with designs containing millions of transistors and wires. Analytical expressions, however, are not sufficiently accurate and do not consider all of interconnect and driver parameters. Different design stages have different requirements for accuracy of modeling interconnect cross talk noise effects. Many studies have been conducted to model metal lines and cross talk effects. For the purpose of the timing verification, ideal ground-based RC or RLC-distributed-circuit models for interconnect lines have been widely used (Sakurai). The conventional ideal RC or RLC transmission line model of the IC interconnects without considering that the following detailed physical phenomena are not accurate enough to verify the picosecond level timing of high performance VLSI circuits (Woojin *et al.*, 2000). These phenomena are the **silicon substrate effect**, the **skin effect**, and the **proximity effect**.

Most of the research efforts have been focused on developing formulas for the peak noise pulse amplitude. The pulse width of the cross talk noise and the peak noise occurring time, which are of similar importance for circuit performance as the peak amplitude, should also be considered. Since digital gates are inherently low-pass filter, they can filter out noise pulses with high amplitude provided that the noise pulse width is sufficiently narrow. Dynamic noise-immunity metrics, such as the noise-immunity curve (NIC), are required (Elgamel *et al.*, 2002).

8.2.1 Lumped Interconnect Model

The interconnect analytical model could be categorized into two categories: lumped and distributed models. In the lumped model, the total capacitance and resistance values are used. Figure 8.4 shows a simple coupling circuit structure with one victim and one aggressor. This is not the practical case because there is always more than one aggressor coupled to the victim line. The lumped capacitance model for this circuit is shown in Figure 8.5. In this model, each pulling resistance, R_v or R_a, is composed of the line resistance and the driver resistance. The load capacitances, C_a and C_v, consist of the line capacitance and the gate capacitance of the load driven by the line. C_m is the coupling capacitance between the two wires. Most of the papers about modeling cross talk noise published in the earlier years belong to this category (Vittal and Marek-Sadowska, 1997; Pillage and Rohrer, 1990; Cong *et al.*, 2000; Kahng *et al.*, 1999). In deep submicron technology, lumped models are no longer capable of satisfying the accuracy requirements and considering some parameters like the coupling location.

8.2.2 Distributed Interconnect Model

Distributed coupled RC(L) tree structure models become necessary even for the early design stages. Many papers addressed this problem (Vittal and Marek-Sadowska, 1999; Davis and Meindl, 2000; Kuhlmann and Sapatnekar, 2001). The distributed π, 3π, and 4π models have been used in (Kahng *et al.*, 1999, 2001; Murat *et al.*, 2002), respectively. We describe the π model in Figure 8.6.

In Figure 8.6, C_{Lv} and C_{La} are the victim and the aggressor load capacitances respectively. R_{dv} and R_{da} are the victim and the aggressor drivers modeled by resistances. C_{c1} and C_{c2} are the coupling capacitance between the two wires distributed to the beginning and the end of the wires. C_{v1}, C_{v2}, C_{a1}, and C_{a2} are the distributed area capacitances for the victim and aggressor wires respectively. In the 4π model found in Becer

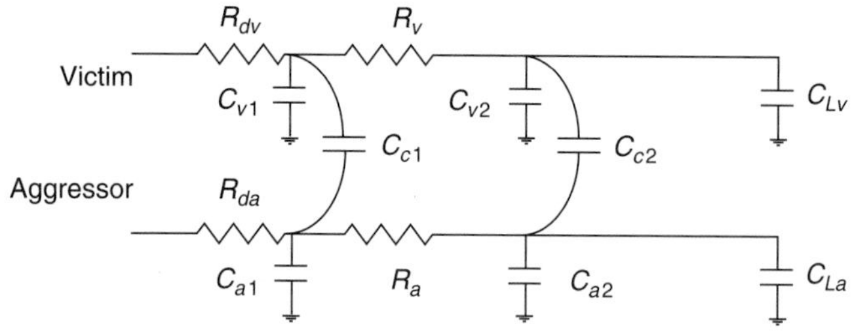

FIGURE 8.6 Interconnect Distributed π Model (for the Configuration in Figure 8.4)

et al. (2002), the users stated that their model is a complete analytical cross talk noise model that incorporates all physical properties including victim and aggressor drivers, distributed RC characteristics of interconnects, and coupling locations in both victim and aggressor lines.

8.2.3 Interconnect Modeling Issues

Some issues need to be considered while modeling the cross talk noise: the **input signal shape**, **number of pins in a net**, **net topology**, and the **driver modeling**. These issues have been discussed in some papers (Sapatnekar, 2000; Vittal and Marek-Sadowska, 1999; Sirichotiyakul *et al.*, 2001).

The nonlinear behavior of the victim driver gate during the transition should be captured (Supamas *et al.*, 2001). The industrial noise analysis tool, ClariNet, has considered this issue. Results on industrial designs were presented to demonstrate the effectiveness of this tool with an 8% error compared to SPICE simulations.

It has been emphasized that the exact functional form used to estimate the capacitance and the delay is not important. The only requirement that the delay model must satisfy is that an increase or decrease in the coupling capacitance should be translated into an increase or reduction in the delay of a net (Sapatnekar, 2000).

Expressions for cross talk amplitude and pulse width in resistive, coactively coupled lines should hold for nets with arbitrary number of pins and topology under any specified input excitation as explained by Vittal and Marek-Sadowska (1999). These authors claim that the magnitude of inductive coupling is small in the presence of good ground return paths close to signal lines. They used their expression in formulations aimed at reducing cross talk. These methods included transistor sizing, wire ordering, wire width optimization, and wire spacing. They failed to get success in some of these formulations. For wire ordering, Vittal and Marek-Sadowska (1997) proposed a new heuristic since they did not report results for this work.

The characteristics that make someone prefer one model over another include the model's operational speed and accuracy. Depending on the design stage, different accuracies and speeds are needed.

8.3 Noise Minimization Techniques

8.3.1 Buffer Insertion

For long interconnects, wire sizing or spacing, explained in following subsections, are not sufficient to limit the interconnect delay. So, buffer insertion is used to trade-off the active device area for reduction of interconnect delays. The additional inserted buffers increase the overall gate delay. Thus, there is an optimal number of repeaters that should be inserted into an RC line to minimize the overall propagation delay (Ismail and Friedman, 2001). This trend is qualitatively illustrated in Figure 8.7. During the routing of global interconnects, macroblocks form useful routing regions that allow wires to go through but forbid buffers to be inserted. They give restrictions on buffer locations. The buffer location restrictions have been taken into consideration and the simultaneous maze routing and buffer insertion problem have been solved in polynomial time (Zhou *et al.*, 2000).

8.3.2 Wire Sizing

It is known that proper wire sizing can effectively reduce the interconnect delay especially in deep submicron or nanometer designs when the wire resistance becomes significant. An optimal wire-sizing algorithm was developed by Cong and Leung (1993) and Cong *et al.* (1993) for a single source RC interconnect tree to minimize the sum of weighted delays from the source to timing-critical sinks under the Elmore delay model. Cong *et al.* (1997) designed an efficient approach to perform global interconnects sizing and spacing (GISS) for multiple nets to minimize interconnect delays with consideration of coupling capacitance as well as area and fringing capacitances. In that paper, they proposed an asymmetric wire-sizing scheme where they could asymmetrically widen or make more narrow the areas above and below the centerline of the original wire. The optimal wire sizing and spacing problem for a single net with fixed surrounding wire segments can be solved by adapting the bottom-up dynamic programming (DP)-based buffer insertion and wire-sizing algorithm proposed by Lillis *et al.* (1996).

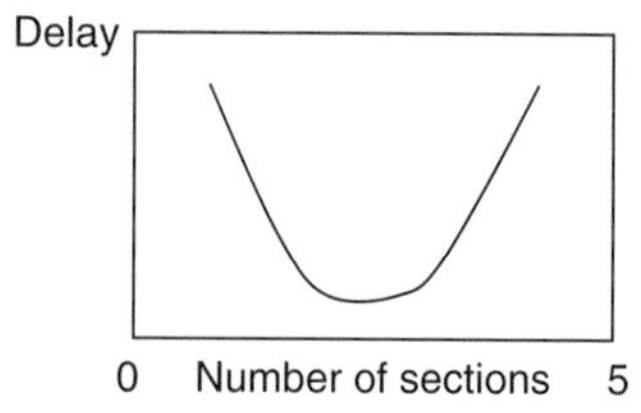

FIGURE 8.7 Relationship Between the Number of Sections of an RC Line and the Total Propagation Delay

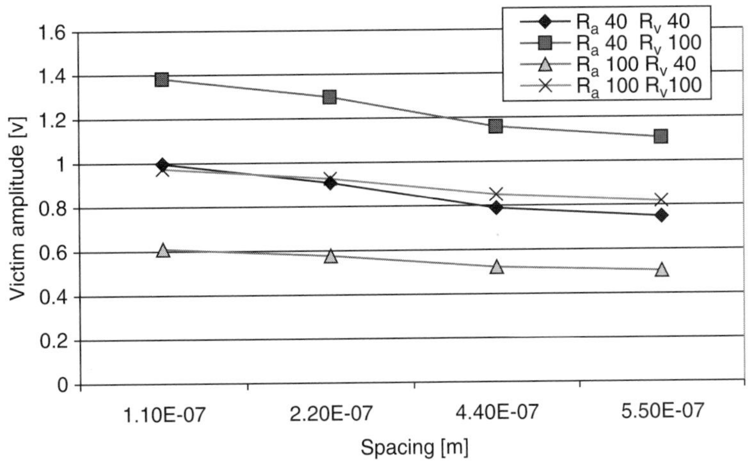

FIGURE 8.8 Victim Amplitude Versus Spacing (Different *Rs*). Adapted from Mohamed *et al.* (2003).

8.3.3 Wire Spacing

Wire spacing can decrease the cross coupling capacitance value effectively (Mohamed *et al.*, 2003). Figure 8.8 shows that as the spacing between wires increases, the peak noise decreases. The noise relatively decreases faster for a small R_a (aggressor driver resistance). Moreover, it is clear that sizing up the victim driver (scaling R_v down) reduces peak noise, where R_v is the victim driver resistance.

For example, the postlayout spacing heuristic in work by Chaudhary *et al.* (1993) proposed a postlayout graph-based spacing algorithm. The framework spaces out wires after detailed routing has been finished. The cross talk effect is based on cross talk voltage glitches because they do not consider the delay. The cross talk effect from driver to sink is simply the superposition of all glitches of wires along the path. The potential disadvantage of this greedy operation is the poor use of space resources around a timing-critical wire.

In Tseng *et al.* (1998), the graph-based optimizer preroutes wires on the global routing grids incrementally in two stages—net order assignment and space relaxation. The timing delay of each critical path is calculated while taking into account interconnect coupling capacitance. The objective is to reduce the delays of critical nets with negative timing slack values by adding extra wire spacing.

8.3.4 Shield Insertion

Shield insertion is another technique for decreasing cross talk. It simply adds a power or a ground line between already existing wires. These power lines act like shields and can isolate between wires. Figure 8.9 shows the effect of adding shields of different sizes on the noise amplitude. As the shield width increases, the noise amplitude decreases. The result is an effective technique specially made to eliminate inductive noise; however, it consumes more area.

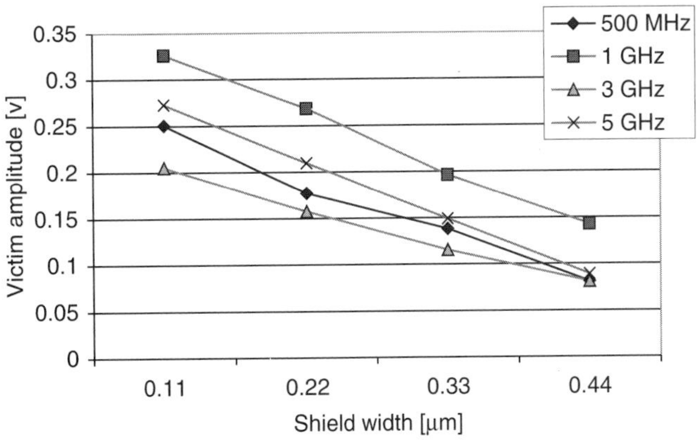

FIGURE 8.9 Noise Amplitude Versus Shield Width at Different Frequencies

The existing net ordering formulations to minimize noise are no longer valid with presence of inductive noise, and shield insertion is needed to minimize inductive noise (He and Lepak, 2000). He and Lepak (2000) formulate two simultaneous shield insertion and net ordering (SINO) problems: (1) the optimal SINO/NF problem to find a min-area SINO solution that is free of capacitive and inductive noise and (2) the optimal SINO/NB problem to find a min-area SINO solution that is free of capacitive noise and is under the given inductive noise bound. They claimed that this work was the first presenting an in-depth study on the simultaneous shield insertion and net ordering problem to minimize both capacitive and inductive noises. The drawbacks of this work are the consideration for having the same wire lengths and the consideration that the appropriate sensitivity matrix is given a priori. In addition, this matrix required an overhead, and He and Lepak (2000) did not mention the complexity or how to build this matrix. They developed four approximate algorithms for solving the SINO/NB problem: greedy-based shield insertion (SI) algorithm, net ordering for minimizing C_x noise followed by SI algorithm (NO+SI algorithm), graph-coloring based SINO algorithm (SINO/GC algorithm), and simulated annealing based SINO algorithm (SINO/SA algorithm). They solved the SINO/NF problem by using the SINO algorithms and setting the noise bound to zero. Note that the SINO/SA always performs the best for any given setting. Lepak *et al.* have extended this work by considering the RLC model.

8.3.5 Network Ordering

It has been shown that signal ordering can effectively reduce the cross talk noise on interconnects. In other words, arranging signals in an order that does not let a wire fight against its neighbor, making a transition in the same direction, can reduce the cross talk noise. For example, two interconnect layout design methodologies for minimizing the cross-coupling capacitance effect in the design of a full-custom data path have been proposed by Yim and Kyung (1999). These are the **control-signal ordering scheme** and the **track assignment algorithm**. Figure 8.10 shows how the control signal for a multiplexer can be ordered to decrease the cross talk effect.

An evolutionary programming approach was used in the track assignment algorithm to minimize the total number of tracks used. Yim and Kyung (1999) proposed an **evolutionary programming-based track assignment** (*EPTA*) algorithm considering both length and switching activity. They showed that the solution quality of EPTA is quite good, and the converging speed is very fast when compared with SA and GA. More than anything else, EPTA does not require an annoying genetic operator design, which makes the implementation of EPTA very easy.

8.4 Interconnect Noise in Early Design Stages

In the current design flow, interconnect optimizations are mainly used in postlayout stages (Chang, 2000). As the global interconnects are largely determined by floor planning, it becomes critical for floor planning engines to be able to handle efficient interconnect planning and optimizations so that the overall timing and design convergence can be achieved. Due to the inherent complexity of the interconnect driven floor planning (IDFP) problem, designers use multistage, adaptive cost functions in IDFP to gradually consider more interconnect optimization, planning, and/or global routing features. Using the adaptive cost functions gives designers flexibility to tune for different design objectives and to trade-off between performance and run time. They use a simulated annealing algorithm to achieve this flow.

The primary goal of Sylvester and Keutzer (2000) was to examine global interconnect effects and determine if there were any significant road blocks that would prevent National Technology Roadmap for Semiconductors (NTRS) performance expectations from being met. Their analysis revealed that due to global RC delays as well as time-of-flight considerations, the global clock would necessarily be slower than the achievable local clock frequency. In addition, they found that cross talk at the global level would not be as significant as at the local level due to the use of large repeaters—their capacitance would dampen the effects of coupling capacitance. Sylvester and Keutzer (2000) proposed a wiring hierarchy that complemented the modular design methodology that they proposed in 1998. This modular methodology proposed the use of 50,000 to 100,000 gate modules of logic to eliminate the impact of interconnect at the local level. These modules are arranged together in isochronous (or locally clocked) regions, which run at a higher clock speed than the global clock. These isochronous regions come together to form the entire design.

Cong *et al.* (2001) presented a multilayer gridless detailed routing system for deep submicrometer physical designs. Their detailed routing system used a hybrid approach consisting of two parts: (1) an efficient variable-width variable-spacing detailed routing engine and (2) a wire-planning algorithm providing high-level guidance as well as rip-up and reroute capabilities. They used a nonuniform grid graph, which has proven to guarantee a gridless connection of the minimum cost in multilayer variable-width and variable-spacing routing problems. They suggested further improving their wire-

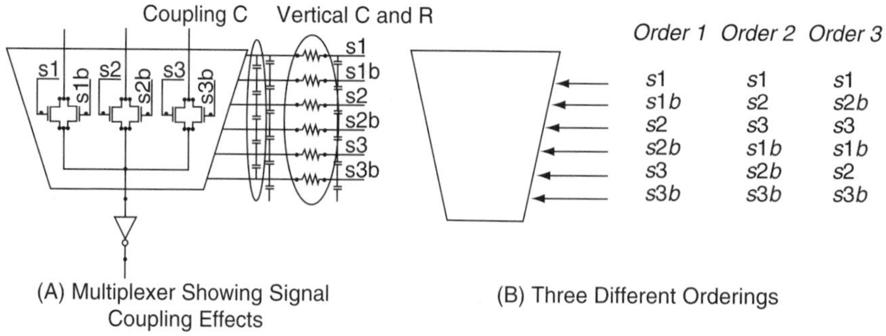

FIGURE 8.10 Decreasing Cross Talk Effects. Adapted from Yim and Kyung (1999). © 2004 IEEE.

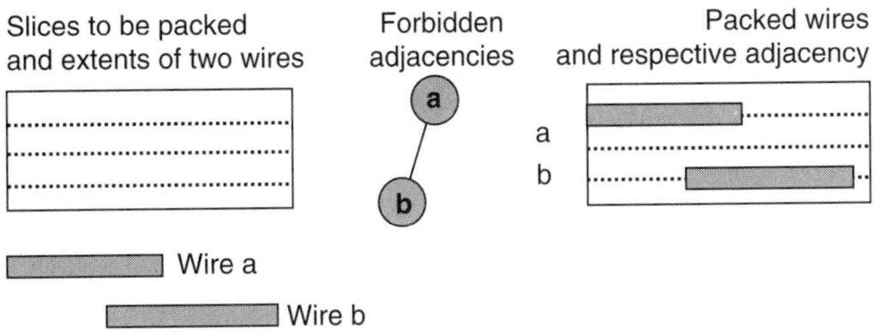

FIGURE 8.11 Wire-Packing Instance. Adapted from Kay and Rutenbar (2001). © 2004 IEEE.

planning algorithm and fine-tuning of rip-up and rerouting algorithm.

An integer linear programming (ILP) formulation that yields tight bounds on the quality of the achievable wire-packing solution was formulated by Kay and Rutenbar (2001). The input to cross talk-aware wire packing is composed of a set of wires, a set of variable tracks (per layer), and a cross talk graph XG that determines forbidden wire adjacencies. The output is a legal assignment such that no forbidden adjacencies and no electrical shorts exist, as shown in Figure 8.11. The key technical insight is to model all constraints (both geometric and cross talk conflicts) as cliques in an appropriate conflict graph; these cliques can be extracted quickly from the interval structure of the slice.

Kastner and Sarrafzadeh (2001) developed methods to reduce interconnect delay and noise caused by coupling. The coupling-free routing (CFR) takes a set of nets and tries to find a one-bend coupling-free routing for a subset of nets. A routed net must not be coupled with any other routed net. These authors defined **coupling** as a Boolean variable, which is appropriate when the coupling is greater than some threshold. They developed an exact algorithm for the CFR decision problem via a transformation to 2-satisfiability (2SAT). This algorithm runs in linear time. Kastner and Sarrafzadeh (2001) also presented the implication graph that models the dependencies associated with CFR. Moreover, they developed an algorithm for the maximum coupling-free layout (MAX–CFL) problem. Given a set of nets, the **MAX–CFL** is defined as finding a subset of nets that are coupling-free routable. The subset should have maximum size and/or criticality. This algorithm is the implication algorithm. In addition, Kastner and Sarrafzadeh (2001) presented the coupling capacitance between two wires i and j as follows:

$$C_c(i, j) = \frac{f_{ij} \cdot l_{ij}}{d_{ij}} \frac{1}{1 - (w_i + w_j/2d_{ij})},$$

where w_i and w_j are the sizes of wires i and j (w_i, $w_j > 0$), f_{ij} is the unit length fringing capacitance between wires i and j, l_{ij} is the overlap length of wires i and j, and d_{ij} is the distance from the center line of wire i to the center of wire j. Because the nets are routed at most one bend, they have minimum wire length. In addition, coupling-free routing minimizes the coupling of the routed nets.

It has been shown that the delay of a wire of length l increases at the rate of $O(l^2)$ without wire sizing, $O(l\sqrt{l})$ with optimal wire sizing, and linearly with proper buffer insertion (Cong and Pan, 1999). The criticality function can easily be changed to incorporate some other functions.

8.5 Case Study Pentium 4

8.5.1 Interconnect Delay and Crosscapacitance Scaling

The interconnect problem has become significant enough to require entire architectural pipe stages in the Pentium IV processor for interconnect communication (Intel, 2001). At the circuit level, widespread use of repeaters has become necessary. To avoid degrading interconnect resistance, the vertical dimension of metals has scaled very weakly compared to the horizontal dimension, leading to extremely high height and width aspect ratios (2-2.2) (see Figure 8.12).

Nowadays, the crosscoupling capacitance between parallel neighboring wires, which can get routed together for long distances, is gaining great importance compared to the self-capacitance wires. The crosscoupling can either lead to a large increase in delay, coupling noise, min delay, or power consumption problems, depending on the switching direction of neighboring wires. Avoiding these delay and noise problems would involve drastically increased wire spacing or extensive shielding. Further, studies on both the Pentium III and Pentium IV processor floor plans have clearly shown that they tend to be interconnect-limited for die area, which increases the penalty for spacing and shielding. Thus, there is a fundamental design trade-off between a simple, robust wiring solution employing extensive spacing and shielding versus an aggressive solution employing short wiring with only judicious shielding leading to high density. The latter requires sophisticated CAD tools, has more risks, but ultimately is much more optimal for a high-volume product. It was therefore the choice for the Intel Pentium IV processor (Intel, 2001).

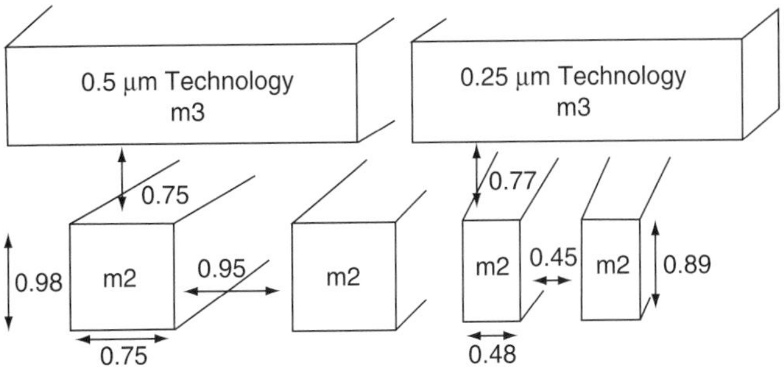

FIGURE 8.12 Wire Aspect Ratio Scaling with Technology. Adapted from Intel (2001).

8.5.2 Wire and Repeater Design Methodology for the Pentium 4 Processor

Delay, noise, slope limits, and gate oxide wear out were all considered by Intel designers when drafting the guidelines for the wire and repeater methodology. Notable features were an increased emphasis on noise robustness and "pushed process" considerations for delay (repeater distance guidelines were made shorter than optimal for delay with the existing process in anticipation of end-of-life process trending when transistors greatly speed up compared to wires). Repeater sizing, rather than best delay optimization for noncoupled wires, was picked to be optimal for noise rejection, for equal rise and fall delays, and for better delay in the presence of coupling. Stringent limitations were put on maximum sizing of repeaters, especially in busses, to reduce collapse of power supply caused by a simultaneously switching bank of repeaters. The methodology and tools allowed Intel developers to use both inverting and noninverting repeaters. Simple length-based design rules were provided for repeaters, and further optimization was possible through internally developed proprietary tools: NoisePad, ROSES, and Visualizer (net routing and timing) analysis. The extensive use of dedicated repeater blocks is evident in the Pentium 4 Processor floor plan. Further, the net length comparison in Figure 8.13 shows that although the Pentium 4 Processor is a much larger chip, it has very few long nets compared to the previous generation of chips, such as the Pentium III Processor. This is even more notable given that the Pentium 4 Processor has more than twice as many full-chip nets as the Pentium III Processor and has architecturally bigger blocks. If we compare the M5 wire segments of the Pentium III and Pentium 4 processors, we note that 90% of the M5 wire segments of the Pentium 4 processor are shorter than 2000 microns, while the same percentage of Pentium III Processor wires are 3500 microns long. These short wires are a key to enabling high-frequency operation.

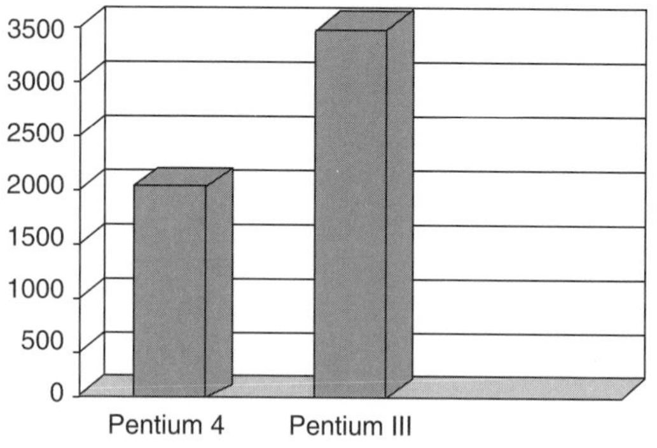

FIGURE 8.13 M5 Length Comparison of Global Wires for Different Processors Using the Same 0.18 μm Technology. Adapted from Intel (2001).

Acknowledgments

The authors acknowledge the support of the U.S. Department of Energy (DoE) EETAPP program (DE97ER12220) and the Governor's Information Technology Initiative.

References

Becer, M., Blaauw, D., Zolotov, V., Panda, R., and Hajj, I.N. (2002). Analysis of noise avoidance techniques in DSM interconnects using a complete cross talk noise model. *Proceedings of the 2002 Design Automation and Test in Europe Conference and Exhibition*, 456–463.

Chang, C.C., Cong, J., Zhigang, D., and Yuan, X. (2000). Interconnect-driven floor planning with fast global wiring planning and optimization. *Proceedings SRC Techcon Conference.*

Chaudhary, K., Onozawa, A., and Kuh, E.S. (1993). A spacing algorithm for performance enhancement and cross talk reduction. *Digital Technology Papers, IEEE/ACM International Conference on Computer-Aided Design*, 697–702.

Cong, J. *et al.* (2001). DUNE—A multilayer gridless routing system. *IEEE Transactions on Computer-Aided Design 20* (5) 633–647.

Cong, J., He, L., and Koh, C.K. (1997). Global interconnect sizing and spacing with consideration of coupling capacitance. *Proceedings ACM/IEEE International Conference on Computer-Aided Design*, 628–633.

Cong, J., and Leung, K.S. (1993). Optimal wire sizing under the distributed Elmore delay model. *Proceedings of the International Conference on Computer-Aided Design*, 634–639.

Cong, J., Leung, K.S., and Zhou, D. (1993). Performance-driven interconnect design based on distributed RC model. *Proceedings on Design Automation Conference*, 606–611.

Cong, J., and Pan, D.Z. (1999). Interconnect delay estimation models for synthesis and design planning. *Proceedings of the Asia and South Pacific Design Automation Conference.* 97–100.

Cong, J., Pan, D.Z., and Srinivas, P.V. (2000). Improved cross talk modeling for noise constrained interconnect optimization. *Proceedings of the ACM/IEEE International Workshop on Timing Issues in the Specification and Synthesis of Digital Systems*, 14–20.

Davis, J.A., and Meindl, J.D. (2000). Compact distributed RLC interconnect models. *IEEE Transactions on Electronic Devices, 47* (11).

Intel (2001). Interconnect and noise immunity design for the Pentium IV processor. http://developer.intel.com/technology/itj/q12001/articles/art_5.htm

Elgamel, M., Darwish, T., and Bayoumi, M. (2002). Noise-Tolerant low-power dynamic TSPCL D flip-flops. *IEEE Annual Symposium on VLSI*, 89–94.

Elgamel, M.A., Tharmalingam, K.S., and Bayoumi, M.A. (2003). Cross talk noise analysis in ultra deep submicrometer technologies. *IEEE Computer Society Annual Symposium on VLSI*, 189–192.

He, L., and Lepak, K.M. (2000). Simultaneous shield insertion and net ordering. *Proceedings of the International Symposium on Physical Design*, 55–60.

Ismail, Y.I., and Friedman, E.G. (2001). *On-chip inductance in high-speed integrated circuits.* Kluwer Academic Publishers.

Jin, W. *et al.* (2000). Experimental characterization and modeling of transmission line effects for high-speed VLSI circuits interconnects. *IEICE Transactions on Electronic Devices E83-c* (5), 728–735.

Kahng, A.B., and Muddu, S. (1997). An analytical delay model for RLC interconnects. *IEEE Transactions on Computer-Aided Design Integrated Circuits and Systems 16* (12), 1507–1514.

Kahng, A., Muddu, S., and Vidhani, D. (1999). Noise and delay uncertainty studies for coupled RC interconnects. *Proceedings of the Twelfth Annual IEEE International ASIC/SOC Conference*, 3–8.

Kahng, A.B., Muddu, S., and Vidhani, D. (1999). Noise and delay uncertainty studies for coupled RC interconnects. *Proceedings of the IEEE International ASIC/SOC Conference* 3–8.

Kahng, A., Muddu, S., and Vidhani, D. (2001). Noise model for multiple segmented RC interconnects. *2001 International Symposium on Quality Electronic Design*, 145–150.

Kastner, R., and Sarrafzadeh, M. (2001). An exact algorithm for coupling-free routing. *International Symposium on Physical Design (ISPD)*, 10–15.

Kay, R., and Rutenbar, R.A. (2001). Wire packing: A strong formulation of cross talk-aware chip-level track/layer assignment with an efficient integer programming. *IEEE Transactions on Computer-Aided Design, 20* (5) 672–679.

Kuhlmann, M., and Sapatnekar, S.S. (2001). Exact and efficient cross talk estimation. *IEEE Transactions on Computer-Aided Design of Integrated Circuits and Systems. 20* (7), 858–866.

Lepak, K.M., Luwandi, I., He, L. (2001). Simultaneous shield insertion and net ordering under explicit RLC noise constraint. *Proceedings on Design Automation Conference*, 199–202.

Lillis, J., Cheng, C.K., and Lin, T.T.Y. (1996). Simultaneous routing and buffer insertion for high performance interconnect. *Proceedings of the Sixth Great Lakes Symposium on VLSI*, 148–153.

Pillage, L.T., and Rohrer, R.A. (1990). Asymptotic waveform evaluation for timing analysis. *IEEE Transactions on Computer-Aided Design of ICs and Systems 9*, 352–366.

Sakurai, T. (1993). Closed-form expression for interconnect delay, coupling, and cross talk in VLSIs. *IEEE Transactions on Electronic Devices 40* (1), 118–124.

Sapatnekar, S.S. (2000). A timing model incorporating the effect of cross talk on delay and its application to optimal channel. *IEEE Transactions on Computer-Aided Design 19* (5).

Sematech (1999). *International technology roadmap for semiconductors 1999 edition: interconnect.* http://public.itrs.net/files/1999_SIA_Roadmap/Int.pdf

Sematech (2001). *International technology roadmap for semiconductors. 2001 edition: interconnect.* http://public.itrs.net/Files/2001ITRS/Interconnect.pdf

Shepard, K.L., and Narayanan, V. (1998). Conquering noise in deep-submicron digital ICs. *IEEE Design & Test of Computers 15* (1), 51–62.

Sirichotiyakul, S., Blaauw, D.Oh, C., Levy, R., Zolotov, V., and Zuo, J. (2001). Driver modeling and alignment for worst-case delay noise. *DAC*, 720–725.

Sylvester, D., and Keutzer, K. (1998). Getting to the bottom of deep submicron. *Proceedings of the ICCAD*, 203–211.

Sylvester, D., and Keutzer, K. (2000). A global wiring paradigm for deep submicron design. *IEEE Transactions on Computer-Aided Design 19* (2) 242–252.

Synopsys. Routing for Complex SOC Designs. (1999, October). http://www.synopsys.com/products/tlr/flexroute_wp.pdf

Tseng, H.P., Scheffer, L., and Sechen, C. (1998). Timing- and cross talk-driven area routing. *Design Automation Conference.* 378–381.

Vittal, A., and Marek-Sadowska, M. (1997). Cross talk reduction for VLSI. *IEEE Transactions on Computer-Aided Design 16.*

Vittal, A., and Marek-Sadowska, M. (1999). Cross talk in VLSI Interconnections. *IEEE Transactions Computer-Aided Design 18* (12), 1817–1824.

Yim, J.S., and Kyung, C.M. (1999). Reducing cross-coupling among interconnect wires in deep-submicron data path design. *The 36th Design Automation Conference (DAC)*, 485–490

Zhou, H., Wong, D.F., Liu, I.M., and Aziz, A. (2000). Simultaneous routing and buffer insertion with restrictions on buffer locations. *IEEE Transactions on Computer-Aided Design 19* (7) 819–824.

IV

DIGITAL SYSTEMS AND COMPUTER ENGINEERING

Sun-Yung Kung
Department of Electrical Engineering, Princeton University, Princeton, New Jersey, USA

Benjamin W. Wah
Computer and Systems Research Laboratory, University of Illinois at Urbana-Champaign, Urbana, Illinois, USA

In the past half century, the world of microelectronic and information techonlogies has experienced a breathtaking and revolutionary evolution—50s semi-conductor, 60s integrated circuits, 70s microprocessor, 80s DSP/FPGA/ASIC processors, 90s multimedia multi-processors, to SoC for ubiquitous computing necessary for the ever-more-connected information world in the 21st century. Subsequently, the field of *digital systems and computer engineering* has also experienced a fast change driven by the devices' technological advances. The performance of digital systems requires integration of system architecture, programming systems, and application. While it is not unusual to have as many as 100 ALU's per chip today, the Moore's law is not directly applicable to performance improvement, which needs also parallelism-friendly software in order to match the hardware speed-up. To this end, chapters related to both computer architecture and sotware programming systems are collected in this section to achieve comprehensive and balanced coverage.

Chapter 1, "Computer Architecture," by Morris Chang, provides an introduction to computer architecture, including microprogramming, memory hierarchy in computer systems, and input and output systems. A computer system consists of processor(s), main memory, clocks, terminals, disks, network interface, and input/output devices. The power of computation can be maximized via a systematic and seamless integration of hardware cores, operating systems, and application softwares. Throughout the 70s, microprogramming computer architecture design was the most dominant approach, and it had a fundamental influence on early development of computing systems. In general, the hierarchy of systems using microprogramming is divided into application softwares, operating systems, machine language, microprogramming, and physical devices. Since the early 80s, the ever-increasing processing power offered by VLSI technology (as governed by the Moore's law) has fundamentally changed the computer design concept. The popularity of reduced instruction set computing (RISC) has virtually eliminated the need for microprogramming. More generally, computer architectures have undergone a rapid change driven by the VLSI, deep-submicron, and nano-scale device technologies.

Chapter 2 is "Multiprocessors," by Peter Y. K. Cheung, G. A. Constantinides, and Wayne Luk. Multiprocessors are categorized by Flynn into SISD (single instruction single data), SIMD (single instruction multiple data), MISD (multiple instruction single data), and MIMD (multiple instruction multiple data) machines. The chapter addresses fundamental issues such as how multiple processors share data, how cache memories on different processors maintain their data correctly, and how multiple processors coordinate with each other. In addition, the following two performance goals of multiprocessor system are explored: (l) increased throughput for independent tasks distributed among a number of processors, and (2) faster execution of a single task on multiple processors. The organization and programming language for the multiprocessor systems vary significantly, depending on the goal.

Chapter 3, "Configurable Computing," by Wayne Luk, Peter Y. K. Cheung, and Nabeel Shirazi, first presents an overview of configurable computing technology and then describes methods for reconfiguration, tools for configurable designs, and the automatic detection of reconfigurable regions. In particular, it covers an efficient approach to exploit run-time reconfigurability. FPGAs have become the favored choice in implementing glue logic, experimental systems, and hardware prototypes due to their clear advantages in user reconfigurability, short turnaround time, and low development costs.

In Chapter 4, "Operating Systems," by Yao-Nan Lien, the hardware core of a basic computer system, which consists of processor(s), memory, and peripheral devices is presented. It is often tedious to write programs that keep track of all these hardware components. The operating system is software, which controls all the computer's resources and provides the base on which the application programs can be written. The operating system puts a layer of software on top of the bare hardware, manages all parts of the system, and presents the user with an interface or virtual machine that is easier to understand and program. The operating system can be viewed as a set of software extensions of primitive hardware, culminating in a virtual machine that serves as a high-level programming environment and manages the flow of work network of computers. Modern operating systems also provide numerous services, such as interprocess communication, file and directory systems, data transfer over networks, and a command language for invoking and controlling programs. In addition to the basic concept on operating systems, the chapter further explores a model operating system as well as two exemplifying operating systems: Unix and MS-DOS.

Chapter 5, "Expert Systems," by Yi Shang, covers four fundamental topics in expert systems—knowledge representation, model- and case-based reasoning, knowledge acquisition, and explanation of solution. An expert system is a computer program that represents and uses knowledge of one or more human experts to provide high-quality performance in a specific domain. Expert systems offer a number of benefits when compared with human experts. They are inexpensive to operate, easy to reproduce and distribute, and can provide permanent documentation of the decision process. Most importantly, expert systems can produce consistent results on the same tasks and handle similar situations consistently.

Two chapters (Chapters 6 and 7) are devoted to emerging topics on multimedia systems. Multimedia data, such as text, audio, images and video, are rapidly evolving as main avenues for the creation, exchange, and storage of information. Rapid advances in VLSI technology have made cost-effective processing of extremely high-volume multimedia data as well as efficient storage and communication over high-bandwidth networks possible. Chapter 6, "Multimedia Systems: Content-Based Indexing and Retrieval," by Faisal Bashir, Shashank Khanvilkar, Ashfaq Khokhar, and Dan Schonfeld, addresses multimedia (image and video) storage and encoding standards (JPEG, MPEG, H.26.X) and explores many multimedia indexing and retrieval techniques. Chapter 7, "Multimedia Networks and Communication," by Shashank Khanvilkar, Faisal Bashir, Dan Schonfeld, and Ashfaq Khokhar, offers comprehesive coverage of many key topics, such as multimedia networks and communication, including multimedia expectations (including real-time characteristics, multicasting support, security, and mobility support, etc.); best-effort internet support for distributed traffic requirements (including service models; integrated and differential services etc.); enhancing the TCP/IP protocol stack to support functional requirements of distributed multimedia applications; and quality of service architecture for third generation cellular systems.

Chapter 8, "Fault Tolerance in Computer Systems from Circuits to Algorithms," by Shantanu Dutt, Federico Rota, and Franco Trovo, starts with a discussion on fundamental issues in fault tolerance and dependability, then it further treats a broad range of fault-detection and fault-tolerance methods with potential applications to field-programmable gate arrays, program control-flow, and processor systems. The ubiquitous use of digital systems in all areas of human endeavor and the decreasing feature sizes of VLSI/ULSI technologies make it important to incorporate fault tolerance mechanisms into modern computer systems.

Chapter 9, "High-Level Petri Nets Extensions, Analysis, and Applications," by Xudong He and Tadao Murata, presents analysis techniques for several extended Petri net models, explores specific system properties, such as performance, reliability, and schedulability, and describes their benefits for many applications in computer science disciplines. Petri nets provide a formal model for concurrent and distributed systems.

Acknowledgments

The Section Editors wish to take this opportunity to sincerely thank all the authors for their outstanding contributions. They also wish to express their profound gratitude to Professor Wai-Kai Chen for his initial invitation to paricipate in this timely project. Moreover, we deeply appreciate his leadership and especially, his persistence thorougout the entire production process.

1

Computer Architecture

Morris Chang
Department of Electrical and Computer Engineering, Iowa State University, Ames, Iowa, USA

1.1 Microprogramming

In 1951, Maurice Wilkes suggested the notion of designing a three-level computer to simplify the hardware. Prior to his suggestion, computers used a two-level approach in which level 1 was for programming (also known as the instruction set architecture [ISA] level) and level 2 was for digital logic. In Wilkes's proposed design, an unchangeable interpreter (the microprogram) was introduced to execute the ISA level program by interpretation. Since microprograms have a smaller instruction set than ISA level programs, the number of hardware components needed was greatly reduced. During the 1970s, the microprogram was the most dominant approach in system design. Nowadays, the popularity of reduced instruction set computing (RISC) has virtually eliminated the need for microprogramming. The basic hierarchy of systems using microprogramming is illustrated in Figure 1.1.

Microprogramming is usually located in read-only memory (ROM), directly controls physical devices, and provides a cleaner interface to the machine language. It is actually an interpreter, fetching the machine language instruction, such as ADD, MOVE, and JUMP, and carrying the instruction out as a series of little steps. To carry out an ADD instruction, for example, the microprogram must determine where the numbers to be added are located, fetch them, add them, and store the result somewhere. The set of instructions that the microprogram interprets defines the machine language, which is not really part of the hard machine at all, but computer manufacturers always describe it in their manuals.

Microprogramming designs the control as a program that implements the machine instructions in terms of simpler microinstructions. Each microinstruction defines the set of data path control signals that must be asserted in a given state. The microprogram is a symbolic representation of the control that will be translated by a program to control logic. In this way, one can choose how many fields a microinstruction should have and what control signals are affected by each field. The format of the microinstruction should be chosen to simplify the representation, making it easier to write and understand the microprogram. For example, it is useful to have one field that controls the ALU and a set of three fields that determine the two sources for the ALU operation as well as the destination of the ALU result. In addition to readability, the microprogram format should make it difficult or impossible to write inconsistent microinstructions. A microinstruction is inconsistent if it requires that a given control signal be set to two different values. To avoid a format that allows inconsistent microinstructions, one can make each field of the microinstruction responsible for specifying a nonoverlapping set of control signals.

Microinstructions are usually placed in a ROM or a programmable logic array (PLA), so one can assign addresses to the microinstructions. The addresses are usually given out sequentially, in the same way that one chooses sequential

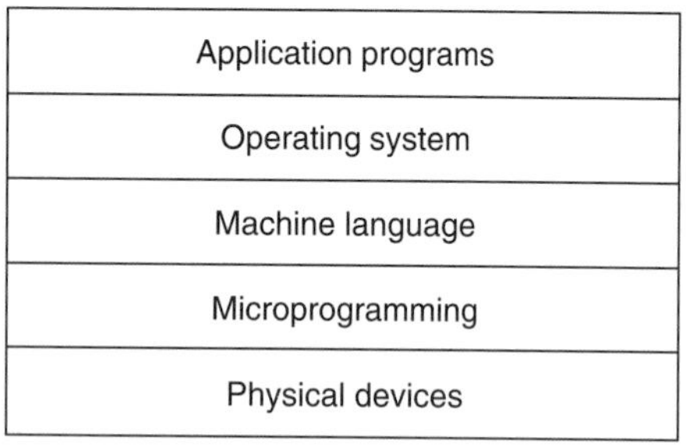

FIGURE 1.1 Basic Hierarchy of Systems Using Microprogramming

numbers for the states in the finite state machine. Three different methods are available to choose the next microinstruction to be executed:

- **Increment the address of the current microinstruction** to obtain the address of the next micro-instruction. This is indicated in the microprogram by putting it in the sequencing field. Because sequential execution is encountered often, many microprogramming systems make this the default and simply leave the entry blank as the default.
- **Branch to the microinstruction** that begins execution of the next instruction. This initial micro-instruction is usually labeled as Fetch and placed in the sequencing field to indicate this action.
- **Choose the next microinstruction** based on control unit input. Choosing the next microinstruction on the basis of some input is called a **dispatch**.

In writing the microprogram, there are two situations in which one may want to leave a field of the microinstruction blank. When a field that controls a functional unit or that causes state to be written (such as the memory field of the ALU destination field) is blank, no control signals should be asserted. When a field only specifies the control of a multiplexor that determines the input to a functional unit, leaving it blank means that one does not care about the input to the functional unit. The easiest way to understand the microprogram is to break in into pieces that deal with each component of instruction execution. The first component of every instruction execution is to fetch the instructions, decode them, and compute both the sequential PC and branch target PC.

1.2 Memory Hierarchy in Computer Systems

Memory is part of the computer used to store information such as instructions and data. Ideally, memory should operate at a speed that is comparable to the CPU so that the CPU can function as close as possible to its maximum speed. Unfortunately, such memory devices are very expensive and only economically feasible in very small systems. For larger systems, the stored information is distributed in a memory hierarchy using different memory devices with various performance and cost. A basic organization of multilevel memory is depicted in Figure 1.2.

As illustrated in Figure 1.2, storage components can be separated by performance into four basic groups: registers, caches, main memory, and secondary memory.

- Registers are small areas of extremely high-speed memory used to store data that will be processed by the CPU. Generally, a register can hold only a few bytes of data, which can be the operands, the results, or the memory addresses that can be accessed in one clock cycle. These registers then form general-purpose register files that can be used by the CPU. In many of today's CPUs, the size of a register file is 32 entries.
- **Caches** are the areas of high-speed memory used to store frequently used data and instructions. Caches allow the CPU to access instructions and data much quicker than if it were to access the data from the main memory. In a nutshell, cache memory works as follows: if the CPU requests a memory access, the memory management unit (MMU) first checks to see if the requested address is in the caches. If the address is in the caches, the data are returned to the CPU (in case of a read request) or updated in the cache (in case of a write request). On the other hand, if the address is not in the cache, the contents of that address (and perhaps its neighbor) must be brought into the cache from the main memory. In today's system, there are often multiple levels of cache memory. For example, Intel's Pentium II uses two levels of cache. In level 1, separate instruction and data caches are used and each with a 16-KB capacity. In level 2, 512 KB of unified

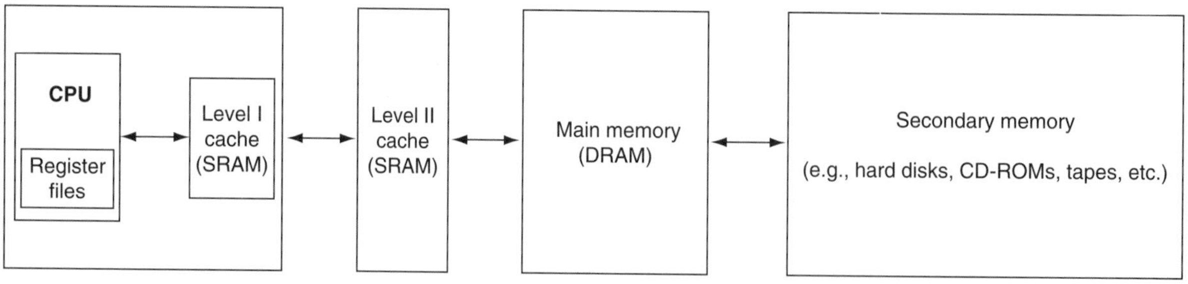

FIGURE 1.2 Basic Organization of Multilevel Memory System

cache is used. Since cache memory is expensive, the cache's capacity is much less than the capacity of main memory (often from 1 to 1000 KB [1 KB is 2^{10} bytes]). On the other hand the common access time of cache memory is only one to three cycles.

- **Main memory** is external memory that is relatively fast and inexpensive. The information stored in the main memory is data and instructions still actively used. The storage capacity of main memory (often from 1 to 1000 MB [1 MB is 2^{20} bytes]) is much larger than that of cache memory. The access time (five or more cycles), however, is also much slower than cache memory.
- **Secondary memory** is much larger in capacity and much slower in access time than main memory. Secondary memory stores information that is not currently used by the CPU. In addition, it also absorbs from the main memory when it is exceedingly used. Examples of secondary memory are a hard disk, a compact disk read-only memory (CD-ROM) disk, a digital video disc (DVD), now called a digital versatile disc, a digital audio tape (DAT), a floppy disk, and a magnetic tape. For hard drives, the typical capacity varies from 0.5 GB to several GB (1 GB is 2^{30} bytes), and the access time is measured in milliseconds. For floppy drives, the capacity varies from one to hundreds of MB (in case of Zip drives and Super drives).

1.2.1 Random-Access Memory

Random-access memory (RAM) allows storage location to be accessed in any order and access time to be fixed regardless of the accessed location. Unlike sequential access memory (e.g., magnetic tape), for which data access can only start from the beginning and the address is searched sequentially, RAM can access any address in a constant time. For example, if the access time of a RAM is stated to be 60 ns, any address location inside that RAM can be accessed within that time. RAM is, therefore, much faster to read from and write to than other storage devices such as hard disks, floppy disks, and magnetic tapes. However, the stored information in RAM is volatile. RAM loses all of its contents as soon as the power to the system is turned off. There are two types of RAM: **static** (SRAM) and **dynamic** (DRAM). It is worth noting that the terms static and dynamic do not refer to mechanical movement in storage device; instead, they refer to the need to refresh the contents. In DRAM, the stored contents have a tendency to decay overtime. Thus, periodic recharging (refreshing) is needed to maintain the content correctness. SRAM, on the other hand, does not require refreshing. A typical refresh rate is 15 ms.

Static RAM

The advantage of **static RAM** SRAM is that it requires no refreshing and often has shorter access time than DRAM. On the other hand, SRAM is more expensive and may require up to four times the size for a given amount of data than DRAM. Moreover, SRAM also consumes more power. SRAM is mainly used in the implementations of level 1 and level 2 caches. Typical capacity of SRAM is from a few kilobytes to 1 MB. In 1998, however, IBM announced the high-performance and high-capacity SRAM that has the capacity of 8 MB and the speed of up to 600 MHz (see Figure 1.3).

FIGURE 1.3 IBM's High-Performance SRAM

Dynamic RAM

Dynamic RAM (DRAM) possesses a characteristic similar to a capacitor; that is, for every read access, a power refresh is required to maintain the data. In addition, a power refresh is also required every 15 ms just to hold the information. Over the years, DRAM has been mainly used to implement the main memory in most computer systems. Some of the most commonly used DRAMs are given in the following list:

- **Enhanced DRAM** (EDRAM) uses combination of SRAM and DRAM. It is mainly used to implement level II cache memory. In this setup, the data are read first from the SRAM. If data are not found there, the data are then read from the DRAM.
- **Fast page mode DRAM** (FPM DRAM) is the most commonly used DRAM for the personal computer from mid-1980s to the early 1990s.
- **Extended data output** (EDO) DRAM is the leading type of DRAM used in mid-1990s. It is faster than FPM DRAM because it allows the CPU to access data while the refresh cycle is being set up. The typical access time for EDO DRAM is 60 ns. As the CPU speed increases beyond 200 MHz, however, the popularity of EDO DRAM gives way to the faster SDRAM.
- **Synchronized DRAM** (SDRAM) is a generic name for any DRAM that is synchronized with the clock speed optimized for the CPU. SDRAM is faster than EDO DRAM because SDRAM chips can synchronize their operations with the processor clock. Moreover, SDRAM also allows new memory access before the preceeding access is completed. The speed of SDRAM is rated in megahertz instead of the traditional nanoseconds because a comparison can easily be made to the system bus speed. A typical speed of the SDRAM is 66 to 125 MHz.

FIGURE 1.4 Typical SDRAM Module for Personal Computers (PCs)

FIGURE 1.5 Illustration of EPROM and EEPROM for Rapid Prototyping. (A) Altera's EPROM-Based Programmable Logic Device (PLD) Family (1988). (B) Altera's EEPROM-Based PLD Family (1992).

- **PC100 SDRAM** is the SDRAM that meets Intel's i440BX specification. The i440BX was designed to use a 100-MHz system bus speed. It is the most commonly used DRAM in personal computer systems (see Figure 1.4).
- **Direct rambus DRAM** (DRDRAM) is a proprietary technology proposed by Rambus in partnership with Intel. It utilizes Rambus DRAM (RDRAM) memory platform. This technology is expected to replace SDRAM as the adopted standard in PCs. In 1999, Rambus reported that its DRDRAM could deliver up to 1.6 GBPS capability. Presently, Rambus DRAM (RDRAM) is used in products ranging from Silicon Graphics workstations to Nintendo-64 video game machines.

1.2.2 Read-Only Memory

Unlike RAM, read-only memory (ROM) cannot have the contents rewritten once they are installed in a system. To read from ROM, the random-addressing method similar to RAM is used. In contrast to RAMs, ROMs have the advantage of being nonvolatile, which means ROMs do not lose their contents when the power to the system is turned off. ROM chips are typically slower than both SRAM and DRAM chips.

One of the first ROM chips was introduced in the late 1960s as masked ROM. This ROM only allowed one-time programming and was used to store unchanged information. Over the years, ROM technology has evolved to include many variations or ROMs. Some of the most commonly used ones are listed in the following list:

- **Erasable programmable ROM** (EPROM) is often used in microcontroller applications. A good example is the basic input output system (BIOS) used to start up the PC. During the start-up process, the EPROM supplies the necessary instructions to the CPU so that the computer's subsystems can be initialized and tested. EPROMs usually have the same organization as the SRAM. EPROMs can be erased by being exposed to ultraviolet light for about 15 min.
- **Electrically erasable programmable ROM** (EEPROM) is user-modifiable ROM that can be erased and reprogrammed repeatedly by applying specific electrical pulses. Unlike for EPROM, an EEPROM chip has to be erased and reprogrammed in its entirety, not selectively. EEPROM also has a limited life span (limited programmability). Moreover, the largest EEPROMs are only 1/64 as large as common EPROMs, and they are only half as fast. At this point, EEPROMs cannot compete with SRAMs and DRAMs because the latter two have much smaller capacity with much less speed and are more expensive. Presently, EEPROMs are mainly used in rapid device prototyping where nonvolatility and easy reprogrammability are crucial. Figure 1.5 illustrates the uses of EPROM and EEPROM technology in high-density programmable logic devices (PLD) for rapid prototyping. Figure 1.5 (A) shows a picture of Altera's MAX 5000 device that is EPROM-based. MAX 5000 family provides logic densities ranging from 600 to 3750 usable gates. This family of devices was launched in 1988. Figure 1.5 (B) shows a picture of Altera's MAX 7000 EEPROM-based device. The MAX 7000 family provides logic densities ranging from 600 to 5000 usable gates. This family of products was launch in 1992.
- **Flash memory** is a more efficient version of EEPROM. Unlike EEPROM, for which the data are byte-erasable and rewritable, flash memory erases and writes data in fixed size block (see Figure 1.6). This effectively improves the performance of flash memory as compared to typical EEPROM. At first, the expectation for flash memory is to replace the hard drive as the adopted mass media storage. This adoption would be a major improvement in mass storage media considering that the access time of flash memory is about 100 ns, whereas the access time of hard disks is in milliseconds. The takeover, however, has not occurred because flash memory is still far too costly to be used as a mass storage device. Presently, flash memory is mainly used in digital photography industry as film for storing digital pictures. The life expectancy of a flash memory is roughly 10,000 erasures.

FIGURE 1.6 Illustration of Flash Memory

1.2.3 Serial-Access Memory

The best way to describe serial-access memory is through an example. Magnetic tapes, hard disk drives, CD-ROM drives, and DVD drives are all serial-access memory devices. To access data on a magnetic tape drive, the tape must be loaded from the beginning. The magnetic tape head is scanned through the tape media until the corresponding location is found and data is accessed. To sum up, the data in serial-access memory must be accessed in a predetermined fashion (in this example, the tape is sequentially scanned from the beginning) using shared read/write devices (e.g., magnetic tape head) for all possible locations on the media. There are three basic mediums to implement serial-access memory devices: **magnetic disk memory**, **magnetic tape memory**, and **optical memory**.

Magnetic Disk Memory

Magnetic disk memory is used to implement hard disks, standard floppy disks, and high-density floppy disks (e.g., Zip drive, Super drive). Hard disks are the most commonly used second memory devices because of their low cost, high speed, and high storage capacity. Hard disk drives are mass storage units that allow read to and write from magnetic media; they consist of one or many thin disks that have magnetic coating, allowing data to be recorded. The recording surface is divided into **concentric tracks**, and each track is divided into segments called **sectors**. The set of tracks at a given radial position is referred to as a **cylinder**. One disk or more are then mount on a spindle and rotate at a constant speed. To access the data, a two-step process is required. First, the read/write head moves across the rotating disk to the locating track. Then the head waits until the right sector is underneath it, and read/write is performed. The descriptions of magnetic disk memory devices are given as follows:

- The **hard disk drive** is the most commonly used mass storage device, as already stated. The size of today's hard drives can vary from 14 inches (used in older mainframe computers) to 1.8 in (used in laptop and portable computers). The most typical size used in a PC is 3.5 in, and the ones used in notebook computers are from 1.8 to 2.5 in. The rotating speed also varies depending on the interface used (discussed more in the bus interface section). For an integrated drive electronics (IDE) interface, the speed varies from 4500 rpm to 7200 rpm. For a small computer systems interface (SCSI), the speed can be as high as 10,800 rpm. The typical capacity varies from one gigabyte to tens of gigabytes (1 GB is 2^{30} bytes).
- The **floppy disk drive**, also known as diskette, is a removable magnetic storage medium that allows recording of data. IBM first introduced it as a 8-in diskette in 1971. In the middle of 1970s, a 5.25 diskette was introduced. Today, the most commonly used floppy disks are 3.5 inches and have the capacity of 800 KB to 2.8 MB (with a standard of 1.44 MB).
- The **high-density floppy disk drive** was first introduced in 1995. High-density floppy disks, while sharing a 3.5-in size with the standard floppy disks, are much faster and have up to one hundred times more capacity than the standard floppy disks. One example is the Zip drive, produced by Iomega. Each Zip disk is capable of storing up to 100 MB of data. Similarly, Imation, a subsidiary of 3 M, also manufactures the Super disk (also known as the LS 120) that can store up to 120 MB of data.
- The **removable hard disk** has been at work in the mainframe computer industry since the 1950s. Back then, the drive mechanism was extremely expensive; therefore, different applications would use different removable disks, during program execution. In the 1980s, removable hard drive was used for backup purposes. The capacity then was 44 MB. Nowadays, removable disks come in various capacities from one gigabyte up to several gigabytes.
- The **redundant array of inexpensive disks** (RAID) was introduced by David Patterson and other researchers at the University of California, Berkeley, in the late 1980s. It is a method where two or more disks are used to store data. Data can be read simultaneously from more than one drive, which improves the performance. Data can also be split among all drives in bits, bytes, or blocks. Typically two or more disks are connected together. A single controller can be used to connect the drives so that they function together as one drive. For extra safety, a second interface controller can be installed to duplex the drives and increase read performance. The major advantages of RAID are improvement in reliability and protection for data in mass storage systems.

Magnetic Tape Memory

The **magnetic-tape memory device** is one of the oldest and cheapest secondary memories. In spite of the popularity of RAM, CD-ROM, and removable hard drives; tape is still used because of its low cost and reliability. In today's systems, tape is mainly used to provide backup data in case of hardware failure to the hard disk system. The major tape formats available for PCs and servers are as follows:

- Quarter-inch cartridges (QIC) are packaged in two formats, 3.5-in mini-cartridges and 5.25-in data-cartridges. Due to its low capacity, QIC is used mainly to back up PCs and small servers.
- Digital linear tape (DLT) was originally developed by Digital Equipment Corporation (DEC) to be used in its midsize systems. Presently, DLT is used to provide backup for midsize local area network (LAN). The current transfer rate of DLT is a 800-KBps backup to tape cartridges that hold up to 40 GB of uncompressed data.
- Digital audio tape (DAT) of 4 mm was originally intended for the recording industry. However, it never caught on, and it was adapted for data storage. It is mainly used to back-up data at work-group level. The transfer rate of DAT is about 1.5 Mbps, and the capacity varies from 4 GB to 24 GB.
- The 8-mm tape was the first high-capacity tape to be widely used in the PC market. The current transfer rate is about 6 Mbps, and the capacity is 20 GB of uncompressed data.

Optical Memory

The popularity of the **optical disk** memory as mass storage media has rapidly grown over the past few years. Optical mediums have much higher recording density than their conventional magnetic counterparts. In 1980, Phillips in partnership with Sony developed the compact disc (CD) that quickly replaced the long-playing record. CDs can be prepared by using high-power infrared laser to burn holes (pits) in a master disk. Once a master disk is made, the successive copies can be made very inexpensively (about 2000 copies for $1.00). To read the disk low-power infrared light is shined through the disk. Since the surface of the CD consists of pits and smooth surfaces (lands), the variations of reflected light are used to translate data into a digital signal. Unlike the hard drive in which the disk is rotated at a constant angular velocity (e.g., 5400 rpm), the CD drive needs to achieve constant linear velocity of 120 cm/sec. This means that the angular velocity when reading on the outside of the disk is 330 rpm, and the velocity when reading the inside is 530 rpm. In 1984, the standard for using CDs to store computer data was precisely defined, and the media was referred to as compact disc read-only memory (CD-ROM). The standard states that CD-ROM is to have physical appearance of CDs. Moreover, it must be optically and mechanically compatible with CDs, which means the manufacturing techniques must be compatible. Since then, there have been many advances in CD-ROM technology, such as creation of the CD-recordable (CD-R), the CD-rewritable (CD-RW), the digital versatile disc (DVD), and the DVD-RAM.

- **CD-ROMs** are the most widely used optical storage. With the capacity of 650 MB, CD-ROMs are ideal for distribution of text, images, and programs in an electronically readable form. Since the mid-1990s, CD-ROMs have become the standard on PC systems.
- **CD-recordable** (CD-R) disks at first glance, appear to be similar to regular CD-ROMs except CD-Rs are either gold or green instead of silver. These green and yellow colors are dye, and they are used to simulate lands and pits. In the initial state, the dye is transparent and allows the laser light to pass through and reflect the inner layer (made out of gold). During the writing stage, the laser power is increased to heat up the dye, which results in a dark spot. During the reading stage, the differences between a dark spot and a transparent spot are interpreted as the differences between pits and lands. Kodak is one of the first manufacturers to produce CD-R disks. One of the first uses of CD-Rs was for Kodak's PhotoCD. CD-Rs are being used for backing up hard disks, and they also make it possible for individuals and small companies to manufacture a small number of CD-ROMs. Unfortunately, this technology also allows many individuals and companies to duplicate CD-ROMs and CDs without any regard for copyright violations.
- **CD-rewritable** (CD-RW) technology allows individuals and companies to write, erase, and rewrite CD-ROMs. Unlike CD-R, CD-RW uses a reflecting alloy that possesses two stable states, crystalline and amorphous. These two states reflect light differently. Inside CD-RW drive, three levels of laser power can be used. At high power, the alloy melts and loses some of its reflectivity (amorphous state). This in effect simulates pits. At medium power, the alloy melts again, but this time returns to its crystalline state and regains its reflectivity (returns lands). At low power, the disk is read but there is no state change. Because CD-RWs still cost considerably more than CD-Rs, they are not as widely used as CD-Rs.
- **Digital versatile disc-ROM** (DVD-ROM) technology is similar to CD technology with only three exceptions. First, the pit size is half of CD's pit size (0.4 mc versus 0.8 mc). Second, the spiral grove for recording is 54% tighter. Last, the laser beam is 17% smaller. With these refinements, the capacity is improved 7-fold. A typical CD-ROM can store 650 MB of data, whereas a DVD-ROM can store 4.7 GB. This translates to the ability to hold up to 133 minutes of high-resolution video, with soundtracks for eight languages and subtitles for additional thirty-two languages. The packaging of DVD-ROM comes in four different formats. The first is single-sided with a single layer, which translates to 4.7 GB. The second is single-sided with a dual layer, which translates to 8.5 GB of storage capacity. The third is double-sided with a single layer, which has the capacity

of 9.4 GB. Finally, a double-sided formate with a dual layer has the capacity of 17 GB.

- **DVD-recordable** (DVD-R) is a write-once system. The original capacity of a single-sided DVD-R disk that has a single-layer was 3.95 GB, which is slightly lower than a single-sided, single-layer DVD-ROM. The capacity has increased to 4.7 GB.
- **DVD-rewritable** (DVD-RAM) is new type of rewritable DVD that provides much greater data storage than today's CD-RW systems. A single-sided, single-layer DVD-RAM has a capacity of 2.6 GB, and a single-sided, single-layer DVD-RAM has a capacity of 5.2 GB. This allows an hour of MPEG-2 video, which is not very practical for home entertainment use when videotapes can have a capacity of up to 5 hr (standard play). This DVD-RAM specification is therefore aimed at the computer market where it can be used to store large amounts of data.

1.3 Bus and Interface

Bus is a set of conductors (wires, PCB tracks, or connections in an integrated circuit) connecting various functional units in a computer. Busses can be either inside or outside the CPU. The outside busses are used to connect the CPU to external memory and peripheral devices. The width of a bus (i.e., the number of parallel connectors) determines the size in bits of the largest data item that it can carry. The bus width and the number of data items per second that it can transmit are factors limiting a computer's performance. Most current microprocessors have 32-bit or 64-bit busses. Some processors have internal busses, which are wider than their external busses (usually twice the width). The width of the internal bus, therefore, affects the speed of all operations and has less effect on the overall system cost than the width of the external bus.

On the other hand, an interface is a communication channel for two or more systems. It can be a hardware connector linking devices or a convention allowing communication between two software systems. There are often some intermediate components between the two systems that connect the interfaces together. For example, two RS-232 interfaces are connected via a serial cable.

When a central processing unit (CPU) sends data to a serial device (receiver) through a RS-232 interface, the data will go through internal busses, external busses, a RS-232 interface, a serial cable, a receiver's RS-232 interface, a receiver's busses, and a receiver's CPU. Normally, a component (e.g., interface card) that is used to transfer signals from busses to some protocol can be recognized by other components or peripheral devices. The interface card will be plugged into an expansion bus slot of a motherboard. Normally, a motherboard has several expansion bus slots allowing components such as a video card, a disk controller, a modem, or a parallel-port card to communicate with the computer. There are three common types of busses: industry standard architecture (ISA), peripheral component interconnect (PCI) local bus, and an accelerated graphics port (AGP). Each (ISA) has its own type of expansion slot and signal definitions. They will be described in the following subsections.

1.3.1 Bus Overview

Bus technology is well-developed in computer systems such as PCs. Most of the busses described here are common in a typical PC-based system. However, some busses such as PCL are also used in Sun Microcomputer systems. In 1982, the first-generation PCs used ISA busses. The ISA bus was an 8-bit bus running at 4.77 MHz on PC XT. Although it was extended to a 16-bit bus with the speed of 8 MHz on PC AT, the low speed led to extended industry standard architecture bus (EISA) design. At the same time, micro channel architecture (MCA) bus was added by IBM to provide users with choices. Newer Intel486TM and Pentium processor-based machines offer ISA/PCI or EISA/PCI. With most vendors now shipping PCI-based machines across their complete product line, it is clear that PCI will be the dominant bus for the future. Furthermore, the transition of corporate PC purchases to Pentium processor-based systems hastens the move to PCI for the adapter interface. This move allows for better response time for PCs, which eliminates the bottlenecks for these high-powered workstations.

ISA expansion slots have existed in PCs for years. Most of the interfaces depending on them such as parallel ports and serial ports, however, have been integrated into motherboards. Moreover, in the PC 99 System Design Guide advocated by Intel and Microsoft, ISA slots have been removed because of their configuration issues.

Terminology

Before introducing bus specification, some terminology needs to be defined so that readers can have better understanding of bus specifications.

- **Auto configurable**: An interface card allows software to identify its requirements and resolve any potential resource conflicts such as IRQ, DMA, I/O address, and BIOS.
- **Bus master support**: This bus master is capable of first-party DMA transfers.
- **Full bus master capability**: This capability allows support of any first-party cycle from any device, including another CPU.
- **Good bus arbitration**: This feature allows fair bus access during conflicts, which means there is no need to back off unless another device needs the bus. This arbitration prevents CPU starvation while allowing a single device to use 100% of the available bandwidth. Other busses let a card use the bus until it decides to release it and attempts

to prevent starvation by having an active card voluntarily release the bus periodically ("bus on time"). This device is to remain off the bus for a period of time ("bus off-time") to give other devices, including the CPU, a chance even if they do not want it.

- **16-Meg addressable**: This limits first-party DMA transfers to the lower 16 Megs of address space. There are various software methods to overcome this problem when more than 16 Megs of main memory are available. This has no effect on the ability of the processor to reach all of the main memory.
- **Backward-compatible with ISA**: This feature allows an ISA card to be placed in the slot of a more advanced bus. It is worth noting that the ISA card does not gain any benefit from being in an advanced slot. Other slots are unaffected.
- **Burst**: This allows the transfer of multiple data items continuously in response to a single request.

Industry Standard Architecture

The **industry standard architecture** (ISA) bus was originally designed for the IBM PC XT with an 8088 (at 4.77 MHz) processor. It is limited to 8 data lines and 20 address lines (1 MB of memory addressing). The PC/AT 80286-based systems maintain the 8-bit PC compatibility and add a 16-bit connector using 16 data lines and 24 address lines. Some features of ISA busses are the following:

- Operating speed of 8 MHz to 8.33 MHz asynchronously
- Maximum throughput of 8 MB ps
- Two CPU cycles used for every read or write
- Bus master support
- Edge-triggered TTL interrupts (IRQs)—no sharing
- Communication with both 8- and 16-bit devices
- DMA capable
- Bus group signals
- 24 address lines
- Low cost

ISA buses are ideal for low- to mid-bandwidth cards, though lack of IRQs can quickly become annoying.

Micro Channel Architecture Bus

The **micro channel architecture** (MCA) **bus** was proprietary and bounded mainly in IBM personal computers. MCA has a better performance than ISA. The data lines and address lines are increased, and the burst data transfer rate can be up to 80 MB/sec. Some of its features are the following:

- Intelligent bus master expansion card support
- 16- or 32-bit data transfers and a 32-bit address bus (4 GB of memory space) support
- CPU clock uncoupled from 10-MHz bus (32-bit data streaming)
- 80 M/s burst, synchronous
- Fully bus master capability
- Good bus arbitration
- Autoconfigurable
- IBM proprietary (not ISA/EISA/VLB compatible)

Extended Industry Standard Architecture Bus

Because MCA was proprietary, the extended industry standard architecture (EISA) was formed to compete with it. EISA gained much more acceptance; MCA has all but disappeared. EISA is an improved version of ISA. It is great for high-bandwidth bus mastering cards, such as SCSI host adapters, but its high cost limits its usefulness for other types of cards. Some of its features are the following:

- Support for intelligent bus master expansion cards
- Support of 8-, 16-, or 32-bit data transfers
- 8–8.33 MHz, synchronous
- Backward-compatible 32-bit address bus (4 GB of memory space)
- Synchronous data transfer protocol
- Cycles synched between CPU, bus master, and DMA
- 33 MB ps data transfer for bursting bus master or DMA
- Automatic configuration of board and expansion cards
- Full bus master capability
- Good bus arbitration
- Sharable IRQs, DMA channels
- Backward compatible with ISA
- Some acceptance outside of the PC architecture
- High cost

Video Electronics Standards Association Local Bus

The **video electronics standards association** (VESA) **local bus** (VLB). VLB is great for video cards, but its lack of a good bus arbiter limits its usefulness for bus mastering cards, and its moderate cost limits its usefulness for low- to mid-bandwidth cards. Since it can coexist with EISA/ISA, a combination of all three types of cards usually works best. Moreover, the VLB has limited electrical integrity at higher speeds. The following list includes some features:

- 32 data bits, 32 address bits
- 25–40 MHz, asynchronous
- Capable of limited bursting (up to 132 MB/sec) in the 386/486 environment
- Bus master capability
- Coexistence with ISA/EISA
- Slot limited to two or three cards typically
- Backward-compatible with ISA
- Moderate cost
- Originally designed to maximize throughput for video graphics
- Memory that adds a local bus for faster access to the processor
- Two types of VL bus

- Type A: no buffering
- Type B: buffering; supports up to three add-in connectors
- Revision 2 adds bus mastering support

Peripheral Component Interconnect Local Bus

The **peripheral component interconnect** (PCI) **local bus** is the newest bus standard accepted by all computer systems such as PC-based systems, Apple's Power Macintosh computers and Workgroup servers, Sun workstations, and PowerPC processor-based computers from IBM and Motorola. The PCI has a high-performance expansion bus architecture that was originally developed by Intel to replace the traditional ISA and EISA busses found in many 80 × 86-based PCs. PCI combines the speed of VLB with the advanced arbitration of EISA. Great for both video cards and bus mastering SCSI/network cards. Some of its features include these:

- 32 data bits (64 bit option), 32 address bits (64-bit option)
- Up to 33 MHz, synchronous
- 132 M/s burst (sustained) (264 M/s with 64-bit option)
- Full bus master capability
- Good bus arbitration
- Slot limited to three or four cards typically
- Autoconfigurable
- Coexistence with ISA/EISA/MCA as well as another PCI bus
- Strong acceptance outside of the PC architecture
- Moderate cost
- Voltage: 3.3 V and 5 V

Accelerated Graphics Port

Because of the large amount of 3-D image processing data, PCI busses limit an animation to be life-like in realism and depth. **Accelerated graphics port** (AGP) technology was invented to accelerate graphics performance by providing a dedicated high-speed port (independent to PCI) for transferring 3-D texture data from system memory to video memory in a graphics controller.

- 32 data bits, 32 address bits (extra 8 lines for sideband addressing)
- Up to 133 MHz
- Up to 1 GBps peak (528 MBps at 66 MHz)
- Slot limited to one card
- Autoconfigurable
- Coexistence with ISA/EISA/MCA as well as another PCI bus
- Voltage: 3.3 V and 1.5 V

1.3.2 Interfaces

Computers communicate with peripherals or each other by using **I/O chips** as interfaces. These chips include universal asynchronous receiver transmitters (UARTs), and parallel input/outputs (PIOs). A video graphics adapter (VGA) interface is used to send signals to a monitor; a parallel port is used to communicate with a printer; and a serial port is used to communicate with a mouse, and so on. First, an interface takes signals from busses according to an *I/O* addresses. For example, a UART chip can read a byte from the data bus and output it in a serial format with parity check. It can be configured as various transmission speeds from 1200 bps to 115,200 bps. Different interfaces have their own specifications and transmission protocols. The following subsections give a brief introduction.

Serial Port and Parallel Port

In general, there are two types of communication ports in a computer: **serial ports** and **parallel ports**. A serial port is an interface through which peripherals can be connected as a communication channel using a serial (bit-stream) protocol. Usually, a serial mouse is connected to COMI (communication port 1). The most common type of serial port is a 25-pin D-type connector carrying RS-232 signals. Smaller connectors (e.g., 9-pin D-type) carrying a subset of RS-232 are often used on personal computers. The serial port is usually connected to an integrated circuit called a UART that handles the conversion between serial and parallel data.

A parallel port is an interface for a computer to communicate peripherals in parallel manner. Data are transferred in or out in parallel, that is, on more than one wire. A parallel port carries 1 bit on each wire, thus multiplying the transfer rate obtainable over a single wire. There will usually be some control signals on the port that indicate when data are ready to be sent or received. The most common type of parallel port is a printer port (e.g., a Centronics port that transfers 8 bits at a time). Disks are also connected via special parallel ports (e.g., SCSI or IDE).

PS/2 Port

A type of port developed by IBM for connecting a mouse or keyboard to a PC is the **PS/2 port**. This port supports a mini-DIN plug containing just six pins. Most PCs have a PS/2 port so that the serial port can be used by another device, such as a modem. The PS/2 port is often called the mouse port. Most laptops have one PS/2 port that is used to hook up a full-sized mouse or keyboard to a laptop. To hook up both a mouse and a keyboard, a PS/2 keyboard and a serial-type mouse are required. The mouse will interface through the laptop's standard serial port.

Integrated Drive Electronics (IDE)

In mid-1980's, the disk controller for a hard drive had integrated with the drive itself. The **integrated drive electronics** (IDE), also called ATA (AT attachment), interface hard drive appeared. However, IDE hard drive has a maximum size limitation (528 MB) due to basic input output system (BIOS) conventions (16 heads, 63 sectors, and 1024 cylinders).

Extended IDE (EIDE) also called ATA-2, drives were later introduced to solve the problem by supporting a second addressing scheme called logical block addressing (LBA) mode. It allows numbering sectors from 0 to 2^24–1 and, consequently, gets beyond the 528-MB limit. The IDE can also control four drives instead of two, has a higher transfer rate, and has the ability to control CD-ROM drives.

AT Attachment Packet Interface (ATAPI)

The enhanced IDE standard functions as an interface for hard drives and PCs. It would be great, however, if other devices such as tape drives and CD-ROMs could be controlled through the same controller. The AT attachment packet interface (ATAPI) is an extension to EIDE that enables the interface to support CD-ROM players and tape drives. Using ATAPI-capable hardware and software drivers, mixing and matching different types of drives on the same EIDE controller, are possible which are both a convenience and savings in hardware costs.

Small Computer System Interface

Small computer system interface (SCSI), standardized in 1986, is a flexible *I/O* bus that has been used in a number of peripherals such as disk drives, CD-ROM drives, tape drives, and scanners. SCSI drives have much higher transfer rates than IDE drives. A SCSI-controlling card plugged into a system expansion slot serves as a host to other SCSI devices. At most, 7 SCSI devices can be linked together to a host (15 for wide SCSI).

There are several versions of SCSI standards: SCSI-1 standard supports transfer rates of up to 3 MBPs, SCSI-2 supports up to 20 MBPs, and SCSI-3 supports up to 40 MBPs. In a chain of SCSI devices, any one can be either an initiator or a target. Normally, the SCSI controller, acting as an initiator, issues commands to other SCSI devices acting as targets. The commands descriptor block (CDB) can be up to 16 bytes and tells the target what to do. These CDB commands are also used to arbitrate the bus when more than two SCSI devices try to use the bus as the same time. This arbitration mechanism improves performance by allowing more than two SCSI devices to be activated at the same time, whereas only one IDE device can be activated at one time. The following are lists of characteristics for different SCSIs:

Major SCSI-1 Characteristics

- Very short bus with only two SCSI connectors or devices
- Single initiator and single target
- Transfer rates of 3 MBPs or less
- No arbitration, no parity, no disconnect
- Single-ended interface
- 6-byte command descriptor block (CDBs)
- 5-MHz bus frequency (transfer rate)

Major SCSI-2 Characteristics

- Higher performance than SCSI-1
- Transfer rates increased to 10 MHz, synchronous mode, routinely using 5 MHz
- Lower overhead, often below 30%
- Single-ended and differential interfaces
- SCSI bus up to 4 bytes wide (32-bit); main use of the 2-byte wide bus, achieving 20-MBPs burst transfer rates
- Diversity of peripheral device types (10 types)
- Greatly improved compatibility
- Improved functionality (e.g., command queuing, disconnect, etc.)
- Improved reliability (e.g., arbitration, parity, error reporting and classification, new commands)

Major SCSI-3 Characteristics

- Much higher performance in respect to burst transfer rates (greater than 40 MBps) and lower overhead
- Distance in kilometers rather than just meters
- Expanded addressability, allowing an almost unlimited number of devices on the same bus
- Universality use of the SCSI system expanded even more
- Higher reliability and robustness of systems
- 20-MHz bus frequency (transfer rate)

Universal Serial Bus

The **universal serial bus** (USB) standard was originally developed by Compaq, Digital Equipment, IBM, Intel, Microsoft, NEC, and Northern Telecom. USB is a two-way serial channel with supports of hot-swapping and a 12-Mbps data transmission rate. It allows up to 127 devices to be connected into a PC through a multilevel tiered star topology using USB hubs. Furthermore, it adopts standard connectors and eliminates add-in interface cards, which results in lower costs for developers and consumers. Features include:

- 12 Mbps (1.5 Mbps for low-speed devices, such as keyboards)
- Up to 127 devices that can be hooked up together
- Cable length less than 5 ms
- Two-way communication channel
- Hot-swapping

IEEE 1394

The **IEEE 1394** is a high-performance serial bus. App developed the bus in early 1990s and called it Fire Wire bus. Sony also markets a similar product under the name i.Link, which is one of many 1394-compliant devices. Because of USB's low speed, it does not provide support for high-speed peripherals, such as digital camcorders and high performance printers. IEEE 1394 can support these high-speed peripherals. IEEE 1394 is the only bus structure that can be used as an internal bus and an external bus in a computer system.

Table 1.1 Illustration of Interface Performance

Bus type	Maximum speed (MBps)
Parallel port (unidirectional)	0.0050–0.015
Serial port (16550 UART)	0.0144
Parallel port (bidirectional)	0.01–0.0375
Serial port (16550/16750 UART)	0.02875–0.0575
USB	12
Parallel port (EPP)	2
SCSI	Regular, 5: wide, 10
ISA (16-bit)	8
SCSI (Fast)	Regular, 10; wide, 20
SCSI (Ultra)	Regular, 20; wide, 40
IEEE 1394a	400
USB 2.0	480
SCSI (Ultra-2)	Regular, 40: wide 80
IEEE 1394b	800
SCSI (Ultra-2)	Regular, 40: wide 80
IEEE 1394a	100
PCI (32-bit)	133
PCI (64-bit)	267
AGP	267
IEEE 1394b	200–400
AGP 2X	533
AGP 4X	1067

The 1394 bus is faster than the USB and most versions of SCSI (it will approach or exceed the speed of PCI in the future). Furthermore, the IEEE 1394 specification is designed to replace the SCSI buses and PCI buses. Features are as follows:

- Up to 400 Mbps (800 Mbps for 1394a and 1394b)
- Up to 63 devices per segment
- Cable length less than 4.5 meters

Comparisons

Table 1.1 illustrates data transfer rates ranking from slowest to fastest.

1.4 Input/Output

Input/output (*I/O*) devices are sometimes called peripheral. They are the parts that link outside world to a computer system. Basically, *I/O* can be divided into three categories: input devices, output devices, and *I/O* devices. The following subsections show a list of input devices, output devices, and *I/O* devices.

1.4.1 Input Devices

- A **bar code reader** is an optical scanning device, usually used to read data of a product such as price tags. Two main technologies are used in reading the tags: charge-coupled devices (CCD) and laser beams. Laser technology offers the advantages of high speed and longer focal lengths, while CCD scanners tend to be more durable since they have no movable parts. CCD scanners are widely used, whereas laser technology is catching up.
- A **digital camera** is similar to a camera except that the image taken is stored in memory instead of film. These image data will usually be uploaded into a computer to be stored in data format.
- **Digitizers/tablets** are drawing tool other than a mouse or a scanner to ease the work of drawing. Whatever is drawn onto the digitizer pad is recorded and displayed onto the video display.
- A **joystick** is usually used to control movement of a cursor or other graphic element for video games and computer graphics.
- A **keyboard** is one of the basic components in a computer system. By definition, it is a set of keys for a user to input data into a computer. American Standard Code for Information Interchange (ASCII) is an encoding system for converting keyboard characters and instructions into the binary number code that the computer understands.
- A **keypad** is a section of a keyboard. It is set up like a calculator to allow users to enter numbers and equations quickly into the computer. Sometimes a keypad is used in embedded systems as the primary input instead of a keyboard to reduce manufacturing cost.
- A **microphone** is a device capable of transforming sound waves into changes in electric currents or voltage. The electric signal is usually interpreted by the computer system through the input port of a sound card.
- A **mouse** is a small handheld device used to control the position of the cursor on the video display. This is the second most important input device other than a keyboard. The movements of a mouse on a desktop correspond to the movements of the cursor on the screen. The most common technology used nowadays is a mechanical mouse due to its reliability and production cost. Other technologies are optical and optomechanical. A mechanical mouse uses a rubber ball to track movements. An optical mouse, on the other hand, uses LED and photo detector to detect changes in gridlines an on the mouse pads an optomechanical mouse combines rubber balls with optical detection that detects rubber ball movements instead of mouse movements. Different manufacturers come up with different designs for a mouse, such as IntelliMouse by Microsoft and NoHands (foot mouse) by Hunter Digital.
- **Scanner** is an electronic device that uses light-sensing equipment to scan paper images (e.g., text, photos, and illustrations) and translate the images into data that the computer can then store, modify, or distribute. Some common data formats are PDF, GIF IPG, TIF and BMP. The most common software interface is called TWAIN. Although TWAIN stands for nothing, some prefer to call it "toolkit without an interesting name". Any software can access the scanner through a TWAIN interface. Software

that supports TWAIN usually has a command option called "Acquire" under their options. Scanners can combine with optical character recognition (OCR) software to convert scanned graphic into text format

- A **touch screen panel** is an input device commonly used in industrial or commercial systems rather than in a personal computer. The information will be directly input into the system by touching the video display. Some of the common technologies used are infrared, resistance, and guided wave.
- A **frackball** is an input device that controls the position of the cursor on the screen, similar to a mouse. The unit is mounted near a keyboard, and moving the ball controls the movement. Most laptop computers come with a trackball because of its small size.
- A **video camera/camcorder** is similar to a video recorder except that the video will be stored in memory instead of film. The video will then be uploaded into a computer.

1.4.2 Output Device

- A **monitor/display** is a video display terminal. It is the most basic output device in a typical computer system. Sometimes this can be an input device if it is modified into a touch screen monitor. Two other common displays are cathode ray tube (CRT) and liquid crystal display (LCD). Other rate technologies are light emitting diode (LED), gas plasma, and other image projection techniques. Most displays use analog signals, thus, they need the video adapter to store them in VRAM, then convert the digital signals to analog signals through DAC. Nowadays, a terminal can display up to 16,777,216 colors or 24-bit long colors. Human beings cannot differentiate this much color, but since it is convenient for programmers to program it in RGB, 1 byte each, it is widely used. For example, colors in HTML are usually expressed in #03F60A using RGB. The sharpness of the display is determined using dots-per-inch (dpi). This depends on the resolution and physical screen size. The screen size's ratio of width to height (aspect ratio) is usually 4:3.
- A **plotter** is a mechanical drafting device similar but more powerful than printers. It incorporates a moving pen whose horizontal and vertical range in two dimensions is limited only by the size of the bed of the device. It is usually used for floor plans, maps, and blueprints.
- A **printer** is a mechanical device for printing the computer's output on paper. The three major types of printers are: (1) dot matrix, in which individual letters are composed of a series of tiny ink dots formed by punching a ribbon with the ends of tiny wires; (2) ink jet, which sprays tiny droplets of ink particles onto paper; and (3) laser, which uses a beam of light to reproduce the image of each page before dry toner is applied to the image and transferred to paper. The laser printer is the most popular. The dot matrix is practically obsolete. The resolution of a printer is measured in dots-per-inch. The common resolution of a laser printer ranges from 600 dpi to 1400 dpi.
- A **projector** for a computer system is the same as a traditional slide projector except that it receives input from a computer instead of a slide.
- **Computer speakers** are similar to normal audio speakers except that they obtain the input from a computer sound card instead of from conventional audio devices.

1.4.3 Input and Output Device

- **Disk drives** include CD-ROM drives, DVD-ROM drives, Zipped drives, floppy-disk drives, and hard drives. These drives are used to read and write information. (Refer to Memory Hierarchy in Computer Systems (Section 1.2) for further detail.)
- A **modem** is a device that connects two computers together over a telephone line by converting the computer's data into an audio signal and vice versa. Modems operate at data rates from 28,800 bps to 57600 bps. Most of them are full duplex although some networks might still need to support half-duplex modems. There is also another device known as cable modem that converts the digital signal into analog, which can be transmitted through coaxial cable.
- **LAN** can be made up of several technologies, such as integrated services digital network (ISDN), fiber optics, twisted pair, and cable.
- **VCRs** can be used as input or output for a computer system through the use of interface cards.
- **Fax machines** are commonly used in offices to send and receive a fax. A fax can be stored in electronic format in a computer.
- **Audio equipment** can be easily connected to a computer system through an interface such as a sound card. Popular audio equipment is MIDI keyboard. MIDI stands for musical instrument digital interface. It stores the information of how the music should be played back instead of all the information about the music, with the help of a wave table in a sound card. Since MIDI skips the storage of actual sound itself, it saves a lot of memory space. However, MIDI might not be accurate without a good wavetable. A wavetable with 32 voices means that the sound card stores 32 voices in its ROM, and MIDI can just call the voices without recording the voices again.

2
Multiprocessors

Peter Y. K. Cheung,
George A. Constantinides,
and Wayne Luk
Department of Electrical and Electronic Engineering, Imperial College of Science, Technology, and Medicine, London, UK

2.1 Introduction

As both the cost and performance of microprocessors continue to improve, it becomes increasingly attractive to build computer systems containing many processors. Performance improvement is one obvious benefit of a multiprocessor system. In the ideal case, a system with N processors can provide N times speedup of compute-bound tasks. Given that the cost of a microprocessor is a small fraction of the total system cost, the cost effectiveness of such an approach is obvious. A different but equally important reason for using multiprocessor systems is for better reliability. The idea here is that even if a single processor fails, the system would continue to work, though at a slower pace. The workload of the failed processor would be automatically taken up by the remaining processors.

Nowadays, multiprocessor systems can be found in many applications. Most file servers and World Wide Web servers are built with machines that can take two or more processors. The data processing industry that requires transaction processing with large databases use, multiprocessor systems as a standard.

There are two ways of looking at how a multiprocessor system improves performance: (1) increased throughput for **independent** tasks distributed among a number of processors and (2) faster execution of a single task on multiple processors. The organization and design of the system differs significantly for the two different goals; some system architectures are discussed in Section 2.2. The programming of the system also depends highly on the architecture and desired functionality, as discussed in Section 2.4.

One common way to categorize all computer systems was originally proposed by Flynn (1972) as shown in the following list:

1. Single instruction, single data (SISD) machines refer to a single processor executing a single instruction stream that operates on data stored in a single memory. All uniprocessor systems belong to this category.
2. Single instruction, multiple data (SIMD) machines allow a single instruction stream to control many processing elements in a lockstep fashion. Each processor has its own data memory so that during each instruction step, many sets of data are processed simultaneously. Both vector processors, such as the CRAY T80, and array processors, such as the Connection Machine (Hillis, 1986), are SIMD machines.
3. Multiple instruction, single data (MISD) machines enable a common sequence of data to be sent to multiple

processors that each operate on data with a different instruction. This type of machine structure is not very useful and has never been implemented in real systems.

4. Multiple instruction, multiple data (MIMD) machines refer to a number of processors executing different instructions on different data streams simultaneously. This is the most common multiprocessor structure; the symmetrical multiprocessor (SMP), clusters, and nonuniform memory access (NUMA) systems are all examples of this category.

In the following sections, we examine the different organization used in multiprocessors. The key issues addressed include:

- How do multiple processors share data?
- How do cache memories on different processors maintain their data correctly?
- How do multiple processors communicate and coordinate with each other?
- How many processors are needed, and what performance improvement can be expected?

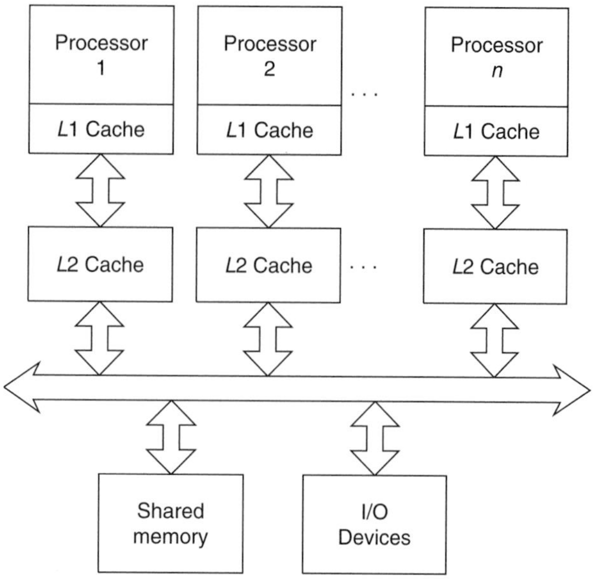

FIGURE 2.1 A Typical SMP Machine

2.2 Architecture of Multiprocessor Systems

2.2.1 Symmetric Multiprocessors

The **symmetric multiprocessor** (SMP) is the most popular form of multiprocessor system available, ranging from low-cost file servers with only two processors to high-performance graphics systems, such as Silicon Graphics's Power Challenge that contains up to 36 processors. In most SMP machines, all processors are connected in a shared backplane. The characteristics of an SMP are the following (Patterson and Hennessy, 1998):

- Two or more similar (or often identical) processors are employed in a stand-alone system.
- All processors share the same memory and *I/O* devices via one or more shared busses with similar access time.
- All processors are capable of performing the same functions.
- The distribution of workload between processors is performed by the operating system in such a way that multiple independent or dependent tasks are shared between processors without any special consideration in the application program.

The distribution workload characteristic is one of the reasons why SMP is widely adopted in commerical products. Most modern operating systems, such as MS-WindowsNT, Linux, and Solaris, support SMP machines. Users no longer need to learn special parallel programming skills to exploit the performance of SMP. Users, however, can also choose to write multithreaded applications to exploit the parallel capability of SMP in a single task.

Figure 2.1 shows a block diagram of a typical SMP machine as would be found in a Pentium-based file server. The common time-shared bus provides the simplest and most cost-effective way to interconnect the processors together. It also, however, limits the number of processors that can be used. Since all memory and *I/O* references pass through the shared bus, the performance of the system is limited by the maximum bandwidth of this bus. One way to alleviate the shared bus bottleneck is the extensive use of cache memory. Most SMP systems have at least two levels of cache memory as shown in Figure 2.1. Typically level 1 cache is on the same chip as the processor, and level 2 cache may or may not be internal to the processor.

Using cache memory in a multiprocessor system introduces the problem of cache coherency: how to guarantee that when a data in a specific cache are altered, this change is reflected in other caches and in main memory. The problem of cache cherency is examined in detail in Section 2.3. Table 2.1 summarizes the key specifications of some SMP systems available in 1999.

2.2.2 Cache Coherent Nonuniform Memory Access

The **cache coherent nonuniform memory access** (CC-NUMA) paradigm, as employed in the Sequent NUMA-Q (Lovett and Clapp, 1996), for example, is a relatively recent idea compared to SMP. CC-NUMA systems strike a balance between the tightly coupled SMP systems and more loosely coupled clusters of communicating computers. The idea behind CC-NUMA systems is that while all processors have access to all portions of the main memory, the memory access time is allowed to vary depending on the part of main memory that is being

Table 2.1 SMP System Specifications

	SGI 1400	SUN Enterprise 10000	HP-V2600
Number of processors	1–4	4–64	2–32
Processors	Pentium III	UltraSPARC	PA-RISC
Clock speed	500 MHz	400 MHz	552 MHz
Memory (max)	4 GB	64 GB	128 GB
Memory bandwidth	800 MB ps	10 GB ps	61.44 GB ps

	IBM RS/6000-SP	Compaq GS60E
Number of processors	2–8	1–8
Processor	Power PC 604e	Alpha 21264a
Clock speed	222 MHz	700 MHz
Memory (max)	1 GB	20 GB
Memory bandwidth	400 MB ps	—

accessed. This allows portions of main memory to be distributed between groups of processors, each group typically arranged in an SMP style or as a uniprocessor. These groups must be able to communicate across a communications network to access nonlocal memory.

The advantage of CC-NUMA over a standard SMP design is that there are several local busses, so each one need not carry all transactions occurring throughout the system. This in turn means that a larger number of processors can be used than in the equivalent SMP design, allowing a greater level of parallelism. The Sequent NUMA-Q system can support over two hundred Pentium II processors. CC-NUMA has a clear advantage over clusters of communicating computers—because the nonlocal memory accesses and cache coherence is maintained automatically, it is a much easier process to port software developed for an SMP to a CC-NUMA system than to a cluster. In addition, the programming paradigm is significantly simpler.

2.3 Cache Coherence

For many years, cache has been used to speed up memory access and overcome bottlenecks in a uniprocessor computer systems. All the reasons for using caches with uniprocessors also exist with multiprocessor systems but exist more acutely. A typical shared-bus SMP system will have its performance severely limited by bus cycle time without using local cache memory.

The existence of caches local to each processor in a multiprocessor system introduces the problem of **cache coherence**. Multiple copies of data could possibly exist in different caches simultaneously due to shared data structures or because of process migration between processors. Each processor must be sure that when it reads a line of memory from its cache, the line has not been previously overwritten in the cache of some other processor or in main memory due to a transaction initiated by another processor. This is the essence of the cache coherence problem for which several techniques have been applied. Techniques to ensure coherence can be divided into **hardware-based** and **software-based approaches**.

Software-based approaches use compiler and operating system techniques to analyze data flow in the processes and act accordingly. Typically, a compiler will mark data as "non-cachable" during periods when the data are both shared by multiple processes and being written to at least one process; the compiler marks the data "cachable" during other periods, such as when one process has exclusive use of the data or read-only sharing is in operation (Lilja, 1993). The advantages of using a software approach are clear—by solving the problem at compile-time, it is not necessary to design complex and expensive coherence-preserving hardware. For software with complex control structures, however, it becomes difficult or impossible to analyze which processes have use of which blocks of memory at compile time. The compiler is therefore forced to produce conservative solutions to preserve correctness, leading to significant lost opportunities for caching.

The type of **hardware-based cache coherence approach** used depends heavily on the hardware architecture. For a shared-bus system, it is relatively easy for each cache controller to monitor, or "snoop," on all bus transactions. This is the popular approach used by so-called snoopy protocols such as MESI, to be described in Section 2.3.1. Under other, more general, forms of system architecture, it becomes difficult or wasteful for all cache controllers to monitor all transactions, therefore, another protocol must be adopted. Directory-based protocols that maintain a directory of which parts of main memory are in which caches eliminate the need for a global broadcast mechanism; these protocols are therefore better suited for complex system architectures.

2.3.1 Snoopy Protocols and MESI

Snoopy protocols distribute the responsibility for cache coherence between all cache controllers. Some mechanism must be incorporated so that each cache knows when a line is also in another cache. If this line is subsequently updated, it must inform the other caches that this update has taken place. There are two approaches in snoopy systems: either the write is broadcasted to all caches so that they can update their line accordingly (**write-broadcast**), or the write is simply used to invalidate the lines in the other caches (**write-invalidate**). The MESI protocol is an example of write-invalidate and is used by the Pentium II processor (Shanley, 1998).

MESI caches include two status bits per tag, indicating the state of the corresponding cache line, which can be described by each letter in the MESI abbreviation:

- **M**odified: The line has been modified and is only in this cache.

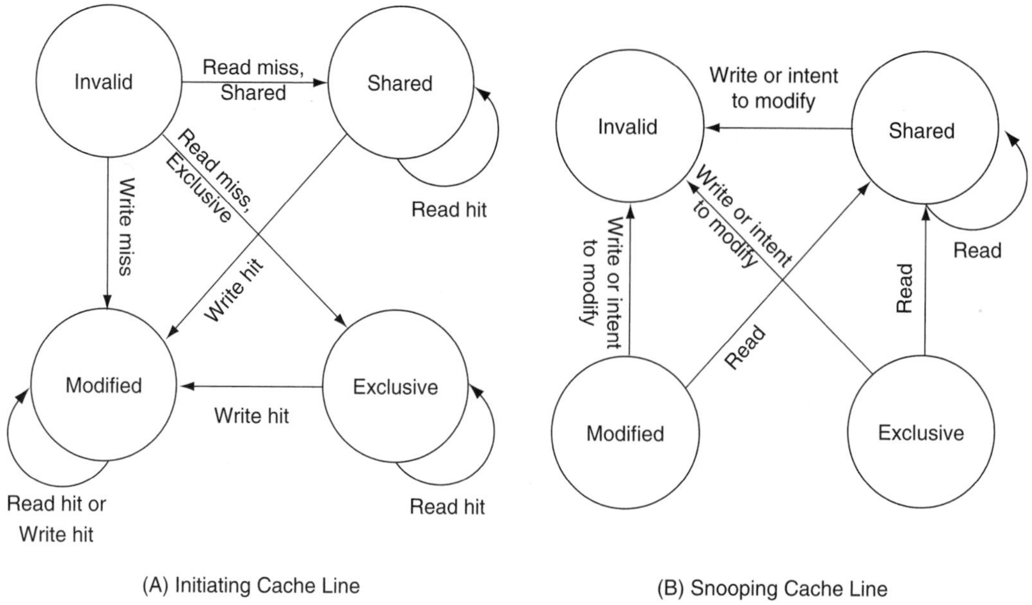

FIGURE 2.2 MESI Protocol State Transition Diagram

- **E**xclusive: The line is "clean" (same as in main-memory) and is only in this cache.
- **S**hared: The line is clean and possibly present in another cache.
- **I**nvalid: The line in the cache does not contain valid data.

The possible states are **read hit**, **read miss**, **write hit**, and **write miss**, as next described; and a state diagram for the MESI-protocol is shown in Figure 2.2 (Stallings, 2000).

1. **Read hit**: On a read hit, the cache controller supplies the data from its cache. No state change occurs.
2. **Read miss**: On a read miss, the cache controller must transfer the relevant line from main memory. A signal is broadcast to all snooping caches that must reply if they have a copy of the cache line.
   - If no signals are returned, the initiating cache controller can set the "exclusive" state.
   - If at least one signal is returned, the initiating cache controller must set the "shared" state.
   - If another cache has an exclusive copy of this line, it too must change the state to shared.
   - If another cache contains a modified copy of the line, this copy is written back to the main memory, and this cache controller must set the shared state before the originating controller can perform the read transfer.
3. **Write hit**: On a write hit to a modified line, the cache controller simply updates the line in the local cache. On a write hit to an exclusive line, the cache controller modifies the line in its local cache, but it must also change the state of the line to "modified." On a write hit to a shared line, the cache controller must inform the other caches so that they can mark their lines as "invalid" (shown as *intent to modify* in the state diagram). After this, the initiating controller can update the line locally and mark it modified.
4. **Write miss**: To write an item of data not in the cache, it is first necessary to read the entire line of memory into the cache. This signals the other caches of the initiating controller's intent to modify.

If another cache contains a modified copy of this line, it is possible that the two modifications are in different positions in the line. It is therefore necessary for the initiating controller to stall while the other cache controller writes its modified line out to main memory before invalidating its line. If no cache has a modified copy of this line, no signals are returned to the initiating processor, which modifies the line in its local cache, marking it modified. If any other caches containing clean copies of this line, their copies are marked invalid.

Some extensions have been proposed to snoopy protocols to reduce the amount of bus traffic generated. In the read-broadcast (Rudolph and Segall, 1984) protocol, a read-miss generated by a cache controller will force an update of that line in every cache. Whether write-invalidate is a better approach than write-broadcast depends on the pattern of data transfer. Karlin *et al.* (1986) proposed a protocol that adaptively decides which approach to take at run time.

2.3.2 Directory-Based Protocols

Directory-based protocols maintain information on which caches contain copies of which memory lines (Agarwal *et al.*,

1988). Most directory-based approaches maintain a single centralized directory, although there have also been proposals to maintain the directory in main memory or to distribute it through each local cache (Stenstrom, 1989).

Maintaining a centralized directory has the least overhead in terms of storage (Stenstrom, 1990), but it suffers from several disadvantages. Every local action that could possibly cause a change in global state must be reported to the controller, causing a bottleneck due to its centralized nature. In addition, a hardware search mechanism must be implemented that allows the controller to search through all duplicates to find which caches have copies of a particular block.

The key advantage of directory schemes is the ability to restrict coherency communications to only those absolutely necessary. There is clearly a trade-off between communication overhead and directory size and complexity. One possibility for exploring the trade-off is to use limited directory schemes (Stenstrom, 1990). In such a scheme, a limit is placed on the number of directory entries pointing to a particular block. The protocol either does not allow more than this number of cache copies of a single line or starts to use broadcasting after this limit has been reached. Another possibility is to use a linked-list style structure in which each cache line points to the next cache to contain a copy of that line. These are known as chained directory schemes, one example being the scalable coherent interface (SFI) as described by IEEE (1989). This approach slows down the coherency communications because they may have to go through several stages of cache controller.

2.4 Software Development and Tools

Two of the goals for multiprocessor software development are the same as those for single-processor systems: (1) the facilities should support correct program development, and (2) the application program should be portable to a wide variety of systems. There is a third goal: performance. The performance goal, however, should satisfy a further requirement of scalability—the speed of the system should increase preferably linearly with the number of processors. This means that the computational load should be optimally distributed among the processors: the task of this optimal distribution is often known as load balancing. While the multiprocessor system architecture usually dictates its scalability, appropriate languages and compilers ensure that the performance potential can be exploited by application programs.

2.4.1 Operating Systems

The two key tasks performed by a multiprocessor operating system are load balancing and failure management (Valiant, 1990), in addition to the tasks required in single-processor operating systems.

For SMP systems, the operating system is required to schedule multiple tasks for execution on multiple processors and to allocate resources to support such execution. The scheduling and resources allocation should take into account the need for load balancing. Since scheduling can be performed by any processor, there should be a mechanism to avoid or resolve conflicts. The same or different parts of the operating system can be executed by many processors, so control information (e.g., operating system tables), storage management (e.g., paging facilities), and synchronization between processors need to be carefully managed to avoid deadlock or invalid operations. Furthermore, the operating system should contribute to failure management—for instance, processor failure should lead to graceful degradation in services and performance.

For cluster systems, load balancing is often achieved by automatically including information about node availability in the scheduling process so that the system can be incrementally scalable. There should be facilities to manage services offered by one or more nodes in the cluster, and such services may also migrate from one node to another.

Failure management in a cluster depends on the cluster method. Two approaches are often adopted: **high-availability clusters** and **fault-tolerant clusters**. A high-availability cluster supports migration of services from failed nodes although partially executed transactions have to be recovered by the application program. A fault-tolerant cluster guarantees service availability by using redundant resources, and there are facilities to complete partially executed transactions.

For CC-NUMA systems, an important task for the operating system is to exploit temporal and spatial locality of data to reduce the need for remote storage access. For instance, spatial locality can be improved by organizing application data into pages to be stored locally to the processor. Another technique is to support page migration: pages in virtual memory can be moved to a node that requires them frequently.

2.4.2 Software Development Methods

This section provides an overview of three methods for parallel programming (Foster, 1995). The first two are based on extending existing sequential languages, C and FORTRAN. The third method, MPI, consists of a library for managing communication in parallel programs and is largely language-independent.

Compositional C++, or CC++, consists of six new keywords in addition to all of the existing ones in C++ (Chandy and Kesselman, 1993). Each new keyword supports a basic abstraction for parallel programming: specification of sequential code execution in a "processor object", linking together processor objects, specification and synchronization of program threads for concurrent execution on one or more processor objects, control of interleave execution of threads in the same processor object, and transfer of data structure between processor objects. These abstractions provide the means for capturing parallelism, locality, communication, and mapping. CC++ is used particularly for applications involving dynamic task creation, irregular communication and computation patterns, and concurrent composition of objects.

High performance FORTRAN, or HPF, is an example of a data parallel language (Foster, 1995). Parallelism is mainly expressed in terms of array operations: data elements of an array are operated on simultaneously. There are data distribution directives to allow programmer control over data partitioning, agglomeration, and mapping. Communication is inferred by the compiler rather than by the programmer. Although HPF offers a high level of abstraction, not all algorithms can be expressed in the form of a data parallel program. Hence, HPF is especially applicable to those programs easily specified using array operations, such as numerical algorithms based on regular domain decomposition.

Message passing interface (MPI) is a library of standard subroutines for sending and receiving messages and performing collective operations (Foster, 1995). A computation consists of processes created at program initialization. Processes can use point-to-point communication to send messages to one another. A group of processes can perform collectively in global operations, such as summation and broadcast. MPI provides support for program modularity: MPI modules can encapsulate internal communication structures. Such modules can be combined by sequential or parallel composition.

A more recent approach for language-independent parallel software development, called BSP, is covered in the next section.

2.4.3 Software Development: The BSP Approach

The **bulk synchronous parallel** (BSP) programming approach is of interest because it promises the achievement of all the goals mentioned at the beginning of this section: software correctness, portability, and scalable performance (McColl, 1996). The approach is applicable to SMP, cluster, and NUMA architectures. Just as for MPI, libraries exist to support BSP programming in standard sequential languages such as FORTRAN and C. There is also a rigorous theoretical foundation (Valiant, 1990) that allows the development of scalable BSP programs.

The key idea in BSP is to decouple communication from synchronization in the programming model. A parallel system is described as a collection of processors, each with its own local memory, and the processors are connected by a communication network capable of efficient barrier synchronization. A **barrier** is a method to separate two phases of a computation to ensure that messages generated in the two phases are not mixed together.

Execution of a BSP program proceeds in phases called supersteps. Global communication between parallel threads, such as access to memory of remote processors, is only allowed between supersteps but not in a superstep. The application program does not need to concern itself about synchronization, which can be implemented in the hardware or software. The same program can be ported to different systems provided that each supports the BSP programming model.

Since communication and synchronization are decoupled in a BSP program, the programmer does not have to worry about problems such as deadlock, which can occur with synchronous message passing. Debugging a BSP program is also simpler: the barrier at the end of a superstep provides an appropriate break point at which the global state of the parallel computation is well-defined and can be interrogated. Debugging and reasoning about the correctness of a BSP program are hence not much more difficult than for a sequential program. In addition, BSP provides an analytical cost model to assess the performance of parallel algorithms, to facilitate analyzing, and to predict their performance.

Machine-specific libraries for process creation, remote data access, and bulk synchronization have been developed to assist BSP programmers. There are native libraries for IBM's SP2, Cray's T3D, and Silicon Graphics's Power Challenge and other parallel systems. There is also a generic version for any homogeneous parallel Unix machine that has access to PARMACS, PVM, TCP/IP, or System V Shared Memory primitives (Pountain, 1996). These libraries simplify parallel program development: they also facilitate performance estimation for specific parallel machines.

2.4.4 Performance Tools

Performance tools are an essential part of the software tool kit for the development of parallel algorithms. These tools allow software developers to capture empirical data on the execution of their algorithms. These data may then be used to calibrate and validate preconceived mathematical performance models, to compare different algorithms, to investigate the scalability of algorithms with a number of processors, and to identify actual or potential bottlenecks in execution.

The type of data collected by performance tools falls into three categories: **profiles**, **counters and timers**, and **traces** (Foster, 1995). Profiling tools typically use sampling techniques, regularly examining the program counters during execution to determine how much time is spent on different procedures and activities. Profiling is extremely simple to implement and can provide an automatic high-level view of process execution. It is particularly useful for identifying unbalanced loading between processors. In addition, if there are several profiling runs for different processor counts, profiling can be used to determine components that do not scale well under the given algorithm. The disadvantage of using sampling techniques is that they are not always accurate, which is of particular concern if the developer is trying to calibrate timing models. The main problem with simple profiling in a multiprocessor system, however, is that it is unable to reveal the complex interprocessor interactions taking place under the surface.

Counters and timers allow precise information to be made available by inserting extra code to count the number of times a portion of source code has been executed or how long is spent executing it. Counters require manual user insertion of codes or special compiler action. They have the advantage of revealing precise information, unlike sampling techniques; however, timers can be complex, and the presence of the counter or timer may itself interfere with the result that is trying to be measured to some extent.

Traces are detailed logs of program execution. Whenever a meaningful event takes place, such as a message passed between two processes or the entry and exit of a procedure, the event is logged and time-stamped. Analysis tools are then heavily relied upon to extract some meaningful information from these logs or to at least reduce the dimensionality of the data such that visualization tools can be used to explore the information present. The main disadvantage to the use of traces is the huge amount of data that can be created; Foster (1995) gives the example of a 20-byte record on every message on a 128-processor system with messages produced every 10 ms. For this system, a volume of 1 GB of data would be produced per hour. In addition to the storage and postprocessing requirements necessary, logging this much data could itself change performance significantly.

2.5 Recent Advances

Multiprocessor systems are being used successfully today to improve performance in systems running multiple programs concurrently. In addition, multiprocessor systems have shown the ability to improve single-program performance significantly for certain applications containing easily parallelized loops. The extraction of coarse-grained parallelism from a software description and, indeed, the study of languages used to describe parallel software are a flourishing area of research.

With the introduction of the single-chip multiprocessor (Olukotun *et al.*, 1996), the dividing line between research into high-performance system architectures and high-performance processors is becoming blurred. Very coarse-grained programmable logic devices such as Chess (Marshall *et al.*, 1999) or the RAW machines proposed by Waingold *et al.* (1997) can be considered radically different forms of multiprocessor architecture. These approaches eliminate traditional instruction-set interfaces and instead rely heavily on compilation to directly customize the hardware to a particular application. Relying on compilation is possible because the hardware consists of a simple regular array of interchangable processing units. Clearly, this blurs the boundary between compiler research and traditional high-level synthesis, another area under intense investigation (Detton and Wawrzynek, 1999) Extending this approach of exposing the inner architecture of a processor to the compiler even further results in the possibility of more fine-grained reconfigurable computing techniques using field programmable gate arrays (Luk *et al.*,).

2.6 Summary

It is becoming increasingly attractive to build computer systems containing several interacting processors. In general, it is significantly more cost-effective to exploit the parallelism inherent in an algorithm by using multiprocessor approaches than it is to design a single faster uniprocessor.

Multiprocessor architectures have been categorized by the existence of single or multiple instruction and data streams. This chapter has examined multiple-instruction, multiple-data, and MIMD architectures in some detail. The most common form of MIMD multiprocessor arrangement is one of symmetric multiprocessors (SMP). The use of multiprocessors creates the problem of maintaining cache coherence over several levels of cache existing at different places in the computer architecture. This chapter has also presented several cache coherence approaches and the common MESI protocol. In addition, the discussion has surveyed approaches to software design for multiprocessors and some of the more recent problems in multiprocessor design.

References

Agarwal, A., Simoni, R., Hennessy, J., and Horowitz, M. (1988). An evaluation of directory schemes for cache coherence. *Proceedings of the 15th International Symposium on Computer Architecture,* 280–289.

Chandy, K.M., and Kesselman, C. (1993) CC++: A declarative concurrent object-oriented programming notation. In *Research Directions in Concurrent Object-Oriented Programming.* Cambridge, MA: MIT Press.

DeHon, A., and Wawrzynek, J. (1999). Reconfigurable computing: What, why, and implications for design automation. *Proceedings of the 36th Design Automation Conference,* 610–615.

Flynn, M.J. (1972). Some computer organizations and their effectiveness. *IEEE Transactions on Computers,* 948–960.

Foster, I. (1995). *Designing and building parallel programs.* Reading, MA: Addison-Wesley.

Hillis, W.D. (1986). *The connections machine.* Cambridge, MA: MIT Press.

IEEE. (1989). Scalable coherent interface IEEE P1596—SCI coherence protocol *Vol #,* pages 000–000.

Karlin, A., Manasse, M., Rudolph, L., and Sleator, D. (1986). Competitive snoopy caching. *Proceedings of the 27th Annual Symposium Foundations of Computer Science,* 244–254.

Lilja, D.J. (1993). Cache coherence in large-scale shared-memory multiprocessors: Issues and comparisons. *ACM Computing Surveys25*(3), 303–338.

Lovett, T., and Clapp, R. (1996). Implementation and performance of a CC-NUMA system. *Proceedings of the 23rd Annual International Symposium on Computer Architecture,* 308–317.

Luk, W., Cheung, P.Y.K., and Shirazi, N. (2004). Configurable computing. New York: Academic Press.

Marshall, A., Stansfield, T., Kostarnov, I., Vuillemin, J., and Hutchings, B. (1999). A reconfigurable arithmetic array for multimedia application. *Proceedings of the ACM International Symposium on Field Programmable Gate Arrays*, 135–143.

McColl, W.F. (1996). Scalable computing. In J. van Leeuwen (Ed.), *Computer Science Today: Recent Trends and Developments, LNCS 1000.* Berlin: Springer-Verlag.

Olukotun, K. *et al.* (1996). The case for a single-chip multiprocessor. *Proceedings of the International Conference for Architectural Support for Programming Languages and Operating Systems VII*, 2–11.

Patterson, D.A., and Hennessy, J.L. (1998). *Computer organization and design.* (2d ed.). City, State Abbrev: Morgan Kaufmann.

Pountain, D. (1996). Parallel processing in bulk. *Byte Magazine*, 71–72.

Rudolph, L. and Segall, Z. (1984). Dynamic decentralized cache schemes for mimd parallel architectures. *Proceedings of the 11th International Symposium on Computer Architecture.* 340–347.

Shanley, T. (1998). *Pentium Pro and Pentium II system architecture.* Reading, MA: Addison-Wesley.

Stallings, W. (2000). *Computer organization and architecture*, Englewood Cliffs, NJ: Prentice-Hall.

Stenstrom, P. (1989). A cache consistency protocol for multiprocessors with multistage networks. *Proceedings of the 16th International Symposium on Computer Architecture*, 407–415.

Stenstrom, P. (1990). A survey of cache coherence schemes for multiprocessors. *IEEE Computer. 23*(6).

Valiant, L. (1990). A bridging model for parallel computation. *Communications ACM.* 103–111.

Waingold, E. *et al.* (1997). Baring it all to software: Raw machines. *IEEE Computer.*

3

Configurable Computing

Wayne Luk
Department of Electrical and Electronic Engineering, Imperial College of Science, Technology, and Medicine, London, UK

Peter Y. K. Cheung
Department of Electrical and Electronic Engineering, Imperial College of Science, Technology, and Medicine, London, UK

Nabeel Shirazi
Xilinx, Inc., San Jose, California, USA

3.1 Introduction

Configurable computing, or **reconfigurable computing**, refers to the use of configurable hardware for computing purposes. The root of configurable computing can be traced back to the early 1960s, when the idea of a "restructurable computer" was first proposed by Estrin *et al.* (1963). The recent interest in configurable computing is mainly due to the rapid advance in technology of field-programmable gate arrays (FPGAs), devices containing a collection of programmable elements connected together by a network that may also be programmable by users. In short, FPGAs combine the flexibility of software with a performance approaching that of custom-developed hardware.

This chapter is intended to provide the reader with a flavor of configurable computing and its current and future trends. Configurable computing covers a wide range of topics and is rapidly growing, so our discussion will focus mainly on the promising technology of run-time reconfiguration. Useful reviews of configurable computing and FPGAs include Amano *et al.* (2000), Buell *et al.* (1996), Hauck (1998), Kean (2000), Mangione-Smith (1997), Schaumont (2001), Tessier and Burleson (2001), Trimberger (1994), Villasenor and Hutchings (1998), Villasenor and Mangione-Smith (1997), and Vuillemin *et al.* 1996. Readers who wish to cover further ground are advised to consult some of these resources.

FPGAs have become the favored choice in implementing glue logic, experimental systems, and hardware prototypes because of advantages such as short turnaround time, user reconfigurability, and low development costs. However, there are area and time overheads for providing configurable logic, configurable storage, and configurable routing resources.

Increasingly, designers realize that the key to minimizing the effects of these overheads is to fully exploit the flexibility of FPGAs, especially those that can be rapidly reconfigured at run time. We shall adopt a broad interpretation for the term **run-time reconfigurability**: it includes devices that support only complete reconfiguration by the user and those that can be partially reconfigured at run time.

Run-time reconfiguration has two main benefits. First, it enables multiplexing nonmutually dependent operations in the time domain: circuit area can be reduced by including only the active parts of a circuit at a particular moment in time. If the tasks or the algorithmic steps in an application can be partitioned into smaller ones to fit into available FPGA resources, run-time reconfiguration can be used to swap between the smaller circuits. This technique will enable the implementation of a large virtual circuit in a small physical FPGA.

The second benefit of run-time reconfiguration is to enable design upgrade. The ability to upgrade product features is a fundamental benefit of reconfigurable technology over application-specific integrated circuits; it has the potential to reshape the electronics industry when combined with a "chipless" business model for providing intellectual properties for reconfigurable devices (Kean, 2000). Design upgradability is particularly important for applications where standards evolve quickly or where there is a need to customize designs to unpredictable run-time conditions.

The frequency of reconfiguration is often dependent on the application. For some applications, there is sufficient time between tasks for reconfiguration to take place without affecting performance; for other applications, careful management of reconfiguration is required to minimize the impact of reconfiguration time.

With the development of partially reconfigurable FPGAs, it is possible to reduce the overhead of reconfiguration: the smaller the circuit to be configured, the shorter time it takes. Existing components are replaced by new ones at run time by partially reconfiguring the appropriate parts of the FPGA, while the rest of the FPGA continues to operate. The same principle applies to systems containing multiple FPGAs; computation and configuration can take place concurrently in different devices.

Although incorporating partial reconfiguration into designs is becoming popular, their development is still largely an art involving tedious and error-prone crafting of the circuits at a very low level. Relatively few studies have been reported that concern methods for describing reconfiguration, compilation tools for configurable designs, and the automatic detection of reconfigurable regions. In the next section, we shall outline an approach to exploit run-time reconfigurability. An overview of configurable computing technology, particularly run-time reconfiguration, will be given in Section 3.3, and current and future trends will be covered in Section 3.4. Finally, concluding remarks will be presented in Section 3.5.

3.2 Approach

This section begins with an outline of an approach for design development for reconfigurable hardware (Luk *et al.*, 1999; Shirazi *et al.*, 1998a, 2000). To enable the high-level description of run-time reconfigurable designs, a model has been developed for specifying, visualizing, and optimizing reconfigurable designs (Luk *et al.*, 1999). Such designs can be compiled into hardware by tools that take into account that reconfigurable designs change dynamically with time (Shirazi *et al.*, 2000). Given information regarding the sequence of reconfiguration, the tools select and refine appropriate components in reconfigurable regions.

To ease the development of reconfigurable designs, it is desirable to automate the identification of reconfigurable regions (Shirazi *et al.*, 1998a). This involves the following steps: (1) identification of identical or similar components between two successive designs and (2) placement of such components to minimize reconfiguration time. An example of this process is shown in Figure 3.1. In this design, blocks A, B, and C are present in both design 1 and design 2, while block D is present only in design 1 and block E only in design 2. Given these two designs, the task of a tool is to produce a run-time reconfigurable (RTR) design such that, when reconfiguring from design 1 to design 2, only block D needs to be reconfigured to block E. In other words, the blocks have to be physically aligned to minimize the reconfigurable region covered by blocks D and E; blocks A, B, and C are outside the reconfigurable region and will not be reconfigured.

Run-time management techniques are required to incorporate the necessary components efficiently based on a sequence for reconfiguration. If run-time conditions determine how reconfiguration should be performed, it may not be feasible to produce all the reconfiguration possibilities at compile time. A solution to this problem is to let the hardware manager build the new configuration as required at run time (Shirazi *et al.*, 1998b); various approaches are discussed in Section 3.3.4.

Since pipelining is a common technique to increase the speed of a circuit, it would be advantageous to minimize reconfiguration time of these types of circuits. Further reductions in reconfiguration overhead can be achieved by avoiding latency associated with emptying and refilling the pipeline during reconfiguration. Figure 3.2 describes a technique for overlapping computation and reconfiguration that can be implemented using either existing FPGAs (Luk *et al.*, 1997) or a

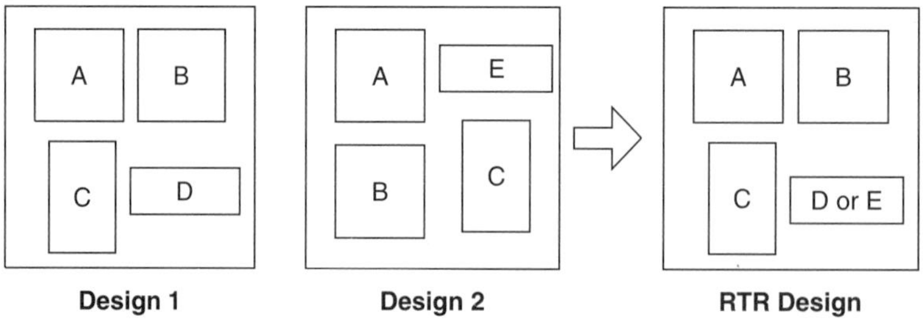

FIGURE 3.1 Example of a Design Process Involving Run-Time Reconfiguration. Blocks A, B, and C are positioned to minimize reconfiguration time. Blocks D and E together constitute the reconfigurable region. RTR design stands for run-time reconfigurable design.

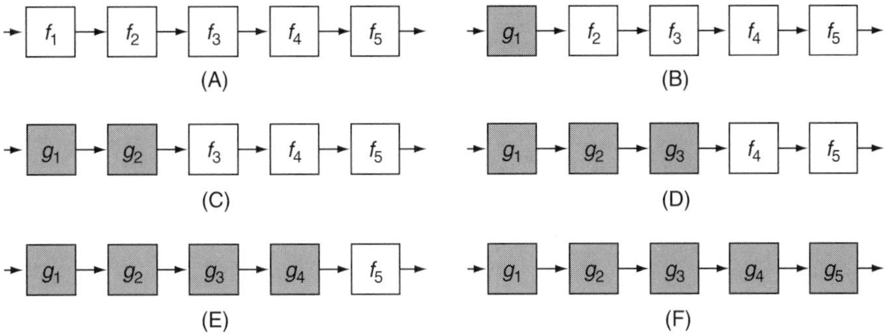

FIGURE 3.2 Overlapping Computation and Reconfiguration. Given the pipeline F with stages $f_1, \ldots, f_5$ and the pipeline G with stages $g_1, \ldots, g_5$, the diagram shows how F can be reconfigured into G in five steps. Only one stage of the pipeline is being reconfigured in each step, while the rest of the pipeline continues to operate.

device specially designed to support such overlapping (Schmit *et al.*, 2000).

3.3 Overview

This section presents an overview of important topics in configurable computing, particularly those involving run-time reconfiguration. The topics covered include run-time reconfigurable devices, run-time reconfiguration methods, tools, run-time management, and hardware platforms and applications.

3.3.1 Run-Time Reconfigurable Devices

FPGAs, by definition, are configurable; most of them are also reconfigurable unless they are based on technologies such as Antifuse, that are one-time programmable. Several commercial devices support partial reconfiguration, including the Virtex (Xilinx, 2001) and 6200 (Churcher *et al.*, 1995) devices from Xilinx, the CLAy chip from National Semiconductor (National Semiconductor, 1993), and the AT 40K devices from Ateml (Atmel, 1997). Useful reviews of FPGA architectures are available (Buell *et al.*, 1996; Hauck, 1998; Kean, 2000; Mangione-Smith, 1997; Trumberger, 1994; Villasenor and Hutchings, 1998). Although some devices such as Xilinx 6200 FPGAs are no longer supported commercially, the ideas in the relevant publications may still inspire future advances.

A simple FPGA model is shown in Figure 3.3. In this figure, processing elements, typically containing configurable logic and storage blocks, are represented by squares. The processing elements are connected to configurable switches, represented as circles, that control data flow by establishing the desired connectivity between the busses. Much of the area in an FPGA is usually taken up by the configurable switches and the busses; local and global busses can also be organized hierarchically. The figure demonstrates the regularity found in most FPGAs; practical FPGAs often contain additional resources, such as

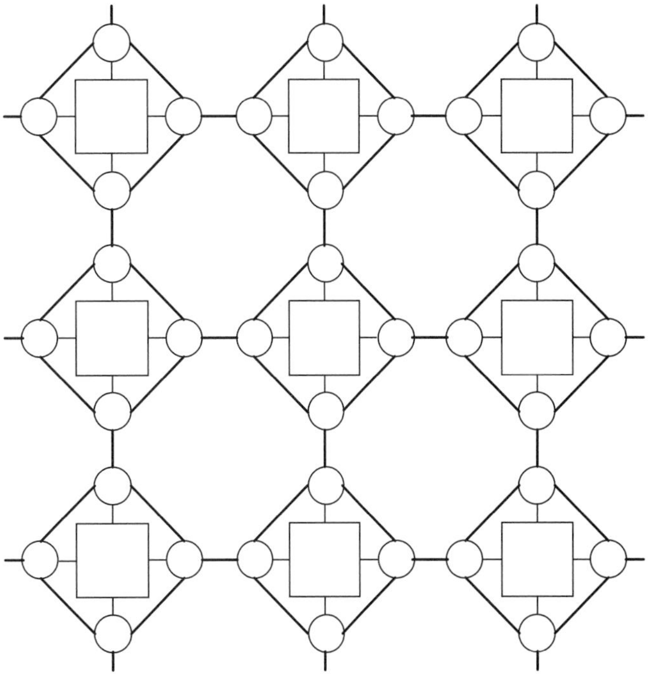

FIGURE 3.3 A Simple Model of an FPGA. Squares represent configurable processing elements, and circles represent configurable switches to control routing.

configurable memory blocks and special-purpose input/output blocks supporting boundary-scan testing (Trimberger, 1994).

Many experimental FPGA architectures support run-time reconfiguration. Tau *et al.*, (1995) have come up with an FPGA that stores multiple configurations in memory banks. In a single clock cycle, which is in the order of tens or hundreds of nanoseconds, the chip can replace configuration by another without erasing partially processed data.

A similar FPGA that can perform a context switch in one cycle has been developed by Trimberger *et al.* (1997). The FPGA can store up to eight configurations in on-chip memory.

This FPGA is based on a Xilinx 4000E device and includes extensions for dealing with saving state from one context to another.

The Colt Group led by Athanas is investigating a run-time reconfiguration technique called Wormhole that lends itself to distributed processing (Bittner and Athanas, 1997). The unit of computing is a stream of data that creates custom logic as it moves through the reconfigurable hardware.

Schmit *et al.* (2000) have developed a reconfigurable FPGA targeted toward pipelined designs. Reconfiguration is performed at the level of individual pipeline stages, similar to that described in Figure 3.2. Others have shown that commercial partially reconfigurable FPGAs can also support efficient reconfiguration of pipelined designs (Luk *et al.*, 1997).

There are also configurable devices based on coarse-grain programmable elements (Conquist *et al.*, 1998), multiple-bit arithmetic units (Marshall et al., 1999), and low-power techniques (Rabaey, 1997). Kean (2000) provides an overview of commercial devices available in the year 2000.

3.3.2 Run-Time Reconfiguration Methods

Hutchings and Wirthin (1995) suggested that two types of run-time reconfiguration exist: global and local. During **global reconfiguration**, all the hardware resources are allocated with a new configuration in each configuration step. This method works for nonpartially reconfigurable FPGAs, such as the Xilinx 4000 series. Applications that use the second type of run-time reconfiguration, local reconfiguration, locally or selectively reconfigure subsets of the FPGA during application execution. Since the larger the size of the region to be reconfigured means the longer the reconfiguration time, Hadley and Hutchings (1995) have proposed techniques to minimize the size of the reconfigurable region to reduce reconfiguration time. Others have developed methods and tools that are able to produce designs both locally reconfigurable as well as globally reconfigurable (Shirazi *et al.*, 2000).

To quantify the advantages of run-time reconfigurable designs, Wirthlin and Hutchings (1998) introduced a metric called **functional density** that measures the computational throughput (operations per second) of unit hardware resources. This metric can be used to identify the conditions under which a run-time reconfigured circuit provides higher functional density than its statically configured alternative. Several applications, including neural nets, template matching, and DNA sequence matching, are used to illustrate the superiority of run-time reconfigurable implementations.

Hutchings and Wirthlin (1995) have built a dynamic instruction set computer (DISC) that emulates a reconfigurable microprocessor by using an FPGA and demonstrates the potential of automatic reconfiguration using stored configurations. As a program runs, the FPGA requests a new reconfiguration if the desired instruction is not resident. DISC allows a designer to create and store a large number of circuit configurations and activate them much as a programmer would initiate a call to a software subroutine in a microprocessor. Partial reconfiguration is used to reconfigure FPGA resources with the instruction, thereby minimizing reconfiguration time.

One feature of DISC is that the instructions are interconnected using a fixed bus structure. Luk, Shirazi, and Cheung also proposed methods (1996) and tools (1998, 2000) for run-time reconfigurable designs. An advantage of their approach is that the reconfigurable designs produced are not confined to a specific bus structure.

On DISC 25% to 71% of the execution time is spent reconfiguring. Hauck *et al.* (1999) have studied techniques for reducing reconfiguration time on the Xilinx 6200 by compressing the configuration data stream. They have been able to reduce reconfiguration time 4-fold by using a device specific feature called **wildcarding**. The approach by Shirazi *et al.* (2000) also takes advantage of wildcarding, and they provide an example in which reconfiguration time has been reduced from linear time with respect to size of the component to constant time at best and logarithmic time at worst. More recently, Li and Hauck (2001) described a method for compressing configuration data for Xilinx Virtex FPGAs that exploits the regularity of the frame-based format of configuration data.

The group led by Woods has developed an efficient technique for fast partial reconfiguration for FPGA-based systems (Sezer *et al.*, 1998). They used a library-based approach to achieve a high degree of commonality between designs. The configuration data are stored efficiently in a **configuration data graph** (CDG). A CDG is a database of designs organized according to their shared components. High-speed configuration is also achieved by using **reconfiguration state graphs** (RSGs) that model reconfiguration as state transitions. The tools developed in this thesis generate representations similar to RSGs automatically from a specified reconfiguration sequence. It is also possible to introduce optimization techniques based on compile time information to optimize the RSGs and decrease the amount of configuration data needed (Shirazi *et al.*, 2000).

McKay and Singh (1998) have used partial evaluation techniques to optimize circuits that have slowly changing inputs. They are able to achieve about a 60% increase in clock speed for a parallel multiplier circuit performing just a single level of partial evaluation. Others have also used partial evaluation techniques in compilation tools (Shirazi *et al.*, 2000) and in application development (Shirazi *et al.*, 1998b).

3.3.3 Tools

The introduction of Xilinx Virtex FPGAs (Xilinx, 2001) offered an opportunity for hardware support of partial run-time reconfiguration in large commercial devices. McMillan and Guccione (2000) describes the JRTR tool in the JBits tool suite that provides software support for partial run-time reconfiguration that makes use of the features of the Virtex architecture. The JBits configuration application programming interface (API) provides a set of Java classes to access and manipulate Virtex configuration data.

The JRTR API includes a caching model in which changes to configuration data are tracked, and only the necessary data are written to or read back from the device. The compilation speed is fast, but the user has to lay out components on the devices carefully. Fortunately, there are tools (Shirazi *et al.*, 1998a, 2000) that can be used in combination with the JRTR tool set to automate the development of run-time reconfigurable designs. Such tools should enhance JRTR by automatically finding the reconfigurable regions between subsequent configurations and calculate the difference between configurations, thus minimizing configuration time. It would also be useful to interface hardware compilers (Page and Luk, 1991; Weinhardt and Luk, 2001) to the JRTR tools.

Research on compilation tools for run-time reconfigurable systems, based on the dbC language, is described by Gokhale and Marks (1995). These tools target the National Semiconductor CLAy FPGA (National Semiconductor, 1993) and provide a path for high-level design compilation. Weinhardt and Luk (2001) have extended vectorization and loop transformation techniques used in super-computers for reconfigurable design; Figure 3.4 summarizes the key steps in this framework.

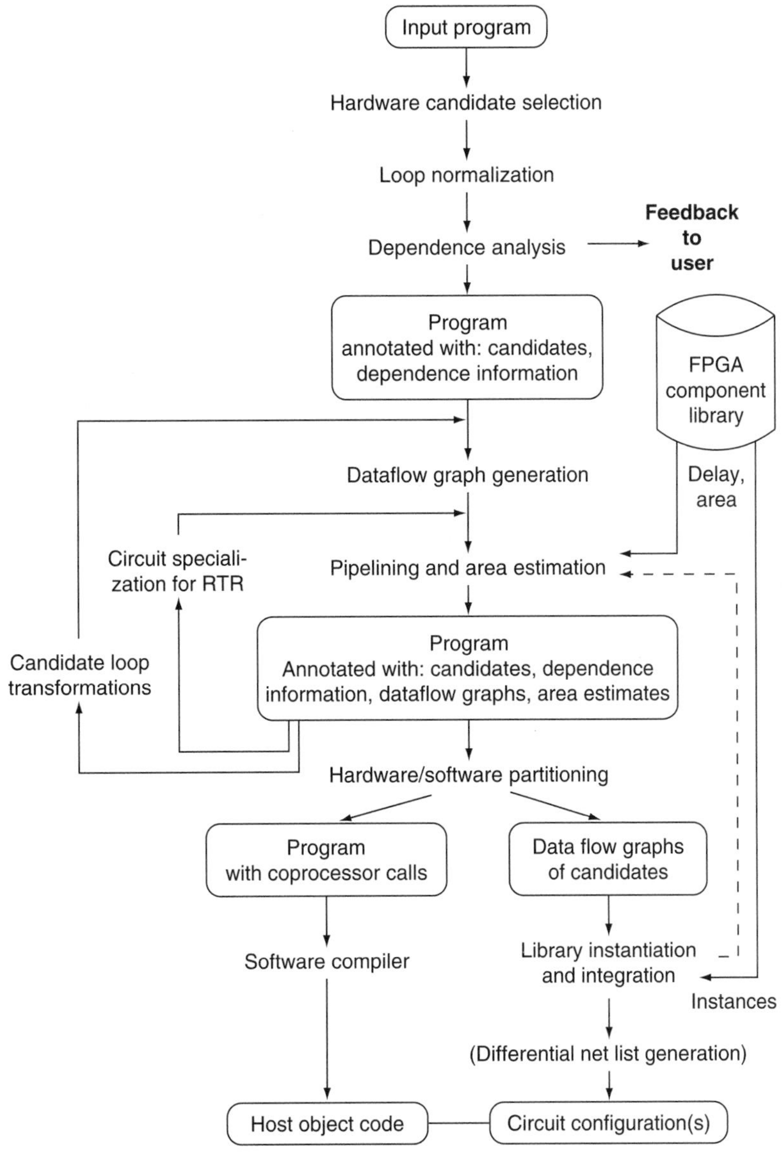

FIGURE 3.4 Design Framework Based on Vectorization Approach. RTR denotes run-time reconfiguration. The dotted line shows possible repetition of some steps that may improve design size and speed.

Promising results have been achieved for mapping loops in a sequential language into reconfigurable hardware.

Page and Luk (1991) describe a hardware compiler for high-level sequential programs augmented with language constructs for parallel operation and communication. This method enables the rapid production of a wide variety of designs, from pipelined video operators to instruction-set processors. The approach is being extended to cover designs involving run-time reconfiguration.

The Trianus system (Gehring and Ludwig, 1996) is an integrated tool set consisting of a layout editor, circuit checker, technology mapper, placer, router, and bit-stream generator and loader for the Xilinx 6200 FPGA. The back ends of the Trianus tool set have been incorporated with the IRIS tools (Woods *et al.*, 1997), allowing designs to be changed at the architectural level in IRIS rather than at the FPGA layout level.

Lysaght and his groups have developed a framework called dynamic circuit switching for specifying, simulating, and debugging reconfigurable designs (Lysaght, 1997; Lysaght and Stockwood, 1996; McGregor *et al.*, 1998; Robinson *et al.*, 1998). Their tools can estimate the reconfiguration time before low-level tools are invoked; this information is used in their compilation process to produce efficient reconfigurable designs.

3.3.4 Run-Time Management

Lysaght and Dunlop (1994) have proposed a self-controlled management technique called **logic caching** that is based on the same principles used by traditional microprocessors. They suggest many improvements to traditional FPGA compilation software to support the development of run-time reconfigurable designs. These include simulation models, automatic design partitioning based on temporal specifications, and support for generating relocatable bit streams. Others have also reported work on temporal partitioning and scheduling of configuration data (Purna and Bhatia, 1999), block-processing based temporal partitioning (Kaul and Vemuri, 2000), identifying and positioning of reconfigurable regions (Shirazi *et al.*, 1998a), and generating relocatable bit streams (Shirazi *et al.*, 1998b).

Bellows and Hutchings (1998) have developed a management technique similar to the method that object-oriented languages use to manage memory. Their management technique is part of their design tool called **JHDL**. A circuit configuration is treated as a distinct object and is downloaded into the FPGA when the constructor of the object is invoked.

The **run-time reconfigurable artificial neural network** developed by Eldridge and Hutchings (1994) uses dynamic reconfiguration to implement three stages of a neural network back-propagation algorithm. Each stage of the algorithm is represented by a circuit module that is swapped in and out of the FPGA hardware as demanded by the application. Only one circuit module is resident on the FPGA at a given time. Some static circuitry is constantly resident which controls data flow and the sequencing of execution of the dynamic modules. Swapping time can be minimized by calculating the difference between the configuration currently resident in the FPGA and the desired new configuration; a tool has been developed for this purpose (Shirazi *et al.*, 2000).

Haug and Rosenstiel (1998) have integrated run-time management techniques into the Linux operating system. Their tools synthesize hardware from user-defined threads in a C program. The operating system's scheduler has been modified to keep track of the configurations in use by the FPGA coprocessor. By incorporating their scheduler into a popular operating system such as Linux, they are one step closer to making run-time reconfiguration a mainstream technology.

Burns *et al.* (1997) have proposed a run-time system for managing the dynamic reconfiguration of FPGAs. The system incorporates operating-system style services that permit high-level operations on circuits, such as circuit transformations and dynamic rerouting of circuits. Shirazi *et al.* (1998b) have also incorporated circuit transformations into their run-time manager.

Brebner (1996) states that the fundamental difference between virtual hardware and virtual memory is that virtual hardware is much more structured. Virtual memory is a collection of independent memory locations and swapping can be done at the single memory location level. In the case of logic circuits, a collection of components operates as a whole; therefore, it is not possible to swap a component unless it is isolated and operates independently from the other components. Brebner defines this type of isolated component as a **swappable logic unit** (SLU). Two different virtual hardware models are defined: **sea of accelerators** and a **parallel harness**. The sea of accelerators model is a collection of SLUs that are independently operating and are accessed via bus-accessible registers. A parallel harness model is a set of SLUs that are cooperating parallel processing elements, interacting with neighbors using perimeter signals, and wiring infrastructure supplied by the operating system. Brebner proposes an operating system to help automate the swapping of SLUs. The swapping of an SLU into an FPGA is broken down into three steps. First, the operating system finds sufficient space to swap the logic into the FPGA by using a two-dimensional recursive bisection technique. If there is not enough available space for the SLU, the operating system removes the least recently used SLUs until enough space is available. Second, the SLU is configured in the FPGA. Third, information about the location of the SLU's *I/O* registers is recorded.

3.3.5 Hardware Platforms and Applications

Luk *et al.* (1998) presented several real-time applications for a hardware platform that includes a daughter board for supplying video data. The architecture of this platform is shown in Figure 3.5. This daughter board is necessary to overcome the

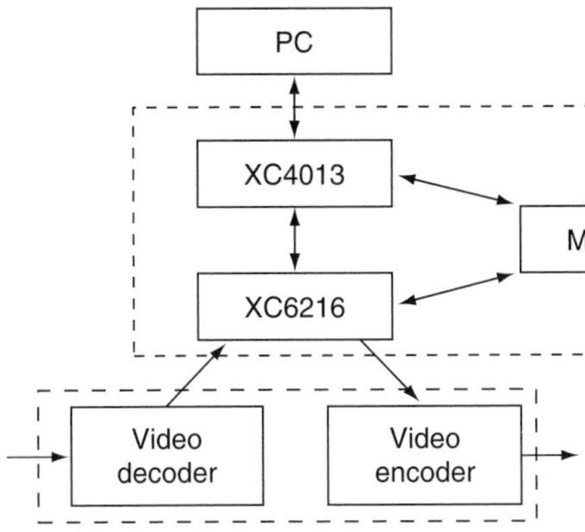

FIGURE 3.5 A Configurable Computing Platform for Video Applications. The upper dashed box contains a commercially available low-cost FPGA board; the lower box contains a video decoder and a video encoder interfaced to the user-programmable FPGA, the XC6216.

I/O bottleneck caused by the PCI bus connecting the platform to a PC host. One use of this platform is in combining real and synthetic video images for augmented reality applications. (Luk et al., 1999).

The FPGA board used in Figure 3.5's hardware platform has been also used to implement an image interpolation engine by Hudson *et al.* (1998). They used run-time reconfiguration to reduce the size of their interpolation engine by four times. To reduce reconfiguration time, they used a tool for incorporating partial reconfiguration into their design (Shirazi *et al.*, 2000).

A further application has been developed for this FPGA board by Singh *et al.* (1998) that involves accelerating Adobe Photoshop computations in a software plug-in framework. As different filters are selected in the application, a new design is downloaded into the platform. The applications implemented include a color to gray scale filter and 1-D and 2-D image convolvers. Partial reconfiguration is not performed when switching between these image processing tasks. By using the same method as Hudson *et al.* (1998), they reduced the time required to reconfigure to a new image processing task. The significance of using this FPGA platform with a mainstream application is that users do not need to know details about the enabling technology to reap its benefits. MacVicar and Singh (1998) have also been able to accelerate the rendering of Bezier curves, a core function in desktop publishing applications.

A different approach to supporting the software plug-in framework was adopted by Haynes *et al.* (2000). They have developed a scalable architecture called Sonic, which consists of multiple modules capable of accommodating different types and sizes of configurable devices, general-purpose processors, and custom-designed circuits. Such modules have a common interface and are connected by high-speed control and data busses. The architecture is particularly effective for video and image processing; the run-time environment includes a multitasking operating system allowing hardware relocation to simplify hardware implementation of plug-in functions.

Although many existing hardware platforms interface to their host through an *I/O* bus such as the PCI bus (Guccione, 2000), the Pilchard (Leong *et al.*, 2001) is unique in interfacing to the host processor through the memory bus. Experiments demonstrate that in read/write benchmarks, the 64-bit Pilchard transfer rates are 6 to 10 times faster than PCI.

Villasenor *et al.* (1996) have implemented an automatic target recognition system using the Xilinx 4000 series FPGAs. Even though the Xilinx 4000 FPGA is not partially reconfigurable, the group was able to achieve an order of magnitude increase in performance over systems built using general purpose processors. In addition, over 75% of the total execution time of their application was spent in reconfiguring the FPGA. This significant reconfiguration overhead can be reduced if a partially reconfigurable FPGA is used.

The **vector dot product**, which involves computing the sum of products of corresponding elements in two vectors, is a common algorithmic kernel in signal processing. Its use ranges from filters and correlators to two-dimensional image transforms, such as the **discrete cosine transform**. Benyamin *et al.* (1999) describes a method for producing optimized reconfigurable hardware implementations of vector dot products. The method is based on a novel representation of common subexpressions in constant data patterns. Whereas traditional techniques such as distributed arithmetic require storage size exponential in the number of vector elements, the storage requirement of the proposed method usually grows linearly with problem size.

Fawcett (1994) describes several applications that have been implemented using reconfigurable logic. He has categorized the applications based on how the FPGAs are reconfigured. These categories are **systems with built-in diagnostics, adaptable systems**, and **multi-purpose hardware**. The examples provided are all implemented using nonpartially reconfigurable FPGAs. By using appropriate tools (2000), designers can retarget these designs on partially reconfigurable FPGAs. For instance, system diagnostics can be performed by partially reconfiguring a portion of the FPGA to perform a boundary scan operation, while the rest of the design continues to operate. An example of an adaptable system that is partially reconfigurable is a network card that changes its back-end functions to support different protocols and that keeps its front-end bus interface functional to service the host's demands.

Other applications involving run-time reconfiguration include data encryption (Dandalis *et al.*, 2000), active networking (Dollas *et al.*, 2001), genomic database search (Lemoine and Merceron, 1995), graph coloring (Rising *et al.*, 1997), continuous speech recognition (Stogiannos *et al.*, 2000), computer graphics (Styles and Luk, 2000), procedural texture mapping (Ye and Lewis, 1999), and solving Boolean satisfiability (Zhong *et al.*, 1998).

3.4 Current and Future Trends

This section discusses various directions of current and future development in configurable computing technology.

3.4.1 Domain-Specific Compilation

Significant productivity gain can be obtained by the use of domain-specific tools. An example is MatLab, a well-known tool for rapid design exploration of digital signal processing applications. Several projects have been undertaken to map MatLab programs, or descriptions in the Simulink block-diagram format, into hardware. Banerjee *et al.* (2001) describe a compiler that maps MatLab programs onto a distributed system containing embedded processors, digital signal processors, and FPGAs. Constantinides *et al.* (2001) present an analytical approach for producing efficient hardware implementations by varying the word length and scaling of each signal in a Simulink block diagram. Hwang *et al.* (2001) demonstrate a visual design flow that addresses issues including the mapping of system parameters, such as sample rates into implementation parameters, and implications of system modeling for testing, such as the test-bench generation. Other domain-specific tools for configurable computing have also been reported: applications include computer graphics, where tools can leverage on the OpenGL interface standard (Styles and Luk, 2000), and image processing, where tools, can build on the Khoros Cantata graphical programming environment (Ong *et al.*, 2001).

The methods and tools described in this subsection are specific to particular application domains. A general approach for producing domain-specific compilers is advocated by Mencer *et al.* (2001). Their approach was based on the use of object-oriented module generators for developing compilers that target domain-specific implementations, such as stream architectures and Boolean satisfiability machines.

3.4.2 Architectural Enhancements

Instead of implementing common operators such as multiplication using fine-grain FPGA resources, it can be more effective to have customized configurable blocks designed specifically for such operators. For instance, Haynes *et al.* (1999) have proposed configurable blocks for efficient multiplication. An array of these blocks are capable of being configured to perform any 8-m by 8-nb signed and unsigned multiplication. The architecture is based on the radix-4 overlapped multiple-bit scanning algorithm. When compared with existing FPGAs for implementing multipliers, this design is shown to be both faster and require fewer transistors.

As another example of architectural enhancement, double buffering techniques can be used for configuration data to support reconfiguration of an FPGA in one clock cycle. This would require only one additional plane of configuration data instead of multiple planes in some context-switchable FPGAs (Tau *et al.*, 1995), hence reducing area and speed overheads. Related ideas on configuration caching algorithms have been explored by Li *et al.* (2000); their algorithms include relocation and defragmentation techniques that allocate and combine configuration data at run time.

3.4.3 Modeling and Optimising Run-Time Reconfiguration

Providing appropriate constructs in hardware description languages is necessary to support run-time reconfiguration. Such constructs are included in the Pebble language (Luk et al., 1997), which can be viewed as a variant of the industrial standard VHDL language. It is also possible to model the reconfiguration mechanism in VHDL (Lysaght and Dunlop, 1994) or in the Circal process algebra (Diessel and Milne, 2000). Currently, most design frameworks require the user to specify the reconfiguration sequence; at compile time, the sequencing information can be extracted and used to guide compilation (Shirazi *et al.*, 2000).

The technique for overlapping configuration and computation (see Section 3.2, Figure 3.2) can be also extended to cover various architectural templates; this extension will enable the reconfiguration of one pipeline to another with a different number of pipeline stages or to change a linear pipeline into a tree-shaped architecture. Frequently, there are multiple ways of reconfiguring designs, and it is important to evaluate the trade-offs involved.

3.4.4 Identification and Mapping of Reconfigurable Regions

Much of current research is focused on automating the optimal partitioning of large circuits into smaller ones to fit into available FPGA resources and the scheduling of the resulting components (Purna and Bhatia, 1999; Kaul and Vemuri, 2000; Liu and Wong, 1999). Run-time reconfiguration can then be performed to swap between the smaller circuits, thus effectively implementing the large circuit in an FPGA, which is physically smaller than the original circuit. If this goal is achieved, circuit designers would not need to worry whether their design fits in the available chip. Multichip partitioning algorithms could be used, however, the partitioning algorithm would need to partition the circuit to maximize the overlap between multiple partitions. Tools for identifying the reconfigurable regions and positioning components (Shirazi, *et al.*, 1998a) can then be applied to maximize physical overlap, thus reducing partial reconfiguration time.

With existing tools, the granularity of the reconfigurable regions is considered to be a logic block (usually the size of a four-input logic gate) or a routing resource. Conceptually, there is no reason why the granularity cannot be expanded to an entire nonreconfigurable FPGA. By doing this, one can develop run-time reconfigurable designs for multi-FPGA systems, such as the Splash 2 platform (Buell *et al.*, 1996) or the Sonic system (Haynes *et al.*, 2000). Although many existing platforms (Guccione, 2000) contain multiple FPGAs that are not partially reconfigurable, the platforms themselves are often partially reconfigurable.

3.4.5 Rapid Compilation and Validation

Fast compilation of configurable computing designs is important. Depending on its complexity, the development cycle of a design often takes a few minutes to many hours. If compilation times for FPGAs were equivalent to today's C++ or Java compilers, productivity would increase tremendously. The JBits tool suite, described in subsection 3.3.3, is capable of rapid compilation provided that users place and route their designs in detail. The JBits tools are being combined with other design tools, such as JHDL (Bellows and Hutchings, 1998) Pebble (Luk and McKeever, 1998) and Lava (Singh and James-Roxby, 2001), so that high-level descriptions can be compiled rapidly into efficient hardware.

The development of intellectual property (IP) cores is another technology that will decrease compilation times. An example is the self-implementing modules (SIMs) methodology (Hwang *et al.*, 2001). A parametrized core is described in SIMs by assigning placement to components in the design; however routing is not specified. Automatic place and route tools are used to route the designs that incorporate SIMs, and these modules are then routed along with user logic. If routing were specified along with placement, compilation times would decrease; however, placement restrictions might be imposed. If at either run time or compile time one of the inputs of the SIM remains constant for a prolonged period, partial evaluation techniques can be used to optimize the SIM. At run time, partial reconfiguration can be used to reconfigure between the constant values. It would be useful to identify automatically the subcomponents of a SIM that remain the same (Shirazi *et al.*, 1998a) when a constant value changes to minimize reconfiguration time.

To speed up the compilation and validation process, Luk et al. (1999) have developed a tool kit for automating the quality assurance of intellectual property cores. Designs can be validated in various ways, including functional simulation, timing analysis, automated theorem proving, and execution on a hardware platform such as the system shown in Figure 3.5. Their work is being extended to cover reconfigurable designs.

3.4.6 Novel Implementation Technology

Faggin (1996) led the design and development of the world's first microprocessor, the Intel 4004. He believes that one day, processors will have some configurable hardware resources incorporated onto the same device. His prediction is being fulfilled by FPGA vendors: Altera, Triscend, and Xilinx are offering devices that contain both reconfigurable logic and a microprocessor. The microprocessor is usually based on an existing architecture from manufacturers such as ARM, MIPS, or IBM to exploit the existing software base. On the research side, the Garp device (Callahan *et al.*, 2000) and the NAPA device (Rupp *et al.*, 1998) are two examples of processor/FPGA hybrid devices.

Perhaps the most radical departure from current silicon technology is to exploit advances in molecular science (Collier *et al.*, 1999) and microfluidic systems (McCaskill and Wagler, 2000) for implementing reconfigurable architectures. If successful, these approaches will have a significant impact on the future of configurable computing.

3.5 Concluding Remarks

Despite the rapid progress made in the last 10 years, many consider configurable computing still in its infancy (Mangione-Smith *et al.*, 1997). As described earlier, there is active research in many aspects of configurable computing: modeling, applications, tools, architectures, platforms, and implementation technology.

In Section 3.1, we explained the two main benefits in configurable computing technology: the ability for time-multiplex resource sharing and ability for hardware upgrades. Many believe that the key to future success of this technology involves techniques and tools that enable designers to gain competitive advantage from these benefits more easily.

We can draw a lesson from history. The invention of virtual memory is a milestone in computing: virtual memory frees programmers from specific machine storage details when developing applications. Run-time reconfiguration can be considered the processing equivalent of virtual memory. Multimillion gate FPGAs are beginning to appear, opening the door to applications that once were never thought to be implementable on FPGAs. The need for more resources, however, will never end. Run-time reconfiguration provides a mechanism for emulating an unlimited amount of gates. For run-time reconfigurable hardware to become as big a success as virtual memory, it needs to be as easy to use as virtual memory—to be completely transparent to the user. By automating run-time reconfiguration, application developers for configurable computing should seldom have to worry about the amount of chip utilization.

The short-term goal in making configurable computing a mainstream technology is to automate the efficient implementation of a configurable design partitioned by hand. The long-term goal is to fully automate the compilation process for specific configurable architectures without compromising design efficiency.

As technology for system-on-a-chip matures, the advantages of configurable computing—the combination of flexibility and performance—can make it increasingly competitive for a wide range of applications.

Acknowledgments

We thank S.P. Seng and M. Weinhardt for producing some of the diagrams in this paper. The support of UK Engineering and Physical Sciences Research Council (Grant numbers GR/L24366, GR/L54356, GR/L59658, and GR/N66599), Xilinx, Incorporated, and a UK Overseas Research Student Award is gratefully acknowledged.

References

Amano, H., Shibata, Y., and Uno, M. (2000). Reconfigurable systems: New activities in asia. *Proceedings of the IEEE Symposium on FPGAs Custom Computing Machines.* 585–594.

Athanas, P.M., and Abbott, A.L. (1995). Real-time image processing on a custom computing platform. *IEEE Computer.* 16–24.

Atmel. (1997). AT40K FPGA preliminary datasheet. http://www.atmel.com/atmel/products/prod99.htm.

Banerjee P. *et al* (2001) A MATLAB compiler for distributed, heterogeneous, reconfigurable computing systems. *Proceedings of the IEEE Symposium on FPGAs Custom Computing Machines.* 39–48.

Bellows, P. and Hutchings, B. (1998). JHDL—An HDL for reconfigurable systems. *Proceedings of the IEEE Symposium on FPGAs for Custom Computing Machines.* 175–184.

Benyamin, D., Luk, W., and Villasenor, J. (1999). Optimising FPGA-based vector product designs. *Proceedings of the IEEE Symposium on FPGAs for Custom Computing Machines.* 188–197.

Bittner, R., and Athanas, P. (1997). Wormhole run-time reconfiguration. *Proceedings of the IEEE Symposium on FPGAs Custom Computing Machines.* 79–85.

Brebner, G. (1996). A virtual hardware operating system for the Xilinx XC6200. *Field Programmable Logic 1142*, 327–336.

Buell, D., Arnold, J., and Kleinfelder, W. (1996). Splash 2. *Proceedings of the IEEE Symposium on FPGAs Custom Computing Machines.*

Burns, J., Donlin, A., Hogg, J., Singh, S., and deWit, M. (1997). A dynamic reconfiguration run-time system. *Proceedings of the IEEE Symposium on FPGAs Custom Computing Machines.* 66–75.

Callahan, T.J., Hauser, J.R., and Wawrzynek, J. (2000). The garp architecture and C compiler. *IEEE Computer.* 50–57.

Churcher, S., Kean, T., and Wilkie, B. (1995). The XC6200 FastMap processor interface. In W. Moore and W. Luk (Eds.), *Field Programmable Logic and Applications 975*, 36–43.

Collier, C.P., Wong, E.W., Belohradsky, M., Raymo, F.J., Stoddart, J.F., Kuekes, P.J., Williams, R.S., and Heath, J.R. (1999). Electronically configurable molecular-based logic gates. *Science 285* (5426), 391.

Conquist *et al.* D.C. (1998). Specifying and compiling applications for RaPid. *Proceedings of the IEEE Symposium on FPGAs Custom Computing Machines.* 115–125.

Constantinides, G.A., Cheung, P.Y.K., and Luk, W. (2001). The multiple wordlength paradigm. *Proceedings of the IEEE Symposium on FPGAs Custom Computing Machines.*

Dandalis, A., Prasanna, V.K., and Rolim, J.D.P. (2000). An adaptive cryptographic engine for IPSec architectures. *Proceedings of the IEEE Symposium on FPGAs Custom Computing Machines.* 132–141.

Diessel, O., and Milne, G. (2000). Behavioral language compilation with virtual hardware management. *Field Programmable Logic and Applications 1896*, 707–717.

Dollas, A., *et al.*, (2001). Architecture and applications of PLATO, a reconfigurable active network platform. Proceedings of the IEEE Symposium on FPGAs Custom Computing Machines. 77–80.

Eldridge, J.G. and Hutchings, B. (1994). RRANN: The run-time reconfiguration artificial neural network. *Proceedings of the IEEE Symposium on FPGAs Custom Computing Machines.* 77–80.

Estrin, G. *et al.* (1963). Parallel processing in a restructurable computer system. *IEEE Transactions on Electronic Computers.* 747–755.

Faggin, F. (1996). The future of the microprocessor. Forbes.

Fawcett, B.K. (1994). Applications of reconfigurable logic. In W. Moore and W. Luk (Eds.), *More FPGAs.* Abingdon EE&CS Books.

Gehring, S., and Ludwig, S. (1996). The trianus system and its application to custom computing. *Field Programmable Logic 1142*, 176–184.

Gokhale, M., and Marks, A. (1995). Automatic synthesis of parallel programs targeted to dynamically reconfigurable logic arrays. *Field Programmable Logic and Applications 975*, 399–408.

Guccione, S. (2000). List of FPGA-based computing machines. http://www.io.com/~guccione/HW_list.html/

Hadley, J., and Hutchings, B. (1995). Design methodologies for partially reconfigured systems. *Proceedings of the IEEE Symposium on FPGAs Custom Computing Machines.* 78–84.

Hauck, S. (1998). The roles of FPGAs in reprogrammable systems. *Proceedings of the IEEE Symposium on FPGAs Custom Computing Machines 86*, 615–638.

Hauck, S. Li, Z., and Schwabe, E. (1999). Configuration compression for the Xilinx XC6200 FPGA. *IEEE Transactions on Computer-Aided Design of Integrated Circuits and Systems 18*(8), 1107–1113.

Haug, G., and Rosenstiel, W. (1998). Reconfigurable hardware as shared resource in multipurpose Computers. *Field Programmable Logic and Applications 1482*, 147–158.

Haynes, S., Ferrari, A.B., and Cheung, P.Y.K. (1999). Flexible reconfigurable multiplier blocks suitable for enhancing the architecture of FPGAs. *Proceedings IEEE of Custom Integrated Circuits Conference.* 191–194.

Haynes, S., Stone, J., Cheung, P.Y.K., and Luk, W. (2000). Video image processing with the sonic architecture. *IEEE Computer.* 50–57.

Hudson, R.D., Lehn, D.I., and Athanas, P. (1998). A run-time reconfigurable engine for image interpolation. *Proceedings of the IEEE Symposium on FPGAs Custom Computing Machines.* 88–95.

Hutchings, B., and Wirthlin, M.J. (1995). Implementation approaches for reconfigurable logic applications. *Field Programmable Logic and Applications 975*, 419–428.

Hwang, J., Milne, B., Shirazi, N., and Stroomer, J.D. (2001). System level tools for DSP in FPGAs. *Field Programmable Logic and Applications 2147*, 534–543.

Hwang, J., Patterson, C., Mohan, S., Dellinger, E., Mitra, S., and Wittig, R. (1998). Generating layouts for self-implementing modules. *Field Programmable Logic and Applications 1482*, 525–529.

Kaul, M., and Vemuri, R. (2000). Design space exploration for block-processing based temporal partitioning of run-Time reconfigurable systems. *Journal of VLSI Signal Processing Systems 24* (2/3), 181–209.

Kean, T., (2000). It's FPL, Jim—But not as we know it! opportunities for the new commercial architectures. *Field Programmable Logic and Applications 1896*, 575–584.

Lemoine, E., and Merceron, O. (1995). Run-time reconfiguration of FPGAs for scanning genomic databases. *Proceedings of the IEEE Symposium on FPGAs Custom Computing Machines.* 90–98.

Leong, P.H.W., Leong, M.P., Cheung, O.Y.H., Tung, T., Kwok, C.M., Wong, M.Y., and Lee, K.H. (2001). Pilchard—A reconfigurable computing platform with memory slot interface. *Proceedings of the IEEE Symposium on Field Programmable Custom Computing Machines.*

Li, Z., Compton, K., and Hauck, S. (2000). Configuration caching management techniques for reconfigurable computing. *Proceedings of the IEEE Symposium on Field Programmable Custom Computing Machines.* 22–36.

Li, Z., and Hauck, S. (2001). Configuration compression for virtex FPGAs. *Proceedings of the IEEE Symposium on Field Programmable Custom Computing Machines.*

Liu, H., and Wong, D.F. (1999). Circuit partitioning for dynamically reconfigurable FPGAs. *Proceedings of the IEEE Symposium on FPGAs Custom Computing Machines.* 187–194.

Luk, W., Andreou, P., Derbyshire, A., Dupont-De-Dinechin, F., Rice, J., Shirazi, N. and Siganos, D. (1998). A reconfigurable engine for real-time video processing. *Field Programmable Logic and Applications 1482,* 169–178.

Luk, W., Lee, T.K., Rice, J.R., Shirazi, N., and Cheung, P.Y.K. (1999). Reconfigurable computing for augmented reality. *Proceedings of the IEEE Symposium on FPGAs Custom Computing Machines.* 136–145.

Luk, W., and McKeever, S. (1998). Pebble: A language for parametrized and reconfigurable hardware design. *Field Programmable Logic and Applications 1482,* 9–18.

Luk, W., Siganos, D., and Fowler, T. (1999). Automating qualification of reconfigurable cores. *Reconfigurable Systems. IEEE Digest 99/061.*

Luk, W., Shirazi, N., and Cheung, P.Y.K. (1996). Modeling and optimizing run-time reconfigurable systems. *Proceedings of the IEEE Symposium on FPGAs Custom Computing Machines.* 167–176.

Luk, W., Shirazi, N., Guo, S.R., and Cheung, P.Y.K. (1997). Pipeline morphing and virtual pipelines. *Field Programmable Logic and Applications 1304,* 111–120.

Lysaght, P. (1997). Towards an expert system for a priori estimation of reconfiguration latency in dynamically reconfigurable logic. *Field Programmable Logic and Applications 1304,* 183–192.

Lysaght, P., and Dunlop, J. (1994). Dynamic reconfiguration of FPGAs. In W. Moore and W. Luk (Eds.), *More FPGAs.* Abingdon EE&CS Books.

Lysaght, P., and Stockwood, J. (1996). A simulation tool for dynamically reconfigurable field programmable gate arrays. *IEEE Transactions on VLSI 4* (3), 381–390.

MacVicar, D., and Singh, S. (1998). Accelerating DTP with reconfigurable computing engines. *Field Programmable Logic and Applications 1482,* 391–395.

Mangione-Smith, W.H., *et al.* (1997). Seeking solutions in configurable computing. *IEEE Computer.* 38–43.

Marshall, A., *et al.* (1999). A reconfigurable arithmetic array for multimedia applications. *FPGA99.* 135–143.

McCaskill, J.S., and Wagler, P. (2000). From reconfigurability to evolution in construction systems: Spanning the electronic, microfluidic, and biomolecular domains. *Field Programmable Logic and Applications 1896,* 286–299.

McKay, N., and Singh, S. (1998). Dynamic specialization of XC6200 FPGAs by partial evaluation. *Field Programmable Logic and Applications 1482,* 298–307.

McGregor, G., Robinson, D., and Lysaght, P. (1998). A hardware/software codesign environment for reconfigurable logic systems. *Field Programmable Logic and Applications 1482,* 258–267.

McMillan, S., and Guccione, S.A. (2000). Partial run-time reconfiguration using JRTR. *Field Programmable Logic and Applications 1896,* 352–360.

Mencer, O., Platzner, M., Morf, M., and Flynn, M.J. (2001). Object-oriented domain specific compilers for programming FPGAs. *IEEE Transactions on VLSI 9* (1), 205–210.

National Semiconductor. (1993). *Configurable Logic Array Data Sheet.*

Ong, S.W., (2001). Automatic mapping of multiple applications to multiple adaptive computing systems. *Proceedings of the IEEE Symposium on FPGAs Custom Computing Machines.*

Page, I., and Luk, W. (1991). Compiling occam into FPGAs. In W. Moore and W. Luk (Eds.), More *FPGAs.* Abingdon EE&CS Books.

Purna, K.G., and Bhatia, D. (1999). Temporal partitioning and scheduling data flow graphs on reconfigurable computers. *IEEE Transactions on Computers 48* (6), 579–590.

Rabaey, J.M. (1997). Reconfigurable computing: The solution to low-power programmable DSP. *Proc. ICASSP.*

Rising, B., *et al.* (1997). Parallel graph coloring using FPGAs. *Field Programmable Logic and Applications 1304,* 121–130.

Robinson, D., McGregor, G., and Lysaght, P. (1998). New CAD framework extends simulation of dynamically reconfigurable logic. *Field Programmable Logic and Applications 1482,* 1–8.

Rupp, C.R., Landguth, M., Garverick, T., Gomersall, E., Holt, H., Arnold, J.M., and Gokhale, M. (1998). The NAPA adaptive processing architecture. *Proceedings of the IEEE Symposium on FPGAs Custom Computing Machines.* 28–37.

Schaumont, P., Verbauwhede, I., Keutzer, K., and Sarrafzadeh, M. (2001). A quick safari through the reconfiguration jungle. *Proceedings of the Design Automation Conference.* 172–177.

Schmit, H., Cadambi, S., Moe, M., and Goldstein, S. (2000). Pipeline reconfigurable FPGAs. *Journal of VLSI Signal Processing Systems, 24* (2/3), 129–146.

Sezer, S., Heron, J., Woods, R., Turner, R., and Marshall, A. (1998). Fast partial reconfiguration for FCCMs. *Proceedings of the IEEE Symposium on FPGAs Custom Computing Machines.* 318–319.

Shirazi, N., Luk, W., and Cheung, P.Y.K. (1998a). Automating production of run-time reconfigurable designs. *Proceedings of the IEEE Symposium on FPGAs Custom Computing Machines.* 147–156.

Shirazi, N., Luk, W., and Cheung, P.Y.K. (2000). Framework and tools for run-time reconfigurable designs. *IEEE Proceedings on Computers and Digital Techniques 147* (3), 147–152.

Shirazi, N., Luk, W., and Cheung, P.Y.K. (1998b). Run-time management of dynamically reconfigurable designs. *Field Programmable Logic and Applications, 1482,* 59–68.

Singh, S., and Slous, R. (1998). Accelerating Adobe Photoshop with reconfigurable logic. *Proceedings of the IEEE Symposium on FPGAs Custom Computing Machines.* 236–244.

Singh, S., and James-Roxby, P. (2001). Lava and Jbits: From HDL to bitstream in seconds. *Proceedings of the IEEE Symposium on Field Programmable Custom Computing Machines.*

Stogiannos, P., Dollas, A., and Digalakis, V. (2000). A configurable logic based architecture for real-time continuous speech recognition using hidden Markov models. *Journal of VLSI Signal Processing Systems 24* (2/3), 223–240.

Styles, H., and Luk, W. (2000). Customizing graphics applications: Techniques and programming interface. *Proceedings of the IEEE Symposium on Field Programmable Custom Computing Machines.* 77–87.

Tau, T., Chen, D., Eslick, I., Brown, J., and DeHon, A. (1995). A first generation DPGA implementation. *Proceedings of the Third Canadian Workshop on Field Programmable Devices.* 138–143.

Tessier, R., and Burleson, W. (2001). Reconfigurable computing for digital signal processing: A Survey. *Journal of VLSI Signal Processing 28* (1/2), 7–27.

Trimberger, S.M. (1994). *Field Programmable Gata Array Technology.* Norwell, MA: Kluwer Academic.

Trimberger, S., Carberry, D., Johnson, A., and Wong, J. (1997). A time-multiplexed FPGA. *Proceedings of the IEEE Symposium on FPGAs Custom Computing Machines.* 22–28.

Villasenor, J., and Hutchings, B. (1998). The flexibility of configurable computing. *IEEE Signal Processing Magazine.* 67–84.

Villasenor, J., and Mangione-Smith, B. (1997). Configurable computing. *Scientific American.* 66–71.

Villasenor, J., Schoner, B., Chia, K., Zapata, C., Kim, H.J., Jones, C., Lansing, S., and Mangione-Smith, B. (1996). Configurable computing solutions for automatic target recognition. *Proceedings of the IEEE Symposium on FPGAs Custom Computing Machines.* 70–79.

Vuillemin, J. (1996). Programmable active memories: Reconfigurable systems come of age. *IEEE Transactions on VLSI 4* (1), 56–69.

Weinhardt, M., and Luk, W. (2001). Pipeline vectorization. *IEEE Transactions on Computer-Aided Design of Integrated Circuits and Systems* 234–248.

Wirthlin, M.J., and Hutchings, B.L., (1995). A dynamic instruction set computer. *Proceedings of the IEEE Symposium on FPGAs Custom Computing Machines.* 99–107.

Wirthlin, M.J., and Hutchings, B.L. (1998). Improving functional density using run-time circuit reconfiguration. *IEEE Transactions on VLSI* 6 (2), 247–256.

Woods, R., Ludwig, S., Heron, J., Trainor, D., and Gehring, S. (1997). FPGA synthesis on the XC6200 using IRIS and Trianus/Hades. *Proceedings of the IEEE Symposium on FPGAs Custom Computing Machines.* 155–164.

Xilinx. (2001). Virtex series FPGAs. http://www.xilinx.com/products/virtex.htm

Ye, A.G., and Lewis, D.M. (1999). Procedural texture mapping on FPGAs. *FPGA99.* 112–120.

Zhong, P., Martonosi, M., Ashar, P., and Malik, S. (1998). Solving Boolean satisfiability with dynamic hardware configurations. *Field Programmable Logic and Applications 1482*, 326–335.

4

Operating Systems

Yao-Nan Lien
Department of Computer Science, National Chengchi University, Taipei, Taiwan

4.1 Introduction

The operating system of a computer is software, which controls all the computer's resources and provides the base upon which the application programs can be written. Modern operating systems also provide numerous services, such as interprocess communication, file and directory systems, data transfer over networks, and a command language for invoking and controlling programs.

A modern computer system consists of one or more processors, some main memory, clocks, terminals, disks, network interfaces, and other input/output devices. Writing programs that keep track of all these components and use them correctly is an extremely difficult job. The operating system puts a layer of software on top of the bare hardware, manages all parts of the system, and presents the user with an interface or **virtual machine** that is easier to understand and program. Thus, an **operating system** can be defined as a set of software extensions of primitive hardware, culminating in a virtual machine that serves as a high-level programming environment and manages the flow of work in a network of computers.

The virtual machine visible to the user is only the outermost of a series of software layers refining the base hardware. The principle behind the layered architecture is called information hiding—confining the details of managing a class of "objects" in a module that has a friendly interface with its users. With information hiding, designers can protect themselves from extensive reprogramming if the hardware or some part of the software changes: the change affects only the small portion of the software interfacing directly with that system component. By nesting these modules, a hierarchy of levels of abstraction is created so that at any level, users can ignore the details at all lower levels. At the highest level is system users who, ideally, are insulated from everything except what they want to accomplish. Operating systems structured in this way can support diverse environments: programming, game playing, real-time processing, office automation, database, and so on. This situation is shown in Figure 4.1.

At the bottom of the hierarchy is the hardware that may be composed of two or more layers. The lowest layer contains physical devices, such as processors, memory, and disks.

On top of the physical devices, there is a layer of system software that directly controls these devices and provides a cleaner interface to the upper layers of software. These software modules are best known as device drivers in the most popular personal computer systems and are sometimes treated as parts of the operating system. The interface to the device drivers is usually in a primitive format, called **machine language**. The

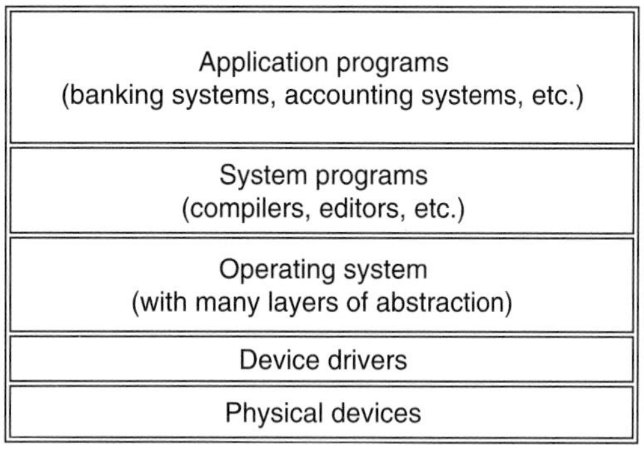

FIGURE 4.1 A Computer System: Hardware, System Programs, and Application Programs

machine language typically has between 50 and 300 instructions, mostly for moving data around the machine, doing arithmetic, and comparing values. Physical devices are controlled by loading values into special device registers. For example, a disk can be commanded to read by loading the values of the disk address, main memory address, byte count, and direction (read/write) into its registers. A major function of the operating system is to hide all this complexity and give the programmer a more convenient set of instructions to work with.

On top of the operating system is the rest of the system software, such as the command interpreter (shell), compilers, editors, and similar application independent programs. Hundreds of various system programs can be found in the Microsoft Windows 95/98/2000.

Finally, above the system programs come the application programs, which are probably written by the users to solve their particular problems regarding commercial data processing, scientific calculation, and game playing, for example.

4.2 Operating System Concepts

4.2.1 Process

A key concept in all operating systems is the process. A **process** is basically a program in execution. It consists of the executable programs; the program's data and stack; its program counter, stack pointer, and other registers; and all the other information needed to run the program. In a time-sharing system, the operating system periodically decides to stop running one process and starts running another.

When a process is temporarily suspended like this, all information about the process must be explicitly saved during the suspension. The process must be resumed (restarted) later in exactly the same state it was in when it was stopped by restoring the previously saved information. In many time-sharing operating systems, all the information about each process is stored in an operating system table called the **process table**; there is one table for each process currently in existence. Thus, a process consists of its address space, usually called the core image and its process table entry.

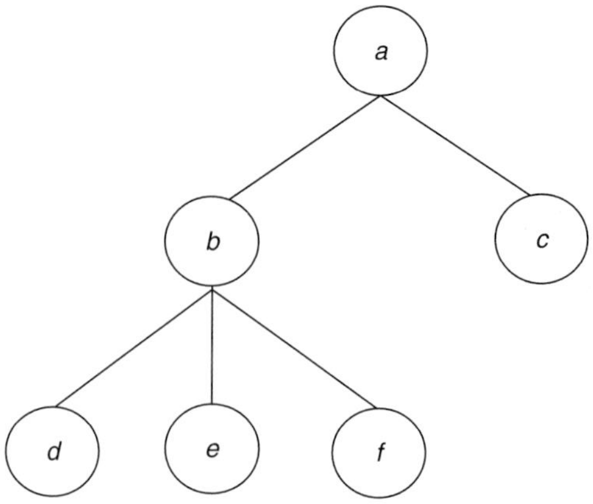

FIGURE 4.2 An Example of Process Tree

A process can be created or terminated by another process using **system calls**, which are extended instructions offered by the operating system. The newly created process is referred to as the **child process** which in turn can create its own child processes. The parent–child relationship among processes forms a process tree structure as shown in Figure 4.2.

Other process system calls are available to request more memory or to release unused memory, wait for a child process to terminate, and overlay its program with a different one.

Operating systems can communicate with a process by sending a signal to the process. The signal causes the process to temporarily suspend whatever it is doing, save its registers on the stack, and start running a special signal handling procedure. Then the signal handler is done, and the running process is resumed. Signals are the software analog of hardware interrupts and can be generated by a variety of causes, such as timers expiring, execution of an illegal instruction, and use of an invalid address. Signals are also used for process-to-process communication, when one process wants to communicate something to another.

In a multiprogramming system, each authorized user is assigned a UID for identification. In addition, each process is assigned a system-wide unique identification, called PID, as well as the UID of its owner for tracking purposes.

4.2.2 Files

Managing various objects, such as information in disks, display monitors, and keyboards, is an important operating system function. To hide peculiarities of these objects, the operating system presents the programmer with a nice, clean abstract model of device-independent files. A file must be opened before it can be read, written, and closed. Therefore, system calls are needed to create, remove, open, read, write, and close files.

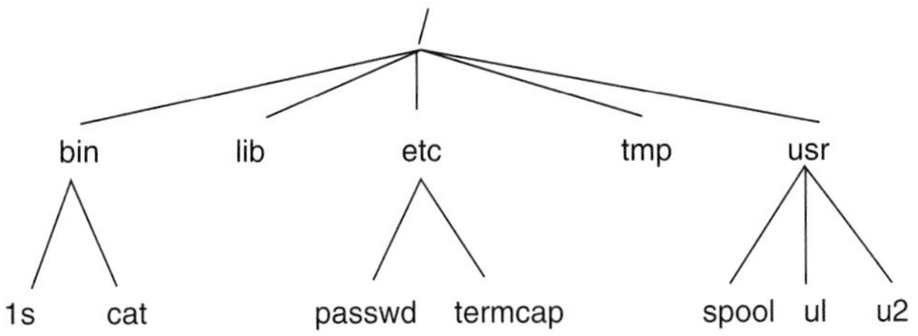

FIGURE 4.3 An Example of Hierarchical File Tree

Most operating systems provide a hierarchical file system for users to manage their files. Files can be grouped together into a directory, which is also a special file and can be grouped into another directory. This model also gives rise to a hierarchical file system, as shown in Figure 4.3.

Every file in the file hierarchy can be specified by its given path name from the top of the hierarchy, the root directory. Such absolute path names consist of the list of directories that must be traversed from the root directory to get to the file, with special symbols such as slashes separating the components. In Figure 4.3, the path for file **passwd** is denoted as **/etc/passwd**. The leading slash indicates that the path is starting from the root directory. As an aside, some systems use a back slash(\) instead of a regular slash(/) as the separator in path names.

At every instant, each process has a current working directory, in which path names not beginning with a slash are looked for. In Figure 4.3, if **etc** is the current directory, then the path name **passwd** denotes the same file as **etc/passwd**. A process can change its working directory by issuing a system call specifying the new working directory.

If several users can access the same computer, the system needs to provide a means for users to protect the privacy of their files. In many systems, a user can specify the type of protection, such as read, write, and execution, against certain groups of users. Before a file can be read or written, it must be opened, at which time the protection is checked. If the access is permitted, the system returns a small integer called a file descriptor or handle to use in subsequent operations. Otherwise, an error code is returned.

Some operating systems such as UNIX and MS-DOS provide a special file abstraction to allow users to perform *I/O* without knowing all details of the hardware. Special files can be read and written in the same way as regular files are read and written. Special files are categorized into two types: **block special files** and **character special files**. Block special files are used to model devices that consist of a collection of randomly addressable blocks, such as disks. By opening a block special file and reading, a program can directly access the desired block on the device without regard to the structure of the file system contained on it.

Character special files are used to model devices that consist of character streams rather than fixed-size randomly addressable blocks. Terminals, line printers, and network interfaces are typical examples of character special files. A program can interact with the user's terminal by reading and writing the corresponding character special file. In UNIX and MS-DOS, when a process is started up, file descriptor 0, called standard input, is normally arranged to refer to the terminal for the purpose of reading. File descriptor 1, called standard output, refers to the terminal for writing. File descriptor 2, called standard error, also refers to the terminal for output but is used only for writing error messages.

In some operating systems, such as UNIX, there is a special concept called **pipe**, which is a pseudo-file that can be used to connect two processes together. When a process wants to send data to another process, it writes on the pipe as though it were an input file for the second process. Thus, communication between processes looks very much like ordinary file reads and writes.

4.2.3 System Calls

In modern operating systems, applications are separated from the operating system itself. The operating system code runs in a privileged processor mode known as **kernel mode** and has access to system data and hardware. Applications run in a nonprivileged processor mode are known as **user mode** and have limited access to system data and hardware by making system calls, which are actually a set of tightly controlled application programming interfaces (APIs).

Corresponding to each system call is a library procedure that user programs can call. This procedure puts the parameters of the system call in a specified place, such as the machine registers; it then issues a TRAP instruction, which is a kind of protected procedure call, to start the operating system. The purpose of the library procedure is to hide the details of the TRAP instruction and make system calls look like ordinary procedure calls.

When the operating system gets control after the TRAP, it examines the parameters to see if they are valid, and if so, performs the work requested. When it is finished, the operating system puts a status code in a register, telling whether it succeeded or failed, and executes a **return from trap** instruction to return control back to the library procedure. The library procedure then returns to the caller in the usual way, returning the status code as a function value. Sometimes additional values are returned in the parameters.

4.2.4 Shell: Command-Based User Interface

A user can interact with the operating system through a special process, **the command interpreter** using a command language. A command language can be either very simple in the early days or as powerful as a regular programming language. Most UNIX systems offer one or more command interpreter with a name called shell or its variations. Unfortunately, the term **shell** may cause some confusion because it is also used by researchers as a generic term to denote the layer of software for users to interact with an operating system. When any user logs in, a shell is started up. The shell has the terminal as standard input and standard output. It starts out by displaying a prompt symbol, such as a dollar sign, indicating that it is waiting to accept a command. For example, if the user types:

```
who,
```

the shell creates a child process and runs the **who** program as the child. The shell is waiting until the child process is terminated. When the child finishes, the shell displays the prompt again and waits for the next command.

UNIX's shell offers a regular programming capability to its users such that its users can request very rich services from the operating system. In addition to the common flow control constructs, shell offers several unique features: *I/O* redirection, pipelining, and background job execution.

A user can specify standard output to be redirected to a file by typing, for example:

```
who > outfile.
```

Similarly, standard input can be redirected from another file, as in:

```
sort < infile > outfile,
```

which invokes the sort program with input taken from file **infile** and output sent to **outfile**.

The output of one program can be used as the input for another program by connecting them with a pipe. Thus, the command:

```
who | wc > /dev/tty 10
```

invokes the **who** program to list all users currently logged in the system and sends the output to **wc** to count the number of entries. The output of **wc** is redirected to a file, **/dev/tty10** which by convention is a special file denoting a terminal.

If a user put an ampersand after a command, the shell does not wait for it to complete. Instead it just gives a prompt immediately. For instance, the command:

```
cc file.c &
```

starts up a compiling job in the background, allowing the user to continue working as the compiler is running.

4.2.5 Graphical User Interface

A graphical user interface (GUI) is another user interface paradigm. A GUI presents the system resources, mostly data files and applications, in graphical objects called icons on the screen and allows users to use a mouse as another input device to express their demands, such as selecting an object, moving an object, and invoking a task. Each task occupies a resizable screen area called a **window** and all objects associated with a task are confined in its window. On the screen, there is a special icon called a **cursor** whose position defines the current focus of the user, and all input given by the user will be delivered to the window where the cursor is located. The user can select any object he or she wants freely by moving the cursor to the top of the object and clicking a button on the mouse. A double click usually means to invoke a task represented by the clicked icon. Because users can intuitively manipulate objects, GUI is also referred to as direct manipulation user interface (DMUI).

4.3 Operating Systems History

Most operatings systems for large mainframes are direct descendants of the third-generation systems, such as Honeywell Multics, IBM VI and VM/370, and CDC Scope. These systems introduced important concepts, such as time sharing, multiprogramming, virtual memory, hierarchical file systems, and device-independent *I/O* (Denning, 1971, 1976).

During the 1960s, many experimental time-sharing systems were constructed to test several new operating systems concepts. These included MIT's Compatible Time Sharing System (CTSS), the University of Cambridge Multiple Access System, IBM TSS/360, and the operating systems for the Manchester University Atlas and the RCA Spectra/70. The most ambitious project of all was Multics (Multiplexed Information and Computing Service) for the General Electric 645 processor (Organick, 1972) Multics simultaneously tested many new concepts such as processes, interprocess communication, segmented virtual memory, hierarchical file system, device independence, *I/O* redirection, a high-level language shell.

Perhaps the most influential current operating system is UNIX, a complete re-engineering of Multics, originally developed at AT&T Bell Laboratories for the DEC PDP computers. Although its size is much smaller than Multics, UNIX retains most of its predecessor's useful characteristics. It also introduces the pipe. The powerful yet simple shell command language together with a large library of utility programs make the UNIX an extremely high productive system for information professionals. Most of UNIX is written in the high-level language C, allowing it to be transported to a wide variety of processors from mainframes to personal computers (Ritchie and Thompson, 1974; Kernighan and Pike, 1984).

Although AT&T developed UNIX, longstanding antitrust provisions prevented AT&T from marketing UNIX or UNIX-based products. Source code was made available to universities for educational purposes. This source code opening policy had a positive effect on the computer industry. It allowed new hardware companies such as SUN, Apollo, and Silicon Graphics to focus on hardware design without having to invest a huge capital in designing operating systems. On the other hand, manufacturers were used to add enhancements and features of their own to differentiate them from other UNIX variants. New versions of UNIX are thus proliferated, resulting in a complex and convoluted history. The most influential two versions, created in the mid-1980s, are AT&T System V and 4.xBSD (Berkeley UNIX) from the Computer Science Department, University of California, Berkeley. Many familiar UNIX features have their origins in academia. Berkeley UNIX was the source of the TCP/IP implementation, sockets, and many common UNIX utilities.

Although compatibility remained a problem among numerous versions, UNIX continued to flourish because it was the only relatively open standards-based operating system not controlled by a single vendor that could be used as an alternative to mainframes.

POSIX is the first serious attempt to reconcile the two flavors of UNIX and was sponsored by the IEEE Standards Board, a highly respected neutral body. The POSIX committee produced a series of standards known as **1003.n-yyyy**, where **1003** is the IEEE POSIX project, **n** denotes the document, and **yyyy** is the year the standard was completed or last amended. It defines a set of library procedures that every conformant UNIX system must supply. There are actually more than 20 standards and draft documents under the POSIX umbrella. The **1003.1–1990** describes interfaces to operating system services at the source-code level. It describes syntax (in the C language) and behavior, not implementation of the interface by the operating system. The **1003.2–1992** describes a programmable shell and related utilities (Walli, 1995).

In the 1980s, a large family of operating systems was developed for personal computers, including MS-DOS, PC-DOS, Apple-DOS, CP/M. Coherent, Xenix, and MINIX. All these systems were of limited function, being initially designed for 8- and 16-bit microprocessor chips with small memories. Processor speeds and memories of personal computers are now sufficient to support full-fledged operating systems such as Linux, FreeBSD, and SUN's Solaris (Torvalds, 1999; Bowman *et al.*, 1999).

With multiprocessors and computer networks in the early 1980s, operating systems began to manage the resources of multiple computers at once. Two early examples are the StarOS and Medusa operating systems for the CM* (pronounced "CM star") machine, a multicomputer consisting of several dozen individual computers linked by a special network (Jones *et al.*, 1979; Ousterhout *et al.*, 1980). Established, single-machine operating systems, such as UNIX and DEC'S VMS, evolved to accommodate networks of computers. Such operating systems typically support standards for accessing files on remote servers from any machine in a network. SUN's Network File System (NFS) was one of the first widely available UNIX-based file systems that provided a single filename space on top of a network of servers and workstations (Sandberg *et al.*, 1985). Carnegie-Mellon's Andrew system provides a UNIX-based network file system that spans more than 5,000 computers around the campus; it allows users to access files without having to know their locations, and it improves performance by caching whole files at individual nodes in the network (Howard *et al.*, 1988).

In addition to a distributed file system, an ideal distributed operating system should give its users an illusion of a single synergistic computer with the collective computing capability extracted from a set of heterogeneous computer systems. The Mach operating system, also developed at Carnegie-Mellon University, is a typical distributed operating system. It handles a variety of distributed system operations, including a uniform file name space, a virtual shared computational memory, and multiprocessing; it is also compatible with UNIX (Accetta *et al.*, 1986; Rashid *et al.*, 1988). Another example is the Amoeba system designed by a team in Vrije Universiteit led by Andrew Tanenbaum (Tanenbaum, 1992).

With massively parallel computers containing thousands of processors, new challenges arise. Operating systems for these machines must support extremely fast synchronization and communication among thousands of processes. Each processor may have its own devices attached, and, hence, the operating system must control thousands of *I/O* channels at once. The concepts of virtual memory and time-sharing must be extended to accommodate massive parallelism. Perhaps the most important challenge is that the programming environment should permit parallel programs to be written with only modest effort beyond what is required for sequential ones (Denning and Tichy, 1990).

As the demand for user mobility increases, a user's need to carry a lightweight handheld personal computer and to be able to access information on the Internet increases as well. Hardware resources, communication bandwidth, and network connectivity available to this type of personal computer are extremely limited so that currently available operating systems running on such machines will not be able to achieve the best performance. Designing a new technology to accommodate this mobile computing environment becomes a new challenge to the operating system research.

The style of contemporary direct manipulation interface evolved largely from a prototype developed at the Xerox Palo Alto Research Center (PARC) in the 1970s. Zerox Star available in 1981 was the first commercial computer system equipped with a GUI. After several failed cases, Apple's Macintosh, introduced in 1984, was a major commercial success. It dominated the GUI personal computer markets for the entire 1980s (Press, 1990). Starting from Windows version 1, version 2, and version 3, Microsoft kept trying to move DOS users away from their command-line interface to a GUI interface and finally achieved its dominant position in the GUI PC market after introducing Windows 95 in 1995. Windows 98 was introduced in 1998 and Windows 2000 was introduced in 2000, as their names imply.

On the UNIX side, the X Window System developed at MIT achieved fairly widespread popularity. Rather than mandate a particular user interface style, X provides primitives to support several policies and styles so that there is no factory standard for any particular interface style. The based system in X is defined by a network protocol so that an application can use windows on any display in a network in a device-independent, network-transparent fashion (Scheifler and Gettys, 1986)

4.4 A Model Operating System

In the hierarchical structure of a model operating system, functions are separated according to their characteristic time scales and their levels of abstraction.

Table 4.1 shows an organization spanning 14 levels. It is not a model of any particular operating system but rather exhibits the relationships among the functions present in most operating systems. Each level encapsulates the system at lower levels into an abstract machine and presents itself to the user, or a program, by a set of objects and the associated operations. A program written at a given level can invoke visible operations at all lower levels, but not operations on higher levels.

4.4.1 Single-Machine Levels 1 through 7

Levels 1 through 7 are called **single-machine levels** because the operating system at these levels deals exclusively with the resources of a single machine. A single machine may contain one or several processors that share a common memory. A multimachine consists of several physically separate computer systems, each of which has its own processor(s) and private memory. There is no shared memory among the separate computers in a multimachine.

The lowest levels include the hardware and firmware of the system. **Level 1** is the electronic circuitry in which objects are such things as registers, gates, and memory cells, and operations are clearing registers, reading memory cells, and other similar items **Level 2** adds the instruction set that can deal with somewhat more abstract entities, such as evaluation stacks and arrays of memory locations. **Level 3** adds the procedure and the operations for call and return. **Level 4** introduces interrupts and a mechanism for invoking special procedures (interrupt-handler routings) when the processor receives an interrupt signal.

The first four levels correspond roughly to the basic machine as it is received from the manufacturer, although there is some interaction with the operating system. For example, interrupts are generated by hardware, but the interrupt-handler routines are part of the operating system.

Level 5 adds primitive processes that are single programs in the course of execution. The information required to represent a primitive process is its state word, which consists of the values of the registers in a processor. This level also provides a context switch operation, which transfers a processor's attention from one process to another by saving the state word of the first and loading the state word of the second. All processes ready to run are queued in a ready list. After a process is switched off its processor, a scheduler in this level selects the next process from the ready list to run. This level also provides semaphores, the special variables used to cause one process to stop and wait until another process has signaled the completion of a task.

Level 6 handles the access to the secondary storage devices. The programs at this level are responsible for operations such as positioning the head of a disk drive and transferring blocks of data. Software at a higher level determines the address of the data on the disk and places requests for data; the requesting process then waits at a semaphore until the transfer has been completed.

Level 7 implements a virtual memory, a scheme that gives the programmer the illusion of having a main memory space large enough to hold the entire program and all its data even if the available real memory is much smaller (Denning, 1970). Software at this level handles the interrupts generated by the hardware when a block of data is addressed that is not in the main memory; this software locates the missing block in the secondary storage, frees space for it in the main storage, and requests

TABLE 4.1 An Operating System Design Hierarchy

Level	Name	Objects	Example operations
14	Shell	User programming environment, data structures	Statements in shell language
13	Directories	Directories	Create, destroy, attach, detach
12	User processes	User process	Fork, suspend, resume, kill
11	Stream *I/O*	Streams	Open, close, read, write
10	Devices	External devices, including printers, displays, keyboards	Create, destroy, open, close, read, write
9	File system	Files	Create, destroy, open, close, read, write
8	Communications	Pipes	Create, destroy, open, close, read, write
7	Virtual memory	Addresses, segments	Create, destroy, map
6	Local secondary	Blocks of data, devices	Read, write, allocate, free store
5	Primitive processes	Primitive process, semaphores, ready list	Fork, suspend, resume, wait, signal, kill
4	Interrupts	Interrupt vectors, fault-handler programs	Invoke, mask, unmask, retry
3	Procedures	Procedure segments, call stacks, displays	Mark stack, call, return
2	Instruction set	Evaluation stacks, memory arrays	Load, store, index, unary and binary operators
1	Electrical circuits	Registers, gates, etc.	Clear, transfer, complement

level 6 to read in the missing block. The event is called a **page fault** and the handling process is called **swapping**.

4.4.2 Multi-Machine Levels 8 through 14

From level 8 and those above, the operating system deals with the resources of a larger world, including peripheral devices such as terminals and printers as well as other computers attached to the network. In this world, pipes, files, devices, user processes, and directories can be shared among all the machines.

Level 8 is concerned with communication between processes, which can be arranged through a single mechanism called a **pipe**. A pipe is a one-way channel: data streams in one end and out the other. A request to read items is delayed until items are actually in the pipe. A pipe can connect two processes of the same machine or of different machines equally well. Pipes are implemented in UNIX, MS-DOS, and many other systems.

Level 9 provides services for long-term storage of files. While level 6 deals with disk storage in terms of tracks and sectors, which are the physical units of the hardware, level 9 deals with more abstract entities of variable length. Indeed, a file may be scattered over many noncontiguous tracks and sectors. To be examined or updated, a file's contents must be copied between virtual memory and the secondary storage system. If a file is kept on a different machine, level 9 software can retrieve it by creating a pipe to level 9 on the file's home machine.

Level 10 provides access to external input and output devices, such as printers, plotters, and the keyboards and display screens of terminals. There is a standard interface with all these devices, and a pipe can be used to get access to a device attached to another machine.

Level 11 provides a means of attaching user processes interchangeably to pipes, files, or *I/O* devices. The idea is to make each fundamental operation of levels 8, 9, and 10 (**open**, **close**, **read**, and **write**) look the same so that the author of a program need not be concerned with the differences in these objects.

Level 12 implements user processes, which are virtual machines executing programs. While the level 5 process is primitive and can be completely defined by a state word that records the contents of the registers in a processor, a level 12 user process is a significant extension. It includes one or more primitive processes, a virtual memory containing the program and its workspace, a list of arguments supplied as parameters when the process was started, a list of objects with which the process can communicate, and certain other information about the context in which the process operates.

Level 13 manages a hierarchy of directories that catalogs the hardware and software objects to which access must be controlled throughout the network: pipes, files, devices, user processes, and the directories themselves. The central concept of a directory is a table that matches external names of objects to internal names. An **external name** is a string of characters having some meaning to users; an **internal name** is a binary code used by the system to locate the object. The user designates external names for objects, and the system generates the mapping from external to internal names automatically. A hierarchy arises because a directory can include among its entries the names of subordinate directories.

Level 14 is the shell that separates the user from the rest of the operating system. The shell interprets a high-level command language through which the user gives instructions to the system. Incorporating a listener program, usually called a command interpreter, that responds to a terminal's keyboard, the shell parses each line of input to identify program names and parameters, creates and invokes a user process for each program, and connects those as needed to pipes, files, and devices. Many shells incorporate window managers that allow users to refer to objects by manipulating icons denoting those objects on the display screen and that allow objects to send and receive data from different regions of the display.

4.5 Case 1: UNIX

4.5.1 Introduction to UNIX

UNIX began life as a minicomputer time-sharing system, but is now used on machines ranging from notebook computers to supercomputers. It offers three interfaces: **the shell, the system call library**, and the **system calls** themselves. The shell allows users to type commands for execution. These may be simple commands or more complex shell scripts. Standard input and output may be redirected.

The key concepts in UNIX include the **process**, the **memory mode**, the **file system**, and **I/O**. Processes may fork off subprocesses, leading to a tree of processes. The memory model consists of a text, data, and stack segment for each process. The file system supports regular files, directories, and two kinds of special files. Directories may contain subdirectories, leading to a hierarchical file system. *I/O* is done using character and block special files that are integrated into the file system. Access to UNIX services is achieved via system calls.

Process management in UNIX uses two key data structures, the **process table** and the **user structure**. The former is always in memory, but the latter can be swapped or paged out. Process creation is done by duplicating the process table entry and then the memory image. Scheduling is done using a priority-based algorithm that favors interactive users.

Memory management used to be done by swapping, but it is now done by paging in most UNIX systems. The core map keeps track of the state of each page, and the page daemon uses a clock algorithm to keep enough free pages around.

The file system uses three main tables: the **file descriptor table**, the **open file description table**, and the ***i*-node table**. Each open file has entries in all three of them.

Block device *I/O* uses a buffer cache to reduce the number of disk accesses. An LRU algorithm is used to manage the cache. Character *I/O* can be done in raw or cooked mode, the latter of which is implemented by a line discipline. Line discipline acts like a filter, taking the raw character stream from the terminal driver, processing it, and producing what is called a **cooked character stream**.

4.5.2 Using UNIX

The UNIX Shell

To use UNIX, a user has to log in first by typing user name and password. After a successful login, the login program starts up the command line interpreter, which is most likely a shell variant such as Bourne Shell, Korn Shell, or Berkeley C Shell that has been designed to make it look like C program. A user can choose whatever shell variant he or she likes. The shell initializes itself and then displays a prompt character, often a dollar sign, on the screen and waits for the user to type a command line.

When the user types a command line. The shell extracts the first word from it, assumes it is the name of a program to be run, searches for this program, and runs the program if it finds it. The shell then suspends itself until the program terminates, at which time it tries to read the next command.

Commands may take arguments, which are passed to the called program as a character string. For example, the command line:

```
rm filea fileb
```

invokes the **rm** program to delete the files specified in the arguments **filea** and **fileb**.

With a few exceptions, an argument started with a character(-)denotes an option specified by the user to the command. In the following command:

```
head -20 testfile,
```

the first arguments, **-20**, tells **head** to print the first 20 lines of **testfile** instead of the default number of lines, **10**. The convention of using a dash to denote an option to a command is strongly recommended to reduce the possible confusion to the user. Most UNIX commands accept multiple options and arguments. With a few exceptions, options are often specified before all other arguments.

To make it easy to specify multiple file names, the shell accepts magic characters, sometimes called **wild cards**. An asterisk matches all possible strings, so the following two commands are equivalent:

```
rm file*
rm filea fileb
```

These commands are equivalent if, in the current working directory, only files named **filea** and **fileb** match to **file***. To execute the first command, the shell first expands **file** into **filea** **fileb** then invokes the **rm** program to delete **filea** and **fileb**. Another wild card is a question mark that matches any one character.

A program like the shell does not have to open the terminal to read from it or write to it. Instead, when it (or any other program) starts up, it automatically has access to a file called **standard input for reading**, a file called **standard output for writing normal output**, and a file called **standard error for writing error messages**. Normally all three default to the terminal, so that reads from standard input come from the keyboard and writes to standard output or standard error go to the screen. Many UNIX programs read from standard input and write to standard output as the default. For example, the command:

```
sort
```

invokes the **sort** program, which reads lines from the terminal (until the user types a CTRL-D, to indicate end of file), sorts them alphabetically, and writes the result to the screen.

It is possible to redirect standard input and standard output, which often is useful. The syntax for redirecting standard input uses a less than sign (<) followed by the input file name. For example, the command:

```
ex File ToBeEdit < exscript
```

invokes **ex** to edit the file **FileToBeEdit** using the editing instructions stored in the file **exscript** as if the editing instructions are given by the user from the keyboard. Similarly, standard output is redirected using a greater than sign. It is permitted to redirect both in the same command. For example, the command:

```
sort < infile > outfile
```

causes **sort** to take its input from the file **infile** and write its output to the file **outfile**. Because standard error has not been redirected, all error messages still go to the screen. A program that reads its input from standard input, does some processing on it, and then writes its output to standard output is called a **filter**.

The UNIX provides pipe capability so that the output of one program can be used as the input for another program by connecting them with a pipe. In the following command:

```
sort < infile | head −30,
```

the vertical bar, called a **pipe symbol**, says to take the output from **sort** and use it as the input to **head**. A collection of commands connected by pipe symbols, called a **pipeline**, may arbitrarily contain many commands. A four-component pipeline is shown by the following two examples. The first one is written as:

```
head −100 infile | cat −n| cut −c1-7
```

and takes the first 100 lines of **infile**, adds line numbers, and removes everything except the first seven characters in each line. This pipeline effectively generates a number sequence from 1 to 100. (Assume the file **infile** has no fewer than 100 lines.) The second one is written as:

```
head -100 infile | cat -n | sort -nr | cut -c8-> outfile
```

and takes the first 100 lines of **infile**, adds line numbers, sorts it in descending numerical order, and finally eliminates line numbers and writes the result into **outfile**. In fact, this example takes the first 100 lines from **infile** and reverses it so that the 100th line becomes the first line and the first line becomes the 100th line in **outfile**. These two examples demonstrate how UNIX provides basic building blocks (numerous filters), each of which does one job, as well as provides a mechanism for them to be put together in almost limitless ways.

UNIX is not only a time-sharing system; it is also a general-purpose multiprogramming system. A single user can run several programs at once, each as a separate process. The shell syntax for running a process in the background is to follow its command with an ampersand (&). Thus, the command:

```
a.out < infile > outfile &
```

runs a user program, **a.out**, from the input file, **infile**, then writes the result to **outfile** but does it in the background. As soon as the command has been typed, the shell displays the prompt and is ready to accept and handle the next command. Pipelines can also run in the background.

It is possible to put a list of shell commands in a file, say **scriptfile** and then start a shell with this file as standard input or as a file argument like following:

```
sh scriptfile
```

The (second) shell just processes them in order, the same as it would with commands typed on the keyboard. Files containing shell commands are called shell scripts. Shell scripts may assign values to shell variables and then read them later. They may also have parameters and use **if, for, while** and **case** constructs. Thus a shell script is really a program written in a shell language.

Files and Directories in UNIX

A UNIX file is a sequence of zero or more bytes containing arbitrary information. No distinction is made between ASCII files, binary files, or any other kinds of files. The meaning of the bits in a file is entirely up to the file's owner. The system does not care.

By convention, many programs expect file names to consist of a base name and an extension separated by a dot (which counts as a character). Thus **prog.c** is typically a C program, **prog.o** is usually a compiled object file, and **a.out** is normally an executable compiled object program. These conventions are not enforced by the operating system.

Files can be protected by assigning each one a 9-bit mode by the owner or the super user, sometimes called the **rights bits.** The first 3 bits refer to the owner's access right to the file. The next 3 bits apply to other members of the owner's group (e.g., department or project). The last 3 bits pertain to everyone else; these 3 bits control reading, writing, and executing the file, respectively. Thus a mode of 640 (octal) means that the owner can read and write the file, other members of the owner's group can read it, and outsiders have no access at all. A mode of 777 (octal) allows everyone to do everything to the file. On the screen, these two example modes are displayed as **rw-r-----** and **rwxrwxrw**, respectively.

Files can be grouped together for convenience in directories. Directories are stored as files and, to a large extent, can be treated like files. They are protected with the same 9 bits as files, except that the 3 **execute bits** grant or deny permission to search the directory rather than execute it.

Directories can contain subdirectories, leading to a hierarchical file system. The root directory is denoted as / and usually contains several subdirectories. The / character is also used to separate directory names so that the name **/usr/man/sort 1** denotes the file **sort. 1** located in the directory **man,** which itself is in the **/usr** directory. Some of the major directories of the average UNIX file system are shown in Table 4.2.

UNIX Utility Programs

The user interface to UNIX consists not only of the shell but also of a large number of standard utility programs. Roughly speaking, these programs can be divided into six categories:

1. File and directory manipulation commands
2. Filters
3. Compilers and program development tools
4. Text processing
5. System administration
6. Miscellaneous

The POSIX 1003.2 standard specifies the syntax and semantics of just under 100 of these programs, primarily in the first three categories. The idea of standardizing them is to make it possible for anyone to write shell scripts that use these programs and work on all POSIX conformant UNIX systems.

A selection of POSIX utility programs is listed in Table 4.3 along with a short description of each one. All UNIX systems have these and many other programs.

TABLE 4.2 Some Important Directories Found in Most UNIX Systems

Directory	Description
/bin	Frequently used system binaries
/dev	Special drivers for I/O devices
/etc	Miscellaneous system administration parameters
/lib	Frequently used libraries
/tmp	Temporary files once stored here
/usr	Contains all user files in this part of the tree
/usr/include	System-provided header files
/usr/man	On-line manuals
/usr/spool	Spooling directories for printers, e-mail, and other daemons

TABLE 4.3 Some Popular UNIX Utility Programs Required by POSIX

Command	Description
awk	A pattern matching language
basename	Strip off prefixes or suffixes from a file name
cat	Link file(s) and write them to standard output
cc	Compile a C program
chmod	Change protection mode for file(s)
comm	Print lines common to two sorted files
cp	Make a copy of a file
cut	Make each column in a document into a separate file
date	Print the date and time
diff	Print all the differences between two files
echo	Print the arguments (used mostly in shell scripts)
find	Find all the files meeting a given condition
grep	Search file(s) for lines containing a given pattern
head	Print the first few lines of file(s)
kill	Send a signal to a process
lp	Print a file on a printer
ls	List files and directories
make	Recompile those parts of a large program that have changed
mkdir	Make a directory
mv	Rename a file or move file(s)
paste	Combine multiple files as columns in a single file
pwd	Print the working directory
rm	Remove file(s)
rmdir	Remove one or more directory
sed	A stream (i.e., noninteractive) editor
stty	Set terminal options such as the characters for line editing
sort	Sort a file consisting of ASCII lines
tail	Print the last few lines of a file
tr	Translate character codes
uniq	Delete consecutive identical lines in a file
wc	Count characters, words, and lines in a file

4.5.3 Linux: An Open Source UNIX Clone for a Personal Computer

Although UNIX has been installed on numerous computer systems ranging from supercomputers to personal computers, it is not suitable for casual personal users for at least two reasons:

1. Its command-based user interface is not friendly enough for casual users.
2. It requires a professional staff for system operation, administration, and maintenance.

The situation did not used to be better for professional personal users either. Not until 1990s was UNIX popular among this type of users for another major reason: there was no standard configuration for the personal computer. Numerous new hardware components, however, are made available on the market everyday. These hardware components are all sold together with necessary system software packages, usually device drivers, so that users can easily **plug-in** these drivers into their operating systems. Although with minimum effort, a casual user can reconfigure the operating system for his or her personal computer, these device drivers usually do not have a version for any UNIX-clone. Furthermore, the companies that sell UNIX-clone for personal computers do not make an operating system source code available, so their users cannot develop required device drivers by themselves. Thus, the popularity of personal computer UNIX is limited by the poor availability of device drivers.

Starting from the early 1990s, a UNIX clone, Linux, was available for free on numerous personal computers with its source code open to the user community. Linux is written from scratch by Linus Torvalds with the assistance of a loosely-knit team of hackers from across the Internet. It aims toward POSIX compliance and has all of the features one would expect of a modern, fully fledged UNIX: true multitasking, virtual memory, shared libraries, TCP/IP networking, and, most importantly, device drivers for most available hardware components.

Linux runs mainly on the IBM-compatible PCs that are equipped with Intel 80386/80486/Pentium CPUs. It requires much less hardware resources than Microsoft Windows. Ports to other architectures are underway. Linux has most standard UNIX utilities including X-Windows and TCP/IP as well as numerous free software that people have compiled or ported from other architecture. There is even a DOS emulator that can run DOS itself and some (but not all) DOS applications. Work has been progressing on an emulator for Microsoft Windows binaries.

The Linux kernel is protected under the GNU General Public License, which basically means that one may freely copy, change, and distribute it, but one may not impose any restrictions on further distribution and must make the source code available to others.

Although Linux popularity has gained great momentum in recent years, it is by no means for casual users because it still requires a certain level of professional skills and some administrative efforts. Professional personal users obtain assistance mainly from various newsgroups on the Internet, especially Usenet. One can look for those newsgroups whose names started with comp.os.linux, comp.unix, and comp.windows.x to obtain more information.

4.6 Case 2: MS-DOS

4.6.1 Introduction to MS-DOS

Microsoft MS-DOS is a single-user operating system for the IBM PC and its successors. It was originally based on CP/M, but many elements from UNIX have been added over the years. Only one process at a time can be active, but a process can create and execute a child process. Doing so, however, suspends the parent until the child is finished.

Memory management in MS-DOS involves four separate regions: **conventional memory** (below 640 K), **upper memory** (between 640 K and 1 M), **high memory** (the 64 K just above 1 M), and **extended memory** (above 1 M). Each of these has different properties and is used in different ways. Ordinary programs are restricted to conventional memory, however, and thus cannot exceed 640 K. Overlays and expanded memory are two of the techniques that are used to get around this limit to some extent.

The MS-DOS file system supports hierarchical directories, absolute and relative path names, and many of the same system calls as the UNIX file system. *I/O* is done using special files, both block and character. It also has the capability to handle primitive shell scripts, called **batch files**.

The implementation of MS-DOS is closely tied to the underlying architecture. MS-DOS makes use of a hardware ROM built into the IBM PC called the basic input output system. (BIOS). The BIOS contains device drivers for the standard devices. Whereas UNIX is self-contained with all device drivers built-in, MS-DOS could just call BIOS procedures to do *I/O*. In addition, users are free to install their own drivers for special devices.

The file system is based on the use of a file allocation table (FAT) for each disk. For each file, a chain of blocks is maintained in the FAT. Directory entries in MS-DOS contain some of the information that in UNIX is in the *I*-nodes.

The character length of file names in MS-DOS has a so-called 8.3 format constraint. Each file can have a name no more than eight characters long with an optional dot extension no more than three characters long. Some examples of valid and invalid file names are listed in Table 4.4.

Although it is not mandatory, file names with certain dot extensions have been associated with certain applications. For example, file names with a **.doc** extension are usually associated with Microsoft word processors, such as MS-Word. The capability of associating file names to applications has been implemented in the Microsoft Windows as a convenient operating system feature. Some examples of such associations are listed in Table 4.5.

Since it was first released along with the IBM PC in August 1981, MS-DOS has kept evolving from version 1.x, 2.x, 3.x, 4.x, up to 5.x with various features added. Its command-based user interface and obsolete programming environment makes it an idiosyncratic and unfriendly operating system for personal computers. In early 1990s, Microsoft developed Windows mouse-driven graphical user interface to replace MS-DOS. Early versions of Windows were executed on the top of MS-DOS so that users had to run MS-DOS first and then invoke Windows to get into a Windows environment. Starting from Windows 95, MS-DOS became a component of Windows so that today's users can start the Windows without running MS-DOS first. After more than 20 years of service, MS-DOS retired from the position of dominant operating system for personal computers although it kept evolving to version 6.x and 7.x.

TABLE 4.4 Examples of Valid and Invalid MS-DOS File Names

Valid file names	Invalid file names
file.txt document.doc hypertxt.htm program.c	file.text mydocument.doc hypertxt.html

TABLE 4.5 Examples of the Association Between Dot File Extensions and Applications

Extensions	Associated applications
.txt	Any plain text viewer (e.g., MS-Notepad)
.doc	MS-Word, MS-Wordpad
.ppt	MS-Powerpoint
.xls	MS-Excel
.wav	MS-MediaPlayer
.mpg	MS-MediaPlayer
.mov	quicktime
.reg	regedit
.zip	pkzip, winzip
.mp3	winamp
.htm	Any HTML viewer (e.g., MS Iexplorer or Netscape Navigator)

4.6.2 Using MS-DOS

Using MS-DOS can be best compared to using a stripped down early version of UNIX. There is a command interpreter or shell (called command.com), a file system, system calls, utility programs, and other features that are often similar to their UNIX counterparts but more primitive.

To use MS-DOS, you just turn on the computer. A few seconds later, the shell prompt appears. Unless the PC hardware itself has a security protection set up, there is no login procedure and no password because MS-DOS is really meant for a personal computer. The underlying assumption is that the machine is used by only one person. For the same reason, files and directories do not have owners, and there are no protection bits. There is also no concept of a superuser. The regular user can do everything.

To run a program, you type its name and arguments into the shell. The shell then forks off a child, passes the arguments to the child, and keeps quiet until the child has finished executing and exists. It is not possible to put an ampersand (**&**) at the end of the command line to start the command off in the background.

Some of the UNIX shell features are also present in MS-DOS, but others are not. The wild card characters * and **?** are present and match all strings and one character, respectively, just as in UNIX. Thus, the command:

```
copy *.c src
```

means copy all files in the current directory ending in **c** to the **src** directory, and the command:

```
del ???
```

means delete from the current directory all files whose names are precisely three characters. Command.com does not distinguish between upper and lower cases. Thus typing:

TABLE 4.6 Some of the MS-DOS Commands Built into the Command.com

MS-DOS	UNIX	Description
Command	Equivalent	
copy	cp	Copy one or more files
date	date	Display or change the current date
del	rm	Remove one or more files
dir	ls	List files and directories
mkdir	mkdir	Create a new directory
rename	mv	Rename a file
rmdir	rmdir	Remove an empty directory
type	cat	Display a file

```
COPY A.TXT B.TXT
```

has exactly the same effect as typing:

```
copy a.txt b.txt.
```

Redirection of standard input and output and filters work the same way as in UNIX and even use the same notation. Pipes are also allowed, but they are implemented differently (using temporary files), which introduces some subtle changes in their behavior. Other features taken from UNIX are shell variables, prompts that can be set by users, and shell scripts.

On the other hand, only one command per line is permitted, there are no background jobs, the use of **[a–z]** for a range is not allowed, and the quoting of strings is not recognized.

Commands in MS-DOS are divided into two categories **internal** and **external.** The are about 40 internal commands that are executed by the shell itself; the external ones are genuine programs, typically in the |**dos** or |**bin** directory. Some of the most common internal commands, along with their UNIX equivalent, are listed in Table 4.6.

References

Accetta, M., *et al.* (1986). Mach: A new kernel foundation for Unix development. *Proceedings of USENIX 1986 Summer Conference Vol,* 93–112.

Bolsky, M.I., and Korn, D.G. (1989). The Kornshell command and programming language. Englewood Cliffs, NJ: Prentice Hall.

Brinch H.P., Bowman, I.T., Holt, R.C., and Brewster, N.V. (1999). Linux as a case study: Its extracted software architecture. *Proceedings of the 1999 International Conference on Software Engineering.* 555–563.

Brinch H.P. (1973). Operating system principles. Englewood Cliffs, NJ: Prentice Hall.

Brown, R.L., Denning, P.J., and Tichy, W.F. (1984). Advanced operating systems. *IEEE Computer 17*(10), 173–190.

Bourne, S.R. (1978). The UNIX shell. *The Bell System Technical Journal 57* (6), 1971–1990.

Comer, D. (1984). Operating system design: The XINU approach. Englewood Cliffs, NJ: Prentice Hall.

Denning, P.J. (1970). Virtual memory. *Computing Surveys 2* (3), 154–216.

Denning, P.J. (1976). Fault-tolerant operating systems. *Computing Surveys 8*(4), 359–389.

Denning, P.J., and Tichy, W.F. (1990). Highly parallel computation. *Science 250,* 1217–1222.

Howard, J.H. *et al.* (1988). Scale and performance in a distributed file system. *ACM Transactions on Computer Systems 6* (1), 51–81.

Jones, A.K. *et al.* (1979). StarOS, A multiprocessor operating system for the support of task forces. *Proceedings of the Seventh Symposium on Operating Systems Principles.* 117–127.

Kernigham, B.W., and Pike, R. (1984). The UNIX programming environment. Englewood Cliffs, NJ: Prentice Hall.

Organick, E.I. (1972). The multics system: An examination of its structure. Cambridge, MA: The MIT Press.

Ousterhout, J.K., *et al.* (1980). Medusa: An experiment in distributed operating system structure. *Communications of the ACM 23* (2), 92–105.

Press, L. (1990). Personal computing: Windows, DOS, and the MAC. *Communications of the ACM 33* (11), 19–26.

Rashid, R. *et al.* (1988). Machine-independent virtual memory management for paged uniprocessor and multiprocessor architectures. *IEEE Transactions on Computers 37* (8), 896–908.

Ritchie, D.M., and Thompson, K.L. (1974). The UNIX time-sharing system. *Communications of the ACM 17* (7), 365–375.

Sandberg, R., *et al.* (1985). Design and implementation of the sun network file system. *Proceedings of USENIX 1985 Summer Conference.* 119–130.

Scheifler, R.W., and Gettys, J. (1986). The X Window system. *ACM Transactions on Graphics 5* (2), 79–109.

Tanenbaum, A.S. (1987). Computer networks. Englewood Cliffs, NJ: Prentice Hall.

Tanenbaum, A.S. (1992). Modern operating system. Englewood Cliffs, NJ: Prentice Hall.

Tanenbaum, A.S., *et al.* Experiences with the amoeba distributed operating system. 33(12), 46–63.

Torvalds, L. (1999). The Linux edge. *Communications of the ACM 42* (4), 38–39.

Walli, S.R. (1995). The POSIX family of standards. *Standard View 3* (1), 11–17.

Wulf, W.A., Levin, R., and Harbison, S.P. (1981). HYDRA/C.mmp, an experimental computer system. New York: McGraw-Hill.

5

Expert Systems

Yi Shang
Department of Computer Science, University of Missouri-Columbia, Columbia, Missouri, USA

5.1 Overview

An expert is a person who has special skills and knowledge about a restricted domain or field. An **expert system** is a computer program that represents and uses skills and knowledge of one or more human experts to provide high-quality performance in a specific domain.

Expert systems offer a number of benefits when compared with human experts. Expertise embodied in an expert system is permanent, like other software, whereas human expertise is perishable. Expert systems can produce consistent results on the same tasks and handle similar situations consistently, whereas humans can get tired or bored and are influenced by various effects (e.g., their judgment can be easily impacted by new information). Expert systems are inexpensive to operate, easy to reproduce and distribute, and can provide permanent documentation of the decision process. In addition, expert systems may contain knowledge from several human experts, giving them more breadth and robustness than a single expert.

Although expert systems enjoy these advantages, they also have many weaknesses that include lack of common sense knowledge, narrow focus and restricted knowledge, inability to respond creatively to unusual situations, and difficulty in adapting to changing environments.

5.1.1 A Brief History

In 1956, a summer workshop at Dartmouth College attended by ten influential computer scientists marked the birth of artificial intelligence (AI). Following this event, researchers pursued different areas to develop various types of intelligent systems including expert systems.

From the mid- to late 1960s, several important expert systems were developed. At Stanford, Edward Feigenbaum, Joshua Ledergerg, Bruce Buchanan, and Georgia Sutherland developed DENDRAL, the first expert system that interpreted the output of a mass spectrometer for analysis of the structure of organic chemical compounds. At MIT, Joseph Weizenbaum built ELIZA, an interactive dialog expert system, and Joel Moses built MACSYMA, an expert system using symbolic reasoning for solving mathematical problems. During the 1960s, LISP was the dominant programming language used in developing expert systems.

Not many expert systems were built in the 1960s because they were developed from scratch and the development time was long (Clancy and Shortliffe, 1984). This situation changed dramatically following the development of the MYCIN expert system. MYCIN was developed at Stanford University from 1972 to 1980 for diagnosing and treating patients with infectious blood diseases caused by bacteria in the blood and meningitis (Shortliffe, 1976). MYCIN is a rule-based expert system that incorporates approximately 500 rules and uses backward-chaining reasoning, which is discussed in more detail later in this chapter.

During the development of MYCIN, the importance of separating a knowledge base and an inference engine was recognized. Shortly afterwards, EMYCIN (Buchanan and Shortliffe, 1984), the domain-independent version of MYCIN, was produced by removing specific domain knowledge about infectious blood disease. It became the forerunner of expert system shells. The impact of EMYCIN was immediate in the development of subsequent expert systems. For example, PUFF (Aikens *et al.*, 1983), a program for analyzing pulmonary problems, was developed based on EMYCIN in about 5 years, a significant improvement over the 20 years required to develop MYCIN.

The value of expert system shells was quickly recognized by the research community and industry. The number of expert systems increased dramatically in the following years. The estimated number of developed expert systems increased from 50 in 1985 to 2200 in 1988 and then to 12,500 in 1992 (Durkin, 1993; Hammond, 1986). The vast majority of them have been developed using a shell, with LISP, PROLOG, and OPS-5 as the other major languages being used. During the past three decades, the field of expert systems has changed from a technology confined to research circles to one being used commercially to aid human decision making in a wide range of applications.

5.1.2 Characteristics and Categories

Compared to conventional computer applications and other AI programs, expert systems have the following distinguished characteristics:

- They deal with complex problems that require a considerable amount of human expertise.
- They encode human knowledge for a specific domain and simulate human reasoning using the knowledge.
- They often solve problems using heuristic or approximate methods, which are not guaranteed to find the best solutions.
- They are able to explain and justify solutions to human users at different knowledge levels.

An expert system is meant to embody the knowledge of human experts in a particular domain so that nonexpert users can use it to solve difficult problems. The knowledge and information handled by the expert system may be inaccurate, fuzzy, or incomplete. Hence, an expert system should be able to reason under approximate and noisy conditions. To improve its performance so that recurring problems can be solved faster and new problems can be solved better, an expert system should learn from its past experience automatically. The learning may be achieved either from previous examples or through analogies between similar problems. In a network environment, an expert system may interact with other systems, such as by obtaining information from a database or by consulting other expert systems. Similar to human experts, an expert system should be able to explain its reasoning process, justify its conclusion, and answer questions about the inference procedure. The explanation ability is also important in the development of the expert system. It enables the designer to validate the encoded knowledge and the reasoning algorithms. Finally, for real-world applications that require real-time response, an expert system should perform well under time and resource constraints.

In the past, expert systems have been built to solve a large range of problems in the fields of engineering, manufacturing, business, medicine, environment, energy, agriculture, telecommunications, government, law, education, and transportation. Generally, they fall into the following broad categories (Waterman, 1986):

- Controlling: Governing the behavior of a complex system
- Designing: Configuring components to meet design objectives and constraints
- Diagnosing: Determining the cause of malfunctions based on observations
- Instructing: Assisting students' learning and understanding of a subject
- Interpretating: Forming high-level conclusions or descriptions from raw data
- Monitoring: Comparing a system's observed behavior to the expected behavior
- Planning: Constructing a sequencing of actions that achieve a set of goals
- Predicting: Projecting probable consequences from given situations
- Scheduling: Assigning resources to a sequence of activities
- Selecting: Identifying the best choice from a list of possibilities
- Simulating: Modeling the behavior of a complex system

Among the large number of developed expert systems, here are some well-known ones:

- DENRAL: A chemical expert system that interprets mass spectra on organic chemical compounds:
- ELIZA: An interactive dialog expert system
- HEARSAY: A speech-understanding expert system
- INTERNIST: One of the largest medical expert systems for internal medicine
- MACSYMA: A symbolic mathematics expert system that solves math problems involving algebra, calculus, and differential equations
- MOLGEN: An object-oriented expert system for planning gene-cloning experiments
- MYCIN: A medical expert system that deals with bacterial infections

- PUFF: A medical diagnostic system for interpreting the results of respiratory tests
- PROSPECTOR: An expert system for predicting potential mineral deposits
- XCON (originally called R1): An expert system used to configure computers that consist of hundreds of modules

5.1.3 Architecture

The architecture of a typical expert system is shown in Figure 5.1. It contains the following major components:

- **A knowledge base** contains specialized knowledge of the problem domain. In a rule-based expert system, the knowledge including facts, concepts, and relationships is represented in the **if...then...** rule form.
- **An inference engine** performs knowledge processing. It is modeled after the reasoning process of human experts and applies knowledge in the knowledge base as well as information on a given problem for its solution.
- **A working memory** holds the data, goal statements, and intermediate results that make up the current state of the inference process.
- **An explanation subsystem** provides explanations of the reasoning process of the system and justifications for the system's actions and conclusions. Adequate explanation facilitates the acceptance of the system by the user community.
- **A user interface** is the window through which the user interacts with the system. Interface styles include question-and-answer, menu-driven, natural language, graphics interfaces, and online help. An expert system needs to be easy to learn and friendly to use.
- **A knowledge-base editor** is the window through which the knowledge engineer develops the system. It also assists in maintaining and updating the knowledge base.
- **A system interface** links the expert system to external programs and information sources. For example, the expert system may consult other expert systems or obtain information over the Internet.

Among these components, the two most important ones are **knowledge base** and **inference engine**. The knowledge base contains formally encoded domain-specific knowledge of a problem, whereas the inference engine solves problems using various reasoning algorithms. To achieve expert-level performance, expert systems need to access a substantial amount of domain knowledge and to apply the knowledge efficiently through one or several reasoning mechanisms.

The separation of the knowledge base and the inference engine is essential for the success of expert systems. It is very important for a number of reasons: First, knowledge can be represented in a more natural and declarative form that is easier to understand and implement. For example, **if... then...** rules represent domain knowledge in a more declarative form than embedding the knowledge in problem-solving procedures. Second, developers can focus on creating and organizing high-level knowledge rather than on the detailed implementation. Third, changes made to the knowledge base do not affect the function of inference engine and vice versa. Finally, the same problem-solving techniques employed by the inference engine can be used in a variety of applications.

As a subfield of artificial intelligence (AI), expert system research has strong links with related topics in AI. The four fundamental topics in expert systems are **representing knowledge, reasoning using knowledge, acquiring knowledge**, and **explaining solutions**. Building a successful expert systems depends on a number of factors, including the nature of the task, the availability of expertise, the amount of knowledge required, and the ability to formally encode the expertise and reason efficiently with it to solve the task.

5.1.4 Languages and Tools

Expert systems have been constructed using various general-purpose programming languages as well as specific tools. LISP and PROLOG have been used widely. OPS-5 has also been

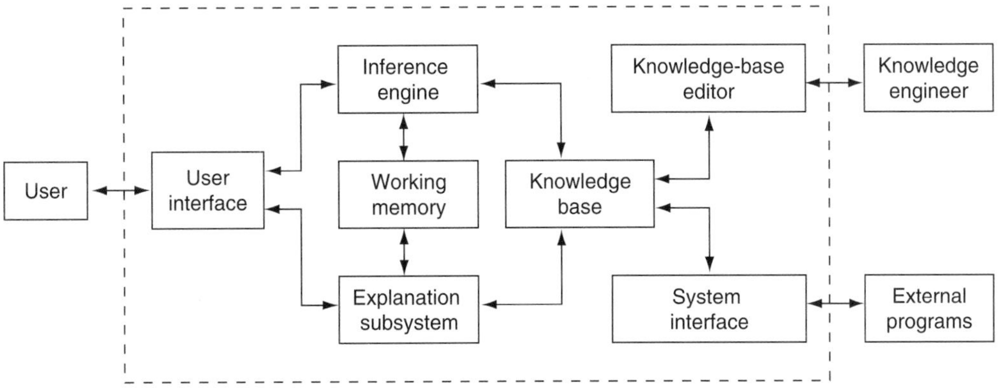

FIGURE 5.1 Architecture of a Typical Expert System

popular among rule-based programmers. OPS is a product of the Instructable Production System Project at CMU in 1975. It was used to create one of the classic expert systems called R1 (later called XCON) to help DEC configure VAX computers. C and C + + have been used in applications when execution speed is of importance. C Language Integrated Production System (CLIPS) is a popular development environment for building rule- and object-based expert systems. CLIPS is written in C for portability and speed. JAVA is relatively new and is gaining popularity in building expert systems mostly due to the desirable attribute of portability on different computer platforms.

Compared to general-purpose programming languages, shells provide better tools for fast design and implementation of expert systems. An expert system **shell** is a programming environment that contains the necessary utilities for both developing and running an expert system. For example, since an expert system shell provides a build-in inference engine, the developer can focus on inputting problem-specific knowledge into the knowledge base. By using expert system shells, the programming skills required to build an expert system become minimal, and the design and implementation time can be greatly reduced.

EMYCIN (Buchanan and Shortliffe, 1984) is an example of shells. The tools it provides include (a) an abbreviated rule language that is easier to read than LISP and more concise than the English subset used by MYCIN; (b) an indexing and grouping scheme for rules; (c) a goal-driven inference engine; (d) a communication interface between the program and the end user; (e) an interface between the system designer and the program, which supports displaying and editing rules, editing knowledge in tables, and running selected rules sets on a problem; and (f) facilities for monitoring the behavior of the program, such as explaining how a conclusion is reached, comparing the results of the current run with correct results and exploring the discrepancies, and reviewing conclusions about a stored library of cases.

General-purpose shells are offered to address a broad range of problems. Many domain-specific tools have also been developed to provide special features to support the development of expert systems for specific domains. The introduction and widespread use of expert system shells promotes the successful applications of expert systems to a wide range of tasks.

5.2 Knowledge Representation

Knowledge representation is a substantial subfield of AI in its own right. Winston defines a **representation** as "a set of syntactic and semantic conventions that make it possible to describe things" (Winston, 1984). The **syntax** of a representation specifies a set of rules for combining symbols to form expressions in the language. The **semantics** of a representation specify the meanings of expressions. In the field of expert systems, knowledge representation is mostly concerned with symbolically coding a large amount of domain knowledge in computer-tractable form such that it can be used to describe the task and the environment unambiguously and to reason with the knowledge efficiently toward certain goals.

Because of the property of unambiguity, mathematical tools are used as the bridges between verbal or mental representation and computer representation of knowledge. These tools include logic, probability theory, fuzzy sets, abstract algebra, and set theories. Major knowledge representation schemes in expert systems include the following:

- **Production rules** are generative rules of intelligent behavior, similar to the grammar rules in automata theory and formal grammars. They encode relationships between facts and concepts as well as associations between data and actions (i.e., the "what to do when" knowledge).
- **Semantic networks and frames** are graphical formalisms for encoding facts, such as objects and events and their properties, and heuristic knowledge for processing information, such as procedures. In these structured representations, nodes stand for concepts and arcs for relationships between them.
- **Logic languages** are derived from mathematical logic; they encode facts and control information in logic formulas.
- **Model-based approaches** use theoretical models to represent knowledge about components and systems.
- **Object-oriented approaches** represent knowledge in terms of interacting objects, which consist of data and procedures.
- **Blackboard architectures** partition domain knowledge into independent knowledge sources and build up solutions on a global data structure, the **blackboard**. The blackboard serves the function of the working memory, but its structure is much more complex.

5.2.1 Rule-Based Expert Systems

Based on predicate logic, probability theory, and fuzzy sets, rule-based expert systems are the most common type of expert systems. Rule-based expert systems use **if-then** rules (or production rules) to represent human expert knowledge, which is often a mix of theoretical knowledge, heuristics derived from experience, and special-purpose rules for dealing with abnormal situations. A production system is an example of rule-based systems: its knowledge base is a rule set, its inference engine is a rule interpreter that decides when to apply which rules, and its working memory contains the data that are examined and modified in the inference process. The rules are triggered by the data, and the rule interpreter controls the selection of the rules and updating of the data.

Generally, the knowledge base of a rule-based expert system consists of a set of rules of the following form:

```
if condition₁ and . . . and conditionₘ are true
then conclusion₁ and . . . and conclusionₙ
```

Each rule contains one or more conditions and one or more conclusions or actions.

MYCIN is a classical rule-based system. It diagnoses certain infectious diseases, prescribes antimicrobial therapy, and can explain its reasoning (Buchanan and Shortliffe, 1984). In a controlled test, its performance equaled that of specialists. Because medical diagnosis often involves a degree of uncertainty, each MYCIN's rule has associated with it a **certainty factor** (CF), which indicates the extent to which the evidence supports the conclusion (i.e., likelihood). A typical MYCIN rule looks like this:

```
If the stain of the organism is gram-positive, and
the morphology of the organism is coccus, and
the growth conformation of the organism is clumps
then there is suggestive evidence (0.7) that the
  identity of the organism is staphylococcus.
```

There are many sources of uncertainty in problem solving. The domain knowledge may be incomplete, error-prone, or approximate. The data may be noisy or unreliable. In the field of expert system, research methods for handling uncertainty include **certainty factor approach, the Bayesian approach, fuzzy logic**, and the **belief theory**.

In the **certainty factor approach** certainty factors are associated with given data, indicating their degree of certainty. Each rule is also associated with a CF representing how strong the conditions of the rule support the conclusions. A certainty factor of a hypothesis h and an evidence e is defined as:

$$CF[h, e] = MB[h, e] - MD[h, e], \tag{5.1}$$

where $MB[h, e]$ measures the extent to which e supports h and where $MD[h, e]$ measures the extent to which e supports the negation of h. Since $MB[h, e]$ and $MD[h, e]$ have values between zero and one, a CF has a value between -1 and 1. As the CF moves toward 1, the evidence increasingly supports the hypotheses. The CF of the conclusion of a rule is computed as the CF of the rule times the minimum of the CFs of its individual premises. In MYCIN, when several different possible diseases are consistent with available evidences, their certainty factors show how likely each is. When there are multiple possible treatments, their certainty factors show the likelihood that each treatment will work.

The certainty factor approach is simple, easy to understand, and efficient to run; it provides a means of estimating belief that is natural to a domain expert. Its combination methods ensure locality, detachment, and modularity (Stefik, 1995). This means that adding or deleting rules does not require rebuilding the entire certainty model.

A more general approach than the certainty factor approach is the **Bayesian approach** based on conditional probabilities. The certainty factor approach is simple because it ignores prior probabilities that are difficult to estimate. It makes a strong assumption, however, that all rules are independent. At the heart of the Bayesian approach is the **Bayes' rule**. The Bayes' rule (also known as Bayes' law or Bayes' theorem) is the following equation:

$$P(Y|X) = \frac{P(X|Y)P(Y)}{P(X)}, \tag{5.2}$$

where $P(A)$ represents the unconditional or prior probability that A is true and where $P(A|B)$ represents the conditional or posterior probability A given that all we know is B. This simple equation underlies modern AI systems for probabilistic inference. The limitations of the Bayesian approach resides in the assumptions that the evidences are independent, prior probabilities are known, and the sets of hypotheses are both inclusive and exhaustive.

Fuzzy logic is also widely used to deal with uncertainty. Probability theory represents uncertainty based on frequency of occurrence, whereas fuzzy logic sets represent imprecision as a graded membership function with a value between zero and one. Fuzzy logic has been applied in various types of expert systems, including rule-based systems, to handle uncertainties. In fuzzy rule-based expert systems, knowledge is represented in fuzzy rules and fuzzy sets. Fuzzy rules are **if-then** statements that contain fuzzy variables. These rules are processed based on the mathematical principles of fuzzy logic.

Belief theory is yet another alternative approach. **Dempster-Shafer theory of evidence** performs reasoning under uncertainty in a structured space of hypotheses. It provides a means for computing a belief function over the hypothesis space and a way for combining belief function derived from different pieces of evidence. This approach is based on better mathematical foundations than the CF approach and is more general than the Bayesian and CF approaches (Durkin, 1993).

5.2.2 Model-Based Systems

Expert systems relying on heuristic rules have a number of limitations. They fail when a problem does not match the heuristics or when the heuristics are applied in inappropriate situations. The **model-based approach** attempts to address these problems by solving problems based on theoretical models of the components and systems. They are less dependent on expert opinion and experience and have higher flexibility and scope for expansion. The downsides are the cost and complexity of building models and the resulting programs that may be large and cumbersome (Stern and Luger, 1997).

The central concept of a model-based system is a model, a description of a device, or a system using an appropriate modeling language. The model specifies the structure, functions, and behaviors of the devices for the purposes of analysis, prediction, diagnosis, and other such procedures. The causal and structural information in the model may be represented using a number of data structures, such as rules and objects. For example, in using objects to represent the component structure in a model, the fields of an object represent a component's state, and the methods define its functionality. A model-based system reasons based on the model.

The types of models employed in model-based systems can be categorized as follows:

- **Declarative versus procedural**: Declarative models describe relationships between entities. They are very general, but can lead to complex and inefficient reasoning algorithms. Procedural models are more appropriate when the knowledge contains a strong element of directionality. They tend to be highly specialized to a certain task or situation and can efficiently derive outputs for inputs.
- **Quantitative versus qualitative**: Quantitative models are numerical models (e.g., the algebraic or differential equation-based models in traditional engineering disciplines). In qualitative models, variables take on qualitative values like *low* and *normal* instead of numerical values. The representation is usually based on fuzzy sets and belief networks. The advantage of qualitative models is that they map the domain knowledge in a more natural way and better reflect a human point of view.
- **Certain versus uncertain**: When the knowledge is inaccurate, incomplete, or ambiguous, the model should still be able to represent and process the knowledge. The major techniques for dealing with uncertainty are probability theory and fuzzy theory.
- **Static versus dynamic**: Static models represent steady-state or equilibrium behaviors of systems, whereas dynamic models represent behaviors in transient states.
- **Continuous versus discontinuous**: In continuous models, behavior trajectories evolve smoothly through adjacent states. Discrete models present discrete state transitions, jumping from one state to another with different properties.

The following simple examples of device and circuit analysis from Davis and Hamscher (1992) illustrate the concept of models. To represent an adder shown in Figure 5.2(A), the model encodes the knowledge that captures the functionality of the adder. It consists of three expressions that represent the relationships between values on the terminals: (1) if we know the values at A and B, then the value at C is $A + B$; (2) if we know the values at A and C, then the value at B is $C - A$; and (3) if we know the values at B and C, then the value at A is $C - B$.

Using this kind of model in fault diagnosis, consider the circuit of two multipliers and two adders shown in Figure 5.2(B). The input values are given at A through D, and the output values are given at E and F with the expected output values in parentheses and the actual ones in brackets. The task is to determine which device is faulty from the fact that we expect 9 at E and instead get 6. Let's assume that only a single device is faulty. Since the value at E depends on Add-1, Mult-1, and Mult-2, one of them must have a fault. Since we get a correct value at F, Mult-2 must be correct. Thus, the fault lies in either Mult-1 or Add-1. Additional tests may be applied to the fewer devices left to identify the faulty one.

5.3 Reasoning

Some reasoning strategies are general, whereas others are specific to particular knowledge representation schemes. Goal-driven and data-driven strategies are general reasoning strategies. In data-driven reasoning, from the conditions that are known to be true, we reason forward to establish new conclusions. In contrast, in goal-driven reasoning, we reason backward from a goal state toward the conditions necessary to make it true. In addition, representation-specific reasoning methods have been developed for the various types of expert systems, including rule-based systems, frame-based systems, logic systems, model-based systems, object-oriented systems, fuzzy logic systems, blackboard systems, and many others.

5.3.1 Forward and Backward Chaining in Rule-Based Systems

In a rule-based system, the inference engine usually goes through a simple recognize-assert cycle. The control scheme is called forward chaining for data-driven reasoning, and backward chaining for goal-driven reasoning. The basic idea of forward chaining is when the premises of a rule (the **if** portion) are satisfied by the data, the expert system asserts the

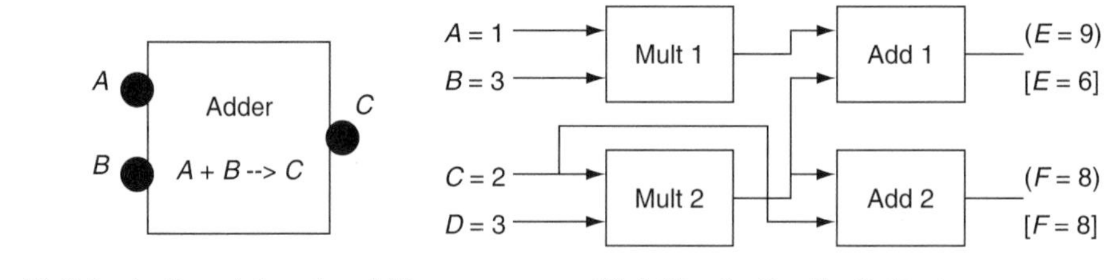

(A) Behavior Description of an Adder (B) A Circuit with a Faulty Device

FIGURE 5.2 Examples of Device and Circuit Analysis

conclusions of the rule (the **then** portion) as true. A **forward-chaining reasoning system** starts by placing initial data in its working memory. Then the system goes through a cycle of matching the premises of rules with the facts in the working memory, selecting one rule, and placing its conclusion in the working memory. This inference process is useful in searching for a goal or an interpretation, given a set of data. For example, XCON is a forward-chaining rule-based system (McDermott, 1982) that contains several thousand rules for designing configurations of computer components for individual customers. It was one of the first clear commercial successes of expert systems. Its underlying technology has been implemented in the general-purpose language OPS-5.

In a **backward-chaining reasoning system** the goal is initially placed in the working memory. The system matches rule conclusions with the goal, selects one rule, and places its premises in the working memory. The process iterates, with these premises becoming new goals to match against rule conclusions. Thus, the system works backward from the original goal until all the subgoals in the working memory are known to be true. Subgoals may also be solved by asking the user for information. For example, MYCIN's inference engine uses a backward-chaining control strategy. From its goal of finding significant disease-causing organisms, MYCIN uses its rules to reason backward to the data available. Once it finds such organisms, it attempts to select a therapy to treat the disease(s). Since it was designed as a consultant for physicians, MYCIN was given the ability to explain both its reasoning and its knowledge (Buchanan and Shortliffe, 1984).

Given a fixed reasoning method, the process of searching through alternative solutions can be affected through the structuring and ordering of the rules in implementations. For example, in production systems, a rule of the form "**if *p* and *q* and *r* then *s***" may be interpreted in backward chaining as a procedure of four steps: **to do s, first do *p*, then do *q*, then do *r*.** Although the procedural interpretation of rules reduces the advantages of declarative representation, it can be used to reflect more efficient heuristic solution strategies. For instance, the premises of a rule may be ordered so that the one that is most likely to fail or is easiest to be satisfied will be tried first.

To illustrate forward and backward chaining, consider a simple example with the following rules in the knowledge base:

```
Rule 1: if a and b, then c
Rule 2: if c and d, then p (s1)
Rule 3: if not d and not e, then p (s2) or p (s3)
Rule 4: if not d and e, then p (s4)
```

The symbols *a* to *e* and *s*1 to *s*4 represents objects, and *p* represents a property of objects.

To perform backward-chaining reasoning, the top-level goal, *p*(*X*), is placed in the working memory as shown in Figure 5.3(A), where *X* is a variable that can match with any object. The conclusions of three rules (rules 2, 3, and 4) match with the expression in the working memory. If we solve conflicts in favor of the lower-numbered rule, then rule 2 will be selected and fire. This causes *X* to be bound to *s*1 and the two premises of rule 2 to be placed in the working memory as in Figure 5.3(B). Then, since the conclusion of rule 1 matches with a fact in the working memory, we then fire rule 1 and place its promises in the working memory as in Figure 5.3(C). At this point, there are three entries in the working memory (*a, b, d*) that do not match with any rule conclusion. The expert system will query the user directly about these subgoals. If the user confirms them as true, the expert system will have successfully determined the causes for the top-level goal *p*(*X*).

The control of the previous backward-chaining process performs a depth-first search, in which each new subgoal is searched exhaustively first before moving onto old subgoals. Other search strategies, such as breadth-first search, can also be applied.

Given the same set of rules, forward chaining can also be applied to derive new conclusions from given data. For example, the algorithm of forward chaining with breadth-first search is as follows: Compare the content of the working memory with the premises of each rule in the rule base using the ordering of the rules. If the data in the working memory match a rule's premises, the conclusion is placed in the working memory, and the control moves to the next rule. Once all rules have been considered, the control starts again from the beginning of the rule sets.

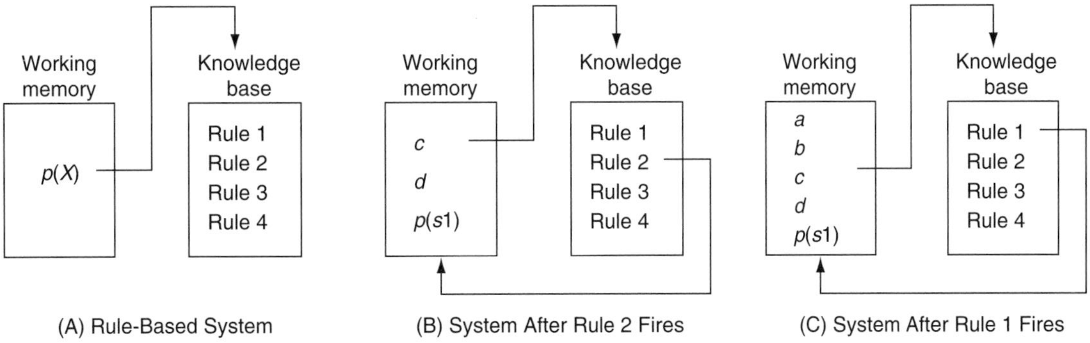

FIGURE 5.3 The Rule-Based System Throughout a Goal-Driven Inference Process

5.3.2 Model-Based Reasoning

Model-based reasoning contains a diverse set of approaches and a collection of loosely connected techniques, with most applications in the areas of monitoring, control, and diagnosis. For diagnosis applications, model-based reasoning usually consists of the following elements: (a) simulate and predict the normal behavior of the system, (b) record dependencies between internal model components and predicted observations, (c) upon detection of abnormal observations, use the dependencies to identify conflicting model assumptions, and (d) in the presence of multiple candidates, apply a measurement strategy to reduce the number of candidates.

Among the various types of models, qualitative models have been the focus of model-based reasoning in AI. They are useful in the situations where the quantitative models are impossible to develop due to lack of knowledge or are prohibitively computationally expensive. They may also correspond to the common sense knowledge better than abstract mathematical models. The major approaches in building qualitative models and related reasoning systems are the following:

- **Constraint-based approach**: A physical system is modeled by a set of qualitative mathematical constraints between the variables that comprise the model (Kuipers, 1986).
- **Component-based approach**: A model represents the topological structure of the target system, consisting of components, connections, and materials that flow through connections (deKleer and Brown, 1984). The behavior of each component is described by a set of local properties (i.e., relationships between inputs, internal states, and outputs).
- **Process-based approach**: A system is modeled based on component models and system topology, similar to the component-based approach. The difference is that the process-based approach uses *processes* to model physical interactions and to define the dynamic characteristics of the target system (Davis and Hamscher, 1992).

5.3.3 Case-Based Reasoning

Knowledge acquisition is a very difficult process in building expert systems. **Case-based reasoning** (CBR) systems simplifies the process by using a collection of past problem solutions (cases) to address new problems (Kolodner, 1993). The basic idea underlining this approach is "what was true yesterday is likely to be true today."

Past cases are either collected from human experts or from previous successes or failures of the system. A case usually contains three components: **problem description**, **solution to the problem**, and **outcome of the solution**. A description of the problem contains all descriptive information necessary for solving the problem. The description of the solution allows the reuse of a previous solution without starting from scratch once a similar case arrives. The outcome description contains explanations of what was carried out, whether it was a success or failure, the repair strategy in the case of failure, and how to avoid the failure. Cases can be represented in many different forms such as rules, logic formulas, frames, and database records. To be successful, it is important for a CBR system to have a sufficiently large collection of cases to reason from and to have a domain that is well understood.

CBR systems select and reason from appropriate past cases to build general rules and are able to learn from experiences because success and failure of previous attempts are retained. The major steps in case-based reasoning are as follows:

1. A new problem is analyzed and represented in the form such that the system can retrieve relevant past cases from its memory. A past case is appropriate if it has the potential to provide a solution to the new problem. Typically, heuristics are used to choose cases similar to the new problem based on certain similarity measures.
2. A retrieved case is modified so that it is applicable to the new problem. Analytic methods or heuristic methods are applied to transform the stored solution into operations suitable to the new problem.
3. In applying the transformed case to the new problem, an initial solution is proposed, tested, and evaluated. If it does not perform well, an explanation of the result is generated.
4. The new case is saved into the case collection with a record of its success or failure for future use. The system goes through an incremental learning process as new cases are added.

To retrieve relevant cases, one technique is to assign indexes (or labels) to all cases. Indexes represent an interpretation of a situation and determine under what conditions a case can be used to make useful inferences (Kolodner, 1993). Assigning indexes is problem dependent, requiring a good understanding of the problem domain. Another technique is to compute the relevance of a case based on similarity measures. For example, in nearest-neighbor matching, each corresponding feature of a new case and a retrieved case is compared by looking at the field type and values. Each feature gets a similarity score, and the weighted sum of the feature scores is the overall similarity score of the two cases.

To apply a retrieved case to the new problem, the previous solution may need to be modified to fit the current problem. There are two general types of modification: **structural** and **derivational**. In structural modification, simple changes, such as substituting values or adjusting and interpolating numerical numbers, are made to the retrieved solution to make it work for the new problem. In derivational modification, procedures that generated the previous solutions are run again to generate a new solution to the new case.

To illustrate how case-based reasoning works, let's look at CHEF, a CBR system that creates cooking recipes (Hammond, 1986). The input to CHEF is a list of goals that specify different types, tastes, and textures of dishes. The output is a recipe that satisfies these constraints. For example, the input may be "create a crisp pork dish; include broccoli; use stir-fry method." CHEF looks in its case base for a recipe that makes a similar meal and then adapts it to solve the new problem. If it already has a recipe for pork with cabbage, it may copy the recipe and substitute broccoli for cabbage. However, broccoli may not remain crisp if cooked like cabbage. Thus, the way of stir-frying may be modified, for example, by using rules from other cases involving crisp broccoli.

CBR has been applied to a variety of applications, including diagnosis, classification, interpretation, instruction, planning, and scheduling. One example is Support Management Automated Reasoning Technology (SMART) of Compaq (Acorn and Walden, 1992), which provides quality technical support by using previous cases to resolve new ones instead of trying to diagnose a new problem from scratch. Another system, CARES (Ong *et al.*, 1997), employs CBR to predict the recurrence of colorectal cancer.

Hybrid systems combining CBR with rule-based reasoning and model-based reasoning have also been developed. In combining CBR and rule-based reasoning, rules may be used to capture broad trends and general strategies in the problem domain, whereas cases are used to support exceptions and explain and justify rules. In combining CBR and model-based reasoning, model-based reasoning is good at handling well-understood components under normal situations, whereas CBR covers the part of the domain that does not have a good model or theory. In addition, by incorporating domain knowledge into CBR, past cases can be more efficiently organized, searched, retrieved, and better adapted to the new cases.

5.4 Knowledge Acquisition

Knowledge acquisition refers to the process of extracting, structuring, and organizing domain knowledge from domain experts into a program. A **knowledge engineer** is an expert in AI language and knowledge representation who investigates a particular problem domain, determines important concepts, and creates correct and efficient representations of the objects and relations in the domain.

Capturing domain knowledge of a problem domain is the first step in building an expert system. In general, the knowledge acquisition process through a knowledge engineer can be divided into four phases:

1. **Planning**: The goal is to understand the problem domain, identify domain experts, analyze various knowledge acquisition techniques, and design proper procedures.
2. **Knowledge extraction**: The goal is to extract knowledge from experts by applying various knowledge acquisition techniques.
3. **Knowledge analysis**: The outputs from the knowledge extraction phase, such as concepts and heuristics, are analyzed and represented in formal forms, including heuristic rules, frames, objects and relations, semantic networks, classification schemes, neural networks, and fuzzy logic sets. These representations are used in implementing a prototype expert system.
4. **Knowledge verification**: The prototype expert system containing the formal representation of the heuristics and concepts is verified by the experts. If the knowledge base is incomplete or insufficient to solve the problem, alternative knowledge acquisition techniques may be applied, and additional knowledge acquisition process may be conducted.

Many knowledge acquisition techniques and tools have been developed with various strengths and limitations. Commonly used techniques include interviewing, protocol analysis, repertory grid analysis, and observation.

Interviewing is a technique used for eliciting knowledge from domain experts and design requirements. The basic form involves free-form or unstructured question–answer sessions between the domain expert and the knowledge engineer. The major problem of this approach results from the inability of domain experts to explicitly describe their reasoning process and the biases involved in human reasoning. A more effective form of interviewing is called **structured interviewing**, which is goal-oriented and directed by a series of clearly stated goals. Here, experts either fill out a set of carefully designed questionnaire cards or answer questions carefully designed based on an established domain model of the problem-solving process. This technique reduces the interpretation problem inherent in the unstructured interviewing as well as the distortion caused by domain expert subjectivity.

As an example, let's look at the interviewing process used in constructing GTE's COMPASS system (Prerau, 1990). COMPASS is an expert system that examines error messages derived from a telephone switch's self-test routines and suggests running of additional tests or replacing a particular component. The interviewing process in building COMPASS has an elicit–document–test cycle as follows:

1. Elicit knowledge from an expert.
2. Document the elicited knowledge in rules and procedures.
3. Test the new knowledge using a set of data:
   (a) Have the expert analyze a new set of data.
   (b) Analyze the same set of data using the documented knowledge.
   (c) Compare the two results.
   (d) If the results differ, find the rules or procedures that lead to the discrepancy and return to step 1 to elicit more knowledge to resolve the problem.

Protocol analysis is another technique of data analysis originated in clinical psychology. In this approach, an expert is asked to talk about his or her thinking process while solving a given problem. The difference from interviewing is that experts find it much easier to talk about specific problem instances than to talk in abstract terms. The problem-solving process being described is then analyzed to produce a structured model of the expert's knowledge, including objects of significance, important attributes of the objects, relationships among the objects, and inferences drawn from the relationships. The advantage of protocol analysis is the accurate description of the specific actions and rationales as the expert solves the problem.

Repertory grid analysis investigates the expert's mental model of the problem domain. First, the expert is asked to identify the objects in the problem domain and the traits that differentiate them. Then, a rating grid is formed by rating the objects according to the traits.

Observation involves observing how an expert solves a problem. It enables the expert to continuously work on a problem without being interrupted while the knowledge is obtained. A major limitation of this technique is that the underlying reasoning process of an expert may not be revealed in his or her actions.

Knowledge acquisition is a difficult and time-consuming task that often becomes the bottleneck in expert system development (Hayes-Roth *et al.*, 1983). Various techniques have been developed to automate the process by using domain-tailored environments containing well-defined domain knowledge and specific problem-solving methods (Rothenfluh *et al.*, 1996). For example, OPAL is a program that expedites knowledge elicitation for the expert system ONCOCIN (Shortliffe *et al.*, 1981) that constructs treatment plans for cancer patients. OPAL uses a model of the cancer domain to acquire knowledge directly from an expert. OPAL's domain model has four main aspects: **entities and relationships**, **domain actions**, **domain predicate**, and **procedural knowledge**. Based on its domain knowledge, OPAL can acquire more knowledge from a human expert and translate it into executable code, such as production rules and finite state tables. Following OPAL, more general-purpose systems called PROTEGE and PROTEGE-II were developed (Musen, 1989). PROTEGE-II contains tools for creating domain ontology and generating OPAL-like knowledge acquisition programs for particular applications PROTEGE-II is a general tool developed by abstraction from a successful application, similar to the process from MYCIN to EMYCIN.

Another example of automated knowledge acquisition is the SALT system (Marcus and McDermott, 1989) associated with an expert system called Vertical Transportation (VT) for designing custom-design elevator systems. SALT assumes a propose-and-revise strategy in the knowledge acquisition process. Domain knowledge is seen as performing one of three roles: (1) proposing an extension to the current design, (2) identifying constraints upon design extension, and (3) repairing constraint violation. SALT automatically acquires these kinds of knowledge by interacting with an expert and then compiles the knowledge into production rules to generate a domain-specific knowledge base. SALT retains the original knowledge in a declarative form as a dependency network, which can be updated and recompiled as necessary.

To build a knowledge base, the knowledge can be either captured through knowledge engineers or be generated automatically by machine learning techniques. For example, rules in rule-based expert systems may be obtained through a knowledge acquisition process involving domain experts and knowledge engineers or may be generated automatically from examples using decision-tree learning algorithms. Case-based reasoning is another example of automated knowledge extraction in which the expert system searches its collection of past cases, finds the ones that are similar to the new problem, and applies the corresponding solutions to the new one. The whole process is fully automatic. An expert system of this type can be built quickly and maintained easily by adding and deleting cases. Automatic knowledge generation is especially good when a large set of examples exist or when no domain expert exists.

In addition to generating knowledge automatically, machine learning methods have also been used to improve the performance of the inference engines by learning the importance of individual rules and better control in reasoning.

5.5 Explanation

Providing explanations that clarify the decision-making process and justify recommendations is an integral component of an expert system. It is one of the primary user requirements (Dhaliwal and Benbasat, 1996; Ye and Johnson, 1995). The explanation facility may be used by different users for different reasons in different contexts. For example, novice users can use the facility to find more about the knowledge being applied to solve a particular problem. Advanced users access it to make sure that the system's knowledge and reasoning process is appropriate. Decision makers use the explanation facility because it aids them in formulating problems and models for analysis. The context may be problem solving by end users, knowledge-base debugging by knowledge engineers, or expert system validation by domain experts and/or knowledge engineers. Although explanations are commonly used by end-users, they also play a significant role in the development of expert systems by offering enhanced debugging and validation abilities. Most current expert system development shells and environments include explanation tools.

The knowledge required for providing explanations may be derived from the knowledge base or may be separate from the knowledge used in solving problems. Depending on the type of problem-solving tasks, explanations may be presented to the users in different ways, such feed-forward and feedback.

Feed-forward explanations focus on the input cues, are not case-specific, and are presented prior to an assessment being performed, whereas feedback explanations focus on the outcome, explain a particular case-specific outcome, and are presented subsequent to the assessment.

The major types of explanations include why, how, what, what-if, and strategic explanations. The **why** explanations provide justification knowledge of the underlying reasons for an action based on causal models. The **how** explanations provide reasoning trace knowledge of the inference process. The **why** and **how** explanations were first introduced in MYCIN. They remain the core of most explanation facilities in current expert system applications and development shells. The **what** explanations provide knowledge about the object definitions or decision variables used by the system. The **what-if** explanations provide direct and explicit information about the sensitivity of decision variables. The **strategic** explanations provide information about the problem-solving strategy and metaknowledge.

In MYCIN, the explanation module is invoked at the end of every consultation. To explain the result, the module retrieves the list of rules that were successfully applied, along with the conclusions drawn. It allows the user to interrogate the system about the conclusions. Inquiries generally fall into two types: why a particular question was put and how a particular conclusion was reached. MYCIN keeps track of the goal–subgoal structure of the computation and uses it to answer a why question by citing the related rules together with other conditions. To answer a how question, MYCIN maintains a record of the decisions it made and cites the rules that it applied as well as the degree of certainty of the decision.

References

Acorn, T., and Walden, S. (1992). SMART: Support-management cultivated reasoning technology for Compaq customer service. In *Proceedings of the 11th National Conference on AI.*

Aikens, J.S., Kunz, J.C., and Shortliffe, E.H. (1983). PUFF: An expert system for interpretation of pulmonary function data. *Computers and Biomedical Research 16,* 199–208.

Buchanan, B.G., and Shortliffe, E.H. (1984). *Rule-based experts programs: The MYCIN experiments of the Stanford heuristic programming project.* Reading, MA: Addison-Wesley.

Clancy, W.J., and Shortliffe, E.H. (1984). *Readings in medical artificial intelligence: The first decade.* Reading, MA: Addison-Wesley.

Davis, R., and Hamscher, W. (1992). Model-based reasoning: Troubleshooting. In W. Hamscher, L. Console, and J. de Kleer (Eds), *Readings in model-based diagnosis.* San Mateo, CA, Morgan Kaufman.

de Kleer, J., and Brown, J.S. (1984). A qualitative physics based on confluences. *Artificial Intelligence 24,* 7–83.

Dhaliwal, J.S., and Benbasat, I. (1996). The use and effects of knowledge-based system explanations: Theoretical foundations and a framework for empirical evaluation. *Information Systems Research 7*(3), 342–362.

Durkin, J. (1993). *Expert systems: Catalog of applications.* Akron, OH: Intelligent Computer Systems.

Gordon, J., and Shortliffe, E.H. (1985). The dempster-shafer theory of evidence. In B. G. Buchanan and E. H. Shortliffe (Eds.), *Rule-based expert systems: The MYCIN experiments of the Stanford heuristic programming project.* Reading, MA: Addison-Wesley.

Hammond, K. (1986). A model of case-based planning. *Proceedings of the 5th National Conference on AI,* 65–95.

Harmon, P., and Sawyer, B. (1990). *Creating expert systems for business and industry.* New York: John Wiley & Sons.

Hayes-Roth, F., Waterman, D.A., and Lenat, D.B. (1983). *Constructing an expert system.* Reading, MA: Addison Wesley.

Kolodner, J.L. (1993). *Case-based reasoning.* San Mateo, CA: Morgan Kaufmann.

Kuipers, B. (1986). Qualitative simulation. *Artificial Intelligence 29,* 289–388.

Marcus, S., and McDermott, J. (1989). SALT: A knowledge acquisition language for propose-and-revise systems. *Artificial Intelligence 39,* 1–37.

McDermott, J. (1982). R1: A rule-based configurer of computer systems. *Artificial Intelligence 19*(1), 39–88.

Musen, M.A. (1989). *Automated generation of model-based knowledge acquisition tools.* San Francisco: Morgan-Kaufmann.

Ong, L.S., Shepherd, S., Tong, L.C., Seow-Choen, F., Ho, Y.H., Tang, C.L., Ho, Y.S., and Tan, K. (1997). The colorectal cancer recurrence support (CARES) system. *Artificial Intelligence in Medicine.* Vol. II, No. 3, 175–188.

Prerau, D.S. (1990). *Developing and managing expert systems,* Reading, MA: Addison-Wesley.

Rothenfluh, T.E., Gennari, J.H., Eriksson, H., Puerta, A.R. Tu, S.W., and Musen, M.A. (1996). Reusable ontologies, knowledge acquisition tools, and performance systems: PROTEGE-II solutions to sisyphus-2. *International Journal of Human-Computer Studies 44,* 303–332.

Shortliffe, E.H. (1976). *Computer-Based Medical Consultation, MYCIN.* New York: American Elsevier.

Shortliffe, E.H., Scott, A.C., Bischoff, M.M., van Melle, W., and Jacobs, C.D. (1981). ONCOCIN: An expert system for oncology protocol management. In *Proceedings of the 7th National Conference on AI* 876–881.

Stefik, M. (1995). *Introduction to knowledge systems.* San Francisco: Morgan Kaufmann.

Stern, C.R., and Luger, G.F. (1997). Abduction and abstraction in diagnosis: A schema-based account. In *Expertise in context.* Cambridge, MA: MIT Press.

Waterman, D.A. (1986). *A guide to expert systems.* Reading, MA: Addison-Wesley.

Winston, P.H. (1984). *Artificial Intelligence* (2nd ed.). Reading, MA: Addison-Wesley.

Ye, R., and Johnson, P.E. (1995). The impact of explanation facilities on user acceptance of expert systems advice. *MIS Quarterly 19*(2), 157–172.

6

Multimedia Systems: Content-Based Indexing and Retrieval

Faisal Bashir,
Shashank Khanvilkar,
Ashfaq Khokhar,
and Dan Schonfeld
Department of Electrical and Computer Engineering, University of Illinois at Chicago, Chicago, Illinois, USA

6.1 Introduction

Multimedia data, such as text, audio, images and video, are rapidly evolving as main avenues for the creation, exchange, and storage of information in the modern era. Primarily, this evolution is attributed to rapid advances in the three major technologies that determine the data's growth: VLSI technology that is producing greater processing power, broadband networks (e.g., ISDN, ATM, etc.) that are providing much higher bandwidth for many practical applications, and multimedia compression standards (e.g., JPEG, H.263, MPEG, MP3, etc.) that enable efficient storage and communication. The combination of these three advances is spurring the creation and processing of increasingly high-volume multimedia data, along with efficient compression and transmission over high-bandwidth networks. This current trend toward the removal of any conceivable bottleneck for those using multimedia data, from advanced research organizations to home users, has led to the explosive growth of visual information available in the form of digital libraries and online multimedia archives. According to a press release by Google Inc. in December 2001, the search engine offers access to over 3 billion Web documents and its Image search comprises more that 330 million images. Alta Vista Inc. has been serving around 25 million search queries per day in more than 25 languages, with its multimedia search featuring over 45 million images, videos, and audio clips.

This explosive growth of multimedia data accessible to users poses a whole new set of challenges relating to data storage and retrieval. The current technology of text-based indexing and retrieval implemented for relational databases does not provide practical solutions for this problem of managing huge multimedia repositories. Most of the commercially available multimedia indexing and search systems index the media based on keyword annotations and use standard text-based indexing and retrieval mechanisms to store and retrieve multimedia data. There are often many limitations with this method of keyword-based indexing and retrieval, especially in the context of multimedia databases. First, it is often difficult to describe with human languages the content of a multimedia object (e.g., an image having complicated texture patterns). Second, a manual annotation of text phrases for a large database is prohibitively laborious in terms of time and effort. Third, since users may have different interests in the same multimedia object, it is difficult to describe it with a complete set of keywords. Finally, even if all relevant object characteristics are annotated, difficulty may still arise due to the use of different indexing languages or vocabularies by different users. As recently as the 1990s, these major drawbacks of searching visual media based on textual annotations were recognized as unavoidable, and this prompted a surging increase in interest in content-based solutions (Goodrum, 2000). In content-based retrieval, manual annotation of visual media is avoided, and indexing and retrieval are instead performed on the basis of media content itself. There have been extensive studies on the design of automatic **content-based indexing and retrieval** (CBIR) systems. For visual media, these contents may include color, shape, texture, and motion. For audio/speech data, contents may include phonemes, pitch, rhythm, and cepstral coefficients. Studies of human visual perception indicate that there exists a gradient of sophistication in human perception, ranging from seemingly primitive inferences (e.g., shapes, textures, and colors), to complex notions of structures (e.g., chairs, buildings, and affordances), and to cognitive processes (e.g., recognition of emotions and feelings). Given the multidisciplinary nature of the techniques for modeling, indexing, and retrieval of multimedia data, efforts from many different communities of engineering, computer science, and psychology have merged in the advancement of CBIR systems. But the field is still in its infancy and calls for more coherent efforts to make practical CBIR systems a reality. In particular, robust techniques are needed to develop semantically rich models to represent data, computationally efficient methods to compress, index, retrieve, and browse the information; and semantic visual interfaces integrating the above components into viable multimedia systems.

This chapter reviews the state-of-the-art research in the area of multimedia systems. Section 6.2 reviews storage and coding techniques for different media types. Section 6.3 studies fundamental issues related to the representation of multimedia data and discusses salient indexing and retrieval approaches introduced in the literature. For the sake of compactness and focus, this chapter reviews only CBIR techniques for visual data (i.e., for images and videos); the review of systems for audio data readers are referred to Foote (1999).

6.2 Multimedia Storage and Encoding

Raw multimedia data require vast amount of storage, and, therefore, they are usually stored in a compressed format. Slow storage devices (e.g., CD-ROMs and hard disk drives) do not support playback/display of uncompressed multimedia data (especially of video and audio) in real-time. The term **compression** refers to removal of **redundancy** from data. The more redundancy in the data is reduced, the higher the **compression ratio** is achieved. The method by which redundancy is eliminated to increase data compression is known as **source coding**. In essence, the same (or nearly the same) information is represented using fewer data bits. There are several other reasons behind the popularity of compression techniques and standards for multimedia data:

- Compression extends the playing time of a storage device. With compression, more data can be stored in the same storage space.
- Compression allows miniaturization of hardware system components. With less data to store, the same playing time is obtained with less hardware.
- Tolerances of system design can be relaxed. With less data to record, storage density can be reduced, making equipment that is more resistant to adverse environments and that requires less maintenance.
- For a given bandwidth, compression allows faster information transmission.
- For a given bandwidth, compression allows a better-quality signal transmission.

The previous bulleted list explains the reasons that compression technologies have helped the development of compressed domain-based modern communication systems and compact and rugged consumer products. Although compression in general is a useful technology, and, in the case of multimedia data, an essential one, it should be used with caution because it comes with some drawbacks as well. By definition, compression removes redundancy from signals. Redundancy is, however, essential in making data resistant to errors. As a result, compressed data is more sensitive to errors than uncompressed data. Thus, transmission systems using compressed data must incorporate more powerful error-correction strategies. Most of the text-compression techniques, such as the Lampel-Ziv-Welch codes, are very sensitive to bit errors: an error in the transmission of the code table value results in bit errors every time the table location is accessed. This phenomenon is known as **error propagation**. Other variable-length coding techniques, such as Huffman coding, are also sensitive to bit errors. In real-time multimedia applications, such as audio and video communications, some error concealment must be used in case of errors.

The applications of multimedia compression are limitless. The **International Standards Organization** ISO has provided standards that are appropriate for a wide range of possible compression products. The video encoding standards by ISO, developed by Motion Pictures Expert Group (MPEG), embrace video pictures from the tiny screen of a videophone to the high-definition images needed for electronic cinema. Audio coding stretches from speech-grade monochannel to multichannel surround sound. Data compression techniques are classified as **lossless** and **lossy** coding. In lossless coding, the data from the decoder is identical bit-for-bit with the original source data. Lossless coding generally provides limited compression ratios. Higher compression is possible only with lossy coding in which data from the decoder is not identical to the original source data and between them minor differences exist. Lossy coding is not suitable for most applications using text data but is used extensively for multimedia data compression as it allows much greater compression ratios. Successful lossy codes are those in which the errors are imperceptible to a human viewer or listener. Thus, lossy codes must be based on an understanding of psychoacoustic and psychovisual perceptions and are often called **perceptive codes**. The following subsections provide a very brief overview of some of the multimedia compression standards. Subsection 6.2.1 presents image encoding standards, and subsection 6.2.2 discusses several video encoding standards.

6.2.1 Image Encoding Standards

As noted earlier, the task of compression schemes is to reduce the redundancy present in raw multimedia data representation. Images contain three forms of redundancy:

- **Coding redundancy**: Consider the case of an 8-bit per pixel image (i.e., each pixel in the image is represented with an 8-bit value ranging between 0 to 255, depending on the local luminosity level in that particular area of the image). Because the gray scale value of some of the pixels may be small (around zero for a darker pixel), representing those pixels with the same number of bits for the ones with a higher pixel value (brighter pixels) is not a good coding scheme. In addition, some of the gray scale values in the image may be occurring more often than others. A more realistic approach would be to assign shorter codes to the more frequent data. Instead of using fixed-length codes as above, the **variable length coding** schemes (e.g., Shannon-Fano, Huffman, arithmetic, etc.) would be used in which the smaller and more frequently occurring gray scale values get shorter codes. If the gray levels of an image are encoded in such a way that uses more code symbols than absolutely necessary to represent each gray level, the resulting image is said to contain the coding redundancy.
- **Interpixel redundancy**: For the case of image data, redundancy will always be present if only the coding redundancy is explored and however rigorously it is minimized by using state-of-the-art variable length coding techniques. The reason for this is that images are typically composed of objects that have a regular and somewhat predictable morphology and reflectance and the pixel values are highly correlated. The value of any pixel can be reasonably predicted from the value of its neighbors; the information carried by each individual pixel is relatively small. To exploit this property of the images, it is often necessary to convert the visual information of the image into somewhat nonvisual format that better reflects the correlation between the pixels. The most effective technique in this regard is to transform the image into frequency domain by taking the discrete fourier transform (DFT), discrete cosine transform (DCT) or any other such transform.
- **Psychovisual redundancy**: This type of redundancy arises from the fact that a human eye's response is not equally sensitive to all the visual information. The information that has less relative importance to the eye is said to be psychovisually redundant. Human perception of the visual information does not involve the quantitative analysis of every pixel; rather, the eye searches for some recognizable groupings to be interpreted as distinguished features in the image. This is the reason that the direct current (dc) component of a small section of the image, which indicates the average luminosity level of that particular section of the image, contains more visually important information than a high frequency alternating current (ac) component, which has the information regarding the difference between luminosity levels of some successive pixels. Psychovisual redundancy can be eliminated by throwing away some of the redundant information. Since the elimination of psychovisually redundant data results in a loss of quantitative information and is an irreversible process, it results in lossy data compression. How coarse or how fine to quantize the data depends on what quality and/or what compression ratio is required at the output. This stage acts as the tuning tap in the whole image compression model. On the same grounds, the human eye's response to color information is not as sharp as it is for the luminosity information. More color information is psychovisually redundant than the gray scale information simply because the eye cannot perceive finer details of colors, whereas it can for gray scale values.

The following paragraphs outline the details of one very popular image compression standard, JPEG, that is a result of a collaborative effort between ITU-T and ISO.

JPEG: Digital Compression and Coding of Continuous-Tone Still Images

The Joint Photographic Experts Group (JPEG) standard is used for compression of continuous-tone still images (ISO 10918-1, 1994). This compression standard is based on the Huffman and run-length encoding of the quantized DCT coefficients of image blocks. The widespread use of the JPEG standard is motivated by the fact that it consistently produces compression ratios in excess of 20:1. The compression algorithm can be operated on both gray scale as well as multichannel color images. The color data, if in psychovisually redundant RGB format, is first converted to a more "compression-friendly" color model like YCbCr or YUV. The image is first broken down into blocks of size 8×8 called **data units**. After that, depending on color model and decimation scheme for chrominance channels involved, **minimum code units** (MCUs) are formed. A minimum code unit is the smallest unit that is processed for DCT, quantization, and variable-length encoding subsequently. One example for the case of the YUV 4:1:1 color model (each of chrominance component being half in width and half in length) is shown in Figure 6.1. Here the MCU consists of four data units from the Y component and one each from the U and V components.

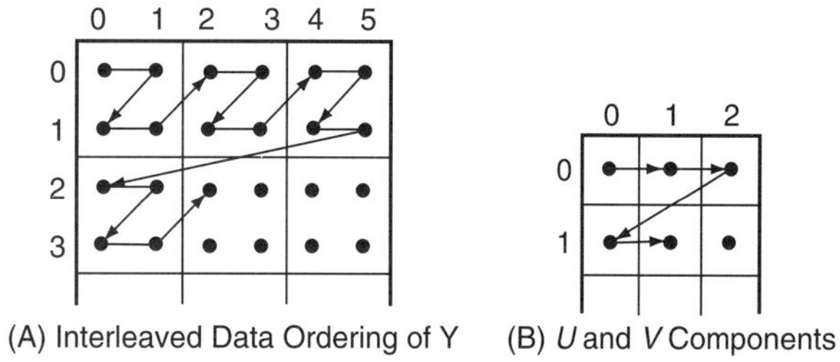

FIGURE 6.1 Minimum Code Unit

Each of the data units in an MCU is then processed separately. First a two-dimensional (2-D) DCT operation is performed that changes the energy distribution of the original image and concentrates more information in low frequencies. The DCT coefficients are then quantized to reduce the magnitude of coefficients to be encoded and to reduce some of the smaller ones to zero. The specs contain two separate quantization tables for luminance (Y) and for chrominance (U and V). The quantized coefficients are then prepared for symbol encoding. This is done by arranging them in a zigzag sequence as shown in the Figure 6.2, with lowest frequencies first and the highest frequencies last, to keep the scores of zero-valued coefficients at the tail of the variable-length bit stream to be encoded in the next phase. The assumption is that low-frequency components tend to occur with higher magnitude, while the high-frequency ones occur with lower magnitudes; placing the high frequency components at the end of the sequence to be encoded is more likely to generate a longer run of zeros yielding a good overall compression ratio.

As far as the symbol encoding is concerned, the standard specifies the use of either Huffman or arithmetic encoding. Realizing the fact that image data applications are computation intensive and performing the Huffman encoding for each block of 8×8 pixels might not be practical in most situations, the specs provide standard Huffman encoding tables for luminance (Y) and for chrominance (U and V) data. The experimentally proven code tables based on the average statistics of a large number of video images with 8 bits per pixel depth yield satisfactory results in most practical situations. For the dc encoding, the difference of each block's dc coefficient with the previous one is encoded. This code is then output in two successive bunches: one for the size of this code and the succeeding one for the most significant bits of the exact code. Since ac coefficients normally contain many zeros scattered between nonzero coefficients, the technique to encode ac coefficients takes into account the run of zeros between current and upcoming ac coefficients.

A direct extension of the JPEG standard to video compression known as motion JPEG (MJPEG) is obtained by the JPEG encoding of each individual picture in a video sequence. This approach is used when random access to each picture is essential, such as in video editing applications. The MJPEG compressed video yields data rates in the range of 8 to 10 Mbps.

6.2.2 Video Encoding Standards

Video data can be thought of as a sequential collection of images. The statistical analysis of video indicates that there is a strong correlation between successive picture frames as well as in the picture elements themselves. Theoretically,

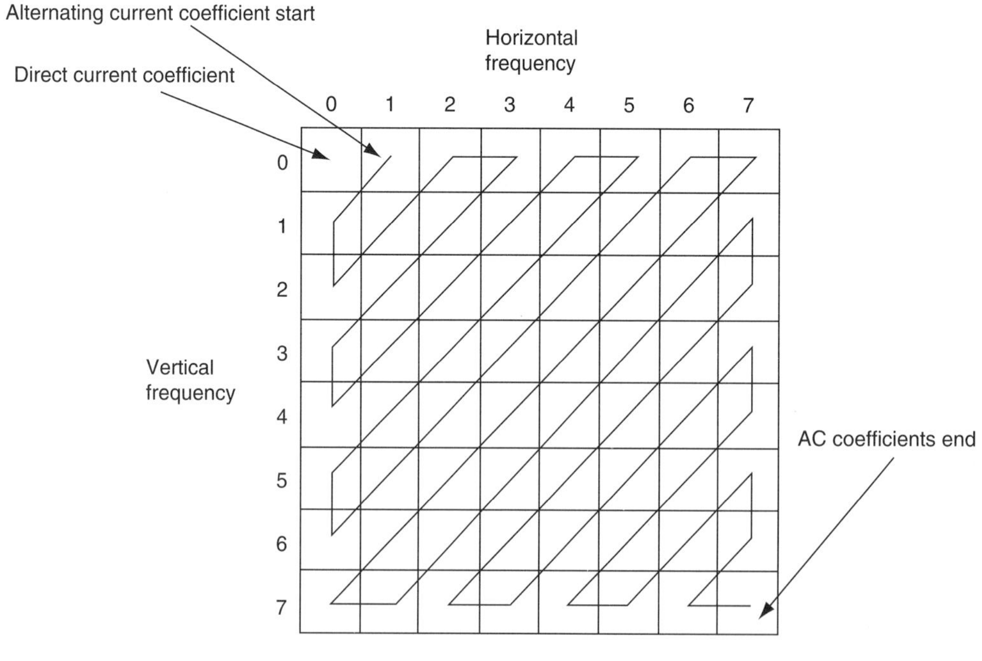

FIGURE 6.2 Zigzag Ordering of the Quantized DCT Coefficients

decorrelation of the temporal information can lead to bandwidth reduction without greatly affecting the video quality. As shown in the previous section, spatial correlation between image pixels is exploited to achieve still image compression. Such a coding technique is called **intraframe coding**. If temporal correlation is exploited as well, this is called **interframe coding**.

Interframe coding is the main coding principal that is used in all standard video codecs. First, a theoretical discussion of temporal redundancy reduction is given and then some of the popular video compression standards are explored in more detail.

Temporal redundancy is removed by using the differences between successive images. For static parts of the image sequence, temporal differences will be close to zero and, hence, are not coded. Those parts that change between frames, either due to illumination variation or motion of objects, result in significant image error that needs to be coded. Image changes due to motion can be significantly reduced if the motion of the object can be estimated and the difference can be taken on a motion-compensated image. To carry out **motion compensation**, the amount and direction of moving objects has to be estimated first. This is called **motion estimation**. The commonly used motion estimation technique in all the standard video codecs is the **block matching algorithm** (BMA). In a typical BMA, a frame is divided into square blocks of N^2 pixels. Then, for a maximum motion displacement of w pixels per frame, the current block of pixels is matched against a corresponding block at the same coordinates but in a previous frame in the square window of width $N + 2w$. The best match on the basis of a matching criterion yields the displacement. Various measures such as cross correlation function (CCF), mean squared error (MSE) and mean absolute error (MAE) can be used in the matching criteria. In practical coders, MSE and MAE are more often used since it is believed that CCF does not give good motion tracking, especially when the displacement is not quite large. MSE and MAE are defined as:

$$MSE(i, j) = \frac{1}{N^2}\sum_{m=1}^{N}\sum_{n=1}^{N}(f(m, n) - g(m+i, n+j))^2, \; -w \leq i, j \leq w.$$
$$MAE(i, j) = \frac{1}{N^2}\sum_{m=1}^{N}\sum_{n=1}^{N}|f(m, n) - g(m+i, n+j)|, \; -w \leq i, j \leq w. \tag{6.1}$$

The $f(m,n)$ variable represents the current block of N^2 pixels at coordinates (m, n), and $g(m+i, n+j)$ represents the corresponding block in the previous frame at new coordinates $(m+i, n+j)$. Motion estimation is one of the most computationally intensive parts of video compression standards, and some fast algorithms for this have been reported. One such algorithm is three-step Search, which is the recommended method for H.261 codecs (to be explained subsequently). It computes motion displacements up to 6 pixels per frame. In this method, all eight positions surrounding the initial location with a step size of $w/2$ are searched first. At each minimum position, the search step size is halved and the next eight positions are searched. The process is outlined in Figure 6.3. This method, for w set as 6 pixels per frame, searches 25 positions to locate the best match.

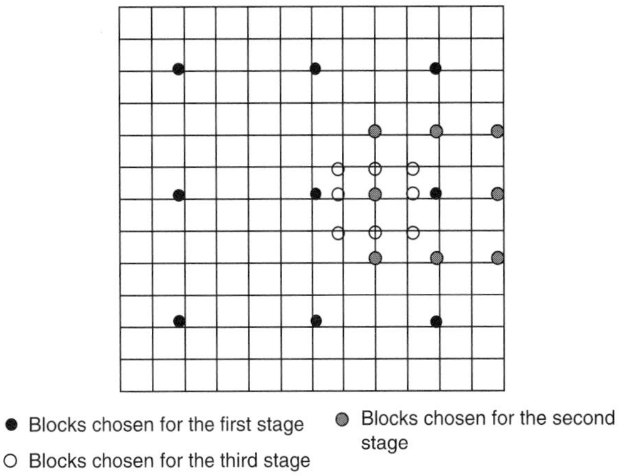

FIGURE 6.3 Example Path for Convergence of a Three-Step Search

Motion Picture Expert Group

The motion picture expert group (MPEG) provides a collection of motion picture compression standards developed by the IEC. Its goal is to introduce standards for movie storage and communication applications. These standards include audio compression representation, video compression representation, and system representation. The following subsection briefly outlines the details of three video compression standards developed by MPEG.

MPEG-1: Coding of Moving Pictures for Digital Storage Media

The goal of MPEG-1 was to produce VCR NTSC (352 × 240) quality video compression to be stored on CD-ROM (CD-I and CD-Video formats) using a data rate of 1.2 Mbps. This approach is based on the arrangement of frame sequences into a group of pictures (GOP) consisting of four types of pictures: I-picture (intra), P-picture (predictive), B-picture (bidirectional), and D-picture (dc). I-pictures are intraframe JPEG-encoded pictures that are inserted at the beginning of the GOP. The P- and B-pictures are interframe motion-compensated JPEG-encoded macroblock difference pictures that are interspersed throughout the GOP.[1] The system level of MPEG-1 provides for the integration and synchronization of the audio and video streams. This is accomplished by multiplexing and including time stamps in both the audio and video streams from a 90-KHz system clock (ISO/IEC, 1991).

[1] MPEG-1 restricts the GOP to sequences of 15 frames in progressive mode.

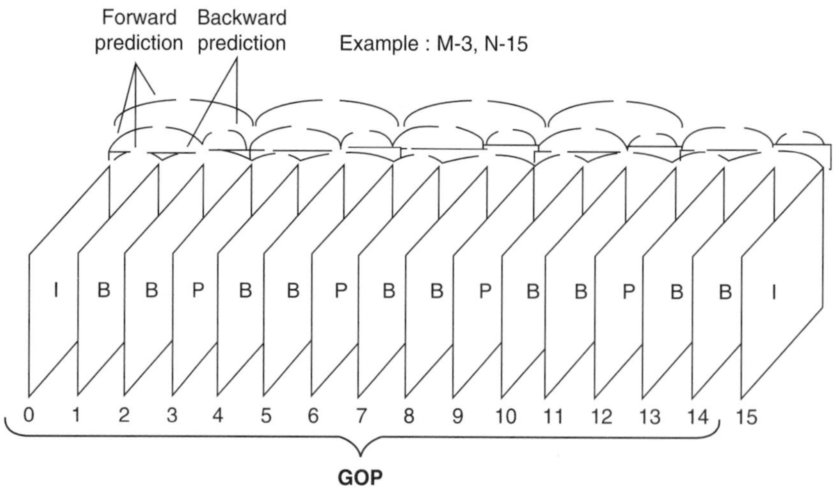

FIGURE 6.4 A Typical MPEG-1 GOP

In MPEG-1, due to the existence of several picture types, GOP is the highest level of hierarchy. The first coded picture in a GOP is an I-picture. It is followed by an arrangement for P- and B-pictures, as shown in Figure 6.4. GOP length is normally defined as the distance N between I-pictures. The distance between anchor I/P to P-pictures is represented by M. The GOP can be any length, but there has to be one I-picture in each GOP. Applications requiring random access, fast-forward play, or fast and reverse play may use short GOPs. GOP may also start at scene cuts; otherwise, motion compensation is not effective. Each picture is further divided into a group of macroblocks, called **slices**. The reason for defining slices is to reset the variable length code to prevent channel error propagation into the picture. Slices can have different sizes in a picture, and the division in one picture need not be the same as in any other picture. The slices can begin and end at any macroblock in a picture, but the first slice has to begin at the top left corner of the picture, and the end of last slice must be the bottom right macroblock of the picture. Each slice starts with a **slice start code** and is followed by a code that defines its position as well as a code that sets the quantization step size. Slices are divided into **macroblocks** of size 16×16, which are further divided into **blocks** of size 8×8 as in JPEG.

The encoding process for the MPEG-1 encoder is as follows. For a given macroblock, the coding mode is first chosen. This depends on the picture type, the effectiveness of motion-compensated prediction in that local region, and the nature of the signal in the block. Next, depending on the coding mode, a motion-compensated prediction of the contents of the block based on past and/or future reference pictures is formed. This prediction is subtracted from the actual data in the current macroblock to form an error signal. After that, this error signal is divided into 8×8 blocks, and a DCT is performed on each block. The resulting 2-D 8×8 block of DCT coefficients is quantized and scanned in zigzag order to convert it into a 1-D string of quantized coefficients like in JPEG. Finally, the side information for the macroblock, including the type, block-pattern and motion vectors, and DCT coefficients are coded. A unique feature of the MPEG-1 standard is the introduction of B-pictures that have access to both past and future anchor points. They can either use past frame, called **forward motion estimation** or the future frame, called **backward motion estimation**, as shown in Figure 6.5.

Such an option increases motion compensation efficiency especially when there are occluded objects in the scene. From the two forward and backward motion vectors, the encoder has a choice of choosing either any of the two or a weighted average of the two, where weights are inversely proportional to the distance of the B-picture with its anchor position.

MPEG-2: Coding of High-Quality Moving Pictures (MPEG-2)

The MPEG-1 standard was targeted for coding of audio and video for storage, in which the media error rate is negligible. Hence, MPEG-1 bitstream was not designed to be robust to bit errors. In addition, MPEG-1 was aimed at software-oriented image processing, where large and variable length packets could reduce the software overhead. The MPEG-2 standard,

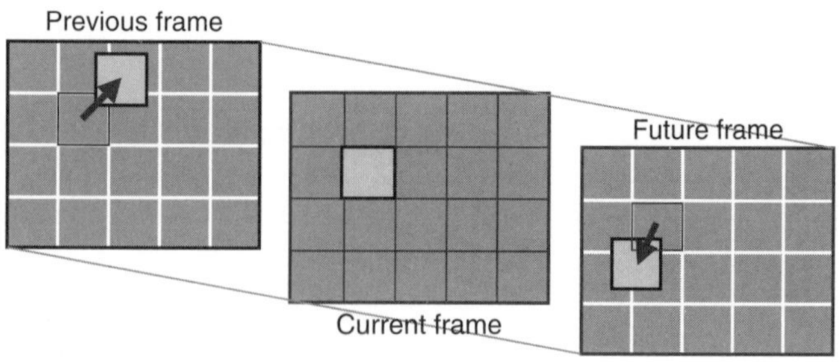

FIGURE 6.5 Motion Estimation in B-Pictures

on the other hand, is more generic for a variety of audio–visual coding applications. It has to have the error resilience for broadcasting and ATM networks. The aim of MPEG-2 is to produce broadcast-quality video compression and support higher resolutions including high definition television (HDTV).[2] MPEG-2 supports four resolution levels: low (352×240), main (720×480), high-1440 (1440×1152), and high (1920×1080) (ISO/IEC, 1994). The MPEG-2 compressed video data rates are in the range of 3 to 100 Mbps.[3] Although the principles used to encode MPEG-2 are very similar to MPEG-1, they provide much greater flexibility by offering several profiles that differ in the presence or absence of B-pictures, chrominance resolution, and coded stream scalability.[4] MPEG-2 supports both progressive and interlaced modes.[5] Significant improvements have also been introduced in the MPEG-2 system level. The MPEG-2 systems layer is responsible for the integration and synchronization of the elementary streams (ES): audio and video streams as well as an unlimited number of data and control streams that can be used for various applications such as subtitles in multiple languages. This is accomplished by first packetizing the ES, thus forming the packetized elementary streams (PES). These PES contain timestamps from a system clock for synchronization. The PES are subsequently multiplexed to form a single output stream for transmission in one of two modes: program stream (PS) and transport stream (TS). The PS is provided for error-free environments, such as storage in a CD-ROM. They are used for multiplexing PES that share a common time base, using long variable-length packets.[6] The TS is designed for noisy environments, such as communication over ATM networks. This mode permits multiplexing streams (PES and PS), which do not necessarily share a common time-base, using fixed-length (188 bytes) packets.

In the MPEG-2 standard, pictures can be interlaced, whereas in MPEG-1, the pictures are progressive only. The dimensions of the units of blocks used for motion estimation/compensation can change. Because the number of lines per field is half the number of lines per frame in the interlaced pictures, for motion estimation it might be appropriate to choose blocks of 16×8 (i.e., 16 pixels over 8 lines) with equal horizontal and vertical resolutions. The second major difference between the two is **scalability**. The scalable modes of MPEG-2 video encoders are intended to offer interoperability among different services or to accommodate the varying capabilities of different receivers and networks upon which a single service may operate. MPEG-2 also has a choice of a different DCT coefficient scanning mode **alternate scan** as well as a zigzag scan.

[2] The HDTV Grand Alliance standard adopted the MPEG-2 video compression and transport stream standards in 1996.

[3] The HDTV Grand Alliance standard video data rate is approximately 18.4 Mbps.

[4] The MPEG-2 video compression standard, however, does not support D-pictures.

[5] The interlaced mode is compatible with the field format used in broadcast television interlaced scanning.

[6] The MPEG-2 program stream is similar to the MPEG-1 systems stream.

MPEG-4: Content-Based Video Coding

The intention of MPEG-4 was to provide low bandwidth video compression at a data rate of 64 Kbps that can be transmitted over a single N-ISDN B channel. This goal has evolved to the development of flexible, scalable, extendable, and interactive compression streams that can be used with any communication network for universal accessibility (e.g., Internet and wireless networks). MPEG-4 is a genuine multimedia compression standard that supports audio and video as well as synthetic and animated images, text, graphics, texture, and speech synthesis (ISO/IEC, 1998). The foundation of MPEG-4 is on the hierarchical representation and composition of audio–visual objects (AVO). MPEG-4 provides a standard for the configuration, communication, and instantiation of object classes: The configuration phase determines the classes of objects required for processing the AVO by the decoder. The communication phase supplements existing classes of objects in the decoder. Finally, the instantiation phase sends the class descriptions to the decoder. A video object at a given point in time is a video object plane (VOP). Each VOP is encoded separately according to its shape, motion, and texture. The shape encoding of a VOP provides a pixel map or a bitmap of the object's shape. The motion and texture encoding of a VOP can be obtained in a manner similar to that used in MPEG-2. A multiplexer is used to integrate and synchronize the VOP data and composition information—position, orientation, and depth—as well as other data associated with the AVOs in a specified bitstream. MPEG-4 provides universal accessibility supported by error robustness and resilience, especially in noisy environments at very low data rates (less than 64 Kbps): bit stream resynchronization, data recovery, and error concealment. These features are particularly important in mobile multimedia communication networks.

H.26X: Video Compression Standards

The H.26X provides a collection of video compression standards developed by the ITU-T. The main focus of this effort is to present standards for videoconferencing applications compatible with the H.310 and H.32X communication network standards. These communication network standards include video compression representation, audio compression representation, multiplexing standards, control standards, and system standards. The H.26X and MPEG standards are very similar with relatively minor differences due to the particular requirements of the intended applications.

H.261: Coding for Video Conferencing

The H.261 standard has been proposed for video communications over ISDN at data rates of $p \times 64$ Kbps. It relies on intraframe and interframe coding for which integer–pixel accuracy motion estimation is required for intermode coding (H. 261, 1990).

H.263 Video Coding for Low Bit-Rate Communications

The H.263 standard is aimed at video communications over POTS and wireless networks at very low data rates (as low as 18 to 64 Kbps). Improvements in this standard result from the incorporation of such features as half-pixel motion estimation, overlapping and variable blocks sizes, bidirectional temporal prediction,[7] and improved variable-length coding options (H. 263, 1995).

H.26L: Video Communications over Wireless Networks

The H.26L standard is designed for video communications over wireless networks at low data rates. It provides features such as fractional pixel resolution and adaptive rectangular block sizes.

6.3 Multimedia Indexing and Retrieval

As discussed in the previous sections, multimedia data poses its distinct challenges for modeling and representation. The huge amount of multimedia information now available makes it all the more important to organize these multimedia repositories in a structured and coherent way to make it more accessible to a large number of users. This section explores the problem of storing multimedia information in a structured form (indexing) and searching the multimedia repositories in an efficient manner (retrieval). Subsection 6.3.1 outlines an image indexing and retrieval paradigm. The subsection first discusses the motivation for using content-based indexing and retrieval for images and then explores several different issues and research directions in this field. Subsection 6.3.2 highlights similar problems in the area of video indexing and retrieval. As with any other emerging field going through intellectual and technical exploration, the domain of content-based access to multimedia repositories causes a number of research issues to surface; these issues cannot be summarized in a single concise presentation. One such issue is query language design for multimedia databases. The interested reader can refer to Catarci *et al.* (1995), Hibino and Rundensteiner (1995), Kaushik and Rundensteiner (1998), and Zhang *et al.* (1997).

6.3.1 Image Indexing and Retrieval

Because of the tremendous growth of visual information available in the form of images, effective management of image archives and storage systems is of great significance and an extremely challenging task indeed. For example, a remote sensing satellite, which generates seven band images including three visible and four infrared spectrum regions, produces around 5000 images per week. Each single spectral image, which corresponds to a 170 km × 185 km of the earth's region, requires 200 mB of storage. The amount of data originated from satellite systems is already reaching a terabyte level per day. Storing, indexing, and retrieving such a huge amount of data by their contents is a very challenging task. Generally speaking, data representation and feature-based content modeling are two basic components required by the management of any multimedia database. As far as the image database is concerned, data representation focuses on image storage, whereas feature-based content modeling is related to image indexing and retrieval. Depending on the background of the research teams, different levels of abstractions have been assumed to model the data. As shown in Figure 6.6, these abstractions can be classified into three categories based on the gradient model of human visual perception. This figure, also captures the mutual interaction of some disciplines of engineering, computer science, and cognitive sciences.

Level 1 represents systems that model raw image data using features such as color histogram, shape, and texture descriptors. This model can be used to serve the queries like "find pictures with dominant red color on a white background." Content-based image indexing and retrieval (CBIR) systems based on these models operate directly on the data, employing techniques from the signal processing domain. Level 2 consists of derived or logical features involving some degree of statistical and logical inference about the identity of objects depicted by visual media. An example query at this level can be "find pictures of Eiffel Tower." Using these models, systems normally operate on low-level feature representation, though they can also use image data directly. Level 3 deals with semantic abstractions involving a significant amount of high-level reasoning about the meaning and purpose of the objects or scenes depicted. An example of a query at this level can be "find pictures of laughing children." As indicated at level 3 of the figure, the artificial intelligence (AI) community has had the leading role in this effort. Systems at this level can take semantic representation based on input generated at level 2.

The following subsections explore some of the major building blocks of CBIR-systems.

Low-Level Feature-Based Indexing

Low-level visual feature extraction to describe image content is at the heart of CBIR systems. The features to be extracted can be categorized into general features and domain-specific features. The latter ones may include human faces, fingerprints, and human skin. Feature extraction in the former context, such as from databases that contain images of wide ranging content with images that do not portray any specific topic or theme and come from various sources, is a very challenging job. One possible approach is to perform segmentation first

[7] Bidirectional temporal prediction, denoted as a PB-picture, is obtained by coding two pictures as a group and avoiding the reordering necessary in the decoding of B-pictures.

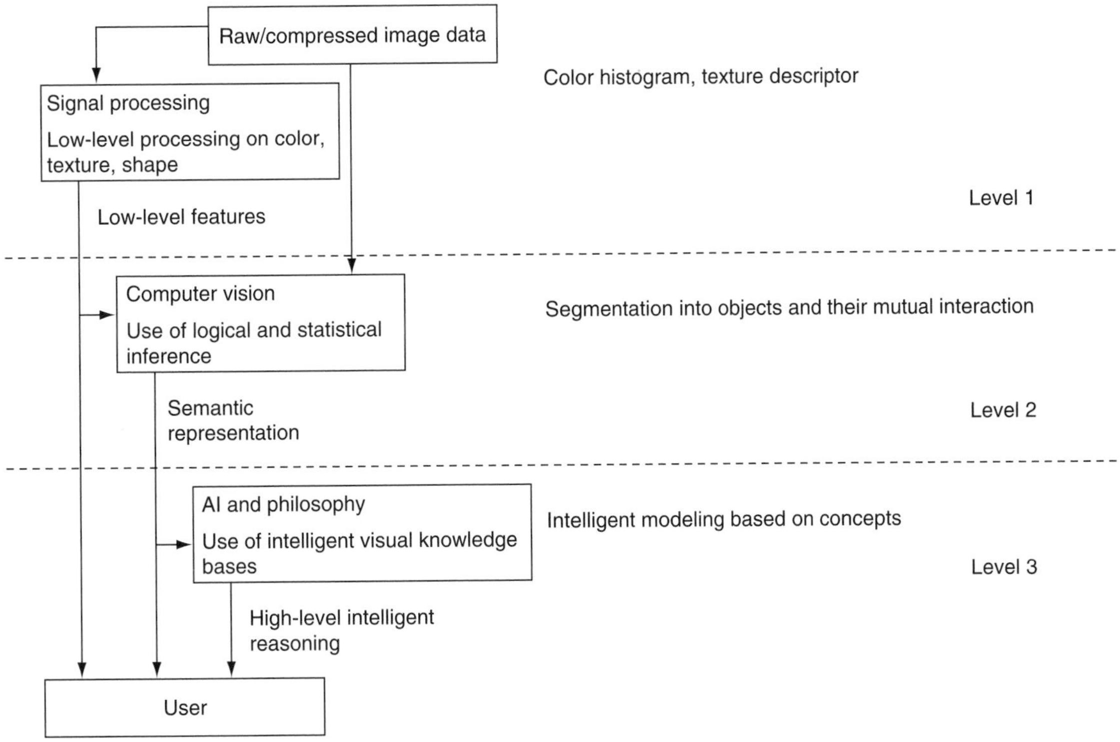

FIGURE 6.6 Classification of CBIR Techniques. Level 1. This diagram shows a low-level physical modeling of raw image data: QBIC (Niblack *et al.*, 1998); PhotoBook (Pentland *et al.*, 1996); VisualSeek (Smith and Chang, 1996b); Seapal (Khokhar *et al.*, 1999, 2000). Level 2. Shown is a representation of derived or logical features: BlobWorld (Carson *et al.*, 1997); Iqbal (Iqbal and Aggarwal, 1999). Level 3. This level illustrates semantic level abstractions.

and then extract visual features from segmented objects. Unconstrained segmentation of an object from the background, however, is often not possible because there generally is no particular object in the image. Therefore, segmentation in such a case is of very limited use as a stage preceding feature extraction. The images thus need to be described as a whole unit, and one should devise feature extraction schemes that do not require segmentation. This restriction excludes a vast number of well-known feature extraction techniques from low-level feature-based representation: all boundary-based methods and many area-based methods. Basic pixel-value-based statistics, possibly combined with edge detection techniques, that reflect the properties of human visual system in discriminating between image patches can be used. Invariance to specific transforms is an issue of interest in feature extraction as well. Feature extraction methods that are global in their nature or perform averaging over the whole image area are often inherently translation invariant. Other types of invariances (e.g., invariance to scaling, rotation, and occlusion) can be obtained with some feature extraction schemes by using proper transformations. Because of the perception subjectivity, there does not exist a single best representation for a given feature. For any given feature, there exist multiple representations that characterize the feature from different perspectives. The main features used in CBIR systems can be categorized into three groups: color features, texture features, and shape features. The following subsections review the importance and implementation of each feature in the context of image content description.

Color

Color is one of the most widely used low-level features in the context of indexing and retrieval based on image content. It is relatively robust to background complication and independent of image size and orientation. Typically, the color of an image is represented through a color model. A color model is specified in terms of a 3-D coordinate system and a subspace in that system where each color is represented by a single point. The more commonly used color models are RGB (red, green, and blue), HSV (hue, saturation, and value), and YIQ (luminance and chrominance). Thus, the color content is characterized by three channels from a color model. One representation of color content of the image is made by using a color histogram. The histogram of a single channel of an image with values in the range $[0, L-1]$ is a discrete function $p(i) = n_i/n$, where i is the value of the pixel in current channel, n_i is the number of pixels in the image with value i, n is the total number of pixels in the image, and $i = 0, 1, 2, \ldots, L-1$. For a three-channel image, there will be three such histograms. The histograms are normally divided into bins in an effort to coarsely represent the content and reduce dimensionality of a subsequent matching phase. A feature vector is then formed by linking the three channel histograms into one vector. For image retrieval, the histogram of a query image is then matched against the histogram of all images in the database using a similarity metric. One similarity metric that can be used in this context is **histogram intersection**. The intersection of histograms h and g is given by:

$$d(h, g) = \frac{\sum_{m=0}^{M-1} \min(h[m], g[m])}{\min\left(\sum_{m0=0}^{M-1} h[m0], \sum_{m1=0}^{M-1} g[m1]\right)}. \tag{6.2}$$

In this metric, colors not present in the user's query image do not contribute to the intersection. Another similarity metric between histograms h and g of two images is **histogram quadratic distance**, which is given by:

$$d(h, g) = \sum_{m0=0}^{M-1} \sum_{m1=0}^{M-1} (h[m0] - g[m0]) a_{m0, m1} \cdot (h[m1] - g[m1]). \tag{6.3}$$

where $a_{m0, m1}$ is the crosscorrelation between histogram bins based on the perceptual similarity of the colors $m0$ and $m1$. One appropriate value for the crosscorrelation is given by:

$$a_{m0, m1} = 1 - d_{m0, m1}, \tag{6.4}$$

where $d_{m0, m1}$ is the distance between colors $m0$ and $m1$ normalized with respect to the maximum distance. **Color moments** have also been applied in image retrieval. The mathematical foundation of this approach is that any color distribution can be characterized by its moments. Furthermore, since most of the information is concentrated in the low-order moments, only the first moment (mean) and the second and third central moments (variance and skewness) can be used for robust and compact color content representation. Weighted Euclidean distance is then used to compute color similarity. To facilitate fast search over large-scale image collections, **color sets** have also been used as approximations to color histograms. The color model used is HSV, and the histograms are further quantized into bins. A color set is defined as a selection of the colors from quantized color space. Because color set feature vectors are binary, a binary search tree is contructed to allow fast search (Smith and Chang, 1996a).

One major drawback of color histogram-based approaches is the lack of explicit spatial information. Specifically, based on global color based representation, it is hard to distinguish between a red car on white background and a bunch of red balloons with a white background. This problem is addressed by Khokhar *et al.* (2000). They have used encoded quadtree spatial data structure to preserve the structural information in the color image. Based on a perceptually uniform color space CIELab*, each image is quantized into k bins to represent k different color groups. A color layout corresponding to pixels of each color in the whole image is formed for each bin and is represented by the corresponding encoded quadtree. This encoded quadtree-based representation not only keeps the spatial information intact but also results in a system that is highly scalable in terms of query search time.

Texture

An image can be considered as a mosaic of regions with different appearances, and the image features associated with these regions can be used for search and retrieval. Although no formal definition of **texture** exists, intuitively this descriptor provides measures of properties such as smoothness, coarseness, and regularity. These properties can generally not be attributed to the presence of any particular color or intensity. Texture corresponds to repetition of basic texture elements called **texels**. A texel consists of several pixels and can be periodic, quasiperiodic, or random in nature. Texture is an innate property of virtually all surfaces, including such elements as clouds, trees, bricks, hair, and fabric. It contains important information about the structural arrangement of surfaces and their relationship to the surrounding environment. The three principal approaches used in practice to describe the texture of a region are statistical, structural, and spectral. **Statistical approaches** yield characterization of textures as smooth, coarse, grainy, and so on. **Structural techniques** deal with the arrangement of image primitives, such as description of texture based on regularly spaced parallel lines. **Spectral techniques** are based on properties of Fourier spectrum and are used primarily to detect global periodicity in an image by identifying high-energy, narrow peaks in the spectrum (Gonzalez, 1992) Haralick *et al.* (1973) proposed the co-occurrence matrix representation of texture feature. This method of texture description is based on the repeated occurrence of some gray-level configuration in the texture; this configuration varies rapidly with distance in fine textures and slowly in coarse textures. This approach explores the gray-level spatial dependence of texture. It first constructs a co-occurrence matrix based on the orientation and distance between image pixels and then extracts meaningful statistics from the matrix as texture representation. Motivated by the psychological studies in human visual perception of texture, Tamura *et al.* (1978) have proposed the texture representation from a different angle. They developed computational approximations to the visual texture properties found to be important in psychology studies. The six visual texture properties are **coarseness, contrast, directionality, linelikeness, regularity, and roughness**. One major distinction between the Tamura texture representation and co-occurrence matrix representation is that all the texture properties in Tamura representation are visually meaningful, whereas some of the texture properties used in the co-occurrence matrix representation may not. This characteristic makes Tamura texture representation very attractive in image retrieval because it can provide a friendlier user interface.

The use of texture feature requires texture segmentation, which remains a challenging and computationally intensive task. In addition, texture-based techniques lack robust texture models and correlation with human perception.

Shape

Shape is an important criterion for matching objects based on their profile and physical structure. In image retrieval applications, shape features can be classified into **global** and **local** features. Global features are the properties derived from the entire shape, such as roundness, circularity, central moments, and eccentricity. Local features are those derived by partial processing of a shape, including size and orientation of consecutive boundary segments, points of curvature, corners, and turning angle. Another categorization of shape representation is **boundary-based** and **region-based**. The former uses only the outer boundary of the shape, while the latter uses the entire shape of the region. Fourier descriptors and moment invariants are the most widely used shape representation schemes. The main idea of a Fourier descriptor is to use the Fourier transformed boundary as the shape feature. Moment invariant technique uses region-based moments, which are invariant to transformations, as the shape features. Hu *et al.* (1962) proposed a set of seven invariant moments derived from second and third moments. This set of moments is invariant to translation, rotation, and scale changes. Finite element method (FEM) (Pentland *et al.*, 1996) has also been used as shape representation tool. FEM defines a stiffness matrix that describes how each point on the object is connected to other points. The eigenvectors of the stiffness matrix are called modes and span a feature space. All the shapes are first mapped into this space, and similarity is then computed based on the eigenvalues. Along the similar lines of Fourier descriptors, Arkin *et al.* (1991) developed a Turning function-based approach for comparing both convex and concave polygons.

Spatial Versus Compressed Domain Processing

Given the huge storage requirements of nontextual data, the vast volumes of images are normally stored in compressed form. One approach in CBIR systems is to first decompress the images and transform them into the format used by the system. The default color format used by majority of the CBIR systems is RGB, which is very redundant and unsuitable for storage but very easy to process and use for display. Once the raw image data has been extracted after decompression, any of the content modeling techniques can be applied to yield a representation of the image content for indexing. This process of decompressing the image before content representation poses an overhead in the most likely scenario of more and more image content being stored in compressed form due to the success of image coding standards like JPEG and JPEG-2000. A better approach toward this issue in many of the modern systems is to process compressed domain images as a first class and default medium (i.e., compressed images should be operated upon directly). Since compressed domain representation either has all (if it is a lossless coding scheme) or most of the important (if it is a lossy scheme, depending on the quantization setting in the encoder) image information intact, indexing based on image content can be performed from minimal decoding of compressed images.

Discrete cosine transform (DCT) is at the heart of the JPEG still-image compression standard and many of the video compression standards, like MPEG-1, MPEG-2, and H.261. Shen and Sethi (1996) have used DCT coefficients of encoded blocks to mark areas of interest in the image. This distinction can be used to give more preference to these areas when processing the image for content representation. Locations of areas of interest are those parts of the image that show sufficiently large intensity changes. This is achieved by computing the variance of pixels in a rectangular window around the current pixel. In the DCT domain, this translates to computing the ac energy according to the relationship:

$$E = \sum_{u=0}^{7} \sum_{v=0}^{7} F_{uv}^2 \quad (u, v) \neq (0, 0), \tag{6.5}$$

where F_{uv} stands for ac coefficients in the block and where the encoded block is 8×8 as in image/video coding standards. Shen and Sethi (1996) also propose fast coarse-edge detection techniques using DCT coefficients. A comparison of their approach with edge detection techniques in spatial domain speaks in favor of DCT coefficients based-edge detection because a coarse representation of edges is quite sufficient for content description purposes.

Because image coding and indexing are quite overlapping processes in terms of storage and searching, one approach is to unify the two problems in a single framework. There has been a recent shift in trends in terms of transformation used for frequency domain processing, from DCT to discrete wavelet transform (DWT) because of DWT's time frequency and multiresolution analysis nature. DWT has been incorporated in modern image and video compression standards like JPEG-2000 and MPEG-4. One such system that uses DWT for compression and indexing of images is proposed in Liang *et al.* (1999). The wavelet-based image encoding techniques depend on successive approximate quantization (SAQ) of wavelet coefficients in different subbands from wavelet decomposition. The image indexing from DWT encoded image is mainly based on significant coefficients in each subband. Significant coefficients in each sub band are recognized as the ones whose magnitude is greater than a certain threshold, which is different at each decomposition level. The initial threshold is chosen to be half the maximum magnitude at first decomposition level, whereas successive thresholds are given by dividing the threshold at a previous decomposition level by 2. During the coding process, a binary map called significant map is maintained so the coder knows the location of significant as well as insignificant coefficients. To index texture, a two-bin histogram of a wavelet coefficient at each sub band is formed with the count of significant and insignificant wavelet coefficients in the two bins. For color content representation, YUV color space is used. A nonuniform histogram of 12 bins containing count of significant coefficients given each of the 12 thresholds is computed for each color channel. This way, three 12-bin histograms are computed for luminance (Y) and chrominance (U and V) channels. For each of the histograms, first-, second-, and third-order moments (mean, variance, and skewness) are computed and used as indexing features for color.

Segmentation

Most of the existing techniques in current CBIR systems depend heavily on a low-level feature-based description of image content. Most existing approaches represent images based only on their composition with little regard to spatial organization of the low-level features. On the other hand, users of the CBIR systems often would like to find images containing particular objects ("things"). This gap between low-level description of the image content and the object images represent can be filled by performing segmentation on images to be indexed. Segmentation subdivides an image into its constituent parts or objects. Segmentation algorithms for monochrome images generally are based on one of two basic properties of gray-level values: **discontinuity** and **similarity**. In the first category, the approach is to partition an image based on abrupt changes in gray level. The principal areas of interest in this category are detection of isolated points and detection of lines and edges in an image. The principal approaches in the second category are based on thresholding, region growing, and region splitting and merging. The BlobWorld system proposed by Carson *et al.* (1997) is based on segmentation using the expectation–maximization algorithm on combined color and texture features. It represents the image as a small set of localized coherent regions in color and texture spaces. After segmenting the image into small regions, a description of each region's color, texture, and spatial characteristics is produced. Each image may be visualized by an ensemble of 2-D ellipses or "blobs," each of which possesses a number of attributes. The number of blobs in an image is not very overwhelming to facilitate fast image retrieval applications and is typically less than ten. Each blob represents a region on the image that is roughly homogeneous with respect to color or texture. A blob is described by its dominant color, mean texture descriptors, spatial centroid, and scatter matrix. The exact retrieval process is then performed on the blobs in a query image. On similar lines but in domain-specific context, Iqbal and Aggarwal (1999) apply perceptual grouping to develop a CBIR system for images containing buildings. In their work, semantic interrelationships between different primitive image features are exploited by perceptual grouping to detect presence of man-made structures. Perceptual grouping uses concepts as grouping by proximity, similarity, continuation, closure, and symmetry to organize primitive image features into meaningful higher level image relations. The approach is based on the observation that the presence of a man-made structure in an image will generate a large number of significant edges, junctions, parallel lines, and groups in comparison with an image of predominantly nonbuilding objects. These structures are generated by the presence of corners, windows, doors, and boundaries of the buildings, for example. The features they extract from an image are hierarchical in nature and include **line segments**, **longer linear lines**, **L junctions**, **U junctions**, **parallel lines**, **parallel groups**, and **significant parallel groups**.

Most of the segmentation methods discussed in image processing and analysis literature are automatic. A major advantage of this type of segmentation algorithms is that it can extract boundaries from a large number of images without occupying the user's time and effort. However, in an unconstrained domain, for nonpreconditioned images, which is the case with image CBIR systems, the automatic segmentation is not always reliable. What an algorithm can segment in this case is only regions and not objects. To obtain high-level objects, human assistance is almost always needed for reliable segmentation.

High-Dimensionality and Dimension Reduction

It is obvious from the discussion thus far that content-based image retrieval is a high-dimensional feature vector-matching problem. To make the systems truly scalable to large size image collections, two factors should be considered. First, the dimensionality of the feature space needs to be reduced to achieve the embedded dimension. Second, efficient and scalable multidimensional indexing techniques need to be adapted to index the reduced but still high-dimensional feature space. In the context of dimensionality reduction, a transformation of the original data set using the Karhunen-Loeve transform (KLT) can be used. KLT features data-dependent basis functions obtained from a given data set and achieves the theoretical ideal in terms of compressing the data set. An approximation to KLT given by principal component analysis (PCA) gives a very practical solution to the computationally intensive process of KLT. PCA, introduced by Pearson in 1901 and developed independently by Hotelling in 1933, is probably the oldest and best known of the techniques of multivariate analysis. The central idea of PCA is to reduce the dimensionality of a data set in which there are a large number of interrelated variables while retaining as much as possible of the variation present in the data set. This reduction is achieved by transforming to a new set of variables, the principal components (PCs), that are uncorrelated and ordered so the first few retain most of the variation present in *all* of the original variables. Computation of the principal components reduces to the solution of an eigenvalue–eigenvector problem for a positive–semidefinite symmetric matrix. Given that x is a vector of p random variables, the first step in a PCA evaluation is to look for a linear function $\alpha_1^{'}x$ of the elements of x that have maximum variance, where α_1 is a vector of p constants $\alpha_{11}, \alpha_{12} \ldots \alpha_{1_p}$. Next, look for a linear function $\alpha_2^{'}x$, uncorrelated with $\alpha_1^{'}x$, which has maximum variance and so on. The kth derived variable $\alpha_k^{'}x$ is the kth PC. Up to p PCs can be found, but in general most of the variation in x can be accounted for by m PCs, where $m \ll p$. If the vector x has known covariance matrix Σ, then the kth PC is given by an orthonormal linear transformation of x as $y_k = \alpha_k^{'}x$, where α_k is an eigenvector of Σ corresponding to its kth largest eigenvalue λ_k. Consider an orthogonal matrix Φ_q with α_k as the kth column and containing $q \ll p$ columns corresponding to q PCs; it can be shown that for the transformation $y = \Phi_q x$, the determinant of covariance matrix for transformed data set y, det (Σ_y) is maximized. The statistical importance of this property follows because the determinant of a covariance matrix, which is called the **generalized variance** can be used as a single measure of spread for a multivariate random variable. The square root of the generalized variance for a multivariate normal distribution is proportional to the *volume* in p-dimensional space, which encloses a fixed proportion of the probability distribution of x. For a multivariate normal x, the first q PCs are therefore q linear functions of x whose joint probability distribution has contours of fixed probability that enclose the maximum volume (Joliffe, 1986). If the data vector x is normalized by its variance and autocorrelation matrix instead of covariance matrix is used, then above mentioned optimality property and derivation of PCs still hold. For the efficient computation of PCs, at least in context of PCA rather than general eigenvalue problems, singular value decomposition (SVD) has been termed as the best approach available (Chambers, 1997).

Even after the dimension of the data set has been reduced, the data set is still almost always fairly high-dimensional. There have been contributions from three major research communities in this direction: computational geometry, database management, and pattern recognition. The history of multidimensional indexing techniques can be tracked back to middle 1970s when cell methods, quadtree and k-d tree were first introduced. However, their performance was far from satisfactory. Pushed by the then urgent demand of spatial indexing from GIS and CAD systems, Guttman proposed the R-tree indexing structure in 1984. Some good reviews of various indexing techniques in the context of image retrieval can be found in (White and Jain, 1996). Khokhar *et al.* (1999) have dealt with the feature vector formation and its efficient indexing as one problem and suggest a solution in which query response time is relatively independent of the database size. They exploited energy compaction properties of the vector wavelets and designed suitable data structures for fast indexing and retrieval mechanisms.

Relevance Feedback

The problem of content-based image retrieval is different than the conventional computer vision-based pattern recognition task. The fundamental difference between the two is that in the latter, we **are looking for exact match for the object to be searched with as small and as accurate a retrieved list as possible**. But in the former, the goal is to extract as many "similar" objects as possible, the notion of similarity being very loose as compared to the notion of exact match. Moreover, the human user is the indispensable part in the former. Early literature and CBIR systems emphasized fully automatic operation. This approach did not take into account the fact that the ultimate end user of the CBIR system is human, and the image is inherently a subjective medium (i.e., the perception of image content is very subjective, and the same content can be interpreted differently by users having different search criteria). This human perception subjectivity has different levels to it: one user might be more interested in a different dominant feature of the image than the other, or two users might be interested in the same feature (e.g., texture), but the perception of a specific texture might be different for the two users. Recent drive is oriented more toward how humans perceive image content and how to integrate such a "human model" into the image retrieval systems. Rui *et al.* (1998) have reported a formal model of a CBIR system with relevance feedback integrated into it. They first initialized a retrieval system with uniformly distributed weights for each feature. Then user's information need is distributed among all the features. The similarity is then computed on the basis of weights by user's input, and retrieval results are displayed to the user. The user marks each retrieved result as **highly relevant, relevant, no-opinion, irrelevant**, and **highly irrelevant** according to his or her information needs and perception subjectivity. The system updates its weights and goes back into the loop.

CBIR Systems: Query By Image Content

The field of content-based indexing and retrieval has been an active area of research for the past few decades. The research effort that has gone into development of techniques for this problem has led to some very successful systems currently available as commercial products as well as other research systems available for the academic community. Some of these CBIR systems include Virage (Bach *et al.*, 1996), Netra (Ma and Manjunath, 1997), PhotoBook (Pentland *et al.*, 1996), VisualSeek (Smith and Chang, 1996b), WebSeek (Smith and Chang, 1997), MARS (Mehrotra *et al.*, 1997), and BlobWorld (Carson *et al.*, 1997). A comparative study of many of the CBIR systems can be found in (Venters and Cooper, 2000). This section, for illustrative purposes, reviews one example of a commercial CBIR system known as Query By Image Content (QBIC).

QBIC is the first commercial content-based image retrieval system. Developed by IBM Almaden Research Centre, it is an open framework technology that can be used for both static and dynamic image retrieval. QBIC has undergone several iterations since it was first reported (Niblack *et al.*, 1998). QBIC allows users to graphically pose and refine queries based on multiple visual properties, including color, shape, and texture. QBIC supports several query types: simple, multifeature, and multipass. A simple query involves only one feature. For example, identifying images that have a color distribution similar to the query image can involve a complex query with more than one feature, which can take the form of a multifeature or a multi-pass query. For identifying images that have similar color and texture features, a multifeature query would be possible and involve the system searching through the different types of feature data in the database to identify similar images. All feature classes would have equal weightings during the search, and all feature tables would be searched in parallel. In contrast, with a multipass query, the output of an initial search would be used as the basis for the next search. The system would reorganize the search results from a previous pass based on the "feature distances" in the current pass. For example, a user could identify images that have a similar color distribution and then reorder the results based on color composition. With multifeature and multipass queries, users can weight features to specify their relative importance. QBIC technology has been incorporated into several IBM software products, including DB2 Image Extender and Digital Library. QBIC supports several matching features, including color, shape, and texture. The global color function computes the average RGB colors in the entire image for both the dominant color and the variation of color throughout the entire image. Similarity is based on the three average color values. The local color function computes the color distribution for both the dominant color and the variation for each image in a predetermined 256 color space. Image similarity is based on the similarity of the color distribution. The shape function analyzes images for combinations of area, circularity, eccentricity, and major axis orientation. All shapes are assumed to be nonoccluded planar shapes, allowing each shape to be represented as a binary image. The texture function analyzes areas for global coarseness, contrast, and directionality features.

6.3.2 Video Indexing and Retrieval

When compared with content-based image indexing and retrieval, which has been an active area of research since the 1970s, the field of content-based access to video repositories is still gaining due attention. As discussed at the beginning of subsection 6.3.1, human visual perception displays a gradient of sophistication, ranging from seemingly primitive inferences (e.g., shapes, textures, and colors) to complex notions of structures (e.g., chairs, trees, and affordances) to cognitive processes (e.g., recognition of emotions and feelings). Given the multidisciplinary nature of the techniques for modeling, indexing, and retrieving visual data, efforts from many different communities have merged in the advancement of content-based video indexing and retrieval (CBVIR) systems. Depending on the background of the research teams, different levels of abstractions have been assumed to model the data. As shown in Figure 6.7, we classify these abstractions into three categories based on the gradient model of human visual perception specifically in the context of CBVIR systems. In this figure, we also capture the mutual interaction of some of the disciplines of engineering, computer science, and cognitive sciences.

Level 1 represents systems that model raw video data using features such as color histogram, shape and texture descriptors, or trajectory of objects. This model can be used to serve a query like "shots of object with dominant red color moving from left corner to right." CBVIR systems based on these models operate directly on the data, employing techniques from signal processing domain. Level 2 consists of derived or logical features involving some degree of statistical and logical inference about the identity of objects depicted by visual media. An example query at this level can be "shots of Sears Tower." Using these models, systems normally operate on low-level feature representation, though they can also use video data directly. Level 3 deals with semantic abstractions involving a significant amount of high-level reasoning about the meaning and purpose of the objects or scenes depicted. An example of a query at this level can be "shots depicting human suffering or sorrow." As indicated at level 3 of the figure, the AI community has had the leading role in this effort. Systems at this level can take semantic representation based on input generated at level 2.

Despite the diversity in modeling and application of CBVIR systems, most systems usually rely on similar video processing modules. The following subsections explore some of the broad classes of modules typically used in a content-based video indexing and retrieval system.

Temporal Segmentation

Video data can be viewed hierarchically. At the lowest level, video data is made up of **frames**; a collection of frames that result from single camera operation depicting one event is called a **shot**, and a complete unit of narration that consists of a series of shots or a single shot taking place in a single location and dealing with a single action defines a **scene** (Monaco, 1977). CBVIR systems rely on the visual content at distinct hierarchical levels of the video data. Although the basic representation of raw video is provided in terms of a sequence of frames, the detection of distinct shots and scenes is a complex task.

Transitions or boundaries between shots can be abrupt (cut) or they can be gradual (fade, dissolve, and wipe). Traditional temporal segmentation techniques have focused on cut detec-

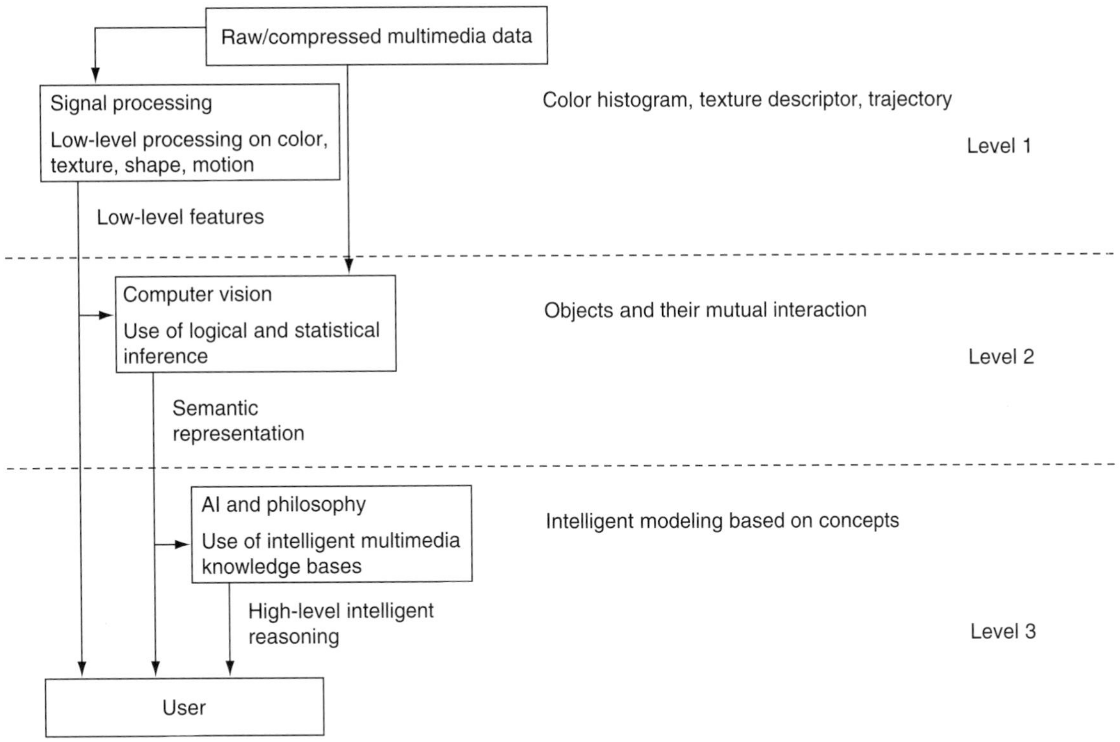

FIGURE 6.7 Classification of Content Modeling Techniques. Level 1: This diagram shows low-level physical modeling of raw video data: ViBE (Chen *et al.*, 2004). Level 2: Shown here is a representation of derived or logical features: VideoQ (Chang *et al.*, 1998); MultiNet (Nephade *et al.*, 2002). Level 3: This figure illustrates semantic level abstraction: MediaNet (Benitez *et al.*, 2000); Purdue (Khokar *et al.*, 1999); OVID (Oomoto and Tanaka, 1993).

tion, but there has been increasing research activity on gradual shot boundary detection as well. Most of the existing techniques reported in the literature detect shot boundary by extracting some form of feature for each frame in the video sequence, then evaluating a similarity measure on features extracted from successive pairs of frames in the video sequence, and finally declaring the detection of a shot boundary if the features difference conveyed by the similarity measure exceeds a threshold. One such approach is presented in (Naphade *et al.*, 1998a) in which two difference metrics, histogram distance metric (HDM) and spatial distance metric (SDM), are computed for every frame pair. HDM is defined in terms of three-channel linearized histograms computed for successive frame pair f_i and f_{i+1} as follows:

$$D_h(f_i, f_{i+1}) = \frac{1}{M \times N} \sum_{j=1}^{256x3} |H_i(j) - H_{i+1}(j)|, \tag{6.6}$$

where H_i represents the histogram of frame f_i and $M \times N$ is the dimension of each frame. For each histogram, 256 uniform quantization levels for each channel are considered. SDM is defined in terms of the difference in intensity levels between successive frames at each pixel location. Let $I_{i,j}(f_k)$ denote the intensity of a pixel at location (i,j) in the frame f_k, and then the spatial distance operator is defined as:

$$d_{i,j}(f_k, f_{k+1}) = \begin{cases} 1 \cdots, |I_{i,j}(f_k) - I_{i,j}(f_{k+1})| > \varepsilon. \\ 0 \cdots, \textit{otherwise}. \end{cases} \tag{6.7}$$

SDM is then computed as follows:

$$D_s(f_k, f_{k+1}) = \frac{1}{M \times N} \sum_{i=1}^{M} \sum_{j=1}^{N} d_{i,j}(f_k, f_{k+1}). \tag{6.8}$$

These two distances are then treated as a 2-D feature vector, and an unsupervised *K*-means clustering algorithm is used to group shot boundaries into one cluster. For a review of major conventional shot boundary detection techniques, refer to Borecsky and Rowe (1996), which also provides a comparison between five different techniques based on pixel difference from raw data, DCT coefficients difference, and motion-compensated difference. Due to the huge amount of data to be processed in the case of full-frame pixel difference-based methods as well as data's susceptibility to intensity differences caused by motion, illumination changes, and noise, many novel techniques (beyond the scope of the review presented in Borecsky and Rowe, 1996), have been proposed in both the compressed as well as uncompressed domain. We shall now present a brief overview of some of the recent advances in shot detection.

Porter *et al.* (2000) proposed a frequency domain correlation approach. This approach relies on motion estimation information obtained by use of template matching; that is, for each 32 × 32 block in a given frame, the best matching block in a corresponding neighborhood in the next frame is sought by calculating the normalized crosscorrelation in the frequency domain as:

$$\rho(\varepsilon) = \frac{F^{-1}\left\{\hat{x}_1(\bar{\omega}) \times \overset{*}{\hat{x}}_2(\bar{\omega})\right\}}{\sqrt{\int |\hat{x}_1(\bar{\omega})|^2 d\bar{\omega} \cdot \int |\hat{x}_2(\bar{\omega})|^2 d\bar{\omega}}}, \quad (6.9)$$

where ε and $\bar{\omega}$ are the spatial and frequency coordinate vectors, respectively, $\hat{x}_i(\bar{\omega})$ denotes the Fourier transform of frame $x_i(\varepsilon)$, F^{-1} denotes the inverse Fourier transform operation, and * is the complex conjugate. Next, the mean and standard deviation of the correlation peaks for each block in the whole image are calculated, and the peaks beyond one standard deviation away from the mean are discarded, thus making the technique more robust to sudden local changes in a small portion of the frame. An average mean is then computed from this pruned data. This average match measure is compared to the average match of the previous pair, and a shot boundary is declared if there is a significant decrease in this similarity match feature.

A novel approach proposed by Liu and Chen (2002) argues that at the shot boundary, the contents of new shot differ from contents of the whole previous shot instead of just the previous frame. They proposed a generic and recursive principal component analysis-based approach that can be built on any feature extracted from frames in a shot and that generates a model of the shot trained from features in previous frames. Features from the current frame are extracted, and a shot boundary is declared if the features from the current frame do not match the existing model by projecting the current feature onto the existing eigenspace.

In an effort to cut back on the huge amount of data available for processing and to emphasize the fact that in video shots, while objects may appear or disappear, the background stays much the same and follows the camera motion in one shot, Oh *et al.* (2000) have proposed a background tracking (BGT) approach. A strip along the top, left, and right borders of the frame, covering around 20% of frame area, is taken as a fixed background area (FBA). A signature 1-D vector called transformed background area (TBA) formed from the Gaussian pyramid representation of the FBA is computed. Background tracking is achieved by a 1-D correlation matching between two TBAs obtained from successive frames. Shot detection is declared if the background tracking fails as characterized by a decrease in the correlation matching parameter. This approach has been reported to detect and classify both abrupt and gradual scene changes.

Observing the fact that single features cannot be used accurately in a wide variety of situations, Chen *et al.* (2004) have proposed to construct a high-dimensional feature vector called generalized trace (GT) by extracting a set of features from each dc frame. For each frame, GT contains the number of intracoded as well as forward- and backward-predicted macroblocks; a histogram intersection of current and previous frames for Y, U, and V color components; and a standard deviation of Y, U, and V components for the current frame. GT is then used in a binary regression tree to determine the probability that each frame is a shot boundary. These probabilities are used to determine the frames that most likely correspond to the shot boundary.

Hanjalic (2002) has put together a nice analysis of the shot boundary detection problem itself, identifying major issues that need to be considered and creating a conceptual solution to the problem in the form of a statistical detector based on minimization of average detection–error probability. The thresholds used in their system are defined at the lower level modules of the detector system. The decision making about the presence of a shot boundary is left solely to a parameter-free detector, where all of the indications coming from different low-level modules are combined and evaluated.

Lelescu and Schonfeld (2000, 2003) presented a scene change detection method using stochastic sequential analysis theory. The dc data from each frame are processed using PCA to generate a very low-dimensional feature vector Y_k corresponding to each frame. These feature vectors are assumed to form an independant identically distributed (*i.i.d*) sequence of multidimensional random vectors having Gaussian distribution. Scene change is then modeled as change in the mean parameter of this distribution. Scene change detection is formulated as a hypothesis testing problem, and the solution is provided in terms of a threshold on a generalized likelihood ratio. Scene change is declared at frame k when the maximum value of the sufficient statistic g_k evaluated over frame interval j to k exceeds the threshold:

$$g_k = \max_{1 \leq j \leq k} \left\{ \frac{k-j+1}{2} (X_j^k)^2 \right\}. \quad (6.10)$$

Here X_j^k is defined as:

$$X_j^k = \left[(\bar{Y}_j^k - \Theta_0)^T \Sigma^{-1} (\bar{Y}_j^k - \Theta_0) \right]^{1/2}. \quad (6.11)$$

In this expression, $\bar{Y}_j^k$ is the mean of feature vectors Y in the current frame interval j to k, and Θ_0 is the mean of Y in an initial training set frame interval consisting of M frames. This approach, which is free from human fine-tuning, has been reported to perform equally well for both abrupt and gradual scene changes.

Video Summarization

Once a video clip has been segmented into atomic units based on visual content coherence, the next step is to compactly represent the individual units. This task is the major block in summarizing video content using a table of contents approach. This step also facilitates efficient matching between two shots at query time for content-based retrieval. Most existing systems represent video content by using one representative frame from the shot called a **keyframe.** Keyframe-based representation has been recognized as an important research issue in content-based video abstraction. The simplest approach for this problem is to use the first frame of each shot as a keyframe (Nagasaka and Tanaka, 1992). Although the approach is simple, it is limited because each shot is allotted only one frame for its representation irrespective of the complexity of the shot content. In addition, the choice of the first frame over other frames in the shot is arbitrary. To have more flexibility in keyframe-based representation of a video shot, Zhang *et al.* (1997) proposed to use multiple frames to represent each shot. They used criteria such as color content change and zoom-in type of effects in shot content to decide on the keyframes for each shot. A technique for shot content representation and similarity measure using subshot extraction and representation is presented in Lin *et al.* (2001). This approach uses two content descriptors, dominant color histogram (DCH) and spatial structure histogram (SSH), to measure content variation and to represent subshots. The quantized hue, saturation, and value (HSV) color histogram are First Computed for each frame. Next, the dominant local maxima positions in each frame's histogram are identified and tracked throughout the shot. After tracking, only the colors with longer durations are retained as dominant colors of the shot. Histogram bins are finally weighted by the duration of each bin in the whole shot. SSH is computed based on spatial information of color blobs. For each blob, histograms are computed for the area, position, and deviation.

Delp *et al.* (2001) represented a shot using a tree structure called a **shot tree**. This tree is formed by an agglomerative clustering technique performed on individual frames in a shot. Starting at the lowest level with each frame representing a cluster, the algorithm iteratively combines the two most similar frames at a particular level into one cluster at the next higher level. The process continues until a single cluster represented by one frame for the whole shot is obtained. This approach unifies the problem of scene content representation for both browsing and similarity matching. For browsing, only the root node of the tree (keyframe) is used, whereas for similarity matching, two or three levels of tree can be used, employing standard tree matching algorithms.

Another approach to video summarization based on a low-resolution video clip has been proposed by Lelescu and Schonfeld (2001). In this approach, a low-resolution video clip is provided by an efficient representation of the dc frames of the video shot using an iterative algorithm for the computation of PCA. Efficient representation of the dc frames is obtained by their projection onto the eigenspace characterized by the dominant eigenvectors for the video shot. The eigenvectors obtained by PCA can also be used for conventional keyframe representation of the video shot by considering the similarity of frames in the video shot to the eigenvectors with the largest eigenvalues.

Compensation for Camera and Background Movement

The motion content in a video sequence is the result of either camera motion (e.g., pan, zoom, or tilt) or object and background motion. **Panning motion** of a camera is defined as rotation along the horizontal axis, whereas **tilt** is rotation along the vertical axis. The major concern in motion content-based video indexing and retrieval is almost always the object's motion and not the camera effects. This is because while querying video indexing and retrieval systems, users tend to be more interested in the maneuvering of the objects in the scene and not the way camera is being tilted, rotated, or zoomed with respect to the object. The problem is that the true motion of the object cannot be assessed unless the camera motion is compensated for. This problem always arises in case of a video of a moving object recorded with a mobile camera. Similarly, quite often there is object motion as well as background movement in the video sequence. In such cases, background movement needs to be differentiated from object movement to give the true object motion. Once these motions have been separated, the object trajectory can be obtained and video data can be indexed based on this motion cue along with other features.

Oh and Chowdary (2002) expressed different motions accurately by estimating motions from the camera and object. First, the total motion (TM) in a shot is measured. This is achieved by computing the accumulated quantized pixel differences on all pairs of frames in the shot. Before computing pixel differences, the color at each pixel location is quantized into 32 bins to reduce the effect of noise. Once the total motion has been estimated, each frame in the shot is checked for the presence of camera motion. If pan and tilt of camera, are present, the amount and direction of camera motion are computed. Object motion (OM) is computed by a technique similar to the computation of TM, such as after compensation of camera motion. Bouthemy *et al.* (1999) addressed the problem of shot change detection as well as camera motion estimation in a single framework. The two objectives are met by computing, at each time instant, the dominant motion in the image sequence represented by a 2-D affine motion model. From each frame pair in the video sequence, a statistical module estimates the motion model parameters and the support for motion (i.e., the area of motion) in the successive frame. A least-squares motion estimation module then computes the confidence of the motion model and maps significant motion parameters onto predefined camera motion classes. These classes include pan, tilt, zoom, and many combinations.

Feature-Based Modeling

Most of the focus on the problems of content-based video indexing from the signal processing community concern the modeling of visual content by using low-level features. Since video is formed by a collection of images, most of the techniques that model visual content rely on extracting image-like features from the video sequence. Visual features can be extracted from keyframes or the sequence of frames after the video sequence has been segmented into shots. This section analyzes different low-level features used to represent the visual content of a video shot.

Temporal Motion Features

Video is a medium that is very rich in dynamic content. Motion stands out as the most distinguishing feature for indexing video data. Motion cue is hard to extract since computation of the motion trail often involves generation of optical flow. The problem of computing optical flow between successive frames of the image sequence is recognized to be computationally intensive, so few systems use motion cue to a full extent. The optical flow represents a 2-D field of instantaneous velocities corresponding to each pixel in the image sequence. Instead of computing the flow directly on image brightness values, it is also possible to first process the raw image sequence for contrast, entropy, or spatial derivatives. The computation of optical flow can then be performed on these transformed pixel brightness values instead of on the original images in an effort to reduce the computational overhead. In either case, a relatively dense flow field is obtained at each pixel in the image sequence. Another approach to estimating object motion in the scene can be performed by using a feature-matching based method. This approach involves computation of relatively sparse but highly discriminatory features in a frame. The features can be points, lines, or curves and are extracted from each frame of the video sequence. Interframe correspondence is then established between these features to compute the motion parameters in the video sequence.

Pioneering work in using motion to describe video object activity has been presented by Dimitrova and Golshani (1995) in their use of macroblock tracing and clustering to derive trajectories and their computation of similarity between these raw trajectories. In Dimitrova and Golshani (1994), a three-level motion analysis methodology has been proposed. Starting from extracting the trajectory of a macroblock in an MPEG video, followed by averaging all trajectories of the macroblocks of objects, and finally estimating the relative position and timing information among objects, a dual hierarchy of spatio-temporal logic is established for representing video. More recently, Schonfeld and Lelescu (2000) developed a video tracking and retrieval system know as *VORTEX*. In this system, a bounding box is used to track an object throughout the compressed video stream. This is accomplished by exploiting the motion vector information embedded in the coded video bit stream. A k-means clustering of the motion vectors is used to avoid occlusions. Schonfeld *et al.* (2003) presented an extension of this approach. After initial segmentation of the object contour, they used an adaptive block matching process to predict the object contour in successive image sequences.

Further research has also been devoted to the indexing and retrieval of object trajectories. One such system that makes use of low-level features extracted from objects in the video sequence with particular emphasis on object motion is VideoQ (Chang *et al.*, 1998). Once the object trajectory has been extracted, modeling of this motion trail is essential for indexing and retrieval applications. A trajectory in this sense is a set of *2-tuples* $\{(x_k, y_k): k = 1, \ldots, N\}$, where (x_k, y_k) is the location of the object's centroid in the kth frame, and the object has been tracked for a total of N frames. The trajectory is treated as separable in its x and y-coordinates, and the two are processed separately as 1-D signals. VideoQ models object trajectory based on physical features like acceleration, velocity, and arc length. In this approach, the trajectory is first segmented into smaller units called subtrajectories. The motivation of this is two-fold. First, modeling of full object trajectories can be very computationally intensive. Second, there might be many scenarios in which a part of the object trajectory is not available due to occlusion. Moreover, the user might be interested in certain partial movements of the objects. Physical feature-based modeling is used to index each subtrajectory using acceleration and velocity, for example. These features are extracted from the original subtrajectory by fitting it with a second-order polynomial as in the following equation:

$$\begin{aligned} r(t) &= (x(t), y(t)) = 0.5at^2 + v_0 t, \\ a &= (a_x, a_y) = \text{acceleration}, \; v_0 = (v_x, v_y) = \text{velocity}, \end{aligned} \tag{6.12}$$

where $r(t)$ is the parametric representation of the object trajectory.

Spatial Image Features

Low-level image representation features can be extracted from keyframes in an effort to efficiently model the visual content. At this level, any of the techniques from representation of image indexing schemes can be used. The obvious candidates for feature space are color, texture, and shape. Thus, features used to represent video data have conventionally been the same ones used for images, extracted from keyframes of the video sequence, with additional motion features used to capture temporal aspects of video data. Nephade *et al.* (1998) first segmented the video spatiotemporally obtaining regions in each shot. According to their experiments, each region is then processed for feature extraction. A linearized HSV histogram is used that has 12 bins per channel as the color feature.

The HSV color space is used because it is perceptually closer to human vision as compared to the RGB space. The three histograms corresponding to the three channels (hue, saturation, and value) are then combined into one vector of dimension 36. Texture is represented by gray-level co-occurrence matrices at four orientations. Shape is captured by moment invariants. A similar approach proposed by Shih-Fu Chang *et al.* (1998) uses quantized CIE–LUV space as the color feature, three Tamura texture measures (coarseness, contrast, and orientation) as texture feature, and shape components and motion vectors. All these features are extracted from objects detected and tracked in video sequence after spatiotemporal segmentation.

High-Level Semantic Modeling

As pointed out earlier, high-level indexing and retrieval of visual information, as depicted at level 2 or level 3 in Figure 6.1, requires semantic analysis that is beyond the scope of many of the low-level feature-based techniques. One important consideration that many existing content modeling schemes overlook is the importance of the multimodal nature of video data composed of a sequence of images along with associated audio and, in many cases, textual captions. Fusing data from multiple modalities improves the overall performance of the system. Many of the content modeling schemes based on low-level features work on a query by example (QBE) paradigm in which the user is required to submit a video clip or an image illustrating the desired visual features. At times, this constraint becomes prohibitive when an example video clip or image depicting what the person is seeking is not at hand. Query by keyword (QBK) offers an alternative to QBE in the high-level semantic representation. In this scenario, a single keyword or a combination of many can be used to search through the video database. This requires more sophisticated indexing, however, because keywords summarizing the video content need to be generated during the indexing stage. This capability can be achieved by incorporating knowledge base into video indexing and retrieval systems. There has been a drive toward incorporating intelligence into CBVIR systems, and intelligence-based ideas and systems will be covered in this section. Modeling video data and designing semantic reasoning-based video database management systems (VDBMSs) facilitate high-level querying and manipulating of video data. A prominent issue associated with this domain is the development of formal techniques for semantic modeling of multimedia information. Another problem in this context is the design of powerful indexing, searching, and organization methods for multimedia data.

Multimodal Probabilistic Frameworks

Multimedia indexing and retrieval presents a challenging task of developing algorithms that fuse information from multiple media to support queries. Content modeling schemes operating in this domain have to bridge the gap between low-level features and high-level semantics often called the **semantic gap**. This effort has to take into account the information from audio as well as from video sources. Naphade *et al.* (1998b) have proposed the concept of **Multiject**, a multimedia object. A Multiject is the high-level representation of a certain object, event, or site having features from audio as well as from video. It has a semantic label, that describes the object in words. It also has associated multimodal features (including both audio and video features) that represent its physical appearance. Multiject also has an associated probability of occurrence in conjunction with other objects in the same domain (shot). Experiments using Multiject concepts from three main categories of objects (e.g., airplane), sites (e.g., indoor), and events (e.g., gunshot) have been conducted. Given the multimodal feature vector $\vec{X}_j$ of the *j*th frame and assuming uniform priors on the presence or absence of any concept in any region, the probability of occurrence of each concept in the *j*th frame is obtained from Bayes' rule as:

$$P(R_{ij} = 1|\vec{X}_j) = \frac{P(\vec{X}_j|R_{ij} = 1)}{P(\vec{X}_j|R_{ij} = 1) + P(\vec{X}_j|R_{ij} = 0)}, \qquad (6.13)$$

where R_{ij} is a binary random variable taking value 1 if the concept *i* is present in frame *j*. During the training phase, the identified concepts are given labels, and the corresponding Multiject consists of a label along with its probability of occurrence and multimodal feature vector. Multijects are then integrated at the frame level by defining frame level features F_i, $i \in \{1 \ldots N\}$ (N is the number of concepts the system is being trained for) in the same way they are for R_{ij}. If M is the number of regions in the current frame, then given $\chi = \{\vec{X}_1, \ldots \vec{X}_M\}$, the conditional probability of Multiject *i* being present in any region in the current frame is:

$$P(F_i = 1|\chi) = \max_{j \in \{1, \ldots, M\}} P(R_{ij} = 1|\vec{X}_j). \qquad (6.14)$$

Observing the fact that semantic concepts in videos do not appear in isolation but rather interact and appear in context, their interaction is modeled explicitly, and a network of multijects, called **Multinet**, was proposed (Naphade *et al.*, 2002). This framework based on Multinet takes into account the fact that presence of some multijects in a scene boosts the detection of other semantically related multijects and reduces the chances for detection of others. Based on this Multinet framework, spatiotemporal constraints can be imposed to enhance detection, support inference, and impose a priori information.

Intelligence-Based Systems

The next step toward future CBVIR systems will be marked by the introduction of intelligence into the systems as they need to be capable of communicating with the user, understanding audio–visual content at a higher semantic level, and reasoning and planning at a human level. **Intelligence** is referred to as the capabilities of the system to build and maintain situational or world models, use dynamic knowledge representation, exploit context, and leverage advanced reasoning and learning capabilities. Insight into human intelligence can help designers to better understand users of CBVIR systems and construct more intelligent systems. Benitez *et al.* (2000) proposed an intelligent information system framework known as **MediaNet** that incorporates both perceptual and conceptual representations of knowledge based on multimedia information in a single framework. MediaNet accomplishes this by augmenting the standard knowledge representation frameworks with the capacity to include data from multiple media. It models the real world by concepts, which are real-world entities and relationships between those concepts that can be either semantic (car Is-A-Subtype-Of vehicles) or perceptual (donkey Is-Similar-To mule). In MediaNet, concepts can be as diverse-natured as living entities (humans), inanimate objects (cars), events in the real world (explosions), or certain property (blue). Media representation of the concepts involves data from heterogeneous sources. Multimodal data from all such sources is combined using the framework that intelligently captures the relationships between its various entities.

Semantic Modeling and Querying of Video Data

Owing to its distinguished characteristics from textual or image data—very rich information content, temporal as well as spatial dimensions, unstructured organization, massive volume, and complex and ill-defined relationship among entities—robust video data modeling is an active area of research. The most important issue that arises in the design of video database management systems (VDBMSs) is the description of video data structure in a form that is appropriate for querying, that is sufficiently easy for updating, and that is compact enough for capturing the rich information content of the video. The process of designing the high-level abstraction of raw video facilitates various information retrieval and manipulation operations is the crux of VDBMSs. To this end, current semantic-based approaches can be classified into **segmentation-based** and **stratification-based**. The drawback of the former approaches is lack of flexibility and incapability of representing semantics residing in overlapping segments. The latter models, however, segment contextual information of video instead of simply partitioning it.

SemVideo (Tran *et al.*, 2000) presents a video model in which semantic content having unrelated time information is modeled as one that does; moreover, not only is the temporal feature used for semantic descriptions but the temporal relationships among the descriptions are components of the model. The model encapsulates information about **videos**, each being represented by a unique identifier; **semantic objects**, relating to the description of knowledge about video having a number of attribute–value pairs; **entities**, which are any of the above two; **relationship**, referring to an association between two entities. Many functions are also defined that help in organizing data and arranging relations between different objects in the video. Tran *et al.* (2000) proposed a graphical model, **VideoGraph** that supports not only the event description but also the interevent description that describes the temporal relationship between two events—a functionality overlooked by most of the existing video data models. Tran *et al.* (2000) also have a provision for exploiting incomplete information by associating the temporal event with a Boolean-like expression. A query language based on their framework is proposed in which query processing involves only simple graph traversal routines.

Khokhar *et al.* (1999) introduced a multilevel architecture for video data in which semantics are shared among various levels. An object-oriented paradigm is proposed for management of information at higher levels of abstraction. For each video sequence to be indexed, they first identified objects inside the video sequence, their sizes and locations, and their relative positions and movements, they encoded this information in a spatiotemporal model. This approach integrates both intraclip and interclip modeling and uses both bottom-up as well as top-down object-oriented data abstraction concepts. Decleir *et al.* (1999) have developed a data model that goes one step beyond the existing stratification-based approaches using **generalized intervals**. Here instead of a time segment to be associated with a description, a set of time segments is associated with a description—an approach that allows handling with a single object all occurrences of an entity in a video document. Also proposed was a declarative, rule-based, constraint query language that can be used to infer relationships from information represented in the model and to intentionally specify relationships among objects.

6.4 Conclusions

This chapter focuses on two major issues pertaining to multimedia systems: (1) storage and encoding standards for image and video data and (2) content-based indexing and retrieval of multimedia data. First, the chapter briefly outlines theoretical foundations of image and video compression and then explores some widely used compression and encoding standards for multimedia. Content-based indexing and retrieval of multimedia is an emerging research area that has received wide attention from the research community over the past decade. This chapter surveys the domain of content-based indexing and retrieval for image and video data in depth. Issues, such as low-level feature-based modeling, dimensionality reduction, and relevance feedback, are discussed. Video is normally thought of as a sequence of images, but the sheer size of the data yields an altogether different set of problems to be addressed. In this context, issues, such as temporal segmentation, video summarization, and high-level semantic modeling, are discussed.

References

Arkin, E.M., Chew, L., Huttenlocher, D., Kedem, K., and Mitchell, J. (1991). An efficiently computable metric for comparing polygonal shapes. *IEEE Transactions on Patt. Recog. and Mach. Intel 13*(3), 209–216.

Bach, J.R., Fuller, C., Gupta, A., Hampapur, A., Horwitz, B., Humphrey, R., Jain, R., and Shu, C.F. (YEAR) The Virage image search engine: An open framework for image management. *Proceedings of the SPIE Storage and Retrieval for Image and Video Databases 2670*, 76–87.

Benitez, A.B., Smith, J.R., and Chang, S.E. MediaNet: A multimedia information network for knowledge representation. *Proceedings of the SPIE 2000 Conference on Internet Multimedia Management Systems 4210*, 1–12.

Borecsky, J.S., and Rowe, L.A. (1996). Comparison of video shot boundary detection techniques. *Proceedings of SPIE 26670*, 170–179.

Bouthemy, P., Gelgon, M., and Ganansia, F. (1997). A unified approach to shot change detection and camera motion characterization. *Research Report IRISA 1148*, 1030–1044.

Carson, C., Belongie, S., Greenspan, H., and Malik, J. (1997). Region-based image querying. CVPR '97 Workshop on Content-Based Access of Image and Video Libraries 42–49.

Catarci, T., Costabile, M.E., Levialdi, S., and Batini, C. (1995). Visual query systems for databases: A survey *Technical Report Rapporto di Ricerca SI/RR*. Universita degli Studi di Roma.

Chambers, J.M. (1997). *Computational methods for data analysis*. New York: John Wiley of Sons.

Chang, S.F., Chen, H., Meng, J., Sundaram, H., and Zhong, D. (1998). A fully automated content-based video search engine supporting spatiotemporal queries. *IEEE Transactions on Circuits and Systems for Video Technology 8* (5), 602–615.

Chen, J.Y., Taskiran, C., Albiol, A., Delp, E.J., and Bouman, C.A. (2001). ViBE: A compressed video database structured for active browsing and search. *IEEE Transactions on Multimedia 6* (1), 103–118.

Decleir, C., Hacid, M.H., and Kouloumdjian, J. (1999). A database approach for modeling and querying video data. *Fifteenth International Conference on Data Engineering 6*.

Dimitrova, N., and Golshani, F. (1994). *Px* for semantic video database Retrieval. *Proceedings of the ACM Multimedia*, 219–226.

Dimitrova, N., and Golshani, F. (1995). Motion recovery for video content classification. *ACM Transactions on Information Systems 13*(4), 408–439.

Foote, J. (1999). An overview of audio information retrieval. *Multimedia systems 7* (1), 2–11.

Gonzalez, R.C., and Woods R.E. Digital image processing. Boston: Addison-Wesley.

Goodrum, A.A. (2000). Image information retrieval: An overview of current research. *Informing Science, Special Issue on Information Sciences 3* (2), 63–66.

H. 261: ITU-T recommendation H. 261, video codec for audiovisual services at p × 64 kbits/sec. (1990).

H. 263: Draft ITU-T Recommendation H.263, video coding for low bit-rate communication. July 1995.

Hanjalic, A. (2002). Shot-boundary detection: Unraveled and resolved? *IEEE Transactions on Circuits and Systems for Video Technology 12* (2), 90–105.

Haralick, R.M., Shanmugam, K., and Dinstein, I. (1973). Texture features for image classification. IEEE Transactions on Sys, Man, and Cyb 3(6), 610.

Hibino, S. and Rundensteiner, E. (1995). A visual query language for identifying temporal trends in video data. *Proceedings of the 1995 International Workshop on Multimedia Database Management Systems*. 74–81.

Hu, M.K. (1962). Visual pattern recognition by moment invariants, computer methods in image analysis. *IRE transactions on Information Theory 8*, 170–187.

Iqbal, Q., and Aggarwal, J.K. (1999). Using structure in content-based image retrieval. *Proceedings of the IASTED International Conference Signal and Image Processing*. 129–133.

ISO 10918–1. (1993). Information technology—*Digital compression and coding of continuous tone still images: Requirements and guidelines. T. 81.*

ISO/IEC. (1991). MPEG-1: Coding of moving pictures and associated audio for digital storage media at up to about 1.5 Mbps. 1117–2: Video.

ISO/IEC. (1994). MPEG-2: Generic coding of moving pictures and associated audio information. 13818–2: Video, Draft International Standard.

ISO/IEC. (1998). MPEG-4 video verification model version-11. JTC1/SC29/WG11, N2171: Video.

Jolliffe, I.T. (1986). Principal component analysis. New York: Springer-Verlag.

Kaushik, S., and Rundensteiner, E.A. (1998). A. SVIQUEL: A spatial visual query and exploration language. *Ninth International Conference on Database and Expert Systems Applications 1460*, 290–299.

Khokhar, A., Albuz, E., and Kocalar, E. (1999). Vector wavelet-based image indexing and retrieval for large color image archives. *IEEE International Conference on Acoustics, Speech, and Signal Processing 4*, 1995–1998.

Khokhar, A., Albuz, E., and Kocalar, E. (2000). Quantized CIELab* space and encoded spatial structure for scalable indexing of large color image archives. *Proceedings of the IEEE International Conference on Acoustics, Speech, and Signal Processing 6*, 3025–3028.

Khokhar, A., Ansari, R., and Malik, H., (2003). Content-based audio indexing and retrieval: An Overview. Multimedia Systems Lab, UIC. Multimedia-2003-1: *Technical Report*.

Khokhar, A., Day, Y.F., and Ghafoor, A. (1999). A framework for semantic modeling of video data for content-based indexing and retrieval. ACM Multimedia.

Lelescu, D., and Schonfeld, D. (2000). Real-time scene change detection on compressed multimedia bit stream based on statistical sequential analysis. *Proceedings of the IEEE International Conference on Multimedia and Expo 2*, 1141–1144.

Lelescu, D., and Schonfeld, D. (2001). Video skimming and summarization based on principal component analysis. In E.S. Aishaer and G. Pacific (Eds.), Management of Multimedia on the Internet, Lecture Notes in Computer Science. Berlin: Springer-Verlag.

Lelescu, D., and Schonfeld, D. (2003). Statistical sequential analysis for real-time scene change detection on compressed multimedia bit stream. *IEEE Transactions on Multimedia*.

Liang, K.C., and JayKuo, C.C. (1999). WageGuide: A joint wavelet-based image representation and description system. *IEEE Transactions on Image Processing 8*(11), 1619–1629.

Lin, T., Zhang, H.J., and Shi, Q.Y. (2001). Video content representation for shot retrieval and scene extraction. *International Journal of Image & Graphics 1* (3), 507–526.

Liu, X.M., and Chen, T. (2002). Shot boundary detection using temporal statistics modeling. *ICASSP 2002 4*, 3389–3392.

Ma, W.Y., and Manjunath, B.S. (1997). Netra: A tool box for navigating large image databases. *Proceedings* IEEE *International Conference on Image Processing.*

Mehrotra, S., Chakrabarti, K., Ortega, M., Rui, Y., and Huang, T.S. (1997). Multimedia analysis and retrieval system. *Proceedings of the third international workshop on information retrieval systems.*

Monaco, J. (1977). How to read a film: The art, technology, language, history, and theory of film and media. New York: Oxford University Press.

Nagasaka, A., and Tanaka, Y. (1992). Automatic video indexing and full-video search for object appearances. *Visual Database Systems II.*

Naphade, M.R., Kozintsev, I.V., and Huang, T.S. (2002). A factor graph framework for semantic video indexing. *IEEE Transactions on Circuits and Systems for Video Technology 12*, (1), 40–52.

Naphade, M.R., Kristjansson, T., Frey, B., and Huang, T.S. (1998b). Probabilistic multimedia objects multijects: A novel approach to indexing and retrieval in multimedia systems. *Proceedings of the IEEE International Conference on Image Processing 3*, 536–540.

Naphade, M.R., Mehrotra, R., Fermant, A.M., Warnick, J., Huang, T. S., and Tekalp A.M. (1998a). A high-performance shot boundary detection algorithm using multiple cues. *Proceedings of the IEEE International Conference on Image Processing 2*, 884–887.

Niblack, W., Zhu, X., Hafner, J.L., Breuel, T., Ponceleon, D.B., Petkovic, D., Flickner, M.D., Upfal, E., Nin, S.I., Sull, S., Dom, B.E., Yeo, B.L., Srinivasan, S., Zivkovic, D., and Penner, M. (1997). Updates to the QBIC system. *Proceedings of Storage and Retrieval for Image and Video Databases VI, 3312*, 150–161.

Oh, J., and Chowdary, T. (2002). An efficient technique for measuring of various motions in video sequences. *Proceedings of 2002 International Conference on Imaging Science, Systems, and Technology.*

Oh, J.H., Hua, K.A., and Liang, N. (2000). A content-based scene change detection and classification technique using background tracking *Proceedings of IS & T/SPIE Conference on Multimedia Computing and Networking 2000.* 254–265.

Oomoto, E., and Tanaka, K. (1993). OVID: Design and implementation of a video–object database system. *IEEE Transactions on Knowledge and Data Engineering 5* (4), 629–643.

Pentland, A., Picard, R.W., and Sclaroff, S. (1996). PhotoBook: Content-based manipulation of image databases. *International Journal of Computer Vision. 18* (3), 233–254.

Porter, S.V., Mirmehdi, M., and Thomas, B.T. (2000). Video cut detection using frequency domain correlation. *Proceedings of the 15th International Conference on Pattern Recognition III*, 413–416.

Rui, Y., Huang, T.S., Ortega, M., and Mehrotra, S. (1998). Relevance feedback: A power tool in interactive content-based image retrieval. IEEE *Transactions on Circuits and Systems for Video Technology 8* (5), 644–655.

Schonfeld, D., and Lelescu, D. (2000). VORTEX: Video retrieval and tracking from compressed multimedia databases—multiple object tracking from MPEG-2 bits tream. *Journal of Visual Communications and Image Representation 11*, 154–182.

Schonfeld, D., Hariharakrishnan, K., Raffy, P., and Yassa, F. (In Press). Object tracking using adaptive block matching. *Proceedings of the IEEE International Conference on Multimedia and Expo.* 65–68.

Shen, B., and Sethi, I.K. (1996). Direct feature extraction from compressed images. *SPIE: Storage and Retrieval for Image and video Databases IV 2670*, 404–414.

Smith, J.R., and Chang S.F. (1997). Visually searching the Web for content. *IEEE Multimedia Magazine 4* (3), 12–20.

Smith, J.R., and Chang, S.F. (1995). Tools and techniques for color image retrieval. *IS & T/SPIE proceedings Storage and Retrieval for Image and Video Databases IV 2670*, 1630–1639.

Smith, J.R., and Chang, S.F. (1996). VisualSeek: A fully automated content-based image query system. *Proceedings of the ACM Multimedia.* 87–98.

Tamura, H., Mori, S., and Yamawaki, T. (1978). Texture features corresponding to visual perception. *IEEE Trans. On Sys, Man, and Cyb SMC 78* (6), 460–473.

Tran, D.A., and Hua, K.A., and Vu, K. (2000). Semantic reasoning-based video database systems. *Proceedings of the 11th International Conference on Database and Expert Systems Applications.* 41–50.

Tran, D.A., Hua, K.A., and Vu, K. (2000). VideoGraph: A graphical object-based model for representing and querying video data. *Proceedings of ACM International Conference on Conceptual Modeling.* 383–396.

Venters, C.C., and Cooper, M. (2000). A review of content-based image retrieval systems. University of Manchester, JISC, Technology Applications Program (JTAP): Report.

White, D., and Jain, R. (1996). Similarity indexing: Algorithms and performance. *Proceedings of the SPIE Storage and Retrieval for Image and Video Databases. 2670*, 62–75.

Zhang, C., Meng, W.E., Zhang, Z., Zhong, U.W. (2000). WebSSQL: A query language for multimedia Web documents. *Proceedings of the IEEE Conference on Advances in Digital Libraries.* 58–67.

Zhang, H., Wu, J., Zhong, D., and Smoliar, S.W. (1997). An integrated system for content-based video retrieval and browsing. *Pattern Recognition 30* (4), 643–658.

7

Multimedia Networks and Communication

Shashank Khanvilkar,
Faisal Bashir,
Dan Schonfeld, and
Ashfaq Khokhar
Department of Electrical and Computer Engineering, University of Illinois at Chicago, Chicago, Illinois, USA

7.1 Preface

Paul Baran from the RAND Corporation first proposed the notion of a distributed communication network in 1964 (Schonfeld, 2000; Tanenbaum, 1996). The aim of the proposal was to provide a communication network that could survive the impact of a nuclear war and employ a new approach to data communication based on packet switching. The Department of Defense (DoD) through the Advanced Research Projects Agency (ARPA) commissioned the ARPANET in 1969. ARPANET was initially an experimental communication network that consisted of only four nodes: UCLA, UCSB, SRI, and the University of Utah. Its popularity grew very rapidly over the next two decades, and by the end of 1989, there were over 100,000 nodes connecting research universities and government organizations around the world. This network later came to be known as the Internet, and a layered protocol architecture (i.e., Transmission Control Protocol/Internet Protocol, or TCP/IP, ref. Model) was adopted to facilitate services such as remote connection, file transfer, electronic mail, and news distribution. The proliferation of the Internet exploded over the past decade to more than 10 million nodes since the release of the World Wide Web.

The current Internet infrastructure, however, behaves as a "best-effort" delivery system. Simply put, it makes an honest attempt to deliver packets from a source to a destination, but it provides no guarantees on the packet either being actually delivered and/or the time it would take to deliver it (Kurose and Ross, 2001). Although this behavior is appropriate for textual data that require correct delivery rather than timely delivery, it is not suitable for time-constraint multimedia data such as video and audio. Recently, there has been a tremendous growth in demand for distributed multimedia applications over the Internet that operate by exchanging multimedia involving a myriad of media types. These applications have shown their value as powerful technologies that can enable remote sharing of resources or interactive work collaborations, thus saving both time and money. Typical applications of distributed multimedia systems include Internet-based radio/television broadcast, video conferencing, video telephony, real-time interactive and collaborative work environments, video/audio on demand, multimedia mail, and distant learning.

The popularity of these applications has highlighted the limitations of the current best-effort Internet service model and viability of its associated networking protocol stack (i.e.,

TCP/IP) for the communication of multimedia data. The different media types exchanged by these applications have significantly different **traffic requirements**—such as **bandwidth**, **delay, jitter**, and **reliability**—from the traditional textual data and demand different constraints or service guarantees from the underlying communication network to deliver an acceptable performance. In networking terminology, such performance guarantees are referred to as **quality of service** (QoS) guarantees that can be provided only by suitable enhancements to the basic Internet service model (Kurose and Ross, 2001). Circuit-switched networks, like the telephony system or plain old telephone service (POTS), have been designed from the ground up to support such QoS guarantees. However, this approach suffers from many shortcomings like scalability, resource wastage, high complexity, and high overhead (Leon-Garcia and Widjaja, 2000). Another approach, known as the asynchronous transfer mode (ATM), relies on cell switching to form virtual circuits that provide some of the QoS guarantees of traditional circuit-switched networks. Although ATM has become very popular as the backbone of high bandwidth and local networks, it has not been widely accepted as a substitute for the protocol stack used on the Internet. Providing QoS in packet-switched Internet, without completely sacrificing the gain of statistical multiplexing, has been a major challenge of multimedia networking. In addition to the QoS guarantees, distributed multimedia applications also demand many **functional requirements**—such as support for **multicasting**, **security**, **session management**, and **mobility**—for effective operation, and these can be provided by introducing new protocols residing above the traditional protocol stack used on the Internet (Wolf *et al.*, 1997). In this chapter, we discuss two popular protocol architectures, H.323 (Thom, 1996; Liu and Mouchtaris, 2000) and SIP (Johnston, 2000; Schulzrinne and Rosenburg, 2000) that have been specifically designed to support distributed multimedia applications.

Apart from the Internet, cellular networks have also seen an unprecedented growth in their usage [13] and consequent demand for multimedia applications. The second generation (2G) cellular systems like GSM, IS-95, IS-136 or PDC, which offered circuit-switched voice services, are now evolving towards third generation (3G) systems that are capable of transmitting high-speed data, video, and multimedia traffic to mobile users. IMT-2000 is composed of several 3G standards under development by the International Telecommunication Union (ITU) that will provide enhanced voice, data, and multimedia services over wireless networks. We will discuss the layered QoS approach adopted by IMT-2000 to provide end-to-end QoS guarantees.

Section 7.2 starts with a general classification of media types from a networking/communication point of view. In this section, we introduce the reader to some common media types like text, audio, images, and video, and we also discuss their **traffic** and **functional** requirements. Section 7.3 discusses the inadequacy of the current best-effort Internet model to satisfy the multimedia **traffic requirements**. We describe three enhanced architectures: Integrated Services (White, 2001), Differentiated Services (Blake *et al.*, 1998) and Multi-Protocol Label Switching (Rosen *et al.*, 2001). These architectures have been proposed to overcome these shortcomings. Section 7.4 presents some standard approaches for meeting the functional requirements posed by multimedia traffic. Later in this section, we present two protocol architectures (H.323 and SIP) that have been introduced for the Internet protocol stack to satisfy these requirements. Section 7.5 describes current efforts to support multimedia traffic over the cellular/wireless networks; we illustrate issues related to internetworking between wired and wireless networks.

7.2 Introduction to Multimedia

The term **multimedia** refers to diverse classess of media employed to represent information. **Multimedia traffic** refers to the transmission of data representing diverse media over communication networks. Figure 7.1 shows the diversity of the media classified into three groups: (1) text, (2) visuals, and (3) sound. As illustrated in this figure symbolic textual material may include not only the traditional unformatted plain text but also formatted text with numerous control characters, mathematical expressions, phonetic transcription of speech, music scores, and other symbolic representations such as

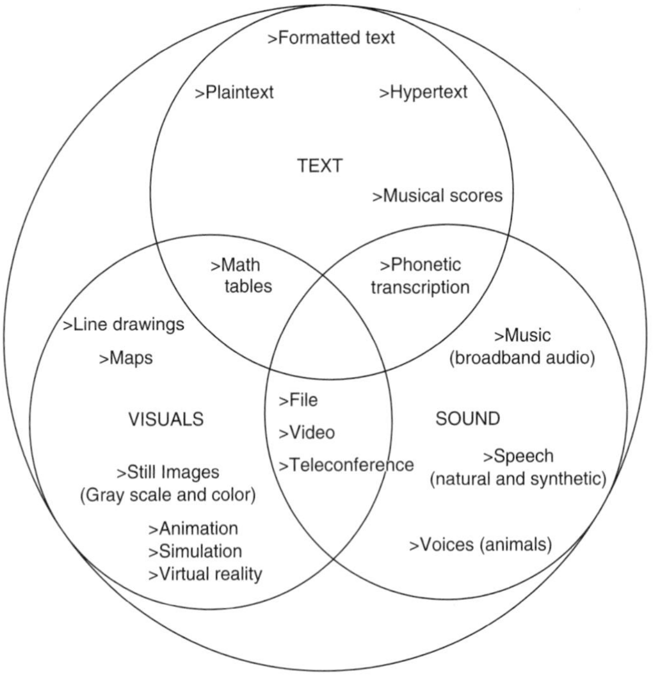

FIGURE 7.1 Diversity of Multimedia Data Signals. Adapted from Kinsner (2002).

hypertext. The visual material may include line drawings, maps, gray-scale or colored images, and photographs as well as animation, simulation, virtual reality objects, and video conferencing and teleconferencing. The sound material may include telephone/broadcast-quality speech to represent voice, wideband audio for music reproduction, and recordings of sounds such as from electrocardiograms or other biomedical signals. Other perceptory senses, such as touch and smell, can very well be considered as part of multimedia but are considered out of scope of this chapter.

Text is inherently digital, whereas other media types like sound and visuals can be analog and need to be converted into digital form using appropriate analog to digital conversion techniques. In this chapter, we assume that all media types have been suitably digitized; the reader is invited to read a book by Chapman and Chapman (2000) that gives an excellent introduction to the principles and standards used to convert many such analog media to digital form. In this chapter, we focus on typical characteristics of different media types when transported over a network. In this regard, multimedia networking deals with the design of networks that can handle multiple media types with ease and deliver scalable performance.

7.2.1 Multimedia Classification

From a networking perspective, all media types can be classified as either real-time (RT) or non real-time (NRT), as shown in Figure 7.2. RT media types require either hard or soft bounds on the end-to-end packet delay/jitter, while NRT media types, like text and image files, do not have any strict delay constraints but may have rigid constraints on error. There are basically two approaches to error control (Leon-Garcia and Widjaja, 2000):

1. **Error detection followed by Automatic Retransmission reQuest** (ARQ): This approach requests retransmission of lost or damaged packets. It is used by Transport Control Protocol (TCP), a transport layer protocol in the TCP/IP protocol stack, to provide reliable connection-oriented service. Applications that require an error-free delivery of NRT media typically use TCP for transport.
2. **Forward error correction** (FEC): This second approach provides sufficient redundancy in packets so that errors can be corrected without the need for retransmissions. It can be used by User Datagram Protocol (UDP), another transport layer protocol in the TCP/IP protocol stack that provides connectionless unreliable service. Applications that exchange error-tolerant media types (both RT and NRT) typically use UDP for transport because it eliminates time lost in retransmissions. Leigh *et al.* (2001) have conducted experiments using FEC along with UDP over a global high-bandwidth communication network, **STARTAP**.

The RT media types are further classified as **discrete media** (DM) or **continuous media** (CM), depending on whether the data is transmitted in discrete quantum as a file or message or continuously as a stream of messages with intermessage dependency. The real-time discrete type of media has recently gained high popularity because of ubiquitous applications like MSN/Yahoo messengers (which are error intolerant) and instant messaging services like stock quote updates (which are error tolerant).

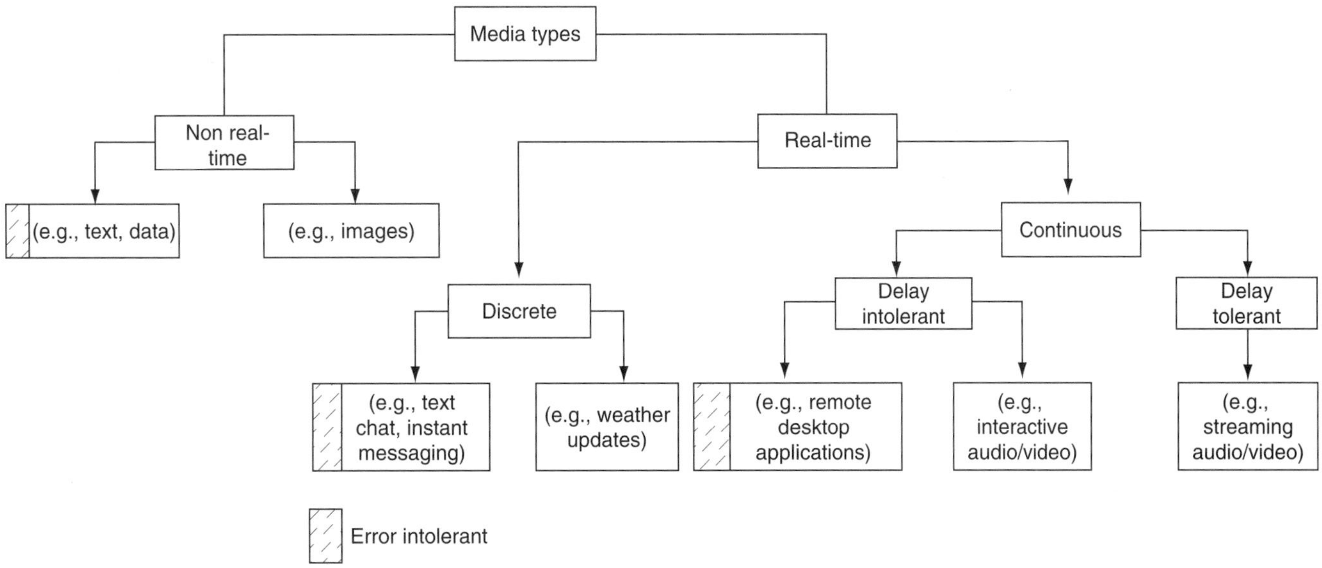

FIGURE 7.2 Network-Oriented Classification of Media Types

The RT continuous type of media can further be classified as **delay tolerant** or **delay intolerant**. We cautiously use the term delay tolerant to signify that such media type can tolerate higher amounts of delay than their delay-intolerant counterparts without significant performance degradation. Examples of RT, continuous, and delay-intolerant media are audio and video streams used in audio or video conferencing systems, and remote desktop applications. Streaming audio/video media used in applications like Internet Webcast are examples of delay-tolerant media types. Their delay dependency is significantly diminished by having an adaptive buffer at the receiver that downloads and stores a certain portion of the media stream before starting playout. The entire classification has been carefully illustrated in Figure 7.2.

We now discuss some common media types and their defining characteristics in terms of bandwidth usage, error requirements, and real-time nature.

7.2.2 Text

Text is the most popular of all the media types. It is distributed over the Internet in many forms, including files or messages using different transfer protocols such as File Transfer Protocol (FTP), which is used to transfer binary and ASCII files over the Internet; Hyper Text Transfer Protocol (HTTP), which is used to transmit HTML pages; or Simple Mail Transfer Protocol (SMTP), which is used for exchanging e-mails. Text is represented in binary as 7-bit US-ASCII, 8-bit ISO-8859, 16-bit Unicode, or 32-bit ISO 10646 character sets, depending on the language of choice and the country of origin. Bandwidth requirements of text media mainly depend on their size, which can be easily reduced using common compression schemes (Salomon, 1998) as detailed by Table 7.1.

The error characteristics of text media depend largely on the application under consideration. Some text applications, such as file transfer, require text communication to be completely loss/error free and therefore use TCP for transport. Other text applications, such as instant messaging, may tolerate some errors as well as losses and therefore can use UDP for transport.

TABLE 7.1 Text Compression Schemes

Compression scheme	Comments
Shannon-Fano coding	This coding uses variable length code words (i.e., symbols with higher probability of occurrence are represented by smaller codes-words).
Huffman coding	This coding is the same as Shannon-Fano coding
Lempel-Ziv-Welch (LZW)	LZW compression replaces strings of characters with single codes. It does analyze the incoming text. Instead, it just adds every new string of characters it sees to a table of strings. Compression occurs when a single code is output instead of a string of characters.
Unix compress	This compression scheme uses LZW with growing dictionary. Initially, the dictionary contains 512 entries and is subsequently doubled until it reaches the maximum value set by the user.

TABLE 7.2 Audio Compression Schemes

Voice/audio codec	Used for	Bit rate (Kbps)
Pulse code modulation (G.711)	Narrowband speech (300–3300 Hz)	64
GSM	Narrowband speech (300–3300 Hz)	13
CS-ACELP (G.729)	Narrowband speech (300–3300 Hz)	8
G.723.3	Narrowband speech (300–3300 Hz)	6.4 and 5.3
Adaptive differential PCM (G.726)	Narrowband speech (300–3300 Hz)	32
SBC (G.722)	Wideband speech (50–7000 Hz)	48/56/64
MPEG layer III (MP3)	CD-quality music wideband audio (10–22 Khz)	128–112 Kbps

Applications that use text as primary media (e.g., Web browsing or e-mail) do not have any real-time constraints, such as bounded delay or jitter. These applications are called **elastic applications**. Applications like instant messaging, however, do require some guarantees on the experienced delay.

Overall, the text media has been around since the birth of the Internet and can be considered as the primary means of information exchange.

7.2.3 Audio

Audio media is sound/speech converted into digital form using sampling and quantization. Digitized audio media is transmitted as a stream of discrete packets over the network. The bandwidth requirements of digitized audio depend on its dynamic range and/or spectrum. For example, telephone-grade voice uses dynamic range reduction using the logarithmic *A*-law (Europe) or μ-law (North America) capable of reducing the linear range of 12 bits to nonlinear range of only 8 bits. This reduces the throughput from 96 kbps to 64 kbps. A number of compression schemes (Garg, 1999) along with their bit rates, as illustrated in Table 7.2, are commonly used for audio media types.

The audio media type has loose requirements on packet loss/errors (or loss/error tolerance) in the sense that it can tolerate up to 1 to 2% packet loss/error without much degradation. Today, most multimedia applications that use audio have built-in mechanisms to deal with the lost packets using advanced interpolation techniques.

The real-time requirements of audio strictly depend on the expected interactivity between the involved parties. Some applications like Internet–telephony, which involves two-way communication, are highly interactive and require shorter response times. The audio media, in this case, requires strong bounds on end-to-end packet delay/jitter to be of acceptable/decipherable quality. Applications that use this media type are called **real-time intolerant** (RTI) applications. In most RTI applications, the end-to-end delay must be limited to ~ 200 msec to get an acceptable performance. Other applications like Internet Webcast, which involves one-way communication, have relatively low interactivity. Interactivity, in this case, is limited to commands that allow the user to change radio channels, for example, which can tolerate higher response times. Consequently, the Webcast requires weaker bounds on delay/jitter, and the applications that use such kind of media are termed as **real-time tolerant** (RTT) applications. **Streaming audio** is also used to refer to this media type.

7.2.4 Graphics and Animation

Graphics and animation include static media types like digital images and dynamic media types like flash presentations. An uncompressed, digitally encoded image consists of an array of pixels, with each pixel encoded in a number of bits to represent luminance and color. Compared to text or digital audio, digital images tend to be large in size. For example, a typical 4 × 6 inch digital image, with a spatial resolution of 480 × 640 pixels and color resolution of 24 bits requires ~ 1 MB of storage. To transmit this image on a 56.6 Kbps line will take at least 2 min. If the image is compressed at the modest 10:1 compression ratio, the storage is reduced to ~ 100 KB, and transmission time drops to ~ 14 sec. Thus, some form of compression schemes is always used that cashes on the property of high spatial redundancy in digital images. Some popular compression schemes (Salomon, 1998) are illustrated in Table 7.3. Most modern image compression schemes are progressive and have important implications for transmission over the communication networks (Kisner, 2002). When such an image is received and decompressed, the receiver can display the image in a low-quality format and then improve the display as subsequent image information is received and decompressed. A user watching the image display on the screen can recognize most of the image features after only 5 to 10% of the information has been decompressed. Progressive compression can be achieved by: (1) encoding spatial frequency data progressively, (2) using vector quantization that starts with a gray image and later adds colors to it, and (3) using pyramid coding that encodes images into layers, in which early layers are low resolution and later layers progressively increase the resolution.

Images are error tolerant and can sustain packet loss, provided the application used to render them knows how to handle lost packets. Moreover, images, like text files, do not have any real-time constraints.

7.2.5 Video

Video is a sequence of images or frames displayed at a certain rate (e.g., 24 or 30 frames per second). Digitized video, like digitized audio, is also transmitted as a stream of discrete packets over the network. The bandwidth requirements for digitized video depend on the spatial redundancy present in every frame as well as the temporal redundancy present in consecutive frames. Both these redundancies can be exploited to achieve efficient compression of video data. Table 7.4, illustrates some common compression schemes that are used in video (Watkinson, 2001).

The error- and real-time requirements of video media are similar to the audio media type. Hence, for the sake of brevity, we do not discuss them here.

TABLE 7.3 Image Compression Schemes

Compression scheme	Comments
Graphics interchange format (GIF)	GIF supports a maximum of 256 colors and is best used on images with sharply defined edges and large, flat areas of color like text and line-based drawings. GIF uses LZW compression to make files small. This is a lossless compression scheme.
Portable network graphics (PNG)	PNG supports any number of colors and works best with almost any type of image. PNG uses the zlib compression scheme, compressing data in blocks dependent on the "filter" of choice (usually *adaptive*). This is a lossless compression scheme and does not support animation.
Joint photographic experts group (JPEG)	JPEG is best suited for images with subtle and smooth color transitions, such as photographs gray-scale, and colored images. This compression standard is based on the Huffman and run-length encoding of the quantized discrete cosine transform (DCT) coefficients of image blocks. JPEG is a lossy compression. Standard JPEG encoding does not allow interlacing, but the progressive JPEG format does. Progressive JPEGs start out with large blocks of color that gradually become more detailed.
JPEG 2000	JPEG 2000 is suitable for a wide range of images, from those produced by portable digital cameras to advanced prepress and medical imaging. JPEG 2000 is a new image coding system that uses state-of-the-art compression techniques based on wavelet technology that stores its information in a data stream instead of blocks as in JPEG. This is a scalable lossy compression scheme.
JPEG-LS	JPEG-LS is suitable for continuous-tone images. The standard is based on the LOCO-I algorithm (Low COmplexity LOssless COmpression for images) developed by HP. This is a lossless/near-lossless compression standard.
Joint bilevel image experts group (JBIG)	JBIG is suitable for compressing black-and-white monochromatic images. It uses multiple arithmetic coding schemes to compress the image. This is a lossless type of compression.

TABLE 7.4 Video Compression Schemes

Compression scheme	Comments
MPEG-I	MPEG-I is used to produce VCR NTSC (352 × 240) quality video compression to be stored on CD-ROM (CD-I and CD-video format) using a data rate of 1.2 Mbps. MPEG-I uses heavy down-sampling of images limits image rate between 24 to 30 Hz to achieve this goal.
MPEG-II	MPEG-II is a generic standard for a variety of audio–visual coding applications and supports error resilience for broadcasting. It supports broadcast-quality video compression (DVB) and high-definition television (HDTV). MPEG-2 supports four resolution levels: low (352 × 240), main (720 × 480), high-1440 (1440 × 1152), and high (1920 × 1080). The MPEG-2 compressed video data rates are in the range of 3 to 100 Mbps.
MPEG-IV	MPEG-IV supports low-bandwidth video compression at a data rate of 64 Kbps that can be transmitted over a single N-ISDN B-channel. MPEG-4 is a genuine multimedia compression standard that supports audio and video as well as synthetic and animated images, text, graphics, texture, and speech synthesis.
H.261	H.261 supports video communications over ISDN at data rates of p × 64 Kbps. It relies on intraframe and interframe coding where integer–pixel accuracy motion estimation is required for intermode coding
H.263	The H.263 standard is aimed at video communications over POTS and wireless networks at very low data rates (as low as 18 to 64 Kbps). Improvements in this standard are due to the incorporation of several features such as half-pixel motion estimation, overlapping and variable block sizes, bidirectional temporal prediction, and improved variable length coding options.

7.2.6 Multimedia Expectations from a Communication Network

In this section, we identify and analyze the requirements that a distributed multimedia application may enforce on the communication network. Due to the vastness of this field, we do not claim that this list is exhaustive, but we have tried to include all the important aspects (from our view point) that have significantly impacted the enhancements to the basic Internet architecture and its associated protocols. In Sections 7.3 and 7.4, we further explore these aspects and give readers a sense of understanding of the efforts made to help the Internet deal with the challenges posed by such applications.

We divide these requirements into two categories (Wolf *et al.*, 1997): **traffic requirements** and **functional requirements**. The traffic requirements include limits on real-time parameters (e.g., delay and jitter), bandwidth, and reliability; functional requirements include support for multimedia services, such as multicasting, security, mobility, and session management. The traffic requirements can be met only by enhancements to the basic Internet architecture, whereas the functional requirements can be met by introducing newer protocols over the TCP/IP networking stack. The functional requirements are not an absolute necessity in the sense that a distributed multimedia application can still operate with high performance by incorporating the necessary functions into the application itself. They represent, however, the most common functionality required among distributed multimedia applications, and it would only help to have standardized protocols operating over the networking protocol stack to satisfy them.

Real-Time Characteristics (Limits on Delay and Jitter)

As discussed in Subsections 7.2.1 to 7.2.5, media types such as audio and video have real-time traffic requirements, and the communication network must honor these requirements. For example, audio and video data must be played back continuously at the rate at which they are sampled. If the data does not arrive in time, the playback process will stop, and human ears and eyes can easily pick up the artifact. In Internet telephony, human beings can tolerate a latency of $\sim$ 200 msec. If the latency exceeds this limit, the voice will sound like a call routed over a long satellite link, which amounts to degradation in quality of the call. Thus real-time traffic enforces strict bounds on end-to-end packet delay (time taken by the packet to travel from the source to the destination) and jitter (variability in the interpacket delay at the receiver). The performance of distributed multimedia applications improves with decrease in both these quantities.

Need for Higher Bandwidth

Multimedia applications require significantly higher bandwidths than conventional textual applications of the past. Moreover, media streams are transmitted using UDP that does not have any mechanism to control congestion. The communication network must be able to handle such high bandwidth requirements without being unfair to other conventional flows. Table 7.5 summarizes the bandwidth requirements of some common audio/video media types. We discussed several compression schemes that take advantage of spatial/temporal redundancy present in audio/video media, but the compressed media still requires significantly higher bandwidth than what is typically required for text-oriented services. Moreover, compression schemes cannot be expected to be used for all multimedia transmissions. There are two types of compression techniques: **lossy** and **lossless**. The lossy compression techniques eliminate redundant information from data and subsequently introduce distortion or noise in the original data. The lossless compression techniques do not loose any information, and data received by the user is exactly identical to the original data. Lossy compression usually yields significantly higher compression ratios than lossless

TABLE 7.5 Sources of Multimedia and Their Effective Bandwidth Requirements

Audio source	Sampling rate	Bits/sample	Bandwidth Requirements
Telephone grade voice (up to 3.4 KHz)	8000 samples/sec	12	96 Kbps
Wideband speech (up to 7 KHz)	1600 samples/sec	14	224 Kbps
Wideband audio two channels (up to 20 KHz)	44.1 K samples/sec	16 per channel	1.412 Mbps for both channels
Image source	Pixels	Bits/Pixel	Bit rate
Color image	512 × 512	24	6.3 Mbps
CCIR TV	720 × 576 × 30	24	300 Mbps
HDTV	1280 × 720 × 60	24	1.327 Gbps

compression. However, lossy compression might not be acceptable for all media types or applications (e.g., medical images such as X-rays, telemedicine, etc.), and it may be necessary to use either lossless compression or not use compression at all.

Error Requirements

As discussed in earlier sections, different media types have vastly different error requirements, ranging from being completely error intolerant to being somewhat error tolerant depending on the application. An error is said to have occurred when a packet is either lost or damaged. Most error-tolerant multimedia applications use error concealment techniques to deal with lost or damaged packets by predicting lost information from correctly received packets. Errors are handled using various Forward Error Correction (FEC) codes that can be used to detect and correct single or multiple errors.

The use of FEC codes implies that extra information has to be added to the packet stream to handle errors. However, if the communication path over which the packets are transmitted introduces additional errors beyond the level of degradation for which the FEC was designed, then some errors will remain undetected or may not be corrected, and the performance will surely degrade. Thus, it is essential for a multimedia application to know the error characteristics of the communication network so that an adequate level of FEC is introduced to supplement the packet stream and protect against data loss or damage. As an example, wireless networks usually rely much more heavily on FEC than wired networks because the latter has a higher probability of packet loss. The minimization of packet retransmission achieved by using FEC can be too costly in wired networks that are characterized by very low probability of packet loss. The cost incurred is attributed to the additional bandwidth required for the representation of FEC information. The use of FEC is also critically dependent on the application. For instance, in real-time applications, some level of FEC is introduced for both wired and wireless communication networks because retransmissions are generally prohibited due to delay constraints.

Multicasting Support

Multicasting refers to a single source of communication with simultaneous multiple receivers. Most popular distributed multimedia applications require multicasting. For example, multiparty audio/video conferencing is one of the most widely used services in Internet telephony. If multicasting is not naturally supported by the communication network (as was the case in some circuit-switched networks) then significant efforts need to be invested in building multimedia applications that support this functionality in an overlaid fashion, which often leads to inefficient bandwidth utilization.

Multicasting is relatively easier to achieve for one-way communication than for two-way communication. For example, in the case of Internet radio, multicasting can be achieved by creating a spanning tree consisting of the sender at the root and the receiver at the leaves as well as replicating packets over all links that reach the receivers. In the case of two-way communication like Internet telephony among multiple parties, however, some form of audio mixing functionality is required that mixes the audios from all participants and only relays the correct information. Without this audio mixer, a two-way communication channel needs to be established between each participant in an all-to-all mesh fashion, which may amount to waste of bandwidth.

Session Management

The session management functionality includes the following features.

- **Media description:** This enables a distributed multimedia application to distribute session information, such as media type (audio, video, or data) used in the session, media encoding schemes (PCM, MPEG-II), session start time, session stop time, and IP addresses of the involved hosts, for example. It is often essential to describe the session before establishment because most participants involved in the session will have different multimedia capabilities.
- **Session announcement**: This allows participants to announce future sessions. For example, there are hundreds of Internet radio stations, each Webcasting different channels. Session announcement allows such radio stations to distribute information regarding their scheduled shows so that a user finds it easier to tune in to the preferred show.

- **Session identification**: A multimedia session often consists of multiple media streams (including continuous media (e.g., audio, video) and discrete media (e.g., text, images) that need to be separately identified. For example, the sender might choose to send the audio and video as two separate streams over the same network connection, and the receiver needs to decode each synchronously. Another example is that the sender might put the audio and video streams together but divide quality into a base layer and some enhancement layers so that low-bandwidth receivers might be able to receive only the base layer, whereas high-bandwidth receivers might also receive the enhancement layers.
- **Session control**: As said previously, a multimedia session involves multiple media streams. The information contained in these data streams is often interrelated, and the multimedia communication network must guarantee to maintain such relationships as the streams are transmitted and presented to the user. This is called **multimedia synchronization** and can be achieved by putting time stamps in every media packet. Moreover, many Internet multimedia users may want to control the playback of continuous media by pausing, playing back, repositioning the playback to a future or past point in time, visual fast-forwarding, or visual rewinding of the playback (Kurose and Ross, 2001). This functionality is similar to what we have in a VCR while watching a video or is similar to what we have in a CD player when listening to a CD.

Security

The security issue has been neglected in almost all discussions of multimedia communication. With the increasing use of online services and issues related to digital asset management, however, it is now apparent that security issues are quite significant. Security provides the following three aspects to multimedia data: **integrity** (data cannot be changed in mid-flight), **authenticity** (data comes from the right source), and **encryption** (data cannot be deciphered by any third party). For example, public broadcasts require data integrity and data authenticity, while private communication requires data encryption. All the above aspects can be provided using different cryptographic techniques like secret key cryptography, public key cryptography, and hash functions (Kessler, 2002).

Another issue is that of protecting the intellectual copyrights for different media components. For example, consider digital movies that are distributed over the Internet using a pay-per-view service. It is possible for any entrepreneur to download such movies and sell them illegally. Digital watermarking techniques (Su *et al.*, 1999), which embed extra information into multimedia data (such information is imperceptible to the normal user (Su *et al.*, 1999) as well as unremovable), can help prevent copyright violations.

Mobility Support

The advent of wireless and cellular networks has also enhanced multimedia applications with mobility. Cellular systems have a large coverage area and hence permit high mobility. Another emerging network is IEEE 802.11x wireless LAN (Crow *et al.*, 1997), which can operate at speeds exceeding 54 Mbps. Wireless LANs (WLAN) typically cover a smaller area and have limited mobility. Their main advantage, however, is that they work in the ISM band (no licensing required, thus eliminating significant investments into license purchase) and are relatively easy to set up; moreover, there is a vast availability of cheap WLAN products in the market that cost much less.

Mobility aspect has added another dimension of complexity to multimedia networks. It opens up questions on a host of complex issues like routing to mobile terminals, maintaining the QoS when the host is in motion, and internetworking between wireless and wired networks.

7.3 Best-Effort Internet Support for Distributed Multimedia Traffic Requirements

In this section, we further analyze why the current Internet, having the best-effort delivery model, is inadequate in supporting **traffic requirements** of multimedia traffic streams and justify the need for enhancements to this basic model. We also point out the research approaches that have been adopted to make best-effort Internet more accommodating to real-time multimedia traffic. To preserve the brevity of this chapter, we do not discuss every such approach at length but provide appropriate references for interested readers.

7.3.1 Best-Effort Internet Support for Real-Time Traffic

Real-time traffic requires strong bounds on packet delay and jitter, and the current Internet cannot provide any such bounds. Packet delay and jitter effects are contributed at different stages of packet transmission, and different techniques are used to reduce them. An analysis of these different components is essential in understanding the overall cause because almost all enhancements to the current Internet architecture aim to reduce one of these components. We explain each of these components (Bertsekas and Gallager, 1987) in more detail in the following subsection.

Packet Processing Delay

Packet processing delay is a constant amount of delay faced at both the source and the destination. At the source, this delay might include the time taken to convert analog data to digital form and packetize them through different layers of protocols until data are handed over to the physical layer for transmission. We can define similar packet processing delay at the receiver in the reverse direction. Usually this delay is the characteristic of the operating system (OS) and the multimedia application under consideration. For a lightly loaded system, this delay can be considered as negligible; however, with increasing load, this delay can become significant. Packet processing delay is independent of the Internet model (whether best-effort or any other enhanced version), and any reductions in this delay would imply software enhancements to the OS kernel—such OSs are called multimedia operating systems (Steinmetz, 1995) that provide enhanced process-, resource-, file-, and memory-management techniques with real-time scheduling—and the application.

Packet Transmission Delay

Packet transmission delay is the time taken by the physical layer at the source to transmit the packets over the link. This delay depends on multiple factors, including the following:

- **Number of active sessions**: The physical layer processes the packets in the FIFO order. Hence, if there are multiple active sessions, this delay becomes quite significant, especially if the OS does not support real-time scheduling algorithms to support multimedia traffic.
- **Transmission capacity of the link**: Increasing the transmission capacity reduces the transmission delay. For example, upgrading from the 10 Mbps ethernet to 100 Mbps fast ethernet will ideally reduce the transmission delay by a factor of 10.
- **Medium access control (MAC) access delay**: If the transmission link is shared, a suitable MAC protocol must be used for accessing the link (Yu and Khanvilkar, 2002). The choice of MAC protocol largely influences this delay. For example, if the transmission capacity is C bps, and the packet length is L bits, time taken to transmit is L/C, assuming a dedicated link. However, if the MAC protocol uses time division multiple access (TDMA) with m slots, this delay becomes mL/C, which is m times larger than the earlier case. The widespread ethernet networks cannot provide any firm guarantees on this access delay (and hence the overall QoS) due to the indeterminism of the carrier sense multiple access/collision detection (CSMA/CD) approach toward sharing of network capacity (Wolf *et al.*, 1997). The reason for this is that the collisions, which occur in the bus-based ethernet when two stations start transmitting at the same time, lead to delayed service time. Fast ethernet exploits the same configuration as 10 Mbps ethernet and increases the bandwidth with the use of new hardware in hubs and end stations to 100 Mbps but provides no QoS guarantees. Isochronous ethernet (integrated voice data LAN, IEEE 802.9) and demand priority ethernet (100Base-VG, AnyLAN, IEEE 802.12) can provide QoS, yet their market potential remains questionable.
- **Context switch in the OS**: Sending or receiving a packet involves context switch in the OS, which takes a finite time. Hence, there exists a theoretical maximum at which a computer can transmit packets. For a 10 Mbps LAN, this delay might seem insignificant; however, for gigabit networks, this delay becomes quite significant. Again, reduction in this delay will require enhancing the device drivers and increasing the operating speed of the computer.

Propagation Delay

Propagation delay is defined as the flight time of packets over the transmission link and is limited by the speed of light. For example, if the source and destination are in the same building at the distance of 200 m, the propagation delay will be ~ 1 μsec. If they are located in different countries at a distance of 20,000 km, however, the delay is in order of 0.1 sec. The above values represent the physical limits and cannot be reduced. This has major implications for interactive multimedia applications that require the response time to be less than $\sim$200 msec. Thus if the one-way propagation delay is greater than this value, then no enhancements can improve the quality of the interactive sessions, and the user will have to settle for a less responsive system.

Routing and Queuing Delay

The routing and queuing delay, is the only delay components that we can reduce (or control) by introducing newer enhanced Internet architecture models. In the best-effort Internet, every packet is treated equally, regardless of whether it is a real-time packet or a non real-time packet. All intermediate routers make independent routing decisions for every incoming packet. Thus, a router can be ideally considered as a First-In First-Out (FIFO) queuing system. When packets arrive at a queue, they have to wait for a random amount of time before they can be serviced, which depends on the current load on the router. This adds up to the queuing delay. The routing and queuing delay is random and, hence, is the major contributor to jitter in the traffic streams. Sometimes when the queuing delay becomes large, the sender application times out and resends the packet. This can lead to an avalanche effect that leads to congestion and thus increase in queuing delays. Different techniques have been adopted to reduce precisely this delay component and thus have given rise to newer Internet service models. For example, in the simplest case, if there is a dedicated virtual circuit connection (with dedicated resources in the form of buffers and bandwidth) from the source to the destination, then this delay will be negligible. The Integrated Services (IntServ) model and Multi-Protocol Label Switching (MPLS) model follow this approach. Another option is to use a combination of traffic policing, admission control, and sophisticated queuing techniques (e.g., priority queuing, weighted fair queuing) to provide a firm upper bound on delay and jitter. The Differentiated Services (DiffServ) model follows this approach. Later, we will discuss in some more detail the principles that need to be followed to reduce this delay component.

7.3.2 High Bandwidth Requirements

Multimedia traffic streams have high bandwidth requirements (refer to Table 7.5). The best-effort Internet model does not provide any mechanism for applications to reserve network resources to meet such high bandwidth requirements and also does not prevent anyone from sending data at such high rates. Uncontrolled transmissions at such high rates can cause heavy congestion in the network, leading to a congestion collapse that can completely halt the Internet. There is no mechanism in the best-effort Internet to prevent this from happening (except using a brute force technique of disconnecting the source of such congestion). It is left to the discretion of the application to dynamically adapt to network congestions. Elastic applications that use TCP utilize a closed-loop feedback mechanism (built into TCP) to prevent congestion (this method of congestion control is called reactive congestion control). However, most multimedia applications use UDP for transmitting media streams; UDP does not have any mechanism to control congestion and has the capability to create a congestion collapse.

To remove these shortcomings, the enhanced Internet service models use admission control, bandwidth reservations, and traffic policing mechanisms. The application must first get permission from some authority to send traffic at a given rate and with some given traffic characteristics. If the authority accepts admission, it will reserve appropriate resources (bandwidth and buffers) along the path for the application to send data at the requested rate. Traffic policing mechanisms are used to ensure that applications do not send at a rate higher than what was initially negotiated.

7.3.3 Error Characteristics

Multimedia streams require some guarantees on the error characteristics of the communication network, and the best-effort Internet cannot provide such guarantees because the path that a packet follows from the source to the destination is not fixed; hence, the network has no idea about the error characteristics of each individual segment. Thus, the sender application has no knowledge of the error characteristics of the network and may end up using an error correction/detection mechanism that may not be optimum.

For the newer Internet service models, the sender application has to go through admission control. At this time, the sender can specify the maximum error that it can tolerate. If the network uses a QoS-based routing algorithm, explained later in this discussion, and is unable to find a path that can satisfy this requirement, it will just reject the connection or make a counteroffer to the sender specifying the error rate at which it is willing to accept the connection. In other words, the sender is made aware of the error characteristics of the network.

7.3.4 Proposed Service Models for the Internet

We now discuss several new architecture models: Integrated Services (IntServ), Differentiated Services (DiffServ), and Multi-Protocol Label Switching (MPLS). These have been proposed for the best-effort Internet to satisfy the traffic requirements of distributed multimedia applications. But, before delving into the discussion of these QoS service models proposed for the Internet, we would like to summarize some of the principles that are common to all of them and are also expected to be seen in any future proposals.

Clearly Defined Service Expectations and Traffic Descriptions

To enhance the current Internet to support service guarantees, it is necessary to define such service guarantees in clear mathematical terms. QoS quantifies the level of service that a distributed multimedia application expects from the communication network. In general, three QoS parameters are of prime interest: **bandwidth**, **delay**, and **reliability**.

Bandwidth, as the most prominent QoS parameter, specifies how much data (maximum or average) are to be transferred in the networked system (Wolf, 1997). In general, it is not sufficient to specify the rate only in terms of bits, as the QoS scheme shall be applicable to various networks as well as to general-purpose end systems. For example, in the context of protocol processing, issues such as buffer management, timer management, and the retrieval of control information play an important role. The costs of these operations are all related to the number of packets processed (and are mostly independent of the packet size), emphasizing the importance of a packet-oriented specification of bandwidth. Information about the packetization can be given by specifying the maximum and the average packet size and the packet rate.

Delay, as the second parameter, specifies the maximum delay observed by a data unit on an end-to-end transmission (Furht, 1994). The delay encountered in transmitting the elements of a multimedia object stream can vary from one element to the next. This delay variance can take two forms: **delay jitter** and **delay skew**. Jitter implies that in an object stream, the actual presentation times of various objects shift with respect to their desired presentation times. The effect of jitter on an object stream is shown in Figure 7.3. In Figure 7.3(A), each arrow represents the position of an object that is equally spaced in time. In Figure 3(B), the dotted arrows represent the desired positions of the objects, and the solid arrows represent their actual positions. It can be seen in Figure 3(B) that these objects are randomly displaced from their original positions. This effect is called jitter in the timing of

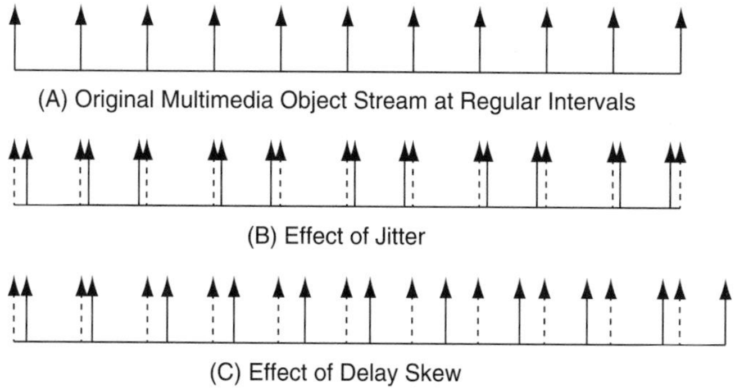

FIGURE 7.3 Effect of Jitter on an Object Stream. Adapted from Furht (1994).

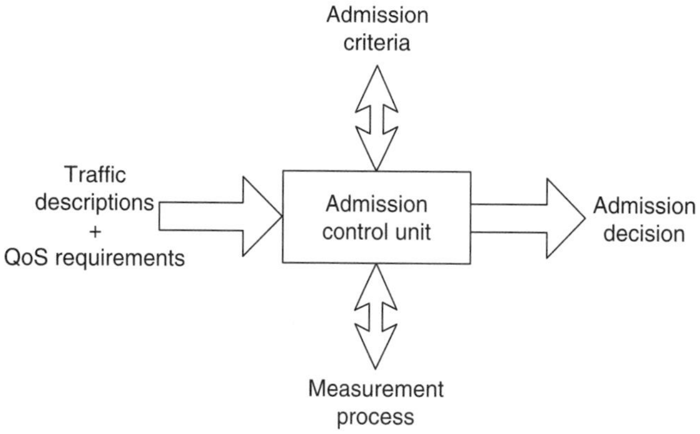

FIGURE 7.4 Admission Control Components. Adapted from Tang *et al.* (1999).

the object stream. The effect of jitter on a video clip is a shaky picture.

Skew implies constantly increasing the difference between the desired presentation times and the actual presentation times of streamed multimedia objects. This effect is shown in Figure 7.3(C). The effect of skew in the presentation times of consecutive frames in a video will be a slow- or fast-moving picture. Jitter can be removed only by buffering at the receiver side.

Reliability pertains to the loss and corruption of data. Loss probability and the method for dealing with erroneous data can also be specified.

It also becomes necessary for every source to mathematically describe the traffic characteristics of the traffic it will be sending. For example, every source can describe its traffic flow characteristics using a **traffic descriptor** that contains the peak rate, average rate, and maximum burst size (Kurose and Ross, 2001). This can be specified in terms of leaky bucket parameters, like the bucket size b, and the token rate r. In this case, the maximum burst size will be equal to the size of the bucket (i.e., b peak rate will be $rT + b$, where T is the time taken to empty the whole bucket, and the average rate over time t is $rt + b$).

Admission Control

Admission control is a *proactive* form of congestion control (as opposed to reactive congestion control used in protocols like TCP) that ensures that demand for network resources never exceeds the supply. Preventing congestions from occurring reduces packet delay and loss, which improves real-time performance.

An admission control module (refer Figure 7.4) takes as input the traffic descriptor and the QoS requirements of the flow, and the module outputs its decision of either accepting the flow at the requested QoS or rejecting it if that QoS is not met (Sun and Jain, 1999). For this it consults **admission criteria** module, which refers to the rules by which an admission control scheme accepts or rejects a flow. Since the network resources are shared by all admitted flows, the decision to accept a new flow may affect the QoS commitments made to the admitted flows. Therefore, an admission control decision is usually made based on an estimation of the effect the new flow will have on other flows and the utilization target of the network.

Another useful component of admission control is the **measurement process** module. If we assume sources can characterize their traffic accurately using traffic descriptors, the admission control unit can simply use parameters in the traffic descriptors. It is observed, however, that real-time traffic sources are very difficult to characterize, and the leaky bucket parameters may only provide a very loose upper bound of the traffic rate. When the real traffic becomes bursty, the network utilization can get very low if admission control is solely based on the parameters provided at call setup time. Therefore, the admission control unit should monitor the network dynamics and use measurements such as instantaneous network load and packet delay to make its admission decisions.

Traffic Shaping and Policing

After a traffic stream gets admitted with a given QoS requirement and a given traffic descriptor, it becomes binding on the source to stick to that profile. If a rogue source breaks its contract and sends more than what it had bargained for, there will be breakdown in the service model. To prevent this possibility, **traffic shaping and policing** becomes essential.

Token bucket algorithm (Tanenbaum, 1996) is almost always used for traffic shaping. Token bucket is synonymous to a bucket with depth b, in which tokens are collected at a rate r. When the bucket becomes full, extra tokens are lost. A source can send data only if it can grab and destroy sufficient tokens from the bucket.

Leaky bucket algorithm (Tanenbaum, 1996) is used for traffic policing, in which excessive traffic is dropped. Leaky bucket is synonymous to a bucket of dept b with a hole at the bottom that allows traffic to flow at a fixed rate *r*. If the bucket is full, the extra packets are just dropped.

Packet Classification

Every packet, regardless of whether it is a real-time packet or a non real-time packet, is treated equally at all routers in the best-effort Internet. However, real-time multimedia traffic demands differential treatment in the network. Thus the newer service models will have to use some mechanism to distinguish between real-time and non real-time packets. In practice, this is usually done by **packet marking**. The type of service (ToS) field in the IP header can be used for this purpose. Some newer Internet architectures like MPLS make use of short labels that are attached to the front of the IP packets for this purpose.

Packet Scheduling

If differential treatment is to be provided in the network, then FIFO scheduling, traditionally used in routers, must be replaced by sophisticated queuing disciplines like **priority queuing** and **weighted fair queuing**. Priority queuing provides different queues for different traffic types. Every queue has an associated priority in which it is served. Queues with lower priority are served only when there are no packets in all the higher priority queues. One disadvantage of priority queuing is that it might lead to starvation of some low-priority flows.

Weighted fair queuing also has different queues for different traffic classes. Every queue, however, is assigned a certain weight *w*, and the packets in that queue always get a fraction w/C of the bandwidth, where C is the total link capacity.

Packet Dropping

Under congestion, some packets need to be dropped by the routers. In the past, this was done at random, leading to inefficient performance for multimedia traffic. For example, an MPEG-encoded packet stream contains *I*, *P*, and *B* frames. The *I* frames are compressed without using any temporal redundancy between frames, while the *P* and *B* frames are constructed using motion vectors from *I* (or *P*) frames. Thus the packets containing *I* frames are more important than those containing *P* or *B* frames. When it comes to packet dropping, the network should give higher dropping priority to the *P* and *B* frames than the *I* frame packets. For a survey on the different packet dropping schemes, refer to Labrador and Banerjee (1999).

QoS-Based Routing

The best-effort Internet uses routing protocols such as Open Shortest Path First (OSPF), Routing Information Protocol (RIP), and Border Gateway Protocol (BGP) (Sun and Jain, 1999). These protocols are called best-effort routing protocols, and they normally use single objective optimization algorithms

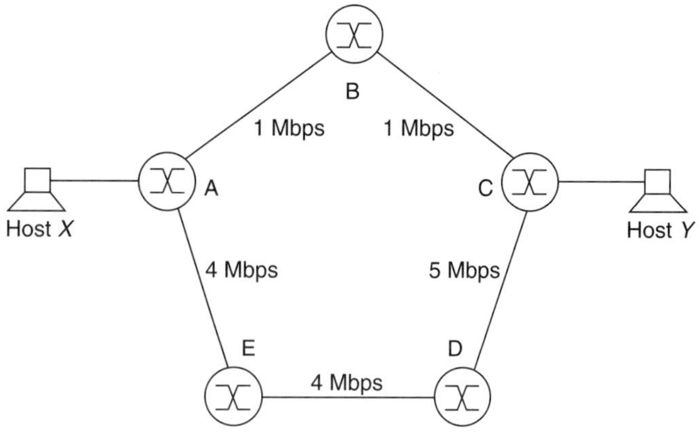

FIGURE 7.5 QoS-Based Routing

that consider only one metric (either hop count or line cost) and minimize it to find the shortest path from the source to the destination. Thus, all traffic is routed along the shortest path leading to congestion on some links, while other links might remain underutilized. Furthermore, if link congestion is used to derive the line cost such that highly congested links have a higher cost, then such algorithms can cause oscillations in the network, where traffic load continuously shifts from heavily congested links to lightly congested links, and this will increase the delay and jitter experienced by end users.

In QoS-based routing, paths for different traffic flows are determined based on some knowledge of resource availability in the network as well as the QoS requirement of the flows. For example, in Figure 7.5, suppose there is a traffic flow from host *X* to host *Y*, which requires 4 Mbps bandwidth. Although path A–B–C is shorter (with just two hops), it will not be selected because it does not have enough bandwidth. Instead, path A–E–D–C is selected because it satisfies the bandwidth requirement.

Besides QoS-based routing, there are two relevant concepts called **policy-based routing** and **constraint-based routing**. Policy-based routing (Avramovic, 1992) commonly means the routing decision is not based on the knowledge of the network topology and metrics but on some administrative policies. For example, a policy may prohibit a traffic flow from using a specific link for security reason even if the link satisfies all the QoS constraints. Constraint-based routing (Kuipers *et al.*, 2002) is another new concept that is derived from QoS-based routing but has a broader sense. In this routing, algorithm routes are computed based on multiple constraints, including both QoS constraints and policy constraints. Both QoS-based routing and policy-based routing can be considered as special cases of constraint-based routing.

7.3.5 Integrated Services

To support multimedia traffic over the Internet, the Integrated Services working group in the Internet Engineering Task Force

(IETF) has developed an enhanced Internet service model called Integrated Services (IntServ) (White, 1997). This model is characterized by resource reservations. It requires applications to know their traffic characteristics and QoS requirements beforehand and signal the intermediate network routers to reserve resources, like bandwidth and buffers, to meet them. Accordingly, if the requested resources are available, the routers reserve them and send back a positive acknowledgment to the source, which allows data to be sent. If, on the other hand, sufficient resources are not available at any router in the path, the request is turned down, and the source has to try again after some time. This model also requires the use of **packet classifiers** to identify flows that are to receive a certain level of service as well as **packet schedulers** to handle the forwarding of different packets in a manner to ensure that the QoS commitments are met.

The core of IntServ is almost exclusively concerned with controlling the queuing component of the end-to-end packet delay. Thus, per-packet delay is the central quantity about which the network makes service commitments.

Intserv introduces three service classes to support RTI, RTT, and elastic multimedia applications. They are **Guaranteed service**, **Controlled Load service**, and the **Best-Effort service**. A **flow descriptor** is used to describe the traffic and QoS requirements of a flow (Leon-Garcia and Widjaja, 2000). The flow descriptor consists of two parts: a **filter specification** (filterspec) and a **flow specification** (flowspec). The filterspec provides the information required by the packet classifier to identify the packets that belong to that flow. The flowspec consists of a **traffic specification** (Tspec) and **service request specification** (Rspec). Tspec specifies the traffic behavior of the flow in terms of token bucket parameters (b,r), while the Rspec specifies the requested QoS requirements in terms of bandwidth, delay, jitter, and packet loss.

Since all network nodes along the path from source to destination must be informed of the requested resources, a signaling protocol is needed. Resource Reservation Protocol (RSVP) is used for this purpose (Braden *et al.*, 1997). The signaling process is illustrated in Figure 7.6. The sender sends a PATH message to the receiver, specifying the characteristics of the traffic. Every intermediate router along the path forwards the PATH message to the next hop determined by the routing protocol. The receiver, upon receiving the PATH message, responds with the RESV message to request resources for the flow. Every intermediate router along the path can reject or accept the request of the RESV message. If the request is rejected, the router will send an error message to the receiver, and the signaling process terminates. If the request is accepted, link buffer and bandwidth are allocated to the flow, and related flow state information will be installed in the router.

The design of RSVP lends itself to be used with a variety of QoS control services. RSVP specification does not define the internal format of the RSVP protocol fields or objects and treats them as opaque; it deals only with the setup mechanism. RSVP was designed to support both unicast and multicast applications. RSVP supports heterogeneous QoS, which means different receivers in the same multicast group can request different QoS. This heterogeneity allows some receivers to have reservations, whereas others could receive the same traffic using the best-effort service. We now discuss the service classes offered by IntServ.

Guaranteed Service Class

The **Guaranteed** service class provides firm end-to-end delay guarantees. Guaranteed service does not control the minimum or average delay of packets; merely the maximal queuing delay. This service guarantees that packets will arrive at the receiver but in a requested delivery time and will not be discarded due to queue overflows, provided the flow's traffic stays in its specified traffic limits, which are controlled using traffic policing. This service is intended for applications that need a firm guarantee that a packet will arrive no later than a certain delay bound.

Using traffic specification (Tspec), the network can compute various parameters describing how it will handle the flow. By combining the parameters, it is possible to compute the maximum queuing and routing delay that a packet can experience. Using the fluid flow model, the queuing delay is approximately a function of two parameters: the token bucket size b and the data rate R that the application requests and gets when admitted.

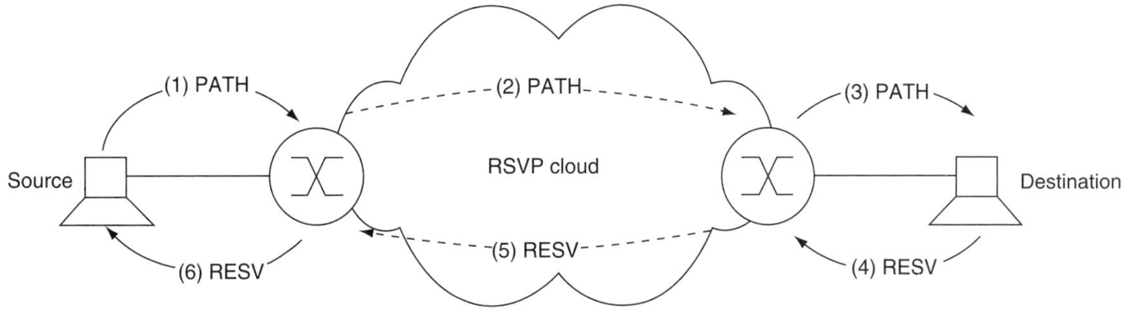

FIGURE 7.6 RSVP Signaling

Controlled Load Service

Controlled Load service is an enhanced quality of service intended to support RTT applications requiring better performance than that provided by the traditional best-effort service. It approximates the end-to-end behavior provided by best effort under unloaded conditions. The assumption here is that under unloaded conditions, a very high percentage of the transmitted packets are successfully delivered to the end nodes, and the transmission delay experienced by a very high percentage of the delivered packets will not vary much from the minimum transit delay.

The network ensures that adequate bandwidth and packet processing resources are available to handle the requested level of traffic. The controlled load service does not make use of specific target values for delay or loss. Acceptance of a controlled-load request is merely a commitment to provide the flow with a service closely equivalent to that provided to uncontrolled traffic under lightly loaded conditions. Over all timescales significantly larger than the burst time, a controlled load service flow may experience little or no average packet queuing delay and little or no congestion loss.

The controlled load service is described only using a Tspec. Since the network does not give any quantitative guarantees, Rspec is not required. The controlled load flows not experiencing excess traffic will get the contracted quality of service, and the network elements will prevent excess controlled load traffic from unfairly impacting the handling of arriving best-effort traffic. The excess traffic will be forwarded on a best-effort basis.

Best-Effort Service

The **Best-Effort** service class does not have a Tspec or an Rspec. There are no guarantees by the network whatsoever. The network does not do any admission control for this class.

Disadvantages of the IntServ Service Model for the Internet

IntServ uses RSVP to make per-flow reservations at routers along a network path. Although this allows the network to provide service guarantees at the flow level, it causes it to suffer from scalability problems. The routers have to maintain per-flow state for every flow that passes through the router, which can lead to huge overhead. Moreover, RSVP is a soft-state protocol, which means that the router state has to be refreshed at regular intervals. This increases traffic overhead.

7.3.6 Differentiated Services

The Differentiated Services working group in the IETF proposed the Differentiated Services (DiffServ) service model for the Internet, which removes some of the shortcomings of the IntServ architecture (Blake *et al.*, 1998). DiffServ divides the network into distinct regions called DS domains, and each DS domain can be controlled by a single entity. For example, an organization's intranet or an ISP can form its own DS domain. One important implication of this is that to provide any service guarantees the entire path between the source and destination must be under some DS domain (possibly multiple). Even if a single hop is not under some DS domain, then service cannot be guaranteed. DiffServ architecture can be extended across multiple domains using a service level agreement (SLA) between them. An SLA specifies rules for traffic remarking, actions to be taken for out-of-profile traffic, and other such specifications.

Every node in a DS domain can be a boundary node or an interior node.

- **Boundary node:** Boundary nodes are the gatekeepers of the DS domain. A boundary node is the first (or last) node that a packet can encounter when entering (or exiting) a DS domain. It performs certain edge functions like admission control, packet classification, and traffic conditioning. The admission control algorithm limits the number of flows that are admitted into the DiffServ domain and is distributed in nature. For example, in the simplest case, the admission control algorithm may maintain a central data structure that contains the current status of all links in the DS domain. When a flow is considered for admission, the corresponding boundary node might check this data structure to verify if all the links of the flow path can satisfy the requested QoS. Every packet belonging to an admitted flow and arriving into the DS domain is classified and marked as belonging to one of the service classes called **behavior aggregates** in DiffServ terminology. Each such behavior aggregate is assigned a distinct 8-bit code word, called the DS code point. Packet marking is achieved by updating the TOS field in the packet's IP header with the appropriate DS code point. Boundary nodes also enforce traffic conditioning agreements (TCA) between their own DS domain and other connected domains, if any.
- **Interior node:** An interior node is completely inside a DS domain and is connected to other interior nodes or boundary nodes in the same DS domain. The interior nodes only perform packet forwarding. When a packet with a particular DS code point arrives at this node, it is forwarded to the next hop according to some predefined rule associated with the packet class. Such predefined rules are called **perhop behaviors** (PHBs), which are discussed next.

Thus unlike IntServ, only the edge routers have to maintain per-flow states, which makes DiffServ relatively more scalable.

Per Hop Behaviors

A **per hop behavior** (PHB) is a predefined rule that influences how the router buffers and link bandwidth are shared among competing behavior aggregates. PHBs can be defined either in terms of router resources (e.g., buffer and bandwidth), in terms of their priority relative to other PHBs, or in terms of their relative traffic properties (e.g., delay and loss). Multiple PHBs can be lumped together to form a PHB group. A particular PHB group can be implemented in a variety of ways because PHBs are defined in terms of behavior characteristics and are not implementation dependent. Thus PHBs can be considered as basic building blocks for creating services. A PHB for a packet is selected at the first node on the basis of its DS code point. The mapping from the DS code point to PHB may be 1 to 1 or *N* to 1. Examples of the parameters of the forwarding behavior that each traffic should receive are **bandwidth partition** and the **drop priority**. Examples of these implementations are weighted fair queuing (WFQ) for bandwidth partition and random early detect (RED) for drop priority. Two commonly used PHBs defined by IETF are the following:

- **Assured forwarding (AF) PHB:** AF PHB divides incoming traffic into four classes; each AF class is guaranteed some minimum bandwidth and buffer space. In each AF class, packets are further assigned one of three drop priorities. By varying the amount of resources allocated to each AF class, different levels of performance can be offered.
- **Expedited forwarding (EF) PHB:** EF PHB dictates that the departure rate of a traffic class from any router must equal or exceed the configured rate. Thus, for a traffic class belonging to EF PHB, during any interval of time, it can be confidently said that departure rate from any router will equal or exceed the aggregate arrival rate at that router. This has strong implications on the queuing delay that is experience by the packet. In this case, the queuing delay can be guaranteed to be bounded and is negligible (limited by the link bandwidth). EF PHB is used to provide premium service (having low delay and jitter) to the customer. However, EF PHB requires a very strong admission control mechanism. The admission control algorithm will basically ensure that the arrival rate of traffic belonging to EF PHB is less than the departure rate configured at any router in its path. Moreover, proper functioning of EF PHB demands strict policing. This job can be carried out by the Ingress routers. If packets are found to be in violation of the contract, they can be either dropped or demoted to a lower traffic class.

7.3.7 Multi-Protocol Label Switching

When an IP packet arrives at a router (Rosen *et al.*, 2001) the next hop for this packet is determined by the routing algorithm in operation, which uses the **longest prefix match** (i.e., matching the longest prefix of an IP destination address to the entries in a routing table) to determine the appropriate outgoing link. This process introduces some latency because the routing tables are very large and table lookups take time. The same process also needs to be repeated independently for every incoming packet, even though all packets may belong to the same flow and may be going toward the same destination. This shortcoming of IP routing can be removed by using IP switching, in which a short label is attached to a packet and updated at every hop. When this modified packet arrives at a switch (*router* in our previous description), this label is used to index into a short switching table [an O(1) operation] to determine the outgoing link and new label for the next hop. The old label is then replaced by new label, and the packet is forwarded to the next hop. All this can be easily done in hardware and results in very high speeds. This concept was applied earlier in ATM for cell switching that used the VPI/VCI field in the packet as the label. MPLS introduces the same switching concept in IP networks. It is called **Multi-Protocol** because this technique can also be used with any network layer protocol other than IP. Label switching provides a low-cost hardware implementation, scalability to very high speeds, and flexibility in the management of traffic flows.

Similar to DiffServ, an MPLS network is divided into domains with boundary nodes called **label edge routers** (LER) and interior nodes called **label switching routers** (LSR). Packets entering an MPLS domain are assigned a label at the ingress LER and are switched inside the domain by a simple label lookup. The labels determine the quality of service that the flow receives in the network. The labels are stripped off the packets at the egress LER, and the packets might be routed in the conventional fashion before they reach their final destination. A sequence of LSRs that is to be followed by a packet in an MPLS domain is called **label switched path** (LSP). Again, similar to DiffServ, to guarantee a certain quality of service to the packet, both the source and destination have to be attached to the same MPLS domain, or if they are attached to different domains, then there should be some service agreement between the two.

MPLS uses the concept of **forward equivalence class** (FEC) to provide differential treatment to different media types. A group of packets that are forwarded in the same manner are said to belong to the same FEC. There is no limit to the number and granularity of FECs that can exists. Thus, it is possible to define a separate FEC for every flow (which is not advisable due to large overheads) or for every media type, each tailored for that media type. One important thing to note here is that labels have only local significance in the sense that two LSRs agree to use a particular label to signify a particular FEC among themselves. The same label can be used to distinguish a different FEC by another pair of LSRs. Thus, it becomes necessary to do label assignments, which includes label allocation and label-to-FEC bindings on every hop of the LSP before the traffic flow can use the LSP. Label assignments can be initiated in one of the following three ways:

(1) **Topology-driven label assignment**: In this assignment, LSPs (for every possible FEC) are automatically set up between every pair of LSR (full mesh). Thus, this scheme can place a heavy demand on the usage of labels at each LSR. However, the main advantage is that there is no latency involved in setting up an LSP before the traffic flow can use it (i.e., zero-call setup delay).
(2) **Request-driven label assignment**. Here, the LSPs are set up based on explicit requests. RSVP can be used to make the request. The advantage of this scheme is that LSPs will be set up only when required, and a full mesh is avoided. However, the disadvantage is the added setup latency, which can dominate short-lived flows.
(3) **Traffic-driven label assignment**. This assignment combines the advantages of the above two methods. LSPs are set up only when the LSR identifies traffic patterns that justify the setup of an LSP. Those that are identified as not needing an established LSP are routed using the normal routing method.

MPLS also supports **label stacking**, which can be very useful for performing tunneling operations. In label stacking, labels are stacked in a FILO order. In any particular domain, only the topmost label can be used to make forwarding decisions. This functionality can be very useful for providing mobility: a home agent can push another label on incoming packets and forward the packet to a foreign agent that pops it off and finally forwards the packet to the destination mobile host.

7.4 Enhancing the TCP/IP Protocol Stack to Support Functional Requirements of Distributed Multimedia Applications

In this section, we illustrate standards/protocols that have been introduced to operate over the basic TCP/IP protocol stack to satisfy the **functional requirements** of multimedia traffic streams. We later describe two protocol architectures: H.323 and Session Initiation Protocol (SIP) that have been standardized to support these functional requirements. Again, to preserve the brevity of this chapter, we do not discuss every such approach at length but provide appropriate references for interested readers.

7.4.1 Supporting Multicasting

The easiest way to achieve multicasting over the Internet is by sending packets to a multicast IP address (Class D IP addresses are multicast IP addresses) (Banikazemi *et al.*, 1997). Hosts willing to receive multicast messages for particular multicast groups inform their immediate-neighboring routers using the Internet Group Management Protocol (IGMP). Multicasting is trivial on a single ethernet segment (where packets can be multicast using the multicast MAC address). For delivering a multicast packet from the source to the destination nodes on other networks, however, multicast routers need to exchange the information they have gathered from the group membership of the hosts directly connected to them. There are many different algorithms such as *flooding, spanning tree, reverse path broadcasting*, and *reverse path multicasting* for exchanging the routing information among the routers. Some of these algorithms have been used in dynamic multicast routing protocols such as Distance Vector Multicast Routing Protocol (DVMRP), Multicast Extension to Open Shortest Path First (MOSPF), and Protocol Independent Multicast (PIM) (Sahasrabuddhe and Mukherjee, 2000). Based on the routing information obtained through one of these protocols, whenever a multicast packet is sent out to a multicast group, multicast routers will decide whether to forward that packet to their network(s) or not.

Another approach is MBone or Multicast Backbone. Mbone is essentially a virtual network implemented on top of some portions of the Internet. In the MBone, islands of multicast-capable networks are connected to each other by virtual links called **tunnels**. Multicast messages are forwarded through these tunnels in non multicast-capable portions of the Internet. For forwarding multicast packets through these tunnels, they are encapsulated as IP-over-IP (with protocol number set to four) such that they look like normal unicast packets to interventing routers.

ITU-T H.323 and IETF Session Initiation Protocol (SIP), as discussed in detail later, support multicasting through the presence of a multipoint control unit that provides both mixing and conferencing functionalies needed for audio/video conferencing. Striegel and Manimaran (2002) offer a survey of QoS multicasting issues.

7.4.2 Session Management

We now discuss the different protocols that have been standardized to meet different aspects of session management.

- **Session description**: Session Description Protocol (Handley and Jacobson, 1998; Garg, 1999), developed by IETF, can be used for providing the session description functionality (to describe media type and media encoding used for that session). It is more of a description syntax than a protocol because it does not provide a full-range media negotiation capability (this is provided by SIP, as discussed in later sections). SDP encodes media descriptions in simple text format. An SDP message is composed of a series of lines, called **fields**, whose names are abbreviated by a single lowercase letter and are in a required order to facilitate parsing. The fields are in the form

Protocol version number	v = 0
Owner/creator of session	o = khanvilkar 8988234542 8988234542 IN IP4 192.168.0.201
Session name	s = Presentation on multimedia
Session information	i = Topics on multimedia communication
URI	u = http://mia.ece.uic.edu/sip
E-mail address	e = shashank@evl.uic.edu
Phone number	p = 1-312-413-5499
Connection information	c = IN IP4 192.168.0.201
Bandwidth information	b = CT:144
Time session starts/stops	t = xxxxxxxxxx xxxxxxxxxx
Media information	m = audio 56718 RTP/AVP 0
Media attributes	a = rtpmap : 0 PCMU/8000
Media information	m = video 67383 RTP/AVP 31
Media attributes	a = rtpmap : 31 H261/90000

FIGURE 7.7 Sample SDP Message

attribute_type = value. A sample SDP message is illustrated in Figure 7.7. The meaning of all attributes is illustrated on the left, while the actual message is illustrated on the right.

- **Session announcement**: Session Announcement Protocol (SAP) (Handley *et al.*, 2000; Arora and Jain, 1999) is used for advertising multicast conferences and other multicast sessions. An SAP announcer periodically multicasts announcement packets to a well-known multicast address and port (port number 9875) with the same scope as the session it is announcing, ensuring that the recipients of the announcement can also be potential recipients of the session being advertised. Multiple announcers may also announce a single session to increase robustness against packet loss or failure of one or more announcers. The time period between repetitions of an announcement is chosen such that the total bandwidth used by all announcements on a single SAP group remains below a preconfigured limit. Each announcer is expected to listen to other announcements to determine the total number of sessions being announced on a particular group. SAP is intended to announce the existence of a long-lived wide area multicast sessions and involves a large startup delay before a complete set of announcements is heard by a listener. SAP also contains mechanisms for ensuring integrity of session announcements, for authenticating the origin of an announcement, and for encrypting such announcements.
- **Session identification**: In the best-effort Internet, every flow (or session) can be identified using the tuple <*Src Ip, Src Port, Dst IP, Dst Port, Protocol*>. Thus, individual transport layer sockets have to be established for every session. If there ever arises a need to bunch sessions together (for cutting costs), however, there is no available mechanism. Hence, there is clearly a need to multiplex different streams into the same transport layer socket. This functionality is similar to that of the **session layer** in the seven-layer OSI model, which has been notably absent in TCP/IP protocol stack used in the Internet. Session identification can be done using RTP and is described in more detail in the next point.
- **Session control**: All the session control functionalities can be satisfied using a combination of RTP, RTCP, and RTSP. Real-Time Protocol (RTP) (Schulzrinne *et al.*, 1996) typically runs on top of UDP. Specifically, chunks of audio/video data that are generated by the sending side of the multimedia application are encapsulated in RTP packets, which in turn are encapsulated in UDP. The RTP functionality needs to be integrated into the application. The functions provided by RTP include (a) **sequence number** field in the RTP header to detect lost packets, (b) a **payload identifier** included in each RTP packet to dynamically change and describe the encoding of the media, (c) a **frame market** bit to indicate the beginning and end of a video or audio frame, (d) a **synchronization source** (SSRC) identifier to determine the originator of the frame because there are many participants in a multicast session and (e) **time stamp** to compensate for the different delay jitter for packets in the same stream, and to assist the play-out buffers.

Additional information pertaining to particular media types and compression standards can also be inserted in the RTP packets by using profile headers and extensions. Cavusoglu *et al.* (2003) have relied on the information contained in the RTP header extensions to provide an adaptive forward error correction (FEC) scheme for MPEG-2 video communications over RTP networks. This approach relies on the group of pictures (GOP) sequence and motion information to assign a dynamic weight to the video sequence. The fluctuations in the weight of the video stream are used to modulate the level of FEC assigned to the GOP. Much higher performance is achieved by the use of the adaptive FEC scheme in comparison to other methods that assign an uneven level of protection to video stream. The basic notion presented by the adaptive weight assignment procedure presented in Cavusoglu *et al.* (2003) can be employed in other applications such as DiffServ networks and selective retransmissions (see discussion on LSP networks later).

RTCP is a control protocol that works in conjunction with RTP and provides participants with useful statistics about the number of packets sent, number of packets lost, interarrival jitter, and round trip time. This information can be used by the sources to adjust their data rate. Other information, such as e-mail address, name, and phone number are included in the RTCP packets and allow all users to know the identities of the other users for that session.

Real-Time Streaming Protocol (RTSP) (Schulzrinne *et al.*, 1998) is an out-of-band control protocol that allows the media player to control the transmission of the media stream, including functions like **pause/resume** and **repositioning playback**. The use of the RTP above UDP provides for additional functionality required for reordering and time stamping the UDP packets. The inherent limitations of the UDP, however, are not completely overcome by the use of RTP. Specifically, the unreliability of UDP persists and, consequently, there is no guarantee of delivery of the transmitted packets at the receiver. On the other hand, the error-free transmission guarantees provided by the TCP pose severe time delays that render it useless for real-time applications. Mulabegovic *et al.* (Mulabegovic *et al.*, 2002) have proposed an alternative to RTP provided by the Lightweight Streaming Protocol (LSP). This protocol also resides on top of the transport layer and relies on the UDP. LSP provides sequence numbers and time stamps—as facilitated by RTP—to reorder the packets and manage the buffer at the receiver. However, unlike UDP and RTP, the LSP allows for limited use of retransmissions to minimize the effects of error over the communication network. This is accomplished by use of negative acknowledgments in the event of lost packets and satisfaction of timing delay constraints required to maintain real-time communication capability.

7.4.3 Security

IpSec (Chapman and Chapman, 2000; Opplinger, 1998) provides a suite of protocols that can be used to carry out secure transactions. IpSec adds security protection to IP packets. This approach allows applications to communicate securely without having to be modified. For example, before IpSec came into picture, applications often used SSH or SSL for having a secure peer-to-peer communication, but this required modifying certain APIs in the application source code and subsequent recompilation. IpSec cleanly separates policy from enforcement. The policy (which traffic flow is secured and how it is secured) is provided by the system administrator, while the enforcement of this policy is carried out by a set of protocols: Authentication Header (AH) and Encapsulated Security Payload (ESP). These policies are placed as rules in a Secure Policy Database (SPD), consulted for every inbound/outbound packet, and tell IpSec how to deal with a particular packet: if a IpSec mechanism needs to be applied to the packet, if the packet should be dropped, or if the packet should be forwarded without placing any security mechanism. If the administrator has configured the SPD to use some security for a particular traffic flow, then IpSec first negotiates the parameters involved in securing the traffic flow with the peer host. These negotiations result in the so-called Security Association (SA) between the two peers. The SA contains the type of IpSec mechanism to be applied to the traffic flow (AH or ESP), the encryption algorithms to be used and the security keys to be applied. SAs are negotiated using the Internet Key Exchange (IKE) protocol.

As said earlier, IpSec provides two protocols to secure traffic flow. The first is the **Authenticated Header** (AH) whose main function is to establish the authenticity of the sender to the receiver. It does not provide data confidentiality. In other words, AH does not encrypt the payload. It may seem, at first, that authentication without confidentiality might not be useful in the industry. However, there are many applications where this does provide a great help. For example, there may be many situations, such as news reports, where the data may not be encrypted, but it may be necessary to establish the authenticity of the sender. AH provides significantly less overhead as compared to ESP. The second protocol is the **Encapsulated Security Payload** (ESP). ESP provides both authentication and encryption services to the IP packets. Since every packet is encrypted, ESP puts a larger load on the processor.

Currently, IpSec can operate in **transport** mode or **tunnel** mode. In transport mode, IpSec takes a packet to be protected, preserves the packet's IP header, and modifies only the upper layer portions by adding IpSec headers and the requested kind of protection between the upper layers and the original IP header. In the tunnel mode, IpSec treats the entire packet as a block of data, adds a new packet header, and protects the data by making it part of the encrypted payload of the new packet.

IpSec can be easily integrated into the current operating system environment, by either changing the native implementation of IP protocol (and subsequent kernel compilation), inserting an additional layer below the IP layer of the TCP/IP protocol stack (also known as bump-in-the-stack [BITS]), or using some external hardware (also known as bump-in-the-wire [BITW]).

7.4.4 Mobility

Mobile IP (Perkins, 1998) is an Internet protocol used to support mobility. Its goal is to provide the ability of a host to stay connected to the Internet regardless of its location. Every site that wants to allow its users to roam has to create a **home agent** (HA) entity, and every site that wants to allow visitors has to create a **foreign agent** (FA) entity. When a mobile host (MH) roams into a foreign site (which allows roaming), it contacts and registers with the FA for that site. The FA, in turn, contacts the HA of that mobile host and gives it a **care-of address**, normally the FA's own IP address. All packets destined for the MH will eventually reach the HA that encapsulates the packets inside another IP packet and forward it to the FA. The FA decapsulates it and ultimately forwards the original packets, to the MH. The HA may also inform the source of the new IP address where the packets must be directly forwarded.

7.4.5 H.323

ITU-T recommendation H.323 (Thom, 1996; Liu and Mouchtaris, 2000; Arora and Jain, 1999) is an umbrella recommendation that specifies the components, protocols, and procedures used to enable voice, video, and data conferencing over a packet-based network like the IP-based Internet or IPX-based local-area networks. H.323 is a very flexible recommendation that can be applied in a variety of ways—audio only (IP telephony); audio and video (video telephony); audio and data; and audio, video, and data—over point-to-point and point-to-multipoint multimedia conferencing. It is not necessary for two different clients to support the same mechanisms to communicate because individual capabilities are exchanged at the beginning of any session, and communication is set up based on the lowest common denominator. Point-to-multipoint conferencing can also be supported without the presence of any specialized hardware or software.

H.323 is part of a family of ITU-T recommendations illustrated in Table 7.6. The family is called H.32x and provides multimedia communication services over a wide variety of networks. Interoperation between these different standards is obtained through a **Gateway** that provides data format translation as well as controls signaling translation, audio and video codec translation, and call setup/termination functionality.

The H.323 standard defines four components that, when networked together, provide the point-to-point and point-to-multipoint multimedia–communication services:

(1) **Terminals:** These are the endpoints of the H.323 conference. A multimedia PC with a H.323 compliant stack can act as a terminal.
(2) **Gateway:** As discussed earlier, gateway is only needed whenever conferencing needs to be done between different H.32X-based clients.
(3) **Gatekeeper**: This provides many functions, including admission control and bandwidth management. Terminals must get permission from the gatekeeper to place any call.
(4) **Multipoint control unit** (MCU): This is an optional component that provides point-to-multipoint conferencing capability to an H.323-enabled network. The MCU consists of a mandatory multipoint controller (MC) and optional multipoint processors (MP). The MC determines the common capabilities of the terminals by using H.245, but it does not perform the multiplexing of audio, video, and data. The multiplexing of media streams is handled by the MP under the control of the MC. Figure 7.8, illustrates an H.323-enabled network with all these different components.

TABLE 7.6 ITU-T Recommendations for Audio/Video/Data Conferencing Standards

ITU-T recommendation	Underlying network over which audio, video, and data conferencing is provided
H.320	ISDN
H.321 and H.310	ATM
H.322	LANs that provide a guaranteed QoS
H.323	LANs and Internet
H.324	PSTN/Wireless

Figure 7.9 illustrates the protocol stack for H.323. RTP and its associated control protocol, RTCP, are employed for timely and orderly delivery of packetized audio/video streams. The operation of RTP and RTCP has been discussed in earlier sections.

The H.225 RAS (registration, admission, and status) is mainly used by H.323 endpoints (terminals and gateways) to discover a gatekeeper, to register/unregister with the gatekeeper, to request call admission and bandwidth allocation, and to clear a call. The gatekeeper can also use this protocol for inquiring on an endpoint and for communicating with other peer gateways.

The Q.931 signaling protocol is used for call setup and teardown between two H.323 endpoints and is a lightweight version of the Q.931 protocol defined for PSTN/ISDN. The H.245 media control protocol is used for negotiating media processing capabilities, such as audio/video codec to be used for each media type between two terminals, and determining master–slave relationships.

Real-time data conferencing capability is required for activities such as application sharing, whiteboard sharing, file transfer, fax transmission, and instant messaging. Recommendation T.120 provides this optional capability to H.323. T.120 is a real-time data communication protocol designed specifically for conferencing needs. Like H.323, Recommendation T.120 is an umbrella for a set of standards that enable the real-time sharing of specific applications data among several clients across different networks.

Table 7.7 illustrates the different phases involved in setting up a point-to-point H.323 conference when a gatekeeper is present in an H.323 network. The first three phases correspond to call setup, whereas the last three correspond to call teardown. When no gatekeeper is involved, phases 1 and 7 are omitted.

Accordingly, two call control models are supported in H.323: **direct call** and **gatekeeper-routed call**. In the direct call model, all Q.931 and H.245 signaling messages are

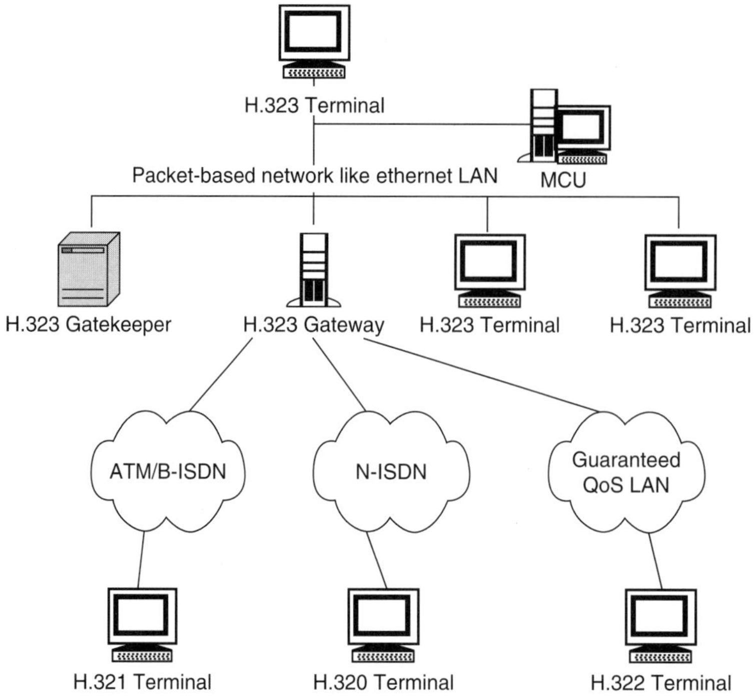

FIGURE 7.8 An H.323-Enabled Network with Different Components

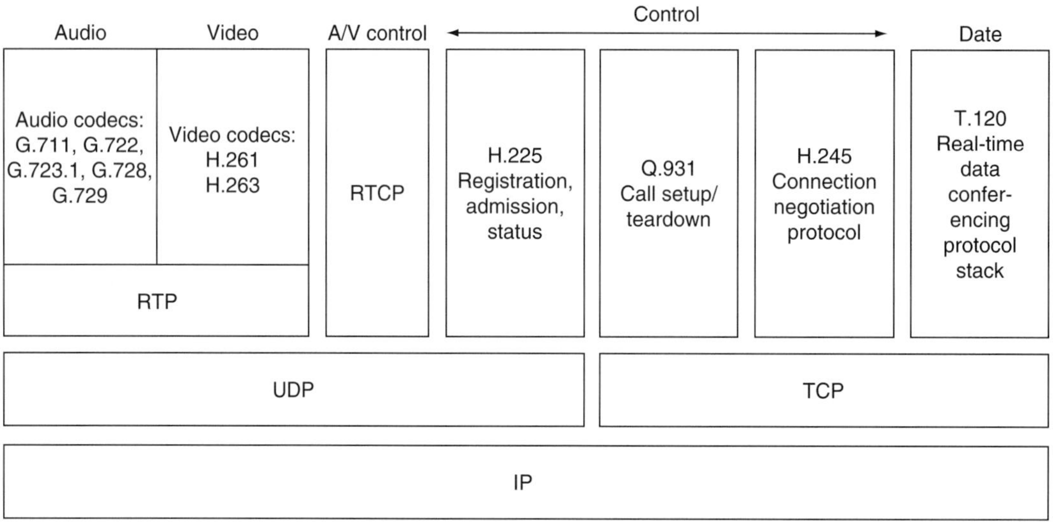

FIGURE 7.9 H.323 Protocol Stack

exchanged directly between the two H.323 endpoints, which is similar to what happens for RTP media streams. As long as the calling endpoint knows the transport address of the called endpoint, it can set up a direct call with the other party. This model is unattractive for large-scale carrier deployments because carriers may be unaware of which calls are being set up, and this may prevent them from providing sufficient resources for the call and charging for it. In the gatekeeper-routed call model, all signaling messages are routed through the gatekeeper. In this case, use of RAS is necessary. This model allows endpoints to access services provided by the gatekeeper, such as address resolution and call routing. It also allows the gatekeepers to enforce admission control and bandwidth allocation over their respective zones. This model is more suitable for IP telephony service providers because they can control the network and exercise accounting and billing functions.

TABLE 7.7 Phases in an H.323 Call

Phase	Protocol	Intended functions
1. Call admission	RAS	Request permission from gatekeeper to make/receive a call. At the end of this phase, the calling endpoint receives the Q.931 transport address of the called endpoint.
2. Call setup	Q.931	Set up a call between the two endpoints. At the end of this phase, the calling endpoint receives the H.245 transport address of the called endpoint.
3. Endpoint capability	H.245	Negotiate capabilities between two endpoints. Determine master-slave relationship. Open logical channels between two endpoints. At the end of this phase, both endpoints know the RTP/RTCP addresses of each other.
4. Stable call	RTP	Two parties in conversation.
5. Channel closing	H.245	Close down the logical channels.
6. Call teardown	Q.931	Tear down the call.
7. Call disengage	RAS	Release the resources used for this call

7.4.6 Session Initiation Protocol

Session Initiation Protocol (SIP) is an application-layer signaling protocol for creating, modifying, and terminating multimedia sessions (voice, video, or data) with either one or more participants (Johnston, 2000; Schulzrinne and Rosenburg, 2000; Arora and Jain, 1999). SIP does not define what a session is; this is defined by the content carried opaquely in SIP messages. To establish a multimedia session, SIP has to go through the following stages.

- **Session initiation:** Initiating a session is perhaps the hardest part because it requires determining where the user to be contacted is residing at the current moment: the user may be at home working on a home PC or may be at work on an office PC. Thus, SIP allows users to be located and addressed by a single global address (usually an e-mail address) irrespective of the user's physical location.
- **Delivery of session description:** Once the user is located, SIP performs the second function of delivering a description of the session that the user is invited to. SIP itself is opaque to the session description in the sense that it does not know anything about the session. It merely notifies the user about the protocol to be used so that the user can understand the session description. Session Description Protocol (SDP) is the most common protocol used for this purpose. SIP can also be used to decide a common format to describe a session so that protocols other than SDP can also be used.
- **Active session management:** Once the session description is delivered, SIP conveys the response (accept or reject) to the session initiation point (the caller). If the response is "accept," the session becomes active. If the session involves multimedia, media streams can now be exchanged between the two users. RTP and RTCP are some common protocols for transporting real-time data. SIP can also be used to change the parameters of an active session, such as removing some video media stream or reducing the quality of the audio stream.
- **Session termination:** Finally, SIP is used to terminate the session.

Thus, SIP is only a signaling protocol and must be used in conjunction with other protocols like SDP, RTP, or RTCP to provide a complete multimedia service architecture as the one provided in H.323. Note that the basic functionality and operation of SIP does not depend on any of these protocols. The SIP signaling system consists of the following components:

- **User agents:** The end system acts on behalf of a user. If the user–agent initiates SIP requests, it is called user–agent client (UAC); a user–agent server (UAS) receives such requests and return responses.
- **Network servers:** There are three types of servers in a network:
  (1) **Registration server (or registrars)**: This server keeps track of the user location (i.e., the current PC or terminal on which the user resides). The user–agent sends a registration message to the SIP registrar, and the registrar stores the registration information in a location service via a non-SIP protocol (e.g., LDAP). Once the information is stored, the registrar sends the appropriate response back to the user–agent.
  (2) **Proxy server**: Proxy servers are application-layer routers that receive SIP requests and forward them to the next hop server that may have more information about the location of the called party.
  (3) **Redirect server**: Redirect servers receive requests and then return the location of another SIP user agent or server where the user might be found.

It is quite common to find proxy, redirect, and registrar servers implemented in the same program.

SIP is based on an HTTP-like request/response transaction model. Each transaction consists of a request that invokes a particular method, or function, on the server and at least one response. Just as for HTTP, all requests and responses use textual encoding for SIP. Some commands and responses of SIP and their use are illustrated in Table 7.8.

Message format for SIP is shown Figure 7.10. The message body is separated from the header by a blank line. The **Via** indicates the host and port at which the caller is expecting a

Table 7.8 Commands and Responses Used in SIP

Method	Used
INVITE	For inviting a user to a call
ACK	For reliable exchange of invitation messages
BYE	For terminating a connection between the two endpoints
CANCEL	For terminating the search for a user
OPTIONS	For getting information about the capabilities of a call
REGISTER	For giving information about the location of a user to the SIP registration server

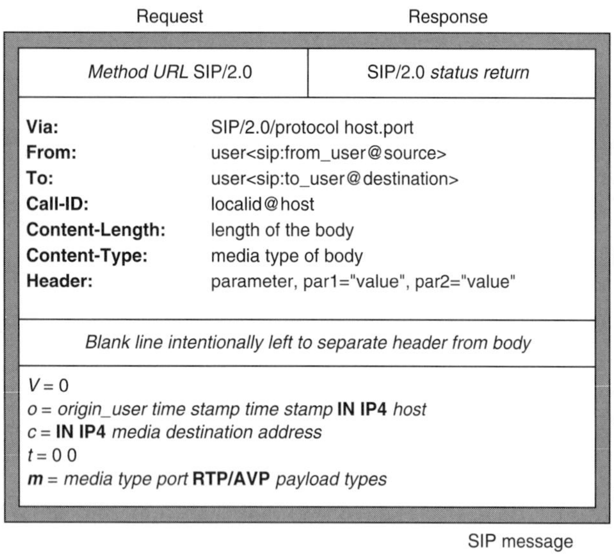

FIGURE 7.10 SIP Message Format

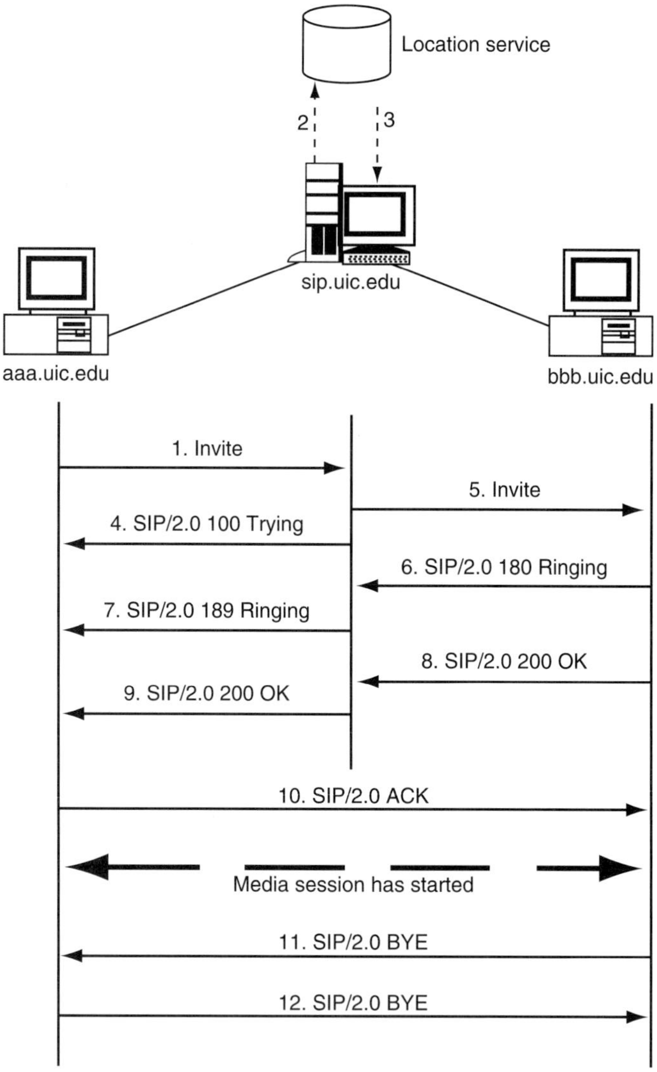

FIGURE 7.11 Timeline for a Typical SIP Session

response. When an SIP message goes through a number of proxies, each such proxy appends to this field with its own address and port. This enables the receiver to send back an acknowledgment through the same set of proxies. The **From** (or **To**) field specifies the SIP URI of the sender (or receiver) of the invitation, which is usually the e-mail address assigned to the user. The **call-ID** contains a globally unique identifier of this call generated by a combination of a random string and IP address. The **Content-Length** and **Content-Type** fields describe the body of the SIP message.

We now illustrate a simple example (refer to Figure 7.11) that captures the essence of SIP operations. Here, a client (caller) is inviting a participant (callee) for a call. The SIP client creates an INVITE message for callee@uic.edu, which is normally sent to a proxy server (step 1). This proxy server tries to obtain the IP address of the SIP server that handles requests for the requested domain. The proxy server consults a location server to determine this next hop server (step 2). The location server is a non-SIP server that stores information about the next hop server for different users and returns the IP address of the machine where callee can be found (step 3). On getting this IP address, the proxy server forwards the INVITE message (step 5) to the host machine. After the UAS has been reached, it sends an OK response back to the proxy server (step 8), assuming that the callee wants to accept the call. The proxy server, in turn, sends back an OK response to the client (step 9). The client then confirms that he or she has received the response by sending an ACK (step 10). A full-fledged multimedia session is now initiated between the two participants. At the end of this session, the callee sends a BYE message to the caller (step 11), which in turn ends the session with another BYE message (step 12). Note that we have skipped the TRYING and RINGING message exchanges (step 4, step 6, and step 7) in the above explanation.

7.5 Quality of Service Architecture for Third-Generation Cellular Systems

Over the past decade, a phenomenal growth in the development of cellular networks around the world has occurred. Wireless communications technology has evolved over these years, from a simple first-generation (1G) analog system supporting only voice (e.g., AMPS, NMT, TACS, etc.) to the current second-generation (2G) digital systems (e.g., GSM, IS-95, IS-136, etc.) supporting voice and low-rate data. We are still evolving towards the third-generation (3G) digital system (e.g., IMT-2000, cdma2000 etc.) supporting multimedia (Garg, 1999).

IMT-2000/UMTS is the 3G specification under development by the ITU that will provide enhanced voice, data, and multimedia services over wireless networks. In its current state, the plan is for IMT-2000 to specify a **family of standards** that will provide at least a 384 kbps data rate at pedestrian speeds, 144 kbps at mobile speeds, and up to 2 Mbps in an indoor environment. Numerous standard bodies throughout the world have submitted proposals to the ITU on UMTS/IMT-2000.

In 1997, Japan's major standards body, the Association for Radio Industry and Business (ARIB), became the driving force behind a 3G radio transmission technology known as wide-band CDMA (WCDMA) (Steinbugl and Jain, 1999). In Europe, the European Telecommunications Standards Institute (ETSI) Special Mobile Group (SMG) technical subcommittee has overall responsibility for UMTS standardization. ETSI and ARIB have managed to merge their technical proposal into one harmonized WCDMA standard air interface. In the United States, the Telecommunications Industry Association (TIA) has proposed two air interface standards for IMT-2000, one based on CDMA and the other based on TDMA. Technical committee TR45.5 in TIA proposed a CDMA-based air interface, referred to as cdma2000, that maintains backward compatibility with existing IS-95 networks. The second proposal comes from TR45.3, which adopted the Universal Wireless Communications Consortium's (UWCC) recommendation for a 3G air interface that builds off existing IS-136 networks. Last but not least, the South Korean Telecommunications Technology Association (TTA) supports two air interface proposals, one similar to WCDMA and the other to cdma2000.

In the discussion that follows, we describe the layered QoS architecture that has been adopted by the UMTS standard (Garg, 1999; Dixit *et al.*, 2001). In UMTS there is a clear separation between the radio access network (called UMTS Terrestrial Radio Access Network or UTRAN), which comprises of all the air interface related functions (like medium access), and the Core Network (CN) that comprises switching- and control-related functions. Such separation allows both the CN and the radio access network to evolve independently of each other. UTRAN and CN exchange information over the Iu air interface.

Network services are end-to-end (i.e., from Terminal Equipment (TE) to another TE). An end-to-end service has certain QoS requirements that are provided to the user by the network (called network bearer service in telecommunication terminology). A network bearer service describes how a given network provides QoS and is set up from the source to the destination. The UMTS bearer service layered architecture is illustrated in Figure 7.12. Each bearer service at layer N offers its service by using the services provided to it by the layer $(N-1)$.

In the UMTS architecture, the end-to-end bearer service has three main components: the **terminal equipment** (mobile terminal) TE/MT local bearer service, the **external local bearer service**, and the **UMTS bearer service**.

The TE/MT local bearer service enables communication between the different components of a mobile station. These

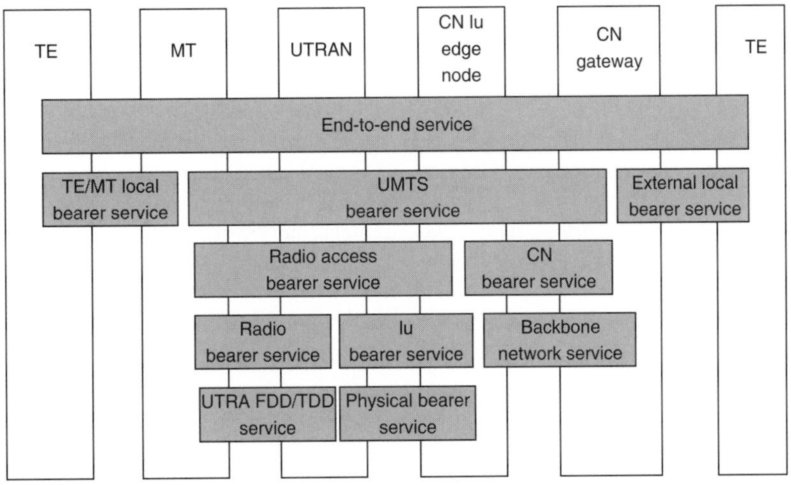

FIGURE 7.12 UMTS QoS Architecture. Adapted from Crow *et al.* (1997).

components make up an MT, mainly responsible for the physical connection to the UTRAN through the air interface, and one or several attached end user devices, also known as TEs. Examples of such devices are communicators, laptops, or traditional mobile phones.

The external bearer service connects the UMTS core network and the destination node located in an external network. This service may use IP transport or other alternatives (like those provided by IntServ, DiffServ, or MPLS).

The UMTS bearer service uses the radio access bearer service (RAB) and the core network bearer service (CN). Both the RAB and CN reflect the optimized way to realize UMTS bearer service over the respective cellular network topology, taking into account aspects such as mobility and mobile subscriber profiles. The RAB provides confidential transport of signaling and user data between the MT and the **CN Iu edge node** with the QoS negotiated by the UMTS bearer service. This service is based on the characteristics of the radio interface and is maintained even for a mobile MT.

The CN service connects the UMTS **CN Iu edge node** with the **CN gateway** to the external network. The role of this service is to efficiently control and use the backbone network to provide the contracted UMTS bearer service. The UMTS packet CN shall support different backbone bearer services for a variety of QoS options.

The RAB is realized by a radio bearer service and an Iu-bearer service. The role of the radio bearer service is to cover all the aspects of the radio interface transport. This bearer service uses the UTRA frequency/time–division duplex (FDD/TDD). The Iu bearer service provides the transport between the UTRAN and CN. Iu bearer services for packet traffic provide different bearer services for different levels of QoS.

The CN service uses a generic backbone network service. The backbone network service covers the layer 1 and layer 2 functionality and is selected according to the operator's choice to fulfill the QoS requirements of the CN bearer service. The backbone network service is not specific to UMTS but may reuse an existing standard.

References

Arora, R., and, Jain, R. (1999). *Voice over IP. Protocols and standards.* ftp://ftp.netlab.ohio-state.edu/pub/jain/courses/cis788-99/voip_protocols/index.html

Avramovic, Z. (1992). *Policy-based routing in the defense information system network. IEEE Military Communications Conference 3,* 1210–1214.

Banikazemi, M., and Jain, R. (1999). *IP multicasting: Concepts, algorithms, and protocols.* http://ftp.netlab.ohio-state.edu/pub/jain/courses/cis788-97/ip_multicast/index.htm

Bertsekas, D., and Gallager, R. (1987). *Data networks.* Englewood Cliffs, NJ: Prentice Hall.

Blake, S., Black, D., Carlson, M., Davies, E. *et al.* (1998). *An architecture for differentiated services* RFC2475.

Braden, R., Zhang, L., Berson, S., Herzog, S., and Jamin, S. (1997). *Resource reservation protocol (RSVP): Version 1 functional specification.* RFC2205.

Cavusoglu, B., Schonfeld, D., and Ansari, R. (2003). Real-time adaptive forward error correction for MPEG-2 video communications over RTP networks. *Proceedings of the IEEE International Conference on Multimedia and Expo. 3,* 261–264.

Chapman, N., and Chapman, J. (2000). *Digital multimedia.* New York: John Wiley & Sons.

Crow, B.P., Widjaja, I., Kim, J.G. and Sakai, P.T. (1997). IEEE 802.11: Wireless local area networks. *IEEE Communications Magazine.* 116–126.

Dixit, S., Guo, Y., and Antoniou, Z. (2001). Resource management and quality of service in third-generation wireless networks. *IEEE Communications Magazine,* 125–133.

Furht, B. (1996). *Multimedia tools and applications.* Norwell, MA: Kluwer Academic Publishers.

Garg, V.K., (1999). *IS-95 CDMA and CDMA2000.* Englewood Cliffs, NJ: Prentice Hall.

Garg, V.K., and Yu, O.T.W. (2000). Integrated QoS support in 3G UMTS networks. *IEEE Wireless Communications and Networking Conference, 3,* 1187–1192.

Handley, M., and Jacobson, V. (1998). SDP: Session description protocol. RFC2327.

Handley, M., Perkins, C., and Whelan, E. (2000). Session announcement protocol. RFC2974.

Johnston, A.B. (2000). *Understanding the session initiation protocol.* Artech House.

Kent, S., and Atkinson, R. (1998). *Security architecture for the internet protocol.*

Kessler, G. (2002). *An overview of cryptography.* http://www.garykessler.net/library/crypto.html

Kinsner, W. (2002). Compression and its metrics for multimedia. *Proceedings of First IEEE International Conference on Cognitive Informatics,* 107–121.

Kuipers, F., Mieghem, P.V., Korkmaz, T., and Krunz, M. (2002). An overview of constraint-based path selection algorithms for QoS routing. *IEEE Communications Magazine,* 50–55.

Kurose, J., and Ross, K. (2001). *Computer networking: A top-down approach featuring the Internet.* Reading, MA: Addision Wesley.

Labrador, M., and Banerjee, S. (1999). Packet dropping policies for ATM and IP networks. *IEEE communication surveys. 2*(3), 2–14.

Leigh, J., Yu, O., Schonfeld, D., Ansari, R. et al., (2001) Adaptive networking for teleimmersion. *Proceedings of the Immersive Projection Technology/Eurographics Virtual Environments Workshop.*

Leon-Garcia, A., and Widjaja, I. (2000). *Communication networks: Fundamental concepts and key architectures.* New York: McGraw-Hill.

Liu, H., and Mouchtaris, P. (2000). Voice over IP signaling: H.323 and beyond. *IEEE Communications Magazine,* 142–148.

Mulabegovic, E., Schonfeld, D., and Ansari, R. (2002). Lightweight streaming protocol (LSP). *ACM Multimedia Conference,* 227–230.

Oppliger, R. (1998). Security at the Internet layer. *Computer 31*(9), 43–47.

Perkins, C.E. (1998). Mobile networking through mobile IP. *IEEE Internet Computing 2*(1), 58–69.

Rosen, E., Viswanathan, A., Callon, R. (2001). Multiprotocol label switching architecture. RFC3031.

Sahasrabuddhe, L.H., and Mukherjee, B.M. (2000). Multicast routing algorithms and protocols: A tutorial. *IEEE Network*, 90–102.

Salomon, D. (1998). *Data compression: The complete reference.* Berlin: Springer.

Schonfeld, D. (2000). Image and video communication networks. In A. Bovik (Ed.), *Handbook of image and video processing.* San Diego, CA: Academic Press.

Schulzrinne, H., and Rosenberg, J. (2000). The session initiation protocol: Internet-centric signaling. *IEEE Communications Magazine*, 134–141.

Schulzrinne, H., Casner, S., Frederick, R., and Jacobson, V. (1996). RTP: A transport protocol for real-time applications. RFC1889.

Schulzrinne, H., Rao, A., and Lanphier, R. (1998). Real-time streaming protocol (RTSP). RFC2326.

Steinbugl, J.J., and Jain, R. (YEAR). *Evolution toward 3G wireless networks.* At ftp://ftp.netlab.ohio-state.edu/pub/jain/courses/cis788-99/3g wireless/index.html

Steinmetz, R. (1995). Analyzing the multimedia operating system. *IEEE Multimedia 2*(1), 68–84.

Striegel, A., and Manimaran, G. (2002). A survey of QoS multicasting issues. *IEEE Communications Magazine*, 82–87.

Su, J., Hartung, F., and Girod, B. (1999). Digital watermarking of text, image, and video documents. *Computers & Graphics 22*(6), 687–695.

Sun, W., and Jain, R. (1999). *QoS/policy/constraint-based routing.* ftp://ftp.netlab.ohio-state.edu/pub/jain/courses/cis788-99/qos_routing/index.html.

Tanenbaum, A. (1996). *Computer networks, 3e.* Eaglewood Cliffs, NJ: Prentice Hall.

Tang, N., Tsui, S., and Wang, L. (1999). *A survey of admission control algorithms.* http://www.cs.ucla.edu/~tang/

Thom, G.A. (1996). H.323: The multimedia communications standard for local area networks. *IEEE Communications Magazine 34*(12), 52–56.

Watkinson, J. (2001). *The MPEG handbook: MPEG-I, MPEG-II, MPEGIV.* Boston: Focal Press.

White, P.P. (1997). RSVP and integrated services in the Internet: A tutorial. *IEEE Communications Magazine, 35*(5), 100–106.

Wolf, L.C., Griwodz, C., and Steinmetz, R. (1997). Multimedia communication. *Proceedings of the IEEE 85*(12), 1915–1933.

Wu, C., and Irwin, J. (1998). *Multimedia computer communication technologies.* Englewood Cliffs, NJ: Prentice Hall.

Yu, O., and Khanvilkar, S. (2002). Dynamic adaptive guaranteed QoS provisioning over GPRS wireless mobile links. *ICC2002 2,* 1100–1104.

8

Fault Tolerance in Computer Systems—From Circuits to Algorithms*

Shantanu Dutt,* Federico Rota,*,† Franco Trovo,*,† and Fran Hanchek‡
*Department of Electrical and Computer Engineering, University of Illinois at Chicago, Chicago, Illinois, USA
†Politecnico di Torina, Italy
‡Intel Corporation, Portland, Oregon, USA

8.1 Introduction

Computer and digital systems are becoming ubiquitous in all aspects of human endeavor, from everyday living (e.g., embedded systems in VCRs/VCDs and microwaves, personal computers) to business transactions (e.g., stock exchanges, efficient business-to-business transactions, economic predictions) to engineering and scientific applications (e.g., design of complex computer chips, mapping the human genome, numerical and non-numerical computations for various aspects of scientific research, weather prediction, modeling and simulation in astronomy and astrophysics, control of complex machine tools) and to life- and mission-critical systems (e.g., avionics, space-mission systems). Simultaneously, the complexity of these computer and digital systems are increasing every few months, and the electronic components (transistors and wires) used to design them are reducing in feature sizes dramatically over the same span. Current feature sizes have reached 90 nanometers, and further decreases are anticipated. Simultaneously, significant efforts are being undertaken in research on nanoelectronics, which will take feature sizes to the next level of compaction: nanometers. Voltage levels for these systems have also dropped significantly from the classical 5 V to its current levels of 1.1 V and lower. These developments augur well for more complex and faster processing on very large-scale integrated (VLSI) chips.

A crucial metric in which these rapidly advancing systems have fallen short is **reliability**. As feature sizes decrease, it

* This work was supported in part by the U.S. Department of Defense under the MURI grant F49620-01-1-0436.

becomes more likely that some latent fabrication defect on a chip shows up during operation (e.g., a transistor goes into permanent conduction mode, two wires placed too close to each other in violation of spacing requirements may get shorted during operation due to electromigration) (**permanent faults**)or some external disturbance such as alpha-particle or electromagnetic radiation can change voltage values in memory cells or interconnects (**soft errors/transient faults**). Reduction of voltage levels also means that digital circuits are more susceptible to external or internal noise, such as cross talk causing logic or delay faults. The study and development of techniques to detect faults in the operation of modern computer and digital systems and to reconfigure around them or correct them in some manner so that either the system performs correctly in the presence of such faults (**fault-tolerant** operation) or shuts down safely in the presence of irrecoverable errors (**fail-safe** operation) are thus exceedingly important. Furthermore, it is also wise to now have a somewhat different mind-set in designing these complex systems, namely, to invest some design time and chip real-estate to improving reliability of these systems, thereby increasing the dependability of a variety of applications that run on these systems.

Another dimension of system complexity, and thus, by definition, lower reliability (more things can fail) stems from multiprocessing. Currently, many critical and very computationally intensive applications (e.g., weather forecasting, environmental monitoring, and providing advanced prosthetic devices for the disabled) employ a number of multiple processors in either general-purpose multicomputer boxes or in embedded computing environments. The reliability problem of such complex computers increases by orders of magnitude. Many such multiprocessor systems have interconnection networks with topologies like meshes, toruses, and hypercubes that are integral to the correct and nondegraded operation of the overall system. In these systems, faulty components need to be reconfigured using spare ones, and structural integrity of the system needs to be preserved; that is, these systems need **structural fault tolerance** (Banerjee *et al.*, 1986; Bruck *et al.*, 1993a, 1993b; Chau and Liestman, 1989; Chung *et al.*, 1983; Dutt and Hayes, 1990, 1991, 1992, 1997; Dutt, 1993; Dutt and Mahapatra, 1997; Hayes, 1973; Mahapatra and Dutt, 2001; Raghavendra *et al.*, 1984; Rennels, 1986; Roy-Chowdhury *et al.*, 1990). A non fault-tolerant system consisting of a large number of highly reliable processors *can* have low system reliability because even a single component failure renders the system faulty or significantly degraded in performability. Thus, a 1024-processor system consisting of highly reliable individual components, each with a mean-time-to-failure (MTTF) of 85.34 years, has a system MTTF of only one month (Dutt and Mahapatra, 1997)! Large multiprocessor servers that power e-commerce and e-business enterprises have 99.9 to 99% availability or 8 to 80 hours of downtime per year (Patterson *et al.*, 2002). This translates to a $200,000 loss per hour for an Internet service like Amazon to a $6,000,000 loss per hour for a stock brokerage firm. It has been estimated that due to such downtime and associated system maintenance and recovery costs, the total cost of ownership is five to ten times or more compared to hardware and software purchase price (Patterson *et al.*, 2002). Consequently, system dependability today is a critical requirement for high-performance multiprocessor systems. This chapter focuses primarily on fault-tolerance techniques for uniprocessor computer systems and will not discuss multiprocessor fault-tolerance methods. The reader is directed to the various works cited previously for structural fault-tolerance methods for multiprocessor systems. There also exists a considerable amount of work on software-based fault-tolerance techniques, primarily **checkpointing** and **rollback recovery** for multiprocessor and distributed-processor systems that the reader can also peruse: Chandy and Ramamoorthy, Elnozahy *et al.* (2002), Huang and Kintala (1993), Koo and Toveg, Li and Fuchs (1990), Long *et al.* (1992), Ssu *et al.* (1999), and Wang *et al.*, (1995). Both classes of fault-tolerance techniques, structural/hardware and software/rollback-recovery are important for multiprocessor dependability.

8.1.1 Dependability and Fault-Tolerance Definitions

The trustworthiness of a computer system is characterized by its **dependability**. A system is dependable if "reliance can justifiably be placed on the service it delivers" (Laprie, 1992). There are many aspects under which a system can be dependable: two of the most important ones are **reliability** and **safety**. The first, reliability, is the ability to guarantee the continuity of the service provided, and the last, safety, is the certainty of avoidance of catastrophic consequences on the environment, either people or the system itself (Yu and Johnson, 2002; Johnson, 1989; Prasad, 1989). The reliability of a system is a time-dependent function and, unfortunately, cannot be always assured. System failures occur when the service of the system does not comply with the specification, and the behavior of the system deviates from that required by the user.

System **failures** are caused by erroneous transitions of the system state, due to either the failure of a system component or erroneous design. An **error** is a defective value in the system state that is liable to lead to failure: an error can lead to a failure depending on the system composition (applied redundancy) or on the activity of the system (an error may be overwritten). The adjudged or hypothesized cause of an error is a **fault**. This relation can be schematized as in Figure 8.1.

The types of faults and relative sources can be classified by nature, origin, and persistence (Yu and Johnson, 2002;

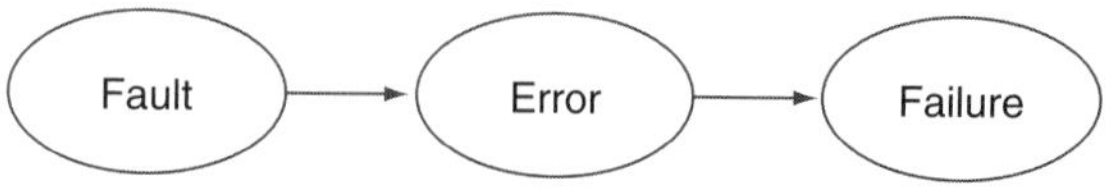

FIGURE 8.1 Fault, Error, and Failure Chain

Gunneflo *et al.*, 1989). The nature of a fault can be accidental or intentional, and the origin can be physically related or human-made. Taking into account the system boundaries, **internal faults** are parts of the system state, whereas **external faults** result from the interaction of the system with its physical environment (temperature, radiation) or human environment (operator).

A **permanent** fault is a fault whose presence is not related to internal or external time conditions and is usually caused by irreversible component damage. **Transient** faults are present for a limited amount of time and are usually triggered by environmental conditions, such as power-line fluctuation, electromagnetic interferences, or other radiation. These faults rarely do any lasting damage to the component affected although they can induce an erroneous state in the system. Finally, **intermittent** faults appear, disappear, and reappear repeatedly and are caused by unstable hardware or varying hardware state.

A fault is **active** when it produces an error and is **dormant** otherwise. An active fault is an external fault or a previously dormant one that was activated by the computation of the system. An error is **latent** prior to its detection and is **detected** if it is recognized by some mechanism.

To increase the dependability of a system, many strategies have been proposed in the literature for the different phases of the computing system development. In the design and implementation phase, **fault avoidance** can be achieved by proven design methodologies and technologies and by formal verification. In the test and debugging phase, **fault removal** can be performed. Both methods can be referred to as fault prevention and can avoid fault occurrence or introduction. In the operational phase, **fault-tolerance** methods are applied to provide a proper system service in spite of faults (Lee and Anderson, 1990).

8.1.2 Some Basics of Error Detection and Fault Tolerance

Detection mechanisms have to find out whether the actual state of the system is erroneous. To this end, the original system is supplemented by additional components that perform checks. Ideal checks are based solely on the specification and check the complete behavior of the system independently from the checked system to avoid the effects of errors to the checks (Lee and Anderson, 1990). The complexity of the system and the need of accessing the information to be checked usually prevents the implementation of an ideal checker. Less strict checks do not guarantee the absence of errors, but the results can still be accepted with a high probability.

The main metrics for error detectors are **hardware overhead** required to implement the check, **error latency** (time period between the occurrence of an error and its detection), and **error coverage** (percentage of errors that are detected).

Fault-tolerant operations and error detection are based on some form of redundancy and use of extra elements:

- **Hardware redundancy**: presence of extra hardware or data components (e.g., replication of the processor)
- **Information redundancy**: complementary information added to the original one (e.g., checksum, parity bits, error detection, and correction codes)
- **Functional or software redundancy**: presence of additional functions (e.g., *N*-version programming, reverse checking) for checking the results of a computation
- **Time redundancy**: additional time that is used to deliver the service of the system (e.g., multiple execution of the operation)

Error detection can be performed off-line, when the system is inactive (or in an idle state), or online, when the system is operational. The latter is referred to as **concurrent error detection**. The detection of transient faults requires continuous and concurrent error detection. Off-line checks usually include the systematic testing of the system by supplying special input test patterns to the components and investigating the outputs. The purpose is either to check whether the component satisfies the specification or to reveal some faults. These types of checks have high coverage but are expensive in resources and time, and so they are rarely implemented as diagnostic checks during the operational period of the system. For high availability-requirement systems, such as Web and database servers, the operational period is almost 100% of the time. Thus, concurrent error detection becomes very important for ensuring the dependability of such systems.

Fault-tolerance (FT) as well as fault-detection (FD) schemes can be implemented at various levels of a computing system. At the highest level is **algorithm/application based fault tolerance** (ABFT), where some property of the computation or algorithm is used to check the correctness of the results and possibly correct erroneous ones (Assaad and Dutt, 1992; Banerjee *et al.*, 1990; Boley and Luk, 1991; Boley *et al.*, 1995; Dutt and Assaad, 1996; Dutt and Boley, 1999; Huang and Abraham, 1984; Jou and Abraham, 1986; Luk and Park, 1988; Roy-Chowdhury and Banerjee, 1993). The most prevalent form of redundancies applied here are information and time redundancies. The next level is the **system level**, where the control and data flow of a computation being executed by a computer are concurrently checked for "reasonableness" by various FD schemes (DeLord and Saucier, 1991; Lu, 1982; Mahmood and McCluskey, 1988; Ohlsson and Rimen, 1995; Rota, 2002; Trovo, 2002). If the computation flow is in the given high-level specifications of the program, it is deemed correct. System level FT techniques include rolling back the state of the computer to the state prior to the occurrence of a detected error and restarting from that correct state, called **microrollback** (Tamir and Tremblay, 1990; Trovo, 2002). System-level FT and FD techniques primarily use hardware redundancy. The third and most detailed level of FT and FD is at the **circuit level**. These include FT and FD techniques for random-access memories (RAMs) (Tanner, 1984; Mazumder and

Chakraborty, 1996; Mazdumer and Yih, 1993) and arithmetic circuits (Al-Arian and Gumusel, 1992; Avizienis, 1973; Chen *et al.*, 1992; Dutt and Hanchek, 1999; Hsu and Swartzlander, 1992; Johnson *et al.*, 1988; Johnson, 1989; Laha and Patel, 1983; Lo *et al.*, 1993; Nicolaidis, 1993; Patel and Fung, 1982, 1983; Rao, 1968, 1974; Sparmann and Reddy, 1994), and general logic circuits (Johnson, 1989; Lala, 1985; Pradhan, 1990; Roth, 1975). Circuit-level FT and FD mechanisms generally involve all three forms of redundancies: hardware, information, and time. A highly dependable computer system is one that will have FT/FD mechanisms implemented at least in the last two levels (system and circuit levels).

8.1.3 Fault Models

We focus here on faults resulting from physical problems, discounting specification, design, implementation, and operator mistakes in the system. The causes of physical faults in computer and digital systems are many and varied. They range from **fabrication defects** (e.g., shorting of the drain and source of an nMOS or pMOS transistor), to **operational faults** caused by extreme operation conditions (e.g., electromigration in adjacent interconnects from high current density resulting in shorting of the interconnects, overheating in a portion of a chip resulting in damage or unpredictable behavior of transistors), and to **external disturbance** (e.g., alpha-particle or electromagnetic radiation that can change voltage values on interconnects and capacitances such as those present in dynamic RAM cells and in environmental extremes of temperature, pressure, and moisture).

To design effective FD and FT techniques, it is necessary to encapsulate the primary effects of such faults in a manageable and general **fault model** that accounts for the behavior for most, if not all, fault types. Mechanisms to tackle faults can then be designed in a general and effective way by addressing the fault model rather than the very large number of faults underlying that model (Hayes, 1985; Johnson, 1989). Two classical fault models that have proven reasonably effective are the **logical stuck-at-fault** model used at the logic circuit level and the **transistor stuck-at-fault** model used at the transistor circuit level.

In the logical stuck-at-fault model, it is assumed that a fault causes a module to respond as if one of its inputs or outputs is physically stuck at a logic 1 (called **stuck-at-1**) or 0 (called **stuck-at-0**) without its basic functionality being altered (Johnson, 1989; Kohavi, 1978). In other words, this fault model assumes that the effect of faults is the same as that of interconnects being stuck-at-0/1. Consider, for example, an AND gate with two inputs A and B and with the A input wire being shorted to a ground wire due to electromigration. This will cause both the A input wire (and the AND gate's output) to be always at logic 0, and thus this physical fault can be modeled as a stuck- at-0 fault of A. Another example is that of a two-input CMOS NAND gate that has two pMOS transistors in parallel connected to the supply voltage and two nMOS transistors in series connected to ground (Johnson, 1989). If the source and drain of one of the pMOS transistors is shorted, then it always conducts irrespective of its gate voltage; the output of the NAND gate is always connected to the supply voltage and thus always at logic 1. Thus, this fault can be modeled by the NAND gate output being stuck-at-1.

This fault model, however, does not cover all types of faults. One example is that of two signal wires X and Y being shorted to each other (called a **bridging fault**). If the driver strength of X is significantly greater then that of Y, then Y's logic value will follow X's, which can take on either 0 or 1 values depending on the input to the circuit. Thus Y is neither stuck-at-0 nor stuck-at-1 permanently, though, of course, it can cause an error in the circuit output. This fault cannot be modeled by the logical stuck-at-fault model. Another example is that of a CMOS NOR gate with inputs A and B. This gate has two pMOS transistors in series connected to the supply voltage and two nMOS transistors in parallel connected to ground. The A input feeds a pMOS and an nMOS transistor. If the branch of the A wire that feeds the nMOS transistor is broken, then the gate of this transistor is at logic 0, and it never conducts. When the inputs are A, $B = 1, 0$, the pMOS network is not conducting, and the nMOS network is not connected to ground (the latter due to the wire break—the correct operation would have been for the nMOS network to conduct). This leaves the output of the NOR gate connected to neither supply voltage nor ground, and it assumes the voltage is stored at its load capacitance, which is its previous value. Thus, this fault does not manifests itself as the NOR gate's output (or for that matter any of its inputs) being either stuck-at-0 or stuck-at-1.

Some of the inadequacies of the logic stuck-at-fault model are overcome by the transistor stuck-at-fault model that assumes a fault's behavior is akin to one or more transistors always conducting (**stuck-at-on**) or always in a nonconducting state (**stuck-at-off**) irrespective of gate inputs. For example, the above wire-break fault in the NOR gate can be modeled as the corresponding nMOS transistor stuck-at-off. Even this model, however, cannot simulate a bridging fault. A clear disadvantage of the transistor stuck-at-fault model vis-a-vis the logic stuck-at-fault model is the additional complexity in terms of the number of components that must be represented, manipulated, and simulated (Johnson, 1989).

In the sequel, we model faults using the logic stuck-at fault model, which, in spite of its limitations, covers a large number of physical fault types.

The rest of this chapter is organized as follows. We discuss various FD and FT techniques at various levels of application, down from the circuit up to the application level. For each of these discussions, we present various important work in that area and then give a more detailed discussion of work done in our group. In Section 8.2, we present FD and FT techniques for arithmetic circuits, whereas in Section 8.3, we explain methods for field-programmable gate arrays that can be considered a mix of circuit-level and system-level techniques. In Section 8.4, we discuss various control-flow checking methods using a watchdog processor, and in Section 8.5, we present processor micro rollback techniques for recovering from detected errors like CFC errors. In Section 8.6, we describe ABFT methods for achieving FD and/or FT at the application level. We conclude in Section 8.7 with a summary of our findings.

8.2 Fault Detection and Tolerance for Arithmetic Circuits

In this section, we discuss various FD and FT techniques for arithmetic circuits. Many previous designs are only able to detect one error in the operation and are classified as one-fault detecting (1-FD). If the error is caused by a transient fault, the arithmetic circuit can still be used. Otherwise, the circuit and possibly the entire processor in which it is embedded become useless and must be replaced. Single error detecting designs include duplication, time-redundancy techniques like recomputation with shifted operands (RESO) (Patel and Fung, 1982, 1995), or recomputation with duplication with comparison (REDWC) (Johnson *et al.* 1988) and information redundancy methods that use residue codes (Avizienis, 1973; Rao, 1968), Berger codes (Lo *et al.*, 1993), *AN* codes (Avizienis, 1973), or parity prediction (Nicolaidis, 1993). While time redundancy techniques have a very high time overhead, duplication has high hardware overhead, and coding methods have appreciable time as well as hardware overhead. Duplication is the classic hardware redundant method for error detection and simply compares the results of two copies of a circuit. It is fast but cannot correct an error. RESO (Patel and Fung, 1982, 1995) is a well-known time redundant error detection method. After an original computation in RESO, operands are shifted left so that a fault, if present, will affect a different output bit, and then a second computation is performed. The result is shifted right, and any difference from the first result indicates an error. The time overhead is 100%, and no error correction is possible without additional recomputations. Another error detecting technique is REDWC (Johnson *et al.*, 1988), which divides a circuit into identical upper and lower partitions. The lower halves of the operands are applied to both partitions, and results are compared. For the upper portions of the operands, the procedure is repeated. If the partitions are linearly dependent, as in a ripple-carry adder, then time overhead is due only to comparisons and can be very low. If the partitions are independent, however, then the time overhead becomes 100%.

Arithmetic codes provide information redundancy, where extra information is encoded with the operands, and this information is processed concurrently with the operands. For residue codes (Avizienis, 1973; Rao, 1968; Sparmann and Reddy, 1994), the additional information is a **residue**, which is generated by performing a **mod** operation on each operand. Processing these residues yields a value that matches the residue of the result if there is no error. Area overhead can be reasonable when applied to a large circuit such as a multiplier (Sparmann and Reddy, 1994) but is generally quite high for an adder. Residues mod $(2^i - 1)$ are considered the easiest to compute, and a tree of $(2^i - 1)$-bit end-around-carry adders is typically used for this purpose, but time overhead is very high, even for a multiplier. In Berger codes (Lo *et al.*, 1993), the extra information encoded is the number of zeros in an operand. This information, combined with information about the internal carries generated during the operation, is used to predict the number of zeros in the result. Berger check prediction methods have been applied to adders and array multipliers with time overheads of less than 100%, but area overhead is very high for low-complexity circuits like ripple-carry adders. *AN* coding (Avizienis, 1973) involves multiplying each data word *N* by a selected constant *A*. Results of arithmetic operations on these codes are checked by dividing by *A*; a nonzero remainder indicates an error. Although *A* is selected to make encoding and decoding easy, these operations are still hardware and time intensive. Parity prediction (Nicolaidis, 1993) uses the parities of the operands and the parity of the internal carries generated during an operation (Nicolaidis, 1993) to predict the parity of the result. A disagreement between predicted and actual result parities indicates an error. Since this method depends on the correctness of the internal carries, it must ensure this by generating them independently of the arithmetic functional unit, thus increasing the area overhead.

There have also been designs proposed that can correct a single error, and these are classified as one-fault tolerant (1-FT). These designs arise by using fault masking as in error-correcting RESO (Laha and Patel, 1983) and triple modular redundancy (TMR) (Johnson, 1989), using a hybrid redundancy scheme as in partitioned TMR (PTMR) (Arian and Gumusel, 1992; Chen *et al.*, 1992; Hsu and Swartzlander, 1992), or using arithmetic codes (Rao, 1971). In such schemes, the circuit can still operate correctly in the presence of a single permanent fault but will become useless when a second permanent fault occurs.

Error-correcting RESO (Laha and Patel, 1983) is essentially triple-time redundancy. The circuit and its operands are divided into at least three partitions. Successive recomputations are performed while shifting the operand portions such that each portion is recomputed into three different circuit partitions. In a circuit with dependent partitions, such as an adder, an erroneous carry may propagate an error between partitions, making it impossible for a given portion of the result to be computed correctly. To avoid this problem, comparison of intermediate results produces error syndromes that can identify the faulty partition and allow correct carries to be saved and then rerouted in successive recomputations. Variations on the technique trade off additional hardware with a fourth computation. Finally, correct result portions are located and assembled into a complete result. Hardware overhead can be quite reasonable for a sufficiently complex original circuit, but the necessity of at least three complete computations makes the time overhead excessively high (≥200%).

In TMR (Johnson, 1989), the classic hardware redundant technique for error correction, three copies of a module perform the same computation simultaneously. A majority voter selects the result upon which at least two of the modules have agreed. Time overhead is very small, but hardware overhead is greater than 200%. A related technique, PTMR, has been developed by three independent groups (Arian and Gumusel, 1992; Chen *et al.*, 1992; Hsu and Swartzlander, 1992), and is also known as hardware partition in time redundancy (HPTR) (Arian and Gumusel, 1992) and recomputing with triplication with voting (RETWV) (Hsu and Swartzlander, 1992). In this technique, three copies of a single partition of the circuit simultaneously process a portion of the operands, and the results are voted on, as in TMR. The remaining portions of the operands are processed similarly by the same triplicated partition. When there are three independent partitions, the time overhead is 200%. For linearly dependent partitions in adders and multipliers, time overhead can be lower, and area overheads range from under 100% to about 150%.

Some arithmetic codes can also provide error correction by generating unique error syndromes for each single-bit error (Rao, 1974). In the case of residue coding, this requires multiple residues, thus increasing the area. For *AN* coding, it precludes the use of "low cost" *A* values, thus increasing the time spent on decoding. Error correcting codes, therefore, have such high time and/or area overheads, making them uncompetitive with other methods.

8.2.1 The Reprocessing with Micro Delays Method

We now discuss the reprocessing with micro delays (REMOD) technique for designing FT arithmetic circuits that can detect errors and reconfigure around the fault for correct operation (Dutt and Hanchek, 1997). REMOD has much smaller time overheads than RESO and REDWC, and its hardware and time overheads are typically less than or comparable to methods that are only 1-FD, such as duplication or information redundancy.

A key feature of REMOD designs is that they are extendible to multiple fault tolerance, with only small increments of hardware and time—in some cases, the designs have virtually no increase in time. On the other hand, PTMR and the other techniques are not as easily extendible, if at all. As a result, the mean time to failure (MTTF) of processors using REMOD-based arithmetic circuits will increase substantially. This is especially crucial in computer systems used in life- and property-critical applications, such as avionics, spacecraft control, military applications, and the control of industrial processes and nuclear reactors. Not only are these environments hostile with regard to increasing fault probabilities, but maintenance is either very difficult or impossible to perform during the operation of these systems. This underscores the need for very high MTTF computer systems in these applications, and REMOD is a means to achieving this in the execution part of a system. It is also a means to increase the availability of more general purpose computer systems and to reduce their maintenance costs.

REMOD Operation

The main idea behind REMOD is to assign a cover v to each cell u of the circuit, where the cover is a functionally identical cell that checks the computation of u after both have finished their original computations. The **original computation** of a cell is the computation it performs in the non-FT circuit. All inputs to cell u are transmitted to its cover v after some delay that depends on the delay of the cells. After cell v repeats u's original computation (called v's **checking computation**), the corresponding outputs of this computation are compared to the original outputs of u. A mismatch indicates a fault in either u or v, and a diagnostic routine can determine which of the two is actually faulty. REMOD circuits are then reconfigured around the faulty cell to resume normal operation.

REMOD can be applied to circuits containing an array of independent cells, such as a carry-save adder, or an array of linearly dependent cells, such as a ripple-carry adder. REMOD has also been applied to partially dependent cells found in a multiplier by partitioning the circuit into sub-arrays of independent cells.

Array of Independent Cells

In Figure 8.2, each cell i is covered by cell $i + 1$. Suppose that all inputs are available to the circuit at time t and that the maximum propagation delay of a cell is δ. Then all original outputs are available from the primary cells at time $t + \delta$. At time $t + \delta$, the outputs are latched, and the inputs are forwarded to the cover cells for the checking computation. At time $t + 2\delta$, the checking outputs are available from the cover cells. These are compared with the latched original outputs, which have also been forwarded to the cover cells, and any error signals are latched.

To provide reconfigurability and multiple fault tolerance, it is necessary to forward inputs and outputs past faulty cells for recomputation and comparison. We refer to such forwarding as **pipelining** and illustrate the concept in Figure 8.2(B), where it is seen that cell 2 has become faulty. Each covering cell $j > 2$ *replaces* its dependent cell $j - 1$ in the sense that it now does the original computation of cell $j - 1$, and a spare cell replaces the last primary cell in the array. In the reconfigured circuit, the second computations (produced at $t + 2\delta$) of all cells $j \geq 2$ become the original outputs, as shown in Figure 8.2(B). The third computation of these cells is now the checking computation, and the pipeline continues to forward the original outputs at the proper time for comparison.

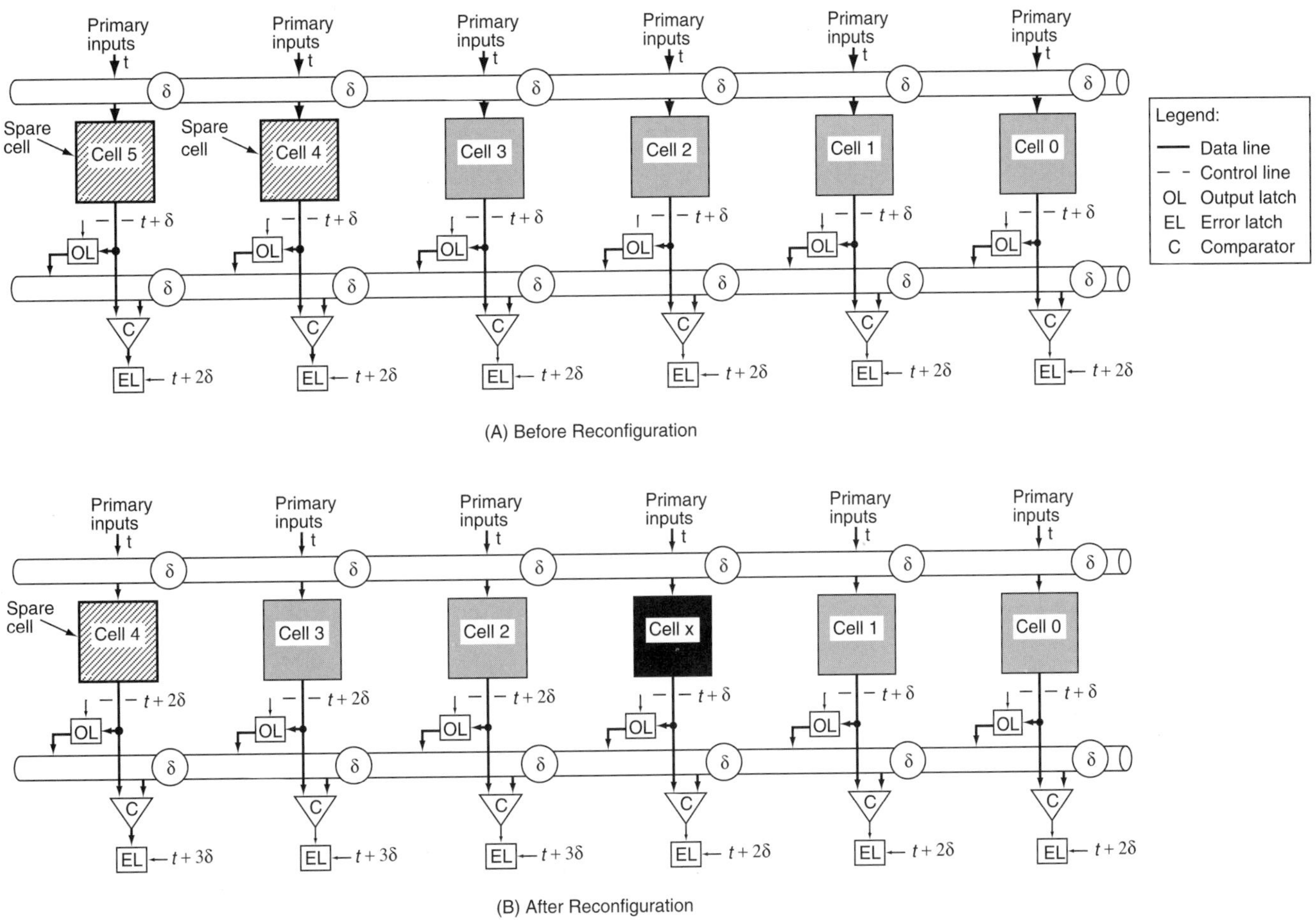

FIGURE 8.2 Array of Independent Cells. (A) This image shows an FT design for independent cells before reconfiguration. (B) Shown here is a reconfiguration around faulty cell 2; cell labels are those after reconfiguration.

Array of Dependent Cells

The FT designs for linearly connected dependent cells can be obtained by some simple modifications to the designs for independent cells presented in the previous paragraphs. Essentially, the fact that the original inputs of all cells are not available at the same time differentiates between arrays of independent and dependent cells.

Consider a ripple-carry adder for which each stage is a full-adder cell dependent on the carry out from the previous stage. Each cell can receive all of its original inputs δ time units only after the previous cell receives its inputs, and so it is not ready to perform a checking operation until after an additional δ time units. Checking inputs must then be forwarded to each cell 2δ time units after they have been input for the primary computation. The last checking computation completes (in the spare cell) 2δ time units after the last original computation, giving a time overhead factor of $2/n$ for error detection in an n-cell adder. This can be reduced to $1/n$ since the spare is not needed for an original computation and can thus perform a checking computation earlier.

Reconfiguration in a circuit of dependent cells is analogous to the method employed for independent cells. Inputs are forwarded past the faulty cell and the computations, original and checking, in all cells beyond the faulty cell are delayed by δ time units compared to those in the circuit that is not faulty.

Multiple FT can be obtained in either independent or dependent cells by adding more spares. The pipelines forward inputs (and outputs used for comparison) past the faulty cells, and the computations in those cells beyond the faults are delayed an additional δ time units each time the pipeline flows past a faulty cell. If there are $f \leq k$ faults in a k-FT REMOD design, then the original computation has a time overhead of $f\delta$ in the worst case. Pipelines implemented with banks of multiplexers rather than delay lines can reduce this overhead to 0, however (Dutt and Hanchek, 1997). In all cases, error detection takes δ more time after the original computation.

Adder and Multiplier Applications

The REMOD method was applied to a ripple-carry adder, which consist of fully dependent linearly connected cells (Dutt and

Hanchek, 1997). The cells are either 1-bit or 4-bit carry look-ahead (CLA) adders. REMOD was also applied to multipliers based on a Wallace tree and on a carry-save array for the summation of partial product terms (Dutt and Hanchek, 1997). The multipliers are partitioned into arrays of independent subcells, but the high time overhead normally associated with independent cells is masked through a clever application of REMOD. Original and checking computations overlap in time between adjacent rows of subcells as the computations proceed down the array or Wallace tree. The designs were laid out using the Berkeley MAGIC tools for CMOS. The multiplier designs were implemented in CMOS dynamic logic. Delay elements were implemented as strings of inverters. Error injection gates were inserted into the layout to simulate stuck-at faults. The layout tools allow geometry information to be extracted and converted to SPICE parameters. Simulations of the circuits using delay element pipelining were then performed using HSPICE. The simulations successfully tested the functions of error detection and reconfiguration, with selected nodes in the SPICE files for the circuits forced to logic *high* or *low* to simulate faults.

Adders

Table 8.1 compares the REMOD method for adders to other methods of fault tolerance. For REMOD adders, multiplexer forwarding was used, and data are presented both for cell output comparison and for output parity comparison (Dutt and Hanchek, 1997). Since most of the available information concerning area overheads for the various fault tolerance methods is given in terms of transistor counts, comparisons are made using these numbers.

Some coding methods can provide error correction, but, as shown in Table 8.1, the area and/or time overheads of several hundred percent are excessively high. Since error-correcting RESO is essentially triple-time redundancy, it is much slower than REMOD although it has a small advantage in area overhead. TMR is very fast, but REMOD has a significant advantage in area, and with similar area overhead REMOD can even tolerate three faults. Partitioned TMR is most comparable to 1-FT REMOD in terms of overhead, with both of these techniques having area overheads of about 100 to 150% and time overheads of 10 to 30%. The big advantage of REMOD, however, is that it can be extended to tolerate multiple sequential faults.

Multipliers

Parallel multipliers have more complex structures than parallel adders. For example, an array multiplier has $\Theta(n)$ diagonals of linearly connected full adders, where n is the input length in bits, whereas the Wallace tree multiplier has a tree structure. The trick to applying REMOD to such complex structures is to bit-slice them and add some components, if needed, to bit-slice $i+1$ so that it can cover bit-slice i. Each bit slice is then a cell of the circuit pertaining to the definition of covering and related concepts discussed earlier. Figure 8.3 shows a

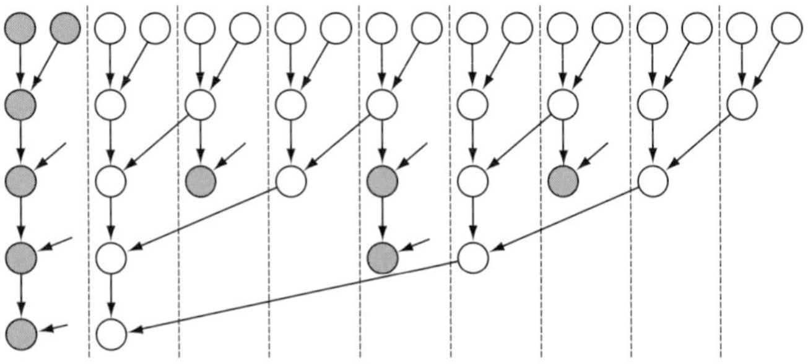

FIGURE 8.3 Bit-Slice-and-Add-(Component) Approach to Deriving a REMOD-Based 1-FD Generic Binary Tree Circuit

TABLE 8.1 Comparison of Area and Time Overheads in FT Adders

ADDER Fault tolerance method	Fault tolerance capability	Area % overhead 32-bit	Area % overhead 64-bit	Time % overhead 32-bit	Time % overhead 64-bit	Derivation notes
AN Coding[1]	1-FT	500+		800+		Functional analysis
Multiresidue Coding[1]	1-FT	600+		150+		Functional analysis
Error-correcting RESO[2]	1-FT	100+		200–300		Functional analysis[2]
TMR[3]	1-FT	250	250	3	2	Transistor count, group CLA[5]
PTMR[4]	1-FT	160	95	16	17	Transistor count, group CLA[5]
REMOD	1-FT	150	135	19	9	Transistor count, group CLA full cell output comparison
	2-FT	225	190	19	9	
	3-FT	300	245	19	9	
REMOD	1-FT	115	105	29	15	Transistor count, group CLA output parity comparison
	2-FT	190	160	29	15	
	3-FT	260	210	29	15	

[1] Rao (1974).
[2] Laha and Patel (1983).
[3] Johnson (1989).
[4] Al-Arian and Gumusel (1992); Chen *et al.* (1992); Hsu and Swartzlander (1992).
[5] Chen *et al.* (1992).

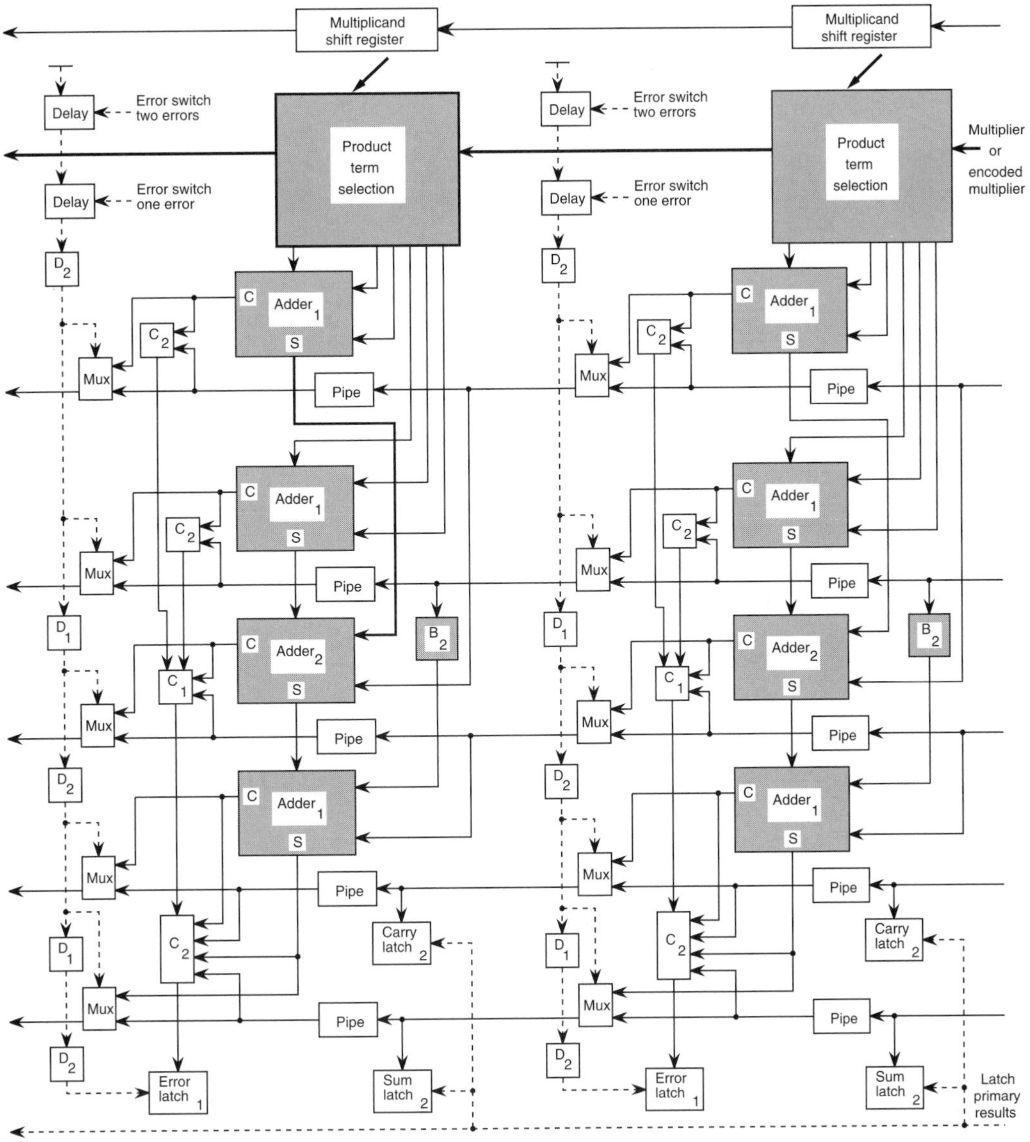

FIGURE 8.4 Two Bit Slices of a Six-Input, Three-Level Wallace Tree for a 6-Bit Multiplier. A subscript in a function block refers to the phase of that block; in two-phase dynamic logic, a circuit cannot drive another circuit of the same phase.

generic circuit with a binary-tree structure that has been bit-sliced with added spare components (shown shaded) in this manner. The last bit slice is composed only of spares. Such a structure with the appropriate input and output pipelines will be 1-FD and, with some slight modifications, would also be 1-FT.

Two bit slices of a 6-bit FT Wallace tree multiplier are obtained using the "bit-slice-and-add" approach, shown in Figure 8.4. A 16-bit version was also laid out. Table 8.2 compares the area and time overheads of REMOD multipliers to those of other FT multipliers. Of the error-correcting multipliers, error-correcting RESO has the lowest area overhead, but its time overhead of up to 300% is extremely high. TMR has by far the lowest time overhead, but its 200% area overhead is extremely high. Comparing REMOD to PTMR for array multipliers shows that the time overheads are much the same, reaching less than 10%. REMOD's area overhead, however, is better than 30% less than that of PTMR. The area overhead fraction for a REMOD Wallace tree is even lower, with a transistor overhead of 48% and a layout area overhead of only 36%. The time overhead fraction *is* higher because the Wallace tree is a faster circuit than a carry-save array. Finally, we note that the hardware overhead for both types of 2-FT REMOD multiplier is also lower than that of a 1-FT PTMR multiplier; for 3-FT REMOD multipliers, the overhead is comparable to that of a 1-FT PTMR and

TABLE 8.2 Comparison of Area and Time Overheads in FT Multipliers

Multiplier fault tolerance method	Fault-tolerance capability	Area % overhead		Time % overhead		Derivation notes
		16-bit	32-bit	16-bit	32-bit	
Error-correcting RESO[1]	1-FT	50	30	200–300		Gate count, array multiplier[2]
TMR[2]	1-FT	200	200	1	1	Transistor count, array multiplier[4]
PTMR[3]	1-FT	125	97	16	8	Transistor count, array multiplier[4]
REMOD	1-FT	83	62	18	9	Transistor count, CSA array only
	2-FT	115	81	29	15	
	3-FT	150	99	41	21	
REMOD	1-FT	69	48	60	43	Transistor count, Wallace tree only
	2-FT	83	54	100	71	
	3-FT	98	60	140	100	

[1] Laha and Patel (1983).
[2] Johnson (1989)
[3] Al-Arian and Gumuse (1992); Chen *et al.*, 1992; Hsu and Swartzlander, (1992).
[4] Produces half-length truncated product of the same length as the operands; Chen *et al.* (1992).

significantly lower than that of a 1-FT TMR multipliers. Thus, REMOD provides a very good balance between time and area efficiencies and provides greater fault tolerance than all the other methods.

Yield Improvement

In addition to increasing the reliability and availability of a circuit, improvement in chip yield is another important benefit of fault tolerance. With the aid of a model relating yield to circuit area and defect density, the yield improvement provided by a REMOD implementation has been analyzed taking into account the possibility of faults due to defects in the redundant portions of the circuitry (Dutt and Hanchek, 1997). Chip yield is defined as the percentage of usable chips. The Poisson yield model (Cunningham, 1990) was used for defect distribution where defects are assumed to occur uniformly and independently.[1] For a k-FT circuit, the yield is the sum of the circuit yields for each number of faulty cells up to k.

For adders, which have small, relatively simple cells, the yield improvement obtained is small. For a circuit with reasonably complex cells, however, very good yield improvement is obtained. This is shown in Figure 8.5 for 32-bit Wallace tree multipliers. It is also apparent in this figure that greater degrees of FT do provide greater yield improvement of well over 50%.

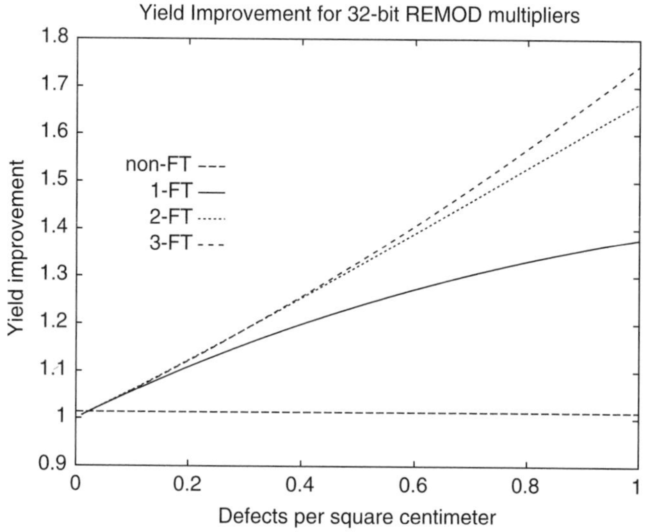

FIGURE 8.5 Yield Improvement for 32-bit REMOD Wallace Tree Multipliers with 2-Bit Cells

8.3 Fault Tolerance in Field-Programmable Gate Arrays

Field-programmable gate arrays (FPGAs) are now being used in production-scale quantities in various commerical and consumer products. Recent increases in speed and size (of the order of 300-MHz and 1-M gates) will expand this trend further. FPGAs have also found use in mission- and life-critical applications (e.g., the 1996 Mars Pathfinder mission) (Lach *et al.*, 1998). FPGAs form the core of adaptive computing platforms used in aircraft avionics, in which the FPGA hardware is rapidly reconfigured in run time to map the desired application to the platform. Current technology trends for FPGA devices are in the deep-submicron (DSM) regime with recent chips using 90 nanometer and seven metal layers. This trend, however, has also resulted in a decrease of fabrication yield (due to the extreme accuracy requirement of the process technology resulting in low error margins) as well as decreased reliability in operation (stemming, for example, from greater risk of electromigration and increased susceptibility to ionized

[1] While the Poisson model tends to underestimate yield when the product of defect density and chip area increases beyond unity, it is perfectly acceptable for the purposes of estimating yield improvement.

particle radiation). The larger die sizes also mean that there is more likelihood of failure. Thus, increasing the reliability of FPGA devices via testing and fault-tolerance techniques is necessary for achieving various goals ranging from increasing device fabrication yield (thus lowering costs) to increasing the availability and reliability of systems ranging from office products to mission- and life-critical systems.

As we mentioned in Section 8.1.1, the generic approach to increasing system reliability is to use some form of redundancy. However, redundancy at the chip level is not cost-effective for FPGAs, especially given the fact that such chips already lag behind in device density to custom ASIC chips. Furthermore, chip-level redundancy is not a feasible solution for various space-based and other systems for which weight, space, and power are at a premium and need to be optimized. Fortunately, FPGAs provide a fine granularity of replicated identical components (logic cells and interconnect segments) that can potentially be exploited to provide both testing and fault tolerance capabilities at low hardware and time overheads. We focus here on reprogrammable SRAM-based FPGAs (with a segmented wiring architecture) that are the most prevalent kind (e.g., Xilinx [Xilinx, 1994], Lucent ORCA [Lucent Technologies, 1998] and Altera Flex 8000 [Altera corporation, 1993] FPGA families) and also offer the most reprogramming flexibility needed for reconfiguration. Such a generic FPGA is shown in Figure 8.6.

Most fault-tolerance (FT) approaches for FPGAs make use of their inherent flexibility, replicability, and homogenity in the logic (PLB) and interconnect architecture. There are two broad classes of techniques: those requiring architectural modifications and those that are software-based.

One technique of the former type is adding spare rows and/or columns of cells and replacing every i-length-segment track with an $(i+1)$-length track and employing an extra such track to retain the original routing flexibility (Hatori, 1993). Another technique that assumes a nonconventional FPGA architecture results from work by Shnidman *et al.* (1998). The architecture required is one in which the entire routing area is composed of busses that run the entire span of rows or columns, with row busses being inputs only and column busses being outputs only—the interconnects are not segmented. Fault tolerance is then simply achieved by transposing configuration bits of a faulty PLB and that of the column busses to which it is connected to a spare PLB in its row and its corresponding column busses. However, bus-based FPGAs have several drawbacks, including inefficient use of interconnect resources, which means that only very sparsely wired designs—not typical of modern circuits—are implementable on such FPGAs, and timing delays for short nets are excessive, leading to slow designs.

One software-based method focuses on noting the locations of faulty cells and/or interconnects and rerouting the circuit to avoid using spares or unused cells/interconnects (McDonald *et al.*, 1991; Narasimhan *et al.*, 1994; Roy and Nag, 1995). This has been proposed for use in factory programmable gate arrays and potentially requires a different routing for every chip to which the user's circuit is mapped (Narasimhan *et al.*, 1994). This software-based method has also been proposed in the context of field reconfiguration for logic cell faults by McDonald *et al.* (1991) and for interconnect faults by Roy and Nag (1995). Requiring the layout tools to perform a new routing of a circuit for each new faulty cell or interconnect location encountered, however, puts a heavy burden on the user who must also keep track of all of the different routings for a given circuit design. Furthermore, in a mission- or life-critical system, it is desirable to perform fault reconfiguration quickly so that system function is suspended for only seconds. Complete rerouting of large FPGA circuits can take hours, which is unacceptable in such systems.

A newer approach of partial rerouting only in faulty tiles—a **tile** is a subarray of logic cells including a spare cell—has been proposed by Lach *et al.* (1998); this is a significant improvement over past full rerouting methods. In this technique, fast reconfiguration is achieved by precomputing all possible configurations of a tile (assuming one fault in it) and storing them for loading into the tile as and when a fault is detected. However, this technique has some of the other disadvantages of full rerouting methods. A rip-up-and-reroute approach for

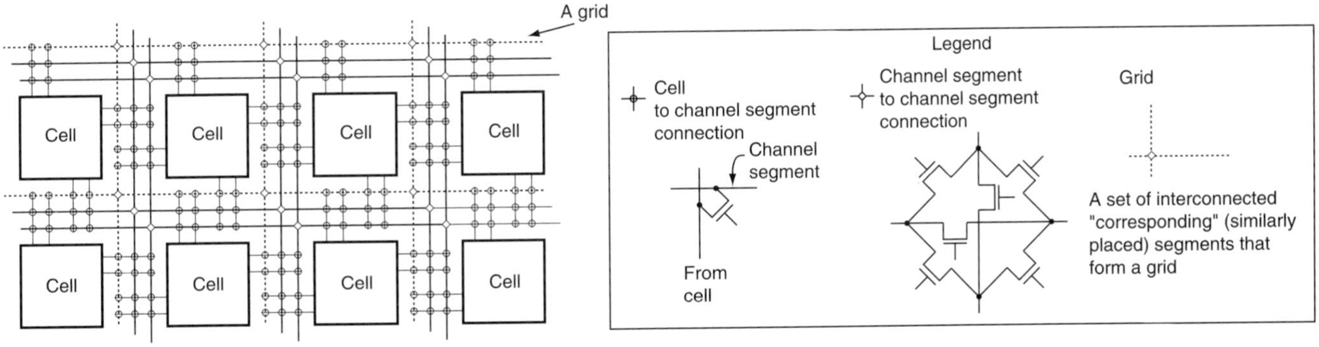

FIGURE 8.6 Segmented FPGA Routing Architecture with i-to-i Connection Switch Boxes

incremental routing was developed by Emmert and Bhatia (1998) in a later work. Some results have been presented on the success of rerouting and computation times that are reasonable.

8.3.1 The Static Node-Covering Approach

Another software-based FT technique is the **node-covering** (NC) method (Hanchek and Dutt, 1998). It has several distinguishing features from other methods of this type. The NC method does not require the factory or the user to generate new routing maps to reconfigure around faulty cells or interconnects, as is required by other techniques (Lach *et al.*, 1998; McDonald *et al.*, 1991; Narasimhan *et al.*, 1994; Roy and Nag, 1995). Instead, the original configuration data can be reused. This FT method involves a routing strategy that requires the use of additional interconnect segments at all branch points of the original (non-FT) circuit routing. On the other hand, no explicit additional tracks are needed in the channels to avoid the loss of connection flexibility as explained by Hatori *et al.* (1993). Further, circuit electrical properties are minimally affected by these additional interconnect segments (called **cover segments** (CSs) or **reserved segments**); reconfiguration does not involve any rerouting and, hence, is very fast. CSs, however, cause some track overhead in the routing. This results in retaining total functionality with somewhat reduced routability or total routability with reduced functionality. We describe this method briefly in the subsequent paragraphs.

Under the NC approach, each logic cell u in the FPGA is assigned a **cover cell** that can be reconfigured to replace the logic cell in the event that cell u becomes faulty. Reconfiguration around a faulty u is completed by a **replacement sequence** in which u is replaced by one of its covers which in turn is replaced by one of its covers, and so on until a spare cell is configured in.

For a cell to cover another cell (the **dependent cell**), the cover cell must be able to duplicate the functionality of the dependent cell. In an FPGA, all cells are identical, so configuration data for the dependent cell can simply be transposed to the cover cell. The cover cell must also be able to duplicate the connectivity of the dependent cell with respect to the rest of the array. This is accomplished by ensuring that each net connected to a cell through a channel segment also includes the corresponding cover segment, CS, bordering the cover cell. Cover segments are included in a net in one of two ways. First, segments in the net may already be in positions to act as covers. In case the above condition does not hold, additional segments should be attached to the net to provide covers. Figure 8.7 illustrates the use of CSs.

There are two approaches for incorporating the required CSs in a circuit mapped to an FPGA: **static** and **dynamic**. These approaches are discussed next.

In the static method, reconfiguration paths are predetermined to conform to a certain pattern, such as from going left to right along rows (as depicted in Figure 8.7). CSs are then inserted wherever needed along these paths (also shown in Figure 8.7) as part of the original routing. This is a static approach in the sense that the CS insertion is oblivious to the presence and location of faults. Whenever faults occur that conform to the predetermined reconfigurable pattern (e.g., one fault per row), the transposition of configuration bits from dependent to cover cells automatically achieves complete reconfiguration—the interconnect "hooks" needed by the cover cells to connect to nets routed to their dependent cells are already in place. Thus, reconfiguration is extremely fast with times determined primarily by transposition of cell configuration bits.

In Hanchek and Dutt (1996a, 1996b, 1998), this technique was implemented by modifying the routing tools obtained from the University of Toronto. It was found that, on the average, for 18 benchmark circuits ranging in size from 79 to

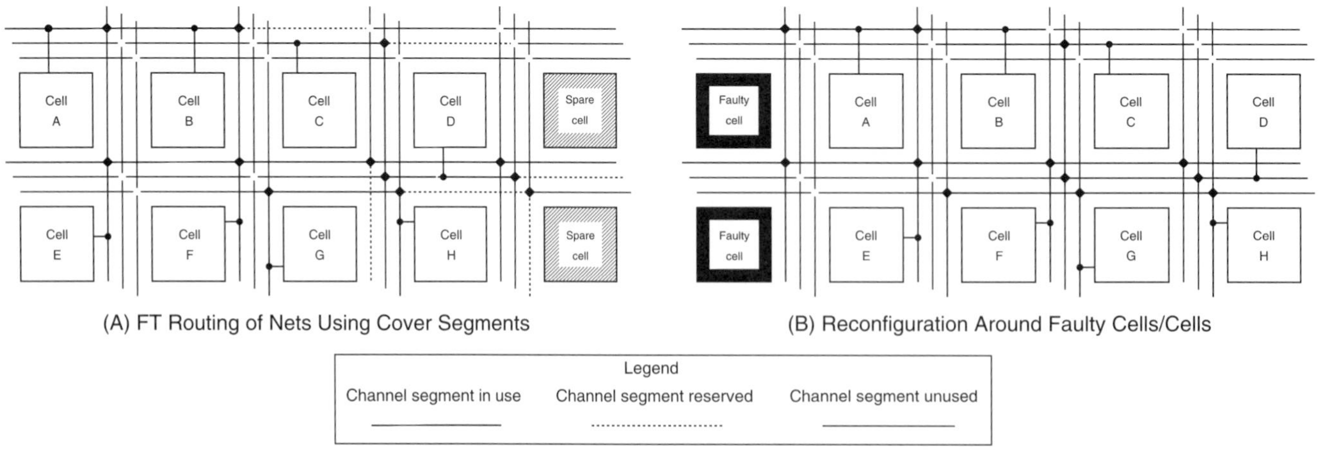

FIGURE 8.7 Using Cover Segments

508 nets, 34% more tracks are required than in a non-FT circuit to put the necessary CSs in place for reconfiguration. Good reliability and yield improvements are afforded by static CS insertion that can tolerate one faulty cell per row; for example, the percentage doubles a 35% non-FT yield for a 16×16 array and increases a non-FT reliability of 0.7 to 0.95. Net delays changed minimally (by at most approximately 7%) after reconfiguration. These results establish the static NC method as one of the best available for software-based reconfiguration of FPGAs with respect to several metrics: track overhead, reliability improvement, reconfiguration time, and circuit delay increase.

An even better-performing technique can potentially be developed by using the dynamic reconfiguration paradigm. Using such an approach will make it much easier to perform online testing and diagnosis, to reduce track overhead significantly, and to increase the number of possible fault patterns (and thus reliability) that can be reconfigured. Such a technique is discussed in Subsection 8.3.2.

8.3.2 The Dynamic Node-Covering Approach

The dynamic NC method has the following advantages: (1) better utilization of track resources—this will mean a higher rate of successful reconfigurations and thus greater fault tolerance; (2) fast on-line reconfiguration; (3) greater flexibility in tolerating fault patterns; (4) minimal changes in a circuit's electrical properties, including timing and power; (5) applicability to both cell and interconnect faults; and (6) applicability to other domains, such as online testing and diagnosis, wherein a "roving tester" can move around the FPGA during FPGA run time to test different parts of the FPGA (this requires run-time reconfiguration of the FPGA for remapping the working circuit and the tester) (Verma *et al.*, 2001).

The primary difference between static and dynamic methods is that, in the latter, CS insertions will not be part of the initial routing but will be made as and when required for fault reconfiguration (or in general, circuit reconfiguration for performing other functions like testing and diagnosis) and only to those nets connected to cells along reconfiguration paths. The dynamic method will thus significantly reduce the number of CSs required for FT, thereby making it easier to retain total functionality with much lower track overheads. Further, it allows reconfiguration paths from faulty cells to spare ones to be chosen dynamically, thus maximizing FT flexibility. The paths can have any arbitrary shape unlike the predetermined straight horizontal paths in the static NC method. The needed CSs can then be put in dynamically along any such chosen path. Since the configuration bits can be transposed along reconfiguration paths of any shape in FPGAs with incremental run-time-reconfiguration (RTR) capabilities, coupled with dynamic CS insertion, this allows reconfiguration around many more fault patterns than possible with the static method (e.g., fault clusters).

The shift from a static method to a dynamic one, however, introduces several interesting algorithmic problems. One of these is the fundamental issue of minimally changing the circuit routing to accommodate the needed CSs for reconfiguration in a fast and (near-)optimal manner with minimal effect on the circuit's electrical properties.

An approach to this problem that has yielded satisfactory results is shifting track assignments of nets (a process called **net bumping**) for accommodating the CSs needed for reconfiguration and "refitting" the bumped nets on other track positions. For simplicity of exposition, we describe this "bump-and-refit" (B&R) algorithm in the context of track segment lengths of one.

The general problem is the following. When a cell u is replaced by another cell v along a reconfiguration path, a net n_i connected to u may need its route to be extended to connect to v [as shown for $u =$ the dark node $B2$ and $v = C2$ in Figure 8.8(A) if the current route of n_i does not include the required track segment. The required extension is the CS. For each CS, if the required interconnect segment where the CS is to be inserted is vacant, the net extension is straightforward, and no other nets are disturbed. If the required interconnect segment is occupied by another net, then the CS insertion will cause a displacement or "bumping" of this net. The net requiring the CS extension is termed the **CS-net** [n_1 in Figure 8.8(A)], and the net occupying the required track segment is termed the **occupying net** or **O-net** [n_2 in Figure 8.8(A)].

The O-net needs to be moved out of its current track to make space for the CS. We use $n_i^{T_j}$ to denote a net n_i on track T_j. Let a **transition** be defined as the movement of net n_i on a track T_j to another track T_k and that is denoted by $n_i^{T_j \rightarrow T_k}$. This transition may result in net n_i bumping into one or more nets on track T_k. These nets will have to move out of their current track T_k, giving rise to a transition for each of them. This transition sequence is shown in Figure 8.8(B) by dark arcs, where net n_2 initiates a set of transitions that finally terminate in **spare track segments**, which are vacant segments of appropriate total lengths into which bumped nets can move without bumping other nets. As seen in the figure, the set of transitions take on a directed-acyclic graph (DAG) structure, termed a **transition DAG (T-DAG)**, with the spares forming the leaf nodes. The CS insertion is successful if a T-DAG rooted at the corresponding O-net can be found whose leaves are spare segments; such a T-DAG is termed a **converging T-DAG.** As stated earlier, because net bumpings only involve nets changing their track positions but not their topologies, their timing and power properties remain intact. Because CS insertions affect CS-net delays minimally (by about 7% [Hanchek and Dutt, 1998]), the B&R incremental rerouting approach only marginally affects FPGA performance.

The concept of an **overlap graph (OG)**, which is a graph representation of the circuit routing on the FPGA, was introduced in Dutt *et al.* (1999). The OG is an undirected graph with the circuit nets represented by the nodes of the graph.

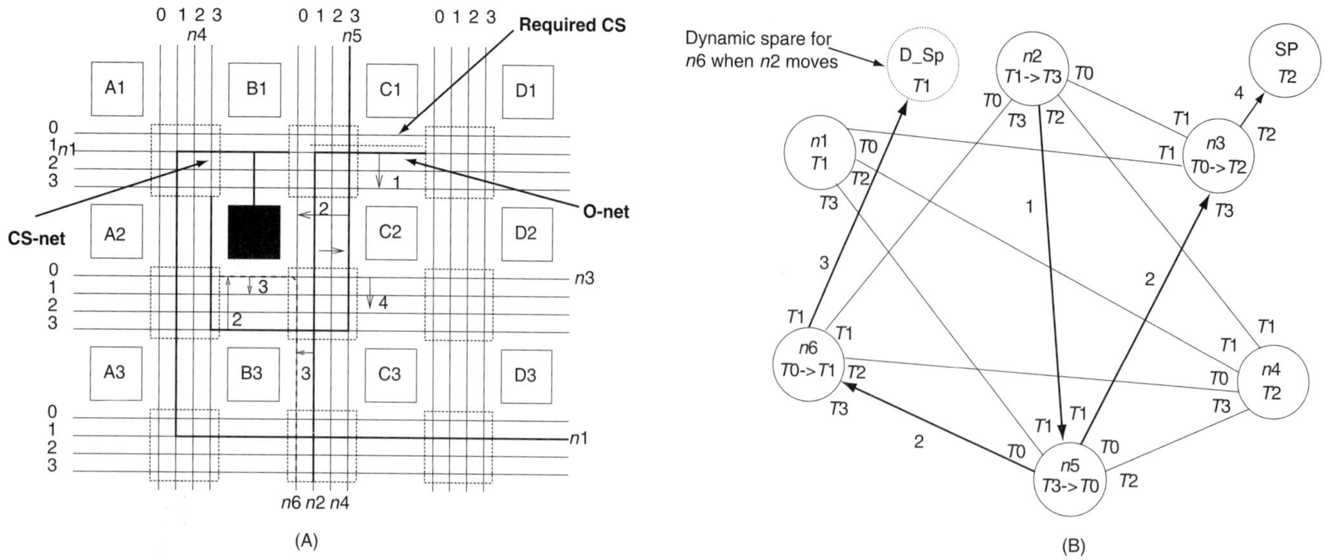

FIGURE 8.8 Searching the OG for a Converging Transition DAG to Determine Feasibility of CS Insertion

In the overlap graph OG (V, E), the set of nodes $V = S \cup \{n_1, n_2, \ldots, n_m\}$, where each n_i is a routed net of the circuit and where S is the set of "spare" track segments as described above. There exists an edge between n_i and n_j in the OG if nets n_i and n_j share a channel[2] in the FPGA. Figure 8.8(B) shows net n_2 and n_6 having an edge between them in the OG since they are routed through a common vertical channel to the right of cell B3.

The OG can be used as an effective model in the evaluation of the required T-DAG. Since it represents the circuit routing in the FPGA, a T-DAG is a DAG embedded in the OG (the undirected edges of the OG become directed arcs in the direction of the transitions 8b), as in Figure 8.8(B). A converging T-DAG rooted at an O-net can be determined by performing a search on the DAG until all leaf nodes of the search DAG are spare nodes.

This process is illustrated in Figure 8.8(B) for a small circuit and for a single CS insertion for extending net n_1, the CS-net. The corresponding O-net n_2 transits from T_1 to T_3 and bumps into n_5. The movement of n_2 from T_1 creates a "dynamic" spare node (labeled by D_Sp in the figure) for net n_6, for which information is added to the OG. The bumped net n_5 then transits from T_3 to T_0 where it bumps n_6 and n_3. Net n_6 then transits to the above dynamically created spare on T_1, while n_3 transits to its spare node on T_2. Thus, a converging T-DAG (here a tree) is determined in the OG. The transition arcs are shown dark in Figure 8.8(B) and numbered chronologically in the order in which they are traversed in the search process. After every transition, the track labels of the nodes and those of their edges need to be updated.

[2] A **channel** is the set of all track segments between two adjacent switch boxes of the FPGA.

The B&R incremental rerouting algorithm Conv_T-DAG that performs a depth-first based T-DAG search in the OG is given in Figure 8.9 (Dutt *et al.*, 1999). It recursively searches for converging T-DAGs rooted at each net bumped by the O-net. A depth-first path terminates in success if a spare node is reached and in failure either when all OG nodes have been visited in that path or a cycle is detected (an ancestor along the current path is revisited). When an $n_i^{T_j \to T_k}$ transition fails in this manner, the search backtracks and tries an unexplored transition $n_i^{T_j \to T_l}$. The algorithm was shown to be optimal—if a converging T-DAG exists among the currently used tracks for a CS, Conv_T-DAG will find it (Dutt *et al.*, 1999). The entire reconfiguration process that invokes Conv_T-DAG for each CS is impervious to the order of insertion of most CSs but not of a certain class of CSs (called "vertical" CSs). Thus, the reconfiguration process is near optimal.

While an existing converging T-DAG will ultimately be found by Conv_T-DAG, it will be time-efficient if some suitable "cost" measure can be used to determine which transitions are more likely to be successful so that fewer T-DAGs are searched and backtracked. A good cost measure will consider both the "magnitude" of bumpings (total length of bumped nets) and the likelihood of convergence of these bumpings.

Two transition cost (TC) measures evaluated are as follows: The first is written as:

$$sum(n_i^{T_j \to T_k}) = \sum\nolimits_{n_j \in adj^{T_k}(n_i)} l(n_j), \qquad (8.1)$$

where $l(n_j)$ is the total length of n_j in terms of the track segments (each of length 1) that it occupies. This heuristic is reasonable but only considers the bumping magnitude.

```
Algorithm Conv_T-DAG(OG,n_i^Tj)/*Find a converging
T-DAG rooted at n_i^Tj*/
for T_k = (T_i1, ..., T_it) in order of increasing TC do begin
    if (adj^Tk (n_i) == Ø)then/*there exists a spare for n_i on T_k
    */
        return (success)
    else begin
        for each n_r ∈ adj^Tk (n_i) do begin
            if (n_r is an ancestor) then break;
            /* this transition to T_k results in a cycle; try the
            next best transition */
        for each n_r ∈ adj^Tk (n_i) do
            result = Conv-T-DAG(OG, n_r^Tk);
            if (result == fail) then break;
            else numb_succ = numb_succ + 1;
        endfor
        if (numb_succ == |adj^Tk(n_i)|)return (success);
        /* converging T-DAGs were found for all nets in
        adj^Tk(n_i) */
    endelse
endfor
return(fail)./* no transition of n_i was successful */
End./* Conv_T-DAG( ) */
```

FIGURE 8.9 The Optimal Depth-First Search Algorithm for Finding a Converging T-DAG

For example, according to it, it is equally costly to bump a net of length 9 as it is to bump three nets each of length 3. The latter case, however, has a higher likelihood of convergence because there is greater flexibility in moving three bumped nets (e.g., each can transit to a different track) than a single net of the same total length. This leads to the second cost function:

$$(2)sqrt(n_i^{T_j \to T_k}) = \left[\sum_{n_j \in ad\ j^{T_k}(n_i)} l(n_j)\right] / \sqrt{|ad\ j^{T_k}(n_i)|}. \quad (8.2)$$

Using such TC functions 8.1 and 8.2 to guide the search results in tremendous savings in computation time compared to a blind depth-first search, as discussed in the following paragraphs.

Experimental Results

The algorithms in Figure 8.9 were coded for the FPGA model of single-length segments and tested on a number of benchmark circuits (obtained from the University of Toronto FPGA group) ranging from about 80 cells and nets to about 500 cells and nets. The input to the incremental routing software of Dutt *et al.* (1999) is the mapped and routed circuits produced by the Toronto's group SEGA tools (Lemieux and Brown, 1993). Two important metrics, track overhead and reconfiguration time (essentially, rerouting time), are the focus of these experiments.

- The dynamic method shows considerable improvement over the static method (Hanchek and Dutt, 1996a, 1996b, 1998) for all the different transition costs (TCs). The static method has a track overhead of 34% with its best combination routing heuristics. The dynamic method's overheads are 25.96% for a worst-case fault pattern and 15.58% for average-case patterns. This represents a 24–54% improvement over the static method.
- The sum TC heuristic is the most time-efficient for both the worst- and average-case simulations with average reconfiguration times over all circuits of about 43 sec and 18 sec, respectively. These times are quite promising.
- The suitability of heuristic TCs for time-efficiency was demonstrated by an experiment in which we randomly chose transitions only for the O-net (root of the T-DAG); the transitions for the other nodes were chosen based on the sum TC[3]. On the average, the partial-random case shows an increase in reconfiguration time over that of the full sum TC case by 65% for the worst-case fault-pattern and by 126% for the average-case pattern. The track overheads remain the same as that of the sum TC used.
- One example considered a limited number of faults; this case occurs mostly in fabrication defects and short-life missions. The track overheads for all cases are very small: 1.75% for one fault to 4.49% for four faults. Except for one circuit, all other circuits require an absolute of one extra track to tolerate four faults; a few do not require any extra tracks.

8.4 Control Flow Checking With a Watchdog Processor

A **watchdog processor** (WD) is a relatively small and simple coprocessor used to perform concurrent system-level error detection by monitoring the behavior of the main processor. Because the WD is much simpler than the processor being tested, this approach is significantly more inexpensive than if system-level redundancy techniques, like duplication, were used. A WD can also be added incrementally to an existing processor system without requiring any significant modification to it. The general configuration of a system with a WD is shown in Figure 8.10.

One behavioral aspect of the main processor that can be concurrently monitored by the WD snooping onto the memory bus is whether control flows of programs being executed are being correctly followed. This section is devoted to discussing various control flow monitoring techniques using a WD as well as presenting some of our recent results on a

[3] The reason for having partial randomness as opposed to randomizing all transition choices was to keep the simulation time tractable.

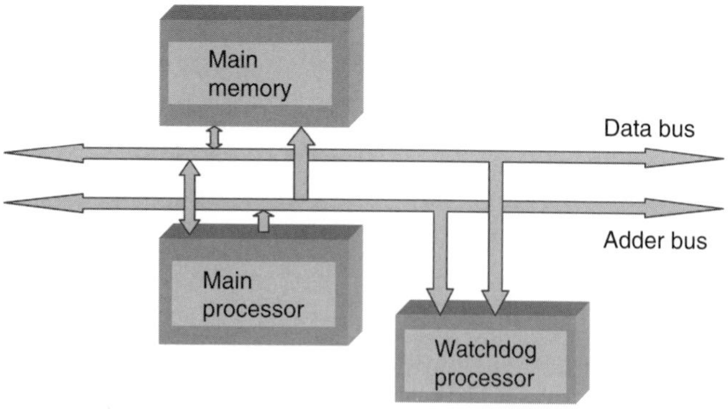

FIGURE 8.10 General Structure of a System with a Watchdog Processor

solution to the problem of monitoring program control flow by an external WD when there is an internal cache in the main processor.

Every application or program consists of processes. A process consists of procedures, and a procedure is composed of statements or instructions. Execution of programs, processes, and a procedures follow well-defined paths of instructions.

The control flow of a program can be represented by a **control flow graph** (CFG) in which the nodes represent some program unit, and edges represent the syntactically allowed flow of control. A node can be a single statement or a block of statements. Formally, a *node* of the CFG is a *block* of the program with exactly one entry point at the top of the block and exactly one exit point at the bottom of the block. The last instruction in a block may be a **branch** or **jump** instruction, although this is not necessary. There is an *arc* in the CFG from node *i* to node *j* if the program execution can correctly flow from the last instruction of block *i* to the first instruction of block *j*. An example of CFG extraction from a program is shown in Figure 8.11.

A correct control flow is a fundamental part of the correct execution of computer programs, and so it should be monitored to ensure the processor/program dependability. Such monitoring is called **control flow checking** (CFC).

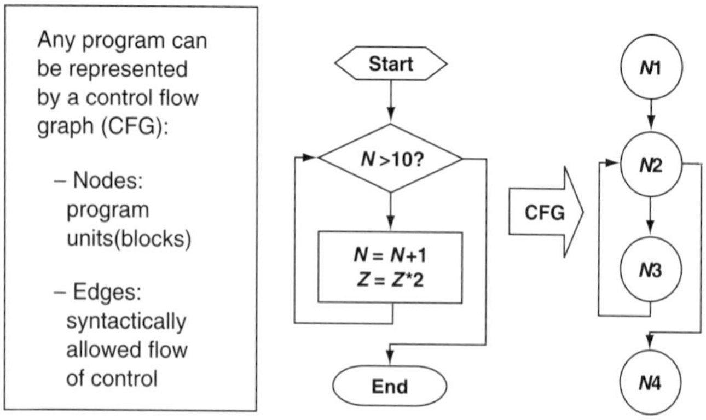

FIGURE 8.11 The Control Flow Graph (CFG) of a Simple Program

The most widely used techniques assign signatures as labels to the nodes of the CFG to determine the control flow. The WD is provided with the reference signatures and the relationships among them (represented by the CFG). During run time, the WD monitors the program run and concurrently computes the signatures that are then compared to the reference ones provided earlier. If a discrepancy is detected, then an error is signaled.

From this basic idea, methods have been developed that differ from each other in the augmentation of a node, by the representation of the reference information, and by the derivation of the run-time information. Figure 8.12 shows a possible augmentation of a basic block of a program for the purpose of CFC. Each "signature instruction" can be one or two NOPS containing the signature in the non-opcode fields.

According to Mahmood and McCluskey (1988), the reference CFG can be provided to the WD in three different ways:

- The control flow graph with the signature of each block is **stored in the local memory of the WD**. If a run-time signature is received or computed by the WD, then the stored reference is searched to determine whether the signature is a valid successor to the previous one.
- The reference information is **stored in the checked processor**, and it is transferred to the WD in run time. No local storage is needed in the WD, and this reduces the hardware overhead significantly. The transfer of the reference information, however, increases the overhead of the checked processor.
- The WD executes a **specialized reference program** that has the same control flow structure as the program executed by the checked processor. Based on the checked run-time information, the WD program performs a compiled graph tracing, which is faster than checking the

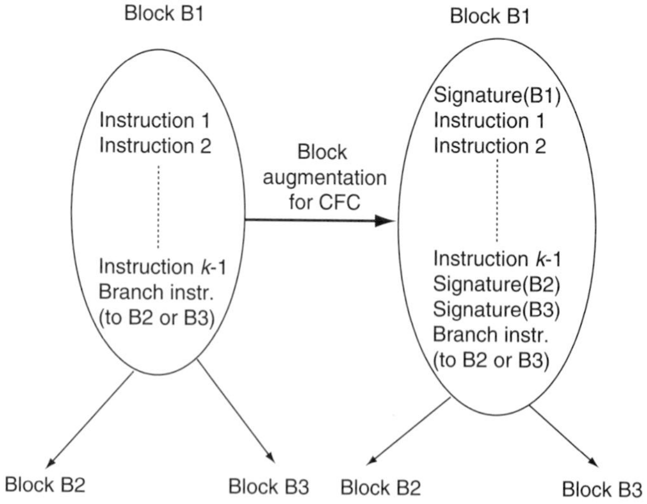

FIGURE 8.12 A Block in a CFG and Its Possible Augmentation for the Purpose of CFC

actual run-time signature by comparing it with the reference one of the WD; the tracing also does not require a separate data structure.

There are also two different ways of assigning the labels (signatures) to each node of the CFG: **assigned signatures** or **derived signatures** (Mahmood and McCluskey, 1988). Various techniques for these two approaches to signature determination are discussed next.

8.4.1 CFC with Assigned Signatures

The signatures labeling the nodes of the CFG are assigned arbitrarily using prime numbers or successive integers. These signatures are transferred to the WD explicitly by the checked processor. For this purpose, signature transfer statements are added into the source of the checked program.

Assigned signature-based methods check only that the nodes are executed in an allowed sequence (i.e., that the main processor traverses the CFG correctly), which is not necessarily the correct sequence of the program execution since the selection itself is unchecked. To check the contents of a node and the type and sequence of the instructions in it, derived signature methods are required. One of the first methods developed to perform CFC using assigned signatures is structural integrity checking (SIC) (Lu, 1982; Mahmood and McCluskey, 1988), which makes use of the control structure embedded in the syntax of the programming language.

To perform CFC, the source code to be executed by the checked system is preprocessed: the CFG is extracted, signatures are assigned to the nodes (blocks), the blocks are augmented with these signatures, and possibly the program of the main processor is modified by inserting the statements that transfer the signatures to the WD.

Methods differ in the way the signatures are assigned to the nodes and in how the reference information is represented. In SIC (Lu, 1982), the nodes are encoded (labeled) by selecting 8-bit numbers randomly and independently from a uniform distribution. The reference information for the WD is represented under the form of a reference program that has the same CFG as the checked program. In place of computations, it contains statements to receive and check the signatures from the main processor.

A newer control flow technique is the extended structural integrity checking (ESIC) (Mahmood and McCluskey, 1988), which extends the checking capabilities of SIC to check run-time computed procedure calls and interrupt handlers. The nodes of the CFG are encoded by successive numbers, and the reference information is extracted in the form of a stored reference signature database in which the valid successors of each signature are given in a sparse matrix format. The first and last nodes of a procedure are tagged by special flags, SOP start of procedure (SOP) and end of procedure (EOP). If an SOP signature is received (procedure call), then the actual state (pointer to the previous signature) is pushed to a WD internal stack and the reference state corresponding to the called procedure is reached. After an EOP signature (return from a procedure call), the original state (pointer to the original reference signature) is popped from the stack.

With this technique, procedure calls that are nondeterministic in the preprocessing phase and interrupt handler procedures can be checked. Whether the correct procedure is called remains unchecked, but the return address is checked using the signature stack.

The relatively easy implementation of the preprocessor and the ability of asynchronous checking are the advantages of the assigned signature-based methods. The disadvantages are the performance degradation of the checked processor due to explicit signature transfer, the need of recompilation of the source code that prevents checking of existing executables, and the fact that the sequence of instructions in a node is unchecked.

8.4.2 CFC with Derived Signatures

In this approach, the signatures labeling the nodes are derived from the opcode sequence of the node by information compaction (mod-2 addition, a checksum, or a linear feedback shift register [LSFR]). The run-time signatures are also computed concurrently by the WD by reading instructions from the memory data bus when they are fetched by the main processor. The WD computed signatures are compared to WD reference information to perform the checking.

A large number of schemes, improvements, and modifications have been proposed in past work, but the methods mainly differ in the augmentation of a node or block and in the representation of the reference information.

A simple method is basic path signature analysis (BPSA), in which the nodes of the CFG are a branch-free sequence of assembly level instructions, and the reference information is given by embedded signatures inserted at the beginning of each block (Mahmood and McCluskey, 1988); see Figure 8.12. The WD monitors the memory data bus of the processor and captures the reference signatures using tag bits to differentiate them from the normal instructions. At the end of each block, this signature is compared to the concurrently computed one (from the read instruction sequence).

Improvements have been proposed to reduce the hardware complexity of the WD: justifying signatures or using the embedded signature for the computation of the concurrent one. These were proposed to make the concurrent computed signature all zeros at the end of a node; thus, a simple gate instead of the comparator can be used for the checking.

Because we can decide during compile time the maximum dimension (number of instructions) of a block by splitting it into two separate but consecutive ones, we can find a trade-off between the overhead caused by the embedded signatures (increased memory usage and reduced performances) and the error coverage and latency.

Another method to reduce memory overhead is generalized path signature analysis (GPSA), which checks paths rather than simple nodes (Mahmood and McCluskey, 1988): the CFG is divided into path sets made of paths starting from the same node. Because there is only one reference signature for each path set, the run-time signature of each path is made the same by inserting justifying signatures in some paths.

To improve the error detection coverage, the universal signature monitor (USM) scheme uses path tags to check the validity of the sequence of nodes at the beginning of the next node and block signatures to check that the block is started at its entry point.

A similar method is implicit signature checking (Patel and Fung, 1983), which assigns unique signatures to each basic block by using the block's start address, and obtaining an implicit signature checking point at the beginning of each block. Justifying signatures are embedded at each branch instruction.

Other methods use advanced techniques of signature encoding, such as hashing of instructions or addresses by reference signatures (Mahmood and McCluskey, 1988).

The overhead caused by the embedded reference signatures can be completely eliminated if the reference information is moved to the environment of the WD. Two main solutions are possible: WD with stored reference information or WD based on a reference program.

The Roving Monitoring Processor implements a scheme based on stored reference signatures. The CFG is stored in a linked list format in the local memory of the WD. Receiving a signature, the WD examines whether it belongs to a node that can be reached from the previous one.

Cerberus–16 is the first synchronous scheme that checks processors by a reference program (Mahmood and McCluskey, 1988). The WD is implemented by a specialized processor that has a restricted instruction set. The program executed by the WD has the same CFG structure as that of the the program being checked. Each instruction of the WD includes the number of instructions in the corresponding node, a justifying signature, and the label of the next instruction in the WD. The reference signature of a path is computed at the end of a node and the justifying signature is fed to this unit to be compared to the run-time one. When a branch is executed at the end of a node, a signal is activated, and the WD takes the same branch as well; otherwise, the next instruction is executed.

The specifications of required performance are sometimes such that it is not possible to build a WD that can fetch all the required signatures or instructions. For these cases, asynchronous versions have been proposed in past work (Mahmood and McCluskey, 1988). These solutions make use of a signature queue allowing an asynchronous checking of the run-time signatures transferred to the WD.

8.4.3 CFC for Modern Processor Architectures

In derived signature methods, run-time signatures can only be computed if the sequence of instructions executed by the checked processor can be monitored by the WD. In the case of traditional processor architectures, signatures can be easily computed by monitoring the memory data bus of the checked processor. The sequence of instructions fetched by the processor is exactly the sequence executed by it. However, modern processor architectures include on-chip instruction caches and prefetch queues (pipelines). The sequence of instructions fetched by the processor can still be monitored by the WD, but it is not exactly the sequence of instructions executed by the processor. Some instructions are executed in a different order or are not executed at all.

The problem of monitoring pipelined RISC processors was addressed in (Delord and Savcier, 1991). The main idea was that the reference signatures were computed on the basis of the instruction sequence fetched by the main processor, and the basic functions of the pipeline were imitated in the WD. Delayed branching and exceptions events, which make a classical signature inconsistent, are handled by some new notions. Delayed branching is resolved by anticipated signatures (i.e., always taking into account in the compaction the delayed instructions following a branch, independent of whether they are executed). Exceptions and the corresponding flush of the pipeline are handled by delayed signatures (i.e., an instruction code is compacted only if a given number of successors are regularly fetched).

The problem of derived signature-based monitoring of processors with built-in caches was, to the best of our knowledge, addressed for the first time by Rota (2002). An external observer cannot guess the instructions fetched by the processor from its cache, and the integration of the WD into the processor core cannot be performed in off-the-shelf commercial systems. Rota (2002) developed a solution to this problem in which the processor explicitly sends information about the instructions that are actually being fetched from the cache. A very simple solution would be to modify the processor in such a way that every instruction fetched from the cache is also asserted on the external bus. Unfortunately, it is not realistic to assume that the external signature transfer time on the external memory bus followed by the concurrent signature computation and error analysis that a WD needs to perform is comparable to the processing speed of a pipelined processor with one or more cache levels. A solution was needed that would allow the watchdog to perform a derived signature control flow checking at a reasonable speed. It was assumed that the external watchdog is able to process instructions at the rate of the external bus clock frequency, which is usually smaller than the processor clock frequency.

The **speed factor** (K) is defined as the ratio of the internal versus the external clock frequency:

$$K = \frac{\text{Processor clock frequency}}{\text{Bus clock frequency}} \quad (8.3)$$

During the execution of the program, the lines of the cache will be filled successively until a branch in the program flow is taken. The design technique of Rota (2002) introduces a small hardware modification in the cache system: a counter that counts the number of accesses by the processor to the cache, and every K that accesses the accessed instruction is sent to an FIFO buffer placed in a bus controller that interfaces with the external memory bus. The WD then performs a derived signature checking of each block based on the instructions of the block that it receives.

Note that to perform control flow checking, each block must have three embedded signatures: the current block signature and the signatures of the two possible blocks following the branch instruction at the end of the block (see Figure 8.12). These three signatures must be sent to the WD (via the buffer) whenever fetched by the processor irrespective of their position in the block and of the counter value.

The buffer will put its entries on the bus as soon as it is available, and if the buffer gets full, the bus controller will stop CPU accesses to the cache until the buffer empties one entry. A block diagram of this approach is shown in Figure 8.13.

According to the technique described above, the WD will always receive the three signatures added to the basic blocks and some other instructions according to the length of the block and its alignment in the memory. A generic example is shown in Figure 8.14.

An assigned signature checking for the program flow was used so that the WD will always know when a new block is started or ending, which block has been taken after the branch, and when a derived signature checking for each block is based on the received instructions. Because which instruction in a block that will be passed first to the WD is not known a priori (except that it will be one of the first K instructions), the derived signature is different for each of the K possibilities of the first instruction of the block seen by the WD. Thus, every block must be assigned K signatures, each one for a different possible alignment. The signature is concurrently computed by the WD and compared with the K references stored in its memory to verify the correctness of the fetches.

The method described in the previous paragraphs for performing a partial derived signature checking CFC (partial CFC, for brevity) was implemented for a processor system with an internal cache for a Motorola 68040 processor. The processor's initial VHDL description was obtained from (Heishman *et al.*). It was augmented greatly and various structural descriptions were added to the original behavioral description for CFC and partial CFC implementations (Rota, 2002; Trovo, 2002). A cache-less 68040 (or equivalently a 68040 with an external cache) was also simulated with the WD then being able to monitor the program flow completely and thus perform a full CFC. The two different WDs were also described in VHDL. Simulations performed show that the error coverage for $K = 2$ of the partial CFC scheme for an internal cache system is almost as good as that for full CFC. This result is shown in Figure 8.15 for two and four faults

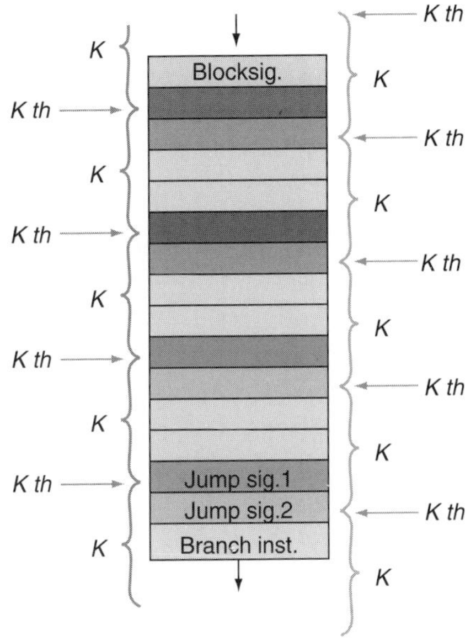

FIGURE 8.14 A Block with K Subsets of Instructions. Instructions in each subset are shaded the same, one of which is sent to the WD based on the alignment of the block on the instruction access pattern of the processor. Each subset has a potentially different derived signature. All K derived signatures are stored in the WD as reference information.

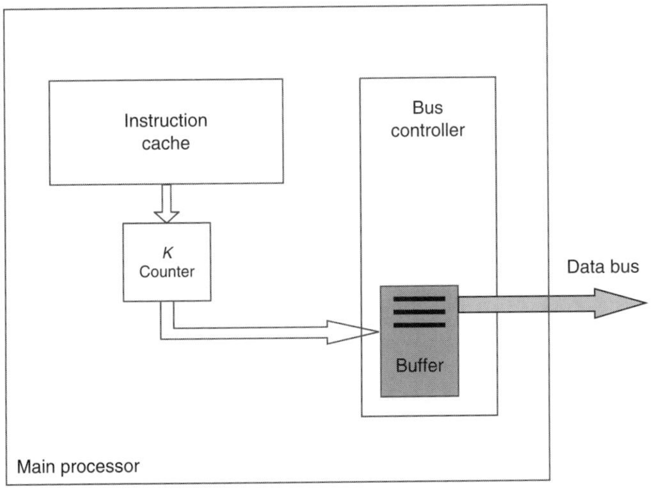

FIGURE 8.13 The Block Diagram of the "1-of-K" Forwarding Technique for Partial Derived Signature Checking. This technique is in an environment with on-chip caches.

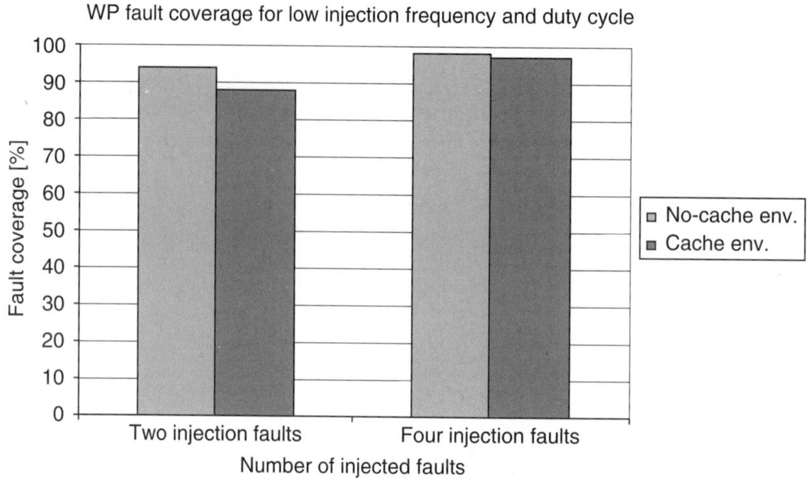

FIGURE 8.15 Fault Coverage Results for Full and Partial CFC Checking. The results are for external-cache/cache-less systems and in internal cache systems with partial CFC respectively, for $K = 2$, and for two and four faults, averaged over a number of patterns (random, semiclustered, fully clustered) for each number of faults.

injected in the memory data bus at a repeat frequency of 100 KHz and at 25% duty cycle.[4]

8.5 Microrollback—A Fault-Tolerance Mechanism for Processor Systems

Microrollback is a technique that is coupled with concurrent error detection mechanisms (such as FD arithmetic circuits [see Section 8.2] and program CFC [see Section 8.4]) and allows a processor system to recover from detected errors by going back to a state prior to the occurrence of these errors and restarting execution from that point. In general, to perform such a rollback to a previous state, it is necessary to have copies of all previous states of the system up to a certain time interval. Because a lot of memory is required to store this kind of information, it is necessary to consider the fact that only a reduced number of snapshots of previous states can be kept. Due to this limitation, it is necessary to determine critical points, called **checkpoints** at which we need to or it is appropriate to store the state of the system (Bowden and Pradhan, 1992; Chandy and Ramamoorthy). This decision process is important for software rollback, but is too complex for a completely hardware solution.

Hardware-based rollback techniques fall in one of two categories: (1) **full checkpointing** and (2) **incremental checkpointing** (Alewine *et al.*, 1992). Full checkpointing maintains snapshots of the entire state of the system (essentially all register, cache, and memory contents) at the checkpoints, whereas incremental checkpointing maintains only the changes to the state from the immediately preceding checkpoint (e.g., contents of registers written between the checkpoints). Rollback in full checkpointing schemes involves setting the current state to be the desired checkpointed state, while in incremental checkpointing, rollback involves undoing or backing out of the state changes made after the desired checkpointed state. We will focus on incremental checkpointing in the discussion of hardware microrollback in the rest of this section.

At the implementation detail level, there are two different approaches to hardware microrollback: clock *cycle microrollback* (state changes saved at every clock cycle), as described by Sparmann and Reddy (1994) and *instruction microrollback*: (state changes saved at the completion of every instruction execution, as described by Trovo (2002)).

Before presenting the hardware solution by Tamir and Tremblay (1990), we define rollback distance and rollback range.

- **Rollback distance** is the number of state change snapshots to be undone. This number is limited by the number of stored snapshots.
- **Rollback range** is the maximum rollback distance that the system can support. This value is determined by the maximum error detection latency of the system, thus allowing it to always roll back to a state prior to that in which the errors were generated.

We next describe the complete hardware solution for clock cycle microrollback discussed by Tamir and Tremblay (1990) for a Reduced Instruction Set Computer (RISC) processor.

[4] This work was done as part of an AFOSR MURI project for determining the effect of deliberate electromagnetic (EM) interference on digital and computer systems; the interference pattern is expected to be a train of EM pulses for some duration followed by a silent period, with this pattern repeating at some frequency that is dependent on the power capability of the interfering source. EM interference is also expected to cause multiple transient faults, especially on off-chip wiring like memory busses.

8.5.1 Support for Microrollback in the Register File

In a RISC processor, a write into a register may be performed every clock cycle. Incremental checkpointing is performed for the register file by using a delayed-write buffer (DWB) to store the written register values for up to *N* clock cycles to realize a rollback range of *N*. Figure 8.16 shows the DWB structure for supporting microrollback in the register file.

The address of the destination register and its new value are stored in the DWB, which is an *N*-level FIFO buffer. This storage structure is composed of an *N*-level data FIFO that contains the values of the registers that have been written and an *N*-level content associative memory (CAM) that stores the addresses of the written registers plus valid bits for the data FIFO entries. In each clock cycle, if a write register operation is executed, a new line of the DWB is filled; otherwise, the line is invalidated by resetting the corresponding valid bit in the CAM. This microrollback structure also accommodates the needed change in the read of the register file so that the latest value of the addressed register is read. During a read, the DWB checks the CAM to determine if the addressed register's content is stored in its data store. If so, a priority circuit chooses the most recent written value of the addressed register to be read out of the data FIFO. A microrollback of *d* clock cycles is implemented simply by invalidating the first (most recent) *d* locations of the DWB.

8.5.2 Support for Microrollback in Individual State Registers

Two different solutions were presented by Tamir and Tremblay (1990) for microrollback in a single-state register: the FIFO and RAM methods. The FIFO solution is shown in Figure 8.17. The storage data structure is a FIFO that contains all previous

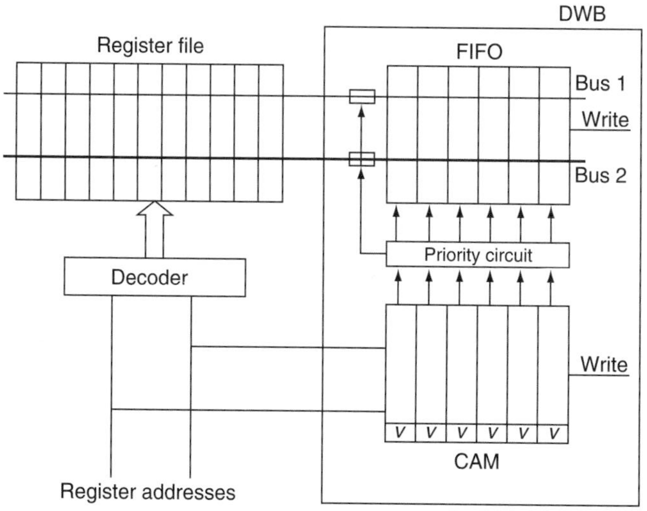

FIGURE 8.16 A Register File with Support for Microrollback

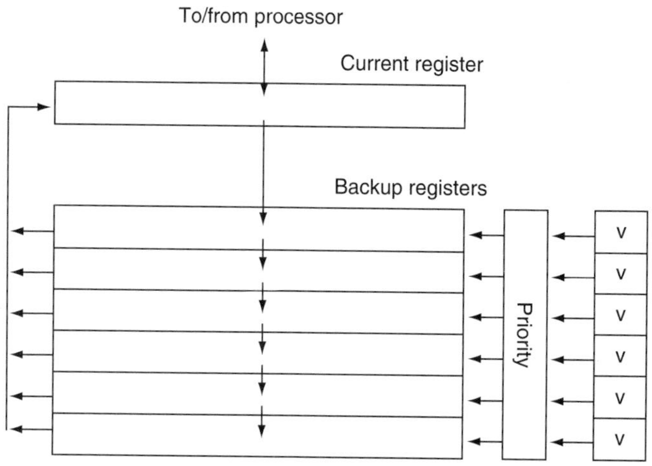

FIGURE 8.17 The FIFO Method for Microrollback in a State Register

values of the register with associated valid bits up to *N* clock cycles. The *current register* is the register at the top of the FIFO structure and always has the most recent value of the register. In case of read or write (R/W) operations, the current register is accessed. After each clock cycle, the data contained in this register is pushed into the FIFO structure, and the corresponding valid bit is set to value 1. The priority circuit ensures that the pushing of an entry into this FIFO results in only the top consecutive sequence of valid entries being pushed down by one entry; entries following and including the first invalid entry do not move. To realize a microrollback of distance *d*, the *d*th valid entry from the top of the FIFO is moved to the current register, and the first *d* entries in the FIFO are invalidated.

The RAM method uses a small RAM and a pointer to the last stored value of the register to create a circular buffer. Like the previous solution, a R/W operation is performed on the current register (see Figure 8.18). During a write operation, the old value of the current register is written into the RAM location that immediately follows the pointer, and the pointer is then updated to point to this location. For microrollback, it is simply necessary to subtract the desired rollback distance from the pointer and move the value in the entry following the newly pointed location to the current register.

8.5.3 Support for Microrollback in Cache Memory

As in the register file, a DWB is used to support microrollback in cache memory. The only modification for a cache, presented by Tamir and Tremblay (1990), is the use of two CAMs instead of one. In this case, each CAM checks only one part of the address, and, thus, the response time of the DWB is smaller. This is done to achieve better performance on cache microrollback support. The cache CAMs' behavior for R/W operations and for rollback is exactly the same as that of the CAM for the register file.

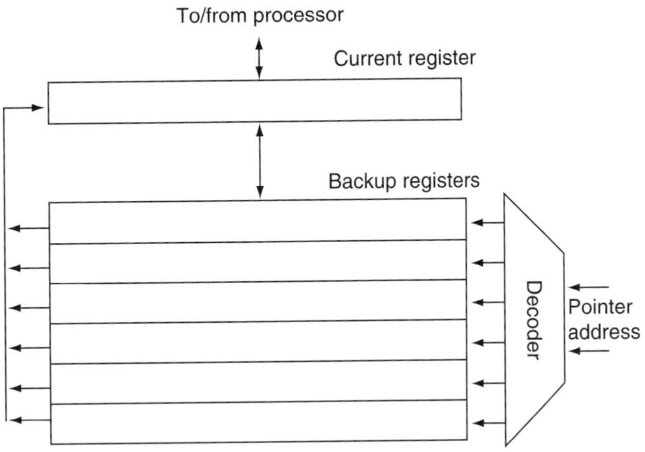

FIGURE 8.18 The RAM Solution for Microrollback of Individual State Registers

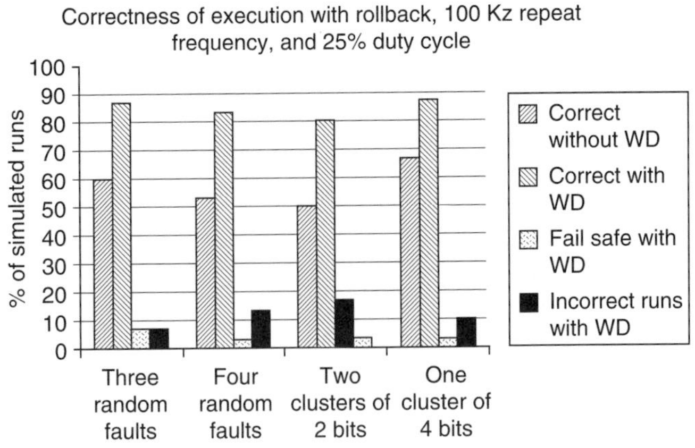

FIGURE 8.19 Execution Correctness of a Matrix Multiplication Program in a Simulated Motorola 68040. Using microrollback and Hamming code on memory and cache busses and WD-based CFC detection schemes for different number of faults and fault patterns.

8.5.4 Microrollback in a CISC Processor and Experimental Results

Trovo (2002) implemented microrollback in VHDL in the Motorola 68040 processor along with concurrent error detection via a watchdog (WD) processor-based control-flow checking (see Section 8.4) and a two-error detecting one-error correcting Hamming code (HC) implemented on the memory and cache data and address busses. Because the 68040 is a CISC processor, an instruction may not be executed every clock cycle, and an instruction microrollback scheme can be implemented for the register file, state registers, cache, and main memory (the main memory rollback support is similar to that of the cache) using hardware similar to that described in Sections 8.5.1–8.5.3.

In simulation experiments, multiple errors of different patterns were injected on the memory and cache busses at various repeat frequencies and duty cycles to simulate the effect of EM interference on a computer system (see footnote 4). These errors can be caught by the HC detector and/or the WD. The maximum error detection latency of the HC error detector is one instruction execution and that of the WD is $b_{\max}$ instruction executions, where $b_{\max}$ is the maximum block size of the program after CFC-based augmentation as described in Section 8.4 (note again that here the rollback distance unit is an instruction execution and not a clock cycle). Whenever either the HC or the WD detects an error, the microrollback unit is signaled and microrollback of the appropriate distance occurs (= error latency of the detector—one instruction execution for HC and b_i instruction executions for the WD if b_i is the size of the current block). When both the HC and WD detect errors simultaneously, the HC is given priority, and a rollback of one instruction is performed.

Sometimes the transient errors injected on the memory bus can become permanent errors if they are not detected by the HC and, if the corresponding instruction or data line fetched is stored in cache. Subsequently, when the instruction is fetched from cache, it will always be erroneous. If this error is subsequently detected by the WD monitoring the cache-processor bus or the HC on that bus, a microrollback occurs. After the microrollback occurs, the same instruction (with permanent errors) will be fetched again, and the microrollback and subsequent fetch will repeat ad infinitum. This type of infinite loop can be detected with high probability if the WD can be augmented to detect the number of microrollbacks of the same type (e.g., defined by just the microrollback distance) and then send a signal to the processor to terminate the program when 10 consecutive microrollbacks of the same type are detected. Such termination is a fail-safe termination of the program as its results are not deemed reliable.

Figure 8.19 shows the efficacy of this microrollback method, HC- and CFC-based error detection, and infinite loop detector at a fault injection repeat frequency of 125 KHz and duty cycle 25% for different patterns of three to four faults injected on the memory and cache busses. In all cases, the matrix multiplication program that was tested ended with incorrect results less than 10% of the time, thus achieving a safety probability of about 90%. This compares to a safety probability in the range of 50 to 68% for microrollback without using concurrent error detection via WD-based CFC, thus also demonstrating the usefulness of CFC.

8.6 Algorithm-Based Fault Tolerance

Algorithm-based fault tolerance (ABFT) refers to techniques that use the property of the algorithm or computation to check the correctness of output data. This is usually done by encoding the input data (e.g., by checksums) and verifying that the output also has a valid encoding. Huang and Abraham

(1984) presented the seminal work in ABFT in which methods of detecting and correcting errors in various matrix computations using checksum-encoded matrices were discussed. These methods use the linear transformation property of matrix computations to detect errors in the result matrix. Subsequently, a number of ABFT schemes were proposed for error detection in different computations like Fourier transforms (Malek and Choi, 1985) and matrix equation solvers (Luk and Park, 1988). Most ABFT schemes use a floating point (FP) checksum equality test that is susceptible to roundoff in FP computations of finite precision. This susceptibility arises from the difference in the sequence of operations used in computing the two sides of the equality test. Roundoff error analysis in Wilkinson (1963) indicates that upper bounds on the (roundoff) error in FP operations are dependent on the quantities involved, the order in which the intermediate operations are performed, and the precision used in the intermediate result's accumulation.

Banerjee *et al.* (1990) presented an evaluation of the error coverage of ABFT schemes for matrix operations on a hypercube multiprocessor. In this work, a threshold Δ was used for the equality test so that a difference of at most Δ between the two FP numbers being compared meant passing the test. The value of Δ was determined arbitrarily using averaging on certain input data sets and did not result in very good error coverage. Nair and Abraham (1990) proposed a generalization of the checksum test in the framework of real number codes. The performances of several weighted checksum schemes that correspond to different encoder vectors were examined. The checksum schemes and the numerical error involved again were found to be data dependent. Roy-Chowdhury and Banerjee (1993) developed a theoretical forward (roundoff) error bounds for some computations; the bounds are used as the threshold values. They found good error coverage for random dense problems. Although this is an innovative method, it has some drawbacks. Although, they have a theoretical basis for their thresholds, their experiments use tighter empirical thresholds to obtain a reasonable error coverage. Second, the inherent nature of threshold methods can miss errors in the low-order bits. Such errors in the low bits could become very large if the system of linear equations in **LU** decomposition is ill-conditioned (Wilkinson, 1965).

Luk and Park (1988) analyzed the effect of roundoff errors on various matrix decomposition methods. They also concluded that small roundoff errors are magnified in the final result, thus necessitating the use of a large threshold for the FP checksum test to prevent false alarms. As Boley and Luk (1991) determined, a good threshold depends on the condition number of the matrix of **checksum coefficients**. Thus, the selection of the threshold is generally a difficult issue. False alarms can occur when the threshold is too small, and relatively large roundoff errors can be mistaken as the presence of an error in the result, which can lead to unnecessary recomputation. On the other hand, a large threshold can cause an actual error to go undetected, and this can lead to substantial inaccuracies in the final result. In addition, if ABFT is being used as a detection mechanism for hardware faults, then transient and, more dangerously, intermittent and permanent faults can go undetected for a long time. Hence, the issue of acceptable accurate error detection of various matrix computations using equality test-based ABFT schemes needed to be satisfactorily solved.

ABFT methods using the mantissas of floating point quantities can potentially be used to check certain floating point operations, such as multiplication using checksums of the input mantissas treated as integers without being susceptible to the problem of roundoff errors (Assaad and Dutt, 1992, 1996). The only drawback to this approach is that it cannot be applied to all floating point operations (e.g., addition). However, a hybrid test was formulated that applies both the floating-point and integer mantissa checksum tests to appropriate parts of a computation (that is amenable to ABFT), thus covering all its operations. In this manner, much higher error coverages are obtained compared to using only the floating-point test.

The floating-point and mantissa-based integer checksum techniques are discussed next.

8.6.1 Floating-Point Checksum Test

Many previous approaches for error detection and correction of linear numerical computations were based on the use of checksum schemes (Banerjee *et al.*, 1990; Huang and Abraham, 1984; Jou and Abraham, 1986; Luk and Park, 1988). A function f is **linear** if $f(\boldsymbol{u}+\boldsymbol{v}) = f(\boldsymbol{u}) + f(\boldsymbol{v})$, where $\boldsymbol{u}$ and $\boldsymbol{v}$ are vectors. We discuss here a commonly used checksum technique for a frequently encountered computation: matrix multiplication.

The floating point checksum technique for matrix multiplication from Huang and Abraham (1984) is as follows. Consider an $n \times m$ matrix A with elements $a_{i,j}$, $1 \le i \le n$, $1 \le j \le m$. The **column checksum matrix** A_c of the matrix A is an $(n+1) \times m$ matrix, whose first n rows are identical to those of A and whose last row **rowsum** (A) consists of elements $a_{n+1,\,j} := \sum_{i=1}^{n} a_{i,j}$ for $1 \le j \le m$. Matrix A_c can also be defined as $A_c := [\frac{A}{e^T A}]$, where $\boldsymbol{e}^T$ is the $1 \times n$ row vector $(1, 1, \ldots, 1)$. Similarly, the **row checksum matrix** A_r of the matrix A is an $n \times (m+1)$ matrix, whose first m columns are identical to those of A and whose last column **colsum** (A) consists of elements $a_{i,\,n+1} := \sum_{j=1}^{n} a_{i,j}$ for $1 \le i \le n$. Matrix A_r can also be defined as $A_r := [A|A\boldsymbol{e}]$, where A$\boldsymbol{e}$ is the column summation vector. Finally, a *full checksum matrix* A_f of A is defined to be the $(n+1) \times (m+1)$ matrix, which is the column checksum matrix of the row checksum matrix A_r. Corresponding to the matrix multiplication $C := A \times B$, the relation $C_f := A_c \times B_r$ was established by Huang and Abraham (1984). This result leads to their ABFT scheme for error detection in matrix multiplication, which can be described as follows.

Algorithm: Mult_Float_Check(A, B)
/*A is an $n \times m$ matrix and B an $m \times l$ matrix. */

1. Compute A_c and B_r.
2. Compute $C_f := A_c \times B_r$.
3. Extract the $n \times l$ submatrix D of C_f consisting of the first n rows and l columns. Compute D_f.
4. Check if $\mathbf{c}_{n+1} \stackrel{?}{=} \mathbf{d}_{n+1}$, where $\mathbf{c}_{n+1}$ and $\mathbf{d}_{n+1}$ are the $(n+1)$th rows of C_f and D_f, respectively.
5. Check if $\mathbf{c}^{n+1} \stackrel{?}{=} \mathbf{d}^{n+1}$, where $\mathbf{c}^{n+1}$ and $\mathbf{d}^{n+1}$ are the $(n+1)$th columns of C_f and D_f, respectively.
6. **If** any of the above equality tests fail **then return** ("error") **else return** ("no error").

The following result was proved indirectly in Theorem 4.6 of work by Huang and Abraham (1984).

Theorem 1: At least three erroneous elements of any full checksum matrix can be detected, and any single erroneous element can be corrected.

Theorem 1 implies that Mult_Float_Check can detect at least three errors and correct a single error in the computation of $C_f = A_c \times B_r$ as long as all operations, especially floating point additions, have a large enough precision such that no roundoff inaccuracies are introduced. Of course, such an 'infinite' precision assumption is unrealistic, and the checksum scheme is susceptible to roundoff introduced by finite precision floating point arithmetic, as described previously. In particular, there can be false alarms in which the checksum test fails due to roundoff in spite of the absence of errors (that can occur due to hardware glitches or failures) in the computation. Alternatively, real errors could be masked or canceled by roundoff leading to nondetection of a potential problem in the hardware.

8.6.2 Integer Checksum Test

The susceptibility of the floating point checksum test to round off inaccuracies can be largely mitigated by applying integer checksums to various (linear) computations that are "mantissa preserving." This results in high error coverage and zero false alarms because integer checksums do not have to contend with the roundoff error problem of floating point checksums. The integers involved are derived from the mantissas of the intermediate floating point results of the floating point computation. To date, we have successfully applied integer checksums (hereafter also called mantissa checksums) to two important matrix computations: matrix–matrix multiplication and **LU** decomposition (using the **GE** algorithm) (Assaad and Dutt, 1992; Dutt and Assaad, 1996). We will briefly discuss the general theory of mantissa checksums and how they are applied to these two computations.

General Theory

In the following discussion, $\boldsymbol{u} = (u_1, \ldots, u_n)^T$ is used to represent column vectors and a,b,c, etc., for scalars. Unless otherwise specified, these variables will denote floating point quantities. The notation mant (a) is used to denote the mantissa of the floating point number a treated as an integers. For example, considering 4-bit mantissas and integers, if 1.100 is the mantissa portion of a, with its implicit binary point shown, then the value of the mantissa is 1.5 in decimal. However, mant $(a) = 1100$ and has value 12 in decimal. Furthermore, for a vector $\boldsymbol{v} := (v_1, \ldots, v_n)^T$, mant $(\boldsymbol{v}) := (\text{mant})(v_1), \ldots, \text{mant}(v_n))^T$, and for a matrix $A := [a_{i,j}]$, $A^{mant} :=$ mant $(A) := [\text{mant}\ (a_{i,j})\ ]$. The $:=$ symbol is used to denote equality by definition, $=$ to denote the standard (derived) equality, and $\stackrel{?}{=}$ to denote an equality test of two quantities that are theoretically supposed to be equal but may not be due to errors and/or round-off.

Let f be any linear function on vectors. The linearity of f allows us to apply the following floating point checksum test on the computation of f on a set S of vectors:

$$f\left(\sum_{\boldsymbol{v}\in S} \boldsymbol{v}\right) \stackrel{?}{=} \sum_{\boldsymbol{v}\in S} f(\boldsymbol{v}). \tag{8.4}$$

Ignoring the round-off problem, the LHS and RHS of the above equation should be equal if there are no errors in the following: (1) computing the $f(\boldsymbol{v})$ for all $\boldsymbol{v} \in S$ (which is the original computation), (2) summing up the $f(\boldsymbol{v})$ to get the RHS, (3) summing up the $\mathbf{v}$, (4) and applying f to the sum to get the LHS. If the sums are not equal, then an error is detected. Unfortunately due to round-off, the test of equation 8.4 often fails to hold in the absence of computation errors. Therefore, an integer version of this test that is not susceptible to round-off problems was sought. Of course, this integer checksum test should involve integers derived from the floating point quantities.

Because f is a linear function, irrespective of whether the vectors are floating points or integers, the following checksum property also holds:

$$f\left(\sum_{\boldsymbol{v}\in S} \text{mant}(\boldsymbol{v})\right) = \sum_{\boldsymbol{v}\in S} f(\text{mant}(\boldsymbol{v})), \tag{8.5}$$

where the multiple mant $(\mathbf{v})$ are integer quantities, as we saw previously. Note that equation 8.5 is not related to the original floating point computation and can be used to check it only if f is **mantissa preserving** [i.e., f (mant $(\boldsymbol{v})$) is equal to mant $(f(\boldsymbol{v}))$]. Then equation 8.5 becomes:

$$f\left(\sum_{\boldsymbol{v}\in S} \text{mant}(\boldsymbol{v})\right) = \sum_{\boldsymbol{v}\in S} \text{mant}(f(\boldsymbol{v})). \tag{8.6}$$

Thus, if there are errors introduced in the mantissas of the $f(\mathbf{v})$, then those errors are also present in the mant $(f(\mathbf{v}))$, and these will be detected by the integer checksum test of equation 8.6. Furthermore, this test is not susceptible to roundoff. Hence, it will not cause any false alarms, and very few computation errors will go undetected vis-a-vis the floating point test of equation 8.4. In practice, since an integer word can store a finite range of numbers, integer arithmetic is effectively done in modulo q, where $q-1$ is the largest integer that can be stored in the computer. Some higher order bits can be lost in a modulo summation. As established shortly, however, a single error on either side of equation 8.6 will always be detected even in the presence of overflow.

The crucial condition that must be satisfied to apply a mantissa-based integer checksum test on f is whether $f(\text{mant}(\boldsymbol{v})) = \text{mant}(f(\boldsymbol{v}))$. To check if f is mantissa preserving, we have to look at the basic floating point operations, such as multiplication, division, addition, subtraction, square-root, and others, that f is composed of and see if they are mantissa preserving. A binary operator $\odot$ is said to be mantissa preserving if mant (a) $\odot$ mant (b) = mant ($a \odot b$). Let a floating point number a be represented as $a_1 \times 2^{a_2}$, where a_1 is the mantissa and a_2 the exponent of a. Ignoring the position of the implicit binary point (i.e., in terms of just the bit pattern of numbers), floating point multiplication is mantissa preserving because:

$$\begin{aligned}\text{mant}(a)\ \text{mant}(b){:} &= a_1 \cdot b_1, \text{ while mant}(a \cdot b) \\ &= \text{mant}(a_1 \cdot b_1 \times 2^{a_2+b_2}){:} = a_1 \cdot b_1.\end{aligned} \tag{8.7}$$

Note that sometimes the mantissa c_1 of the product $c = a \cdot b$ is "forcibly" normalized by the floating point hardware when the "natural" mantissa of the resulting product is not normalized (e.g., $1.100 \times 1.110 = 10.101000$; the product mantissa is not normalized and is normalized to 1.010100, assuming 6 bits of precision after the binary point, and the exponent is incremented by 1). In such a case, c_1 is either equal to $(a_1 \cdot b_1)/2$ as in the previous example or is equal to $(a_1 \cdot b_1)/2 - 1$ when the mantissa that is not normalized has a 1 in its least-significant bit. When normalization is performed, the exponent of c becomes $a_2 + b_2 + 1$. This normalization done by the floating point multiplication unit is easy to detect and reverse in c (a process called **denormalization**) so that the multiplication is effectively mantissa preserving. Similarly, floating point division is also mantissa preserving. Floating point addition and subtraction, however, are not mantissa preserving.

Thus, if f is composed of only floating point multiplications and/or divisions, it is mantissa preserving, and the integer checksum test can be applied to it. On the other hand, if f has floating point additions as well, and there is no guarantee that the exponents of all numbers involved are equal, then f is not mantissa preserving. All is not lost in such a case because it might be possible to formulate f as a composition $g \circ h[g \circ h(\boldsymbol{u}){:} = g(h(\boldsymbol{u}))]$ of two (or more) linear functions g and h, where, without loss of generality, h is mantissa preserving and g is not. In such a case, one can apply an integer checksum test to the h portion of f [i.e., after computing $h(\boldsymbol{u})$] and a floating point checksum test to f [i.e., after computing $g(h(\mathbf{u})){:} = f(\mathbf{u})$]. Since errors in $h(\boldsymbol{u})$ are caught precisely, the error coverage will still increase and the false alarm rate will reduce in checking f vis-a-vis just by applying the floating point checksum test to f. This type of a combined mantissa and floating point checksum is called a **hybrid** checksum.

Application to Matrix Multiplication

We discuss here the application of integer mantissa checksums to matrix multiplication; the description of this test for **LU** decomposition can be found in work by Dutt and Assaad (1996). Matrix multiplication is not mantissa preserving, since it contains floating point additions. However, matrix multiplication can be formulated as a composition of two functions, one mantissa-preserving and the other not, as shown below.

First of all, matrix multiplication can be thought of as a sequence of vector–matrix multiplications:

$$A_{n\times m} \cdot B_{m\times l}{:} = \begin{pmatrix} \boldsymbol{a}_1^T \cdot B \\ \boldsymbol{a}_2^T \cdot B \\ \vdots \\ \boldsymbol{a}_n^T \cdot B \end{pmatrix}, \tag{8.8}$$

where $\boldsymbol{a}_i^T$ is the ith row of A, and $\boldsymbol{a}_i^T \cdot B$ is a vector–matrix multiplication. Note that $f_B(\boldsymbol{a}_i^T){:} = \boldsymbol{a}_i^T \cdot B$ is a linear function. This property leads to the **floating point row checksum test** for matrix multiplication. In terms of f_B, the row checksum test is:

$$f_B\left(\sum_{i=1}^{n} \boldsymbol{a}_i^T\right) \stackrel{?}{=} \sum_{i=1}^{n} f_B(\boldsymbol{a}_i^T). \tag{8.9}$$

Matrix multiplication can also be thought of as a sequence of matrix vector products $A \cdot B = (A \cdot \boldsymbol{b}_1, A \cdot \boldsymbol{b}_2, \ldots, A \cdot \boldsymbol{b}_l)$. This leads to a similar **column checksum test**.

A vector–vector component-wise product $\diamond$ for two vectors $\boldsymbol{u}$ and $\boldsymbol{v}$ is defined as the vector:

$$\boldsymbol{u} \diamond \boldsymbol{v}{:} = (u_1 \cdot v_1, u_2 \cdot v_2, \ldots, u_n \cdot v_n)^T. \tag{8.10}$$

For a matrix $B_{m\times l}$, and a m-vector $\boldsymbol{u}$, $\boldsymbol{u}^T \diamond B$ is defined as:

$$\boldsymbol{u}^T \diamond B{:} = (\boldsymbol{u}^T \diamond \boldsymbol{b}_1, \boldsymbol{u}^T \diamond \boldsymbol{b}_2, \ldots, \boldsymbol{u}^T \diamond \boldsymbol{b}_l), \tag{8.11}$$

where $\boldsymbol{b}_i$ denotes the ith column of B. Thus $\boldsymbol{u}^T \diamond B$ is an $m \times l$ matrix. For example:

$$(5, 2) \diamond \begin{pmatrix} 2 & 3 \\ 1 & 4 \end{pmatrix}{:} = \left((5, 2) \diamond (2, 1)^T, (5, 2) \diamond (3, 4)^T\right) = \begin{pmatrix} 10 & 15 \\ 2 & 8 \end{pmatrix} \tag{8.12}$$

It is easy to see that h_B defined by $h_B(\boldsymbol{u}){:} = \boldsymbol{u}^T \diamond B$ is linear and mantissa preserving.

Finally, defining function rowsum (C) for a matrix $C = (c_1, \ldots, c_m)$ as rowsum $(C) := (\mp(c_1), \ldots, \mp(c_m))$ where $\mp(v) := \sum_{j=1}^{m} v_i$, we obtain the decomposition of the vector–matrix product as stated in theorem 2 (Dutt and Assaad, 1996):

Theorem 2: The vector–matrix product $\boldsymbol{u}^T \cdot B := f_B(\boldsymbol{u}) =$ rowsum $\circ\, h_B(\boldsymbol{u})$.

Since matrix multiplication $A \cdot B$ is a sequence of $f_B(\boldsymbol{a}_i)$ computations, one for each row of A, one can apply a mantissa-based integer row checksum test to the $h_B(\boldsymbol{a}_i)$ components to precisely check for errors in the floating point multiplicant in $A \cdot B$. This integer row checksum test is as follows:

$$h_B^{\text{mant}}\left(\sum_{i=1}^{n} \text{mant}(a_i)\right) \stackrel{?}{=} \sum_{i=1}^{n} \text{mant}(h_B(\boldsymbol{a}_i)), \qquad (8.13)$$

or in other words

$$\text{rowsum}(\text{mant}(A)) \diamond \text{mant}(B) \stackrel{?}{=} \sum_{i=1}^{n} \text{mant}(\boldsymbol{a}_i^T \diamond B). \qquad (8.14)$$

Note that the RHS of (Banerjee, 1986) is obtained almost for free from the floating point computations $\boldsymbol{a}_i^T \diamond B$ that are computed as part of the entire floating point vector matrix product $\boldsymbol{a}_i^T \times B$. A similar derivation can be made for an integer column checksum test.

The floating point additions have to be tested by applying the floating point checksum tests to rowsum $\circ h_B(\boldsymbol{u}) := f_B(\boldsymbol{u})$ (i.e., to the final matrix product $A \cdot B$) to give rise to the hybrid test for matrix multiplication.

Error Coverage Results

Analytical Results

Two noteworthy results that have been obtained regarding the error coverage of the mantissa checksum method are given in the two theorems in this subsection.

Theorem 3: If either modulo or extended-precision integer arithmetic is used in a mantissa checksum test of the form of equation 8.6 shown again here:

$$f\left(\sum_{v \in s} \text{mant}(v)\right) \stackrel{?}{=} \sum_{v \in S} \text{mant}(f(v)),$$

then any single bit error in each of the scalar component of this test will be detected even in the presence of overflow in modulo (or single-precision) integer arithmetic (Dutt and Assaad, 1996).

In equation 8.6, scalars a_i and b_i are compared, where $a := (a_1, \ldots, a_n)^T$ and $b := (b_1, \ldots, b_n)^T$ are the LHS and RHS, respectively of equation 8.6. The above result means that single bit errors in either a_i or b_i can be detected for each i even when single-precision integer arithmetic is used. The following result is regarding the maximum number of arbitrarily distributed errors (i.e., not necessarily restricted to one error per scalar component of the check) that can be detected by the mantissa checksum test.

Theorem 4: The row and column mantissa checksums for matrix multiplication can detect errors in any three elements of the product matrix $C = A \cdot B$ that are due to errors in the floating point multiplications used to compute these elements (Assaad and Dutt, 1992).

The mantissa checksum test also implicitly detects errors in the exponents of the floating point products. This is done during the denormalization process by checking if $\exp(a) + \exp(b) = \exp(a \cdot b)$, which occurs when the floating point multiplier did not need to normalize the product $a \cdot b$, or if $\exp(a) + \exp(b) = \exp(a \cdot b + 1)$, meaning that a normalization was needed and the mantissa of $a \cdot b$ needs to be denormalized for use in the mantissa checksum test. If neither of these conditions hold, then an error is detected in the exponent of $a \cdot b$.

Empirical Results

A **dynamic range** of x means that the exponents of the input data lie in the interval $[-x, x]$. In Figure 8.20, the error coverage or the number of detection events (for single errors) is plotted against different dynamic ranges of the input data for the following tests.

(1) The **thresholded floating point checksum test** is used with the lower 24 bits masked in the checksum comparison for matrix multiplication and 12 bits for **LU** decomposition. The threshold of the floating point checksum test component of the hybrid checksum test was chosen to correspond to masking the lower 24 (12) bits, which guarantees almost zero false alarm in matrix multiplication (**LU** decomposition).

(2) The **mantissa checksum test** is used alone, as described previously.

(3) The **hybrid checksum test** uses both the thresholded floating point test and the mantissa checksum test—an error is detected in the hybrid test if an error is detected in its mantissa checksum test or in its floating point checksum test.

The plots in Figure 8.20 clearly show the significant improvements in coverage of the hybrid checksum test with respect to both the mantissa and the floating point checksum tests. For the low false alarm case, the mantissa checksum test has superior coverage compared to the floating point checksum test. An important point to be noted that is not apparent from the plots of Figure 8.20 is that the mantissa checksum test detects 100% of all multiplication errors for both matrix multiplication and **LU** decomposition.

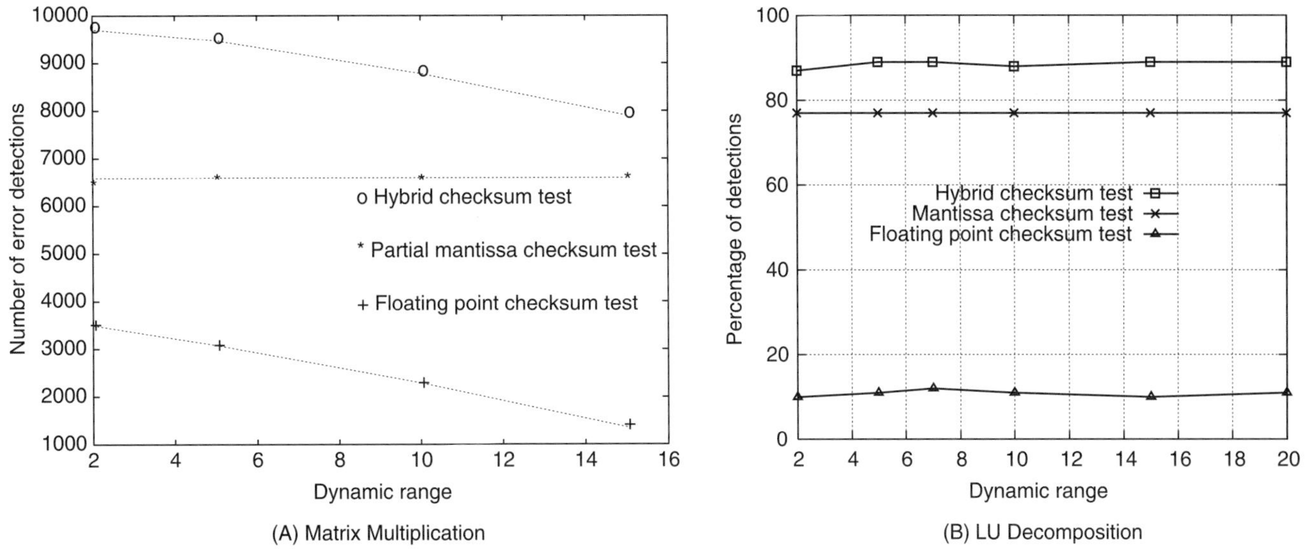

FIGURE 8.20 Error Coverage Versus Dynamic Range of Data. These data are for the mantissa checksum test, a properly thresholded floating point checksum test, and the hybrid checksum test.

Note that for matrix multiplication, the error coverage of the hybrid test is as high as 97% for a dynamic range of 2 and is 80% for a dynamic range of 15; this is a much higher error coverage than the technique of forward-error propagation used with the floating point checksum test in work by Roy-Chowdhury and Banerjee (1993). For **LU** decomposition, error coverage of 90% is obtained for a dynamic range of 7, which is similar to Roy-Chowdhury and Banerjee's results.

Timing Results

Note that part of the overhead of a mantissa checksum test is extracting the mantissas of the input matrices or vectors and also extracting and denormalizing the mantissas of the intermediate multiplications $a_{i,j} \cdot b_{j,k}$. The latter overhead can be eliminated by a very simple modification to the floating point multiplication unit that is shown in Figure 8.21. With this modification, the mantissa that is not normalized is also available (along with the normalized mantissa) as an output of the floating point multiplier. In many computers, the floating point product is also available in the form that is not normalized by using the appropriate multiply instruction—this requires tinkering with the compiler in such a machine to use the non-normalized multiply instruction where appropriate. No hardware modification is needed in this case to extract the mantissa for free.

Figure 8.22 shows the plots of the times of the fault-tolerant computations that use the hybrid checksum test and that use only the floating point checksum test. The average overhead of the hybrid checksum for matrix multiplication is 15%, whereas for **LU** decomposition, the overhead is only 9.5%. Thus, the significantly higher error coverages yielded by the mantissa checksum test are obtained at only nominal time overheads that are lower than those of previous techniques (such as by Roy-Chowdhury and Banerjee, 1993) developed for combatting the susceptibility of the floating point checksum test to round-off.

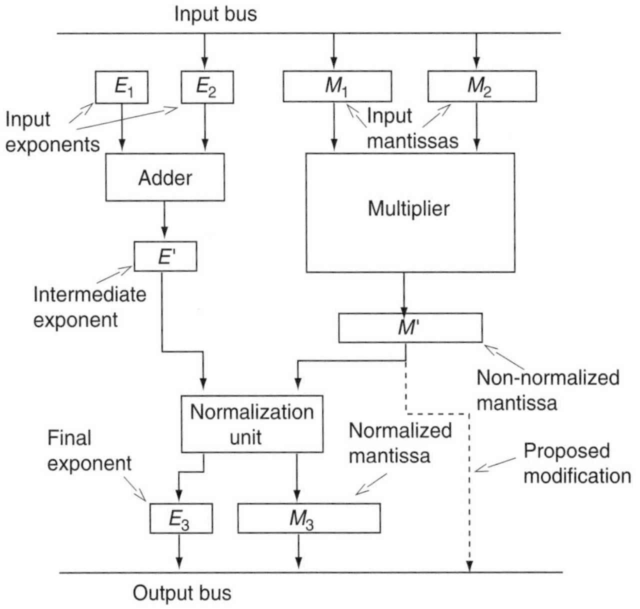

FIGURE 8.21 A Simple Modification of a Floating Point Multiplier. This multiplier is shown by the dashed line from internal register M' to the output bus to make the non-normalized mantissa of the product available at no extra time penalty.

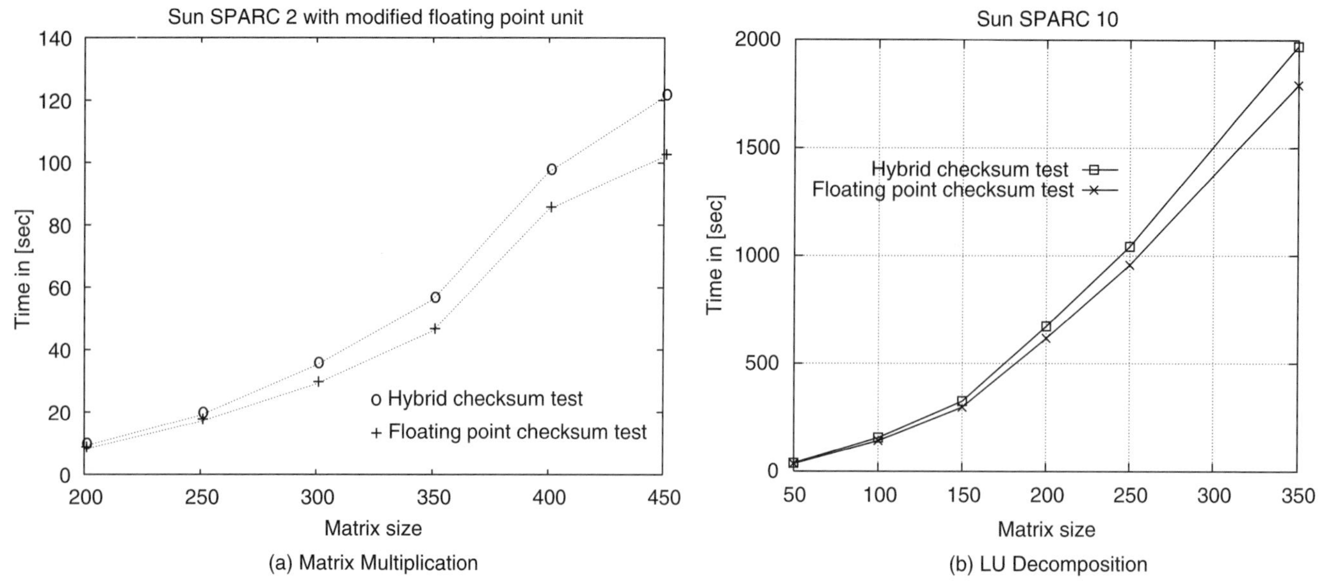

FIGURE 8.22 Timing Results with a Simulated Modification of the Floating Point Multiplier

8.7 Conclusions

We have presented a summary of various fault-detection and fault-tolerance methods at various levels of a computing system: arithmetic circuit, FPGA, program control flow, processor system and algorithm. We have also presented some of our works in these areas. Results show that the techniques discussed produce significant increases in system reliability. As computer chips become more complex and denser, with smaller devices, they and systems incorporating them become more and more vulnerable to faults arising from a myriad of sources like fabrication errors, operation extremes, and external disturbances. Computer systems are also being increasingly used in various economic- and life-critical environments. It is thus hoped that microprocessor and computer manufacturing will undergo a paradigm shift whereby reliability becomes an important metric, and fault tolerance capabilities of different degrees such as those discussed here are explicitly designed in computer chips and systems to provide different levels of dependability and reliability needed in various application environments.

References

Al-Arian, S., and Gumusel, M. (1992). HPTR: Hardware partition in time redundancy technique for fault tolerance. *Proceedings IEEE SOUTHEASTCON 2*, 630–633.

Alewine, N.S., Chen, S.K., Fuchs, K.W., and Hwu, W.M.W. (1995). Compiler-assisted multiple instruction rollback recovery using a read buffer. *IEEE Transactions on Computers 44*(9), 1096–1107.

Altera Corporation. (1993). *Flex 8000 programmable logic device family.*

Assaad, F.T., and Dutt, S. (1992). More robust tests in algorithm-based fault-tolerant matrix multiplication. *The Twenty-Second Fault-Tolerant Computing Symposium*, 430–439.

Avizienis, A. (1973). Arithmetic algorithms for error-coded operands. *IEEE Transactions on Computers. C-22*, 567–572.

Banerjee, P., Kuo, S.Y., and Fuchs, W.K. (1986). Reconfigurable cube-connected cycles architecture. *Proceedings of the Sixteenth Fault-Tolerant Computer Symposium.* 286–291.

Banerjee, P., Rahmeh, J.T., Stunkel, C., Nair, V.S., Roy, K., and Abraham, J.A. (1990). Algorithm-based fault tolerance on a hypercube multiprocessor. *IEEE Transactions on Computers 39*, 1132–1145.

Boley, D.L., Golub, G.H., Makar, S., Saxena, N., and McCluskey, E.J. (1995). Floating point fault tolerance using backward error assertions. *IEEE Transactions on Computers 44*(2), 302–311.

Boley, D.L., and Luk, F.T. (1991). A well-conditioned checksum scheme for algorithmic fault tolerance. *Integration, the VLSI Journal 12*, 21–32.

Bowden, N.S., and Pradhan, D.K. (1992). Virtual checkpoints: Architecture and performance. *IEEE Transactions on Computers 41*(5), 516–525.

Bruck, J., Cypher, R., and Ho, C.T. (1993a). Wildcard dimensions coding theory, and fault-tolerant meshes and hypercubes. *Proceedings of the 23rd Fault-Tolerant Computer Symposium.* 260–267.

Bruck, J., Cypher, R., and Ho, C.T. (1993b). Fault-tolerant meshes and hypercubes with a minimal number of spares. *IEEE Transactions on Computers 42*(9), 1089–1104.

Chandy, M., and Ramamoorthy, C.V. Rollback and recovery strategies for computer programs. *IEEE Transactions on Computers 21*(6), 546–556.

Chau, S.C., and Liestman, A.L. (1989). A proposal for a fault-tolerant binary hypercube. *Proceedings of the Nineteenth Fault-Tolerant Computer Symposium*, 323–330.

Chen, T.H., Chen, L.G., and Jehng, Y.S. (1992). Design and analysis of VLSI-based arithmetic arrays with error correction. *International Journal of Electronics 72*(2), 253–271.

Chung, F.R.K., Leighton, F.T., and Rosenberg, A.L. (1983). Diogenes: A methodology for designing fault-tolerant VLSI processor arrays. *Proceedings of the Thirteenth Fault-Tolerant Computer Symposium.* 26–31.

Cliff, R., Ahanin, B., Cope, L.T., Heile, F., Ho, R., Huang, J., Lytle, C., Mashruwala, S., Pederson, B., Ranian, R., Reddy, S., Singhal, V., Sung, C.K., Veenstra, K., Gupta, A. (1993). A dual granularity and globally interconnected architecture for a programmable logic device. *Proceedings of the IEEE CICC.* 7.3.1–7.3.5.

Cunningham, J. (1990). The use and evaluation of yield models in integrated circuit manufacturing. *IEEE Transactions of Semiconducter manufacturing 3*(2), 60–71.

Delord, X., and Saucier, G. (1991). Formalizing signature analysis for control flow testing of pipelined RISC microprocessors. *Proceedings of the International Test Conference.* 936–945.

Dutt, S., and Hayes, J.P. (1990). On designing and reconfiguring *k*-fault-tolerant tree architectures. *IEEE Transactions on Computers C-39,* 490–503.

Dutt, S., and Hayes, J.P. (1991). Designing fault-tolerant systems using automorphisms. *Journal of Parallel and Distributed Computing.* 249–268.

Dutt, S., and Hayes, J.P. (1992). Some practical issues in the design of fault-tolerant multiprocessors. *IEEE Transactions on Computers 41,* 588–598.

Dutt, S. (1993). Fast polylog-time reconfiguration of structurally fault-tolerant multiprocessors. *Proceedings of the Fifth IEEE Symposium on Parallel and Distributed Processings.* 762–770.

Dutt, S., and Assaad, F. (1996). Mantissa-preserving operations and robust algorithm-based fault tolerance for matrix computations. *IEEE Transactions on Computers 45*(4), 408–424.

Dutt, S., and Hanchek, F. (1999). REMOD: A new hardware- and time-efficient methodology for designing fault-tolerant arithmetic circuits. *IEEE Transactions on VLSI Systems 4*(1). 34–56.

Dutt, S., and Mahapatra, N.R. (1997). Node covering, error correcting codes, and multiprocessors with high average fault tolerance. *IEEE Transactions on Computers.* 997–1015.

Dutt, S., and Hayes, J.P. (1997). A local-sparing design methodology for fault-tolerant multiprocessors. *Computers and Mathematics with Applications 34*(11), 25–50.

Dutt, S., and Boley, D. (1999). Roundoff errors. *Wiley Encyclopedia of Electrical and Electronics Engineering 18.* 617–627.

Dutt, S., Shanmugavel, V., and Trimberger, S. (1999). Efficient incremental rerouting for fault reconfiguration in field programmable gate arrays. *Proceedings of the IEEE International Conference on Computer-Aided Design.* 173–177.

Elnozahy, E.N., Alvisi, L., Wang, Y.M., and Johnson, D.B. (2002). A survey of rollback-recovery protocols in message-passing system. *ACM Computing Surveys 34*(3).

Emmert, J.M., and Bhatia, D.K. (1998). Incremental routing in FPGAs. *Proceedings of 11th Annual IEEE International Application Specific Integrated Circuit Conference.*

Gray, C.T., Liu, N., and Cavin, R.K., III. (1994). *Wave pipelining: Theory and CMOS implementation.* Norwell, MA: Kluwer Academic Publishers.

Gunneflo, U., and Karlsson, J.T. (1989). Evaluation of error-detection schemes using fault injection by heavy-ion radiation. *Nineteenth International Symposium in Fault-Tolerant Computing.*

Hanchek, F., and Dutt, S. (1996a). Node-covering based defect and fault-tolerance methods for increased yield in FPGAs. *Proceedings of the International Conference on VLSI Design.* 225–229.

Hanchek, F., and Dutt, S. (1996b). Design methodologies for tolerating cell and interconnect faults in FPGAs. *Proceedings of the International Conference on Computer Design.* 326–331.

Hanchek, F., and Dutt, S. (1998). Methodologies for tolerating logic and interconnect faults in FPGAs. *IEEE Transactions on Computers.* 15–33.

Hastie, N., and Cliff, R. (1990). The implementation of hardware subroutines on field programmable gate arrays. *Proceedings of the IEEE Custom Integrated Circuits Conference.* 31.4.1–31.4.4.

Hatori, F. *et al.,* (1993). Introducing redundancy in field programmable gate arrays. *Proceedings of the IEEE Custom Integrated Circuits Conference.* 7.1.1–7.1.4.

Hayes, J.P. (1976). A graph model for fault-tolerant computing systems. *IEEE Transactions on Computers C-25,* 875–883.

Hayes, J.P. (1985). Fault modeling, *IEEE Design and Test 2*(2), 88–95.

Heishman, R., Hollinden, D., K.J. and T., M. Model of the mc 68000 developed at the laboratory for digital design environments of the university of cincinnati. http://www.uc.edu.

Hsu, Y.M., and Swartzlander, E.E., Jr., (1992). Time-redundant error correcting adders and multipliers. *Proceedings of the IEEE International Workshop on Defect and Fault Tolerance in VLSI Systems.* 247–256.

Huang, K.H., and Abraham, J.A. (1984). Algorithm-based fault tolerance for matrix operations. *IEEE Transactions on Computers C-33*(6), 518–528.

Huang, A., and Kintala, C. (1993). Software-implemented fault tolerance: Technologies and experience. *Proceedings of the Twenty-Third Annual International Symposium on Fault-Tolerant Computing,* 2–9.

Johnson, B., Aylor, J., and Hana, H. (1988). Efficient use of time and hardware redundancy for concurrent error detection in a 32-bit VLSI adder. *IEEE Journal of the Solid-State Circuits 23*(1), 208–215.

Johnson, B. (1989). *Design and analysis of fault-tolerant digital systems.* Reading, MA: Addison-Wesley.

Jou, J.Y., and Abraham, J.A. (1986). Fault-tolerant matrix arithmetic and signal processing on highly concurrent computing structures. *Proceedings of the IEEE 74*(5) 732–741.

Kohavi, Z. (1978). *Switching and finite automata theory.* New York: McGraw-Hill.

Koo, R., and Toueg, S. Checkpointing and rollback-recovery for distributed systems. *IEEE Transactions on Software Engineering, 13*(1), 23–31.

Kumar, V., Dahbura, A., Fisher, F., and Juola, P. (1989). An approach for the yield enhancement of programmable gate arrays. *Proceedings of the IEEE International Conference on Computer–Aided Design.* 226–229.

Lach, J., Mangione-Smith, W.H., and Potkonjak, M. (1998). Low overhead fault-tolerant FPGA systems. *IEEE Transactions on VLSI Systems 6*(2).

Laha, S., and Patel, J. (1983). Error correction in arithmetic operations using time redundancy. *Proceedings of the Thirteenth International Symposium on Fault-Tolerant Computing,* 298–305.

Lala, P.K. (1985). *Fault-tolerant and fault-testable hardware design.* London: Prentice-Hall International.

Laprie, J.C. (1992). Dependability: Basic concepts and terminology. *Dependable Computing and Fault-Tolerant Systems 5.*

Lee, P.A., and Anderson, T. (1990). Fault tolerance: Principles and practice. *Dependable Computing and Fault-Tolerant Systems 3.*

Lemieux, G., and Brown, S. (1993). A detailed router for allocating wire segments in FPGAs. *ACM Physical Design Workshop.* 215–226.

Li, C.C.J., and Fuchs, W.K. CATCH (1990). Compiler-assisted techniques for checkpointing. *Proceedings of the 20th Annual IEEE International Fault-Tolerant Computing Symposium.* 74–81.

Lo, J.C., Thanawastien, S., and Rao, J.R.N. (1993). Berger check prediction for array multipliers and array dividers. *IEEE Transactions on Computers C-42*(7), 892–896.

Long, J., Fuchs, W.K., and Abraham, J.A. (1992). Implementing forward recovery using checkpoints in distributed systems. In Meyer and Schlichting (Eds.), *Dependable Computing and Fault-Tolerant Systems—Dependable Computing for Critical Applications* Vol. 6, New York: Springer-Verlag.

Lu, D.J. (1982). Watchdog processors and structural integrity checking. *IEEE Transactions on Computers 31,* 681–685.

Lucent Technologies. (1998). *Field programmable GateArrays data book.*

Luk, F.T. (1985). Algorithm-based fault tolerance for parallel matrix solvers. *Proceedings of SPIE Real-Time Signal Processing VIII 564,* 49–53.

Luk, F.T., and Park, H. (1988). An analysis of algorithm-based fault-tolerance techniques. *Journal of Parallel and Distributed Computing 15,* 172–184.

Mahapatra, N.R., and Dutt, S. (2001). Hardware efficient and highly reconfigurable 4- and 2-track fault-tolerant designs for mesh-connected arrays. *Journal of Parallel and Distribution Computing 61*(10), 1391–1411.

Mahmood, A., and McCluskey, E.J. (1988). Concurrent error detection using watchdog processors—a survey. In *IEEE Transactions on Computers 37*(2), 160–174.

Malek, M., and Choi, Y.H. (1985). A fault-tolerant FFT processor. *Proceedings of the Fifteenth Fault-Tolerant Computer Symposium.* 266–271.

McDonald, J., Philhower, B., and Greub, H.A. (1991). A fine-grained, highly fault-tolerant system based on WSI and FPGA technology. W. Moore and W. Luk (Eds.), *FPGAs.* Abingdon, England: Abingdon EE & CS Books.

Mazumder, P., and Chakraborty, K. (1996). *Testing and testable design of random-access memories.* Norwell, MA: Kluwer Academic.

Mazumder, P., and Yih, J. (1993). A new built-in self-repair approach to VLSI memory yield enhancement by using neural-type circuits. *IEEE Transactions on Computer-Aided Design of Integrated Circuits and Systems 12*(1), 124–136.

Nair, V.S.S., and Abraham, J.A. (1990). Real-number codes for fault-tolerant matrix operations on processor arrays. *IEEE Transactions on Computers 39*(4), 426–435.

Narasimhan, J., Nakajima, K., Rim, C., and Dahbura, A. (1994). Yield enhancement of programmable ASIC arrays by reconfiguration of circuit placements. *IEEE Transactions on Computer–Aided Design of Integrated Circuits and Systems 13*(8), 976–986.

Nicolaidis, M. (1993). Efficient implementations of self-checking adders and ALUs. *Proceedings of the Twenty-Third International Symposium on Fault-Tolerant Computing.* 586–595.

Ohlsson, J., and Rimen, M. (1995). Implicit signature checking. *Proceedings of the 22nd International Symposium on Fault-Tolerant Computing.* 218–228.

Patel, J.H., and Fung, L.H. (1982). Concurrent error detection in ALU's by recomputing with shifted operands. *IEEE Transactions on Computers C-31*(7), 589–595.

Patel, J.H., and Fung, L.Y. (1983). Concurrent error detection in multiply and divide arrays. *IEEE Transactions on Computers, 32*(4), 417–422.

Patterson, D.A., Brown, A., Broadwell, P., Candea, G., Chen, M., Cutler, J., Enriquez, P., Fox, A., Kiciman, E., Merzbacher, M., Oppenheimer, P., Sastry, N., Tetzlaff, W., Traupman, J., and Treuhaft, N. (2002). Recovery-oriented computing (ROC): Motivation, definition, techniques, and case studies. *UC Berkeley Computer Science Technical Report UCB//CSD-02-1175.*

Pradhan, D.K. (Ed.) (1990). *Fault-tolerant computing—Theory and techniques, Vol. 1,* Englewood Cliffs, NJ: Prentice-Hall.

Prasad, V.B. (1989). Fault-tolerant digital systems. *IEEE Potentials 8*(1), 17–21.

Raghavendra, C.S., Avizienis, A., and Ercegovac, M.D. (1984). Fault tolerance in binary tree architectures. *IEEE Transactions on Computers C-33,* 568–572.

Rao, T.R.N. (1968). Error-checking logic for arithmetic-type operations of a processor. *IEEE Transactions on Computers, C-17,* 845–849.

Rao, T.R.N. (1974). *Error coding for arithmetic processors.* New York: Academic Press.

Rennels, D.A. (1986). On implementing fault-tolerance in binary hypercubes. *Proceedings of the Sixteenth Fault-Tolerant Computer Symposium.* 344–349.

Rota, F. (2002). *Control flow checking using main memory bus monitoring in an internal cache environment.* M.S. Thesis, University of Illinois at Chicago.

Roth, C.H. Jr., (1975). *Fundamentals of logic design.* St. Paul, MN: West Publishing.

Roy, K., and Nag, S. (1995). On routability for FPGAs under faulty conditions. *IEEE Transactions on Computers 44,* 1296–1305.

Roy-Chowdhury, A., and Banerjee, P. (1993). Tolerance determination for algorithm-based checks using simple error analysis techniques. *Fault-Tolerant Computing Symposium FTCS-23,* 290–298.

Roy-Chowdhury, V.P., Bruck, J., and Kailath, T. (1990). Efficient algorithms for reconfiguration in VLSI/WSI arrays. *IEEE Transactions on Computers 39*(4), 480–489.

Shnidman, N.R., Mangione-Smith, W.H., and Potkonjak, M. (1998). On-line fault detection for bus-based field programmable gate arrays. *IEEE Transactions on VLSI Systems 6*(4), pages 656–666.

Sparmann, U., and Reddy, S.M. (1994). On the effectiveness of residue code checking for parallel two's complement multipliers. *Proceedings of the Twenty-Fourth International Symposium on Fault-Tolerant Computing.* 219–228.

Ssu, K.F., Yao, B., Fuchs, W.K., and Neves, N.F. (1999). Adaptive checkpointing with storage management for mobile environments. *IEEE Transactions on Reliability 48*(4), 315–324.

Tamir, Y., and Tremblay, M. (1990). High-performance fault-tolerant VLSI system using micro rollback. *IEEE Transactions on Computers 39*, 548–554.

Tanner, R.M. (1984). Fault-tolerant 256K memory designs. *IEEE Transactions on Computers C-33*(4), 314–322.

Trovo, F. (2002). *Concurrent control flow checking with microrollback in a CISC processor.* M.S. Thesis. University of Illinois at Chicago.

Verma, V., Dutt, S., and Suthar, V. (2004). Efficient on-line testing of FPGA's with provable diagnostabilities. Accepted for publication, Proc. IEEE/ACM Design Automation Conference, June 2004. 498–503. (Nominated for Best Paper Award).

Wang, Y.M., Chung, P.Y., Lin, J.J., and Fuchs, W.K. (1995). Checkpoint space reclamation for uncoordinated checkpointing in message-passing systems. *IEEE Transactions on Parallel and Distributed Systems 6*(5), 546–554.

Wilkinson, J.H. (1963). *Rounding errors in algebraic processes.* Englewood Cliffs, NJ: Prentice Hall.

Wilkinson, J.H. (1965). *The algebraic eigenvalue problem.* Oxford: Clarendon Press.

Xilinx. (1994). *The programmable logic data book.*

Yu, Y., and Johnson, B.W. (2002). A perspective on the state of research in fault injection techniques. University of Virginia, Center for Safety-Critical Systems, Department of Electrical and Computer Engineering: *Technical Report.*

9

High-Level Petri Nets—Extensions, Analysis, and Applications

Xudong He
School of Computer Science, Florida International University, Miami, Florida, USA

Tadao Murata
Department of Computer Science, University of Illinois at Chicago, Chicago, Illinois, USA

9.1 Introduction

Petri nets are an excellent formal model for studying concurrent and distributed systems and have been widely applied in many different areas of computer science and other disciplines (Murata, 1989). There have been over 8000 publications on Petri nets (refer to Website *http://www.daimi.au.dk/PetriNets/*). Since Carl Adam Petri originally developed Petri nets in 1962, Petri nets have evolved through four generations: the first-generation low-level Petri nets primarily used for modeling system control (Reisig, 1985a), the second-generation high-level Petri nets for describing both system data and control (Jensen and Rozenberg, 1991), the third-generation hierarchical Petri nets for abstracting system structures (He and Lee, 1991; He, 1996; Jensen, 1992), and the fourth-generation object-oriented Petri nets for supporting modern system development approaches (Agha, 2001). Petri nets have also been extended in many different ways to study specific system properties, such as performance, reliability, and schedulability. Well-known examples of extended Petri nets include timed Petri nets (Wang, 1998) and stochastic Petri nets (Marsan *et al.*, 1994; Haas, 2002). In this article, we present several extensions to Petri nets based on our own research work and provide analysis techniques for these extended Petri net models. We also discuss the intended applications of these extended Petri nets and their potential benefits.

9.2 High-Level Petri Nets

In the past few years, a concerted effort by the worldwide Petri net community has resulted in an international standard on defining high-level Petri nets (HLPN) (ISO/IEC, 2002) that will profoundly help to facilitate and promote the research and applications of these nets. In the following sections, we follow as closely as possible the notations, concepts, and definitions given in the standard documentation to introduce high-level Petri nets, which are used later to define our extensions.

9.2.1 The Syntax and Static Semantics of High-Level Petri Nets

An **HLPN** is a structure:

$$N = (NG, \text{Sig}, V, H, Type, AN, M_0).$$

Where:

- $NG = (P, T; F)$ is a net graph, with:
  - P a finite set of nodes, called places;
  - T a finite set of nodes, called transitions, disjoint from P ($P \cap T = \emptyset$); and
  - $F \subseteq (P \times T) \cup (T \times P)$ a set of directed edges called arcs, known as the flow relation.
- $Sig = (S, O)$ is a Boolean signature, where S is a set of sorts and O is a set of operators defined in the Annex A of ISO/IEC (2002).
- V is an S-indexed set of variables, disjoint from O.
- $H = (S_H, O_H)$ is a many-sorted algebra for the signature Sig, defined earlier in this list:
- $Type$: $P \rightarrow S_H$ is a function that assigns types to places.
- $AN = (A, TC)$ is a pair of net annotations.
  - A: $F \rightarrow TERM(O \cup V)$ such that for all (p, t), $(t, p) \in F$ and all bindings α, $Val_\alpha(A(p, t))$, $Val_\alpha(A(t', p)) \in \mu Type(p)$. $TERM(O \cup V)$, α, Val_α and $\mu Type(p)$ are defined in Annex A of ISO/IEC (2002). A is a function that annotates each arc with a term that when evaluated (for some binding) results in a multiset over the associated place's type.
  - TC: $T \rightarrow TERM(O \cup V)_{\text{Bool}}$ is a function that annotates transitions with Boolean expressions.
- $M_0: P \rightarrow \cup_{p \in P}\ \mu Type(p)$ such that $\forall p \in P$, $M_0(p) \in \mu Type(p)$ is the initial marking function that associates a multiset of tokens (of the correct type) with each place.

The above definitions are directly from ISO/IEC 2002. Basically, an HLPN has three essential parts: (1) a net graph NG defining the syntax, (2) an underlying algebraic specification (Sig, V, H) defining the semantic domain, and (3) a net inscription ($Type$, AN, M_0) mapping syntactic entities to their semantic denotations. By restricting the underlying algebraic specification and/or the net inscription, different variations of high-level Petri nets can be obtained.

9.2.2 Dynamic Semantics

Marking

The marking M of the HLPN is defined in the same way as the initial marking:

$M: P \rightarrow \cup_{p \in P}\ \mu Type(p)$ such that $\forall p \in P$, $M(p) \in \mu Type(p)$.

Enabling

- **Enabling of a Single Transition Mode**

A transition $t \in T$ is enabled in a marking M for a particular assignment of α_t to its variables and satisfies the transition condition, $Val_{\text{bool}}(TC(t)) = \text{true}$, known as a **mode** of t, if:

$$\forall p \in P\ Val_{\alpha_t}(\overline{p, t}) \leq M(p),$$
$$\text{where for } (u, v) \in (P \times T) \cup (T \times P),$$
$$\overline{u, v} = A(u, v) \text{ for } (u, v) \in F, \text{ or } \overline{u, v} = \emptyset \text{ for } (u, v) \notin F.$$

- **Concurrent Enabling of Transition Modes**

Let α_t be an assignment for the variables of transition $t \in T$ that satisfies its transition condition, and then denote the set of all assignments for transition t, by Assign_t. Define the set of all transition modes to be $TM = \{(t, \alpha_t) | t \in T, \alpha_t \in \text{Assign}_t\}$ and a step to be a finite nonempty multiset over TM.

A step X of transition modes is enabled in a marking, M, if:

$$\forall p \in P \sum_{(t, \alpha_t) \in X} Val_{\alpha_t}(\overline{p, t}) \leq M(p).$$

Thus, all of the transition modes in X are concurrently enabled if X is enabled. Enabling of a single transition mode is a special case of concurrent enabling of transition modes.

The enabling condition of a transition specifies that enough right tokens are available to make the transition fire.

Transition Rule

If $t \in T$ is enabled in mode α_t for marking M, t may occur in mode α_t. When t occurs in mode α_t, the marking of the net is transformed to a new marking M', denoted in $M[t, \alpha_t > M'$, according to the following rule:

$$\forall p \in P(M'(p) = M(p) - Val_{\alpha_t}(\overline{p, t}) + Val_{\alpha_t}(\overline{t, p})).$$

If a step X is enabled in marking M, then the step can occur, resulting in a new marking of M' denoted by $M[X > M'$, where M' is given by:

$$\forall p \in P(M'(p) = M(p) - \sum_{(t, \alpha_t) \in X} Val_{\alpha_t}(\overline{p, t}) + \sum_{(t, \alpha_t) \in X} Val_{\alpha_t}(\overline{t, p}))$$

The firing rule defines the effect of firing a transition, which consumes specific tokens according to the mode in the preset (input places) of the transition and generates new tokens in the postset (output places) of the transition.

Behavior of A HLPN N

An **execution sequence** $M_0[X_0> M_1[X_1> \ldots$ of N is either finite when the last marking is terminal (no more enabled transition in the last marking) or infinite, in which each X_i is a step. The **behavior** of N is the set of all execution sequences starting from the initial marking. The set of all reachable markings from the initial marking is denoted by $[M_0>$.

Although ISO/IEC (2002) has defined the concepts of markings, transition enabling, and firing rules, it stopped short in defining the dynamic semantics of an HLPN. In the existing Petri net literature, there have been several semantic models of Petri nets, such as interleaving, concurrent, and causal executions (Reisig, 1985b). From our own experience, we feel that the interleaving set semantics model defined above is adequate for studying many useful system behavioral properties.

9.3 Temporal Predicate Transition Nets

Petri nets are a model-oriented formal method and are well suited for modeling the dynamic behaviors of concurrent and distributed systems. Petri net specifications reveal system design structures and thus provide guidelines for system implementation. Furthermore, Petri net specifications are executable and support system simulation and testing in addition to formal analysis. It is not easy, however, to directly and explicitly express behavioral properties using Petri nets. On the other hand, property-oriented formal methods, such as temporal logic, are ideal for specifying and analyzing behavioral properties. It would be great to have a hybrid formal method by integrating the strengths of model-oriented and property-oriented formal methods. In recent years, integrating formal methods has become a major research area (Clarke and Wing, 1996). There have been several published results on integrating Petri nets and temporal logic (Anttila *et al.*, 1983; Diaz *et al.*, 1983; He and Lee, 1990; He, 1992; Mandrioli *et al.*, 1996; Suzuki and Lu, 1989). Most of the earlier work was either lacking a systematic approach or using a low-level Petri net model. In the following subsections, we describe our results on integrating predicate transition nets (Genrich and Lautenbach, 1981) and first-order linear time temporal logic (Manna and Pnueli, 1992).

9.3.1 Definition of Temporal Logic

In this section, we define a linear time first-order temporal logic (LTFOTL) in the style of Lamport (1994a) based on a given PrTN. Let $N = (NG, Sig, V, H, Type, AN, M_0)$ be a PrTN.

Values, State Variables, and States

The **set of values** is the multiset of tokens defined by the ground terms TERM(O) of N. Multisets can be viewed as partial functions. For example, multiset $\{3a, 2b\}$ can be represented as $\{a \mapsto 3, b \mapsto 2\}$.

The **set of state variables** is the set P of places of NG, which change their meanings during the executions of N. The arity of a place p is determined by its type $Type(p)$.

The **set of states** ***St*** is the set of all reachable markings $[M_0>$ of N. A marking is a mapping from the set of state variables into the set of values. We use $M[\![x]\!]$ to denote the value of x under state (marking) M.

Since state variables take partial functions as values, they are *flexible* function symbols. We can access a particular component value of a state variable. There is a problem, however, associated with partial functions (i.e., many values are undefined). The above problem can easily be solved by extending state variables into total functions in the following way: for any n-ary state variable p, any tuple $c \in$ TERM(O), and any state M, if $p(c)$ is undefined under M, then let $M[\![p(c)]\!] = 0$. The above extension is consistent with the semantics of PrTN. Furthermore we can consider the meaning $[\![p(c)]\!]$ of the function application $p(c)$ as a mapping from states to **Nat** using a postfix notation for function application $M[\![p(c)]\!]$.

Rigid Variables, Rigid Function, and Predicate Symbols

Rigid variables are individual variables that do not change their meanings during the executions of N. All rigid variables occurring in our temporal logic formulas are bound (quantified), and they are the only variables that can be quantified. Rigid variables are variables appearing in the label expressions and constraints of N.

Rigid function and **predicate symbols** do not change their meanings during the executions of N. The set of rigid function and predicate symbols is defined in V of N.

State Functions, Predicates, and Transitions

A **state function** is an expression built from values, state variables, rigid function, and predicate symbols. For example, $[\![p(c) + 1]\!]$ is a state function, where c and 1 are values, p is a state variable, and $+$ is a rigid function symbol. Since the meanings of rigid symbols are not affected by any state, for any given state M, $M[\![p(c) + 1]\!] = M[\![p(c)]\!] + 1$.

A **predicate** is a Boolean-valued state function. A predicate p is satisfied by a state M if and only if $M[\![p]\!]$ is true.

A **transition** is a particular kind of predicate that contains primed state variables (e.g., $[\![p'(c) = p(c) + 1]\!]$). A transition relates two states (an old state and a new state), where the unprimed state variables refer to the old state and where the primed state variables refer to the new state. Therefore, the meaning of a transition is a relation between states. The term **transition** used here is a temporal logic entity. Although it reflects the nature of a transition in a PrTN net N, it is not a transition in N. For example, given a pair of states M and M': $M[\![p'(c) = p(c) + 1]\!]M'$ is defined by $M'[\![p(c)]\!] = M[\![p(c)]\!] + 1$. Given a transition t, a pair of states M and M' is called a "transition step" if $M[\![p]\!]M'$ equals true.

We can easily generalize any predicate p without primed state variables into a relation between states by replacing all unprimed state variables with their primed versions such that $M[\![p]\!]M'$ equals $M'[\![p]\!]$ for any states M and M'.

Temporal Formulas

Temporal formulas are built from elementary formulas (predicates and transitions) using logical connectives $\neg$ and $\wedge$ (and derived logical connectives $\vee$, $\Rightarrow$, and $\Leftrightarrow$), universal quantifier $\forall$ (and derived existential quantifier $\exists$), and temporal operator **always** $\Box$ (and derived temporal operator sometimes $\Diamond$).

The semantics of temporal logic is defined on infinite sequences of states that are extracted from the execution sequences of PrTNs, where the last marking of a finite execution sequence is repeated infinitely many times at the end of the execution sequence. For example, for an execution sequence $M_0[X_0> M_1[X_1> \ldots M_n$, the following infinite sequence behavior $\sigma = << M_0 \ldots M_n, M_n, \ldots >>$ is obtained. We denote the set of all possible behaviors obtained from a given PrTN as St^{∞}. Let u and v be two arbitrary temporal formulas; p be an m-ary place; t be a transition; $x, x_1 \ldots, x_n$ be rigid variables; $\sigma = << M_0, M_1, \ldots >>$ be a behavior; and $\sigma^k = << M_k, M_{k+1}, \ldots >>$ be a k-step-shifted behavior sequence. We define the semantics of temporal formulas recursively as follows:

(1) $\sigma[\![p(x_1, \ldots, x_n)]\!] \equiv M_0[\![p(x_1, \ldots, x_n)]\!]$
(2) $\sigma[\![t]\!] \equiv M_0[\![t]\!]M_1$
(3) $\sigma[\![\neg u]\!] \equiv \neg\sigma[\![u]\!]$
(4) $\sigma[\![u \wedge v]\!] \equiv \sigma[\![u]\!] \wedge \sigma[\![v]\!]$
(5) $\sigma[\![\forall xu]\!] \equiv \forall x \cdot \sigma[\![u]\!]$
(6) $\sigma[\![\Box u]\!] \equiv \forall n \in \textbf{Nat}\ \sigma^n[\![u]\!]$

Intuitively temporal operator always $\Box$ means every state in a state sequence, and its dual operator sometimes $\Diamond$ means some future state in a state sequence. The relationship between these two temporal operators is:

$$\neg\Box \text{W} \Diamond.$$

A temporal formula u is said to be **satisfiable** denoted as $\sigma \models u$, if there is an execution σ such that $\sigma[\![u]\!]$ is true (i.e., $\sigma \models u \Leftrightarrow \exists\sigma \in ST^{\infty} \cdot \sigma[\![u]\!]$). A temporal formula u is **valid** with regard to N, denoted as $N \models u$, if it is satisfied by all possible behaviors St^{∞} from N: $N \models u \Leftrightarrow \forall\sigma \in St^{\infty} \cdot \sigma[\![u]\!]$.

9.3.2 Temporal Predicate Transition Net

A temporal predicate transition net (TPrTN) is a tuple $TN = (NG, \text{Sig}, V, H, \text{Type}, AN, M_0, f)$, where $N = (NG, Sig, V, H, Type, AN, M_0)$ is a PrTN, and f is a LTFOTL formula that constrains the execution of TN. The semantics of TN are defined to be the set of execution sequences $\Sigma = \{\sigma | \sigma \in St^{\infty} \wedge \sigma[\![f]\!]\}$.

It is easy to see that a TPrTN TN is a PrTN N with a temporal logic formula f. The temporal logic formula is defined according to the net graph NG using logical connectives and operators, which can be viewed as a part of the underlying algebraic specification (Sig, V, H). The temporal logic formula f is evaluated using the dynamic behaviors of N. By incorporating a temporal formula into the definition of a PrTN, we are able to explicitly specify and verify many system properties such as fairness.

9.3.3 An Example of TPrTN

The **Five Dining Philosophers** problem is a classical example for studying behavioral properties of concurrent processes. There are five philosophers sitting around a round table, and there is one chopstick between two adjacent philosophers. Each philosopher is either thinking or eating. In order for a philosopher to eat, he needs to pick up two chopsticks next to him. There are several interesting issues with regard to a system model of this simple problem:

- Is the system deadlock free? A deadlock occurs when the system cannot progress anymore due to a formation of a waiting cycle.
- Has the system enforced mutual exclusion? Mutual exclusion ensures that no two neighboring philosophers can eat at the same time.
- Is the system live-lock free? A live lock occurs when some philosopher has no chance to eat anymore from a certain point and thus starves to death.

The following TPrTN models the Five Dining Philosophers problem, shown in Figure 9.1, in which five philosophers are denoted by five integer tokens 0 to 4 and five chopsticks are also denoted by five integer tokens 0 to 4. Places p_1 and p_3 stand for the **Thinking** and **Eating** states of philosophers, respectively. Place p_2 defines the availability of chopsticks. Transitions t_1 and t_2 stand for the actions of **Pick up** and **Put down** chopsticks, respectively.

The algebraic definition of the net $TN = (NG, Sig, V, H, Type, AN, M_0, f)$ is as follows:

$P = \{p_1, p_2, p_3\}$, $T = \{t_1, t_2\}$,
$F = \{(p_1, t_1), (p_2, t_1), (t_1, p_3), (p_3, t_2), (t_2, p_1), (t_2, p_2)\}$,
$\text{Type}(p_1) = \text{Type}(p_2) = \text{Type}(p_3) = \text{INT}$,
$A(p_1, t_1) = x, \quad A(p_2, t_1) = \{x, y\}, \quad A(t_1, p_3) = x,$
$A(p_3, t_2) = x, \quad A(t_2, p_1) = x, \quad A(t_2, p_2) = \{x, y\},$
$TC(t_1) = TC(t_2) = y = x \oplus 1,$
$M_0(p_1) = M_0(p_2) = \{0, 1, 2, 3, 4\}$, and $M_0(p_1) = \{\ \}$,

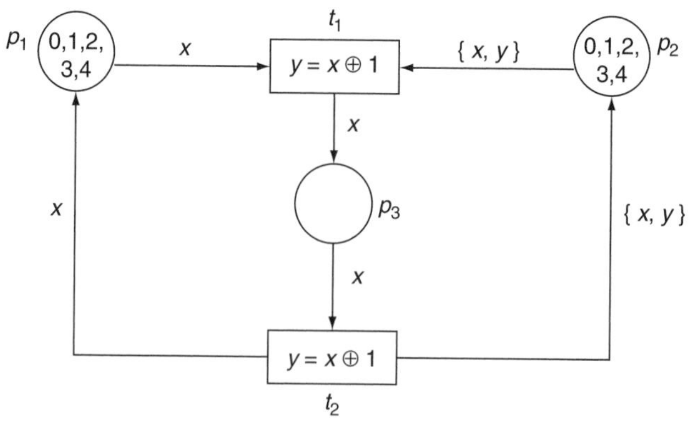

FIGURE 9.1 A TPrTN Specification of the Five Dining Philosophers Problem

where INT is a sort defined in *Sig*, x and y are variables in V, $\oplus$ is a modulo 5 addition operation given in *Sig* and defined in the algebra H, and $y = x \oplus 1$ is a term in algebra H.

Temporal formula f is defined as follows:

$$f = \forall x(\Box\Diamond(p_1(x) = 1 \wedge p_2(x) = 1 \wedge p_2(x \oplus 1) = 1) \Rightarrow \Box\Diamond(p_3'(x) = 1 \wedge p_1'(x) = 0 \wedge p_2'(x) = 0 \wedge p_2'(x \oplus 1) = 0)).$$

Temporal formula f defines the strong fairness (Manna and Pnueli, 1992) that captures the enabling condition and the firing result of transition t_1 with every mode (defined by $\forall x$). Intuitively, it states that any philosopher who wants to eat **infinitely many times** (defined by the temporal operator sequence $\Box\Diamond$) will eat **infinitely many times**.

9.3.4 Analysis of TPrTNs

System behavioral properties can be divided into two major categories: **safety** properties and **liveness** properties (Manna and Pnueli, 1992). Widely accepted formal definitions of safety and liveness properties were given in (Alpern and Schneider, 1985). A safety property, which is different from the concept **safeness** in Petri net literature (Murata, 1999) needs to hold in every state and in every state sequence. The canonical form of a safety property is characterized by a temporal formula $\Box w$. A liveness property, which is different from the concept transition **liveness** in Petri net literature (Murata, 1989), needs to hold in some future state in every state sequence. The canonical form of a liveness property is characterized by a temporal formula $\Diamond w$.

In He and Lee (1990) and He and Ding (1992), we developed an axiomatic approach of using temporal logic to analyze PrTNs. The essential ideas are to derive system-dependent temporal inference rules from PrTN transitions. These inference rules capture the causal relationships of transition firings. Thus, executions of a PrTN become temporal logic deductions in the derived axiomatic system. We have proved that safety properties (Manna and Pnueli, 1992) can be effectively analyzed using this axiomatic approach. Furthermore, we have shown how to analyze liveness properties (Manna and Pnueli, 1992) using the above temporal formula (He, 1991).

Given $TN = (NG, Sig, V, H, Type, AN, M_0, f)$, we can obtain the following system-dependent temporal axiom system L:

(1) A system-dependent axiom captures the initial marking M_0: $\wedge_{p\in P}(IC_p)$, (IC),

where IC_p represents the initial condition of place p under M_0

(2) For each place p of arity k, let $y_1, \ldots, y_k$ be new variables not used in any label and t be any transition in the postset $p\bullet$ with label $L(p, t) = \{m_1(x_{11}, \ldots, x_{1k}), \ldots, m_n(x_{n1}, \ldots, x_{nk})\}$. We construct an inference rule as follows: $p(y_1, \ldots, y_k) > p'(y_1, \ldots, y_k) \mid - \vee_{t\in p\bullet}((ET_t) \wedge (\vee_{1\le i\le n} (\wedge_{1\le j\le k} (y_j = x_{ij})))),\quad (PR_p)$

where ET_t represents the causal relationship of firing transition t. Intuitively, each inference rule states if a place loses tokens from one marking to another, then some relevant transition must have fired (He and Ding, 1992). The variables are universally quantified and are omitted for simplicity. The above inference rules only reflect the local state changes. Together with system independent axioms and inference rules (Manna and Pnueli, 1992, 1995), we can use IC and PR_p to prove a variety of safety properties.

The following is the derived system-dependent temporal axiom system of the Five Dining Philosophers problem shown in Figure 9.1:

$$p_1(0) = 1 \wedge p_1(1) = 1 \wedge p_1(2) = 1 \wedge p_1(3) = 1 \wedge p_1(4) = 1 \wedge p_2(0) = 1 \wedge p_2(1) = 1 \wedge p_2(2) = 1 \wedge p_2(3) = 1 \wedge p_2(4) = 1. \quad (IP)$$

$$p_1(x) > p_1'(x) | - p_1(x) \ge 1 \wedge p_2(x) \ge 1 \wedge p_2(y) \ge 1 \wedge y = x \oplus 1 \wedge p_1'(x) = p_1(x) - 1 \wedge p_2'(x) = p_2(x) - 1 \wedge p_2'(y) = p_2(y) - 1 \wedge p_3'(x) = p_3(x) + 1. \quad (PR_{p_1})$$

$$p_2(x) > p_2'(x) | - p_1(x) \ge 1 \wedge p_2(x) \ge 1 \wedge p_2(y) \ge 1 \wedge y = x \oplus 1 \wedge p_1'(x) = p_1(x) - 1 \wedge p_2'(x) = p_2(x) - 1 \wedge p_2'(y) = p_2(y) - 1 p_3'(x) = p_3(x) + 1. \quad (PR_{p_2})$$

$$p_3(x) > p_3'(x) | - p_3(x) \ge 1 \wedge p_1'(x) = p_1(x) + 1 \wedge p_2'(x) = p_2(x) + 1 \wedge p_2'(y) = p_2(y) + 1 \wedge p_3'(x) = p_3(x) - 1 \wedge y = x \oplus 1. \quad (PR_{p_3})$$

A TPrTN is **deadlock free** if and only if there is at least one transition being enabled in any reachable state or if a normal terminal state has been reached. Deadlock freedom is a safety property. The deadlock freedom property of the Five Dining Philosophers problem is formulated as follows:

$$\Box\exists x,\ y(p_3(x) \ge 1 \vee (p_1(x) \ge 1 \wedge p_2(x) \ge 1 \wedge p_2(y) \ge 1 \wedge y = x \oplus 1)). \quad (*)$$

A TPrTN net is **live lock free** if and only if for any transition enabled infinitely often with a mode α, the same transition with mode α will fire infinitely often. Livelock freedom is a liveness property. One of the liveness properties of the Five Dining Philosophers problem is that every individual philosopher will eventually get a piece of a meal (starvation free), which can be expressed by the following temporal formula:

$$\forall x\Box\Diamond(p_3(x) = 1). \quad (**)$$

The proofs of the above formulas involve logical deductions using inference rules of ordinary first-order logic and temporal logic as well as system-dependent inference rules listed previously. Because logical deductions are machine-oriented and not suitable for human comprehension, we omit the proofs of the above properties (*) and (**). Interested readers can find the proofs of (*) and (**) in He (1991) and He and Ding (1992).

9.4 *PZ* Nets

A widely accepted formal notation for specifying the functionality of sequential systems is *Z* (Spivey, 1992). The *Z* is based on typed set theory and first-order logic and, thus, offers rich type definition facility and supports formal reasoning. However, *Z* does not support an effective definition of concurrent and distributed systems, and *Z* specifications do not have explicit operational semantics. Many researchers have attempted to combine *Z* with other concurrent models, including CSP (Benjamin, 1990), CCS (Taguchi and Araki, 1997), and temporal logic (Duke and Smith, 1989; Clarke and Wing, 1996; Evans, 1997; Lamport, 1994b) in recent years. Several works attempted to integrate Petri nets and *Z* (van Hee *et al.*, 1991; He, 2001). In van Hee *et al.* (1991), *Z* was used to specify (1) the metamodel of restricted hierarchical colored Petri nets (that allows super transitions but not super places) and (2) the transitions of specific colored Petri nets. A complete specification consists of a hierarchical net structure, one global state schema for defining all places, and one operation schema for each transition. Schemas of transitions are piped to obtain an operational semantics through the use of input and output variables. In He (2001), a formal method of integrating Petri nets with *Z* was presented with the objectives to extend Petri nets with data and function definition capabilities through an underlying *Z* specification and to extend *Z* with an explicit operational semantics and concurrency mechanisms through Petri nets. In the following subsections, we briefly describe the basics of *Z* notations and introduce the results given in He (2001).

9.4.1 A Brief Overview of *Z*

Besides elementary data types, *Z* allows the user to introduce other primitive types, called given types. Given types are in capital letters and enclosed by a pair of brackets that do not require further definitions. For example, [IDEN] introduces a given type IDEN.

Part of *Z* is an essential notation called *schema* from which new types and their properties can be defined. A *Z* schema has a boxed structure often consisting of two major parts: the **declaration part** and the **predicate part** (optional) as follows:

Name
Declaration part
Predicate part

The name of a schema is global and can be used in the declaration part of other schemas to reuse the variable definitions.

The declaration part defines local variables of the schema in the form: **var: Type** (the collection of these variable definitions forms the **signature** of the schema), and thus a reference of this variable in another schema needs to be prefixed by its schema name followed by a period (i.e., **Name.var**). A variable name can be a simple name (defining the old value), a name with a prime′ (defining the new value of the same name without the prime), a name with a question mark ? (an input from the external environment), or a name with an exclamation point ! (an output to the external environment). If a schema includes both the name of a schema S and its primed version S', the following notation is used: ΔS when the value of some variable defined in S has been changed by the required processing or ΞS when nothing in S is changed by the required processing.

The predicate part specifies the constraint (invariant) of a definition or the precondition (subformulas without dashed variables) and postcondition (subformulas containing dashed variables) of some processing in terms of first-order logic formulas. Subformulas on separate lines in the predicate part are conjoined by default. The predicate thus defines the **property** of the schema.

State Schemas and Operation Schemas

A schema can be used to define the abstract state of a system, called a **state schema** in the sequel, when the state machine model is used. The predicate part in this case defines the data invariant among involved variables.

Each state schema S needs an initial value that is defined by an initialization schema that enumerates the initial values of state variables in the predicate part.

A schema SC can be used to define an operation, called an **operation schema** in the sequel, which includes both a state schema name S and its version S' in its declaration part. However, ΔS or ΞS is often used instead of S and S'. The predicate part can be divided into the precondition part (denoted by pre-SC in the sequel) and the postcondition part (denoted by post-SC in the sequel).

Operations on Schemas

New schemas can be built using existing schemas, in addition to the inclusion mentioned in the previous paragraphs.

(1) **Schema disjunction**: Schema disjunction has the form New $\hat{=}$ Old1 $\vee$ Old2, where New, Old1, and Old2 are schema names. The declaration part of new schema New is obtained by merging the declaration parts of Old1 and Old2, and the predicate part of New is obtained by making a disjunction of the predicate part of Old1 with that of Old2.

(2) **Schema conjunction**: Schema conjunction has the form New $\hat{=}$ Old1 $\wedge$ Old2, where New, Old1, and Old2 are schema names. The declaration part of new schema New is obtained by merging the declaration parts of Old1 and Old2, and the predicate part of New is obtained by making a conjunction of the predicate part of Old1 with that of Old2.

(3) **Schema composition**: Let Old1 and Old2 be two schemas; a new schema of the form New $\hat{=}$ Old1; Old2 can be defined and has the following meaning:
- The signature of New is the inputs and outputs of both Old1 and Old2, together with the nondashed variables in Old1 and dashed variables of Old2;
- The property of Old1 is included in New, but all the dashed names are redecorated with a decorator not used in either Old1 or Old2;
- The property of Old2 is included in New, but all the nondashed names are decorated with the same decorator as was used in Old1; and
- The newly decorated names are hidden with an existential quantifier.

Other operations related to schemas can be found in work by Spivey (1992), including using a schema as a type in the declaration part of other schemas, hiding declarations in a schema, and reusing the operations defined in one schema in other related schemas through promotion.

9.4.2 Definition of *PZ* Nets

Let $Z = (Z_P, Z_T, Z_I)$ be a collection of Z schemas. If we let z be a Z schema, we use **name**(z), **sig**(z), and **prop**(z) to denote the name, the signature part (as a typed set of mappings), and property part of z, respectively in the sequel. The Z schemas in Z satisfy the following conditions:

- $\forall z1, z2 \in Z_P \cdot (z1 \neq z2 \Rightarrow \mathbf{sig}(z1) \cap \mathbf{sig}(z2 = \varnothing)$;
- $\forall z1, z2 \in Z_I \cdot (z1 \neq z2 \Rightarrow \mathbf{sig}(z1) \cap \mathbf{sig}(z2) = \varnothing)$; and
- $|Z_P| = |Z_I|$.

The first two conditions state that the signatures of z schemas are pair-wise disjoint, and the last condition states that the number of z schemas in Z_P is the same as that in Z_I (i.e., one-to-one correspondence).

A *PZ* net is a tuple $ZN = (NG, Sig, V, H, Type, AN, M_0, Z)$.

Here, Z is a collection of Z schemas satisfying the above conditions, $AN = (Type, A, TC)$, and:

- *Type*: $P \rightarrow Z_P$ is one-to-one mapping, giving the data definition of *ZN*. For each place $p \in P$, *Type* maps p to a unique Z schema $z \in Z_P$ such that $p = \mathbf{name}(z)$. The type of p is defined by the signature of z.
- *TC*: $T \rightarrow Z_T$ is one-to-one mapping, providing the functionality definition of *ZN*. For each transition $t \in T$, C maps t to a Z schema $z \in Z_T$ such that $t = \mathbf{name}(z)$. The functional processing of t is defined by the property of z. Furthermore, the following constraint is satisfied. For any $p \in P$, $t \in T$:
  (1) If $(p, t) \in F$, then $\mathbf{sig}(Type\ (p)) \cap \mathbf{sig}(TC(t)) \neq \varnothing$.
  (2) If $(t, p) \in F$, then $\mathbf{sig}(Type\ (p)) \cap \mathbf{sig}(TC(t)) \neq \varnothing$.
- A: $F \rightarrow \wp Var$ is the control flow definition of *ZN*, where *Var* is the set of all hidden variables (through quantification) in Z_T. Let $Var(t)$ denote the set of hidden variables in $TC(t)$ for any transition t, let $S(v)$ denote the sort (or type) of a variable v, and let $S(V)$ denote the set of the sorts of the variables in the sorted set V. Then, the following constraints are satisfied for any $p \in P$:
  (1) $\bar{A}(p, t) \subseteq Var(t)$ and $\bar{A}(t, p) \subseteq Var(t)$, where
  $$\bar{A}(x, y) = \begin{cases} A(x, y) & \text{if } (x, y) \in F. \\ \varnothing & \text{otherwise} \end{cases}$$
  (2) For any $x \in \bar{A}(p, t)(\text{or } \bar{A}(t, p)) \cdot S(x) \in S(\mathbf{sig}(Type\ (p)))$.
  (3) $\mathbf{prop}(TC(t)) \Rightarrow \mathbf{sig}(Type(p)') = \mathbf{sig}(Type(p)) - \bar{A}(p, t) \cup \bar{A}(t, p)$.
- M_0: $P \rightarrow Z_I$ is a *Type*-respecting (i.e., $\mathbf{sig}(Type(p)) = \mathbf{sig}(M_0(p))$ initial marking of *ZN*, where Z_I is a set of Z schemas defining the initial state of the system. $M_0(p)$ is the Z schema in Z_I, defining the initial marking of place p.

The following is a simple example of a *PZ* net specification of the familiar Five Dining Philosophers problem. The overall system structure is shown in Figure 9.2, which is the same as Figure 9.1 except for the renaming.

The Z schemas (Z_P, Z_T, Z_I) are the following:

(1) Z_P = {thinking, chopstick, eating}:

— Thinking —
tphil: $\wp$ PHIL

— Chopstick —
chop: $\wp$ CHOP

— Eating —
ephil: $\wp$ (PHIL × CHOP × CHOP)
left: PHIL → CHOP
right: PHIL → CHOP

(2) Z_T = {pickup, putdown}:

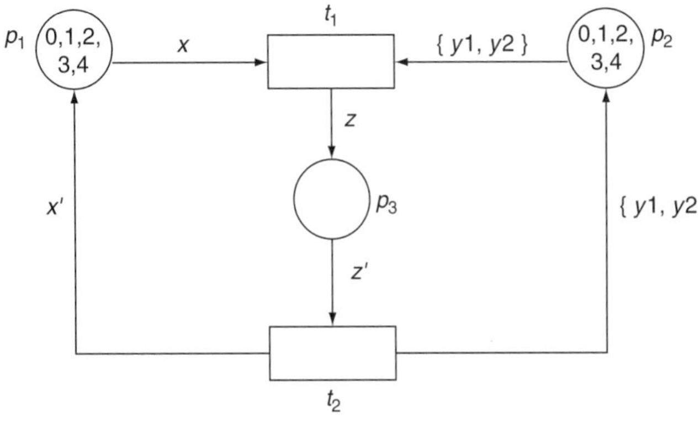

FIGURE 9.2 A *PZ* Net Specification of the Five Dining Philosophers Problem

(3) $Z_1 = \{$Init_Thinking, Init_Chopstick, Init_Eating$\}$:

Pick up

Δ Thinking
Δ Chopstick
Δ Eating

$\exists x$: PHIL • $\exists y_1, y_2$: CHOP • $\exists z'$: PHIL × CHOP × CHOP•
$(x \in \text{tphil} \wedge y_1 \in \text{chop} \wedge y_2 \in \text{chop} \wedge y_1 = \text{left}(x) \wedge$
$y_2 = \text{right}(x) \wedge$
$(z' = < x, y_1, y_2 >$
$\text{tphil}' = \text{tphil} \setminus \{x\}$
$\text{chop}' = \text{chop} \setminus \{y_1, y_2\}$
$\text{ephil}' = \text{ephil} \cup \{z'\}$
$\text{left}' = \text{left} \wedge \text{right} = \text{right}))$

Put down

Δ Thinking
Δ Chopstick
Δ Eating

$\exists x'$: PHIL • $\exists y_1', y_2'$: CHOP• $\exists z$: PHIL × CHOP × CHOP•
$(z \in \text{ephil} \wedge z = < x', y_1', y_2' >$
$\text{tphil}' = \text{tphil} \cup \{x'\}$
$\text{chop}' = \text{chop} \cup \{y_1', y_2'\}$
$\text{ephil}' = \text{ephil} \setminus \{z\}$
$\text{left}' = \text{left} \wedge \text{right} = \text{right})$

Init_Thinking

$\text{tphil} = \{0, 1, 2, 3, 4\}$

Init_Chopstick

$\text{chop} = \{0, 1, 2, 3, 4\}$

Init_Eating

Eating

$\text{ephil} = \varnothing$
$\text{left} = \{0 \mapsto 4, 1 \mapsto 0, 2 \mapsto 1, 3 \mapsto 2, 4 \mapsto 3\}$
$\text{right} = \{0 \mapsto 0, 1 \mapsto 1, 2 \mapsto 2, 3 \mapsto 3, 4 \mapsto 4\}$

$\text{Type} = \{p_1 \mapsto \text{Thinking}, p_2 \mapsto \text{Chopstick}, p_3 \mapsto \text{Eating}\}$
$TC = \{t1 \mapsto \text{Pick up}, t2 \mapsto \text{Put down}\}$,
$A = \{(p1, t1) \mapsto \{x\}, (p2, t1) \mapsto \{y1, y2\}, (t1, p3) \mapsto \{z'\}$,
$(p3, t2) \mapsto \{z\}, (t2, p1) \mapsto \{x'\}, (t2, p2) \mapsto \{y1', y2'\}\}$,
$M_0 = \{p1 \mapsto \text{Init\_Thinking}, p2 \mapsto \text{Init\_Chopstick}$,
$p3 \mapsto \text{Init\_Eating}\}$.

9.4.3 *PZ* Net Analysis

In He (1995, 2001), a structural induction technique was proposed for analyzing safety properties of *PZ* nets, which is based on an invariance inference rule from work by Manna and Pnueli (1995). This technique is briefly discussed in the following subsections.

Formalizing Invariant Properties

Let $[M_0>^{\omega}$ denote the set of all valid marking sequences extracted from all execution sequences of a given *PZ* net, and let σ be a valid marking sequence with $|\sigma|$ as the length of the sequence and $\sigma(i)$ as the *i*th state (marking) in σ. W (a first order logic formula) is an invariant property if and only if the following holds: $\forall \sigma: \sigma \in [M_0>^{\omega} \cdot (\forall i: 0 \leq i \leq |\sigma| \cdot (\sigma(i) \models W))$, where $\sigma(i) \models W$ denotes that marking $\sigma(i)$ satisfies W (i.e., the evaluation of W under marking $\sigma(i)$ yields true). Thus, a safety property holds in every state (marking) of every valid marking sequence. The above formulation can be simplified to the following equivalent version in terms of the set of all reachable markings only:

$$\forall M: M \in [M_0> \cdot (M \models W).$$

Proving Invariant Properties

In Manna and Pnueli (1995), several temporal logic-based inference rules for invariance (or safety properties) were given. Among the rules, the following basic invariance rule (in its state validity form) is reformulated in terms of *PZ* nets and is essential.

The Basic Invariance Rule

B1. $M_0 \Rightarrow W$

B2. $\dfrac{C(t): \alpha \wedge W \Rightarrow W' \text{ for every } t \in T \text{ and occurrence } \alpha}{M \models W \text{ for every } M \in [M_0>}$

In the basic invariance rule, premise B1 requires that the initial marking M_0 imply property W, and premise B2 requires that all transitions preserve W. $C(t)$ is the Z schema associated with t that defines the enabling condition (precondition) and the firing result (postcondition) of t. W' is obtained from W by changing the names of variables to their dashed version. Based on premises B1 and B2, we conclude that W is valid or is satisfied under any reachable marking from M_0.

We provide the following procedure to use the above inference rule:

Step 1. Prove the initial marking satisfies a system property formula.

Step 2. Assume the system property formula holds after k events in a state M.

Step 3. Prove the system property formula holds after $k+1$ events in any directly reachable state M' from M.

It is easy to see that steps 2 and 3 in the temporal induction proof technique fulfill premise B2 of the invariance rule.

Furthermore, we only need to consider the firing of a relevant transition and the associated Z schema with regard to the given property under the guidance of the net structure during step 3. For example, a system deadlock may occur when a particular place has a special marking. To show the system does not have the deadlock, we only need to show that transitions connected to this place cannot result in this special marking. Therefore, the proof is in general local in the sense that only a subset of transitions needs to be considered. Similar ideas have been also explored in other temporal analysis techniques (He and Lee, 1990; He and Ding, 1992). In general, logical, net structural, and net behavioral reasonings are needed to prove an invariant property. The above temporal induction technique is demonstrated in the following example.

The deadlock freedom property of the Five Dining Philosophers problem is formulated as follows:

$$\Box(\exists x \in \text{ephil} \cdot (p_3(x) \geq 1 \vee \exists x \in \text{tphil} \cdot (\text{right}(x) \in p_2 \cdot \wedge \text{left}(x) \in p_2)).$$

This problem explains that at any state, a philosopher is eating, which ensures the enabled condition of transition Put down, or a philosopher is thinking, which ensures the enabled condition transition Pick up.

The above formula can be rewritten as follows without using the temporal operator:

$$\forall M \in [M_0> \cdot (\exists x \in \text{ephil} \cdot (p_3(x) \geq 1 \vee \exists x \in \text{tphil}\ (\text{right}(x) \in p_2 \wedge \text{left}(x) \in p_2)).$$

Thus, we need to prove the following formula using the structural induction technique:

$$\exists x \in \text{ephil} \cdot (p_3(x) \geq 1 \vee \exists x \in \text{tphil} \cdot (\text{right}(x) \in p_2 \wedge \text{left}(x) \in p_2) \quad (*)$$

Here is the proof outline of the formula (*):

Step 1: Under the initial marking M_0, Init_Thinking, Init_Chopstick, and Init_Eating endures:

$$\exists x \in \text{tphil} \cdot (\text{right}(x) \in p_2 \wedge \text{left}(x) \in p_2) \text{ and thus } (*).$$

Step 2: Assume (*) holds after k transitions in a state M.

Step 3: Prove (*) holds after $k+1$ transitions in a state M' such that $M[t, \alpha >M'$.

Case 1: Firing transition Pick up with $\alpha = \{x/ph_1, y_1/ch_1, y_2/ch_2\}$ and from the postcondition of schema Pick up with $ch_1 \in \text{ephil}'$ in M'. Thus, $\exists x \in \text{ephil} \cdot (p_3(x) \geq 1)$ is true in M', and hence (*) holds in M'.

Case 2: Firing transition put down $\alpha = \{z/< ph_1, ch_1, ch_2 >\}$: From the precondition of put down with left $(ph_1) = ch_1$ and right $(ph_1) = ch_2$; from the postcondition of schema put down with $ph_1 \in \text{tphil}$, $ch_1 \in \text{chop}'$, and $ch_2 \in \text{chop}'$. Thus, $\exists x \in \text{tphil} \cdot (\text{right}(x) \in p_2 \wedge \text{left}(x) \in p_2)$ is true in M', and hence (*) holds in M'.

Therefore, (*) has been proven.

9.5 Hierarchical Predicate Transition Nets

The development of hierarchical predicate transition nets (HPrTNs) was motivated by the need to construct specifications for large systems using Petri nets (Reisig, 1987) and inspired by the development of modern high-level programming languages and other hierarchical and graphical notations, such as data flow diagrams (Yourdon, 1989) and statecharts (Harel, 1988). Similar work on introducing hierarchies into colored Petri nets was given in Jensen (1992, 1995). With the introduction of hierarchical structures into predicate transition nets, the resulting net specifications are more understandable, and the specification construction process becomes more manageable. HPrTNs were used in specifying several systems, including an elevator system (He and Lee, 1991), a library system (He and Yang, 1992), and a hurried dining philosophers system (He and Ding, 2001). HPrTNs can be analyzed directly by using a structural induction technique combining structural, behavioral, and logical reasoning (He, 2001) and can be translated into program skeletons in a concurrent and parallel object-oriented programming language CC++ (He, 2000c) and Java (Lewandowski and He, 1998, 2000). A complete formal definition of HPrTNs was given in He (1996). In the following subsections, basic concepts and notation of HPrTNs are briefly introduced.

9.5.1 Definition of HPrTNs

An HPrTN is a structure:

$$HN = (NG, Sig, V, H, Type, AN, M_0, \rho).$$

Where:

- $NG = (P, T, F)$ is a net graph. P and T are finite sets of places and transitions such that $P \cap T = \emptyset$. Elements in P are represented by solid and dotted circles. Similarly, elements in T are represented by solid and dotted boxes. Solid circles or boxes are elementary nodes, and dotted circles and boxes are super nodes. In particular, we identify two subsets $IN \subseteq P \cup T$ and OUT $\subseteq P \cup T$ such that IN contains the heads of all incoming **nonterminating arcs** (an arc inside a super node is a nonterminating arc if one of its ends is connected to the boundary of the super node) and OUT contains the tails of all outgoing nonterminating arcs. Nodes in IN $\cup$ OUT are called **interface** nodes. We use •IN to denote the set of the presets of all elements in IN (i.e., •IN $= \{\bullet\, n | n \in \text{IN}\}$), and OUT• to denote the set of the postsets of all elements in OUT. F is the set of arcs and is called the **flow relation**, satisfying the conditions: $P \cap F = \emptyset$, $F \cap T = \emptyset$, and $F \subseteq (\bullet\, \text{IN} \times \text{IN} \cup P \times T \cup T \times P \cup \text{OUT} \times \text{OUT}\bullet)$. An arc f can be uniquely identified by a pair of nodes (n_1, n_2) denoting its source and sink, in which $n_1(n_2)$ may denote the preset (postset) of $n_2(n_1)$ when f is a nonterminating arc.

- *Sig, V, H, Type* are defined as in HLPN in Section 9.2.
- $AN = (A, TC)$ is a pair of net annotations.

  A is a function that annotates each arc with a term that when evaluated (for some binding) results in a multiset over the associated place's type. Furthermore, all simple labels of a compound label must have distinct identifiers, and all simple labels of arcs connected to the same node must have distinct identifiers. Because compound labels define data flows as well as control flows, the following basic control flow patterns (He and Lee, 1991) must be correctly labeled: (1) data flowing into and out of an elementary transition must take place concurrently, and (2) data flowing into and out of an elementary predicate can occur at different times. Furthermore, data flows between different levels of hierarchies must be balanced (i.e., a simple label occurs in a nonterminating arc if and only if it also appears in an arc with the same direction connected to the enclosing super node).
- *TC* is a function that annotates transitions with Boolean expressions. A super transition is an abstraction of low-level actions, and its meaning is thus completely defined by the low-level refinement. Therefore, the constraint of a super transition is true by default (it is conceivable that a nontrivial constraint for a super transition might be useful; however, in general it is very difficult to define such a constraint and also very difficult to interpret the constraint with regard to the operational (dynamic) semantics of the super transition).
- M_0 is the initial marking function.
- $\rho: P \cup T \rightarrow \wp(P \cup T)$ is a hierarchical mapping that defines the hierarchical relationships among the nodes in P and T; this mapping also satisfies the constraint that the interface nodes $\in$ IN $\cup$ OUT be all predicates if their parent node is a predicate or all transitions if their parent node is a transition. For any node n, $\rho(n)$ defines the immediate descendant nodes of n. The ancestor and descendants of any node can be easily expressed by using well-known relations, such as transitive closure on ρ. A node in an HPrTN is local to its parent and can be uniquely identified by prefixing its ancesters' names separated with periods to its own name; however, often its own name is referred whenever a no name clash occurs.

The enabling condition of an elementary transition is defined exactly the same as that of an *HLPN*'s in Section 9.2. A super transition is enabled if at least one of its interface child transitions in IN is enabled and its firing is defined by an execution sequence of its child transitions; thus, its behavior is fully defined by its child transitions. The firing rule of a transition is formally defined in He (1996). Two transitions (including the same transition with two different occurrence modes) can fire concurrently if they are not in conflict (the firing of one of them disables the other). Conflicts are resolved nondeterministically. The firing of an elementary transition is atomic, and the firing of a super transition implies the firing of some elementary transition and may not be atomic. We define the behavior of an HPrTN to be the set of all possible maximal execution sequences containing only elementary transitions. Each execution sequence represents consecutively reachable markings from the initial marking in which a successor marking is obtained through a step (firing of some enabled transitions) from the predecessor marking.

Figure 9.3 shows an HPrTN specification of the Five Dining Philosophers problem.

Type(Thinking) $= \varphi$(Eating) $= \wp$(PHIL), Type(Avail) $=$ Type(Used) $= \wp$(CHOP),

Type(Chop)) $= \varphi$(Avail) $\cup\ \varphi$(Used), Type(Relation) $= \wp$(PHIL $\times$ CHOP $\times$ CHOP),

$A(f_3) = <1, re>$, $A(f_4) = <2, re>$, $A(f_5) = <3, ph>$,
$A(f_6) = <4, ph>$, $A(f_7) = <5, ph>$, $A(f_8) = <6, ph>$,
$A(f_{13}) = <7, \{ch_1, ch_2\}>$, $A(f_{14}) = <8, \{ch_1, ch_2\}>$,
$A(f_{15}) = <9, \{ch_1, ch_2\}>$, $A(f_{16}) = <10, \{ch_1, ch_2\}>$,
$A(f_9) = A(f_3) \times A(f_{13})$, $A(f_{10}) = A(f_4) \times A(f_{16})$,
$A(f_{11}) = A(f_3) \times A(f_{15})$, $A(f_{12}) = A(f_4) \times A(f_{14})$,
$A(f_1) = A(f_{13}) \times A(f_{15})$, $A(f_2) = A(f_{14}) \times A(f_{16})$,
$TC(\text{Pick up}) = (ph = re[1]) \wedge (ch1 = re[2]) \wedge (ch2 = re[3])$,
$TC(\text{Put down}) = (ph = re[1]) \wedge (ch1 = re[2]) \wedge (ch2 = re[3])$,
$TC(\text{Phil}) = \text{True}$,
$M_0(\text{Thinking}) = \{1, 2, \ldots, k\}$, $M_0(\text{Eating}) = \{\}$,
$M_0(\text{Avail}) = \{1, 2, \ldots, k\}$, $M_0(\text{Used}) = \{\}$,
$M_0(\text{Chop}) = M_0(\text{Avail}) \cup M_0(\text{Used}) = \{1, 2, \ldots, k\}$,
$M_0(\text{Relation}) = \{(1, 1, 2), (2, 2, 3), \ldots, (k, k, 1)\}$

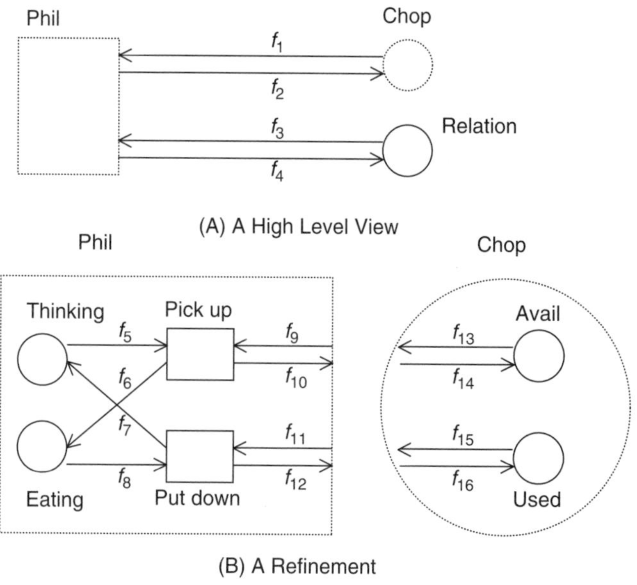

FIGURE 9.3 An HPrTN Specification of the Five Dining Philosophers Problem

9.5.2 System Modeling Using HPrTNs

HPrTNs can be used to model systems using the traditional structured approach (Yourdon, 1989) or modern object-oriented approach (Booch, 1994). He and Ding (2001) proposed an approach to realize various object-oriented (OO) concepts in HPrTNs. Furthermore, He (2000a, 2000b) and Dong and He (2001) applied HPrTNs to formalize several diagrams in the Unified Modeling Language (UML)—a software industry standard second-generation object-oriented modeling language. In the following sections, we briefly describe how to model several key OO concepts in HPrTNs.

Classes

One of the central ideas of OO paradigm is data encapsulation captured by the **class** concept. A class is essentially an abstract data type with a name, a set of related data fields (or attributes), and a set of operations on the data fields. It is straightforward to use a predicate to denote a data field (structure) and a transition to represent an operation in Petri nets. The current value of a data field is determined by the tokens of the denoting predicate under the current marking. The meaning or definition of an operation is specified by the constraint associated with the denoting transition.

HPrTNs were originally developed for structured analysis that provides separate mechanisms for data abstraction and processing abstraction through super predicates and super transitions, respectively. Therefore, we can use a super predicate and super transition pair in an HPrTN to capture the notion of a class although it is adequate to define a class by using a super predicate when there is no externally visible operation or using a super transition when there is no externally visible attribute. This view of class is a major improvement over the view in (He and Ding, 1996), where a class was represented by a super predicate only. In this view, the interface of the class is defined by the super predicate and the super transition. The super predicate defines data and internal operations of the class, while the super transition mainly defines the externally visible operations of the class. The corresponding subnets further define the internal structures of the data and the operations. The net inscription defines the meanings of net components through predicate types, token values, and transition constraints. When the resulting HPrTN is simple enough, there is no need to separate the super nodes from their subnets (i.e., the subnets are directly embedded inside the super nodes). An attribute or operation is externally visible if the corresponding denoting predicate or transition is an interface node (i.e., connected with a nonterminating arc). It should be noted, however, that not every super predicate or transition needs to be considered as a class. A super predicate or transition may simply denote a data abstraction or operation abstraction as originally intended; for example, a super predicate can be used to hide the internal states of an attribute that is defined by several related predicates, and a super transition can be used to define alternative implementations of an operation to realize operation overloading or overriding. Thus, our approach supports the coexistence of various modeling paradigms.

Based on the above analysis, we use the following C + +-like class schema to document a class defined by the super node(s) in an HPrTN (it is worth noting that the class schema is only used for understanding purpose and does not add functionality to the given HPrTN:

```
class Name [:superclass(es)]
{  public:
        predicates and transitions
   [private:
        predicates and transitions],
}
```

where brackets [...] denote optional items. Predicates and transitions listed in both public and private sections are those contained in the super node(s). The name(s) of the super node(s) are used to form the class name.

In the HPrTN shown in Figure 9.3, both super transition Phil and super predicate Chop can be viewed as classes. Thus, the following class definitions can be obtained:

```
class Chop
{  public:
        Avail, Used
},
class Phil
{  public:
        Pick up, Put down
   private:
        Thinking, Eating
}.
```

Objects

An instance or object of a class has its own copy of data while sharing operations with other objects of the same class. To distinguish an object from other objects, a unique identifier is needed for each object.

In an HPrTN, an object is essentially defined by a set of tokens related through the same identifier; thus, the sort of any predicate p needs to contain a component sort of relevant identifiers (i.e., $\text{Type}(p) = \wp(ID \times \ldots)$. Different objects of a class share the same class data structure (i.e., tokens with different identifiers can reside in a predicate at the same time in an HPrTN). In general, however, objects of the same class cannot interact with each other directly. The above problem can be easily solved by defining a subexpression comparing token identifiers in the constraint of each transition. Movements of tokens and/or changes of token values while maintaining the object identifier indicate state changes of the object.

In the HPrTN shown in Figure 9.3, there are k philosopher objects with identifiers of sort PHIL, and there are k chopstick objects with identifiers of sort CHOP.

Class Reference Relation

Classes work together to fulfill the functionality of the underlying system. A class can use the operations and/or data provided by other classes.

In HPrTNs, the interface of a transition includes a box with a name and the labels of relevant arcs (the label identifiers determining the calling context and the flow expressions specifying parameters). The meaning of an elementary transition is defined by its constraint, and the meaning of a super transition is defined through its corresponding subnet.

It is easy to model a class reference by adding some arc when a class needs to access some public attribute of another class. For example, Figure 9.4 illustrates simple class reference relationships where some operation in class C_1 (p_1 and t_1) uses some public attributes defined in class C_2 (p_2 and t_2), and some operation in class C_2 uses some public attributes defined in class C_1. Figure 9.3 contains simple reference relationships between classes Phil and Chop.

To define an operation in one class using another operation in a different class, we cannot simply add an arc since Petri nets do not allow direct connections between transitions. As discussed earlier, there are two main ways to handle class reference relationships in the existing research works: (1) to fuse the two operations in two classes into one such that only synchronized communication is allowed (Biberstein and Buchs, 1995; Battison *et al.*, 1995; Lakos, 1995b) and (2) to create some places inside one class to hold parameters to simulate message passing and function calls (Bastida, 1995; Lakos, 1995b), which supports asynchronous communication. HPrTNs support both synchronous and asynchronous communications through **reference places** to model different communication protocols. These reference predicates do not belong to any class and can be viewed as connectors in software architecture languages (Shaw and Garlan, 1996). It is quite easy to model a function call through passing two messages by using one reference predicate to hold the input values and another to hold output results; another easy way to model a function call is by defining the calling operation (function) as a super transition whose subnet has at least two transitions to handle sending and receiving values.

Figure 9.5 shows the general pattern of a function call from class C_1 containing t_1 to class C_2 containing t_2, in which p_1 and p_2 are reference places. The above pattern defines a one-way **synchronized communication** (i.e., the caller must wait for the callee to continue its execution), whereas a simple message passing from an operation in one class to an operation in another class in general defines an **asynchronous communication**.

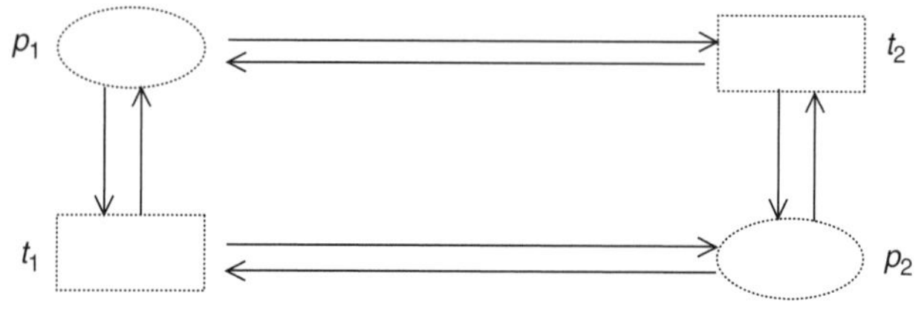

FIGURE 9.4 Simple Public Attributes Access

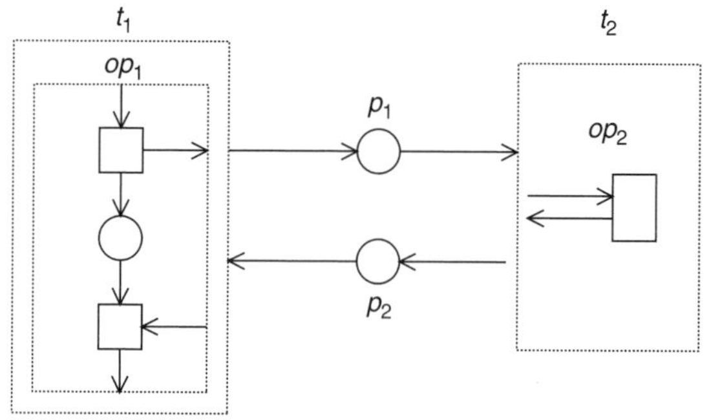

FIGURE 9.5 Reference Through Function Call

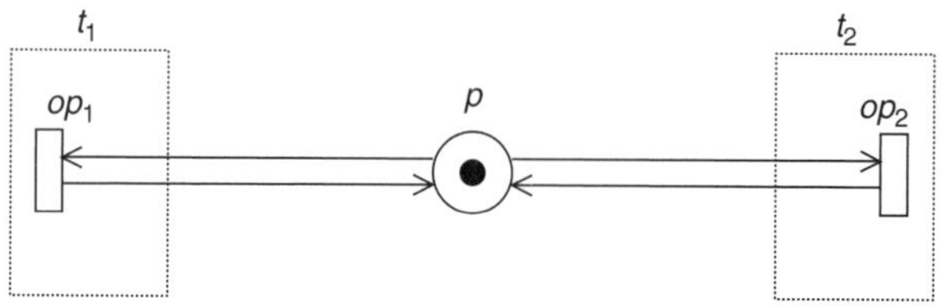

FIGURE 9.6 Synchronized Communication

Figure 9.6 defines a general synchronization pattern such that two operations in class C_1 containing t_1 and C_2 containing t_2 must execute mutual exclusivity, where p is a reference predicate with an initial dummy token.

It is quite natural and easy to define class reference relationships by using the decomposition and synthesis techniques of HPrTNs discussed in He and Lee (1991).

Class Inheritance Relation

Another major feature of the OO paradigm is **class inheritance relation** that captures the generalization–specialization relationships in the real world (Coad and Yourdon, 1991). A class inheritance relationship exists between a superclass and a subclass such that the subclass inherits data structures as well as operation definitions from the superclass without the need to define them again. Thus, class inheritance relation supports a flexible and manageable way to reuse existing data structures and operations.

A class inheritance relation is realized in HPrTNs through the reuse of the net structures of inherited super nodes and the net inscription of inherited elementary nodes (the sorts of predicates, the label expressions of relevant arcs, and the constraints of transitions) defined in an existing HPrTN denoting a class. The inherited predicates and transitions, however, are explicitly represented or embedded in the subclass to clearly define its role, the same convention was used by Lakos (1995b). An inherited element in a subclass has a name of the form super_node.element_name, where super_node is the partial name of the superclass and element_name is the internal name of the element in the superclass. Renaming of relevant arcs are also necessary to reflect the current context and to ensure flow balance. It is clear that inheritance does not reduce the size of an HPrTN specification since inherited elements are embedded; an alternative way to embedding is through delegation, as explained by Abadi and Cardelli (1996). However, the advantages are obvious since the meaning or structure (the most difficult part in writing an HPrTN specification) of an inherited element is already available and is obtained without any additional effort; furthermore, many known properties of the inherited element might be maintained through inheritance (structural properties are surely kept, but behavioral properties may need additional validation). It is worth noting that (1) only public components of a superclass can be inherited; (2) inheritances from multiple superclasses are supported, and an element can be inherited by multiple subclasses because no ambiguity will occur due to the naming convention; and (3) a redefined (overriding) operation is considered as a new operation in a subclass and is distinguished from an inherited operation such that an overriding operation in a subclass has the same name as the overriden operation in the superclass; this distinction between inheritance and overriding was also made by Abadi and Cardelli (1996).

Figure 9.7 shows a class inheritance relationship defined as follows:

class p_1 and t_1
{ public:
  p_2, t_2, t_3
};
class p_3 and t_3 : p_1 and t_1
{ public:
  t_3, $p_1 \cdot p_2$, $t_1 \cdot t_2$
  private:
  p_4, p_5, t_6
}.

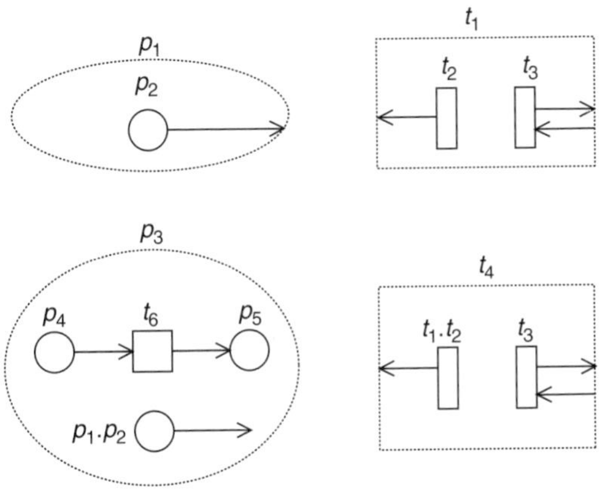

FIGURE 9.7 An Example of Class Inheritance Relation

Polymorphism

OO paradigm also supports **polymorphism** such that an operation's name (with possibly different signatures) may have different meanings or behaviors (implementations) through inheritance or overriding.

Polymorphism can be achieved in HPrTNs in two different yet related ways. First, polymorphism is a major feature of the underlying many-sorted algebra H of an HPrTN; detailed discussions of algebraic specifications and polymorphism can be found in work by Ehrig and Mahr (1985). The same operation symbol in H is used for many derived sorts. A simple example is the overloaded equality (=) operator when an algebraic specification H contains two elementary data types (or classes) INT and CHAR with a single parameterized definition of the equality (=). Second, polymorphism can be accomplished through net structure and inscription. An operation provides overriding capability if its constraint distinguishes a superclass object and a subclass object (or two objects from two different subclasses with the same superclass) and processes them differently. To realize polymorphism in an HPrTN, a shared predicate can be used to hold tokens of a superclass as well as tokens of subclasses, and the shared predicate is connected to the transition defined in the superclass and its inherited (or overriding) versions in the subclasses. The constraint of the original transition is only satisfied by the tokens of the superclass, and the constraint of each inherited (or overriding) transition is only satisfied with tokens of the subclass containing the transition.

Figure 9.8 shows the general pattern of realizing polymorphism through net structure in HPrTNs, in which p is a shared place, t_1 is a part of the superclass, and t_2 is a part of the subclass, and op_1^* is either an inherited or an overriding version of op_1, as shown in Figure 9.8.

Furthermore, operation overriding can be achieved through a super transition in an HPrTN such that the firing of a particular component transition is determined partly by the (dynamic) instantiation of an object identifier (and thus its sort). The use of the above case-like net structure in realizing polymorphism can be avoided in the implementation of an HPrTN.

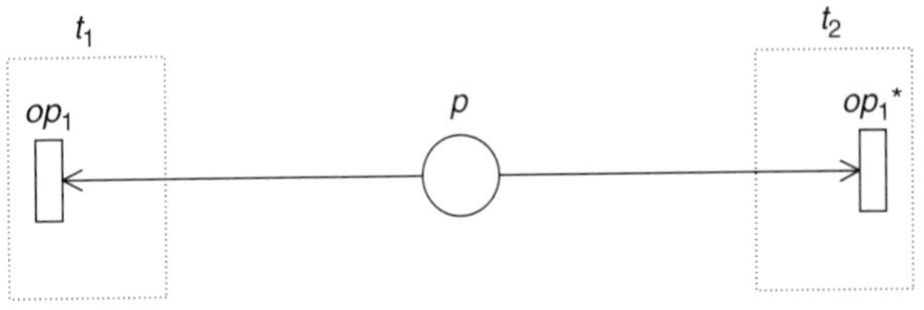

FIGURE 9.8 Polymorphism Through Choice Structure

9.6 Fuzzy-Timing High-Level Petri Nets

9.6.1 Definition of FTHN

A Fuzzy-timing high-level Petri net (FTHN) is a structure:

$$FN = (N, D, FT),$$

where N is an HLPN defined in Section 9.2; D is the set of all fuzzy delays, $d_{tp}(\tau)$, associated with arcs (t, p) from each transition $t \in T$ to its output place p; and FT is the set of all **fuzzy time stamps**. A **fuzzy time stamp** $\pi(\tau) \in FT$ is associated with each token and each place. A fuzzy time stamp $\pi(\tau)$ is a **fuzzy time function** or **possibility distribution** giving the numerical estimate of the possibility that a particular token arrives at time τ in a particular place. Any type of possibility distribution can be represented or approximated by using a number of trapezoidal distributions. Thus, we use the trapezoidal **possibility distribution** specified by the five parameters, $h(a, b, c, d)$ as shown in Figure 9.9, where h is the height having the following properties: $0 \leq h \leq 1$, $h = 1$ for an event (arrival of a token) that has occurred or will occur and $h < 1$ for an event that will not necessarily occur. A so-called **fuzzy number** is represented by the triangular distribution $h(a, b, b, d)$, which is a special case of the trapezoidal form with $b = c$. A deterministic interval between a and d denoted $[a, d]$ can be represented by (a, a, d, d), a special case of trapezoidal form with $a = b$, $c = d$, and $h = 1$. In addition, given an arbitrary-shaped possibility distribution, we can approximate it with the union of a number of trapezoidal distributions (Zhou and Murata, 1999).

In Zhou *et al.* (2000), *FTHN* is extended by adding a time interval with a possibility value p in the form of $p[\alpha, \beta]$ to each transition. That is, each transition is associated with a firing interval denoted $p[\alpha, \beta]$, where the default interval is $1[0, 0]$ (a transition definitely fires as soon as it is enabled). If a transition t is enabled at time instant τ, it may not fire before time instant $\tau + \alpha$ and must fire before or at time instant $\tau + \beta$. Possibility p is a value in the interval $[0, 1]$, where p is 1 if transition t is not in structural conflict with any other transition, and p can be less than 1 when we want to assign different chances to transitions in conflict. Here, **structural conflict** means that a transition t shares some input place with another transition that can be enabled simultaneously with transition t; firing one transition will disable the other transition.

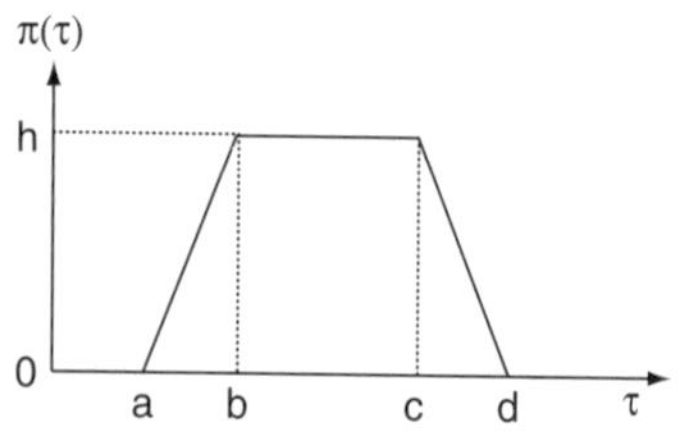

FIGURE 9.9 Trapezoidal Possibility Distribution

9.6.2 Computation for Updating Fuzzy Time Stamps

Suppose that a transition t is enabled by n tokens and the fuzzy enabling time $e_t(\tau)$ of transition t is computed by $e_t(\tau) = \text{latest}\ \{\pi_i(\tau), i = 1, 2, \ldots, n\}$, where **latest** is the operator that constructs the "latest-arrival-lowest-possibility distribution" from n distributions (Murata, 1996; Murata *et al.*, 1999). The $\pi_i(\tau)$ is the fuzzy time stamp to which the enabling token arrives at the ith input place of transition t. When there are m transitions in structural conflict that are enabled with their fuzzy enabling times, $e_i(\tau)$, $i = 1, 2, \ldots, t, \ldots, m$, and with their possibility intervals, $p_i[\alpha_i, \beta_i]$, we compute the fuzzy occurrence time $o_t(\tau)$ of transition t whose fuzzy enabling time is $e_t(\tau)$ as follows:

$$o_t(\tau) = \min\{e_t(\tau) \oplus p_i(\alpha_t, \alpha_t, \beta_t, \beta_t),$$
$$\text{earliest}\{e_i(\tau) \oplus p_i(\alpha_i, \alpha_i, \beta_i, \beta_i), i = 1, 2, \ldots, t, \ldots, m\}\},$$

where **earliest** is the operator that constructs the "earliest-arrival-highest-possibility distribution" from m distributions, (Murata, 1996; Murata *et al.*, 1999), **min** denotes the minimum or intersection operation, and $\oplus$ is the extended addition (Dubois and Prade, 1989). We compute the fuzzy time stamp $\pi_{tp}(\tau)$, which is the fuzzy time distribution at which a token arrives at the transition t output place p, as follows:

$$\pi_{tp}(\tau) = o_t(\tau) \oplus d_{tp}(\tau) = h_1(o_1, o_2, o_3, o_4) \oplus h_2(d_1, d_2, d_3, d_4)$$
$$= \min\{h_1, h_2\}(o_1 + d_1, o_2 + d_2, o_3 + d_3, o_4 + d_4),$$

where $d_{tp}(\tau)$ is the fuzzy delay associated with the arc (t, p) (Murata, 1996). When there are no transitions in conflict with transition t with its possibility interval $p_t[\alpha t, \beta_t]$, the fuzzy occurrence time is given by $o_t(\tau) = e_t(\tau) \oplus (\alpha_t, \alpha_t, \beta_t, \beta_t)$, where we set $p_t = 1$ since no conflict exists. We use the following formulas as approximate computations of the earliest and latest operations:

$$\text{earliest}\{e_i(\tau), i = 1, 2, \ldots, n\} = \text{earliest}\{h_i(e_{i1}, e_{i2}, e_{i3}, e_{i4}), i = 1, 2, \ldots, n\}$$
$$= \max\{h_i\}(\min\{e_{i1}\}, \min(e_{i2}, \min\{e_{i3}, \min\{e_{i4}\}), i = 1, 2, \ldots, n$$

$$\text{latest}\{\pi_i(\tau), i = 1, 2, \ldots, n\} = \text{latest}\{h_i(\pi_{i1}, \pi_{i2}, \pi_{i3}, \pi_{i4}), i = 1, 2, \ldots, n\}$$
$$= \min\{h_i\}(\max\{\pi_{i1}\}, \max\{\pi_{i2}\}, \max\{\pi_{i3}\}, \max\{\pi_{i4}\}), i = 1, 2, \ldots, n.$$

Using the above procedure, we compute and update fuzzy time stamps $\pi(\tau)$, fuzzy enabling times $e(\tau)$, and fuzzy occurrence times $o(\tau)$ each time a transition firing (atomic action) occurs, starting from the initial (given) fuzzy time stamps of tokens in the initial marking M_0 and initially specified fuzzy delays. $d_{tp}(\tau)$. (See Zhou *et al.* [2000].)

9.6.3 Intended Application Areas and Application Examples of FTHNs

As seen in Subsections 9.6.1 and 9.6.2, the essence of FTHNs is the computation of updating fuzzy time stamps, and this computation involves only additions and comparisons of real numbers. Thus, computation can be done very fast, and FTHNs are suitable for performance analysis in real-time applications. The notion of fuzzy timing is highly flexible in that it can capture imprecise or incomplete knowledge regarding time behavior with specified or given possibility distributions, as well as the more conventional deterministic and probabilistic knowledge. However, it is expected that the traditional probabilistic approach and the FTHN method using possibility theory are complementary, rather than competitive, as possibility theory is considered to be complementary to probability theory (Zadeh, 1995). We expect that the FTHN method is more scalable than the traditional stochastic approaches because the FTHN computations are done using real arithmetic operations.

Some examples of FTHN applications include the following. A real-time network protocol used in local area networks (LANs) is modeled using FTHNs in (Murata, *et al.*, 1999). Here we are interested in evaluating the worst-case performance in a given network, where propagation delays, times for processing message frames, and other such processes are specified as trapezoidal fuzzy-time functions. The fuzziness is due to the uncertain length (thus uncertain delay) of each message. In addition, FTHN models are used for performance evaluation of manufacturing systems, where the abilities of machines and/or workers involved in a manufacturing process are fuzzily known (Watanuki, 1999). Another area in which FTHN models are applied is the synchronization of multimedia systems (Zhou and Murata, 1998, 2001). Multimedia systems integrate a variety of media with different temporal characteristics (e.g., time-dependent media, such as video, audio, or animation) and time-independent media (e.g., text, graphics, and images). Thus, synchronization is a critical issue in these systems. The temporal specification has to be properly represented for presentation reviewing and planning by the user as well as for storing purposes. Multimedia synchronization has a time-critical problem because it must guarantee the temporal constraints for presenting the media items. Thus, there is a benefit in using FTHN methods to specify and analyze the temporal relations and specifications. The FTHN method has been used to present a new fine-grained temporal FTHN model for distributed multimedia synchronization (Zhou and Murata, 2001). The FTHN method has also been applied to a video-on-demand (VOD) system, which is a continuous playback multimedia application in which constant real-time media data are required for smooth real-time presentation. Thus, timing is a critical issue, and the response time is the only timing that has direct interaction with the subscribers and also impacts the quality-of-service (QoS), as discussed by Murata and Chen (2000). A nontrivial example of FTHN application is found in Zhou *et al.* (2000), where the FTHN method has been used to model networked virtual-reality systems, such as Cave Automatic Virtual Environment (CAVE) and Narrative Immersive Constructionist/Collaborative Environments (NICE) projects at the University of Illinois at Chicago. Using the FTHN models, various simulations have been conducted to study real-time behaviors, network effects, and performance (latencies and jitters) of the NICE. The simulation results are consistent with data and measurements obtained experimentally. This study shows how powerful FTHN models are in specifying and analyzing performance and how helpful the performance analysis is in improving a real-time networked-virtual-environment system design process.

Acknowledgements

Xudong He's research was supported in part by the NSF under grant HRD-0317692, and by the NASA under grant NAG2-1440. Tadao Murata's research was supported by the NSF under grant CCR-9988326.

References

Abadi, M., and Cardelli, L. (1996). *A theory of objects.* Springer-Verlag.

Agha, G., De Cindio, F., Rozenberg, G. (Eds.). (2001). Concurrent object-oriented programming and Petri nets—Advances in Petri nets. *Lecture Notes in Computer Science.* Berlin: Springer Verlag.

Anttila, M., Eriksson, H., and Ikonen, J. (1983). Tools and studies of formal techniques—Petri nets and temporal "logic." In H. Rudin and C.H. West (Eds.). *Protocol specification, testing, verification.* New York: Elsevier Science.

Alpern, B., and Schneider, F.B. (1985). Defining liveness. *Information Processing Letters 21,* 181–185.

Bastide, R. (1995). Approaches in unifying Petri nets and the object-oriented approach. *Proceedings of the 1st Workshop on Object-Oriented Programming and Models of Concurrency.*

Biberstein, O., and Buchs, D. (1995). Structured algebraic nets with object-orientation. *Proceedings of the 1st Workshop on Object-Oriented Programming and Models of Concurrency.*

Battiston, E., Chizzoni, A., and Cindio, F.D. Inheritance and concurrency in CLOWN. *Proceedings of the 1st Workshop on Object-Oriented Programming and Models of Concurrency.*

Benjamin, M. (1990). A message passing System: An example of combining CSP and Z. *Proceedings of the 5th Annual Z Users Workshop.* 221–228.

Booch, G. (1994). *Object-oriented analysis and design with applications.* (2d. ed.). MA: Benjamin/Cummings.

Booch, G., Rumbaugh, J., and Jacobson, I. (1997). *Unified modeling language user guide.* Reading, MA: Addison-Wesley.

Clarke, E., and Wing, J. (1996). Formal methods: State of the art and future. *ACM Computing Surveys 28*(4), 626–643.

Coad, P., and Yourdon, E. (1991). *Object-oriented analysis.* NJ: Yourdon Press.

Diaz, M., Guidacci, G., and Silverira, D. (1983). On the specification and validation of protocols by temporal logic and nets. *Information Processing 83,* 47–52.

Dong, Z., and He, X. (2001). Integrating UML statechart and collaboration diagrams using hierarchical predicate transition nets. *Lecture Notes in Informatics, P-7*, 99–112.

Dubois, D., and Prade, H. (1989). Processing fuzzy temporal knowledge. *IEEE Transactions On Systems, Man and Cybernetics, 19*(4), 729–744.

Duke, R., and Smith, G. (1989). Temporal logic and Z specifications. *Australian Computer Journal 21*(2), 62–69.

Ehrig, H., and Mahr, B. (1985). *Fundamentals of algebraic specification 1—Equations and initial semantics.* Berlin: Springer-Verlag.

Evans, A.S. (1997). An improved recipe for specifying reactive systems in Z. *Proceedings of the 10th International Conference of Z Users (Lecture Notes in Computer Science) 1212*, 273–294.

Genrich, H.J., and Lautenbach, K. (1981). System modeling with high-level Petri nets. *Theoretical Computer Science 13*, 109–136.

Haas, P. (2002). *Stochastic Petri nets: Modeling, stability, simulation.* Berlin: Springer-Verlag.

Harel, D. (1988). On visual formalisms. *Communications of the ACM 31*, 514–530.

He, X., and Ding, Y. (1992). A temporal logic approach for analyzing safety properties of predicate transition nets. *Proceedings of the 12th IFIP World Computer Congress (Information Processing '92).* 127–133.

He, X., and Ding, Y. (1996). Object-oriented specification using hierarchical predicate transition nets. *Proceedings of the 2nd International Workshop on Object-Oriented Programming and Models of Concurrency.* 72–79.

He, X., and Ding, Y. (2001). Object orientation in hierarchical predicate transition nets. *Lecture Notes in Computer Science.* 196–215.

He, X. (1991). Specifying and verifying real-time systems using time Petri nets and real-time temporal logic. *Proceedings of the 6th Annual Conference on Computer Assurance.* 135–140.

He, X. (1992). Temporal predicate transition nets—A new formalism for specifying and verifying concurrent systems. *International Journal of Computer Mathematics 45*(1/2), 171–184.

He, X. (1995). A method for analyzing properties of hierarchical predicate transition nets. *Proceedings of the 19th Annual International Computer Software and Applications Conference (COMPSAC'95).* 50–55.

He, X. (1996). A formal definition of hierarchical predicate transition nets. *Proceedings of the 17th International Conference on Application and Theory of Petri Nets (ICATPN'96), Lecture Notes in Computer Science 1091*, 212–229.

He, X. (1998). Transformations on hierarchical predicate transition nets: Abstractions and refinements. *Proceedings of the 22nd International Computer Software and Application Conference (COMPSAC '98).* 164–169.

He, X. (2000a). Formalizing use case diagrams in hierarchical predicate transition nets. *Proceedings of the IFIP 16th World Computer Congress.* 484–491.

He, X. (2000b). Formalizing class diagrams using hierarchical predicate transition nets. *Proceedings of the 24th International Computer Software and Application Conference (COMPSAC'2000).*

He, X. (2000c). Translating hierarchical predicate transition nets into CC++ programs. *Information and Software Technology 42*(7), 475–488.

He, X. (2001). PZ nets—A formal method integrating Petri nets with Z. *Information and Software Technology 43*, 1–18.

He, X., and Lee, J.A.N. (1990). Integrating predicate transition nets and first-order temporal logic in the specification of concurrent systems. *Formal Aspects of Computing 2*(3), 226–246.

He, X., and Lee, J.A.N. (1991). A methodology for constructing predicate transition net specifications. *Software—Practice & Experience 21*(8), 845–875.

He, X., and Yang, C.H., Structured analysis using hierarchical predicate transition nets. *Proceedings of the 16th International Computer Software and Applications Conference (COMPSAC'92).* 212–217.

ISO/IEC. (2002). High-level Petri nets—Concepts, definitions, and graphical notation. *Final Draft International Standard* 15909, version 4.7.1.

Jensen, K. (1992). *Coloured Petri nets—Basic concepts, analysis methods, and practical use.* Berlin: Springer-Verlag.

Jensen, K. (1995). *Coloured Petri nets—Basic concepts, analysis methods, and practical use.* Berlin: Springer-Verlag.

Jensen, K., and Rozenberg, G. (Eds.). (1991). *High-level Petri nets—Theory and applications.* Berlin: Springer-Verlag.

Kan, C., and He, X. (1995). High-level algebraic Petri nets. *Information and Software Technology 37*(1), 23–30.

Kan, C., and He, X. (1996). A method for constructing algebraic Petri nets. *Journal of Systems and Software 35*, 12–27.

Kappel, G., and Schrefl, M. (1991). Using an object-oriented diagram technique for the design of information systems. In *Dynamic modeling of information systems.* New York: Elsevier Science.

Lewandowski, S., and He, X. (1998). A Java framework for implementing hierarchical predicate transition nets. *Proceedings of the 10th International Conference on Software Engineering and Knowledge Engineering (SEKE'98).* 261–268.

Lewandowski, S., and He, X. (2000). Automating the generation of code for a hierarchical predicate transition net-based design. *Proceedings of the 12th International Conference on Software Engineering and Knowledge Engineering.*

Lakos, C. (1995a). From colored Petri nets to object Petri nets: *Proceedings of the 16th International Conference on the Application and Theory of Petri Nets.*

Lakos, C. (1995b). The object orientation of object Petri nets. *Proceedings of the 1st Workshop on Object-Oriented Programming and Models of Concurrency.*

Lamport, L. (1994a). The temporal logic of actions. *ACM Transactions on Programming Languages and Systems 16*, 872–923.

Lamport, L. (1994b). TLZ. *Proceedings of the 1994 Z User Workshop.* 267–268.

Lee, Y.K., and Park, S.J. (1993). OPNets: An object-oriented high-level Petri net model for real-time system modeling. *Journal of Systems and Software 20*, 69–86.

Mandrioli, D., Morzenti, A., Pezze, M., Pietro, P., and Silva, S. (1996). A Petri net and logic approach to the specification and verification of real-time systems. *Formal Methods for Real-time Computing.*

Manna, Z., and Pnueli, A. (1995). *The temporal verification of reactive systems—safety.* Berlin: Springer-Verlag.

Manna, Z., and Pnueli, A. (1992). *The temporal logic of reactive and concurrent systems—specification.* Berlin: Springer-Verlag.

Marsan, M., Balbo, G., Conte, G., Donatelli, S., and Franceschinis, G. (1994). *Modeling with generalized stochastic Petri nets.* New York: John Wiley & Sons.

Matsuoka, S., and Yonezawa, A. (1993). Analysis of inheritance anomaly in object-oriented concurrent programming languages. In G. Agha, P. Wegner, and A. Yonezawa (Eds.). *Research directions in concurrent object-oriented programming.* Boston: MIT Press.

Medvidovic, N., and Taylor, R. (2000). A classification and comparison framework for software architecture description languages. *IEEE Transaction on Software Engineering 26*(1), 70–93.

Murata, T. (1996). Temporal uncertainty and fuzzy-timing high-level Petri nets. *Application and theory of Petri nets: Lecture Notes in Computer Science, 1091*, 11–28.

Murata, T. (1989). Petri nets, properties, analysis and applications. *Proceedings of IEEE, 77*(4), 541–580.

Murata, T., and Chen, C.P. (2000). Fuzzy-timing Petri-net modeling and analysis of video-on-demand system response times. *Proceedings of the 5th World Conference on Integrated Design & Process Technology*, 298–306.

Murata, T., Suzuki, T., and Shatz, S. (1999). Fuzzy-timing high-level Petri nets (FTHNs) for time-critical systems. In J. Cardoso and H. Camargo (Eds.). *Fuzziness in Petri nets*, Vol. 22: *Studies in Fuzziness and Soft Computing.* Berlin: Springer-Verlag.

Queille, J.P., and Sifakis, J. (1982). Specification and verification of concurrent systems in CESAR. *Lecture Notes in Computer Sciences 137*, 337–351.

Reisig, W. (1987). Petri nets in software engineering. *Lecture Notes in Computer Science 255*, 63–96.

Reisig, W. (1985a). Petri nets—An introduction. *EATCS Monographs on Theoretical Computer Science 4.*

Reisig, W. (1985b). On the semantics of Petri nets. In J. Neuhold and G. Chroust (Eds.), *Formal models in programming.* North-Holland.

Shaw, M., and Garlan, D. (1996). *Software architecture.* Englewood Cliffs, NJ: Prentice-Hall.

Spivey, J.M. (1992). *The Z notation: A reference manual.* Englewood Cliffs, NJ: Prentice-Hall.

Stroustrup, B. (1991). *The C++ Programming Language.* (2d ed.). Reading, MA: Addison-Wesley.

Suzuki, I., and Lu, H. (1989). Temporal Petri nets and their application to modeling and analysis of a handshake daisy chain arbiter. *IEEE Transactions on Computer 38*(5), 696–704.

Taguchi, K., and Araki, K. (1997). The state-based CCS semantics for concurrent Z specification. *Proceedings of the 1st International Conference on Formal Engineering Methods.* 283–292.

van Hee, K.M., Somers, L.J., and Voorhoeve, M. (1991). Z and high-level Petri nets. *Lecture Notes in Computer Science 551*, 204–219.

Wang, J. (1998). *Timed Petri nets, theory and application.* Norwell, MA: Kluwer Academic Publisher.

Watanuki, K., and Murata, T. (1999). Evaluation method for assembly/disassembly by Petri nets. *Proceedings of the International Conference on Engineering Design* (ICED'99) *1*, 519–522.

Yourdon, E. (1989). *Modern structured analysis.* Englewood Cliffs, NJ: Prentice Hall.

Zadeh, L.A. (1995). Discussion: Probability theory and fuzzy logic are complementary rather than competitive. *Technometrics of American Statistical Association and American Society for Quality Control 37*(3).

Zhou, Y., and Murata, T. (2001). Modeling and analysis of distributed multimedia synchronization by extended fuzzy-timing Petri nets. *Journal of Integrated Design and Process Science 4*(4), 23–38.

Zhou, Y., and Murata, T. (1998). Fuzzy-timing Petri net model for distributed multimedia synchronization. *Proceedings of the 1998 IEEE International Conference on Systems, Man, and Cybernetics*, 244–249.

Zhou, Y., and Murata, T. (1999). Petri net model with fuzzy-timing and fuzzy-metric temporal logic. *International Journal of Intelligent Systems 14*(8), 719–746.

Zhou, Y., Murata, T., and DeFanti, T. (2000). Modeling and performance analysis using extended fuzzy-timing Petri nets for networked virtual environments. *IEEE Transactions on Systems, Man, and Cybernetics 30*(5), 737–756.

V

ELECTROMAGNETICS

Hung-Yu David Yang
Department of Electrical and Computer Engineering, University of Illinois at Chicago, Chicago, Illinois, USA

Electromagnetics is fundamental in electrical and electronic engineering. Electromagnetic theory based on Maxwell's equations establishes the basic principle of electrical and electronic circuits over the entire frequency spectrum from dc to optics. It is the basis of Kirchhoff's current and voltage laws for low-frequency circuits and Snell's law of reflection in optics. For low-frequency applications, the physics of electricity and magnetism are uncoupled. Coulomb's law for electric field and potential and Ampere's law for magnetic field govern the physical principles. Infrared and optical applications are usually described in the content of photonics or optics as separate subjects. This section emphasizes the engineering applications of electromagnetic field theory that relate directly to the coupling of space and time-dependent vector electric and magnetic fields, and, therefore, most of the subjects focus on microwave and millimeter-wave regimes. The eleven chapters in this section cover a broad area of applied electromagnetics, including fundamental electromagnetic field theory, guided waves, antennas and radiation, microwave components, numerical methods, and radar and inverse scattering.

Chapter 1 discusses the basic theory of magnetostatics. Magnetic field and energy due to a direct current is defined based on Ampère's law and the Biot-Savart law. Macroscopic properties of magnetic material are described. In addition, domains and hysteresis are introduced. Inductance relating the magnetic flux to the current is defined. The concept of a magnetic circuit, which finds important applications in power transformers, is also introduced.

Chapter 2 is devoted to the fundamental theory of electrostatics. The concept of electric field and potential based on Coulomb's law and Gauss's law is introduced. Electric energy and force based on the field and potential are described. Boundary value problem based on the Poisson's equation and Laplace's equation for electric potential is formulated and canonical examples are given.

Waves propagating in a homogeneous isotropic region are usually in the form of plane-waves. Chapter 3 describes the basic properties of plane waves for both lossless and lossy media. These basic properties include the nature of the electric and magnetic fields, the properties of the wave number vector, and the power flow of the plane wave. Special attention is given to the specific case of a homogenous (uniform) plane wave, i.e., one having real direction angles because this case is most often met in practice. Properties such as wavelength, phase and group velocity, penetration depth, and polarization are discussed for these plane waves.

Chapter 4 describes the theory of transmission lines. Transmission equations for voltage and current are derived based on lumped-element circuit models. The propagation characteristics of both lossless and lossy lines are discussed with the latter emphasized on low-loss cases and cases lacking distortion. Useful parameters of a terminated transmission line, including impedance, reflection coefficient, voltage, and current, at various locations on the line are discussed in detail. The basic operation of the Smith chart to relate the reflection coefficient to the input impedance at the transmission line is explained and examples are given.

Distributing electromagnetic power from one point to another in a prescribed way usually requires transmission lines or waveguides. Chapter 5 discusses the properties of a class of guided wave structures. The emphasis is on non-TEM structures where the guided waves are dispersive. The mode characteristics of rectangular metallic waveguides, circular metallic waveguides, microstrip lines, slot lines, coplanar waveguides, and the circular dielectric waveguides are summarized.

In wireless communication systems, it is necessary to send signals in the form of electromagnetic waves through air, such as in radio or television broadcasting, or via point-to-point microwave links. An antenna is a device for transmission or reception of electromagnetic signals. Chapter 6 describes the basic theory of antennas and their arrays. This chapter presents the fundamental properties of electromagnetic waves emanating from any antenna as well as the antenna parameters, including polarization, radiation patterns, beam width, side lobe level, efficiency, gain, bandwidth, input impedance, directivity, and receiving cross section. Chapter 6 discusses the radiation/reception properties of selected antenna structures, including a dipole, a monopole, a wire-loop, a slot, and a microstrip. The theory of antenna arrays made of a number of individual antenna elements at different locations is also discussed.

Active and passive components are essential building blocks of microwave circuits and systems that have become increasingly important due to the booming of next-generation wireless communications. Chapter 7 describes the basic characteristics of a class of microwave passive components, including tuning stubs, lumped elements, impedance transformers and matching network, couplers, power dividers/combiners, resonators, and filters. Scattering parameters are usually used to characterize the frequency-dependent components. Examples are given mostly for rectangular metallic waveguides and microstrip circuits. Active microwave components will also be discussed in this chapter.

The engineering applications of electromagnetic fields and waves usually require accurate solutions of Maxwell's equations subject to proper boundary conditions. Due to the increasing capability of computers, many of the complicated electromagnetic problems are now becoming solvable. Numerical computation has become an indispensable subject in electromagnetics. The two most widely applied numerical methods are discussed in Chapter 8. Chapter 8 also describes a frequency-domain integral-equation based approach known as the method of moments. This method is particularly useful when the Green's function (the kernel) can be found analytically. This section also discusses the recent progress of using a fast algorithm to deal with large electromagnetic systems. Chapter 9 describes a time-domain differential equation-based approach known as the finite-difference time-domain (FDTD) method. This method is particularly useful for complicate noncanonical or nonlinear structures with impulsive sources. The volume of the structure usually dictates computer time and memory required. Perfect absorbing boundary facilitates fast computation with minimum required memory.

An early application of electromagnetics is on radio detection and ranging known as radar, which is an interdisciplinary subject involving communications, signal processing, and propagation. Chapter 10 discusses the principle of both pulsed and CW radar systems and radar parameters. Specialized radars for various applications are also addressed, including MTI radar, Doppler radar, tracking radar, high cross-range resolution radar, and synthetic aperture radar. An inverse scattering problem is to reconstruct or recover physical or geometric properties of an object from measured electromagnetic fields. The basic principles of inverse scattering are also discussed in this chapter. The approach based on an integral equation formulation in conjunction with an iterative scheme to solve the inverse scattering is also discussed.

Trends in compact communication systems to involve the integration of antenna (including matching network) and active circuits (such amplifiers) together as one component. Chapter 11 discusses the basic principles of active integrated circuit antennas. The basic operation of microwave transistors (both BJT and FET) is discussed. Active circuits of amplifiers, oscillators, and detectors/mixers are described. Active antennas using microstrip patches or printed slots are also described.

The section 5 editor would like to thank all the authors and reviewers for their volunteer effort and cooperation to make this chapter possible. It is our hope that this chapter will be valuable to electrical and electronic engineers who are interested in the subject of electromagnetics.

1

Magnetostatics

Keith W. Whites
Department of Electrical and Computer Engineering, South Dakota School of Mines and Technology, Rapid City, South Dakota, USA

1.1 Introduction

Magnetostatics involves the computation of magnetic forces and fields produced by direct (i.e., time-stationary) currents and from materials with permanent magnetization (**magnets**). Only magnetic forces and fields that do not change with time are **magnetostatic**.

There are many applications of magnetostatics and even a few industries that are almost wholly based upon it. The magnetic recording and electric power industries both apply principles from magnetostatics. Other applications include magnetic resonance imaging (MRI) (Inan and Inan, 1999), magnetic brush applicators in electrophotography (laser printers) (Schein, 1992), and aurora in the earth's atmosphere (Paul *et al.*, 1998), to name a few.

1.2 Direct Current

1.2.1 Current and Current Density

Current is the flow of charge. By convention, the direction of this flow is with the movement of positive charge. The amount of charge δQ flowing *through* (i.e., perpendicular) to a surface in time δt is defined as $\delta Q = I\delta t$, where I is the **current**. In the limit of infinitesimally small time increments, the current I through the surface can be defined as:

$$I = \frac{dQ}{dt} \text{ [A]}, \tag{1.1}$$

where the units are coulombs per second [C/s] or amperes [A].

Magnetostatics is a field theory and, consequently, the quantities of interest are usually distributed throughout space. As such, the **volume current density,** $\boldsymbol{J}(x, y, z)$ is often employed. In terms of the charge carriers, the current density is given by

$$\boldsymbol{J} = Nq\boldsymbol{v}[\text{A/m}^2], \tag{1.2}$$

where N is the number of charge carriers per unit volume (i.e., the number density), q is the charge, and $\boldsymbol{v}$ is the average (or drift) velocity. In addition, a current density through an open surface S is related to the current as:

$$I = \int_s \boldsymbol{J} \cdot d\boldsymbol{s}[\text{A}]. \tag{1.3}$$

A **surface current density,** $\boldsymbol{J}_s[\text{A/m}]$, is an approximation for $\boldsymbol{J}$ in a very thin layer.

Example 1.1

One end of a copper wire (diameter = 2 mm) is attached to one end of an aluminum wire (diameter = 4 mm). A direct current $I = 10$ mA is passing through the wires. To determine the current density in each wire, equation 1.3 is applied assuming the current density is uniformly distributed over the cross-section:

$$J_{Cu} = \frac{I}{\text{Area}_{Cu}} = \frac{0.01\text{A}}{\pi(0.002)^2\text{m}^2} = 795.8\text{A/m}^2. \tag{1.4}$$

$$J_{Al} = \frac{I}{\text{Area}_{Al}} = \frac{0.01\text{A}}{\pi(0.004)^2\text{m}^2} = 198.9\text{A/m}^2. \tag{1.5}$$

The much smaller current density in Al is due solely to the larger diameter and not because of the material type.

The drift speed of the conduction electrons in each wire can be determined using equation 1.2 and by knowing that there are approximately 8.49×10^{28} conduction charges/m^3 in Cu and 6.02×10^{28} conduction charges/m^3 in Al (Halliday *et al.*, 2002). Therefore, the following two equations result:

$$v_{Cu} = \frac{J_{Cu}}{N_{Cu}e} = \frac{795.8\ \text{A/m}^2}{(8.49 \times 10^{28}/\text{m}^3)(1.6022 \times 10^{-19}\text{C})} = 5.850 \times 10^{-8}\text{m/s} \tag{1.6}$$

$$v_{Al} = \frac{J_{Al}}{N_{Cu}e} = \frac{198.9\text{A/m}^2}{(6.02 \times 10^{28}/\text{m}^3)(1.6022 \times 10^{-19}\text{C})} = 2.062 \times 10^{-8}\text{m/s}. \tag{1.7}$$

1.2.2 Ohm's Law

At each point in an ohmic material, such as in a conductor, the volume current density $\boldsymbol{J}$ and electric field $\boldsymbol{E}$ are related by **Ohm's law:**

$$\boldsymbol{J} = \sigma\boldsymbol{E}\ [\text{A/m}^2], \tag{1.8}$$

through the **electrical conductivity** σ of the material. The units of σ are siemens per meter [S/m]. The conductivities for various materials are listed in Table 1.1. It is apparent that σ varies enormously for different materials. The materials near the top of the table are called conductors, whereas those near the bottom are called insulators.

The electrical conductivity σ of metals varies with temperature. As a simple estimate of this variation, the conductivity can be assumed to change linearly with temperature (Halliday *et al.*, 2002):

$$\frac{1}{\sigma} - \frac{1}{\sigma_0} = \frac{\alpha}{\sigma_0}(T - T_0). \tag{1.9}$$

In this linear equation, σ_0 is the conductivity at temperature T_0, α is the temperature coefficient of the conductor (see Table 1.1), and σ is the conductivity at temperature T. For metals with a positive α, the conductivity decreases with increasing T (or resistivity $\equiv 1/\sigma$ increases with increasing T).

Example 1.2

From Table 1.1, the conductivity of copper at 20°C is $\sigma_0 = 5.8 \times 10^7$. We can use equation 1.9 to determine the temperature at which the conductivity is half that at 20°C. Solving for T in equation 1.9 using $T_0 = 20°\text{C}$, $\sigma = \sigma_0/2$, and $\alpha = 0.00393/°\text{C}$ gives the following:

$$T = T_0 + \frac{1}{\alpha} = 274.5°\text{C}. \tag{1.10}$$

TABLE 1.1 Electrical Conductivity σ and Temperature Coefficient α (Near 20°C) for Selected Materials at dc.

Material	σ(S/m) (20°C)	α (per degree Celsius)
Silver[1]	6.29×10^7	0.0038
Copper (annealed)[1]	5.8001×10^7	0.00393
Gold[1]	4.10×10^7	0.0034
Aluminum[1]	3.541×10^7	0.0039
Tungsten	1.90×10^7	0.0045
Iron (99.98% pure)[1]	1.0×10^7	0.005
Tin[1]	8.70×10^6	0.0042
Constantan[1]	2.0×10^6	0.00001
Nichrome[1]	1.0×10^6	0.0004
Carbon (graphite)	7.1×10^4	−0.0005
Seawater	4	—
Silicon (pure)	4×10^{-4}	−0.07
Distilled water	$\approx 10^{-4}$	—
Glass	$\approx 10^{-10} - 10^{-14}$	—
Polystyrene	$> 10^{-14}$	—
Hard rubber	$\approx 10^{-15}$	—
Quartz (fused)	$\approx 10^{-16}$	—

[1]Weast (1984).

1.2.3 Resistance

The ratio of the potential difference along a conductor to the current through the conductor is called the **resistance** R with units of ohms (Ω). Referring to an arbitrary conductor as in Figure 1.1, this ratio of voltage to current can be expressed as:

$$R = \frac{V}{I} = \frac{\int_c \boldsymbol{E} \cdot \boldsymbol{dl}}{\int_s \boldsymbol{J} \cdot \boldsymbol{ds}} = \frac{\int_c \boldsymbol{E} \cdot \boldsymbol{dl}}{\int_s \sigma \boldsymbol{E} \cdot \boldsymbol{ds}} [\Omega]. \quad (1.11)$$

To conform to the convention that $R \geq 0$ for passive conductors, the path of integration c is from the surfaces of higher to lower potential through the conductor, and $\boldsymbol{ds}$ is in the direction of current, as shown in Figure 1.1. If the conductor in this Figure is homogeneous with a cross-sectional area A, then from equation 1.11:

$$R = \frac{1}{\sigma} \frac{V}{A\left(\frac{V}{L}\right)} = \frac{L}{\sigma A} [\Omega]. \quad (1.12)$$

Equation 1.12 can be used to compute R for any straight, homogeneous conductor with a uniform cross-sectional area A at zero frequency. Conversely, if the conductor is inhomogeneous or has a nonuniform cross section, R must be computed using equation 1.11.

The resistance R and capacitance C of two perfect conductors (or, simply, two constant potential surfaces) at zero frequency are related as (Cheng, 1981):

$$RC = \frac{\varepsilon}{\sigma}, \quad (1.13)$$

where ε (permittivity) and σ are the material parameters of the otherwise homogeneous space between the perfect conductors.

Example 1.3

Suppose you wish to determine the resistance of a 1-m length of 12-gauge copper wire at 20°C and at a temperature of 274.5°C. At 20°C, using equation 1.12 and σ from Table 1.1, you get:

$$R = \frac{1}{5.8001 \times 10^7 \,\mathrm{S/m}(\pi \cdot 0.001025^2 \,\mathrm{m}^2)} = 5.224 \,\mathrm{m}\Omega. \quad (1.14)$$

As mentioned previously, the conductivity of copper decreases with increasing temperature. At $T = 274.5$°C, $\sigma = 2.9001 \times 10^7$ S/m from equation 1.9. Hence, the following results

$$R = \frac{1}{2.9001 \times 10^7 \,\mathrm{S/m}(\pi \cdot 0.001025^2 \mathrm{m}^2)} = 10.45 \,\mathrm{m}\Omega. \quad (1.15)$$

This resistance is twice that at 20°C as expected, because the conductivity has decreased by half.

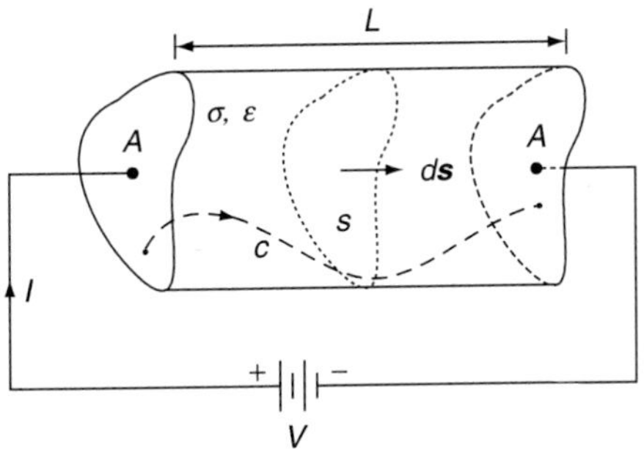

FIGURE 1.1 Conventions Used in the Computation of Resistance Using Equation 1.11

1.2.4 Power and Joule's Law

Ohm's law in equation 1.8 relates the conduction current $\boldsymbol{J}$ to the electric field $\boldsymbol{E}$ at every point in conductive material. Because of collisions between the charge carriers (electrons) comprising the current with the lattice of atoms forming the conductive material, there will be a loss of electrical energy. The **power** P delivered to electrical charges in a volume v is given by **Joule's law**:

$$P = \int_v \boldsymbol{E} \cdot \boldsymbol{J} dv [\mathrm{W}], \quad (1.16)$$

where P has units of joules per second [J/s] or watts [W]. This power is dissipated as heat in the conductive material through an irreversible process since P is unchanged when the direction of $\boldsymbol{E}$ in equation 1.16 is reversed with $\boldsymbol{J}$ given in equation 1.8.

Considering a conductor with a uniform cross section and a length L, if both $\boldsymbol{E}$ and $\boldsymbol{J}$ are directed along the conductor's length at all points, then from equation 1.16:

$$P = \int_L E dl \int_s J ds = VI [\mathrm{W}]. \quad (1.17)$$

This familiar expression for power in electrical circuits can be expressed in two alternative forms using Ohm's law for resistors (in equation 1.11) as:

$$P = \frac{V^2}{R} = I^2 R [\mathrm{W}]. \quad (1.18)$$

This power is dissipated in the resistor and transferred to its surroundings through **joule heating**.

Example 1.4

If a 75-W incandescent light bulb is connected to a 120-V electrical outlet, equation 1.18 can be used to compute the resistance of the light bulb provided V and I are root–mean–square values. In this case:

$$R = \frac{V_{RMS}^2}{P} = \frac{120^2}{75} = 19\Omega, \quad (1.19)$$

which is the resistance when the light bulb is turned on. If the cost of electricity is 5 ¢(kW·h), it would cost \$2.70 ($= 75 \times 24 \times 30 \times 0.05/1000$) to light this bulb continuously for 1 month. The resistance in equation 1.19 will likely be very different than when the light bulb is off and cold. From equation 1.9 using equation 1.12:

$$R - R_0 = \alpha R_0 (T - T_0)[\Omega] \quad (1.20)$$

Assuming the tungsten filament is approximately 3000°C when lit and using $\alpha = 0.0045/^\circ C$ from Table 1.1, then using this last equation gives $R_0 \approx 13\ \Omega$ for the resistance of a room-temperature 75-W light bulb. This agrees closely with a measured value of 13.1 Ω even though using α at such a large temperature was not justified.

1.2.5 Conservation of Charge and Kirchhoff's Current Law

A basic postulate of physics is that electrical charge can neither be created nor destroyed. This fact is manifested in electromagnetics through the **continuity equation**:

$$\nabla \cdot \boldsymbol{J} = -\frac{\partial \rho}{\partial t}. \quad (1.21)$$

This equation relates the net outward flux of $\boldsymbol{J}$ per unit volume to the time rate of change of the **volume electric charge density** ρ at every point. When there is no time variation (which is the situation for magnetostatics), the conservation of charge equation 1.21 becomes:

$$\nabla \cdot \boldsymbol{J} = 0. \quad (1.22)$$

Physically, this equation tells us that the net outward flux of $\boldsymbol{J}$ per unit volume at every point must vanish. In other words, the electric current density $\boldsymbol{J}$ acts like an incompressible fluid.

Applying the divergence theorem (Paul *et al.*, 1998) to equation 1.22 gives the integral form of the static continuity equation as:

$$\oint_s \boldsymbol{J} \cdot d\boldsymbol{s} = 0. \quad (1.23)$$

This result is **Kirchhoff's current law** (KCL) expressed in integral form. Using equation 1.23 at a junction of N conducting wires in a nonconducting space (such as air), the currents I_j in all wires satisfy:

$$\sum_j I_j = 0, \quad (1.24)$$

which is the circuit form of KCL.

Example 1.5

Conductors are intrinsically charge neutral even if a current exists in the conductor. If an excess charge is somehow introduced into the conductor, it will redistribute itself until the electric field due to the excess charge is zero. For example, suppose an excess charge is introduced into an isolated conductor. Beginning with equation 1.21, substituting equation 1.8, and using Gauss's law $\nabla \cdot (\varepsilon \boldsymbol{E}) = \rho$, we can determine that the free volume charge ρ must satisfy the equation (Cheng, 1989):

$$\frac{\partial \rho}{\partial t} + \frac{\sigma}{\varepsilon}\rho = 0, \quad (1.25)$$

with solution

$$\rho = \rho_0 e^{-t/\tau} \quad [\mathrm{C/m^3}], \quad (1.26)$$

where

$$\tau = \frac{\varepsilon}{\sigma} \ [\mathrm{s}]. \quad (1.27)$$

In these results, ρ_0 is the charge density distribution at an initial time and τ is the **relaxation time constant**. For all practical purposes, the charge is redistributed in a time equal to 5τ. This time is very brief for metals and much longer for good dielectrics. At these two extremes, copper has a relaxation time $\tau \approx 1.5 \times 10^{-19}$ s, while for fused quartz, $\tau \approx 4$ days.

1.3 Governing Equations of Magnetostatics

1.3.1 Postulates of Magnetostatics

The natural phenomenon of magnetostatics is governed by a short and succinct set of equations. The circulation of the **magnetic field intensity** $\boldsymbol{H}$ [A/m] is governed by **Ampère's law**:

$$\nabla \times \boldsymbol{H} = \boldsymbol{J} \text{ (point form).} \quad (1.28)$$

$$\oint_c \boldsymbol{H} \cdot d\boldsymbol{l} = I_{net} \text{ (integral form).} \quad (1.29)$$

The I_{net} is the net current passing through the open surface bounded by the closed contour c (Cheng, 1989). Furthermore, the net outward flux of the **magnetic flux density** $\boldsymbol{B}$ [Wb/m^2 or T] is governed by **Gauss's law** for magnetic fields:

$$\nabla \cdot \boldsymbol{B} = 0 \text{ (point form).} \quad (1.30)$$

$$\oint_s \boldsymbol{B} \cdot d\boldsymbol{s} = 0 \text{ (integral form).} \tag{1.31}$$

The $\boldsymbol{B}$ and $\boldsymbol{H}$ vector fields are related through the constitutive equation:

$$\boldsymbol{B} = \mu \boldsymbol{H}[\mathrm{T}], \tag{1.32}$$

where μ is the **permeability** [H/m] of the material.

All magnetostatic fields must satisfy the equations 1.28 through 1.31. Very few magnetostatic problems, however, have simple and analytical solutions for the vector fields $\boldsymbol{B}$ and $\boldsymbol{H}$. Table 1.2 contains a representative list of problems that do have simple analytical solutions. These problems all contain much symmetry, which is the key to arriving at these simple solutions.

Example 1.6

In rare instances, the integral forms of Ampère's law equation 1.29 and Gauss's law equation 1.31 can be used to *solve* for $\boldsymbol{H}$ and $\boldsymbol{B}$. This occurs when the problem contains sufficient symmetry so that $\boldsymbol{H}$ and $\boldsymbol{B}$ can be factored out of the integrals in these two equations (Paul *et al.*, 1998). As an example consider the toroid shown in Table 1.2. Assuming that the wire (carrying current I) is "tightly wound" on the toroid, then using Ampere's law around a circular contour inside the toroid gives:

$$\oint_c \boldsymbol{H} \cdot d\boldsymbol{l} = NI, \tag{1.42}$$

because the current pierces the open surface (bounded by contour c) N times. By symmetry and using a cylindrical coordinate system with $d\boldsymbol{l} = \hat{\boldsymbol{\varphi}} r d\phi$, in this case, gives:

$$2\pi r\, H_\phi = NI, \tag{1.43}$$

$$B_\phi = \frac{\mu NI}{2\pi r}[\mathrm{T}]. \tag{1.44}$$

Equation 1.44 is the solution given in Table 1.2. Outside the toroid, $I_{net} = 0$, and consequently $B_\phi = 0$.

1.3.2 Biot-Savart Law and Vector Magnetic Potential *A*

Ampère's law (equations 1.28 and 1.29) and Gauss's law (equations 1.30 and 1.31) are the rules that all magnetostatic fields must obey. Except in very limited situations, as discussed in the previous section, these laws cannot be used to directly compute $\boldsymbol{B}$ or $\boldsymbol{H}$. Instead, if a given current density $\boldsymbol{J}$ is prescribed at source coordinates $\boldsymbol{r}'$, the $\boldsymbol{B}$ field can be directly computed at any observation coordinate $\boldsymbol{r}$ using the **Biot-Savart law** (Paul *et al.*, 1998):

$$\boldsymbol{B}(\boldsymbol{r}) = \frac{\mu}{4\pi} \int_{r^1} \frac{\boldsymbol{J}(\boldsymbol{r}') \times \boldsymbol{R}}{R^3} dv'[\mathrm{T}]. \tag{1.45}$$

The vector $\boldsymbol{R} = \boldsymbol{r} - \boldsymbol{r}'$ points from the source point to the observation point. For a surface current density $\boldsymbol{J}_s$, the Biot-Savart law reads:

$$\boldsymbol{B}(\boldsymbol{r}) = \frac{\mu}{4\pi} \int_{s'} \frac{\boldsymbol{J}_s(\boldsymbol{r}') \times \boldsymbol{R}}{R^3} ds'\ [\mathrm{T}], \tag{1.46}$$

whereas for a filamentary current I (pointing in the direction of $d\boldsymbol{l}'$ at $\boldsymbol{r}'$), the Biot-Savart law is written as:

$$\boldsymbol{B}(\boldsymbol{r}) = \frac{\mu}{4\pi} \oint_{c'} \frac{I(\boldsymbol{r}')d\boldsymbol{l}' \times \boldsymbol{R}}{R^3}\ [\mathrm{T}]. \tag{1.47}$$

The $\boldsymbol{B}$ field can also be computed from the **magnetic vector potential *A*** as:

$$\boldsymbol{B} = \nabla \times \boldsymbol{A}\ [\mathrm{T}] \tag{1.48}$$

where for a volume current density $\boldsymbol{J}$ at source coordinates $\mathbf{r}'$:

$$\boldsymbol{A}(\boldsymbol{r}) = \frac{\mu}{4\pi} \int_{v'} \frac{\boldsymbol{J}(\boldsymbol{r}')}{R} dv'\ [\mathrm{Wb/m}]. \tag{1.49}$$

Expressions for $\boldsymbol{A}$ produced by $\boldsymbol{J}_s$ and I are similar to equations 1.46 and 1.47, respectively (Paul *et al.*, 1998). In rare instances, computation of $\boldsymbol{A}$ first and then $\boldsymbol{B}$ is simpler than computing $\boldsymbol{B}$ directly using the Biot-Savart law.

Example 1.7

A direct current I exists in the circular loop shown in Table 1.2. You can use the Biot-Savart law (equation 1.47) to find a simple analytical solution for $\boldsymbol{B}$ along the z-axis. (This is not true for observation points off the z-axis. Instead, numerical integration of (equation 1.47) would be required) (Whites, 1998, Example 4.10). For this geometry, referring to equation 1.47 and the circular loop figure in Table 1.2, $d\boldsymbol{l}' = \phi' a d\phi'$ and $\boldsymbol{R} = \boldsymbol{r} - \boldsymbol{r}' = \hat{\boldsymbol{z}}z - \hat{\boldsymbol{r}}' a$. Consequently, from equation 1.47:

$$\boldsymbol{B} = \frac{\mu I}{4\pi} \int_0^{2\pi} \frac{(\boldsymbol{\phi}' a d\phi') \times (\hat{\boldsymbol{z}}z - \hat{\boldsymbol{r}}' a)}{(a^2 + z^2)^{3/2}} = \frac{\mu I}{4\pi(a^2 + z^2)^{3/2}} \int_0^{2\pi} (\hat{\boldsymbol{r}}' az + \hat{\boldsymbol{z}} a^2)\, d\phi'$$
$$= \frac{\mu I}{4\pi(a^2 + z^2)^{3/2}} \left[\hat{\boldsymbol{x}} az \int_0^{2\pi} \cos\phi' d\phi' + \hat{\boldsymbol{y}} az \int_0^{2\pi} \sin\phi' d\phi' + \hat{\boldsymbol{z}} a^2 \int_0^{2\pi} d\phi' \right]. \tag{1.50}$$

TABLE 1.2 Examples of Problems with Simple Analytical Solutions for $\boldsymbol{B}$ [T]

Name	Geometry	Solution	
Infinite sheet[1]	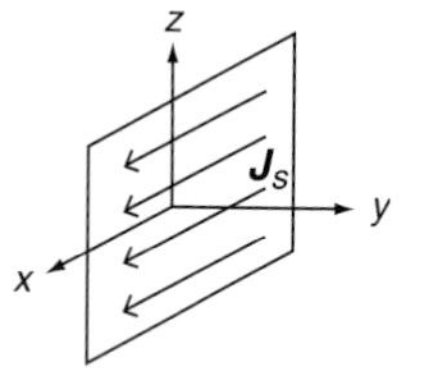	$\boldsymbol{B} = \pm\hat{\boldsymbol{z}}\frac{\mu J_s}{2}\ y \gtrless 0$	(1.33)
Straight wire[2]	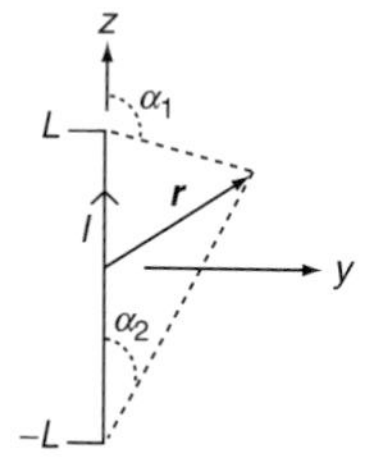	Finite: $\boldsymbol{B} = \boldsymbol{\phi}\frac{\mu I}{4\pi r}(\cos\alpha_2 - \cos\alpha_1)$	(1.34)
		Infinite: $\boldsymbol{B} = \boldsymbol{\phi}\frac{\mu I}{2\pi r}\ (z = 0,\ L \to \infty)$	(1.35)
Circular loop (along z axis)[3]	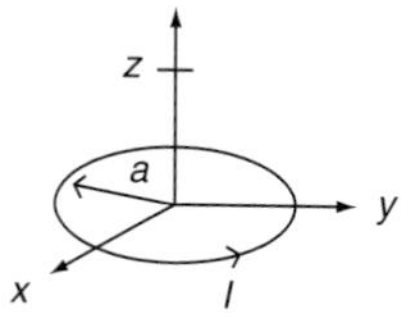	$\boldsymbol{B} = \hat{\boldsymbol{z}}\frac{\mu I a^2}{2(a^2+z^2)^{3/2}}\ (x = y = 0)$	(1.36)
Magnetic dipole[1]	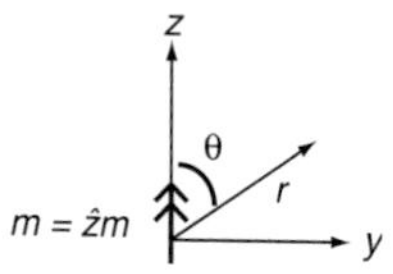	$\boldsymbol{B} = \frac{\mu m}{4\pi r^3}(\hat{\boldsymbol{r}}2\cos\theta + \hat{\boldsymbol{\theta}}\sin\theta)$	(1.37)
Long solenoid ($L \gg a$)[1]	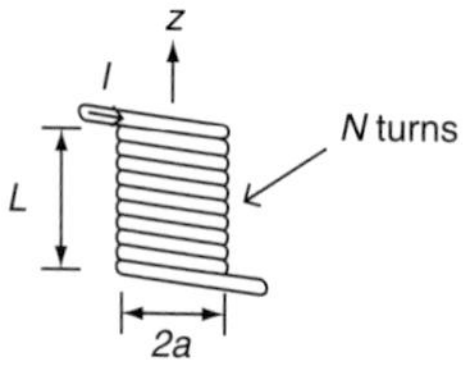	$\boldsymbol{B} \approx \hat{\boldsymbol{z}}\mu\left(\frac{N}{L}\right)I$ inside	(1.38)
Toroid[1]	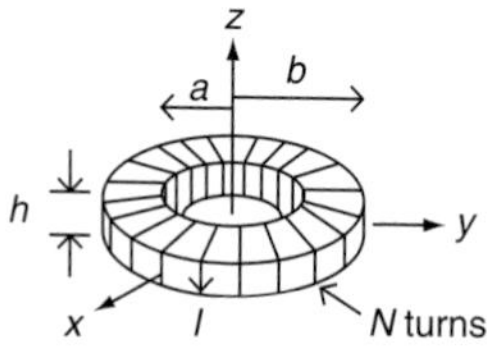	$\boldsymbol{B} = \begin{cases} \boldsymbol{\phi}\frac{\mu NI}{2\pi r} & \text{inside} \\ 0 & \text{outside} \end{cases}$	(1.39)
Coaxial cable[1]	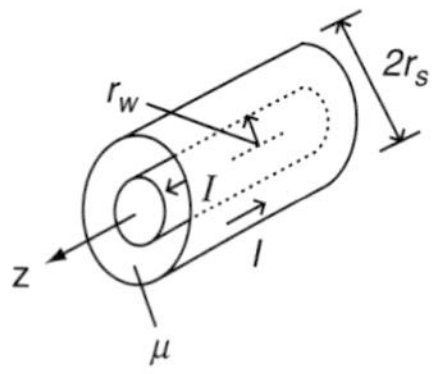	$\boldsymbol{B} = \begin{cases} \boldsymbol{\phi}\frac{\mu_0 I r}{2\pi r_w^2} & r < r_w \\ \boldsymbol{\phi}\frac{\mu I}{2\pi r} & r_w < r < r_s \\ 0 & r > r_s \end{cases}$	(1.40)
Helmholtz coils (along z axis)[4]	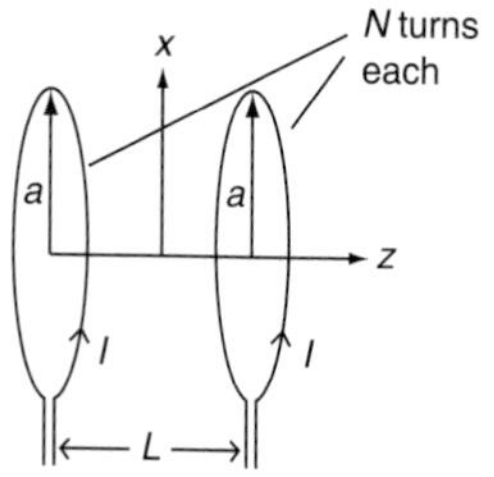	$\boldsymbol{B} = \hat{\boldsymbol{z}}\frac{\mu N I a^2}{2}\left\{\left[a^2 + \left(z + \frac{L}{2}\right)^2\right]^{-3/2} + \left[a^2 + \left(z - \frac{L}{2}\right)^2\right]^{-3/2}\right\}$ (along z axis)	(1.41)

[1]Paul *et al.* (1998). [2]Inan and Inan (1999). [3]Cheng (1989). [4]Whites (1998), Sec 4.3, Prob. 4.3.4.

Therefore,

$$B(z) = \hat{z}\frac{\mu I a^2}{2(a^2 + z^2)^{3/2}} \text{ [T]}, \quad (1.51)$$

which is the result of equation 1.36 listed in Table 1.2.

1.3.3 Boundary Conditions for *B*, *H*, and *J*

The vector field quantities ***B*** and ***H*** behave in a prescribed manner at the interface between two different magnetic materials. The component of ***B*** perpendicular to the interface (in the direction of the unit vector $\hat{n}$) is continuous across the interface. That is:

$$\hat{n} \cdot (B_2 - B_1) = 0, \quad (1.52)$$

or equivalently,

$$B_{n2} = B_{n1}. \quad (1.53)$$

The B_1 (or B_2) is ***B*** at the interface but just inside material 1 (or 2) and $\hat{n}$ is a unit vector pointing from region 1 toward region 2.

Conversely, the tangential components of ***H*** are discontinuous across a boundary that has a surface current density J_s as:

$$\hat{n} \times (H_2 - H_1) = J_s. \quad (1.54)$$

In scalar form, using the right-handed coordinate system $\{n, t, u\}$, the tangential components of $H(H_{t2}$, and $H_{t1})$ are discontinuous across the material interface by an amount equal to $J_{s,u}$, which is the component of J_s perpendicular to H_{t2} and H_{t1}. That is:

$$H_{t2} - H_{t1} = J_{s,u}. \quad (1.55)$$

A surface current density J_s exists at an interface only in certain situations such as an impressed source layer, on the surface of superconductors, and, for time-varying fields, on the surface of perfect electrical conductors ($\sigma \to \infty$) (Paul *et al.*, 1998). When J_s does not exist at the interface, then from equation 1.55:

$$H_{t2} = H_{t1}, \quad (1.56)$$

so that the tangential components of ***H*** are continuous across the interface.

At the interface between two materials with conductivities σ_1 and σ_2, the components of ***J*** that are normal (i.e., perpendicular) to the interface are continuous across the interface:

$$J_{n2} = J_{n1}, \quad (1.57)$$

while the tangential components are discontinuous according to the relationship:

$$\frac{J_{t2}}{\sigma_2} = \frac{J_{t1}}{\sigma}. \quad (1.58)$$

1.4 Magnetic Force and Torque

1.4.1 Ampère's Force Law

One of the most fundamental principles in magnetostatics is that a current immersed in a magnetic field experiences a force. This magnetic force behaves very differently than the electrostatic force. Considering two closed loops of current shown in Figure 1.2, the total net magnetic force on the current in loop 1 due to the current in loop 2 is given by **Ampère's force law** (Paul *et al.*, 1998):

$$F_{12} = \frac{\mu I_1 I_2}{4\pi}\oint_{c_1}\oint_{c_2}\frac{dl_1 \times (dl_2 \times R_{12})}{R_{12}^3} \text{ [N]}. \quad (1.59)$$

This force law can also be expressed in the slightly different form:

$$F_{12} = -\frac{\mu I_1 I_2}{4\pi}\oint_{c_1}\oint_{c_2}\frac{R_{12}(dl_1 \cdot dl_2)}{R_{12}^3} \text{ [N]}. \quad (1.60)$$

By interchanging indices 1 and 2 in equation 1.60, it is easy to see that:

$$F_{21} = -F_{12}, \quad (1.61)$$

as required by Newton's third law (Whites, 1998).

Substituting equation 1.47 into equation 1.59, the net force on loop 1 can be expressed as:

$$F_{12} = \oint_{c_1} I_1 dl_1 \times B_2(r_1) \text{ [N]}. \quad (1.62)$$

It is very clear from this result that a current immersed in a magnetic field experiences a force. Specifically, the magnetic field is produced by the current in loop 2, and the force F_{12} in equation 1.62 is that experienced by the current in loop 1.

Example 1.8

Using equation 1.62 it can be shown that the net magnetic force on any closed loop of current immersed in a uniform ***B***

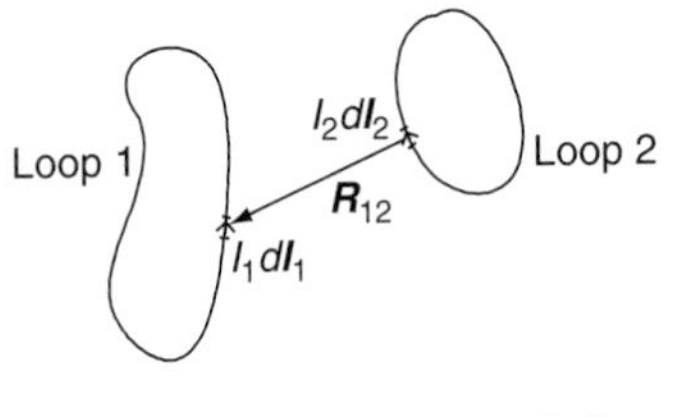

FIGURE 1.2 Differential Current Elements $I_1 dl_1$ and $I_2 dl_2$ Located in Two Current Loops

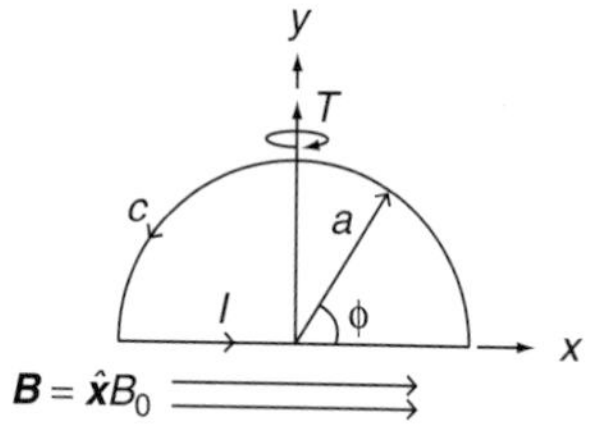

FIGURE 1.3 Closed Loop of Current I Immersed in a Uniform **B** Field

field is zero (Ulaby, 2001). When $\boldsymbol{B}$ is uniform, then from equation 1.62:

$$\boldsymbol{F} = I\oint_c d\boldsymbol{l} \times \boldsymbol{B} = I\underbrace{\left(\oint_c d\boldsymbol{l}\right)}_{=0} \times \boldsymbol{B} = 0. \qquad (1.63)$$

This result is true regardless of the shape of the current loop so long as the $\boldsymbol{B}$ field is uniform. As an example, the net magnetic force on the loop in Figure 1.3 can be computed using equation 1.62 as:

$$\boldsymbol{F} = I\oint_c d\boldsymbol{l} \times \boldsymbol{B} = I\left[\int_{-a}^{a} (\hat{\boldsymbol{x}}dx) \times (\hat{\boldsymbol{x}}B_0) + \int_0^{\pi} (\boldsymbol{\phi} a d\phi) \times (\hat{\boldsymbol{x}}B_0)\right]$$

$$= I\int_0^{\pi} (-\hat{\boldsymbol{x}}\sin\phi + \hat{\boldsymbol{y}}\cos\phi)a d\phi \times (\hat{\boldsymbol{x}}B_0) = -\hat{\boldsymbol{z}}IaB_0 \sin\phi|_0^{\pi} = 0, \qquad (1.64)$$

which is zero, as expected in light of equation 1.63.

1.4.2 Lorentz Force Equation

Moving charges in a magnetic field also experience a magnetic force, similar to current in the previous section. The total force experienced by a charge q moving with velocity $\boldsymbol{v}$ is given by the celebrated **Lorentz force equation:**

$$\boldsymbol{F} = q(\boldsymbol{E} + \boldsymbol{v} \times \boldsymbol{B}) \ [\text{N}], \qquad (1.65)$$

which is the sum of an electric force ($\boldsymbol{F}_e = q\boldsymbol{E}$) and a magnetic force ($\boldsymbol{F}_m = q\boldsymbol{v} \times \boldsymbol{B}$).

There are two important distinctions concerning the behavior of the electric and magnetic forces in the Lorentz force equation. First, it is apparent from equation 1.65 that the electric force $\boldsymbol{F}_e$ acts on both moving and stationary charges, whereas the magnetic force $\boldsymbol{F}_m$ acts only on moving charges. Second, energy is transferred from the electric field to the charged particle, whereas no energy is transferred from the magnetic field to a moving (or stationary) charged particle.

Because of this second fact, a magnetic force can only change the direction of a moving charge and not its speed. This can be easily understood by considering the differential work dW performed by the magnetic field when the particle is displaced a small distance $d\boldsymbol{l} = \boldsymbol{v}dt$ in time dt (Ulaby, 2001):

$$dW = \boldsymbol{F}_m \cdot d\boldsymbol{l} = (\boldsymbol{F}_m \cdot \boldsymbol{v})dt = 0, \qquad (1.66)$$

since from equation 1.65, $\boldsymbol{F}_m \cdot \boldsymbol{v} = 0$.

Example 1.9

The Lorentz force equation can be applied to charged particle motion through solids and gases. An example of the former is the **Hall effect** in a conductor as illustrated in Figure 1.4. In steady-state, the force of equation 1.65 on an electron q_e passing through this material is zero such that:

$$\boldsymbol{E} = -\boldsymbol{v} \times \boldsymbol{B} = -\hat{\boldsymbol{y}}vB_0 \ [\text{V/m}]. \qquad (1.67)$$

The difference in potentials at the $y = w$ and $y = 0$ faces of the material is as follows:

$$V_H = -\int_0^w E_y dy = vwB_0 \ [\text{V}], \qquad (1.68)$$

because of the charge accumulation on these two faces. This V_H is called the **Hall voltage.**

From the sign of V_H, the polarity of the charge can be determined; from the magnitude of V_H, the number density of charge carriers can be computed. A Hall device can also be used to measure conductivities of materials and to measure magnetic field strengths (Inan and Inan, 1999; Halliday, 1992). Imagine that a copper foil of thickness $d = 100\,\mu\text{m}$ with current $I = 20$ A is placed in magnetic field with $B_0 = 0.5$ T. Substituting equation 1.2 in equation 1.68 and rearranging gives the following:

$$V_H = \frac{IB_0}{dNe} = \frac{(20\text{A})(0.5\ \text{T})}{(100\times10^{-6}\ \text{m})(8.49\times10^{28}\ \text{m}^{-3})(1.6022\times10^{-19}\ \text{C})} = 7.35\,\mu\text{V}. \qquad (1.69)$$

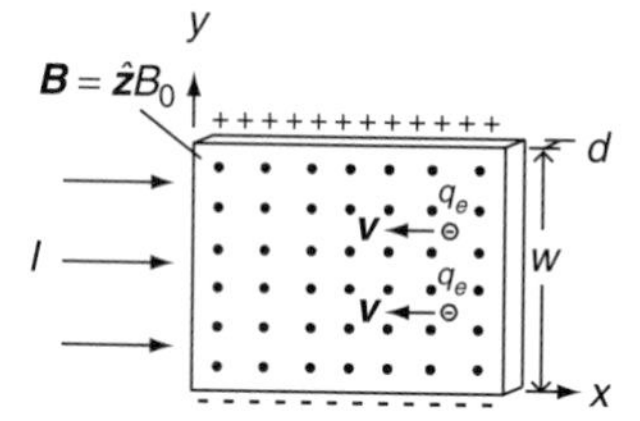

FIGURE 1.4 Hall Effect and a Conducting Material Placed in a Uniform **B** Field

1.4.3 Torque and Magnetic Dipole Moment

In example 1.8, it was mentioned that the net magnetic force on a closed loop of current in a uniform $\boldsymbol{B}$ field is zero. However, this closed loop of current does experience a torque and will rotate if it is free to do so. The **torque** $\boldsymbol{T}$ exerted on any planar loop of current immersed in a uniform $\boldsymbol{B}$ field is given as (Cheng, 1989):

$$\boldsymbol{T} = \boldsymbol{m} \times \boldsymbol{B} \text{ [N-m]}, \tag{1.70}$$

where the **magnetic dipole moment** $\boldsymbol{m}$ of a planar current loop of area A is as follows:

$$\boldsymbol{m} = \hat{\boldsymbol{n}}m = \hat{\boldsymbol{n}}IA \text{ [A-m}^2\text{]}. \tag{1.71}$$

The unit normal vector $\hat{\boldsymbol{n}}$ is determined using the right-hand rule. That is, with the fingers pointing in the direction of the current, the thumb points in the direction of $\hat{\boldsymbol{n}}$. The magnetic dipole moment $\boldsymbol{m}$ of any planar loop can be computed using equation 1.71. (Note, however, that the $\boldsymbol{B}$ field from equation 1.37 produced by $\boldsymbol{m}$ is valid only at distances "far" from the loop [White, 1998, Example 4.10].)

Example 1.10

The object shown in Figure 1.3 is an example of a planar current loop in a uniform $\boldsymbol{B}$ field that experiences a torque. The magnetic dipole moment of this loop can be found using equation 1.71:

$$\boldsymbol{m} = \hat{\boldsymbol{z}}\frac{I}{2}\pi a^2 \text{ [A-m}^2\text{]}, \tag{1.72}$$

and the torque can be determined from equation 1.70 as:

$$\boldsymbol{T} = \left(\hat{\boldsymbol{z}}\frac{I}{2}\pi a^2\right) \times (\hat{\boldsymbol{x}}B_0) = \hat{\boldsymbol{y}}\frac{I}{2}B_0\pi a^2 \text{ [N-m]}. \tag{1.73}$$

The directions of $\boldsymbol{T}$ and the loop rotation—if it were free to move—are both indicated in Figure 1.3. The direction of rotation is determined from $\boldsymbol{T}$ using the right-hand rule: With the thumb in the direction of $\boldsymbol{T}$, the fingers give the sense of rotation. The loop will not experience a torque when the unit normal vector $\hat{\boldsymbol{n}}$ is pointing in the direction of $\boldsymbol{B}$.

1.5 Magnetic Materials

1.5.1 Magnetization Vector and Permeability

When magnetic material is placed in an incident (or external) magnetic field, a secondary magnetic field is produced by the material. Similar to a dielectric material, the magnetic material is said to have become polarized by the alignment of microscopic magnetic dipole moments $\boldsymbol{m}$. These magnetic dipole moments can be used to develop a phenomenological model for the classical effects and many quantum mechanical aspects of magnetization (Paul *et al.*, 1998; Plonus, 1978).

The macroscopic effects of magnetization are described by a **magnetization vector** field $\boldsymbol{M}$ as a vector sum of $\boldsymbol{m}_i$ in a small volume $\Delta v'$:

$$\boldsymbol{M} = \lim_{\Delta v' \to 0} \frac{\sum_{i=1}^{N} \boldsymbol{m}_i}{\Delta v'} \text{ [A/m]}, \tag{1.74}$$

where N moments are assumed contained in $\Delta v'$. By definition, the **magnetic field intensity** $\boldsymbol{H}$ is then given as:

$$\boldsymbol{H} = \frac{\boldsymbol{B}}{\mu_0} - \boldsymbol{M} \text{ [A/m]}. \tag{1.75}$$

The **permeability of free space**, μ_0, is identically equal to $4\pi \times 10^{-7}$ [H/m].

From experimentation, it has been found that for many materials, $\boldsymbol{M}$ and $\boldsymbol{H}$ are simply related as:

$$\boldsymbol{M} = \chi_m \boldsymbol{H} \text{ [A/m]}. \tag{1.76}$$

In this expression, χ_m is the **magnetic susceptibility** of the material and is a dimensionless quantity. The values of χ_m are typically found through experimental measurement.

Substituting equation 1.76 into 1.75 and rearranging gives the following equations:

$$\boldsymbol{B} = \mu_0(1 + \chi_m)\boldsymbol{H} \text{ [T]} \tag{1.77}$$

or

$$\boldsymbol{B} = \mu \boldsymbol{H} \text{ [T]}, \tag{1.78}$$

where

$$\mu = \mu_0 \mu_r = \mu_0(1 + \chi_m) \text{ [H/m]} \tag{1.79}$$

is the **permeability** and μ_r is the **relative permeability** of the material. For free space, $\mu_r = 1$. Equation 1.78 is the **constitutive relationship** for magnetic fields.

Example 1.11

A magnetic sphere of radius a and permeability μ is placed in a uniform incident field $\boldsymbol{H}^{\text{inc}} = \hat{\boldsymbol{z}}H_0$. It can be shown that the secondary field inside and outside the sphere is the following (Plonus, 1978; Ramo *et al.*, 1999):

$$\boldsymbol{H}^{\text{secondary}} = \begin{cases} -\hat{\boldsymbol{z}}\frac{\mu-\mu_0}{\mu+2\mu_0}H_0 & r < a \\ \frac{\mu-\mu_0}{\mu+2\mu_0}H_0\frac{a^3}{r^3}\left(\hat{\boldsymbol{r}}2\cos\theta + \hat{\boldsymbol{\theta}}\sin\theta\right) & r > a \end{cases} \text{ [A/m]}. \tag{1.80}$$

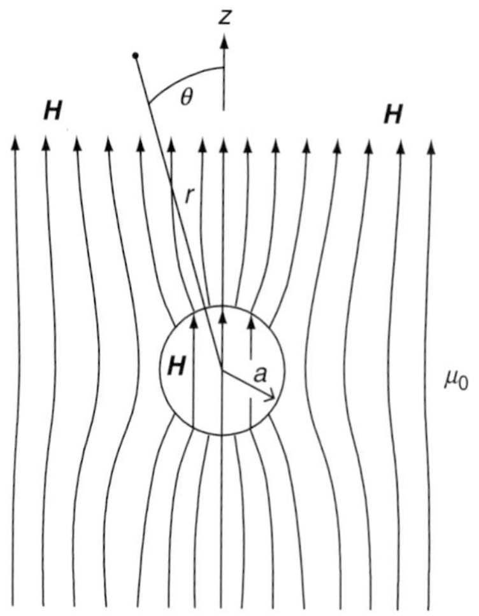

FIGURE 1.5 Magnetic Sphere of Permeability μ Immersed in a Uniform *H* Field. The field lines become strictly vertical when the sphere is removed.

The total $\boldsymbol{H}$ at any point is the sum $\boldsymbol{H} = \boldsymbol{H}^{inc} + \boldsymbol{H}^{secondary}$. A plot of the $\boldsymbol{H}$ field lines is shown in Figure 1.5. The magnetic sphere causes $\boldsymbol{H}$ to be disturbed from its original uniform nature. Inside the sphere, however, $\boldsymbol{H}$ is uniform and pointed along the same axis as the incident field, though its amplitude is now less than H_0 (and the amplitude of $\boldsymbol{B}$ is *greater* than B_0) when $\mu > \mu_0$. This disturbance in $\boldsymbol{H}$ is due to the secondary $\boldsymbol{H}$ produced by the induced magnetization $\boldsymbol{M}$ of the sphere. Using equation 1.76:

$$\boldsymbol{M} = \begin{cases} \hat{\boldsymbol{z}}\frac{3\chi_m\mu_0}{\mu+2\mu_0}H_0 = \hat{\boldsymbol{z}}\frac{3(\mu-\mu_0)}{\mu+2\mu_0}H_0 & r < a \\ 0 & r > a \end{cases} \text{[A/m]}, \qquad (1.81)$$

which is zero outside the sphere because there is nothing to magnetize in free space.

1.5.2 Magnetic Materials

There are five major classifications for magnetic materials as briefly discussed in this subsection (Inan and Inan, 1999; Paul *et al.*, 1998; Halliday *et al.*, 1992). Diamagnetic materials have $\chi_m < 1$, and the remaining four types all have $\chi_m > 1$. Because of this, diamagnetic materials are repelled by a strong magnet, whereas specimens of the other four categories are attracted to a strong magnet.

- **Diamagnetic** materials have a small negative χ_m so that $\mu_r \lesssim 1$ as shown in Table 1.3. All materials are diamagnetic to some extent although this behavior may be superceded by a more dominant effect, such as ferromagnetism. Diamagnetism is a classical (*versus* quantum mechanical) effect produced by moving charges. The induced magnetization $\boldsymbol{M}$ is opposed to the applied $\boldsymbol{B}$, thus reducing the total $\boldsymbol{B}$ in such a material sample. This effect is directly analogous to the polarization effects in ordinary dielectrics.
- **Paramagnetic** materials have a small positive χ_m so that $\mu_r \gtrsim 1$ as shown in Table 1.3. Paramagnetism is a quantum mechanical effect largely due to the spin magnetic moment of the electron. While these permanent magnetic moments are usually randomly oriented (so that $\boldsymbol{M} = 0$), they become partially aligned in an applied $\boldsymbol{B}$. In this latter state, the magnetization $\boldsymbol{M}$ is aligned with the applied $\boldsymbol{B}$, thus increasing the total $\boldsymbol{B}$ in the sample (and, consequently, decreasing $\boldsymbol{H}$).
- **Ferromagnetic** materials also have a positive χ_m, but the resulting μ_r is usually much greater than 1. There is a much stronger quantum mechanical interaction between neighboring spin moments than with paramagnetic materials that can also lead to magnetization $\boldsymbol{M}$ without an applied $\boldsymbol{B}$. Ferromagnetism is strongly temperature dependent. Only the elements iron, nickel, and cobalt are ferromagnetic at room temperature. Above the **Curie temperature** T_c, a ferromagnetic material becomes paramagnetic, and the magnetic susceptibility decreases with increasing temperature. The T_c of the room temperature ferromagnetic materials are 770°C for iron, 354°C for nickel, and 1115°C for cobalt.
- **Antiferromagnetic** materials are quite similar to ferromagnetic materials in that there is a strong quantum mechanical interaction between neighboring atomic molecules. However, this strong interaction causes an **anti-parallel** alignment of magnetic moments yielding a zero

TABLE 1.3 Relative Permeability μ of Selected Materials at Zero Frequency and at Room Temperature[1]

Material	μ_r
Diamagnetic ($\mu_r \approx 1$)	
Water	0.99999
Copper	0.99999
Silver	0.99998
Gold	0.99996
Bismuth	0.99983
Paramagnetic ($\mu_r \approx 1$)	
Air	1.000004
Magnesium	1.000012
Aluminum	1.000021
Titanium	1.00018
FeO_2	1.0014
Ferromagnetic ($\mu_{r,\,max}$)	
Cobalt	250
Nickel	600
Mild steel	2,000
Iron	5,000
Mumetal	100,000
Supermalloy	1,000,000

[1]Paul *et al.* (1998).

magnetization $\boldsymbol{M}$. The elements chromium and manganese are examples of antiferromagnetic materials. This effect is highly temperature dependent. Below the **Néel temperature** T_N, the magnetic susceptibility increases with increasing temperature but decreases for temperature greater than T_N (Kittel, 1996).

- **Ferrimagnetic** materials possess characteristics between those of ferromagnetic and antiferromagnetic materials. There is an incomplete antiparallel alignment of magnetic moments as in the antiferromagnetic materials, and, consequently, there is a net magnetization $\boldsymbol{M}$, though typically less than ferromagnetic materials. **Ferrites** are a special class of ferrimagnetic material that have a low conductivity at high frequencies. These ceramic-like materials are extremely useful in high-frequency applications due to their small conduction losses. Examples include MnZn with $\mu_{r,\,\max} \approx 2000$ and NiZn with $\mu_{r,\,\max} \approx 100$.

1.5.3 Domains and Hysteresis

The origin of the extremely large permeabilities of ferromagnetic materials is the existence of magnetized domains in these materials (Plonus, 1978; Kittel, 1996). A **magnetized domain** is a microscopic region (on the order of 0.1 to 1 mm^3) where the magnetization is uniform and saturated. Between adjacent domains are **domain walls.** Without an external field (at point O in Figure 1.6), these domains are randomly orientated so that the net magnetization of the material is zero. In the presence of external magnetic fields, these domains align and produce an enormous secondary $\boldsymbol{B}$ in the same direction as the applied field, thus giving a possibly enormous permeability.

If the applied fields become strong enough, the domains rotate and the material leaves the initial linear region of operation and enters the nonlinear and multivalued region indicated by the **hysteresis curve** in Figure 1.6. The ratio B/H at any point on the curves is the permeability μ. Around point O, the slope is nearly constant, whereas on the hysteresis curve it is not; this indicates that the ferromagnetic material has become nonlinear.

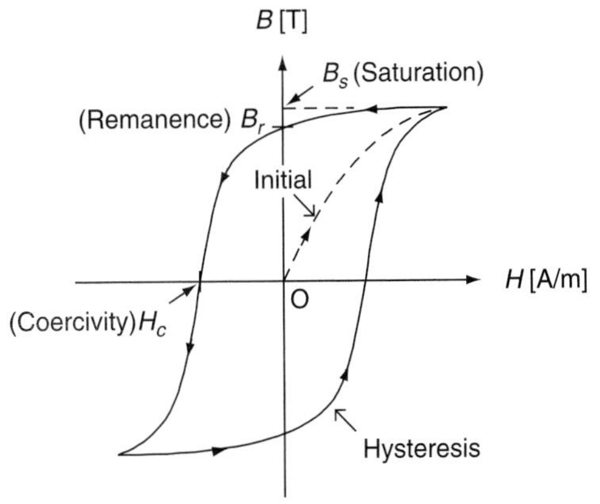

FIGURE 1.6 Magnetization Curves for a Typical Ferromagnetic Material. The initial magnetization curve is indicated by the dashed line, while the hysteresis curve is indicated by the solid line.

The maximum B that the material attains for any H is called the **saturated** B, B_s in Figure 1.6, corresponding to total alignment of domains. When H is then reduced from this point to zero, B reduces to B_r, which is the remanent B field or **remanance.** The negative (or antiparallel) H required to further reduce B to zero is H_c, the **coercivity.** Materials with a large H_c (called **hard ferromagnetic materials**) expend more energy per complete transversal of the hysteresis curve and consequently find applications as permanent magnets (see the next section). Conversely, materials with small H_c (called **soft ferromagnetic materials**) expend much less energy per hysteresis-curve cycle and find uses in transformers, relays, and generators.

The properties of a few selected soft ferromagnetic materials are listed in Table 1.4, including μ along the initial magnetization curve in Figure 1.6 and the maximum μ found anywhere along the hysteresis curve.

1.5.4 Permanent Magnets

Permanent magnets are materials that possess a magnetization $\boldsymbol{M}$ in the absence of an applied magnetic field. These magnets

TABLE 1.4 Properties of Selected Soft Ferromagnetic Materials[1]

Material (% by weight; remainder is Fe)	Initial μ_r	Max μ_r	B_s [T][2]	B_r [T][2]	H_c [A/m][3]
Commercial iron (0.2 impurities)	250	9,000	2.15	0.77	≈ 80
Purified iron (0.05 impurities)	10,000	200,000	2.15	—	4
Silicon-iron (4 Si)	1,500	7,000	1.95	0.5	20
Silicon-iron (3 Si)	7,500	55,000	2	0.95	8
Mumetal (5 Cu, 2 Cr, 77 Ni)	20,000	100,000	0.65	0.23	4
78 Permalloy (78.5 Ni)	8,000	100,000	1.08	0.6	4
Supermalloy (79 Ni, 5 Mo)	100,000	1,000,000	0.79	0.5	0.16

[1]Plonus (1978).
[2]Multiply by 10,000 for cgs unit of gauss [G].
[3]Multiply by $4\pi \times 10^{-3}$ for cgs unit of oersted [Oe].

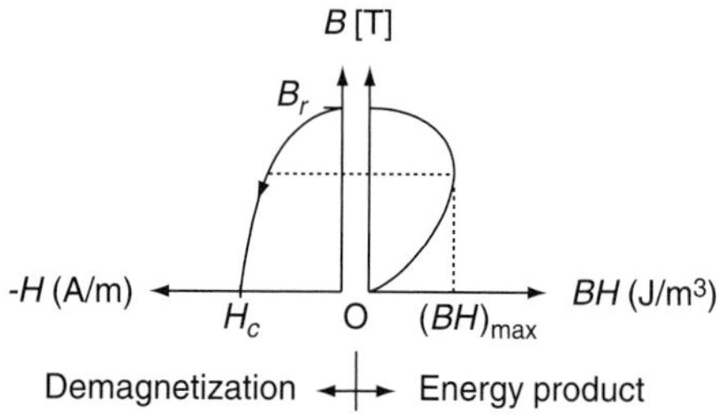

FIGURE 1.7 Demagnetization and Energy Product Curves for a Permanent Magnet

can be constructed from ferromagnetic or other materials that are nonlinear with magnetization curves containing hysteresis, as shown in Figure 1.6. If the material has been properly magnetized, then a remanent B field B_r exists in the material even when $H = 0$. Consequently, this material acts as a **permanent magnet**.

In many applications of permanent magnets—such as in electrical motors, generators, and relays—the permanent magnet is part of a magnetic circuit containing an air gap. In these situations, as derived in example, 1.12, the permanent magnet generally operates in the second (or fourth) quadrant of the hysteresis curve (Bozorth, 1978). This second-quadrant portion of the hysteresis curve is called the **demagnetization curve** and is illustrated in Figure 1.7.

One quality measure of a permanent magnet is the "size" of this demagnetization curve. A large B_r and H_c in Figure 1.7 ensures that the intrinsic magnetization of the magnet is large and remains "permanent" even for large H. Another selection criterion used for permanent magnets is based on the **energy product** BH. This quantity is appropriately named since energy density is proportional to this product (see subsection 1.7.2), which is also plotted in Figure 1.7 by multiplying B and H at each point along the demagnetization curve. The **maximum energy product**$(BH)_{max}$ is yet another measure of the quality of a permanent magnet since H in the air gap produced by the permanent magnet shown in Figure 1.8 is maximum (with $I = 0$) when the energy product is maximum. This maximum energy product $(BH)_{max}$ for a selected number of steels, alloys, and other materials is listed in Table 1.5 together with the remanance B_r and coercivity H_c.

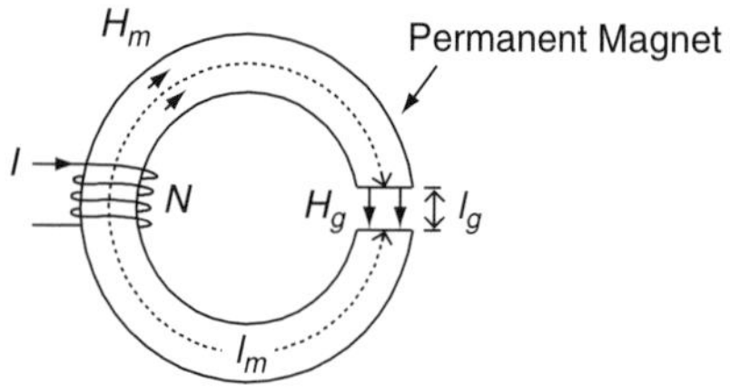

FIGURE 1.8 Ring of Permanent Magnet Material and Air Gap Excited by a Coil of Wire with N Turns and Current I

Example 1.12

The geometry shown in Figure 1.8 is an excellent canonical problem to illustrate the salient features of permanent magnet circuits. Assume a small cross-section and ignore all leakage flux. The first topic is to show the steps necessary to determine B in this problem, which is the same in the gap and the magnet because of the boundary condition shown in equation 1.52. To do this, two equations will be solved simultaneously. The first equation is the hysteresis curve for the permanent magnet, which is shown in Figure 1.9.

The second equation comes from an application of Ampère's law of equation 1.29 around the permanent magnet circuit in Figure 1.8:

$$\oint_c \mathbf{H} \cdot d\mathbf{l} = NI. \tag{1.82}$$

Assuming no "flux leakage" from the permanent magnet or "flux fringing" in the air gap (see Section 1.8), then:

$$H_m l_m + H_g l_g = NI, \tag{1.83}$$

where H_m and H_g are the magnetic fields in the magnet and air gap, respectively. In the air gap, $H_g = B/\mu_0$. Substituting this into equation 1.82 and rearranging gives:

$$B = -\mu_0 \frac{l_m}{l_g} H_m + \mu_0 \frac{NI}{l_g} [\mathrm{T}]. \tag{1.84}$$

This straight-line equation 1.84 is also drawn in Figure 1.9. The intersections of this "load line" (Bozorth, 1978, Ch. 9) with the hysteresis curve give the two possible solutions for B in the permanent-magnet circuit of Figure 1.8. The actual operating point depends on the previous time history because this is a nonlinear circuit. Nevertheless, note from this example that the second quadrant of the hysteresis curve (which is symmetrical with the fourth) is the important quadrant. This illustrates why the second quadrant is used as a measure of comparison for magnets, as shown in Figure 1.7.

The second topic considered in this example is the relationship between H_g in the air gap and the maximum energy product $(BH_m)_{max}$ of the permanent magnet in Figure 1.8. With $I = 0$ in equation 1.82, then:

$$H_g = -\frac{l_m}{l_g} H_m [\mathrm{A/m}]. \tag{1.85}$$

Multiplying this result by H_g and using the constitutive relationship $B = \mu_0 H_g$ yields (Bozorth, 1978, Ch. 9):

$$H_g^2 = -\frac{l_m}{l_g} \frac{BH_m}{\mu_0}. \tag{1.86}$$

TABLE 1.5 Properties of Selected Permanent Magnet Materials[1]

Material	B_r [T][2]	H_c [A/m][3]	$(BH)_{max}$[J/m^3][4]
Barium ferrite (Ferroxdure)	0.2	120,000	8,000
Iron powder (100% Fe)	0.57	61,000	12,800
Steel (% by weight; remainder is Fe):			
Carbon steel (0.9 C, 1 Mn)	1	4,000	1,600
Chromium steel (1 C, 0.5 Mn, 3.5 Cr)	0.95	5,200	2,200
Tungsten steel (0.7 C, 0.5 Mn, 0.5 Cr, 6 W)	0.95	5,900	2,600
Cobalt steel (0.7 C, 0.35 Mn, 2.5 Cr, 8.25 W, 17 Co)	0.95	13,500	5,200
Alnico (% by weight; remainder is Fe):			
I (12 Al, 20 Ni, 5 Co)	0.71	33,800	10,700
II (10 Al, 17 Ni, 12.5 Co, 6 Cu)	0.72	43,400	13,100
III (12 Al, 25 Ni)	0.68	36,600	10,700
IV (12 Al, 28 Ni, 5 Co)	0.55	55,700	10,300
V (8 Al, 14 Ni, 24 Co, 3 Cu)	1.25	47,700	39,800
VI (8 Al, 15 Ni, 24 Co, 3 Cu, 1.25 Ti)	1.03	59,700	29,000
VIII (7 Al, 15 Ni, 35 Co, 4 Cu, 5 Ti)	1.04	126,000	44,000
Samarium Cobalt, SmCO (rare earth):			
18	0.86	573,000	140,000
22	0.985	696,000	180,000
26 H	1.06	736,000	210,000
27 H	1.1	820,000	220,000
32 H	1.16	756,000	250,000
Neodymium Iron Boron, NdFeB (rare earth):			
27	1.085	768,000	210,000
30 H	1.12	851,000	240,000
35	1.23	899,000	280,000
39 H	1.28	979,000	320,000
45	1.355	935,000	350,000
48	1.41	1,030,000	380,000

[1]Plonus (1978); Pollock (1993); Magnet Sales + Manufacturing (1995).
[2]Multiply by 10,000 for cgs unit of gauss [G].
[3]Multiply by $4\pi \times 10^{-3}$ for cgs unit of oersted [Oe].
[4]Multiply by $4\pi \times 10^{-5}$ for cgs unit of mega-gauss-oersted [MG-Oe].

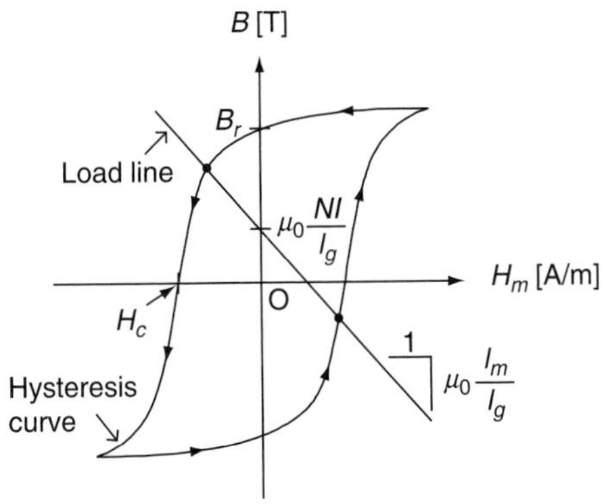

FIGURE 1.9 Solution for B in the Permanent-Magnet Circuit of Figure 1.8

That is, the square of the magnetic field in the air gap is directly proportional to the maximum energy product of the magnet. In other words, choosing a magnet with a larger maximum energy product will produce a larger magnetic field in the air gap shown in Figure 1.8.

1.6 Inductance

1.6.1 Magnetic Flux and Flux Linkage

Magnetic flux and flux linkage are two qualities of a spatially distributed magnetic field that are closely related to inductance. The flux of the **B** field, ψ_m, through a loop is the integral of the component of **B** normal to the loop over the loop cross-section. Mathematically, the **magnetic flux** through loop i produced by current in loop j is defined as (Paul *et al.*, 1998):

$$\psi_{m,ij} = \int_{s_i} \mathbf{B}_j \cdot d\mathbf{s}_i \text{ [Wb]}. \qquad (1.87)$$

In the case that $i = 1$ and $j = 2$, for example, then $\psi_{m,12}$ is the magnetic flux through loop 1 produced by the current in loop 2. Alternatively, magnetic flux can also be computed as (Paul *et al.*, 1998):

$$\psi_{m,ij} = \oint_{c_i} \mathbf{A}_j \cdot d\mathbf{l}_i \text{ [Wb]}, \qquad (1.88)$$

where $\boldsymbol{A}_j$ is the magnetic vector potential of current loop j.

Inductors are commonly constructed by wrapping many turns of wire around a core. This is an efficient method of increasing the inductance since for the same amount of current, the flux through (or "linking") the open surface bounded by the contour is approximately proportional to the number of wire turns. Accordingly, for loop i with N_i identical and "tightly wound" turns, the **flux linkage** is defined as:

$$\Lambda_{ij} = N_i \psi_{m,ij} [\text{Wb}]. \quad (1.89)$$

This flux linkage represents the total magnetic flux that passes through (or is "linked" by) the open surface bounded by the multiturn contour.

Example 1.13

The toroid in Figure 1.10 will be used to illustrate the computation of magnetic flux and flux linkage. Assume that μ is so large that all of the magnetic field is trapped in the toroid. In other words, ignore all "flux leakage." With a current I_1 in coil 1, applying Ampère's law of equation 1.29 along a closed contour in the toroid yields:

$$B_{\phi,1} = \mu \frac{N_1 I_1}{2\pi r} \ [\text{T}]. \quad (1.90)$$

Using the definition in equation 1.87, the magnetic flux through coil 1 due to I_1 (the "self flux") is as follows:

$$\psi_{m,11} = \int_{s_1} \boldsymbol{B}_1 \cdot d\boldsymbol{s}_1 = \int_{-h/2}^{h/2} \int_a^b \boldsymbol{\phi} \frac{\mu N_1 I_1}{2\pi_r} \cdot \boldsymbol{\phi} dr dz = \frac{\mu N_1 I_1 h}{2\pi} \ln\left(\frac{b}{a}\right) \ [\text{Wb}], \quad (1.91)$$

which is also equal to $\psi_{m,21}$ (the "mutual flux") since the cross sections of the two coils are identical.

The flux linkage through coil 1 is from equations 1.89 and 1.91:

$$\Lambda_{11} = N\psi_{m,11} = \frac{\mu N_1^2 I_1 h}{2\pi} \ln\left(\frac{b}{a}\right) \ [\text{Wb}], \quad (1.92)$$

whereas the flux linkage through coil 2 is as written here:

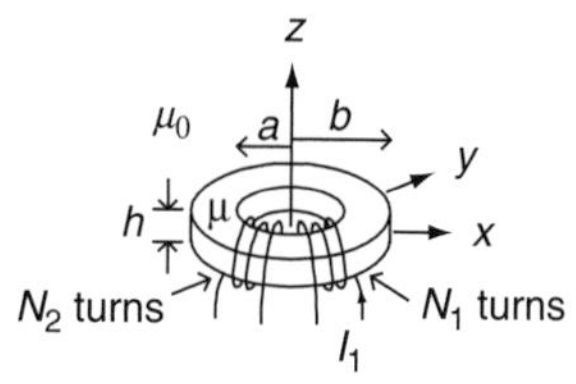

FIGURE 1.10 Two Wire Coils Wrapped Around a Toroid with Large μ

$$\Lambda_{21} = N_2 \psi_{m,21} = \frac{\mu N_2 N_1 I_1 h}{2\pi} \ln\left(\frac{b}{a}\right) \ [\text{Wb}]. \quad (1.93)$$

1.6.2 Definition of Inductance

An inductor is a device that can store energy in a magnetic field. Coils, solenoids, and toroids are all examples of inductors. For an ideal two-terminal inductor with inductance L, the voltage–current relationship is:

$$v_L(t) = L \frac{di_L(t)}{dt} \ [\text{V}]. \quad (1.94)$$

For practical inductors, however, nonideal effects may alter this voltage–current relationship through an additional series resistance and possible lead capacitance and inductance (Paul, 1992, Ch. 6).

Assuming an ideal inductor, the computation of inductance becomes a strictly magnetostatic problem.

Considering one or more loops of current in space, it can be shown that the magnetic flux through a loop is proportional to the current that produces the magnetic flux (Paul *et al.*, 1998). (This behavior is evident in equation 1.91 for the toroid.) The constant of proportionality is called **inductance** with units of henry (H). Specifically, inductance is defined in terms of the flux linkage and current as:

$$L_{ij} = \frac{\text{flux linkage through } ith \text{ coil due to current in } jth \text{ coil}}{\text{current in } jth \text{ coil}} \ [\text{H}] \quad (1.95)$$

or

$$L_{ij} = \frac{\Lambda_{ij}}{I_j} = \frac{N_i \psi_{m,ij}}{I_j} \ [\text{H}], \quad (1.96)$$

assuming from equation 1.89 that all N_i turns of the inductor are identical. The **Neumann formula** (Cheng, 1989):

$$L_{ij} = \frac{\mu}{4\pi} \oint_{c_i} \oint_{c_j} \frac{d\boldsymbol{l}_i \cdot d\boldsymbol{l}_j}{R_{ij}} \ [\text{H}] \quad (1.97)$$

is a useful alternative form to equation 1.96, especially for filamentary currents.

If $i = j$, $L_{ii}(= L)$ is called a **self-inductance**, whereas if $i \neq j$, L_{ij} is called a **mutual inductance.** A list of self-inductances for a few selected geometries is shown in Table 1.6. Inductance will not depend on current if the inductor is constructed from linear materials. Conversely, if ferromagnetic materials are used to fabricate the inductor, the inductance may be dependent on current since, as discussed previously, ferromagnetic materials can behave nonlinearly.

Other than for very simple and ideal geometries, neither equations 1.96 and 1.97 can be analytically evaluated to give

TABLE 1.6 Self-Inductances for a Selected Set of Geometries

Name	Geometry	Inductance [H]	
Toroid[1]	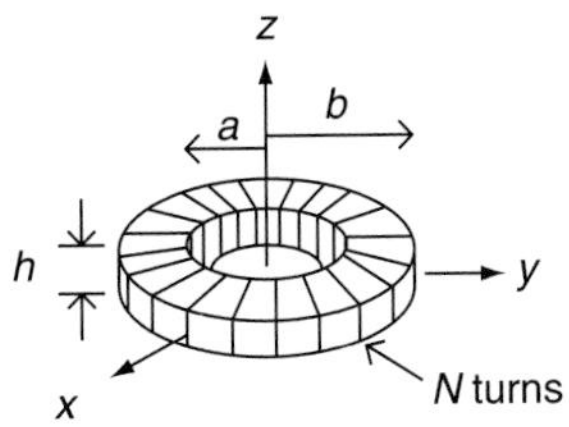	$L = \frac{\mu N^2 h}{2\pi} \ln\left(\frac{b}{a}\right)$	(1.98)
Solenoid (radius = a)	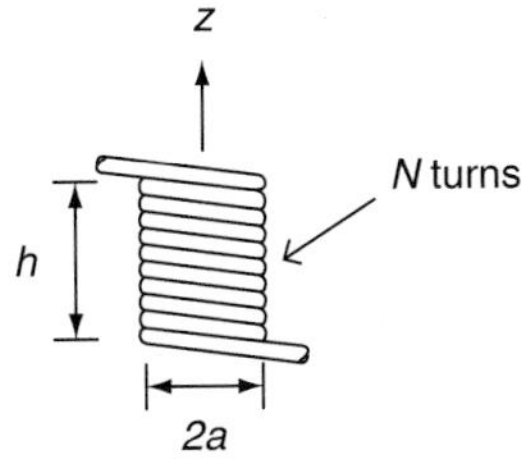	$h \gg a_1 3$: $L \approx \frac{\mu N^2 \pi a^2}{h}$	(1.99)
		$h \gtrsim 0.4a$ (Wheeler formula)[3]: $L \approx \frac{10\mu N^2 \pi a^2}{10h + 9a}$	(1.100)
Coaxial cable[1] ($h \gg r_s$)	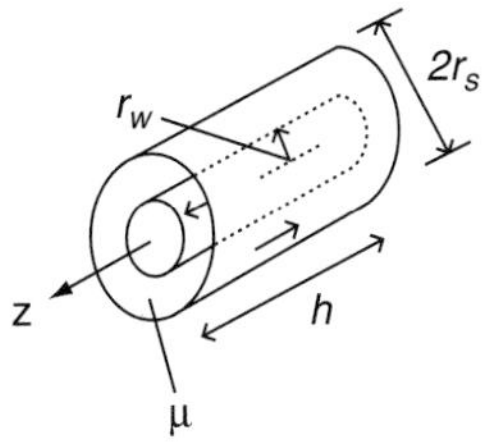	$L \approx \frac{\mu h}{2\pi} \ln\left(\frac{r_s}{r_w}\right) + \frac{\mu_0 h}{8\pi}$	(1.101)
Two-wire line[2]	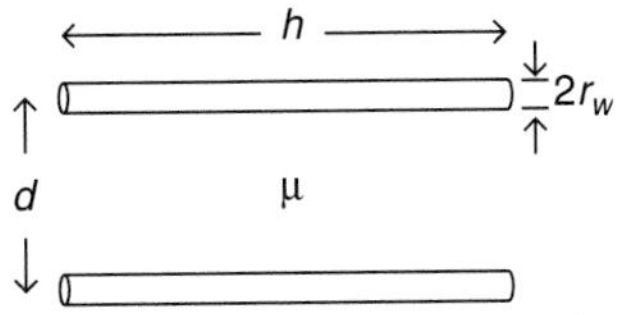	$L \approx \frac{\mu h}{\pi} \ln\left(\frac{d}{r_w}\right)$	(1.102)

[1]Paul *et al.* (1998). [2]Plonus (1978). [3]Ramo *et al.* (1994).

simple formulas for inductance. This is only possible for inductors with high degrees of symmetry, such as the toroid or infinitely long coaxial cable shown in Table 1.6. Alternatively, numerical integration can be used instead to evaluate equations 1.96 and 1.97; the latter is particularly suited for coil-type inductors (Whites, 1998, Example 4.15).

Example 1.14

The self- and mutual inductances of the wire coils on the toroid in Example 1.13 can be computed using equation 1.96 together with the flux linkages in equations 1.92 and 1.93. For coil 1, the self-inductance is the following:

$$L_{11} = \frac{\Lambda_{11}}{I_1} = \frac{\mu N_1^2 h}{2\pi} \ln\left(\frac{b}{a}\right) \text{ [H]}, \qquad (1.103)$$

which is the first entry in Table 1.6, while the mutual inductance is the following:

$$L_{21} = \frac{\Lambda_{21}}{I_1} = \frac{\mu N_2 N_1 h}{2\pi} \ln\left(\frac{b}{a}\right) \text{ [H]}. \qquad (1.104)$$

In both cases, the inductance is proportional to the square of the number of wire turns. This behavior is common to all coil-type inductors.

The self-inductance for coil 2 and the mutual inductance L_{12} can also be computed using equation 1.96. A current I_2 is assumed to exist in coil 2 for the purposes of these inductance calculations. Since the inductance of inductors formed from linear materials does not depend on the current in the wire, assuming this current I_2 is only a construction step. From equation 1.96:

$$L_{22} = \frac{\Lambda_{22}}{I_2} = \frac{\mu N_2^2 h}{2\pi} \ln\left(\frac{b}{a}\right) \text{ [H]}, \qquad (1.105)$$

and the mutual inductance is

$$L_{12} = \frac{\Lambda_{12}}{I_2} = \frac{\mu N_1 N_2 h}{2\pi} \ln\left(\frac{b}{a}\right) \text{ [H]}. \quad (1.106)$$

Comparing equations 104 and 106, it is apparent that $L_{12} = L_{21}$. This is a general result. In particular, by interchanging indices i and j in equation 1.97 it can be shown that:

$$L_{ij} = L_{ji} \quad i \neq j \quad (1.107)$$

1.7 Stored Energy

1.7.1 Energy Stored in a Magnetic Field

As described by Ampère's force law discussed in Section 1.4.1, a magnetic force exists between loops of constant current. Were the loops free to move, they would migrate toward or away from each other depending on their orientations. The work done to keep the current loops stationary is stored as energy in the magnetic field around the loops (Halliday *et al.*, 2002). Energy is also stored in the magnetic field of a single current loop as well as other situations where a magnetic field exists.

Magnetic energy W_m can be computed by integrating the dot product of $\boldsymbol{B}$ and $\boldsymbol{H}$ through space as (Cheng, 1989):

$$W_m = \frac{1}{2} \int_{\text{all space}} \boldsymbol{B} \cdot \boldsymbol{H} dv \quad \text{[J]}. \quad (1.108)$$

Any *change* to this stored magnetic energy occurs when the magnetic field changes with time according to Faraday's law. Once $\boldsymbol{B}$ and $\boldsymbol{H}$ have reached a steady-state, equation 1.108 provides a method to compute the time-stationary stored magnetic energy.

A **magnetic energy density** w_m can be defined at every point $\boldsymbol{r}$ in space from the integrand of equation 1.108 as:

$$w_m(\boldsymbol{r}) = \frac{1}{2} \boldsymbol{B}(\boldsymbol{r}) \cdot \boldsymbol{H}(\boldsymbol{r}) \quad [\text{J/m}^3]. \quad (1.109)$$

The total stored magnetic energy can be computed by integrating w_m throughout space as:

$$W_m = \int_{\text{all space}} w_m(\boldsymbol{r}) dv \quad \text{[J]}. \quad (1.110)$$

Example 1.15

We will compute the magnetic energy stored in a 1-m section of a long coaxial cable shown in Figure 1.11. Using Ampère's law, the magnetic field inside and outside the center conductor are, respectively, (Whites, 1998, Example 4.2):

$$\boldsymbol{H}_{\text{in}} = \boldsymbol{\phi} \frac{Ir}{2\pi r_w^2} \quad r < r_w \text{ [A/m]} \quad (1.111)$$

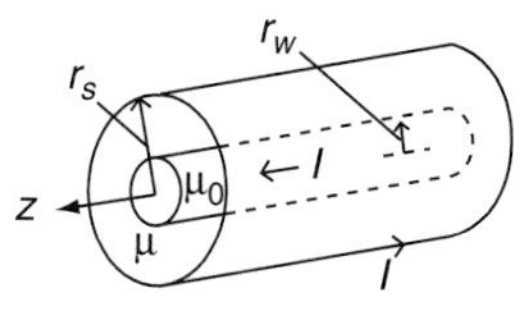

FIGURE 1.11 Coaxial Cable with a Nonmagnetic Center Conductor. All current I in the center conductor is assumed to return on the outer conductor.

and

$$\boldsymbol{H}_{\text{out}} = \boldsymbol{\phi} \frac{I}{2\pi r} \quad r_w < r < r_s \text{ [A/m]}. \quad (1.112)$$

From equations 1.108 and 1.78:

$$\begin{aligned} W_m &= \frac{1}{2} \int_{\text{coax}} \mu |\boldsymbol{H}|^2 dv = \frac{1}{2} \int_{z_0}^{z_0+1} \int_0^{2\pi} \int_0^{r_s} \mu H_\phi^2 r dr d\phi dz \\ &= \pi \left\{ \mu_0 \int_0^{r_s} H_{\phi,\text{in}}^2 r dr + \mu \int_{r_w}^{r_s} H_{\phi,\text{out}}^2 r dr \right\} \text{[J]}. \end{aligned} \quad (1.113)$$

Substituting H_ϕ inside and outside the center conductor from equations 1.111 and 1.112, respectively, and integrating gives

$$W_m = I^2 \left[\frac{\mu_0}{16\pi} + \frac{\mu}{4\pi} \ln\left(\frac{r_s}{r_w}\right) \right] \text{ [J]}, \quad (1.114)$$

for the 1-m section of coax.

1.7.2 Energy Stored in an Inductor

Inductors are the primary circuit elements for storage of magnetic energy. Ideally, they are the only circuit elements with this property. The **magnetic energy** W_m stored by an inductor with current I is as follows:

$$W_m = \frac{1}{2} L I^2 \text{ [J]}. \quad (1.115)$$

In this expression, L is the self-inductance of the inductor as discussed in Subsection 1.6.2.

With coupled inductors—such as when two or more coils are wrapped around a common high permeability core—the energy expression becomes more complicated. In the case of two coupled inductors with currents I_1 and I_2 in coils 1 and 2, respectively, the energy stored in the magnetic field is (Paul *et al.*, 1998):

$$W_m = \frac{1}{2} L_{11} I_1^2 + \frac{1}{2} L_{22} I_2^2 + L_{12} I_1 I_2 \text{ [J]}, \quad (1.116)$$

where L_{11} and L_{22} are the self-inductances of coils 1 and 2, respectively, and $L_{12}(=L_{21})$ is the mutual inductance between the two coils. As expected, the first two terms in equation 1.116 represent the magnetic energies stored in coils 1 and 2, respectively, while the last term is the energy stored in the mutual inductances of the two-coil system. One potential subtlety is that these self-inductances in equation 1.116 are those computed for a coil with the other one present but having a current equal to zero (Paul *et al.*, 1998).

Example 1.16

The energy stored in a 1-m section of the long coaxial cable in Figure 1.11 was computed to be equation 1.114 in example 1.15. This magnetic energy calculation will be repeated here but using the inductance expression of equation 1.115. The inductance for a 1-m section of coaxial cable is given from the third entry in Table 1.6 as:

$$L = \frac{\mu}{2\pi}\ln\left(\frac{r_s}{r_w}\right) + \frac{\mu_0}{8\pi}\ \text{[H]}. \tag{1.117}$$

Using this inductance in equation 1.115 with a current I in the coaxial cable gives:

$$W_m = \frac{1}{2}\left[\frac{\mu}{2\pi}\ln\left(\frac{r_s}{r_w}\right) + \frac{\mu_0}{8\pi}\right] I^2\ \text{[J]}, \tag{1.118}$$

which is identical to equation 1.114 as expected.

In some situations, applying this procedure in reverse can be a convenient method for computing inductance. That is, the stored magnetic energy is computed first, and from this, the inductance is determined using equation 1.115.

1.8 Magnetic Circuits

Magnetic circuit analysis is a technique that can be applied to certain magnetic field problems to greatly simplify the solution. In short, this technique is a lumped-element approximation used to solve for magnetic fields and magnetic fluxes (Cheng, 1989; Paul *et al.*, 1998; Planus, 1978). Applications for this method include the design of electrical generators, motors, transformers, and actuators. Magnetic circuit analysis is quite similar to electrical circuit analysis. Quantities that serve analogous purposes in both types of circuit analyses are listed side by side in Table 1.7.

Magnetic circuit analysis is illustrated with the specific example of the toroid shown in Figure 1.12(A). There are two primary assumptions made in magnetic circuit analysis (Paul *et al.*, 1998). First, it is assumed that the permeability of the structure is very large ($\mu \gg \mu_0$). Second, since $\mu \gg \mu_0$, it is further assumed that all of the magnetic field lines flow along the magnetic core and do not deviate outside of the structure; that is, it is assumed there is no **flux leakage** out of the toroid in Figure 1.12(A). Applying Ampère's law of equation 1.29 gives:

$$\oint_c \boldsymbol{H} \cdot d\boldsymbol{l} = NI, \tag{1.119}$$

such that inside a linear magnetic core:

$$B_\phi = \frac{\mu NI}{2\pi}\ \text{[T]}. \tag{1.120}$$

Another common assumption in magnetic circuit analysis – though not required – is that the cross-sectional dimensions

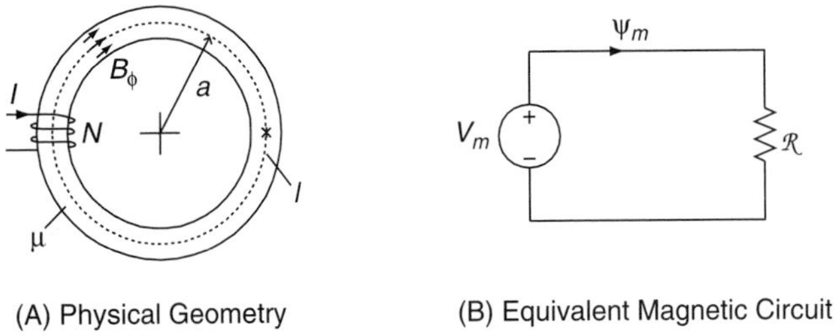

FIGURE 1.12 Toroid with High Permeability Core of Circular Cross-Sectional Area A

TABLE 1.7 Analogous Quantities in Electrical and Magnetic Circuit Analysis[1]

Electrical circuits		Magnetic circuits	
Conductivity [S/m]	σ	Permeability [H/m]	μ
EMF [V]	V	MMF [A]	V_m
Current [A]	I	Magnetic flux [Wb]	ψ_m
Resistance [Ω]	$R = \frac{l}{\sigma A}$	Reluctance [H^{-1}]	$\mathcal{R} = \frac{l}{\mu A}$
Ohm's law	$V = IR$	—	$V_m = \psi_m \mathcal{R}$
KVL (around a loop)	$\sum_i V_i = \sum_j R_j I_j$	—	$\sum_i V_{m,i} = \sum_j \mathcal{R}_j \psi_{m,j}$
KCL (at a junction)	$\sum_j I_j = 0$	—	$\sum_j \psi_{m,j} = 0$

[1]Paul *et al.*, (1998); Cheng (1989).

are small with respect to the length (the "large aspect ratio" assumption). In these situations, $\boldsymbol{B}$ will be approximately uniform over the cross section. Assuming that is the case here, then from equation 1.120:

$$B_\phi \approx \frac{\mu NI}{2\pi a} \equiv B \text{ [T]}. \quad (1.121)$$

Consequently, the magnetic flux ψ_m in a core with cross-sectional area A will be approximately

$$\psi_m \approx BA = \frac{\mu ANI}{2\pi a} \text{ [Wb]}, \quad (1.122)$$

and the flux will have this value all around the toroid.

The equivalent magnetic circuit for the toroid in Fig. 1.12(A) is developed by expressing equation 1.122 as the ratio of two quantities:

$$\psi_m = \frac{NI}{\left(\frac{2\pi a}{\mu A}\right)} \equiv \frac{V_m}{\mathcal{R}} \text{ [Wb]}. \quad (1.123)$$

In the numerator,

$$V_m = NI \quad \text{[A]}. \quad (1.124)$$

Equation 1.124 shows what is called the **magnetomotive force (MMF)** that serves an analogous function in this circuit as a **voltage source (EMF)** in an electrical circuit. In the denominator of equation 1.123:

$$\mathcal{R} = \frac{l}{\mu A} \quad [\text{H}^{-1}]. \quad (1.125)$$

Equation 1.125 shows the **reluctance** of the core (with mean length l), which is analogous to resistance as shown in Table 1.7. The equivalent magnetic circuit for the toroid is shown in Figure 1.12(B).

Other magnetic field problems can be solved using this magnetic circuit analysis by dividing the geometry into different magnetic flux paths, as appropriate, then computing the reluctances of the paths using equation 1.125. The last two entries in Table 1.7 list the laws used to solve magnetic circuit problems. In particular, the sum of mmfs around a closed loop is equal to the sum of "reluctance drops" (KVL analogy), while the sum of magnetic fluxes into a junction equals zero (KCL analogy) (Cheng, 1989; Paul *et al.*, 1998)

Example 1.17

The geometry shown in Figure 1.13(A) is used to illustrate the solution of magnetic field problems using magnetic circuit analysis. The structure is assumed to have a square cross section of area 10^{-6} m^2, a core with $\mu_r = 1{,}000$, and dimensions $l_1 = 1$ cm, $l_3 = 3$ cm, and $l_4 = 2$ cm.

One distinguishing characteristic of this structure is the **air gap** at the bottom of the center section shown in Figure 1.13(A). If the length of the air gap is small with respect to the cross-sectional dimensions, we can ignore **flux fringing** (or "spreading out" of the $\boldsymbol{B}$ field lines) in the air gap (Plonus, 1978). Consequently, this gap can be simply modeled as another reluctance as indicated by $\mathcal{R}_g$ in the equivalent magnetic circuit of Figure 1.13(B).

The goal here will be to solve for the magnetic flux density in the air gap. Using equation 1.125 the four reluctances in Figure 1.13(B) can be calculated as:

$$\mathcal{R}_1 = \frac{2l_1 + l_4}{1000\mu_0 A} = 31.83 \times 10^6 \quad [\text{H}^{-1}]. \quad (1.126)$$

$$\mathcal{R}_2 = \frac{l_4 - l_g - 0.001/2}{1000\mu_0 A} = 15.44 \times 10^6 \quad [\text{H}^{-1}]. \quad (1.127)$$

$$\mathcal{R}_3 = \frac{2l_3 + l_4}{1000\mu_0 A} = 63.66 \times 10^6 \quad [\text{H}^{-1}]. \quad (1.128)$$

$$\mathcal{R}_g = \frac{l_g}{\mu_0 A} = 79.58 \times 10^6 \quad [\text{H}^{-1}]. \quad (1.129)$$

The magnetic flux through the source coil, $\psi_{m,1}$, can be calculated as the source mmf divided by the total reluctance seen by the source as:

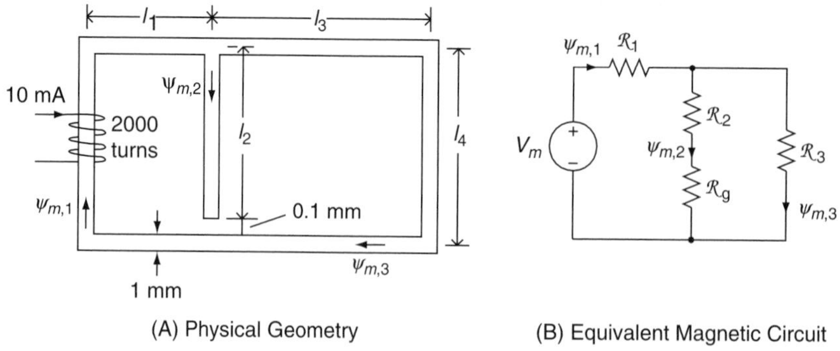

FIGURE 1.13 Magnetic Circuit Geometry with a Small Air Gap

$$\psi_{m,1} = \frac{V_m}{\mathcal{R}_{\text{total}}} = \frac{NI}{\mathcal{R}_1 + (\mathcal{R}_2 + \mathcal{R}_g)\|\mathcal{R}_3} = 0.286 \times 10^{-6}[\text{Wb}]. \quad (1.130)$$

Using "flux division" (which is analogous to current division in electrical circuits), the magnetic flux through the air gap is then:

$$\psi_{m,2} = \frac{\mathcal{R}_3}{\mathcal{R}_3 + \mathcal{R}_2 + \mathcal{R}_g}\psi_{m,1} = 0.115 \times 10^{-6}[\text{Wb}]. \quad (1.131)$$

References

Bozorth, R.M. (1978). *Ferromagnetism.* New York: IEEE Press.

Cheng, D.K. (1989). *Field and wave electromagnetics.* (2d ed.). Reading, MA: Addison-Wesley.

Inan, U.S., and Inan, A.S. (1999). *Engineering electromagnetics.* Menlo Park, CA: Addison Wesley.

Halliday, D., Resnick, R., and Krane, K.S. (2002). *Physics.* (5th ed.). New York: John Wiley & Sons.

Kittel, C. (1996). *Introduction to solid state physics.* (7th ed.). New York: John Wiley & Sons.

Magnet Sales & Manufacturing. (1995). *Catalog 7, High-performance permanent magnets.*

Paul, C.R., Whites, W., and Nasar, S.A. (1998). *Introduction to electromagnetic fields.* (3d ed.). New York: McGraw-Hill.

Paul, C.R. (1992). *Introduction to electromagnetic compatibility.* New York: John Wiley & Sons.

Plonus, M.A. (1978). *Applied electromagnetics.* New York: McGraw-Hill.

Pollock, D.D. (1993). *Physical properties of materials for engineers.* (2d ed.). Boca Raton: CRC Press.

Ramo, S., Whinnery, J.R., and Van Duzer, T. (1994). *Fields and waves in communication electronics,* (3d ed.). New York: John Wiley & Sons.

Schein, L.B. (1992). *Electrophotography and development physics.* (2d ed.). Berlin: Springer-Verlag.

Ulaby, F.T. (2001). *Fundamentals of applied electromagnetics.* Upper Saddle River, NJ: Prentice Hall.

Weast, R.C. (Ed.). (1984). *CRC handbook of chemistry and physics.* (65th ed.). Boca Raton, FL: CRC Press.

Whites, K.W. (1998). *Visual electromagnetics for mathcad.* New York: McGraw-Hill.

2

Electrostatics

Rodolfo E. Diaz
Department of Electrical Engineering, Ira A. Fulton School of Engineering, Arizona State University, Tempe, Arizona, USA

2.1 Introduction

Electrostatics in its most restrictive sense is the specialization of Maxwell's equations to a system whose sources are steady-state, time-invariant electric charges. Because the conservation of charge is implicit in this definition, the unifying principle of all the equations is the conservation of total **electric flux**. Therefore, electrostatics also properly includes steady-state conduction current problems.

In this chapter, the fundamental relationship between source and field is between the **electric charge q** (measured in coulombs) and the **electric flux density D** (measured in coulombs per meter squared) because that relationship has the form of a conservation law. The **electric field E** (measured in volts per meter) is introduced with the concept of the **electrostatic potential Φ** (measured in volts) as the quantity involved in the dynamics of electrostatic systems (i.e., their interaction forces and energies). In this way, ε (measured in farads per meter), the permittivity of the material medium through which the flux traverses, appears as a proportionality constant that gauges the amount of energy stored in a given electrostatic system. Capacitance C (measured in farads) is then a purely geometric expression of the arrangement of that energy inside the system. The concepts of electrostatics are extended to the case of current flow in resistive environments by the recognition of the formal analogy between the electrostatic flux and the current density flux J (in amperes per meter^2). Wherever possible, the method of derivation of the important results is indicated.

2.1.1 Conventions

Although the presentation in this chapter will follow the steps outlined above, many other methods of exposition are possible. The reader should be aware of this whenever consulting the various textbooks available on electromagnetic theory. It must be emphasized that apart from the fundamental governing Maxwell equations and the constitutive relations connecting the various field quantities, almost anything that the author of this chapter adds is a matter of opinion or philosophy. This is most evident when the issue of the "fundamentality" of a field is brought up. The viewpoint adopted here is that electrostatics is the specialization of Maxwell's equations to elucidate the behavior of the electric fields in the case of no time variation or zero frequency. This idea is exemplified when equations 2.1 (written in their most conventional form and using the Lorentz Gauge) become equations 2.2.

$$\nabla \times \vec{E} = -\frac{\partial \vec{B}}{\partial t}. \quad \nabla \times \vec{H} = \frac{\partial \vec{D}}{\partial t} + \vec{J}. \quad \text{(a) The curl equations}$$
$$\vec{B} = \mu(t)\vec{H}. \quad \vec{D} = \varepsilon(t)\vec{E}. \quad \text{(b) The constitutive relations} \quad (2.1)$$
$$\nabla \cdot \vec{D} = \rho. \quad \nabla \cdot \vec{B} = 0. \quad \nabla \cdot \vec{J} = -\frac{\partial \rho}{\partial t}. \quad \text{(c) The divergence equations}$$

$$\nabla \times \vec{E} = 0. \quad \text{(a) The irrotational nature of the electric field}$$
$$\vec{D} = \varepsilon\vec{E}. \quad \text{(b) The constitutive relation} \quad (2.2)$$
$$\nabla \cdot \vec{D} = \rho. \quad \text{(c) The divergence equation}$$

No field in equations 2.1 and 2.2 can be considered more "fundamental" than the others; each plays a unique part in the dynamics of electromagnetic and electrostatic systems. In particular, when dealing with electrostatics, the distinction between D and E should never be blurred. The practical fact that the constitutive properties of dielectric matter vary in space prevents the moving of the permittivity function, ε, through the vector differential operators of equations 2.2a and 2.2c as if it were a constant. Therefore, the following two expressions are, in general, inequalities. They only become equalities **inside homogeneous regions of space**.

$$\frac{\nabla \times \vec{D}}{\varepsilon} \neq 0, \nabla \cdot \vec{E} \neq \frac{\rho}{\varepsilon}. \quad (2.3)$$

Equation 2.2c defines the D field as the flux produced in space by its source, the charge density ρ, a quantity that is always conserved. Because of equation 2.2a, the E field is shown as the gradient of a potential function; therefore, it is a measure of the force exerted on charges by a given electrostatic field. Equation 2.2b then shows that the strength of this force depends on the environment in which the charges in question are located. The interested reader is referred to the book by Hammond (1981) for an illuminating proof that neither D nor E is more fundamental.

2.2 Sources and Fields

Electric charges are the sources of a vector field called the **electric flux density**. The relationship between a source and its flux is always expressed in the form of a conservation law as in equation 2.2c. The fundamental analysis problem of electrostatics is to calculate the vector fields resulting from a specific charge distribution. In the following section, this is done directly from equation 2.2c by exploiting symmetry. To solve for the fields due to more general asymmetric charge distributions requires a source-field relationship, as derived in Subsection 2.2.2. Finally, a simplification of the problem can be obtained by considering the vector fields to be derivable from potential functions. This is presented in Section 2.2.3.

2.2.1 *D* Fields of Charge Distributions Using Gauss's Law

The following derivations require the calculus of vector differential operators. Most textbooks in electromagnetic theory contain the relevant theorems and their application to electromagnetic fields. A particularly complete and concise presentation of the same can be found in Chapter 2 of the textbook by Jefimenko (1996). The book by Schey (1996) is also recommended. Helmholtz's theorem states that any vector field that is continuous and regular at infinity can be completely specified by its divergence and curl. Thus, including the mediation of the permittivity function, it is clear that the electrostatic field is completely defined by equations 2.2. Integrating both sides of equation 2.2c over all of space and applying the divergence theorem yields for the left-hand side:

$$\int_V \nabla \cdot \vec{D} dv = \oint_S \vec{D} \cdot d\vec{S} \quad (2.4)$$

Integration of the right-hand side over all of space encompasses the volume containing the charge density and simply gives the total charge enclosed. The result is **Gauss's Law** stating that the total electric flux crossing any surface bounding the charge is equal to the total charge enclosed in that surface, clearly a conservation law:

$$\oint \vec{D} \cdot d\vec{S} = \int \rho dv = Q_{\text{enclosed}}, \quad \text{(Gauss's law)} \quad (2.5)$$

where S is the bounding surface of the integration volume V and where the vector differential surface element is given by the differential surface element, dS, multiplied by the unit vector normal, $\hat{n}$, to the surface at the point of integration as Figure 2.1 illustrates.

With equation 2.5, the electric flux density, or D field, from **highly symmetric charge distributions** can be obtained. The method of approach is to guess at the form of the field and pick a bounding surface over which the D field is constant and

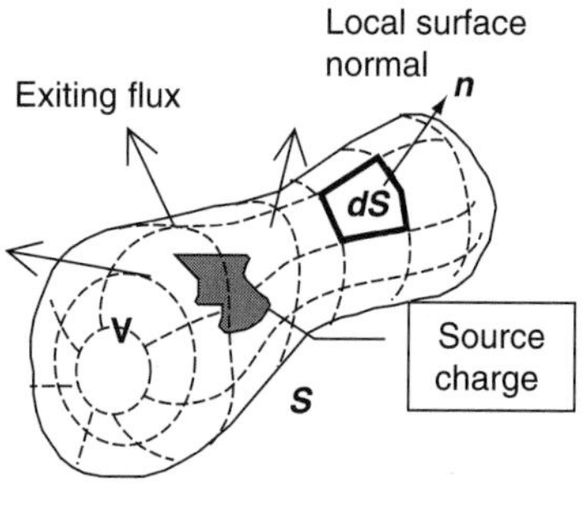

FIGURE 2.1 Gauss's Law Connects the Total Source of Flux Contained Inside the Volume V to the Flux Density that Exits its Bounding Surface S.

perpendicular to the surface, thus allowing it to be pulled out of the integral on the left-hand side of equation 2.5. The bounding surface is known as a **Gaussian surface.**

For instance, assume $\vec{D} = \hat{r}f(r)$. The left-hand side of equation 2.5 in spherical coordinates becomes $4\pi D(r_0)r_0^2$, where r_0 is the radius of the spherical surface of integration enclosing the charge distribution. This result must equal Q, the total charge enclosed. In particular, if r_0 is arbitrarily small, the field surrounding a **point charge** is obtained.

$$\vec{D}(r) = \frac{Q}{4\pi r^2}\hat{r}. \quad \text{(Point source)} \tag{2.6}$$

This result holds in general *outside* any spherically symmetric charge distribution. By considering such distributions and evaluating the integral on the right-hand side of equation 2.5 for each one, the following classic cases are obtained:

(a) The infinitesimally thin spherical shell of **surface charge density** σ_0
(b) The uniformly charged spherical region of **volumetric charge density** ρ_0
(c) Concentric shells of radii b and a, $b > a$ carrying equal and opposite charges

The results are shown in the left-hand column of Table 2.1. Similarly, by looking for charge distributions that yield a purely radial D field in **two dimensions** the solutions for cylindrically symmetric charge distributions are obtained. The field of the line charge equivalent to equation 2.6 is:

$$\vec{D}(r) = \frac{\lambda}{2\pi r_c}\hat{r}, \quad \text{(Line source)} \tag{2.7}$$

where λ is the lineal charge density in coulombs per meter. When the charge distribution is symmetrically spread in a

TABLE 2.1 Common Symmetric Charge Distributions with Exact Solutions Via Gauss's Law

Spherical coordinates		Cylindrical coordinates
$\vec{D}(r) = \frac{Q}{4\pi r^2}\hat{r}$	Arbitrarily small source 	$\vec{D}(r) = \frac{\lambda}{2\pi r}\hat{r}$
$\rho_0 \cdot \delta r = \sigma_0$, where σ_0 = surface charge density in coulombs per meter squared. Total charge $Q = 4\pi a^2\sigma_0$. $D(r) = \sigma_0\frac{a^2}{r^2}$ for $r \geq a$. $D(r) = 0$ for $r < a$.	The infinitesimally thin shell 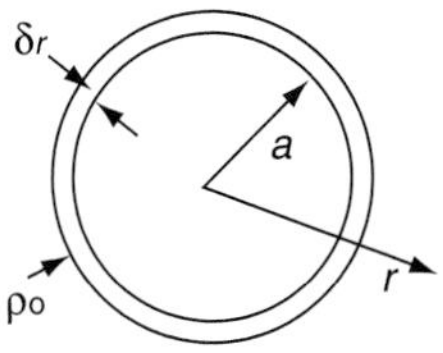	$\sigma_0 \cdot 2\pi a = \lambda_0$, where σ_0 = surface charge density in coulombs per meter squared. Total lineal charge density is $\lambda_0 = 2\pi a\sigma_0$. $D(r_c) = \frac{\lambda_0}{2\pi r_c}$ for $r_c \geq a$. $D(r_c) = 0$ for $r_c < a$.
ρ_0 = volumetric charge density in coulombs per meter squared. Total charge $Q = \rho_0 4\pi a^3/3$. $D(r) = \rho_0\frac{a^3}{3r^2}$ for $r \geq a$. $D(r) = \rho_0\frac{r}{3}$ for $r < a$.	The uniformly charged region 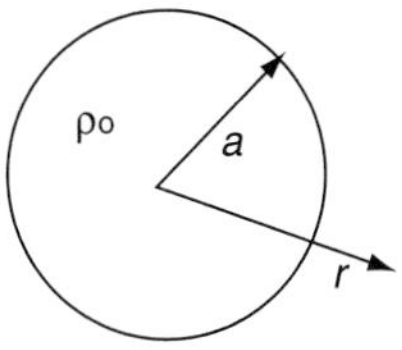	Total lineal charge density is $\lambda_0 = \rho_0\pi a^2$. $D(r_c) = \frac{\lambda_0}{2\pi r_c}$ for $r_c \geq a$. $D(r_c) = \frac{\lambda_0}{2\pi a}\left(\frac{r_c}{a}\right)$ for $r_c < a$.
$-\sigma = -Q/4\pi b^2$, $+\sigma = +Q/4\pi a^2$. $D(r) = \frac{Q}{4\pi r^2}$ for $a < r < b$. $D(r) = 0$ for $r < a$ and $r > b$.	Concentric shells of radii b and a, $b > a$ carrying equal and opposite charges 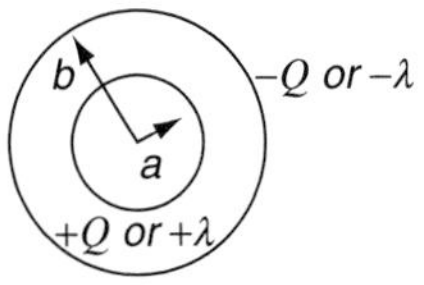	$-\sigma = -\lambda/2\pi b$, $+\sigma = +\lambda/2\pi a$. $D(r_c) = \frac{\lambda_0}{2\pi r_c}$ for $a < r_c < b$. $D(r_c) = 0$ for $r_c < a$ and $r_c > b$.
Rectangular coordinates $\vec{D} = +\hat{z}\frac{\sigma_0}{2}$ for $z > 0$. $\vec{D} = -\hat{z}\frac{\sigma_0}{2}$ for $z < 0$.	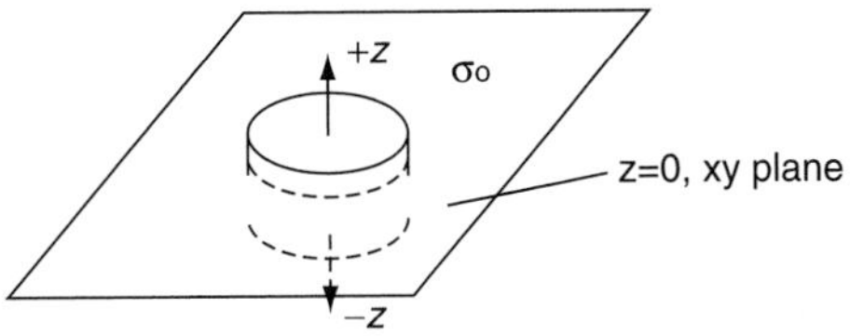	

two-dimensional space, cases corresponding to those of the spherical distribution are obtained. Table 2.2 summarizes these results along the right-hand column. Note that in the right-hand column of Table 2.1, the variable r_c is the radial variable in cylindrical coordinates and, by the definition of a two-dimensional field, the solution is for an infinitely long line charge. In all these cases, the field vanishes identically in those regions where the Gaussian surfaces surround zero net charge, and it becomes identical to the field of a point (line) source when the Gaussian surface surrounds a region with nonzero net charge. When the field inside a charge-filled region is being calculated, the Gaussian surface penetrates the region, as suggested in Figure 2.2. As long as flux is perpendicular to the surface, which is known from symmetry, Gauss's law always applies.

Apart from linear combinations of the above cases (such as the shell of finite thickness), the last case that can be derived with ease from Gauss's law is the case of the infinite charged plane. If a Gaussian surface in the form of a pillbox is drawn cutting through the surface of a plane of charge, symmetry dictates that equal amounts of flux must exit the pillbox above and below the plane. Thus, the result at the bottom of Table 2.1 is obtained.

This last example of the charged plane highlights a very important restatement of Gauss's law, relating the charge density on a surface to the total flux density cutting through that surface: Because the electric flux is conserved, the electric flux density normal to a surface is expected to be continuous across that surface. Any discontinuity of that normal flux implies that there is a net surface charge density stored on that surface numerically equal to that discontinuity. As the surface $z = 0$ is crossed in the last case of Table 2.1, the D field changes from $-\hat{z}\sigma_0/2$ to $+\hat{z}\sigma_0/2$, for a total jump discontinuity of magnitude σ_0. This is an example of the fundamental boundary condition on the D field as it crosses any (material) boundary surface:

$$D_{n1} - D_{n2} = \sigma_f, \quad \text{(Boundary condition on normal } D) \quad (2.8)$$

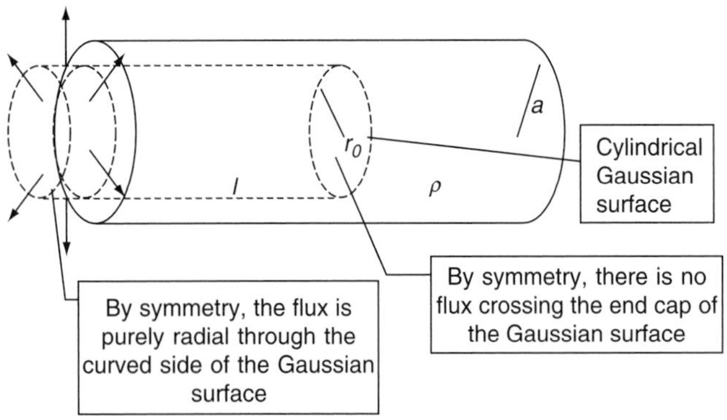

FIGURE 2.2 A Gaussian Surface of Radius r_0. The surface is used to derive the field inside and outside a uniformly charged infinitely long cylinder.

where D_{n1} is the normal component of the D field on side 1 of the surface, D_{n2} is the normal component of the D field on the other side of the boundary, and σ_f is the **free** surface charge on that boundary.

The simply symmetric charge distributions examined here are the only type of source distributions that can be solved by direct application of Gauss's law, equation 2.5. To find the field of more general (and realistic) charge distributions, a connection between a source charge and the D field it produces is needed.

2.2.2 First Alternative to Gauss's Law: Integration over Charge Distributions

From the result of equation 2.6 and the principle of linear superposition, it follows that an arbitrary charge distribution can be built up from individual differential (or point) charges and yield a total D field given by:

$$\vec{D} = \frac{1}{4\pi}\int \frac{dq}{r^2}\hat{r}. \quad (2.9)$$

In this equation, the vector $\vec{r}$ (and therefore $\hat{r}$ and r^2) is measured from the particular elementary charge being integrated, dq, to the **observation point.** Equation 2.9 can be proven as a consequence of Poisson's theorem for the field vector D in an infinite homogeneous space. Denoting the charge distribution by $\rho(x', y', z')$ and the differential volume element by dv', equation 2.9 is rewritten in the more common form:

$$\vec{D}(\vec{r}) = \int_{\text{all-space}} \frac{\rho(r')dv'(\vec{r} - \vec{r}')}{4\pi|\vec{r} - \vec{r}'|^3}, \quad \text{(The source – field relationship)} \quad (2.10)$$

where the primed coordinates refer to the position of the differential element of charge being integrated, and the unprimed coordinates refer to the position of the observation point, as suggested in Figure 2.3.

When the charge distribution is volumetric, the elementary differential element of charge is $dq = \rho dv'$. For surface charge distributions, $dq = \sigma dS'$; for lineal charge distributions, the element is $\lambda dl'$. The meaning and usefulness of equation 2.10 can be illustrated with the following two examples.

Consider the field along the axis of a **charged ring** of constant lineal charge density $\lambda C/m$ and radius a, as shown in Figure 2.4(A). For the sake of convenience, the ring is placed on the plane $z = 0$ of the coordinate system. Then, the differential element of charge in the primed (source) coordinate system is a differential arc of the ring with charge $dq = \lambda dl' = \lambda a d\varphi'$ located at the primed (cylindrical-r'_c, ϕ', z') coordinates $(a, \phi', 0)$. For an observation point on the axis a distance h from the center of the ring, the unprimed

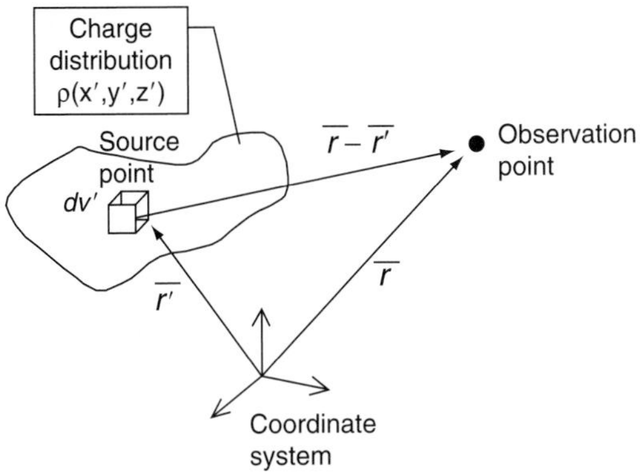

FIGURE 2.3 Definition of the Prime and Unprimed Coordinates for Source–Field Relationships

coordinates are (0,0,h). At this observation point, each particular differential element exerts a vector D field with components in the z and r_c directions. By the symmetry of the problem, the r_c components all cancel out when integrating around the ring, leaving only the z-directed components. Because the charge density is constant around the ring and every differential charge is at the same distance $|\vec{r} - \vec{r}'| = \sqrt{a^2 + h^2}$, the integration over the ring merely contributes a factor $\int dl' = 2\pi a$, giving the result shown in the Figure 2.4(A).

For the second example, consider the field along the axis of a uniformly charged disk of charge density σ and radius b, as shown in Figure 2.4(B). In this special case, the results of Figure 2.4(A) are used to provide us with a new kind of differential elementary charge, namely the ring of radius r_c' and thickness dr_c', for which the z-directed vector D field along the axis is known. Every such ring is located at the primed coordinate (0,0,0), whereas the observation point is at (0,0,h). Then, the differential charge is $dq = \sigma 2\pi r_c' dr_c'$, and every element is again at the same distance h from the observation point. Because each elementary ring charge has a different radius, the following integral must be performed:

$$\vec{D}(r = 0,\, z = h) = \hat{z} \int_0^b \frac{\sigma 2\pi r_c' dr_c' h}{4\pi \left(r_c'^{\,2} + h^2\right)^{3/2}} \tag{2.11}$$

The final result is given in Figure 2.4(B). The reader can verify that for observation distances $h \ll b$, the equation in this figure reduces to the case of the infinite plane of charge seen in Table 2.1.

When equation 2.10 is applied to a delta function distribution of charge, equation 2.6 results. This is the field of a **monopole**, or point charge. The fields of ensembles of point charges are obtained by linear superposition of this result. Equation 2.6 states that the D field of a monopole drops as $1/r^2$ with distance. When two or more oppositely charged sources, however, are located in a finite region of space, the external field may drop faster than this. In particular, when the total sum of all charges inside a given volume of space is equal to zero, the far field must drop as $1/r^n$, with $n \geq 3$. Consider the configuration of Figure 2.5, where a negative charge of strength $-q$ is located on the z axis at the point $z = -h/2$, and an equal and opposite charge $+q$ is located a distance h above it on the z axis.

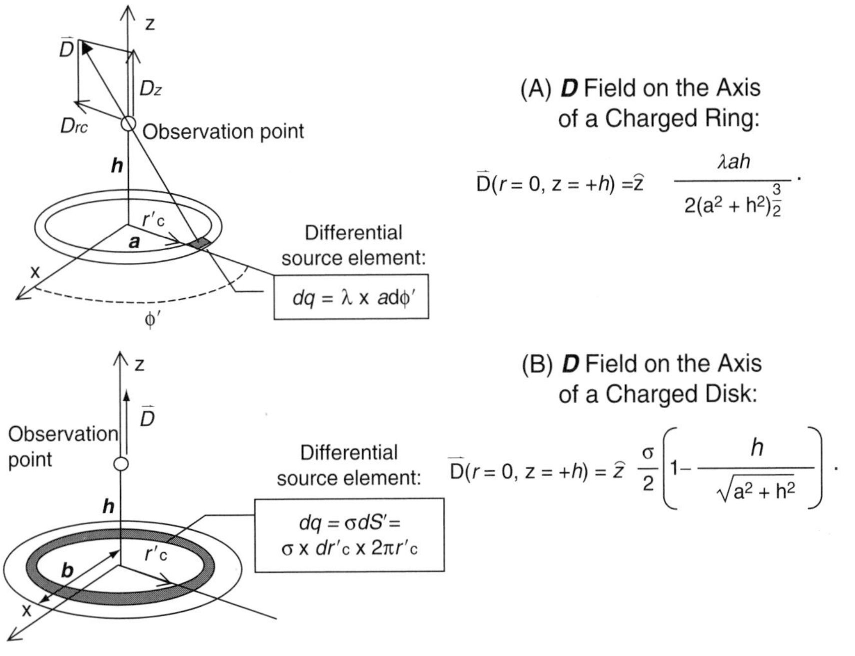

FIGURE 2.4 Examples of the Use of the Source–Field Relationship for Calculating the D-Field of a General Charge Distribution

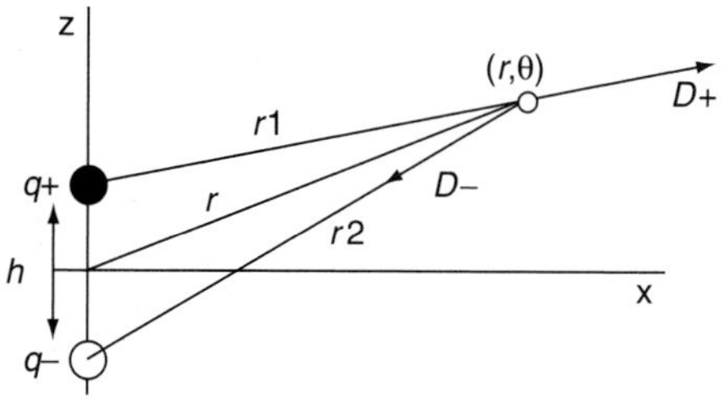

FIGURE 2.5 The Near Cancellation of the Fields of Equal and Opposite Charges Leading to a Far Field that Drops as $1/r^3$

When the observation point is far away, $r_1, r_2 \gg h$. Then $r_1 \approx r_2 - h\cos\theta$. In the **far field**, the resultant field is known as a **dipole field**, given by:

$$\vec{D}_{\text{dipole}} = \hat{r}\frac{2p\cos\theta}{4\pi r^3} + \hat{\theta}\frac{p\sin\theta}{4\pi r^3}, \quad (D \text{ field of a point dipole}) \tag{2.12}$$

where the quantity $p = |q|h$ is known as the dipole moment. The dipole moment **vector** points from the negative charge to the positive charge. This equation holds for a dipole of *any orientation*, with θ being the angle subtended to the observation point from the direction of p.

Similarly, the far field of a two-dimensional dipole, consisting of two line charges of equal and opposite strength λ and separated by a distance h, is given by:

$$\vec{D}_{\text{line-dipole}} = \hat{r}\frac{\lambda h\cos\theta}{2\pi r^2} + \hat{\theta}\frac{\lambda h\sin\theta}{2\pi r^2}. \tag{2.13}$$

(D field of a two-dimensional "line" dipole)

Just as a dipole results from the combination of equal and opposite monopoles, a **quadrupole** results from the combination of equal and opposite dipoles. The far field of quadrupoles drops as $1/r^4$. Higher order 2^n poles result from linear combinations of $2^{(n-1)}$ poles.

2.2.3 Second Alternative to Gauss's Law: The Potential Function

The calculation of vector fields from arbitrary charge distributions using equation 2.10 is made straightforward by modern computational machines and vector field visualization software. When simple expressions, however, are desired to aid intuition, or when memory storage is at a premium during numerical computations, a scalar field is much more convenient.

At this point, the second electrostatic vector field, the **electric field E**, measured in Volts per meter must be introduced. It is related to the electric flux density D through the **constitutive relation**:

$$\vec{D} = \varepsilon\vec{E}, \tag{2.14}$$

where the quantity ε is known as the **permittivity** of the medium. Its value for free space is $\varepsilon_0 = 8.854 \times 10^{-12}$ F/m. When the permittivity of a material medium is normalized to this value, it is known as the **relative permittivity** ε_r, or the **dielectric constant**. For most materials, the value is a scalar quantity. However, for anisotropic materials, the value depends on the orientation of the electric field. In such cases, the quantity ε in equation 2.14 is understood to be a tensor.

While the D field is a measure of the amount of flux produced by an electric charge, the E field is a measure of the force that the flux can exert on another charge through the equation:

$$\vec{F} = q\vec{E}. \quad \text{(The electrostatic force equation)} \tag{2.15}$$

Thus, the permittivity can be understood as a proportionality constant that determines the force a charge can exert on neighboring charges in a given material environment. Further discussion on the permittivity function is given in Subsection 2.3.4. All results derived up to this point for the D fields of sources in an infinite homogeneous space translate directly to the E fields of the same sources by simply dividing by ε. When boundaries exist across material properties that change, the field must be obtained as the solution of a boundary value problem (see Subsection 2.3.3). Equation 2.2a reveals that the E field is irrotational, so it must be derivable from the gradient of a scalar potential function because the curl of a gradient is zero. Therefore, the following is true:

$$\vec{E} = -\nabla\Phi,$$

where it follows that:

$$\Phi = \frac{1}{4\pi}\int_{\text{all-space}} \frac{\nabla\cdot\vec{E}}{r}\,dv' + \Phi_0, \tag{2.16}$$

and where Φ_0, as a constant of integration, is an arbitrary fixed potential.

The electrostatic potential Φ constitutes a **conservative field**. To prove this, take the line integral of both sides of the first equation in 2.16, traversing the field E along an arbitrary path from position a to position b:

$$\int_a^b \vec{E}\cdot d\vec{l} = -\int_a^b \nabla\Phi\cdot d\vec{l} = -\int_a^b d\Phi = \Phi(a) - \Phi(b). \tag{2.17}$$

The result is independent of the path of integration followed in going from a to b. The potential function is thus a single-valued point-wise continuous function everywhere in space, even in the presence of material boundaries. Therefore:

TABLE 2.2 Electrostatic Potential of Selected Charge Distributions

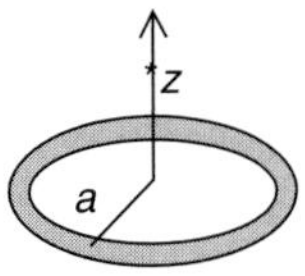

On axis above charged ring

$q_{\text{ring}} = \lambda 2\pi a.$

$$\Phi_{\text{ring}}(Z) = \frac{q_{\text{ring}}}{4\pi\varepsilon\sqrt{a^2+z^2}}.$$

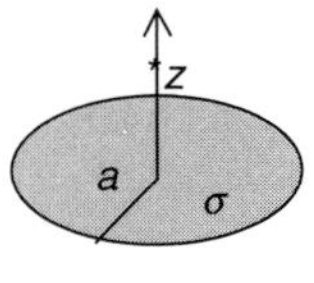

On axis above a charged disk

$$\Phi_{\text{disk}}(z) = \frac{\sigma}{2\varepsilon}\left[\sqrt{a^2+z^2} - z\right].$$

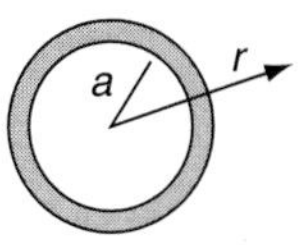

A charged spherical shell

$$\Phi_{\text{shell}}(r) = \frac{q}{4\pi\varepsilon r} \text{ for } r \geq a.$$

$$\Phi_{\text{shell}}(r) = \frac{q}{4\pi\varepsilon a} \text{ for } r < a.$$

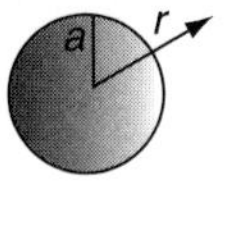

A uniformly charged sphere

$$\Phi_{\text{sphere}}(r) = \frac{q}{4\pi\varepsilon r} \text{ for } r \geq a.$$

$$\Phi_{\text{sphere}}(r) = \frac{q}{4\pi\varepsilon a}\left(\frac{3}{2}\right) - \frac{q}{4\pi\varepsilon r}\left(\frac{r^2}{2a^2}\right) \text{ for } r < a.$$

A dipole of moment $p = qh$

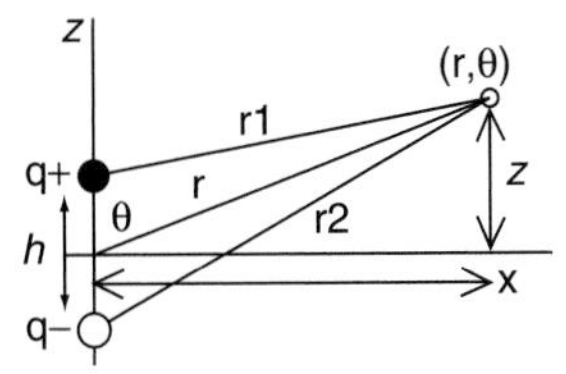

A two-dimensional line dipole

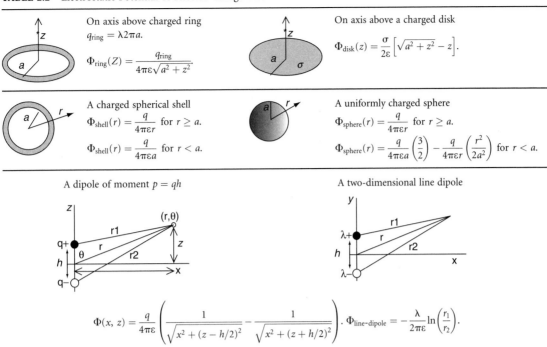

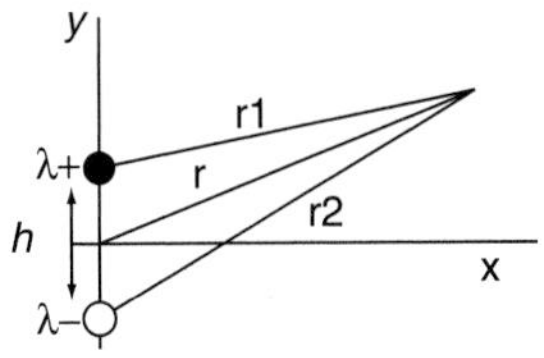

$$\Phi(x, z) = \frac{q}{4\pi\varepsilon}\left(\frac{1}{\sqrt{x^2+(z-h/2)^2}} - \frac{1}{\sqrt{x^2+(z+h/2)^2}}\right).$$

$$\Phi_{\text{line-dipole}} = -\frac{\lambda}{2\pi\varepsilon}\ln\left(\frac{r_1}{r_2}\right).$$

$$\Phi_{\text{dipole}}(r, \theta) = \frac{p}{4\pi\varepsilon r^2}\cos\theta. \text{ (In the far field)}$$

$$\lim_{\vec{\delta}\to 0}\Phi(\vec{r}+\vec{\delta}) = \Phi(\vec{r}) \text{ for all } \vec{r}, \vec{\delta}. \text{ (The boundary condition for } \Phi) \quad (2.18)$$

The quantity measured by equation 2.17 is called the **electromotive force** (emf) in the path *a–b*, or the **potential difference** or **voltage difference** between the points *a* and *b* measured in volts.

The second equation of equations 2.16 can be used to derive the relationship between the electrostatic potential and the electric charges inside an infinite homogeneous space by recognizing that:

$$\nabla\cdot\vec{E} = \frac{1}{\varepsilon}\nabla\cdot\vec{D}.$$

Thus:

$$\Phi = \frac{1}{4\pi\varepsilon}\int\frac{\rho}{|\vec{r}-\vec{r}'|}dv' + \Phi_0 = \frac{1}{4\pi\varepsilon}\int\frac{dq}{r}, \quad (2.19)$$

(The source – potential relationship)

where, as is customary, the potential at infinity = 0 has been defined; thus, $\Phi_0 = 0$, and $\Phi(r)$ is the *absolute* potential.

Thus, the potential of a point charge, or monopole, is simply:

$$\Phi_{\text{point}}(r) = \frac{q}{4\pi\varepsilon r}. \quad \text{(The potential of a point source)} \quad (2.20)$$

To calculate the electrostatic field of any charge distribution, equation 2.19 can be used first to calculate the electrostatic potential by integration over the source, and then the vector electric field can be derived from its gradient. By this process, the potential on the axis of the **charged ring** of Figure 2.4(A) is easily obtained. Then, the on-axis potential of the disk follows by integration over a continuum of rings, as before. The results are given in Table 2.2. Constructing a spherical shell by integrating over a series of rings spanning the surface of a sphere ($\theta = 0$ to π) yields the potential of a spherical shell of surface charge density σ and radius a. A similar derivation yields the potential of a **uniformly charged sphere**. These are also included in Table 2.2.

The potential function for ensembles of point charges is particularly simple, even in the near field. Thus, for an observation point anywhere near the dipole of Figure 2.5, the total potential is given as the scalar sum of the two point source potentials, shown at the bottom of Table 2.2. The latter discussion naturally applies to the case of two-dimensional charge distributions. In two dimensions, however, the electrostatic potential suffers an ambiguity from the infinite length of the structures in that the potential at infinity becomes undefined. The solution is to always calculate the potential of **pairs** of equal and oppositely charged line charges so that the net charge is zero and so that the potential at infinity is exactly zero. Thus, at an observation point a distance r_1 from a positive line charge and a distance r_2 from an equal and opposite line charge, the potential is as given at the end of Table 2.2.

2.2.4 Energy and Force in the Electrostatic Field

The potential function is positive near positive charges, negative near negative charges, and in general tends to zero at infinity. The function is directly related to the energy stored in the electrostatic system. The **self-energy** of a system of charges in their own potential is contained in all of space surrounding the system and is given by:

$$U_e = \frac{1}{2}\int \rho\Phi dv = \frac{1}{2}\int \vec{E}\cdot\vec{D}dv. \tag{2.21}$$

Thus, for a spherical shell of radius a:

$$U_{shell} = \frac{q^2}{8\pi\varepsilon a}. \tag{2.22}$$

The *interaction energy* between two systems of charges (a and b) is defined as the potential of charge b integrated over charge a or as the potential of charge a integrated over charge b. The equality of these two expressions is an instance of the **principle of reciprocity**:

$$U_{\text{int}} = \int_{Va} \rho_a \Phi_b dv = \int_{Vb} \rho_b \Phi_a dv = \frac{\varepsilon}{2}\int \vec{E}_a \cdot \vec{E}_b dv. \quad (2.22)$$

For a monopole in an external field, the interaction energy is then:

$$U = q_a \Phi_b. \tag{2.23}$$

For an infinitesimal dipole, the interaction energy in the presence of an external field is given by:

$$U = -\vec{p}_a \cdot \vec{E}_b. \tag{2.24}$$

The total energy of a system is the sum of all these individual energies. These expressions are useful to determine the forces in electrostatic systems in the same way that the potential is useful for obtaining the electric fields of those systems. For any system of charges where the *individual* charges are conserved, that is, *when the flux is constant,* the force is given by the negative gradient of the energy:

$$\vec{F}_{\textbf{const-flux}} = -\nabla U. \tag{2.25}$$

Similarly, in situations where the forces can lead to rotations of charged structures (e.g., a dipole in an external field), the torque is given by:

$$T_\theta = -\frac{\partial U}{\partial \theta}. \tag{2.26}$$

When equations 2.23 and 2.25 are applied to a point charge in the presence of the field caused by another point charge, the result is *Coulomb's law*:

$$\vec{F} = q_a \vec{E}_b = \frac{q_a q_b}{4\pi\varepsilon r^2}\hat{r} \quad \text{(Coulomb's law)} \tag{2.27}$$

By equation 2.26, a dipole in an external field experiences a torque:

$$\vec{T} = \vec{p} \times \vec{E}. \tag{2.28}$$

2.3 Boundary Conditions and Laplace's Equation

The definition of an electric field as the negative gradient of the electrostatic potential implies that the direction of the electric field at a given point in space is always perpendicular to the local equipotential line (or surface) at that point in space. When the electric field vectors are drawn throughout an electrostatic field, they take on the character of streamlines in a classic fluid flow field. Those lines are called **electric field lines**, and the tube-like volumes formed by bundles of such lines are called Faraday's tubes of electric flux or **lines of force.** The intensity of the electric field is represented by the density of these lines, being greatest where they are closest to each other. The lines of force and the equipotential surfaces form a curvilinear orthogonal coordinate system with important properties. Figure 2.6 illustrates such a system for the case of two line charges.

2.3.1 The Fields of Charged Conductors and the Method of Images

An idealized conductor, also known as a **perfect electric conductor** (PEC), is defined as a material that allows charges to move freely through its body and on its surface so that (a) the charges move to the configuration of least energy and (b) the surface of the conductor is an equipotential. At zero frequency, all conductors behave as perfect conductors because all charges placed or induced upon them have infinite time available to migrate to the surface. Therefore, for the purposes of electrostatics, all conductors are PECs and will be labeled as such in the figures to follow. When the conductor's surface is an equipotential, by the orthogonality of the system of lines described previously, all the electric field lines enter its surface at 90°. Therefore, another definition of a conductor is a statement of a **boundary condition**: a conductor is a surface on which the tangential component of the E field must vanish.

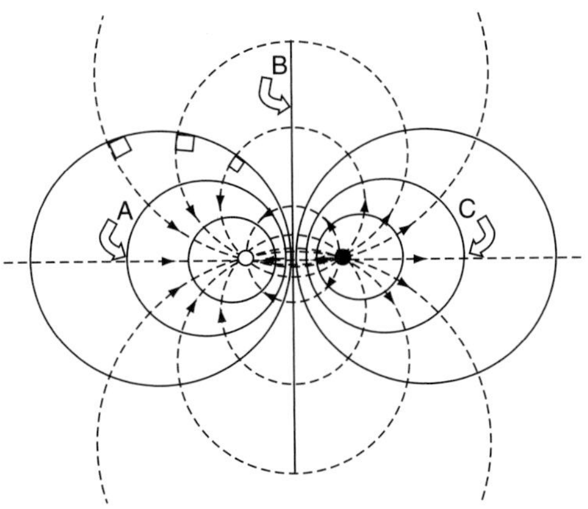

FIGURE 2.6 The Electric Field Lines (dashed) and Equipotentials (solid). These are in a system of two equal and opposite line charges that form a curvilinear orthogonal coordinate system in which the lines intersect each other at 90°.

Because of these properties, a conductor can be slipped into an existing electrostatic field and be left unchanged as long as its surface conforms to an equipotential and it is raised to the same potential level (voltage) as that equipotential relative to infinity (ground). Of course, an arbitrary constant potential can always be added to the entire system without changing the fields. Thus, a system of two line charges, such as the system in Figure 2.6, can also describe a system of one line charge (the black one) in the presence of a grounded PEC cylinder (the circle A in Figure 2.7), of a grounded plane (the symmetry line B in Figure 2.6), inside a grounded cylinder (the circle C in Figure 2.6), or even of the field of two cylindrical conductors

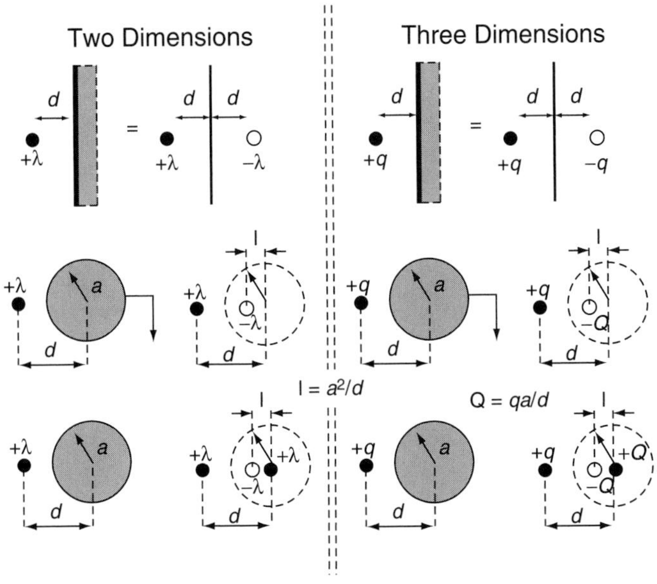

FIGURE 2.7 Image Systems for 2-D and 3-D Charge–Conductor Configurations

raised to different potentials (A and C in Figure 2.6). When a conductor replaces an equipotential surface, all the fields that existed in the space that were replaced, or surrounded, by the conductor vanish. Therefore, the surface of a conductor is a surface across which the normal D field undergoes a jump discontinuity from the value just outside it to zero inside it. Accordingly, by Gauss's law, there must be a surface charge everywhere a D field line terminates on a conductor that is numerically equal to the strength of the D field. Where the lines are crowded closer together, the charge density is greatest. Therefore on the surface of a conductor, the following special boundary conditions hold:

$$D_n = \sigma, \; E_{\tan} = 0. \text{ (Boundary conditions on the surface of a conductor)} \tag{2.29}$$

Problems involving grounded or floating conductors in the presence of sources can be solved by the ***method of images***. In this method, the total field (potential and vector) of a charge in the presence of a PEC object is obtained by finding the position of a fictitious source of appropriate strength that, when combined with the original source, would make the surface of the PEC an equipotential. When the PEC is grounded, there is usually only one oppositely charged image. When the PEC is **floating**, conservation of charge demands that the total charge over its surface remain zero. In that case, the image charges form a dipole. Figure 2.7 summarizes the most common source–conductor–image systems for both two-dimensional and three-dimensional problems.

Note that for the three-dimensional case of a grounded PEC sphere in front of a point charge, the image charge induced on the sphere is smaller than the source charge, whereas in a two-dimensional case the image charge is equal in strength to the source charge. Problems involving combinations of the boundaries shown in Figure 2.7 can be solved by first obtaining the image in one of the boundaries and then reflecting both source and image on the other boundary.

Whenever a two-dimensional configuration of conductor surfaces is not one of the simple planar or cylindrical geometries given in the figures above, they may still be derived from these results through a conformal mapping (Binns *et al.*, 1992).

2.3.2 Laplace's Equation and Boundary Value Problems

Inside any homogeneous region of space, equations 2.2c and 2.16 combine to yield a new equation:

$$\nabla \cdot \vec{E} = \frac{\rho}{\varepsilon} \Rightarrow \nabla \cdot (-\nabla \Phi) = \frac{\rho}{\varepsilon} \Rightarrow \nabla^2 \Phi = -\frac{\rho}{\varepsilon}. \tag{2.30}$$

This last equation is known as **Poisson's equation**. At points in space where there is no charge, it becomes **Laplace's equation**:

$$\nabla^2 \Phi = 0 \tag{2.31}$$

The most general method for the solution of electrostatic problems is to solve the partial differential equations (Poisson's or Laplace's) as boundary value problems. Since Poisson's equation is the inhomogeneous version of Laplace's equation, the foundation of all solutions is the solution of Laplace's equation.

Exact solutions only exist for a few coordinate systems; however, these solutions are important both as useful approximations to real problems and as **canonical** problems to use as standards for testing numerical methods of solution. Works by Boyce and DiPrima (1969) and Sommerfeld (1949) are good introductions to the subject, whereas the work by Abramowitz and Stegun (1972) is one of the standard reference handbooks for the functions commonly appearing in boundary value problems. The most common of these problems are those in which the boundaries across which the material properties change either fit Cartesian, cylindrical, or spherical coordinates. Since Laplace's equation is a second-order partial differential equation, two boundary conditions are needed at each boundary to solve the problem in addition to conditions at infinity if the problem space includes the point at infinity.

The first of the two boundary conditions is equation 2.18: the potential must be continuous across every boundary. This condition is equivalent to stating that the tangential E field must be continuous across every boundary. **The second condition is the conservation of flux**, equation 2.8: the jump discontinuity of normal D field across a boundary must equal the free surface charge density on that boundary. In the absence of free boundary charges, this means D_n must be continuous across the boundary. **The third boundary condition** is the total field must be regular at infinity.

Solutions by separation of variables in the standard coordinate systems are expressed in the form of a sum of **eigenfunctions** or **Harmonics**. The most common are the following:

- In two-dimensional Cartesian (x,y) coordinates:

$$\Phi = \sum_{n=1}^{\infty} (A_n \sin(\alpha_n x) + B_n \cos(\alpha_n x))(C_n e^{\alpha_n y} + D_n e^{-\alpha_n y}) + K. \tag{2.32}$$

- In two-dimensional cylindrical (r, ϕ) coordinates for z-independent solutions:

$$\Phi = \sum_{n=1}^{\infty} (A_n r^n + B_n r^{-n})(C_n \cos(n\phi) + D_n \sin(n\phi)) + (F \ln r + G)(H\phi + K). \tag{2.33}$$

- In spherical coordinates for azimuthally (ϕ) independent solutions:

$$\Phi = \sum_{n=0}^{\infty} (A_n r^n + B_n r^{-n-1})(C_n P_n(\cos\theta) + D_n Q_n(\sin\theta)) + K. \tag{2.34}$$

In equations 2.32, 2.33, and 2.34, K is a constant, and $P_n(\cos\theta)$ and $Q_n(\cos\theta)$ are the Legendre polynomials of the first and second kind respectively. (Note that the Q_n functions go to infinity on the z-axis, that is $\theta = 0, \pi$).

All solutions in these coordinate systems must be expressible as sums of these harmonics. The method of constructing the solution is (1) assume the solution has the form of the infinite sum, independently, in every region in which the material properties are constant; (2) eliminate the terms in the sums of harmonics that would cause infinities to appear in the given regions; (3) enforce satisfaction of the boundary conditions across every boundary connecting the regions by setting the potentials on either side and the normal D fields equal to each other; and (4) solve for the arbitrary constants. Once the constants are found, according to the **uniqueness theorem**, the solution has been found.

As an example, consider the case of **the dielectric sphere in a uniform field**. A uniform applied field exists in all of space directed along the z direction and of strength E. A material sphere of permittivity ε_1 different from the permittivity ε_2 of the surrounding medium is introduced into the field.

TABLE 2.3 The Boundary Conditions of Electrostatics

D field	E field	Potential Φ
Across any boundary: $D_{n1} - D_{n2} = \sigma_f$.	Across any boundary: $E_{\tan 1} = E_{\tan 2}$.	Across any boundary: $\Phi_1 = \Phi_2$.
In the absence of free charge: $D_{n1} = D_{n2}$ or $\varepsilon_1 \frac{\partial \Phi_1}{\partial n} = \varepsilon_2 \frac{\partial \Phi_2}{\partial n}$.		

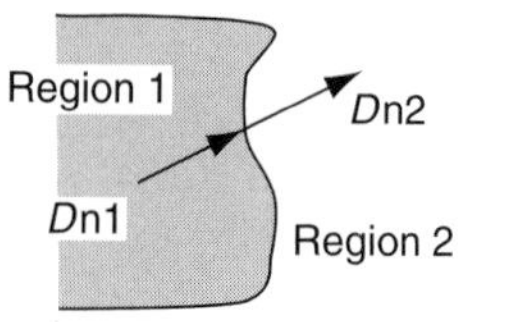

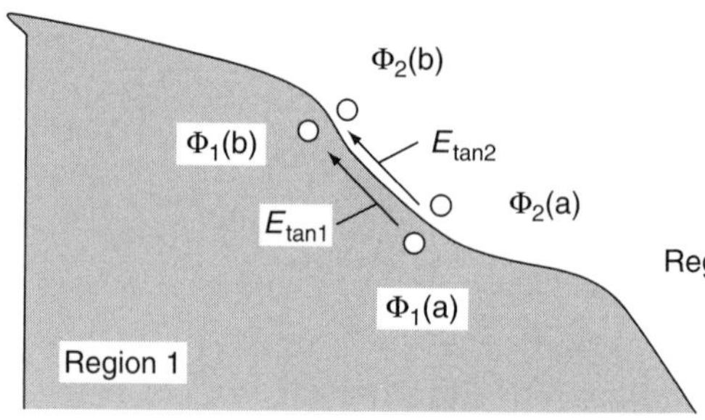

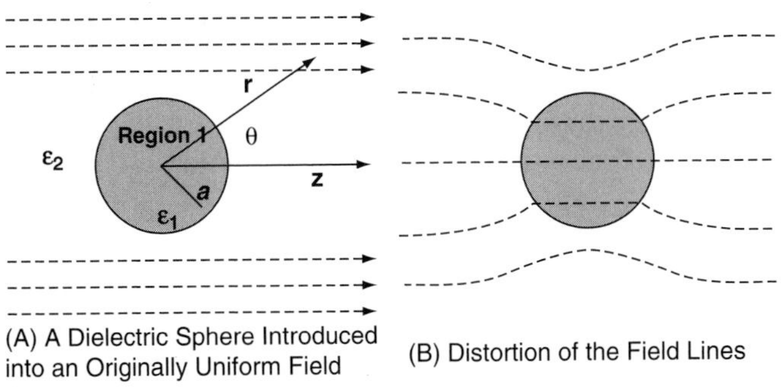

FIGURE 2.8 The Field Distribution About a Dielectric Sphere Immersed in a Constant Field can be Obtained as a Sum of Spherical Harmonics.

Figure 2.8(A) shows the orientation of the coordinate axes, and Figure 2.8(B) shows the distortion in the field lines caused when $\varepsilon_1 > \varepsilon_2$.

The potential function of the original applied field is as written here:

$$\Phi = -Ez = -Er\cos\theta. \tag{2.35}$$

The third boundary condition requires the final total field in region 2 to approach this value as r goes to infinity. The first and second boundary conditions require that:

$$\Phi_1(a, \theta) = \Phi_2(a, \theta). \tag{2.36a}$$

$$\varepsilon_1 \frac{\partial \Phi_1}{\partial r}(a, \theta) = \varepsilon_2 \frac{\partial \Phi_2}{\partial r}(a, \theta). \tag{2.36b}$$

Assuming solutions of the form of equation 2.34 in each region, the results are the following:

$$\Phi_1 = -\frac{3\varepsilon_2}{\varepsilon_1 + 2\varepsilon_2} Er\cos\theta. \text{ (Inside the sphere)} \tag{2.37a}$$

$$\Phi_2 = -Er\cos\theta + Ea^3 \frac{\varepsilon_1 - \varepsilon_2}{\varepsilon_1 + 2\varepsilon_2} \frac{\cos\theta}{r^2}. \text{ (Outside the sphere)} \tag{2.37b}$$

Two important features of this result are:

(1) The field inside the sphere is a scaled version of the original applied field. It is identical in form but weaker for the case $\varepsilon_1 > \varepsilon_2$ and stronger for the case $\varepsilon_1 < \varepsilon_2$. This is true in general for any ellipsoidal body inserted into a uniform field.

(2) A comparison of equation 1.37b with the far field form of the potential of a dipole in Table 2.2 shows that the external field is a linear superposition of the original applied field and a dipole field. The dielectric sphere has an **induced dipole moment** given by:

$$p = \varepsilon_2 E 4\pi a^3 \frac{\varepsilon_1 - \varepsilon_2}{\varepsilon_1 + 2\varepsilon_2}. \tag{2.38}$$

We say that the sphere has been **polarized**. The dipole moment is positive when $\varepsilon_1 > \varepsilon_2$ (e.g., a liquid droplet in air), meaning that the dipole is collinear with the applied field, and is negative for $\varepsilon_1 < \varepsilon_2$ (e.g., a bubble in a liquid), meaning that the dipole moment is antilinear.

The case of a metal sphere can be simulated by setting the permittivity of the sphere equal to infinity. Then the dipole moment is:

$$p = 4\pi a^3 \varepsilon_2 E. \tag{2.39}$$

2.3.3 The Connection Between *D*, *P*, *E*, and ϵ

In addition to serving as the prototypical example of the boundary value problem for Laplace's equation, this solution of the sphere immersed in the uniform field can be used to show the relationship between the D field and the phenomenon of polarization. The polarization, P, is defined as the dipole moment per unit volume in a dielectric. In the example, the sphere has an induced dipole moment per unit volume given by:

$$\frac{\vec{p}}{(4\pi a^3/3)} = \varepsilon_2 \vec{E}\left(3\frac{\varepsilon_1 - \varepsilon_2}{\varepsilon_1 + 2\varepsilon_2}\right) = \vec{P}, \tag{2.40}$$

which has the units of a flux density, like the D field. Now, taking the gradient of equation 2.37a, the D field inside the sphere can be obtained:

$$\vec{D}_1 = -\varepsilon_1 \vec{E}_1 = -\varepsilon_1 \nabla \Phi_1 = \varepsilon_1 \left(\frac{3\varepsilon_2}{\varepsilon_1 + 2\varepsilon_2}\right)\vec{E}. \tag{2.41}$$

It can be shown that:

$$\varepsilon_2 \vec{E}_1 + \vec{P} = \vec{D}_1. \tag{2.42}$$

In words: The D field can always be interpreted as the sum of the D field that would be assigned to the region if its material were the same as the external medium (ε_2 in this case) plus a polarization term due to the induced dipole moments arising from the contrast in properties across the boundary of the region.

This statement is true in general even when the dipoles are not induced but intrinsic. **Electrets** are permanently polarized dielectrics in complete analogy to magnets. They are characterized by an internal P. Since P is a flux density, its discontinuity at the faces of an electret has the appearance of surface charges on those faces to external observers. In fact, given any region containing a distribution of P, its electric field E over all of space can be calculated by finding the equivalent charges and replacing the electret by those charges. The charges are found by:

$$\sigma_{\text{equiv}} = \vec{P} \cdot \hat{n} \quad \text{and} \quad \rho_{\text{equiv}} = -\nabla \cdot \vec{P}. \tag{2.43}$$

Thus, for a uniformly polarized cylindrical electret found in free space, the total E field can be obtained by replacing the electret with two equivalent surface charge disks at its ends with $\varepsilon = \varepsilon_0$ everywhere. Then, the D field everywhere is obtained by applying equation 2.42.

When in equation 2.42, the permittivity of the surrounding medium is set to that of free space, and the D field only differs from the E field by the polarization P. In particular, if $\vec{D} = \varepsilon_0\vec{E} + \vec{P} = \varepsilon\vec{E}$, a proportionality constant χ can be defined, such that $\vec{P} = \varepsilon_0\chi\vec{E}$. The relative permittivity is given by:

$$\varepsilon_r = \frac{\varepsilon}{\varepsilon_0} = 1 + \chi. \tag{2.44}$$

From the example of the dielectric sphere and equation 2.44, it can be seen that the permittivity function is an expression of the fact that an applied E field induces dipole moments inside material objects. There are two common physical mechanisms for the creation of these induced dipole moments inside matter. The first is the rotation of existing permanently polarized subregions ("dipoles") inside the material into partial alignment with the applied field through the torque of equation 2.28. These subregions can be individual molecules, as in the **polar** molecules of water, or **domains** consisting of thousands of atoms, as in **ferroelectric** ceramics. The second mechanism is the creation of dipoles by charge separation in initially **nonpolar** atoms or molecules. In this case, the electron clouds are pulled in one direction by the E-field, while the positive nuclei are pushed in the opposite direction. Lorrain and Corson (1970), Clemmow (1973), and Bunget and Popescu (1984) provide excellent introductions to the subject of the physics of dielectrics. Work by von Hippel (1995) is the standard engineering reference on dielectric materials.

Table 2.4 gives typical values of the dielectric constant for various materials. It must be emphasized that these values are for electrostatics and will vary dramatically at nonzero frequencies.

2.4 Capacitance

Capacitance is a purely geometric quantity of the form ε times the area through which the flux flows, divided by the distance over which it flows. For instance, a parallel plate capacitor is a structure where two equal metal plates of area A are oppositely charged and brought into close proximity at a distance d from each other. Ignoring the fringing fields the geometric definition just given yields as the capacitance of this structure:

$$C_{par-plate} = \varepsilon\frac{A}{d} \text{ in Farads.} \tag{2.45}$$

Capacitance is a measure of the capacity of a material configuration for storing electrostatic energy.

TABLE 2.4 Typical Dielectric Properties of Materials

Gases and liquids		Organic solids		Inorganic solids	
Hydrogen	1.0003	Vaseline	2.2	Quartz	4–5
Air	1.0006	Paraffin	2–2.5	Glass	4–10
Gasoline	1.9	Rubber	3	Mica	5–7
Oils	2.2–4.3	Plexiglass	3.4	Silicon	12
Glycerine	10	Paper	3.7	Rutile TiO_2	89
Ethyl alcohol	28	Wood	2.5–8	Barium titanate $BaTiO_3$	1,200
Distilled water	81	Nylon	8	Barium strontium titanate $2BaTiO_3{:}\,1SrTiO_3$	10,000

2.4.1 Capacitance of Various Configurations and Energy

Capacitance is defined as the ratio of the charge stored (or separated) to the potential difference between the conductors. (In the case of only one conductor, the point at infinity serves as the second conductor.) Ignoring fringing in a parallel plate capacitor means the charge distribution is uniform throughout the plates. Since $\sigma = Q/A$, and the D field near a charged plane is $\sigma/2$, the total D field between the two plates is σ. Therefore, E is σ/ε and is practically uniform. The potential difference is then the integral of this nearly uniform E from one plate to the other, or $V = Ed$. Dividing Q by V, the equation 2.45 is obtained.

A pair of concentric spherical shells that are oppositely charged constitutes a spherical capacitor. The inner shell has radius a, and the outer shell has radius b. The capacitance is the following:

$$C_{\text{sph-cap}} = \varepsilon\frac{4\pi ab}{(b-a)}. \tag{2.46}$$

Letting the outer sphere go to infinity, the capacitance of the isolated spherical conductor relative to infinity is obtained:

$$C_{\text{sphere}} = \varepsilon 4\pi a. \tag{2.47}$$

For two-dimensional structures (i.e., structures that are infinitely long along the z-axis), the same method is employed except the charge used in the expression is the charge per unit length, and the resulting capacitance is a capacitance per unit length in Farads per meter (which coincidentally is the same as the units of the permittivity.) Thus, for coaxial cylindrical shells of inner radius a and outer radius b, the result is:

$$C_{cyl-cap} = \varepsilon\frac{2\pi}{\ln\frac{a}{b}}[\text{F/m}]. \tag{2.48}$$

The capacitance of the system is connected to its electrostatic energy by:

$$U_e = \frac{1}{2}CV^2 = \frac{1}{2}\frac{Q^2}{C}. \tag{2.49}$$

The first of these expressions is used in problems where the voltage between the conductors is constant, whereas the second is used when the charge is constant. When the charge is constant, flux is conserved. The force experienced by the system is then:

$$\vec{F}_{\text{const-flux}} = -\nabla U_e. \tag{2.50a}$$

But when voltage is constant, the force equation is:

$$\vec{F}_{\text{const-volt}} = +\nabla U_e. \tag{2.50b}$$

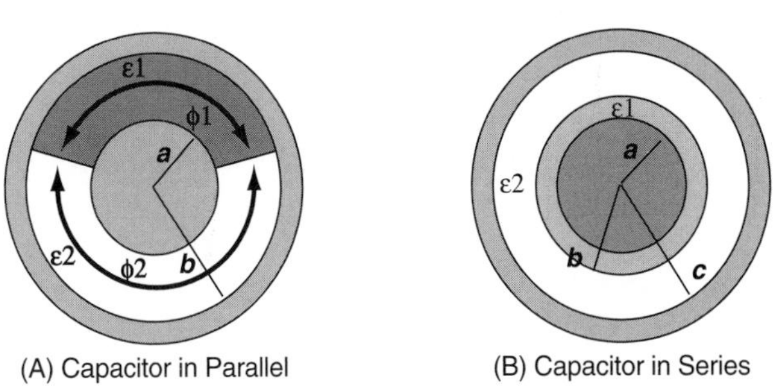

FIGURE 2.10 Partially Filled Cylindrical Capacitors

2.4.2 Partially Filled Capacitors

The boundary conditions of continuity of normal D and tangential E allow the determination of the capacitance of capacitors filled with an inhomogeneous dielectric. Consider the parallel plate capacitor filled with two materials. The two basic ways to fill it are shown in Figure 2.9.

Assuming fringing is negligible, the charges on the plates of Figure 2.9(A) are uniformly distributed. In this **series** arrangement, continuity of normal D guarantees that the same uniform D field crosses from the top charges to the bottom charges. Therefore, the E fields in the two regions must be different, with $E_1 = D/\varepsilon_1$ and $E_2 = D/\varepsilon_2$. The total voltage drop from the top plate to the bottom plate is then the sum of the integrals of these or $V = E_1 d_1 + E_2 d_2 = (D/\varepsilon_1)d_1 + (D/\varepsilon_2)d_2$. Because the charge on the plates must be $Q = \sigma A = DA$, the capacitance Q/V is the following:

$$C_{\text{series}} = \left(\frac{d_1}{A\varepsilon_1} + \frac{d_2}{A\varepsilon_2}\right)^{-1} = \left(\frac{1}{C_1} + \frac{1}{C_2}\right)^{-1}. \tag{2.51}$$

Note that in this case, the charge (that is the flux) was treated as the constant parameter. For the case of Figure 2.9(B), the voltage difference between the plates is the constant parameter because continuity of tangential E guarantees that the E field is the same everywhere between the plates. As the figure indicates, the charges on the plates are not uniformly distributed. They are denser where the flux density is greater. Now, $E = V/d$, and so the D fields ($D_1 = \varepsilon_1 E$ and $D_2 = \varepsilon_2 E$) are different. The charges Q_1 and Q_2 are given by $Q_1 = D_1 A_1 = \varepsilon_1 V A_1/d$ and $Q_2 = D_2 A_2 = \varepsilon_2 V A_2/d$, so the total capacitance $(Q_1 + Q_2)/V$ is given by:

$$C_{\text{parallel}} = \frac{\varepsilon_1 A_1}{d} + \frac{\varepsilon_2 A_2}{d} = C_1 + C_2. \tag{2.52}$$

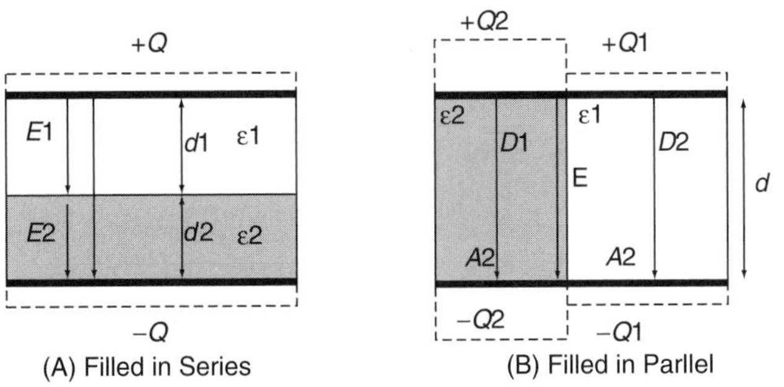

FIGURE 2.9 A Parallel Capacitor Filled with Two Different Dielectrics. The capacitor can be filled in series or in parallel.

These two fundamental ways of adding dielectric regions inside a capacitor can be used to determine the capacitance of arbitrarily filled capacitors, as long as the path taken by the lines of flux can be recognized. Thus, the partially filled coaxial cylindrical capacitors of Figure 2.10 have the following capacitances **per unit length**

$$C_a = \frac{\phi_1 \varepsilon_1}{\ln(b/a)} + \frac{\phi_2 \varepsilon_2}{\ln(b/a)}. \tag{2.53a}$$

$$C_b = \left(\frac{\ln(b/a)}{2\pi\varepsilon_1} + \frac{\ln(c/b)}{2\pi\varepsilon_2}\right)^{-1}. \tag{2.53b}$$

In a problem involving both series and parallel combinations of regions, a reasonable estimate can be obtained by partitioning the capacitor into individual regions and combining them appropriately.

In general, a partially filled capacitor exerts a force on its dielectric sections. A dielectric plunger partly inserted into a parallel plate capacitor will experience a force that tends to pull it inside the capacitor.

2.4.3 Static Current Fields

There is one more field that obeys all the laws of electrostatics: the static conduction current field. The last divergence equation of equations 2.1c also known as the equation of continuity is a conservation law, just like the equation for the D field. Invoking Ohm's law:

$$\vec{J} = \sigma\vec{E}, \tag{2.54}$$

as the fundamental relationship between the electric field in conducting media and the current density J and specializing to the case of statics, reduces that equation to:

$$\nabla \cdot \vec{J} = \nabla \cdot \sigma\vec{E} = 0. \tag{2.55}$$

In these equations, σ is the conductivity of the material measured in siemens/meter. By the same argument employed in the derivation of equations 2.30 and 2.31, it is seen that the problem of current flow in conducting bodies also satisfies Laplace's equation. This means that replacing D by J and ε by σ translates every solution of Laplace's equation for the electric field in the presence of perfect conductors and dielectric objects into a solution of an equivalent electric current problem involving perfect conductors and conducting objects. The total flux of current density crossing a cross-sectional area is called the current I measured in amperes. The **conductance** $G = I/V$, of a system of perfectly conducting electrodes immersed in a system of interconnected conducting regions is the direct analog of the capacitance. Its inverse is the **resistance**, $R = 1/G$, from which the more familiar form of Ohm's law is obtained:

$$V = IR. \tag{2.56}$$

The resistivity in ohm–meters is defined as the inverse of the conductivity: $\rho = 1/\sigma$. Therefore, from equations 2.51 and 2.52 follow the rules for adding resistors in parallel or in series.

Finally, it is pointed out that the power dissipated in a conductor is given by:

$$\int \vec{J} \cdot \vec{E} dv = IV[\mathrm{J/sec}] \text{ or } [\mathrm{w}]. \tag{2.62}$$

References

Abramowitz, M., and Stegun, J. (Eds). (1972). *Handbook of mathematical functions.* New York: Dover Publications.

Binns, K.J., Lawrenson, P.J., and Trowbridge, C.W. (1992). *The analytical and numerical solution of electric and magnetic fields.* New York: John Wiley & Sons.

Boyce, W.E., and DiPrima, R.C. (1969). *Elementary differential equations and boundary value problems.* New York: John Wiley & Sons.

Bunget, I., and Popescu, M. (1984). *Physics of solid dielectrics.* New York: Elsevier.

Clemmow, P.C. (1973). *An introduction to electromagnetic theory.* Cambridge: Cambridge University.

Hammond, P. (1981). *Energy methods in electromagnetism.* Oxford: Clarendon Press.

Jefimenko, O. (1966). *Electricity and magnetism.* New York: Appleton–Century–Crofts.

Lorrain P., and Corson, D. (1970). *Electromagnetic fields and waves.* San Francisco: W. H. Freeman and Company.

Schey, H.M. (1996). *Div, grad, curl, and all that.* New York: W. W. Norton & Company.

Sommerfeld, A. (1949). *Partial differential equations in physics.* New York: Academic Press.

von Hippel, A. (1995). *Dielectric materials and applications.* Boston: Artech House.

Weeks, W.L. (1964). *Electromagnetic theory for engineering applications.* New York: John Wiley & Sons.

3

Plane Wave Propagation and Reflection

David R. Jackson
Department of Electrical and Computer Engineering, University of Houston, Houston, Texas, USA

3.1 Introduction

Plane waves are the simplest solution of Maxwell's equations in a homogeneous region of space, such as free space (vacuum). In spite of their simplicity, plane waves have played an important role throughout the development of electromagnetics, starting from the time of the earliest radio transmissions through the development of modern communications systems. Plane waves are important for several reasons. First, the far-field radiation from any transmitting antenna has the characteristics of a plane wave sufficiently far from the antenna. The incoming wave field impinging on a receiving antenna can therefore usually be approximated as a plane wave. Second, the exact field radiated by any source in a region of space can be constructed in terms of a continuous spectrum of plane waves via the Fourier transform. Understanding the nature of plane waves is thus important for understanding both the far-field and the exact radiation from sources.

The theory of plane wave reflection from layered media is also a well-developed area, and relatively simple expressions suffice for understanding reflection and transmission effects when layers are present. Problems involving reflections from the earth or sea, for example, are easily treated using plane wave theory. Even when the incident wave front is actually spherical in shape, as from a transmitting antenna, plane wave theory may often be approximately used with accurate results.

This chapter's discussion assumes article that the regions of interest are homogeneous (the material properties are constant) and isotropic, which covers most cases of practical interest.

3.2 Basic Properties of a Plane Wave

3.2.1 Definition of a Plane Wave

The most general definition of a **plane wave** is an electromagnetic field having the form:

$$\boldsymbol{E} = \boldsymbol{E}_0 \psi(x, y, z) \tag{3.1}$$

and

$$\boldsymbol{H} = \boldsymbol{H}_0 \psi(x, y, z). \tag{3.2}$$

In equations 3.1 and 3.2, $\boldsymbol{E}_0$ and $\boldsymbol{H}_0$ are constant vectors, and the wave function ψ is defined as:

$$\psi(x, y, z) = e^{-j(k_x x + k_y y + k_z z)} = e^{-j\mathbf{k}\cdot\mathbf{r}}, \quad (3.3)$$

where

$$\boldsymbol{k} = \hat{\boldsymbol{x}}k_x + \hat{\boldsymbol{y}}k_y + \hat{\boldsymbol{z}}k_z = \beta - j\alpha \quad (3.4)$$

$$\boldsymbol{r} = \hat{\boldsymbol{x}}x + \hat{\boldsymbol{y}}y + \hat{\boldsymbol{z}}z. \quad (3.5)$$

The k_x, k_y, and k_z are complex constants that define a wave number vector k. (A time-harmonic dependence of $e^{j\omega t}$ is assumed and suppressed.) The vector $\boldsymbol{E}_0$ defines the polarization of the plane wave. The real and imaginary parts of the wave number vector $\boldsymbol{k}$ define the phase vector $\boldsymbol{\beta}$ and an attenuation vector $\boldsymbol{\alpha}$. The phase vector has units of radians per meter and gives the direction of most rapid phase change, whereas the attenuation vector has units of nepers per meter and gives the direction of most rapid attenuation. The magnitude of the phase vector gives the phase change per unit length along the direction of the phase vector, and the magnitude of the attenuation vector determines the attenuation rate along the direction of the attenuation vector.

Basic Properties

Wave Number Vector

In a homogeneous lossless space, a plane wave must satisfy Maxwell's equations, which in the time-harmonic form are the following (Harrington, 1961):

$$\nabla \times \boldsymbol{H} = j\omega\varepsilon \boldsymbol{E}. \quad (3.6)$$

$$\nabla \times \boldsymbol{E} = -j\omega\mu \boldsymbol{H}. \quad (3.7)$$

$$\nabla \cdot \boldsymbol{E} = \boldsymbol{0}. \quad (3.8)$$

$$\nabla \cdot \boldsymbol{H} = \boldsymbol{0}. \quad (3.9)$$

In these equations, ε_c and μ are the permittivity and permeability of the space. For free space, $\varepsilon = \varepsilon_0$ and $\mu = \mu_0$, where μ_0 is defined as $4\pi \times 10^{-7}$ H/m, and ε_0 is determined from the **defined** velocity of light (or any plane wave) in a vacuum (Hayt, 1989): $c = 2.99792458 \times 10^8$ m/sec because $c = 1/\sqrt{\varepsilon_0\mu_0}$. This gives the approximate value $\varepsilon_0 = 8.85418781762039 \times 10^{-12}$ F/m. A lossy medium can be modeled using a complex effective permittivity, accounting for conduction loss and/or polarization loss (Harrington, 1961). The complex effective permittivity is expressed as:

$$\varepsilon = \varepsilon - j\left(\frac{\sigma}{\omega}\right). \quad (3.10)$$

In this equation, ε is the complex permittivity of the material, accounting for polarization loss (if any), and σ is the conductivity of the medium. Henceforth, $\varepsilon_e = \varepsilon' - j\varepsilon''$ will be denoted as ε for simplicity. Therefore, equations (3.6)–(3.9) will remain valid in the general lossy case.

Taking the curl of equation 3.7 and then substituting in equation 3.6 gives the vector wave equation:

$$\nabla \times (\nabla \times \boldsymbol{E}) - k^2\boldsymbol{E} = \boldsymbol{0}, \quad (3.11)$$

where k is the (possibly complex) **wave number** defined by:

$$k = k' - jk'' = \omega\sqrt{\mu\varepsilon}, \quad (3.12)$$

with the square root chosen so that k lies in the fourth quadrant on the complex plane. The definition of the vector Laplacian is:

$$\nabla^2\boldsymbol{E} \equiv \nabla(\nabla \cdot \boldsymbol{E}) - \nabla \times (\nabla \times \boldsymbol{E}). \quad (3.13)$$

The fact that the divergence of the electric field is zero for a time-harmonic field in a homogeneous region (Harrington, 1961) results in the vector Helmholtz equation:

$$\nabla^2\boldsymbol{E} + k^2\boldsymbol{E} = \boldsymbol{0}. \quad (3.14)$$

In rectangular coordinates, the vector Laplacian is expressed as:

$$\nabla^2\boldsymbol{E} = \hat{\boldsymbol{x}}\nabla^2 E_x + \hat{\boldsymbol{y}}\nabla^2 E_y + \hat{\boldsymbol{z}}\nabla^2 E_z. \quad (3.15)$$

Hence, all three rectangular components of the electric field satisfy the scalar Helmholtz equation in a homogeneous region:

$$\nabla^2\Psi + k^2\Psi = 0. \quad (3.16)$$

Substituting equation 3.3 into equation 3.16 gives the result:

$$k_x^2 + k_y^2 + k_z^2 = k^2$$

or

$$\boldsymbol{k} \cdot \boldsymbol{k} = k^2. \quad (3.17)$$

This is the **separation equation** that relates the components of the wave number vector $\boldsymbol{k}$ defined in equation 3.4. Note that the term on the left side of equation 3.17 is not in general equal to $|\boldsymbol{k}|^2$, because $\boldsymbol{k}$ may be complex.

Orthogonality of Vectors and Impedance Relations

Other fundamental relations for a plane wave may be found by substituting equations 3.1 and 3.2 into Maxwell's equations 3.6 through 3.9. Noting that $\nabla \rightarrow -j\boldsymbol{k}$ for a plane wave, Maxwell's equations reduce to:

$$\boldsymbol{k} \times \boldsymbol{H} = -\omega\varepsilon \boldsymbol{E}. \quad (3.18)$$

$$\boldsymbol{k} \times \boldsymbol{E} = \omega\mu \boldsymbol{H}. \quad (3.19)$$

$$\boldsymbol{k} \cdot \boldsymbol{E} = 0. \quad (3.20)$$

$$\boldsymbol{k} \cdot \boldsymbol{H} = 0. \quad (3.21)$$

Equations 3.18 and 3.19 each imply that $\boldsymbol{E} \cdot \boldsymbol{H} = 0$, and together, they also imply that $\boldsymbol{k} \cdot \boldsymbol{E} = 0$ and $\boldsymbol{k} \cdot \boldsymbol{H} = 0$; that is, all three vectors $\boldsymbol{k}$, $\boldsymbol{E}$, and $\boldsymbol{H}$ are **mutually orthogonal**. Another interesting property that is true for any plane wave, which may be derived directly from equations 3.18, or 3.19, is that:

$$\boldsymbol{E} \cdot \boldsymbol{E} = \eta^2 \boldsymbol{H} \cdot \boldsymbol{H}, \tag{3.22}$$

where η is the **intrinsic impedance** of the space (possibly complex) defined from:

$$\eta = \sqrt{\frac{\mu}{\varepsilon}}. \tag{3.23}$$

The principle branch of the square root is chosen so that the real part of η is non-negative. For a vacuum, the intrinsic impedance is often denoted as η_0 and has a value of approximately 376.7303 Ω.

Power Flow

The complex Poynting vector for a plane wave, giving the **complex power flow**, is (assuming peak notation for phasors):

$$\boldsymbol{S} = \frac{1}{2}\boldsymbol{E} \times \boldsymbol{H}^*. \tag{3.24}$$

Using equation 3.19, the Poynting vector for a plane wave can be written as:

$$\boldsymbol{S} = \frac{1}{2\omega\mu^*}|\Psi|^2 \boldsymbol{E}_0 \times \left(\boldsymbol{k}^* \times \boldsymbol{E}_0^*\right). \tag{3.25}$$

Using the triple product rule $\boldsymbol{A} \times (\boldsymbol{B} \times \boldsymbol{C}) = (\boldsymbol{A} \cdot \boldsymbol{C})\boldsymbol{B} - (\boldsymbol{A} \cdot \boldsymbol{B})\boldsymbol{C}$, this can be rewritten as:

$$\boldsymbol{S} = \frac{1}{2\omega\mu^*}|\Psi|^2|\boldsymbol{E}_0|^2\boldsymbol{k}^* - \frac{1}{2\omega\mu^*}|\Psi|^2(\boldsymbol{E}_0 \cdot \boldsymbol{k}^*)\boldsymbol{E}_0^*. \tag{3.26}$$

The second term in equation 3.26 is not always zero for an arbitrary plane wave, even though $\boldsymbol{E}_0 \cdot \boldsymbol{k} = 0$, since $\boldsymbol{k}$ may be complex in the most general case. If either $\boldsymbol{k}$ or $\boldsymbol{E}_0$ is proportional to a real vector (i.e., all of the components of the vector have the same phase angle), then it is easily demonstrated that the second term vanishes. In this case, the Poynting vector becomes:

$$\boldsymbol{S} = \frac{1}{2\omega\mu^*}|\Psi|^2|\boldsymbol{E}_0|^2\boldsymbol{k}^*. \tag{3.27}$$

If the medium is also lossless (μ is real), the time-average power flow (coming from the real part of the complex Poynting vector) is in the direction of the phase vector $\boldsymbol{\beta}$. A similar derivation, casting the Poynting vector in terms of the $\boldsymbol{H}_0$ vector, yields:

$$\boldsymbol{S} = \frac{1}{2\omega\varepsilon}|\Psi|^2|\boldsymbol{H}_0|^2\boldsymbol{k}, \tag{3.28}$$

provided either $\boldsymbol{k}$ or $\boldsymbol{H}_0$ is proportional to a real vector. If the region is lossless (ε is real), the time-average power flow is then in the direction of the phase vector. Hence, for a lossless region, the time-average power flow is in the direction of the phase vector $\boldsymbol{\beta}$ if one of the three vectors $\boldsymbol{k}$, $\boldsymbol{E}_0$, or $\boldsymbol{H}_0$ is proportional to a real vector. In many practical cases of interest, one of the three vectors will be proportional to a real vector, and hence, the conclusion will be valid. However, it is always possible to find exceptions, even for free space. One such example is the plane wave defined by the vectors $\boldsymbol{k} = (1, j, 1)$, $\boldsymbol{E}_0 = (2, 1 - 2 - j)$, and $\boldsymbol{H}_0 = (1/(\omega\mu))\,(-2j, 4 + j, 1 - 2j)$ at a frequency $\omega = c$ ($k_0 = 1$). This plane wave satisfies Maxwell's equations 3.18 through 3.21. For this plane wave, however, the vectors $\boldsymbol{\beta}$, $\boldsymbol{\alpha}$, and the power-flow vector $\boldsymbol{p} = \mathrm{Re}(\boldsymbol{S})$ are all in different directions because the power flow is in the direction of the vector (5,0,4). One must then be careful to define what is meant by the "direction of propagation" for such a plane wave.

Direction Angles

One way to characterize a plane wave is through **direction angles** (θ, ϕ) in spherical coordinates. The direction angles (which are in general complex) are defined from the relations:

$$k_x = k \sin\theta \cos\phi. \tag{3.29}$$

$$k_y = k \sin\theta \sin\phi. \tag{3.30}$$

$$k_z = k \cos\theta. \tag{3.31}$$

Homogeneous (Uniform) Plane Wave

One important class of plane waves is the class of **homogeneous** or **uniform** plane waves. A homogeneous plane wave is one for which the direction angles are, by definition, **real**. A homogeneous plane wave enjoys certain special properties that are not true in general for all plane waves. For such a plane wave, the wave number vector can be written as:

$$\boldsymbol{k} = k\hat{\boldsymbol{R}} = \hat{\boldsymbol{R}}(k' - jk''), \tag{3.32}$$

where $\hat{\boldsymbol{R}}$ is a real unit vector defined from:

$$\hat{\boldsymbol{R}} = \hat{\boldsymbol{x}} \sin\theta \cos\phi + \hat{\boldsymbol{y}} \sin\theta \sin\phi + \hat{\boldsymbol{z}} \cos\theta. \tag{3.33}$$

In this case, the $\boldsymbol{k}$ vector is proportional to the real vector $\hat{\boldsymbol{R}}$, so the result of equation 3.27 applies. The unit vector $\hat{\boldsymbol{R}}$ then gives the direction of time-average power flow and also points in the direction of the phase and attenuation vectors. That is, all three vectors point in the same direction for a homogeneous plane wave. This direction is, unambiguously, the direction of propagation of the plane wave. The planes of constant phase are also then the planes of constant amplitude, being the planes perpendicular to the $\hat{\boldsymbol{R}}$ vector. That is, the plane wave has a uniform amplitude across the plane perpendicular to the direction of propagation. If the plane wave is not homogeneous, corresponding to complex direction angles, then the physical interpretation of the direction angles is not clear.

From equations 3.18 or 3.19, it may be easily proven that the fields of a homogeneous plane wave obey the relation:

$$|\boldsymbol{E}| = |\eta||\boldsymbol{H}|. \quad (3.34)$$

(Recall that all plane waves obey equation 3.22 but not in general equation 3.34).

Any homogeneous plane wave can be broken up into a sum of two plane waves, with real electric field vectors to within multiplicative constants polarized perpendicular to each other. This follows from a simple rotation of coordinates (the direction of propagation is then z', with electric field vectors in the x' and y' directions). The two plane waves have the fields (E_x, H_y) and (E_y, H_x), with $E_x/H_y = -E_y/H_x = \eta$. This decomposition is used later in the discussion of polarization.

Lossless Media: Relation Between Phase and Attenuation Vectors

Another important special case for a plane wave concerns a lossless medium, so that $k'' = 0$. If equation 3.4 is substituted into the separation equation 3.17, the imaginary part of this equation immediately yields the relation:

$$\boldsymbol{\beta} \cdot \boldsymbol{\alpha} = 0. \quad (3.35)$$

Hence, for a lossless region, **the phase and attenuation vectors** are always perpendicular. In some applications (e.g., a Fourier transform solution of radiation from an aperture in a ground plane at $z = 0$ or from a planar current source at $z = 0$ [Clemmow, 1996]), a plane wave propagating in a lossless region has the characteristic of that two wave numbers (e.g., k_x and k_y) are real (corresponding to the transform variables). In this case, the third wave number k_z will be real if $k_x^2 + k_y^2 < k^2$ and will be imaginary if $k_x^2 + k_y^2 > k^2$; that is, all transverse wave numbers (k_x, k_y) that lie in a circle of radius k in the wave number plane will be propagating, while all wave numbers outside the circle will be evanescent. In the first case, the power flow is in the direction of the (real) $\boldsymbol{k}$ vector (k_x, k_y, k_z), so power leaves the aperture from this plane wave. In the second case, the power flow is in the direction of the transverse wave number vector $(k_x, k_y, 0)$, so no power leaves the aperture for this plane wave component. If the medium is lossy, there is no sharp distinction between propagating and evanescent plane waves. In this case, all plane waves carry power in the direction of the vector $(k_x, k_y, \mathrm{Re}k_z)$.

Finally, it can be noted that if a homogeneous plane wave is propagating in a lossless region, then the attenuation vector α must be zero, and all wave number components k_x, k_y, and k_z are real.

Summary of Basic Properties

A summary of the basic properties of a plane wave, discussed in the previous subsections, is given in Table 3.1.

3.3 Propagation of a Homogeneous Plane Wave

The case of a homogeneous plane wave is important enough to warrant further attention. Far away from a transmitting antenna, the radiation field always behaves as a homogeneous plane wave (if the spherical wave front is approximated as

TABLE 3.1 Summary of the Basic Properties of a Plane Wave

	Inhomogeneous		Homogeneous	
Property	Lossy	Lossless	Lossy	Lossless
Direction angles	Complex	Complex	Real	Real
Relation between phase and attenuation vectors	$\boldsymbol{\beta} \cdot \boldsymbol{\alpha} = k', k''$	$\boldsymbol{\beta} \cdot \boldsymbol{\alpha} = 0$	$\boldsymbol{\beta} \parallel \boldsymbol{\alpha}$	$\boldsymbol{\alpha} = \boldsymbol{0}$
Impedance relations	$\boldsymbol{E} \cdot \boldsymbol{E} = \eta^2 \boldsymbol{H} \cdot \boldsymbol{H}$	$\boldsymbol{E} \cdot \boldsymbol{E} = \eta^2 \boldsymbol{H} \cdot \boldsymbol{H}$	$\boldsymbol{E} \cdot \boldsymbol{E} = \eta^2 \boldsymbol{H} \cdot \boldsymbol{H}$ $\lvert\boldsymbol{E}\rvert = \lvert\eta\rvert\lvert\boldsymbol{H}\rvert$	$\boldsymbol{E} \cdot \boldsymbol{E} = \eta^2 \boldsymbol{H} \cdot \boldsymbol{H}$ $\lvert\boldsymbol{E}\rvert = \eta\lvert\boldsymbol{H}\rvert$
Direction of power flow vector	Not necessarily in $\boldsymbol{\beta}$ direction	$\boldsymbol{\beta}$ If $\boldsymbol{k}$, $\boldsymbol{E}_0$, or $\boldsymbol{H}_0$ is proportional to a real vector	$\boldsymbol{\beta}$	$\boldsymbol{\beta}$
Wave number vector	$\boldsymbol{k} \cdot \boldsymbol{k} = k^2$	$\boldsymbol{k} \cdot \boldsymbol{k} = k^2$	$\boldsymbol{k} \cdot \boldsymbol{k} = k^2$ $\boldsymbol{k} \cdot \boldsymbol{k}^* = \lvert k^2\rvert$	$\boldsymbol{k} \cdot \boldsymbol{k} = k^2$ $\boldsymbol{k} \cdot \boldsymbol{k}^* = k^2$
Orthogonality of vectors	$\boldsymbol{E} \cdot \boldsymbol{H} = 0$ $\boldsymbol{E} \cdot \boldsymbol{k} = 0$ $\boldsymbol{H} \cdot \boldsymbol{k} = 0$	$\boldsymbol{E} \cdot \boldsymbol{H} = 0$ $\boldsymbol{E} \cdot \boldsymbol{k} = 0$ $\boldsymbol{H} \cdot \boldsymbol{k} = 0$	$\boldsymbol{E} \cdot \boldsymbol{H} = 0$ $\boldsymbol{E} \cdot \boldsymbol{k} = 0$ $\boldsymbol{H} \cdot \boldsymbol{k} = 0$	$\boldsymbol{E} \cdot \boldsymbol{H} = 0$ $\boldsymbol{E} \cdot \boldsymbol{k} = 0$ $\boldsymbol{H} \cdot \boldsymbol{k} = 0$

planar). As mentioned, a homogeneous plane wave propagates with real direction angles in space, with a direction of propagation described by a real unit vector. By a suitable rotation of coordinate axis, the direction of propagation may be assumed to be along the z-axis. Furthermore, the plane wave can always be represented as a sum of two separate plane waves, one with the electric field polarized in the x direction and the other polarized in the y direction. Considering the wave polarized in the x direction, the fields are as follows:

$$E_x = E_0 e^{-jkz}. \tag{3.36}$$

$$H_y = \frac{E_0}{\eta} e^{-jkz}. \tag{3.37}$$

The wave number k and the intrinsic impedance η of the space are given by equations 3.12 and 3.23, respectively.

3.3.1 Wavelength, Phase Velocity, and Group Velocity

The **wavelength** λ is defined at the distance required for the plane wave to change phase by 2π radians and is given by:

$$\lambda = \frac{2\pi}{k}. \tag{3.38}$$

The **phase velocity** of the plane wave is defined as the velocity at which a point of constant phase travels (such as the crest of the wave where the electric field is the maximum). The phase velocity is given by:

$$v_p = \frac{\omega}{k'} = \frac{\omega}{\mathrm{Re}(\omega\sqrt{\mu\varepsilon})}. \tag{3.39}$$

If the permittivity ε is not a function of frequency (true for a vacuum and often approximately true for nonconducting media over a certain frequency range), then the phase velocity is constant with frequency (the plane wave propagates without dispersion). For a vacuum, the phase velocity is $c = 2.99792458 \times 10^8$ m/sec. The group velocity (velocity of energy flow and often a good approximation to the velocity of signal propagation [Collin, 1991], is given by:

$$v_g = \frac{d\omega}{dk'} = \frac{1}{\frac{d}{d\omega}\mathrm{Re}(\omega\sqrt{\mu\varepsilon})}. \tag{3.40}$$

For a nondispersive lossless media (constant real-valued permittivity), the group velocity is equal to the phase velocity, and both are equal to:

$$v_p = v_g = \frac{1}{\sqrt{\mu\varepsilon}}. \tag{3.41}$$

3.3.2 Depth of Penetration (Skin Depth)

For a lossy medium, the plane wave decays as it propagates because $k'' > 0$. The **depth of penetration**, d_p, is defined as the distance required to reduce the field level by a factor of $e \approx 2.71828$, so the field is 36.788% of the starting value (the power is reduced by a factor of e^2 or is at a level of 13.534% of the starting value). The depth of penetration is given as:

$$d_p = \frac{1}{k''} = \frac{-1}{\mathrm{Im}(\omega\sqrt{\mu\varepsilon})}. \tag{3.42}$$

The permittivity is complex for lossy media and is represented by equation 3.10. Special cases are of particular interest. For low-loss media, $\varepsilon'' \ll \varepsilon'$, a simple binomial approximation of the square root in equation 3.42 gives:

$$d_p = \frac{1}{k''} \approx \frac{1}{0.5\, k' \tan\delta}, \tag{3.43}$$

where the following is true:

$$k' \approx \omega\sqrt{\mu\varepsilon'}, \tag{3.44}$$

and the loss tangent is defined as:

$$\tan\delta = \frac{\varepsilon''}{\varepsilon'}. \tag{3.45}$$

If the loss is purely due to medium conductivity (no polarization loss so that $\varepsilon_c = \varepsilon'$ is a real number), the penetration depth formula 3.43 for the low-loss case becomes:

$$d_p \approx \frac{2}{\sigma}\sqrt{\frac{\varepsilon'}{\mu}}. \tag{3.46}$$

For a highly lossy medium, $\varepsilon' \ll \varepsilon''$, in which case:

$$k' \approx k'' \approx \omega\sqrt{\frac{\mu\varepsilon''}{2}}. \tag{3.47}$$

For a good conductor, the conductivity is very large, and $\varepsilon'' \approx \sigma/\omega$ from equation 3.10. In this case, the penetration depth is more commonly referred to as the **skin depth** δ because the value may be very small. (There is no relation to the δ symbol appearing in the loss tangent symbol.) For a good conductor:

$$\delta = d_p \approx \sqrt{\frac{2}{\omega\mu\sigma}}. \tag{3.48}$$

For pure copper ($\sigma = 5.8 \times 10^7$ S/m, $\mu \approx \mu_0$) at 2.45 GHz, $\delta = 1.335\ \mu$m.

Whether a particular material falls into the low-loss or highly lossy categories may depend on frequency. For example, consider typical seawater, which has roughly $\varepsilon_r = 78$ (ignoring polarization losses) and $\sigma = 4.0\,\text{S/m}$. (In actuality, polarization loss also becomes important at microwave and higher frequencies.) Table 3.2 shows the loss tangent versus frequency. The loss tangent is on the order of unity at microwave frequencies. For much lower frequencies the seawater is highly lossy; for much higher frequencies, it is a low-loss media. Note that the depth of penetration continues to decrease as the frequency is raised, even though the medium becomes more of a low-loss material at higher frequencies according to the definition used (a low-loss tangent). At low frequencies, the penetration depth is increasing inversely as the square root of frequency according to equation 3.48. At high frequencies, the penetration depth approaches a constant according to equation 3.46.

3.3.3 Summary of Homogeneous Plane Wave Properties

A summary of the properties relating to wavelength, velocity, and depth of penetration for a homogeneous plane wave is given in Table 3.3.

3.3.4 Polarization

The previous discussion assumed a homogeneous plane wave propagating in the z direction, polarized with the electric field in the x direction. The most general polarization of a wave propagating in the z direction is one having both x and y components of the field:

$$\boldsymbol{E} = \left(\hat{\boldsymbol{x}}E_{x0} + \hat{\boldsymbol{y}}E_{y0}\right)e^{-j(kz)}. \tag{3.49}$$

The field components are represented in polar form as:

$$E_{x0} = |E_{x0}|e^{j\phi_x}. \tag{3.50}$$

$$E_{y0} = |E_{y0}|e^{j\phi_y}. \tag{3.51}$$

The phase difference between the two components is defined as:

$$\phi = \phi_y - \phi_x. \tag{3.52}$$

Without any real loss of generality, ϕ_x may be chosen as zero, so that $\phi = \phi_y$. In the time domain, the field components are then:

$$\varepsilon_x = |E_{x0}|\cos(\omega t) \tag{3.53}$$

$$\varepsilon_y = |E_{y0}|\cos(\omega t + \phi). \tag{3.54}$$

Using trigonometric identities, equation 3.54 may be expanded into a sum of $\sin(\omega t)$ and $\cos(\omega t)$ functions, and then equation 3.53 may be used to put both $\cos(\omega t)$ and $\sin(\omega t)$ in terms of ε_x (using $\sin^2 = 1 - \cos^2$). After simplification, the result is as follows:

$$A\varepsilon_x^2 + B\varepsilon_x\varepsilon_y + C\varepsilon_y^2 = D, \tag{3.55}$$

where the following holds:

$$A = \left|\frac{E_{y0}}{E_{x0}}\right|^2. \tag{3.56}$$

$$B = -2\left|\frac{E_{y0}}{E_{x0}}\right|\cos\phi. \tag{3.57}$$

$$C = 1. \tag{3.58}$$

$$D = |E_{y0}|^2\sin^2\phi. \tag{3.59}$$

After simplification, the discriminant of this quadratic curve is:

$$\Delta = B^2 - 4AC = -4\left|\frac{E_{y0}}{E_{x0}}\right|^2\sin^2\phi, \tag{3.60}$$

which is always negative. This curve thus always represents an ellipse. The general form of the ellipse is shown in Figure 3.1. The tilt angle of the ellipse is τ, and the **axial ratio** (AR) is defined as the ratio of the major axis of the ellipse to the

TABLE 3.2 Loss Tangent and Penetration Depth for Typical Seawater Versus Frequency

Frequency [GHz]	Loss tangent ($\varepsilon''/\varepsilon'$)	Penetration depth d_p [meters]
0.000001	921800	7.958
0.00001	92180	2.5165
0.0001	9218	0.7958
0.001	921.8	0.2518
0.01	92.18	0.0800
0.1	9.218	0.0266
1.0	0.9218	0.0127
10.0	0.09218	0.01173
100.0	0.009218	0.01172

TABLE 3.3 Propagation Properties of a Homogeneous Plane Wave

	General	Highly conducting	Low loss	Lossless
Wavelength	$\frac{2\pi}{k'}$	$2\pi\sqrt{\frac{2}{\omega\mu\sigma}}$	$\frac{2\pi}{\omega\sqrt{\mu\varepsilon'}}$	$\frac{2\pi}{\omega\sqrt{\mu\varepsilon}}$
Phase velocity	$\frac{\omega}{k'}$	$\sqrt{\frac{2\omega}{\mu\sigma}}$	$\frac{1}{\sqrt{\mu\varepsilon'}}$	$\frac{1}{\sqrt{\mu\varepsilon}}$
Group velocity	$\frac{d\omega}{dk}$	$2\sqrt{\frac{2\omega}{\mu\sigma}}$	$\frac{1}{\sqrt{\mu\varepsilon'}}$	$\frac{1}{\sqrt{\mu\varepsilon}}$
Depth of penetration	$1/k''$	$\sqrt{\frac{2}{\omega\mu\sigma}}$	$\frac{2}{\sigma}\sqrt{\frac{\varepsilon'}{\mu}}$	∞

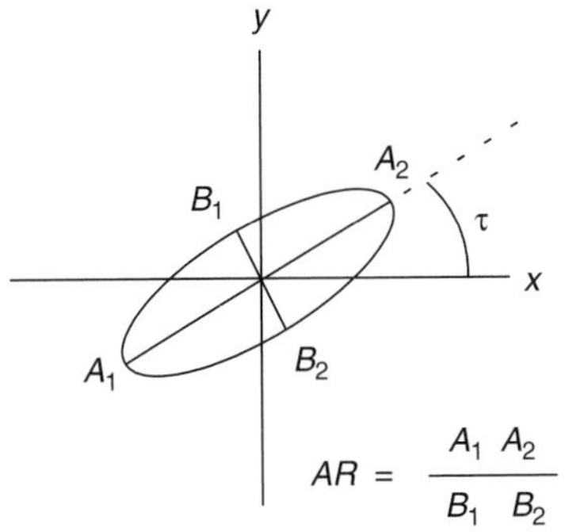

FIGURE 3.1 Geometry of the Polarization Ellipse

minor axis ($AR \geq 1$). In the time domain, the electric field vector rotates with the tip of the vector lying on the ellipse. **Right-handed elliptical polarization** (RHEP) corresponds to counterclockwise rotation (the thumb of the right hand aligns with the direction of propagation, and the fingers of the right hand align with the direction of rotation in time; this is the IEEE definition, which is opposite to the usual optics convention.) LHEP corresponds to rotation in the opposite direction.

A convenient way to represent the polarization state is with the Poincaré sphere (Kraus, 1988). Using spherical trigonometric relations, the following results may be derived for the tilt angle of the ellipse and the axial ratio:

$$\tan 2\tau = \tan 2\gamma \cos \phi, \tag{3.61}$$

$$\sin 2\xi = \sin 2\gamma \sin \phi. \tag{3.62}$$

The parameter ξ is related to the axial ratio as:

$$\xi = \cot^{-1}(\pm AR), \quad -45^\circ \leq \xi \leq +45^\circ. \quad \begin{matrix} +\text{for LHEP} \\ -\text{for RHEP} \end{matrix} \tag{3.63}$$

The phase angle ϕ is defined in equation 3.52. The parameter γ characterizes the ratio of the fields along the x- and y-axes, and is defined from:

$$\gamma = \tan^{-1}\left|\frac{E_{y0}}{E_{x0}}\right|. \tag{3.64}$$

There is no ambiguity in equation 3.62 for ξ because $-90^\circ \leq 2\xi \leq +90^\circ$. Equation 3.61, however, gives an ambiguity for τ because adding multiples of 180° does not change the tangent. To resolve this ambiguity, Table 3.4 may be used to determine the appropriate quadrant (1, 2, 3, or 4) the angle 2τ is in based on the Poincaré sphere (Kraus, 1988).

The following special cases are important.

(1) **Linear polarization:** $\phi = 0$ or 180°, or either E_{x0} or E_{y0} is zero.

TABLE 3.4 Quadrant Containing the Angle 2τ

	$\cos\phi > 0$	$\cos\phi < 0$
$\cos 2\gamma > 0$	1	4
$\cos 2\gamma < 0$	2	3

In this case, $AR = \infty$ (the ellipse degenerates to a straight line).

(2) **Circular polarization:** $E_{y0} = E_{x0}$ and $\phi = \pm 90^\circ$ (to within any multiple of 180°).

In this case, the ellipse becomes a circle (either RHCP or LHCP), and $AR = 1$.

It can also be noted that a wave of arbitrary polarization can be represented as a sum of RHCP and LHCP waves, by noting that the arbitrarily polarized wave in equation 3.49 can be written as:

$$\mathbf{E} = \hat{\mathbf{r}}\left[\frac{1}{\sqrt{2}}(E_{x0} + jE_{y0})\right] + \hat{\mathbf{l}}\left[\frac{1}{\sqrt{2}}(E_{x0} - jE_{y0})\right], \tag{3.65}$$

where the unit-amplitude RHCP and LHCP waves are the following:

$$\hat{\mathbf{r}} = \frac{1}{\sqrt{2}}(\hat{\mathbf{x}} - j\hat{\mathbf{y}})e^{-jkz}. \tag{3.66}$$

$$\hat{\mathbf{l}} = \frac{1}{\sqrt{2}}(\hat{\mathbf{x}} + j\hat{\mathbf{y}})e^{-jkz}. \tag{3.67}$$

3.4 Plane Wave Reflection and Transmission

A general plane wave **reflection and transmission problem** consists of an incident plane wave impinging on a multilayer structure, as shown in Figure 3.2. An important special case is the two-region problem shown in Figure 3.3. The incident plane wave may be either homogeneous or inhomogeneous, and any of the regions may have an arbitrary amount of loss. The vector representing the direction of propagation for the incident wave in Figures 3.2 and 3.3 is a real vector (as shown) if the incident wave is homogeneous, but the analysis is valid for the general case.

The key to obtaining a simple solution to such reflection and transmission problems is the use of a transmission line model of the layered structured, which is based on TE–TM decomposition of the plane waves, discussed next.

3.4.1 Transverse Electric and Transverse Magnetic Decomposition

According to a basic electromagnetic theorem (Harrington, 1961) the fields in a source-free homogeneous region can be

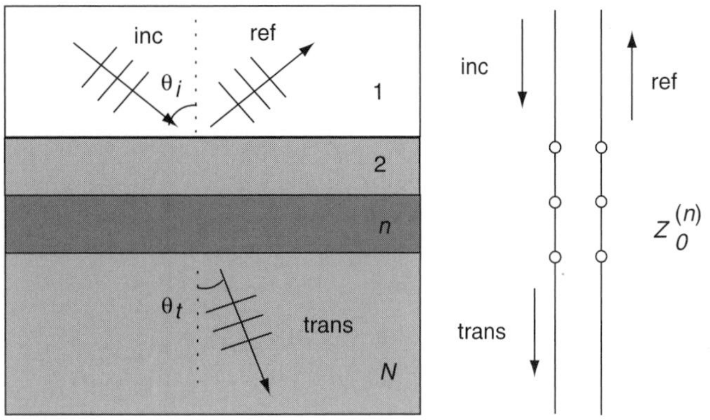

FIGURE 3.2 A Plane Wave Reflecting from a Multilayer Stack of Different Materials. Each layer has a uniform (constant) set of material parameters. On the right side, the equivalent transmission-line model (transverse equivalent network) is shown.

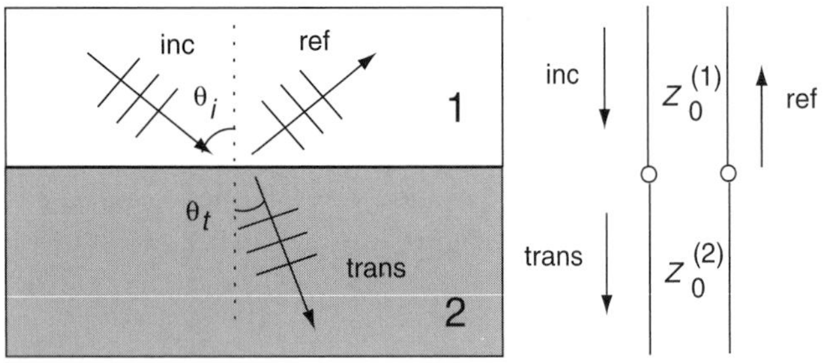

FIGURE 3.3 Reflection and Transmission from a Single Interface Between Two Different Media. The transmission-line model (transverse equivalent network) for the two-region reflection problem is shown on the right.

represented as the sum of two types of fields: a field that is **transverse magnetic** to z (TM_z) and a field that is **transverse electric** to z (TE_z). The TM_z field has $H_z = 0$, and the TE_z field by definition has $E_z = 0$. A general plane wave field may therefore be written as the sum of a TM_z plane wave and a TE_z plane wave. In free space, the direction z is rather arbitrary. When a layered media is present, the preferred direction z is perpendicular to the layers because this allows for the TM_z and TE_z plane waves in each region to be modeled independently as waves on a transmission line. Standard transmission-line theory may then be conveniently used to solve plane wave reflection and transmission problems in a relatively simple manner, without having to solve the electromagnetic boundary-value problem of matching fields at the interfaces. The TM_z plane wave is commonly referred to as one that is polarized with the electric field "in the place of incidence," whereas the TE_z place wave is polarized with the electric field "perpendicular to the plane of incidence." (The plane of incidence is the $y-z$ plane.) For a homogeneous plane wave with an incident wave vector in the $y-z$ plane, the polarizations are shown in Figure 3.4.

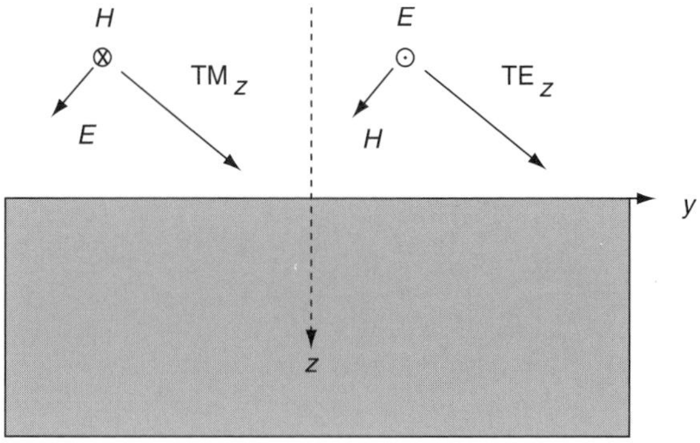

FIGURE 3.4 Polarizations of Incident TM_z and TE_z Plane Waves

The field components may be found by using Maxwell's equations to write the transverse (x and y) field components in terms of the longitudinal components E_z and H_z. The nonzero longitudinal component is assumed to be proportional to the wave function:

$$\psi = \exp\big(-j(k_x x + k_y y \pm k_z z)\big). \tag{3.68}$$

The plus sign in this equation is chosen for plane waves propagating or decaying in the positive z direction, while the negative sign is for propagation or decay in the minus z direction. This representation is convenient for plane wave reflection problems, since the characteristic impedance of a transmission line that models a particular region will then have a positive characteristic impedance for both upward and downward propagating plane waves, in agreement with the usual transmission line convention.

For the TM_z plane wave, the normalized field components may then be written as:

$$\begin{aligned} E_x &= \mu\ k_x k_z \Psi(x, y, z). & H_x &= \omega\varepsilon\ k_y \Psi(x, y, z). \\ E_y &= \mu\ k_y k_z \Psi(x, y, z). & H_y &= -\omega\varepsilon\ k_x \Psi(x, y, z). \\ E_z &= (k^2 - k_z^2)\Psi(x, y, z). & H_z &= 0. \end{aligned} \tag{3.69}$$

For a TE_z, plane wave the corresponding results are the following:

$$\begin{aligned} E_x &= -\omega\mu\ k_y \Psi(x, y, z). & H_x &= \mu\ k_x k_z \Psi(x, y, z). \\ E_y &= \omega\mu\ k_x \Psi(x, y, z). & H_y &= \mu\ k_y k_z \Psi(x, y, z). \\ E_z &= 0. & H_z &= (k^2 - k_z^2)\Psi(x, y, z). \end{aligned} \tag{3.70}$$

Note that a plane wave propagating in the z direction ($k_z = k$) has both longitudinal field components that are zero. Such a plane wave is transverse electric and magnetic (TEM) to the z direction.

It may be seen from the above equations that the transverse fields obey the relations:

$$\boldsymbol{E}_t^{TM} = \mu Z_0^{TM}(\hat{\boldsymbol{z}} \times \boldsymbol{H}_t^{TM}), \tag{3.71}$$

and

$$E_t^{TE} = \mu Z_0^{TE}(\hat{z} \times H_t^{TE}), \qquad (3.72)$$

where the following equations apply:

$$Z_0^{TM} = \frac{k_z}{\omega\varepsilon}. \qquad (3.73)$$

$$Z_0^{TE} = \frac{\omega\mu}{k_z}. \qquad (3.74)$$

3.4.2 Transverse Equivalent Network

From the equations 3.71 and 3.76, the transverse (perpendicular to z) field components behave as voltages and currents on a transmission line model called the **transverse equivalent network**. In particular, the correspondence is given through the relations ($i =$ TM or TE):

$$E_t^i = \hat{e}\psi_t(x, y)V^i(z). \qquad (3.75)$$

$$H_t^i = \hat{h}\psi_t(x, y)I^i(z). \qquad (3.76)$$

In these equations, the unit vectors are chosen so that:

$$\hat{e} \times \hat{h} = \pm\hat{z}, \qquad (3.77)$$

for a plane wave propagating in the $\pm\hat{z}$ direction. The transverse wave function is written as:

$$\psi_t(x, y) = e^{-j(k_x x + k_y y)}, \qquad (3.78)$$

which is a common transverse phase term that must be the same for all regions (if the transverse wave numbers k_x or k_y were different between two regions, a matching of transverse fields at the boundary would not be possible). The fact that the transverse wave numbers are the same in all regions leads to the **law of reflection** that states the direction angle θ for a reflected plane wave must equal the direction angle for the incident plane wave. This fact also leads to Snell's law that states the direction angles θ inside each of the regions are related to each other, through the following:

$$n_i \sin\theta_i = n_1 \sin\theta_1 \text{ and } i = 1, 2, \ldots N, \qquad (3.79)$$

where n_i is the index of refraction (possibly complex) of region i, defined as $n_i = k_i/k_0 = \sqrt{\varepsilon_{ri}\mu_{ri}}$. It is often convenient to express the characteristic impedance for region i from equations 3.73 and 3.74 in terms of the medium intrinsic impedance η_i as:

$$Z_0^{TM} = \eta_i \cos\theta_i, \qquad (3.80)$$

$$Z_0^{TE} = \eta_i \sec\theta_i. \qquad (3.81)$$

Using Snell's law, the impedances then become:

$$Z_0^{TM} = \eta_i\sqrt{1 - \left(\frac{n_1}{n_i}\right)^2 \sin^2\theta_1}. \qquad (3.82)$$

$$Z_0^{TE} = \frac{\eta_i}{\sqrt{1 - \left(\frac{n_1}{n_i}\right)^2 \sin^2\theta_1}} \qquad (3.83)$$

The above expressions remain valid for lossy media. The square roots are chosen so that the real part of the characteristic impedances are positive.

The functions $V(z)$ and $I(z)$ behave as voltage and current on a transmission line, with characteristic impedance Z_0^{TM} or Z_0^{TE}, depending on the case. Hence, any plane wave reflection and transmission problem reduces to a transmission line problem, giving the exact solution that satisfies all boundary conditions. One consequence of this is that TM_z and TE_z plane waves do not couple at a boundary. If the incident plane wave is TM_z, for example, the waves in all regions will remain TM_z plane waves. Hence, this situation creates the motivation for the TM_z–TE_z decomposition. The transverse equivalent network for the multilayer and two-region problems are shown on the right sides of Figures 3.2 and 3.3, respectively. The network model is the same for either TM_z or TE_z polarization, except that the characteristic impedances are different. If an incident plane wave is a combination of both TM_z and TE_z waves, the two parts are solved separately and then summed to get the total reflected or transmitted field.

3.4.3 Special Case: Two-Region Problem

For the simple **two-region problem** of Figure 3.3, the reflected and transmitted voltages (modeling the transverse electric fields) are represented as:

$$V^{\text{ref}}(z) = \Gamma V^{\text{inc}}(0)e^{+jk_z^{(1)}z} \qquad (3.84)$$

and

$$V^{\text{trans}}(z) = TV^{\text{inc}}(0)e^{-jk_z^{(2)}z}, \qquad (3.85)$$

where the reflection and transmission coefficients are given by the standard transmission line equations:

$$\Gamma = \frac{Z_0^{(2)} - Z_0^{(1)}}{Z_0^{(2)} + Z_0^{(1)}}. \qquad (3.86)$$

$$T = 1 + \Gamma. \qquad (3.87)$$

Note that a plus sign is used in the exponent of equation 3.84 to account for upward propagation of the reflected wave.

Critical Angle

If regions 1 and 2 are lossless, and region 1 is more dense than region $2 (n_1 > n_2)$, an incident angle $\theta_1 = \theta_c$ will exist for which $k_z^{(2)} = 0$. From Snell's law, this **critical angle** is the following:

$$\theta_c = \sin^{-1}\left(\frac{n_2}{n_1}\right). \qquad (3.88)$$

When $\theta_1 > \theta_c$, the wave number $k_z^{(2)}$ will be purely imaginary, of the form $k_z^{(2)} = -j\alpha_z^{(2)}$. In this case, there is no power flow into the second region because the phase vector in region 2 has no z component. In this case, 100% of the incident power is reflected back from the interface. There are, however, fields still present in region 2, decaying exponentially with distance z. In region 2, the power flow is in the horizontal direction only.

Brewster Angle

For lossless layers, it is possible to have 100% of the incident power transmitted into region 2 with no reflection. This corresponds to a matched transmission line circuit, with:

$$Z_0^{(1)} = Z_0^{(2)}. \qquad (3.89)$$

For nonmagnetic layers, it may be easily shown that this matching equation can only be satisfied in the TM_z case (there is always a nonzero reflection coefficient in the TE_z case, unless the trivial case of identical medium is considered). The angle θ_b, at which no reflection occurs, is called the **Brewster angle**. For the TM_z case, a simple algebraic manipulation of equation 3.89 yields the result:

$$\tan\theta_b = \sqrt{\frac{\varepsilon_2}{\varepsilon_1}}. \qquad (3.90)$$

3.4.4 Orthogonality

When treating reflection problems, there is more than one plane wave in at least one of the regions: a wave traveling in the positive z direction (focusing on the z variation) and a wave traveling in the negative z direction. Often, calculating power flow in such a region is desired. For a single plane wave, the complex power density in the z direction (the z component of the complex Poynting vector) is equal to the complex power flowing on the corresponding transmission line of the transverse equivalent network. When both an incident and a reflected wave are present, or both TM_z and TE_z waves are present, the following orthogonality results are useful. These theorems related to power flow in the z direction may be proven by direct calculation using the field components in rectangular coordinates.

1. An **orthogonality** exists between a homogeneous TM_z plane wave and a homogeneous TE_z plane wave propagating in the same direction in the sense that the complex power density in the z direction, S_z, is the sum of the two complex power densities S_{1z} and S_{2z}. This is true for a lossy or lossless media.

2. An **orthogonality** exists between an incident wave and a reflected wave in a lossless medium (both are either TE_z or TM_z), provided the wave number component k_z of the two waves is real. The two waves are orthogonal in the sense that the time-average power density in the z direction, $\mathrm{Re}S_z$, is the sum of the two time-average power densities $\mathrm{Re}S_{1z}$ and $\mathrm{Re}S_{2z}$.

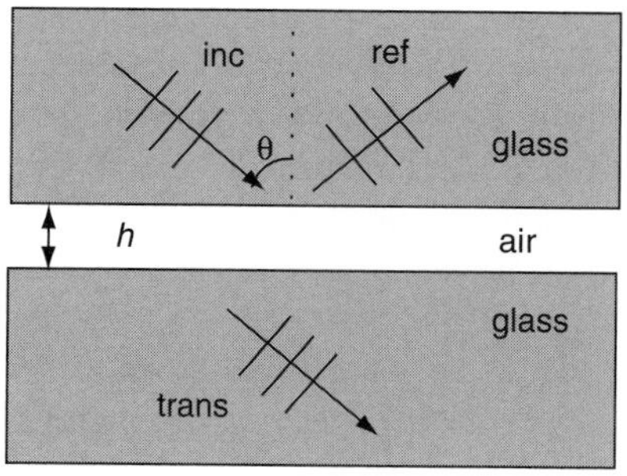

FIGURE 3.5 An Incident Wave Traveling in a Semi-Infinite Region of Lossless Glass. The wave impinges on an air gap separating the glass region from an identical region below. Incident, reflected, and transmitted plane waves are shown.

To illustrate the second orthogonality property, consider an incident plane wave traveling in a glass region, impinging on an air gap that separates the glass region from another identical glass region, as shown in Figure 3.5.

If the incident plane wave is beyond the critical angle, the plane waves in the air region will be evanescent with an imaginary vertical wave number k_z. Each of the two plane waves in the air region (upward and downward) that constitute a standing wave field do not, individually, have a time-average power flow in the z direction because k_z is imaginary. There is an overall power flow in the z direction inside the air region, however, because there is a transmitted field in the lower region. The total power flow in the z direction is thus not the sum of the two individual power flows. The two waves in the air region are not orthogonal, and the second orthogonal property does not apply since the wave number k_z is not real.

3.5 Example: Reflection of an RHCP Wave

An RHCP plane wave at a frequency of 1.0 GHz is incident on the surface of the ocean at an angle of $\theta = 30^\circ$. Determine the percentage of power that is reflected from the ocean, and characterize the polarization of the reflected wave (axial ratio, tilt angle, and handedness). In addition, determine the field of the inhomogeneous transmitted plane wave. The parameters of the ocean water are assumed to be $\varepsilon' = 78$ (ignoring polarization loss) and $\sigma = 4$ S/m.

3.5.1 Solution

The geometry of the incident and reflected waves is shown in Figure 3.6. The incident plane wave is represented as:

$$\boldsymbol{E}^{\text{inc}} = E_0[\hat{\boldsymbol{x}} + \hat{\boldsymbol{u}}(-j)]e^{-j(k_y y + k_{z1} z)}, \tag{3.91}$$

where the following equations apply:

$$k_y = k_0 \sin\theta = \frac{1}{2} k_0. \tag{3.92}$$

$$k_{z1} = k_0 \cos\theta = \frac{\sqrt{3}}{2} k_0. \tag{3.93}$$

$$\hat{\boldsymbol{u}} = \hat{\boldsymbol{y}} \cos\theta - \hat{\boldsymbol{z}} \sin\theta. \tag{3.94}$$

The reflected wave is represented as:

$$\boldsymbol{E}^{\text{ref}} = E_0[A\hat{\boldsymbol{x}} + B\hat{\boldsymbol{v}}]e^{-j(k_y y - k_{z1} z)}, \tag{3.95}$$

where:

$$\hat{\boldsymbol{v}} = -\hat{\boldsymbol{y}} \cos\theta - \hat{\boldsymbol{z}} \sin\theta. \tag{3.96}$$

The x component of the incident and reflected waves corresponds to TE_z waves, whereas the u and v components correspond to TM_z waves. (The u and v directions substitute for the y direction in the previous discussion on polarization.) The transverse equivalent network is shown in Figure 3.3. From equations 3.82 and 3.83, the impedances are $Z_1^{TM} = 326.258\Omega$, $Z_1^{TE} = 435.011\Omega$, $Z_2^{TM} = 34.051 + j(13.269)\Omega$, and $Z_2^{TE} = 34.089 + j(13.346)\Omega$. The reflection coefficients are then $\Gamma^{TM} = -0.808535 + j(0.066600)$ and $\Gamma^{TE} = -0.853161 + j(0.052724)$.

The coefficient A in equation 3.95 is determined directly from:

$$A = \Gamma^{TE} = -0.853161 + j(0.052724). \tag{3.97}$$

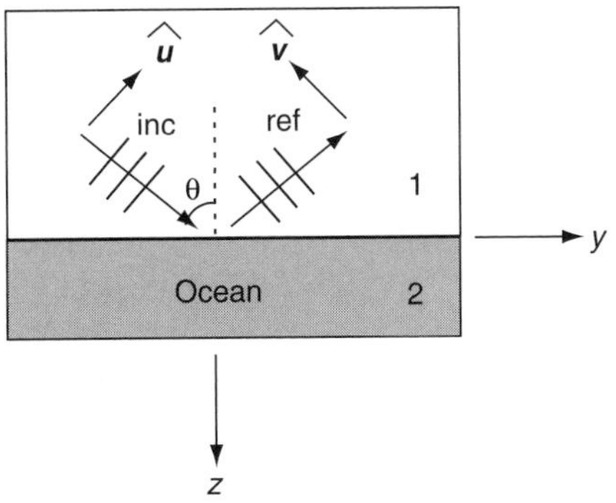

FIGURE 3.6 Geometry that Defines the Coordinate System for the Example of a Plane Wave Reflecting from the Surface of the Ocean

To determine the coefficient B, recall that the voltages in the transverse equivalent network model only the transverse (horizontal) component of the electric field (the y component in the TM case). Hence, the following equation results:

$$-BE_0 \cos\theta^{\text{ref}} = \Gamma^{TM} E_0(-j)\cos\theta^{\text{inc}}. \tag{3.98}$$

Since $\theta^{\text{inc}} = \theta^{\text{ref}} = \theta$, B is determined directly as:

$$B = \Gamma^{TM}(j) = -0.066600 - j(0.808535). \tag{3.99}$$

The percent power reflected is calculated from:

$$P_{\%}^{\text{ref}} = 100\left(\frac{\frac{1}{2}\left[\frac{|A|^2}{Z_1^{TE}} + \frac{|B\cos\theta|^2}{Z_1^{TM}}\right]}{\frac{1}{2}\left[\frac{1}{Z_1^{TE}} + \frac{|-j\cos\theta|^2}{Z_1^{TM}}\right]}\right), \tag{3.100}$$

which yields:

$$P_{\%}^{\text{ref}} = 69.441. \tag{3.101}$$

Hence, 30.559% of the incident power is transmitted into the ocean.

The phase angle between the v and x components of the reflected field is as written here:

$$\phi = \angle B - \angle A = 88.827^\circ. \tag{3.102}$$

Hence, the reflected wave is a left-handed elliptically polarized wave.

The parameter γ is determined from the ratio of the magnitudes of the v and x components of the reflected field as:

$$\gamma = \tan^{-1}\left(\frac{|B|}{|A|}\right) = 43.504^\circ. \tag{3.103}$$

From the formulas in the polarization section, Subsection 3.34, the tilt angle τ of the polarization ellipse from the x-axis (toward the positive v-axis) is then:

$$\tau = 10.690^\circ, \tag{3.104}$$

and the polarization parameter ξ is the following:

$$\xi = 43.393^\circ. \tag{3.105}$$

The axial ratio of the reflected wave is then:

$$AR = +\cot\xi = 1.058. \tag{3.106}$$

The reflected wave is nearly CP because the ocean is highly reflecting (if the ocean were replaced by a perfect conductor, the reflected wave would be LHCP).

The transmitted field in the ocean is represented as:

$$\boldsymbol{E}^{\text{trans}} = E_0[\hat{\boldsymbol{x}}C + \hat{\boldsymbol{y}}D + \hat{\boldsymbol{z}}E]e^{-j(k_y y - k_{z2}z)}, \tag{3.107}$$

where k_{z2} is the vertical wave number in the ocean found from equation 3.31 as:

$$\begin{aligned} k_{z2} = k_2 \cos\theta_2 &= k_0 n_2 \sqrt{1 - \left(\frac{1}{n_2}\right)^2 \sin^2\theta} \\ &= k_0(9.582527 - j3.751642). \end{aligned} \tag{3.108}$$

From the transverse equivalent network:

$$C = T^{TE} = 1 + \Gamma^{TE} = 0.146839 + j(0.052724), \tag{3.109}$$

and

$$\begin{aligned} D = T^{TM}(-j\cos\theta) &= (1 + \Gamma^{TM})(-j\cos\theta) \\ &= 0.057678 - j(0.165813). \end{aligned} \tag{3.110}$$

The z component of the transmitted field can be determined from the electric Gauss's law equation 3.20, which yields:

$$E = -\frac{1}{k_{z2}}(k_y D). \tag{3.111}$$

Hence, the following result is obtained:

$$E = -0.0055466 + j(0.00648030). \tag{3.112}$$

From the wave numbers k_y and k_{z2}, the direction angles of the transmitted wave may be determined from equations 3.29 through 3.31. The results are as follows:

$$\phi = 90^\circ, \tag{3.113}$$

$$\theta_2 = 0.045227 + j(0.017679)\text{radians}. \tag{3.114}$$

The transmitted plane wave is inhomogeneous because the angle θ_2 is complex.

References

Clemmow, P. C. (1996). *The plane wave spectrum representation of electromagnetic fields.* New York: *IEEE Press.*

Collin, R. E. (1991). *Field theory of guided waves.* (2d ed.). New York: *IEEE Press.*

Kraus, J. D. (1988). *Antennas.* New York: McGraw-Hill.

Harrington, R. F. (1961). *Time-harmonic electromagnetic fields.* New York: McGraw-Hill.

Hayt, W. H. (1989). *Engineering electromagnetics.* (5th ed.). New York: McGraw-Hill.

4

Transmission Lines

Krishna Naishadham
Massachusetts Institute of Technology, Lincoln Laboratory, Lexington, Massachusetts, USA

4.1 Introduction

A transmission line is used to transfer a signal from the generator to the load, as in Figure 4.1, by guiding an electromagnetic (EM) wave between two conductors. In microwave communications, the signal is usually some form of modulation on a high-frequency carrier. The generator circuit represents the Thevenin equivalent (a voltage V_g in series with an impedance Z_g) of the EM wave's source, such as a radar transmitter, a continuous tone source, or a high-speed data terminal. The load Z_L represents the equivalent impedance at the terminals of the receiver (e.g., the transmitting antenna's input impedance).

Figure 4.2 illustrates a cross section of some common transmission lines. The transmission lines are assumed to be uniform along the direction of propagation, so discontinuity effects are not considered in the field distribution. All these lines, except the microstrip, support electric and magnetic field distributions, which are entirely **transverse** to the direction of propagation. Thus, they can be described mathematically in terms of **transverse electromagnetic** (TEM) waves. The microstrip line has a small longitudinal field component relative to the transverse component because of flux leakage across the air–dielectric interface. Such a transmission line can be considered as **quasi-TEM**. The coaxial line and the stripline are completely shielded and do not have any leakage under ideal conditions.

A TEM wave has the same phase constant and intrinsic impedance as a plane wave propagating in an unbounded medium and can be uniquely represented by voltages and currents of the form:

$$v(z, t) = -\int_{C_V} \boldsymbol{E}(r, z) \cdot \boldsymbol{dl}. \tag{4.1}$$

$$i(z, t) = \int_{C_I} \boldsymbol{H}(r, z) \cdot \boldsymbol{dl}. \tag{4.2}$$

In equations 4.1 and 4.2, ***E*** and ***H*** are the electric and magnetic field intensities. C_V and C_I are arbitrary integration contours, parallel to electric and magnetic flux lines, respectively, as displayed in Figure 4.3 for a coaxial line. The TEM field distribution can be entirely calculated by considering static electric and magnetic fields, namely, an irrotational (or curl-free) electric field and the absence of displacement currents. Hence, the voltages and currents defined in equations 4.1 and 4.2 are independent of the integration contour's orientation.

With reference to Figure 4.3, assuming that the current (I) flows out of the center conductor and returns into the outer conductor, the contour C_V terminates radially on the conductors, while C_I follows a closed circular path along a magnetic flux line.

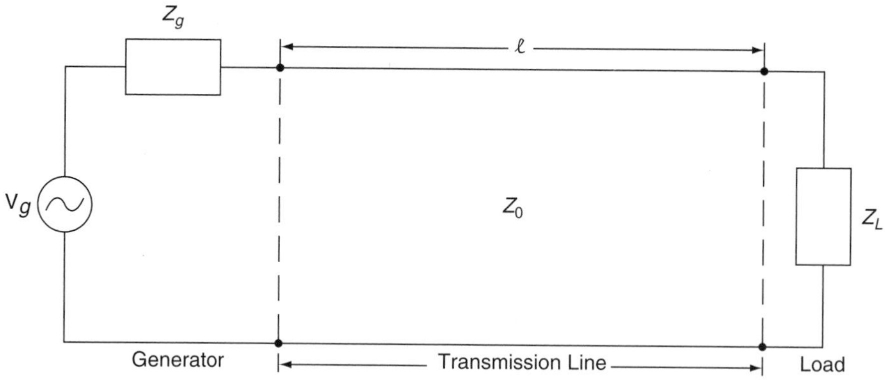

FIGURE 4.1 A Transmission Line of Length, ℓ, and Characteristic Impedance, Z_0, Connected Between the Generator and the Load

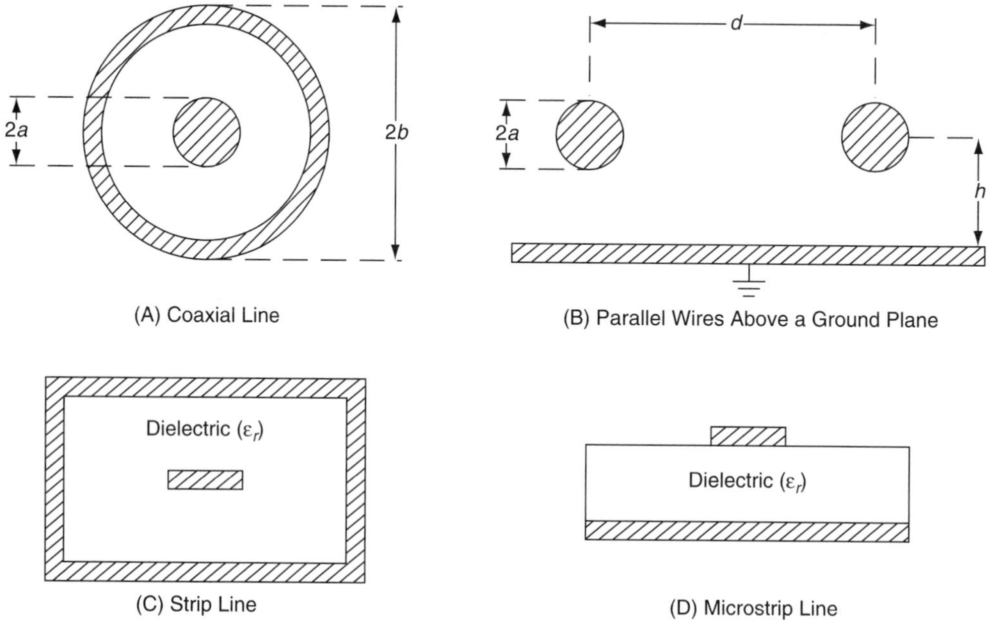

FIGURE 4.2 Examples of Transmission Lines. The dielectric medium has relative permittivity ε_r.

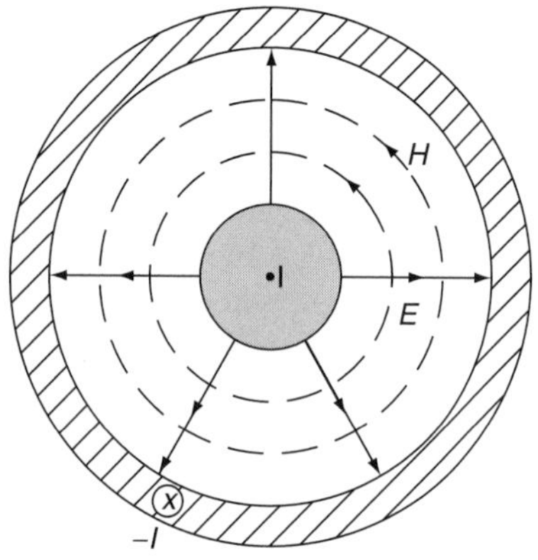

FIGURE 4.3 Transverse Electric and Magnetic Field Distribution in a Coaxial Line

4.1.1 What Is a Transmission Line?

A **transmission line** is a distributed circuit element. Unlike a conventional low-frequency circuit, the voltages and currents on a transmission line vary with longitudinal position because they experience a phase (or time) delay as the wave propagates from one end of the line to the other. This effect becomes important when the line length becomes an appreciable fraction of the wavelength at the operating frequency. Consider the circuit shown in Figure 4.4, where a lossless transmission line of length ℓ is connected between an ideal generator and a load resistance, R_L. For simplicity, the load is assumed to be matched to Z_0, the characteristic impedance of the line, so that there is no reflected wave.

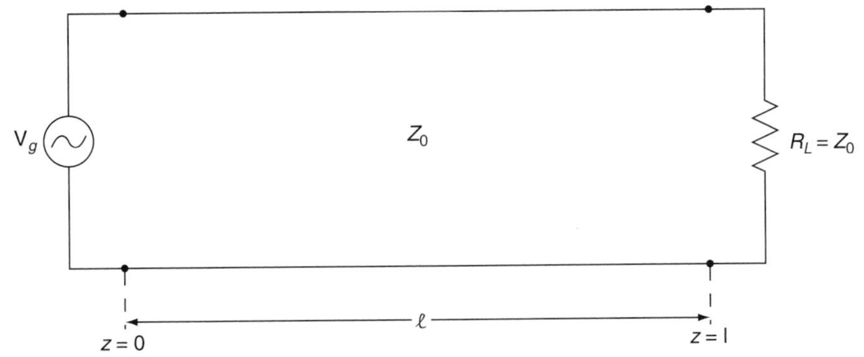

FIGURE 4.4 A Matched Transmission Line of Characteristic Impedance, Z_0

At time $t = 0$, let a sinusoidal signal be given by:

$$v_g(t) = V_0 \cos(\omega t), \tag{4.3}$$

where $\omega = 2\pi f$ is the radian frequency. Let this signal be applied to the generator located at $z = 0$. If the line is an air line, the signal travels at a velocity $v_p = 3 \times 10^8$ m/sec and reaches the load at time $\tau = \ell / v_p$. Thus, the signal experiences a time delay proportional to the distance that the wave has traveled along the line. At the load, the signal waveform is given by:

$$v_L(t) = V_0 \cos[\omega(t - \tau)] = V_0 \cos\left[\omega t - \frac{2\pi\ell}{\lambda}\right], \tag{4.4}$$

where $v_p = \lambda f$ is used to obtain the second equality. It is evident from equation 4.4 that the delay tracks with the distance traveled, normalized to the wavelength λ. The larger the fraction ℓ/λ gets, the longer is the delay. For conventional circuits operated at a low frequency, the fraction ℓ/λ is small; hence, the phase delay along the wire can be neglected. At higher frequencies, the wavelength gets shorter; hence, ℓ/λ becomes larger for a fixed ℓ. An important effect of the phase delay is that the line voltage varies with position along the line because the wave requires a finite time to travel from the source location to the measurement location. At low frequencies, these local potential differences are negligible because of very short transit times. This is not valid, however, when the line length becomes an appreciable fraction of the operating wavelength. Equation 4.4 can be generalized to any position z along the line by writing:

$$v(z, t) = V_0 \cos[\omega(t - \tau_z)] = V_0 \cos\left[\omega t - 2\pi\frac{z}{\lambda}\right], \tag{4.5}$$

where $\omega\tau_z$ is the phase delay incurred over a transit distance z, measured relative to the origin $z = 0$ at the generator-end of the line. The electrical distance may be written in terms of the phase constant:

$$\beta = \frac{2\pi}{\lambda} (\text{rad/m}), \tag{4.6}$$

so that the voltage wave becomes:

$$v(z, t) = V_0 \cos[\omega t - \beta z)]. \tag{4.7}$$

Equation 4.7 represents an incident wave traveling from the generator toward the load, and it remains valid on an infinitely long line, in which case there is no reflected wave. We can obtain some insight into wave propagation along the line by plotting the voltage as a function of phase delay βz for a fixed time variable ωt. Figure 4.5 shows three progression stages of the voltage waveform.

As time progresses from $t = 0$ to $t = t_1$ and then to $t = t_2$, the reference voltage sample at the waveform's peak moves along the positive z direction from location $z = 0$ to z_1 to z_2, for example. This is true for every point on the waveform. In fact, there is a constant phase delay, $\beta\Delta z = \omega\Delta t$, between successive snapshots. Since ω and β are constant, it means that as time increases by Δt, the position of the reference sample is proportionately displaced along positive z direction

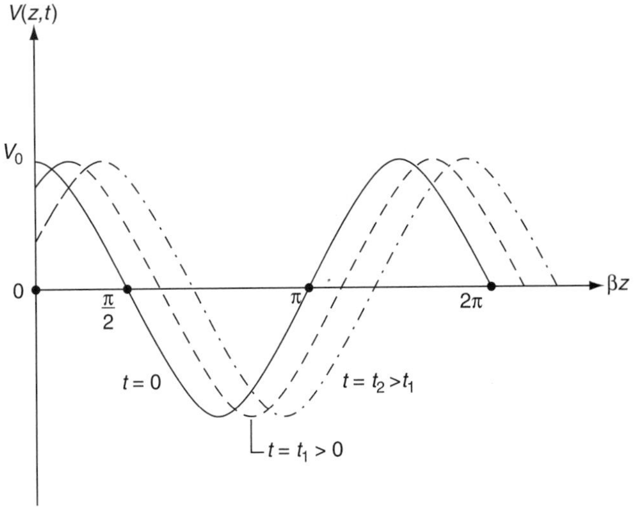

FIGURE 4.5 Progression of the Voltage Waveform on a Lossless Transmission Line

by a distance $(\omega/\beta)\Delta t$. The displacement per unit time is defined as the velocity of propagation or the **phase velocity**:

$$v_p = \frac{\omega}{\beta} = f\lambda. \tag{4.8}$$

The current on the incident wave is given by:

$$i(z, t) = \frac{v(z, t)}{Z_0} = \frac{V_0}{Z_0}\cos[\omega t - \beta z]. \tag{4.9}$$

Equation 4.9 depicts only a change in amplitude relative to the voltage in equation 4.7. Since the phase is unchanged, the conclusions on the distributed nature, discussed earlier for the voltage, are also valid for the current.

4.1.2 Lossy Transmission Line

On a **lossy transmission line** the voltage and current waveforms for a wave traveling along the z direction are given by:

$$v(z, t) = V_0 e^{-\alpha z}\cos[\omega t - \beta z], \tag{4.10}$$

$$i(z, t) = \frac{v(z, t)}{Z_0} = \frac{V_0}{Z_0} e^{-\alpha z}\cos[\omega t - \beta z]. \tag{4.11}$$

In addition to the phase delay linearly proportional to the distance traveled, the envelope of the wave pattern attenuates in amplitude exponentially according to $e^{-\alpha z}$, as shown in Figure 4.6. In this case, the propagation constant is $\gamma = \alpha + j\beta$, where α is the attenuation constant (measured in nepers per meter) and β is the phase constant. In general, the characteristic impedance Z_0 is also complex. For practical low-loss lines, however, Z_0 can be considered approximately real.

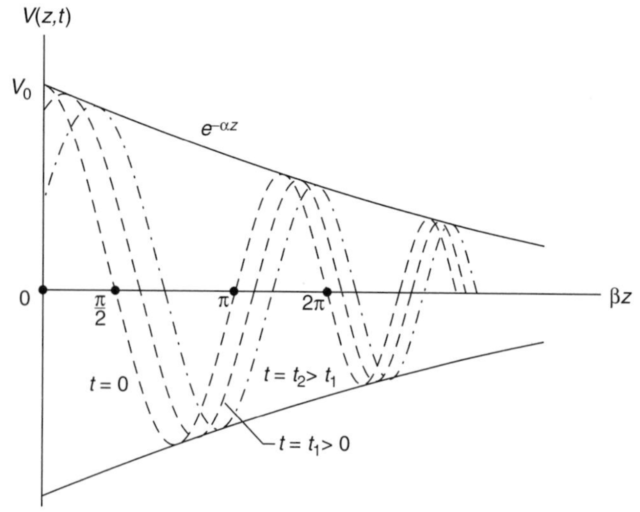

FIGURE 4.6 Progression of the Voltage Waveform on a Lossy Transmission Line

The next section shows that these two parameters, Z_0 and γ, are complex because of series resistive loss in the conductors and shunt conductive loss in the dielectric medium.

4.2 Equivalent Circuit

The TEM mode on a transmission line can be represented by the **equivalent circuit**, shown in Figure 4.7, for a length Δz. Physically, this is a **distributed circuit** because the circuit elements are not lumped at discrete locations as in a conventional low-frequency circuit but distributed uniformly along the length of the line. Thus, the elements are defined on **per-unit-length** basis. The series resistance per unit length, R, is the combined resistance of all the conductors in the line, and it accounts for power dissipation in the conductors. As exemplified in Figure 4.3 for a coaxial line, current flow along the conductor's surface is accompanied by a circumferential magnetic field. The linkage of this magnetic flux with the current produces a combined inductance per unit length, L, which accounts for flux linkages both internal and external to the conductors. Positive and negative charges of equal magnitude are deposited on the two conductors (Figure 4.3). This charge separation creates a potential difference between the conductors, which accounts for the capacitance per unit length, C. The shunt conductance per unit length, G, accounts for the power dissipation in the non-ideal dielectric medium between the conductors.

Application of Kirchhoff's voltage and current laws to the circuit in Figure 4.7 leads to the system of coupled differential equations for the phasor voltage and current in the limit $\Delta z \to o$:

$$\begin{aligned}\frac{dV(z)}{dz} &= -(R + j\omega L)I(z),\\ \frac{dI(z)}{dz} &= -(G + j\omega C)V(z).\end{aligned} \tag{4.12}$$

Eliminating the current variable in the former and the voltage variable in the latter of equation 4.12 leads to the uncoupled system of second-order differential equations:

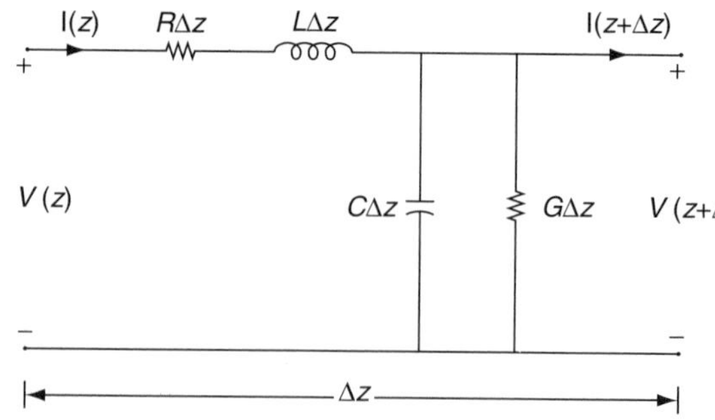

FIGURE 4.7 Lumped Parameter Equivalent Circuit of a General Transmission Line

$$\frac{d^2V(z)}{dz^2} - \gamma^2 V(z) = 0,$$
$$\frac{d^2I(z)}{dz^2} - \gamma^2 I(z) = 0. \tag{4.13}$$

The propagation constant is defined by:

$$\gamma = \sqrt{(R + j\omega L)(G + j\omega C)}. \tag{4.14}$$

Solution of the differential equations 4.13 results in the time-instantaneous traveling wave representation:

$$v(z,\ t) = V_0^+ e^{-\alpha z} \cos[\omega t - \beta z] + V_0^- e^{\alpha z} \cos[\omega t + \beta z], \tag{4.15}$$

$$i(z,\ t) = \frac{V_0^+}{Z_0} e^{-\alpha z} \cos[\omega t - \beta z] - \frac{V_0^-}{Z_0} e^{\alpha z} \cos[\omega t + \beta z]. \tag{4.16}$$

Equations 4.15 and 4.16 use the following equation:

$$f(z,\ t) = \mathcal{R}e[F(z)e^{j\omega t}], \tag{4.17}$$

to convert the voltage and current phasors, $V(z)$ and $I(z)$, into the time-domain. The first term in each equation corresponds to a wave traveling along the $+z$ direction, while the second corresponds to a wave traveling along the opposite direction. Section 4.3 analyzes the transmission line circuit shown in Figure 4.1 using equations 4.15 and 4.16 and also determines the wave amplitudes V_0^+ and V_0^-. The characteristic impedance Z_0 can be calculated using:

$$Z_0 = \sqrt{\frac{R + j\omega L}{G + j\omega C}}. \tag{4.18}$$

The per-unit-length elements R, L, G, and C depend on the resistivity of the conductors, cross-sectional dimensions and shape of the transmission line, and the permittivity of the dielectric medium. Expressions of these elements for some practical transmission lines can be found in any standard electromagnetics text, such as in Ulaby (1999), listed in Appendix A. In summary, the propagation constant γ and the characteristic impedance Z_0 for the TEM mode depend only on the transmission line's geometrical and physical parameters.

4.2.1 Lossless Line

For the **lossless line** $R = 0 = G$; hence, the attenuation constant $\alpha = 0$, and the characteristic impedance Z_0 is real. In this case, these equations apply:

$$\gamma = j\beta = j\omega\sqrt{LC}. \tag{4.19}$$

$$Z_0 = \sqrt{\frac{L}{C}}. \tag{4.20}$$

A lossless line has these properties: (a) it does not dissipate any power, (b) it is non-dispersive (i.e., the phase constant varies linearly with frequency ω, or the velocity $v_p = \omega/\beta$ is independent of frequency), and (c) its characteristic impedance Z_0 is real.

4.2.2 Low-Loss Line

A **low-loss line** is one for which $R \ll \omega L$ and $G \ll \omega C$. In this case, γ and Z_0 can be approximated from equations 4.14 and 4.18, respectively, as:

$$\alpha \cong \frac{R}{2}\sqrt{\frac{C}{L}} + \frac{G}{2}\sqrt{\frac{L}{C}}. \tag{4.21}$$

$$\beta \cong \omega\sqrt{LC}. \tag{4.22}$$

$$Z_0 \cong \sqrt{\frac{L}{C}},\ 2\omega \gg \frac{R}{L} - \frac{G}{C}. \tag{4.23}$$

It is emphasized that a low-loss line is non-dispersive subject to the approximations used to derive equation 4.22.

4.2.3 Distortionless Line

A **distortionless line** is truly non-dispersive, exhibits a real characteristic impedance Z_0, and satisfies the condition $RC = LG$. Thus, despite being dissipative, it possesses two of the three desirable characteristics of the lossless line. In this case, the propagation parameters are *exactly* given by:

$$\alpha = R\sqrt{\frac{C}{L}}. \tag{4.24}$$

$$\beta = \omega\sqrt{LC}. \tag{4.25}$$

$$Z_0 = \sqrt{\frac{L}{C}}. \tag{4.26}$$

4.3 Alternating Current Analysis

This section focuses on the steady-state alternating current (ac) analysis of transmission lines using the basic configuration in Figure 4.8, which shows a line connected between a generator of impedance Z_g and a load of impedance Z_L. This discussion begins with the general solution in equations 4.15 and 4.16, which contain two unknown voltage amplitudes, V_0^+ and V_0^-. The boundary condition is applied at the load to introduce the reflection coefficient, which relates V_0^+ to V_0^-. The condition at the generator-end of the line gives an expression for V_0^-. The standing wave properties of terminated lossless lines as well as lossy transmission lines are presented with examples.

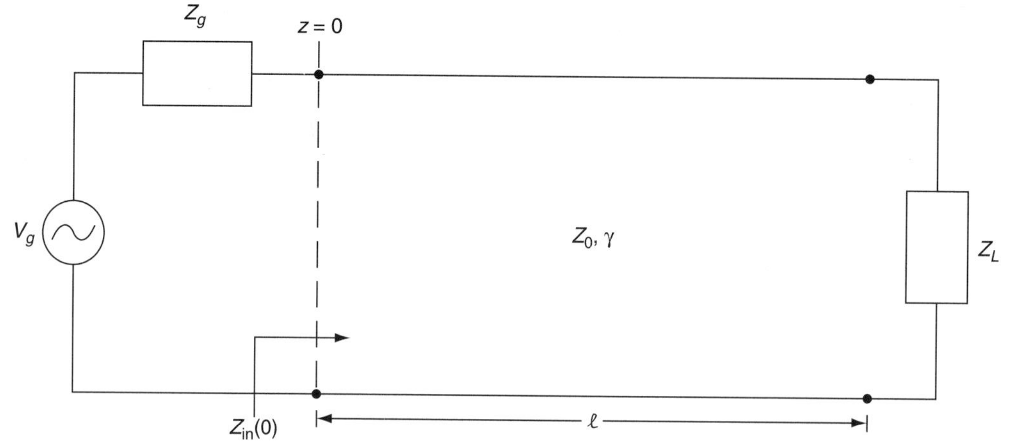

FIGURE 4.8 A Terminated Transmission Line of Characteristic Impedance, Z_0, and Propagation Constant, γ

4.3.1 Terminated Lossless Line

Reflection Coefficient

For a lossless transmission line, the voltage and current phasors can be calculated from equations 4.15 and 4.16 subject to $\alpha = 0$ and are given by:

$$V(z) = V_0^+ e^{-j\beta z} + V_0^- e^{j\beta z} = V_0^+ e^{-j\beta z}[1 + \Gamma(z)], \tag{4.27}$$

$$I(z) = \frac{1}{Z_0}\left[V_0^+ e^{-j\beta z} - V_0^- e^{j\beta z}\right] = \frac{V_0^+ e^{-j\beta z}}{Z_0}[1 - \Gamma(z)]. \tag{4.28}$$

In these equations, the voltage reflection coefficient has been introduced as:

$$\Gamma(z) = \frac{V_0^-}{V_0^+} e^{j2\beta z}. \tag{4.29}$$

Specializing equations 4.27 and 4.28 to the load position $z = \ell$ and taking the ratio of voltage to current yields:

$$\frac{V(\ell)}{I(\ell)} = Z_L = Z_0 \frac{1 + \Gamma(\ell)}{1 - \Gamma(\ell)}. \tag{4.30}$$

Equation 4.30 defines the **load reflection coefficient**:

$$\Gamma_L \equiv \Gamma(\ell) = \frac{Z_L - Z_0}{Z_L + Z_0}. \tag{4.31}$$

We can eliminate V_0^-/V_0^+ from equation 4.29 by specializing it to $z = \ell$ and the **translation formula**, which relates the reflection coefficient for any z to that at the load:

$$\Gamma(z) = \Gamma_L e^{j2\beta(z-\ell)}. \tag{4.32}$$

Input Impedance

The voltage to current ratio for any z, called the **input impedance** of the line, can be calculated using equations 4.27 and 4.28 as:

$$Z_{in}(z) = Z_0 \frac{1 + \Gamma(z)}{1 - \Gamma(z)}. \tag{4.33}$$

After substituting the translation formula 4.32 for $\Gamma(z)$ into equation 4.33, using equation 4.31 and the identity $e^{j\theta} = \cos\theta + j\sin\theta$, equation 4.33 can be rewritten as:

$$Z_{in}(z) = Z_0 \frac{Z_L + jZ_0 \tan[\beta(\ell - z)]}{Z_0 + jZ_L \tan[\beta(\ell - z)]}. \tag{4.34}$$

At the input terminal $z = 0$, the following equation is obtained:

$$Z_{in}(0) = Z_0 \frac{Z_L + jZ_0 \tan\beta\ell}{Z_0 + jZ_L \tan\beta\ell}. \tag{4.35}$$

In particular, for a short circuit ($Z_L = 0$), $Z_{in}(0) = jZ_0 \tan\beta\ell$; for an open circuit ($Z_L \to \infty$), $Z_{in}(0) = -jZ_0 \cot\beta\ell$. Clearly, the input impedance is periodic in position with a period of a **half wavelength**. It can be seen from equation 4.32 that the reflection coefficient also has the same periodicity. Thus, both the input impedance and the reflection coefficient attain identical values at points separated by a distance of $\lambda/2$. This property, known as the **replication property**, has been used in designing a graphical tool called the **Smith chart** and considerably simplifies the analysis of lossless transmission lines (see Appendix A, Smith [1969], for more information and further reading).

We conclude this discussion now with some examples. Figure 4.9 shows the input impedance of a reactively termin-

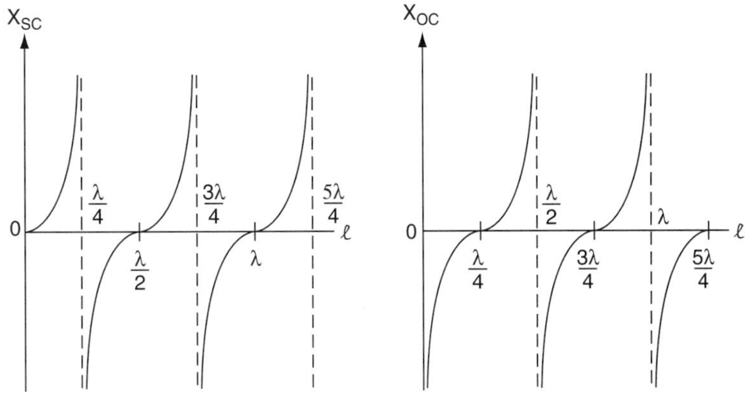

(A) Short-Circuited Line (B) Open-Circuited Line

FIGURE 4.9 The Input Reactance of Lossless Lines

ated line (open or short) plotted as a function of its length. The half wavelength periodicity is evident. Depending on the line's electrical length, the input reactance is either capacitive ($X < 0$) or inductive ($X > 0$). This suggests that a discrete capacitor or inductor can be constructed from an appropriate length of shorted or open-circuited transmission line. In practice, at microwave frequencies, it is necessary to have a small footprint and reduced parasitic effects for the element. This precludes the use of conventional transmission lines. Instead, for example, an RF inductor is made of a planar spiral etched on a printed circuit board, and a miniature capacitor is constructed in parallel plate configuration using a material with high ε_r and metallized pads for the contacts (see Wadell [1991] in Appendix A for further reading).

Quarter Wavelength Transformer

The input impedance of a $\lambda/4$ long line can be calculated from equation 4.35 as:

$$Z_{\text{in}}(0) = \frac{Z_0^2}{Z_L}, \tag{4.36}$$

which indicates that a $\lambda/4$ line can be used as an impedance inverter, (i.e., a high impedance at the load can be converted to a low impedance at the input and vice versa). Using equation 4.36, it follows that a $\lambda/4$ line with characteristic impedance given by $Z_0 = \sqrt{Z_{01}Z_{02}}$ provides matched impedance when connected between two transmission lines of characteristic impedances Z_{01} and Z_{02}. Such a line, known as a **quarter-wave impedance transformer**, is frequency-selective (i.e., the mismatch becomes significant as the operating frequency deviates from the narrow band around the center frequency at which the line length $\ell = \lambda/4$).

Standing Waves

From equations 4.27, 4.28, and 4.32 the voltage and current on a lossless line are given by:

$$V(z) = V_0^+ e^{-j\beta z}[1 + \Gamma_L e^{j2\beta(z-\ell)}] \tag{4.37}$$

$$I(z) = \frac{V_0^+}{Z_0} e^{-j\beta z}[1 - \Gamma_L e^{j2\beta(z-\ell)}]. \tag{4.38}$$

At the generator end $z = 0$, this equation results:

$$V(0) = V_0^+\left[1 + \Gamma_L e^{-j2\beta\ell}\right]. \tag{4.39}$$

Applying voltage division at the generator-end of the line (see Figure 4.8), the following is calculated:

$$V(0) = V_g \frac{Z_{\text{in}}(0)}{Z_g + Z_{\text{in}}(0)}, \tag{4.40}$$

where, using equations 4.32 and 4.33:

$$Z_{\text{in}}(0) = Z_0 \frac{1 + \Gamma_L e^{-j2\beta\ell}}{1 - \Gamma_L e^{-j2\beta\ell}}. \tag{4.41}$$

From equations 4.39 through 4.41, the unknown amplitude V_0^+ is calculated as:

$$V_0^+ = V_g \frac{Z_0}{Z_g + Z_0} \frac{1}{1 - \Gamma_g \Gamma_L e^{-j2\beta\ell}}. \tag{4.42}$$

Therefore, the voltage and current on a lossless transmission line are given by:

$$V(z) = V_g \frac{Z_0}{Z_g + Z_0} \frac{e^{-j\beta z}[1 + \Gamma_L e^{j2\beta(z-\ell)}]}{1 - \Gamma_g \Gamma_L e^{-j2\beta\ell}}, \tag{4.43}$$

$$I(z) = V_g \frac{1}{Z_g + Z_0} \frac{e^{-j\beta z}[1 - \Gamma_L e^{j2\beta(z-\ell)}]}{1 - \Gamma_g \Gamma_L e^{-j2\beta\ell}}. \quad (4.44)$$

The voltage or current magnitude describes the standing wave nature of the field on a terminated transmission line. Standing waves result from reflections along the line, and they describe mismatch between load or source termination and the line. From equations 4.37 and 4.42, these equations are calculated:

$$\begin{aligned} |V(z)| &= \left|V_0^+\right| \left|1 + |\Gamma_L| e^{j[\theta_L + 2\beta(z-\ell)]}\right| \\ &= \left|V_0^+\right| \left[1 + |\Gamma_L|^2 + 2|\Gamma_L| \cos\theta\right]^{1/2}, \end{aligned} \quad (4.45)$$

and

$$\left|V_0^+\right| = \left|V_g\right| \left|\frac{Z_0}{Z_g + Z_0}\right| \left[1 + |\Gamma_g \Gamma_L|^2 - 2|\Gamma_g \Gamma_L| \cos\phi\right]^{-1/2}, \quad (4.46)$$

where $\theta = \theta_L \; 2\beta(z - \ell)$, $\phi = \theta_g + \theta_L - 2\beta\ell$, $\Gamma_L = |\Gamma_L|\angle\theta_L$, and $\Gamma_g = |\Gamma_g|\angle\theta_g$. The voltage magnitude in equation 4.45 depends on the position as $2\beta z$ and, thus, replicates every half wavelength. The phase can be calculated from equation 4.37 and shown to replicate at full wavelength spacing. The magnitude of the current is derived from equation 4.37 as:

$$\begin{aligned} |I(z)| &= \frac{|V_0^+|}{Z_0} \left|1 - |\Gamma_L| e^{[\theta_L + 2\beta(z-\ell)]}\right| \\ &= \frac{|V_0^+|}{Z_0} \left[1 + |\Gamma_L|^2 - 2|\Gamma_L| \cos\theta\right]^{1/2}. \end{aligned} \quad (4.47)$$

It can be observed from equations 4.45 and 4.47 that the voltage is maximum when the current is minimum and vice versa. The maximum voltage is given by:

$$|V|_{\max} = \left|V_0^+\right| (1 + |\Gamma_L|), \quad (4.48)$$

and occurs for $\cos\theta = 1$ in equation 4.45 or at:

$$z_{\max} = \ell + \frac{n\lambda}{2} - \frac{\theta_L \lambda}{4\pi}, n = 0, 1, \ldots N. \quad (4.49)$$

Likewise, the voltage minimum of:

$$|V|_{\min} = \left|V_0^+\right| (1 - |\Gamma_L|), \quad (4.50)$$

occurs for $\cos\theta = -1$ in (4.45) or at:

$$z_{\min} = z_{\max} \pm \frac{\lambda}{4}. \quad (4.51)$$

The ratio of maximum to minimum voltage (or current) is called the **standing wave ratio** (SWR) and is given by:

$$S = \frac{|V|_{\max}}{|V|_{\min}} = \frac{1 + |\Gamma_L|}{1 - |\Gamma_L|}. \quad (4.52)$$

This factor defines the mismatch along the line and can range from unity for a matched load to infinity for an ideal short or open termination. When the load termination is resistive, equation 4.52 yields:

$$S = \begin{cases} R_L/Z_0, & R_L > Z_0. \\ Z_0/R_L, & R_L < Z_0. \end{cases} \quad (4.53)$$

When $R_L > Z_0$ ($R_L < Z_0$), a voltage maximum (minimum) occurs at the load position.

Figure 4.10 displays the magnitude of the voltage and current for a terminated transmission line with $V_g = 10\angle 45° V$, $Z_0 = 50\Omega$, $Z_g = 75\Omega$, $Z_L = 100\Omega$, and $\ell = \lambda$. The voltage and current magnitudes replicate every half wavelength. The

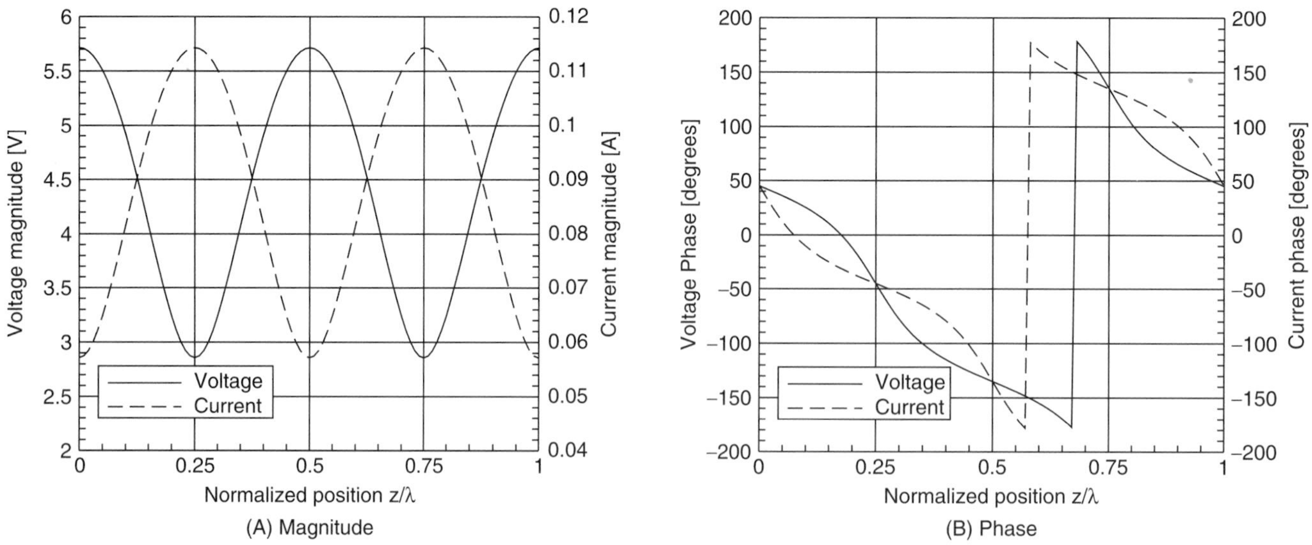

FIGURE 4.10 Voltage and Current on a Resistively Terminated Lossless Line

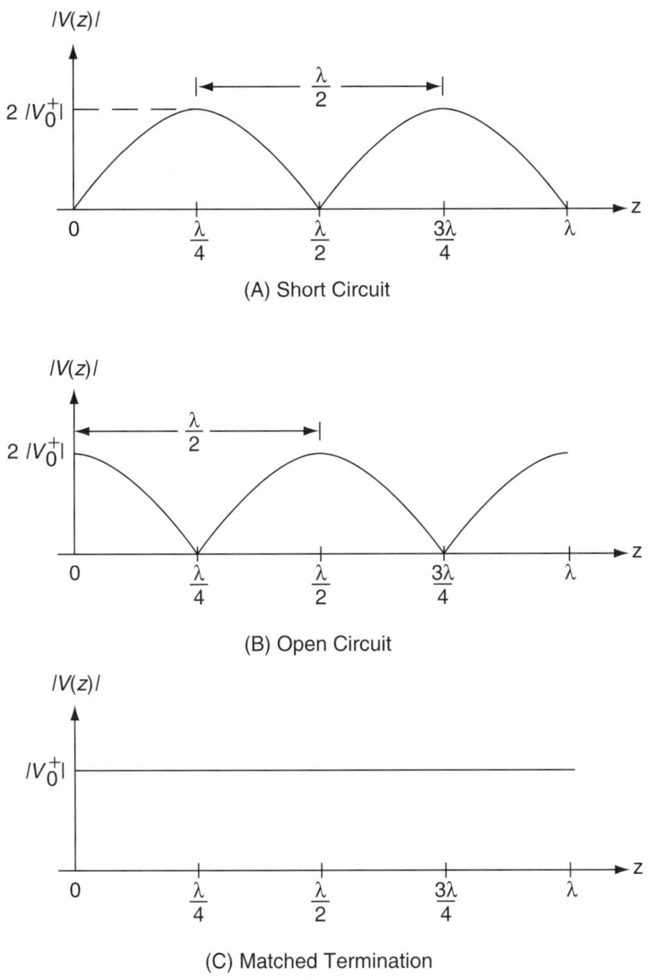

FIGURE 4.11 The Voltage Standing Wave Patterns on a Lossless Line

standing wave ratio is observed to be $5.7/2.85 = 2$, which agrees with equation 4.53. Likewise, Figure 4.11 shows the voltage standing wave patterns for a short circuit, an open circuit, and a matched termination, with $\ell = \lambda$. For the short circuit, equation 4.45 yields:

$$|V(z)| = 2\left|V_0^+\right|\left|\sin\left[\frac{2\pi}{\lambda}(z - \ell\right]\right|, \tag{4.54}$$

and for the open circuit, the pattern is as follows:

$$|V(z)| = 2\left|V_0^+\right|\left|\cos\left[\frac{2\pi}{\lambda}(z - \ell)\right]\right|. \tag{4.56}$$

These waves are known as **complete** standing waves. The minimum value is zero, the nulls are sharper than those in Figure 4.10 for the resistive load, and the SWR becomes infinite. For a matched load $Z_L = Z_0$, equation 4.52 shows that there are no standing waves ($S = 1$) because $\Gamma_L = 0$. In this case, the standing wave pattern looks flat as shown in Figure 4.11.

Power Flow

A transmission line is intended for power transfer between the generator and the load. The average power flow along a lossless line can be calculated using:

$$P_{av} = \frac{1}{2}\text{Re}[V(z)I(z)^*], \tag{4.57}$$

where the voltage and current are given by equations 4.37 and 4.38, respectively, and the asterisk denotes a complex conjugate. On a lossless line, since there is no dissipation, the average power is independent of position z. Therefore, the voltage and current in equation 4.57 can be calculated at any point along the line. Using the conditions at the load in equations 4.37 and 4.38, it follows that:

$$P_{av} = \frac{1}{2Z_0}|V_0^+|^2\left(1 - |\Gamma_L|^2\right). \tag{4.58}$$

The first term in equation 4.58 corresponds to the incident wave, and the second corresponds to the reflected wave. The magnitude $|V_0^+|$ can be computed using equation 4.46.

For a matched load, $\Gamma_L = 0$ ($Z_L = Z_0$); therefore, the power transferred to the load is just the first term in equation 4.58. This is not the maximum power transfer, however, for there is still a mismatch at the generator-end of the line because $Z_g \neq Z_{in} = Z_0$ ($\Gamma_g \neq 0$). The maximum power transfer on a lossless line occurs for a conjugate match at the generator-end, $Z_{in} = Z_g^*$, irrespective of the load impedance, and is given by:

$$P_{av} = \frac{|V_g|^2}{4R_g}. \tag{4.59}$$

4.3.2 Terminated Lossy Transmission Line

The voltage and current on a lossy transmission line are given by:

$$\begin{aligned} V(z) &= V_0^+ e^{-\alpha z} e^{-j\beta z} + V_0^- e^{\alpha z} e^{j\beta z} \\ &= V_0^+ e^{-\alpha z} e^{-j\beta z}[1 + \Gamma(z)]. \end{aligned} \tag{4.60}$$

$$\begin{aligned} I(z) &= \frac{1}{Z_0}\left[V_0^+ e^{-\alpha z} e^{-j\beta z} - V_0^- e^{\alpha z} e^{j\beta z}\right] \\ &= \frac{V_0^+ e^{-\alpha z} e^{-j\beta z}}{Z_0}[1 - \Gamma(z)]. \end{aligned} \tag{4.61}$$

The translation formula for the reflection coefficient follows from:

$$\Gamma(z) = \Gamma_L e^{2\alpha(z-\ell)} e^{j2\beta(z-\ell)} \tag{4.62}$$

The magnitude of the reflection coefficient is given by:

$$|\Gamma(z)| = |\Gamma_L| e^{2\alpha(z-\ell)}, \tag{4.63}$$

which is a function of both frequency and position, unlike the constant reflection coefficient magnitude:

$$|\Gamma(z)| = |\Gamma_L|. \tag{4.64}$$

Equation 4.64 is for a lossless line. The input impedance at $z = 0$ is as follows:

$$Z_{\text{in}}(0) = Z_0 \frac{Z_L + Z_0 \tanh \gamma\ell}{Z_0 + Z_L \tanh \gamma\ell}. \tag{4.65}$$

The impedance at any position z is obtained by replacing $\ell \rightarrow (\ell - z)$. It should be noted that the characteristic impedance Z_0 for a lossy line is complex. For practical low-loss lines, however Z_0 may be considered as real. Because of attenuation, it is observed from equations 4.63 and 4.65 that neither the reflection coefficient nor the input impedance exhibit the replication property. Also varying with position along the line is the standing wave ratio:

$$S = \frac{1 + |\Gamma_L| e^{2\alpha(z-\ell)}}{1 - |\Gamma_L| e^{2\alpha(z-\ell)}}. \tag{4.66}$$

The power flow is calculated from equation 4.57 using the voltage and current from equations 4.60 and 4.61, respectively. Denoting $\Gamma_L = |\Gamma_L| \angle \theta_L$ and $Z_0 = |Z_0| \angle \theta_0$, the following is obtained:

$$\begin{aligned} P_{av}(z) = {} & \frac{|V_0^+|^2}{2|Z_0|} e^{-2\alpha z} \cos\theta_0 \left[1 - |\Gamma_L|^2 e^{4\alpha(z-\ell)}\right] \\ & - \frac{|V_0^+|^2}{|Z_0|} |\Gamma_L| e^{-2\alpha\ell} \sin\theta_0 \sin[\theta_L + 2\beta(z-\ell)]. \end{aligned} \tag{4.67}$$

The first term in 4.67 computes the incident power, the second term corresponds to the reflected power, and the third term represents the interference between these two waves. Note that for a lossless line, since $\theta_0 = 0$, the interaction term will vanish, and the power transferred to the load reduces to the difference between incident and reflected contributions.

The power dissipated along the line can be computed by subtracting the power at the load, $P_{av}(\ell)$, from the power at the input, $P_{av}(0)$, and is given by:

$$\begin{aligned} P_{diss} = P_{av}(0) - P_{av}(\ell) = {} & \frac{|V_0^+|^2}{2|Z_0|} \cos\theta_0 \left[\left(1 - e^{-2\alpha\ell}\right)\right. \\ & \left. + |\Gamma_L|^2 e^{-2\alpha\ell}\left(1 - e^{-2\alpha\ell}\right)\right] \\ & + \frac{2|V_0^+|^2}{|Z_0|} |\Gamma_L| e^{-2\alpha\ell} \sin\theta_0 \sin(\beta\ell) \cos(\theta_L - \beta\ell). \end{aligned} \tag{4.68}$$

Again, the first term denotes the power loss in the incident wave, the second term denotes the power loss in the reflected wave, and the third term represents power loss due to interaction between the two. As α increases, the power loss in the incident and reflected waves as well as in the interaction term increases. In the limiting case of a lossless line, there should not be any dissipation. Clearly, the interaction term vanishes because $\theta_0 = 0$, and the power dissipation in the first two terms reduces to zero.

4.4 Smith Chart

The Smith chart is a graphical tool for determination of the reflection coefficient and impedance along a transmission line. It is an integral part of microwave circuit performance visualization, modern computer-aided design (CAD) tools, and RF/microwave test instrumentation. Basically, a Smith chart is a polar graph of normalized line impedance in the complex reflection coefficient plane. Let $Z = R + jX$ be the impedance at some location along a lossless line. The reflection coefficient is given by:

$$\Gamma = \frac{Z - Z_0}{Z + Z_0} = \Gamma_r + j\Gamma_i. \tag{4.69}$$

If the impedance Z is normalized with respect to Z_0 and $z \equiv Z/Z_0 = r + jx$ is written in terms of the reflection coefficient (z in this section should not be confused with the position variable z used elsewhere), the following equation is obtained:

$$r + jx = \frac{(1 + \Gamma_r) + j\Gamma_i}{(1 - \Gamma_r) - j\Gamma_i}. \tag{4.70}$$

The key to understanding the Smith chart is the realization that equation 4.70 corresponds to two families of circles given by:

$$r = \frac{1 - \Gamma_r^2 - \Gamma_i^2}{(1 - \Gamma_r)^2 + \Gamma_i^2}, \tag{4.71}$$

$$x = \frac{2\Gamma_i}{(1 - \Gamma_r)^2 + \Gamma_i^2}. \tag{4.72}$$

For a given r, 4.71 represents a circle in the complex Γ-plane that is centered at the point $[r/(1 + r), 0]$ and has a radius of $1/(1 + r)$. For a given x, equation 4.72 yields a circle centered at $[1, 1/x]$ with a radius $1/|x|$. A few of the former circles are shown as solid lines and the latter circles as dashed lines in Figure 4.12. The reactance is inductive in the upper half of the plane and capacitive in the lower half. Viewed in terms of impedance coordinates, the Smith chart displays the normalized impedance $z = r + jx$ at the intersection of an $r - x$ circle

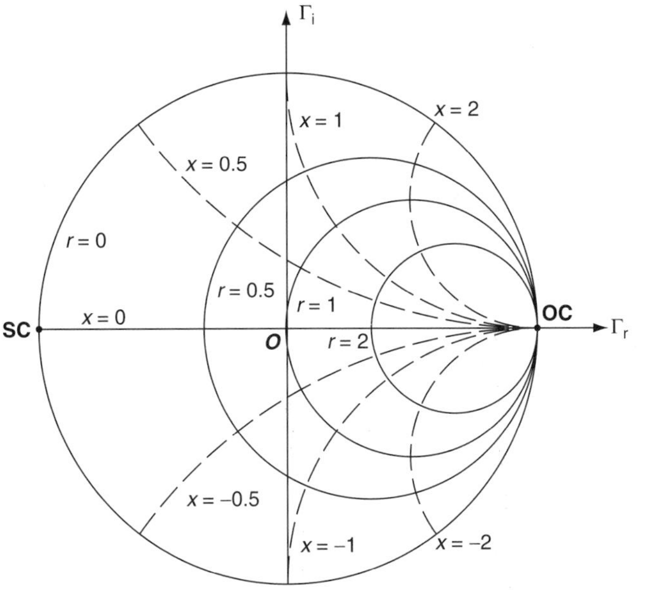

FIGURE 4.12 Basic Construction of the Smith Chart

pair. Because the underlying polar grid is a reflection coefficient plane, at any point on the chart the reflection coefficient can be read by measuring the radius and angle of the point (polar coordinates) on an appropriate scale. The outermost circle (OC) corresponds to the unity reflection coefficient. The open circuit lies at (1,0) in the reflection coefficient plane, and the short circuit (SC) is at (− 1, 0).

Figure 4.13 displays a typical commercially available Smith chart. The basic operations of the chart can be understood by an example to be discussed next. For better clarity and practice, it is highly recommended to follow this example by repeating the graphical solution taps on a new Smith chart. Suppose that a $(1/3)\lambda$-long, 50Ω line is connected to a load impedance of $(25 + j25)\ \Omega$. The input impedance and reflection coefficient can be determined by using the Smith chart. The load impedance is normalized, and the point $(0.5 + j0.5)$ is plotted and shown as point A in Figure 4.13. This point can be translated to the input by moving $(1/3)\lambda$ *toward the generator.* Two scales on the Smith chart's outer periphery indicate movement in wavelengths either **toward the generator** (clockwise) or **toward the load** (counterclockwise). One complete rotation around the

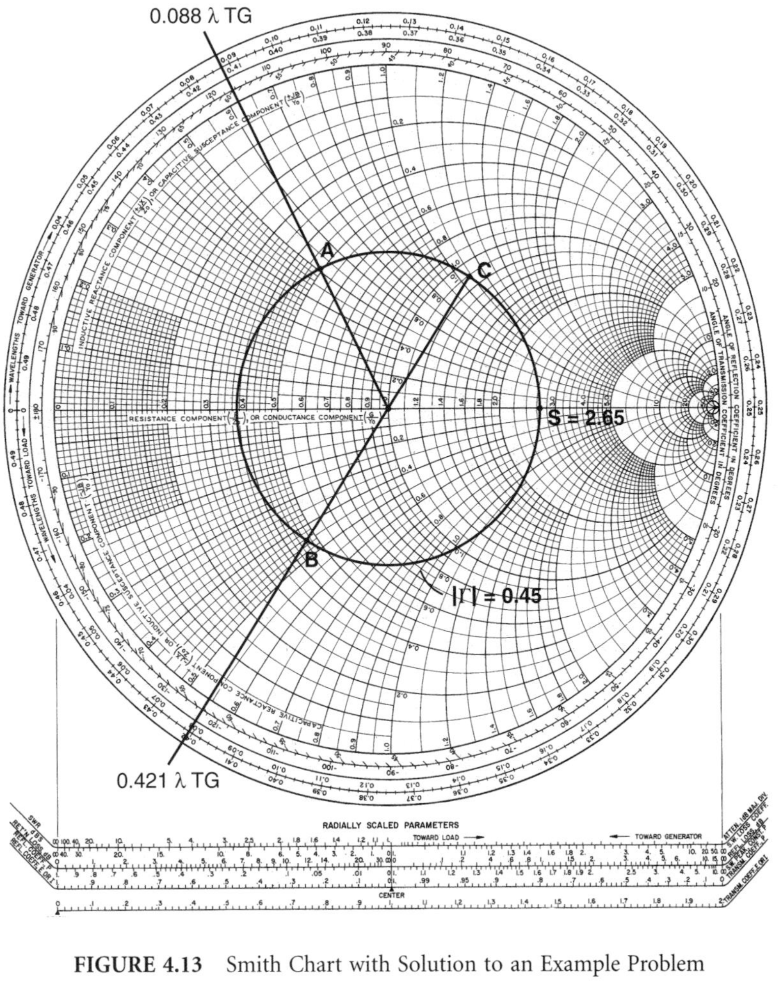

FIGURE 4.13 Smith Chart with Solution to an Example Problem

chart amounts to a half wavelength traversal. The relative position of a point on the transmission line can be determined by extending the radius of the point to intersect one of these two scales and reading its value in λ. Thus, point A reads 0.088λ toward generator (TG). If the radius of point A is projected onto the **reflection coefficient, E or I,** scale at the bottom of the chart, the measurement is $|\Gamma_L| = 0.45$. The angle of Γ_L is measured using the corresponding **angle of reflection coefficient** scale on the periphery of the unit circle as 116.5°. Therefore, $\Gamma_L = 0.45\angle 116.5°$. The input terminal is at a distance of 0.333λ TG from the load, located at 0.088λ. Thus, the position of the input terminal is at $0.088 + 0.333 = 0.421\lambda$ TG. Since $|\Gamma|$ is constant for a lossless transmission line, the reflection coefficient at the input, denoted by point B on the 0.421λ TG radial line, has the same radius as point A (= 0.45). Thus, we obtain the input reflection coefficient $\Gamma_{in} = 0.45\angle -123°$. For convenience, $|\Gamma| = 0.45$ circle is drawn in Figure 4.13 and allows the determination of the reflection coefficient at any point along the line. The normalized impedance at point B is read as $0.47 - j0.44$, which, after denormalization (multiplication by 50), yields $(23.5 - j22)\ \Omega$.

The standing wave ratio can be directly read from the scale at the bottom of the chart by projecting the radius of the reflection coefficient circle. It is instructive, however, to determine the point at which the normalized impedance equals SWR. From equation 4.48 there is a voltage maximum of:

$$|V|_{max} = |V_0^+|(1 + |\Gamma_L|), \tag{4.73}$$

which occurs at points along the line where the incident and reflected waves add *in* phase (**constructive interference**). Likewise, from equation 4.50 there is a voltage minimum of:

$$|V|_{min} = |V_0^+|(1 - |\Gamma_L|), \tag{4.74}$$

which occurs when the incident and reflected waves add *out of* phase (**destructive interference**). Because the current reflection coefficient is negative of the voltage reflection coefficient, at a voltage maximum location the current is a minimum and is given by:

$$|I|_{min} = \frac{|V_0^+|}{Z_0}(1 - |\Gamma_L|). \tag{4.75}$$

Dividing equation 4.75 into 4.73 and normalizing with respect to Z_0 results in the impedance:

$$z_{max} = \left|\frac{V_{max}}{I_{min}}\right| = \frac{1 + |\Gamma_L|}{1 - |\Gamma_L|} \stackrel{\Delta}{=} S. \tag{4.76}$$

Thus, at the position of a voltage maximum (or a current minimum), the normalized impedance is a maximum and equals S. Similarly, it can be shown that at the position of a voltage minimum (or a current maximum), the normalized impedance is a minimum and equals:

$$z_{min} = \left|\frac{V_{min}}{I_{max}}\right| = \frac{1 - |\Gamma_L|}{1 + |\Gamma_L|} \stackrel{\Delta}{=} \frac{1}{S}. \tag{4.77}$$

These two points can be easily obtained on the Smith chart as the two points where the real axis intersects the constant $|\Gamma|$ circle. From Figure 4.13, $S = 2.65$ at the intersection of the $|\Gamma|$ circle and the positive real axis. This agrees with the value read from the SWR scale at the bottom of the chart. In addition, the voltage maximum nearest to the load is at a distance of $0.25 - 0.088 = 0.162\lambda$ from the load.

The Smith chart can be used to calculate admittances, a feature very useful in designing impedance matching circuits. The normalized admittance occurs on the reflection coefficient circle, diametrically opposite to the normalized impedance. Corresponding to the r and x circles, respectively, the admittance coordinates are the constant conductance and constant susceptance circles. The upper half of the plane yields positive susceptances, and the lower half of the plane yields negative susceptances. The normalized input admittance for the example in Figure 4.13 is observed (at point C) as $(1.1 + j1.08)$. The net admittance follows from:

$$\mathbb{Y}_n = \frac{y}{Z_0} = \frac{1.1 + j1.08}{50}\ \text{(mhos)}. \tag{4.78}$$

The Smith chart is used in several applications, including microwave measurements, instrument displays, and computer-aided design of microwave circuits. For example, impedance matching networks such as single-stub and double-stub tuners or quarter-wave transformers can be conveniently designed using the Smith chart (see Cheng [1989] in Appendix A for further reading). For space limitations, these applications are not presented in this chapter. The reader is referred to Appendix A at the end of the chapter for several useful references illustrating microwave circuit design utilizing the Smith chart.

4.5 Summary

This chapter was devoted to the fundamentals of frequency-domain analysis of transmission lines with an emphasis on physical concepts rather than detailed mathematical derivations. A distributed circuit model was employed to derive the transmission line equations, whose solution describes the wave behavior of voltages and currents along the line. Initially, we considered an infinite line to emphasize the concepts of phase delay and spatial dependence, which account for the distributed nature of voltages and currents along a transmission line. The propagation characteristics of lossless and lossy lines were discussed, with the latter specialized to low-loss and distortionless cases. Properties of the standing waves along a terminated transmission line, such as impedance, reflection coefficient, voltage and current distribution, were discussed in detail. Mathematical expressions were derived for power flow, including power loss due to interference between incident and reflected waves on a lossy line and the condition of conjugate matching for maximum power transfer. Finally, the basic operations of the Smith chart to characterize standing waves on a terminated line were explained with an example.

Appendix A: References

The technical literature on transmission lines is pervasive with several articles scattered in books and journals, ranging from introductory material to advanced topics and covering both theoretical and practical aspects. The basic background material introduced in this chapter can be supplemented with further reading from the references listed below that are arranged by topics. A few books are devoted exclusively to transmission lines. Transmission line theory, however, is covered exhaustively in several electromagnetics textbooks. The list is not intended to be exhaustive.

Transmission Lines

Dworsky, L.N. (1979). *Modern transmission line theory and applications.* New York: John Wiley & Sons.

Freeman, J.C. (1996). *Fundamentals of microwave transmission lines.* New York: John Wiley & Sons.

Gardiol, F.E. (1987). *Lossy transmission lines.* Norwood, MA: Artech House.

Granzow, K.D. (1998). *Digital transmission lines: Computer modelling and analysis.* New York: Oxford University.

Johnson, W.C. (1950). *Transmission lines and networks.* New York: McGraw-Hill.

Magnusson, P.C., Alexander, G.C., and Tripathi, V.K. (1992). *Transmission lines and wave propagation.* Boca Raton, FL: CRC Press.

Matick, R.E. (1995). *Transmission lines for digital and communication networks.* New York: IEEE Press.

Smith, P.H. (1969). *Electronic applications of the smith chart in waveguide: Circuit and Component Analysis.* New York: McGraw-Hill.

Wadell, B.C. (1991). *Transmission line design handbook.* Norwood, MA: Artech House.

Introductory Electromagnetics

Cheng, D.K. (1989). *Field and wave electromagnetics.* Reading, MA: Addison-Wesley.

Hayt, W.H. Jr. (1989). *Engineering electromagnetics.* New York: McGraw-Hill.

Iskander, M.F. (1992). *Electromagnetic fields and waves.* Upper Saddle River, NJ: Prentice Hall.

Johnk, C.T.A. (1988). *Engineering electromagnetic fields and waves.* New York: John Wiley & Sons.

Paul, C.R., Whites, K.W., and Nasar, S.A. (1988). *Introduction to electromagnetic fields.* New York: McGraw-Hill.

Rao, N.N. (1987). *Elements of engineering electromagnetics.* Upper Saddle River, NJ: Prentice Hall.

Ulaby, F. (1999). *Fundamentals of applied electromagnetics.* Upper Saddle River, NJ: Prentice Hall.

Advanced Electromagnetics

Balanis, C.A. (1989). *Advanced engineering electromagnetics.* New York: John Wiley & Sons.

Harrington, R.F. (1961). *Time-harmonic electromagnetic fields.* New York: McGraw-Hill.

King, R.W.P., and Prasad, S. (1986). *Fundamental electromagnetic theory and applications.* Englewood Cliffs, NJ: Prentice Hall.

Kong, J.A. (1990). *Electromagnetic wave theory.* New York: John Wiley & Sons.

Kraus, J.D., and Fleisch, D.A. (1999). *Electromagnetics with applications.* New York: McGraw-Hill.

Wait, J.R. (1985). *Electromagnetic theory.* New York: Harper and Row.

Microwave Circuits

Collin, R.E. (1992). *Foundations for microwave engineering.* New York: McGraw-Hill.

Edwards, T. (1992). *Foundations for microstrip circuit design.* New York: John Wiley & Sons.

Gupta, K.C., Garg, R., and Chadha, R. (1981). *Computer-aided design of microwave circuits.* Norwood, MA: Artech House.

Howe, H. (1973). *Stripline circuit design.* Dedham, MA: Artech House.

Matthaei, G.L., Young, L., and Jones, E.M.T. (1965). *Microwave filters, impedance matching networks, and coupling structures.* New York: McGraw-Hill.

Pozar, D.M. (1998). *Microwave engineering.* New York: John Wiley & Sons.

Ramo, S., Whinnery, J.R., and van Duzer, T. (1994). *Fields and waves in communication electronics.* New York: John Wiley & Sons.

Sander, K.F., and Reed, G.A.L. (1978). *Transmission and propagation of electromagnetic waves.* Cambridge: Cambridge University Press.

5

Guided Waves

Franco De Flaviis
Department of Electrical and Computer Engineering, University of California, Irvine, California, USA

5.1 Definition of Guiding Structure or Waveguide

A **waveguide** can be defined as a geometrical structure capable of propagating electromagnetic energy in a preferred direction in space within a certain frequency range. Usually, this direction is referred to as the **direction of propagation**, and the frequency range is referred as the **waveguide operating bandwidth**. The lower operating frequency is determined by the electrical property of the guiding structure and does not necessarily start from zero (dc field) unless the guide is capable of supporting transverse electromagnetic modes (TEM) as in the case of a transmission line. It has been proven (Balanis, 1989) that to support a TEM, a structure must be also capable of supporting static field distribution (multiconductor topology), so structures made of a single conductor are not capable of supporting TEMs. The higher operating frequency is usually determined by the **single-mode guide operation**, which is the highest frequency at which the waveguide operates while supporting a single field distribution usually referred as the **fundamental mode**. The propagation of an electromagnetic wave inside a waveguide is quite different from the propagation of a TEM wave, which usually occurs in a traditional transmission line. This is because when a wave is introduced at one end of the waveguide, it is reflected from the sides of the waveguide whenever it hits them. These various reflected waves interact with each other to produce an infinite number of discrete characteristic patterns called **modes**. The existence of a discrete mode depends on the shape of and size of the waveguide, the medium in the waveguide, and the operating frequency. Unlike the TEM, which can be excited at any frequency, the transverse magnetic (TM) and transverse electric (TE) modes can only propagate when the wave frequency is higher than a certain frequency, called the **cutoff frequency**. The cutoff frequency is different for each mode. When the operating frequency of the wave is lower than the cutoff frequency of the lowest mode, the wave experiences exponential attenuation and disappears after traveling a very short distance. On the other hand, those modes with cutoff frequencies lower than the operating frequency can exist simultaneously inside the waveguide. To avoid existence of multiple modes, the waveguide is operated at a frequency that lies between the cutoff frequencies of the lowest and the next lowest modes. Guided waves can have different shapes and be made for different materials, such as metal or dielectrics or a combination of them. The shape and the type of waveguide used in a specific application depends on the desired performance, geometrical constraint, and cost.

5.2 Classification and Definitions

There are two major types of waveguides, namely of the metallic or dielectric type. **Metallic waveguides** are closed structures surrounded by metal and confine the electric field inside the guide. **Dielectric waveguides** are open structures where the field confinement is obtained from the dielectric constant discontinuities between the dielectric waveguiding structure and free space. In addition, a third category of waveguide can be identified under the name of **planar waveguide**, whose structure is a combination of metal and dielectric assembled in a planar geometry. Figure 5.1 gives a schematic diagram of the most common type of waveguide.

Several parameters are used to describe waveguide structures. The following list provides a description of some of them and their corresponding meanings.

- The **operating frequency** is the frequency range at which the guide operates for each field distribution, called **modes.** Some modes have overlapping frequency ranges for some geometrical dimensions of the waveguide. Such modes are called **degenerate modes.**

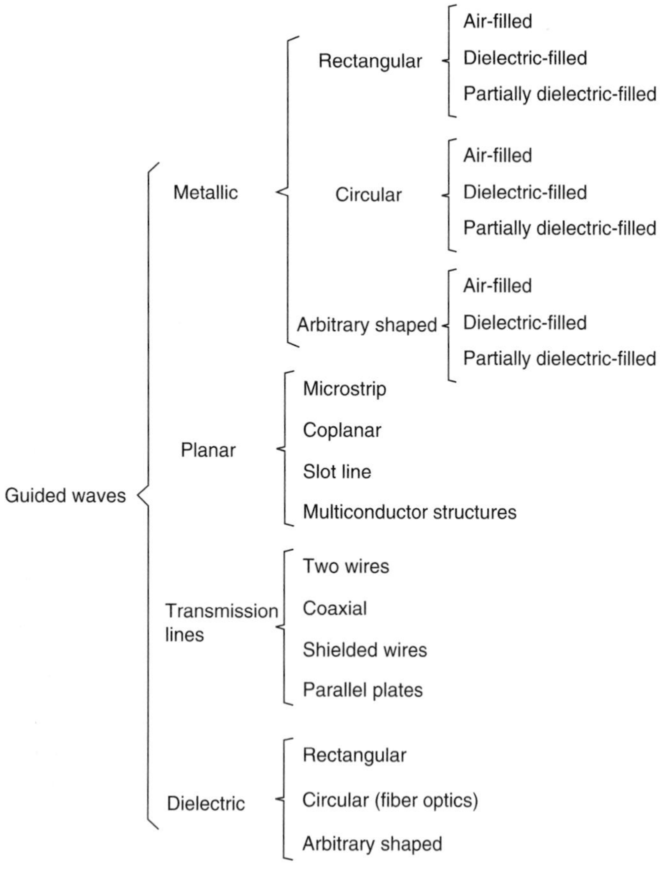

FIGURE 5.1 Schematic Classification of Waveguides

- The **operation mode** represents the field distribution of the guide and, more specifically, the type of field distribution. If both the electric and magnetic field components are zero in the direction of propagation, the mode is called **transverse electromagnetic** (TEM waves). If only the electric field is zero in the direction of propagation of the guide, then the mode is called **transverse electric** (TE wave); it is called **transverse magnetic** (TM) if the magnetic field is zero in the direction of propagation. In some cases, the field is a combination of the two (TE and TM). For such a case, the mode is referred to as a **hybrid mode.**
- The **propagation constant** (β) of a waveguide is defined as the rate of change (radians per second) of the phase angle of the signal and is directly related to the wavelength (λ) and its frequency (f):

$$\beta = \frac{2\pi}{\lambda} = \frac{2\pi f}{v_p}, \tag{5.1}$$

where v_p is the phase velocity.

- The **phase velocity** is the speed v_p, of the constant phase points on the wave as the wave travels and is usually represented by the symbol v_p and expressed in meters per second (MKS) units.
- The **group velocity** is the speed at which the energy propagates in the waveguide, usually referred to as v_g, and is expressed in meters per second.
- The **wavelength** is the distance in space (usually called λ_0 of free space or λ_g if inside the waveguide and expressed in meters) between two points having the same value in terms of field amplitude, phase, and correspondent derivatives.
- The **characteristic impedance** quantity is borrowed from transmission line theory and is also used for a waveguide to define the ratio between the transverse electric field and the corresponding transverse magnetic field. It is usually represented by the letter Z and is expressed in ohms. In case a structure supports more than one mode (for example TE and TM), two types of impedances corresponding to each mode (Z^{TE} and Z^{TM}) need to be identified.

5.3 Rectangular Waveguide

A rectangular waveguide is among the most commonly used waveguides. This is due to its manifacturing simplicity, low cost, and low-loss operation up to 100 GHz. Rectangular waveguides are also capable of handling very high-power microwave signals, and they are suitable for operations in outdoor environments. Because these waveguides are made of metals such as brass, copper, silver, or iron, they are heavy. Their

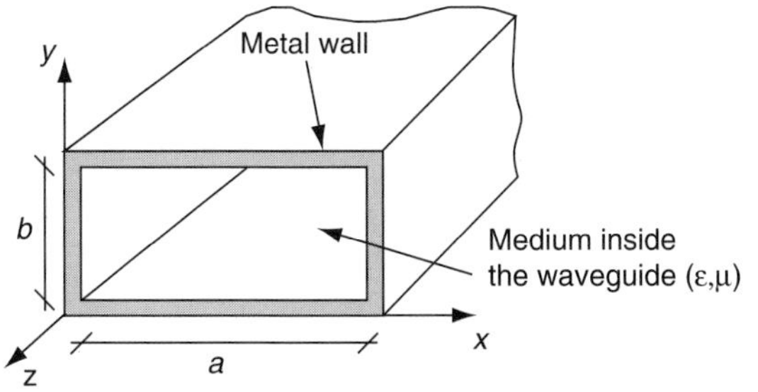

FIGURE 5.2 Schematic View of a Metallic Rectangular Waveguide

dimensions become large at low frequency (below X-band). A schematic view of a rectangular wave guide is shown in Figure 5.2.

5.3.1 Field Configuration

Because of its electrical topology (single conductor), a structure cannot support a static electric field, so it will not support TEM modes. The modes supported are TM. As mentioned earlier for these modes, either the electric or the magnetic field is zero along the direction of propagation (z direction in the figure). Starting from Maxwell's equations and imposing the proper boundary conditions for the field components on the four sides of the rectangular waveguide (Balanis, 1989), it is possible to derive the exact field expressions for all the field components in the guide. The reason there are more possible modes in a rectangular section waveguide ($TE_{mn}TM_{mn}$) is because imposing boundary conditions on the waveguide walls produce multiple solutions due to the periodicity of the trigonometric functions in which the fields are expressed:

TE_{mn}

$$E_x = A_{mn}\frac{\beta_v}{\varepsilon}\cos(\beta_x x)\sin(\beta_y y)e^{-j\beta_z z} \tag{5.2}$$

$$E_y = -A_{mn}\frac{\beta_x}{\varepsilon}\sin(\beta_x x)\cos(\beta_y y)e^{-j\beta_z z} \tag{5.3}$$

$$E_z = 0 \tag{5.4}$$

$$H_x = -A_{mn}\frac{\beta_x\beta_z}{\omega\mu\varepsilon}\sin(\beta_x x)\cos(\beta_y y)e^{-j\beta_z z} \tag{5.5}$$

$$H_y = -A_{mn}\frac{\beta_y\beta_z}{\omega\mu\varepsilon}\cos(\beta_x x)\sin(\beta_y y)e^{-j\beta_z z} \tag{5.6}$$

$$H_z = -A_{mn}\frac{\beta_c^2}{\omega\mu\varepsilon}\cos(\beta_x x)\cos(\beta_y y)e^{-j\beta_z z} \tag{5.7}$$

$$E_x = -B_{mn}\frac{\beta_x\beta_z}{\omega\varepsilon\mu}\cos(\beta_x x)\sin(\beta_y y)e^{-j\beta_z z} \tag{5.8}$$

$$E_y = -B_{mn}\frac{\beta_y\beta_z}{\omega\varepsilon\mu}\sin(\beta_x x)\cos(\beta_y y)e^{-j\beta_z z} \tag{5.9}$$

$$E_z = -jB_{mn}\frac{\beta_c^2}{\omega\varepsilon\mu}\sin(\beta_x x)\sin(\beta_y y)e^{-j\beta_z z} \tag{5.10}$$

$$H_x = B_{mn}\frac{\beta_y}{\mu}\sin(\beta_x x)\cos(\beta_y y)e^{-j\beta_z z} \tag{5.11}$$

$$H_y = -B_{mn}\frac{\beta_x}{\mu}\cos(\beta_x x)\sin(\beta_y y)e^{-j\beta_z z} \tag{5.12}$$

$$H_z = 0 \tag{5.13}$$

In all the equations listed under TE_{mn} and TM_{mn}, A_{mn} and B_{mn} are the constant coefficients proportional to the amplitude of the TE_{mn} and TM_{mn} wave, respectively. $\beta_i^s (i \equiv x,\ y,\ or\ z)$ are the propagation constants along the three Cartesian directions, and they satisfy the following relationship:

$$\beta_x^2 + \beta_y^2 + \beta_z^2 = \beta^2 = \omega^2\varepsilon\mu, \tag{5.14}$$

where β is the propagation constant of the signal in the particular mudium and is linked to the operating frequency ($\omega = 2\pi f$) and material property (ε and μ). The propagation constants along each direction ($\beta_{x/y/z}$) are linked to the guide dimension and operating frequency using the following equations:

$$\beta_x = \frac{m\pi}{a}. \quad m = 0, 1, 2, \ldots \tag{5.15}$$

$$\beta_y = \frac{n\pi}{b}. \quad n = 0, 1, 2, \ldots \tag{5.16}$$

$$\beta_z = \sqrt{\beta^2 - \beta_c^2} = \sqrt{\beta^2 - \beta_x^2 - \beta_y^2};\ \beta_c = \sqrt{\beta_x^2 - \beta_y^2}. \tag{5.17}$$

The constant β_c represents the value of the propagation, for which there is no propagation along the z direction ($\beta_z = 0$). It is also referred to as the **cutoff wave number**. The cutoff frequency from a given mode mn (either TE or TM) is given by:

$$f_{mn} = \frac{1}{2\pi\sqrt{\varepsilon\mu}}\sqrt{\left(\frac{m\pi}{a}\right)^2 + \left(\frac{n\pi}{b}\right)^2} = \frac{\beta_c}{2\pi\sqrt{\varepsilon\mu}}. \tag{5.18}$$

The phase velocity v_p of the wave along the direction of propagation is derived from equations 5.14 and 5.18 and is expressed as:

$$v_{p_{mn}} = \frac{\omega}{\beta_z} = \frac{1}{\sqrt{\varepsilon\mu}\sqrt{1 - (f_{mn}/f)^2}}. \tag{5.19}$$

Equation 5.19 provides the $\omega - \beta$ diagram for a waveguide operating with the TE_{mn} or TM_{mn} mode. The $\omega - \beta$ diagram is also known as a **dispersion diagram**. From equation 5.19, observe that when $f = f_c$, $v_p = \infty$ and $\beta_z = 0$. This is defined as the **cutoff frequency** of the waveguide (no propagation along the z direction). Also note that for a particular mode, the phase velocity of propagation along the guide varies with frequency. As a consequence of this characteristic of the guided wave propagation, the field patterns of different frequency components of a signal composing a band of frequencies do not maintain the same relationship as they propagate down the guide. This phenomenon is known as **dispersion**, so termed after the phenomenon of dispersion of colors by prism. The phase velocity is actually the speed of the constant phase points on the wave as it travels. So it does not represent the velocity of the propagating energy in the waveguide. The energy propagates at a speed equal to the group velocity (v_g) of the wave. The group velocity is related to the cutoff frequency as the following:

$$v_g = \frac{d\beta_z}{d\omega} = v_p\sqrt{1 - [(f_c)_{mn}/f]^2}. \tag{5.14}$$

Figure 5.3 shows the variations of the phase and group velocities versus the operating frequency.

From what has been illustrated so far, a signal of a given frequency can propagate in a rectangular waveguide in several modes for which the cutoff frequencies are less than the signal frequencies. Waveguides, are however, designed so that only one mode, the one with the lowest cutoff frequency, propagates. This is known as the **dominant mode**. The reason for using a single-mode waveguide is because in a multimode waveguide, the total energy is distributed among the propagating modes. For such a case, the instrumentation (e.g., detectors, exitation probes, etc.) required to detect the total power of multimode is more complex compared to instrumentation needed for single-mode operation.

From equation 5.18, the dominant mode is the $TE_{1,0}$ mode or the $TE_{0,1}$ mode depending on whether the dimension a or the dimension b is the larger of the two. By convention, the larger dimension is designed to be a, hence, the $TE_{1,0}$ mode is the dominant mode. For example, consider a rectangular waveguide with dimensions $a/b = 3.0$, $a/b = 2.0$, and $a/b = 1.0$, and find the propagation modes that occur in the guide. From equation 5.18, the cutoff frequency of the $TE_{1,0}$ case is the following:

$$[f_c]_{TE_{1,0}} = \frac{1}{2a\sqrt{\varepsilon\mu}}. \tag{5.15}$$

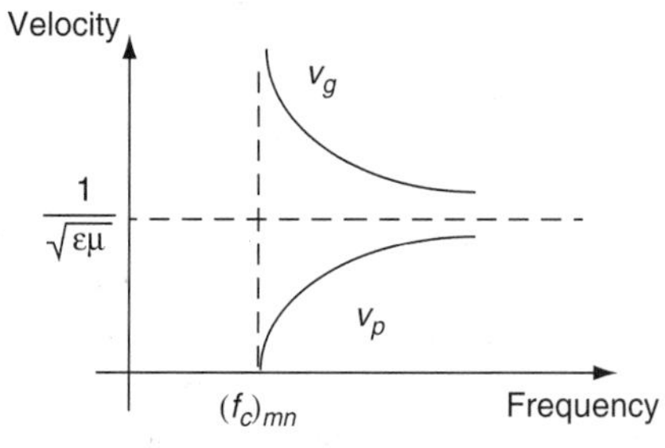

FIGURE 5.3 Variation of v_p and v_g with Frequency

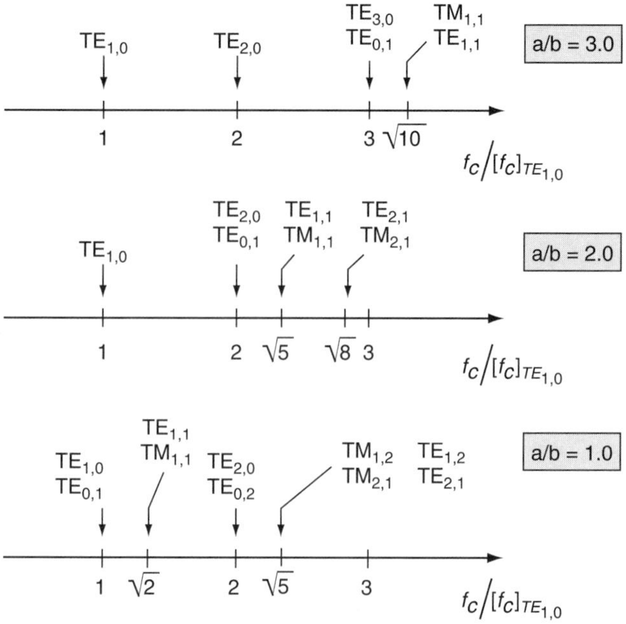

FIGURE 5.4 Cutoff Frequency of Different Modes in a Rectangular Waveguide for Different Ratios *a/b*.

Hence:

$$\frac{f_c}{[f_c]_{TE_{1,0}}} = \sqrt{m^2 + \left(n\frac{a}{b}\right)^2}. \tag{5.16}$$

By assigning different pairs of values for m and n, the lowest four values of $f_c/[f_c]_{TE_{1,0}}$ can be computed for different ratios of a/b as reported in Figure 5.4.

From Figure 5.4 the second lowest cutoff frequency, for $b/a \leq 1/2$, which corresponds to that of the $TE_{2,0}$ mode, is twice the cutoff frequency of the dominant mode $TE_{1,0}$. For this reason, the dimension b of a rectangular waveguide is generally chosen to be less than or equal to $a/2$ to achieve a single-mode transmission over a complete octave (factor 2:1) range of frequencies.

5.3.2 Power

The field inside a waveguide has power associated with each mode propagating in the guide. The total **power** propagating can be calculated using the **Poynting vector formulation** to find the power associated with each mode in the waveguide. The result is reported as (Balanis, 1989).

$$P_{m,n}^{TE} = |A_{m,n}|^2 \frac{\beta_c^2}{2\eta\varepsilon}\left(\frac{a}{\varepsilon_{0,m}}\right)\left(\frac{b}{\varepsilon_{0,n}}\right)\sqrt{1 - \left(\frac{f_{c,m,n}}{f}\right)^2}. \tag{5.17}$$

$$P_{m,n}^{TM} = |B_{m,n}|^2 \frac{\beta_c^2\eta}{2\mu^2}\left(\frac{a}{2}\right)\left(\frac{b}{2}\right)\sqrt{1 - \left(\frac{f_{c,m,n}}{f}\right)^2}. \tag{5.18}$$

In equations 5.17 and 5.18, η is the wave medium impedance defined as $\eta = \sqrt{\mu/\varepsilon}$. The total power associated with the wave is the sum of all the power components associated with each mode that exist inside the waveguide.

5.3.3 Attenuation

An important parameter in a rectangular section waveguide is the **attenuation per unit length**. The attenuation for propagating modes in a waveguide results from two types of losses: those due to the dielectric imperfection (dielectric loss α_d) and those due to the finite conductivity of the waveguide metal walls (conductor loss α_c). The calculation of those losses is performed considering the time average power lost in the conducting walls and in the dielectric per unit length. The resulting expressions for this calculation are reported, respectively, for TE and TM modes as (Collin, 1991):

$$(\alpha_c)_{TE_{m,n}} = \frac{2R_s}{b\eta\sqrt{1-(f_c/f)}}\left\{\left(1+\frac{b}{a}\right)\left(\frac{f_c}{f}\right)^2 + \left[1-\left(\frac{f_c}{f}\right)^2\right]\left[\frac{\frac{b}{a}\left(\frac{b}{a}m^2+n^2\right)}{\frac{b^2m^2}{a^2}+n^2}\right]\right\}(np/m). \tag{5.19}$$

$$(\alpha_c)_{TM_{m,n}} = \frac{2R_s}{b\eta\sqrt{1-(f_c/f)}}\frac{m^2(b/a)^3+n^2}{m^2(b/a)^2+n^2}(np/m). \tag{5.20}$$

$$\alpha_D = \frac{k\varepsilon''/\varepsilon'}{2\sqrt{1-(f_c/f)^2}}(np/m). \tag{5.21}$$

In equations 5.19 through 5.21, η is the medium impedance, k is the propagation constant for the lossy medium, and R_s the surface resistivity due to the skin effect of the walls. The R_s value is obtained from the metal conductivity (σ_m) and its permeability (μ_m) by:

$$R_s = \sqrt{\pi f \mu_m/\sigma_m} \tag{5.22}$$

Note that the dielectric loss is independent from the type of mode (TE or TM), but it depends on the frequency of the mode. A complete reference table for a commercially available rigid rectangular waveguide and its fundamental characteristics are reported in Table 5.1.

The first two columns of Table 5.1 represent the waveguide designation name and the corresponding designated band

TABLE 5.1 Reference Table for a Commercially Available Rigid Rectangular Waveguide

Designation: WR (_)	Designation (_) Band	Dimensions inside (in)	Dimensions outside (in)	Operating frequency $TE_{1,0}$ [Gigahertz]	Cutoff frequency $TE_{1,0}$ [Gigahertz]	Att. high to low frequency [decibels per 100 ft] ‡Al, *Cu, †Ag
2300	2300	23.00–11.50	23.25–11.75	0.32–0.49	0.256	0.051–0.031‡
2100	2100	21.00–10.50	21.25–10.75	0.35–0.53	0.281	0.054–0.034‡
1800	1800	18.00–9.00	18.25–9.25	0.41–0.625	0.328	0.056–0.038‡
1500	1500	15.00–7.50	15.25–7.75	0.49–0.75	0.393	0.069–0.050‡
1150	1150	11.50–5.75	11.75–6.00	0.64–0.96	0.513	0.128–0.075‡
975	975	9.75–4.875	10.00–5.125	0.75–1.12	0.605	0.137–0.095‡
770	770	7.70–3.85	7.95–4.10	0.96–1.45	0.766	0.201–0.136‡
650	L	6.50–3.25	6.66–3.41	1.12–1.70	0.908	0.317–0.212*
510	510	5.10–2.55	5.26–2.71	1.45–2.20	1.157	
430	W	4.30–2.15	4.46–2.31	1.70–2.60	1.372	0.588–0.385*
340	340	3.40–1.70	3.56–1.86	2.20–3.30	1.736	0.877–0.572*
284	S	2.84–1.34	3.00–1.50	2.60–3.95	2.078	1.102–0.752*
229	229	2.29–1.145	2.418–1.273	3.30–4.90	2.577	
187	C	1.872–0.872	2.00–1.00	3.95–5.85	3.152	2.08–1.44*
159	159	1.59–0.795	1.718–0.923	4.90–7.05	3.711	
137	X_B	1.372–0.622	1.50–0.75	5.85–8.20	4.301	2.87–2.30*
112	X_L	1.122–0.497	1.25–0.625	7.05–10.00	5.259	4.12–3.21*
90	X	0.90–0.40	1.00–0.50	8.20–12.40	6.557	6.45–4.48*
75	75	0.75–0.375	0.85–0.475	10.00–15.00	7.868	
62	K_u	0.622–0.311	0.702–0.391	12.40–18.00	9.486	9.51–8.31*
51	51		0.59–0.335	15.00–22.00	11.574	
42	K	0.42–0.17	0.50–0.25	18.00–26.50	14.047	20.7–14.8*
34	34	0.34–0.17	0.42–0.25	22.00–33.00	17.328	
28	K_A	0.28–0.14	0.36–0.22	26.50–40.00	21.081	21.90–15.00†
22	Q	0.224–0.112	0.304–0.192	33.00–50.00	26.342	31.0–20.90†
19	19	0.188–0.094	0.268–0.174	40.00–60.00	31.357	
15	V	0.148–0.074	0.228–0.154	50.00–75.00	39.863	52.9–39.1†
12	12	0.122–0.061	0.202–0.141	60.00–90.00	48.35	93.3–52.2†
10	10	0.10–0.05	0.18–0.13	75.00–110.00	59.01	

symbol. The third and fourth columns are the waveguide physical dimensions, and the fifth and sixth columns are the operating frequency and the cutoff frequency, respectively. The last column gives the loss per unit length for different plating materials in the waveguide internal walls, such as aluminum (Al), silver (Ag), and copper (Cu).

5.4 Partially Filled Metallic Rectangular Waveguide

A **dielectric-filled metallic waveguide** has the same property as an air-filled metallic waveguide with the only exception: the cutoff frequency for the modes is lower due to the presence of the dielectric ($\varepsilon > 1$ or $\mu > 1$) as seen from equation 5.18. A schematic view of this type of guide is given in Figure 5.5.

Partially filled waveguides are a classic type of waveguide with a dielectric slab placed on one wall of the guide, either the broad wall or the narrow one, as illustrated in Figure 5.6.

For this type of structure, the field distribution is a combination of TE and TM modes. Those modes are referred to as hybrid modes or longitudinal section electric (LSE) and longitudinal section magnetic (LSM) modes (Balanis, 1989). Using perturbation technique shows that, in general, the cutoff frequency of the partially filled waveguide is between the same air-filled waveguide and the one that is fully loaded:

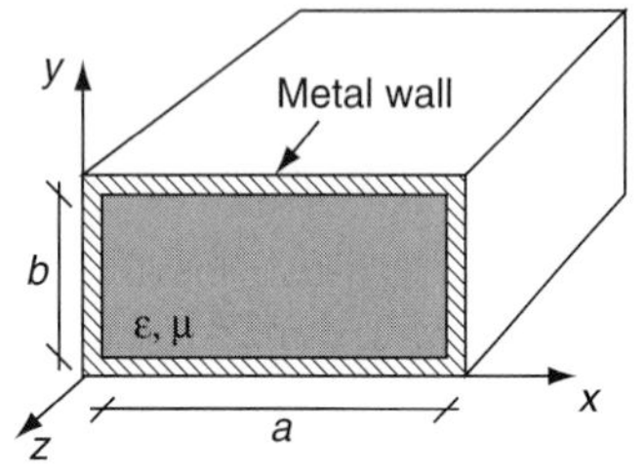

FIGURE 5.5 Cross Section of Dielectric-Filled Metallic Rectangular Waveguide

$$\frac{1}{2b\sqrt{\mu\varepsilon}} \leq (f_c)_{01}^{TE} \leq \frac{1}{2b\sqrt{\mu_0\varepsilon_0}}. \quad (5.23)$$

$$\frac{1}{2a\sqrt{\mu\varepsilon}} \leq (f_c)_{01}^{TM} \leq \frac{1}{2a\sqrt{\mu_0\varepsilon_0}}. \quad (5.24)$$

5.5 Circular Metal Waveguide

The **circular waveguide** is also very popular due to its simplicity in manufacturing and low attenuation of the fundamental mode. The circular waveguide is made of a cylindrical structure that maintains a uniform cross section along the length as illustrated in Figure 5.7. As in the case of a rectangular waveguide, the inside of the waveguide is either filled with air or with dielectric material.

The field configurations (modes) that can be supported by such a structure are TE and TM modes along the longitudinal direction (z-axis in the figure). Starting from Maxwell's equations after proper expansion in cylindrical coordinates and imposing the boundary conditions, it is possible to obtain the propagation constant and the field distribution for each TE mode and TM mode in the waveguide. The propagation constant for the TE modes can be written as a function of the roots (χ'_{mn}) of the derivative of the Bessel function of the first kind (Balanis, 1989). The result is the following equation:

$$\beta_{mn} = \begin{cases} \sqrt{\beta^2 - \beta_\rho^2}, & \beta > \beta_\rho \\ 0, & \beta = \beta_\rho = \beta_c, \\ -j\sqrt{\beta_\rho^2 - \beta^2}, & \beta < \beta_\rho \end{cases} \quad (5.25)$$

where the following is true:

$$\beta_\rho = \frac{\chi'_{mn}}{a}. \quad (5.26)$$

In equation 5.25, χ'_{mn} is the mth root of the derivative of the nth order Bessel function, and β_c is called the **cutoff**

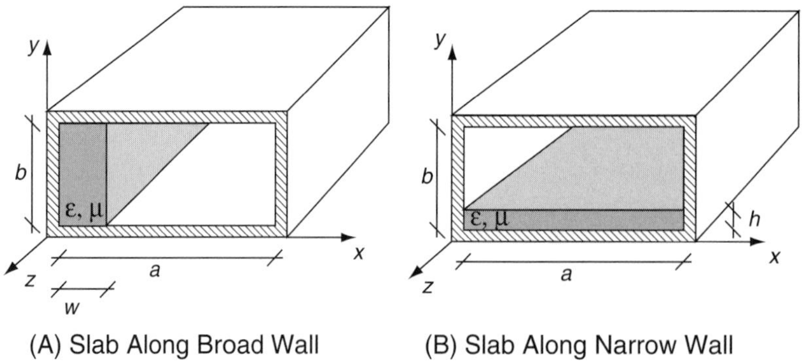

FIGURE 5.6 Schematic View of Dielectric Loaded Waveguide

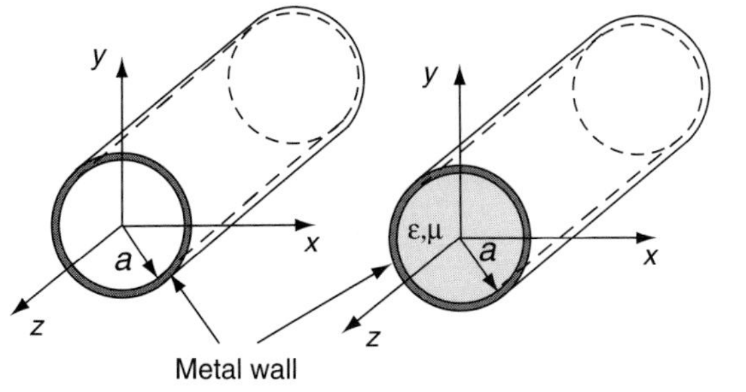

FIGURE 5.7 Circular Waveguide Structure

propagation constant of the *mn* mode. Since $\beta_c = \omega_c\sqrt{\varepsilon\mu}$, the result for the TE_{mn} mode is as written here:

$$(f_c)_{mn} = \frac{\chi'_{mn}}{2\pi a\sqrt{\varepsilon\mu}}. \quad (5.27)$$

The corresponding guide wavelength is defined as $(\lambda_g)_{mn} = \frac{2\pi}{(\beta_z)_{mn}}$ and is given by:

$$(\lambda_g)_{mn} = \begin{cases} \frac{2\pi}{\beta\sqrt{1-\left(\frac{f_c}{f}\right)^2}} & \text{for } f > (f_c)_{mn}. \\ \infty & \text{for } f = (f_c)_{mn}. \end{cases} \quad (5.28)$$

In equation 5.28, λ_g is the wavelength of the wave in an infinite medium having the same electrical property as the one inside the waveguide. Below the cutoff, there is no expression for the wavelength because the wave decays exponentially. The values of χ'_{mn} for the first five values of m and n are reported in Table 5.2.

The electric and magnetic field components for the TE_{mn} can be expressed in cylindrical coordinate as:

$$E_\rho = -A_{mn}\frac{m}{\varepsilon\rho}J_m(\beta_\rho\rho)[-C_2\sin(m\phi) + D_2\cos(m\phi)]e^{-j\beta_z z}. \quad (5.29)$$

$$E_\phi = A_{mn}\frac{\beta_\rho}{\varepsilon}J'_m(\beta_\rho\rho)[C_2\cos(m\phi) + D_2\sin(m\phi)]e^{-j\beta_z z}. \quad (5.30)$$

$$E_Z = 0. \quad (5.31)$$

$$H_\rho = -A_{mn}\frac{\beta_\rho\beta_z}{\omega\varepsilon\mu}J'_m(\beta_\rho\rho)[C_2\cos(m\phi) + D_2\sin(m\phi)]e^{-j\beta_z z}. \quad (5.32)$$

$$H_\phi = -A_{mn}\frac{m\beta_z}{\omega\mu\varepsilon}\frac{1}{\rho}J'_m(\beta_\rho\rho)[C_2\sin(m\phi) + D_2\cos(m\phi)]e^{-j\beta_z z}. \quad (5.33)$$

$$H_z = -jA_{mn}\frac{\beta_\rho^2}{\omega\varepsilon\mu}J_m(\beta_\rho\rho)[C_2\cos(m\phi) + D_2\sin(m\phi)]e^{-j\beta_z z}. \quad (5.34)$$

According to the values of χ'_{mn} in Table 5.2, the order in which the TE modes propagate along the z direction is TE_{11}, TE_{21}, TE_{01}, and so forth. The bandwidth of the first single-mode TE_{11} operation is 1.6588:1 (less than 2:1 obtained with rectangular waveguide obtained for $a/b \geq 2$).

The expression of the propagation constant for the TM modes can be written as function of the derivative (χ_{mn}), the Bessel function of the first kind, as:

$$\beta_{mn} = \begin{cases} \sqrt{\beta^2 - \beta_\rho^2}, & \beta > \beta_\rho \\ 0, & \beta = \beta_\rho = \beta_c, \\ -j\sqrt{\beta_\rho^2 - \beta^2}, & \beta < \beta_\rho \end{cases} \quad (5.35)$$

where the following is true:

$$\beta_\rho = \frac{\chi_{mn}}{a}. \quad (5.36)$$

The β_c is the cutoff propagation constant of the *mn* mode. Since $\beta_c = \omega_c\sqrt{\varepsilon\mu}$, the result for the TE_{mn} mode is as written here:

$$(f_c)_{mn} = \frac{\chi_{mn}}{2\pi a\sqrt{\varepsilon\mu}}. \quad (5.37)$$

The corresponding guide wavelength is defined as $(\lambda_g)_{mn} = 2\pi/(\beta_2)_{mn}$ and is given by:

$$(\lambda_g)_{mn} = \begin{cases} \frac{2\pi}{\beta\sqrt{1-\left(\frac{f_c}{f}\right)^2}} & \text{for } f > (f_c)_{mn}. \\ \infty & \text{for } f = (f_c)_{mn}. \end{cases} \quad (5.40)$$

In equation 5.40, as in the case of TE waves, λ_g is the wavelength of the wave in an infinite medium having the same electrical property as the one inside the waveguide. The values of χ_{mn} for the first five values of m and n are reported in Table 5.3.

The electric and magnetic field components for the TM_{mn} modes can be expressed in cylindrical coordinates as:

TABLE 5.2 χ'_{mn} for the First Five values of m and n

	$m=0$	$m=1$	$m=2$	$m=3$	$m=4$	$m=5$
$n=1$	3.8318	1.8412	3.0542	4.2012	5.3175	6.4155
$n=2$	7.0156	5.3315	6.7062	8.0153	9.2824	10.5199
$n=3$	10.1735	8.5363	9.9695	11.3459	12.6819	13.9872
$n=4$	13.3237	11.7060	13.1704	14.5859	15.9641	17.3129
$n=5$	16.4706	14.8636	16.3475	17.7888	19.1960	20.5755

TABLE 5.3 χ_{mn} for the First Five Values of m and n

	$m=0$	$m=1$	$m=2$	$m=3$	$m=4$	$m=5$
$n=1$	2.409	3.8318	5.1357	6.3802	7.5884	8.7715
$n=2$	5.5201	7.0156	8.4173	9.7610	11.0647	12.3386
$n=3$	8.6537	10.1735	11.6199	13.0152	14.3726	15.7002
$n=4$	11.7915	13.3237	14.7960	16.2235	17.6160	18.9801
$n=5$	14.9309	16.4706	17.9598	19.4094	20.8269	22.2178

$$E_\rho = -B_{mn}\frac{\beta_\rho\beta_z m}{\omega\varepsilon\mu}J_m(\beta_\rho\rho)[C_2\cos(m\phi) + D_2\sin(m\phi)]e^{-j\beta_z z}. \quad (5.41)$$

$$E_\phi = B_{mn}\frac{m\beta_z}{\omega\mu\varepsilon\rho}J_m(\beta_\rho\rho)[-C_2\sin(m\phi) + D_2\cos(m\phi)]e^{-j\beta_z z}. \quad (5.42)$$

$$E_z = -jB_{mn}\frac{\beta_\rho^2}{\omega\varepsilon\mu}J_m(\beta_\rho\rho)[C_2\cos(m\phi) + D_2\sin(m\phi)]e^{-j\beta_z z}. \quad (5.43)$$

$$H_\rho = -B_{mn}\frac{m}{\mu}\frac{1}{\rho}J_m(\beta_\rho\rho)[-C_2\sin(m\phi) + D_2\cos(m\phi)]e^{-j\beta_z z}. \quad (5.44)$$

$$H_\phi = -B_{mn}\frac{\beta_\rho}{\mu}J_m(\beta_\rho\rho)[C_2\cos(m\phi) + D_2\sin(m\phi)]e^{-j\beta_z z}. \quad (5.45)$$

$$H_z = 0. \quad (5.46)$$

According to the values of χ_{mn} of Table 5.3, the order in which the TM modes propagate along the z direction is TM_{01}, TM_{11}, TM_{21}, and so forth. The bandwidth of the first single-mode TM_{01} operation is 1.5933:1 (less than 2:1 of rectangular waveguide). From the results in the Tables 5.2 and 5.3, the order of the modes for circular waveguide is associated with the values of χ_{mn} and χ'_{mn} and the sequence of modes is as follows: TE_{11}, TM_{01}, TE_{21}, TE_{10}, TM_{11}, TE_{31}, and so forth. Because the zero of the Bessel function of the first kind is the same as the derivative of the Bessel function of order zero, the TE_{0n} and TM_{0n} modes have same cutoff frequency and are *degenerate*.

5.6 Microstrip Line

A **microstrip line** is a guided wave structure consisting of a strip conductor and a ground plane separated by a dielectric substrate as shown in Figure 5.8.

The electromagnetic field in the microstrip is not contained entirely in the substrate. Therefore, the propagating mode in the microstrip is not a pure TEM but a hybrid one. Assuming a quasi-TEM mode (valid for lower frequencies) of propagation in the microstrip line, the phase velocity is given by:

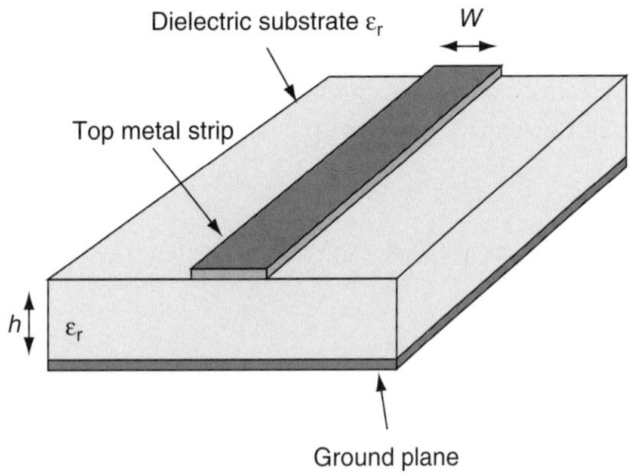

FIGURE 5.8 Microstrip Line Geometry

$$v_p = \frac{c}{\sqrt{\varepsilon_{ff}}}, \quad (5.47)$$

where c is the speed of light (3×10^8 m/sec), and ε_{ff} is the effective relative dielectric constant of the dielectric substrate.

5.6.1 Characteristic Impedance and Wavelength

The effective relative dielectric constant of the microstrip is related to the relative dielectric constant of the dielectric substrate, and the constant takes into account the effect of the external electromagnetic fields. The **characteristic impedance** of the microstrip line is given by:

$$Z_0 = \frac{1}{v_p C}, \quad (5.48)$$

where C is the capacitance per unit length of the microstrip. The **wavelength** in the microstrip is given by:

$$\lambda = \frac{v_p}{f} = \frac{c}{f\sqrt{\varepsilon_{ff}}} = \frac{\lambda_0}{\sqrt{\varepsilon_{ff}}}, \quad (5.49)$$

where λ_0 is the free space wavelength ($\lambda_0 = c/f$). There are different methods for determining ε_{ff} and C. Of course, closed-form expressions are of great importance in microstrip line design. The evaluation of ε_{ff} and C based on a quasi-TEM approximation is accurate for design at lower microwave frequencies. At higher microwave frequencies, however, the longitudinal components of the electromagnetic fields are significant, and the quasi-TEM assumption is no longer valid. A useful set of relations for the characteristic impedance, assuming zero or negligible thickness of the strip conductor (i.e., $t/h < 0.005$), is provided here (Bahl, 1977):

$$\begin{cases} Z_0 = \frac{60}{\sqrt{\varepsilon_{ff}}}\ln\left(8\frac{h}{W} + 0.25\frac{W}{h}\right) & \text{for } W/h \le 1.0. \\ Z_0 = \frac{120\pi/\sqrt{\varepsilon_{ff}}}{W/h + 1.393 + 0.667\ln(W/h + 1.444)} & \text{for } W/h \ge 1.0. \end{cases} \quad (5.50)$$

For equations 5.50:

$$\begin{cases} \varepsilon_{ff} = \frac{\varepsilon_r+1}{2} + \frac{\varepsilon_r-1}{2}\left[\left(1 + 12\frac{h}{W}\right)^{-1/2} + 0.04\left(1 - \frac{W}{h}\right)^2\right] & \text{for } W/h \le 1.0. \\ \varepsilon_{ff} = \frac{\varepsilon_r+1}{2} + \frac{\varepsilon_r-1}{2}\left[\left(1 + 12\frac{h}{W}\right)^{-1/2}\right] & \text{for } W/h \ge 1.0. \end{cases} \quad (5.51)$$

Based on the results for equations 5.49 through 5.51 and/or in experimental data, the wavelength in the microstrip line, assuming zero or negligible thickness (i.e., $t/h \le 0.005$) for the strip conductor, is given by the relations (Sobol, 1967):

$$\begin{cases} \lambda = \frac{\lambda_0}{\sqrt{\varepsilon_r}}\left[\frac{\varepsilon_r}{1 + 0.63(\varepsilon_r - 1)(W/h)^{0.1255}}\right]^{1/2} & \text{for } W/h \ge 0.6. \\ \lambda = \frac{\lambda_0}{\sqrt{\varepsilon_r}}\left[\frac{\varepsilon_r}{1 + 0.63(\varepsilon_r - 1)(W/h)^{0.1255}}\right]^{1/2} & \text{for } W/h \le 0.6. \end{cases} \quad (5.52)$$

For design purposes, a set of equations relating Z_0 and ε_r to the ratio *W/h* of the microstrip line is desirable. Assuming zero or negligible thickness of the strip conductor (i.e., $t/h < 0.005$), the expressions are as follows (Sobol, 1971):

$$\begin{cases} \frac{W}{h} = \frac{8e^A}{e^{2A}-2} & \text{for } W/h \leq 2.0. \\ \frac{W}{h} = \frac{2}{\pi}\left\{B - 1 - \ln(2B-1) + \frac{\varepsilon_r - 1}{2\varepsilon_r}\right. & \\ \left.\left[\ln(B-1) + 0.39 - \frac{0.61}{\varepsilon_r}\right]\right\} & \text{for } W/h \geq 2.0. \end{cases} \tag{5.53}$$

For equations 5.53:

$$A = \frac{Z_0}{60}\sqrt{\frac{\varepsilon_r+1}{2}} + \frac{\varepsilon_r - 1}{\varepsilon_r + 1}\left(0.23 + \frac{0.11}{\varepsilon_r}\right), \tag{5.54}$$

and

$$B = \frac{377\pi}{2Z_0\sqrt{\varepsilon_r}}. \tag{5.55}$$

The zero or negligible thickness formulas given in equations 5.50 to 5.55 can be modified to include the thickness of the strip conductor. The first-order effect of a strip conductor of finite thickness t is to increase the capacitance. Therefore, an approximate correction is made by replacing the strip width W by the effective width W_{eff}. The following relations for W_{eff}/h are useful when $t < h$ and $t < W/2$:

$$\begin{cases} \frac{W_{eff}}{h} = \frac{W}{h} + \frac{t}{\pi h}\left(1 + \ln\frac{2h}{t}\right) & \text{for } W/h \geq 0.5\pi. \\ \frac{W_{eff}}{h} = \frac{W}{h} + \frac{t}{\pi h}\left(1 + \ln\frac{4\pi W}{t}\right) & \text{for } W/h \leq 0.5\pi. \end{cases} \tag{5.56}$$

The restrictions $t < h$ and $t < W/2$ are usually satisfied because for dielectric substrates, a typical thickness is $t = 0.002$ in. The formulas presented thus far are valid at frequencies where the quasi-TEM assumption can be made. When the quasi-TEM assumption is not valid, ε_{ff} and Z_0 are functions of frequency and, therefore, the microstrip line becomes dispersive. The phase velocity of the microstrip line decreases with increasing frequency. Therefore, $\varepsilon_{ff}(f)$ increases with frequency. Moreover, the characteristic impedance of the microstrip line increases with frequency, and it follows that the effective width $W_{eff}(f)$ decreases. The frequency in equation 5.57, in which dispersion may be neglected, is given by:

$$f_0(GH_z) = 0.3\sqrt{\frac{Z_0}{h\sqrt{\varepsilon_r - 1}}}, \tag{5.57}$$

where h must be expressed in centimeters. An analytical expression that shows the effect of dispersion in $\varepsilon_{ff}(f)$ is the following (Bahl and Trivedi, 1977):

$$\varepsilon_{ff}(f) = \varepsilon_r - \frac{\varepsilon_r - \varepsilon_{ff}}{1 + G(f/f_p)^2} \quad [f \text{ in gigahertz}], \tag{5.58}$$

where:

$$f_p = \frac{Z_0}{8\pi h} \quad [h \text{ in centimeters}], \tag{5.59}$$

and

$$G = 0.6 + 0.009Z_0. \tag{5.60}$$

Observe that when $f_p \gg f$, then $\varepsilon_{ff}(f) = \varepsilon_{ff}$. In other words, high-impedance lines on thin substrates are less dispersive. The expression for the dispersion in Z_0 is (Bahl and Trivedi, 1977):

$$Z_0(f) = \frac{377h}{W_{eff}(f)\sqrt{\varepsilon_{ff}}}, \tag{5.61}$$

where the following is true:

$$W_{eff}(f) = W + \frac{W_{eff}(0) - W}{1 + (f/f_p)^2}, \tag{5.62}$$

and

$$W_{eff}(0) = \frac{377h}{Z_0(0)\sqrt{\varepsilon_{ff}(0)}}. \tag{5.63}$$

5.6.2 Attenuation

The **attenuation constant** is a function of the microstrip geometry, the electrical properties of the dielectric substrate and the conductors, and frequency. There are two types of losses in a microstrip line: a dielectric substrate loss and than ohmic skin loss in the conductors. The losses can be expressed as a loss per unit length along the microstrip line in terms of the attenuation factor α. The power carried by a wave traveling in the positive direction in a quasi-TEM mode is given by:

$$P^+(z) = \frac{1}{2}(V^+e^{-\alpha z}I^+e^{-\alpha z}) = \frac{1}{2}\frac{|V^+|^2}{Z_0}e^{-2\alpha z} = P_0e^{-2\alpha z}, \tag{5.64}$$

where $P_0 = |V^+|^2/2Z_0$ is the power at $z = 0$. Then, from equation 5.64:

$$\alpha = \frac{-dP(z)/dz}{2P(z)} = \alpha_d + \alpha_c, \tag{5.65}$$

where α_d is the dielectric loss factor and where α_c the conduction loss factor. A useful set of expressions for calculating α_d (Bahl and Trivedi, 1977) for a dielectric with low losses is written as:

$$\alpha_d = 27.3\frac{\varepsilon_r}{\sqrt{\varepsilon_{ff}}}\frac{\varepsilon_{ff} - 1}{\varepsilon_r - 1}\frac{\tan\delta}{\lambda_0} \quad [\text{decibels/centimeter}], \tag{5.66}$$

where the loss tangent δ is given by:

$$\tan\delta = \frac{\sigma}{\omega\varepsilon}. \tag{5.67}$$

For a dielectric with $\sigma \neq 0$:

$$\alpha_d = 4.34\frac{\varepsilon_{ff}-1}{\sqrt{\varepsilon_{ff}}(\varepsilon_r-1)}\left(\frac{\mu_0}{\varepsilon_0}\right)^{1/2}\sigma[\text{decibels/centimeter}]. \tag{5.68}$$

In (38) and (40) σ is the conductivity of the dielectric and $\mu_0 = 4\pi \times 10^{-7}$ H/m. A set of expressions for calculating α_c is as follows (Bahl and Trivedi, 1977)

$$\alpha_c = \frac{8.68}{Z_0 W}R_s \quad \text{for } W/h \to \infty, \tag{5.69}$$

where these equations apply:

$$R_s = \sqrt{\frac{\pi f \mu_0}{\sigma}}. \tag{5.70}$$

$$\alpha_c = \frac{8.68 R_s P}{2\pi Z_0 h}\left[1 + \frac{h}{W_{eff}} + \frac{h}{\pi W_{eff}}\left(\ln\frac{4\pi W}{t} + \frac{t}{W}\right)\right] \quad \text{for } W/h \leq 1/2\pi. \tag{5.71}$$

$$\alpha_c = \frac{8.68 R_s}{2\pi Z_0 h}PQ \qquad \text{for} \frac{1}{2}\pi < W/h \leq 2. \tag{5.72}$$

$$\alpha_c = \frac{8.68 R_s Q}{Z_0 h}\left\{\frac{W_{eff}}{h} + \frac{2}{\pi}\ln\left[2\pi e\left(\frac{W_{eff}}{2h} + 0.94\right)\right]\right\}^{-2} \left[\frac{W_{eff}}{h} + \frac{W_{eff}/\pi h}{(W_{eff}/2h) + 0.94}\right] \text{for } W/h \geq 2. \tag{5.73}$$

In equations 5.69 through 5.73:

$$P = 1 - \left(\frac{W_{eff}}{4h}\right)^2 \tag{5.74}$$

and

$$Q = 1 + \frac{h}{W_{eff}} + \frac{h}{\pi W_{eff}}\left(\ln\frac{2h}{t} - \frac{t}{h}\right). \tag{5.75}$$

In dielectric substrates, the dielectric losses are normally smaller than conductor losses. Dielectric losses in silicon substrates, however, can be of the same order or larger than conductor losses.

5.7 Slot Line

A **slot line** is another planar waveguide that is used above X-band. The basic configuration of a slot line is shown in Figure 5.9.

The line consists of a dielectric substrate and a narrow slot on one side of the substrate, the other side is without metallization. In a slot line, the wave propagates along the slot with major electric field component oriented across the slot in the plane of metallization. The mode of propagation is non-TEM and almost transverse electric in nature. However, unlike the conventional waveguide, there is no low frequency cutoff because the slotline is a two-conductor structure. The most widely used method of slotline analysis is the one given by Cohn and employs a transverse resonance approach. Most methods used to find the wavelength and impedance for this type of waveguide are based on numerical formulation, and they do not provide closed form expressions. This becomes a serious handicap for circuit analysis and design. Closed form expression has been obtained by curve-fitting the results based on Cohn's analysis. These expressions have an accuracy of the order of 2% for the following set of parameters:

$$\begin{aligned} &9.7 \leq \varepsilon_r \leq 20 \\ &0.02 \leq W/h \leq 1.0 \\ &0.01 \leq h/\lambda_0 \leq (h/\lambda_0)_c \end{aligned} \tag{5.76}$$

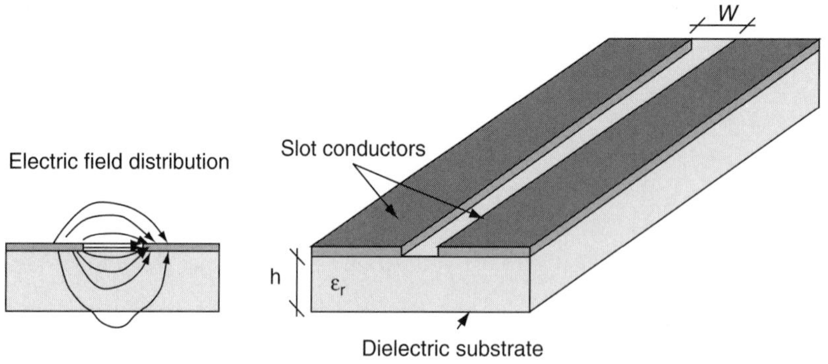

FIGURE 5.9 Slot Line Geometry

The $(h/\lambda_0)c$ is equal to the cutoff value for the TE_{10} surface wave mode on the slot line and is given by:

For $0.02 \leq \varepsilon_r \leq 0.2$:

$$\begin{aligned}\lambda_s/\lambda_0 &= 0.923 - 0.448 \log \varepsilon_r \\ &\quad + 0.2W/h - (0.29W/h + 0.047) \log (h/\lambda_0 10^2).\end{aligned} \quad (5.77)$$

$$\begin{aligned}Z_{0s} &= 72.62 - 35.19 \log \varepsilon_r + 50 \frac{(W/h - 0.02)(W/h - 0.1)}{W/h} \\ &\quad + \log (W/h - 10^2) \times (44.28 - 19.58 \log \varepsilon_r) + [0.32 \log \varepsilon_r - 0.11 \\ &\quad + W/h(1.07 \log \varepsilon_r + 1.44)] \times (11.4 - 6.07 \log \varepsilon_r - h/\lambda_0 10^{-2})^2.\end{aligned} \quad (5.78)$$

for $0.2 \leq \varepsilon_r \leq 1.0$:

$$\begin{aligned}\lambda_s/\lambda_0 &= 0.987 - 0.4483 \log \varepsilon_r + W/h(0.111 - 0.0022\varepsilon_r) \\ &\quad - (0.121 - 0.094W/h + -0.0032\varepsilon_r) \times \log (h/\lambda_0 10^2).\end{aligned} \quad (5.79)$$

$$\begin{aligned}Z_{0S} &= 113.19 - 353.55 \log \varepsilon_r + 1.25W/h(114.59 - 51.88 \log \varepsilon_r) \\ &\quad + 20(W/h - 0.2) \times (1 - W/h) - [0.15 + 0.23 \log \varepsilon_r \\ &\quad + W/h(-0.79 + 2.07 \log \varepsilon_r)] \times [10.25 - 5 \log \varepsilon_r \\ &\quad + W/h(2.1 - 1.42 \log \varepsilon_r) - h/\lambda_0 10^2]^2.\end{aligned} \quad (5.80)$$

Note the all the logarithms are assumed to be with respect to base 10.

5.8 Coplanar Waveguide

A **coplanar waveguide** consists of a center strip with two ground planes located parallel to and in the plane of the strip, as shown in Figure 5.10.

Quasistatic analysis formulation for CPW is available (Gupta, 1979). The synthesis formulas take the following closed-form expressions.

When the following equations result:

$$\frac{S}{H} \leq \frac{10}{3(1 + \ln \varepsilon_r)} \text{ and } \frac{W}{H} \leq \frac{80}{3(1 + \ln \varepsilon_r)},$$

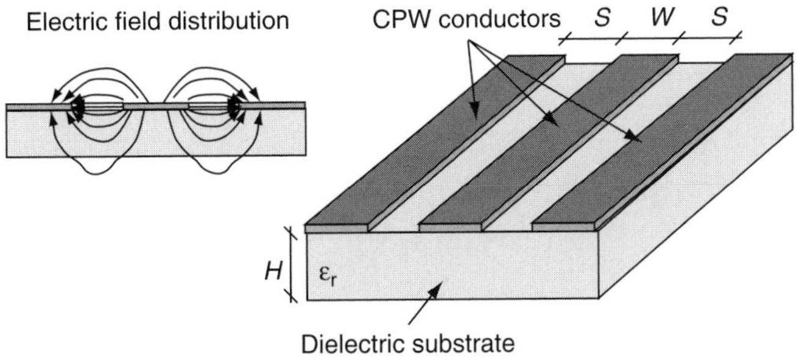

FIGURE 5.10 Coplanar Line Geometry

Then:

$$W = S \cdot G(\varepsilon_r, H, Z_0, S), \quad (5.81)$$

with:

$$G = \begin{cases} \left[\frac{1}{4}\exp\left(\frac{\pi}{4\varepsilon_{re}}\frac{\eta_0}{Z_0}\right) + \exp\left(-\frac{\pi}{4\varepsilon_{re}}\frac{\eta_0}{Z_0}\right) - 1\right] & \text{for } Z_0 < \frac{\eta_0}{\sqrt{2(\varepsilon_r+1)}}. \\ \left[\frac{1}{8}\exp\left(2\pi\sqrt{\varepsilon_{re}}\frac{Z_0}{\eta_0}\right) - \frac{1}{2}\right]^{-1} & \text{for } Z_0 \geq \frac{\eta_0}{\sqrt{2(\varepsilon_r+1)}}. \end{cases} \quad (5.82)$$

The ε_{re} is a relative effective dielectric constant:

$$\varepsilon_{re} = \varepsilon_{re}(\varepsilon_r, H, Z_0, S) = AB, \quad (5.83)$$

with these equations:

$$A = 1 + \sqrt{2}(\varepsilon_r - 1)\sqrt{\varepsilon_r + 1}\frac{Z_0}{\eta_0}\frac{K(k)}{K(k')}. \quad (5.84)$$

$$\begin{aligned}B = \operatorname{sech}\Bigg\{&\frac{\varepsilon_r^5}{4\pi(\varepsilon_r + 1)^6}\left(\frac{\eta_0}{Z_0}\right)^2. \\ &\exp\left[\left(1 + 0.0016\varepsilon_r Z_0\frac{S}{H}\right)\ln\left(0.6 + \frac{S}{H}\right)\right]\Bigg\}.\end{aligned} \quad (5.85)$$

$$k = \frac{\exp\left(\frac{\pi(1+g)S}{2H}\right) - \exp\left(\frac{\pi S}{2H}\right)}{\exp\left(\frac{\pi(2+g)S}{2H}\right) - 1}. \quad (5.86)$$

$$k' = \sqrt{1 - k^2}. \quad (5.87)$$

$$g = G|_{\varepsilon_{re}} = \frac{\varepsilon_r + 1}{2}. \quad (5.88)$$

$$\eta_0 = 120\,\pi\ \Omega. \quad (5.89)$$

The $K(k)/K(k')$ is the ratio of complete elliptic integrals of the first kind (Abramowitz and Stegun, 1972). Therefore, in a CPW wave guide design, for a given substrate (ε_r, H) and a required characteristic impedance Z_0, the central conductor width W can be evaluated immediately by choosing an appropriate gap width S.

5.9 Dielectric Circular Waveguide and Optical Fiber

Optical fiber is a guiding structure consisting of a core and a cladding that have a circular cross section, as illustrated in Figure 5.11.

The core (radius a) is made of a material with permittivity greater than that of the cladding (radius b) so that a critical angle exists for waves inside the core incident on the interface between the core and the cladding. Hence, **waveguiding** is made possible in the core by total internal reflection. Although

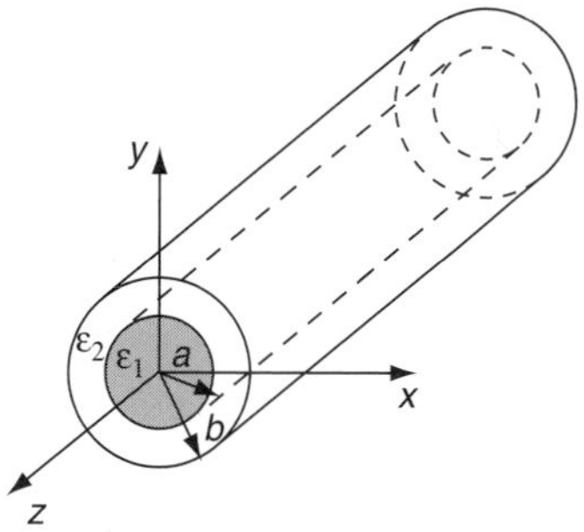

FIGURE 5.11 Circular Optical Fiber Waveguide Structure

the cladding is not necessary for the purpose of waveguiding, since the permittivity of the air is smaller than the permittivity of the fiber, it avoids scattering by the supporting structure of the fiber because the field decays exponentially outside the core, hence, it is very small outside the cladding. Moreover, the cladding allows a single-mode propagation for a larger value of the radius of the core than permitted in the absence of it. A common approach to simplify the analysis of this type of structure is to consider the cladding region to extend to infinity, having relative dielectric constant equal to one (air). From the application of Maxwell's equations to the problem illustrated, it turns out that TE and TM modes are not supported by this structure because there is no metal to separate the core from the outside dielectric. It is important to identify two modes for the structure, the ones inside the rod ($\rho < a$) and the ones outside the rod ($\rho > a$). For this structure to act as a waveguide, the fields outside the rod must vanish and exhibit a decay in the radial direction. The diameter of the rod in this structure is very important: an increase in the diameter reduces the attenuation of the field inside the rod, increases the distance to which the fields outside the rod extend, while producing a propagation constant slightly higher to the one of free space. The field distribution inside a dielectric waveguide is a combination of TE (or H) and TM (or E) modes. One way to designate hybrid modes is to denote them as HE when the TE modes predominate or EH when TM modes predominate. Pure TE or TM modes exhibit cutoff frequencies below which attenuating modes cannot propagate. The cutoff frequency is determined by the minimum electrical radius (a/λ) of the rod. For small values of a/λ, the modes cannot propagate unattenuated in the rod. There is, however, one hybrid mode called the HE_{11} that does not have a cutoff frequency. This mode is called the **dominant mode** or **dipole mode** because it is used in end-fire antennas. The field components of the modes in the internal rod are given in these equations:

$$E_\rho = -j\frac{1}{\beta_\rho^2}\left[m\omega\mu A_m\frac{J_m(\beta_\rho\rho)}{\rho} + \beta_z\beta_\rho B_m J_m(\beta_\rho\rho)\right]\cos(m\phi)e^{-j\beta_z z}. \quad (5.90)$$

$$E_\phi = j\frac{1}{\beta_\rho^2}\left[\omega\beta_\rho\mu A_m J_m(\beta_\rho\rho) + m\beta_z B_m\frac{1}{\rho}J_m(\beta_\rho\rho)\right]\sin(m\phi)e^{-j\beta_z z}. \quad (5.91)$$

$$E_z = B_m J_m(\beta_\rho\rho)\cos(m\phi)e^{-j\beta_z z} \quad (5.92)$$

$$H_\rho = -j\frac{1}{\beta_\rho^2}\left[m\omega\varepsilon B_m\frac{J_m(\beta_\rho\rho)}{\rho} + \beta_z\beta_\rho A_m J_m(\beta_\rho\rho)\right]\sin(m\phi)e^{-j\beta_z z}. \quad (5.93)$$

$$H_\phi = -j\frac{1}{\beta_\rho^2}\left[\omega\beta_\rho\varepsilon B_m J_m(\beta_\rho\rho) + m\beta_z A_m\frac{1}{\rho}J_m(\beta_\rho\rho)\right]\cos(m\phi)e^{-j\beta_z z}. \quad (5.94)$$

$$H_z = A_m J_m(\beta_\rho\rho)\sin(m\phi)e^{-j\beta_z z}. \quad (5.95)$$

In equations 5.90 through 5.95:

$$\beta_\rho^2 + \beta_z^2 = \beta^2 = \omega^2\varepsilon\mu. \quad (5.96)$$

$$A_m = -j\frac{\beta_\rho^2}{\omega\varepsilon\mu}A_{mn}D_2. \quad (5.97)$$

$$B_m = -j\frac{\beta_\rho^2}{\omega\varepsilon\mu}B_{mn}C_2$$

$$= \frac{\partial}{\partial(\beta_\rho\rho)} \quad (5.98)$$

Note that in equations 5.97 and 5.98, the coefficients A_m and B_m are dependent of each other, and their dependence can be found by applying the proper boundary conditions. To find the propagating mode in the rod for a given radius of the rod, it is necessary to write the field expression outside the rod and impose that this field decays for increasing radius. Upon application of this procedure, it is possible to determine an eigenvalue equation for the rod that gives a relationship between the fiber parameters and the minimum radius of the rod capable of supporting guided modes. The relationship is expressed as:

$$(\beta_\rho a)_{max} = \omega a\sqrt{\varepsilon_0\mu_0(\varepsilon_r - 1)}. \quad (5.99)$$

5.10 Line-Type Waveguide

A transmission line can also be considered as a guiding structure because it forces propagation of the electromagnetic field in a specific direction. The fundamental mode in the transmission lines illustrated in Figure 5.12 is the transverse electric and magnetic mode (TEM).

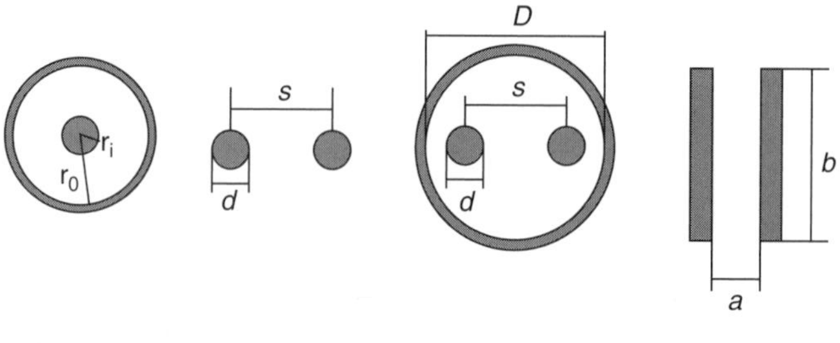

FIGURE 5.12 Fundamental Mode TEM: Transmission Lines

TABLE 5.4 Basic Parameters of Line-Type Waveguide

Capacitance [farad/meter]	$\frac{2\pi}{\ln(r_0/r_i)}$	$\frac{\pi\varepsilon}{\cosh^{-1}(s/d)}$	*Not defined*	$\frac{\varepsilon b}{a}$
Inductance (external) [henry/meter]	$\frac{\mu}{2\pi}\ln(r_0/r_i)$	$\frac{\mu}{\pi}\cosh^{-1}(s/d)$	*Not defined*	$\frac{\mu a}{b}$
Conductance [siemens/meter]	$\frac{2\pi\omega\varepsilon''}{\ln(r_0/r_i)}$	$\frac{\pi\omega\varepsilon''}{\cosh^{-1}(s/d)}$	*Not defined*	$\frac{\omega\varepsilon'' b}{a}$
Resistivity [ohms per meter]	$\frac{R_s}{2\pi}\left(\frac{1}{r_0}+\frac{1}{r_i}\right)$	$\frac{2R_s}{\pi d}\left[\frac{s/d}{\sqrt{(s/d)^2-1}}\right]$	*See note*	$\frac{2R_s}{b}$
High frequency internal inductance [henry/meter]	R/ω	R/ω	R/ω	R/ω
Characteristic impedance [ohms]	$\frac{\eta}{2\pi}\ln(r_0/r_i)$	$\frac{\eta}{\pi}\cosh^{-1}(s/d)$	*See note*	$\eta\frac{a}{b}$
Conductor attenuation α_c	$R/2Z_0$	$R/2Z_0$	$R/2Z_0$	$R/2Z_0$
Dielectric attenuation α_d	$\frac{\pi}{\lambda}\frac{\varepsilon''}{\varepsilon'}$	$\frac{\pi}{\lambda}\frac{\varepsilon''}{\varepsilon'}$	$\frac{\pi}{\lambda}\frac{\varepsilon''}{\varepsilon'}$	$\frac{\pi}{\lambda}\frac{\varepsilon''}{\varepsilon'}$
Total attenuation [decibels/meter]	$8.686(\alpha_c+\alpha_d)$	$8.686(\alpha_c+\alpha_d)$	$8.686(\alpha_c+\alpha_d)$	$8.686(\alpha_c+\alpha_d)$
Propagation constant β (for low-loss case)	$\omega\sqrt{\mu\varepsilon'}=2\pi/\lambda$	$\omega\sqrt{\mu\varepsilon'}=2\pi/\lambda$	$\omega\sqrt{\mu\varepsilon'}=2\pi/\lambda$	$\omega\sqrt{\mu\varepsilon'}=2\pi/\lambda$

Note: Assume $p = s/d$ and $q = s/D$.
Note: All the units in Table 5.4 are in the MKS system, and R_S is the surface resistance of the conductor due to the skin effect.

For this type of waveguide, it is important to evaluate the capacitance and the inductance per unit length, respectively. These two parameters enable the computation of the characteristic impedance of the line. To characterize the propagation, the propagation constant can be identified for the line, whereas the loss is divided in dielectric loss (α_d) and conductor loss (α_c). Table 5.4 illustrates the fundamental parameters of four different types of transmission line that are used in practice. The **first type of line** is the coaxial transmission line. This line is made out of a central conductor surrounded by a grounded one. The separation between the two conductors can be simple air or dielectric. Air-filled coaxial lines are very attractive because of very low dielectric loss. They are, however, difficult to manufacture since the central conductor must be suspended in air.

The **second line** presented in Table 5.4 is made out of two wires (signal-ground) running parallel to each other. The diameter of the conductor is indicated with d and the separation with s. There are several variations of this simple transmission line, among them the most common is called '**twisted pair**', and it consists of two wires twisted with each other, forming a helix.

The **third line** is a combination of the first two and consists of two parallel conductors shielded by a circular shaped electric shell that is usually grounded.

The **last line** is made of two parallel thin plates facing each other, having air or dielectric as the separation medium.

$$\text{Resistivity[ohm]} = \frac{2R_{s2}}{\pi d}\left[1+\frac{1+2p^2}{4p^4}\left(1-4q^2\right)\right] + \frac{8R_{s3}}{\pi D}q^2\left[1+q^2-\frac{1+4p^2}{8p^4}\right]. \quad (5.100)$$

$$\text{Char impedance[ohm]} = \frac{\eta}{\pi}\left\{\ln\left[2p\left(\frac{1-q^2}{1+q^2}\right)\right] - \frac{1+4p^2}{16p^4}\left(1-4q^2\right)\right\}. \quad (5.101)$$

References

Abramowitz, M. and Stegun, A. (1972). *Handbook of mathematical functions.* Dover.

Bahl, I.J., and Trivedi, D.K. (1977). *A designer's guide to microstrip line.*

Balanis, C.A. (1989). *Advanced engineering electromagnetics.* New York: John Wiley & Sons.

Collin, R.E. (1991). *Guided waves* (2d ed.). New York.

Gupta, K.C., Garg, R., and Bahl, I.J. (1979). *Microstrip lines and slot Lines.* Dedham, MA: Artech House.

Sobol, H. (1971). Application of integrated circuit technology to microwave frequencies. *Proceeding of IEEE.*

Sobol, H. (1967). Extending IC technology to microwave equipment. *Electronics.*

6

Antennas and Radiation

Nirod K. Das
Department of Electrical and Computer Engineering, Polytechnic University, Brooklyn, New York, USA

I

Antenna Fundamentals

6.1 Introduction

An **antenna** is a device used for transmission or reception of **electromagnetic waves**, allowing **radio-frequency** communication between distant locations across empty space or other material media. Antennas are fabricated using conducting and/or dielectric materials, are designed with desirable radiation or reception characteristics, and employ suitable coupling arrangements to an external port for an input/output (I/O) connection. The **I/O** may be used for exciting the antenna from a radio-frequency **signal source** or for coupling any radio-frequency signal received by the antenna to an output load, respectively, when operated as a **transmitter** or a **receiver**. Depending on the particular application and technical requirements, an antenna can be a very simple form, like a piece of metal wire or rod used in a **car or home radio**, or can be much more complex in shape and size for more sophisticated applications in **wireless communication** and **radar**. This chapter studies the fundamental operation of a general antenna structure, independent of its specific configuration and application, and device basic performance parameters commonly used in antenna technology.

Specifically, this discussion analyzes basic antenna parameters under four operating conditions: (1) antenna as a transmitter, (2) antenna as a receiver, (3) antenna in a transmit–receive communication link, and (4) antenna as a scatterer. For related studies on fundamental antenna parameters or for information on more advanced concepts of antenna theory, one may refer to some of the standard antenna books such as those by Balanis (1997), Kraus and Fleisch (1999), Kraus (1950), Jasik (1961), Collin and Zucker (1969), and Schelkunoff and Friis (1952).

6.2 Antenna as a Transmitter

6.2.1 Basic Far-Field Relations

Figure 6.1 shows the geometry of a general radiating antenna placed at the origin of a spherical co-ordinate system (r, θ, ϕ). The electric field $\bar{E}$ and the magnetic field $\bar{H}$ of any arbitrary antenna, as observed in the far-field region, can be expressed in a general form as follows. It is assumed that the antenna is excited from a time-harmonic source and that the field expressions represent the corresponding complex phasors:

$$\bar{E}(r, \theta, \phi) = \hat{\theta}E_\theta(r, \theta, \phi) + \hat{\phi}E_\phi(r, \theta, \phi). \tag{6.1}$$

$$\bar{H}(r, \theta, \phi) = \hat{\theta}H_\theta(r, \theta, \phi) + \hat{\phi}H_\phi(r, \theta, \phi). \tag{6.2}$$

$$|\bar{E}(r, \theta, \phi)| = \sqrt{|E_\theta(r, \theta, \phi)|^2 + |E_\phi(r, \theta, \phi)|^2}. \tag{6.3}$$

$$\bar{H} = \frac{\hat{r} \times \bar{E}}{\eta_0}, \quad \frac{|\bar{E}|}{|\bar{H}|} = \frac{E_\theta}{H_\phi} = -\frac{E_\phi}{H_\theta} = \eta_0 = \sqrt{\frac{\mu_0}{\varepsilon_0}} = 120\pi \tag{6.4}$$

The far-field region may be loosely defined to be sufficiently far away from the antenna (i.e., $r \to \infty$). In this region, the radiating fields behave like a uniform plane wave, having no field component along the radial direction ($\hat{r}$). Using the basic principles of a plane wave (Cheng, 1993; Rao, 2000; Ramo *et al.*, 1993), one can relate the magnetic field $\bar{H}$ in the far-field region to the corresponding electric field $\bar{E}$ in a quite simple manner, as given in equation 6.4. Therefore, it may be convenient and sufficient to use only the electric field $\bar{E}$ to model various far-field antenna parameters.

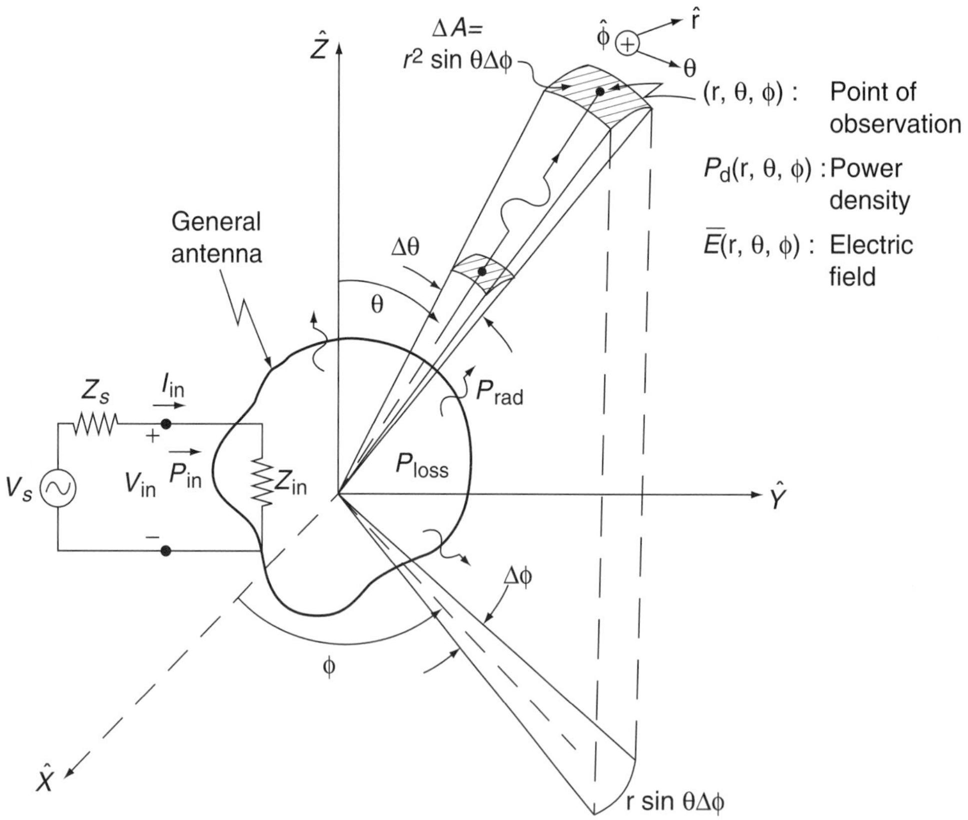

FIGURE 6.1 A General Antenna Geometry in a Spherical Coordinate System, Showing its Radiation Fields and Input Excitation

6.2.2 Radiation Power Density

The electromagnetic power flows in radial direction $\hat{r}$. The power density of radiation, P_d, is defined as the time-average power flow per unit area, ΔA, perpendicular to the power flow:

$$\Delta P = \bar{S}(r, \theta, \phi) \cdot \hat{n}\Delta A = P_d(r, \theta, \phi)\hat{r} \cdot \hat{n}\Delta A, \quad (6.5)$$

where the vector $\bar{S}(r, \theta, \phi) = \hat{r}P_d(r, \theta, \phi)$ is called the power-flow vector or the Poynting vector and where $\hat{n}$ is normal to the surface ΔA. The power density, P_d, is related to the electric-field magnitude:

$$P_d(r, \theta, \phi) = P_{d\theta}(r, \theta, \phi) + P_{d\phi}(r, \theta, \phi) = \frac{|\bar{E}|^2}{2\eta_0}. \quad (6.6)$$

$$P_{d\theta} = \frac{|E_\theta|^2}{2\eta_0}, \; P_{d\phi} = \frac{|E_\phi|^2}{2\eta_0}. \quad (6.7)$$

It is assumed that the magnitude of a complex phasor quantity, such as $|E_\theta|$ or $|E_\phi|$, refers to the peak value of the respective time-harmonic quantity.

6.2.3 Power per Solid Angle

The **solid angle**, $\Delta\Omega$, sometimes referred to as the **cone angle**, is produced by a surface element $\overline{\Delta A} = \Delta A\hat{n}$ and is defined as follows:

$$\Delta\Omega = \frac{\overline{\Delta A} \cdot \hat{r}}{r^2} = \frac{(\Delta A\hat{n}) \cdot \hat{r}}{r^2}. \quad (6.8)$$

The unit of a solid angle is **Steradian** or *Sr* for short. The solid angle formed by the angle segments $(\Delta\theta, \Delta\phi)$, shown in Figure 6.1, can be expressed as:

$$\Delta\Omega = \frac{\Delta A}{r^2} = \frac{r^2 \sin\theta\Delta\theta\Delta\phi}{r^2} = \sin\theta\Delta\theta\Delta\phi. \quad (6.9)$$

Using equations 6.8 and 6.9 in 6.5, one may rewrite the power flow ΔP through the cone angle $\Delta\Omega$ formed by the angle segments $(\Delta\theta, \Delta\phi)$ (see Figure 6.1):

$$\Delta P = P_d(r, \theta, \phi)r^2\Delta\Omega = P_d(r, \theta, \phi)r^2 \sin\theta\Delta\theta\Delta\phi. \quad (6.10)$$

$$\frac{\Delta P}{\Delta\Omega} = P_d(r, \theta, \phi)r^2 = [P_{d\theta}(r, \theta, \phi) + P_{d\phi}(r, \theta, \phi)]r^2. \quad (6.11)$$

The power ΔP is conserved while propagating in empty space through the cone angle $\Delta\Omega$. Therefore, from equation 6.10, $P_d(r, \theta, \phi)r^2$ should be independent of *r*. In other words, $P_d(r, \theta, \phi)$ and its constituent parts $P_{d\theta}(r, \theta, \phi)$ and $P_{d\phi}(r, \theta, \phi)$ should be inversely proportional to r^2:

$$P_{d\theta}(r, \theta, \phi) = \frac{p_{d\theta}(\theta, \phi)}{r^2}, \; P_{d\phi}(r, \theta, \phi) = \frac{p_{d\phi}(\theta, \phi)}{r^2},$$
$$P_d(r, \theta, \phi) = \frac{p_d(\theta, \phi)}{r^2}. \quad (6.12)$$

$$p_d(\theta, \phi) = p_{d\theta}(\theta, \phi) + p_{d\phi}(\theta, \phi). \quad (6.13)$$

$$\frac{\Delta P}{\Delta\Omega} = p_d(\theta, \phi). \quad (6.14)$$

The $p_d(\theta, \phi)$ is called the **power per solid angle**, with W/Sr as its unit. The $p_d(\theta, \phi)$ consists of two parts, $p_{d\theta}(\theta, \phi)$ and $p_{d\phi}(\theta, \phi)$, that correspond to the $\hat{\theta}$ and $\hat{\phi}$ components of the radiated electric field, respectively.

Now, from equations 6.12, 6.6, and 6.7, the magnitudes of the total electric field $\bar{E}$ and its components E_θ and E_ϕ should be inversely proportional to *r*:

$$|E_\theta(r, \theta, \phi)| = \frac{f_\theta(\theta, \phi)}{r}, \; |E_\phi(r, \theta, \phi)| = \frac{f_\phi(\theta, \phi)}{r},$$
$$|\bar{E}(r, \theta, \phi)| = \frac{f(\theta, \phi)}{r}. \quad (6.15)$$

$$p_{d\theta} = \frac{f_\theta^2}{2\eta_0}, \; p_{d\phi} = \frac{f_\phi^2}{2\eta_0}, \; p_d = \frac{f^2}{2\eta_0}; \; f^2 = f_\theta^2 + f_\phi^2. \quad (6.16)$$

6.2.4 Radiation Pattern

The r-independent parts of the corresponding power density functions are p_d, $p_{d\theta}$, and $p_{d\phi}$. The r-independent parts of the corresponding electric field functions are f, f_θ, f_ϕ. These power density and field functions may be normalized with respect to their maximum total values $p_{d\max}$ and $f_{\max}$ respectively:

$$\underline{p}_d(\theta, \phi) = \frac{p_d(\theta, \phi)}{p_{d\max}}, \; \underline{p}_{d\theta}(\theta, \phi) = \frac{p_{d\theta}(\theta, \phi)}{p_{d\max}},$$
$$\underline{p}_{d\theta}(\theta, \phi) = \frac{p_{d\phi}(\theta, \phi)}{p_{d\max}}. \quad (6.17)$$

$$\underline{f}(\theta, \phi) = \frac{f(\theta, \phi)}{f_{\max}}, \; \underline{f}_\theta(\theta, \phi) = \frac{f_\theta(\theta, \phi)}{f_{\max}}, \; \underline{f}_\phi(\theta, \phi) = \frac{f_\theta(\theta, \phi)}{f_{\max}}. \quad (6.18)$$

$$\underline{p}_d = \underline{p}_{d\theta} + \underline{p}_{d\phi} \le 1; \; \underline{f} = \sqrt{\underline{f}_\theta^2 + \underline{f}_\phi^2} \le 1;$$
$$\underline{p}_d = \underline{f}^2, \; \underline{p}_{d\theta} = \underline{f}_\theta^2, \; \underline{p}_{d\phi} = \underline{f}_\phi^2. \quad (6.19)$$

The normalized functions $\underline{p}_d$, $\underline{p}_{d\theta}$, and $\underline{p}_{d\phi}$, are called the total-, theta-, and phi-power radiation patterns, respectively. Similarly, $\underline{f}$, $\underline{f}_\theta$, and $\underline{f}_\phi$ are called the total-, theta-, and phi-field radiation patterns, respectively. An example of a total power-radiation pattern, $\underline{p}_d(\theta, \phi = \phi_0)$, is shown in Figure 6.2 in both rectangular and polar forms. Radiation patterns are normally plotted with respect to one angle variable, θ or ϕ, keeping the other variable fixed. Alternatively, the radiation pattern may be plotted using a three-dimensional (3-D) graph, including both θ and ϕ variables.

6.2.5 Beam Width and Side Lobe Level

As shown in Figure 6.2, an antenna has a direction of maximum radiation, such that $\underline{p}_d(\theta_{\max}, \phi = \phi_0) = 1$ (in Figure 6.2, $\theta_{\max}$ is assumed 0^0). The lobe surrounding the maximum

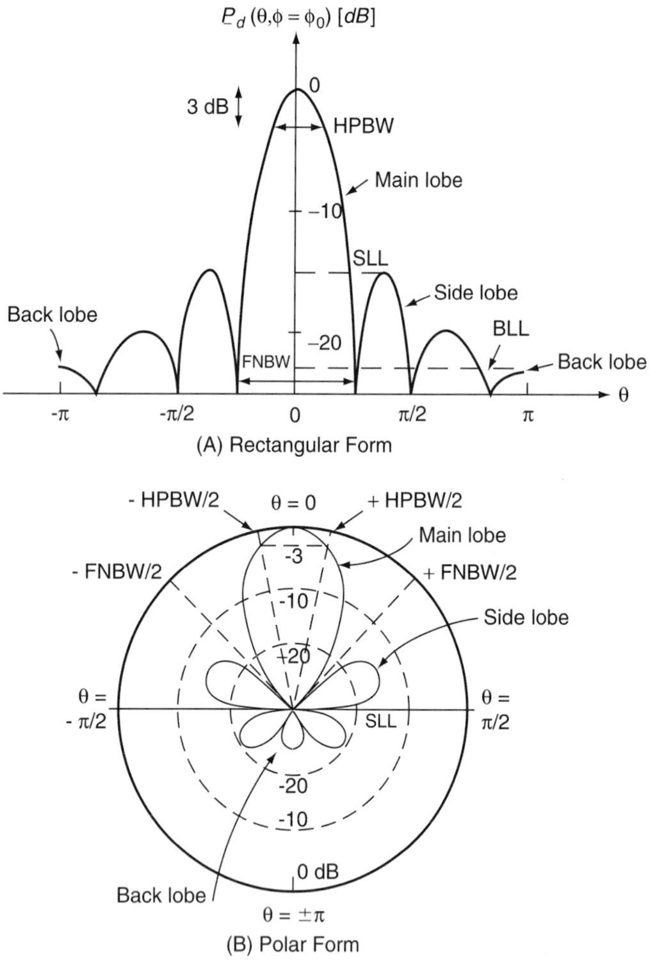

FIGURE 6.2 Examples of a Power Radiation Pattern $\underline{P}_d$, are Plotted in Rectangular and Polar Forms as a Function of θ for a Fixed Value of $\phi = \phi_0$. The half-power beam width (HPBW), first-null beam width (FNBW), side lobe level (SLL), and back lobe level (BLL) are shown, indicating the main lobe, side lobe, and back lobe of radiation.

radiation is called the **main lobe** or the **main beam**. The width of the main beam, defined using the angles of first nulls in both sides of θ_{max}, is called the **first-null beam width** (FNBW). Sometimes it is more useful to define the beam width using the angles of half-power radiation relative to the maximum radiation. The beam width defined in such manner is called the **half-power beam width** (HPBW). The lobes of the radiation pattern in both sides of the main lobe are called the **side lobes**, and the maximum level of radiation in the side lobes is called the **side lobe level** (SLL). In many applications, it is desirable to have a sufficiently low SLL to minimize unwanted interference away from the main lobe of communication. Low SLL also helps in efficient use of radiated power by using most of the radiated power in the desired main lobe of communication. Antennas also often radiate to the direction opposite the main beam, and the corresponding lobe is called the **back lobe.** The level of maximum radiation in the back lobe is called the **back lobe level** (BLL). A low BLL may be desirable to minimize interference to the back side to allow for efficient use of the radiated power.

It may be noted, depending on the specific application in mind, that the desired radiation pattern may take many different forms. In addition to the total power radiation pattern, one may be interested in the theta- or phi-power patterns to evaluate the polarization properties of radiation. The polarization properties are discussed separately in the following section.

6.2.6 Polarization Unit Vector

Equation 6.15 indicates that the magnitude of the electric field is inversely proportional to *r*. Assuming linearity, the electric field should be proportional to the antenna input current I_{in}. In addition, the phase of the electric field must vary as $e^{-jk_o r}$ to account for the wave propagation in the $\hat{r}$ direction, where $k_0 = 2\pi/\lambda_0 = 2\pi f/c$ is called the wave number in free space, λ_0 is the free-space wavelength, $c = 3 \times 10^8$ m/s is the velocity of the electromagnetic wave in free space, and *f* is the frequency of operation. Accordingly, the complete phasor of the *E*-field vector may be expressed in the following form:

$$\bar{E}(r, \theta, \phi) = I_{in}\frac{e^{-jk_0 r}}{r}\bar{e}(\theta, \phi) = I_{in}\frac{e^{-jk_0 r}}{r}[\hat{\theta}e_\theta(\theta, \phi) + \hat{\phi}e_\phi(\theta, \phi)] \quad (6.20)$$

$$E_\theta(r, \theta, \phi) = I_{in}\frac{e^{-jk_0 r}}{r}e_\theta(\theta, \phi), \; E_\phi(r, \theta, \phi) = I_{in}\frac{e^{-jk_0 r}}{r}e_\phi(\theta, \phi) \quad (6.21)$$

$$p_d = \frac{|I_{in}|^2|\bar{e}|^2}{2\eta_0}, \; p_{d\theta} = \frac{|I_{in}|^2|\bar{e}_\theta|^2}{2\eta_0}, \; p_{d\phi} = \frac{|I_{in}|^2|\bar{e}_\phi|^2}{2\eta_0}. \quad (6.22)$$

$$k_0 = \frac{2\pi}{\lambda_0}, \; \lambda_0 = \frac{c}{f}, \; c = 3 \times 10^8 \text{ m/s} \quad (6.23)$$

The (θ, ϕ)-dependent part, $\bar{e}(\theta, \phi)$, of the *E*-field is a complex vector called the **transmit vector,** which is separated into its magnitude and a complex unit vector $\hat{\rho}$:

$$\bar{e}(\theta, \phi) = |\bar{e}(\theta, \phi)|\hat{\rho}(\theta, \phi). \quad (6.24)$$

$$|\bar{e}(\theta, \phi)| = \sqrt{[|e_\theta(\theta, \phi)|^2 + |e_\phi(\theta, \phi)|^2]}. \quad (6.25)$$

The complex unit vector $\hat{\rho}$ is called the **transmit polarization unit vector** or simply the **polarization unit vector,** $\hat{\rho}$ contains information about the phase and direction of the transmit radiation. The ρ has $\hat{\theta}$ and $\hat{\phi}$ components and has a unit magnitude:

$$\hat{\rho}(\theta, \phi) = \hat{\theta}\rho_\theta(\theta, \phi) + \hat{\phi}\rho_\phi(\theta, \phi); \; |\hat{\rho}(\theta, \phi)| = 1. \quad (6.26)$$

$$\rho_\theta = \frac{e_\theta}{|\bar{e}|} = \frac{E_\theta}{|\bar{E}|}, \rho_\phi = \frac{e_\phi}{|\bar{e}|} = \frac{E_\phi}{|\bar{E}|}. \quad (6.27)$$

$$\arg(\rho_\theta) = \arg(e_\theta) = \alpha_\theta, \; \arg(\rho_\phi) = \arg(e_\phi) = \alpha_\phi,$$
$$\alpha_\phi\text{-}\alpha_\theta = \Delta\phi. \quad (6.28)$$

6.2.7 Polarization Ellipse and Axial Ratio

The polarization unit vector, $\hat{\rho}$, is originally expressed in the (θ, ϕ) coordinates. The locus of the vector, with time, is in general an ellipse with its major and minor axes oriented along $\hat{\theta}'$ and $\hat{\phi}'$, respectively, as shown in Figure 6.3. The polarization vector $\hat{\rho}(\theta, \phi)$ may be transformed into the new coordinate system (θ', ϕ'):

$$\hat{\rho}(\theta, \phi) = e^{j\alpha(\theta, \phi)}\hat{\rho}'(\theta, \phi) = e^{j\alpha(\theta, \phi)}[\hat{\theta}'\hat{\alpha}(\theta, \phi) + \zeta(\theta, \phi)\hat{\phi}' jb(\theta, \phi)]; \quad a, b \geq 0, \; a \geq b, \; \hat{\theta}' \times \hat{\phi}' = \hat{r}. \tag{6.29}$$

$$\zeta = +1 \text{ for CW wave}; \quad \zeta = -1 \text{ for CCW wave}. \tag{6.30}$$

The ζ is the rotational parameter of the wave, indicating in which manner the polarization ellipse rotates with time. In Figure 6.3, $\zeta = +1$ when the polarization ellipse of the particular wave turns clockwise (CW), and $\zeta = -1$ when the polarization ellipse turns counterclockwise (CCW). The CW and CCW waves are also referred to, respectively, as left-hand polarized (LHP) and right-hand polarized (RHP) when the wave propagates in the $\hat{r} = \hat{\theta}' \times \hat{\phi}$ direction (which is the case here). A more standardized convention may be followed to describe the rotation of the polarization vector, which is independent of the coordinate system one chooses to use. A wave is called an RHP wave when the direction of wave propagation corresponds to the thumb of the right hand as the wave polarization vector spins around in the directions of the fingers of the right hand (with fingers curved in a rotational orientation). Alternately, the wave is called an LHP wave when the left hand matches the directions and orientations just described.

Using simple trigonometry for rotating the coordinates by an angle γ (see Figure 6.3), the following transformation equations can be written as:

$$e^{j\alpha}a = \rho_\theta \cos\gamma + \rho_\phi \sin\gamma. \tag{6.31}$$

$$e^{j\alpha}jb\zeta = -\rho_\theta \sin\gamma + \rho_\phi \cos\gamma. \tag{6.32}$$

The unknown parameters a, b, γ, and α, as well as the rotational parameter ζ, can be solved in terms of ρ_θ and ρ_ϕ, using the pair of complex equations 6.31 and 6.32. The γ is called the tilt angle, a the major axis, b the minor axis, and α the phase of the ellipse. The ratio of the major and minor axes is called the axial ratio (AR):

$$AR = \frac{a}{b} \equiv 20\log\frac{a}{b}.[\text{dB}] \tag{6.33}$$

The parameters of the polarization ellipse are expressed in the following and are conditional upon the phase difference between ρ_θ and $\rho_\phi(\arg(\rho_\phi)\text{-}\arg(\rho_\theta) = \Delta\phi$. In the following, it is assumed that $-\pi/2 \leq \gamma \leq \pi/2$, and $a \geq b$. It is also assumed that one represents the phase difference, $\Delta\phi$, in the domain $-\pi < \Delta\phi \leq \pi$.

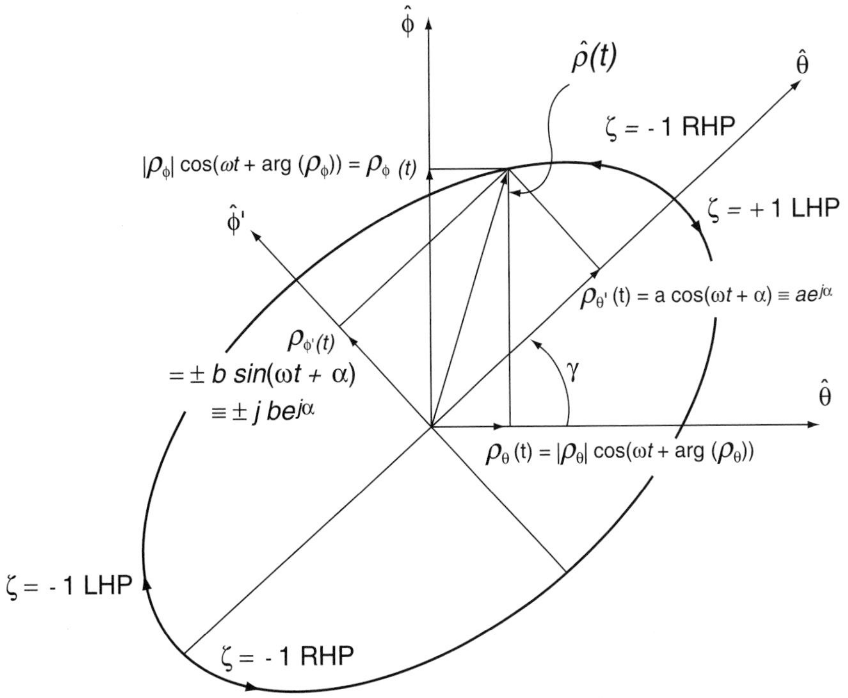

FIGURE 6.3 Vector $\hat{\rho}$ is Shown in Two Coordinate Systems: the Original Coordinate System (θ, ϕ) and a Rotated Coordinate System (θ', ϕ'). The locus of the unit vector $\hat{\rho}(t)$ is in general an ellipse, with its major axis aligned along $\hat{\theta}'$, which is in general tilted by an angle γ with respect to the original axis $\hat{\theta}$.

Case A: Linearly Polarized Wave

In case A, $\Delta\phi = 0$ or π, or $Im(\rho_\phi \rho_\theta^*) = 0$:

$$a = \sqrt{|\rho_\theta|^2 + |\rho_\phi|^2} = 1,\ b = 0,\ AR = \infty,\ \alpha = \alpha_\theta,$$
$$\gamma = \arctan\left(\frac{|\rho_\phi|}{|\rho_\theta|}\right) \text{ if } \Delta\phi = 0,$$
$$\gamma = \text{-}\arctan\left(\frac{|\rho_\phi|}{|\rho_\theta|}\right) \text{ if } \Delta\phi = \pi, \tag{6.34}$$
$$\zeta = \text{undefined}.$$

In this case, the polarization ellipse becomes a straight line, with $b = 0$ and $AR = \infty$, and the rotational parameter ζ becomes irrelevant.

Case B: Circularly Polarized Wave

In case B, $\Delta\phi = \pm\pi/2$ (or equivalently $Re(\rho_\phi \rho_\theta^*) = 0$) and $|\rho_\theta| = |\rho_\phi|$:

$$a = b = |\rho_\theta| = |\rho_\phi|,\ \gamma = 0,\ \alpha = \alpha_\theta,\ \zeta = Im\left(\frac{\rho_\phi}{\rho_\theta}\right),\ AR = 1 \equiv 0 \text{ dB}. \tag{6.35}$$

In this case, the polarization ellipse becomes a circle with $a = b$, and axial ratio $AR = 1(\equiv 0\text{dB})$. The wave is called a **circularly polarized** wave (CP). The tilt angle is arbitrary and is chosen to be zero. When the wave is RHP ($\zeta = -1$), it is called **right-hand circularly polarized** (RHCP), and on the other hand, when the wave is LHP ($\zeta = +1$), it is called **left-hand circularly polarized** (LHCP).

Case C: Elliptically Polarized Wave with $\gamma = 0$ or $\pi/2$

In case C, $\Delta\phi = \pm\pi/2$ (or equivalently $Re(\rho_\phi \rho_\theta^*) = 0$) but $|\rho_\theta| \neq |\rho_\phi|$:

$$\text{if} |\rho_\theta| > |\rho_\phi| \begin{cases} a = |\rho_\theta|,\ b = |\rho_\phi|,\ \gamma = 0,\ \alpha = \alpha_\theta, \\ \zeta = +1 \text{ if } \Delta\phi = +\pi/2;\ \zeta = \text{-}1 \text{ if } \Delta\phi = \text{-}\pi/2. \end{cases} \tag{6.36}$$

$$\text{if} |\rho_\theta| < |\rho_\phi| \begin{cases} a = |\rho_\phi|,\ b = |\rho_\theta|,\ \gamma = +\pi/2,\ \alpha = \alpha_\phi, \\ \zeta = +1 \text{ if } \Delta\phi = +\pi/2;\ \ \zeta = \text{-}1 \text{ if } \Delta\phi = \text{-}\pi/2. \end{cases} \tag{6.37}$$

Case C is the special situation of an elliptical wave with the major and minor axes coinciding with the original coordinates. Depending on the relative magnitudes of ρ_θ and ρ_ϕ however, one needs to choose the tilt angle as 0 or $\pi/2$ degrees, as indicated.

Case D: General Elliptically Polarized Wave

Case D includes all cases other than A, B, and C, when $\Delta\phi \neq 0,\ \pm\pi/2,\ \pi$, (irrespective of $|\rho_\theta|$ and $|\rho_\phi|$):

$$\zeta = +1(\text{LHP}) \text{ if } Im(\rho_\phi \rho_\theta^*) > 0 (\text{or} \sin\Delta\phi > 0);$$
$$\zeta = \text{-}1(\text{RHP}) \text{ if } Im(\rho_\phi \rho_\theta^*) < 0 (\text{or} \sin\Delta\phi < 0). \tag{6.38}$$

When $\zeta = +1$, the wave is called **left-hand elliptically polarized** (LHEP), and when $\zeta = -1$, the wave is called **right-hand elliptically polarized** (RHEP):

$$\tan 2\gamma = \frac{2Re(\rho_\phi \rho_\theta^*)}{|\rho_\theta|^2 \text{-} |\rho_\phi|^2} = \frac{2|\rho_\theta||\rho_\phi|\cos\Delta\phi}{|\rho_\theta|^2 \text{-} |\rho_\phi|^2};\ \text{-}\frac{\pi}{2} \le \gamma \le +\frac{\pi}{2}. \tag{6.39}$$

Note that there can be two values of $\text{-}\pi/2 \le \gamma \le \pi/2$ satisfying equation 6.39. One needs to eliminate one value as follows:

$$0 < \gamma < \pi/2 \text{ if } Re(\rho_\phi \rho_\theta^*) > 0 (\text{or} \cos\Delta\phi > 0);$$
$$\text{-}\pi/2 < \gamma < 0 \text{ if } Re(\rho_\phi \rho_\theta^*) < 0 (\text{or} \cos\Delta\phi < 0). \tag{6.40}$$

$$a = \sqrt{|\rho_\theta|^2 \cos^2\gamma + |\rho_\phi|^2 \sin^2\gamma + 2|\rho_\theta||\rho_\phi| \sin\gamma\cos\gamma\cos\Delta\phi}$$
$$= \sqrt{\frac{1}{2}[|\rho_\theta|^2 + |\rho_\phi|^2 + \{|\rho_\theta|^4 + |\rho_\phi|^4 + 2|\rho_\theta|^2|\rho_\phi|^2 \cos(2\Delta\phi)\}^{1/2}}. \tag{6.41}$$

$$b = \sqrt{|\rho_\theta|^2 \sin^2\gamma + |\rho_\phi|^2 \cos^2\gamma \text{-} 2|\rho_\theta||\rho_\phi| \sin\gamma\cos\gamma\cos\Delta\phi}$$
$$= \sqrt{\frac{1}{2}[|\rho_\theta|^2 + |\rho_\phi|^2 \text{-} \{|\rho_\theta|^4 + |\rho_\phi|^4 + 2|\rho_\theta|^2|\rho_\phi|^2 \cos(2\Delta\phi)\}^{1/2}}. \tag{6.42}$$

$$\alpha = \arg(\rho_\theta \cos\gamma + \rho_\phi \sin\gamma)$$
$$= \alpha_\theta + \arctan\left[\frac{|\rho_\phi|\sin\gamma\sin\Delta\phi}{|\rho_\phi|\sin\gamma\cos\Delta\phi + |\rho_\theta|\cos\gamma}\right]. \tag{6.43}$$

It should be remembered that all parameters a, b, γ, α, ζ, and AR as well as the unit vectors $\hat{\theta}'$ and $\hat{\phi}'$ in equations 6.40 to 6.43 are functions of (θ, ϕ). Figure 6.4 shows a four-quadrant chart for $\Delta\phi$ that may be used to determine the type of polarization (linear, circular or elliptical, left-handed or right-handed) and the corresponding tilt angle γ. The other parameters a, b, AR, and α can then be determined from cases A through D as appropriate.

6.2.8 Total Radiated Power and Radiation Efficiency

From equations 6.10 and 6.12, the total radiated power can be determined by integrating the power density function $P_d(r, \theta, \phi)$ over a closed spherical surface.

$$P_{\text{rad}} = \int_{\theta=0}^{\pi}\int_{\phi=0}^{2\pi} P_d(r, \theta, \phi) r^2 \sin\theta d\phi d\theta = \int_{\theta=0}^{\pi}\int_{\phi=0}^{2\pi} P_d(\theta, \phi) \sin\theta\, d\phi\, d\theta. \tag{6.44}$$

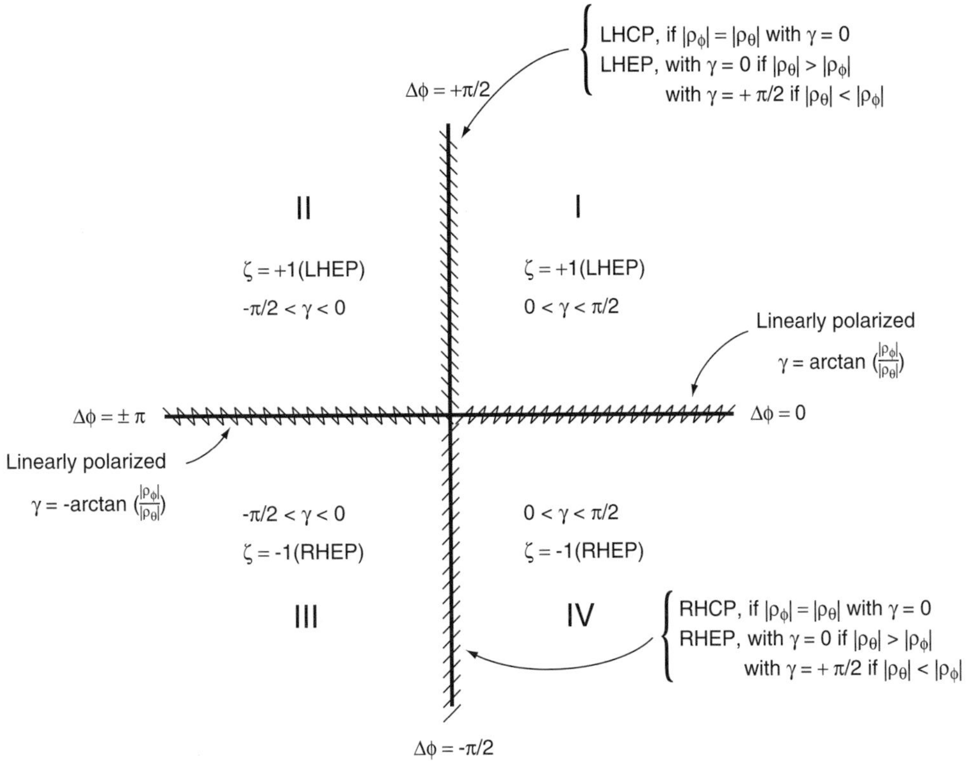

FIGURE 6.4 Different Quadrants of the Phase Difference $\Delta\phi = \arg(\rho_\phi) - \arg(\rho_\theta)$ Determine the Tilt Angle γ and Rotational Parameter ζ

The total radiated power, P_{rad}, is less than or equal to the input power, P_{in}, supplied at the input of the antenna. A part of the input power, P_{loss}, is dissipated as heat in the material (conductor or dielectric) of the antenna itself.

$$P_{\text{rad}} = P_{\text{in}} - P_{\text{loss}} = P_{\text{in}}\left(1 - \frac{P_{\text{loss}}}{P_{\text{in}}}\right) \tag{6.45}$$

$$\eta_r = 1 - \frac{P_{\text{loss}}}{P_{\text{in}}}. \tag{6.46}$$

The parameter η_r, is called the radiation efficiency. A part of the P_{loss} is dissipated as dielectric loss (P_{ld}), a part as conductor loss (P_{lc}), and a part is sometimes lost as unwanted surface wave in the antenna substrate material (P_{sw}), with efficiencies associated with the individual loss mechanisms as η_d, η_c, and η_{sw}, respectively:

$$P_{\text{loss}} = P_{ld} + P_{lc} + P_{sw}. \tag{6.47}$$

$$\eta_d = 1 - \frac{P_{ld}}{P_{\text{in}}},\ \eta_c = 1 - \frac{P_{lc}}{P_{\text{in}}},\ \eta_{sw} = 1 - \frac{P_{sw}}{P_{\text{in}}}. \tag{6.48}$$

If the individual losses are small, the total radiation efficiency can be approximated as the product of individual parts:

$$\eta_r = 1 - \frac{P_{ld} + P_{lc} + P_{sw}}{P_{\text{in}}} \simeq \left(1 - \frac{P_{ld}}{P_{\text{in}}}\right)\left(1 - \frac{P_{lc}}{P_{\text{in}}}\right)\left(1 - \frac{P_{sw}}{P_{\text{in}}}\right) = \eta_c \eta_d \eta_{sw}. \tag{6.49}$$

6.2.9 Antenna Input Impedance and Radiation Resistance

An antenna is seen by the input source as an equivalent impedance Z_{in} (see Figure 6.1) with real and imaginary parts R_{in} and X_{in}, respectively. The imaginary part is produced due to stored energy in the antenna field. The real part R_{in} consists of two parts, R_{rad} and R_{loss} respectively contributed due to the radiated power P_{rad} and the power loss P_{loss} in the antenna:

$$Z_{\text{in}} = R_{\text{in}} + jX_{\text{in}} = R_{\text{rad}} + R_{\text{loss}} + jX_{\text{in}},\ R_{\text{in}} = R_{\text{rad}} + R_{\text{loss}}. \tag{6.50}$$

$$P_{\text{rad}} = \frac{|I_{\text{in}}|^2 R_{\text{rad}}}{2},\ P_{\text{loss}} = \frac{|I_{\text{in}}|^2 R_{\text{loss}}}{2},$$
$$P_{\text{in}} = \frac{|I_{\text{in}}|^2 (R_{\text{rad}} + R_{\text{loss}})}{2},\ \eta_r = \frac{R_{\text{rad}}}{R_{\text{rad}} + R_{\text{loss}}}. \tag{6.51}$$

The R_{rad} is called the **radiation resistance**, and R_{loss} is called the **loss resistance** of the antenna.

6.2.10 Resonant Frequency and Bandwidth

The input impedance Z_{in} of the antenna is a function of the frequency of operation. Figure 6.5 shows the magnitude of the input impedance of an example antenna as a function of

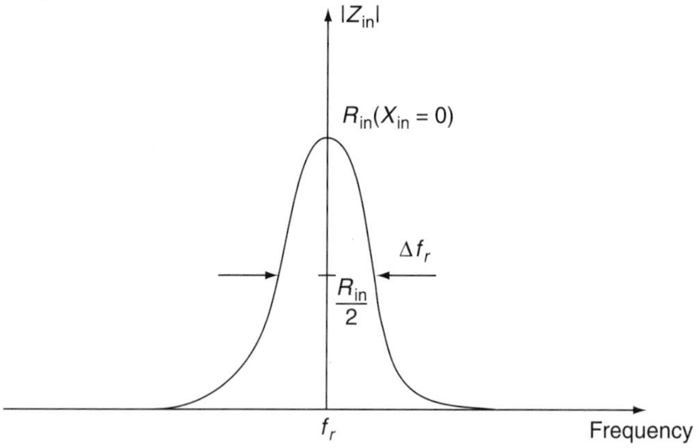

FIGURE 6.5 The frequency dependence of impedance magnitude of an example antenna is shown with the resonant frequency f_r and the bandwidth Δf_r of the antenna

frequency. In this case, the antenna impedance looks like a parallel RLC resonant circuit. The frequency, f_r, for which the impedance magnitude is maximum, or equivalently the reactance is zero, is often defined as the **resonant frequency**. It is desirable to operate the antenna at this resonant frequency so that it can be easily matched to an input transmission line with a real characteristic impedance. The frequency span Δf_r, beyond which the impedance magnitude falls below half of the resonant value, may be defined as the **impedance bandwidth** of the antenna. The antenna may be usable in this frequency band. Outside the antenna's bandwidth, the input power to the antenna would be significantly reflected due to impedance mismatch, resulting in poor radiation.

It may be noted that Figure 6.5 shows only an example case. Depending on the type of the antenna and its application, the resonance frequency and bandwidth may be defined based on the performance parameter one desires for the particular application, such as the gain, directivity, axial ratio. In such cases, the resonant frequency is defined as the frequency of best performance and the bandwidth as the frequency span over which the performance can be tolerated, as dictated by the application. It is meaningful to call the resonant frequency and the bandwidth of an antenna with an appropriate parameter prefix, such as the impedance–resonant frequency, the impedance–bandwidth, gain–resonant frequency, the gain–bandwidth, the axial–ratio–resonant frequency, or the axial–ratio–bandwidth. Often, the definitions based on the different performance parameters have strong correlations with each other.

6.2.11 Directivity

How well an antenna directs the radiation in a particular angle is quantified by comparing the radiation intensity with that of an ideal isotropic radiator. An isotropic radiator is defined as an antenna that radiates uniformly in all angles. Let the total radiated power from a given antenna, P_{rad}, also be radiated from the ideal isotropic radiator. The resulting power density, $P_{d(iso)}$, at a given distance r is independent of θ and ϕ:

$$P_{d(\text{iso})} = \frac{P_{\text{rad}}}{4\pi r^2}. \tag{6.52}$$

The ratio of the power density $P_d(r, \theta, \phi)$ from the actual antenna, in a given direction (θ, ϕ) with that from the isotropic radiator $P_{d(\text{iso})}$, is called the **directive gain function** $D_g(\theta, \phi)$:

$$D_g(\theta, \phi) = \frac{P_d(r, \theta, \phi)}{P_{d(iso)}} = \frac{4\pi r^2 P_d(r, \theta, \phi)}{P_{\text{rad}}} = \frac{4\pi p_d(\theta, \phi)}{P_{\text{rad}}}. \tag{6.53}$$

$$P_d(r, \theta, \phi) = \frac{P_{\text{rad}}}{4\pi r^2} D_g(\theta, \phi), \; p_d(\theta, \phi) = \frac{P_{\text{rad}}}{4\pi} D_g(\theta, \phi). \tag{6.54}$$

The maximum value of $D_g(\theta, \phi)$, occurring along the main beam direction $(\theta_{\max}, \phi_{\max})$, is called the **directivity** (D).

$$\text{Directivity} = D = D_g(\theta, \phi)_{\max} = D_g(\theta_{\max}, \phi_{\max}). \tag{6.55}$$

Now, using equation 6.44 in equation 6.53 $D_g(\theta, \phi)$ can be expressed using either $p_d(\theta, \phi)$ or the power pattern $\underline{P}_d(\theta, \phi)$:

$$\begin{aligned} D_g(\theta, \phi) &= \frac{4\pi p_d(\theta, \phi)}{\int_{\phi=0}^{2\pi} \int_{\theta=0}^{\pi} \underline{p}_d(\theta, \phi) \sin\theta d\theta d\phi} \\ &= \frac{4\pi p_d(\theta, \phi)}{\int_{\phi=0}^{2\pi} \int_{\theta=0}^{\pi} \underline{P}_d(\theta, \phi) \sin\theta d\theta d\phi}. \end{aligned} \tag{6.56}$$

- ***D_g, R_{rad}, and $\bar{e}$ Relationship***: Combining the equations (6.22, 6.51, and 6.54), a useful relationship results between the field vector $\bar{e}(\theta, \phi)$, directive gain function $D_g(\theta, \phi)$, and the radiation resistance R_{rad}:

$$\frac{|I_{\text{in}}|^2 |\bar{e}(\theta, \phi)|^2}{2\eta_0} = \frac{|I_{\text{in}}|^2 R_{\text{rad}}}{8\pi} D_g(\theta, \phi), \; |\bar{e}(\theta, \phi)| = \sqrt{\frac{\eta_0 R_{\text{rad}}}{4\pi}} D_g(\theta, \phi). \tag{6.57}$$

6.2.12 Gain

It may also be useful to find the ratio of the power density $P_d(r, \theta, \phi)$ from a given antenna to that ($= P'_{d(\text{iso})}$) from an ideal isotropic radiator when supplied by the same input power P_{in} in both cases (unlike the same P_{rad} considered in the last section.) The ratio of $P_d(r, \theta, \phi)$ to $P'_{d(\text{iso})}$ is called the **gain function** $G(\theta, \phi)$:

$$P'_{d(\text{iso})} = \frac{P_{\text{in}}}{4\pi r^2}. \tag{6.58}$$

$$G(\theta, \phi) = \frac{P_d(r, \theta, \phi)}{P'_{d(\text{iso})}} = \frac{4\pi r^2 P_d(r, \theta, \phi)}{P_{\text{in}}} = \frac{4\pi p_d(\theta, \phi)}{P_{\text{in}}}. \tag{6.59}$$

$$P_d(r, \theta, \phi) = \frac{P_{in}}{4\pi r^2} G(\theta, \phi) = \frac{P_{rad}}{4\pi r^2 \eta_r} G(\theta, \phi);$$
$$p_d(\theta, \phi) = \frac{P_{in}}{4\pi} G(\theta, \phi). \quad (6.60)$$

Comparing equations 6.59 and 6.53, and using the efficiency relationship of equation 6.45 a simple relationship is obtained between $G(\theta, \phi)$ and $D_g(\theta, \phi)$:

$$G(\theta, \phi) = \eta_r D_g(\theta, \phi). \quad (6.61)$$

The maximum value of the gain function is referred to as the maximum gain or simply the **gain G**:

$$\text{Gain} = G = G(\theta, \phi)_{max} = G(\theta_{max}, \phi_{max}) = \eta_r D. \quad (6.62)$$

- ***G*, R_{in}, and $\bar{e}$ Relationship**: Combining equations (6.22, 6.51, 6.60) leads to a useful relationship between the field vector $\bar{e}(\theta, \phi)$, gain function $G(\theta, \phi)$, and the antenna input resistance R_{in}:

$$\frac{|I_{in}|^2 |\bar{e}(\theta, \phi)|^2}{2\eta_0} = \frac{|I_{in}|^2 R_{in}}{8\pi} G(\theta, \phi), \; |\bar{e}(\theta, \phi)| = \sqrt{\frac{\eta_0 R_{in}}{4\pi} G(\theta, \phi)}. \quad (6.63)$$

6.2.13 Beam Efficiency

The **beam efficiency η_{beam}** of an antenna is defined as the ratio of the radiated power, $P_{mainbeam}$, confined in the antenna main lobe to the total radiated power P_{rad}. This quantifies how well the antenna uses the total radiated power in the desired main beam.

$$\eta_{beam} = \frac{P_{mainbeam}}{P_{rad}} = \frac{\int\int_{mainbeam} p_d(\theta, \phi) \sin\theta \, d\theta \, d\phi}{\int_{\phi=0}^{2\pi} \int_{\theta=0}^{\pi} p_d(\theta, \phi) \sin\theta \, d\theta \, d\phi}. \quad (6.64)$$

6.3 Antenna as a Receiver

This section treats an antenna as a receiving element. Before a receiving antenna can be modeled, a uniform plane wave must first be modeled incident on the antenna placed at the origin of a spherical coordinate system (see Figure 6.6).

6.3.1 Incident Plane Wave

The incident plane wave has both $\hat{\theta}$ and $\hat{\phi}$ components with arbitrary amplitudes and phases. Unlike a radiating wave propagating radially away from the origin, the incident plane wave is propagating toward the origin with a $e^{+jk_0 r}$ phase factor:

$$\bar{E}_i = \hat{\theta} E_{i\theta} + \hat{\phi} E_{i\phi} = \bar{e}_i e^{+jk_0 r} = (\hat{\theta} e_{i\theta} + \hat{\phi} e_{i\phi}) e^{+jk_0 r}. \quad (6.65)$$

$$\bar{e}_i = \hat{\theta} e_{i\theta} + \hat{\phi} e_{i\phi} = |e_i| \hat{\rho}_i = |\bar{e}_i| (\hat{\theta} \rho_{i\theta} + \hat{\phi} \rho_{i\phi}). \quad (6.66)$$

$$|\bar{e}_i| = \sqrt{|e_{i\theta}|^2 + |e_{i\phi}|^2}, \; \rho_{i\theta} = \frac{e_{i\theta}}{|\bar{e}_i|}, \; \rho_{i\phi} = \frac{e_{i\phi}}{|\bar{e}_i|}. \quad (6.67)$$

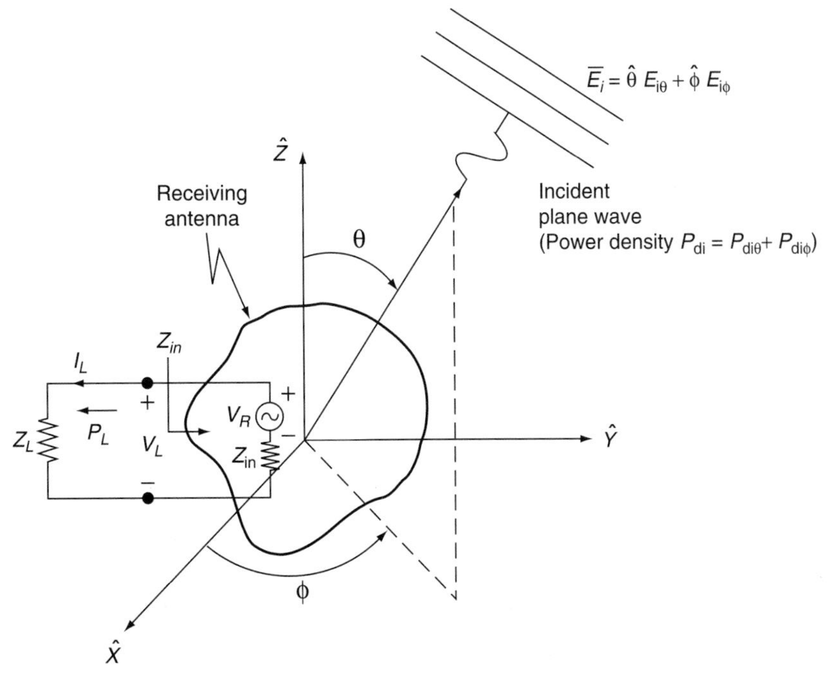

FIGURE 6.6 A General Antenna System in a Receiving Mode of Operation Shows an Incident Plane Wave and a Thevenin's Equivalent Model at the Input Terminals

$$|\hat{\rho}_i| = \sqrt{|\rho_{i\theta}|^2 + |\rho_{i\phi}|^2} = \frac{\sqrt{|e_{i\theta}|^2 + |e_{i\phi}|^2}}{|\bar{e}_i|} = 1. \quad (6.68)$$

$$P_{dinc} = \frac{|\bar{E}_i|^2}{2\eta_0} = \frac{|\bar{e}_i|^2}{2\eta_0}. \quad (6.69)$$

The $\hat{\rho}_i$ is the polarization unit vector of the incident wave. Like the radiating polarization unit vector $\hat{\rho}$, the incident polarization unit vector $\hat{\rho}_i$ may be transformed to a new coordinate system $(\hat{\theta}'_i, \hat{\phi}'_i)$ aligned with the major and minor axes of the incident polarization ellipse, respectively. The angle between the $\hat{\theta}$ and $\hat{\theta}'_i$ axes is the tilt angle γ_i:

$$\hat{\rho}_i = e^{j\alpha_i}\hat{\rho}'_i = e^{j\alpha_i}(a_i\hat{\theta}'_i - \zeta_i j b_i \hat{\phi}'_i); \quad a_i, b_i \geq 0, \ a_i \geq b_i, \ \theta'_i \times \hat{\phi}'_i = \hat{r}. \quad (6.70)$$

$$\zeta_i = +1 \text{ for LHP wave}; \quad \zeta_i = -1 \text{ for RHP wave}. \quad (6.71)$$

The equations in Section 6.2.7 may be used here to determine the various parameters of the incident wave. The major and minor axes a_i and b_i, axial ratio AR_i, tilt angle γ_i, the rotation parameter ζ_i, and the phase angle α_i should be replaced for the corresponding variables in Section 6.27: a, b, AR, γ, ζ, and α, respectively. However, it may be noted that, in contrast with equation 6.29, equation 6.70 has a negative sign before ζ_i. This takes into account the propagation of the incident wave in the -$\hat{r}$ direction instead of the +$\hat{r}$ direction of propagation for the outgoing radiated field. The change in the direction of propagation, by definition, automatically switches the "handedness" (right- or left-hand) of the wave.

6.3.2 Receive Voltage Vector

As shown in Figure 6.6, the receiving antenna is modeled as a Thevenin's equivalent circuit at its output terminals. The open-circuit voltage V_R can be written as a linear combination of the θ and the ϕ components of the incident fields:

$$V_R(\theta, \phi) = V_{R\theta}(\theta, \phi) + V_{R\phi}(\theta, \phi) = e_{\theta i} v_{R\theta}(\theta, \phi) + e_{\phi i} v_{R\phi}(\theta, \phi) = \bar{e}_i \cdot \bar{v}_R(\theta, \phi). \quad (6.72)$$

$$\bar{v}_R(\theta, \phi) = \hat{\theta} v_{R\theta}(\theta, \phi) + \hat{\phi} v_{R\phi}(\theta, \phi) = |\bar{v}_R(\theta, \phi)| \hat{\rho}_R(\theta, \phi). \quad (6.73)$$

$$\hat{\rho}_R(\theta, \phi) = \hat{\theta}\rho_{R\theta}(\theta, \phi) + \hat{\phi}\rho_{R\phi}(\theta, \phi). \quad (6.74)$$

$$|\bar{v}_R| = \sqrt{|v_{R\theta}|^2 + |v_{R\phi}|^2}, \ \rho_{R\theta} = \frac{v_{R\theta}}{|\bar{v}_R|}, \ \rho_{R\phi} = \frac{v_{R\phi}}{|\bar{v}_R|}. \quad (6.75)$$

$$|\hat{\rho}_R| = \sqrt{|\rho_{R\theta}|^2 + |\rho_{R\phi}|^2} = \frac{\sqrt{|v_{R\theta}|^2 + |v_{R\phi}|^2}}{|\bar{v}_R|} = 1. \quad (6.76)$$

The $\bar{v}_R(\theta, \phi)$ is called the **receive vector**, and the unit vector $\hat{\rho}_R(\theta, \phi)$ is called the **receive polarization unit vector.** Like the radiation and incident polarization unit vectors $\hat{\rho}$ and $\hat{\rho}_i$, the receive polarization unit vector may be transformed to a new coordinate system $(\hat{\theta}_R, \hat{\phi}_R)$ aligned with the major and minor axes of the receive polarization ellipse. The angle between the $\hat{\theta}$ and $\hat{\theta}_R$ axes is the tilt angle γ_R:

$$\hat{\rho}_R = e^{j\alpha_R}\hat{\rho}'_R = e^{j\alpha_R}(a_R\hat{\theta}'_R + \zeta_R j b_R \hat{\phi}'_R); \quad a_R, b_R \geq 0, \ a_R \geq b_R, \ \hat{\theta}'_R \times \hat{\phi}'_R = \hat{r}. \quad (6.77)$$

$$\zeta_R = +1 \text{ for LHP receiver}; \quad \zeta_R = -1 \text{ for RHP receiver}. \quad (6.78)$$

Like the incident polarization unit vector in equation 6.70, the equations in Section 6.2.7 may also be used here to determine the various parameters for the receive polarization unit vector $\hat{\rho}_R$. It may be noted that, in contrast with equation 6.70, equation 6.77 has a positive sign before ζ_R. The rotational parameter ζ_R as a rule follows the same convention as for ζ, which is different from the rule for the incident wave ζ_i

6.3.3 Received Power and Effective Area

Once the Thevenin's voltage V_R of the receiving antenna is known, the load power P_L delivered to a load impedance Z_L can be easily obtained (see Figure 6.6):

$$V_R = \bar{e}_i \cdot \bar{v}_R = |\bar{e}_i||\bar{v}_R|\hat{\rho}_i \cdot \hat{\rho}_R. \quad (6.79)$$

$$V_L = \frac{V_R Z_L}{Z_L + Z_{\text{in}}}, \ I_L = \frac{V_R}{Z_L + Z_{\text{in}}}, \ P_L = \frac{1}{2}\left|\frac{V_R}{Z_L + Z_{\text{in}}}\right|^2 R_L. \quad (6.80)$$

The effective area A_e of a receiving antenna is defined as the ratio of the load power P_L and the incident power density P_{dinc}. The receiving antenna is seen as an equivalent capture area for the incident wave. The equivalent area A_e of an antenna may be clearly distinguished from the physical area of an antenna because, in general, they can be significantly different from each other. For example, a commonly used wire antenna has a nonzero effective area allowing a fairly good reception, whereas its physical area is very small—practically zero:

$$A_e = \frac{P_L}{P_{\text{dinc}}} = \frac{R_L\eta_0}{|\bar{e}_i|^2}\left|\frac{V_R}{Z_L + Z_{\text{in}}}\right|^2. \quad (6.81)$$

$$A_e = \frac{R_L\eta_0}{|\bar{e}_i|^2}\frac{|\bar{v}_R|^2|\bar{e}_i|^2|\hat{\rho}_i \cdot \hat{\rho}_R|^2}{|Z_L + Z_{\text{in}}|^2} = \frac{R_L\eta_0|\bar{v}_R|^2|\hat{\rho}_i \cdot \hat{\rho}_R|^2}{|Z_L + Z_{\text{in}}|^2}. \quad (6.82)$$

The effective area can be maximized in two ways: (1) by matching the load impedance for maximum power transfer, which requires $Z_L = Z_{\text{in}}^*$ or (2) by adjusting the incident and receive polarization vectors such that $|\hat{\rho}_i \cdot \hat{\rho}_R| = 1 = |\hat{\rho}_i \cdot \hat{\rho}_R|$ whenever possible. Under these conditions, the maximum effective area A_{em} can be derived from equation 6.82:

$$A_{em}(\theta, \phi) = \frac{\eta_0|\bar{v}_R(\theta, \phi)|^2}{4R_{\text{in}}}; Z_L = Z^*_{\text{in}}, |\hat{\rho}_i \cdot \hat{\rho}_R| = 1 = |\hat{\rho}'_i \cdot \hat{\rho}'_R|. \quad (6.83)$$

The element $|\hat{\rho}_i \cdot \hat{\rho}_R|$ is called the **polarization match factor** between the incident wave and the receiving antenna. It can be shown that the polarization match factor is at the maximum possible (unity) only when $\gamma_i = \gamma_r$, $\zeta_i = \zeta_r$, and $AR_i = AR_r$ ($a_i = a_R$, $b_i = b_R$). From equations 6.70 and 6.77, one can see that the above condition established when $\hat{\rho}_i = \hat{\rho}^*_R$. Other possible practical situations may be considered. First, for a given receiving antenna (given ζ_r and AR_r), a given incident plane wave (given ζ_i and AR_i), and a given load impedance Z_L, the receive power P_L can be maximized by just turning the antenna about the $\hat{r}$ axis, such that the major axes of the incident and receive polarization vectors are aligned ($\gamma_i = \gamma_R$). Second, every other parameter remaining the same, the received power is larger if the rotational parameters are matched ($\zeta_i = \zeta_r$) compared to when they are mismatched ($\zeta_i \neq \zeta_r$).

6.3.4 Received Noise Temperature

Figure 6.7(A) shows the noise-equivalent circuit model of an antenna. Like an ohmic resistance, the input resistance of the antenna, R_{in} is associated with a noise source v_{na}, with equivalent noise temperature T_{ea} (Pozar, 1998). Unlike an ohmic resistance, however, T_{ea} is not the physical temperature of the antenna. The v_{na} consists of two parts, v_{nL} and v_{nr}, with noise temperatures, T_a and T_{er}, respectively associated with the loss resistance R_{loss} and radiation resistance R_{rad} of the antenna. The v_{nL} is generated due to the thermal process in the antenna itself, and therefore T_a is equal to the physical temperature of

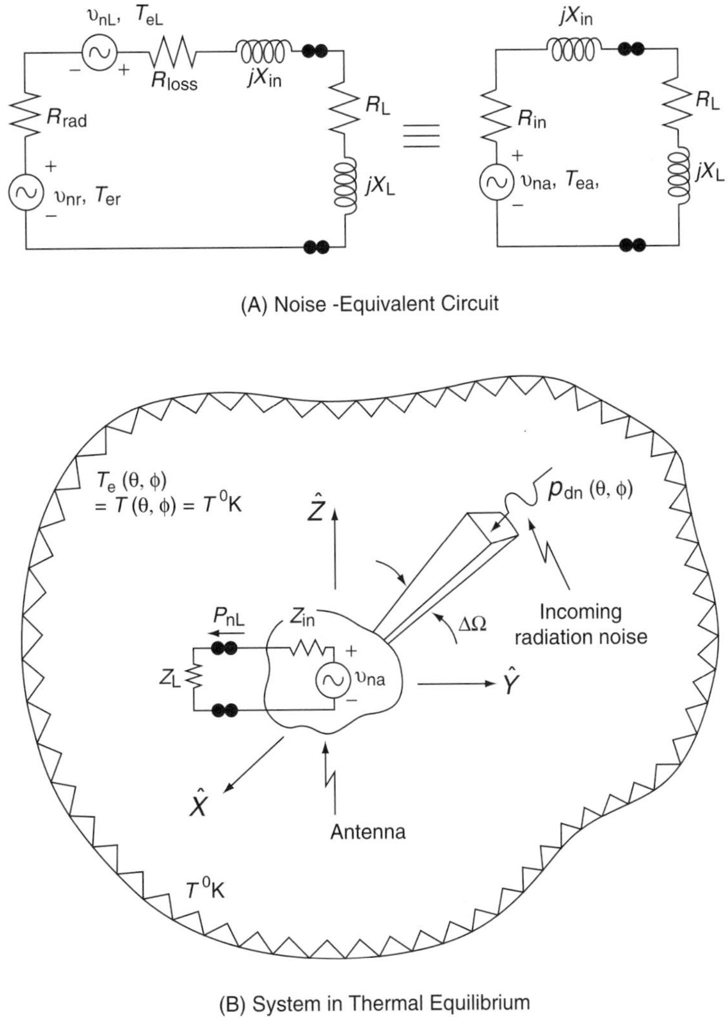

FIGURE 6.7 Noise Model for a Receiving Antenna. (A) This noise-equivalent circuit of an antenna shows noise voltage v_{nr} received through radiation and v_{nl} produced by the antenna itself due to material-loss resistance R_{loss}. (B) The antenna is placed in a closed "perfectly absorbing" chamber at temperature T^0K (p_{dn} = noise power per unit solid angle, as seen by an ideal isotropic antenna).

the antenna. In comparison, T_{er} is associated with the incoming radiation noise that depends on the temperature distribution of the surrounding objects that generate it:

$$\bar{v}_{nL}^2 = 4kT_a R_{\text{loss}},\ \bar{v}_{nr}^2 = 4kT_{er}R_{\text{rad}},\ \bar{v}_{na}^2 = 4kT_{ea}R_{\text{in}};\ \bar{v}_{na}^2 = \bar{v}_{nL}^2 + \bar{v}_{nr}^2. \quad (6.84)$$

The bar over the v represents root-mean-square (rms) values per unit bandwidth. Let the total noise–power per unit bandwidth, delivered by the antenna to a conjugate-matched load $Z_L = Z_{\text{in}}^*$ be P_{na}. The respective constituent parts delivered by v_{nL} and v_{nR} are P_{nL} and P_{nr}. These noise powers can be expressed using the respective noise temperatures and antenna radiation efficiency η_r (from equation 6.51), which would then provide a relationship between T_{ea}, T_a, and T_{er}:

$$P_{nL} = \frac{\bar{v}_{nL}^2}{4R_{\text{in}}} = kT_a\frac{R_{\text{loss}}}{R_{\text{in}}} = kT_a(1\text{-}\eta_r). \quad (6.85)$$

$$P_{nr} = \frac{\bar{v}_{nr}^2}{4R_{\text{in}}} = kT_{er}\frac{R_{\text{rad}}}{R_{\text{in}}} = kT_{er}\eta_r. \quad (6.86)$$

$$P_{na} = \frac{\bar{v}_{na}^2}{4R_{\text{in}}} = kT_{ea} = P_{nL} + P_{nr};\ T_{ea} = \eta_r T_{er} + (1\text{-}\eta_r)T_a. \quad (6.87)$$

Relationship Between T_{er} and Radiation–Noise Density

Let $p_{dn}(\theta, \phi)$ be the noise–power density of the incoming radiation per unit solid angle and per unit bandwidth, as seen by an ideal, matched "isotropic" antenna (gain function $G_{iso}(\theta, \phi) = 1$, matched effective area $= A_{em(iso)}$). The matched radiation noise P_{nr} received by an arbitrary antenna can be expressed in terms of $p_{dn}(\theta, \phi)$ and the antenna's effective area $A_{em}(\theta, \phi)$ relative to $A_{em(iso)}$:

$$P_{nr} = \int_\Omega p_{dn}(\theta, \phi)\frac{A_{em}(\theta, \phi)}{A_{em(iso)}}d\Omega = \int_\Omega p_{dn}(\theta, \phi)\frac{G(\theta, \phi)}{G_{iso}}d\Omega$$
$$= \int_\Omega p_{dn}(\theta, \phi)G(\theta, \phi)d\Omega. \quad (6.88)$$

It may be noted that equation 6.88 uses the $G(\theta, \phi) \sim A_{em}(\theta, \phi)$ relationship, to be developed in 6.110. Then, using equations 6.86 and 6.61 in 6.88, we can relate the radiation noise–temperature T_{er} with $p_{dn}(\theta, \phi)$ and the directive gain function $D_g(\theta, \phi)$:

$$r_r kT_{er} = \int_\Omega p_{dn}(\theta, \phi)G(\theta, \phi)d\Omega;$$
$$T_{er} = \int_\Omega \frac{p_{dn}(\theta, \phi)}{k}D_g(\theta, \phi)d\Omega. \quad (6.89)$$

Relationship Between T_{er} and Spatial Temperature Distribution $T_e(\theta, \phi)$

Consider first the special situation when the antenna is surrounded by perfect absorbers, and the entire system is in thermal equilibrium at the temperature $T = T_a$, as shown in Figure 6.7(B). Under this condition, the noise–temperature of the antenna should be equal to the surrounding temperature. Consequently, using equation 6.87, the radiation–noise temperature of the antenna can be shown to be the temperature of the surrounding objects T:

$$T_{ea} = T = T_a = \eta_r T_{er} + (1\text{-}\eta_r)T_a;\ T_{er} = T. \quad (6.90)$$

Due to symmetry of the situation, $p_{dn}(\theta, \phi)$ is independent of the angle of arrival (θ, ϕ). With this assumption, and using equations 6.56 and 6.90 in equation 6.89 a relationship between p_{dn} and the absorber temperature T results:

$$T_{er} = T = \int_\Omega \frac{p_{dn}(\theta, \phi)}{k}D_g(\theta, \phi)d\Omega$$
$$= \frac{p_{dn}}{k}\int_\Omega D_g(\theta, \phi)d\Omega = \frac{4\pi p_{dn}}{k};\ p_{dn} = \frac{kT}{4\pi}. \quad (6.91)$$

Equation 6.91 can be extended to arbitrary situations when the surrounding objects are at different physical temperatures $T(\theta, \phi)$, or the objects can be modeled with noise-equivalent temperatures $T_e(\theta, \phi)$. This will relate $p_{dn}(\theta, \phi)$ to $T(\theta, \phi)$ or $T_e(\theta, \phi)$. Then, using equation 6.89 T_{er} can be related to the surrounding temperature profile $T_e(\theta, \phi)$ via $D_g(\theta, \phi)$:

$$p_{dn}(\theta, \phi) = \frac{kT_e(\theta, \phi)}{4\pi},\ T_{er} = \frac{1}{4\pi}\int_\Omega T_e(\theta, \phi)D_g(\theta, \phi)d\Omega. \quad (6.92)$$

6.4 Transmit–Receive Communication Link

Figure 6.8(A) shows a communication link between two antennas, with their antenna parameters indicated by a subscript 1 or 2. Each antenna is described with respect to its own spherical coordinate system, represented with subscripts 1 or 2, having an origin at the center of the corresponding antenna. First, antenna 1 can be treated as the transmit antenna. The transmit field, $\bar{E}$, radiated from antenna 1 is viewed as the incident wave, $\bar{e}_i$, for the receiving antenna 2:

$$\bar{E}_1(r_1, \theta_1, \phi_1) = I_1\frac{e^{-jk_0 r_1}}{r_1}\bar{e}_1(\theta_1, \phi_1) = \bar{e}_i. \quad (6.93)$$

$$|\bar{e}_i| = \frac{|I_1||\bar{e}_1(\theta_1, \phi_1)|}{r_1},\ \hat{\rho}_i = e^{j(\arg(I_1)-k_0 r_1)}\hat{\rho}_1(\theta_1, \phi_1). \quad (6.94)$$

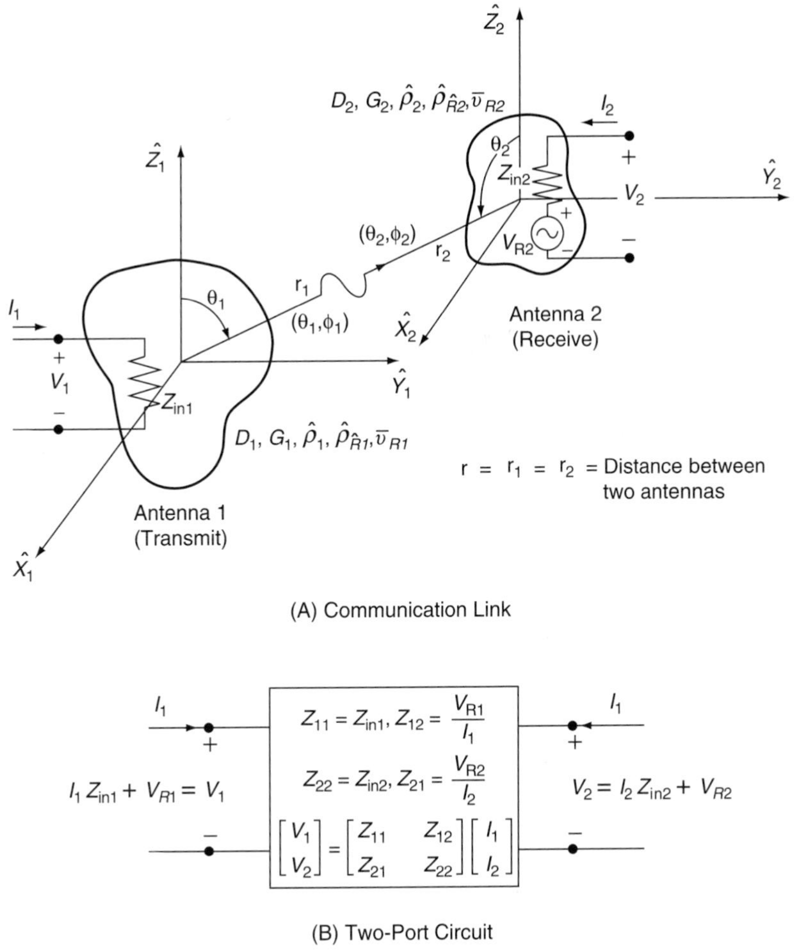

FIGURE 6.8 A Communication Link Between Two Antennas, Shown with Antenna 1 Used as the Transmitter and Antenna 2 as the Receiver. The Link can be Viewed as a Two-Port Circuit as Shown in (B).

Note that the incident polarization unit vector $\hat{\rho}_i$, as seen by receiving antenna 2, is equal to the transmit polarization unit vector $\hat{\rho}_1$ of antenna 1 multiplied by the phase factor $e^{j(\arg(I_1)-k_0 r_1)}$ due to propagation over distance r_1 and the phase of the input current I_1. The magnitude of the incident wave in equation 6.94 may be expressed in terms of the input impedance R_{in1} of antenna 1 (or alternately the input power P_{in1} of antenna 1) using equation 6.63:

$$|\bar{e}_i| = |I_1|\sqrt{\frac{\eta_0 R_{in1}}{4\pi r_1^2} G(\theta_1, \phi_1)} = \sqrt{\frac{2\eta_0 P_{in1}}{4\pi r_1^2} G(\theta_1, \phi_1)}. \quad (6.95)$$

The Thevenin's voltage, or the open-circuit voltage V_{R2}, at the terminals of antenna 2 can now be written in different forms using equations 6.79, 6.93, 6.94, and 6.95:

$$V_{R2} = I_1 \frac{e^{-jk_0 r_1}}{r_1} \bar{e}_1(\theta_1, \phi_1) \cdot \bar{v}_{R2}(\theta_2, \phi_2). \quad (6.96)$$

$$V_{R2} = \frac{|I_1||\bar{e}_1(\theta_1, \phi_1)||\bar{v}_{R2}(\theta_2, \phi_2)|}{r_1} e^{j(\arg(I_1)-k_0 r_1)} \hat{\rho}_1(\theta_1, \phi_1) \cdot \hat{\rho}_{R2}(\theta_2, \phi_2). \quad (6.97)$$

$$V_{R2} = \left[\sqrt{\frac{2\eta_0 P_{in1}}{4\pi r_1^2} G_1(\theta_1, \phi_1)}\right] |\bar{v}_{R2}(\theta_2, \phi_2)| e^{j(\arg(I_1)-k_0 r_1)} \hat{\rho}_1(\theta_1, \phi_1) \cdot \hat{\rho}_{R2}(\theta_2, \phi_2). \quad (6.98)$$

6.4.1 Transmit–Receive Link as a Two-Port Circuit

If antenna 2 is also excited by an input current I_2, antenna 1 will receive a Thevenin's terminal voltage V_{R1}. Repeating the procedure described in the previous paragraphs, V_{R1} may be expressed by interchanging the subscripts 1 and 2:

$$V_{R1} = I_2 \frac{e^{-jk_0 r_2}}{r_2} \bar{e}_2(\theta_2, \phi_2) \cdot \bar{v}_{R1}(\theta_1, \phi_1). \quad (6.99)$$

The link between antennas 1 and 2 may be seen as a two-port circuit, as illustrated in Figure 6.8(B). The total input voltages V_1 and V_2, respectively across the terminals of antennas 1 and 2 may be expressed as a superposition of the receive voltage and the transmit voltage:

$$V_1 = I_1 Z_{in1} + V_{R1},\ V_2 = I_2 Z_{in2} + V_{R2}. \quad (6.100)$$

$$Z_{12} = \frac{V_{R1}}{I_2} = \frac{e^{-jk_0 r_2}}{r_2} \bar{e}_2(\theta_2, \phi_2) \cdot \bar{v}_{R1}(\theta_1, \phi_1). \quad (6.101)$$

$$Z_{21} = \frac{V_{R2}}{I_1} = \frac{e^{-jk_0 r_1}}{r_1} \bar{e}_1(\theta_1, \phi_1) \cdot \bar{v}_{R2}(\theta_2, \phi_2);\ r_1 = r_2. \quad (6.102)$$

If the circuit is assumed to be reciprocal, then $Z_{12} = Z_{21}$ (Pozar, 1998). Hence, the following is true:

$$\bar{e}_1(\theta_1, \phi_1) \cdot \bar{v}_{R2}(\theta_2, \phi_2) = \bar{e}_2(\theta_2, \phi_2) \cdot \bar{v}_{R1}(\theta_1, \phi_1). \quad (6.103)$$

Equation 6.103 shows an important relationship in antenna theory called the **antenna reciprocity relationship**.

6.4.2 Relationship Between Transmit and Receive Vectors of an Arbitrary Antenna

Using equation 6.103, one can relate the transmit and receive vectors of any arbitrary antenna (replaced for antenna 1), once the transmit and receive vectors of one specific known antenna can be substituted for antenna 2.

A small dipole antenna with unit input current and length l can be used as the reference antenna 2. Let this dipole current be oriented in the -$\hat{\theta}_1$ direction, such that the radiation from the dipole antenna 2 is maximum toward the antenna 1. Alternately, one may visualize the current of the dipole antenna 2 to be oriented along the $\hat{Z}_2$ axis (see Figure 6.8(A)). Then, let the $\hat{z}_2$ axis be turned to align in the $-\hat{\theta}_1$ direction so that in the line of communication $\hat{\theta}_1 = \hat{\theta}_2$ and $\hat{\theta}_2 = 90°$. Under this situation, using the results from Das (2004) in Section 6.2.1, the transmit and receive vectors of the dipole antenna 2 are found as:

$$\bar{I} = I_2 \hat{z}_2 = -\hat{\theta}_1, I_2 = 1. \quad (6.104)$$

$$\bar{e}_2(\theta_2, \phi_2) = \frac{j\eta_0 l}{2\lambda_0} \hat{\theta}_2 = \frac{j\eta_0 l}{2\lambda_0} \hat{\theta}_1,\ \bar{v}_{R2}(\theta_2, \phi_2) = l\hat{\theta}_2 = l\theta_1,\ \theta_1 = \theta_2. \quad (6.105)$$

Now, using equation 6.105 in equation 6.103, the relationship between the $\hat{\theta}$ components of the transmit and receive vectors of the arbitrary antenna 1 is established as follows:

$$\bar{e}_1(\theta_1, \phi_1) \cdot \hat{\theta}_1 l = \frac{j\eta_0 l}{2\lambda_0} \hat{\theta}_1 \cdot \bar{v}_{R1}(\theta_1, \phi_1), e_{1\theta}(\theta_1, \phi_1) = \frac{j\eta_0}{2\lambda_0} v_{R1\theta}(\theta_1, \phi_1). \quad (6.106)$$

One may repeat the above procedure by reorienting the dipole antenna 2 such that the current is in the -$\hat{\phi}_1$ direction. This would provide the relationship between the $\hat{\phi}$ components of the transmit and receive vectors:

$$e_{1\phi}(\theta_1, \phi_1) = \frac{j\eta_0}{2\lambda_0} v_{R1\phi}(\theta_1, \phi_1). \quad (6.107)$$

Combining equations 6.106 and 6.107, the relationship between the complete transmit and receive vectors becomes:

$$\bar{e}_1(\theta_1, \phi_1) = \frac{j\eta_0}{2\lambda_0} \bar{v}_{R1}(\theta_1, \phi_1). \quad (6.108)$$

Because an arbitrary antenna is being considered for antenna 1, and the direction of consideration (θ_1, ϕ_1) is arbitrary, it can be generalized that for any antenna and any direction of radiation-reception the following is true:

$$\bar{e}(\theta, \phi) = \frac{j\eta_0}{2\lambda_0} \bar{v}_R(\theta, \phi),$$
$$|\bar{e}(\theta, \phi)| = \frac{\eta_0}{2\lambda_0} |\bar{v}_R(\theta, \phi)|,\ \hat{\rho}(\theta, \phi) = j\hat{\rho}_R(\theta, \phi). \quad (6.109)$$

Equation 6.109 is a useful universal relationship between the transmit and receive vectors of any antenna.

$G \sim A_{em}$ Relationship

Consider now the relationship between the magnitudes of the transmit and receive vectors in equation 6.109. Then use equation 6.63 to relate $|\bar{e}|$ to G and equation 6.83 to relate $|\bar{v}_R|$ to maximum effective area A_{em}. Thus, an equivalent universal relationship between $G(\theta, \phi)$ and $A_{em}(\theta, \phi)$ results:

$$\frac{|\bar{e}(\theta, \phi)|^2}{|\bar{v}_R(\theta, \phi)|^2} = (\frac{\eta_0}{2\lambda_0})^2 = \frac{G(\theta, \phi)}{A_{em}(\theta, \phi)} \frac{\eta_0^2}{16\pi}, \frac{G(\theta, \phi)}{A_{em}(\theta, \phi)} = \frac{4\pi}{\lambda_0^2}. \quad (6.110)$$

6.4.3 Friis Transmission Formula

Using the universal $\bar{e} \sim \bar{v}_R$ relationship of equation 6.109 in the equation 6.96, the received Thevenin's voltage at the antenna 2 terminals can be expressed using transmit vectors of the two antennas:

$$V_{R2} = I_1 \frac{e^{-jk_0 r_1}}{r_1} \bar{e}_1(\theta_1, \phi_1) \cdot \bar{e}_2(\theta_2, \phi_2) \frac{2\lambda_0}{j\eta_0}. \quad (6.111)$$

$$|V_{R2}| = \frac{|I_1|}{r_1} |\bar{e}_1(\theta_1, \phi_1)||\bar{e}_2(\theta_2, \phi_2)| \frac{2\lambda_0}{\eta_0} |\hat{\rho}_1(\theta_1, \phi_1) \cdot \hat{\rho}_2(\theta_2, \phi_2)|. \quad (6.112)$$

Equation 6.63 may be used to relate the multiple $|\bar{e}|$s of equation 6.112 to corresponding Gs:

$$|V_{R2}|^2 = \left(\frac{\lambda_0}{4\pi r_1}\right)^2 4|I_1|^2 R_{in1} R_{in2} G_1(\theta_1, \phi_1) G_2(\theta_2, \phi_2) |\hat{\rho}_1(\theta_1, \phi_1) \cdot \hat{\rho}_2(\theta_2, \phi_2)|^2. \quad (6.113)$$

$$|V_{R2}|^2 = \left(\frac{\lambda_0}{4\pi r_1}\right)^2 8P_{in1} R_{in2} G_1(\theta_1, \phi_1) G_2(\theta_2, \phi_2) |\hat{\rho}_1(\theta_1, \phi_1) \cdot \hat{\rho}_2(\theta_2, \phi_2)|^2. \quad (6.114)$$

Once V_{R2} is known, the load voltage V_{L2} and load power P_{L2}, when the antenna 2 is loaded by an arbitrary load impedance Z_{L2}, are found from equation 6.80:

$$P_{L2} = \frac{8R_{\text{in}2}R_{L2}}{2|Z_{L2}+Z_{\text{in}2}|^2}\left(\frac{\lambda_0}{4\pi r_1}\right)^2 P_{\text{in}1}G_1(\theta_1,\phi_1)G_2(\theta_2,\phi_2)|\hat{\rho}_1(\theta_1,\phi_1)\cdot\hat{\rho}_2(\theta_2,\phi_2)|^2. \tag{6.115}$$

$$P_{L2\max} = \left(\frac{\lambda_0}{4\pi r_1}\right)^2 P_{\text{in}1}G_1(\theta_1,\phi_1)G_2(\theta_2,\phi_2);\ Z_{L2} = Z_{\text{in}2}^*, |\hat{\rho}_1\cdot\hat{\rho}_2| = 1. \tag{6.116}$$

$$P_{L2} = P_{L2\max}\left(\frac{4R_{\text{in}2}R_{L2}}{|Z_{L2}+Z_{\text{in}2}|^2}\right)|\hat{\rho}_1(\theta_1,\phi_1)\cdot\hat{\rho}_2(\theta_2,\phi_2)|^2. \tag{6.117}$$

Equations 6.115, 6.116, and 6.117 are called the **Friis transmission formula**, which relates the received load power to the input power through the antenna and load parameters.

6.4.4 Polarization Match Factor

The element $\hat{\rho}_1(\theta_1,\phi_1)\cdot\hat{\rho}_2(\theta_2,\phi_2)$ in the above Friis formulas is the **polarization match factor** M_p.

$$M_p = \hat{\rho}_1(\theta_1,\phi_1)\cdot\hat{\rho}_2(\theta_2,\phi_2). \tag{6.118}$$

Using equation 6.29 (also see Figure 6.3), M_p may be alternately represented in the coordinates of the respective polarization ellipses:

$$M_p = e^{j\alpha_1}\hat{\rho}_1'(\theta_1,\phi_1)\cdot\hat{\rho}_2'(\theta_2,\phi_2)e^{j\alpha_2}. \tag{6.119}$$

It should be remembered, however, that equation 6.118 applies for general conditions, where the polarization unit vectors are referenced with respect to the coordinate systems of the corresponding antennas (see Figure 6.8) that in general can have arbitrary relative orientation with respect to each other. It may be useful to express M_p for a simpler situation, when the axes $(\hat{x}_1,\hat{y}_1,\hat{z}_1)$ and $(\hat{x}_2,\hat{y}_2,\hat{z}_2)$ are respectively aligned with each other (with different origins). With this situation, in the direction of communication between the antennas, $\hat{\theta}_1\cdot\hat{\theta}_2 = 1$, but $\hat{\phi}_1\cdot\hat{\phi}_2 = -1$:

$$M_p = \hat{\rho}_1\cdot\hat{\rho}_2 = (\rho_{1\theta}\hat{\theta}_1+\rho_{1\phi}\hat{\phi}_1)\cdot(\rho_{2\theta}\hat{\theta}_2+\rho_{2\phi}\hat{\phi}_2) = \rho_{1\theta}\rho_{2\theta}-\rho_{1\phi}\rho_{2\phi}. \tag{6.120}$$

M_p Using $\hat{\rho}$ Vectors in Common Coordinates

Consider again the just described situation of aligned coordinates for the two antennas of a communication link. However, consider that the polarization unit vectors of the two antennas, $\hat{\rho}_1$ and $\hat{\rho}_2$, are expressed while the two antennas are placed in a common coordinate system (r,θ,ϕ) with origin at the antenna center:

$$\hat{\rho}_1 = \rho_{1\theta}\hat{\theta}+\rho_{1\phi}\hat{\phi}, \hat{\rho}_2 = \rho_{2\theta}\hat{\theta}+\rho_{2\phi}\hat{\phi}. \tag{6.121}$$

With this situation, one may see that the polarization mismatch factor, $M_p = \rho_{1\theta}\rho_{2\theta}-\rho_{1\phi}\rho_{2\phi}$, is no longer equal to $\hat{\rho}_1\cdot\hat{\rho}_2$. The correct expression for M_p is as follows:

$$M_p = \rho_{1\theta}\rho_{2\theta}-\rho_{1\phi}\rho_{2\phi} = \hat{\rho}_{1s}\cdot\hat{\rho}_2 = \hat{\rho}_1\cdot\hat{\rho}_{2s};\quad \hat{\rho}_{1s} = \hat{\rho}_1|_{\hat{\phi}\to-\hat{\phi}}, \hat{\rho}_{2s} = \hat{\rho}_2|_{\hat{\phi}\to-\hat{\phi}}. \tag{6.122}$$

$$|M_p| = |\hat{\rho}_{1s}\cdot\hat{\rho}_2| = |\hat{\rho}_1\cdot\hat{\rho}_{2s}|, \tag{6.123}$$

where the unit vectors with subscript s refer to original vectors with $-\hat{\phi}$ switched for $\hat{\phi}$. As may be noted, this essentially switches the "handedness" of the coordinates.

6.5 Antenna as a Scatterer

Figure 6.9 shows a plane wave, $\bar{E}_i$, being scattered from an arbitrary antenna. $\bar{E}_i$ is incident from angle (θ_i,ϕ_i), which in general has both $\hat{\theta}$ and $\hat{\phi}$ components of the electric field:

$$\bar{E}_i = e^{jk_0r}(e_{i\theta}\hat{\theta}+e_{i\phi}\hat{\phi}), P_{di} = P_{di}^{\theta}+P_{di}^{\phi};\ P_{di}^{\theta} = \frac{|e_{i\theta}|^2}{2\eta_0}, P_{di}^{\phi} = \frac{|e_{i\phi}|^2}{2\eta_0}. \tag{6.124}$$

The scattered field $\bar{E}_s(r,\theta,\phi;\ \theta_i,\phi_i)$ in the far-field region can be decomposed into two parts. The first part, $\bar{E}_{so}(r,\theta,\phi;\ \theta_i,\phi_i)$, is the scattering with the antenna input terminals open-circuited ($I_L = 0$). The second part, $\bar{E}_r(r,\theta,\phi;\ \theta_i,\phi_i)$, is the reradiated field due to the load current I_L = -I_{in} when the antenna is loaded with a finite impedance:

$$\bar{E}_s = \bar{E}_{so}+\bar{E}_r = \frac{e^{-jk_0r}}{r}[e_{i\theta}\bar{e}_s^{\theta}(\theta,\ \phi;\ \ \theta_i,\ \phi_i)+e_{i\phi}\bar{e}_s^{\phi}(\theta,\ \phi;\ \theta_i,\ \phi_i)]. \tag{6.125}$$

The scattered field is a linear combination of the incident field components $e_{i\theta}$ and $e_{i\phi}$. As for a radiating antenna, the scattered field also has the e^{-jk_0r}-dependence, in the far-field region, that account for the propagation phase and the free-space loss. The $\bar{e}_s^{\theta}$ and $\bar{e}_s^{\phi}$ may respectively be referred to as the θ- and ϕ-scattering vectors and are the r-independent part of the scattered field due to one unit of $e_{i\theta}$ and $e_{i\phi}$, respectively.

The part due to the open-circuited antenna may be separately expressed in the following form, where the subscript refers to the corresponding values in the open-circuited condition:

$$\bar{E}_{so} = \frac{e^{-jk_0r}}{r}[e_{i\theta}\bar{e}_{so}^{\theta}(\theta,\ \phi;\ \ \theta_i,\ \phi_i)+e_{i\phi}\bar{e}_{so}^{\phi}(\theta,\ \phi;\ \theta_i,\ \phi_i)]. \tag{6.126}$$

Using equations 6.20, 6.72, and 6.80, the reradiated part can be expressed in terms of the input current I_{in} = -I_L, the transmit vector $\bar{e}(\theta,\ \phi)$, and the receive vector $\bar{v}_R(\theta_i,\ \phi_i)$. The

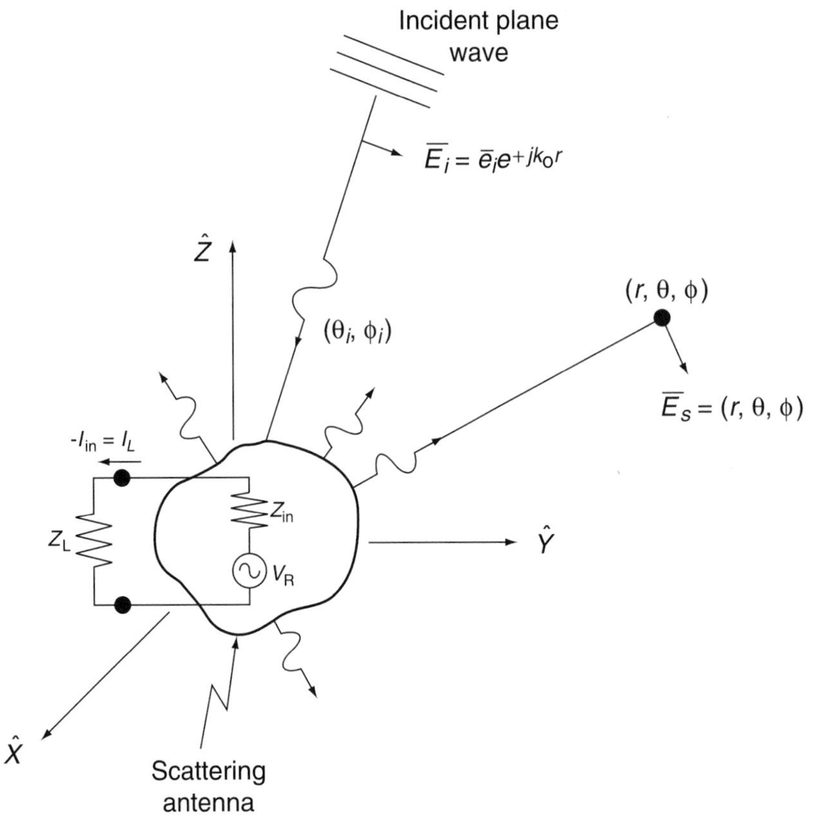

FIGURE 6.9 A Plane Wave Incident From Angle (θ_i, ϕ_i) is Scattered from an Antenna. The scattered signal $\overline{E}_s(r, \theta, \phi)$ is observed at a point $(r,\ \theta,\ \phi)$. The scattered signal is affected by the load impedance Z_L.

input/load current is induced by the incident field and related by the receive vector $\bar{v}_R$:

$$\bar{E}_r = \frac{e^{-jk_o r}}{r} I_{\text{in}}(\theta_i, \phi_i)\bar{e}(\theta, \phi) = -\frac{e^{-jk_o r}}{r}\left[\frac{\bar{e}_i \cdot \bar{v}_R(\theta_i, \phi_i)}{Z_{\text{in}} + Z_L}\bar{e}(\theta, \phi)\right]. \quad (6.127)$$

Now the individual parts in equations 6.126 and 6.127 may be combined with equation 6.125 to obtain the following relationships:

$$\bar{e}_s^\theta = \bar{e}_{so}^\theta - \frac{v_{R\theta}(\theta_i, \phi_i)}{Z_{\text{in}} + Z_L}\bar{e}(\theta, \phi), \bar{e}_s^\phi = \bar{e}_{so}^\phi - \frac{v_{R\phi}(\theta_i, \phi_i)}{Z_{\text{in}} + Z_L}\bar{e}(\theta, \phi). \quad (6.128)$$

6.5.1 Intercepted Power and Radar Cross Section

First consider a θ-polarized incident field ($e_{i\phi} = 0$). With this condition, the power–density, P_{ds}^θ, of the scattered field at a given location $(r,\ \theta, \phi)$ can be expressed using the resulting scattered E-field (from equations 6.124 and 6.125). For brevity of presentation, the angle of incidence (θ_i, ϕ_i) and the observation point $(r,\ \theta,\ \phi)$ can be considered as implicit variables, as appropriate:

$$P_{ds}^\theta(r,\ \theta,\ \phi;\ \theta_i,\ \phi_i) = \frac{|e_{i\theta}|^2|\bar{e}_s^\theta|^2}{2\eta_o r^2} = \frac{P_{di}^\theta|\bar{e}_s^\theta|^2}{r^2} = \frac{P_{di}^\theta[|e_{s\theta}^\theta|^2 + |e_{s\phi}^\theta|^2]}{r^2}. \quad (6.129)$$

An **effective intercept power**, P_{int}^θ, is defined as associated with a scatterer. The P_{int}^θ is the equivalent power intercepted by the scattering antenna that when reradiated by an isotropic antenna provides the same power density at the observation location as the scattering wave. The P_{int}^θ is a function of the angle of observation $(\theta,\ \phi)$. The corresponding **radar cross section**, σ^θ, is defined as the ratio of the intercept power and the incident power density P_{di}^θ:

$$P_{ds}^\theta = \frac{P_{int}^\theta(\theta,\ \phi;\ \theta_i, \phi_i)}{4\pi r^2} = \frac{P_{di}^\theta \sigma^\theta}{4\pi r^2};$$
$$\sigma^\theta(\theta,\ \phi;\ \ \theta_i,\ \phi_i) = \frac{P_{int}^\theta}{P_{di}^\theta} = \frac{4\pi r^2 P_{ds}^\theta}{P_{di}^\theta}. \quad (6.130)$$

Using equations 6.129 in 6.130, the relationship is created between the θ-radar cross section σ^θ and the θ-scattering vector $\bar{e}_s^\theta$. σ^θ is further decomposed into σ_θ^θ and σ_ϕ^θ which respectively correspond to the θ and ϕ components of $\bar{e}_s^\theta = \hat{\theta}e_{s\theta}^\theta + \hat{\phi}e_{s\phi}^\theta$:

$$\sigma^\theta(\theta, \phi;\ \ \theta_i, \phi_i) = 4\pi|\bar{e}_s^\theta|^2 = 4\pi[|e_{s\theta}^\theta|^2 + |e_{s\phi}^\theta|^2]$$
$$= \sigma_\theta^\theta(\theta,\ \phi;\ \ \theta_i, \phi_i) + \sigma_\phi^\theta(\theta,\ \phi;\ \ \theta_i, \phi_i). \quad (6.131)$$

Similarly, the radar cross section may be defined for the ϕ-polarized incident field, with the θ-superscripts exchanged for ϕ:

$$P_{ds}^{\phi}(r, \theta, \phi; \theta_i, \phi_i) = \frac{P_{int}^{\phi}(\theta, \phi; \ \theta_i, \phi_i)}{4\pi r^2} = \frac{P_{di}^{\phi}|\bar{e}_s^{\phi}|^2}{r^2} = \frac{P_{di}^{\phi}[|e_{s\theta}^{\phi}|^2 + |e_{s\phi}^{\phi}|^2]}{r^2}. \tag{6.132}$$

$$\sigma^{\phi}(\theta, \phi; \ \theta_i, \phi_i) = \frac{4\pi r^2 P_{ds}^{\phi}}{P_{di}^{\phi}} = 4\pi[|e_{s\theta}^{\phi}|^2 + |e_{s\theta}^{\phi}|^2] = \sigma_{\theta}^{\phi}(\theta, \phi; \ \theta_i, \phi_i) + \sigma_{\phi}^{\phi}(\theta, \phi; \ \theta_i, \phi_i). \tag{6.133}$$

References

Balanis, C.A. (1997). *Antenna theory: Analysis and design.* (2d ed.). John Wiley & Sons.

Cheng, D.K. (1993). *Fundamentals of engineering electromagnetics.* Reading, MA: Addison-Wesley.

Collin, R.E., and Zucker, F.J. (Eds.) (1969). *Antenna theory: Parts I, II.* New York: McGraw-Hill.

Das, N.K. (2004). Antennas and radiation: Antenna elements and arrays. *EE Handbook.* Boston: Academic Press.

Jasik, J. (Ed.). (1961). *Antenna engineering handbook.* New York: McGraw-Hill.

Kraus, J.D. (1950). *Antennas.* New York: McGraw-Hill.

Kraus, J.D., and Fleisch, D.A. (1999). *Electromagnetics with applications.* New York: McGraw-Hill.

Pozar, D.M. (1998). *Microwave engineering.* New York: John Wiley & Sons.

Ramo, S., Whinnery, J.R., and van Duzer, T. (1993). *Fields and waves in communication electronics.* (3rd ed.) New York: John Wiley & Sons.

Rao, N.N. (2000). *Elements of engineering electromagnetics.* New Jersey: Prentice Hall.

Schelkunoff, S.A., and Friis, H.T. (1952). *Antennas, theory, and practice.* New York: John Wiley & Sons.

II Antenna Elements and Arrays

6.6 Introduction

One can envision literally infinite possible antenna geometries that can be designed to achieve a variety of performance features. Among these diverse configurations, one may identify a few **basic antenna elements** that are often used for general-purpose wireless applications. The following sections analyze the radiation/reception mechanism in selected basic antenna geometries: (1) dipole antenna, (2) monopole antenna, (3) wire-loop antenna, (4) slot antenna, and (5) microstrip antenna. Selected other types of antennas in use will also be introduced.

An **antenna array** is made of a number of individual antenna elements each located at a different position in space, with independent excitation to each input. Such an array may provide many practical advantages over a single antenna element. By properly designing for the individual locations and input excitations, one can achieve radiation characteristics that might not be normally feasible using a single antenna. If one can also control the input excitations to the individual elements in the real-time, the radiation characteristics of the array, such as the pointing direction or radiation pattern, can be changed in a dynamic manner. This will allow significant flexibility for adjusting to a dynamically changing situation. This chapter will analyze the basic theory of a general antenna array and derive results for specific situations of a one-dimensional array and a two-dimensional array.

6.7 Antenna Elements

This section presents a number of basic antenna elements, starting with a small dipole antenna that is considered the most fundamental radiating structure. Behaviors only in the far-field region are emphasized, and various radiation and impedance parameters of the specific antennas are derived from them. For general definitions and notations used in this study, the reader is referred to Das (2004). Some useful parameters of the antenna elements covered in this section are summarized in Table 6.1.

6.7.1 Dipole Antennas

Small Dipole Antenna

A **small dipole antenna** may be visualized as a small length of current $\bar{I} = \hat{z}I_{in}$, excited by connecting a pair of small metal wires or rods to a radio frequency source. This is shown in Figure 6.10. It is considered small when the length l is significantly smaller than the wavelength ($l \ll \lambda_0$) of operation. This basic element is often used in many radio communication systems as a **whip antenna**. For communication in a relatively

TABLE 6.1 Summary of Antenna Parameters of Basic Antenna Types

Antenna type		Polarization	Power pattern $P_{-d}(\theta, \pi)$	Directivity D	Radiation resistance $R_{rod}(\Omega)$
Dipole antennas (current along $\hat{z}$)	Short dipole (length = l)	$E_\theta(TM_z)$	$\sin^2\theta$	1.5	$80\pi^2(l/\lambda_0)^2$
	Half-wave dipole (length = $\lambda_0/2$)	$E_\theta(TM_z)$	$\frac{\cos^2(\pi/2\cos\theta)}{\sin^2\theta}$	1.64	73
	Half-wave folded dipole (length = $\lambda_0/2$)	$E_\theta(TM_z)$	$\frac{\cos^2(\pi/2\cos\theta)}{\sin^2\theta}$	1.64	292
Monopole antennas (current along $\hat{z}$)	Short monopole (length = l)	$E_\theta(TM_z)$	$\sin^2\theta$ (only half space, $0 \le \theta \le \pi/2$)	3.0	$160\pi^2(l/\lambda_0)^2$
	Half-wave monopole (length = $\lambda_0/4$)	$E_\theta(TM_z)$	$\frac{\cos^2(\pi/2\cos\theta)}{\sin^2\theta}$ (only half space, $0 \le \theta \le \pi/2$)	3.28	36.5
	Half-wave folded monopole (length = $\lambda_0/4$)	$E_\theta(TM_z)$	$\frac{\cos^2(\pi/2\cos\theta)}{\sin^2\theta}$ (only half space, $0 \le \theta \le \pi/2$)	3.28	146
Loop antenna (loop on x–y plane)	Small loop antenna (loop area = S)	$E_\phi(TE_z)$	$\sin^2\theta$	1.5	$320\pi^4(S/\lambda_0^2)^2$
Slot antennas (slot length along $\hat{z}$)	Short slot antenna (length = l)	$E_\phi(TE_z)$	$\sin^2\theta$	1.5	$45(\lambda_0/l)^2$
	Half-wave slot antenna (length = $\lambda_0/2$)	$E_\phi(TE_z)$	$\frac{\cos^2(\pi/2\cos\theta)}{\sin^2\theta}$	1.64	485
Microstrip antenna (antenna current along $\hat{x}$, patch on the x–y plane)	Rectangular, edge fed (length = $\lambda_0/(2\sqrt{\varepsilon_r})$	Both E_θ and $E_\phi(TM_x)$	See Section 6.2.5	No analytical results, depending on several parameters: d, L, W, ε_r, λ_0.	

Refer to appropriate sections in the text for the geometry of the specific antennas.

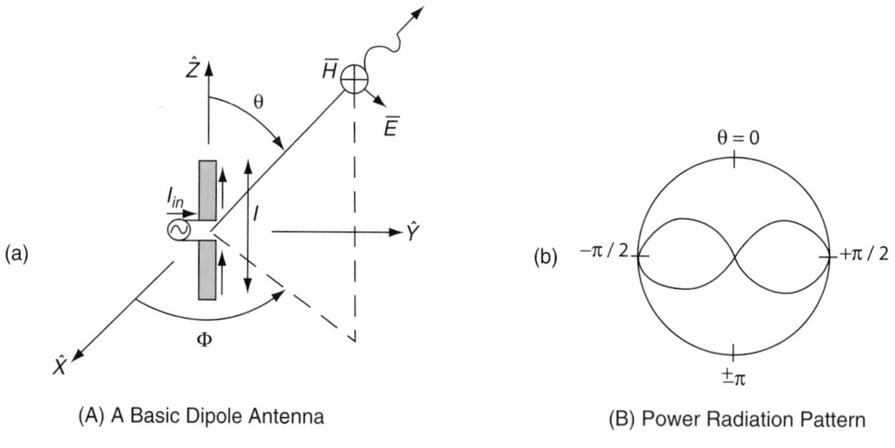

FIGURE 6.10 (A) This Basic Dipole antenna is excited by an Input Current I_{in} and the Figure Shows the Directions of the Electric $\bar{E}$ and Magnetic $\bar{H}$ Fields in the Far-Field Region. (B) The Power Radiation Pattern $\underline{p}_d(\theta, \phi)$ in the Elevation Plane (θ) for a Given ϕ is Shown Here. The pattern is uniform in the azimuth (ϕ) plane.

low frequency range (such as a short-wave radio), where the wavelength of radiation is quite long, a whip antenna of any reasonable length can be safely approximated as a small dipole antenna. The metal wire or rod is ideally made of a perfect metal when the ohmic loss in the metal is zero. In practical situations when the wire is made from a very good conductor, unless the wire is too thin (which results in very high ohmic resistance), only a small fraction of the input power is dissipated as ohmic loss. Most of the input power is thus radiated into the free space, resulting in a good radiation efficiency.

Transmitting Mode of Operation

The rigorous electric and magnetic fields of this dipole antenna can be derived using Maxwell's equations (Cheng, 1993; Ramo *et al.*, 1993). The final expressions in the far-field region are provided here and are used to derive the various basic parameters. The radiated far-fields of the dipole antenna can be expressed in the following form:

$$\bar{E}(r, \theta, \phi) = \hat{\theta} I_{\text{in}} \frac{e^{-jk_0 r}}{r} \frac{j\eta_0 l}{2\lambda_0} \sin\theta = I_{\text{in}} \frac{e^{-jk_0 r}}{r} \bar{e}(\theta, \phi). \quad (6.134)$$

$$\bar{H}(r, \theta, \phi) = \frac{\hat{r} \times \bar{E}(r, \theta, \phi)}{\eta_0} = \hat{\phi} I_{\text{in}} \frac{e^{-jk_0 r}}{r} \frac{jl}{2\lambda_0} \sin\theta$$
$$= I_{\text{in}} \frac{e^{-jk_0 r}}{r} \frac{jl}{2\lambda_0} (\hat{z} \times \hat{r}). \quad (6.135)$$

$$\bar{e}(\theta, \phi) = \hat{\theta} \frac{j\eta_0 l}{2\lambda_0} \sin\theta, \; \hat{\rho}(\theta, \phi) = j\hat{\theta}. \quad (6.136)$$

The dipole does not radiate in the vertical directions ($\theta = 0, \pi$) and produces maximum radiation perpendicular to the current ($\theta = \pi/2$) with uniform variation in the azimuth. There is a $\sin\theta$ factor for variation of the radiation fields in the elevation in proportion to the projection of the input current in the particular direction. The magnetic field $\bar{H}$ is in the $\hat{\phi}$ direction and, thus, revolves around the wire in analogy to a magnetostatic field, due to a linear current. The electric field $\bar{E}$ and, accordingly, the polarization vector of the dipole antenna, are in the $\hat{\theta}$ direction. Once the far-field expressions are known, then the power density P_d and power per unit solid angle p_d of the radiation can be derived using the theory of Das (2004):

$$P_d(r, \theta, \phi) = \frac{|\bar{E}|^2}{2\eta_0} = \frac{|I_{\text{in}}|^2 \eta_0}{8r^2} \left(\frac{l}{\lambda_0}\right)^2 \sin^2\theta, \; p_d(\theta, \phi)$$
$$= r^2 P_d(r, \theta, \phi) = \frac{|I_{in}|^2 \eta_0}{8} \left(\frac{l}{\lambda_0}\right)^2 \sin^2\theta. \quad (6.137)$$

The power radiation pattern $\underline{p}_d(\theta, \phi)$ of the dipole antenna is a $\sin^2\theta$ function in elevation and uniform in the azimuth:

$$\underline{p}_d(\theta, \phi) = \frac{p_d(\theta, \phi)}{p_{d\max}} = \sin^2\theta. \quad (6.138)$$

The total radiated power from the dipole antenna for a given input current I_{in} can be obtained by integrating $p_d(\theta, \phi)$ over the entire sphere:

$$P_{\text{rad}} = \int_{\phi=0}^{2\pi} \int_{\theta=0}^{\pi} p_d(\theta, \phi) \sin\theta \, d\theta \, d\phi$$
$$= \frac{|I_{\text{in}}|^2 \eta_0 \pi}{3} \left(\frac{l}{\lambda_0}\right)^2 = \frac{|I_{\text{in}}|^2 R_{\text{rad}}}{2}. \quad (6.139)$$

$$R_{\text{rad}} = \frac{2\eta_0 \pi}{3} \left(\frac{l}{\lambda_0}\right)^2 = 80\pi^2 \left(\frac{l}{\lambda_0}\right)^2, \; R_{\text{in}} = R_{\text{rad}}; \; \eta_r = 1. \quad (6.140)$$

The directive gain and directivity of the dipole antenna are obtained from $p_d(\theta, \phi)$ or $\underline{p}_d(\theta, \phi)$ Das (2004):

$$D_g(\theta, \phi) = \frac{4\pi p_d(\theta, \phi)}{\int_{\phi=0}^{2\pi} \int_{\theta=0}^{\pi} p_d(\theta, \phi) \sin\theta \, d\theta \, d\phi}$$
$$= \frac{4\pi \underline{p}_d(\theta, \phi)}{\int_{\phi=0}^{2\pi} \int_{\theta=0}^{\pi} \underline{P}_d(\theta, \phi) \sin\theta \, d\theta \, d\phi}. \quad (6.141)$$

$$D_g(\theta, \phi) = \frac{4\pi \sin^2\theta}{\int_{\phi=0}^{2\pi} \int_{\theta=0}^{\pi} \sin^3\theta \, d\theta \, d\phi} = \frac{3\sin^2\theta}{2}; \; D = 1.5. \quad (6.142)$$

In an ideal situation, when the ohmic loss is zero or radiation efficiency is one, the antenna gain is equal to the directivity. In practical cases using good conductors, the ohmic loss can be calculated using the metal conductivity, shape of the wire, and the current distribution:

$$G(\theta, \phi) = \eta_r \frac{3\sin^2\theta}{2} \simeq \frac{3\sin^2\theta}{2}, \; G \simeq 1.5; \; \eta_r \simeq 1. \quad (6.143)$$

Receiving Mode of Operation

Figure 6.11 shows the dipole antenna in the **receiving mode** of operation. The Thevenin's voltage excited at the antenna terminals can be obtained by taking the tangential component of the incident electric field and multiplying it with the antenna length:

$$V_R = -lE_{z\text{inc}} = le_{i\theta} \sin\theta = \bar{v}_R(\theta, \phi) \cdot \bar{e}_i$$
$$= v_{R\theta}(\theta, \phi) e_{i\theta} + v_{R\phi}(\theta, \phi) e_{i\phi}. \quad (6.144)$$

$$v_{R\theta}(\theta, \phi) = l \sin\theta, \; v_{R\phi}(\theta, \phi) = 0;$$
$$\bar{v}_R(\theta, \phi) = \hat{\theta} l \sin\theta, \; \hat{\rho}_R(\theta, \phi) = \hat{\theta}. \quad (6.145)$$

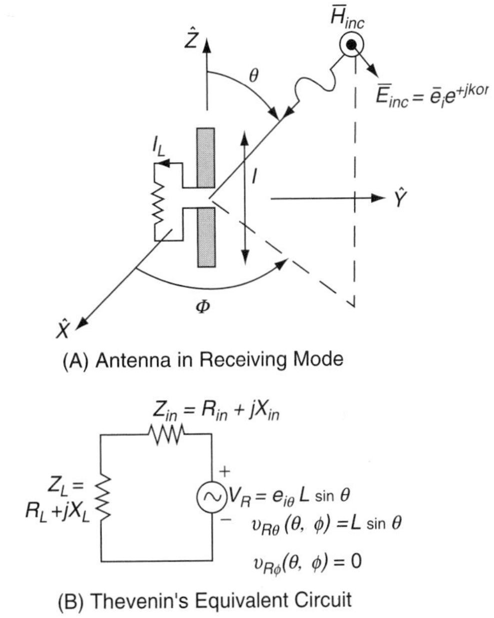

FIGURE 6.11 (A) A Small Dipole Antenna in its Receiving Mode of Operation (B) Thevenin's Equivalent Circuit of the Receiving Dipole Antenna

The receiving vector $\bar{v}_R$ is linearly polarized, with the receiving polarization unit vector in the $\hat{\theta}$ direction. From the magnitude of the receiving vector, the maximum effective area A_{em} can be found using the theory of Das (2004):

$$A_{em}(\theta, \phi) = \frac{\eta_0 |\bar{v}_R(\theta, \phi)|^2}{4R_{in}} = \frac{3\lambda_0^2 \sin^2\theta}{8\pi}. \qquad (6.146)$$

Notice that the ratio of $A_{em}(\theta, \phi)$ and $G(\theta, \phi)$ is equal to $\frac{\lambda_0^2}{4\pi}$, which is a constant independent of (θ, ϕ). This is the universal constant applicable for any arbitrary antenna, as derived in Das (2004):

$$\frac{A_{em}(\theta, \phi)}{G(\theta, \phi)} = \frac{3\lambda_0^2 \sin^2\theta}{8\pi(1.5\sin^2\theta)} = \frac{\lambda_0^2}{4\pi}. \qquad (6.147)$$

Dipole Antenna of Arbitrary Length

If the length of the dipole antenna is not electrically small, then one can derive the radiation fields by slicing it into small pieces and then superposing (integrating) the individual fields, if the current distribution is known. This is shown in Figure 6.12. Each piece can be modeled as a small dipole with a different current of excitation and a different location.

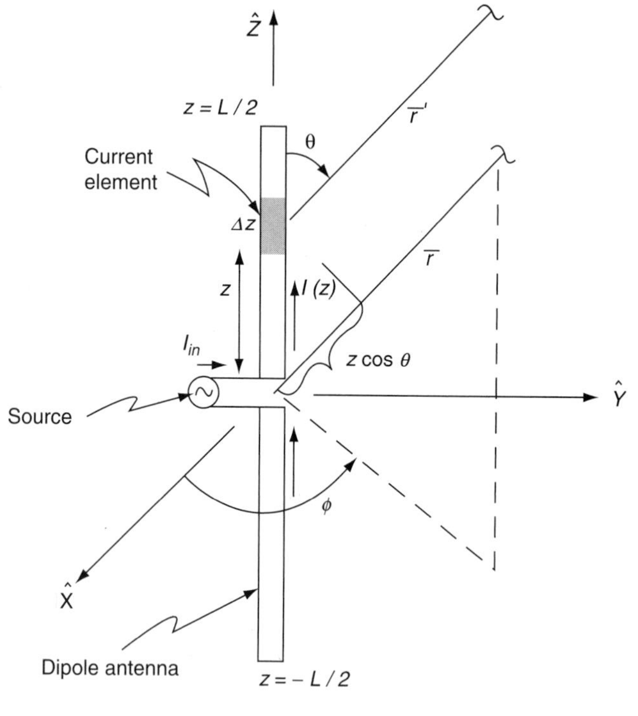

FIGURE 6.12 A Linear Dipole Antenna of Arbitrary Length L. This antenna can be treated as a number of "slices" of small dipole elements. The total field can be obtained by superposing the fields due to the individual small dipole slices having current $I(z)$, having length Δz, and being located at distance z from the center.

The elemental field $\Delta\bar{E}$ produced by a current element of length Δz, located at distance z from the center, with a current I_z, can be written using equations 6.134 through 6.136:

$$\Delta\bar{E} = \hat{\theta} I(z) \frac{e^{-jk_0 r'}}{r'} \frac{j\eta_0}{2\lambda_0} \sin\theta \Delta z; \; r' = r - z\cos\theta. \qquad (6.148)$$

Assuming that $z \ll r$ in the far-field, equation 6.148 can be simplified by approximating the r' in the denominator as r:

$$\Delta\bar{E} \simeq \hat{\theta} I(z) \frac{e^{-jk_0 r} e^{jk_0 z\cos\theta}}{r} \frac{j\eta_0}{2\lambda_0} \sin\theta \Delta z. \qquad (6.149)$$

The total field can be obtained by integrating equation 6.149 with respect to z, from $z = -L/2$ to $z = +L/2$:

$$\bar{E} = \hat{\theta} \frac{e^{-jk_0 r}}{r} \frac{j\eta_0}{2\lambda_0} \sin\theta \int_{z=-L/2}^{L/2} I(z) e^{jk_0 z\cos\theta} dz. \qquad (6.150)$$

If the current distribution $I(z)$ is known or can be approximated by a known function, then the far-field of the total antenna can be obtained using equation 6.150. One would expect the current distribution to exhibit a standing wave pattern, with zero current at the ends of the line ($z = \pm L/2$) and $I = I_{in}$ at the center. This behavior is similar to the current distribution along an open-circuited transmission line, having propagation constant k_0:

$$I(z) = I_{in} \frac{\sin[k_0(L/2 - |z|)]}{\sin(k_0 L/2)}; \; I(z = 0) = I_{in}. \qquad (6.151)$$

Using the current distribution of equation 6.151 in 6.150 and then performing the integration over z, one can obtain a close-form expression for the far-field distribution:

$$\bar{E}(r, \theta, \phi) = \hat{\theta} \frac{e^{-jk_0 r}}{r} \frac{j\eta_0}{2\lambda_0} \sin\theta \int_{z=-L/2}^{L/2} I_{in} \frac{\sin[k_0(L/2 - |z|)}{\sin(k_0 L/2)} e^{jk_0 z\cos\theta} dz$$
$$= \hat{\theta} \frac{e^{-jk_0 r}}{r} \frac{j\eta_0 I_{in}}{2\pi} \left(\frac{\cos(k_0 L/2 \cos\theta) - \cos(k_0 L/2)}{\sin(k_0 L/2)\sin\theta} \right). \qquad (6.152)$$

Once the electric field is obtained, then basic antenna parameters can be derived using steps similar to those used for a small dipole antenna (Das, 2004). The radiated power is obtained by integrating the power-density function over the sphere, as usual, from which the radiation resistance can be derived. The necessary spherical integration involving equation 6.152 cannot be easily expressed in closed form, requiring numerical integration.

Half-Wave Dipole Antenna

A dipole antenna with $L = \lambda_0/2$, called a **half-wave dipole**, is often useful because it operates close to a resonant condition of its input impedance. This results in an almost real value of the input impedance, which can be easily matched to an input source having a real source impedance for maximum power transfer. For this case, the equations 6.151 and 6.152 can be simplified:

$$I(z) = I_{\text{in}} \cos(k_0 z) = I_{\text{in}} \cos \frac{2\pi z}{\lambda_0}. \tag{6.153}$$

$$\bar{E}(r, \theta, \phi) = \hat{\theta} \frac{e^{-jk_0 r}}{r} \frac{j\eta_0 I_{\text{in}}}{2\pi} \left(\frac{\cos\left(\frac{\pi}{2}\cos\theta\right)}{\sin\theta} \right)$$
$$= \hat{\theta} \frac{e^{-jk_0 r}}{r} j60 I_{\text{in}} \left(\frac{\cos\left(\frac{\pi}{2}\cos\theta\right)}{\sin\theta} \right). \tag{6.154}$$

The polarization vector is in the $\hat{\theta}$ direction, which is the same as that for a small dipole. The power density P_d and power per unit solid angle p_d can be expressed using equation 6.154 (Das, 2004):

$$P_d(r, \theta, \phi) = \frac{|E_\theta|^2}{2\eta_0} = \frac{15|I_{\text{in}}|^2}{\pi r^2} \left(\frac{\cos\left(\frac{\pi}{2}\cos\theta\right)}{\sin\theta} \right)^2,$$
$$p_d(\theta, \phi) = \frac{15|I_{\text{in}}|^2}{\pi} \left(\frac{\cos\left(\frac{\pi}{2}\cos\theta\right)}{\sin\theta} \right)^2. \tag{6.155}$$

The direction of maximum radiation is along $\theta = \frac{\pi}{2}$, which is perpendicular to the direction of the antenna current. The power radiation pattern $\underline{p}_d(\theta, \phi)$ has the following form with respect to θ, having a uniform variation in the azimuth:

$$\underline{p}_d(\theta, \phi) = \frac{p_d(\theta, \phi)}{p_{d\max}} = \left(\frac{\cos\left(\frac{\pi}{2}\cos\theta\right)}{\sin\theta} \right)^2. \tag{6.156}$$

Now, by following steps similar to those used for a small dipole, the radiated power P_{rad}, radiation resistance R_{rad}, directive gain $D_g(\theta, \phi)$, and directivity D for a half-wave dipole antenna can be derived. This will require integrations involving the pattern function of equation 6.156:

$$P_{\text{rad}} = \int_{\phi=0}^{2\pi} \int_{\theta=0}^{\pi} p_d(\theta, \phi) \sin\theta \, d\theta \, d\phi = 30|I_{\text{in}}|^2 \int_{\theta=0}^{\pi} \frac{\cos^2\left(\frac{\pi}{2}\cos\theta\right)}{\sin\theta} d\theta. \tag{6.157}$$

$$R_{\text{rad}} = \frac{2P_{\text{rad}}}{|I_{\text{in}}|^2} = 60 \int_{\theta=0}^{\pi} \frac{\cos^2\left(\frac{\pi}{2}\cos\theta\right)}{\sin\theta} d\theta. \tag{6.158}$$

$$D_g(\theta, \phi) = \frac{4\pi \underline{p}_d(\theta, \phi)}{\int_{\phi=0}^{2\pi} \int_{\theta=0}^{\pi} \underline{p}_d(\theta, \phi) \sin\theta \, d\theta \, d\phi} = \frac{2\left(\frac{\cos\left(\frac{\pi}{2}\cos\theta\right)}{\sin\theta} \right)^2}{\int_{\theta=0}^{\pi} \frac{\cos^2\left(\frac{\pi}{2}\cos\theta\right)}{\sin\theta} d\theta}. \tag{6.159}$$

$$D = D_g(\theta, \phi)_{\max} = \frac{2}{\int_{\theta=0}^{\pi} \frac{\cos^2\left(\frac{\pi}{2}\cos\theta\right)}{\sin\theta} d\theta}. \tag{6.160}$$

The θ integral that identically appears in the above equations 6.157 through 6.160 may be computed numerically and has an approximate value of 1.22. Using this value, the antenna parameters can be approximated as follows:

$$P_{\text{rad}} = 36.6|I_{\text{in}}|^2, \; D_g(\theta, \phi) = 1.64 \left(\frac{\cos\left(\frac{\pi}{2}\cos\theta\right)}{\sin\theta} \right)^2, \tag{6.161}$$
$$D = 1.64, \; R_{\text{rad}} = 73.2 \; \Omega$$

Notice that the directivity D of a half-wave dipole ($= 1.64$) is not significantly different from that of a small dipole ($= 1.5$).

Folded Dipole Antenna

Figure 6.13 shows the geometry of a **folded dipole antenna.** It consists of a source (V_{in}) connected to the ends of a wire of total length $2L$ and folded once as a rectangular loop of height L and width $\delta \ll L$. The length L can have any value but is often designed to be about $\lambda_0/2$ for a resonant operation. This antenna configuration can be viewed as a superposition of an **odd mode** and an **even mode**, having dual excitations at the center of the two arms of the antenna with odd and even symmetry. The odd mode represents a transmission line mode of operation without any radiation. Here, the top and bottom sections are two identical transmission-line stubs of length $L/4$ connected in series, as shown in Figure 6.13. The input current I_t in this mode can then be related to V_{in} as follows:

$$I_t = \frac{V_{\text{in}}}{2Z_t} = \frac{V_{\text{in}}}{2jZ_0 \tan(k_0 L/2)}, \tag{6.162}$$

where Z_0 is the characteristic impedance of a two-wire transmission line having the same wire diameter as that of the folded dipole and having a line-to-line separation δ. In contrast, the even mode represents an antenna mode of operation with $I = 0$ at the top and bottom ends to satisfy the symmetry condition. The antenna mode operates very much as two dipole antennas of length L that are placed close to each other. The current distribution along the wire is the same as that for an equivalent dipole antenna and, therefore, has the same radiation pattern and gain. The input current I_a for the antenna mode can be related to the input voltage V_{in} using the input impedance Z_{dipole} of an equivalent dipole antenna of length L:

$$Z_{\text{dipole}} = \frac{V_{\text{in}}/2}{2I_a}, \; I_a = \frac{V_{\text{in}}}{4Z_{\text{dipole}}}. \tag{6.163}$$

Now, using superposition, the input impedance of the folded dipole can be expressed as a parallel combination of two parts:

$$I_{\text{in}} = I_a + I_t = \frac{V_{\text{in}}}{4Z_{\text{dipole}}} + \frac{V_{\text{in}}}{2jZ_0 \tan(k_0 L/27)}. \tag{6.164}$$

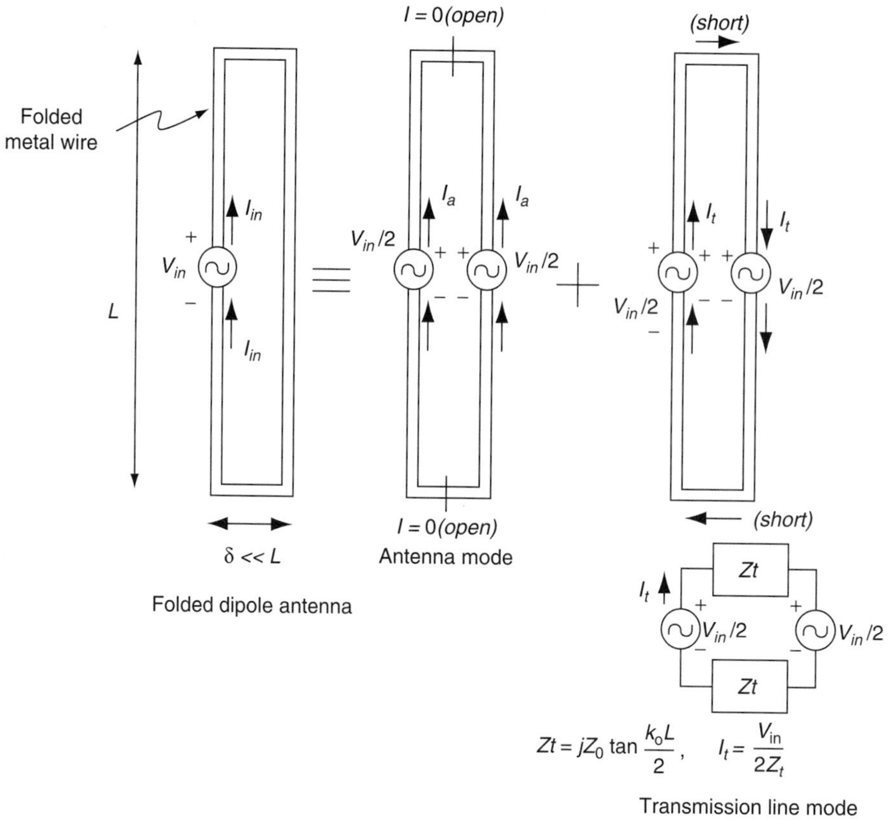

FIGURE 6.13 A Folded Dipole Antenna of Vertical Length L. The antenna can be treated as a superposition of an **even mode** that radiates as a regular dipole antenna of length L and an **odd mode** that can be viewed as two transmission line stubs of length $L/2$, each connected in series.

$$Z_{\text{in}} = \frac{V_{\text{in}}}{I_{\text{in}}} = \frac{1}{\frac{1}{4Z_{\text{dipole}}} + \frac{1}{2jZ_0 \tan(k_0 L/2)}} = (4Z_{\text{dipole}}) \,\|\, (2jZ_0 \tan(k_0 L/2)). \tag{6.165}$$

If the length L is a half wavelength, or $k_0 L/2 = \pi/2$, then the input impedance of the folded dipole antenna is four times that of a half-wave dipole antenna:

$$Z_{\text{in}}(L = \lambda_0/2) = 4Z_{\text{dipole}}(L = \lambda_0/2) = 4 \times 73 = 292\,\Omega. \tag{6.166}$$

So, a folded-dipole antenna is a derivative of a regular dipole antenna with the same radiation pattern, polarization, and directivity, except with a different input impedance. For the half-wave case, the folded-dipole antenna simply produces an impedance transformation by a factor of four. The resulting $292\,\Omega$ impedance is convenient to match to $300\,\Omega$ systems used in some broadcast applications.

6.7.2 Monopole Antennas

Monopole antennas, as shown in Figure 6.14, constitute a group of derivatives of dipole antennas. Here, only half of the dipole antenna is needed for operation. A metal ground plane (ideally of infinite size) is used, with respect to which the excitation voltage is applied to the half structure. The half structure for a regular dipole antenna is called a monopole antenna, in reference to the presence of only one physical side. A similar half structure for a folded dipole antenna is called a folded monopole antenna. The presence of the ground plane allows the monopole antenna to operate as electrically equivalent to a dipole antenna. The ground plane equivalently replaces the lower half by an imaging principle, similar to creating an optical image through a mirror. Notice in Figure 6.14 that for the currents in the monopole and dipole structures to be the same, one needs the source voltage of the equivalent dipole antenna to be twice that of the monopole antenna. As a result, the input impedance of the monopole structure is half that of the corresponding dipole structure:

$$Z_{\text{monopole}} = \frac{1}{2} Z_{\text{dipole}};\ Z_{\text{monopole}} = \frac{V_{\text{in}}}{I_{\text{in}}},\ Z_{\text{dipole}} = \frac{2V_{\text{in}}}{I_{\text{in}}}. \tag{6.167}$$

For example, if the length of the monopole antenna $L = \lambda_0/4$, such that the corresponding length of the equivalent dipole antenna is $\lambda_0/2$, the following values of the radiation impedances result:

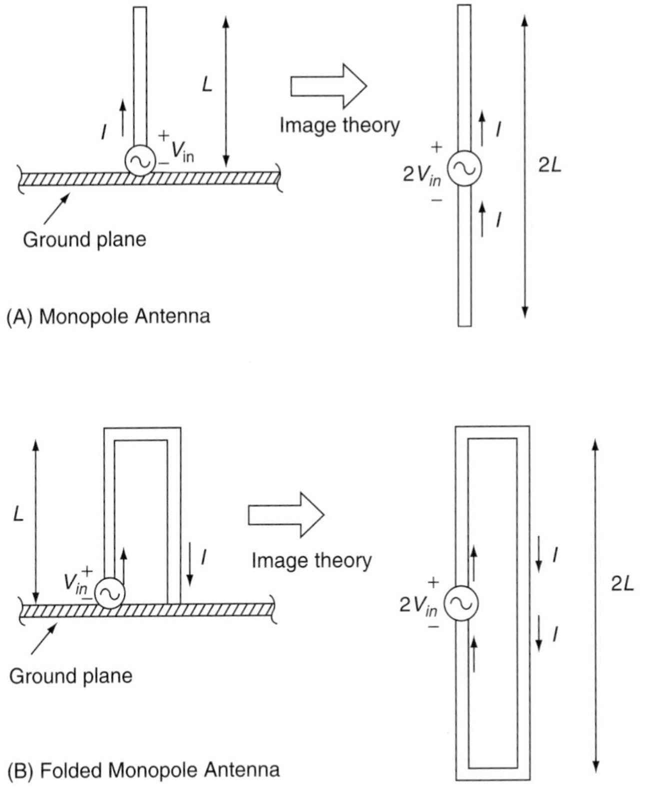

FIGURE 6.14 The Corresponding Equivalent Dipole Structures Employing Image Theory

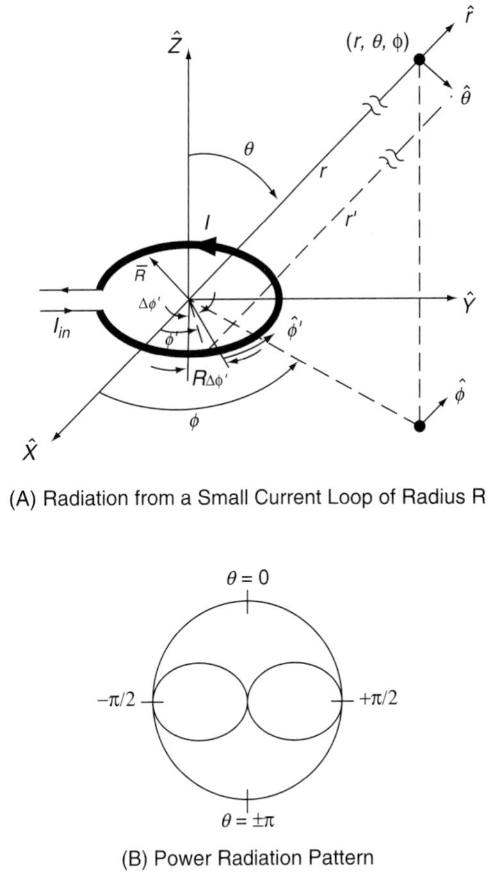

FIGURE 6.15 Circular Loop Antenna

$$Z_{\text{monopole}} = \frac{1}{2} Z_{\text{dipole}} = 36.5\,\Omega;$$
$$Z_{\text{foldedmonopole}} = \frac{1}{2} Z_{\text{foldeddipole}} = 146\,\Omega. \quad (6.168)$$

Due to the imaging principle, the polarization of radiation and radiation patterns of a monopole antenna is the same as that of its equivalent dipole antenna. However: the monopole antenna has a field only in the top half of the space, having zero radiation below the ground plane. In contrast, the equivalent dipole structure has fields in both sides, with the radiation to the bottom side symmetric to that above. In this situation, the expression of the directive gain functions in equations 6.141, 6.142, 6.159, and 6.160 (Das, 2004) can be used to show that the directivity D of a monopole antenna is twice that of its equivalent dipole structure.

$$D_{\text{monopole}} = 2D_{\text{dipole}};\ D_{\text{short monopole}} = 3.0,$$
$$D_{\text{half-wave monopole}} = 3.28. \quad (6.169)$$

6.7.3 Loop Antenna

A **loop antenna** consists of a wire loop of arbitrary shape fed by an input current I_{in}. A circular loop, as shown in Figure 6.15, is the shape most commonly used. The radius of the loop, R, could be electrically small or large. If the current distribution along the wire loop is known, or can at least be closely approximated based on some physical insight, then the radiated fields of the loop antenna can be derived analytically or computed numerically. This is possible by treating the loop as a superposition of small current elements with known amplitudes, physical location, and orientation. The far-field quantities for a small dipole antenna, as presented in Section 6.7.1, with suitable transformation of its coordinates to account for different physical location and orientation, can be used here to model the elemental radiation.

If the current loop is small ($R \ll \lambda_0$), the far-field quantities can be derived into a simple final form. In this case, the current along the loop $I(\phi')\hat{\phi}'$ can be assumed to be independent of $\hat{\phi}'$, with a magnitude equal to the input current I_{in}. With reference to Figure 6.15, consider the small current element of length $R\Delta\phi'$ carrying a current along the direction $\hat{\phi}'$. The radiation fields due to this current element can be expressed using results from Section 6.7.1, equations 6.134 and 6.135. One must properly adjust these equations, however, to account for the orientation of the current element along $\hat{\phi}'$ and its location at (R, ϕ'), defined by a position vector $\bar{R} = R\hat{R}$.

For simplicity of analysis, one should first use equation 6.135 to express the radiating magnetic field $\bar{H}$ due to the current element:

$$\Delta\bar{H}(r,\theta,\phi;\phi') = \frac{jI_{\text{in}}R\Delta\phi'}{2\lambda_0}\frac{e^{-jk_0r'}}{r'}(\hat{\phi}'\times\hat{r}). \tag{6.170}$$

$$\hat{\phi}' = \cos(\phi-\phi')\hat{\phi} + \sin\theta\sin(\phi-\phi')\hat{\theta}. \tag{6.171}$$

$$\bar{R} = R\hat{R} = R[\sin\theta\cos(\phi-\phi')\hat{r} + \cos\theta\cos(\phi-\phi')\hat{\theta} - \sin(\phi-\phi')\hat{\phi}]. \tag{6.172}$$

$$r' = r - \bar{R}\cdot\hat{r} = r - R\sin\theta\cos(\phi-\phi'). \tag{6.173}$$

$$(\hat{\phi}'\times\hat{r}) = \cos(\phi-\phi')\hat{\theta} - \sin\theta\sin(\phi-\phi')\hat{\phi}. \tag{6.174}$$

Using the coordinate relationships of equations 6.171 through 6.174, equation 6.170 can be rewritten in the following form:

$$\Delta\bar{H}(r,\theta,\phi;\phi') \simeq \frac{jI_{\text{in}}R\Delta\phi'}{2\lambda_0}\frac{e^{-jk_0r}e^{jk_0R\sin\theta\cos(\phi-\phi')}}{r} \times [\cos(\phi-\phi')\hat{\theta} - \sin\theta\sin(\phi-\phi')\hat{\phi}]. \tag{6.175}$$

In the far-field region, $r \gg R$. This condition has been used in equation 6.175 to approximate r' in the denominator by r. In addition, for a small loop, $R \ll \lambda_0 = \frac{2\pi}{k_0}$, which is equivalent to $Rk_0 \ll 1$. This condition can be used to further approximate equation 6.175 by approximating one of the exponential functions in the numerator:

$$\Delta\bar{H}(r,\theta,\phi;\phi') \simeq \frac{jI_{in}R\Delta\phi'}{2\lambda_0}\frac{e^{-jk_0r}[1 + jk_0R\sin\theta\cos(\phi-\phi')]}{r} \times [\cos(\phi-\phi')\hat{\theta} - \sin\theta\sin(\phi-\phi')\hat{\phi}]. \tag{6.176}$$

The total magnetic field can be obtained by integrating equation 6.176 over $0 < \hat{\phi}' < 2\pi$ to cover the entire current loop. The $\hat{\phi}$ component of equation 6.176 can be shown to integrate to zero, whereas the $\hat{\theta}$ component will be nonzero:

$$\bar{H}(r,\theta,\phi) = \int_{\phi'=0}^{2\pi}\Delta\bar{H}(r,\theta,\phi;\hat{\phi}') = \frac{jI_{\text{in}}R}{2\lambda_0}\frac{e^{-jk_0r}}{r}(jk_0R\sin\theta)(\pi\hat{\theta})$$
$$= -\frac{I_{\text{in}}(\pi R^2)}{\lambda_0^2}\frac{e^{-jk_0r}}{r}(\pi\sin\theta)\hat{\theta} = -\frac{e^{-jk_0r}}{4\pi r}I_{\text{in}}Sk_0^2\sin\theta\hat{\theta};\ S = \pi R^2. \tag{6.177}$$

Once the total radiating magnetic field is known, the radiating electric field can be obtained from it using their far-field relationship:

$$\bar{E}(r,\theta,\phi) = \hat{\phi}\frac{\eta_0k_0^2(I_{\text{in}}S)\sin\theta e^{-jk_0r}}{4\pi r} = \eta_0\bar{H}(r,\theta,\phi)\times\hat{r}. \tag{6.178}$$

If there are N number of loops in the form of a coil instead of a single loop shown in Figure 6.15, then the corresponding field quantities would have to be multiplied by N. One may view the far-field of a small loop antenna as **dual** to that of a small dipole antenna. This is because the loop antenna has H_θ and E_ϕ far-field components, as shown in equations 6.177 and 6.178, in contrast to which a small dipole antenna has E_θ and H_ϕ components. Both the small loop and the small dipole antennas have the same $\sin\theta$ field pattern, thus having the same directive gain $D_g(\theta,\phi) = 1.5\sin^2\theta$ and directivity $D = 1.5$. Once the radiation fields are known, using a procedure similar to that used for a small dipole antenna in Section 6.7.1, one can derive the total radiated power, P_{rad}, and the radiation resistance, R_{rad}.

$$P_{\text{rad}} = \frac{\eta_0k_0^4(S|I_{\text{in}}|)^2}{12\pi},\ R_{\text{rad}} = \frac{2P_{\text{rad}}}{|I_{\text{in}}|^2} = \frac{\eta_0k_0^4S^2}{6\pi} = 20k_0^4S^2. \tag{6.179}$$

It may be noted, that equation 6.179 needs to be multiplied by N^2 for a N-loop coil antenna.

6.7.4 Slot Antenna

A **slot antenna**, as shown in Figure 6.16, is considered another basic radiating element. It is realized by removing a small area of metal from an infinite ground plane. For a practical design, however, the ground plane may be finite but large in size. The open area, or the 'slot,' can be of any shape and size, but is shown in Figure 6.16 to be rectangular, as is often done for simplicity. As shown in this figure, a slot antenna may be excited by applying a voltage source across the slot.

Babinet's Principle

One may view a slot antenna to be physically *complementary* to a **strip dipole antenna** consisting of a metal strip of the same

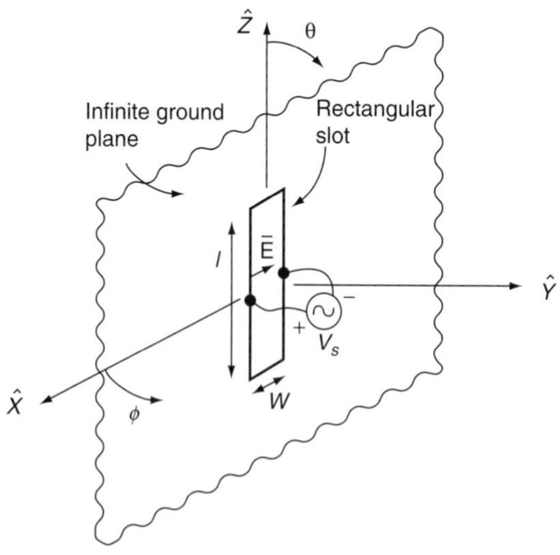

FIGURE 6.16 A Slot Antenna. This antenna is produced by a rectangular opening at the center of an infinite ground plane.

shape and size as that of the slot. The complementary strip dipole is to be excited by a voltage gap, breaking across the strip at the same position where the slot antenna is excited. If the slot antenna is superimposed on its complementary dipole antenna structure, one would obtain an infinite conducting plane. Interestingly, employing certain symmetry in Maxwell's equations and realizing the complementary nature of the above two structures, it can be shown that the fields of a slot antenna and that of a complementary dipole antenna are related to each other in a simple manner. This theory is called **Babinet's principle** (Balanis, 1997), which consequently relates the various transmit/receive characteristics, such as the pattern, gain, and input impedance of the two complementary antennas.

According to Babinet's principle, the fields produced by a slot antenna can be obtained directly from the corresponding complementary strip dipole antenna using the following substitutions:

- For the front side of the slot antenna of ($0 < \phi < \pi$; see Figure 6.16), replace $\bar{E}$ of the dipole with $-\bar{H}$ for the slot antenna, $\bar{H}$ with $\bar{E}$, ε_0 with μ_0, μ_0 with ε_0, and η_0 with $1/\eta_0$, but replace the input current I_{in} of the strip dipole with twice the source voltage V_s for the slot antenna.
- For the back side ($\pi < \phi < 2\pi$), one needs the same substitutions as for the front side, except the signs for the field expressions are reversed.

The above substitutions also lead to a simple relationship between the complex input impedances of the complementary antennas. If Z_{sl} is the complex input impedance as seen across the input of an arbitrary slot antenna and Z_{cd} is that of its complementary dipole antenna, then using the substitutions prescribed in Babinet's principle, one can show:

$$4\frac{Z_{cd}}{\eta_0} = \frac{\eta_0}{Z_{sl}},\; Z_{cd}Z_{sl} = \frac{\eta_0^2}{4}. \tag{6.180}$$

Small Slot Antenna

As an example, let us apply Babinet's principle to a **small rectangular slot antenna**, whose complementary structure is a small dipole antenna studied in Section 6.7.1. Employing the substitutions prescribed in the previous subsection, the far-field expressions for the small slot antenna can be obtained from those in equations 6.134 and 6.135 for a short dipole antenna:

$$\bar{H}(r, \theta, \phi) = \begin{cases} -\hat{\theta} V_s \frac{e^{-jk_0 r}}{r}\frac{jl}{\eta_0\lambda_0}\sin\theta; & 0 < \phi < \pi. \\ \hat{\theta} V_s \frac{e^{-jk_0 r}}{r}\frac{jl}{\eta_0\lambda_0}\sin\theta; & \pi < \phi < 2\pi. \end{cases} \tag{6.181}$$

$$\bar{E}(r, \theta, \phi) = \begin{cases} \hat{\phi} V_s \frac{e^{-jk_0 r}}{r}\frac{jl}{\lambda_0}\sin\theta; & 0 < \phi < \pi. \\ -\hat{\phi} V_s \frac{e^{-jk_0 r}}{r}\frac{jl}{\lambda_0}\sin\theta; & \pi < \phi < 2\pi. \end{cases} \tag{6.182}$$

Using these field expressions described previously, one can obtain the antenna parameters for the small slot antenna following similar steps to Section 6.7.1 for a complementary dipole antenna:

$$P_d(r, \theta, \phi) = \frac{|V_s|^2}{2\eta_0 r^2}\left(\frac{l}{\lambda_0}\right)^2 \sin^2\theta,\; p_d(\theta, \phi) = \frac{|V_s|^2}{2\eta_0}\left(\frac{l}{\lambda_0}\right)^2 \sin^2\theta. \tag{6.183}$$

$$\underline{p}_d(\theta, \phi) = \frac{p_d(\theta, \phi)}{p_{d\max}} = \sin^2\theta. \tag{6.184}$$

$$P_{\text{rad}} = \int_{\phi=0}^{2\pi}\int_{\theta=0}^{\pi} p_d(\theta, \phi)\sin\theta\, d\theta\, d\phi = \frac{|V_s|^2 4\pi}{3\eta_0}\left(\frac{l}{\lambda_0}\right)^2 = \frac{|V_s|^2}{2R_{\text{rad}}}. \tag{6.185}$$

$$G_{\text{rad}} = \frac{1}{R_{\text{rad}}} = \frac{8\pi}{3n_0}\left(\frac{l}{\lambda_0}\right)^2 = \frac{1}{45}\left(\frac{l}{\lambda_0}\right)^2. \tag{6.186}$$

$$D_g(\theta, \phi) = \frac{3\sin^2\theta}{2};\; D = 1.5. \tag{6.187}$$

Comparing the radiation resistances of a small slot in equation 6.186 to that of a small dipole antenna in equation 6.140, one sees that $R_{\text{rad}}(\text{dipole}) \times R_{\text{rad}}(\text{slot}) = \frac{\eta_0^2}{4}$, which is consistent with the general impedance relationship of equation 6.180. As shown in equation 6.184, a narrow slot antenna (see Figure 6.16) has the same $\sin^2\theta$ power pattern as that of a small dipole antenna. The small slot antenna has E_ϕ and H_θ radiation fields, in contrast with the E_θ and H_ϕ radiation fields of its complementary small dipole antenna.

Similar procedures can be extended to a half-wave slot antenna or, in general, to a rectangular slot antenna of any arbitrary length. This will allow derivation of the far fields, as well as radiation/impedance parameters of a long slot antenna, by relating to those of the complementary dipole structures in Section 6.2.1 under Dipole Antenna of Arbitrary Length.

6.7.5 Microstrip Antenna

The **microstrip antenna** is a relatively modern invention. It was invented to allow convenient integration of an antenna and other driving circuitry of a communication system on a common printed-circuit board or a semiconductor chip (Carver and Mink, 1981; Pozar, 1992). Besides other resulting advantages, the integrated-circuit technology for the antenna fabrication allowed high dimensional accuracy, which was otherwise difficult to achieve in traditional fabrication methods. The geometry of a microstrip antenna consists of a dielectric substrate of certain thickness d, having a complete metalization on one of its surfaces and of a metal "patch" on the other side. The substrate is usually thin ($d \ll \lambda$). The metal patch on the front surface can have various shapes, although a rectangular shape, as shown in Figure 6.17, is commonly used. The antenna may be excited using various methods (Pozar, 1992; Pozar and Schawbert, 1995). One common approach is to feed from a microstrip line, connecting the microstrip antenna at the center of one of its edges. The microstrip line

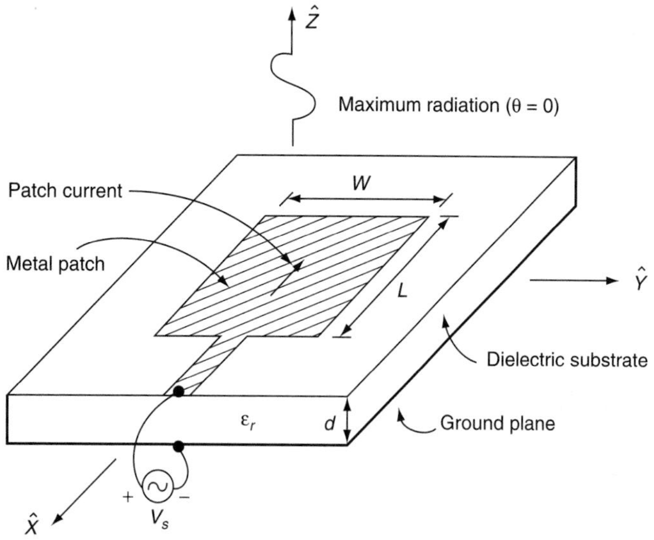

FIGURE 6.17 Geometry of a Rectangular Microstrip Antenna. It consists of a rectangular metal patch on a dielectric substrate and is excited by a voltage source across the metal patch and the bottom ground plane of the substrate. The microstrip antenna produces maximum radiation in the broadside direction ($\theta = 0$), with ideally no radiation along the substrate edges ($\theta = 90°$).

may be connected to a feeding circuitry or directly fed by connecting a signal source across the microstrip line and the ground plane.

The microstrip antenna produces maximum radiation in the broadside (perpendicular to the substrate) direction and ideally no radiation in the end-fire (along the surface of the substrate) direction. The size of the antenna is usually designed such that the antenna resonates at the operating frequency, producing a real input impedance. For a rectangular microstrip antenna, this requires the length of the antenna, L, to be about half a wavelength in the dielectric medium. The width of the antenna, W, on the other hand, determines the level of the input impedance. The microstrip antenna can be thought of as a rectangular cavity with open sidewalls. The fringing fields through the open sidewalls are responsible for the radiation. However, the structure is principally a resonant cavity, with only limited fringing radiation. Therefore, the bandwidth of the radiation is poor compared to the bandwidth of antennas discussed earlier. The small bandwidth, however, is adequate in a large class of communication applications. Readers may refer to Balanis (1997) and Carver and Mink (1981) for some analytical modeling of a microstrip antenna. Simple and approximate expressions for the radiated electric field components of a microstrip antenna are given by Carver and Mink (1981):

$$E_\theta = \frac{-jV_o k_0 W e^{-jk_0 r}}{\pi r} \cos\phi \cos(k_0 \sqrt{\varepsilon_r} d \cos\theta) \times \frac{\sin(k_0 \sin\theta \sin\phi W/2)}{k_0 \sin\theta \sin\phi W/2} \cos(k_0 L/2 \sin\theta \cos\phi);\ 0 \le \theta \le \pi/2. \quad (6.188)$$

$$E_\phi = \frac{jV_o k_0 W e^{-jk_0 r}}{\pi r} \sin\phi \cos\theta \cos(k_0 \sqrt{\varepsilon_r} d \cos\theta) \times \frac{\sin(k_0 \sin\theta \sin\phi W/2)}{k_0 \sin\theta \sin\phi W/2} \cos(k_0 L/2 \sin\theta \cos\phi);\quad 0 \le \theta \le \pi/2. \quad (6.189)$$

More accurate and complete modeling of various characteristics of a microstrip antenna often requires numerical methods (Pozar and Schawbert, 1995). Sophisticated designs, including techniques for achieving broad bandwidths, are also found in Pozar and Schawbert (1995).

6.7.6 Other Common Antenna Geometries

There are many possible antenna geometries in practical use today, each having some desirable properties suited for specific applications. This chapter has only discussed a few of the most basic radiating elements. The reader may be referred to antenna textbooks (Balanis, 1997; Kraus, 1950; Collin and Zucker, 1969; Jasik, 1961; Schelkunoff and Friis, 1952; Kraus and Fleisch, 1999) for a more elaborate set of practical antenna geometries. A few of the antenna geometries in common use are shown in Figure 6.18: (A) a **horn antenna**, (B) a **Yagi-Uda array**, and (C) a **spiral plate antenna**. A horn antenna is distinguished by its feeding from a input waveguide. The Yagi-Uda array is popularly used as a television antenna, having a high-gain radiation along the end-fire direction over a broad bandwidth. A much broader bandwidth of radiation can be achieved by a class of **frequency-independent** antennas, such as a spiral plate antenna. In principle, one needs an infinite-sized spiral structure for an ideal frequency-independent operation. However, a practical spiral antenna will have to be truncated at a finite spiral length. This would result in a finite, but large bandwidth. Other frequency-independent antennas include conical spiral and log-periodic structures.

Another common class of antennas make use of additional material structures, such as a focusing reflector or a dielectric lens structure, surrounding a basic source antenna to achieve focusing of the antenna radiation. This results in increasing the antenna gain. Although these structures are not really a part of the fundamental radiation mechanism, their designs are often critical for the overall performance of the total antenna system.

6.8 Antenna Array

An arbitrary configuration of an array of N identical antennas is shown in Figure 6.19. The center of the ith antenna is defined by a position vector $\bar{R}_i = \hat{x}x_i + \hat{y}y_i + \hat{z}z_i$, and the input current to the ith antenna is I_i. The electric field of the array in the far-field region can be expressed as the superposition of the individual contributions:

$$\bar{E}_i = I_i \frac{e^{-jk_0 r_i}}{r_i} \bar{e}_i(\theta, \phi),\ \bar{E} = \sum_{i=1}^{N} \bar{E}_i = \sum_{i=1}^{N} I_i \frac{e^{-jk_0 r_i}}{r_i} \bar{e}_i(\theta, \phi). \quad (6.190)$$

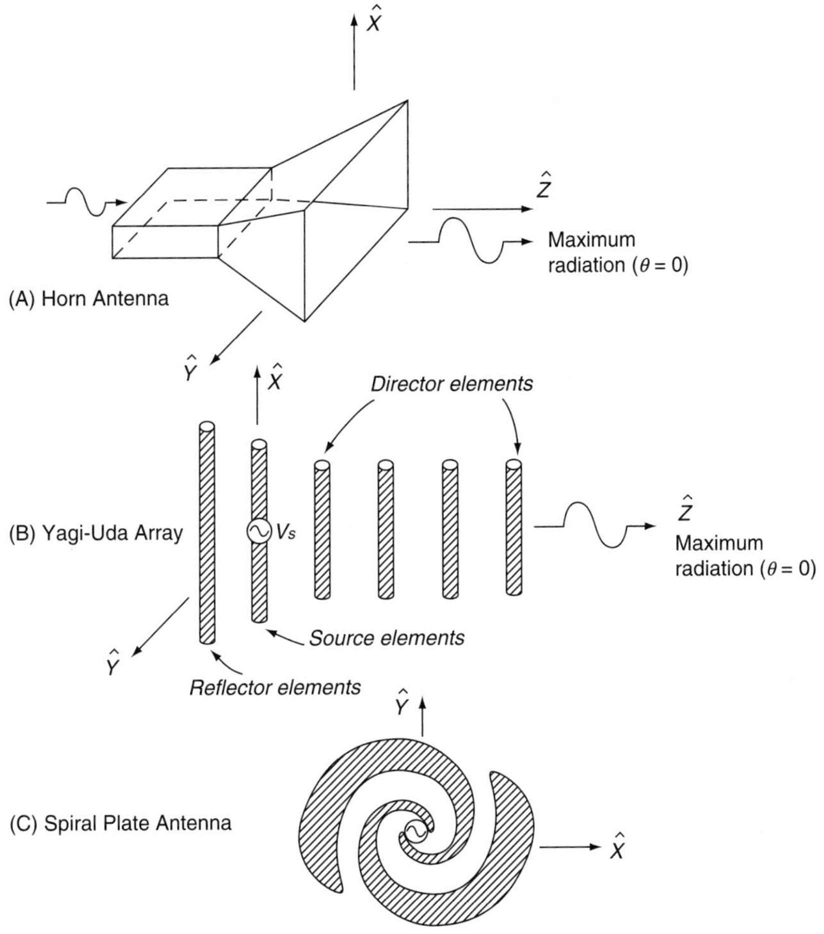

FIGURE 6.18 A Few Practical Antenna Geometries. (A) A waveguide horn antenna is excited from a waveguide input. (B) A Yagi-Uda array is commonly used in television receivers, having an end fire and high-gain beam. (c) A spiral plate antenna is shown here with broad band radiation along the broadside (z) direction.

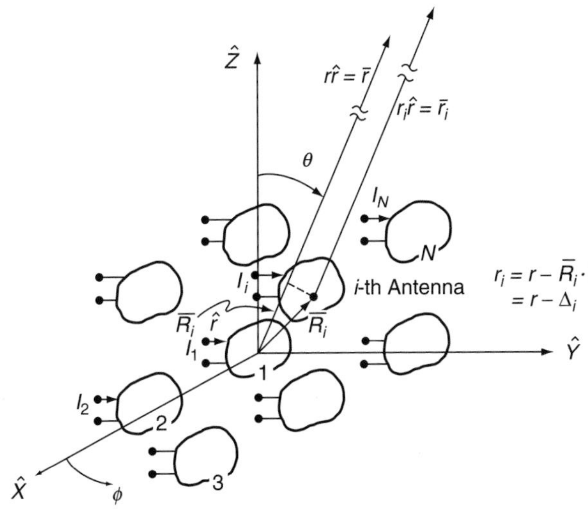

FIGURE 6.19 A General Configuration of an Array of N Identical Antennas. Each antenna is excited by a different input current I_i and placed at a different location $\bar{R}_i = x_i\hat{x} + y_i\hat{y} + z_i\hat{z}$.

$$\bar{r}_i = \bar{r} - \bar{R}_i = \hat{r}r_i, r - \bar{R}_i \cdot \hat{r} = r - \Delta_i. \qquad (6.191)$$

in the far-field region, $r \gg \Delta_i$. Under this situation, the r_i in the denominator of equation 6.190 may be approximated as $r_i = r - \Delta_i \simeq r$. Such approximation, however, is not appropriate for the propagation factor in the numerator because of the oscillating nature of the propagation factor, which is sensitive to the small difference Δ_i between r_i and r.

$$\bar{E} = \sum_{i=1}^{N} I_i \frac{e^{-jk_0(r-\Delta_i)}}{r - \Delta_i}\bar{e}_i(\theta, \phi) \simeq \frac{e^{-jk_0 r}}{r}\sum_{i=1}^{N} I_i e^{jk_0\Delta_i}\bar{e}_i(\theta, \phi). \quad (6.192)$$

Often it may be assumed, with good approximation, that the transmit properties of each antenna are unaffected by the presence of the other elements. This assumption is valid when the **mutual-coupling effects** between the antennas have negligible effects on the radiation characteristics of the individual elements. Under this practical approximation, the total radiation from the array can be further simplified:

$$\bar{E} \simeq \bar{e}(\theta, \phi)\frac{e^{-jk_0 r}}{r}\sum_{i=1}^{N} I_i e^{jk_0\Delta_i}; \ \bar{e}_i(\theta, \phi) = \bar{e}(\theta, \phi). \quad (6.193)$$

$$\bar{E} = (EF)\times(AF); \ EF = \bar{e}(\theta, \phi)\frac{e^{-jk_0 r}}{r}, \ AF = \sum_{i=1}^{N} I_i e^{jk_0\Delta_i}. \quad (6.194)$$

The total radiation is thus separated into two independent factors. The element factor (EF) is the contribution due to the individual antenna elements, whereas the array factor (AF) may be seen as the effect due to the array configuration. Such factoring into EF and AF allows the study and design of the array to be used, independent of the specific antenna elements. When the antenna input currents are allowed to have different phases α_i as well as different amplitudes $|I_i|$, the AF may be expressed in the following compact form:

$$AF = \sum_{i=1}^{N} |I_i| e^{j\alpha_i} e^{jk_0\Delta_i} = \sum_{i=1}^{N} |I_i| e^{j\psi_i}; \ \psi_i = \alpha_i + k_0\Delta_i$$
$$= \alpha_i + k_0(\hat{r}\cdot\bar{R}_i). \quad (6.195)$$

6.8.1 One-Dimensional Array

As an example, consider an array of N antenna elements, equally spaced along the $\hat{y}$ axis, as shown in Figure 6.20. Also assume that each antenna is excited with the same amplitude, but phased in a linearly progressive manner. The array factor for this case can be expressed as follows:

$$I_i = |I_i| e^{j(i-1)\alpha} = I e^{j(i-1)\alpha}. \quad (6.196)$$

$$\bar{R}_i = \hat{y}(i-1)d, \ \Delta_i = \hat{r}\cdot\bar{R}_i = (i-1)d\sin\theta\sin\phi = (i-1)\Delta. \quad (6.197)$$

$$AF = I\sum_{i=1}^{N} e^{j\psi_i}; \ \psi_i = (i-1)\psi = (i-1)(\alpha + k_0\Delta)$$
$$= (i-1)(\alpha + k_0 d\sin\theta\sin\phi). \quad (6.198)$$

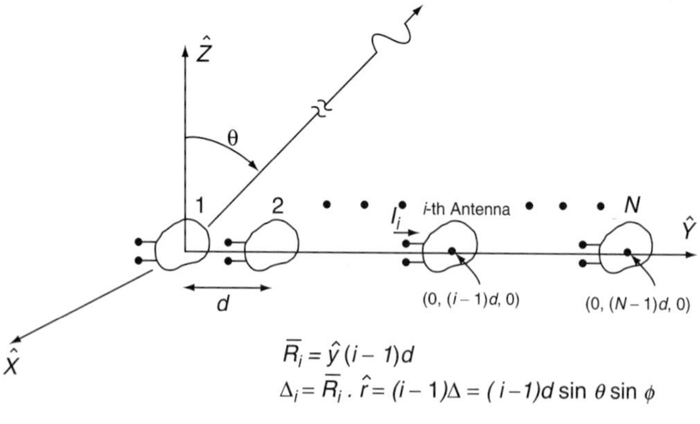

FIGURE 6.20 A One-Dimensional Array of N Antenna Elements. Elements are placed along the $\hat{y}$ axis, with interelement spacing d. The input currents are of equal magnitude I, but are progressively phased with interelement phase shift α.

$$|AF| = I\left|\sum_{i=1}^{N} e^{j(i-1)\psi}\right| = I\left|\frac{1-e^{jN\psi}}{1-e^{j\psi}}\right| = I\left|\frac{\sin\frac{N\psi}{2}}{\sin\frac{\psi}{2}}\right|. \quad (6.199)$$

It may be useful to normalize the array factor with respect its maximum value NI:

$$AF_n(\theta, \phi) = \frac{|AF(\theta, \phi)|}{|AF(\theta, \phi)|_{max}} = \frac{\sin\frac{N\psi}{2}}{N\sin\frac{\psi}{2}}; \ AF_n(\theta, \phi) \le 1. \quad (6.200)$$

Array Side Lobe Level

The peak value of the normalized array factor $AF_n(\theta, \phi)$ is unity, which occurs at $\psi = 0$ corresponding to the main direction of radiation. The next peak of AF_n occurs approximately at $N\psi/2 = 1.5\pi$, where $AF_n = \frac{1}{N\sin\frac{1.5\pi}{N}}$. This corresponds to the first side lobe level (SLL) of the array factor. For an array with large number of elements N, this level can be shown to be independent of N:

$$SLL = \frac{1}{N\sin\frac{1.5\pi}{N} \simeq \frac{1}{1.5\pi} \cong 13\text{dB}.} \quad (6.201)$$

Assuming that the element factor EF does not change significantly from the main beam to the side lobe of the array, the SLL of equation 6.201 is approximately equal to the SLL of the total radiation pattern. Remember that this SLL is derived for any large array ($N \to \infty$) while assuming that the amplitudes of excitation are uniform over the array. One can achieve a lower value of SLL, however, by using a nonuniform amplitude distribution.

Graphical Method of Sketching Array Factor

Notice that the array factor in equation 6.199 is represented as a function of $\psi = \alpha + k_0 d\sin\theta\sin\phi$. This should be fairly easy to sketch on the ψ plane. Ultimately one needs to sketch the array factor as a function of the space angles (θ, ϕ). This mapping may be done in a graphical manner, as demonstrated in Figure 6.21 for a four-element array. In Figure 6.21, the mapping from the ψ (upper sketch) to the θ plane (lower polar sketch) is demonstrated for the specific case of $\phi = \pi/2$ and $\alpha = \pi/2$ when $d = \lambda_0/2$. The reasoning for the method, as illustrated in this figure, should be self-explanatory and accordingly extended for other values of ϕ and α.

Notice that the array factor will be maximum ($= NI$) when ψ is zero or even multiples of π. This will determine the angle of the maximum radiation θ_{max} for a given value of ϕ, or vice versa:

$$|AF(\theta, \phi)|_{max} = I\left|\frac{\sin\frac{N\psi_{max}}{2}}{\sin\frac{\psi_{max}}{2}}\right| = NI, \quad (6.202)$$

$$\psi_{max} = \alpha + k_0 d\sin\theta_{max}\sin\phi = 2n\pi,$$
$$n = 0, \pm 1, \pm 2, \ldots \quad (6.203)$$

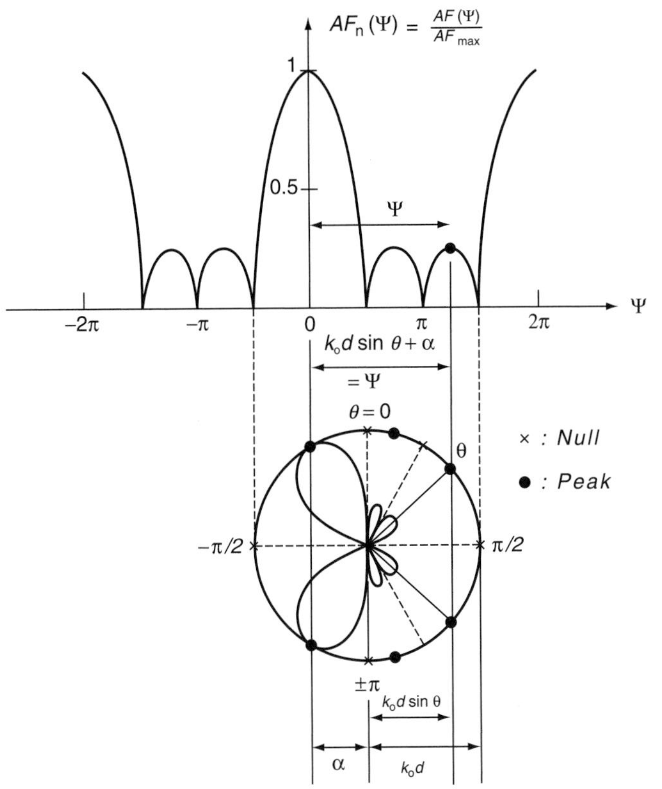

FIGURE 6.21 Graphical Method of Sketching the Array Factor of a Four-Element Array. Sketching is done as a function of θ in the polar form. (Top) Normalized array factor as a function of ψ, plotted in the rectangular coordinates. (Bottom) Mapping from Ψ to θ in polar coordinates, with $\phi = \pi/2$, where $\Psi = k_0 d \sin\theta \sin\phi + \alpha = k_0 d \sin\theta + \alpha$.

For example, if on the $\phi = \pi/2$ plane:

$$\psi_{max} = \alpha + k_0 d \sin\theta_{max} = 2n\pi, \quad n = 0, \pm 1, \pm 2, \ldots; \phi = \frac{\pi}{2}. \tag{6.204}$$

Grating Lobes

In general, depending on the spacing d, there can be more than one value of n for which one can find a solution for θ_{max}. The radiation lobe corresponding to $n = 0$ is called the **main lobe**, whereas all other lobes corresponding to all n other than $n = 0$ are called the **grating lobes** (borrowing the name from a similar situation in the study of periodic gratings). For the main-lobe radiation:

$$\theta_{max} = \arcsin\frac{-\alpha}{k_0 d}; \; n = 0, \; \phi = \pi/2. \tag{6.205}$$

Accordingly, the value of θ_{max} can be steered by controlling the interelement phasing α. However, it may be desirable to design d such that no grating lobe is allowed to exist for all possible values of α. As can be shown from equation 6.205, this is achieved if $k_0 d \leq \pi$, requiring the interelement spacing α to be less than half a wavelength.

6.8.2 Two-Dimensional Array

The one-dimensional array has the freedom of steering the main beam only in one plane: θ or ϕ. The concept may be extended to an array in two-dimensions, as shown in Figure 6.22, to allow beam steering both in θ and ϕ. As shown in this figure, consider an array of $(m \times n)$ elements placed on the x–y plane, with uniform amplitudes of excitation. The interelement spacing is d_x along the x axis and d_y along the y axis, which are in general different from each other. The interelement phasing is α_x and α_y, respectively, along the x and y axes. The interelement phasing can be independently controlled to allow steering in both θ and ϕ.

The derivation of the array factor for the two-dimensional array, as presented below, follows the same procedure used for a one-dimensional array. The same terminology is also used for equivalent parameters. Needed are double indexing of the elements along x and y as well as double summations for the total fields, instead of a single summation for a one-dimensional array:

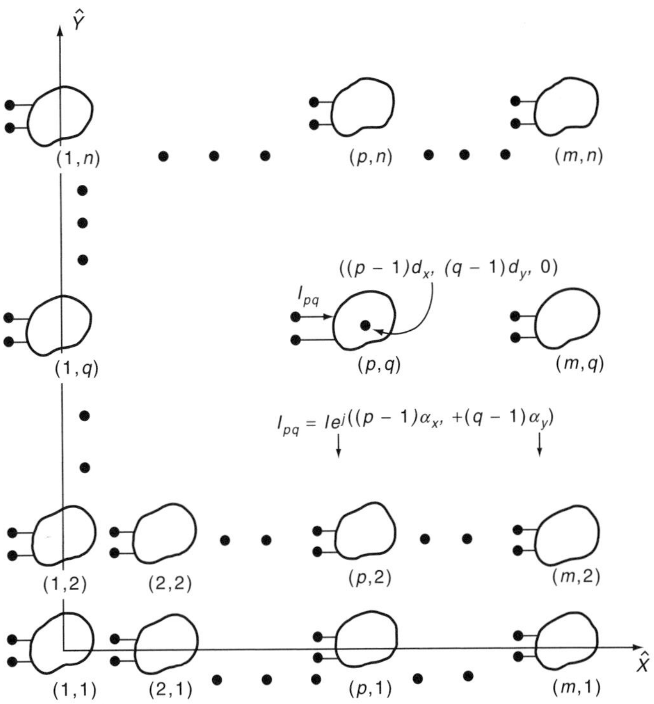

FIGURE 6.22 A Two-Dimensional Array of $m \times n$ Antennas. These antennas are placed on the $x-y$ plane. The interelement spacings are d_x and d_y, respectively, along the $\hat{x}$ and $\hat{y}$ directions. The input currents to the elements are assumed here to have the same magnitude ($= I$), with linearly progressing phases along the $\hat{x}$ and $\hat{y}$ directions.

$$I_{pq} = |I_{pq}|e^{j[(p-1)\alpha_x+(q-1)\alpha_y]} = Ie^{j[(p-1)\alpha_x+(q-1)\alpha_y]}. \quad (6.206)$$

$$\bar{R}_{pq} = \hat{x}(p-1)d_x + \hat{y}(q-1)d_y. \quad (6.207)$$

$$\begin{aligned}\Delta_{pq} &= \hat{r}\cdot\bar{R}_{pq} = (p-1)d_x \sin\theta\cos\phi \\ &+ (q-1)d_y\sin\theta\sin\phi = (p-1)\Delta_x + (q-1)\Delta_y.\end{aligned} \quad (6.208)$$

$$\begin{aligned}AF &= I\sum_{p=1}^{m}\sum_{q=1}^{n} e^{j\psi_{pq}} = I\sum_{p=1}^{m} e^{j(p-1)\psi_x}\sum_{q=1}^{n} e^{j(q-1)\psi_y} \\ &= I(AF_x)(AF_y).\end{aligned} \quad (6.209)$$

$$\begin{aligned}\psi_{pq} &= (p-1)(\alpha_x + k_0\Delta_x) + (q-1)(\alpha_x + k_0\Delta_y) \\ &= (p-1)\psi_x + (q-1)\psi_y.\end{aligned} \quad (6.210)$$

$$|AF_x| = \left|\sum_{p=1}^{m} e^{j(p-1)\psi_x}\right| = \left|\frac{1-e^{jm\psi_x}}{1-e^{j\psi_x}}\right| = \left|\frac{\sin(m\psi_x/2)}{\sin(\psi_x/2)}\right|. \quad (6.211)$$

$$|AF_y| = \left|\sum_{q=1}^{n} e^{j(q-1)\psi_y}\right| = \left|\frac{1-e^{jn\psi_y}}{1-e^{j\psi_y}}\right| = \left|\frac{\sin(n\psi_y/2)}{\sin(\psi_y/2)}\right|. \quad (6.212)$$

$$|AF| = I|AF_x||AF_x| = I\left|\frac{\sin(m\psi_x/2)}{\sin(\psi_x/2)}\right|\left|\frac{\sin(n\psi_y/2)}{\sin(\psi_y/2)}\right|. \quad (6.213)$$

$$|AF(\theta,\phi)|_{\max} = I\left|\frac{\sin(m\psi_{x(\max)}/2)}{\sin(\psi_{x(\max)}/2)}\right|\left|\frac{\sin(n\psi_{y(\max)}/2)}{\sin(\psi_{y(\max)}/2)}\right| = mnI. \quad (6.214)$$

$$AF_n(\theta,\phi) = \frac{AF(\theta,\phi)}{|AF(\theta,\phi)|_{\max}} = \left(\frac{\sin(m\psi_x/2)}{m\sin(\psi_x/2)}\right)\left(\frac{\sin(n\psi_y/2)}{n\sin(\psi_y/2)}\right). \quad (6.215)$$

In general, it may be possible to have more than one value of $\psi_{x(\max)}$ and $\psi_{y(\max)}$ for which the array factor is maximum. All such values other than the dominant ones ($\psi_{x(\max)} = 0$, $\psi_{y(\max)} = 0$) correspond to the grating lobes of the array structure:

$$\begin{aligned}\psi_{x(\max)} &= 0, \pm 2\pi, \pm 4\pi \ldots; \psi_{y(\max)} \\ &= 0, \pm 2\pi, \pm 4\pi \ldots.\end{aligned} \quad (6.216)$$

As for a one-dimensional array, no grating lobe will occur when one designs with $d_x, d_y \le \lambda_0/2$. For the dominant lobe of radiation, which is of particular practical interest, the equation is the following:

$$\begin{aligned}\alpha_x + k_0 d_x \sin\theta_{\max}\cos\phi_{\max} &= 0, \\ \alpha_y + k_0 d_y \sin\theta_{\max}\sin\phi_{\max} &= 0.\end{aligned} \quad (6.217)$$

$$\begin{aligned}\theta_{\max} &= \arcsin\left(\sqrt{\left(\frac{\alpha_x}{k_0 d_x}\right)^2 + \left(\frac{\alpha_y}{k_0 d_y}\right)^2}\right), \\ \phi_{\max} &= \arccos\left(\frac{-\alpha_x}{k_0 d_x \sin\theta_{\max}}\right).\end{aligned} \quad (6.218)$$

Accordingly, for given spacings d_x and d_y, the maximum directions $\theta_{\max}$ and $\phi_{\max}$ may be varied by controlling the interelement phasing α_x and α_y.

6.8.3 Mutual Coupling

As mentioned before, the analyses discussed earlier in this section are based on a critical assumption—there are negligible mutual interactions between the antenna elements. This allowed the key simplification of equation 6.193. In situations when this assumption is no longer valid, one needs to use the rigorous equation 6.192. In this rigorous formulation, $\bar{e}_i(\theta, \phi)$ is different from that of an isolated antenna. Instead, $\bar{e}_i(\theta, \phi)$ is the transmit vector when only the *i*th element is excited (other elements are open circuited with zero input currents), while all other elements are still physically present in their respective locations. In other words, $\bar{e}_i(\theta, \phi)$ is the superposition of the primary radiation from the *i*th antenna (when isolated), with secondary radiations from all the antenna elements passively excited through mutual coupling.

In addition to the radiated fields $\bar{e}_i(\theta, \phi)$, the voltages at the antenna inputs will also change due to the mutual coupling effects. An N-element array needs to be rigorously modeled as a N-port network using a $[Z]_{N\times N}$ impedance matrix:

$$[V] = [Z][I]; \; V_i = \sum_{k=1}^{N} I_k Z_{ik}, \; i = 1, N. \quad (6.219)$$

The input voltage of an antenna element operating in an array environment is dependent on the input currents of all other elements. The linearity constant relating V_i and I_k is called the mutual impedance Z_{ik}. When one excites only the *i*th element, with all other currents zero (but all other elements physically present) from equation 6.219, the result is $V_i = I_i Z_{ii}$. The Z_{ii} is called the self-impedance of the *i*th element, operating in the array environment. Note that the corresponding value, when the *i*th antenna is allowed to operate in an isolated manner, is $V_i = I_i Z_{in}$. The Z_{in} is, in general, different from Z_{ii}, and they would be equal to each other only in the limit of zero mutual coupling.

References

Balanis, C.A. (1997). *Antenna theory: Analysis and design.* (2d ed.). New York: John Wiley & Sons.

Carver, K.R., and Mink, J.W. (1981). Microstrip antenna technology. *IEEE Transactions on Antennas and Propagation* AP-29 (1), 2–24.

Cheng, D.K. (1993). *Fundamentals of engineering electromagnetics.* Reading, MA: Addison-Wesley.

Collin, R.E., and Zucker, F.J. (Eds.). (1969). *Antenna theory, parts I, II.* New York: McGraw-Hill.

Das, N.K. (2004). *Antennas and radiation, part I: Antenna fundamentals.* In *EE Handbook.* Boston: Academic Press.

Jasik, J. (Ed.). (1950). *Antenna engineering handbook.* New York: McGraw-Hill.

Kraus, J.D. (1961). *Antennas.* New York: McGraw-Hill.

Kraus, J.D., and Fleisch, D.A. (1999). *Electromagnetics with applications.* New York: McGraw-Hill.

Pozar, D.M., and Schaubert, D.H. (1995). *Microstrip antennas.* New York: IEEE Press.

Pozar, D.M. (1992). Microstrip antennas. *IEEE Proceedings 80,* 79–91.

Ramo, S., Whinnery, J.R., and van Duzer, T. (1993). *Fields and waves in communication electronics.* (3d ed.). New York: John Wiley & Sons.

Schelkunoft, S.A., and Friis, H.T. (1952). *Antennas, theory, and practice.* New York: John Wiley & Sons.

7

Microwave Passive Components

Ke Wu,* Lei Zhu,*
and Ruediger Vahldieck†
*Department of Electrical and Computer Engineering, Ecole Polytechnique, Montreal, Quebec, Canada
†Laboratory for Electromagnetic Fields and Microwave Electronics, Swiss Federal Institute of Technology, Zurich, Switzerland

7.1 General Concepts and Basic Definitions

Active and passive microwave devices and components are the essential building blocks of microwave circuits and systems that operate in the frequency range from 300 MHz to 300 GHz (corresponding to wavelengths of 1 m to 1 mm in free space). This chapter only presents passive components that are composed of lumped or distributed elements or a combination of both. At lower frequencies (e.g., 300 MHz–10 GHz), the emphasis is on lumped elements to keep circuit dimensions small. At higher frequencies, the quality of lumped elements deteriorates rapidly, and mostly distributed elements are employed. Microwave circuits are a combination of passive and active components, whereby the passive part easily makes up 75% or more of the circuit real estate area. Without passive components (e.g., filters, matching circuits, circulators, isolators resistors, etc.), active components (e.g., transistors, tubes) cannot be operated.

A **passive component** is a physical structure or circuit layout that performs one or multiple linear electronic functions without resorting to and consuming external biasing or controlling electric and/or magnetic sources. In other words, the circuit functionality of a passive component is self-contained and self-manifested without any consumable and external third-party intervention. In practice, a microwave signal passing through a passive component always experiences losses and dissipates power. A passive component may

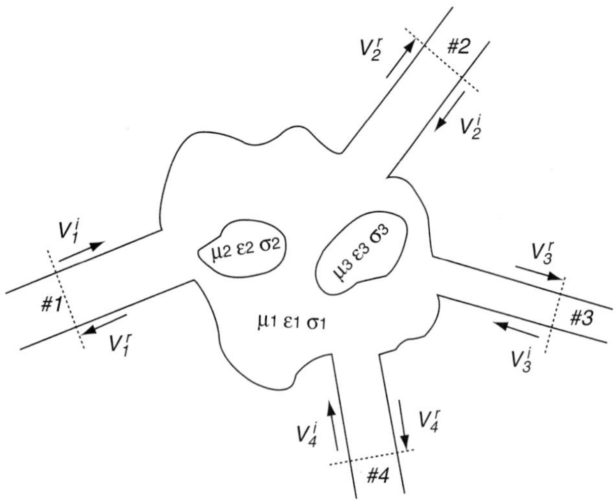

FIGURE 7.1 Graphic Description of a Generalized Four-Port Microwave Passive Component that may Involve Different Segmented Materials Within its Arbitrarily Shaped Circuit Domain.

be described as a closed-form network of one or multiple ports.

An arbitrarily shaped multiport (four ports in the following example) microwave passive component is depicted in Figure 7.1, in which the inbound (or incident) power is always equal to the outbound (or reflected) power plus the possible dissipated power in the circuit. In most practical applications, a microwave circuit is modeled by scattering or S-parameters that are defined through the incident and reflected measurable voltage (or alternatively current) waves V^i and V^r (or I^i and I^r) at particular planes of reference for a given mode of propagation, such as a transverse electric mode (TEM). Circuit properties considering multiple modes of operation (propagating or evanescent) are described with the generalized S-parameter matrix. The difference is that for only one mode of propagation, the voltages and S-parameters per port are single-valued, whereas for multimode propagation or evanescent modes, the voltages and S-parameters become vectors and submatrices, respectively. For single-mode propagation, the S-matrix of the four-port example in Figure 7.1 is given as a 4×4 matrix with complex elements as follows:

$$\begin{bmatrix} V_1^r \\ V_2^r \\ V_3^r \\ V_4^r \end{bmatrix} = \begin{bmatrix} S_{11} & S_{12} & S_{13} & S_{14} \\ S_{21} & S_{22} & S_{23} & S_{24} \\ S_{31} & S_{32} & S_{33} & S_{34} \\ S_{41} & S_{42} & S_{43} & S_{44} \end{bmatrix} \begin{bmatrix} V_1^i \\ V_2^i \\ V_3^i \\ V_4^i \end{bmatrix} \quad (7.1)$$

This matrix establishes a port-to-port signal flow relationship, such as transmission and reflection, at and between ports. The $S_{ii}(i = 1, 2, 3, 4)$ is referred to as the **reflection coefficient** (or Γ) at the corresponding ith port. This value is often expressed in decibels $[-20\log(S)]$ and is called the return loss. The S_{ii}(i, j = 1, 2, 3, 4 and i $\neq$ j) is defined as the transmission coefficient (or T) from the jth port (inbound) to the ith port (outbound) according to equation 7.1. This value expressed in decibels is defined as the insertion loss between two specific ports. One of the fundamental properties of the S-parameter matrix for a linear and lossless passive network is the unitary relationship:

$$[S][S]^{t^*} = I, \quad (7.2)$$

in which the superscripts t and * stand for transpose and conjugate complex, respectively. This equation is derived from the principle of energy conservation. The magnitude of S-parameters of a passive network is always equal to or smaller than one, depending on whether it is lossless or lossy. From the unitary condition, the correct relationship between the entries of the S-matrix can be established.

The S-parameters fully describe the electrical properties of a lumped or distributed microwave network. S-parameters can be converted to and from other network parameter representations such as Y, Z, and ABCD matrices (Collin, 1992). The ABCD (chain) matrix is popular in the analysis and design of microwave circuits when the voltage and current concepts can be applied. Smith chart techniques also provide a straightforward tool for most practical problems of design and analysis (Collin, 1992).

Microwave passive components may be composed of **lumped elements** (inductors, capacitors, and resistors) or **distributed elements** (transmission line sections and discontinuities) or both. The circuit is arranged in a particular configuration (topology) so that selected signal properties (amplitude, frequency, and phase) are processed or controlled in a desired manner. The lumped elements may also be regarded as a special class of transmission line structures whose geometrical size is always within a small fraction of the operating wavelength and thus, the embedded electromagnetic fields behave in a stationary manner. In that sense, all microwave passive components can be considered as transmission line structures consisting of certain transmission line discontinuities and connecting lines commensurate with wavelengths.

Judging from their circuit functionality, passive components can be classified into the following categories: **impedance matching/transformer, power dividing/combining, signal rerouting/coupling, frequency filter/resonator,** and **phase delay/shifter.** Among them, filters are a special and very important class of passive components because of their technical scope and design complexity. Hybrids and directional couplers are no less important but their realization spans a much smaller range than that of filters. Impedance matching or transformer networks are considered the most fundamental building blocks for nearly any microwave circuit. Possible realizations are strongly dependent on the circuit environment, and stand-alone components are not possible unless they are tunable. Ferrite components and some special material-related circuits (e.g., ferroelectrics) constitute another class of passive microwave components. They may be used to design tunable circuits by applying external magnetic or electric fields. Ideally, no external energy consumption takes place even in this case. Ferrite components have been widely used for isolators, phase shifters, and other nonreciprocal circuit functions like circulators and others.

In the design of a passive component, function-specific parameters, such as frequency bandwidth, insertion/return losses, input/output impedance levels, group/phase delays, and isolation and transient response, are usually the most important. Other parameters like temperature stability, vibration resistance, low passive intermodulation, or multipacting effects are secondary although, depending on the application, may have a direct influence on circuit design.

7.2 Basic Passive Elements and Circuits

The simplest form of a passive component is a section of transmission line bounded by certain impedance conditions, including short and open (electrically). The transmission line may be a **planar structure,** a **quasiplanar structure,** or a **nonplanar waveguide**. These three basic categories of transmission lines form the foundation for designing a microwave passive component.

Planar structures include strip line, microstrip, slot line, coplanar waveguide (CPW), and coplanar strip line (CPS) as well as other variants. Nonplanar structures usually refer to classical metallic waveguides, coaxial (coax) lines, and dielectric waveguides. Hybrid or quasiplanar structures are a combination of planar and nonplanar (waveguides) platforms such as slot lines suspended in the E-plane of a waveguide (finline) or strip lines suspended in the H-plane of a wave guide. From the manufacturing point of view, planar and hybrid structures offer easy integration of passive and active components. This is an advantage over classical waveguide or coax circuits for which integration of active and passive circuit parts is much more difficult and usually leads to much larger circuit layouts. Figure 7.2 shows a cross-sectional view of commonly used transmission lines that have already been discussed in Gupta *et al.* (1996) and Itoh (1987) with respect to their electrical and mechanical properties.

In the following sections, unless otherwise specified, we refer to the microstrip line or the rectangular waveguide as the basic building block for passive components. Important aspects of commonly used passive circuits are highlighted. Limited space does not permit to introduce such useful principles and theories like the even–odd mode decoupling scheme, Kuroda identities, and Richards's theorem (Pozar, 1998) that have been developed for modeling, designing, and synthesizing passive networks. For the relevant details, the reader is referred to the appropriate literature.

7.2.1 Transmission-Line Stubs

A **stub** is a typical one-port passive element in which one of the two ports of a transmission line is terminated by a certain reactive and/or resistive load denoted by Z_L or $Y_L(Y_L = 1/Z_L)$, as shown in Figure 7.3(A). A stub is an attachment to the main transmission line path, either at its end or somewhere along the line in parallel or in series. The input terminal of a stub (the part attached to the main transmission line), depending on its load condition and length, can appear as a capacitive or inductive impedance, as a series or parallel resonance circuit, or as purely inductive or purely capacitive. A transmission line stub is essentially an impedance transformer that must be mismatched at its end terminal to become effective or operative. Because of a wide range of input impedance possible, stubs are widely used as impedance transformers and matching networks, as tuning elements or dc-bias networks (a kind of low-pass filter), and as baluns or filters. The most commonly used load conditions of a stub are either a short load ($Z_L = 0$) or an open load ($Y_L = 0$). Their effect, however, is typically only narrow band. If broadband stubs are needed, so-called radial microstrip open-ended stubs (Figure 7.3), for example, are utilized. Other loads may be in the form of a lumped capacitor or a lumped inductor or a mixture of both. From transmission line theory (Collin, 1992), the input impedance Z_i of a stub can simply be formulated by:

$$Z_i = Z_0 \frac{Z_L + Z_0 \tanh \beta l}{Z_0 + Z_L \tanh \beta l}\Bigg|_{\text{if } \beta=\alpha+j\beta}^{\text{lossy line}} = Z_0 \frac{Z_L + Z_0 \tan \beta l}{Z_0 + Z_L \tan \beta l}\Bigg|_{\text{if } \beta=j\beta}^{\text{lossless line}}, \quad (7.2)$$

in which Z_L and Z_i are in general complex quantities. The Z_0 represents the characteristic impedance of the stub line or the main transmission line that, for a microstrip line, can be calculated at low frequency by knowing the line width, the substrate thickness, and the dielectric constant. Calculating Z_0 of a waveguide also requires knowledge of the frequency because waveguides are very dispersive (frequency dependent). Equation 7.2 clearly illustrates the important fact that impedance transformation on a transmission line is only possible if the line is mismatched ($Z_0 \neq Z_L$). Examples of short-circuited and open-ended transmission lines with an electrical length smaller than $\lambda/4$ are shown in Figure 7.4. Also shown is their possible realization in a microstrip circuit as well as the equivalent network.

7.2.2 Transmission Line Discontinuities

Passive circuits always consist of uniform sections of transmission lines interconnected by discontinuities (i.e., impedance steps). **Transmission-line discontinuities** can also appear as gaps, transmission line bends, waveguide apertures, and posts. The discussed short and open stubs may be regarded as part of the circuit discontinuities. Generally, discontinuities are

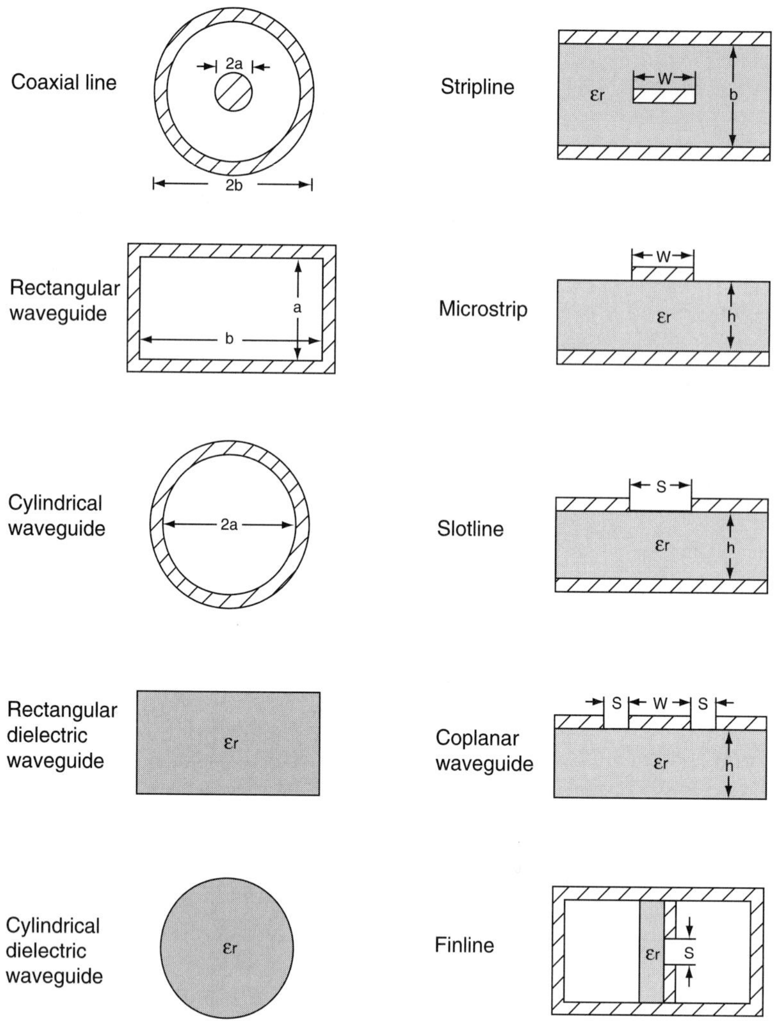

FIGURE 7.2 Cross-Sectional View of Commonly Used Planar and Nonplanar Transmission Lines that are Used in the Construction of Passive Microwave Components

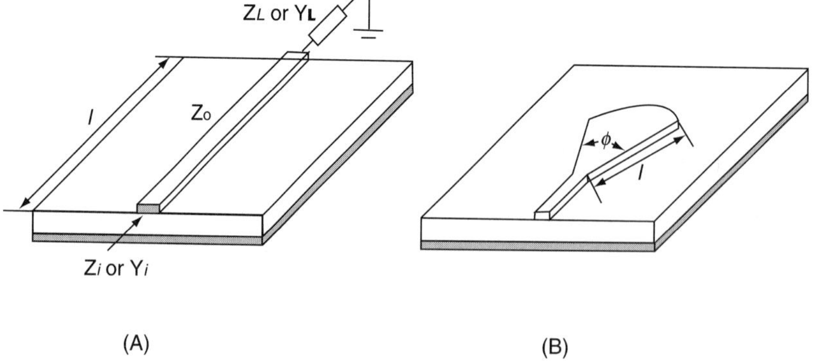

FIGURE 7.3 Examples of One-Port Microwave Passive Circuits. (A) Microstrip line loaded with reactive and/or resistive element; (B) Microstrip radial open-ended stub element.

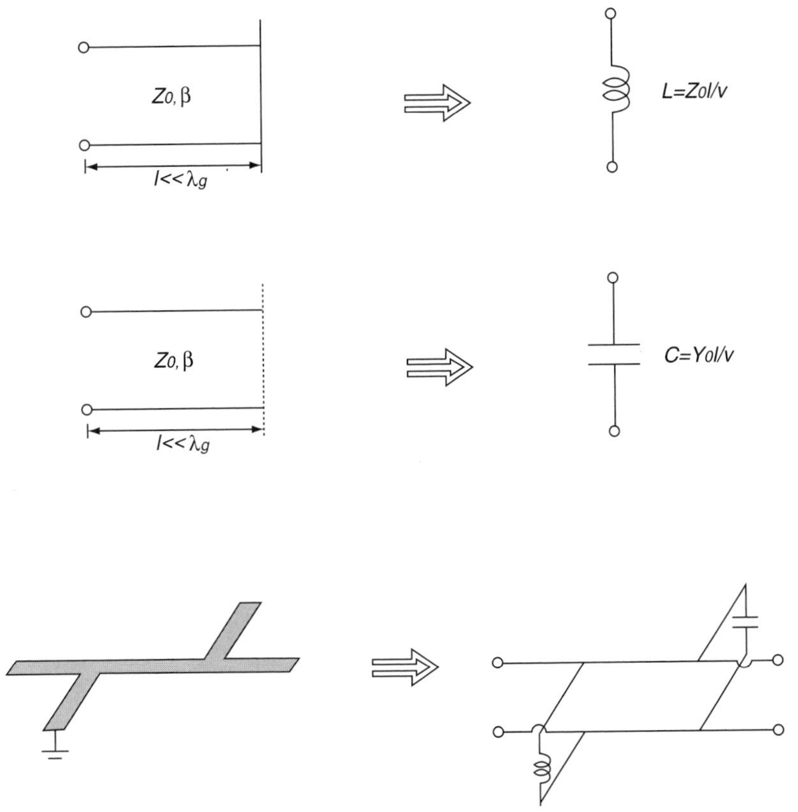

FIGURE 7.4 Short- and Open-Ended Transmission Line Elements Having Electrical Length Shorter than a Quarter-Wavelength and Their Equivalent Circuit Representations.

electrically small with respect to the wavelength. Lumped-element equivalent circuits can be developed to represent these discontinuities. A lumped-element representation of a transmission line discontinuity, such as bend or gap, is important in the design and analysis of passive microwave circuits and components. Values for the lumped elements can be obtained from either field theoretical computations, measurements, or empirical formulas.

Transmission-line discontinuities may be represented by a combination of lumped elements of circuit. Figure 7.5 presents a class of typical microstrip line discontinuities and its equivalent network parameters. Resistive elements may be included (series resistance and/or shunt conductance) to represent radiation and leakage as well as ohmic effects. Closed-form analytical formulas on the basis of quasistatic or simplified models are provided in Gupta (1996) and Itoh (1987). These useful tools allow approximate calculations of the lumped-element parameters for some commonly used discontinuities. Accurate parameter extractions using field theory-based full-wave techniques are preferred for higher frequency design. Figure 7.6 gives extracted lumped-element values from a self-calibrated method of moments (MoM) (Zhu and Wu, 1999) for asymmetrical microstrip gap and step discontinuities.

7.2.3 Lumped Circuit Elements

A **lumped element** or a **discrete circuit part** can normally be realized with a much smaller size than its distributed or transmission-line parameter counterpart. At lower microwave frequencies, a size reduction to $\lambda/10$ or less is possible. Therefore, lumped elements are used to miniaturize passive circuit components and also reduce losses as well as cost. Almost all of the low (usually megahertz to lower gigahertz range) radio frequency (RF) electronic components are in lumped-element form. At higher microwave frequencies and, in particular, at millimeter wavelengths, the loss of lumped elements becomes prohibitive for discrete circuit design and is therefore typically limited to special applications, such as in monolithic microwave integrated circuits (MMICS) (Robertson, 1995). In these applications, geometrical factors, dispersion, and, of course, losses must be taken into account. In addition, the variation of lumped-element values is often limited by structural topology, material, operating frequency, and manufacturing process.

Example topologies for the three basic lumped-element types (**inductor, capacitor,** and **resistor**) and their realization in layered planar geometry for higher frequency applications in MMIC and miniature hybrid microwave integrated circuit (MHMIC) technologies are shown in Figure 7.7. These elements

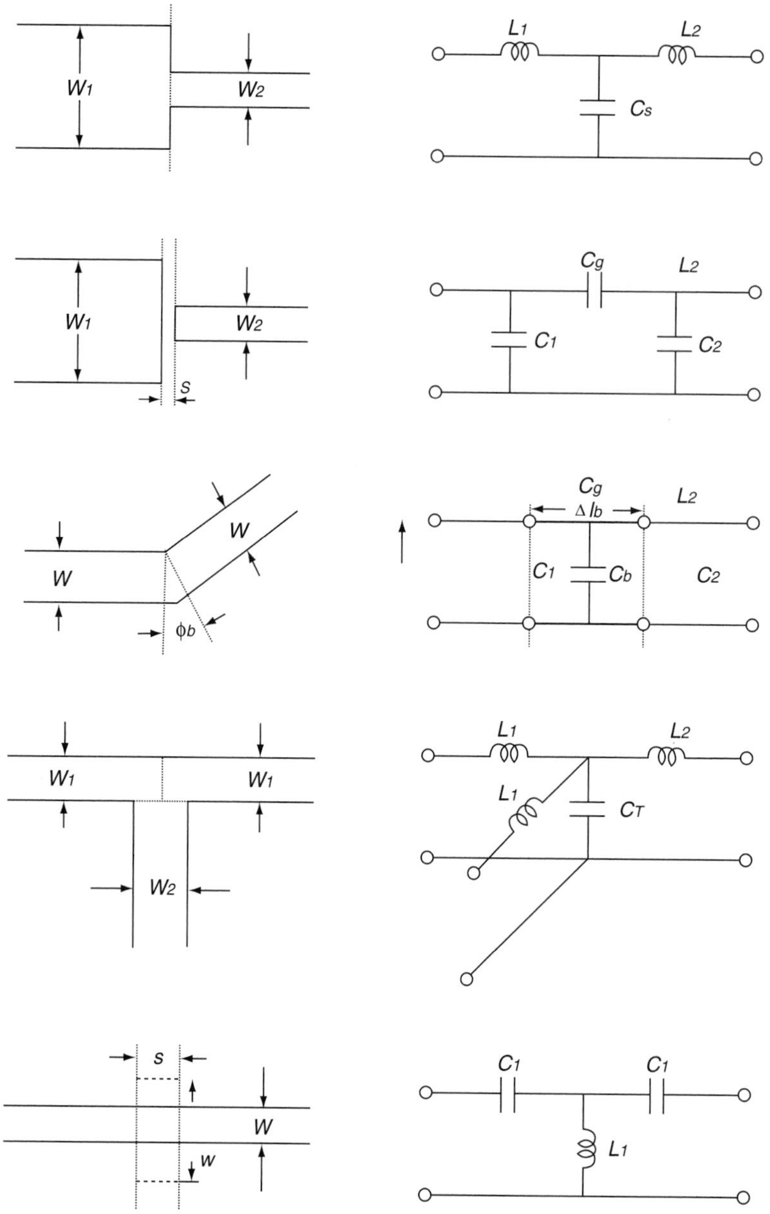

FIGURE 7.5 Physical Layout and Equivalent Circuit of Typical Transmission Line Discontinuities Used in the Design of Microwave Circuits

are commercially available in the form of surface-mountable chips for hybrid integration or in standard MMIC processes. For inductors and capacitors, the most important parameter is the quality (Q-) factor in addition to the characteristic value of the element. The Q-factor, which is used to measure the loss effects, depends on conductor and dielectric dissipation properties of the structure at a given frequency and can be calculated from the equivalent RGLC (resistance conductance inductance capacitance) model of the element.

A high-density current flowing over a transmission line with relatively narrow line width generates a strong magnetic field and thus creates the lumped inductance effect. Multiturn planar spiral inductors easily produce inductance values higher than 1 nH and are frequently used in MMIC design, either in the form of squared or circular topologies. Meander-line inductors may also be considered as a special class of spiral inductors, but they generate lower inductance values. The difficulty of the spiral layout is the electrical crossover from the center turn back to the outside circuit. This may be achieved by means of an air bridge or dielectrically spaced underpasses. If a silicon substrate is used, the resulting Q-factor is low, which is a major design concern for application at higher frequencies.

Two tightly spaced but physically insulated metallic plates or ends form a capacitor, in which a strong electric field is

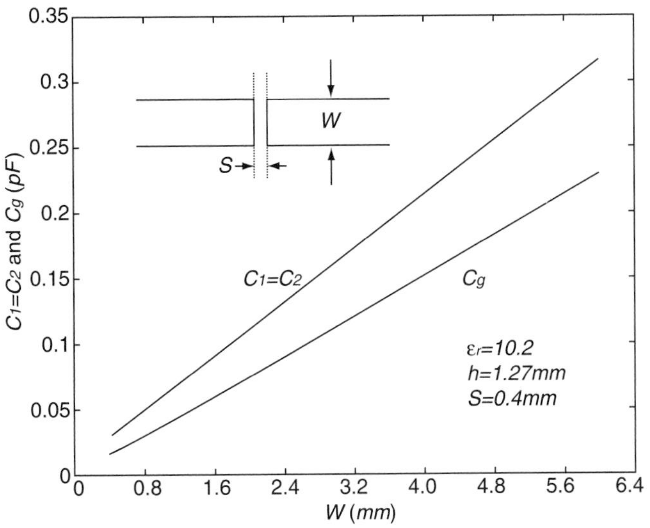

(A) Gap Discontinuity

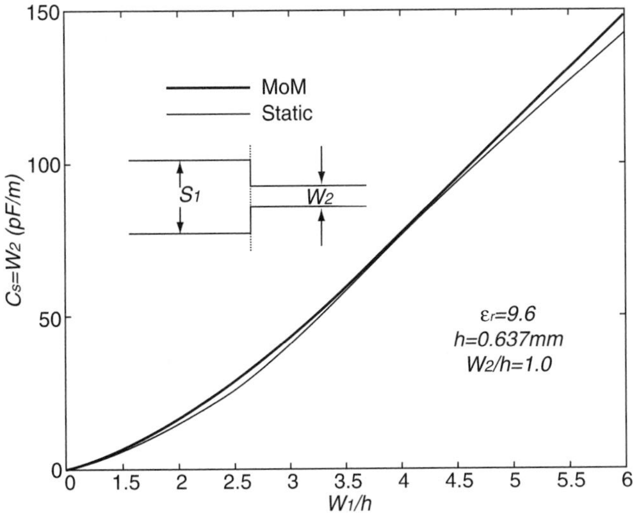

(B) Step Discontinuity

FIGURE 7.6 Extracted Equivalent Circuit Parameter Values of Asymmetrical Microstrip Gap and Step Discontinuity Formed on a Dielectric Substrate whose Thickness and Dielectric Constant Are Denoted by h and ε_r, Respectively.

induced once a voltage is applied between them. Such a capacitor is easily realizable in a multilayered platform. Two widely used structures are capacitors with overlay and interdigital topologies. The overlay capacitor is usually referred to as metal-insulator-metal (MIM) with insulators made of low-loss and temperature-stable thin-film materials, such as ceramic, silicon nitride, silicon dioxide, and polyimide. The interdigital capacitor consists of a number of interleaved microstrip fingers coupled together as indicated in Figure 7.7. Usually, the interdigital capacitor cannot be used to produce a capacitance value above 1 pF and tends to saturate at high frequency even with a large number of fingers, whereas the MIM technique can yield high capacitances. The tolerance of interdigital capacitors can be well controlled, and they are ideal for tuning, coupling, and matching element design where a small capacitance and a high precision are required. The Q-factors for the MIM and interdigital capacitors are mainly limited by insulator dielectric loss and ohmic effects of the fingers. Figure 7.8 illustrates some typical lumped element values for the interdigital capacitor (Zhu and Wu, 2000).

Resistors, such as the isolating resistor used in the design of Wilkinson's power dividers and combiners, are another fundamental building block for matching loads or special applications. Resistors can be made from deposited thin-film resistive layers or thin-metal films, whereby the film thickness is much less than its skin depth. Resistors can also be made from doped semiconductor layers (mesa resistors or implanted planar resistors). In either case, the layer or film thickness t is fixed, and the resistivity is conveniently quoted in terms of an ohms-per-square figure or simply square resistance $R_0 = (\sigma t)^{-1}$. As such, the film or layer resistance can be calculated by a simple formula:

$$R = R_0 \frac{l}{w}, \tag{7.3}$$

in which l/w is often called aspect ratio. The geometrical parameters l and w stand for the film length and width in view of the connecting pads of a resistor. The required tolerances and the expected power dissipation also determine the absolute size of a resistor. In addition, contact pads and grounding blocks may contribute to its parasitic effects (capacitance) that may be severe at high frequency. Other properties such as linearity and temperature coefficients are also critical for certain solid-state power device applications. In practice, $50 - \Omega$ resistors are the most important and popular standard reference for circuit design.

7.3 Impedance Transformers and Matching Networks

The purpose of an **impedance transformer** is to transform a given impedance to a specific value. An impedance **matching network** may consist of more than one impedance transformer. Its main goal is to match a given impedance to a prescribed value over a frequency range of interest to ensure maximum power transfer from a source to a load. Typical applications include returning loss optimization between an antenna and low-noise amplifier, achieving low-noise characteristics in active circuits, and providing conjugate complex matching in amplifier design. Electrical and/or mechanical impedance tuners are widely used in RF and microwave measurement setups, such as load-pull nonlinear characterization systems. Impedance transformers and matching networks are perhaps the most important and most widely used passive microwave circuit component.

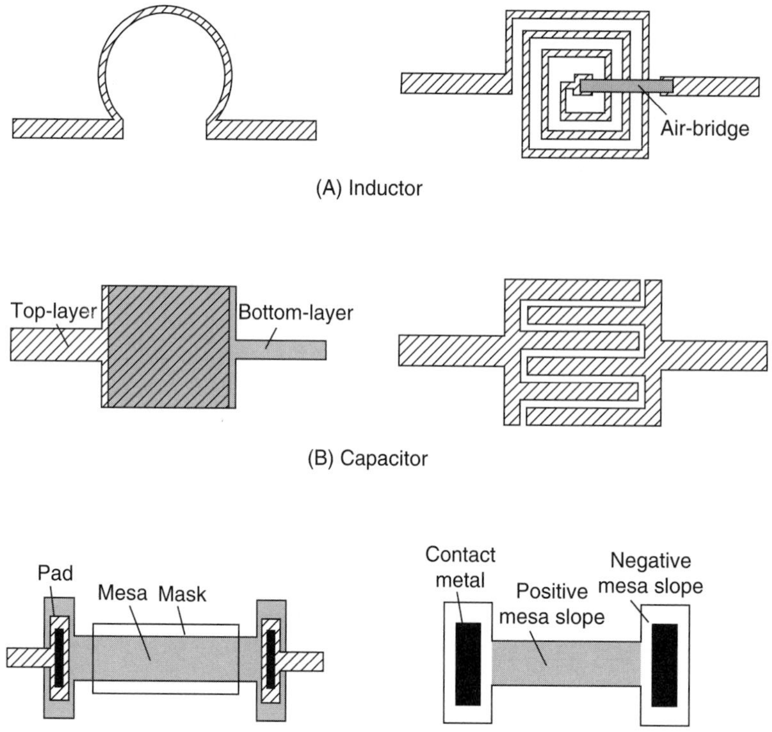

FIGURE 7.7 Three Basic Lumped Circuit Elements for Microwave IC Design: Inductor, Capacitor, and Resistor

In a practical design, the characteristic impedance of the transformer and matching network depends on the type of transmission line that dictates its achievable impedance range. Further, the type of transmission line used gives preference to either shunt (common voltage) or series (common current) branch connection. In a microstrip line, for example, stubs are best used in shunt–parallel connections.

Techniques based on the Smith chart are popular in the design of impedance transformers and matching networks, a basic block diagram of which is shown in Figure 7.9. Matching networks may be built from lumped or distributed elements or a combination of both. The quarter-wavelength section of a transmission line may be the easiest to realize. Classical shunt or series branch lines are widely used for more sophisticated matching networks. Commonly used topologies include single or multiple stubs and one or more quarter-wavelength sections. In addition, taped or nonuniform transmission line sections are also often found in broadband design.

7.3.1 Matching Sections and Stubs

Figure 7.10 shows a class of commonly used series cascaded sections and stubs for the design of impedance matching networks. A simple series connection of transmission lines may be used to match a complex load $Z_L(=R_L+jX_L)$ to a real resistance R, as shown in Figure 7.10(A). Note that this is a narrow-band matching scheme with very limited use because of a low degree of flexibility in the impedance choice and matchable area on the Smith chart. The two-section transformer as shown in Figure 7.10(B) can overcome this problem to some degree. Still, the useful bandwidth is relatively narrow for most practical applications.

The stubs attached to the main section of the transmission line provide a useful alternative if there is a requirement to reduce the circuit area. In addition, the use of stub impedance tuners gives an extra degree of freedom in the choice of matching strategies. This is because the lossless stub (open or short) represents just a pure reactance to the main line (see Subsection 7.2.1 for more details) and can effectively be used for canceling the load's imaginary part. In general, two stub tuners together with one line section are always enough to generate a complete matching condition for any complex load. The position, characteristic impedance, and line length of the stubs and the connecting lines are conveniently designed through the use of the Smith chart. Once again, the bandwidth remains inherently narrow because of the line length dependence of the impedance transformer or the matching network in question.

7.3.2 Quarter-Wavelength Transformers

An important class of impedance-matching networks is the **quarter-wavelength transformer;** these transformers are used

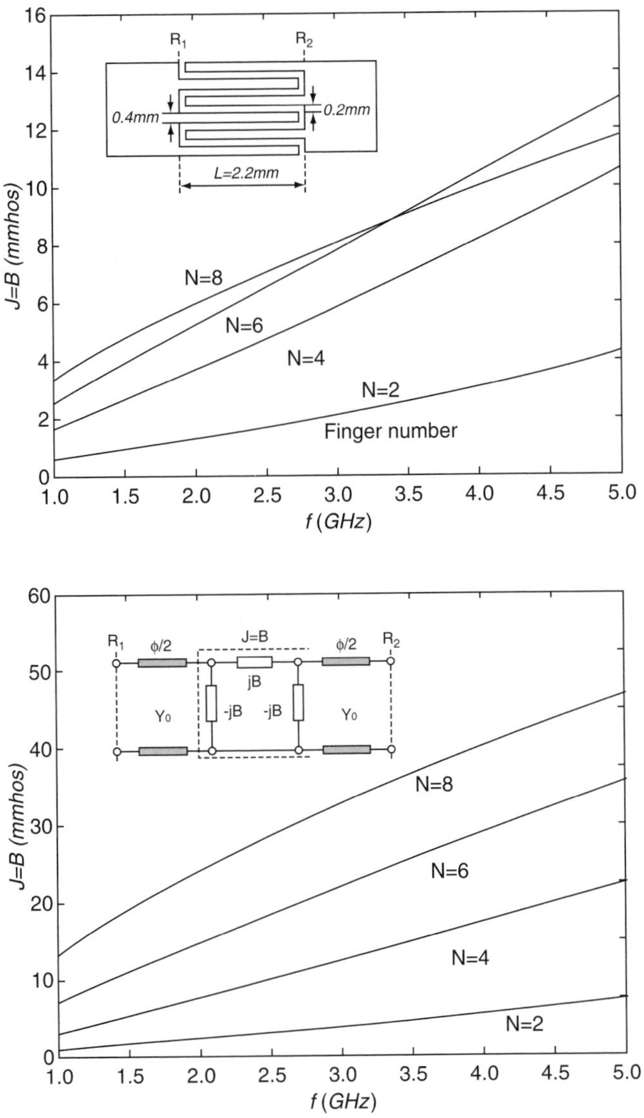

FIGURE 7.8 Typical Extracted Lumped Values of a Microstrip Interdigital Multifinger Capacitor

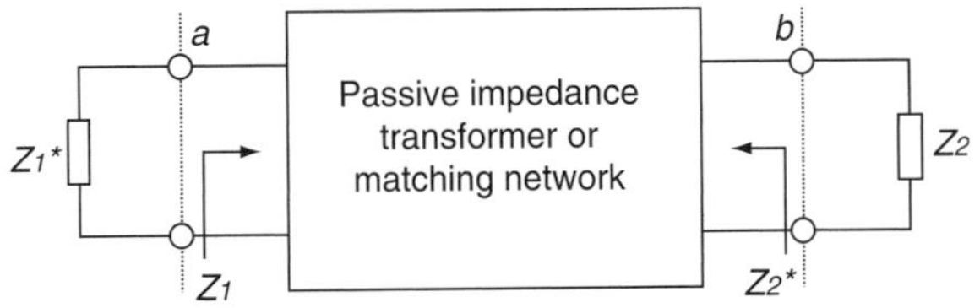

FIGURE 7.9 Schematic Description of a Two-Port Passive Impedance Transformer or Matching Network

to match a real-valued impedance load to another real-valued impedance at the input. If complex impedances are involved, the stub-tuning technique may be used to cancel out the transformer's imaginary part. Therefore, the quarter-wavelength transformer is widely used in practice in conjunction with impedance tuning stubs. For narrow band applications, a single section is adequate. Otherwise, multiple cascaded sections are required.

With reference to Figure 7.10(A), a quarter-wavelength section ($\beta l = \theta = \pi/2$) of a lossless network with the input impedance denoted by R gives the input impedance as follows (from equation 7.2):

$$Z_i = \frac{Z_0^2}{Z_L}. \tag{7.4}$$

The load impedance Z_L is matched to R if $Z_i = R$ and $Z_0 = \sqrt{RZ_L}$. Obviously, the prescribed quarter-wavelength condition yields a narrow band operation, which is also further limited by the impedance ratio of R/Z_0 and Z_0/Z_L.

A multisection impedance matching network is described in Figure 7.11, in which the sections may be equal or not equal to quarter-wavelengths. Broadband matching condition can be achieved with adequate design schemes, such as Chebyshev synthesis techniques. The matching bandwidth can be enlarged if the number of sections increases. Of course, in most practical cases, the limit is given by the maximum circuit area and loss considerations.

7.3.3 Tapered and Nonuniform Impedance Transformers

The nonuniform impedance transformer uses transmission lines with gradually varying line geometry. The **tapered impedance transformer** is a special case of nonuniform lines, and its impedance changes smoothly and monotonically in the propagation direction of the wave. This change of impedance is either linear or exponential or follows rules derived from Chebyshev or other designs. The synthesis of a tapered-line transformer is usually made at its lowest frequency of interest over the predesignated bandwidth because the derivative of the reflection coefficient decreases rapidly with frequency. The **nonuniform line** may be used not only for a broadband impedance transformer but also for other purposes, such as the realization of low-pass and stop-band filtering functions. Figure 7.12 presents two typical layout examples of a tapered impedance transformer and a nonuniform transmission line transformer.

7.3.4 Coupled-Line Transformers

In the design of coupling and filtering elements, impedance transformation may also be required because of the potential difference in impedance between the source at the input and load at the output. Such transformers not only match the input to the output, but, they function at the same time as building blocks for the couplers and the filters. Quarter-wavelength coupled lines with various terminal conditions are among the most popular structures, as shown in Figure 7.13 **Coupled-line**

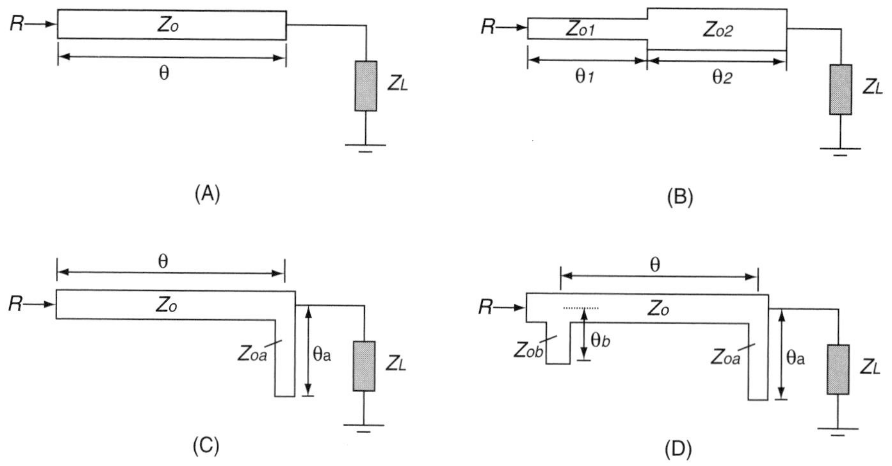

FIGURE 7.10 A Class of Series Cascaded Sections and Stubs for Design of Impedance Matching Network

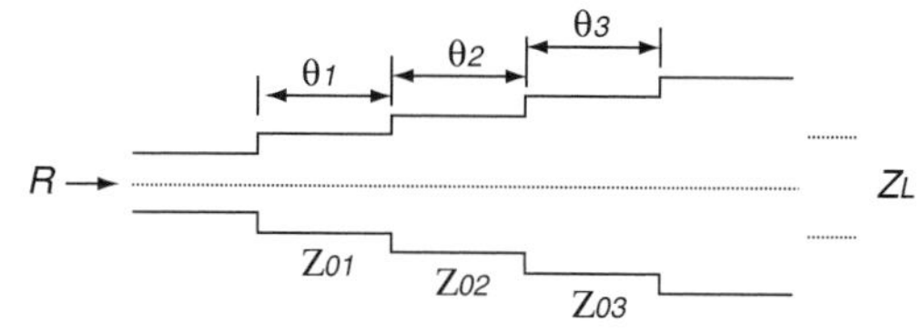

FIGURE 7.11 Multi-Section Impedance Transformer

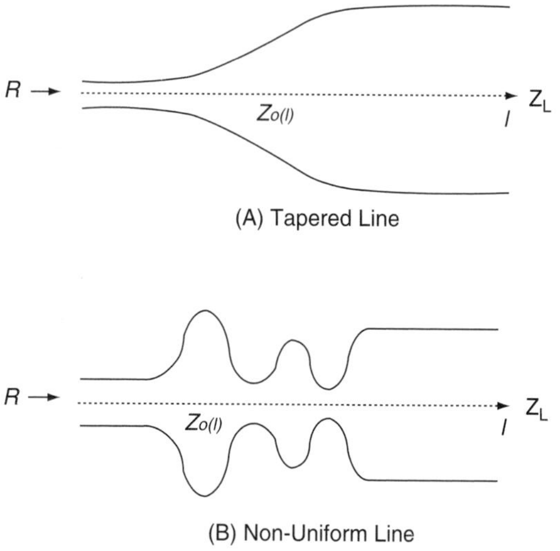

FIGURE 7.12 Tapered Impedance and Non-Uniform Transmission Line Impedance Transformers

elements have no direct electrical contact, and this is why they are frequently used as dc bias blocks for active devices. Symmetrical coupling structures are most often used as coupled-line impedance transformers, but asymmetrical designs may offer more design freedom, for example, in

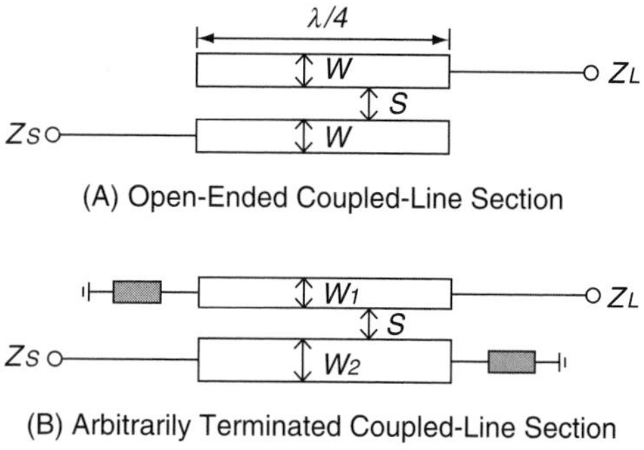

FIGURE 7.13 Two Representative Quarter-Wavelength Coupled-Line Sections for Impedance Transformation and Matching Purposes

enhancing the impedance transformation ratio. Usually, the coupled-line schemes may provide an impedance ratio up to 10, depending on the geometry and the substrate used in the design. For symmetrical structures, even- and odd-mode parameters are widely used for the analysis and synthesis of impedance transformers.

7.3.5 Lumped-Element Transformers

Transmission-line-based impedance transformation is made possible through distributed effects, which may also be realized by using lumped-element-based networks in terms of capacitors and inductors. Capacitors and inductors connected in L-/π-shaped configurations are widely used as impedance-matching networks. There are eight possible arrangements of the LC network topology, as shown in Figure 7.14. They are popular in the design of miniaturized ICs, such as monolithic (MMIC) and high-density hybrid circuits (MHMIC) as well as low-frequency ICs.

The ratio of impedance transformation depends on the value of the inductor and capacitor used in the network. It is

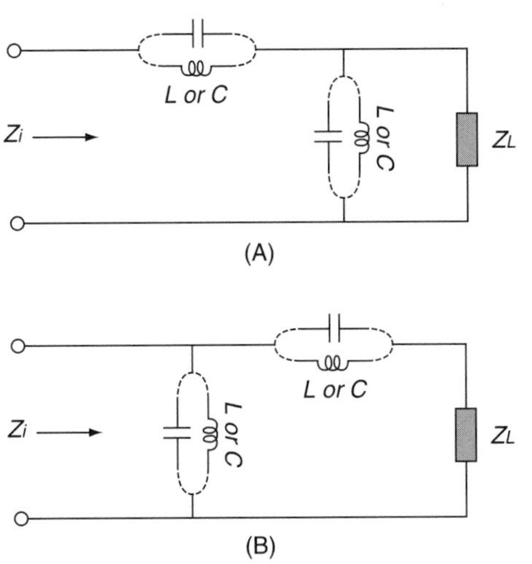

FIGURE 7.14 Generalized Lumped-Element Network Transformers with Various L-C Combinations

also possible to design broadband impedance matching networks by cascading lumped-element circuits. The design and synthesis of such lumped-element circuits can easily be made through the Smith chart by normalizing LC values with respect to an impedance of reference (normally the impedance of external connection lines).

7.3.6 Balun Transformers

This is a special class of microwave passive components that converts the fundamental guided mode from balanced to unbalanced or vice versa. This is why it is called **balun** (**bal**ance-to-**un**balance). In this case, the characteristic impedance may also be transformed in this line-to-line modal conversion. Usually, a balun structure deals with two different types of transmission line geometry that are used as input and output of the circuit.

Although the early version of a balun was proposed for coaxial lines and other nonplanar structures, **planar baluns** have received much attention and may be realized by two-sided and uniplanar MIC technologies. Generally, the baluns may be divided into two categories (Trifunovic and Jokanovic, 1994); **Marchand baluns** (band-pass networks) and **double-*Y* baluns** (all-pass networks, ideally), as shown in Figure 7.15. This figure presents three basic transformations: microstrip-slot line, CPW-slot line, and finite-ground CPW–CPS line.

The input and output lines are orthogonally crossed over in the Marchand baluns, and two ports are electrically terminated with either a short or open end that transforms either open or short effects at the junction through two quarter-wavelength line transformers. The modal conversion is thus achieved over the junction together with a potential line-to-line impedance transformation. Generally, the Marchand baluns are simple in geometry but have a narrow operating bandwidth (filter behavior). On the other hand, the double-*Y* baluns may be used for broadband design. They are based on a six-port double-*Y* junction that consists of three balanced and three unbalanced lines placed alternatively around the center of the junction. The input and output lines are usually aligned along the same axes, and the other four branch lines are short- or open-circuited and perform in-phase and out-of-phase signal dividing/combining procedures. This is very similar to the well-known low-frequency electrical bridge transformer.

7.4 Hybrids, Couplers, and Power Dividers/Combiners

Hybrids and **couplers** (directional couplers) are important components in microwave and millimeter-wave systems for rerouting or redistribution of the signal (Cohn and Levy, 1984). They are widely used in high-frequency signal measurements, amplitude, phase and frequency discriminators, balanced amplifiers, balanced mixers, feeding networks of antennas, and many other applications. In principle, a hybrid or a coupler may simply be described as a reciprocal multiport (usually four-port) network as shown in Figure 7.16. Hybrids are special classes of directional couplers that use commensurate line sections or junctions to connect ports or arms. Branch-line couplers, hybrid-rings (rat-race), and magic-tees are the most widely used designs of the hybrids. Their performance depends critically on the transmission line properties. Couplers rated at a 3-dB power splitting ratio are also often called hybrids. The designation of hybrid is not consistently used in the literature. In the following paragraphs, a hybrid describes a directional coupler consisting of several line sections that are physically connected to each other via junctions. Hybrids may be derived from all types of couplers, and they may be classified as 90° (quadratic) hybrids or 180° hybrids, depending on their circuit properties and application features.

Other types of couplers are based on distributed electromagnetic coupling between lines. In such cases, the mechanism of line-to-line coupling is important. 1-to-N way **dividers** or N-to-1 way **combiners** present a class of alternative structures to directional couplers or hybrids for achieving equal or non-equal power dividing/combining functions. They are widely used in the design of multicell power amplifiers and feeding networks of antenna arrays. Among them, the **Wilkinson power divider/combiner** is the most prominent.

The four ports of a hybrid or a coupler are labeled as **input, direct, coupled,** and **isolated.** Suppose that P_1 is the power fed into port *l*, while P_2, P_3, and P_4 denote the power available at the other corresponding ports. Then, four important parameters that describe the electrical performance of this passive circuit are defined as follows:

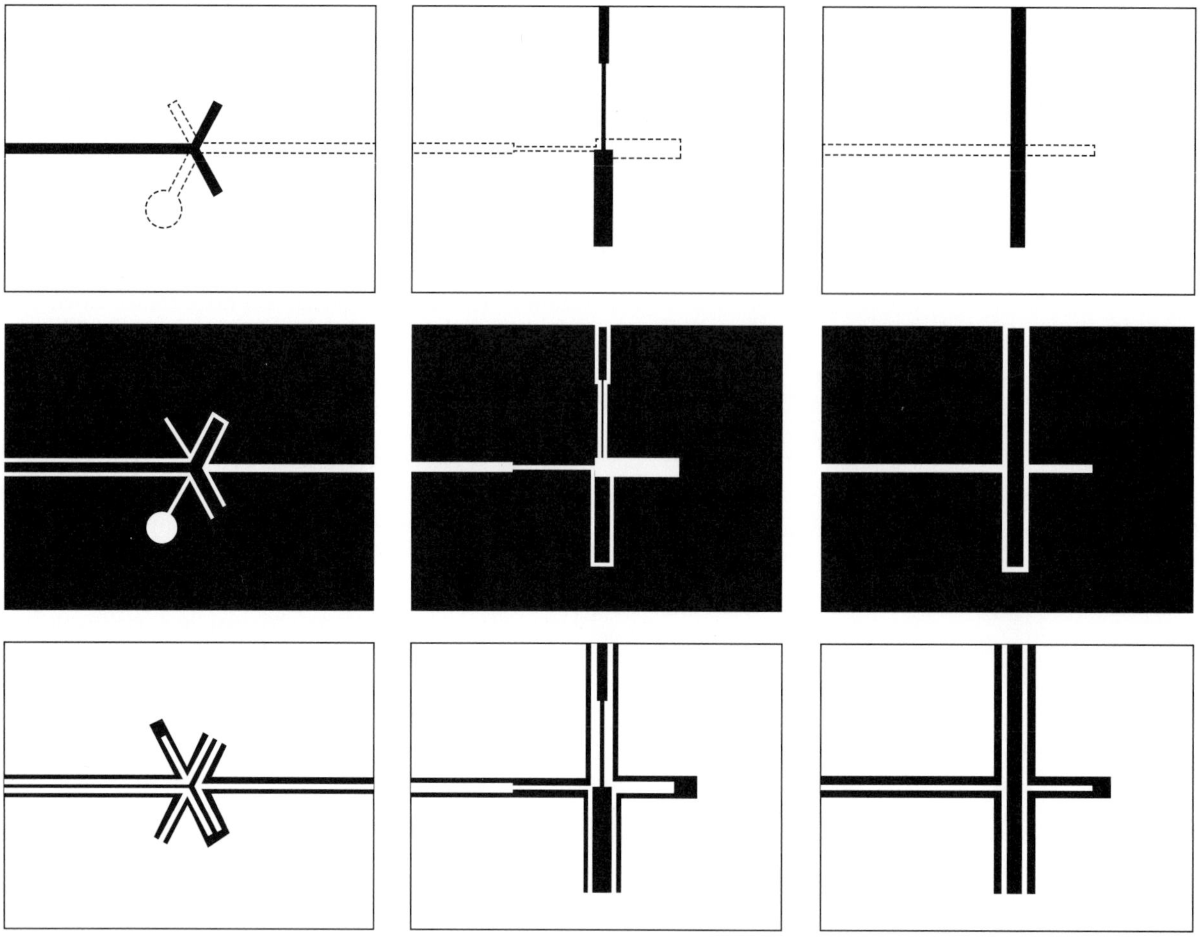

FIGURE 7.15 Three Basic Planar Marchand and Double-Y Balun Transformers: Microstrip-Slot Line, CPW-Slot Line, and Finite-Ground CPW-CPS Line

- **Coupling factor** = $C_{dB} = 10\log(P_1/P_3) = -20\log S_{31}$. (7.5)
- **Isolation** = $I_{dB} = 10\log(P_1/P_4) = -20\log S_{41}$. (7.6)
- **Directivity** is sometimes also used to characterize the isolation properties of a hybrid or directional coupler, which is defined as $D_{dB} = 10\log(P_3/P_4) = -20\log S_{43} = I_{dB} - C_{dB}$. In addition, the transmitted power at port 2 can be calculated through $T_{dB} = 10\log(P_1/P_2) = -20\log(S_{21})$.

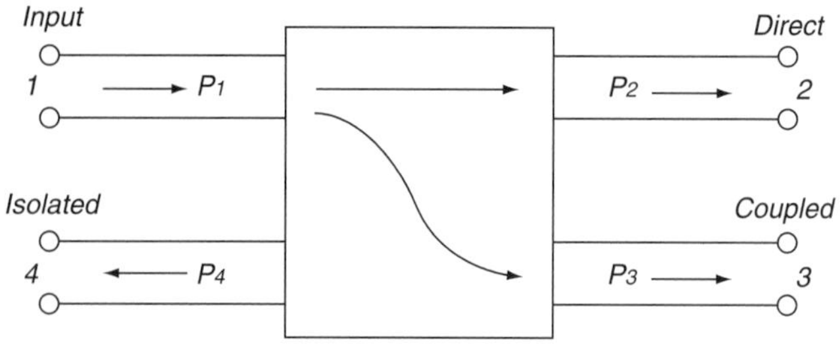

FIGURE 7.16 Reciprocal four-port representation of a microwave hybrid or directional coupler.

- **Input VSWR** (voltage standing wave ratio) is usually used to define the matching condition at port 1 because the other ports are matched to the corresponding external loads.
- **Bandwidth** refers to the operating bandwidth of a hybrid or directional coupler when its coupling factor, isolation, and input VSWR all satisfy the prescribed specifications. Generally, the single-section coupler (typically on the order of 10% for a branch-line hybrid) does not provide sufficient bandwidth. Bandwidth limitations, however, can easily be overcome by multisection designs that combine a number of single-section coupling cells.

7.4.1 Quadrature 90° Hybrids (Branch-Line Couplers)

The structure of a **quadrature 90° hybrid** has two series quarter-wavelength sections that are connected by two shunt quarter-wavelength sections, as shown in Figure 7.17 for a microstrip realization. In most practical cases, the equal power dividing (3-dB) structure is most popular. For a lossless hybrid,

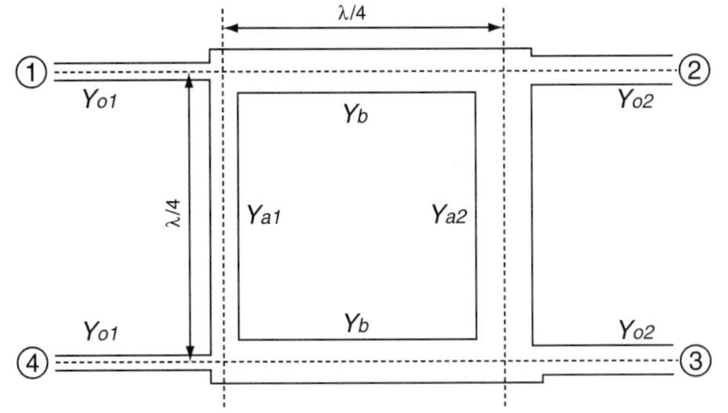

FIGURE 7.17 Physical Layout of a Microstrip Quadrature 90° Hybrid Branch-Line Coupler

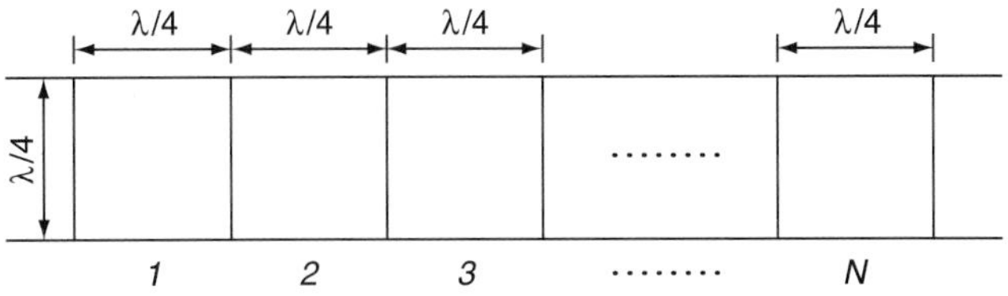

FIGURE 7.18 Schematic View of an N-Section Cascaded Hybrid Branch-Line Coupler with N + 1 Branches

the input signal at port 1 is divided into two parts. Some of the power is combined at port 4 but out of phase by 180°. Thus, if the signal amplitude is equal (depending on adequate choice of branch-line admittance Y_{a1}, Y_{a2}, and Y_b between both ports), no signal appears at port 4, the signals at ports 2 and 3 may be equal (3-dB) but out of phase by 90°. In practice, the isolated port 4 is terminated with a matched load. The power-dividing ratio at ports 2 and 3 also depends on an appropriate choice of the branch-line admittances. The direct and the coupling port (2 and 3) may be connected to admittance Y_{02} that is different from Y_{01} of the input port, which suggests that this type of hybrid has impedance transformation properties.

Table 7.1 presents a list of closed-form design equations for this type of hybrid, which are derived from transmission line theory such as the even–odd mode decoupling technique. As an example, it shows a specific selection of branch-line admittances for the design of a 3-dB hybrid with various input-to-output impedance transformations. Figure 7.18 gives a schematic view of an N-section cascaded branch-line coupler with $N + 1$ branches. The aim of this arrangement is to enlarge the 10% limiting bandwidth of a single-section hybrid to more than one octave.

7.4.2 The 180° Hybrid Rings (Rat-Race Couplers)

Hybrid rings also nicknamed as **rat-race couplers,** are made of an annular line of one-and-half wavelengths, (1.5λ) in circumference with four arms connected at appropriate

TABLE 7.1 3dB 90° Hybrid Design Data ($C_{dB} = 3\ dB$)

$R = \dfrac{Y_{o1}}{Y_{o2}}$	1	0.75	0.5	1/3
$\lvert V\rvert = \sqrt{\dfrac{R}{\log^{-1}(C_{dB}/10)}}$	$1/\sqrt{2}$	0.614	0.5	$1/\sqrt{6}$
$\dfrac{Y_b}{Y_{o1}} = \dfrac{1}{\sqrt{R - \lvert V\rvert^2}}$	$\sqrt{2}$	1.61	2	$\sqrt{6}$
$\dfrac{Y_{a1}}{Y_{o1}} = \dfrac{Y_b}{Y_{o1}}\lvert V\rvert$	1	1	1	1
$\dfrac{Y_{a2}}{Y_{o1}} = \dfrac{Y_b}{RY_{o1}}\lvert V\rvert$	1	1.34	2	3

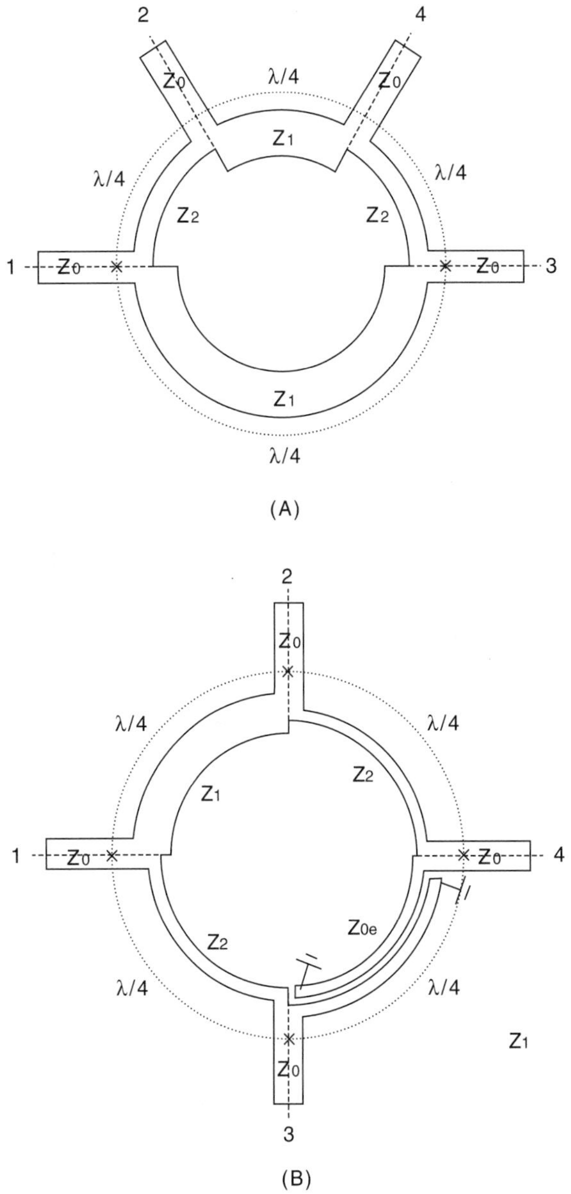

FIGURE 7.19 Two Structure Representations of 180° Hybrid Microstrip Ring Rat-Race Coupler

positions along the annular ring as shown in Figure 7.19(A). The electrical characteristics of hybrid rings are quite similar to those of the 90° branch-line hybrids except that the output ports 2 and 3 have a phase difference of 180° instead of 90°. With reference to Figure 7.19, the design of a 3-dB coupler requires $Z_1 = Z_2 = \sqrt{2}Z_0$. Generally, the rat-race hybrid has a broader bandwidth than its branch-line counterpart, but it is also limited by resonance of the ring or the commensurate line sections.

A substantial increase in bandwidth for this type of coupler is made possible by replacing the $3\lambda/4$-section of the original ring geometry by a shorted $\lambda/4$ parallel coupled-line section as shown in Figure 7.19(B). In this case, the ring is reduced to one wavelength, and its bandwidth can be increased by roughly one octave. The effective bandwidth of ring-based hybrids can be further extended by using a generalized design theory developed by Wang and Wu (1999) that requires a small modification of the impedance ratio of the ring with respect to the externally connected branch lines. In addition, innovative structures can be found in the literature for the design of a 3-dB coupler on the basis of some bandwidth-broadening and size-reduced schemes, such as microstrip disc or waveguide cavity geometry.

7.4.3 Magic Tees (Matched Hybrid Tees)

A **matched hybrid tee** (also called a **magic tee** or **T** consists of a combined *E*- and *H*-plane waveguide *T*-junction, as shown in Figure 7.20. A magic tee is a 3-dB hybrid with directional coupling properties quite similar to the above-discussed 180° hybrid rate-race ring. The most critical issue in the design of a magic tee is a broadband matching post or rod located in the center of the waveguide junction. The magic tee is a popular component for low-loss and/or high-power circuit applications.

Referring to Figure 7.20, if a fundamental TE_{10} wave is incident at port 4, the electric field shows even symmetry at the mid-plane and, hence, no power is directed into port 3 because the electric field is in the opposite direction and cancels out. The coplanar arms 1 and 2 receive equal powers that appear in phase at both ports if the arm lengths are equal.

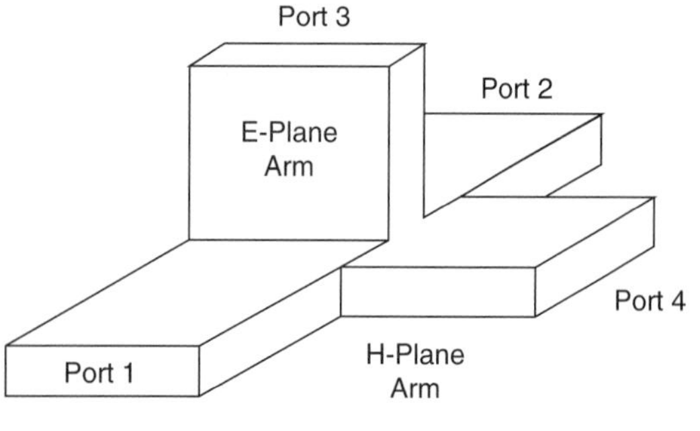

FIGURE 7.20 Waveguide Hybrid Tee Structure or Magic T

When a TE_{10} wave becomes incident at port 3, its electric field is parallel to the broadwall of arm 4 in which it is below cutoff. Hence, no power is directed into that arm. Arms 1 and 2, however, receive once again equal powers that are 180° out of phase in this case (provided both arms are of equal length). If ports 3 and 4 are both matched, the amplitude of all four phasors is the same. Provided all arms are of equal length, the signals add at port 1 and cancel at port 2. Arms 3 and 4 are decoupled. The magic tee is described by the following symmetrical *S*-matrix:

$$[S] = \frac{1}{\sqrt{2}} \begin{bmatrix} 0 & 0 & 1 & 1 \\ 0 & 0 & -1 & 1 \\ 1 & -1 & 0 & 0 \\ 1 & 1 & 0 & 0 \end{bmatrix} \tag{7.7}$$

The magic tees may also be realized through planar approaches similar to the balun design. In practice, a planar magic tee is made of a hybrid strip/slot geometry, where slot and strip are usually located on opposite sides of a dielectric substrate.

7.4.4 Parallel-Line Coupler

The **parallel-line coupler** is a well-known class of directional couplers. When two parallel TEM lines are brought in close proximity of each other, electrical and magnetic field coupling takes place, and a certain portion of the signal of one line is coupled onto the other. With an appropriate choice of coupling length, broadband directional couplers can be built; a typical example is described in Figure 7.21. Symmetrical geometries are most often used even though nonsymmetry of the coupled lines can be considered as well. Geometrical symmetry of the coupled lines leads to a simplification of design models, and even–odd mode analysis can be applied. Otherwise, a modified $C - \pi$ model (Tripathi, 1975) must be employed to decouple the two-line system.

The parallel-line coupler is not restricted to TEM-mode structures although its earliest versions were made from coaxial- or strip-coupled lines, which support pure TEM-modes. The microstrip parallel-line coupler supports quasi-TEM-mode propagation, whereas non-TEM mode couplers may be in the form of a dielectric waveguide, a rectangular waveguide, and slot lines that are known to be dispersive. Design equations, obtained from even–odd mode analysis,

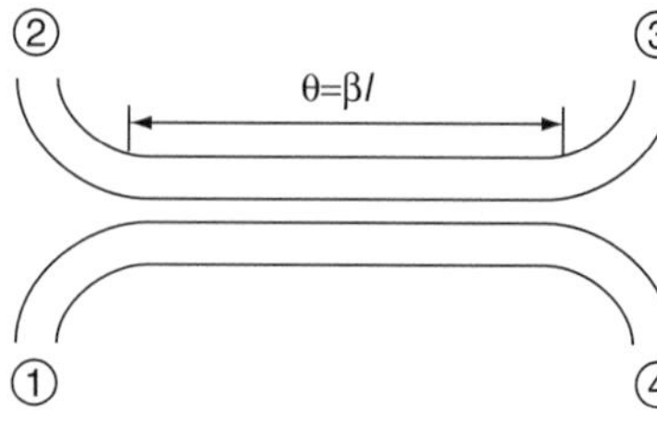

FIGURE 7.21 Typical Sketch of Parallel-Line Directional Coupler.

can be summarized as follows (with quarter-wavelength coupling sections):

$$Z_0^2 = Z_{0e} Z_{0o}. \tag{7.8}$$

$$C = \frac{Z_{0_e} - Z_{0_o}}{Z_{0_e} + Z_{0_o}} = -20 \log \left| \frac{Z_{0_e} - Z_{0_o}}{Z_{0_e} + Z_{0_o}} \right| \mathrm{dB}. \tag{7.9}$$

The characteristic impedances for the even- and odd-modes of the coupled line system can be easily found from a given external line impedance Z_0 and a specified coupling coefficient C required for the coupler design. In the TEM mode case, the even- and odd-mode velocities are always equal and given by:

$$\begin{cases} Z_{0_e} = Z_0 \sqrt{\dfrac{1+C}{1-C}}. \\ Z_{0_o} = Z_0 \sqrt{\dfrac{1-C}{1+C}}. \end{cases} \tag{7.10}$$

As indicated in Figure 7.21, this type of coupler presents a backward coupling property because the coupled energy is directed at port 2 instead of port 3 as it is in the branch-line hybrid that is a forward coupler. For microstrip quasi-TEM mode couplers and non-TEM mode structures, the above design equations are no longer valid. As a general rule, equal even- and odd-mode phase velocities are required for achieving the best directivity performance, which may be difficult to achieve in quasi- and non-TEM mode coupled lines unless a compensation technique is applied.

The type of coupler shown in Figure 7.21 is best suited for weak couplings (e.g., 10- or 20-dB couplers for microwave instrumentations and measurements). Tight coupling requires the lines to be very close together, which may not be realizable because of fabrication tolerances.

Some practical solutions have been developed to remedy these shortcomings of the parallel-line coupler. Figure 7.22 shows a class of commonly used parallel-line couplers including edge, broadside, and interdigital multiconductor couplings. The **De Ronde coupler** may be regarded as a special type of coupler that uses both sides of a substrate (Hoffmann, 1987) employing a combination of coupled microstrip and slot lines to ease the tight coupling problem. On the other hand, the interdigital **Lange coupler** is most noteworthy since it combines the features of both a 90° hybrid and a parallel-line coupler. It usually uses $\lambda/4$ coupled microstrip-line sections for a 3-dB design. The Lange coupler has the advantage of small size, relatively large coupled line spacing, and wide bandwidth.

Ultrabroadband components can be designed with long or multisectioned inhomogeneous and nonuniform tapered coupled lines as described by the example of Figure 7.22(E). In the case of a continuously and gradually tapered line coupler, its operating bandwidth can be easily extended by several octaves. The design is usually made for the upper frequency range in the bandwidth of the coupler.

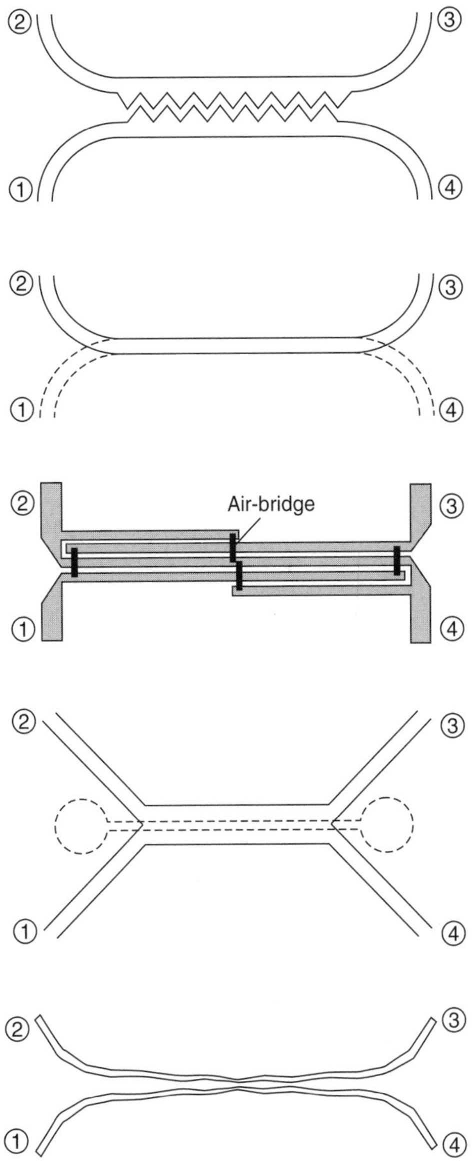

FIGURE 7.22 Various Physical Realizations of Parallel-Line Directional Coupler

7.4.5 Hole- and Aperture-Based Waveguide Couplers

Some waveguide couplers are made of a series of **coupling holes** or some well-placed short slots in the common broad or sidewall of two parallel waveguide sections (shown in Figure 7.23). This design principle is also applicable to multilayered planar structures. In this case, **coupling slots** or **apertures** are etched into the common ground plane(s). Hybrid planar/nonplanar design approaches can also be used for the coupling from planar lines to nonplanar lines or vice versa.

Figure 7.23 presents four well-known couplers: (a) **Bethe-hole coupler** (b) **Moreno cross-guide coupler** (c) **Schwinger reversed-phase coupler** and (d) **Riblet short-slot coupler.**

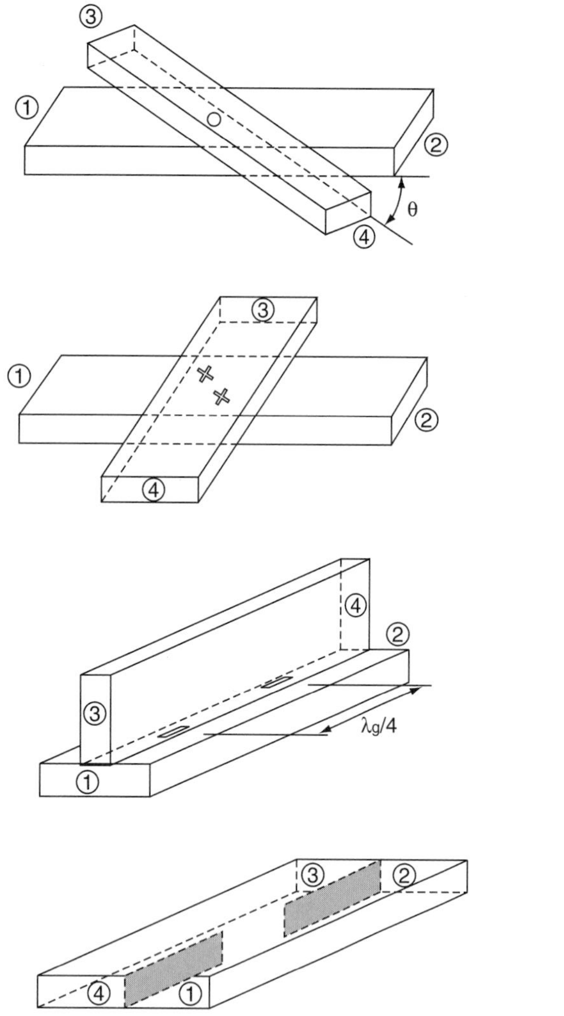

FIGURE 7.23 Typical Hole- and Aperture-Based Waveguide Couplers

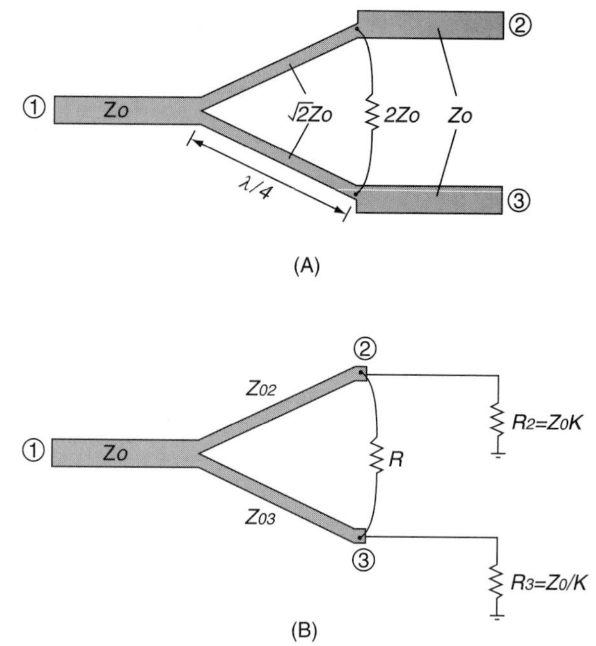

FIGURE 7.24 Wilkinson Microstrip Line Power Divider and Combiner with Equal (A) and Unequal (B) Power Dividing and Combining Schemes

The Bethe-hole coupler consists of two stacked waveguides with small-hole dipole coupling on the common broad wall or sidewall of the guides, the cross-guide coupler consists of two joint waveguides that are coupled through the crossed slots on the broadside of the guides. The cross-guide coupler has an added-up forward coupling due to a backward cancellation of the two excited waves generated by the crossed slots. The short-slot coupler couples electromagnetic energy of the TE_{10} (even) and TE_{20} (odd) modes along the common cross section of the two guides. This coupler design is actually derived from the parallel-line principle but results in forward coupling instead of backward coupling. This difference is caused by the different nature of the mode coupling.

7.4.6 Wilkinson Power Dividers/Combiners

The original three-port **Wilkinson strip-like** (microstrip and strip line) **divider/combiner** offers a one-to-two equal power dividing or two-to-one equal power combining function over a broad bandwidth. Two Wilkinson dividers are shown in Figure 7.24. This type of divider can be designed to give an arbitrary ratio of power division at the outputs. It is a well-known fact that a lossless reciprocal passive three-port junction cannot be matched at all ports simultaneously. In the Wilkinson power divider/combiner, a resistor is introduced to bridge the two output ports for improving their isolation, which is usually called "resistor of isolation." The resistor is placed $\lambda/4$ away from the junction and can be considered as the fourth port.

The requirements for power dividing on the component design may be quite different from power combining even though the same component may posses both functions. This is because only a single input signal is concerned in the divider while two signals are simultaneously injected at the parallel-ports of the combiner, and these signals are not necessarily equal in magnitude and in phase. This may cause a serious problem of power or signal unbalance in the combiner junction. The added resistor, however, ensures that the two output/input ports are electrically isolated.

The two splitting branch lines of the strip-line equal-power Wilkinson divider/combiner have the characteristic impedances of $\sqrt{2}Z_0$, which is obtained from the input-matching requirement (characteristic impedance $= Z_0$ of the three external ports is assumed). The resistor of isolation is $2Z_0$. A Wilkinson divider may offer a bandwidth of about one octave, which can be further improved by using a cascaded multi-section topology.

Wilkinson dividers can also be made with unequal power splits $K^2 = P_3/P_2$, as shown in Figure 7.24(B). The design parameters are then obtained as follows:

$$R = Z_0\left(K + \frac{1}{K}\right). \qquad (7.11)$$

$$Z_{02} = Z_0 K\sqrt{\left(K + \frac{1}{K}\right)}. \qquad (7.12)$$

$$Z_{03} = \frac{Z_0}{K}\sqrt{\left(K + \frac{1}{K}\right)}. \qquad (7.13)$$

Generalized N-way Wilkinson power divider/combiner can be easily constructed in a similar manner. The only problem is that it requires crossover-bridged resistors if $N > 2$, which makes realization difficult in planar form (Maurin and Wu, 1996). In some cases, the design of Wilkinson power divider or combiner may not need a resistor of isolation. This is particularly the case with binary (corporate) power dividers or combiners for antenna feeding networks and may require a symmetrical and balanced topology. The bandwidth in this application, however, is relatively narrow.

7.4.7 Non-Wilkinson Power Dividers/Combiners

Although the Wilkinson structure is probably the most popular choice in the design of a power divider/combiner, other solutions may be useful alternatives when certain design requirements such as phase, power level, bandwidth, operating frequency, combining efficiency, and compactness may not be satisfied with the Wilkinson structure. These alternatives may be classed as **resonant** and **nonresonant** structures and are detailed in Chang (1989). In general, the nonresonant type is more useful, providing broader operating bandwidth. Among them, the corporate (binary-tree type) network may use hybrids, directional couplers, and Wilkinson topology that depend on the required phase relationship of the inputs and the outputs. The N-way network may use spatial or quasi-optical techniques or radial lines and generalized Wilkinson topology. Innovative dividing/combining approaches, have also been developed, such as multihole planar patch dividers/combiners (Kobeissi and Wu, 1999).

7.5 Resonators and Cavities

Resonators are fundamental building blocks for microwave filters, diplexers, frequency discriminators, and tuners. Resonators are used for frequency stabilization in oscillators or as control or tuning elements for amplifiers. A resonator is generally a structure that confines electromagnetic energy either inside a block of dielectric material, inside a section of metallic waveguide enclosure (Figure 7.25), or between the open or shorted ends of a piece of planar transmission line (i.e., microstrip, Figure 7.27). A combination of these may also be used as a resonator. The electrical and magnetic energies stored in the resonator determine its equivalent inductance and capacitance. Any energy dissipation in the resonator due to ohmic, dielectric, or radiation losses can be expressed by an equivalent resistance or a conductance, which directly determines the quality factor (Q-factor) of the resonator. The three electrical parameters RLC constitute a complete equivalent circuit network for characterizing a resonator in terms of bandwidth, center frequency, and Q-factor.

Resonant cavities are an important class of resonators that refer to almost completely closed metallic enclosures that have

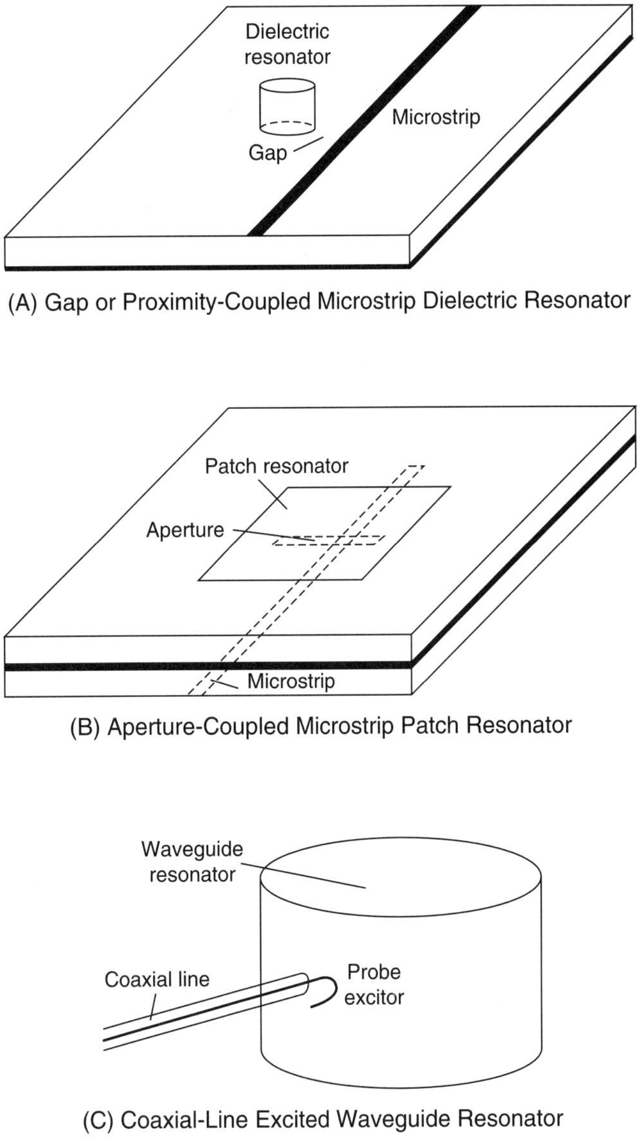

FIGURE 7.25 Different Mechanisms of Excitation on the Basis of a Transmission Line for Dielectric, Planar Patch, and Waveguide Resonators

openings only for input and/or output coupling of electromagnetic energy. Cavity resonators are usually in the form of empty or dielectric-filled rectangular or circular waveguide bodies as well as coaxial line sections. The **Fabry-Perot cavity** is a special case of a quasioptical resonator. The resonance takes place between two face-to-face mirrors while other boundaries are open space. Therefore, it is sometimes called **open or quasioptical resonator** and is useful at very high frequency.

A resonator has an infinite number of resonant modes that satisfy the boundary conditions inside the cavity (eigenvalues of the cavity). Each mode corresponds to a resonant frequency that is characteristic for the cavity dimensions. At resonance, the electric W_e and magnetic W_m energies are equal, and the lowest resonant frequency is characterized by the fundamental (or dominant) mode. Regardless of its geometry or shape, a resonator can usually be modeled by either a series or parallel RLC lumped-element equivalent circuit, and the choice of an equivalent circuit may also depend on a resonant mode in question.

The resonant frequency can be found by:

$$\omega_0 = \frac{1}{\sqrt{L_0 C_0}}. \tag{7.14}$$

The quality factor Q, which is a measure of the resonator loss, is defined as:

$$Q = \omega \frac{\text{average energy stored}}{\text{energy dissipation/second}} = \omega \frac{W_e + W_m}{P_{\text{loss}}}. \tag{7.15}$$

A lossless resonator yields an infinite Q. If the resonator is completely isolated from external loading or coupling, the resulting Q is called the unloaded Q_u. Otherwise, the term **loaded Q_L** is used to account for external loading effects. Q_L is always lower than Q_u as long as the external coupling or load is passive. External loading effects are also extracted by a term called **external Q_e** that can be defined in the same way as equation 7.15, except that in this case the loss P_{loss} is due to the energy transferred to the external circuitry. The loaded Q_L can be expressed as:

$$\frac{1}{Q_L} = \frac{1}{Q_u} + \frac{1}{Q_e}. \tag{7.16}$$

In practice, the unloaded Q(or Q_u) value is closely related to the half-power or 3-dB fractional bandwidth (BW) of the resonator by.

$$Q = \frac{1}{BW_{3\text{dB}}}. \tag{7.17}$$

Because the resonator is always coupled to an external circuity, the coupling scheme is another important aspect in the design of a resonant circuit. Usually, a gap or an aperture is used as the coupling structure. The choice of an adequate coupling mechanism depends on the type of mode one wants to excite as well as the coupling strength required. To measure the coupling or energy transfer between a resonator and the external circuitry, a coupling coefficient κ is defined as:

$$\kappa = \frac{Q}{Q_e}. \tag{7.18}$$

Three types of coupling can be observed:

- $\kappa < 1$ if the feed line is overcoupled to the resonator.
- $\kappa = 1$ if the feed line is critically coupled to the resonator.
- $\kappa > 1$ if the feed line is undercoupled to the resonator.

Critical coupling occurs when the resonator is matched to the feed line. In most filter designs, this is the desired case.

Microwave resonators can be classified into three basic categories: **metallic cavities**, **planar resonators,** and **dielectric resonators**. In addition, open resonators of the Fabry-Perot type and other quasioptical geometries are gaining interest in millimeter-wave applications.

7.5.1 Metallic Cavity Resonators

A **cavity resonator** can be regarded as a section of metallic waveguide including coaxial lines that is usually short-circuited at both ends, thus forming a closed box or cavity. Electric and magnetic fields are confined in the structure, and currents flowing on the interior's imperfect conducting walls cause power dissipation. Electrical and/or magnetic coupling to this type of resonator can be made through a small aperture or a small loop or probe.

Resonant frequencies and mode profiles can be obtained through the solution of Maxwell's satisfying the metallic boundary conditions. The current distribution over the surface of the cavity depends on the profile of magnetic fields that are in turn determined by a particular mode. Generally, the existence of a mode also depends on the excitation mechanism, such as the shape of aperture or probe. Consequently, the power dissipation is expected to be different from mode to mode, leading to different Q values. In practice, the selection of an adequate resonant mode is of critical importance for high-Q operation.

Figure 7.26 illustrates three basic types of cavity resonators: a **rectangular cavity** of height b, width a, and length d; a **cylindrical cavity** of length d with radius a; and a **coaxial line cavity** of length d with an inner and outer radius denoted by a and b. The coaxial cavity resonator operators in the TEM mode, which is much easier to control than other (waveguide) modes. For simplicity, the coaxial line cavity will not be discussed because its information can be found widely in the literature.

For the rectangular cavity resonator, the resonant frequency of the TE_{mnl} or TM_{mnl} modes is given by:

$$f_r^{mnl} = \frac{c}{2}\sqrt{\left(\frac{m}{a}\right)^2 + \left(\frac{n}{b}\right)^2 + \left(\frac{l}{d}\right)^2}, \tag{7.19}$$

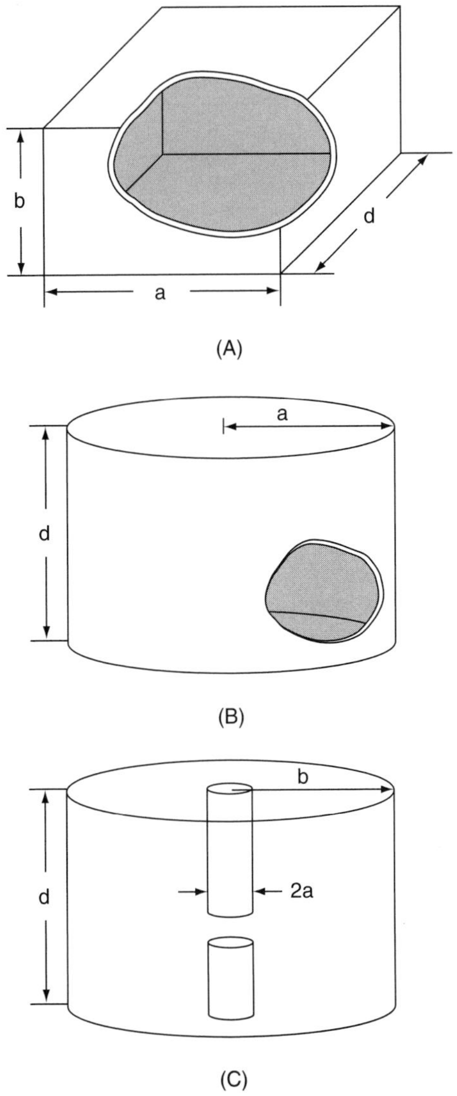

FIGURE 7.26 Representative Waveguide and Coaxial Line Cavity Resonators

in which c is the speed of light. The mode subscript mnl refers to the number of half-sinusoidal variations in the standing-wave pattern (resonance) along the x, y, and z directions, respectively. If $b < a < d$ is valid, the cavity will support the dominant mode TE_{101}, which has the lowest resonant frequency. The unloaded Q value of this mode can range from 5,000 to 15,000 depending on the cavity surface and the metal conductivity. Obviously, the smoother the surface is, the lower the losses. The dominant TM resonant mode is TM_{110}.

The cylindrical cavity resonator is a closed section of circular waveguide similar to its rectangular waveguide counterpart. The resonant frequency of the TE_{mnl} or TM_{mnl} modes is given by:

$$f_r^{mnl} = \frac{c}{2}\sqrt{\left(\frac{p_{nm}^{TE,TM}}{a\pi}\right)^2 + \left(\frac{l}{d}\right)^2}. \quad (7.20)$$

The integer number $l = 0, 1, 2, 3$ refers to the number of half-guided-wavelengths that fit into a longitudinal section of a waveguide. The values of p can be obtained from the zeros of the Bessel functions and their derivatives for TM and TE modes, respectively. In view of the resonant frequency, the dominant mode is TM_{010} for $d/a < 2$, while the TE_{111} mode becomes dominant for $d/a > 2$.

Of all the modes, the TE_{011} is perhaps the most interesting for practical applications because it has only a circumferential current in both the cylindrical wall and the end metallic plates. This means that the end plate is free to be moved to adjust the length d for frequency tuning purpose, which hardly affects the unloaded Q. The maximum Q can be obtained with $d/a = 2$, and a range of 20,000 to 60,000 is possible. This value is significantly higher than the achievable Q values of other modes even though the TE_{011} is not the dominant mode. This is why this mode is used in the design of precision frequency meters that function on the basis of a circular cavity resonator with a movable length of the cavity.

To control the coupling mechanisms in connection with the above cavity resonators, in particular for input/output coupling, can become quite involved, especially if modes other than the wanted one can resonate in the cavity as well. The design of an appropriate coupling topology is also important in view of achieving the highest Q possible for the wanted mode. Two modes are sometimes possible to resonate at the same resonant frequency. These are so-called **degenerated modes** which can be used quite favorably as dual-mode filters, for example, in very narrowband applications. The advantage is that with only one physical cavity, two electrical resonators can be realized.

7.5.2 Planar Resonators

Planar resonators may conveniently be classified as **transmission line resonators** and **patch resonators**. Figure 7.27 shows three typical planar transmission line resonators. Their line length is always commensurate with wavelength. Figure 7.28 illustrates two examples of patch resonators. The transmission line resonators consist of sections of line having the physical length equal to a certain fraction of the guided-wave wavelength. A closed ring geometry where the electrical length equals an integer multiple of the wavelength also acts as a resonator. For open-ended planar resonators, the fringing field effects at the open end introduce an effective lengthening of the line and, thus, lower the resonant frequency. This effect must be taken into account in the calculation. Hence, resonant modes of such resonators may be considered to have approximately a one-dimensional dependency. This is quite different for the resonance condition of a patch resonator, which depends on both length and width of the patch (two-dimensional problem).

The electric energy is predominately confined in the dielectric substrate under the strip, while the magnetic energy

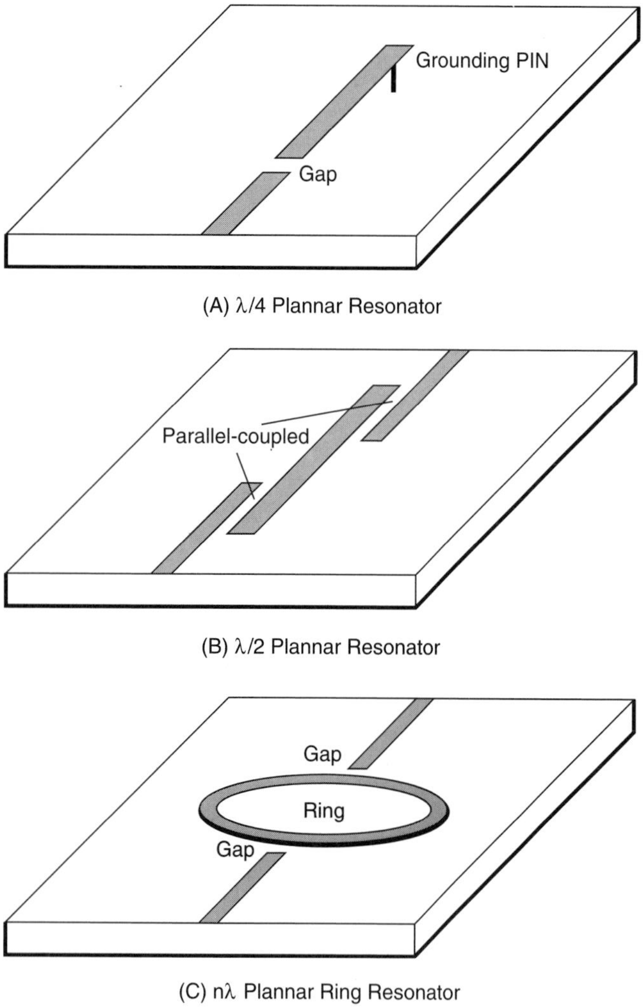

FIGURE 7.27 Gap-Excited Typical Microstrip Line Resonators

extends into both the substrate and air regions around the line or patch. Simple lossy transmission line theory can be used to calculate the resonant frequencies and Q values of the transmission line resonators. Their unloaded Q values are evaluated by:

$$Q = \frac{\beta}{2\alpha} \frac{v_p}{v_g}, \tag{7.21}$$

in which α and β are the attenuation and phase constants of the resonator line, and v_p and v_g are the phase and group velocities, respectively (Konishi, 1991). As for the TEM-mode transmission line resonators, the condition of $v_p = v_g$ is always valid. The resonant characteristics of an arbitrarily shaped planar patch resonator can be obtained by applying a field modeling technique (two-dimensional problem). For a regularly shaped resonator, such as the circular disk or radial resonator shown in Figure 7.28(A), the resonant frequencies can be approximately calculated by simple closed-form formulas. TM_{nm0} modes are

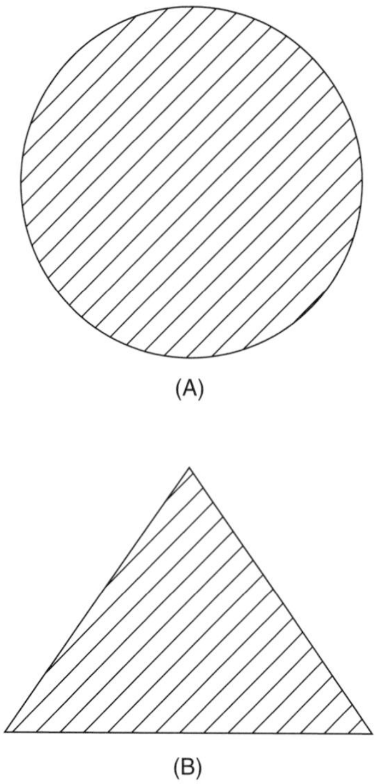

FIGURE 7.28 Layout View of Circular and Triangular Microstrip Patch Resonators.

dominant in those structures where the subscripts n and m indicate the number of half-wave variations of the fields along two (planar) orthogonal directions of the coordinate, respectively. The subscript 0 stands for no variation of fields in the vertical direction of substrate.

Generally speaking, the unloaded Q of the planar resonators is basically determined by the conductor loss if the dielectric substrate has a loss tangent $\tan\delta$ small than 10^{-3}. Practical values of Q are in the range of 100 to 500. A patch resonator usually has a slightly higher Q than its transmission line counterpart. On the other hand, the Q value of a patch resonator also depends on its operating mode because ohmic losses are different from mode to mode. Planar resonators find wide-ranging applications in particular in the lower microwave regime. Coupling between planar resonators is easily accomplished by a capacitive gap. The resulting loaded Q_L value is usually low and may range from 10 to 100. Patch resonators are good candidates for antenna design.

A planar resonator may also be constructed by lumped elements such as parallel or series LC networks, which are popular choices in MMICs. This is in particular adequate for some active circuit designs, such as voltage-controlled oscillators (VCO).

7.5.3 Dielectric Resonators

A **dielectric resonator** (DR) is a small piece (disc, cube, or rectangle) of ceramic or other low-loss dielectric material with a high relative dielectric constant ε_r, normally in the range of

10 to 100. The reflections at the dielectric–air interface confine the electromagnetic fields mainly inside the dielectric region and its immediate vicinity. The operating principle of a dielectric resonator is quite similar to that of rectangular or cylindrical cavity resonators. Calculation of the resonant frequencies is, however, more involved because of the inhomogeneous medium in which the electromagnetic wave resonates.

Dielectric resonators can serve two basic functions. One is to keep the electromagnetic energy away from any metallic boundary and thus increase the Q factor significantly, provided the dielectric is of low loss (specialized ceramic). The other function is to reduce the resonator size by approximately $1/\sqrt{\varepsilon_r}$. Specially designed ceramic materials offer excellent temperature stability and high Q value. Typical values for Q are in the range of 1,000 to 60,000. In a practical realization, a dielectric resonator is either enclosed in a metallic cavity (placed sufficiently away from the housing) or, for lower Q applications, metallized at its surface. A dielectric resonator can also be used as a drop-in resonator in a planar circuit (surface-mounted on a metallic surface) or even sandwiched between two parallel metallic plates. The latter application prevents to some degree radiation loss and unwanted interference with other circuits.

Figure 7.29 presents four different uses of the dielectric resonator, including two applications for millimeter-wave circuit applications (for a nonradiative dielectric waveguide [NRD] resonator and for a planar dielectric resonator (Ishikawa *et al.*, 1996). A **cylindrical dielectric resonator** as shown in Figure 7.29(A) is probably the most commonly used structure in which the low-permittivity dielectric substrate may be used to keep the resonator in place. In this resonator structure, the dominant resonant modes are usually the $HE_{11\delta}$ and $TE_{01\delta}$ modes, respectively. The subscript δ denotes the number of resonating waves in the longitudinal direction of the resonator. The lowest number is $\delta = 1$. A complete field solution for the resonance properties of cylindrical resonators can be obtained through field modeling and matching the boundary conditions at the air–dielectric interfaces.

Figure 7.30 shows some field profiles of the principal modes for the cylindrical dielectric resonator suspended between two metallic plates and surrounded by air. In this figure, two classes of hybrid modes are illustrated: the HE_{nm} and EH_{nm} modes. The former describes a mode in which the H_l component of the mode is dominant, whereas in the latter case, it is the E_z component. Figure 7.31 depicts three practical techniques commonly used to achieve magnetic coupling between a microstrip line and a $TE_{01\delta}$ mode dielectric resonator.

In general, applications of dielectric resonators are limited to lower frequencies only since the loss of tangents at millimeter-wave frequencies may become prohibitive. Furthermore, size reductions at millimeter wavelengths may be too great to be practical. An exception may be the **whispering-gallery** (WG) **mode dielectric resonator** that can be generated in a hollow dielectric cylinder as shown in Figure 7.29(B). This resonator exhibits a large number of azimuthal field variation around the hollow cylinder, which is quite similar to the higher order resonances of a microstrip ring resonator. The WG resonator is essentially "oversized" for millimeter wavelengths and thus enables the designer to use larger structures.

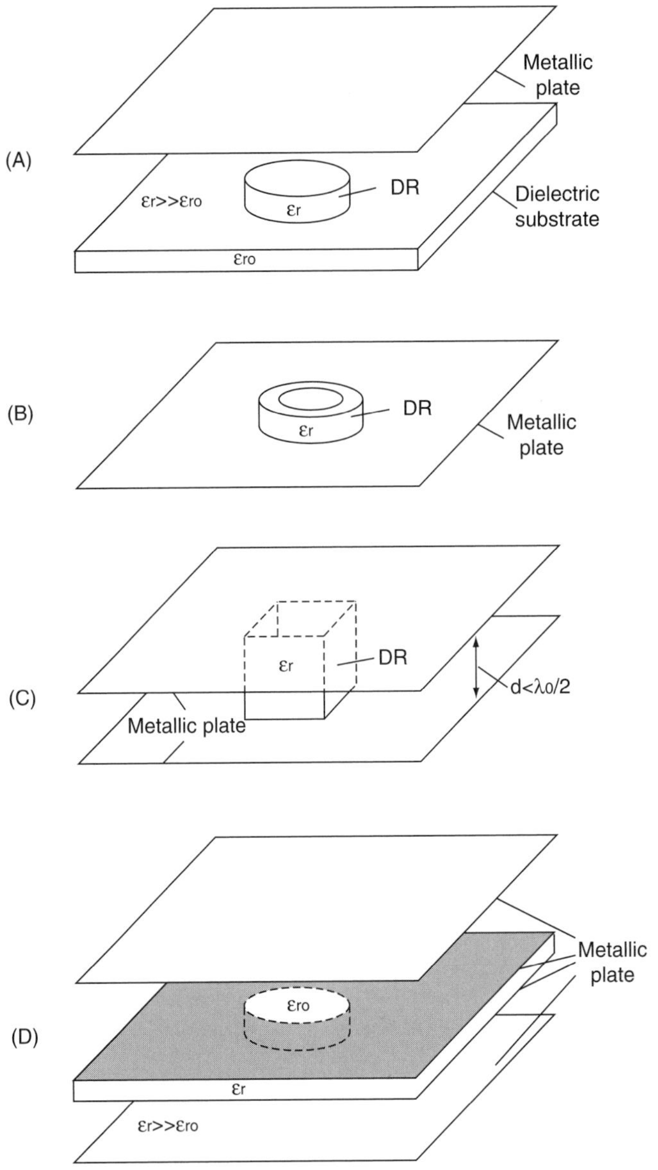

FIGURE 7.29 Four Different Dielectric Resonators (DR): (A) Surface-Mounted DR; (B) Image DR; (C) Non-Radiative Dielectric (NRD) resonator; and (D) Planar DR

In case of an **NRD-guide resonator,** the dielectric block may be in the form of a rectangle made of a relatively low permittivity and low-cost dielectric material. Thus, size is not a problem. The intrinsic nonradiating condition of the NRD-guide will give rise to a high Q value of the resonator since the fields are well confined inside the dielectric region and pose no leakage problem. The NRD-guide resonator is also a promising

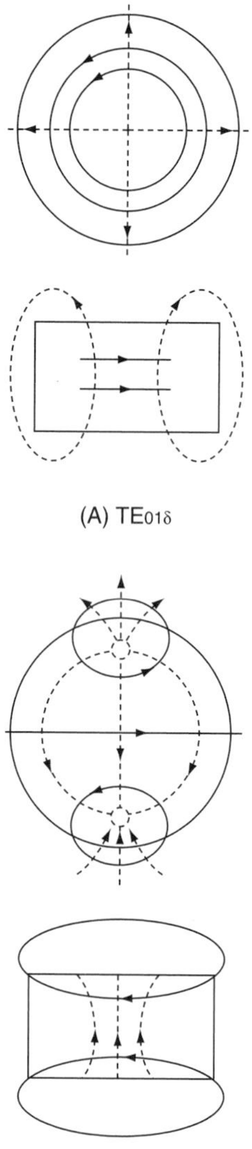

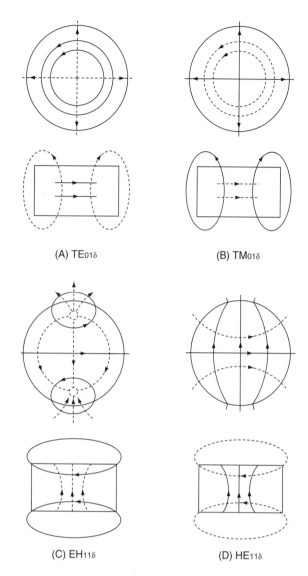

FIGURE 7.30 Illustration of Field Profiles of Some Principal Modes for a Cylindrical DR (Solid Line for E-Fields and Dotted Line for H-Fields)

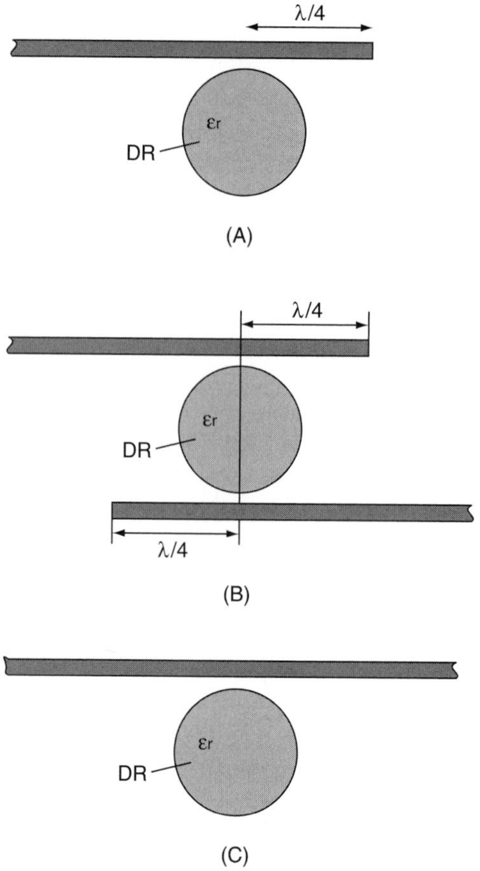

FIGURE 7.31 Basic Coupling Schemes Between $TE_{01\delta}$ Dielectric Resonator and Microstrip Line: (A) Reflection Type, (B) Transmission Type, and (C) Reaction Type

solution for millimeter-wave circuits. Figure 7.29(D) shows a planar dielectric resonator that is compatible with a conventional lithographic fabrication process for planar circuits. The resonance takes place around the circular aperture area of the dielectric substrate while the remaining dielectric surfaces are metallized. This structure offers a very good Q value at millimeter-wave frequencies while it is compatible with coplanar waveguide or slot lines realized on the same substrate.

7.6 Filter Circuits

Filters may be constructed by active and passive elements, and they are indispensable in almost all microwave electronics and communication systems that need to control, channelize, or combine signals. A passive filter may be designed by either lumped LC or distributed resonators (planar, dielectric resonators, or metallic cavities) or both and arranged in a cascaded transfer network.

A passive filter is usually regarded as a two-port linear network that allows a signal to pass over desired frequency range(s) with minimum possible attenuation, whereas undesired frequency bands are attenuated as much as possible in the given constraints. In the design of a filter, the most important specifications are related to bandwidth, insertion loss, stopband attenuation or out-of-band rejection, spurious response, return loss, input/output impedance levels, VSWR, group delay, topological size, temperature range, power handling capability, and transient response. In practice, a selected group of those specifications is most dominant and determines the design approach.

A filter is usually synthesized and designed through some well-established network theories for a given frequency response or a designated frequency-related network transfer function. A reciprocal two-port filter network is schematically

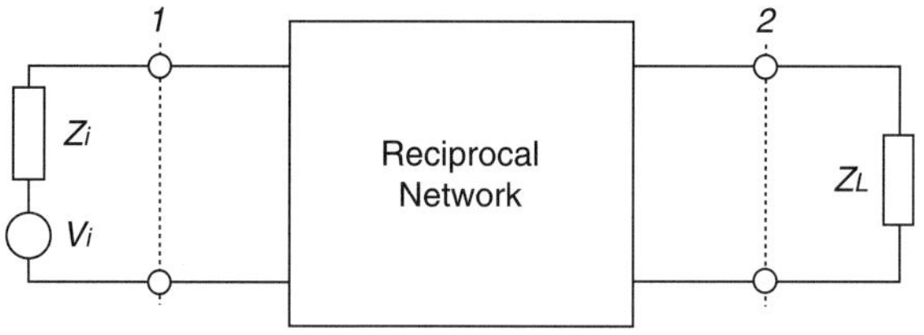

FIGURE 7.32 Schematic Reciprocal Two-Port Filtering Network

described in Figure 7.32. Note that the source and load terminal conditions are part of the design parameters and are included in the network synthesis. In this case, two-port S-parameters can effectively be used to characterize a filter in terms of S_{11} and S_{21} for its return and insertion losses, respectively. In a low-loss filtering system, magnitudes of both S-parameters are directly related to each other, thus allowing to derive one if the other is known. Phase characteristics of a filter are important for certain practical applications, such as phase and frequency modulation schemes. The group delay is also usually used to measure how long it takes for the signal to pass through the filter (in seconds). The group delay is also usually frequency dependent (dispersion). In addition, temperature sensitivity may be a delicate issue of design considerations, in particular for narrow band applications in all areas (e.g., commercial, military, and satellite).

There are four basic types of filters classified according to the frequency response: **low-pass**, **band-pass**, **band-stop or stopband** (also called band rejection or notch), and **high-pass** filters. They are illustrated in Figure 7.33. An ideally lossless filter has zero transmission loss and infinite out-of-band rejection as well as abrupt band response. Practically, a filter always presents a loss-related degradation of performance, and it also exhibits unwanted spurious responses beyond a certain limit of the frequency range of interest.

Filter design is usually based on a prototype network for which the transfer function can be described mathematically. The transfer function is usually given in the form of a

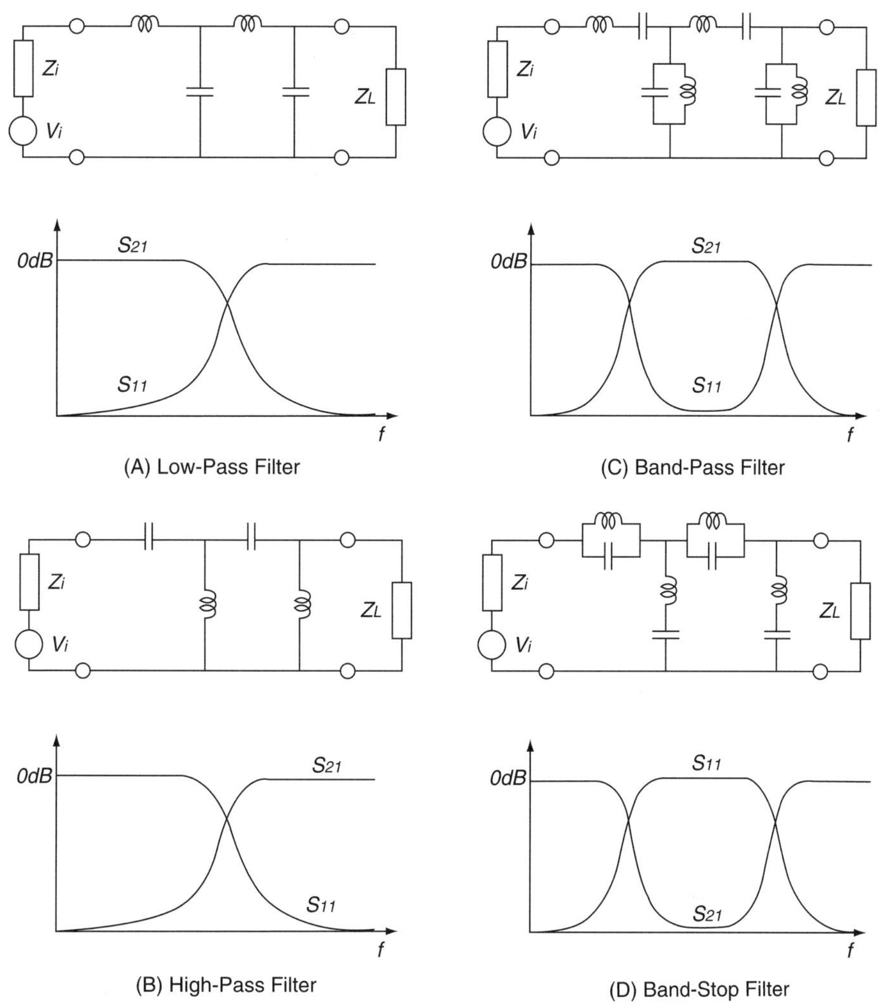

FIGURE 7.33 Simplified Equivalent Circuit Representation and Typical Frequency Response of Four Basic Filters

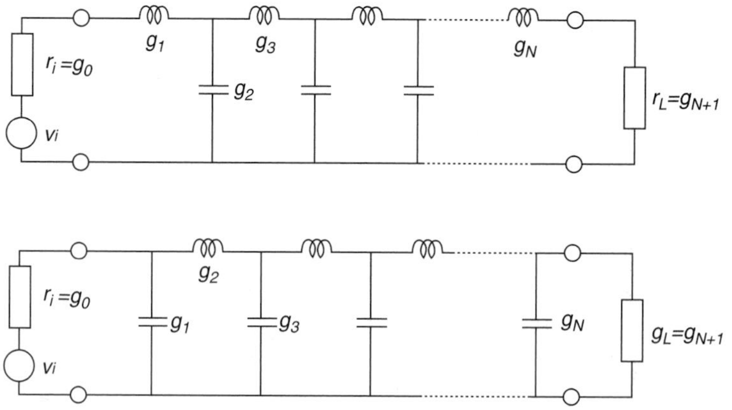

FIGURE 7.34 Two Possible Prototype Low-Pass Filter Networks

frequency-dependent polynomial or other mathematical functions. In general, the design of a filter requires three steps, starting with the network synthesis of a low-pass LC prototype (using normalized frequency, normalized bandwidth, and specified slope selectivity) as shown in Figure 7.34. The insertion–loss method and the image–parameter method are most frequently used to calculate network parameters of the prototype transfer function. Then, the prototype network and parameters are transformed or frequency-scaled to the actual frequency specifications. Finally, the synthesized network is physically designed and produced by either lumped LC elements, commensurate line planar microstrip, waveguide sections, or other building blocks. Special techniques to transform LC parameters into corresponding line dimensions or waveguide geometries have been developed. Most of them are based on the well-established impedance and admittance inverters technique (so-called *K*- and *J*-inverters) (Mattaei *et al.* 1980). In practice, postfabrication tuning elements or other similar corrective measures are required to account for interelement coupling, inaccurate material parameters, mechanical tolerances, and parasitic effects.

Most **microwave filters** are designed as either **maximally flat** (Butterworth), **equal-ripple** (Chebyshev), or **elliptic function response** (Figure 7.35). Other types of transfer functions are possible and depend on specific design requirements. The degree of the low-pass prototype function (the order of the

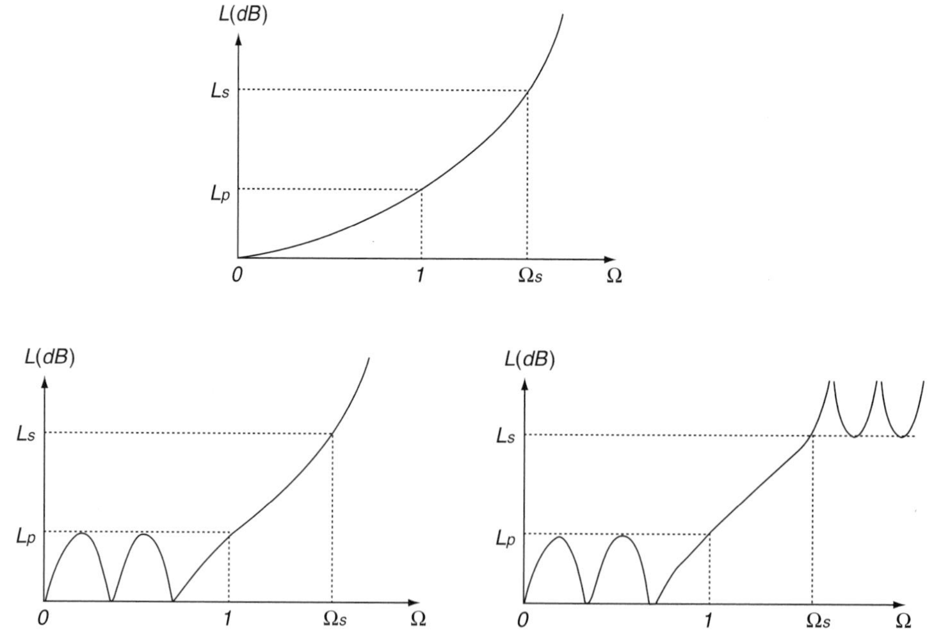

FIGURE 7.35 Three Popular Frequency Responses or Transfer Functions of the Prototype Filters Selected in the Design: Maximally Flat (Butterworth), Equal-Ripple (Chebyshev), and Elliptic Function

polynomial) corresponds to the number of reactive LC elements required to obtain the desired response. This is called the **order of the filter**. In the case of band-pass and stopband filter designs, the order is also related to the number of poles or zeros of the frequency response (resonance-related effects). A practical filter design usually requires an order ranging from 2 to 10.

A filter may be realized in any one of the following technologies: **metallic waveguides** (including coaxial lines), **dielectric waveguides,** and **planar transmission lines**. Planar filters are small in size, lightweight, and low in cost but have usually a low *Q* value compared to waveguide and dielectric resonator filters. In practice, high-quality band-pass and stopband filters that require high *O* resonators are made of waveguide and/or dielectric resonators. For narrow band realization (less than 1.5%) with low insertion loss as well as broadband application, extremely low-loss resonators are needed. Broadband applications (greater than 15%) with high slope-selectivity are possible. Generally, the broader the bandwidth is, the higher the order of the filter. Moreover, the higher the required slope-selectivity is, the higher the order of the filter. Single-mode resonators arranged as direct coupled (Chebyshev or Butterworth) or crosscoupled filters with elliptic transfer function response are possible. Dual-mode or multimode resonators allow the realization of more than one electrically equivalent resonator in a single physical cavity. These filters are attractive for space applications due to their small size and lightweight features.

7.6.1 Low-Pass Filters

The equivalent circuit of a typical **low-pass filter** consists of a cascade of series inductances and parallel capacitors. It is usually realized with TEM or quasi-TEM mode transmission lines like strip line, microstrip line, coaxial line, and coplanar waveguide because they have no cutoff frequency. The equivalent LC circuit elements can be designed with short line sections of high and low impedances (stepped impedance), respectively. In some cases, LC lumped elements like spiral inductors and MIM capacitors are used to realize a filter on an MMIC chip or a high-density hybrid integrated circuit. Although using LC lumped elements reduces the size of the passive filter structure, the Q value, insertion loss, and slope selectivity may deteriorate significantly.

Figure 7.36 illustrates examples of low-pass filters that can be made out of strip line, microstrip line, and coaxial line sections. The simplest example of a low-pass filter may be a dc-biasing block filter used to isolate the RF from the dc source, for example, in a microwave amplifier circuit. The operating bandwidth of a low-pass filter can range widely depending on its lowest cutoff frequency point. Note that a waveguide filter can also be designed as pseudo low-pass filter in a waveguide band, but below the waveguide cutoff frequency, no signal propagation will take place.

7.6.2 Band-Pass Filters

A **band-pass filter** may also be called a **band-select filter** as it selects a specific frequency range to pass a signal unattenuated. This type of filter is the most frequently used. Band-pass filters may be built from all common transmission line media, ranging from waveguide to microstrip line. A basic feature of the geometry of a band-pass filter is that it may consist of resonators coupled along the transmission path of signal, and its number of poles is related to the number of resonant modes of the filter.

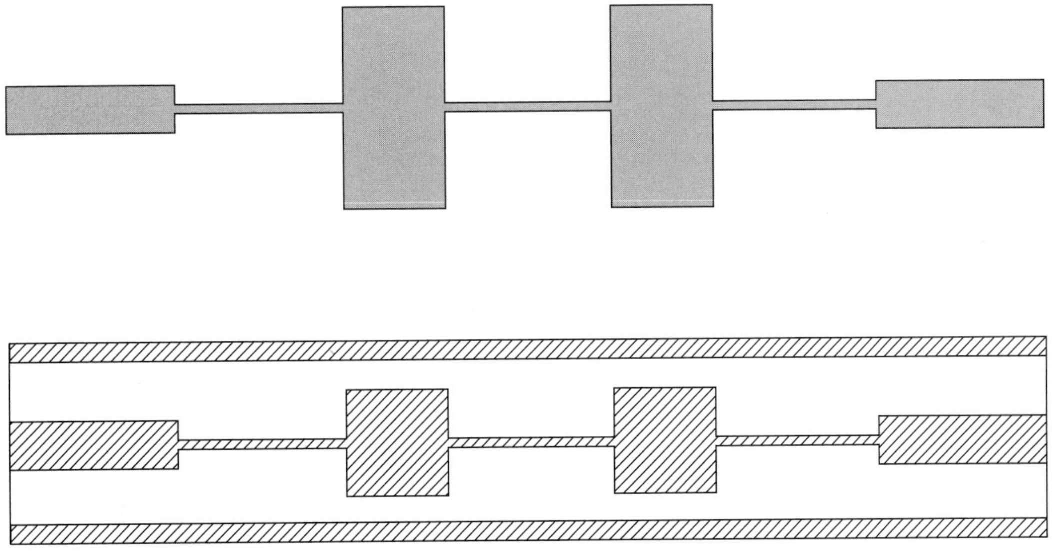

FIGURE 7.36 Examples of Low-Pass Filter Made of Microstrip, Stripline, or Coaxial Line Sections

Planar resonators are known to suffer from high ohmic losses, in particular at high frequency, which excludes them from the construction of high-performance, narrow band filters. Usually, they are limited to the lower microwave domain (up to 15 GHz) and only if insertion loss and slope-selectivity is not a major concern. Some *Q*-enhanced planar filters are possible in conjunction with strip line and suspended microstrip structures (up to 30 GHz). In the higher microwave domain and well into the millimeter wave region, waveguide filters are most commonly used. With the emergence of commercial broadband systems in the millimeter-wave range, quasiplanar filters have made a comeback. Coaxial line and dielectric resonator filters are most frequently used in the lower microwave range if high *Q* values or small size are needed. In addition, low-loss and lightweight dielectric filters are constructed for millimeter-wave applications.

One of the most difficult problems for band-pass filters (especially for narrow-band types) is the need for postfabrication tuning. Fabrication tolerances and material uncertainties as well as inaccurate design techniques may all contribute to the need for tuning. Field theory-based design techniques may alleviate this problem to some degree.

Most frequently used configurations for planar band-pass filters are **direct-coupled, parallel-coupled, interdigital, comb-line, hairpin-line, dual-mode ring,** and **square-patch resonators,** some of which are shown in Figure 7.37. Direct-coupled resonator filters are of excessive length that can be reduced by using a parallel-coupled geometry. The parallel coupling can be made stronger to achieve a larger bandwidth. The interdigital combline, and hairpin-line have side-to-side inter-resonator coupling schemes, and filters can be made compact if spurious responses are suppressed. In addition,

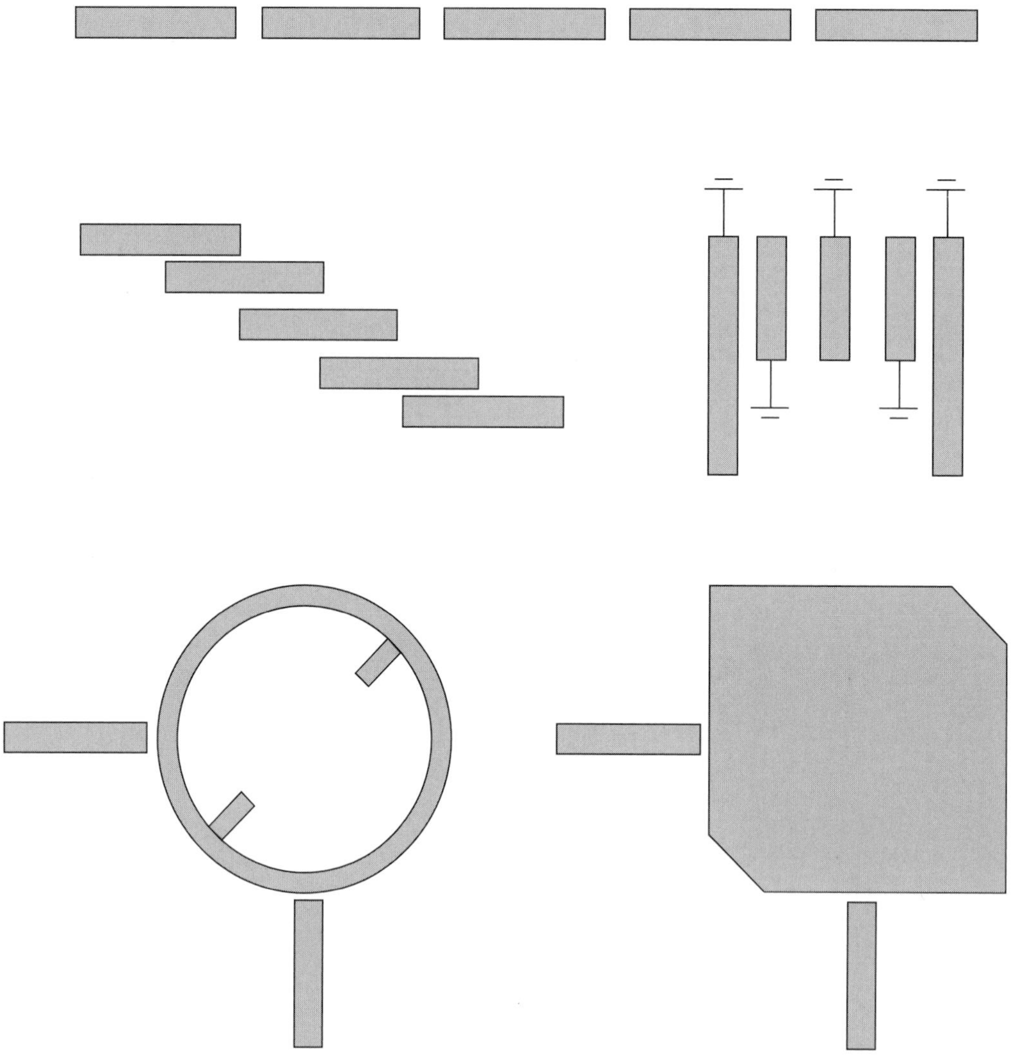

FIGURE 7.37 Some Popular Configurations of Planar Transmission-Line Band-Pass Filter Including Dual-Mode Filters

they are good candidates for narrow band designs. The dual-mode ring and square-patch resonators can simultaneously induce two resonant modes that are orthogonal in space and are excited by the two orthogonally arranged input and output lines. The coupling between the two modes is accomplished by a topological perturbation that takes place along the symmetrical axes with respect to the input and output lines.

Figure 7.38 presents a number of nonplanar band-pass filters, including waveguide, dielectric, and coaxial line filters. Various waveguide filters have been developed over the frequency range from 1 GHz to more than 100 GHz. They are in the form of cascaded rectangular or circular cavities in which low-loss resonant modes are excited. Waveguide filters may be categorized as **stub type, *E*- or *H*-plane type, dual-mode type,** or **filters with corrugated geometry**. Most of the resonator sections are coupled through evanescent mode waveguide sections. Among them are *E*-plane filters that are most suitable for mass-fabrication. In contrast, popular dual-mode and crosscoupled filters are very special designs requiring very particular knowledge in design and tuning.

E-plane filters are usually referred to as **finline** and metal insert filters that consist of ladder-shaped inserts in the *E*-plane of metallic waveguide (Vahldriek, 1989). The inserts can either be made from pure metal or they consist of a ladder-shaped metal pattern etched on a thin supporting low-permittivity substrate. The latter structure is called a quasiplanar filter.

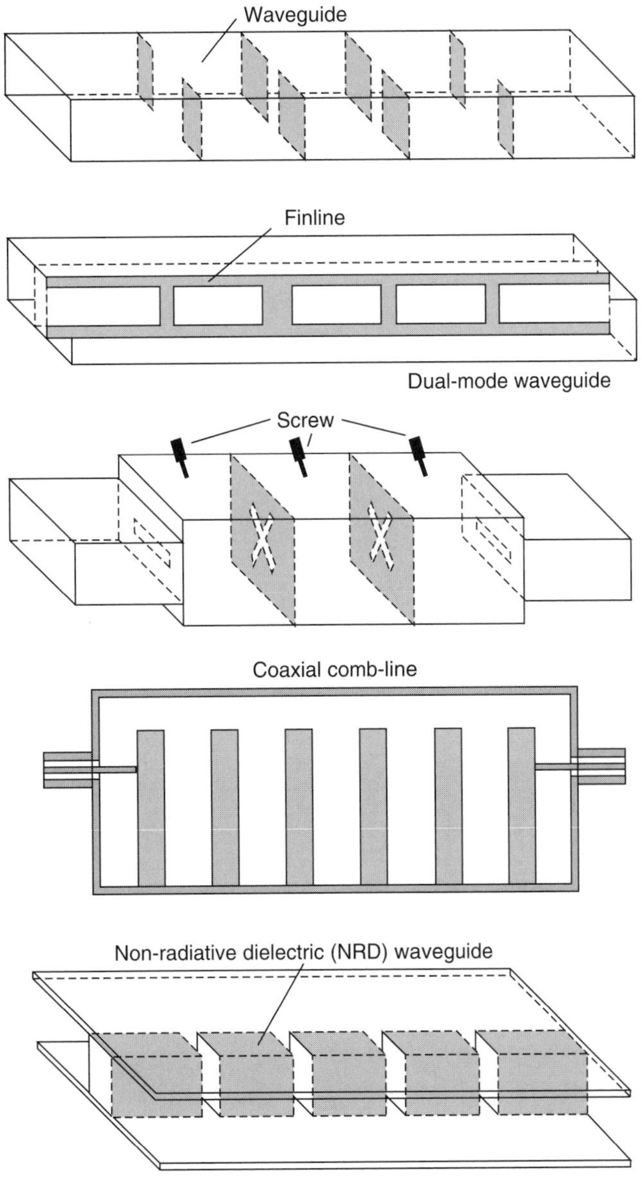

FIGURE 7.38 Selected Design Choices of Band-Pass Filter with Nonplanar Metallic Waveguide, Finline, Coaxial Line, and Dielectric Waveguide

The filter function is mostly determined by the ladder-shaped pattern. The pure metal insert structure is used if insertion loss is of paramount importance. The quasiplanar structure is used when the following components are printed on the same substrate, allowing high integration density.

Dual-mode filters (canonical symmetric design), as described in Figure 7.38, are based on square or circular waveguide cavities that support two orthogonals in a single cavity (Zaki *et al.*, 1987). Thus, the filter structure can be miniaturized, and elliptic function and linear phase performance can also be obtained. Dual-mode filters can be based on low-loss high-permittivity square or cylindrical dielectric blocks. The idea is to maintain the electromagnetic field inside the dielectric material and away from the lossy metallic wall to improve the Q factor. These design approaches are found in very narrow band applications (less than 0.1%) at low frequencies (Less than 3 GHz). Even smaller bandwidth is only obtainable with superconductor filters. Depending on how much the dielectric fills the metal cavity, the size reduction effect sets in.

Coax filters are found as combline and interdigital filters and are mainly used in the lower microwave domain. Interdigital filters are based on side-to-side coupled $\lambda/4$ resonators, whereas the combline filters (Figure 7.38) consist of coupled resonators that are loaded with lumped capacitors at the open ends to reduce the lengths of the resonators and thus help to shrink the complete filter volume. Dielectric waveguide filters use high Q dielectric resonator building blocks, and the input and output lines are usually in the form of waveguide or coaxial lines. The dielectric resonators are canonical with high dielectric constants, and they are coupled with each other through irises or adjacent air interfaces. Generally, the dielectric band-pass filters can be used over the frequency range from 300 MHz to 100 GHz. For high-frequency applications, NRD waveguide filters (Figure 7.38) gain interests because of the extremely low-loss and low dielectric constant materials that can be used in the design.

7.6.3 Band-Stop Filters

The band-stop class of filters is designed to reject signals over the selected bandwidth and pass all other frequencies. Therefore, these filters are also called band rejection filters. They can be used, for example, to reject only the harmonics generated by a nonlinear circuit or to cancel out interferences at selected frequencies. The basic strategy of *band-stop* filter design is to place resonators in parallel with the main transmission line. When these resonators are in resonance, they short-circuit the signal at that particular frequency. At other frequencies, they do not load the main transmission line and are therefore invisible to the signal.

Figure 7.39 describes two typical *band-stop* filters. One of them is a **waveguide stub filter** with stubs attached to the broad wall of the main line. A signal will be rejected if its operating frequency falls into the resonant frequency range of the stubs. The other example is a simple **microstrip line** coupled with a series of dielectric resonators that observe the similar operating mechanism. On the other hand, a periodic structure is used that exhibits alternated pass and stopbands. The stopband is called **bandgap** in quantum physics and optics for periodic patterned structures or simply bandgap structures.

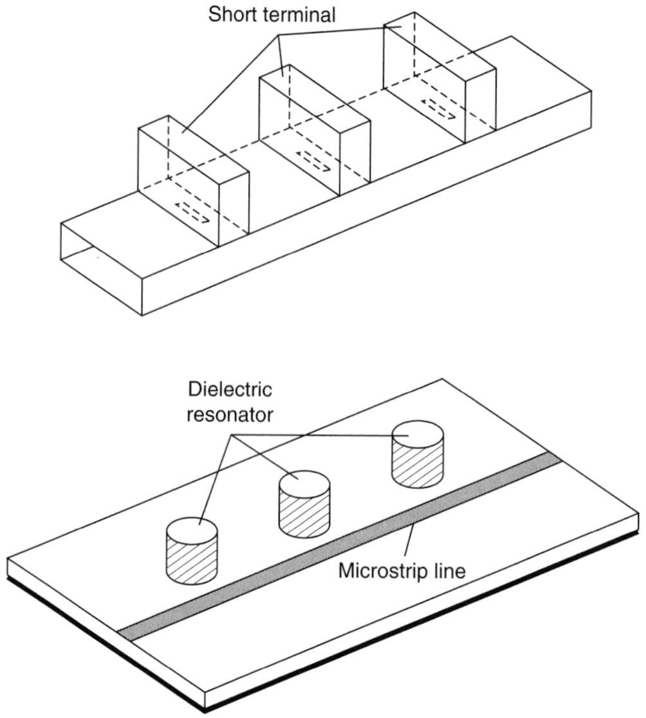

FIGURE 7.39 Two Typical Waveguide and Planar Line/DR Stopband Filters

7.6.4 High-Pass Filters

High-pass filters are complementary to low-pass filters. From an equivalent network point of view, the design of a high-pass network is quite straightforward as it is sufficient to interchange the topological position of inductors and capacitors of the low-pass filter. Thus, high-pass filters consist of a series capacitor elements with joint shunt inductors. As a matter of fact, a metallic waveguide itself is a natural high-pass filter because the signal is completely rejected below the cutoff frequency of the guide. Figure 7.40 shows two representative examples of stopband filters made of planar elements and waveguide sections, respectively.

7.6.5 Multiplexers

Multiplexers (Rhodes and Levy, 1979) are important components for channel combination or separation in multichannel communication systems. They are usually part of the antenna feed systems in front-end modules. Generally, there are four

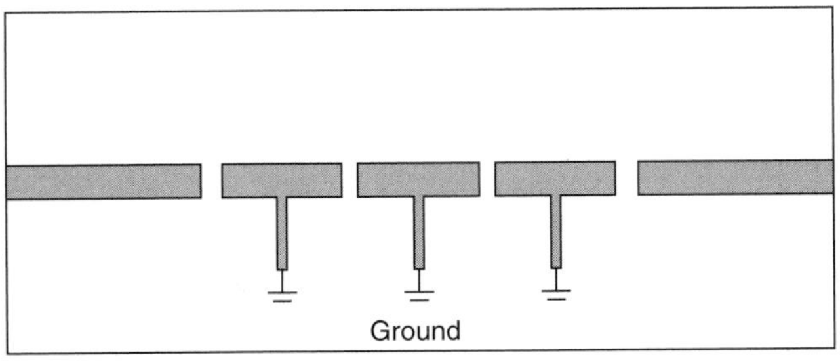

FIGURE 7.40 Representative Microstrip Line High-Pass Filter

different multiplexing methods: (1) the **circulator/filter chain** (2) the **directional filter technique**, (3) the **manifold multiplexing approaches**, and (4) the **branching filter concepts**. Each of them has its particular merits and applications (Uher *et al.*, 1993).

The **diplexer** (Menzel *et al.*, 1997) is a special case of a multiplexer. Its purpose is to separate two channels, the receiving channel and the transmit channel by means of two band-pass filters whose paths are terminated on the same antenna. The diplexer is a low-loss alternative to a circulator-based solution. In the design of the band-pass filters, the slope-selectivity between both filters must be sufficient to ensure a certain level of isolation between the transmit and the receiving paths. This isolation is necessary to protect the input of the receiver from the output power of the transmitter. The isolation bandwidth between the transmitting and the receiving bandwidths is known as guard band. Figure 7.41 illustrates two possible diplexer arrangements based on waveguide and planar circuit technology, respectively.

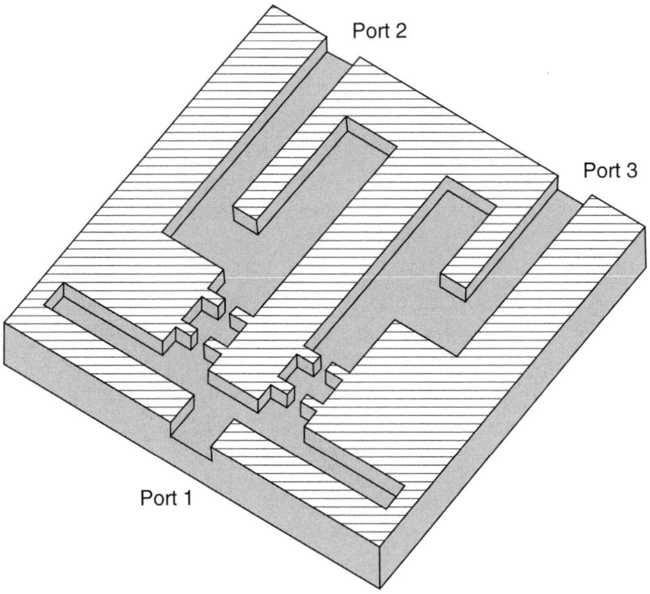

FIGURE 7.41 Transparent Half-Plane View of a Metallic Waveguide Diplexer

7.7 Ferrite Components

This discussion has only presented reciprocal passive components, and they were realized with isotropic materials (air or any other dielectric). Nonreciprocal components, dealt with in the following, show different properties in different propagation directions of the wave (electrically and/or magnetically). They are realized with anisotropic materials, such as ferrimagnetic and ferromagnetic compounds, that are generically termed as **ferrites**. Ferrites are widely used in the design of microwave passive components to control the amplitude (e.g., attenuators, limiters), the phase (e.g., phase shifters), frequency (e.g., YIG or yttrium iron garnet), the transmission paths (e.g., circulators, isolators, switches), and frequency bandwidth (e.g., filters) of microwave signals. Ferrites possess low-loss and high-anisotropy properties for microwave applications.

Ferrite materials are generally categorized as garnets, spinels, and hexagonal ferrites. The latter are composed of iron oxides (a kind of ceramics) and many other elements. The magnetic anisotropy of a ferrite is controlled by applying an external static (dc) magnetic field. This field aligns the inherent magnetic dipoles in the material to produce a net (nonzero) magnetic dipole moment, which causes the magnetic dipoles to precess at a frequency controlled by the strength of the bias field. The ideally lossless macroscopic description of the anisotropic properties of ferrites can be described by a Hermitian permeability tensor as follows:

$$[\mu] = \mu_0 \begin{bmatrix} \mu & j\kappa & 0 \\ -j\kappa & \mu & 0 \\ 0 & 0 & 1 \end{bmatrix}. \quad (\vec{z}\text{-biased with } \vec{H}_0 = \vec{z}H_0) \quad (7.21)$$

Similar forms of the permeability tensor can be derived for the x- and y-oriented static magnetic fields. In equation 7.21, the dependency of the applied external magnetic field strength H_0 and the material parameters are involved in the matrix elements μ and κ (Chang, 1989).

The nonreciprocal behavior of ferrite microwave components is basically described by two different scalar permeabilities for oppositely rotating microwave magnetic fields that are perpendicular to the static magnetic field. Thus, a microwave signal will propagate through a ferrite differently in different directions. The propagation characteristics are controlled by the strength of the applied magnetic field, which is also related to the power handling capability. Once the biasing field exceeds a threshold value, the transmission loss increases rapidly. The dielectric properties of ferrite materials are described by the relative permittivity and the loss tangent $\tan\delta = \varepsilon''/\varepsilon'$, with ε' being the real part of the dielectric constant and ε'' being the imaginary part. High-quality commercial ferrites have values of the loss tangent lower than 0.001 for microwave applications.

In the following paragraphs, a number of widely used ferrite components are presented, including isolators, phase shifters, and circulators.

7.7.1 Isolators

The **isolator** is a two-port device with a very low insertion loss (e.g., less than 0.5 dB) in one direction (forward) and very high insertion loss (e.g., greater than 20 dB) in the other (reverse) direction. The forward and reverse transmission losses depend on the choice of ferrite materials and design topology as well as operating frequency. Isolators are used to provide protection from high-power reflection and to stabilize signal sources or power sensitive devices from frequency shifts caused by variations in load impedance. Isolators are also important devices for eliminating unwanted interactions between components. An ideal isolator is shown in Figure 7.42 with matched ports and lossless transmission from port 1 to port 2. The corresponding *S*-matrix is as follows:

$$[S] = \begin{bmatrix} 0 & 0 \\ 1 & 0 \end{bmatrix}. \tag{7.22}$$

Basically, there are four types of ferrites isolators: **terminated circulators, Faraday rotation isolators, resonance isolators,** and **field displacement isolators**. They may be designed with different transmission line media, such as waveguide or microstrip. Important design specifications are forward and reverse insertion loss, frequency bandwidth, and power handling capability. The ratio of reverse (isolation) to forward insertion loss may also be used and is called the **figure of merit**.

The most commonly used isolators are related to the resonance and the field displacement principle and will be discussed here.

Resonance Isolators

Resonance isolators make direct use of the ferrimagnetic resonance phenomenon to achieve the nonreciprocal operation. A circularly polarized wave rotating in the same direction as the precessing magnetic dipoles is known to have a strong interaction with the material, while its opposite counterpart has a weaker interaction. Thus, the attenuation (absorption) of the strong interacting wave is very large near the gyromagnetic resonance of the ferrite, while it is very small for the wave propagation in the other direction. This resonance effect is used to construct the resonance isolators, which usually consist of a ferrite slab or strip mounted at a certain position along a transmission line.

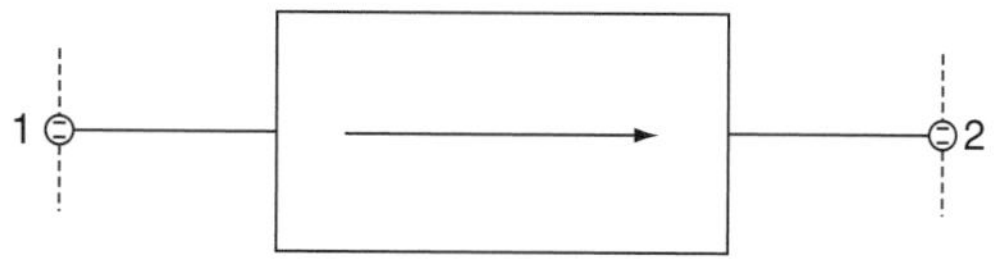

FIGURE 7.42 Network Representation of an Ideal Isolator

In principle, resonance isolators can be designed for any transmission line in which circularly polarized electromagnetic waves exist. Most often, isolators are realized in rectangular waveguides. Figure 7.43 shows two typical waveguide isolators in the *E*-plane and *H*-plane. The *E*-plane geometry has the advantages of requiring a lower biasing field to drive the isolator, while its *H*-plane counterpart is normally used for high-power applications because of easy heat dissipation.

Field Displacement Isolators

The **field displacement isolators** are based on the fact that field profiles of forward and reverse waves can be quite different in transmission lines, such as waveguides, when loaded with a biased ferrite. Such a distinct nonreciprocal behavior can be used to construct field displacement isolators. As shown in Figure 7.44 for a TE_{10} waveguide field displacement isolator, the largest electric field of the reverse wave is located at the ferrite slab where the electric field of the forward wave has a negligible level. If a thin resistive sheet is introduced in this position, the forward wave will be essentially unaffected while the reverse wave will be greatly attenuated. This is the operating principle of the field displacement isolators.

The field displacement isolators show excellent performance with reverse isolation better than 30 dB and forward insertion loss lower than 0.3 dB. Compared to the resonance isolators, they require a much smaller bias field because they operate below resonance. In addition to the waveguide geometry, other forms of the field displacement isolators have been developed, which include strip line and finline types.

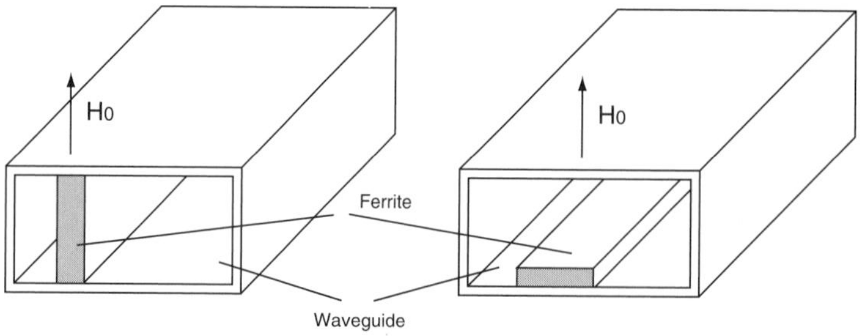

FIGURE 7.43 Two Typical Waveguide Isolators Loaded with Ferrite Slab

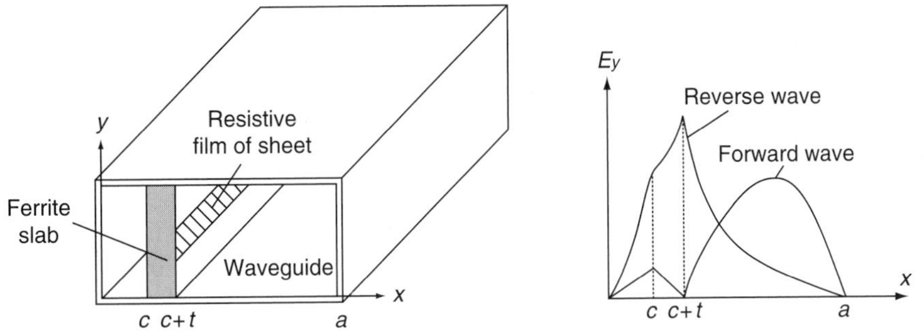

FIGURE 7.44 Geometrical View and Operating Principle of a Field Displacement Waveguide Isolator

7.7.2 Phase Shifters

Ferrite phase shifters are usually two-port devices that provide a variable phase shift of the transmission path by changing the bias field of the ferrite. There are many types that are characterized by their microwave performances, their switching characteristics, and their control characteristics. These include waveguides and planar circuits, transverse and longitudinal biasing, latching and continuous phase variation, and reciprocal and nonreciprocal operations. Special devices are also developed with the ferrite phase shifters, such as gyrators that have a 180° differential phase shift in the forward direction but no phase shift in the reverse direction.

One example of a ferrite phase shifter is a **nonreciprocal waveguide latching phase shifter** whose geometry is described in Figure 7.45. It consists of twin-toroids symmetrically located in the waveguide with bias wires passing through their centers, and the two toroids are separated by a dielectric spacer. If β_+ designates the propagation constant in one direction when a positive bias field saturates the ferrite, then β_- denotes the propagation constant for a negative bias field. The maximum amount of variable phase shift per unit length is the difference of $\beta_+ - \beta_-$. This differential phase shift may be achieved over a very broad bandwidth with good switching speed and phase accuracy.

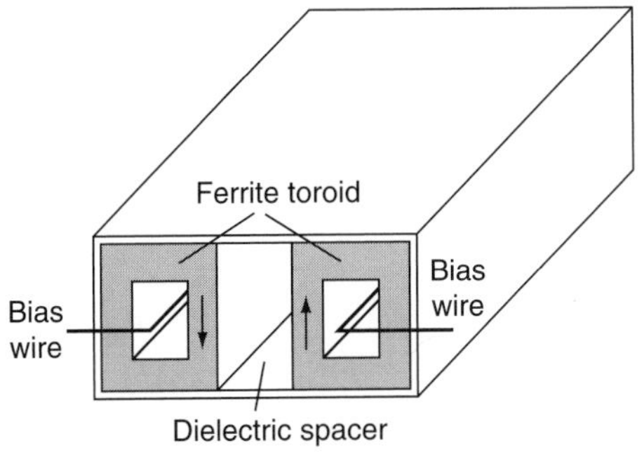

FIGURE 7.45 Topological Sketch of a Ferrite-Based Waveguide Phase Shifter

7.7.3 Circulators

Circulators find wide applications in microwave systems particularly in separating transmitter and receiver paths of a microwave front-end. Circulators are three-port devices that can ideally be lossless and simultaneously matched at all ports because of their nonreciprocity. An ideal circulator has the following *S*-matrix form:

$$[S] = \begin{bmatrix} 0 & 0 & 1 \\ 1 & 0 & 0 \\ 0 & 1 & 0 \end{bmatrix}. \tag{7.23}$$

Equation 7.23 indicates that signal flow can occur from ports 1 to 2, 2 to 3, and 3 to 1, but flow cannot occur in the reverse direction. The opposite path may be obtained by changing the polarity of the ferrite bias field. In practice, permanent magnets are used to generate the bias field. A circulator may also be used as an isolator by terminating one of its ports with a matched load. Design and performance specifications include required insertion loss, isolation, power handling capability, and bandwidth.

There are three different types of circulators: **Y-junction circulators, differential phase-shift circulators,** and **Faraday rotation circulators**. The most popular configurations are the junction circulators that can be constructed in waveguide or strip line and microstrip, for example.

Figure 7.46 presents typical geometries of strip line and waveguide *H*-plane junction circulators. In the strip line junction geometry, two ferrite disks are stacked between the two ground planes and the center metallic disk, forming a resonant cavity. Three external strip lines are symmetrically connected to the periphery of the center disk at 120° intervals and form three ports of the circulators. The static magnetic field is applied normal to the ground planes. In the waveguide junction, a cylindrical ferrite or other shaped ferrite is placed in a usually ridged or stepped *Y*-junction. This ridge or step form is also used to provide impedance match to the three external waveguide ports.

In the absence of a magnetic bias field for the strip line circulator, the resonant cavity supports two contrarotating

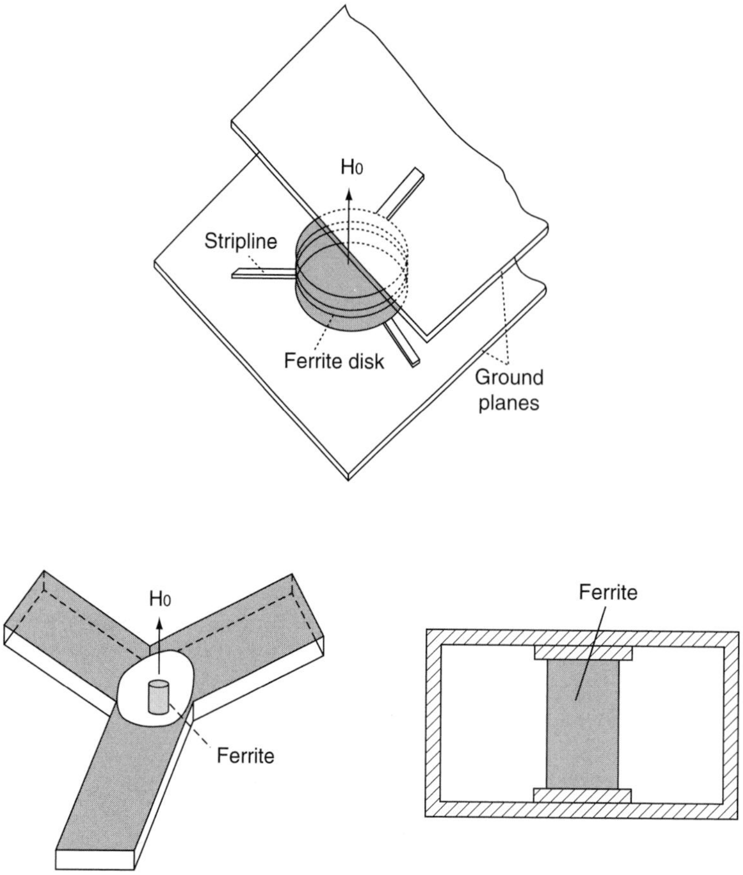

FIGURE 7.46 Three-Port Stripline and Waveguide Circulators with the Ferrite Film or Post-Loaded Within the Junction of Structure

modes that are degenerated (have the same resonant frequency). Application of a bias field will remove this mode degeneracy with two resonant modes having slightly different resonant frequencies, which are referred to as **mode splitting**. The operating frequency of the circulator can then be chosen between the two resonant frequencies so that the superposition of those two modes add at the output port and cancel out at the isolated port. A similar operating principle is also applied to the waveguide *Y*-junction circulator.

7.8 Other Passive Components

In this section, several classes of passive components are briefly described that are related to special materials and special techniques for designing and realizing certain classes of passive microwave components (as described in previous sections). Without getting into details, these components include **acoustic microwave devices, microwave ferroelectric devices, micromachined components,** and **superconducting components**. No doubt, new and innovative passive components are emerging, owing to the advancement of new materials (e.g., low-loss RF ferroelectrics), new processing techniques (e.g., nanotechnologies), and new design schemes (e.g., multilayers).

7.8.1 Acoustic Microwave Devices

The class of passive components that is usually constructed on piezoelectric crystals or similar materials is called **acoustic microwave devices**. These devices allow the conversion of electromagnetic and mechanical energies within the bulk or surface structure. The most popular microwave devices based on this technology are generally known as surface acoustic wave (SAW) devices. Because the velocity of ultrasonic waves in crystals is in the order of 10^5 times lower than that of electromagnetic waves, the microwave SAW circuits can be made extremely compact. Usually, the SAW components are loss-limited at higher frequency by characteristics of the transducers, and they are used for microwave frequency up to several gigahertz ($\sim 2\,\text{GHz}$). Higher frequency applications are still subject of ongoing research. Well-known passive SAW devices include filters, delay lines, and resonators. Technical details on the microwave SAW devices and useful references can be found in Chang (1989), Konishi (1991), and Webster (1999).

7.8.2 Microwave Ferroelectric Devices

Analog to the ferrite materials, the **ferroelectric materials** (a kind of ceramics) provide a variable dielectric constant ranging from a dozen to several thousands by applying an electric field or voltage over the ferroelectric substrate. This controllable dielectric constant allows the design of electrically tunable devices. In addition, attractive properties of ferroelectric materials, such as barium strontium titanate ($Ba_{1-x}Sr_xTiO_3$), are able to provide reciprocal and broadband operations. Fundamental design considerations are related to the tunability with biasing voltage levels and transmission losses to be as low as possible. The design of the geometry of such circuits is very involved to reduce the potential bias voltage and to have an appreciable tunability of the dielectric constant or other electrical properties. Ferroelectric thin films are compatible with integrated circuits. Usually, a high-quality ferroelectric substrate has a negligible current leakage over the biasing direction and thus has virtually no power consumption. This is why some ferroelectric components are considered passive.

Passive components based on ferroelectric materials include **phase shifters**, **delay lines**, **tunable filters**, **resonators**, and many others. Useful references are given in (Webster, 1999).

7.8.3 Micromachined Components

The micromachining process was developed in the study of the micro-electro-mechanical system (MEMS) in the 1960s for the miniaturization of mechanical machines for sensors and other applications. It can be used to fabricate a number of high-quality microwave components with the existing silicon (Si) or GaAs-based process. The Si-based micromachined components are attractive because they are relatively low cost. There are two basic micromachining techniques: **bulky** and **surface micromachining**. In certain microwave designs, nearly lossless SiO_2 or Si_3N_4-based membranes with low dielectric constant are used. Micromachined passive components offer attractive performance, namely ultrabroadband nonuniform line couplers, high-quality filters, and ultra-rapid MEMS switches. Further details can be found in De Los Santos (1999) and Nguyen *et al.* (1999).

7.8.4 Superconducting Components

The **superconducting class** of passive components is fabricated by using both low- and high-temperature superconductors, such as NbN and YBCO materials, that are able to reduce the conductor loss to a negligible level. Therefore, extremely high Q superconductive planar transmission lines and waveguides can be used to design various loss- and noise-sensible passive microwave components that include filters, diplexers, low-loss antenna feeders, delay lines, and phase shifters. In particular, the extremely narrow band-pass filters and very long time delay lines can be realized, for which the ordinary planar lines are not amenable in practice. Generally, the superconducting technologies are expensive, and they are widely studied for planar passive components that require special dielectric substrates in support of superconducting thin films, such as MgO and $LaAlO_3$.

Superconducting passive components (Shen, 1994) can be used in space-and military-oriented systems. Limited commercial applications are also being exploited in competition with other enabling technologies.

References

Chang, K. (1989). *Handbook of microwave and optical components—Volume 1: Microwave passive and antenna components.* New York: John Wiley & Sons.

Cohn, S. B., and Levy, R. (1984). History of microwave passive components with particular attention to directional couplers. *IEEE Transactions on Microwave Theory Technology 32,* 1046–1054.

Collin, R. E. (1992). *Foundations for microwave engineering.* New York: McGraw-Hill.

de los Santos, H. J. (1999). *Introduction to microelectromechanical microwave systems (MEMS).* Norwood, MA: Artech House.

Gupta, K. C., Garg, R., and Bahl, I. J. (1996). *Microstrip lines and slotlines.* (2d, ed.). Norwood, MA: Artech House.

Hoffmann, R. K. (1987). *Handbook of microwave integrated circuits.* Norwood, MA: Artech House.

Ishikawa, Y., Hiratsuka, T., Yamashita, S., and Iio, K. (1996). Planar-type dielectric resonator filter at millimeter-wave frequency. IEICE Transactions, *E79-C* (5), 679–684.

Itoh, T. (Ed.). (1987). *Planar transmission line structures.* New York: IEEE Press.

Kobeissi, H., and Wu, K. (1999). Design technique and performance assessment of new multiport multihole power divider suitable for M(H)MICs. *IEEE Transactions on Microwave Theory Technology 47,* 499–505.

Konishi, Y. (Ed.) (1991). *Microwave integrated circuits.* New York: Marcel Dekker.

Mattaei, G., Young, L., and Jones, E. M. T. (1980). *Microwave filters, impedance-matching networks, and coupling structures.* Dedham, MA: Artech House.

Maurin, D., and Wu, K. (1996). A compact 1.7–2.1 GHz three-way power combiner using microstrip technology with better than 93.8% combining efficiency. *IEEE Microwave and Guided-Wave Letters 6,* 106–108.

Menzel, W., Alessandri, F., Plattner, A., and Bornemann, J. (1997). Planar integrated waveguide diplexer for low-loss millimeter-wave applications. *Proceedings of 27th European Microwave Conference Vol* 6, 676–680.

Nguyen, C. T. C., Katehi, L. P. B., and Rebeiz, G. M. (1988). Micromachined devices for wireless communications. *Proceedings of the IEEE 86,* 1756–1768.

Pozar, D. M. (1998). *Microwave engineering.* (2d. ed.). New York: John Wiley & Sons.

Rhodes, J. D., and Levy, R. (1979). A generalized multiplexer theory. *IEEE Transactions on Microwave Theory Technology 27,* 99–111.

Robertson, I. D. (1995). *MMIC Design.* London: Institute of Electrical Engineering.

Shen, Z. Y. (1994). *High-temperature superconducting microwave circuits.* Dedham, MA: Artech House.

Trifunovic, V., and Jokanovic, B. (1994). Review of printed Marchand and double-Y baluns: Characteristics and application. *IEEE Transactions on Microwave Theory Technology 42,* 1454–1462.

Tripathi, V. K. (1975). Asymmetric coupled transmission lines in an inhomogeneous medium. *IEEE Transactions on Microwave Theory Technology 23,* 734–739.

Uher, J., Bornemann, J., and Rosenberg, U. (1993). *Waveguide components for antenna feed systems: Theory and CAD.* Dedham, MA: Artech House.

Vahldieck, R. (1989). Quasi-planar filters for millimeter-wave applications. *IEEE Transactions on Microwave Theory Technology 37,* 324–334.

Wang, T. Q., and Wu, K. (1999). Size-reduction and band-broadening design technique of uniplanar hybrid ring coupler using phase inverter for M(H)MICs. *IEEE Transactions on Microwave Theory Technology 47,* 198–206.

Webster, J. G. (1999). *John Wiley's encyclopedia of electrical and electronics engineering.* New York: John Wiley & Sons.

Zaki, K. A., Chen, C., and Atia, A. E. (1987). Canonical and longitudinal dual mode dielectric resonator without iris. *IEEE Transactions on Microwave Theory Technology 35,* 1130–1135.

Zhu, L., and Wu, K. (1999). Unified equivalent circuit model of planar discontinuities suitable for field theory-based CAD and optimization of M(H)MIC. *IEEE Transactions on Microwave Theory Technology 47,* 1589–1602.

Zhu, L., and Wu, K. (2000). Accurate circuit model of interdigital capacitor and its application to design of new quasi-lumped miniaturized filters with suppression of harmonic resonance. *IEEE Transactions on Microwave Theory Technology 48,* 347–356.

8

Computational Electromagnetics: The Method of Moments

Jian-Ming Jin and
Weng Cho Chew
Center for Computational Electromagnetics, Department of Electrical and Computer Engineering, University of Illinois at Urbana-Champaign, Urbana, Illinois, USA

8.1 Introduction

Due to the rapidly increasing capability of computers, computational electromagnetics has become a very important analysis tool in electromagnetics. This is partly due to the predictive power of Maxwell's theory as proven over the years—**Maxwell's theory** can predict the design performances or experimental outcomes if Maxwell's equations are solved correctly. Moreover, Maxwell's theory, which governs the basic principles behind the manipulation of electricity, is also extremely pertinent in many electrical engineering and scientific technologies. Examples of these technologies are radar, remote sensing, geoelectromagnetics, bioelectromagnetics, antennas, wireless communication, optics, and high-frequency circuits. Furthermore, Maxwell's theory is valid over a broad range of frequencies spanning static to optics and over a large dynamic range of length scales, from subatomic to intergalactic. In view of this, there is always a quest to solve Maxwell's equations accurately from first principles using numerical methods so that increasingly complex structures can be handled.

After the establishment of Maxwell's theory in 1873, the early electromagnetics analyses were associated with such simple shapes as spheres, cylinders, and planes. As the scientific and engineering demand for sophistication rose, solutions to more complex geometries were needed. As a result, approximate techniques were developed to solve Maxwell's equations. One can view circuit theory as a reduced form of Maxwell's theory in the low-frequency limit where approximate analyses of many complex geometries have been obtained with astounding success. At the other end of the spectrum, high-frequency ray theory, diffraction theory, and perturbation theory were developed to provide approximate solutions to Maxwell's equations. Recently, with the advent of computer technology, numerical methods were developed in the 1960s to allow more versatility and accuracy in the solution methods.

The **method of moments**, also known as the **moment method**, is one of the numerical methods developed to rise up to the challenge of solving increasingly complex problems in electromagnetics. It transforms the governing equation of a boundary-value problem into a matrix equation to enable its solution on a digital computer. Although the basic mathematical concepts of the moment method were in existence during early 20th century, true interest in it did not arise until the mid-1960s, with the publication of pioneering work by Mei and Van Bladel (1963), Andreasen (1964), Oshiro (1965), and Richmond (1968). The unified formulation of the method was presented by Harrington in his seminal book in 1968. Since then, the method has been further developed and applied to a variety of important electromagnetic problems. It has become one of the dominant methods in computational electromagnetics. It is not our intent here to present a complete history and list of references on the method. The reader is referred to Miller (1992) for a 35-page bibliography of the journal articles from 1960 to 1990. This book also presents a selection of key articles for the analytical formulation, numerical implementation, and practical application of the method. Other books on the topic include the ones by Mittra (1973), Hansen (1990), Wang (1991), and Peterson *et al.* (1997).

8.2 Basic Principle

The solutions to Maxwell's equations can be sought directly by solving related differential equations. Alternatively, they can be obtained by solving an integral equation derived from Maxwell's equations. For example, in the electrostatic case, an integral equation can be derived using the **Green's function approach**. A Green's function is a point source response and, in this case, is the potential produced by a point charge. As an example, in the capacitance problem, where one is interested in finding the total capacitance of a piece of metal, one can formulate an integral equation to solve for the charge in the metal. Using the Green's function approach, one can write, by the principle of linear superposition, the total potential due to the charges on the surface of a metallic conductor as:

$$\phi(\boldsymbol{r}) = \int_S G(\boldsymbol{r}, \boldsymbol{r}')\sigma(\boldsymbol{r}')dS', \tag{8.1}$$

where $\sigma(\boldsymbol{r}')$ is the surface charge density on the metallic surface, $G(\boldsymbol{r}, \boldsymbol{r}')$ is the Green's function or the point source response, S is the surface of the metallic object, and $\phi(\boldsymbol{r})$ is the potential generated by the surface charge. On the metallic surface S, $\phi(\boldsymbol{r})$ must be a constant. As a consequence, one has:

$$\Phi = \int_S G(\boldsymbol{r}, \boldsymbol{r}')\sigma(\boldsymbol{r}')dS' \quad \boldsymbol{r} \in S. \tag{8.2}$$

Equation 8.2 is an integral equation where Φ is a known constant, $G(\boldsymbol{r}, \boldsymbol{r}')$ is a known function, and $\sigma(\boldsymbol{r}')$ is the unknown surface charge density on the metal surface. The integral equation can be solved for the unknown $\sigma(r')$.

Equation 8.2 typifies an integral equation in electromagnetics. Integral equations can also be derived for dielectric and metallic bodies for electrostatics and electrodynamics. These integral equations are generally solved using the moment method. Notice that the function $\sigma(\boldsymbol{r}')$ in equation 8.2 has infinite degrees of freedom. Such infinite degrees of freedom cannot be handled by a computer. We say that $\sigma(\boldsymbol{r}')$ is a function representing a vector in an infinite dimensional space. To make equation 8.2 solvable by a computer, we approximate its solution in a finite dimension subspace using the moment method.

To illustrate this procedure, we choose a set of functions called basis functions that can be used to approximate $\sigma(\boldsymbol{r}')$. For example, we let:

$$\sigma(\boldsymbol{r}') = \sum_{n=1}^{N} c_n \upsilon_n(\boldsymbol{r}'), \tag{8.3}$$

where $\upsilon_n(\boldsymbol{r}')$ are the basis functions that can be used to approximate $\sigma(\boldsymbol{r}')$ and c_n are the unknown coefficients yet to be determined. In this manner, we have given $\sigma(\boldsymbol{r}')$ only N degrees of freedom. Substituting equation 8.3 into equation 8.2, we have:

$$\sum_{n=1}^{N} c_n \int_S G(\boldsymbol{r}, \boldsymbol{r}')\upsilon_n(\boldsymbol{r}')dS' = \Phi \quad \boldsymbol{r} \in S. \tag{8.4}$$

However, equation 8.4 still does not allow us to solve for c_n—we have to convert equation 8.4 into a matrix equation. To this end, we choose a set of functions $w_1(\boldsymbol{r}), w_2(\boldsymbol{r}), \ldots, w_N(\boldsymbol{r})$ and multiply equation 8.4 by these functions individually. We then integrate to obtain:

$$\begin{aligned} \sum_{n=1}^{N} c_n \int_S w_m(\boldsymbol{r}) \int_S (\boldsymbol{r}, \boldsymbol{r}')\upsilon_n(\boldsymbol{r}')dS'dS \\ = \int_S w_m(\boldsymbol{r})\Phi dS \quad m = 1, 2, \ldots, N. \end{aligned} \tag{8.5}$$

Notice that the double integrals above produce a number A_{mn} that depends only on m and n, and the integral on the right-hand side produces a number b_m that depends only on the index m. Consequently, equation 8.5 becomes:

$$\sum_{n=1}^{N} A_{mn} c_n = b_m \quad m = 1, 2, \ldots, N. \tag{8.6}$$

The 8.6 matrix equation from which we can solve for c_n in turn can be used in equation 8.3 to provide an approximate solution $\sigma(\boldsymbol{r}')$.

In a more abstract level, the moment method can be presented as follows. All integral equations can be expressed as:

$$\mathcal{L}\phi = f, \tag{8.7}$$

where $\mathcal{L}$ represents an integral operator, ϕ is the unknown to be sought, and f is a known driving term for the integral equation. We pick a finite basis set to approximate the solution:

$$\phi = \sum_{n=1}^{N} c_n \phi_n, \tag{8.8}$$

to arrive at:

$$\sum_{n=1}^{N} c_n \mathcal{L}\phi_n = f. \tag{8.9}$$

As mentioned, ϕ_n are called **basis** or **expansion functions**. We next convert the above equation into a matrix equation by testing the above with w_m, $m = 1, 2, \ldots, N$, and integrate over the solution domain. This corresponds to equation 8.5 and can be abstractly written as:

$$\sum_{n=1}^{N} c_n \langle w_m, \mathcal{L}\phi_n \rangle = \langle w_m, f \rangle \quad m = 1, 2, \ldots, N, \tag{8.10}$$

where $\langle w_m, f \rangle$ is an abbreviation for an integration; in particular, it is the following:

$$\langle w_m, f \rangle = \int_S w_m(\mathbf{r}) f(\mathbf{r}) dS, \tag{8.11}$$

in the electrostatic example. The functions w_m are usually referred to as **testing** or **weighting functions**. Equation 8.10 can be written as a matrix equation:

$$[A]\{c\} = \{b\}, \tag{8.12}$$

where $[A]$ is called the **system matrix** whose elements are given by:

$$A_{mn} = \langle w_m, \mathcal{L}\phi_n \rangle, \tag{8.13}$$

and $\{b\}$ and $\{c\}$ are called the **source** and **unknown** vectors, respectively, with the elements of $\{b\}$ given by:

$$b_m = \langle w_m, f \rangle. \tag{8.14}$$

The solution procedure described above is called the **method of moments** because equation 8.10 is equivalent to taking the moments of equation 8.9. It is also known as the **weighted residual method** because equation 8.10 can be interpreted as setting the weighted residual of equation 8.9 to zero. The solution procedure works for both differential and integral operators. In fact, it is a widely used procedure to formulate the well-known finite element method (Jin, 1993). In the computational electromagnetics community, the name of the moment method is reserved for the case when $\mathcal{L}$ is an integral operator involving a Green's function.

It is clear that there are four basic steps for solving an electromagnetic boundary-value problem using the moment method:

- To formulate the problem in terms of an integral equation
- To represent the unknown quantity using a set of basis functions
- To convert the integral equation into a matrix equation using a set of testing functions
- To solve the matrix equation and calculate the desired quantities

These steps are discussed in more detail in the following sections.

8.3 Integral Equations

This section presents some fundamental integral equations in two and three dimensions that are found in most electromagnetic problems.

8.3.1 Scalar Equations

Consider the problem of a **scalar wave** produced by a source $f(\rho)$ in the presence of an arbitrarily shaped object immersed in an infinite medium, as illustrated in Figure 8.1. Assume that both the source and the object have no variation along the z-axis; thus, we only have to consider a plane perpendicular to the z-axis. Exterior to the object, the wave function $\phi(\boldsymbol{\rho})$ satisfies the inhomogeneous Helmholtz equation:

$$\nabla^2\phi(\boldsymbol{\rho}) + k^2\phi(\rho) = -f(\boldsymbol{\rho}) \quad \boldsymbol{\rho} \in \Omega_\infty, \tag{8.15}$$

where k is the wavenumber and Ω_∞ denotes the exterior region. The wave function should also satisfy the radiation condition:

$$\sqrt{\rho}\left[\frac{\partial\phi(\boldsymbol{\rho})}{\partial\rho} + jk\phi(\boldsymbol{\rho})\right] = 0 \quad \rho \to \infty, \tag{8.16}$$

which simply indicates that the wave propagates outward into infinity without reflection.

Using Green's identity (Tai, 1994; Poggio and Miller, 1973; Morita *et al.*, 1990), we can derive an integral representation for the wave function everywhere:

$$\phi^{\text{inc}}(\boldsymbol{\rho}) + \int_{\Gamma_o}\left[\phi(\boldsymbol{\rho}')\frac{\partial G_0(\boldsymbol{\rho}, \boldsymbol{\rho}')}{\partial n'} - G_0(\boldsymbol{\rho}, \boldsymbol{\rho}')\frac{\partial\phi(\boldsymbol{\rho}')}{\partial n'}\right] d\Gamma' = \begin{cases} \phi(\boldsymbol{\rho}) & \boldsymbol{\rho} \in \Omega_\infty \\ 0 & \boldsymbol{\rho} \in \Omega_o, \end{cases} \tag{8.17}$$

where $\phi^{\text{inc}}(\boldsymbol{\rho})$ denotes the incident wave generated by the source $f(\boldsymbol{\rho})$ in the absence of the object and where G_0 is called the free space **Green's function** given by:

$$G_0(\boldsymbol{\rho}, \boldsymbol{\rho}') = \frac{1}{4j}H_0^{(2)}(k|\boldsymbol{\rho} - \boldsymbol{\rho}'|), \tag{8.18}$$

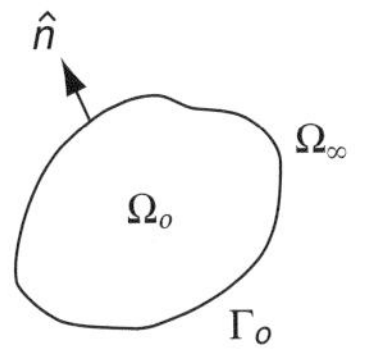

FIGURE 8.1 A Two-Dimensional Object Ω_o Having a Boundary Γ_o Immersed in an Infinite Medium Ω_∞

in which $H_0^{(2)}$ denotes the zero-order Hankel function of the second kind.

Equation 8.17 provides the foundation to establish an integral equation for ϕ and $\partial\phi/\partial n$ on the surface of the object. We now consider four cases.

Impenetrable with a Hard Surface

If the object is impenetrable with a hard surface where ϕ satisfies the boundary condition:

$$\phi(\boldsymbol{\rho}) = 0 \quad \boldsymbol{\rho} \in \Gamma_o, \tag{8.19}$$

equation 8.17 becomes:

$$\phi^{\text{inc}}(\boldsymbol{\rho}) - \int_{\Gamma_o} G_0(\boldsymbol{\rho}, \boldsymbol{\rho}') \frac{\partial\phi(\boldsymbol{\rho}')}{\partial n'} d\Gamma' = \begin{cases} \phi(\boldsymbol{\rho}) & \boldsymbol{\rho} \in \Omega_\infty. \\ 0 & \boldsymbol{\rho} \in \Omega_o. \end{cases} \tag{8.20}$$

Applying this equation on Γ_o, we obtain:

$$\int_{\Gamma_0} G_0(\boldsymbol{\rho}, \boldsymbol{\rho}') \frac{\partial\phi(\boldsymbol{\rho}')}{\partial n'} d\Gamma' = \phi^{\text{inc}}(\boldsymbol{\rho}) \quad \boldsymbol{\rho} \in \Gamma_o, \tag{8.21}$$

which is the integral equation for $\partial\phi/\partial n$ on Γ_o.

Impenetrable with a Soft Surface

If the object is impenetrable with a soft surface where ϕ satisfies the boundary condition:

$$\frac{\partial\phi(\boldsymbol{\rho})}{\partial n} = 0 \quad \boldsymbol{\rho} \in \Gamma_o, \tag{8.22}$$

equation 8.17 becomes:

$$\phi^{\text{inc}}(\boldsymbol{\rho}) + \int_{\Gamma_o} \phi(\boldsymbol{\rho}') \frac{\partial G_0(\boldsymbol{\rho}, \boldsymbol{\rho}')}{\partial n'} d\Gamma' = \begin{cases} \phi(\boldsymbol{\rho}) & \boldsymbol{\rho} \in \Omega_\infty. \\ 0 & \boldsymbol{\rho} \in \Omega_o. \end{cases} \tag{8.23}$$

Applying this equation on Γ_o, we obtain:

$$\frac{1}{2}\phi(\boldsymbol{\rho}) - ⨍_{\Gamma_o} \phi(\boldsymbol{\rho}') \frac{\partial G_0(\boldsymbol{\rho}, \boldsymbol{\rho}')}{\partial n'} d\Gamma' = \phi^{\text{inc}}(\boldsymbol{\rho}') \quad \boldsymbol{\rho} \in \Gamma_o, \tag{8.24}$$

where $⨍$ denotes the integral excluding the contribution from the singular point $\boldsymbol{\rho} = \boldsymbol{\rho}'$, which is known as the **principal value integral**. Equation 8.24 is the integral equation for ϕ on Γ_o.

Impenetrable with a Surface Satisfying Mixed Boundary

If the object is impenetrable with a surface where ϕ satisfies the mixed boundary condition:

$$\frac{\partial\phi(\boldsymbol{\rho})}{\partial n} + \gamma\phi(\boldsymbol{\rho}) = 0 \quad \boldsymbol{\rho} \in \Gamma_o, \tag{8.25}$$

Equation 8.17 becomes:

$$\begin{aligned} \phi^{\text{inc}}(\boldsymbol{\rho}) - \int_{\Gamma_o} \phi(\boldsymbol{\rho}') \left[\gamma G_0(\boldsymbol{\rho}, \boldsymbol{\rho}') - \frac{\partial G_0(\boldsymbol{\rho}, \boldsymbol{\rho}')}{\partial n'}\right] d\Gamma' \\ = \begin{cases} \phi(\boldsymbol{\rho}) & \boldsymbol{\rho} \in \Omega_\infty. \\ 0 & \boldsymbol{\rho} \in \Omega_o. \end{cases} \end{aligned} \tag{8.26}$$

Applying this equation on Γ_o, we obtain:

$$\begin{aligned} \frac{1}{2}\phi(\boldsymbol{\rho}) - ⨍_{\Gamma_o} \phi(\boldsymbol{\rho}') \left[\gamma G_0(\boldsymbol{\rho}, \boldsymbol{\rho}') + \frac{\partial G_0(\boldsymbol{\rho}, \boldsymbol{\rho}')}{\partial n'}\right] d\Gamma' \\ = \phi^{\text{inc}}(\boldsymbol{\rho}) \quad \boldsymbol{\rho} \in \Gamma_o, \end{aligned} \tag{8.27}$$

which is the integral equation for ϕ on Γ_o.

Penetrable and Homogeneous Object

If the object is penetrable and homogeneous, we apply equation 8.17 on Γ_o to obtain:

$$\begin{aligned} \frac{1}{2}\phi(\boldsymbol{\rho}) - ⨍_{\Gamma_o} \left[\phi(\boldsymbol{\rho}') \frac{\partial G_0(\boldsymbol{\rho}, \boldsymbol{\rho}')}{\partial n'} - G_0(\boldsymbol{\rho}, \boldsymbol{\rho}') \frac{\partial\phi(\boldsymbol{\rho}')}{\partial n'}\right] d\Gamma' \\ = \phi^{\text{inc}}(\boldsymbol{\rho}) \quad \boldsymbol{\rho} \in \Gamma_o. \end{aligned} \tag{8.28}$$

To solve for ϕ and $\partial\phi/\partial n$ on Γ_o, we need another equation that can be derived by considering the interior of the object. This is given by:

$$\begin{aligned} \frac{1}{2}\phi(\boldsymbol{\rho}) - ⨍_{\Gamma_o} \left[\tilde{G}_0(\boldsymbol{\rho}, \boldsymbol{\rho}') \frac{\partial\phi(\boldsymbol{\rho}')}{\partial n'} - \phi(\boldsymbol{\rho}') \frac{\partial \tilde{G}_0(\boldsymbol{\rho}, \boldsymbol{\rho}')}{\partial n'}\right] d\Gamma' \\ = 0 \quad \boldsymbol{\rho} \in \Gamma_o, \end{aligned} \tag{8.29}$$

where:

$$\tilde{G}_0(\boldsymbol{\rho}, \boldsymbol{\rho}') = \frac{1}{4j} H_0^{(2)}(\tilde{k}|\boldsymbol{\rho} - \boldsymbol{\rho}'|). \tag{8.30}$$

The $\tilde{k}$ in equation 8.30 is the wave number inside the object. Equation 8.29 can be used together with equation 8.28 for a numerical solution of ϕ and $\partial\phi/\partial n$ on Γ_o.

For an inhomogeneous object, we can follow the basic approach outlined above to derive a so-called **volume-surface integral equation** (Jin *et al.*, 1988), which contains an integral over Ω_o in addition to the integral over Γ_o. As a result, the unknown function is distributed over Ω_o and Γ_o.

8.3.2 Vector Equations

Consider the problem of electromagnetic fields ($\boldsymbol{E}$, $\boldsymbol{H}$) produced by an electric current source of density $\boldsymbol{J}_i$ in the presence of an arbitrarily shaped object in an infinite homogeneous space, as illustrated in Figure 8.2. The $\boldsymbol{E}$ and $\boldsymbol{H}$ satisfy the **vector wave equations**:

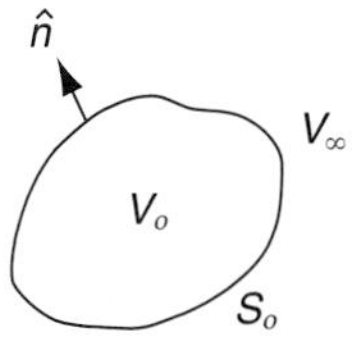

FIGURE 8.2 A Three-Dimensional Object V_o Bounded by Surface S_o in an Infinite Space V_∞

$$\nabla \times \nabla \times \boldsymbol{E}(\boldsymbol{r}) - k^2\boldsymbol{E}(\boldsymbol{r}) = -j\omega\mu \boldsymbol{J}_i(\boldsymbol{r}) \quad \boldsymbol{r} \in V_\infty. \tag{8.31}$$

$$\nabla \times \nabla \times \boldsymbol{H}(\boldsymbol{r}) - k^2\boldsymbol{H}(\boldsymbol{r}) = \nabla \times \boldsymbol{J}_i(\boldsymbol{r}) \quad \boldsymbol{r} \in V_\infty. \tag{8.32}$$

In these equations, V_∞ denotes the exterior region.

Using Green's identity (Tai, 1994; Poggio and Miller, 1973; Morita, *et al.*, 1990), we can derive the following integral representations:

$$\boldsymbol{E}^{\text{inc}}(\boldsymbol{r}) - \int_{S_o} [(\hat{n}' \cdot \boldsymbol{E})\nabla G_0 + (\hat{n}' \times \boldsymbol{E}) \times \nabla G_0 + j\omega\mu(\hat{n}' \times \boldsymbol{H})G_0]dS'. = \begin{cases} \boldsymbol{E}(\boldsymbol{r}) & \boldsymbol{r} \in V_\infty. \\ 0 & \boldsymbol{r} \in V_o. \end{cases} \tag{8.33}$$

$$\mathbf{H}^{\text{inc}}(\boldsymbol{r}) - \int_{S_o} [(\hat{n}' \cdot \boldsymbol{H})\nabla G_0 + (\hat{n}' \times \boldsymbol{H}) \times \nabla G_0 - j\omega\varepsilon(\hat{n}' \times \boldsymbol{E})G_0]dS'. = \begin{cases} \mathbf{H}(\mathbf{r}) & \boldsymbol{r} \in V_\infty. \\ 0 & \boldsymbol{r} \in V_o. \end{cases} \tag{8.34}$$

In these equations:

$$G_0(\boldsymbol{r}, \boldsymbol{r}') = \frac{e^{-jk|\boldsymbol{r}-\boldsymbol{r}'|}}{4\pi|\boldsymbol{r}-\boldsymbol{r}'|}, \tag{8.35}$$

and $\boldsymbol{E}^{\text{inc}}(\boldsymbol{r})$ and $\boldsymbol{H}^{\text{inc}}(\boldsymbol{r})$ represent the electric and magnetic fields produced by $\boldsymbol{J}_i$ in the infinite space without the object.

To write equations 8.33 and 8.34 in compact form, we define the operators:

$$\boldsymbol{L}(\boldsymbol{X}) = jk\int_{S_o} \left(\boldsymbol{X}G_0 + \frac{1}{k^2}\nabla' \cdot \boldsymbol{X}\nabla G_0\right) dS'. \tag{8.36}$$

$$\boldsymbol{K}(\boldsymbol{X}) = \int_{S_o} \boldsymbol{X} \times \nabla G_0 dS'. \tag{8.37}$$

We also introduce the equivalent surface currents:

$$\bar{\boldsymbol{J}}_s = \hat{n} \times \bar{\boldsymbol{H}} = \sqrt{\frac{\mu}{\varepsilon}}\hat{n} \times \boldsymbol{H}, \quad \boldsymbol{M}_s = \boldsymbol{E} \times \hat{n}. \tag{8.38}$$

As a result, equations 8.33 and 8.34 can be written as:

$$\boldsymbol{E}^{\text{inc}}(\boldsymbol{r}) - \boldsymbol{L}(\bar{\boldsymbol{J}}_s) + \boldsymbol{K}(\boldsymbol{M}_s) = \begin{cases} \boldsymbol{E}(\boldsymbol{r}) & \boldsymbol{r} \in V_\infty. \\ 0 & \boldsymbol{r} \in V_o. \end{cases} \tag{8.39}$$

$$\bar{\boldsymbol{H}}^{\text{inc}}(\boldsymbol{r}) - \boldsymbol{K}(\bar{\boldsymbol{J}}_s) - \boldsymbol{L}(\boldsymbol{M}_s) = \begin{cases} \bar{\boldsymbol{H}}(\boldsymbol{r}) & \boldsymbol{r} \in V_\infty. \\ 0 & \boldsymbol{r} \in V_o. \end{cases} \tag{8.40}$$

Equations 8.39 and 8.40 provide the foundation to derive integral equations for $\boldsymbol{J}_s$ and $\boldsymbol{M}_s$. Taking the cross product of these with $\hat{n}$ and letting $\boldsymbol{r}$ approach S_o, we have:

$$\frac{1}{2}\boldsymbol{M}_s(\boldsymbol{r}) - \hat{n} \times \boldsymbol{L}(\bar{\boldsymbol{J}}_s) + \hat{n} \times \tilde{\boldsymbol{K}}(\boldsymbol{M}_s) = -\hat{n} \times \boldsymbol{E}^{\text{inc}}(\boldsymbol{r}) \quad \boldsymbol{r} \in S_o. \tag{8.41}$$

$$\frac{1}{2}\bar{\boldsymbol{J}}_s(\boldsymbol{r}) + \hat{n} \times \tilde{\boldsymbol{K}}(\bar{\boldsymbol{J}}_s) + \hat{n} \times \boldsymbol{L}(\boldsymbol{M}_s) = \hat{n} \times \bar{\boldsymbol{H}}^{\text{inc}}(\boldsymbol{r}) \quad \boldsymbol{r} \in S_o. \tag{8.42}$$

The $\tilde{\boldsymbol{K}}$ is the same integral as the one in equation 8.37, except that the singular point $\boldsymbol{r} = \boldsymbol{r}'$ is now removed by using a principal value integral. Equation 4.41 is known as the **electric field integral equation** (EFIE) and equation 8.42 is called the **magnetic field integral equation** (MFIE).

If the object is a perfect electric conductor, $\boldsymbol{E}$ satisfies the boundary condition:

$$\hat{n} \times \boldsymbol{E} = 0 \quad \boldsymbol{r} \in S_o. \tag{8.43}$$

Hence, $\boldsymbol{M}_s = 0$ and equations 8.41 and 8.42 reduce to:

$$\hat{n} \times \boldsymbol{L}(\bar{\boldsymbol{J}}_s) = \hat{n} \times \boldsymbol{E}^{\text{inc}}(\boldsymbol{r}) \quad \boldsymbol{r} \in S_o. \tag{8.44}$$

$$\frac{1}{2}\bar{\boldsymbol{J}}_s(\boldsymbol{r}) + \hat{n} \times \tilde{\boldsymbol{K}}(\bar{\boldsymbol{J}}_s) = \hat{n} \times \bar{\boldsymbol{H}}^{\text{inc}}(\boldsymbol{r}) \quad \boldsymbol{r} \in S_o. \tag{8.45}$$

Any of these two equations can be used to solve for $\bar{\boldsymbol{J}}_s$. However, for a given S_o, $\boldsymbol{L}$ can be singular at certain frequencies when the exterior medium is lossless. Consequently, equation 8.44 may give an erroneous solution at these frequencies. This is known as the **problem of interior resonance**, and the singular frequencies correspond to the resonant frequencies of a cavity formed by filling the interior of S_o with the exterior medium. A similar problem occurs in equation 8.45 as well. To eliminate this problem, we can combine equations 8.44 and 8.45 to find:

$$\alpha n \times \hat{n} \times \boldsymbol{L}(\bar{\boldsymbol{J}}_s) + (1-\alpha)\left[\frac{1}{2}\bar{\boldsymbol{J}}_s(\boldsymbol{r}) + \hat{n} \times \tilde{\boldsymbol{K}}(\bar{\boldsymbol{J}}_s)\right] = \alpha\hat{n} \times \hat{n} \times \boldsymbol{E}^{\text{inc}}(\boldsymbol{r}) + (1-\alpha)\hat{n} \times \bar{\boldsymbol{H}}^{\text{inc}}(\boldsymbol{r}) \quad \boldsymbol{r} \in S_o, \tag{8.46}$$

which is known as the **combined field integral equation** (CFIE) (Mautz and Harrington, 1978). This combination results in an integral operator corresponding to the corresponding operator for a cavity with a resistive wall whose resonant frequencies are complex. As a result, the operator cannot be singular for a real frequency. The combination parameter α is usually chosen between zero and one.

If the object is an impedance body, there exists a relation between $\bar{\boldsymbol{J}}_s$ and $\boldsymbol{M}_s$, which is known as the **impedance boundary condition**. Substituting this relation into either equation

8.41 or equation 8.42 yields an integral equation containing only one unknown function.

If the object is a homogeneous body, we can apply Green's theorem to the interior of the object to formulate another two equations: EFIE and MFIE. One of these two equations can be used with either equations 8.41 or 8.42 to form a complete system for the solution of $\bar{\boldsymbol{J}}_s$ and $\boldsymbol{M}_s$. Again, to overcome the problem of interior resonance, one can combine the EFIE and MFIE to form the CFIE for the interior field. This CFIE can then be used with the CFIE for the exterior field to form a complete system (Rao and Wilton, 1990). A better approach is, however, to add the EFIEs for the interior and exterior fields to form a new integral equation. This is then used with the integral equation obtained by adding the MFIEs for the interior and exterior fields. This formulation results from work by Poggio and Miller (1973), Chang and Harrington (1977), and Wu and Tsai (1977); the formulation is found to be free of interior resonances and yields accurate and stable solutions (Mautz and Harrington, 1979).

If the object is inhomogeneous, one can employ the volume equivalence principle to derive a volume integral equation (Livesay and Chen, 1974; Schaubert *et al.*, 1984), which contains an integral over V_o, in contrast to the surface integral equations described previously. As a result, the unknown function is distributed over the entire V_o, and its numerical solution is very computationally intensive unless some special techniques with certain approximations are used (Borup and Gandhi, 1984; Zwamborn and van der Berg, 1994; Gan and Chew, 1995; Wang and Jin, 1998). Problems involving inhomogeneous media can usually be handled more efficiently using the finite element method (Jin, 1993).

8.4 Basis Functions

The second step in the moment method is to expand the unknown function of the integral equation in terms of **basis functions**. There are two classes of basis functions. The first class is defined over the entire solution domain and hence called the **entire-domain basis functions**. For most electromagnetic problems, the solution domain is complicated, and it is difficult to find entire-domain basis functions that can form approximately a complete set over the domain. For this reason, the entire-domain basis functions have limited use. Contrasted with the entire-domain basis functions is the second class defined over small parts of the solution domain. This class is called the **subdomain basis functions**. Although the unknown function over the entire solution domain can be complicated and cannot be represented by simple functions, its behavior over a sufficiently small region can be rather simple and hence representable with simple functions. For this reason, the subdomain basis functions are widely used in the moment method for a variety of electromagnetic problems. In the following subsections, we describe some commonly used subdomain basis functions.

8.4.1 Line Basis Functions

If the solution domain is a contour, it can first be divided into small segments denoted by $C_1, C_2, \ldots, C_N$. The simplest basis function is the **zero-order**, widely known as the pulse function in the electromagnetics community. This function is defined as:

$$P_n(s) = \begin{cases} 1 & s \in C_n. \\ 0 & \text{elsewhere.} \end{cases} \quad (8.47)$$

With this basis function, the unknown ϕ is approximated by a piecewise constant function, as illustrated in Figure 8.3. Pulse functions are widely used in the moment method because they lead to simple integrals that can be evaluated easily.

The next basis function is the **first-order**, commonly known as the triangle function. It is defined over two neighboring segments with a value of unity at the joint:

$$\Lambda_n(s) = \begin{cases} \dfrac{s - s_{n-1}}{s_n - s_{n-1}} & s \in C_n. \\ \dfrac{s_{n+1} - s}{s_{n+1} - s_n} & s \in C_{n+1}. \\ 0 & \text{elsewhere.} \end{cases} \quad (8.48)$$

The s_n denotes the joint between the segment of C_n and C_{n+1}. With this choice, the unknown ϕ is represented by a piecewise linear function (Figure 8.4).

The third basis function is the so-called piecewise sinusoidal, defined as:

$$S_n(s) = \begin{cases} \dfrac{\sin[k(s - s_{n-1})]}{\sin[k(s_n - s_{n-1})]} & s \in C_n. \\ \dfrac{\sin[k(s_{n+1} - s)]}{\sin[k(s_{n+1} - s_n)]} & s \in C_{n+1}. \\ 0 & \text{elsewhere.} \end{cases} \quad (8.49)$$

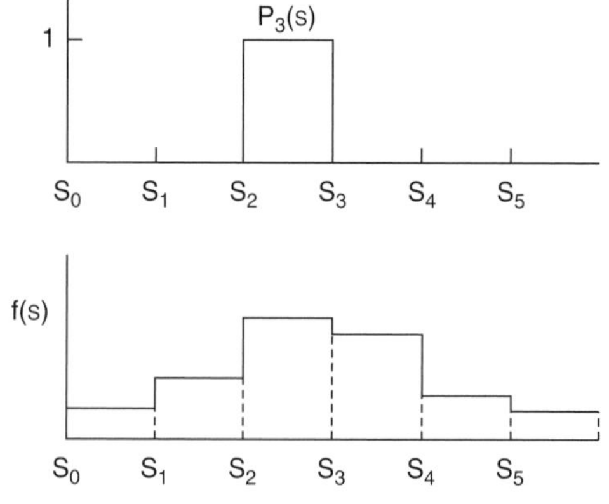

FIGURE 8.3 (A) A Pulse Function and (B) A Piecewise Constant Approximation

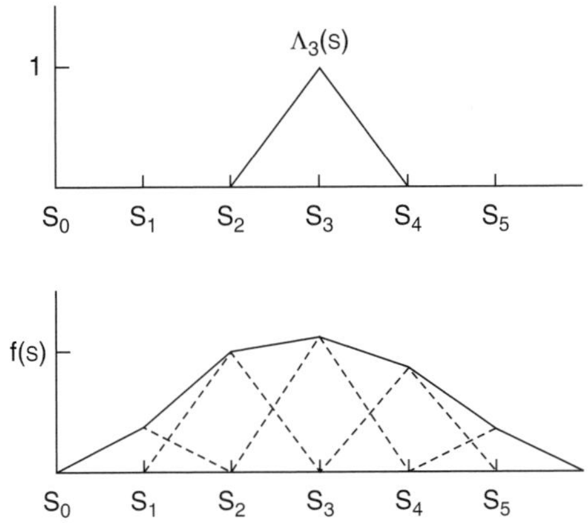

FIGURE 8.4 (A) A Triangle Function and (B) A Piecewise Linear Approximation

The k is the wave number. Obviously, when kC_n and kC_{n+1} are sufficiently small, the piecewise sinusoidal functions reduce to triangle ones.

It is also possible to introduce high-order basis functions such as the quadratic and cubic ones, in a manner similar to one used in the finite element method (Jin, 1993). The high-order functions can yield higher accuracy without increasing the number of unknowns.

8.4.2 Surface Basis Functions

If the solution domain is a surface, it can be divided into small triangular cells. In some special cases, a surface can also be divided into small rectangular cells. Triangular cells, however, are more flexible and accurate when representing an arbitrary surface. As in the case for one-dimensional segments, we can have the zeroth-order (pulse) basis function, which is a constant on one cell and zero elsewhere, and the first-order (linear) basis function, which is one at a node and decreases linearly to zero at the neighboring nodes (Figure 8.5).

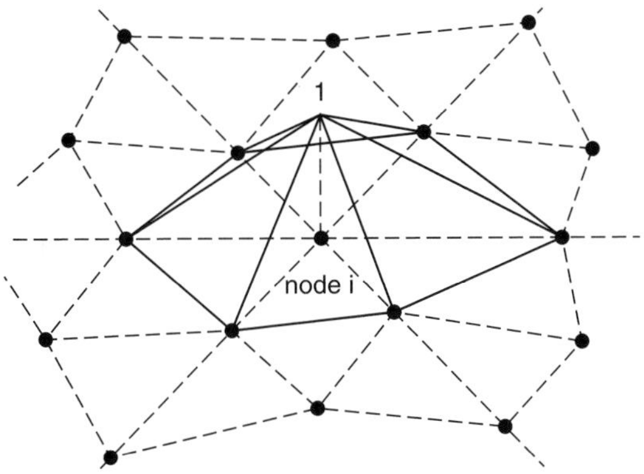

FIGURE 8.5 A First-Order Basis Function for Node i on a Surface

In most three-dimensional electrodynamic problems, the unknown function of an integral equation is the electric surface current density, which is a vector function. For such an unknown, basis functions must be constructed properly based on the properties of the current density. One of the most popular basis functions for this purpose is the so-called **rooftop function.** For a rectangular mesh in the x–y plane, the basis function for the x component of the current density is defined as:

$$f_n^x = \begin{cases} \dfrac{x - x_{i-1}}{x_i - x_{i-1}} & x_{i-1} \le x \le x_i,\ y_{j-1} \le y \le y_j. \\ \dfrac{x_{i+1} - x}{x_{i+1} - x_i} & x_i \le x \le x_{i+1},\ y_{j-1} \le y \le y_j. \\ 0 & \text{elsewhere.} \end{cases} \tag{8.50}$$

The basis function for the y component is defined as:

$$f_n^y = \begin{cases} \dfrac{y - y_{j-1}}{y_j - y_{j-1}} & x_{i-1} \le x \le x_i,\ y_{j-1} \le y \le y_j. \\ \dfrac{y_{j+1} - y}{y_{j+1} - y_j} & x_{i-1} \le x \le x_i,\ y_j \le y \le y_{j+1}. \\ 0 & \text{elsewhere.} \end{cases} \tag{8.51}$$

It is easy to see that the rooftop function is a product of a pulse and a triangle function: $f_n^x = \Lambda_i(x)P_j(y)$ and $f_n^y = P_i(x)\Lambda_j(y)$. The rooftop functions are sketched in Figure 8.6. They guarantee the continuity of current flow and permit the edge conditions to be satisfied approximately.

When a surface is divided into triangular elements, one may use the triangular rooftop function (Rao *et al.*, 1982) as the basis function to expand the surface current. This function is defined over two triangular elements joined at a common edge ℓ_n:

$$\mathbf{f}_n(\mathbf{r}) = \begin{cases} \frac{\ell_n}{2A_n^+}\boldsymbol{\rho}_n^+ & \mathbf{r} \in T_n^+. \\ \frac{\ell_n}{2A_n^-}\boldsymbol{\rho}_n^- & \mathbf{r} \in T_n^-. \\ 0 & \text{otherwise.} \end{cases} \tag{8.52}$$

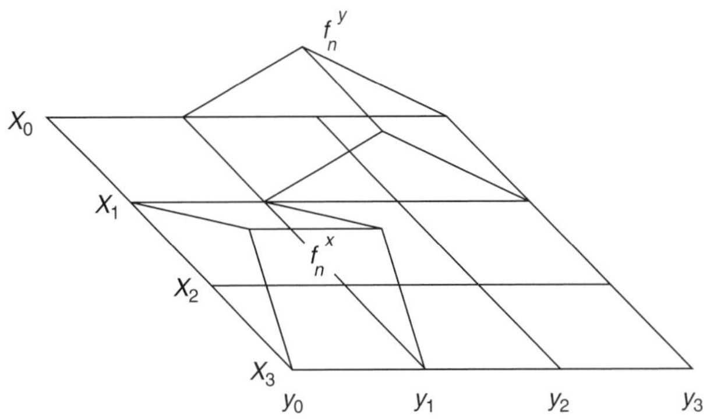

FIGURE 8.6 Rooftop Functions Defined on a Rectangular Mesh

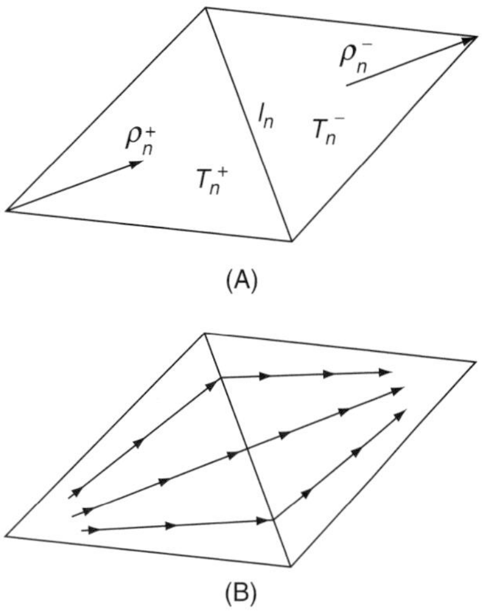

FIGURE 8.7 (A) Illustration of Two Joint Triangular Elements (B) Vector Plot of the Triangular Rooftop Function $\boldsymbol{f}_n$

the $T_n^{\pm}$ denote the two triangles associated with the nth edge, $A_n^{\pm}$ are the areas of triangles $T_n^{\pm}$, ℓ_n is the length of the nth edge, and $\boldsymbol{\rho}_n^{\pm}$ are the vectors defined in Figure 8.7(A). The vector plot of $\boldsymbol{f}_n(\boldsymbol{r})$ is illustrated in Figure 8.7(B). The most important feature of this basis function is that its normal component to edge ℓ_n is a constant (normalized to 1), whereas the normal components to other edges are zero. This feature guarantees the continuity of current flow over all edges, as in the case of the rooftop functions.

8.5 Testing Functions

To convert an integral equation into a matrix equation, we have to choose a set of **testing functions.** This set of functions should be approximately complete in the range of the integral operator. There are several commonly used testing functions that lead to several different formulations of the moment method.

The **first kind of testing** is the Dirac delta function:

$$w_m(\boldsymbol{r}) = \delta(\boldsymbol{r} - \boldsymbol{r}_m), \qquad (8.53)$$

where $\boldsymbol{r}_m$ denotes a set of points in the solution domain. This testing is equivalent to satisfying the integral equation at a set of specific points. The resulting formulation is called **point collocation** which is better known as **point matching** in the electromagnetics community.

The **second kind of testing** is a constant over a subdomain and zero elsewhere:

$$w_m(\boldsymbol{r}) = \begin{cases} 1 & \boldsymbol{r} \in \Omega_m. \\ 0 & \text{elsewhere.} \end{cases} \qquad (8.54)$$

The Ω_m denotes the mth subdomain. This testing is equivalent to satisfying the integral equation over each subdomain in the average sense. The resulting formulation is called **subdomain collocation.**

The **third kind of testing** is to use the basis functions as the testing functions, that is:

$$w_m(\boldsymbol{r}) = v_m(\boldsymbol{r}). \qquad (8.55)$$

With this choice, the system matrix becomes symmetric when the integral operator is symmetric. This is known as **Galerkin's formulation.**

The **fourth kind of testing** is to use $\mathcal{L}v_m$ as the testing functions:

$$w_m(\boldsymbol{r}) = [\mathcal{L}v_m(\boldsymbol{r})]. \qquad (8.56)$$

It can be shown that this is equivalent to minimizing the integral of the square of the residual of the governing equation. For this reason, it is called the **least squares formulation.**

Among the four kinds of testing functions described, the point-matching and Galerkin's formulations are widely used in the moment method. It was a common belief that Galerkin's formulation yields best accuracy; however, recent research indicates that the solution accuracy mainly depends on the basis functions and much less on the choice of testing functions (Peterson *et al.*, 1997).

8.6 Solution of Matrix Equations

The final step in the moment method is to solve the resulting matrix equation that represents a set of linear equations. Typical solution methods include **direct matrix inversion, factorization**, and **iteration.**

Both matrix inversion and factorization make use of Gaussian elimination. The matrix inversion finds the inverse of the matrix $[A]$, from which the solution to equation 8.12 is obtained as:

$$\{c\} = [A]^{-1}\{b\}. \qquad (8.57)$$

This method requires $O(N^3)$ operations, where O stands for *on the order of.* The factorization method decomposes the matrix $[A]$ into the product of a lower triangular matrix, denoted as $[L]$, and an upper triangular matrix, denoted as $[U]$. The solution is then obtained by a forward substitution:

$$[L]\{y\} = \{b\}. \qquad (8.58)$$

The solution is also obtained by a backward substitution:

$$[U]\{c\} = \{y\}. \qquad (8.59)$$

The decomposition method also requires $O(N^3)$ operations. It is about three times faster, however, than the direct inversion. Once the system matrix is inverted or factorized, the solution for a new source vector $\{b\}$ can be obtained easily. Therefore, the inversion and factorization methods are suitable for cases when the solution is sought for many right-hand sides. This is their major advantage.

The third method is to solve the matrix equation through iteration by minimizing the residual vector:

$$\{r\} = \{b\} - [A]\{\tilde{c}\}, \quad (8.60)$$

where $\{\tilde{c}\}$ denotes an approximate solution. Commonly used iterative methods include the **conjugate gradient** (Hestenes and Stiefel, 1952), the **biconjugate gradient** (Fletcher, 1975), the **quasiminimal residual** (Freund and Nachtigal, 1991), the **generalized minimal residual** (Saad, 1986), and their variations, such as the **conjugate gradient stabilized** (Sonneveld, 1989), the **conjugate gradient normal residual** (Greenbaum, 1997), the **biconjugate gradient stabilized** (van der Vorst, 1992), and the **transpose-free quasiminimal residual** (Freund, 1993). The algorithms implementing these methods can be found in public literature and software packages of Barret (1993) and Saad (1995). For positive definite systems, the conjugate gradient method is the optimum iterative solver. For indefinite systems, such as most matrix equations for electromagnetic problems, it is not clear which is optimum. An iterative solver calculates one or two matrix-vector products for each iteration. Hence, it requires $O(N^2)$ operations per iteration, and the total computing time is proportional to $N_{\text{iter}}N^2$, where N_{iter} denotes the total number of iterations for convergence. If N_{iter} is small, an iterative solver can be faster than the LU decomposition method for one right-hand side. The interative solver, however, must repeat the process for every new right-hand side.

To achieve an efficient solution through iteration, it is critical to reduce the number of iterations and/or to reduce the operation count for computing the matrix–vector product. The first can be achieved by **preconditioning**, and the second can be achieved by using **fast Fourier transform** (Borup and Gandhi, 1984; Zwamborn and van der Berg, 1994; Gan and Chew, 1995; Wang and Jin, 1988), the **adaptive integral method** (Wang *et al.*, 1988), the **fast multipole method** (Rokhlin, 1990; Coifman *et al.*, 1993; Lu and Chew, 1993), and the **multilevel fast multipole algorithm** (Song and Chew, 1993; Dembart and Yip, 1995; Song *et al.*, 1997).

References

Andreasen, M.G. (1964). Scattering from parallel metallic cylinders with arbitrary cross section. *IEEE Transactions of Antennas and Propagations 12*, 746–754.

Barret, R. (1993). *Templates for the solution of linear systems.* Philadelphia: SIAM.

Bleszynski, E., Bleszynski, M., and Jaroszewicz, T. (1996). AIM: Adaptive integral method for solving large-scale electromagnetic scattering and radiation problems. *Radio Science, 31*, 1225–1251.

Bojarski, N.N. (1971). *k*-space formulation of the electromagnetic scattering problem. Air Force Avionics Lab. *Technical Report* AFAL-TR-71-75.

Borup, D.T., and Gandhi, O.P. (1984). Fast-Fourier transform method for calculation of SAR distributions in finely discretized inhomogeneous models of biological bodies. *IEEE Transactions on Microwave Theory Technology 32*, 355–360.

Catedra, M.F., Cuevas, J.G., and Nuno, L. (1988). A scheme to analyze conducting plates of resonant size using the conjugate gradient method and fast Fourier transform. *IEEE Transactions of Antennas and Propagations 37*, 528–537.

Chang, Y., and Harrington, R.F. (1977). A surface formulation for characteristic modes of material bodies. *IEEE Transactions of Antennas and Propagations 25*, 789–795.

Coifman, R., Rokhlin, V., and Wandzura, S. (1993). The fast multipole method for the wave equation: A pedestrian prescription. *IEEE Transactions of Antennas and Propogations 35*, 7–12.

Dembart, B., and Yip, E. (1995). A 3-D fast multipole method for electromagnetics with multiple levels. *11th Annual Review Progress Appl. Computat. Electromag.* 621–628.

Fletcher, R. (1975). Conjugate gradient methods for indefinite systems. In *Proc. Dundee Conf. Numer. Anal. Syst.* New York: Springer.

Freund, R.W. (1993). A transpose-free quasi-minimal residual algorithm for non-Hermitian linear systems. *SIAM Journal of Scientific and Statistical Computations 14*, 470–482.

Freund, R.W., and Nachtigal, N.M. (1991). QMR: A quasi-minimal residual method for non-Hermitian linear systems. *Numerische Mathematik 60*, 315–339.

Gan, H., and Chew, W.C. (1995). A discrete BCG-FFT algorithm for solving 3-D inhomogeneous scattering problems. *Journal of Electromagnetic Waves Applications 9*, 1339–1357.

Greenbaum, A. (1997). *Iterative methods for solving linear systems.* Philadelphia: SIAM.

Hansen, R.C. (Ed.). (1990). *Moment methods in antennas and scattering.* Norwood, MA: Artech House.

Harrington, R.F. (1968). *Field computation by moment methods.* New York: Macmillan.

Hestenes, M.R., and Stiefel, E. (1952). Method of conjugate gradients for solving linear systems. *Journal of Research of the National Bureau of Standards 49*, 409–435.

Jin, J.M. (1993). *The finite element method in electromagnetics.* New York: John Wiley & Sons.

Jin, J.M., Liepa, V.V., and Tai, C.T. (1988). A volume-surface integral equation for electro-magnetic scattering by inhomogeneous cylinders. *Journal of electromagnetic waves and Applications 2*, 573–588.

Jin, J.M., and Volakis, J.L. (1992). A biconjugate gradient FFT solution for scattering by planar plates. *Electromagnetics 12*, 105–119.

Livesay, D.E., and Chen, K.M. (1974). Electromagnetic fields induced inside arbitrarily shaped biological bodies. *IEEE Transactions of Microwave Theory Technology 22*, 1273–1280.

Lu, C.C., and Chew, W.C. (1993). Fast algorithm for solving hybrid integral equations. *IEEE Proceedings-H 140*, 455–460.

Mautz, J.R., and Harrington, R.F. (1978). H-field, E-field, and combined-field solutions for conducting body of revolution. *Archiv für Elektronik und Übertragungestechnik 32*, 157–164.

Mautz, J.R., and Harrington, R.F. (1979). Electromagnetic scattering from a homogeneous material body of revolution. *Archiv für Elektronik und Übertragungestechnik 33, 71–80.*

Mei, K.K., and Van Bladel, J. (1963). Scattering by perfectly conducting rectangular cylinders. *IEEE Transactions of Antennas and Propagations 11,* 185–192.

Miller, E.K., Medgyesi-Mitschang, L., and Newman, E.H. (1992). (Ed.). *Computational electromagnetics: Frequency-domain method of moments.* New York: IEEE Press.

Mittra, R. (Ed.). (1973). *Computer techniques for electromagnetics.* Elmsford, NY: Permagon.

Morita, N., Kumagai, N., and Mautz, J.R. (1990). *Integral equation methods for electromagnetics.* Norwood, MA: Artech House.

Oshiro, F.K. (1965). Source distribution techniques for the solution of general electromagnetic scattering problems. *Proceedings of the First GISAT Symposium on Mitre Corporation 1,* 83–107.

Peterson, A.F., Ray, S.L., and Mittra, R. (1997). *Computational methods for electromagnetics.* New York: IEEE Press.

Poggio, A.J., and Miller, E.K. (1973). Integral equation solutions of three-dimensional scattering problems. In *Computer techniques for electromagnetics.* Elmsford, NY: Permagon.

Rao, S.M., and Wilton, D.R. (1990). E-field, H-field, and combined field solution for arbitrarily shaped three-dimensional dielectric bodies. *Electromagnetics 10*(4), 407–421.

Rao, S.M., Wilton, D.R., and Glisson, A.W. (1982). Electromagnetic scattering by surfaces of arbitrary shape. *IEEE Transactions of Antennas and Propagations 30,* 409–418.

Richmond, J.H. (1965). Scattering by a dielectric cylinder of arbitrary cross section shape. *IEEE Transactions of Antennas and Propagations 13,* 334–341.

Rokhlin, V. (1990). Rapid solution of integral equations of scattering theory in two dimensions. *Journal of Computer Physics 86,* 414–439.

Saad, Y. (1986). GMRES: A generalized minimal residual algorithm for solving nonsymmetric linear systems. *SIAM Journal of Scientific and Statistical Computations 7,* 856–869.

Saad, Y. (1995). *Iterative method for sparse linear systems.* New York: PWS Publishing.

Schaubert, D.H., Wilton, D.R., and Glisson, A.W. (1984). A tetrahedral modeling method for electromagnetic scattering by arbitrarily shaped inhomogeneous dielectric bodies. *IEEE Transactions of Antennas and Propagations 32,* 77–85.

Song, J.M., and Chew, W.C. (1995). Multilevel fast-multipole algorithm for solving combined field integral equations of electromagnetic scattering. *Microwave Optical Technology Letters 10*(1), 14–19.

Song, J.M., Lu, C.C., and Chew, W.C. (1997). MLFMA for electromagnetic scattering by large complex objects. *IEEE Transactions of Antennas and Propagations 45,* 1488–1493.

Sonneveld, P. (1989). CGS: A fast Lanczos-type solver for nonsymmetric linear systems. *SIAM Journal of Scientific and Statistical Computations 10,* 36–52.

Tai, C.T. (1994). *Dyadic Green functions in electromagnetic theory* (2d ed.). New York: IEEE Press.

van der Vorst, H.A. (1992). BI-CGSTAB: A fast and smoothly converging variant of BI-CG for the solution of nonsymmetric linear systems. *SIAM Journal of Scientific and Statistical Computations 13,* 631–644.

Wang, C.F., and Jin, J.M. (1998). Simple and efficient computation of electromagnetic fields in arbitrarily-shaped, inhomogeneous dielectric diodes using transpose-free QMR and FFT. *IEEE Transactions on Microwave Theory Technology 46,* 553–558.

Wang, C.F., Ling, F., and Jin, J.M. (1998). A fast full-wave analysis of scattering and radiation from large finite arrays of microstrip antennas. *IEEE Transactions of Antennas and Propagations 46,* 1467–1474.

Wang, J.J.H. (1991). *Generalized moment methods in electromagnetics.* New York: John Wiley & Sons.

Wu, T.K., and Tsai, L.L. (1977). Scattering from arbitrarily shaped lossy dielectric bodies of revolution. *Radio Science 12,* 709–718.

Zwamborn, A.P.M., and van der Berg, P.M. (1991). A weak form of the conjugate gradient FFT method for plate problems. *IEEE Transactions of Antennas and Propagations 39,* 224–228.

Zwamborn, A.P.M., and van der Berg, P.M. (1994). Computation of electromagnetic fields inside strongly inhomogeneous objects by the weak-conjugate-gradient fast-Fourier-transform method. *Journal of the Optical Society of America A, 11,* 1414–1420.

9

Computational Electromagnetics: The Finite-Difference Time-Domain Method

Allen Taflove
Department of Electrical and Computer Engineering, Northwestern University, Chicago, Illinois, USA

Susan C. Hagness
Department of Electrical and Computer Engineering, University of Wisconsin, Madison, Wisconsin, USA

Melinda Piket-May
Department of Electrical and Computer Engineering, University of Colorado, Boulder, Colorado, USA

9.1 Introduction

9.1.1 Background

Prior to about 1990, the modeling of electromagnetic engineering systems was primarily implemented using solution techniques for the sinusoidal steady-state Maxwell's equations. Before about 1960, the principal approaches in this area involved closed-form and infinite-series analytical solutions, with numerical results from these analyses obtained using mechanical calculators. After 1960, the increasing availability of programmable electronic digital computers permitted such frequency-domain approaches to rise markedly in sophistication. Researchers were able to take advantage of the capabilities afforded by powerful new high-level programming languages such as Fortran, rapid random-access storage of large arrays of numbers, and computational speeds that were orders of magnitude faster than possible with mechanical calculators. In this period, the principal computational approaches for Maxwell's equations included the high-frequency asymptotic methods of Keller (1962) as well as Kouyoumjian and Pathak (1974) and the integral equation techniques of Harrington (1968).

However, these frequency-domain techniques have difficulties and trade-offs. For example, while asymptotic analyses are well suited for modeling the scattering properties of electrically large complex shapes, such analyses have difficulty treating nonmetallic material composition and volumetric complexity of a structure. While integral equation methods can deal with material and structural complexity, their need to construct and solve systems of linear equations limits the electrical size of possible models, especially those requiring detailed treatment of geometric details in a volume as opposed to just the surface shape.

While significant progress has been made in solving the ultra-large systems of equations generated by frequency-domain integral equations (Song and Chew 1998), the capabilities of even the latest such technologies are exhausted by many volumetrically complex structures of engineering interest. This also holds true for frequency-domain finite element techniques that generate sparse rather than dense matrices. Further, the very difficult incorporation of material and device nonlinearities into frequency-domain solutions of Maxwell's equations poses a significant problem as engineers seek to design active electromagnetic/electronic and electromagnetic/quantum-optical systems such as high-speed digital circuits, microwave and millimeter-wave amplifiers, and lasers.

9.1.2 Rise of Finite-Difference Time-Domain Methods

During the 1970s and 1980s, a number of researchers realized the limitations of frequency-domain integral equation solutions of Maxwell's equations. This led to early explorations of a novel alternative approach: direct time-domain solutions of Maxwell's differential (curl) equations on spatial grids or lattices. The **finite-difference time-domain** (FDTD) method, introduced by Yee (1966), was the first technique in this class and has remained the subject of continuous development (Taflove and Hagness [2000]).

There are seven primary reasons for the expansion of interest in FDTD and related computational solution approaches for Maxwell's equations:

1. FDTD uses no linear algebra. Being a fully explicit computation, FDTD avoids the difficulties with linear algebra that limit the size of a frequency-domain integral equation and finite-element electromagnetics models to generally fewer than 10^6 field unknowns. FDTD models with as many as 10^9 field unknowns have been run. There is no intrinsic upper bound to this number.
2. FDTD is accurate and robust. The sources of error in FDTD calculations are well understood and can be bounded to permit accurate models for a very large variety of electromagnetic wave interaction problems.
3. FDTD treats impulsive behavior naturally. Being a time-domain technique, FDTD directly calculates the impulse response of an electromagnetic system. Therefore, a single FDTD simulation can provide either ultrawide-band temporal waveforms or the sinusoidal steady-state response at any frequency in the excitation spectrum.
4. FDTD treats nonlinear behavior naturally. Being a time-domain technique, FDTD directly calculates the nonlinear response of an electromagnetic system.
5. FDTD is a systematic approach. With FDTD, specifying a new structure to be modeled is reduced to a problem of mesh generation rather than the potentially complex reformulation of an integral equation. For example, FDTD requires no calculation of structure-dependent Green's functions.
6. Computer memory capacities are increasing rapidly. Although this trend positively influences all numerical techniques, it is of particular advantage to FDTD methods that are founded on discretizing space over a volume and that inherently require a large random access memory.
7. Computer visualization capabilities are increasing rapidly. Although this trend positively influences all numerical techniques, it is of particular advantage to FDTD methods that generate time-marched arrays of field quantities suitable for use in color videos to illustrate field dynamics.

An indication of the expanding level of interest in FDTD Maxwell's equations solvers is the hundreds of papers currently published in this area worldwide each year, as opposed to fewer than ten as recently as 1985 and prior to that year (Shlager and Schneider, 1998). This expansion continues as engineers and scientists in nontraditional electromagnetics-related areas such as digital systems and integrated optics become aware of the power of such direct solution techniques for Maxwell's equations.

9.1.3 Characteristics of FDTD and Related Space-Grid Time-Domain Techniques

FDTD and related space-grid time-domain techniques are direct solution methods for Maxwell's curl equations. These methods employ no potentials. Rather, they are based on volumetric sampling of the unknown electric and magnetic fields in and surrounding the structure of interest over a period of time. The sampling in space is at subwavelength resolution set by the user to properly sample the highest near-field spatial frequencies thought to be important in the physics of the problem. Typically, 10 to 20 samples per wavelength are needed. The sampling time is selected to ensure numerical stability of the algorithm.

Overall, FDTD and related techniques are marching-in-time procedures that simulate the continuous actual electromagnetic waves in a finite spatial region by sampled-data numerical analogs propagating in a computer data space. Time-stepping continues as the numerical wave analogs propagate in the space lattice to causally connect the physics of the modeled region. For simulations where the modeled region must extend to infinity, absorbing boundary conditions (ABCs) are employed at the outer lattice truncation planes, which ideally permit, all outgoing wave analogs to exit the region with negligible reflection. Phenomena such as induction of surface currents, scattering and multiple scattering, aperture penetration, and cavity excitation are modeled time-step by time-step by the action of the numerical analog to the curl equations. Self-consistency of these modeled phenomena is generally ensured if their spatial and temporal variations are well resolved by the space and time sampling process. In fact, the goal is to provide a self-consistent model of the mutual coupling of all of the electrically small volume cells constituting the structure and its near field, even if the structure spans tens of wavelengths in three dimensions and there are hundreds of millions of space cells.

Time-stepping is continued until the desired late-time pulse response is observed at the field points of interest. For linear wave interaction problems, the sinusoidal response at these field points can be obtained over a wide band of frequencies by discrete Fourier transformation of the computed field versus time waveforms at these points. Prolonged "ringing" of the computed field waveforms due to a high Q-factor or large electrical size of the structure being modeled requires a combination of extending the computational window in time and extrapolating the windowed data before Fourier transformation.

9.1.4 Classes of Algorithms

Current FDTD and related space-grid time-domain algorithms are fully explicit solvers employing highly vectorizable and parallel schemes for time-marching the six components of the electric and magnetic field vectors at each of the space cells. The explicit nature of the solvers is usually maintained by employing a leapfrog time-stepping scheme. Current methods differ primarily in how the space lattice is set up. In fact, gridding methods can be categorized according to the degree of structure or regularity in the mesh cells.

Almost Completely Structured

In this case, the space lattice is organized so that its unit cells are congruent wherever possible. The most basic example of such a mesh is the pioneering work of Yee (1966), who employed a uniform Cartesian grid having rectangular cells. Staircasing was used to approximate the surface of structural features not parallel to the grid coordinate axes. Later work showed that it is possible to modify the size and shape of the space cells located immediately adjacent to a structural feature to conformally fit its surface (Jurgens *et al.*, 1992; Dey and Mittra, 1997). This result is accurate and computationally efficient for large structures because the number of modified cells is proportional to the surface area of the structure. Thus, the number of modified cells becomes progressively smaller relative to the number of regular cells filling the structure volume as its size increases. As a result, the computer resources needed to implement a fully conformal model approximate those required for a staircased model. A key disadvantage of this technique, however, is that special mesh-generation software must be constructed.

Surface-Fitted

In this case, the space lattice is globally distorted to fit the shape of the structure of interest. The lattice can be divided into multiple zones to accommodate a set of distinct surface features (Shankar *et al.*, 1990). The major advantage of this approach is that well-developed mesh-generation software of this type is available. The major disadvantage, relative to the Yee algorithm, is the substantial added computer burden due to:

- Memory allocations for the position and stretching factors of each cell
- Extra computer operations to implement Maxwell's equations at each cell and to enforce field continuity at the interfaces of adjacent cells

Another disadvantage is the possible presence of numerical dissipation in the time-stepping algorithm used for such meshes. This can limit the range of electrical size of the structure being modeled due to numerical wave-attenuation artifacts.

Completely Unstructured

In an unstructured case, the space containing the structure of interest is completely filled with a collection of lattice cells of varying sizes and shapes but conforms to the structure surface (Madsen and Ziolkowski, 1990). As for the case of surface-fitted lattices, mesh-generation software is available and capable of modeling complicated three-dimensional shapes possibly having volumetric inhomogeneities. A key disadvantage of this approach is its potential for numerical inaccuracy and instability due to the unwanted generation of highly skewed space cells at random points in the lattice. A second disadvantage is the difficulty in mapping the unstructured mesh computations onto the architecture of either parallel vector computers or massively parallel machines. The structure-specific irregularity of the mesh mandates a robust preprocessing algorithm that optimally assigns specific mesh cells to specific processors.

At present, the best choice of a computational algorithm and mesh remains unclear. For the next several years, progress should continue in this area as various groups develop their favored approaches and perform validations.

9.1.5 Predictive Dynamic Range

For computational modeling of electromagnetic wave interaction structures using FDTD and related space-grid time-domain techniques, it is useful to consider the concept of a predictive dynamic range. Let the power density of the primary (incident) wave in the space grid be $P_0 \mathrm{W/m}^2$. Further, let the minimum observable power density of a secondary (scattered) wave be $P_S \mathrm{W/m}^2$, where "minimum observable" means that the accuracy of the field computation degrades due to numerical artifacts to poorer than n decibels (some desired figure of merit) at lower levels than P_S. Then, the predictive dynamic range can be defined as $10\log_{10}(P_0/P_S)$dB.

This definition is well suited for FDTD and other space-grid time-domain codes for two reasons:

- It squares nicely with the concept of a "quiet zone" in an experimental anechoic chamber, which is intuitive to most electromagnetics engineers.
- It succinctly quantifies the fact that the desired numerical wave analogs propagating in the lattice exist in an additive noise environment due to nonphysical propagating wave analogs caused by the imperfect ABCs.

In addition to additive noise, the desired physical wave analogs undergo gradual progressive deterioration while propagating due to accumulating numerical dispersion artifacts, including phase velocity anisotropies and inhomogeneities in the mesh.

In the 1980s, researchers accumulated solid evidence for a predictive dynamic range on the order of 40 to 50 dB for FDTD codes. This value is reasonable if one considers the additive noise due to imperfect ABCs to be the primary limiting factor, because the analytical ABCs of this era (Mur, 1981) provided outer-boundary reflection coefficients in the range of about 0.3 to 3% (-30 to -50dB).

The 1990s saw the emergence of powerful, entirely new classes of ABCs, including the perfectly matched layer (PML) of Berenger (1994), the uniaxial anisotropic PML (UPML) of Sacks *et al.* (1995) and Gedney (1996), and the complementary operator methods (COM) of Ramah (1997, 1998). These ABCs were shown to have effective outerboundary reflection coefficients of better than -80dB for impinging pulsed electromagnetic waves having ultrawideband spectra. Solid capabilities were demonstrated to terminate free space lattices, multimoding and dispersive waveguiding structures, and lossy and dispersive materials.

However, for electrically large problems, the overall dynamic range was shown not to reach the maximum permitted by these new ABCs because of inaccuracies due to accumulating numerical dispersion artifacts generated by the basic grid-based solution of the curl equations. Fortunately, by the end of the 1990s, this problem was being attacked by a new generation of low-dispersion algorithms. Examples include the wavelet-based multiresolution time-domain (MRTD) technique introduced by Krumpholz and Katehi (1996) and the pseudo-spectral time-domain (PSTD) technique introduced by Liu, (1996, 1997). As a result of these advances, there is the possible emergence of FDTD and related space-grid time-domain methods, demonstrating predictive dynamic ranges of 80 dB or more in the first decade of the 21st century.

9.1.6 Scaling to Very Large Problem Sizes

Using FDTD and related methods, electromagnetic wave interaction problems can be modeled requiring the solution of considerably more than 10^8 field–vector unknowns. At this level of complexity, it is possible to develop detailed, three-dimensional models of complete engineering systems, including the following:

- Entire aircraft and missiles illuminated by radar at 1 GHz and above
- Entire multilayer circuit boards and multichip modules for digital signal propagation, cross talk, and radiation
- Entire microwave and millimeter-wave amplifiers, including the active and passive circuit components and packaging
- Entire integrated-optical structures, including lasers, waveguides, couplers, and resonators

A key goal for such large models is to achieve algorithm/computer-architecture scaling such that for N field unknowns to be solved on M processors, we approach an order (N/M) scaling of the required computational resources.

We now consider the factors involved in determining the computational burden for the class of FDTD and related space-grid time-domain solvers.

1. **Number of volumetric grid cells, N:** The six vector electromagnetic field components located at each lattice cell must be updated at every time-step. This yields by itself an order (N) scaling.
2. **Number of time steps, $n_{\max}$:** A self-consistent solution in the time domain mandates that the numerical wave analogs propagate over time scales sufficient to causally connect each portion of the structure of interest. Therefore, $n_{\max}$ must increase as the maximum electrical size of the structure. In three dimensions, it can be argued that $n_{\max}$ is a fractional power function of N such as $N^{1/3}$. Further, $n_{\max}$ must be adequate to step through "ring-up" and "ring-down" times of energy storage features such as cavities. These features vary from problem to problem and cannot be ascribed a dependence relative to N.

3. **Cumulative propagation errors**: Additional computational burdens may arise due to the need for either progressive mesh refinement or progressively higher accuracy algorithms to bound cumulative positional or phase errors for propagating numerical modes in progressively enlarged meshes. Any need for progressive mesh refinement would feed back to factor 1.

For most free space problems, factors 2 and 3 are weaker functions of the size of the modeled structure than factor 1. This is because geometrical features at increasing electrical distances from each other become decoupled due to radiative losses by the electromagnetic waves propagating between these features. Further, it can be shown that replacing second-order accurate algorithms by high-order versions sufficiently reduces numerical dispersion error to avoid the need for progressive mesh refinement for object sizes up to the order of 100 wavelengths. Overall, a computational burden of order $(N \cdot n_{\max}) = \text{order}\ (N^{4/3})$ is estimated for very large FDTD and related models.

9.2 Maxwell's Equations

In this section, we establish the fundamental equations and notation for the electromagnetic fields used in the remainder of this chapter.

9.2.1 Three-Dimensional Case

Using meter per second (MKS) units, the time-dependent Maxwell's equations in three dimensions are given in differential and integral form by:

Faraday's law:

$$\frac{\partial \vec{B}}{\partial t} = -\nabla \times \vec{E} - \vec{M}. \tag{9.1a}$$

$$\frac{\partial}{\partial t}\iint_A \vec{B} \cdot d\vec{A} = -\oint_\ell \vec{E} \cdot d\vec{\ell} - \iint_A \vec{M} \cdot d\vec{A}. \tag{9.1b}$$

Ampere's law:

$$\frac{\partial \vec{D}}{\partial t} = \nabla \times \vec{H} - \vec{J}. \tag{9.2a}$$

$$\frac{\partial}{\partial t}\iint_A \vec{D} \cdot d\vec{A} = \oint_\ell \vec{H} \cdot d\vec{\ell} - \iint_A \vec{J} \cdot d\vec{A}. \tag{9.2b}$$

Gauss' law for the electric field:

$$\nabla \cdot \vec{D} = 0. \tag{9.3a}$$

$$\oint\!\!\oint_A \vec{D} \cdot d\vec{A} = 0. \tag{9.3b}$$

Gauss' law for the magnetic field:

$$\nabla \cdot \vec{B} = 0. \tag{9.4a}$$

$$\oint\!\!\oint_A \vec{B} \cdot d\vec{A} = 0. \tag{9.4b}$$

For equations 9.1 through 9.4, the following symbols (and their MKS units) are defined:

$\vec{E}$: Electric field [volts/meter]
$\vec{D}$: Electric flux density [coulombs/meter^2]
$\vec{H}$: Magnetic field [amperes/meter]
$\vec{B}$: Magnetic flux density [webers/meter^2]
A : Arbitrary three-dimensional surface
$d\vec{A}$: Differential normal vector that characterizes surface A [meter^2]
ℓ : Closed contour that bounds surface A
$d\vec{\ell}$: Differential length vector that characterizes contour ℓ [meters]
$\vec{J}$: Electrical current density [amperes/meter^2]
$\vec{M}$: Equivalent magnetic current density [volts/meter^2]

In linear, isotropic, and nondispersive materials (i.e., materials having field-independent, direction-independent, and frequency-independent electric and magnetic properties), we can relate $\vec{D}$ to $\vec{E}$ and $\vec{B}$ to $\vec{H}$ using simple proportions:

$$\vec{D} = \varepsilon\vec{E} = \varepsilon_r\varepsilon_0\vec{E}; \quad \vec{B} = \mu\vec{H} = \mu_r\mu_0\vec{H}, \tag{9.5}$$

where the following applies:

ε : Electrical permittivity [farads/meter]
ε_r : Relative permittivity [dimensionless scalar]
ε_0 : Free-space permittivity [8.854×10^{-12} farads/meter]
μ : Magnetic permeability [henry/meter]
μ_r : Relative permeability [dimensionless scalar]
μ_0 : Free-space permeability [$4\pi \times 10^{-7}$ henry/meter]

Note that $\vec{J}$ and $\vec{M}$ can act as **independent sources** of E- and H-field energy, $\vec{J}_{\text{source}}$, and $\vec{M}_{\text{source}}$. We also allow for materials with isotropic, nondispersive electric, and magnetic losses that attenuate E- and H-fields via conversion to heat energy. This yields:

$$\vec{J} = \vec{J}_{\text{source}} + \sigma\vec{E}; \ \vec{M} = \vec{M}_{\text{source}} + \sigma^*\vec{H}, \tag{9.6}$$

where the following applies:

σ : Electric conductivity [siemens/meter]
σ^* : Equivalent magnetic loss [ohms/meter]

Finally, we substitute equations 9.5 and 9.6 into (9.1a) and (9.2a). This yields Maxwell's curl equations in linear, isotropic, nondispersive, and lossy materials:

$$\frac{\partial \vec{H}}{\partial t} = -\frac{1}{\mu}\nabla \times \ddot{E} - \frac{1}{\mu}\left(\vec{M}_{source} + \sigma^* \vec{H}\right). \quad (9.7)$$

$$\frac{\partial \vec{E}}{\partial t} = \frac{1}{\varepsilon}\nabla \times \vec{H} - \frac{1}{\varepsilon}\left(\vec{J}_{source} + \sigma \vec{E}\right). \quad (9.8)$$

We now write out the vector components of the curl operators of equations 9.7 and 9.8 in Cartesian coordinates. This yields the following system of six coupled scalar equations:

$$\frac{\partial H_x}{\partial t} = \frac{1}{\mu}\left[\frac{\partial E_y}{\partial z} - \frac{\partial E_z}{\partial y} - \left(M_{source_x} + \sigma^* H_x\right)\right]. \quad (9.9a)$$

$$\frac{\partial H_y}{\partial t} = \frac{1}{\mu}\left[\frac{\partial E_z}{\partial x} - \frac{\partial E_x}{\partial z} - \left(M_{source_y} + \sigma^* H_y\right)\right]. \quad (9.9b)$$

$$\frac{\partial H_z}{\partial t} = \frac{1}{\mu}\left[\frac{\partial E_x}{\partial y} - \frac{\partial E_y}{\partial x} - \left(M_{source_z} + \sigma^* H_z\right)\right]. \quad (9.9c)$$

$$\frac{\partial E_x}{\partial t} = \frac{1}{\varepsilon}\left[\frac{\partial H_z}{\partial y} - \frac{\partial H_y}{\partial z} - \left(J_{source_x} + \sigma E_x\right)\right]. \quad (9.10a)$$

$$\frac{\partial E_y}{\partial t} = \frac{1}{\varepsilon}\left[\frac{\partial H_x}{\partial z} - \frac{\partial H_z}{\partial x} - \left(J_{source_y} + \sigma E_y\right)\right]. \quad (9.10b)$$

$$\frac{\partial E_z}{\partial t} = \frac{1}{\varepsilon}\left[\frac{\partial H_y}{\partial x} - \frac{\partial H_x}{\partial y} - \left(J_{source_z} + \sigma E_z\right)\right]. \quad (9.10c)$$

The system of six coupled partial differential equations of (9.9) and (9.10) forms the basis of the FDTD numerical algorithm for electromagnetic wave interactions with general three-dimensional objects. The FDTD algorithm need not explicitly enforce the Gauss's law relations indicating zero free electric and magnetic charge shown in equations 9.3 and 9.4. This is because these relations are theoretically a direct consequence of the curl equations, as can be readily shown. However, the FDTD space grid must be structured so that the Gauss's law relations are *implicit* in the positions of the *E*- and *H*-field vector components in the grid and in the numerical space-derivative operations upon these components that model the action of the curl operator. This will be discussed later in the context of the Yee mesh.

Before proceeding with the introduction of the Yee algorithm, it is instructive to consider simplified two-dimensional cases for Maxwell's equations. These cases demonstrate important electromagnetic wave phenomena and can yield insight into the analytical and algorithmic features of the general three-dimensional case.

9.2.2 Reduction to Two Dimensions

Let us assume that the structure being modeled extends to infinity in the *z* direction with no change in the shape or position of its transverse cross section. If the incident wave is also uniform in the *z* direction, then all partial derivatives of the fields with respect to *z* must equal zero. Under these conditions, the full set of Maxwell's curl equations given by equations 9.9 and 9.10 reduces to two modes: the **transverse-magnetic mode with respect to z** (TM_z) and the **transverse-electric mode with respect to z** (TE_z). The reduced sets of Maxwell's equations for these modes are as follows.

TM_z Mode (Involving only H_x, H_y and E_z)

$$\frac{\partial H_x}{\partial t} = \frac{1}{\mu}\left[-\frac{\partial E_z}{\partial_y} - \left(M_{\text{source}_x} + \sigma^* H_x\right)\right]. \quad (9.11a)$$

$$\frac{\partial H_y}{\partial t} = \frac{1}{\mu}\left[\frac{\partial E_z}{\partial x} - \left(M_{\text{source}_y} + \sigma^* H_y\right)\right]. \quad (9.11b)$$

$$\frac{\partial E_z}{\partial t} = \frac{1}{\varepsilon}\left[\frac{\partial H_y}{\partial x} - \frac{\partial H_x}{\partial y} - \left(J_{\text{source}_z} + \sigma E_z\right)\right]. \quad (9.11c)$$

TE_z Mode (Involving only E_x, E_y and H_z)

$$\frac{\partial E_x}{\partial t} = \frac{1}{\varepsilon}\left[\frac{\partial H_z}{\partial y} - \left(J_{\text{source}_x} + \sigma E_x\right)\right]. \quad (9.12a)$$

$$\frac{\partial E_y}{\partial t} = \frac{1}{\varepsilon}\left[-\frac{\partial H_z}{\partial x} - \left(J_{\text{source}_y} + \sigma E_y\right)\right]. \quad (9.12b)$$

$$\frac{\partial H_z}{\partial t} = \frac{1}{\mu}\left[\frac{\partial E_x}{\partial y} - \frac{\partial E_y}{\partial x} - \left(M_{\text{source}_z} + \sigma^* H_z\right)\right]. \quad (9.12c)$$

The TM_z and TE_z modes contain no common field vector components. Thus, these modes can exist simultaneously with *no* mutual interactions for structures composed of isotropic materials or anisotropic materials having no off-diagonal components in the constitutive tensors.

Physical phenomena associated with these two modes can be very different. The TE_z mode can support propagating electromagnetic fields bound closely to, or guided by, the surface of a metal structure (the "creeping wave" being a classic example for curved metal surfaces). On the other hand, the TM_z mode sets up an *E*-field which must be negligible at a metal surface. This diminishes or eliminates bound or guided near-surface propagating waves for metal surfaces. The presence or absence of surface-type waves can have important implications for scattering and radiation problems.

9.3 The Yee Algorithm

9.3.1 Basic Ideas

Yee (1966) originated a set of finite-difference equations for the time-dependent Maxwell's curl equations 9.9 and 9.10 for the lossless materials case $\sigma = 0$ and $\sigma^* = 0$. This section summarizes Yee's algorithm, which forms the basis of the

FDTD technique. Key ideas underlying the robust nature of the **Yee algorithm** are as follows:

1. The Yee algorithm solves for both electric and magnetic fields in time and space using the coupled Maxwell's curl equations rather than solving for the electric field alone (or the magnetic field alone) with a wave equation.
   - This is analogous to the combined-field integral equation formulation of the method of moments, wherein both $\vec{E}$ and $\vec{H}$ boundary conditions are enforced on the surface of a material structure.
   - Using both $\vec{E}$ and $\vec{H}$ information, the solution is more robust than using either alone (i.e., it is accurate for a wider class of structures). Both electric and magnetic material properties can be modeled in a straightforward manner. This is especially important when modeling radar cross section mitigation.
   - Features unique to each field (e.g., tangential $\vec{H}$ singularities near edges and corners; azimuthal [looping] $\vec{H}$ singularities near thin wires; and radial $\vec{E}$ singularities near points, edges; and thin wires) can be individually modeled if both electric and magnetic fields are available.
2. As illustrated in Figure 9.1, the Yee algorithm centers its $\vec{E}$ and $\vec{H}$ components in a three-dimensional space so that every $\vec{E}$ component is surrounded by four circulating $\vec{H}$ components and so that every $\vec{H}$ component is surrounded by four circulating $\vec{E}$ components.

   This provides a beautifully simple picture of a three-dimensional space being filled by an interlinked array of Faraday's law and Ampere's law contours. For example, it is possible to identify Yee $\vec{E}$ components associated with displacement current flux linking $\vec{H}$ loops as well as $\vec{H}$ components associated with magnetic flux linking $\vec{E}$ loops. In effect, the Yee algorithm simultaneously simulates the pointwise differential form *and* the macroscopic integral form of Maxwell's equations. The latter possibility is extremely useful in specifying field boundary conditions and singularities.

   In addition, we have the following attributes of the Yee space lattice:
   - The finite-difference expressions for the space derivatives used in the curl operators are central different in nature and second-order accurate.
   - Continuity of tangential $\vec{E}$ and $\vec{H}$ is naturally maintained across an interface of dissimilar materials if the interface is parallel to one of the lattice coordinate axes. For this case, there is no need to specially enforce field boundary conditions at the interface. At the beginning of the problem, we simply specify the material permittivity and permeability at each field component location. This yields a stepped or "staircase" approximation of the surface and internal geometry of the structure, with a space resolution set by the size of the lattice unit cell.
   - The location of the $\vec{E}$ and $\vec{H}$ components in the Yee space lattice and the central-different operations on these components implicitly enforce the two Gauss's Law relations (Taflove and Hagness, 2000). Thus, the Yee mesh is divergence-free with respect to its E and H fields in the absence of free electric and magnetic charges.
3. As illustrated in Figure 9.2, the Yee algorithm also centers its $\vec{E}$ and $\vec{H}$ components in time in what is termed a **leapfrog arrangement**. All of the $\vec{E}$ computations in the modeled space are completed and stored in memory for a particular time point using previously stored $\vec{H}$ data. Then, all of the $\vec{H}$ computations in the space are completed and stored in memory using the $\vec{E}$ data just computed. The cycle begins again with the recomputation of the $\vec{E}$ components based on the newly obtained $\vec{H}$. This process continues until time-stepping is concluded.
   - Leapfrog time-stepping is fully explicit, thereby avoiding problems involved with simultaneous equations and matrix inversion.
   - The finite-difference expressions for the time derivatives are central different in nature and second-order accurate.
   - The time-stepping algorithm is nondissipative. That is, numerical wave modes propagating in the mesh do not spuriously decay due to a nonphysical artifact of the time-stepping algorithm.

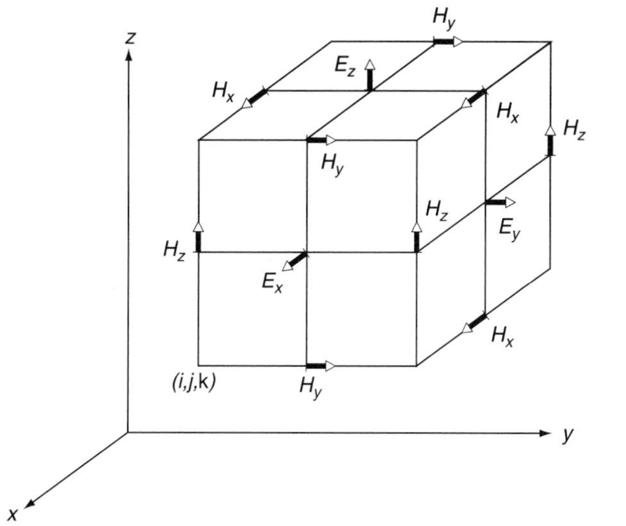

FIGURE 9.1 Position of the Electric and Magnetic Field Vector Components about a Cubic Unit Cell of the Yee Space Lattice. Adapted from Yee (1966).

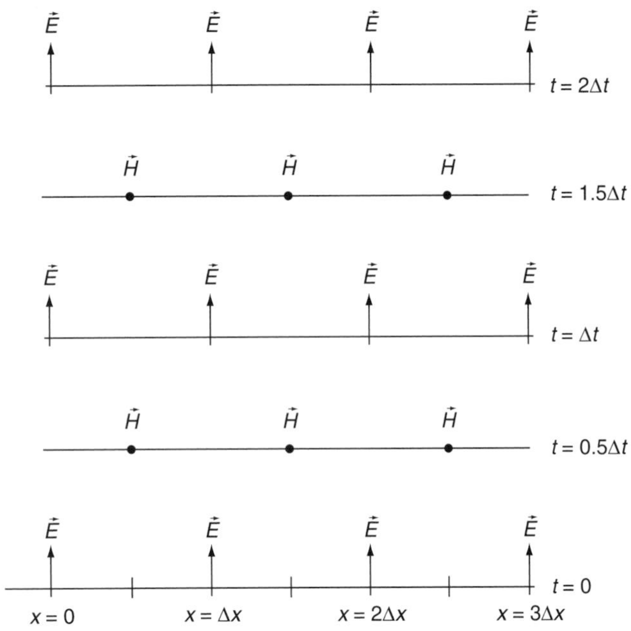

FIGURE 9.2 Space-Time Chart of the Yee Algorithm for a One-Dimensional Wave Propagation. This example shows the use of central differences for the space derivatives and leapfrog for the time derivatives. Initial conditions for both electric and magnetic fields are zero everywhere in the grid.

9.3.2 Finite Differences and Notation

Yee (1966) introduced the following notation for space points and functions of space and time. A space point in a uniform, rectangular lattice is denoted as:

$$(i, j, k) = (i\Delta x, j\Delta y, k\Delta z). \quad (9.13)$$

In equation 9.13, Δx, Δy, and Δz are, respectively, the lattice space increments in the x, y, and z coordinate directions, and i, j, and k are integers. Further, we denote any function u of space and time evaluated at a discrete point in the grid and at a discrete point in time as:

$$u(i\Delta x, j\Delta y, k\Delta z, n\Delta t) = u^n_{i,j,k}, \quad (9.14)$$

where Δt is the time increment that is assumed uniform over the observation interval, and n is an integer.

Yee (1966) used centered finite-difference (central-difference) expressions for the space and time derivatives that are both simply programmed and second-order accurate in the space and time increments. Consider his expression for the first partial space derivative of u in the x direction, evaluated at the fixed time $t_n = n\Delta t$:

$$\frac{\partial u}{\partial x}(i\Delta x, j\Delta y, k\Delta z, n\Delta t) = \frac{u^n_{i+1/2,j,k} - u^n_{i-1/2,j,k}}{\Delta x} + O\left[(\Delta x)^2\right]. \quad (9.15)$$

We note the $\pm 1/2$ increment in the i subscript (x coordinate) of u, denoting a space finite difference over $\pm\Delta x/2$. Yee's goal (1966) was second-order accurate central differencing, but it is apparent that he desired to take data for his central differences to the right and left of his observation point by only $\Delta x/2$, rather than a full Δx.

Yee chose this notation because he interleaved his $\vec{E}$ and $\vec{H}$ components in the space lattice at intervals of $\Delta x/2$. For example, the difference of two adjacent $\vec{E}$ components, separated by Δx and located $\pm\Delta x/2$ on either side of an $\vec{H}$ component, would be used to provide a numerical approximation for $\partial E/\partial x$ to permit stepping the $\vec{H}$ component in time. For completeness, it should be added that a numerical approximation analogous to equation 9.15 for $\partial u/\partial y$ or $\partial u/\partial z$ can be written simply by incrementing the j or k subscript of u by $\pm\Delta y/2$ or $\pm\Delta z/2$, respectively.

Yee's expression for the first time partial derivative of u, evaluated at the fixed space point (i, j, k), follows by analogy:

$$\frac{\partial u}{\partial t}(i\Delta x, j\Delta y, k\Delta z, n\Delta t) = \frac{u^{n+1/2}_{i,j,k} - u^{n-1/2}_{i,j,k}}{\Delta t} + O\left[(\Delta t)^2\right]. \quad (9.16)$$

Now the $\pm 1/2$ increment is in the n superscript (time coordinate) of u, denoting a time finite difference over $\pm\Delta t/2$. Yee (1996) chose this notation because he wished to interleave his $\vec{E}$ and $\vec{H}$ components in time at intervals of $\Delta t/2$ for purposes of implementing a leapfrog algorithm.

9.3.3 Finite-Difference Expressions for Maxwell's Equations in Three Dimensions

We now apply the above ideas and notation to achieve a finite-difference numerical approximation of the Maxwell's curl equations in three dimensions given by equations 9.9 and 9.10. We begin by considering as an example the E_x field-component equation 9.10a. Referring to Figures 9.1 and 9.2, a typical substitution of central differences for the time and space derivatives in equation 9.10a at $E_x(i, j+1/2, k+1/2, n)$ yields the following expression:

$$\frac{E_x|^{n+1/2}_{i,j+1/2,k+1/2} - E_x|^{n-1/2}_{i,j+1/2,k+1/2}}{\Delta t} = \frac{1}{\varepsilon_{i,j+1/2,k+1/2}} \times \left(\begin{array}{c} \frac{H_z|^n_{i,j+1,k+1/2} - H_z|^n_{i,j,k+1/2}}{\Delta y} - \frac{H_y|^n_{i,j+1/2,k+1} - H_y|^n_{i,j+1/2,k}}{\Delta z} \\ - J_{\text{source}_x}|^n_{i,j+1/2,k+1/2} - \sigma_{i,j+1/2,k+1/2} E_x|^n_{i,j+1/2,k+1/2} \end{array} \right). \quad (9.17)$$

Note that all field quantities on the right-hand side are evaluated at time-step n, including the electric field E_x appearing due to the material conductivity σ. Since E_x values at time-step n are not assumed to be stored in the computer's memory (only the previous values of E_x at time-step $n - 1/2$ are assumed to be in memory), we need some way to estimate such terms. A very good way is as follows, using what we call a **semi-implicit approximation**:

$$E_x\big|^{n}_{i,j+1/2,k+1/2} = \frac{E_x\big|^{n+1/2}_{i,j+1/2,k+1/2} + E_x\big|^{n-1/2}_{i,j+1/2,k+1/2}}{2}. \quad (9.18)$$

Here E_x values at time-step n are assumed to be simply the arithmetic average of the stored values of E_x at time-step $n-1/2$ and the yet-to-be computed new values of E_x at time-step $n+1/2$. Substituting equation 9.18 into 9.17 and collecting terms yields the following explicit time-stepping relation for E_x (which is numerically stable for values of σ from zero to infinity):

$$E_x\big|^{n+1/2}_{i,j+1/2,k+1/2} = \left(\frac{1-\frac{\sigma_{i,j+1/2,k+1/2}\Delta t}{2\varepsilon_{i,j+1/2,k+1/2}}}{1+\frac{\sigma_{i,j+1/2,k+1/2}\Delta t}{2\varepsilon_{i,j+1/2,k+1/2}}}\right) E_x\big|^{n-1/2}_{i,j+1/2,k+1/2} + \left(\frac{\frac{\Delta t}{\varepsilon_{i,j+1/2,k+1/2}}}{1+\frac{\sigma_{i,j+1/2,k+1/2}\Delta t}{2\varepsilon_{i,j+1/2,k+1/2}}}\right) \times \left(\begin{array}{l}\frac{H_z|^{n}_{i,j+1,k+1/2}-H_z|^{n}_{i,j,k+1/2}}{\Delta y} \\ -\frac{H_y|^{n}_{i,j+1/2,k+1}-H_y|^{n}_{i,j+1/2,k}}{\Delta z} \\ -J_{\text{source}_x}\big|^{n}_{i,j+1/2,k+1/2}\end{array}\right). \quad (9.19a)$$

Similarly, we can derive finite-difference expressions based on Yee's algorithm for the E_y and E_z field components given by Maxwell's equations 9.10b and 9.10c. Referring again to Figure 9.1, we have:

$$E_y\big|^{n+1/2}_{i-1/2,j+1,k+1/2} = \left(\frac{1-\frac{\sigma_{i-1/2,j+1,k+1/2}\Delta t}{2\varepsilon_{i-1/2,j+1,k+1/2}}}{1+\frac{\sigma_{i-1/2,j+1,k+1/2}\Delta t}{2\varepsilon_{i-1/2,j+1,k+1/2}}}\right) E_y\big|^{n-1/2}_{i-1/2,j+1,k+1/2} + \left(\frac{\frac{\Delta t}{\varepsilon_{i-1/2,j+1,k+1/2}}}{1+\frac{\sigma_{i-1/2,j+1,k+1/2}\Delta t}{2\varepsilon_{i-1/2,j+1,k+1/2}}}\right) \times \left(\begin{array}{l}\frac{H_x|^{n}_{i-1/2,j+1,k+1}-H_x|^{n}_{i-1/2,j+1,k}}{\Delta z} \\ -\frac{H_z|^{n}_{i,j+1,k+1/2}-H_z|^{n}_{i-1,j+1,k+1/2}}{\Delta x} \\ -J_{\text{source}_y}\big|^{n}_{i-1/2,j+1,k+1/2}\end{array}\right). \quad (9.19b)$$

$$E_z\big|^{n+1/2}_{i-1/2,j+1/2,k+1} = \left(\frac{1-\frac{\sigma_{i-1/2,j+1/2,k+1}\Delta t}{2\varepsilon_{i-1/2,j+1/2,k+1}}}{1+\frac{\sigma_{i-1/2,j+1/2,k+1}\Delta t}{2\varepsilon_{i-1/2,j+1/2,k+1}}}\right) E_z\big|^{n-1/2}_{i-1/2,j+1/2,k+1} + \left(\frac{\frac{\Delta t}{\varepsilon_{i-1/2,j+1/2,k+1}}}{1+\frac{\sigma_{i-1/2,j+1/2,k+1}\Delta t}{2\varepsilon_{i-1/2,j+1/2,k+1}}}\right) \times \left(\begin{array}{l}\frac{H_y|^{n}_{i,j+1/2,k+1}-H_y|^{n}_{i-1,j+1/2,k+1}}{\Delta x} \\ -\frac{H_x|^{n}_{i-1/2,j+1,k+1}-H_x|^{n}_{i-1/2,j,k+1}}{\Delta y} \\ -J_{\text{source}_z}\big|^{n}_{i-1/2,j+1/2,k+1}\end{array}\right). \quad (9.19c)$$

By analogy, we can derive finite-difference equations for equations 9.9a to 9.9c to time-step H_x, H_y, and H_z. Here $\sigma^* H$ represents a magnetic loss term on the right-hand side of each equation, which is estimated using a semi-implicit procedure analogous to equation 9.18. Referring again to Figures 9.1 and 9.2, we have, for example, the following time-stepping expressions for the H components located about the unit cell:

$$H_x\big|^{n+1}_{i-1/2,j+1,k+1} = \left(\frac{1-\frac{\sigma^*_{i-1/2,j+1,k+1}\Delta t}{2\mu_{i-1/2,j+1,k+1}}}{1+\frac{\sigma^*_{i-1/2,j+1,k+1}\Delta t}{2\mu_{i-1/2,j+1,k+1}}}\right) H_x\big|^{n}_{i-1/2,j+1,k+1} + \left(\frac{\frac{\Delta t}{\mu_{i-1/2,j+1,k+1}}}{1+\frac{\sigma^*_{i-1/2,j+1,k+1}\Delta t}{2\mu_{i-1/2,j+1,k+1}}}\right) \times \left(\begin{array}{l}\frac{E_y|^{n+1/2}_{i-1/2,j+1,k+3/2}-E_y|^{n+1/2}_{i-1/2,j+1,k+1/2}}{\Delta z} \\ -\frac{E_z|^{n+1/2}_{i-1/2,j+3/2,k+1}-E_z|^{n+1/2}_{i-1/2,j+1/2,k+1}}{\Delta y} \\ -M_{\text{source}_x}\big|^{n+1/2}_{i-1/2,j+1,k+1}\end{array}\right). \quad (9.20a)$$

$$H_y\big|^{n+1}_{i,j+1/2,k+1} = \left(\frac{1-\frac{\sigma^*_{i,j+1/2,k+1}\Delta t}{2\mu_{i,j+1/2,k+1}}}{1+\frac{\sigma^*_{i,j+1/2,k+1}\Delta t}{2\mu_{i,j+1/2,k+1}}}\right) H_y\big|^{n}_{i,j+1/2,k+1} + \left(\frac{\frac{\Delta t}{\mu_{i,j+1/2,k+1}}}{1+\frac{\sigma^*_{i,j+1/2,k+1}\Delta t}{2\mu_{i,j+1/2,k+1}}}\right) \times \left(\begin{array}{l}\frac{E_z|^{n+1/2}_{i+1/2,j+1/2,k+1}-E_z|^{n+1/2}_{i-1/2,j+1/2,k+1}}{\Delta x} \\ -\frac{E_x|^{n+1/2}_{i,j+1/2,k+3/2}-E_x|^{n+1/2}_{i,j+1/2,k+1/2}}{\Delta z} \\ -M_{\text{source}_y}\big|^{n+1/2}_{i,j+1/2,k+1}\end{array}\right). \quad (9.20b)$$

$$H_z\big|^{n+1}_{i,j+1,k+1/2} = \left(\frac{1-\frac{\sigma^*_{i,j+1,k+1/2}\Delta t}{2\mu_{i,j+1,k+1/2}}}{1+\frac{\sigma^*_{i,j+1,k+1/2}\Delta t}{2\mu_{i,j+1,k+1/2}}}\right) H_z\big|^{n}_{i,j+1,k+1/2} + \left(\frac{\frac{\Delta t}{\mu_{i,j+1,k+1/2}}}{1+\frac{\sigma^*_{i,j+1,k+1/2}\Delta t}{2\mu_{i,j+1,k+1/2}}}\right) \times \left(\begin{array}{l}\frac{E_x|^{n+1/2}_{i,j+3/2,k+1/2}-E_x|^{n+1/2}_{i,j+1/2,k+1/2}}{\Delta y} \\ -\frac{E_y|^{n+1/2}_{i+1/2,j+1,k+1/2}-E_y|^{n+1/2}_{i-1/2,j+1,k+1/2}}{\Delta x} \\ -M_{\text{source}_z}\big|^{n+1/2}_{i,j+1,k+1/2}\end{array}\right). \quad (9.20c)$$

With the systems of finite-difference expressions of equations 9.19 and 9.20, the new value of an electromagnetic field vector component at any space lattice point depends only on its previous value, the previous values of the components of the other field vector at adjacent points, and the known electric and magnetic current sources. Therefore, at any given time step, the computation of a field vector can proceed either one point at a time, or, if p parallel processors are employed concurrently, p points at a time.

9.3.4 Reduction to the Two-Dimensional TM$_z$ and TE$_z$ Modes

The finite-difference systems of equations 9.19 and 9.20 can be reduced for the two-dimensional TM$_z$ and TE$_z$ modes of

Subsection 9.2.2. For consistency, we again consider the field vector components of the unit cell of Figure 9.1. Assuming now that all partial derivatives of the fields with respect to z are equal to zero, the following conditions hold:

1. The sets of (E_z, H_x, H_y) components located in each lattice-cut plane k, $k+1$, and so on, are identical and can be completely represented by any one of these sets, which we designate as the **TM$_z$ mode.**
2. The sets of (H_z, E_x, E_y) components located in each lattice-cut plane $k+1/2$, $k+3/2$, and so on, are identical and can be completely represented by any one of these sets, which we designate as the **TE$_z$ mode.**

The resulting finite-difference systems for the TM$_z$ and TE$_z$ modes are as follows:

TM$_z$ Mode, Corresponding to the System of Equation 9.11

$$H_x\Big|_{i-1/2,j+1}^{n+1} = \left(\frac{1-\frac{\sigma^*_{i-1/2,j+1}\Delta t}{2\mu_{i-1/2,j+1}}}{1+\frac{\sigma^*_{i-1/2,j+1}\Delta t}{2\mu_{i-1/2,j+1}}}\right) H_x\Big|_{i-1/2,j+1}^{n} + \left(\frac{\frac{\Delta t}{\mu_{i-1/2,j+1}}}{1+\frac{\sigma^*_{i-1/2,j+1}\Delta t}{2\mu_{i-1/2,j+1}}}\right) \times \left(\frac{E_z\big|_{i-1/2,j+1/2}^{n+1/2} - E_z\big|_{i-1/2,j+3/2}^{n+1/2}}{\Delta y} - M_{\text{source}_x}\Big|_{i-1/2,j+1}^{n+1/2}\right). \tag{9.21a}$$

$$H_y\Big|_{i,j+1/2}^{n+1} = \left(\frac{1-\frac{\sigma^*_{i,j+1/2}\Delta t}{2\mu_{i,j+1/2}}}{1+\frac{\sigma^*_{i,j+1/2}\Delta t}{2\mu_{i,j+1/2}}}\right) H_y\Big|_{i,j+1/2}^{n} + \left(\frac{\frac{\Delta t}{\mu_{i,j+1/2}}}{1+\frac{\sigma^*_{i,j+1/2}\Delta t}{2\mu_{i,j+1/2}}}\right) \times \left(\frac{E_z\big|_{i+1/2,j+1/2}^{n+1/2} - E_z\big|_{i-1/2,j+1/2}^{n+1/2}}{\Delta x} - M_{\text{source}_y}\Big|_{i,j+1/2}^{n+1/2}\right). \tag{9.21b}$$

$$E_z\Big|_{i-1/2,j+1/2}^{n+1/2} = \left(\frac{1-\frac{\sigma_{i-1/2,j+1/2}\Delta t}{2\varepsilon_{i-1/2,j+1/2}}}{1+\frac{\sigma_{i-1/2,j+1/2}\Delta t}{2\varepsilon_{i-1/2,j+1/2}}}\right) E_z\Big|_{i-1/2,j+1/2}^{n-1/2} + \left(\frac{\frac{\Delta t}{\varepsilon_{i-1/2,j+1/2}}}{1+\frac{\sigma_{i-1/2,j+1/2}\Delta t}{2\varepsilon_{i-1/2,j+1/2}}}\right) \times \left(\frac{H_y\big|_{i,j+1/2}^{n} - H_y\big|_{i-1,j+1/2}^{n}}{\Delta x} + \frac{H_x\big|_{i-1/2,j}^{n} - H_x\big|_{i-1/2,j+1}^{n}}{\Delta y} - J_{\text{source}_z}\Big|_{i-1/2,j+1/2}^{n}\right). \tag{9.21c}$$

TE$_z$ Mode, Corresponding to the System of Equation 9.12

$$E_x\Big|_{i,j+1/2}^{n+1/2} = \left(\frac{1-\frac{\sigma_{i,j+1/2}\Delta t}{2\varepsilon_{i,j+1/2}}}{1+\frac{\sigma_{i,j+1/2}\Delta t}{2\varepsilon_{i,j+1/2}}}\right) E_x\Big|_{i,j+1/2}^{n-1/2} + \left(\frac{\frac{\Delta t}{\varepsilon_{i,j+1/2}}}{1+\frac{\sigma_{i,j+1/2}\Delta t}{2\varepsilon_{i,j+1/2}}}\right) \times \left(\frac{H_z\big|_{i,j+1}^{n} - H_z\big|_{i,j}^{n}}{\Delta y} - J_{\text{source}_x}\Big|_{i,j+1/2}^{n}\right). \tag{9.22a}$$

$$E_y\Big|_{i-1/2,j+1}^{n+1/2} = \left(\frac{1-\frac{\sigma_{i-1/2,j+1}\Delta t}{2\varepsilon_{i-1/2,j+1}}}{1+\frac{\sigma_{i-1/2,j+1}\Delta t}{2\varepsilon_{i-1/2,j+1}}}\right) E_y\Big|_{i-1/2,j+1}^{n-1/2} + \left(\frac{\frac{\Delta t}{\varepsilon_{i-1/2,j+1}}}{1+\frac{\sigma_{i-1/2,j+1}\Delta t}{2\varepsilon_{i-1/2,j+1}}}\right) \times \left(\frac{H_z\big|_{i-1,j+1}^{n} - H_z\big|_{i,j+1}^{n}}{\Delta x} - J_{\text{source}_y}\Big|_{i-1/2,j+1}^{n}\right). \tag{9.22b}$$

$$H_z\Big|_{i,j+1}^{n+1} = \left(\frac{1-\frac{\sigma^*_{i,j+1}\Delta t}{2\mu_{i,j+1}}}{1+\frac{\sigma^*_{i,j+1}\Delta t}{2\mu_{i,j+1}}}\right) H_z\Big|_{i,j+1}^{n} + \left(\frac{\frac{\Delta t}{\mu_{i,j+1}}}{1+\frac{\sigma^*_{i,j+1}\Delta t}{2\mu_{i,j+1}}}\right) \times \left(\frac{E_x\big|_{i,j+3/2}^{n+1/2} - E_x\big|_{i,j+1/2}^{n+1/2}}{\Delta y} + \frac{E_y\big|_{i-1/2,j+1}^{n+1/2} - E_y\big|_{i+1/2,j+1}^{n+1/2}}{\Delta x} - M_{\text{source}_z}\Big|_{i,j+1}^{n+1/2}\right) \tag{9.22c}$$

9.4 Numerical Dispersion

9.4.1 Introduction

The FDTD algorithm for Maxwell's curl equations reviewed in Section 9.3 causes nonphysical results such as dispersion of the simulated waves in a free space computational lattice. That is, the phase velocity of numerical wave modes can differ from c by an amount varying with the wavelength, direction of propagation in the grid, and grid discretization. This artifact causes propagating numerical waves to accumulate delay or phase errors that can lead to nonphysical results, such as broadening and ringing of pulsed waveforms, imprecise cancellation of multiple scattered waves, anisotropy, and pseudorefraction. Numerical dispersion is a factor that must be accounted to understand the operation of FDTD algorithms and their accuracy limits, especially for electrically large structures.

9.4.2 Two-Dimensional Wave Propagation

We begin our discussion of **numerical dispersion** with an analysis of the two-dimensional TM$_z$ mode, assuming for simplicity no electric or magnetic loss. (It can easily be shown that the same results are obtained for the TE$_z$ mode.) The analysis procedure involves substitution of a plane, monochromatic, and sinusoidal traveling-wave mode into equations 9.21a through 9.21c. After algebraic manipulation, an equation is derived that relates the numerical wavevector components, the wave frequency, the time step, and the grid space increments. This equation, the numerical dispersion relation, can be solved for a variety of grid discretizations, wavevectors, and wave frequencies to illustrate the principal nonphysical results associated with numerical dispersion.

Initiating this procedure, we assume the following plane, monochromatic, and sinusoidal traveling wave for the TM$_z$ mode:

$$E_z|_{I,J}^n = E_{z_0} e^{j(\omega n\Delta t - \tilde{k}_x I\Delta x - \tilde{k}_y J\Delta y)}. \quad (9.23a)$$

$$H_x|_{I,J}^n = H_{x_0} e^{j(\omega n\Delta t - \tilde{k}_x I\Delta x - \tilde{k}_y J\Delta y)}. \quad (9.23b)$$

$$H_y|_{I,J}^n = H_{y_0} e^{j(\omega n\Delta t - \tilde{k}_x I\Delta x - \tilde{k}_y J\Delta y)}. \quad (9.23c)$$

The $\tilde{k}_x$ and $\tilde{k}_y$ in these equations are the x and y components of the numerical wavevector, and ω is the wave angular frequency. Substituting the traveling-wave expressions of equations 9.23 into the finite-difference equations of 9.21 yields, after simplification, the following relations for the lossless material case:

$$H_{x_0} = \frac{\Delta t E_{z_0}}{\mu \Delta y} \cdot \frac{\sin(\tilde{k}_y \Delta y/2)}{\sin(\omega \Delta t/2)}. \quad (9.24a)$$

$$H_{y_0} = -\frac{\Delta t E_{z_0}}{\mu \Delta x} \cdot \frac{\sin(\tilde{k}_x \Delta x/2)}{\sin(\omega \Delta t/2)}. \quad (9.24b)$$

$$E_{z_0} \sin\left(\frac{\omega \Delta t}{2}\right) = \frac{\Delta t}{\varepsilon}\left[\frac{H_{x_0}}{\Delta y}\sin\left(\frac{\tilde{k}_y \Delta y}{2}\right) - \frac{H_{y_0}}{\Delta x}\sin\left(\frac{\tilde{k}_x \Delta x}{2}\right)\right]. \quad (9.24c)$$

Upon substituting H_{x_0} of equation 9.24a and H_{y_0} of equation 9.24b into 9.24c we obtain:

$$\left[\frac{1}{c\Delta t}\sin\left(\frac{\omega \Delta t}{2}\right)\right]^2 = \left[\frac{1}{\Delta x}\sin\left(\frac{\tilde{k}_x \Delta x}{2}\right)\right]^2 + \left[\frac{1}{\Delta y}\sin\left(\frac{\tilde{k}_y \Delta y}{2}\right)\right]^2, \quad (9.25)$$

where $c = 1/\sqrt{\mu\varepsilon}$ is the speed of light in the material being modeled. Equation 9.25 is the general numerical dispersion relation of the Yee (1966) algorithm for the TM_z mode.

We shall consider the important special case of a square-cell grid having $\Delta x = \Delta y = \Delta$. Then, defining the **Courant stability factor** $S = c\Delta t/\Delta$ and the **grid sampling density** $N_\lambda = \lambda_0/\Delta$, we rewrite equation 9.25 in a more useful form:

$$\frac{1}{S^2}\sin^2\left(\frac{\pi S}{N_\lambda}\right) = \sin^2\left(\frac{\Delta \cdot \tilde{k}\cos\phi}{2}\right) + \sin^2\left(\frac{\Delta \cdot \tilde{k}\sin\phi}{2}\right), \quad (9.26)$$

where ϕ is the propagation direction of the numerical wave with respect to the grid's x-axis. To obtain the numerical dispersion relation for the one-dimensional wave-propagation case, we can assume without loss of generality that $\phi = 0$ in equation 9.26, yielding:

$$\frac{1}{S}\sin\left(\frac{\pi S}{N_\lambda}\right) = \sin\left(\frac{\tilde{k}\Delta}{2}\right), \quad (9.27)$$

or equivalently:

$$\tilde{k} = \frac{2}{\Delta}\sin^{-1}\left[\frac{1}{S}\sin\left(\frac{\pi S}{N_\lambda}\right)\right]. \quad (9.28)$$

9.4.3 Extension to Three Dimensions: Cartesian Yee Lattice

The dispersion analysis presented in the previous subsection is now extended to the full three-dimensional case, following the analysis presented by Taflove and Brodwin (1975a). We consider a normalized, lossless region of space with $\mu = 1$, $\varepsilon = 1$, $\sigma = 0$, $\sigma^* = 0$, and $c = 1$. Letting $j = \sqrt{-1}$, we rewrite Maxwell's equations in compact form as:

$$j\nabla \times (\vec{H} + j\vec{E}) = \frac{\partial}{\partial t}(\vec{H} + j\vec{E}), \quad (9.29a)$$

or more simply as:

$$j\nabla \times \vec{V} = \frac{\partial \vec{V}}{\partial t}. \quad (9.29b)$$

In these equations, $\vec{V} = \vec{H} + j\vec{E}$. Substituting the vector–field traveling-wave expression:

$$\vec{V}\Big|_{I,J,K}^n = \vec{V}_0 e^{j(\omega n\Delta t - \tilde{k}_x I\Delta x - \tilde{k}_y J\Delta y - \tilde{k}_z K\Delta z)} \quad (9.30)$$

into the Yee space-time central-differencing realization of equation 9.29b, we obtain:

$$\left[\frac{\hat{x}}{\Delta x}\sin\left(\frac{\tilde{k}_x \Delta x}{2}\right) + \frac{\hat{y}}{\Delta y}\sin\left(\frac{\tilde{k}_y \Delta y}{2}\right) + \frac{\hat{z}}{\Delta z}\sin\left(\frac{\tilde{k}_z \Delta z}{2}\right)\right] \times \vec{V}\Big|_{I,J,K}^n = \frac{-j}{\Delta t}\vec{V}\Big|_{I,J,K}^n \sin\left(\frac{\omega \Delta t}{2}\right), \quad (9.31)$$

where $\hat{x}$, $\hat{y}$, and $\hat{z}$ are unit vectors in the x, y, and z coordinate directions. After performing the vector cross product in equation 9.31 and writing out the x, y, and z vector component equations, we obtain a homogeneous system (zero right-hand side) of three equations in the unknowns V_x, V_y, and V_z. Setting the determinant of this system equal to zero results in:

$$\left[\frac{1}{\Delta t}\sin\left(\frac{\omega \Delta t}{2}\right)\right]^2 = \left[\frac{1}{\Delta x}\sin\left(\frac{\tilde{k}_x \Delta x}{2}\right)\right]^2 + \left[\frac{1}{\Delta y}\sin\left(\frac{\tilde{k}_y \Delta y}{2}\right)\right]^2 + \left[\frac{1}{\Delta z}\sin\left(\frac{\tilde{k}_z \Delta z}{2}\right)\right]^2. \quad (9.32)$$

Finally, we denormalize to a nonunity c and obtain the general form of the numerical dispersion relation for the full vector–field Yee algorithm in three dimensions:

$$\left[\frac{1}{c\Delta t}\sin\left(\frac{\omega \Delta t}{2}\right)\right]^2 = \left[\frac{1}{\Delta x}\sin\left(\frac{\tilde{k}_x \Delta x}{2}\right)\right]^2 + \left[\frac{1}{\Delta y}\sin\left(\frac{\tilde{k}_y \Delta y}{2}\right)\right]^2 + \left[\frac{1}{\Delta z}\sin\left(\frac{\tilde{k}_z \Delta z}{2}\right)\right]^2. \quad (9.33)$$

This equation can reduce to equation 9.25, the numerical dispersion relation for the two-dimensional TM_z mode, simply by letting $\tilde{k}_z = 0$.

9.4.4 Comparison with the Ideal Dispersion Case

In contrast to equation 9.33, the **analytical (ideal) dispersion relation** for a physical plane wave propagating in three dimensions in a homogeneous lossless medium is simply:

$$\left(\frac{\omega}{c}\right)^2 = (k_x)^2 + (k_y)^2 + (k_z)^2. \tag{9.34}$$

Although at first glance equation 9.33 bears little resemblance to the ideal case of equation 9.34, we can easily show that the two dispersion relations are identical in the limit as Δx, Δy, Δz, and Δt approach zero. Qualitatively, this suggests that numerical dispersion can be reduced to any degree that is desired if we only use fine enough FDTD gridding.

It can also be shown that equation 9.33 reduces to equation 9.34 if the Courant factor S and the wave-propagation direction are suitably chosen. For example, reduction to the ideal dispersion case can be demonstrated for a numerical plane wave propagating along a diagonal of a three-dimensional cubic lattice ($\tilde{k}_x = \tilde{k}_y = \tilde{k}_z = \tilde{k}/\sqrt{3}$) if $S = 1/\sqrt{3}$. Similarly, ideal dispersion results for a numerical plane wave propagating along a diagonal of a two-dimensional square grid ($\tilde{k}_x = \tilde{k}_y = \tilde{k}/\sqrt{2}$) if $S = 1/\sqrt{2}$. Finally, ideal dispersion results for any numerical wave in a one-dimensional grid if $S = 1$. These reductions to the ideal case have little practical value for two- and three-dimensional simulations, occurring only for diagonal propagation. However, the reduction to ideal dispersion in one dimension is very interesting, since it implies that the Yee algorithm (based on numerical finite-difference approximations) yields an *exact* solution for wave propagation.

9.4.5 Anisotropy of the Numerical Phase Velocity

This section probes a key implication of numerical dispersion relations of equations 9.26 and 9.33. Namely, numerical waves in a two- or three-dimensional Yee space lattice have a propagation velocity that is dependent on the direction of wave propagation. The space lattice thus represents an anisotropic medium.

Our strategy in developing an understanding of this phenomenon is to first calculate sample values of the numerical phase velocity $\tilde{v}_p$ versus wave-propagation direction ϕ to estimate the magnitude of the problem. Then, we will conduct an appropriate analysis to examine the issue more deeply.

Sample Values of Numerical Phase Velocity

We start with the simplest possible situation where numerical phase-velocity anisotropy arises: two-dimensional TM_z modes propagating in a square-cell grid. Dispersion relation 9.26 can be solved directly for $\tilde{k}$ for propagation along the major axes of the grid: $\phi = 0°$, 90°, 180°, and 270°. For this case, the solution for $\tilde{k}$ is given by equation 9.28, which is repeated here for convenience:

$$\tilde{k} = \frac{2}{\Delta}\sin^{-1}\left[\frac{1}{S}\sin\left(\frac{\pi S}{N_\lambda}\right)\right]. \quad \text{(propagation along major grid axes)} \tag{9.35a}$$

The corresponding numerical phase velocity is given by:

$$\tilde{v}_p = \frac{\omega}{\tilde{k}} = \frac{\pi}{N_\lambda \sin^{-1}\left[\frac{1}{S}\sin\left(\frac{\pi S}{N_\lambda}\right)\right]} c. \quad \text{(propagation along major grid axes)} \tag{9.35b}$$

Dispersion relation 9.27 can also be solved directly for $\tilde{k}$ for propagation along the diagonals of the grid $\phi = 45°$, 135°, 225°, and 315°, yielding:

$$\tilde{k} = \frac{2\sqrt{2}}{\Delta}\sin^{-1}\left[\frac{1}{S\sqrt{2}}\sin\left(\frac{\pi S}{N_\lambda}\right)\right]. \quad \text{(propagation along grid diagonals)} \tag{9.36a}$$

$$\tilde{v}_p = \frac{\pi}{N_\lambda \sqrt{2}\sin^{-1}\left[\frac{1}{S\sqrt{2}}\sin\left(\frac{\pi S}{N_\lambda}\right)\right]} c. \quad \text{(propagation along grid diagonals)} \tag{9.36b}$$

As an example, assume a grid having $S = 0.5$ and $N_\lambda = 20$. Then equation 9.35b and 9.36b provide unequal $\tilde{v}_p$ values of $0.996892c$ and $0.998968c$, respectively. The implication is that a sinusoidal numerical wave propagating obliquely in this grid has a speed that is $0.998968/0.996892 = 1.00208$ times that of a wave propagating along the major grid axes. This represents a velocity anisotropy of about 0.2% between oblique and along-axis numerical wave propagation.

Taflove and Hagness (2000) demonstrated that this theoretical anisotropy of the numerical phase velocity appears in FDTD simulations. Figure 9.3 presents their modeling results for a radially outward-propagating sinusoidal cylindrical wave in a two-dimensional TM_z grid. The Taflove and Hagness, grid was configured with 360 × 360 square cells with $\Delta x = \Delta y = \Delta = 1.0$. A unity-amplitude sinusoidal excitation was provided to a single E_z component at the center of the grid. Choosing a grid-sampling density of $N_\lambda = 20$ and a Courant factor $S = 0.5$ permitted direct comparison of the FDTD modeling results with the theoretical results for the anisotropy of $\tilde{v}_p$, discussed immediately above.

Figure 9.3(A) illustrates snapshots of the E_z field distribution versus radial distance from the source at the center of the grid. Here, field observations are made along cuts through the grid passing through the source point and are either parallel to the principal grid axes $\phi = 0°$ and 90° or parallel to the grid diagonal $\phi = 45°$. (Note that, by the 90° rotational symmetry of the Cartesian grid geometry, identical field distributions are obtained along $\phi = 0°$ and $\phi = 90°$.) The snapshots are taken $328\Delta t$ after the beginning of time-stepping. At this time, the

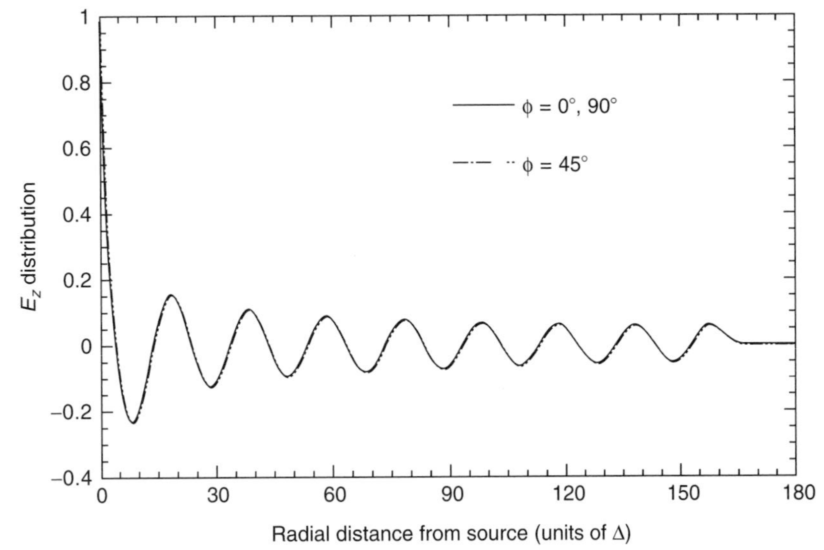

(A) Comparison of Calculated Wave Propagation Along the Grid Axes and Along a Grid Diagonal.

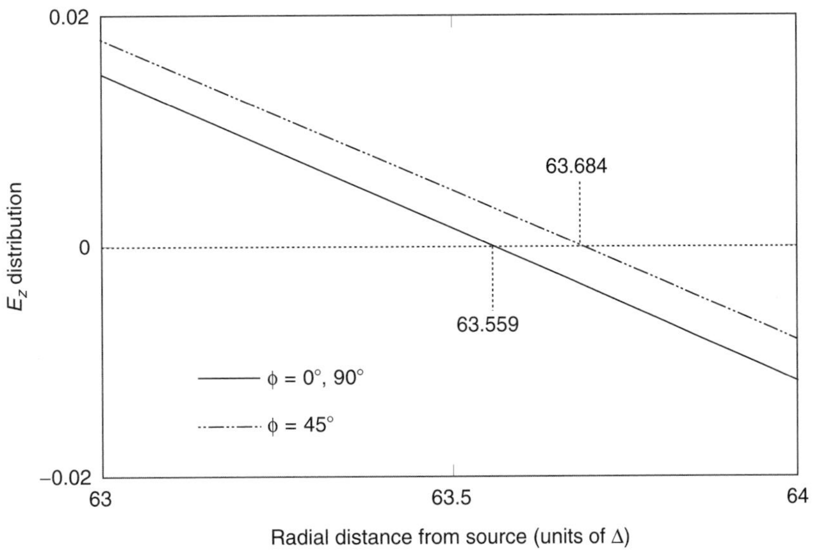

(B) Expanded View of (A) at Distances Between 63 and 64 Grid Cells from the Source.

FIGURE 9.3 Effect of Numerical Dispersion on a Radially Propagating Cylindrical Wave in a 2-D TM Yee Grid. The grid is excited at its center by applying a unity-amplitude sinusoidal time function to a single E_z field component. The value of $S = 0.5$, and the grid sampling density is $N_\lambda = 20$.

wave has not yet reached the outer grid boundary, and the calculated E_z field distribution is free of error due to outer-boundary reflections.

Figure 9.3(B) is an expanded view of Figure 9.3(A) at radial distances between 63Δ and 64Δ from the source. This enables evaluation (with three-decimal-place precision) of the locations of the zero-crossings of the E_z distributions along the two observation cuts through the grid. From the data shown in Figure 9.3(B) the sinusoidal wave along the $\phi = 45^\circ$ cut passes through zero at 63.684Δ cells, whereas the wave along the $\phi = 0^\circ, 90^\circ$ cut passes through zero at 63.559Δ cells. Taking the difference, we see that the obliquely propagating wave "leads" the on-axis wave by 0.125Δ cells. This yields a numerical phase-velocity anisotropy $\Delta\tilde{v}_p/\tilde{v}_p \cong 0.125/63.6 = 0.197\%$. This number is only about 5% less than the 0.208% value obtained using equations 9.35b and 9.36b.

To permit determination of $\tilde{k}$ and $\tilde{v}_p$ for any wave-propagation direction ϕ, it would be very useful to derive closed-form equations analogous to equations 9.35 and 9.36. However, for this general case, the underlying dispersion relation 9.26 is a transcendental equation. Taflove (1995) provided a useful alternative approach for obtaining sample values of $\tilde{v}_p$ by applying the following Newton's method iterative procedure to equation 9.26:

$$\tilde{k}_{\text{icount}+1} = \tilde{k}_{\text{icount}} - \frac{\sin^2(A\tilde{k}_{\text{icount}}) + \sin^2(B\tilde{k}_{\text{icount}}) - C}{A\sin(2A\tilde{k}_{\text{icount}}) + B\sin(2B\tilde{k}_{\text{icount}})}. \quad (9.37)$$

Here, $\tilde{k}_{\text{icount}+1}$ is the improved estimate of $\tilde{k}$, and $\tilde{k}_{\text{icount}}$ is the previous estimate of $\tilde{k}$. The A, B, and C are coefficients given by:

$$A = \frac{\Delta \cdot \cos\phi}{2}, \quad B = \frac{\Delta \cdot \sin\phi}{2}, \quad C = \frac{1}{S^2}\sin^2\left(\frac{\pi S}{N_\lambda}\right). \quad (9.38)$$

Additional simplicity results if Δ is normalized to the free space wavelength, λ_0. This is equivalent to setting $\lambda_0 = 1$. Then, a very good starting guess for the iterative process is simply 2π. For this case, $\tilde{v}_p$ is given by:

$$\frac{\tilde{v}_p}{c} = \frac{2\pi}{\tilde{k}_{\text{final icount}}}. \quad (9.39)$$

Usually, only two or three iterations are required for convergence.

Figure 9.4 graphs results obtained using this procedure that illustrate the variation of $\tilde{v}_p$ with propagation direction ϕ.

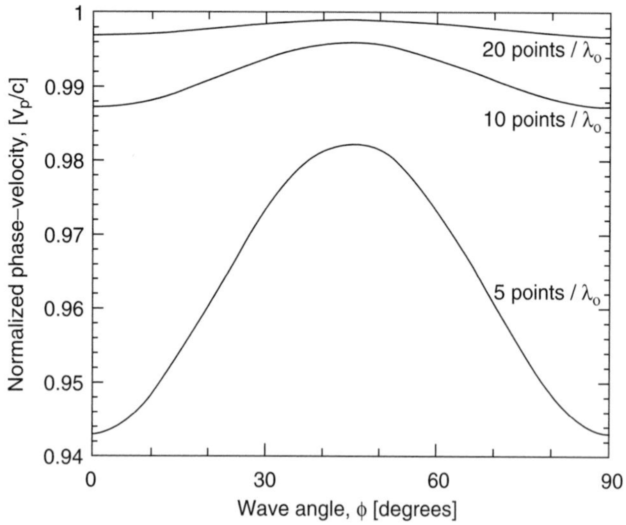

FIGURE 9.4 Variation of the Numerical Phase–Velocity with Wave Propagation Angle. The angle is in a 2-D FDTD grid for three sampling densities of the square unit cells. For all cases, $S = c\Delta t/\Delta = 0.5$.

Here, for the Courant factor fixed at $S = 0.5$, three different grid sampling densities N_λ are examined: $N_\lambda = 5$ points per λ_0, $N_\lambda = 10$, and $N_\lambda = 20$. We see that $\tilde{v}_p < c$ and is a function of both ϕ and N_λ. The $\tilde{v}_p$ is maximum for waves propagating obliquely within the grid ($\phi = 45°$) and is minimum for waves propagating along either grid axis ($\phi = 0°, 90°$).

It is useful to summarize the algorithmic dispersive-error performance by defining two normalized error measures: (1) the physical phase–velocity error $\Delta\tilde{v}_{\text{physical}}$, and (2) the velocity–anisotropy error $\Delta\tilde{v}_{\text{aniso}}$. These are given by:

$$\Delta\tilde{v}_{\text{physical}}\big|_{N_\lambda} = \frac{\min\left[\tilde{v}_p(\phi)\right] - c}{c} \times 100\%. \quad (9.40)$$

$$\Delta\tilde{v}_{\text{aniso}}\big|_{N_\lambda} = \frac{\max\left[\tilde{v}_p(\phi)\right] - \min\left[\tilde{v}_p(\phi)\right]}{\min\left[\tilde{v}_p(\phi)\right]} \times 100\%. \quad (9.41)$$

The $\Delta\tilde{v}_{\text{physical}}$ is useful in quantifying the phase lead or lag that numerical modes suffer relative to physical modes propagating at c. For example, from Figure 9.4 and equation 9.35b, $\Delta\tilde{v}_{\text{physical}} = -0.31\%$ for $N_\lambda = 20$. This means that a sinusoidal numerical wave traveling over a $10\lambda_0$ distance in the grid (200 cells) could develop a lagging phase error up to 11°. We note that $\Delta\tilde{v}_{\text{physical}}$ is a function of N_λ. Because the grid cell size Δ is fixed, for an impulsive wave-propagation problem there exists a spread of effective N_λ values for the spectral components comprising the pulse. This causes a spread of $\Delta\tilde{v}_{\text{physical}}$ over the pulse spectrum, which in turn yields a temporal dispersion of the pulse evidenced in the spreading and distortion of its waveform as it propagates.

The $\Delta\tilde{v}_{\text{aniso}}$ is useful in quantifying wavefront distortion. For example, a circular cylindrical wave would suffer progressive distortion of its wavefront since the portions propagating along the grid diagonals would travel slightly faster than the portions traveling along the major grid axes. For example, from Figure 9.4 and equations 9.35b and 9.36b, $\Delta\tilde{v}_{\text{aniso}} = 0.208\%$ for $N_\lambda = 20$. The wavefront distortion due to this anisotropy would total about 2.1 cells for each 1,000 cells of propagation distance.

It is clear that errors due to inaccurate numerical velocities are cumulative (i.e., they increase linearly with the wave-propagation distance). These errors represent a fundamental limitation of *all* grid-based Maxwell's equations' algorithms and can be troublesome when modeling electrically large structures. A positive aspect seen in Figure 9.4 is that both $\Delta\tilde{v}_{\text{physical}}$ and $\Delta\tilde{v}_{\text{aniso}}$ decrease by approximately a 4:1 factor each time the grid-sampling density doubles, indicative of the second-order accuracy of the Yee algorithm. Therefore, finer meshing is one way to control the dispersion error.

As discussed in Taflove and Hagness (2000), there are proposed means to improve the accuracy of FDTD algorithms to allow much larger structures to be modeled. Specifically, using pseudo-spectral techniques (Y. Liu, 1996, 1997), $\Delta\tilde{v}_{\text{aniso}}$ can be reduced to very low levels approaching zero. In this case, residual errors involve primarily the dispersion of $\Delta\tilde{v}_{physical}$ with N_λ, which can be optimized by the proper choice of Δt. The new approaches presently have limitations regarding their ability to model material discontinuities, however, and require more research.

Intrinsic Grid Velocity Anisotropy

Following Taflove and Hagness (2000), this section provides a deeper discussion of the numerical phase–velocity errors of the Yee algorithm. We will find that the nature of the grid discretization, in a manner virtually independent of the time-stepping scheme, determines the velocity anisotropy $\Delta\tilde{v}_{aniso}$.

Relation of the Time and Space Discretizations in Generating Numerical Velocity Error

In the previous section we determined that $\Delta\tilde{v}_{aniso} = 0.208\%$ for a two-dimensional Yee algorithm having $N_\lambda = 20$ and $S = 0.5$. An important and revealing question is how is $\Delta\tilde{v}_{aniso}$ affected by the choice of S, assuming that N_λ is fixed at 20?

To begin to answer this question, we first choose (what will later be shown to be) the largest possible value of S for numerical stability in two dimensions, $S = 1/\sqrt{2}$. Substituting this value of S into equations 9.35b and 9.36b yields:

$$\left.\begin{array}{l}\tilde{v}_p(\phi = 0^\circ) = 0.997926\,c\\ \tilde{v}_p(\phi = 45^\circ) = c\end{array}\right\}\Delta\tilde{v}_{aniso} = \frac{c - 0.997926\,c}{0.997926\,c} \times 100\% = 0.208\%.$$

To three decimal places, there is no change in $\Delta\tilde{v}_{aniso}$ from the previous value, $S = 0.5$. We next choose a very small value $S = 0.01$ for substitution into equations 9.35b and 9.36c:

$$\left.\begin{array}{l}\tilde{v}_p(\phi = 0^\circ) = 0.995859\,c\\ \tilde{v}_p(\phi = 45^\circ) = 0.997937\,c\end{array}\right\}\Delta\tilde{v}_{aniso} = \frac{0.997937\,c - 0.995859\,c}{0.995859\,c} \times 100\% = 0.208\%.$$

Again, there is no change in $\Delta\tilde{v}_{aniso}$ to three decimal places.

We now suspect that, for a given N_λ, $\Delta\tilde{v}_{aniso}$ is at most a weak function of S and, therefore, is only weakly dependent on Δt. In fact, this is the case. More generally, Y. Liu, (1996) has shown that $\Delta\tilde{v}_{aniso}$ is only weakly dependent on the specific type of time-marching scheme used, whether leapfrog, Runge-Kutta, or other schemes. Thus, we can say that $\Delta\tilde{v}_{aniso}$ is virtually an intrinsic characteristic of the space–lattice discretization. Following Y. Liu (1996), three key points should be made in this regard:

- Numerical-dispersion errors associated with the time discretization are isotropic relative to the propagation direction of the wave.
- The choice of time discretization has little effect on the phase–velocity anisotropy $\Delta\tilde{v}_{aniso}$ for $N_\lambda > 10$.
- The choice of time discretization does influence $\Delta\tilde{v}_{physical}$. However, it is not always true that higher order time-marching schemes, such as Runge-Kutta, yield less $\Delta\tilde{v}_{physical}$ than simple Yee leapfrogging. Errors in $\Delta\tilde{v}_{physical}$ are caused separately by the space and time discretizations and can either partially reinforce or cancel each other. Thus, the use of fourth-order Runge-Kutta may actually shift the $\tilde{v}_p(\phi)$ profile away from c, representing an increased $\Delta\tilde{v}_{physical}$ relative to ordinary leapfrogging.

Approximate Analytical Expression

Y. Liu (1996) has shown that, to determine the relative velocity anisotropy characteristic intrinsic to a space grid, it is useful to set up an eigenvalue problem for the matrix that delineates the spatial derivatives used in the numerical algorithm. Consider as an example the case of two-dimensional FDTD modeling of electromagnetic wave propagation. Following the development in Taflove and Hagness (2000), it can be shown that Y. Liu's (1996) technique leads to a simple approximate analytical expression for $\Delta\tilde{v}_{aniso}$ that is useful for $N_\lambda > 10$:

$$\Delta\tilde{v}_{aniso} \cong \frac{\pi^2}{12(N_\lambda)^2} \times 100\%. \tag{9.42}$$

For example, equation 9.42 provides $\Delta\tilde{v}_{aniso} \cong 0.206\%$ for $N_\lambda = 20$. This is very close to the 0.208% value previously obtained using equations 9.35b and 9.36b, the exact solutions of the full numerical dispersion relation for $\phi = 0^\circ$ and $\phi = 45^\circ$, respectively.

9.4.6 Complex-Valued Numerical Wave Numbers

Schneider and Wagner (1999) found that the Yee algorithm has a low sampling-density regime that allows complex-valued numerical wavenumbers. In this regime, spatially decaying numerical waves can propagate faster than light, causing a weak, nonphysical signal to appear ahead of the nominal leading edges of sharply defined pulses. This subsection reviews the theory underlying this phenomenon as related in Taflove and Hagness (2000).

Case 1: Numerical Wave Propagation Along the Principal Lattice Axes

Consider again numerical wave propagation along the major axes of a Yee space grid. For convenience, we rewrite equation 9.35a, the corresponding numerical dispersion relation:

$$\tilde{k} = \frac{2}{\Delta}\sin^{-1}\left[\frac{1}{S}\sin\left(\frac{\pi S}{N_\lambda}\right)\right] \equiv \frac{2}{\Delta}\sin^{-1}(\zeta), \tag{9.43}$$

where:

$$\zeta = \frac{1}{S}\sin\left(\frac{\pi S}{N_\lambda}\right). \tag{9.44}$$

Schneider and Wagner (1999) realized that, in evaluating numerical dispersion relations such as equation 9.43, it is possible to choose S and N_λ such that $\tilde{k}$ is complex. In the case of this equation, it can be shown that the transition between real and complex values of $\tilde{k}$ occurs when $\zeta = 1$. Solving for N_λ at this transition results in:

$$N_\lambda\big|_{\text{transition}} = \frac{\pi S}{\sin^{-1}(S)}. \quad (9.45)$$

For a grid sampling density greater than this value (i.e., $N_\lambda > N_\lambda|_{\text{transition}}$, $\tilde{k}$ is a real number, and the numerical wave undergoes no attenuation while propagating in the grid). Here, $\tilde{v}_p < c$. For a coarser grid-sampling density $N_\lambda < N_\lambda|_{\text{transition}}$, $\tilde{k}$ is a complex number, and the numerical wave undergoes a nonphysical exponential decay while propagating. Further, in this coarse-resolution regime, $\tilde{v}_p$ can exceed c.

We now discuss how $\tilde{k}$ and $\tilde{v}_p$ vary with grid sampling N_λ, both above and below the transition between real and complex numerical wavenumbers.

Real Numerical Wave Number Regime

For $N_\lambda > N_\lambda|_{\text{transition}}$, we have from equation 9.43:

$$\tilde{k}_{\text{real}} = \frac{2}{\Delta}\sin^{-1}\left[\frac{1}{S}\sin\left(\frac{\pi S}{N_\lambda}\right)\right]; \quad \tilde{k}_{\text{imag}} = 0 \quad (9.46\text{a, b})$$

The numerical phase–velocity is given by:

$$\tilde{v}_p = \frac{\omega}{\tilde{k}_{\text{real}}} = \frac{\pi}{N_\lambda \sin^{-1}\left[\frac{1}{S}\sin\left(\frac{\pi S}{N_\lambda}\right)\right]}\,c. \quad (9.47)$$

This is exactly the expression from equation 9.35b. The wave-amplitude multiplier per grid cell of propagation is given by:

$$e^{\tilde{k}_{\text{imag}}\Delta} \equiv e^{-\alpha\Delta} = e^{0} = 1. \quad (9.48)$$

Thus, there is a constant wave amplitude with spatial position for this range of N_λ.

Complex Numerical Wave Number Regime

For $N_\lambda < N_\lambda|_{\text{transition}}$, we observe that $\zeta > 1$ in equation 9.43. Here, the following relation for the complex-valued arc–sine function given by Churchill *et al.* (1976) is useful:

$$\sin^{-1}(\zeta) = -j\ln\left(j\zeta + \sqrt{1-\zeta^2}\right). \quad (9.49)$$

Substituting equation 9.49 into 9.43 yields after some algebraic manipulation:

$$\tilde{k}_{\text{real}} = \frac{\pi}{\Delta}; \quad \tilde{k}_{\text{imag}} = -\frac{2}{\Delta}\ln\left(\zeta + \sqrt{\zeta^2 - 1}\right). \quad (9.50\text{a, b})$$

The numerical phase–velocity is then:

$$\tilde{v}_p = \frac{\omega}{\tilde{k}_{\text{real}}} = \frac{\omega}{(\pi/\Delta)} = \frac{2\pi f\Delta}{\pi} = \frac{2f\lambda_0}{N_\lambda} = \frac{2}{N_\lambda}c. \quad (9.51)$$

The wave-amplitude multiplier per grid cell of propagation is given by:

$$e^{\tilde{k}_{\text{imag}}\Delta} \equiv e^{-\alpha\Delta} = e^{-2\ln\left(\zeta+\sqrt{\zeta^2-1}\right)} = \frac{1}{\left(\zeta + \sqrt{\zeta^2-1}\right)^2}. \quad (9.52)$$

Because $\zeta > 1$, the numerical wave amplitude decays exponentially with spatial position.

We now consider the possibility of $\tilde{v}_p$ exceeding c in this situation. Nyquist theory states that any physical or numerical process that obtains samples of a time waveform every Δt seconds can reproduce the original waveform without aliasing for spectral content up to $f_{\max} = 1/(2\Delta t)$. In the present case, the corresponding minimum free space wavelength that can be sampled without aliasing is therefore:

$$\lambda_{0,\,\min} = c/f_{\max} = 2c\Delta t. \quad (9.53\text{a})$$

The corresponding minimum spatial-sampling density is as follows:

$$N_{\lambda,\,\min} = \lambda_{0,\,\min}/\Delta = 2c\Delta t/\Delta = 2S. \quad (9.53\text{b})$$

Then, from equation 9.51, the maximum numerical phase velocity is given by:

$$\tilde{v}_{p,\,\max} = \frac{2}{N_{\lambda,\,\min}}c = \frac{2}{2S}c = \frac{c}{S}. \quad (9.54\text{a})$$

From the definition of S, this maximum phase velocity can also be expressed as:

$$\tilde{v}_{p,\,\max} = \frac{1}{S}c = \left(\frac{\Delta}{c\Delta t}\right)c = \frac{\Delta}{\Delta t}. \quad (9.54\text{b})$$

This relation tells us that in one Δt, a numerical value can propagate at most one Δ. This is intuitively correct given the local nature of the spatial differences used in the Yee algorithm. That is, a field point more than one Δ away from a source point that undergoes a sudden change cannot possibly "feel" the effect of that change during the next Δt. Note that $\tilde{v}_{p,\,\max}$ is independent of material parameters and is an inherent property of the grid and its method of obtaining space derivatives.

Case 2: Numerical Wave Propagation Along a Grid Diagonal

We next explore the possibility of complex-valued wavenumbers arising for oblique numerical wave propagation in a square-cell grid. For convenience, we rewrite equation 9.36a, the corresponding numerical dispersion relation:

$$\tilde{k} = \frac{2\sqrt{2}}{\Delta}\sin^{-1}\left[\frac{1}{S\sqrt{2}}\sin\left(\frac{\pi S}{N_\lambda}\right)\right] \equiv \frac{2\sqrt{2}}{\Delta}\sin^{-1}(\zeta), \tag{9.55}$$

where:

$$\zeta = \frac{1}{S\sqrt{2}}\sin\left(\frac{\pi S}{N_\lambda}\right), \tag{9.56}$$

Similar to the previous case of numerical wave propagation along the principal lattice axes, it is possible to choose S and N_λ such that $\tilde{k}$ is complex. In the specific case of equation 9.55, the transition between real and complex values of $\tilde{k}$ occurs when $\zeta = 1$. Solving for N_λ at this transition results in:

$$N_\lambda|_{\text{transition}} = \frac{\pi S}{\sin^{-1}(S\sqrt{2})}. \tag{9.57}$$

We now discuss how $\tilde{k}$ and $\tilde{v}_p$ vary with grid sampling N_λ, both above and below the transition between real and complex numerical wavenumbers.

Real Numerical Wave Number Regime

For $N_\lambda \geq N_\lambda|_{\text{transition}}$, we have from equation 9.55:

$$\tilde{k}_{\text{real}} = \frac{2\sqrt{2}}{\Delta}\sin^{-1}\left[\frac{1}{S\sqrt{2}}\sin\left(\frac{\pi S}{N_\lambda}\right)\right]; \quad \tilde{k}_{\text{imag}} = 0. \tag{9.58a, b}$$

The numerical phase velocity is given by:

$$\tilde{v}_p = \frac{\omega}{\tilde{k}_{\text{real}}} = \frac{\pi}{N_\lambda\sqrt{2}\sin^{-1}\left[\frac{1}{S\sqrt{2}}\sin\left(\frac{\pi S}{N_\lambda}\right)\right]}c. \tag{9.59}$$

This is exactly expression 9.36b. The wave-amplitude multiplier per grid cell of propagation is given by:

$$e^{\tilde{k}_{\text{imag}}\Delta} \equiv e^{-\alpha\Delta} = e^0 = 1. \tag{9.60}$$

Thus, there is a constant wave amplitude with spatial position for this range of N_λ.

Complex Numerical Wave Number Regime

For $N_\lambda < N_\lambda|_{\text{transition}}$, we observe that $\zeta > 1$ in equation 9.55. Substituting the complex-valued arc–sine function of equation 9.49 into 9.55 into yields after some algebraic manipulation gives the following:

$$\tilde{k}_{\text{real}} = \frac{\pi\sqrt{2}}{\Delta}; \quad \tilde{k}_{\text{imag}} = -\frac{2\sqrt{2}}{\Delta}\ln\left(\zeta + \sqrt{\zeta^2 - 1}\right). \tag{9.61a, b}$$

The numerical phase velocity for this case is as follows:

$$\tilde{v}_p = \frac{\omega}{\tilde{k}_{\text{real}}} = \frac{\omega}{(\pi\sqrt{2}/\Delta)} = \frac{\sqrt{2}f\lambda_0}{N_\lambda} = \frac{\sqrt{2}}{N_\lambda}c, \tag{9.62}$$

and the wave-amplitude multiplier per grid cell of propagation is as follows:

$$e^{\tilde{k}_{\text{imag}}\Delta} \equiv e^{-\alpha\Delta} = e^{-2\sqrt{2}\ln\left(\zeta+\sqrt{\zeta^2-1}\right)} = \frac{1}{\left(\zeta + \sqrt{\zeta^2 - 1}\right)^{2\sqrt{2}}}. \tag{9.63}$$

Since $\zeta > 1$, the numerical wave amplitude decays exponentially with spatial position.

We again consider the possibility of $\tilde{v}_p$ exceeding c. From our previous discussion of equations 9.53a and 9.53b, the minimum free space wavelength that can be sampled without aliasing is $\lambda_{0,\min} = c/f_{\max} = 2c\Delta t$, and the corresponding minimum spatial-sampling density is $N_{\lambda,\min} = \lambda_{0,\min}/\Delta = 2S$. Then from equation 9.62, the maximum numerical phase velocity is given by:

$$\tilde{v}_{p,\ \max} = \frac{\sqrt{2}}{N_{\lambda,\ \min}}c = \frac{\sqrt{2}}{2S}c. \tag{9.64a}$$

From the definition of S, this maximum phase velocity can also be expressed as:

$$\tilde{v}_{p,\ \max} = \frac{\sqrt{2}}{2}\left(\frac{\Delta}{c\Delta t}\right)c = \frac{\sqrt{2}\Delta}{2\Delta t}. \tag{9.64b}$$

This relation tells us that in $2\Delta t$, a numerical value can propagate at most $\sqrt{2}\Delta$ along the grid diagonal. We can show that this upper bound on $\tilde{v}_p$ is intuitively correct given the local nature of the spatial differences used in the Yee algorithm. Consider two nearest neighbor field points $P_{i,j}$ and $P_{i+1,j+1}$ along a grid diagonal and how a sudden change at $P_{i,j}$ could be communicated to $P_{i+1,j+1}$. Now, a basic principle is that the Yee algorithm can communicate field data only along Cartesian (x and y) grid lines and not along grid diagonals. Thus, at the minimum, $1\Delta t$ would be needed to transfer any part of the field perturbation at $P_{i,j}$ over a distance of 1Δ in the x direction to $P_{i+1,j}$. Then, a second Δt would be needed, at the minimum, to transfer any part of the resulting field perturbation at $P_{i+1,j}$ over a distance of 1Δ in the y direction to reach $P_{i+1,j+1}$. Because the distance between $P_{i,j}$ and $P_{i+1,j+1}$ is $\sqrt{2}\Delta$, the maximum effective velocity of signal transmission between the two points is $\sqrt{2}\Delta/2\Delta t$. By this reasoning, we see that $\tilde{v}_{p,\ \max}$ is an inherent property of the FDTD grid.

Example of Calculation of Numerical Phase–Velocity and Attenuation

This subsection provides sample calculations of values of the numerical phase–velocity and the exponential attenuation

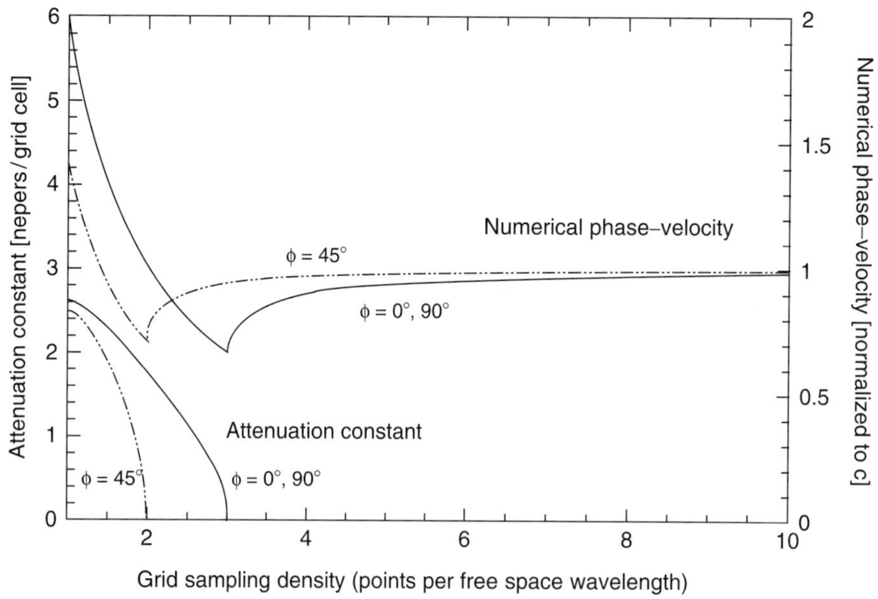

FIGURE 9.5 Normalized Numerical Phase–Velocity and Exponential Attenuation Constant per Grid Cell. These quantities are plotted versus the grid sampling density for on-axis and oblique wave propagation. The value of $S = 0.5$.

constant for the case of a two-dimensional square-cell Yee grid. These calculations are based on the numerical dispersion analyses of the previous two sections.

Figure 9.5 graphs the normalized numerical phase–velocity and the exponential attenuation constant per grid cell as a function of grid sampling density N_λ. A Courant factor $S = 0.5$ is assumed. From this figure, we note that:

- For propagation along the principal grid axes $\phi = 0°$, $90°$, a minimum value of $\tilde{v}_p = (2/3)c$ is reached at $N_\lambda = 3$. This sampling density is also the onset of attenuation. As N_λ is reduced below 3, $\tilde{v}_p$ increases inversely with N_λ. Eventually, $\tilde{v}_p$ exceeds c for $N_\lambda < 2$ and reaches a limiting velocity of $2c$ as $N_\lambda \to 1$. In this limit, as well, the attenuation constant approaches a value of 2.634 nepers/cell.
- For propagation along the grid diagonal at $\phi = 45°$, a minimum value of $\tilde{v}_p = (\sqrt{2}/2)c$ is reached at $N_\lambda = 2$. This point is also the onset of exponential attenuation. As N_λ is reduced below 2, $\tilde{v}_p$ increases inversely with N_λ. Eventually, $\tilde{v}_p$ exceeds c for $N_\lambda < \sqrt{2}$ and reaches a limiting velocity of $\sqrt{2}c$ as $N_\lambda \to 1$. In this limit, as well, the attenuation constant approaches a value of 2.493 nepers/cell.

Overall, for both the on-axis and oblique cases of numerical wave propagation, we see that very coarsely resolved wave modes in the grid can propagate at superluminal speeds but are rapidly attenuated.

Figure 9.6 graphs the percent error in the numerical phase–velocity relative to c for lossless wave propagation along the principal grid axes $\phi = 0°, 90°$. In the present example wherein $S = 0.5$, this lossless propagation regime exists for $N_\lambda \geq 3$. Figure 9.6 also graphs the percent velocity error for lossless wave propagation along the grid diagonal $\phi = 45°$. This lossless regime exists for $N_\lambda \geq 2$ for $S = 0.5$. As $N_\lambda \gg 10$, we see that the numerical phase–velocity error at each wave-propagation angle diminishes as the inverse square of N_λ. This is indicative of the second-order-accurate nature of the Yee algorithm.

Examples of Calculations of Pulse Propagation in a One-Dimensional Grid

Figure 9.7(A) graphs examples of the calculated propagation of a 40-cell-wide rectangular pulse in free space for two cases of the Courant factor: $S = 1$ (i.e., Δt is equal to the value for dispersionless propagation in a one-dimensional grid) and $S = 0.99$. To permit a direct comparison of these results, both "snapshots" are taken at the same absolute time after the onset of time-stepping. There are three key observations:

1. When $S = 1$, the rectangular shape and spatial width of the pulse are completely preserved. For this case, the abrupt step discontinuities of the propagating pulse are modeled perfectly. In fact, this is expected since $\tilde{v}_p \equiv c$ for all numerical modes in the grid.
2. When $S = 0.99$, there is appreciable "ringing" located behind the leading and trailing edges of the pulse. This is due to short-wavelength numerical modes in the grid generated at the step discontinuities of the wave. These numerical modes are poorly sampled in space and,

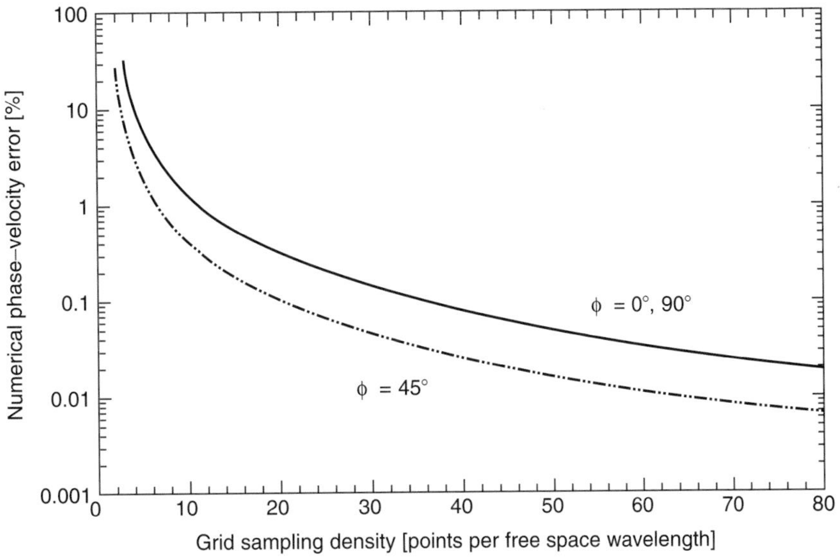

FIGURE 9.6 Percent Numerical Phase–Velocity Error Relative to the Free Space Speed of Light. This quantity is plotted as a function of the grid sampling density for on-axis and oblique wave propagation. The value of $S = 0.5$.

hence, travel slower than c, thereby lagging behind the causative discontinuities.

3. When $S = 0.99$, a weak superluminal response propagates just ahead of the leading edge of the pulse. This is again due to short-wavelength numerical modes in the grid generated at the step-function wavefront. These modes, however have spatial wavelengths even shorter than those noted in point 2. These are in fact so short that their grid sampling density drops below the upper bound for complex wavenumbers and the modes appear in the superluminal and exponentially decaying regime.

Figure 9.7(B) repeats the examples of Figure 9.7(A), but for the Courant factors $S = 1$ and $S = 0.5$. We see that the duration and periodicity of the ringing is greater than that for $S = 0.99$. Further, the superluminal response is more pronounced and less damped.

Figures 9.8(A) and 9.8(B) repeat the above examples but for a Gaussian pulse having a 40-grid-cell spatial width between its $1/e$ points. We see that the calculated pulse propagation for $S = 0.99$ shows no observable difference relative to the perfect propagation case of $S = 1$. Even for $S = 0.5$, the calculated propagation shows only a slight retardation relative to the exact solution, as expected, because $\tilde{v}_p < c$ for virtually all modes in the grid. Further, there is no observable superluminal precursor. All of these phenomena result from the fact that, for this case, virtually the entire spatial spectrum of propagating wavelengths in the grid is well resolved by the grid's sampling process. As a result, almost all numerical phase-velocity errors relative to c are well below 1%. This allows the Gaussian pulse to "hold together" while propagating over significant distances in the grid.

Example of Calculation of Pulse Propagation in a Two-Dimensional Grid

Figure 9.9 presents an example of the calculation of a radially outward-propagating cylindrical wave in a two-dimensional TM_z grid. A 360 × 360 – cell square grid with $\Delta x = \Delta y = \Delta = 1.0$ is used in this example. The grid is numerically excited at its center point by applying a unit-step time-function to a single E_z field component. We assume the Courant factor $S = \sqrt{2}/2$, which yields dispersionless propagation for numerical plane-wave modes propagating along the grid diagonals $\phi = 45°$, 135°, 225°, and 315°. In Figure 9.9(A), we graph snapshots of the E_z distribution versus radial distance from the source. Here, field observations are made along cuts through the grid passing through the source and either parallel to the principal grid axes $\phi = 0°$, 90° or parallel to the grid diagonal $\phi = 45°$. The snapshots are taken $232\Delta t$ after the beginning of time-stepping. At this time, the wave has not yet reached the outer grid boundary.

Figure 9.9(A) illustrates two nonphysical artifacts arising from numerical dispersion. First, for both observation cuts, the leading edge of the wave exhibits an oscillatory spatial jitter superimposed upon the normal field falloff profile. Second, for the observation cuts along the grid axes, the leading edge of the wave exhibits a small, spatially decaying, superluminal component.

To see these artifacts more easily, Figure 9.9(B) shows an expanded view of the leading edge of the wave. Consider first the oscillatory jitter. Similar to the results shown in Figure 9.7, the jitter is due to short-wavelength numerical modes in the grid generated at the step-function wave front. According to our dispersion theory, these numerical modes are poorly

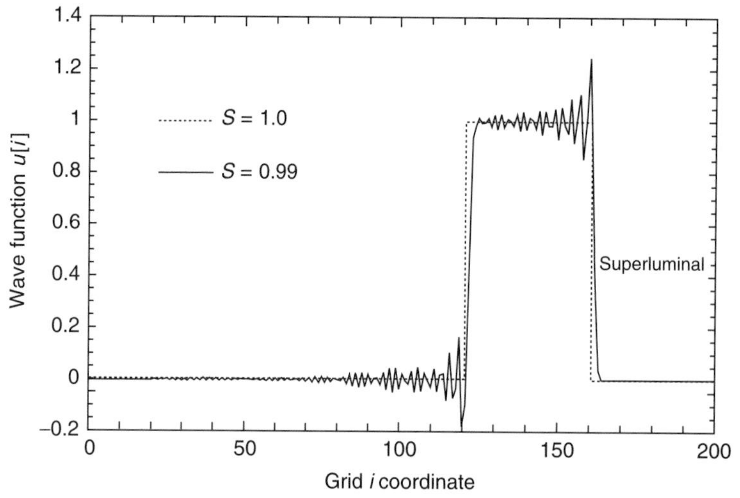

(A) Comparison of Calculated Pulse Propagation for $S = 1$ and $S = 0.99$

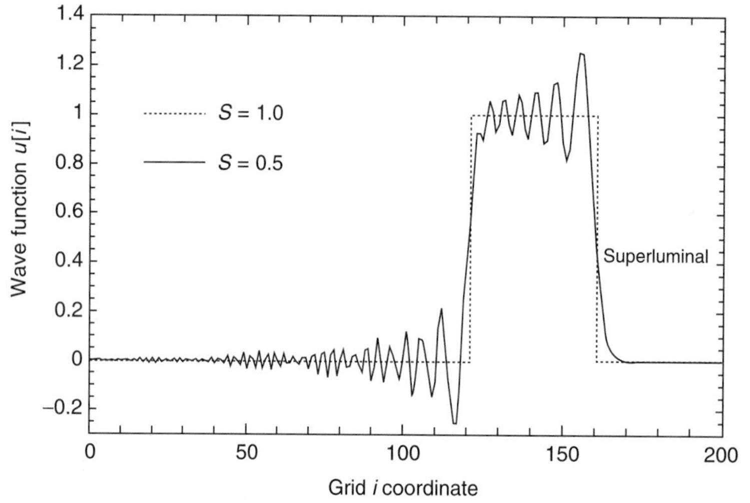

(B) Comparison of Calculated Pulse Propagation for $S = 1$ and $S = 0.5$

FIGURE 9.7 Effect of Numerical Dispersion on a Rectangular Pulse Propagating in Free Space. This occurs in a 1-D grid for three different Courant factors: $S = 1$, $S = 0.99$, and $S = 0.5$.

sampled in space and, hence, travel slower than c, thereby lagging behind the actual wavefront. While the jitter is most pronounced along the grid axes $\phi = 0°, 90°$, it is nonetheless finite along $\phi = 45°$ despite the choice of $S = \sqrt{2}/2$ (which implies dispersionless propagation along grid diagonals).

In Figure 9.9(B), the apparent anomaly of observing dispersion-induced jitter along what should be a dispersionless wave path at $\phi = 45°$ is resolved by noting that numerical dispersion introduces a slightly anisotropic propagation characteristic of the background free space in the grid versus azimuth angle ϕ. The resulting inhomogeneity of the free space background scatters part of the radially propagating numerical wave into the ϕ direction. Thus, no point behind the wavefront can avoid the short-wavelength numerical jitter.

Consider next the superluminal artifact present at the leading edge of the wave shown in Figure 9.9(B) for $\phi = 0°, 90°$ but not for $\phi = 45°$. This is again due to short-wavelength numerical modes in the grid generated at the leading edge of the outgoing step-function wave. However, these modes have spatial wavelengths so short that their grid sampling density drops below the threshold delineated in equation 9.45, and the modes appear in the superluminal and exponentially decaying regime. With $S = \sqrt{2}/2$ in the present example, we conclude that the lack of a superluminal artifact along the $\phi = 45°$ cut (and the consequent exact modeling of the step discontinuity at the leading edge of the wave) is due to a dispersionless numerical wave propagation along grid diagonals.

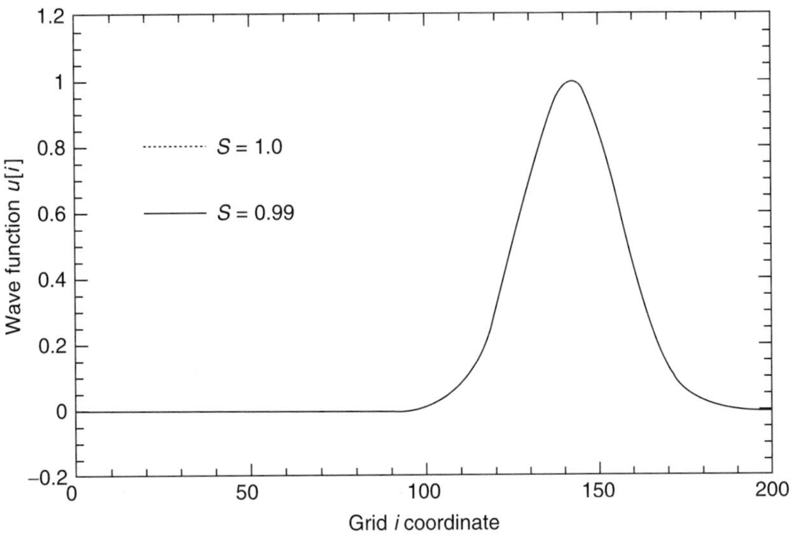

(A) Comparison of Calculated Pulse Propagation for $S = 1$ and $S = 0.99$

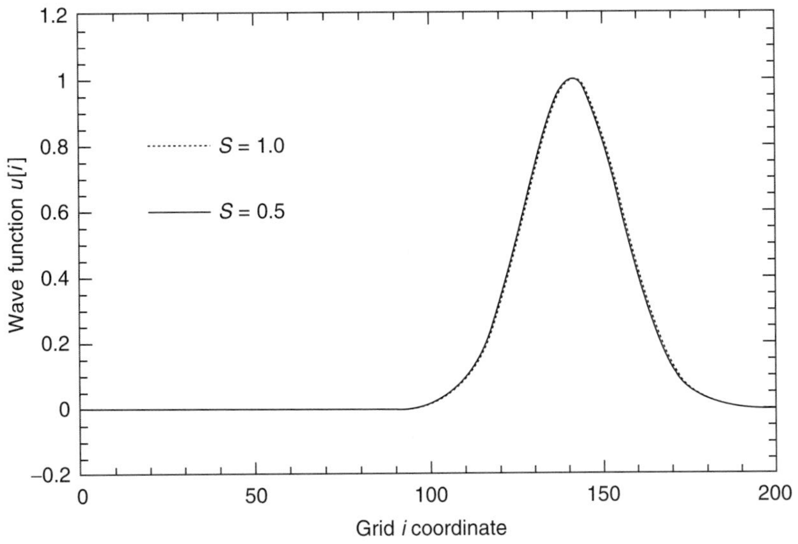

(B) Comparison of Calculated Pulse Propagation for $S = 1$ and $S = 0.5$

FIGURE 9.8 Effect of Numerical Dispersion on a Gaussian Pulse Propagating in Free Space. This occurs in a 1-D grid for three different Courant factors: $S = 1$, $S = 0.99$, and $S = 0.5$.

9.5 Numerical Stability

9.5.1 Introduction

In Section 9.4, we saw that the choice of Δ and Δt can affect the propagation characteristics of numerical waves in the FDTD space lattice and, therefore, the numerical error. In this section, we show that, in addition, Δt must be bounded to ensure numerical stability. Our approach to determine the upper bound on Δt is based on the complex-frequency analysis reported by Taflove and Hagness (2000). This technique allows straightforward estimates of the exponential growth rate of unstable numerical solutions.

9.5.2 Complex-Frequency Analysis

We first postulate a sinusoidal traveling wave present in the three-dimensional FDTD space lattice and discretely sampled at (x_I, y_J, z_K, t_n), allowing for the possibility of a complex-valued numerical angular frequency, $\tilde{\omega} = \tilde{\omega}_{real} + j\tilde{\omega}_{imag}$. A field vector in this wave can be written as:

$$\begin{aligned} \vec{V}\big|^{n}_{I,\,J,\,K} &= \vec{V}_0 e^{j\left[(\tilde{\omega}_{real} + j\tilde{\omega}_{imag})n\Delta t - \tilde{k}_x I\Delta x - \tilde{k}_y J\Delta y - \tilde{k}_z K\Delta z\right]} \\ &= \vec{V}_0 e^{-\tilde{\omega}_{imag} n\Delta t} e^{j\left(\tilde{\omega}_{real} n\Delta t - \tilde{k}_x I\Delta x - \tilde{k}_y J\Delta y - \tilde{k}_z K\Delta z\right)}, \end{aligned} \tag{9.65}$$

where $\tilde{k}$ is the wavenumber of the numerical sinusoidal traveling wave. We note that equation 9.65 permits either a constant

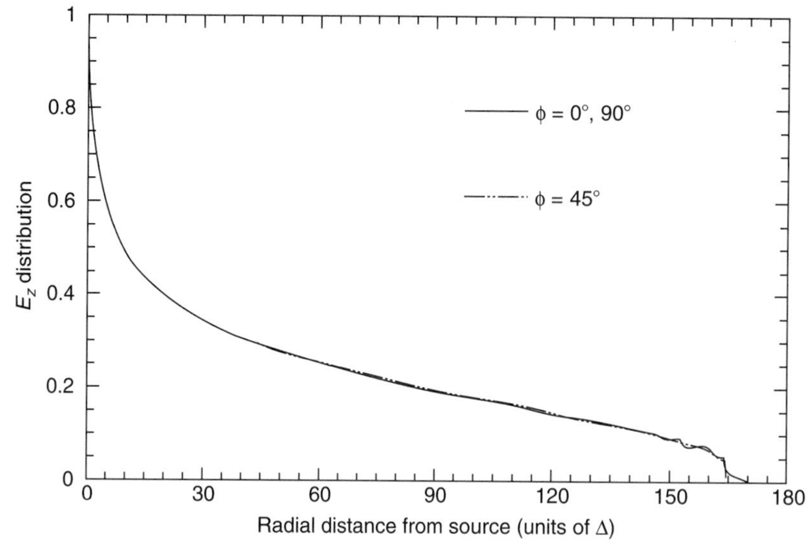

(A) Comparison of Calculated Wave Propagation Along the Grid Axes and Along a Grid Diagonal

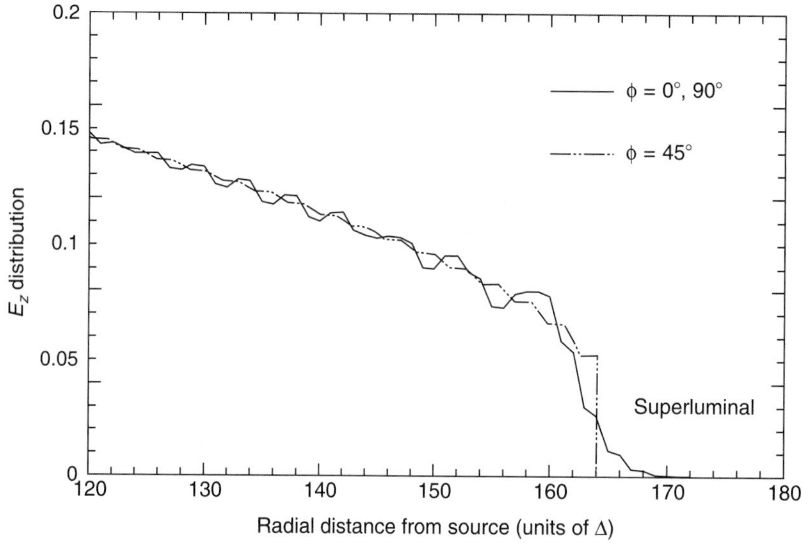

(B) Expanded View of (A) at Distances Between 120Δ and 180Δ Grid Cells from the Source

FIGURE 9.9 Effect of Numerical Dispersion on a Radially Propagating Cylindrical Wave in a 2-D Square Cell TM_z Yee Grid. The grid is excited at its center point by applying a unit-step time-function to a single E_z field component. The Courant factor is $S = \sqrt{2}/2$.

wave amplitude with time ($\tilde{\omega}_{imag} = 0$), an exponentially decreasing amplitude with time ($\tilde{\omega}_{imag} > 0$), or an exponentially increasing amplitude with time ($\tilde{\omega}_{imag} < 0$).

Given this basis, we proceed to analyze the numerical dispersion relation of equation 9.33, allowing for a complex-valued angular frequency:

$$\left[\frac{1}{c\Delta t}\sin\left(\frac{\tilde{\omega}\Delta t}{2}\right)\right]^2 = \left[\frac{1}{\Delta x}\sin\left(\frac{\tilde{k}_x\Delta x}{2}\right)\right]^2 + \left[\frac{1}{\Delta y}\sin\left(\frac{\tilde{k}_y\Delta y}{2}\right)\right]^2 + \left[\frac{1}{\Delta z}\sin\left(\frac{\tilde{k}_z\Delta z}{2}\right)\right]^2. \tag{9.66}$$

We first solve equation 9.66 for $\tilde{\omega}$. This yields:

$$\tilde{\omega} = \frac{2}{\Delta t}\sin^{-1}(\xi), \tag{9.67}$$

where:

$$\xi = c\Delta t\sqrt{\frac{1}{(\Delta x)^2}\sin^2\left(\frac{\tilde{k}_x\Delta x}{2}\right) + \frac{1}{(\Delta y)^2}\sin^2\left(\frac{\tilde{k}_y\Delta y}{2}\right) + \frac{1}{(\Delta z)^2}\sin^2\left(\frac{\tilde{k}_z\Delta z}{2}\right)}. \tag{9.68}$$

We observe from equation 9.68 that:

$$0 \le \xi \le c\Delta t\sqrt{\frac{1}{(\Delta x)^2}+\frac{1}{(\Delta y)^2}+\frac{1}{(\Delta z)^2}} \equiv \xi_{\text{upper bound}}, \quad (9.69)$$

for all possible real values of $\tilde{k}$ (i.e., those numerical waves having zero exponential attenuation per grid space cell). The $\xi_{\text{upper bound}}$ is obtained when each $\sin^2$ term under the square root of 9.68 simultaneously reaches a value of 1. This occurs for the propagating numerical wave having the wave vector components:

$$\tilde{k}_x = \pm\frac{\pi}{\Delta x}; \quad \tilde{k}_y = \pm\frac{\pi}{\Delta y}; \quad \tilde{k}_z = \pm\frac{\pi}{\Delta z}. \quad (9.70\text{a, b, c})$$

It is clear that $\xi_{\text{upper bound}}$ can exceed 1 depending upon the choice of Δt. This yields complex values of $\sin^{-1}(\xi)$ in equation 9.67 and, therefore, complex values of $\tilde{\omega}$. To investigate further, we divide the range of ξ given in equation 9.69 into two subranges.

Stable Range: $0 \le \xi \le 1$

Here, $\sin^{-1}(\xi)$ is real-valued, and real values of $\tilde{\omega}$ are obtained in equation 9.67. With $\tilde{\omega}_{\text{imag}} = 0$, equation 9.65 yields a constant wave amplitude with time.

Unstable Range: $1 < \xi < \xi_{\text{upper bound}}$

This subrange exists only if:

$$\xi_{\text{upper bound}} = c\Delta t\sqrt{\frac{1}{(\Delta x)^2}+\frac{1}{(\Delta y)^2}+\frac{1}{(\Delta z)^2}} > 1. \quad (9.71)$$

The unstable range is defined in an equivalent manner by:

$$\Delta t > \frac{1}{c\sqrt{\frac{1}{(\Delta x)^2}+\frac{1}{(\Delta y)^2}+\frac{1}{(\Delta z)^2}}} \equiv \Delta t_{\substack{\text{stable}\\ \text{limit/3-D}}} \quad (9.72)$$

To prove the claim of instability for the range $\xi > 1$, we substitute the complex-valued $\sin^{-1}(\xi)$ function of equation 9.49 into 9.67 and solve for $\tilde{\omega}$. This yields:

$$\tilde{\omega} = \frac{-j2}{\Delta t}\ln\left(j\xi + \sqrt{1-\xi^2}\right). \quad (9.73)$$

Upon taking the natural logarithm, we obtain:

$$\tilde{\omega}_{\text{real}} = \frac{\pi}{\Delta t}; \quad \tilde{\omega}_{\text{imag}} = -\frac{2}{\Delta t}\ln\left(\xi + \sqrt{\xi^2-1}\right). \quad (9.74)$$

Substituting equation 9.74 into 9.65 yields:

$$\begin{aligned}\vec{V}\big|^n_{I,J,K} &= \vec{V}_0 e^{2n\ln\left(\xi+\sqrt{\xi^2-1}\right)} e^{j[(\pi/\Delta t)(n\Delta t)-\tilde{k}_x I\Delta x-\tilde{k}_y J\Delta y-\tilde{k}_z K\Delta z]}\\ &= \vec{V}_0\left(\xi+\sqrt{\xi^2-1}\right)^{**2n} e^{j[(\pi/\Delta t)(n\Delta t)-\tilde{k}_x I\Delta x-\tilde{k}_y J\Delta y-\tilde{k}_z K\Delta z]},\end{aligned} \quad (9.75)$$

where $^{**}2n$ denotes the $2n$'th power. From equation 9.75, we define the following multiplicative factor greater than 1 that amplifies the numerical wave every time-step:

$$q_{\text{growth}} \equiv \left(\xi + \sqrt{\xi^2-1}\right)^2. \quad (9.76)$$

Equations 9.75 and 9.76 define an exponential growth of the numerical wave with time-step number n. We see that the dominant exponential growth occurs for the most positive possible value of ξ (i.e., $\xi_{\text{upper bound}}$ defined in equation 9.69).

Example of Calculating a Stability Bound: Three-Dimensional Cubic-Cell Lattice

Consider the practical case of a three-dimensional (3-D) cubic-cell space lattice with $\Delta x = \Delta y = \Delta z = \Delta$. From equation 9.72, numerical instability arises when:

$$\Delta t > \frac{1}{c\sqrt{\frac{1}{(\Delta)^2}+\frac{1}{(\Delta)^2}+\frac{1}{(\Delta)^2}}} = \frac{1}{c\sqrt{\frac{3}{(\Delta)^2}}} = \frac{\Delta}{c\sqrt{3}}. \quad (9.77)$$

We define an equivalent Courant stability limit for the cubic-cell lattice case:

$$S_{\substack{\text{stability}\\ \text{limit/3-D}}} = \frac{1}{\sqrt{3}}. \quad (9.78)$$

From equation 9.70, the dominant exponential growth is seen to occur for numerical waves propagating along the lattice diagonals. The relevant wave vectors are as follows:

$$\tilde{\vec{k}} = \frac{\pi}{\Delta}(\pm\hat{x}\pm\hat{y}\pm\hat{z}) \rightarrow \left|\tilde{\vec{k}}\right| = \frac{\pi\sqrt{3}}{\Delta} \rightarrow \tilde{\lambda} = \left(\frac{2\sqrt{3}}{3}\right)\Delta, \quad (9.79)$$

where $\hat{x}$, $\hat{y}$, and $\hat{z}$ are unit vectors defining the major lattice axes. Further, equation 9.69 yields:

$$\xi_{\substack{\text{upper}\\ \text{bound}}} = c\Delta t\sqrt{\frac{1}{(\Delta)^2}+\frac{1}{(\Delta)^2}+\frac{1}{(\Delta)^2}} = \left(\frac{c\Delta t}{\Delta}\right)\sqrt{3} = S\sqrt{3}. \quad (9.80)$$

From equation 9.76, this implies the following maximum possible growth factor per time step under conditions of numerical instability:

$$q_{\text{growth}} \equiv \left[S\sqrt{3} + \sqrt{\left(S\sqrt{3}\right)^2 - 1}\right]^2 \quad \Big\} \text{ for } S \ge \frac{1}{\sqrt{3}}. \quad (9.81)$$

Courant Factor Normalization and Extension to Two- and One-Dimensional Grids

It is instructive to use the results of equation 9.78 to normalize the Courant factor S in equation 9.81. This will permit us to generalize the three-dimensional results for the maximum growth factor q to two-dimensional and one-dimensional (1-D) Yee (1966) grids. In this spirit, we define:

$$S_{\text{norm 3-D}} \equiv \frac{S}{S_{\substack{\text{stability}\\ \text{limit 3-D}}}} = \frac{S}{(1/\sqrt{3})} = S\sqrt{3}. \tag{9.82}$$

Then, equation 9.81 can be written as:

$$q_{\text{growth}} = \left[S_{\text{norm 3-D}} + \sqrt{(S_{\text{norm 3-D}})^2 - 1} \right]^2 \Bigg\} \text{for } S_{\text{norm 3-D}} \geq 1. \tag{9.83}$$

Given this notation, it can be shown that analogous expressions for the Courant stability limit and the growth factor under conditions of numerical instability are given by:

Two-Dimensional Square Yee Grid

$$S_{\substack{\text{stability}\\ \text{limit 2-D}}} = \frac{1}{\sqrt{2}}. \tag{9.84}$$

$$S_{\text{norm 2-D}} \equiv \frac{S}{S_{\substack{\text{stability}\\ \text{limit 2-D}}}} = \frac{S}{(1/\sqrt{2})} = S\sqrt{2}. \tag{9.85}$$

Here, dominant exponential growth occurs for numerical waves propagating along the grid diagonals. The relevant wave-vectors are the following:

$$\tilde{\vec{k}} = \frac{\pi}{\Delta}(\pm\hat{x} \pm \hat{y}) \rightarrow \left|\tilde{\vec{k}}\right| = \frac{\pi\sqrt{2}}{\Delta} \rightarrow \tilde{\lambda} = \sqrt{2}\Delta. \tag{9.86}$$

This yields the following solution growth factor per time-step:

$$q_{\text{growth}} = \left[S_{\text{norm 2-D}} + \sqrt{(S_{\text{norm 2-D}})^2 - 1} \right]^2 \Bigg\} \text{ for } S_{\text{norm 2-D}} \geq 1. \tag{9.87}$$

One-Dimensional Uniform Yee Grid

$$S_{\substack{\text{stability}\\ \text{limit 1-D}}} = 1. \tag{9.88}$$

$$S_{\text{norm1-D}} \equiv \frac{S}{S_{\substack{\text{stability}\\ \text{limit 1-D}}}} = \frac{S}{1} = S. \tag{9.89}$$

Dominant exponential growth occurs for the wave vectors:

$$\tilde{\vec{k}} = \pm\frac{\pi}{\Delta}\hat{x} \rightarrow \left|\tilde{\vec{k}}\right| = \frac{\pi}{\Delta} \rightarrow \tilde{\lambda} = 2\Delta. \tag{9.90}$$

This yields the following solution growth factor per time-step:

$$q_{\text{growth}} = \left(S + \sqrt{S^2 - 1} \right)^2 \Bigg\} \text{ for } S \geq 1. \tag{9.91}$$

We see from the above discussion that the solution growth factor q under conditions of numerical instability is the same, regardless of the dimensionality of the FDTD space lattice if the same normalized Courant factor is used. A normalized Courant factor equal to 1 yields no exponential solution growth for any dimensionality grid. However, a normalized Courant factor only 0.05% larger, yields a multiplicative solution growth of 1.0653 every time-step for each dimensionality grid:

- $S = 1.0005$ for a uniform, one-dimensional grid
- $S = 1.0005 \times (1/\sqrt{2}) = 0.707460$ for a uniform, square, two-dimensional grid
- $S = 1.0005 \times (1/\sqrt{3}) = 0.577639$ for a uniform, cubic, three-dimensional grid

This is equivalent to a solution growth of 1.8822 every 10 time-steps, 558.7 every 100 time-steps, and 2.96×10^{27} every 1,000 time-steps.

9.5.3 Examples of Calculations Involving Numerical Instability in a 1-D Grid

We first consider an example of the beginning of a numerical instability arising because the Courant stability condition is violated equally at *every* point in a uniform one-dimensional grid. Figure 9.10(A) graphs three snapshots of the free space propagation of a Gaussian pulse in a grid having the Courant factor $S = 1.0005$. The exciting pulse waveform has a $40\Delta t$ temporal width between its $1/e$ points and reaches its peak value of 1.0 at time-step $n = 60$. Graphs of the wavefunction $u(i)$ versus the grid coordinate i are shown at time-steps $n = 200$, $n = 210$, and $n = 220$.

From Figure 9.10(A), we see that the trailing edge of the Gaussian pulse is contaminated by a rapidly oscillating and growing noise component that does not exist in Figure 9.8(A), which shows the same Gaussian pulse at the same time but with $S \leq 1.0$. In fact, the noise component in Figure 9.10(A) results from the onset of numerical instability in the grid due to $S = 1.0005 > 1.0$. Because this noise grows exponentially with time-step number n, it quickly overwhelms the desired numerical results for the propagating Gaussian pulse. Shortly thereafter, the exponential growth of the noise increases the calculated field values beyond the dynamic range of the computer being used, resulting in run-time floating-point overflows and errors.

Figure 9.10(B) is an expanded view of Figure 9.10(A) between grid points $i = 1$ and $i = 20$, showing a segment of the numerical noise on the trailing edge of the Gaussian pulse. We see that the noise oscillates with a spatial period of 2 grid cells

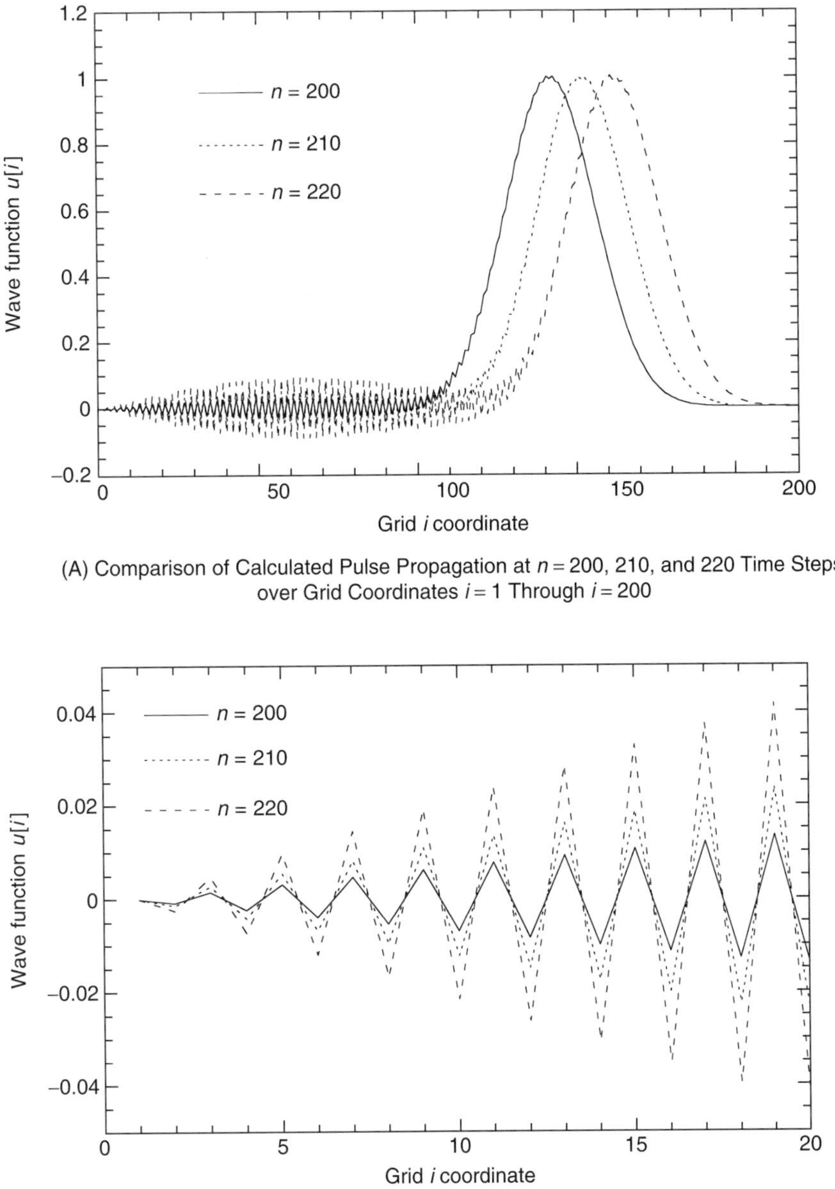

(A) Comparison of Calculated Pulse Propagation at $n = 200$, 210, and 220 Time Steps over Grid Coordinates $i = 1$ Through $i = 200$

(B) Expanded View of (A) over Grid Coordinates $i = 1$ Through $i = 20$

FIGURE 9.10 The Beginning of Numerical Instability for a Gaussian Pulse Propagating in a Uniform, Free Space 1-D Grid. The Courant factor is $S = 1.0005$ at each grid point.

(i.e., $\tilde{\lambda} = 2\Delta x$) in accordance with equation 9.90. In addition, when analyzing the raw data underlying Figure 9.10(B), it is observed that the growth factor q is in the range 1.058 to 1.072 per time-step. This compares favorably with the theoretical value of 1.0653 determined using equation 9.91.

We next consider an example of the beginning of a numerical instability arising because the Courant stability condition is violated at only a *single* point in a uniform one-dimensional grid. Figure 9.11(A) graphs two snapshots of the free space propagation of a narrow Gaussian pulse in a grid having the Courant factor $S = 1.0$ at all points except at $i = 90$, where $S = 1.2075$. The exciting pulse has a $10\Delta t$ temporal width between its $1/e$ points and reaches its peak value of 1.0 at time-step $n = 60$. Graphs of the wave function $u(i)$ versus the grid coordinate i are shown at time-steps $n = 190$ and $n = 200$. In contrast to Figure 9.10(A), the rapidly oscillating and growing noise component due to numerical instability originates at just a single grid point along the trailing edge of

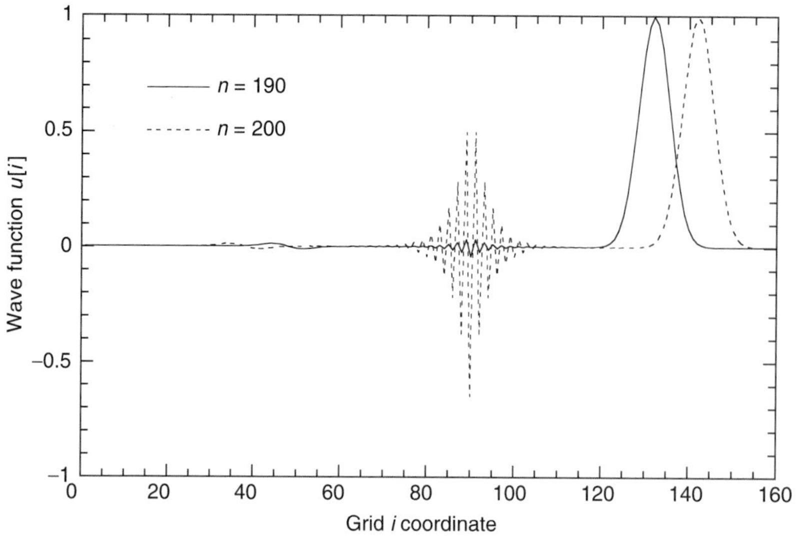

(A) Comparison of Calculated Pulse Propagation at $n = 190$ and $n = 200$ Time-Steps. These exist over grid coordinates $i = 1$ through $i = 160$.

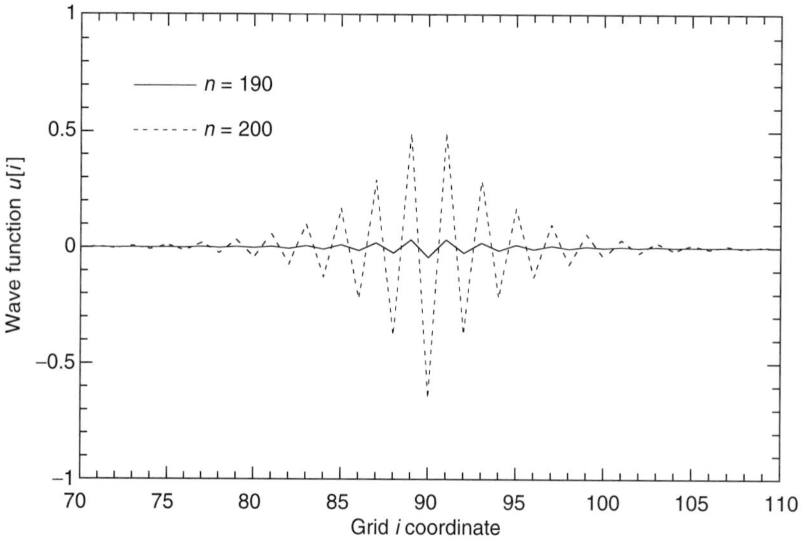

(B) Expanded View of (A) over Grid Coordinates $i = 70$ Through $i = 110$

FIGURE 9.11 The Beginning of Numerical Instability for a Gaussian Pulse Propagating in a Uniform, Free Space 1-D Grid. The Courant factor is $S = 1$ at all grid points but $i = 90$, where $S = 1.2075$.

the Gaussian pulse ($i = 90$), where S exceeds 1.0, rather than along the entirety of the trailing edge. Despite this localization of the source of the instability, the noise again grows exponentially with time-step number n. In this case, the noise propagates symmetrically in both directions from the unstable point. Ultimately, the noise again fills the entire grid, overwhelms the desired numerical results for the propagating Gaussian pulse, and causes run-time floating-point overflows.

Figure 9.11(B) is an expanded view of Figure 9.11(A) between grid points $i = 70$ and $i = 110$, showing how the calculated noise due to the numerical instability originates at $i = 90$. Again, in accordance with equation 9.90, the noise oscillates with a spatial period of two grid cells (i.e. $\tilde{\lambda} = 2\Delta x$). However, the rate of exponential growth here is much less than that predicted by equation 9.91, wherein *all* grid points were assumed to violate Courant stability. Upon analyzing the raw data underlying Figure 9.11(B), a growth factor of $q \cong 1.31$ is observed per time-step. This compares to $q \cong 3.55$ per time-step determined by substituting $S = 1.2075$ into equation 9.91. Thus, it is clear that a grid having one or just a few localized points of numerical instability can "blow up" much more slowly than a uniformly unstable grid having a comparable or even smaller Courant factor S.

9.5.4 Example of Calculation Involving Numerical Instability in a Two-Dimensional Grid

We next consider an FDTD modeling example where the Courant stability condition is violated equally at every point in a uniform two-dimensional TM_z grid. To allow direct comparison with a previous example of stable pulse propagation, the same grid in Figure 9.9 is used. The overall grid size is again 360×360 square cells with $\Delta x = \Delta y = 1.0 \equiv \Delta$. Numerical excitation to the grid is again provided by specifying a unit-step time-function for the center E_z component. The only condition that differs from those assumed in the original grid is that the Courant factor S is increased above the threshold for numerical instability given by equation 9.84.

Figure 9.12(A) visualizes the two-dimensional E_z distribution at $n = 40$ time-steps for $S = 1.005 \times (1/\sqrt{2})$. This value of S quickly generates a region of numerical instability spreading out radially from the source, where the field amplitudes are large enough to mask the normal wave propagation. Individual E_z components in the grid are depicted as square pixels. We see that the unstable field pattern has the form of a checkerboard, wherein dark and gray pixels denote positive and negative E_z values, respectively. Here, pixel saturation denotes the relative amplitude of its positive or negative value.

Figure 9.12(B) graphs the variation of E_z versus radial distance from the source at $n = 200$ time-steps for $S = 1.0005 \times (1/\sqrt{2})$. Two distinct plots are shown. The solid line graph exhibits a rapid spatial oscillation with the period 2Δ. This is the E_z behavior along the $\phi = 0°, 90°$ (and similar on-axis) cuts through the grid. The smooth dashed-dotted curve with no spatial oscillation represents the E_z behavior along the $\phi = 45°$ (and similar oblique) cuts through the grid. Analysis of the underlying data reveals growth factors in the range 1.060 to 1.069 per time-step along the leading edge of the instability region. This agrees very well with $q_{\text{growth}} = 1.0653$ calculated using equation 9.87 and is an excellent validation of the Courant-factor-normalization theory.

An interesting observation in Figure 9.12(B) is that the smooth E_z variation along $\phi = 45°$ forms the envelope of the oscillatory E_z distribution observed along the grid's major axes. This difference in behavior is confirmed in Figure 9.12(A), which shows that the $\phi = 45°$ cut lies entirely within a diagonal string of dark (positive) pixels, whereas the $\phi = 0°$ cut passes through alternating dark (positive) and gray (negative) pixels. We attribute this behavior to equation 9.86, which states that the exponential growth along the grid diagonal has $\tilde{\lambda} = \sqrt{2}\Delta$. That is, the numerical wavelength along the 45° observation cut for the unstable mode is exactly the diagonal length across one $\Delta \times \Delta$ grid cell. Thus, there exists 2π (or equivalently, 0) phase shift of the unstable mode between adjacent observation points along the $\phi = 45°$ cut. Adjacent E_z values along $\phi = 45°$ cannot change sign. In contrast, equation 9.86 reduces to $\tilde{k} = \pi/\Delta$, (i.e., $\tilde{\lambda} = 2\Delta$) for the unstable mode along the $\phi = 0°, 90°$ cuts. Therefore, there is π phase shift of the unstable mode between adjacent observation points along $\phi = 0°, 90°$, yielding the point-by-point sign reversals (rapid spatial oscillations) seen in Figure 9.12(B).

9.5.5 Linear Instability When the Normalized Courant Factor Equals 1

The general field vector postulated in equation 9.65 permits a numerical wave amplitude that is either constant, exponentially growing, or exponentially decaying as time-stepping progresses. Min and Teng (2001) have identified a linear growth mode, (i.e., a linear instability) that can occur if the normalized Courant factor equals exactly 1. While this growth mode is much slower than the exponential instability discussed previously, the analyst should proceed with caution when using $S_{\text{norm}} = 1$.

9.6 Perfectly Matched Layer Absorbing Boundary Conditions

9.6.1 Introduction

Many electromagnetic wave interaction problems are defined on "open" regions where the spatial domain is partially or completely unbounded. Since no computer can store an unlimited data set, FDTD space lattices for such problems must somehow be truncated without introducing error due to spurious reflection of outward-propagating numerical waves. Grid truncation conditions of this type are popularly called **absorbing boundary conditions** (ABCs) because of their requirement to absorb outward-propagating numerical modes with negligible reflection.

ABCs cannot be directly obtained from the Yee algorithm (1966) because it requires field data on both sides of an observation point and, hence, cannot be implemented at the outermost planes of a space lattice. Although backward finite differences could conceivably be used here, these are generally of lower accuracy for a given space discretization and have not been used in any major FDTD software.

Research in this area since 1970 has resulted in two principal categories of ABCs for FDTD simulations:

1. The first category is special analytical boundary conditions imposed on the electromagnetic field at the outermost planes of the space lattice. This category was reviewed by Taflove and Hagness (2000).
2. The second category is incorporation of **perfectly matched layer** (PML) absorbing media adjacent to the

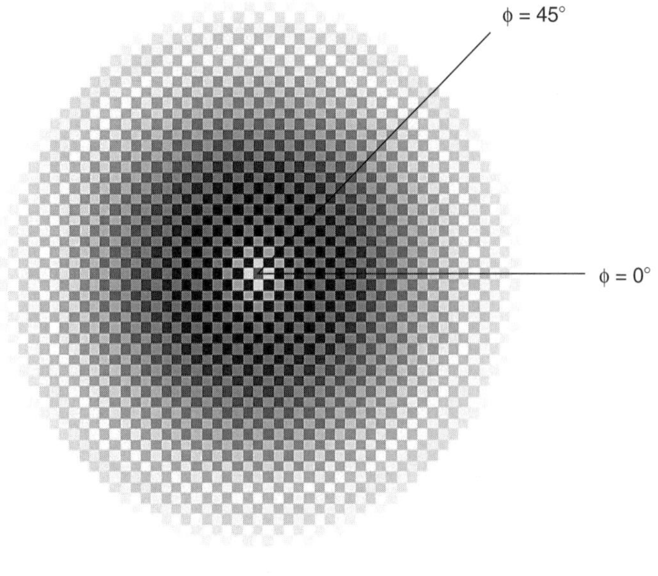

(A) Visualization of the 2-D E_z Distribution at $n = 40$ for $S = 1005 \times (1/\sqrt{2})$

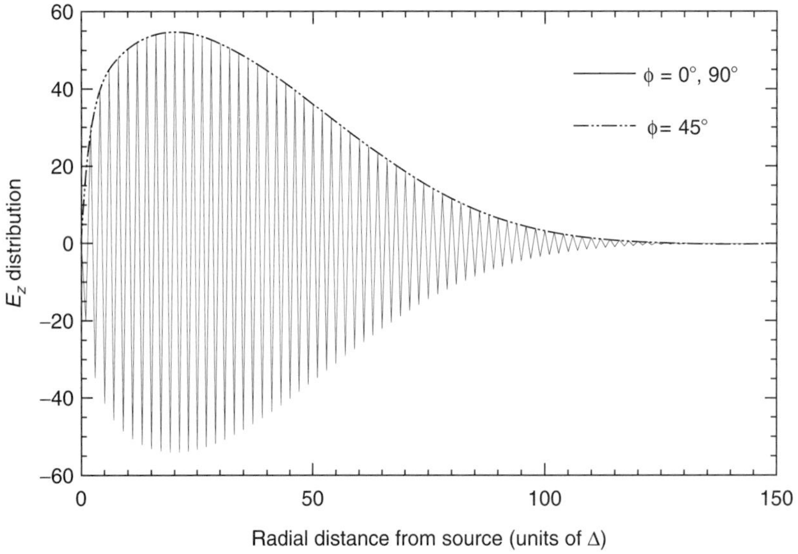

(B) E_z Distributions Along the Grid Axes and Grid Diagonal at $n = 200$ for $S = 1.005 \times (1/\sqrt{2})$. The theoretical and measured growth factor is $q_{\text{growth}} \cong 1.065$ per time-step.

FIGURE 9.12 Effect of Numerical Instability on a 2-D Pulse-Propagation Model.

outer planes of the space lattice (by analogy with the treatment of the walls of an anechoic chamber). ABCs of this type have excellent capabilities for truncation of FDTD lattices in free space, in lossy or dispersive materials, or in metal or dielectric waveguides. Extremely small numerical wave reflection coefficients in the order of 10^{-4} to 10^{-6} can be attained with an acceptable computational burden, allowing the possibility of achieving FDTD simulations having a dynamic range of 70 dB or more.

This section reviews the formulation and application of PML ABCs. The review is based on the publication by Gedney and Taflove (2000).

9.6.2 Brief History

Consider implementing an ABC by using an impedance-matched electromagnetic wave absorbing layer adjacent to the outer planes of the FDTD space lattice. Ideally, the absorbing medium is only a few lattice cells thick, reflectionless to all impinging waves over their full frequency spectrum, highly absorbing, and effective in the near field of a source or a scatterer. An early attempt at implementing such an absorbing material boundary condition was reported by Holland and Williams (1983) who utilized a conventional lossy, dispersionless, and absorbing medium. The difficulty with this tactic is that such an absorbing layer is matched only to normally incident plane waves.

Berenger (1994) provided the seminal insight that a nonphysical PML could be implemented at the outer boundary of an FDTD grid to completely absorb outgoing waves regardless of their frequency, angle of incidence, and polarization. The key was to exploit additional degrees of freedom arising from a novel split-field formulation of Maxwell's equations. Here, each field vector component is split into two orthogonal components, and the 12 resulting field components are expressed as satisfying two coupled first-order partial differential equations. By choosing loss parameters consistent with a dispersionless medium, a perfectly matched planar interface is derived.

Following Berenger's work, many papers appeared validating his technique as well as applying FDTD with the PML medium. An important advance was made by Chew and Weedon (1994), who restated the original split-field PML concept in a stretched-coordinate form. Subsequently, this allowed Teixeira and Chew (1997) to extend PML to cylindrical and spherical coordinate systems. A second important advance was made by Sacks *et al.* (1995) and Gedney (1995, 1996), who reposed the split-field PML as a lossy, uniaxial anisotropic medium having both magnetic permeability and electric permittivity tensors. The uniaxial PML, or UPML, is intriguing because it is based on a potentially physically realizable material formulation rather than Berenger's nonphysical mathematical model.

9.6.3 Berenger's Perfectly Matched Layer

Two-Dimensional TE$_z$ Case

This section reviews the theoretical basis of Berenger's PML for the case of a TE$_z$-polarized plane wave incident from region 1, the lossless material half-space $x < 0$, onto region 2, the PML half-space $x > 0$.

Field-Splitting Modification of Maxwell's Equations

Within region 2, Maxwell's curl equations (9.12a through 9.12c) as modified by Berenger are expressed in their time-dependent form as:

$$\varepsilon_2 \frac{\partial E_x}{\partial t} + \sigma_y E_x = \frac{\partial H_z}{\partial y}. \qquad (9.92a)$$

$$\varepsilon_2 \frac{\partial E_y}{\partial t} + \sigma_x E_y = -\frac{\partial H_z}{\partial x}. \qquad (9.92b)$$

$$\mu_2 \frac{\partial H_{zx}}{\partial t} + \sigma_x^* H_{zx} = -\frac{\partial E_y}{\partial x}. \qquad (9.92c)$$

$$\mu_2 \frac{\partial H_{zy}}{\partial t} + \sigma_y^* H_{zy} = \frac{\partial E_x}{\partial y}. \qquad (9.92d)$$

Here, H_z is assumed to be split into two additive subcomponents:

$$H_z = H_{zx} + H_{zy}. \qquad (9.93)$$

Further, the parameters σ_x and σ_y denote electric conductivities, and the parameters σ_x^* and σ_y^* denote magnetic losses.

Berenger's formulation represents a generalization of normally modeled physical media. If $\sigma_x = \sigma_y = 0$ and $\sigma_x^* = \sigma_y^* = 0$, equations 9.92a through 9.92d reduce to Maxwell's equations in a lossless medium. If $\sigma_x = \sigma_y = \sigma$ and $\sigma_x^* = \sigma_y^* = 0$, equations 9.92a through 9.92d describe an electrically conductive medium. If $\varepsilon_2 = \varepsilon_1$, $\mu_2 = \mu_1$, $\sigma_x = \sigma_y = \sigma$, $\sigma_x^* = \sigma_y^* = \sigma^*$, and:

$$\sigma^*/\mu_1 = \sigma/\varepsilon_1 \quad \rightarrow \quad \sigma^* = \sigma\mu_1/\varepsilon_1 = \sigma(\eta_1)^2, \qquad (9.94)$$

then equations 9.92a through 9.92d describe an absorbing medium that is impedance-matched to region 1 for normally incident plane waves.

Additional possibilities present themselves, however. If $\sigma_y = \sigma_y^* = 0$, the medium can absorb a plane wave having field components (E_y, H_{zx}) propagating along x, but it does not absorb a wave having field components (E_x, H_{zy}) propagating along y, since in the first case propagation is governed by equations 9.92b and 9.92c and in the second case by equations 9.92a and 9.92d. The converse situation is true for waves (E_y, H_{zx}) and (E_x, H_{zy}) if $\sigma_x = \sigma_x^* = 0$. These properties of particular Berenger media characterized by the pairwise parameter sets $(\sigma_x, \sigma_x^*, 0, 0)$ and $(0, 0, \sigma_y, \sigma_y^*)$ are closely related to the fundamental premise of this novel ABC, as proved later in this chapter. That is, if the pairwise electric and magnetic losses satisfy equation 9.94, then at interfaces normal to x and y, respectively, the Berenger media have zero reflection of electromagnetic waves.

Now consider equations 9.92a through 9.92d expressed in their time-harmonic form in the Berenger medium. Letting the hat symbol denote a phasor quantity, we write:

$$j\omega\varepsilon_2\left(1 + \frac{\sigma_y}{j\omega\varepsilon_2}\right)\breve{E}_x = \frac{\partial}{\partial y}\left(\breve{H}_{zx} + \breve{H}_{zy}\right). \qquad (9.95a)$$

$$j\omega\varepsilon_2\left(1 + \frac{\sigma_x}{j\omega\varepsilon_2}\right)\breve{E}_y = -\frac{\partial}{\partial x}\left(\breve{H}_{zx} + \breve{H}_{zy}\right). \qquad (9.95b)$$

$$j\omega\mu_2\left(1+\frac{\sigma_x^*}{j\omega\mu_2}\right)\breve{H}_{zx} = -\frac{\partial \breve{E}_y}{\partial x}. \quad (9.95c)$$

$$j\omega\mu_2\left(1+\frac{\sigma_y^*}{j\omega\mu_2}\right)\breve{H}_{zy} = \frac{\partial \breve{E}_x}{\partial y}. \quad (9.95d)$$

The notation is simplified by introducting the variables:

$$s_w = \left(1+\frac{\sigma_w}{j\omega\varepsilon_2}\right); \quad s_w^* = \left(1+\frac{\sigma_w^*}{j\omega\mu_2}\right) : w = x, y. \quad (9.96)$$

Then, equations 9.95a and 9.95b are rewritten as:

$$j\omega\varepsilon_2 s_y \breve{E}_x = \frac{\partial}{\partial y}\left(\breve{H}_{zx}+\breve{H}_{zy}\right). \quad (9.97a)$$

$$j\omega\varepsilon_2 s_x \breve{E}_y = -\frac{\partial}{\partial x}\left(\breve{H}_{zx}+\breve{H}_{zy}\right). \quad (9.97b)$$

Plane-Wave Solution in the Berenger Medium

The next step is to derive the plane-wave solution in the Berenger medium. To this end, equation 9.97a is differentiated with respect to y and equation 9.97b with respect to x. Substituting the expressions for $\partial\breve{E}_y/\partial x$ and $\partial\breve{E}_x/\partial y$ from equations 9.95c and 9.95d leads to:

$$-\omega^2\mu_2\varepsilon_2\breve{H}_{zx} = -\frac{1}{s_x^*}\frac{\partial}{\partial x}\frac{1}{s_x}\frac{\partial}{\partial x}\left(\breve{H}_{zx}+\breve{H}_{zy}\right). \quad (9.98a)$$

$$-\omega^2\mu_2\varepsilon_2\breve{H}_{zy} = -\frac{1}{s_y^*}\frac{\partial}{\partial y}\frac{1}{s_y}\frac{\partial}{\partial y}\left(\breve{H}_{zx}+\breve{H}_{zy}\right). \quad (9.98b)$$

Adding these together and using equation 9.93 leads to the representative wave equation:

$$\frac{1}{s_x^*}\frac{\partial}{\partial x}\frac{1}{s_x}\frac{\partial}{\partial x}\breve{H}_z + \frac{1}{s_y^*}\frac{\partial}{\partial y}\frac{1}{s_y}\frac{\partial}{\partial y}\breve{H}_z + \omega^2\mu_2\varepsilon_2\breve{H}_z = 0. \quad (9.99)$$

This wave equation supports the solutions:

$$\breve{H}_z = H_0\tau e^{-j\sqrt{s_x s_x^*}\beta_{2_x}x - j\sqrt{s_y s_y^*}\beta_{2_y}y}, \quad (9.100)$$

with the dispersion relationship:

$$(\beta_{2_x})^2+(\beta_{2_y})^2 = (k_2)^2 \rightarrow \beta_{2_x} = \left[(k_2)^2-(\beta_{2_y})^2\right]^{1/2}. \quad (9.101)$$

Then, from equations 9.97a, 9.97b, and 9.93, we have:

$$\breve{E}_x = -H_0\tau\frac{\beta_{2_y}}{\omega\varepsilon_2}\sqrt{\frac{s_y^*}{s_y}}e^{-j\sqrt{s_x s_x^*}\beta_{2_x}x - j\sqrt{s_y s_y^*}\beta_{2_y}y}. \quad (9.102)$$

$$\breve{E}_y = H_0\tau\frac{\beta_{2_x}}{\omega\varepsilon_2}\sqrt{\frac{s_x^*}{s_x}}e^{-j\sqrt{s_x s_x^*}\beta_{2_x}x - j\sqrt{s_y s_y^*}\beta_{2_y}y}. \quad (9.103)$$

Despite the field splitting, continuity of the tangential electric and magnetic fields must be preserved across the $x = 0$ interface. To enforce this field continuity, we have $s_y = s_y^* = 1$ or, equivalently, $\sigma_y = 0 = \sigma_y^*$. This yields the phase-matching condition $\beta_{2_y} = \beta_{1_y} = k_1 \sin\theta$. Further, we derive the H-field reflection and transmission coefficients as:

$$\Gamma = \left(\frac{\beta_{1_x}}{\omega\varepsilon_1} - \frac{\beta_{2_x}}{\omega\varepsilon_2}\sqrt{\frac{s_x^*}{s_x}}\right) \times \left(\frac{\beta_{1_x}}{\omega\varepsilon_1} + \frac{\beta_{2_x}}{\omega\varepsilon_2}\sqrt{\frac{s_x^*}{s_x}}\right)^{-1}; \tau = 1+\Gamma. \quad (9.104a, b)$$

Reflectionless Matching Condition

Now, assume $\varepsilon_1 = \varepsilon_2$, $\mu_1 = \mu_2$, and $s_x = s_x^*$. This is equivalent to $k_1 = k_2$, $\eta_1 = \sqrt{\mu_1/\varepsilon_1} = \sqrt{\mu_2/\varepsilon_2}$, and $\sigma_x/\varepsilon_1 = \sigma_x^*/\mu_1$ (i.e., σ_x and σ_x^* satisfying equation 9.94 in a pairwise manner). With $\beta_{2_y} = \beta_{1_y}$, equation 9.101 now yields $\beta_{2_x} = \beta_{1_x}$. Substituting into 9.104a gives the reflectionless condition $\Gamma = 0$ for *all* incident angles regardless of frequency ω. For this case, equations 9.100, 9.102, 9.103, and 9.104b specify the following transmitted fields in the Berenger medium:

$$\breve{H}_z = H_0 e^{-js_x\beta_{1_x}x - j\beta_{1_y}y} = H_0 e^{-j\beta_{1_x}x - j\beta_{1_y}y}e^{-\sigma_x x\eta_1\cos\theta}. \quad (9.105)$$

$$\breve{E}_x = -H_0\eta_1\sin\theta e^{-j\beta_{1_x}x - j\beta_{1_y}y}e^{-\sigma_x x\eta_1\cos\theta}. \quad (9.106)$$

$$\breve{E}_y = H_0\eta_1\cos\theta e^{-j\beta_{1_x}x - j\beta_{1_y}y}e^{-\sigma_x x\eta_1\cos\theta}. \quad (9.107)$$

Within the matched Berenger medium, the transmitted wave propagates with the same speed and direction as the impinging wave while simultaneously undergoing exponential decay along the x-axis normal to the interface between regions 1 and 2. Further, the attenuation factor $\sigma_x\eta_1\cos\theta$ is independent of frequency. These desirable properties apply to all angles of incidence. Hence, Berenger's coining of the term "perfectly matched layer" makes excellent sense.

Structure of an FDTD Grid Employing Berenger's PML ABC

The analysis in the previous paragraphs can be repeated for PMLs that are normal to the y direction. This permitted Berenger to propose the two-dimensional TE_z FDTD grid shown in Figure 9.13, which uses PMLs to greatly reduce outer-boundary reflections. Here, a free-space computation zone is surrounded by PML backed by perfect electric conductor (PEC) walls. At the left and right sides of the grid (x_1 and x_2), each PML has σ_x and σ_x^* matched according to equation 9.94 along with $\sigma_y = 0 = \sigma_y^*$ to permit reflectionless transmission across the interface between the free space and PML regions. At the lower and upper sides of the grid (y_1 and y_2), each PML has σ_y and σ_y^* matched according to

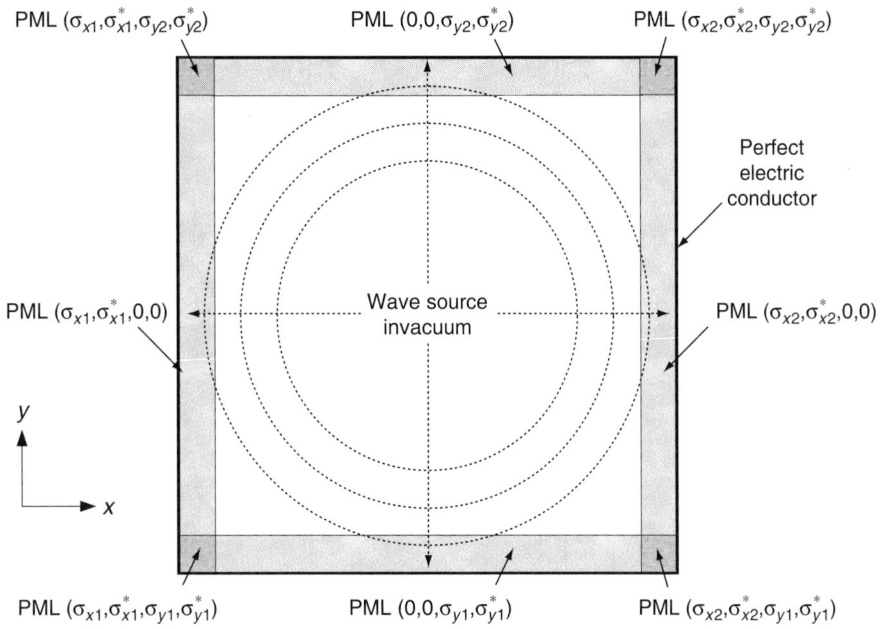

FIGURE 9.13 Structure of a 2-D TE_zFDTD Grid Employing the Berenger PML ABC. Figure adapted from Berenger (1994).

equation 9.94 along with $\sigma_x = 0 = \sigma^*_x$. At the four corners of the grid where there is overlap of two PMLs, all four losses (σ_x, σ^*_x, σ_y, and σ^*_y) are present and set equal to those of the adjacent PMLs.

Two-Dimensional TM_z Case

The analysis of the previous section can be repeated for the case of a TM_z-polarized incident wave wherein we implement the field splitting $E_z = E_{zx} + E_{zy}$. Analogous to equations 9.92, Maxwell's curl equations 9.11a to 9.11c as modified by Berenger are expressed in their time-dependent form as:

$$\mu_2 \frac{\partial H_x}{\partial t} + \sigma^*_y H_x = -\frac{\partial E_z}{\partial y}. \tag{9.108a}$$

$$\mu_2 \frac{\partial H_y}{\partial t} + \sigma^*_x H_y = \frac{\partial E_z}{\partial x}. \tag{9.108b}$$

$$\varepsilon_2 \frac{\partial E_{zx}}{\partial t} + \sigma_x E_{zx} = \frac{\partial H_y}{\partial x}. \tag{9.108c}$$

$$\varepsilon_2 \frac{\partial E_{zy}}{\partial t} + \sigma_y E_{zy} = -\frac{\partial H_x}{\partial y}. \tag{9.108d}$$

A derivation of the PML properties conducted in a manner analogous to that of the TE_z case yields slightly changed results. In most of the equations, the change is only a permutation of ε_2 with μ_2 and of σ with σ^*. However, the PML matching conditions are unchanged. This permits an absorbing reflectionless layer to be constructed adjacent to the outer grid boundary, as in the TE_z case.

Three-Dimensional Case

Katz *et al.* (1994) showed that Berenger's PML can be realized in three dimensions by splitting all six Cartesian field vector components. For example, the modified Ampere's law is given by:

$$\left(\varepsilon \frac{\partial}{\partial t} + \sigma_y\right) E_{xy} = \frac{\partial}{\partial y}(H_{zx} + H_{zy}). \tag{9.109a}$$

$$\left(\varepsilon \frac{\partial}{\partial t} + \sigma_z\right) E_{xz} = -\frac{\partial}{\partial z}(H_{yx} + H_{yz}). \tag{9.109b}$$

$$\left(\varepsilon \frac{\partial}{\partial t} + \sigma_z\right) E_{yz} = \frac{\partial}{\partial z}(H_{xy} + H_{xz}). \tag{9.109c}$$

$$\left(\varepsilon \frac{\partial}{\partial t} + \sigma_x\right) E_{yx} = -\frac{\partial}{\partial x}(H_{zx} + H_{zy}). \tag{9.109d}$$

$$\left(\varepsilon \frac{\partial}{\partial t} + \sigma_x\right) E_{zx} = \frac{\partial}{\partial x}(H_{yx} + H_{yz}). \tag{9.109e}$$

$$\left(\varepsilon \frac{\partial}{\partial t} + \sigma_y\right) E_{zy} = -\frac{\partial}{\partial y}(H_{xy} + H_{xz}). \tag{9.109f}$$

Similarly, the modified Faraday's law is given by:

$$\left(\mu \frac{\partial}{\partial t} + \sigma^*_y\right) H_{xy} = -\frac{\partial}{\partial y}(E_{zx} + E_{zy}). \tag{9.110a}$$

$$\left(\mu \frac{\partial}{\partial t} + \sigma^*_z\right) H_{xz} = \frac{\partial}{\partial z}(E_{yx} + E_{yz}). \tag{9.110b}$$

$$\left(\mu\frac{\partial}{\partial t}+\sigma_z^*\right)H_{yz} = -\frac{\partial}{\partial z}(E_{xy}+E_{xz}). \qquad (9.110c)$$

$$\left(\mu\frac{\partial}{\partial t}+\sigma_x^*\right)H_{yx} = \frac{\partial}{\partial x}(E_{zx}+E_{zy}). \qquad (9.110d)$$

$$\left(\mu\frac{\partial}{\partial t}+\sigma_x^*\right)H_{zx} = -\frac{\partial}{\partial x}(E_{yx}+E_{yz}). \qquad (9.110e)$$

$$\left(\mu\frac{\partial}{\partial t}+\sigma_y^*\right)H_{zy} = \frac{\partial}{\partial y}(E_{xy}+E_{xz}). \qquad (9.110f)$$

PML matching conditions analogous to the two-dimensional cases discussed previously are used. Specifically, if we denote $w = x, y, z$, the matching condition at a normal-to-w PML interface has the parameter pair (σ_w, σ_w^*) to satisfy equation 9.94. This causes the transmitted wave in the PML to undergo exponential decay in the $\pm w$ directions. All other (σ_w, σ_w^*) pairs in this PML are zero. In a corner region, the PML is provided with each matched (σ_w, σ_w^*) pair that is assigned to the overlapping PMLs forming the corner. Thus, PML media located in dihedral corner overlapping regions have two nonzero and one zero (σ_w, σ_w^*) pairs. PML media located in trihedral corner overlapping regions have three nonzero (σ_w, σ_w^*) pairs.

9.6.4 An Anisotropic PML Absorbing Medium

The split-field PML introduced by Berenger is a hypothetical, nonphysical medium based on a mathematical model. Due to the coordinate dependence of the loss terms, if such a physical medium exists, it must be anisotropic.

Indeed, a physical model based on an anisotropic, perfectly matched medium can be formulated. This was first discussed by Sacks *et al.* (1995). For a single interface, the anisotropic medium is uniaxial and is composed of both electric and magnetic constitutive tensors. This uniaxial PML, or UPML, performs as well as Berenger's PML while avoiding its nonphysical field splitting. Further, as summarized in Gedney and Taflove (2000), UPML has capabilities exceeding Berenger's PML by providing additional degrees of freedom which permit it to attenuate evanescent waves and terminate conductive and dispersive materials. The next subsection summarizes the frequency- and time-domain realizations of the UPML ABC.

Generalized Three-Dimensional Formulation: Frequency Domain

Following Gedney and Taflove (2000), we can write the time-harmonic Maxwell's curl equations in a three-dimensional UPML medium as:

$$\nabla\times\breve{\vec{H}} = j\omega\varepsilon\bar{\bar{s}}\breve{\vec{E}}; \quad \nabla\times\breve{\vec{E}} = -j\omega\mu\bar{\bar{s}}\breve{\vec{H}}, \qquad (9.111a, b)$$

where $\bar{\bar{s}}$ is the diagonal tensor defined by:

$$\bar{\bar{s}} = \begin{bmatrix} s_x^{-1} & 0 & 0 \\ 0 & s_x & 0 \\ 0 & 0 & s_x \end{bmatrix}\begin{bmatrix} s_y & 0 & 0 \\ 0 & s_y^{-1} & 0 \\ 0 & 0 & s_y \end{bmatrix}\begin{bmatrix} s_z & 0 & 0 \\ 0 & s_z & 0 \\ 0 & 0 & s_z^{-1} \end{bmatrix} = \begin{bmatrix} s_ys_zs_x^{-1} & 0 & 0 \\ 0 & s_xs_zs_y^{-1} & 0 \\ 0 & 0 & s_xs_ys_z^{-1} \end{bmatrix} \qquad (9.112)$$

Allowing for a nonunity real part κ, the multiplicative components of the diagonal elements of $\bar{\bar{s}}$ are given by:

$$s_x = \kappa_x + \frac{\sigma_x}{j\omega\varepsilon}; \; s_y = \kappa_y + \frac{\sigma_y}{j\omega\varepsilon}; \; s_z = \kappa_z + \frac{\sigma_z}{j\omega\varepsilon}. \qquad (9.113a, b, c)$$

If desired, the UPML medium can be specialized to apply throughout the entire FDTD space lattice. The following lists all of the possible special cases.

Lossless, Isotropic Interior Zone

The tensor $\bar{\bar{s}}$ is an identity tensor realized by setting $s_x = s_y = s_z = 1$ in equation 9.112. This requires $\sigma_x = \sigma_y = \sigma_z = 0$ and $\kappa_x = \kappa_y = \kappa_z = 1$ in equations 9.113.

UPML absorbers at x_{min} and x_{max} outer-boundary planes:

We set $s_y = s_z = 1$ in equation 9.112. This requires $\sigma_y = \sigma_z = 0$ and $\kappa_y = \kappa_z = 1$ in equations 9.113

UPML absorbers at y_{min} and y_{max} outer-boundary planes:

We set $s_x = s_z = 1$ in equation 9.112. This requires $\sigma_x = \sigma_z = 0$ and $\kappa_x = \kappa_z = 1$ in equations 9.113.

UPML absorbers at z_{min} and z_{max} outer-boundary planes:

We set $s_x = s_y = 1$ in equation 9.112. This requires $\sigma_x = \sigma_y = 0$ and $\kappa_x = \kappa_y = 1$ in equations 9.113.

Overlapping UPML absorbers at x_{min}, x_{max} and y_{min}, y_{max} dihedral corners:

We set $s_z = 1$ in equation 9.112. This requires $\sigma_z = 0$ and $\kappa_z = 1$ in equations 9.113.

Overlapping UPML absorbers at x_{min}, x_{max} and z_{min}, z_{max} dihedral corners:

We set $s_y = 1$ in equation 9.112. This requires $\sigma_y = 0$ and $\kappa_y = 1$ in equations 9.113.

Overlapping UPML absorbers at y_{min}, y_{max} and z_{min}, z_{max} dihedral corners:

We set $s_x = 1$ in equation 9.112. This requires $\sigma_x = 0$ and $\kappa_x = 1$ in equations 9.113.

Overlapping UPML absorbers at all trihedral corners:

We use the complete general tensor in equation 9.112.

Note that the generalized constitutive tensor defined in equation 9.112 is not uniaxial by strict definition but rather is anisotropic. The anisotropic PML, however, is still referenced as uniaxial since it is uniaxial in the nonoverlapping PML regions.

Efficient Implementation of UPML in FDTD

Starting with equations 9.111a and 9.112, Ampere's law in a matched UPML is expressed as:

$$\begin{bmatrix} \frac{\partial \breve{H}_z}{\partial y} - \frac{\partial \breve{H}_y}{\partial z} \\ \frac{\partial \breve{H}_x}{\partial z} - \frac{\partial \breve{H}_z}{\partial x} \\ \frac{\partial \breve{H}_y}{\partial x} - \frac{\partial \breve{H}_x}{\partial y} \end{bmatrix} = j\omega\varepsilon \begin{bmatrix} \frac{s_y s_z}{s_x} & 0 & 0 \\ 0 & \frac{s_x s_z}{s_y} & 0 \\ 0 & 0 & \frac{s_x s_y}{s_z} \end{bmatrix} \begin{bmatrix} \breve{E}_x \\ \breve{E}_y \\ \breve{E}_z \end{bmatrix}, \tag{9.114}$$

where s_x, s_y, and s_z are defined in equations 9.113. Directly inserting equation 9.113 into equation 9.114 and then transforming into the time domain would lead to a convolution between the tensor coefficients and the E-field. This is not advisable because implementing this convolution would be computationally intensive. As shown by Gedney (1995, 1996), a much more efficient approach is to define the proper constitutive relationship to decouple the frequency-dependent terms. Specifically, let:

$$\breve{D}_x = \varepsilon \frac{s_z}{s_x} \breve{E}_x; \quad \breve{D}_y = \varepsilon \frac{s_x}{s_y} \breve{E}_y; \quad \breve{D}_z = \varepsilon \frac{s_y}{s_z} \breve{E}_z. \tag{9.115a, b, c}$$

Then, equation 9.114 is rewritten as:

$$\begin{bmatrix} \frac{\partial \breve{H}_z}{\partial y} - \frac{\partial \breve{H}_y}{\partial z} \\ \frac{\partial \breve{H}_x}{\partial z} - \frac{\partial \breve{H}_z}{\partial x} \\ \frac{\partial \breve{H}_y}{\partial x} - \frac{\partial \breve{H}_x}{\partial y} \end{bmatrix} = j\omega \begin{bmatrix} s_y & 0 & 0 \\ 0 & s_z & 0 \\ 0 & 0 & s_x \end{bmatrix} \begin{bmatrix} \breve{D}_x \\ \breve{D}_y \\ \breve{D}_z \end{bmatrix}. \tag{9.116}$$

Now, we substitute s_x, s_y, and s_z from equations 9.113 into 9.116, and then apply the inverse Fourier transform using the identity $j\omega f(\omega) \to (\partial/\partial t) f(t)$. This yields an equivalent system of time-domain differential equations for equation 9.116:

$$\begin{bmatrix} \frac{\partial H_z}{\partial y} - \frac{\partial H_y}{\partial z} \\ \frac{\partial H_x}{\partial z} - \frac{\partial H_z}{\partial x} \\ \frac{\partial H_y}{\partial x} - \frac{\partial H_x}{\partial y} \end{bmatrix} = \frac{\partial}{\partial t} \begin{bmatrix} \kappa_y & 0 & 0 \\ 0 & \kappa_z & 0 \\ 0 & 0 & \kappa_x \end{bmatrix} \begin{bmatrix} D_x \\ D_y \\ D_z \end{bmatrix} + \frac{1}{\varepsilon} \begin{bmatrix} \sigma_y & 0 & 0 \\ 0 & \sigma_z & 0 \\ 0 & 0 & \sigma_x \end{bmatrix} \begin{bmatrix} D_x \\ D_y \\ D_z \end{bmatrix}. \tag{9.117}$$

The system of 9.117 equations can be discretized on the standard Yee lattice. It is suitable to use normal leapfrogging in time wherein the loss terms are time-averaged according to the semi-implicit scheme. This leads to explicit time-stepping expressions for D_x, D_y, and D_z. For example, the D_x update is given by:

$$D_x\big|_{i+1/2,\,j,\,k}^{n+1} = \left(\frac{2\varepsilon\kappa_y - \sigma_y\Delta t}{2\varepsilon\kappa_y + \sigma_y\Delta t}\right) D_x\big|_{i+1/2,\,j,\,k}^{n} + \left(\frac{2\varepsilon\Delta t}{2\varepsilon\kappa_y + \sigma_y\Delta t}\right) \times \left(\frac{H_z\big|_{i+1/2,j+1/2,k}^{n+1/2} - H_z\big|_{i+1/2,j-1/2,k}^{n+1/2}}{\Delta y} - \frac{H_y\big|_{i+1/2,j,k+1/2}^{n+1/2} - H_y\big|_{i+1/2,j,k-1/2}^{n+1/2}}{\Delta z}\right) \tag{9.118}$$

Next, we focus on equations 9.115a through 9.115c. For example, we consider 9.115a. After multiplying both sides by s_x and substituting for s_x and s_z from 9.113a and 9.113c, we have:

$$\left(\kappa_x + \frac{\sigma_x}{j\omega\varepsilon}\right)\breve{D}_x = \varepsilon\left(\kappa_z + \frac{\sigma_z}{j\omega\varepsilon}\right)\breve{E}_x. \tag{9.119}$$

Multiplying both sides by $j\omega$ and transforming into the time domain leads to:

$$\frac{\partial}{\partial t}(\kappa_x D_x) + \frac{\sigma_x}{\varepsilon} D_x = \varepsilon\left[\frac{\partial}{\partial t}(\kappa_z E_x) + \frac{\sigma_z}{\varepsilon} E_x\right]. \tag{9.120a}$$

Similarly, from equations 9.115b and 9.115c, we obtain:

$$\frac{\partial}{\partial t}(\kappa_y D_y) + \frac{\sigma_y}{\varepsilon} D_y = \varepsilon\left[\frac{\partial}{\partial t}(\kappa_x E_y) + \frac{\sigma_x}{\varepsilon} E_y\right]. \tag{9.120b}$$

$$\frac{\partial}{\partial t}(\kappa_z D_z) + \frac{\sigma_z}{\varepsilon} D_z = \varepsilon\left[\frac{\partial}{\partial t}(\kappa_y E_z) + \frac{\sigma_y}{\varepsilon} E_z\right]. \tag{9.120c}$$

The time derivatives in equations 9.120 are discretized using standard Yee leapfrogging and time-averaging the loss terms. This yields explicit time-stepping expressions for E_x, E_y, and E_z. For example, the E_x update is given by:

$$E_x\big|_{i+1/2,j,k}^{n+1} = \left(\frac{2\varepsilon\kappa_z - \sigma_z\Delta t}{2\varepsilon\kappa_z + \sigma_z\Delta t}\right) E_x\big|_{i+1/2,j,k}^{n} + \left[\frac{1}{(2\varepsilon\kappa_z + \sigma_z\Delta t)\varepsilon}\right] \times \left[(2\varepsilon\kappa_x + \sigma_x\Delta t) D_x\big|_{i+1/2,j,k}^{n+1} - (2\varepsilon\kappa_x - \sigma_x\Delta t) D_x\big|_{i+1/2,j,k}^{n}\right] \tag{9.121}$$

Overall, updating the components of $\vec{E}$ in the UPML requires two steps in sequence: (1) obtaining the new values of the components of $\vec{D}$ according to equation 9.118, and (2) using these new $\vec{D}$ components to obtain new values of the $\vec{E}$ components according to equation 9.121.

A similar two-step procedure is required to update the components of $\vec{H}$ in the UPML. Starting with Faraday's law in equations 9.111b and 9.112, the first step involves developing the updates for the components of $\vec{B}$. A procedure analogous to that followed in obtaining equation 9.118 yields, for example, the following update for B_x:

$$B_x\Big|_{i,j+1/2,k+1/2}^{n+3/2} = \left(\frac{2\varepsilon\kappa_y - \sigma_y\Delta t}{2\varepsilon\kappa_y + \sigma_y\Delta t}\right) B_x\Big|_{i,j+1/2,k+1/2}^{n+1/2} - \left(\frac{2\varepsilon\Delta t}{2\varepsilon\kappa_y + \sigma_y\Delta t}\right) \times \left(\frac{E_z\big|_{i,j+1,k+1/2}^{n+1} - E_z\big|_{i,j,k+1/2}^{n+1}}{\Delta y} - \frac{E_y\big|_{i,j+1/2,k+1}^{n+1} - E_y\big|_{i,j+1/2,k}^{n+1}}{\Delta z}\right) \tag{9.122}$$

The second step involves updating the $\vec{H}$ components in the UPML using the values of the $\vec{B}$ components just obtained with equation 9.122 and similar expressions for B_y and B_z. For example, employing the dual constitutive relation $\breve{B}_x = \mu(s_z/s_x)\breve{H}_x$, a procedure analogous to that followed in obtaining equation 9.121 yields the following update for H_x:

$$H_x\Big|_{i,j+1/2,k+1/2}^{n+3/2} = \left(\frac{2\varepsilon\kappa_z - \sigma_z\Delta t}{2\varepsilon\kappa_z + \sigma_z\Delta t}\right) H_x\Big|_{i,j+1/2,k+1/2}^{n+1/2} + \left[\frac{1}{(2\varepsilon\kappa_z + \sigma_z\Delta t)\mu}\right] \times \left[(2\varepsilon\kappa_x + \sigma_x\Delta t) B_x\Big|_{i,j+1/2,k+1/2}^{n+3/2} - (2\varepsilon\kappa_x - \sigma_x\Delta t) B_x\Big|_{i,j+1/2,k+1/2}^{n+1/2}\right] \tag{9.123}$$

Similar expressions can be derived for H_y and H_z.

Nehrbass *et al.* (1996) showed that such an algorithm is numerically stable within the Courant limit. Further, Abarbanel and Gottlieb (1997) showed that the resulting discrete fields satisfy Gauss's law, and the UPML is well posed. Finally, Gedney and Taflove (2000) discussed how the UPML formulation in both the frequency domain and the time domain can be extended to terminate material regions in the FDTD space grid that are either electrically lossy or have frequency-dispersive dielectric parameters.

9.6.5 Theoretical Performance of the PML

The Continuous Space

When used to truncate an FDTD lattice, the PML has a thickness d and is terminated by the outer boundary of the lattice. If the outer boundary is assumed to be a PEC wall, finite power reflects back into the primary computation zone. For a wave impinging on the PML at angle θ relative to the w-directed surface normal, this reflection can be computed using transmission line analysis, yielding:

$$R(\theta) = e^{-2\sigma_w \eta d \cos\theta} \tag{9.124}$$

Here, η and σ_w are, respectively, the PML's characteristic wave impedance and its conductivity, referred to propagation in the w direction. In the context of an FDTD simulation, $R(\theta)$ is referred to as the "reflection error" since it is a nonphysical reflection due to the PEC wall that backs the PML. We note that the reflection error is the same for both the split-field PML and the UPML, because both support the same wave equation. This error decreases exponentially with σ_w and d. However, the reflection error increases as $\exp(\cos\theta)$, reaching the worst case for $\theta = 90°$. At this grazing angle of incidence, $R = 1$ and the PML is completely ineffective. To be useful in an FDTD simulation, we want $R(\theta)$ to be as small as possible. Clearly, for a thin PML, we must have σ_w as large as possible to reduce $R(\theta)$ to acceptably small levels, especially for θ approaching 90°.

The Discrete Space

Grading of the PML Loss Parameters

Theoretically, reflectionless wave transmission can take place across a PML interface regardless of the local step-discontinuity in σ and σ^* presented to the continuous impinging electromagnetic field. However, in FDTD or any discrete representation of Maxwell's equations, numerical artifacts arise due to the finite spatial sampling. Consequently, implementing PML as a single-step discontinuity of σ and σ^* in the FDTD lattice leads to significant spurious wave reflection at the PML surface.

To reduce this reflection error, Berenger (1994) proposed that the PML losses gradually rise from zero along the direction normal to the interface. Assuming such a grading, the PML remains matched. Pursuing this idea, we consider as an example an x-directed plane wave impinging at angle θ upon a PEC-backed PML slab of thickness d, with the front planar interface located in the $x = 0$ plane. Assuming the graded PML conductivity profile $\sigma_x(x)$, we can obtain:

$$R(\theta) = e^{-2\eta\cos\theta \int_0^d \sigma_x(x)dx}. \tag{9.125}$$

Several profiles have been suggested for grading $\sigma_x(x)$ (and $\kappa_x(x)$ in the context of the UPML). One of the most successful uses a polynomial variation of the PML loss with depth x. Polynomial grading is simply:

$$\sigma_x(x) = (x/d)^m \sigma_{x,\max}\ ;\ \kappa_x(x) = 1 + (\kappa_{x,\max} - 1)\cdot(x/d)^m. \tag{9.126a, b}$$

This increases the value of the PML σ_x from zero at $x = 0$ (the surface of the PML) to $\sigma_{x,\max}$ at $x = d$ (the PEC outer boundary). Similarly, for the UPML, κ_x increases from one at $x = 0$ to $\kappa_{x,\max}$ at $x = d$. Substituting equation 9.126a into 9.125 yields:

$$R(\theta) = e^{-2\eta\sigma_{x,\max} d\cos\theta/(m+1)}. \tag{9.127}$$

For a fixed d, polynomial grading provides two parameters: $\sigma_{x,\max}$ and m. A large m yields a $\sigma_x(x)$ distribution that is relatively flat near the PML surface. However, deeper in the

PML, σ_x increases more rapidly than for small m. In this region, the field amplitudes are substantially decayed and reflections due to the discretization error contribute less. Typically, $3 \leq m \leq 4$ has been found to be nearly optimal for many FDTD simulations (Berenger, 1996).

For polynomial grading, the PML parameters can be readily determined for a given error estimate. For example, let m, d, and the desired reflection error $R(0)$ be known. Then, from equation 9.127, $\sigma_{x,\,\max}$ is computed as:

$$\sigma_{x,\,\max} = -\frac{(m+1)\ln[R(0)]}{2\eta d}. \qquad (9.128)$$

Discretization Error

The design of an effective PML requires balancing the theoretical reflection error $R(\theta)$ and the numerical discretization error. For example, equation 9.128 provides $\sigma_{x,\,\max}$ for a polynomial-graded conductivity given a predetermined $R(0)$ and m. If $\sigma_{x,\,\max}$ is too small, the primary reflection from the PML is due to its PEC backing, and equation 9.125 provides a fairly accurate approximation of the reflection error. However, if $\sigma_{x,\,\max}$ is too large, the discretization error due to the FDTD approximation dominates, and the actual reflection error is potentially much higher than what equation 9.125 predicts. Consequently, there is an optimal choice for $\sigma_{x,\,\max}$ that balances reflection from the PEC outer boundary and discretization error.

Through extensive numerical experimentation, Gedney (1996) and He (1997) found that, for a broad range of applications, an optimal choice for a 10-cell-thick, polynomial-graded PML is $R(0) \approx e^{-16}$. For a 5-cell-thick PML, $R(0) \approx e^{-8}$ is optimal. From equation 9.128, this leads to an optimal $\sigma_{x,\,\max}$ for polynomial grading:

$$\sigma_{x,\,\text{opt}} \approx -\frac{(m+1)\cdot(-16)}{(2\eta)\cdot(10\Delta)} = \frac{0.8(m+1)}{\eta\Delta}. \qquad (9.129)$$

This expression has proven to be adequate for many applications. Its value, however, may be too large when the PML terminates highly elongated resonant structures or sources with a very long time duration. For detailed discussions and examples of PML performance over a wide range of possible loss profiles, see Gedney (1998) and Gedney and Taflove (2000).

9.7 Examples of FDTD Modeling Applications

Current examples of FDTD modeling applications span much of the electromagnetic spectrum. A worldwide FDTD developer/user community is involved in numerical simulation of electrodynamic phenomena ranging from extremely low-frequency (ELF) propagation about the entire earth to the lasing behavior of aggregates of micron-scale particles exhibiting four-level quantum-system characteristics. Such applications of FDTD vividly illustrate the principle that "Maxwell's equations work from dc to light." The goal of this section is to provide examples of current and emerging FDTD applications that literally span this range.

9.7.1 Global ELF Propagation in the Earth's Ionosphere Waveguide

Global propagation of extremely low-frequency (ELF: 3 Hz – 3 *kHz*) and very low-frequency (VLF: 3–30 kHz) electromagnetic waves in the earth-ionosphere waveguide is a problem having a rich history of theoretical investigation extending over many years (Wait, 1965). ELF/VLF propagation phenomena form the physics basis of submarine communications and remote-sensing investigations of lightning and sprites, global temperature change, subsurface structures, and potential earthquake precursors.

Most theoretical techniques for modeling ELF/VLF propagation in the earth-ionosphere waveguide are based on frequency-domain waveguide mode theory. However, these techniques cannot account for arbitrary horizontal as well as vertical geometrical and electrical inhomogeneities of the ionosphere, continents, and oceans. Recently, Simpson and Taflove (2002, 2004) and Otsuyama *et al.* (2003) have developed two- and three-dimensional FDTD models of impulsive ELF propagation around the entire surface of the earth. Periodic boundary conditions are used in conjunction with a latitude-longitude grid that wraps around the complete earth-sphere. These models have been verified by numerical studies of frequency-dependent propagation attenuation with distance, antipodal propagation, and the Schumann resonance. Figure 9.14 visualizes the results of one such model (Simpson and Taflove [2004]) which illustrates the propagation completely around the Earth of an ELF electromagnetic pulse generated by a vertical lightning stroke off the coast of South America. In this model, all features of the lithosphere and ionosphere located within ±100 km of sea level are modeled with a resolution of approximately 40 × 40 × 5 km.

9.7.2 High-Speed Electronics

High-speed electronic circuits have been traditionally grouped into two classes: **analog microwave circuits** and **digital logic circuits**.

(1) Microwave circuits typically process band-pass signals at frequencies above 3 GHz. Common circuit features include microstrip transmission lines, directional couplers, circulators, filters, matching networks, and individual transistors. Circuit operation is fundamentally based on electromagnetic wave phenomena.

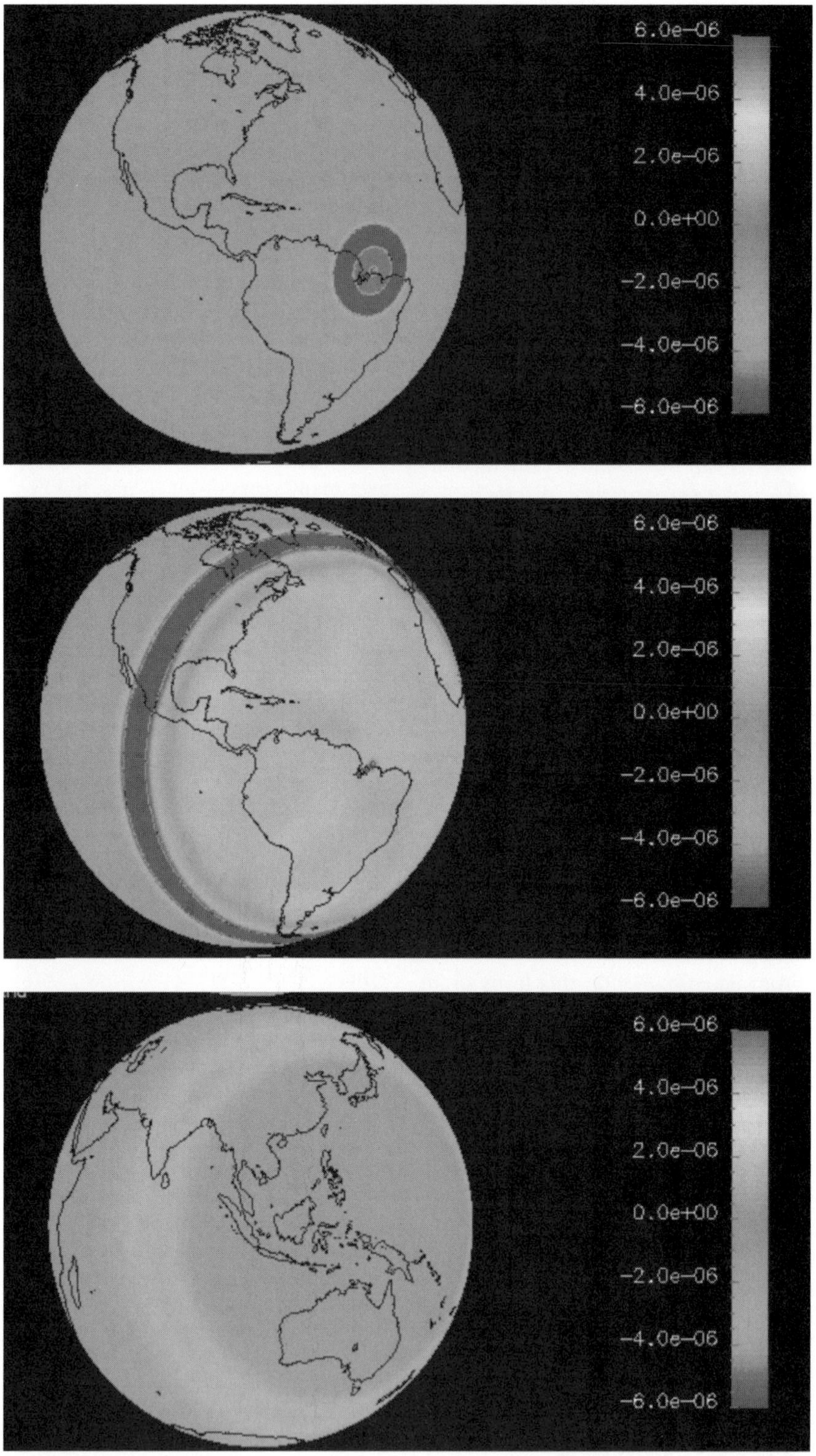

FIGURE 9.14 Snapshot Visualizations of the FDTD-Computed Global Propagation of an ELF Electromagnetic Pulse Generated by a Lightning Strike Off the Coast of South America (Simpson and Taflove [2004]). All features of the lithosphere and atmosphere located within ± 100 km of sea level are modeled in three dimensions with a resolution of approximately $40 \times 40 \times 5$ km.

(2) Digital circuits typically process low-pass pulses having clock rates below 3 GHz. Typical circuits include densely packed, multiple planes of metal traces providing flow paths for the signals, dc power feeds, and ground returns. Via pins provide electrical connections between the planes. Circuit operation is nominally *not* based on electromagnetic wave effects.

The distinction between the design of these two classes, however, is blurring. Microwave circuits are becoming very complex systems comprised of densely packed elements. On the digital-circuit side, the rise of everyday clock speeds to 3 GHz implies low-pass signal bandwidths above 10 GHz, which are well into the microwave range. Electromagnetic wave effects that, until now, were in the domain of the microwave engineer are becoming a limiting factor in digital-circuit operation. For example, hard-won experience has shown that high-speed digital signals can spuriously:

- Distort as they propagate along the metal circuit paths
- Couple (cross talk) from one circuit path to another
- Radiate and create interference to other circuits and systems

An example of electromagnetic field effects in a digital circuit is shown in Figure 9.15, which illustrates the results of applying FDTD modeling to calculate the coupling and cross talk of a high-speed logic pulse entering and leaving a microchip embedded in a conventional dual in-line integrated-circuit package. The fields associated with the logic pulse are not confined to the metal circuit paths and, in fact, smear out and couple to all adjacent circuit paths.

9.7.3 Microwave Penetration and Coupling

Computational electromagnetics has played an important role in helping to assess and mitigate the effects of high-level electromagnetic wave coupling into sensitive electrical and electronic equipment. The primary sources of such waves include **lightning, nuclear electromagnetic pulse** (NEMP), and **high-power microwaves** (HPM).

NEMP can burn out electrical and electronic equipment on the earth's surface located many hundreds of miles away from the detonation of a nuclear bomb above the earth's atmosphere. Equipment failures on this geographical scale could leave a nation largely defenseless against subsequent attack. HPM neutralize electronics in the same manner as NEMP but can be applied on a more selective basis for either tactical or strategic applications. For both NEMP and HPM, computational solutions of Maxwell's equations have been used to understand the complex electromagnetic wave penetration and coupling mechanisms into potential targets and create means to mitigate these mechanisms. FDTD is particularly useful for these purposes since it can model extremely complicated three-dimensional structures in a straightforward manner. Figure 9.16 illustrates the results of

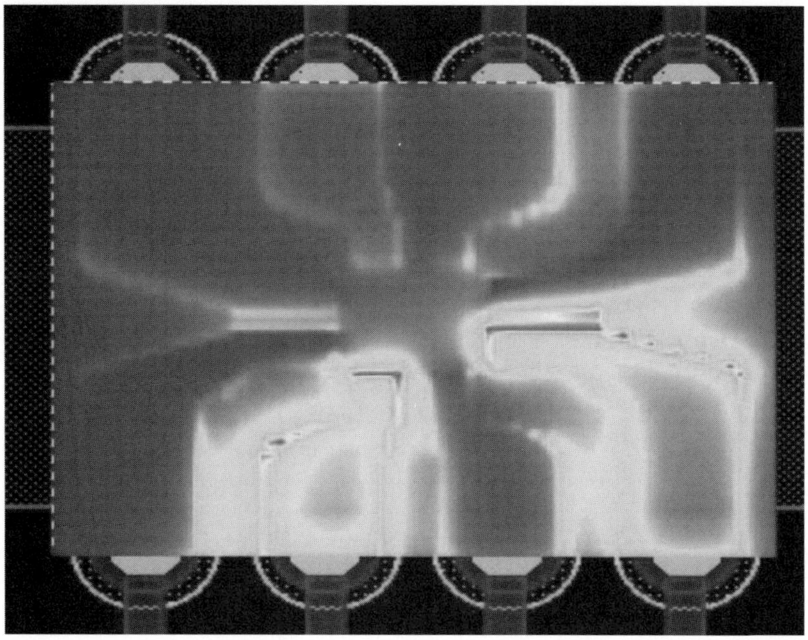

FIGURE 9.15 Snapshot Visualization of FDTD-Computed Coupling and Crosstalk of a High-Speed Logic Pulse. The pulse is entering and leaving a microchip embedded in a conventional dual in-line integrated-circuit package. Courtesy of Prof. M. Picket-May, University of Colorado-Boulder.

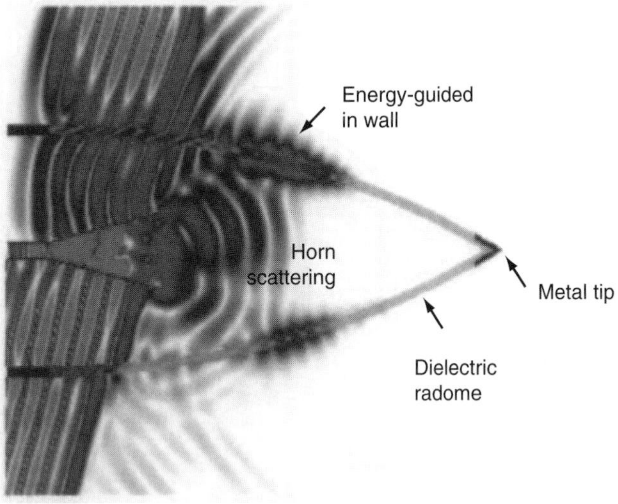

FIGURE 9.16 Snapshot Visualization of the FDTD-Computed Penetration of a Microwave Pulse into a Missile Radome Containing a Horn Antenna. The impinging plane wave propagates from right to left and is obliquely incident at 15° from boresight. Courtesy of Maloney *et al.* (2000).

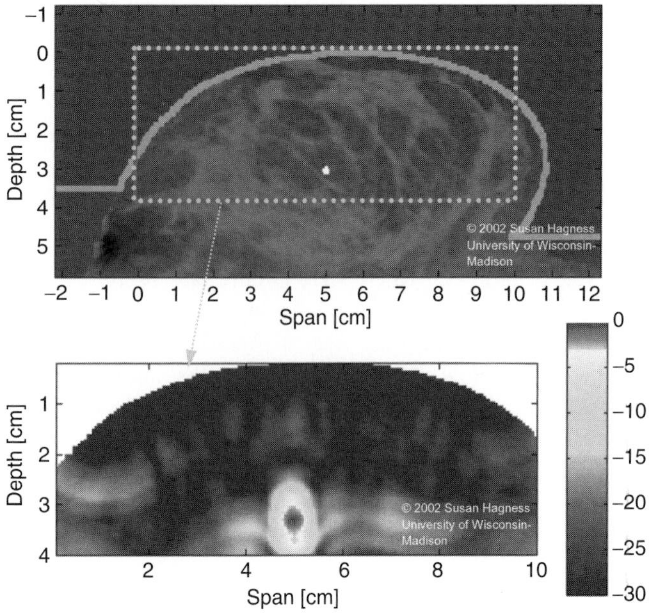

FIGURE 9.17 Top: MRI-derived Breast Model. This image shows the location of a 2-mm diameter malignant breast tumor at a depth of 3 cm. Bottom: image reconstructed from backscattered waveforms obtained by FDTD modeling. The scale is in decibels. Note that the tumor signature is 15–30 dB (30 to 1,000 times) stronger than the backscattering clutter due to the surrounding normal tissues. Courtesy of Bond *et al.* (2003).

applying FDTD modeling to calculate the interaction of a 10-GHz radar beam with a missile radome containing a horn antenna.

9.7.4 Ultrawideband Microwave Imaging for Early Stage Breast Cancer Detection

Recently, several researchers have conducted theoretical investigations of the use of ultrawideband microwave pulses for early stage breast cancer detection. In principle, this technique could detect smaller tumors over larger regions of the breast, a detection that is not currently possible using X-ray mammography. This technique would also help prevent a patient's exposure to potentially hazardous ionizing radiation. In this proposed technique, an array of small antennas would be placed on the surface of the breast to emit and then receive a short electromagnetic pulse lasting less than 100 ps. Signal-processing techniques would then be applied to the received pulses at each antenna element to form the breast image. In work to date, FDTD modeling has provided simulated test data and allowed optimization of the imaging algorithms. As shown in Figure 9.17, these simulations indicate promise for imaging small, deeply embedded malignant tumors in the presence of the background clutter due to complicated surrounding normal tissues.

9.7.5 Photonic Integrated Circuits

Microcavity ring and disk resonators are proposed components for filtering, routing, switching, modulation, and multiplexing/demultiplexing tasks in ultrahigh-speed photonic integrated circuits. Figure 9.18 is a scanning electron microscope image of a portion of a prototype photonic circuit comprised of 5.0-μm diameter aluminum gallium arsenide (AlGaAs) microcavity disk resonators coupled to 0.3-μm wide optical waveguides across air gaps spanning 0.1 to 0.3 μm.

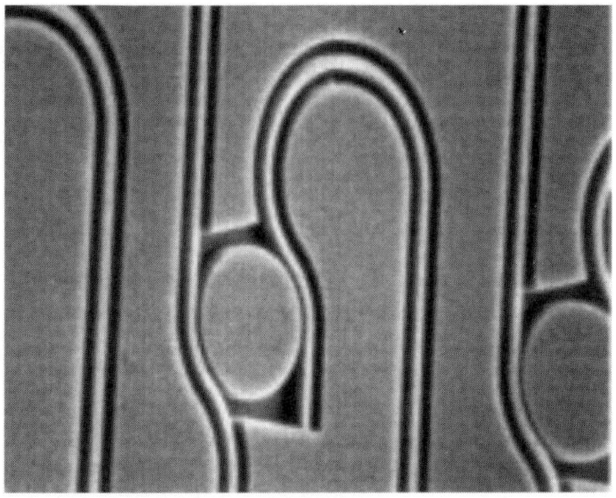

FIGURE 9.18 Scanning Electron Microscope Image of a Prototype Photonic Integrated Circuit. The photonic circuit is comprised of 5.0-μm diameter AlGaAs microcavity disk resonators coupled to 0.3-μm-wide AlGaAs optical waveguides across air gaps spanning as little as 0.1 μm. Courtesy of Prof. S.-T. Ho, Northwestern University.

By computationally solving Maxwell's equations, the coupling, transmission, and resonance behavior of the micro-optical structures in Figure 9.18 can be determined. This permits effective engineering design. For example, Figure 9.19 shows visualizations of the FDTD-calculated sinusoidal steady-state optical electric field distributions for a typical microdisk in Figure 9.18. In the upper left panel, the optical excitation is at a nonresonant frequency, 193.4 THz (an optical wavelength, λ, of 1.55 μm). Here, 99.98% of the rightward-directed power in the incident signal remains in the lower waveguide. In the upper right panel, the excitation is at the resonant frequency of the first-order radial whispering-gallery mode of the microdisk, 189.2 THz ($\lambda = 1.585\,\mu$m). Here, there is a large field enhancement in the microdisk, and 99.79% of the incident power switches to the upper waveguide in the reverse (leftward) direction. This yields the action of a passive, wavelength-selective switch.

The lower left and lower right panels of Figure 9.19 are visualizations at, respectively, the resonant frequencies of the second- and third-order whispering-gallery modes: 191.3 THz

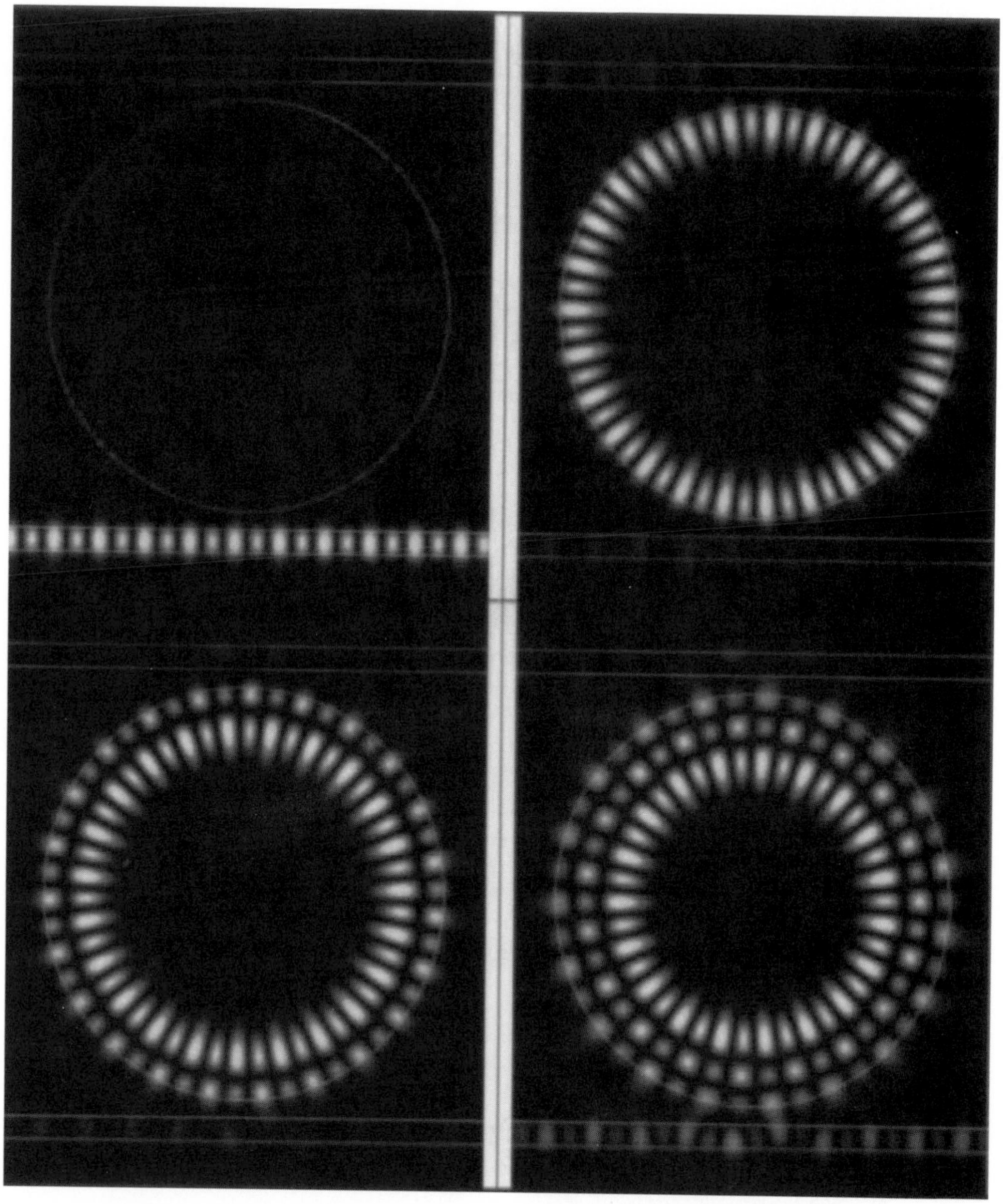

FIGURE 9.19 Visualizations Illustrating the FDTD-Calculated Sinusoidal Steady-State Optical Electric Field Distributions in a 5.0-μm Diameter GaAlAs Microdisk Resonator. The resonator is coupled to straight 0.3-μm wide GaAlAs optical waveguides for single-frequency excitations propagating to the right in the lower waveguide. Upper left: Off-resonance signal; Upper right: On-resonance signal, first-order radial mode; Lower left: Second-order radial-mode resonance; Lower right: Third-order radial mode resonance. Courtesy of Hagness *et al.* (1997).

($\lambda = 1.567\,\mu m$) and 187.8 THz ($\lambda = 1.596\,\mu m$). A goal of current design efforts is to suppress such higher order modes to allow the use of microdisks as passive wavelength-division multiplexing devices having low cross talk across a wide spectrum or as active single-mode laser sources.

9.7.6 Light Switching: Light in Femtoseconds

Recent optical systems can generate laser pulses down to 6 fs in duration. From a technology standpoint, it is clear that controlling or processing these short pulses involves understanding the nature of their interactions with materials over enormous bandwidths and very likely in beam-intensity regimes where material nonlinearity plays an important role. A key factor here is material dispersion, having two components: (1) **linear dispersion** (i.e., the variation of the material's index of refraction with frequency at low optical power levels) and (2) **nonlinear dispersion** (i.e., the variation of the frequency-dependent refractive index with optical power).

Advances in FDTD computational techniques for solving Maxwell's equations have provided the basis for modeling both linear and nonlinear dispersions over ultrawide bandwidths. An example of the possibilities for such modeling is shown in Figure 9.20, which is a sequence of snapshot visualizations of the dynamics of a potential femtosecond all-optical switch. This switch uses the collision and subsequent "bouncing" of pulsed optical spatial **solitons**, which are pulsed beams of laser light that are prevented from spreading in their transverse directions by the optical focusing effect of the nonlinear material in which they propagate.

Referring to the top panel of Figure 9.20, this switch would inject an in-phase, 100-fs pulsed signal and control solitons

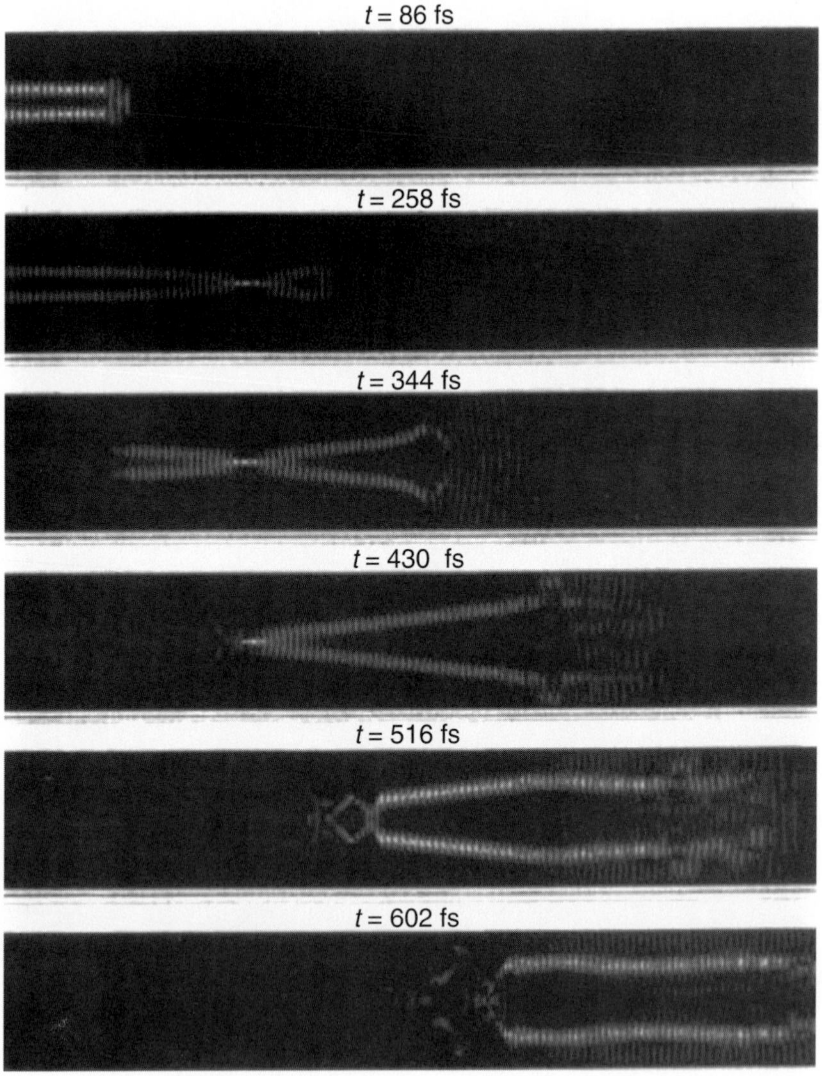

FIGURE 9.20 Sequential Snapshot Visualizations of the FDTD-Computed Electric Field of Equal-Amplitude, In-Phase, 100-fs Optical Spatial Solitons Copropagating in Glass. The optical pulses propagate from left to right at the speed of light in glass. Courtesy of Joseph and Taflove (1994).

into ordinary glass (a Kerr-type nonlinear material) from a pair of optical waveguides on the left side. Each pulsed beam has a 0.65-μm transverse width, while the beam-to-beam spacing is in the order of 1 μm. In the absence of the control soliton, the signal soliton would propagate to the right at the speed of light in glass with zero deflection and then be collected by a receiving waveguide. However, as shown in the remaining panels of this figure, a copropagating control soliton would first merge with and then laterally deflect the signal soliton to an alternate collecting waveguide. (Curiously, the location of the two beams' merger point would remain stationary in space while the beams would propagate at light-speed through this point.) Overall, the optical pulse dynamics shown in Figure 9.20 could provide the action of an all-optical "AND" gate working on a time scale about 1/10,000th that of existing electronic digital logic.

Application of FDTD solutions of Maxwell's equations to problems such as the one illustrated in this figure shows great promise in allowing the detailed study of a variety of novel nonlinear optical wave species that may one day be used to implement all-optical switching circuits (light switching light), attaining speeds 10,000 times faster than those of the best semiconductor circuits today. The implications may be profound for the realization of "optonics," a proposed successor technology to electronics that would integrate optical-fiber interconnects and optical microchips into systems of unimaginable information-processing capability.

9.8 Summary and Conclusions

This chapter reviewed key elements of the theoretical foundation and numerical implementation of finite-difference time-domain (FDTD) solutions of Maxwell's equations. The chapter included:

- Introduction and background
- Review of Maxwell's equations
- The Yee algorithm
- Theory of numerical dispersion
- Theory of numerical stability
- Perfectly matched layer (PML) absorbing boundary conditions
- Examples of FDTD modeling applications

With literally hundreds of papers on FDTD methods and applications published each year, it is clear that FDTD is one of the most powerful and widely used numerical modeling approaches for electromagnetic wave interaction problems. With expanding developer and user communities in an increasing number of disciplines in science and engineering, FDTD technology is continually evolving in terms of its theoretical basis, numerical implementation, and technological applications, which now span the spectral range from ELF to daylight.

Bibliography

Abarbanel, S., and Gottlieb, D. (1997). A mathematical analysis of the PML method. *Journal of Computational Physics 134*, 357–363.

Berenger, J.P. (1994). A perfectly matched layer for the absorption of electromagnetic waves. *Journal of Computational Physics 114*, 185–200.

Berenger, J.P. (1996). Perfectly matched layer for the FDTD solution of wave-structure interaction problems. *IEEE Transactions Antennas Propagat.* 44, 110–117.

Berenger, J.P. (1997). Improved PML for the FDTD solution of wave-structure interaction problems. *IEEE Transactions Antennas Propagat. 45*, 466–473.

Bond, E.J., Li, X., Hagness, S.C., and Van Veen, B.D. (2003). Microwave imaging via space-time beamforming for early detection of breast cancer. *IEEE Transactions on Antennas and Propagation* 51, 1690–1705.

Chew, W.C., and Jin, J.M. (1996). Perfectly matched layers in the discretized space: An analysis and optimization. *Electromagnetics 16*, 325–340.

Chew, W.C., and Weedon, W.H. (1994). A 3-D perfectly matched medium from modified Maxwell's equations with stretched coordinates. *IEEE Microwave and Guided Wave Letters 4*, 599–604.

Churchill, R.V., Brown, J.W., and Verhey, R.F. (1976). *Complex variables and applications.* New York: McGraw-Hill.

Dey, S., and Mittra, R. (1997). A locally conformal finite-difference time-domain algorithm for modeling three-dimensional perfectly conducting objects. *IEEE Microwave and Guided Wave Letters 7*, 273–275.

Engquist, B., and Majda, A. (1977). Absorbing boundary conditions for the numerical simulation of waves. *Mathematics of Computation 31*, 629–651.

Fang, J. (1989). Time-domain finite difference computations for Maxwell's equations. Ph.D. dissertation. University of California, Berkeley, CA.

Gedney, S.D. (1995). An anisotropic perfectly matched layer absorbing medium for the truncation of FDTD lattices. University of Kentucky, Lexington, KY. *Report EMG-95-006.*

Gedney, S.D. (1996). An anisotropic perfectly matched layer absorbing medium for the truncation of FDTD lattices. *IEEE Transactions on Antennas and Propagations 44*, 1630–1639.

Gedney, S.D. (1998). The perfectly matched layer absorbing medium. In A. Taflove (Ed.), *Advances in computational electrodynamics: The finite-difference time-domain method.* Norwood, MA: Artech House.

Gedney, S.D., and Lansing, F. (1995). Nonuniform orthogonal grids. In A. Taflove (Ed.), *Computational electrodynamics: The finite-difference time-domain method.* Norwood, MA: Artech House.

Gedney, S.D., and Taflove, A. (2000). Perfectly matched layer absorbing boundary conditions. In A. Taflove and S.C. Hagness (Eds.), *Computational electrodynamics: The finite-difference time-domain method.* (2nd ed.). Norwood, MA: Artech House.

Hagness, S.C., Rafizadeh, D., Ho, S.T., and Taflove, A. (1997). FDTD microcavity simulations: Design and experimental realization of waveguide-coupled single-mode ring and whispering-gallery-mode disk resonators. *Journal of Lightwave Technology 15*, 2154–2165.

Harrington, R.F. (1968). *Field computation by moment methods.* New York: Macmillan.

He, L. (1997). FDTD—Advances in subsampling methods, UPML, and higher order boundary conditions. M. S. thesis. University of Kentucky, Lexington, KY.

Higdon, R.L. (1986). Absorbing boundary conditions for difference approximations to the multidimensional wave equation. *Mathematics of Computation 47*, 437–459.

Holland, R., and Williams, J. (1983). Total-field versus scattered-field finite-difference. *IEEE Transactions on Nuclear Science 30*, 4583–4587.

Joseph, R.M., and Taflove, A. (1994). Spatial soliton deflection mechanism indicated by FDTD Maxwell's equations modeling. *IEEE Photonics Technology Letters 2*, 1251–1254.

Jurgens, T.G., Taflove, A., Umashankar, K.R., and Moore, T.G. (1992). Finite-difference time-domain modeling of curved surfaces. *IEEE Transactions on Antennas and Propagations 40*, 357–366.

Katz, D.S., Thiele, E.T., and Taflove, A. (1994). Validation and extension to three dimensions of the Berenger PML absorbing boundary condition for FDTD meshes. *IEEE Microwave and Guided Wave Letters 4*, 268–270.

Keller, J.B. (1962). Geometrical theory of diffraction. *Journal of Optical Society of America 52*, 116–130.

Kouyoumjian, R.G., and Pathak, P.H. (1974). A uniform geometrical theory of diffraction for an edge in a perfectly conducting surface. *Proceedings of the IEEE 62*, 1448–1461.

Kreiss, H., Manteuffel, T., Schwartz, B., Wendroff, B., and White, J. A. B. (1986). Supraconvergent schemes on irregular meshes. *Mathematics of Computation 47*, 537–554.

Krumpholz, M., and Katehi, L.P.B. (1996). MRTD: New time-domain schemes based on multiresolution analysis. *IEEE Transactions on Microwave Theory Technology 44*, 555–572.

Liao, Z.P., Wong, H.L., Yang, B.P., and Yuan, Y.F. (1984). A transmitting boundary for transient wave analyses. *Scientia Sinica (Series A) XXVII*, 1063–1076.

Liu, Q.H. (1996). The PSTD algorithm: A time-domain method requiring only two grids per wavelength. New Mexico State University Las Cruces, NM. *Report NMSU-ECE96-013.*

Liu, Q.H. (1997). The pseudospectral time-domain (PSTD) method: A new algorithm for solutions of Maxwell's equations. *Proceedings of the IEEE Antennas and Propagation Society 1*, 122–125.

Liu, Y. (1996). Fourier analysis of numerical algorithms for the Maxwell's equations. *Journal of Computational Physics 124*, 396–416.

Madsen, N.K., and Ziolkowski, R.W. (1990). A three-dimensional modified finite volume technique for Maxwell's equations. *Electromagnetics 10*, 147–161.

Maloney, J.G., Smith, G.S., Thiele, E.T., and Gandhi, O.P. (2000). Modeling of antennas. In A. Taflove and S.C. Hagness (Eds.). *Computational electrodynamics: The finite-difference time-domain method.* (2d ed.). Norwood, MA: Artech House.

Min, M.S., and Teng, C.H. (2001). The instability of the Yee scheme for the "magic time step." *Journal of Computational Physics 166*, 418–424.

Monk, P. (1994). Error estimates for Yee's method on nonuniform grids. *IEEE Transactions on Magnetics 30*, 3200–3203.

Mur, G. (1981). Absorbing boundary conditions for the finite-difference approximation of the time-domain electromagnetic field equations. *IEEE Transactions on Electromagnetic Compatibility 23*, 377–382.

Nehrbass, J.W., Lee, J.F., and Lee, R. (1996). Stability analysis for perfectly matched layered absorbers. *Electromagnetics 16*, 385–389.

Otsuyama, T., Sakuma, D., and Hayakawa, M. (2003). FDTD analysis of ELF wave propagation and Schumann resonances for a subionospheric waveguide model. *Radio Science* 38, 1103.

Pozar, D.M. (1998). *Microwave engineering.* (2d ed.). New York: John Wiley & Sons.

Ramahi, O.M. (1997). The complementary operators method in FDTD simulations. *IEEE Antennas Propagation Magazine 39/6*, 33–45.

Ramahi, O.M. (1998). The concurrent complementary operators method for FDTD mesh truncation. *IEEE Transactions on Antennas and Propagations 46*, 1475–1482.

Rappaport, C.M. (1995). Perfectly matched absorbing boundary conditions based on anisotropic lossy mapping of space. *IEEE Microwave and Guided Wave Letters 5*, 90–92.

Reuter, C.E., Joseph, R.M., Thiele, E.T., Katz, D.S., and Taflove, A. (1994). Ultrawideband absorbing boundary condition for termination of waveguiding structures in FDTD simulations. *IEEE Microwave and Guided Wave Letters 4*, 344–346.

Sacks, Z.S., Kingsland, D.M., Lee, R., and Lee, J.F. (1995). A perfectly matched anisotropic absorber for use as an absorbing boundary condition. *IEEE Transactions on Antennas and Propagations 43*, 1460–1463.

Shankar, V., Mohammadian, A.H., and Hall, W.F. (1990). A time-domain finite-volume treatment for the Maxwell equations. *Electromagnetics 10*, 127–145.

Schneider, J.B., and Wagner, C.L. (1999). FDTD dispersion revisited: Faster-than-light propagation. *IEEE Microwave and Guided Wave Letters 9*, 54–56.

Sheen, D. (1991). Numerical modeling of microstrip circuits and antennas. Ph.D. thesis. Massachusetts Institute of Technology, Cambridge, MA.

Shlager, K.L., and Schneider, J.B. (1998). A survey of the finite-difference time-domain literature. In A. Taflove (Ed.), *Advances in computational electrodynamics: The finite-difference time-domain method.* Norwood, MA: Artech House.

Simpson, J. J., and Taflove, A. (2004). Two-dimensional FDTD model of antipodal ELF propagation and Schumann resonance of the earth. *IEEE Antennas and Wireless Propagation Letters 1*, 53–56.

Simpson, J.J., and Taflove, A. (2004). Three-dimensional FDTD modeling of impulsive ELF propagation about the earth-sphere. *IEEE Transactions on Antennas and Propagation 52*, 443–451.

Song, J., and Chew, W.C. (1998). The fast Illinois solver code: Requirements and scaling properties. *IEEE Computer Science and Engineering 5*, 19–23.

Taflove, A., and Hagness, S.C. (2000). *Computational electrodynamics: The finite-difference time-domain method.* (2d ed.). Norwood, MA: Artech House.

Taflove, A., Umashankar, K.R., Beker, B., Harfoush, F.A., and Yee, K. S. (1988). Detailed FDTD analysis of electromagnetic fields penetrating narrow slots and lapped joints in thick conducting screens. *IEEE Transactions on Antennas and Propagation 36*, 247–257.

Teixeira, F.L., and Chew, W.C. (1997). PML-FDTD in cylindrical and spherical coordinates. *IEEE Microwave and Guided Wave Letters 7*, 285–287.

Tulintseff, A. (1992). The finite-difference time-domain method and computer program description applied to multilayered microstrip antenna and circuit configurations. Jet Propulsion Laboratory, Pasadena, CA. *Technical Report AMT: 336.5-92-041.*

Wait, J.R. (1965). Earth-ionosphere cavity resonances and the propagation of ELF radio waves. *Radio Science 69D*, 1057.

Yee, K.S. (1966). Numerical solution of initial boundary value problems involving Maxwell's equations in isotropic media. *IEEE Transactions on Antennas and Propagations 14*, 302–307.

10

Radar and Inverse Scattering

Hsueh-Jyh Li and
Yean-Woei Kiang
*Department of Electrical Engineering,
National Taiwan University,
Taipei, Taiwan*

10.1 Introduction

Radar is constructed from the words **radio detection and ranging**. The early purpose of a radar was to detect the presence of a target and measure its range by transmitting radio waves. Modern radars not only detect target and measure distances, but they also have the capability of locating, imaging, and identifying targets. A typical radar consists of a transmitter, an antenna, a receiver, a signal processor, and a display.

There are different types of radars. According to the physical relationship between the transmitting and receiving antennas, they can be classified into **monostatic radars** and **bistatic radars**. The monostatic radar has a common antenna used for both transmitting and receiving, while the bistatic radar has transmitting and receiving antennas separated by a considerable distance. According to the waveforms transmitted, radars can be classified into **continuous wave** (CW) radars or **pulsed** radars. A CW radar's transmitter operates continuously.

A pulsed radar transmits a relatively short burst of pulses, and after each pulse, the receiver is turned on to receive the echo. According to the primary missions, radars can be classified into **search radars** and **tracking radars.** Search radars continuously scan a volume of space without dwelling at any location. Their primary missions are detecting targets and determining a target's range and direction. Tracking radars dwell on individual targets and track their motion in range, azimuth, elevation, and/or Doppler.

10.2 Parameters of a Pulsed Radar

The most popular type of radar is the **pulsed radar**. The waveform of a pulsed radar is usually represented by the **pulse duration** τ, the **pulse repetition interval** (PRI) T, or the **pulse repetition frequency** (PRF) f_p. PRI is the time interval between two adjacent pulses. PRF is the rate that pulses repeat per second and is equal to the inverse of PRI. The fraction of the total time that the transmitter is on is called the **duty cycle** d_c. For a fixed PRI, duty cycle is given by:

$$d_c = \tau T. \tag{10.1}$$

Target range is the distance between the target and the radar. It can be determined by measuring the time required by the pulse to travel to the target and return. The target range is given by:

$$R = c\tau_d/2, \tag{10.2}$$

where c is the speed of light, τ_d is the time delay, and the factor 2 accounts for the round-trip.

If the time delay is smaller than the PRI, the range determined by equation 10.2 is unambiguous. If the PRF is too high or the PRI is too short, the echo pulses from the target might arrive after the transmission of the next pulse, and then the range measured might be ambiguous. The maximum unambiguous range is given by:

$$R_{\text{unamb}} = cT/2 = c/(2f_p), \tag{10.3}$$

where T is the PRI and where f_p is the PRF.

10.3 Radar Equation

A **radar equation** relates the range of a radar to the characteristics of the transmitter, receiver, antenna, target, and distance. Consider a monostatic radar. Let P_t be the transmitted power, G the antenna gain, R the distance of the target. Then, the power density at the target is given by:

$$P_d = \frac{P_t G}{4\pi R^2}. \tag{10.4}$$

Assume the target receives the incident power with an area σ, called the **radar cross section** (RCS), and radiates isotropically. Hence, the power density at the radar is given by:

$$P'_d = \frac{P_d\sigma}{4\pi R^2} = \frac{P_t G_t \sigma}{(4\pi R^2)^2}. \tag{10.5}$$

Assume the radar antenna has an effective area A_e, which is related to the antenna gain by $A_e = G_r\lambda^2/4\pi$. Then, the power received by the radar is as written here:

$$P_r = P_d A_e = \frac{P_t \cdot G_t \sigma G_r \lambda^2}{(4\pi R^2)^2 \cdot 4\pi} = \frac{P_t G^2 \sigma \lambda^2}{(4\pi)^3 R^4}, \tag{10.6}$$

where the same antenna has been used both for transmitting and receiving. Let $S_{\min}$ be the minimum detectable signal, and let the maximum range that a target can be detected for a given RCS σ be the following:

$$R_{\max} = \left[\frac{P_t G^2 \sigma \lambda^2}{(4\pi)^3 S_{\min}}\right]^{1/4}. \tag{10.7}$$

10.4 Radar Cross Section

The **radar cross section** (RCS) of a target is the equivalent area seen by a radar. It is the fictitious area intercepting that amount of power which, when scattered equally in all directions, produces an echo at the radar equal to that from the target. Mathematically, it is written as:

$$\sigma = \lim_{R\to\infty} 4\pi R^2 \frac{|E_s|^2}{|E_i|^2}, \tag{10.8}$$

where:

- R = distance between radar and target
- E_s = scattered field strength at radar
- E_i = incident field strength at target

10.4.1 RCS of Simple Objects

In theory, the RCS of a target can be determined by solving Maxwell's equations with proper boundary conditions. Only objects with simple geometries, however, can be determined in this way. The RCS of a simple conducting sphere as a function of the normalized circumference ($2\pi a/\lambda$, where a is the radius of the sphere and λ is the wavelength) is shown in Figure 10.1. The RCS of this figure can be divided into the **Rayleigh region**, the **Mie** or the **resonance region**, and the **optical region**. In the Rayleigh region, where the size of the sphere is small compared with the wavelength, the RCS varies with λ^4. In the optical region, where the dimensions of the sphere are

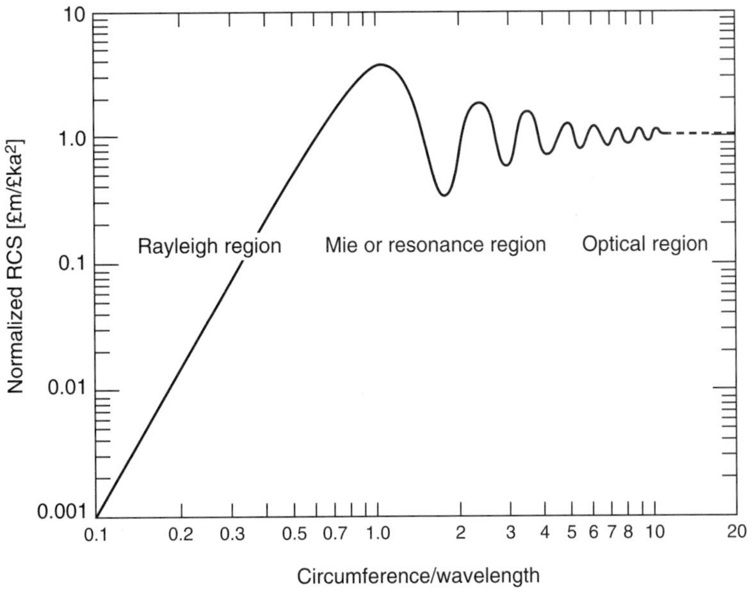

FIGURE 10.1 The RCS of a Conducting Sphere as a Function of the Normalized Circumference

large compared with the wavelength, the RCS approaches a constant value πa^2. In the Mie or resonance region, which is between the optical and the Rayleigh region, the RCS is oscillatory with frequency.

The RCS of a target depends on the aspect angle, frequency, and polarization. It is not directly related to the physical area. When the object dimension is much greater than the wavelength, the RCS of a conducting plate with a physical area A observed at the normal direction can be approximated by the product of the effective gain of the plate and the physical area. The product is given by:

$$\sigma = G_e \cdot A = \frac{4\pi A}{\lambda^2} \cdot A = \frac{4\pi A^2}{\lambda^2}. \tag{10.9}$$

As an example, if the conducting plate has an area of 1 m^2, and the wavelength is 1 cm, then the RCS at the normal direction is 12, 560 m^2, which is a hundred thousand times the physical area!

The approximate formulas for the RCS of some simple objects are shown in Table 10.1.

The RCS of a complex target that is large compared to a wavelength is highly dependent on the aspect angle. A small

TABLE 10.1 Approximate Formulas for the RCS of Some Simple Objects

Object	Aspect	RCS	Symbol
Sphere	Any	πa^2	a: Radius
Cone	Axial	$\frac{\lambda^2}{16\pi}\tan^4\theta$	θ: Cone half angle
Paraboloidal	Axial	πa^2	a: Apex radius of curvature
Cylinder	Normal to axis	$\frac{2\pi a L^2}{\lambda}$	a: Radius L: Length
Large flat plate	Normal	$\frac{4\pi A^2}{\lambda^2}$	A: Plate area
Square plate	Angle to normal	$\frac{4\pi a^4}{\lambda^2}\left[\frac{\sin(ka\sin\theta)}{ka\sin\theta}\right]$	a: Length of side
Circular plate	Angle θ to normal	$\frac{\pi a^2}{\tan^2\theta} J_1^2(2ka\sin\theta)$	a: Radius of plate J_1 (): Bessel function of the first order
Dihedral	Maximum direction	$\frac{8\pi a^2 b^2}{\lambda^2}$	a,b: Length of side
Trihedral	Maximum direction	$\frac{4\pi a^4}{3\lambda^2}$	a: Edge length
Square trihedral	Maximum direction	$\frac{12\pi a^4}{\lambda^2}$	a: Edge length

change in the aspect angle can cause the RCS to fluctuate several tens of decibels.

10.4.2 RCS Enhancement and Reduction

In some civilian applications, such as navigation, it is desirable to enhance the radar echo of the object so that it can be easily detected or tracked. The metal corner reflectors are usually used for this purpose because their RCSs have large values over a wide angular coverage. On the contrary, for electronic warfare and electronic countermeasure, it is desirable to reduce the RCS of military targets so that they will not be detected by the radar. There are several ways to reduce the RCS of a target: **shaping, absorbing,** and **cancelling**. Shaping includes specific design configuration, such as placing engine intakes where they can be shielded by other parts of the object. The purpose of the radar absorber is to absorb incident energy and thereby reduce the energy scattered back to the radar. Cancellation methods require loading the object with suitable impedance to cancel the returns from other parts of the body. Cancellation methods are too complicated to implement and are seldom used.

10.5 Radar Transmitters

The function of the **radar transmitter** is to generate an electromagnetic signal to illuminate the target. The transmitters can be classified into the oscillators and power amplifiers. The power amplifiers can be divided into three groups: the **linear beam O-type**, the **cross-field M-type**, and the **solid-state type** amplifiers. The O-type power amplifiers are characterized by high gain, high power capability, low noise, and large size. Typical O-type tubes include klystrons, travelling wave tubes, and twystrons. The cross-field M-type amplifiers are characterized by low gain, moderate to high power capability, moderate noise, wide bandwidth, and small size. Typical M-type tubes include cross-field amplifiers and magnetrons. The solid-state transmitters have lower voltage requirements and poor operation at higher frequency bands. Solid-state devices for amplifiers include bipolar junction transistors, field effect transistors, transferred-electron devices, and avalanche transit-time devices.

Two types of radar transmitters exist: **coherent** and **incoherent**. Coherent transmitters produce a signal whose phase is known prior to transmission. Coherent-on-receive and incoherent transmitters produce a signal whose phase is unknown prior to the start of the transmit output.

10.6 Radar Receivers and Displays

The function of the radar receiver is to detect desired echo signals in the presence of noise, interference, or clutter. In general, a radar receiver has four functional blocks: the **amplification block**, the **band-select filter block**, the **matched filter block**, and the **demodulation block**. The amplification block increases the amplitude of the incoming signal to levels usable in the post stages—band-select filter block, the matched filter block, and the demodulation block. The band-select filter block rejects out-of–band interference. The matched filter block shapes the signals and noise to gain the maximum signal-to-noise ratio. The demodulation function block removes the carrier and translates the signal to its information frequencies or baseband.

The **superheterodyne** is the most popular receiver type. In a superheterodyne receiver, the signals are offset downward by a local oscillator. Amplification and filtering take place at a lower frequency or intermediate frequency. This type of receiver has the advantage of good sensitivity, high gain selectivity, and reliability.

10.6.1 Receiver Noise

Noise is the chief limiting factor of the sensitivity of a receiver. If the receiver itself were perfect and did not generate any excess noise, there would still exist noise generated by the thermal motion of the conduction electrons. This noise is called **thermal noise**. The thermal noise power is proportional to the temperature T (degrees Kelvin) and the receiver bandwidth B_n (hertz) and is given by:

$$P_n = kTB_n. \tag{10.10}$$

Where k is Boltzmann's constant $= 1.38 \times 10^{-23}$ J/deg. The B_n is the equivalent noise bandwidth defined as the width of a rectangular filter whose peak response equals that of the actual filter and whose area equals that under the actual filter. Mathematically, B_n is given by:

$$B_n = \frac{\int_{-\infty}^{\infty} |H(f)|^2 df}{|H(f_0)|^2}, \tag{10.11}$$

where $H(f)$ is the frequency response of the filter and where $H(f_0)$ is the peak response of the filter.

10.6.2 Noise Figure and Temperature

For a practical receiver, the noise power is often greater than the thermal noise power. The total noise power at the receiver output may be considered to be equal to the thermal noise power multiplied by a factor called the **noise figure**. The noise figure is a measure of the degradation of signal-to-noise-ratio as the signal passes through the receiver. It measures how much noise is added by the receiver and can be expressed by:

$$F_n = \frac{(S/N)_i}{(S/N)_o}, \tag{10.12}$$

where $(S/N)_i$ is the signal-to-noise-ratio at the input to the receiver and where $(S/N)_o$ is the signal-to-noise-ratio at the output to the receiver.

If there are N amplifiers in cascade, the equivalent noise figure is given by:

$$F_e = F_1 + \frac{F_2 - 1}{G_1} + \frac{F_3 - 1}{G_1 G_2} + \cdots \frac{F_N - 1}{G_1 G_2 \ldots G_{N-1}}, \quad (10.13)$$

where F_i and G_i are the noise figure and the gain of the ith stage amplifier, respectively.

The **noise temperature** is the noise introduced by an amplifier that may be expressed as an effective temperature:

$$T_e = (F_n - 1)T_0, \quad (10.14)$$

where T_0 is the room temperature.

10.6.3 Mixers

The function of a **mixer** is to convert RF energy to intermediate frequency (IF) energy with minimum loss and without spurious responses. The inputs to the mixer are the signal at the radio frequency and the sinusoid from a local oscillator (LO). The output is the signal at the intermediate frequency. Mixing is the process of multiplying the RF signal by the local oscillator and then low-pass filtering the product.

The conversion process of a mixer will cause a loss known as **conversion loss**, which is defined as:

$$Lc = \frac{\text{available RF power}}{\text{available IF power}}. \quad (10.15)$$

The available power means that power can be obtained when the impedences are matched. Conversion loss is a measure of the efficiency of the mixer in converting RF signal power into IF.

Balanced mixer

Noise that accompanies the LO signal can appear at the IF. A method of eliminating LO noise is via the **balanced mixer**. A balanced mixer uses a four-port hybrid junction as shown in Figure 10.2. The LO and RF signals are applied to the two ports. Diode mixers are in each of the remaining two ports. At one diode, the sum of the RF and LO signals appears, and at the other diode is the difference of the two signals. The two diode mixers should have the same characteristics and be well matched. The IF signal is the subtraction of the two diode outputs. Because the LO noise at the two diode mixers are in phase, it will be canceled at the output.

10.6.4 IF Amplifiers

Most gain and filtering are obtained in the IF amplifier. Typical IF amplifiers have gains of up to 120 dB. The design of the IF filters involves the matched filter and the Doppler shift.

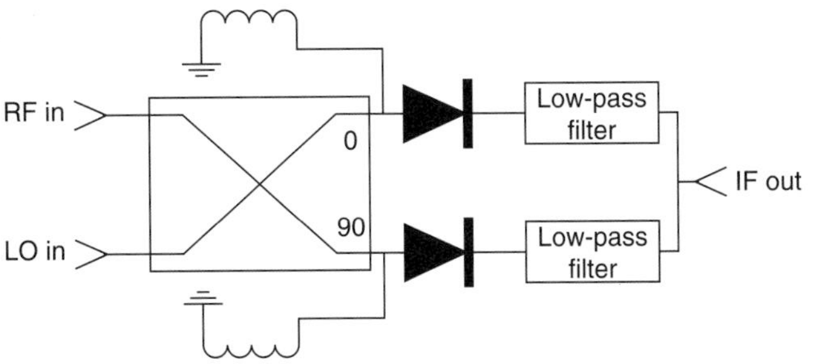

FIGURE 10.2 A Balanced Mixer Using a Four-Port Hybrid Junction

The purpose of a **matched filter** is to maximize the signal-to-noise ratio. If the noise is a white noise, the matched filter's characteristics are given by:

$$H(f) = G_0 S^*(f) \exp(-j2\pi f t_1), \quad (10.16)$$

where $H(f)$ is the transfer function of the matched filter, G_0 is a constant, $S^*(f)$ is the complex conjugate of the spectrum of the received signal, and t_1 is the time of maximum response of the filter.

Most IF amplifiers have high gains and **gain control**. The automatic gain control (AGC) is to hold the amplitude at the output of the receiver constant. Multichannel tracking receivers should have the same gain control voltage applied to all receivers, and they must maintain the same gain over their gain control range (90 dB or more). Maintaining identical gains in multiple channels over wide gain-control excursions is called **gain tracking**.

The amplitude of radar signals has a very wide dynamic range. To avoid the saturation of the amplifier output, the **log receiver** is usually used. In a log receiver, the amplitude of the output is proportional to the logarithm of the amplitude of the input. It can compress the output dynamic range so that weak signals can be enhanced and observed.

10.6.5 Displays

The function of the **display** is to visually present the information contained in the radar echo signal to the operator so that he or she can easily interpret and take action. There are various types of displays. Some of the more usual displays adapted from the IEEE standard definitions are listed below:

- An **A-scope** is a deflection-modulated display in which the vertical deflection is proportional to target echo strength, and the horizontal coordinate is proportional to the range.
- A **B-scope** is an intensity-modulated rectangular display with an azimuth angle indicated by the horizontal coordinate and range by the vertical coordinate.
- A **C-scope** is an intensity-modulated rectangular display with an azimuth angle indicated by the horizontal coordinate and elevation angle by the vertical coordinate.

- A **plan position indicator (PPI) or P-scope** is an intensity-modulated circular display where echo signals are shown in plan position with range and azimuth angle displayed in polar coordinates.
- A **range-height indicator** (RHI) is an intensity—modulated display with height as the vertical axis and range as the horizontal axis.
- A **duplexer** is a device that allows the transmitter and the receiver to share the same antenna. On transmission, it protects the receiver from burnout; on reception, it directs the echo signal to the receiver. The duplexer must provide a very large isolation (greater than 60 dB) between the transmitter and receiver with a very small loss of the desired signal. Most radar duplexers are in one of the two types: **circulator** and **TR tube**.

10.7 Radar Antennas

The basic role of the **radar antenna** is to act as a transducer between the free space and the electromagnetic wave sources or receivers. During transmission, it is used to concentrate the radiated energy into a shaped beam or in a desired direction. During reception, it is used to collect the echo signal and deliver it to the receiver.

Most radar systems employ a single antenna for both transmitting and receiving. If the antenna has identical performance parameters (such as impedance, pattern, and gain) for both functions, it is called **reciprocal.**

Directivity of an antenna is defined as the ratio of the maximum radiation power intensity to the average intensity and is expressed by:

$$G_D = \frac{\text{maximum radiation intensity}}{\text{average radiation intensity}} = \frac{\text{maximum power per steradian}}{\text{total power radiated}/4\pi} \qquad (10.17)$$

Gain of an antenna involves antenna losses and is defined in terms of power accepted by the antenna at its input port. The relation between gain and directivity is written as:

$$G = \eta G_D, \qquad (10.18)$$

where η is called the efficiency factor, which counts for antenna losses, such as impedance mismatch loss or ohmic loss. An approximate relationship between directivity and the antenna half-power beamwidths can be expressed by:

$$G_D = \frac{40,000}{B_{az} \cdot B_{el}}, \qquad (10.19)$$

where B_{az} and B_{el} are the azimuth and elevation half-power beamwidths in degrees, respectively.

The **effective aperture** A_e of an antenna is the portion of its total projected area that effectively captures the electromagnetic waves. It is related to the physical aperture by:

$$A_e = \eta_A A, \qquad (10.20)$$

where η_A is called the aperture efficiency. A_e is related to the antenna gain by:

$$G = \frac{4\pi A_e}{\lambda^2}, \qquad (10.21)$$

where λ is the operating wavelength. One can estimate the antenna gain if the physical area, the efficiency, and the operating frequency are known.

Radar antennas can be classified into two broad categories: Aperture antennas and array antennas.

10.7.1 Aperture Antennas

Cosecant-Squared and Cassegrain Antennas

The **cosecant-squared antenna** is a popular type of aperture antenna. In the cosecant-squared antenna pattern, the gain as a function of elevation angle θ is given by:

$$G(\theta) = G(\theta_0) \cdot \frac{\csc^2\theta}{\csc^2\theta_0} \quad \text{for } \theta_0 < \theta < \theta_m, \qquad (10.22)$$

where θ_0 and θ_m are the angular limits that the beam follows in a csc^2 shape. The cosecant-squared antenna has the important property of the echo power received from a target of constant RCS at a constant altitude being independent of the target range.

The **cassegrain antenna** is the most common antenna using multiple reflectors. In the cassegrain antenna, the feed, placed at or near the paraboloid main reflector, illuminates the hyperboloidal subreflector, which in turn illuminates the paraboloidal main reflector.

Aperture blockage and **spillover** are two main problems of multiple reflectors. An obstacle, such as a feed or feed support, in front of a reflector is called aperture blockage. Aperture blockage degrades the antenna performance by lowering the gain, increasing the side lobes, and filling in the nulls. If the feed beamwidth is excessive and illuminates over the reflector aperture, it is called spillover. Spillover will cause an undesired antenna response in the direction of the spillover.

Antennas for ground-based or airplane radars are usually covered by a **radome** for protection. A radome is a dielectric shield covering to protect an antenna operated under severe weather conditions. Radomes must be mechanically strong and not interfere with the normal operation of the antenna.

10.7.2 Array Antennas

An **array antenna** is a directive antenna made of individual elements. Its radiation pattern can be adjusted by controlling the amplitude and phase of the currents at the individual elements. By changing the relative phase shift between elements, the beam of an antenna array can be scanned. The beam of an array antenna can be steered by properly varying the phase of the signals applied to each element.

Consider a uniformly spaced array with a number of elements N and spacing d. Assume the progressive phase shift between elements is ϕ. The radiation pattern of the array can be expressed by:

$$G(\theta) = \frac{\sin^2[N/2(kd\sin\theta - \phi)]}{N^2\sin^2[1/2(kd\sin\theta - \phi)]}. \qquad (10.23)$$

The direction of the main beam will be at angle θ_0 when the progressive phase shift ϕ is equal to $kd\sin\theta_0$. The following is a list of different arrays.

- **Frequency-scan array:** If the adjacent antenna elements are connected by a transmission line with length ℓ, the progressive phase shift will be $k\ell = 2\pi\ell/\lambda$. By changing the frequency, the progressive phase shift will be varied, and the main beam of the array can be controlled by changing the operating frequency. This type of array is called the frequency-scan array. If the beam is to point in a direction θ_0, then the phase difference ϕ should be:

$$\phi = kd\sin\theta_0 + 2m\pi = 2\pi\,\ell/\lambda, \qquad (10.24)$$

  where m is an integer.
- **Beamforming:** If the array is only used for reception, the antenna pattern can be digitally "beamformed." The output from each array element may be amplified and sampled. The sampled signals are then transferred to a computer for processing. The beam pattern can be formed digitally by applying suitable beamforming algorithms.
- **Multiple beam array:** In principle, an N-element array can generate N-independent beams. Multiple beams can be simultaneously obtained by attaching additional phase shifters to each element. The advantage of using multiple beams is that they allow parallel operation and allow a higher data rate than can be achieved from a single beam. A popular multiple beam array is the **Butler beamforming array.** It uses a lossless network consisting of 3-dB 90 hybrids along with fixed-phase shifters to form N contiguous beams from an N-element array, where N is some power of 2. The number of 3-dB hybrids required for an N element is equal to $(N/2)\log_2 N$, and the number of fixed phase shifters is $(N/2)(\log_2 N - 1)$.
- **Adaptive array:** An adaptive array adjusts the weights (amplitude and phase) of elements to achieve some desired performance, such as maximizing the received signal to interference plus noise ratio. Interference or jamming can be greatly reduced by forming a beam pattern with nulls pointing to the jammers. Adaptive arrays require some a priori knowledge of the desired signal, such as the direction, waveform, or statistical properties to train the weights.
- **Coherent side lobe canceller:** The coherent side lobe canceller uses a small number of auxiliary elements to adaptively place nulls in the directions of jammers. The auxiliary elements have a side lobe structure matched to that of the main antenna, but the antenna's main lobe is much smaller. The response of the main and auxiliary antennas are subtracted, thus cancelling the response of the radar antenna's side lobes but not the main lobe.

10.8 Clutter

Clutter is the unwanted signal echo from such sources as the sea, land, and atmosphere. It is real echo signal rather than random noise. The ground effects on radar signals can be divided into forward scattering and backscattering. A smooth surface provides strong forward scattering but weak backscattering, whereas a rough surface enables poor forward scattering but good backscattering.

The distinction between rough and smooth surfaces can be determined by the **Rayleigh criterion.** The critical height h_c of surface irregularities for a given angle of incidence θ_i can be defined as:

$$h_c = \frac{\lambda}{8\sin\theta_i}. \qquad (10.25)$$

A surface is considered smooth or rough if its irregularity height is less than or greater than h_c. At a small incident angle θ_i, a surface with large irregularities can still be viewed as a smooth surface.

A flat surface can be assumed to have a reflection coefficient. For rough surfaces, the flat surface reflection coefficient should be multiplied by a **roughness loss factor** to account for the diminished reflected field. The roughness loss factor ρ_s is given by:

$$\rho_s = \exp\left[-\left(\frac{\pi\sigma_h\sin\theta_i}{\lambda}\right)^2\right], \qquad (10.26)$$

where σ_h is the standard deviation of the surface height about the mean surface height, and θ_i is the incident angle.

The RCS of an extended clutter surface is usually expressed by a parameter called the **normalized RCS,** which is the RCS per unit area. It is a dimensionless quantity and is strongly dependent on the incident angle, the surface material, and roughness. To find the total RCS of a ground surface illuminated by a pulse, integration should be performed on all the area elements that return to the radar at the same instant. The area that should be integrated is determined by the radar–surface geometry, the antenna's illuminating pattern, and the pulse duration.

Consider an airborne pulse radar pointing its antenna toward the ground or sea surface at a depression angle. If the area illuminated by the pulse is smaller than the area covered by the antenna beamwidth, then the illumination is called **pulse width-limited illumination.** If the area covered by the antenna beamwidth is smaller than that covered by the pulse, the illumination is called **beamwidth-limited illumination.** With pulse width-limited illumination, the received clutter power is inversely proportional to R^3, whereas with beam-width-limited illumination, the received clutter power is inversely proportional to R^2.

10.9 Radar Detection

Detection is the process of deciding whether a target is present. The received field consists of a signal, interference, and noise. After the processing unit, the receiver output is compared with a predetermined threshold. If the receiver output is greater than the threshold, a detection is declared. Otherwise, no detection occurs. There are four possible states:

1. No target is present, and the receiver output is smaller than the threshold.
2. The target is present, and the receiver output crosses the threshold.
3. The target is present, but the receiver output is smaller than the threshold.
4. No target is present, but the receiver output crosses the threshold.

The first two states are called **correct decision**, and the last two states are called **detection errors**. The third case is called **missing**, and the fourth case is called **false alarm**.

Because noise is a random process and the target signal may fluctuate, detection performance is treated by probability. The following quantities are important in radar detection:

- **Probability of detection:** The $\mathbf{P_d}$ probability that the receiver output exceeds the threshold and the target is present.
- **Probability of missing P_m:** The probability that the received voltage is below the threshold but the target is present. $P_m = 1 - P_d$.
- **Probability of false alarm P_f:** The probability that the received voltage exceeds the threshold but the target is not present.

In radar detection, the threshold is usually determined by the allowable probability of false alarm. The probability of detection for a sine wave in white Gaussian noise as a function of the signal-to-noise ratio and the probability of false alarm are shown in Figure 10.3.

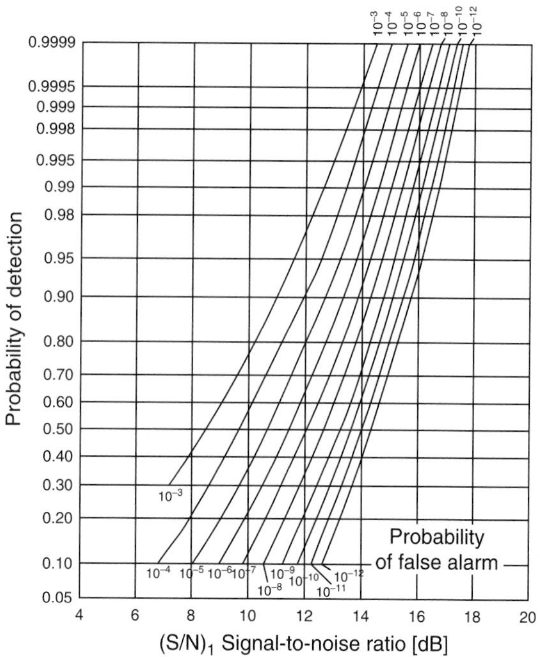

FIGURE 10.3 The Probability of Detection in White Gaussian Noise and the Probability of False Alarm. The probability is shown as a function of the signal-to-noise ratio.

10.9.1 Pulse Integration

Radar detection can be based on a single pulse echo. It can also be based on the integration of many pulses returned from a particular target. Integration can be accomplished either before or after the detector. Integration before the detector is called **predetection integration** or **coherent integration** because both amplitude and phase of each pulse echo can be preserved in the integration. Integration after the detector is called **postdetection** or **incoherent integration**, indicating the phase is not preserved after detector.

If n pulses (all having the same signal-to-noise ratio and phase) were coherently integrated, the resultant signal-to-noise ratio would be n times that of a single pulse. If the same pulses are incoherently integrated, the resultant signal-to-noise ratio would be less than n times that of a single pulse.

Let $(S/N)_1$ be the value of signal-to-noise ratio of a single pulse required to produce given probability of detection, and let $(S/N)_n$ be the value of signal-to-noise ratio per pulse required to produce the same probability of detection when n pulses are integrated. The integration improvement factor is defined as the ratio of $(S/N)_1$ to $(S/N)_n$, which is the improvement in the signal-to-noise ratio when n pulses are postdetection integrated.

10.9.2 RCS Fluctuation

The radar cross section of a complex target is highly dependent on the aspect angle. The echo signal from a target in motion usually fluctuates with time. Swerling (1960) has postulated four fluctuation models:

- **Case 1**: Scan-to-scan fluctuation exists, and the **probability density function** (pdf) for the cross section is given by:

$$P(\sigma) = \frac{1}{\sigma_{av}} \exp(-\sigma/\sigma_{av}), \quad (10.27)$$

where σ_{av} is the average cross section over all target fluctuation.

- **Case 2**: Pulse-to-pulse fluctuation exists, and the pdf is the same as case 1.
- **Case 3**: Scan-to-scan fluctuation exists, and the pdf for the cross section is given by:

$$P(\sigma) = \frac{4\sigma}{\sigma_{av}^2} \exp(-2\sigma/\sigma_{av}). \quad (10.28)$$

- **Case 4**: Pulse-to-pulse fluctuation exists, and the pdf is the same as case 3.

The pdf assumed in equation 10.27 applies to complex targets consisting of many independent scatterers of equal RCS, and the pdf assumed in equation 10.28 applies to targets consisting of a dominant reflector and many small scatterers. Pulse-to-pulse fluctuation corresponds to fast fluctuation, and scan-to-scan fluctuation corresponds to slow fluctuation.

The detection of radar signals depends on the distribution of noise and interference. The white Gaussian noise has a flat spectrum and has a Gaussian distribution of noise voltage with pdf given by:

$$P(v) = \frac{1}{\sqrt{2\pi\psi_0}} \exp(-v^2/2\psi_0), \quad (10.29)$$

where ψ_0 is the variance of the noise, and the mean value of the noise is taken to be zero.

If the Gaussian noise is passed through a narrow band IF filter, the probability density function of the noise voltage output is the following:

$$P(R) = \frac{R}{\psi_0} \exp(-R^2/2\psi_0), \quad (10.30)$$

if a sine wave signal of amplitude A along with noise is at the input to the IF filter. The probability density function of the envelope detector output is known as the Rician distribution and is given by:

$$P(R) = \frac{R}{\psi_0} \exp\left(-\frac{R^2 + A^2}{2\psi_0}\right) I_0\left(\frac{RA}{\psi_0}\right), \quad (10.31)$$

where $I_0(x)$ is the modified Bessel function of zero order.

With white Gaussian noise, given a threshold V_T, the probability of false alarm is as follows:

$$P_f = P_{rob}(V_T < R < \infty) = \int_{V_T}^{\infty} \frac{R}{\psi_0} \exp\left(-\frac{R^2}{2\psi_0}\right) dR = \exp\left(-\frac{V_T^2}{2\psi_0}\right) \quad (10.32)$$

The probability of detection is as written here:

$$P_d = \int_{V_T}^{\infty} \frac{R}{\psi_0} \exp\left(-\frac{R^2 + A^2}{2\psi_0}\right) I_0\left(\frac{RA}{\psi_0}\right) dR \quad (10.33)$$

Constant False Alarm Rate

The threshold for detection is usually determined to achieve a desired false alarm rate. The false alarm rate is quite sensitive to the threshold level. A **constant false alarm rate** (CFAR) can be obtained by adjusting the threshold in accordance with the measured background. There are different schemes for obtaining CFAR. The cell-average CFAR uses a tapped delay line to sample the range cells to either side of the test cell. The number of taps used in the cell-average CFAR might vary from 16 to 20. Clutter-map CFAR is to divide the radar's space into range and azimuth bins. Over several scans, a moving average of the clutter residue in each range-azimuth cell is calculated and is used to determine the threshold level. In the guard-band CFAR scheme, the interference level is determined by examining the frequency bands adjacent to the band containing the signal. The threshold is determined by this interference level. This type of CFAR is effective against wideband interference but not effective against clutter. The limiting CFAR scheme is used when the bandwidth of the interference is much greater than that of the target echo signal, such as with wideband noise jamming and with impulse jamming. In this scheme, except the wideband interference, signal plus high amplitude impulse interface is amplified and then pass through a limiter. After limiting, the signal and the interference have the same amplitude. Because the interference has much wider spectral bandwidth in the frequency domain, the power distributed to the signal band is much smaller than the signal. After passing through a band pass filter matched to the signal, the signal amplitude is thus much greater than the interference amplitude.

10.10 Continuous Wave Radars

10.10.1 Continuous Wave Radar

For a continuous wave (CW) **radar**, the transmitter is operated continuously rather than pulsed. A CW radar can obtain the target speed by measuring the Doppler frequency. Consider a CW radar with frequency f_0 and a target with a relative velocity v_r with respect to the radar. The Doppler shift of the received signal is given by:

$$f_d = 2v_r/\lambda = 2v_r f_0/c. \tag{10.34}$$

In the IF stage, the CW Doppler radar has a Doppler filter bank, which is a bank of narrow band filters centered at different frequencies. The center frequencies of the filters are staggered to cover the entire range of Doppler frequencies. The Doppler frequency is determined by the filter that has the largest output.

10.10.2 Frequency-Modulated CW Radar

The CW radar can measure the Doppler frequency of the target, but it cannot measure the target range. The **frequency-modulated CW radar** (FM–CW) can measure both the range and Doppler frequency of the target. In the FM–CW radar, the transmitted frequency changed as a function of time in a known manner.

Figure 10.4(A) shows the principle of triangular FM–CW ranging on a single target with no Doppler shift. The range information is contained in the frequency difference between the signal echo and the radar's present transmitting frequency. If there is no Doppler frequency, the difference frequency is a measure of the target range, which is given by:

$$R = \frac{cTf_r}{2B}, \tag{10.35}$$

where B is the bandwidth of the transmitted signal, T is the period of the modulation wave, f_r is the frequency difference between the signal echo and the present transmitting signal, and c is the light speed.

If there is a Doppler shift, there is a received frequency–time relationship, as shown in Figure 10.4(B). There are two difference frequencies: the upper beat frequency, $f_b(\text{up})$, and the down beat frequency, $f_b(\text{down})$. The range frequency f_r and the Doppler frequency f_d can be extracted by:

$$\begin{aligned} f_r &= \frac{1}{2}[f_b(\text{up}) + f_b(\text{down})]. \\ f_d &= \frac{1}{2}[f_b(\text{down}) - f_b(\text{up})]. \end{aligned} \tag{10.36}$$

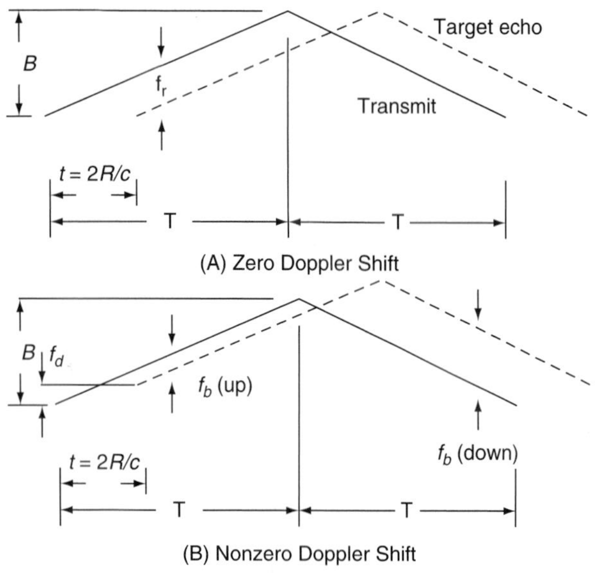

FIGURE 10.4 Principle of Triangular FM–CW Ranging on a Single Target with Different Doppler Shifts

FM–CW radars can be used in airborne applications. For example, an FM–CW altimeter can be placed in the aircraft to measure height above the surface of the earth. A Doppler navigation radar can measure the vector velocity relative to the frame of reference of the antenna assembly. A Doppler navigation radar having forward and rearward beams is called a **Janus system**. With the Janus system, the angular displacement of the aircraft heading and the speed along the ground track can be measured.

10.11 Moving Target Indicator and Pulse Doppler Radars

10.11.1 Moving Target Indicator Radar

In some radar applications, moving targets' signals are embedded in strong stationary clutter. Moving targets will produce a Doppler frequency shift, while the stationary clutter has very small spectral spreading around zero frequency. The **moving target indicator** (MTI) radar is a pulsed radar that uses the Doppler frequency shift as a means for discriminating moving targets from stationary clutter.

Delay-Line Canceller

A basic MTI single **delay-line canceller filter** is shown in Figure 10.5(A). The delay line has a time delay equal to the pulse repetition interval. The received signal and the reference signal are fed to a mixer called the phase detector, whose output is proportional to the phase difference of the two input signals.

The canceller output in the time domain is as follows:

$$y(t) = x(t) - x(t - T). \tag{10.37}$$

In the frequency domain, the output is shown by:

$$Y(\omega) = X(\omega)[1 - \exp(-j\omega T)]. \tag{10.38}$$

The transfer function of the filter is as follows:

$$\begin{aligned} H(\omega) &= \frac{Y(\omega)}{X(\omega)} = [1 - \exp(-j\omega T)] \\ &= 2j \sin\frac{\omega T}{2} \exp(-j\omega T/2), \end{aligned} \tag{10.39}$$

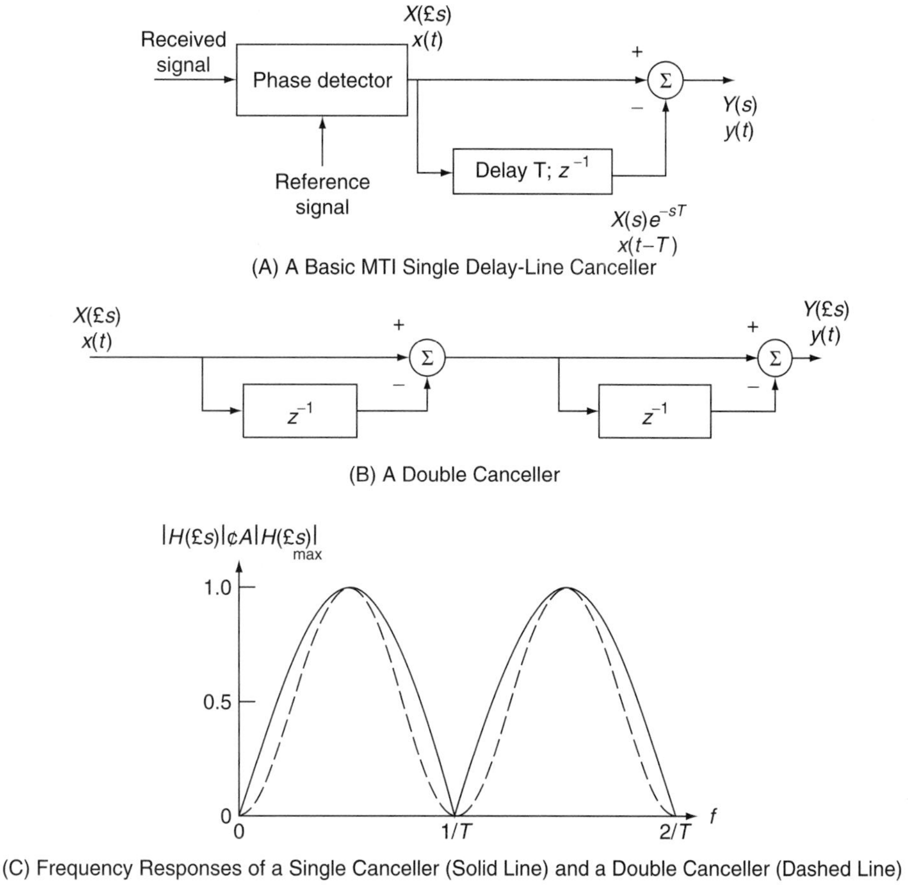

FIGURE 10.5 Delay-Line Canceller Filter

and

$$|H(\omega)| = 2\left|\sin\frac{\omega T}{2}\right|. \tag{10.40}$$

Two single cancellers can be cascaded to become a double canceller as shown in Figure 10.5(B). The transfer function of the double canceller is written as:

$$|H(\omega)| = 4\sin^3\frac{\omega T}{2}. \tag{10.41}$$

The frequency responses of a single canceller and a double canceller are shown in Figure 10.5(C).

Blind Speeds

It is noted that in Figure 10.5(C), the frequency responses have nulls at multiples of the repetition frequency f_R. This indicates that Doppler frequencies equal to multiples of f_R will subject to total attenuation. Returns from targets with corresponding velocities will be highly attenuated and cannot be detected by the radar. These velocities are called **blind speeds**. The blind speed problem can be alleviated by the staggered PRF technique, which modulates the interpulse period with more than one PRF and uses the corresponding delays in the canceller filter. With the staggered PRF, the first null response or the blind speed will be increased.

Consider a double canceller with stagger pulse repetition intervals, aT and $(2-a)T$, which has an average delay T. Define the stagger ratio as $a/(2-a)$. If $a = (m-1)/m$, the stagger ratio $= (m-1)/(m+1)$, where m is a large and even integer; then, the first blind Doppler frequency will be at $f_{\text{blind}} = m/2T$.

Several figures of merit can be used to compare performance between various MTI filters. Some definitions are given here:

- **Clutter attenuation (CA):** The ratio of the clutter power at the filter *input* to the clutter power at the filter *output*
- **MTI improvement factor:** The *output* target-to-clutter ratio divided by the *input* target-to-clutter ratio
- **Subclutter visibility (SCV):** The ratio between the *input clutter power* and the *input target power* that yields equal powers at the output of the canceller filter

If $S_c(\omega)$ is the clutter power spectral density, and $|H(\omega)|$ is the magnitude of the filter response, the clutter attenuation is given by:

$$CA = \frac{\int_{-\infty}^{\infty} S_c(\omega)d\omega}{\int_{-\infty}^{\infty} S_c(\omega)|H(\omega)|^2 d\omega}. \quad (10.42)$$

The filter attenuates the clutter, and it may also attenuate the signal. Assume that all Doppler frequencies are equally likely between 0 and the highest Doppler angular frequency $\omega_M = 4\pi V_m/\lambda$, where V_m is the highest target velocity. Define the average power gain as:

$$\bar{G} = \frac{1}{\omega_M}\int_0^{\omega_M} |H(\omega)|^2 d\omega. \quad (10.43)$$

Then the output target power P_{Tout} is equal to $\bar{G}P_T$, where P_T is the input target power. By the definition of clutter attenuation, the output clutter power P_{cout} is equal to P_c/CA, where P_c is the input clutter power. The improvement factor I is related to $\bar{G}$ and CA by:

$$I = \frac{P_{Tout}/P_{cout}}{P_T/P_c} = CA \cdot \bar{G}. \quad (10.44)$$

The expressions of CA, I, $\bar{G}$, and SCV depend on the target spectrum, the clutter spectrum, and the types of canceller filters. Assume that the clutter spectrum is normally distributed around zero Doppler with a variance of σ_ω^2, and assume that the Doppler spectrum is uniformly distributed between 0 and $1/T$. For a single canceller, we then have:

$$\begin{aligned}
CA &\cong \frac{1}{2\left[1-\exp\left(-(\mathbf{T}\sigma_\omega)^2/2\right)\right]} \cong \frac{1}{(T\sigma_\omega)^2} \quad \textit{if } T\sigma_\omega \ll 1. \\
\bar{G} &= 2. \\
I &\cong CA \cdot \bar{G} = \frac{1}{\left[1-\exp\left(-(\mathbf{T}\sigma_\omega)^2/2\right]\right.} \cong \frac{2}{(T\sigma_\omega)} \quad \textit{if } T\sigma_\omega \ll 1.
\end{aligned} \quad (10.45)$$

For a double canceller, the corresponding parameters are the following:

$$\begin{aligned}
CA &\cong \frac{1}{6\left[1-4/3\exp\left[-(T\sigma_\omega)^2/2\right]+\frac{1}{3}\exp\left[-(2T\sigma_\omega)^2/2\right]\right]} \\
&\cong \frac{1}{3(T\sigma_\omega)^4} \quad \text{for } T\sigma_\omega \ll 1. \\
\bar{G} &= 6. \\
I &= CA \cdot \bar{G} \\
&\cong \frac{2}{(T\sigma_\omega)^4} \quad \text{for } T\sigma_\omega \ll 1.
\end{aligned} \quad (10.46)$$

10.11.2 Pulse Doppler Radar

A **pulse Doppler radar** uses the Doppler shift to discriminate moving targets from stationary clutter. A low PRF radar has a long unambiguous range but results in blind speeds. On the contrary, a high PRF radar can avoid blind speeds but experiences ambiguity in range. MTI usually refers to a radar in which the PRF is low enough to avoid ambiguity in range but results in blind speeds. The pulse Doppler radar operates with high PRF to avoid blind speeds but at the expense of range ambiguity.

The pulse Doppler radar usually uses filter banks, which are implemented with a discrete Fourier transform rather than delay cancellers to remove the clutter. The improvement factor is a function of the size of the Fourier transform and the window function used.

10.12 Tracking Radar

Tracking refers to a radar following the position of one or multiple targets in space. Before the tracking process, the radar has to detect targets and find their range, angular location, and sometimes velocity. The requirements on the accuracy of angle measurement for a tracking radar are more strict than those for a search radar.

The principle of tracking radar is to use the error signal to adjust the antenna's pointing direction. The difference between the target direction and the reference direction, usually the axis of the antenna, is defined as the **angular error**. The tracking radar attempts to position the antenna with zero angular error (i.e., to locate the target along the reference direction).

Tracking radars are classified by how the tracking errors are developed. The principal tracking schemes include **lobe switching, conical scan** and **monopulse tracking**.

- **Lobe switching:** In the **lobe switching** mode, the antenna beam is switched alternately between two positions. The difference in amplitude between the voltages obtained in the two positions is a measure of the angular displacement of the target from the switching axis. The sign of the difference determines the direction that the antenna must be moved. When the switching axis is in the direction of the target, the voltages in the two positions are equal. Another two additional positions can be used to determine the direction of the target in the orthogonal plane.
- **Conical scan:** In the **conical scan** mode, an offset antenna beam is rotated continuously about an axis. The angle between the axis of rotation and the axis of the beam is called the **squint angle**. The echo from a target at the rotation axis will be constant all along the rotation period, while the echo from a target off the rotation axis will be sinusoidally modulated at the rotation frequency. The amplitude of the modulation indicates how far the

target is located off the axis. The phase of the modulation gives information about the direction of the target.

- **Monopulse system:** In the lobe switching and conical scan, tracking radars require several pulses to extract the error signal. During this interval, if the pulse echoes contain additional modulation components, such as fluctuating RCS, the tracking accuracy will be degraded. In fact, the angle of arrival of the target echo can be determined with a single pulse by measuring the relative phase or relative amplitude of the received echo pulse. Because information of the angular error is obtained on the basis of a single pulse, the system is called a **monopulse system.**

The monopulse system employs two overlapping antenna patterns to form the **sum pattern** and the **difference pattern.** The sum pattern is used for transmission, and both the sum pattern and difference pattern are used for reception. The sum pattern provides target detection and range measurement. The difference pattern provides the magnitude of the angular error. The monopulse antenna should have a sum pattern with high gain in the foresight and have a difference pattern with a large value of slope at the crossover of the offset beams.

10.12.1 Glint Error

Amplitude fluctuation, **angle fluctuation**, **receiver thermal noise**, and **servo-noise** limit the tracking accuracy. Amplitude fluctuation of an echo signal is an important limitation factor to the lobe switching and conical scan radars but not an important factor to the monopulse radar. In a tracking radar, the apparent angle of arrival (AOA) of a target is at the direction of the antenna with zero error signal. The apparent AOA may not be confined to the physical extent of the target and may be off the target. Changes in the target aspect with respect to the radar may cause the apparent AOA to wander from one direction to another direction. This random wandering of the apparent AOA is called angle fluctuation or **target glint**. The glint phenomenon is more pronounced when a relatively large target is at short range, which may make the radar unable to track the target and may be the chief factor limiting tracking accuracy. The angular error due to glint varies inversely with distance. Different carrier frequencies may produce different glint errors. The glint error can be reduced by averaging the independent measurements obtained with frequency agility over a bandwidth greater than $c/2D$, where D is the target depth. The qualitative angular error as a function of range is shown in Figure 10.6. The thermal noise term is inversely proportional to R^2, the glint error is inversely proportional to R, and the servo-noise is independent of range.

10.12.2 Track-While-Scan Radar

The track of a target or tracks of multiple targets can be determined with a surveillance radar. As the surveillance radar passes each target, its position is reported to the radar processor. After consecutive scans and reports, the targets' positions can be smoothed, and their future locations can be predicted. A surveillance radar that develops tracks on targets is sometimes called a **track-while-scan** (TWS) **radar.** In a modern radar, the target detection and tracking can be automatically processed by a data processor called **automatic detection and tracking** (ADS). An ADS system can perform the functions of target detection, track initiation, track association, track update, track smoothing, and track termination.

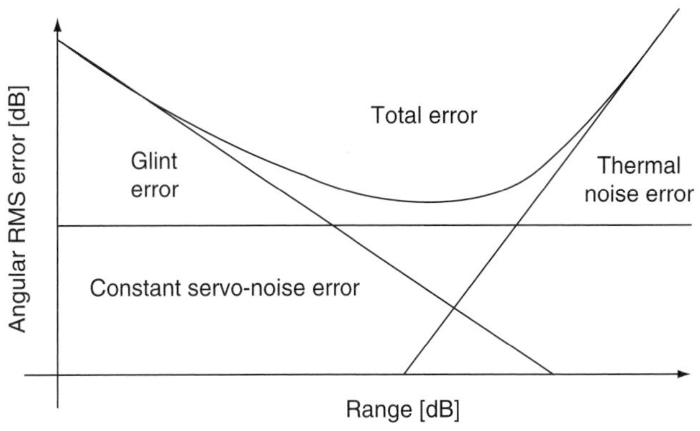

FIGURE 10.6 The Qualitative Angular Error as a Function of Range

10.13 High-Resolution Radar

A low-resolution radar views a target as a single point. Therefore, most targets to the low-resolution radar are point targets with different RCS or echo powers. When a high-resolution radar views the same target, it sees many scattering points and information about the targets, such as sizes, features, or images, that can be extracted from the received echoes.

10.13.1 Resolution

Targets can be resolved in range, azimuth cross-range, elevation cross-range, and Doppler. Resolution represents the smallest separation between two quantities that the radar can distinguish. Range resolution is a function of the transmitted waveform and is inversely proportional to the signal bandwidth. The resolution can be expressed by:

$$\Delta R \cong c/(2B), \tag{10.47}$$

where c is the speed of light and where B is the bandwidth. The factor 2 accounts for the round-trip. The cross-range resolution is proportional to the antenna 3-dB beamwidth $\Delta\theta$ and range R and can be expressed by:

$$\Delta X \cong R \cdot \Delta\theta. \tag{10.48}$$

The Doppler resolution is inversely proportional to the time T_D for collecting the target signal and can be expressed by:

$$\Delta f_D \cong 1/T_D \quad (10.49)$$

10.13.2 Pulse Compression

To obtain a high-range resolution, the pulse is usually compressed. Pulse compression is the process of transmitting a wide pulse to achieve large radiated energy and processing it to a narrow pulse to obtain fine-range resolution. The transmitted pulse is called the **expanded pulse**, and the processed pulse is called the **compressed pulse**. The received signal is processed by passing it through a matched filter. The ratio of the transmitted pulse width to the compressed (or processed) pulse width is called the **pulse compression ratio**.

There are two primary classes of pulse compression: **analog** and **digital**. For analog compression, the transmitted waveform is a **linear FM pulse**. For digital compression, the transmitted waveform is a **phase-coded pulse**.

- **FM pulse compression:** In the linear FM pulse compression, the transmitted waveform consists of a rectangular pulse of constant amplitude. The frequency increases linearly over the duration of the pulse. On reception, the echo is passed through a pulse compression filter, or a matched filter. The output is the autocorrelation of the modulated pulse. The peak power of the correlator output is increased by the pulse compression ratio.
- **Phase-coded pulse compression:** In the phase-coded pulse compression, a long pulse with duration T_p is divided into N subpulses, called chips, of width T_c. The phase of each subpulse is chosen to be either 0 or π radian. The output of the matched filter has a spike of width T_c with an amplitude N times greater than that of the long pulse. The pulse compression ratio is $T_p/T_c = N$.

Two popular code sequences are used for the phase-coded pulse compression: the **Barker code** and the **pseudorandom code**. A binary phase-coded sequence that results in equal side lobes after passing through the matched filter is called a Barker code. The longest Barker code is of length 13. Pseudorandom codes are generated using shift registers with feedback and a modulo-two adder. An n-stage shift register has a total of 2^n different possible states. When the output sequence is of period $2^n - 1$ (the state of all zeros is excluded), it is called a maximal length sequence or a pseudonoise (PN) sequence.

Ambiguity Function

The ambiguity function is the two-dimensional (2-D) autocorrelation of waveform in time and frequency. It can be described as:

$$X(t, f) = \int_{-\infty}^{\infty} \mu(\tau)\mu^*(t + \tau)e^{j2\pi f\tau}d\tau, \quad (10.50)$$

where $X(t, f)$ is the 2-D ambiguity function in time and frequency, and $\mu(\tau)$ is the waveform described in the time domain.

The diagram that represents the ambiguity function is called the ambiguity diagram. It indicates the limitations and utility of particular classes of waveforms and gives a general guideline for the selection of suitable waveforms for various applications.

10.14 High Cross-Range Resolution Radar

The cross-range resolution of a conventional radar is proportional to the product of the 3-dB antenna beamwidth and range. Consider a conventional imaging radar having an antenna with 2 m in length and transmitting linear FM pulses with a bandwidth of 10 MHz. Assume the radar is operated at 2 GHz; then, the 3-dB beamwidth is $\Delta\theta \approx \lambda/2L \cong 0.0375$ rad. If the range from the radar to the ground surface is 20 km, the cross-range resolution will be $\Delta x = R \times \Delta\theta = 20000 \times 0.0375 = 750$ m, much greater than the range resolution, which is equal to $\Delta R \cong c/2B = 15$ m. The cross-range resolution can be greatly improved by using the synthetic aperture technique.

10.14.1 Synthetic Aperture Radar

Synthetic aperture radar (SAR) is a radar carried by a moving vehicle that provides a high-resolution image in both range and cross-range. The fine range is obtained by transmitting waveforms with wide bandwidth, and the cross-range resolution is enhanced by moving the antenna to synthesize a large effective length.

Consider a side-looking SAR geometry as shown in Figure 10.7. The effective length of the synthetic antenna L_{eff} is the distance the radar moves while a scatterer remains in the beam. The beamwidths for a real array with length ℓ and a synthetic

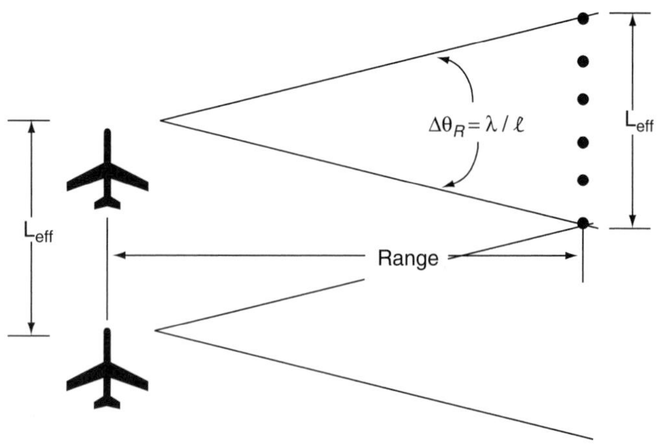

FIGURE 10.7 A Side-Looking SAR Geometry

array having an effective length L_{eff} are shown via these equations, respectively:

$$\Delta\theta_R = \lambda/\ell. \tag{10.51}$$

$$\Delta\theta_S = \lambda/2L_{eff}. \tag{10.52}$$

It is noted that there is a factor 2 in the denominator of equation 10.52. The effective synthetic length is related to the real array length by:

$$L_{eff} = R \cdot \Delta\theta_R = R\lambda/\ell,$$

where R is the range between the scatterer and the radar. The cross-range resolution of SAR at range R is as written here:

$$\Delta X_s = R \cdot \Delta\theta_S = R \cdot \lambda/2L_{eff} = \frac{\ell}{2} \tag{10.53}$$

The cross-range resolution is independent of range and wavelength, and the resolution is better with smaller real antenna length.

Unfocused and Focused SAR

The ranges between the scatterer and the antenna at different flying positions along the synthetic length are not a constant. The range when the antenna is at both ends is greater than the range when the antenna is at the normal to the flight path. When the range difference for the antenna at different positions exceeds a range resolution cell, it is called **range walk**. A range walk will cause the reconstructed image to be unfocused or obscure. To avoid the range walk, the maximum synthetic length should be limited to a length that allows the scatterers to always be in the far-field of the synthetic antenna. The synthetic length should be:

$$L_{eff}^2 < \lambda R/2.$$

The SAR constrained by the range walk is called the **unfocused SAR**. The best resolution for an unfocused SAR is the following:

$$\Delta X = R \cdot \frac{\lambda}{2L_{eff}} \cong (\lambda R/2)^{1/2}. \tag{10.54}$$

A focused SAR refers to data received that are phase-compensated so the range walks are effectively removed. After focusing, the cross-range resolution of equation 10.53 can be obtained.

Motion Compensation

In generating the synthetic aperture, the signal processing assumes that the radar flies along a straight line at constant speed. Practically, the vehicle carrying the antenna may be subjected to deviation from linear motion. Motion compensation is to compensate the phase of the received signal for the displacement of real antenna motion from the position of the ideal SAR being generated.

10.14.2 Inverse Synthetic Aperture Radar

In SAR, the synthetic aperture is obtained by the motion of radar antenna. For inverse synthetic aperture radar (ISAR), the radar is stationary, and the aperture is synthesized by the motion of the object. A basic type of ISAR is the synthetic aperture with rotating objects.

ISAR Imaging of Rotating Objects

The procedure of ISAR imaging of rotating objects is as follows: The object to be imaged is mounted on a rotation pedestal. At each aspect, wideband complex frequency responses (amplitude and phase) or high-resolution range profiles of the backscattered fields are measured and recorded. The object is then rotated by a small angle, and another set of frequency responses or range profiles are taken. This process is repeated until data are gathered for a specified total rotation angle. If the measured quantities are the frequency responses, the data collected are arranged in a polar format as shown in Figure 10.8(A), where the radial line represents the frequency and where the azimuth represents the rotation angle. The point (f_i, ϕ_m) represents the data measured at frequency f_i and

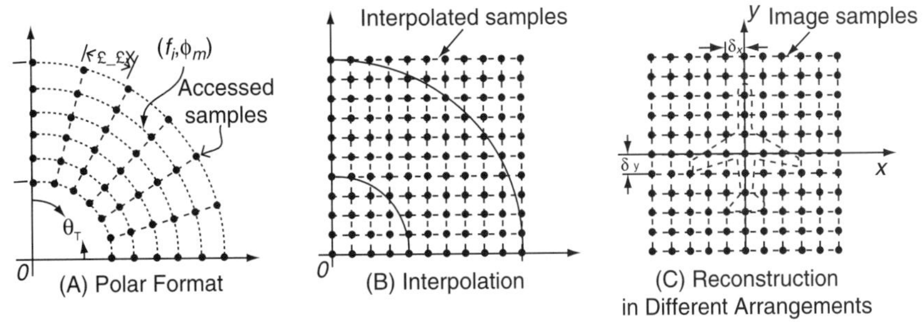

FIGURE 10.8 The Data Collected by Rotation Object Imaging in Different Arrangements

aspect ϕ_m. The image can be reconstructed by taking a 2-D Fourier transform over the collected frequency responses. To use the efficient 2-D fast Fourier transform (FFT), the data sampled in polar grid format should be converted to data sampled in rectangular grid format through an interpolation process as shown in Figure 10.8(B). The image is then processed by taking the 2-D FFT on the interpolated samples as shown in Figure 10.8(C). If the data measured are the complex range profiles for each aspect angle, we can convert the complex range profile into the complex frequency response through a 1-D FFT and arrange the frequency response data in polar format. The image can be reconstructed by the same procedure shown in Figure 10.8. The cross-range resolution ΔX of rotating object imaging is related to the total rotation angle ϕ_T and wavelength λ by:

$$\Delta X = \frac{\lambda}{4\sin(\phi_T/2)}. \qquad (10.55)$$

If the total rotation angle is small, equation 10.55 can be simplified to $\Delta X \cong \lambda/2\phi_T$. The sampling criterion for the angular rotation is to satisfy the Nyquist criterion and avoid the aliasing effect. If the maximum extent of the object is $D_{\max}$, the angular increment $\delta\phi$ should be:

$$\delta\phi < \frac{\lambda}{2D_{\max}} \qquad (10.56)$$

10.15 Inverse Scattering

10.15.1 Definition and Applications

The inverse scattering problem is to reconstruct or recover some physical and/or geometric properties of an object from the measured scattered field under the illumination of an incident wave. The reconstructed information of interest includes, for instance, the dielectric constant distribution and the shape or structure. The interrogating or probing radiation can be an electromagnetic wave (e.g., microwave, optical wave, and X-ray), an acoustic wave, or some other waves. The problem of inverse scattering is important when details about the structure and composition of an object are required but can not be ascertained from measurements made *in situ*.

Due to its noninvasiveness, inverse scattering has wide applications in nondestructive evaluation, medical imaging, remote sensing, seismic exploration, target identification, geophysics, optics, atmospheric sciences, and other such fields.

10.15.2 Properties of the Inverse Scattering Problem

From a mathematical point of view, the existence, uniqueness, and stability of the solution are important concerns of the inverse scattering problem. Several conditions have been suggested to ensure the existence and uniqueness of the solution. It is quite difficult, however, to deal with the instability, which is inherent to the problem. The ill-posed problem, which means that a small error on the data (i.e., the measured scattered field) may cause a large error on the solution (i.e., the reconstructed result), usually makes the inverse scattering problem rather intractable. This phenomenon is even more significant from the viewpoint of applied science or engineering where the incompleteness of measured data and the noise contamination are not uncommon. In addition, it can be shown that the measured data depend nonlinearly on the property function of the object. This nonlinearity also increases the complexity of the inverse scattering problem.

Categories of Electromagnetic Inverse Scattering Problems

Here, the concern is with electromagnetic inverse scattering or electromagnetic imaging problem for which the electromagnetic wave is used as a probing tool. From different aspects, we can classify the electromagnetic inverse scattering problems in various ways. According to the physical properties of the material, the object may be a conductor (for which the information to be reconstructed is the shape function) or a dielectric (for which the dielectric constant distribution and hence also the shape function are to be reconstructed). Moreover, the dielectric object can be homogeneous or inhomogeneous, and different inversion algorithms may apply. According to the environment in which the object is embedded, there may be whole-space problem or half-space problem. The latter is more difficult than the former due to the complexity of Green's function and the geometric limitation on the data measurement for the buried object. According to the dimensionality, the inverse scattering problem may be of one, two, or three dimensions. The one-dimensional problems, which are related to the remote sensing of layered media, have been intensively investigated. There still remain difficulties, however, in the two- or three-dimensional case.

10.15.3 Integral Equation Formulations: Three-Dimensional Case

This discussion now presents some rigorous integral equation formulations for inverse scattering of a conductor as well as a dielectric in the general three-dimensional case. Consider an object occupying the volume V bounded by the surface S with unit outer normal $\hat{n}$, as shown in Figure 10.9.

This object of interest is embedded in a homogeneous background with dielectric constant ε_{rb}. When an incident time-harmonic electromagnetic wave with electric field $\vec{E}^i(\vec{r})$ and angular frequency ω impinges on the object, one can measure the scattered field $\vec{E}^s(\vec{r})$ outside it. The purpose of inverse scattering is to reconstruct the shape of the object (for a conductor) or the dielectric constant distribution $\varepsilon_r(\vec{r})$ of the object (for a dielectric) based on the measured scattered field.

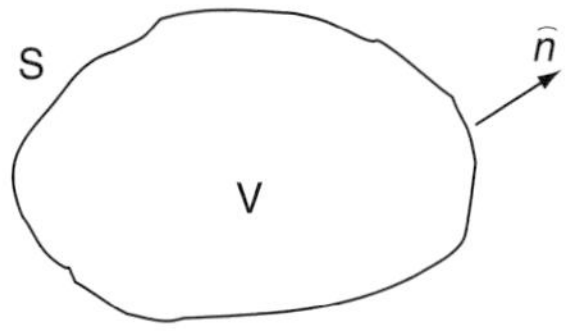

FIGURE 10.9 Object Embedded in a Homogeneous Background

Starting with Maxwell's equations as well as appropriate boundary conditions and using the Green's function technique, one can derive the following integral equations relating the scattered field to the property function of the object:

For a conductor:

$$\begin{cases} \vec{E}^s(\vec{r}) = j\omega\mu \int\int_S \bar{\bar{G}}(\vec{r},\vec{r}') \cdot \vec{J}_s(\vec{r}')dS', & \text{some } \vec{r} \notin V. \\ -\hat{n} \times \vec{E}^i(\vec{r}_0) = j\omega\mu\hat{n} \times \int\int_S \bar{\bar{G}}(\vec{r}_0,\vec{r}') \cdot \vec{J}_s(\vec{r}')dS', & \vec{r}_0 \in S. \end{cases} \tag{10.57}$$

For a dielectric:

$$\begin{cases} \vec{E}^s(\vec{r}) = -k^2 \int\int\int_V \bar{\bar{G}}(\vec{r},\vec{r}') \cdot \left[\frac{\varepsilon_r(\vec{r})}{\varepsilon_{rb}} - 1\right] \cdot \vec{E}(\vec{r}')dV', & \text{some } \vec{r} \notin V. \\ \vec{E}(\vec{r}_0) = \vec{E}^i(\vec{r}_0) - k^2 \int\int\int_V \bar{\bar{G}}(\vec{r}_0,\vec{r}') \cdot \left[\frac{\varepsilon_r(\vec{r}')}{\varepsilon_{rb}} - 1\right] \cdot \vec{E}(\vec{r}')dV', & \vec{r}_0 \in V. \end{cases} \tag{10.58}$$

In equations 10.57 and 10.58 $\vec{J}_s$ is the induced surface current density on the conductor, $\bar{\bar{G}}$ is the dyadic Green's function, $\vec{E}$ denotes the total electric field in the dielectric object, μ is the permeability (a constant), and k is the wave number of the background medium. For the conductor case, the induced surface current density $\vec{J}_s$ is related to shape function through the second part of equation 10.57. Then, from the first part of equation 10.57, it is apparent that the unknown shape function is nonlinearly related to the known (measured) scattered field. In a similar manner, the total field $\vec{E}$ in the dielectric object is related to the dielectric constant distribution ε_r through the second part of equation 10.58. Therefore, the first part of equation 10.58 implies that the unknown dielectric constant distribution depends nonlinearly on the known (measured) scattered field. In other words, the inverse scattering problem is a nonlinear ill-posed problem and various numerical techniques have been developed to tackle it. Note also that for the cases of lower dimensionality or under some circumstances, integral equations 10.57 and 10.58 can be simplified or transformed into other equivalent forms.

10.15.4 Methods for Solving the Inverse Scattering Problem

A variety of inversion algorithms have been proposed to solve the aforementioned inverse scattering problems under various conditions. The numerical techniques can be roughly classified into two general categories: the **approximate approach** and the **rigorous method**. The former is usually based on the linearized model in which the scattered field is approximated as a linear function of the object function. That is, the theoretical formulation is approximated at the beginning, and hence, the problem can be simplified and easily treated. On the other hand, the rigorous approach deals directly with the nonlinear problem based on the rigorous formulation. It is almost impossible to have a closed-form or exact solution, however, and some numerical techniques, including how to discretize the problem, are still needed to obtain a reconstructed result.

Approximate Approaches

Bojarski's Identity

Based on the physical optics (PO) approximation, Bojarski proposed in 1967 the so-called Bojarski's identity that can be used to determine the shape function of a conducting object from the measured back-scattered far-field over all frequencies and all aspect angles. This approach is illustrated in the following. Consider a conductor of volume V bounded by the surface S. When this conductor is illuminated by an electromagnetic plane wave $\vec{E}_o^i e^{-j\vec{k}^i \cdot \vec{r}}$, the back-scattered far-field can be expressed as:

$$\vec{E}^s \approx -\frac{e^{-jkR}}{\sqrt{2\pi}R}\vec{E}_0^i \cdot \rho(\vec{K}), \tag{10.59}$$

with

$$\rho(\vec{K}) = \frac{-j}{2\sqrt{\pi}}\int\int_{\vec{K}\cdot\hat{n}>0} e^{j\vec{k}\cdot\vec{r}'}(\vec{K}\cdot\hat{n})dS'. \tag{10.60}$$

In these equations, $\vec{K}$ is $-2\vec{k}^i$, R is the distance from the reference point on the object, and $\vec{K}\cdot\hat{n} > 0$ refers to the illuminated part of the surface. Now consider the scattered field when illuminated from the opposite direction ($\vec{K} \to -\vec{K}$). Take the complex conjugate and obtain:

$$\rho^*(-\vec{K}) = \frac{-j}{2\sqrt{\pi}}\int\int_{\vec{K}\cdot\hat{n}<0} e^{j\vec{k}\cdot\vec{r}'}(\vec{K}\cdot\hat{n})dS'. \tag{10.61}$$

If the conducting object is convex, then:

$$\rho(\vec{K}) + \rho^*(-\vec{K}) = \frac{-j}{2\sqrt{\pi}}\oint\!\!\oint_S e^{j\vec{k}\cdot\vec{r}'}(\vec{K}\cdot\hat{n})dS' = \frac{K^2}{2\sqrt{\pi}}\int\int\int_V e^{j\vec{k}\cdot\vec{r}'}dV'. \tag{10.62}$$

Here, the divergence theorem has been used. The complex scattering amplitude $\Gamma(\vec{K})$ and the characteristic function $\gamma(\vec{r}')$ of the object can be defined as:

$$\Gamma(\vec{K}) = \frac{2\sqrt{\pi}}{K^2}[\rho(\vec{K}) + \rho^*(-\vec{K})]. \tag{10.63}$$

$$\gamma(\vec{r}') = \begin{cases} 1 & \text{inside } V \\ 0 & \text{outside } V \end{cases}. \quad (10.64)$$

The result is the following three-dimensional Fourier transform relationship:

$$\Gamma(\vec{K}) = \iiint \gamma(\vec{r}')e^{j\vec{k}\cdot\vec{r}'}dV'. \quad (10.65)$$

Taking the inverse Fourier transform, yields:

$$\gamma(\vec{r}) = \frac{1}{(2\pi)^3}\iiint \Gamma(\vec{K})e^{j\vec{k}\cdot\vec{r}}d\vec{K}. \quad (10.66)$$

Equation 10.66 is Bojarski's identity and shows that the object shape can be reconstructed by measuring the backscattered far-field over all $\vec{K}$. Since the PO approximation is valid only at high frequencies and for smooth convex conducting surfaces, Bojarski's identity has its own theoretical limitation. Another practical difficulty is that the measured far-field data should be complete over "all" frequencies and "all" aspect angles. However, this difficulty is common to almost all the inversion algorithms. In spite of these difficulties, Bojarski's identity is useful for reconstructing the shape of a large, smooth convex conductor for which the multiple scattering effect can be neglected. Moreover, the algorithm is also efficient in computation because it makes use of the **fast Fourier transform** (FFT).

Diffraction Tomography

The X-ray **computerized tomography** (CT) has been widely used for medical imaging and diagnosis since the 1970s. The basic idea of it is that the total attenuation of the X-ray traversing the object along some direction is related to the integrated absorption by the object along that direction. Hence, the attenuation distribution of the transmitted X-ray forms a Radon transform of the absorption distribution of the object. The projection–slice theorem is used to collect the data in the Fourier domain, and then the inverse Fourier transform is performed (by FFT) to reconstruct the absorption distribution of the object. Now, if the electromagnetic wave is used to reconstruct dielectric constant distribution of the object in a similar manner, the diffraction effect due to finite wavelength should be taken into account. It is quite complicated to do in general. For a weakly inhomogeneous object, however, it is possible to formulate the diffraction tomography in a manner similar to CT, and the general solution can be obtained. To be specific, here the dielectric object of interest is assumed to be weakly inhomogeneous such that the total field $\vec{E}$ in the first part of equation 10.58 can be approximated by the incident field $\vec{E}^i$. This is usually referred to as the **Born approximation**. Consider a two-dimensional dielectric object illuminated by a plane wave incident from the direction $\hat{i}$, as shown in Figure 10.10.

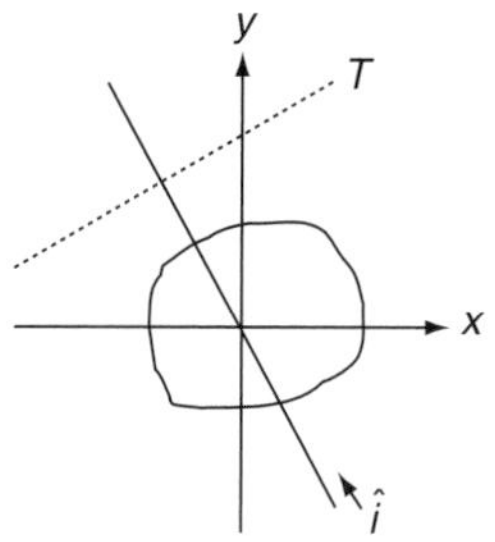

FIGURE 10.10 A 2-D Dielectric Illuminated by a Wave

The forward scattered field measured along the line T perpendicular to $\hat{i}$ can be expressed as the integral of the product of the Green's function, incident field and the dielectric constant of the object. Taking the one-dimensional Fourier transform of the scattered field along the line perpendicular to $\hat{i}$, one sees that the transformed field is related to the values of the two-dimensional Fourier transform of the dielectric constant contrast evaluated on a semicircle. The orientation of the semicircle depends on the incident direction $\hat{i}$, and its radius is frequency-dependent. Note that this semicircle corresponds to the slice for the conventional X-ray CT. If we change the direction of the incident wave (and hence the corresponding direction of receiving line) over all aspect angles and/or vary the frequency of the incident wave, sufficient transformed data will be available in the two-dimensional spatial frequency domain. After proper interpolation of the data to fit the rectangular coordinates, one can take the two-dimensional inverse Fourier transform (by FFT) to obtain the reconstructed dielectric constant distribution. This is the basic principle of the forward-type diffraction tomography. For the reflection-type, one can measure the back scattered field along the line perpendicular to $\hat{i}$ and perform similar processes (except that the transformed data are now on another semicircle away from the origin). This discussion is based on the Born approximation that holds only for the case of very weak inhomogeneity of the dielectric object. However, one can enlarge the range of inhomogeneity a little further such that the Rytov approximation is valid. By properly defining the measured quantity (taking the logarithm of the total field divided by the incident field), it can be shown that all the aforementioned processes are still valid. Therefore, the diffraction tomography can be applied to a larger range of inhomogeneity of the dielectric object as long as the Rytov approximation is satisfied. Note that the idea of diffraction tomography can also be applied to inverse scattering of the conducting object. Under the physical optics approximation, the scattered field can be approximately expressed as a linear function of a properly defined shape function of the object. Therefore, the one-dimensional Fourier transform of the scattered field will be related to the values of the two-dimensional Fourier transform of the shape function evaluated on a semicircle. This is indeed the basic scheme of the diffraction tomography. However, the shape function defined here is not rigorous since it depends on the incident wave vector. Note that this type of diffraction tomography is applicable only to the large, smooth convex conductor because of the limitation on the PO approximation. Compared to the inversion algorithm of Bojarski's identity for which the far-field measured data are required, the measurement for the diffraction tomography can be made in the near zone. This may be convenient in practical applications.

Rigorous Approaches

Iterative Methods

As mentioned previously, the inverse scattering problem is in general a nonlinear problem in the sense that the unknown object function is nonlinearly related to the measured scattered field. To solve it rigorously, iterative techniques are usually applied. Although various algorithms have been proposed, they are based on the same principle—taking the variation of the nonlinear equation and solving the resulting linearized problem at each iterative step to update the solution. For illustration, consider a dielectric object of dielectric constant distribution $\varepsilon_r(\vec{r})$ illuminated by the incident field $\vec{E}^i(\vec{r})$. The scattered field $\vec{E}^s(\vec{r})$ is related to the dielectric constant $\varepsilon_r(\vec{r})$ through equation 10.58. First discretize the object into N small cells and assume that the dielectric constant and electric field are constants in each cell. By the moment method, transform the integral equation 10.58 into the following algebraic matrix equation:

$$[G_2][\tau]([I] - [G_1][\tau])^{-1}[E^i] = [E^s]. \tag{10.67}$$

Here, matrix $[G_2]$ corresponds to the Green's function relating an interior point in the object to an exterior point outside the object, $[G_1]$ corresponds to the Green's function associated with two interior points, $[\tau]$ is a diagonal matrix with elements being the contrast of the object's dielectric constant with respect to the surroundings, $[E^i]$ lists the incident field in N cells, $[E^s]$ lists the measured scattered field outside the object, and $[I]$ is an identity matrix. The nonlinear relation between $[\tau]$ and $[E^s]$ is apparent in equation 10.67. For solving this nonlinear algebraic equation, first take the variation or differential of equation 10.67 to yield:

$$[G_2]([I] - [\tau][G_1])^{-1}[\delta\tau]([I] - [G_1][\tau])^{-1}[E^i] = [\delta E^s], \tag{10.68}$$

which describes a variation $[\delta E^s]$ due to a variation $[\delta\tau]$. The linear relation between $[\delta E^s]$ and $[\delta\tau]$ in equation 10.68 can also be rewritten in the following familiar form:

$$[A_\tau][x] = [b], \tag{10.69}$$

where matrix $[A_\tau]$ depends on the contrast $[\tau]$ of the dielectric constant and is related to matrices $[G_2]([I] - [\tau][G_1])^{-1}$ and $([I] - [G_1][\tau])^{-1}[E^i]$. All the unknowns in $[\delta\tau]$ are listed in the column matrix $[x]$, and the column matrix $[b]$ contains the elements in $[\delta E^s]$. Note that equation 10.69 is the basic equation used at each iterative step. The whole iteration scheme proceeds by first choosing an initial guess $[\tau_{(0)}]$ of the true contrast of the dielectric constant $[\tau_{\text{true}}]$. The resulting scattered field $[E^s_{(0)}$ due to $[\tau_{(0)}]$ can be calculated by equation 10.67, and the difference in scattered field $[\delta E^s_{(0)}]$ between the measured value and the calculated one is evaluated. To the first-order approximation, $[\delta\tau_{(0)}]$ obtained from solving equation 10.69 is used to update the initial guess to $[\tau_{(1)}] = [\tau_{(0)}] + [\delta\tau_{(0)}]$. This new $[\tau_{(1)}]$ is then taken as the initial guess at the second step, and a similar procedure is repeated to obtain further updated $[\tau_{(2)}]$. In other words, the above procedure is repeatedly executed, using $[\tau]$ updated at the previous step, until the variation $[\delta\tau]$ between consecutive steps is smaller than some prescribed value. This illustrates the basic principle of the iterative methods. Based on the idea of iteration, other algorithms have been created, such as the Newton-Kantorovitch technique, Born iterative method, and distorted Born iterative method.

Other Methods

Note that after discretization, the inverse scattering problem can be cast into a multivariable nonlinear problem. There exist, of course, miscellaneous methods for solving this problem. Apart from those mentioned previously, other available methods include genetic algorithm and simulated annealing.

Regularization

Inverse scattering is an ill-posed problem in the sense that its solution is unstable with respect to the perturbation of the measured data. Explicit evidence is the occurrence of inverting an ill-conditioned matrix during the inversion process. For example, matrix $[A_\tau]$ in equation 10.69 is usually of high condition-number inducing errors at each iterative step. To address the ill-posed problem the regularization process is usually used. The basic idea of regularization is to compute a meaningful "smooth" solution or to filter out the high-frequency components associated with the small singular values. If the inverse scattering problem is so formulated that the convolution or Fourier transform is involved, the regularization is realized by introducing a low-pass filter during the deconvolution or inverse Fourier transform process. If the matrix inversion is involved in the inverse scattering problem, one can take the pseudoinverse transformation or, equivalently, singular value decomposition method to regularize the problem. In this approach, the effect of singular values smaller than some threshold is neglected to obtain the well-behaved solution. Another equivalent technique, known as **Tikhonov method**, has also been proposed. It incorporates a properly chosen well-posed problem into the original ill-posed one to increase the stability of the resulting solution. All of the discussions in this chapter are from the mathematical point of view. In practice, one can incorporate prior knowledge or some constraints in the reconstruction operation to limit the range of possible solutions. In addition, more measured data at multiple frequencies and/or from multiview illumination will be helpful for the stabilization of the solution.

References

Baltes, H.P. (1980). *Inverse scattering problems in optics.* New York: Springer-Verlag.

Barton, D.K. (1976). *Radar system analysis.* Norwood, MA: Artech House.

Barton, D.K. (1988). *Modern radar system analysis.* Norwood, MA: Artech House.

Barton, D.K., Cook, C.E., and Hamilton, P. (1991). *Radar evaluation handbook.* Norwood, MA: Artech House.

Berkowitz, R. (1965). *Modern radar.* New York: John Wiley & Sons.

Chew, W.C. (1990). *Waves and Fields in Inhomogeneous Media.* New York: Van Nostrand.

Chew, W.C., and Wang, Y.M. (1990). Reconstruction of two-dimensional permittivity distribution using the distorted born iterative method. *IEEE Transactions on Medical Imaging 9,* 218–225.

Chu, T.H., and Lin, D.B. (1991). Microwave diversity imaging of perfectly conducting objects in the near-field region. *IEEE Transactions on Microwave Theory Technology MTT-39,* 480–487.

Davis, L. (1987). *Genetic algorithms and simulated annealing.* Los Altos, CA: Morgan Kaufmann.

Devaney, A.J. (1982). A filtered back propagation algorithm for diffraction tomography. *Ultrasonic Imaging 4,* 336–350.

Edde, B. (1993). *Radar principles, technology, applications.* Englewood Cliffs, NJ: Prentice Hall.

Elachi, C., Bicknell, T., Jordan, R.L., and Wu, C. (1982). Spaceborne synthetic aperture imaging radars: Applications, technology, and technique. *Proceedings of the IEEE 70,* 1174–1209.

Hopcraft, K.I., and Smith, P.R. (1992). *An introduction to electromagnetic inverse scattering.* Dordrecht: Kluwer Academic.

Institute of Electrical and Electronics Engineers. (1982). IEEE Standard 686-1982, *IEEE standard radar definition.* New York: IEEE Press.

Ishimaru, A. (1991). *Electromagnetic wave propagation, radiation, and scattering.* Englewood Cliffs, NJ: Prentice Hall.

Joachimowicz, N., Pichot, C., and Hugonin, J.P. (1991). Inverse scattering: An iterative numerical method for electromagnetic imaging. *IEEE Transactions on Antennas Propagation 39,* 1742–1752.

Kak, A.C. (1979). Computerized tomography with X-ray, emission, and ultrasound sources. *Proceedings of the IEEE 67,* 1245–1272.

Knott, E.F., Shaeffer, J.F., and Tuley, M.T. (1985). *Radar cross-section.* Norwood, MA: Artech House.

Larsen, L.E., and Jacobi, J.H. (1986). *Medical applications of microwave imaging.* New York: IEEE Press.

Levanon, N. (1988). *Radar principles.* New York: John Wiley & Sons.

Lewis, R.M. (1969). Physical optics inverse diffraction. *IEEE Transactions on Antennas Propagation, AP-17,* 308–314.

Li, H.J., Farhat, N.H., Shen, Y., and Werner, C.L. (1989). Image understanding and interpretation in microwave diversity imaging. *IEEE Transactions on Antennas Propagation, 37*(8), 1048–1057.

Li, H.J., Lin, F.L., Shen, Y., and Farhat, N.H. (1990). A generalized interpretation and prediction in microwave imaging involving frequency and angular diversity. *Journal of Electromagnetic Waves and Applications 4*(5), 415–430.

Liao, S.Y. (1980). *Microwave devices and circuits.* Englewood Cliffs, NJ: Prentice Hall.

Marcum, J.I. (1960). A statistical theory of detection by pulsed radar and mathematical appendix. *IRE Transaction IT-6,* 59–267.

Mensa, D.L. (1991). *High-resolution radar cross section imaging.* (2d. ed.). Norwood, MA: Artech House.

Nathanson, F.E. (1969). *Radar design principles.* New York: McGraw-Hill.

Ney, M.M., Smith, A.M., and Stuchly, S.S. (1984). A solution of electromagnetic imaging using pseudoinverse transformation. *IEEE Transactions on Medical Imaging MI-3,* 155–162.

Roger, A. (1981). Newton-Kantorovitch algorithm applied to an electromagnetic inverse problem. *IEEE Transactions on Antennas Propagation AP-29,* 232–238.

Ruck, G.T. Barrick, D.E., Stuart, W.D., and Krichbaum, C.K. (1970). *Radar cross-section handbook.* New York: Plenum Press.

Skolnik, M.I. (1980). *Introduction to radar systems.* (2d ed.). New York: McGraw-Hill.

Skolnik, M.I. (1988). *Radar applications.* New York: IEEE Press.

Skolnik, M.I. (1990). *Radar handbook.* (2d ed.). New York: McGraw-Hill.

Slaney, M., Kak, A.C., and Larsen, L.E. (1984). Limitations of imaging with first-order diffraction tomography. *IEEE Transactions on Microwave Theory Technology MTT-32,* 860–874.

Steinberg, B.D., Carlson, D.L., and Lee, W. (1989). Experimental localized radar cross sections of aircraft. *Proceedings of the IEEE 77*(5), 663–669.

Stuzman, W.L., and Thiele, G.A. (1998). *Antenna theory and design.* (2d. ed.). New York: John Wiley & Sons.

Swerling, R. (1960). Probability of detection for fluctuating targets. *IRE Transaction IT-6,* 269–2308.

Ulaby, F.T., Moore, R.K., and Fung, A.K. (1981). *Microwave remote sensing.* Reading, MA: Addison-Wesley.

Walker, J.L. (1980). Range-Doppler imaging of rotating objects. *IEEE Transactions on Aerospace and Electronic Systems AES-16* (1), 23–53.

Wang, Y.M., and Chew, W.C. (1989). An iterative solution of two-dimensional electromagnetic inverse scattering problem. *International Journal on Imaging Systems Technology 1,* 100–108.

Wehner, D.R. (1987). *High-resolution radar.* Norwood, MA: Artech House.

11

Microwave Active Circuits and Integrated Antennas

William R. Deal,*
Vesna Radisic,[†]
Yongxi Qian,[‡]
and Tatsuo Itoh[‡]

*Northrup Grumman Space Technologies, Redondo Beach, California, USA
[†]Microsemi Corporation, Los Angeles, California, USA
[‡]Department of Electrical Engineering, University of California, Los Angeles, Los Angeles, California, USA

11.1 Introduction

In this chapter, a variety of topics concerning communication systems at microwave- and millimeter-wave frequencies are discussed.

Section 11.2, Device Technology and Concepts, begins with a discussion on noise at these frequencies. Because wireless applications are the primary drive for development of these frequency bands, understanding the basic concepts of noise at these frequencies becomes of fundamental importance in understanding the sensitivity of high-frequency components, such as low-noise amplifiers and mixers. The section continues with an introduction to the most common microwave devices, including bipolar transistors, field-effect transistors, high electron mobility transistors, and two-terminal microwave devices. Noise mechanisms and most common applications as well as their most common small-signal models are discussed. Heterojunction bipolar transistors are briefly mentioned in the section on bipolar transistors.

In Section 11.3, Active Microwave Circuits, issues relevant to the design of the most common types of active microwave circuits are discussed. The section first presents the design of microwave amplifiers using an S-parameter approach. Next, microwave oscillator design is briefly outlined. The stability conditions for both one- and two-port microwave oscillators are presented. The section concludes with a discussion on microwave detectors and mixers. Detectors are often used in control circuitry. Mixers are used in up-conversion and down-conversion circuitry of wireless systems. Fundamental operation of mixing phenomena is described and figures of merits for mixers are defined.

Section 11.4, Planar Antennas, presents a unique and practical class of antennas for application at microwave- and millimeter-wave frequencies. These antennas, often referred to as printed antennas, are completely compatible with many types of high-frequency transmission lines and can be constructed directly on a printed circuit board. Their low cost, ease of fabrication, and control over their radiation properties makes these antennas very useful in commercial wireless systems at these frequencies. Four types of planar antennas are discussed, including the patch antenna, slot antenna, the tapered slot antenna (TSA), and the newly developed quasi-Yagiantenna. The section concludes with an antenna selection guide, containing geometry and radiation characteristics of a wide variety of planar antennas in one compact table for easy reference.

Last, this chapter briefly discusses active integrated antennas (AIAs). In this approach, the compatibility of planar antennas with popular microwave frequency transmission lines and active devices is exploited by directly integrating the active circuitry with the antenna platform. This results in a compact and highly functional system with many interesting characteristics.

11.2 Device Technology and Concepts

This section deals with fundamental basics of active microwave circuits: **noise in a microwave system** and **noise in active microwave devices.** As this section first explains, at microwave frequencies, noise is dominated by thermal and shot noise, not $1/f$ noise. Then, the bipolar transistor, heterojunction bipolar transistor, field effects transistor, high electron mobility transistor, and two-terminal microwave devices are discussed. Excellent references on these topics can be found in Fukui (1981) and Yngresson (1991).

11.2.1 Noise in a Microwave System

In a microwave system, a small signal can be measured even when no applied signal is present. This is due to noise that can come from several sources. The sensitivity of receiver components, such as the low noise amplifier (LNA) and mixer, is fundamentally limited by the noise figures of their components and background noise. For these reasons, noise is of fundamental importance in many active microwave frequency designs and will be covered briefly in this section. Dominant mechanisms are typically thermal noise and shot noise at microwave frequencies. If a microwave device is used to directly convert a signal to baseband, $1/f$ noise may become an important noise factor as well since $1/f$ noise determines oscillator phase noise.

Thermal (Johnson) noise is one of the dominant types of noise present in microwave circuits, and it is caused by the random fluctuations of the electrons due to thermal agitation. At microwave frequencies and for a temperature, T, close to room temperature, an expression for the rms voltage produced by a resistor of value R is the following:

$$v_n = \sqrt{4kTRB}. \tag{11.1}$$

Note that k is Boltzmann's constant (1.374×10^{-23} J/°K), and B is the bandwidth that the noise is measured over. Therefore, the thermal noise depends on bandwidth and not on the center frequency. For this reason, it is also referred to as **white noise.**

Shot noise is noise due to electron and hole fluctuations in any device with an average dc current flow. **Flicker noise** (*1/f noise*) is noise due to a variety of sources that exhibits a $1/f^{\alpha}$ dependence, where α is close to one. At microwave frequencies, the contribution of flicker noise is usually negligible with respect to other noise mechanisms. Its contribution can be significant for some microwave devices operated below 50 MHz.

Typically, flicker noise will be dominant at lower frequencies until the **corner frequency** is reached, above which thermal noise dominates. For particular devices, shot noise will also have a contributive effect. As with thermal noise, however, shot noise is also approximately independent of frequency.

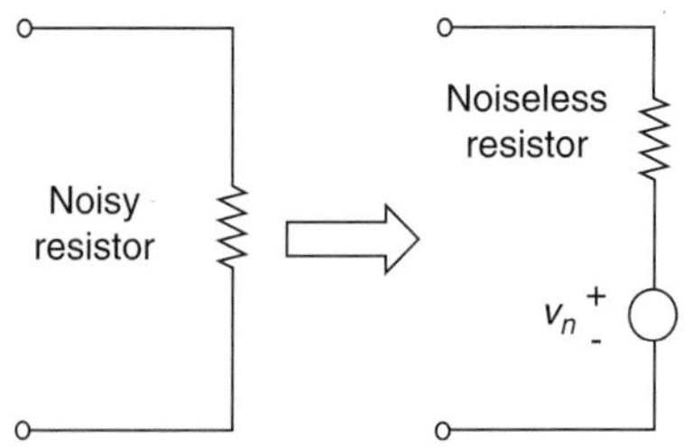

FIGURE 11.1 Model of a Noisy Resistor

Therefore, from a macroscopic point of view, both of these mechanisms can be lumped together when characterizing a device or component in terms of its noise characteristics. The question then becomes how to best describe a noise source. The simplest way is to consider a simple resistor. As shown in Figure 11.1, this may be a **noisy resistor** or a **noiseless resistor.**

The noise power available from an arbitrary 1-port network with an input impedance $Z(f) = R(f) + jX(f)$ can be expressed in terms of a noise temperature. If the available noise power, P_a, is constant over the frequency range of interest, then a noise temperature can be defined as:

$$T_s = \frac{P_a}{kB}. \tag{11.2}$$

Note that this is not the physical temperature at which the network is measured but the temperature required for a resistor to give the proper available noise power. This can be determined by measurements with a calibrated noise source.

Noise of two-port networks, such as from mixers or amplifiers, is also of interest in receiver design. For a noisy two-port network with known gain G_a when both ports are terminated in a simultaneous conjugate match, the available noise power at the output will be given by:

$$P_{no} = P_{ni}G_a + P_{ne} = kT_sBG_a + P_{ne}. \tag{11.3}$$

The first variable represents the noise amplified (or attenuated) by the network. The second, P_{ne}, represents the excess noise generated by the network itself. In addition, if the input noise temperature, T_s, is held at 0° Kelvin, the output noise power will consist of only P_{ne}. The noisy two-port can be replaced by a noiseless two-port giving the same noise power output. This is done by defining an effective noise temperature of the two-port network itself, T_e:

$$T_e = \frac{P_{ne}}{kBG_a}. \tag{11.4}$$

The total noise power output of the equivalent noiseless two-port is then:

$$P_{no} = kT_s\left[1 + \frac{T_e}{T_s}\right]BG_a. \tag{11.5}$$

In equation 11.5, the calculation in brackets is of particular interest. By simple manipulation, it can be shown that this part of the calculation is equivalent to the signal-to-noise ratio (SNR) at the input to that at the output. This term is defined to be the **actual noise figure** of the two-port:

$$F_a = 1 + \frac{T_e}{T_s} = \frac{SNR_i}{SNR_o}. \quad (11.6)$$

By convention, the noise figure is often cited for a two-port network. The noise figure, F, is defined to be the actual noise, F_a, with T_s held at room temperature ($T_s = 290°K$).

Often, two-port networks will be cascaded in a receiver. For instance, several LNAs may be required to boost a received signal to a sufficient level that it can be down-converted. In this case, the designer may be interested in the overall effect of cascading each two-port network. The total effective noise temperature and noise figure of n cascaded two-ports can be found using:

$$F = F_1 + \frac{F_2 - 1}{G_{a1}} + \cdots + \frac{F_n - 1}{G_{a1}G_{a2}\cdots G_{a(n-1)}}. \quad (11.7)$$

$$T_e = T_{e1} + \frac{T_{e2}}{G_{a1}} + \cdots + \frac{T_{en}}{G_{a1}G_{a2}\cdots G_{a(n-1)}}. \quad (11.8)$$

Bipolar Junction Transistor

Typically made from silicon in the NPN configuration, the microwave bipolar junction transistor (BJT) is historically the dominant device below 4 GHz but does find some application above this frequency. Currently, silicon CMOS devices are increasingly competitive at these frequencies for many applications.

Conceptually, operation of the microwave BJT is similar to its lower frequency counterparts, but it requires scaled-down base widths (commonly 0.1 μm). While the gate voltage controls the FET, the base current controls the BJT. A cross-sectional view of an idealized BJT with an interdigital design is shown in Figure 11.2.

Several commonly referred to characteristics for BJT transistors are useful for choosing a device. The first of these is f_T: the gain-bandwidth frequency. This is the frequency for which the **short-circuit current gain** of the device is unitary (0 dB). The second is $f_{\max}$, the maximum frequency of oscillation, or the maximum frequency that the device can produce negative resistance. Typically, the parameter h_{fe}, the low frequency short-circuit current gain, is not used with microwave frequency devices. The meaning of these parameters is illustrated in Figure 11.3. Note that gain rolls off at the rate of −6dB/Octave.

A standard hybrid-π model for the BJT transistor is shown in Figure 11.4. At microwave frequencies, the resistance of $r_{b'c}$ is typically much larger than the reactance of $C_{b'c}$, which it is in parallel with, allowing it to be replaced with an open circuit. The resistance r_{ce} may also be large enough with respect to $C_{b'e}$ and $r_{b'e}$ so that it may be omitted as well, resulting in a simpler model at microwave frequencies. Extrinsic components of the package, however, are also important at microwave frequencies. A typical equivalent model of the package is shown to the right of the intrinsic BJT model.

Due to their low cost and reliable performance, Si BJTs are found in many applications below 4 GHz. Packaged low-power (general purpose), medium-power, and low-noise BJTs are available from many manufacturers. Power versions are also available. Figures of merit for microwave BJTs are f_T and $f_{\max}$. In terms of the equivalent circuit model, these are given by:

$$f_T = \frac{g_m}{2\pi Cb'e}. \quad (11.9)$$

$$f_{\max} = \sqrt{\frac{f_T}{8\pi rb'eCb'c}}. \quad (11.10)$$

Noise mechanisms in microwave BJTs are thermal noise caused by thermal agitation of the carriers in the ohmic resistance of the emitter and base, collector, and shot noise caused

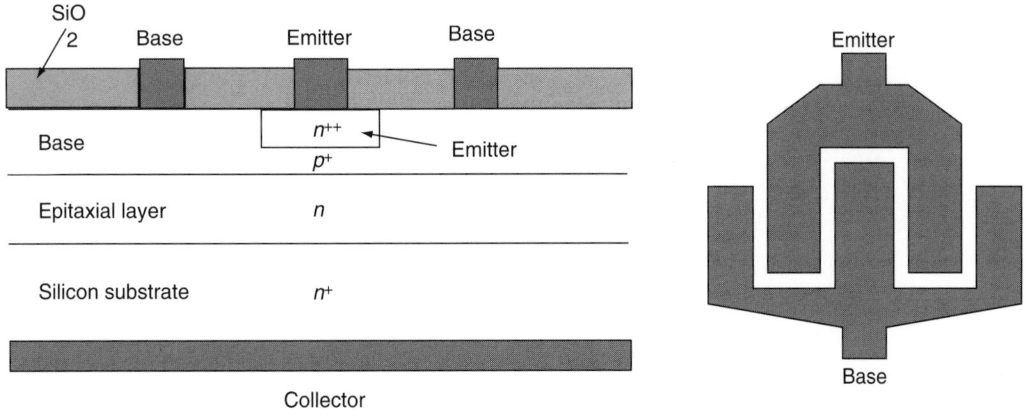

FIGURE 11.2 Cross-Sectional and Top Views of Interdigitated BJT

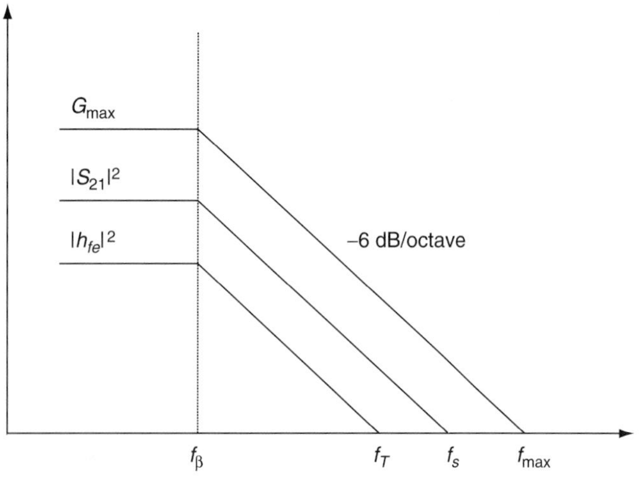

FIGURE 11.3 Frequency Characteristics of G_{max}, $|S_{21}|^2$, and $|h_{fe}|^2$

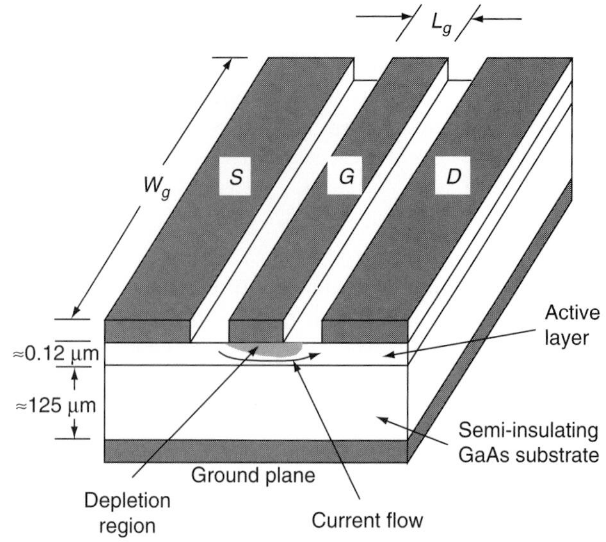

FIGURE 11.5 Physical Structure of Idealized GaAs MOSFET

by fluctuation in the electron and hole currents due to biasing conditions.

Inclusion of a heterojunction in the BJT has been demonstrated to yield a device with superior microwave properties, including higher gains and superior noise figure. The device is commonly referred to as the heterojunction bipolar transistor (HBT). Unlike conventional microwave frequency BJT transistors that are Si-based, the HBT typically uses GaAs technology with a significantly higher velocity of electrons in the n-doped GaAs. Commercial development of the HBT is not nearly as mature as competing technologies, such as development of the HEMT. It is therefore expected that significant improvements will be made in these devices, which already show respectable performance. SiGe HBT transistors have also been successfully developed.

Metal-Oxide-Semiconductor Field Effect Transistor (MOSFET)

The MOSFET, typically made of gallium arsenide (GaAs FET) because of its high electron mobility, is the most common device used for amplification (low noise and power) in the regions between 4 to 60 GHz. It is also commonly used in oscillators, mixers, and switches. Due to the excellent noise properties of the MOSFET, it is often used for sensitive receivers below 4 GHz. GaAs FETs typically exhibit higher power gain, lower noise figure, and higher output capabilities than Si BJTs above 4 GHz. The idealized geometry of a GaAs FET is shown in Figure 11.5. Note that the depletion region below the gate controls the amount of current flowing between the source and drain. With zero applied bias, the depletion region will consist of only the built-in potential of the Schottky barrier. As the gate potential is made negative relative to the source, the depth of the depletion region increases, with the effect of increasing channel resistance and reducing current. Pinch-off occurs when the depletion region reaches the semi-insulating GaAs substrate.

The GaAs FET is the primary building block of monolithic microwave-integrated circuits (MMICs). Generally, smaller gate length (L_g) corresponds to higher f_{max}. Longer gate width (W_g) corresponds to higher output power. A gate

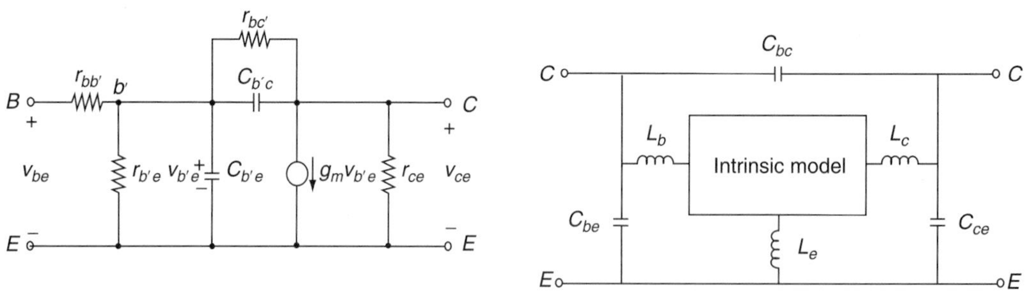

FIGURE 11.4 Hybrid-π Model for the BJT Transistor. Note that simplifications to this model can often be made at microwave frequencies. The BJT (intrinsic) model (right) is incorporated into an equivalent circuit model for packaging parasitics.

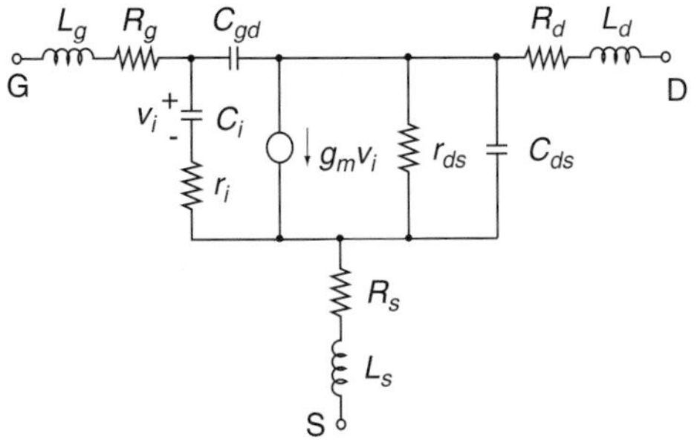

C_i:	Gate-to-source capacitance	(0.3 pF)
r_i:	Gate-to-source channel resistance	(2.5 Ω)
C_{gd}:	Gate-to-drain capacitance	(0.02 pF)
C_{ds}:	Drain-to-source capacitance	(0.05 pF)
r_{ds}:	Drain-to-source resistance	(600 Ω)
g_m:	Transconductance	(40 mS)
R_g:	Gate parasitic resistance	(0.1 Ω)
L_g:	Gate parasitic inductance	(0.1–0.9 nH)
R_s:	Source parasitic resistance	(0.1 Ω)
L_s:	Source parasitic inductance	(0.1–0.9 nH)
R_d:	Drain parasitic resistance	(0.1 Ω)
L_d:	Drain parasitic inductance	(0.1–0.9 nH)

FIGURE 11.6 Equivalent Common-Source Model of Typical MOSFET. This model includes packaging effects.

width that is too long, however, can result in degraded performance. This is often overcome by using an interdigitated design with additional "fingers." Conceptually, this is like connecting several lower power devices in parallel.

A small signal equivalent model, including external parasitics, for the MOSFET transistor in common-source mode is shown in Figure 11.6. Each component is defined to the right of the equivalent circuit, with typical values for a GaAs MOSFET with a gate length of 1 μm and a gate width of 250 μm given in parentheses. There are many minor variations of this model. Often, the drain-to-source resistance, r_{ds}, is omitted. In addition, when the gate-to-drain feedback capacitance is small enough to be neglected, it is replaced with an open circuit, resulting in the simplified unilateral model for a common-source GaAs FET.

In terms of the equivalent model, the frequencies f_T and $f_{\max}$ are given by:

$$f_T = \frac{g_m}{2\pi C_i}. \tag{11.11}$$

$$f_{\max} = \frac{f_T}{2}\sqrt{\frac{r_{ds}}{r_i}}. \tag{11.12}$$

Intrinsic noise sources in a GaAs FET are thermal channel noise and noise at the gate induced by the channel noise voltage. Flicker noise is insignificant above 50 MHz. Extrinsic noise sources in the equivalent model are R_g and R_s as well as the gate bonding pad resistance.

High Electron Mobility Transistor

Also known as a modulation-doped field effect transistor (MODFET), the high electron mobility transistor (HEMT) structure is similar to that of a typical MOSFET, with the difference being the inclusion of a heterojunction formed by growing an additional layer of AlGaAs under the gate contact, as shown in Figure 11.7. This results in very high electron mobilities and consequently much higher f_T and $f_{\max}$ for a given feature size. HEMT devices typically exhibit much better noise performance than MESFET devices and are used well into the millimeter-wave frequency range.

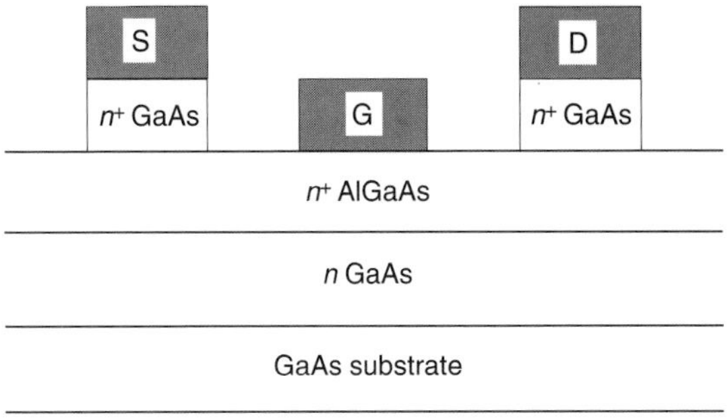

FIGURE 11.7 Cross Section of Typical HEMT Transistor

Two-Terminal Devices

While microwave and millimeter-wave transistor technology has matured considerably in last decade, many applications still rely on two-terminal devices for a number of reasons, including frequency limitations of transistors and noise or niche applications, such as P-i-N diodes in switching circuitry. Microwave and millimeter-wave diodes are still commonly used in oscillators and mixer circuitry. Two of the most commonly used two-terminal devices at these frequencies are the **transferred electron device** (TED), which relies on the Gunn effect, and the **impact ionization transit-time** (IMPATT) diode. Both of these devices may be used in reflection-type oscillators and amplifiers. Schottky diodes are used well into the millimeter-wave region and beyond.

The TED consists of a bulk semiconductor, typically GaAs, that demonstrates the Gunn effect. Contacts are attached at each terminal. Under specific biasing conditions, a region of negative differential resistance (NDR) is formed in a region of the device. If the electric field jumps above the threshold value, additional carries (both positive and negative) will be generated, forming a dipole domain. This domain will propagate to the terminal of the device, where a current spike will be observed. At this point, the field again passes threshold and a new domain is formed that will again propagate to the terminals. The result is a periodic spike in the terminal current. The period of the spike is $f = L/v_d$, where L is the length of the device and v_d is the velocity of the domain in the material.

The IMPATT diode relies on avalanche breakdown in the device to generate an external current. The nonlinear device provides negative resistance that is a function of the amplitude of the RF current flowing through the device. At higher RF current amplitude, the negative resistance is depleted, which is essential for stable operation as an oscillator.

Although IMPATT and TED are the most common two-terminal microwave- and millimeter-wave devices, a variety of other devices also exist, depending on the application. However, the trend of increasing monolithic integration (which uses three-terminal devices) in microwave- and millimeter-wave is reducing the prevalence of these devices.

11.3 Active Microwave Circuits

Design for microwave- and millimeter-wave frequency circuits are fundamentally different from the approach used at lower frequencies. Since measurements at these frequencies most naturally consist of S-parameter measurements, components are most commonly characterized by their S-parameter response. Design techniques also naturally reflect this. This section briefly outlines microwave amplifier design, oscillator design, and detectors and mixers. Because these circuits are most commonly used in communications with tight EMI and linearity constraints, issues such as intermodulation distortion, harmonic generation, saturation, and nonlinear models are briefly discussed in applicable contexts. An excellent reference on microwave amplifier design is Gonzalez (1996). A good reference dealing with many practical aspects of nonlinear circuits Maas (1998).

11.3.1 Amplifiers

Amplification is one of the primary functions in an active microwave circuit. Historically, microwave tubes or diodes in a negative resistance configuration were used to achieve gain. Currently, transistors (usually GaAs FETs) are almost universally used for this purpose. Design of microwave transistor amplifiers takes two specific paths: small signal amplifiers, such as the low noise amplifiers (LNA), use small signal S-parameters for an adequate design, whereas large signal amplifiers, such as power amplifiers, rely on more advanced techniques including load-pull and nonlinear modeling for an adequate design. This section discusses the basics of transistor amplifier design in terms of S-parameters. The basic concepts will be the same for both small-signal and large-signal amplifier design. However, large-signal amplifier design requires that large-signal S-parameters be available.

Historically, amplifiers have been designed with a graphical approach using Smith charts and simultaneously plotting circles for the parameters of interest, such as gain, stability, voltage standing wave ratio (VSWR), or noise figure at a particular frequency. A treatment of this technique is not possible in this chapter due to space limitations. Most modern microwave design CAD tools have built these functions into their interface, and the user requires only a physical understanding of the technique rather than a detailed knowledge of the equations for generating the circles that are involved. This discussion, hence, focuses on presenting the most important parameters used for microwave amplifiers.

Figures of merit for an amplifier depend on a particular application. Gain and stability are important for all amplifiers. Another figure of merit is noise figure (NF), which is of primary importance for a low-noise amplifier in a typical amplifier. In a power amplifier, efficiency may be of primary concern if the system has a limited supply of dc power, such as in satellite transmitters or handheld phones. The first topic of discussion is general amplifier concepts. Issues relevant to LNA and PA design will be discussed separately.

A general block diagram of a transistor amplifier is shown in Figure 11.8. It consists of three blocks with associated gains: input matching (G_s), output matching (G_L), and transistor S-parameters (G_0). Using transmission line theory, it can be easily shown that the reflection coefficients on the input and output sides of the transistors are given as:

$$\Gamma_{in} = S_{11} + \frac{S_{12}S_{21}\Gamma_L}{1 - S_{22}\Gamma_L}. \tag{11.13}$$

$$\Gamma_{out} = S_{22} + \frac{S_{12}S_{21}\Gamma_S}{1 - S_{11}\Gamma_S}. \tag{11.14}$$

One important case is the **unilateral transistor**, which occurs if $S_{12} = 0$ or is small enough to be neglected. In this case, the effective gain factors for each component in terms of these reflection coefficients are as follows:

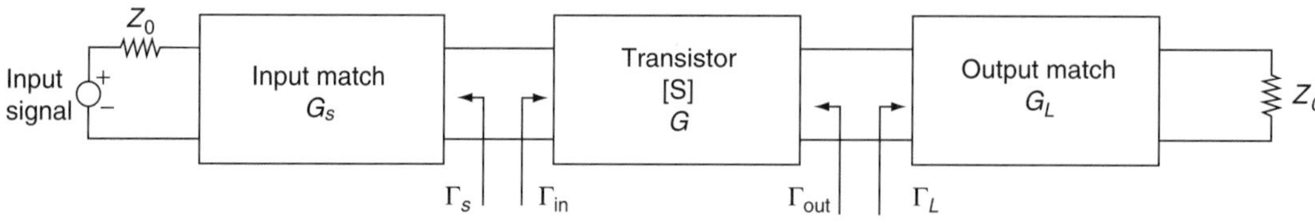

FIGURE 11.8 Block Diagram of a Transistor Amplifier

$$G_s = \frac{1 - |\Gamma_s|^2}{|1 - \Gamma_{in}\Gamma_s|^2}. \quad (11.15)$$

$$G_0 = |S_{21}|^2. \quad (11.16)$$

$$G_L = \frac{1 - |\Gamma_L|^2}{|1 - S_{22}\Gamma_L|^2}. \quad (11.17)$$

One of the most useful figures for an amplifier is the **transducer gain**, which is the ratio of the power delivered to the load to the power delivered by the source and depends on the input and output match. In terms of the gain coefficients, the transducer gain is $G_T = G_S G_0 G_L$. In terms of reflection coefficients, the transducer gain is given as:

$$G_T = \frac{1 - |\Gamma_s|^2}{|1 - \Gamma_{in}\Gamma_s|^2} |S_{21}|^2 \frac{1 - |\Gamma_L|^2}{|1 - S_{22}\Gamma_L|^2}. \quad (11.18)$$

Note that maximum gain occurs when both the input and output provide conjugate matches (i.e., $\Gamma_s = S_{11}^*$ and $\Gamma_L = S_{22}^*$). Moreover, the amplifier must be operated in a stable configuration to ensure that it will not oscillate. Oscillation may occur if the input or output port impedance has a negative real part, leading to $|\Gamma_{in}|$ or $|\Gamma_{out}|$ greater than the value of 1. Since Γ_{in} and Γ_{out} rely on Γ_L and Γ_S, stability can be ensured by proper matching conditions with possibly reduced gain. Two types of stability are typically defined: **unconditionally stable** and **conditionally stable**. An unconditionally stable network must have both $|\Gamma_{in}| < 1$ and $|\Gamma_{out}| < 1$ for *all* passive source and load terminations. A conditionally stable network will have both $|\Gamma_{in}| < 1$ and $|\Gamma_{out}| < 1$ for a range of passive source and load terminations. Note that in a unilateral device, it is sufficient that $|S_{11}| < 1$ and $|S_{22}| < 1$ for the device to be unconditionally stable. In addition, the maximum stable gain of a transistor is $G_{MSG} = |S_{21}|/|S_{12}|$. This is an important parameter when choosing a transistor. A convenient way of expressing the necessary and sufficient conditions for unconditional stability in terms of the S-parameters is through these equations:

$$|\Delta| = |S_{11}S_{22} - S_{12}S_{21}| < 1. \quad (11.19)$$

$$K = \frac{1 - |S_{11}|^2 - |S_{22}|^2 + |\Delta|^2}{2|S_{12}S_{21}|} > 1. \quad (11.20)$$

Low-Noise Amplifier

A low-noise amplifier (LNA) is commonly found in all receivers. Its role is to boost the received signal a sufficient level above the noise floor so that it can be used for additional processing. The noise figure of the LNA therefore directly limits the sensitivity of the receiver. Minimum noise performance, F_{min}, occurs with a source termination with reflection coefficient Γ_{opt}. The noise figure of a two-port amplifier is given as:

$$F = F_{min} + \frac{4r_n|\Gamma_s - \Gamma_{opt}|^2}{\{1 - |\Gamma_s|^2)|1 + \Gamma_{opt}|^2}. \quad (11.21)$$

Note that r_n is the equivalent normalized noise resistance of the two-port, $r_n = R_n/Z_0$. Moreover, r_n, F_{min}, and Γ_{opt} are known as noise parameters and given by the transistor manufacturer. They can also be determined experimentally.

Design of an LNA typically consists of trade-off between noise figure and gain while designing at the required stability. This can often be a difficult task. One technique to make a potentially unstable transistor an unconditionally stable transistor is to use resistive loading or feedback at the expense of reduced power gain and degraded noise figure.

Power Amplifier

Although the primary concern in an LNA is noise figure, **power amplifiers** (PA) have linearity and often efficiency requirements. Among other places, power amplifiers are found in all transmitters and are a significant user of dc power.

Since small-signal S-parameters are almost always operated in large-signal mode, they are typically not sufficient to ensure an accurate design. Other techniques include load-pull to determine optimal terminations or large-signal models, which are often available from the manufacturer or built into popular microwave CAD tools. The most common figure of merit for a microwave power transistor is the 1-dB **compression point**, a measure of the power-handling capabilities of the transistor. This is defined as the power where the saturation of the transistor reduces the power gain of the transistor by 1 dB from the small-signal power gain. This is graphically illustrated in Figure 11.9. The dynamic range is the range of linear performance that the amplifier provides measured from the minimal detectable output signal at the noise power level to the 1-dB compression point, $P_{1\ dB}$.

Typically, power amplifiers are operated in the nonlinear regime, producing significant harmonics. These undesired har-

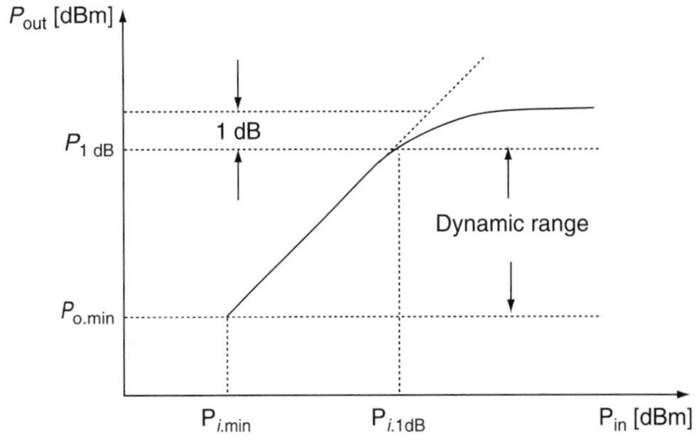

FIGURE 11.9 Graphical Illustration of the 1-dB Compression Point of a Microwave Amplifier

monics must be either "tuned" or filtered at the output. In the case of harmonic tuning, a resonant circuit at harmonic frequencies is formed at the output from either transmission line stubs or chip capacitors, where the capacitors possess a self-resonance near the harmonic frequencies. The power in the harmonics is then reflected back to the device with an appropriate phase to add to the fundamental power, therefore increasing efficiency. In practice, it is difficult to tune more than the first two or three harmonics. Additional filtering will often be required at the output.

Another concern in a microwave power amplifier is **intermodulation distortion**, which occurs when more than one input signal is injected into the amplifier. In particular, if two sinusoidal signals are simultaneously injected, the output will contain additional frequency components known as **intermodulation products** at *dc*, f_1, f_2, $2f_1$, $2f_2$, $3f_1$, $3f_2$, $f_1 \pm f_2$, $2f_1 \pm f_2$, and $f_1 \pm 2f_2$. Higher products may also be observable, depending on the amplifier. The frequencies $f_1 \pm f_2$ are known as second-order intermodulation products, and $2f_1 \pm f_2$ and $f_1 \pm 2f_2$ are third-order intermodulation products. The frequency of the third-order intermodulation products may be quite close to the input signal f_1 and f_2 and, therefore, fall into the bandwidth of the amplifier, causing distortion at the output. Of particular interest is the third-order intercept point, P_{IP3}. This is the point where the output power of P_f1 and P_{2f1-f2} intercepts extrapolated from linear operation of the amplifier. This quantity is useful for estimating third-order intermodulation products for different input power levels. A typical PA will exhibit P_{IP3} about 10 dB higher than the 1-dB compression point.

Because power amplifiers are often operated with a limited power supply, such as in portable handsets or satellite communications, efficiency can be an extremely important figure of merit. Due to limited gain in high-power microwave devices, power-added efficiency (PAE) is the most important type of efficiency and is given by:

$$PAE = \frac{P_{\text{out}} - P_{\text{in}}}{P_{DC}} \times 100\,\%. \qquad (11.22)$$

11.3.2 Oscillators

The desired output of an **oscillator** is a stable, single frequency tone. Microwave frequency oscillators typically use loop-feedback or reflection-type feedback for stabilization. Microwave frequency oscillators may use any device that exhibits negative resistance when configured properly. Within their operating frequency range, bipolar junction transistor (BJT) devices as well as FET and HEMT devices are often used. At millimeter-wave frequencies, Gunn diodes or IMPATT diodes are often used. Accurate prediction of the output power and precise frequency of oscillation requires a nonlinear model of the device used in the oscillator. Use of high-Q resonators, however, allows reliable prediction of oscillation frequency even

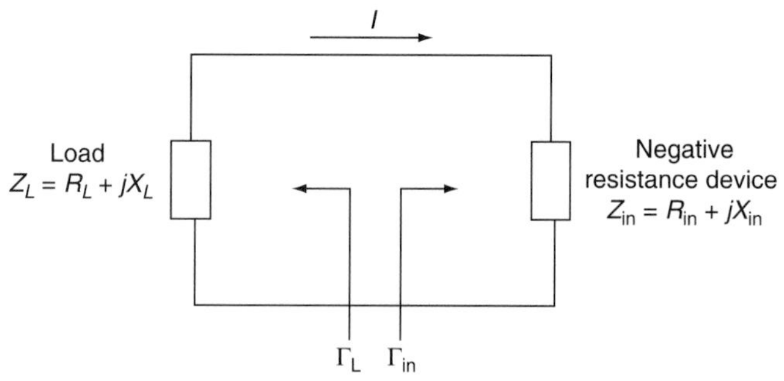

FIGURE 11.10 One-Port Negative Resistance Oscillator

when small signal *S*-parameters are used. In practice, dielectric resonators (DR) are often used because of their high-*Q* value and stability. This section briefly outlines a few of the fundamentals of microwave oscillators.

Oscillator circuits that are 1-port use diodes, commonly the Gunn diode, to provide negative resistance for oscillation. A typical circuit for a 1-port negative resistance oscillator is shown in Figure 11.10. The frequency-dependant input impedance of the active device is $Z_{in}(\omega)$, with real and imaginary components $R_{in}(\omega)$ and $X_{in}(\omega)$. Note that Z_{in} is also typically a function of the bias voltage and current. Applying Kirchhoff's voltage law gives:

$$(Z_L + Z_{in})I = 0. \qquad (11.23)$$

For nonzero *I*, this implies that oscillation can occur at any frequency if the following is true:

$$R_L + R_{in} = 0. \qquad (11.24)$$

$$X_L + X_{in} = 0. \qquad (11.25)$$

For a passive load, it is apparent that the device must provide negative resistance. For single-frequency operation, it is necessary to design the circuit so that the previous condition is met at only one frequency. Also note that because the impedance of the negative resistance device is affected by its operating mode, oscillation almost invariably undergoes some frequency shift. Moreover, conditions are (11.24) and (11.25) necessary for oscillation but do not guarantee stable steady-state oscillation. It can be shown that stable oscillation will occur if the following is satisfied:

$$\frac{\partial R_{in}}{\partial I}\frac{\partial}{\partial \omega}(X_L + X_{in}) - \frac{\partial X_{in}}{\partial I}\frac{\partial}{\partial \omega}R_{in} > 0. \qquad (11.26)$$

The first factor, $\partial R_{in}/\partial I$, is positive for a typical device. The condition can then be satisfied if $\partial(X_L + X_{in})/\partial\omega \gg 0$. A high-*Q* circuit, such as a dielectric resonator, will ensure this.

Transistors are used in a two-port oscillator configuration, as shown in Figure 11.11. The circuit is designed so that the

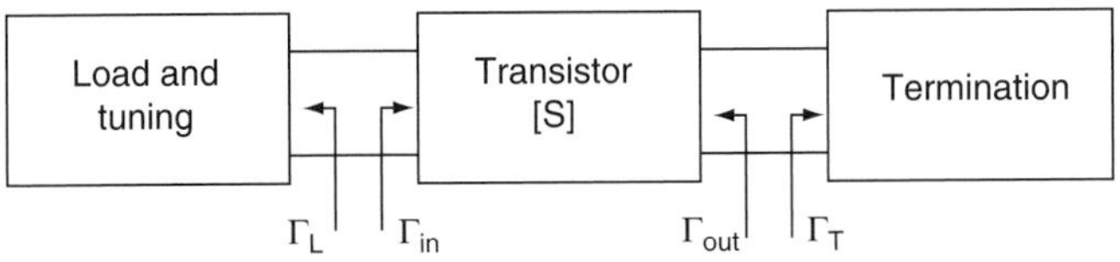

FIGURE 11.11 Two-Port Transistor Oscillator

transistor provides negative impedance at one port. Terminating one port of the transistor with impedance chosen to make the device unstable can cause this impedance. Then, to initiate oscillation, the load is chosen so that $R_L + R_{in} < 0$ and $X_L = -X_{in}$. Note that this concerns refer the small-signal impedance of the device. In practice, the value $R_L = -R_{in}/3$ is often used. When oscillation is initiated, the negative impedance provided by the device will decrease until a steady-state is reached when the large signal S-parameters are such that $\Gamma_L\Gamma_{in} = 1, 0$. As with the 1-port oscillator, high-Q resonators are often used to ensure stable oscillation and minimize phase noise.

11.3.3 Detectors and Mixers

Detectors and mixers use the nonlinear property of devices to achieve frequency conversion. Either diodes or transistors can be used for this purpose. A **detector** produces a dc signal whose amplitude varies in proportion to the amplitude of an applied RF signal. They are often used in power sensors and test equipment. A detector can also be used to demodulate information on a modulated RF signal. Mixers are used for both up-conversion in transmitters and down-conversion in receivers. A **mixer** is a circuit with two input ports for the RF and local oscillator (LO) signals and an output port for the intermediate frequency (IF) signal.

The basic concept of mixing and rectification is easily observed by considering a diode with the following I–V dependence:

$$i(V) = I_s[\exp(V\alpha) - 1]. \tag{11.27}$$

Note that $\alpha = q/nkT$, with q the charge of an electron, n as the ideality factor, I_s the saturation current, and k as Boltzmann's constant. Assume that V contains a dc voltage term, V_0, and a small signal ac signal, v, which may contain one or more frequency components. The current can be expanded in a Taylor series. Considering only the first three terms, the expansion becomes:

$$i = I_0\left(1 + \alpha v + \frac{\alpha^2 v^2}{2}\right). \tag{11.28}$$

If the diode is operated as a detector, the small signal ac signal, $v = A\cos(\omega t)$, may be inserted in the preceding equation:

$$\frac{i}{I_0} = 1 + \frac{\alpha^2 A^2}{4} + \alpha A\cos(\omega t) + \frac{\alpha^2 A^2}{4}\cos(\omega t) \tag{11.29}$$

Of particular interest for a detector is the small-signal current responsivity, β_0, defined to be the change in dc current due to application of the RF signal divided by time-average power. The change in dc current will be given by the second term on the right-hand side of the preceding equation ($\Delta i/I_0 = \alpha^2 A^2/4$). Since the change in current is proportional to the square of the amplitude of the input voltage, the small-signal operation is known as **square-law** detection. Taking the product of the diode current and voltage and integrating over one period, the small-signal current responsivity is simply:

$$\beta_0 = \alpha/2 \tag{11.30}$$

Operation of the detector is depicted in Figure 11.12. Note that at higher input levels, violating small-signal requirement conditions causes the device to saturate, and the three-term expansion of current is no longer sufficient to describe the behavior of the detector. When two signals are present, **mixing** occurs. This is most common in receivers, where the amplification is used to boost the signal at the intermediate frequency, thus increasing the sensitivity over direct detection schemes. With mixing, both an RF and a LO signal will be present. Expression of the small-signal ac is shown in this equation:

$$v = v_{RF}\cos(\omega_{RF}t) + v_{LO}\cos(\omega_{LO}t). \tag{11.31}$$

Note that, in general, *all* mixing products will be present in the spectrum of the output current (i.e., the current will contain components at $|m\omega_{RF} \pm n\omega_{LO}|$ with m and n integers ranging from zero to infinity. The first few mixing products are typically of primary interest. Using the three-term expansion of the output current, yields the following:

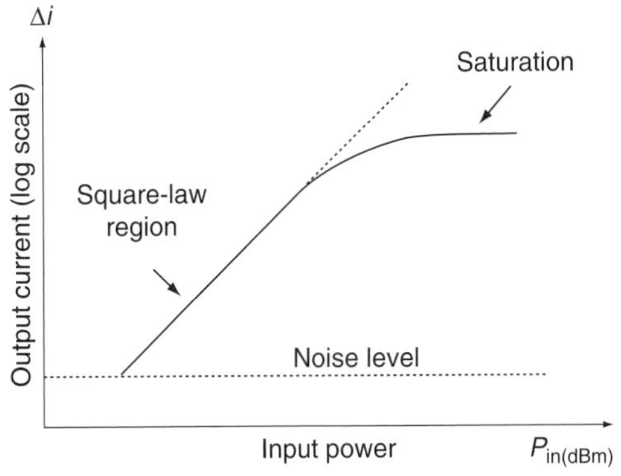

FIGURE 11.12 Square-Law Region for a Typical Diode Detector

$$\frac{i}{I_0} = 1 + \frac{\alpha^2}{4}\left(v_{RF}^2 + v_{LO}^2\right) + \alpha v_{RF}\cos(\omega_{RF}t) + \alpha v_{LO}\cos(\omega_{LO}t)$$
$$+ \frac{\alpha^2}{4} v_{RF}\cos(2\omega_{RF}t) + \frac{\alpha^2}{4} v_{LO}\cos(2\omega_{LO}t)$$
$$+ \frac{\alpha^2}{2} v_{RF}v_{LO}\cos[(\omega_{RF} - \omega_{LO})t] + \frac{\alpha^2}{2} v_{RF}v_{LO}\cos[(\omega_{RF} + \omega_{LO})t]. \quad (11.32)$$

The last two terms of the right-hand side of the preceding equation correspond to down-conversion and up-conversion, respectively. Moreover, the mixing products are also proportional to both RF and LO voltage. Mixing therefore exhibits linear behavior rather than the square-law behavior found with detection.

Specific issues with mixer design include matching for optimal performance, isolation, and rejection of image frequencies that can cause spurious responses near the desired intermediate frequency. A block diagram of a typical down-conversion mixer is shown in Figure 11.13. With applied RF and LO signals, the frequencies $f_{RF} \pm f_{LO}$ are present at the output of the mixer. Note that in practice, all mixing products will be present at the output. Then a low-pass filter is used to eliminate all but the desired intermediate frequency, $f_{RF} \pm f_{LO}$, for improved isolation. In addition to isolation, conversion loss is an important figure of merit for a mixer. In decibels, this is given as:

$$CL = 10\log\left(\frac{P_{RF}}{P_{IF}}\right). \quad (11.33)$$

Because FETs are commonly used in mixer circuits, this number can be positive, resulting in a mixer with gain. In fact, FET mixers with up to 10 dB of gain can be produced at X-band frequencies, with lower LO requirements than with diode mixers. Typically, this comes at the expense of degraded noise performance.

11.4 Planar Antenna Technology

Planar antennas are printed on a dielectric circuit board, typically with ground metallization on one side. These antennas are typically compatible with conventional planar transmission lines, such as microstrip, coplanar waveguide (CPW), and coplanar stripline (CPS), and may be integrated with other microwave circuitry. For this reason, planar antennas are often referred to as **integrated antennas**. These antennas are low profile and low cost. At microwave- and millimeter-wave frequencies, they are also typically quite compact. As the operating frequency of commercial wireless systems is consistently pushed higher, it becomes more feasible and economical to integrate the antenna and system on a single printed circuit board (PCB) or even on a die at millimeter-wave frequencies. In this section, four planar antennas are described, including the **microstrip patch antenna**, the **slot antenna**, the **tapered slot antenna** and the **quasi-Yagi antenna**. The section concludes with a brief discussion about the active integrated antenna approach.

11.4.1 Microstrip Patch Antenna

The patch antenna has several desirable qualities, including a broadside radiation pattern that allows it to be integrated into two-dimensional arrays. The antenna is also low profile and low cost, has good conformability, and has ease of manufacturing. It is readily integrated with microstrip or coaxial probe feeding. Multilayer schemes have been used for other types of feeding, including CPW and strip line. With microstrip feeding, it is also a relatively simple task to implement either linear or circular polarization excitation of the antenna.

The patch antenna, shown in Figure 11.14 with microstrip feeding, is one of the most widely used planar antennas. Feeding is extremely important with the patch antenna, and it contributes to bandwidth, crosspolarization levels, and

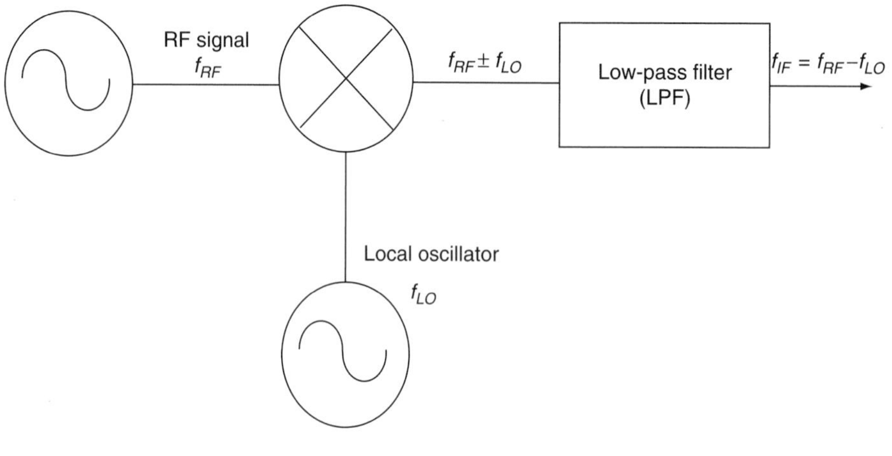

FIGURE 11.13 Block Diagram for a Down-Conversion Mixer Configuration

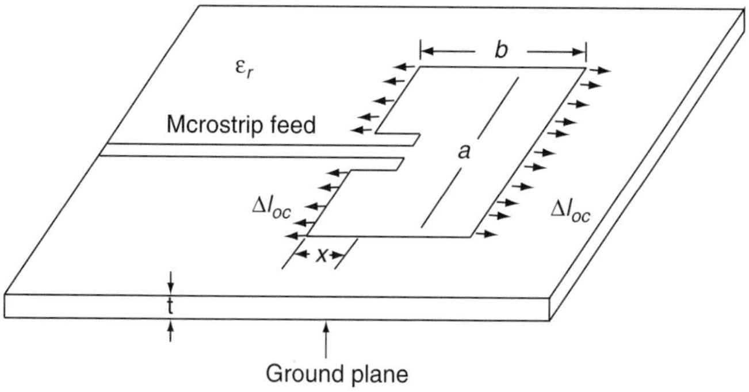

FIGURE 11.14 Top View of Microstrip-Fed Patch Antenna

ripple. Microstrip-fed patches have very narrow bandwidths, almost invariably less than 5%. Other feed mechanisms have been used to increased bandwidth, including proximity coupling and aperture coupling, both of which require multilayer fabrication. A review of this technology is discussed in Pozar (1992). Alternatively, matched bandwidth of the antenna can be increased by making the antenna substrate electrically thicker, effectively lowering the *Q*-factor of the antenna cavity for increased bandwidth. High levels of TM_0 surface waves, however, can result and therefore reduce the radiation efficiency as well as degrade the radiation pattern if the surface wave generates radiation (which can occur at the edge of finite ground antennas). The problem of electrically thick substrate is also a common one for high-frequency antennas on high-permittivity substrates, and high amounts of TM surface waves can result.

Returning to Figure 11.14, the microstrip feed is inset into the antenna a distance x to obtain an input match. The dimension b is chosen so that the cavity formed by the conductor on the top plane of the structure is resonant. This causes radiation at the two edges of the antenna, as shown by the fringing fields in the diagram. A simple and intuitive technique for modeling this antenna is the **transmission line model**. This model provides a reasonable estimate for the resonant frequency and a fairly accurate estimate of the input impedance close to resonance. Due to the narrow bandwidth of the patch antenna, it is typically not accurate enough to guarantee first-pass design success. The bandwidth, however, does provide a useful starting point as well as useful insight into the operation of the antenna.

A cross section of the patch antenna is shown in Figure 11.15. In this model, it is assumed that the patch antenna consists of a perfect magnetic conductor (PMC) walls on the sides of the patch antenna, giving rise to standing wave type modes inside the patch antenna cavity. The total length of the cavity is the length of the patch antenna (dimension b) and an effective length at each edge due to the microstrip open-end effect. The fundamental resonance of the cavity formed by the microstrip patch antenna will occur at the frequency where the

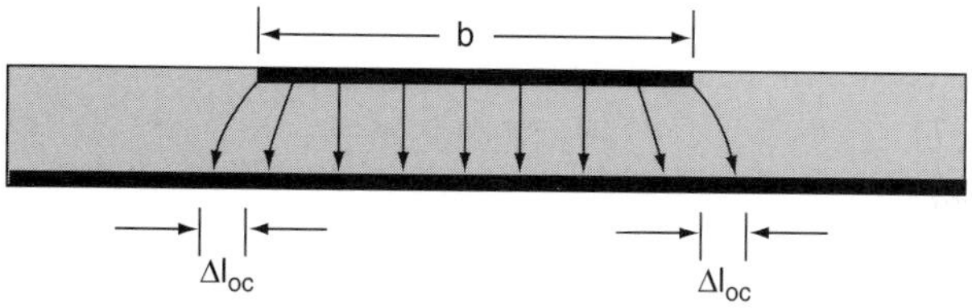

FIGURE 11.15 Cross Section of a Microstrip Patch Antenna

total effective length of the patch antenna, $b + 2\Delta l_{oc}$, is equal to one-half a guided wavelength in the microstrip cavity. The equation representing this concept is as follows:

$$f_r = \frac{c}{2\sqrt{\varepsilon_{re}}}\frac{1}{b + 2\Delta l_{oc}}. \tag{11.34}$$

Note that c is the speed of light in a vacuum ($c = 3^* 10^8$ m/s), and ε_{re} is the effective permittivity of the microstrip. In terms of the features in Figure 11.14, an approximate expression for the effective permittivity is as written here:

$$\varepsilon_{re} = \frac{\varepsilon_r + 1}{2} + \frac{\varepsilon_r - 1}{2}\left(1 + \frac{10t}{a}\right)^{-\frac{1}{2}}. \tag{11.35}$$

It is also necessary to have an estimate of the effective length due to the fringing effects. The following is a commonly used formula for the effective length of the fringing field:

$$\frac{\Delta l_{oc}}{t} = 0.412\frac{(\varepsilon_{re} + 0.3)(a/t + 0.264)}{(\varepsilon_{re} - 0.258)(a/t + 0.813)}. \tag{11.36}$$

By using these equations together, dimensions of the patch antenna can be chosen to achieve a particular resonant frequency. The accuracy of the model is typically a few percent. Because the bandwidth of a microstrip-fed patch is on the same order, this may not be accurate enough for first pass design.

It is also desirable to have an estimate of the input impedance of the antenna. The simplest way for estimating this is the transmission line model for the patch antenna. In this case, the antenna is modeled as two radiating slots of width Δl_{oc} and length a separated by a microstrip transmission line with dimensions corresponding to the dimensions of the patch antenna. Note that the feeding can be placed at one end of the antenna or at some point a distance x inside the patch, either by the use of an inset feed or a coaxial probe. The equivalent structure for interior feeding is shown in Figure 11.16. At resonance, the impedance of the radiating slot will be pure real. To first-order for $a \ll \lambda_0$ (which will be true on high-permittivity substrates), the radiation resistance of each slot may be approximated as:

$$R_a = \frac{90\lambda_0^2}{a^2}. \tag{11.37}$$

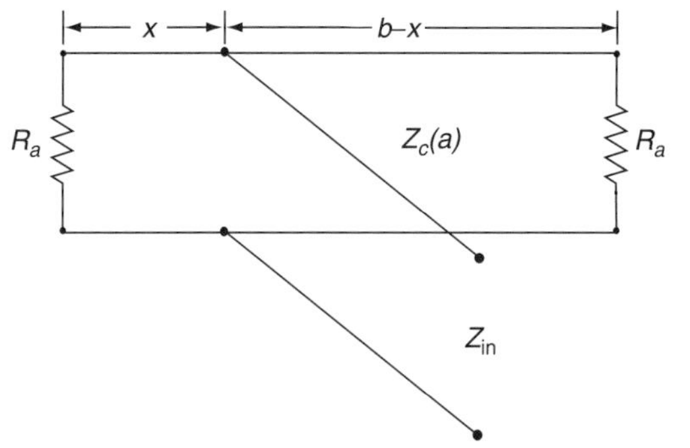

FIGURE 11.16 Transmission Line Model of a Patch Antenna. This model is used for determining the input impedance as a function of feed location.

Equation 11.37 provides a reasonable first-order estimate of the input impedance of the patch antenna near resonance. Note that there are many other simple models that also provide a first-order estimate of the input impedance. Often, the designer must either fabricate and perform measurements or obtain a full-wave solution to the structure using an electromagnetic (EM) simulator to achieve accurate design data for the antenna. This simple model, however, provides a useful starting place for the design.

Because the antenna's length determines the resonant frequency of the patch, higher antenna resonances will coincide with frequencies that are multiples of the fundamental resonance. If active circuitry, which may generate harmonic frequencies, is integrated with the patch antenna, harmonic radiation leading to co-site interference may occur. A circular geometry patch antenna may be used to reduce this problem. In this case, higher resonances of the antenna will be determined by circular harmonics (Bessel functions) and can be designed to occur away from circuit harmonics. A plot of the input impedance of one particular type of circular geometry patch, the circular segment patch antenna, is shown in Figure 11.17 along with its geometry. By looking at the frequency scale of the plot, it is apparent that higher resonances do not correspond to harmonic frequency of active devices that may be integrated with the antenna. Note that the input impedance was obtained by full-wave analysis. Measured radiation patterns of these types of antenna are comparable to that of a standard rectangular geometry patch antenna. In addition, circular geometry patch antennas are often more compact than rectangular geometry patch antennas. Note that the antenna is fabricated on a standard RT/Duroid of permittivity 2.33 and a thickness of 31 mils. A 120° sector of the antenna has been removed for optimal impedance. A microstrip feed is placed 30° from the edge of the voided sector. The radius of the antenna is 740 mils.

11.4.2 Slot Antenna

The **slot antenna**, consisting of a narrow slit in a ground plane, is a very versatile antenna. With modification, it is amenable to waveguide, coplanar waveguide (CPW), coaxial, slot line, or microstrip feeding schemes and has been used in all aspects of wireless and radar applications. The planar form of the slot antenna with CPW feeding is shown in Figure 11.18. Unlike the waveguide slot antenna that is excited from one side, the planar slot antenna demonstrates bidirectional radiation, which limits the application of this antenna. Alternatively, the slot may be either cavity- or reflector-backed to eliminate backside radiation at the cost of additional bulk and complexity.

As seen in Figure 11.18, CPW feeding is very natural to implement. Microstrip feeding may also be used, either by using vias to connect the microstrip transmission line or by electromagnetic coupling. With electromagnetic coupling, a quarter-wave open circuited microstrip transmission line is used to form a virtual short at the far side of the slot. If designed properly, all of the energy will couple to the slot antenna.

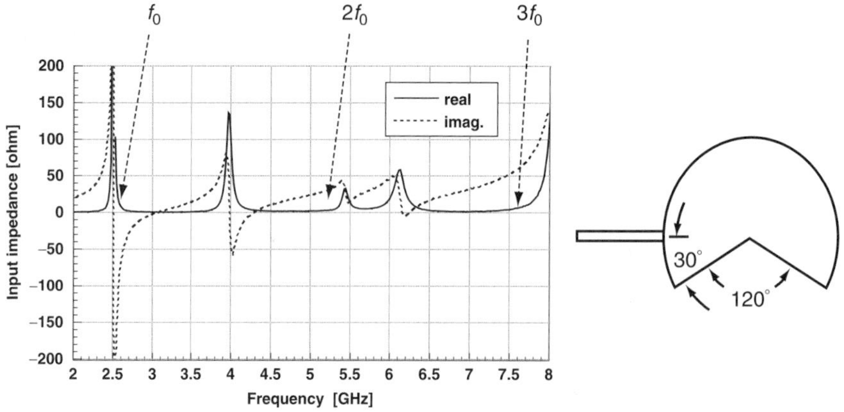

FIGURE 11.17 Input Impedance as a Function of Frequency for a Circular-Segment Patch Antenna (Right)

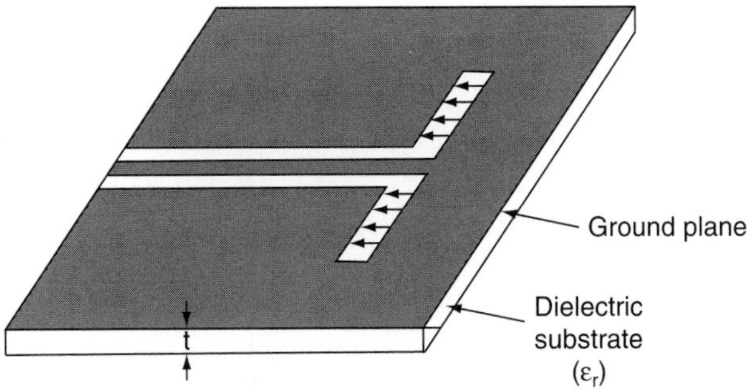

FIGURE 11.18 Geometry of a CPW-Fed Slot Antenna

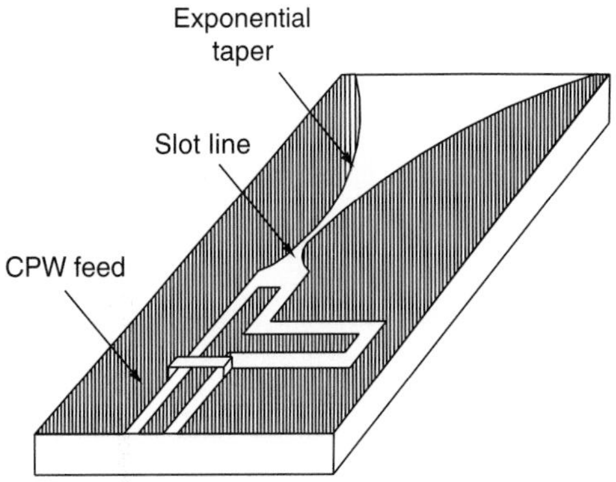

FIGURE 11.19 Tapered Slot Featuring CPW Feeding Via a CPW Slot Line Transition and Exponential Taper

The length and width of the slot, as well as material characteristics, determine the operating characteristics of the antenna. The planar slot antenna is typically operated at the first or second resonance. To estimate the proper length of the antenna, it is necessary to know the effective permittivity of the slot. For very thick dielectric substrates, the average of the dielectric permittivity and air is used (i.e., $\varepsilon_{eff} = (\varepsilon_r + 1)/2$). This may be inaccurate for thin substrates, however. In this case, it may be more suitable to use the effective permittivity of a slot line transmission line with proper width. The width of the antenna is often chosen as 10% of the length, allowing fairly large input return loss bandwidths.

The resonant half-wavelength slot antenna is desirable because of its compact size; however, it has large input impedance, typically larger than 300 Ω, which makes it difficult to match to. This can be circumvented by using an offset microstrip feed or the folded slot antenna, which stems directly from the folded dipole by Booker's relation. In this case, the slot is *folded* in upon itself. The overall length of the antenna remains approximately a half-wavelength, but increasing the number of folds reduces the radiation resistance. A folded slot with one fold will have one-quarter the radiation resistance of a standard slot antenna. The one-wavelength slot is also used with center-feeding because it is fairly simple to achieve an input match.

11.4.3 Tapered Slot Antenna

Like the slot antenna, the **tapered slot antenna** (TSA) is etched on a grounded dielectric slab. A TSA with a CPW feed via a CPW slot line transition is shown in Figure 11.19. This particular TSA with its exponential taper is known as a **Vivaldi antenna**. Other configurations of different taper profiles are also commonly used, including the linear taper (LTSA) and the constant width (CWSA). These antennas are completely planar, have an end-fire pattern, and can obtain high directivity and/or bandwidth. Proposed applications include millimeter-wave imaging, power combining, and use as an active integrated antenna element.

There are several important design concerns with TSAs. Radiation of a particular frequency will occur where the slot is a certain width. Slot width should reach at least one-half a wavelength for efficient radiation to occur. Therefore, maximum and minimum widths roughly determine the bandwidth of the structure. TSAs are usually built on thin, low-permittivity substrates if they are to achieve good radiation patterns and maintain good radiation efficiency that will be reduced by TM_0 surface wave losses. More about the tapered slot antenna may be found in Yngvesson *et al.* (1985, 1989).

11.4.4 Quasi-Yagi Antenna

Printed wire antennas are planar adaptations of wire antennas, such as dipole, loop, or spiral antennas. Due to a number of reasons, only limited success has been achieved at developing working versions of truly planar wire antennas. This is in part due to the popularity of microstrip and coplanar waveguide transmission lines at these frequencies, which are difficult to integrate with wire-based antennas. For instance, a printed microstrip dipole will have very small radiation resistance due to the shorting effect of the microstrip ground plane and consequently will have low efficiency when realistic estimates of losses are taken. This can be overcome by using an electrically thick substrate at the cost of increased losses due to substrate waves and increased weight and cost.

Alternatively, the microstrip ground plane can be modified to accommodate the printed wire antenna. Recently, a new type of microstrip-compatible planar end-fire antenna based on the well-known Yagi-Uda antenna has been developed. The layout and an X-band prototype of the quasi-Yagi antenna are shown in Figure 11.20. The antenna uses the truncated microstrip ground plane as the reflector and uses a microstrip-CPS transition as balanced feed on a single-layer high-dielectric substrate (Duroid, $\varepsilon_r = 10.2$). The X-band antenna shown in this figure uses a 0.635-mm thick Duroid.

The antenna radiates an end-fire beam, with a front-to-back ratio greater than 15 dB and a crosspolarization level below −12 dB across the entire frequency band. Figure 11.21 plots input return loss and radiation patterns for the antenna. Both

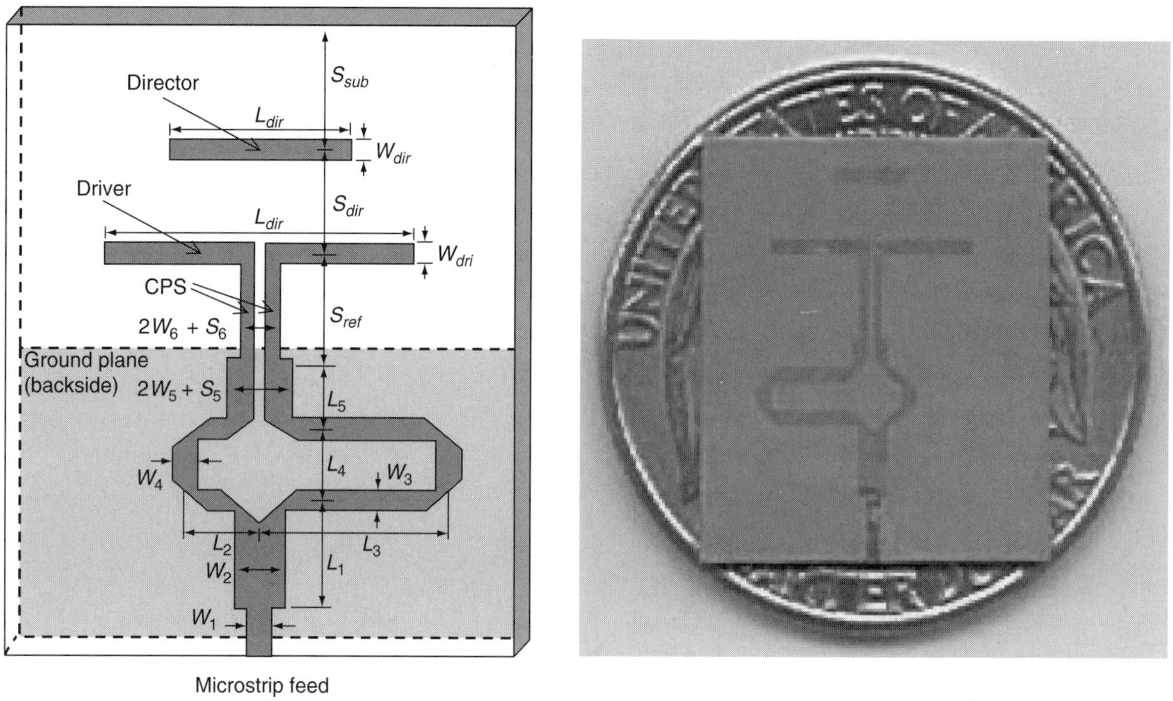

FIGURE 11.20 Layout of Quasi-Yagi Antenna (Left) and Photograph of X-Band Prototype (Right)

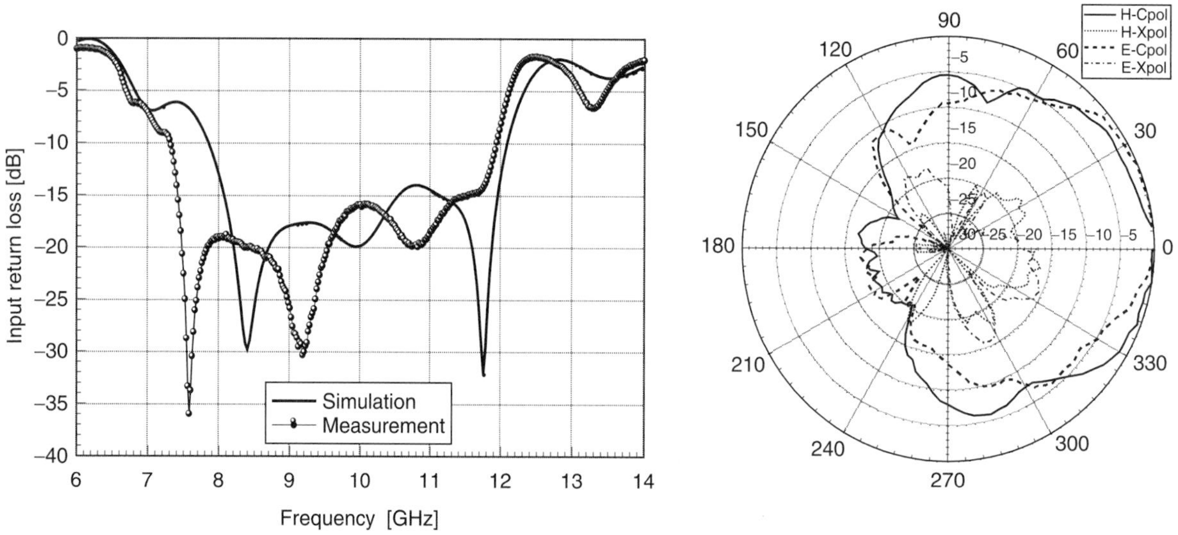

FIGURE 11.21 Input Return Loss (Left) and Measured Radiation Characteristics (Right) of X-Band Quasi-Yagi Antenna

finite-difference time-domain (FDTD) simulation and measurement results for the input return loss are shown. The simulated and measured bandwidths (VSWR < 2) are 43 and 48%, respectively, covering the entire X-band. The antenna is at least two orders smaller in volume than a standard horn antenna for the same frequency coverage. The quasi-Yagi antenna can be scaled linearly to any frequency band of interest while retaining the wideband characteristics. In fact, a *C*-band prototype simulated and fabricated on 1.27-mm thick Duroid ($\varepsilon_r = 10.2$) measured 50% frequency bandwidth (4.17 to 6.94 GHz). FDTD simulation for a millimeter-wave version indicates that a single quasi-Yagi antenna works from 41.6 to 70.1 GHz (51% bandwidth), which covers part of *Q*- and most of the *V*-band.

The quasi-Yagi antenna should prove to be an excellent array element and should find application in any architecture that

TABLE 11.1 Planar Antenna Selection Guide

		Pattern	Directivity	Polarization	Bandwidth	Comments
Patch	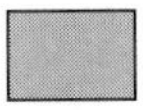	Broadside	Medium	Linear/circular	Narrow	Easiest design
Slot	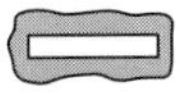	Broadside	Low/medium	Linear	Medium	Bidirectional
Ring		Broadside	Medium	Linear/circular	Narrow	Feeding complicated
Spiral		Broadside	Medium	Linear/circular	Wide	Balun and absorber
Bow-tie	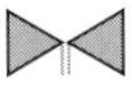	Broadside	Medium	Linear	Wide	Same as spiral
TSA (Vivaldi)	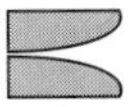	End-fire	Medium/high	Linear	Wide	Feed transition
LPDA	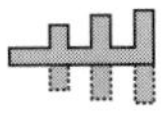	End-fire	Medium	Linear	Wide	Balun, Two-layer
Leaky-wave		Scannable	High	Linear	Medium	Beam-steering, beam-tilting
Quasi-Yagi	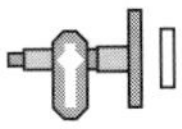	End-fire	Medium/high	Linear	Wide	Uniplanar, compact

has been proposed for tapered-slot antennas, such as imaging arrays and power combining. The quasi-Yagi antenna has several additional advantages over the tapered slot including narrower width, direct microstrip feeding, and the ability to design on a high-permittivity substrate. Moreover, radiation efficiency may be improved because of the truncated ground plane. This will eliminate losses due to TM_0 mode in the antenna region that may be a significant loss mechanism with the TSA antenna. More about the quasi-Yagi antenna may be found in Qian *et al.* (1998) and Deal *et al.* (1999).

11.4.5 Planar Antenna Selection Guide

Due to space limitations, this chapter has presented four types of planar antennas. A wide variety of other types of planar antennas also exist, with varying radiation characteristics and applications. For convenience, many of these antennas are included in Table 11.1.

11.5 Active Integrated Antennas

The compatibility of planar antennas with various types of printed circuit transmission lines allows the possibility of active circuitry and antenna being fabricated on a single printed circuit board, which allows for several attractive possibilities. First, minimizing losses between the antenna and receiver will increase sensitivity with no additional engineering of the receiver. In transmit mode systems, placing the power amplifier directly at the antenna minimizes output losses to realize high-efficiency performance by eliminating interconnects. This is possible if proper matching can be maintained. This approach is limited at lower frequency by the physical size of the planar antenna; typically, resonant-type antennas even on high-permittivity substrates will be a few fractions of a free space wavelength. However, at microwave- and millimeter-wave frequencies, smaller wavelength causes this approach to be attractive for an integrated system with maximum performance.

In this approach, the antenna often takes on additional functionality. For instance, the circular-segment antenna of Figure 11.17 has been used to perform harmonic tuning and filtering in addition to its role as radiator. This is possible because the higher resonances of the antenna do not correspond to harmonic frequencies of the amplifier because of its circular nature (Radisic *et al.*, 1997, 1998). Various other tasks have been assigned to the planar antenna, including using a two-port antenna for power combining (Deal *et al.*, 1998, 1999) and noise matching. Reviews of AIA technology can be found in Lin and Itoh (1994, 1998).

References

Deal, W.R., Radisic, V., Qian, Y., and Itoh, T. (1998). Novel push–pull integrated antenna transmitter front-end. *IEEE Microwave and Guided Wave Letters 8*, 405–407.

Deal, W.R., Radisic, V., Qian, Y., and Itoh, T. (1999). Integrated-antenna push–pull power amplifiers. *IEEE Transactions on Microwave Theory and Techniques 47*(8), 1901–1909.

Deal, W.R., Sor, J., Qian, Y., and Itoh, T. (1999). A broadband uniplanar quasi-Yagi array for power combining. *Proceedings of RAWCON*, 231–234.

Fukui, H. (1981). *Low-noise microwave transistors and amplifiers. IEEE Press.*

Gonzalez, G. (1996). *Microwave transistor amplifiers.* Englewood Cliffs, NJ: Prentice Hall.

Lin, J., and Itoh, T. (1994). Active integrated antennas. *IEEE Transactions on Microwave Theory Techniques MTT-42, 33*(12) 2186–2194.

Maas, S. (1988). *Nonlinear microwave circuits.* IEEE Press.

Pozar, D.M. (1992). Microstrip antennas. *Proceedings of the IEEE 80*: 79–91.

Qian, Y., Radisic, V., and Itoh, T. (1998). Novel architectures for high-efficiency amplifiers for wireless applications. *IEEE Transactions on Microwave Theory and Techniques 46*, 1901–1909.

Qian, Y., Deal, W.R., Kaneda, N., and Itoh, T. (1998). Microstrip-fed quasi-Yagi antenna with broadband characteristics. *Electronic Letters 34*(23), 2194–2196.

Radisic, V., Qian, Y., and Itoh, T. (1997). Class F power amplifier integrated with circular sector microstrip antenna. *IEEE MTT-S International Microwave Dig. 2*, 687–690.

Radisic, V., Qian, Y., and Itoh, T. (1998) Novel architectures for high-efficiency amplifiers for wireless applications. *IEEE Transactions on Microwave Theory Techniques 46*, 1901–1909.

Yngvesson, K.S. (1994). *Microwave semiconductor devices.* Norwell, MA: Kluwer Academic.

Yngvesson, K.S., Korzeniowski, T.L., Kim, Y.K., Kollberg, E.L., and Johansson, J.F. (1989). The tapered slot antenna—A new integrated element for millimeter-wave applications. *IEEE Transactions on Microwave Theory and Techniques 37*(2), 365–374.

Yngvesson, K.S., Schaubert, D.H., Korzeniowski, T.L., Kollberg, E.L., Thungren, T., and Johansson, J.F. (1985). End-fire tapered slot antennas on dielectric substrate. *IEEE Transactions on Antennas and Propagation 33*(12), 1392–1400.

VI

ELECTRIC POWER SYSTEMS

Anjan Bose
College of Engineering and Architecture, Washington State University, Pullman, Washington, USA

The Importance of the Electric System

In the list of the greatest engineering achievement of the 20th century, the National Academy of Engineering ranked electrification at the very top. The availability of electricity for industrial, commercial, and domestic uses has affected human society more profoundly than any other technology in the history of mankind. From the modest start in 1882 of Edison's Pearl Street electric generating station in New York City that could provide electric lights in a few buildings in lower Manhattan, to the highly interconnected power grids of today that span continents to bring reliable and affordable electric power to most of human habitation, the development of the technology of electric power systems is a remarkable story of not only engineering innovation but also of business practices, governmental regulations, and societal changes.

This section covers, albeit briefly and in a sweeping overview, the electrical engineering aspects of the **electric power system**. The electric power system, often referred to as the electric power grid, is made up of electric generation, transmission, and distribution, all aspects of which are touched on in the following chapters. The generation of electricity requires the conversion of fossil (e.g., coal, oil, and gas), nuclear (e.g., fission), or renewable fuels (e.g., hydro, solar, wind, and fusion) into electricity, and the considerable engineering needed on mechanical, chemical, and nuclear aspects is an explanation that is outside the scope of this textbook. Similarly, the various use of electricity to produce light, heat, and mechanical work is also not covered here. The electrical portion of the power system, however, is covered in some detail.

Chapters 1 and 2 cover the general principles of three-phase alternating current systems and the various electrical components that make up the electric power system. Of these components, electric machines are covered in the next two chapters, with Chapter 3 covering power transformers and Chapter 4 covering rotating machines that generate electricity (generators) or use electricity to do mechanical work (motors). Chapters 5 and 6 cover high-voltage transmission and lower voltage distribution, respectively. The rest of the section is then devoted to system aspects, with Chapter 7 covering the analytical tools needed to study and design the power system and Chapter 8 explaining operation and control. Techniques and equipment to protect the power system against short circuits are covered in Chapter 9. The concluding Chapter 10 briefly covers power quality issues that have become more important today as electricity is used for more precise applications like the production of integrated circuit chips.

1

Three-Phase Alternating Current Systems

Anjan Bose
College of Engineering and Architecture, Washington State University, Pullman, Washington, USA

1.1 Introduction

Although the first electric power system, Edison's Pearl Street system, was based on direct current (dc), the advantages of alternating current (ac) systems were obvious by the turn of the 20th century. The voltage drop in an electrical circuit limited the distance from the source of electricity to where it was consumed. The Westinghouse transformer made it possible to boost and lower voltage levels in ac systems, making it possible early on to bring electricity into Buffalo, NY, and Portland, OR, from generators at waterfalls many miles away. Further, the Tesla induction motor replaced all steam-driven manufacturing machinery because it was more clean and flexible, thus ensuring the usage of ac as the preferred technology. Despite this, some pockets of dc power systems survived until after World War II.

The choice of a three-phase transmission and distribution system over a single-phase system also came very early because of the increased efficiency of transmitting power. Although the use of electricity at the consuming end is in one phase low voltage (except for very large industrial use), transmission and distribution are always done in a three-phase system. To understand the efficiency of transmitting power, consider the discussion following section.

1.2 Two-Wire and Three-Wire Systems: Current

For a two-wire, single-phase system, let the root-mean-square (rms) voltage between the wires be V, the rms current be I, and the phase angle between the voltage and current be ϕ. The power transmitted over this line is then given by:

$$P = VI\cos\phi. \tag{1.1}$$

In a three-wire, three-phase system, let the rms voltage between each wire and the ground (the ground is often a grounded wire, making it a four-wire system) be V', the current in each wire be I, and the phase angle between the two be ϕ. Because there are three phases, the total power transmitted is as follows:

$$P = 3V'I\cos\phi. \tag{1.2}$$

However, V' in equation 1.2 is the voltage of each wire to ground, whereas V in equation 1.1 is the voltage between the two wires. In the three-phase system, the voltage in each wire is 120° out of phase with each other. Thus, the voltage V between any two wires of the three-phase system is $\sqrt{3}V'$. Hence, $V' = V/\sqrt{3}$, and:

$$P = \sqrt{3}VI\cos\phi. \tag{1.3}$$

A three-wire, three-phase system can then transmit 73% more power than a two-wire, single-phase system by just the addition of one wire. A three-phase system also has some major advantages in the generation and use of electricity by rotating machines as will be explained later.

The three-phase ac systems have been adopted worldwide. The frequency of the ac and the voltage levels chosen around the world vary. The frequency of 60 Hz was adopted in North America (and a few other places), and 50 Hz is used in all other parts of the world. Because the frequency must be the same to interconnect power systems, only these two frequencies have become standards (exceptions to this are a few isolated systems like those used in rail transportation).

1.3 Voltages

Voltages, unlike current, can vary a lot, and transformers can always be used to match voltages. Moreover, several levels of voltages are used in one geographic area from the 110- or 220-V single phase inside a residence to maybe a 765-KV three phase for transmission, with five or six intermediate levels in between.

In general, generators will produce electricity at low voltages, up to about 20 KV for large generators, but this voltage will be boosted immediately outside the generating stations to high transmission level voltages by transformers. The advantages of transmitting power at higher voltages are very easy to see from equation 1.3. The same power transmitted at a higher voltage requires less current. Less current implies that the voltage drop IR and the power losses I^2R are both lower. Less current also means that the wire size needed will be smaller, which is a savings not just in the wire but in the whole transmission structure.

The siting of generators at hydro dams and mine mouths required the transmission of power over long distances to population centers, and the economies of scale encouraged larger power plants. The concentration of generation in a limited number of locations required that alternate transmission paths be available between the generator locations and the consumption areas. Thus, transmission developed as a meshed network of lines so that the loss of a line did not disrupt the flow of power to consumers.

The main difference between **transmission** and **distribution** used to be that transmission lines transferred larger amounts of power at higher voltages over longer distances, while distribution lines transferred smaller amounts of power at lower voltages locally. However, distribution lines are radial, and the major distinction between transmission lines and distribution lines today is that the transmission system is a meshed network, while the distribution system is radial. In terms of voltages, all lines over 100 KV are always meshed and part of the transmission system, and all voltages below 30 KV are always radial and are classified as distribution. The in-between voltage levels (e.g., 69 KV or 34.5 KV) are sometimes radial in sparsely populated rural areas while meshed in densely populated downtown urban areas; when they are in a mesh network, the term **subtransmission** is sometimes used for these voltage levels.

1.3.1 Industry Business Structure

The business structure for this industry for decades has been either a state-owned monopoly, a model used in most countries, or a state-regulated monopoly; the latter is a model used in the United States. Thus, the generation, transmission, and distribution in one contiguous geographic region would be owned by one corporation as a vertically integrated monopoly. Such a configuration is shown in Figure 1.1 with the generators (circles), transmission, and distribution shown in one corporate box. The transmission system of one corporation is often

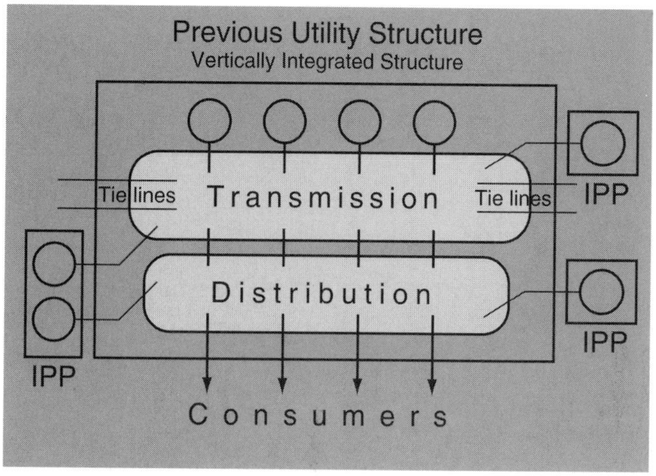

FIGURE 1.1 The Old Vertically Integrated Power Industry Structure

interconnected to the transmission of the neighboring corporations, thus creating the large electric power grid enabling the exchange of power between corporations.

In some countries during the last couple of decades, to encourage more entrepreneurial building of generation plants, governments started to allow independent power producers (IPPs) to build generating plants that could be connected to the grid. Rules were set up to require the monopoly power companies to buy this generation from the IPPs at certain rates. These IPPs are shown in separate boxes in Figure 1.1.

In most countries, there is a move to "deregulate" the power industry by introducing competition in a region. The main restructuring has been the separation and privatization of the generation into separate companies that compete with each other to sell electricity to the distribution companies or directly to the customers. This breaking up of the vertically integrated corporation is shown by the many boxes in Figure 1.2. The transmission system, being the main pathway for the generation

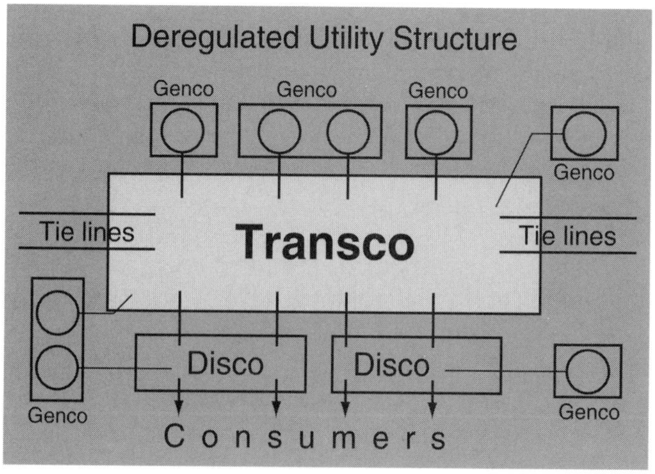

FIGURE 1.2 The New Deregulated Power Industry Structure

to reach the customers, remains under strict regulation with the main rule being the nondiscriminatory availability of electricity to all sellers and buyers of electric power. In most cases, the distribution companies are also regulated to ensure that the retail customer is not affected by big swings in electricity prices.

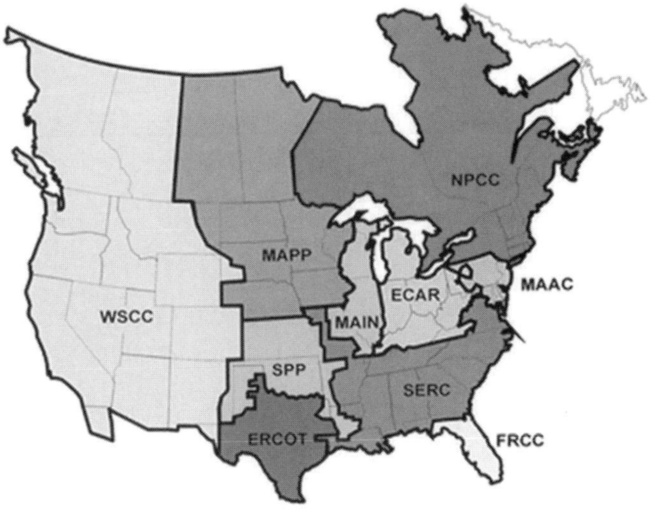

FIGURE 1.3 The Reliability Regions of the North American Power Grid

The breakup of the old vertically integrated monopolies has resulted in more corporate entities that are involved. In addition, new companies providing new services, like brokering power sales, have also entered the picture. Active markets in wholesale power are operating.

The technical operation of the power grid, however, has not changed much. The interconnected transmission network is still the same. Some of these interconnected transmission networks are very large, spanning vast geographical areas. The North American power grid is shown in Figure 1.3 and divided up into **reliability regions** that plan and operate the regional portion of the grid in a coordinated fashion. The Western region (WSCC) and Texas (ERCOT) are only connected to the rest of the grid with dc ties instead of ac. The Eastern Interconnection is synchronously connected through the ac transmission and is the largest synchronously connected power grid in the world with 687 GW of installed generation and 128,000 miles of transmission.

The West European Power Grid is shown in Figure 1.4. The separate regions are again connected with dc ties. The largest synchronously operating grid is that of continental Europe recently enlarged by the interconnecting of some of the Eastern European countries that disconnected from the old Soviet grid to interconnect with the West. The installed generation capacity is 512 GW with 125,000 miles of transmission.

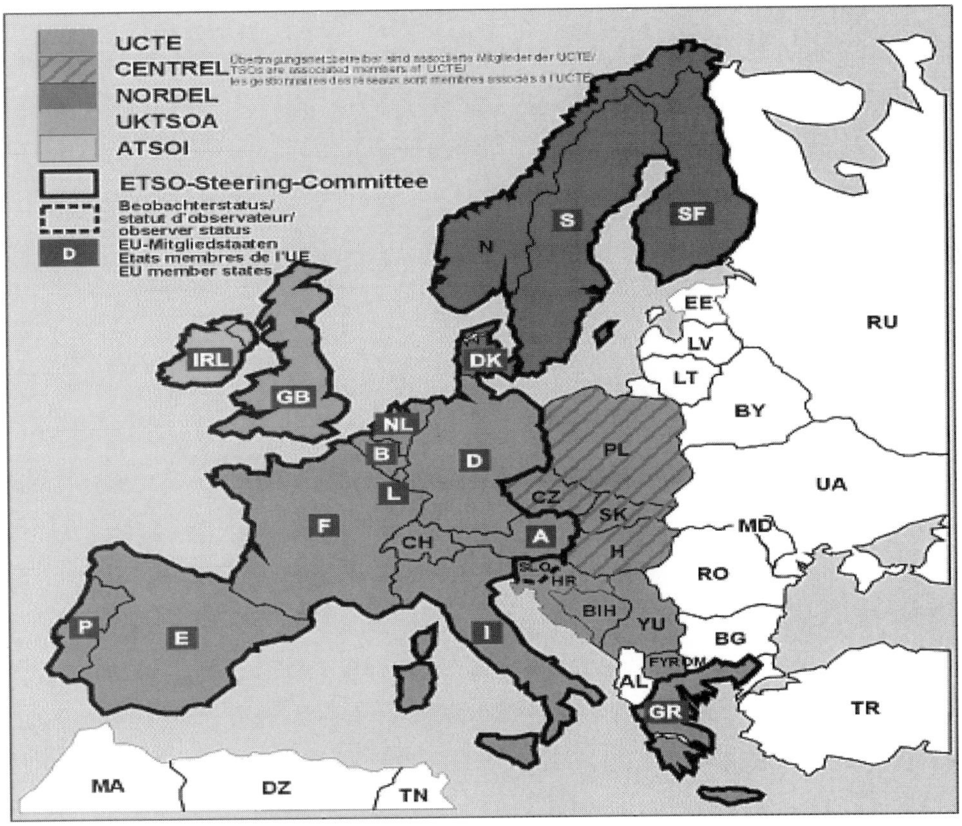

FIGURE 1.4 The West European Power Grid

2

Electric Power System Components

Anjan Bose
College of Engineering and Architecture, Washington State University, Pullman, Washington, USA

2.1 Introduction

The **electric power system**, in very general terms, is made up of generators, transformers, transmission and distribution lines, and loads. Although these are called components in this chapter, each of these is a complex system on its own and has many components. All of these are three-phase, 60-Hz components except for the smallest loads, such as residences, where the three phases are split into single-phase, low-voltage supply.

2.2 Generators and Transformers

Almost all **generators** turn mechanical power into electrical power through a 60-Hz synchronously rotating machine. The only exceptions are the new generators that convert solar, wind, or chemical power into electrical energy, but the amount of generation available in these modes is yet a small fraction; however, any nonsynchronous generator has to be connected to the grid through interfacing equipment that converts to 60 Hz.

For a synchronous machine driven by a turbine, the turbine is controlled by a governor that regulates its speed to maintain 60 Hz while generating the power required. The electrical generator has a voltage regulator and exciter that control its output voltage. The synchronous generator is covered in more detail in Chapter 4, Electric Machines.

The main purpose of a **transformer** is to transform one ac voltage level to another by a fixed ratio. A transformer has no moving parts except that some have the capability of changing the transformation ratio by small percentages through moving taps; this enables better control of the output voltage. Transformers can be single-phase or three-phase, and three single-phase transformers can sometimes be used instead of one three-phase transformer. Although one three-phase transformer is cheaper than three one-phase transformers for the same purpose, if a spare is needed for reliability purposes, two three-phase transformers are more expensive than four one-phase transformers!

A special type of transformer is the **phase-shifting transformer**. This type of transformer may or may not transform the voltage magnitude, but it does change the phase relationship between the output and input voltages. The power transfer through the transformer can be controlled by changing this phase angle. Details on transformers are provided in Chapter 3, Power Transformers.

2.2.1 Transmission and Distribution

The main difference between **transmission** and **distribution** lines is the voltage level. They always have three wires for the three phases insulated for the voltage level and can be with or without a fourth grounded (uninsulated) wire. They can be strung overhead on wooden or steel towers, or they can be laid underground in conduits. **Overhead lines** are bare and are hung from a string of insulators that can insulate the high voltage from the tower. **Underground cables** have to be covered in insulation throughout.

The design and installation of overhead lines and underground cables require very specialized engineering that gets quite complex for higher and higher voltages. For high-voltage overhead lines, the design of the wires, towers, and insulators and their installation over hundreds of miles of different terrain require the resolution of electrical, mechanical, material, and civil engineering issues. Similarly, for high-voltage cables, a different set of equally complex issues arise in design, especially in the insulation that may even require pressurized oil; installation underground in crowded downtown areas or under water has its own challenges. Transmission is covered in Chapter 5 and distribution in Chapter 6.

Transmission that is dc is not covered in this section although its usage is increasing around the world. The main use of dc transmission is to interconnect two separate areas where some power transfer is desirable but a synchronous (ac) connection is not. The main features of such an interconnection are the ac–dc conversion and the dc–ac inversion that allows the connection of two ac systems with a dc line. The design of the line itself—two wires with or without a ground wire—is not particularly different from ac transmission. In some cases, where the geographic distance between the two separate ac areas is small, the interconnection may not even require a dc transmission line but only back to back ac–dc and dc–ac converters.

The locations where transmission and distribution lines are connected through switches (circuit breakers) and transformers are known as **switchyards** or **substations.** The design of these substations becomes more complex with higher voltages and the numbers of lines and transformers. In less crowded areas these substations can take up many acres of open space, but it becomes more difficult to fit equipment into limited urban spaces. In addition, special insulating techniques for high voltages, like gas insulation, may have to be used to reduce sizes. Other than transformers and circuit breakers, such substations may also include reactors and capacitors that are used to control the voltage levels.

2.2.2 Protection and Control

Protecting all the equipment—generators, transformers, transmission lines, and distribution feeders—against short circuits is essential. The general principle is to detect the fault (the short circuit) and isolate the equipment. In its simplest form, a fuse detects the fault by burning out and thus isolating the line. In its most sophisticated form, microprocessor-based relays can detect the fault and analyze which circuit breakers need to be opened to isolate the equipment. Protection system details are covered in Chapter 9, Power System Protection.

Many other types of control are also used for the normal and emergency operations of the power system. For example, remote operation of circuit breakers can be manually initiated by operators through the supervisory control and data acquisition (SCADA) system at the control center. Voltage control is done automatically by generators, transformers, reactors, and capacitors, but the operator can remotely set the target voltages. Control and operation are covered in Chapter 8, Power System Operation and Control.

Hence, all of these subjects can be explored in more detail in the upcoming chapters in this section.

3

Power Transformers

Bob C. Degeneff
Department of Computer, Electrical, and Systems Engineering, Rensselaer Polytechnic Institute, Troy, New York, USA

3.1 Introduction

This chapter presents a very brief overview of the characteristics of a transformer. The basic theory and principles of application are introduced. A simple lumped-parameter equivalent circuit is presented. Magnetizing current and leakage reactance are introduced, and cooling methods are presented and compared. In addition, transformer applications are presented and discussed. The basic core and winding arrangements are also presented. Finally, transformer sound, overcurrent, and losses/efficiency are presented, and acceptance tests are outlined.

3.2 Transformers: Description and Use

A **transformer** is a static device used for transforming electric energy from one circuit to another magnetically (e.g., by induction rather than conduction). Normally, this transformation is accomplished between circuits of different voltages of the same frequency. A power transformer will have a magnetic core surrounded by two or more windings. These windings are insulated from each other and from the ground. The windings are connected together in a manner to achieve the desired voltage transformation ratio. The overall assembly of core and coils is generally insulated and cooled by immersion in mineral oil or other suitable liquid in an enclosing tank. Connection to the windings is usually by means of insulated bushings.

In a modern utility, electrical energy may undergo four or five voltage transformations between the generation site and the point of utilization. As such, a given system possesses approximately five times the kilovoltamperes [kVA] of transformers compared to that of generators.

3.3 Transformers: Theory and Principle

3.3.1 Equivalent Circuits

Figure 3.1 is a cross section of a two-winding transformer. Figure 3.1 illustrates how both windings are linked with the same magnetic circuit. A sinusoidal voltage is induced in the windings by a sinusoidal variation of flux $E = 4.44\ a_c BfN$, with a_c as the area of the core in square meters, B as the peak

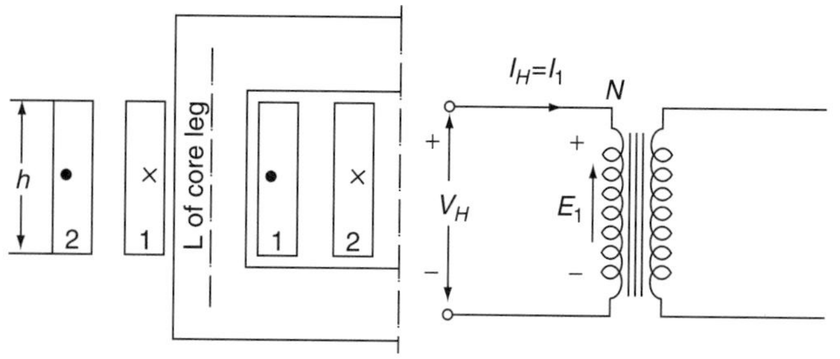

FIGURE 3.1 Cross Section of a Two-Winding Transformer

flux density in tesla, E as the rms-induced voltage, f as the frequency in hertz, and N as the numbers of turns in the winding. With no load on the secondary circuit, a small exciting current flows in the primary, which produces the alternating flux in the core. This flux links both windings and induces essentially the same volts per turn in each winding. To a close approximation:

$$\frac{E_1}{N_1} = \frac{E_2}{N_2},$$

where E_1 and E_2 are the primary and secondary voltages and where N_1 and N_2 are the primary and secondary turns. If a load is applied to the secondary of the transformer and the exciting current is small with respect to the load current, the currents in the primary and secondary are related by:

$$I_1 N_1 = I_2 N_2.$$

3.3.2 Magnetizing Current

Figure 3.2 is a lumped-parameter model of the transformer in Figure 3.1, with both the primary and secondary solidly grounded. If the secondary is open-circuited, the only current that flows is an exciting current in the primary winding. This exciting current contains only odd harmonics and is composed of a core loss component and a **magnetizing component**. The core loss component is in phase with the induced voltage, and the magnetizing component is lagging the induced voltage by 90° (i.e., in phase with the magnetizing flux). The core loss is represented in Figure 3.2 by b_m, and it is composed of both hysteresis and *eddy current* losses. *Hysteresis* losses are directly proportional to the frequency, and eddy current losses vary proportionally to the square of frequency.

3.3.3 Leakage Reactance

If the secondary of the transformer is shorted, the impedance of the transformer presented to the system is approximately $(R_1 + jX_{l1}) + (R_2 + jX_{l2})$. This is generally referred to as the **impedance of a transformer** or its leakage impedance. The R is the effective resistance of the primary and secondary winding and is composed of both dc and skin effect losses. Generally, R is much smaller than X and normally can be ignored in practical calculation.

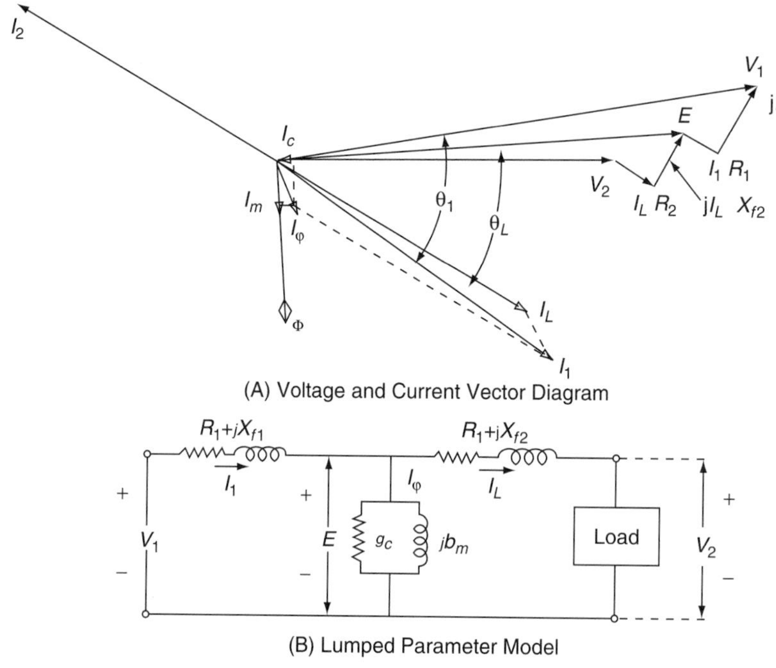

FIGURE 3.2 Lumped-Parameter Equivalent

3.3.4 Magnetizing Inrush

When a transformer is energized, the amount of current that flows will depend on the amount of residual flux that resides in the core. This residual is typically between 50 and 90% of the peak operating flux. In extreme cases, the peak flux in the core upon energization will be greater than twice the peak design limit, and, as such, the **inrush current** can be 3 to 40 times the load current (Grigsby, 2001). The actual inrush experienced depends on the point on the voltage wave during energization, the residual flux, air core inductance of the winding, and losses of the transformer and system.

3.4 Cooling Methods

3.4.1 Liquid-Filled Transformers

A **liquid-immersed transformer** consists of a magnetic core and coil assembly immersed in a fluid, normally mineral oil. The fluid must possess both good heat transfer characteristics and electrical insulating characteristics. An advantage of a liquid-immersed transformer is that it permits compact design. Generally, transformers above 10 MVA and/or 34.5 kV are liquid-filled. Since oils are flammable, liquid-immersed transformers must be applied with adequate precautions recognizing the flammability of oil. This has presented challenges for indoor applications.

Mineral oil used in transformer insulation systems degrades with prolonged exposure to oxygen. Moreover, mineral oil has a fairly large thermal expansion coefficient. In recognition of this disadvantage, several systems are used to minimize degradation. Liquid-filled transformers normally protect insulating oil with one of three types of preservation systems: **sealed tank system**, **nitrogen blanked**, or **conservator**.

3.4.2 Dry-Type Transformers

Dry-type transformers are generally more expensive per kVa than liquid-filled transformers but address the concern for indoor applications. Dry-type transformers use solid insulation systems or film coatings and/or paper tapes. A variation is the resin-encapsulated system (or cast coil). Dry-type transformers have less of an ability to withstand impulse voltages than corresponding liquid-filled designs. Dry-type transformers are offered in ventilated and nonventilated, totally enclosed systems for use in hostile industrial environments. Dry-type transformers are routinely offered up to 34.5 kV and 10 MVA.

3.5 Transformer Applications

The most common arrangement of a power transformer is two or more isolated windings wound around a common core. The advantage is complete electrical isolation between electrical circuits. These units can be of the **step-down** or **step-up** configuration. A step-down is designed to decrease the incoming voltage to a lower level more suitable for distribution. Power transformers of this type are frequently located in substations or at large industrial consumer locations. Transformers that perform this reduction in voltage from a primary feeder to the utilization level are referred to as **distribution transformers**. A step-up transformer takes the input voltage and raises it to a higher lever suitable for transmission. American National Standards Institute (ANSI) standards require the nameplate to indicate the transformer is suitable for this type of operation. A generator step-up transformer is an example of this type of power transformer. The majority of these type power transformers are three-phase units.

It is often more economical to transfer energy between two circuits of similar voltage using an autotransformer rather than a conventional two-winding transformer. Figure 3.3 provides an illustration of this arrangement. Most autotransformers are *Y*-connected with a delta tertiary (used to provide a path for third harmonic currents required for excitation). The autotransformer does not afford electrical separation between the two circuits but is a more economical transformer on a $/kVA basis. In addition, the autotransformer is not inherently self-protecting and typically will be subjected to higher short circuit currents and forces than a corresponding two-winding transformer. The autotransformer can also be used to increase or decrease voltage in a system. By using suitable on-load tap-changing equipment, it is possible to regulate the voltage of a system within desired limits. An autotransformer used in this manner is referred to as a voltage-regulating transformer. Typically, the range of the regulation is $\pm 10\%$ of the system voltage in 5/8% steps.

It may be desirable to create an electrical connection between systems that have different phase angles. These systems may also be at different voltages. The function of the phase–angle regulator is to provide this interconnection between systems possessing different phase angles and voltages. Figure 3.4 illustrates one method to accomplish this transformation.

In power system applications, there are two types of instrument transformers: **current transformers** and **voltage trans-**

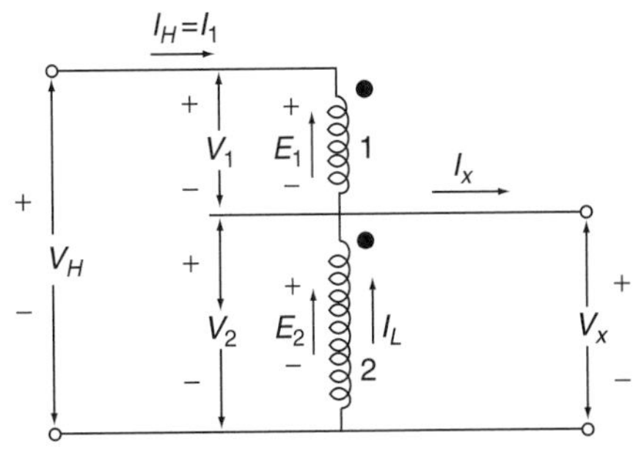

FIGURE 3.3 Autotransformer

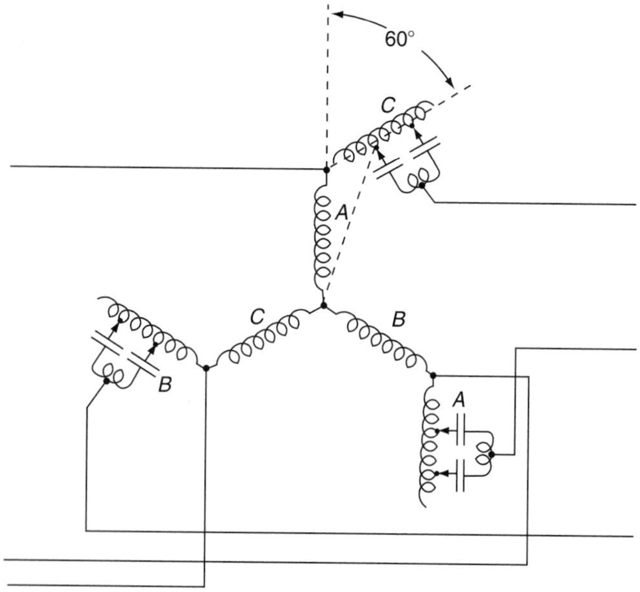

FIGURE 3.4 Phase-Angle Regulator

formers. Both are used to secure information about the condition and operation of the system. Typically, this information is used in relaying information and in operational control of the system. The current transformer is a series transformer. Its primary winding is connected in series with the circuit in which the current is to be measured, and the secondary winding supplies a current output that is proportional to the primary current (Figure 3.5). The voltage transformer primary winding is connected to the terminals of the circuit where the voltage is to be measured, and the secondary winding supplies a voltage proportional to the primary voltage.

3.6 Cores and Windings

The successful design of a commercial transformer requires the selection of a simple structure so that the core and coils are easy to manufacture. At the same time, the structure should be as compact as possible to reduce materials, shipping concerns, and footprints. The form should also allow convenient removal of heat, sufficient mechanical strength to withstand forces generated during system faults, acceptable noise characteristics, and an electrical insulation system that meets both the system steady-state and transient requirements. There are two common transformer structures in use today for power transformers. When the magnetic circuit is encircled by two or more windings, the transformer is referred to as a **core-type transformer**. When the primary and secondary windings are encircled by the magnetic material, the transformer is referred to as a **shell-type transformer**. Refer to Figure 3.6 for an illustration of each.

3.6.1 Core Form Transformer

Characteristics of the **core-form transformer** are a long magnetic path and a shorter mean length of turn. Commonly used core-form magnetic circuits are single-phase transformers with a two-legged magnetic path with turns wound around each leg, a three-legged magnetic path with the center leg wound with conductor, or a four-legged magnetic path with the two interior legs wound with conductors (Bean *et al.*, 1959; Massachusetts Institute of Technology, 1943). Three-phase core-form designs are generally three-legged magnetic cores with all three legs possessing windings or a five-legged core arrangement with the three center legs possessing windings. The simplest winding arrangement has the low-voltage winding nearest the core and the high-voltage winding on top of the low. Normally, in the core form construction, the winding system is constructed from helical, layer, or disk-type windings (Massachusetts Institute of Technology, 1943; Franklin, 1998). The disk-type winding itself can be constructed on several different winding configurations (i.e., continuous, interleaved, and internally shielded), all of which affect the transient voltage response and this insulation design. Often the design requirements, such as impedance or shipping size limitations, call for a core and winding arrangement that is a more complex arrangement (e.g., interleaving high- and low-voltage windings, interwound taps, and entry and exit points other than the top or bottom of the coil). All of these variations have an effect on the transformer's transient voltage

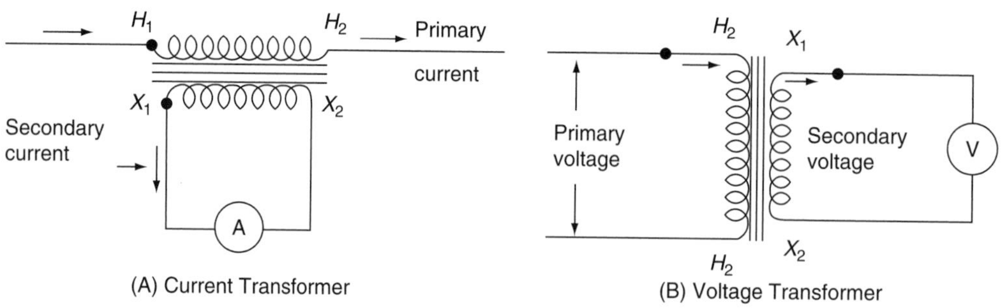

FIGURE 3.5 Instrument Transformers

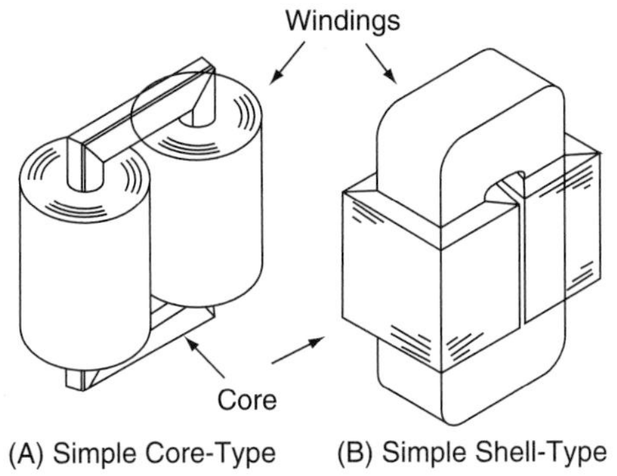

FIGURE 3.6 Core-Form and Shell-Form Transformers

response. To ensure an adequate insulation structure, each possible variation must be explored during the design stage to evaluate the variation's effect on the transient overvoltages.

3.6.2 Shell-Form Transformer

The **shell-form transformer** construction features a short magnetic path and a longer mean length of electrical turn. Bean *et al.* (1959) point out that this results in the shell-form transformer having a larger core area and a smaller number of winding turns than the core-form of the same output and performance. In addition, the shell-form will generally have a larger ratio of steel to copper than an equivalently rated core-form transformer. The most common winding structure for shell-form windings are the primary–secondary–primary (P–S–P), but it is common to encounter a shell form winding of P–S–P–S–P. The winding structure for both the primary and secondary windings are normally of the pancake-type winding structure (Bean *et al.*, 1959).

3.7 Transformer Performance

3.7.1 Losses and Efficiency

Total losses for transformer performance are made up of **no-load losses** and **load losses**. No-load losses are primarily hysteresis and eddy-current loss in the core. This loss is independent of load and exists whenever the transformer is connected to the system. The no-load loss is a function of the voltage of the system. The core's eddy-current loss is proportional to the square of the system frequency, and the hysteresis loss is proportional to the frequency. Load losses consist of I^2R loss in the windings, stray losses in the windings, and structural clamps and fittings. These losses are caused by the load current. On large well-designed transformers, the total rated losses are on the order of 0.3 to 0.5% of the rated kVA of the transformer. The cost per kilovoltampere is different for load and no-load losses and depends on a number of factors (Massachusetts Institute of Technology, 1943). Typically the no-load losses are 25 to 35% of the total losses and are 2.5 to 3.5 times the $/kVA of load losses.

3.7.2 Sound

Transformers are static devices; however, they vibrate due to a number of causes and because of the radiate sound. There are two distinct sources of sound: **auxiliary cooling equipment** and **magnetostriction**. The sound produced by the fans and pumps of a cooling system generally possesses a broadband frequency spectrum of approximately equal magnitude (white noise). The major source of a transformer's audible noise is caused by the dimensional change of the core laminations from the variation of flux density in each voltage cycle, referred to as **magnetostriction**. The core will radiate noise at the even harmonics of the voltage (e.g., for a 60-Hz system, the noise radiates at 120-, 240-, 360-, 480-Hz, and so on). The magnitude of this noise is measured by determining the sound-pressure level given in decibels:

$$P = 20 \log \frac{F}{0.0002},$$

with P the sound pressure in decibels and F the sound pressure in rms dynes. IEEE C57.12.90, Test Code for Liquid-Immersed Transformers, specifies the method for measuring the average sound level of a transformer. The goal is to avoid a transformer design and installation that produces noise at a level that is an annoyance. Sound reduction can be accomplished at the transformer design stage and/or by judicious application. During design, the most basic modification effecting noise production is to reduce the core's flux density. In addition, barriers or structural changes in the tank enclosure will reduce emitted noise. Upon installation, the erection of sound barriers is very effective in reducing noise levels. With proper design, reductions of 10 to 30 dB are possible; with the application of sound barriers at a site, an additional 10-dB reduction is achievable.

3.7.3 Overcurrent

A transformer may at some point be subjected to current in excess of its rated load current. Current up to approximately 2% of the rated load current can be anticipated from overload conditions on the system. These currents may last only a few minutes or several hours. Faults on the system will subject the transformer to peak overcurrents that may be up to 40 times the rated load current. These currents may produce either mechanical or thermal damage to the transformer. Mechanical forces are proportional to the square of the current and manifest themselves in a number of different failure mechanisms. These failure mechanisms include loosening of coils, conductor tipping, radial and axial conductor buckling, winding deformation, core deformation, and telescoping of windings.

3.8 Acceptance Tests

The desire of the purchaser is to obtain a transformer at a reasonable price that will achieve the required performance for an extended period of time. The desire of the manufacturer is to construct and sell a product, at a profit, that meets the customer's goals. The specification and purchase contract combines both the purchaser's requirements and manufacturer's commitment in a legal format. The specification will typically address the transformer's service condition, rating, general construction, control and protection, design and performance review, testing requirements, and transportation and handling. Since it is impossible to address all issues in a specification, the industry uses standards that are acceptable to purchaser and supplier. In the case of power transformers, the applicable standard would include IEEE C57, IEC 76, and NEMA TR-1.

ANSI/IEEE C57.12.00 defines routine and optional test and testing procedures for power transformers. The following are listed as routine tests for transformers larger than 501 kVA: winding resistance, winding turns ratio, phase–relationship tests, polarity, angular displacements, phase sequence, no-load loss and exciting current, load loss and impedance voltage, low-frequency dielectric tests (applied voltage and induced voltage), and leak test on a transformer tank.

The following are listed as type tests to be performed on only one of a number of similar design units for transformers 501 kVA and larger: temperature rise tests, lightning-impulse tests (full and chopped wave), audible sound tests, mechanical test from lifting and moving of transformer, and pressure tests on tank.

Other tests include short circuit forces and switching surge impulse tests.

The variety of transient voltages a transformer may experience in its normal useful lifetime are virtually unlimited. It is impractical to test each transformer for every conceivable combination of transient voltage. However, the electrical industry has found that it is possible, in most instances, to assess the integrity of the transformer's insulation systems to withstand transient voltages with the application of a few specific aperiodic voltage waveforms (Massachusetts Institute of Technology, 1943), illustrating the full, chopped, and switching surge waveforms. Each of these tests is designed to test the insulation structure for a different transient condition. The purpose of applying this variety of tests is to substantiate adequate performance of the total insulation system for all the various transient voltages a transformer may see in service.

References

Bean, R.L., Crackan, N., Moore, H.R., and Wentz, E. (1959). *Transformers for the electric power industry.* New York: McGraw-Hill.

Blume, L.F., Boyajian, A., Camilli, G., Lennox, T.C., Minneci, S., and Montsinger, V.M. (1951). *Transformer Engineering.* (2d ed.). New York: John Wiley & Sons.

Fink, D.G., and Beaty, H.W. (1987). *Standard handbook for electrical engineers.* (12th ed.). New York: McGraw-Hill.

Grigsby, L.L. (2001). The Electric Power Engineering Handbook. CRC Press.

Heathcote, M., Franklin, A., Franklin, D. (1998). *J&P Transformer Book.* London: Butterworth-Heinemann.

IEEE guide and standards for distribution, power, and regulating transformers. New York: IEEE Press, 1998.

Massachusetts Institute of Technology, Department of Electrical Engineering (1943). *Magnetic circuits and transformers.* New York: John Wiley & Sons.

4

Introduction to Electric Machines

Sheppard Joel Salon
Department of Electrical Power Engineering, Renssalaer Polytechnic Institute, Troy, New York, USA

4.1 Introduction

Essentially all electric energy is generated in a rotating machine, the **synchronous generator**, and most of it is consumed by electric motors. In many ways, the world's entire technology is based on these devices. The study of the behavior of electric machines is based on three fundamental principles: Ampère's law, Faraday's law and Newton's Law. Various configurations result and are classified generally by the type of electrical system to which the machine is connected: **direct current** (dc) machines or **alternating current** (ac) machines. Machines with a dc supply are further divided into permanent magnet and wound field types, as shown in Figure 4.1. The wound motors are further classified according to the connections used. The field and armature may have separate sources (separately excited), they may be connected in parallel (shunt connected), or they may be series (series connected).

Machines that have ac are usually single-phase or three-phase machines and may be synchronous or asynchronous. Several variations are shown in Figure 4.2.

The following sections state very briefly the basic principles on which all standard electric machines operate and then, for the most common devices, explain the principles of operation and the parameters for predicting and understanding their behavior.

4.1.1 Basic Electromagnetic Laws: Ampère's Law and Faraday's Law

The two principles that describe the electromagnetic behavior of electric machines are **Ampère's law** and **Faraday's law.** These are two of Maxwell's equations. Most electric machines operate by attraction or repulsion of electromagnets and/or permanent magnets. Ampère's law describes the magnetic field that can be produced by currents or magnets. In an electric machine, there will always be at least one set of coils with currents. A motor cannot be produced with permanent magnets alone.

Ampère's law states that the line integral of the component of the magnetic field along the path of integration is equal to the current enclosed by the path. This is exactly true for static fields and is a very good approximation for the low-frequency fields dealt with in electric machines:

$$\oint \vec{H} \cdot d\vec{\ell} = I_{\text{enclosed}}. \tag{4.1}$$

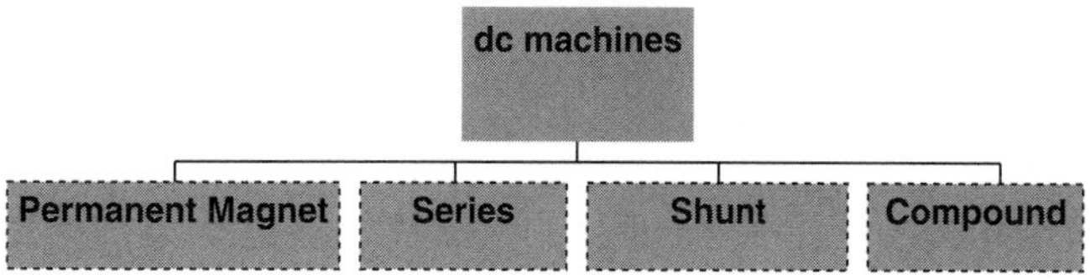

FIGURE 4.1 Direct Current Machine Classifications

The right-hand side of equation 4.1 represents the current enclosed by the integration path and is called the **magnetomotive force** (MMF). In electric machines, currents are frequently placed in slots surrounded by ferromagnetic teeth. An example illustrating the determination of the MMF is shown in Figure 4.3, where different integration paths are shown by dotted lines. The MMF corresponding to each path is the total current enclosed by the path. If the slots contain currents that are approximately sinusoidally distributed in space, then the MMF will be cosinusoidally distributed. In this way, the magnetic field or flux density in the air gaps of the machine will often have a sinusoidal or cosinusoidal distribution.

Faraday's law relates the induced voltage or electromotive force (EMF) to the time rate of change of the magnetic flux linkage:

$$\oint \vec{E} \cdot d\vec{\ell} = -\frac{d}{dt}\oint \vec{B} \cdot d\vec{S}, \tag{4.2}$$

where $\vec{E}$ is the electric field and $\vec{B}$ is the magnetic flux density. This law states that the voltage induced in a loop is equal to the time rate of change of the flux linking the loop. The negative sign indicates that the voltage is induced such that the current would oppose the change in flux linkage. The change in flux linkage can be caused by a change in flux density and/or a change in geometry.

4.1.2 MMF and Traveling Waves

Presented now is the important concept of rotating MMF and fields in machines. Consider the following equation:

$$\mathfrak{F} = F_1 \sin(\omega t)\cos(P\theta). \tag{4.3}$$

Equation 4.3 represents a standing wave. The sine term describes the time variation, while the cosine term describes the space variation. P is the number of pole pairs around the machine. At a fixed point on the wave θ is constant, and the amplitude is sinusoidally time varying. The peak of the wave and the nodes or zero crossings, however, are always located at the same position. This equation is contrasted with an expression of the form:

$$\mathfrak{F} = F_1 \sin(\omega t - P\theta). \tag{4.4}$$

This equation represents a traveling wave. If the behavior of a fixed point in space is considered, as before, the amplitude varies sinusoidally in time. However, it is no longer true that the peak or any point on the wave remains in the same position. To see this, consider the argument of the sine function. The peak will be located at $\omega t - P\theta = \frac{\pi}{2}$. To remain at the peak of the wave as time moves forward, movement must be made in the positive θ direction. The expression therefore represents a wave traveling in the positive θ direction. If the minus sign in the argument is replaced by a plus sign, the wave travels backward or in the negative θ direction. The speed of the traveling wave can be determined by considering the argument of the sine function. In one cycle, the ωt term increases by 2π. To remain at the same point on the wave, the space term must also advance by 2π. This means that θ must increase by $\frac{2\pi}{P}$. Thus, the wave progresses one wavelength or two poles in one cycle.

A standing wave can be decomposed into two traveling waves, one going forward and one backward, each of half the amplitude of the standing wave. See Figure 4.4. This is seen by using a trigonometric identity:

$$\sin \omega t \cos P\theta = \frac{1}{2}(\sin \omega t + P\theta) + \sin(\omega t - P\theta)). \tag{4.5}$$

The first term on the right-hand side of equation 4.5 represents a backward traveling wave, and the second term represents a forward traveling wave.

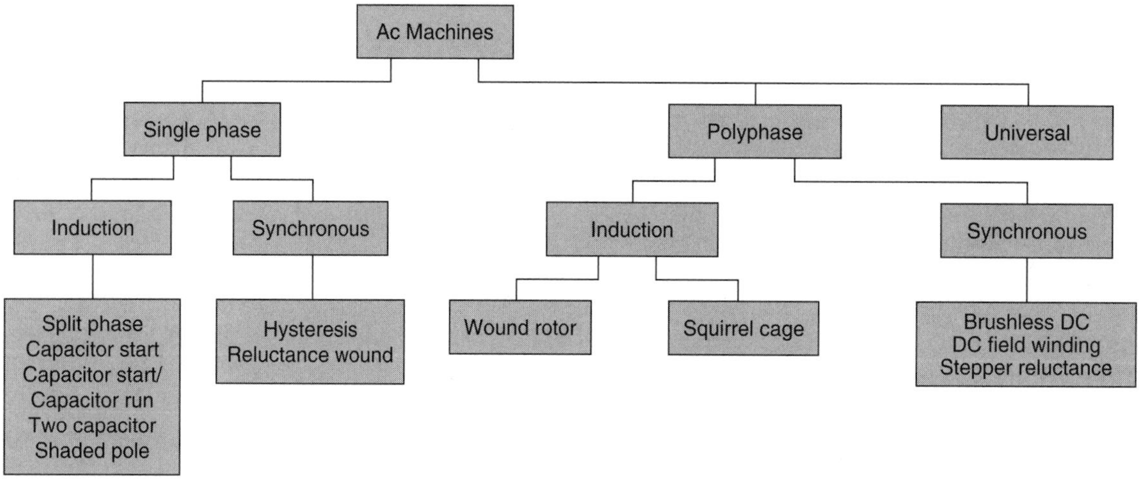

FIGURE 4.2 Alternating Current Machine Classifications

Let us consider the case of a three-phase winding in which each of the phases are displaced by 120° in space and the winding currents are displaced by 120° in time. For the first phase, phase a, the fundamental space component of the MMF has the form:

$$\mathfrak{F}_a = K\sin(\omega t)\cos(P\theta) = \frac{K}{2}(\sin(\omega t - P\theta) + \sin(\omega t + P\theta)), \tag{4.6}$$

where K is a constant containing information on the winding geometry. Similarly, for phases b and c:

$$\begin{aligned}\mathfrak{F}_b &= K\sin(\omega t - \frac{2\pi}{3})\cos(P\theta - \frac{2\pi}{3})\\ &= \frac{K}{2}(\sin(\omega t - P\theta) + \sin(\omega t + P\theta - \frac{4\pi}{3}).\end{aligned} \tag{4.7}$$

$$\begin{aligned}\mathfrak{F}_c &= K\sin(\omega t - \frac{4\pi}{3})\cos(P\theta - \frac{4\pi}{3})\\ &= \frac{K}{2}(\sin(\omega t - P\theta) + \sin(\omega t + P\theta - \frac{2\pi}{3}).\end{aligned} \tag{4.8}$$

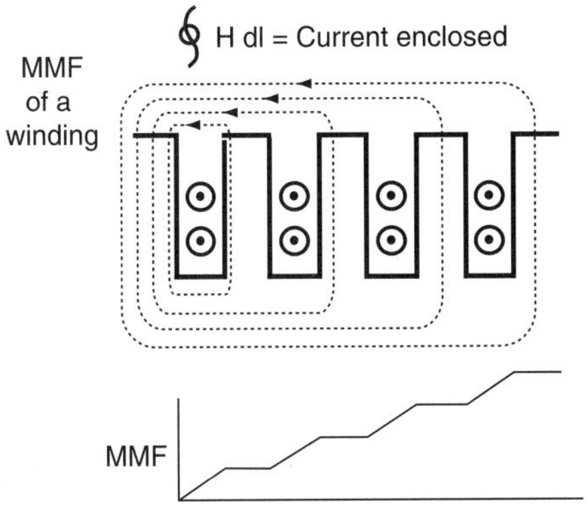

FIGURE 4.3 Slots with Current and MMF

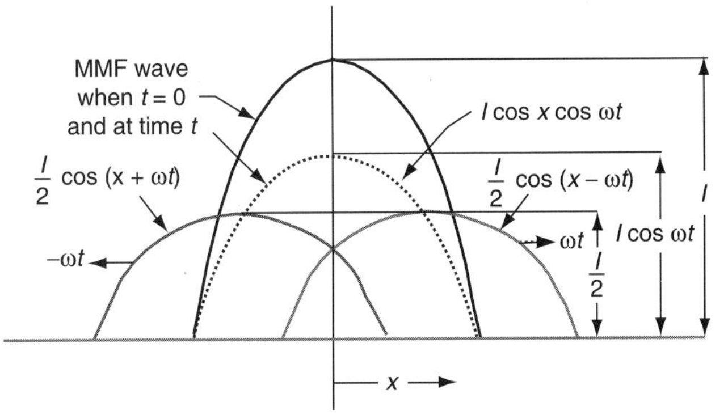

FIGURE 4.4 Two Equal Traveling Waves Equivalent to a Standing Wave

The forward waves of all three phases are the same, while the backward waves are 120° out of phase. By adding the MMFs of the three phases, a resultant wave is obtained in which the fundamental space component travels in the forward direction and has a magnitude 3/2 times the peak MMF of one phase alone. The backward waves cancel. This is the most common method of producing traveling magnetic fields in motors. Figure 4.5 shows a simplified 18-slot machine with three-phase sinusoidal currents. Each phase is placed into three consecutive slots with the return nine slots away. The MMF at different instants of time is plotted, and the wave moves to the right at a rate of one wavelength (two poles) per cycle. The wave is distorted due to the space harmonics included. These have little influence on the energy conversion process and are not dealt with here.

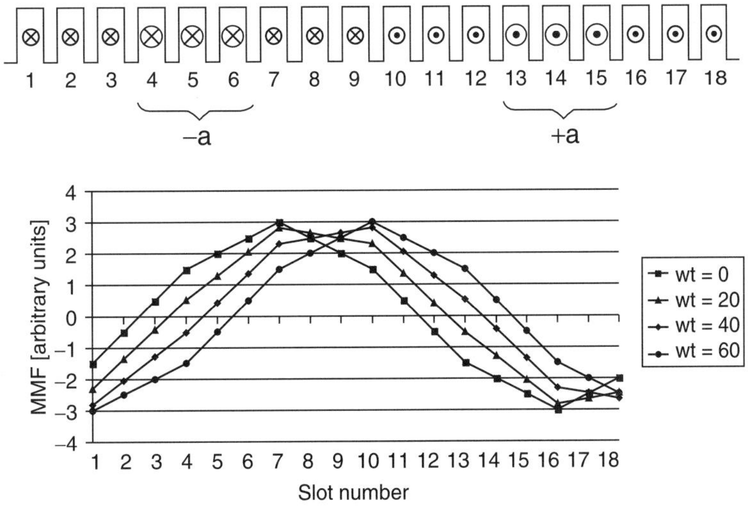

FIGURE 4.5 An 18-Slot Machine with MMF at 4 Times: The Movement of the Wave

4.1.3 Mechanical Description of a Drive System

An electrical machine will operate at a speed determined by the electric torque or force produced and the mechanical load (Figure 4.6). If these are equal and opposite, the machine will operate at a constant speed; if not, the machine will accelerate or decelerate at a rate proportional to the net accelerating force or torque. Newton's law describes this process. Writing this for a rotational system:

$$T_{em} - T_{\text{load}} = J\frac{d\omega_m}{dt}, \tag{4.9}$$

where T_{em} is the electromagnetic torque delivered to the load, T_{load} is the load torque, J is the moment of inertial of load and motor, and ω_m is the mechanical angular frequency [radians per second]. Figure 4.7 illustrates a typical motor and load torque characteristic. At point a, the motor torque is greater than the load torque and the system accelerates. At point b, the load torque is greater than the motor torque and the system decelerates. Point c is a stable equilibrium point, and the system will operate at this speed.

The electric machine may operate as a motor or as a generator. Only the direction of energy transfer is affected. In a motor, the electrical energy is transformed into mechanical energy, and in a generator, the mechanical energy is transformed into electrical energy. In either case, the machine may

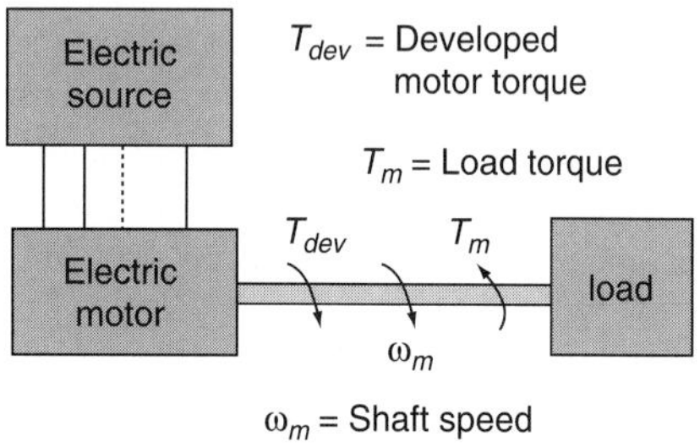

FIGURE 4.6 Drive System

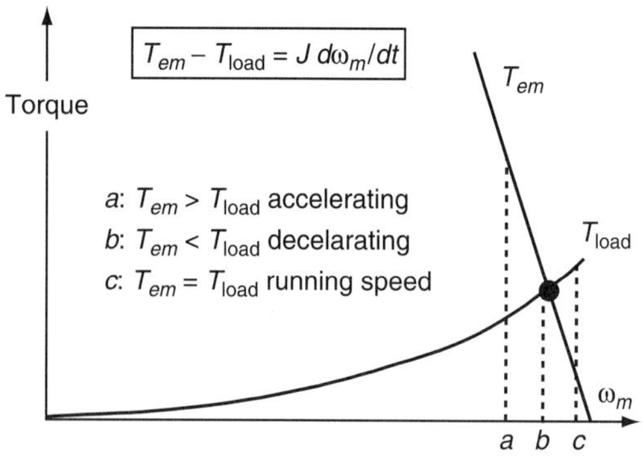

FIGURE 4.7 Motor and Load Torque

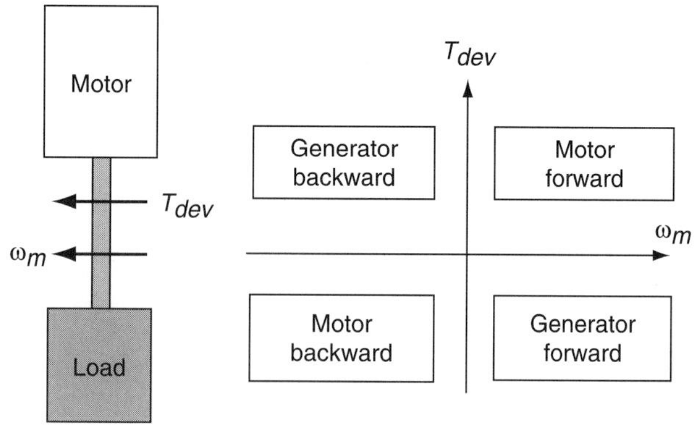

FIGURE 4.8 Four-Quadrant Operation

run forward or backward. These combinations are illustrated in Figure 4.8, where a four-quadrant operation is shown. Depending on the type of drive, an electromechanical system may have the ability to operate in any one or perhaps all four quadrants.

4.1.4 Energy Conversion

All of the devices considered here are **magnetic** (i.e., they use the energy in a magnetic field to produce force or torque). It is possible to construct electrostatic motors and actuators, but in standard sizes, the energy density is too low to make them competitive. Consider a simple linear system consisting of a resistance and a position-dependent inductance connected to an electrical supply. The circuit equation is as follows:

$$v = iR + \frac{d\psi}{dt} = iR + \frac{d}{dt}(L(x)i), \tag{4.10}$$

Where $\psi = Li$ is the flux linkage. Performing the differentiation yields:

$$v = Ri + L(x)\frac{di}{dt} + i\frac{\partial L(x)}{\partial x}\frac{dx}{dt}. \tag{4.11}$$

Note how the velocity dx/dt appears due to the position dependence of the inductance. The energy input from the source in time dt is as written here:

$$vi\ dt = i^2Rdt + iL(x)di + i^2\frac{\partial L(x)}{\partial x}dx. \tag{4.12}$$

The first term on the right-hand side of equation 4.12 is the joule heating. To understand the physical significance of the second and third terms, consider the change in stored magnetic energy. The energy stored in an inductor is as follows:

$$W_m = \frac{1}{2}L(x)i^2. \tag{4.13}$$

If both i and $L(x)$ change, the change in stored energy becomes:

$$dW_m = \frac{\partial W_m}{\partial i} di + \frac{\partial W_m}{\partial x} dx = iL(x)di + \frac{1}{2} i^2 \frac{\partial L(x)}{\partial x} dx. \quad (4.14)$$

The energy input from the circuit cannot completely be explained by losses and a change in stored energy. There is a missing $\frac{1}{2} i^2 (\partial L(x)/\partial x) dx$. This is the energy turned into mechanical work. This term is the one associated with motion. As a general rule, a situation must occur in which there is a change of inductance (self and/or mutual) with position.

Now consider the case of two coupled circuits. The losses can be treated as above. The total energy stored in the magnetic field is as written here:

$$W_m = \frac{1}{2} L_{11} i_1^2 + \frac{1}{2} L_{22} i_2^2 + L_{12} i_1 i_2. \quad (4.15)$$

Using the virtual work principle, the mechanical force in any direction is the partial derivative of the magnetic energy with respect to an incremental motion in that direction with the current held constant. To find torque, differentiate with respect to the rotational angle. Now look at the expression for torque in this two circuit system:

$$T = \frac{1}{2} i_1^2 \frac{\partial L_{11}}{\partial \theta} + \frac{1}{2} i_2^2 \frac{\partial L_{22}}{\partial \theta} + i_1 i_2 \frac{\partial L_{12}}{\partial \theta}. \quad (4.16)$$

There are three possible sources of torque: the change in the self-inductance of winding 1 with position, the change of the self-inductance of winding 2 with position, and the change of the mutual inductance with position. In an electric machine, one, two, or all three of these effects may be involved. Figure 4.9(A) shows a double salient structure in which all three effects are in operation. With winding 1 excited alone, there is a torque due to the position dependence of the winding 1 self-inductance. The torque will be in a direction to maximize L_{11} and thus make the air gap smaller. Similarly, if winding 2 is excited alone, a torque is obtained due to the variation of L_{22}. If both windings are excited together, there is also a torque due to the variation of the mutual inductance L_{12}. Figures 4.9(B) and 4.9(C) show the case of a single salient structure (one self and one mutual term) and a nonsalient structure (only the mutual term).

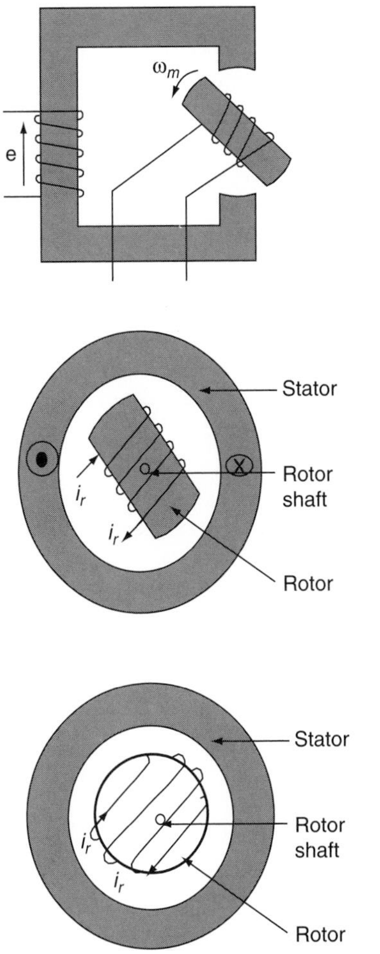

FIGURE 4.9 Double Salient Structure (Top), Single Salient Structure (Middle), and Nonsalient Structure (Bottom)

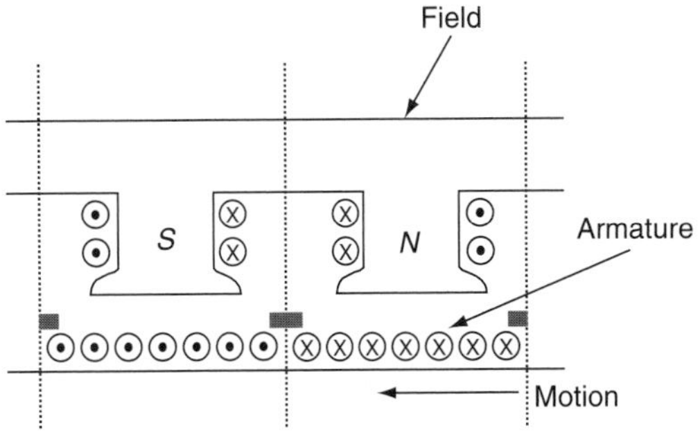

FIGURE 4.10 Typical Direct Current Motor

4.2 Direct Current Machines

The dc machine usually has a dc winding or a set of permanent magnets on the stator (the field) and a winding connected to brushes on the rotor (the armature). This configuration is shown in Figure 4.10. The armature currents produce a flux density that peaks approximately 90° from the peak of the flux density produced by the field. As the rotor turns, the currents in the armature conductors are switched by a commutator so that the field pattern remains the same. The dc machine is inherently a variable speed machine. The speed and torque can be controlled by the field current and by the armature voltage. Machines that are ac run at constant speed or near-constant speed, the speed being determined by the supply frequency.

4.2.1 Induced EMF in a Full-Pitched Winding

The expression for the induced voltage or back EMF in a full-pitched armature conductor is now formed (see Figure 4.11). **Full-pitched** means that the coil spans one pole. Assume that the flux density in the air gap has a periodic spatial relationship, and it may be expressed as:

$$B(\alpha) = B_1 \sin(\alpha) + B_3 \sin(3\alpha) + B_5 \sin(5\alpha) + \dots. \quad (4.17)$$

Integrating to find the flux linkage yields:

$$\psi = N \cdot D \cdot L\left(B_1 \cos(\theta) + \frac{1}{3}B_3 \cos(3\theta) + \frac{1}{5}B_5 \cos(5\theta) + \dots\right), \quad (4.18)$$

where N is the number of turns in a coil, D is the diameter, L is the length, and $\theta = \omega t$.

From Faraday's law:

$$e_a = \frac{d\psi}{dt} = \frac{d\psi}{d\theta} \cdot \frac{d\theta}{dt}. \quad (4.19)$$

So, for each armature coil:

$$e_a = \omega N \cdot D \cdot L(B_1 \sin(\theta) + B_3 \sin(3\theta) + B_5 \sin(5\theta) + \dots). \quad (4.20)$$

Note that the induced voltage waveform has the same harmonic content as the flux density waveform.

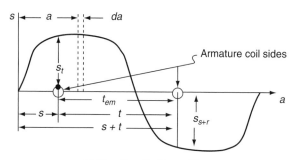

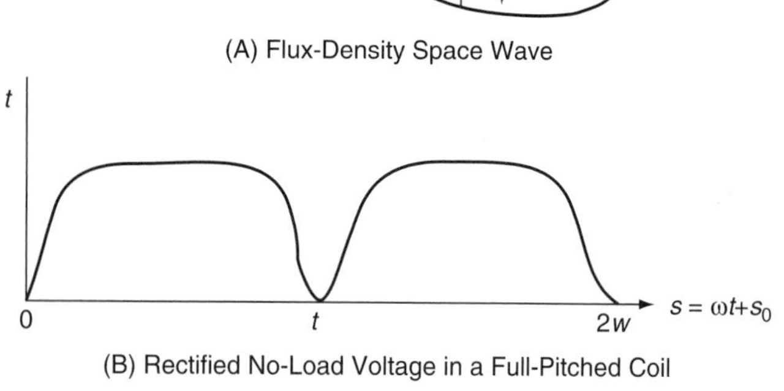

FIGURE 4.11 Direct Current Motor Field Distribution

The average value of the EMF is the following:

$$E_a = \frac{2}{T}\int_0^{T/2} e_a \, dt = \frac{1}{\pi}\int_0^{\pi} e_a \, d\theta, \quad (4.21)$$

where T is the period. Using symmetry to integrate over one pole pitch:

$$E_a = \frac{\omega}{\pi} N \int_0^{\pi} B(\theta) DL \, d\theta. \quad (4.22)$$

Note here that $D \cdot L \cdot d\theta = 2dA$, where dA is the incremental area shown in Figure 4.11. Now, the flux per pole is written as:

$$\Phi = \int_0^{\pi} B(\theta) \, dA. \quad (4.23)$$

Hence, the following is true:

$$E_a = \frac{2\omega_m}{\pi} N\Phi. \quad (4.24)$$

This is the average EMF per pole, per coil. To generalize, there may be P poles. Then:

$$E_a = \frac{P}{2} \cdot \frac{2}{\pi} \cdot N\Phi\omega_m. \quad (4.25)$$

If there are a parallel paths in the armature, and N_a is the total number of turns in the armature, then:[1]

$$E_a = \frac{P}{a\pi} N_a \Phi\omega_m. \quad (4.26)$$

In terms of the total number of active conductors in the armature, $Z = 2N_a$:

$$E_a = \frac{PZ}{2a\pi}\Phi\omega_m = K_i\Phi\omega_m, \quad (4.27)$$

where:

$$K_t = \frac{PZ}{2\pi a} = \frac{PN_a}{\pi a}. \quad (4.28)$$

4.2.2 Speed–Torque Relationships

Consider the circuit of Figure 4.12, where R_a is the armature resistance and V_a is the applied voltage. E_a represents the back EMF. Assume that there is no change in the energy stored in

[1] Direct current (dc) machines use two types of windings. In a lap winding, the number of parallels is equal to the number of poles. In a wave winding, there are two parallels.

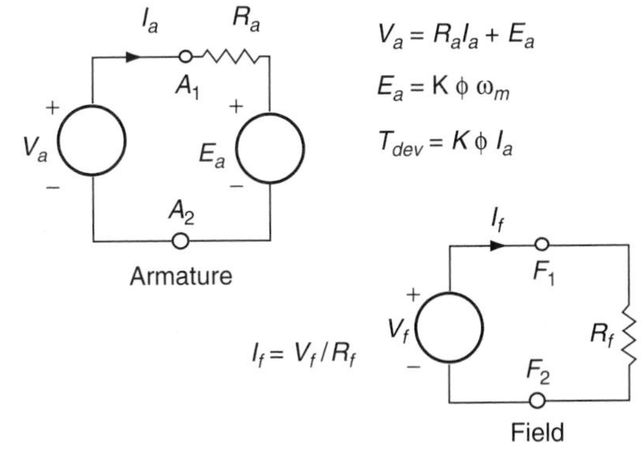

FIGURE 4.12 Direct Current Machine Diagram

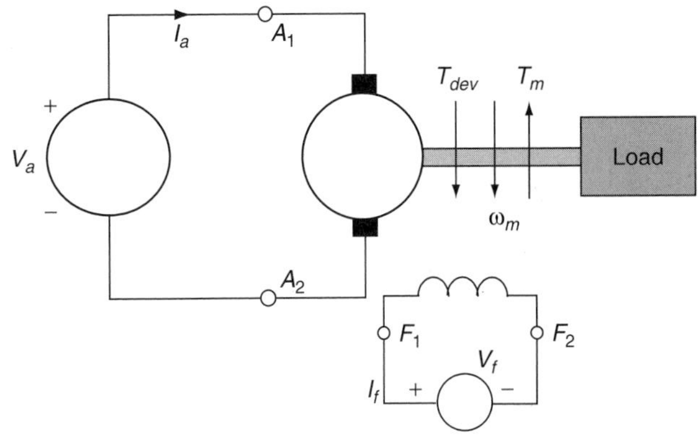

FIGURE 4.13 Separately Excited Direct Current Motor

the field. Then, the input power P_{in} must go into joule heating, P_L, and mechanical power, P_m:

$$P_{\text{in}} = P_L + P_m. \tag{4.29}$$

Therefore:

$$P_m = V_aI_a - R_aI_a^2 = I_a(V_a - R_aI_a) = E_aI_a. \tag{4.30}$$

So the mechanical power is the product of the armature current and the back EMF. The mechanical power is related to the torque by:

$$P_m = E_aI_a = \omega T_m. \tag{4.31}$$

Solving for the torque:

$$T_m = \frac{E_aI_a}{\omega} = I_a\frac{K_t\Phi\omega}{\omega} = K_t\Phi I_a. \tag{4.32}$$

The voltage equation is as written here:

$$V_a = E_a + R_aI_a = K_t\Phi\omega + R_aI_a, \tag{4.33}$$

which gives this equation:

$$\omega = \frac{V_a - R_aI_a}{K_t\Phi}. \tag{4.34}$$

Equation 4.34 gives a relationship between the motor speed and torque. Notice that the speed is inversely proportional to the field flux.

4.2.3 Direct Current Machine Connections

The dc machine can be connected in several ways to control the speed–torque characteristic, as explained in the following subsections.

Separately Excited Machine

When the field and the armature are excited by separate sources, as indicated in Figure 4.13, the machine is **separately excited**. The field flux is controlled by I_f, and the armature current depends on the applied voltage V_a and the back EMF.

The rotor speed is given by:

$$\omega_m = \frac{V_a - R_aI_a}{K_t\Phi}. \tag{4.35}$$

Equation 4.35 can be broken into two terms, one that depends on the applied armature voltage and one that depends on the armature current:

$$\omega_m = \omega_0 - \frac{R_aI_a}{K_t\Phi}, \tag{4.36}$$

where $\omega_0 = \frac{V}{K_t\Phi}$ is the no-load speed. There are two mechanisms for speed control. As the armature voltage increases while the field current holds constant, the velocity ω_0 increases. A number of parallel curves are obtained of ω_m versus I_a. If I_f is increased (and therefore Φ), both terms of ω_0 and $R_a/K_t\Phi$ decrease.

Shunt-Excited Machine

If both windings of a dc machine are connected in parallel, then the voltage to the armature and the field are the same. This is indicated in Figure 4.14 as a **shunt = excited machine**. Ignoring saturation, the result is the following:

$$K_t\Phi = K_1I_f = KV_a. \tag{4.37}$$

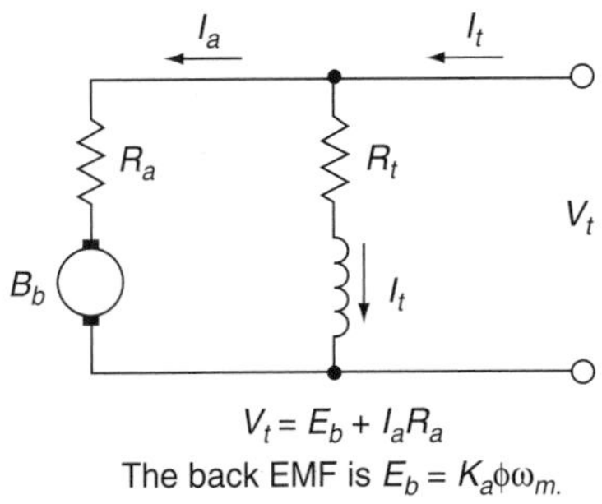

FIGURE 4.14 Shunt-Excited Direct Current Machine

So:

$$\omega_m = \frac{V_a - I_a R_a}{K_t \Phi} = \frac{V_a - I_a R_a}{K V_a} = \frac{1}{K} - \frac{R_a}{K V_a} I_a. \quad (4.38)$$

Therefore, at no load ($I_a = 0$):

$$\omega_m = \frac{1}{K}. \quad (4.39)$$

The no load speed is independent of the armature voltage. Speed control is achieved by controlling field current through a variable resistance connected to the field.

Series-Excited Machine

In the **series-excited machine**, the current through the field winding and armature winding are the same. Neglecting saturation yields:

$$K_t \Phi = K_2 I_f = K_2 I_a. \quad (4.40)$$

$$\omega_m = \frac{V - R_a I_a}{K_t \Phi} = \frac{V - R_a I_a}{K_2 I_a} = \frac{V}{K_2 I_a} - \frac{R_a}{K_2}. \quad (4.41)$$

If an external resistance is used for speed control, then:

$$\omega_m = \frac{V}{K_2 I_a} - \frac{R_a + R_{ext}}{K_2}. \quad (4.42)$$

For the torque, the result is as follows:

$$T = K_t \Phi I_a = K_2 I_a^2. \quad (4.43)$$

If the load torque increases, the current increases only as the square root of the torque, and the speed drops. This makes the series-excited dc machine a popular choice for traction applications.

4.3 Three-Phase Induction Motor

Induction motors are by far the most common type of motor used by industry. The most common **polyphase induction motor** has a laminated stator core and a three-phase winding contained in slots. The rotor has two standard configurations: **squirrel cage** and **wound rotor**. The squirrel cage is the most common and consists of solid rotor bars that are contained in slots. The bars are typically made of a good conductor, such as copper or aluminum. The rotor is usually laminated. The bars are connected together at each end by a short-circuiting ring. The squirrel cage has the advantage of being inexpensive and very rugged, having no insulation. In the wound rotor induction machine, the rotor winding is an insulated three-phase winding. The terminals are brought out through slip rings. The advantage of the wound rotor is that the resistance of the rotor can be controlled by inserting or removing stationary resistance through the slip rings. A squirrel cage induction motor cross section is shown in Figure 4.15 along with a plot of the flux lines for rated conditions.

At steady state, the three-phase stator winding produces a magnetic field rotating at synchronous speed. The rotor rotates at a speed slightly less than synchronous speed. The low-frequency currents induced on the rotor produce a field that also travels at synchronous speed. There is a fixed angle between these two fields. They react with each other to produce a constant torque. This is illustrated schematically in Figure 4.16.

4.3.1 Equivalent Circuit

A typical single-phase equivalent circuit for the induction machine is shown in Figure 4.17(A). This circuit is valid for steady-state sinusoidal conditions.

This circuit resembles a two winding transformer equivalent circuit, with the secondary (rotor) short-circuited. Here, V_1 is the phase voltage, r_1 the stator winding resistance per phase, X_1

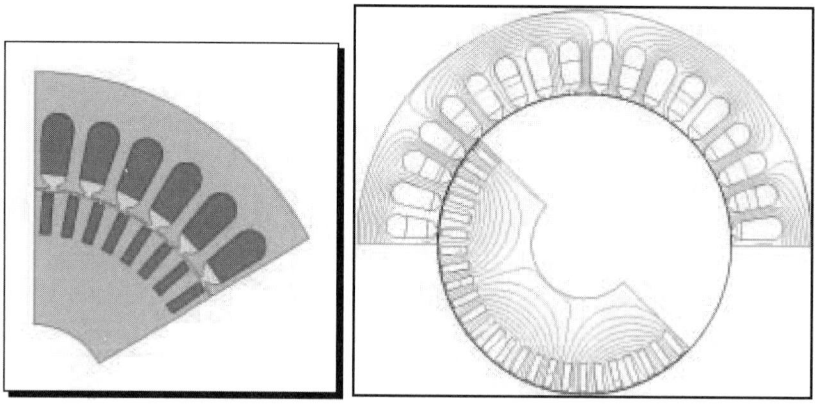

FIGURE 4.15 Squirrel Cage Induction Motor

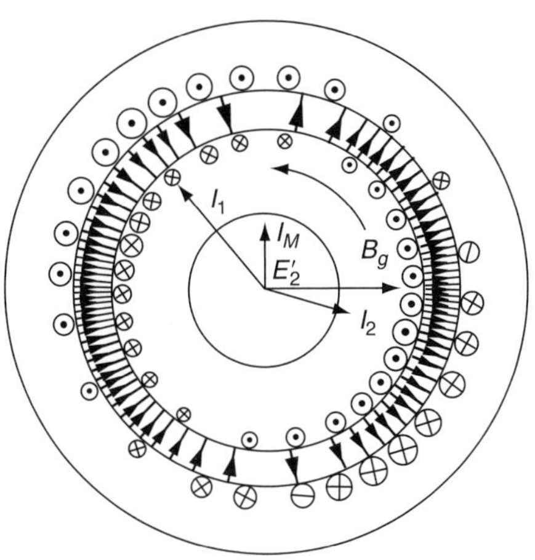

FIGURE 4.16 Induction Motor Rotor and Stator Currents and Fields

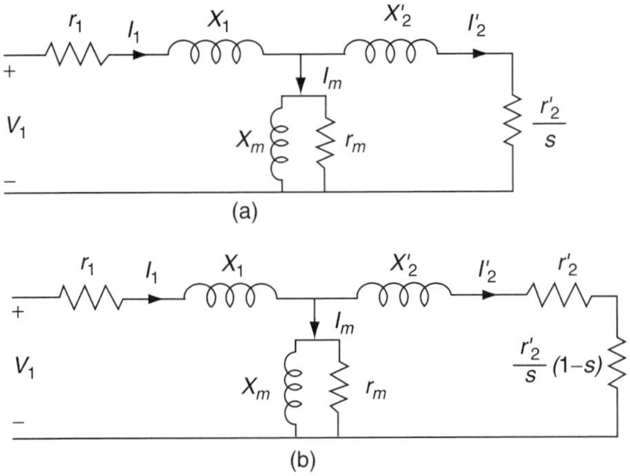

FIGURE 4.17 Phase-Equivalent Circuit of Polyphase Induction Motor

the stator winding leakage reactance per phase, X_m the magnetizing reactance, r_m the equivalent core resistance representing eddy and hysteresis losses, X_2' the rotor reactance referred to the stator, and r_2' the rotor resistance referred to the stator.

To understand the rotor circuit, consider the concept of slip. The application of a balanced three-phase supply to a three-phase winding produces a rotating flux distribution at synchronous frequency ω_s. The rotor of the induction machine will be rotating at frequency ω_m. The slip s can be defined as:

$$s = \frac{\omega_s - \omega_m}{\omega_s}. \tag{4.44}$$

Equation 4.44 shows that if the rotor is moving at synchronous speed, the slip is zero. If the rotor is stationary, the slip is one. Since the stator produces a rotating field at ω_s, an observer on the rotor sees a time-varying magnetic field at frequency $s\omega_s$. Applying Faraday's law to the rotor winding, the EMF is multiplied by the slip for a given mutual flux. The secondary parameters are then r_2 and sX_2. The rotor reactance per phase has been corrected by s due to the frequency change. Call the effective turns ratio between the rotor and the stator n. Using this value, first reflect the rotor quantities to the stator. Having done this, we connect the rotor and stator as is done in a normal transformer equivalent circuit if it were not for the frequency difference. The rotor current is then:

$$I_r = \frac{sE_2}{r_2' + jsX_2'}. \tag{4.45}$$

To connect the two circuits together, divide the voltage sE_2 by the slip. To keep the current correct, also divide the denominator by s to give:

$$\frac{r_2'}{s} + jX_2'. \tag{4.46}$$

The reactance appears to be at stator frequency, while the resistance is divided by the slip. The two circuits may now be connected together. The rotor quantities have been reflected to the stator by both the turns ratio and the frequency. All current and voltage is at applied (stator) frequency. Now drop the primed notation and assume all quantities are referred to the stator. The significance of dividing the rotor resistance by the slip can be seen by considering an alternative form of the circuit shown in Figure 4.17(B). The real power crossing the air gap is:

$$P_g = I_2^2 \frac{r_2}{s}. \tag{4.47}$$

The heat dissipated in the rotor circuit is the following:

$$P_L = I_2^2 r_2, \tag{4.48}$$

where the power is in the first rotor resistance. The remaining power is as follows:

$$P_m = I_2^2 r_2 \frac{1-s}{s}, \tag{4.49}$$

which is the mechanical power delivered to the rotor. The output torque per phase is then:

$$T = \frac{P_m}{\omega_m}. \tag{4.50}$$

4.3.2 Torque–Speed Calculations

We obtain an expression for the output torque and power from the equivalent circuit by applying Thévenin's theorem.

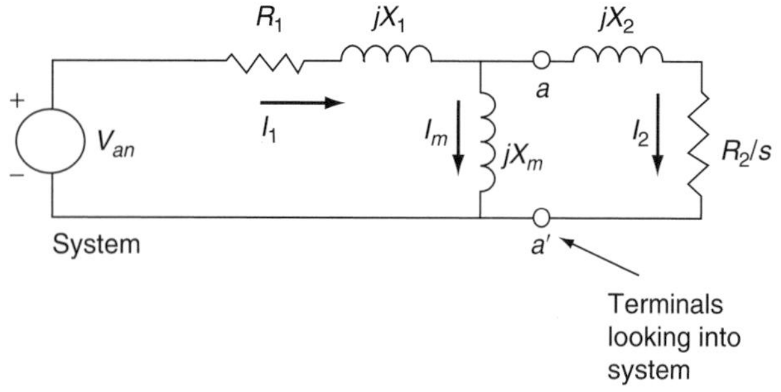

FIGURE 4.18 Induction Motor and System

Considering the circuit of Figure 4.18, we look back into the system from terminals $a - a'$ and find that the source voltage becomes the voltage at these terminals when the rotor is open circuited:

$$V_{th} = V_1 \frac{jX_m}{R_1 + j(X_1 + X_m)}. \tag{4.51}$$

The **Thévenin impedance** is the impedance seen from $a - a'$ with the voltage source shorted. This is the parallel combination of jX_m and $r_1 + jX_1$:

$$r_{th} + jx_{th} = \frac{(R_1 + jX_1)jX_m}{R_1 + j(X_1 + X_m)}. \tag{4.52}$$

Equation 4.52 gives a simple series circuit for the induction machine. The torque is now found as:

$$T = \frac{1}{\omega_s} \frac{3V_{th}^2 (r_2/s)}{\left(r_{th} + \frac{r_2}{s}\right)^2 + (x_{th} + X_2)^2}. \tag{4.53}$$

For a constant voltage, the familiar torque versus slip characteristic is obtained and shown in Figure 4.19.

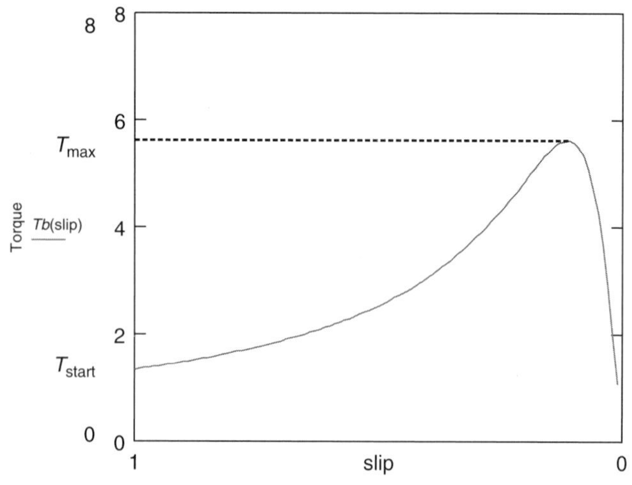

FIGURE 4.19 Torque Versus Slip

It is useful to have an expression for the maximum torque (pull-out torque) and the slip at which it occurs. This can be obtained from equation 4.53 by first differentiating with respect to the slip and setting the result to zero. This gives the slip at maximum torque as:

$$S_{max} = \frac{r_2}{\sqrt{r_{th}^2 + (x_{th} + X_2)^2}}. \tag{4.54}$$

Substituting equation 4.54 into equation 4.53 yields:

$$T_m = \frac{1}{\omega_s} \frac{1.5V_{th}^2}{r_{th} + \sqrt{r_{th}^2 + (x_{th} + X_2)^2}}. \tag{4.55}$$

These expressions tell us that the slip at maximum torque is proportional to the rotor resistance but that the pull-out torque is independent of the rotor resistance. Figure 4.20 shows the torque slip curves for an induction motor with different values of rotor resistance. As the rotor resistance increases, the starting torque increases, but the running speed decreases and the losses increase. The increase of starting torque is obtained at the expense of lower running efficiency.

4.3.3 Performance Calculation Method

Given the equivalent circuit values, calculate the machine steady state performance at a fixed speed as follows:

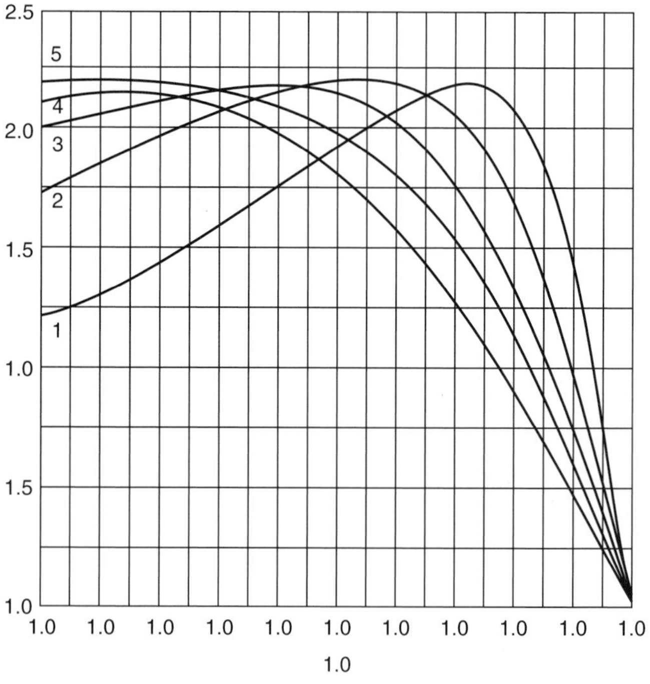

FIGURE 4.20 Torque–Slip Curves at Different r_2

1. Find synchronous speed as:

$$\omega_s = \frac{4\pi f_0}{P}. \quad (4.56)$$

2. Find the slip as:

$$s = \frac{\omega_s - \omega_m}{\omega_s}. \quad (4.57)$$

3. Find the rotor impedance as:

$$Z_2 = \frac{r_2}{s} + jX_2. \quad (4.58)$$

4. Find the equivalent secondary impedance as:

$$Z_{eq} = R_{eq} + jX_{eq} = jX_m \frac{r_2/s + jX_2}{r_2/s + j(X_2 + X_m)}. \quad (4.59)$$

5. Find the total impedance per phase as:

$$Z_{\text{in}} = r_1 + R_{eq} + j(X_1 + X_{eq}). \quad (4.60)$$

6. Find the phase current as:

$$I_1 = \frac{V_1}{Z_{\text{in}}}. \quad (4.61)$$

7. Find the power factor of the phase current as the cosine of the angle of the current.
8. Find the total input power as:

$$P_{\text{in}} = 3V_1 I_1 \cos\theta. \quad (4.62)$$

9. Find the stator copper loss as:

$$P_{scl} = 3I_1^2 r_1. \quad (4.63)$$

10. Find the air-gap power as:

$$P_g = 3I_1^2 R_{eq}. \quad (4.64)$$

11. Find the rotor copper loss as:

$$P_{rcl} = sP_g. \quad (4.65)$$

12. Find the developed mechanical power as:

$$P_{\text{mech dev}} = (1 - s)P_g. \quad (4.66)$$

13. Find the output torque as:

$$T = \frac{P_g}{\omega_s}. \quad (4.67)$$

14. Find the output power as:

$$P_{\text{out}} = P_{\text{mech dev}} - P_{\text{rot}}, \quad (4.68)$$

where P_{rot} are the rotational losses (windage and friction).
15. Find the output torque as:

$$T = \frac{P_{\text{out}}}{\omega}. \quad (4.69)$$

16. Find the efficiency as:

$$\eta = \frac{P_{\text{out}}}{P_{\text{in}}}. \quad (4.70)$$

4.3.4 NEMA Motor Types

NEMA, the National Electrical Manufacturers' Association, has developed standard design classifications for motors with par-

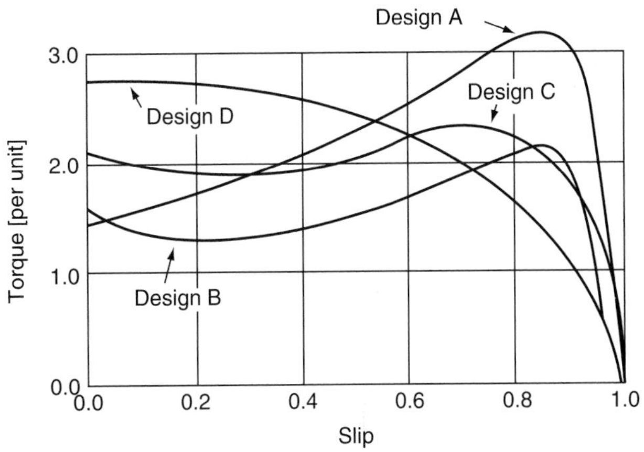

FIGURE 4.21 Typical Torque Speed Curves for NEMA Motors

ticular speed–torque characteristics. Classes A through D are common squirrel cage designs. Typical torque–slip curves are shown in Figure 4.21. The motor designs differ mainly in the design of the rotor bars. As explained, by increasing the rotor resistance, the starting torque is increased and the slip at pull-out is reduced. By changing the rotor reactance, the starting current can be regulated. Class A machines have rotors with large bar cross sections, which give large starting currents, high efficiency, and relatively low starting torque. A class D motor, on the other hand, will have a smaller bar cross section and, therefore, has high starting torque, lower efficiency, and lower starting current.

4.3.5 Motor Tests

There are two basic tests used to obtain machine parameters. These are the **no-load test** and the **locked rotor test**. In the no-load test, the motor is run at rated voltage with no mechanical load attached. In this case, the slip is approximately zero, and very little current circulates on the rotor. There is a small amount of mechanical power required, enough to overcome friction and windage losses. Assuming that the rotor current is negligible, then, referring to the equivalent circuit, the stator current circulates in the mutual branch representing the excitation current. The stator resistance is measured independently with a bridge. The stator voltage, current, and power are measured. Since the stator resistance is known, the stator winding losses, $3I_1^2 r_s$, can be subtracted from the input power. The remaining losses are associated with the core loss and mechanical losses. The total reactance $X_{NL} = X_1 + X_m$ is now found as follows:

$$Z_{NL} = \frac{V_1}{I_1}. \quad (4.71)$$

$$R_{NL} = \frac{P_{NL}}{3I_{NL}}. \quad (4.72)$$

$$X_{NL} = \sqrt{Z_1^2 - R_{NL}^2}. \quad (4.73)$$

The second test is the locked (or blocked) rotor test. In this case, a balanced three-phase voltage is applied to the stator while the rotor is constrained so as not to turn. This test is done at low voltage to keep the torque low and reduce the losses. Because the voltage is low, the core flux density is reduced, and the core loss is also very low. At zero speed, the input voltage, current, and power are measured. The stator winding losses are subtracted as for the previous no-load test. There is no mechanical power and little core loss. The series reactances are much smaller than the magnetizing reactance, and it can be assumed the magnetizing current is negligible. Then, the impedance is found as $Z_{LR} = \frac{V_1}{I_1}$. After subtracting the stator losses from the measured power, the rotor resistance is found as:

$$r_2 = \frac{P_{LR} - P_{scl}}{3I_1^2}. \quad (4.74)$$

The series reactance is then found as:

$$X_{LR} = X_1 + X_2 = \sqrt{Z_{LR}^2 - (r_1 + r_2)^2}. \quad (4.75)$$

The total reactance is then divided into a stator and a rotor part by using the following empirical distributions. If the total reactance is X_{LR}, then for class A motors, $X_1 = X_2 = 0.5X_{LR}$. For class B motors, $X_1 = 0.4X_{LR}$ and $X_2 = 0.6X_{LR}$. For class C motors, $X_1 = 0.3X_{LR}$ and $X_2 = 0.7X_{LR}$. For class D motors, $X_1 = X_2 = 0.5X_{LR}$. For wound rotor machines, $X_1 = X_2 = 0.5X_{LR}$. Using these values of leakage reactance, the magnetizing reactance is found as:

$$X_m = X_{NL} - X_1. \quad (4.76)$$

4.4 Synchronous Machines

Essentially all of the world's electric power is generated by **synchronous machines**, the device in the power system where mechanical energy is converted into electrical energy. Further, much of the power system control is done via the synchronous machine. Synchronous motors have their applications as well, especially when constant speed operation is important. In terms of energy conversion, though, the synchronous machine is far more important as a generator, and this section describes the generating aspect of synchronous machines.

In its most common configuration, the synchronous machine stator has a three-phase winding in a laminated core. In this sense, the stator is the same as the stator in the three-phase induction motor. The rotor of the synchronous machine has a dc winding (the field) or permanent magnets. In steady-state, the rotor is moving at synchronous speed so an observer on the rotor will see a constant field. For this reason, the rotor may or may not be laminated. For high-speed machines, the rotor is often made of a solid steel forging in which slots are machined to hold the field winding. These are called round rotor machines. For lower speed machines, the rotors are usually built from laminations. A salient pole is constructed, and the field winding places around the pole.

4.4.1 Three-Phase Round Rotor Generator

A simplified analysis of a round rotor generator illustrates the important points of synchronous machine operation (see Figure 4.22). The machine consists of a balanced three-phase stator winding and a dc rotor winding. Assuming a steady-state balanced operation, it is not necessary to consider the effect of damper windings. The rotor to stator mutual inductances are a function of the rotor angle. With a round rotor machine, the stator to stator mutual inductances and the stator self-inductances are constant:

$$\begin{aligned} L_{af} &= L_{afm}\cos(\theta). \\ L_{bf} &= L_{afm}\cos\left(\theta - \frac{2\pi}{3}\right). \\ L_{cf} &= L_{afm}\cos\left(\theta - \frac{4\pi}{3}\right). \end{aligned} \quad (4.77)$$

The flux linkages of the windings are as follows:

$$\begin{aligned} \psi_a &= L_{aa}i_a - L_{ab}(i_b + i_c) + L_{af}i_f. \\ \psi_b &= L_{aa}i_b - L_{ab}(i_c + i_a) + L_{bf}i_f. \\ \psi_c &= L_{aa}i_c - L_{ab}(i_a + i_b) + L_{cf}i_f. \\ \psi_f &= L_{ff}i_f + L_{af}i_a + L_{bf}i_b + L_{cf}i_c. \end{aligned} \quad (4.78)$$

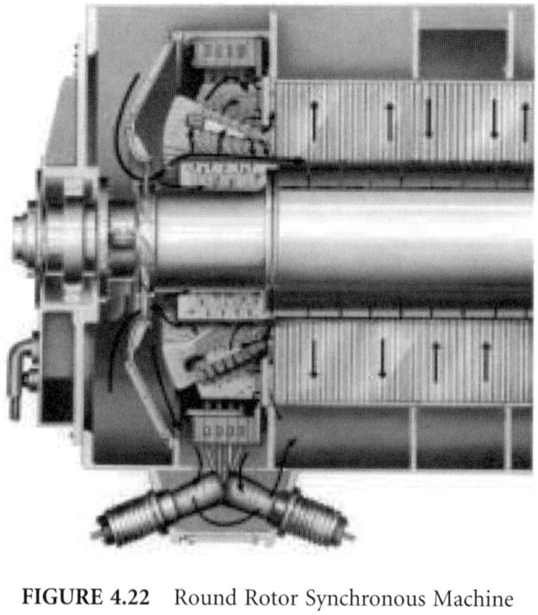

FIGURE 4.22 Round Rotor Synchronous Machine

Substituting the values of rotor to stator mutual inductances yields the following equations:

$$\psi_a = L_{aa}i_a - L_{ab}(i_b + i_c) + L_{afm}i_f \cos(\theta).$$
$$\psi_b = L_{aa}i_b - L_{ab}(i_c + i_a) + L_{afm}i_f \cos(\theta - \frac{2\pi}{3}).$$
$$\psi_c = L_{aa}i_c - L_{ab}(i_a + i_b) + L_{afm}i_f \cos(\theta - \frac{4\pi}{3}).$$
$$\psi_f = L_{ff}i_f + L_{afm}(i_a \cos(\theta) + i_b \cos(\theta - \frac{2\pi}{3}) + i_c \cos(\theta - \frac{4\pi}{3})). \quad (4.79)$$

The terminal voltages are written as:

$$e_{an} = -R_a i_a - \frac{d\psi_a}{dt}.$$
$$e_{bn} = -R_a i_b - \frac{d\psi_b}{dt}.$$
$$e_{cn} = -R_a i_c - \frac{d\psi_c}{dt}. \quad (4.80)$$
$$e_f = R_f i_f + \frac{d\psi_f}{dt}.$$

For steady-state operation, the rotor velocity is constant as:

$$\theta = \omega t + \theta_0, \quad (4.81)$$

and the currents are balanced:

$$i_a + i_b + i_c = 0. \quad (4.82)$$

Using these, yields the following:

$$e_{an} = \sqrt{2}E_i \sin(\omega t + \theta_0) - R_a i_a - (L_{aa} + L_{ab})\frac{di_a}{dt}.$$
$$e_{bn} = \sqrt{2}E_i \sin(\omega t + \theta_0 - \frac{2\pi}{3}) - R_a i_b - (L_{aa} + L_{ab})\frac{di_b}{dt}.$$
$$e_{cn} = \sqrt{2}E_i \sin(\omega t + \theta_0 - \frac{4\pi}{3}) - R_a i_c - (L_{aa} + L_{ab})\frac{di_c}{dt}. \quad (4.83)$$

In these equations, E_i, the internal voltage, is the component of voltage induced by the field current.

The Phasor Diagram

In phasor form, the voltage equation can be written as:

$$E_{an} = E_i - (R_a + j\omega(L_{aa} + L_{ab}))I_a. \quad (4.84)$$

The synchronous reactance can be defined as:

$$X_s = \omega(L_{aa} + L_{ab}). \quad (4.85)$$

Hence, the result is as written here:

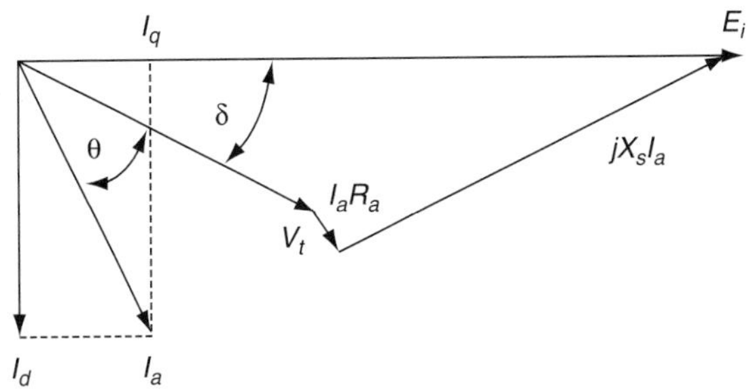

FIGURE 4.23 Phasor Diagram

$$E_{an} = E_i - (R_a + jX_s)I_a. \quad (4.86)$$

The phasor diagram is illustrated in Figure 4.23.

V Curve and Capability Curve

Two standard curves that describe the performance and operating limits of the synchronous machine are the ***V* curve** and the **reactive capability curve**. The *V* curve is a plot of the armature current versus the field current at constant real power. Considering the voltage to be fixed at rated conditions or 1.0 per unit, then the family of currents whose projection onto the voltage vector is the same all produce the same real power. The power factor is varying and the minimum current occurs at a unity power factor. In the phasor diagram in Figure 4.23, the field current is high in the overexcited (generator lagging power factor) condition and low in the underexcited (generator leading power factor) cases. Plotting the armature current versus the field current in a small range around unity power factor, which is the normal operating range of the machine, the curve is *V* shaped. A typical curve is shown in Figure 4.24 for different values of real power.

The reactive capability curve is used to indicate the safe thermal operating range of the synchronous machine. The machine is assumed to be operating at rated voltage. First consider stator heating, and operate at the maximum allowable stator current (1.0 per unit). The power generated is then:

$$VI^* = P + jQ. \quad (4.87)$$

Squaring both sides yields:

$$(VI)^2 = P^2 + Q^2. \quad (4.88)$$

Equation 4.88 represents a circle on the *P*, *Q* plane. If the voltage is constant, then the radius is proportional to the stator current. Power can be safely generated operating anywhere in the semicircle centered on the origin.

Supplying more reactive power to the system causes the rotor current to increase. Eventually, a heating limit of the rotor is reached. This condition can be analyzed by considering the locus of operating points having constant field current and

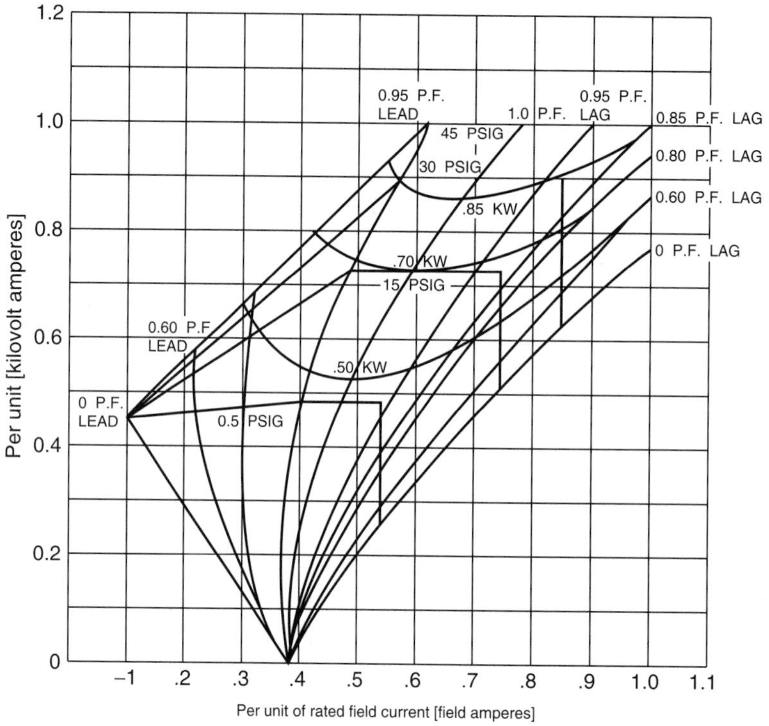

FIGURE 4.24 *V* Curve for a Turbine Generator

therefore constant internal voltage, E_i. The internal voltage transmits power to the terminals through an impedance X_s. The angle between these voltages is δ. Then:

$$P + jQ = VI^* = V\angle 0\left(\frac{E_i\angle\delta - V\angle 0}{jX_s}\right)^* \tag{4.89}$$

$$= V\angle 0 \frac{(E_i \cos\delta - V\angle 0 + jE_i \sin\delta)^*}{jX_s}$$

$$= \frac{VE_i \sin\delta}{X_s} + j\frac{VE_i \cos\angle\delta - V^2}{X_s}. \tag{4.90}$$

Rearranging these equations results in:

$$P^2 + \left(Q + \frac{V^2}{X_S}\right)^2 = \left(\frac{VE_i}{X_s}\right)^2 = K. \tag{4.91}$$

The locus of constant field current is also a circle whose center is displaced on the negative Q axis by V^2/X_s and whose radius is VE_i/X_s. The region of safe operation of the rotor is inside this second circle. This rotor limit circle intersects the stator limit circle so that the operator must cut back power when moving to more lagging power factor to protect the rotor.

There is another limitation placed on operation. When moving to the leading power factor, the rotor and stator fields are reinforcing each other at the end of the core. Therefore the operation is limited in the leading power factor regime to keep the end of the stator core from overheating. The capability curve in this region does not represent a locus of maximum loss but is an approximation derived from temperature measurements on generators under test. The complete capability curve is shown in Figure 4.25.

4.4.2 Machine Tests

There are two standard tests that are performed on synchronous machines to determine their steady state parameters and verify the design. These are the **open circuit saturation test** and the **steady-state short circuit test** (also known as the synchronous impedance test). In the open circuit test, the machine is run at rated speed with the stator winding open circuited. The field current is varied, and the terminal voltage is plotted as a function of the field current. A typical test is shown in Figure 4.26.

As the field current rises the stator voltage increases proportionally. For most machines at about 80% of rated voltage, the curve becomes nonlinear, showing the effects of magnetic saturation. If the straight line characteristic is continued, the so-called **air-gap line** results, which is shown dotted in the figure.

In the short circuit test, the rotor is turned at rated speed with a three-phase short circuit on the stator. As before, increasing the field current allows the plotting of the stator current versus the field current. A typical short circuit characteristic is also shown in Figure 4.26. The short circuit characteristic is linear. The flux density in the machine is

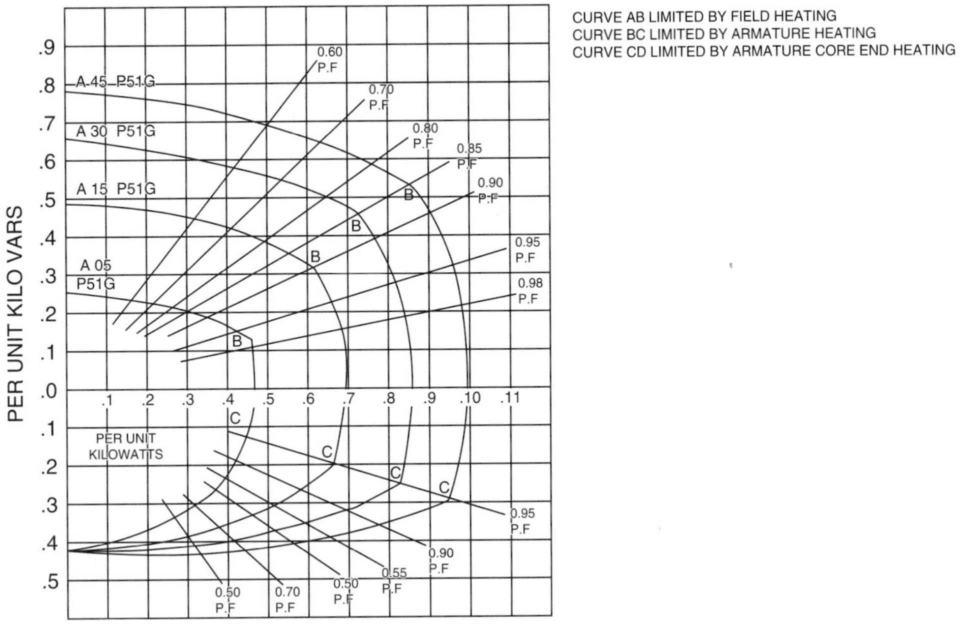

FIGURE 4.25 Reactive Capability Curve

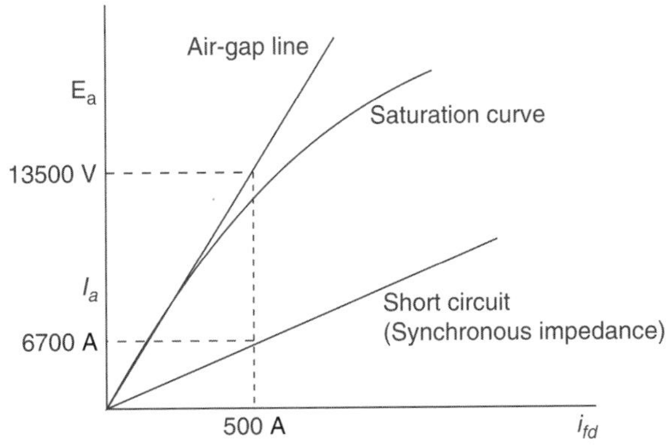

FIGURE 4.26 Open Circuit Test

very low since the terminal voltage is zero. The air-gap flux is sufficient to produce a voltage to overcome the leakage reactance drop.

From these two curves, the synchronous reactance can be found. For the open circuit condition, the terminal voltage is equal to the internal voltage E_i.

For the same field current, take the stator current from the short circuit test and find X_s as $X_s = E_i/I_a$. Referring to the figure, the synchronous reactance is $X_s = 13{,}500/6{,}700 = 2.01\ \Omega$.

4.5 Single-Phase Induction Machines

The electrical supply to most residences and many businesses is single phase, and, therefore, most appliances use **single-phase motors**. A single-phase winding produces a standing wave, which may be resolved into a forward and a backward rotating wave. These waves each produce a torque versus slip curve as shown in Figure 4.27. The net torque is the sum of the forward and backward components. As indicated by the figure, there is no starting torque. If we can manage, however, to start the motor (in either direction), it will continue to accelerate in that direction. In single-phase induction motors, a starting torque can be produced by using two stator windings situated 90° electrical distance apart. The power factor of these windings is different, producing a phase shift between the two currents and therefore a traveling wave. Once the motor begins to turn, one of the windings may be disconnected.

4.5.1 Split-Phase Motors

The **split-phase motor** has a main winding and an auxiliary winding that are situated 90° apart. The auxiliary winding has smaller wire and fewer turns than the main winding. When

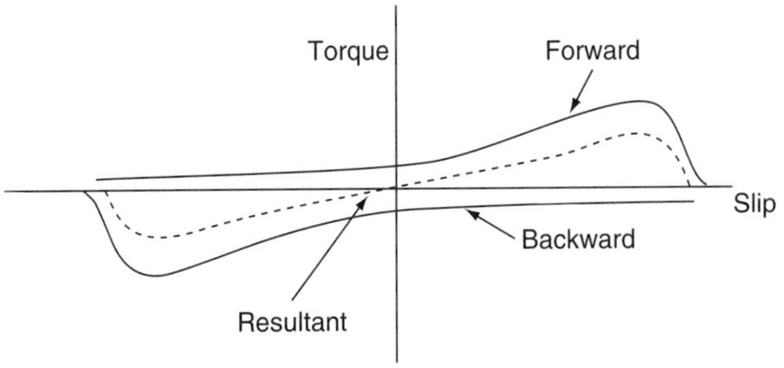

FIGURE 4.27 Single-Phase Induction Motor Torque Characteristic

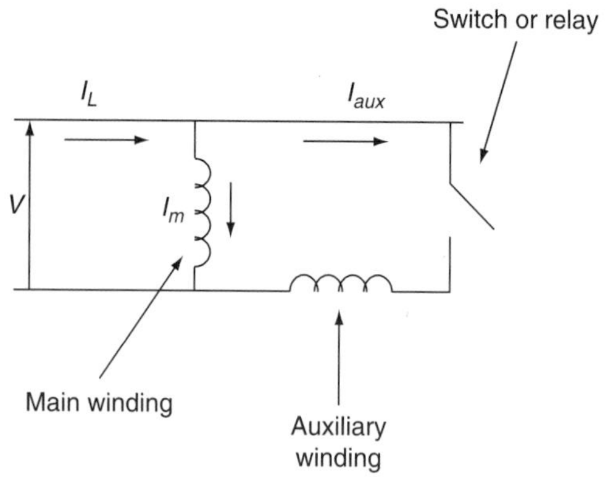

FIGURE 4.28 Split-Phase Motor

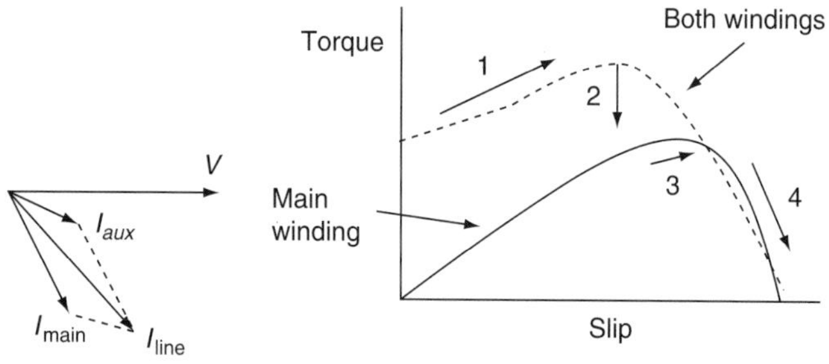

FIGURE 4.29 Split-Phase Motor Performance

connected in parallel, the current in the auxiliary winding leads the current in the main winding and produces a traveling wave. Once the motor comes up to speed, the auxiliary winding is switched out (Figure 4.28). Figure 4.29 shows the phasor diagram and torque speed curve. With both windings connected, the torque–slip characteristic follows path 1. When the auxiliary winding is disconnected, the characteristic moves via path 2 to the main winding curve and then along paths 3 and 4.

4.5.2 Capacitor Start Motor

In the capacitor start motor, a capacitor is used in the auxiliary winding to change the power factor to leading so that the auxiliary current can lead the main winding current by approximately 90°. The starting torque is higher and the starting current is lower than in the case of the split-phase motor.

4.5.3 Two-Value Capacitor and Permanent Capacitor Motors

The **two-value capacitor motor** is similar to the capacitor start motor, but the auxiliary winding is designed for continuous operation. The impedance of the auxiliary winding depends on

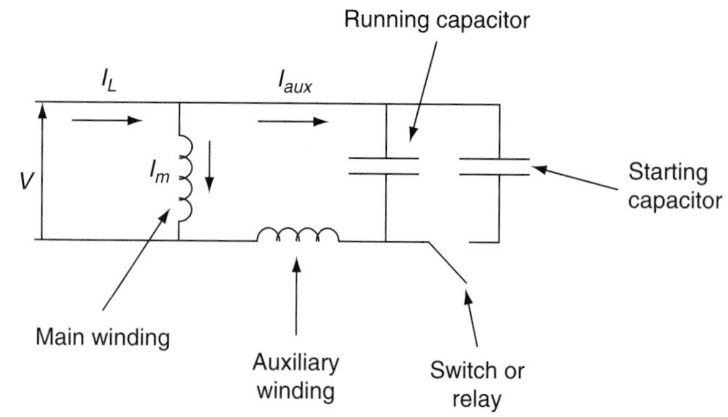

FIGURE 4.30 Two-Value Capacitor Motor

the slip, so a smaller capacitor is needed for optimal running performance. Suitable values for starting and running are achieved by using a short-time rated capacitor in parallel with the running capacitor. The short-time rated capacitor is removed when the motor comes up to speed. The configuration of the two-value capacitor motor is illustrated in Figure 4.30. Because of the running capacitor, this motor basically has a two-phase supply and, therefore, runs more efficiently and quietly.

To avoid the extra expense of two capacitors and a switch, another type of motor, the **permanent capacitor motor**, uses a single capacitor in series with the auxiliary winding. This permanent split capacitor motor does not give optimum starting or running performance.

4.5.4 Shaded-Pole Motor

The shaded-pole induction motor does not use an auxiliary winding. Rather, there is a shorted turn or turns on part of the stator pole. The portion of the flux crossing the air gap through this turn produces an eddy current reaction, which effectively delays the flux. This produces a traveling wave that enables the motor to start. The starting torque is not as great as in the capacitor motors, but these motors are very reliable and inexpensive. They are often used in appliances where efficiency and torque requirements are not great.

4.5.5 Universal Motors

The **universal motor** is configured like a series dc motor. In fact, the universal motor can be run from a dc or ac supply. If the motor is connected to an ac supply, the armature current and the field flux change direction at the same time, and the torque is always in the same direction. The universal motor typically has a laminated rotor and stator core due to the ac operation. These motors are useful for high-speed, low-torque operations.

5

High-Voltage Transmission

Ravi S. Gorur
Department of Electrical Engineering, Arizona State University, Tempe, Arizona, USA

5.1 Introduction

Reliable and economical transport of power from generating stations to the load centers is accomplished by overhead lines operating at a high voltage. Transmission by underground cables is limited to special cases and for short spans as in water crossings and in metropolitan areas. Currently in the United States, there exists more than 200,000 miles of overhead transmission lines and about 3500 miles of underground transmission cables. Voltage in excess of 69 kV is considered **transmission**, and voltage below 69 kV is considered **distribution**.

The conductor current for a given Kilovoltampere [kVA] rating is inversely proportional to the voltage used; hence the higher the voltage, lower is the current and even lower is the power losses (I^2R) occurring during the transmission process. In the United States, the highest transmission voltage used is 765 kV; in Canada, it is 735 kV; in Western Europe, it is 400 kV; and in Japan, a 1000-kV line has been recently built. In most countries, the transmission voltage varies between 220 and 500 kV.

It is interesting to note that even though extensive research on higher voltages such as 1500 kV was performed in the 1970s in the United States, Canada, and Russia, there are no commercial lines operating at such voltages simply because of the exorbitant cost of equipment required. The electric stress occurring on the insulation varies nonlinearily with the voltage; hence, design requirements at such high voltages are not a simple extrapolation of the values used at lower voltages.

The demand for energy use is largely driven by increased urbanization. It is estimated that for a 2-fold increase in the energy use (usually expressed in terms of GWH or TWH), the transmission line capacity will increased by about 50%. This manifests either as an upgrade in the voltage or as new line construction.

The use of energy has increased slightly for industrialized countries but significantly for developing countries in the last 10 to 15 years. This is shown in Table 5.1. It can therefore be expected that the bulk of new activity, with respect to new transmission line construction, will be in the developing countries.

TABLE 5.1 Expected Increase in Energy for Various Regions of the World

Geographic region	Expected increase in energy use by 2010	Comments
North America	7.5%	Lower in Canada and Mexico, compared to USA
Western Europe	13%	
Industrialized Pacific	8%	Japan, Australia, New Zealand
Eastern Europe and former Soviet Union	10%	Difficult to get reliable information
China	40%	
India	25%	
Other parts of developing Asia	21.5%	S. Korea, Thailand, Taiwan, Phillipines, Indonesia, Malaysia
Africa	20%	
Middle East	10%	
Central and South America	15.5%	Largest growth to come in Brazil and Argentina

TABLE 5.2 Major Transmission Projects (by 2010)

Country or region	Voltage and length	Comments
Vietnam	500 kV, 500 km	500 kV line of 2400 km by 2010, 1.5 billion/year USD investment needed
India	400 kV, 436 km 220 kV, 600 km	
Brazil	500 kV, 2800 km	A 25 billion (USD) total investment anticipated
Southern Africa	400 kV, 400 km	Operated by ESKOM
Northern Africa	400 kV, Few thousand km (not clear)	Mediterranean Power Grid to be completed by 2015
China	220 kV, 6600 km (and above)	Some dc lines also created
Nigeria	330 kV, 144 km	Estimated as a $40 million USD project
Philippines	Build 24 new lines, rebuild 13 existing lines 110–230 kV (length not known but probably a few thousand kilometers)	Estimated as a $165 million USD project
North America	About 230 kV, 1000 km and 500 kV, 1000 km	A 500-kV line linking several states (in the planning stages)

From Table 5.1, it should be obvious that majority of transmission line construction should occur in China, India, and Brazil, followed by Africa and the rest of South America.

A listing of new major construction projects planned in the next 5 years in several of these regions is shown in Table 5.2. Based on Table 5.2, a cumulative length of about 15,000 km of transmission lines is to be built by 2010.

5.2 Design Considerations for Overhead Lines

The most popular conductor type used for transmission lines is the aluminum conductor steel reinforced (ACSR). Other conductor types such as all aluminum conductor (AAC), and all aluminum alloy conductor (AAAC) are also used and offer advantages such as higher operating temperature and low weight.

It is desirable to increase current carrying capacity of a conductor without increasing its weight (and hence sag). This has been possible by the substitution of the steel core of the conductor with composite materials, such as carbon fibers and fiberglass. There are claims that this permits a 2- to 3-fold increase in the electrical power transmitted. This increase is even more impressive when considering that it can be achieved without changing the towers. But it remains to be seen at what cost. Obviously, the new composite conductors are more expensive than the conventional conductors.

The conductor size is determined by the current. The electric field on the conductor's surface and at any other point in space is determined by the voltage. For overhead lines, clearances to ground are determined both by the voltage and current. Heating produced by resistive heating of the conductor causes it to sag between the support structures. The electric field on the surface of the conductor is inversely proportional to the conductor diameter.

For transmission lines below 230 kV, conductor size is determined by thermal heat dissipation of the conductor, whereas for lines operating at voltages greater than 230 kV, the conductor size is determined primarily by electric field considerations. Higher conductor diameters are required to keep the electric field on the surface of the conductor below the inception values that trigger corona. This is most effectively achieved by the use of bundled conductors. It is desirable to keep the maximum surface electric stress below 15 kV/cm under dry conditions.

A **corona** is the phenomenon of air breakdown when the electric stress at the surface of a conductor exceeds a certain value. At higher values, the stress results in luminous discharge. Corona produces undesirable effects like radio and television interference, ozone, audible noise, and power loss. The use of grading rings on dead-end towers is a common practice to reduce corona. When nonceramic insulators are used, the use of corona rings at the terminals of the insulators is a common practice for voltages above 230 kV. These along with ceramic or glass insulators are normally not used.

If the conductor diameter is inadequate to prevent corona, the remedial measures to fix the situation, although available, are very expensive. These include increasing the number of conductors per phase, applying special surface treatment of conductors, and isolating line sections with filters. Therefore, it is important to ensure that the conductor details are chosen after a careful review.

The clearance to ground is determined by the induced electric field with a magnitude that reduces with distance from the conductor. The clearances of the line from the tower are dependent on overvoltage produced by lightning surges and switching events. Transmission lines at 230 kV and below are governed by lightning overvoltages, whereas for lines at 345 kV and above, overvoltage produced by switching surges is the determining factor.

The magnitude of lightning produced overvoltage is mainly dependent on the lightning stroke current and impedence of the object struck. For a stroke to a tower, the lightning strike creates a traveling wave that travels down the tower and is reflected at the footing and is reflected back. The sign of the reflected wave depends on the relative tower footing resistance. If the footing resistance is high, a voltage wave will be reflected, which will be about double. This could raise the tower voltage to a level that a flashover could occur across the insulator. This is referred to as **back flashover**.

It is important to keep the tower footing resistance as small as possible as the reflected wave then will have the opposite sign and will reduce the tower voltage. Tower footing resistance can be reduced by using driven rods and counterpoises.

The magnitude of switching overvoltage can be reduced by the use of circuit breakers with preinsertion resistors and surge arresters. It has been realized that the preinsertion resistors are maintenance intensive and actually can be dispensed when using metal–oxide surge arresters.

The appropriate clearances of the conductor to the tower and ground are achieved by proper selection of insulators. More details about the different insulator types and important aspects of insulators are provided below.

5.2.1 Types of Outdoor Insulators

An insulator is actually a system of components consisting of the dielectric, terminal electrodes or end-fittings, and internal parts that help attach the dielectric to the electrodes. There are many details involved in these components and constructional methods. However, insulators are recognized largely by the classification of the dielectric employed. There are three main classes of dielectrics that have been used for outdoor HV insulator construction: **porcelain, glass**, and **polymer**. Hence, the nomenclature of porcelain, glass and polymeric insulators is well-known. Polymeric insulators are also commonly known by other names, such as composite (in Europe) and non-ceramic (in North America).

Porcelain and glass insulators have been used for more than 100 years. The widespread use of composite insulators began in North America during the 1970s. Figure 5.2 shows a schematic of these insulators for use on outdoor lines (Gorur *et al.*, 1999).

The most widely used type of porcelain and glass insulator, worldwide, is the **cap and pin insulator** where each unit or bell is connected to each other by metal hardware. The number of bells is dependent on the voltage class. In Europe, the use of porcelain long-rod insulators, where intermediate metal electrodes are significantly reduced or eliminated, is equally popular. In comparison, composite insulators have a single piece of dielectric, which is the fiberglass rod that is covered by a polymer housing and is attached to the metal terminals.

The following dimensional details are relevant to insulators for transmission. Figure 5.1 shows an illustration of the important dimensions.

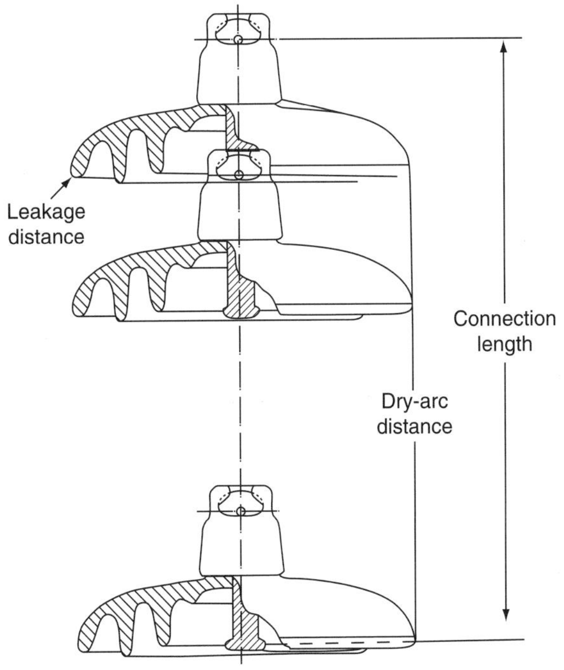

FIGURE 5.1 Illustration of Important Dimensions for Insulators

- **Dry-Arcing or strike distance**: The shortest distance through the surrounding medium between terminal electrodes, or the sum of the distances between intermediate electrodes, whichever is shorter
- **Connection length**: The shortest distance between the conductor and the support structure, including the strike distance plus the hardware dimensions
- **Leakage or creepage distance**: The sum of the shortest distances measured along the insulating surfaces between the conductive parts
- **Protected leakage or creepage distance:** Parts of the insulator surface that are not directly exposed to natural elements like sun and rain. (In porcelain and glass insulators, the ribs on the underside of the insulator contribute to the protected leakage distance.)

5.3 Stresses Encountered in Service

Outdoor insulators are subjected to a variety of stresses in service, such as mechanical, electrical, and environmental stresses. These stresses act in unison. The exact nature and magnitude of these stresses varies significantly and depends on the details of insulator design, application, and location. For example, suspension and dead-end insulators encounter a tensile load due to the weight and tension of the conductor. Wind and ice impose additional loading. Insulators used for a supporting station apparatus encounter a compressive mechanical load. Line-post insulators are subjected to a cantilever or bending load in supporting the conductor. These loads are to be expected under steady-state conditions. In additional, transient loading conditions can be generated. A torsional or a

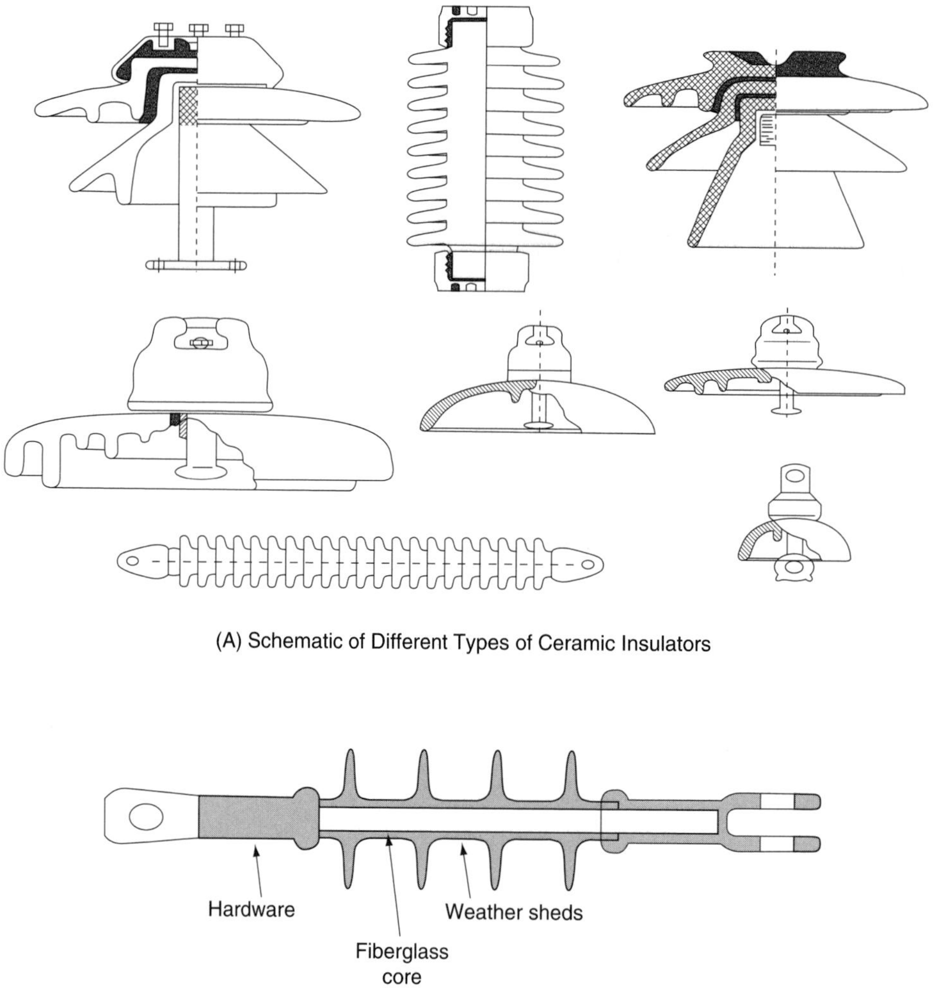

(A) Schematic of Different Types of Ceramic Insulators

(B) Schematic of Nonceramic Insulator

FIGURE 5.2 Schematic Diagram of Insulators for Outdoor Lines twisting type of load can be experienced by line insulators during line construction. Vibrational loads are imposed on insulators due to conductor vibration and movement. Shock (or impact) loading is also possible during natural events like earthquakes, ice-shedding, or man-made events, like vehicular impact on poles and vandalism (gun shots).

Electrical stresses include those imposed by power frequency nominal operating voltage, which is a steady-state stress. Voltage surges generated by lightning or switching operations impose a higher albeit momentary stress on the insulator. In the event of an insulator flashover, the insulator is subjected to a large fault current (several kiloamperes) at power frequency in the form of an arc, called **power arc**, until the protection isolates the fault.

Outdoor environmental conditions vary over a wide range. Temperature impacts the insulation properties of all materials as the conductivity increases with temperature. For polymers that are organic materials, ultraviolet (UV) radiation from sunlight can break certain chemical bonds and cause cross-linking on the surface, resulting in surface degradation. Moisture in any form (e.g., rain, dew, fog, melting ice and snow) lowers the surface insulation resistance significantly from the dry state. In the presence of contamination, the surface resistance is reduced even more drastically. Altitude or the elevation above sea level affects the insulation properties. Higher altitudes reduce the air density, hence, weakening the surface insulation strength.

All of these stresses could be acting on the insulators in various combinations. Hence, it is clear that the insulators have to perform under a wide range of service conditions. Needless to say, only insulators that are designed while suitably considering all these stresses will work satisfactorily in the field for many years.

5.4 Insulator Performance

5.4.1 Electrical Performance

Dielectric material is largely responsible for the electrical performance of an insulator. It is important to distinguish between bulk or volume properties and surface properties. The volume dielectric strength is determined by defects in the form of impurities and voids in the system. The presence of these defects serves to provide sites for electrical stress concentration, which could lead to the formation of a permanent failure path in the system. Failures along the bulk of the material, normally called **punctures**, are permanent in nature. The surface dielectric strength is determined largely by surface deposits and moisture. Resistivity values, which are indicative of the dielectric strength, are typically greater than $10^{10}\ \Omega$ per square for the bulk material (Looms, 1988). Under dry conditions, such high surface resistance is also obtained. In the presence of humidity, however, surface resistance values are lowered by several orders of magnitude and are even further lowered in the presence of ionic contaminants on the surface.

For dielectric materials that do not contain large voids or impurities, which is typical of a well-made insulator, the electric stress required for failure via the surrounding air medium is much lower than for bulk failure. Such failures are called **flashovers**. The arc produced by the flashover is usually away from the surface of the dielectric. Therefore, as far as flashovers are concerned, porcelain, glass, and composite insulators **are self-restoring types of insulation.**

Whether an insulator fails by surface flashover or a puncture depends on the magnitude and duration of the electric stress applied, insulator dimensions, and defects in the material. Breakdown requires the formation of an ionized channel, and this channel has to be established within the duration of the applied voltage. If the insulator is defective, (i.e., has large voids or impurities), puncture can be caused by extremely short duration and large magnitude pulses. Lightning surges, which have a rise time in the microsecond range, do not cause puncture if the insulator is sound. Similarly, switching surges normally do not cause puncture. Both lightning and switching surges can cause flashover if they have adequate magnitude, even under dry conditions. Longer duration stresses, such as those imposed by power frequency, do not result in a puncture. A flashover is, however, possible in a wet and contaminated environment at the nominal operating voltage, as illustrated in Figure 5.3.

5.4.2 Mechanical Performance

The mechanical performance of outdoor insulators is determined by all the main components of the insulator, namely, dielectric and end-fittings, as well as the details of internal attachment of the dielectric with the end-fittings. Failure of the insulator to perform mechanically results in dropping of the conductors. This type of insulator failure is not acceptable as it leads to a prolonged interruption of power, possible personnel injury and equipment damage.

5.4.3 Role of Insulators on Power System Reliability

Power system apparatuses are subjected to faults from both man-made and natural causes. In a system that uses many types of apparatus with costs that vary widely, it is important that expensive apparatuses are protected from faults, even if it is at the expense of a cheaper device. Outdoor insulators, especially the line insulators, are one of the least expensive devices in the power system. Therefore, it is understandable that to protect a more expensive apparatus, such as a transformer or circuit breaker in a station, a line insulator must fail first under *overvoltages*. When an insulator fails, it is required that the failure mode is a flashover rather than a puncture.

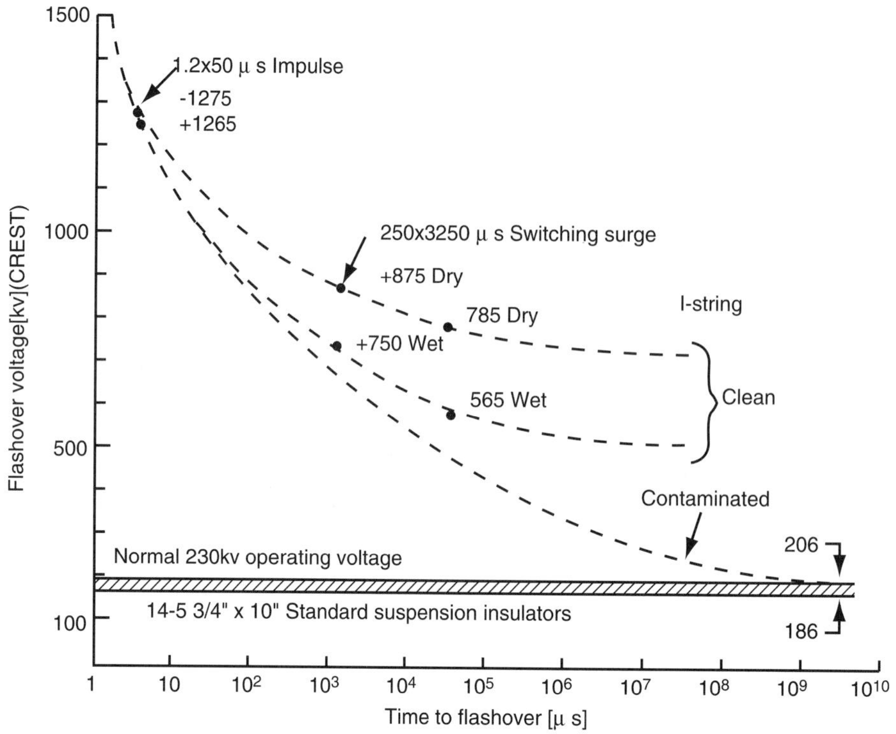

FIGURE 5.3 Illustration of Reduction in Flashover Voltage with Increasing Duration of Voltage Application. Note that the contaminated conditions represent the most onerous conditions for outdoor insulator operation.

Voltage surges produced by lightning and/or switching operations are a major factor to be considered in the electrical design of the insulator. Insulators should be designed so that they do not fail at surge voltage magnitudes that are lower than the value for which the power system apparatus has been designed. For higher magnitude surges, the insulator should be a flashover. The determining values of these surges are defined by the **basic lightning impulse insulation level** (BIL) and **basic switching impulse insulation level** (BSL).

Insulator design for voltage 230 kV and below is largely dependent on lightning-produced surges, whereas for 345 kV and above, the design is dependent on switching surges. However, regardless of the voltage class, in locations where contamination is a serious problem, insulator design is determined by the required contamination performance.

Flashovers usually cause momentary interruption of power flow in the circuit. Interruption may cause customer interruption depending on the system configuration. Such interruptions, although a nuisance, can be tolerated in rural locations. In urban areas with high-tech industries that use semiconductors and in pharmaceutical and automotive fields even temporary interruptions are not acceptable because they lead to huge financial losses from lost production, jammed machinery, and loss of process control.

Therefore, today's power system reliability is to a large measure linked with the avoidance of insulator failure.

Flashover due to surges is determined by the shortest spacing in air between the electrodes of an insulator, called the **dry-arcing distance**. Moisture has little effect on the flashover voltage due to lightning surges. The flashover voltage under surges is dependent on the dry-arcing distance. The flashover under contaminated conditions, however, is dependent on the **leakage** or **creepage distance**.

If the dielectric material is not altered during service, electrical characteristics like power frequency wet withstand or flashover, lightning and switching surges are defined by dry arcing distance. The shape or profile of the dielectric, which determines the leakage distance, is important for the contamination performance.

To increase the leakage distance and to help maintain certain parts of the insulator dry, it is common to see corrugations on the underside of porcelain and glass cap and pin insulators. The required dry-arcing and leakage distance is obtained by stacking several units, the number being dependent on the voltage level, contamination severity, and the profile of the insulator. Many shapes of porcelain and glass insulators have been developed, as shown in Figure 5.2.

For composite insulators, because there are no intermediate electrodes, the required leakage distance can be obtained with relatively simple shapes. The long-rod porcelain insulator, used commonly in Europe, also has a simple shape, with no corrugations or under-ribs.

5.4.4 Mechanical and Electrical Ratings of Insulators

Because insulators perform both mechanical and electrical functions, they are identified by a combined **mechanical and electrical** (M and E) **rating**. The M and E rating is defined as the mechanical load at which the insulator fails to perform its function either electrically or mechanically and when voltage and stress are applied simultaneously. It is normal to design the insulator such that the daily and the maximum loads experienced by the insulator are 20% and 50% of the M and E rating, respectively. International standards require that every suspension insulator unit is routinely test with a 50% (per ANSI) or 60% (per IEC) of the M- and E-rated load for 10 sec.

Composite insulators are rated by their **specified mechanical load**, which is defined as the tensile load specified by the manufacturer that has to be verified during a mechanical load test. International standards require that every suspension insulator unit is routinely tested with a 50% M- and E-rated load for 10 secs. The relationship between the M and E rating or SML rating of insulators, everyday load, maximum load, and routine test load that every insulator is subjected to is shown in Figure 5.4. It is clear that there is a comfortable safety factor in the mechanical rating of the insulator.

Comparison of Porcelain, Glass, and Composite Insulators

Porcelain and glass are inorganic materials, which historically are known to resist degradation from natural elements for many hundreds of years. They have melting points in excess of 1500°C. They are inert to most chemicals. Consequently, insulators made from porcelain and glass are known to possess extremely high resistance to heat from electrical discharge activity in the form of arcs or sparks, corona, and chemicals encountered during service. Their high stability is related to the strong electrostatic bonds between the various atoms in the material. This attribute also imparts a high value of **surface free energy**, a thermodynamic quantity that determines the strength of adhesion of the surface with water. In simple terms, porcelain and glass are easily rendered wettable by water. Hence, insulators made from these materials need to have sufficient leakage distance and complicated shapes to retain a sufficiently high surface resistance under wet and contaminated conditions.

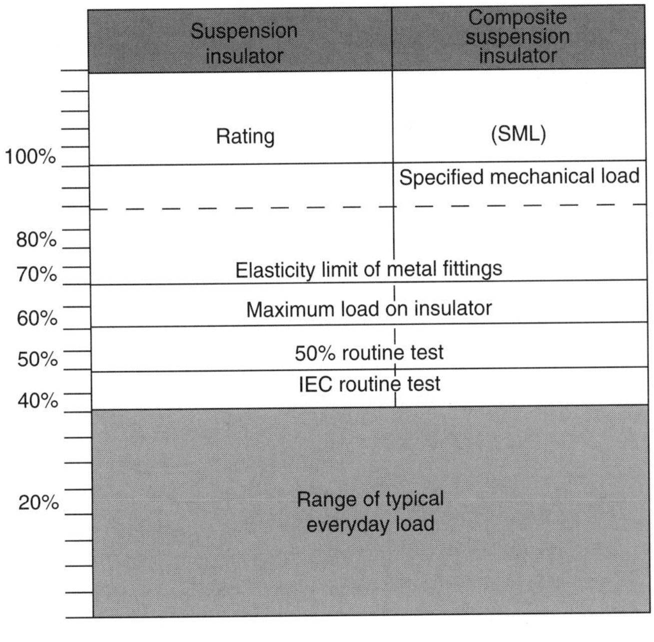

FIGURE 5.4 Relationship of M and E/SML Ratings with Daily, Maximum, and Routine Test Loads

Porcelain and glass are also dense materials. Hence, insulators made from these materials are heavy. They possess a compressive strength that is at least an order of magnitude higher than the tensile strength. They are brittle materials, and such insulators need to be handled with great care to avoid breakages.

The housings or weathersheds of composite insulators are made by organic materials, mainly hydrocarbons. Organic materials, in comparison with inorganic porcelain and glass, have weaker electrostatic bonds that can be broken more easily. Therefore, composite insulators are more easily prone to deterioration by heat from electrical discharge activity, chemicals (including water), and natural elements like sunlight, humidity, and temperature. Hence, composite insulators can undergo a reduction in the electrical and mechanical properties with time, and these changes are irreversible.

Aging is the term used to describe irreversible permanent changes. All materials undergo aging at different rates. Because organic materials are more susceptible to changes from the environment and applied stress, it is more common to talk of aging with composite insulators than it is with porcelain and glass insulators. However, it should be mentioned that even porcelain and glass insulators are subjected to aging.

When chemical bonds in hydrocarbons are broken, free carbon is generated, which is conducted even in the dry state. This is obviously not desirable. The formation of a carbonaceous conducting path on the surface of composite materials is called **tracking**. Materials can be formulated such that the carbon is physically dislodged from the surface or removed in the form of gaseous products. Such modes of carbon removal will lead to a loss of material, which is termed **erosion**. Erosion is a much slower mode of degradation than tracking. Hence, the resistance to tracking and erosion of materials is an extremely important aspect of composite insulators.

Owing to the weak nature of the electrostatic bonds, organic materials have a much lower surface free energy; hence, they are not easily wetted by water. In fact, on new composite insulators, water usually forms small individual beads instead of a continuous film. This property of water repellency is called hydrophobicity. This resistance to water wetting is desirable because it increases the surface resistance of the insulator under wet and contaminated conditions.

Composite insulators are much lighter in weight than equivalent porcelain and glass insulators. Also, the materials are nonbrittle. Therefore, composite insulators are much easier to handle and install as well as more resistant to breakages from handling and acts of vandalism.

5.4.5 Methods to Improve Performance Under Contaminated Conditions

Making insulating systems work satisfactorily under contaminated or polluted conditions is critical for maintaining high reliability of power delivery. This has been accomplished by a combination of design and a variety of periodic maintenance procedures, retrofits, and product innovation. It is worthwhile to note that while many situations in service can and have been dealt with satisfactorily by relatively simple means, others need a more in-depth analysis and require research and development.

Even a momentary loss of power can cause huge losses to industrial productivity. Losses incurred depend on the type of customer affected and the socioeconomic conditions. For example, in the United States, a 0.25 sec interruption to a paper plant can result in a loss of revenue in excess of $100,000, and this figure does not include legal liabilities. Insulators used on overhead lines and in stations are responsible for a significant part of factors that cause power outages. Hence, such interruptions, especially if originating from manmade causes, should be eliminated.

Contamination-related failures of traditional porcelain and glass insulators usually manifest as flashovers, a self-restoring failure mode. This is true provided the quality of insulators is good, which means that there will be no internal failure. Flashovers mostly result in short-term interruptions in power delivery. For nonceramic/composite insulators, failure could be in the form of a flashover and/or degradation of the insulator, in which case the insulator has to be replaced, leading to a prolonged outage.

Contamination Flashover Process

The processes that lead to flashover under contaminated conditions have been extensively researched and documented. A simplified model, originally developed around 1930, is shown in Figure 5.5 and is adequate to explain the process. Leakage current that is promoted on a wet surface in the combined presence of moisture and contamination initiates surface discharges, often called **dry band arcing**. The existence of a highly nonlinear distribution of the applied voltage along the insulator and the insulator shape that has surface areas with widely differing diameters, contribute significantly to the initiation of surface discharges. Normally, these discharges are self-limiting and localized to the narrow parts of the insulator. But, under conditions usually associated with increased contamination (or a decreased resistance in series with the arc), the discharges

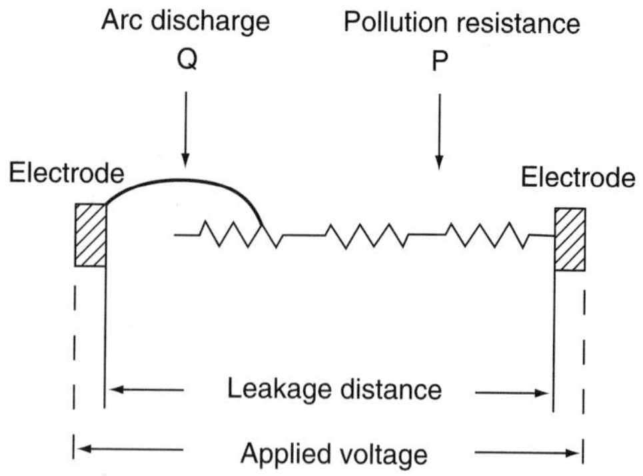

FIGURE 5.5 Simplified Schematic for Modeling Contaminated Insulators

TABLE 5.3 Principle of Improving Contamination Performance and Methods Used

Method	Principle
Additional insulators or longer leakage distance insulators	Increase leakage/creepage distance
Complicated shapes	Protect part of insulator from getting wet
Creepage extenders	Increase leakage distance
Cleaning	Reduce contamination buildup
Semiconducting or resistance glazed (RG) insulators	Improve voltage distribution and maintain dry surface due to heating produced by leakage current
High-voltage insulator coatings (HVIC), such as greases and RTV silicone rubbers	Maintain hydrophobic surface
Nonceramic insulators	Maintain hydrophobic surface and smaller diameter sheds

elongate and can bridge the air gap between the electrodes causing a flashover.

Making insulators work under contamination usually revolves around increasing the surface resistance in series with the arc (P in Figure 5.5). Linearizing the voltage distribution is also an effective method for improving the contamination performance. The different techniques and the underlying principle that helps achieve improvement in contamination performance are listed in Table 5.3.

5.5 Established Methods Employed for Installations In-Service

5.5.1 Insulator Cleaning

A frequent occurrence involves the contamination severity changing significantly in a region after the station or a section of the line has been built. Construction of new roads or industrial sites cannot always be predicted at the design stage. Under such situations, insulator cleaning is a method commonly used by utilities. Insulator cleaning is covered by the IEEE Standard 957. **High-pressure water washing** is routinely employed and is effective for most types of contamination. However, there are certain types of contaminants (e.g., cement, greasy deposits, certain fertilizers) that adhere intimately with the insulator surface and cannot be easily removed easily by high-pressure water. In such cases, cleaning is performed by using dry-type abrasive agents, such as crushed corncob mixed with limestone. Water washing can be done online (without de-energizing the circuit) or off-line (after de-energization).

The major limitation with this method is that it is labor intensive and expensive (more so when using dry-type cleaning agents). In addition, there is no reliable method other than past service experience to determine when, and if, insulator cleaning is necessary. Caution should be exercised in cleaning nonceramic insulators because not all designs and materials can be washed at high-pressure with water.

Grease Coatings

Another technique employed by utilities for servicing existing porcelain insulators is the application of **high-voltage insulator coatings**. Grease coatings have been used for many years. Greases used fall into two main categories: **hydrocarbon greases** (or petroleum jellies) and **silicone greases**. Petroleum jellies are cheaper than silicone greases, which explain their popularity. But they are not as stable as silicone greases at both extremes of operating temperature. In addition, silicone greases offer superior resistance to tracking than petroleum jellies. IEEE Standard 957 also addresses grease coatings for insulators. This practice is used mainly for station insulation. Most greasing applications are done off-line, although online application is possible.

The big disadvantage of grease coatings is the short time interval (typically less than 1 to 2 years) between subsequent applications to prevent any contamination-related outages. During this time, the grease becomes saturated with dirt and loses its water repellency. Removal of spent grease is labor intensive, involving the use of dry cleaning agents and followed by water washing. Failure to remove the spent grease can lead to heavy surface arcing that can cause the porcelain insulator to crack. In addition, in some countries like the United States, disposal of spent grease is becoming increasingly difficult due to environmental issues

5.6 Newer Developments to Improve Performance of Installations In-Service

5.6.1 Silicone Rubber Coatings

Room temperature vulcanized (RTV) **coatings** were developed with the intention of making a high-voltage insulator coating that would be effective for a much longer time than grease coatings. These coatings consist of a polydimethylsiloxane (PDMS) polymer, an alumina trihydrate (ATH) or alternate filler for increased tracking and erosion resistance, a catalyst, and a cross-linking agent. Coating formulations may also contain a condensation catalyst, adhesion promoter, reinforcing filler, and pigments (for color). The coating systems are dispersed in an organic solvent. Naphtha and 1-1-1 trichloroethane are the solvents commonly used. The solvent merely acts as a carrier medium to transfer the RTV coating to the ceramic insulator surface and serves the same purpose as a thinner for paints. As the solvent evaporates from the surface, moisture from the air triggers vulcanization, forming a solid rubber coating. The speed at which this takes place depends on the type of solvent, cure system chemistry, and relative humidity. Recommended coating thickness is about 0.5 mm (20 ml).

RTV coatings can be applied online (energized application) or off-line (after de-energizing the circuit). Coatings that use a combustible solvent like naphtha, however, are not suitable for energized applications.

RTV coatings offer superior contamination performance by virtue of their inherent hydrophobicity. The hydrophobic surface breaks up the water into discrete droplets, resulting in suppression of leakage current even in the presence of contamination. This is illustrated in Figure 5.6.

Under conditions of long-term corona or dry-band discharge activity promoted by wetting and contamination deposition, the RTV-coated insulator surface can lose its hydrophobicity in those areas subjected to the discharges. The hydrophobic loss is usually temporary, and the surface will recover its hydrophobicity if the surface is dry for a few hours. This is due to the dynamic nature of the silicone polymer that allows polymer chains to migrate and reorient

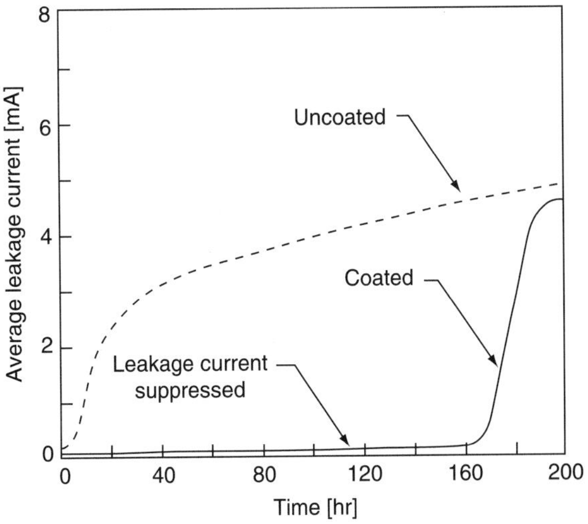

FIGURE 5.6 Leakage Current Suppression Capability of Uncoated and RTV-Coated Porcelain Insulator. The results were obtained in the laboratory using a salt-fog test.

themselves to the surface in the least surface energy configuration (Gorur *et al.*, 1999).

Details of application of coatings to ceramic insulators are partially covered in the IEEE Standard 957. Important aspects are insulator cleaning, coating preparation, and details of equipment to be used for application of the coating. The interested reader is referred to the documents listed in the reference section. Proper adhesion of the coating to the ceramic surface is critical for successful operation of the coated insulators.

5.6.2 Creepage Extenders

Retrofitting existing insulators with attachments called **creepage extenders** has been adopted successfully. Figure 5.7 shows a photograph of a field installation in a transmission line.

Presently, creepage attachments are made from an ethylene vinyl acetate (EVA) polymer compound and are formulated for properties desirable for outdoor insulation. These devices increase the leakage distance, prevent cascading of rain, and can also maintain a certain part of the insulator surface dry by the umbrella effect. They could also be effective in reducing bird excrement-related flashover problems. Reports from field experience suggest that they provide long-term improvement in contamination performance. Adhesion of the creepage extender to the porcelain surface is critical for satisfactory performance. Some creepage extenders require the use of a heat source (flame torch) to ensure good bonding with the ceramic insulator surface. Obviously, such systems should be installed off-line.

5.7 Methods for Improving Contamination Performance of New Installations

5.7.1 Additional Units and Longer Leakage Units

If the line or station is in the design stage, increasing **leakage distance** by adding additional bells or using longer leakage distance bells (such as the fog-type insulator) in a string is a practice followed routinely by users for both line and station insulation and is quite successful. It is common to see insulators normally rated for the next higher voltage to be used for lower voltage-rated systems in areas with contamination problems. This practice works well for most cases, except in those with severe contamination. In such regions, this method has only reduced the frequency of outages but not eliminated it, thus calling for additional measures such as water washing to prevent contamination-related power interruptions.

5.7.2 Resistive (or Semiconducting) Glazed Porcelain Insulators

The use of **resistive** or **semiconducting glazed** porcelain insulators, in place of normal porcelain insulators for contaminated locations (RG) is a well-known practice that was first tried more than 50 years ago. The semiconducting glaze offers a lower resistance than the normal insulating glaze and promotes a steady current (less than 1 mA) at the nominal operating voltage. This current ensures a more linear voltage distribution, and the resulting I^2R heating keeps the surface warm, ensuring that large areas of the insulator are dry. This combination of factors significantly reduces dry-band discharges and improves the contamination performance. The ability to curtail discharges in the presence of wet contamination is shown schematically in Figure 5.8.

RG insulators have demonstrated their success for station- and line-post applications. However, for cap and pin suspension insulators, judging from previous service experience, their record has not been as good. One problem for line insulators is the corrosion of the glaze near the pin where the current density is the highest, which ultimately causes a discontinuity or "opening" of the glaze in that region. Once the glaze has been opened, the effectiveness of the glaze is lost as surface discharges can develop in areas similar to those for a regular glazed porcelain insulator. Another problem experienced with RG cap and pin suspension insulators is related to the expan-

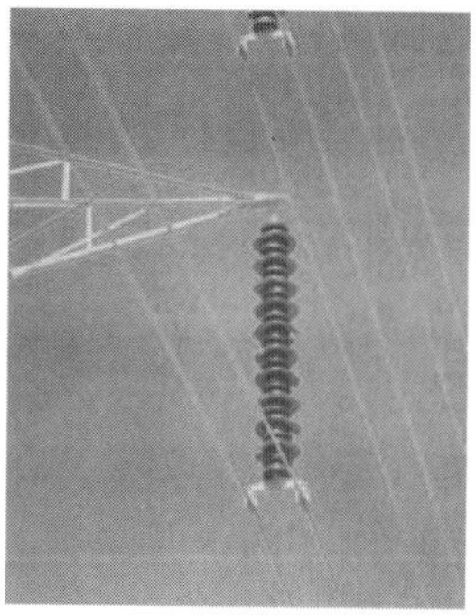

FIGURE 5.7 Creepage Extenders Installed in a Transmission Line. Courtesy of Tyco Electronics.

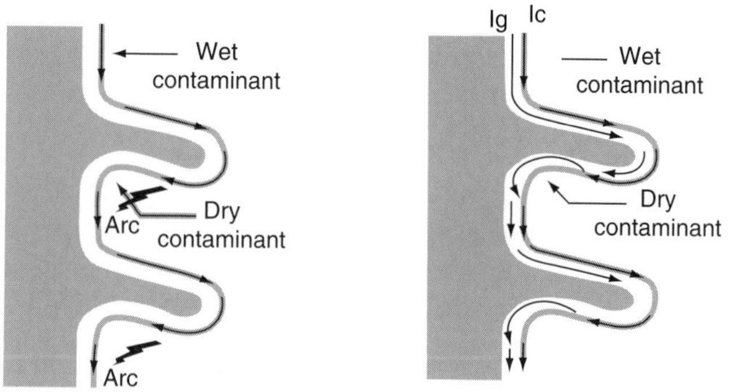

FIGURE 5.8 Schematic of Discharge Formation on Conventional Glazed Porcelain and Suppression with the Use of RG Porcelain

sion of the cement in the pin area that ultimately shatters the porcelain dielectric in the event of a flashover. These problems, however, are not insurmountable, and RG cap and pin insulators are available presently. New products have hopefully addressed the issues experienced by earlier products.

5.7.3 Nonceramic and Composite Insulators

The use of **nonceramic and composite insulators** in place of traditional porcelain and glass insulators for line insulation has become widespread in the last 20 years. Such insulators have several advantages over porcelain and glass insulators, such as lighter weight, easier handling, better resistance to damage from vandals, lower cost (in some countries), and superior contamination performance. Different material families have been used for the exposed part of the insulator (hereafter called the housing). High temperature vulcanized (HTV) silicone rubber, ethylene propylene rubber (EPR), cycloaliphatic epoxy, and EVA are among the materials proven suitable for outdoor use, with the first two varieties dominating for transmission voltages.

A nonceramic and composite insulator's performance under polluted conditions merits careful consideration. For porcelain or glass insulators, flashover resulting in a temporary outage is the end result of a contamination event. This does not usually cause any major permanent damage to the insulator string although burning of the glaze and/or fragmentation of the bells in contact with the power-follow fault current can occur. There is little risk of the insulator failing mechanically. It is also unlikely that the inorganic dielectric is degraded due to surface discharge activity, which could be long-lasting.

For nonceramic insulators, there is actually less risk than with the porcelain or glass insulators from the flashover event itself due to the elastic nature of the material. But it is from the surface discharge activity that the exposed insulation can be subjected to degradation, and this can be a major concern. In addition, the organic nature of the insulating materials can make them vulnerable to degradation from natural elements, such as heat, UV from sunlight, moisture, and chemicals. A permanent reduction of their desirable properties under service conditions can occur with time, referred to as aging. It is also important to note that some degradation modes may actually occur even in clean conditions, such as from exposure to corona activity. In fact, mechanical failure of nonceramic insulators from a mode of failure called **brittle fracture** has been experienced in relatively benign outdoor conditions. Users should be aware of all these possibilities.

Despite these concerns, it should be said that judicious selection and application of nonceramic insulators has resulted in improved reliability and lower installation costs for both transmission and distribution lines. Progress at all fronts, namely research, development, testing, manufacturing, and usage, has made this possible.

Just from geometrical considerations alone, nonceramic insulators should offer superior performance under contaminated conditions when compared to their porcelain and glass counterparts due to their smaller diameter. Additional improvement in contamination performance can be obtained by using materials that are hydrophobic, hence suppressing leakage current and discharge activity; they can remain in this state for a long time in service. Silicone rubber is one type of material that fits into this category. Within a particular material family, the leakage current suppression capability is dependent on the formulation, but in general, silicone polymers have better leakage current suppression capability than other outdoor insulating materials.

The typical practice is to use a leakage distance similar to porcelain for silicone rubber insulators; and for nonceramic insulators employing materials other than silicone rubber, the leakage distance is about 20 to 30% higher than the distance for porcelain insulators.

5.8 Underground Transmission Cables

Power cable technology can be traced back to 1880, just shortly after the introduction of incandescent lighting. Prior to World War I, oil-impregnated paper cables were extensively used. These had a three-conductor belted core made of copper and were used up to 25 kV. Due to the nonuniform electric field distribution inside the cable, this method of cable construction was extremely prone to partial discharges. This problem was solved by shielding the individual conductors with thin copper tapes, and it was possible to raise the voltage to as high as 69 kV.

At higher voltages, partial discharges were found to occur in voids in the insulation. This problem was solved by the use of a low-viscosity oil-impregnated paper insulation system. Voltages as high as 132 kV could be reached. Present day self-contained oil-filled cables have not changed much from this design that was first introduced in 1917. Self-contained oil-filled cables are in operation in Canada and the United States at 230 kV, 315 kV, and 500 kV.

In North America, starting around 1932, the first installation of oil-filled pipe-type cables occurred. The pipe-type cables proved to be more economical, and by 1954, the total circuit mileage of installed pipe-type cable surpassed that of the self-contained oil-filled cable in the United States. The first 230-kV pipe-type cables were installed in 1956. With increasing loads, neither increasing cooling nor operating voltage can cope up with the increased dielectric losses in the oil-impregnated paper dielectric. Thus, for higher operating voltage, synthetic dielectrics, such as cross-linked polyethylene (XLPE) and paper-polypropylene laminates (PPL), were introduced and have inherently lower dielectric losses.

PPL has lower dielectric losses and significantly higher dielectric strength. However, the widely used material is cross-linked polyethylene (XLPE) due to a combination of cost and performance considerations.

Transmission cables use a hermetic seal over the core to protect it from water exposure. Extruded lead, extruded- or seam-welded aluminum, or seam-welded copper sheath provides both the hermetic barrier and path for short circuit currents. Due to the presence of a good seal, problems such as **water treeing** widely experienced with distribution XLPE cables has not been an issue with transmission cables. Presently, extruded XLPE cables are being designed for operation at 500 kV. The use of thermoplastic-type low-density polyethylene (LDPE) is widespread in France on 225 kV cables. These cables have been in service since 1969. In 1985, the voltage of the LDPE cables was extended to 400 kV and subsequently to 500 kV in 1995.

The operating temperature allowable for LDPE cables has increased to 200° C. Much of the work on power transmission cables has been done in Norway, Japan, Sweden, and the United States. Work done in the United States has resulted in improved cleanliness of the insulation and semiconducting shield interfaces, a more uniform distribution of the cross-linking agent, and reduced void formation by increasing the extrusion pressure and replacing the wet curing process by a dry curing process. In Japan, the present practice is to use XLPE cables that are being used at 275 kV and 500 kV.

References

ANSI C 29.1. (1992). *Electrical power insulators—test methods.*

ANSI C 29.11. (1989). *Composite suspension insulators for overhead transmission lines—tests.*

ANSI C2. (1992). *National electric safety code.*

IEC 1109. (1992). *Composite insulators for ac overhead lines with a nominal voltage greater than 1000 V: Definitions, test methods, and acceptance criteria.*

IEC 60815. (1986). *Guide for the selection of insulators in respect of polluted conditions.*

IEEE Std. 100. (1992). *The new IEEE standard dictionary of electrical and electronic terms.* New York: IEEE Press.

IEEE Std. 957. (1995). *IEEE guide for insulator cleaning.* NERC Web site: *http://www.nerc.com.*

EPRI. (1982). *EPRI transmission lines reference book: 345 kV and above.*

Gorur, R.S, Cherney, E.A., and Burnham, J.T. (1999). *Outdoor insulators.*

Looms, J.S.T. (1998). *Insulators for high voltages.* Peter Peregrinus Ltd.

6
Power Distribution

Turan Gönen
College of Engineering and Computer Science, California State University, Sacramento, Sacramento, California, USA

6.1 Distribution System

In a broad definition, the **distribution system** is the part of the electric utility system between the bulk power source and the consumers' service switches. It includes a subtransmission system; distribution substations; primary distribution feeders; distribution transformers; secondary circuits, including the services to the consumer (called **service drops**); and appropriate protective and control equipment. Some distribution engineers, however, define the distribution system as that part of the power system between the distribution substations and the consumers' service entrance. Figure 6.1 shows a one-line diagram of a typical distribution system.

6.1.1 Subtransmission

The **subtransmission system** is the part of the power system that delivers power from bulk-power sources, such as large transmission substations. The subtransmission may be composed of overhead open-wire construction on wood poles or underground cables. Although the voltages of these circuits can occasionally range from 12.47 to 345 kV, the majority are at 69-, 115-, and 138-kV voltage levels. There is a trend toward the use of the higher voltages as a result of increasing use of the higher transmission voltages. The subtransmission-system designs vary from simple radial-type and/or loop-type systems to a grid- or network-type subtransmission. In the radial system, the circuits radiate from the bulk-power stations to the distribution substations. The radial system is simple and has a low cost but it also has a low service continuity. In general, due to higher service reliability, the transmission system is designed as loop circuits or multiple circuits forming a subtransmission grid or network. In the loop-type subtransmission system, a single circuit originating from a bulk-power bus runs through a number of substations and returns to the same bus. The major considerations affecting the design are cost and reliability.

6.1.2 Substation

The distribution **substation** is an assemblage of equipment for the purpose of switching, regulating, and changing the level of supply voltage from subtransmission level to primary distribution level. A typical substation may have:

- Power transformers
- Circuit breakers
- Disconnecting switches
- Station busses and insulators
- Current-limiting reactors
- Shunt reactors
- Current transformers
- Potential transformers
- Capacitor voltage transformers

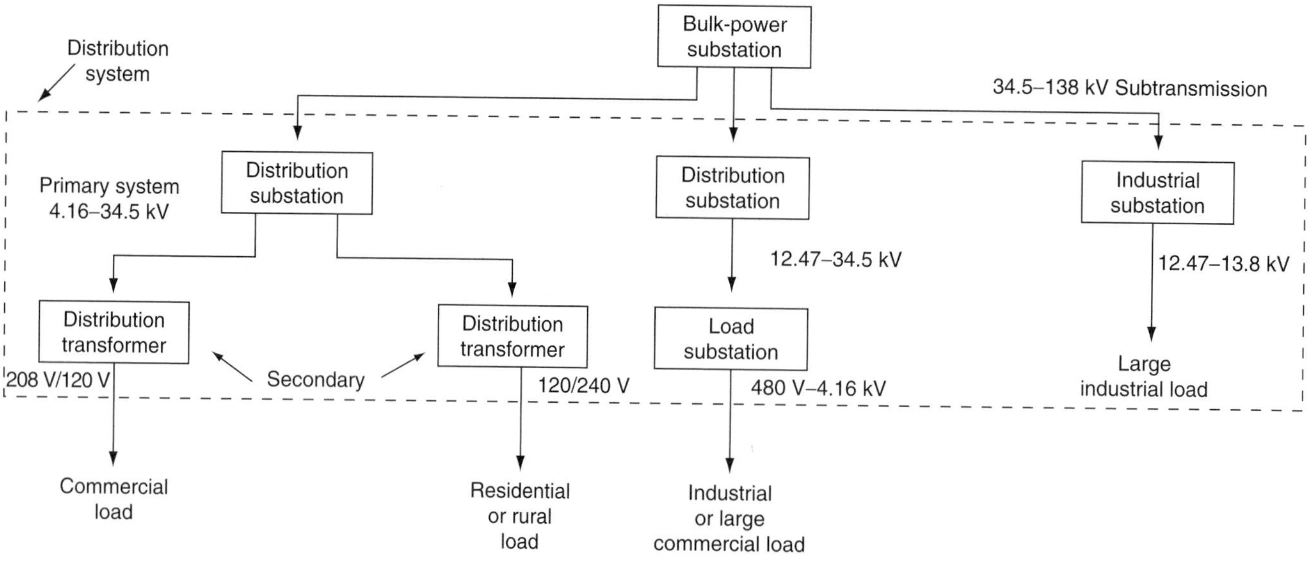

FIGURE 6.1 Overview of the Power Distribution System

- Coupling capacitors
- Series capacitors
- Shunt capacitors
- Grounding systems
- Lightning arresters and/or gaps
- Line traps
- Protective relays
- Station batteries
- Other equipment

More important substations are designed so that the failure of a piece of equipment in the substation or one of the subtransmission lines to the substation will not cause an interruption of power to the consumer. The location of a substation is dictated by the voltage levels, voltage regulation considerations, subtransmission costs, substation costs, and the primary feeders, mains, and distribution transformers. Location is also restricted by esthetic considerations.

The electrical and physical arrangements of the switching and bussing at the subtransmission voltage level are determined by the selected substation scheme (or diagram). On the other hand, the selection of a particular substation scheme is based on safety, reliability, economy, simplicity, and other considerations. The most commonly used substation bus schemes include (1) single bus scheme, (2) double bus–double breaker (or double main) scheme, (3) main-and-transfer bus scheme, (4) double bus–single breaker scheme, (5) ring bus scheme, and (6) breaker-and-a-half scheme. Figure 6.2 shows various bus schemes.

Each scheme has some advantages and disadvantages depending on economical justification of a specific degree of reliability. For example, the advantage of the **single bus scheme** is its lowest cost; its disadvantages include (a) the shutdown of the entire substation because of the failure of bus or any circuit breaker, (b) the difficulty in performing any maintenance work, (c) its limited use where loads can be interrupted or have other supply arrangements, and (d) the inability to be extended without completely de-energizing the substation.

The advantages of the **double bus–double breaker scheme** include (a) two dedicated breakers in each circuit, (b) breakers that can be taken out of service for maintenance, (c) high reliability, and (d) flexibility in allowing feeder circuits to be connected to either bus. Its disadvantages include high cost and the loss of half the circuits for breaker failure if circuits are not connected to both busses.

The advantages of the **main-and-transfer scheme** include (a) its low initial and ultimate cost, (b) breakers that can be taken out of service for maintenance, and (c) potential devices that may be used on the main bus for relaying. Disadvantages include (a) its requirement of one extra breaker for the tie bus, (b) switching that is somewhat complicated when maintaining a breaker and (c) the failure of bus or any circuit breaker resulting in the shutdown of the entire substation.

The advantages of the **double bus–single breaker scheme** include (a) some flexibility with two operating busses, (b) the ability of either main bus to be isolated for maintenance and (c) transferring the circuit readily from one bus to the other by using bus-tie breaker and bus selector disconnect switches. Disadvantages include (a) one extra breaker required for the bus tie, (b) four switches required per circuit, (c) a bus protection scheme that may cause loss of substation when it operates if all circuits are connected to that bus, (d) high exposure to bus faults, (e) line breaker failure taking all circuits connected to that bus out of service, and (f) bus-tie breaker failure taking the entire substation out of service.

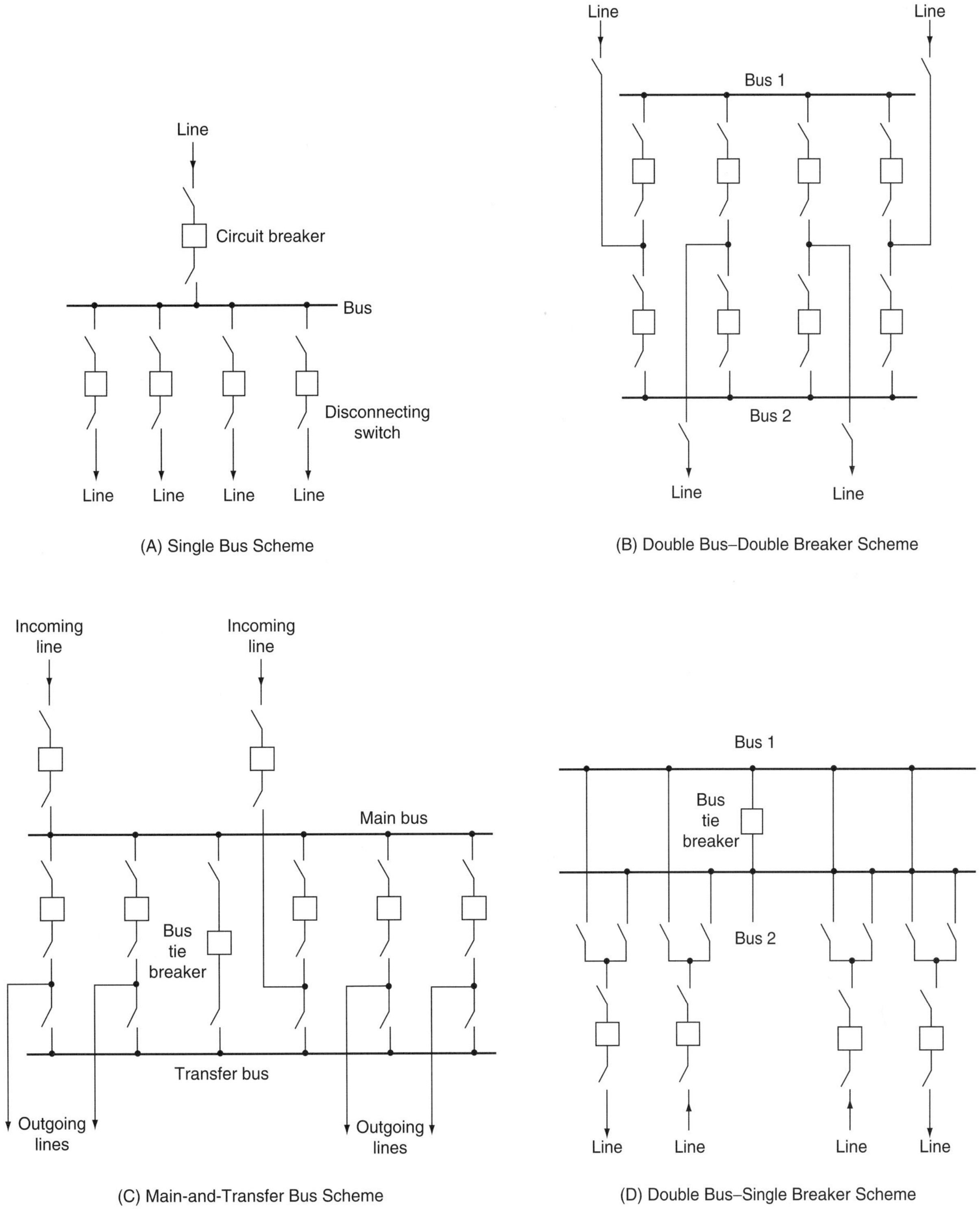

FIGURE 6.2 Typical Substation Bus Schemes

The advantages of the **ring bus scheme** include (a) low initial and ultimate cost, (b) flexible operation for breaker maintenance, (c) the ability of any breaker to be removed for maintenance without interrupting the load, (d) the requirement of only one breaker per circuit, (e) its ability to run without using the main bus, (f) all circuits that are fed by two breakers, and (g) all switching that is performed with breakers. Disadvantages include (a) the ability of the ring to

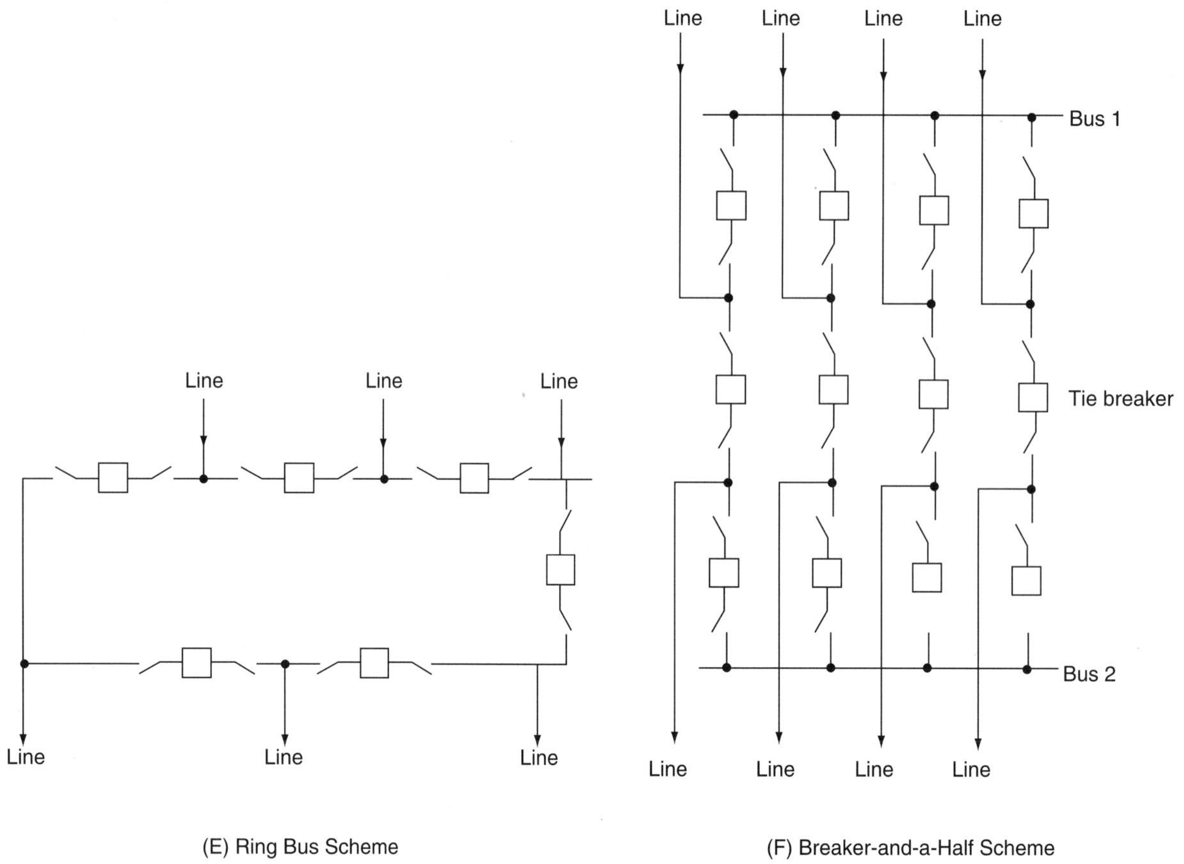

FIGURE 6.2 Typical Substation Bus Schemes (*continued*)

be separated into two sections if a fault occurs during a breaker maintenance period, (b) complex automatic reclosing and protective relaying circuitry, (c) taking the circuit out of service to maintain the relays (common on all schemes if a single set of relays is used), (d) the requirement of potential devices on all circuits since there is no definite potential reference point (these devices may be required in all cases for synchronizing, live-line, or voltage indication), and (e) the breaker failure during a fault on one of the circuits that causes loss of one additional circuit (owing to operation of breaker-failure relaying).

The advantages of the **breaker-and-a-half scheme** include (a) most flexible operation, (b) high reliability, (c) breaker failure of bus side breakers removing only one circuit from service, (d) all switching being performed with breakers, (e) simple operation without disconnect switching for normal operation, (f) the ability of either main bus to be taken out of service at any time for maintenance, and (g) no removal of any feeder circuits, from service during bus failure. Disadvantages include 1.5 breakers per circuit and relaying and automatic reclosing that are somewhat involved because the middle breaker must be responsive to either of its associated circuits.

6.1.3 Primary System

The **primary system** is made up of circuits known as primary feeders or distribution feeders. A feeder includes the main or main feeder (which usually is a three-phase four-wire circuit) and branches or laterals (which usually are single-phase or three-phase circuits) tapped off the main, as shown in Figure 6.3. A feeder is usually sectioned by means of reclosing devices at various locations to remove as little as possible of a faulted circuit and hinder service to as few consumers as possible. This is accomplished through the coordination of all the fuses and reclosers.

The factors that affect the design of the primary system include the cost, importance of the load it serves, and the required level of service continuity and reliability. The simplest and least expensive and, therefore, the most common design is **radial design**. The low reliability of this design is improved by a modification called the **loop-type primary design**, in which the feeder loops through the feeder load area and returns to the substation bus. The most expensive and reliable design is the **primary-network type**, in which a system of interconnected feeders is supplied by a number of substations. Such a system supplies a load from several directions and consequently provides the maximum reliability and quality of service.

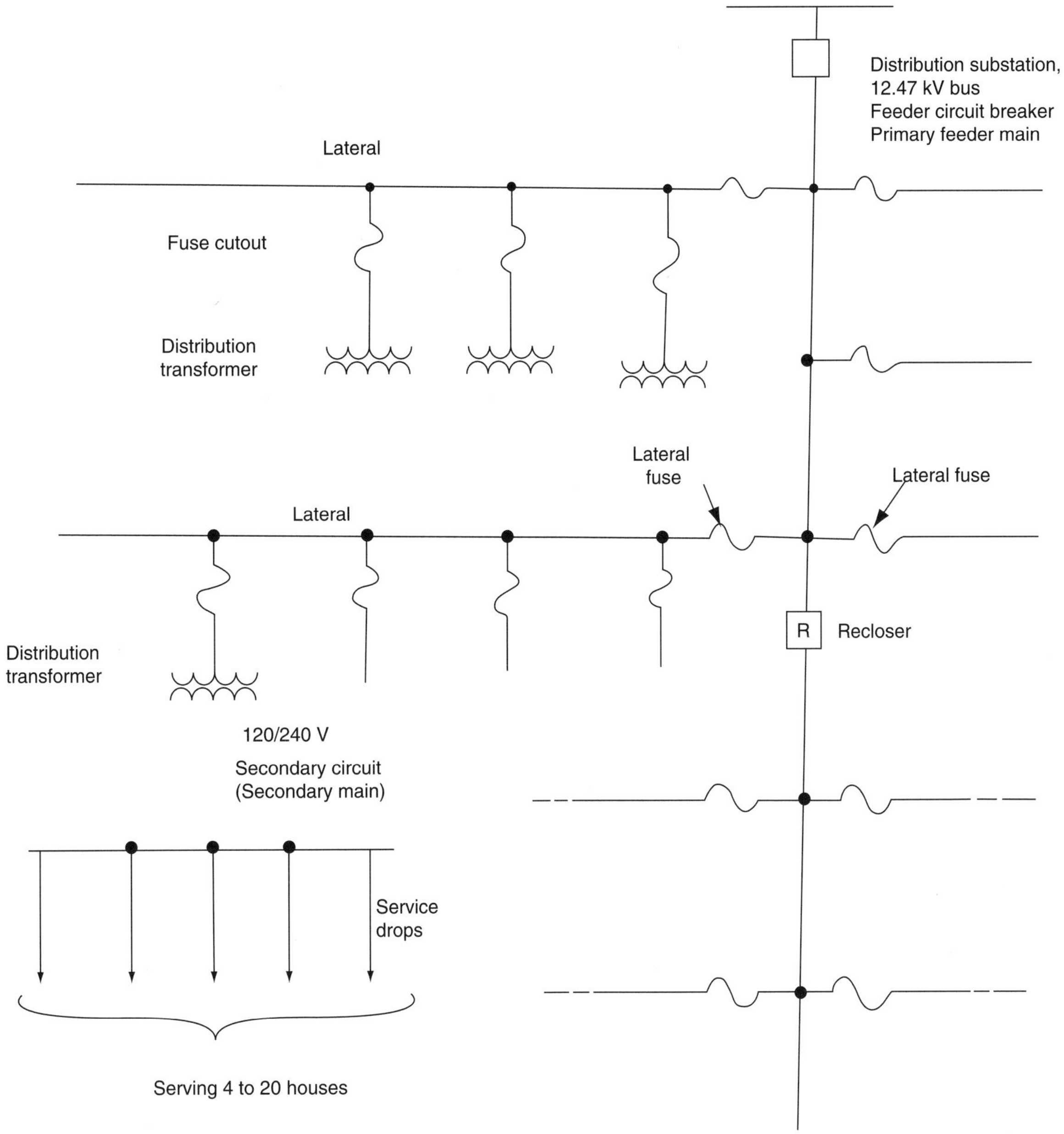

FIGURE 6.3 Typical Power Distribution Feeder

The primary system leaving the substation is most often in the 11,000 to 15,000-V range. A specific voltage used is 12,470-V line-to-line and 7,200-V line-to-neutral (conventionally written 12,470Y/7,200 V). Some utilities still use a lower voltage, such as 4,160Y/2,400 V. In recent years, however, the use of primary distribution circuits in the 25- and 35-kV classes is increasing; all of these circuits are four-wire systems. Single-phase loads are connected line-to-neutral on the four-wire systems. Congested and heavy-load locations in metropolitan areas as well as new residential areas are customarily served by underground primary feeders made up of radial three-conductor cables.

6.1.4 Secondary System

The **secondary system** is the part of the electric power system between the primary system and the consumer's property. The secondary distribution system includes distribution transformers, secondary circuits (secondary mains), consumer services (or service drops), and meters to measure consumer energy consumption. Usually, the secondary systems are designed in single phase for areas of residential consumers and in three phase for areas of industrial or commercial consumers with high load densities. Secondary voltages are provided by distribution transformers that are connected to the primary system, which are usually associated with utilization voltages. Residential and most rural consumers are supplied by 120Y/240-V single-phase, three-wire systems. Commercial and small industrial consumers are supplied by 208Y/120-V or 480Y/277-V three-phase, four-wire systems. The secondary voltage also supplies multiple street lights.

6.1.5 Spot Networks

Spot networks are used in downtown areas for high-rise buildings with extremely high load densities and sometimes also for areas of industrial or commercial consumers. Often, large shopping centers are supplied by spot networks. In such a network, all transformers and protecting equipment are placed at the same location.

6.1.6 Good Voltage and Continuity

Basically, **good voltage** means that the average voltage level is correct, that variations are within acceptable voltage limits, and that sudden and momentary changes in voltage level do not cause objectionable light flicker. Naturally, the utilization voltage of the consumer varies with changing load on the system as well as the location of the consumer on the system. Voltage variation, however, is usually less then 5% at the consumer's meter.

Service continuity is the provision of uninterrupted electric power to the consumer. **Good continuity** means that such service is almost always provided.

6.1.7 Backup Systems and Overhead Construction

To increase the service reliability for critical loads, such as hospitals, computer centers, and crucial industrial loads, **backup systems**, such as emergency generators or batteries, with automatic switching devices are provided. Computer installations are backed up by uninterruptible power supplies (UPS).

Most existing distribution systems in residential, industrial, and rural areas are of **overhead construction**. The distribution transformers are mounted near the tops of poles, and bare primary and secondary conductors are strung from pole to pole.

6.1.8 Underground Construction

Almost all residential developments now being built are served by **underground residential distribution** (URD) **systems** using underground cables for esthetic and safety reasons. In general, underground distribution systems cost more than comparable overhead systems. An URD system is free from service outages and accompanying repair expenses caused by lightning, rain, ice, sleet, snow, and wind storms as well as by any man-made hazards. Some other costs, such as the cost of tree trimming, is totally eliminated by using the URD systems. The URD systems used in residential areas do not require all the refinements and operating advantages found in conventional downtown commercial underground systems. Extensive duct banks, transformer vaults, manholes, and submersible apparatuses are not required in residential systems.

In the past, the favored underground cable was stranded copper with an oil-impregnated paper insulation wrapped in a lead sheath. It was expensive to buy, install, and splice, but it had a long service once in service. For example, it is common to see a 50-year-old paper-insulated cable still in service in duct vaults. Today, URD cables are made up of a relatively inexpensive plastic insulation, typically cross-linked polyethylene (XLPE), and practical elbow connectors. They are buried directly into ground or placed in narrow and shallow trenches, which naturally reduces installation and maintenance costs. In general, the heavy three-phase feeders are overhead along the periphery of a residential development, and the laterals to the pad-mount transformers are buried about 40 in deep. The lateral cable conductors come up from underground into the high voltage compartment of the transformer. The secondary service lines then run to the individual houses at a depth of about 24 in and come up into a meter through a conduit. The service conductors are run along easement and do not cross adjacent property lines.

The frequency of faults is lower on underground systems than on overhead systems. Faults, however, are much more difficult and time-consuming to find, isolate, and repair on underground systems. Service-restoration requirements dictate that primary-system designs should operate as a normally open loop. In the event of a cable fault, such design facilitates location and isolation of the failure and faster service restoration to all consumers on the unfaulted segment of the primary loop. A well-designed URD system should permit ease of sectionalizing to isolate a faulted portion; it should be flexible to accept load growth with a minimum number of changes. The looped primary and single-phase banked secondary system provide these requirements at about the lowest cost. The banked secondary system takes better advantage of diversity among loads and permits a smaller installed kilovoltampere of transformer capacity and possibly fewer transformers. Compared to the simple radial system, the secondary system also provides better voltage conditions at the loads and less voltage dip due to motor starting for a given transformer rating, conductor size, and spacing among transformers.

Transformers used for the URD system are hermetically sealed against moisture, including their bushings and terminals. The terminals may have one or more insulated disconnecting elbows, enabling the disconnection of the transformer and the sectioning of the primary circuits. The transformer may be completely buried or installed on ground-level pads or semiburied pads. **Pad-mounted transformers** are the predominant type of transformer being used for URD. They are usually installed on concrete slabs or pads. Pad-mounted transformers may use **dead-front** configuration with separable insulated connectors or elbows. There are many other combinations of pad-mount construction and accessory equipment available including **live-front** primary connections with stress cones for the cables, internal or external primary fuses and switches, and secondary breakers. **Residential subsurface transformers** (RSTs) are used much less frequently. RSTs are installed in relatively tight-fitting vaults with the cover grating of the vault at ground level. Cooling is usually accomplished by natural convection of air. RSTs are submersible and, therefore, have dead-front primary cable terminations. Switches, fuses, and circuit breakers are located on the cover of the transformer so that they can be operated by a lineman standing on the surface of the ground. The last type of URD transformer is a **direct-buried transformer**. This type of transformer is not popular because of the problem of accessory equipment and its poor operation under adverse weather conditions, such as with snow and ice. There also are concerns about galvanic corrosion under certain conditions.

6.2 Quality of Service and Voltage Standards

In general, performance of distribution systems and quality of the service provided are measured in terms of freedom from interruptions and maintenance of satisfactory voltage levels at the consumer's premises that are within limits appropriate for this type of service. Due to economic considerations, an electric utility company cannot provide each consumer with a constant voltage matching exactly the nameplate voltage on the consumer's utilization equipment. Thus, a common practice among the utilities is to stay with preferred voltage levels and ranges of variation for satisfactory operation of equipments set forth by the ANSI standard.

The nominal voltage standards for a majority of the electric utilities in the United States to serve residential and commercial consumers are as follows:

1. 120/240-V three-wire, single-phase
2. 240/120-V four-wire, three-phase delta
3. 208Y/120-V four-wire, three-phase wye
4. 480Y/277-V four-wire, three-phase wye

The voltage on a distribution circuit varies from a maximum value at the consumer nearest to the source (first consumer) to a minimum value at the end of the circuit (last consumer). For any given voltage level, the actual operating values can vary over a large range. This range has been segmented into three zones: (1) the **favorable zone** or preferred zone, (2) the **tolerable zone**, and (3) the **extreme zone**. The favorable zone includes the majority of the existing operating voltages that produce satisfactory operation of the customer's equipment. The tolerable zone contains a band of operating voltages slightly above and below the favorable zone. The operating voltages in the tolerable zone are usually acceptable for most purposes. The extreme or emergency zone includes voltages on the fringes of the tolerable zone, usually within 2 or 3% above or below the tolerable zone. Usually, the maximum voltage drop in the customer's wiring between the point of delivery and the point of utilization is accepted as 4 V based on 120 V.

6.2.1 Voltage Control

To keep distribution–circuit voltages within permissible limits, means must be provided to control the voltage. There are numerous ways to improve the overall voltage regulation of a distribution system. The complete list includes the following:

1. Use of generator voltage regulators
2. Application of voltage-regulating equipment in the distribution substations
3. Application of capacitors in the distribution substation
4. Balancing of the loads on the primary feeders
5. Increasing of conductor size
6. Changing of feeder sections from single-phase to multiphase
7. Transferring of loads to new feeders
8. Installing of new substations and primary feeders
9. Increasing of primary voltage level
10. Application of voltage regulators out on the primary feeders
11. Application of shunt capacitors on the primary feeders
12. Application of series capacitors on the primary feeders

The selection of a technique or techniques depends on the particular system requirement. However, automatic voltage regulation is always provided by (1) bus regulation at the substation, (2) individual feeder regulation in the substation, and (3) supplementary regulation along the main by regulators mounted on poles. Distribution substations are equipped with **load-tap-changing** (LTC) transformers that operate automatically under load or with separate voltage regulators that provide bus regulation.

Voltage-regulating equipment is designed to maintain a predetermined level of voltage automatically that would otherwise vary with the load. As the load increases, the regulating equipment boosts the voltage at the substation to compensate for the increased voltage drop in the distribution feeder. In cases where consumers are located at long distances from the substation or where voltage along the primary circuit is excessive, additional regulators or capacitors located at selected points on the feeder provide supplementary regulation. Many utilities have experienced that the most economical way of regulating the voltage within the required limits is to apply both step voltage regulators and shunt capacitors. Capacitors are installed out on the feeders and on the substation bus in adequate quantities to accomplish the economic power factor. Many of these installations have sophisticated controls designed to perform automatic switching. Of course, a fixed capacitor is not a voltage regulator and cannot be directly compared to regulators, but, in some cases, automatically switched capacitors can replace conventional step-type voltage regulators for voltage control on distribution feeders.

6.2.2 Feeder Voltage Regulators

Feeder voltage regulators are used extensively to regulate the voltage of each feeder separately to maintain a reasonable constant voltage at the point of utilization. They are either the induction-type or the step-type regulators. However, today's modern step-type voltage regulators have practically replaced induction-type regulators.

Step-type voltage regulators can be either (1) **station-type,** which can be single- or three-phase regulators and can be used in substations for bus voltage regulation or individual feeder voltage regulation or (2) **distribution type,** which can be only single-phase regulators and are used pole-mounted out on overhead primary feeders. Single-phase step-type voltage regulators are available in sizes from 25 to 833 kVA, whereas three-phase step-type voltage regulators are available in sizes from 500 to 2,000 kVA. For some units, the standard capacity ratings can be increased by 25 to 33% by forced-air cooling. Standard voltage ratings are available from 2,400 to 19,920 V, allowing regulators to be used on distribution circuits from 2,400 to 34,500 V grounded wye/19,920 V multigrounded wye. Station-type voltage regulators for bus voltage can be up to 69 kV.

A step-type voltage regulator is basically an autotransformer with a common winding and a tapped (stepped) series winding. A reversing switch enables the series winding to be connected for either raising or lowering of the line voltage. The number of steps depends on the percentage of increments in voltage regulation desired per step. Typically, step voltage regulators have 16 steps of 5/8% per step, for a total regulation range of $\pm 10\%$.

6.2.3 Line-Drop Compensation

Voltage regulators located in the substation or on a feeder are used to keep the voltage constant at a fictitious regulation or regulating point without regard to the magnitude or power factor of the load. The regulating point is usually selected to be somewhere between the regulator and the end of the feeder. The automatic voltage maintenance is achieved by dial settings of the adjustable resistance and reactance elements of a unit called the **line-drop compensator** (LDC) located on the control panel of the voltage regulator. Determination of the appropriate dial settings depends on whether any load is tapped off the feeder between the regulator and the regulating point.

In the event that no load is tapped off the feeder between the regulator and the regulating point, the R dial setting of the line-drop compensator can be determined from:

$$R_{set} = \frac{CT_P}{VT_N} \times R_{eff}, \tag{6.1}$$

where CT_P is the rating of the current transformer's primary; VT_N is the voltage transformer's turns ratio, which is V_{pri}/V_{sec}; and R_{eff} is the effective resistance of a feeder conductor from regulator station to regulating point. The *Reff* can be determined from:

$$R_{eff} = r_a \times \frac{l - s_1}{2}. \tag{6.2}$$

In equation 6.2, the r_a is the resistance of a feeder conductor from regulator station to regulation point, Ω/mi per conductor; s_1 is the length of three-phase feeder between regulator station and substation, mi (multiply length by 2 if feeder is in single-phase); and l is the primary feeder length, mi.

In addition, the X dial setting of the line-drop compensator can be determined from:

$$X_{est} = \frac{CT_P}{VT_N} \times X_{eff}, \tag{6.3}$$

where X_{eff} is the effective reactance of a feeder conductor from regulator to regulation point:

$$X_{eff} = x_L \times \frac{l - s_1}{2}, \tag{6.4}$$

and

$$x_L = x_a + x_d. \tag{6.5}$$

In equation 6.5, the x_a is the inductive reactance of an individual phase conductor of feeder at 12-in spacing, Ω/mi; x_d is the inductive-reactance spacing factor, Ω/mi; and x_L is the inductive reactance of feeder conductor, Ω/mi.

Note that since the R and X settings are determined for the total connected load, rather than for a small group of customers, the resistance and reactance values of the transformers are not included in the effective resistance and reactance calculations.

On the other hand, in the event that load is tapped off between the regulator station and the regulating point, the R setting of the line-drop compensator can still be determined from equation 6.2, but the determination of the R_{eff} is somewhat more involved. In this case, the effective resistance is calculated from:

$$R_{eff} = \frac{\sum_{i=1}^{n} |VD_R|_i}{|I_L|}, \tag{6.6}$$

and

$$\sum_{i=1}^{n} |VD_R|_i = |I_{L,1}| \times r_{a,1} \times l_1 + |I_{L,2}| \times r_{a,2} \times l_2 + \ldots + |I_{L,n}| \times r_{a,n} \times l_n. \tag{6.7}$$

In equation 6.7, $|VD_R|_i$ is the voltage drop due to line resistance of the ith section of feeder between regulator station and regulating point, V/section; $\left|\sum_{i=1}^{n} VD_R\right|_i$ is the total voltage drop due to line resistance of feeder between regulating station and regulation point; $|I_L|$ is the magnitude of load current at regulator location; $|I_{L,i}|$ is the magnitude of load current in the ith feeder section; $r_{a,i}$ is the resistance of a feeder conductor in the ith section of feeder, Ω/mi; and l_i = length of ith feeder section, mi.

Also, the X dial setting of the line-drop compensation can be found from equation 6.4, but the determination of the X_{eff} is again somewhat more involved. The following equations need to be used:

$$X_{eff} = \frac{\sum_{i=1}^{n} |VD_x|_i}{|I_L|}, \tag{6.8}$$

and

$$\sum_{i=1}^{n} |VD_X|_i = |I_{L,1}| \times X_{L,1} \times l_1 + |I_{L,2}| \times X_{L,2} \times l_2 + \ldots + |I_{L,n}| \times X_{L,n} \times l_n. \tag{6.9}$$

In this equation, $|VD_X|_i$ is the voltage drop due to line reactance of the ith section of a feeder between the regulator station and regulating point, V/section; $\left|\sum_{i=1}^{n} VD_X\right|_i$ is the total voltage drop due to line reactance of a feeder between the regulating station and regulation point; and $X_{L,1}$ is the inductive reactance as defined in equation 6.6, of ith section of feeder, Ω/mi.

Since the methods just described to find the effective R and X are rather involved, an alternative and practical method can be used to measure the current (I_L) and voltage at the regulator location and the voltage at the regulating point. The difference between the two voltage values is the total voltage drop between the regulator and the regulation point, which can be found as:

$$VD = |I_L| \times R_{eff} \times \cos\theta + |I_L| \times X_{eff} \times \sin\theta, \tag{6.10}$$

from which the R_{eff} and X_{eff} values can be found easily if the load power factor of the feeder and average r/x ratio of the feeder conductors between the regulator and the regulating point are known.

6.2.4 Voltage Regulation

Customer taps along the feeder draw current through the line, causing the voltage to drop as one gets further from the substation. If the voltage at the substation is set at the nominal voltage, the consumers at the end of the feeder have too low a voltage under peak load. If the voltage is set so that consumers at the end of the feeder receive the nominal voltage under heavy load, the consumers near the substation have too high a voltage, and the voltage is too high for all of the consumers at light load. Therefore, a compromise voltage setting must be chosen so that the voltage is at an acceptable level for all consumers regardless of the load, and the line drop must be acceptably low under all load conditions. However, a favorable compromise of voltage drop and voltage settings is not always possible for all load conditions, so other means of voltage regulation have been devised for such conditions.

In voltage regulation studies, it is customary to express the primary feeder voltage values in a 120-V base. The voltage-regulating relay (VRR) of a voltage regulator is adjustable within the approximate range from 110 to 125 V. The VRR measures the voltage at the regulating point by means of LDC. The LDC has R and X settings that are both adjustable within the approximate range from 0 to 24 Ω (often called volts because the current transformers used with regulators have 1-A secondaries). The bandwidth (BW) of the VRR is adjustable within the approximate range from ±0.75 to ±1.5 V based on 120 V. The time delay is adjustable between about 10 and 120 sec. The location of the regulating point is controlled by the R and X settings of the LDC. If the R and X settings are set at zero, the regulator adjusts the voltage at its local terminal to the setting of the VRR ± BW. Good advantage sometimes can be taken of this designed "overload" type of limited-range operation. If load growth occurs, however, both a larger range of regulation and larger regulator size may be required.

The peak-load voltage profile is not in linear but in parabolic shape. The voltage-drop value for any given point s between the substation and the regulator station can be found from:

$$VD_x = K\left(S_{3\phi} - \frac{S_{3\phi} \times s}{l}\right)s + K\left(\frac{S_{3\phi} \times s}{l}\right)\frac{s}{2} \quad \text{[pu]}, \tag{6.11}$$

where K is the percentage of voltage drop per kilovoltampere–mile [kVA] characteristic of feeder; $S_{3\phi}$ is the uniformly distributed three-phase annual peak load, kVA; l is the primary feeder length, mi; and s is the distance from substation, mi.

Alternatively, the voltage-drop value for any given point s between the substation and the regulator station can be found from:

$$VD_s = I(r \times \cos\theta + x \times \sin\theta)s\left(1 - \frac{s}{2l}\right), \tag{6.12}$$

where:

- I_L = load current in feeder at substation end

$$= \left(\frac{S_{3\phi}}{\sqrt{3} \times V_{L-L}}\right) \quad (6.13)$$

- r = resistance of feeder main, Ω/mi per phase
- x = reactance of feeder main, Ω/mi per phase

Thus, the voltage drop in per unit can be found as:

$$VD_s = \frac{VD_s}{V_{L-N}}. \quad (6.14)$$

The voltage-drop value for any given point s between the end of the feeder and the regulator station can be found from:

$$VD_s = K\left(S_{3\phi} - \left(\frac{S_{3\phi} \times s}{1-s}\right)\right)s + K\left(\frac{S_{3\phi} \times s}{1-s}\right)\frac{s}{2}, \quad (6.15)$$

where:

- $S_{3\phi}$ = uniformly distributed three-phase annual peak load at distance s_1, kVA

$$= S_{3\phi}\left(1 - \frac{s_1}{l}\right) \quad (6.16)$$

- s_1 = distance of feeder regulator station from substation, mi

The voltage profiles for annual peak load can be found by plotting the primary voltage values, $V_{P,pu}$, against the s distance on the feeder. Because there is no voltage drop at zero load, the primary voltage values, $V_{P,pu}$, remain constant. Thus, the voltage profile for the zero load is a horizontal line.

6.2.5 Use of Capacitors for Voltage Regulation

Because most loads lag and the line reactance is much greater than the line resistance, switching shunt capacitors across a line will increase the voltage by reducing the inductive vars drawn. If so much capacitance is switched across the line that the current becomes leading, the method is not cost-effective because the vars begin to rise again. Shunt capacitors are only useful with lagging load power factors.

In general, capacitor banks are installed near load points. If only fixed-type capacitors are installed, the utility will experience an excessive leading power factor and voltage rise at that feeder. Therefore, some of the capacitors are installed as switched-capacitor banks so they can be switched off during light-load conditions. Because of this, the fixed capacitors are sized for light load and connected permanently. The switched capacitors can be switched as a block or in several consecutive steps as the reactive load becomes greater from light-load level to peak load and sized accordingly.

A system survey is required in choosing the type of capacitor installation. As a result of power flow program runs or manual load studies on feeders or distribution substations, the system's lagging reactive loads (i.e., power demands) can be determined and the results can be plotted on a curve. This curve is called the **reactive load duration curve** and is the cumulative sum of the consumer's reactive loads and the system's reactive power requirements. Once the daily reactive load curve is obtained, then, by visual inspection of the curve, the size of the fixed capacitors can be found to meet the minimum reactive load. All capacitor banks require protection. The entire bank will be protected with a fuse or circuit breaker, and each capacitor will be fused. The capacitors can be connected via either delta or wye. The ungrounded wye is a preferred configuration because if a capacitor should short out in one leg of the wye, the other two legs will limit the fault current.

6.2.6 Voltage Regulation by Use of Tap Changers

Another method of line voltage regulation is **tap changing.** Taps are connections on a transformer winding that change the turns ratio slightly. The ratio change is normally ±10% (0.625% per step). The taps are usually located on the primary because less current has to be switched by the tap changing connections than would be necessary on the secondary. Tap changers can be manual or automatic. Some distribution transformers and distribution substation transformers have manual tap changers so that added load can be compensated for. Motor-driven automatic tap changers are necessary for voltage regulation with widely fluctuating loads. This is called **tap changing under load** (TCUL) or **load tap changing** (LTC). LTC is used in distribution and subtransmission substations to keep the secondary line voltage at the proper level in response to load and primary voltage changes.

6.2.7 Voltage Regulation by Use of Voltage Regulating Transformers

Regulating transformers are designed to provide a boost in voltage magnitude along a line or a change in phase. They are used basically to control the flow of power between two systems with different sources, or they are used along a tie feeder between two load centers that are fed by the same bulk power substation. Phase-regulating transformers are employed to control the power flow around loops with two or more sources. They are almost always motor driven and have extensive control and protection systems.

6.2.8 Dispersed Storage and Generation

Electric distribution systems are undergoing major changes to accommodate the development of alternative sources for generating electric energy and provide enhanced opportunities for small power producers and cogenerators. In general, these generators are small (typically ranging from 10 kW to 10 MW and connectable to either side of the meter) and can be

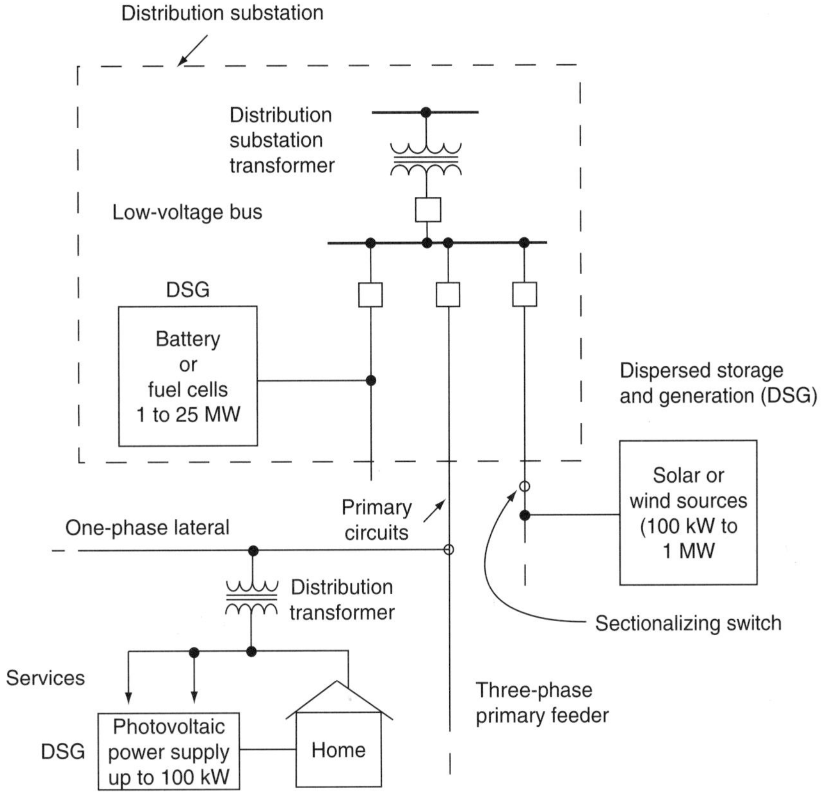

FIGURE 6.4 Small Dispersed Storage and Generation Units. These example units are attached to a distribution system

economically connected only to the distribution system, as shown in Figure 6.4.

If properly planned and operated, **dispersed storage and generation** may provide benefits to distribution systems by reducing capacity requirements, improving reliability, and reducing losses. Dispersed-storage-and-generation technologies include hydroelectric systems, diesel generators, wind-electric systems, solar-electric systems, batteries, storage space and water heaters, storage air conditioners, hydroelectric pumped storage, photovoltaics, and fuel cells.

References

Gönen, T. (1986). *Electric power distribution system engineering.* New York: McGraw-Hill.

Westinghouse Electric Corp. (1964). *Electric transmission and distribution reference book.* East Pittsburgh, PA: Westinghouse Electric Corporation.

Westinghouse Electric Corp. (1965). *Electric utility engineering reference book: Distribution systems.* East Pittsburgh, PA: Westinghouse Electric Corporation.

7

Power System Analysis

Mani Venkatasubramanian
and Kevin Tomsovic
*School of Electrical Engineering
and Computer Science,
Washington State University,
Pullman, Washington, USA*

7.1 Introduction

The **interconnected power system** is often referred to as the largest and most complex machine ever built by humankind. This may be hyperbole, but it does emphasize an inherent truth: there is a complex interdependency between different parts of the system. That is, events in geographically distant parts of the system may interact strongly and in unexpected ways. **Power system analysis** is concerned with understanding the operation of the system as a whole. Generally, the system is analyzed either under steady-state operating conditions or under dynamic conditions during disturbances.

Electric power is primarily transmitted as a three-phase signal. Three ac current currents are sent that are out of phase by 120° but of equal magnitude. Such balanced currents sum to zero and, thus, obviate the need for a return line. If the voltages are balanced as well, then the total power transmitted will be constant in time, which is a more efficient use of equipment capacity. For large scale systems analysis, the assumption is usually made that the system is balanced. Each phase can be then analyzed independently, greatly simplifying computations. In the following paragraphs, the implicit assumption is that three-phase systems are being used.

7.2 Steady-State Analysis

In **steady-state analysis**, any transients from disturbances are assumed to have settled down, and the system state is assumed as unchanging. Specifically, system loads including transmission system losses, are precisely matched with power generation so that the system frequency is constant (e.g., 60 Hz in North America). Perhaps the foremost concern during steady-state analysis is economic operation of the system; reliability is also important as the system must be operated to avoid outages should disturbances occur. The primary analysis tool for steady-state operation is the so-called **power flow analysis**, where the voltages and power flow through the system are determined. This analysis is used for both operation and planning studies and throughout the system at both the high and low transmission voltages.

The power system can be roughly separated into three subcomponents: **generation, transmission and distribution**, and **load**. The transmission and distribution network consists of power transformers, transmission lines, capacitors, reactors, and protection devices. The vast majority of generation is produced by synchronous generators. Loads consist of a large number and a diverse assortment of devices, from home appliances and lighting to heavy industrial equipment and sophisticated electronics. As such, modeling the aggregate effect is a challenging problem in power system analysis. In the following sections, the appropriate models for these components in the steady-state are introduced.

7.2.1 Modeling

Transformers

A transformer is a device used to convert voltage levels in an alternating current (ac) circuit. The device has numerous uses in power systems. To begin, it is more efficient to transmit power at high voltages and low current than low voltages and high current. Conversely, lower voltages are safer and more economic for end use. Thus, transformers are used to step-up voltages from generators and then used to stepdown the voltage for end use. Another wide use of transformers is instrumentation: sensitive equipment can be isolated from the high voltages and currents of the transmission system. Transformers may also be used as means of controlling real power flow by phase shifting.

Transformers function by the linkage of magnetic flux through a core of ferromagnetic material. Figure 7.1(A) illustrates a magnetic core with a single winding. When a current I is supplied to the first set of windings, called the **primary windings**, a magnetic field, H, will develop, and magnetic flux, ϕ, will flow in the core. Ampère's law relates the enclosed current to the magnetic field encountered on a closed path. If H is constant throughout the path, then:

$$Hl = NI, \tag{7.1}$$

where l is the path length through the core; N is the number of turns of the winding on the core so that NI is the enclosed current by the path referred to as the **magnetomotive force** (mmf).

The **magnetic field** is related to the **magnetic flux** by the properties of the material, specifically the permeability. If a linear relationship is assumed (i.e., neglecting hysteresis and saturation effects), then the flux density, B, or the flux, ϕ, is as follows:

$$B = \mu H = \mu\frac{NI}{l} \text{ or } \phi = \mu A\frac{NI}{l}, \tag{7.2}$$

where A is the cross-sectional area of the core. This relationship between the flux flow in the core and the mmf is called the **reluctance**, R, of the core so that:

$$R\phi = NI. \tag{7.3}$$

Now, if a second set of windings, the secondary windings, is wrapped around the core as shown in Figure 7.1(B), the two currents will be linked by magnetic induction. Assuming that no flux flows outside the core, then the two windings will see the exact same flux, ϕ. Because the two windings also see the same core reluctance, the two mmfs are identical:

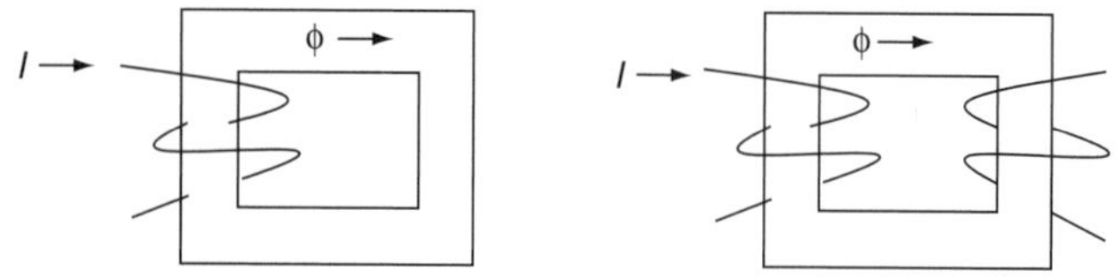

(A) Flux Flowing Through (B) Flux as Linked

FIGURE 7.1 (A) The Flux Flows Through Core from First Winding; (B) The Flux is Linked to a Second Set of Windings.

$$N_1 I_1 = N_2 I_2. \tag{7.4}$$

If the flux ϕ or, equivalently, the current I are changing in time, then according to Faraday's law, a voltage will be induced. Assuming this ideal transformer has no losses, the power input will be the same as the power output, so:

$$V_1 I_1 = V_2 I_2, \tag{7.5}$$

where V_1 and V_2 are the primary and secondary voltages, respectively. Substituting (7.4) and rearranging shows:

$$\frac{V_2}{V_1} = \frac{N_2}{N_1}. \tag{7.6}$$

Thus, the voltage gain in an ideal transformer is simply the ratio of the primary and secondary windings' turns. A practical transformer experiences several non-ideal effects. Specifically, these include nonzero winding resistance, finite permeability of the core, eddy currents that flow in the core, hysteresis (the effect arising from the energy required to reorient the magnetic dipoles as the magnetic polarity changes), and magnetic saturation. For steady-state studies of the large system, linear circuit models are desired. These effects are typically modeled as a combination of series and parallel impedances in the following way:

- **Series impedances**: Because the transformer core has a finite permeability, some of the magnetic flux flows outside the core. This leakage flux will not link the primary and secondary windings. Thus, the voltage at the input sees not only the voltage that links the primary and secondary windings but also a voltage drop caused by this leakage inductance. Similarly, the finite winding resistance causes an additional voltage drop to be seen at the terminals.
- **Shunt impedances**: Finite permeability implies nonzero core reluctance and requires current to magnetize the core (i.e., a nonzero mmf). This difference between the primary and secondary mmfs can be modeled as a shunt inductance. Hysteresis and eddy currents lead to energy losses in the core that can be approximately modeled by a shunt resistor. Saturation is an important nonlinear effect that results in additional losses and the creation of odd order harmonics in the current and voltage signals. Because in steady-state system analysis only the 60-Hz component of the currents and voltages is considered, saturation effects are typically ignored.

An equivalent circuit for the transformer model just described is shown in Figure 7.2.

The main difficulty with the model in Figure 7.2 as it now stands concerns the ideal transformer component. Carrying this component around in the calculations creates unnecessary complexity. Further, from an engineering point of view, the

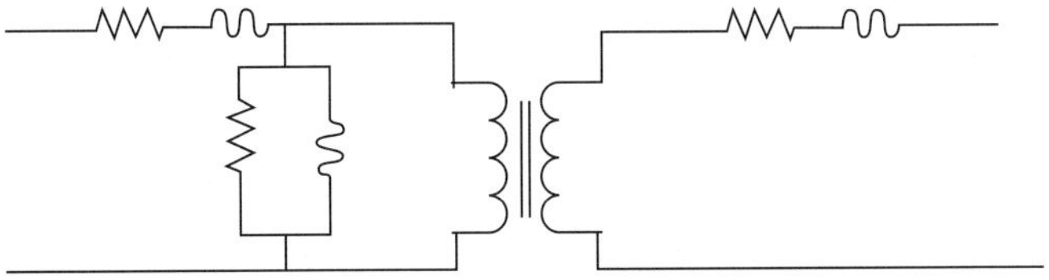

FIGURE 7.2 Transformer Circuit Model

voltages and currents in the system are most easily seen relative to their rated values. Thus, most system analysis is done on a normalization called the per unit system. In the per unit system, a system power base is established, and the rated voltages at each point in the network are determined. All system variables are then given relative to this value. These base quantities for the currents can be found as:

$$I_B = \frac{S_B}{V_B}, \tag{7.7}$$

and for impedances:

$$Z_B = \frac{V_B}{I_B} = \frac{V_B^2}{S_B}. \tag{7.8}$$

This normalization has the great added advantage of reducing the need to represent the ideal transformer in the circuit. One must simply keep track of the nominal base voltage in each part of the network. In this way, the equivalent transformer model is as given in Figure 7.3. Note that phase shifting and off-nominal transformer ratios result in asymmetric circuits and require some additional manipulation in the per unit framework. Those details are omitted here for brevity.

Transmission Line Parameters

As mentioned previously, electric power is transmitted in three phases. This accounts for the common site of three lines or for dual circuits of six lines seen strung between transmission towers. Typically, a high-voltage transmission line has several feet of spacing between the three conductors. The conductors are stranded wire for improved mechanical as well as electrical properties. If the currents are expected to be large, several conductors may be strung per phase. This improves cooling

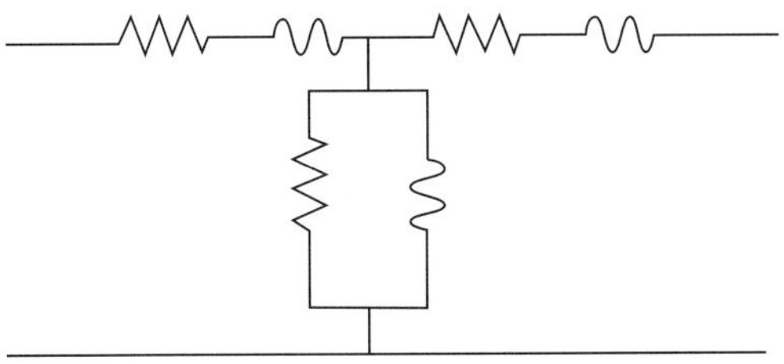

FIGURE 7.3 Simplified Transformer Circuit Model Under Per Unit System

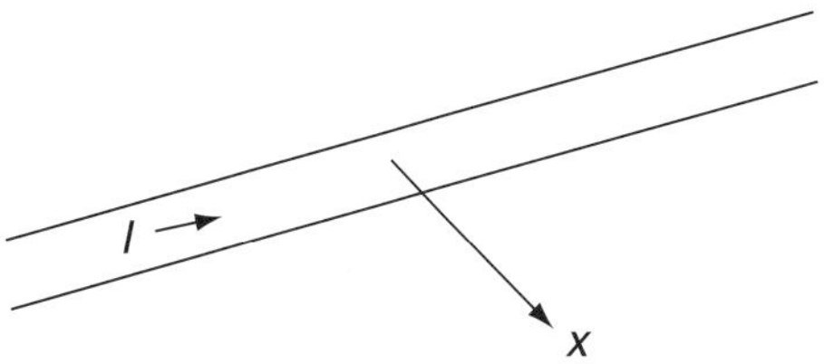

FIGURE 7.4 Infinite Transmission Line

compared to using one large conductor. This geometry is important as it impacts the electrical properties of the line.

As current flows in each conductor, a magnetic field develops. Adjacent lines then may induce voltages in nearby conductors through mutual induction (as seen for transformers, only now with coupling that is not as tight). This interaction largely determines the inductance seen by the respective phase currents. To understand this phenomenon, consider a single line of radius r and infinite length with some current flow, I, as sketched in Figure 7.4. Similar to what is done for the transformer development, apply Ampère's law to characterize the magnetic field. The magnetic field at some distance x from the line can be found by assuming that the field is constant at all points equal distance from the line. The closed path is then a circle with circumference $2\pi x$, which gives:

$$2\pi x H = I \text{ or } H = \frac{I}{2\pi x}. \tag{7.9}$$

If x is less than the line radius, the closed path will not link all of the current. Assuming an equal distribution of current throughout the wire:

$$2\pi x H = I\frac{x^2}{r^2} \text{ or } H = \frac{Ix}{2\pi r^2}. \tag{7.10}$$

Once again, the flux density is determined by the permeability of the material, which in this case is either the conductor itself if $x < r$ or that of free space for $x > r$. Then, the flux relationship is the following:

$$d\phi = \mu \frac{Ix}{2\pi r^2} dx. \tag{7.11}$$

For $x > r$, the flux linked up to some radial distance R per unit of length is simply:

$$\lambda_{\text{external}} = \int_r^R \mu_0 \frac{I}{2\pi x} dx = \frac{\mu_0 I}{2\pi} \ln \frac{R}{r}. \tag{7.12}$$

For $x < r$, only the enclosed current will be linked. Continuing with the even distribution of current assumption, the flux linked is as written here:

$$\lambda_{\text{internal}} = \int_0^r \mu_c \frac{Ix^3}{2\pi r^4} dx = \mu_c \frac{I}{8\pi}. \tag{7.13}$$

For simplicity, assume the permeability of the conductor is that of free space, and then the total flux linkage is as follows:

$$\lambda = \frac{\mu_0 I}{8\pi} + \frac{\mu_0 I}{2\pi} \ln\frac{R}{r} = \frac{\mu_0 I}{2\pi} \ln\frac{R}{re^{-1/4}}. \tag{7.14}$$

Typically, $re^{-1/4}$ is written as r'. Consider a three-phase transmission line of phase currents I_a, I_b, and I_c, with each line spaced equally by the distance D. Flux from each of the currents will link with each of the other conductors. The flux linkage for phase a out to some point R a far distance away from the conductors is approximately:

$$\lambda_a = \frac{\mu_0}{2\pi}\left(I_a \ln\frac{R}{r'} + I_b \ln\frac{R}{D} + I_c \ln\frac{R}{D}\right). \tag{7.15}$$

Assuming balanced currents (i.e., $I_a + I_b + I_c = 0$) and recalling that inductance is simply the ratio of flux linkage to current, the series inductance in phase a per unit length of line will be the following:

$$\tilde{L}_a = \frac{\mu_0}{2\pi} \ln\frac{D}{r'}. \tag{7.16}$$

In practice, the phase conductors may not be equally spaced as they are in Figure 7.5. This results in unbalanced conditions due to the imbalance in mutual inductance. High-voltage transmission lines with such a layout can be transposed so that, on average, the distance between phases is equal canceling out the imbalance. The equivalent distance of separation between phases can then be found as the geometric mean of this spacing. Similarly, if several conductors are used per phase, an equivalent conductor radius can be found as the geometric mean.

Transmission lines also exhibit capacitive effects. That is, whenever a voltage is applied to a pair of conductors separated by a nonconducting medium, charge accumulates, which leads to capacitance. Similar to the previous development for inductance, the capacitance can be determined based on Gauss's law. For a point P at a distance x from a conductor with charge q, the electric flux density D is:

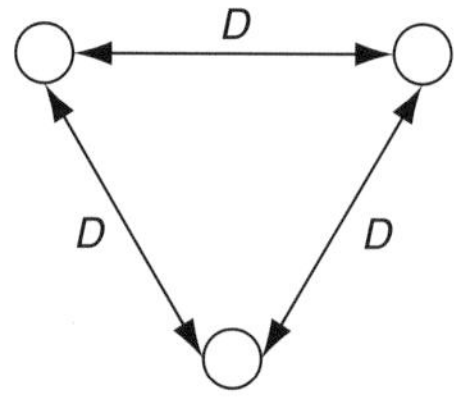

FIGURE 7.5 End View of Equally Spaced Phase Conductors

$$D = \frac{q}{2\pi x}. \tag{7.17}$$

Assuming a homogeneous medium, the electric field density E is related to D by the permitivity ε of the dielectric, which, in this case, will be assumed to be that of free space:

$$E = \frac{q}{2\pi\varepsilon_0 x}. \tag{7.18}$$

Integrating E over some path (a radial path is chosen for simplicity) yields the voltage difference between the two end points:

$$V_{12} = \int_{R_1}^{R_2} \frac{q}{2\pi\varepsilon_0 x} dx = \frac{q}{2\pi\varepsilon_0} \ln\frac{R_1}{R_2}. \tag{7.19}$$

Now consider a three-phase transmission line again with each line spaced equally by the distance D. Superposition holds so that the voltage arising from each of the charges can be added. To find the voltage from phase to ground arising from each of the conductors, assume a balanced system with $q_a + q_b + q_c = 0$ and a neutral system located at some far distance R from phase a:

$$V_{an} = \frac{1}{2\pi\varepsilon_0}\left(q_a \ln\frac{R}{r} + q_b \ln\frac{R}{D} + q_c \ln\frac{R}{D}\right) = \frac{1}{2\pi\varepsilon_0} q_a \ln\frac{D}{r}. \tag{7.20}$$

Now since the capacitance is the ratio of charge to voltage, the capacitance from phase a to ground per unit length of line will be:

$$\tilde{C}_{an} = \frac{q_a}{V_{an}} = \frac{2\pi\varepsilon_0}{\ln D/r}. \tag{7.21}$$

If the conductors are not evenly spaced, transposition results in an equivalent geometric mean distance, and using bundled conductors per phase can also be accommodated by using a geometric mean.

Finally, conductors have finite resistances that depend on the temperature, the frequency of the current, the conductor material, and other such factors. For most systems analysis problems, these can be based on values provided by manufacturers or from tables for commonly used conductors and typical ambient conditions.

Transmission Line Circuit Models

Transmission lines may be classified according to their total length. If the line is around 50 miles or less, a so-called **short line**, capacitance can be neglected, and the series inductance

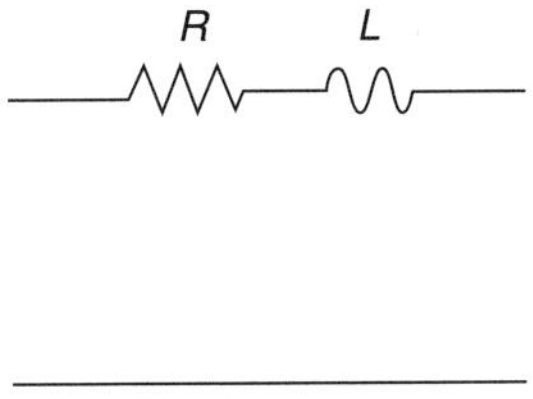

FIGURE 7.6 Short Line Model

and resistance can be modeled as lumped parameters. Figure 7.6 depicts the short line model per phase. The series resistance and inductance are simply found by calculating the per unit distance parameters times the line length; therefore, at 60 Hz for line length l, the line impedance is the following:

$$Z = \tilde{R}l + j120\pi\tilde{L}l = R + jX. \quad (7.22)$$

For lines longer than 50 miles, up to around 150 miles, capacitance can no longer be neglected. A reasonable circuit model is to simply split the total capacitance evenly with each half represented as a shunt capacitor at each end of the line. This is depicted as the π–circuit model in Figure 7.7. Again, the total capacitance is simply the per unit distance capacitance times the line length:

$$Y = j120\pi\tilde{C}_{an}l = jB. \quad (7.23)$$

For line lengths longer than 150 miles, the lumped parameter model may not provide sufficient accuracy. To see this, note that at 60 Hz for a low-loss line, the wavelength is around 3000 miles. Thus, a 150-mile line begins to cover a significant portion of the wave, and the well-known wave equations must be used. The relationship between voltage and current at a point x (i.e., distance along the line) to the receiving end voltage, V_r, and current, I_r, is seen through the following equations:

$$V(x) = V_r\cosh\,\gamma x + I_r Z_c \sinh\,\gamma x. \quad (7.24)$$

$$I(x) = I_r\cosh\,\gamma x + \frac{V_r}{Z_c}\sinh\,\gamma x. \quad (7.25)$$

The $Z_c = \sqrt{z/y}$ is the **characteristic impedance** of the line, and $\gamma = \sqrt{zy}$ is the **propagation constant**. It would be useful,

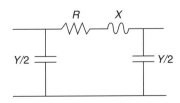

FIGURE 7.7 Medium Line Model

and it turns out to be possible, to continue to use the π–model for the transmission line and simply modify the circuit parameters to represent the distributed parameter effects. The relationship between the sending end voltage, V_s, and current, I_s, to the receiving end voltage and current in a π–model can be found as:

$$V_s = V_r\left(1 + \frac{YZ}{2}\right) + I_r Z. \quad (7.26)$$

$$I_s = V_r Y\left(1 + \frac{YZ}{4}\right) + I_r\left(1 + \frac{YZ}{2}\right). \quad (7.27)$$

Now equating 7.24 and 7.25 for a line of length l to equations 7.26 and 7.27 and solving shows the equivalent shunt admittance and series impedance for a long line as:

$$Z' = Z_c\sinh\,\gamma l = Z\frac{\sinh\,\gamma l}{\gamma l}. \quad (7.28)$$

$$\frac{Y'}{2} = \frac{1}{Z_c}\tanh\,\frac{\gamma l}{2} = \frac{Y}{2}\frac{\tanh\,\gamma l/2}{\gamma l/2}. \quad (7.29)$$

In these equations, the prime indicates the modified circuit values arising from a long line.

Generators

Three-phase synchronous generators produce the overwhelming majority of electricity in modern power systems. Synchronous machines operate by applying a dc excitation to a rotor that, when mechanically rotated, induces a voltage in the armature windings due to changing flux linkage. The per phase flux for a balanced connection can be written as:

$$\lambda = K_f I_f \sin\theta_m, \quad (7.30)$$

where I_f is the field current, θ_m is the angle of the rotor relative to the armature, and K_f is a constant that depends on the number of windings and the physical properties of the machine. The machine may have several poles so that the armature will "see" multiple rotations for each turn of the rotor. So, for example, a four pole machine appears electrically to be rotating twice as fast as two pole machine. For a machine rotating at ω_m radions per second with p poles, the electric frequency is as follows:

$$\omega_s = \omega_m\frac{p}{2}, \quad (7.31)$$

with ω_s as the desired synchronous frequency. If the machine is rotated at a constant speed Faraday's law indicates that the induced voltage can be written as:

$$V = \frac{d\lambda}{dt} = K_f I_f \omega_s \sin(\omega_s t + \theta_0). \quad (7.32)$$

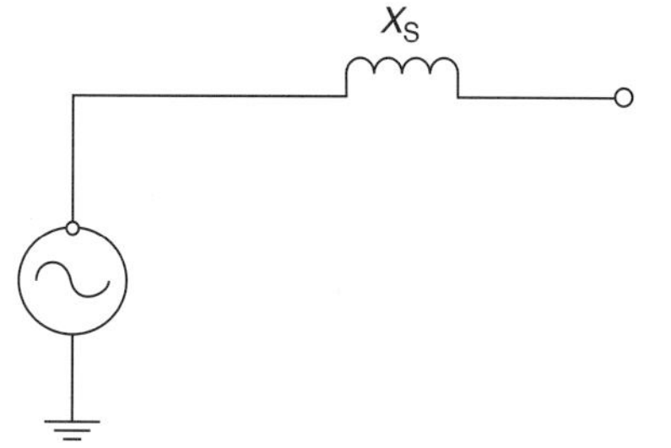

FIGURE 7.8 Simple Synchronous Generator Model

If a load is applied to the armature windings, then current will flow and the armature flux will link with the field. This effectively puts a mechanical load on the rotor, and power input must be matched to this load to maintain the desired constant frequency. Some of the armature flux "leaks" and does not link with the field. In addition, there are winding resistive losses, but those are commonly neglected. The circuit model shown in Figure 7.8 is a good representation for the synchronous generator in the steady-state. Note that most generators are operated at some fixed terminal voltage with a constant power output. Thus, for steady-state studies, the generator is often referred to as a *PV* bus since the terminal node has fixed power P and voltage V.

Loads

Modeling power system loads remains a difficult problem. The large number of different devices that could be connected to the network at any given time renders precise modeling intractable. Broadly speaking, loads may vary with voltage and frequency. In the steady-state, frequency is constant, so the only concern is voltage. For most steady-state analysis, a fixed (i.e., constant over an allowable voltage range) power consumption model can be used. Still, some analysis requires consideration of voltage effects to be useful, and then the traditional exponential model can be used to represent real power consumption P and reactive power consumption Q as:

$$P = P_0 V^a. \tag{7.33}$$

$$Q = Q_0 V^b. \tag{7.34}$$

In these equations, the voltage V is normalized to some rated voltage. The exponents a and b can be 0, 1, or 2 where they could represent constant power, current, or impedance loads, respectively. Alternatively, they can represent composite loads with a generally ranging between 0.5 and 1.8 and b ranging between 1.5 and 6.0.

7.2.2 Power Flow Analysis

Power flow equations represent the fundamental balancing of power as it flows from the generators to the loads through the transmission network. Both real and reactive power flows play equally important roles in determining the power flow properties of the system. Power flow studies are among the most significant computational studies carried out in power system planning and operations in the industry. Power flow equations allow the computation of the bus voltage magnitudes and their phase angles as well as the transmission line current magnitudes. In actual system operation, both the voltage and current magnitudes need to be maintained within strict tolerances for meeting consumer power quality requirements and for preventing overheating of the transmission lines, respectively. The difficulty in computing the power flow solutions arises from the fact that the equations are inherently nonlinear because of the balancing of power quantities. Moreover, the large size of the power network implies that power flow studies involve solving a very large number of simultaneous nonlinear equations. Fortunately, the sparse interconnected nature of the power network reflects itself in the computational process, facilitating the computational algorithms.

In this section, we first study a simple power flow problem to gain insight into the nonlinear nature of the power flow equations. We then formulate the power flow problem for the large power system. A classical power flow solution method based on the Gauss-Seidel algorithm is studied. The popular Newton-Raphson algorithm, which is the most commonly used power flow method in the industry today, is introduced. We then briefly consider the fast decoupled power flow algorithm, which is a heuristic method that has proved quite effective for quick power flow computations. Finally, we will discuss the dc power flow solution that is a highly simplified algorithm for computing approximate linear solutions of the power flow problem and is becoming widely used for electricity market calculations.

Simple Example of a Power Flow Problem

Let us consider a single generator delivering the load $P + jQ$ through the transmission line with the reactance x. The generator bus voltage is assumed to be at the rated voltage, and it is at 1 per unit (pu). The generator bus angle is defined as the phasor reference, and hence, the generator bus voltage phase angle is set to be zero. The load bus voltage has magnitude V and phase angle δ. Because the line has been assumed to be lossless, note that the generator real power output must be equal to the real

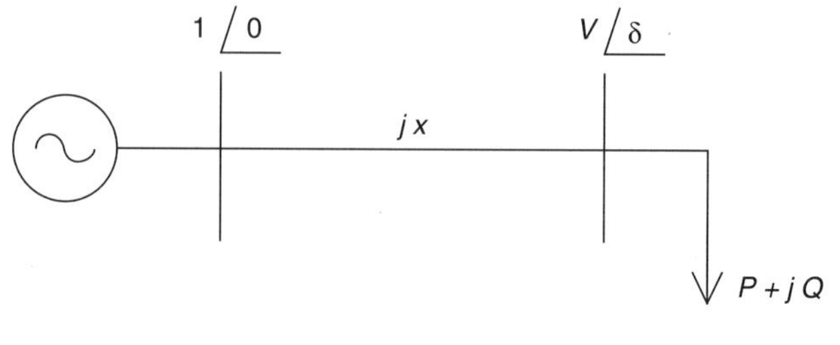

FIGURE 7.9 A Simple Power System

power load P. However, the reactive power output of the generator will be the sum of the reactive load Q and the reactive power "consumed" by the transmission line reactance x.

Let us write down the power flow equations for this problem. Given a loading condition $P + jQ$, we want to solve for the unknown variables, namely, the bus voltage magnitude V and the phase angle δ. For simplicity, we will assume that the load is at unity power factor of $Q = 0$. The line current phasor $\boldsymbol{I}$ from the generator bus to the load bus is easily calculated as:

$$\boldsymbol{I} = \frac{1\angle 0 - V\angle\delta}{jx}. \tag{7.35}$$

Next, the complex power $\boldsymbol{S}$ delivered to the load bus can be calculated as:

$$\boldsymbol{S} = \boldsymbol{V}\boldsymbol{I}^* = \frac{V\angle(\delta + \pi/2)}{x} - \frac{V^2\angle(\pi/2)}{x}. \tag{7.36}$$

Therefore, we get the real and reactive power balance equations:

$$P = \frac{-V\sin\delta}{x} \text{ and } Q = \frac{-V^2 + V\cos\delta}{x}. \tag{7.37}$$

After setting $Q = 0$ in equation 7.37, we can simplify equation 7.37 into a quadratic equation in V^2 as follows:

$$V^4 - V^2 + x^2P^2 = 0. \tag{7.38}$$

Therefore, given any real power load P, the corresponding power flow solution for the bus voltage V can be solved from equation 7.38. We note that for nominal load values, there are two solutions for the bus voltage V, and they are the positive roots of V^2 in the next equation:

$$V^2 = \frac{1 \pm \sqrt{1 - 4x^2P^2}}{2}. \tag{7.39}$$

Equation 7.39 implies that there exist two power flow solutions for load values $P < P_{max}$ where $P_{max} = 1/(2x)$, and there exist no power-flow solutions for $P > P_{max}$. A qualitative plot of the power flow solutions for the bus voltage V in terms of different real power loads P is shown in Figure 7.10.

From the plot and from the analysis thus far, we can make the following observations:

1. The dependence of the bus voltage V on the load P is very much nonlinear. It has been possible for us to compute the power flow solutions analytically for this simple system. In the large power system with hundreds of generators delivering power to thousands of loads, we have to solve for thousands of bus voltages and their phase angles from large coupled sets of nonlinear power flow equations, and the computation is a nontrivial task.

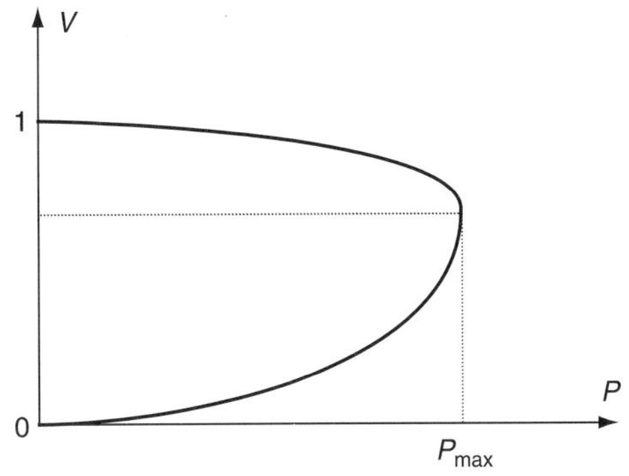

FIGURE 7.10 Qualitative Plot of the Power-Flow Solutions

2. Multiple power flow solutions can exist for a specified loading condition. In Figure 7.10, there exist two solutions for any load $P < P_{max}$. Among the two solutions, the solution on the upper locus with voltage V near 1 pu is considered the nominal solution. For the solutions on the lower locus, the bus voltage V may be unacceptably low for normal operation. The lower voltage solution also requires higher line current to deliver the specified load P, and the line current values can become unacceptably high. In general, for any specified loading condition, we would like to locate the power flow solution that has the most acceptable values of voltages and currents among the multiple power flow solutions. In this example of a single generator delivering power to a single load, there exist two power flow solutions. In a large power system, there may exist a very large number of possible power flow solutions.
3. Once the bus voltage V has been computed from equation 7.39, the bus voltage phase angle δ can be computed from equation 7.37. Then, the line current phasor $\boldsymbol{I}$ can be solved from equation 7.35. Specifically, we would like to ensure that the magnitude of the line current I stays below the thermal limit of the transmission line for preventing potential damage to the expensive transmission line.
4. Power flow solutions may fail to exist at high loading conditions, such as when $P > P_{max}$ in Figure 7.10. The loading value P_{max} beyond which power flow solutions do not exist is called the **static limit** in the power literature. Because power flow solutions denote the steady-state operating conditions in our formulation, lack of power flow solutions implies that it is not possible to transfer power from the generator to the load in a steady-state fashion, and the dynamic interactions of the generators and the loads become significant. Operating the power system at loading conditions

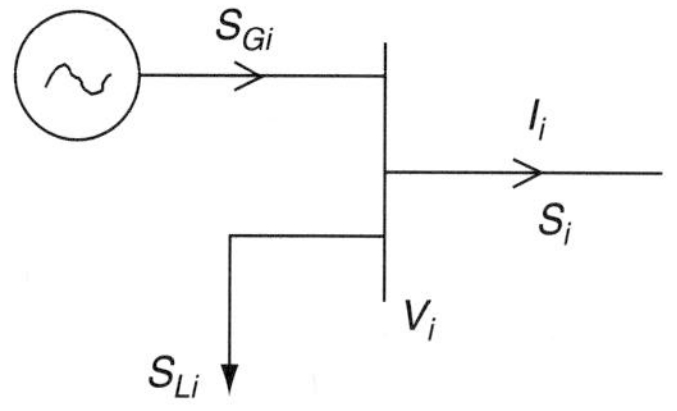

FIGURE 7.11 Complex Power Balance at Bus i

beyond the static limit may lead to catastrophic failure of the system.

Power flow Problem Formulation

In this subsection, we will construct the power flow equations in a structured manner using the admittance matrix $\boldsymbol{Y}_{\text{bus}}$ representation of the transmission network. The admittance matrix $\boldsymbol{Y}_{\text{bus}}$ is assumed to be known for the system under consideration. Let us first look at the complex power balance at any bus, say bus i, in the network.

The power balance equation is given by:

$$\boldsymbol{S}_i = \boldsymbol{V}_i \boldsymbol{I}_i^* = \boldsymbol{S}_{G_i} - \boldsymbol{S}_{L_i}. \tag{7.40}$$

Let us denote the vector of bus voltages as $\underline{\boldsymbol{V}}_{\text{bus}}$ and the vector of bus injection currents as $\underline{\boldsymbol{I}}_{\text{bus}}$. By definition, the admittance matrix $\underline{\boldsymbol{Y}}_{\text{bus}}$ provides the relationship $\underline{\boldsymbol{I}}_{\text{bus}} = \underline{\boldsymbol{Y}}_{\text{bus}} \underline{\boldsymbol{V}}_{\text{bus}}$. Suppose the ith or jth entry $\boldsymbol{Y}_{ij}$ of the $\underline{\boldsymbol{Y}}_{\text{bus}}$ matrix has the magnitude Y_{ij} and the phase γ_{ij}. Then, we can simplify the current injection $\boldsymbol{I}_i$ as:

$$\boldsymbol{I}_i = \sum_j \boldsymbol{Y}_{ij} \boldsymbol{V}_j = \sum_j Y_{ij} V_j \angle(\delta_j + \gamma_{ij}). \tag{7.41}$$

Then, combining equations 7.40 and 7.41, we get the complex power balance equations for the network as:

$$\boldsymbol{S}_i = \boldsymbol{S}_{G_i} - \boldsymbol{S}_{L_i} = \sum_j Y_{ij} V_i V_j \angle(\delta_i - \delta_j - \gamma_{ij}). \tag{7.42}$$

Taking the real and imaginary parts of the complex equation 7.42 gives us the real and reactive power flow equations for the network:

$$P_i = P_{G_i} - P_{L_i} = \sum_j Y_{ij} V_i V_j \cos(\delta_i - \delta_j - \gamma_{ij}). \tag{7.43}$$

$$Q_i = Q_{G_i} - Q_{L_i} = \sum_j Y_{ij} V_i V_j \sin(\delta_i - \delta_j - \gamma_{ij}). \tag{7.44}$$

Generally speaking, our objective in this section is to solve for the bus voltage magnitudes V_i and the phase angles δ_i when the power generations and loads are specified. For a power system with N buses, there are $2N$ number of power flow equations. At each bus, there are six variables: P_{Gi}, Q_{Gi}, P_{Li}, Q_{Li}, V_i, and δ_i. Depending on the nature of the bus, four of these variables will be specified at each bus, leaving two unknown variables at each bus. We will end up with $2N$ unknown variables related by $2N$ equations 7.43 and 7.44, and our aim in the rest of this section is to develop algorithms for solving this problem.

Let us consider a purely load bus first, that is, with $P_{Gi} = Q_{Gi} = 0$. In this case, the loads P_{Li} and Q_{Li} are assumed to be known either from measurements or from load estimates, and the bus voltage variables V_i and δ_i are the unknown variables. Purely load busses with no generation support are called **PQ busses** in power flow studies because both real-power injection P_i and reactive power Q_i have been specified at these busses.

Typically, every generator in the system consists of two types of internal controls, one for maintaining the real power output of the generator and the other for regulating the bus voltage magnitude. In power flow studies, we usually assume that both these control mechanisms are operating perfectly and so the real power output P_{Gi} and V_i are maintained at their specified values. Again, the load variables P_{Li} and Q_{Li} are also assumed to be known. This leaves the generator reactive output Q_{Gi} and the voltage phase angle δ_i as the two unknown variables for the bus. In terms of injections, the real power injection P_i and the bus voltage V_i are then the specified variables; thus, the generator busses are normally denoted **PV busses** in power flow studies.

In reality, the generator voltage control for keeping the bus voltage magnitude at a specified value becomes inactive when the control is pushed to the extremes, such as when the reactive output of the generator becomes either too high or too low. This voltage control limitation of the generator can be represented in power flow studies by keeping track of the reactive output Q_{Gi}. When the reactive generation Q_{Gi} becomes larger than a prespecified maximum value of $Q_{Gi,\,\max}$ or goes lower than a prespecified minimum value $Q_{Gi,\,\min}$, the reactive output is assumed to be fixed at the limiting value $Q_{Gi,\,\max}$ or $Q_{Gi,\,\min}$, respectively, and the voltage control is disabled in the formulation; that is, the reactive power Q_{Gi} becomes a known variable, either at $Q_{Gi,\,\max}$ or $Q_{Gi,\,\min}$, and the voltage V_i then becomes the unknown variable for bus i. In power flow terminology, we say that the generator at bus i has "reached its reactive limits" and, hence, bus i has changed from a PV bus to a PQ bus. Owing to space limitations, we will not discuss generator reactive limits in any more detail in this section.

In addition to PQ busses and PV busses, we also need to introduce the notion of a **slack bus** in the power flow formulation. Note that power conservation demands that the real power generated from all the generators in the network must equal the sum of the total real power loads and the line losses on the transmission network:

$$\sum_i P_{G_i} = \sum_i P_{L_i} + \sum_i \sum_j P_{\text{losses}_{ij}}. \quad (7.45)$$

The line losses associated with any transmission line in turn depend on the line resistance and the line current magnitude. As stated earlier, one of the main objectives of power flow studies is to compute the line currents, and as such, the line current values are not known at the beginning of a power flow computation. Therefore, we do not know the actual values for the line losses in the transmission network. Looking at equation 7.45, we need to assume that at least of one of the variables P_{Gi} or P_{Li} should be a free variable for satisfying the real power conservation. Traditionally, we assume that one of the generations is a "slack" variable, and such a generator bus is denoted the slack bus. At the slack bus, we specify both the voltage V_i and the angle δ_i. The power injections P_i and Q_i are the unknown variables. Again, by tradition, we set the voltage at slack bus to be the rated voltage or at 1 pu and the phase angle to be at zero.

Like in standard textbooks, slack bus is defined in this section to be the first bus in the network with $V_1 = 1$ and the angle $\delta_1 = 0$. Assuming the number of generators to be N_G, the busses 2 through $N_G + 1$ are set to be the PV busses. The remaining buses $N_G + 2$ through N are then the PQ busses.

Gauss-Seidel Algorithm

Let us consider a set of simultaneous linear equations of the form $\underline{A}\underline{x} = \underline{b}$, where $\underline{A}$ is an $n \times n$ matrix and $\underline{x}$ and $\underline{b}$ are $n \times 1$ vectors. Clearly, there exists a unique solution to the problem when the matrix $\underline{A}$ is invertible, and the solution is given by $\underline{x} = \underline{A}^{-1}\underline{b}$. When the matrix size is very large, it may not be possible to compute the inverse of the matrix $\underline{A}$ for finding the solution; there exist other numerical techniques. **Gauss-Seidel algorithm** is one such classical algorithm that tries to arrive at the solution $\underline{x} = \underline{A}^{-1}\underline{b}$ iteratively by starting from an approximate initial condition of $\underline{x}^0$. The iteration for the solution $\underline{x}^{k+1}$ from the previous iterate $\underline{x}^k$ proceeds as follows.

$$x_i^{k+1} = \frac{1}{a_{ii}}\left(b_i - \sum_{j<i} a_{ij}x_j^{k+1} - \sum_{j>i} a_{ij}x_j^k\right) \text{ for } i = 1, 2, \ldots, n \quad (7.46)$$

Here, a_{ij} denotes the ith or jth entry of the matrix $\underline{A}$ as usual. It can be shown that the iterative solution $\underline{x}^k$ converges to the exact solution $\underline{A}^{-1}\underline{b}$ for any initial condition $\underline{x}^0$, provided the matrix $\underline{A}$ satisfies certain "diagonal dominance" properties. The details are limited here to save space, and they can be found in standard numerical analysis textbooks.

In the previous section, we formulated the power flow problem as a set of simultaneous nonlinear equations of 7.42 and as such, it is not obvious how the Gauss-Seidel algorithm can be applied for solving these equations. The trick here is to visualize the power balance equations to be arising from the network admittance equations $\underline{Y}_{\text{bus}}\ \underline{V}_{\text{bus}} = \underline{I}_{\text{bus}}$. The matrix $\underline{Y}_{\text{bus}}$ takes over the role of the matrix $\underline{A}$ in the linear equations. We will be solving for the bus voltage vector $\underline{V}_{\text{bus}}$. The current injections $\underline{I}_{\text{bus}}$ are not known per se. The current injections are in fact dependent on the bus voltages. As we see next, they can also be computed iteratively from the power injections $\boldsymbol{S}_i$ by using the relationship $\boldsymbol{I}_i = \boldsymbol{S}_i^*/\boldsymbol{V}_i^*$. For a PQ bus, the injection $\boldsymbol{S}_i$ is a specified variable and, hence, is known. For PV busses, only the real power injection P_i is known, while the reactive injection Q_i is evaluated first using the latest estimate of bus voltages $\underline{V}_{\text{bus}}$.

An outline of the Gauss-Seidel algorithm for solving the power flow equations of 7.42 is presented next. Let us start with an initial condition for the bus voltages $\underline{V}_{\text{bus}}^0$, and we would like to compute the iterate $\underline{V}_{\text{bus}}^{k+1}$ from the previous iterate $\underline{V}_{\text{bus}}^k$. Recall that bus 1 is a slack bus and, hence, $\boldsymbol{V}_1^k = 1\angle 0$ for all iterations. Also, busses 2 through $N_G + 1$ are PV busses; therefore we need to keep V_i^k at specified values $V_{i,\text{ specified}} = V_i^0$ for all iterations for the PV busses.

Gauss-Seidel Iterations

- Slack bus ($i = 1$): $\boldsymbol{V}_1^{k+1} = 1\angle 0$
- PV busses ($i = 2, \ldots, N_G + 1$):
  (1) Compute the reactive power generation at bus i.

$$Q_i^{k+1} = \sum_{j<i} \boldsymbol{Y}_{ij}\boldsymbol{V}_i^k\boldsymbol{V}_j^{k+1}\sin\left(\delta_i^k - \delta_j^{k+1} - \gamma_{ij}\right) + \sum_{j\geq i} \boldsymbol{Y}_{ij}\boldsymbol{V}_i^k\boldsymbol{V}_j^k\sin\left(\delta_i^k - \delta_j^k - \gamma_{ij}\right). \quad (7.47)$$

  (2) Update the bus voltage phasor $\boldsymbol{V}_i^{k+1}$.

$$\boldsymbol{V}_i^{k+1} = \frac{1}{\boldsymbol{Y}_{ii}}\left(\frac{P_i - jQ_i^{k+1}}{\boldsymbol{V}_i^{k*}} - \sum_{j<i}\boldsymbol{Y}_{ij}\boldsymbol{V}_j^{k+1} - \sum_{j>i}\boldsymbol{Y}_{ij}\boldsymbol{V}_j^k\right). \quad (7.48)$$

  (3) Normalize the magnitude $\boldsymbol{V}_i^{k+1}$ to be V_i^0.

$$\boldsymbol{V}_i^{k+1} = \frac{\boldsymbol{V}_i^{k+1}}{\left|\boldsymbol{V}_i^{k+1}\right|}V_i^0. \quad (7.49)$$

- PQ busses ($i = N_G + 2, \ldots, N$):
  (4) Update the bus voltage phasor $\boldsymbol{V}_i^{k+1}$.

$$\boldsymbol{V}_i^{k+1} = \frac{1}{\boldsymbol{Y}_{ii}}\left(\frac{P_i - jQ_i}{\boldsymbol{V}_i^{k*}} - \sum_{j<i}\boldsymbol{Y}_{ij}\boldsymbol{V}_j^{k+1} - \sum_{j>i}\boldsymbol{Y}_{ij}\boldsymbol{V}_j^k\right). \quad (7.50)$$

At the end of the $(k+1)$th iteration, we have the updated values of the bus voltages $\underline{V}_{\text{bus}}^{k+1}$. The values of $\underline{V}_{\text{bus}}^{k+1}$ can be compared with the previous set of values $\underline{V}_{\text{bus}}^k$ to check whether the solution has converged.

For the power flow problems, the Gauss-Seidel algorithm has been known to converge even from poor initial conditions, which is one of its main strengths. The algorithm is typically used for "flat starts" when all the initial voltage magnitudes are set to be at their rated values ($V_i^0 = 1$ pu for all the PV busses), and the bus voltage angles are set to be zero ($\delta_i^0 = 0$ for all buses). The relative simplicity of the computations involved in each iteration step in equations 7.46 to 7.49 implies that the algorithm is very fast to implement. On the other hand, the error convergence rate is typically linear in the sense that the ratios of the error norm from one iteration to the previous iteration tend to be constant. Therefore, while the Gauss-Seidel algorithm can converge to the proximity of the actual solution in tens of iterations, it typically takes a large number of iterations to get to an accurate solution estimate.

Newton-Raphson Algorithm

Unlike the Gauss-Seidel algorithm, which was originally developed for solving simultaneous linear equations, the **Newton-Raphson** (NR) algorithm is specifically designed for solving nonlinear equations. The algorithm proceeds iteratively by linearizing the nonlinear equations into linear equations at each step and by solving the linearized equations exactly.

Suppose we want to solve the nonlinear equations $\underline{F}(\underline{x}) = \underline{0}$, where $\underline{x}$ is a $n \times 1$ vector and where $\underline{F}: \Re^n \rightarrow \Re^n$ is a smooth nonlinear function. We have been given an initial condition $\underline{x}^0$. Then, for computing the estimate $\underline{x}^{k+1}$ from x^k, we first linearize the functions $\underline{F}(\underline{x})$ at $\underline{x}^k$ as follows:

$$0 = \underline{F}(\underline{x}) \approx \underline{F}(\underline{x}^k) + \left.\frac{\partial \underline{F}}{\partial \underline{x}}\right|_{\underline{x}^k} (\underline{x} - \underline{x}^k). \tag{7.51}$$

The solution to the linearized equations 7.17 and 7.18 is defined as the iteration estimate $\underline{x}^{k+1}$:

$$\underline{x}^{k+1} = \underline{x}^k - \left(\left.\frac{\partial \underline{F}}{\partial \underline{x}}\right|_{\underline{x}^k}\right)^{-1} \underline{F}(\underline{x}^k). \tag{7.52}$$

Note that the linearization of equation 7.51 will be a good approximation if the estimate $\underline{x}^k$ is close to the true solution of $\underline{x}^*$ since $\underline{F}(\underline{x}^*) = 0$. The NR algorithm stated in equation 7.52 can be proved to converge to the true solution $\underline{x}^*$ when the initial condition $\underline{x}^0$ is sufficiently close to $\underline{x}^*$. On the other hand, for initial conditions away from $\underline{x}^*$, the approximation of equation 7.51 becomes poorly justified, and the iterations can quickly diverge away from $\underline{x}^*$. When the iterations converge, owing to the linearized nature of the algorithm, the norm of the error decreases to zero in a "quadratic" fashion. Roughly speaking, the ratios of the error norm from one iteration to the square of the error norm in the previous iteration tend to be a constant. An example would be that the error norms decrease from 0.1 in one iteration, to 0.01 in the next iteration, to 0.0001 in the following iteration. Therefore, given good initial conditions, the NR algorithm can typically get to an accurate solution estimate within a few iterations.

Let us apply the NR algorithm for solving the power flow equations 7.43 and 7.44. We will solve for the unknown variables among the bus voltage magnitudes V_i and angles δ_i first. That is, we define the vector $\underline{x}$ as consisting of all the PV and PQ bus angles and all the PQ bus voltages. PV bus voltages are known and, hence, they are not included in $\underline{x}$:

$$\underline{x} = (\delta_2, \ldots, \delta_{N_G+1}, \delta_{N_G+2}, \ldots, \delta_N, V_{N_G+2}, \ldots, V_N)^T. \tag{7.53}$$

The corresponding power flow equations are as follows:

$$\underline{F}(\underline{x}) = \begin{pmatrix} P_2 - p_2(\underline{x}) \\ \cdots\cdots\cdots \\ P_{N_G+1} - p_{N_G+1}(\underline{x}) \\ P_{N_G+2} - p_{N_G+2}(\underline{x}) \\ \cdots\cdots\cdots \\ P_N - p_N(\underline{x}) \\ Q_{N_G+2} - q_{N_G+2}(\underline{x}) \\ \cdots\cdots\cdots \\ Q_N - q_N(\underline{x}) \end{pmatrix} = \begin{pmatrix} P_2 - \sum_j Y_{2,j} V_2 V_j \cos\left(\delta_2 - \delta_j - \gamma_{2,j}\right) \\ \cdots\cdots\cdots \\ P_{N_G+1} - \sum_j Y_{N_G+1,j} V_{N+1} V_j \cos\left(\delta_{N_G+1} - \delta_j - \gamma_{N_G+1,j}\right) \\ P_{N_G+2} - \sum_j Y_{N_G+2,j} V_{N+2} V_j \cos\left(\delta_{N_G+2} - \delta_j - \gamma_{N_G+2,j}\right) \\ \cdots\cdots\cdots \\ P_N - \sum_j Y_{N,j} V_N V_j \cos\left(\delta_N - \delta_j - \gamma_{N,j}\right) \\ Q_{N_G+2} - \sum_j Y_{N_G+2,j} V_{N+2} V_j \sin\left(\delta_{N_G+2} - \delta_j - \gamma_{N_G+2,j}\right) \\ \cdots\cdots\cdots \\ Q_N - \sum_j Y_{N,j} V_N V_j \sin\left(\delta_N - \delta_j - \gamma_{N,j}\right) \end{pmatrix}. \tag{7.54}$$

The entries of the $\underline{F}$ function in equation 7.54 are the differences between the specified power injections and the computed power injections from the current power flow solutions, and these are usually denoted as the **real** and **reactive power mismatches** at the different busses. In the power flow problem, we want to find a solution that makes the power mismatches in equation 7.54 equal zero.

Suppose an initial condition $\underline{x}^0$ has been specified. Then, the NR algorithm for solving the power flow equations 7.54 proceeds iteratively as follows.

Newton-Raphson Iterations

(1) Compute the power mismatches $\underline{F}(\underline{x}^k)$ for step k from equation 7.54. If the mismatches are within desired tolerance values, the iterations stop.

(2) Compute the power flow Jacobian:

$$\underline{J}^k = \left.\frac{\partial \underline{F}}{\partial \underline{x}}\right|_{\underline{x}^k}. \tag{7.55}$$

Owing to the nice structure of the equations in 7.54, explicit formulas can be derived for the entries of the Jacobian matrix, and the Jacobian for step k can be evaluated by substituting the current values of $\underline{x}^k$ into these formulas.

(3) Compute the correction factors $\Delta\underline{x}^k$ from equation 7.52 by solving a set of simultaneous linear equations:

$$\underline{J}^k \Delta\underline{x}^k = -\underline{F}(\underline{x}^k). \tag{7.56}$$

The Jacobian matrix $\underline{J}^k$ is extremely sparse even for very large power systems, which facilitates the solution of $\Delta\underline{x}^k$ in equation 7.55.

(4) Evaluate $\underline{x}^{k+1}$ from $\underline{x}^k$ by adding the correction factors $\Delta\underline{x}^k$:

$$\underline{x}^{k+1} = \underline{x}^k + \Delta\underline{x}^k. \tag{7.57}$$

As compared with the Gauss-Seidel algorithm, each iteration step in the NR algorithm is computationally much more intensive because of (a) evaluating the Jacobian and (b) solving the linear equations 7.55. On the other hand, the error convergence rate of the NR algorithm is spectacularly faster, and hence, the NR algorithm requires much fewer iterations to reach comparable solution accuracies. In usual practice, the Gauss-Seidel algorithm is used only for flat starts with poorly known initial conditions. In most other situations, the NR algorithm is the preferred choice.

Fast Decoupled Power Flow Algorithm

Both the Gauss-Seidel algorithm and the Newton-Raphson algorithm are general methods for solving linear and nonlinear equations respectively, and they were tailored toward solving the power flow problem. On the other hand, we study a specific method for power systems in this section called the **fast decoupled power flow algorithm**, which is a heuristic method that is derived by exploiting specific properties for a power system.

The fast decoupled power flow algorithm is essentially a highly simplified and approximated version of the Newton-Raphson algorithm of the previous subsection. We recall that the NR iteration steps are computation intensive because of evaluating $\underline{J}^k$ and solving the Jacobian equation 7.55. In this subsection, we will proceed to simplify this by replacing the iteration specific $\underline{J}^k$ with a constant matrix.

It is a well-known property of power systems that variations in the bus voltage magnitudes mostly affect the reactive power injections under nominal operating conditions. Similarly, the variations in the bus voltage phase angles mostly influence the real power injections. By idealizing this property, we assume that all the Jacobian entries of the form $\frac{\partial p_i}{\partial V_j}$ and $\frac{\partial q_i}{\partial \delta_j}$ are all identically zero. Next, if we assume that the bus voltage magnitudes are all close to one pu, and the voltage phase angle differences on the two ends of any transmission line are all close to zero, the NR algorithm greatly simplifies to the fast decoupled algorithm stated in equation 7.58.

Let us split the power flow state vector $\underline{x}$ in equation 7.53 into the voltage magnitude $\underline{V}$ and angle δ counterparts:

$$\underline{\delta} = (\delta_2, \ldots, \delta_{N_G+1}, \delta_{N_G+2}, \ldots, \delta_N)^T, \quad \underline{V} = (V_{N_G+2}, \ldots, V_N)^T. \tag{7.58}$$

Similarly, we separate the real and reactive power mismatches in equation 7.54:

$$\Delta\underline{P} = \begin{pmatrix} P_2 - p_2(\underline{x}) \\ \cdots\cdots\cdots \\ P_{N_G+1} - p_{N_G+1}(\underline{x}) \\ P_{N_G+2} - p_{N_G+2}(\underline{x}) \\ \cdots\cdots\cdots \\ P_N - p_N(\underline{x}) \end{pmatrix} \text{ and } \Delta\underline{Q} = \begin{pmatrix} Q_{N_G+2} - q_{N_G+2}(\underline{x}) \\ \cdots\cdots\cdots \\ Q_N - q_N(\underline{x}) \end{pmatrix}. \tag{7.59}$$

Suppose initial conditions for the voltage magnitude vector $\underline{V}^0$ and angle vector δ^0 have been specified. The fast decoupled algorithm then proceeds iteratively as follows.

Fast Decoupled Iterations

(1) Compute the real and reactive power mismatches $\Delta\underline{P}(\underline{x}^k)$ and $\Delta\underline{Q}(\underline{x}^k)$. If the mismatches are within desirable tolerance, the iterations end.

(2) Normalize the mismatches by dividing each entry by its respective bus voltage magnitude:

$$\Delta\underline{\tilde{P}}^k = \begin{pmatrix} \Delta P_2^k / V_2^k \\ \cdots\cdots\cdots \\ \Delta P_{N_G+1}^k / V_{N_G+1}^k \\ \Delta P_{N_G+2}^k / V_{N_G+2}^k \\ \cdots\cdots\cdots \\ \Delta P_N^k / V_N^k \end{pmatrix} \text{ and } \Delta\underline{\tilde{Q}}^k = \begin{pmatrix} \Delta Q_{N_G+2}^k / V_{N_G+2}^k \\ \cdots\cdots\cdots \\ \Delta Q_N^k / V_N^k \end{pmatrix}. \tag{7.60}$$

(3) Solve for the voltage magnitude and angle correction factors $\Delta\underline{V}^k$ and $\Delta\underline{\delta}^k$ by using the constant matrices $\overline{B}$ and $\overline{\overline{B}}$, which are extracted from the bus admittance matrix $\underline{\boldsymbol{Y}}_{\text{bus}}$:

$$\overline{B}\Delta\underline{\delta}^k = \Delta\underline{\tilde{P}}^k \text{ and } \overline{\overline{B}}\Delta\underline{V}^k = \Delta\underline{\tilde{Q}}^k. \tag{7.61}$$

Define $B_{ij} = \text{imag}(\boldsymbol{Y}_{ij})$. The matrices $\overline{B}$ and $\overline{\overline{B}}$ are constructed as follows:

$$\overline{B} = -\begin{bmatrix} B_{2,2} & \cdots & B_{2,N} \\ \cdots & \cdots & \cdots \\ B_{N,2} & \cdots & B_{N,N} \end{bmatrix} \text{ and}$$
$$\overline{\overline{B}} = -\begin{bmatrix} B_{N_G+2,\,N_G+2} & \cdots & B_{N_G+2,\,N} \\ \cdots & \cdots & \cdots \\ B_{N,\,N_G+2} & \cdots & B_{N,\,N} \end{bmatrix} \tag{7.62}$$

(4) Update the voltage magnitude and angle vectors.

$$\underline{\delta}^{k+1} = \underline{\delta}^k + \Delta\underline{\delta}^k, \; \underline{V}^{k+1} = \underline{V}^k + \Delta\underline{V}^k. \tag{7.63}$$

By using the constant matrices $\overline{B}$ and $\overline{\overline{B}}$ in equation 7.60, the time taken for evaluating one iteration for the fast decoupled algorithm is considerably less than that of the Newton-Raphson algorithm. In repeated power flow runs of the same power system, the inverses of the matrices $\overline{B}$ and $\overline{\overline{B}}$ can directly be stored, which enables the implementation to be very fast. However, the convergence speed of the fast decoupled algorithm is not quadratic, and it takes considerably more iterations to converge to an accurate power flow solution. Fast decoupled algorithm is used in applications where quick approximate estimates of power flow solutions are required.

Direct Current Power Flow Algorithm

A further simplification of the fast decoupled algorithm is the highly approximate dc power flow algorithm, which completely transforms the nonlinear power flow equations into linear equations by using drastic assumptions. In addition to the assumptions used in deriving the fast decoupled method, we also assume that all the voltage magnitudes are at 1 pu and all the transmission lines are lossless.

With these assumptions, the voltage correction factors $\Delta \underline{V}^k$ become irrelevant in equation 7.60. Moreover, the angle variables can be solved explicitly by the linear equations:

$$\overline{B}\underline{\delta} = \underline{P}, \tag{7.64}$$

where $\underline{P}$ is the vector of bus power injections. The resulting solution for the bus voltage phase angles is called the **dc power flow model**. It gives approximate values for the phase angles across the power system. The phase angles can be used to approximate the real power flow on any transmission line by dividing the phase angle difference between the two ends of the transmission lines by the line reactance. The advantage of the dc power flow model is its extreme simplicity in finding a power flow solution. The limitations of the solution need to be kept in mind, however, in light of the drastic assumptions that were used to simplify the nonlinear power flow equations into linear equations.

7.3 Dynamic Analysis

The power system in practice is constantly undergoing changes because of such factors as changing loads, planned outages of equipment for maintenance or other disturbances (e.g., equipment failures, line faults, lightning strikes etc), or any number of other events that cause outages. During disturbances, the precise balance between generation and load is not maintained. These disturbances may lead to oscillations, and the system must be able to dampen these and reach a viable steady-state operating condition. Extremely fast electromagnetic transients, such as those that arise from lightning strikes or switching actions, are not considered from a system point of view, hence, the network models introduced in the previous sections are still valid. Dynamic models for generator units models still must be introduced to understand system response. Load dynamics are also important, particularly from large induction motors, but such details are beyond the scope here. This section focuses on the transient response of the system over fractions of a second to oscillations over several seconds and then up to slow dynamics that occur with voltage problems over several minutes.

7.3.1 Modeling

Electric Generators

To understand modeling generators for dynamic analysis, more details on the physical construction are needed. Most **generators** are three-phase synchronous machines, which means they are designed to operate at a constant frequency. The machine rotor is driven mechanically by the prime mover governing system to control power output and speed. Synchronous machine rotors can be classified as either **cylindrical** or **salient pole**. Cylindrical rotors have an even air gap at all points between the stator, i.e., the stationary part of the machine, and the rotor. This construction is used for machines that rotate at high speed, typically steam-driven generators. Steam generators generally are two-pole or four-pole machines and so rotate, in North America, at 1800 RPM or 3600 RPM to produce the desired 60 Hz signal. Hydro generators, conversely, may have numerous pole pairs, as it is more efficient to drive them at lower speeds. These generators have a salient pole construction that leads to a variable air gap between the stator and the rotor.

For modeling purposes, the pole construction is important because it represents the mechanical speed of rotations and also because the transfer of power from the rotor to the stator through mutual inductance depends on the size of the air gap (i.e., the reluctance of the air gap and, thus, the effective coupling). During disturbances, these variations are most evident, and the effective circuit inductances must be modeled accurately. In modern power system modeling for dynamic analysis, a rotating frame of reference is chosen to represent these effects. The circuit equations are then written in terms of direct and quadrature axes. For simplicity, the more involved rotating frame of reference models are not developed here but instead a simple model approximating the variable inductances is presented. To begin, there are primarily two sets of windings of concern:

- **Armature windings**: The windings on the stator are referred to as the armature windings. The armature is the source from which the power generated will be drawn. A voltage is induced in the armature from the rotation of the field generated by currents on the rotor. For purposes of modeling, the self-inductance, including leakage, and mutual inductance between phases of the armature windings describe the circuit performance as current flows from the terminals. The armature windings

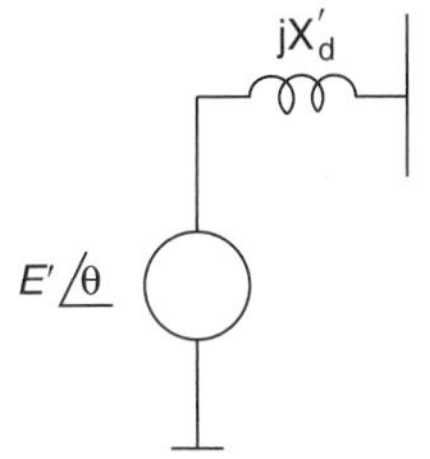

FIGURE 7.12 A Simple Model of a Synchronous Generator Armature for Dynamic Studies

are distributed in slots around the stator to produce a high-quality sinusoidal signal. Winding resistance may also be included but is small and often neglected in simple studies.

- **Field windings**: These windings reside on the rotor and provide the primary excitation for the machine. The windings are supplied with a dc current so that with rotation, they induce a voltage in the armature windings. The current is controlled by the exciter to provide the desired voltage across the armature windings. This current must be constrained to avoid overheating of the windings; the modeling of these limits and excitation of the control circuit are critical for analysis.

In addition, in salient pole machines, there are solid conducting bars in the pole faces called **damper windings**, or amortisseurs windings, that influence the effective inductance. These windings serve to damp out higher frequency oscillations. Similar effects are also seen in the cylindrical case through eddy current flows in the rotor even though damper windings may not be present. These details are not pursued further here.

The armature will be modeled as a voltage controlled by the rotor current behind a simple transient reactance as illustrated in Figure 7.12. The induced voltage $E'\angle\theta$, referred to as the voltage behind the reactance, is connected to the generator terminal through the transient reactance x_d', where the subscript indicates a direct axis quantity and the superscript a transient quantity.

To control the terminal voltage, the field current is controlled by the exciter, which in turn varies the voltage behind the reactance. A simple high gain exciter with limits can be modeled as in Figure 7.13. For the field circuit, there is a time constant associated with the winding inductance and resistance. Together, these describe the basic electromagnetic time constants for a simple generator model. There is often a supplementary stabilization control on large units referred to as a power system stabilizer (PSS). The interested reader is referred to the literature.

Mechanically, the generator is similar to a mass spring system with some frictional damping. If there is a net imbalance of torque acting on the rotor, then, neglecting damping, the machine will accelerate according to the well-known swing equations:

$$\frac{d\omega_m}{dt} = \dot{\omega}_m = \frac{1}{J}(T_m - T_e), \tag{7.65}$$

where J is the rotational inertia, T_m is the mechanical torque input, and T_e the electromagnetic torque that arises from producing an electric power output. Recall that power equals the torque times angular velocity, so multiplying both sides of equation 7.65 by ω_m yields:

$$\dot{\omega}_m \omega_m = \frac{1}{J}(P_m - P_e). \tag{7.66}$$

Typically, engineers normalize the machine inertia based on the machine rating and use the per unit inertia constant H as:

$$H = \frac{1}{2}\frac{J(\omega_m^o)^2}{S_B}, \tag{7.67}$$

with ω_m^0 the synchronous mechanical speed. If in addition the speed and the real powers are expressed on a per unit basis, then substituting equation 7.67 into equation 7.66 gives:

$$\dot{\omega} = \frac{\omega_s}{2H}(P_m - P_e) \text{ or } \dot{\omega} = \frac{\pi f_s}{H}(P_m - P_e), \tag{7.68}$$

where $\dot{\omega}$ is now the acceleration relative to synchronous speed. The electrical power output is a function of the rotor angle θ, and so we need to write the simple differential equation that relates rotor angle to speed:

$$\dot{\theta} = \omega - \omega_s. \tag{7.69}$$

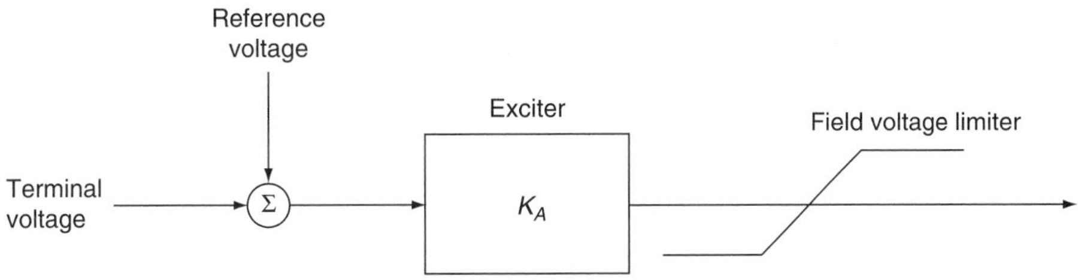

FIGURE 7.13 Basic Components of Simple Voltage Regulation and Excitation System

Thus, the mechanical equations can be expressed as a second-order dynamic system. In the next section, the analysis of this in a connected system is discussed.

7.3.2 Power System Stability Analysis

The power system is a nonlinear dynamic system consisting of generators, loads, and control devices, which are coupled together by the transmission network. The dynamic equations of the generator have been discussed in the previous section. The interactions of the generators with the loads and the control devices can result in diverse nonlinear phenomena. Specifically, we are interested in understanding whether the system response can return to an acceptable operating condition following disturbances. Normally, we distinguish between two types of disturbances in the power system context:

- **Minor disturbances** such as normal random load fluctuations, are denoted small-disturbances, and the ability of the power system to damp out such small disturbances is called **small-signal stability**. In the next section, we will learn that small-signal stability can be understood by computing the eigenvalues of the system Jacobian matrix that is evaluated at an equilibrium point.
- **Major disturbances**, such as transmission line trippings and generator outages, are denoted large disturbances, and the capability of the power system to return to acceptable operating conditions following a large disturbance is called the **transient stability**. In the following paragraphs, we will also study techniques for verifying the transient stability of a power system following a specific large disturbance.

Small-Signal Stability

Consider a nonlinear system described by the following ordinary differential equation:

$$\frac{d\underline{x}}{dt} = \underline{f}(\underline{x}),\ \underline{x} \in \Re^n,\ \underline{f}: \Re^n \rightarrow \Re^n,\ \underline{f} \text{ is smooth.} \qquad (7.70)$$

Suppose that $\underline{x}^*$ is an equilibrium point for the system. That is, $\underline{f}(\underline{x}^*) = \underline{0}$. We define the Jacobian matrix $\underline{A}$ for the equilibrium to be the matrix of partial derivatives $\frac{\partial \underline{f}}{\partial \underline{x}}$ evaluated at $\underline{x}^*$. Then, classical analysis in nonlinear dynamical system theory tells us that the equilibrium $\underline{x}^*$ is locally stable if all the eigenvalues of the matrix $\underline{A}$ have negative real parts. Recall that the eigenvalues of a matrix $\underline{A}$ are defined as the solutions λ_i of the polynomial matrix characteristic equation det $(\lambda \underline{I}_n - \underline{A}) = 0$, where $\underline{I}_n$ denotes the n–X–n identity matrix. In addition, the equilibrium will be locally unstable if any one of the eigenvalues has a positive real part.

In the power system context, the concept of local stability is known as the small-signal stability or the small-disturbance stability. We will look at a simple power system example next for studying the concept in more detail.

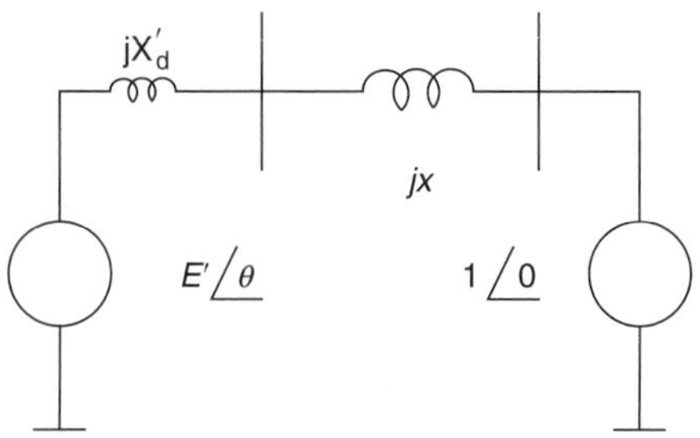

FIGURE 7.14 A Simple Power System

A single generator that is connected to an ideal generator bus through a transmission line is shown in Figure 7.14. The ideal generator maintains its bus voltage at 1 pu irrespective of the external dynamics and is referred to as an infinite bus. It also defines the reference angle for this power system, and the angle is set at zero. The other generator is represented by the classical machine model with a voltage source of $E' \angle \theta$ connected to the generator terminal through the transient reactance x_d'. In the classical machine model, the internal induced voltage E' is assumed to remain constant, and the rotor angle θ follows the second-order swing equations as in equation 7.68 but here includes a damping component P_d:

$$\frac{2H}{\omega_s}\ddot{\theta} = P_m - P_e - P_d. \qquad (7.71)$$

This can be rewritten into the standard form of:

$$\begin{pmatrix} \dot{\theta} \\ \dot{\omega} \end{pmatrix} = \begin{pmatrix} \omega - \omega_s \\ \frac{\omega_s}{2H}(P_m - P_e - P_d) \end{pmatrix}. \qquad (7.72)$$

For the power system in Figure 7.14, the electrical power output P_e is easily calculated as:

$$P_e = \frac{E' \sin\theta}{x_d' + x} = P_{\max} \sin\theta, \qquad (7.73)$$

where $P_{\max} = E'/(x_d' + x)$ is a constant. The remaining term, the damping power P_d, is usually defined as $P_d = D(\omega' - \omega_s)$, where D is known as the damping constant for the generator, which is a positive constant. Substituting the entries for P_e and P_d into equation 7.72, we get the dynamic equations for the generator as:

$$\begin{pmatrix} \dot{\theta} \\ \dot{\omega} \end{pmatrix} = \begin{pmatrix} \omega - \omega_s \\ \frac{\omega_s}{2H}(P_m - P_{\max}\sin\theta - D(\omega - \omega_s)) \end{pmatrix}. \qquad (7.74)$$

The equilibrium points for equation 7.74 can be easily solved by setting the derivatives to be zero. The $(\theta^*, \omega_s)^T$ is an equilibrium point for this system if $P_m = P_{\max}\sin\theta^*$. The equilibrium points can be identified visually by plotting P_e and P_m as shown in Figure 7.15. The intersection of the

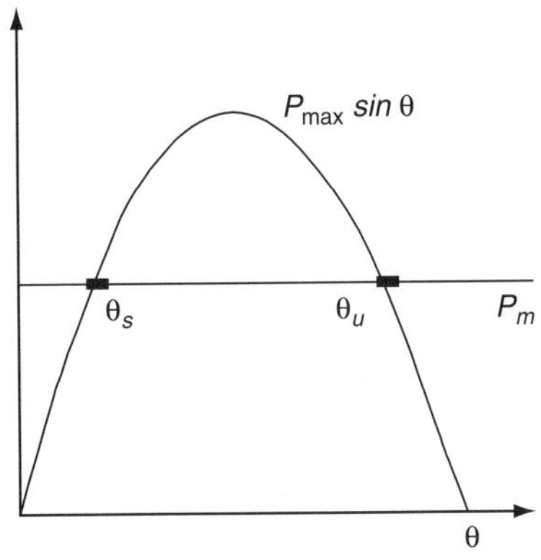

FIGURE 7.15 Plot of P_e and P_m

constant line P_m with the curve $P_{\max} \sin\theta$ highlighted by the black marks depict the two possible equilibrium points when $P_m < P_{\max}$. Let us denote the equilibrium point between 0 and $\pi/2$ by θ_s and the other equilibrium between $\pi/2$ and π by θ_u.

We will show that the equilibrium $(\theta_s, \omega_s)^T$ is small-signal stable, while the other equilibrium $(\theta_u, \omega_s)^T$ is small-signal unstable. For assessing the small-signal stability, we need to evaluate the Jacobian matrix $\underline{A}$ at the respective equilibrium point. The Jacobian is first derived as:

$$\frac{\partial f}{\partial \underline{x}} = \begin{bmatrix} 0 & 1 \\ \frac{\omega_s}{2H}(-P_{\max}\cos\theta) & \frac{\omega_s}{2H}(-D) \end{bmatrix}. \tag{7.75}$$

The eigenvalues of the system matrix $\underline{A}$ can easily be computed at the equilibrium point $(\theta_s, \omega_s)^T$ as the solutions of the second-order polynomial:

$$\lambda^2 + \frac{\omega_s}{2H} D\lambda + \frac{\omega_s}{2H} P_{\max}\cos\theta_s = 0. \tag{7.76}$$

Since θ_s by definition lies between 0 and $\pi/2$, *cos* θ_s is positive. Therefore, the two roots of the characteristic equation 7.76 have negative real parts, and the equilibrium $(\theta_s, \omega_s)^T$ is small-signal stable. Therefore, the system will return to the equilibrium condition following any small perturbations away from this equilibrium point. On the other hand, when the equilibrium point $(\theta_u, \omega_s)^T$ is considered, note that $\cos\theta_u$ is negative because θ_u lies between $\pi/2$ and π. Then, the characteristic equation 7.76 will have one positive real eigenvalue and one negative real eigenvalue when θ_s is replaced by θ_u. Therefore, the equilibrium $(\theta_u, \omega_s)^T$ is small-signal unstable.

Normally, the power system is operated only at the stable equilibrium point $(\theta_s, \omega_s)^T$. The unstable equilibrium point $(\theta_u, \omega_s)^T$ also plays an important role in determining the large disturbance response of the system that we study in the next subsection. Unlike this simple system, assessing the small-signal stability properties of a realistic large power system is an extremely challenging task computationally.

Transient Stability

Assume that the power system is operating at a small-signal stable equilibrium point. Suppose the system is suddenly subject to a large disturbance such as a line fault. Then, the power system protective relays will detect the fault, and they will isolate the faulty portion of the network by possibly tripping some lines. The occurrence of fault and the subsequent line trippings will cause the system response to move away from the equilibrium condition. After the fault has been cleared, the ability of the system to return to nominal equilibrium condition is called the **transient stability** for that fault scenario. We will study a transient stability example for the simple system in Figure 7.16.

Suppose we want to study the occurrence of a solid three-phase-to-ground fault at the middle point of the lower transmission line in Figure 7.16. We usually assume that the system is operating at a nominal equilibrium point before the fault occurs. For this prefault system, the effective transmission line reactance is $x/2$, since there are two transmission lines in parallel each with reactance x. The dynamic equations for the prefault system are then given by:

$$\begin{pmatrix} \dot\theta \\ \dot\omega \end{pmatrix} = \begin{pmatrix} \omega - \omega_s \\ \frac{\omega_s}{2H}\left(P_m - P_{\max}^{\text{pre}}\sin\theta - D(\omega - \omega_s)\right) \end{pmatrix}, \tag{7.77}$$

where $P_{\max}^{\text{pre}} = E'/x'_d + x/2$. The equilibrium points can be solved by setting the derivatives to zero in equation 7.77. Let us denote the stable equilibrium by $(\theta_s^{pre}, \omega_s)^T$ where θ_s^{pre} is the equilibrium solution of the rotor angle between 0 and $\pi/2$.

Next, let us say that the fault occurs at time t = 0. When the solid fault is present at the middle of the lower transmission line, the system in Figure 7.14 changes to the configuration shown in Figure 7.17.

The computation of the generator electrical power output P_e for the fault-on system in Figure 7.15 requires a little more work. Looking from the generator terminal bus, the effective Thevenin voltage is $1\angle 0[x/(\frac{x}{2} + x)] = \frac{1}{3}\angle 0$, which is $1/3\angle 0$. Next, the effective Thevenin reactance is the parallel equivalent of reactance x and the reactance $x/2$, which is $x/3$. Therefore, the electrical power P_e during the fault-on period is given by:

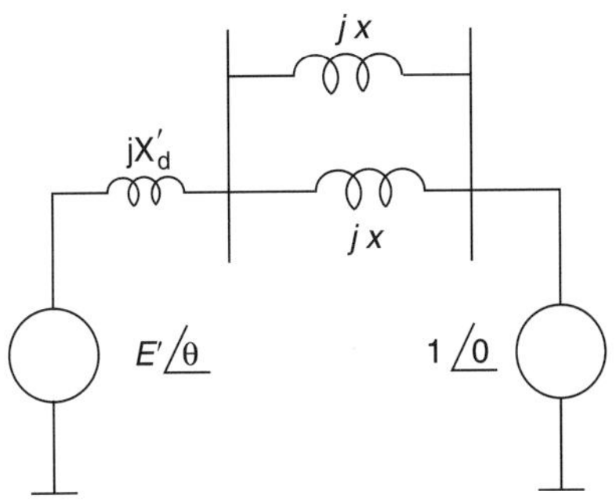

FIGURE 7.16 Prefault Power System

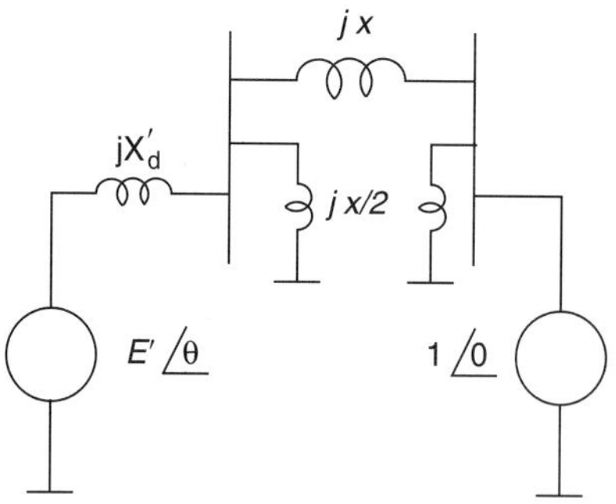

FIGURE 7.17 Fault-On Power System

$$P_e^{fault} = P_{max}^{fault} \sin\theta = \frac{E'/3}{x_d' + x/3} \sin\theta. \quad (7.78)$$

The dynamic equations for the fault-on system are then given by:

$$\begin{pmatrix} \dot{\theta} \\ \dot{\omega} \end{pmatrix} = \begin{pmatrix} \omega - \omega_s \\ \frac{\omega_s}{2H}(P_m - P_{max}^{fault} \sin\theta - D(\omega - \omega_s)) \end{pmatrix}. \quad (7.79)$$

Let us assume that the relays clear the fault at time t = t$_c$ by opening the lower transmission line at both the sending and receiving end. Then, the system configuration becomes as shown in Figure 7.14, and the dynamic equations for the postfault system are given by:

$$\begin{pmatrix} \dot{\theta} \\ \dot{\omega} \end{pmatrix} = \begin{pmatrix} \omega - \omega_s \\ \frac{\omega_s}{2H}(P_m - P_{max}^{post} \sin\theta - D(\omega - \omega_s)) \end{pmatrix}, \quad (7.80)$$

where $P_{max}^{post} = E'/xd' + x$. If the system settles down to the stable equilibrium of the postfault system of equation 7.80 after the fault is cleared, we say that the system is transient stable for the fault under study. To verify this numerically, we proceed as follows.

Transient Stability Study by Numerical Integration

(1) The system is at $(\theta_s^{pre}, \omega_s)$ before the fault occurs. Therefore, we start the simulation with $(\theta, \omega)^T = (\theta_s^{pre}, \omega_s)^T$ at time t = 0.

(2) During the fault-on period, the system dynamics is governed by equation 7.79, and the fault-on period is from time t = 0 through time t = t$_c$. Therefore, we integrate the equation 7.79 starting from the initial condition $(\theta_s^{pre}, \omega_s)^T$ at time t = 0 for a time period from t = 0 to t = t$_c$. Let us denote the system state at the end of fault-on period numerical integration at time t = t$_c$ by the clearing state $(\theta_c, \omega_c)^T$.

(3) At time t = t$_c$, the system equations change to the postfault equation 7.80. Therefore, we integrate the equation 7.80 starting from the clearing state $(\theta_c, \omega_c)^T$ at time t = t$_c$ for a period of several seconds to assess whether the system response converges to the stable equilibrium $(\theta_s^{post}, \omega_s)^T$. If the postfault response settles down to the nominal equilibrium, we can determine the system to be transient stable for the disturbance. If the postfault system response diverges away, the system is not transient stable.

The steps involved in the numerical integration procedure for a realistic large system are similar to the outline above for the simple system. The models are of very large dimensions for real-size power systems. Hence, the formulation of the prefault, fault-on, and postfault models becomes a nontrivial task, and the numerical integration becomes highly time-consuming.

Equal Area Criterion

To verify the transient stability, it is possible to derive analytical conditions for the power system in Figure 7.15. For simplicity, we assume that the damping constant $D = 0$ and that the fault clearing is instantaneous (t$_c$ = 0). We recall that the dynamics of the prefault system are described by:

$$\frac{2H}{\omega_s}\ddot{\theta} = P_m - P_{max}^{pre} \sin\theta, \quad (7.81)$$

where $P_{max}^{pre} = E'/xd' + x/2$. Similarly, the equations for the postfault system are described by:

$$\frac{2H}{\omega_s}\ddot{\theta} = P_m - P_{max}^{post} \sin\theta, \quad (7.82)$$

where $P_{max}^{post} = E'/xd' + x$. Clearly, it follows that $P_{max}^{pre} > P_{max}^{post}$ because $x/2 < x$. Intuitively, more power can be transferred in the prefault configuration with two parallel lines compared to one transmission line present in the postfault configuration.

As stated earlier, the system is operating at the prefault equilibrium $(\theta_s^{pre}, \omega_s)^T$ before the fault occurs where $\sin\theta_s^{pre} = P_m/P_{max}^{pre}$, and θ_s^{pre} lies between 0 and $\pi/2$. When the fault is cleared instantaneously at time t = 0, we would like to know whether the transient starting from $(\theta_s^{pre}, \omega_s)^T$ will settle to the postfault system equilibrium $(\theta_s^{post}, \omega_s)^T$ or whether it will diverge away. Convergence to $(\theta_s^{post}, \omega_s)^T$ or divergence implies transient stability or instability, respectively.

Let us start with the analysis. First, we note that $\theta_s^{pre} < \theta_s^{post}$ because $P_{max}^{pre} > P_{max}^{post}$. For visualization, let us plot P_e and P_m for the postfault system as shown in Figure 7.18.

The dynamics of the transient starting from $(\theta_s^{pre}, \omega_s)^T$ are governed by the second-order equation:

$$\frac{2H}{\omega_s}\ddot{\theta} = \frac{2H}{\omega_s}\dot{\omega} = P_m - P_{max}^{post} \sin\theta. \quad (7.83)$$

Therefore, the sign of the term $P_m - P_{max}^{post} \sin\theta$ determines whether the speed derivative $\dot{\omega}$ is positive or negative. Inspecting the plot Figure 7.18, we conclude that the rotor frequency ω increases whenever the rotor angle θ is below the P_m line because then $P_m - P_{max}^{post} \sin\theta$ will be positive. Similarly, the

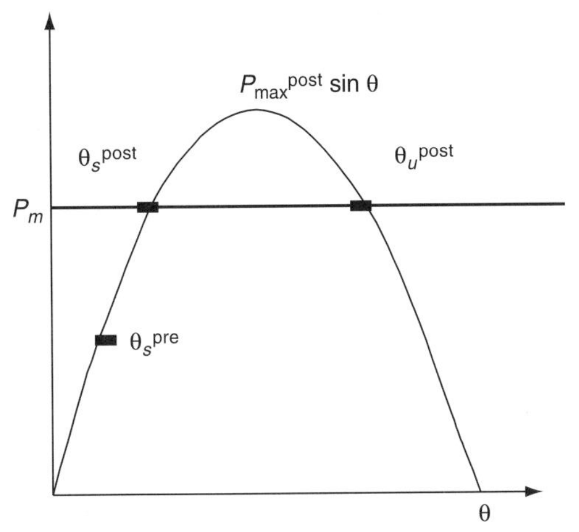

FIGURE 7.18 Power-Angle Curve for the Postfault System

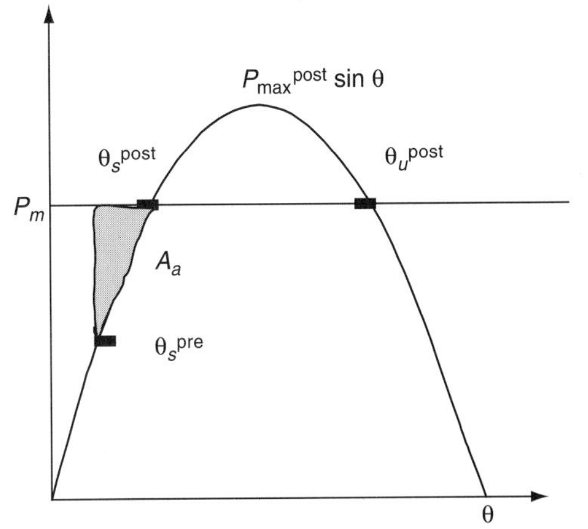

FIGURE 7.19 Acceleration Area for the Postfault System

rotor speed decreases in value when the angle θ is above the P_m line because then $P_m - P_{\max}^{\text{post}} \sin\theta$ will be negative.

Let us recall that the postfault system response starts at $(\theta_s^{\text{pre}}, \omega_s)^T$ at time t = 0. As noted earlier, the initial rotor angle θ_s^{pre} lies beneath the P_m line in Figure 7.18, hence, the speed ω increases from the initial value $\omega = \omega_s$ at time t = 0 as soon as the fault is cleared. When ω increases above ω_s, the rotor angle starts to increase from θ_s^{pre} because $\dot{\theta} = \omega - \omega_s$ for the machine dynamics. Therefore, the rotor angle moves up on the power-angle curve in Figure 7.18, and the rotor speed keeps increasing until the rotor angle reaches the value θ_s^{post} at the intersection point with the P_m line in Figure 7.18. Let us say that the angle θ takes time t_1 seconds to increase from the initial value θ_s^{pre} at time t = 0 to the value θ_s^{post}. During this time period from t = 0 to t = t_1, the speed ω has increased from ω_s to some higher value, such as ω_1. The dynamic state of the postfault system at time t = t_1 is then given by $(\theta_s^{\text{post}}, \omega_1)^T$. Note that even though the rotor angle θ equals θ_s^{post} at time t = t_1, the system is not in the equilibrium condition because the speed ω equals ω_1 at time t_1, and ω_1 is greater than the equilibrium speed value ω_s.

By construction, the speed value ω_1 is defined as follows by the dynamics of equation 7.83:

$$\omega_1 - \omega_s = \int_{t=0}^{t=t_1} \dot{\omega}\, dt = \int_{\theta_s^{pre}}^{\theta_s^{post}} \frac{\omega_s}{2H}\left(P_m - P_{\max}^{\text{post}} \sin\theta\right) d\theta = \frac{\omega_s}{2H} A_a, \quad (7.84)$$

where A_a is the shaded area shown in Figure 7.19. Because the rotor acceleration $\ddot{\theta}$ has been positive during this time period from t = 0 to t = t_1, the angle has been accelerating in the area shown, and hence, the area A_a is called the **acceleration area.**

When the transient reaches $(\theta_s^{\text{post}}, \omega_1)^T$ at time $t = t_1$, the rotor angle keeps on increasing because $\dot{\theta} = \omega_1 - \omega_s > 0$ at time t_1. However, for time $t > t_1$, the rotor angle moves above the P_m line in Figure 7.18, and hence, the derivative of speed becomes negative. That is, the speed ω starts to decrease from the value ω1 as the time increases from $t = t_1$. Only after the speed ω has decreased below the synchronous speed ω_s can the rotor angle start to decrease. Until then, the rotor angle will keep increasing, and the rotor speed keeps decreasing as time increases from $t = t_1$.

Looking at the power-angle plot in Figure 7.19, the rotor angle stays above the P_m line only up to the unstable equilibrium value θ_u^{post}. If the rotor angle were to increase above θ_s^{post}, then the speed derivative $\dot{\omega}$ becomes positive again, and the speed will start to increase. In this case, there is no scope for speed ω to decrease to the synchronous speed ω_s, and transient instability results. Therefore, for any chance of transient stability, and for response to settle down around $(\theta_s^{\text{post}}, \omega_s)^T$, we need the speed ω to decrease below ω_s before the rotor angle reaches the critical value θ_u^{post}. Graphically, this implies that the maximum deceleration area $A_d^{\max}$ shown in Figure 7.20 needs to be larger than the acceleration area A_a shown in Figure 7.19.

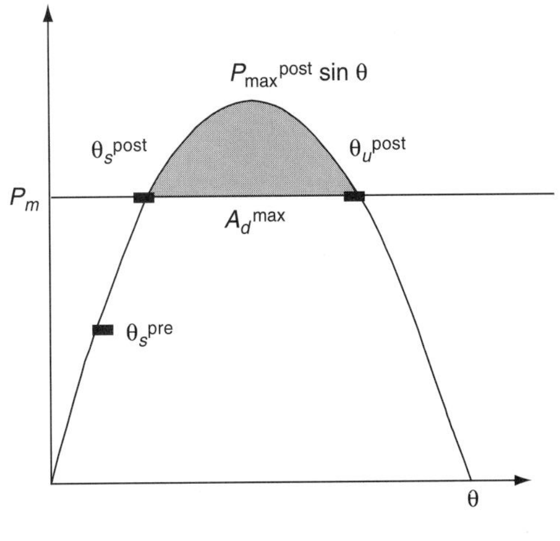

FIGURE 7.20 Maximum Deceleration Area for the Postfault System

When $A_d^{\max} > A_a$, as the rotor angle decelerates past θ_s^{post} from time $t = t_1$, the deceleration area will become equal to the accelaration area at some intermediate value of θ between θ_s^{post} and θ_u^{post} at some time of $t = t_2$. For time $t > t_2$, the speed ω falls below ω_s, and the rotor angle θ starts to decrease back toward θ_s^{post}. The alternating scenarios of rotor angle acceleration and deceleration will continue before the angle swings are damped out eventually by the rotor damping effects that have been ignored thus far. Therefore, we say that the system is transient stable whenever $A_d^{\max} > A_a$.

On the contrary, when $A_d^{\max} < A_a$, the rotor speed stays above ω_s, the rotor angle reaches θ_u^{post}. Then, the rotor speed starts to increase away from ω_s monotonously. In this case, the rotor speed never recovers below ω_s, and the rotor angle continuously keeps increasing. The transient diverges away, thus resulting in transient instability.

The analytical criterion presented in this section for the simple system can be extended to multimachine models using Lyapunov theory and based on the concepts of energy functions. Development of analytical criteria for checking the transient stability of large representative dynamic models remains a research area. Numerical integration procedures outlined in the previous sections are commonly used by the power industry for studying the transient stability properties of large power systems.

7.4 Conclusion

This chapter has introduced the readers to the basic concepts in power system analysis, namely modeling issues, power flow studies, and dynamic stability analysis. The concepts have been illustrated on simple power system representations. In real power systems, power-flow studies and system stability studies are routinely carried out for enduring the reliability and security of the electric grid separation. While the basic concepts here have been summarized in this chapter on simple examples, the real power systems are large-scale, nonlinear systems. The large interconnected nature of electric networks makes the computation aspects highly challenging. We have highlighted some of these issues in this section, and the readers are encouraged to refer to advanced power system analysis textbooks for additional details.

References

Bergen, A., and Vittal, V. (2000). *Power systems analysis, 2nd Ed.* New Jersey: Prentice Hall.

Chapman, S. (2002). *Electric machinery and power system fundamentals.* New York: McGraw Hill.

Glover, J., and Sarma, M. (1994). *Power system analysis and design, 2nd Ed.* Boston: PWD Publishing.

Grainger, J.J., and Stephenson, W.D. (1994). *Power system analysis.* New York: McGraw Hill.

Kundur, P. (1994). *Power system stability and control.* New York: McGraw Hill.

Saadat, H. (2002). *Power system analysis, 2nd Ed.* New York: McGraw Hill.

8

Power System Operation and Control

Mani Venkatasubramanian
and Kevin Tomsovic
School of Electrical Engineering and Computer Science, Washington State University, Pullman, Washington, USA

8.1 Introduction

The primary objective of power system operation is delivering power to consumers meeting strict tolerances on voltage magnitude and frequency. Accordingly, the operation control problems naturally divide into the control of voltage magnitudes or the **voltage control** issues and the control of system frequency or the **frequency control** problems. Because a power system is an interconnected, large system spread over a geographically wide network, operation of the large system is complex. The controls are built to exploit the inherent timescale and structural properties of the system. In this chapter, we focus on the frequency control problem as an example of power system controls. In fact, the **automatic frequency control** in the North American electric power grid was the first instance of a successful implementation of a large-scale network-based control scheme.

The frequency control includes two subproblems. First, we need to determine optimal values of generations that minimize the total generation costs while meeting the load demands. This problem is denoted the **economic dispatch problem** and is discussed in Section 8.2.

It can be shown that differences between total active power that is generated and the total active power that is consumed lead to frequency drifting. Because the load fluctuations themselves are random, it is not possible to exactly match the total generation with the power consumption at all times. Therefore, the system frequency will tend to drift around on its own. In North American grids, there exists a central closed-loop controller that samples system-wide power flows and frequency to maintain the system frequency within tight tolerances while also maintaining economically dispatched generations. A brief introduction to the control called the **automatic generation control** or the **load frequency control** is presented in Section 8.3.

In addition to the topics in Sections 8.2 and 8.3, the voltage control problem is another important topic because it is a complex problem. It is mostly done by distributed automatic local controls in the North American grid. However, some European countries such as France do have automatic coordinated voltage control schemes for the large network.

The operation of the power system also has to meet regulations on security and reliability. Roughly speaking, the system is required to continue normal operation even with the loss of any one component. These studies are grouped under the framework of **power system security**, which is a broad topic in itself.

8.2 Generation Dispatch

A power system must generate sufficient power at all times to meet the load demand from the grid. The amount of load connected to the system varies significantly based on seasonal and time-of-day considerations. The cost of producing power at different generators also varies from plant to plant, depending on the efficiency of plant design and fuel costs at the generator location. Therefore, it is not economical to divide the required generation capacity arbitrarily among the available generators. The problem of determining how the total load requirement is to be divided among the generators in service is denoted the **generation dispatch** problem. This problem clearly optimizes the total generation costs in producing the required amount of active power. We discuss this problem first in Secton 8.2.1 under a classical formulation. Moreover, there are also active power losses involved in transmitting real power from generators to loads. Some generators, such as those near coal mines with low fuel costs, may incur large transmission losses in transferring the generated power to load centers. An optimization formulation that includes some simple consideration of transmission losses together with generation costs is presented in Section 8.2.2. Section 8.2.3 discusses a general framework for posing detailed optimization problems in the form of optimal power flow formulation.

8.2.1 Classical Lossless Generation Dispatch

The cost C of generating power P at a thermal plant can be roughly stated by the nonlinear function:

$$C(P) = \alpha + \beta P + \gamma P^2$$

Let us assume that there are N thermal generators in the system that share the total load demand of P_D. The economic dispatch problem tries to minimize the total generation costs for generating power at the N generators while meeting the load demand P_D. In the classical lossless formulation, we also assume that there are no transmission losses involved, which simplifies the optimization considerably. For the lossless case, the power conservation equation or the power balance equation is simply stated as:

$$\sum_{i=1}^{N} P_i = P_D,$$

where P_i denotes the power generated at plant i. Then, the economic dispatch reduces to the constrained optimization problem for minimizing the total generation cost C_T:

$$\underset{P_i}{\text{Min}} \quad C_T = \sum_{i=1}^{N} C_i(P_i),$$

$$\text{subject to} \sum_{i=1}^{N} P_i = P_D.$$

The problem can be easily solved using the Lagrangian formulation by defining the Lagrangian $L(P_1, P_2, \ldots, P_N, \lambda)$ as:

$$L(P_1, P_2, \ldots, P_N, \lambda) = \sum_{i=1}^{N} C_i(P_i) + \lambda\left(P_D - \sum_{i=1}^{N} P_i\right).$$

By setting the partial derivatives of the Lagrangian L with respect to P_i and λ equal to zero, the optimal solution can be described by the conditions:

$$IC_i = \frac{dC_i}{dP_i} = \lambda \text{ for } i = 1, 2, \ldots N, \tag{8.1}$$

and

$$\sum_{i=1}^{N} P_i = P_D. \tag{8.2}$$

In other words, the generation dispatch becomes optimal when the incremental generation costs IC_i become equal at all the generators. Since the generation cost C_i is a quadratic function of P_i, the incremental cost IC_i is a linear function of P_i. Therefore, the optimality problem reduces to solving $N+1$ linear equations 8.1 and 8.2 stated above in terms of $N+1$ unknown variables P_1 through P_N and λ. The unique solution is easily solved as:

$$\lambda = \frac{P_D + \sum_{i=1}^{N} \beta_i/2\gamma_i}{\sum_{i=1}^{N} 1/2\gamma_i} \text{ and } P_i = \frac{\lambda - \beta_i}{2\gamma_i}. \tag{8.3}$$

Thus far, we have considered no limits on the generation capacity at individual plants. In reality, there are lower and upper limits, $P_{\min, i}$ and $P_{\max, i}$, on the generation output P_i. These limits can be easily incorporated into the optimization problem above by appending the inequality constraints:

$$P_{\min, i} \leq P_i \leq P_{\max, i}, \tag{8.4}$$

with the conditions of equations 8.1 and 8.2. Since these equations are linear, the optimal solution can be computed by simply freezing P_i at either $P_{\min, i}$ or $P_{\max, I}$ whenever the limit is reached while finding the optimal solution. Details can be found in any standard textbook on power system analysis.

The discussion in this section has thus far been focused on thermal generation plants. The operation costs of a hydroelectric plant are fundamentally different because there are no fuel costs involved. Here, the concern is to maintain an adequate level of water storage while maintaining required water flow through the generator turbines. Clearly, the stored water capacity depends on water in-flow and out-flow rates, and these issues are typically studied over longer time horizons as compared to the dispatch of thermal power plants. The problem of coordinating the generation outputs of hydro generators and thermal plants for meeting load demands while minimizing generation costs is called the **hydrothermal coordination**. Again, the formulation and solution details can be found in standard power system textbooks.

8.2.2 Lossy Economic Dispatch

In the real system, there will always be transmission losses associated with sending power from the generation facilities to the load busses. In the previous section, these losses were ignored, which resulted in simple solutions for the generation dispatch problem. In this section, we modify the power balance equation 8.2 to include the power losses P_L. For a given set of generations P_i, the line losses have to be computed from the line currents, which in turn require solving the corresponding power flow equations. Such a problem of optimizing the total generation costs while solving nonlinear power flow equations is called the **optimal power flow formulation**, and it will be discussed in the next section. In this section, we approximate the total line losses P_L as a direct function of the generations P_i using the equation:

$$P_L = \sum_{i=1}^{N}\sum_{j=1}^{N} B_{ij}P_iP_j + \sum_{i=1}^{N} B_{i0}P_i + B_{00}. \tag{8.5}$$

Here, the term B_{00} is constant while the coefficients B_{ij} and B_{i0} summarize the quadratic and linear dependence of line losses P_L on the generations. Equation 8.5 is called the **B matrix loss formula**, and the B coefficients in equation 8.5 are called the **loss factors.** There exists a rich history in the literature on the computation of loss factors for a given economic dispatch problem.

In this section, let us address the economic dispatch problem when the losses are represented by equation 8.5. Then, the optimization changes to:

$$\underset{P_i}{\text{Min}}\ C_T = \sum_{i=1}^{N} C_i(P_i), \tag{8.6}$$

subject to:

$$\sum_{i=1}^{N} P_i = P_D + P_L(P_1,\ P_2,\ \ldots P_N).$$

We redefine the Lagrangian $L(P_1, P_2, \ldots, P_N, \lambda)$ as:

$$L(P_1,\ P_2,\ \ldots,\ P_N,\lambda) = \sum_{i=1}^{N} C_i(P_i) + \lambda\left(P_D + P_L(P_1,\ P_2,\ \ldots P_N) - \sum_{i=1}^{N} P_i\right). \tag{8.7}$$

The optimal solution can be determined by solving these two equations:

$$IC_i = \frac{dC_i}{dP_i} = \lambda\left(1 - \frac{\partial P_L}{\partial P_i}\right) \text{ for } i = 1, 2, \ldots N. \tag{8.8}$$

$$\sum_{i=1}^{N} P_i = P_D + P_L(P_1,\ P_2, \ldots P_N). \tag{8.9}$$

Whereas equations 8.1 and 8.2 for the lossless case were linear in the variables P_i and λ, the equations 8.8 and 8.9 for the lossy case are quadratic. The solution from the lossless formulation provides an excellent initial condition, and the lossy economic dispatch solution can be computed by iterative solution strategies such as the Newton-Raphson algorithm discussed in the power system analysis chapter.

Equation 8.8 provides the relationship between the power generations P_i and the Lagrangian multiplier λ. That is, we can restate 8.8 as $\lambda = IC_i/(1 - \partial P_L/\partial P_i)$. The product term $1/(1 - \partial P_L/\partial P_i)$ denotes the penalty factor that is being multiplied to the incremental cost IC_i from the contribution of generation P_i to the line losses P_L.

Moreover, once the multiplier λ is specified, the power generations P_i can be uniquely determined from equation 8.8 since this equation is linear in P_i. Therefore, it follows that the optimal solution of the equations 8.8 and 8.9 essentially reduces to the problem of finding the Lagrangian multiplier λ. There exist excellent iterative techniques in the literature for finding the optimal solution by iterating on λ.

8.2.3 Optimal Power Flow Formulation

In previous subsection, we simplified the network power balance equations into a single power conservation equation by either ignoring the losses (equation 8.2) or by approximating the line losses (equation 8.9). However, in both earlier approaches, the network nature of the power transmission was completely ignored. In this section, we treat the power transmission in earnest by stating the power balance equations in full, which summarizes the transfer of power from the generators to the loads through the transmission network. The optimization itself has the same objective of minimizing the total cost of generation.

Let us assume that the system has N generators like before, and the total number of busses including generator and load busses is M. To simplify the notation, bus number 1 is assumed to be the slack bus. Busses 2 to N are the PV busses, and busses numbered $N+1$ through M are PQ busses. The slack bus generation P_{GI} becomes a dependent variable, while the remaining generations P_{G2} through P_{GN} are the control variables for the minimization. Real and reactive loads are denoted by P_{Di} and Q_{Di}, respectively. Suppose the ith or jth entry $\boldsymbol{Y}_{ij}$ of the $\boldsymbol{Y}_{\mathbf{bus}}$ matrix has the magnitude Y_{ij} and the phase γ_{ij}.

The basic optimal power flow problem can then be stated as:

$$\underset{P_{G_i}}{\text{Min}}\ C_T = \sum_{i=1}^{N} C_i(P_{G_i}), \tag{8.10}$$

which is subject to the real power flow equations of:

$$P_i = P_{G_i} - P_{D_i} = \sum_j Y_{ij}V_iV_j\cos\left(\delta_i - \delta_j - \gamma_{ij}\right), \tag{8.11}$$

for $i = 2, \ldots, M$, and the reactive power flow equations of:

$$Q_i = Q_{G_i} - Q_{D_i} = \sum_j Y_{ij} V_i V_j \sin(\delta_i - \delta_j - \gamma_{ij}), \quad (8.12)$$

for $i = N+1, \ldots, M$. The power flow variables, namely the bus voltages $V_{N+1}, \ldots, V_M$ and phase angles $\delta_2, \ldots, \delta_N$ become the dependent variables in the optimization procedure. In general, there may exist inequality constraints on the control variables as well as the state variables in the optimal power flow formulation.

The optimal power flow problem can be formally stated as:

$$\min_u f(x, u) \quad (8.13)$$

$$\text{subject to} \quad g(x, u) = 0 \quad (8.14)$$

$$\text{and} \quad h_i(x, u) \leq 0 \quad \text{for } i = 1, 2, \ldots, k. \quad (8.15)$$

Here u denotes the optimization control variables, and x denotes the network state variables that are dependent on u through equation 8.14. The function $f(x, u)$ is the objective function to be minimized. The equations 8.14 denote the equality constraints, and the number of equations of n in equation 8.14 matches the dimension of x. The equation 8.15 represents k different inequality constraints.

A solution to the general constrained optimization problem in equations 8.13 to 8.15 can be found by using the celebrated Kuhn-Tucker theorem. We first define a generalized Lagrangian $L(x, u, \lambda, \mu)$ for the problem as:

$$L(x, u, \lambda, \mu) = f(x, u) + \lambda^T g(x, u) + \mu^T h(x, u), \quad (8.16)$$

where $\lambda \in \mathcal{R}^n$ and $\mu \in \mathcal{R}^k$. The Kuhn-Tucker theorem provides the conditions that must be satisfied at the optimal solution as:

$$\frac{\partial f}{\partial x} + \lambda^T \frac{\partial g}{\partial x} + \mu^T \frac{\partial h}{\partial x} = 0. \quad (8.17)$$

$$\frac{\partial f}{\partial u} + \lambda^T \frac{\partial g}{\partial u} + \mu^T \frac{\partial h}{\partial u} = 0. \quad (8.18)$$

$$g(x, u) = 0. \quad (8.19)$$

$$h_i(x, u)\mu_i = 0 \text{ with either } \mu_i = 0 \text{ and } h_i(x, u) < 0 \text{ or } \mu_i > 0 \text{ and } h_i(x, u) = 0. \quad (8.20)$$

Therefore, the optimal power flow problem becomes the solution of equations 8.17 through 8.20. In large-scale power system formulations, it is very difficult to find an exact solution to the Kuhn-Tucker conditions. It is quite often sufficient to find a suboptimal solution using heuristic optimization algorithms. In recent history, excellent progress has been made in the development of such heuristic algorithms.

8.3 Frequency Control

The power system load is continually undergoing changes as individual loads fluctuate while others are energized or de-energized. Generation must precisely match these changes to maintain system frequency, a function called **load frequency control** (LFC), and at the same time must follow an appropriate economic dispatch of the units as discussed in the previous section. Together, these functions are referred to as **automatic generation control** (AGC). Accordingly, all modern power systems have centralized control centers that run software called **Energy Management Systems** (EMS) that, in addition to other functions, monitor frequency and generator outputs. Units that are on AGC will receive raise and lower signals to adjust their set points.

8.3.1 AGC

Ensuring the power balance is commonly referred to as regulation and can be sensed by changes in the system frequency. If load (including losses) exceeds the generation input, then energy must be leaving the system over time. This energy will be drawn from the kinetic energy stored in the rotating masses of the generators. Hence, the generators will begin to rotate more slowly and the system frequency will decrease. Conversely, if generation input exceeds the load, then frequency will increase. It is the responsibility of the governor on a generator to sense these speed deviations and adjust the power input (say through the opening or closing of valves on a steam unit) as appropriate.

Specifically, the governor will change the power input in proportion to the speed deviation. This is referred to as the **droop** or **speed regulation**, R, and can be expressed as:

$$\Delta P = -\frac{1}{R}\Delta f, \quad (8.21)$$

where Δf is the change in frequency and where ΔP is the resulting change in power input. The R is measured in hertz per megawatts or alternatively as a percentage of the rated capacity of a unit. Typical droops in practice are on the order of 5 to 10%. If no further action is taken, the system will then operate at this new frequency.

Figure 8.1(A) illustrates a standard scenario. Assume a simple system with a nominal frequency of 60 Hz and a 100 MW unit is operating with a 5% droop. The regulation constant is calculated to be $R = 0.05(60/100) = 0.03$ Hz/MW. If the generator unit senses a drop in frequency of 0.6 Hz, then this corresponds to a 20 MW increase in power output. In addition to governor actions, many loads are frequency sensitive (e.g., motors). For these loads, the load will decrease as the frequency drops so that the needed increase from the generators is less. This can be expressed as:

$$\Delta P = -\left(\frac{1}{R} + D\right)\Delta f, \quad (8.22)$$

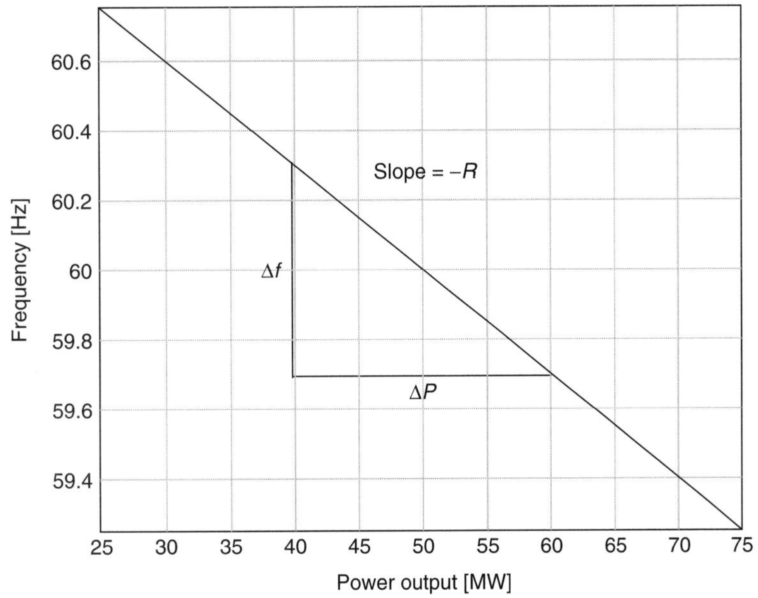

(A) Illustration of 5% Speed Droop on a 100-MW Unit

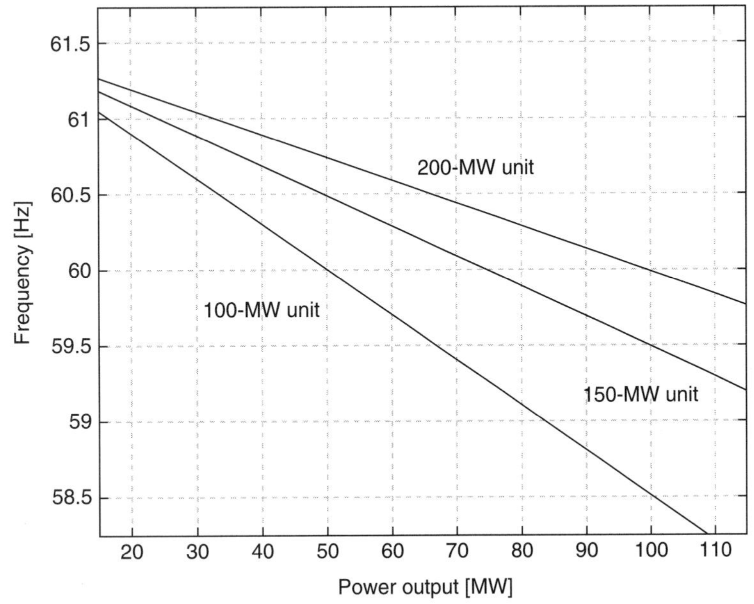

(B) Illustration of Three Units with 5% Droop

FIGURE 8.1 Speed Droop Characteristics

where D is the damping. Thus, the effective droop is slightly less (i.e., the frequency drop will be less) when load damping is considered. This load damping is often approximated as a 1% decrease in load for every 1% decrease in frequency.

Now, as would normally be true, if there are several generators interconnected and on regulation, then each will see the same frequency change (after any initial transients die out) and respond. Assuming the same percentage droop on each unit, the load will be picked up according to the relative capacities. This is depicted in Figure 8.1(B) with units of 100 MW, 150 MW and 200 MW, respectively, all operating with a 5% droop. Analytically, with R on a per unit basis, we can express this for n units as:

$$\Delta P = -\left(\frac{1}{R_1} + \frac{1}{R_2} + \cdots + \frac{1}{R_n} + D\right)\Delta f. \tag{8.23}$$

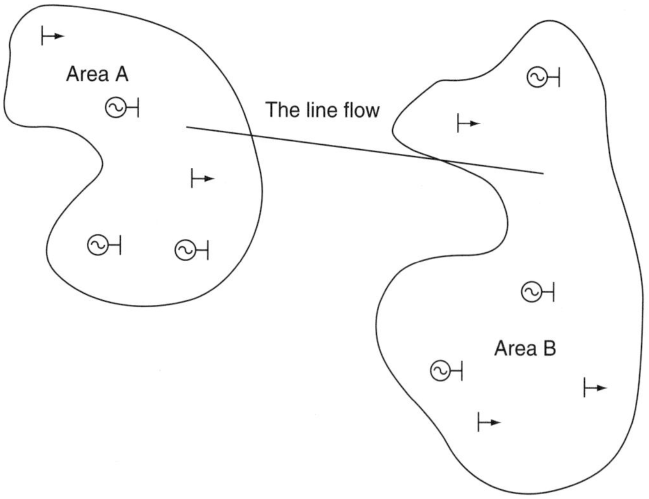

FIGURE 8.2 Multiple Control Areas Interconnected by Tie Lines

The control system as described in the previous paragraphs suffers from a serious drawback. Frequency will never return to the nominal (i.e., desired point). Supplemental control is needed to make the appropriate adjustments. One may think of this as adjusting for the initial loss of rotating kinetic energy when the load change first occurred by modifying the generator set points. This action must be coordinated among the units.

An additional issue arises in the coordination for this supplemental control. With relatively few exceptions (e.g., some islands), utilities are interconnected with their neighboring systems. Not only does this provide additional security and reliability in the case of outages, but it allows for more economic operation by judicious energy trades. Still, the utilities must coordinate their operation to maintain energy schedules and meet demand. If a load change occurs in a neighboring system, both systems will see the frequency change and respond. There is no difference from the viewpoint of the generator between the loads in the different areas. Clearly, each utility wishes only to supply loads for which it is responsible and will receive compensation.

Systems are generally broken into separate control areas reflecting the different responsibilities of the utilities as shown in Figure 8.2. The flow on tie lines between these control areas is monitored. The generator set point adjustments are made to maintain these scheduled tie flows. Note that while the first function of AGC is load following, these adjustments to generator set points may over time lead the units away from the most economic dispatch. Thus, the supplemental control may include additional adjustments for economic operation of the units. As summarized in the block diagram of Figure 8.3, the supplemental control serves several functions, including the following:

- Restoration of the nominal frequency
- Maintenance of the scheduled interchanges
- Provision for the economic dispatch of units

The coordination among areas is achieved by defining the so-called **area control error** (ACE) as a measure of the needed adjustments to the generator set points. Let the following be true:

$$\text{ACE} = \Delta P_{Tie} - 10\beta_f \Delta f, \tag{8.24}$$

where ΔP_{Tie} is the deviation in the tie line from the scheduled exchange and $\beta_f \Delta f$ is called the frequency bias, which is by tradition negative and measured in MW/0.1 Hz (thus, the multiplier 10 in equation 8.4). If the ACE for an area is

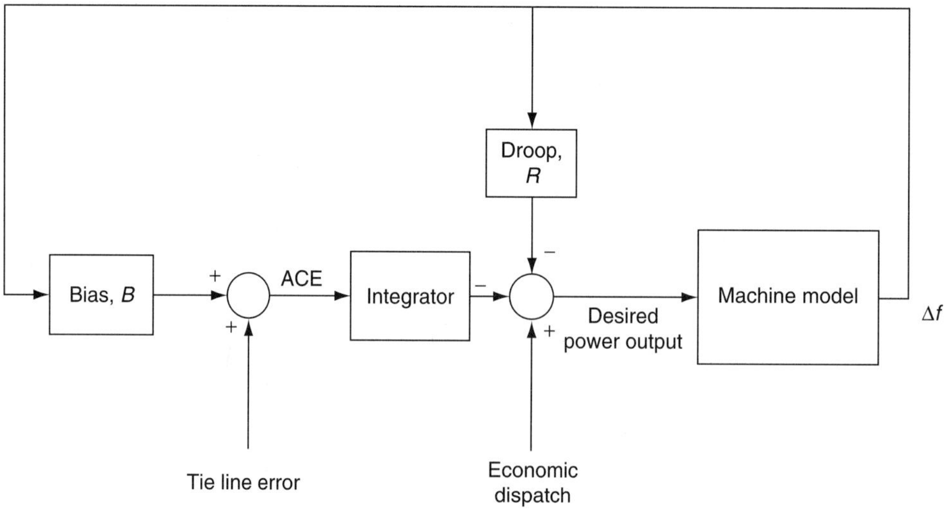

FIGURE 8.3 Block Diagram of AGC System

negative, then the area generation is too small and the unit set points should be increased.

Considering the system of Figure 8.2, notice the effect of a sudden load increase in area A on the ACE in both areas A and B. First, the frequency in both areas will decrease and, accordingly, the power output for all regulated units will increase according to their respective droop settings. Since the load change occurs in area A, there will be an additional unscheduled in-flow from B. ACE in area A will have two negative terms: **schedule error** ΔP_{Tie} and the **frequency bias**, or control, error. Conversely, ACE in area B will have a positive schedule error but a negative frequency bias.

It may seem that ideally the ACE in B would be zero and that in A it would precisely match the load change; however, this is not practical or particularly necessary. Instead, the ACE is integrated over time, and this signal is used to determine the generator set points. Integrating the error ensures that the actual energy exchange between areas is precisely maintained over time. Since the control center must calculate the ACE after gathering data from the system, the set point controls are discrete. In North America, the AGC signals are fed to the units typically, about once every 2–4 sec. In many parts of the world, such frequent adjustment is considered excessive, and supplementary control signals are sent over much less frequent time intervals. In each country, regulating agencies determine the required performance of the AGC so that utilities do not "lean" too hard on their neighbors. For example, the traditional performance criteria in North America is that the ACE has to return to zero or its predisturbance level within 15 min following the start of the disturbance.

8.4 Conclusion: Contemporary Issues

The supplementary control system as described above reflects AGC operation more or less as it has existed since the advent of computer controls in the late 1960s. Recent moves to deregulate the power system and open up the system to competition greatly complicate the traditional control philosophy. In an open market, the load responsibilities are not so clearly defined or geographically restricted. For example, an independent generator may have a contract to serve a load in a neighboring control area, and this impacts the scheduled flows on the intertie. A number of methods have been proposed to facilitate this control, including the establishment of a separate market for a "load-following" service. Consumers would pay for this service along with their energy fee. No regions in the world have fully opened AGC control to market competition, but many incremental steps have been taken. So while the supply of energy has proven to be amenable to economic competition, the vagaries of control are more difficult to formulate as an open transparent market.

9

Fundamentals of Power System Protection

Mladen Kezunovic
*Department of Electrical Engineering,
Texas A & M University,
College Station, Texas, USA*

9.1 Fundamentals of Power System Protection

This chapter defines the power system faults, the role of protective relaying, and the basic concepts of relaying. The discussion is a rather general overview. More specific issues are discussed in several excellent textbooks (Horowitz and Phadke, 1992; Blackburn, 1998; Ungrad *et al.*, 1995).

9.1.1 Power System Faults

Power systems are built to allow continuous generation, transmission, and consumption of energy. Most of the power system operation is based on a three-phase system that operates in a balanced mode, often described with a set of symmetrical phasors of currents and voltages being equal in magnitude and having the phase shifts between phases equal to 120° (Blackburn, 1993). The voltages and current behave according to Kirchhoff's laws of the electrical circuits stating that the sum of all currents entering and leaving a network node is equal to zero, and the sum of all voltage drops and gains in a given loop is also equal to zero. In addition, the voltage and currents generate electric power that integrated over a period of time produce energy. For the three-phase systems, a variety of definitions for the delivered, consumed, or transmitted power may be established as follows: **instantaneous**, **average**, **active**, **reactive**, and **complex**. The power system consists of components that are put together with a goal of matching each other regarding the power ratings, dielectric insulation levels, galvanic isolation, and number of other design goals. In the normal operating conditions, currents, voltages, power, and energy are matched to meet the design constraints.

As such, the system is capable of sustaining a variety of environmental and operating impacts that resemble normal operating conditions.

The abnormal operating conditions that the system may experience are rare but do happen. They include lightning striking the transmission lines during severe weather storms, excessive loading and environmental conditions, deterioration or breakdown of the equipment insulation, and intrusions by humans and/or animals. As a result, power systems may experience occasional faults. The **faults** may be defined as events that have contributed to a violation of the design limits for the power system components regarding insulation, galvanic isolation, voltage and current level, power rating, and other such requirements. The faults occur randomly and may be associated with any component of the power system. As a result, the power component experiences an exceptional stress, and unless disconnected or de-energized, the component may be damaged beyond repair. In general, the longer the duration of a fault, the larger is the damage. The fault conditions may affect the overall power system operation since the faulted component needs to be removed, which in turn may contribute to violation of the stability and/or loading limits. Last, but not least, the faults may present a life threat to humans and animals since the damage caused by the faults may reduce safety limits otherwise satisfied for normal operating conditions. Protective relaying was introduced in practice as early as the first power systems were invented to make sure that faults are detected and damaged components are taken out of service quickly.

To facilitate graphic representation of different types of faults, circuit diagrams shown in Figure 9.1 are used. The example is related to the faults on transmission lines and covers eleven types of most common transmission line ground and/or phase faults. To facilitate the presentation, multiple fault types are shown on the same diagram.

The rest of the discussion in this section describes the basic power system components and how different power system components are used to facilitate implementation of the protection concept. The basic requirements for the protection system solution are outlined pointing out the most critical implementation criteria.

9.1.2 Power System Components

The most basic **power system components** are generators, transformers, transmission lines, busses, and loads. They allow for power to be generated (generators), transformed from one voltage level to another (transformers), transmitted from one location to another (transmission lines), distributed among a number of transmission lines and power transformers (busses), and used by consumers (loads). In the course of doing this, the power system components are being switched or connected in a variety of different configurations using circuit breakers and associated switches (Horowitz and Phadke, 1992; Blackburn, 1998; Ungrad *et al.*, 1995). The circuit breakers are capable of interrupting the flow of power at a high energy level and, hence, may also be used to disconnect the system components on an emergency basis, such as in the case when the component experiences a fault (Flurscheim, 1985). Because the power systems are built to cover a large geographical area, the power system components are scattered across the area and interconnected with transmission lines. The grouping of the components associated with generation, switching, transformation, or consumption are called **power plants** (generation and transformation), **substations** (transformation and switching), and **load centers** (switching, transformation, and consumption). In turn, the related monitoring, control, protection, and communication gear is also located at the mentioned facilities.

To facilitate the description of power systems, a graphical representation of the power system components as shown in Figure 9.2 is used. Such representation is called a **one-line diagram**. It is reducing the presentation complexity of the three-phase connections into a single-line connection. This is sufficiently detailed when the normal system operation is considered since the solutions of voltages and currents are symmetrical and one-line representations resemble very closely the single-phase system representation used to obtain the solution. The solution for the faulted systems requires more detailed three-phase representation, but the one-line diagram is still sufficient to discuss the basic relaying concepts. In that case, a detailed representation of the faults shown in Figure 9.1 is not used, but a single symbol representing all fault types is used instead.

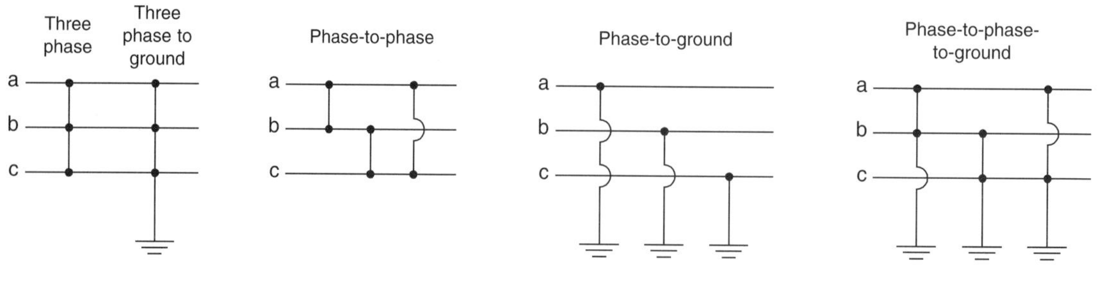

FIGURE 9.1 Eleven Types of Most Common Transmission Line Faults

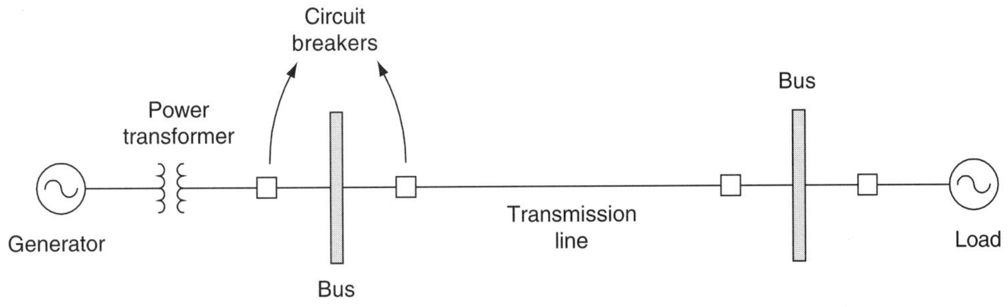

FIGURE 9.2 One-Line Representation of the Power System Components and Connection

The next level of detail is the **representation of the points** where the components merge, shown in Figure 9.2 as busses. A good example of such a point is a substation where a number of lines may come together, and a transformation of the voltage level may also take place. Figure 9.3 shows a one-line representation of a substation.

Substations come in a variety of configurations, and the one selected in Figure 9.3 is called a breaker-and-a-half. This configuration is used in high-voltage substations containing a number of transmission lines and transformers as well as different voltage levels and associated busses. This representation also includes circuit breakers and busses as the principal means of switching and/or connecting the power system components in a substation. The protective relaying role is to disconnect the components located or terminated in the substation when a fault occurs. In the case shown in Figure 9.3, the transmission line is connected to the rest of the system through two breakers marked up as "L," the bus is surrounded with several breakers connected to the bus and marked up as "B," and the power transformer is connected between the two voltage level busses with four breakers marked up as "T." In the common relaying terminology, all the breakers associated with a given relaying function are referred to as a **bay**, hence, the terminology exists of "protection bays" for a transmission line, a bus, and a transformer. It may be observed in the high-voltage substation example, given in Figure 9.3, that each breaker serves at least two protection bays. In Figure 9.3, each breaker box designated as "L" or "T" also acts as the breaker designated with "B." This property will be used later when introducing the concept of overlapping protection zones.

9.1.3 Relay Connections and Zones of Protection

Protective relays are devices that are connected to instrument transformers to receive input signals and to circuit breakers to issue control commands for opening or closing. In some instances, the relays are also connected to the communication channels to exchange information with other relays. The electronic relays always require a power supply, which is commonly provided through a connection to the station dc battery. Often, relays are connected to some auxiliary monitoring and control equipment to allow for coordination with other similar equipment and supervision by the operators. In the high-voltage power systems, relays are located in substations and, most frequently, in a control house. The connections to the instrument transformers and circuit breakers located in the substation switchyard are done through standard wiring originating from the substation switchyard and terminating in the control house.

To achieve effective protection solutions, the entire relaying problem is built around the concept of **relaying zones**. The zone is defined to include the power system component that

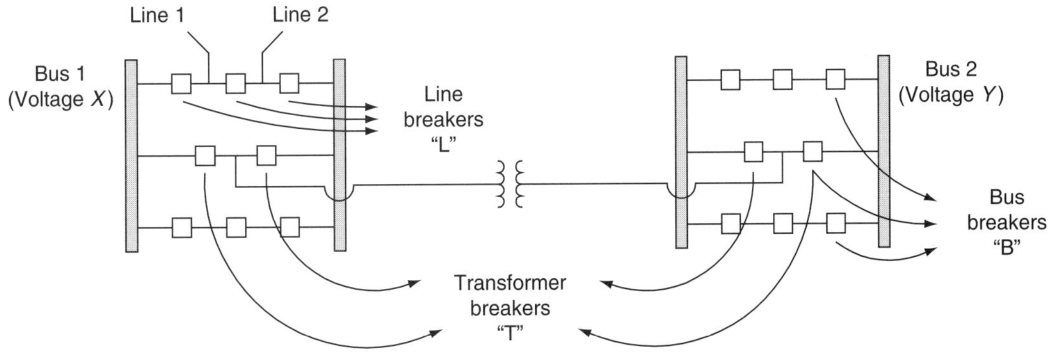

FIGURE 9.3 Breaker-and-a-Half Substation Connection

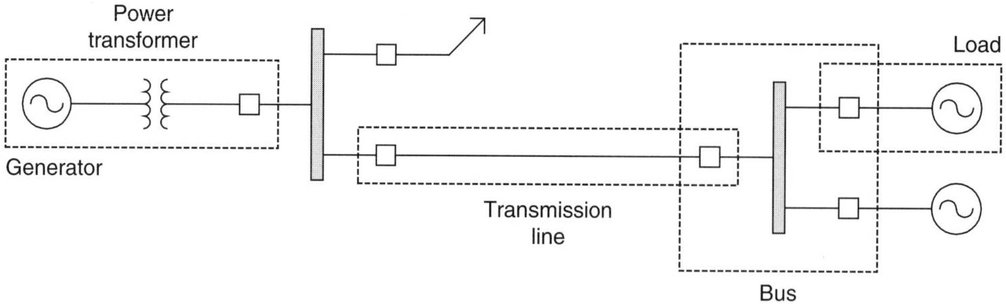

FIGURE 9.4 Allocation of Zones of Protection for Different Power System Components

has to be protected and include the circuit breakers needed to disconnect the component from the rest of the system. A typical allocation of protective relaying zones for the power system shown in Figure 9.2 is outlined in Figure 9.4.

Several points regarding the zone selection and allocation are important. First, the zones are selected to ensure that a multiple usage of the breakers associated with each power system component is achieved. Each circuit breaker may be serving at least two protection functions. This enables separation of the neighboring components in the case either one is faulted. At the same time, the number of breakers used to connect the components is minimized. By overlapping at least two zones of protection around each circuit breaker, it is important to make sure that there is no part of the power system left unprotected. This includes the short connection between circuit breakers or between circuit breakers and busses. Such an overlap is only possible if instrument transformers exist on both sides of the breaker, which is the case with so-called dead-tank breakers that have current transformers located in the breaker bushings.

Another important notion of the zones is to define a **backup coverage**. This is typical for the transmission line protection where multiple zones of protection are used for different sections of the transmission line. Figure 9.5 shows how the zones may be selected in case three zones of protection are used by each relay. The zones of protection are selected by determining the **settings** of the relay reach and the time associated with relay operation.

Each zone of protection is set to cover specific length of the transmission line, which is termed the **relay reach**. Typical selection of the zones in the transmission line protection is to cover 80 to 90% of the line in zone 1, 120–130% in zone 2, and 240–250% in zone 3. This protection is selected by locating a relay at a given line terminal and determining the length corresponding to the relay coverage as a percentage of the line length between the relay terminal and adjacent relay terminals. When doing this, the selected direction is down the transmission line starting from the terminal where the relay is located. The length of the transmission line originating from the location of the relay and ending at the next terminal is assumed to be 100%. The meaning of 120% is that the entire transmission line is covered as well as the additional 20% of the line originating from the adjacent terminal. The **times of operation** associated with zones are different: zone 1 operation is instantaneous, zone 2 is delayed to allow zone 1 relays to operate first, and zone 3 times allow the corresponding relays closer to the fault to operate first in either the zone 1 or zone 2. With this time-step approach selected for different zones of protection, the relays closest to the fault are allowed to operate first. If they fail to operate, the relays located at the remote terminals, that "see" the same fault in zone 2, will still disconnect the failed component. If zone 2 relay operation fails, relays located further away from the faulted line will operate next with the zone 3 settings. The advantage of this approach is a redundant coverage of each line section. They are also covered with multiple relay zones of the relay located on the adjacent lines, ensuring that the faulted component will be eventually removed even if the relay closest to the fault fails. The disadvantage is that each time a backup relay operates, a larger section of the system is removed from service because the relays operating in zone 2 (sometimes) or zone 3 (always) are connected to the circuit breakers that are remote from the ends of the transmission line experiencing the fault. In addition, the time to remove faulted sections from service increases as the

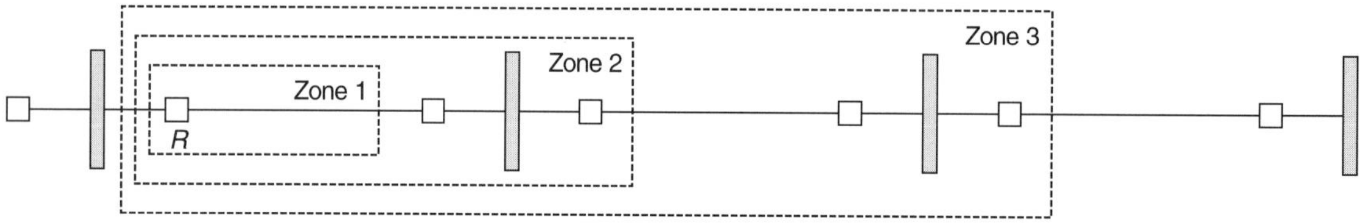

FIGURE 9.5 Selection of the Overlapping Zones for Transmission Line Protection

zone coverage responsible for the relay action increases due to the time delays associated with the zone 2 and zone 3 settings.

9.2 Relaying Systems, Principles, and Criteria of Operation

This section describes elements of a relaying system and defines the basic concept of a relaying scheme. In addition, the basic principles of protective relaying operation are discussed. More detailed discussions of each of the relaying solutions using the mentioned principles aimed at protecting different power system components are outlined in subsequent sections.

9.2.1 Components of a Relaying System

Each **relaying system** consists, as a minimum, of an instrument transformer, relay, and a circuit breaker. A typical connection for protection of high-voltage transmission lines using distance relays is shown in Figure 9.6.

In the case of Figure 9.6, since the relay measures the impedance (which is proportional to distance), both current and voltage instrument transformers are used. The relay is used to protect the transmission line, and it is connected to a circuit breaker at one end of the line. The other end of the line has another relay protecting the same line by operating the breaker at that end. In a case of a fault, both relays need to operate, causing the corresponding breakers to open and resulting in the transmission line being removed from service.

The role of **instrument transformers** is to provide galvanic isolation and transformation of the signal energy levels between the relay connected to the secondary side and the voltages and currents connected to the primary side. The original current and voltage signal levels experienced at the terminals of the power system components are typically much higher than the levels used at the input of a relay. To accommodate the needed transformation, instrument transformers with different ratios are used (IEEE, 1996; Ungrad, 1995). Next, a brief discussion of the options and characteristics of most typical instrument transformer types is given.

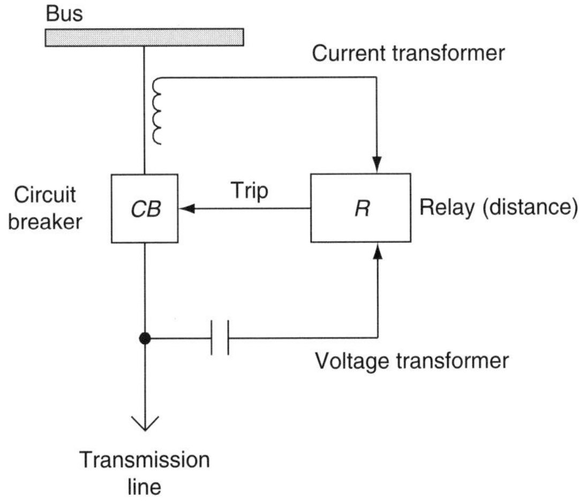

FIGURE 9.6 Protective Relaying System Consisting of Instrument Transformers, a Relay, and a Breaker

Current Transformer

Current transformers (CTs) are used to reduce the current levels from thousands of amperes down to a standard output of either 5 A or 1 A for normal operation. During faults, the current levels at the transformer terminals can go up several orders of magnitude. Most of the current transformers in use today are simple magnetically coupled iron-core transformers. They are input/output devices operating with a hysteresis of the magnetic circuit and, as such, are prone to saturation. The selection of instrument transformers is critical for ensuring a correct protective relaying operation. They need to be sized appropriately to prevent saturation. If there is no saturation, instrument transformers will operate in a linear region, and their basic function may be represented via a simple turns ratio. Even though this is an ideal situation, it can be assumed to be true for computing simple relaying interfacing requirements. If a remanent magnetism is present in an instrument transformer core, then the hysteresis may affect the time needed to saturate next time the transformer gets exposed to excessive fault signals. The current transformers come as free-standing solutions or as a part of the circuit breaker or power transformer design. If they come preinstalled with the power system apparatus, they are located in the bushings of that piece of equipment.

Voltage Transformer

Voltage transformers come in two basic solutions: **potential transformer** (PT) with iron-core construction and **capacitor coupling** voltage transformers (CVTs) that use a capacitor coupling principle to lower the voltage level first and then use the iron-core transformer to get further reduction in voltage. Both transformer types are typically free-standing. PTs are used frequently to measure voltages at substation busses, whereas CVTs may be used for the same measurement purpose on individual transmission lines. Since the voltage levels in the power system range well beyond kilovolt values, the transformers are used to bring the voltages down to an acceptable level used by protective relays. They come in standard solutions regarding the secondary voltage, typically 69.3 V or 120 V, depending if either the line-to-ground or line-to-line quantity is measured respectively. In an ideal case, both types of instrument transformers are assumed to be operating as voltage dividers, and the transformation is proportional to their turns ratio. In practice, both designs may experience specific deviations from the ideal case. In PTs, this may manifest as a nonlinear behavior caused by the effects of the hysteresis. In CVTs, the abnormalities include various ringing effects at the output when a voltage is collapsed at the input due to a close-in fault as well as impacts of the stray capacitances in the inductive transformer, which may affect the frequency response.

Relays

Another component shown in Figure 9.6 is the **relay** itself. Relays are controllers that measure input quantities and compare them to thresholds, commonly called relay settings, which in turn define operating characteristics. The relay characteristics may be quite different depending on the relaying quantity used and the relaying principle employed. A few different **operating characteristics** of various relays are shown in Figure 9.7. The first one is for the relay operating using an overcurrent principle with an inverse time-delay applied for different levels of input current. The second one is used by transmission line relays that operate using an impedance (distance) principle with the time-step zone implementation.

Further discussion of the specific relaying principles will be given later. In general, the relay action is based on a comparison between the measured quantity and the operating characteristic. Once the characteristic thresholds (**settings**) are exceeded, the relay assumes that this is caused by the faults affecting the measuring quantity, and it issues a command to operate associated circuit breaker(s). This action is commonly termed as a **relay tripping**, meaning opening a circuit breaker.

The relays may come in different designs and implementation technologies. The number of different designs at the early days when the relaying was invented was rather small, and the main technology was the electromechanical one. Today's design options are much wider with a number of different relay implementation approaches being possible. This is all due to a great flexibility of the microprocessor-based technology almost exclusively used to build relays today. Since the microprocessor-based relays use very low-level voltage signal at inputs to the signal measurement circuitry, all these relays have **auxiliary transformers** at the front-end to scale down the input signal levels even further from what is available at the secondary of an instrument transformer. To accommodate specific needs and provide different levels of the relay input quantities, some of the relay designs come with multiple connections of the auxiliary transformers called **taps**. Selecting appropriate tap determines a specific turns ratio for the auxiliary transformer that allows more precise selection of the specific level of the relay input signal. Besides the voltage and/or current signals as inputs and trip signals as outputs, the relays have a number of input/output connections aimed at other functions: coordination with other relays, communication with relays and operators, and monitoring. For an electronic relay, a connection to the power supply also exists.

Circuit Breaker

The last component in the basic relaying system is the **circuit breaker**. The breakers allow interruption of the current flow, which is needed if the fault is detected and a tripping command is issued by the relay. Circuit breakers operate based on different principles associated with physical means of interrupting the flow of power (Flurscheim, 1985). As a result, vacuum, air-blast, and oil-field breakers are commonly used depending on the voltage level and required speed of operation. All breakers try to detect the zero crossing of the current and interrupt the flow at that time since the energy level to be

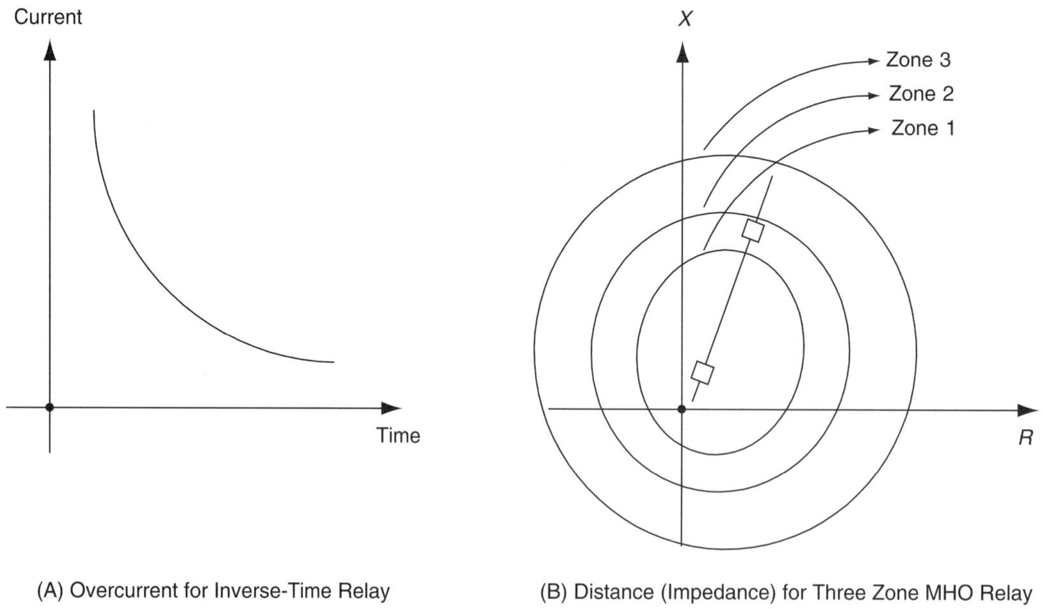

FIGURE 9.7 Typical Relay-Operating Characteristic

interrupted is at a minimum. The breakers often do not succeed in making the interruption during the first attempt and, as a result, several cycles of the fundamental frequency current signal may be needed to completely interrupt the current flow. This affects the **speed of the breaker operation**. The fastest breakers used at the high-voltage levels are one-cycle breakers, whereas a typical breaker used at the lower voltage levels may take 20 to 50 cycles to open. Circuit breakers are initiated by the relays to disconnect the power system component in the case a fault is present on the component. In the case of the transmission line faults, many faults are temporary in nature. To distinguish between permanent and temporary faults on transmission lines, the concept of **breaker autoreclosing** is used. It assumes that once the breaker is tripped (opened) by the relay, it will stay open for a while, and then it will automatically reclose. This action allows the relays to verify if the fault is still present and, if so, to trip the breaker again. In the case the fault has disappeared, the relays will not act, and the transmission line will stay in service. The autoreclosing function may be implemented quite differently depending on the particular needs. The main options are to have a single or multiple reclosing attempts and to operate either a single pole or all three poles of the breaker. Circuit breakers are also quite often equipped with auxiliary relays called **breaker failure** (BF) relays. If the breaker fails to open when called upon, the BF relay will initiate operation of other circuit breakers that will disconnect the faulted element, quite often at the expense of disconnecting some additional healthy components. This may be observed in Figure 9.3: once a transmission line relay operates the two breakers in the transmission line bay and one of the breakers fails to operate, the BF relay will disconnect all the breakers on the bus side where the failed breaker is connected, making sure the faulted line is disconnected from the bus.

9.2.2 Basic Relaying Principles

When considering protection of the most common power system components, namely generators, power transformers, transmission lines, busses, and motors, only a few basic relaying principles are used. They include **overcurrent**, **distance**, **directional**, and **differential**. In the case of transmission line relaying, communication channels may also be used to provide exchange of information between relays located at two ends of the line. The following discussion is aimed at explaining the generic properties of the above-mentioned relaying principles. Many other relaying principles are also in use today. The details may be found in a number of excellent references on the subject (Horowitz and Phadke, 1992; Blackburn, 1998; Ungrad *et al.*, 1995; Blackburn, 1993).

Overcurrent protection is based on a very simple premise that in most instances of a fault, the level of fault current dramatically increases from the prefault value. If one establishes a threshold well above the nominal load current, as soon as the current exceeds the threshold, it may be assumed that a fault has occurred and a trip signal may be issued. The relay based on this principle is called an **instantaneous overcurrent relay**, and it is in wide use for protection of radial low-voltage distribution lines, ground protection of high-voltage transmission lines, and protection of machines (motors and generators). The main issue in applying this relaying principle is to understand the behavior of the fault current well, in particular when compared to the variation in the load current caused by significant changes in the connected load. A typical example where it may become difficult to distinguish the fault levels from the normal operating levels is the overcurrent protection of distribution lines with heavy fluctuations of the load. To accommodate the mentioned difficulty, a variety of overcurrent protection applications are developed using the basic principle as described previously combined with a **time delay**. One approach is to provide a fixed time delay, and in some instances, the time delay is proportional to the current level. One possible relationship is an inverse one where the time delay is small for high currents and long for smaller ones. The example shown earlier in Figure 9.7 describes an inverse time characteristic, which may also be a very or extremely inverse type. Further variations of the overcurrent relay are associated with the use of the **directional element**, which is discussed later. The issues of coordinating overcurrent relays and protecting various segments of a distribution line are also discussed later.

Distance relaying belongs to the principle of **ratio comparison**. The ratio is between the voltage and current, which in turn produces **impedance**. The impedance is proportional to the distance in transmission lines, hence the "distance relaying" designation for the principle. This principle is primarily used for protection of high-voltage transmission lines. In this case, the overcurrent principle cannot easily cope with the change in the direction of the flow of power, simultaneous with variations in the level of the current flow, which is common in the transmission but not so common in the radial distribution lines. Computing the impedance in a three-phase system is a bit involved because each type of fault produces a different impedance expression (Lewis, 1947). Because of these differences the settings of a distance relay need to be selected to distinguish between the ground and phase faults. In addition, fault resistance may create problems for distance measurements because the value of the fault resistance may be difficult to predict. It is particularly challenging for distance relays to measure correct fault impedance when a current in-feed from the other end of the line creates an unknown voltage drop on the fault resistance. This may contribute to erroneous computation of the impedance, called apparent impedence, "seen" by the relay located at one end of the line and using the current and voltage measurement just from that end. Once the impedance is computed, it is compared to the settings that define the operating characteristic of a relay. Based on the comparison, a decision is made if a fault has occurred and, if so, in what **zone**. As mentioned earlier, the

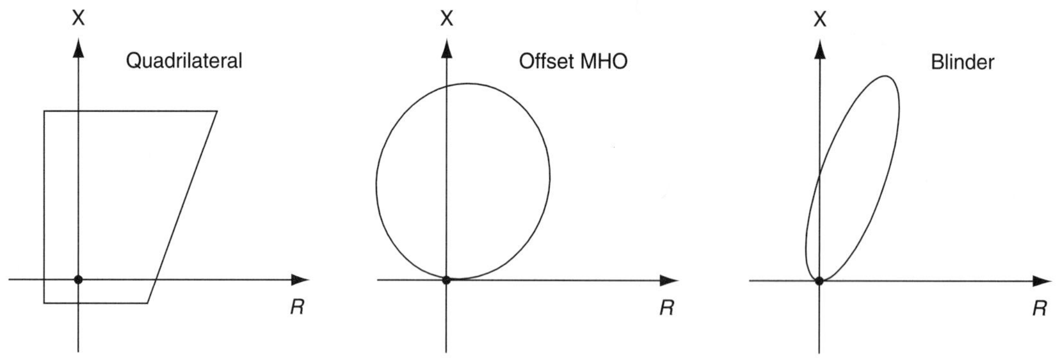

FIGURE 9.8 Operating Characteristics of a Distance Relay

impedance relay may be set to recognize multiple zones of protection. Due to variety of application reasons, the operating characteristics of a distance relay may have different shapes, the quadrilateral and MHO being the most common. The different operating characteristic shapes are shown in Figure 9.8 (Blackburn, 1998). The characteristics dictate relay performance for specific application conditions, such as the changes in the loading levels, different values of fault resistance, effects of power swings, presence of mutual coupling, and reversals of fault direction.

Distance relays may be used to protect a transmission line by taking the input measurements from only one end of the line. Another approach is to connect two distance relays to perform the relaying as a system by exchanging the data between the relays from two ends through a communication link. In this case, it is important to decide what information gets exchanged and how is it used. The logic that describes the approach is called a **relaying scheme**. Most common relaying schemes are based on the signals sent by the relay from one end causing either blocking and unblocking or facilitating and accelerating of the trip from the relay at the other end. In all of the mentioned applications, **directionality** of a distance relay is an important feature. It will be discussed in more detail later on.

As an example of the relaying scheme operation, Figure 9.9 shows relaying zones used for implementation of a blocking scheme. The zones for each relay are **forward overreaching** (FO) and **backward reverse** (BR). The FO setting is selected so that the relay can "see" the faults occurring in a forward direction, looking from the relay position toward the adjacent line terminal and beyond. The BR setting is selected so that the relay can "see" the faults occurring in the backward direction, causing a reversal of the power flow. If the relay *R*1 (at location A) has "seen" the fault (at location *X*1) in zone FO behind the relay *R*2 positioned at the other end (at location B) of the line, the relay is blocked by relay *R*2 from operating. Relay *R*2 has "seen" the fault at location *X*1 in a BR zone and, hence, can "tell" the relay *R*1 not to trip by sending a blocking signal. Should a fault occur in between the two relays (location *X*2), the blocking signal is not sent, and both relays operate instantaneously since both relays "see" the fault in zone FO and neither in zone BR.

A relaying concept widely used for protection of busses, generators, power transformers, and transmission lines is the **current differential**. It assumes that the currents entering and leaving the power system component are measured and compared to each other. If the input and output currents are the same, then it means that the protected component is "healthy," and no relaying action is taken. If the current comparison indicates that there is a deference, which means that a difference is caused by a fault, the relay action is called upon. The difference has to be significant enough to be attributed to a fault since some normal operating conditions and inaccuracies in the instrument transformers may also indicate a difference that is not attributed to a fault. More discussion about the

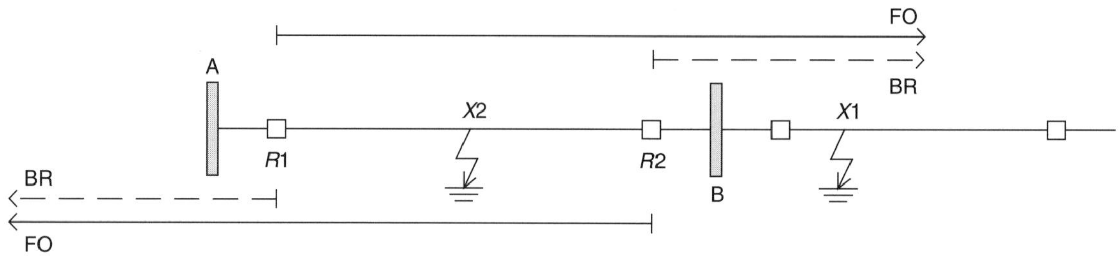

FIGURE 9.9 The Blocking Principle of Relaying with Fault Directionality Discrimination

criteria for establishing the type and level of the difference that are typical for a fault event are presented at a later time when this relaying principle is discussed in more detail. In the case of a differential protection of transmission lines, the measured quantities need to be sent over a communication channel to compute the difference. Since the speed and reliability of communication channels may play a role in the overall performance of the concept, slightly different philosophy is established regarding the type of measurements taken and the comparison criteria for the transmission lines versus other, more local, applications.

Directionality of a relay operation is quite important in many applications. It is established based on the angle of the fault current with the respect to the line voltage. It can be established for current alone or for impedance. In the latter case, the directionality is detected by looking at the angle between the reference voltages and fault current. In the impedance plane, the directionality is detected by the quadrant where the impedance falls. Whether the calculated fault impedance is in a forward (first-quadrant) or reverse (third-quadrant) zone determines the directions of the fault. Typically, the zone associated with the reverse direction is called a **reverse zone** and is numbered in a sequence after the **forward zones** are numbered first. The directionality may also be based on the power calculation, which in turn determines the direction of the power flow to/from a given power system component. Besides being used for implementation of the transmission line relaying schemes, directionality is quite important when applying overcurrent principle and is often used when implementing various approaches to ground protection.

9.2.3 Criteria for Operation

A number of different criteria for operation may be established, but the three most common ones are **speed**, **dependability/security**, and **selectivity** (WD G5, 1997a, b). All the criteria need to be combined in a sound engineering solution to produce the desirable performance, but for the sake of clarity, each of the criteria is now discussed separately.

Speed

The **speed** of operation is the most critical protective relay operating criterion. The relays have to be fast enough to allow clearing of a fault in the minimum time needed to ensure reliable and safe power system operation. The minimum operating time of a relay is achieved when the relay operates without any intentional time delay settings. Such an example is the time of operation of a distance relay in a direct (instantaneous) trip in Zone 1. The operating time may vary from the theoretical minimum possible to the time that a practical solution of a relaying algorithm may take to produce a decision. The operating time is dependent on the algorithm and technology used to implement the relay design. Because the relays respond to the fault transients, the relay operating time may vary slightly for the same relay if subjected to the transients coming from different types of fault. The minimum acceptable operating time is often established to make sure that the relay will operate fast enough to meet other time-critical criteria. For the transmission line protection example, the overall time budget for clearing faults in a system is expressed based on the number of cycles of the fundamental frequency voltage and current signals. This time is computed from the worst-case fault type persisting and potentially causing an instability in the overall power system. To prevent the instability from occurring, the fault needs to be cleared well before this critical time is reached, hence the definition of the minimum **fault clearance** time. The relay operating time is only a portion of this time-budget allocation. The rest is related to the operation of circuit breakers and a possible multiple reclosing action that needs to be taken. The consideration also includes the breaker failure action taken in the case a breaker fails to open and other breakers get involved in clearing the fault. The relay operating time is a pretty critical criterion even though it is allocated a very small portion of the mentioned fault-clearing time-budget criteria. As an example, the expected average operating time of transmission line relays in zone 1 is around one to two fundamental frequency cycles, where one cycle duration is 16.666 ms.

Dependability/Security

Another important operating criterion for protective relays is dependability/security. It is often mentioned as a pair since dependability and security are selected in a trade-off mode. **Dependability** is defined as the relay ability to respond to a fault by recognizing it each time it occurs. **Security** is defined as an ability of a relay not to act if a disturbance is not a fault. In almost all the relay approaches used today, the relays are selected with a bias toward dependability or security in such a way that one affects the other. A more dependable approach will cause the relays to **overtrip**, the term used to designate that the relay *will* operate whenever there is a fault but at the expense of possibly tripping even for nonfault events. The security emphasis will cause relays ***not* to trip** for nofault conditions but at a risk of not operating correctly when the fault occurs. The mentioned trade-off when selecting the relaying approach is made by choosing different types of relaying schemes and related settings to support one or the other aspect of the relay operation.

Selectivity

One criterion often used to describe how reliable a relaying scheme is relates to the relay ability to differentiate between a variety of operating options it is designed for. This criterion is called relay selectivity. It may be attributed to the relay accuracy, relay settings, or, in some instances, the measuring capabilities of the relay. In all cases, it designates how well the relay has recognized the fault conditions that it is designed or set to operate for. An example of the selectivity problem is an inability of a relay to correctly decide if it should operate in zone 1 or zone 2 for a fault that occurs in the region close to the set point between the zones. In this case, the relay operation may be termed as "overreaching" or "underreaching," depending if the relay has mistaken that the fault was inside of the selected zone while it was actually outside and vice versa.

9.3 Protection of Transmission Lines

The protection of transmission lines varies in the principle used and the implementation approaches taken, depending on the voltage level. The main principles and associated implementation issues are discussed for the following most frequent transmission line protective relaying cases: overcurrent protection of distribution radial feeders, distance protection of transmission lines and associated relaying schemes, and differential protection of transmission lines. Detailed treatment of the subject may be found in an excellent recent IEEE survey of the subject (IEEE, 1999).

9.3.1 Overcurrent Protection of Distribution Feeders

Protection of distribution feeders is most adequately accomplished using the overcurrent relaying principle. In a radial configuration of the feeder, shown in Figure 9.10, the fault current distribution is such that the fault currents are the highest for the faults closest to the source and the current decays as the fault gets farther away from the source. This property allows the use of the fault current magnitudes as the main criterion for the relaying action. The **overcurrent relaying** principle is combined with the inverse-time operating characteristic, as shown in Figure 9.11, and this represents the relaying solution for radial distribution feeders in most cases. The **inverse-time property** designates the relationship between the current magnitude and the relay operating time (the higher the fault current magnitude, the shorter the relay operation time). Further discussion is related to the setting procedures for the overcurrent relays with inverse-time characteristics.

To embark on determination of **relay settings**, the following data have to be made available for an overcurrent relay either through a calculation or through simple selection of design options: time dial, tap (pick-up setting), and operating characteristic type; current transformer ratio for each relay location; and extreme (minimum and maximum) short circuit values for fault currents. The mentioned data applied to the relaying system shown in Figure 9.10 are provided in Table 9.1. The values are determined only for relays *R*2 and *R*3 as an example. Similar approaches will yield corresponding values for relay *R*1.

The next step is to establish the criteria for **setting coordination**. The criteria selected as examples for relays *R*3 and *R*2 shown in Figure 9.10 are as these rules state:

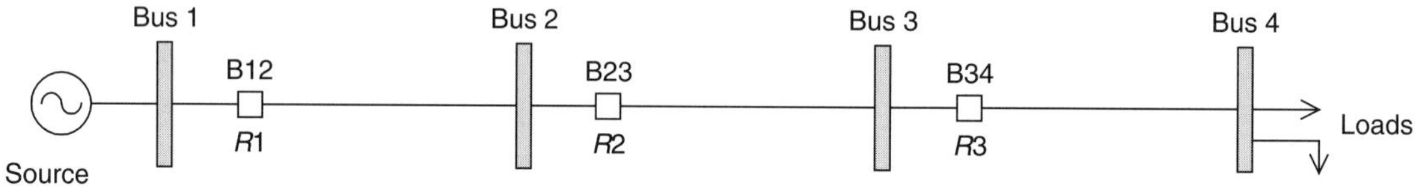

FIGURE 9.10 Protection of a Radial Distribution Feeder

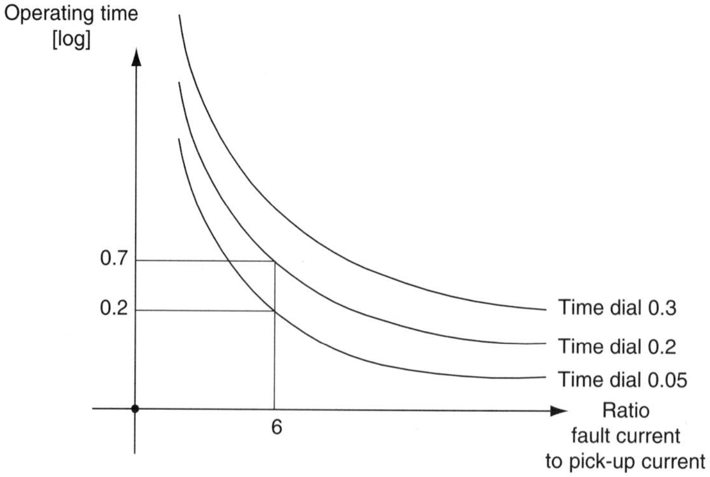

FIGURE 9.11 Inverse-Time Operating Characteristic of an Overcurrent Relay

TABLE 9.1 Data Needed for Setting Determination for the Case in Figure 9.10

	Bus/relay	
Max and min fault current [A]	*R2*	*R3*
Max fault current	306	200
Min fault current	250	156
CT ratio	50:5	50:5
Pick-up setting	5	5
Time-dial setting	0.2	0.05

1. *R2* must pick up for a value exceeding one-third of the minimum fault current (rule of thumb) seen by relay *R3* (assuming this value will never be below the maximum load current)
2. *R2* must pick up for the maximum fault current seen by *R3* but no sooner than 0.5 sec (rule of thumb) after *R3* should have picked up for that current.

Based on the mentioned criteria and data provided in Table 9.1, the following are the setting calculation steps.

Step 1: Settings for Relay *R3*

The relay has to operate for all currents above 156 A. For reliability, one-third of the minimum fault current is selected. This yields a primary fault current of 156/3 = 52 A. Based on this, a CT ratio of 50/5 = 10 is selected. This yields a secondary current of 52/10 = 5.2 A. To match this, the relay tap (pick-up value) is selected to be 5.0 A. To ensure the fastest tripping for faults downstream from relay *R3*, the time dial of 0.05 is selected (see Figure 9.11).

Step 2: Settings for Relay *R2*

- Selection of CT ratio and relay tap
  The relay *R2* must act as a backup for relay *R3*, and, hence, it has to operate for the smallest fault current seen by relay *R3*, which is 156 A. Therefore, the selection of the CT ratio and relay tap is the same for relay *R2* as it was for relay *R3*.
- Time dial selection
  Based on the rule 2, relay *R2* acting as a backup for relay *R3* has to operate 0.5 sec after relay *R3* should have operated. This means that relay *R2* has to have a delay of 0.5 sec for the highest fault current seen by relay *R3* to meet the above mentioned criteria. Let us assume that the highest fault current seen by relay *R3* is at location next to breaker B34 looking downstream from breaker B34, and this current is equal to 306 A. In that case, the primary fault current is equal to 306/10 = 30.6 A. The selected relay tap setting will produce a pick-up current of 30.6/5 = 6.12 A. From Figure 9.11, the time delay corresponding to this pick-up value is 0.2 sec. Hence, if relay *R3* fails to operate, the relay *R2* will operate as a backup with a time delay of 0.2 sec + 0.5 sec = 0.7 sec. According to Figure 9.11, the time delay of 0.7 sec requires the selection of time dial of 0.2. It also becomes obvious that selection of a smaller current for calculation of the time delay of relay *R3* will not allow the criteria in rule 2 to be met.

The overcurrent relaying of distribution feeders is a very reliable relaying principle as long as the time coordination can be achieved for the selected levels of fault current as well as the given circuit breaker operating times and instrument transformer ratios. The problems start occurring if the feeder loading changes significantly during a given period of time and/or the level of fault currents are rather small. This may affect a proper selection of settings that will accommodate a wide-range fluctuation in the load and fault currents. Some other, less likely, phenomena that may affect the relaying performance are as follows: current transformer saturation, selection of inadequate auxiliary transformer taps, and large variation in the circuit breaker opening times.

9.3.2 Distance Protection of Transmission Lines

As explained earlier, the distance protection principle is based on calculation of the impedance "seen" by the relay. This impedance is defined as the **apparent impedance** of the protected line calculated using the voltages and currents measured at the relaying point. Once the impedance is calculated, it is compared to the relay **operating characteristic**. If identified as falling into one of the zones, the impedance is considered as corresponding to a fault; as a result, the relay issues a **trip**. The concept of the impedance being proportional to the length of the transmission line and the idea of using the relay settings to correspond to the line length lead to the reason for calling this relaying principle **distance relaying**.

To illustrate the process of selecting distance relay **settings**, a simple network configuration, with data given in Table 9.2, is considered in Figure 9.12. Table 9.2 also contains additional information about the instrument transformer ratios.

The setting selection and coordination for the example given in Figure 9.12 can be formulated as follows.

Step 1: Determination of Maximum Load Current and Selection of CT and CVT Ratios

From the data given in Table 9.2, the maximum load current is computed as:

TABLE 9.2 Data Needed for Calculation of Settings for a Distance Relay

	Line impedance	
Data	Line 1–2	Line 2–3
Line impedance	$1.0 + j1.0$	$1.0 + j1.0$
Max load current	50 MVA	50 MVA
Line length	50 miles	50 miles
Line voltages	138 kV	138 kV

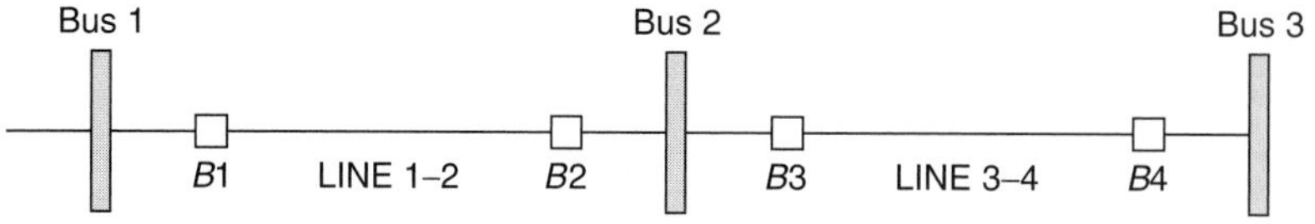

FIGURE 9.12 A Sample System for Distance Relaying Application

$$\frac{50(10^6)}{(\sqrt{3})(138)(10^3)} = 418.4 \text{ A} \qquad (9.1)$$

The CT ratio is 400/5 = 80, which produces about 5 A in the secondary winding. If one assumes that the CVT secondary phase-to-ground voltage needs to be close to 69 V, then computing the actual primary voltage and selecting the ratio to produce the secondary voltage close to 69 V allows one to calculate the CVT ratio. The primary phase-to-ground voltage is equal to $138\sqrt{3}(10,000) = 79.67(10,000)$ V. If we allow the secondary voltage to be 69.3 V, then the CVT ratio can be selected as $79.67(10,000)/69.3 = 1150/1$.

Step 2: Determination of the Secondary Impedance "Seen" by the Relay

The CT and CVT ratios are used to compute the impedance as follows:

$$\frac{V_p/1150}{I_p/80} = \frac{V_p}{I_p}(0.07) = Z_{\text{line}}(0.07). \qquad (9.2)$$

Therefore, the secondary impedance "seen" by the relay is for both lines equal to $0.07 + j0.07$.

Step 3: Computation of Apparent Impedance

Apparent impedance is the impedance of the relay seen under specific loading conditions. If we select the power factor of 0.8 lagging for the selected CT and CVT ratios as well as the selected fault current, the apparent impedance is equal to:

$$Z_{\text{load}} = \frac{69.3}{418.4\left(\frac{5}{400}\right)}(0.8 + j0.6) = 10.6 + 8.0. \qquad (9.3)$$

Step 4: Selection of Zone Settings

Finally, the zone settings can now be selected by multiplying each zone's impedance by a safety factor. This factor is arbitrarily determined to be 0.8 for zone 1 and 1.2 for zone 2. As a result, the following settings for zone 1 and zone 2, respectively, are calculated as:

$$\begin{aligned} &\text{Zone 1} \quad 0.8(0.007 + j0.7) = (0.056 + 0.56)\Omega \\ &\text{Zone 2} \quad 1.2(0.007 + j0.7) = (0.084 + 0.84)\Omega. \end{aligned} \qquad (9.4)$$

Distance relaying of transmission lines is not free from inherent limitations and application ambiguities. The most relevant inherent limitation is the influence of the current in-feed from the other end as described earlier. The other source of possible error is the fault resistance, which cannot be measured online. Therefore, it has to be assumed at the time of the relay setting computation, when a value different from what is actually present during the fault may be picked up. The anticipated value of the fault resistance is selected arbitrarily and may be a cause for a gross error when computing the impedance for a ground fault. Yet another source of error is the mutual coupling between adjacent transmission line or phases, which if significant, can adversely affect relay operation if not taken adequately into account when computing the fault impedance. The distance relaying becomes particularly difficult and sometimes unreliable if applied to some special protection case, such as multiterminal lines, lines with series compensation, very short lines, and parallel lines. In all of the cases, selecting relay settings is quite involved and prone to errors due to some arbitrary assumptions about the value of the fault impedance.

9.3.3 Directional Relaying Schemes for High-Voltage Transmission Lines

Due to the inherent shortcoming of distance relays not being able to recognize the effect of the current in-feed from the other end of the line and because of this possibly leading to a wrong decision, the concept of a relaying scheme is developed. The concept assumes that the relays at the two ends of a transmission line will coordinate their actions by exchanging the information that they detect for the same fault. To be able to perform the coordination, the relays have to use a **communication channel** to exchange the information. In an earlier discussion, illustrated in Figure 9.9, one of the principles for scheme protection of transmission lines was explained. A more comprehensive summary is given now.

The choices among basic relaying scheme principles are influenced by two factors: the **approach** in using the communication channels and the **type** of information sent over the channel. Regarding the approach for using the channels, one option is for the channels to be active during a fault, which means sending the information all the time and making sure that the communication did not fail. The other option is for the communication channels to be activated only when a command is to be transmitted and not being activated during other intervals or cases related to the fault. Regarding the type of information sent, the channels are used for issuing either a

TABLE 9.3 Summary of Basic Characteristics of Most Common Relaying Schemes

Scheme type (Directional Comparison)	The use of communication channel	Type of signal sent
Blocking	Blocking a signal (use of power line carrier)	Block
Unblocking	Sending either block or unblock signal all the time (use of frequency shift keying channel)	Block/unblock
Overreaching transfer trip Underreaching transfer trip	Sending a trip signal (use of audio tones)	Trip/guard

blocking or a tripping signal. In the case a **blocking signal** is issued, the relay from one end, after making a decision about a fault, sends a blocking signal to the other relay. If a **trip signal** is used, the relay that first detects the fault sends a signal to the other end. This will make the other relay perform a trip action immediately after it has detected a fault as well. Additional consideration in selecting the scheme relates to the type of **detectors** used in the relays for making the decisions about the existence of a fault and the location of the fault with the respect to the zones. Specific choices in setting up the zones are made for each scheme operation. Typical choices for the detector types are overcurrent and/or directional, with the zone I and / or zone II settings being involved in the decision making. A summary of the above considerations is given in Table 9.3.

As with any other relaying approach, the scheme implementations also have different performance criteria established. If one takes the dependability/security criterion (WG D5, 1997; WG, 1981) as the guiding factor in making the decisions, then the property of the various relaying scheme solutions can be classified as follows:

1. **Blocking/unblocking:** The blocking solution tends to provide higher dependability than security. Failure to establish a blocking signal from a remote end can result in overtripping for external faults. The unblocking solution offers a good compromise of both high dependability (channel not required to trip) and high security (blocking is continuous).
2. **Transfer trip (overreaching/underreaching):** The transfer trip solution offers higher security than dependability. A failure to receive the channel signal results in a failure to trip for internal faults. The transfer trip systems require extra logic for internal-trip operation at a local terminal when the remote terminal breaker is open or for a "weak infeed" when the fault contribution is too low to send a trip signal. This is not a problem with blocking and unblocking systems.

The options for scheme protection implementation are much more involved than what has been discussed here. Further details may be found in Blackburn (1993, 1998).

9.3.4 Differential Relaying of High-Voltage Transmission Lines

In the cases when the above-mentioned relaying schemes are not sufficiently effective, a differential scheme may be used to protect high-voltage (HV) transmission lines. In this type of relaying, the measurements from two (or multiple) ends of a transmission line are compared. Transmitting the measurement from one end to the other enables the comparison. The decision to trip is made once the comparison indicates a significant difference between the measurements taken at the transmission line end(s).

The main reason for introducing differential relaying is the ability to provide **100% protection** of transmission lines; at the same time, the influence of the rest of the system is minimal. The scheme has primarily been used for high-voltage transmission lines that have a strategic importance or have some difficult application requirements such as series compensation or multiterminal configuration. The use of a **communication system** is needed for implementation of this approach. This may be considered a disadvantage due to the increased cost and possibility for the channel malfunctioning.

The classification of the existing approaches can be based on two main **design properties**: the type of the communication **media** used and the type of the **measurements** compared. The communication media most commonly used are metallic wire, also known as pilot-wire; leased telephone lines; microwave radio; and fiber-optic cables. The measurements typically used for the scheme are composite values obtained by combining several signal measurements at a given end (IEEE, 1999) and sample-by-sample values of the phase currents (IEEE, 1999). A summary of the most common approaches for the differential relaying principle for transmission lines is given in Table 9.4.

In the past, the most common approach was to use metallic wire to compare the sequence values. The sequence values are a particular representation of the three-phase original values obtained through a symmetrical component transformation (Blackburn, 1993). Due to a number of practical problems caused by the ground potential rise and limited length of the physical wire experienced with the metallic wire use, these schemes have been substituted by other approaches where different media such as fiber-optic or microwave links are used. Most recently, as the wideband communication channels have become more affordable, the differential schemes are being implemented using either dedicated fiber-optic cables or a high-speed leased wideband communication system.

9.4 Protection of Power Transformers

For power transformers, the **current differential** relaying principle is the most common one used. In addition, other types of protection are implemented, such as a sudden pressure relay on the units with large ratings. As much as the current

TABLE 9.4 Summary of Most Common Differential Relaying Principles

Differential principle	Signals used for comparison	Properties
Pilot-wire	Three-phase currents converted into sequence voltage	Use of sequence filters due to direct use of wires prone to transients caused by interference
Phase comparison	Three-phase currents converted into a single current waveform	Composite waveform converted into binary string used for phase comparison
Segregated phase comparison	Phase currents in each phase directly compared at two ends	Currents converted into square waves used for phase comparison
Current differential	Samples of each current transmitted to the other end for comparison	Currents from both ends directly compared
Composite–waveform differential	Each current converted into sequence current and a combined composite signal is transmitted	Composite signals from both ends directly compared

differential relaying has been a powerful approach in the past, it has some inherent limitations that pose difficulties for special applications or practical design considerations. Further discussion concentrates on the mentioned limitations and their impact on the current differential approaches. The other relaying principles applied for protecting power transformers are not discussed here. Further details can be found in references (Horowitz and Phadke, 1992; Blackburn, 1998; Ungrad *et al.*, 1995).

9.4.1 Operating Conditions: Misleading Behavior of Differential Current

Power transformers are energy storage devices that experience transient behavior of the terminal conditions when the stored energy is abruptly changed. Such conditions may be seen during transformer energization, energization of a parallel transformer, removal of a nearby external fault, and a sudden increase in the terminal voltage. The following is a discussion of the mentioned phenomena and the impact they have on the terminal currents.

Energizing a transformer causes a transient behavior of the currents at the transformer primary due to so-called **magnetizing inrush** current. As a voltage is applied on an unloaded transformer, the nonlinear nature of the magnetizing inductance of the transformer causes the magnetizing current to experience an initial value as high as 8–30 times the full-load current, which may appear to the differential scheme as a difference caused by a fault. An example of the harmonic inrush wave shape for the magnetizing current is given in Figure 9.13. Fortunately, the inrush current has a rich harmonic content, which can be used as the basis for distinguishing between the high currents caused by a fault and the ones caused by the inrush. Since the magnetizing inrush is a function of both the prior history of remanent magnetism as well as the type of the transformer connection, selecting the scheme for recognizing proper levels of the harmonics needs to be carried out carefully.

Similar transient behavior of the primary current is seen in a transformer connected in parallel to a transformer that is being energized. The change in the magnetizing current is affecting the primary current of the parallel transformer due to an inrush created on the transformer being energized. This phenomenon is called **sympathetic inrush**. Sudden removal of an external fault and a sudden increase in the transformer voltage also cause the inrush phenomenon, again well recognized by an occurrence of particular harmonics in the primary current.

9.4.2 Implementation Impacts Causing Misleading Behavior of Differential Currents

The current differential relayed for power transformer applications may be affected by practical implementation constraints.

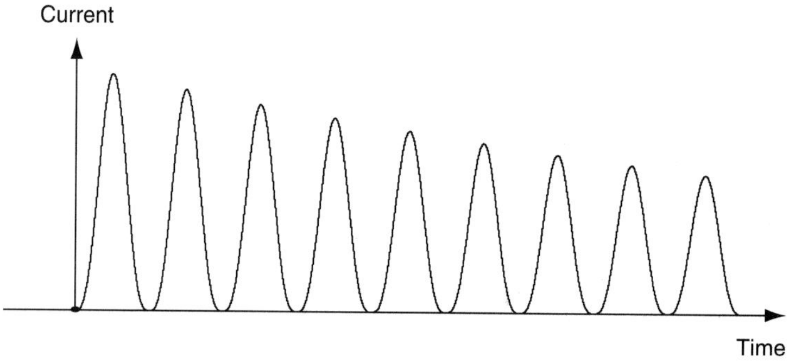

FIGURE 9.13 Magnetizing Inrush Affecting Primary Current

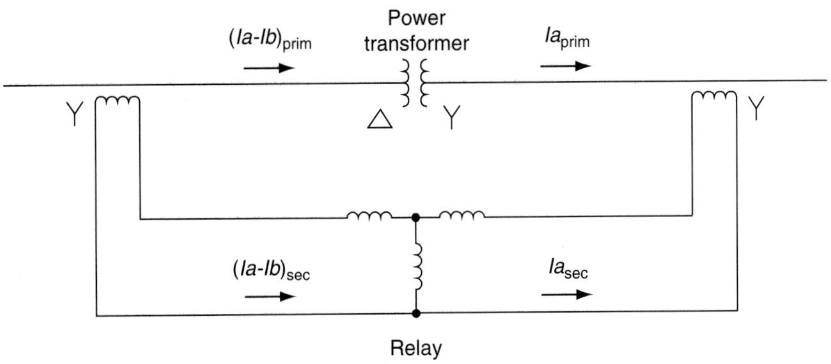

FIGURE 9.14 Phase Mismatch

One of the common problems is to have a mismatch between ratios of the instrument transformers located at the two power transformer terminals. This is called a **ratio mismatch**, and it is corrected by selecting appropriate taps on the auxiliary transformers located at the inputs of the transformer differential relay. Yet another obstacle may be a **phase mismatch**, where the instrument transformer connection may cause a phase shift between the two currents seen at the transformer terminals. This is because the connection of the power transformer may introduce a phase shift, and if the instrument transformer does not correct for this, a phase mismatch will occur at the terminals of the instrument transformer. The phase mismatch is illustrated in Figure 9.14, where the currents in the relaying circuit are not correctly selected. The mismatch can be avoided if the instrument transformer at the Y side of the power transformer is of a Δ type.

Besides the mentioned constraints, other constraints include the mismatch due to a changing tap position on the load tap changer as well as the mismatch caused by the errors in current transformers located at two terminals.

9.4.3 Current Differential Relaying Solutions

The straightforward solution for differential relaying is to take a difference of currents I_1 and I_2 at two ends and compare it to a threshold I_T as shown in equation:

$$|(I_1 - I_2)| \geq I_T. \tag{9.5}$$

This solution will have a problem to accommodate an error due to a mismatch discussed earlier. Hence, a different equation may be used to make sure that a higher error is allowed for higher current levels:

$$|(I_1 - I_2)| \geq \frac{|(I_1 - I_2)|}{2}. \tag{9.6}$$

Finally, to distinguish the case of the inrush condition mentioned earlier, a harmonic restraint scheme is used as represented by equation:

$$|(I_1 - I_2)| \geq k \cdot \frac{|(I_1 - I_2)|}{2}. \tag{9.7}$$

This type of operating characteristic will recognize that the current difference is caused by an event that is not an internal fault, and it will block the relay from operating. The criterion for recognizing a nonfault event is the presence of a particular harmonic content in the differential current. This knowledge is used to restrain the relay operation by relating the factor k to the presence of the harmonic content, hence the "harmonic restraint" terminology.

9.5 Protection of Synchronous Generators

Synchronous generators are commonly used in high-voltage power systems to generate electric power. They are also protected using the **current differential** relaying principle. In addition, the generators require a number of other special operating conditions to be met. This leads to the use of a score of other relaying principles. This section reviews some basic requirements for generator protection and discusses the basic relaying principles used. A much more comprehensive coverage of the subject may be found in an IEEE Tutorial (1995).

9.5.1 Requirements for Synchronous Generator Protection

Generators need to be protected from the **internal faults** as well as the abnormal operating conditions. Since generators consist of two parts, namely the **stator** and the **rotor**, protection of both is required. The stator is protected from both phase and ground faults, while the rotor is protected against ground faults and loss of field excitation. Due to the particular conditions required for the synchronous machine to operate, a number of operating conditions that represent either a power system disturbance or operational hazard need to be avoided. The conditions associated with **network disturbances** are overvoltage or undervoltage, unbalanced currents, network frequency deviation, and subsynchronous oscillations. The conditions of **hazardous operation** are loss of prime mover (better known as generator motoring), inadvertent energization causing a nonsynchronized connection, overload, out-of-step or loss of synchronism, and operation at unallowed frequencies.

9.5.2 Protection Principles Used for Synchronous Generators

The **current differential** protection principle is most commonly used to protect against phase faults on the **stator**, which are the most common faults. Other conditions require other principles to be used. The loss of field and the ground protection of the rotor are quite complex and depend on the type of the grounding and current sensing arrangement used. This subject is well beyond the basic considerations and is treated in a variety of specialized literature (IEEE, 1995). Protection from the **abnormal** generator operating **condition** requires the use of relaying principles based on detection of the changes in voltage, current, power, or frequency. A reverse power relay is used to protect against loss of a prime mover known as generator motoring, which is a dangerous condition since it can cause damage of the turbine and turbine blades. In addition, synchronous generators should not be subjected to an overvoltage. With normal operation near the knee of the iron saturation curve, small overvoltages result in excessive flux densities and abnormal flux patterns, which can cause extensive structural damage in the machine. A relaying principle based on the ratio between voltage and frequency, called a volt-per-hertz principle, is used to detect this condition. An inadvertent connection of the generator to the power system not meeting the synchronization requirements can also cause damage. The overcurrent relaying principle in combination with the reverse power principle are used to detect such conditions. The overload conditions as well as over- and under-frequency operations can cause damage from overheating, and thermal relays in combination with frequency relays are used. Undervoltage and overvoltage protections are used for detecting loss of synchronism and overvoltage conditions.

9.6 Bus Protection

Protecting **substation busses** is a very important task because operation of the entire substation depends on availability of the busses. The bus faults are rare, but in the open-air substations, they occasionally happen. They have to be cleared very fast due to high-fault current flow that can damage the bus. This section briefly discusses the requirements and basic principles used for bus protection. Detailed treatment of the subject is given in several classical references (Horowitz and Phadke, 1992; Blackburn, 1998; Ungrad *et al.*, 1995).

9.6.1 Requirements for Bus Protection

High-voltage substations typically have at least two busses: one at one voltage level and the other at a different voltage level with power transformer(s) connecting them. In high-voltage substations, the breaker-and-a-half arrangement, shown earlier in Figure 9.3, is used to provide a redundant bus connection at each voltage level. In addition, in some substation arrangements, there is a provision for separating one portion of the bus from another, allowing for independent operation of the two segments. All of those configurations are important when defining the bus protection requirements (Blackburn, 1998).

The first and foremost requirement for bus protection is **the speed of relay operation**. There are several important reasons for that all relate to the fact that the fault currents are pretty high. First, the **current transformers** used to measure the currents may get **saturated**; hence, a fast operation will allow for the relaying decision to be made before this happens. Next, due to the high currents, the equipment damage caused by a sustained fault may be pretty severe. Hence, the need to accomplish an isolation of the faulted bus from the rest of the system in the shortest time possible is paramount. Last, but not least, due to the various possibilities in reconfiguring busses, it is very important that the bus protection scheme is developed and implemented in a flexible way. It needs to allow for various parts of the bus to be isolated for maintenance purposes related to the breakers on the connected lines, and yet the protection for the rest of the bus needs to stay intact.

9.6.2 Protection Principles Used for Bus Protection

The most common bus protection principle is the **current differential approach**. All connections to the bus are monitored through current transformers to detect current imbalance. If an internal bus fault occurs, the balance between incoming and outgoing currents is drastically disturbed, and that becomes a criterion for tripping the bus breakers and isolating the bus from the rest of the system. This relaying principle would be very simple to implement if there were no problems with **CT saturation**. Due to high-fault currents, one of the CTs may see extraordinary high currents during a close-in external fault. This may cause the CT to saturate, making it very difficult to distinguish if the CT was saturated due to high currents from an internal or external fault. If **air-gap** or **air-core** CTs are used, the saturation may not be an issue, and the current differential relaying is easily applied. This solution is more costly and is used far less frequently than the solution using standard iron-core CTs.

To cope with the CT saturation, two types of relaying solutions are most commonly used (Blackburn, 1998). The first one is a **multirestraint current differential**, and the other one is a **high-impedance voltage differential**. The multirestraint current differential scheme provides the restraint winding connection to each circuit that is a major source of the fault current. These schemes are designed to restrain correctly for heavy faults just outside the differential zone, with maximum offset current, as long as the CTs do not saturate for the maximum symmetrical current. These schemes are more difficult to apply and require use of auxiliary CTs to match the current transformer ratios. The high-impedance voltage differential principle uses a property that CTs, when loaded with high impedance, will be forced to take the error differential current, and, hence, this current will not flow through the relay operating coil. This principle translates the current differential sensitivity problem is to the voltage differential sensitivity probem for distinguishing the close-in faults from the bus faults. The voltage differential scheme is easier to implement since the worst case voltage for external faults can be determined much more precisely knowing the open voltage of the CT.

9.7 Protection of Induction Motors

Induction motors are a very common type of load. The main behavioral patterns of the induction motor are representative of the patterns of many loads; their operation depends on the conditions of the network that is supplying the power as well as the **loading conditions**. This section gives a brief discussion on the most important relaying requirements as well as the most common relaying principles that apply to the induction motor protection. Further details can be found in (Horowitz and Phadke, 1992; Blackburn, 1998; Ungrad *et al.*, 1995).

9.7.1 Requirements for Induction Motor Protection

The requirements may be divided in three categories: **protecting the motor from faults**, **avoiding thermal damage**, and **sustaining abnormal operating conditions**. The protection from faults has to detect both phase and ground faults. The thermal damage may come from an overload or locked rotor and has to be detected by correlating the rise in the temperature to the occurrence of the excessive currents. The abnormal operating conditions that need to be detected are unbalanced operation, under voltage or overvoltage, reversed phases, high-speed reclosing, unusual ambient temperature, voltage and incomplete starting sequence.

The protection principle used has to differentiate the causes of problems that may result in a damage to the motor. Once the causes are detected and linked to potential problems, the motor needs to be quickly disconnected to avoid any damage.

9.7.2 Protection Principles Used for Induction Motor Protection

The most common relaying principles used for induction motor protection are the **overcurrent protection** and **thermal protection**. The overcurrent protection needs to be properly set to differentiate between various changes in the currents caused either by faults or excessive starting conditions. A variety of current-based principles can be used to implement different protection tasks that fit the motor design properties. Both the phase and ground overcurrents, as well as the differential current relays, are commonly used for detecting the phase and ground faults. The thermal relays are available in several forms: a "**replica**" **type**, where the motor-heating characteristics are approximated closely with a bimetallic element behavior; resistance temperature detectors embedded in the motor winding; and relays that operate on a combination of current and temperature changes.

References

Blackburn, J.L. (1993). *Symmetrical components for power systems engineering.* Marcel-Dekker.

Blackburn, J.L. (1998). *Protective relaying—Principles and applications.* Marcel-Dekker.

Flurscheim, C.H. (1985). *Power circuit breaker theory and design.* Peter Peregrinus.

Horowitz, S.H., and Phadke, A.G. (1992). *Power system relaying.* City, State Abbreviation: Research Studies Press.

IEEE Standard, IEEE guide for generator protection C37.102-1995.

IEEE Standard, IEEE guide for the application of current transformers used for protective relaying C37.110-1996.

IEEE Standard, IEEE guide for protective relay applications of transmission lines C37.113-1999.

Lewis, W.A., Tippett, L.S. (1947). Fundamental basis for distance relaying on three-phase systems. *AIEE Transactions 66*, 694–708.

Ungrad, H., Winkler, W., Wiszniewski, A. (1995). *Protection techniques in electrical energy systems.* Marcel-Dekker.

WG D5, Line Protection Subcommittee, Power System Relaying Committee. (1997a). Proposed statistical performance measures for microprocessor-based transmission-line protective relays: Part I—Explanation of the statistics. *IEEE Transactions on Power Delivery 12*(1).

WG D5, Line Protection Subcommittee, Power System Relaying Committee. (1997b). Proposed statistical performance measures for microprocessor-based transmission-line protective relays: Part II—Collection and uses of data. *IEEE Transactions on Power Delivery 12*(1).

WG of the Relay Input Sources Subcommittee, Power System Relaying Committee. (1981). Transient response of coupling capacitor voltage transformers. *IEEE Transactions on Power Apparatus and Systems PAS-100*(2).

10

Electric Power Quality

Gerald T. Heydt
*Department of Electrical Engineering,
Arizona State University,
Tempe, Arizona, USA*

10.1 Definition

There is no single acceptable definition of the term **electric power quality**. The term generally applies to the goodness of the electric power supply, its voltage regulation, its frequency, voltage wave shape, current wave shape, level of impulses and noise, and the absence of momentary outages. Some engineers include reliability considerations in electric power quality studies, some consider electromagnetic compatibility, and some perform generation supply studies. Narrower definitions of electric power quality generally focus on the bus voltage wave shape.

Electric power quality studies generally span the entire electrical system, but the main points of emphasis are in the primary and secondary distribution systems. This is the case because loads generally cause distortion of bus voltage wave shape, and this distortion is mainly noted near the source of the difficulty—namely near the load—in the secondary distribution system. Because the primary distribution system is closely coupled to the secondary system and the load, which is sometimes served directly by the primary distribution system, the primary network is also a point of focus. Transmission and generation systems are also studied in certain types of power quality evaluation and analyses. Table 10.1 shows the main points in electric power quality.

10.2 Types of Disturbances

There are two main classes of electric power quality disturbances: the **steady-state disturbance** that lasts for a long period of time (and is often periodic) and the **transient**. The latter generally exists for a few milliseconds, and then decays to zero. Controversy surrounds which is more important, and a case could be made for either type of disturbance being more problematic as far as cost. The steady-state type of disturbance is generally less evident in its appearance, often at lower voltage and current levels, and less harmful to the operation of the system. Because steady-state phenomena last for a long period of time, the integrated effects of active power losses (low or high voltage) and inaccurate timing signals may be quite costly. Transient effects tend to be higher level in amplitude and are often quite apparent in harmful effects as well as occasionally spectacular in cost (e.g., causing loss of a manufactured product or causing long-term outages). The cost of transient power quality problems has been estimated in the 100 million to 3 billion dollar range annually in the United States. Table 10.2 lists some types of steady-state and transient power quality problems. The transient problems are often termed **events**.

10.3 Measurement of Electric Power Quality

Many indices have been developed for the specification of electric power quality. A few are listed in Table 10.3. The most widely used index of power quality is the total harmonic distortion, which is an index that compares the intensity of harmonic signals in voltages and currents to the fundamental component. The main indices are as follows:

TABLE 10.1 Electric Power Quality Considerations

Consideration	Focus	Comments
Region of analysis	• Distribution systems • Points of utilization of electric power • Transmission systems • Electromagnetic compatibility	The main region of analysis is the primary and secondary distribution system. This is where nonsinusoidal waves are most prevalent and of greatest amplitude.
Types of problems	• Harmonics • Momentary outages and low voltages (sags)	There is a controversy concerning whether momentary low voltages (sags) or harmonics are the most problematic in terms of cost of the problem.
Analysis methods	• Circuit analysis programs • Harmonic power flow studies • Focused studies on particular events using circuit theory to obtain solution • Pspice	A range of commercial software is available for both smaller and larger studies. Many software tools are linked to elaborate graphics. Most methods are data intensive and approximate.
Mitigation techniques	• Filters • Capacitors • Problematic loads • Higher pulse order (e.g., twelve-pulse rather than six-pulse systems)	Mitigation techniques are often customized to the particular problem and application. In general, higher pulse order systems give much less problem than single-phase and six-pulse, three-phase systems.

TABLE 10.2 Power Quality Problems

Type	Problem	Appearance	Causes
Transient system problems	Impulses (surges, pulses)	High-voltage impulse for a short time, typically in the microsecond to 1 ms range	• Lightning • Switching surges • Rejection of inductive loads
	Momentary outages	Collapse of ac supply voltage for up to a few (e.g., 20) cycles	Circuit breaker operations
	Phase shift	Sinusoidal supply voltage proportional to a sine function whose phase angle suddenly shifts by an angle φ	Faults
	Sags (low voltage)	Momentary low voltage caused by faults in the supply	Faults
	Ringing	Damped sinusoidal voltages impressed on the ac wave	Capacitor switching
Steady-state system problems	Harmonics	Integer multiples of the ac supply frequency (e.g., 60 Hz) of (usually) lower amplitude signals impressed on the power frequency wave	• Nonlinear loads • Adjustable speed drives • Rectifiers • Inverters • Fluorescent lamps
	Voltage notches	Momentary low voltages of duration much shorter than one cycle caused by commutated loads	Adjustable speed drives
	Noise	Noise impressed on the power frequency	• Static discharge and corona • Arc furnaces
	Radio frequency	High-frequency (e.g., $f > 500$ kHz) sinusoidal signals of typically low amplitude impressed on the power frequency	Radio transmitters
	Interharmonics and fractional harmonics	Components of noninteger multiples of the power frequency	• Cycloconverters • Kramer drives • Certain types of adjustable speed drives

- **Power factor**: The relationship of power factor to the IEEE Standard 519-1992, especially with regard to the failure of the power factor to register harmful and undesirable effects of high-frequency harmonics in power distribution systems, is well known. Many electric utilities have limits to the power factor of consumer loads, but there may not be a clear definition of power factor for the nonsinusoidal case. In particular, many electric utilities

TABLE 10.3 Voltage Measurement

Instrument	Configuration	Application considerations
Potential transformer (PT)	Energized at full line potential on the high side and typically in the 100-V range on the low side	• Bandwidth of the PT • Safety (e.g., upon failure of the PT) • Turns ratio, accuracy
Voltmeters	Energized directly or through a probe	• Isolation from the line • Bandwidth • True RMS reading questioned • Accuracy
Voltage divider	Energized at full line potential at the upper resistor; resistive voltage divider used to give low voltage at the low end of the string	• Safety (e.g., opening of low end of string) • Heat loss
Capacitively coupled voltage transformer	Capacitive voltage divider energized at full line potential at upper end; instrument transformer used to obtain some isolation	• Bandwidth • Resonance of the capacitors and transformer • Accuracy • Frequency dependence

do not distinguish between displacement of the power factor (i.e., the cosine between the voltage and current phasors at the fundamental frequency) and the power factor defined in Table 10.4. The term **true power factor** has been used by some to refer to $P/|V||I|$, but the IEEE Standard 100 used the term **power factor** for this ratio, and this simpler term is used in this chapter.

- **Total harmonic distortion**: The use of the total harmonic distortion (THD) is perhaps the most widespread power quality index, and many electric utilities have adopted a THD-based measure of the limits of customer load currents.
- ***K*-factor**: Explanation of the use of the *K*-factor to derate transformers that are expected to carry nonsinusoidal load currents has been used, and an alternative calculation of this index in the time domain has been shown.
- **Flicker factor**: This flicker factor index has been used in connection with electric arc furnaces for the purpose of quantifying the load impact on the power system.

Perhaps the main application of power quality indices has been in guides, recommended practices, and standards. As an example, the IEEE Standard 519-1992 contains an often cited limit to harmonic load currents and voltages. The ANSI

TABLE 10.4 Common Power Quality Indices

Index	Definition	Main applications
Total harmonic distortion (THD)	$\left(\sqrt{\sum_{i=2}^{\infty} I_i}\right)/I_1$	General purpose; standards
Power factor (PF)	$P_{tot}/\lvert V_{\text{rms}}\rvert\lvert I_{\text{rms}}\rvert$	Potentially in revenue metering
Telephone influence factor	$\left(\sqrt{\sum_{i=2}^{\infty} w_i^2 I_i}\right)/I_{\text{rms}}$	Audio circuit interference
C message index	$\left(\sqrt{\sum_{i=2}^{\infty} c_i^2 I_i}\right)/I_{\text{rms}}$	Communications interference
IT product	$\sqrt{\sum_{i=1}^{\infty} w_i^2 I_i^2}$	Audio circuit interference, shunt capacitor stress
VT product	$\sqrt{\sum_{i=1}^{\infty} w_i^2 V_i^2}$	Voltage distortion index
K factor	$\left(\sum_{h=1}^{\infty} h^2 I_h^2\right)/\sum_{h=1}^{\infty} I_h^2$	Transformer derating
Crest factor	$V_{\text{peak}}/V_{\text{rms}}$	Dielectric stress
Unbalance factor	$\lvert V_-\rvert/\lvert V_+\rvert$	Three-phase circuit balance
Flicker factor	$\Delta V/\lvert V\rvert$	Incandescent lamp operation, bus voltage regulation, sufficiency of short circuit capacity
Total demand distortion (TDD)	THD*(fundamental current/circuit rating)	In IEEE Standard 519

Standard 368 contains a well-quoted guide on limits of the IT product. The Underwriters Laboratories applies the *K*-factor to the specification of transformers that carry nonsinusoidal load currents.

10.4 Instrumentation Considerations

Because power quality is often stated in terms of voltages and currents, the main instrumentation needed to assess power quality relates to bus voltages and line and load currents. In terms of voltages, usually bus voltages are measured using potential transformers because isolation from the power circuit is desirable and because power system voltages are usually too high to measure directly. Typical potential transformers are capable of bringing circuit voltages (e.g., in the 440–7200 V range for distribution circuits and up to 40 kV for subtransmission circuits) to about 110 V. These potential transformers must have the proper bandwidth to "see" the desired voltages to be instrumented (e.g., harmonic voltages), they must have the proper dynamic range to allow measurement of the voltage, and they must have isolation suitable for safety. Some voltage measurement instruments are listed in Table 10.3.

Current instrumentation is analogous to voltage measurement with the replacement of the current transformer for the potential transformer. Main considerations are appropriateness of the current ratio in the current transformer, isolation for safety, loading with the proper current transformer burden, and bandwidth and dynamic range of the current transformer. Table 10.5 shows some of the basic instruments used in current instrumentation.

Both current and voltage transformers are available in several grades. General purpose transformers are not generally usable for power quality measurements because the bandwidth of these devices is often not much larger than 60 Hz (in 60-Hz systems). Harmonics may be attenuated by the current or potential transformer, and this adds intolerable error to the measurement. Relaying grade current and relaying potential transformers are usually not much better because of limited bandwidth. Revenue meter grade current and voltage transformers are generally not designed for high bandwidth applications. Laboratory grade transformers are usually the best choice. Modern power quality instruments generally have companion current and voltage transformers for use with a given instrument, and the manual for the instrument will contain the bandwidth measurement of the transformers. It is noted that for many voltage and current instruments used in power quality tests, current and voltage transformers are the most costly components of the instrumentation system.

Modern power quality assessment often involves more than voltage and current measurement. The following are also part of the test regimen in most cases:

- Event measurement (i.e., measurement of three-phase voltages and currents, plus neutral ground voltage and neutral current versus time)
- On-board evaluation of harmonics (of voltages and currents, often plotted versus time)
- Measurement of active and reactive power
- Measurement of total harmonic distortion
- Measurement of active power loss in a system component
- Assessment of high-frequency effects
- Measurement of rise time
- Oscillograph capability (often written digitally to a disk for subsequent analysis and report writing)
- Energy measurement.

Table 10.6 illustrates a few of these capabilities in commercial instruments in use for power quality assessment.

TABLE 10.5 Current Instrumentation

Instrument	Configuration	Application considerations
Current transformer (CT)	Placed around the conductor to be instrumented	• Bandwidth • Safety • Operation with correct CT burden
Resistive shunt for current measurement	Resistor in series with load to be instrumented	• No ohmic isolation • Accuracy • Heating of the shunt
Ammeter	Placed in series with circuit to be instrumented	• Accuracy • Bandwidth • Safety—no ohmic isolation for circuit
Optical instruments	Rotation of plane of polarized light in a fiber-optic cable around or near the instrumented circuit	• Accuracy • Vibration sensitivity • Cost
Hall-effect device	Measurement of magnetic field near instrumented conductor	• Accuracy • Bandwidth • Vibration and mechanical placement sensitivity • Linearity

TABLE 10.6 Power Quality Assessment Instruments

Instrument	Typical capability	Bandwidth	Dynamic range
Digital fault recorder (DFR)	Measure three-phase voltage and current; perform basic analysis on these data; measure digital output	Typically to about 3000 Hz	At least 80 db, possibly 100 db
Power quality node	Commercialized event recorder for distribution system instrumentation and field rugged, and telephone interrogations	Typically to about 3000 Hz	About 80 db
Event recorder, digital oscilloscope	Oscilloscope functions, digital readout, basic analysis functions, signal triggers recording	Varies widely—possibly into the megahertz range	About 100 db

10.4.1 Advanced Instrumentation

In the preceding section, several basic voltage and current measurement instruments were discussed. Because power quality measurements are often demanding in bandwidth and dynamic range, several types of advanced instrumentation techniques have been studied and developed for this application. A few are listed in Table 10.7.

10.5 Analysis Techniques

The main analytical techniques for power quality studies are the following:

- Power flow studies
- Injection current analysis
- Simulation methods, such as Pspice and EMTP
- Direct circuit analysis

Power flow studies are software tools that rely on steady-state operation of the system to be studied. These software tools use input data such as load type, load level (P, Q), and circuit data (e.g., impedance, connection diagram). The output is typically all nodal voltages and all line currents at all frequencies.

Injection current analysis is the analysis of how currents injected into a system propagate. This is a simplified form of power flow study.

Power quality engineering may be taken to be a specialized branch of signal analysis. In this regard, there are many useful formulas that may be applied to solve problems. One particularly useful set of formulas is the ideal rectifier set of formulas shown in Table 10.8. Rectifier loads are a main source of harmonic load currents. The ideal rectifier formulas give the various interrelationships of voltage and current parameters for the idealized case of no rectifier power loss and very high dc side inductance. These formulas, while useful, should be used with caution because the assumptions made in arriving at these formulas are all idealizations. Note that the displacement factor is the cosine of the angle between fundamental voltage and fundamental current. The power factor is the ratio $P/V_{\text{rms}}I_{\text{rms}}$.

10.6 Nomenclature

c_h	*C*-message weight
CBEMA	Computer Business Equipment Manufacturers Association
CT	current transformer
DF	displacement factor
DPF	displacement power factor
EPRI	Electric Power Research Institute
IEC	International Electrotechnical Commission
IT	current—telephone influence factor product, read as IT product
I_h, V_h	harmonic components of current $i(t)$ and $v(t)$
K	*K*-factor
kIT	thousands of IT units
kVT	thousands of VT units

TABLE 10.7 Advanced and Unconventional Instrumentation for Electric Power Quality Assessment

Basic instrumentation technique	Basis	Application
Poeckels effect	Rotation of the plane of polarized light in a medium due to electric field strength	Wideband measurement of voltage
Faraday rotation effect	Rotation of the plane of polarized light in a medium due to magnetic field strength	Wideband measurement of current
Hall effect	Variation of the resistance of a material in a magnetic field	Wideband measurement of current
Global Positioning System (GPS) applications	Triangular spotting of four or more artificial earth satellites to obtain time and position of a receiver	Accurate time tagging of measurements, phasor measurements

TABLE 10.8 The Ideal Rectifier Formulas

Single-phase bridge rectifiers	
Line commutated: Infinite dc inductance, Zero supply inductance	$V_{dc} = \frac{2\sqrt{2}}{\pi} V_{ac}$ $I_{ac,\,\text{fundamental}} = \frac{2\sqrt{2}}{\pi} I_{dc}$ $I_{\text{supply},\,h} = I_{ac,\,\text{fundamental}/h}$ $\text{THD}_I = 48.4\%$ Displacement factor = DF = 1.0 Displacement power factor = $\text{DPF}\frac{2\sqrt{2}}{\pi}$ $P_{dc} = P_{ac} = \frac{2\sqrt{2}}{\pi} V_{s,\,\text{rms}}\, I_{dc}$
Line commutated: Infinite dc inductance, Nonzero supply inductance	$\cos(u) = 1 - \frac{2\omega I_{dc} L_s}{\sqrt{2} V_s}$ $V_{dc} = \frac{2\sqrt{2}}{\pi} V_s - \frac{2\omega L_s I_{dc}}{\pi}$ $\text{DF} \approx \cos(u/2)$ $P = V_s I_{ac,\,\text{fundamental}} {}^{*}\text{DF} = V_{dc} I_{dc}$
Forced commutated: Infinite dc inductance, Nonzero supply inductance	$\cos(\alpha + u) = \cos(\alpha) - \frac{2\omega L_s I_{dc}}{V_s\sqrt{2}}$ $\text{DPF} = \cos\left(\alpha + \frac{u}{2}\right)$ $I_{\text{supply, fundamental}} = \frac{\frac{2\sqrt{2}}{\pi} V_s I_{dc}\cos(\alpha) - \frac{2\omega L_s}{\pi} I_{dc}^2}{V_s \cos\left(\alpha + \frac{u}{2}\right)}$ $V_{dc} = \frac{2\sqrt{2}}{\pi} V_s \cos(\alpha) - \frac{2}{\pi}\omega L_s I_{dc}$
Three-phase bridge rectifiers (six pulse)	
Line commutated: Infinite dc inductance, Zero supply inductance	$V_{dc} = \frac{3\sqrt{2}}{\pi} V_{LL}$ $I_{s,\,\text{rms}} = \sqrt{2/3} I_{dc}$ $I_{\text{supply, fundamental}} = \frac{\sqrt{6}}{\pi} I_{dc}$ DF = 1.00 $\text{DPF} = 3/\pi$ $\text{THD}_\text{I} = 31.1\%$ (harmonics 5, 7, 11, ...)
Line commutated: Infinite dc inductance, Nonzero supply inductance	$V_{dc} = \frac{3\sqrt{2}}{\pi} V_{LL} - \frac{3\omega L_S}{\pi} I_{dc}$ $\cos(u) = 1 - \frac{2\omega L_s I_{dc}}{\sqrt{2} V_{LL}}$ $\text{DF} \approx \cos(u/2)$ $P = \sqrt{3} V_{LL} I_{\text{supply, fundamental}}$ ${}^{*}\cos(u/2) = V_{dc} I_{dc}$
Forced commutated: Infinite dc inductance, Nonzero supply inductance	$V_{dc} = \frac{3\sqrt{2}}{\pi} V_{LL}\cos(\alpha) - \frac{3\omega L_S}{\pi} I_{dc}$ $\cos(\alpha + u) = \cos(\alpha) - \frac{2\omega L_s I_{dc}}{\sqrt{2} V_{LL}}$ $\text{DPF} \approx \cos\left(\alpha + \frac{u}{2}\right)$

P_{tot}	total active power (for all harmonics)
PCC	point of common coupling
PF	power factor
PT	potential transformer
SCR	short circuit ratio
rms	root mean square value $\sqrt{\frac{1}{T}\int_T f^2(t)dt}$
T	period of a periodic wave; time horizon under study for an aperiodic wave
TDD	total demand distortion
THD	total harmonic distortion
TIF	telephone influence factor
V_+, V_-	positive and negative sequence components of a sinusoidal three-phase voltage
VT	voltage—telephone influence factor product, read as VT product
w_h	telephone influence factor weight

References

Arrillaga, J. (1985). *Power system harmonics.* New York: John Wiley & Sons.

Bollen, M.H.J. (2000). *Understanding power quality problems: Voltage sags and interruptions.* New York: IEEE.

Dugan, R.C., McGranaghan, M.F., and Beaty, H.W. (1996). *Electrical power systems quality.* New York: McGraw-Hill.

Heydt, G. (1995). *Electric power quality.* Scottsdale, AZ: Stars in a Circle.

IEEE. (1986). IEEE Standard C57.110-1986, IEEE recommended practice for establishing transformer capability when supplying nonsinusoidal load currents. New York.

IEEE. (1997a). IEEE Standard 368-1977, IEEE recommended practice for measurement of electrical noise and harmonic filter performance of high-voltage direct-current systems. New York.

IEEE. (1997b).

IEEE. (1992a). IEEE Standard 1100, IEEE recommended practice for powering and grounding sensitive electronic equipment. New York.

IEEE. (1992b). IEEE Standard 519-1992, IEEE recommended practices and requirements for harmonic control in electrical power systems. New York.

Kennedy, B.W. (2000). *Power quality primer.* New York: McGraw-Hill.

Porter, G., and Van Sciver, J.A. (1998). *Power quality solutions: Case studies for troubleshooters.* Lilburn, GA: Fairmont Press.

Power Quality Assurance Magazine, http://industryclick.com/magazine.asp?magazineid=286&siteid=13

Shepherd, W., and Zand, P. (1979). *Energy flow and power factor in nonsinusoidal circuits.* Cambridge: Cambridge University.

VII

SIGNAL PROCESSING

Yih-Fang Huang
Department of Electrical Engineering, University of Notre Dame, Notre Dame, Indiana, USA

This section includes seven chapters that address various topics of *Signal Processing*, written by leading experts on the respective topics. Signal processing is one of the most important and rapidly growing fields in Electrical Engineering. Over the last few decades, it has played an important role in many different fields in science and engineering.

Information is most effectively represented by electronic signals or optical signals. Both types of signals can then be processed conveniently by electronic devices and systems. In the second half of the twentieth century, the successes in semiconductor device fabrication and integrated circuits technologies continued to fuel the development of powerful signal processing tools that benefit further development of various technological fields.

The first chapter of this section provides an overview of basic principles of signals and systems. It presents important fundamental concepts in transforming time-domain signals into frequency domain signals (like Fourier transform and z-transform and their properties) and linear shit-invariant systems and their properties.

The presentation of Chapter 1 lays down a nice foundation for the discussion of digital filters, with some emphasis on finite impulse response (FIR) and infinite impulse response (IIR) filters. The FIR filter and IIR filters are two most commonly used digital filters and they are realizations of linear shift-invariant systems. Properties and applications of those filters are addressed.

Chapters 3 to 5 present principles of the processing of various forms of signals, namely, speech signals, image signals and multimedia signals. Each form of signals has specific properties and applications that require distinctive processing techniques.

Chapter 6 presents fundamental concepts of statistical signal processing, which is an important aspect of signal processing. It provides a brief summary of the principles of mathematical statistics on which signal detection and estimation are based. Signal detection and estimation are critical in extracting information from signals in communication and control systems.

The Section ends with a chapter that addresses issues related to implementation of signal processing algorithms. It is concerned with application-specific VLSI architecture, including programmable digital signal processors, and dedicated signal processors implemented with VLSI technology. It also provides a good survey of recent development in VLSI signal processing technologies.

On behalf of all the authors in this section, I thank you for your interests in signal processing and hope you enjoy reading these chapters.

1

Signals and Systems

Rashid Ansari and
Lucia Valbonesi
Department of Electrical and Computer Engineering, University of Illinois at Chicago, Chicago, Illinois, USA

1.1 Introduction

Signal processing deals with operations performed on a signal with the intent of either modifying it in some desirable way or extracting useful information from it. Signal processing encompasses the theory, design, and practice of software and hardware for converting signals produced by different sources into a form that allows a more effective utilization of the signal content. The signals might be speech, music, image, video, multimedia data, biological signals, sensor data, telemetry, electrocardiograms, or seismic data. Typical objectives of processing include manipulation, enhancement, transmission, storage, visualization, detection, estimation, diagnosis, classification, segmentation, or interpretation. Signal processing finds applications in many areas, such as in telecommunications, consumer electronics, medical imaging and instrumentation, remote sensing, space research, oil exploration, radar, sonar, and biometrics (Anastassiou, 2001; Baillet *et al.*, 2001; Ebrahimi *et al.*, 2003; Hanzo *et al.*, 2000; Molina *et al.*, 2001; Ohm, 1999; Podilchuck and Delp, 2001; Strintzis and Malassiotis, 1999; Strobel *et al.*, 2001; Wang *et al.*, 2000; National Science Foundation, 1994). Mathematical foundations of signal processing are mainly drawn from transform theory, probability, statistical inference, optimization, and linear algebra.

The field of signal processing has grown phenomenally in the past years. This growth has been fueled by theoretical and algorithmic developments, technological advances in hardware and software, and ever widening applications that have generated an insatiable demand for signal processing tools. Theory and practice in a number of fields have been greatly influenced by signal processing (e.g., communications, controls, acoustics, music, seismology, medicine, and biomedical engineering).

Signal processing finds widespread use in everyday life, though the ubiquity of signal processing may not be evident to most users. Users may be familiar with JPEG, MPEG, and MP3 without recognizing that these tools are rooted in signal processing (Ansari and Memon, 2000; Wang *et al.*, 2000).

Recording of movies on a digital video disc (DVD), one of the most successful consumer electronics products, is made possible by compression techniques based on signal processing. DVD video is usually encoded from digital studio master tapes to MPEG-2 format, where MPEG refers to Moving Pictures Expert Group. The encoding process uses lossy compression that removes information that is either redundant or not readily perceptible by the human eye. The ultimate objective in compression is to reduce the bit rate for storage and transmission of video. The signal processing used here consists of applying suitable transforms to the data. Typically, a reduction in bit-rate requirements by a factor of 30 to 40 permits acceptable video quality.

The reduced memory requirements of audio coded in MP3 format has made it a popular vehicle for exchanging and storing music by individual users. MP3 refers to the MPEG Layer 3 audio compression format. Raw digital audio signals typically consist of 16-bit samples acquired at a sampling rate that is at least twice the bandwidth of the audio signal. In compact discs (CDs), a sampling rate of 44.1 kHz is used for each channel of stereo music. As a result, more than 1.4 Mbits is needed to represent just 1 second of music in raw CD recordings. By using MP3 audio coding, the storage requirement is reduced by a factor of 12 to 14 without significant loss in sound quality. This saving in storage is realized by using a signal processing tool, called a filter bank, in conjunction with exploiting the nature of perception of sound waves by the human ear.

In some applications, both speech and visual information may be jointly processed (Chen, 2001; Strobel *et al.*, 2001). This joint processing is referred to as multimedia or multimodal signal processing where data of multiple media like text, audio, images, and video are processed in an integrated manner, and interaction among the different media or modalities is investigated and exploited. Person recognition and authentication systems that have recently acquired increased importance in security and surveillance applications can be made more reliable by combining inputs from different modalities, such as voice and visual information.

Progress in the signal processing field has in recent years been largely driven by specific applications. Signal processing often figures in multidisciplinary research efforts. A noteworthy success in the application of signal processing has been in the fields of medicine and healthcare (Baillet *et al.*, 2001; Ebrahimi *et al.*, 2003). A number of imaging tools have been developed to aid diagnosis and recognition of disease. Computer tomography (CT), magnetic resonance imaging (MRI), and positron emission tomography (PET) are being widely used. These imaging technologies are testimony to the success of multidisciplinary interaction among researchers in fields such as signal processing, medicine, biology, and physics. The multidisciplinary effort can be witnessed in the collaboration on new challenges in modeling and processing in the emerging modalities for medical imaging such as electro/magneto-encephalography (E/MEG) and electrical impedance tomography (EIT).

In this chapter, we provide an overview of signals and systems that is largely based on signal representation framework, linear system theory, and transform theory. We begin by describing some basic notions of **discrete-time** and **continuous-time** signals and systems. The emphasis is on discrete-time signals and systems, but notions for the two cases are presented in parallel. The exposition of the continuous-time case is deliberately brief because the similarity in the notions should be evident to the reader. The concepts of sampling and quantization that relate continuous- and discrete-time signals are briefly discussed toward the end of the chapter.

Fundamentals of discrete-time systems and properties such as linearity, time-invariance, causality, and stability are reviewed first. Frequency domain analysis of signals and Fourier transforms are then presented followed by a discussion of frequency-selective filtering. Basics of z-transforms are covered next with a discussion of the z-domain interpretation of system properties such as causality and stability. Operations of sampling and quantization are then presented. Finally, the discrete Fourier transform is presented, and its importance in implementing signal processing algorithms is described. Our discussion is confined to the description and processing of **deterministic signals**.

1.2 Signals

Notions of signals and systems are encountered in a wide variety of disciplines. **Signal processing** formalizes these notions in a unified framework. It enables us to deal with diverse phenomena whose behavior is captured in variations of functions that characterize the phenomena. Signals are an abstraction for representing these variations. The signal may represent amplitude changes in speech or audio, concentration of a solute, biologically generated voltages, temperature inside an engine, pressure in a urinary bladder, noise in a communication receiver, and other such variables.

A **signal** is a mathematical function of one or more independent variables that usually represent time and/or space. An example of a signal as a function of time is a speech signal that represents a variation of acoustic pressure with time. An example of a signal as a function of space is an **image**: it is a function of light intensity and color in two spatial coordinates.

Some signals may depend both on space and time. Examples are video signals or the measurements of temperature in a room that depend on the three spatial coordinates defining each point in the room and that also vary with time.

In this section, we focus on signals that are functions of a single independent variable, and we will generally refer to this variable as **time**.

The independent variable of a signal can be either continuous or discrete, depending on the set of values for which the function is defined. In the first case, the signal is **continuous-time**, and the function is defined over a continuum of values of the independent variable; in the second case, the signal is **discrete-time**, and the function is defined over a countable set of values of the independent variable.

To distinguish between continuous-time and discrete-time signals, we will denote the independent variable as t in the continuous-time case and n in the discrete-time case.

The amplitude of a signal may also assume either continuous or discrete set of values. Thus, the signals can be classified into four categories depending on the nature of their independent and dependent variables.

It is convenient to introduce some notation here: $\mathcal{Z}$ denotes the set of integers, $\mathcal{C}$ the set of complex numbers, and $\mathcal{R}$ the set of real numbers:

- **Continuous-time or analog signals**: A continuous-time or an analog signal x_c is a mapping $x_c\colon \mathcal{R} \to \mathcal{R}$, where the domain is $\mathcal{R}$ and the range is a subset of the real line. A continuous-time signal will be denoted as $x_c(t)$ and not x_c, using the engineering convention of attaching the time variable as argument. This may be a source of confusion at times, but this aspect is clarified later. An example of a continuous-time signal is a speech signal: the acoustic pressure assumes a continuum of values, and pressure is defined at values of the independent variable, time, on the entire real line. Figure 1.1 shows an example of a continuous-time or an analog signal.
- **Quantized continuous-time signals**: A quantized continuous-time (boxcar) signal is a special case of a continuous-time signal, where the range of the signal is finite or countably infinite (i.e. it is a function defined over a continuum of values of the independent variable, while the range of values the signal assumes is discrete). An example of a quantized continuous-time (boxcar) signal is the control signal of a switch. At every instant of time, the switch is either open or closed depending on the control signal: if it is positive, the switch is open, whereas if it is negative, the switch is closed. Figure 1.2 shows an example of the control signal.

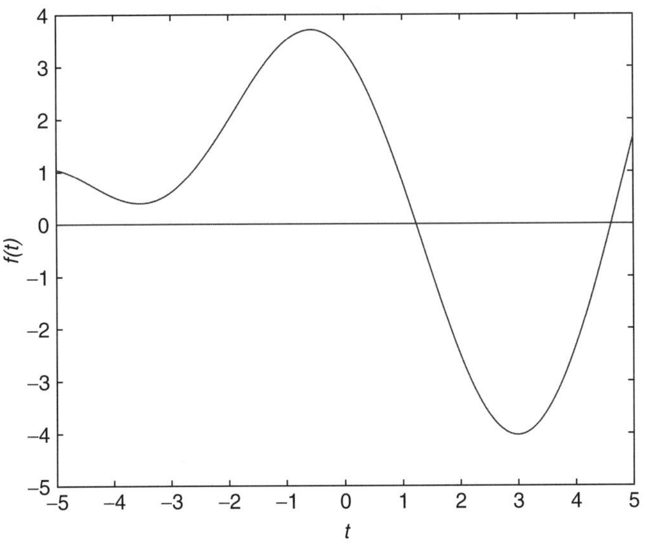

FIGURE 1.1 Example of **Continuous-Time** or **Analog Signal**. Both the independent and the dependent variables assume a continuum of values.

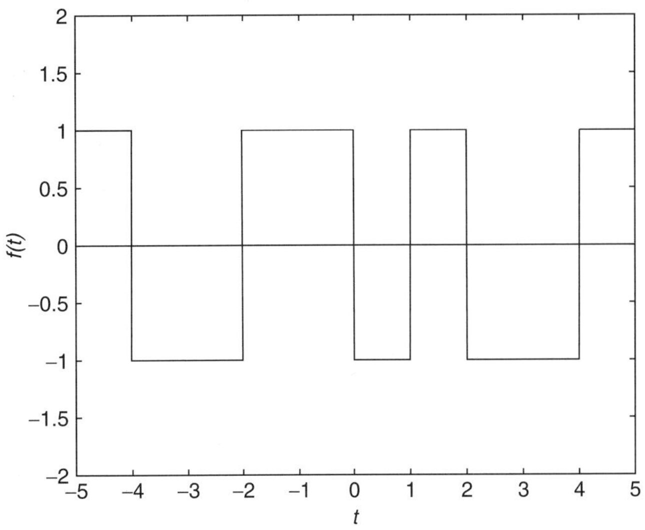

FIGURE 1.2 Example of **Quantized Continuous-Time (or Boxcar)** *Signal*. The independent variable is "continuous," whereas the dependent variable is "discrete".

- **Discrete-time or sampled data signals**: A discrete-time signal x is a mapping $x\colon\ \mathcal{Z} \to \mathcal{R}$, where the domain is $\mathcal{Z}$ and the range is a subset of the real line. A discrete-time signal will be denoted as $x[n]$, and not x, again using the engineering convention of attaching the time variable as argument. Note that the independent variable is integer-valued. Sampling a continuous-time signal at equally spaced instants is commonly used to produce a discrete-time signal. The distance, T, in time between two adjacent sampling instants is called the **sampling interval**, and its inverse is called **sampling frequency**. Upon sampling, $x[n] = x_c(nT)$.

  Figure 1.3 shows an example of a discrete-time signal. This is the sampled but not necessarily quantized version of the signal in Figure 1.1.
- **Digital signals**: A digital signal is a special case of a discrete-time signal, where the range of the signal is finite or countably infinite (i.e., it is a function defined over integer values of the independent variable and the range of values the signal assumes is also discrete).

  Figure 1.4 shows an example of a digital signal. This is the same signal shown in Figure 1.1, sampled and quantized. In this case, we note that the function assumes only integer values (i.e., it is discrete-valued).

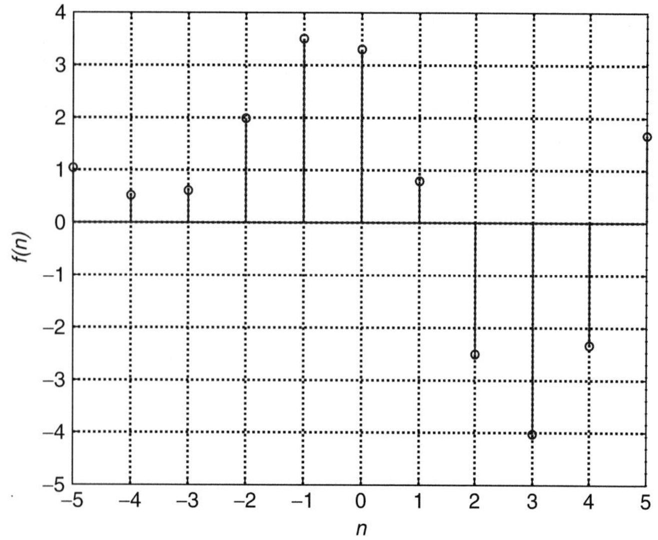

FIGURE 1.3 Example of a **Sampled Signal**. The independent variable is discrete, whereas the dependent variable is continuous.

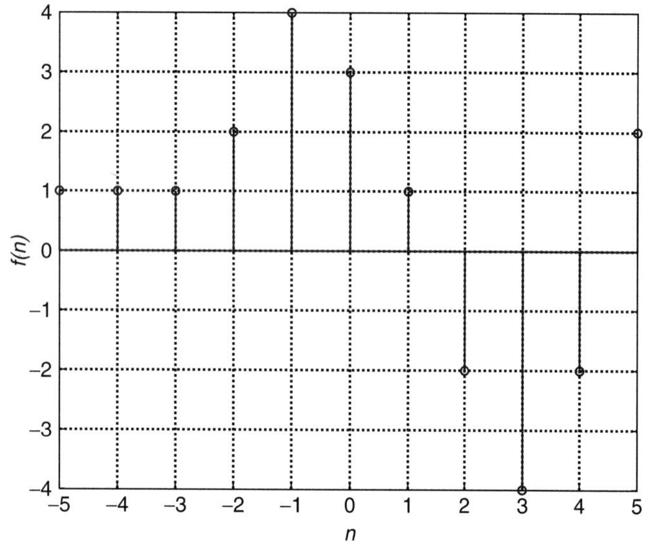

FIGURE 1.4 Example of **Digital Signal**. Both the independent and the dependent variables are discrete.

1.2.1 Basic Signals

Some basic signals are worthy of special attention because they play a particularly important role in the representation of other more general signals, either because other signals can be written as their weighted sum or integral or because they have useful properties. Signals in this category include the **unit sample** or **unit impulse**, the **unit step**, and the **exponential signal**. Another signal that plays an important role is a time-shifted version of a given signal. In the discrete-time case, this may be viewed as a transformation of a signal $x[n]$ given by $y = D_{n_0}x$ and defined by:

$$y[n] = D_{n_0}x[n] = x[n - n_0]. \tag{1.1}$$

In the case of a continuous-time signal $x_c(t)$, the time-shifted signal is defined as:

$$y_c(t) = D_{t_0}x(t) = x(t - t_0). \tag{1.2}$$

We will now define some common basic signals in continuous-time and discrete-time cases.

Unit Sample or Unit Impulse

The discrete **unit sample** or **unit impulse** is defined as:

$$\delta[n] = \begin{cases} 1, & n = 0. \\ 0, & n \neq 0. \end{cases} \tag{1.3}$$

The continuous-time counterpart of $\delta[n]$ is the unit impulse or the **Dirac delta-function** $\boldsymbol{\delta_c(t)}$. This function is zero-valued for $t \neq 0$ and $\int_{-\infty}^{\infty} \delta_c(t)dt = 1$.

The discrete-time and the continuous-time unit impulses are shown in Figure 1.5.

The unit impulse has the property that:

$$x[n_0] = \sum_{n=-\infty}^{\infty} x[n]\delta[n - n_0], \tag{1.4a}$$

$$x_c(t_0) = \int_{-\infty}^{\infty} x_c(t)\delta_c(t - t_0)dt. \tag{1.4b}$$

Moreover, every signal can be written as a weighted sum or integral of unit impulses:

$$x[n] = \sum_{k=-\infty}^{\infty} x[k]\delta[n - k]. \tag{1.5a}$$

$$x_c(t) = \int_{-\infty}^{\infty} x_c(\tau)\delta_c(t - \tau)d\tau. \tag{1.5b}$$

Unit Step

The **discrete-time unit step** is defined as:

$$u[n] = \begin{cases} 1, & n \geq 0. \\ 0, & n < 0. \end{cases} \tag{1.6}$$

The **continuous-time unit step** is defined as:

$$u_c(t) = \begin{cases} 1, & t \geq 0. \\ 0, & t < 0. \end{cases} \tag{1.7}$$

The discrete-time and the continuous-time unit steps are shown in Figure 1.6.

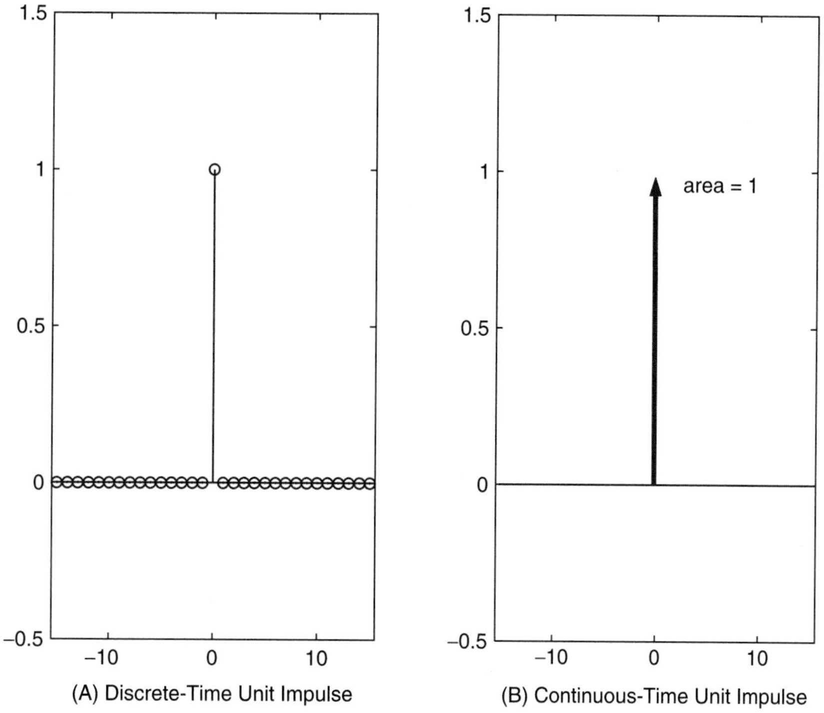

FIGURE 1.5 Discrete-Time and Continuous-Time Unit Impulses

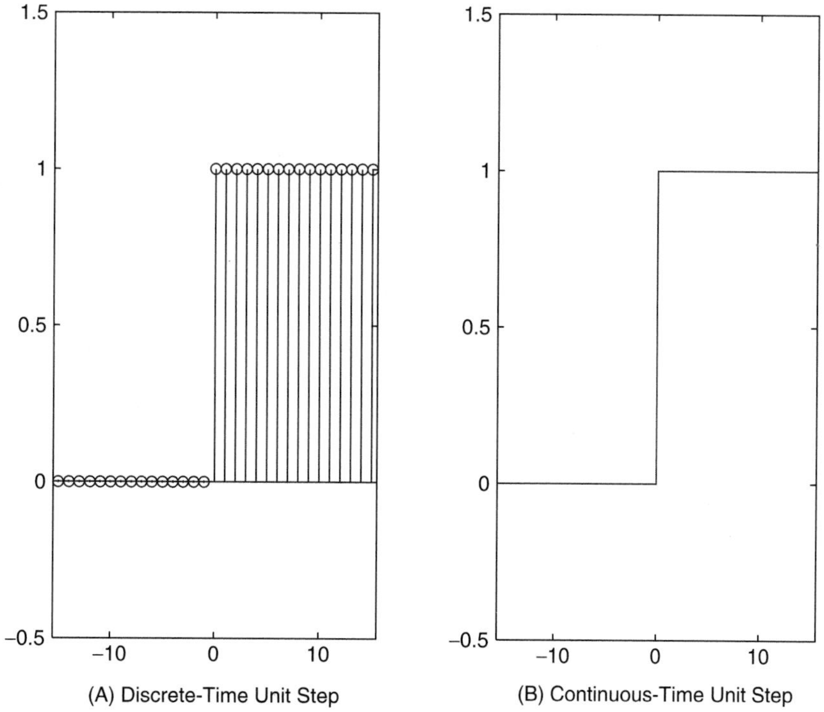

FIGURE 1.6 Discrete-Time and Continuous-Time Unit Steps

The unit impulse and the unit step functions are related by:

$$u[n] = \sum_{k=-\infty}^{n} \delta[k], \quad \delta[n] = u[n] - u[n-1]. \tag{1.8a}$$

$$u_c(t) = \int_{-\infty}^{t} \delta_c(\tau)d\tau, \quad \delta_c(t) = \frac{d}{dt}u_c(t). \tag{1.8b}$$

Exponential Signal

The continuous-time exponential signal is defined as:

$$f(t) = Ke^{st}, \tag{1.9}$$

where K is a constant, t is the independent variable, and s is in general a complex number with $s = \sigma + j\omega$.

When $\sigma = 0$ and $K = 1$, the real part of the signal is $\cos(\omega t)$, and the imaginary part is $\sin(\omega t)$.

Figure 1.7 shows the real part of complex exponential functions for different values of σ.

This is a periodic signal with period Δt given by:

$$\omega \Delta t = 2\pi, \tag{1.10}$$

from which we have:

$$\Delta t = \frac{2\pi}{\omega}. \tag{1.11}$$

The parameter ω is expressed in radians per second, and the *frequency* $f = \omega/2\pi$ is expressed in cycles per second or Hertz.

The role of the parameter σ is to amplify ($\sigma > 0$) or reduce ($\sigma < 0$) the amplitude of the signal when the independent variable increases.

In a similar way, the discrete-time exponential signal is defined as:

$$x[n] = Kz^n, \tag{1.12}$$

where K is a constant and z is in general a complex variable $z = \gamma e^{j\omega}$.

The discrete-time exponential signal is shown in Figure 1.8, with $z = 3/4$ and k = 1.

1.2.2 Signal Classification

Signals can be classified according to their support, energy, power, range bounds, and other characteristics. Here, we focus

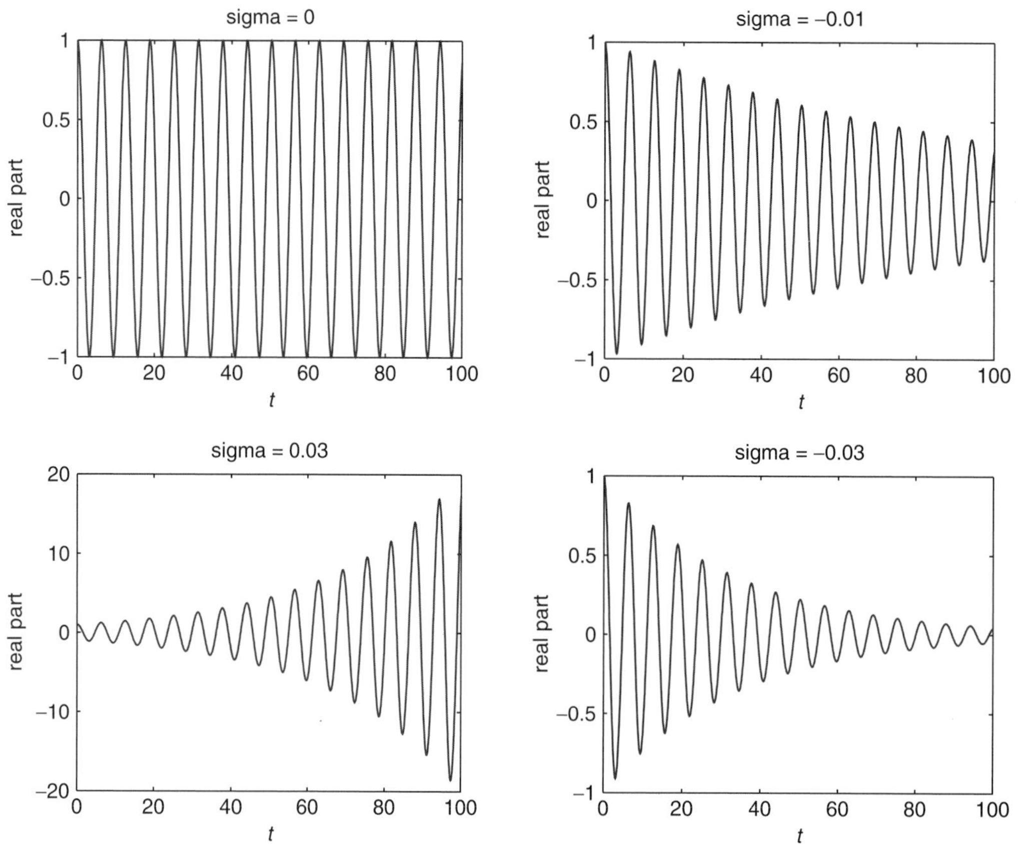

FIGURE 1.7 Real Part of the Complex Exponential Function for Different Values of the Parameter σ

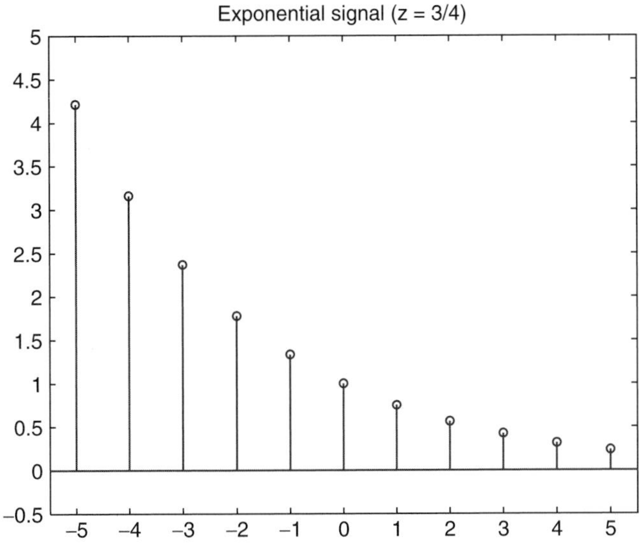

FIGURE 1.8 Discrete-Time Exponential Signal

on classification according to support. Support denotes the subset of the domain where the signal is nonzero (i.e., the extent of the independent variable or "time-axis" where the signal is nonzero). The nature of support bears a relation to types of convergence regions of the z-transform or Laplace transform of a signal as seen later.

Signal Duration

Signals can be classified according to their **duration** or support (i.e., the extent of domain over which the signal is nonzero).

- **Finite-extent signal**: A signal is of finite extent or finite duration if its support is bounded. In the discrete-time case, there exist integers, N_1 and N_2, such that the signal $x[n]$ satisfies the constraint:

$$x[n] = \begin{cases} 0, & n < N_1. \\ 0, & n > N_2. \end{cases} \tag{1.13}$$

  The corresponding requirement on a finite-extent continuous-time signal $x_c(t)$ is that there exist real numbers T_1 and T_2, such that:

$$x_c(t) = \begin{cases} 0, & t < T_1. \\ 0, & t > T_2. \end{cases} \tag{1.14}$$

- **Infinite-extent signal**. A signal is of infinite extent or infinite duration if its support is not bounded. Infinite-extent signals are further categorized into:
  (1) **Left-sided infinite-extent** signal: if there exists an integer N_1, such that $x[n] = 0$ for $n > N_1$
  (2) **Right-sided infinite-extent** signal: if there exists an integer N_2, such that $x[n] = 0$ for $n < N_2$
  (3) **Two-sided infinite-extent** signal: if the infinite-extent signal is neither left-sided no right sided

Causal Support

- **causal Signal**
  Causality is a system property that is described later in this chapter. For linear-time invariant systems, causality constrains the support of the system impulse response to non-negative values of the domain. The notion is extended to define a general "causal" signal. A discrete-time signal, $x[n]$, is causal if:

$$x[n] = 0, \quad n < 0. \tag{1.15}$$

  A continuous-time signal is causal if:

$$x_c(t) = 0, \quad t < 0. \tag{1.16}$$

  A signal is **noncausal** if it is *not* causal. An infinite-duration causal signal is a special case of an infinite-duration right-sided signal.
- **Anticausal Signal**
  A discrete-time signal is anticausal if:

$$x[n] = 0, \quad n > 0. \tag{1.17}$$

  A continuous-time signal is anticausal if:

$$x_c(t) = 0, \quad t > 0. \tag{1.18}$$

  An infinite-duration anticausal signal is a special case of a left-sided infinite-duration signal.

1.3 Systems

A **system** is an entity that transforms an **input signal** into an **output signal** according to a specified rule. Systems can be classified into **discrete-time systems** or **continuous-time systems** if the pair of input and output signals is discrete-time or continuous time, respectively, as shown in Figure 1.9. Systems can be hybrid (e.g., analog-to-digital converters) where the input is continuous-time and the output discrete-time.

A discrete-time or continuous-time input signal is denoted by $x[n]$ or $x_c(t)$, respectively, while the corresponding output signal is denoted as $y[n]$ or $y_c(t)$. The system (i.e., the transformation that maps that input signal into the output signal) is denoted by T so that $y[n] = T(x[n])$ or $y_c(t) = T(x_c(t))$. We will often refer to the system itself as T.

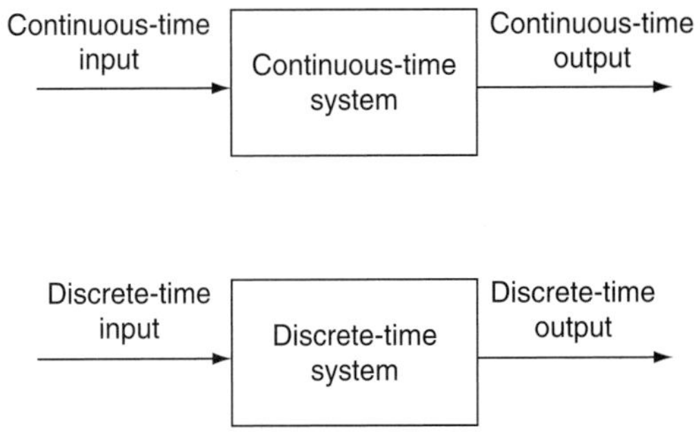

FIGURE 1.9 Continuous-Time and Discrete-Time Systems

Unfortunately the engineering notation here could be misleading. The relation $y_c(t) = T(x_c(t))$ does not mean that the output at instant t depends on the input only at time t. Rather, the equation should be understood in the mathematical sense of $y_c = T(x_c)$.

Time-Shift Operator

We had noted earlier that a signal that plays an important role in signal processing is a time-shifted version of a given signal. This signal can be viewed as the output of a **time-shifting system**. In the discrete-time case, this may be viewed as a transformation of a signal $x[n]$ given by $y = D_{n_0}x$ and defined by:

$$y[n] = D_{n_0}x[n] = x[n - n_0]. \tag{1.19}$$

In the case of a continuous-time signal $x_c(t)$, the time-shifted signal is as follows:

$$y_c(t) = D_{t_0}x_c(t) = x_c(t - t_0). \tag{1.20}$$

1.3.1 Common System Properties

To characterize system behavior, we need to impose some structure on a system. This structure takes the form of system properties that provide useful mathematical descriptions as well as insights into analysis and design of systems. We consider a few of these properties with a focus on linear shift-invariant systems.

Linearity

A **linear system** is one where the response (output) of a system to the sum of scaled stimuli (inputs) equals the sum of its scaled responses to each individual stimulus. Consider a discrete-time system T such that the output corresponding to an input $x_1[n]$ is $y_1[n] = T(x_1[n])$, and the output corresponding to an input $x_2[n]$ is $y_2[n] = T(x_2[n])$. The system is said to be *linear* if:

$$T(ax_1[n] + bx_2[n]) = ay_1[n] + by_2[n], \tag{1.21}$$

where a and b are **scalars** (real or complex numbers depending on the mapping T). The above rule can be extended to sum of a finite set of inputs by induction. We will assume that the system allows the extension of this property to a countably infinite sum of inputs.

Example

The system:

$$y[n] = \frac{1}{2}(x[n] + x[n-1]),$$

is linear, since:

$$y_1[n] = \frac{1}{2}(x_1[n] + x_1[n-1])$$

$$y_2[n] = \frac{1}{2}(x_2[n] + x_2[n-1])$$

$$\Rightarrow T(ax_1[n] + bx_2[n])$$

$$= \frac{1}{2}(ax_1[n] + bx_2[n] + ax_1[n-1] + bx_2[n-1])$$

$$= a\frac{1}{2}(x_1[n] + x_1[n-1]) + b\frac{1}{2}(x_2[n] + x_2[n-1])$$

$$= ay_1[n] + by_2[n].$$

Shift Invariance (or Time Invariance)

A **shift-invariant** or **time-invariant** system is one where the response of a system to a shifted version of an original input is the same as the shifted version of the system's response to the original input. Consider a discrete-time system T such that the output corresponding to an input $x[n]$ is $y[n] = T(x[n])$. The system T is **shift invariant** if:

$$T(x[n - n_0]) = y[n - n_0]. \tag{1.22}$$

Example

The system of the previous example with output given by:

$$y[n] = \frac{1}{2}(x[n] + x[n-1]),$$

is shift invariant. If the input to a system is $x_1[n] = x[n - n_0]$, then:

$$y_1[n] = \frac{1}{2}(x_1[n] + x_1[n-1])$$
$$= \frac{1}{2}(x[n - n_0] + x[n - n_0 - 1]) = y[n - n_0].$$

On the other hand, the system with output $y[n] = nx[n]$ is not shift invariant.

If $x_1[n] = x[n - n_0]$, then:

$$y_1[n] = nx_1[n] = nx[n - n_0] \neq y[n - n_0] = (n - n_0)x[n - n_0].$$

Causality and Stability

A discrete-time system T is **causal** if its output at any time instant depends only on the input samples at that time or prior to it. In other words, the output value $y[n_0]$ depends only on input samples $(x[n_0], x[n_0 - 1], x[n_0 - 2], \ldots)$ and not on future input samples $(x[n_0 + 1], x[n_0 + 2], \ldots)$.

Stability of a system is defined in many ways, but a common characterization is in terms of bounds on system response when the input range is confined to some specified bounds. A discrete-time system T is said to be **stable in the bounded-input**

bounded-output (BIBO) **sense** if, given any input $x[n]$ bounded for all n (i.e., $|x[n]| \leq B_1 < \infty$) the output is also bounded for all n (i.e., $|y[n]| \leq B_O < \infty$) for some B_O.

Examples

The system $y[n] = e^{x^2[n]}$ is BIBO stable. If $|x[n]| \leq 10 = B_I$, then $|y[n]| \leq e^{100} = B_O$.

The two systems with outputs $y[n] = nx[n]$ and $y[n] = \sum_{k=-\infty}^{\infty} x[n-k]$ are not BIBO stable because their outputs are not bounded (i.e., one cannot find a finite positive number B_O such that $|y[n]| \leq B_O$ for all n).

Invertibility

An **invertible system** is one in which an arbitrary input to the system can be uniquely inferred from the corresponding output of the system. The cascade of the system and its inverse, if it exists, is an identity system (i.e., the output of the cascade is identical to the input).

All properties so far were defined for discrete-time systems. Equivalent properties can be defined for continuous-time systems by requiring output to satisfy the following conditions, with notation analogous to that in the discrete-time case:

- **Linearity**: $T(ax_1(t) + bx_2(t)) = ay_1(t) + by_2(t)$.
- **Shift invariance**: $T(x(t - t_0)) = y(t - t_0)$.
- **Causality**: $y(t)$ is function of $x(\tau)$ for $-\infty < \tau \leq t$.
- **BIBO stability**: $|x(t)| \leq B_I$ for all $t \Rightarrow |y(t)| \leq B_O$ for all t.

1.3.2 Linear Shift-Invariant Systems

Linearity and shift invariance are often used in combination to model the behavior of practical systems. This chapter is largely devoted to developing descriptions and tools for the analysis and design of **linear shift-invariant** (LSI) **systems**. These are often referred to as linear time-invariant (LTI) systems because the independent variable is commonly **time**.

As pointed out before, a discrete-time signal can be expressed as a weighted sum of shifted discrete-time unit impulses. Similarly, a continuous-time signal can be expressed as the integral of weighted and shifted continuous-time unit impulses. This representation, together with system properties of linearity and shift invariance, allows us to create a practical and useful characterization of a system in terms of a function called the **impulse response** of the system.

To arrive at the characterization of a linear time-invariant discrete-time system in terms of its response to an impulse input, the input signal, $x[n]$, to a digital system is expressed as a weighted sum of discrete-time unit impulses:

$$x[n] = \sum_{k=-\infty}^{\infty} x[k]\delta[n-k]. \tag{1.23}$$

Here, a comment on the engineering notation is in order. Equation 1.23 should be understood as $x = \sum_{-\infty}^{\infty} x[k]D_k\delta$, so that $x[k]$ in this equation is a scalar and $\delta[n-k]$ is a discrete-time signal.

The output $y[n]$ for the input in equation 1.23 is expressed as:

$$y[n] = T(x[n]) = T\left(\sum_{k=-\infty}^{\infty} x[k]\delta[n-k]\right). \tag{1.24}$$

The linearity of the system allows us to rewrite equation 1.24 as:

$$y[n] = \sum_{k=-\infty}^{\infty} x[k]T(\delta[n-k]), \tag{1.25}$$

where we have assumed the system behavior allows us to extend the linearity property to the case of an infinite sum of inputs. Here, $T(\delta[n])$ denotes the output signal when a unit impulse is applied as input. This signal is called the impulse response of the system, and it is denoted by $h[n]$.

Next, we invoke the shift invariance property to assert that $T(\delta[n-k]) = h[n-k]$. The output signal can therefore be rewritten as:

$$y[n] = \sum_{k=-\infty}^{\infty} x[k]h[n-k]. \tag{1.26}$$

The last expression in equation 1.26 is called the **convolution sum** of the signals $x[n]$ and $h[n]$, which is denoted by $x[n] * h[n]$. The convolution sum of $h[n]$ and $x[n]$ can also be expressed as:

$$\begin{aligned} y[n] = x[n] * h[n] &= \sum_{k=-\infty}^{\infty} x[k]h[n-k] \\ &= \sum_{m=-\infty}^{\infty} x[n-m]h[m] = h[n] * x[n]. \end{aligned} \tag{1.27}$$

By proceeding in an analogous manner in the case of continuous-time systems, one arrives at equivalent relations. As pointed out before, each continuous-time signal can be written as the integral of weighted and shifted unit impulses:

$$x_c(t) = \int_{-\infty}^{\infty} x_c(\tau)\delta_c(t-\tau)d\tau. \tag{1.28}$$

We assume that $h_c(t)$ is the **impulse response** of the continuous-time system (i.e., it is the output of the system when a unit impulse is applied at the input).

By invoking linearity and shift invariance property of the system, the output signal $y(t)$ can be written as:

$$\begin{aligned} y_c(t) &= T(x_c(t)) = T\left(\int_{-\infty}^{\infty} x_c(\tau)\delta_c(t-\tau)d\tau\right) \\ &= \int_{-\infty}^{\infty} x_c(\tau)T(\delta_c(t-\tau))d\tau \\ &= \int_{-\infty}^{\infty} x_c(\tau)h_c(t-\tau)d\tau = x_c(t) * h_c(t). \end{aligned} \tag{1.29}$$

Therefore, for a continuous-time system, the output signal is given by the convolution integral of the input signal and the system impulse response.

An important observation is that both discrete-time and continuous-time systems that are linear and time invariant are completely characterized by their impulse responses under suitable assumptions.

Example

An example of the convolution for two continuous-time signals is considered now. It illustrates the property that the duration of **convolved** signal is equal to the sum of the durations of the two signals that are convolved. For discrete-time convolution, the duration of the sum obtained as a convolution sum is equal to the sum of the durations of the signals less one.

Consider a signal $x_c(t)$:

$$x_c(t) = \begin{cases} 1/2, & -1 \leq t \leq 1, \\ 0, & \text{otherwise,} \end{cases}$$

which is applied as input to a system with impulse response $h_c(t)$:

$$h_c(t) = \begin{cases} 1, & 0 \leq t \leq 2 \\ 0, & \text{otherwise.} \end{cases}$$

Figure 1.10 shows the two signals $x_c(t)$ and $h_c(t)$ and their convolution $y_c(t)$. We notice that the duration of the convolution signal is $L_y = L_x + L_h = 2 + 2 = 4$.

For discrete-time convolution, the duration of the signal obtained as a convolution sum is equal to the sum of the durations of the convolved signals less one. Let $x[n]$ and $h[n]$ be two finite-duration discrete-time signals such that $x[n] = 0$ for $n < N_1$ or $n > N_2 \geq N_1$, and $h[n] = 0$ for $n < M_1$ or $n > M_2 \geq M_1$. We assume that $x[N_1]$, $x[N_2]$, $h[M_1]$, and $h[M_2]$, are nonzero, so that the duration of $x[n]$ is $N = N_2 - N_1 + 1$ and the duration of $h[n]$ is $M = M_2 - M_1 + 1$. Then, the convolved signal $y[n] = x[n] * h[n]$ has duration $L = M + N - 1$. The signal $y[n]$ is zero for $n < L_1 = M_1 + N_1$ or $n > L_2 = M_2 + N_2$. Note that $L = L_2 - L_1 + 1$.

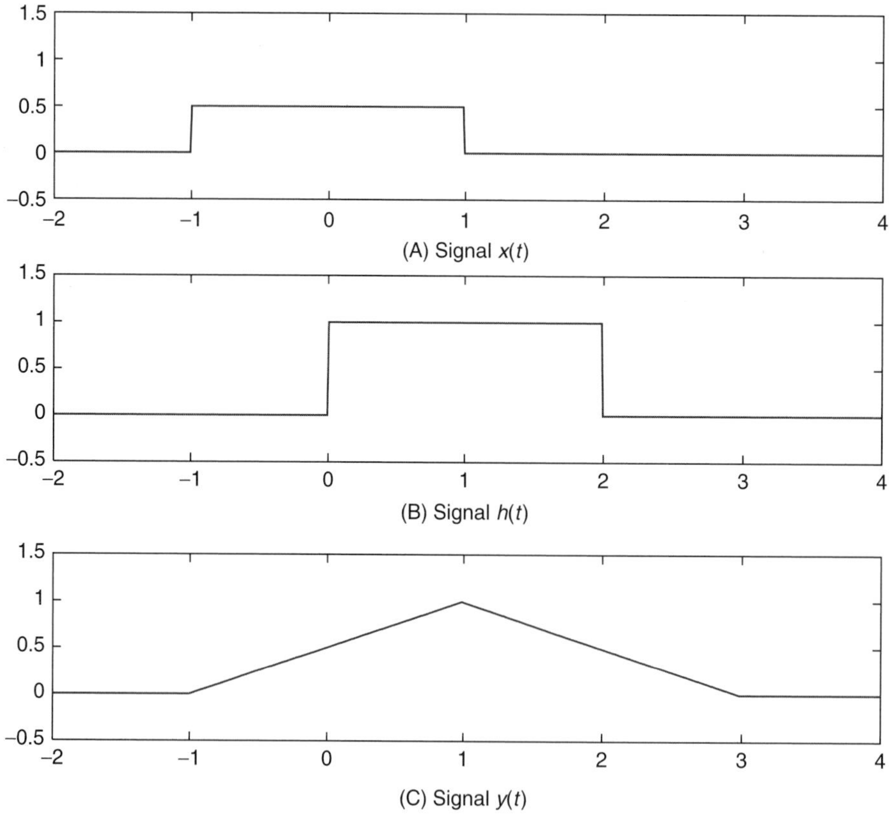

FIGURE 1.10 Two signals and their convolution

To conclude this section, we list three properties of the convolution operator that are valid both in the continuous-time and in the discrete-time case. Their proofs follow directly from properties of the convolution sum and integral operators.

1. **Commutative property**: $x_1 * x_2 = x_2 * x_1$.
2. **Associative property**: $x_1 * [x_2 * x_3] = [x_1 * x_2] * x_3$.
3. **Distributive property**: $x_1 * [x_2 + x_3] = x_1 * x_2 + x_1 * x_3$.

In expressions 1 through 3, if x_1 is assumed to represent system input and x_2 and x_3 represent system impulse responses, then these have useful interpretations for an interconnected system. The associative property implies that a cascade connection of two LSI systems is equivalent to a single system with impulse response equal to the convolution of the two individual system impulse responses. The distributive property implies that a parallel connection of two LSI systems can be represented by a single system with impulse response equal to the sum of the two individual system impulse responses.

1.3.3 Properties of LSI Systems

An important observation is that both discrete-time and continuous-time systems that are **linear** and **time invariant** are completely characterized by their impulse responses. Other properties of LSI systems, such as causality, stability, and invertibility, can then be characterized in terms of a system's impulse response.

The characterization of the LSI system properties of causality, stability, and invertibility will first be examined for the case of discrete-time systems.

Causality

If a discrete-time system is causal, then the current output sample depends only on the current and past input samples and not on future samples. This definition is now examined in the case of an LSI system with an impulse response $h[n]$.

The output $y[n]$ of an LSI system can be expressed as:

$$\begin{aligned} y[n] = h[n] * x[n] &= \sum_{k=-\infty}^{\infty} x[n-k]h[k] \\ &= \ldots + x[n+2]h[-2] + x[n+1]h[-1] \\ &\quad + x[n]h[0] + x[n-1]h[1] + x[n-2]h[2] + \ldots. \end{aligned} \tag{1.30}$$

The expression of equation 1.30 immediately suggests the constraint on the impulse response of a causal LSI system. If the output is required to be independent of future samples of an arbitrary input, then the coefficients $h[-1], h[-2], \ldots$ in the convolution sum should each be equal to zero.

Therefore, an LSI system is causal if and only if $h[n] = 0$, $n < 0$.

BIBO Stability

A system was previously defined to be BIBO stable if a bounded input produces a bounded output. It will be shown that an LSI system is BIBO stable if and only if $\sum_{n=-\infty}^{\infty} |h[n]| < \infty$.

Proof

We first assume that $S_h = \sum_{n=-\infty}^{\infty} |h[n]| < \infty$. Now suppose the system input $x[n]$ is bounded (i.e., $|x[n]| < B_I$) for all n. The output satisfies:

$$|y[n]| \leq \sum_{k=-\infty}^{\infty} |x[n-k]||h[k]| \leq B_I \sum_{k=-\infty}^{\infty} |h[k]| = B_I S_h. \tag{1.31}$$

We conclude that $|y[n]| \leq B_O = B_I S_h$, and therefore the system is stable.

We now assume that the LSI system is BIBO stable with an impulse response that is not identically zero. Consider a bounded input $x[n]$ defined by:

$$x[n] = \begin{cases} B_I \frac{h^*[-n]}{|h[-n]|}, & h[n] \neq 0, \\ 0, & h[n] = 0, \end{cases} \tag{1.32}$$

where $h^*[n]$ is the complex conjugate of $h[n]$. The output $y[n]$ for $n = 0$ is given by:

$$|y[0]| = \sum_{k=-\infty}^{\infty} x[-k]h[k] = B_I \sum_{k=-\infty}^{\infty} |h[k]| = B_I S_h. \tag{1.33}$$

For the system to be stable, we require that $y[n]$ be bounded for all n. In particular, $|y[0]|$ above should be finite. This implies that $S_h = \sum_{n=-\infty}^{\infty} |h[n]| < \infty$.

Invertibility

Let an LSI system with impulse response $h[n]$ be invertible. It can then be shown that the inverse system is LSI. If the inverse system has impulse response $h^i[n]$, then $h[n] * h^i[n] = \delta[n]$.

In the case of continuous-time, we can obtain similar conditions on the impulse response, $h_c(t)$, of an LSI system as summarized here:

- An LSI system is causal $\Leftrightarrow h_c(t) = 0,\ t < 0$.
- An LSI system is BIBO stable $\Leftrightarrow \int_{-\infty}^{\infty} |h_c(t)| dt < \infty$.
- An Inverse system exists and has impulse response $h_c^i(t)$; then $h_c(t) * h_c^i(t) = \delta_c(t)$.

A special class of LSI systems commonly used in modeling and implementation is the class that is described by a linear constant-coefficient difference or differential equation in the case of discrete and continuous time, respectively. It should be pointed out that depending on the specified auxiliary (or "initial") conditions, a system described by a linear constant-coefficient difference or differential equation may not be LSI

(Oppenheim *et al.*, 1999; oppenheim and Willsky, 1997). We will assume that the auxiliary conditions are consistent with the LSI requirement. A condition of initial rest is assumed for the systems, so that we will be dealing only with linear shift-invariant systems described by the difference or differential equation.

The LSI systems can be further classified according to the duration of the impulse response. An LSI system is called **finite-duration impulse response** (FIR) if its impulse response has finite support. An **infinite-duration impulse response** (IIR) LSI system has an impulse response with infinite support. LSI systems are often referred to as filters in a generalized sense, though the term filters is commonly understood in terms of passing or rejecting signal content in a **frequency-selective** manner. Hence, the labels FIR and IIR filters are commonly used even when frequency selectivity is not the objective. We now consider frequency domain analysis of signals and systems.

1.4 Analysis in Frequency Domain

So far, we have examined discrete-time and continuous-time signals as well as LSI systems in the time domain. We now consider signal and LSI system descriptions in the frequency domain motivated by the fact that analysis and design may be more conveniently done in the frequency domain. Central to frequency-domain description and analysis is the fact that signals (and impulse responses) can be conveniently represented in terms of sinusoids and complex exponentials that are eigenfunctions of LSI systems.

Let T denote the system operator for an LSI system. We are interested in finding the eigenfunctions of this operator, that is the input signals $x[n]$ or $x(t)$, such that:

$$Tx = \lambda x, \tag{1.34}$$

where λ is a scalar. Equation 1.34 implies that for certain input signals x, the system output is a scaled version of the input. The scalar multiplier λ is called the **eigenvalue**.

It has already been mentioned that the **exponential signal** plays an important role in signal representation. We now consider the response of a linear shift-invariant system to an exponential signal applied as input.

We again focus on the discrete-time case first. Consider an LSI discrete-time system with impulse response $h[n]$. If the input to the system is $x[n] = e^{j\omega n}$ for $-\infty < n < \infty$, then the output signal is as written here:

$$\begin{aligned} y[n] &= \sum_{k=-\infty}^{\infty} x[n-k]h[k] = \sum_{k=-\infty}^{\infty} e^{j\omega(n-k)}h[k] \\ &= e^{j\omega n}\sum_{k=-\infty}^{\infty} e^{-j\omega k}h[k] = H(e^{j\omega})e^{j\omega n}. \end{aligned} \tag{1.35}$$

From equation 1.35, we observe that e^{jwn} is an eigenfunction of the linear shift invariant system. The LSI system output is equal to the input multiplied by a scalar factor $H(e^{j\omega})$, which is the eigenvalue given by:

$$H(e^{j\omega}) = \sum_{k=-\infty}^{\infty} e^{-j\omega k}h[k]. \tag{1.36}$$

The $H(e^{j\omega})$ represents the complex gain as a function of the frequency variable ω and is called the **frequency response** of the system and also the **discrete-time Fourier transform** (DTFT) of $h[n]$.

In general, the DTFT of a signal $x[n]$ is defined as:

$$X(e^{j\omega}) = \sum_{n=-\infty}^{\infty} e^{-j\omega n}x[n]. \tag{1.37}$$

The DTFT of a signal is a complex function that can be expressed as $|X(e^{j\omega})|e^{\phi(\omega)}$. Here $|X(e^{j\omega})|$ is the **magnitude**, and $\phi(\omega)$ is the **phase** of the DTFT. The magnitude $|X(e^{j\omega})|$ can be viewed as a measure of the strength of the signal content at different frequencies, and $\phi(\omega)$ represents the phase of the frequency components.

Example

Consider the computation of the DTFT of the rectangular window signal shown in Figure 1.11 and expressed as:

$$x[n] = \begin{cases} A, & -n_0 \le n \le n_0 \\ 0, & \text{otherwise.} \end{cases}$$

For $\omega = 0$, we note that $X(e^{j\omega}) = A(2n_0 + 1)$. For $\omega \neq 0$:

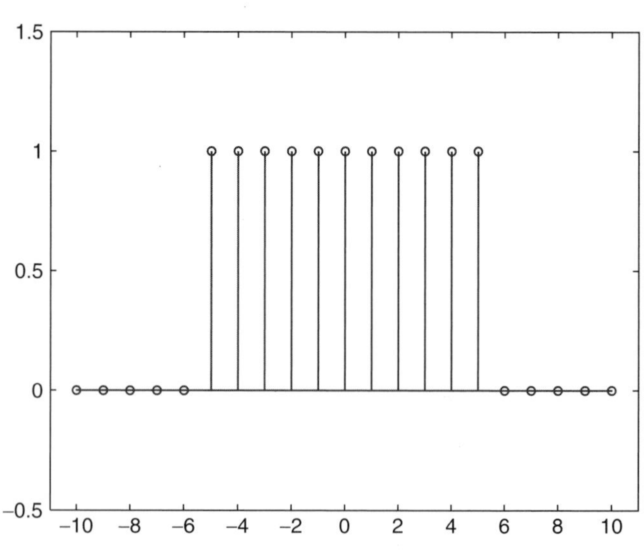

FIGURE 1.11 Rectangular Window

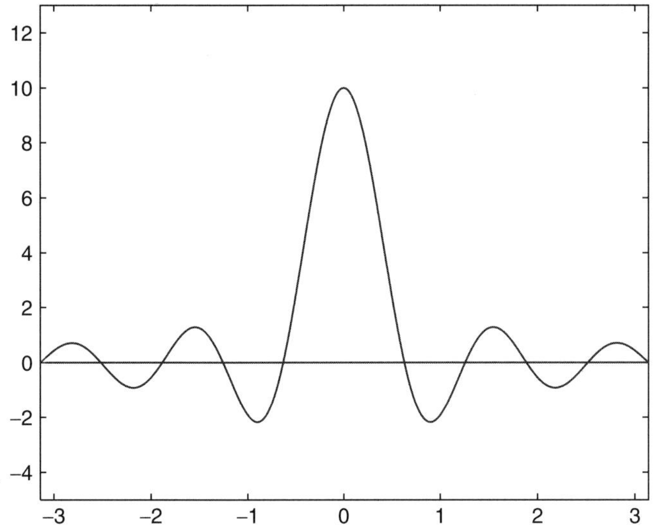

FIGURE 1.12 Discrete-Time Fourier Transform of the Rectangular Window.

$$X(e^{j\omega}) = \sum_{n=-\infty}^{\infty} x[n]e^{-j\omega n} = \sum_{n=-n_0}^{n=n_0} Ae^{-j\omega n}$$
$$= A\frac{e^{j\omega(n_0+1)} - e^{-j\omega n_0}}{e^{j\omega} - 1} = A\frac{\sin((2n_0+1)\omega/2)}{\sin(\omega/2)}.$$

The DTFT is shown in Figure 1.12.

1.4.1 Properties of the Discrete-Time Fourier Transform

The discrete-time Fourier transform (DTFT) was defined a complex series for an arbitrary discrete-time signal. The issue of its convergence is briefly examined. In the description here, a **sufficient condition** for the existence of DTFT is assumed. This is done by considering the signal $x[n]$ to be **absolutely summable** (i.e., $\sum_{n=-\infty}^{\infty} |x[n]| < \infty$). In this case:

$$|X(e^{j\omega})| = \left|\sum_{n=-\infty}^{\infty} e^{-j\omega n}x[k]\right| \leq \sum_{n=-\infty}^{\infty} |e^{-j\omega n}|\,|x[k]|$$
$$= \sum_{n=-\infty}^{\infty} |x[n]| < \infty.$$

Therefore, if the signal $x[n]$ is absolutely summable, the series converges and the DTFT exists.

Inverse DTFT

Given the DTFT of a signal, the signal can be recovered from it using the so-called inverse DTFT given by:

$$x[n] = \frac{1}{2\pi}\int_0^{2\pi} X(e^{j\omega})e^{j\omega n}d\omega. \tag{1.38}$$

The validity of the above relation can be seen from:

$$\frac{1}{2\pi}\int_0^{2\pi} X(e^{j\omega})e^{j\omega n}d\omega = \frac{1}{2\pi}\int_0^{2\pi}\left(\sum_{m=-\infty}^{\infty} e^{-j\omega m}x[m]\right)e^{j\omega n}d\omega$$
$$= \sum_{n=-\infty}^{\infty} x[m]\frac{1}{2\pi}\int_0^{2\pi} e^{j\omega(n-m)}d\omega = \sum_{m=-\infty}^{\infty} x[m]\delta[n-m] = x[n],$$

where we have used the fact that:

$$\frac{1}{2\pi}\int_0^{2\pi} e^{j\omega(n-m)}d\omega = \begin{cases} 1, & n = m. \\ 0, & n \neq m. \end{cases}$$

Some useful properties of DTFT are now examined.

Periodicity

The DTFT is a periodic function in ω with period 2π, since:

$$X(e^{j(\omega+2\pi)}) = \sum_{n=-\infty}^{\infty} x[n]e^{-j(\omega+2\pi)n} = \sum_{n=-\infty}^{\infty} x[n]e^{-j\omega n}e^{-j2\pi n} = X(e^{j\omega}).$$

Linearity

A property that follows from the definition is that the DTFT of a linear combination of two or more signals is equal to the same linear combination of the DTFTs of each signal:

$$y[n] = ax_1[n] + bx_2[n] \Leftrightarrow Y(e^{j\omega}) = aX_1(e^{j\omega}) + bX_2(e^{j\omega}).$$

Symmetry of DTFT for Real Signals

If the signal $x[n]$ is real ($x[n] = x^*[n]$), then $X(e^{j\omega}) = X^*(e^{-j\omega})$. This follows from:

$$X(e^{-j\omega}) = \sum_{n=-\infty}^{\infty} x[n]e^{j\omega n}.$$
$$X^*(e^{-j\omega}) = \left(\sum_{n=-\infty}^{\infty} x[n]e^{j\omega n}\right)^* = \sum_{n=-\infty}^{\infty} x^*[n]e^{-j\omega n}$$
$$= \sum_{n=-\infty}^{\infty} x[n]e^{-j\omega n} = X(e^{j\omega}).$$

If $X(e^{j\omega}) = |X(e^{j\omega})|e^{\phi(\omega)}$, then $X^*(e^{-j\omega}) = |X(e^{-j\omega})|e^{\phi(-\omega)}$. Therefore, for real $x[n]$, the magnitude of the DTFT is an even function of ω and the phase of the DTFT is an odd function of ω.

Time Shift of Signals

A shift of the signal $x[n]$ in the time domain does not affect the magnitude of the DTFT but produces a phase shift that is linear in ω and in proportion to the time shift:

$$y[n] = x[n - n_0] \Leftrightarrow Y(e^{j\omega}) = X(e^{j\omega})e^{-j\omega n_0}.$$

This follows from:

$$Y(e^{j\omega}) = \sum_{n=-\infty}^{\infty} y[n]e^{-j\omega n} = \sum_{n=-\infty}^{\infty} x[n - n_0]e^{-j\omega n}$$

$$= e^{-j\omega n_0} \sum_{m=-\infty}^{\infty} x[m]e^{-j\omega n} = e^{-j\omega n_0} X(e^{j\omega}).$$

Convolution

The DTFT of the convolution sum of two signals $x_1[n]$ and $x_2[n]$ is the product of their DTFTs, $X_1(e^{j\omega})$ and $X_2(e^{j\omega})$. That is:

$$y[n] = x_1[n] * x_2[n] \Leftrightarrow Y(e^{j\omega}) = X_1(e^{j\omega})X_2(e^{j\omega}).$$

The DTFT of $y[n] = x_1[n] * x_2[n] = \sum_{k=-\infty}^{\infty} x_1[k]x_2[n-k]$ is given by:

$$Y(e^{j\omega}) = \sum_{n=-\infty}^{\infty} y[n]e^{-j\omega n} = \sum_{n=-\infty}^{\infty}\sum_{k=-\infty}^{\infty} x_1[k]x_2[n-k]e^{-j\omega n}$$

$$= \sum_{k=-\infty}^{\infty} x_1[k]e^{-j\omega k} \sum_{n=-\infty}^{\infty} x_2[n-k]e^{-j\omega(n-k)} = X_1(e^{j\omega})X_2(e^{j\omega}).$$

Modulation

A modulation of the signal in the time domain by a complex exponential signal corresponds to a shift in the frequency domain. That is:

$$y[n] = e^{j\omega_0 n}x[n] \Leftrightarrow Y(e^{j\omega}) = X(e^{j(\omega-\omega_0)}).$$

This is seen from:

$$Y(e^{j\omega}) = \sum_{n=-\infty}^{\infty} y[n]e^{-j\omega n} = \sum_{n=-\infty}^{\infty} x[n]e^{j\omega_0 n}e^{j\omega n}$$

$$= \sum_{n=-\infty}^{\infty} x[n]e^{-j(\omega-\omega_0)n} = X(e^{j(\omega-\omega_0)}).$$

Product of Signals

The DTFT of the product of two signals, $x_1[n]$ and $x_2[n]$ as well as the **periodic convolution** of the two DTFTs, $X_1(e^{j\omega})$ and $X_2(e^{j\omega})$, are shown on the right-hand side of the equation:

$$y[n] = x_1[n]x_2[n] \Leftrightarrow Y(e^{j\omega}) = \frac{1}{2\pi}\int_{-\pi}^{\pi} X_1(e^{j\theta})X_2(e^{j(\omega-\theta)})d\theta.$$

Computing the DTFT of the product:

$$Y(e^{j\omega}) = \sum_{n=-\infty}^{\infty} x_1[n]x_2[n]e^{-j\omega n} = \sum_{n=-\infty}^{\infty}\left[\frac{1}{2\pi}\int_{-\pi}^{\pi} X_1(e^{j\theta})e^{j\theta n}d\theta\right]x_2[n]e^{-j\omega n}$$

$$= \frac{1}{2\pi}\int_{-\pi}^{\pi} X_1(e^{j\theta})\left[\sum_{n=-\infty}^{\infty} x_2[n]e^{-jn(\omega-\theta)}\right]d\theta = \frac{1}{2\pi}\int_{-\pi}^{\pi} X_1(e^{j\theta})X_2(e^{j(\omega-\theta)})d\theta.$$

Parseval's Relation

The signal energy can equivalently be expressed in the time or in the frequency domain as:

$$\sum_{n=-\infty}^{\infty} |x[n]|^2 = \frac{1}{2\pi}\int_{-\pi}^{\pi} |X(e^{j\theta})|^2 d\theta.$$

Defining $x_1[n] = x[n]$, $x_2[n] = x^*[n]$ and $y[n] = x_1[n]x_2[n] = |x^2[n]|$ and then using the product relation, we obtain:

$$Y(e^{j0}) \sum_{n=-\infty}^{\infty} x[n]x^*[n] = \sum_{n=-\infty}^{\infty} |x^2[n]|$$

$$= \frac{1}{2\pi}\int_{-\pi}^{\pi} X(e^{j\theta})X(e^{-j\theta})d\theta = \frac{1}{2\pi}\int_{-\pi}^{\pi} |X(e^{j\theta})|^2 d\theta.$$

For continuous-time signals, the Fourier transform and its inverse are given by:

$$X_c(j\Omega) = \int_{-\infty}^{\infty} x_c(t)e^{j\Omega t}, \tag{1.39}$$

$$x_c(t) = \frac{1}{2\pi}\int_{-\infty}^{\infty} X_c(j\Omega)e^{j\Omega t}. \tag{1.40}$$

Properties analogous to those for DTFT can be obtained for the continuous-time Fourier transform (CTFT), and these are summarized in Table 1.1. The subscript c used for continuous-time signals is dropped in the table.

1.4.2 Frequency Response of LSI System

An LSI system is characterized in the time domain by its impulse response. It can be equivalently characterized in the frequency domain by the system frequency response, which is the Fourier transform of the impulse response. We first examine the discrete-time case.

The frequency response of a discrete-time system is defined as the discrete-time Fourier transform (DTFT) of its impulse response $h[n]$, as shown in equation 1.36. If an input $x[n]$ is applied to the system, the output is the convolution sum $y[n] = x[n] * h[n]$. From the property of the DTFT of convolution sums discussed above, the DTFT, $Y(e^{j\omega})$, of the output,

TABLE 1.1 Properties of the Continuous-Time Fourier Transform

Property	Continuous-time signal	Continuous-time Fourier transform
Linearity	$y(t) = ax_1(t) + bx_2(t)$.	$Y(j\Omega) = aX_1(j\Omega) + bX_2(j\Omega)$.
Time shift	$y(t) = x(t - t_0)$.	$Y(j\Omega) = X(j\Omega)e^{-j\Omega t_0}$.
Frequency shift	$y(t) = x(t)e^{j\Omega_0 t}$.	$Y(j\Omega) = X(j(\Omega - \Omega_0))$.
Convolution	$y(t) = x_1(t) * x_2(t)$.	$Y(j\Omega) = X_1(j\Omega)X_2(j\Omega)$.
Product	$y(t) = x_1(t)x_2(t)$.	$Y(j\Omega) = X_1(j\Omega) * X_2(j\Omega) = \int_{-\infty}^{\infty} X_1(j\Theta)X_2(j(\Omega - \Theta))d\Theta$.
Area under $X(j\Omega)$		$\int_{-\infty}^{\infty} X(j\Omega) = x(0)$.
Parseval's relation		$\int_{-\infty}^{\infty} \lvert x(t)\rvert^2 dt = \int_{-\infty}^{\infty} \lvert X(j\Omega)\rvert^2 d\Omega$.
Conjugate function	$y(t) = x^*(t)$.	$Y(j\Omega) = X^*(-j\Omega)$.
Real part	$Re[x(t)]$	$\frac{1}{2}[X(j\Omega) + X^*(-j\Omega)]$.
Imaginary part	$Im[x(t)]$	$\frac{1}{2j}[X(j\Omega) - X^*(-j\Omega)]$.

$y[n]$, is the product of the DTFTs $H(e^{j\omega})$ and $X(e^{j\omega})$ of $h[n]$ and $x[n]$, respectively:

$$Y(e^{j\omega}) = H(e^{j\omega})X(e^{j\omega}). \tag{1.41}$$

If the DTFTs are expressed in terms of magnitude and phase, (i.e., $Y(e^{j\omega}) = |Y(e^{j\omega})|e^{j\phi_Y(\omega)}$, $X(e^{j\omega}) = |X(e^{j\omega})|e^{j\phi_X(\omega)}$, and $H(e^{j\omega}) = |H(e^{j\omega})|e^{j\phi_H(\omega)}$, then the following two equation apply:

$$\left|Y(e^{j\omega})\right| = \left|H(e^{j\omega})\right|.\left|X(e^{j\omega})\right|. \tag{1.42}$$

$$\phi_Y(\omega) = \phi_H(\omega) + \phi_X(\omega). \tag{1.43}$$

The $\left|H(e^{j\omega})\right|$ and $\phi_H(\omega)$ are referred to as the magnitude response and the phase response of the system, respectively. In equations 1.42 and 1.43, the system magnitude response represents the gain or attenuation that the input signal content at frequency ω experiences when the system maps it to the output. The system phase response represents the corresponding phase shift at frequency ω.

If the system magnitude response, $\left|H(e^{j\omega})\right|$, is unity, then the system is said to ideally pass the input signal content at frequency ω_0 to the output. The set of all such frequencies is called the **passband** of the system. If $\left|H(e^{j\omega_0})\right|$ is zero, then the system is said to ideally stop the input signal content at frequency ω_0 from appearing at the output. The set of all such frequencies is called the **stopband** of the system.

If the set of all frequencies $|\omega| \leq \pi$ is partitioned into passbands and stopbands, then the system is called an ideal filter that passes or stops the signal content in a **frequency-selective** manner. A filter is said to be ideal low-pass if its frequency response is the following:

$$H(e^{j\omega}) = \begin{cases} 1, & |\omega| < \omega_c. \\ 0, & |\omega_c| < |\omega| \leq \pi. \end{cases} \tag{1.44}$$

The filter passes all low-frequency input content in the passband $|\omega| < \omega_c$ and stops all high-frequency input content in the stopband $|\omega_c| < |\omega| \leq \pi$. The impulse response of the ideal low-pass filter can be computed as the inverse DTFT of $H(e^{j\omega})$:

$$h[n] = \begin{cases} \omega_c/\pi, & n = 0. \\ \sin \omega_c n/\pi n, & n \neq 0. \end{cases} \tag{1.45}$$

This filter can be shown to be unstable in the BIBO sense.

A practical low-pass filter can only approximate the ideal low-pass response. In practice, filters have passband magnitude within a tolerance of unity and have stopband magnitude within a tolerance of zero. When practical filters are specified, transition bands are allowed between adjacent passbands and stopbands so that the magnitude response in the transition bands may swing between the allowed passband and stopband magnitudes. The impulse response $h[n]$ of a practical filter can vary depending on given filter specifications. The task of determining $h[n]$ to satisfy the specifications is called **filter design**.

The frequency response of a continuous-time LSI system is the continuous-time Fourier transform $H_c(j\Omega)$ of the impulse response $h_c(t)$:

$$H_c(j\Omega) = \int_{-\infty}^{\infty} h_c(t)e^{j\Omega t}, \tag{1.46}$$

$$h_c(t) = \frac{1}{2\pi}\int_{-\infty}^{\infty} H_c(j\Omega)e^{j\Omega t}. \tag{1.47}$$

Frequency response of continuous-time filters is specified for frequencies over the range $-\infty < \Omega < \infty$.

1.5 The z-Transform and Laplace Transform

Fourier transforms are useful, but they exist for a restricted class of signals. They can be generalized so that the generalization exists for a larger class of signals. These generalizations provide additional insights into system analysis and design. If time is discrete, the generalization is called the z-transform. The definition of the DTFT is modified by replacing $e^{j\omega}$ by a complex variable $z = re^{j\omega}$. In the case of continuous time, the definition of the CTFT is modified by replacing $j\Omega$ by a complex variable $s = \sigma + j\Omega$.

1.5.1 z-Transform and Its Properties

The discrete-time Fourier transform of a signal $x[n]$, expressed as:

$$X(e^{j\omega}) = \sum_{n=-\infty}^{\infty} x[n]e^{-j\omega n}, \tag{1.48}$$

can be viewed as a power series in the complex variable $z = e^{j\omega}$ but with the variable restricted to values on the unit circle in the z-plane. Consider allowing z to assume any complex value of $z = re^{j\omega}$, r real, and ≥ 0. The z-transform of a discrete-time signal $x[n]$ is defined as:

$$X(z) = \sum_{n=-\infty}^{\infty} x[n]z^{-n}. \tag{1.49}$$

Note the relation of the z-transform to a Laurent's series. The DTFT of $x[n]$ is identical to values that the z-transform assumes on the unit circle (i.e., the circle with unit radius centered at the origin in the z-plane). With $z = re^{j\omega}$ viewed as a position vector, r is the distance from the origin, and ω is the angle between the vector z and the real axis.

The z-transform may converge only over some region in the complex z-plane. In general, the region of convergence of z-transform, if not empty, is either the interior, $|z| < R$, of a circle or an annular region in the z-plane:

$$R_1 < |z| < R_2, \tag{1.50}$$

where $0 < R_1 < R_2 \leq \infty$.

If the region of convergence includes the unit circle, then the DTFT of the signal exists.

Example 1

Consider the discrete-time signal:

$$x[n] = a^n u[n] = \begin{cases} a^n, & n \geq 0. \\ 0, & n < 0. \end{cases} \tag{1.51}$$

The z-transform of $x[n]$ is the following:

$$X(z) = \sum_{n=-\infty}^{\infty} x[n]z^{-n} = \sum_{n=0}^{\infty} a^n z^{-n} = \sum_{n=0}^{\infty} (az^{-1})^n = \frac{1}{1 - az^{-1}}, \tag{1.52}$$

provided $|az^{-1}| < 1$ (i.e., $|z| > |a|$). For the DTFT of the signal to exist, we require that $|a| < 1$.

Example 2

Consider a signal $x[n]$ given by:

$$x[n] = b^n u[-n] = \begin{cases} b^n, & n < 0. \\ 0, & n \geq 0. \end{cases}$$

The z-transform of $x[n]$ is expressed as:

$$X(z) = \sum_{n=-\infty}^{-1} b^n z^{-n} = \sum_{m=1}^{\infty} (bz^{-1})^{-m} = \sum_{m=1}^{\infty} (b^{-1}z)^m = \frac{b^{-1}z}{1 - b^{-1}z},$$

provided $|b^{-1}z| < 1 \Rightarrow |z| < |b|$. In this case, the existence of the DTFT requires $|b| > |1|$.

Example 3

The z-transform of the signal:

$$x[n] = a^{|n|} = \begin{cases} a^n, & n \geq 0, \\ a^{-n}, & n < 0, \end{cases}$$

is expressed as:

$$X(z) = \sum_{n=0}^{\infty} a^n z^{-n} + \sum_{n=-\infty}^{-1} a^{-n} z^{-n} = \frac{z}{z-a} - \frac{z}{z-\frac{1}{a}}.$$

The region of convergence $ROC = \{[|z| > |a|] \cap [|z| < |1/a|]\}$ is $\{|1/a| < |z| < |a|\}$ provided $|a| < 1$. One notices that in these examples, the z-transforms are rational functions in z, with pole locations defining the boundaries of the region of convergence. In example 3, the two poles are located at $z = a$ and $z = 1/a$.

When signals are sums of left-sided or right-sided complex exponential sequences, of the form $a^n u[n - n_a]$ or $b^n u[-(n - n_b)]$, then the z-transforms are rational functions in z.

Given a z-transform of a signal $x[n]$, the signal can be recovered with the *inverse z-transform*:

$$x[n] = \frac{1}{2\pi j} \oint_C X(z) z^{n-1} dz, \tag{1.53}$$

where the integral is evaluated in the counterclockwise direction over a contour within the region of convergence that encloses the origin. To compute the inverse z-transform, we need information not only about the functional form of the z-transform but also about the region of convergence. In fact, two identical functions with different regions of convergence have different inverse transforms.

Let us consider the case of a z-transform $X(z)$ with two poles located at $z = a$ and $z = b$, $|a| < |b|$. There are three possible regions of convergence:

- ROC_1—the region $|z| < |a|$
- ROC_2—the annulus $|a| < |z| < |b|$
- ROC_3—the region $|z| > |b|$

Figure 1.13 shows the three possible regions of convergence for $a = 1/2$ and $b = 2$. To select the correct region of convergence, we need further information about the signal.

If for example we know that the DTFT of the signal exists (i.e., signal is absolutely summable), then we know that the unit circle is included in the region of convergence. Among the three previous regions, we choose the one that includes the unit circle. In our example, $b = 1/2$ and $a = 2$, and therefore, the ROC is the annulus $|1/2| < |z| < |2|$ (i.e., region ROC_2).

If the system is causal, then the z-transform converges in the region outside the largest pole, which is region ROC_3 in the example.

Depending on the selected region of convergence, the signal obtained through the inverse transform is different. For example, the inverse z-transform of $X(z) = z/z - a$ is equal to $x[n] = (a)^n u[n]$ if the ROC is $|z| > |a|$, whereas it is equal to $x[n] = -(a)^n u[-n-1]$ if the ROC is $|z| < |a|$.

Table 1.2 summarizes the properties of the z-transform. The derivation of these properties is similar to that carried out for the discrete-time Fourier transform and is not presented here.

Example

Consider the z-transform of a finite-duration sequence:

$$x[n] = a^n,\ 0 \leq n \leq N-1.$$

The z-transform of this sequence is as written here:

$$X(z) = \sum_{n=0}^{N-1} x[n]z^{-n} = \begin{cases} \frac{1-(az^{-1})^N}{1-az^{-1}} = \frac{1}{z^{N-1}}\frac{z^N - a^N}{z-a}, & z \neq a \\ N, & z = a. \end{cases}$$

Zeros of the z-transform are located at:

$$z_k = ae^{j2\pi k/N},\ 1 \leq k \leq N-1.$$

There are $N-1$ poles at $z = 0$ and $N-1$ zeros as shown in Figure 1.14. In general, the z-transform of a finite-duration sequence can possibly have poles only at $z = 0$ or $z = \infty$, while the zeros may be located anywhere in the z-plane. This means that for finite sequences, the ROC is the whole z-plane except perhaps the origin and/or $z = \infty$.

TABLE 1.2 Properties of the z-Transform

Property	Discrete-time signal	z-Transform
Linearity	$y[n] = ax_1[n] + bx_2[n]$.	$Y(z) = aX_1(z) + bX_2(z)$.
Time shift	$y[n] = x[n - n_0]$.	$Y(z) = X(z)z^{-zn_0}$.
Exponential weighting	$y[n] = x[n]a^n$.	$Y(z) = X(a^{-1}z)$.
Linear weighting	$y[n] = nx[n]$.	$Y(z) = z\frac{dX(z)}{dz}$
Convolution	$y[n] = x_1[n] * x_2[n]$.	$Y(z) = X_1(z)X_2(z)$.
Product	$y[n] = x_1[n]x_2[n]$.	$Y(z) = \frac{1}{2\pi j}\int_C X_1(v)X_2(\frac{z}{v})\frac{dv}{v}$.

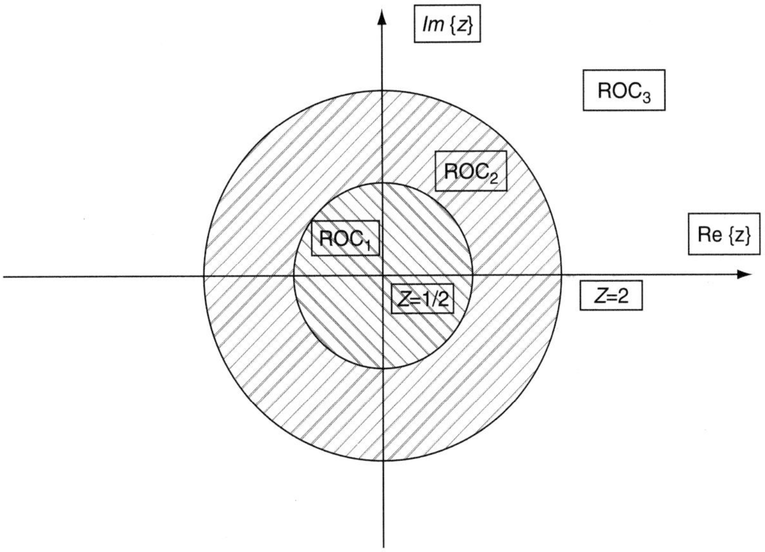

FIGURE 1.13 Region of Convergence with Poles at $z = 2$ and $z = 1/2$

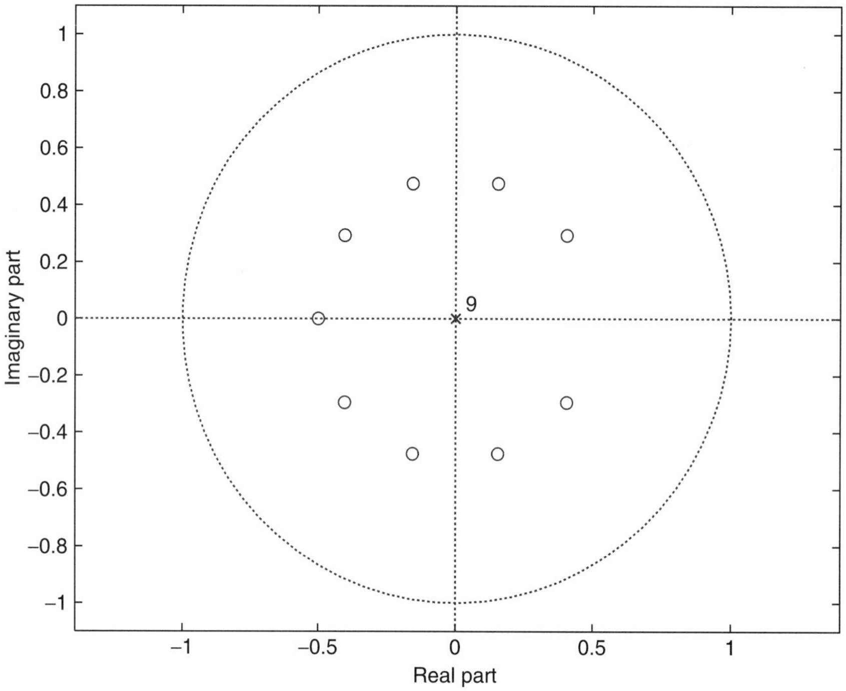

FIGURE 1.14 Location of Zeros and Poles for a Finite Sequence with $N = 10$ and $a = 0.5$

The magnitude and the phase of the z-transform and the DTFT of the finite-duration sequence are shown in Figure 1.15.

1.5.2 The Laplace Transform and Its Properties

For continuous-time signals, the Fourier transform integral of equation 1.39 can be generalized as:

$$X(s) = \int_{-\infty}^{\infty} x(t)e^{-st}dt, \qquad (1.54)$$

where $s = \sigma + j\Omega$. Equation 1.54 is the **Laplace transform** of the signal $x(t)$.

The Laplace transform can be represented in the two-dimensional s-plane, with σ along the real axis, and the frequency Ω on the imaginary axis. The Laplace transform evaluated for $\sigma = 0$ (i.e., for $s = j\Omega$) is identical to the Fourier transform $X(\Omega)$.

If a continuous-time signal $x(t)$ is bounded by:

$$|x(t)| \leq K_1 e^{\sigma_1 t} u_c(t) + K_2 e^{\sigma_2 t} u_c(-t),$$

then one can determine that the Laplace transform converges for:

$$\sigma_1 < Re\{s\} < \sigma_2.$$

The signal $x(t)$ can be recovered using the inverse Laplace transform:

$$x(t) = \frac{1}{2\pi j} \int_{\Gamma} X(s)e^{st}ds, \qquad (1.55)$$

where Γ is a contour within the region of convergence that goes from $s = \sigma - j\infty$ to $s = \sigma + j\infty$. The following rules are useful in determining the region of convergence:

- If $x(t)$ is causal and of infinite duration, then the ROC lies to the right of all poles.
- If $x(t)$ is anticausal and of infinite duration, then the ROC lies to the left of all the poles.
- If $x(t)$ is a two-sided infinite-duration signal, then the ROC is the section of plane included between the poles of the causal part and the poles of the anticausal part.
- If $x(t)$ is of finite duration and there exists at least one value of σ for which the Laplace transform exists, then the ROC is the whole plane.

The properties of the Laplace transform are similar to the properties of the z-transform and are summarized in Table 1.3.

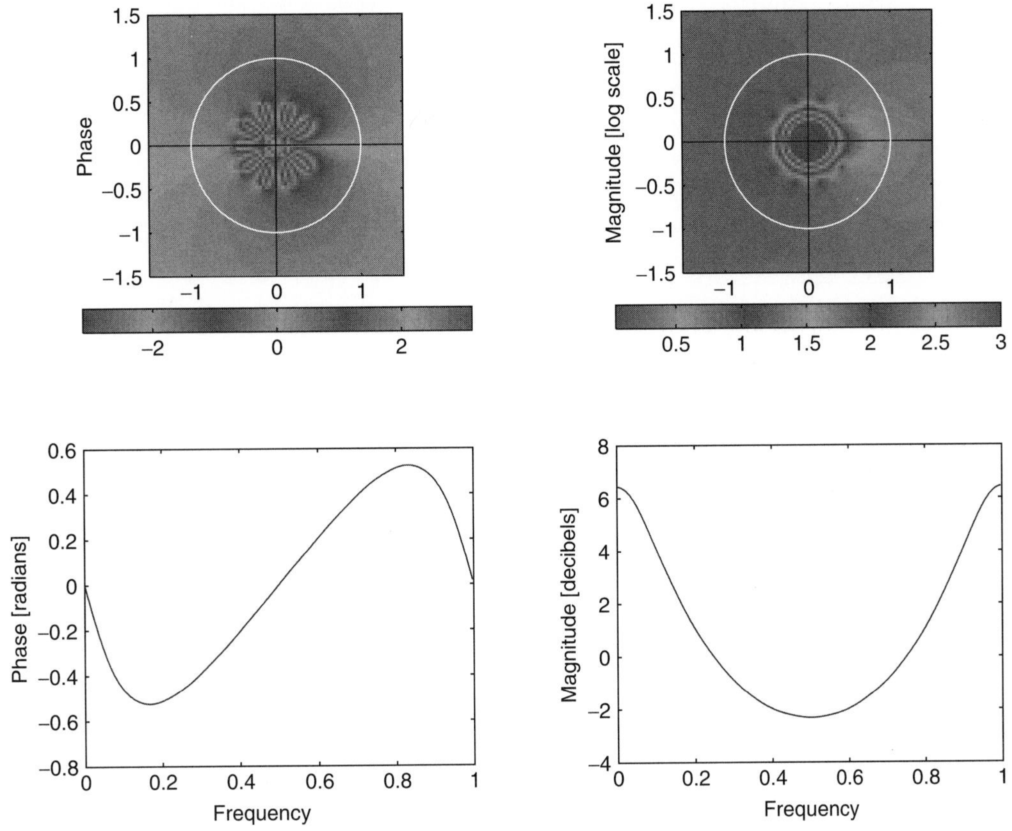

FIGURE 1.15 Phase and Log-Magnitude of z-Transform and of DTFT of Finite Sequence With $N = 10$ and $a = 0.5$

1.6 Sampling and Quantization

Until this point in our discussion, continuous-time and discrete-time signals have been considered separately. We did not consider how the signals arise in practice and where they are linked in some cases. Now we examine a mechanism—the sampling process—by which a discrete-time signal is obtained from a continuous-time signal.

1.6.1 Sampling

Sampling refers to the process of obtaining a discrete-time signal by extracting values of a continuous-time signal at predetermined instants that are usually equally spaced. This process maps a function with an uncountably infinite domain to a function with a countable domain. One would like the mapping to be such as to allow an exact recovery of the continuous-time signal from the sampled signal. The conditions that allow a continuous-time signal to be reconstructed from its samples can be obtained from examining the signal Fourier transform. The main requirements are that the signal be bandlimited and that an adequate sampling rate be used.

Consider a signal sampled at equally spaced instants. The time separation between two adjacent samples is called **sampling period (T)**, and its inverse is the **sampling rate** *or* **sampling frequency** (f_s). Because signal values are retained only at sampling instants, the signal information can be captured in the product of the continuous-time signal with a train, $s(t)$, of impulses spaced apart by T:

$$s(t) = \sum_{n=-\infty}^{\infty} \delta(t - nT). \tag{1.56}$$

A sampled **continuous-time** signal is just the product of the original signal $x_c(t)$ and the sampling signal $s(t)$:

$$x_s(t) = s(t)x_c(t) = \sum_{n=-\infty}^{\infty} x_c(nT)\delta(t - nT). \tag{1.57}$$

The Fourier transform of the signal $x_s(t)$ is as follows:

TABLE 1.3 Properties of the Laplace Transform

Property	Continuous-time signal	Laplace transform
Linearity	$y(t) = ax_1(t) + bx_2(t).$	$Y(s) = aX_1(s) + bX_2(s).$
Time shift	$y(t) = x(t - t_0).$	$Y(s) = X(s)e^{-st_0}.$
Exponential weighting	$y(t) = x(t)e^{at}.$	$Y(s) = X(s - a).$
Differentiation	$tx(t)$	$-\frac{dX(s)}{ds}$
Convolution	$y(t) = x_1(t) * x_2(t).$	$Y(s) = X_1(s)X_2(s).$
Product	$y(t) = x_1(t)x_2(t).$	$Y(s) = X_1(s) * X_2(s).$

$$X_s(j\Omega) = \int_{-\infty}^{\infty} \left(\sum_{n=-\infty}^{\infty} x_c(nT)\delta(t - nT) \right) e^{-j\Omega t dt} = \sum_{n=-\infty}^{\infty} x_c(nT)e^{-j\Omega nT}. \quad (1.58)$$

The samples define a discrete-time signal $x[n] = x_c(nT)$, so that equation 1.58 becomes:

$$X_s(j\Omega) = \sum_{n=-\infty}^{\infty} x[n]e^{-j\Omega nT} = X(e^{j\Omega T}) = X(e^{j\omega}), \quad (1.59)$$

where $\omega = \Omega T$ is the **normalized frequency**.

In Figure 1.16, a continuous-time signal $x_c(t)$ is shown, and we assume that its Fourier transform is bandlimited. If $x_c(t)$ is multiplied with an impulse train, $s(t)$, we obtain the sampled signal $x_s(t)$. Multiplication in the time domain yields convolution in the frequency domain. This is another way of expressing the Fourier transform of the sampled signal $x_s(t)$ as a convolution integral of the Fourier transforms $X_c(e^{j\Omega})$ and $S(e^{j\Omega})$.

If $s(t)$ is a train of impulses with spacing T, its Fourier transform is a train of impulses with spacing $f = 1/T$

$$S(j\Omega) = \frac{2\pi}{T} \sum_{k=-\infty}^{\infty} \delta\left(\Omega - k\frac{2\pi}{T}\right). \quad (1.60)$$

The convolution of this train of impulses with the $X(e^{j\Omega})$ results in periodic repetition of the Fourier transform image at the locations of the impulses.

$$X_s(j\Omega) = \frac{1}{2\pi} X_c(j\Omega) * S(j\Omega) = \frac{1}{T} X_c(j\Omega) * \sum_{k=-\infty}^{\infty} \delta\left(\Omega - k\frac{2\pi}{T}\right) = \frac{1}{T} \sum_{k=-\infty}^{\infty} X_c\left(j\left(\Omega - k\frac{2\pi}{T}\right)\right). \quad (1.61)$$

Equating the Fourier transform of $x_s(t)$ in equation 1.59 to 1.61, we get:

$$X(e^{j\Omega T}) = \frac{1}{T} \sum_{k=-\infty}^{\infty} X_c\left(j\left(\Omega - k\frac{2\pi}{T}\right)\right). \quad (1.62)$$

If we substitute the discrete frequency $\omega = \Omega T$, equation 1.62 becomes:

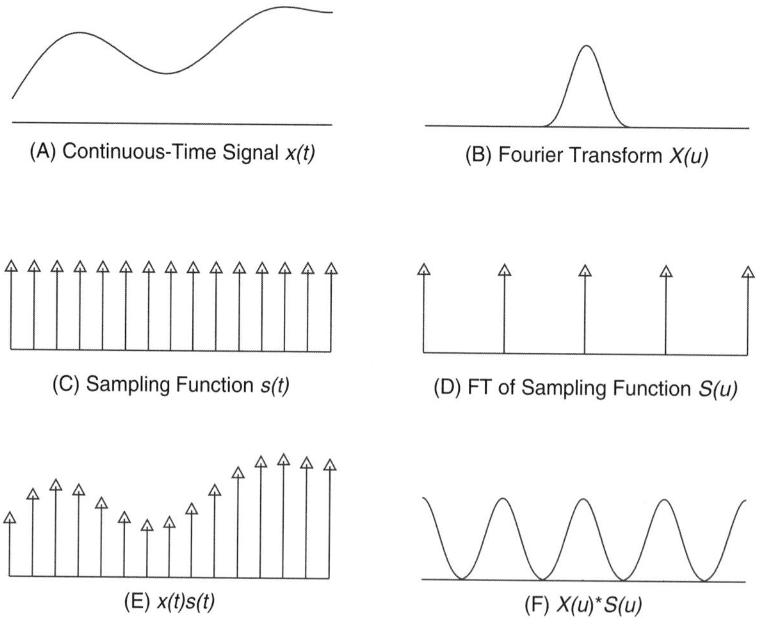

FIGURE 1.16 Graphic Development of Sampling Concepts

$$X(e^{j\omega}) = \frac{1}{T}\sum_{k=-\infty}^{\infty} X_c\left(j\frac{\omega - k2\pi}{T}\right). \qquad (1.63)$$

Therefore, sampling in the time domain corresponds to adding shifted replicas of the Fourier transform in the frequency domain. To avoid losing information, we need to avoid any overlap in the replicas in the frequency domain.

1.6.2 Signal Reconstruction

To reconstruct the original signal (i.e., to extract the central replica from $X_s(j\Omega)$), we need to apply an ideal low-pass filter. In the time domain, this means that we need to convolve the signal with a sinc function, shown in Figure 1.17, and given by:

$$h(t) = \frac{\sin(\pi \pm / T)}{\pi t / T} = \mathrm{sinc}\left(\frac{\pi t}{T}\right). \qquad (1.64)$$

The reconstructed signal, $x_r(t)$, is the sum of shifted sinc functions weighted by the signal samples:

$$\begin{aligned} x_r(t) &= \sum_{n=-\infty}^{\infty} x[n] \frac{\sin(\pi/T(t - nT))}{\pi/T(t - nT)} \\ &= \sum_{n=-\infty}^{\infty} x[n]\, \mathrm{sinc}\left(\frac{\pi}{T}(t - nT)\right). \end{aligned} \qquad (1.65)$$

The reconstructed signal $x_r(t)$ equals the original signal $x(t)$ only if there is no overlapping of the replicas in the frequency domain.

In Figure 1.16 replicas of the Fourier transform do not overlap, so the original signal can be recovered by simply using a low-pass filter. The sampling rate has been chosen adequately to avoid loss of information.

Suppose that the Fourier transform of the signal $x_c(t)$ is bandlimited to $|\Omega| \leq \Omega_0 = 2\pi W$ so that:

$$X_c(j\Omega) = 0. \quad |\Omega| \geq \Omega_0 = 2\pi W. \qquad (1.66)$$

We need to choose a sampling rate such that:

$$f_s = \frac{1}{T} \geq 2W. \qquad (1.67)$$

In the example of Figure 1.17, the sampling period is $T = 1/2W$.

Figure 1.18 shows the case of **undersampling** for the signal with Fourier transforms shown in Figure 1.16. Undersampling means that the sampling rate is below the so-called Nyquist rate of $2W$, resulting in a frequency content above $\Omega > 2\pi W$ that cannot be recovered by applying the low-pass reconstruction filter. The content is folded back to lower frequencies, a phenomenon called **aliasing**.

Finally, Figure 1.19 illustrates the case of **oversampling** where $f_s > 2W$. The requirements on the low-pass reconstruction filter can be more relaxed.

In many applications, there is a need to increase or decrease the sampling rate of a discrete-time signal directly in the discrete-time domain without reconstructing the continuous-time signal. These tasks are referred to as **interpolation** and **decimation**. These are basic operations needed in multirate signal processing, filter banks, and sub-band/wavelet analysis (Akansu *et al.*, 1996; Strang and Nguyen, 1996; Vaidyanathan, 1993; Vetterli and Kovacevic, 1995).

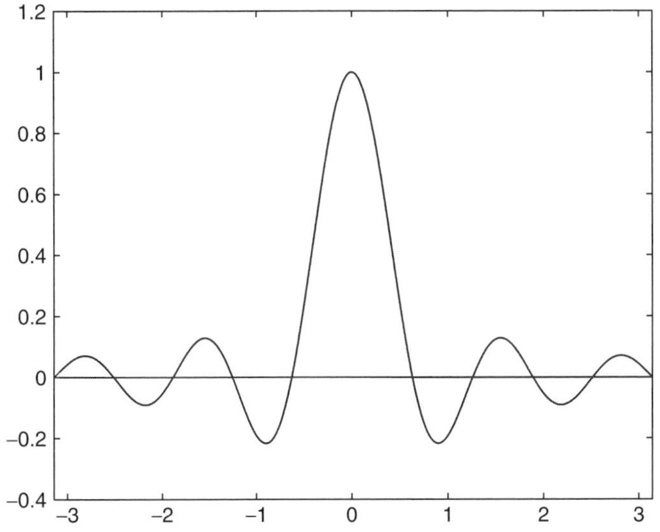

FIGURE 1.17 Impulse Response of an Ideal Low-Pass Filter

1.6.3 Quantization

At the beginning of this chapter, signals were classified into four categories according to the values assumed by the independent and the dependent variables. So far, we have focused our attention only on the independent variable, namely time, to distinguish signals between continuous-time and discrete-time signals, and we assumed that the dependent variable can assume a continuum of values.

However, when processing data with a computer or with a digital signal processor, both variables can assume only discrete values. This means that not only is the signal discrete in time but also in amplitude. These signals are referred to as **digital signals**. Therefore, if a signal has to be processed in digital form, first its independent variable and then its dependent variable are converted from continuous to discrete form through **sampling** and **quantization** respectively. Quantization is the process that converts data from infinite or high precision to finite or lower precision.

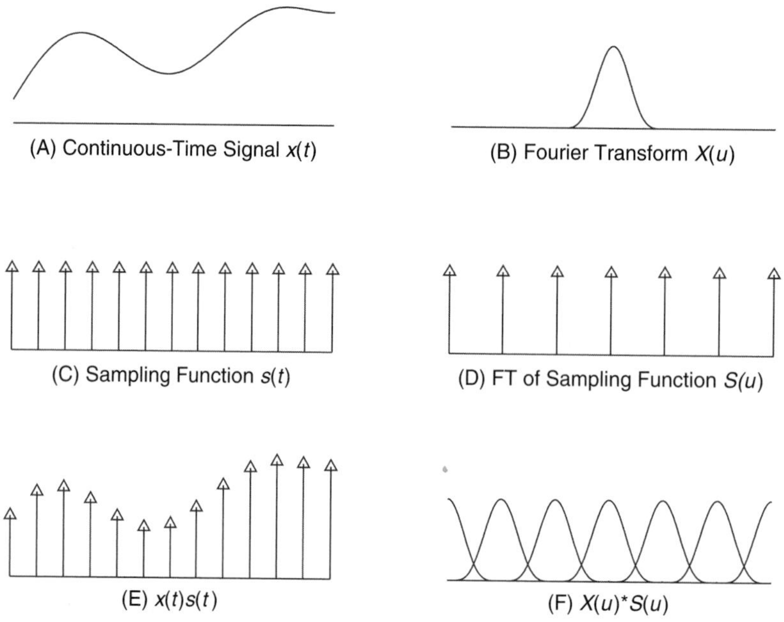

FIGURE 1.18 Graphic Development of Undersampling Concepts

Each digital variable is stored in a finite length register and is represented by a finite number of bits, B. There are two possible ways of representing data in binary form: **fixed point** and **floating point**. In fixed point representation, the binary point is fixed at a specific location, and the implementation of arithmetic operations takes into account this property. On the other hand, in the floating point representation, each number is represented by two parameters: a **mantissa**, represented with B_M bits, and an **exponent** E, represented as B_E bits, with $B = B_M + B_E$.

The error introduced in the processing of digital signals due to quantization depends on the format of the numbers, the quantization method (**truncation** or **rounding**), the number of bits used to represent signal values and filter coefficients, and so on. Details can be found in Mitra (1998) and Oppenheim *et al.* (1999).

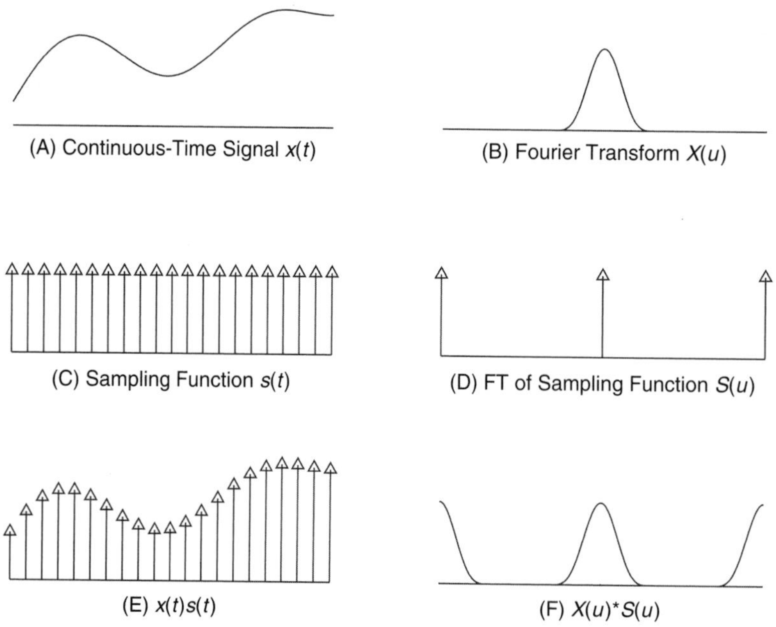

FIGURE 1.19 Graphic Development of Oversampling Concepts

1.7 Discrete Fourier Transform

For general discrete-time signals, we examined the discrete-time Fourier transform (DTFT) as a useful analysis and design tool for frequency domain representation. We now consider Fourier analysis that is expressly formulated for finite-duration discrete-time and that serves as a computational aid in many signal processing tasks.

Consider a sequence $x[n]$ that is zero for n other than $0 \leq n \leq N-1$. The **discrete Fourier transform** (DFT) of the signal $x[n]$ is defined as:

$$X[k] = \begin{cases} \sum_{n=0}^{N-1} x[n]e^{-j2\pi nk/N}, & 0 \leq k \leq N. \\ 0, & \text{otherwise.} \end{cases} \quad (1.68)$$

With the following two equations:

$$W_N = e^{-j2\pi/N}, \quad (1.69)$$

$$R_N[n] = \sum_{m=0}^{N-1} \delta[n-m], \quad (1.70)$$

the DFT can be rewritten as:

$$X[k] = \sum_{n=0}^{N-1} x[n] W_N^{nk} R_N[k]. \quad (1.71)$$

Note that the DFT maps sequences to sequences. In other words, the frequency variable (k) is discrete, unlike that for the DTFT, where ω assumes all real values. The signal $x[n]$ can be recovered from X[k] through the **inverse discrete Fourier transform** (IDFT) relation:

$$x[n] = \left(\frac{1}{N}\sum_{k=0}^{N-1} X[k]e^{j\frac{2\pi}{N}nk}\right) R_N[n] = \frac{1}{N}\sum_{k=0}^{N-1} X[k] W_N^{-nk} R_N[n], \quad (1.72)$$

which differs from the forward transform in the normalization factor $\frac{1}{N}$ and in the sign of the exponent.

1.7.1 Relation Between the DFT and the Samples of DTFT

In equation 1.37, we defined the DTFT of a signal $x[n]$ as:

$$X(e^{j\omega}) = \sum_{n=-\infty}^{\infty} e^{-j\omega n} x[n]. \quad (1.73)$$

If the signal $x[n]$ has finite duration and it is limited to the interval $0 \leq n \leq N-1$, then the previous expression becomes:

$$X(e^{j\omega}) = \sum_{n=0}^{N-1} e^{-j\omega n} x[n]. \quad (1.74)$$

Clearly $X[k]$, the DFT of $x[n]$, consists of samples of $X(e^{j\omega})$ at $\omega = 2\pi k/N$ for $0 \leq k \leq N-1$.

Therefore:

$$X[k] = X(e^{j\omega})|_{\omega=\frac{2\pi k}{N}} R_N[k] = X(e^{j2\pi k/2}) R_N[k]. \quad (1.75)$$

Because a sequence of duration N can be recovered from its DFT $X[k]$, then $x[n]$ of duration N samples can be recovered from N samples of $X(e^{j\omega})$ at $\omega = 2\pi k/N$, $0 \leq k \leq N-1$, by computing the inverse DFT of these samples.

We conclude that a sequence $x[n]$ of duration N can be completely recovered from N equally spaced samples of its DTFT by computing the inverse DFT.

But what happens in the case of a general sequence that is not necessarily of duration N? That is, what is the inverse DFT of the samples of $X(e^{j\omega})$ at $\omega = 2\pi k/N$, $0 \leq k \leq N-1$, when the sequence $x[n]$ is not necessarily of duration N?

Let $X(e^{j\omega})$ be the DTFT of $x[n]$ that may be of duration $> N$. Then the following applies.

- Denote samples of $X(e^{j\omega})$ as $X_N[k] = X(e^{j2\pi k/N})$ for $k = 0, 1, \ldots.\ N-1$.
- Compute the inverse DFT of $X_N[k]$ and call the result $x_N[n]$.
- Determine $x_N[n]$ such that:

$$X_N[k] = \sum_{n=0}^{N-1} x_N[n]e^{-j2\pi kn/N}. \quad (1.76)$$

It can be shown that:

$$x_N[n] = \sum_{r=-\infty}^{\infty} x[n+rN],\ n = 0, 1, \ldots, N-1. \quad (1.77)$$

Proof: $X_N[k] = X(e^{j2\pi k/N}) = \sum_{n=-\infty}^{\infty} x[n]e^{-j2\pi kn/N}$

$$= \sum_{n=0}^{N-1} \sum_{r=-\infty}^{\infty} x[n+rN]e^{-j2\pi k(n+rN)/N}$$

$$= \sum_{n=0}^{N-1}\left(\sum_{r=-\infty}^{\infty} x[n+rN]\right)e^{-j2\pi kn/N}.$$

1.7.2 Linear and Circular Convolution

In implementing discrete-time LSI systems, we need to compute the convolution sum, otherwise called linear convolution, of the input signal $x[n]$ and the impulse response $h[n]$ of the system. For finite duration sequences, this convolution can be carried out using DFT computation.

Let $x[n]$ and $h[n]$ be of finite duration. Assume $x[n]$ is zero outside the interval $0 \leq n \leq N-1$ and $h[n]$ is zero outside the interval $0 \leq n \leq M-1$.

The sequence $y[n]$ is the **linear convolution** between $x[n]$ and $h[n]$:

$$y[n] = x[n] * h[n] = \sum_{k=-\infty}^{\infty} x[k]h[n-k], \qquad (1.78)$$

and it is of duration $M+N-1$.

If we wish to recover $y[n]$ from its DTFT $Y(e^{j\omega})$, we need samples of it at $(M+N-1)$ points $\omega = 2\pi/Nk$ and $k = 0, 1, \ldots, M+N-2$.

To get these samples, we observe that $Y(e^{j\omega}) = X(e^{j\omega})H(e^{j\omega})$. Therefore, if we know the values of the samples of $X(e^{j\omega})$ and $H(e^{j\omega})$ at $\omega = 2\pi/Nk$ for $k = 0, 1, \ldots, M+N-2$, then $Y(e^{j2\pi/Nk})$ can be obtained as a product of these samples.

But the samples of $X(e^{j\omega})$ at these points are the same as the $(M+N-1)$-point DFT of $x[n]$. Similarly the samples of $H(e^{j\omega})$ at these points are obtained by computing the $(M+N-1)$-point DFT of $h[n]$.

The product $X[k]\ H[k]$ is therefore equal to $Y[k]$, the DFT of $y[n]$.

However the relationship $Y[k] = X[k]H[k]$ is valid only if $Y[k]$ is of length $\geq (M+N-1)$, which is the length of the linear convolution $y[n]$. To obtain this result, we need to zero-pad the sequences to the desired length before computing the DFTs and multiplying them together.

Let $L \geq \max\{N, M\}$. If we compute the product $X[k]H[k]$ using an L-point DFT $\max\{N, M\}$, possibly without the necessary zero-padding, then the resulting sequence can be denoted as $Y_L[k]$, with inverse DFT given by:

$$y_L[n] = \sum_{r=-\infty}^{\infty} y[n+rL] = x[n] \ o \ h[n]. \qquad (1.79)$$

Here o denotes **circular convolution** of the sequences $x[n]$ and $h[n]$.

The samples of circular convolution, $y_L[n]$, are obtained from the samples of linear convolution, $y[n]$, by wrapping around all samples that exceed the index $n = L-1$ as shown in equation 1.79.

From the definitions of linear and circular convolution, we observe that if $L \geq (N+M-1)$, then the two expressions coincide and $y_L[n] = y[n]$ as determined previously.

1.7.3 Use of DFT

Many fast algorithms have been developed to implement the DFT, and these are referred to as **fast Fourier transform** (FFT) algorithms (Oppenheim *et al.*, 1997). FFT allows an efficient computation of the DFT with a number of operations proportional to $N \log N$ instead of N^2.

Algorithms are commonly used in implementing signal processing tasks:

- Convolution of two finite duration sequences can be performed using DFTs of adequate length. The use of FFT enables an efficient implementation of convolution and FIR filters.
- The DFT of a sequence consists of samples of the DTFT of the sequence. It thus provides information about the frequency content of signals, and it is used in estimating the spectra of signals.

1.8 Summary

In this chapter, we provided an overview of discrete-time and continuous-time signals and systems. The emphasis was on discrete time, but notions for the two cases were presented in parallel. Fundamentals of discrete-time systems and properties such as linearity, time-invariance, causality, and stability were presented. Fourier analysis of signals was presented followed by a discussion of frequency-selective filtering. Basics of z-transforms were covered next along with a discussion of the z-domain interpretation of system properties, such as causality and stability. The operations of sampling and quantization that relate continuous and discrete time signals were briefly discussed. Finally, the discrete Fourier transform was presented and its importance in implementing signal processing algorithms described. Our discussion was largely confined to the description and processing of deterministic signals. A discussion of issues in implementation and the use and design of digital signal processors (Lapsley *et al.*, 1997) is not within the scope of this chapter.

Many excellent books cover the topic of signal processing, such as those by Haykin and Van Veen (1999), Lathi (1998), Mitra (1998), Oppenheim (1999, 1997), Proakis and Manolakis (1995) Strang and Nguyen (1996), Vaidyanathan (1993), and Vetterli and Kovacevic (1995). Introductory articles on new developments and research in signal processing appear in *IEEE Signal Processing Magazine.* Most of the research articles describing the advances in signal processing appear in *IEEE Transactions on Signal Processing, IEEE Transactions on Image Processing, IEEE Transactions on Circuits and Systems, Electronics Letters; International Conference on Acoustics, Speech, and Signal Processing (ICASSP)*, and *International Symposium on Circuits and Systems.* Several information resources are available on the World Wide Web. The impact of signal processing on information technology is described in a workshop report by the National Science Foundation (1994).

References

Akansu, A.N., and Smith, M.J.T. (1996). *Sub-band and wavelet transforms: Design and applications.* Norwell, MA: Kluwer Academic.

Anastassiou, D. (2001). Genomic signal processing. *IEEE Signal Processing Magazine 18* (4), 8–20.

Ansari, R., and Memon, N. (2000). The JPEG lossy image compression standard. In A. Bovik (Ed), *The handbook of image and video processing.* Burlington, MA: Academic Press.

Baillet, S., Mosher, J.C., and Leahy, R.M. (2001). Electromagnetic brain mapping. *IEEE Signal Processing Magazine 18* (6).

Chen, T. (2001). Audiovisual speech processing. *IEEE Signal Processing Magazine 18* (1) 9–21.

Ebrahimi, T., Vesin, J.M., and Garcia, G. (2003). Brain-computer interface multimedia communication. *IEEE Signal Processing Magazine 20* (1), 14–29.

Hanzo, L., Wong, C.H., and Cherriman, P. (2000). Channel-adaptive wideband wireless video telephony. *IEEE Signal Processing Magazine 17* (4), 10–30.

Haykin, S., and Van Veen, B. (1999). *Signals and systems.* New York: John Wiley & Sons.

Lapsley, P., Bier, J., Shoham, A., and Lee, E.A. (1997). *DSP Processor fundamentals—Architectures and features.* New York: IEEE Press.

Lathi, B.P. (1998). *Signal processing and linear systems.* New York: Oxford University.

Mitra, S.K. (1998). *Digital signal processing, A computer-based approach.* New York: McGraw-Hill.

Molina, R., Nunez, J., Cortijo, F.J., and Mateos, J. (2001). Image restoration in astronomy. *IEEE Signal Processing Magazine 18* (2), 11–29.

Ohm, J.R. (1999). Encoding and reconstruction of multiview video objects. *IEEE Signal Processing Magazine 16* (3), 47–54.

Oppenheim, A.V., Schafer, R.W., and Buck, J.R. (1999). *Discrete-time signal processing.* Englewood Cliffs, NJ: Prentice Hall.

Oppenheim, A.V., and Willsky, A.S. (1997). *Signals and systems.* Englewood Cliffs, NJ: Prentice Hall.

Podilchuk, C.I., and Delp, E.J. (2001). Digital watermarking, algorithms, and applications. *IEEE Signal Processing Magazine 18* (4), 33–46.

Proakis, J., and Manolakis, M. (1995). *Digital signal processing.* Englewood Cliffs, NJ: Prentice Hall.

Strang, G., and Nguyen, T. (1996). *Wavelets and filter banks.* Wellesley, MA: Wellesley-Cambridge.

Strintzis, M.G., and Malassiotis, S. (1999). Object-based coding of stereoscopic and 3D image sequences. *IEEE Signal Processing Magazine 16* (3), 14–28.

Strobel, N., Spors, S., and Rabenstein, R. (2001). Joint audio-video object localization and tracking. *IEEE Signal Processing Magazine 18* (1), 22–31.

Vaidyanathan, P.P. (1993). *Multirate systems and filter banks.* Englewood Cliffs, NJ: Prentice Hall.

Vetterli, M., and Kovacevic, J. (1995). *Wavelets and sub-band coding.* Englewood Cliffs, NJ: Prentice-Hall.

Wang, Y., Wenger, S., Wen, J., and Katsaggelos, A.K. (2000). Error resilient video coding techniques. *IEEE Signal Processing Magazine 18* (4), 61–82.

National Science Foundation. (1994). Signal processing and the national information infrastructure. Report of workshop organized by the National Science Foundation, Ballston, Virginia. http://www-isl.stanford.edu/gray/iii.pdf

2

Digital Filters

Marcio G. Siqueira
Cisco Systems, Sunnyvale, California, USA

Paulo S.R. Diniz
Program of Electrical Engineering, Federal University of Rio de Janeiro, Rio de Janeiro, Brazil,

2.1 Introduction

The rapid development of integrated circuit technology led to the development of very powerful digital machines able to perform a very high number of computations in a very short period of time. In addition, digital machines are flexible, reliable, reproducible, and relatively cheap. As a consequence, several signal processing tasks originally performed in the analog domain have been implemented in the digital domain. In fact, several signal processing tools are only feasible to be implemented in the digital domain.

As most real life signals are continuous functions of one or more independent variables, such as time, temperature, and position, it would be natural to represent and manipulate these signals in their original analog domain. In this chapter we will consider time as the single independent variable. If sources of information are originally continuous in time, which is usually the case, and they occupy a limited range of frequencies, it is possible to transform continuous-time signals into discrete-time signals without losing any significant information. By interpolation, the discrete-time signal can be mapped back into the original continuous-time signal.

It is worth mentioning that signals exist that are originally discrete in time, such as the monthly earnings of a worker or the average yearly temperature of a city.

As a result, the current technological trend is to sample and quantize the signals as early as possible so that most of the operations are performed in the digital domain. The sampling process consists of transforming a continuous signal into a discrete signal (a sequence). This transformation is necessary because digital machines are suitable to manipulate sequences of quantized numbers at high speed. Therefore, quantization of the continuous amplitude samples generate digital signals tailored to be processed by digital machines.

A **digital filter** is one of the basic building blocks in digital signal processing systems. Digital filtering consists of mapping a discrete-time sequence into another discrete-time sequence that highlights the desired information while reducing the importance of the undesired information.

Digital filters are present in various digital signal processing applications related to speech, audio, image, video, and multirate processing systems as well as in communication systems, CD players, digital radio, television, and control systems. Digital filters can have either fixed (Antoniou, 1993; Jackson, 1996; Oppenheim and Schafer, 1989; Mitra, 2001; Diniz, 2002) or adaptive coefficients (Diniz, 2002).

In this chapter, we will describe:

- The most widely used types of digital filter transfer functions;
- How to design these transfer functions (i.e., how to calculate coefficients of digital filter transfer functions);
- How to map a transfer function into a digital filter structure;
- The main concerns in the actual implementation in a finite precision digital machine; and
- The current trends in the hardware and software implementations of digital filters

In particular, we focus on digital filters that implement linear systems represented by difference equations. We will show how to design digital filters based on magnitude specifications. In addition, we present efficient digital filter structures that are often required for implementation with minimal cost, taking into consideration that real-time implementation leads to nonlinear distortions caused by the finite word length representations of internal signals and coefficients. Finally, the mostly widely used approaches to implement a digital filter are briefly discussed.

2.2 Digital Signal Processing Systems

A typical digital signal processing (DSP) system comprises the following modules, shown in Figure 2.1.

- **Analog-to-digital (A/D) converter**: Assuming that the input signal is band-limited to frequency components below half of the sampling frequency, the A/D converter obtains signal samples at equally spaced time intervals and converts the level of these samples into a numeric representation that can be used by a digital filter.
- **Digital filter**: The digital filter uses the discrete-time numeric representation obtained by the A/D converter to perform arithmetic operations that will lead to the filtered output signal.
- **Digital-to-analog (D/A) converter**: The D/A converter converts the digital filter output into analog samples that are equally spaced in time.
- **Low-pass filter**: The low-pass filter converts the analog samples of the digital filter output into a continuous-time signal.

2.3 Sampling of Analog Signals

Figure 2.2 shows the mechanism behind sampling of continuous time signals and its effect on the frequency representation. Figure 2.2(A) shows that a continuous signal $x(t)$ is multiplied in the time domain by a train of unitary impulses $x_i(t)$. The resulting signal from the multiplication operation is $x^*(t)$, which contains samples of $x(t)$. Figure 2.2(B) shows the frequency representation of the original signal $x(t)$ before sampling. Figure 2.2(C) shows the spectrum of the impulse train $x_i(t)$. Finally, Figure 2.3 shows the spectrum of the sampled signal $x^*(t)$. According to the convolution theorem, the Fourier transform of the sampled signal $x^*(t)$ should be the convolution, in the frequency domain, of the Fourier transform of the impulse train and the Fourier transform of the original signal before sampling. Therefore, it can be shown that:

$$\begin{aligned} X_D(e^{j\omega}) &= \frac{1}{T}\sum_{l=-\infty}^{\infty} X\left(j\frac{\omega}{T} + j\frac{2\pi l}{T}\right) \\ &= \frac{1}{T}\sum_{l=-\infty}^{\infty} X(j\omega_a + j\omega_s l). \end{aligned} \tag{2.1}$$

From the above equation and from Figure 2.2, it can be seen that the spectrum of the sampled signal will be repeated at every $\omega_s = 2\pi/T$ interval. An important consequence of the spectrum repetition is the sampling theorem shown next.

2.3.1 Sampling Theorem

If a continuous signal is band-limited (i.e., $X(j\omega_a) = 0$ for $\omega_a > \omega_c$), $x(t)$ can be reconstructed from $x_D(n)$ for

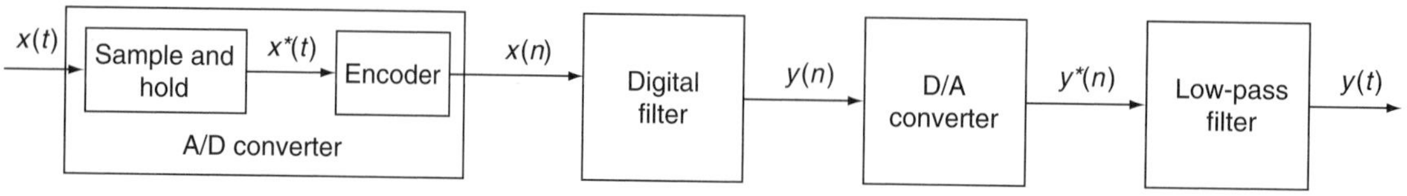

FIGURE 2.1 Architecture of a Complete DSP System Used to Filter an Analog Signal

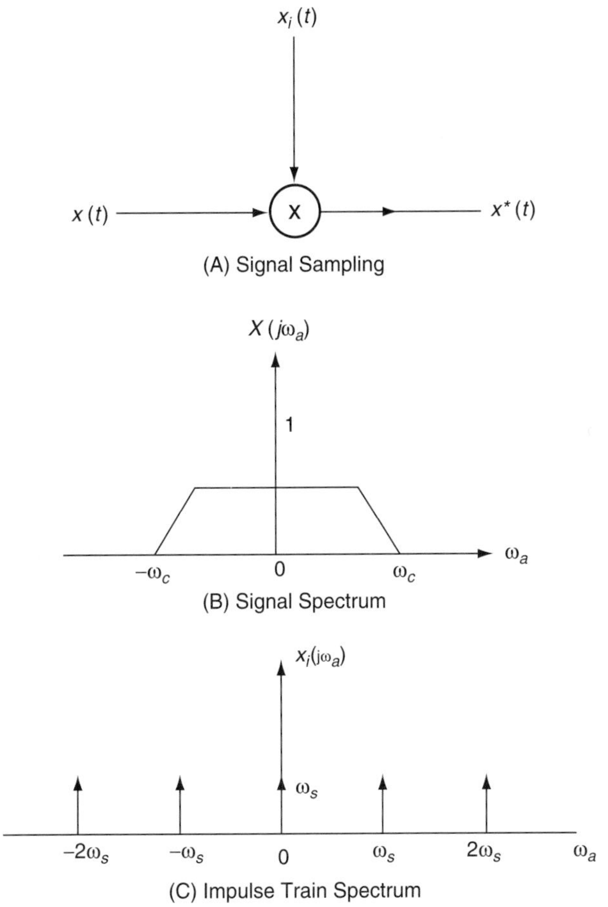

FIGURE 2.2 Sampling of Continuous-Time Signals

$-\infty < n < \infty$ if $\omega_s > 2\omega_c$, where ω_s is the sampling frequency in radians per second.

Figure 2.4 shows sampling effects on a continuous-time signal. Figure 2.4(A) shows the continuous-time signal spectrum, and Figures 2.4(B), 2.4(C), and 2.4(D) show the sampled signal spectra for the cases $\omega_c = \omega_s/2$, $\omega_c > \omega_s/2$, and $\omega_c < \omega_s/2$. It can be seen from these pictures that only Figure 2.4(D) does not produce distortion in the frequencies between $-\omega_c$ and ω_c. Figure 2.4(C) shows that when the sampled signal has frequency components above $\omega_s/2$ (half of the sampling frequency), the sampled signal spectrum will be distorted in the frequencies around $\omega_s/2$. In the case of Figure 2.4(B), no distortion occurred only because the signal had no power at $\omega_s/2$. This distortion is known as aliasing and is undesirable in most cases. As a consequence, only case 2.4(D)[1] can be used to recover the spectrum of the original continuous signal shown in Figure 2.4(A).

The recovery of the continuous signal spectrum from the sampled signal can be theoretically accomplished by using an analog filter with flat frequency response to retain only the components between $-\omega_s/2$ and $\omega_s/2$ of the spectra shown in Figure 2.4(A). The frequency response of the analog filter with these characteristics is shown in Figure 2.5. The impulse response of this filter is as follows:

$$h(t) = \frac{T \sin(\omega_{LP} t)}{\pi t}. \tag{2.2}$$

In equation 2.2, ω_{LP} is the cutoff frequency of the designed low-pass filter. The ω_{LP} shall be chosen to guarantee no aliasing (i.e., $\omega_c < \omega_{LP} < \omega_s/2$). The impulse response in the above equation is depicted in Figure 2.5. It is easy to verify that a causal filter with this exact impulse response cannot be implemented. In practice, it is possible to design analog filters that are good approximations for the frequency response in Figure 2.5.

2.4 Digital Filters and Linear Systems

An example of a discrete-time signal is shown in Figure 2.6. In a digital filter, the input signal $x(n)$ is a sequence of numbers indexed by the integer n that can assume only a finite number of amplitude values. Such a sequence might originate from a continuous-time signal $x(t)$ by periodically sampling it at the time instants $t = nT$, where T is the sampling interval. The output sequence $y(n)$ arises by applying $x(n)$ to the input of the digital filter, with the relationship between $x(n)$ and $y(n)$ represented by the operator $\mathcal{F}$ as:

$$y(n) \equiv \mathcal{F}[x(n)]. \tag{2.3}$$

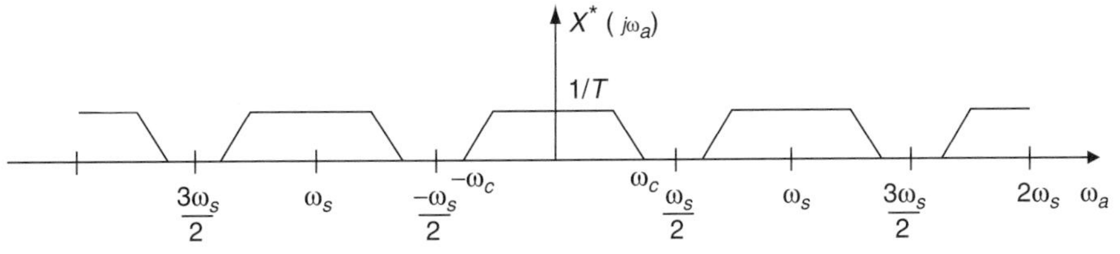

FIGURE 2.3 Spectrum of Sampled Signal

[1] In the case of Figure 2.4(B), distortion could occur only at $\omega_s/2$.

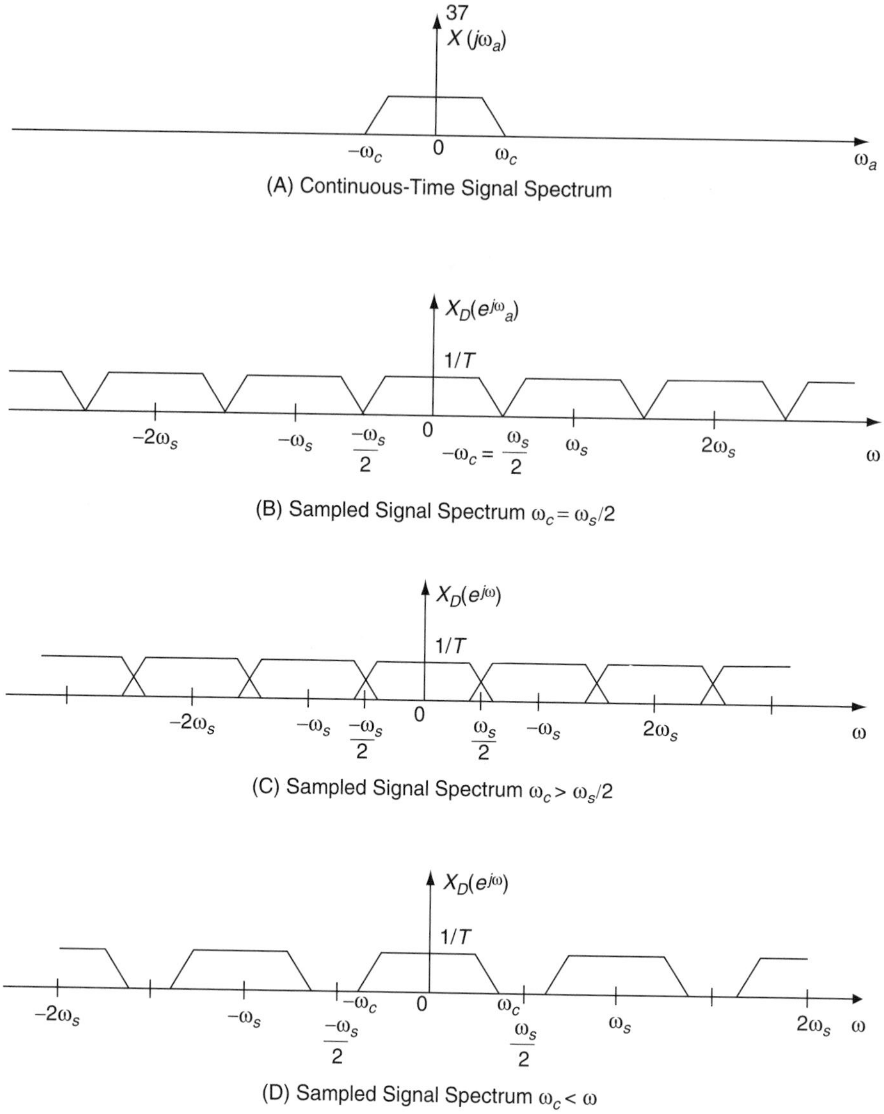

FIGURE 2.4 Sampling Effects

2.4.1 Linear Time-Invariant Systems

The main classes of digital filters are the **linear, time-invariant** (LTI) and **causal** filters. A linear digital filter responds to a weighted sum of input signals with the same weighted sum of the corresponding individual responses; that is:

$$\mathcal{F}[\alpha_1 x_1(n) + \alpha_2 x_2(n)] = \alpha_1 \mathcal{F}[x_1(n)] + \alpha_2 \mathcal{F}[x_2(n)], \quad (2.4)$$

for any sequences $x_1(n)$ and $x_2(n)$ and for any arbitrary constants α_1 and α_2. A digital filter is said to be time-invariant when its response to an input sequence remains the same, irrespective of the time instant that the input is applied to the filter. That is, $\mathcal{F}[x(n)] = y(n)$, and then:

$$\mathcal{F}[x(n - n_0)] = y(n - n_0), \quad (2.5)$$

for all integers n and n_0. A causal digital filter is one whose response does not anticipate the behavior of the excitation signal. Therefore, for any two input sequences $x_1(n)$ and $x_2(n)$ such that $x_1(n) = x_2(n)$ for $n \leq n_0$, the corresponding responses of the digital filter are identical for $n \leq n_0$; that is:

$$\mathcal{F}[x_1(n)] = \mathcal{F}[x_2(n)], \text{ for } n \leq n_0. \quad (2.6)$$

An initially relaxed linear-time invariant digital filter is characterized by its response to the unit sample or impulse sequence $\delta(n)$. The filter response when excited by such a sequence is denoted by $h(n)$, and it is referred to as **impulse response** of the digital filter. Observe that if the digital filter is causal, then $h(n) = 0$ for $n < 0$. An arbitrary input sequence

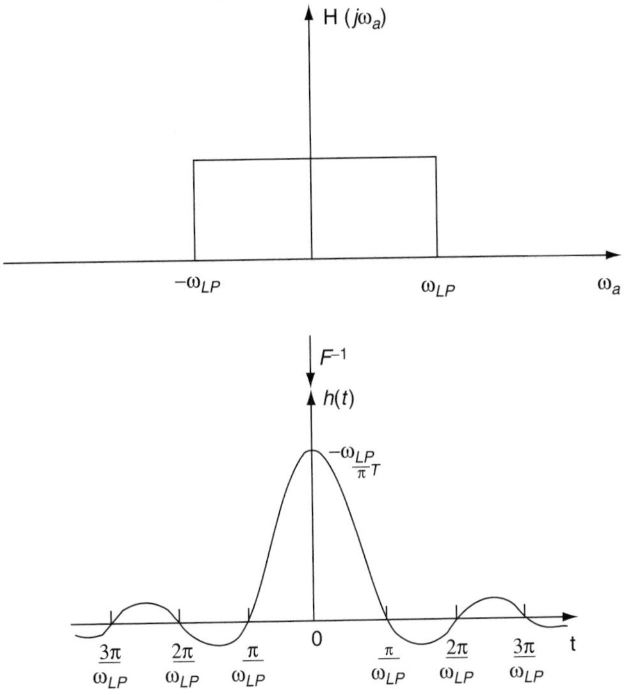

FIGURE 2.5 Ideal Low-Pass Filter

can be expressed as a sum of delayed and weighted impulse sequences:

$$x(n) = \sum_{k=-\infty}^{\infty} x(k)\delta(n-k), \tag{2.7}$$

and the response of an LTI digital filter to $x(n)$ can then be expressed by:

$$\begin{aligned} y(n) &= \mathcal{F}\left[\sum_{k=-\infty}^{\infty} x(k)\delta(n-k)\right] \\ &= \sum_{k=-\infty}^{\infty} x(k)\mathcal{F}[\delta(n-k)] \\ &= \sum_{k=-\infty}^{\infty} x(k)h(n-k) \\ &= x(n) * h(n). \end{aligned} \tag{2.8}$$

The summation in the last two lines of the above expression, called the **convolution sum**, relates the output sequence of a digital filter to its impulse response $h(n)$ and to the input sequence $x(n)$. Equation 2.8 can be rewritten as:

$$y(n) = \sum_{k=-\infty}^{\infty} h(k)x(n-k). \tag{2.9}$$

Defining the z-transform of a sequence $x(n)$ is written as:

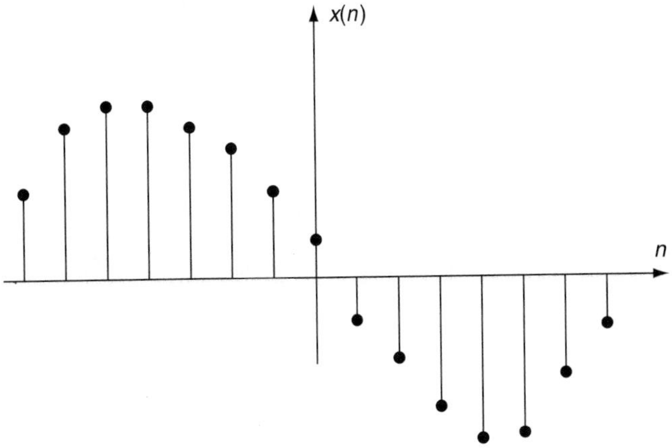

FIGURE 2.6 Discrete-Time Signal Representation

$$X(z) = \sum_{n=-\infty}^{\infty} x(n)z^{-n}. \tag{2.10}$$

The **transfer function** of a digital filter is the ratio of the z transform of the output sequence to the z-transform of the input signal:

$$H(z) = \frac{Y(z)}{X(z)}. \tag{2.11}$$

Taking the z-transform of both sides of the convolution expression of equation 2.9 yields:

$$\sum_{n=-\infty}^{\infty} y(n)z^{-n} = \sum_{n=-\infty}^{\infty}\sum_{k=-\infty}^{\infty} h(k)x(n-k)z^{-n}, \tag{2.12}$$

and substituting variables ($l = n - k$):

$$\sum_{n=-\infty}^{\infty} y(n)z^{-n} = \sum_{k=-\infty}^{\infty} h(k)z^{-k} \sum_{l=-\infty}^{\infty} x(l)z^{-l}. \tag{2.13}$$

The following relation among the z-transforms of the output $Y(z)$, of the input $X(z)$, and of the impulse response $H(z)$ of a digital filter is obtained as:

$$Y(z) = H(z)X(z). \tag{2.14}$$

Hence, the transfer function of an LTI digital filter is the z-transform of its impulse response.

2.4.2 Difference Equation

A general digital filter can be described according to the following **difference equation**:

$$\sum_{i=0}^{N} a_i y(n-i) - \sum_{l=0}^{M} b_l x(n-l) = 0. \tag{2.15}$$

Most digital filters can be described by difference equations that are suitable for implementation in digital machines. It is important to guarantee that the difference equation represents a system that is linear, time-invariant, and causal. This is guaranteed if the auxiliary conditions for the underlying difference equation correspond to its initial conditions, and the system is initially relaxed.

Nonrecursive digital filters are a function of past input samples. In general, they can be described by equation 2.15 if $a_0 = 1$ and $a_i = 0$ for $i = 1, \ldots, N$:

$$y(n) = \sum_{l=0}^{M} b_l x(n-l). \tag{2.16}$$

It can be easily shown that these filters have finite impulse response and are known as FIR filters. For $a_0 = 1$, equation 2.15 can be rewritten as:

$$y(n) = -\sum_{i=1}^{N} a_i y(n-i) + \sum_{l=0}^{M} b_l x(n-l). \tag{2.17}$$

It can be shown that these filters have, in most cases, infinite impulse response [i.e., $y(n) \neq 0$ when $n \to \infty$] and are therefore known as IIR filters.

2.5 Finite Impulse Response (FIR) Filters

The general transfer function of an **finite impulse response** (FIR) filter is given by:

$$H(z) = \sum_{l=0}^{M} b_l z^{-l} = H_0 z^{-M} \prod_{l=0}^{M} (z - z_l). \tag{2.18}$$

A filter with the transfer functions in equation 2.18 will always be stable because it does not have poles outside the unit circle. FIR filters can be designed to have a linear phase. A linear phase is a desirable feature in a number of signal processing applications (e.g., image processing). FIR filters can be designed by using optimization packages or by using approximations based on different types of windows. The main drawback is that to satisfy demanding magnitude specifications, the FIR filter requires a relatively high number of multiplications, additions, and storage elements. These facts make FIR filters potentially more expensive than IIR filters in applications where the number of arithmetic operations or number of storage elements are expensive or limited. However, FIR filters are widely used because they are suitable for designing linear-phase filters.

An FIR filter has a linear phase if and only if its impulse response is symmetric or antisymmetric; that is $h(n) = \pm h(M-n)$.

Let us consider in detail a particular case of linear-phase FIR filters:

For M even:

$$\begin{aligned} H(z) &= \sum_{n=0}^{M/2-1} h(n) z^{-n} + h\left(\frac{M}{2}\right) z^{-M/2} + \sum_{n=M/2+1}^{M} h(n) z^{-n} \\ &= \sum_{n=0}^{M/2-1} h(n)(z^{-n} + z^{-(M-n)}) + h\left(\frac{M}{2}\right) z^{-M/2}. \end{aligned} \tag{2.19}$$

At unity circle:

Evaluating equation 2.19 (M even) at $z = e^{j\omega}$, we have:

$$H(e^{j\omega}) = \sum_{n=0}^{M/2-1} h(n) e^{-j\omega n} + h\left(\frac{M}{2}\right) e^{-j\omega M/2} + \sum_{n=M/2+1}^{M} h(n) e^{-j\omega n}.$$

Considering the case of symmetric impulse response (i.e., $h(n) = h(M-n)$ for $n = 0, 1, \ldots, M/2$), we have:

$$H(e^{j\omega}) = e^{-j\omega M/2}\left[h\left(\frac{M}{2}\right) + \sum_{n=0}^{M/2-1} 2h(n)\cos\left[\omega\left(n - \frac{M}{2}\right)\right]\right].$$

By replacing n by $(M/2 - m)$, the equation above becomes:

$$H(e^{j\omega}) = e^{-j\omega M/2}\left[h\left(\frac{M}{2}\right) + \sum_{m=1}^{M/2} 2h\left(\frac{M}{2} - m\right)\cos\left[\omega(m)\right]\right].$$

Defining $a_0 = h(M/2)$ and $a_m = 2h(M/2 - m)$, for $m = 1, 2, \ldots, M/2$, we have:

$$H(e^{j\omega}) = e^{-j\omega M/2} \sum_{m=0}^{M/2} a_m \cos(\omega m). \tag{2.20}$$

In summary, we have four possible cases of FIR filters with a linear phase:

- *Case A*: M Even and Symmetric Impulse Responses
  (1) Characteristics:
  $h(n) = h(M-n)$ for $n = 0, 1, \ldots, M/2$
  $H(e^{j\omega}) = e^{-j\omega M/2} \sum_{m=0}^{M/2} a_m \cos(\omega m)$, where $a_0 = h(M/2)$ and $a_m = 2h(M/2 - m)$ for $m = 1, 2, \ldots, M/2$
  (2) Phase: $\Theta(\omega) = \tan^{-1}\left\{\frac{\mathrm{Im}[H(e^{j\omega})]}{\mathrm{Re}[H(e^{j\omega})]}\right\} = -\omega \frac{M}{2}$.
  (3) Group delay: $\tau = -\frac{\partial \Theta(\omega)}{\partial \omega} = \frac{M}{2}$.
- *Case B*: M Odd and Symmetric Impulse Responses
  (1) Characteristics:
  $h(n) = h(M-n)$ for $n = 0, 1, \ldots, (M/2 - 0.5)$.
  $H(e^{j\omega}) = e^{-j\omega M/2} \sum_{m=1}^{M/2+0.5} b_m \cos[\omega(m - \frac{1}{2})]$, where $b_m = 2h(M/2 + 0.5 - m)$ for $m = 1, 2, \ldots, (M/2 + 0.5)$.

(2) Phase: $\Theta(\omega) = -\omega\frac{M}{2}$.
(3) Group delay: $\tau = -\frac{\partial\Theta(\omega)}{\partial\omega} = \frac{M}{2}$.

- *Case C*: M Even and Antisymmetric Impulse Response
  (1) Characteristics:
  $h(n) = -h(M-n)$ for
  $n = 0, 1, \ldots, (M/2-1)$ and $h(M/2) = 0$.
  $H(e^{j\omega}) = e^{-j(\omega M/2-\pi/2}\sum_{m=1}^{M/2} c_m \sin(\omega m)$, where $c_m = 2h(M/2-m)$ for $m = 1, 2, \ldots, M/2$.
  (2) Phase: $\Theta(\omega) = -\omega\frac{M}{2} + \frac{\pi}{2}$.
  (3) Group delay: $\tau = -\frac{\partial\Theta(\omega)}{\partial\omega} = \frac{M}{2}$.
- *Case D*: M Odd and Antisymmetric Impulse Response
  (1) Characteristics:
  $h(n) = -h(M-n)$, for $n = 0, 1, \ldots, (M/2-0.5)$.
  $H(e^{j\omega}) = e^{-j(\omega M/2-\pi/2)}\sum_{m=1}^{M/2+0.5} d_m \sin[\omega(m-\frac{1}{2})]$, where $d_m = 2h(\frac{M}{2}+0.5-m)$, for $m = 1, 2, \ldots, (M/2+0.5)$.
  (2) Phase: $\Theta(\omega) = -\omega\frac{M}{2} + \frac{\pi}{2}$.
  (3) Group delay: $\tau = -\frac{\partial\Theta(\omega)}{\partial\omega} = \frac{M}{2}$.

As an illustration, Figure 2.7 depicts the typical impulse responses for the linear-phase FIR filter.

Equation 2.19 for linear-phase FIR filters can be rewritten as:

$$H(z) = z^{-M/2}\sum_{l=0}^{L} g_l(z^l \pm z^{-l}), \qquad (2.21)$$

where the g_l's are related to $h(l)$, where $L = M/2 - 0.5$ for M odd and $L = M/2$ for M even. If there is a zero of $H(z)$ at z_0, z_0^{-1} is also a zero of $H(z)$. This means that all complex zeros not located on the unit circle occur in conjugate and reciprocal quadruples. Zeros on the real axis outside the unit circle occur in reciprocal pairs. Zeros on the unit circle occur in conjugate pairs.

It is possible to verify that the case B linear-phase transfer function has a zero at $\omega = \pi$, such that high-pass and band-stop filters cannot be approximated using this case. For case C, there are zeros at $\omega = \pi$ and $\omega = 0$; therefore they are not suitable for low-pass, high-pass, and band-stop filter designs. Case D has zero at $\omega = 0$ and is not suitable for low-pass and band-stop design.

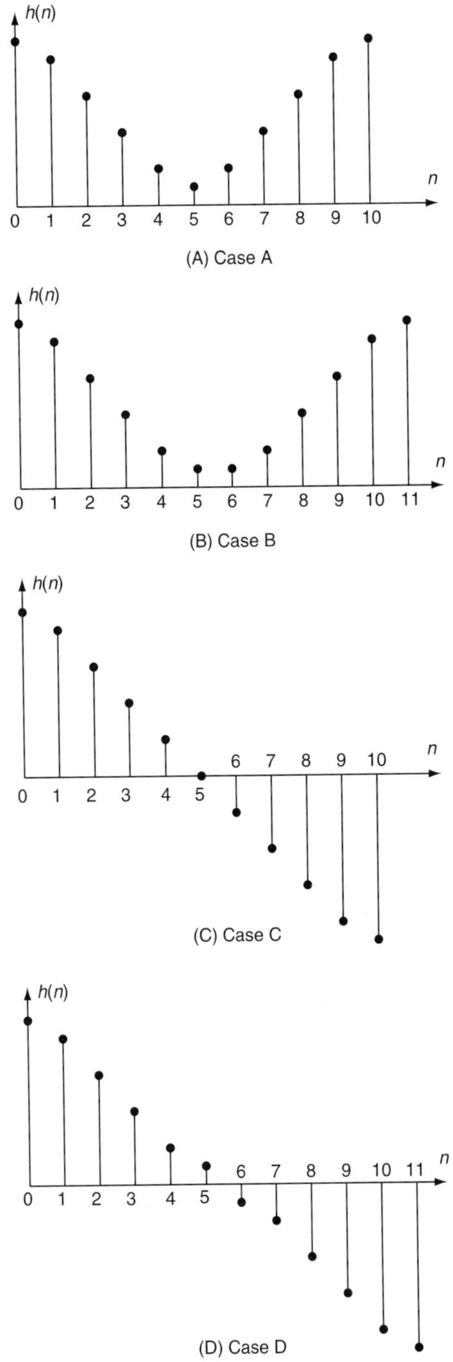

FIGURE 2.7 Linear-Phase FIR Filter: Typical Impulse Response

2.6 Infinite Impulse Response Filters

The general transfer function for **infinite impulse response** (IIR) **filters** is shown below:

$$H(z) = \frac{Y(z)}{X(z)} = \frac{\sum_{l=0}^{M} b_l z^{-l}}{1 + \sum_{i=1}^{N} a_i z^{-i}}. \qquad (2.22)$$

The above equation can be rewritten in the following alternative forms:

$$H(z) = H_0\frac{\prod_{l=0}^{M}(1 - z^{-1}z_l)}{\prod_{i=0}^{N}(1 - z^{-1}p_i)} = H_0 z^{N-M}\frac{\prod_{l=0}^{M}(z - z_l)}{\prod_{i=0}^{N}(z - p_i)}. \qquad (2.23)$$

The above equation shows that, unlike FIR filters, IIR filters have poles located at p_i. Therefore, it is required that $|p_i| < 1$ to guarantee stability for the IIR transfer functions.

2.7 Digital Filter Realizations

From equation 2.15, it can be seen that digital filters use three basic types of operations:

- **Delay**: Used to store previous samples of the input and output signals, besides internal states of the filter
- **Multiplier**: Used to implement multiplication operations of a particular digital filter structure; also used to implement divisions
- **Adder**: Used to implement additions and subtractions of the digital filter

2.7.1 FIR Filter Structures

General FIR and IIR filter structures are shown in Figures 2.8 and 2.9. Figure 2.8 shows the direct-form nonrecursive structure for FIR filters, while Figure 2.9 depicts an alternative transposed[2] structure. These structures are equivalent when implemented in software.

Since the impulse responses of linear-phase FIR filters are either symmetric or antisymmetric [e.g., $h(n) = \pm h(M - n)$], this property can be exploited to reduce the overall number of multiplications of the FIR filter realization. In Figure 2.10 the resulting realizations for the case of symmetric $h(n)$ are shown. For the antisymmetric case, the structures are straightforward to obtain.

2.7.2 IIR Filter Realizations

A general IIR transfer function can be written as in equation 2.22. The numerator in this transfer function can be implemented by using an FIR filter. The denominator entails the use of a recursive structure. The cascade of these realizations for the numerator and denominator is shown in Figure 2.11. If the recursive part is implemented first, the realization in Figure 2.11 can be transformed into the structure in Figure 2.12 for the case $N = M$. The transpose of the structure in Figure 2.12 is shown in Figure 2.13.

In general, IIR transfer functions can be implemented as a cascade or as a summation of lower order IIR structures. In case an IIR transfer function is implemented as a cascade of second-order sections of order m, it is possible to write:

$$H(z) = \prod_{k=1}^{m} \frac{\gamma_{0k} + \gamma_{1k}z^{-1} + \gamma_{2k}z^{-2}}{1 + m_{1k}z^{-1} + m_{2k}z^{-2}} = \prod_{k=1}^{m} \frac{\gamma_{0k}z^2 + \gamma_{1k}z + \gamma_{2k}}{z^2 + m_{1k}z + m_{2k}}$$
$$= H_0 \prod_{k=1}^{m} \frac{z^2 + \gamma'_{1k}z + \gamma'_{2k}}{z^2 + m_{1k}z + m_{2k}} = \prod_{k=1}^{m} H_k(z). \tag{2.24}$$

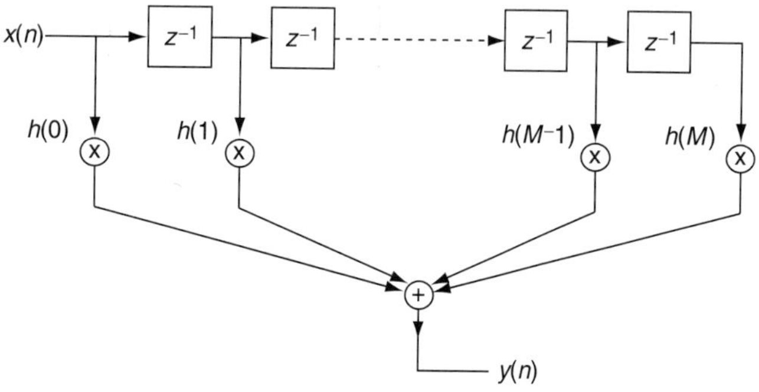

FIGURE 2.8 Direct-Form Nonrecursive Structure

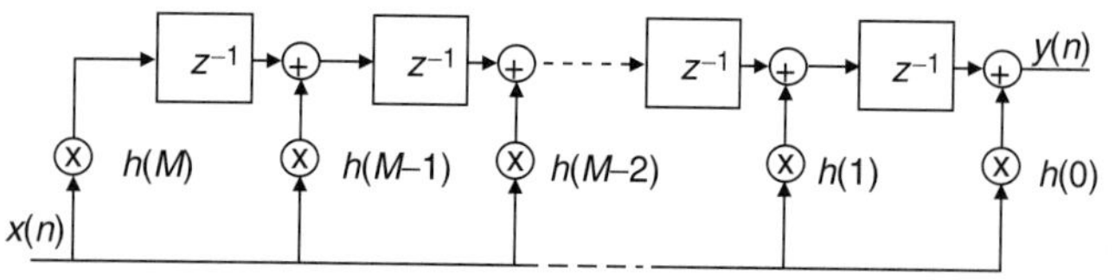

FIGURE 2.9 Alternative Form Nonrecursive Structure

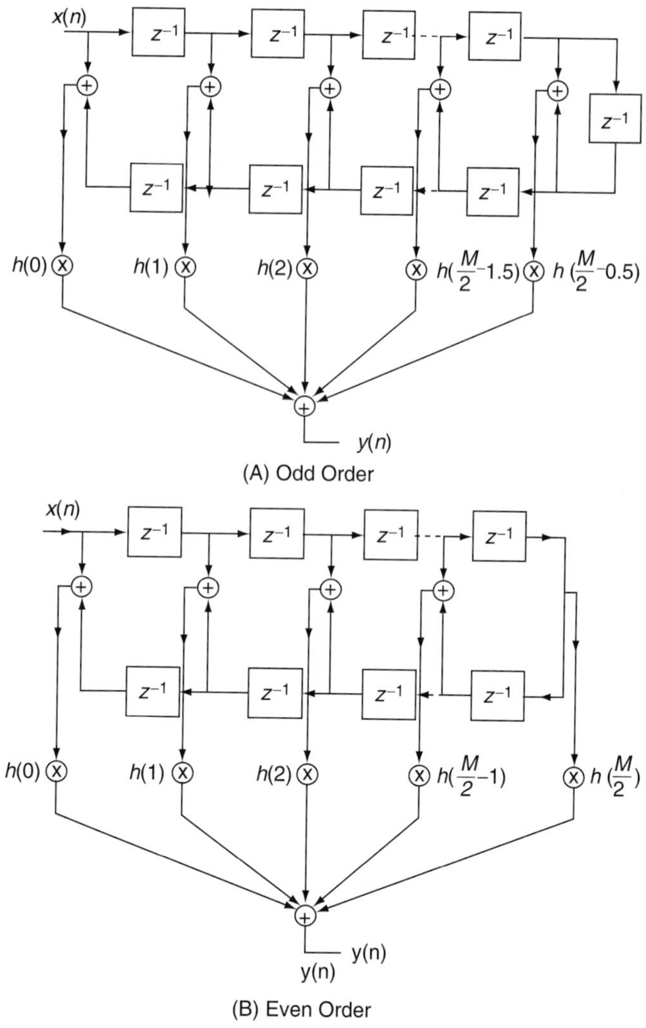

FIGURE 2.10 Linear-Phase FIR Filter Realization With Symmetric Impulse Response

[2] This structure is characterized by the branches being reversed, such that the sinks turn to summations, and the summations become distribution nodes.

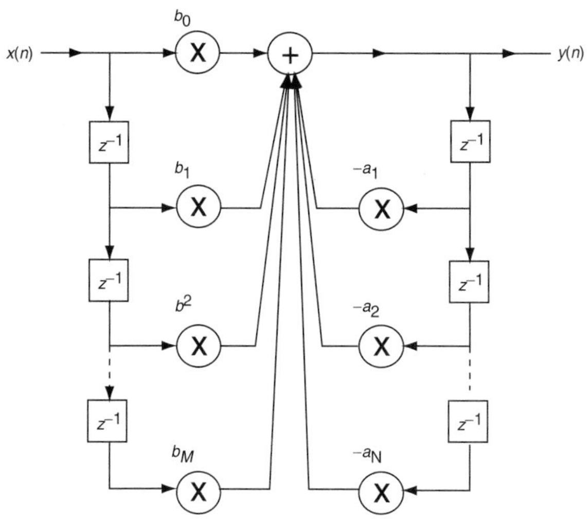

FIGURE 2.11 Recursive Structure in Direct Form

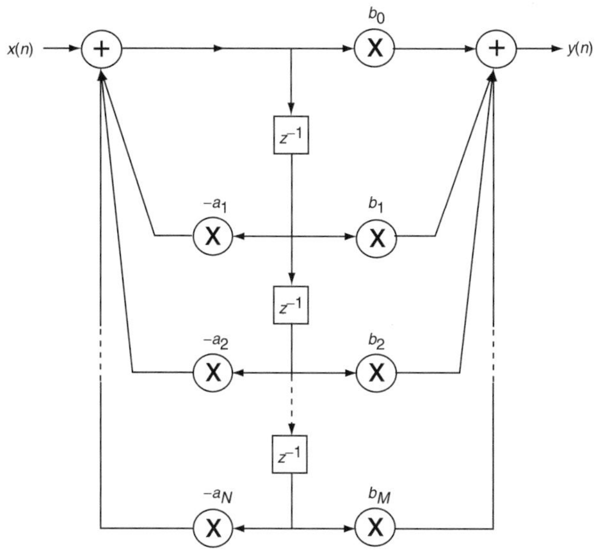

FIGURE 2.12 Direct Form Structure for $N = M$

Equation 2.24 is depicted in Figure 2.14(A).

In the case an IIR transfer function is implemented as the addition of lower order sections of order m, it is possible to write:

$$H(z) = \sum_{k=1}^{m} \frac{\gamma_{0k}^{p} z^2 + \gamma_{1k}^{p} z + \gamma_{2k}^{p}}{z^2 + m_{1k} z + m_{2k}} = h_0 + \sum_{k=1}^{m} \frac{\gamma_{0k}^{p'} z^2 + \gamma_{1k}^{p'} z}{z^2 + m_{1k} z + m_{2k}}$$
$$= h_0' + \sum_{k=1}^{m} \frac{\gamma_{1k}^{p''} z + \gamma_{2k}^{p''}}{z^2 + m_{1k} z + m_{2k}} = \sum_{k=1}^{m} H_k(z). \quad (2.25)$$

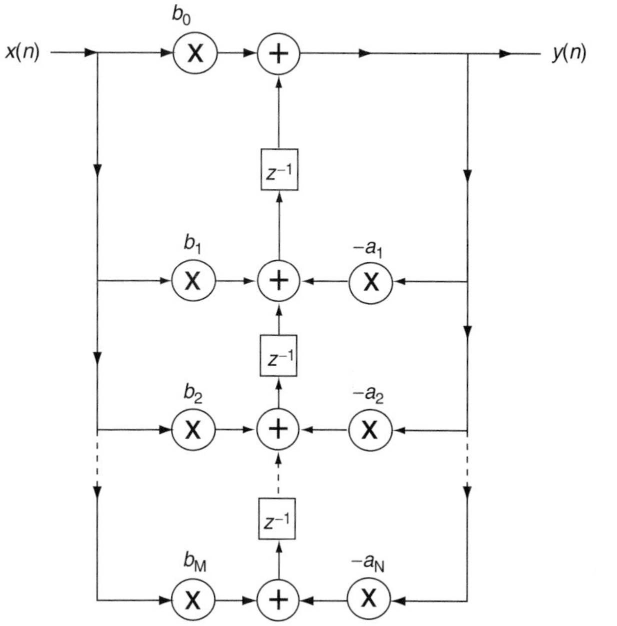

FIGURE 2.13 Alternative Direct-Form Structure for $N = M$

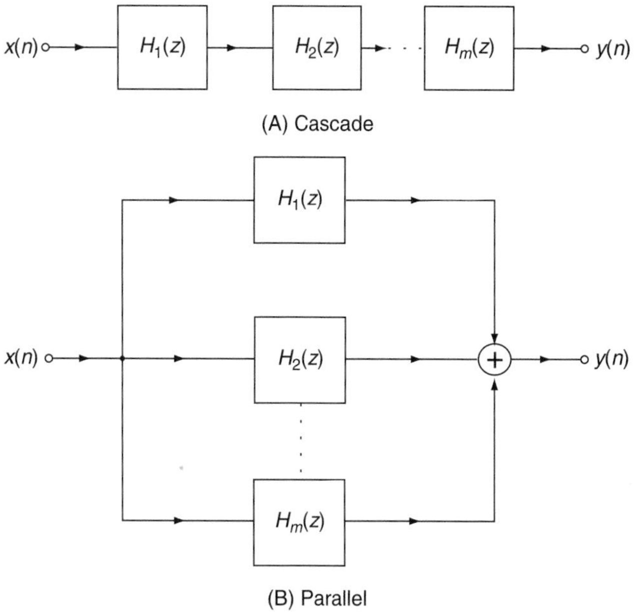

FIGURE 2.14 Realization with Second-Order Sections

Equation 2.25 depicted in Figure 2.14(B).

In general, for implementations with high signal-to-quantization noise ratios, sections of order two are preferred. Different types of structures can be used for second-order sections, such as the direct-form structures shown in Figures 2.15(A) and 2.15(B) or alternative biquadratic structures like state-space (Diniz and Antoniou, 1986).

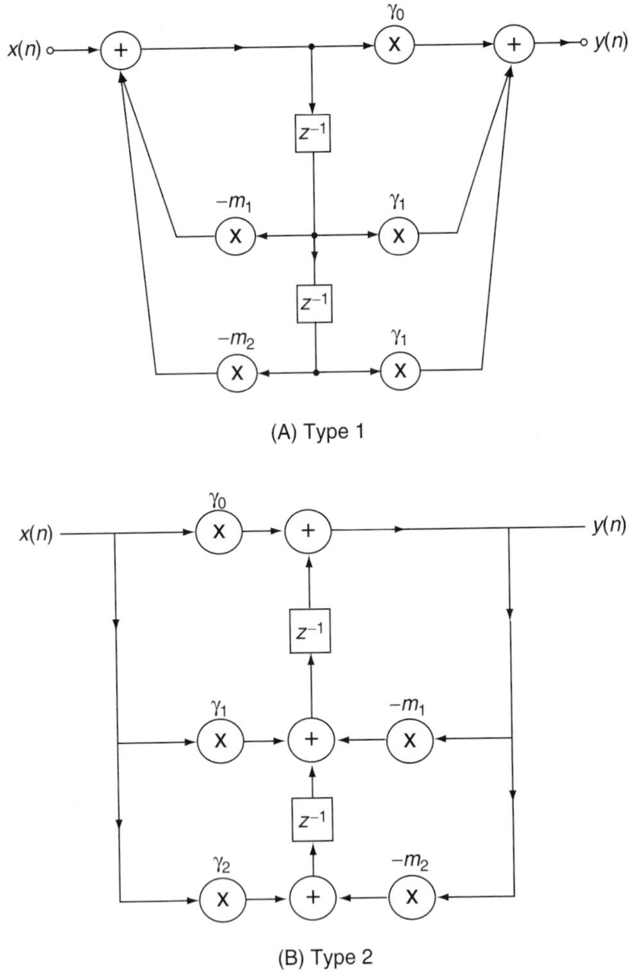

FIGURE 2.15 Basic Second-Order Sections

2.8 FIR Filter Approximation Methods

There are three basic approximation methods for FIR filters satisfying given specifications. The first one, the **window method**, is very simple and well established but generally leads to transfer functions that normally have a higher order than those obtained by optimization methods. The second method, known as **minimax method**, calculates a transfer function that has minimum order to satisfy prescribed specifications. The minimax method is an iterative method and is available in most filter design packages. The third method, known as **weighted least squares-Chebyshev**[3] (WLS-Chebyshev) is a reliable but not widely known method. The WLS-Chebyshev method is particularly suitable for multirate systems because the resulting transfer functions might have decreasing energy at a prescribed range of frequencies.

[3] The WLS-Chebyshev method is also known as the peak constrained method.

2.8.1 Window Method

The **window method** starts by obtaining the impulse response of ideal prototype filters. These responses for standard filters are shown in Figure 2.16. In general, the impulse response is calculated according to:

$$h(n) = \frac{1}{2\pi}\int_{-\pi}^{\pi} H(e^{j\omega})e^{j\omega n}d\omega, \quad (2.26)$$

for a desired filter.

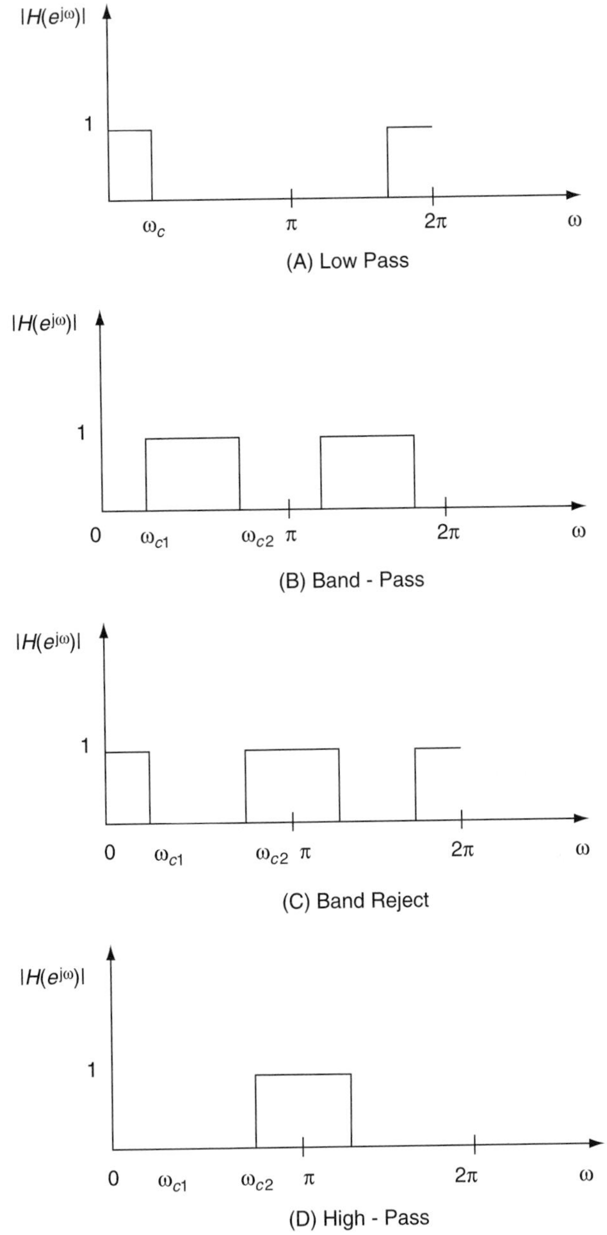

FIGURE 2.16 Ideal Frequency Responses

In the case of a band-pass filter, the prototype filter can be described by following transfer function:

$$|H(e^{j\omega})| = \begin{cases} 0 & \text{for } 0 < |\omega| < \omega_{c1}. \\ 1 & \text{for } \omega_{c1} \leq |\omega| \leq \omega_{c2}. \\ 0 & \text{for } \omega_{c2} < |\omega| \leq \pi. \end{cases} \quad (2.27)$$

Using the inverse Fourier transform, the impulse response of the filter with the response in equation 2.27 can be obtained as follows:

$$\begin{aligned} h(n) &= \frac{1}{2\pi}\int_{-\pi}^{\pi} H(e^{j\omega})e^{j\omega n}d\omega \\ &= \frac{1}{2\pi}\left[\int_{\omega_{c1}}^{\omega_{c1}} 2\cos(\omega n)d\omega\right] \\ &= \frac{1}{\pi n}[\sin(\omega_{c2}n) - \sin(\omega_{c1}n)], \end{aligned} \quad (2.28)$$

for $n \neq 0$, for $n = 0$, $h(0) = \frac{1}{\pi}(\omega_{c2} - \omega_{c1})$.

Different window sequences can be multiplied to $h(n)$ to limit its length, generating the impulse response for the designed FIR filter as follows:

$$h'(n) = h(n)w(n). \quad (2.29)$$

In equation 2.29, $w(n)$ is the window sequence, and $h'(n)$ is the impulse response of the obtained filter. The rectangular window consists of truncating the sequence $h(n)$; that is, $w(n) = 1$ for $n = -M/2, -M/2+1, \ldots, 0, \ldots, M/2-1, M/2$.

2.8.2 Sarämaki Window

The abrupt truncation of the impulse response performed by the rectangular window leads to oscillations at frequencies close to the resulting filter band edges. The satisfaction of prescribed specification is difficult because these ripples have uncontrollable maximum magnitude.

As alternatives, there are more flexible windows incorporating design parameters that allow a trade-off between the transition bandwidth and the ripples magnitude (Sarämaki, 1992).

The frequency response of the unscaled **Sarämaki window** is known to be the following:

$$\begin{aligned} \Psi(\omega) &= \sum_{n=-\frac{M}{2}}^{\frac{M}{2}} \bar{\mathbf{w}}(n)e^{-j\omega n} = 1 + \sum_{k=1}^{\frac{M}{2}} 2T_k[\gamma\cos\omega + (\gamma - 1)] \\ &= \frac{\sin\left[\frac{M+1}{2}\cos^{-1}\{\gamma\cos\omega + (\gamma-1)\}\right]}{\sin\left[\frac{1}{2}\cos^{-1}\{\gamma\cos\omega + (\gamma-1)\}\right]}, \end{aligned} \quad (2.30)$$

where T_k is the kth degree Chebyshev polynomial and:

$$\gamma = \frac{1 + \cos(2\pi/M + 1)}{1 + \cos(2\beta\pi/M + 1)}. \quad (2.31)$$

In the above equation, β is a variable parameter that adjusts the Sarämaki window main lobe width given by $4\beta\pi/M + 1$. In the special case of $\beta = 1$, the Sarämaki window becomes identical to the rectangular window.

The desired normalized window function is as written here:

$$\mathbf{w}(n) = \bar{\mathbf{w}}(n)/\bar{w}(0), \quad (2.32)$$

for $-M/2 \leq n \leq M/2$ and $w(0) = 1$.

The unscaled coefficients $\mathbf{w}(n)$ can be expressed as:

$$w(n) = v_0(n) + 2\sum_{k=1}^{M/2} v_k(n), \quad (2.33)$$

where $v_k(n)$ can be calculated according to the recursive equations below:

$$v_0(n) = \begin{cases} 1 & n = 0. \\ 0, & \text{otherwise}. \end{cases} \quad (2.34)$$

$$v_1(n) = \begin{cases} \gamma - 1, & n = 0. \\ \gamma/2, & |n| = 1. \\ 0, & \text{otherwise}. \end{cases} \quad (2.35)$$

$$v_k(n) = \begin{cases} 2(\gamma-1)v_{k-1}(n) - v_{k-2}(n) + \gamma[v_{k-1}(n-1) + v_{k-1}(n+1)], & -k \leq n \leq k. \\ 0, & \text{otherwise}. \end{cases} \quad (2.36)$$

The causal transfer function is given by:

$$H(z) = z^{-M/2}\sum_{n=-M/2}^{M/2} h'(n)z^{-n} \quad (2.37)$$

Example

Design an FIR filter satisfying the following specifications: $A_p = 0.2$ dB, $A_r = 32$ dB, $\omega_p = 2\pi \times 3400$ rad/sec, $\omega_r = 2\pi \times 4600$ rad/sec, and $\omega_s = 2\pi \times 20{,}000$ rad/sec

Solution

Figure 2.17 shows different responses obtained by using the rectangular and Sarämaki windows on a low-pass filter. It can be seen that the Sarämaki window showed improvements with respect to the rectangular window response. In fact, the rectangular window has a high order of 248; to satisfy the specifications, the transition band had to be reduced due to the high ripples close to the band edges. The Sarämaki window required order 48 to meet the specifications.

2.9 FIR Filter Design by Optimization

In the minimax approach, our aim is to design filters that minimize the peak (maximum) of the error with respect to

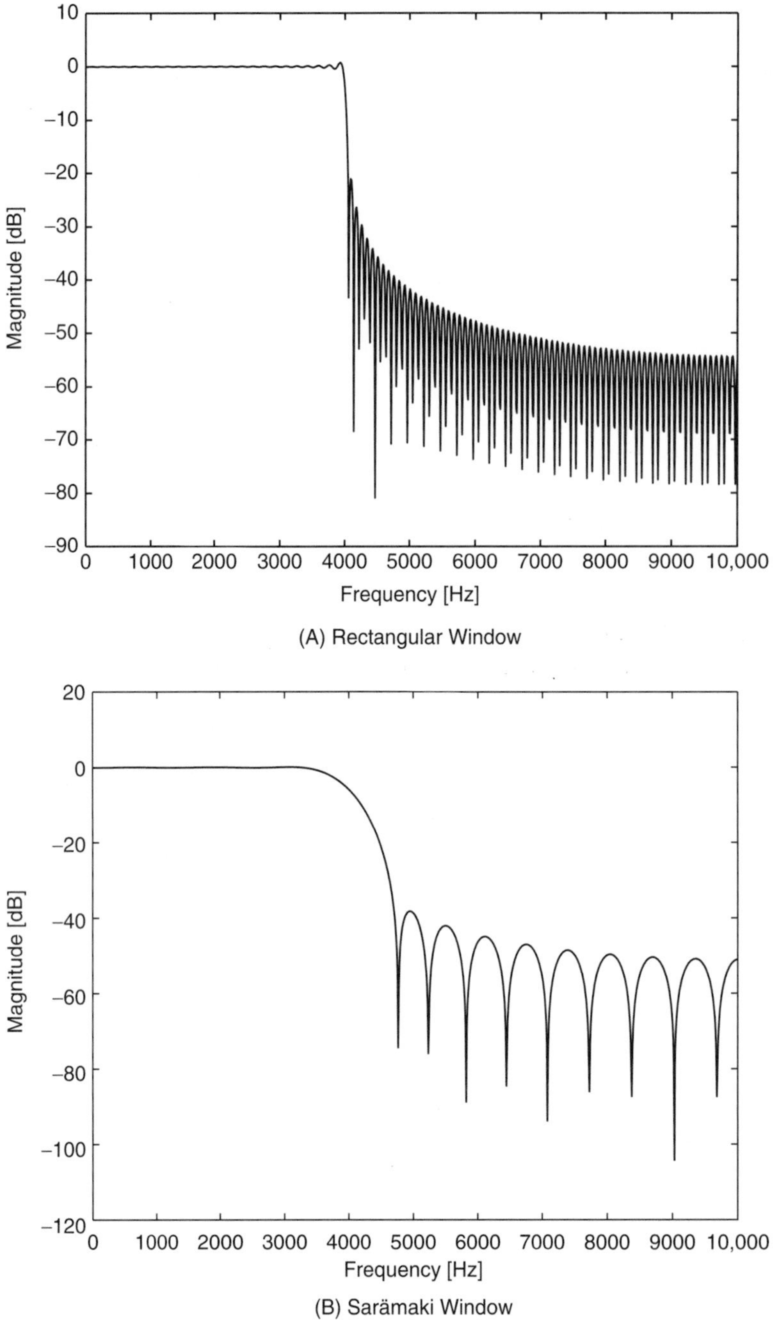

FIGURE 2.17 Low-Pass Filter Using Window Method

desired characteristics without directly considering the error energy. On the other hand, the least squares approach aims to minimize the error energy. As can be observed in Figure 2.18, there are solutions between these two extremal types of objective functions that provide good trade-off between peak error and error energy. In this section, we present a very flexible approach to designing FIR digital filters satisfying prescribed specifications in terms of the maximum deviation the passband and in part of the stopband, while minimizing the energy in the stopband frequency range. This type of approximation is very useful in modern DSP systems involving subsystems with different sampling periods. The approximation method discussed here includes the least squares and minimax approaches as special cases.

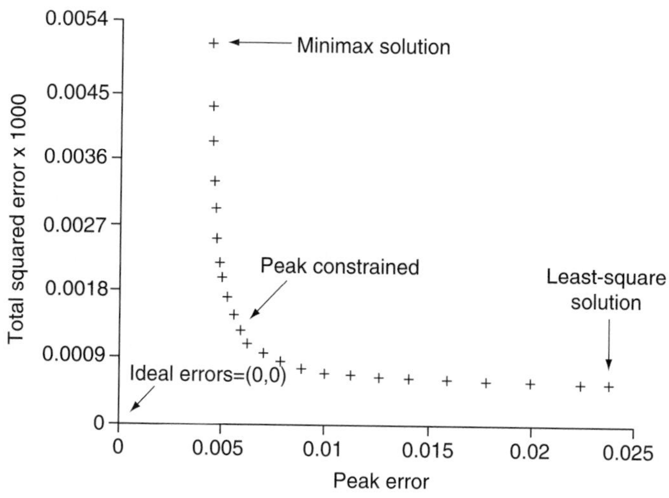

FIGURE 2.18 Minimax Versus Least Squares

2.9.1 Problem Formulation

Consider a nonrecursive filter of length $M+1$ described by the transfer function:

$$H(z) = \sum_{n=0}^{M} h(n) z^{-n}, \qquad (2.38)$$

assuming that $\omega_s = 2\pi$. The frequency response of such filter is then given by:

$$H(e^{j\omega}) = \sum_{n=0}^{M} h(n) e^{-j\omega n} = e^{j\Theta(\omega)} \hat{H}(\omega), \qquad (2.39)$$

where $\Theta(\omega)$ and $\hat{H}(\omega)$ are the phase and magnitude responses of $H(e^{j\omega})$, respectively defined as:

$$\Theta(\omega) = \tan^{-1}\left\{\frac{\mathrm{Im}[H(e^{j\omega})]}{\mathrm{Re}[H(e^{j\omega})]}\right\}. \qquad (2.40)$$

$$\hat{H}(\omega) = |H(e^{j\omega})|. \qquad (2.41)$$

Assume that the phase response $\Theta(\omega)$ is linear on ω, that M is even, and that $h(n)$ is symmetrical. The cases of M odd and/or $h(n)$ antisymmetrical can be easily derived based on the proposed formulation. The frequency response of such filter thus becomes:

$$H(e^{j\omega}) = e^{-jM/2\omega} \sum_{n=0}^{M/2} a_n \cos(\omega n), \qquad (2.42)$$

with $a_0 = h(M/2)$ and $a_n = 2h(\frac{M}{2} - n)$ for $n = 1, \ldots, M/2$.

The desired frequency response is given by $e^{-jM/2\omega} D(\omega)$, and $W(\omega)$ is a strictly positive weighting function. The weighted error function $E(\omega)$ is defined in the frequency domain as:

$$E(\omega) = W(\omega)[D(\omega) - \hat{H}(\omega)]. \qquad (2.43)$$

The approximation problem for FIR linear-phase digital filters is equivalent to the minimization of some objective function of $E(\omega)$ in such way that $|E(\omega)| \le \delta$. Then:

$$|H(\omega) - \hat{H}(\omega)| \le \frac{\delta}{W(\omega)}. \qquad (2.44)$$

A good discrete approximation of $E(\omega)$ can be obtained by evaluating the weighted error function on a dense frequency grid with $0 \le \omega_i \le \pi$ for $i = 1, \ldots, L(M+1)$. For practical purposes, for a filter of length $M+1$, we use $8 \le L \le 16$. Points placed at the transition band are disregarded, whereas the remaining frequencies should be linearly redistributed in the passband and stopband including their corresponding edges. Thus, the following vector equation results:

$$\boldsymbol{e} = \boldsymbol{W}(\boldsymbol{d} - \boldsymbol{U}\boldsymbol{b}), \qquad (2.45)$$

where:

$$\boldsymbol{e} = [E(\omega_1)E(\omega_2) \ \ldots \ E(\omega_{\bar{L}(M+1)})]^T. \qquad (2.46)$$

$$\boldsymbol{W} = \mathrm{diag}[W(\omega_1)W(\omega_2) \ \ldots \ W(\omega_{\bar{L}(M+1)})]. \qquad (2.47)$$

$$\boldsymbol{d} = [D(\omega_1)D(\omega_2) \ \ldots \ D(\omega_{\bar{L}(M+1)})]^T. \qquad (2.48)$$

$$\boldsymbol{U} = \begin{bmatrix} 1 & \cos(\omega_1) & \cos(2\omega_1) & \ldots & \cos(\frac{M}{2}\omega_1) \\ 1 & \cos(\omega_2) & \cos(2\omega_2) & \ldots & \cos(\frac{M}{2}\omega_2) \\ \vdots & \vdots & \vdots & \ddots & \vdots \\ 1 & \cos(\omega_{\bar{L}(M+1)}) & \cos(2\omega_{\bar{L}(M+1)}) & \ldots & \cos(\frac{M}{2}\omega_{\bar{L}(M+1)}) \end{bmatrix}. \qquad (2.49)$$

$$\boldsymbol{b} = \left[a_0 \ a_1 \ \ldots \ a_{\frac{M}{2}}\right]^T. \qquad (2.50)$$

In these equations, $\bar{L} < L$ because the transition frequencies were discarded.

A low-pass filter specification includes δ_p as the passband maximum ripple and δ_r as the stopband minimum attenuation, with ω_p and ω_r being the passband and stopband edges, respectively.

Based on these values, the following can be defined:

$$DB_p = 20\log_{10}\left(\frac{1+\delta_p}{1-\delta_p}\right)[\mathrm{dB}]. \qquad (2.51)$$

$$DB_r = 20\log_{10}(\delta_r)[\mathrm{dB}]. \qquad (2.52)$$

The design of a low-pass digital filter using either the minimax method or the WLS approach is performed by choosing the ideal response and weight functions, respectively, as:

$$D(\omega) = \begin{cases} 1, & 0 \le \omega \le \omega_p. \\ 0, & \omega_r \le \omega \le \pi. \end{cases} \quad (2.53)$$

$$W(\omega) = \begin{cases} 1, & 0 \le \omega \le \omega_p. \\ \delta_p/\delta_r, & \omega_r \le \omega \le \pi. \end{cases} \quad (2.54)$$

Other types of transfer functions can be similarly defined.

2.9.2 Chebyshev Method

Chebyshev filter design consists of the minimization of the maximum absolute value of $E(\omega)$ with respect to the set of filter coefficients:

$$\|E(\omega)\|_\infty = \min_{\boldsymbol{b}} \max_{0 \le \omega \le \pi} [W(\omega)|D(\omega) - \hat{H}(\omega)|]. \quad (2.55)$$

Using equations 2.46 to 2.50, the Chebyshev method attempts to minimize:

$$\|E(\omega)\|_\infty \approx \min_{\mathbf{b}} \max_{0 \le \omega_i \le \pi} [\boldsymbol{W}|\boldsymbol{d} - \boldsymbol{Ub}|], \quad (2.56)$$

at the discrete set of frequencies. The Chebyshev method minimizes:

$$\mathrm{DB}_\delta = 20 \log_{10}(\delta)[\mathrm{dB}], \quad (2.57)$$

where $\delta = \max[\delta_p, \delta_r]$.

2.9.3 Weighted Least-Squares Method

The weighted least-squares (WLS) approach minimizes:

$$\|E(\omega)\|_2^2 = \int_0^\pi |E(\omega)|^2 \, d\omega = \int_0^\pi W^2(\omega)|D(\omega) - \hat{H}(\omega)|^2 \, d\omega. \quad (2.58)$$

This objective function at a discrete set of frequencies is estimated by:

$$\|E(\omega)\|_2^2 \approx \boldsymbol{e}^T \boldsymbol{e}, \quad (2.59)$$

the minimization of which is achieved with:

$$\boldsymbol{b}^* = (\boldsymbol{U}^T \boldsymbol{W}^2 \boldsymbol{U})^{-1} \boldsymbol{U}^T \boldsymbol{W}^2 \boldsymbol{d} \quad (2.60)$$

The WLS objective is to maximize the passband-to-stopband ratio (PSR) of energies:

$$\mathrm{PSR} = 10 \log_{10} \left(\frac{\int_0^{\omega_p} |\hat{H}(\omega)|^2 d\omega}{\int_{\omega_r}^{\pi} |\hat{H}(\omega)|^2 \, d\omega} \right) [\mathrm{dB}]. \quad (2.61)$$

2.9.4 Minimax Filter Design Employing WLS

This section describes a scheme that performs Chebyshev approximation as a limit of a special sequence of weighted least-squares approximations. The algorithm is implemented by a series of WLS approximations using a varying weight matrix $\boldsymbol{W}_k$, whose elements are calculated by:

$$W_{k+1}^2(\omega) = W_k^2(\omega) B_k(\omega), \quad (2.62)$$

where:

$$B_k(\omega) = |E_k(\omega)|. \quad (2.63)$$

This approach entails increasing the weight where the error is smaller.

The convergence of this algorithm is slow because usually 10 to 15 WLS designs are required in practice to approximate the Chebyshev solution. An efficiently accelerated algorithm was presented by Lim *et al.* (1992) that is characterized by the weight matrix $\boldsymbol{W}_k$ whose elements are recurrently updated by:

$$W_{k+1}^2(\omega) = W_k^2(\omega) \hat{B}_k(\omega), \quad (2.64)$$

where $\hat{B}_k(\omega)$ is the envelope function of $B_k(\omega)$ formed by a set of piecewise linear segments that start and end at consecutive extremes of $B_k(\omega)$. The band edges are considered extreme frequencies, and the edges from different bands are not connected. By denoting the extreme frequencies at a particular iteration k as ω_l', for $l = 1, 2, \ldots,$ the envelope function is formed as:

$$\hat{B}_k(\omega) = \frac{(\omega - \omega_l') B_k(\omega_{l+1}') + (\omega_{l+1}' - \omega) B_k(\omega_l')}{(\omega_{l+1}' - \omega_l')}; \; \omega_l' \le \omega \le \omega_{l+1}'. \quad (2.65)$$

Figure 2.19 depicts typical cases of the absolute value of the error function (dash-dotted curve) used by the algorithm of equation 2.62 to update the weighting function; the figure also depicts a corresponding envelope (solid curve) used by the algorithm of equation 2.65.

2.9.5 The WLS-Chebyshev Method

Comparing the adjustments used by the algorithms described by equations 2.62 through 2.65 and illustrated by Figure 2.19, with the piecewise-constant weight function used by the WLS method, it is possible to derive a very simple approach for designing digital filters with some compromises on minimax and WLS constraints (Diniz and Netto, 1999). This approach consists of a modification on the weight-function updating procedure in such way that it becomes constant after a particular extreme of the stopband of $B_k(\omega)$:

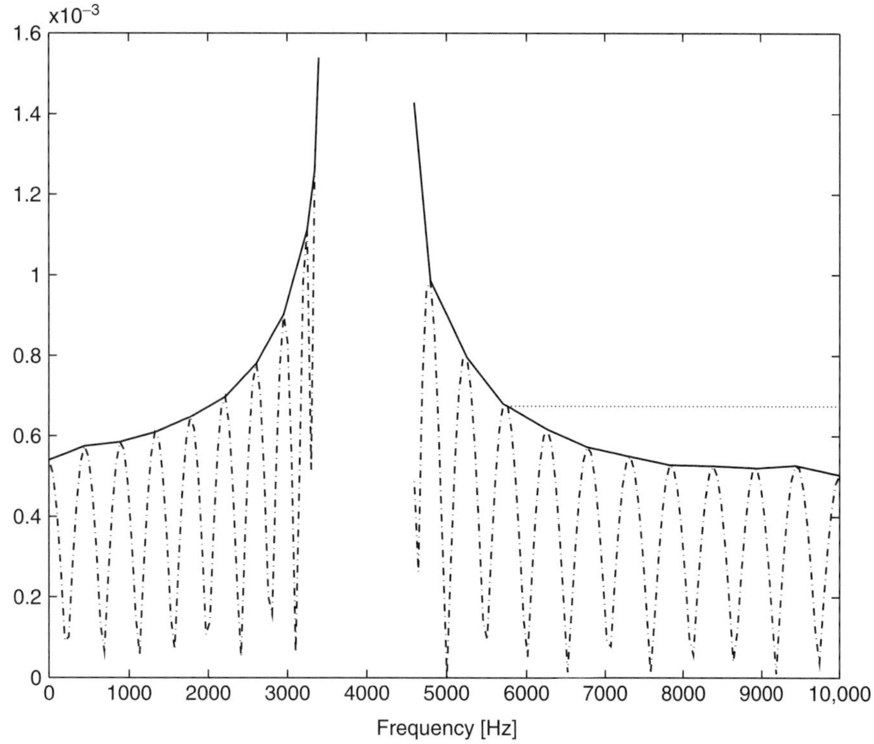

FIGURE 2.19 Typical Absolute Error Function $B(\omega)$ (Dash-Dotted Line) and Corresponding Envelope $\hat{B}(\omega)$ (Solid Curve). The $\beta(\omega)$ corresponds to the WLS-Chebyshev method (darker dotted curve).

$$W_{k+1}^2(\omega) = W_k^2(\omega)\beta_k(\omega). \quad (2.66)$$

For the algorithm of equation 2.65, $\beta_k(\omega)$ is given by:

$$\beta_k(\omega) \equiv \bar{B}_k(\omega) = \begin{cases} \hat{B}_k(\omega), & 0 \leq \omega \leq \omega_L'. \\ \hat{B}_k(\omega_L'), & \omega_L' < \omega \leq \pi. \end{cases} \quad (2.67)$$

The ω_L' is the Lth extreme value of the stopband of $B(\omega) = |E(\omega)|$. The passband values of $B(\omega)$ and $\hat{B}(\omega)$ are left unchanged in equation 2.67 to preserve the equiripple property of the minimax method, although it does not have to be that way. The parameter L is the single design parameter for the proposed scheme. Choosing $L = 1$ makes the new scheme similar to an equiripple-passband WLS design, whereas choosing L to be as large as possible (i.e., making $\omega_L' = \pi$), we return to the original minimax algorithms.

The computational complexity of WLS-based algorithms, like the algorithms here described, is of the order of $(M+1)^3$, where $M+1$ is the length of the filter. This complexity can be greatly reduced by taking advantage of the Toeplitz-plus-Hankel internal structure of the matrix $(\boldsymbol{U}^T \boldsymbol{W}^2 \boldsymbol{U})$ in equation 2.60 and by using an efficient grid scheme to minimize the number of frequency values. These simplifications make the computational complexity of WLS-based algorithms comparable to the classic minimax approach (Antoniou, 1993). The WLS-based methods, however, do have the additional advantage of being easily coded into computer routines.

Example

Design FIR using some WLS-Chebyshev filters by satisfying the following specifications: $A_p = 0.02$ dB, $A_r = 32$ dB, $\omega_p = 2\pi \times 3400$ rad/sec, $\omega_r = 2\pi \times 4600$ rad/sec, $\omega_s = 2\pi \times 20{,}000$ rad/sec. These are specifications established by industry standards for prefiltering speech signals before these signals are subsampled to 8kHz.

Solution

This example applied to the functions seen in Figure 2.19 is depicted in Figure 2.20. The results are for the least-squares and minimax approaches and for the WLS-Chebyshev filter where ω_L' was chosen as the fourth extreme in the filter's stopband. The order of all filters is 40. Compared with the filters designed by the window methods, the WLS, WLS-Chebyshev, and the minimax filters are much more economical. The reader should note that the passband ripple specified here is 10 times smaller than the value used in the window example.

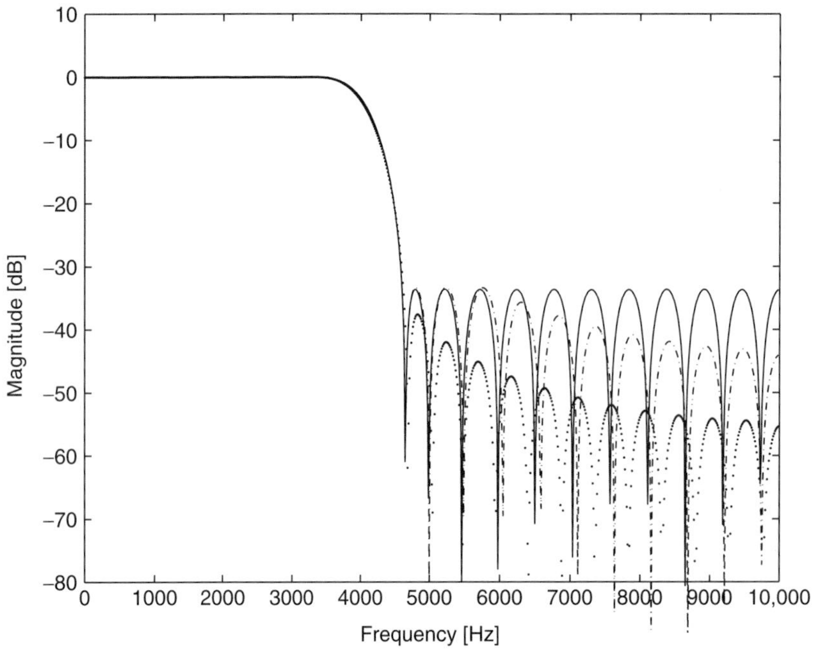

FIGURE 2.20 Frequency Responses of the WLS, WLS (Dotted Line), WLS-Chebyshev (Dashes and Dotted Line), and the Minimax (Solid Line) Filters

2.10 IIR Filter Approximations

IIR filters are usually designed by using analog filter approximations that are easy to apply and are well established. IIR filters can also be designed by using optimization methods. It is known, however, that simultaneous approximation of magnitude and phase is a difficult problem, and no reliable iterative design methods are available. The use of optimization methods in IIR filter design is more common in equalization problems when specifications for both magnitude and phase are given.

2.10.1 Analog Filter Transformation Methods

There are two popular methods for designing digital filters starting from analog approximations: the **impulse invariance method** and the **bilinear transformation method**. These methods are discussed in the following sections.

Impulse Invariance Method

In the impulse invariance method, a digital filter is designed so that its impulse response is close to the sampled analog impulse response. This method is useful for low-pass and some band-pass filter approximations. In general:

$$h_d(n) = T h_a(nT), \tag{2.68}$$

where $h_d(n)$ is the digital filter impulse response and $h_a(t)$ is the analog filter impulse response with desired specifications. The above equation implies:

$$H_d(z) = \sum_{k=1}^{N} T \frac{r_k z}{z - e^{p_k T}}, \tag{2.69}$$

where r_k and p_k are the zeros and poles of the analog transfer function approximation.

Bilinear Transformation Method

The **bilinear transformation method** is based on applying the transformation $s = 2(z-1)/T(z+1)$ to the designed analog transfer function, so that:

$$H_d(z) = H_a(s)|_{s=\frac{2z-1}{Tz+1}}. \tag{2.70}$$

As a consequence, the bilinear transformation maps the analog frequencies into the digital frequencies as follows:

$$\omega_a = \frac{2}{T} \tan \frac{\omega T}{2}. \tag{2.71}$$

Therefore, to design a digital filter with predefined specifications, it is necessary to first design an analog transfer function with adjusted specifications (prewarped) according to the above equation.

Figure 2.21(A) shows how the frequencies in the digital filter are mapped into the designed analog filter frequencies. Figure 2.21(B) shows how the frequency response is affected by the bilinear transformation frequency mapping.

The prewarped specifications of the low-pass analog filter prototype are given by:

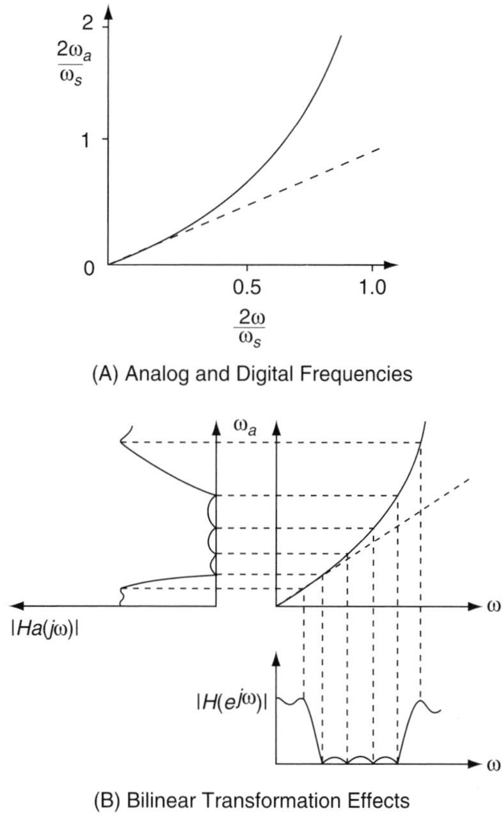

FIGURE 2.21 Bilinear Transformation Warping Effects

$$\omega_{a_p} = \frac{2}{T} \tan \frac{\omega_p T}{2}. \quad (2.72)$$

$$\omega_{a_r} = \frac{2}{T} \tan \frac{\omega_r T}{2}. \quad (2.73)$$

We can apply prewarping to as many frequencies of interest as desired. These frequencies are given by ω_i for $i = 1, 2, \ldots, n$, such that the analog filter specifications are written as:

$$\omega_{a_i} = \frac{2}{T} \tan \frac{\omega_i T}{2} \quad \text{for } i = 1, 2, \ldots, n. \quad (2.74)$$

Design Procedure

- Prewarp the prescribed frequency specifications ω_i to obtain ω_{a_i}.
- Generate $H_a(s)$, satisfying the specifications at the frequencies ω_{a_i}.
- Generate $H_d(s)$ by replacing s by $2(z-1)/T(z+1)$ at $H_a(s)$.

2.10.2 Bilinear Transformation by Pascal Matrix

Consider an analog transfer function $H(s)$ shown below:

$$H(s) = \frac{\sum_{l=0}^{N} \hat{b}_{N-l} s^l}{\sum_{i=0}^{N} \hat{a}_{N-i} s^i}. \quad (2.75)$$

When the bilinear transformation in equation 2.70 is applied to the transfer function shown above, the resulting transfer function in the z-domain is shown below:

$$H(z) = \frac{\sum_{l=0}^{N} b_{N-l} z^l}{\sum_{i=0}^{N} a_{N-i} z^i}. \quad (2.76)$$

It can be shown (Psenicka *et al.*, 2002) that it is possible to create a transformation that maps the coefficients $\hat{a}_i$ into a_i and $\hat{b}_l$ into b_l as shown below.

$$\boldsymbol{a} = \boldsymbol{P}_N \times \hat{\boldsymbol{a}}. \quad (2.77)$$

$$\boldsymbol{b} = \boldsymbol{P}_N \times \hat{\boldsymbol{b}}. \quad (2.78)$$

The $\boldsymbol{a}$ and $\boldsymbol{b}$ are vectors for the coefficients a_i and b_l defined by:

$$\begin{aligned} \boldsymbol{a} &= [a_N \; a_{N-1} \; a_{N-2} \ldots a_0]^T. \\ \boldsymbol{b} &= [b_N \; b_{N-1} \; b_{N-2} \ldots b_0]^T. \end{aligned} \quad (2.79)$$

The vectors $\hat{\boldsymbol{a}}$ and $\hat{\boldsymbol{b}}$ contain, respectively, the coefficients $\hat{a}_i$ and $\hat{b}_l$ as follows:

$$\hat{\boldsymbol{a}} = \left[\hat{a}_N \; \hat{a}_{N-1} \left[\frac{2}{T}\right] \; \hat{a}_{N-2} \left[\frac{2}{T}\right]^2 \ldots \hat{a}_0 \left[\frac{2}{T}\right]^N \right]^T. \quad (2.80)$$

$$\hat{\boldsymbol{b}} = \left[\hat{b}_N \; \hat{b}_{N-1} \left[\frac{2}{T}\right] \; \hat{b}_{N-2} \left[\frac{2}{T}\right]^2 \ldots \hat{b}_0 \left[\frac{2}{T}\right]^N \right]^T. \quad (2.81)$$

It can be shown that $\boldsymbol{P}_N$ is a Pascal matrix with the following properties:

- All elements in the first row must be ones.
- The elements of the last column can be computed by:

$$P_{i,N+1} = (-1)^{i-1} \frac{N!}{(N-i+1)!(i-1)!}, \quad (2.82)$$

where $i = 1, 2, \ldots, N+1$.

- The remaining elements can be calculated by:

$$P_{i,j} = P_{i-1,j} + P_{i-1,j+1} + P_{i,j+1}, \quad (2.83)$$

where $i = 2, 3, \ldots, N, N+1$ and $j = N, N-1, \ldots, 2, 1$.

Example

Design an elliptic filter satisfying the following specifications: $A_p = 0.02$ dB, $A_r = 32$ dB, $\omega_p = 2\pi \times 3400$ rad/sec, $\omega_r = 2\pi \times 4600$ rad/sec, and $\omega_s = 2\pi \times 20{,}000$ rad/sec

Solution

According to equations 2.72 and 2.73, it is possible to calculate ω_{a_p} and ω_{a_r} as follows:

$$\omega_{a_p} = 8.8162 \times 10^{-5}\text{rad/sec.}$$
$$\omega_{a_r} = 5.9140 \times 10^{-5}\text{rad/sec.}$$

An analog elliptic filter that meets the above specifications for ω_{a_p} and ω_{a_r} has the following transfer function:

$$H(s) = \frac{6.8799 \times 10^{-6}s^4 + 1.4569 \times 10^{-26}s^3 + 1.1041 \times 10^{-13}s^2 - 1.5498 \times 10^{-34}s + 3.9597 \times 10^{-22}}{s^5 + 8.6175 \times 10^{-5}s^4 + 8.6791 \times 10^{-9}s^3 + 4.3333 \times 10^{-13}s^2 + 1.6932 \times 10^{-17}s + 3.9597 \times 10^{-22}}.$$

In this case, the vectors $\hat{\boldsymbol{a}}$ and $\hat{\boldsymbol{b}}$ are as follows:

$$\hat{\boldsymbol{a}} = [3.9597 \times 10^{-26}, 1.5498 \times 10^{-42}, 1.1041 \times 10^{-25}, 1.4569 \times 10^{-42}, 6.8799 \times 10^{-26}, 0]^T.$$
$$\hat{\boldsymbol{b}} = 10^{-23}[0.0040\ 0.0169\ 0.0433\ 0.0868\ 0.0862\ 0.1000]^T.$$

and the matrix $\boldsymbol{P}_5$ is as written here:

$$\boldsymbol{P}_5 = \begin{bmatrix} 1 & 1 & 1 & 1 & 1 & 1 \\ 5 & 3 & 1 & -1 & -3 & -5 \\ 10 & 2 & -2 & -2 & 2 & 10 \\ 10 & -2 & -2 & 2 & 2 & -10 \\ 5 & -3 & 1 & 1 & -3 & 5 \\ 1 & -1 & 1 & -1 & 1 & -1 \end{bmatrix}.$$

The resulting transfer function $H(z)$ is as follows:

$$H(z) = 10^{-1}\frac{0.2188 + 0.1020z^{-1} + 0.3127z^{-2} + 0.3127z^{-3} + 0.1020z^{-4} + 0.2188z^{-5}}{0.3372 - 0.7314z^{-1} + 0.9856z^{-2} - 0.7350z^{-3} + 0.3406z^{-4} - 0.0703z^{-5}}.$$

The magnitude response for the transfer function above is shown in Figure 2.22.

2.11 Quantization in Digital Filters

Quantization errors in digital filters can be classified as:

- Round-off errors derived from internal signals that are quantized before or after more down additions;
- Deviations in the filter response due to finite word length representation of multiplier coefficients; and
- Errors due to representation of the input signal with a set of discrete levels.

A general, digital filter structure with quantizers before delay elements can be represented as in Figure 2.23, with the quantizers implementing rounding for the granular quantization and saturation arithmetic for the overflow nonlinearity.

The criterion to choose a digital filter structure for a given application entails evaluating known structures with respect to the effects of finite word length arithmetic and choosing the most suitable one.

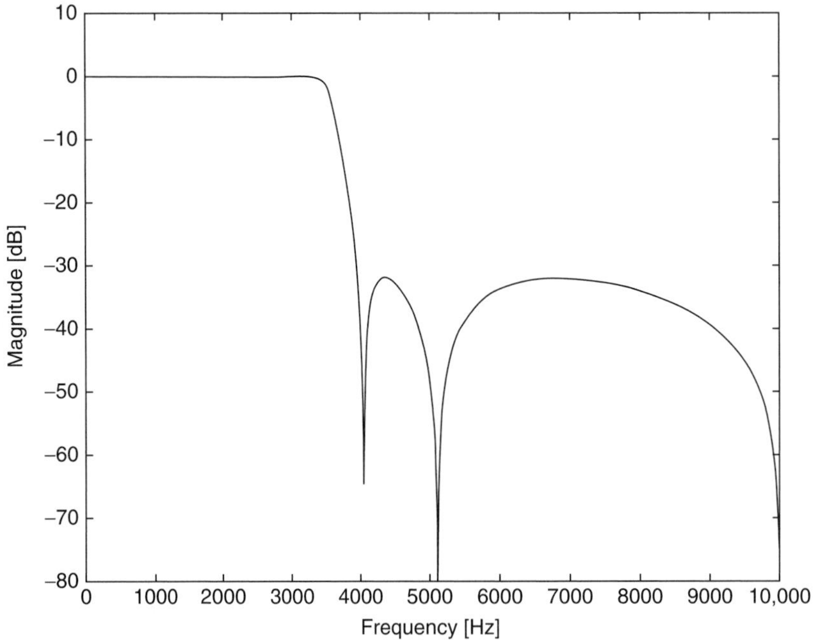

FIGURE 2.22 Magnitude Response of the Designed Low-Pass Elliptic Digital Filter

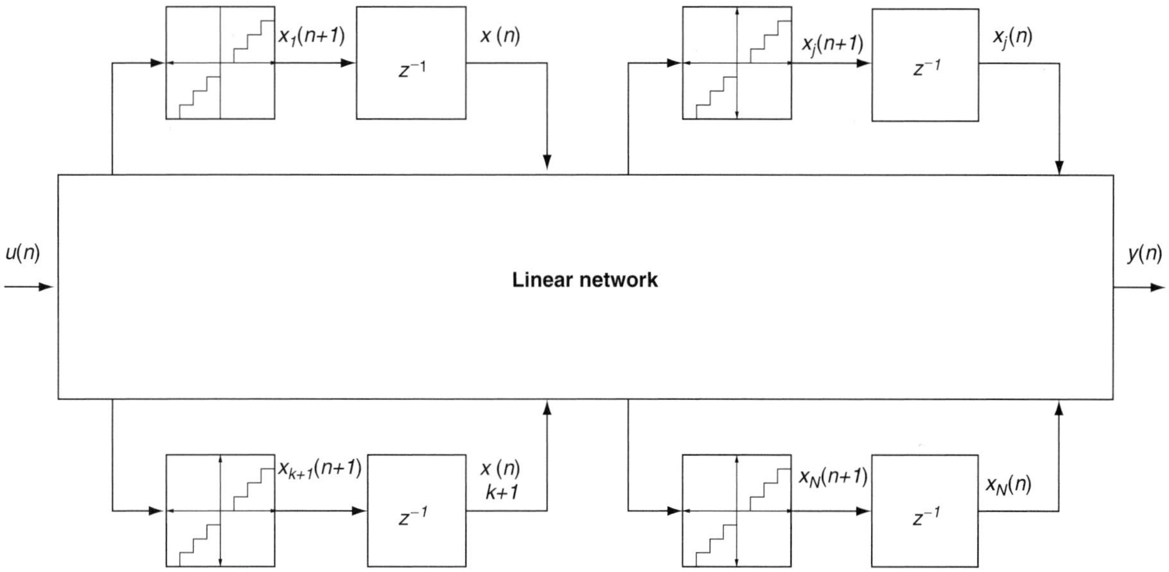

FIGURE 2.23 Digital Filter Including Quantizers at the Delay Inputs

2.11.1 Coefficient Quantization

Approximations are known to generate digital filter coefficients with high accuracy. After coefficient quantization, the frequency response of the realized digital filter will deviate from the ideal response and eventually fail to meet the prescribed specifications. Because the sensitivity of the filter response to coefficient quantization varies with the structure, the development of low-sensitivity digital filter realizations has raised significant interest (Antoniou, 1993; Diniz *et al.*, 2002).

A common procedure is to design the digital filter with infinite coefficient word length satisfying tighter specifications than required, to quantize the coefficients, and to check if the prescribed specifications are still met.

2.11.2 Quantization Noise

In fixed-point arithmetic, a number with a modulus less than one can be represented as follows:

$$x = b_0 b_1 b_2 b_3 \ldots b_b, \tag{2.84}$$

where b_0 is the sign bit and where $b_1 b_2 b_3 \ldots b_b$ represent the modulus using a binary code. For digital filtering, the most widely used binary code is the **two's complement** representation, where for positive numbers $b_0 = 0$ and for negative numbers $b_0 = 1$. The fractionary part of the number, called x_2 here, is represented as:

$$x_2 = \begin{cases} x & \text{if } b_0 = 0. \\ 2 - |x| & \text{if } b_0 = 1. \end{cases} \tag{2.85}$$

The discussion here concentrates in the fixed-point implementation.

A finite word length multiplier can be modeled in terms of an ideal multiplier followed by a single noise source $e(n)$ as shown in Figure 2.24.

For product quantization performed by rounding and for signal levels throughout the filter much larger than the quantization step $q = 2^{-b}$, it can be shown that the power spectral density of the noise source $e_i(n)$ is given by:

$$P_{e_i}(z) = \frac{q^2}{12} = \frac{2^{-2b}}{12}. \tag{2.86}$$

In this case, $e_i(n)$ represents a zero mean white noise process. We can consider that in practice, $e_i(n)$ and $e_k(n + l)$ are statistically independent for any value of n or l (for $i \neq k$). As a result, the contributions of different noise sources can be taken into consideration separately by using the principle of superposition.

The power spectral density of the output noise, in a fixed-point digital-filter implementation, is given by:

$$P_y(z) = \sigma_e^2 \sum_{i=1}^{K} G_i(z) G_i(z^{-1}), \tag{2.87}$$

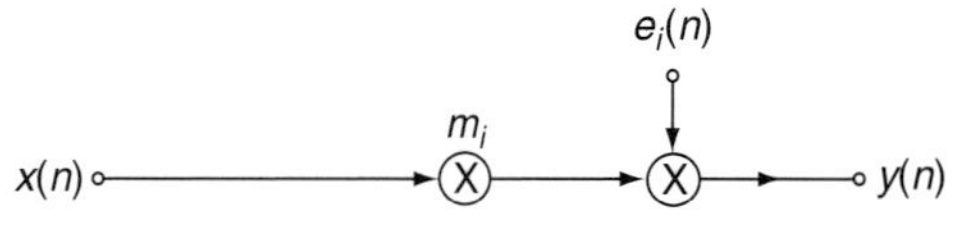

FIGURE 2.24 Model for the Noise Generated after a Multiplication

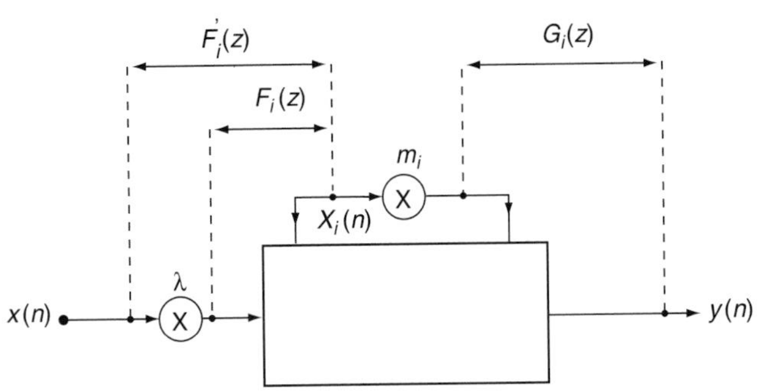

FIGURE 2.25 Digital Filter Including Scaling and Noise Transfer Functions.

where $P_{e_i}(e^{jw}) = \sigma_e^2$, for all i, and each $G_i(z)$ is a transfer function from multiplier output ($g_i(n)$) to the output of the filter as shown in Figure 2.25. The word length, including sign, is $b+1$ bits, and K is the number of multipliers of the filter.

2.11.3 Overflow Limit Cycles

Overflow nonlinearities influence the most significant bits of the signal and cause severe distortion. An overflow can give rise to self-sustained, high-amplitude oscillations known as **overflow limit cycles**. Digital filters, which are free of zero-input limit cycles, are also free of overflow oscillations if the overflow nonlinearities are implemented with saturation arithmetic, that is, by replacing the number in overflow by a number with the same sign and with maximum magnitude that fits the available wordlength.

When there is an input signal applied to a digital filter, overflow might occur. As a result, input signal scaling is required to reduce the probability of overflow to an acceptable level. Ideally, signal scaling should be applied to ensure that the probability of overflow is the same at each internal node of the digital filter. This way, the signal-to-noise ratio is maximized in fixed-point implementations.

In two's complement arithmetic, the addition of more than two numbers will be correct independently of the order in which they are added even if overflow occurs in a partial summation as long as the overall sum is within the available range to represent the numbers. As a result, a simplified scaling technique can be used where only the multiplier inputs require scaling. To perform scaling, a multiplier is used at the input of the filter section as illustrated in Figure 2.25.

It is possible to show that the signal at the multiplier input is given by:

$$x_i(n) = \frac{1}{2\pi j}\oint_c X_i(z)z^{n-1}dz = \frac{1}{2\pi}\int_0^{2\pi} F_i(e^{j\omega})X(e^{j\omega})e^{j\omega n}d\omega, \quad (2.88)$$

where c is the convergence region common to $F_i(z)$ and $X(z)$.

The constant λ is usually calculated by using L_p norm of the transfer function from the filter input to the multiplier input $F_i(z)$, depending on the known properties of the input signal. The L_p norm of $F_i(z)$ is defined as:

$$\|F_i(e^{j\omega})\|_p = \left[\frac{1}{2\pi}\int_0^{2\pi} |F_i(e^{j\omega})|^p d\omega\right]^{\frac{1}{p}}, \quad (2.89)$$

for each $p \geq 1$, such that $\int_0^{2\pi} |F_i(e^{j\omega})|^p d\omega \leq \infty$. In general, the following inequality is valid:

$$|x_i(n)| \leq \|F_i\|_p \|X\|_q, \left(\frac{1}{p} + \frac{1}{q} = 1\right), \quad (2.90)$$

for p, $q = 1$, 2 and ∞.

The scaling guarantees that the magnitudes of multiplier inputs are bounded by a number M_{max} when $|x(n)| \leq M_{max}$. Then, to ensure that all multiplier inputs are bounded by M_{max} we must choose λ as follows:

$$\lambda = \frac{1}{Max\{\|F_1\|_p, \ldots, \|F_i\|_p, \ldots, \|F_K\|_p\}}, \quad (2.91)$$

which means that:

$$\|F_i'(e^{j\omega})\|_p \leq 1, \text{ for} \|X(e^{j\omega})\|_q \leq M_{max}. \quad (2.92)$$

The K is the number of multipliers in the filter.

The norm p is usually chosen to be infinity or 2. The L_∞ norm is used for input signals that have some dominating frequency component, whereas the L_2 norm is more suitable for a random input signal. Scaling coefficients can be implemented by simple shift operations provided they satisfy the overflow constraints.

In case of modular realizations, such as cascade or parallel realizations of digital filters, optimum scaling is accomplished by applying one scaling multiplier per section.

As an illustration, we present the equation to compute the scaling factor for the cascade realization with direct-form second-order sections:

$$\lambda_i = \frac{1}{\|\prod_{j=1}^{i-1} H_j(z)F_i(z)\|_p}, \quad (2.93)$$

where:

$$F_i(z) = \frac{1}{z^2 + m_{1i}z + m_{2i}}.$$

The noise power spectral density is computed as:

$$P_y(z) = \sigma_e^2\left[3 + \frac{3}{\lambda_1^2}\prod_{i=1}^{m} H_i(z)H_i(z^{-1}) + 5\sum_{j=2}^{m}\frac{1}{\lambda_j^2}\prod_{i=j}^{m} H_i(z)H_i(z^{-1})\right], \quad (2.94)$$

whereas the output noise variance is given by:

$$\sigma_o^2 = \sigma_e^2 \left[3 + \frac{3}{\lambda_1^2} \| \prod_{i=1}^{m} H_i(e^{j\omega}) \|_2^2 + 5 \sum_{j=2}^{m} \frac{1}{\lambda_j^2} \| \prod_{i=j}^{m} H_i(e^{j\omega}) \|_2^2 \right]. \tag{2.95}$$

As a design rule, the pairing of poles and zeros is performed as explained here: poles closer to the unit circle pair with closer zeros to themselves, such that $\|H_i(z)\|_p$ is minimized for $p = 2$ or $p = \infty$.

For ordering, we define the following:

$$P_i = \frac{\|H_i(z)\|_\infty}{\|H_i(z)\|_2}. \tag{2.96}$$

For L_2 scaling, we order the section such that P_i is decreasing. For L_∞ scaling, P_i should be increasing.

2.11.4 Granularity Limit Cycles

The quantization noise signals become highly correlated from sample to sample and from source to source when signal levels in a digital filter become constant or very low, at least for short periods of time. This correlation can cause autonomous oscillations called granularity limit cycles.

In recursive digital filters implemented with rounding, magnitude truncation,[4] and other types of quantization, limit-cycles oscillations might occur.

In many applications, the presence of limit cycles can be harmful. Therefore, it is desirable to eliminate limit cycles or to keep their amplitude bounds low.

If magnitude truncation is used to quantize particular signals in some filter structures, it can be shown that it is possible to eliminate zero-input limit cycles. As a consequence, these digital filters are free of overflow limit cycles when overflow nonlinearities, such as saturation arithmetic, are used.

In general, the referred methodology can be applied to the following class of structures:

- State-space structures: Cascade and parallel realization of second-order state-space structures includes design constraints to control nonlinear oscillations (Diniz and Antoniou, 1986).
- Wave digital filters: These filters emulate doubly terminated lossless filters and have inherent stability under linear conditions as well as in the nonlinear case where the signals are subjected to quantization (Fettweis, 1986).
- Lattice realization: Modular structures allowing easy limit cycles elimination (Gray and Markel, 1975).

[4] Truncation here refers to the magnitude of the number is reduced, which leads to the decrease of its energy.

2.12 Real-Time Implementation of Digital Filters

There are many distinct means to implement a digital filter. The detailed description of these implementation methods is beyond the scope of this chapter. The interested reader can refer to Wanhammar (1990) and Diniz (2002) and the references therein.

The most straightforward way to implement digital filters relies on general purpose computers by programming their central processing units (CPUs) to execute the operations related to a particular digital filter structure. This type of implementation is very flexible because it consists of writing software, allowing fast prototyping and testing. This solution, however, might not be acceptable in applications requiring high-processing speed, fast data input/output interfaces, or large-scale production.

Efficient software implementations of digital filters are usually based on special-purpose CPUs known as Digital Signal Processors (DSPs). These processors are capable of implementing a sum of product operations, also referred to as multiply-and-accumulate (MAC) operations, in a very efficient manner.

Another implementation alternative is to employ programmable logic devices (PLD) that include a large number of logic functions on a single chip. An advanced version of PLD is the field-programmable gate array (FPGA). An FPGA is an array of logic macro-cells that are interconnected through a number of communication channels configured in horizontal and vertical directions. The FPGAs allow the system designer to configure very complex digital logic that can implement digital signal processing tasks at low cost with reduced power consumption. Usually, high-level software tools are available for the design of digital systems using FPGAs.

Special purpose hardware is also possible for implementing a digital filter. Hardware implementations consist of designing and possibly integrating a digital circuit with logical gates to perform the basic operating blocks inherent to any digital filter structure, namely **multiplications**, **additions**, and **storage elements**. Multiplications and additions can be implemented using bit-serial or bit-parallel architectures. In general, hardware implementations of digital filters are less flexible than software (CPU-based) implementations. Special purpose hardware, however, is usually necessary when high cost of development is offset by a large production and a DSP-based solution is too expensive or is incapable of meeting sampling frequency specifications.

Consider a heuristic related to the computational complexity of digital filters in a full custom design (Dempster, 1995). Figure 2.26 shows an intuitive plot of complexity as a function of filter order for all digital filter implementations meant to satisfy a given set of specifications. The complexity of a digital filter is measured in terms of number of bits used in the

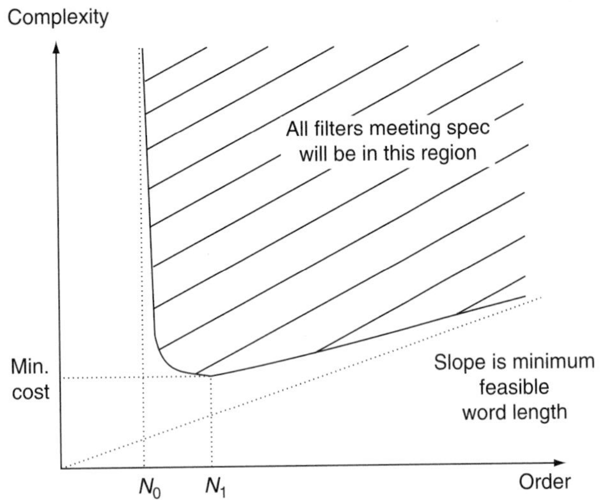

FIGURE 2.26 Complexity of a Digital Filter

coefficients and number of multiplier operations.[5] For a prescribed set of frequency specifications and a particular filter structure, a certain order N_0 may be capable of meeting the specifications with infinite precision. For filters with an order lower than N_0, it is not possible to meet the prescribed specifications. Infinite precision is translated as very high complexity because an infinite number of bits is required. If higher order is used, the number of bits necessary for multiplier coefficients may be reduced, implying lower complexity. An order N_1 in Figure 2.26 is the order of the implementation with the lowest possible complexity. Orders higher than N_1 will necessarily imply higher complexity. The heuristic introduced here, when applied to the case of FIR filter designs using direct-form realizations, leads to complexity curves that are very flat around their minimum point, indicating that almost minimum complexity can be achieved for a wide range of values for the filter order. On the other hand, for IIR filters, the complexity curve is rather sharp, implying that a more careful choice for the filter order should be made.

2.13 Conclusion

In this chapter the main steps for the design and implementation of linear and time-invariant digital filters are introduced. These filters are very important building blocks for signal processing systems implemented in discrete-time domain. In particular, the widely available digital technology allows the implementation of very fast and sophisticated filters in a cheap and reliable manner. As a result, the digital filers are found in numerous commercial products such as audio systems, biomedical equipment, digital radio, and TV just to mention a few.

References

Antoniou, A. (1993). *Digital filters: Analysis, design, and applications.* (2d ed). New York: McGraw-Hill.

Dempster, A. (1995). *Digital filter design for low-complexity implementation.* Ph.D. Thesis. University of Cambridge, Cambridge, UK.

Diniz, P.S.R. (2002). *Adaptive filtering: Algorithms and practical implementation.* (2d Ed.). Boston, MA: Kluwer Academic.

Diniz, P.S.R., and Antoniou, A. (1986). More economical state-space digital filter structures which are free of constant-input limit cycles. *IEEE Transactions on Acoustic Speech Signal Processing ASSP-34,* 807–815.

Diniz, P.S.R., da Silva, E.A.B., and Netto, S.L. (2002). *Digital signal processing: System analysis and design.* Cambridge, UK: Cambridge University.

Diniz, P.S.R., and Netto, S.L. (1999). On WLS-Chebyshev FIR digital filters. *Journal of Circuits, Systems, and Computers 9,* 155–168.

Fettweis, A. (1986). Wave digital filters: Theory and practice. *Proceedings of the IEEE 74,* 270–327.

Gray, A.H. Jr., and Markel, J.D. (1975). A normalized digital filter structure. *IEEE Transactions on Acoustic Speech Signal Processing ASSP-23,* 268–277.

Jackson, L.B. (1996). *Digital filters and signal processing.* (3rd ed.). Boston, MA: Kluwer Academic.

Lim, Y.C., Lee, J.H., Chen, C.K., and Yang, R.H. (1992). A weight least-squares algorithm for quasi-equiripple FIR and IIR digital filter design. *IEEE Transactions on Signal Processing 40,* 551–558.

Mitra, S.K., (2001). *Digital signal processing: A computer-based approach.* (2d ed). New York: McGraw-Hill.

Oppenheim, A.V., and Schafer, R.W. (1989). Schafer, *Discrete-time signal processing.* Englewood Cliffs, NJ: Prentice Hall.

Pšenička, B., Garcia-Ugalde, F., and Herrera-Camacho, A. (2002). The bilinear Z transform by Pascal matrix and its application in the design of digital filters. *IEEE Signal Processing Letters 9,* 368–370.

Sarämaki, T. (1992). Finite-impulse response filter design. In S.K. Mitra and J.F. Kaiser, (Eds.) *Handbook of digital signal processing.* New York: John Wiley & Sons.

Wanhamman, L. (1999). *DSP integrated circuits.* New York: Academic Press.

[5] For fixed-point implementation, computational complexity is mainly determined by the number of adders used in the implementation of multipliers.

3

Methods, Models, and Algorithms for Modern Speech Processing

John R. Deller, Jr.* and
John Hansen
Department of Electrical and Computer Engineering, Michigan State University, East Lansing, Michigan, USA

3.1 Introduction

In this chapter, we address digital speech processing, one of the most important applications of modern signal processing (SP). Speech technologies have become fundamental to the development of many products and services that influence the way we live and work. The enormous commercial potential for speech-based products has been a major impetus for research and development of two broad application areas: **speech communications** and **speech recognition**. Public awareness of speech technologies is based largely on direct experience with speech recognition (and related) products and through marketing of communications services with novel speech features. Internet-based telephony and other audio services have also brought speech processing into the homes and offices of consumers. Underlying the more recent commercial manifestations of speech technology are several decades' worth of ongoing research into fundamental aspects of speech processing at government, military, industry, and university laboratories. These efforts address a broad interdisciplinary span of issues, ranging from basic physiology and speech science, to acoustic modeling and processing algorithms, to higher level language concepts probing questions of intelligence and meaning.

The purpose of this chapter is to present an overview of some salient methods and tools of modern speech processing with the aim of familiarizing the reader with the key vocabulary and main concepts and the most important models and algorithms. In many instances, we will necessarily direct the reader to literature sources for a deeper study of topics. Several current and comprehensive textbooks on speech processing are available (Deller *et al.*, 2000; Rabiner and Juang, 1993), and these texts contain extensive references lists to original citations.[1] The most current developments in the field are often found in conference proceedings, notably in *Proceedings of the IEEE International Conference on Acoustic, Speech, and Signal Processing*, *Proceedings of the International Conference on Spoken Language Processing*, and *Proceedings of Eurospeech*. Many relevant background topics are treated elsewhere in this book.

* This work was supported in part by the National Science Foundation (NSF) under Cooperative Agreement No. IBIS-9817485. Opinions, findings, or recommendations expressed are those of the author and do not necessarily reflect the views of the NSF.

[1] Space constraints, rather than any lack of deference to the original researchers, make it impossible to cite all relevant original research papers and to give detailed historical accounts.

3.2 Modeling Speech Production

3.2.1 Essential Interdisciplinary Aspects of Speech Production

Basic Anatomy and Physiology

To fully appreciate the engineering systems used to model speech, it is necessary to have a basic understanding of the living system that produces it. We first give a very brief overview of the anatomical structures that comprise the human **speech production system**. Then, with the aid of some primitive processing of a speech waveform, we discuss the physiology of speech production. The principal aim of this cursory treatment of speech "biology" is to provide justification for the discrete-time (DT) model used pervasively in contemporary speech processing.

Figure 3.1 portrays the speech system anatomy midway through the human upper torso as viewed from the left. The gross components of the system are the **lungs**, **trachea** ("windpipe"), **larynx** (organ of voice production), **pharyngeal cavity** (throat), **oral** (or **buccal**) **cavity** (mouth), and **nasal cavity** (nose). In technical discussions, the pharyngeal and oral cavities are grouped into one unit referred to as the **vocal tract** and the nasal cavity is often called the **nasal tract**[2]. Accordingly, the vocal tract begins at the output of the larynx and terminates at the input to the lips. The nasal tract begins at the velum and ends at the nostrils of the nose. Finer anatomical features that are critical to speech production include the **vocal folds** or **vocal cords**, **soft palate** or **velum**, **tongue**, **teeth** and **lips**. The **uvula** is the soft tip of the velum that hangs down in the back of the oral cavity. The **glottis** is the name given to the opening between the vocal folds, and the adjective *glottal* is frequently used to refer to laryngeal or vocal phenomena (e.g., "glottal waveform"). These finer anatomical components move to different positions to produce various speech sounds and are known as **articulators** by speech scientists. The **mandible** (jaw) is also considered to be an articulator because it is responsible for both gross and fine movements that affect the size and shape of the vocal tract as well as the positions of the other articulators.

Speech-processing engineers view speech production in terms of an acoustic filtering operation. Let us associate the anatomy with such a technical model. The three main cavities of the speech production system (vocal plus nasal tracts) compose the main acoustic filter. Much of the information in speech is encoded in the spectral peaks associated with given sounds. The size and shape of the vocal tract (pharyngeal and oral cavities) at a given instant are the principal determiners of the resonant structures in the short-term speech spectrum. In particular, the resonances associated with a speech sound are a consequence of the articulators having formed various acoustical cavities and subcavities out of the vocal tract cavities, much like concatenating different lengths of organ pipe in various orders. Thus, the center frequencies of these resonances depend on the shape and physical dimensions of the vocal tract. Conversely, each vocal-tract shape is characterized by a set of resonant frequencies. From a modeling point of view, the vocal-tract articulators determine the important "pole" properties of the speech system. Since these resonances tend to "form" (impose an envelope upon) the spectrum, speech scientists refer to them as **formants**. This term is often used to refer to the nominal center frequencies of the resonances, so formant may be used interchangeably with **formant frequency**. The formants in the spectrum are denoted[3] as $F_1, F_2, F_3, \ldots, F_N$, beginning with the lowest frequency. In practice, there are usually three to five discernable formants in the Nyquist band for typical sampling rates (8–16 kHz).

Spectral "valleys" (antiresonances) in speech spectra are not nearly as prevalent or prominent (i.e., they tend to be of wider bandwidth) as spectral peaks (resonances). Nevertheless, there are speech sounds for which zeros in the production model seem warranted both empirically (by spectral analysis) and physically (by analysis of the anatomy and physiology). The degree to which the nasal cavity is coupled to the vocal tract (as a "side cavity") determines, in part, the antiresonance structure. Generally, any component of the vocal and/or nasal tract systems that creates a cavity with the potential to "trap" energy (which is nominally on a direct path from "lungs to lips") will manifest some degree of antiresonant spectral character. The spectral dip will appear in the frequency region of the resonance of the trapping cavity.

The articulators are used to change the properties of the acoustic filter over time. The filter is loaded at its main output by a radiation impedance ("flow-to-pressure converter") due to the lips. The vocal tract (and/or nasal tract, when appropriate) is excited in two principal ways in most natural languages.[4] The main function of the larynx in speech is to provide a periodic flow waveform (called the **voice** or **phonation**) as input to the acoustic system in the production of **voiced** speech sounds. The voice waveform is a periodic lowpass pulse train whose fundamental frequency, usually denoted F_0, reflects the frequency of glottal vibration.[5] Voiced speech

[2] The term **vocal tract** is often used in imprecise ways by engineers. Sometimes it is used to refer to the combination of all three cavities and even more often to the entire speech production system. We will be careful in this chapter not to use vocal tract when we mean speech production system, but, as is conventionally done in the literature, we will sometimes use the term vocal tract when we actually mean vocal tract *plus* nasal tract. Whether vocal tract implicitly includes the nasal tract will be clear from context, or explicitly stated.

[3] Uppercase F and Ω are used throughout to denote real frequencies (in hertz and radians per second, respectively), while lowercase f and ω are used to represent "normalized" frequencies (dimensionless and in radians, respectively); that is, frequencies normalized to the sample frequency are $f = F/F_s$ and $\omega = 2\pi/F_s$.

[4] The term **natural language** refers to any of the world's human languages. This will be contrasted with a technical meaning of the term **formal language** in the material on language modeling in Section 3.4.2 and beyond.

[5] The improper use of the term *voice* to mean *speech* seems irrevocably entrenched in the technical lexicon (e.g., voice channel) just as it is in common use (e.g., a singer with a "beautiful voice").

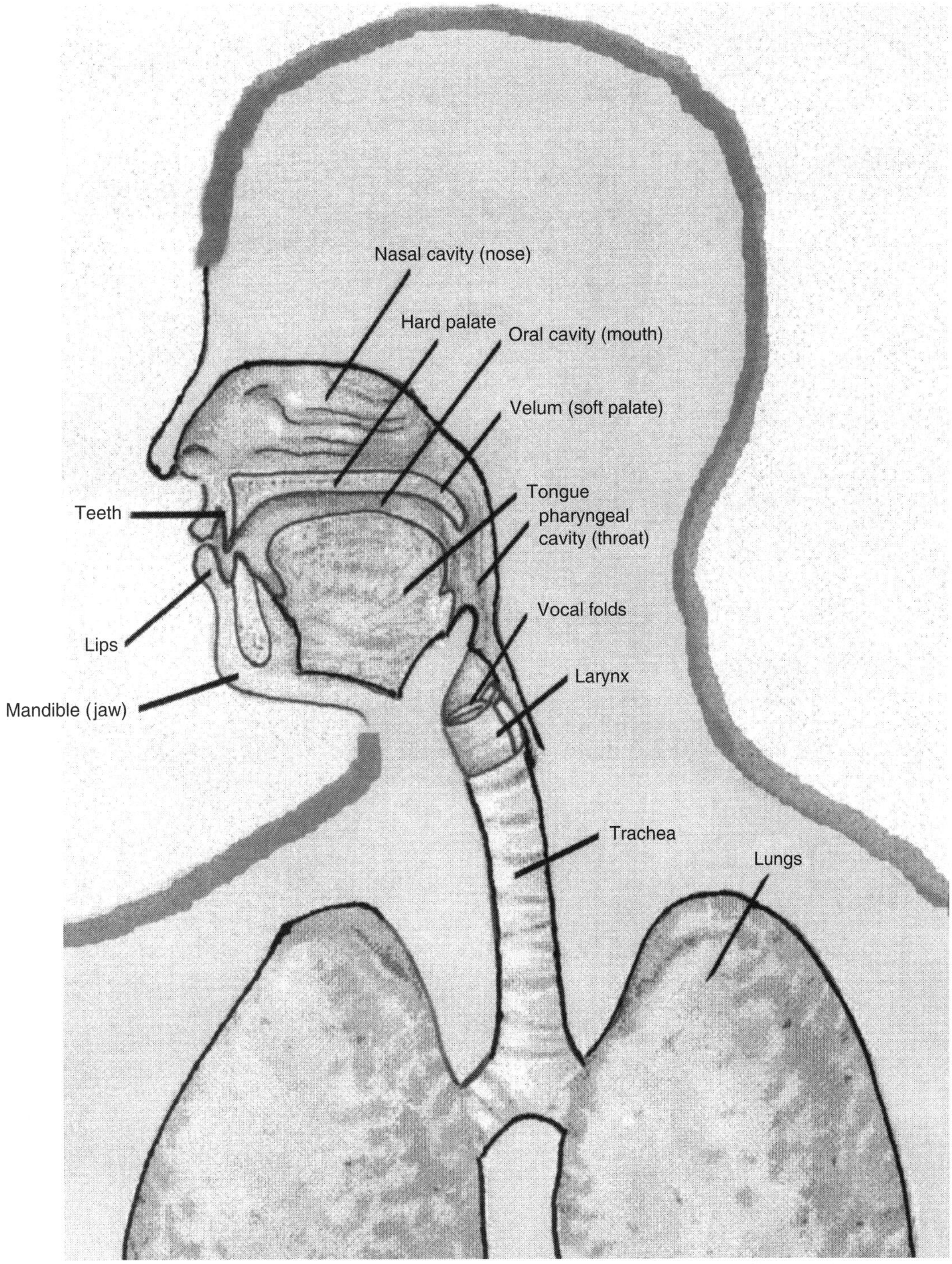

FIGURE 3.1 Gross Anatomy of the Human Speech Production System

waveforms are short-term quasiperiodic, with the period determined by the frequency of vocal-fold vibration—the **pitch** frequency.[6] Accordingly, voice speech is modeled as a deterministic process. **Unvoiced** speech sounds are produced by exciting the vocal tract by turbulent airflow at some constriction in the vocal tract. The excitation source is often modeled

[6] **Pitch** is actually a psychoacoustic rather than a waveform-domain concept. Pitch is precisely defined as the *perceived* fundamental frequency of an audible stimulus, whether or not that fundamental frequency is actually present in the acoustic waveform. In a telephone conversation, each conversant perceives the fundamental voicing frequency of the other even though that frequency is ordinarily not in the telephone channel bandwidth.

as white noise. Unvoiced waveforms are usually lower in energy than voiced utterances, and their noisy character necessitates modeling them as stochastic processes. In addition, some sounds are created using **mixed** voiced and unvoiced excitations. Examples of voiced and unvoiced sounds are given below. The sound made by the letter *z* in *zoo* (denoted /*z*/ by phoneticians) is the result of mixed excitation.

Some of the differences between voiced and unvoiced speech segments are evident in the time waveform for the utterance of the letter *X* in Figure 3.2. The leading voiced sound *eh* (written /*E*/ by phoneticians—see discussion concerning Table 3.1) is higher in amplitude and contains strong periodic structure. The final unvoiced sound /*s*/ is low in amplitude and noise-like. Figure 3.3 shows corresponding magnitude spectra of the steady-state portions of /*E*/ and /*s*/ in Figure 3.2. In Figure 3.3, we note that in each spectrum (and especially for the vowel sound), there are well-defined frequency ranges of emphasis (resonances) and de-emphasis (antiresonances).

Other forms of input to the vocal tract include **plosive** excitation and **whisper**. Because there are pauses[7] in speech, we also include **silence** as a form of "excitation" for modeling completeness. Plosive sounds are formed by making a complete closure (normally toward the front of the vocal tract), building air pressure behind the closure, and suddenly releasing it. The word *pat* begins with the **voiced plosive** /*p*/ (silence, then voiced), and ends with the **unvoiced plosive** /*t*/ (silence, then unvoiced). The whisper excitation is formed by forcing air through a partially constricted glottis.

One of the main challenges in speech processing is that the dynamics of the speech system are constantly changing. This is evident in Figure 3.2. A conservative rule of thumb is that speech frames remain quasistationary for intervals of 10 to 20 ms. Moreover, speech is not a string of discrete, well-formed sounds but rather a series of steady-state or target sounds (sometimes quite brief, if achieved at all) with intermediate transitions. The preceding and/or succeeding sound in a string can affect whether a target is reached, how long it is held, and other finer acoustic details. This interplay among sounds in an utterance is called **coarticulation**. Changes witnessed in the speech waveform are a direct consequence of movements of the speech system articulators that rarely remain fixed for any sustained period of time. On the other hand, the articulators have mass and cannot be moved instantaneously from one target to the next without a period of transition.

Since the vocal-tract shape is varied as a function of time to produce desired speech sounds, so must the spectral properties of the speech signal be time-varying. In contemporary SP, it is possible to exhibit these changes using a three-dimensional plot of magnitude spectra over time (sometimes called a "waterfall plot") This information was historically gleaned from the **speech spectrograph** (Flanagan, 1972; Koenig *et al.*, 1946; Potter *et al.*, 1966).

[7] In addition to "long" pauses that can occur, for example, as a person stops to think, there are also very short regions of naturally occurring silence in continuous speech. Read on, for example, to the discussion of plosives.

Phonetics and Linguistics

Phonemes and Phones

Speech processing engineers rely heavily on research into the complex sounds of speech (phonetics) and into the characterization of the higher-level "rules" by which those sounds are combined to transmit information (linguistics).

The basic theoretical unit for describing how speech conveys linguistic meaning is called a **phoneme**. For American English, there are about 42 phonemes which are made up of **vowels**, **semivowels**, **diphthongs**, and **consonants**, including **nasals**, **stops**, **fricatives**, and **affricates**. Each phoneme can be considered to be a code that consists of a unique set of **articulatory gestures**. These articulatory gestures include the type and location of sound excitation as well as the position or movement of the vocal-tract articulators. We can view a phoneme as an ideal sound unit with a set of corresponding articulatory gestures. If speakers could exactly and consistently produce (in the case of English) these 42 phonemes, then speech would amount to a stream of discrete codes. Many different factors, including accents, gender, and coarticulatory effects, ensure that a given phoneme will have a variety of acoustic manifestations in the course of flowing speech. From an acoustical point of view, therefore, the phoneme represents a **class** of sounds whose elements convey the same meaning. Accordingly, the phonemes of a language compose a minimal theoretical set of distinct units which are sufficient to convey all meaning in the language. This is to be juxtaposed with the actual *sounds* that can be produced which speech scientists call **phones**. Associated with each phoneme is a collection of **allophones** (phone variations) that represent slight acoustic variations of the basic unit.

The study of the abstract units and their relationships in a language is called **phonemics**, while the study of the actual sounds of the language is called **phonetics**. Moreover, each of three branches of phonetics approaches the subject somewhat differently. **Articulatory phonetics** is concerned with the manner in which speech sounds are produced by the articulators of the vocal system; **acoustic phonetics** research explores the sounds of speech through analysis of acoustic waveforms; and, **auditory phonetics** research focuses on perceptual responses to speech sounds as reflected in listener trials. Our cursory explorations into phonetics below will represent a blend of articulatory and acoustic analysis.

A string of symbols representing the phonemes is called **phonemic transcription**. When the transcription also includes **diacritical marks** on the phonemic symbols that indicate allophonic variations, then the result is a **phonetic transcription**. Although the term phonetic transcription is commonly used, speech processing engineers tend to think about speech more at the level of phonemic transcriptions even though allophonic variations are inherently modeled in many systems.

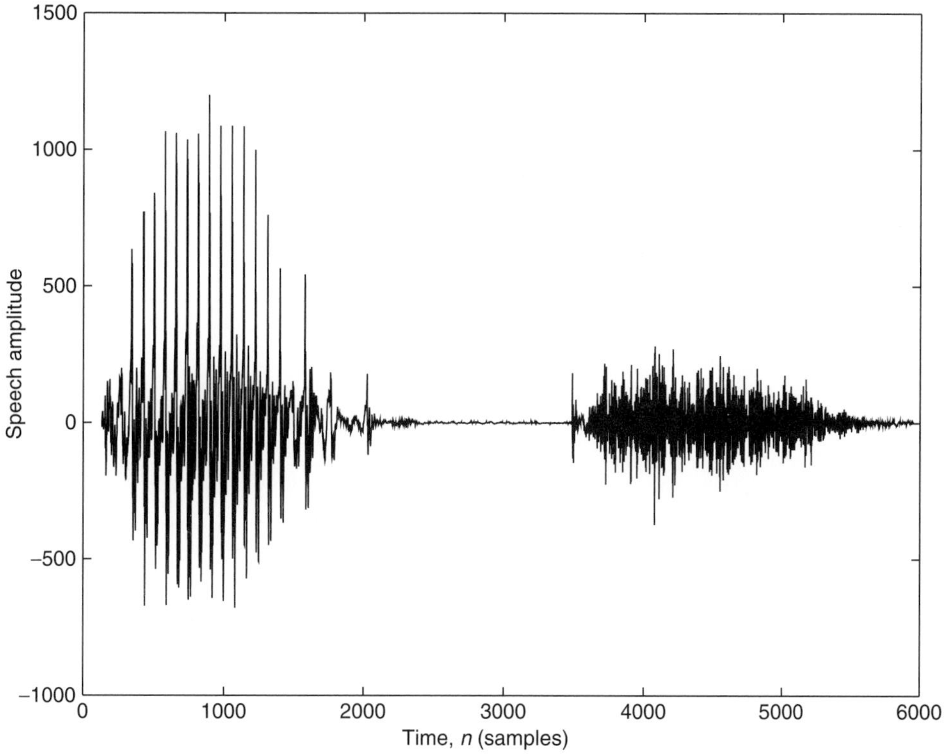

(A) Utterance of Letter X by Male Speaker

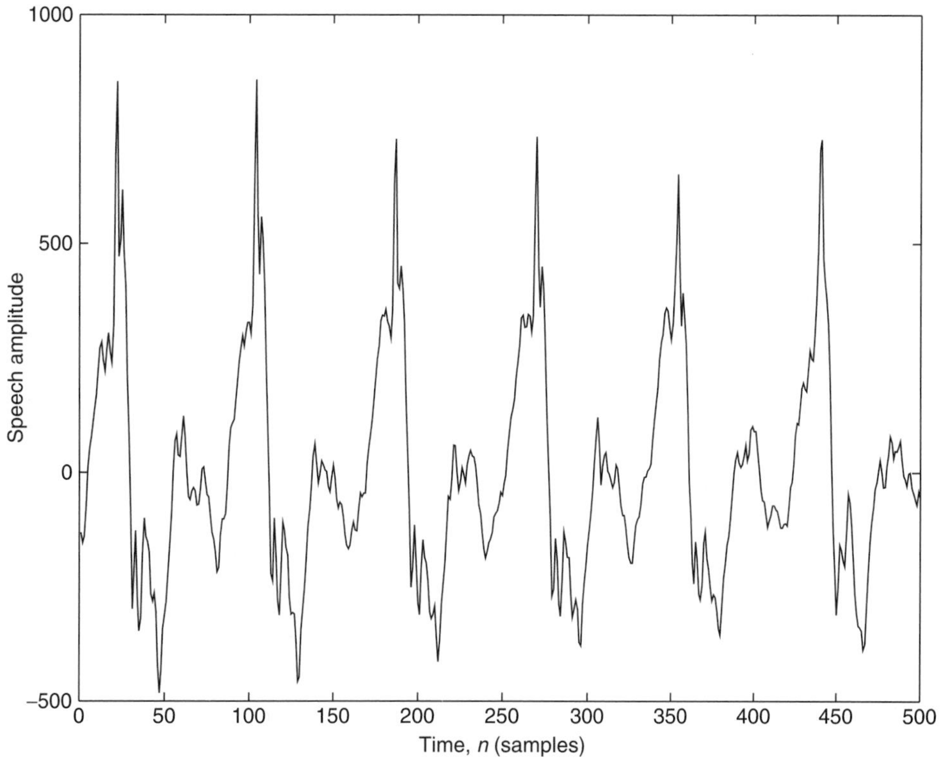

(B) Blowup of Vowel Sound /E/ in Utterance Shown in (A)

FIGURE 3.2 Differences Between Voiced and Unvoiced Speech Segments

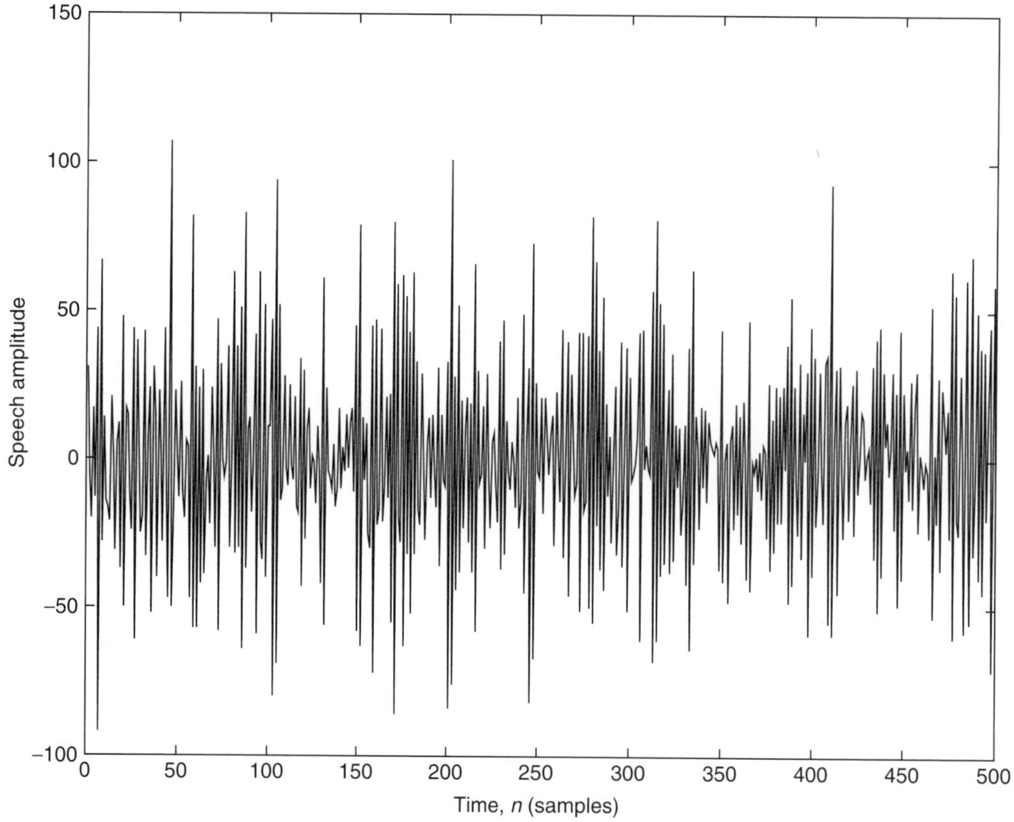

(C) Blowup of Fricative Portion of Fricative /s/ in Utterance of (A)

FIGURE 3.2 Difference Between Voiced and Unvoiced Speech Segments (*Continued*)

Three "phonetic alphabets" are widely used; the two noted in this paragraph are preferred by most speech engineers. The **International Phonetic Alphabet** (IPA) was developed by European phoneticians in the late 19th century, and it remains the standard for basic science writing on phonetics. The complete IPA has entries covering phonemes in all the world's languages (Ladefoged, 1975). To facilitate keyboard entry of phonetic transcriptions, a more recent phonetic alphabet was developed under the auspices of the United States Defense Advanced Research Projects Agency (DARPA) and is accordingly called the **ARPAbet**. There are two versions of the ARPAbet, one that uses single letter symbols and one that uses all uppercase symbols. The use of all uppercase necessitates some double letter designators. When transcription is necessary in this discussion, we will employ the single-letter ARPAbet symbols. In general, phonemic or phonetic transcriptions are shown between two forward slashes as we did when denoting the phonemes /*E*/ and /*s*/ for Figures 3.2 and 3.3.

Classification of Phonemes

There are many methods for classifying phonemes. Phonemes can be categorized based on properties related to the time waveform, frequency characteristics, manner of articulation, place of articulation, type of excitation, and the stationarity of the phoneme. We mention only a few salient facts about the various of phoneme classes, leaving detailed study to the reader's further pursuit (Deller *et al.*, 2000; Ladefoged, 1975).

A phoneme is **continuant** if the speech sound is produced by a steady-state vocal-tract configuration and **noncontinuant** if a change in the vocal-tract configuration is required. The phonemes in Table 3.1 are classified based on the continuant/noncontinuant property. Noncontinuant phonemes are generally more difficult to characterize and model.

Vowels are voiced speech sounds involving a constant vocal-tract shape. There are twelve principle vowels in American English. A thirteenth vowel, called a **schwa**, is a sort of "degenerate vowel" to which many others gravitate when articulated hastily in the course of flowing speech. The initial vowel in "ahead" is a schwa vowel. All vowels are voiced and are normally among the phonemes of largest amplitude. Vowels can vary widely in duration (typically from 40 to 400 ms). From an articulatory point-of-view, vowels are differentiated by the tongue-hump position and the degree of constriction at that position. The tongue position is largely responsible for the variation of cross-sectional area along the vocal tract which, in turn, determines the formants of the vowel. From an acoustic point-of-view, therefore, the sets of formant frequencies are fairly reliable discriminators of the various vowels.

TABLE 3.1 The "ARPAbet" for Phonemes Used in American English

Single-symbol version	Uppercase version	Example	Single-symbol version	Uppercase version	Example
I	IY	heed	v	V	vice
I	IH	hid	T	TH	thing
e	EY	made	D	DH	then
E	EH	head	s	S	see
@	AE	had	z	Z	zoo
a	AA	odd	S	SH	shy
c	AO	thaw	Z	ZH	measure
o	OW	hoed	h	HH	hope
U	UH	hood	m	M	mom
u	UW	who'd	n	N	nine
R	ER	heard	G	NX	sing
x	AX	ago	l	L	love
A	AH	mud	L	EL	cattle†
Y	AY	hide	M	EM	sum†
W	AW	how'd	N	EN	son†
O	OY	boy	F	DX	batter‡
X	IX	roses	Q	Q	§
p	P	pea	w	W	want
b	B	bat	y	Y	yard
t	T	tea	r	R	ride
d	D	deed	C	CH	church
k	K	kick	J	JH	just
g	G	go	H	WH	when
f	F	five			

† *vocalic l,m,n* ‡ *flapped t* § *glottal stop*

There are two further categories of "vowel-like" phonemes. A **diphthong** involves an intentional movement from one vowel toward another vowel. The three diphthongs that are universally accepted are /*Y*/, /*W*/, and /*O*/ (corresponding examples occur in *pie*, *out*, and *toy*). The other group of vowel-like phonemes consists of /*w*/, /*l*/, /*r*/, and /*y*/, which are called **semivowels**. Semivowels are classified as either **liquids** (/*w*/ in *wet* and /*l*/ in *lawn*) or **glides** (/*r*/ in *ran* and /*y*/ in *yam*).

The second broad class of phonemes, the **consonants**, consists of speech sounds that are produced with a constricted vocal tract. Some consonants require precise dynamic movement of the articulators for their production. Other consonants, however, may not require articulator motion, so their sounds, like vowels, are sustained (hence, continuants). Consonants are subclassified as fricatives, affricates, stops, and nasals.

The **fricatives** are produced by exciting the vocal tract with a steady airstream that becomes turbulent at some point of constriction. The constriction in the vocal tract or glottis results in an unvoiced excitation. However, some fricatives also have a simultaneous voicing component (hence, mixed excitation). Those with simple unvoiced excitation are usually called **unvoiced fricatives** whereas those of mixed excitation are called **voiced fricatives**. In the unvoiced case, the constriction causes a noise source anterior to the constriction. The location of the

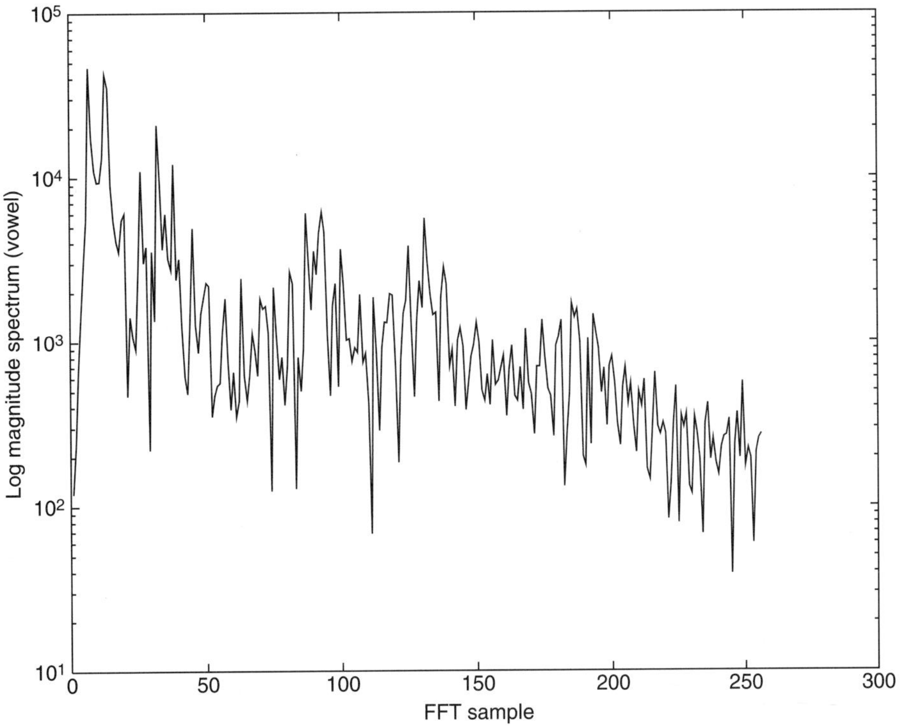

(A) Magnitude Spectrum of Vowel /*E*/ Portion of Utterance *X*

FIGURE 3.3 Corresponding Magnitude Spectra

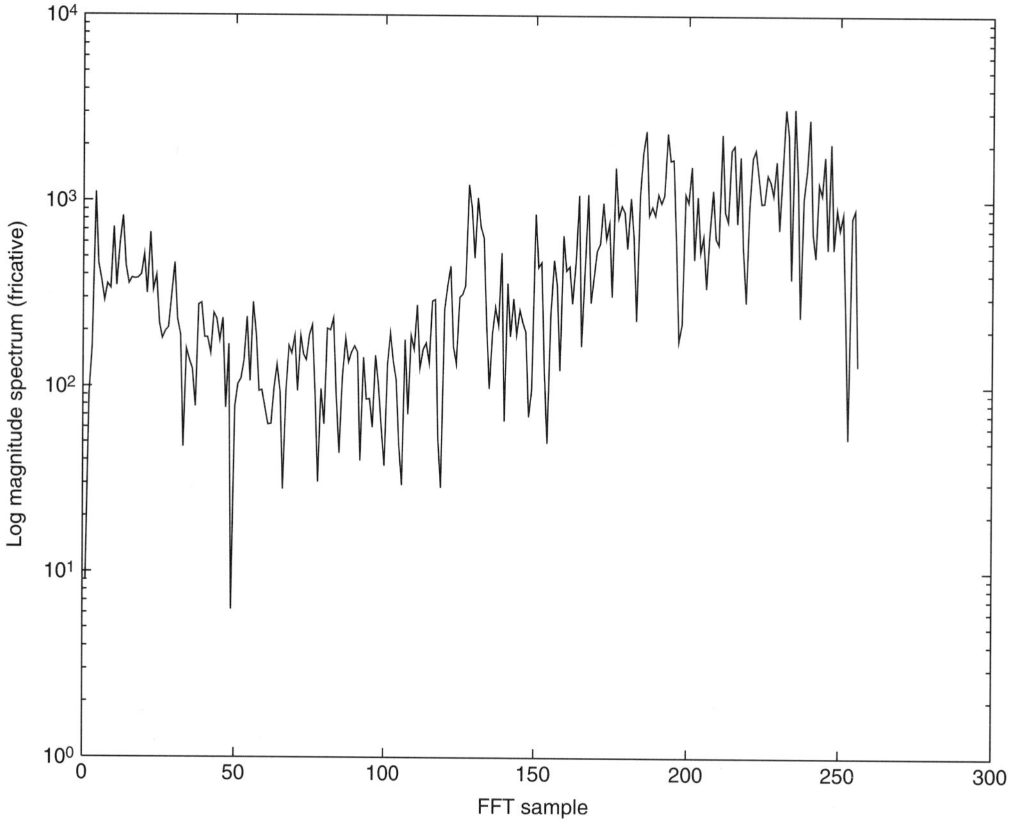

(B) Magnitude Spectrum of Final Fricative /s/ Portion of Utterance *X*

FIGURE 3.3 Corresponding Magnitude Spectra (*Continued*)

constriction serves to determine the fricative sound produced. Unvoiced fricatives include /*f*/ (free), /*T*/ (thick), /*s*/ (cease), /*S*/ (mesh), and /*h*/[8] (heat). The voiced fricatives /*v*/ (vice), /*D*/ (then), /*z*/ (zephyr), and /*Z*/ (measure) are the voiced counterparts of the unvoiced fricatives /*f*/, /*T*/, /*s*/, /*S*/, respectively. Voiced fricatives possess the usual frication noise source at the point of major constriction but also have periodic glottal pulses exciting the vocal tract.

Similarly to glides, liquids, and diphthongs, **affricates** are formed by transitions from a stop to a fricative. The two affricates found in American English are the unvoiced affricate /*C*/ (stop /*t*/, followed by unvoiced fricative /*S*/) as in *change* and the voiced affricate /*J*/ (stop /*d*/, followed by the voiced fricative /*Z*/) as in jam.

The **stops**, or **plosives**, are transient, noncontinuant consonants produced by building up pressure behind a total constriction of the vocal tract and suddenly releasing this pressure. The constriction can be at the lips (/*b*/ in *be* and /*p*/ in *pea*), at the gum ridge (/*d*/ in *day*, and /*t*/ in tea), or at the soft or hard palate (/*g*/ in *go*, /*k*/ in *key*). The consonants /p, t, k/ are unvoiced stops; /b, d, g/ are voiced.

[8] The *whisper* sound /h/ is normally referred to as a unvoiced glottal fricative.

The **nasals** are voiced consonants produced with the glottal waveform exciting an open nasal cavity. Their waveforms resemble vowels but are normally weaker in energy due to limited ability of the nasal cavity to radiate sound. In forming a nasal, a complete closure is made toward the front of the vocal tract, either at the lips (/*m*/ as in *more*), with the tongue resting on the gum ridge (/*n*/ as in *noon*), or by the tongue pressing at the soft or hard palate (/*G*/ as in *sing*).

Prosodic Features

Expressive uses of speech depend on tonal patterns of pitch, syllable stresses, and timing to form rhythmic patterns. Timing and rhythm contribute significantly to the formal linguistic structure of speech. The tonal and rhythmic aspects of speech, called **prosodic features**, are created by special manipulations of the speech production system during the normal sequence of phoneme production. The glottal source excitation conveys important prosodic cues such as **intonation** and **stress**. Stress is used to distinguish similar phonetic sequences or to highlight a syllable or word against a background of unstressed syllables. Intonation refers to the distinctive use of patterns of pitch. The same utterance, for example, can often be changed from a statement into a question by use of intonation.

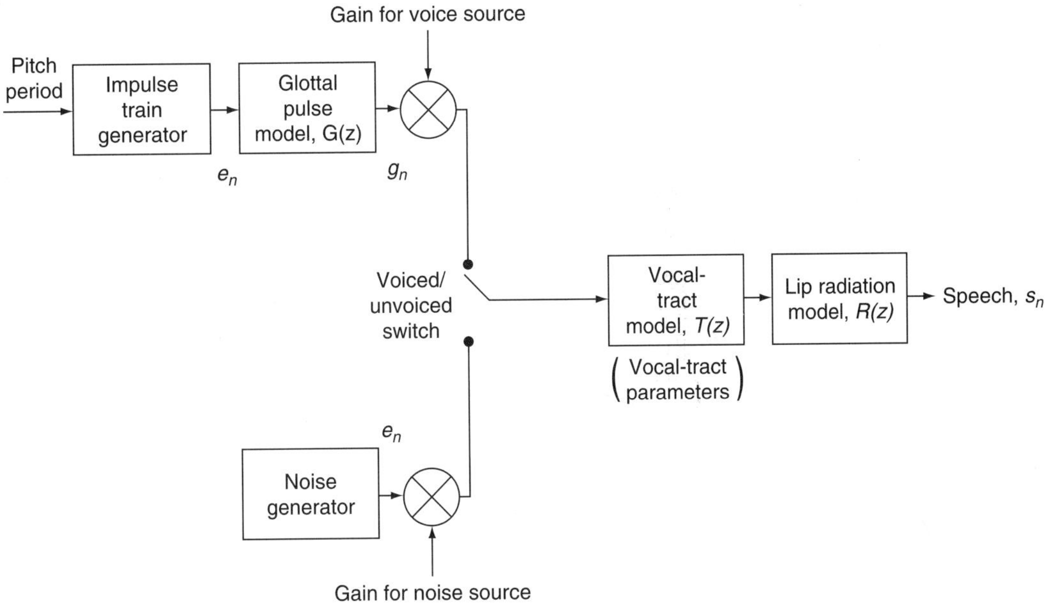

FIGURE 3.4 A General Discrete-Time Model for Speech Production

A detailed discussion of prosodics is found, for example, in Chomsky and Halle (1968) and Lieberman (1967).

3.2.2 Discrete-Time Model of Speech Production

Efforts to model speech production have focused principally on the "source–filter" view of the physiologic system in which, over short stationary intervals of an utterance, speech is modeled as a linear, time-invariant, acoustic filtering operation on an appropriate excitation (the "source"). The process of producing the source input is assumed to be decoupled from the dynamics of the acoustic filter. The prominent discrete-time (DT) source-filter model of speech, known variously as the "all-pole" or "linear prediction" model, is essentially a DT version of classic analog acoustic models. The linear-prediction model underlies a majority of the modern speech products and services.

The history of the developments leading up to modern DT models for speech production is interesting and informative, and it has great value in illuminating the present models (Flanagan, 1972; Klatt, 1987; Morgan and Gold, 1999; Shafer and Markel, 1979; Dudley, 1939). Circuit models developed early in the 20th century, consisting mainly of band-pass filters (resonators) to simulate formant spectra, represent the first attempts of the modern era to model speech production. The primary goal of this work was speech synthesis, and there was no assertion that the models bore any internal similarity to the physiologic system. Such models are often called **terminal–analog** models because they are analogous to the real system only at the terminus. Work on acoustic tube models in the 1960s (Fant, 1960) represents an attempt to model the internal physics of the speech system. Tube models and their differential equations of wave propagation proved to be remarkably useful tools in understanding the speech production system. Although the prevailing DT "all-pole" model of speech production was not derived directly from the acoustic tube theory, important insights and interpretations of the DT model have been obtained from classical theory (Deller, 2000; Rabiner and Schafer, 1978).

A general linear, time-invariant DT model for speech production is shown in the block-diagram form in Figure 3.4. The model is a terminal analog, and it is understood to represent only short intervals of speech over which signal dynamics are assumed to be stationary. Frequent updating or adaptation of model parameters is necessary to properly model the quickly time-varying nature of speech. With this caveat, such a system produces reasonable quality speech for coding and synthesis purposes and serves as a useful analysis model for recognition and related tasks, in spite of the fact that there are no provisions for coupling or nonlinear effects among subsystems in the model.

In the general DT model, a vocal-tract system $T(z)$ is cascaded with a radiation model $R(z)$. [We henceforth include any nasal-tract dynamics in the vocal-tract model $T(z)$.] This system is excited by an uncorrelated excitation signal, say $\{e_n\}$, or by a filtered version of $\{e_n\}$. During unvoiced speech activity, the excitation $\{e_n\}$ is a flat-spectrum **noise** source that excites the vocal tract model directly. During periods of voiced speech activity, the excitation is a periodic DT impulse train whose period corresponds to the desired fundamental frequency.[9] A more realistic voicing signal is created by using

[9] The $\{e_n\}$ is uncorrelated as assumed only as the pitch period becomes infinite (i.e., only when $e_n = \delta_n$).

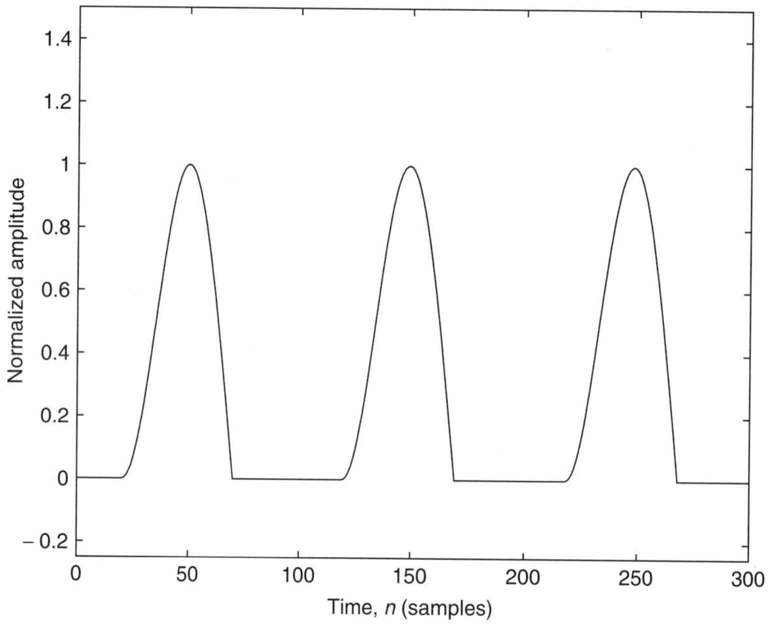

FIGURE 3.5 Idealized Low-Pass Pulse Train. This figure shows three cycles of the glottal volume–velocity waveform during voiced speech.

the pulse train as input to a glottal pulse shaping filter, $G(z)$. In principle, we would like $G(z)$ to produce a low-pass pulse waveform similar to that shown in Figure 3.5. The filter $G(z)$ is further discussed in upcoming paragraphs.

Assuming that the vocal-tract system has a spectrum that is well-modeled by pure resonances, its model is taken to be all pole:[10]

$$T(z) = \frac{T_o}{1 - \sum_{k=1}^{N} b_k z^{-k}} = \frac{T_o}{\prod_{k=1}^{N} (1 - \rho_k z^{-1})}, \qquad (3.1)$$

where T_o represents an overall gain term and ρ_k the (generally complex) pole locations for the model. Based on our knowledge of phonetics, we might suspect that the model is inadequate for certain classes of phonemes. We have noted that some phonemes (e.g., the nasals) exhibit spectral nulls, strongly suggesting the need for zeros in the model. We will return to the issue of zeros in the model below. A strong impetus to justify an all-pole model is the availability of powerful analytical methods that follow from its use.

Each pair of poles in the z-plane at complex conjugate locations (ρ_i, ρ_i^*) roughly corresponds to a formant in the spectrum of $T(z)$. Since the $T(z)$ should be a stable system, all poles are inside the unit circle in the z-plane. If the poles in the z-plane are well-separated, estimates for formant frequencies and bandwidths can be obtained from the pole locations.

[10] Describing a system function as "all pole" is strictly improper and can be misleading. The term means that all zeros of the system function are at $z = 0$ or $z = \infty$ in the z-plane. In other words, the system function has a "denominator polynomial (in z^{-1}) only."

Now let us further characterize the glottal filter $G(z)$. In keeping with the desire to have an all-pole speech model, it is sometimes suggested that the two-pole signal:

$$g_n = [\alpha^n - \beta^n] u_n, \quad \beta < \alpha < 1, \alpha \approx 1, \qquad (3.2)$$

in which $\{u_n\}$ is the unit step sequence; the signal is an appropriate impulse response for the filter. An all-pole impulse–response model for $\{g_n\}$ (regardless of the number of poles used), however, is incapable of producing the realistic pulse shapes that are observed in many experiments because an all-pole model is constrained to be of minimum phase (otherwise it would be unstable). More realistic pulse signals have been suggested in the literature. The Rosenberg pulse [18] is given by:

$$g_n = \begin{cases} \frac{1}{2}[1 - \cos(\pi n/P)], & 0 \le n \le P, \quad P = \text{time of pulse peak.} \\ \cos[\pi(n-P)/2(K-P)], & P \le n \le K, \quad K = \text{final sample before complete closure.} \\ 0, & \text{otherwise.} \end{cases} \qquad (3.3)$$

Equation 3.3 is popular because it can flexibly represent many realistic pulse shapes by adjusting its parameters. The Rosenberg pulse, however, cannot be approximated well using a model with only poles, so we add another concern for the discussion of zeros below.

The radiation component, $R(z)$, is a low-impedance load that terminates the vocal-tract and converts the volume velocity at the lips to a pressure wave in the far field. The radiation load has been observed to have a high-pass filtering effect that is well-modeled by a simple differencer:

$$R(z) = Z_{\text{lips}}(z) = 1 - \zeta_o z^{-1}, \quad \zeta_o < 1, \zeta_o \approx 1. \tag{3.4}$$

We must finally come to terms with the apparent contradiction between the need for zeros in the model and the desire to avoid them. Indeed, $R(z)$ itself is composed of a single zero and no poles. Some of the early writings on this subject argued that if equation 3.2 is a good model for the glottal dynamics, then one of the poles of $G(z)$ will approximately cancel the zero ζ_o in $R(z)$. This does not, however, resolve the question of whether $G(z)$ should contain zeros, or whether, in certain cases, $T(z)$ should include zeros to appropriately model a phoneme. The answer to these questions depends largely on what aspect of speech we are trying to model. In most applications, correct *spectral magnitude* information is all that is required of the model. Generally speaking, this is because the human auditory system is "phase deaf" (Milner, 1970), so information gleaned from the speech is extracted from its magnitude spectrum. While a detailed discussion is beyond our current scope (Deller *et al.*, 2000), the critical fact that justifies the all-pole model is that a magnitude (but not a phase) spectrum for any rational system function can be modeled to an arbitrary degree of accuracy with a sufficient number of stable poles. Therefore, the all-pole model can exactly preserve the magnitude spectral dynamics (the "information") in the speech but might not retain the phase characteristics. In fact, for the speech model to be stable and all pole, it must necessarily have a *minimum phase* spectrum, regardless of the true characteristics of the signal being encoded. Ideally, the all-pole model will have the correct magnitude spectrum but minimum phase characteristic with respect to the "true" model. If the objective is to code, store, resynthesize, and perform other such tasks on the magnitude spectral characteristics but not necessarily on the temporal dynamics, the all-pole model is perfectly adequate. One should not, however, anticipate that the all-pole model can preserve time-domain features of speech waveforms because such features depend explicitly on the phase spectrum.

Let us summarize these important results. Ignoring the technicalities of z-transform existence, we assume that the output (pressure wave) of the speech production system is the result of filtering the appropriate excitation by two (in the unvoiced case) or three (voiced) linear, separable filters. Ignoring the developments above momentarily, let us suppose that we know "exact" or "true" linear models of the various components. By this we mean that we (somehow) know models that will exactly produce the speech waveform under consideration. These models are only constrained to be linear and stable and are otherwise unrestricted. In the unvoiced case $S(z) = E(z)T(z)R(z)$, where $E(z)$ represents a partial realization of a white noise process. In the voiced case, $S(z) = E(z)G(z)T(z)R(z)$, where $E(z)$ represents a DT impulse train of period P, the pitch period of the utterance. Accordingly, the true overall system function is as follows:

$$H(z) = \frac{S(z)}{E(z)} = \begin{cases} T(z)R(z), & \text{unvoiced case.} \\ G(z)T(z)R(z), & \text{voiced case.} \end{cases} \tag{3.5}$$

With enough painstaking experimental work, we could probably deduce reasonable "true" models for any stationary utterance of interest. In general, we would expect these models to require zeros as well as poles in their system functions. Yet, an all-pole model exists that will at least produce a model/speech waveform with the correct magnitude spectrum (Deller *et al.*, 2000), and a waveform with correct spectral magnitude is frequently sufficient for coding, recognition, and synthesis.

We henceforth assume, therefore, that during a stationary frame of speech, the speech production system can be characterized by a z-domain system function of the form:

$$H(z) = \frac{H_0}{1 - \sum_{i=1}^{M} a_i z^{-i}} \text{ with } H_0 > 0, \tag{3.6}$$

which is driven by an excitation sequence:

$$e_n = \begin{cases} \sum_{q=-\infty}^{\infty} \delta_{n-qP}, & \text{voiced case.} \\ \text{zero mean, unity variance, uncorrelated noise,} & \text{unvoiced case.} \end{cases} \tag{3.7}$$

In equation 3.6, $0 < M < \infty$. A block diagram for this system is shown in Figure 3.6.

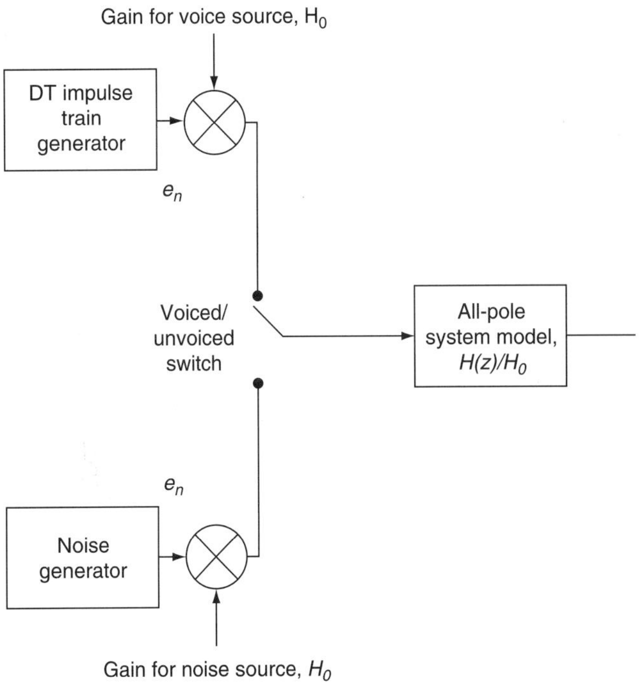

FIGURE 3.6 Block Diagram for z-Domain System Function

3.3 Fundamental Methods and Algorithms Used in Speech Processing

3.3.1 Parametric and Feature-Based Representations of the Speech Signal

Speech processing algorithms rarely work directly with the (sampled) speech waveform but rather with a sequence of quantified **features** that are extracted from the signal, usually at regularly-spaced intervals. In a typical scenario, features are computed periodically in time over 256-point **frames** that are intentionally overlapped by 128 samples. For a 10-kHz sampling rate, this represents feature computations using 25.6 ms segments of speech that overlap by 12.8 ms as the processing moves through time. Using the rule of thumb that speech signal dynamics remain stationary for blocks of 10 to 20 ms, the frame duration is chosen with this in mind, balanced against the need to represent the waveform with the smallest possible number of feature computations (for economy of computation and storage, adherence to bandwidth requirements, and other factors). Increased duration of the frame results in enhanced **spectral resolution** of the feature being computed, while sequences of features computed over shorter windows provide better **time resolution** (Deller *et al.*, 2000). Overlapping frames are used to smooth the transitions of features from one stationary region to the next.

Speech features are often **vector-valued**. An example would be (estimates of) the set of filter coefficients $\{a_i\}$ associated with the DT model of speech in equation 3.6. As we detail in Section 3.3.2, these coefficient estimates, typically 14 in number, are the result of "linear prediction" analysis of speech. Such a set of features is called a **parametric** representation of the frame because it consists of values for the parameters associated with a predetermined model. In the present example, the set of 14 model parameters might be complemented by an estimate of the system gain (H_0 in equation 3.6) and a binary valued feature to indicate a voiced or unvoiced frame. Each frame, therefore, would be represented by a real vector of dimension 16.

Features that have been used to characterize speech usually convey information about spectral properties. These include direct spectral measures (FFT based), mappings or transformations of spectral information (see the cepstrum in Section 3.3.3.), correlation information, zero-crossing measures (crude frequency content information), model parameter estimates (as in the linear-prediction example, or the "reflection coefficients," discussed in Section 3.3.2.), and various transformations of such parameters. Energy measures are also used as conveyors of information about, for example, the presence of voicing or for "silence/speech" decisions.

3.3.2 Linear Prediction Analysis

LP Model and the Normal Equations

The time-domain implication of the all-pole system function is that the difference equation governing the system includes memorized values for the output only; given these output lags, only the present sample of the input is needed to determine s_n to within a gain factor at each n. In particular:

$$s_n = a_1 s_{n-1} + a_2 s_{n-2} + \cdots + a_M s_{n-M} + H_0 e_n = \boldsymbol{a}^T \boldsymbol{s}_n + H_0 e_n, \tag{3.8}$$

where we have defined vectors $\boldsymbol{a} \stackrel{\text{def}}{=} [a_1\ a_2 \cdots a_M]^T$ and $\boldsymbol{s}_n \stackrel{\text{def}}{=} [s_{n-1}\ s_{n-2}\ \cdots\ s_{n-M}]^T$.

A model of form equation 3.8 driven by a zero-mean, uncorrelated (white) sequence, it is called an **autoregressive** (AR) **model** by time-series analysts. The output of an AR model of order M, denoted AR(M), is said to "(linearly) regress on itself," meaning that the present value s_n is a linear combination of (or a **prediction** based on) its past M values plus a component, $H_0 e_n$, which is not predictable from any data, past or future.

Let us assume for the moment that $\{e_n\}$ is indeed stationary white noise, so $\{s_n\}$ is a stationary random process. In light of the difference equation 3.8, a natural way to estimate the parameters of a signal believed to follow an AR(M) model is to design an FIR filter that attempts to "predict" s_n from its past M values. Let us denote the estimated filter parameters by $\{\hat{a}_i\}_{i=1}^{M}$ and the prediction at time n by $\hat{s}_n$:

$$\hat{s}_n = \hat{a}_1 s_{n-1} + \hat{a}_2 s_{n-2} + \cdots + \hat{a}_M s_{n-M} = \hat{\boldsymbol{a}}^T \boldsymbol{s}_n, \tag{3.9}$$

in which the vector $\hat{\boldsymbol{a}} \in \mathbb{R}^M$ is defined similarly to $\boldsymbol{a}$ in equation 3.8. In accordance with the AR model, the parameters, $\hat{\boldsymbol{a}}$, should be chosen to account for all but an unpredictable component, $\{H_0 e_n\}$, in $\{s_n\}$. That is, we seek $\hat{\boldsymbol{a}}$ such that:[11]

$$\hat{e}_n = s_n - \hat{\boldsymbol{a}}^T \boldsymbol{s}_n \approx H_0 e_n \quad \text{(a white-noise sequence of variance } H_0^2\text{)}. \tag{3.10}$$

The sequence $\{\hat{e}_n\}$, which represents the error in prediction at each n, is called the **prediction residual**.

Using the **orthogonality principle** from optimization theory (Haykin, 1996), it can be shown that selecting parameters $\{\hat{a}_i\}$ such that $\{\hat{e}_n\}$ has minimum power (i.e., minimum $\mathcal{E}\{\hat{e}_n^2\}$ in which $\mathcal{E}$ denotes the statistical expectation) will also result in the desired spectral whiteness of $\{\hat{e}_n\}$. Therefore, we seek $\hat{\boldsymbol{a}}$ that minimizes the **mean squared error** (MSE) in prediction; that is:

$$\hat{\boldsymbol{a}} = \underset{\boldsymbol{a}}{\operatorname{argmin}}\ \mathcal{E}\{\hat{e}_n^2\} = \underset{\boldsymbol{a}}{\operatorname{argmin}}\ \mathcal{E}(s_n - \boldsymbol{a}^T \boldsymbol{s}_n)^2$$
$$= \underset{\{a_i\}}{\operatorname{argmin}}\ \mathcal{E}\left(s_n - \sum_{i=1}^{M} a_i s_{n-i}\right)^2 \tag{3.11}$$

[11] The approximation should be taken to mean that $\{\hat{e}_n\}$ and $\{H_0 e_n\}$ are approximately the "same stochastic process."

The well-known solution is the unique parameter vector solving the system of **normal equations**:

$$\bar{\boldsymbol{R}}_s\hat{\boldsymbol{a}} = \bar{\boldsymbol{r}}_s, \tag{3.12}$$

in which $\bar{\boldsymbol{R}}_s \in \mathbb{R}^{M<M}$ is the **autocorrelation matrix** for the process $\{s_n\}$ [i.e., a matrix whose (i, k) element is $\bar{r}_{s;k-i} = \mathcal{E}\{s_{n-i}s_{n-k}\}$] and $\bar{\boldsymbol{r}}_s \in \mathbb{R}^M$ is an auxiliary vector of autocorrelations $\bar{\boldsymbol{r}}_s = \left[\bar{r}_{s;1} \; \cdots \; \bar{r}_{s;M}\right]^T$. The overbars are to distinguish the correlation quantities as statistical expectations rather than time averages.

In the voiced case, where $\{e_n\}$ is modeled as a periodic DT impulse train, the normal equations are used without modification except that the statistical autocorrelations must be replaced by long-term time averages because the speech waveform is deterministic. This approach can be shown to produce unbiased results as the pitch period becomes infinite. Generally speaking, the analysis works satisfactorily for voiced speech for most pitch ranges encountered in applications although performance predictably degrades for very high-pitched utterances that are more typical of women's and children's voices (Deller *et al.*, 2000).

Practical Solutions of the Normal Equations

Frame-Wise Normal Equations

In practice, the parameters must be identified using finite signal records with nominally constant dynamics—typically 10 to 20 ms. Let us denote the block of samples over which the analysis takes place by $n \in \{0, N-1\}$. There are two methods conventionally used to estimate the LP parameters in speech processing.

In the **autocorrelation method**, the solution takes the form of equation 3.12, with the statistical autocorrelations estimated by values of the **autocorrelation sequence**:

$$r_{s;k} = \sum_{n=0}^{N-1} s_n^w s_{n-k}^w, \quad k = 0, 1, \ldots, M, \tag{3.13}$$

in which $s_n^w = s_n w_n$, where w_n is a data window (e.g., Hamming) of duration $N \geq M$ beginning at the time origin, $n = 0$. For future purposes, let us write the resulting system of equations in vector–matrix form as:

$$\boldsymbol{R}_s\hat{\boldsymbol{a}} = \boldsymbol{r}_s, \tag{3.14}$$

in which the overbars have been dropped with respect to equation 3.12 to indicate the short-term temporal solution.

In the so-called **covariance method** the same solution form applies, but the statistical autocorrelations are estimated by values of the **covariance sequence**:

$$\phi_{s;i,k} = \sum_{n=0}^{N-1} s_{n-1}s_{n-k}, \quad i, k = 0, 1, \ldots, M. \tag{3.15}$$

The $\bar{r}_{s;k-i}$ is estimated by $\phi_{s;i,k}$. The data are *not windowed* in this case. Let us denote the corresponding vector–matrix equation by:

$$\boldsymbol{\Phi}_s\hat{\boldsymbol{a}} = \boldsymbol{\phi}_s. \tag{3.16}$$

In addition to its interpretation as a temporally estimated version of the stochastic normal equations, each LP result can be derived as the solution to a short-term optimization problem (Deller *et al.*, 2000).

Autocorrelation Method and Lattice Structures

In either the autocorrelation or covariance solution, obtaining the final LP parameter estimate, $\hat{\boldsymbol{a}}$, ultimately requires the solution of a matrix–vector equation of the form $\boldsymbol{Ax} = \boldsymbol{b}$, in which $\boldsymbol{A}$ is a square matrix that is reasonably assumed to be of full rank, M. This is a widely studied problem for which numerous methods exist (Golub and van Loan, 1989). Solutions to general problems of this form tend to be of $\mathcal{O}(M^3)$ computational complexity.

In the autocorrelation case, the symmetric and Toeplitz structure of the correlation matrix $\boldsymbol{R}_s$ is exploited to obtain numerical and computational $[\mathcal{O}(M^2)]$ benefits as well as an alternative view of the solution that leads to the so-called **lattice** methods. The basis for the more efficient solution is the **Levinson-Durbin** (L-D) **recursion** (Deller *et al.*, 2000; Haykin, 1996; Golub and van Loan, 1989), a **recursive-in-model-order** solution (see Figure 3.7). By this we mean that, using the fixed block of windowed data at each iteration, $\{s_n^w\}_{n=0}^{N-1}$, the solution for the desired order M model is successively built up from lower order models, beginning with the "zero order predictor," which is no predictor at all ($\hat{s}_n = 0$ for every n). Associated with the ith iteration for each $i \in [0, M]$ is a sequence, say $\{\hat{e}_n^i\}$, that can be interpreted as the error in prediction associated with the ith order filter as well as a "backward" prediction error, say $(\hat{b}_n^i)$, corresponding to an attempt to predict the waveform on a reversed time axis (i.e., "backward prediction" in time). Each stage is also characterized by a **parcor** (short for *partial correlation*) **coefficient** or **reflection coefficient**, the ith one of which, say k_i, can be interpreted as a proportion of the amount of "forward error" from stage i that is reflected back as backward error as the forward error propagates into the $(i+1)$st stage. Figure 3.8 illustrates the **analysis lattice**, so-named because it takes the speech as input, sequentially analyzes its dynamics to return only the unpredictable error sequence as output. It is a simple matter to reverse the flow of the lattice to create a **synthesis lattice** that reconstructs the speech from the (forward) error signal (Deller *et al.*, 2000).

As the name suggests, reflection coefficients, $\{\kappa_i\}_{i=1}^M$, bear a strong formal relationship to acoustic reflection coefficients associated with the vocal-tract acoustic-tube model briefly discussed in Section 3.2.2, with the forward and backward errors analogous to similar flow quantities in the model. This

Initialization: For $\ell = 0$,
ξ^0 = scaled total energy in the "error" from an "order-0" predictor
= average energy in the speech frame $\{s_n^w\}_{n=0}^{N-1} = r_{s;0}$

Recursion: For $\ell = 1, 2, \ldots, M$:

1. Compute the ℓth reflection coefficient, $\kappa_\ell = \frac{1}{\xi^{\ell-1}}\left\{r_{s;\ell} - \sum_{i=1}^{\ell-1} \hat{a}_i^{\ell-1} r_{s;\ell-i}\right\}$.
2. Generate the order-ℓ set of LP parameters:

$$\hat{a}_\ell^\ell = \kappa_\ell.$$
$$\hat{a}_i^\ell = \hat{a}_i^{\ell-1} - \kappa_\ell \hat{a}_{\ell-i}^{\ell-1},\; i = 1, \ldots, \ell - 1.$$

3. Compute the error energy associated with the order-ℓ solution, $\xi^\ell = \xi^{\ell-1}\{1 - \kappa_\ell^2\}$.
4. Return to step 1 with ℓ replaced by $\ell + 1$ if $\ell < M$.

FIGURE 3.7 The Levinson-Durbin Recursion. The recursion is applied to the time frame $n = 0, 1, \ldots, N - 1$.

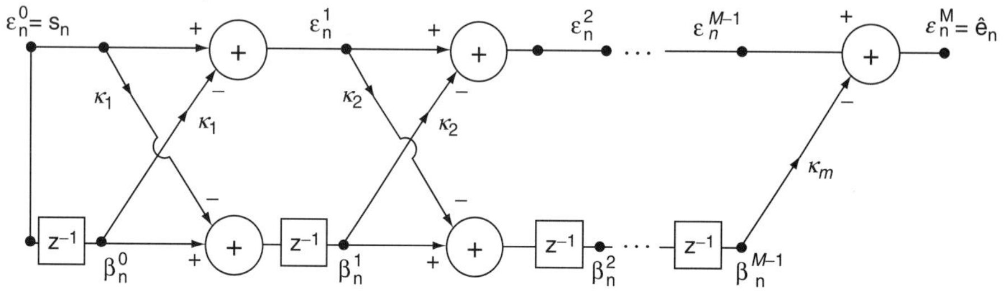

FIGURE 3.8 Analysis Lattice Structure

is a manifestation of the physics of the all-pole system in the mathematics of the DT model. The set $\{\kappa_i\}_{i=1}^M$ contains equivalent information to the set of LP coefficients, as is evident from the fact that the reflection coefficients can be used to reconstruct the speech from the error signal $\{\hat{e}_n\}$. The reflection coefficients are often used as alternatives to the LP parameters in compression and coding schemes because of their desirable numerical property of having a restricted dynamic range (in theory), $|\kappa_i| < 1$ for all i. (Deller *et al.*, 2000). Checking the reflection coefficients for this property on an iteration-by-iteration basis in the L-D recursion offers a convenient check for numerical stability.

Further topics related to lattice structures and reflection coefficients include formulations that compute reflection coefficients independently of the LP parameters (Itakura and Saito, 1969; Burg, 1967) sets of parameters that are related to, or are transformations of, reflection coefficients (Deller *et al.*, 2000; Viswanathan and Makhoul, 1975; and Itakura, 1975); and more efficient lattice structures (Makhoul, 1977; Strobach, 1991).

Decomposition Methods for the Covariance Solution

The covariance method equations to be solved are of the form of equation 3.16. Similar to the autocorrelation matrix $\boldsymbol{R}_s$, the covariance matrix $\boldsymbol{\Phi}_s$ is symmetric and positive definite. Unlike $\boldsymbol{R}_s$, however, $\boldsymbol{\Phi}_s$ it is not Toeplitz, so there is less structure to exploit the solution. The most common solution methods are based on the decomposition of the covariance matrix into lower and upper triangular matrices, say $\boldsymbol{L}$ and $\boldsymbol{U}$, such that $\boldsymbol{\Phi}_s = \boldsymbol{LU}$. Given this decomposition, equation 3.16 can be solved by sequentially solving $\boldsymbol{L}\boldsymbol{y} = \phi_s$ and $\boldsymbol{U}\hat{\boldsymbol{a}} = y$ in each case using simple algorithms (Golub and van Loan, 1989). The most efficient algorithms for accomplishing the $\boldsymbol{LU}$ decomposition are based on two methods from linear algebra (for symmetric matrices): the $\boldsymbol{LDL}^T$ **decomposition** and the **Cholesky** or **square root decomposition**. For details, see Golub and van Loan (1989). Specific algorithms are found in Deller *et al.* (2000) and Golub and van Loan (1989).

Comparisons and Further Notes

When approached as "batch" (i.e., nonlattice) solutions to the $\boldsymbol{Ax} = \boldsymbol{b}$ problem according to the methods described above, the solution methods are of $\mathcal{O}(M)$ complexity, with the autocorrelation method being slightly, but insignificantly, more efficient. Variations exist on the general algorithms discussed above, some being more efficient than others (Markel and Gray, 1976), but all commonly used methods are of $\mathcal{O}(M)$ complexity.

Factors other than computational burden must be considered in choosing between the two general methods. Stability of the solution, for example, can be an important consideration. The autocorrelation solution is theoretically guaranteed to represent a stable filter if infinite precision arithmetic is used. In practice, the L-D recursion contains a built-in check

($|\kappa_i| \leq 1$) for instabilities arising from finite-precision computations. No such theoretical guarantee nor numerical check is available in the covariance case.

The lattice approaches are the most computationally expensive of the conventional methods but are among the most popular because they offer the inherent ability to directly generate the parcor coefficients to monitor stability and to deduce appropriate model orders "online".

Finally, the popularity of the autocorrelation method notwithstanding, the covariance method has an important property that can be very significant in certain applications. In fact, the covariance approach is precisely equivalent to the classical least square error (LSE) solution of an overdetermined system of equations (Golub and van Loan, 1989). This fact can be exploited to develop alternative solutions (Deller *et al.*, 2000; McWhiter, 1983; Deller and Odeh, 1993).

Estimating the Model Order and Gain

In practice, the "true" order of the system of equation 3.6 is unknown. A generally accepted rule of thumb is to allocate one pole pair per kilohertz of Nyquist bandwidth; then, if the segment is voiced, one can add an additional four or five (whichever makes the total even) (Markel and Gray, 1976). Since the order of the model would ordinarily not be changed from frame to frame, the "extra" four or five would be included for unvoiced speech frames. Hence, we choose:

$$M = F_s + \{4 \text{ or } 5\} \quad (F_s = \text{sampling frequency in kHz}). \tag{3.17}$$

The pole-pair per kHz rule is based on the fact that speech spectra tend to have about one formant (which can ordinarily be modeled by a resonant pole pair) per kHz (Fant, 1956). The supplemental poles are needed in the voiced case to account for the glottal shaping filter and to generally allow some flexibility in fitting the model to the speech spectrum.

An experimental technique for determining an appropriate value for M is to study the sum of squared errors, either $\xi_{\text{autoc}} = \sum_{n=0}^{N+M-1} (\hat{s}_n^w - s_n^w)^2$ (autocorrelation method) or $\xi_{\text{cov}} = \sum_{n=0}^{N-1} (\hat{s}_n - s_n)^2$ (covariance), as a function of M. Each is a nonincreasing function of M, and each will tend to reach a point of diminishing returns in the neighborhood of the value of M suggested in equation 3.17. For unvoiced speech, the asymptote will be reached before the four or five supplemental poles in this equation are added. In the unvoiced case, ξ_{autoc} and ξ_{cov} will also tend to level off at a higher final value than the respective values in the voiced case, suggesting that the LP model is not as accurate for the former. Note that ξ_{autoc} as a function of M is a byproduct of the L-D recursion so that the solution method itself can be used to ascertain an appropriate model size.

Finally, to have a completely specified model, we need a means for estimating the gain parameter, H_0. The relative gain from frame to frame is clearly important for proper resynthesis of a speech waveform, for example. By some straightforward manipulations involving the difference equations 3.8 and 3.9, two sets of approximations are obtained (Deller *et al.*, 2000). These are as follows:

$$\hat{H}_0 = \begin{cases} \sqrt{r_{s;0} - \sum_{i=1}^{M} \hat{a}_i r_{s;i}}, & \text{unvoiced case.} \\ \sqrt{P[r_{s;0} - \sum_{i=1}^{M} \hat{a}_i r_{s;i}]}, & \text{voiced case, } (P = \text{pitch period}). \end{cases} \tag{3.18}$$

$$\hat{H}_0 = \begin{cases} \sqrt{r_{\hat{e};0}} = \sqrt{\xi_{\text{autoc}}}, & \text{unvoiced case.} \\ \sqrt{P r_{\hat{e};0}} = \sqrt{P\xi_{\text{autoc}}}, & \text{voiced case, } (P = \text{pitch period}). \end{cases} \tag{3.19}$$

Each of these expressions assumes that the autocorrelation method is used. For the covariance method, the values $\phi_{s;o,i}$, are inserted in place of the $r_{s;i}$ in equation 3.18. In equation 3.19, $\phi_{\hat{e};0,0}$ and ξ_{cov} are similarly substituted.

3.3.3 Cepstral Analysis

The Cepstrum

As its name suggests, **cepstral analysis** represents a variation on speech **spectral** analysis. The nominal goal of cepstral processing is a representation of a voiced speech signal, the **cepstrum**, in which the excitation information is separated, or **deconvolved** from the dynamics of the speech system impulse response. With such a representation, the speech system dynamics and the excitation signal can be studied, coded, and manipulated separately. Applications of the cepstrum have included pitch detection (using the voice component) and "homomorphic" vocoding (using the speech production system component of the cepstrum to derive an efficient representation of the speech dynamics) (Deller *et al.*, 2000; Rabiner and Huang, 1993). The history of the cepstrum is interesting and the details of the theory are extensive. The objective of this treatment is to present an overview of the main concepts and then focus on a modern application of the cepstrum in which most of the connections to the original motivations and theoretical underpinnings are obscured.

Cepstral analysis is a special case within a class of methods collectively known as **homomorphic signal processing**. Historically, the discovery and use of the cepstrum by Bogert, Healy, and Tukey (1963) and Noll (1967) predate the formulation of the more general homomorphic SP approach by Oppenheim and Schafer (1968). In fact, the earlier cepstrum is a special case of a more general cepstrum, which, in turn, is a special instance of homomorphic processing. When it is not clear from context, the cepstrum derived from homomorphic processing is called the **complex cepstrum** (even though the version of it used is ordinarily *real*-valued!), while the original Bogert–Tukey–Healy cepstrum is called the **real cepstrum**.

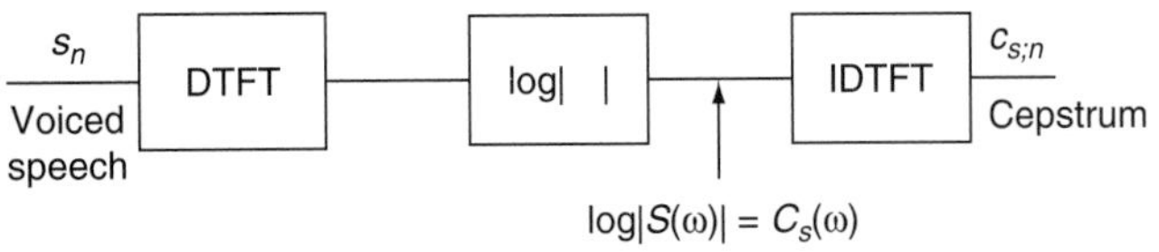

FIGURE 3.9 Computation of the Real Cepstrum

We use the term *cepstrum* and restrict attention to the simpler *real cepstrum*. In fact, there are various definitions of the real cepstrum, but all are equivalent to the even part of the complex cepstrum to within a scale factor.

In spite of its connection to the spectrum, the cepstrum is actually a sequence indexed by time, say $\{c_{s;n}\}$, representing the speech sequence $\{s_n\}$. The set of operations leading to $\{c_{s;n}\}$ is depicted in Figure 3.9. The first two operations can be interpreted as an attempt to transform the signal $\{s_n\}$ into a "linear" domain in which the two parts of the signal that are convolved in the unaltered signal have representatives that are added in the new domain. Let us denote by $\mathcal{Q}$ the operation corresponding to the first two boxes in the figure. Then:

$$\begin{aligned} C_s(\omega) &= \mathcal{Q}\{s_n\} = \log|S(\omega)| = \log|E(\omega)H(\omega)| \\ &= \log|E(\omega)| + \log|H(\omega)| = C_e(\omega) + C_h(\omega). \end{aligned} \tag{3.20}$$

In the linear domain, we are able to apply familiar linear techniques to the new signal, $C_s(\omega)$. In particular, we might wish to apply Fourier analysis to view the frequency domain properties of the new signal. Because $C_s(\omega)$ is a *periodic* function of ω, it is appropriate to compute the Fourier series coefficients for the "harmonics" of the signal. These would take the form:[12]

$$\alpha_n = \frac{1}{2\pi}\int_{-\pi}^{\pi} C_s(\omega)e^{-j\omega n}d\omega. \tag{3.21}$$

Now, according to the definition:

$$c_{s;n} = \frac{1}{2\pi}\int_{-\pi}^{\pi} C_s(\omega)e^{j\omega n}d\omega, \tag{3.22}$$

but $C_s(\omega)$ is a real, even function of ω, so that equations 3.21 and 3.22 produce equivalent results.[13] Therefore, the cepstrum can be interpreted as the Fourier series "line spectrum" of the "signal" $C_s(\omega)$.

[12] Note that function $C_s(\omega)$ is periodic on the ω axis where its fundamental "period" is $\omega_s = 2\pi$. The role of the fundamental "frequency" in this formulation is the quantity $T_s = \omega_s/2\pi = 1$. Hence, the harmonic frequencies are the numbers $nT_s = n$, $n \geq 0$.

[13] In fact, since $C_s(\omega)$ is even, the IDTFT in the definition can be replaced by a **cosine transform**:

$$c_{s;n} = \frac{1}{2\pi}\int_{-\pi}^{\pi} C_s(\omega)\cos(\omega n)d\omega = \frac{1}{\pi}\int_{0}^{\pi} C_s(\omega)\cos(\omega n)d\omega.$$

In practice, the cepstrum would be computed over a frame of, say, length *N*, using a DFT–IDFT pair. Suppose that we seek the cepstrum of the frame (possibly windowed) $\{s_n\}_{n=0}^{N-1}$. Then, we would compute:

$$C_{s;k} = \log\left|\sum_{n=0}^{N-1} s_n e^{-j\frac{2\pi}{N}kn}\right|, \quad \text{for } k = 0, 1, \ldots, N-1$$

(log magnitude spectrum). (3.23)

$$c_{s;n} = \sum_{k=0}^{N-1} C_{s;k}e^{j\frac{2\pi}{N}kn}, \quad \text{for } n = 0, 1, \ldots, N-1 \quad \text{(cepstrum)}. \tag{3.24}$$

Because the cepstrum is an infinitely long sequence, care must be taken to avoid aliasing in the cepstral domain (by zero padding, often unnecessary) and to account for the artificial periodicity of the result.

In the introduction above, quotation marks appear around terms that are used in an unusual manner. The signal that is being transformed into the frequency domain is, in fact, *already* in what we ordinarily consider the frequency domain. Therefore the "new" frequency domain was dubbed the **quefrency domain** by Tukey in the earlier work on the cepstrum, and the *cepstrum* was so named because it plays the role of a *spectrum* in the quefrency domain. The index of the cepstrum (which actually represents DT) is called the **quefrency axis.** The harmonic frequencies of $C_s(\omega)$, which are actually time indices of the cepstrum, are called **rahmonics**. There is an entire vocabulary of amusing terms that accompanies cepstral analysis [see the title of the paper by Bogert *et al.* (1996)], but only the terms *cepstrum* and *quefrency* (and sometimes *liftering*, a concept discussed next, are widely used by speech processing engineers.

The stated purpose of the cepstrum is to resolve the two convolved pieces of the speech, $\{e_n\}$ and $\{h_n\}$, into two additive components, then to analyze those components with spectral (cepstral) analysis. From equation 3.20:

$$c_{s;n} = c_{e;n} + c_{h;n}, \tag{3.25}$$

and if the nonzero parts of $c_{e;n}$ and $c_{h;n}$, occupy different parts of the quefrency axis, we should be able to examine them as separate entities—something that we are unable to do when they are convolved in $\{s_n\}$. In fact, the excitation component, $\{c_{e;n}\}$, tends to have most of its energy at relatively large (and "rahmonically spaced") values of *n*, since its quefrency domain counterpart manifests fast, periodic variations. The use of a time window to select out the component due to the excitation, $\{c_{e;n}\}$, has been called **high-time liftering** to stress the analogy to high-pass filtering. For similar reasons, the vocal system component, $\{c_{h;n}\}$, appears at very low quefrencies and can be extracted using a **low-time lifter**.

Mel Cepstrum

In the 1980s, the cepstrum began to supplant the direct use of the LP parameters as the premier features in speech recognition because of two convenient enhancements that improve performance (Davis and Mermelstein, 1980). The first is the ability to easily smooth the spectrum using the liftering process described above. This process removes the inherent variability of the LP-based spectrum due to the excitation. A second improvement over direct use of the LP parameters can be obtained by use of the so-called **mel-cepstrum**.

In principle, the mel cepstrum represents a set of features that are more similar (than those derived uniformly over the frequency spectrum) to those used by the human auditory system in sound perception. A **mel** is a unit of measure of **perceived pitch** or **frequency** of a tone. It does not correspond linearly to the physical frequency of the tone because the human auditory system does not perceive pitch in this linear manner (Stevens and Volkman, 1940; Koenig, 1949). The mapping between real frequencies, say F_{Hz}, and perceived frequency, say F_{mel}, is approximately (Fant, 1959):

$$F_{\text{mel}} \simeq \frac{1000}{\log 2}\left[1 + \frac{F_{\text{Hz}}}{1000}\right], \tag{3.26}$$

in which F_{mel} (F_{Hz}) is the perceived (real) frequency in mels or Hz.

The cepstrum is particularly well-suited to working with a warped frequency axis since it is based directly on the magnitude spectrum of the speech. Suppose one desires to compute the cepstral parameters at frequencies distributed linearly on the range 0–1 kHz and logarithmically above 1 kHz. One approach is to "oversample" the frequency axis with the DFT–IDFT pair by using, say, a $N' = 1024$ or 2048 point DFT, then selecting those frequency components in the DFT that represent the appropriate distribution. The remaining components can be set to zero, or, more commonly, a second psychoacoustic principle is invoked.

Loosely speaking, it has been found that the perception of a particular frequency by the auditory system, say F', is influenced by energy in a **critical band** of frequencies around F' (O'Shaughnessy, 1987). Further, the bandwidth of these critical bands varies with frequency, beginning at about 100 Hz for frequencies below 1 kHz, then increasing logarithmically above 1 kHz. Therefore, rather than simply using the mel-distributed log magnitude frequency components to compute the cepstrum, it has become common to use the **log total energy** in critical bands around the mel frequencies as inputs to the final DFT. The energy components in each critical band are weighted by a **critical band filter** to account for their relative influences in the band (Davis and Mermelstein, 1980; Dautrich *et al.*, 1983). A detailed description of this computastion is found in Chapter 6 of Deller *et al.* (2000).

Delta Cepstrum

Another popular feature in speech recognition is the delta cepstrum. If $\{c^m_{s;n}\}$ denotes the (mel-) cepstrum or for the mth frame of the signal $\{s_n\}$, then the **delta** or **differenced cepstrum** at frame m is defined as:

$$\Delta c^m_{s;n} \stackrel{\text{def}}{=} c^{m+\delta}_{s;n} - c^{m-\delta}_{s;n}, \tag{3.27}$$

for all n. The parameter δ is chosen to smooth the estimate and typically takes a value of 1 or 2 (look forward and backward one or two frames). A vector of such features at relatively low ns (quefrencies) intuitively provides information about spectral changes that have occured since the previous frame although the precise meaning of $\Delta c^m_{s;n}$ for a particular n and m is difficult to ascertain.

A generalization and enhancement on the delta cepstrum is represented by the **shifted delta cepstrum** (SDC) (Bielefeld, 1994) in which a set of flexible parameters is used to construct enhanced cepstral vectors. The SDC has been shown to be beneficial in limited study (Bielefeld, 1994; Torres-Carrasquillo *et al.*, 2002).

Typically, 8 to 14 cepstral coefficients and their differences are used as features for speech recognition (Deller *et al.*, 2000; Rabiner and Juang, 1993; Juang *et al.*, 1987).

Log Energy

The measures $c_{s;0}$ and $\Delta c_{s;0}$ are often used as relative measures of spectral energy and its change. For the cepstrum, inserting $\log|S(\omega)|$ for $C_s(\omega)$ in equation 3.22, then evaluating at $n = 0$, we see that $c_{s;0}$ differs by a factor of two from the area under the log energy density spectrum. In the practical case in which the IDFT is used in the final operation, this becomes $c_{s;0} = N^{-1}\sum_{k=0}^{N-1}\log|S_k|$ that has a similar interpretation. In the case of the mel-cepstrum, the sum of all critical band energies provides a measure of total spectral energy.

3.3.4 Traditional Search and Pattern Matching Algorithms

Introduction

Pattern recognition and matching and related problems in optimal search play key roles in speech processing. In this subsection, we examine some critical algorithms related to these tasks. The main objective of this material is to introduce the **hidden Markov model** (HMM), a critical technology that serves as a foundation for many aspects of modern speech recognition systems. The HMM can be considered to be a stochastic form of the widely used (across many disciplines and applications) search method known as **dynamic programming** (DP). After treating some general issues in pattern recognition, we begin by studying the DP foundation, then progress to the specific DP-based speech algorithms.

Similarity Measures and Clustering

Underlying all statistical pattern matching problems is the need to have objective measures of similarity among objects and the related problem of grouping objects into clusters whose elements are similar in some sense. Here, we touch on four interrelated classes of pattern-matching problems that occur in specific contexts in speech processing. We return to the clustering problem at the end of the section.

The first problem involves the relatively straightforward task of determining a measure of similarity between sets of features resulting from the analysis of two frames of speech. When features composing the feature vectors represent uncorrelated pieces of information with similar scales (e.g., outcomes of uncorrelated random variables with equal variances), then the widely used Euclidean distance is an appropriate measure of similarity between the vectors. Such is the case, for example, with cepstral coefficients because the parameters are coefficients of a Fourier series and, therefore, represent an expansion on an orthogonal set of basis functions. Determining the "distance" between (models represented by) two sets of LP parameters, however, is a common task for which the Euclidean distance is not appropriate because the LP parameters are highly correlated. For this problem, some variation of the **Itakura distance** measure (Itakura, 1975) (discussed next) is often used.

The LP parameter-matching problem can be considered as an example of the first class of pattern-matching problem—that of assessing the similarity between two feature vectors. It also, however, belongs to a second class of problems important to speech processing, that of assessing similarity between two *models*.[14] The Itakura distance is used to measure intermodel distance reflected in LP vectors. To understand the Itakura distance, consider two LP models, say $\hat{\boldsymbol{a}}$ and $\hat{\boldsymbol{b}}$, derived using two different windowed sequences of data, say $\mathcal{X} = \{x_n^{\omega}\}_{n=0}^{N-1}$ and $\mathcal{Y} = \{y_n^{\omega}\}_{n=0}^{N-1}$. The Itakura distance is given by:

$$d_1(\hat{\boldsymbol{a}}, \hat{\boldsymbol{b}}) \stackrel{\text{def}}{=} \log \frac{\tilde{\boldsymbol{b}}^T \tilde{\boldsymbol{R}}_x \tilde{\boldsymbol{b}}}{\tilde{\boldsymbol{a}}^T \tilde{\boldsymbol{R}}_x \tilde{\boldsymbol{a}}}, \tag{3.28}$$

where

$$\tilde{\boldsymbol{R}}_x \stackrel{\text{def}}{=} \begin{bmatrix} r_{x;0} & \boldsymbol{r}_x^T \\ \boldsymbol{r}_x & \boldsymbol{R}_x \end{bmatrix}, \quad \tilde{\boldsymbol{a}} \stackrel{\text{def}}{=} \begin{bmatrix} 1 \\ -\hat{\mathbf{a}} \end{bmatrix}, \quad \text{and } \tilde{\boldsymbol{b}} \stackrel{\text{def}}{=} \begin{bmatrix} 1 \\ -\hat{\boldsymbol{b}} \end{bmatrix}, \tag{3.29}$$

in which $\boldsymbol{R}_x$ and $\boldsymbol{r}_x$ have as elements the short-term autocorrelations of the data (i.e., $\mathcal{X}$) used to estimate $\hat{\boldsymbol{a}}$ (Schafer and Markel, 1979). If we let $\xi_{\hat{a}}$ be the total squared error incurred in predicting $\mathcal{X}$ using $\hat{\boldsymbol{a}}$ and $\xi_{\hat{b}}$ be the total squared error in predicting $\mathcal{X}$ using $\hat{\boldsymbol{b}}$, then it is not difficult to show that $d_I(\hat{\boldsymbol{a}}, \hat{\boldsymbol{b}}) = \log(\xi_{\hat{b}}/\xi_{\hat{a}})$; hence, the measure gives an indication of how "far" from optimal is the estimate of $\hat{\boldsymbol{b}}$. The Itakura distance is always positive because the denominator yields the smallest possible value of ξ.

The sense of the Itakura distance is to measure similarity between models, but it is based, in fact, on the measurements of the distances from *sequences* to models. These measurements indirectly provide the intermodel distance. Assessment of **sequence-to-model** similarity is a third important class of pattern-matching problems occurring in speech processing. Such problems figure prominently in an array of modern speech-processing technologies, including speech and speaker recognition, speaker identification, speaker verification, language identification, and accent recognition (Rabiner and Juang, 1993). In nearly every case, the model is a statistical representation of the production of sequences associated with that model's class or entity (e.g., a word, a language, a speaker). The "distance" between a sequence consists of some likelihood measure that the model in question is coherent with the given sequence. The discussion later in this chapter of speech recognition using the HMM provides a prime example of this strategy.

A fourth important class of pattern-matching problems occurring in speech processing is that of assessing similarity between two **sequences**. The line between this class and the one immediately preceeding is not distinct because one may consider a particular sequence as a "model" for a given class or entity (e.g., a word), or several sequences may be combined in some manner to produce a "model sequence" for a particular entity. Older techniques in speech recognition and related problems in particular, "dynamic time warping" (discussed later) rely on such direct measures of intersequence distance. Coding and decoding problems in speech processing routinely deploy such techniques as well. The upcoming discussions will provide ample illustration.

Finally, we return to the clustering problem. **Clustering** refers to the act of categorizing J objects into $K \leq J$ classes or categories according to strategies that maximize intraclass similarity and minimize interclass similarity. Clustering methods are numerous and varied depending on the application, types of measurements available, similarity measures used, as well as other factors (Devijver and Kittler, 1982). Although cluster formation often employs feature-to-feature matching (i.e., "objects" represented by features or feature vectors), the "objects" can be sequences or models as well. Thus, any of the pattern-matching problems already discussed may figure into the clustering procedure.

The need for clustering arises in many speech processing applications. Again, the HMM provides two important examples. These include clustering procedures for "codebook" formation for the so-called discrete-observation HMM and

---

[14] A second important example of model-to-model distance occurs in the assessment of the similarity of HMMs. These models are discussed later in this chapter.

clustering for Gaussian mixture model formation for continuous-observation HMMs. In each case, the **K-means algorithm** (Deller *et al.*, 2000; Devijver and Kittler, 1982) is ordinarily used. In this popular method, the initial measurements are subdivided into K distinct groups. The average feature vector, or **centroid** in each group is used as the exemplar feature for that group. The entire population of measurements is then assigned to the group to which it is closest according to the similarity measure used. The centroids are recomputed. The process is repeated until no measurement changes classes.

Dynamic Programming

Dynamic programming (DP) has a rich and varied history in mathematics (Silverman and Morgan, 1990; Bellman, 1957). DP as we discuss it here is actually a special class of DP problems that is concerned with discrete sequential decisions.

Let us first view DP in a general framework. Consider a "grid" in the plane where discrete points or **nodes** of interest are, for convenience, indexed by ordered pairs of non-negative integers as though they are points in the first quadrant of the Cartesian plane. The basic problem is to find a "shortest distance" or "least cost" path through the grid that begins at a designated **original node**, (0,0), and ends at a designated **terminal node**, (I, J). A **path** from node (i_0, j_0) to node (i_N, j_N) is an ordered set of nodes (index pairs) of the form $(i_0, j_0), (i_1, j_1), (i_2, j_2), (i_3, j_3), \ldots, (i_N, j_N)$, where the intermediate (i_k, j_k) pairs are not, in general, restricted.

Distances, or **costs**, may be assigned to nodes or transitions (arcs connecting nodes) along a path in the grid, or both. Suppose that we focus on a node with indices (i_k, j_k). Let us define the notation as:

$$\begin{aligned} D[(i_0, j_0), (i_1, j_1), \ldots, (i_N, j_N)] &\stackrel{\text{def}}{=} \text{distance associated with the path } (i_0, j_0), (i_1, j_1), \ldots, (i_N, j_N); \\ D_{\min}(i_k, j_k) &\stackrel{\text{def}}{=} \text{distance from}(0,0) \text{ to } (i_k, j_k) \text{ over the best path; and} \\ d[(i_k, j_k)|(i_{k-1}, j_{k-1})] &\stackrel{\text{def}}{=} \textbf{transition cost} \text{ from node } (i_{k-1}, j_{k-1}) \text{ to node } (i_k, j_k) \odot \textit{node cost} \text{ incurred upon entering node } (i_k, j_k), \end{aligned} \tag{3.30}$$

where "$\odot$" indicates the rule (usually addition or multiplication) for combining these costs. This total cost is first-order Markovian in its dependence on the immediate predecessor node only. The distance measure d is ordinarily a non-negative quantity, and any transition originating at (0,0) is usually costless. Let us further define the notation:

$$(i_0, j_0) \rightarrow (i_N, j_N): \quad \text{best path (in the sense of optimal cost) from } (i_0, j_0) \text{ to } (i_N, j_N) \tag{3.31}$$

$$(i_0, j_0) \xrightarrow{(i', j')} (i_N, j_N): \quad \text{best path}(i_0, j_0) \text{ to } (i_N, j_N) \text{ that also passes through } (i', j'). \tag{3.32}$$

In these terms, the **Bellman Optimality Principle** (BOP) implies the following (Deller *et al.*, 2000; Bellman, 1957).

$$\left[(i_0, j_0) \xrightarrow{(i', j')} (i_N, j_N)\right] = \left[(i_0, j_0) \rightarrow (i', j')\right] \oplus \left[(i', j') \rightarrow (i_N, j_N)\right], \tag{3.33}$$

for any i_0, j_0, i', j', i_N, and j_N, such that $0 \leq i_0, i', i_N \leq I$ and $0 \leq j_0, j', j_N \leq J$; the $\oplus$ denotes concatenation of the path segments. In words, the BOP asserts that the best path from node (i_0, j_0) to node (i_N, j_N) that includes node (i', j') is obtained by concatenating the best paths from (i_0, j_0) to (i', j') and from (i', j') to (i_N, j_N).

Because it is consistent with most path searches encountered in speech processing, let us assume that a viable search path is always "eastbound" in the sense that for sequential pairs of nodes in the path, say $(i_{k-1}, j_{k-1}), (i_k, j_k)$, it is true $i_k = i_{k-1} + 1$; that is, each transition involves a move by one positive unit along the abscissa in the grid. DP searches generally have similar **local path constraints** to this assumption (Deller *et al.*, 2000). This constraint, in conjunction with the BOP, implies a simple, sequential update algorithm for searching the grid for the optimal path. To find the best path to a node (i, j) in the grid, it is simply necessary to try extensions of all paths ending at nodes with the previous abscissa index, that is, extensions of nodes $(i-1, p)$ for $p = 1, 2, \ldots, J$ and then choose the extension to (i, j) with the least cost. The cost of the best path to (i, j) is:

$$D_{\min}(i, j) = \min_{p \in [1, j]} \{D_{\min}(i-1, p) \odot d[(i, j)|(i-1, p)]\}. \tag{3.34}$$

Ordinarily, there will be some restriction on the allowable transitions in the vertical direction so that the index p above will be restricted to some subset of indices $[1, J]$. For example, if the optimal path is not permitted to go "southward" in the grid, then the restriction $p \nless j$ applies.

Ultimately, node (I, J) is reached in the search process through the sequential updates, and the cost of the globally optimal path is $D_{\min}(I, J)$. To actually locate the optimal path, it is necessary to use a **backtracking** procedure. Once the optimal path to (i, j) is found, we record the immediate predecessor node, nominally at a memory location attached to (i, j). Let us define:

$$\psi(i, j) \stackrel{\text{def}}{=} \text{index of the predecessor node to } (i, j) \text{ on the minimum cost path from } (0, 0) \text{ to } (i, j). \tag{3.35}$$

Initialization: "Origin" of all paths is node (0,0).
For $j = 1, 2, \ldots, J$.
$D_{\min}(1, j) = d[(1, j)|(0, 0)]$.
$\psi(1, j) = (0, 0)$.

Recursion: For $i = 2, 3, \ldots, I$.
For $j = 1, 2, \ldots, J$.
Compute $D_{\min}(i, j)$ according to equation 3.34.
Record $\psi(i, j)$ according to equation 3.35.
Next j
Next i

Termination: Distance of optimal path from (0,0) to (I, J) is $D_{\min}(I, J)$.
Best path is found by backtracking.

FIGURE 3.10 Example Dynamic Programming Algorithm for the Eastbound Salesperson Problem.

If we know the predecessor to any node on the path from (0, 0) to (i, j), then the entire path segment can be reconstructed by recursive backtracking beginning at (i, j).

As an example of how structured paths can yield simple DP algorithms, consider the problem of a salesperson trying to traverse a shortest path through a grid of cities. The salesperson is required to drive eastward (in the positive i direction) by exactly one unit with each city transition. To locate the best route into a city (i, j), only knowledge of optimal paths ending in the column just to the west of (i, j), that is, those ending at $\{(i-1, p)\}_{p=1}^{J}$, is required. A DP algorithm that finds the optimal path for this problem is shown in Figure 3.10.

There are many application-dependent constraints that govern the path search region in the DP grid. We mentioned the possibility of local path constraints that govern the local trajectory of a path extension. These local constraints imply **global constraints** on the allowable region in the grid through which the optimal path may traverse. Further, in searching the DP grid, it is often the case that relatively few partial paths sustain sufficiently low costs to be considered candidates for extension to the optimal path. To cut down on what can be an extraordinary number of paths and computations, a **pruning** procedure is frequently employed that terminates consideration of unlikely paths. This procedure is called a **beam search** (Deller *et al.*, 2000; Rabiner and Juang, 1993; Lowerre and Reddy, 1980).

3.4 Specialized Speech Processing Methods and Algorithms

3.4.1 Applied Dynamic Programming Methods

Dynamic Time Warping

Time alignment is the process by which temporal regions of a test utterance are locally matched with appropriate regions of a reference utterance. The need for time alignment arises not only because different utterances of the same word generally have different durations but also because phones within words are also of different durations across utterances. In this section, we will focus on discrete utterances as the unit to be recognized. For the sake of discussion, we will assume the usual case in which these discrete utterances are words although there is nothing that precludes the use of these principles on subword or superword units.

Dynamic time warping (DTW) is a feature-matching scheme that uses DP to achieve time alignment between sequences of reference and test features. The brief description of DTW presented here is mostly of historical interest and for its value in introducing the HMM (Deller *et al.*, 2000; Rabiner and Juang, 1993; Silverman and Morgan, 1990) DTW has been successfully employed in simple applications requiring relatively straightforward algorithms and minimal hardware. The technique had its genesis in **isolated-word recognition** (IWR) in which words are deliberately spoken in isolation but has also been applied in a limited way to **continuous-speech recognition** (CSR) in which speech is produced naturally without concern for the recognition process. The approach to CSR using DTW has been called **connected-speech recognition** a sort of hybrid approach in which the continuous speech (spoken naturally or with minimal "cooperation" by the talker), is assumed to have been uttered as isolated words. The DTW process essentially concatenates isolated-word templates against which to match the flowing speech feature stream. Details and algorithms used in connected-speech recognition are found in Chapter 11 of Deller *et al.* (2000). Because DTW requires a template (or concatenation of templates) to be available for any utterance to be recognized, the method does not generalize well to accommodate the numerous sources of variation in speech. Therefore, DTW is not used for complex tasks involving large vocabularies.

Generally speaking, DTW is used to match an incoming **test word** (represented by a string of features) with numerous **reference words** (also represented by feature strings). The reference word with the best match score is declared the "recognized" word. In fairness to the correct reference string, the test features should be aligned with it in the manner that gives the best matching score to prevent time differences from unduly influencing the match. On the other hand, constraints on the matching process prevent unreasonable alignments from taking place.

The procedure by which a test word is assigned an optimal match score with respect to a reference string is straightforward. Both the test and reference utterances are reduced to strings of features extracted from the acoustic speech waveform. Let us denote the test utterance feature vectors by $\{\mathbf{t}_i\}_{i=1}^{I}$ and those of the reference utterance by $\{\mathbf{r}_j\}_{j=1}^{J}$, where i and j index the frames as we progress in time. The test and reference features are (conceptually, at least) laid out on the abscissa and ordinate of a grid much like that discussed previously in Section 3.3.4. Although the indices i and j denumerate frame numbers rather than absolute sample times, it is customary to refer to the i and j axes as "time" axes. Costs are assigned to each node in the grid, say node (i, j), according to some measure of distance between the feature vectors $\mathbf{t}_i$ and $\mathbf{r}_j$. With an appropriate set of transition costs for the grid, the feature mapping problem can clearly be posed as a minimum-cost path search through the $i - j$ plane, and we can solve the problem using DP.

Transition costs are sometimes explicitly or implicitly used to control the general trend of the candidate paths as well as the local trajectories at each step along a path.[15] For example, a horizontal transition into a node corresponds to the association of two sequential test frames with a single reference frame, whereas a unity slope transition into a node corresponds to the pairing of sequential frames in both the test and reference strings. The latter may be given a lower cost, for example, to promote a more linear path. Such local costs are often dependent on the short-term history of the path (e.g., one or two steps back) so that, for example, excessive numbers of test string features are not associated with a single reference feature vector. In addition to the local trajectory costs, hard constraints (effectively infinite transition costs) are placed on the search. These may include, for example, strict endpoint matching constraints [nodes (1, 1) and (I, J) must appear on any valid path] and a **monotonicity** requirement that a path may never take a "westward" trajectory at any step. The latter requirement ensures proper time sequencing of events in the utterance. These "soft" and "hard" **local constraints** on path trajectories imply **global constraints** on the overall search space. In turn, the restricted search space implies computational savings as certain node costs need not be computed. For further details on local and global constraints and conditions, see Deller *et al.* (2000), Rabiner and Juang (1993), and Myers *et al.* (1980).

A final issue of significance in the use of DTW is the development of an appropriate set of reference strings—the so-called **training problem**. DTW training is a problem to which there is no simple or well-formulated solution, and this fact represents one of the central contrasts between the DTW method and the pervasive HMM approach. Unlike the DTW reference models, the HMM can be trained in a "supervised" paradigm in which the model can learn the statistical makeup of the exemplars. The application of the HMM to speech recognition in the 1980s was a revolutionary development, largely because of the supervised training aspect of the model. For more information on DTW training, see Deller *et al.* (2000).

Viterbi Decoding

Occurring in many important engineering tasks is a problem that can be solved by the DP search procedure when modified to accommodate stochastic node or transition costs or both. Such problems arise when the test or reference sequence, or both, are modeled by stochastic process(es) so that the search costs are random variables whose values depend on the outcome(s) of the stochastic process(es) to be matched. Given random test and reference sequences, the search costs can only be modeled as random variables. Once realizations of the test and reference sequences are available, however, the search can proceed as in the deterministic case with costs determined "online." Such a modified search procedure, called the **Viterbi algorithm** (Viterbi, 1967), plays a key role in the training and decoding of the HMM, to which we now turn.

3.4.2 The Hidden Markov Model

Introduction

In the 1970s and 1980s, speech researchers began to turn to stochastic approaches to modeling of speech in an effort to address the problem of variability, particularly in large-scale systems (Baker, 1975; Jelinek *et al.*, 1975). The term **stochastic approach** is used to indicate that models are employed that inherently characterize some of the variability in the speech. This is to be contrasted with the straightforward "deterministic" use of the speech data in template-matching (e.g., DTW) approaches in which no probabilistic modeling of variability is present.

Two very different types of stochastic methods have been researched. The first, the **hidden Markov model** (HMM), is amenable to computation on conventional sequential computing machines. Speech research has driven a majority of the engineering interest in the HMM in the last three decades, and the HMM has been the basis for several successful large-scale laboratory and commercial speech recognition systems. In contrast, the second class of stochastic techniques in speech recognition based on the **artificial neural network** (ANN) has been a small part of a much more general research effort to explore alternative computing architectures with some superficial resemblances to the massively parallel computing of biologic neural systems. ANNs have had relatively little impact on the speech field, and we will not pursue this topic here. Interested readers can refer to Morgan and Scofield (1991).

HMM Structure and Formulation

The HMM as an Automaton

An HMM can be viewed as a **stochastic finite-state automaton**—a type of abstract "machine" used to model the generation of a sequence of events or observations (Hopcroft and Ullman, 1979), in this case representing a speech utterance. The utterance may be a word, a subword unit, or, in principle, a sentence or paragraph. In small vocabulary systems, the HMM tends to be used to model words, whereas in larger vocabulary systems, the HMM is used for subword units like phonemes and for more abstract language information (see Section 3.4.3.). To introduce the operation of the HMM,

[15] Normalization of **path lengths** (i.e., sum of the transition costs along a path) is required to prevent undue penalty to longer reference strings. In a DP algorithm, where optimization is done locally, a path-dependent normalization is clearly inappropriate, and we must resort to the use of an arbitrary normalization factor (Myers *et al.*, 1980).

however, it is sufficient to assume that the speech unit is a word.

The HMM models an utterance in the form of a string of features (or feature vectors) to which the corresponding analog waveform has been reduced. In HMM literature, it is customary to refer to the string of test features as the **observations** or **observables** because these features represent information that is "observed" from the speech utterance to be modeled. Let us denote a string of observations by $\{y_t\}_{t=1}^{T}$.

During the **training phase** an HMM is "taught" the statistical makeup of the observation strings for its dedicated word. Training an HMM can be thought of as an attempt to create the automaton (the HMM) that "produces" a given word, in the sense that the HMM will tend to generate strings of features that are similar (statistically) to those produced by a human speaker uttering the given word. In producing the word, the human's vocal tract, roughly speaking, progresses through a sequence of "states" (corresponding to phonemes, for example) that are not directly observable to a listener who must recognize the utterance. The listener can observe the speech features but generally does not know the exact state of the speaker's vocal system that produced a given observation. The HMM is likewise composed of a set of interconnected states, a sequence through which the HMM progresses in the production of a feature string. As in the case of the human generator, the states of the HMM are "hidden" from the observer of the feature string, and the number of discrete states composing the HMM is usually very small compared with the number of feature vectors in a string of observables. In modeling a word, for example, the HMM might have six states corresponding to a like number of phonemes that compose the word being modeled, whereas an acoustic utterance of the word could result in a hundred or more feature vectors.

During the **recognition phase** given an incoming observation string, it is imagined that one of the trained HMMs produced the observation string. Then, for each HMM, the question is asked: How likely (in some sense) is it that *this* HMM produced this incoming observation string? The word associated with the HMM of highest likelihood is declared to be the recognized word. Note carefully that it is *not* the purpose of an HMM to generate observation strings. We *imagine*, however, that one of the HMMs *did* generate the observation string to be recognized as a gimmick for performing the recognition.

The "Hidden" and Observable Random Processes Associated with an HMM

A diagram of a typical HMM with six states is shown in Figure 3.11. The states are labeled by integers. The **structure, or topology**, of the HMM is determined by its allowable state transitions, and speech-related HMMs almost always have a left-to-right or **Bakis** (Bakis, 1976) topology in keeping with the natural temporal ordering of events in an utterance. Two slightly different forms of the HMM are used. The one usually (but not always) discussed for acoustic processing (modeling the signal, as we are doing here) emits an observation upon arrival at each successive state. (It also emits one from the initial state at the outset.) The alternative form, generally employed in language processing, emits an observation *during* the transition. The state emitter form of the model is called a **Moore machine** in automata theory, while the transition emitter form is a **Mealy machine** (Hopcroft and Ullman, 1979). Throughout the rest of this discussion, we will discuss the Moore form. The differences in algorithms associated with the two forms are largely a matter of interpretations of quantities.

The HMM is imagined to "produce" strings of features as follows. At each observation time (corresponding to the times at which we extract observations), a state transition is assumed to occur in the HMM. Upon entering the next state, an observation is emitted. Accordingly, there are two main stochastic processes for which the characterizations must be learned (inferred from data) during training. The first random process models the (hidden) sequence of states that occurs en route to generating a given observation sequence. The second models the generation of observations by the various states.

The state sequence random process is governed by the **state-transition probabilities**, which appear as labels on the arcs connecting the states. Suppose that we call the state random process $\underline{x}$. The associated random variables are $\{\underline{x}_t\}_{t=-\infty}^{\infty}$. We then have that:

$$a(i|j) \stackrel{\text{def}}{=} P(\underline{x}_t = i | \underline{x}_{t-1} = j), \tag{3.36}$$

for arbitrary t. By assumption, the state transition at time t does not depend on the history of the state sequence prior to time $t-1$; hence, the random sequence is called a (first-order) *Markov process* (Leon-Garcia, 1989). The matrix of state transition probabilities is the **state-transition matrix** $\boldsymbol{A} \in \mathbb{R}^{S \times S}$,

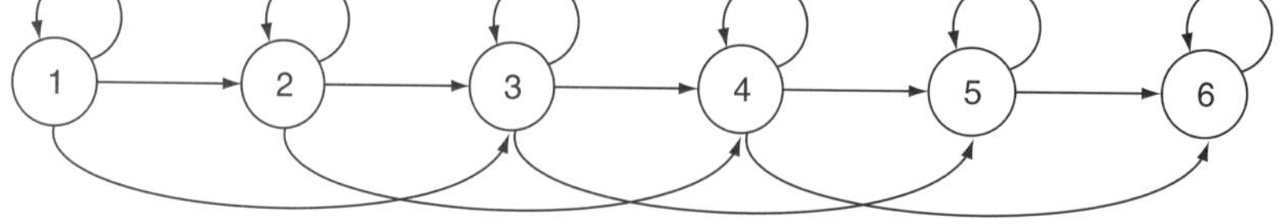

FIGURE 3.11 Typical HMM with Six States

that has $a(i|j)$ as its (i,j) element, where S is the total number of states in the model. Since the random variables of the Markov process take only discrete values (integers), the state process is called a **Markov chain**. Further, since the state-transition probabilities do not depend on t, the Markov chain is said to be homogeneous in time.

The second stochastic process characterized in the learning phase models the generation of observations by the HMM. Let us denote this (generally vector-valued) random process by $\underline{y} = \left\{\underline{y}_t\right\}_{t=-\infty}^{\infty}$, and let $\mathcal{Y}$ be the assumed set of all possible observations (e.g., features, feature vectors, or some mapping of features) that can be produced by a human talker. If $\mathcal{Y}$ is a finite discrete set, then the model is called a **discrete-observation** HMM, whereas if $\mathcal{Y}$ is a continuum, then the model is a **continuous-distribution** HMM. Associated with state i in the HMM, for $i \in [1, S]$, is a probability distribution (discrete) or density (continuous) over the set $\mathcal{Y}$, $f_{y_t|\underline{x}_t}(\xi|i)$. (We henceforth use the abbrevation pdf for this function). Both y_t and ξ are, in general, M-dimensional vectors, where M is the dimension of the feature vector extracted from the speech. For mathematical tractability, it is customary to make the unrealistic assumption that the random process $\boldsymbol{y}$ has conditionally independent and identically distributed random variables, $\{y_t\}$, where the conditioning is upon the state sequence. This means that $f_{y_t|\underline{x}_t(\xi|i)}$ is not dependent on t, and we write:

$$f_{y}|_{\underline{x}}(\xi|i) \stackrel{\text{def}}{=} f_{y_t|\underline{x}_t(\xi|i)} \quad \text{for arbitrary } t. \tag{3.37}$$

Two auxiliary pieces of information are needed to complete the characterization of the HMM—the initial probability vector, say π_1, and the number of states, S. The ith element of $\pi_1 \in \mathbb{R}^S$ is the probability that state i is the initial state in any path that produces an observation string. In the left-to-right model, the starting state is naturally taken to be the "leftmost" one, ordinarily numbered "state 1." Accordingly, $\pi_1 = [1 \quad 0 \quad 0 \ \dots \ 0]$. Formally, then, a Moore-form HMM, say $\mathcal{M}$, is comprised of the set of elements:

$$\mathcal{M} = \left\{S, \pi_1, \boldsymbol{A}, \left\{f_{y|\underline{x}}(\xi|i), 1 \le i \le S\right\}\right\}. \tag{3.38}$$

The Discrete-Observation Model

The discrete-observation HMM, a special case of model set 3.38, is restricted to the production of a finite set of observations. In this case, the naturally occurring observation vectors are quantized into one of the permissible sets using a technique known as **vector quantization** (VQ). In the VQ process, a large population of continuous-observation vectors, assumed to be statistically representative of all the features to be encountered in a speech-recognition task, is partitioned into a fixed number of clusters, say K, typically using the K-means algorithm (Deller *et al.*, 2000; Devijver and Kittler, 1982). The centroids of the clusters are assumed to be representative of the sounds associated with the respective clusters. Only the centroids are kept, and collectively they are called a **codebook** for the vector quantizer. Subsequently, any observation vector used for either training or recognition is quantized (assigned to the nearest code) using this codebook; hence, each feature vector suffers some degree of quantization in the feature space. If there are K possible vectors (observations) in the codebook, then it is sufficient to assign to an observation a single integer, say k, where $1 \le k \le K$. Formally, the vector random process $\underline{\boldsymbol{y}}$ is replaced by a scalar random process, say $\underline{y}$, where each of the random variables $\underline{y}_t$ may take only integer values in $[1, K]$.

For the discrete-observation HMM, the quantized observation pdf for state i takes the form of K impulses on the real line. In this case, it is sufficient to know the probability distribution over the K symbols for each state (weights on the impulses), which we shall denote as:

$$b(k|i) \stackrel{\text{def}}{=} P(\underline{y}_t = k|\underline{x}_t = i), \quad \text{for } k \in [1, K], \ i \in [1, S]. \tag{3.39}$$

These **observation probabilities** are clearly defined to be dependent on the state but are assumed independent of time t. In general, we will not know the value assumed by a particular observation, $\underline{y}_t$, and we will write:

$$b(y_t|i) \stackrel{\text{def}}{=} P(\underline{y}_t = y_t|\underline{x}_t = i). \tag{3.40}$$

Similar to the definition of the transition probability matrix, $\boldsymbol{A}$, we define the **observation probability matrix,** $\boldsymbol{B} \in \mathbb{R}^{K \times S}$, with (k, i) element $b(k|i)$. The general mathematical specification of the HMM of equation 3.38 can be modified to reflect the discrete observations:

$$\mathcal{M} = \{S, \pi_1, \boldsymbol{A}, \boldsymbol{B}, \mathcal{Y}_{\text{VQ}}\}, \tag{3.41}$$

where $\mathcal{Y}_{\text{VQ}} \stackrel{\text{def}}{=} \{y_k, 1 \le k \le K\}$, the set of K vectors in the VQ codebook.

Recognition Using the HMM

Discrete-Observation Case

We assume that we have a population of M HMMs, say $\{\mathcal{M}_\ell\}_{\ell=1}^{M}$, each of which represents a word and each of which has been trained (we will discuss how in subsequent paragraphs) with appropriate probabilities. Every HMM represents a word in a vocabulary, $\mathcal{V}$. Given a (vector-quantized) observation sequence, say $y \stackrel{\text{def}}{=} \{y_1, y_2, \dots, y_t, \dots, y_T\}$, which is a realization of random process $\underline{y}$, we seek to deduce which of the words in $\mathcal{V}$ is represented by y. That is, we want to determine the *likelihood* with which each of the $\mathcal{M}_\ell$ produces y.

Available data will generally not allow us to characterize the more intuitive *a posteriori* probability $P(\mathcal{M}|\underline{y} = y)$ to use as a likelihood measure, so the measure $P(\underline{y} = y|\overline{\mathcal{M}})$, the probability that the observation sequence y is produced given the model $\mathcal{M}$, is used instead. If the model probabilities are equal, then Bayes rule implies that the two approaches will produce the same result.

First, we consider an approach based on the likelihood that the observations could have been produced using *any* state sequence (path) through a given model. The most common technique applied to this problem is the so-called the **forward–backward** (F–B) **algorithm,** also called **Baum-Welch re-estimation.** The full F–B algorithm was developed in the series of papers by Baum and colleagues (Baum, 1972). The underlying principles described in these papers are well beyond the scope of this treatment. The interested reader is referred, for example, to Deller *et al.* (2000) and Rabiner and Juang (1993). We simply note that the method represents an enormous savings in computation with respect to the effort required by exhaustive evaluation of all paths in each model. For $S = 5$ and $T = 100$, for example, the F–B method yields an improvement of 69 orders of magnitude with respect to the same computation carried out directly.

The second approach evaluates the likelihood be based on only the *best* state sequence through $\mathcal{M}$. A Viterbi search provides a slightly more cost-efficient alternative to the F–B method, with typically 10 to 25% fewer computations required. Recalling that the Viterbi algorithm amounts to a DP search with stochastic costs, it is not surprising that Viterbi search returns a total cost based on the single best state path through the HMM. Formally, given observation string y of length T and HMM $\mathcal{M}$, the Viterbi search returns the likelihood $P(y, \mathcal{I}^*|\mathcal{M})$, where $\mathcal{I}^* \stackrel{\text{def}}{=} \arg\max_{\mathcal{I}\in\Sigma_T} P(y, \mathcal{I}|\mathcal{M})$ in which Σ_T represents the set of legal state sequences (paths) of duration T.

The formulation of the Viterbi problem as a DP grid search is straightforward. In principle, the observations (indexed by time, t) are laid out along the abscissa of the search grid and the states along the ordinate. Each point in the grid is indexed by a time, state pair (t,i). Two natural restrictions are imposed on the search. The first involves **sequential grid points** along any path that must be of the form (t, i), $(t + 1, j)$, where $1 \le i, j \le S$. This says that every path must advance in time by one, and only one, time-step for each path segment. The second refers to **final grid points** on any path that must be of the form (T, i_f), where i_f is a legal final state in the model. For a left-to-right HMM, this means that the ordinate of the grid point at time T must be the final ("rightmost") state in the model.

The stochastic cost assigned to node (t,i) in the grid is simply $b(y_t|i)$, and the transition cost, given that the state occupied at time $t-1$ is j, is taken to be $a(i|j)$ for any i and j and for arbitrary $t > 1$. In addition, to account for initial state probabilities, all paths are assumed to originate at a fictitious and costless node (0,0), which makes a transition of cost $P(\underline{x}_1 = i)$ to any initial node of the form $(1,i)$. Upon arriving at the initial node, the path will also incur a node cost of the form $b(y_1|i)$. The total cost associated with any

Initialization: "Origin" of all paths is node (0,0).
For $i = 1, 2, \ldots, S$.
$D_{\min}(1, i) = -\ln[a(i|0)b(y_1|i)] = -[\ln P(\underline{x}_1 = i) + \ln b(y_1|i)]$.
$\psi(1, i) = 0$.
Next i

Recursion: For $t = 2, 3, \ldots, T$
For $i_t = 1, 2, \ldots, S$.
Compute $D_{\min}(t, i_t) = \min_{p\in[1,S]}\{D_{\min}(t-1, p) - \ln a(i_t|p) - \ln b(y_t|i_t)\}$.
Record backtracking information at the present node, $\psi(t, i_t)$.
Next i_t
Next t

Termination: Distance of optimal path from (0,0) to (T, i_T^*) is $D_{\min}(T, i_t^*)$.
Best state sequence, $\mathcal{I}^*$, is found as follows:
Terminal state is i_T^*, then:
For $t = T - 1, I - 2, \ldots, 0$.
$i_t^* = \psi(t + 1, i_{t+1}^*)$.
Next t

FIGURE 3.12 Computation of $P(y, \mathcal{I}^*|\mathcal{M})$ Using the Viterbi Algorithm.

transition, say $(t-1, j)$ to (t, i), is therefore $a(i|j)b(y_t|i)$ for $t > 1$ and $P(\underline{x}_1 = i)b(y_1|i)$ when $t = 1$. We note that multiplication is the natural operation ($\odot$ in equation 3.30) by which node and transition costs are combined in this case. Also note that a *large* "distance" (i.e., probability) is desirable, so we will apparently want to seek a *maximum* cost path if such measures are used. With these cost assignments, and a minor adjustment to seek the maximum rather than minimum cost path, the Viterbi procedure is applied in the usual way.

To reduce the computational load and to mitigate the numerical problems arising from many products of probabilities, it is common to *sum* negative log probabilities rather than to *multiply* probabilities directly. If this strategy is used, the problem reverts to a *minimal* cost path search because high probabilities result in small negative logs. The steps of a Viterbi search algorithm based on this cost strategy are shown in Figure 3.12. Note that a provision for state backtracking is included in the algorithm.

Continuous-Observation Cases

The F–B and Viterbi methods already described are applicable to the continuous-observation case with a simple modification. We define the likelihood of generating observation $\mathbf{y}_t$ in state i as $b(\mathbf{y}_t|i) \stackrel{\text{def}}{=} f_y|x(\mathbf{y}_t|i)$. With this substitution for the observation probabilities, the methods developed for the discrete-observation case can be applied directly. The resulting measure computed for a model, say $\mathcal{M}$, and an observation string, say $\mathbf{y} \stackrel{\text{def}}{=} \{\mathbf{y}_1, \ldots, \mathbf{y}_T\}$, will be $P(\mathbf{y}|\mathcal{M})$ for the F–B approach and $P(y, \mathcal{I}_{\min}|\mathcal{M})$ in the Viterbi case. These measures will no longer be proper probabilities, but they provide meaningful likelihood measures with which appropriate relative comparisons across models can be made.

HMM Training

Discrete-Observation Case

Next we address the question of training a particular HMM to correctly represent its designated word or other utterance.[16] We assume that we have one or more feature strings of the form $y = \{y_1, \ldots, y_t\}$ extracted from training utterances of the word (this assumes that we already have the codebook that will be used to deduce symbols), and the problem is to use these strings to find an appropriate model of form 3.38. In particular, we must find the matrices $\boldsymbol{A}$ and $\boldsymbol{B}$ and the initial state probability vector π_1. There is no known way to analytically compute these quantities from the observations in any optimal sense. However, an extension to the F–B algorithm provides an iterative estimation procedure for computing a model, $\mathcal{M}$, corresponding to a local maximum of the likelihood $P(y|\mathcal{M})$. This method is most widely used to estimate the HMM parameters, and for algorithm details, the reader is referred to Deller *et al.* (2000) and Rabiner and Juang (1993). Repeated re-estimation of the model (by repeated iterations of the F–B algorithm) is guaranteed to converge to an HMM corresponding to a local maximum of $P(y|\mathcal{M})$ (Baum and Sell, 1968). The model improves with each iteration unless its parameters already represent a local maximum. F–B training, therefore, does not necessarily produce the globally best possible model. Accordingly, it is common practice to run the algorithm several times with different sets of initial parameters and to take as the trained model the $\mathcal{M}$ that yields the largest value of $P(y|\mathcal{M})$.

Training using the Viterbi algorithm is much simpler and more computationally efficient, though equally effective. Beginning with an initial model estimate, we implement the recognition procedure on the training sequence. Once the optimal path is discovered, backtracking information is used to determine the number of transitions into and out of each state and the number of times each symbol is produced in each state. The tallies are used to reestimate the state and observation probabilities, and then the process is repeated. The algorithm can be shown to converge to a proper characterization of the underlying probabilities (Fu, 1982).

Continuous-Observation Case

The state-transition and initial-state probabilities in the continuous-observation case have exactly the same meaning and structure as in the discrete-observation case and can be found in exactly the same manner.

The pdf most widely used to model continuous observation outcomes is the **Gaussian mixture density**, which is of the form:

[16] In addition to the two most prominent training methods described here, gradient search techniques (Levinson *et al.*, 1983) and techniques based on the concept of statistical discrimination (Ephraim *et al.*, 1989; Bahl *et al.*, 1988), have been used for training. The latter works includes algorithms for *corrective training* which reinforce model parameters when they achieve successful detection and "negatively train" models with data that cause false recognition.

$$f_{y|\mathbf{x}}(\xi|i) = \sum_{m=1}^{M} c_{im}\mathcal{N}(\xi;\, \mu_{im},\, C_{im}), \tag{3.42}$$

in which c_{im} is the **mixture coefficient** for the mth component for state i, and $\mathcal{N}(\cdot)$ denotes a multivariate Gaussian pdf with mean vector μ_{im} and covariance matrix $\boldsymbol{C}_{im}$. The mixture coefficients must be non-negative and satisfy the constraint $\sum_{m=1}^{M} c_{im} = 1,\ 1 \leq i \leq S$. For a sufficiently large number of mixture densities, M, equation 3.42 can be used to approximate arbitrarily accurately any continuous pdf.

Modified Baum-Welch reestimation formulas have been derived for the three quantities c_{im}, μ_{im}, and $\boldsymbol{C}_{im}$ for mixture density m in state i (Liporace, 1982; Juang *et al.*, 1986). As in the discrete-observation case, the use of these formulas will ultimately lead to a model, $\mathcal{M}$, that represents a *local* maximum of the likelihood $P(\boldsymbol{y}|\mathcal{M})$. However, finding a *good* local maximum depends rather critically on a reasonable initial estimate of the vector and matrix parameters μ_{im} and $\boldsymbol{C}_{im}$ for each i and m. We now describe a convenient procedure for deriving good separation of the mixture components for use with a Viterbi procedure.

In the Viterbi approach, the mean vectors and covariance matrices for the observation densities are reestimated by simple averaging, similarly to what is done in the discrete-observation procedure. This is easily described when there is only one mixture component per state. In this case, for a given $\mathcal{M}$, the recognition experiment is performed on the observation sequence. Each observation vector is then assigned to the state that produced it on the optimal path by examining the backtracking information. The vectors assigned to each state are then used to update the mean vector and covariance matrix for that state, and the process is repeated.

When $M > 1$ mixture components appear in a state, then the observation vectors assigned to that state must be subdivided into M subsets prior to averaging. This is done by clustering, using, for example, the K-means algorithm with $K = M$. If there are N_{im} vectors assigned to the mth mixture in state i, then the mixture coefficient c_{im} is reestimated as $c_{im} = N_{im}/N_i$.

Practical Considerations in the Use of the HMM

Features

Thus far, we have discussed the HMM in principally abstract terms. Here, we briefly consider some of the practical aspects of the use of the HMM, turning first to the fundamental acoustical analysis. Many features have been used as observations, but the most prevalent are the LP parameters, cepstral parameters, and related quantities. These are frequently supplemented by short-term time differences that capture the dynamics of the signal as well as supplemented by energy measures, such as short-term energy and differenced energy. A comparative discussion is found in Bocchieri and Doddington (1986) and Nocerino *et al.* (1985).

In a typical application, the speech would be sampled at 8 kHz and analyzed on frames of 256 points with a 156-point overlap. The analysis frame end-times, say 100, 200, ..., M, become observation times $t = 1, 2, \ldots, T$. For a typical word utterance lasting 1 sec, $T = 80$ observations. On each frame, 8 to 10 LP coefficients are computed, which are then converted to 12 cepstral coefficients. Alternatively, 12 mel-cepstral coefficients might be computed directly from the data. To add dynamic information, 12 differenced cepstral coefficients are also included in the vector. Finally a short-term energy measure and a differenced energy measure are included for each frame, for a total of 26 features in the observation.

A related issue is to decide upon the number of symbols (centroids) in the VQ codebook if a discrete-observation model is used. One measure of the quality of a codebook is the average distance of a vector observation in the training data from its corresponding symbol. This figure is often called the **codebook distortion**. The smaller the distortion, the more accurately the centroids represent the vectors they replace. Typically, recognizers use codebooks of size 32–256 with the larger sizes being more common. Since a larger codebook implies increased computation, there is an incentive to keep the codebook as small as possible without decreasing performance. Like the other parametric decisions that must be made about an HMM, this issue is principally guided by experimental evidence.

Model Topology and Size

Some variation of the Bakis topology is almost uniformly used in speech applications because it most appropriately represents the acoustic process. Interestingly, and perhaps not surprisingly, when used with speech, an HMM with no state-transition constraints will often train so that it essentially represents a sequential structure (backward transition probabilities turn out zero).

Experimental evidence suggests that states frequently represent identifiable acoustic phenomena.[17] Therefore, the number of states is often chosen to roughly correspond to the expected number of such phenomena in the utterance. If words are being modeled with discrete observations, for example, 5 to 10 states are typically used to capture the phones in the utterances. Continuous-observation HMMs sometimes use more states. If HMMs are used to model discrete phones or phonemes, three states are often used—one each for onset and exit transitions and one for the steady-state portion of the phone.

The Generality of the HMM and a First View of Language Models

We have generally discussed the HMM above as though it were a model for a single word but stressed that much of what was developed applies to other uses of the model. For large vocabulary systems, it is necessary to use HMMs to model sub-word acoustic units (e.g., phones or phonemes) since appropriate training of thousands of isolated-word models is prohibitive. Regardless of whether we view speech as constructed of phones or words (or some other acoustic unit), the structure of human language is, of course, much more complex than that conveyed by isolated acoustic units. A major part of the information in the speech code is conveyed by the way in which the units are ordered. Still more information is carried in other acoustic cues such as intonation, nonspeech cues like hand gestures and facial expressions, and in abstract notions like the physical environment of the conversation or the meaning of a phrase in the context of the overall message. Generally speaking, a set of rules by which fundamental acoustic units of speech ultimately become complete utterances is called a **language model** (LM). From the earliest efforts to do large vocabulary speech recognition, it has been clear to researchers that a good LM is critical to a successful performance (Lowerre and Reddy, 1980). For large vocabulary systems, acoustic processing will almost never provide satisfactory recognition as a stand-alone technology. LMs have been a very active area of research on speech processing for many years.

We raise the issue of LMs in the present context because the HMM is, in fact, formally equivalent to one of the most prominent LMs used in practice—the so-called "regular" or "finite state" grammar (Hopcroft and Ullman, 1979; Fu, 1982). Because of its simplicity, regular grammar is often used to model the speech production code at several levels of linguistic and acoustic processing. This is why the HMM figures so prominently in the continuous speech recognition (CSR) problem. The general idea is straightforward: an HMM for **syntax** (word ordering) in a speech recognizer would have dynamics (states and transitions) very much like the HMMs we described for the word model. The difference would be that the observations "emitted" by the model would be words, which, in turn, would be represented by an acoustic HMM (probably further decomposed into phonetic HMMs). It is possible, therefore, to envision this type of speech recognizer as a large graph (or grid) consisting of HMMs embedded inside of HMMs several levels "deep."

Not unexpectedly, the training procedures for such an embedded system of models require some enhancement and reconsideration of methods applied to the simple word models. Nevertheless, the basic ideas are the same, and the same F–B and Viterbi reestimation algorithms are applicable. One remarkable property of the HMM is evident in the training process. In working with continuous speech, we must face the problem of not knowing where the temporal boundaries (i.e., phonetic or word) are in the acoustic observation strings used for training (or recognition). The creation of sufficiently large databases marked according to phonetic time boundaries is generally impractical. An important

[17] Perhaps more accurately, assuming that states represent acoustic phenomena seems to provide a good rule of thumb for creating HMMs with successful performance. However, the relationship of the number of states to the performance of the HMM is very imprecise, and in practice, it is often necessary to experiment with different model sizes.

property of the HMM comes to our aid in this situation. Researchers have discovered that HMMs can be trained *in context* (i.e., hooked together in the language graph) as long as reasonable "seed" models are used to initiate the estimation procedure. This means that the entire observation sequence for a sentence can be presented to the appropriate string of HMMs, and the models will tend to "soak up" the parts of the observation sequence corresponding to their words or phones, for example. This ability of the HMM has revolutionized the CSR field because it obviates the time-consuming procedure of temporally marking a database.

3.4.3 Language Modeling

Introduction

The concept of a LM[18] was introduced in the just-completed material on HMMs because these stochastic models play such a prominent role in representing language structure in contemporary speech recognition systems. More generally, however, LM research has a long and interesting history in which numerous and varied systems have been produced. The field remains very active with many challenging unsolved problems. In the scope of this vast history and level of current activity, the goal of this section is very modest—to provide an overview of some of the basic operating principles and terminology of LM'ing[18] particularly those related to the HMM-like formalisms discussed above.

Inasmuch as LMs ultimately imply structure that assists in pattern-matching and search, some of the techniques and algorithms discussed in earlier sections are directly applicable to the LM'ing problem. The field of LM'ing, however, has developed largely outside of the SP discipline—principally in certain areas of computer science, linguistics, and related fields—so the vocabulary and concepts of LM'ing can seem unfamiliar to the traditional SP engineer, even when the techniques are similar to SP methods. Further, the abstraction of some of the problems encountered in LM'ing (e.g., the *meaning* of an utterance) defy the construction of rigorously quantified models, which is another break from the SP engineer's typical approach to problem solving.

The Purpose and Structure of a Language Model

Whether we view phones, phonemes, syllables, or words as the basic unit of speech, LMs are generally concerned with how fundamental speech units may be concatenated: in what order, in what context, and with what intended meaning. This characterization immediately suggests that LMs have both similarities to, and differences from, the sets of "grammatical rules" of a natural language that are typically studied in primary education. "Primary school" grammar is indeed concerned with the rules by which speech components are combined. Such studies rarely, if ever, are concerned with basic sub-word phonetic elements, and seldom do they deal with or higher level abstractions like subtle nuances of meaning. In contrast, an LM often reaches deep into the acoustic structure of speech and sometimes high into the more abstract aspects of intention and meaning. In essence, what primary school students know as "grammar" is the natural-language version of only one of the ingredients of a formal LM used in speech recognition. A LM is a broader framework, consisting of the (often quantified) formalization of some or all of the structure that grammarians, linguists, psychologists, and others who study language can ascribe to a natural language. Indeed, an LM need not conform to the grammatical rules of the natural language it is modeling.

Peirce (Liszka, 1996) identifies four components of the natural-language code: symbolic, grammatical,[19] semantic, and pragmatic. Implicit or explicit banks of linguistic knowledge resident in speech recognizers, sometimes called **knowledge sources** (Reddy, 1976), can usually be associated with a component of Peirce's model. The **symbols** of a language are defined to be the most fundamental units from which all messages are ultimately composed. In the spoken form of a language, for example, the symbols might be words or phonemes, whereas in the written form, the alphabet might serve as the symbols. For the purposes of this discussion, let us consider the phonemes to be the symbols of a natural language. The **grammar** of the language is concerned with how symbols are related to one another to form message units. If we consider the sentence to be the ultimate message unit, then how words are formed from phonemes are part of Peirce's grammar as well as the manner in which words form sentences. How phonemes form words is governed by **lexical** constraints and how words form sentences by **syntactic** constraints. Lexical and syntactic constraints are both components of the grammar.

The grammar of a language is, in principle, arbitrary in the sense that any rule for combining symbols may be posed. On the other hand, **semantics** is concerned with the way in which symbols are combined to form *meaningful* communication. Systems imbued with semantic knowledge sources straddle the line between speech *recognition* and speech *understanding* and draw heavily on artificial intelligence research. Beyond binary unmeaningful or meaningful decisions about symbol strings, semantic processors can be used to impose meaning upon incomplete, ambiguous, noisy, or otherwise hard-to-understand speech.

Finally, the **pragmatics** component of the LM is concerned with the relationship of the symbols to their users and the environment of the discourse. This aspect of language is very difficult to formalize. To understand the nature of pragmatic knowledge, consider the phrase "rocking chair." Depending on the nature of the conversation, the word "rocking" could

[18] In the following please read "LM" as "language model" and "LM'ing" as "language modeling."

[19] We have replaced Peirce's word *syntax* with *grammar* for more consistency with our "engineering" formalisms in the following. We will reserve the word *syntax* to refer to the rules that govern how *words* may combine.

describe either a *type* of chair or the *motion* of a chair or both. A pragmatic knowledge source in a recognizer must be able to discern among the various meanings of symbol strings and, hence, find the correct decoding.

In the 1950s and 1960s, Chomsky further formalized the structure of an LM (Chomsky, 1959), and he and others (Fu, 1982) generalized the concept of *language* and *grammar* to any phenomenon or process that can be viewed as generating structured entities by building them from primitive patterns ("symbols") according to certain "production" rules. The most restrictive type of grammar in Chomsky's hierarchy, and the one that is of most interest in practical application, is the **regular,** or **finite-state, grammar**. In the parlance of the HMM, Chomsky's regular grammar contains production rules that allow any "state" to produce a "symbol" plus a new "state," or to produce a sole "symbol."

In Chomsky's formalism, a **stochastic grammar** is one whose production rules have probabilities associated with them, so each of the strings in the **stochastic language** occurs with a probability that depends on the probabilities of its productions. Clearly, when probabilities are added to the production rules of the regular grammar, the regular grammar can be used as an alternative description of the operation of the Mealy-form HMM. In each "production" (observation time), the HMM produces a symbol and a new state with a certain probability. It is very important to generalize our thinking about the HMM, however, in the present context. Our fundamental descriptions of the HMM involved *acoustic* modeling in which states nominally represented physiologic acoustic states of the vocal systems and in which symbols represented acoustic features. In the LM context, the "HMM" can represent higher level structures as well. For example, the states may represent words while the observations are phonemes, so the model embodies syntactical and lexical information. As noted earlier, the HMM is usually known as a **finite-state automaton** (FSA) in the formal language theory. "HMM" is almost invariantly used to refer to the model at the acoustic level, while "FSA" is more likely to be used for the linguistic models. Regardless of what we call the model, the critical point is the one-to-one correspondence between the model and a regular stochastic grammar.

n-Gram and Other Language Models

While FSA and FSA-like models are prominent among those deployed in practice, other important models have been applied. Notable among these are the *n-gram* LMs in which the occurrence of a string element (usually a word) is characterized by an $(n-1)$-order Markov dependence on past string elements. For example, in a **2-gram** or **bi-gram** LM for syntax, the probability that a particular word appears as the tth word in a string is conditioned on the $(t-1)$st word. In the **3-gram** or **tri-gram** LM, the dependency is on two previous elements. For most vocabularies, the use of n-gram models for $n > 3$ is prohibitive.

Clearly, the more constrained the rules of language in the recognizer, the less freedom of expression the user has in constructing spoken messages. The challenge of LM'ing is to balance the need for maximally constraining the "pathways" that messages may take in the recognizer while minimizing the degree to which freedom of expression is diminished. A measure of the extent to which an LM constrains permissible discourse is given by the **perplexity**. This term roughly means the average number of branches at any decision point when the decoding of messages is viewed as the search through a graph of permissible utterances.

Training and Searching the Language Model

Training methods akin to those studied in previous sections are available for constructing LMs (e.g., see Chapter 13 of Deller *et al.*, 2000). Both F–B-like and Viterbi-like approaches exist for the inference of the probabilities of the production rules of a stochastic grammar, given the characteristic grammar (rules without probabilities) (Fu, 1982; Levinson, 1985). For a regular grammar, this task is equivalent to the problem of finding the probabilities associated with an HMM or an FSA with a fixed structure. Consequently, the fact that these training algorithms exist is not surprising. In fact, however, any stochastic grammar may be shown to have a correlative doubly stochastic (HMM-like) process. A discussion of this issue and related references are found in Levinson (1985).

3.5 Summary and Conclusions

The focus of this chapter has been on fundamental ideas and algorithms underlying the DT processing of speech. Indeed, it is difficult to find a modern application of speech processing that does not employ one or more of the algorithms discussed here, and virtually all contemporary speech technology begins with some variation of the linear DT model to which we devoted much initial attention in this chapter. As a technology with vast economic significance, the amount of research and development that has been, and continues to be, devoted to speech processing is extraordinary. The commercial, military, forensic, and other driving forces for speech technologies are centered primarily in the broad areas of communications and recognition. Modern communication networks, in their many traditional and emerging forms, have created vast interest in speech processing technologies, particularly with regard to coding, compression, and security. At the same time, modern communications systems have engendered a merging of subspecialties in signal processing as speech technologists find themselves increasingly working with image (and other data) processing engineers, specialists in detection and estimation theory, and computer hardware and software systems developers. Further, the global nature of communications has caused some aspects of speech communications and recognition to merge as well. "Recognition" was once envisioned as a quest for a typewriter that takes dictation, and the field has now broadened to include such issues as **language** and **accent recognition**—motivated, at least in part, as a facilitating technology for multilanguage communications systems and other computer-assisted telephony applications (Waibel *et al.*, 2000). Both as stand-alone technologies and also as part of the communications revolution, security and forensics applications such as **speaker identification** and **speaker verification** have also expanded the activities of "recognition" engineers (Campbell, 1997).

Although the commercial interests have played dominant roles in setting the speech processing agenda, many other applications of speech processing continue to be explored. Biomedicine represents one area, for example, in which speech engineering is playing an important role in the development of diagnostic software, techniques for assessment of speech and hearing therapies, and assistive aids for persons with speech, hearing, and motor disabilities. At the same time, funding agencies continue to support new theories and applications of speech processing, including models of human discourse, theories of hearing and psychoacoustics, and other basic research activities. To explore these emerging technologies and well as the wealth of information about modern applications, the reader is encouraged to explore the literature cited here, the further citations to which it leads, and the ever-increasing amount of information available on the World Wide Web.

References

Bahl, L.R., Brown, P.F., DeSouza, P.V., and Mercer, R.L. (1988). A new algorithm for the estimation of hidden Markov model parameters. *Proceedings of the IEEE International Conference on Acoustics, Speech, and Signal Processing 1*, 493–496.

Baker, J.K. (1975). Stochastic modeling for automatic speech understanding. In D.R. Reddy (Ed.), *Speech recognition.* New York: Academic Press.

Bakis, R. (1976). Continuous speech word recognition via centisecond acoustic states. *Proceedings of the 91st Annual Meeting of the Acoustics Society of America.*

Baum, L.E. (1972). An inequality and associated maximization technique in statistical estimation for probabilistic functions of Markov processes. *Inequalities 3*, 1–8.

Baum, L.E., and Sell, G.R. (1968). Growth functions for transformations on manifolds. *Pacific Journal of Mathematics 27*, 211–227.

Bellman, R.E. (1975). *Dynamic programming.* Princeton, NJ: Princeton University Press.

Bielefeld, B. (1994). Language identification using shifted delta cepstrum. *Proceedings of the 14th Annual Speech Research Symposium.*

Bocchieri, E.L., and Doddington, G.R. (1986). Frame-specific statistical features for speaker-independent speech recognition. *IEEE Transactions on Acoustics Speech, and Signal Processing 34*, 755–764.

Bogert, B.P., Healy, M.J.R., and Tukey, J.W. (1963). The quefrency alanysis of time series for echoes: Cepstrum, pseudo-autocovariance, cross-cepstrum, and saphe cracking. In M. Rosenblatt (Ed.), *Proceedings of the Symposium on Time Series Analysis.* New York: John Wiley & Sons.

Burg, J.P. (1967). Maximum entropy spectral analysis. *Proceedings of the 37th Meeting of the Society of Exploration Geophysicists.*

Campbell, Jr., J.P. (1997). Speaker recognition: A tutorial. *Proceedings of the IEEE 85*, 1437–1462.

Chomsky, N. (1959). On certain formal properties of grammars. *Information and Control 2*, 137–167.

Chomsky, N., and Halle, M. (1968). *The sound pattern of English.* New York: Harper and Row. Reprinted (1991). Boston: MIT Press.

Dautrich, B., Rabiner, L.R., and Martin, T.B. (1983). On the effects of varying filter bank parameters on isolated word recognition. *IEEE Transactions on Acoustics, Speech, and Signal Processing 31*, 793–807.

Davis, S.B., and Mermelstein, P. (1980). Comparison of parametric representations for monosyllabic word recognition in continuously spoken sentences. *IEEE Transactions on Acoustics, Speech, and Signal Processing 28*, 357–366.

Deller, Jr., J.R., and Odeh, S.F. (1993). Adaptive set-membership identification in $\mathcal{O}$ (m) time for linear-in-parameters models. *IEEE Transactions on Signal Processing 41*, 1906–1924.

Deller, Jr., J.R., Hansen, J.H.L., and Proakis, J.G. (2000). *Discrete-time processing of speech signals.* (2d ed.). New York: IEEE Press.

Devijver, P.A., and Kittler, J. (1982). *Pattern recognition: A statistical approach.* London: Prentice Hall.

Dudley, H. (1939). The vocoder. *Bell Labs Record. 17*, 122–126.

Ephraim, Y., Dembo, A., and Rabiner, L.R. (1989). A minimum discrimination information approach for hidden Markov modeling. *IEEE Transactions on Information Theory 35*, 1001–1013.

Fant, C.G.M. (1956). On the predictability of formant levels and spectrum envelopes from formant frequencies. In Halle, M., Lunt, H., MacLean, H. (eds.). *For Roman Jakobson.* The Hague: Mouton.

Fant, C.G.M. (1960). *Acoustic theory of speech production.* The Hague: Mouton.

Fant, C.G.M. (1973). *Speech sounds and features.* Cambridge, MA: MIT Press.

Flanagan, J.L. (1972). *Speech analysis, synthesis, and perception.* (2nd. ed.). New York: Springer-Verlag.

Fu, K.S. (1982). *Syntactic pattern recognition and applications.* Englewood Cliffs, NJ: Prentice Hall.

Golub, G.H., and van Loan, C.F. (1989). *Matrix computations.* (2nd. ed.). Baltimore: Johns-Hopkins University.

Haykin, S. (1996). *Adaptive filter theory.* (3d. ed.). Englewood Cliffs, NJ: Prentice Hall.

Hopcroft, J.E., and Ullman, J.D. (1979). *Introduction to automata theory, languages, and computation.* Reading, MA: Addison-Wesley.

Itakura, F. (1975). Line spectrum representation of linear prediction coefficients of speech signals. *Journal of the Acoustics Society of America 57*, 535.

Itakura, F. (1975). Minimum prediction residual principle applied to speech recognition. *IEEE Transactions on Acoustics Speech, and Signal Processing 23*, 67–72.

Itakura, F., and Saito, S. (1969). Speech analysis-synthesis system based on the partial autocorrelation coefficient. *Proceedings of the Acoustic Society of Japan Meeting.*

Jelinek, F., Bahl, L.R., and Mercer, R.L. (1975). Design of a linguistic statistical decoder for the recognition of continuous speech. *IEEE Transactions on Information Theory 21*, 250–256.

Juang, B.H., Levinson, S.E., and Sondhi, M.M. (1986). Maximum likelihood estimation for multivariate mixture observations of Markov chains. *IEEE Transactions on Information Theory 32*, 307–309.

Juang, B.H., Rabiner, L.R., and Wilpon, J.G. (1987). On the use of bandpass liftering in speech recognition. *IEEE Transactions on Acoustics Speech, and Signal Processing 35*, 947–954.

Klatt, D. (1987). Review of text-to-speech conversion for English. *Journal of Acoustic Society of America 82*, 737–793.

Koenig, W. (1949). A new frequency scale for acoustic measurements. *Bell Telephone Laboratory Record 27*, 299–301.

Koenig, W., Dunn, H.K., and Lacy, L.Y. (1946). The sound spectrograph. *Journal of Acoustic Society of America 17*, 19–49.

Ladefoged, P. (1975). *A course in phonetics.* New York: Harcourt-Brace-Javanovich.

Leon-Garcia, A. (1989). *Probability and random processes for electrical engineering.* Reading, MA: Addison-Wesley.

Levinson, S.E. (1985). Structural methods in automatic speech recognition. *Proceedings of the IEEE 73*, 1625–1650.

Levinson, S.E., Rabiner, L.R., and Sondhi, M.M. (1983). An introduction to the application of the theory of probabilistic functions of a Markov process to automatic speech recognition. *Bell Systems Technology Journal 62*, 1035–1074.

Lieberman, P. (1967). *Intonation, perception, and language.* Cambridge, MA: MIT Press.

Liporace, L.A. (1982). Maximum likelihood estimation for multivariate observations of Markov sources. *IEEE Transactions on Information Theory, 28*, 729–734.

Liszka, J.J. (1996). *A general introduction to the semeiotic of Charles Sanders Peirce.* Bloomington: Indiana University Press.

Lowerre, B.T., and Reddy, D.R. (1980). The HARPY speech understanding system. In W.A. Lea (Ed.), *Trends in speech recognition.* Englewood Cliffs, NJ: Prentice Hall.

Makhoul, J. (1977). Stable and efficient lattice methods for linear prediction. *IEEE Transactions on Acoustics Speech, and Signal Processing 25*, 423–428.

Markel, J.D., and Gray, A.H. (1976). *Linear prediction of speech.* New York: Springer-Verlag.

McWhiter, J.G. (1983). Recursive least squares solution using a systolic array. *Proceedings of the Society of Photooptical Instrumentation Engineers (Real-Time Signal Processing IV) 431*, 105–112.

Milner, P.M. (1970). *Physiological psychology.* New York: Holt–Rinehart–Winston.

Morgan, D.P., and Scofield, C.L. (1991). *Neural networks and speech processing.* Boston: Kluwer Academic.

Morgan, N., and Gold, B. (1999). *Speech and audio signal processing: Processing and perception of speech and music.* New York: John Wiley & Sons.

Myers, C.S., Rabiner, L.R., and Rosenberg, A.E. (1980). Performance trade-offs in dynamic time warping algorithms for isolated word recognition. *IEEE Transactions on Acoustics, Speech, and Signal Processing 28*, 622–635.

Nocerino, N., Soong, F.K., Rabiner, L.R., and Klatt, D.H. (1985). Comparative study of several distortion measures for speech recognition. *Proceedings of the IEEE International Conference on Acoustics, Speech, and Signal Processing 1*, 25–28.

Noll, A.M. (1967). Cepstrum pitch determination. *Journal of the Acoustic Society of America 41*, 293–309.

O'Shaughnessy, D. (1987). *Speech communication: Human and machine.* Reading, MA: Addison-Wesley.

Oppenheim, A.W., and Schafer, R.W. (1968). Homomorphic analysis of speech. *IEEE Transactions on Audio and Electroacoustics 16*, 221–226.

Potter, R.G., Kopp, G.A., and Kopp, H.G. (1966). *Visible speech.* New York: Van Nostrand.

Rabiner, L.R., and Juang, B.-H. (1993). *Fundamentals of speech recognition.* Englewood Cliffs, NJ: Prentice Hall.

Rabiner, L.R., and Schafer, R.W. (1978). *Digital processing of speech signals.* Englewood Cliffs, NJ: Prentice Hall.

Reddy, D.R. (1976). Speech recognition by machine: A review. *Proceedings of the IEEE 64*, 501–531.

Rosenberg, A.E. (1971). Effects of glottal pulse shape on the quality of natural vowels. *Journal of Acoustic Society of America 49*, 583–590.

Schafer, R.W., and Markel, J.D. (Eds.). (1979). *Speech analysis.* New York: John Wiley & Sons.

Silverman, H.F., and Morgan, D.P. (1990). The application of dynamic programming to connected speech recognition. *IEEE Acoustics, Speech, and Signal Processing Magazine 7*, 6–25.

Stevens, S.S., and Volkman, J. (1940). The relation of pitch to frequency. *American Journal of Psychology 53*, 329.

Strobach, P. (1991). New forms of Levinson and Schur algorithms. *IEEE Signal Processing Magazine*, 12–36.

Torres-Carrasquillo, P.A. *et al.*, (2002). An approach to language identification using Gaussian mixture models and shifted delta cepstra. *Proceedings of the International Conference on Spoken Language Processing.*

Viswanathan, R., and Makhoul, J. (1975). Quantization properties of the transmission parameters in linear predictive systems. *IEEE Transactions on Acoustics Speech, and Signal Processing 23*, 309–321.

Viterbi, A.J. (1967). Error bounds for convolutional codes and an asymptotically optimal decoding algorithm. *IEEE Transactions on Information Theory 13*, 260–269.

Waibel, A., Geutner, P., Tomokiyo, L.M., Schultz, T., and Woszczyna, M. (2000). Multilinguality in speech and spoken language systems. *Proceedings of the IEEE 88*, 1181–1200.

4

Digital Image Processing

Eduardo A.B. da Silva
Program of Electrical Engineering, Federal University of Rio de Janeiro, Rio de Janeiro, Brazil

Gelson V. Mendonça
Department of Electronics, COPPE/EE/Federal University of Rio de Janeiro, Rio de Janeiro, Brazil

4.1 Introduction

Digital image processing consists of the manipulation of images using digital computers. Its use has been increasing exponentially in the last decades. Its applications range from medicine to entertainment, passing by geological processing and remote sensing. Multimedia systems, one of the pillars of the modern information society, rely heavily on digital image processing.

The discipline of digital image processing is a vast one, encompassing digital signal processing techniques as well as techniques that are specific to images. An image can be regarded as a function $f(x, y)$ of two continuous variables x and y. To be processed digitally, it has to be **sampled** and transformed into a matrix of numbers. Since a computer represents the numbers using finite precision, these numbers have to be **quantized** to be represented digitally. Digital image processing consists of the manipulation of those finite precision numbers. The processing of digital images can be divided into several classes: **image enhancement**, **image restoration**, **image analysis**, and **image compression**. In image enhancement, an image is manipulated, mostly by heuristic techniques, so that a human viewer can extract useful information from it. Image restoration techniques aim at processing corrupted images from which there is a statistical or mathematical description of the degradation so that it can be reverted. Image analysis techniques permit that an image be processed so that information can be automatically extracted from it. Examples of image analysis are image segmentation, edge extraction, and texture and motion analysis. An important characteristic of images is the huge amount of information required to represent them. Even a gray-scale image of moderate resolution, say 512×512, needs $512 \times 512 \times 8 \approx 2 \times 10^6$ bits for its representation. Therefore, to be practical to store and transmit digital images, one needs to perform some sort of image compression, whereby the redundancy of the images is exploited for reducing the number of bits needed in their representation.

In what follows, we provide a brief description of digital image processing techniques. Section 4.1 deals with image sampling, and Section 4.2 describes image quantization. In Section 4.3, some image enhancement techniques are given. Section 4.4 analyzes image restoration. Image compression, or **coding**, is presented in Section 4.5. Finally, Section 4.6 introduces the main issues involved in image analysis.

4.2 Image Sampling

An analog image can be seen as a function $f(x, y)$ of two continuous space variables: x and y. The function $f(x, y)$ represents a variation of gray level along the spatial coordinates. Since each color can be represented as a combination of three primaries (R, G, and B), a color image can be represented by the three functions $f_R(x, y)$, $f_G(x, y)$, and $f_B(x, y)$, one for each color component.

A natural way to translate an analog image into the discrete domain is to **sample** each variable; that is, we make $x = n_x T_x$ and $y = n_y T_y$. This way, we get a digital image $f_D(nx, ny)$ such that:

$$f_D(n_x, n_y) = f(n_x T_x, n_y T_y). \tag{4.1}$$

Likewise, for the one-dimensional case, one has to devise conditions under which an analog image can be recovered from its samples. We will first generalize the notions of frequency content and Fourier transform for two-dimensional signals.

One can represent an image in the frequency domain by using a two-dimensional Fourier transform. It is a straightforward extension of the one-dimensional Fourier transform, as can be seen in these equations:

$$F(\Omega_x, \Omega_y) = \int_{-\infty}^{\infty}\int_{-\infty}^{\infty} f(x, y)e^{-j(\Omega_x x + \Omega_y y)}dxdy. \tag{4.2}$$

$$f(x, y) = \frac{1}{4\pi^2}\int_{-\infty}^{\infty}\int_{-\infty}^{\infty} F(\Omega_x, \Omega_y)e^{j(\Omega_x x + \Omega_y y)}d\Omega_x d\Omega_y. \tag{4.3}$$

The function $e^{j(\Omega_x x + \Omega_y y)}$ is a complex sinusoid in two dimensions. It represents a two-dimensional pattern that is sinusoidal both along the x (horizontal) and y (vertical) directions. The Ω_x and Ω_y are the spatial frequencies along the x and y directions, respectively. An example can be seen in Figure 4.1, where a two-dimensional sinusoid is depicted as $\cos(\Omega_x x + \Omega_y y)$, with $\Omega_x = 2\Omega_y$. One can note that there are twice as many cycles in the x direction than in the y direction. The period in the x direction is $2\pi/\Omega_x$ and in the y direction is $2\pi/\Omega_y$.

Therefore, equation 4.3 means that any two-dimensional signal is equivalent to an infinite sum of two-dimensional complex sinusoids. The term that multiplies the sinusoid of frequency (Ω_x, Ω_y) is its Fourier transform, $F(\Omega_x, \Omega_y)$, and is given by equation 4.2.

Having defined the meaning of frequency content of two-dimensional signals, we are now ready to state conditions under which an image can be recovered from its samples. It is well known that most useful digital images have the significant part of the amplitude spectrum concentrated at low frequencies. Therefore, an image in general can be modeled as a bandlimited two-dimensional signal; that is, its Fourier transform $F(\Omega_x, \Omega_y)$ is approximately zero outside a bounded region, as shown in Figure 4.2.

FIGURE 4.1 Two-Dimensional Sinusoid with $\Omega_x = 2\Omega_y$. Its expression is $f(x, y) = 1/2(e^{j(\Omega_x x + \Omega_y y)} + e^{-j(\Omega_x x + \Omega_y y)})$.

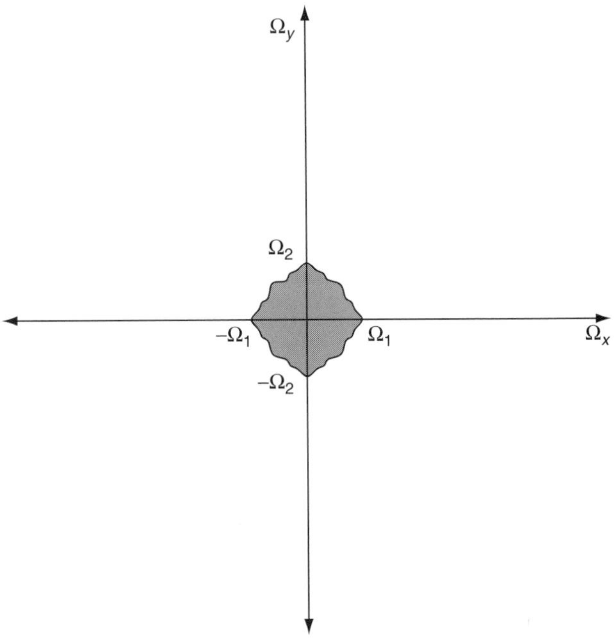

FIGURE 4.2 Support Region of the Fourier Transform of a Typical Image

When the sampling process is performed according to equation 4.1, it is equivalent to sampling the continuous signal using a rectangular grid as in Figure 4.3. As in the one-dimensional case (Oppenheim and Schaffer, 1989), one can show that the spectrum of the digital image is composed of replicas of the analog image spectrum, repeated with a period of $2\pi/T_x$

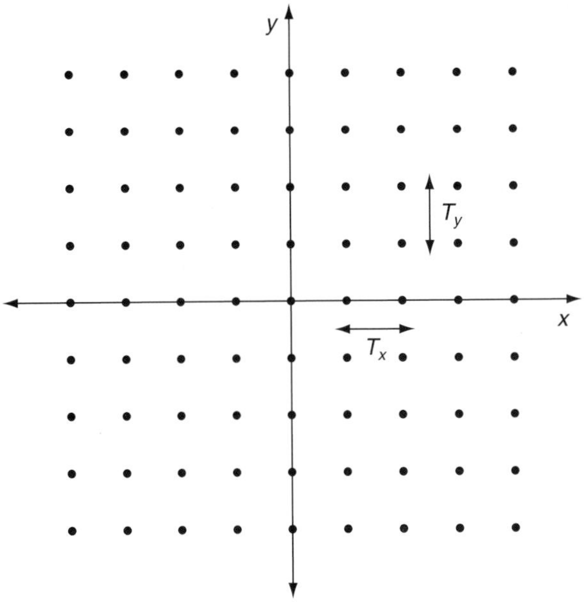

FIGURE 4.3 Sampling a Two-Dimensional Signal Using a Rectangular Grid

along the horizontal frequency axis and $2\pi/T_y$ along the vertical frequency axis, as depicted in Figure 4.4.

The mathematical expression for the spectrum of the sampled signal is then:

$$F_s(\Omega_x, \Omega_y) = \frac{1}{T_x T_y} \sum_{k_x=-\infty}^{\infty} \sum_{k_y=-\infty}^{\infty} F\left(\Omega_x - \frac{2\pi k_x}{T_x}, \Omega_y - \frac{2\pi k_y}{T_y}\right). \quad (4.4)$$

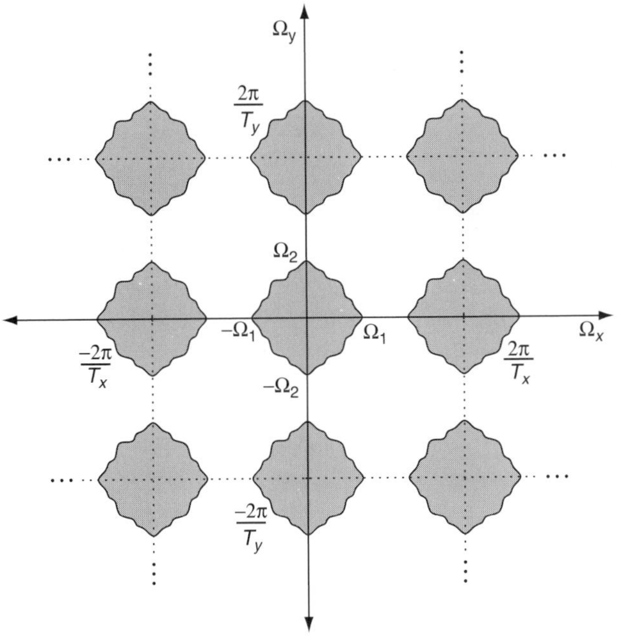

FIGURE 4.4 Spectrum of the Two-Dimensional Discrete Signal Generated by Sampling Using a Rectangular Grid

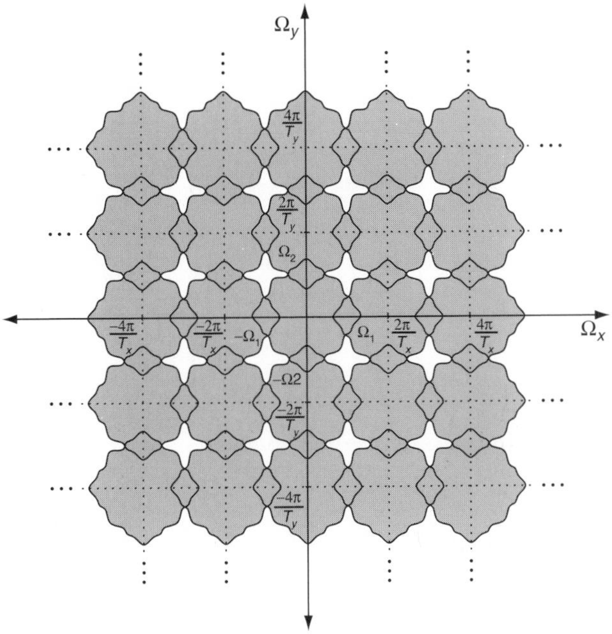

FIGURE 4.5 Aliased Spectrum of the Two-Dimensional Discrete Signal Generated by Sampling Using a Rectangular Grid

From the above equation and Figure 4.4, we note that if the sampling intervals do not satisfy the Nyquist conditions $\Omega_1 \leq 2\pi/T_x$ and $\Omega_2 \leq 2\pi/T_y$ then the analog signal cannot be reconstructed from its samples. This is due to the overlap of the replicated frequency spectra, which, likewise in the one-dimensional case, is known as aliasing. Aliasing is illustrated in Figure 4.5.

Provided that aliasing is avoided, the sampled signal has to be processed by a two-dimensional low-pass filter to obtain the image signal back into its analog form. This filter should ideally have a transfer function with an amplitude spectrum equal to one in the support region of the Fourier transform of the original image and zero outside that region.

In Figure 4.6(A), we see the original LENA image. It has been sampled without aliasing. Figure 4.6(B) shows the LENA image sampled at a rate much smaller than the minimum sampling frequencies in the horizontal and vertical directions necessary to avoid aliasing. Aliasing can be clearly noted. In Figure 4.6(C), the bandwidth of the image LENA has been reduced prior to sampling so that aliasing is avoided after sampling. Figure 4.6(D) shows the filtered image after sampling. We can notice that despite the blur on the image due to the lower bandwidth, there is indeed less aliasing indicated by the smoother edges.

4.2.1 Nonrectangular Grid Sampling

A two-dimensional signal does not necessarily need to be sampled using a rectangular grid like the one described by equation 4.1 and Figure 4.3.

(A) Original

(B) Sampled at Below the Sampling Frequencies

(C) Low-Pass Filtered

(D) Low-Pass Filtered and Sampled

FIGURE 4.6 LENA Image

For example, a nonrectangular grid as in Figure 4.7 can be used. The nonrectangular grid sampling can be best described by using vector and matrix notation. Let:

$$\boldsymbol{t} = \begin{pmatrix} t_x \\ t_y \end{pmatrix} \; \boldsymbol{n} = \begin{pmatrix} n_x \\ n_y \end{pmatrix} \; \boldsymbol{V} = \begin{pmatrix} v_{xx} & v_{xy} \\ v_{yx} & v_{yy} \end{pmatrix} = \begin{pmatrix} \boldsymbol{v}_x & \boldsymbol{v}_y \end{pmatrix} \; \Omega = \begin{pmatrix} \Omega_x \\ \Omega_y \end{pmatrix}. \tag{4.5}$$

The sampling process then becomes:

$$f_D(\boldsymbol{n}) = f_D(n_x, n_y) = f(n_x \boldsymbol{v}_x + n_y \boldsymbol{v}_y) = f(\boldsymbol{V}\boldsymbol{n}). \tag{4.6}$$

In the case of Figure 4.7:

$$\boldsymbol{V} = \begin{pmatrix} T_x & T_x \\ T_y & -T_y \end{pmatrix}, \tag{4.7}$$

and it can be shown that the spectrum of the sampled signal becomes (Mersereau and Dudgeon, 1984):

$$F_s(\Omega) = \frac{1}{|\det(\boldsymbol{V})|} \sum_{\boldsymbol{k}} F(\Omega - 2\pi(\boldsymbol{V}^{-1})^t \boldsymbol{k}). \tag{4.8}$$

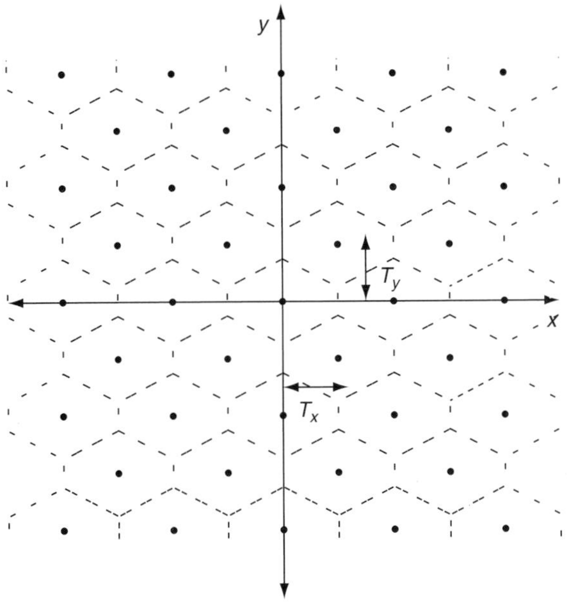

FIGURE 4.7 Sampling a Two-Dimensional Signal Using a Nonrectangular Grid

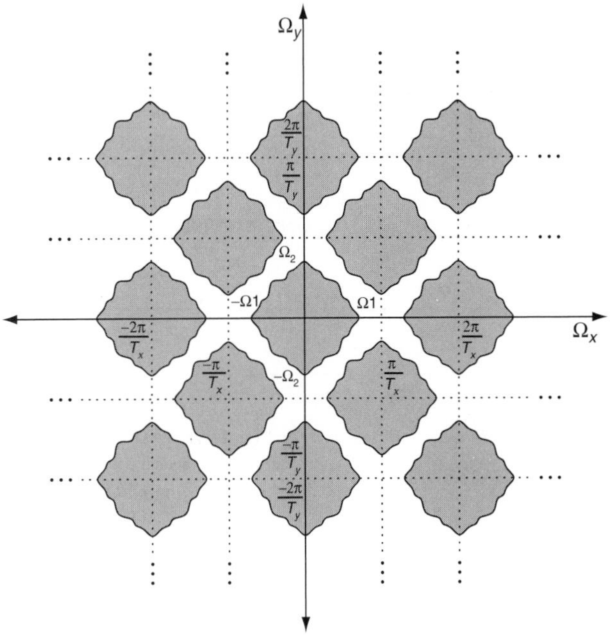

FIGURE 4.8 Spectrum of the Two-Dimensional Discrete Signal Generated by Sampling Using a Nonrectangular Grid

The matrix $2\pi(\boldsymbol{V}^{-1})^t = \boldsymbol{U}$ gives the periodicity, in the two-dimensional frequency plane, of the spectrum repetitions.

For the sampling grid defined in equation 4.7, we have that the periodicity matrix in the frequency plane is the following:

$$\boldsymbol{U} = \begin{pmatrix} \dfrac{\pi}{T_x} & \dfrac{\pi}{T_x} \\ \dfrac{\pi}{T_y} & -\dfrac{\pi}{T_y} \end{pmatrix}. \tag{4.9}$$

Then, the spectrum becomes like the one shown in Figure 4.8.

One particular nonrectangular grid of interest arises when we make $T_y = \sqrt{3}T_x$. We refer to it as **hexagonal sampling**. One advantage of using such a sampling grid instead of the rectangular one is that, for signals with isotropic frequency response, hexagonal sampling needs 13.4% fewer samples than the rectangular one for representing them without aliasing (Mersereaw and Dudgeon, 1984). Such sampling grid is also useful when the continuous signals have a baseband as in Figure 4.2, as can be observed from Figures 4.4, 4.5, and 4.8. In fact, given the support region of a baseband signal, one can always determine the best sampling grid so that aliasing is avoided with the smallest possible density of samples.

4.3 Image Quantization

To be represented in a digital computer, every signal must be **quantized** once sampled. A digital image, for example, is a matrix of numbers represented with finite precision in the machine being used to process it. More generally, a quantizer is a function $Q(\cdot)$ such that:

$$Q(x) = r_l \text{ for } t_{l-1} < x \le t_l \text{ for } l = 1, \ldots, L. \tag{4.10}$$

The numbers r_l, $l = 1, \ldots, L$ are referred to as the **reconstruction levels** of the quantizer, and t_l, $l = 0, \ldots, L$ are referred to as its **decision levels**. A typical quantizer is depicted in Figure 4.9.

In many cases, the decision and reconstruction levels of a quantizer are optimized such that the mean-squared error between the quantizer's input and output is minimized. This is usually referred to as a **Lloyd-Max quantizer** (Jain, 1989). In digital image processing, one often uses linear quantizers. For a linear quantizer, the reconstruction and decision levels satisfy the following relations:

$$t_0 = -\infty. \tag{4.11}$$

$$t_{l+1} - t_l = q, \quad \text{for } l = 1, \ldots, L. \tag{4.12}$$

$$r_{l+1} = t_l + \frac{q}{2} \quad \text{for } l = 1, \ldots, L. \tag{4.13}$$

$$t_L = \infty. \tag{4.14}$$

The parameter q is referred to as the **quantizer step size**.

The mean square error of the linear quantizer, assuming that the quantization error $e(n) = x(n) - Q[x(n)]$ is uniformly distributed, is given by (Jain, 1989):

$$E[e_2(n)] = \frac{q^2}{12}. \tag{4.15}$$

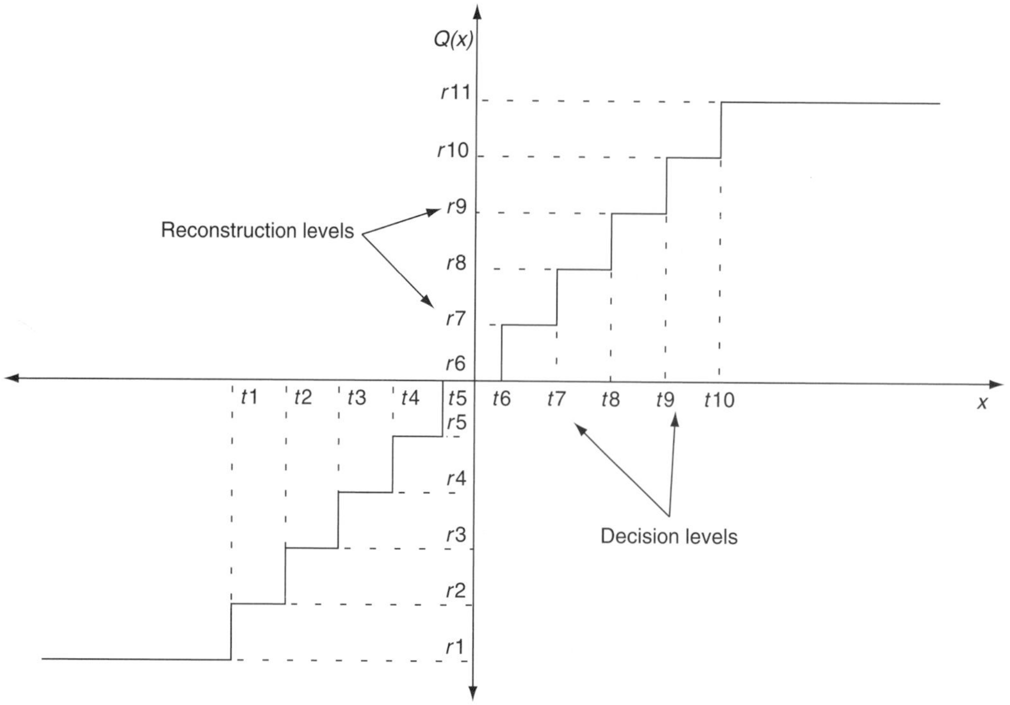

FIGURE 4.9 Typical Quantizer

Since the human eye cannot in general recognize more than around 256 gray levels, usually 8 bits per pixel are enough for representing a gray-scale image. In the case of color images, one uses 8 bits for each color component. For RGB images, this gives 24 bits/pixel.

In many occasions, one needs to quantize an image with less than 8 bits/pixel. Figure 4.10(A) shows the image LENA quantized with 4 bits/pixel. One can notice the effect of false contours. One way to overcome this problem is to perform **dithering** (Jain, 1989). It consists in adding pseudo-random noise to the image prior to quantization. Before display, this noise can be either subtracted from the image or not subtracted. This is illustrated in Figures 4.10(A) to 4.10(C). Figure 4.10(B) shows the LENA image with a pseudo-random noise uniformly distributed in the interval $[-25, 25]$ quantized with 8 levels. Figure 4.10(C) shows the image in Figure 4.10(B) with the pseudo-random noise subtracted from it. One can notice that in both cases there is a great improvement in visual quality when comparing it with Figure 4.10(A). The false contours almost disappear.

(A) Quantized with 8 Levels

(B) Quantized with 8 Levels with Noise Added Prior to Quantization

(C) The Image in (B) with the Noise also Subtrated after Quntization

FIGURE 4.10 LENA Image

4.4 Image Enhancement

Image enhancement is the name given to a class of procedures by which an image is processed so that it can be better displayed or improved for a specific analysis. This can be obtained at the expense of other aspects of the image. There are many enhancement techniques that can be applied to an image.

To provide a flavor of the image enhancement techniques, we will give two examples namely, **histogram manipulation** and **image filtering**.

4.4.1 Histogram Manipulation

The histogram of an image is a function that maps each gray level of an image to the number of times it occurs in the image. For example, the image in Figure 4.11(A) has the histogram shown in Figure 4.11(B).

One should note that the pixels have, in general, gray levels in the integer range [0,255].

Histogram Equalization

By looking at Figure 4.11(A), one notices that the image is too dark. This can be confirmed by the image's histogram in Figure 4.11(B), where one can see that the most frequent gray levels have low values. To enhance the appearance of the image, one would need to re-map the image's gray levels so that they become more uniformly distributed. Ideally, one would need to apply a transformation that would make the histogram of the image look uniform. If $F_U(u) = \int_0^u p_U(x)dx$ is the distribution function of the image, then this transformation would

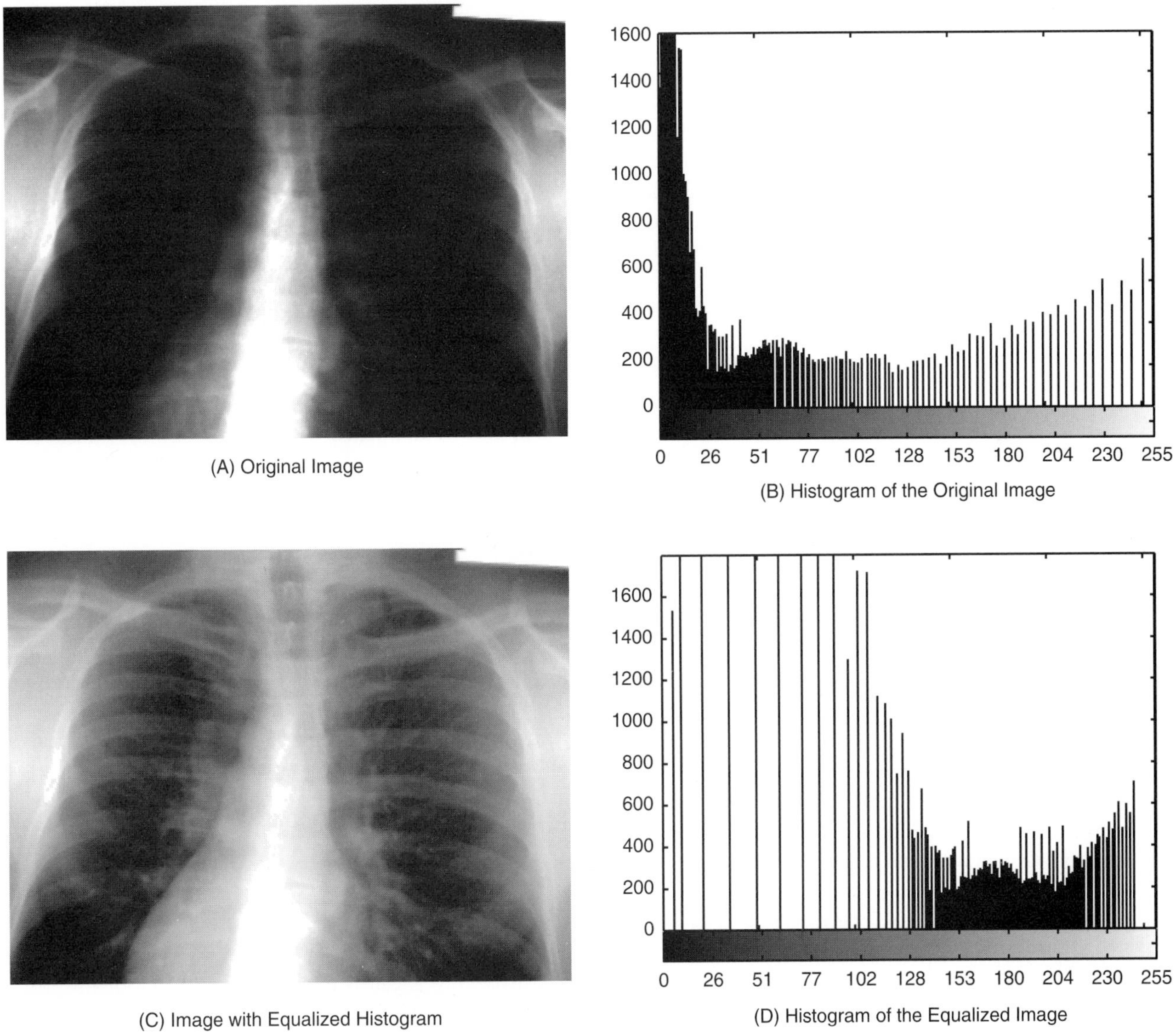

FIGURE 4.11 Histogram of an Image

be $y = F^{-1}(x)$ (Gonzalez and Wintz, 1977). In practice, since the pixels can attain only integer values, this operation cannot be performed exactly, and some sort of quantization must be carried out (Jain, 1989).

Figure 4.11(C) shows the image with equalized histogram, and Figure 4.11(D) shows its histogram. Note that the quality of the image is far superior to the original one, and the histogram is much more uniform then the one in Figure 4.11(B).

Another similar histogram manipulation technique is **histogram specification**, where we try to make the histogram of an image as similar as possible to a given one (Gonzalez and Wintz, 1977; Jain, 1989). Some texts refer to **histogram matching**. It is a kind of histogram specification technique in which the histogram of an image is matched to the one of another image. It can be used, for example, when there are two images of the same scene taken from two different sensors.

4.4.2 Linear Filtering

Linear filtering is one of the most powerful image enhancement methods. It is a process in which part of the signal frequency spectrum is modified by the transfer function of the filter. In general, the filters under consideration are linear and shift-invariant, and thus, the output images are characterized by the convolution sum between the input image and the filter impulse response; that is:

$$y(m, n) = \sum_{i=0}^{M}\sum_{j=0}^{N} h(m-i, n-j)x(i, j) = h(m, n) ** x(m, n), \quad (4.16)$$

where the following is true:

- The $y(m, n)$ is the output image.
- The $h(m, n)$ is the filter impulse response.
- The $x(m, n)$ is the input image.

For example, low-pass filtering has the effect of smoothing an image, as can be observed from Figure 4.6(C). On the other hand, high-pass filtering usually sharpens the edges of an image. They can even be used for edge detection, which is used in image analysis algorithms.

The image filtering can be carried out either in the spatial domain, as in equation 4.16, or in the frequency domain, using the discrete Fourier transform (DFT) (Mersereau and Dudgeon, 1984; Oppenheim and Schaffer, 1989). For filtering using the DFT, we use the well known property that the DFT of the circular convolution of two sequences is equal to the product of the DFTs of the two sequences. That is, for $y(m,n)$ defined as in equation 4.16, provided that a DFT of sufficient size is used, we have that:

$$\text{DFT}\{y(m, n)\} = \text{DFT}\{h(m, n)\}\ \text{DFT}\{x(m, n)\}. \quad (4.17)$$

Therefore, one can perform image filtering in the frequency domain by modifying conveniently the DFT of the image and taking the inverse transformation. This is exemplified in Figures 4.12(A) through 4.12(D).

In Figure 4.12(A), we see an image contaminated with periodic noise. Figure 4.12(B) shows the DFT of this image. One can clearly see the periodic noise as two well-defined points on the DFT of the image. For convenience, arrows are pointing to them. In Figure 4.12(C), one can see the DFT of the image with the periodic noise removed; the frequency locations corresponding to the periodic noise were made equal to zero. Figure 4.12(D) shows the inverse DFT of the image in Figure 4.12(C). One can notice that the noise has been effectively removed, improving the quality of the image a great deal.

Image enhancement is an extensive topic. In this text, we aim at giving a flavor of the topic by providing typical examples. For a more detailed treatment see works by Gonzalez and Wintz (1977) and Jain (1989).

4.5 Image Restoration

In **image restoration**, one is usually interested in recovering an image from a degraded version of it. It is essentially different from image enhancement, which is concerned with accentuation or extraction of image features. Moreover, in image restoration, one usually has mathematical models of the degradation and a statistical description of the ensemble of images used. In this section, we will study one type of restoration problem, namely, the recovery of images degraded by a linear system and contaminated with additive noise. A linear model for degradations is depicted in Figure 4.13. The matrix $u(m, n)$ is the original image, $h(m, n)$ is a linear system representing the degradation, $\eta(m, n)$ is an additive noise, and $v(m, n)$ is the observed degraded image. We can write:

$$v(m, n) = u(m, n) ** h(m, n) + \eta(m, n), \quad (4.18)$$

where $**$ represents a two-dimensional convolution (see equation 4.16).

An illustrative case is when $\eta(m, n) = 0$. In this situation, we can recover $u(m, n)$ from $v(m, n)$ by using a linear filter $g(m, n)$ such that $g(m, n) ** h(m, n) = \delta(m, n)$. In the frequency domain, this is equivalent to having:

$$G(\omega_x, \omega_y) = \frac{1}{H(\omega_x, \omega_y)}. \quad (4.19)$$

This procedure is called **inverse filtering**. The main problem with the inverse filter is that it is not defined in the cases that there is a pair (ω_1, ω_2) such that $H(\omega_1, \omega_2) = 0$. A solution to this problem is the **pseudo-inverse filter** defined as:

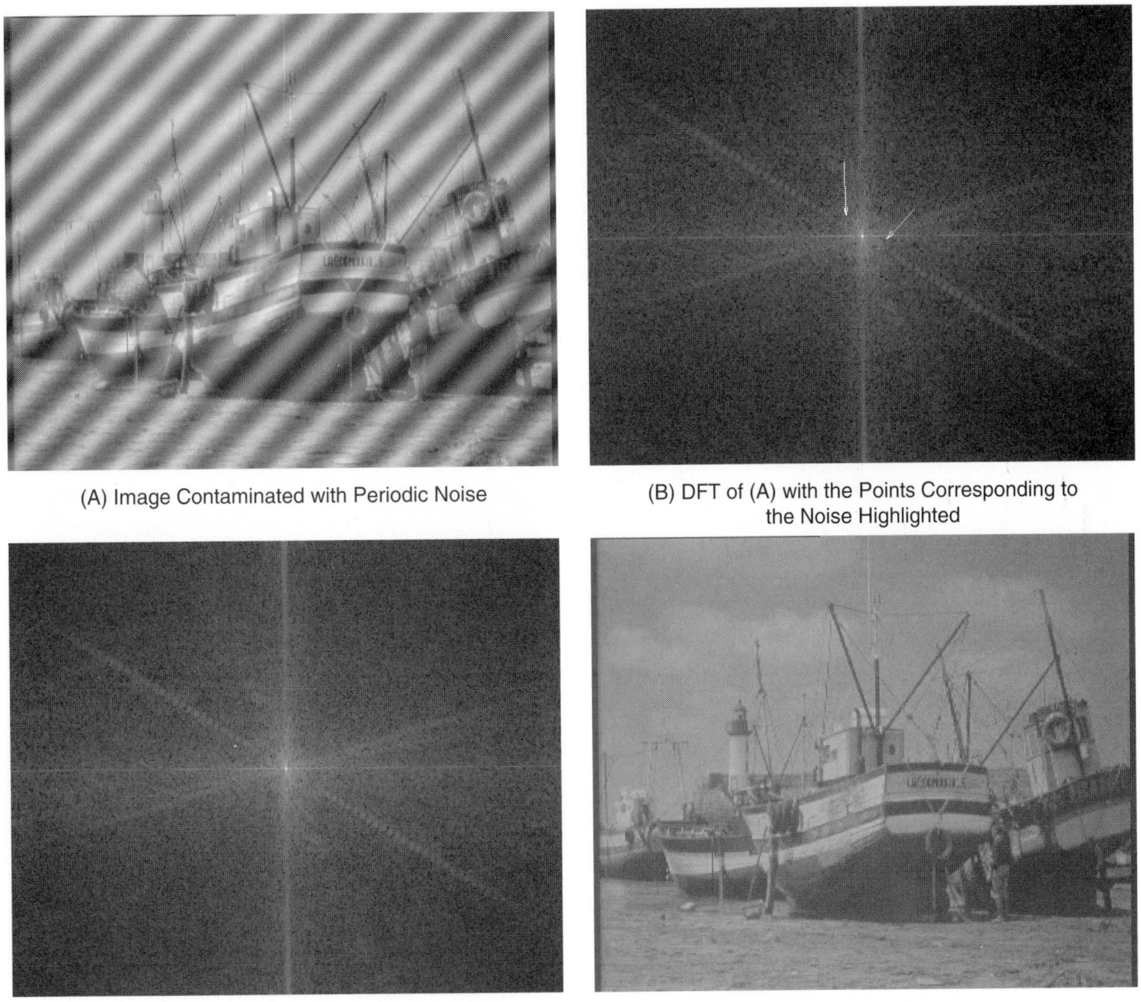

(A) Image Contaminated with Periodic Noise

(B) DFT of (A) with the Points Corresponding to the Noise Highlighted

(C) DFT (A) with the Periodic Noise Remove

(D) Recovered Image

FIGURE 4.12 Image Filtering

$$G(\omega_x, \omega_y) = \begin{cases} \frac{1}{H(\omega_x, \omega_y)}, & H(\omega_x, \omega_y) \neq 0. \\ 0, & H(\omega_x, \omega_y) = 0. \end{cases} \quad (4.20)$$

As an example, we analyze now the restoration of images degraded by blur. For image blur caused by excessive film exposure time in the presence of camera motion in the horizontal direction, this degradation can be modeled, in the case of absence of additive noise, as:

$$v(x, y) = \frac{1}{T}\int_0^T u(x - ct, y)dt, \quad (4.21)$$

where the relative speed between the camera and the scene is c and the camera exposure time is T. In the digital domain, this is equivalent to the following operation:

$$v(m, n) = \frac{1}{L}\sum_{t=0}^{L-1} u(m - l, n). \quad (4.22)$$

Figure 4.14(A) shows the ZELDA image, of dimensions 360×288, blurred using equation 4.22 with $L = 48$. Applying the pseudo-inverse filter (see equation 4.20) in the frequency

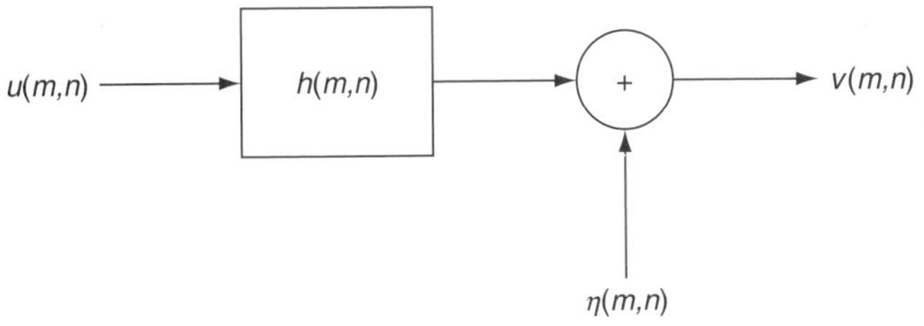

FIGURE 4.13 Image Degradation Model

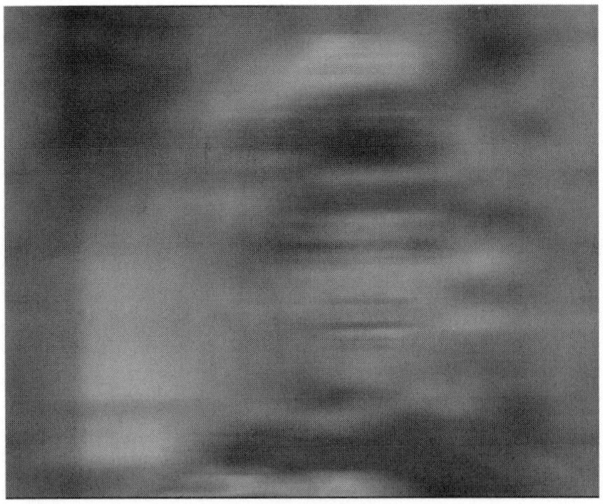

(A) Image Degraded by Blur

(B) Image Restored Using the Pseudo Inverse Filter

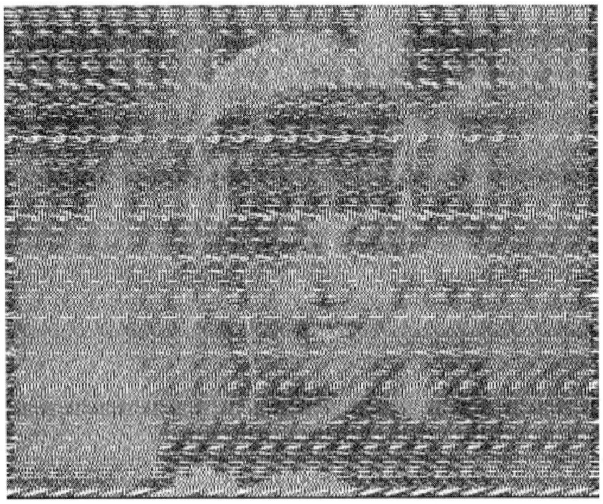

(C) Image in (A), Quantized with 256 Levels, after Application of the Pseudo-Inverse Filter

(D) Image in (A), Quantized with 256 Levels, after Application of the Wiener Filter

FIGURE 4.14

domain (using a DFT), we obtain Figure 4.14(B). We notice that the recovery can be quite good. It is important to note that, in this case, since the process is supposed to be noiseless, the degradation had to be performed using full machine precision. If we quantize the degraded image using 256 levels, as is usual in image processing applications, a small quantization noise is superimposed on the degraded image. As we explained in Section 4.2, this quantization noise is invisible to the human eye. On the other hand, applying the pseudo-inverse filter to the quantized degraded image, we obtain the image in Figure 4.14(C). We can clearly see that the restoration process has failed completely. The excessive noise present in the restored image results from the fact that, since the quantization noise is relatively white, there is noise information even around the frequencies where $H(\omega_1, \omega_2)$ is very small; as a consequence, $G(\omega_1, \omega_2)$ is very large. This greatly amplifies the noise around these frequencies, thus producing an image like the one in Figure 4.14(C). In fact, one would need a filter that is more or less equivalent to the pseudo-inverse in frequencies where the signal to noise ratio is high but has little effect in regions where the signal to noise ratio is low. One way to accomplish this is using the **Wiener filter** which is optimum for a given image and noise statistics. We describe it in what follows.

4.5.1 Wiener Filter

Given the image degradation model in Figure 4.13, suppose that $x(m, n)$ and $\eta(m, n)$ are zero mean, stationary random processes (Papoulis, 1984). The image restoration problem can be formally stated as: given $v(m, n)$, find the best estimate of $u(m, n)$ and $\hat{u}(m, n)$ such that the mean squared estimation error $E[e^2(m, n)]$ is minimum, where:

$$e(m, n) = u(m, n) - \hat{u}(m, n). \tag{4.23}$$

Such best estimate $\hat{u}(m, n)$ is known to be (Papoulis, 1984; Jain, 1989):

$$\hat{u}(m, n) = E[u(m, n)|v(k, l), \forall(k, l)]. \quad (4.24)$$

In our case, we are looking for linear estimates, that is, $\hat{u}(m, n)$ of the form:

$$\hat{u}(m, n) = v(m, n) ** g(m, n). \quad (4.25)$$

According to the orthogonality principle (Papoulis, 1984), we have the best estimate when:

$$E[e(m_1, n_1)v^*(m_2, n_2)] = 0, \quad \forall m_1, n_1, m_2, n_2, \quad (4.26)$$

where $*$ represents the complex conjugation.

From equation 4.23, we have that the above equation is equivalent to:

$$E[\{u(m_1, n_1) - \hat{u}(m_1, n_1)\}v^*(m_2, n_2)] = 0, \quad \forall m_1, n_1, m_2, n_2. \quad (4.27)$$

For stationary processes, equation 4.27 implies that:

$$r_{uv}(m, n) = r_{\hat{u}v}(m, n), \quad (4.28)$$

where $r_{ab}(m, n)$ is the crosscorrelation (Papoulis, 1984) between signals $a(m, n)$ and $b(m, n)$.

If $S_{ab}(\omega_x, \omega_y)$ is the cross power spectral density (Papoulis, 1984) between signals $a(m, n)$ and $b(m, n)$, the above equation implies that:

$$S_{uv}(\omega_x, \omega_y) = S_{\hat{u}v}(\omega_x, \omega_y). \quad (4.29)$$

Since, from equation 4.25, $\hat{u}(m, n) = v(m, n) ** g(m, n)$, equation 4.29 is equivalent to:

$$S_{\hat{u}v}(\omega_x, \omega_y) = G(\omega_x, \omega_y)S_{vv}(\omega_x, \omega_y). \quad (4.30)$$

From equation 4.29, this implies that:

$$G(\omega_x, \omega_y) = \frac{S_{uv}(\omega_x, \omega_y)}{S_{vv}(\omega_x, \omega_y)}. \quad (4.31)$$

Since, from Figure 4.13 and equation 4.18:

$$v(m, n) = u(m, n) ** h(m, n) + \eta(m, n),$$

and $u(m, n)$ and $v(m, n)$ are considered uncorrelated, we have that:

$$S_{vv}(\omega_x, \omega_y) = S_{uu}(\omega_x, \omega_y)|H(\omega_x, \omega_y)|^2 + S_{\eta\eta}(\omega_x, \omega_y). \quad (4.32)$$

$$S_{uv}(\omega_x, \omega_y) = S_{uu}(\omega_x, \omega_y)H^*(\omega_x, \omega_y). \quad (4.33)$$

Then, from equation 4.31, the above equations imply that the optimum linear estimation filter is as follows:

$$G(\omega_x, \omega_y) = \frac{S_{uu}(\omega_x, \omega_y)H^*(\omega_x, \omega_y)}{S_{uu}(\omega_x, \omega_y)|H(\omega_x, \omega_y)|^2 + S_{\eta\eta}(\omega_x, \omega_y)}. \quad (4.34)$$

This is known as the Wiener filter. Note that the implementation of the Wiener filter requires, besides the knowledge of the degradation process $H(\omega_x, \omega_y)$, the knowledge of the power spectral densities of the ensemble of the original inputs, $S_{uu}(\omega_x, \omega_y)$ and of the noise, $S_{\eta\eta}(\omega_x, \omega_y)$.

Figure 4.14(D) shows the image recovered from the blurred and quantized image. In the implementation of the Wiener filter, it was supposed that the ensemble of the input could be modeled by a first-order Gauss-Markov process [AR(1)] with correlation coefficient equal to 0.95. Since the image has a dynamic range of 256 and was quantized with quantization interval 1, the noise was supposed to be white and uniform with density 1/12 (see equation 4.15). We note that the recovery is far better than the one obtained with the pseudo-inverse filter.

We can rewrite equation 4.34 as:

$$G(\omega_x, \omega_y) = \frac{H^*(\omega_x, \omega_y)}{|H(\omega_x, \omega_y)|^2 + S_{\eta\eta}(\omega_x, \omega_y)/S_{uu}(\omega_x, \omega_y)} \quad (4.35)$$

A couple of interesting conclusions can be drawn from equation 4.35. The first one is that, in regions of the frequency plane where the signal-to-noise ratio is high $\left(S_{\eta\eta}(\omega_x, \omega_y)/S_{uu}(\omega_x, \omega_y) \to 0\right)$, then equation 4.35 is equivalent to equation 4.20, the one for the pseudo-inverse filter. On the other hand, in regions of the frequency plane where the signal-to-noise ratio is low, then the Wiener filter depends on the power spectral densities of both the noise and the input.

In this section, we attempted to provide the reader with a general notion of the image restoration problem, which is still an active area of research (Banham and Katsaggetos, 1997). For a deeper treatment of the subject, the reader is referred to Gonzalez and Wintz (1977), Jain (1989), Lim (1990), Pratt (1978), and Rosenfeld and Kak (1982).

4.6 Image Coding

One of the main difficulties when it comes to image processing applications concerns the large amount of data required to represent an image. For example, one frame of a standard definition digital television image has dimensions 720×480 pixels. Therefore, using 256 gray levels (8 bits) for each color component, it requires $3 \times 8 \times 720 \times 480 = 8,294,400$ bits to be represented. Considering that in digital video applications, one uses 30 frames/sec, one second of video needs more than 240 Mbits! Thus, for the processing, storage, and transmission

FIGURE 4.15 Illustration of the Image Compression Problem

of images to be feasible, one needs to reduce somehow this number of bits. **Image compression** or **coding**, is the discipline that deals with forms of achieving this. To better understand the image compression problem, consider the 256×256 image in Figure 4.15. If we need to represent this image as it is, one would need 8 bits/pixel, giving a total of $8 \times 256 \times 256 = 524{,}288$ bits. Alternatively, if we know that the image in Figure 4.15 is a circle, it would be enough to say that it represents a circle of radius 64, it has a gray level of 102, it has a center coordinates $x = 102$ and $y = 80$, and it is superimposed on a background of gray level 255. Since the radius, x coordinate, and y coordinate are between 0 and 255, one would need 8 bits for specifying each one; since both the gray level of the circle and of the background need 8 bits each, the whole image in Figure 4.15 would only need 40 bits to be specified. This is equivalent to a compression ratio of more than 13,000:1!

The vast majority of images, however, is not as simple as the one in Figure 4.15, and such amount of compression cannot be easily obtained. Fortunately, most useful images are redundant; that is, their pixels carry a lot of superfluous information. This can be better understood by resorting to the images in Figure 4.16.

By looking at Figure 4.16(A), we note that by taking a pixel, for example, at the girl's face, the pixels around it have on average similar values. Certainly there are pixels for which this does not hold, such as at the boundary of the girl's skin and hair. However, this is true of most of the image's pixels. Examining the images in Figures 4.16(B) and 4.16(C), we see that the same reasoning applies to them. In other words, if we know something about one pixel, it is likely that we can guess, with a good probability of success, many things about the pixels around it. This implies that we do not need all the pixels to represent the images, only the ones carrying information that cannot be guessed.

At this point, a question arises: are all images redundant? The answer is no, as one can see in Figure 4.16(D). This image represents uniformly distributed white noise between 0 and 255. To know something about a given pixel does not imply knowing anything about the pixels around it. Such an image, however, is not useful for carrying information and is not of the kind frequently present in nature. Indeed, one can state that most natural images are redundant. Therefore, they are amenable to compression. How to exploit this redundancy to create a compact representation of an image is the main objective of image compression techniques.

In general, an image compression method can be split into three basic steps, as depicted in Figure 4.17: **transformation, quantization**, and **coding**. In the transformation step, a mathematical transformation is applied to the image pixels $x(m, n)$, generating a set of coefficients $c(k, l)$ with less correlation among themselves. Its main objective is to exploit the correlation between the image pixels. For these coefficients $c(k, l)$ to be represented with a limited number of bits, they have to be mapped to a finite number of symbols, $\hat{c}(k, l)$. This mapping is commonly referred to as quantization. In Section 4.2, we introduced some forms of quantization. After the coefficients are mapped to a finite number of symbols, these symbols $\hat{c}(k, l)$ must be mapped to a string of bits $b(s)$ to be either transmitted or stored. This operation is commonly referred to as coding. Actually, we have compression only when the number of bits in $b(s)$, $s = 0, 1, \ldots N$ is smaller than the number of bits needed to represent the pixels $x(m, n)$ themselves. These three steps can be clearly illustrated in the next example, **differential pulse-coded modulation** (DPCM).

4.6.1 DPCM

When looking at the redundant images in Figures 4.16(A) through 4.16(C) one notices, for example, that if we know the value of a pixel $x(m, n-1)$, a good guess for the value of the pixel on its right, $\bar{x}(m, n)$, would be the value of the pixel $x(m, n-1)$. A way to reduce the redundancy on those images would be to transmit or encode only the "part" of the pixel that cannot be guessed. This is represented by the difference between the actual value of the pixel and the guess; that is, one would encode $c(m, n)$ as:

$$c(m, n) = x(m, n) - \bar{x}(m, n) = x(m, n) - x(m, n-1). \quad (4.36)$$

Because the images are redundant, the values of $c(m, n)$ tend to be small. This is illustrated in Figure 4.18, where we can see the histograms of the image LENA [see Figure 4.6(A)] and the histogram of the coefficients generated according to equation 4.36. One can see that the histogram of the original image in Figure 4.18(A) has nothing special about it; it has a more or less random shape. On the other hand, one can see that the histogram of the coefficients $c(m, n)$ has a well-defined shape, tending to a Laplacian distribution. One can also see that indeed the small differences $c(m, n)$ are much more likely than the smaller differences, confirming the redundancy of the LENA

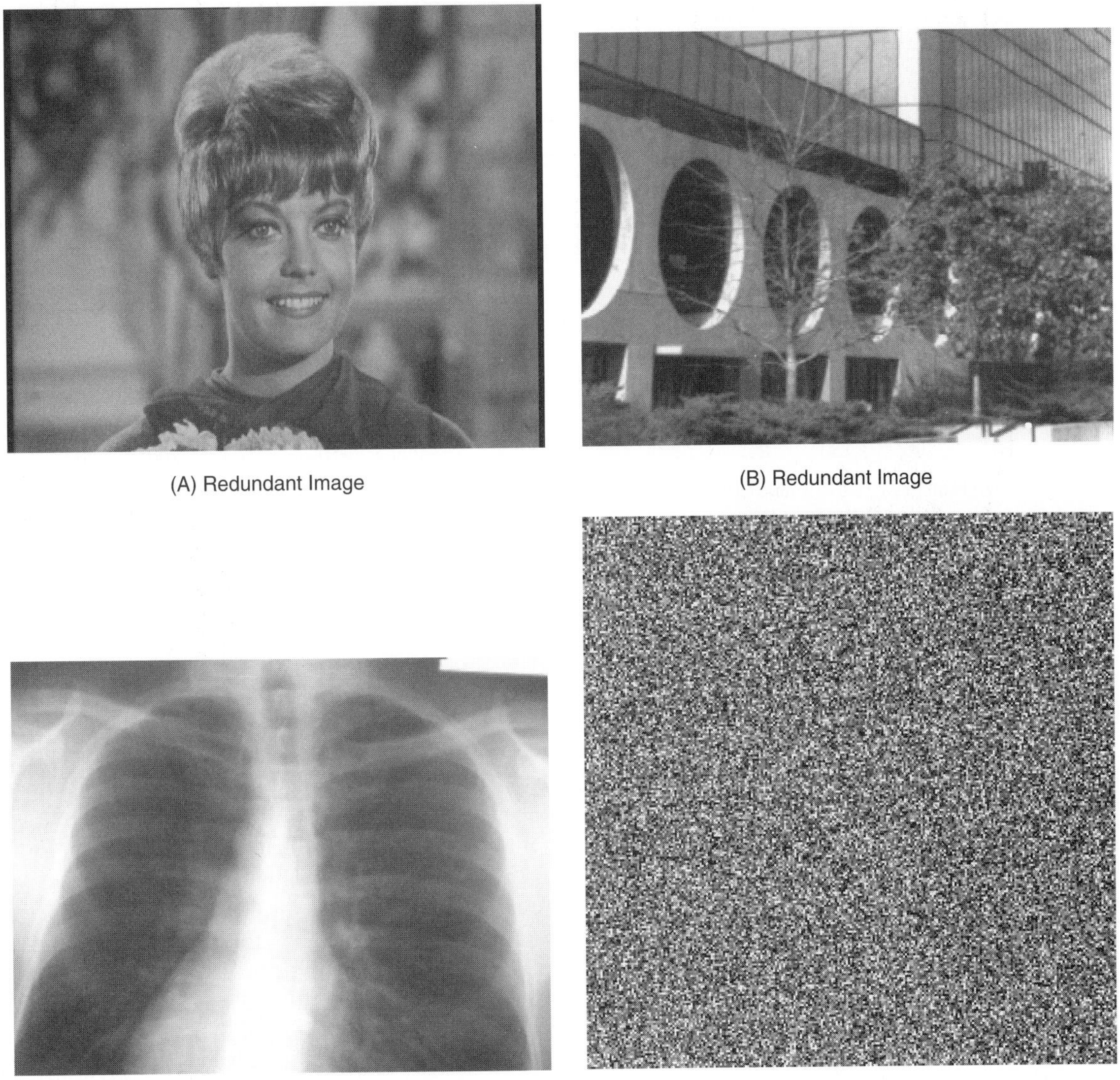

(A) Redundant Image (B) Redundant Image

(C) Redundant Image (D) Nonredundant Image

FIGURE 4.16 Redundant and Nonredundant Images: Differences in Pixels

image. A remarkable fact is that the histogram of the differences has a similar distribution for almost any type of natural images. Then, in fact, the transformation we performed, consisting of taking differences between adjacent pixels, has effectively reduced the redundancy of the original image. In addition, because the smaller values are much more probable than the larger ones, if we use a variable-length code to represent the coefficients, where the more likely values are represented with a smaller number of bits and the less likely values are represented with a larger number of bits, we can effectively obtain a reduction on the number of bits used. It is well known from information theory (Thomas and Joy, 1991) that, by properly choosing a code, one can encode a source using a number of bits arbitrarily close to its **entropy**, defined as:

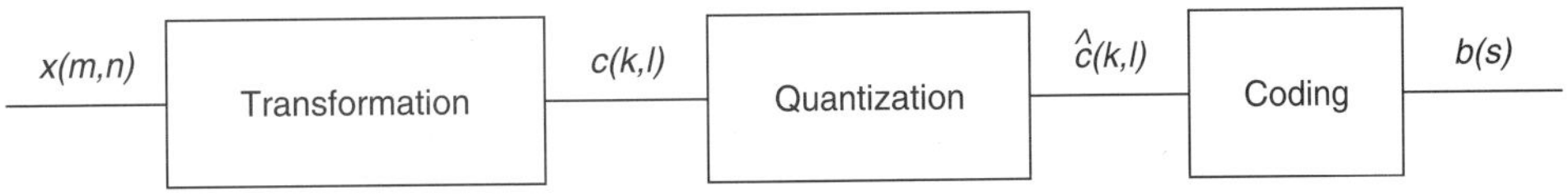

FIGURE 4.17 Three Basic Steps Involved in Image Compression

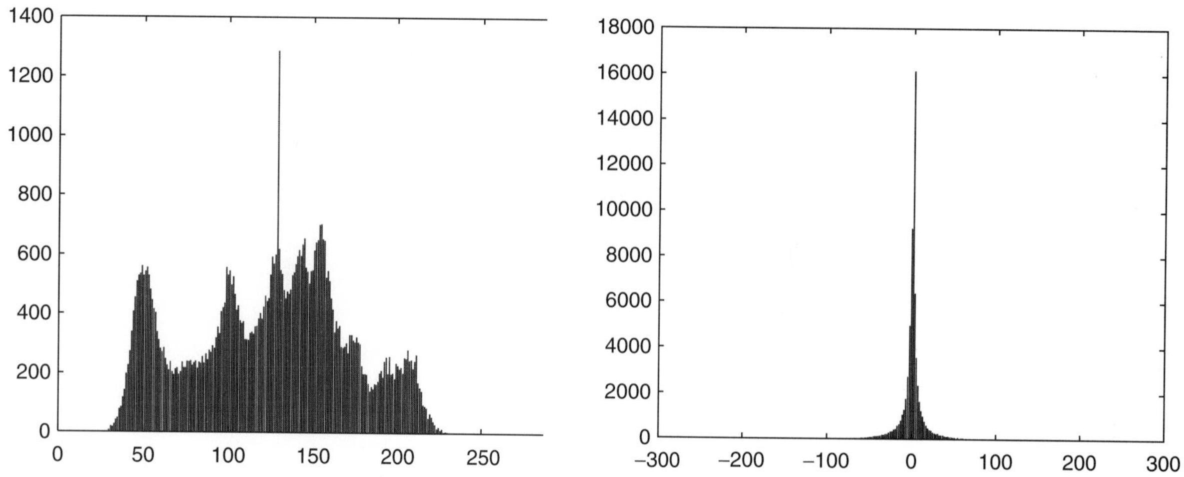

(A) Histogram of the Image LENA

(B) Histogram of the Coefficients $c(m, n)$ Generated According to Equation 4.36

FIGURE 4.18 Redundant Images with Small Coefficient Values

$$H(S) = -\sum_{i=1}^{L} p_i \log_2 p_i, \qquad (4.37)$$

where the source S consists of L symbols $\{s_1, \ldots, S_L\}$, with symbol s_i having probability of occurrence p_i. Examples of such codes are **Huffman codes** (Cover and Thomas, 1991) and **arithmetic codes** (Bell *et al.*, 1990). A discussion of them is beyond the scope of this article. For a thorough treatment of these codes, the reader is referred to Bell *et al.* (1990) and Cover and Thomas (1991).

For example, the image LENA has the histogram in Figure 4.18(A). It corresponds to an entropy of 7.40 bits/pixel, and thus, one needs at least 8 bits to represent each pixel of the image LENA. On the other hand, the entropy of the coefficients, whose histogram is in Figure 4.18(B), is 3.18 bits/coefficient. Therefore, one can find a Huffman code that just needs 4 bits to represent each of them. This is equivalent to a compression ratio of 2:1. This compression was achieved by the combination of the transformation (taking differences) and coding (Huffman codes) operation.

It is noteworthy that the compression obtained in the above example was achieved without any loss; that is, the input image could be recovered, without error, from its compressed version. This is referred to as **lossless compression**. In some circumstances, one is willing to give up the error-free recovery to obtain larger compression rates. This is referred to as **lossy compression**. In the case of images, lossy compression is used a great deal, especially when one can introduce distortions with just a low degree of visibility (Watson, 1993).

These higher compression rates can be obtained through quantization, as described in Section 4.2. In the example above, the differences $c(m, n)$ have a dynamic range from $0 - 255 = -255$ to $255 - 0 = 255$. Therefore, there are 512 such differences. By applying quantization as in Figure 4.9, the number of possible differences can be decreased at the expense of some distortion being introduced. Figure 4.19 describes a practical way of achieving a lossy compression using DPCM. Notice that, in this case, due to quantization, the decoder cannot recover the original pixels without loss. Therefore, instead of $c(m, n)$ being computed as in equation 4.36, with $x(m, n-1)$ being the prediction for the pixel $x(m, n)$, the prediction for $x(m, n)$ is the "recoverable" value of $x(m, n-1)$ at the decoder. Note that to achieve this, there must be a perfect copy of the decoder in the encoder. Details of this kind of coder/decoder can be found in Jayant and Noll (1984).

For the LENA image, if one uses a 16-level quantizer of the same type as the one in Figure 4.9, then the histogram of the quantized differences $Q[c(m, n)]$ is as in Figure 4.20(A). The bit rate necessary to represent them can be made arbitrarily close to the entropy by means of conveniently designed Huffman codes. In this case, the entropy is 1.12 bits/pixel. Therefore, the reconstructed image in Figure 4.20(B) is the result of a compression ratio of more than 7:1. As can be seen, despite the high compression ratio achieved, the quality of the reconstructed image is quite acceptable.

This example highlights the nature and effect of the three basic operations described in Figure 4.17: **transformation** for reducing the redundancy between the pixels; **quantization** for reducing the number of symbols to encode; and **coding** to associate to the generated symbols, the smallest average number of bits possible.

4.6.2 Transform Coding

In the DPCM encoder/decoder already discussed, the transformation consisted of taking the differences between a pixel

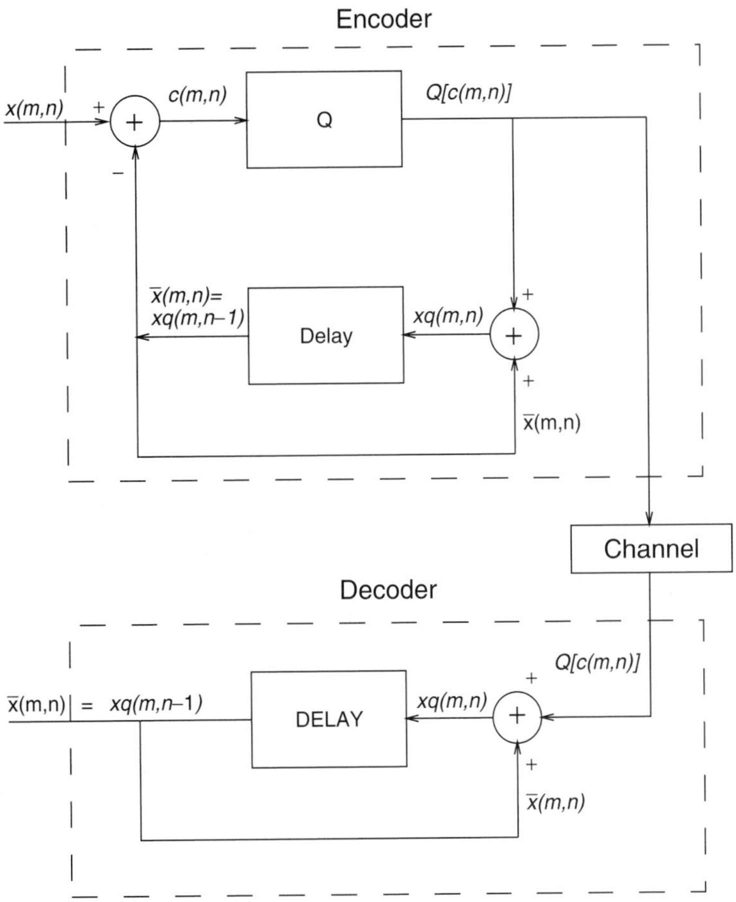

FIGURE 4.19 DPCM Encoder/Decoder

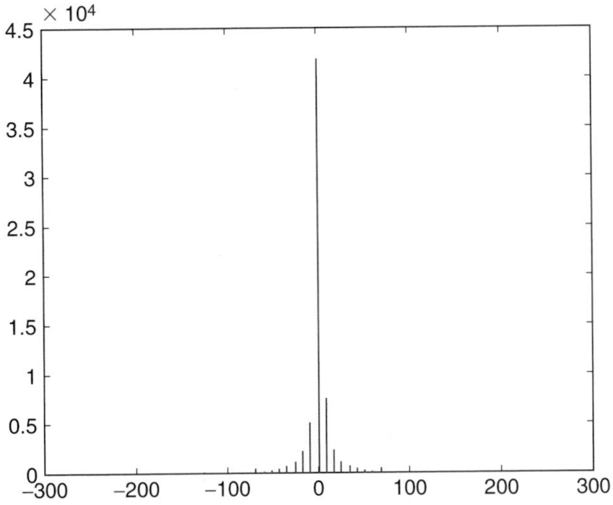

(A) Histogram of the Quantized Differences $Q[c(m, n)]$ in Figure 4.19 for the Case of the LENA Image

(B) Recovered Image $x(m,n)$

FIGURE 4.20

and the adjacent pixel to its left. It indeed reduced the redundancy existing in the image. However, only the redundancy between two adjacent pixels was exploited. By looking at images like the ones in Figures 4.16(A) through 4.16(B), we notice that a pixel has a high correlation with many pixels surrounding it. This implies that not only the redundancy between adjacent pixels but also the redundancy in an area of the image can be exploited. By exploiting this, one expects to obtain even higher compression ratios. **Transform coding** is the generic name given to compression methods in which this type of redundancy reduction is carried out by means of a linear transform (Clarke, 1985).

By far, the most widely used linear transform in compression schemes up to date is the **discrete cosine transform** (DCT). It is part of the old JPEG standard (Pennebaker and Mitchell, 1993) as well as most modern video compression standards, like H.261 (ITU-T, 1990), H.263 (ITU, 1996), MPEG-1 (ISO/IEC, 1994), MPEG-2 (ISO/IEC, 1994) and MPEG-4 (ISO/IEC, 1997). The DCT of a length N signal $x(n)$ consists of the coefficients $c(k)$ such that (Ahmed *et al.*, 1974):

$$c(k) = \alpha(k) \sum_{n=0}^{N-1} x(n) \cos\left[\frac{\pi(2n+1)k}{2N}\right], \tag{4.38}$$

where $\alpha(0) = \sqrt{\frac{1}{N}}$, $\alpha(k) = \sqrt{\frac{2}{N}}$, and $1 \leq k \leq N-1$.

Given the DCT coefficients, the original signal can be recovered using:

$$x(n) = \sum_{k=0}^{N-1} c(k)\alpha(k) \cos\left[\frac{\pi(2n+1)k}{2N}\right]. \tag{4.39}$$

For images, the DCT can be calculated by first computing the one-dimensional DCT of all the rows of the image and then computing the one-dimensional DCT of all the columns of the result.

The DCT is very effective in reducing the redundancy of real images. It tends to concentrate energy in very few transform coefficients. In other words, the coefficients with most of the energy are in general the ones with low values of k in equation 4.38. Since, in this equation, k is proportional to the frequency of the cosine function, usually one refers to them as low-frequency coefficients. To illustrate this, Figure 4.21 shows the 256×256 DCT of the image LENA 256×256. Actually, the logarithm of the absolute value of the coefficients has been plotted. Black corresponds to zero, and the whiter the pixel, the higher its value.

FIGURE 4.21 A 256×256 DCT of the 256×256 LENA Image

We can see clearly that the energy is highly concentrated in the low-frequency coefficients. Therefore, it is natural to think of a lossy image compression system in which only the coefficients having higher energies are encoded. To get a feeling of this, one can look at Figure 4.22. There, the LENA image has been divided into 8×8 blocks, and the DCT of each block was computed. In Figures 4.22(A) to 4.22(D), we can see the image recovered, respectively, using 1, 3, 15, and all 64 coefficients of each block. We note, for example, that the difference between the original image and the one recovered using 15 coefficients is almost imperceptible.

Because the DCT highly concentrates the energy in few coefficients, a large number of the DCT coefficients are zero after quantization. Therefore, a coding method that avoids explicitly encoding the zeros can be very efficient. In practice, this is achieved by scanning the quantized DCT coefficients as in Figure 4.23 prior to encoding. Then, instead of encoding each coefficient, one encodes the pairs (run, level); **level** indicates the quantized value of the nonzero coefficient, while **run** indicates the number of zeros that comes before them. The encoding is performed using an entropy coder, like a Huffman or arithmetic coder. High compression rates can be achieved using such schemes. For example, the JPEG standard (Pennebaker and Mitchell, 1993) is essentially based on the above scheme. The same applies to the H.261, H.263, MPEG-1, MPEG-2, and MPEG-4 standards. Figure 4.24 shows the LENA 256×256 image encoded using JPEG with 0.5 bits/pixel and a compression ratio of 16:1.

One of the main drawbacks of DCT-based image compression methods is the blocking effect that occurs at low rates (see Figure 4.22). One way to solve this is to use a **wavelet transform** in the transformation stage.

Essentially, a wavelet transform is an octave band decomposition in which each band is subsampled according to its bandwidth (Mallat, 1998; Vetterli and Kovačević, 1995). A way of achieving this is to first divide a signal into low- and high-pass bands, with each band being subsampled by two. Then, only the low-pass channel is again low- and high-pass filtered, and each band is subsampled by two. This process is recursively repeated until a predetermined number of stages is reached. For an image, the same process is performed for

(A) Using 1 Coefficient

(B) Using 3 Coefficients

(C) Using 15 Coefficients

(D) Using 64 Coefficients

FIGURE 4.22 LENA Image Recovered Using, from a 8 × 8 DCT, Various Coefficients

both its rows and columns. Figure 4.25 shows an image and its wavelet transform. Figure 4.25(B) shows that the wavelet transform contains mostly coefficients with small magnitudes. In addition, the transform's bands are directional; that is, besides the low-frequency band, one can see horizontal, vertical, and diagonal bands. From an image compression point of view, it is important to notice that the coefficients with small value, which are put to zero after quantization, tend to be clustered. This facilitates their efficient coding, thereby producing high compression ratios. Image coding methods based on the wavelet transform have the advantage of the absence of blocking effects in low rates. Figure 4.24(B) shows a LENA 256 × 256 image compressed at a rate of 16:1 using a wavelet-based encoder. One can notice the superior image quality compared to the DCT-coded one in Figure 4.24(A). Indeed, the JPEG 2000 standard (ISO/IEC, 2000) is based on the wavelet transform. References to state-of-the-art wavelet image encoding methods can be found in works by Andrew (1997), Lan and Tewfik (1999), Said and Pearlman (1996), Shapiro (1993), and Taubman (1999).

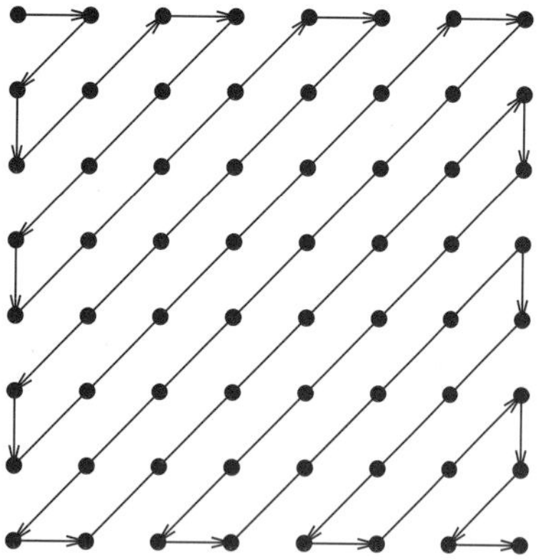

FIGURE 4.23 Scanning Order of the Quantized DCT Coefficients

4.7 Image Analysis

The discipline of image analysis is a very vast one. It deals with the automatic extraction of information from an image or sequence of images. It is also an essential part of computer vision systems. In this section, we provide a brief overview of its main issues. For a more detailed treatment, the reader is referred to some excellent texts on image processing and computer vision, such as Castleman (1979), Faugeras (1993), Gonzales and Wintz (1977), Haralick and Shapiro (1992, 1993), Jain (1989), Law (1991), Pratt (1978), Rosenfeld and Kak (1982), and Russ (1995).

For an image to be analyzed, its main **features** have to be isolated. **Spatial features** like edges, central moments, and entropy are used a great deal. However, **frequency domain** features obtained in the Fourier transform domain may also be useful. Since, for analysis purposes, a scene must be segmented in objects, there must be ways of extracting the objects and representing them. Edge extraction systems are some of the first steps for isolating the different image objects. They are usually based on some form of gradient-based approach, where the edges are associated to large gray level variations. Once the edges are extracted, one can use contour following algorithms to identify the different objects. **Texture** is also a useful feature to be used.

Once the features are extracted, one uses image **segmentation algorithms** to separate the images into different regions. The simplest are the ones based on amplitude thresholding, in which the features used for region determination are the amplitudes of the pixels themselves; each region is defined by a range of amplitudes. If the features used are the edges, one can use **boundary representation** algorithms to define the regions in an image. One can also use **clustering algorithms** to define a region as the set of pixels having similar features. **Quad-trees** are a popular approach for defining regions. **Template** or **texture matching algorithms** can be used to determine which pixels represent a same shape or pattern, thereby defining regions based on them. These belong to the class of **pattern recognition** algorithms that have, by themselves, a large number of applications. In the case of image sequences **motion detection** can be used to define the regions as the pixels of the image having the same kind of movement. For each region, one can also perform **measurements**, such as **perimeter area** and **Euler number**. **Mathematical morphology** (Serra, 1982) is a discipline commonly employed for performing such measurements.

(A) Using the JPEG Standard

(B) Using Wavelet Transforms

FIGURE 4.24 LENA 256 × 256 Image with a Compression Ratio of 16:1

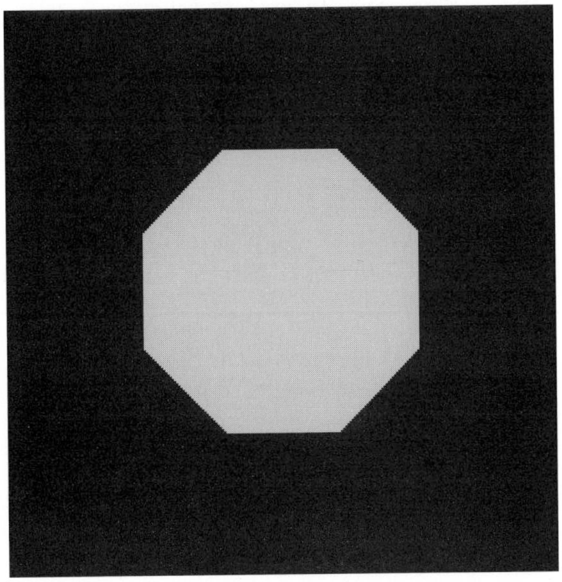

(A) Octagon Image

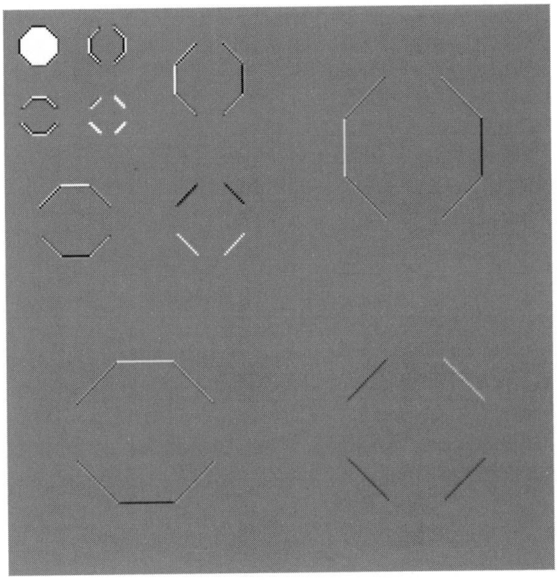

(B) Wavelet Transform

FIGURE 4.25 An Image and Its Wavelet Transform. White corresponds to positive values, black to negative values, and gray to zero values.

Having the different image regions defined, one must group them into objects by labeling the regions so that all the regions belonging to an object have the same label and regions belonging to different objects have different labels. To achieve this, one can use some kind of clustering algorithm.

Once the objects are identified, they must be classified in different categories. The classification can be based, for example, on their shapes. Then, image understanding techniques can be used to identify high-level relations among the different objects.

The image analysis tools mentioned above can be exemplified by Figure 4.26. It shows one bolt and two nuts. The features to be extracted are, for example, the contours of the nuts and the bolt. The regions can be defined as the pixels inside the different contours; alternatively, they can be represented as the pixels having gray level within predetermined ranges. Depending on the application, we may know the shape of one object (e.g., the bolt) and perform pattern matching to determine the position of the bolt on the image. One could also use mathematical morphology to compute the Euler number to count the "number of holes" of each object and consider the nuts as the ones having just one hole. Then, the inner perimeter of the nuts can be determined to check whether they comply to manufacturing specifications. By closely observing the bolt, one can notice that it is composed of several different regions (e.g., its head clearly belongs to a region different from its body). Therefore, after the segmentation algorithm, one should perform region labeling to determine which regions should be united to compose the bolt. Once having established that the image is composed of three objects, image classification and understanding techniques should be used to figure out what kind of objects they are. In this case, one must determine that there are two nuts and one bolt from the data extracted in previous steps. Finally, a robot could use this information, extracted automatically, to screw the bolt into the round nut.

FIGURE 4.26 Hardware Image

As mentioned previously, this section provides only a very brief overview of this vast subject, but the open literature provides a vast treatment of it as first described in this section.

4.8 Summary

This chapter presented an introduction to image processing techniques. First, the issues of image sampling and quantization were dealt with. Then, image enhancement and restoration techniques were analyzed. Image compression methods were described, and the chapter finished with a very brief description of image analysis techniques.

References

Ahmed, N., Natarajan, T., and Rao, K.R. (1974). Discrete cosine transform. *IEEE Transactions on Computers C-23*, 90–93.

Andrew, J. (1997). A simple and efficient hierarchical image coder. *IEEE International Conference on Image Processing 3*, 658–664.

Banham, M.R., and Katsaggelos, A.K. (1997). Digital image restoration. *IEEE Signal Processing Magazine 14*(2), 24–41.

Bell, T.C., Cleary, J.G., and Witten, I.H. (1990). *Text compression.* Englewood Cliffs, NJ: Prentice Hall.

Castleman, K.R. (1979). *Digital image processing.* Englewood Cliffs, NJ: Prentice Hall.

Clarke, R.J. (1985). *Transform coding of images.* London: Academic Press.

Cover, T.M., and Thomas, J.A. (1991). *Elements of information theory.* New York: John Wiley & Sons.

Faugeras, O. (1993). *Three-dimensional computer vision—A geometric viewpoint.* Cambridge, MA: MIT Press.

Gonzalez, R.C., and Wintz, P. (1977). *Digital image processing.* London: Addison-Wesley.

Haralick, R.M., and Shapiro, L.G. (1992). *Computer and robot vision, Vol. 1*, Reading, MA: Addison-Wesley.

Haralick, R.M., and Shapiro, L.G. (1993). *Computer and robot vision, Vol. 2.* Reading, MA: Addison-Wesley.

International Telecommunication Union (ITU). (1996). *Video codec for low bitrate communication.*

International Telecommunication Union-T (ITU-T). (1990). *Recommendation H.261, Video codec for audio visual services at p×64 kbit/s.*

(ISO/IEC). (1992). *JTC1/CD 11172, Coding of moving pictures and associated audio for digital storage media up to 1.5 mbit/s.*

(ISO/IEC). (1994). *JTC1/CD 13818, Generic coding of moving pictures and associated audio.*

(ISO/IEC). (1999). *JTC1/SC29/WG1 (ITU/T SG28), JPEG 2000 verification model* 5.3.

(ISO/IEC). (1997). *JTC1/SC29/WG11, MPEG-4 video verification model version 8.0*, July.

Jain, A.K. (1989). *Fundamentals of digital image processing.* Englewood Cliffs, NJ: Prentice Hall.

Jayant, N.S. and Noll, P. (1984). *Digital coding of waveforms.* Englewood Cliffs, NJ: Prentice Hall.

Lan, T.H., and Tewfik, A.H. (1999). Multigrid embedding (MGE) image coding. *Proceedings of the 1999 International Conference on Image Processing 3*, 24–28.

Lim, J.S. (1990). *Two-dimensional Signal and Image Processing.* Englewood Cliffs, NJ: Prentice Hall.

Low, A. (1991). *Introductory computer vision and image processing.* Berkshire, England: McGraw-Hill.

Mallat, S.G. (1998). *A wavelet tour of signal processing.* San Diego, CA: Academic Press.

Mersereau, R.M., and Dudgeon, D.E. (1984). *Multidimensional digital signal processing.* Englewood Cliffs, NJ: Prentice Hall.

Oppenheim, A.V., and Schaffer, R.W. (1989). *Discrete time signal processing.* Englewood Cliffs, NJ: Prentice Hall.

Papoulis, A. (1984). *Probability, random variables, and stochastic processes.* New York: McGraw-Hill.

Pennebaker, W.B., and Mitchell, J.L. (1993). *JPEG still image data compression standard.* New York: Van Nostrand Reinhold.

Pratt, W.K. *Digital image processing.* New York: John Wiley & Sons.

Rosenfeld, A. and Kak, A.C. (1982). *Digital picture processing, Vol. 1.* New York: Academic Press.

Rosenfeld, A., and Kak, A.C. (1982). *Digital picture processing, Vol. 2*, New York: Academic Press.

Russ, J.C. (1995). *The image processing handbook.* Boca Raton, FL: CRC Press.

Said, A., and Pearlman, W.A. (1996). A new, fast, and efficient image codec based on set partitioning in hierarchical trees. *IEEE Transactions on Circuits and Systems for Video Technology 6*(3), 243–250.

Serra, J. (1982). *Image analysis and mathematical morphology.* New York: Academic Press.

Shapiro, J.M. (1993). Embedded image coding using zero-trees of wavelet coefficients. *IEEE Transactions on Acoustics, Speech, and Signal Processing 41*(12), 3445–3462.

Taubman, D. (1999). High performance scalable image compression with EBCOT. *1999 IEEE International Conference on Image Processing.*

Vetterli, M., and Kovačević, J. (1995). *Wavelets and sub-band coding.* Englewood Cliffs, NJ: Prentice Hall.

Watson, A.B. (Ed.) (1993). *Digital images and human vision.* Cambridge, MA: MIT Press.

5

Multimedia Systems and Signal Processing

John R. Smith
IBM, T. J. Watson Research Center, Hawthorne, New York, USA

5.1 Introduction

With the growing ubiquity and portability of multimedia-enabled devices, **universal multimedia access** (UMA) is emerging as one of the important applications for the next generation of multimedia systems (Smith, 2000). The basic concept of UMA is the adaptation, summarization, and personalization of multimedia content according to usage environment. The different dimensions for adaptation include rate and quality reduction (Kuhn *et al.*, 2001), adaptive spatial and temporal sampling (Smith, 1999), and semantic summarization of the multimedia content (Tseng *et al.*, 2002). The different relevant dimensions of the user environment include device capabilities, bandwidth, user preferences, usage context, and spatial and temporal awareness.

UMA facilitates the scalable or adaptive delivery of multimedia to users and terminals regardless of bandwidth or capabilities of terminal devices and their support for media formats. In addition, UMA allows the content to be customized for users at a high-semantic level through summarization and personalization according to user preferences and usage context (Tseng *et al.*, 2002). UMA is relevant for emerging applications that involve delivery of multimedia for pervasive computing (e.g., handheld computers, palm devices, portable media players), consumer electronics (e.g., television set-top boxes, digital video recorders, television browsers, Internet appliances), and mobile applications (e.g., cell phones, wireless computers) (Bickmore and Schlit, 1997; Smith *et al.*, 1999, 1998b; Han and Smith, 2000). UMA is partially addressed by scalable or layered encoding, progressive data representation, and object- and scene-based encodings (such as MPEG-4) that inherently provide different embedded levels of content quality (Vetro *et al.*, 1999). From the network perspective, UMA involves important concepts related to the growing variety of communication channels, dynamic bandwidth variation, and perceptual quality of service (QoS) (Mohan and Smith, 1999, 1998). UMA involves different preferences of the user (recipients of the content) and the content publisher in choosing the form, quality, and personalization of the content. UMA promises to integrate these different perspectives into a new class of content-adaptive applications that allows users to access multimedia content without concern for specific encodings, terminal capabilities, or network conditions.

5.1.1 MPEG-7/-21 and UMA

The emerging MPEG-7 and MPEG-21 standards address UMA in a number of ways. The overall goal of MPEG-7 is to enable fast and efficient searching, filtering, and adaptation of multimedia content (Salembier and Smith, 2001; Smith, 2003). In MPEG-7, the application of UMA was conceived to allow the scalable or adaptive delivery of multimedia by providing tools for describing transcoding hints and content variations (van Beek *et al.*, 2003; Smith and Reddy, 2001). MPEG-21 addresses the description of user environment, which includes terminals and networks (Smith, 2002). Furthermore, in MPEG-21, digital item adaptation facilitates the adaptation of digital items and media resources for usage environment. The MPEG-21 digital item adaptation provides tools for describing user preferences and usage history in addition to tools for digital item and media resource adaptation (Salembier and Smith, 2001).

5.1.2 UMA Architecture

The basic problem of UMA concerns allowing users to access multimedia content anytime, anywhere, and without concern for particular formats, devices, networks, and so forth. Consider the server–proxy–client network architecture shown in Figure 5.1. UMA can be enabled in a number of ways. First, it is possible to store multiple variations of the content at the server and select the best variation for each particular request (Li *et al.*, 1998). Given multiple media resources in a presentation, it may be necessary to make the selection of multiple variants jointly to meet the overall delivery constraints (Smith *et al.*, 1999). As an alternative, it is possible to perform content adaptation at the server or in the network, such as at the proxy device. For example, considering a scenario in which a user allocates 10 min for receiving a video summary of a sports game, the server or proxy can manipulate the video content to produce a highlighted video that is personalized for the user's particular preference, such as a focus on a favorite team or player or preferred types of plays in the game (e.g., for a baseball game, this could be strikeouts, hits, homeruns, etc.).

5.1.3 Outline

This chapter describes UMA as one of the emerging applications of multimedia systems and signal processing. In particular, it describes how different multimedia systems and signal processing tools use MPEG-7 and MPEG-21 for enabling UMA. Section 5.2 describes the MPEG-7 support for UMA, which includes transcoding hints and variations descriptions. Section 5.3 describes the MPEG-21 support for UMA, which includes digital item adaptation. Section 5.4 describes how transcoding of images can be optimized using MPEG-7 transcoding hints. Section 5.5 describes a method for the optimized selection of media resources, which applies to MPEG-7 variations and MPEG-21 digital items.

5.2 MPEG-7 UMA

MPEG-7 addresses the requirements of UMA by providing a number of different tools for content adaptation (Smith and Reddy, 2001; Kuhn and Suzuki, 2001). For one, MPEG-7 defines **description schemes** that describe different abstraction

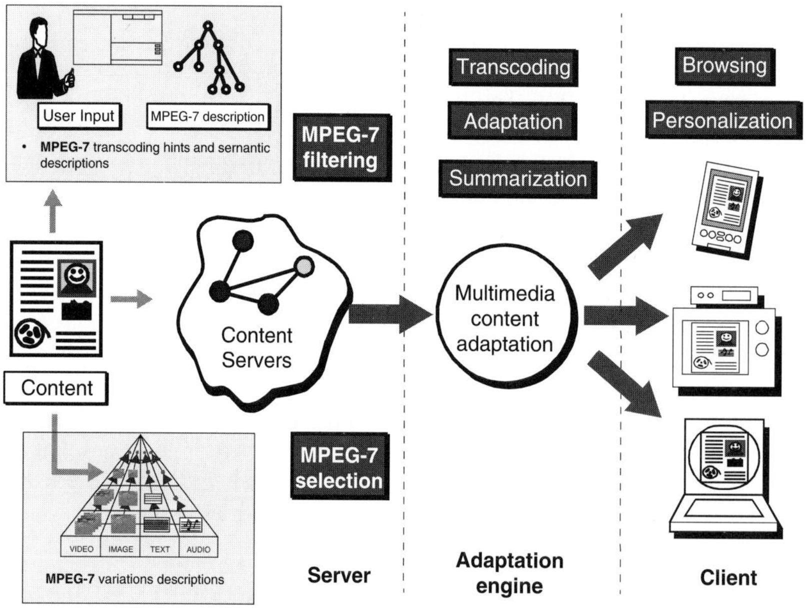

FIGURE 5.1 Universal Multimedia Access. UMA involves the adaptive delivery of multimedia content according to usage environment, which includes devices and networks, user preferences, and usage context.

levels and variations of multimedia content. For example, the different abstraction levels include the composition of objects from sub-objects, the extraction of plot structure of a video, summaries of multimedia programs, and different variations of the multimedia data with different resource requirements. In addition, MPEG-7 supports the transcoding, translation, summarization, and adaptation of multimedia material according to the capabilities of the client devices, network resources, and user and author preferences. For example, adaptation hints may be provided that indicate how a photograph should be compressed to adapt it for a handheld computer or how a video should be summarized to speed up browsing over a low-bandwidth network (Smith *et al.*, 1998a).

5.2.1 MPEG-7 Transcoding Hints

The MPEG-7 **Media Transcoding Hint Description Scheme** gives information that can be used to guide the transcoding of multimedia, including description of importance, priority, and content value and description of transcoding behavior based on transcoding utility functions and network scaling profile, as shown in Figure 5.1. In addition, media coding tools give information about multimedia data including the image and video frame size (width and height), frame rates of video, data sizes for image, video and audio download and storage, formats, and MIME-types.

The MPEG-7 media transcoding hints allow content servers, proxies, or gateways to adapt image, video, audio, and multimedia content to different network conditions, user and publisher preferences, and capabilities of terminal devices with limited communication, processing, storage, and display capabilities. MPEG-7 provides the following types of transcoding hint information as part of its Media Transcoding Hints:

- **Importance** MPEG-7 specifies the relative importance of resources, segments, regions, objects, or media resources. The importance takes values from 0.0 to 1.0, where 0.0 indicates the lowest importance and 1.0 indicates the highest importance.
- **Spatial resolution** MPEG-7 specifies the maximum allowable spatial resolution reduction factor for perceptability. This feature takes values from 0.0 to 1.0, where 0.5 indicates that the resolution can be reduced by half, and 1.0 indicates the resolution cannot be reduced

Importance Hints

The **Importance Hints** can be used to annotate different regions with information that denotes the importance segment hint type of each region. The Importance Hint descriptions can then be used to transcode the images for adaptive delivery according to constraints of client devices and bandwidth limitations. For example, text regions and face regions can be compressed with a lower compression factor than the factors for the remaining part of the image. The other parts of the image are blurred and compressed with a higher factor to greatly reduce the overall size of the compressed image. MPEG-7 UMA enables more intelligent transcoding capability using the MPEG-7 Importance Hint information than otherwise would be provided. The MPEG-7 Importance Hint information has advantages over methods for automatic extraction of regions from images in the transcoder in that the Importance Hints can be provided by the content authors or publishers directly. This allows greater control in the adaptation and delivery of content.

Spatial Resolution Hints

The **Spatial Resolution Hints** can be used to denote the minimum resolution with which a segment should be displayed. For example, if an image region contains a face or textual information, then it may be necessary to preserve a minimum resolution for the contents to be discernible.

Example Transcoding Hints Annotation

The following example shows an MPEG-7 description of transcoding hints for an image in which three regions have been annotated (*R*1, *R*2, and *R*3). Values are assigned for the importance of Spatial Resolution Hint attributes of each region.


```
<Mpeg7 xmlns= ''urn:mpeg:mpeg7:schema:2001'' xml:lang= ''en''>
 <ContentDescription xsi:type= ''ContentEntityType''>
  <MultimediaContent xsi:type= ''ImageType''>
   <Image>
    <SpatialDecomposition gap= ''true'' overlap=''false''>
     <StillRegion id=''R1''>
      <MediaInformation>
       <MediaProfile>
        <MediaTranscodingHints importance= ''1.0''
        spatialResolutionHint= ''0.5''/>
       </MediaProfile>
      </MediaInformation>
      <SpatialLocator>
       <Box>
        <Coords mpeg7:dim=''4 1''>16 34 64 64</Coords>
       </Box>
      </SpatialLocator>
     </StillRegion>
     <StillRegion id=''R2''>
      <MediaInformation>
       <MediaProfile>
        <MediaTranscodingHints importance= ''0.5''
        spatialResolutionHint= ''0.35''/>
       </MediaProfile>
      </MediaInformation>
      <SpatialLocator>
       <Box>
        <Coords mpeg7:dim=''4 1''>384 256 128 128</Coords>
       </Box>
      </SpatialLocator>
     </StillRegion>
     <StillRegion id=''R3''>
      <MediaInformation>
       <MediaProfile>
        <MediaTranscodingHints importance= ''0.35''
        spatialResolutionHint= ''0.25''/>
       </MediaProfile>
      </MediaInformation>
      <SpatialLocator>
       <Box>
```

```
          <Coords mpeg7:dim=''4 1''>512 384 64 128</Coords>
        </Box>
      </SpatialLocator>
    </StillRegion>
   </SpatialDecomposition>
  </Image>
 </MultimediaContent>
</ContentDescription>
</Mpeg7>
```

5.2.2 MPEG-7 Variations

The **MPEG-7 Variation Description** addresses a general framework for managing and selecting multimedia content for adapting delivery. As depicted in Figure 5.2, the Variation DS describes different variations of media resources with different **modalities** (e.g., video, image, text, and audio) and **fidelities** (e.g., summarized, compressed, scaled, etc.) (Li *et al.*, 1998).

The MPEG-7 Variation DS describes the relationships among the different variations of a media resource. For example, Figure 5.2 shows the variations of a video resource *A*. The variations differ in fidelity and modality. The vertical axis shows lower fidelity variations of each resource. The horizontal axis shows different modality variations (e.g., image, text, and audio) of each resource.

```
<Mpeg7>
  <Description xsi:type= ''VariationDescriptionType''>
    <VariationSet>
      <Source xsi:type= ''VideoType''>
        <Video>
          <MediaLocator>
            <MediaUri> file://video-high-res.mpg< /MediaUri>
          </MediaLocator>
        </Video>
      </Source>
      <Variation fidelity=''0.75'' priority=''1''>
        <Content xsi:type= ''ImageType''>
          <Image>
            <MediaLocator>
              < MediaUri>file://key-frame.jpg</MediaUri>
            </MediaLocator>
          </Image>
        </Content>
        <VariationRelationship> extraction</VariationRelationship>
      </Variation>
      <Variation fidelity=''0.65'' priority=''2'' timeOffset=
      ''PT10S'' timeScale=''0.5''>
        <Content xsi:type= ''VideoType''>
          <Video>
            <MediaLocator>
              <MediaUri> file://video-low-res.mpg< /MediaUri>
            </MediaLocator>
          </Video>
        </Content>
        <VariationRelationship> spatialReduction</
        VariationRelationship>
          <VariationRelationship> temporalReduction</
          VariationRelationship>
          <VariationRelationship> rateReduction</
          VariationRelationship>
      </Variation>
      <Variation fidelity=''0.5'' priority=''3''>
        <Content xsi:type= ''AudioType''>
          <Audio>
            <MediaLocator>
              <MediaUri> file://audio-track.mp3< /MediaUri>
            </MediaLocator>
          </Audio>
        </Content>
        <VariationRelationship> extraction</VariationRelationship>
      </Variation>
    </VariationSet>
  </Description>
</Mpeg7>
```

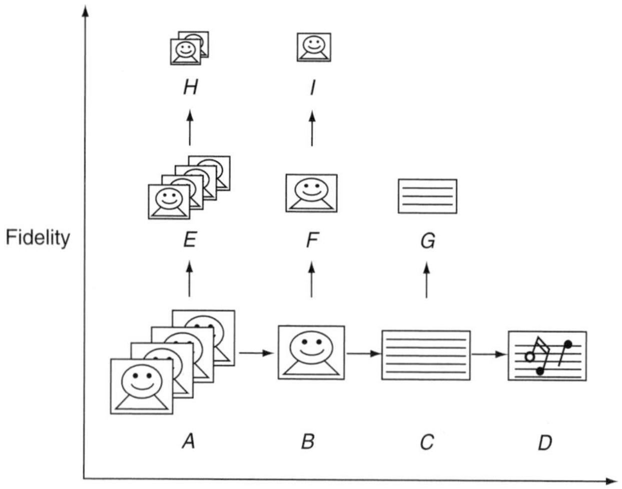

FIGURE 5.2 The MPEG-7 Variation DS. This feature allows the selection and adaptive delivery of different variations of multimedia content.

5.3 MPEG-21 Digital Item Adaptation

MPEG-21 specifies an XML Schema-based language for describing media resource adaptability, user environment, and other concepts related to adapting and configuring digital items and adapting media resources to user environments. While MPEG-7 has an important role in describing media resource adaptability, user preferences, and controlled terms, MPEG-21 further describes media resource adaptability, usage environment, and adaptation rights. MPEG-21 is also investigating a language for describing and manipulating bit streams (BSDL).

The Digital Item Declaration part of MPEG-21 specifies an XML Schema-based language for declaring digital items, which are packages of media resources and metadata. The following example shows a digital item that contains music plus images and gives its digital item declaration.

Digital Item Example (Music and Images)

Figure 5.3 shows an example digital item that contains two descriptors and three media resource items. One of the descriptors gives creation information about the digital item and defines a title. The other descriptor gives a purely text-based description of the digital item. The digital item contains one

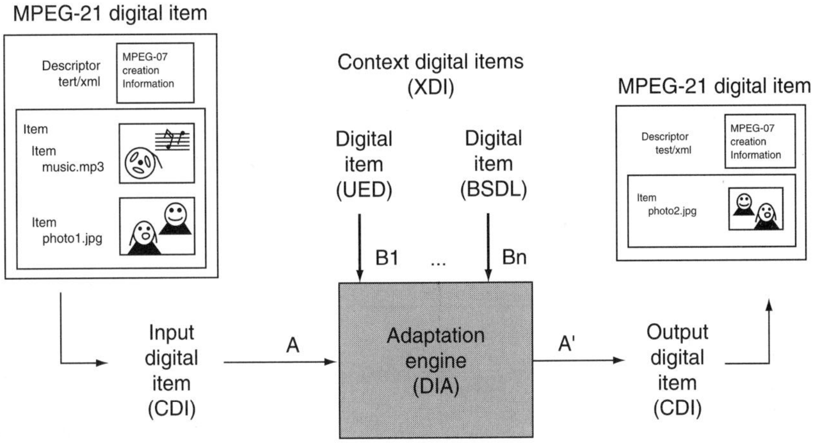

FIGURE 5.3 MPEG-21 Digital Item Adaptation. This feature addresses the configuration of digital items and adaptation of their media resources.

music file and two photos. The media resources are referenced by a locator.

The following MPEG-21 XML description gives the declaration for the digital item depicted in Figure 5.3. As shown next, the digital item declaration provides a container at the top level, which forms the logical package for the digital item. It contains a descriptor that holds the XML statement giving an MPEG-7 description and an item that contains further description and sub-item elements. The description contained by the item holds a plain text statement giving a textual description of the item. The three sub-items each contain one component that gives a media resource. The first one gives an audio resource (music.mp3); the latter two each give an image (photo1.jpg and photo2.jpg).

```
<DIDL>
 <Container>
  <Descriptor>
   <Statement type= ``text/xml''>
    <mpeg7:Mpeg7>
     <mpeg7:DescriptionUnit xsi:type=
     ``CreationInformationType''>
      <mpeg7:Creation>
       <mpeg7:Title>Musical experience package</mpeg7:Title>
      </mpeg7:Creation>
     </mpeg7:DescriptionUnit>
    </mpeg7:Mpeg7>
   </Statement>
  </Descriptor>
  <Item>
   <Descriptor>
    <Statement type= ``text/plain''>
     Music package (one song plus two photos)
    </Statement>
   </Descriptor>
   <Item>
    <Component>
     <Resource ref=``music.mp3'' type=``audio/mp3''/>
    </Component>
   </Item>
   <Item>
    <Component>
     <Resource ref=``photo1.jpg'' type=``image/jpg''/>
    </Component>
   </Item>
   <Item>
    <Component>
     <Resource ref=``photo2.jpg'' type=``image/jpg''/>
    </Component>
   </Item>
  </Item>
 </Container>
</DIDL>
```

5.3.2 Digital Item Configuration

MPEG-21 accommodates digital item adaptation in a number of ways, such as through digital item configuration or media resource adaptation or transcoding as shown in Figure 5.3. **Digital item configuration** is the mechanism by which selections are made among the elements of the digital items. The digital item declaration includes a **choice** element that describes a set of related selections that configure an item. A **selection** describes a specific decision affecting the inclusion of conditional elements from the item. The result is the digital item declaration can specify that items with the package that are optional or subject to conditions. For example, the declaration can describe two video files and indicate that one is conditionally included subject to display on a high-resolution terminal, whereas the other is included subject to display on a low-resolution terminal.

5.3.3 Media Resource Adaptation

Alternatively, MPEG-21 supports digital item adaptation through **media resource adaptation**. In this case, the media resources themselves are adapted or transcoded. The media resources in the digital item declarations can be manipulated in an adaptation engine according to usage environment. The adaptation engine can take as input a user environment description that includes user preferences, usage context, device capabilities, and so forth.

5.4 Transcoding Optimization

Since media-rich content often contains multiple media resources, there is much opportunity for optimizing the adaptation of this content by trading off the different transcoding operation for each of the component media resources. This chapter now investigates how this transcoding optimization can be applied to images consisting of multiple annotated regions.

5.4.1 Image Transcoding

Given that multiple regions in the image can be annotated, the **transcoding of images** needs to consider the individual importance and spatial resolution hint for each region (Smith and Reddy, 2001). Overall, this can be approached as an optimization problem in which the image needs to be manipulated, such as through cropping and rescaling, to produce an output image that maximizes the overall content value given the constraints on its delivery. The optimization problem is expressed as follows: The device imposes a constraint on the size of the image (i.e., size of the screen). The transcoding engine seeks to maximize the benefit derived from the content value of the transcoded image. The problem, thus, is to maximize the total content value given the constraints.

Following the particular structure provided by the MPEG-7 transcoding hints, consider that each region R_i has an importance I_i and spatial resolution hint S_i. Consider also that each region has a content value score V_i after rescaling the image globally using rescaling factor L, which is a function of its importance I_i and spatial resolution hint S_i. Rescaling goes as follows:

- **Importance Hint** (I_i): Indicates relative importance of region R_i, where $0 \leq I_i \leq 1$
- **Spatial Resolution Hint** (S_i): Indicates the minimum resolution of region for preserving details, where
- **Rescaling factor** (L): Indicates the global rescaling of the image, where $0 \leq S_i \leq 1$
- **Content Value** (V_i): Indicates value of the transcoded region $V_i = f(I_i, S_i, L)$as follows:

$$V_i = \begin{cases} I_i & \text{if } 1 \leq L. \\ L\frac{I_i}{S_i} & \text{if } 0 \leq L < 1. \\ 0 & \text{otherwise.} \end{cases}$$

Then, the problem can be stated as $\max\left(\sum_i V_i\right)$ such that $F_x(R_i^*) \leq D_s$, where:

- D_s gives the size of the device screen:
- R_i^* gives a selected rescaled region; and
- $F_s(R_i^*)$ gives the spatial size of the minimum bounding rectangle that encapsulates the selected rescaled regions.

One way to solve this problem is by using an exhaustive search for all possible combinations of the rescaled regions. In this case, for each unique combination of regions, the image is cropped to include only those selected regions, and the cropped image is then rescaled to fit the device screen size. Then, the regions are evaluated in terms of the image's content value under the rescaling and selection, and the total content value is computed. Then, the combination of regions with maximum benefit is selected as the optimal transcoding of the image. The complexity of this search is not great considering that each image will typically only have a handful of annotated regions.

5.4.2 Image Transcoding Examples

The following examples illustrate the transcoding of an image (shown in Figure 5.4) both with and without the MPEG-7 transcoding hints for importance and spatial resolution. This example considers four different display devices: personal computer (PC), television (TV) browser, handheld computer, and personal digital assistant (PDA) or mobile phone.

Figure 5.5 shows the transcoding of the image without using transcoding hints. In this case, the original image is adapted to the screen sizes by globally rescaling the image to fit the screen. The drawback of this type of adaptation is that the image details are lost when the size of the output image is small. For example, it is difficult to discern any details from the image displayed on the PDA screen (far right). The result is an equal loss of detail for all regions including those important regions, which results in a *lower* overall content value of the image for a display size.

Figure 5.6 shows the transcoding of the image using transcoding hints. In this case, the original image is adapted to the screen sizes by using a combination of cropping and rescaling to fit a set of selected image regions on the screen. The advantage of this type of adaptation is that the important image details are preserved when the size of the output image is small. For example, it is possible to discern important details from

FIGURE 5.4 Image with Four Regions Annotated Using MPEG-7 Transcoding Hints

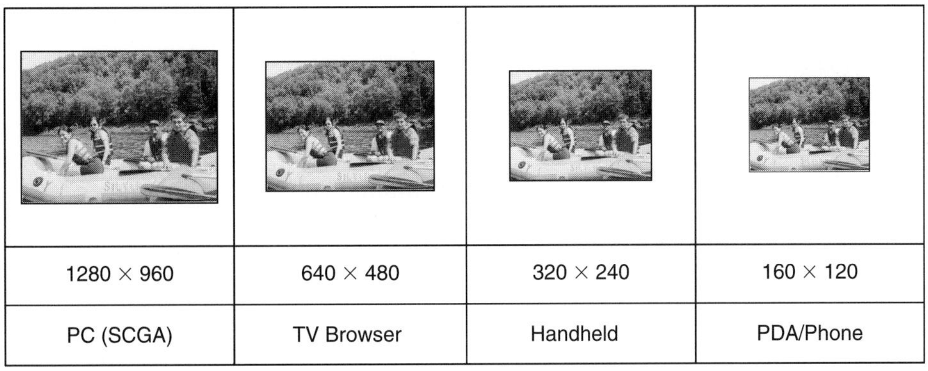

FIGURE 5.5 Example Transcoded Image Output. The image was globally rescaled to fit the screen.

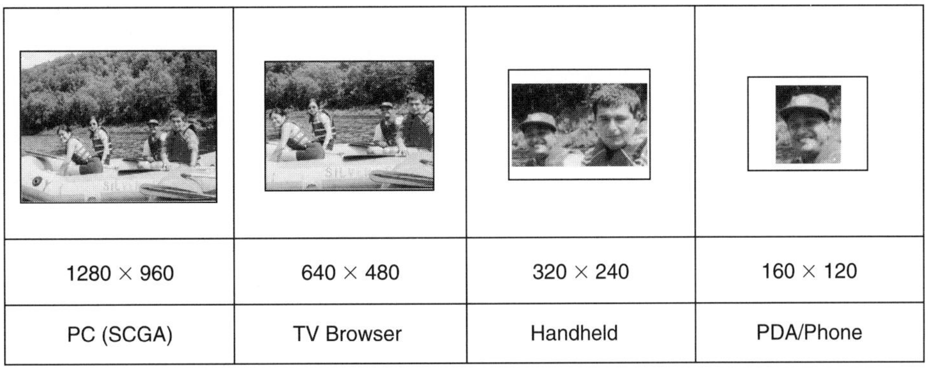

FIGURE 5.6 Example Transcoded Image Output Using the MPEG-7 Transcoding Hints. The hints are for image region importance and minimum allowable spatial resolution. The transcoding uses a combination of cropping and rescaling to maximize the overall content value.

the image displayed on the handheld and PDA screens. The result is an adaptive loss of detail for different regions by cropping those important regions, which results in a *higher* overall content value of the image for a display size.

5.5 Multimedia Content Selection

Considering that media-rich content is composed of multiple constituent media resources, this section investigates how different variations of the media resources can be selected based on the overall device and delivery constraints.

5.5.1 Resource Selection

A procedure is now described for optimizing the selection of the different variations of media resources in a multimedia document that maximizes the total **content value** given the constraints of the client devices (Smith *et al.*, 1999). A multimedia presentation $M = [D, L]$ can be defined as a tuple consisting of a multimedia document D and a document layout L. The multimedia document D is the set of media resources O_{ij}, written as follows:

$$D = \{(O_{ij})_n\} \text{ (multimedia document)},$$

where $(O_{ij})_n$ gives the nth media resource, which has modality i and fidelity j. The document layout L gives the relative spatial and temporal location and size of each media resource. The variation description V of a media resource describes the collection of the different variations of the media resource O_{ij}, as follows:

$$V = \{O_{ij}\} \text{ (variation description)}.$$

We then define a variation document VD as a multimedia document consisting of resources described by variation description $\{O_{ij}\}$, written as follows:

$$VD = \{V_n\} = \{\{O_{ij}\}_n\} \text{ (variation document)}.$$

Content Value Scores

To optimize the selection, the variation descriptions include content value scores $V((O_{ij}))$ for each of the media resources O_{ij}, as shown in Figure 5.7. The content value scores can be based on automatic measures, such as entropy, or on loss in fidelity that results from translating or summarizing the con-

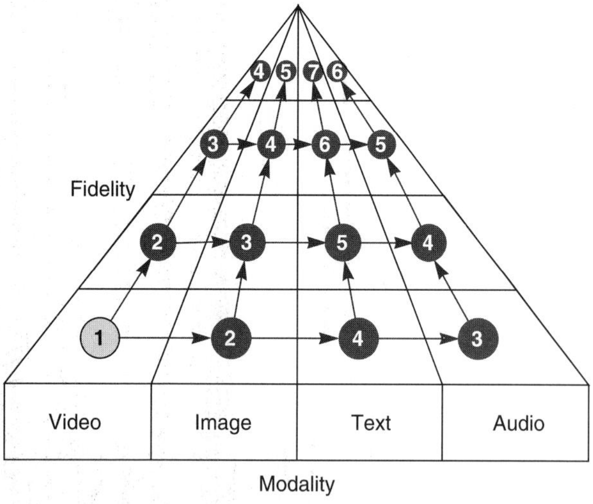

FIGURE 5.7 Example of the Reciprocal Content Value Scores. These scores are assigned for different video media resource variations.

tent. For example, the content value scores can be linked to the distortion introduced from compressing the images or audio. Otherwise, the content value scores can be tied directly to the methods that manipulate the content or be assigned manually.

Figure 5.7 illustrates examples of the relative reciprocal content value scores of different variations of a video media resource. In this example, the original video (lower left) has the highest content value. The manipulation of the video along the dimension of fidelity or modality reduces the content value. For example, converting the video to a sequence of images results in a small reduction in content value. Converting the video to a highly compressed audio track produces a higher reduction in the content value.

5.5.2 Variation Selection

Given a multimedia document with N media resources, let $\{O_{ij}\}_n$ give the variation description of the nth media resource. Let $V((O_{ij})_n)$ give the relative content value score of the variation of the nth media resource with modality i and fidelity j, and let $D((O_{ij})_n)$ give its data size.

Let D_T give the total maximum data size allocated for the multimedia document by the client device. The total maximum data size may, in practice, be derived from the user's specified maximum load time and the network conditions or from the device constraints in storage or processing.

Maximum Content Value

The content selection process selects media resource variations O^*_{ij} from each variation description to maximize the total content value for a target data size D_T as follows:

$$\sum_n V((O^*_{ij})_n) = \max\left(\sum_n V((O_{ij})_n)\right)$$

and

$$\sum_n D((O^*_{ij})_n) \leq D_T, \tag{5.1}$$

where $(O^*_{ij})_n$ gives for each n the optimal variation of the media resource, which has fidelity i and modality j.

Minimum Load Time

Alternatively, given a minimum acceptable total content value V_T, the content select process selects media resource variations O^*_{ij} from each variation description to minimize the total data size as follows:

$$\sum_n D((O^*_{ij})_n) = \min\left(\sum_n D((O_{ij})_n)\right),$$

and

$$\sum_n V((O^*_{ij})_n) \geq V_T, \tag{5.2}$$

where, as above, $(O^*_{ij})_n$ gives for each n the optimal variation of the media resource, which has fidelity i and modality j.

Device Constraints and Preferences

By extending the selection process, other constraints of client devices can be considered. For example, the content selection system can incorporate device screen size S_T as follows: let $S((O_{ij})_n)$ give the spatial size of the variation of the nth media resource with modality i and fidelity j. Then, we add the following constraint to the optimization process:

$$\sum_n S((O^*_{ij})_n) \leq S_T. \tag{5.3}$$

In the same way, we can include additional device constraints such as color depth, streaming bandwidth, and processing power.

5.5.3 Selection Optimization

Given the variation descriptions for describing different variations of the media resources, the overall number of different variations of each variation document is combinatorial in the number of media resources (N) and number of variations of each media resource (M) and is given by M^N. To solve the optimization problems of equations 5.1 and 5.2 convert the constrained optimization problems into the equivalent Lagrangian unconstrained problems, as described in Mohan *et al.* (1998).

The optimization solution is based on the resource allocation technique proposed in Shoham and Gersho (1990) for arbitrary discrete functions. This is illustrated by converting the problem in equation 5.1 to the following unconstrained problem:

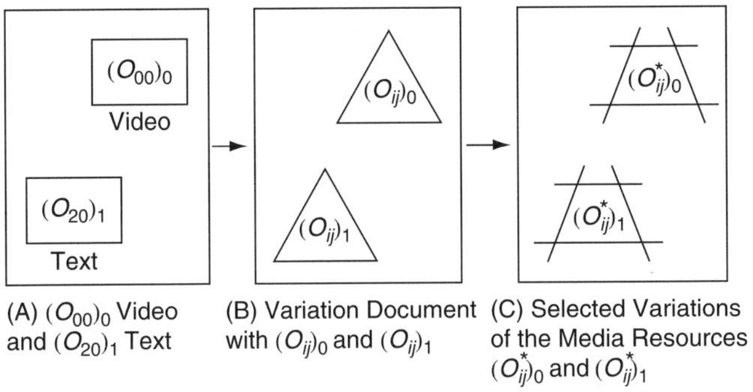

FIGURE 5.8 Example Content Selection for a Multimedia Document **D** Consisting of Two Media Resources

$$\min\left\{\sum_n D((O_{ij})_n) - \lambda(V_T - V((O_{ij})_n))\right\}. \qquad (5.4)$$

the optimal solution gives that for all n, the selected variations of the media resources $(O^*_{ij})_n$ operate at the same constant trade-off λ in content value $V((O_{ij})_n)$ versus data size $D((O_{ij})_n)$. To solve the optimization problem, it is only necessary to search over values of λ.

5.5.4 Example Selection

Illustrated here is the content selection in an example multimedia document, as shown in Figure 5.8. The multimedia document has two media resources: a video resource $= (O_{00})_0$ and a text resource $= (O_{20})_1$. For each media resource $(O_{ij})_n$, where $n \in \{0, 1\}$, a variation document $\{O_{ij}\}_n$ can be constructed, which gives the different variations of the media resource. The selection process selects the variations $(O^*_{ij})_0$ and $(O^*_{ij})_1$, respectively, to maximize the total content value.

Consider the four variations of each media resource with content values and data sizes given in Table 5.1. By iterating over values for the trade-off λ in content value and data size, the content selection table of Table 5.2 is obtained which shows the media resource variations that maximize the total content value $\max(\Sigma_n V((O_{ij})_n))$ for different total maximum data sizes D_T, as given in equation 5.1.

5.6 Summary

Universal Multimedia Access (UMA) is one of the important emerging applications for multimedia systems and signal processing. The basic idea of UMA is to adapt media-rich content according to usage environment. The emerging MPEG-7 and MPEG-21 standards address different aspects of UMA by standardizing metadata descriptions for adapting this content. In this chapter, we described how UMA is supported in multimedia systems through media resource transcoding and multimedia content selection.

TABLE 5.1 Summary of Different Variations of Two Media Resources: $(O_{ij})_0$ and $(O_{ij})_1$

$(O_{ij})_n$	$V((O_{ij})_n)$	$D((O_{ij})_n)$	Modality
$(O_{00})_0$	1.0	1.0	Video
$(O_{01})_0$	0.75	0.25	Video
$(O_{10})_0$	0.5	0.10	Image
$(O_{11})_0$	0.25	0.05	Image
$(O_{20})_1$	1.0	0.5	Text
$(O_{21})_1$	0.5	0.10	Text
$(O_{32})_1$	0.75	0.25	Audio
$(O_{33})_1$	0.25	0.05	Audio

TABLE 5.2 Summary of the Selected Variations of the Two Media Resources: $(O_{ij})_0$ and $(O_{ij})_1$*

D_T	$(O^*_{ij})_0$	$(O^*_{ij})_1$	$\Sigma_n V((O^*_{ij})_n)$	$\Sigma_n D((O^*_{ij})_n)$
1.5	$(O_{00})_0$	$(O_{20})_1$	2.0	1.5
1.25	$(O_{00})_0$	$(O_{32})_1$	1.75	1.25
1.0	$(O_{01})_0$	$(O_{20})_1$	1.75	0.75
0.6	$(O_{10})_0$	$(O_{20})_1$	1.5	0.6
0.35	$(O_{01})_0$	$(O_{21})_1$	1.25	0.35
0.35	$(O_{10})_0$	$(O_{32})_1$	1.25	0.35
0.2	$(O_{10})_0$	$(O_{21})_1$	1.0	0.2
0.1	$(O_{11})_0$	$(O_{33})_1$	0.5	0.1

*These resources are under different total maximum data size constraints D_T.

References

Bickmore, T.W., and Schlitt, B.N., (1997). Digestor: Device-independent access to the World Wide Web. *Proceedings of the Sixth International WWW Conference.*

Bjrk, N., and Christopoulos, C. (2000). Video transcoding for universal multimedia access. *Proceedings of the ACM International Conference on* Multimedia (ACMMM). 75–79.

Han, R., and Smith, J.R. (2000). Transcoding of Internet multimedia for universal access. *Multimedia communications: Directions and innovations.* San Diego, CA: Academic Press.

Li, C.S., Mohan, R., and Smith, J.R. (1998). Multimedia content description in the InfoPyramid. *IEEE Proceedings of the International Conference on Acoustics, Speech, and Signal Processing.*

Kuhn, P.M., and Suzuki, T. (2001). MPEG-7 metadata for video transcoding: Motion and difficulty hints and *Storage and Retrieval for Media Database 4315*, 352–361.

Kuhn, P.M., Suzuki, T., and Vetro, A. (2001). MPEG-7 transcoding hints for reduced complexity and improved quality. *International Packet Video Workshop.* 276–285.

Mohan, R., Smith, J.R., and Li, C.S. (1999). Adapting multimedia Internet content for universal access. *IEEE Transactions on Multimedia 1*(1), 104–114.

Mohan, R., Smith, J.R., and Li, C.S. (1998). Multimedia content customization for universal access. *SPIE, Photonics east—Multimedia storage and archiving systems III.*

Salembier, P., and Smith, J.R. (2001). MPEG-7 multimedia description schemes. *IEEE Transactions on Circuits and Systems for Video Technology, 11*(6), 748–759.

Shoham, Y., and Gersho, A. (1990). Efficient bit allocation for an arbitrary set of quantizers. *IEEE Transactions on Acoustics, Speech, and Signal Processing 36*(9), 289–296.

Smith, J.R. (2003). MPEG-7 multimedia content description standard. *Multimedia Information Retrieval and Management.* Springer.

Smith, J.R. (2002). Multimedia content management in the MPEG-21 framework. *SPIE ITCOM, Internet Multimedia Management Systems III.*

Smith, J.R. (2000). Universal multimedia access. *Proceedings of the SPIE Multimedia Networking Systems III.*

Smith, J.R. (1999). VideoZoom spatio-temporal video browser. IEEE *Transactions on Multimedia 1*(2), 157–171.

Smith, J.R., Mohan, R., and Li, C.S. (1999). Scalable multimedia delivery for pervasive computing. *Proceedings of the ACM International Conference on Multimedia* (ACMMM).

Smith, J.R., Mohan, R., and Li, C.S. (1998a) Content-based transcoding of images in the Internet. *IEEE Proceedings of the International Conference on Image Processing.*

Smith, J.R., Mohan, R., and Li, C.S. (1998b). Transcoding Internet content for heterogenous client devices. *Proceedings of the IEEE International Symposium on Circuits and Systems.*

Smith, J.R., and Reddy, V. (2001). An application-based perspective on universal multimedia access using MPEG-7. *Proceedings of the SPIE Multimedia Networking Systems IV.*

Tseng, B., Lin, C.Y., and Smith, J.R. (2002). Video personalization and summarization system. *Proceedings of the IEEE Multimedia Signal Processing Conference.*

van Beek, P., Smith, J.R., Ebrahimi, T., Suzuki, T., and Askelof, J. (2003). Metadata driven multimedia access. *IEEE Signal Processing Magazine 52*(11), 969–979.

Vetro, A., Sun, H., and Wang, Y. (1999). MPEG-4 rate control for multiple video objects. *IEEE Transactions on Circuits and Systems for Video Technology 9*(1), 186–199.

6

Statistical Signal Processing

Yih-Fang Huang
Department of Electrical Engineering, University of Notre Dame, Notre Dame, Indiana, USA

6.1 Introduction

Statistical signal processing is an important subject in signal processing that enjoys a wide range of applications, including communications, control systems, medical signal processing, and seismology. It plays an important role in the design, analysis, and implementation of adaptive filters, such as adaptive equalizers in digital communication systems. It is also the foundation of multiuser detection that is considered an effective means for mitigating multiple access interferences in spread-spectrum based wireless communications.

The fundamental problems of statistical signal processing are those of signal detection and estimation that aim to extract information from the received signals and help make decisions. The received signals (especially those in communication systems) are typically modeled as random processes due to the existence of uncertainties. The uncertainties usually arise in the process of signal propagation or due to noise in measurements. Since signals are considered random, statistical tools are needed. As such, techniques derived to solve those problems are derived from the principles of **mathematical statistics**, namely, hypothesis testing and estimation.

This chapter presents a brief introduction to some basic concepts critical to statistical signal processing. To begin with, the principles of Bayesian and Fisher statistics relevant to estimation are presented. In the discussion of Bayesian statistics, emphasis is placed on two types of estimators: the minimum mean-squared error (MMSE) estimator and the maximum a posteriori (MAP) estimator. An important class of linear estimators, namely the Wiener filter, is also presented as a linear MMSE estimator. In the discussion of Fisher statistics, emphasis will be placed on the maximum likelihood estimator (MLE), the concept of sufficient statistics, and information inequality that can be used to measure the quality of an estimator.

This chapter also includes a brief discussion of signal detection. The signal detection problem is presented as one of binary hypothesis testing. Both Bayesian and Neyman-Pearson optimum criteria are presented and shown to be implemented with likelihood ratio tests. The principles of hypothesis testing also find a wide range of applications that involve decision making.

6.2 Bayesian Estimation

Bayesian estimation methods are generally employed to estimate a random variable (or parameter) based on another random variable that is usually observable. Derivations of Bayesian estimation algorithms depend critically on the a posteriori distribution of the underlying signals (or signal parameters) to be estimated. Those a posteriori distributions are obtained by employing Bayes' rules, thus the name Bayesian estimation.

Consider a random variable X that is some function of another random variable S. In practice, X is what is observed and S is the **signal** (or signal parameter) that is to be estimated. Denote the estimate as $\hat{S}(X)$, and the error as $\varepsilon \triangleq S - \hat{S}(X)$. Generally, the **cost** is defined as a function of the estimation error that clearly depends on both X and S, thus $J(\varepsilon) = J(S, X)$.

The objective of Bayesian estimation is to minimize the **Bayes' risk** $\mathcal{R}$, which is defined as the ensemble average (i.e., the expected value) of the cost. In particular, the Bayes' risk is defined by:

$$\mathcal{R} \triangleq E\{J(\varepsilon)\} = \int_{-\infty}^{\infty}\int_{-\infty}^{\infty} J(s, x) f_{S,X}(s, x)\,dx\,ds,$$

where $f_{S,X}(s, x)$ is the joint probability density function (pdf) of the random variables S and X. In practice, the joint pdf is not directly obtainable. However, by Bayes' rule:

$$f_{S,X}(s, x) = f_{S|X}(s|x) f_X(x), \tag{6.1}$$

the **a posteriori** pdf can be used to facilitate the derivation of Bayesian estimates. With the **a posteriori** pdf, the Bayes' risk can now be expressed as:

$$\mathcal{R} = \int_{-\infty}^{\infty}\left\{\int_{-\infty}^{\infty} J(s, x) f_{S|X}(s|x)\,ds\right\} f_X(x)\,dx. \tag{6.2}$$

Because the cost function is, in general, non-negative and so are the pdfs, minimizing the Bayes' risk is equivalent to minimizing:

$$\int_{-\infty}^{\infty} J(s, x) f_{S|X}(s|x)\,ds. \tag{6.3}$$

Depending on how the cost function is defined, the Bayesian estimation principle leads to different kinds of estimation algorithms. Two of the most commonly used cost functions are the following:

$$J_{MS}(\varepsilon) = |\varepsilon|^2. \tag{6.4a}$$

$$J_{MAP}(\varepsilon) = \begin{cases} 0 & \text{if } |\varepsilon| \le \frac{\Delta}{2} \\ 1 & \text{if } |\varepsilon| > \frac{\Delta}{2} \quad \text{where } \Delta \ll 1. \end{cases} \tag{6.4b}$$

These example cost functions result in two popular estimators, namely, the minimum mean-squared error (MMSE) estimator and the maximum a posteriori (MAP) estimator. These two estimators are discussed in more detail below.

6.2.1 Minimum Mean-Squared Error Estimation

When the cost function is defined to be the mean-squared error (MSE), as in equation 6.4a, the Bayesian estimate can be derived by substituting $J(\varepsilon) = |\varepsilon|^2 = (s - \hat{s}(x))^2$ into equation 6.2. Hence the following is true:

$$\mathcal{R}_{MS} = \int_{-\infty}^{\infty}\left\{\int_{-\infty}^{\infty} (s - \hat{s}(x))^2 f_{S|X}(s|x)\,ds\right\} f_X(x)\,dx. \tag{6.5}$$

Denote the resulting estimate as $\hat{s}_{MS}(x)$, and use the same argument that leads to equation 6.3. A necessary condition for minimizing $\mathcal{R}_{MS}$ results as:

$$\frac{d}{d\hat{s}}\int_{-\infty}^{\infty} (s - \hat{s}(x))^2 f_{S|X}(s|x)\,ds = 0. \tag{6.6}$$

The differentiation in equation 6.6 is evaluated at $\hat{s} = \hat{s}_{MS}(x)$. Consequently:

$$\int_{-\infty}^{\infty} (s - \hat{s}_{MS}(x)) f_{S|X}(s|x)\,ds = 0;$$

and

$$\hat{s}_{MS}(x) = \int_{-\infty}^{\infty} s f_{S|X}(s|x)\,ds = E\{s|x\}. \tag{6.7}$$

In essence, the estimate that minimizes the MSE is the conditional-mean estimate. If the a posteriori distribution is Gaussian, then the conditional-mean estimator is a linear function of X, regardless of the functional relation between S and X. The following theorem summarizes a very familiar result.

Theorem

Let $\boldsymbol{X} = [X_1, X_2, \ldots, X_K]^T$ and S be jointly Gaussian with zero means. The MMSE estimate of S based on $\boldsymbol{X}$ is $E[S|\boldsymbol{X}]$ and $E[S|\boldsymbol{X}] = \sum_{i=1}^{K} a_i x_i$, where a_i is chosen such that $E[(S - \sum_{i=1}^{K} a_i X_i)X_j] = 0$ for any $j = 1, 2, \ldots, K$.

6.2.2 Maximum a Posteriori Estimation

If the cost function is defined as in equation 6.4b, the Bayes' risk is as follows:

$$\mathcal{R}_{MAP} = \int_{-\infty}^{\infty} f_X(x)\left[1 - \int_{\hat{s}_{MAP}-\frac{\Delta}{2}}^{\hat{s}_{MAP}+\frac{\Delta}{2}} f_{S|X}(s|x)\,ds\right]dx.$$

To minimize $\mathcal{R}_{MAP}$, we maximize $\int_{\hat{s}_{MAP}-\frac{\Delta}{2}}^{\hat{s}_{MAP}+\frac{\Delta}{2}} f_{S|X}(s|x)\,ds$. When Δ is extremely small (as is required), this is equivalent to maximizing $f_{S|X}(s|x)$. Thus, $\hat{s}_{\text{MAP}}(x)$ is the value of s that maximizes $f_{S|X}(s|x)$.

Normally, it is often more convenient (for algebraic manipulation) to consider $\ln f_{S|X}(s|x)$, especially since $\ln(x)$ is a monotone nondecreasing function of x. A necessary condition for maximizing $\ln f_{S|X}(s|x)$ is as written here:

$$\frac{\partial}{\partial s}\ln f_{S|X}(s|x)\bigg|_{s=\hat{s}_{\text{MAP}}(x)} = 0. \tag{6.8}$$

Equation 6.8 is often referred to as the **MAP equation.** Employing Bayes' rule, the MAP equation can also be written as:

$$\frac{\partial}{\partial s}\ln f_{X|S}(x|s) + \frac{\partial}{\partial s}\ln f_S(s)\bigg|_{s=\hat{s}_{\text{MAP}}(x)} = 0. \tag{6.9}$$

Example (van Trees, 1968)

Let X_i, $i = 1, 2, \ldots, K$ be a sequence of random variables modeled as follows:

$$X_i = S + N_i \quad i = 1; 2, \cdots, K,$$

where S is a zero-mean Gaussian random variable with variance σ_s^2 and $\{N_i\}$ is a sequence of independent and identically distributed (iid) zero-mean Gaussian random variables with variance σ_n^2. Denote $\boldsymbol{X} = [X_1 X_2 \ldots X_K]^T$.

$$f_{X|S}(x|s) = \prod_{i=1}^{K}\frac{1}{\sqrt{2\pi}\sigma_n}\exp\left[-\frac{(x_i - s)^2}{2\sigma_n^2}\right]$$

$$f_S(s) = \frac{1}{\sqrt{2\pi}\sigma_s}\exp\left[-\frac{1}{2}\frac{s^2}{\sigma_s^2}\right]$$

$$f_{S|X}(s|x) = \frac{f_{x|s}(x|s)f_S(s)}{f_x(x)}$$

$$= \frac{1}{f_X(x)}\frac{1}{\sqrt{2\pi}\sigma_s}\left[\prod_{i=1}^{K}\frac{1}{\sqrt{2\pi}\sigma_n}\right]\exp\left[-\frac{1}{2}\left(\frac{\sum_{i=1}^{K}(x_i - s)^2}{\sigma_n^2} + \frac{s^2}{\sigma_s^2}\right)\right]$$

$$f_{S|x}(s|x) = C(x)\exp\left[-\frac{1}{2\sigma_p^2}\left(s - \frac{\sigma_s^2}{\sigma_s^2 + \sigma_n^2/K}\left(\frac{1}{K}\sum_{i=1}^{K}x_i\right)\right)^2\right], \tag{6.10}$$

where $\sigma_p^2 \triangleq \left(\frac{1}{\sigma_s^2} + \frac{K}{\sigma_n^2}\right)^{-1} = \frac{\sigma_s^2\sigma_n^2}{K\sigma_s^2 + \sigma_n^2}$.

From equation 6.10, it can be seen clearly that the conditional mean estimate and the MAP estimate are equivalent. In particular:

$$\hat{s}_{ms}(x) = \hat{s}_{MAP}(x) = \frac{\sigma_s^2}{\sigma_s^2 + \sigma_n^2/K}\left(\frac{1}{K}\sum_{i=1}^{K}x_i\right).$$

In this example, the MMSE estimate and the MAP estimate are equivalent because the a posteriori distribution is Gaussian. Some useful insights into Bayesian estimation can be gained through this example (van Trees, 1968).

Remarks

1. If $\sigma_s^2 \ll \frac{\sigma_n^2}{K}$, the a priori knowledge is more useful than the observed data, and the estimate is very close to the a priori mean (i.e., 0). In this case, the a posteriori distribution almost has no effect on the value of the estimate.
2. If $\sigma_s^2 \gg \frac{\sigma_n^2}{K}$, the estimate is directly related to the observed data as it is the sample mean, while the a priori knowledge is of little value.
3. The equivalence of $\hat{s}_{\text{MAP}}$ to $\hat{s}_{ms}(x)$ is not restricted to the case of Gaussian a posteriori pdf. In fact, if the cost function is symmetric and nondecreasing and if the a posteriori pdf is symmetric and unimodal and satisfies $\lim_{s\to\infty} J(s, x)f_{S|X}(s|x) = 0$, then the resulting Bayesian estimation (e.g., MAP estimate) is equivalent to $\hat{s}_{MS}(x)$.

6.3 Linear Estimation

Since the Bayesian estimators (MMSE and MAP estimates) presented in the previous section are usually not linear, it may be impractical to implement them. An alternative is to restrict the consideration to only the class of linear estimators and then find the **optimum estimatior** in that class. As such, the notion of optimality deviates from that of Bayesian estimation, which minimizes the Bayes' risk.

One of the most commonly used optimization criteria is the MSE (i.e., minimizing the error variance). This approach leads to the class of **Wiener filters** that includes the **Kalman filter** as a special case. In essence, the Kalman filter is a realizable Wiener filter as it is derived with a realizable (state-space) model. Those linear estimators are much more appealing in practice due to reduced implementational complexity and relative simplicity of performance analysis.

The problem can be described as follows. Given a set of zero-mean random variables, $X_1, X_2, \ldots, X_K$, it is desired to estimate a random variable S (also zero-mean). The objective here is to find an estimator $\hat{S}$ that is linear in X_i and that is optimum in some sense, like MMSE.

Clearly, if $\hat{S}$ is constrained to be linear in X_i, it can be expressed as $\hat{S} = \sum_{i=1}^{K} a_i X_i$. This expression can be used independently of the model that governs the relation between $\boldsymbol{X}$ and S. One can see that once the coefficients a_i, $i = 1, 2, \ldots, K$ are determined for all i, $\hat{S}$ is unambiguously (uniquely) specified. As such, the problem of finding an optimum estimator becomes one of finding the optimum set of coefficients, and estimation of a random signal becomes estimation of a set of deterministic parameters.

If the objective is to minimize the MSE, namely:

$$E\{\|S - \hat{S}\|^2\} = E\left\{ \|S - \sum_{i=1}^{K} a_i X_i\|^2 \right\}, \tag{6.11}$$

a necessary condition is that:

$$\frac{\partial}{\partial a_i} E\{\|S - \hat{S}\|^2\} = 0 \quad \text{for all } i. \tag{6.12}$$

Equation 6.12 is equivalent to:

$$E\left\{ \left(S - \sum_{j=1}^{K} a_j X_j \right) X_i^* \right\} = 0 \quad \text{for all } i. \tag{6.13}$$

In other words, a necessary condition for obtaining the linear MMSE estimate is the **uncorrelatedness** between the estimation error and the observed random variables. In the context of vector space, equation 6.13 is the well-known **orthogonality principle**. Intuitively, the equation states that the linear MMSE estimate of S is the projection of S onto the subspace spanned by the set of random variables $\{X_i\}$. In this framework, the norm of the vector space is the mean-square value while the inner product between two vectors is the correlation between two random variables.

Let the autocorrelation coefficients of x_i be $E\{X_j X_i^*\} = r_{ji}$ and the crosscorrelation coefficients of X_i and S be $E\{SX_i^*\} = \rho_i$. Then, equation 6.13 is simply:

$$\rho_i = \sum_{j=1}^{K} a_j r_{ji} \quad \text{for all } i, \tag{6.14}$$

which is essentially the celebrated **Wiener-Hopf** equation. Assume that r_{ij} and ρ_j are known for all i, j, then the coefficients $\{a_i\}$ can be solved by equation 6.14. In fact, this equation can be stacked up and put in the following matrix form:

$$\begin{bmatrix} r_{11} & r_{21} & \cdots & r_{K1} \\ r_{12} & r_{22} & \cdots & r_{K2} \\ \vdots & \vdots & \ddots & \vdots \\ r_{1K} & r_{2K} & \cdots & r_{KK} \end{bmatrix} \begin{bmatrix} a_1 \\ a_2 \\ \vdots \\ a_K \end{bmatrix} = \begin{bmatrix} \rho_1 \\ \rho_2 \\ \vdots \\ \rho_K \end{bmatrix} \tag{6.15}$$

or simply:

$$R\underline{a} = \underline{\rho}. \tag{6.16}$$

Thus, the coefficient vector can be solved by:

$$\underline{a} = R^{-1}\underline{\rho}, \tag{6.17}$$

which is sometimes termed the **normal equation**. The orthogonality principle of equation 6.13 and the normal equation 6.17 are the basis of Wiener filter, which is one of the most studied and commonly used linear adaptive filters with many applications (Haykin, 1996, 1991).

The matrix inversion in equation 6.17 could present a formidable numerical difficulty, especially when the vector dimension is high. In those cases, some computationally efficient algorithms, like Levinson recursion, can be employed to mitigate the impact of numerical difficulty.

6.4 Fisher Statistics

Generally speaking, there are two schools of thoughts in statistics: **Bayes** and **Fisher**. Bayesian statistics and estimation were presented in the previous section, where the emphasis was on estimation of random signals and parameters. This section is focused on estimation of a deterministic parameter (or signal). A natural question that one may ask is if the Bayesian approach presented in the previous section is applicable here or if the estimation of deterministic signals can be treated as a special case of estimating random signals. A closer examination shows that an alternative approach needs to be taken (van Trees, 1968) because the essential issues that govern the performance of estimators differ significantly.

The fundamental concept underlying the Fisher school of statistics is that of **likelihood function**. In contrast, Bayesian statistics is derived from conditional distributions, namely, the a posteriori distributions. This section begins with an introduction of the likelihood function and a derivation of the maximum likelihood estimation method. These are followed by the notion of sufficient statistics, which plays an important role in Fisherian statistics. Optimality properties of maximum likelihood estimates are then examined with the definition of Fisher information. Cramér-Rao lower bound and minimum variance unbiased estimators are then discussed.

6.4.1 Likelihood Functions

Fisher's approach to estimation centers around the concept of likelihood function (Fisher, 1992). Consider a random variable X that has a probability distribution $F_X(x)$ with probability density function (pdf) $f_X(x)$ parameterized by a parameter θ.

The likelihood function (with respect to the parameter θ) is defined as:

$$L(x; \theta) = f_X(x|\theta). \tag{6.18}$$

It may appear, at the first sight, that the likelihood function is nothing but the pdf. It is important, however, to note that the likelihood function is really a function of the parameter θ for a fixed value of x, whereas the pdf is a function of the realization of the random variable x for a fixed value of θ. Therefore, in a likelihood function, the variable is θ, while in a pdf the variable is x. The likelihood function is a quantitative indication of how *likely* that a particular realization (observation) of the random variable would have been produced from a particular distribution. The higher the value the likelihood function, the more likely the particular value of the parameter will have produced that realization of x. Hence, the cost function in the Fisherian estimation paradigm is the likelihood function, and the objective is to find a value of the parameter that maximizes the likelihood function, resulting in the **maximum likelihood estimate** (MLE).

The MLE can be derived as follows. Let there be a random variable whose pdf, $f_X(x)$, is parameterized by a parameter θ. Define the objective function:

$$L(x;\,\theta) = f_X(x|\theta), \tag{6.19}$$

where $f_X(x|\theta)$ is the pdf of X for a given θ. Then, the MLE is obtained by:

$$\hat{\theta}_{MLE}(x) = Arg\{\max_{\theta} L(x;\,\theta)\}. \tag{6.20}$$

Clearly, a necessary condition for $L(x;\,\theta)$ to be maximized is that:

$$\frac{\partial}{\partial\theta} L(x;\theta)\Big|_{\theta=\hat{\theta}_{MLE}} = 0. \tag{6.21}$$

This equation is sometimes referred to as the **likelihood equation**. In general, it is more convenient to consider the **log-likelihood function** defined as:

$$l(x;\,\theta) = \ln L(x;\,\theta) \tag{6.22}$$

and the **log-likelihood equation** as:

$$\frac{\partial}{\partial\theta} l(x;\theta)\Big|_{\theta=\hat{\theta}_{MLE}} = 0. \tag{6.23}$$

Example

Consider a sequence of random variables X_i, $i = 1, 2, \cdots, K$ modeled as:

$$X_i = \theta + N_i,$$

where N_i is a sequence of iid zero-mean Gaussian random variables with variance σ^2. In practice, this formulation can be considered as one of estimating a DC signal embedded in white Gaussian noise. The issue here is to estimate the strength, i.e., the magnitude, of the DC signal based on the a set of observations x_i, i = 1, 2, $\cdots$, K. The log-likelihood function as defined in equation 6.22 is given by:

$$l(\underline{x};\theta) = -K\ln(\sqrt{2\pi}\sigma) - \frac{1}{2\sigma^2}\sum_{i=1}^{K}(x_i-\theta)^2,$$

where $\underline{x} = [x_1, x_2, \ldots, x_k]^T$.

Solving the log-likelihood equation, 6.23, yields:

$$\hat{\theta}_{MLE}(x) = \frac{1}{K}\sum_{i=1}^{K} x_i.$$

Thus, the MLE for a DC signal embedded in additive zero mean white Gaussian noise is the **sample mean.** As it turns out, this sample mean is the **sufficient statistic** for estimating θ. The concept of the sufficient statistic is critical to the optimum properties of MLE and, in general, to Fisherian statistics. Generally speaking, the likelihood function is directly related to sufficient statistics, and the MLE is usually a function of sufficient statistics.

6.4.2 Sufficient Statistics

Sufficient statistics is a concept defined in reference to a particular parameter (or signal) to be estimated. Roughly speaking, a sufficient statistic is a function of the set of observations that contains all the information possibly obtainable for the estimation of a particular parameter. Given a parameter θ to be estimated, assume that $\underline{x}$ is the vector consisting of the observed variables. A statistic $T(\underline{x})$ is said to be a **sufficient statistic** if the probability distribution of $\underline{X}$ given $T(\underline{x}) = t$ is independent of θ. In essence, if $T(\underline{x})$ is a sufficient statistic, then all the information regarding estimation of θ that can be extracted from the observation is contained in $T(\underline{x})$. The Fisher factorization theorem stated below is sometimes used as a definition for the sufficient statistic.

Fisher Factorization Theorem

A function of the observation set $T(\underline{X})$ is a sufficient statistic if the likelihood function of $\underline{X}$ can be expressed as:

$$L(\underline{x};\theta) = h(T(\underline{x}),\,\theta)g(\underline{x}). \tag{6.24}$$

In the example shown in Section 6.4.1, $\sum_{i=1}^{K} x_i$ is a sufficient statistic, and the sample mean $\frac{1}{K}\sum_{i=1}^{K} x_i$ is the MLE for θ. The fact that $\sum_{i=1}^{K} x_i$ is a sufficient statistic can be easily seen by using the Fisher factorization theorem. In particular:

$$L(\underline{x};\theta) = \left(\frac{1}{\sqrt{2\pi}\sigma}\right)^K \exp\left[-\frac{1}{2\sigma^2}\sum_{i=1}^{K}(x_i-\theta)^2\right]. \tag{6.25}$$

The above equation can be expressed as:

$$L(\underline{x};\theta) = \left\{\exp\left[\frac{\theta}{\sigma^2}\left(\sum_{i=1}^{K} x_i\right) - \frac{K\theta^2}{2\sigma^2}\right]\right\}\left(\frac{1}{\sqrt{2\pi}\sigma}\right)^K \exp\left[-\frac{1}{2\sigma^2}\sum_{i=1}^{K} x_i^2\right]. \tag{6.26}$$

Identifying equation 6.26 with 6.24, it can be easily seen that, if the following is defined:

$$h(T((x), \theta) = \exp\left[\frac{\theta}{\sigma^2}\left(\sum_{i=1}^{K} x_i\right) - \frac{K\theta^2}{2\sigma^2}\right]$$

and

$$g(x) = \left(\frac{1}{\sqrt{2\pi}\sigma}\right)^K \exp\left[-\frac{1}{2\sigma^2}\sum_{i=1}^{K} x_i^2\right],$$

then $T(x) = \sum_{i=1}^{K} x_i$ is clearly a sufficient statistic for estimating θ. It should be noted that the sufficient statistic may not be unique. In fact, it is always subject to a scaling factor.

In equation 6.26, it is seen that the pdf can be expressed as:

$$f_X(x|\theta) = \{\exp[c(\theta)T(x) + d(\theta) + S(x)]\}I(x), \qquad (6.27)$$

where $c(\theta) = \frac{\theta}{\sigma^2}$, $T(x) = \sum_{i=1}^{K} x_i$, $d(\theta) = -\frac{1}{2\sigma^2}\sum_{i=1}^{K} x_i^2$, $S(x) = -\frac{K}{2}\ln(2\pi\sigma^2) - \frac{1}{2\sigma^2}\sum_{i=1}^{K} x_i^2$, and $I(x)$ is an indicator function that is valued at 1 wherever the value of the pdf is nonzero and zero otherwise. A pdf that can be expressed in the form given in equation 6.27 is said to belong to the **exponential family of distributions** (EFOD).

It can be verified easily that Gaussian, Laplace (or two-sided exponential), binomial, and Poisson distributions all belong to the EFOD. When a pdf belongs to the EFOD, one can easily identify the sufficient statistic. In fact, the mean and variance of the sufficient statistic can be easily calculated (Bickel and Doksum, 1977).

As stated previously, the fact that the MLE is a function of the sufficient statistic helps to ensure its quality as an estimator. This will be discussed in more detail in the subsequent sections.

6.4.3 Information Inequality and Cramér-Rao Lower Bound

There are several criteria that can be used to measure the quality of an estimator. If $\theta(X)$ is an estimator for the parameter θ based on the observation X, three criteria are typically used to evaluate its quality:

1. **Bias**: If $E[\hat{\theta}(X)] = \theta, \theta(\hat{X})$ is said to be an **unbiased estimator**.
2. **Variance**: For estimation of deterministic parameters and signals, variance of the estimator is the same as variance of the estimation error. It is a commonly used performance measure, for it is practically easy to evaluate.
3. **Consistency**: This is an asymptotic property that is examined when the sample size (i.e., the dimension of the vector X) approaches infinity. If $\theta(\hat{X})$ converges with probability one to θ, then it is said to be **strong consistent**. If it converges in probability, it is **weak consistent**.

In this section, the discussion will be focused on the second criterion, namely the variance, which is one of the most commonly used performance measures in statistical signal processing. In the estimation of deterministic parameters, **Fisher's information** is imminently related to variance of the estimators.

The Fisher's information is defined as:

$$I(\theta) = E\left\{\left[\frac{\partial}{\partial\theta} l(x; \theta)\right]^2\right\}. \qquad (6.28)$$

Note that the Fisher's information is non-negative, and it is additive if the set of random variables is independent.

Theorem (Information Inequality)

Let $T(X)$ be a statistic such that its variance $V_\theta(T(X)) < \infty$, for all θ in the parameter space Θ. Define $\psi(\theta) = E_\theta\{T(X)\}$. Assume that the following regularity condition holds:

$$E\left\{\left[\frac{\partial}{\partial\theta} l(x; \theta)\right]\right\} = 0 \quad \text{for all} \quad \theta \in \Theta, \qquad (6.29)$$

where $l(x, \theta)$ has been defined in equation 6.22. Assume further that $\psi(\theta)$ is differentiable for all θ. Then:

$$V_\theta(T(X)) \leq \frac{[\psi'(\theta)]^2}{I(\theta)}. \qquad (6.30)$$

Remarks

1. The regularity condition defined in equation 6.29 is a restriction imposed on the likelihood function to guarantee that the order of expectation operation and differentiation is interchangeable.
2. If the regularity condition holds also for second order derivatives, $I(\theta)$ as defined in equation 6.28 can also be evaluated as:

$$I(\theta) = -E\left\{\frac{\partial^2}{\partial\theta^2} l(x; \theta)\right\}.$$

3. The subscript θ of the expectation (E_θ) and of the variance (V_θ) indicates the dependence of expectation and variance on θ.
4. The information inequality gives a lower bound on the variance that any estimate can achieve. It thus reveals the best that an estimator can do as measured by the error variance.
5. If $T(X)$ is an unbiased estimator for θ, (i.e., $E_\theta\{T(X)\} = \theta$), then the information inequality, equation 6.30, reduces to:

$$V_\theta(T(X)) \leq \frac{1}{I(\theta)}, \qquad (6.31)$$

which is the well-known **Cramér-Rao lower bound** (CRLB).

6. In statistical signal processing, closeness to the CRLB is often used as a measure of *efficiency* of an (unbiased) estimator. In particular, one may define the **Fisherian efficiency** of an unbiased estimator $\hat{\theta}(\boldsymbol{X})$ as:

$$\eta(\hat{\theta}(X)) = \frac{I^{-1}(\theta)}{V_\theta\{\hat{\theta}(X)\}}. \tag{6.32}$$

Note that η is always less than one, and the larger η, the more efficient that estimator is. In fact, when $\eta = 1$, the estimator achieves the CRLB and is said to be an efficient estimator in the Fisherian sense.

7. If an unbiased estimator has a variance that achieves the CRLB for all $\theta \in \Theta$, it is called a **uniformly minimum variance unbiased estimator** (UMVUE). It can be shown easily that UMVUE is a consistent estimator and MLE is usually the UMVUE.

6.4.4 Properties of MLE

The MLE has many interesting properties, and it is the purpose of this section to enlist some of those properties.

1. It is easy to see that the MLE may be biased. This is because being unbiased was not part of the objective in seeking MLE. MLE, however, is always asymptotically unbiased.
2. As shown in the previous section, MLE is a function of the sufficient statistic. This can also be seen from the Fisher factorization theorem.
3. MLE is asymptotically a minimum variance unbiased estimator. In other words, its variance asymptotically achieves the CRLB.
4. MLE is consistent. In particular, it converges to the parameter with probability one (or in probability).
5. Under the **regularity condition** of equation 6.29, if there exists an unbiased estimator whose variance attains the CRLB, it is the MLE.
6. Generally speaking, the MLE of a transformation of a parameter (or signal) is the transformation of the MLE of that parameter (or signal). This is referred to as the **invariance properly** of the MLE.

The fact that MLE is a function of sufficient statistics means that it depends on **relevant information**. This does not necessarily mean that it will always be the *best* estimator (in the sense of, say, minimum variance), for it may not make the best use of the information. However, when the sample size is large, all the information relevant to the unknown parameter is essentially available to the MLE estimator. This explains why MLE is **asymptotically unbiased**, it is consistent, and its variance achieves asymptotically the CRLB.

Example (Kay, 1993)

Consider the problem of estimating the phase of a sinusoidal signal received with additive Gaussian noise. The problem is formulated as:

$$X_i = A\cos(\omega_o i + \phi) + N_i \quad i = 0, 1, \ldots, K-1$$

where $\{N_i\}$ is an iid sequence of zero-mean Gaussian random variables with variance σ^2. Employing equation 6.23, MLE can be obtained by minimizing:

$$J(\phi) = \sum_{i=1}^{K-1} (x_i - A\cos(\omega_o i + \phi))^2.$$

Differentiating $J(\theta)$ with respect to θ and setting it equal to zero yields:

$$\sum_{i=0}^{K-1} x_i \sin(\omega_o i + \hat{\phi}_{MLE}) = A\sum_{i=0}^{K-1} x_i \sin(\omega_o i + \hat{\phi}_{MLE})\cos(\omega_o i + \hat{\phi}_{MLE}).$$

Assume that:

$$\frac{1}{K}\sum_{i=0}^{K-1} \cos(2\omega_o i + 2\phi) = 0 \quad \text{for all } \phi.$$

Then the MLE for the phase can be approximated as:

$$\hat{\phi}_{MLE} \approx -\arctan\frac{\sum_{i=0}^{K-1} x_i \sin(\omega_o i)}{\sum_{i=0}^{K-1} x_i \cos(\omega_o i)}.$$

Perceptive readers may see that implementation of MLE can easily become complex and numerically difficult, especially when the underlying distribution is non-Gaussian. If the parameter to be estimated is a simple scalar, and its admissible values are limited to a finite interval, search algorithms can be employed to guarantee satisfactory results. If this is not the case, more sophisticated numerical optimization algorithms will be needed to render good estimation results. Among others, iterative algorithms such as the Newton-Raphson method and the expectation maximization method are often employed.

6.5 Signal Detection

The problem of signal detection can be formulated mathematically as one of binary hypothesis testing. Let X be the random variable that represents the observation, and let the observation space be denoted by Ω (i.e., $x \in \Omega$). There are basically two hypotheses:

$$\textbf{Null hypothesis } H_0 : X \sim F_0. \tag{6.33a}$$

$$\textbf{Alternative } H_1 : X \sim F_1. \tag{6.33b}$$

The F_0 is the probability distribution of X given that H_0 is true, and F_1 is the probability distribution of X given that H_1 is true. In some applications (e.g., a simple radar communication system), it may be desirable to detect a signal of constant amplitude, and then F_1 will simply be F_0 shifted by a mean value equal to the signal amplitude. In general, this formulation also assumes that, with probability one, either the null hypothesis or the alternative is true. Specifically, let π_0 and π_1 be the prior probabilities of H_0 and H_1 being true, respectively. Then:

$$\pi_0 + \pi_1 = 1. \tag{6.34}$$

The objective here is to decide whether H_0 or H_1 is true based on the observation of $X = x$. A decision rule, namely a detection scheme, $d(x)$ essentially partitions Ω into two subspaces Ω_0 and Ω_1. The subspace Ω_0 consists of all observations that lead to the decision that H_0 is true, while Ω_1 consists of all observations that lead to the decision that H_1 is true. For notational convenience, one may also define $d(x)$ as follows:

$$d(x) = \begin{cases} 0 & \text{if } x \in \Omega_0 \\ 1 & \text{if } x \in \Omega_1 \end{cases}. \tag{6.35}$$

For any decision rule $d(x)$, there are clearly four possible outcomes:

1. Decide H_0 when H_0 is true
2. Decide H_0 when H_1 is true
3. Decide H_1 when H_0 is true
4. Decide H_1 when H_1 is true

Two types of error can occur:

1. **Type-I error** (**false alarm**) Decide H_1 when H_0 is true. The probability of type-I error is as follows:

$$\alpha = \int_{\Omega_1} f_0(x)dx. \tag{6.36}$$

2. **Type-II error** (**miss**): Decide H_0 when H_1 is true. The probability of type-II error is as follows:

$$P_m = \int_{\Omega_0} f_1(x)dx. \tag{6.37}$$

Another quantity of concern is the probability of detection, which is often termed the **power** in signal detection literature, defined by:

$$\beta = 1 - P_m = \int_{\Omega_1} f_1(x)dx. \tag{6.38}$$

Among the various decision criteria, Bayes and Neyman-Pearson are most popular. Detection schemes may also be classified into parametric and nonparametric detectors. The discussion here focuses on Bayes and Neyman-Pearson criteria that are considered parametric. Before any decision criteria can be derived, the following constraints need be stated first:

$$Prob\{d(x) = 0|H_0\} + Prob\{d(x) = 1|H_0\} = 1 \tag{6.39a}$$

$$Prob\{d(x) = 0|H_1\} + Prob\{d(x) = 1|H_1\} = 1. \tag{6.39b}$$

The above constraints of equations 6.39a and 6.39b are simply outcomes of the assumptions that $\Omega_0 \cup \Omega_1 = \Omega$ and $\Omega_0 \cap \Omega_1 = \emptyset$ (i.e., for every observation, an unambiguous decision must be made).

6.5.1 Bayesian Detection

The objective of a Bayes criterion is to minimize the so-called Bayes' risk which is, again, defined as the expected value of the cost. To derive a Bayes' detection rule, the costs of making decisions need to be defined first. Let the costs be denoted by C_{ij} being cost of choosing i when j is true. In particular:

$$C_{01} = \text{cost of choosing } H_0 \text{ when } H_1 \text{ is true.}$$

$$C_{10} = \text{cost of choosing } H_1 \text{ when } H_0 \text{ is true.}$$

In addition, assume that the prior probabilities π_0 and π_1 are known. The Bayes' risk is then evaluated as follows:

$$\begin{aligned} \mathcal{R} \triangleq E\{C\} = \pi_0\{Prob\{d(x) = 0|H_0\}C_{00} + Prob\{d(x) = 1|H_0\}C_{10}\} \\ + \pi_1\{Prob(d(x) = 0|H_1)C_{10} + Prob(d(x) = 1|H_1)C_{11}\}. \end{aligned} \tag{6.40}$$

Substituting equations 6.34, 6.39a and 6.39b into equation 6.40 yields:

$$\begin{aligned} \mathcal{R} &= \pi_0 C_{00} \int_{\Omega_0} f_0(x)dx + \pi_0 C_{10} \int_{\Omega_1} f_0(x)dx \\ &+ \pi_1 C_{01} \int_{\Omega_0} f_1(x)dx + \pi_1 C_{11} \int_{\Omega_1} f_1(x)dx \\ &= \pi_0 C_{10} + (1 - \pi_0)C_{11} + \int_{\Omega_0} \{(1 - \pi_0)(C_{01} - C_{11})f_1(x) \\ &- \pi_0(C_{10} - C_{00})f_0(x)\}dx. \end{aligned}$$

Note that the sum of the first two terms is a constant. In general, it is reasonable to assume that:

$$C_{01} - C_{11} > 0 \quad \text{and} \quad C_{10} - C_{00} > 0.$$

In other words, the costs of making correct decisions are less than those of making incorrect decisions:

$$I_1(x) \triangleq (1 - \pi_0)(C_{01} - C_{11})f_1(x).$$

$$I_2(x) \triangleq \pi_0(C_{10} - C_{00})f_0(x).$$

It can be seen easily that $I_1(x) \geq 0$ and $I_2(x) \geq 0$. Thus, $\mathcal{R}$ can be rewritten as:

$$\mathcal{R} = \text{constant} + \int_{\Omega_0} [I_1(x) - I_2(x)]dx.$$

To minimize $\mathcal{R}$, the observation space Ω need be partitioned such that $x \in \Omega_1$ whenever:

$$I_1(x) \geq I_2(x).$$

In other words, decide H_1 if:

$$(1 - \pi_0)(C_{01} - C_{11})f_1(x) \geq \pi_0(C_{10} - C_{00})f_0(x). \quad (6.41)$$

So, the Bayes' detection rule is essentially evaluating the likelihood ratio defined by:

$$L(x) \stackrel{\Delta}{=} \frac{f_1(x)}{f_0(x)}. \quad (6.42)$$

Comparing it to the threshold yields:

$$\lambda \stackrel{\Delta}{=} \frac{\pi_0}{1 - \pi_0} \frac{C_{10} - C_{00}}{C_{01} - C_{11}}. \quad (6.43)$$

In particular:

$$L(x) = \frac{f_1(x)}{f_0(x)} \begin{cases} \geq \lambda \Rightarrow H_1 \\ < \lambda \Rightarrow H_0 \end{cases}. \quad (6.44)$$

A decision rule characterized by the likelihood ratio and a threshold as in equation 6.44 is referred to as a likelihood ratio test (LRT). A Bayes' detection scheme is always an LRT.

Depending on how the a posteriori probabilities of the two hypotheses are defined, the Bayes' detector can be realized in different ways. One typical example is the so-called MAP detector that renders the minimum probability of error by choosing the hypothesis with the maximum a posteriori probability. Another class of detectors is the minimax detectors, which can be considered an extension of Bayes' detectors. The minimax detector is also an LRT. It assumes no knowledge of the prior probabilities (i.e., π_0 and π_1) and selects the threshold by choosing the prior probability that renders the maximum Bayes' risk. The minimax detector is a *robust* detector because its performance does not vary with the prior probabilities.

6.5.2 Neyman-Pearson Detection

The principle of the Neyman-Pearson criterion is founded on the **Neyman-Pearson Lemma** stated below:

Neyman-Pearson Lemma

Let $d_{\lambda^*}(x)$ be a likelihood ratio test with a threshold λ^* as defined in equation 6.44. Let α^* and β^* be the false-alarm rate and power, respectively, of the test $d_\lambda * (x)$. Let $d_\lambda(x)$ be another arbitrary likelihood ratio test with threshold λ and false-alarm rate and power as, respectively, α and β. If $\alpha \leq \alpha^*$, then $\beta \leq \beta^*$.

The Neyman-Pearson Lemma showed that if one desires to increase the power of an LRT, one must also accept the consequence of an increased false-alarm rate. As such, the Neyman-Pearson detection criterion is aimed to maximize the power under the constraint that the false-alarm rate be upper bounded by, say, α_0. The Neyman-Pearson detector can be derived by first defining a cost function:

$$J = (1 - \beta) + \lambda(\alpha - \alpha_0). \quad (6.45)$$

It can be shown that:

$$J = \lambda(1 - \alpha_0) + \int_{\Omega_0} [f_1(x)dx - \lambda f_0(x)]dx, \quad (6.46)$$

and an LRT will minimize J for any positive λ. In particular:

$$L(x) \stackrel{\Delta}{=} \frac{f_1(x)}{f_0(x)} \begin{cases} \geq \lambda \Rightarrow H_1 \\ < \lambda \Rightarrow H_0 \end{cases}.$$

To satisfy the constraint and to maximize the power, choose λ so that $\alpha = \alpha_0$, namely:

$$\alpha = \int_\lambda^\infty f_{L|H_0}(l|H_0)dl = \alpha_0,$$

where $f_{L|H_0}(l|H_0)$ is the pdf of the likelihood ratio. The threshold is determined by solving the above equation.

The Neyman-Pearson detector is known to be the most powerful detector for the problem of detecting a constant signal in noise. One advantage of the Neyman-Pearson detector is that its implementation does not require explicit knowledge of the prior probabilities and costs of decisions. However, as is the case for Bayes' detector, evaluation of the likelihood ratio still requires exact knowledge of the pdf of X under both hypotheses.

6.5.3 Detection of a Known Signal in Gaussian Noise

Consider a signal detection problem formulated as follows:

$$\begin{aligned} H_0 &: X_i = N_i \\ H_1 &: X_i = N_i + S, \end{aligned} \quad (6.47)$$

$i = 1, 2, \ldots, K$. Assume that S is a deterministic constant and that X_i, $i = 1, 2, \ldots, K$ are iid zero-mean Gaussian random variables with a known variance[2]. The likelihood ratio is then as written here:

$$L(x) = \frac{f_1(x_1, x_2, \ldots, x_K)}{f_0(x_1, x_2, \ldots, x_K)} = \prod_{i=1}^{K} \frac{f_1(x_i)}{f_0(x_i)}.$$

Taking the logarithm of $L(x)$ yields the log-likelihood ratio of:

$$lnL(x) = \sum_{i=1}^{K} \frac{(2x_i S - S^2)}{2\sigma^2}. \quad (6.48)$$

Straightforward algebraic manipulations show that the LRT is characterized by comparing a test statistic:

$$T(x) \triangleq \sum_{i=1}^{K} x_i \quad (6.49)$$

with the threshold λ. If $T(x) \geq \lambda$, then H_1 is said to be true; otherwise, H_0 is said to be true.

It is interesting to note that $T(x) = \sum_{i=1}^{K} x_i$ turns out to be the sufficient statistic for estimating S, as shown in Section 6.4.2. This should not be surprising as detection of a constant signal in noise is a dual problem of estimating the mean of the observations.

Under the iid Gaussian assumption, the test statistic is clearly a Gaussian random variable. Furthermore, $E\{T(X)|H_0\} = 0$, $E\{T(X)|H_1\} = KS$, and $Var\{T(X)|H_0\} = Var\{T(X)|H_1\} = K\sigma^2$. The pdf of $T(x)$ plotted in Figure 6.1, can be written as:

$$f_{T|H_0}(t|H_0) = \frac{1}{\sqrt{2\pi K\sigma^2}} e^{-\frac{t^2}{2K\sigma^2}}$$
$$f_{T|H_1}(t|H_1) = \frac{1}{\sqrt{2\pi K\sigma^2}} e^{\frac{-(t-KS)^2}{2K\sigma^2}}.$$

The false alarm rate and power are given by:

$$\alpha = \int_{\lambda_0}^{\infty} f_{T|H_0}(t|H_0)dt.$$
$$\beta = \int_{\lambda_0}^{\infty} f_{T|H_1}(t|H_1)dt.$$

The above two equations can be further specified as:

$$\alpha = \int_{\lambda 0}^{\infty} \frac{1}{\sqrt{2\pi K\sigma^2}} e^{-t^2/2K\sigma^2} \mathrm{dt} = 1 - \Phi\left(\frac{\lambda_0}{\sigma\sqrt{K}}\right). \quad (6.50)$$

$$\beta = \int_{\lambda 0}^{\infty} \frac{1}{\sqrt{2\pi K\sigma^2}} e^{-(t-Ks)^2/2K\sigma^2} \mathrm{dt} = 1 - \Phi\left(\frac{\lambda 0 - KS}{\sigma\sqrt{K}}\right). \quad (6.51)$$

In equations 6.50 and 6.51, this is true:

$$\Phi(x) \triangleq \int_{-\infty}^{x} \frac{1}{\sqrt{2\pi}} e^{-t^2/2} dt.$$

For the Neyman-Pearson detector, the threshold λ_0 is determined by the constraint on the false alarm rate:

$$\lambda_0 = \sigma\sqrt{K}\Phi^{-1}(1 - \alpha).$$

Combining equations 6.50 and 6.51 yields a relation between the false alarm rate and power, namely:

$$\beta = 1 - \Phi\left(\Phi^{-1}(1 - \alpha) - \sqrt{K}\frac{S}{\sigma}\right). \quad (6.52)$$

Figure 6.1 is a good illustration of the Neyman-Pearson Lemma. The shaded area under $f_{T|H_1}(t|H_1)$ is the value of power, and the shaded area under $f_{T|H_0}(t|H_0)$ is the false alarm rate. It is seen that if the threshold is moved to the left, both the power and the false alarm rate increase.

Remarks

1. It can be seen from equation 6.51 that as the sample size K increases, α decreases and β increases. In fact, $\lim_{K-\infty} \beta = 1$.
2. Define $d^2 \triangleq \frac{s^2}{\sigma^2}$, which can be taken as the signal-to-noise ratio (SNR). From equation 6.52, one can see that $\lim_{d\to\infty} \beta = 1$.
3. If the test statistic is defined with a scaling factor, namely $T(x) = \sum_{i=1}^{K} \frac{S}{\sigma^2} X_i$, the detector remains unchanged as long as the threshold is also scaled accordingly. Let $h_i \triangleq \frac{s}{\sigma^2}$, and then $T(x) = \sum_{i=1}^{K} h_i x_i$. The detector is essentially a discrete-time **matched filter.** This detector is also known as the correlation detector as it correlates signal with the observation. When the output of the detector is of a large numerical value, the correlation between the observation and the signal is high, and H_1 is (likely to be) true.

The Neyman-Pearson detector can be further characterized by the receiver–operation curve (ROC) shown in Figure 6.2, which is a plot of equation 6.52, parameterized by d, the SNR.

It should be noted that all continuous LRTs have ROCs that are above the line of $\alpha = \beta$ and are concave downward. In addition, the slope of a curve in a ROC curve at a particular point is the value of the threshold λ required to achieve the prescribed value of α and β.

6.6 Suggested Readings

There is a rich body of literature on the subjects of statistical signal processing and mathematical statistics. A classic textbook on detection and estimation is by van Trees (1968). This book provides a good basic treatment of the subject, and it is easy to read. Since then, many books have been written. Textbooks written by Poor (1988) and Kay (1993, 1998) are the more popular ones. Poor's book provides a fairly complete coverage of the subject of signal detection and estimation. Its presentation is built on the principles of mathematical statistics and includes some brief discussions of nonparametric and robust detection theory. Kay's books are more relevant to signal processing applications though they also include a good deal of theoretical treatment in statistics. In addition, Kassam (1988)

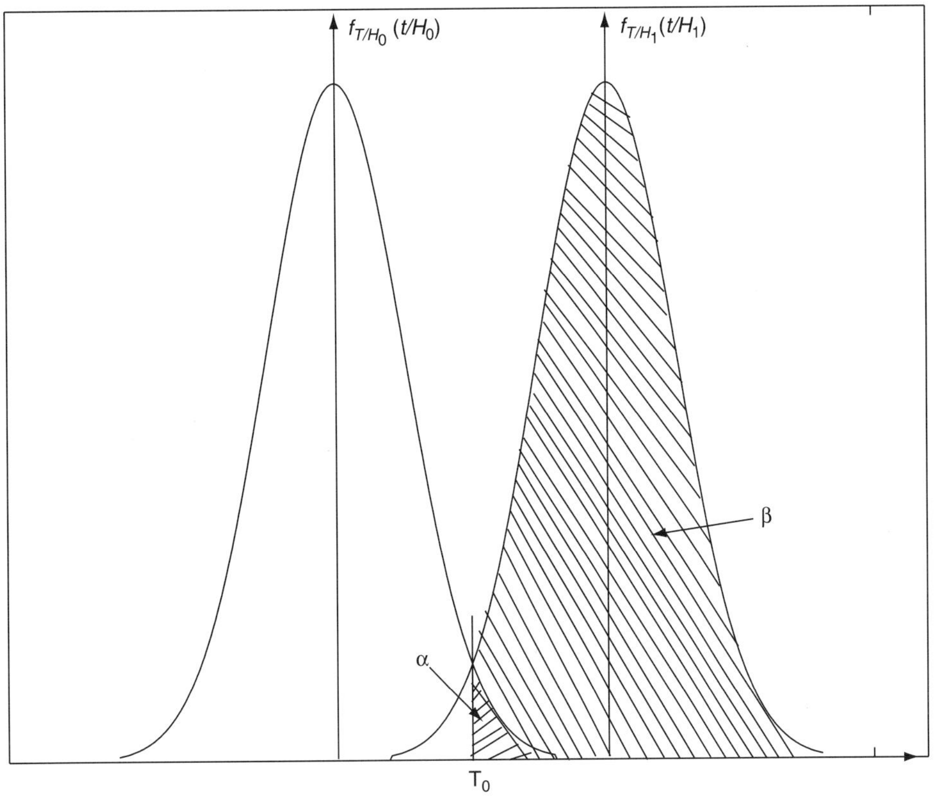

FIGURE 6.1 Probability Density Function of the Test Statistic

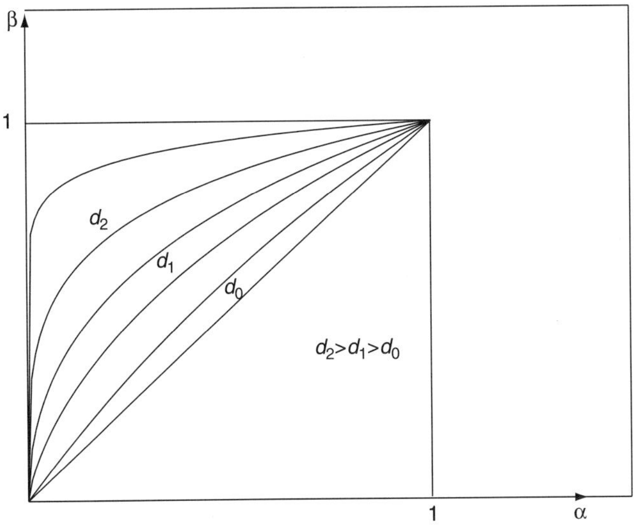

FIGURE 6.2 An Example of Receiver Operation Curves

offers a good understanding of the subject of signal detection in non-Gaussian noise, and Weber (1987) offers useful insights into signal design for both coherent and incoherent digital communication systems. If any reader is interested in learning more about mathematical statistics, Bickel and Doksum (1977) would be a good reference. It contains in-depth coverage of the subject. For quick references to the subject, however, a book by Silvey (1975) is a very useful one.

It should be noted that studies of statistical signal processing cannot be effective without proper background in probability theory and random processes. Many reference books on that subject, such as Billingsley (1979), Kendall and Stuart (1977), Papoulis and Pillai (2002), and Stark and Woods (2002), are available.

References

Bickel, P.J., and Doksum, K.A. (1977). *Mathematical statistics: Basic ideas and selected topics.* San Francisco: Holden-Day.

Billingsley, P. (1979). *Probability and measure.* New York: John Wiley & Sons.

Fisher, R.A. (1950). On the mathematical foundations of theoretical statistics. In R.A. Fisher, *Contributions to mathematical statistics.* New York: John Wiley & Sons.

Haykin, S. (2001). *Adaptive filter theory.* (4th ed.) Englewood-Cliffs, NJ: Prentice Hall.

Haykin, S. (Ed.). (1991). *Advances in spectrum analysis and array processing.* Vols. 1 and 2. Englewood Cliffs, NJ: Prentice Hall.

Kassam, S.A. (1988). *Signal detection in non-Gaussian noise.* New York: Springer-Verlag.

Kay, S.M. (1993). *Fundamentals of statistical signal processing: Estimation theory.* Upper Saddle River, NJ: Prentice Hall.

Kay, S.M. (1998). *Fundamentals of Statistical Signal Processing: Detection Theory.* Upper Saddle River, New Jersey: Prentice-Hall.

Kendall, M.G., and Stuart, A. (1977). *The advanced theory of statistics,* Vol. 2. New York: Macmillan Publishing.

Papoulis, A., and Pillai, S.U. (2002). *Probability, random variables, and stochastic processes.* (4th Ed.) New York: McGraw-Hill.

Poor, H.V. (1988). *An introduction to signal detection and estimation.* New York: Springer-Verlag.

Silvey, S.D. (1975). *Statistical inference.* London: Chapman and Hall.

Stark, H., and Woods, J.W. (2002). *Probability, random processes, and estimation theory.* (3rd ed.). Upper Saddle River, NJ: Prentice Hall.

van Trees, H.L. (1968). *Detection, estimation, and modulation theory.* New York: John Wiley & Sons.

Weber, C.L. (1987). *Elements of detection and signal design.* New York: Springer-Verlag.

7

VLSI Signal Processing

Surin Kittitornkun
King Mongkut's Institute of Technology Ladkrabang, Bangkok, Thailand

Yu-Hen Hu
Department of Electrical and Computer Engineering, University of Wisconsin-Madison, Madison, Wisconsin, USA

7.1 Introduction

The field of very large scale integrated (VLSI) signal processing concerns the design and implementation of signal processing algorithms using application-specific VLSI architecture, including programmable digital signal processors and dedicated signal processors implemented with VLSI technology. In this chapter, we survey important developments in this field, including algorithm design, architecture development, and design methodology.

An implementation of a digital signal processing (DSP) algorithm consists of the computer program of that algorithm and the hardware on which the program is executed. In many signal processing applications, real-time processing is an essential requirement. Real-time implies that the results of a signal processing algorithm must be computed by a predefined deadline after the inputs are sampled. For example, in a cellular phone, the speech coding signal processing algorithm must be executed to match the speed of normal conversation. An implementation of a real-time signal processing application has three special characteristics:

(1) Input signal samples are made available while the program is being executed. The computation cannot be started early until the input signal samples are received.
(2) Results must be computed before the prespecified deadline. When real-time constraint is not met, the quality of services will be dramatically compromised.
(3) Vast amount of operations must be computed. The data rate of a single Moving Picture Experts Group MPEG-II encoded video signal stream can easily exceed 20 million samples per second. On average, each sample will require tens or even hundreds of fixed-point or floating point arithmetic operations to process. Multiple streams of video and audio signals often need to be processed simultaneously. Hence, signal processing algorithms are always computation-intensive.

An efficient implementation of a real-time signal processing algorithm must be able to perform an extremely large amount of arithmetic operations within a short duration. In other words, it must sustain high throughput rate.

Signal processing is often found in embedded systems such as electrical appliances where the user interacts with the system's main function instead of specific signal processing algorithms. For example, speech coding is regularly performed in cellular phones while users may never be aware of its existence.

A signal processing algorithm can be implemented on a general purpose computer, a special purpose programmable digital signal processor, or even dedicated hardware. The tasks of implementation involve algorithm design, code generation (programming), and architecture synthesis. With the same integrated circuit technology, a specialized hardware platform may offer better performance than general-purpose hardware by eliminating redundant operations and components. However, the design and manufacturing cost will be higher.

7.1.1 VLSI Application Specific Processors

In the early 1960s, most DSP algorithms, such as the fast Fourier transform (FFT), were implemented in Fortran programs running on a general-purpose mainframe computer. It could take hours to process a short 30-second speech. Obviously, general purpose computing systems were insufficient to meet the high throughput rate demanded by a real-time signal processing algorithm. Dedicated application-specific computing systems, however, were too expensive to be a realistic solution for most commercial signal processing applications.

This situation changed in mid-1970s. Quantum leaps in integrated circuit manufacturing technology led to the era of VLSI systems. By 1980, hundreds of thousands transistors could be reliably and economically fabricated on a single silicon chip. With the transistor count per chip growing exponentially, it became quite clear that to manage the design complexity of VLSI circuits, IC design methodologies must be revolutionized. In 1980, Mead and Conway championed the notion of structured VLSI design. In Mead and Conway's (1980) seminal book Introduction to VLSI Systems, the authors argued that a hierarchical design style exhibiting regularity and locality must be adopted to design millions of transistors on a single chip. A novel architecture called **systolic array** was used as an example that satisfies all these requirements.

This idea of structured VLSI design further inspired the concept of a silicon compiler, which, in analogy with the software compiler, would automatically generate a silicon implementation starting from a high-level description. These pioneering ideas stimulated many important developments in the IC industry, such as the proliferation of electronic design automation (EDA) tools, the popularity of semi-custom design styles (e.g., gate array and standard cell layout), and the availability of silicon foundry services. By the time of mid-1980s, a new industry known as application specific IC (ASIC) design started to thrive. Numerous chip sets for video coding, three-dimensional (3-D) audio processing, and graphic rendering have been available on the market at appealing costs.

7.1.2 VLSI and Signal Processing

The VLSI revolution impacted on signal processing system architecture in a number of important ways:

- **High speed**: As the IC manufacturing technology evolves, the feature dimensions of transistors continue to shrink. Smaller transistors means faster switching speed and, hence, higher clock rate. Faster processing speed means more demanding signal processing algorithms can now be implemented for real-time processing.
- **Parallelism**: Higher device density and larger chip area promise to pack millions of transistors on a single chip. This makes it feasible to exploit parallel processing to achieve an even higher throughput rate by processing multiple data streams concurrently. To fully exploit the benefit of parallel processing, however, the formulation of signal processing algorithms must be reexamined. Algorithm transformation techniques are also developed to exploit maximum parallelism from a given DSP algorithm formulation.
- **Local communication**: As device dimensions continue to decrease and chip area continues to increase, the cost of intercommunication becomes significant in terms of both chip real estate and transmission delay. Hence, pipelined operation with a local bus is preferred to broadcasting using global interconnection links. Compiler and code generation methods need to be updated to maximize the efficiency of pipelining.
- **Low-power architecture**: Smaller transistor feature size makes it possible to reduce the operating voltage and, thereby, significantly reduces the power consumption of an IC chip. This trend makes it possible to develop digital signal processing systems on portable or handheld mobile computers.

On the other hand, the stringent performance requirement and regular deterministic formulation of signal processing applications also profoundly influenced the VLSI design methodology.

- **High-level synthesis design methodology**: The quest to streamline the process of translating a complex algorithm into a functional piece of silicon that meets the stringent performance and costs constraints has led to significant progress in the area of high-level synthesis, system compilation, and optimal code generation. Ideas such as data-flow modeling, loop unrolling, software pipelining, which were originally developed for general purpose computing systems, have enjoyed great success when applied to aiding the synthesis of an application-specific signal processing system from a high-level behavioral description.
- **Multimedia processing architecture**: With the maturity and popularity of multimedia signal processing applications, general purpose microprocessors have incorporated special-purpose architecture, such as the multimedia extension instruction set (e.g., MMX). Signal processors also led the wave of a novel architectural concept such as very long instruction word (VLIW) architecture. In fact, it is argued that incorporating multimedia features is the only way to sustain the exponential growth in performance through the next decade.

7.1.3 Chapter Overview

This chapter puts more emphasis on DSP algorithm to hardware synthesis and its hardware implementation. First, a DSP algorithm can be expressed as an n-level nested Do-loop, a recurrent equation, and a data flow graph (DFG). Next, one of these representations gets synthesized to its hardware counterpart. The hardware architecture is not only driven by the algorithm representation but also the sampling rate of input/output signals. Due to limited hardware resources, operations and data of a particular algorithm can be scheduled at the right time and assigned at the right execution unit basis provided that the precedence and semantics are preserved. For a high-throughput application such as image/video processing, synthesis of a regular network of processing elements (PEs) is discussed in detail. Eventually, implementation technology and tools of a particular architecture are surveyed.

7.2 Algorithm to Hardware Synthesis

Other than a mathematical equation, a DSP algorithm has a variety of different representations including a nested Do-loop algorithm, recurrent algorithm, and data/signal flow graph. The DSP algorithm can also be classified as a terminating and a nonterminating program. Unlike the terminating program, a nonterminating one is specified to execute for an infinite amount of time.

7.2.1 DSP Common Characteristics

DSP systems implemented in either hardware or software-programmable DSP processors are characterized by the following common characteristics. First, data flow/real-time computing corresponds to its analog counterpart in that an output signal is a convolution of an input signal and the impulse response of a given analog linear time-invariant (LTI) system. The convolution is assumed to take infinitesimal time to compute. In a practical DSP system, a stream of output data is a discrete convolution sum of another stream of sampled/discretized input data and the impulse response of a discrete LTI system. Furthermore, the discrete convolution sum takes a finite amount of time to compute a useful datum (sampling time period). Hence, the notion of real-time computing is imposed. Such DSP systems may involve human interface, such as speech recognition for voice dialing or audio and video signal processing in Dolby Digital/DVD home theaters. Second, some DSP systems have an infinite run time. For instance, a particular communication system on either the transmitting or receiving end must process the transmitted/received signal to maintain the continuous data/audio/video stream such as the backbone network of the Internet. Third, some DSP algorithms can be described in a nested Do-loop construct. This includes any vector/matrix multiplication (inner/outer product), correlation, and the like. A number of DSP algorithms can be written in a recursive equation: infinite impulse response (IIR) filter, least mean square/recursive least square (LMS/RLS) adaptive filter, and so on. Finally, DSP is applied in a variety of applications ranging from seismogram to the communication system in the space shuttle/station. As a result, a variety of data types and sampling rates must be taken into consideration.

Because of these DSP common characteristics, many DSP algorithms can be conveniently formulated into the form of nested Do-loops perhaps with an infinite loop bound.

7.2.2 Nested Do-Loop Algorithm

The regular structure of **nested Do-loop algorithms** has been explored extensively for a compiler of general-purpose parallel computing (Banerjee, 1993; Wolfe, 1996). With the help of the compiler, the algorithms can be executed according to particular scheduling and assignment to match the underlying architecture such as SIMD, MIMD, and many others. In general, without the regular structure of loop indices, such scheduling and assignment are NP-hard.

Similarly, the implementation of a nested Do-loop formulation of DSP algorithms is actually a "mapping" process consisting of task scheduling and assignment from abstract operations to a piece of hardware. In contrast to general-purpose computing, the underlying hardware is of the form programmable digital signal processor or ASIC in array-like structure.

Since the introduction of the temporal hyperplane by Lamport (1974), the scheduling and assignment problem of nested Do-loop algorithms has been explored and developed to the so-called **systolic array** (Kung, 1982), exploiting regular array structure, temporal pipelining, and local communication. Along with the advance of VLSI technology, certain nested Do-loop algorithms can be scaled down and realizable in a single chip. Thus, the nested Do-loop scheduling and assignment problem can be formulated as an algebraic projection. Each loop index in the index space will be projected onto the index of a regular array of processing elements (PEs), and their execution time is determined by a family of equitemporal hyperplanes, where all indices on the same hyperplane are executed simultaneously.

Two different but closely related methodologies to processor array design are the **dependence method** (Moldovan, 1982; Rao and Kailath, 1988; Kurg, 1988) and the **parameter method** (Li and Wah, 1985). Both the dependence method and parameter method are based on the assumption that an algorithm is written in a uniform recurrence nested Do-loop, which can be depicted as a shift-invariant dependence graph (DG).

In this section, we mainly focus on the dependence method design methodology because it is easy to visualize in multidimensional representation. Despite the fact that the parameter method establishes the relationship between velocity, distance, and period of each variable throughout the course of parallel execution, it is not as popular as the dependence method. Nonetheless, the relationship between these two design methodologies has been elaborated by O'Keefe *et al.* (1992).

Space-Time Mapping

The dependence method can be illustrated by the notion of space-time mapping of this matrix–matrix multiplication. Multiplication of two 3×3 matrices, a and b, results in a 3×3 matrix c.

Example 1: Matrix–Matrix Multiplication

$$c = a \times b \tag{7.1}$$

First, the multiplication is formulated as a three-level nested Do-loop as follows:

Do $i = 1$ to 3.
  Do $j = 1$ to 3.
    $c[i, j] = 0.$
    Do $k = 1$ to 3.
      $c[i, j] = c[i, j] + a[i, k] \times b[k, j].$
    EndDo k
  EndDo j
EndDo i

It can then be reformulated in single assignment format (Kung, 1982) as the following:

Do $i = 1$ to 3.
  Do $j = 1$ to 3.
    Do $k = 1$ to 3.

$$a_3[i, j, k] = \begin{cases} a[i, k], & j = 0. \\ a_3[i, j-1, k], & j > 0. \end{cases}$$

$$b_3[i, j, k] = \begin{cases} b[j, k], & i = 0. \\ b_3[i-1, j, k], & i > 0. \end{cases}$$

$$c_3[i, j, k] = \begin{cases} 0, & k = 0. \\ c_3[i, j, k-1] + a_3[i, j, k] \times b_3[i, j, k], & k > 0. \end{cases}$$

$$c[i, j] = \{c_3[i, j, k], k = 3.$$

    EndDo k
  EndDo j
EndDo i

The a_3, b_3, and c_3 are the 3-D version a, b, and c, respectively. □

From this example, a three-level nested Do-loop algorithm can be written in a single assignment format and a perfect loop nest. The innermost loop body is composed of a set of input/propagation, computing/initialization, and output statements as shown in Figure 7.1. We denote $\boldsymbol{J}^n$ as the computing index space.

Do $i_1 = l_1$ to u_1
  ....
  Do $i_n = l_n$ to u_n
    Input/Propagation Statements

$$v_j^n[\vec{I}] = \begin{cases} v_j[\mathcal{G}_{v_j}(\vec{I})] & , \vec{I} \in \mathbf{I}_{v_j}^I \\ v_j^n[\vec{I} - \vec{d}_{v_j^n}] & , \vec{I} \in \mathbf{I}_{v_j}^P \end{cases}$$

    ....
    Computation/Initialization and Statements

$$v_k^n[\vec{I}] = \begin{cases} v_{k,\,Init}^n & , \vec{I} \in \mathbf{I}_{v_k}^N \\ \mathcal{F}_{v_k}(v_k^n[\vec{I} - \vec{d}_{v_k^n}], v_j^n[\vec{I}], \ldots & , \vec{I} \in \mathbf{I}_{v_k}^C \end{cases}$$

    ....
    Output Statements

$$v_k[\mathcal{G}_{v_k}(\vec{I})] = v_k^n[\vec{I}] \;, \vec{I} \in \mathbf{I}_{v_k}^O$$

    ....
  EndDo i_n
  ....
EndDo i_1

FIGURE 7.1 n-Level Nested Do-Loop Algorithm.

Definition 1 Computing Index Space. $\boldsymbol{J}^n$

The computing index space is a set of n-dimensional integer-valued vectors such that $\boldsymbol{J}^n = \{\vec{I} = (i_1, i_2, \ldots, i_n)^t | i_1, i_2, \ldots, i_n \in \boldsymbol{Z}. l_1 \le i_1 \le u_1, l_2 \le i_2 \le u_2, \ldots ln \le i_n \le u_n\}$, where $\vec{a}^t$ denotes the transpose of $\vec{a}$. In other words, $\boldsymbol{J}^n \subset \boldsymbol{Z}^n$ is an n-D polyhedron with integer-valued vertices.

$\boldsymbol{J}^3 = \{(i, j, k)^t | 1 \le i, j, k \le 3\}$ in the matrix–matrix multiplication example.

Definition 2 Generic n-Level Nested Do-Loop

An n-level nested Do-loop model is shown in Figure 7.1. Each iteration is indexed by a computing index (vector) $\vec{I} = (i_1, i_2, \ldots i_n)^t$. The loop body consists of three different kinds of statements:

- Input/Propagation
  Let v_j be an input variable where v_j^n is its corresponding $n-D$ variable. At a particular input, index $\vec{I} \in \boldsymbol{I}_{v_i}^I$, $v_j[\mathcal{G}_{v_j}(\vec{I})]$ is assigned to v_j^n by the index transformation function $\mathcal{G}_{v_j}$. Other than for this exception, v_j^n is propagated along the direction of the propagation dependence vector $(DV) d_{v_j}^n$ at any index $\vec{I} \in \boldsymbol{I}_{v_j}^P$.
- Computation/Initialization
  The v_k^n an n-D version of variable v_k, is computed as a recurrent function $\mathcal{F}_{v_k}$ of itself and other input variables (e.g., v_j^n and so on). The vector $\vec{d}_{v_k^n}$ represents the operation dependence of current of v_k^n. Most of the time, the computation is carried on at index $\vec{I} \in \boldsymbol{I}_{v_k}^C \subseteq \boldsymbol{J}^n$. During the initialization phase, v_k^n is assigned to $v_{k,\,Unit}^n$ at index $I \in \boldsymbol{I}_{v_k}^V$

- Output v_k^n is output as v_k by the index transformation function $\mathcal{G}_{v_k}$ at index $\vec{I} \in \boldsymbol{I}_{v_k}^{O}$.

Note that each variable in the loop body follows the single assignment format. □

Based on the definition of an n-level nested Do-loop, there are three different kinds of variables: **input**, **output**, and **intermediate**.

Definition 3 Input Variable, v

For each input variable v_j its n-D correspondence is v_j^n. It is inputted at $\vec{I} \in \boldsymbol{I}_{v_j}^{I}$ and propagated at $\vec{I} \in \boldsymbol{I}_{v_j}^{P}$, where $\boldsymbol{I}_{v_j}^{I} \cap \boldsymbol{I}_{v_j}^{P} = \phi$ and ϕ is the empty setted.

Definition 4 Output Variable, v_k

For each output variable v_k, its n-D correspondence is v_k^n. It is initialized at $\vec{I} \in \boldsymbol{I}_{v_k}^{N}$ and computed at $\vec{I} \in \boldsymbol{I}_{v_k}^{C}$, where $\boldsymbol{I}_{v_k}^{C} \cap \boldsymbol{I}_{v_k}^{N} = \phi$. Its output value is defined at $\vec{I} \in \boldsymbol{I}_{u_k}^{U}$. □

Definition 5 Intermediate Variable v_k

For each intermediate variable v_k, its $n - D$ correspondence is v_k^n. It is intialized at $\vec{I} \in \boldsymbol{I}_{v_k}^{I}$ and computed at $\vec{I} \in I_{v_k}^{C}$, where $I_{v_k}^{C} \cap I_{v_k}^{N} = \phi$. Its output value is consumed only at $\vec{I} \in I_{v-k}^{U}$ □

In matrix–matrix multiplication, a and b are input variables, while c is the output variable. The propagation dependence vectors of a and b are $\vec{d}_{a_3} = (0, 1, 0)^t$ and $\vec{d}_{b_3} = (1, 0, 0)^t$, respectively. On the other hand, the true dependence vector of c is $(0, 0.1)^t$. If all dependence vectors are plotted in the index space $\boldsymbol{J}^3$, a 3-D DG is yielded in Figure 7.2(A). Each node represents a data flow graph (DFG) of multiplication and addition as shown in Figure 7.2(B), respectively.

Definition 6 Dependence Graph

A dependence graph (DG) (Kung, 1988) is a directed graph $G_{DG} = (\Omega_{DG}, L_{DG})$, where Ω_{DG} is a set of nodes in index space $\boldsymbol{J}^n$ and where L_{DG} is a set of edges connecting a pair of nodes, i and j and $i, j \in \Omega_{DG}$ if and only if computing of node j depends on the result from node i. Note that these edges are also called dependence vectors either due to propagation or computation. In other words, a DG is a graphical representation of an n-level nested Do-loop defined in definition 2. Furthermore, any *DG* is computable if and only if it contains no loops or cycles. □

A DG with its detailed DFG in each node is considered a data flow graph as it is defined here.

Definition 7 Data Flow Graph

A data flow graph is a directed graph $G_{DFG} = (\Omega_{DFG}, L_{DFG})$, where Ω_{DFG} is a set of nodes representing operations or functions and where L_{DFG} is a set of edges connecting a pair of nodes,

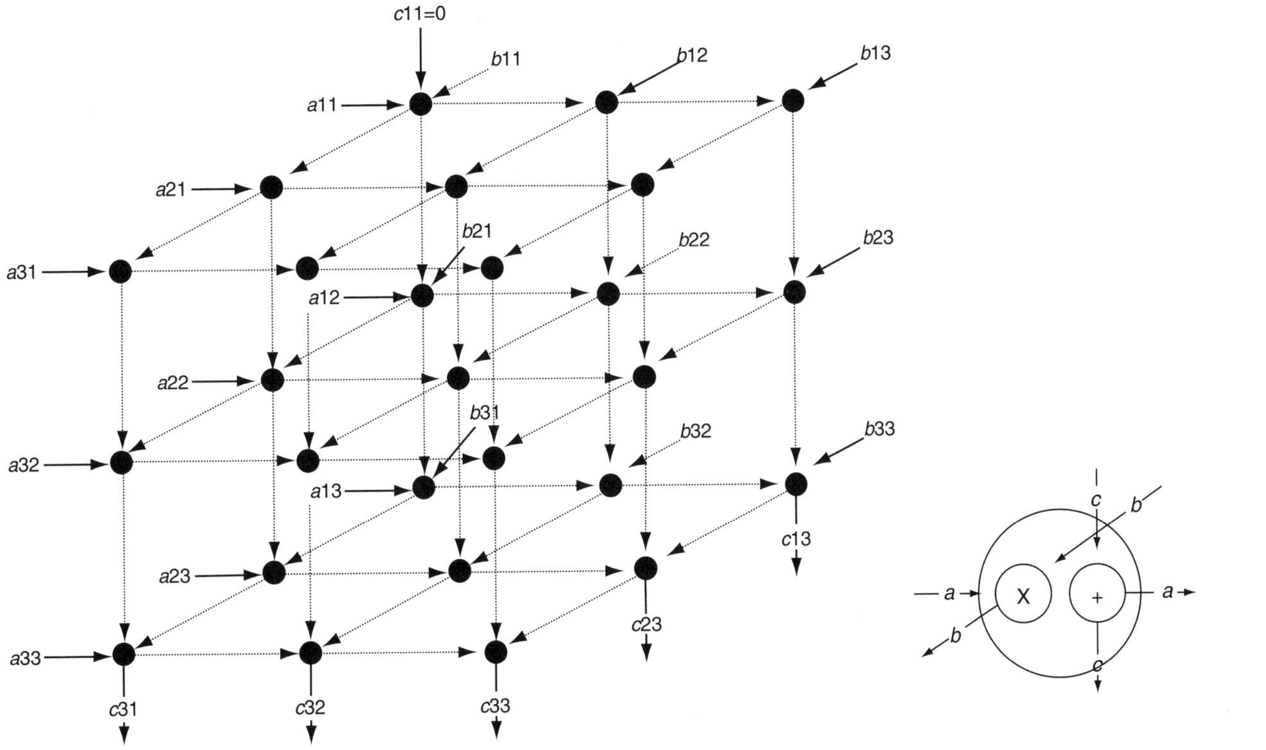

(A) A 3 x 3 Matrix-Matrix Multiplication in a 3-D Dependence Graph

(B) Node (Loop Body) as a Data Flow Graph

FIGURE 7.2

i and j and $i, j \in \Omega_{DFG}$. This indicates the flow of data with or without delay element(s) as an output from node i to node j. □

The notion of linear space-time mapping is a transformation of index space J^n onto an $(n\text{-}1)\text{-}D$ space and 1-D time index spaces. The space-time mapping matrix T consists of **scheduling vector** $s \in Z^n$ and **processor allocation matrix** $P \in Z^{n \times n-1}$, where Z is the integer number space. In other words, an algorithm described as a multiple perfect loop nest can be projected onto an $(n\text{-}1)\text{-}D$ array of PEs and a scalar valued execution schedule with respect to the dependencies. The computing takes place in parallel at the allocated PEs according to the time (clock) schedule.

Definition 8 Space-Time Mapping
The space-time mapping matrix $T \in Z^{n \times n}$ is given by:

$$T = \begin{bmatrix} s^{-t} \\ P^t \end{bmatrix}, \tag{7.2}$$

where $\vec{s}$ and P are linearly independent. In other words, Rank $(T) = n$.

The **schedule-allocation matrix** is another representation of the original index space in the computing schedule (time) and processor allocation (space) domains. The computing schedule $t(J^n) \subset Z^n$ and processor allocation $J^1 \subset Z$ can be obtained by:

$$\begin{bmatrix} t(J^n) \\ J^{n-1} \end{bmatrix} = \begin{bmatrix} \vec{s}^t J^n \\ P^t J^n \end{bmatrix} = TJ^n, \tag{7.3}$$

where J^{n-1} is the $(n-1)-D$ index spaces, respectively. In other words, the matrix T maps the execution index $\vec{q} \in J^n$ to processor $P^t\vec{q}$ at time step (clock) $t(\vec{q})$.

Due to the shift-invariant DG_j all dependence vectors associated with any DG node can be represented in a **dependence matrix**, D_V:

$$D_V = [D_{v_1} \quad D_{v_2} \quad \ldots], \tag{7.4}$$

where variables $v_i \in V$ and V represent a set of all variables in the algorithm. The D_{v_i} is recursively defined as a set of dependence vectors associated with variable v_i:

$$D_{v_i} = \begin{bmatrix} \vec{d}_{i1} & \vec{d}_{i2} & \ldots \end{bmatrix}, \tag{7.5}$$

where $\vec{d}_{ij} \in Z^n$ is a dependence vector j of variable v_i. It is obvious that all the dependencies in every node in Figure 7.2(A) are captured by this dependence matrix:

$$D_V = \begin{bmatrix} \vec{d}_{b3} & \vec{d}_{a3} & \vec{d}_{c3} \end{bmatrix} = \begin{bmatrix} 1 & 0 & 0 \\ 0 & 1 & 0 \\ 0 & 0 & 1 \end{bmatrix}. \tag{7.6}$$

Similarly, the mapping can be applied to the dependence matrix to obtain the **delay-edge matrix**. The D_V is transformed via space–time mapping to interprocessor links or edges and registers or delays, respectively. Interprocessor communication is carried out via an $\vec{e}_{v_i}$ edge that is pipelined with a number of delay elements or shift registers, $r(\vec{e}_{v_i})$:

$$\begin{bmatrix} r(e_V) \\ e_V \end{bmatrix} = \begin{bmatrix} \vec{s}^t D_V \\ P^t D_V \end{bmatrix} = TD_V. \tag{7.7}$$

In equation 7.7, the **delay vector** is as follows:

$$r(e_V) = [r(\vec{e}_{v_1}) \quad r(\vec{e}_{v_2}) \quad \ldots]. \tag{7.8}$$

The **edge matrix** is as follows:

$$e_V = [\vec{e}_{v_1} \quad \vec{e}_{v_2} \quad \ldots]. \tag{7.9}$$

The delay-edge matrix describes the global structure of the resulting PE array regardless of the functionality inside each PE. This structure is called a **signal flow graph** and is mainly used to explain algorithms in DSP.

Definition 9 Signal Flow Graph
A signal flow graph (SFG) (Kung, 1988) shown by $G = (\Omega, L)$ is a directed graph consisting of a set of vertices, (PEs) Ω and a set of edges (links) L. Each variable $v \in V$ has its own set of edges L_v such that an edge $l_v \in L_v$ associated with $r(e_v)$ delay elements of direction e_v connect two PEs ω_i and ω_j, where $\omega_i, \omega_j \in \Omega$. Based on shift-invariant DG, the resulting SFG after space–time mapping is considered structurally time-invariant in which only edge structures remain invariant. □

Partitioning

If the number of processors is limited and much less than the number of PEs required for a particular algorithm, the DG must be partitioned and mapped to fit that fixed size array. Hwang and Hu (1992) have proposed a two-level locally sequential globally parallel (LSGP) scheduling to reduce the overall computing time. The DG is first partitioned into a set of disjoint blocks. Those blocks will be allocated to available processors and scheduled with a two-level schedule. The first-level sequential schedule is obtained from applying permissible loop transformations. The second-level parallel schedule is formulated as an integer programming problem subject to the total computation latency. An interblock schedule is applied to determine the relative timing of each block's computation and satisfy the precedence constraints imposed by dependence vectors across the block boundaries.

Partitioning the same DG to several blocks is considered a special case of multistage systolic mapping MSSM (Hwang and Hu, 1992), where each stage is characterized by a different DG of a different nested Do-loop algorithm. It can be viewed as an locally parallel globally parallel (LPGP). Scheduling problem for parallel processing of a complex operation chain. After mapping consecutive stages, it may result in mismatched I/O operations.

Therefore, a systematic approach is proposed to reduce data communication overhead and overlap part of the computation in successive stages to minimize performance degradation.

Lower-*D* Mapping

In this section, the dependence method or space–time mapping will be extended to map an *n-D* DG to the final 1-D or 2-D SFG. The earliest attempt to map an *n-D* DG to an $(n-2)$-*D* clock schedule and a 2-D PE array using only a single space–time mapping matrix was proposed by Wong and Delosme (1985). The resulting multidimensional clock must be linearized to a scalar value to reduce implementation cost. This idea of mapping to a lower dimension is similar to **multiprojection** (Kung, 1988), which is equivalent to a sequence of consecutive space–time mappings. Later efforts to extend space–time mapping to *k-D* array, where $1 \le k < n$, are reported by Lee and Kedem (1988, 1990), Shang and Fortes (1992), and Zimmerman (1996). Similarly, the generalization of the parameter method, called the **general parameter method** (Ganapathy and Wah, 1992), is formulated for a lower-*D* array as well.

Definition 10 One-Dimensional Space–Time Mapping

Contrary to the traditional space–time mapping, the transformation matrix consists of a scheduling vector $\vec{s}$ and a linear allocation vector P_1 as follows:

$$T_1 = \begin{bmatrix} \vec{s}^t \\ P_1^t \end{bmatrix} = \begin{bmatrix} s_1 & s_2 & \cdots & s_n \\ p_1 & p_2 & \cdots & p_n \end{bmatrix}, \tag{7.10}$$

where $T_1 \in Z^{2\times n}$ and s_i, $p_i \in Z$ are relatively prime. Furthermore, $\vec{s}$ and P_1 must be linearly independent so that $Rank(T_1) = 2$. □

Similar to traditional mapping, the schedule–allocation matrix can also be obtained from low-*D* space–time mapping matrix T_1. The computation schedule, $t(\boldsymbol{J}^n) \subset \boldsymbol{Z}$ and processor allocation $\boldsymbol{J}^1 \subset \boldsymbol{Z}$ are as follows:

$$\begin{bmatrix} t(\boldsymbol{J}^n) \\ \boldsymbol{J}^1 \end{bmatrix} = \begin{bmatrix} \vec{s}^t \boldsymbol{J}^n \\ P_1^t \boldsymbol{J}^n \end{bmatrix} = T_1 \boldsymbol{J}^n. \tag{7.11}$$

In other words, the matrix T_1 maps the execution of index $\vec{q} \in \boldsymbol{J}^n$ at processor $P_1^t \vec{q}$ at time step (clock tick) $t(\vec{q})$. This 1-D array mapping can be applied to obtain the delay-edge matrix as shown earlier in equation 7.7.

Design Objectives

Unlike the parameter method and the general parameter method, the space–time mapping has close form relationships of design objective functions. For instance, the total execution time, τ_e, can be obtained from:

$$\tau_e = t_{\text{cycle}} \times N_{\text{cycle}}, \tag{7.12}$$

where t_{cycle} is the propagation delay of the critical path in each SFG node and where N_{cycle} is the number of clock cycles. The N_{cycle} is the number of steps between the first and the last computation indices (Kung, 1988) which is given by:

$$N_{\text{cycle}} = \max_{\vec{p},\, \vec{q} \in J^n} \{\vec{s}(\vec{p} - \vec{q})\} + 1. \tag{7.13}$$

In other words, N_{cycle} is the number of cuts by the temporal hyperplane. In addition, to minimizing only, the N_{cycle}. Wong and Delosme (1992) has proposed to minimize both t_{cycle} and N_{cycle}.

Although N_{cycle} seems to depend on the scheduling vector $\vec{s}$ only, the question on how many PEs are used and how the data are delivered to the right PE still remain. In the 1-D or linear array mapping, the number of PEs, N_{PE}, can be obtained explicitly as follows:

$$PE_{\max} = \max\{P_1^t q \mid \vec{q} \in \boldsymbol{J}^n\} \tag{7.14}$$

$$PE_{\min} = \min\{P_1^t q \mid \vec{q} \in \boldsymbol{J}^n\}. \tag{7.15}$$

As a result, the number of PEs, N_{PE}, is expressed as:

$$N_{PE} = PE_{\max} - PE_{\min} + 1. \tag{7.16}$$

In other words, N_{PE} is the number of distinct projections of the index space $\boldsymbol{J}^n$ on the vector P_1.

Example 2 4 × 4 Matrix–Matrix Multiplication

A 4 × 4 matrix–matrix multiplication (Pee and Kedem, 1988) DG is scheduled and allocated to a linear or 1-D array of PEs by this 1-D space–time mapping matrix:

$$T_1 = \begin{bmatrix} \vec{s}^t \\ P_1^t \end{bmatrix} = \begin{bmatrix} 2 & 1 & 3 \\ 1 & 1 & -1 \end{bmatrix}. \tag{7.17}$$

As a result, the array is shown in Figure 7.3. It takes 10 cycles for the data to fill in the pipeline and $N_{\text{cycle}} = 19$ cycles to finish the execution. The maximum PE utilization is 60%. This is due to the constraint that I/O is allowed on either end of the array. □

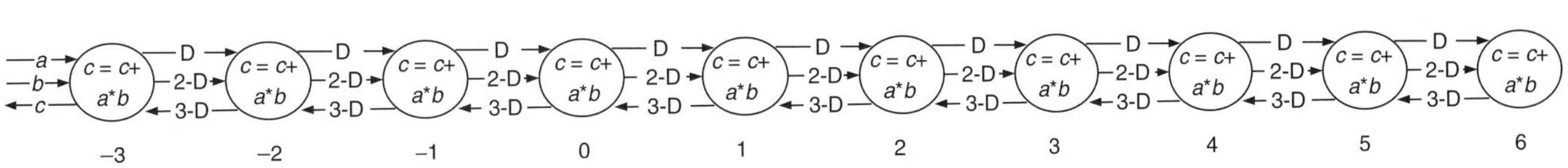

FIGURE 7.3 Signal Flow Graph of 4 × 4 Matrix–Matrix Multiplication. PE number ranges are from −3 to 6 (Lee and Kedem, 1988).

7.2.3 Recurrent Algorithm

In the previous section, an n-level nested Do-loop can be regarded as a terminating program. In this section, a nonterminating recurrent program is taken into consideration. Examples of this kind of algorithm include IIR and adaptive filters. These algorithms are characterized by a current output that is defined by a given function of the previous outputs and the current input. As such, the validity of the output depends on how often its computation is initiated, called the **initiation period**. Two optimization techniques **loop unfolding** and **look-ahead transformation** are discussed next.

Loop Unfolding

Loop unfolding exploits the interiteration precedence or the so-called **loop-carried dependence**. After the loop is unfolded, the new initiation period may be reduced. For example, a nonterminating program is shown in algorithm 1.

Algorithm 1 A Nonterminating Recurrent Program

Do $n = 1$ *to* ∞.

$aa(n) = f_{aa}[aa(n-1),\ ca(n-1),\ ba(n)].$

$ab(n) = f_{ab}[aa(n-1),\ ca(n-1),\ ba(n)].$

$ba(n) = f_{ba}[ab(n-1)].$

$bc(n) = f_{bc}[ab(n-1)].$

$ca(n) = f_{ca}[bc(n)].$

EndDo n

The $xy(n)$ denotes an arc (data) flowing from node x to node y at the iteration n, and f_{xy} denotes a definition of $xy(n)$. Its corresponding DFG is illustrated in Figure 7.4. After the loop body is unfolded by a factor of two, the resulting program is listed here:

Do $n = 1$ *to* ∞.

$aa(2n+1) = f_{aa}[aa(2n),\ ca(2n),\ ba(2n+1)].$

$ab(2n+1) = f_{ab}[aa(2n),\ ca(2n),\ ba(2n+1)].$

$ba(2n+1) = f_{ba}[ab(2n)].$

$bc(2n+1) = f_{bc}[ab(2n)].$

$ca(2n+1) = f_{ca}[bc(2n+1)].$

$aa(2n+2) = f_{aa}[aa(2n+1),\ ca(2n+1),\ ba(2n+2)].$

$ab(2n+2) = f_{ab}[aa(2n+1),\ ca(2n+1),\ ba(2n+2)].$

$ba(2n+2) = f_{ba}[ab(2n+1)].$

$bc(2n+2) = f_{bc}[ab(2n+1)].$

$ca(2n+2) = f_{ca}[bc(2n+2)].$

EndDo n

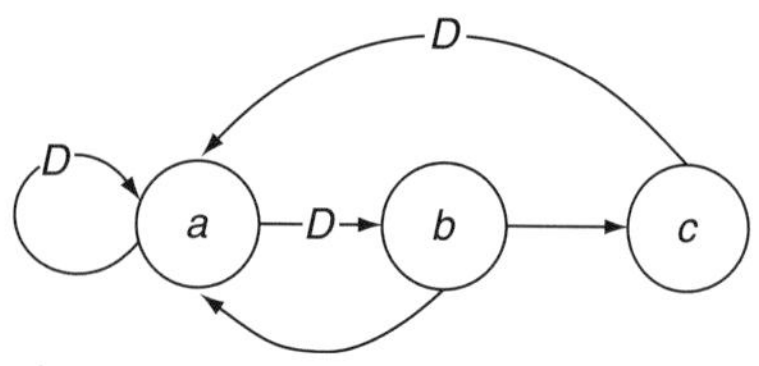

FIGURE 7.4 A Data Flow Graph of Algorithm 1.

As discussed by Parhi (Parhi, 1989), loop unfolding can reduce the iteration period in a multiprocessor implementation. For a perfect-rate DFG, however, its iteration period is bounded and cannot be reduced further. The schedule associated with a perfect-rate DFG is called a rate-optimal schedule.

Look-Ahead Transformation

Loop unfolding can only improve the iteration period but cannot achieve an iteration period less than the iteration bound of the algorithm. To lower the iteration bound, the look-ahead transformation is devised such that the nth iteration's output can be expressed as a function of previous $n-k$ iteration outputs and the necessary inputs. The program shown here is a terminating first-order recurrent program.

Do $n = 0$ to 7

$y(n+1) = a(n)y(n) + x(n)$

EndDo n

The **look-ahead transformation** can be illustrated by a back substitution as follows:

$$\begin{aligned} y(8) &= a(7)y(7) + x(7). \\ &= a(7)[a(6)y(6) + x(6)] + x(7). \\ &= y(6)a(7)a(6) + x(6)a(7) + x(7). \end{aligned} \tag{7.18}$$

For example, the last element $y(8)$ can be obtained directly from $y(6)$, $x(6)$, and $x(7)$. The term $y(6)a(7)a(6) + x(6)a(7)$ is called the *overhead* of this transformation. As we keep substituting, the final result becomes:

$$\begin{aligned} y(8) = y(0)\prod_{i=0}^{8} a(8-i) + x(0)\prod_{i=0}^{6} a(8-i) \\ + x(5)\prod_{i=0}^{1} a(8-i) + \ldots + x(6)a(8) + x(7). \end{aligned} \tag{7.19}$$

After the look-ahead transformation, a new DFG of this algorithm is obtained. The iteration period of the new DFG may be less than that of the one that is not look-ahead.

7.2.4 SFG/DFG Optimization

In this subsection, the relationship between multiprocessor implementation of a recurrent algorithm and VLSI implementation of the space–time mapping is established. After the SFG/DFG is obtained, some optimization techniques such as cut-set/retiming and geometric transformation can be applied.

DG/SFG Versus DFG

As definition 6, the DG with its functional DFG in each DG node can be perceived as a complete DG (Kung, 1988) or a DFG of a nested Do-loop algorithm. On the other hand, each DFG node describing a recurrent algorithm in Section 7.2.3 corresponds to a set of operations or functions in a processor.

During the space–time mapping, each DG node or computation index is allocated and scheduled to execute at a specific SFG node. In parallel, multiprocessor implementation of recurrent DFG can be obtained by scheduling and assignment strategies, such as the one proposed by Wang and Hu (1995). As a result, several DG nodes correspond to an SFG node (PE), while several DFG nodes can be assigned to the same processor.

In a fully systolic SFG, each edge is associated with register(s). The clock cycle time of each SFG node is determined by its critical path. The delay of the critical path is determined by the maximum computation delay on a zero-register path. On the other hand, a recurrent algorithm is constrained by its iteration period analogous to a clock period. Other than minimizing the iteration period or cycle time, both space–time mapping and multiprocessor implementation try to minimize the number of PEs/processors, respectively.

Due to some undesirable properties of SFG (e.g., long zero-delay link, broadcasting link, dimensionality or geometry) the following strategies have been exploited to reorganize the SFG in such a way that the result is more suitable for VLSI implementation.

Cut-Set and Retiming

The primary methods of SFG optimization employed are cut-set and retiming (Kung, 1988). The objective is to convert an SFG into a fully pipelined form so that all the edges between modular sections have at least one delay element. A cut-set is a minimal set of edges including the target edge, non-zero-delay edges going in either direction, and zero-delay edges going in the same direction as the target edge. The procedure consists of these three steps:

- Partitioning cuts the DFG/SFG nodes to two sets to achieve a cut-set.
- Delay elements associated with a cut-set are all scaled by a positive integer.
- A number of delay elements are transferred between inbound and outbound edges with respect to the partition in such a way that there remains at least one delay element.

The retiming was first proposed by Leiserson *et al.* (1983). It is equivalent to moving around the delays in the DFG such that the total number of delays in any loop remains unaltered. The retiming can be applied to a single-rate DFG to reduce the iteration period in a programmable multiprocessor implementation. A rate-optimal schedule cannot be guaranteed, however, and cannot reduce the iteration period in perfect-rate DFG (Parhi *et al.*, 1989). Very similar to the retiming, delay distribution methodology is presented for the FIR and IIR digital filter (Chen and Moricz, 1991). The optimization is performed on an DFG/SFG, which represents a 1-D algorithm and then is generalized to *n-D* one. The strategies consist of balancing, rescaling, distribution, factorization, and inversion.

Geometric Transformation

The recent effort to transform a 1-D SFG to 2-D SFG was exhibited in Yeo and Hu (1995) and Kittitornkum and Hu (2001) by making use of a spiral and long interconnect. Both arrays were proposed for full-search block matching motion estimation, which is one of the most computation-intensive tasks in real-time digital video coding.

7.2.5 Bit-Level Algorithm

In digital communication, a huge amount of information is transformed into a series of data bits by source coding (e.g., audio and video coding) and various forms of channel coding (e.g., error correction) prior to the transmission. Therefore, every bit of received data is so important that many algorithms have been proposed and developed to save the communication bandwidth and error resilience. Such algorithms include **variable length coding** at the source coding level and **Viterbi decoding** at the channel coding level.

Variable Length Coding

The most well-known and widely used **variable-length coding** (VLC) is the Huffman code. The coding part can be easily implemented as a look-up table to achieve a high output rate. On the other hand, its decoding end can be implemented in a number of VLSI architectures to satisfy different input/output data rates as surveyed by Chang and Messerchmitt (1992). The VLSI architecture can be classified as tree-based and programmable logic array-based (PLA-based). The tree-based architecture is a direct map of the Huffman binary tree. The average decoding throughput of the pipelined version is in the range of $[L_{min}, L_{max}]$ bits/cycle, where L_{min} and L_{max} are minimum and maximum code lengths, respectively. This rate is apparently higher than a 1 bit/cycle of the sequential tree-based architecture.

A PLA in the PLA-based architecture is functioning as a look-up table storing the states of an finite state machine (FSM). Each state corresponds to either an internal node of the Huffman tree or a decoded output symbol. The design can be constrained to the following conditions.

- Constant input rate at N bits/cycle:

  An $N < L_{max}$-bit input is buffered and used to look up the next state, the decoded symbol, as well as the shift amount if it is found from a PLA.

- Constant output rate of one symbol per clock cycle:
  The architecture resembles a constant input rate. A longer bit string $N \geq L_{max}$, however, is used to look up a decoded symbol and its shift amount. The bit string will be shifted by a barrel shifter at the corresponding shift amount.

Viterbi Algorithm

The convolutional coding has been one of the most widely used error corrections in digital wireless communication. Therefore, the **Viterbi decoding algorithm** must be implemented efficiently in a pipelined/systolic fashion. A (K, R) convolutional coding scheme can be described by an FSM, where K is the constraint length and R is the code rate. Similar to a mealy FSM, there are a total of 2^K coding states where output bits and the next state depend on the current state and an input bit.

Among many VLSI implementations, a modified trace-back Viterbi algorithm (Truong, 1992) has been proposed to save register storage. The survivor path is traced back through the trellis diagram and used to determine the decoded information bit in the window of length $5K$. Besides, a 1-Gb/s $K = 2$ convolutional code sliding block Viterbi decoder has been recently reported (Black and Menge, 1997). It is designed for a high throughput magnetic and optical storage channel. Both backward and forward processing exist, with total window length of $M = 2L$ where $L = 5K$. However, $L = 3$ is chosen for performance/complexity trade-off. Thus, each processing unit uses an L-stage trace-back decoding unit. Its fully systolic architecture results in a very high throughput of $2Lf_{clk} = 2 \times 6 \times 83 \simeq 1$ Gb/s, where $f_{clk} = 83$ MHz. Each pipeline stage is an add-compare-select (ACS) unit. The fabricated chip size is $9.21 \times 8.77\,\text{nm}^2$ with a 1.2-µm double-metal CMOS technology.

7.3 Hardware Implementation

DSP application-specific integrated circuits (ASICs) are mostly implemented in digital CMOS circuit technology due to several advantages and maturity of fabrication technology. Other than an ASIC design in a large volume, field-programmable gate array (FPGA) and reconfigurable computing (RC) have been invented to amortize the cost in a low to medium volume. As the application advances toward multiprotocol mobile/wireless computing/communication. RC may emerge as an alternative to hard-wired ASIC. In this section, we put the emphasis on current state-of-the-art FPGA and RC implementation technology and tools.

7.3.1 Implementation Technology

FPGA

A custom computing system (CCS) is an array of interconnected FPGA chips hosted by a general purpose computer. CCS can be perceived as a dynamically reconfigurable data path for special-purpose tasks that cannot be efficiently implemented by software. There have been several examples of CCSs, including CHAMP (Patriguin and Gurevich, 1995), national adaptive processing architecture (NAPA) (Rupp *et al.*, 1998), and Splash-2 (Buell *et al.*, 1996).

- CHAMP is a system of 8 PEs interconnected in a 32-bit ring topology. Each PE consists of two Xilinx XC4013 FPGAs, which is a dual-port memory system.
- NAPA's aim is to develop a teraoperations/second-class computing system. Its emphasis is on algorithmic data path and function generation.
- Splash-2 is a large system of up to 16 array boards each containing 17 Xilinx XC4010 FPGA chips. Splash-2 is hosted by SPARCstation 2 and connected through an SBus interface board.

Besides custom computing systems, the following are some DSP algorithms implemented in FPGA:

- A 1-D discrete fourier transform (DFT) (Dick, 1996) systolic array of PE with coordinate rotation digital computer (CORDIC) arithmetic is reported. Each PE consumes about 44 configurable logic blocks (CLBs) of the Xilinx XC4010PG191-4. As a result, each FPGA chip can accommodate 10 PEs. With the operating 15.3-MHz clock frequency, 1000-point 1-D DFT only takes 51.45 ms to compute.
- A CORDIC arithmetic processor performing polar to Cartesian coordinate transformation is reported by Andraka (1998). The 52-MHz CORDIC processor features pipelined 14-bit 5 iterations per transformation and occupies only 50% of a Xilinx XC4013E-2 FPGA.
- An 8×8-pixel 2-D discrete cosine transform (DCT) implementation is reported (Woods *et al.*, 1996). The 2-D DCT is separated into two passes of 1-D DCT, each consisting of bit-serial arithmetic 12 multipliers and 32 adders/subtractors. The 2-D DCT consumes about 70% of a Xilinx XC6216 FPGA. At maximum clock frequency of 38 MHz, it can achieve 30 frames/second at 720×480 pixels.
- Direct sequence spread spectrum (DSSS) RAKE Receiver (Shankiti and Leeser, 2000) has been attempted on a Xilinx XC4028EX–3HQ240 FPGA chip. Each finger occupies 1000 CLBs, which is approximately one FPGA chip. With a 1-MHz clock, the system throughput can reach up to 4000 symbols/sec.
- A 1-D discrete wavelet transform (DWT) realization on FPGA is presented in (Stone and Manalokos, 1998). Both high-pass and low-pass filters are a six-tap bi-orthogonal binary filter. The system consumes 90% of an Altera EPF10K100GC503 whose capacity is equivalent to 100.000 gates. The operating clock frequency is 20 MHz.

TABLE 7.1 Summary of DSP Algorithms on FPGA Implementation

Design	f_{clk} (MHz)	Space Util. (%)	Device
CORDIC	52.0	50	Xilinx XC4013
1-D DFT	15.3	10	Xilinx XC4010
2-D 8 × 8 DCT	38.0	70	Xilinx XC6216
1-D six-tap DWT	20.0	90	Altera EPF10K100
RAKE Receiver	1.0	90	Xilinx XC4028

In addition to these examples summarized in Table 7.1, several efficient formats of finite impulse response (FIR) filter implementation using distributed arithmetic (White, 1989) were reported by Goslin (1995). As discussed earlier, LUTs are one of the fundamental constructs in FPGA. There are a few more advantages of implementing DSP algorithms on FPGA:

- Efficient bit-level logic operations
- Pipelining using flip/flops and shift registers
- ROM- or LUT-based algorithms, such as the Huffman code or Viterbi decoder
- Bit-parallel arithmetic support and dedicated multiplier unit in Xilinx Virtex II FGPA (Xilinx, 2000)

Reconfigurable Computing

Reconfigurable computing (RC) has a variety of hardware architecture as well as applications. In the next two sections, existing RC architecture and its algorithm to hardware mapping are surveyed.

Inspired by the 90/10 locality rule, in which a program executes about 90% of its instructions in 10% of its code, HW/SW partitioning is influenced by the following facts:

1. The higher percentage of run time of specialized computation, the more improvement of cost/performance if it is implemented in hardware.
2. The more specialized computation dominates the application, the more closely the specialized processor should be coupled with the host processor.

This rules of thumb have been successfully applied to the floating-point unit as well as RC.

The basic principle is that the hardware programmability of RC is combined with a general-purpose processing capability of the RISC microprocessor. The interface between these two can be either closely or loosely coupled depending on its applications. How frequently an RC's functionality should be reconfigured must be determined based on the performance criteria of particular HW/SW architecture. Unlike the long configuration latency of the FPGA, RC is attributed by its low-latency dynamic configurability.

Existing RC architecture models are mostly of the form 1-D or 2-D array of configurable logic cells interconnected by programmable links and switches. The array communicates with the outside world through peripheral I/O cells. The architecture can be classified by its processing capacity of each logic cell at different granularity levels.

TABLE 7.2 Granularity Versus Architecture of Reconfigurable Computing

Granularity	Reference
Fine-grained	DRLE
Multigrain	Pleiades
Coarse-grained	Garp, MATRIX, PipeRench, MorphoSys, REMARC

The architecture of a configurable cell, either fine-grained or coarse-grained, is mainly driven by either general-purpose or signal/image processing computing. Its configuration or *context* can be stored locally like SRAM-based FPGA. Instead of being off-chip and loosely coupled to the host processor, RC is going toward an on-chip coprocessor as the technology advances to SOC.

The following are some of the exiting RC systems as they are classified according to their granularity and application in Table 7.2.

- A **fine-grained configurable cell** performs simple logic functions with more complex function (e.g., fast carry look-ahead adder). A perfect example of this architecture is the dynamically reconfigurable logic engine (DRLE) (Nishitani, 1999). It is capable of real-time reconfiguration with several layers of configuration tables. An experimental chip is composed of a 4 × 12 array of configurable cells. Each cell can realize two different logic operations equivalent to 4-bit input to 1-bit output or 3-bit input to 2-bit output. Up to eight different configurations can be locally stored in each memory cell. With 0.25-micron CMOS technology, the chip contains 5.1 million transistors in a 10 × 10-mm^2 die area and consumes 500 mW at 70 MHz.
- A **multigrain configurable cell**, such as Pleiades (Zhang *et al.*, 2000; Rabaey *et al.*, 1997; Abnovs and Rabaey, 1996; Wan *et al.*, 1999), is a multigrain heterogeneous system partitioned by control-flow computing on microprocessor and data flow computing on RC for future wireless embedded devices. An RC array is composed of satellite (configurable) processors and a programmable interconnection to the main microprocessor. Data-flow computing is implemented using global asynchronous and local synchronous clocking to reduce overhead. Therefore, operation starts only when all input data are ready.
- A **coarse-grained configurable cell** performs more complex word parallel arithmetic functions such as, addition and multiplication, with simple bit-level logic functions. Mostly, this RC architecture is organized in a 2-D array of nanoprocessors with an arithmetic and logic unit (ALU), a multiplier, an instruction RAM, a data RAM, data registers, input registers, and an operand register. Reported architecture examples include Garp (Callahan

et al., 2000), PipeRench (Goldstein *et al.*, 1999), REMARC Reconfigurable Multimedia Array Coprocessor (REMARC) (Miyamori and Olukotun, 1998) MATRIX (Mirsky and Detton, 1996) and MorphoSys (Lu *et al.*, 1999).

7.3.2 Implementation Tools

In this subsection, we are focused more on an architecture-driven synthesis, in particular the regular array synthesis. Such array architecture is suitable to both FPGA and reconfigurable computing platforms.

Architecture-Driven Synthesis

One of the well-known DSP synthesis tools is the Cathedral I/II/III (De Man *et al.*, 1990, 1988; Note *et al.*, 1991). A DSP algorithm is described as a graph with operators as vertices and with arcs representing the flow of data in a language called SILAGE. The bit-true specification of all signal formats is specified up to bit-level including quantization and overflow characteristics. Since Cathedral is characterized as an architecture-driven synthesis tool, it supports the following architectural styles. Each architecture style is categorized by sample rate.

1. **Hard-wired bit serial architecture** supported in Cathedral I is suitable to a low-sample rate filter in audio, speech, and telecommunication applications. Because signals are processed bit by bit, least significant bit (LSB) first, the amount of logic is independent of the word length.
2. **Microcoded processors architecture** supported by Cathedral II is appropriate for low- to medium-rate algorithms with heavy decision making and intensive computations. Each processor is a dedicated interconnection of bit-parallel execution units controlled by branch controllers (e.g., highly complex, block-oriented DSP algorithms in the audio- to near-video-frequency range).
3. **Bit-sliced multiplexed data paths** supported in Cathedral III are suitable to a medium-level image, video, and radar processing ranging 1–10 MHz data rates with irregular and recursive high-speed algorithms. The data path consists of interconnected dedicated application-specific units such as registers, adders, multipliers, and so on.
4. **Regular array architecture**, which would be supported in Cathedral IV, is suited for front-end modules of image, video, and radar processing. A regular network of PEs features mostly localized communication and distributed storage. If it is fully pipelined and modular, the array corresponds to the so-called systolic architecture.

In the next subsection, the focus is shifted to a more specific architecture (i.e., the regular array architecture or systolic array mapped from *n*-level nested Do-loop).

Regular Array Architecture Synthesis

The design flow of traditional systolic synthesis is constituted of the following steps as illustrated in Figure 7.5.

1. **DG Capturing**: A DG is captured by means of either description language or graphical input. It is composed of nodes representing computation and edges representing dependencies between connected nodes. Nodes are organized in such a way that they are located on integer grids. In addition, information such as bit width and variable type can be associated with the graph.
2. **Space–time mapping**: It is a process of allocating and scheduling a number of DG nodes to a (usually smaller) number of PEs with respect to their dependencies. The allocation and scheduling can be either a linear or a nonlinear function of node indices. The result of space–time mapping is an SFG with desired objective cost function subject to design criteria.
3. **SFG optimization**: SFG is composed of nodes representing PEs and edges representing wires or busses associated with delay elements or shift registers. SFG optimization includes cut-set and retiming (Kung, 1988) according to additional design criteria and better VLSI realization.
4. **Intermediate representation generation (IR)**: An IR is generated from a given SFG with the help of a predesigned component library. The bit width of each variable is used to determine the size of register and bus. The final result is usually in HDL behavioral description for further behavioral-to-register transfer HDL synthesis.

As summarized in Table 7.3, these synthesis tools are discussed in alphabetical order:

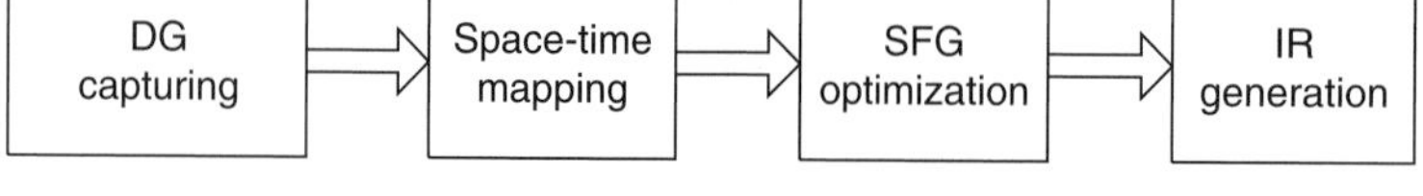

FIGURE 7.5 Systolic Space–Time Mapping Design Flow.

TABLE 7.3 Summary of Systolic Synthesis Tools

	DG capturing	Space–time lapping	SFG optimization	IR generation
ARREST	√	√	√	√
DG2VHDL	√	√	√	√
IRIS[1]			√	√
Khoros[2]	√	√	√	√
VACS	√	√	√	√
VASS	√	√	√	√

[1] Woods *et al.* (1997).
[2] Moorman and Crates (1999).

- ARREST (Burleson and Jung, 1992) extracts dependency vectors from an affine recurrence equation (ARE) description and generates the corresponding DG. Users are allowed to explore design space using a number of provided DG transformations. An SFG is obtained from affine transformation according to user-specified scheduling and projection vectors (Kung, 1988). ARREST permits the user to verify the functionality via a Verilog simulator and cost/performance tradeoff of different design decisions.
- The DG2VHDL (Stone and Manalokos, 2000, 1998) is proposed for general systolic mapping from a net list of DG to a synthesizable behavioral VHDL description of the input DSP algorithm. In addition, a corresponding test bench is generated for design validation.
- The VLSI array compiler system (VACS) (Kung and Jean, 1988) is developed for systolic and wave front array design. DG is captured and displayed in graphical format. Other than scheduling and projection vectors required in the space–time mapping process, processor axes and local memory size in each processor are optional inputs from the user. The optimal SFG is obtained by enumerating different criteria over a bounded search space.
- VASS captures DG in graphical format. It features multiprojection and the two-level pipeline technique for space–time mapping. An approach to transport all interior I/O ports to the array boundary is implemented. The output is the register transfer (RT) level data path synthesis of PE and associated control signals in net list format.

Even though systolic array design methodology was explored in the early 1980s, a deficit of automatic systolic design tools have inhibited traditional systolic design methodology to become useful for RC. To the best of both authors' knowledge, no such automatic design tool yet, exists which may be due to the following reasons.

- Design space of systolic arrays must be explored with human interaction (Kung and Jean, 1988; Burleson and Jung, 1992; Stone and Manalokos, 2000) to achieve an optimal trade-off between different design objectives and to satisfy design constraints.
- The exploration of the final design space is limited because the array dimension is decreased by one after each projection according to the multiprojection procedure.
- I/O port allocation is cumbersome because it is allocated to only array boundaries.
- The resultant array may not be applicable due to its large number of I/O ports and a massive amount of memory bandwidth required.

Regular Array to FPGA/RC Synthesis

RC emerges to exploit inherent and distributed storage, regularity and local communication of VLSI layout, dynamic reconfiguration, and spatial parallelism. As a consequence, more computational power as well as data flow computing with high-throughput demand for multimedia and communication applications can be realized at low-power consumption (Zhang *et al.*, 2000). Meanwhile, advances in semiconductor technology allow faster and larger FPGA/RC to become common practice. This leads to demand of efficient and automatic algorithm mapping tools.

Prior to algorithm-to-hardware synthesis, a system specification undergoes HW/SW partitioning. The success of FPGA/RC lies in how effective the algorithm is mapped to its hardware counterpart and how both hardware and software interact with one another. The latter part, however, is beyond the scope of this chapter.

Despite different levels of granularity of logic cells, the algorithm-to-hardware mapping procedures are performed similarly. Started from algorithm compilation as shown in Figure 7.6, an algorithm mapping methodology of FPGA/RC can be decomposed into the following phases.

1. **Algorithm compilation**: Compilation takes one of several forms of high-level languages (HLLs) to graph representation and then converts it suitable for algorithm transformation and optimization. Such, HLLs include C/C + +. NAPA C (Rupp, 1998), single assignment (SA)-C (Moorman and Cates, 1999), and Matlab. The tool to accomplish the task above is in various forms of compiler such as GNU C/C + + (Isshiki

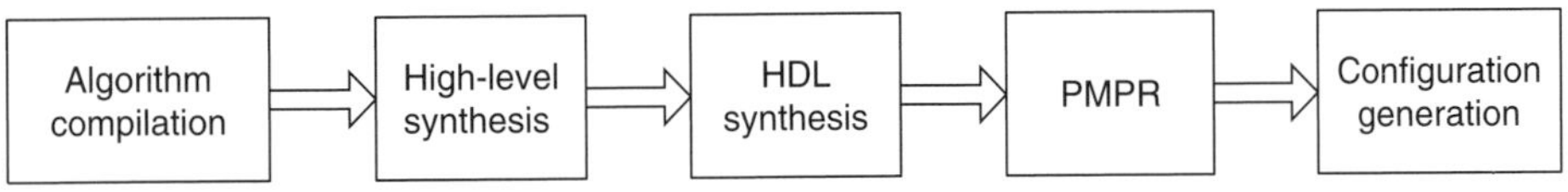

FIGURE 7.6 Algorithm-to-Hardware Mapping of Reconfigurable Computing

and Dai, 1996), NAPA C (Rupp, 1988). SUIF (Callahan *et al.*, 2000), and Cameron (Hammes *et al.*, 1999; Moorman and Cates, 1999). The most recent one is the System Generator from Xilinx and Mathworks (Xilinx, 2000). The suite of tools for algorithm development to hardware prototyping consists of an algorithm mapper, a net lister, a test bench generator, a synthesizer, a control designer a core generator an FPGA placer and router, and a logic simulator.

2. **High-level synthesis**: Synthesis converts the graph representation from the algorithm compilation phase into a behavioral description of hardware description language (HDL), such as VHDL. Verilog HDL, and Lola (Woods *et al.*, 1997). The available tools include ARREST (Burleson and Jung, 1992). DG2VHDL (Stone and Manalokos, 2000). IRIS (Woods *et al.*, 1997), Khoros (Moorman and Cates, 1999), VACS (Kung and Jean, 1988), and VASS (Yen *et al.*, 1996).
3. **HDL synthesis**: This synthesis is responsible for synthesizing a behavioral description of a given algorithm to its corresponding structural or RT level representation using, for example, the Synopsys Behavioral Compiler, Design Compiler, and other such compilers.
4. **Partitioning, mapping, placing, and routing** (PMPR): Partitioning is required for multichip design if it is also necessary for multichip implementation. After design is partitioned for an individual FPGA chip, logic formulation is mapped to underlying logic cells and placed according to the desired floor plan. Logic cells are interconnected by an automatic routing tool. A mapping, placing, and routing tool is provided by the manufacturer or locally implemented and integrated to the design flow such as IRIS (Woods *et al.*, 1997) for faster execution time.
5. **Configuration generation**. A configuration word (bit stream) is generated according to both the logic cell formulation and the routing information.

Table 7.4 summarizes some existing tools/methodologies and their corresponding design phases.

7.4 Conclusion

DSP is attributed by its data flow and real-time computing nature. This dictates the implementation of a high-throughput DSP application as a dedicated chip called ASIC. A broad range of DSP applications include seismic/radar/sonar/radio signal processing, human/computer/machine interface, storage and communication, and security/military operations. As such, a variety of data types, algorithm constructs, and sampling rates are involved. Data type can be either bit-level or bit-parallel (fixed-point/floating-point number). The algorithm can be described as a nonterminating/terminating nested Do-loop or recurrent equation. The sampling rate of the input signals influences the architectural style of algorithm-to-hardware synthesis. As the sampling rate increases, more parallellism/computation power is demanded from the hardware architecture. Consequently, processing power can be obtained from an implementation in either a network of interconnected execution units or a regular array of PEs.

Other than CMOS ASIC realization that needs high product volume to amortize the prototyping cost, FPGA is one of the alternatives provided that allows the timing requirement to be satisfied. In addition, reconfigurable computing has emerged with similar architecture to FPGA where a 2-D array of reconfigurable logic cells are interconnected by a programmable interconnect and surrounded by input/output cells. To satisfy dynamic reconfigurability on the same platform, the context or configuration is stored locally in each cell. This emergence of the dynamically configurable hardware may soon influence algorithm-to-hardware synthesis. Hence, the time to market can be shortened.

TABLE 7.4 Summary of Algorithm Mapping Methodologies

	Algorithm compilation	High-level synthesis	HDL synthesis	PMPR	Configuration generation
CHAMP[1]	√	√	√	√	√
Garp[2]	√	√	√	√	√
IRIS[3]		√	√	√	√
MorphoSys[4]	√	√	√	√	√
NAPA[5]	√	√	√	√	√
Pleiades[6]	√	√	√	√	√
Xilinx[7]	√	√	√	√	√

[1] Patriquin and Gurevich (1995).
[2] Callahan *et al.* (2000).
[3] Woods *et al.* (1997).
[4] Lu *et al.* (1999).
[5] Rupp *et al.* (1998).
[6] Wan *et al.* (1999)
[7] Turney *et al.* (2000).

References

Abnous, A., and Rabaey, J. (1996). Ultra-low-power domain-specific multimedia processors. *VLSI Signal Processing IX*, 461–470.

Andraka, R. (1998). A survey of cordic algorithms for FPGA-based computers. *Proceedings of the ACM/SIGDA Sixth International Symposium on Field-Programmable Gate Arrays* 191–200.

Banerjee, U. (1993). *Loop transformations for restructuring compilers: The foundations.* Norwell, MA: Kluwer Academic.

Black, P.J., and Meng, T.H.Y. (1997). A 1-gb/s. four-state, sliding block viterbi decoder. *IEEE Journal of Solid-State Circuits 32*(6), 797–805.

Buell, D., Arnold, J.M., and Kleinfelder, W.J. (1996). *Splash 2 FPGAs in a custom computing machine.* Los Alamitos, CA: IEEE Computer Society.

Burleson, W., and Jung, B. (1992). Arrest: An interactive graphic analysis tool for VLSI arrays. *Proceedings of the International Conference on Application Specific Array Processors* 149–162.

Callahan, T. J., Hauser, J. R., and Wawrzynek, J. (2000). The garp architecture and C compiler. *IEEE Computer 33*(4), 62–69.

Chang, S. F., and Messerchmitt, D. G. (1992). Designing a high-throughput VLC decoder part *i*-concurrent VLSI architectures. *IEEE Transactions on Circuits and Systems for Video Technology 2*(2), 187–196.

Chen, C. Y. R., and Moricz, M. Z. (1991). A delay distribution methodology for the optimal systolic synthesis of linear recurrence algorithms. *IEEE Transactions on Computer-Aided Design of Integrated Circuits and Systems 10*(6), 685–697.

Dick, C. (1996). Computing the discrete fourier transform on FPGA-base systolic arrays. *Proceedings of the 1996 ACM Fourth International Symposium on Field-Programmable gate arrays* 129–135.

Ganapathy, K. G., and Wah, B. (1992). Optimal design of lower dimensional processor arrays for uniform recurrences. *International Conference on Application Specific Array Processors*, 636–648.

Goldstein, S. C., Schmit, H., Moe, M., Budiu, M., Cadambi, S., Taylor, R. R., and Laufer, R. (1999). Piperench: A coprocessor for streaming multimedia acceleration. *Proceedings of the 26th International Symposium on Computer Architecture* 28–39.

Goslin, G. R. (1995). A guide to using field-programmable gate arrays (FPGAs) for application-specific digital signal processing performance. San Jose, CA: Xilinx.

Hammes, J., Rinker, B., Bohm, W., Najjar, W., Draper, B., and Beveridge, R. (1999). *Cameron: High-level language compilation for reconfigurable system*. 236–244.

Hwang, Y. T., and Hu, Y. H. (1992). MSSM-a design aid for multi-stage systolic mapping. *Journal of VLSI Signal Processing 4*(2/3), 125–146.

Hwang, Y. T., and Hu, Y. H. (1992). Novel scheduling scheme for systolic array partitioning problem. *VLSI Signal Processing 5* 355–64.

Xilinx. (2000). Virtex ii datasheet. San Jose, CA: Xilinx.

Isshiki, T., and Dai, W. W. M. (1996). Bit-serial pipeline synthesis for multi-FPGA systems with C + + design capture. *Proceedings of the IEEE Symposium on FPGAs for Custom Computing Machines* 38–47.

Kittitornkun, S., and Hu, Y. H. (2001). Frame-level pipelined motion estimation array processor. *IEEE Transactions on Circuit and System for Video Technology 11*(2), 248–251.

Kung, H. T. (1982). Why sysytolic architectures? *IEEE Computer 15*(1), 37–45.

Kung, S. Y. (1988). *VLSI Array Processors*. Englewood Cliffs, NJ: Prentice Hall.

Kung, S. Y., and Jean, S. N. (1988). A VLSI array compiler system. *VLSI Signal Processing 3*, 495–508.

Lamport, L. (1974). The parallel execution of Do loops. *Communications of the ACM 17*(2), 83–93.

Lee, P., and Kedem, Z. M. (1988). Synthesizing linear array algorithms from nested for loop algorithms. *IEEE Transactions on Computers 37*(12), 1578–98.

Lee, P., and Kedem, Z. M. (1990). Mapping nested loop algorithms into multidimensional systolics arrays. *IEEE Transactions on Parallel and Distributed Systems 1*(1), 64–76.

Leiserson, C. E., Rose, F., and Saxe, J. (1983). Optimizing synchronous circuitry for retiming. *Third Caltech Conference on VLSI* 87–116.

Li, G. J., and Wah, B. W. (1985). The design of optimal systolic arrays. *IEEE Transactions on Computers 34*(1), 66–77.

Lu, G., Singh, H., Lee, M. H., Bagherzadeh, N., Kurdahi, F. J., Filho, E. M. C., and Castro-Alves, V. (1999). The morphosys dynamically reconfigurable system-on-chip. *Proceedings of the First NASA/DoD Workshop on Evolvable Hardware* 152–160.

De Man, H., Catthoor, F., Goossens, G., Vanhoof, J., Meerbergen, J. V., Note, S., and Huisken, J. (1990). Architecture-driven synthesis techniques or VLSI implementation of DSP algorithms. *Proceedings of IEEE 78*(2), 319–334.

De Man, H., Rabaey, J., Vanhoof, J., Goosens, G., Siz, P., and Claesen, L. (1988). *Cathedral-II-A computer-aided synthesis system for digital signal processing VLSI systems* 55–66.

Mead, C., and Conway, L. (1980). *Introduction to VLSI Systems*. Reading, MA: Addison-Wesley.

Mirsky, E., and DeHon, A. (1996). Matrix: A reconfigurable computing architecture with configurable instruction distribution and deployable resources. *Proceedings of the IEEE Symposium on FPGAs for Custom Computing Machines* 157–166.

Miyamori, T., and Olukotun, U. (1998). A quantitative analysis of reconfigurable co-processors for multimedia applications. *Proceedings of the IEEE Symposium on FPGAs for Custom Computing Machines* 2–11.

Moldovan, D. I. (1982). On the analysis and synthesis of VLSI algorithms. *IEEE Transactions on Computers 31*(11), 1121–1126.

Moorman, A. C., and Cates, D. M., Jr. (1999). A complete development environment for image processing applications on adaptive computing systems. *IEEE ICASSP 4*, 2159–62.

Nishitani, T. (1999). An approach to a multimedia system on a chip. *IEEE Workshop on Signal Processing Systems* 13–21.

Note, S., Geurts, W., Catthoor, F., and De Man, H. (1991). Cathedral-III: Architecture-driven high-level synthesis for high throughput DSP applications. *Proceedings of the 28th Conference on ACM/IEEE Design Automation Conference* 597–602.

O'Keefe, M. T., Fortes, J. A. B., and Wah, B. W. (1992). On the relationship between two systolic array design methodologies. *IEEE Transactions on Computers 41*(12), 1589–93.

Parhi, K. K. (1989). Algorithm transformation techniques for concurrent processors. *Proceedings of the IEEE 77*(12), 1879–95.

Patriquin, R., and Gurevich, I. (1995). An automated design process for the champ module. *Proceedings of the IEEE National Aerospace and Electronics Conference 1*, 417–424.

Rabaey, J., Abnolus, A., Ichikawa, Y., Seno, K., and Wan, M. (1997). Heterogeneous reconfigurable systems. *IEEE Workshop on Signal Processing Systems* 24–34.

Rao, S. K., and Kailath, T. (1989). Regular iterative algorithms and their implementation on processor arrays. *Proceedings of the IEEE 76*(3), 259–269.

Rupp, C. R., Landguth, M., Garverick, T., Gomersall, E., Holt, H., Arnold, J. M., and Gokhale, M. (1988). The NAPA adaptive processing architecture. *IEEE Symposium on FPGAs for Custom Computing Machines* 28–37.

Shang, W., and Fortes, J. A. B. (1992). On-time mapping of uniform dependence algorithms into lower dimensional processor arrays. *IEEE Transactions on Parallel and Distributed Systems 3*(3), 350–363.

Shankiti, A. M., and Leeser, M. (2000). Implementing a rake receiver for wireless communications on an FPGA-based computer system. *Proceedings of the ACM/SIGDA International Symposium on Field-Programmable Gate Arrays* 145–151.

Stone, A., and Manalokos, E. S. (1998). Using DG2VHDL to synthesize an FPGA 1-D discrete wavelet transform. *IEEE Workshop on Signal Processing Systems* 489–498.

Stone, A., and Manalokos, E. S. (2000). DG2YHDL: To facilitate the high-level synthesize of parallel processing array architectures. *Journal of VLSI Signal Processing 24*(1), 99–120.

Truong, T. K., Shih, M. T., Reed, I. S., and Satorius, E. H. (1992). A VLSI design for a trace-back Viterbi decoder. *IEEE Transactions on Communications 40*(3), 616–624.

Turney, R. D., Dick, C., Parlour, D. B., and Hwang, J. (2000). *Modeling and implementation of DSP FPGA solutions.* San Jose, CA: Xilinx.

Wan, M., Zhang, H., Benes, M., and Rabaey, J. (1998). A low-power reconfigurable data-flow driven DSP system. *IEEE Workshop on Signal Processing Systems* 191–200.

Wang, D. J., and Hu, Y. H. (1995). Multiprocessor implementation of real-time DSP algorithms. *IEEE Transactions on Very Large Scale Integration (VLSI) Systems 3*(3), 393–403.

White, S. A. (1989). Applications of distributed arithmetic to digital signal processing: A tutorial review. *IEEE ASSP Magazine* 4–19.

Wolfe, M. (1996). *High performance compilers for parallel computing.* Redwood City, CA: Addison-Wesley.

Wong, Y., and Delosme, J. M. (1985). Optimization of computation time for systolic arrays. *IEEE International Conference on Computer Design: VLSI in Computers* 618–621.

Wong, Y., and Delosme, J. M. (1992). Optimization of computation time for systolic arrays. *IEEE Transactions on Computers 41*(2), 159–177.

Woods, R., Cassidy, A., and Gray, J. (1996). VLSI architectures for field programmable gate arrays: A case study. *Proceedings on IEEE Symposium on FPGAs for Custom Computing Machines* 2–9.

Woods, R., Ludwig, S., Heron, J., Trainor, D., and Gehring, S. (1997). FPGA synthesis on the xc6200 using IRIS and TRIANUS/HADES (or from heaven to hell and back again). *Proceedings of the 5th Annual IEEE Symposium on Field-Programmable Custom Computing Machines* 155–164.

Xilinx. (2000). *System generator v.1.0 product datasheet.*

Yeh, J. W., Cheng, W. J., and Jen, C. W. (1996). VASS-A VLSI array system synthesizer. *Journal of VLSI Signal Processing 12*(2), 135–158.

Yeo, H., and Hu, Y. H. (1995). A novel modular systolic array architecture for full-search block matching motion estimation. *IEEE Transactions on Circuit and Systems for Video Technology 5*(5), 407–416.

Zhang, H., Prabhu, V., George, V., Wan, M., Benes, M., Abnolus, A., and Rabaey, J. M. (2000). A 1-V heterogeneous reconfigurable processor IC for baseband wireless applications. *Digest of Technical Papers IEEE International Solid-State Circuits Conference*, 68–69.

Zimmermann, K. H. (1996). Linear mapping of *n*-dimensional uniform recurrences onto *k*-dimensional sysytolic arrays. *Journal of VLSI Signal Processing 12*, 187–202.

VIII

DIGITAL COMMUNICATION AND COMMUNICATION NETWORKS

Vijay K. Garg
Department of Electrical and Computer Engineering, University of Illinois at Chicago, Chicago, Illinois, USA

Yih-Chen Wang
Lucent Technologies, Naperville, Illinois, USA

This section is concerned with *digital communication* and *data communication networks.* In the first part of the section we focus on the essentials of digital communication and in the second part we discuss important aspects of data communication networks.

In a digital communication system, the information is processed so that it can be represented by a sequence of discrete messages. The digital source may be the result of sampling and quantizing an analog source, or it may represent a digital source such as the contents of a computer memory. When the source is a binary source, the two possible values 0 and 1 are called bits for binary source.

A digital signal can be processed independently of whether it represents a discrete data source or a digitized analog source. This implies that an essentially unlimited range of signal conditioning and processing options is available to the designer. Depending on the origination and destination of the information being conveyed, it may include *source coding, encryption, pulse shaping* for spectral control, *forward error correction (FEC) coding, special modulation* to spread the signal spectrum, and *equalization* to compensate for channel distortion.

In most communication system designs, a general objective is to use as efficiently as possible the resources of bandwidth and transmitted power. Thus, we are interested in both *bandwidth efficiency,* defined as the ratio of data rate to signal bandwidth, and *power efficiency,* characterized by the probability of making a reception error as a function of signal-to-noise ratio (SNR).

Modulation produces a continuous-time waveform suitable for transmission through the communication channel, whereas *demodulation* is to extract the data from the received signal. We discuss various modulation schemes.

Spread spectrum refers to any modulation scheme that produces a spectrum for the transmitted signal much wider than bandwidth of the information to be transmitted independently of the bandwidth of the information-bearing signal. We discuss the *code division multiple access (CDMA)* spread spectrum scheme.

In the second part of the section, our focus is on data communications network usually involving a collection of computers and other devices operating autonomously and interconnected to provide computing or data services. The interconnection of computers and devices can be established with either wired or wireless transmission media. This part starts with an introduction to fundamental concepts of data communications and network architecture. These concepts would help readers to understand how different pieces of system components are working together to build a data communication network. Many data communication and networking terminologies are explained so that they would also help readers to understand advanced networking technologies. Two important network architectures are described. They are Open System Interconnection (OSI) and TCP/IP architecture. OSI is a network reference model that is referenced by all types of networks while the TCP/IP architecture provides a framework where the Internet has been built upon. The network evolution cannot be so successful without the widespread use of local networking. The local network technology including the wireless Local Area Network (WLAN) was then introduced followed by various wireless network access technologies, like FDMA, TDMA, CDMA, and WCDMA. The convergence of varieties of networks under the Internet has begun. One important driver has been the introduction of Session Initiation Protocol (SIP) that could provide a framework for Integrated Multimedia Network Service (IMS). The convergence won't be possible without extensive use of a Softswitch, the creation of ALL-IP architecture and providing broadband access utilizing optical networking technologies.

1

Signal Types, Properties, and Processes

Vijay K. Garg
Department of Electrical and Computer Engineering, University of Illinois at Chicago, Chicago, Illinois, USA

Yih-Chen Wang
Lucent Technologies, Naperville, Illinois, USA

1.1 Signal Types

A **signal** is a function that carries some information. Mathematically, a signal can be classified as deterministic or random. For a **deterministic** signal, there is no uncertainty with respect to its value at any time. Deterministic signals are modeled by explicit mathematical expressions. For a **random** signal, there is some degree of uncertainty before the signal occurs. A probabilistic model is used to characterize a random signal. Noise in a communication system is an example of the random signal.

A signal $x(t)$ is **periodic** in time if there exists a constant $T_0 > 0$, such that:

$$x(t) = x(t + T_0) \text{ for } -\infty < t < \infty, \quad (1.1)$$

where t is the time.

The smallest value of T_0 that satisfies equation 1.1 is called the **period of $x(t)$**. The period T_0 defines the duration of one complete cycle of $x(t)$. A signal for which there is no such value of T_0 that satisfies equation 1.1 is referred to as a **nonperiodic signal**.

1.2 Energy and Power of a Signal

Instantaneous power of an electrical signal is given as:

$$p(t) = x^2(t), \quad (1.2)$$

where $x(t)$ is either a voltage or current of the signal.

The energy dissipated during the time interval $(-T/2, T/2)$ by a real signal with instantaneous power given by equation 1.2 is written as:

$$E_x^T = \int_{-T/2}^{T/2} x^2(t)dt. \quad (1.3)$$

The average power dissipated by the signal during interval T will be as follows:

$$P_x^T = \frac{1}{T} \cdot \int_{-T/2}^{T/2} x^2(t)dt. \quad (1.4)$$

In analyzing communication signals, it is often desirable to deal with **waveform energy**. We refer to $x(t)$ as an energy signal only if it has nonzero but finite energy $0 < E_x < \infty$ for all time, where:

$$E_x = \lim_{T\to\infty} \int_{-T/2}^{T/2} x^2(t)dt = \int_{-\infty}^{\infty} x^2(t)dt. \quad (1.5)$$

A signal is defined to be a power signal only if it has finite but nonzero power $0 < P_x < \infty$ for all time, where:

$$P_x = \lim_{T \to \infty} \frac{1}{T} \int_{-T/2}^{T/2} x^2(t)dt. \quad (1.6)$$

1.3 Random Processes

A **random process** $X(t, s)$ can be viewed as a function of two variables: **sample space** s, and **time** t. We show N sample functions of time, $\{X_k(t)\}$ (see Figure 1.1). Each of the sample functions can be considered as the output of a different noise generator. For a fixed sample point s_k, the graph of the function $X(t, s_k)$ versus time t is called a **realization** or **sample function** of the random process. To simplify the notation, we denote this sample function as:

$$X_k(t) = X(t, s_k). \quad (1.7)$$

We have an indexed ensemble (family) of random variables $\{X(t, s)\}$, which is called a random process. To simplify the notation, we suppress the s and use $X(t)$ to denote a random process.

A **random process** $X(t)$ is an ensemble of time functions together with a probability rule that assigns a probability to any meaningful event associated with an observation of one sample function of the random process.

1.3.1 Statistical Average of a Random Process

A random process whose distribution functions are continuous can be described statistically with a **probability density function** (pdf). We define the mean of the random process $X(t)$ as:

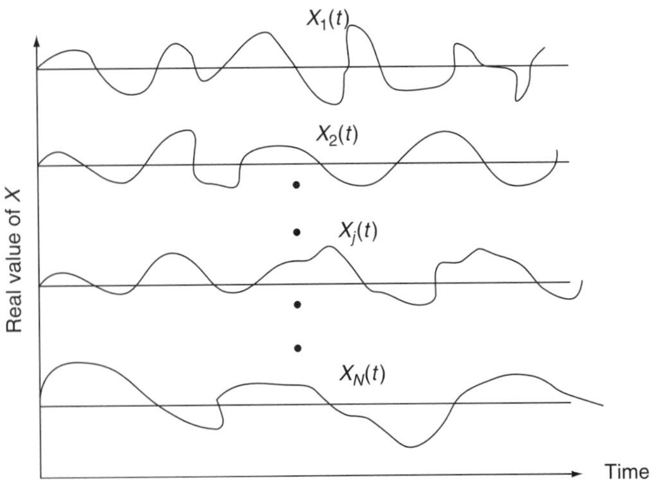

FIGURE 1.1 Random Noise Process

$$E\{X(t_k)\} = \int_{-\infty}^{\infty} (x \cdot px_k)dx = \mu_x(t_k), \quad (1.8)$$

where $X(t_k)$ is the random variable and $px_k(x)$ is the pdf of $X(t_k)$ at time t_k.

We define the autocorrelation function of the random process $X(t)$ to be a function of two variables, t_1 and t_2, as:

$$R_x(t_1, t_2) = E\{X(t_1)X(t_2)\}, \quad (1.9)$$

where $X(t_1)$ and $X(t_2)$ are random variables obtained by observing $X(t)$ at times t_1 and t_2, respectively.

The **autocorrelation function** is a measure of the degree to which two time samples of the same random process are related.

1.3.2 Stationary Process

A random process $X(t)$ is said to be **stationary** in the strict sense if none of its statistics is affected by a shift in time origin. A random process is said to be **wide-sense stationary** (WSS) if two of its statistics (mean and autocorrelation function) do not vary with shift in time origin. Therefore, a random process is WSS if:

$$E\{X(t)\} = \mu_x = \text{constant}, \quad (1.10)$$

and

$$R_x(t_1, t_2) = R_x(t_1 - t_2) = R_x(\tau), \quad (1.11)$$

where $\tau = t_1 - t_2$.

In a strict sense, *stationary* implies WSS, but WSS does *not* imply stationary.

1.3.3 Properties of an Autocorrelation Function

An **auto correlation function** reads as:

$$R_x(\tau) = E[X(t + \tau)X(t)] \text{ for all } t. \quad (1.12)$$

- The mean square value of a random process can be obtained from $R_x(\tau)$ by substituting $\tau = 0$ in equation 1.12:

$$R_x(0) = E[X^2(t)]. \quad (1.13)$$

- The autocorrelation function $R_x(\tau)$ is an even function of τ; that is:

$$R_x(\tau) = R_x(-\tau). \quad (1.14)$$

- The autocorrelation function $R_x(\tau)$ has its maximum magnitude at $\tau = 0$; that is:

$$|R_x(\tau)| \leq R_x(0). \quad (1.15)$$

To prove this property, we consider the non-negative quantity:

$$E[\{X(t+\tau) \pm X(t)\}^2] \geq 0. \quad (1.16)$$

Or we can consider the following:

$$E[X^2(t+\tau)] \pm 2E[X(t+\tau)X(t)] + E[X^2(t)] \geq 0. \quad (1.17)$$

$$2R_x(0) \pm 2R_x(\tau) \geq 0. \quad (1.18)$$

$$-R_x(0) \leq R_x(\tau) \leq R_x(0). \quad (1.19)$$

$$|R_x(\tau)| \leq R_x(0). \quad (1.20)$$

1.3.4 Crosscorrelation Functions

We consider two random processes $X(t)$ and $Y(t)$ with autocorrelation function $R_X(t, u)$ and $R_Y(t, u)$, respectively. The two **crosscorrelation functions** of $X(t)$ and $Y(t)$ are defined as:

$$R_{XY}(t, u) = E[X(t)Y(u)], \quad (1.21)$$

and

$$R_{YX}(t, u) = E[Y(t)X(u)]. \quad (1.22)$$

$$R(t, u) = \begin{bmatrix} R_X(t, u) & R_{XY}(t, u) \\ R_{YX}(t, u) & R_Y(t, u) \end{bmatrix}, \quad (1.23)$$

or

$$R(\tau) = \begin{bmatrix} R_X(\tau) & R_{XY}(\tau) \\ R_{YX}(\tau) & R_Y(\tau) \end{bmatrix}. \quad (1.24)$$

where $\tau = t - u$.

The crosscorrelation function is not generally an even function of τ, and it does not have a maximum value at the origin. However, it obeys a symmetry relationship:

$$R_{XY}(\tau) = R_{YX}(-\tau). \quad (1.25)$$

1.3.5 Ergodic Processes

We consider the sample function $x(t)$ of a stationary process $X(t)$, with observation interval $-T/2 \leq t \leq T/2$. The time average will be as written here:

$$\mu_x(T) = \frac{1}{T}\int_{-T/2}^{T/2} x(t)dt. \quad (1.26)$$

The time average $\mu_x(T)$ is a random variable, and its value depends on the observation interval and which particular sample function of the random process $X(t)$ is selected for use in equation 1.26. Since $X(t)$ is assumed to be stationary, the mean of the time average $\mu_x(T)$ is given as:

$$E[\mu_x(T)] = \frac{1}{T}\int_{-T/2}^{T/2} E[x(t)]dt = \frac{1}{T}\int_{-T/2}^{T/2} \mu_x dt = \mu_x, \quad (1.27)$$

where μ_x is the mean of process $X(t)$.

We say that the process $X(t)$ is **ergodic** in mean if two conditions are satisfied:

(1) The time average $\mu_x(T)$ approaches the ensemble average μ_x in limit as the observation time T goes to infinity:

$$\lim_{T\to\infty} \mu_x(T) = \mu_x. \quad (1.28)$$

(2) The variance of $\mu_x(T)$, treated as a random variable, approaches zero in limit as the observation interval T goes to infinity.

$$\lim_{T\to\infty} var[\mu_x(T)] = 0. \quad (1.29)$$

Likewise, the process $X(t)$ is ergodic in the autocorrelation function if the following two limiting conditions are fulfilled:

$$(1)\ \lim_{T\to\infty} R_x(\tau, T) = R_x(\tau). \quad (1.30)$$

$$(2)\ \lim_{T\to\infty} var[R_x(\tau, T)] = 0. \quad (1.31)$$

Example 1

A random process is described by $x(t) = A\cos(2\pi ft + \phi)$, where ϕ is a random variable uniformly distributed on $(0, 2\pi)$. Find the autocorrelation function and mean square value.

Solutions:

$$R_x(t_1, t_2) = E[A\cos(2\pi ft_1 + \phi)A\cos(2\pi ft_2 + \phi)]$$

$$= A^2E\left\{\frac{1}{2}\cos 2\pi f(t_1 - t_2) + \frac{1}{2}\cos[2\pi f(t_1 - t_2) + 2\phi]\right\}$$

$$= \frac{A^2}{2}\cos 2\pi f(t_1 - t_2).$$

$$R_x(\tau) = \frac{A^2}{2}\cos 2\pi f(t_1 - t_2) = \frac{A^2}{2}\cos 2\pi f\tau.$$

$$R_x(0) = \frac{A^2}{2}.$$

1.4 Transmission of a Random Signal Through a Linear Time-Invariant Filter

Figure 1.2 shows a random input signal $X(t)$ passing through a linear time-invariant filter. The output signal $Y(t)$ can be

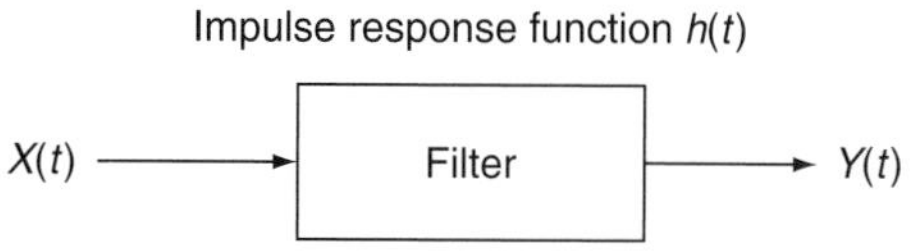

FIGURE 1.2 Random Signal Through a Linear Time-invariant Filter

related to the input through an impulse response function $h(\tau_1)$ as:

$$Y(t) = \int_{-\infty}^{\infty} h(\tau_1) X(t - \tau_1) d\tau_1. \quad (1.32)$$

The mean of $Y(t)$ is as follows:

$$\mu_Y(t) = E[Y(t)] = E\left[\int_{-\infty}^{\infty} h(\tau_1) X(t - \tau_1) d\tau_1\right]. \quad (1.33)$$

The system is stable provided that the expectation $E[X(t)]$ is finite for all t. We may interchange the order of expectation and integration in equation 1.33, and the following results:

$$\mu_Y(t) = \int_{-\infty}^{\infty} h(\tau_1) E[X(t - \tau_1)] d\tau_1 = \int_{-\infty}^{\infty} h(\tau_1) \mu_X(t - \tau_1) d\tau_1. \quad (1.34)$$

For a stationary process $X(t)$, the mean $\mu_X(t)$ is constant μ_X, so we may simplify equation 1.34 as:

$$\mu_Y = \mu_X \cdot \int_{-\infty}^{\infty} h(\tau_1) d\tau_1 = \mu_X \cdot H(0), \quad (1.35)$$

where $H(0)$ is the zero-frequency (dc) response of the system.

Equation 1.35 states that the mean of the random process $Y(t)$ produced at the output of a linear time-invariant system in response to $X(t)$ acting as the input process is equal to the mean of $X(t)$ multiplied by the dc response of the system, which is intuitively satisfying.

The autocorrelation function of the output random process $Y(t)$ is as follows:

$$R_Y(t, u) = E\left\{\left[\int_{-\infty}^{\infty} h(\tau_1) X(t - \tau_1) d\tau_1\right] \cdot \left[\int_{-\infty}^{\infty} h(\tau_2) X(u - \tau_2) d\tau_2\right]\right\}. \quad (1.36)$$

Assuming that mean square $E[X^2(t)]$ is finite for all t and the system is stable, then:

$$R_Y(t, u) = \int_{-\infty}^{\infty} h(\tau_1) d\tau_1 \int_{-\infty}^{\infty} h(\tau_2) d\tau_2 E[X(t - \tau_1) \cdot X(u - \tau_2)].$$

$$R_Y(t, u) = \int_{-\infty}^{\infty} h(\tau_1) d\tau_1 \int_{-\infty}^{\infty} h(\tau_2) d\tau_2 R_X(t - \tau_1, u - \tau_2). \quad (1.37)$$

When the input $X(t)$ is a stationary process, the autocorrelation function of $X(t)$ is only a function of difference between the observation time $t - \tau_1$ and $u - \tau_2$. Thus, with $\tau = t - u$, we have:

$$R_Y(\tau) = \int_{-\infty}^{\infty}\int_{-\infty}^{\infty} [h(\tau_1) h(\tau_2) \cdot R_X(\tau - \tau_1 + \tau_2)](d\tau_1)(d\tau_2) \quad (1.38)$$

$$R_Y(0) = E[Y^2(t)] = \int_{-\infty}^{\infty}\int_{-\infty}^{\infty} h(\tau_1) h(\tau_2) R_X(\tau_2 - \tau_1)(d\tau_1)(d\tau_2) \quad (1.39)$$

$R_Y(0)$ is a constant.

1.5 Power Spectral Density

Let $H(f)$ denote the frequency response of the system; then:

$$h(\tau_1) = \int_{-\infty}^{\infty} [H(f) \cdot e^{j2\pi f \tau_1}] df. \quad (1.40)$$

The following also applies:

$$E[Y^2(t)] = \int_{-\infty}^{\infty}\int_{-\infty}^{\infty}\left(\left[\int_{-\infty}^{\infty} [H(f) \cdot e^{j2\pi f \tau_1}] df\right] \cdot h(\tau_2) R_X(\tau_2 - \tau_1)\right)(d\tau_1)(d\tau_2)$$
$$= \int_{-\infty}^{\infty} H(f) df \int_{-\infty}^{\infty} h(\tau_2) d\tau_2 \int_{-\infty}^{\infty} [R_X(\tau_2 - \tau_1) e^{j2\pi f \tau_1}] d\tau_1. \quad (1.41)$$

Let $\tau = \tau_2 - \tau_1$, and the following results:

$$E[Y^2(t)] = \int_{-\infty}^{\infty} H(f) df \cdot \int_{-\infty}^{\infty} [h(\tau_2) d\tau_2 e^{j2\pi f \tau_2}] d\tau_2 \cdot \int_{-\infty}^{\infty} [R_X(\tau) e^{-j2\pi f \tau}] d\tau$$
$$= \int_{-\infty}^{\infty} |H(f)|^2 df \cdot \int_{-\infty}^{\infty} [R_X(\tau) e^{-j2\pi f \tau}] d\tau, \quad (1.42)$$

where $|H(f)|$ is the magnitude response of the filter.

We define:

$$S_X(f) = \int_{-\infty}^{\infty} [R_X(\tau) e^{-j2\pi f \tau}] d\tau.$$

$S_X(f)$ is called the **power spectral density** or **power spectrum** of the stationary process $X(t)$,

Now, we can rewrite equation 1.42 as:

$$E[Y^2(t)] = \int_{-\infty}^{\infty} [|H(f)|^2 \cdot S_X(f)]df. \quad (1.43)$$

The mean square value of a stable linear time-invariant filter is output in response to a stationary process is equal to the integral over all frequencies of the **power spectral density** of the input process multiplied by the squared magnitude response of the filter.

1.5.1 Properties of psd

We use the following definitions:

$$S_X(f) = \int_{-\infty}^{\infty} [R_X(\tau)e^{-j2\pi f\tau}]d\tau,$$

and

$$R_X(\tau) = \int_{-\infty}^{\infty} [S_X(f)e^{j2\pi f\tau}]df.$$

Using above expressions, we derive the following properties:

- **Property 1**
  For $f = 0$, we get:
  $$S_X(0) = \int_{-\infty}^{\infty} R_X(\tau)d\tau. \quad (1.44)$$
- **Property 2**
  The mean square value of a stationary process equals the total area under the graph of psd:
  $$R_X(0) = E[X^2(t)] = \int_{-\infty}^{\infty} S_X(f)df. \quad (1.45)$$
- **Property 3**
  The psd of a stationary process is always non-negative:
  $$S_X(f) \geq 0 \text{ for all } f. \quad (1.46)$$
- **Property 4**
  The psd of a real-valued random process is an even function of frequency:
  $$S_X(-f) = S_X(f). \quad (1.47)$$
  $$S_X(-f) = \int_{-\infty}^{\infty} [R_X(\tau)e^{j2\pi f\tau}]d\tau.$$
  Because $R_X(\tau)$ is equal to $R_X(-\tau)$, we can write:
  $$S_X(-f) = \int_{-\infty}^{\infty} [R_X(\tau)e^{j2\pi f\tau}]d\tau = S_X(f).$$
- **Property 5**
  The psd, appropriately normalized, has the properties usually associated with a probability density function:
  $$p_x(f) = \frac{S_X(f)}{\int_{-\infty}^{\infty} S_X(f)df} \geq 0 \text{ for all } f. \quad (1.48)$$

1.6 Relation Between the psd of Input Versus the psd of Output

$$S_Y(f) = \int_{-\infty}^{\infty} [R_Y(\tau)e^{-j2\pi f\tau}]d\tau.$$

$$S_Y(f) = \int_{-\infty}^{\infty}\int_{-\infty}^{\infty}\int_{-\infty}^{\infty} [h(\tau_1)h(\tau_2)R_X(\tau - \tau_1 + \tau_2)e^{-2j\pi f\tau}](d\tau_1)(d\tau_2)d\tau.$$

Let $\tau_0 = \tau + \tau_1 - \tau_2$:

$$S_Y(f) = H(f)H^*(f)S_X(f) = |H(f)|^2 \cdot S_X(f). \quad (1.49)$$

The psd of the output process $Y(t)$ equals the power spectral density of the input $X(t)$ multiplied by the squared magnitude response of the filter.

Example 2

Numbers 1 and 0 are represented by pulse of amplitude A and $-A$ volts, respectively, and duration T sec. The pulses are not synchronized, so the starting time t_d of the first complete pulse for positive time is equally likely to lie anywhere between 0 and T sec (see Figures 1.3 and 1.4). That is, t_d is the sample value of a uniformly distributed random variable T_d with its probability density function defined by:

$$f_{T_d}(t_d) = \begin{cases} \frac{1}{T} & 0 \leq t_d \leq T. \\ 0 & \text{elsewhere.} \end{cases}$$

$$R_X(\tau_1) = \frac{A^2}{T}\int_{-T/2}^{T/2} X(t)X(t - \tau_1)d\tau.$$

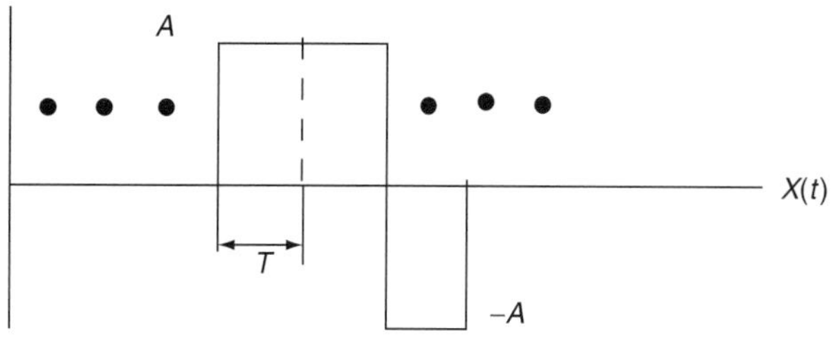

FIGURE 1.3 Pulse of Amplitude A

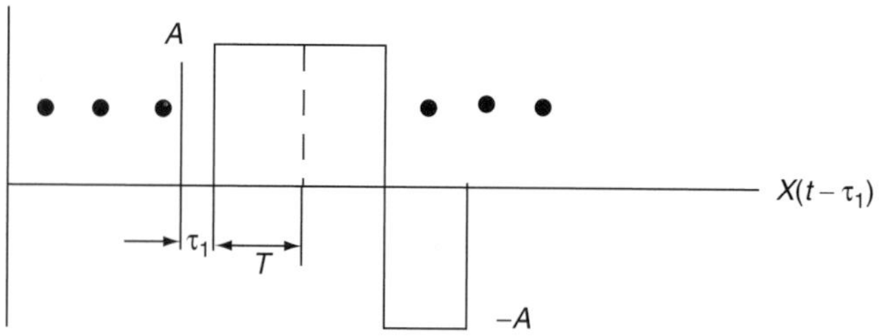

FIGURE 1.4 Shifted Pulse of Amplitude A

$$R_X(0) = \text{Total average power.}$$

$$R_X(\tau) = A^2\left(1 - \frac{|\tau|}{T}\right) \quad |\tau| < T$$

$$= 0 \quad |\tau| \geq T.$$

$$S_X(f) = A^2 \cdot T\left(\frac{\sin \pi f T}{\pi f T}\right)^2.$$

$$\int_{-\infty}^{\infty} S_X(f)df = \text{Total average power}$$

$$S_X(f) = \int_{-T}^{T} A^2\left(1 - \frac{|\tau|}{T}\right)e^{-j2\pi f\tau}d\tau = A^2 T\left(\frac{\sin \pi fT}{\pi fT}\right)^2$$

$$= \frac{A^2T^2}{T}\text{sinc}^2(fT) = \frac{\xi_g(f)}{T}.$$

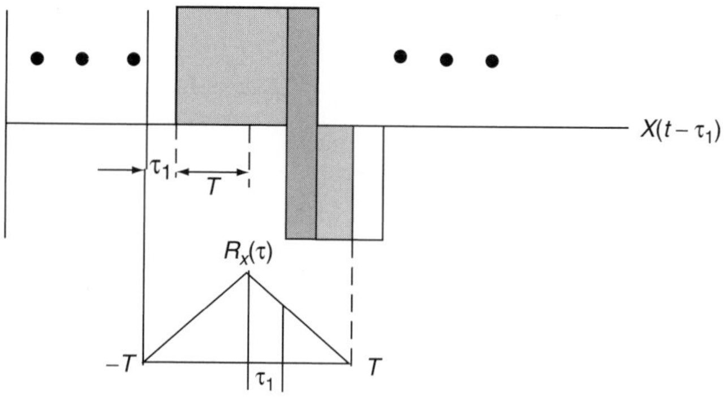

FIGURE 1.5 Autocorrelation Function

These example equations show that for a random binary wave in which binary numbers 1 and 0 are represented by pulse $g(t)$ and $-g(t)$, respectively, the psd $S_X(f)$ is equal to the energy spectral density $\xi_g(f)$ of the symbol shaping pulse $g(t)$, divided by the symbol duration T. Note the autocorrelation function is shown in Figure 1.5.

2

Digital Communication System Concepts

Vijay K. Garg
Department of Electrical and Computer Engineering, University of Illinois at Chicago, Chicago, Illinois, USA

Yih-Chen Wang
Lucent Technologies, Naperville, Illinois, USA

2.1 Digital Communication System

Figure 2.1 shows a block diagram of a typical **digital communication system**. We focus primarily on formatting and transmission of baseband signal. Data already in a digital format would bypass the formatting procedure. Textual information is transformed into binary digits by use of a coder. Analog information is formatted using three separate processes:

- Sampling
- Quantization
- Encoding

In all cases, the formatting steps result in a sequence of **binary digits.** These digits are transmitted through a baseband channel, such as a pair of wires or a coaxial cable. However, before we transmit the digits, we must transform the digits into **waveforms** that are compatible with the channel. For baseband channels, compatible waveforms are pulses. The conversion from binary digits to pulse waveform takes place in a **wave encoder** also called a **baseband modulator.** The output of the waveform encoder is typically a sequence of pulses with characteristics that correspond to the binary digits being sent. After transmission through the channel, the received waveforms are detected to produce an estimate of the transmitted digits, and then the final step is (reverse) formatting to recover an estimate of the source information.

2.2 Messages, Characters, and Symbols

When digitally transmitted, the characters are first encoded into a sequence of bits, called a **bit stream** or **baseband signal.** Groups of n bits can be combined to form a finite symbol set or **word** of $M = 2^n$ for such **symbols.** A system using a symbol set size of M is called an ***M*-ary system.** The value of n or M represents an important initial choice in the design of any digital communication system. For $n = 1$, the system is referred to as **binary**, the size of symbol set is $M = 2$, and the modulator uses two different waveforms to represent the binary 1 and the binary 0. In this case, the symbol rate and the bit rate are the same. For $n = 2$, the system is called **quaternary** or **4-ary** ($M = 4$). At each symbol time, the modulator uses one of the four different waveforms to represent the symbol (see Figure 2.2).

2.3 Sampling Process

Analog information must be transformed into a digital format. The process starts with **sampling** the waveform to produce a **discrete pulse-amplitude-modulated waveform** (see Figure 2.3). The **sampling process** is usually described in a time domain. This is an operation that is basic to digital signal processing and digital communication. Using the sampling

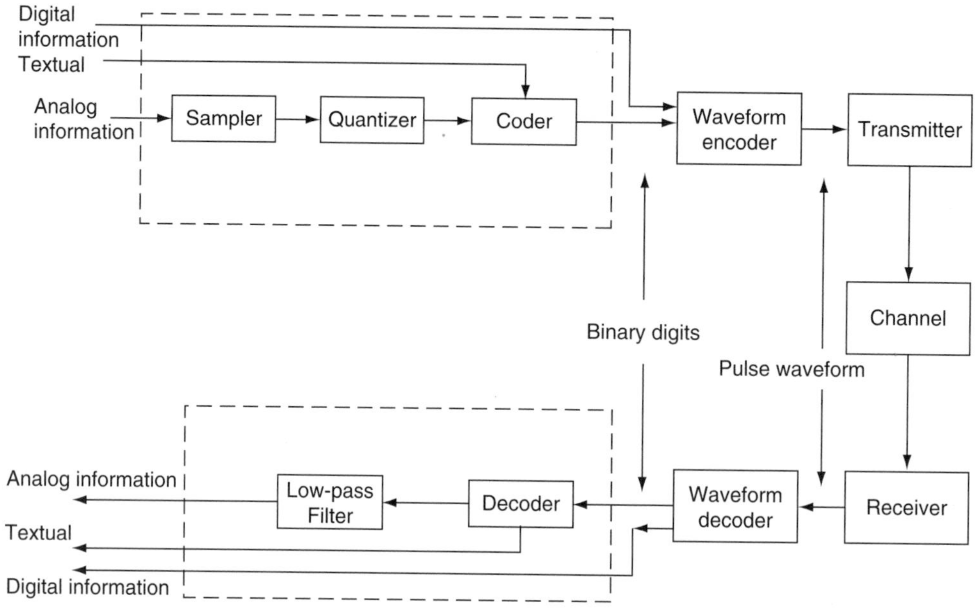

FIGURE 2.1 Block Diagram of a Typical Digital Communication System

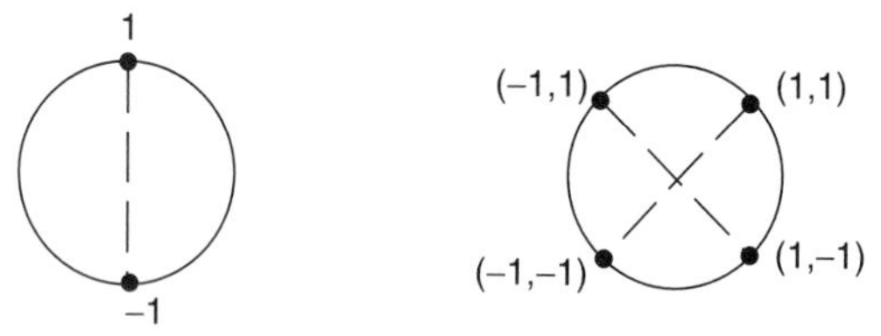

FIGURE 2.2 Binary and Quaternary Systems

process, we convert the analog signal in a corresponding sequence of samples that are usually spaced uniformly in time. The sampling process can be implemented in several ways, the most popular being the **sample-and-hold operation.** In this operation, a switch and storage mechanism (such as a transistor and a capacitor, or shutter and a film strip) form a sequence of samples of the continuous input waveform. The output of the sampling process is called **pulse amplitude modulation (PAM)** because the successive output intervals can be described as a sequence of pulses with amplitudes derived from the input waveform samples. The analog waveform can be approximately retrieved from a PAM waveform by simple low-pass filtering, provided we choose the sampling rate properly. The ideal form of sampling is called **instantaneous sampling.**

We sample the signal $g(t)$ instantaneously at a uniform rate of f_s once every T_s sec. Thus, we can write:

$$g_\delta(t) = \sum_{n=-\infty}^{\infty} g(nT_s)\delta(t - nT_s), \tag{2.1}$$

where $g_\delta(t)$ is the ideal sampled signal and where $\delta(t - nT_s)$ is the delta function positioned at time $t = nT_s$.

A delta function is closely approximated by a rectangular pulse of duration Δt and amplitude $g(nT_s)/\Delta t$; the smaller we make Δt, the better will be the approximation:

$$g_\delta(t) = f_s \sum_{m=-\infty}^{\infty} G(f - mf_s), \tag{2.2}$$

where $G(f)$ is the Fourier transform of the original signal $g(t)$ and f_s is sampling rate.

Equation 2.2 states that the process of uniformly sampling a continuous-time signal of finite energy results in a periodic spectrum with a period equal to the sampling rate.

Taking the Fourier transform of both side, of Equation 2.1 and noting that the Fourier transform of the delta function $\delta(t - nT_s)$ is equal to $e^{-j2\pi nfT_s}$:

$$G_\delta(f) = \sum_{n=-\infty}^{\infty} g(nT_s)e^{-j2\pi nfT_s}. \tag{2.3}$$

Equation 2.3 is called the **discrete-time Fourier transform.** It is the complex Fourier series representation of the periodic frequency function $G_\delta(t)$, with the sequence of samples $g(nT_s)$ defining the coefficients of the expansion.

We consider any continuous-time signal $g(t)$ of finite energy and infinite duration. The signal is strictly band-limited with no frequency component higher than W Hz. This implies that the Fourier transform $G(f)$ of the signal $g(t)$ has the property that $G(f)$ is zero for $|f| \geq W$. If we choose the

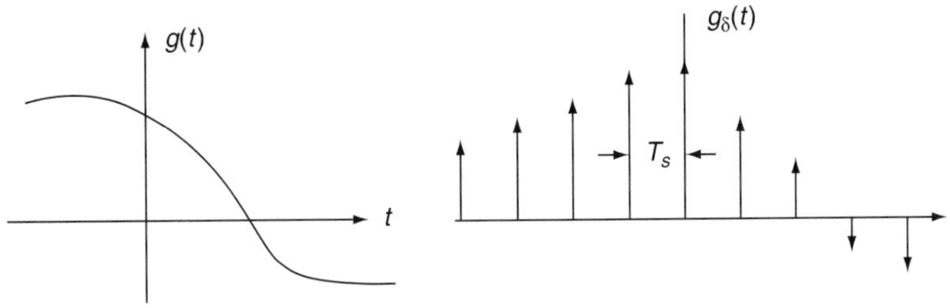

FIGURE 2.3 Sampling Process

sampling period $T_s = 1/2W$, then the corresponding spectrum is given as:

$$G_\delta(f) = \sum_{n=-\infty}^{\infty} g\left(\frac{n}{2W}\right) e^{-\frac{j\pi nf}{W}} = f_s G(f) + f_s \sum_{m=-\infty,\ m\neq 0}^{\infty} G(f - mf_s) \quad (2.4)$$

Consider the following two conditions:

(1) $G(f) = 0$ for $|f| \geq W$.
(2) $f_s = 2W$.

We find from equation 2.4 by applying these conditions,

$$G(f) = \frac{1}{2W} G_\delta(f) \quad -W < f < W.$$
$$\therefore G(f) = \frac{1}{2W} \sum_{n=-\infty}^{\infty} g\left(\frac{n}{2W}\right) e^{-\left(\frac{j\pi nf}{W}\right)} \quad -W < f < W. \quad (2.5)$$

Thus, if the sample value $g(n/2W)$ of a signal $g(t)$ is specified for all n, then the Fourier transform $G(f)$ of the signal is uniquely determined by using the discrete-time Fourier transform of equation 2.5. Because $g(t)$ is related to $G(f)$ by the inverse Fourier transform, it follows that the signal $g(t)$ is itself uniquely determined by the sample values $g(n/2W)$ for $-\infty < n < \infty$. In other words, the sequence $\{g(n/2W)\}$ has all the information contained in $g(t)$.

We state the sampling theorem for band-limited signals of finite energy in two parts that apply to the transmitter and receiver of a pulse modulation system, respectively.

(1) A band-limited signal of finite energy with no frequency components higher than W Hz is completely described by specifying the values of signals at instants of time separated by $1/2\ W$ sec.
(2) A band-limited signal of finite energy with no frequency components higher than W Hz may be completely recovered from a knowledge of its samples taken at the rate of $2\ W$ samples/sec.

This is also known as the **uniform sampling theorem**. The sampling rate of $2\ W$ samples per second for a signal bandwidth W Hz is called the **Nyquist rate** and $1/2\ W$ sec is called the **Nyquist interval**.

We discuss the sampling theorem by assuming that signal $g(t)$ is strictly band-limited. In practice, however, an information-bearing signal is not strictly band-limited, with the result that some degree of under sampling is encountered. Consequently, some aliasing is produced by the sampling process. Aliasing refers to the phenomenon of a high-frequency component in the spectrum of the signal seemingly taking on the identity of a lower frequency in the spectrum of its sampled version.

2.4 Aliasing

Figure 2.4 shows the part of the spectrum that is aliased due to **under sampling**. The aliased spectral components represent ambiguous data that can be retrieved only under special conditions. In general, the ambiguity is not resolved and ambiguous data appear in the frequency band between $(f_s - f_m)$ and f_m.

In Figure 2.5, we show a higher sampling rate f_s' to eliminate the aliasing by separating the spectral replicas.

Figures 2.6 and 2.7 show two ways to eliminate aliasing using antialiasing filters. The analog signal is **prefiltered** so that the new maximum frequency f_m is less than or equal to $f_s/2$. Thus, there are no aliasing components seen in Figure 2.6 because $f_s > 2f_m'$. Eliminating aliasing terms prior to sampling is a good engineering practice. When the signal structure is

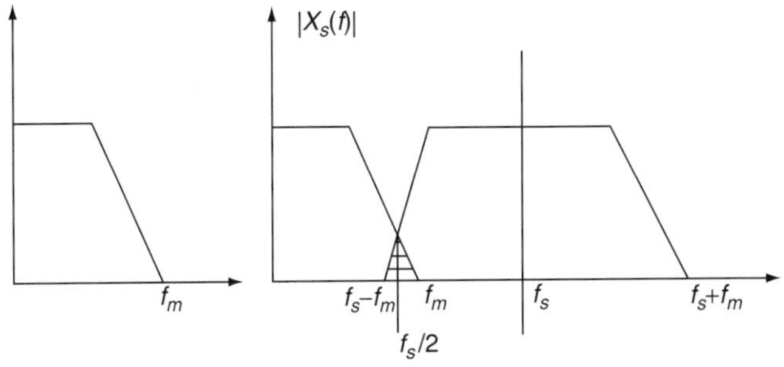

FIGURE 2.4 Sampled Signal Spectrum

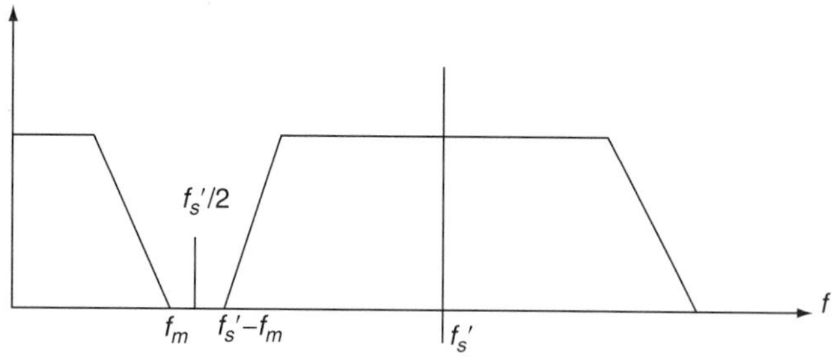

FIGURE 2.5 Higher Sampling Rate to Eliminate Aliasing

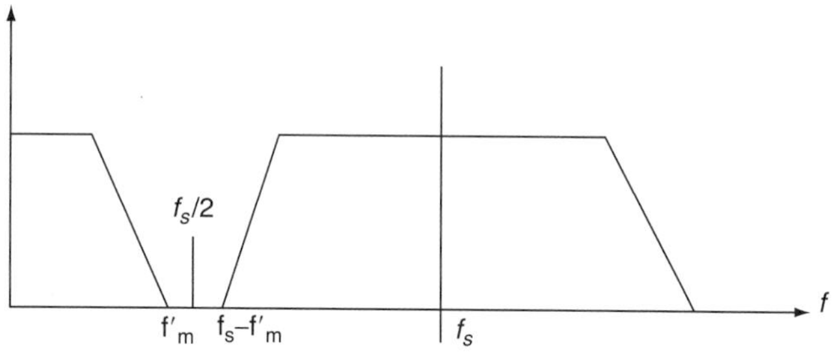

FIGURE 2.6 Prefiltering to Eliminate Aliasing

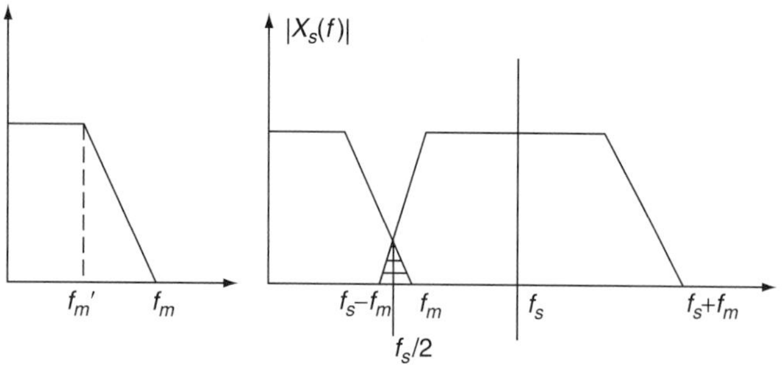

FIGURE 2.7 Postfiltering to Eliminate Aliasing Portion of the Spectrum

well known, the aliased terms can be eliminated after sampling with a linear pass filter (LPF) operating on the sampled data. In this case, the aliased components are removed by **postfiltering** after sampling. The filter cutoff frequency f_m' removes the aliased components; f_m' needs to be less than $(f_s - f_m)$. It should be noted that filtering techniques for eliminating the aliased portion of the spectrum will result in a loss of some signal information. For this reason, the sample rate, cutoff bandwidth, and filter type selected for a particular signal bandwidth are all interrelated.

Realizable filters require a nonzero bandwidth for the transition between the passband and the required out-of-band attenuation. This is called the **transition bandwidth**. To minimize the system sample rate, we desire that the **antialiasing** filter has a small transition bandwidth. Filter complexity and cost rise sharply with narrower transition bandwidth, so a trade-off is needed between the cost of a small transition bandwidth and costs of the higher sampling rate, which are those of more storage and higher transition rates.

In many systems, the answer has been to make the transition bandwidth 10 and 20% of the signal bandwidth. If we account for the 20% transition bandwidth of the antialiasing filter, we have an engineering version of Nyquist sampling rate: $f_s \geq 2.2 f_m$.

Example 3

We want to produce a high-quality digitalization of a 20-kHz bandwidth music signal. The sampling rate of greater than or equal to 22 ksps should be used.

The sampling rate for compact disc digital audio player is 44.1 ksps, and the standard sampling rate for studio-quality audio player is 48 ksps.

2.5 Quantization

In Figure 2.8, each pulse is expressed as a level from a finite number of predetermined levels; each such level can be represented by a symbol from a finite alphabet. The pulses in Figure 2.8 are called **quantized samples.** When the sample values are quantized to a finite set, this format can interface with a digital system. After quantization, the analog waveform can still be recovered but not precisely; improved reconstruction fidelity of the analog waveform can be achieved by increasing the number of quantization levels (requiring increased system bandwidth).

2.6 Pulse Amplitude Modulation

There are two operations involved in the generation of the pulse amplitude modulation (PAM) signal:

(1) Instantaneous sampling of the message signal $m(t)$ every T_s sec, where $f_s = 1/T_s$ is selected according to the sampling theorem
(2) Lengthening the duration of each sample obtained to some constant value T

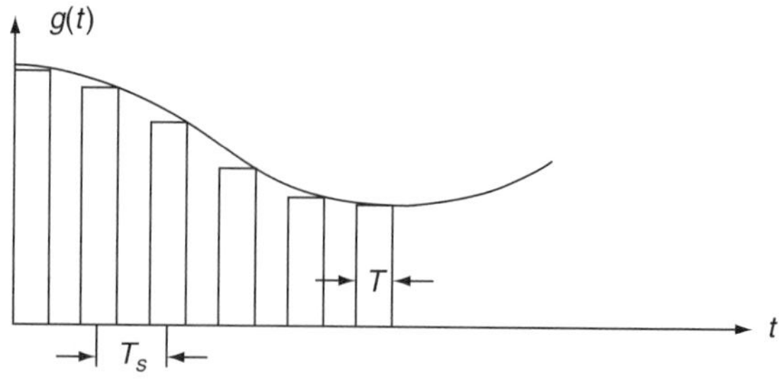

FIGURE 2.8 Flattop Quantization

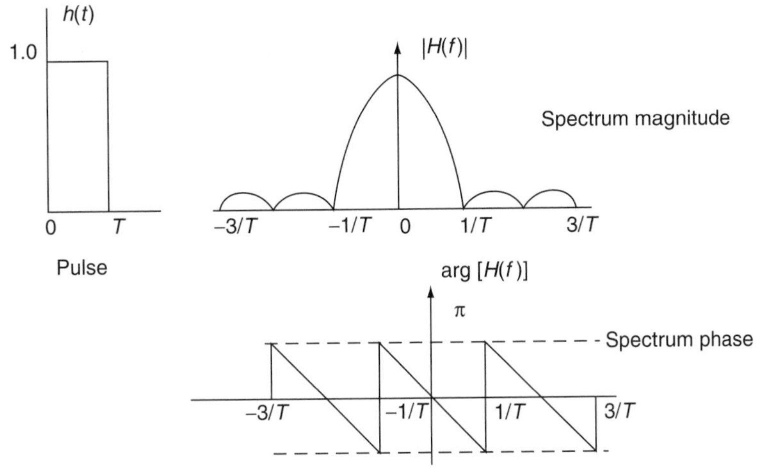

FIGURE 2.9 Rectangular Pulse and Its Spectrum

These two operations are jointly referred to as sample and hold. One important reason for intentionally lengthening the duration of each sample is to avoid the use of an excessive channel bandwidth because bandwidth is inversely proportional to pulse duration.

The Fourier transform of the rectangular pulse $h(t)$ is given as (see Figure 2.9):

$$H(f) = T\text{sinc}(fT)e^{-j2\pi fT}. \tag{2.6}$$

We observe that by using flattop samples to generate a PAM signal, we introduce **amplitude distortion** as well as a **delay** of $T/2$. This effect is similar to the variation in transmission frequency that is caused by the finite size of the scanning aperture in television. The distortion caused by the use of PAM to transmit an analog signal is called the **aperture affect.** This distortion may be corrected by using an **equalizer** (see Figure 2.10). The equalizer has the effect of decreasing the in-band loss of the filter as the frequency increases in such a manner to compensate for the aperture effect. For $T/T_s \leq 0.1$, the amplitude distortion is less than 0.5%, in which case the need of equalization may be omitted altogether.

Example 4

Sampled uniformly and then time-division multiplexed are 24 voice signals. The sampling operation involved flattop samples with 1 μs duration. The multiplexing operation includes provision for synchronization by adding an extra pulse of sufficient amplitude and also 1 μs duration. The highest frequency component of each voice signal is 3.4 kHz.

(1) Assuming a sampling rate of 8 kHz, calculate the spacing between successive pulses of the multiplexed signal.
(2) Repeat your calculations using Nyquist rate sampling.

$$T_s = \frac{10^6}{8000} = 125\ \mu\text{s}.$$

For 25 channels (24 voice channels +1 sync), time allocated for each channel is $125/25 = 5\ \mu$s. Since the pulse duration is 1 μs, the time between pulses is $(5 - 1) = 4\ \mu$s.

The Nyquist rate is 7.48 Hz (2.2 × 3.4).

In addition:

$$T_s = \frac{10^6}{7480} = 134\ \mu\text{s}.$$

$$T_c = \frac{134}{25} = 5.36\ \mu\text{s}.$$

The time between pulses is 4.36 μs.

2.7 Sources of Corruption

The sources of corruption include sampling and quantization effects as well as channel effects, as described in the following bulleted list.

- **Quantization noise:** The distortion inherent in quantization is a roundoff or truncation error. The process of

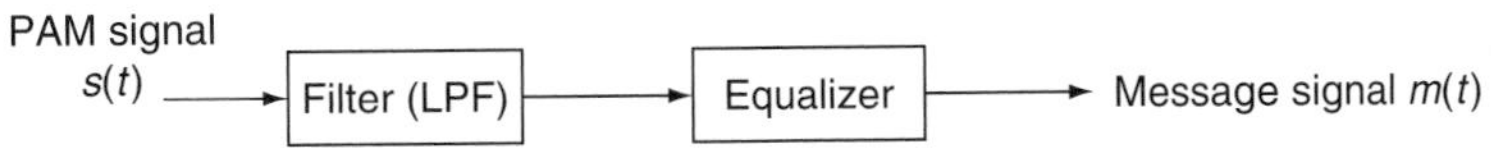

FIGURE 2.10 An Equalizer Application

encoding the PAM waveform into a quantized waveform involves discarding some of the original analog information. This distortion is called quantization noise; the amount of such noise is inversely proportional to the number of levels used in the quantization process.

- **Quantizer saturation:** The quantizer allocates L levels to the task of approximating the continuous range of inputs with a finite set of outputs (see Figure 2.11). The range of inputs for which the difference between the input and output is small is called the **operating range** of the converter. If the input exceeds this range, the difference between the input and output becomes large, and we say that the converter is operating in **saturation**. Saturation errors are more objectionable than quantizing noise. Generally, saturation is avoided by use of automatic gain control (AGC), which effectively extends the operating range of the converter.
- **Timing jitter:** If there is a slight jitter in the position of the sample, the sampling is no longer uniform. The effect of the jitter is equivalent to frequency modulation (FM) of the baseband signal. If the jitter is random, a low-level wideband spectral contribution is induced whose properties are very close to those of the quantizing noise. Timing jitter can be controlled with very good power supply isolation and stable clock reference.
- **Channel noise:** Thermal noise, interference from other users, and interference from circuit switching transients can cause errors in detecting the pulses carrying the digitized samples. Channel-induced errors can degrade the reconstructed signal quality quite quickly. The rapid degradation of the output signal quality with channel-induced errors is called a **threshold effect.**
- **Intersymbol interference:** The channel is always band-limited. A band-limited channel spreads a pulse waveform passing through it. When the channel bandwidth is much greater than pulse bandwidth, the spreading of the pulse will be slight. When the channel bandwidth is close to the signal bandwidth, the spreading will exceed a symbol duration and cause signal pulses to overlap. This overlapping is called **inter-symbol interference** ISI), ISI causes system degradation (higher error rates); it is a particularly insidious form of interference because raising the signal power to overcome interference will not improve the error performance:

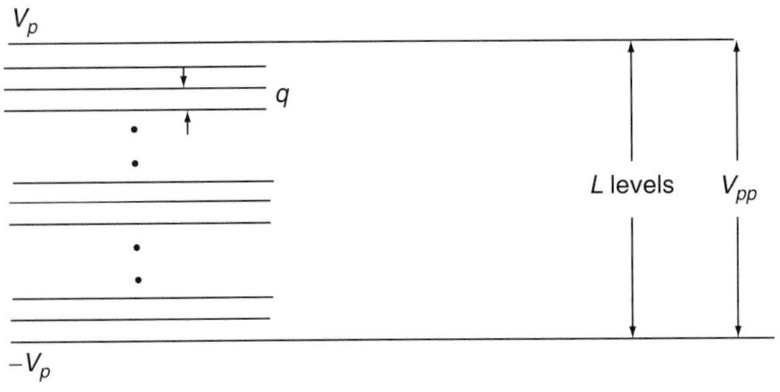

FIGURE 2.11 Uniform Quantization

$$\sigma^2 = \int_{-q/2}^{q/2} e^2 p(e)de = \int_{-q/2}^{q/2} e^2 \frac{1}{q} de = \frac{q^2}{12}$$

$$= \text{average quantization noise power.}$$

$$V_p^2 = \left(\frac{V_{pp}}{2}\right)^2 = \left(\frac{Lq}{2}\right)^2 = \frac{L^2 q^2}{4}. \tag{2.7}$$

$$\left(\frac{S}{N}\right)_q = \frac{(L^2 q^2)/4}{q^2/12} = 3L^2. \tag{2.8}$$

In the limit as $L \rightarrow \infty$, the signal approaches the PAM format (with no quantization error) and signal-to-quantization noise ratio is infinite. In other words, with an infinite number of quantization levels, there is zero quantization error.

Typically $L = 2^R$, $R = Log_2 L$, and $q = \frac{2V_p}{L} = (2V_p)/2^R$.

$$\therefore \sigma^2 = \left(\frac{2V_p}{2^R}\right)^2 /12 = \frac{1}{3} V_p^2 2^{-2R}.$$

Let P denote the average power of the message signal $m(t)$, and then:

$$(SNR)_o = \frac{P}{\sigma^2} = \left(\frac{3P}{V_p^2}\right) 2^{2R}. \tag{2.9}$$

The output SNR of the quantizer increases exponentially with increasing number of bits per sample, R. An increase in R requires a proportionate increase in the channel bandwidth.

Example 5

We consider a full-load sinusodial modulating signal of amplitude A that uses all representation levels provided. The average signal power is (assuming a load of 1 Ω):

The equations are written and solved as follows:

$$P = \frac{A^2}{2}.$$

$$\sigma^2 = \frac{1}{3} A^2 2^{-2R}.$$

$$(SNR)_o = \frac{\frac{A^2}{2}}{(1/3A^2 2^{-2R})} = \frac{3}{2}(2^{2R}) = 1.8 + 6R \text{ dB}.$$

L	R [bits]	SNR [decibels]
32	5	31.8
64	6	37.8
128	7	43.8
256	8	49.8

2.8 Voice Communication

For most **voice communication**, very low speech volumes predominate; about 50% of the time, the voltage characterizing detected speech energy is less than 1/4 of the root-mean-square (rms) value. Large amplitude values are relatively rare; only 15% of the time does the voltage exceed the rms value. The quantization noise depends on the step size. When the steps are uniform in size, the quantization is called the uniform **quantization**. Such a system would be wasteful for speech signals; many of the quantizing steps would rarely be used. In a system that uses equally spaced quantization levels, the quantization noise is same for all signal magnitudes. Thus, with uniform quantization, the signal-to-noise ratio (SNR) is worse for low-level signals than for high-level signals. Non-**uniform quantization** can provide fine quantization of the weak signals and coarse quantization of the strong signals. Thus, in the case of nonuniform quantization, quantization noise can be made proportional to signal size. Improving the overall SNR by reducing the noise for predominant weak signals, at the expense of an increase in noise, can be done for rarely occurring signals. The nonuniform quantization can be used to make the SNR a constant for all signals within the input range. For voice, the signal dynamic range is 40 dB.

Nonuniform quantization is achieved by first distorting the original signal with logarithmic compression characteristics and then using a uniform quantizer. For small magnitude signals, the compression characteristics have a much steeper slope than the slope for large magnitude signals. Thus, a given signal change at small magnitudes will carry the uniform quantizer through more steps than the same change at large magnitudes. The compression characteristic effectively changes the distribution of the input signal magnitude so there is no preponderance of low-magnitude signals at the output of the compressor. After compression, the distorted signal is used as an input to a uniform quantizer. At the receiver, an inverse compression characteristic, called expansion, is used so that the overall transmission is not distorted. The whole process (compression and expansion) is called **companding**.

- The **μ**-Law, used in North America, is as follows:

$$\frac{y}{y_{\max}} = \frac{\ln[1 + |x|/x_{\max}]}{\ln[1+\mu]} \operatorname{sgn} x. \tag{2.10}$$

$$\operatorname{sgn} x = 1,\ x \geq 0.$$

$$\operatorname{sgn} x = -1,\ x < 0.$$

In equation 2.10, μ is constant, x and y are the input and output voltages, $\mu = 0$ represents uniform quantization, and $\mu = 255$ is the standard value used in North America.

- **A-Law**, used in Europe, is as follows:

$$\frac{y}{y_{\max}} = \frac{A(|x|/x_{\max})}{1+\ln A}\operatorname{sgn} x, \quad 0 < \frac{|x|}{x_{\max}} \leq \frac{1}{A}. \tag{2.11a}$$

$$= \frac{1+\ln(|x|/x_{\max})}{1+\ln A}\operatorname{sgn} x, \quad \frac{1}{A} < \frac{|x|}{x_{\max}} < 1. \tag{2.11b}$$

The A is the positive constant, and $A = 87.6$ is the standard value used in Europe.

Example 6

The information in an analog waveform with maximum frequency $f_m = 3$ kHz is transmitted over an M-ary PCM system, where the number of pulse levels is $M = 32$. The quantization distortion is specified not to exceed ± 1% of the peak-to-peak analog signal.

(1) What is minimum number of bits/sample or bits/PCM word that should be used?
(2) What is minimum sampling rate, and what is the resulting transmission rate?
(3) What is the PCM pulse or symbol transmission rate?

Solutions:

$$|e| \leq pV_{pp},$$

where p is fraction of the peak-to-peak analog voltage.

- $|e_{\max}| = \dfrac{V_{pp}}{2L}$,

$$\therefore \left(\frac{V_{pp}}{2L} \leq pV_{pp}\right).$$

- $2^R = L \geq \dfrac{1}{2p}$.

$$2^R \geq \frac{1}{2 \times 0.01} = 50,$$

$$\therefore (R \geq 5.64) \quad \text{use } R = 6.$$

- $f_s = 2f_m = 6000$ samples/sec.

$$f_s = 6 \times 6000 = 36 \text{ kbps}.$$

- $M = 2^b = 32$.

$$b = 5 \text{ bits/symbol}.$$

$$R_s = \frac{36000}{5} = 7200 \text{ symbols/sec}.$$

2.9 Encoding

Codeword timeslots are shown in Figure 2.12 in which the codeword is 4-bit representation of each quantized sample. In the bit duration portion of Figure 2.12, each binary 1 is represented by a pulse, and each binary 0 is represented by the absence of a pulse.

If we increase the pulse width to the maximum possible (equal to bit duration, t), we have the waveform shown in the +V and −V bottom portion of Figure 2.12. Rather than describe this waveform as a sequence of present or absent pulses, we can describe it as a sequence of transitions between two levels. When the waveform occupies the upper voltage level, it represents a binary 1; when it occupies the lower voltage, it represents a binary 0.

We need an **encoding** process to translate the discrete sets of sample value to a more appropriate form of signal. Any plan to represent each of the discrete sets of value as a particular arrangement of discrete events is called a **code**. One of the discrete events in a code is called a **code symbol** or **symbol**. A particular arrangement of symbols used in a code to represent a single value of the discrete set is called a **codeword** or **character**.

Most commonly used pulse code modulation (PCM) waveforms are classified into the following groups:

- Nonreturn to zero (NRZ)
- Return to zero (RZ)
- Phase-encoded
- Multilevel binary

The reason for the large selection relates to the differences in performance that characterize each waveform. In selecting a coding scheme for a particular application, some of the parameters worth examining are:

- The dc component
- Self-clocking
- Error detection
- Bandwidth compression
- Noise immunity
- Biphase level (Manchester code)

2.9.1 Encoding Schemes

The following encoding schemes are often used.

(1) **Nonreturn to Zero-Level (NRZ-L)**
- 1 = high level
- 0 = low level

(2) **Nonreturn to Zero-Mark (NRZ-M)**
- 1 = transition at the beginning of interval
- 0 = no transition

(3) **Nonreturn to Zero-Space (NRZ-S)**
- 1 = no transition
- 0 = transition at the beginning of the interval

(4) **Return to Zero (RZ)**
- 1 = pulse in first half of bit interval
- 0 = no pulse

(5) **Biphase-level (Manchester)**
- 1 = transition from high to low in middle of interval
- 0 = transition from low to high in middle of interval

(6) **Biphase–Mark**
- Always a transition at the beginning of interval
- 1 = transition in middle of interval
- 0 = no transition in middle of interval

(7) **Biphase–space**
- Always a transition at the beginning of interval
- 1 = no transition in middle of interval
- 0 = transition in middle of interval

(8) **Differential Manchester**
- Always a transition in middle of interval
- 1 = no transition at the beginning of interval
- 0 = transition at beginning of interval

(9) **Delay modulation (Miller)**
- 1 = transition in middle of interval
- 0 = no transition if followed by 1, or transition at end of interval if followed by 0

(10) **Bipolar**
- 1 = pulse in first half of interval, alternating polarity from pulse to pulse
- 0 = no pulse

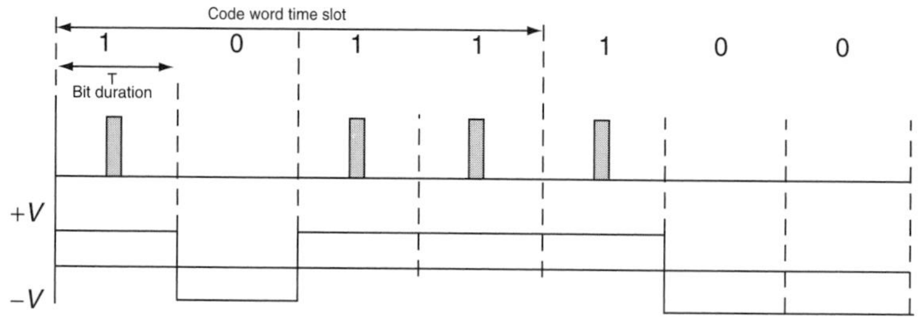

FIGURE 2.12 Bit Sequence and Waveform

3

Transmission of Digital Signals

Vijay K. Garg
Department of Electrical and Computer Engineering, University of Illinois at Chicago, Chicago, Illinois, USA

Yih-Chen Wang
Lucent Technologies, Naperville, Illinois, USA

3.1 Transmission of Digital Data

We now focus on the transmission of digital data over a baseband channel. Digital data have a broad spectrum with a significant low-frequency content. Baseband transmission of digital data requires the use of a low-pass channel with a bandwidth large enough to include the essential frequency content of the data stream. The channel is dispersive in that its frequency response deviates from that of an ideal low-pass filter. The result of data transmission over such a channel is that each received pulse is affected somewhat by adjacent pulses, thereby giving rise to intersymbol interference (ISI). ISI is a major source of bit errors in the reconstructed data stream at the receiver output. Another source of bit errors in a baseband data transmission system is the ubiquitous channel noise. Noise and ISI arise in the system simultaneously.

3.2 Detection of Binary Signals in Gaussian Noise

After the digital symbols are converted into electrical waveforms, they are transmitted through a channel. During a given signaling interval, T, a binary system transmits one of two waveforms, denoted $s_1(t)$ and $s_2(t)$. The signal, $r(t)$, received by the receiver is represented by these two equations:

$$r(t) = s_1(t) + n(t) \text{ for a binary 1} \quad 0 \leq t \leq T. \tag{3.1}$$

$$r(t) = s_2(t) + n(t) \text{ for a binary 0} \quad 0 \leq t \leq T. \tag{3.2}$$

In equations 3.1 and 3.2, $n(t)$ is a zero-mean additive Gaussian white noise, and T is the symbol duration.

We assume that the receiver has knowledge of the starting and ending times of each transmitted pulse; in other words, the receiver has prior knowledge of the pulse shape but not its polarity. Given the noisy signal, the receiver has to make a decision in each signaling interval as to whether the transmitted symbol is 1 or 0.

We refer to Figure 3.1 **step 1** involves reducing the received waveform to a single number $z(t = T)$. This operation can be performed by a linear filter followed by a sampler or optimally by a matched filter. The initial conditions of the filter are set to zero just before the arrival of each new symbol. At the end of a symbol duration T, the output of step 1 yields the sample $z(T)$:

$$z(T) = a_i(T) + n_0(T) \quad i = 1, 2, \tag{3.3}$$

where $a_i(T)$ is the signal component of $z(T)$ and where $n_0(T)$ is the noise component.

Since noise component $n_0(t)$ is a zero-mean Gaussian random variable, $z(T)$ is also a Gaussian random variable with a mean of either a_1 or a_2 depending on whether a binary 1 or 0 was sent. The probability density function (pdf) of the Gaussian random noise, n_0, can be expressed as:

$$p(n_0) = \frac{1}{\sigma_0\sqrt{2\pi}} e^{-1/2(n_0/\sigma_0)^2}, \tag{3.4}$$

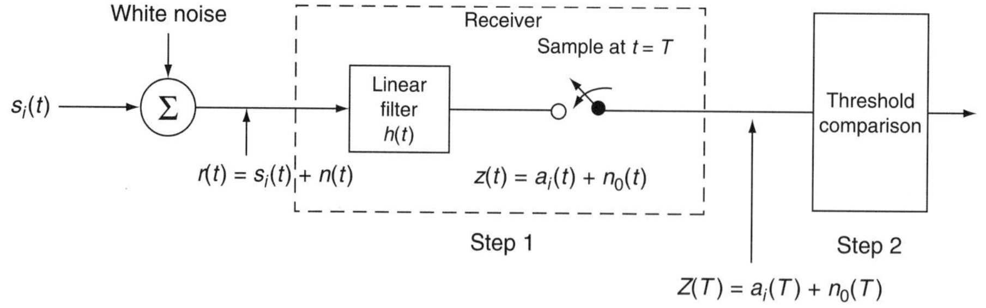

FIGURE 3.1 Detection of Binary Signals in Gaussian Noise

where, σ_0^2 is the noise variance.

The conditional pdfs, p(z s$_1$) and p(z s$_2$), can be given as:

$$p(z|s_1) = \frac{1}{\sigma_0\sqrt{2\pi}} e^{-\frac{1}{2}\left(\frac{z-a_1}{\sigma_0}\right)^2}. \quad (3.5)$$

$$p(z|s_2) = \frac{1}{\sigma_0\sqrt{2\pi}} e^{-\frac{1}{2}\left(\frac{z-a_2}{\sigma_0}\right)^2}. \quad (3.6)$$

These conditional pdfs are shown in Figure 3.2. The rightmost conditional pdf $p(z|s_1)$, shows the probability density of the detector output, $z(T)$, given that $s_1(t)$ was transmitted. Similarly, the leftmost conditional pdf $p(z|s_2)$, shows the probability density of the detector output $z(T)$, given $s_2(t)$ was transmitted.

Step 2 in signal detection process consists of comparing the $z(T)$ to a threshold level λ in block 2 of Figure 3.1 to estimate which signal, $s_1(t)$ or $s_2(t)$, has been transmitted. The filtering operation in block 1 does not depend on the decision criterion in block 2. Thus, the choice of how best to implement block 1 can be independent of the particular decision choice of the threshold setting, λ. After a received waveform, $r(t)$, is transformed to $z(T)$, the actual shape of the waveform is no longer important. The signal energy (not its shape) is the important parameter in the detection process. Thus, the detection analysis for **baseband signals** is the same as that for **bandpass signals.** The final step in block II is to make the detection.

A popular criterion for choosing the threshold level λ for the binary decision is based on minimizing the probability of error. It can be shown that if $p(s_1) = p(s_2)$, and if the likelihoods $p(z|s_i)$ $(i = 1, 2)$ are symmetrical, the optimum value of λ is given as:

$$\lambda_0 = \frac{a_1 + a_2}{2}, \quad (3.7)$$

where a_1 is the signal component of $z(T)$ when $s_1(t)$ is transmitted and where a_2 is the signal component of $z(T)$ when $s_2(t)$ is transmitted.

The threshold level, λ_0, represented by $(a_1 + a_2)/2$, is the optimum threshold to minimize the probability of making an incorrect decision for this important special case. The strategy is known as the **minimum error criterion.**

3.3 Error Probability

For the binary example in Figure 3.2, there are two ways in which errors can occur. An error, e, will occur when $s_1(t)$ is sent, and the channel noise results in the receiver output signal $z(T)$ being less than λ. The probability of such an occurrence is given as:

$$P(e|s_1) = \int_{-\infty}^{\lambda} p(z|s_1)dz. \quad (3.8)$$

Similarly, an error occurs when $s_2(t)$ is sent, and the channel noise results in the receiver output signal $z(T)$ being greater than λ. The probability of such an occurrence is given as:

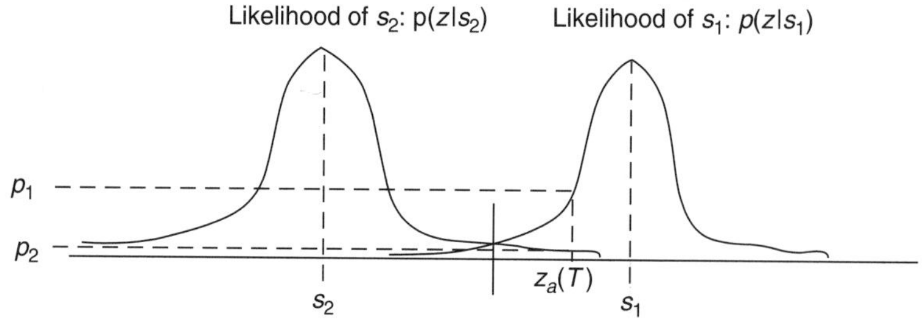

FIGURE 3.2 Conditional Probability Density Functions

$$P(e|s_2) = \int_{\lambda}^{\infty} p(z|s_2)dz. \tag{3.9}$$

Let α and $1-\alpha$ denote the a prior probabilities of transmitting 0 and 1, respectively, and then the average probability of symbol error P_e in the receiver is given by:

$$P_e = \alpha P(e|s_1) + (1-\alpha)P(e|s_2). \tag{3.10}$$

For the special case when 1 and 0 are equiprobable, we have $\alpha = 1/2$:

$$P_e = \frac{1}{2}[P(e|s_1) + P(e|s_2)]. \tag{3.11}$$

Because of the symmetry of pdf:

$$P_e = P(e|s_1) = P(e|s_2). \tag{3.12}$$

The average probability of symbol error with optimum λ_0:

$$P_e = \int_{(a_1+a_2)/2}^{\infty} \frac{1}{\sigma_0\sqrt{2\pi}} e^{-\frac{1}{2}\left(\frac{z-a_2}{\sigma_0}\right)^2} \tag{3.13}$$

Let:

$$u = (z-a_2)/\sigma_0,$$
$$\therefore \sigma_0 du = dz.$$

$$\therefore P_e = \int_{(a_1-a_2)/2\sigma_0}^{\infty} \frac{1}{\sqrt{2\pi}} e^{-u^2/2} du = Q\left(\frac{a_1-a_2}{2\sigma_0}\right). \tag{3.14}$$

In equation 3.14 $(a_1 - a_2)$ is the difference of signal components at the filter output at time $t = T$, and the square of this difference signal is the instantaneous power of difference signal. The $Q(x)$ is the complementary error function, and it is defined as:

$$Q(x) = \frac{1}{\sqrt{2\pi}} \int_{x}^{\infty} e^{-u^2/2} du \approx \frac{1}{x\sqrt{2\pi}} e^{-x^2/2}. \tag{3.15}$$

$$\left(\frac{S}{N}\right) = \frac{(a_1-a_2)^2}{\sigma_0^2} = \frac{E_d}{\left(\frac{N_0}{2}\right)}. \tag{3.16}$$

The E_d is the energy of a difference signal at the filter input:

$$E_d = \int_{0}^{T} [s_1(t) - s_2(t)]^2 dt, \tag{3.17}$$

$$\therefore P_e = Q\left(\sqrt{\frac{E_d}{2N_0}}\right). \tag{3.18}$$

3.4 The Matched Filter

A **matched filter** is a linear filter designed to provide the maximum SNR at its output for a given transmitted symbol waveform. We refer to Figure 3.1 for the ratio of instantaneous signal power to average noise power at time $t = T$; out of the receiver block 1, the following results:

$$\left(\frac{S}{N}\right)_T = \frac{a_i^2}{\sigma_0^2}. \tag{3.19}$$

We want to find the filter transfer function $H_0(f)$ to *maximize* equation 3.19. We express the signal, $a(t)$, at the filter output in terms of the filter transfer function, $H(f)$. The Fourier transform of the input signal will be:

$$a(t) = \int_{-\infty}^{\infty} H(f)S(f)e^{j2\pi ft} dj, \tag{3.20}$$

where $S(f)$ is the Fourier transform of the input signal $s(t)$. With power spectral density of the input noise equal to $N_0/2$, we can express the output noise power as:

$$\sigma_0^2 = \frac{N_0}{2} \int_{-\infty}^{\infty} |H(f)|^2 df. \tag{3.21}$$

Using equations 3.20 and 3.21, we rewrite equation 3.19 as:

$$\left(\frac{S}{N}\right)_T = \frac{\left|\int_{-\infty}^{\infty} H(f)S(f)e^{j2\pi fT} df\right|^2}{N_0/2 \int_{-\infty}^{\infty} |H(f)|^2 df}. \tag{3.22}$$

We must find that value of $H(f) = H_0(f)$ for which the maximum $(S/N)_T$ is achieved, by using **Schwarz's inequality**. One form of the inequality can be stated as:

$$\left|\int_{-\infty}^{\infty} f_1(x)f_2(x)dx\right|^2 \le \int_{-\infty}^{\infty} |f_1(x)|^2 dx \int_{-\infty}^{\infty} |f_2(x)|^2 d.$$

The equality holds if $f_1(x) = k_2(x)^*$, where k is an arbitrary constant and * indicates complex conjugate. If we identify $H(f)$ with $f_1(x)$ and $S(f)e^{j2\pi ft}$ with $f_2(x)$, we can write:

$$\left|\int_{-\infty}^{\infty} H(f)S(f)e^{j2\pi fT} df\right|^2 \le \int_{-\infty}^{\infty} |H(f)|^2 df \int_{-\infty}^{\infty} |S(f)|^2 d(f). \tag{3.23}$$

Using equation 3.23 in equation 3.22, we get:

$$\left(\frac{S}{N}\right)_T \leq \frac{2}{N_0} \int_{-\infty}^{\infty} |S(f)|^2 df. \quad (3.24)$$

$$\max \cdot \left(\frac{S}{N}\right)_T = \frac{2}{N_0} E. \quad (3.25)$$

The energy E of the input signal $s(t)$ is the following:

$$E = \int_{-\infty}^{\infty} |S(f)|^2 df \quad (3.26)$$

Thus, the maximum output $(S/N)_T$ depends on the input **signal energy** and power spectral density of the noise, *not* on the *particular shape* of the waveform that is used.

Therefore, the impulse response of a filter that produces the maximum output signal-to-noise ratio is the mirror image of the message signal $s(t)$, delayed by the symbol time duration, T.

3.5 Error Probability Performance of Binary Signaling

In the following sections, we cover the error probability performance of binary signaling.

3.5.1 Unipolar Signaling

$s_1(t) = A \quad 0 \leq t \leq T$ for binary 1.

$s_2(t) = 0 \quad 0 \leq t \leq T$ for binary 0.

$$E_d = \int_0^T A^2 dt = A^2 T.$$

$$\therefore P_e = Q\left(\sqrt{\frac{A^2 T}{2N_0}}\right) = Q\left(\sqrt{\frac{E_b}{N_0}}\right),$$

where $E_b = A^2 T/2$, the average energy per bit (see Figure 3.3).

3.5.2 Bipolar Signaling

$s_1(t) = A \quad 0 \leq t \leq T$ for binary 1.

$s_2(t) = -A \quad 0 \leq t \leq T$ for binary 0.

$$E_d = \int_0^T 4A^2 dt = 4A^2 T.$$

$$\therefore P_e = Q\left(\sqrt{\frac{4A^2 T}{2N_0}}\right) = Q\left(\sqrt{\frac{2A^2 T}{N_0}}\right) = Q\left(\sqrt{\frac{2E_b}{N_0}}\right),$$

where the average energy per bit is $E_b = A^2 T$ (see Figure 3.3).

3.6 Equalizer

In practical systems, the frequency response of the channel is not known with sufficient accuracy to allow for a receiver design that will compensate for ISI for all time. The filter for handling ISI at the receiver contains various parameters that are adjusted on the basis of measurements of the channel

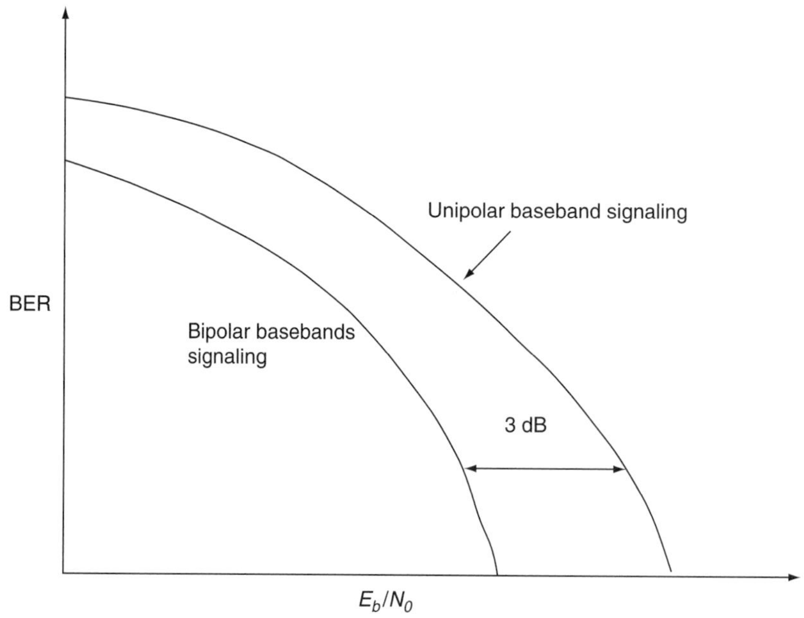

FIGURE 3.3 Bit Error Performance of Unipolar and Bipolar Signaling

characteristics. The process to correct channel-induced distortion is referred to as **equalization**. The adjustable filter is called an **equalizer**. Equalizers may be preset or adaptive. The parameters of a **preset equalizer** are adjusted by making measurements of channel impulse response and solving a set of equations for the parameters using these measurements. An **adaptive equalizer** is automatically adjusted by sending a known signal through the channel and allowing the equalizer to adjust its own parameters in response to this known signal.

A transversal filter—a delay line with T-second taps (where T is the symbol duration)—is a common choice for the equalizer. The outputs of the taps are amplified, summed, and fed to a decision device. The tap coefficients C_n are set to subtract the effects of interference from symbols that are adjacent in time to the desired symbol. The output samples y_k of the equalizer are written as:

$$y_k = \sum_{n=-N}^{N} C_n x_{k-n} \quad k = -2N, \ldots, 2N, \tag{3.27}$$

where:

$$\begin{pmatrix} y_{-2N} \\ \cdot \\ \cdot \\ y_0 \\ \cdot \\ \cdot \\ y_{2N} \end{pmatrix} = \{y_k\};\; C = \begin{pmatrix} C_{-N} \\ \cdot \\ C_0 \\ \cdot \\ C_N \end{pmatrix}, \text{ and}$$

$x = (2N+1) \times (2N+1)$ is the channel response matrix.

We can write the equation in matrix notation as:

$$y = Cx.$$

The impulse response is as follows:

$$h_E(t) = \sum_{n=-N}^{N} C_n \delta(t - nT). \tag{3.28}$$

The frequency response is as follows:

$$H_E(f) = \sum_{n=-N}^{N} C_n e^{-j2\pi nTf}. \tag{3.29}$$

Since there are only $2N + 1$ unknown coefficients, it follows that only a finite number of interfering symbols can be nulled or forced to be zero:

$$y_k = 1 \text{ for } k = 0:$$
$$y_k = 0 \text{ for } k = \pm 1, \pm 2 \ldots \pm N.$$

The channel response matrix is given as:

$$[x] = \begin{bmatrix} x_{-N} & 0 & 0\cdot\cdot & 0 & 0 \\ x_{-N+1} & x_{-N} & 0\cdot\cdot & 0 & 0 \\ \cdot & \cdot & \cdots & \cdot & \cdot \\ \cdot & \cdot & \cdots & \cdot & \cdot \\ x_N & x_{N-1} & \cdots & x_{-N+1} & x_{-N} \\ 0 & 0 & 0\cdot\cdot & x_N & x_{N-1} \\ 0 & 0 & 0\cdot\cdot & 0 & x_N \end{bmatrix}. \tag{3.30}$$

The zero-forcing equations do not account for the effects of noise. In addition, the finite-length transversal filter equalizer can minimize worst-case ISI only if the peak distortion is less than 100% of the eye opening. Another type of equalizer is the minimum mean square error (MMSE) equalizer. In these equalizers, coefficients are selected to minimize the mean square error that consists of the sum of the square of all the ISI terms plus noise power at the equalizer output. The MMSE equalizer maximizes the signal-to-distortion ratio at its output within the constraints of equalizer length and delay.

Example 7

Consider a channel that uses a five-tap equalizer (see Figure 3.4) to correct ISI. The following measurements were made: $x(0) = 1.0$, $x(-1) = 0.2$, $x(-2) = 0.1$, $x(-3) = 0.05$, $x(-4) = -0.02$, $x(-5) = 0.01$, $x(1) = -0.1$, $x(2) = 0.1$, $x(3) = -0.05$, $x(4) = 0.02$, and $x(5) = 0.005$.

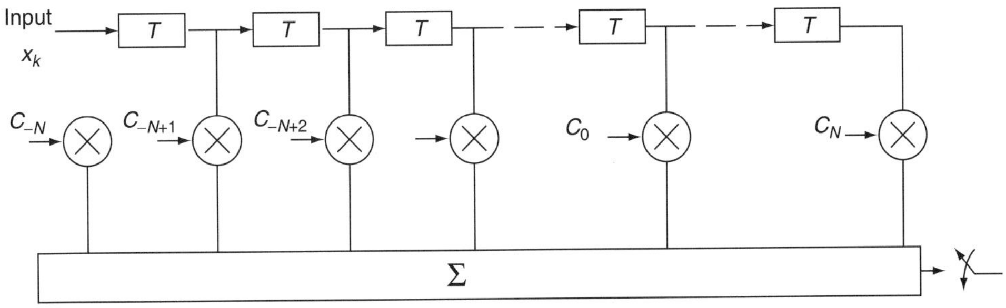

FIGURE 3.4 Tap Equalizer

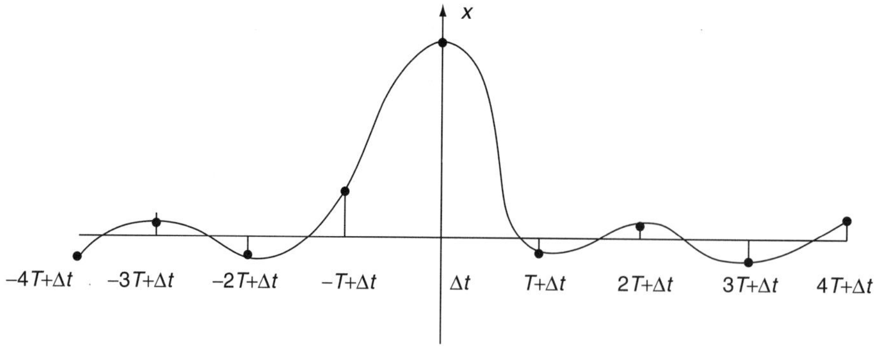

FIGURE 3.5

In addition, note the following results and equations:

$$\begin{bmatrix} x(0) & x(-1) & x(-2) & x(-3) & x(-4) \\ x(1) & x(0) & x(-1) & x(-2) & x(-3) \\ x(2) & x(1) & x(0) & x(-1) & x(-2) \\ x(3) & x(2) & x(1) & x(0) & x(-1) \\ x(4) & x(3) & x(2) & x(1) & x(0) \end{bmatrix} =$$

$$\begin{bmatrix} 1.0 & 0.2 & -0.1 & 0.05 & -0.02 \\ -0.1 & 1.0 & 0.2 & -0.1 & 0.05 \\ 0.1 & -0.1 & 1.0 & 0.2 & -0.1 \\ -0.05 & 0.1 & -0.1 & 1.0 & 0.2 \\ 0.02 & -0.05 & 0.1 & -0.01 & 1.0 \end{bmatrix} = [x].$$

$$[x]^{-1} = \begin{bmatrix} 0.966 & -0.170 & 0.117 & -0.083 & 0.056 \\ 0.118 & 0.945 & -0.158 & 0.112 & -0.083 \\ -0.091 & 0.133 & 0.937 & -0.158 & 0.117 \\ 0.028 & -0.095 & 0.133 & 0.945 & -0.170 \\ -0.002 & 0.028 & -0.091 & 0.118 & 0.966 \end{bmatrix}.$$

$$\{C\} = \begin{bmatrix} 0.117 \\ -0.158 \\ 0.945 \\ 0.133 \\ -0.091 \end{bmatrix} = \begin{bmatrix} C_{-2} \\ C_{-1} \\ C_0 \\ C_1 \\ C_2 \end{bmatrix}.$$

$$\begin{aligned} y_0 &= C_{-2} \cdot x_2 + C_{-1} \cdot x_1 + C_0 \cdot x_0 + C_1 \cdot x_{-1} + C_2 \cdot x_{-2} \\ &= 0.117 \times 0.1 + (-0.158) \times (-0.1) + 0.937 \times 1 \\ &\quad + 0.133 \times 0.2 - 0.091 \times -0.1 = 1.0. \end{aligned}$$

4

Modulation and Demodulation Technologies

Vijay K. Garg
Department of Electrical and Computer Engineering, University of Illinois at Chicago, Chicago, Illinois, USA

Yih-Chen Wang
Lucent Technologies, Naperville, Illinois, USA

4.1 Modulation and Demodulation

The digital signals that are generated in the process of transmitting voice, data, and signaling information are generated at low data rates. These data rates, typically 1 to 50 kbps, are low enough in frequency that their transmission directly from the transmitter to the receiver would require antennas that are thousands of meters long. Furthermore, the signals from one transmitter would interfere with the signals from another transmitter if they all used the same frequency band. Therefore, the baseband signals are modulated onto a radio frequency carrier for transmission from the transmitter to the receiver. The radio environment at 800 to 2000 MHz is hostile. We must therefore choose modulation methods that are robust in this hostile environment. In addition to the modulation methods, we must also choose encoding algorithms that improve the performance of the system.

In this section, we study three modulation methods: minimum shift keying (MSK), **Gaussian minimum shift keying** (GMSK), and **π/4-differential quadrature phase shift keying** ($\pi/4$-DQPSK). GMSK is the modulation used by Global System of Mobile communications (GSM), GSM-1800, and GSM-1900, and Digital European Cordless Telephone (DECT). MSK is introduced as a first step toward GMSK. PWT and PWT-E, the variations of DECT for the licensed and unlicensed 1900-MHz band in North America, use $\pi/4$-DQPSK. Because each of these methods descend from phase shift keying, we will first study PSK and then show its relationship to the other modulation methods.

4.2 Introduction to Modulation

The baseband data rates of a wireless transmitter are usually a few kilobits per second [kbps] to as high as several hundred kilobits. The wavelengths for those signals vary from a thousand meters to several hundred thousand meters. If we attempted to send these signals directly, the antennas would be very long, and multiple transmitters would interfere with each other.

Therefore, when we want to send signals over any distance, baseband signaling is not sufficient. We must therefore modulate the signals onto a radio frequency (RF) carrier. When we transmit the digital bit stream, we convert the bit stream into the analog signal: $a(t)\cos(\omega t + \theta)$. The characteristic of this signal has amplitude $a(t)$, frequency $\omega/2\pi$, and phase θ; thus, we can change any of the three characteristics to formulate the modulation method. The basic form of the three modulation methods used for transmitting digital signals are these:

- Amplitude shift keying (ASK)
- Frequency shift keying (FSK)
- Phase shift keying (PSK)

When ω and θ remain unchanged, we have ASK. When $A(t)$ and θ remain unchanged, we have binary (or M-ary) FSK. When $A(t)$ and ω remain unchanged, we have binary (or M-ary) PSK (these methods are covered in more detail in Section 2.6). Hybrid systems exist where two characteristics are changed with each new symbol transmitted. The most common method is to fix ω and change $A(t)$ and θ. This method is known as quadrature amplitude modulation (QAM). Each of the modulation methods results in a different transmitter and receiver design, different occupied bandwidth, and different error rates. In the remainder of this section, we examine the methods used for GSM and DECT and calculate their error rates. Since all signals have a theoretical bandwidth that is infinite, all modulation methods must be band-limited. The band-limiting introduces detection errors, and the filter bandwidths must be chosen to optimize trade-offs between bandwidth and error rates.

The baseband outputs of the data transmitters are a series of binary data that cannot be sent directly over a radio link. The communications designer must choose radio signals that represent the binary data and permit the receiver to decode the data with minimal errors. For the simplest binary signaling system, we choose two signals denoted by $s_0(t)$ and $s_1(t)$ to represent the binary values of 0 and 1, respectively. Since no channel is perfect the receiver will also have additive Gaussian white noise, $n(t)$. The data receiver (see Figure 4.1) will then process the signal and noise through a filter, $h(t)$, and at the end of the signaling interval, T, it is possible to make a determination of whether the transmitter sent a 0 or a 1.

If the transmitted pulses are allowed to take on any of M transmitted levels with equal probability, then the information rate per transmitted pulse is $\log_2 M$ bits. For a constant information rate, the bandwidth of the transmitted system can be reduced by the same factor. With M-ary transmission, we will show that the error rates are higher, but if we have sufficient signal-to-noise ratio, then the higher errors rates will not matter. Thus, we are using excess signal-to-noise ratio to code the signal and reduce its bandwidth.

When we add additional levels to a baseband system, we are reducing the distance between detection levels in the receiver output. Thus, the error rate of a multilevel baseband system can be determined by calculating the appropriate reduction in the error distance. If the maximum amplitude is V, the error distance d_e between equally spaced levels at the detector is as follows:

$$d_e = \frac{V}{M-1}, \tag{4.1}$$

where M is the number of levels.

Setting the error distance V of a binary system to that defined in equation 4.1 provides the error probability of the multilevel system:

$$P_e = \frac{1}{\log_2 M}\left[\frac{M-1}{M}\right] erfc\left[\frac{V}{(M-1)\sqrt{2}\sigma}\right], \tag{4.2}$$

where the factor [M − 1/M] reflects that the interior signal levels are vulnerable to both positive and negative noise and where the factor $1/\log_2 M$ arises because the multilevel system is assumed to be coded so symbol errors produce single bit errors ($\log_2 M$ is the number of bits per symbol). The probability of multiple bit errors is assumed to be small and can be neglected.

Equation 4.2 relates error probability to the peak signal power V^2. To determine the P_e with respect to average power, the average power of an M-level system is determined by averaging the power associated with the various pulse amplitude levels:

$$[V^2]_{avg} = \frac{2}{M}\left[\left(\frac{V}{M-1}\right)^2 + \left(\frac{3V}{M-1}\right)^2 + \ldots + V^2\right]. \tag{4.3}$$

$$[V^2]_{avg} = \frac{2V^2}{M(M-1)^2}\sum_{j=1}^{M/2}(2j-1)^2. \tag{4.4}$$

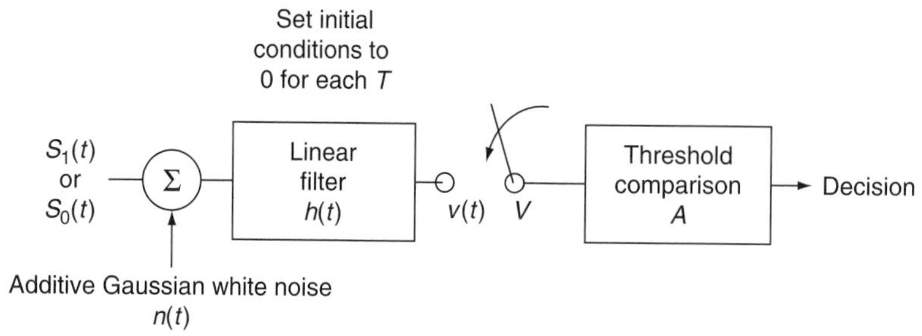

FIGURE 4.1 Receiver Structure to Detect Binary Signals in Gaussian White Noise

In equations 4.3 and 4.4, the levels $V/M - 1[\pm 1, \pm 3, \pm 5, \ldots, \pm (M-1)]$ are assumed to be equally likely.

If T is the signaling interval for a two-level system, the signaling interval T_M for an M-level system providing the same data rate is determined as:

$$T_M = T \log_2 M \quad (4.5)$$

For a raised cosine filter, the noise bandwidth is $B_N = 1/2T_M$. We can write:

$$\sigma^2 = \frac{N_o}{2T_M}. \quad (4.6)$$

$$\sigma = \frac{1}{\sqrt{2}}\left[\frac{N_o}{T_M}\right]^{1/2}. \quad (4.7)$$

Substituting equation 4.7 into equation 4.2, we get:

$$P_e = \left[\frac{1}{\log_2 M}\right]\left[\frac{M-1}{M}\right] erfc\left[\frac{V}{(M-1)(N_o/T_M)^{1/2}}\right]. \quad (4.8)$$

The energy per symbol is $E_s = E_b \log_2 M = V^2 T_M$, where E_b is the energy per bit:

$$\therefore V^2 = \frac{E_b \log_2 M}{T_M}. \quad (4.9)$$

Substituting for V from equation 4.9 into equation 4.8, we get:

$$P_e = \left[\frac{1}{\log_2 M}\right]\left[\frac{M-1}{M}\right] erfc\left[\left(\frac{E_b}{N_o}\right)^{1/2}\frac{(\log_2 M)^{1/2}}{M-1}\right]. \quad (4.10)$$

$$\mathrm{SNR} = \frac{\text{signal power}}{\text{noise power}} = \frac{E_b(\log_2 M)(1/T_M)}{N_o(1/2T_M)}. \quad (4.11)$$

$$\therefore \mathrm{SNR} = 2\log_2 M\left(\frac{E_b}{N_o}\right). \quad (4.12)$$

Another variation of baseband signaling is antipodal baseband signaling (APBBS), where two signals of opposite polarities are sent. If $s_0(t) = -V$, and $s_1(t) = V$ for $0 \le t \le T$, then $s_1(t) - s_0(t) = 2V$.

We then calculate the value of z as:

$$z = \frac{1}{4N_o}\int_0^T (2V)^2 dt = \frac{V^2 T}{N_o} = \frac{E_b}{N_o}, \quad (4.13)$$

where E_b is the energy in either $s_0(t)$ or $s_1(t)$, that is, the bit energy:

$$P_e = \frac{1}{2} erfc\left[\sqrt{\frac{E_b}{N_o}}\right] = Q\left[\sqrt{\frac{2E_b}{N_o}}\right]. \quad (4.14)$$

APBBS is used to modulate some signals, and we will compare its signal-to-noise ratio with other modulation methods.

4.3 Phase Shift Keying

4.3.1 Binary Phase Shift Keying

For binary phase shift keying (BPSK), we transmit two different signals. If the baseband signal is a binary 0, we transmit:

$$A\cos(\omega t + \pi) = -A\cos(\omega t), \quad (4.15)$$

and for binary $1'$, we transmit:

$$A\cos(\omega t). \quad (4.16)$$

BPSK can be considered as a form of amplitude shift keying where each nonreturn to zero (NRZ) data bit of value 0 is mapped into a -1, and each NRZ 1 is mapped into a $+1$. The resulting signal is then passed through a filter to limit its bandwidth and then multiplied by the carrier signal $\cos\omega t$ (see Figure 4.2).

We can also define PSK where there are M phases rather than two phases. In M-ary PSK, every n (where $M = 2^n$) bits of the binary bit stream are coded as a signal that is transmitted as $A\sin(\omega t + \theta_j)$, $j = 1, M$.

The error distance of a PSK system with M phases is $V\sin(\pi/M)$, where V is the signal amplitude at the detector. A detection error occurs if noise of the proper polarity is present at the output of either of the two phase detectors. The probability of error is (Garg and Wilkes, 1996):

$$P_e = \left(\frac{1}{\log_2 M}\right) erfc\left[\sin\left(\frac{\pi}{M}\right)(\log_2 M)^{1/2}\left(\frac{E_b}{N_o}\right)^{1/2}\right]. \quad (4.17)$$

The signal-to-noise ratio is given as:

$$\mathrm{SNR} = \log_2 M\left(\frac{E_b}{N_o}\right)(\text{For } M > 2). \quad (4.18)$$

Each symbol has length T. Therefore, the following is true:

$$E_b = V^2 T/\log_2 M, \quad (4.19)$$

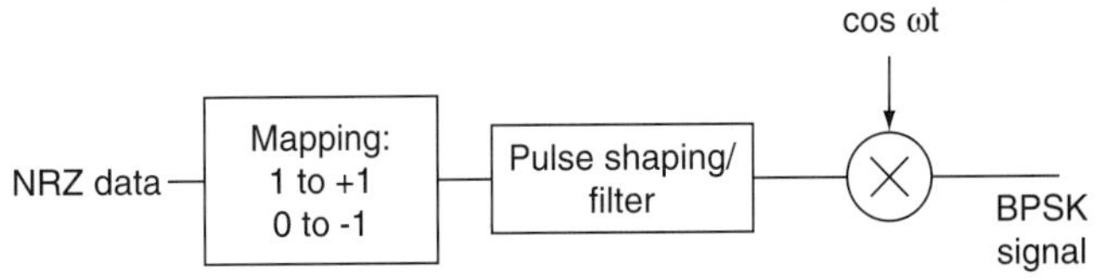

FIGURE 4.2 BPSK Modulator

and the RMS noise σ as:

$$No = \sigma^2 2T, \quad (4.20)$$

for noise in a Nyquist bandwidth.

Also, as shown in the bandwidth efficiency of the M-ary PSK is given as:

$$\frac{R}{B_w} = \frac{\log_2 M}{2},$$

where R is the data rate and B_w is the bandwidth.

We will now examine several variations of Phase Shift Keying.

4.4 Quadrature Phase Shift Keying

If we define four signals, each with a phase shift differing by 90° then we have quadrature phase shift keying (**QPSK**). We have previously calculated the error rates for a general phase shift keying signal with M signal points. For QPSK, $M = 4$, so substituting $M = 4$ in equation 4.17, we get:

$$P_e = \left(\frac{1}{\log_2 4}\right) erfc\left[\sin\left(\frac{\pi}{4}\right)(\log_2 4)^{1/2}\left(\frac{E_b}{N_o}\right)^{1/2}\right].$$

$$P_e = \frac{1}{2} erfc\sqrt{\frac{E_b}{N_o}} = Q\left[\sqrt{\frac{2E_b}{N_o}}\right]. \quad (4.21)$$

The input binary bit stream $\{b_k\}$, $b_k = \pm 1$; k = 0, 1, 2, ..., arrives at the modulator input at a rate $1/T$ bits/sec and is separated into two data streams, a_I(t) and a_Q(t), containing even and odd bits, respectively (Figure 4.3). The modulated QPSK signal $s(t)$ is given as:

$$s(t) = \frac{1}{\sqrt{2}} a_I(t)\cos\left(2\pi ft + \frac{\pi}{4}\right) + \frac{1}{\sqrt{2}} a_Q(t)\sin\left(2\pi ft + \frac{\pi}{4}\right). \quad (4.22)$$

$$s(t) = A\cos\left[2\pi ft + \frac{\pi}{4} + \theta(t)\right]. \quad (4.23)$$

In equations 4.22 and 4.23, the following apply:

$$A = \sqrt{(1/2)(a_I^2 + a_Q^2)} = 1.$$

$$\theta(t) = -a\tan\frac{a_Q(t)}{a_I(t)}.$$

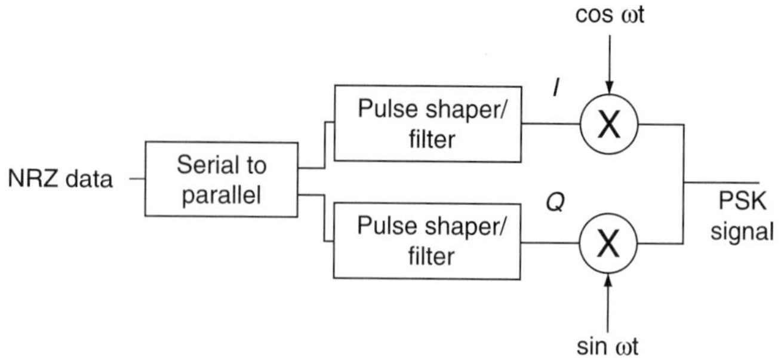

FIGURE 4.3 QPSK Modulator

In Figure 4.3, we show an NRZ data stream of 01110001. We then show the I(0100) and Q (1101) signals that are generated from the NRZ data stream. Notice that the I and Q signals have bit lengths that are twice as long as the NRZ data bits. In the figure, there is no delay between the NRZ data and the I and Q data. In a real implementation, there would be a 1- to 2-bit delay before the I and Q signals were generated. This delay accounts for the time for two bits to be received and decoded into the I and Q signals. Finally, we show the QPSK signal that is generated. To make the figure clearer, we chose a carrier frequency that is four times higher than the data rate. In real systems, the carrier frequency would be many times that data rate.

The values of $\theta(t) = 0, -\pi/2, \pi/2, \pi$ represent the four values of $a_I(t)$ and $a_Q(t)$. On the I/Q plane, QPSK represents four equally spaced points separated by $\pi/2$ (see Figure 4.5). Each of the four possible phases of carriers represents two bits of data. Thus, there are two bits per symbol. Since the symbol rate for QPSK is half of the bit rate, twice the information can be carried in the same amount of channel bandwidth as compared to binary phase shift keying. This is possible because the two signals I and Q are orthogonal to each other and can be transmitted without interfering with each other.

In QPSK, the carrier phase can change only once every $2T$ seconds. If from one $2T$ interval to the next, neither bit stream changes sign, the carrier phase remains the same. If one component $a_I(t)$ or $a_Q(t)$ changes sign, a phase shift of $\pi/2$ occurs. If both components, I and Q change sign, however then a phase shift of π or 180° occurs. When this 180° phase shift is filtered by the transmitter and receiver filters, it generates a change in amplitude of the detected signal and causes additional errors. Notice the 180° shift at the end of bit interval 4 in Figure 4.4.

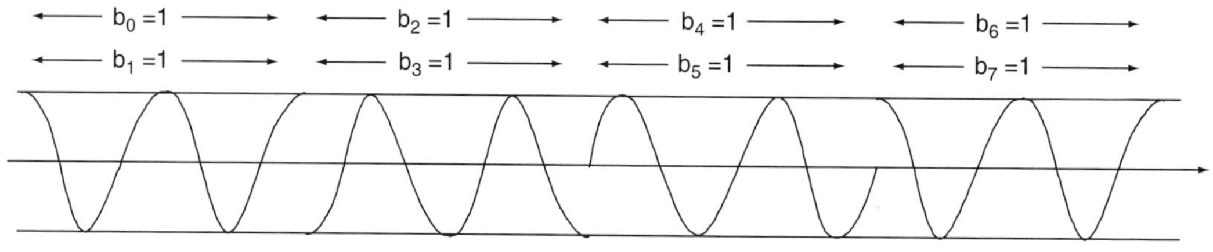

FIGURE 4.4 QPSK Signals

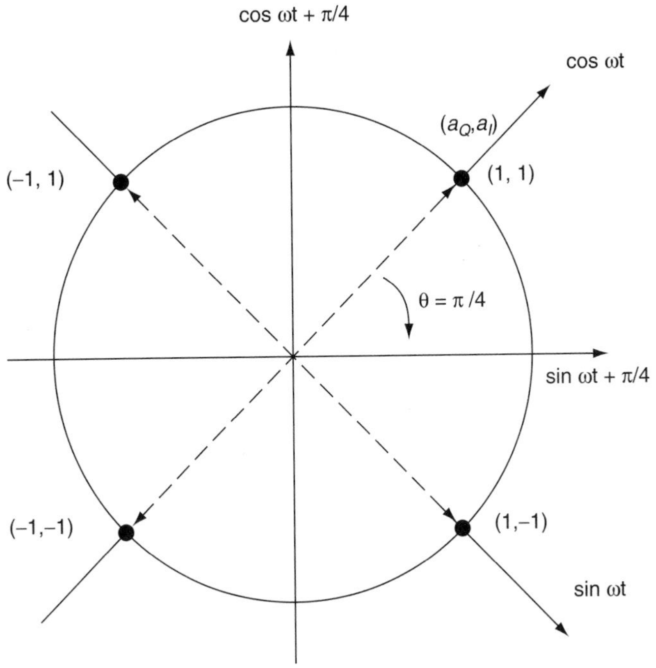

FIGURE 4.5 Signal Constellation for QPSK

If the two bit streams, I and Q, are offset by 1/2 a bit interval, then the amplitude fluctuations are minimized because the phase never changes by 180° (see Figures 4.6 and 4.7). This modulation scheme, **offset quadrature phase shift keying** (OQPSK), is obtained from the conventional quadrature phase shift keying by delaying the odd-bit stream by a half-bit interval with respect to the even bit stream. Thus, the range of phase transition is 0° and 90° and occurs twice as often but with half the intensity of the quadrature phase shift keying system. Comparing Figure 4.4 with Figure 4.7, notice that the Q signal is the same for both QPSK and OQPSK, but the I signal is delayed by 1/2 bit. Thus, the 180° phase change at the end of bit interval 4 of the QPSK signal is replaced by a 90° phase change at the end of bit interval 4 of the OQPSK signal. Also notice that phase changes occur more frequently with OQPSK. Although the phase changes will still cause amplitude fluctuations to occur in the transmitter and receiver, they have a smaller magnitude. The bit error rate and bandwidth efficiency of QPSK and OQPSK is the same as for BPSK.

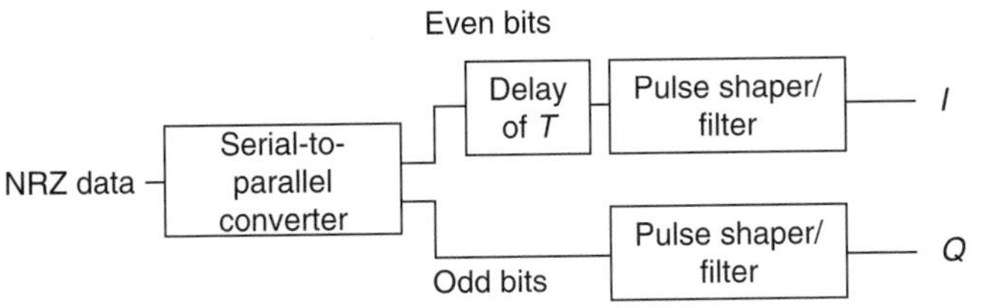

FIGURE 4.6 OQPSK Encoding

In theory, quadrature (or offset quadrature) phase shift keying systems can improve the spectral efficiency of mobile communication. They do, however, require a coherent detector, and in a multipath fading environment, the use of coherent detection is difficult and often results in poor performance over noncoherently based systems. The coherent detection problem can be overcome by using a differential detector, but then OQPSK is subject to intersymbol interference that results in poor system performance. The spectrum of offset QPSK (OQPSK) is the following (Proakis, 1989):

$$P_{QPSK}(f) = T\left[\frac{\sin \pi fT}{\pi fT}\right]^2. \quad (4.24)$$

4.5 The π/4 Differential Phase Shift Keying

We can design a phase shift keying system to be inherently differential and thus solve the detection problems. The **π/4 differential quadrature phase shift keying** (π/4-DQPSK) is a compromise modulation method because the phase is restricted to fluctuate between ±π/4, and ±3π/4 rather than the ±π/2 phase changes for OQPSK. The method has a spectral efficiency of about 20% more than the Gaussian minimum shift keying (see upcoming section) modulation used for DECT and GSM.

The π/4-DQPSK is essentially a π/4-shifted QPSK with differential encoding of symbol phases. The differential encod-

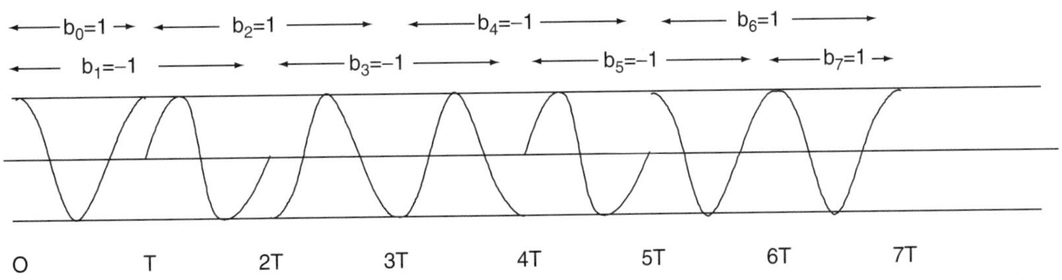

FIGURE 4.7 OQPSK Signals

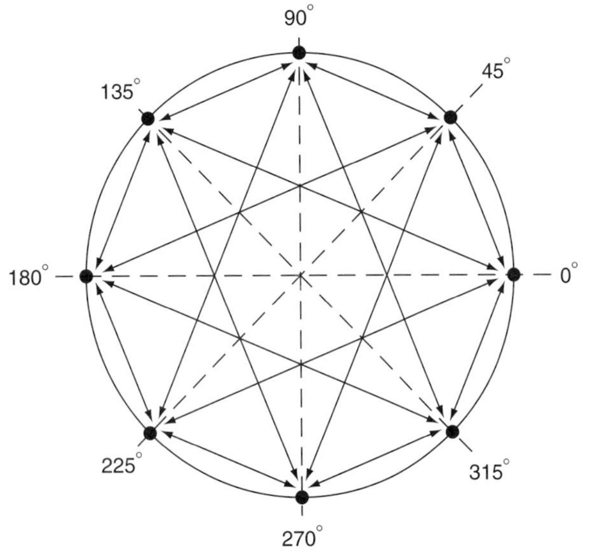

FIGURE 4.8 The $\pi/4$-DQPSK Modulation

ing mitigates loss of data due to phase slips. However, differential encoding results in the loss of a pair of symbols when channel errors occur. This can be translated to approximately a 3-dB loss in E_b/N_o relative to coherent $\pi/4$-QPSK.

A $\pi/4$-shifted QPSK signal constellation in Figure 4.8 consists of symbols corresponding to eight phases. These eight phase points can be considered to be formed by superimposing two QPSK signal constellations, offset by 45 degrees relative to each other. During each symbol period, a phase angle from only one of the two QPSK constellations is transmitted. The two constellations are used alternately to transmit every pair of bits (di-bits). Thus, successive symbols have a relative phase difference that is one of the four phases shown in Table 4.1.

Figure 4.8 shows the $\pi/4$-shifted QPSK signal constellation. When the phase angles of $\pi/4$-shifted QPSK symbols are differentially encoded, the resulting modulation is $\pi/4$-shifted DQPSK. This can be done either by differential encoding of the source bits and mapping them onto absolute phase angles or, alternately, by directly mapping the pairs of input bits onto relative phase $(\pm\pi/4 \pm 3\pi/4)$ as shown in Figure 4.8. The binary data stream entering the modulator $b_M(t)$ is converted by a serial-to-parallel converter into two binary streams $b_o(t)$ and $b_e(t)$ before the bits are differentially encoded (see Figure 4.9):

TABLE 4.1 Phase Transitions of $\pi/4$-DQPSK

Symbol	$\pi/4$-DQPSK phase transition
00	45°
01	135°
10	−45°
11	−135°

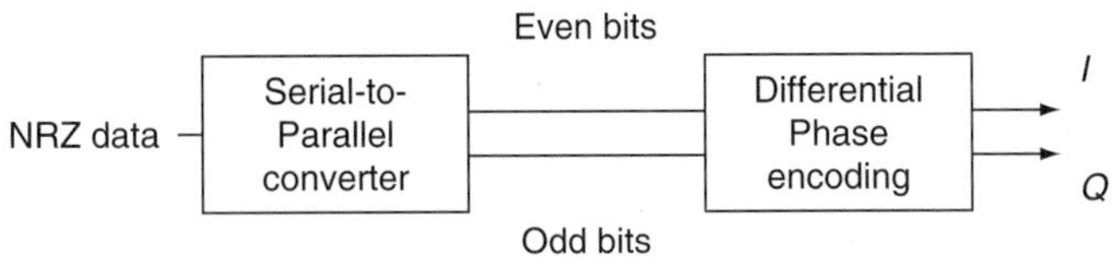

FIGURE 4.9 Differential Encoding of $\pi/4$-DQPSK

$$I_k = I_{k-1}\cos\Delta\phi_k - Q_{k-1}\sin\Delta\phi_k. \qquad (4.25)$$

$$Q_k = I_{k-1}\sin\Delta\phi_k + Q_{k-1}\cos\Delta\phi_k. \qquad (4.26)$$

The I_k and Q_k are the in-phase and quadrature components of the $\pi/4$-shifted DQPSK signal corresponding to the kth symbol. The amplitudes of I_k and Q_k are ± 1, 0, $\pm 1/\sqrt{2}$. Because the absolute phase of $(k-1)$th symbol is ϕ_{k-1}, the in-phase and quadrature components can be expressed as:

$$I_k = \cos\phi_{k-1}\cos\Delta\phi_k - \sin\phi_{k-1}\sin\Delta\phi_k = \cos(\phi_{k-1} + \Delta\phi_k). \qquad (4.27)$$

$$Q_k = \cos\phi_{k-1}\sin\Delta\phi_k + \sin\phi_{k-1}\cos\Delta\phi_k = \sin(\phi_{k-1} + \Delta\phi_k). \qquad (4.28)$$

These component signals (I_k, Q_k) are then passed through baseband filters having a raised cosine frequency response as:

$$|H(f)| = \begin{cases} 1 & 0 \le |f| \le \dfrac{1-\alpha}{2T_s}. \\ \sqrt{\frac{1}{2}\left\{1 - \sin\left[\dfrac{\pi T_s}{\alpha}\left(|f| - \dfrac{1}{2T_s}\right)\right]\right\}} & \dfrac{1-\alpha}{2T_s} \le |f| \le \dfrac{1+\alpha}{2T_s}. \\ 0 & |f| \ge \dfrac{1+\alpha}{2T_s}. \end{cases} \qquad (4.29)$$

In equation 4.29 α is the roll-off factor, and T_s is the symbol duration.

If $g(t)$ is the response to pulses I_k and Q_k at the filter input, then the resultant transmitted signal is given as:

$$s(t) = \sum_k g(t - kT_s)\cos\phi_k\cos\omega t - \sum_k g(t - kT_s)\sin\phi_k\sin\omega t, \qquad (4.30)$$

$$s(t) = \sum_k g(t - kT_s)\cos(\omega t + \phi_k), \qquad (4.31)$$

where $2\pi\omega$ is the carrier frequency of transmission.

The component ϕ_k results from differential encoding (i.e., $\phi_k = \phi_{k-1} + \Delta\phi_k$).

Depending on the detection method (coherent detection or differential detection), the error performance of $\pi/4$-DQPSK can either be the same or 3 dB worse than QPSK.

4.6 Minimum Shift Keying

We previously showed that OQPSK is derived from QPSK by delaying the Q data stream by 1 bit or Tsec with respect to the corresponding I data stream. This delay has no effect on the error rate or bandwidth.

Minimum shift keying (MSK) is derived from OQPSK by replacing the rectangular pulse in amplitude with a half-cycle sinusoidal pulse. Notice that the in-phase and quadrature signals are delayed by interval T from each other.

The MSK signal is defined as:

$$s(t) = a_I(t)\left|\cos\left(\frac{\pi(t-2nT)}{2T}\right)\right|\cos 2\pi ft + a_Q(t)\left|\sin\left(\frac{\pi(t-2nT)}{2T}\right)\right|\sin 2\pi ft. \quad (4.32)$$

$$s(t) = \cos\left[2\pi ft + b_k(t)\frac{\pi(t-2nT)}{2T} + \phi_k\right]. \quad (4.33)$$

In equations 4.32 and 4.33, the following apply:

- $n = 0, 1, 2, 3, \ldots$
- $b_k = +1$ for $a_I \cdot a_Q = -1$.
- $b_k = -1$ for $a_I \cdot a_Q = 1$.
- $\phi_k = 0$ for $a_I = 1$.
- $\phi_k = \pi$ for $a_I = -1$.

Note that since the I and Q signals are delayed by one bit interval, the cosine and sine pulse shape in equation 4.32 are actually both in the shape of a sin pulse.

Minimal shift keying has the following properties:

(1) For a modulation bit rate of R, the high tone is $f_H = f + 0.25R$ when $b_k = 1$, and the low tone is $f_L = f - 0.25\,R$ when $b_k = -1$.
(2) The difference between the high tone and the low tone is $\Delta f = f_H - f_L = 0.5R = 1/(2T)$, where T is the bit interval of the NRZ signal.
(3) The signal has a constant envelope.

The error probability for an ideal minimal shift keying system is:

$$P_e = \frac{1}{2}\,erfc\sqrt{\frac{E_b}{N_o}} = Q\left[\sqrt{\frac{2E_b}{N_o}}\right], \quad (4.34)$$

which is the same as for QPSK/OQPSK.

The minimal shift keying modulation makes the phase change linear and limited to $\pm\pi/2$ over a bit interval T. This enables MSK to provide a significant improvement over QPSK. Because of the effect of the linear phase change, the power spectral density has low side lobes that help to control adjacent channel interference. However, the main lobe becomes wider than the quadrature shift keying (see Figure 4.11). Thus, it becomes difficult to satisfy the CCIR recommended value of −60 dB-side lobe power levels. The power spectral density for MSK can be shown (Proakis, 1989) to be as:

$$P_{MSK}(f) = \frac{16T}{\pi^2}\left[\frac{\cos 2\pi fT}{1-16f^2T^2}\right]^2. \quad (4.35)$$

Figure 4.11 shows the spectral density for both MSK and QPSK. Notice that while the first null in the side lobes occurs at a data rate of R for MSK and $R/2$ for QPSK, the overall side lobes are lower for MSK. Thus, MSK is more spectrally efficient than QPSK.

The detector for MSK (Figure 4.10) is slightly different than for PSK. We must generate the matched filter equivalent to the two transmitted in-phase and quadrature signals. These two reference signals are as follows:

$$i(t) = \cos\left(\frac{\pi t}{2T}\right)\cos\omega t. \quad (4.36)$$

$$q(t) = \sin\left(\frac{\pi t}{2T}\right)\sin\omega t. \quad (4.37)$$

We multiply the received signal by $i(t)$ and $q(t)$ and perform an integration with detection at the end of the bit interval; we then dump the integrator output. This is the standard integrate-and-dump matched filter with the reference signal $i(t)$ and $q(t)$ match to the received waveform. At the end of the bit interval, we make a decision on the state of the bit (+ 1 or − 1) and output the decision as our detected bit. We do this for both the I channel and the Q channel with I and Q out of phase by Tsec to account for the differential nature of MSK.

4.7 Gaussian Minimum Shift Keying

In minimal shift keying, we replace the rectangular data pulse with a sinusoidal pulse. Obviously, other pulse shapes are possible. A Gaussian-shaped impulse response filter generates a signal with low side lobes and a narrower main lobe than the rectangular pulse. Since the filter theoretically has output before input, it can only be approximated by a delayed and shaped impulse response that has a Gaussian-like shape. This modulation is called **Gaussian minimum shift keying** (GMSK).

The relationship between the premodulation filter bandwidth B and the bit period T defines the bandwidth of the system. If $B > 1/T$, then the waveform is essentially minimal shift keying. When $B < 1/T$, the intersymbol interference occurs because the signal cannot react to its next position in the symbol time. However, intersymbol interference can be traded for bandwidth reduction if the system has sufficient

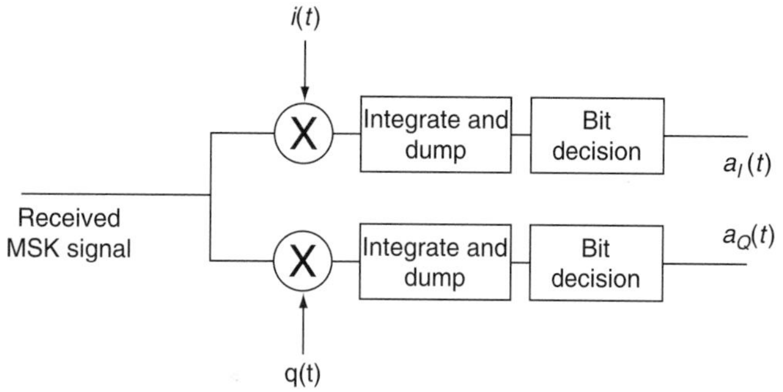

FIGURE 4.10 Optimum MSK Detector

signal-to-noise ratio. GSM designers use a BT of 0.3 with a channel data rate of 270.8 kbps. DECT designers adopt $BT = 0.5$ with a data rate of 1.152 Mbs. A choice of $BT = 0.3$ in GSM is a compromise between bit error rate and out-of-band interference since the narrow filter increases the intersymbol interference and reduces the signal power.

When GMSK was first proposed (Murota and Hirade, 1981), the modulator was based on using frequency modulation (FM), illustrated by Figure 4.12. Since newer integrated circuits are available that enable an I and Q modulator to be easily constructed, a more modern method to generate the GSMK signal is shown in Figure 4.13.

Filtering of the NRZ data by a Gaussian low-pass filter generates a signal that is no longer constrained to one bit interval. Intersymbol interference is generated by the modulator.

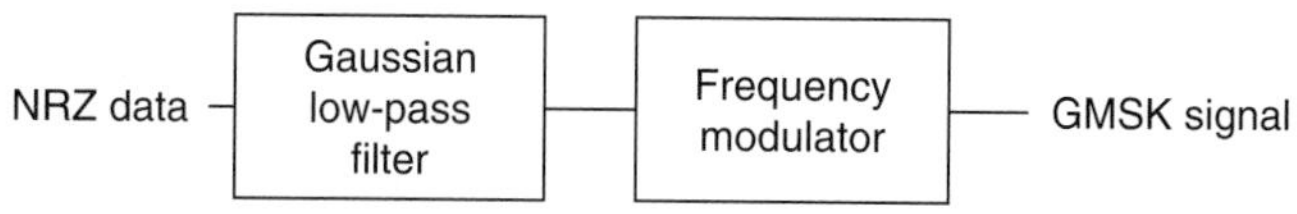

FIGURE 4.12 GMSK Modulator Using Frequency Modulator

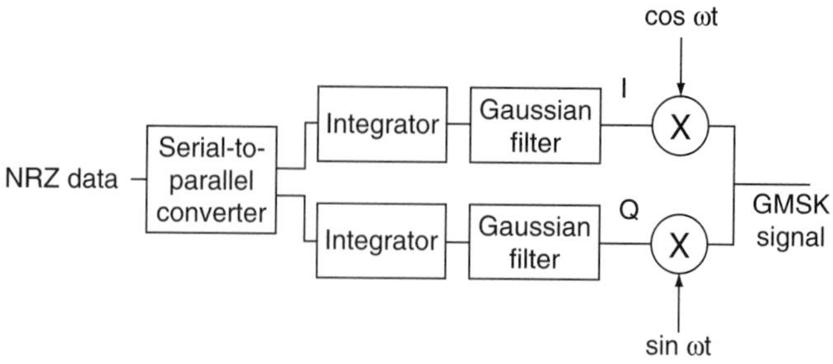

FIGURE 4.13 GMSK Modulator Using Phase Modulator

The bit error rate performance of GMSK can be expressed (Murota and Hirade, 1981) as:

$$P_e = \frac{1}{2} erfc\sqrt{\frac{\alpha E_b}{2N_o}} = Q\left[\sqrt{\frac{\alpha E_b}{N_o}}\right], \tag{4.38}$$

where α depends on the BT product. For GSM where $BT = 0.3$, $\alpha \approx 0.9$; for DECT where $BT = 0.5$, $\alpha \approx 0.97$. For $BT = \infty$, $\alpha = 2$, which is the case of MSK.

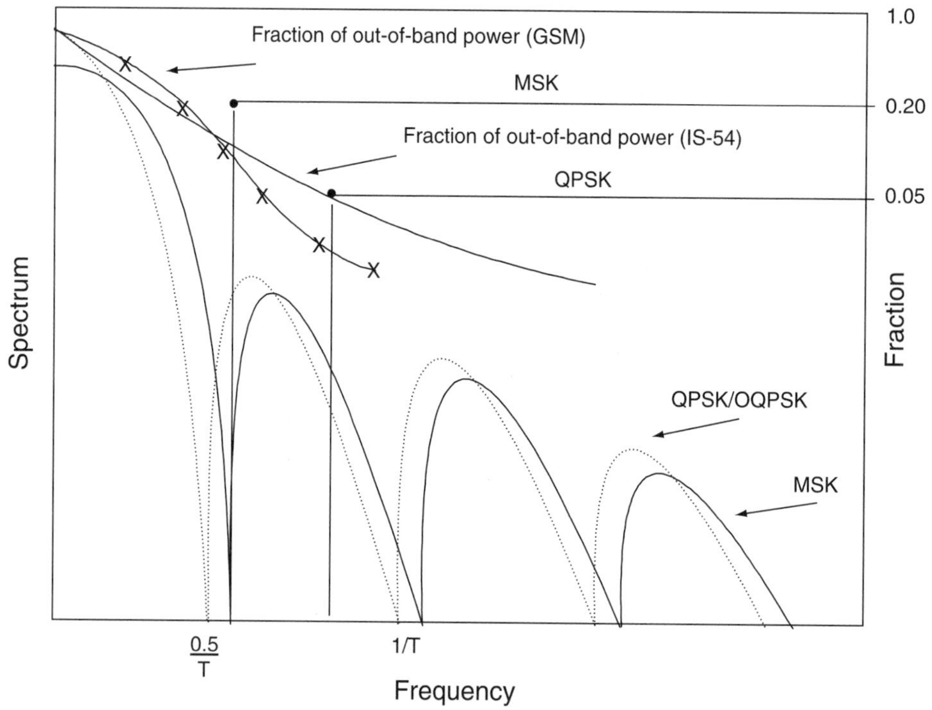

FIGURE 4.11 Spectral Density of QPSK and MSK

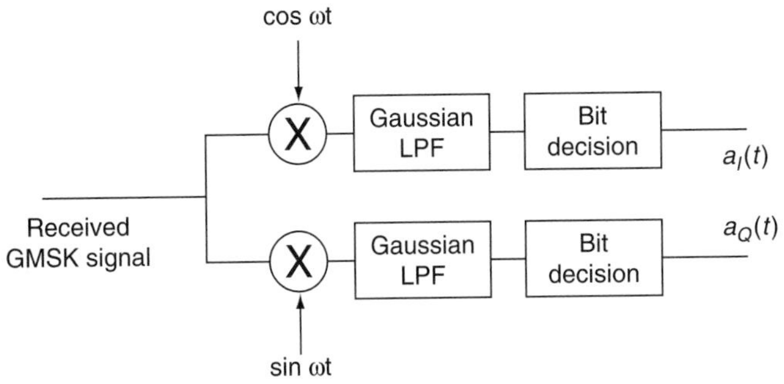

FIGURE 4.14 GMSK Demodulator

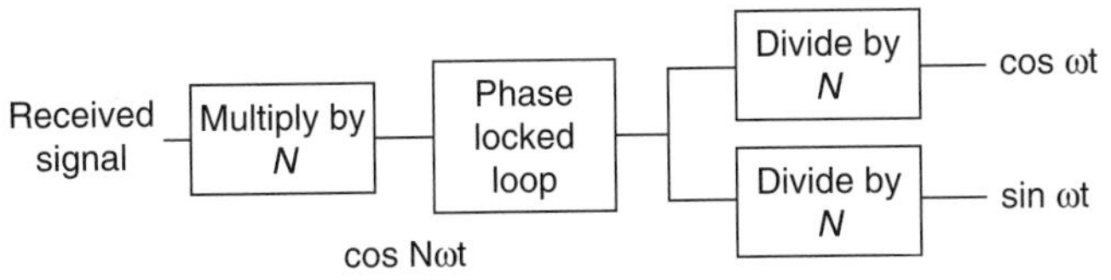

FIGURE 4.15 Carrier Recovery for PSK

Demodulation of GMSK, shown in Figure 4.14, requires multiplication by the in-phase and quadrature carrier signals followed by a low-pass filter with Gaussian shape. At the end of the bit interval, we make a decision on the state of the bit (+1 or −1) and output the decision as our detected bit. As with MSK, we do this for both the *I* channel and the *Q* channel with *I* and *Q* out of phase by *T* seconds to account for the differential nature of GMSK. We ignore the intersymbol interference caused by the longer than 1-bit interval nature of the Gaussian transmitted pulse.

4.8 Synchronization

The demodulation of a signal requires that the receiver be synchronized with the transmitted signal, as received at the input of the receiver. The synchronization must be for:

- **Carrier synchronization:** The receiver is exactly on the same frequency that was transmitted and adjusted for the effects of Doppler shifts.
- **Bit synchronization:** The receiver is aligned with the beginning and end of each bit interval.
- **Word synchronization:** The receiver is aligned with the beginning and end of each word in the transmitted signal.

If the synchronization in the receiver is not precise for any of the above operations, then the bit error rate of the receiver will not be the same as described by the equations in the previous sections. The design of a receiver is an area that standards have traditionally not specified. It is usually an art that enables one company to offer better performance in its equipment compared to a competitor's equipment. The methods of achieving synchronization discussed in this section are the traditional methods. A particular receiver may or may not use any of these methods. Many companies use proprietary methods.

For PSK the carrier signal changes phase every bit interval. If we multiply the received signal by a factor *N*, an integer, we can convert all of the phase changes in the multiplied signal to multiples of 360°. The new signal then has no phase changes, and we can recover it using a narrow band phase locked loop (PLL). After the PLL recovers, the multiplied carrier signal it is divided by *N* to recover the carrier at the proper frequency. By the suitable choice of digital dividing circuits, it is possible to get a precise 90° difference in the output of two dividers and thus generate both the cos ωt and the sin ωt signals needed by the receiver. There are also some down-converter integrated circuits that contain a precise phase shift network. The carrier recovery is typically performed at some lower intermediate frequency rather than directly at the received frequency. For BPSK, we would need an *N* or 2, but an *N* of 4 would be used to enable the sine and cosine term to be generated. For QPSK and its derivatives, an *N* of 4 is necessary, and for π/4-DQPSK, an *N* of 8 would be needed.

After we recover the carrier, we must reestablish the carrier phase to determine the values of the received bits. Somewhere in the transmitted signal must be a known bit pattern that we can use to determine the carrier phase. The bit pattern can be alternating zeros and ones that we use to determine bit timing, or it could be some other known pattern.

The advantage of differential keying (e.g., π/4-DQPSK) is that the knowledge of the absolute carrier phase is not important. Only the change in carrier phase from one symbol to the next is important.

MSK is a form of frequency modulation; therefore, a different method of carrier recovery is needed. In Figure 4.16, the MSK signal has frequency *f* and deviation $\Delta f = 1/2T$. We first multiply the signal by 2, thus doubling the deviation and generating strong frequency components (Pasupathy, 1979) at $2f + 2\Delta f$ and $2f - 2\Delta f$. We use two PLLs to recover these two signals:

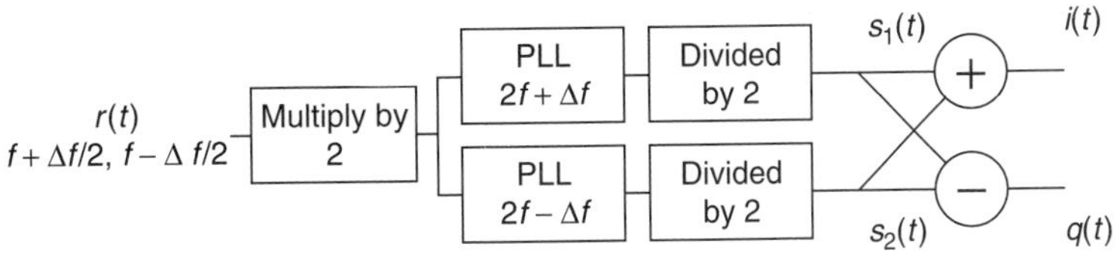

FIGURE 4.16 Carrier Recovery for MSK

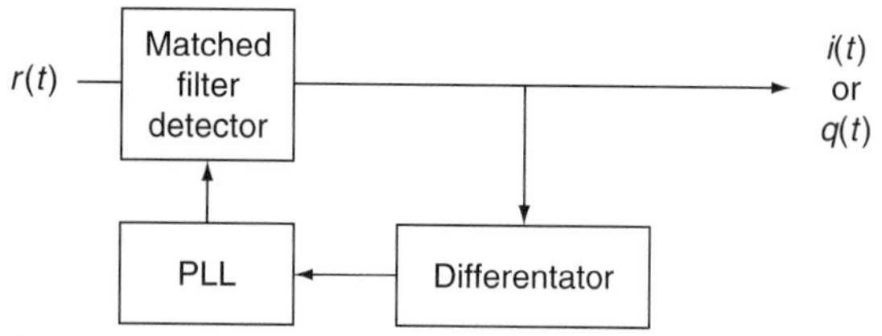

FIGURE 4.17 Generalized Data Timing Recovery Circuit

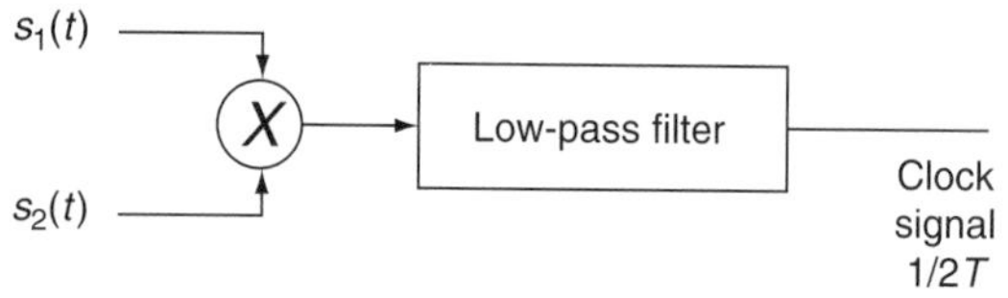

FIGURE 4.18 MSK Data Timing Recovery Circuit

$$s_1(t) = \cos(2\pi ft + 2\pi\Delta ft). \tag{4.39}$$

$$s_2(t) = \cos(2\pi ft - 2\pi\Delta ft). \tag{4.40}$$

We then take the sum and difference of $s_1(t)$ and $s_2(t)$ to generate the desired $i(t)$ and $q(t)$ signals:

$$i(t) = s_1(t) + s_2(t) = \cos(2\pi ft + \pi\Delta ft) + \cos(2\pi ft - \pi\Delta ft) = 2\cos 2\pi ft \cos \pi\Delta ft. \tag{4.41}$$

$$q(t) = s_1(t) - s_2(t) = \cos(2\pi ft + \pi\Delta ft) - \cos(2\pi ft - \pi\Delta ft) = 2\sin 2\pi ft \sin \pi\Delta ft. \tag{4.42}$$

The identical circuit can also be used to recover a carrier for a GMSK system. (de Buda, 1972; Murota and Hirade, 1981; Pasupathy, 1979). The next step is to recover data time or bit synchronization. Most communication systems transmit a sequence of zeros and ones in an alternating pattern to enable the receiver to maintain bit synchronization. A PLL operating at the bit timing is used to maintain timing. Once the PLL is synchronized on the received 101010... pattern (Figure 4.17), it will remain synchronized on any other patterns except for long sequences of all zeros or all ones.

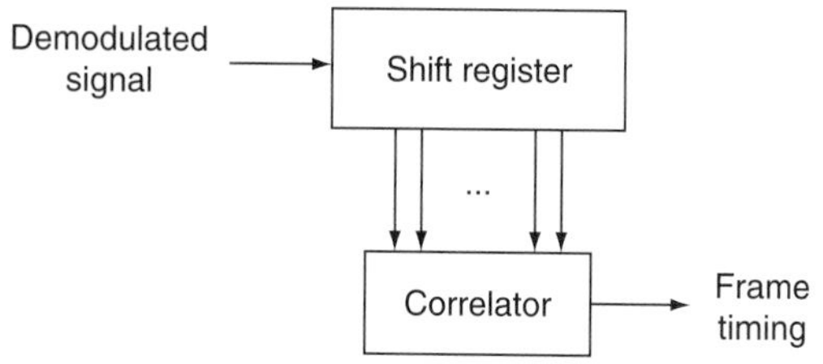

FIGURE 4.19 Generalized Framing Recovery Circuit

MSK uses an additional circuit to achieve bit timing (Figure 4.18). The $s_1(t)$ and $s_2(t)$ signals are multiplied together and low-pass filtered:

$$\begin{aligned} s_1(t)s_2(t) &= \cos(2\pi ft + \pi\Delta ft) \times \cos(2\pi ft - \pi\Delta ft) \\ &= 0.5\cos 4\pi ft + 0.5\cos 2\pi\Delta ft. \end{aligned} \tag{4.43}$$

$$\text{low-pass filtered}[s_1(t)s_2(t)] = 0.5\cos 2\pi\Delta ft = 0.5\cos\frac{\pi t}{T}. \tag{4.44}$$

The output of the low-pass filter is a clock signal at one-half the bit rate, which is the correct rate for demodulation of the signal because the I and Q signals are at one-half of the bit rate.

Word synchronization or framing is determined by correlating on a known bit pattern being transmitted. The receiver then performs an autocorrelation function to determine when the bit pattern is received and outputs a framing pulse (Figure 4.20).

4.9 Equalization

The received signal in a mobile radio environment travels from the transmitter to the receiver over many paths. The signal thus fades in and out and undergoes distortion because of the multipath nature of the channel. For a transmitted signal $s(t) = a(t)\cos(\omega t + \theta(t))$, we can represent the received signal, $r(t)$, as:

$$\begin{aligned} r(t) = \sum_{i=0}^{n} & x_i(t-\tau_i)a(t-\tau_i)\cos(\omega(t-\tau_i) + \theta(t-\tau_i)) \\ & + y_i(t-\tau_i)a(t-\tau_i)\sin(\omega(t-\tau_i) + \theta(t-\tau_i)). \end{aligned} \tag{4.45}$$

The received signal has Rayleigh fading statistics. But what are the characteristics of the x and y terms in equation 4.45? If the transmitter signal is narrow enough compared to the fine multipath structure of the channel, then the individual fading components, $x_i(t)$ and $y_i(t)$ will also have Rayleigh statistics. If a particular path is dominated by a reflection off of a mountain, hill, building, or similar structure, then the statistics of that path may be Rician rather than Raleigh. If the range of τ_i is small compared to the bit interval, then little distortion of the received signal occurs. If the range of τ_i is greater than a bit interval, then the transmissions from one bit will interfere with the transmissions of another bit. This effect is called **intersymbol interference**.

Spread spectrum systems transmit wideband width signals and attempt to recover the signals in each of the paths and add them together in a diversity receiver. In the discussion in this section, we fowson the transmission of narrow band signals. Therefore, the multipath signals are interference to the desired signal. We need a receiver that removes the effects of the multipath signal or cancels the undesired multipaths.

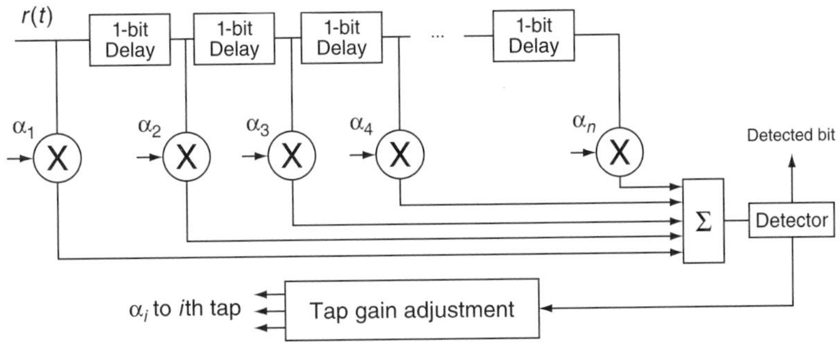

FIGURE 4.20 Block Diagram of Equalizer

Another method to describe the multipath channel is to describe the channel as having an impulse response $h(t)$. The received signal is then written as:

$$r(t) = \int_{-\infty}^{\infty} s(t)h(t-\tau)d\tau. \tag{4.46}$$

We can then recover $s(t)$ if we can determine a transfer function $h^{-1}(t)$, the inverse of $h(t)$. One difficulty in performing the inverse function is that it is time varying. The circuit that performs the inverse transfer function is called an **equalizer**, as shown in Figure 4.20.

Generally, we are interested in minimizing the intersymbol interference at the time when we do our detection (the sample time in a sample and hold circuit). Thus, we can model the equalizer as a series of equal time delays (rather than random as in the general case) with the shortest delay interval being a bit interval. We then construct a receiver that determines r_{eq}, the equalized signal:

$$r_{eq}(t) = \sum_{i=0}^{n} \alpha_i(t-\tau_n)r(t-\tau_n). \tag{4.47}$$

We use our equalized signal r_{eq} as the input to our detector to determine the value of the kth transmitted bit. We must adjust the values of α_i to achieve some measure of performance of the receiver. A typical measure is to minimize the mean square error between the value of the detected bit at the output of the summer and the output of the detector. Other measures for the equalizer are possible. For more details, see Proakis (1989).

4.10 Summary of Modulation and Demodulation Processes

In this section, we studied the modulation and demodulation processes that are applicable to GSM and DECT (and its North American variant PWT). Since baseband signals are can only be transmitted over short distances with wires and require very long antennas to transmit them without wires, the baseband signals are modulated onto radio frequency carriers. First, we studied amplitude shift keying (ASK) to determine its probability of error and methods for demodulation. From ASK, we studied phase modulation and the specific form ($\pi/4$-DQPSK) used for PWT. We showed that $\pi/4$-DQPSK has the same shaped error rate curve as ASK, but the signal-to-noise ratio definition is different. We then studied MSK, a form of frequency modulation, as a first step toward GMSK, which is used for DECT and GSM. Based on the literature, the error performance for GMSK also has the same shaped curve as ASK with the proper definition of a correcting factor α. For GSM where $BT = 0.3, \alpha \approx 0.9$; for DECT where $BT = 0.5$, $\alpha \approx 0.97$.

We then presented methods and block diagrams for recovery of clocks used for the carrier frequency and phase, the bit timing, and the framing for various modulation methods. While our block diagrams are based on hardware approaches, they could just as easily be implemented in software in many cases. We also explained that many manufacturers have proprietary methods for determining clock recovery techniques. Finally, we studied equalization techniques.

5

Data Communication Concepts

Vijay K. Garg
Department of Electrical and Computer Engineering, University of Illinois at Chicago, Chicago, Illinois, USA

Yih-Chen Wang
Lucent Technologies, Naperville, Illinois, USA

5.1 Introduction to Data Networking

The explosive growth of Internet and its applications have made the use of data networks essential for many people's daily lives. The advanced data networking technologies will soon combine data, voice, and multimedia applications on a single platform for an economic reason. We introduce fundamental concepts of **data communication and networking** and also describe important data networking technologies that will shape the future of communication networks.

5.1.1 Fundamental Concepts and Architecture of Data Communication and Networking

The changing face of the communication industry was started with the marriage of computer and communication technologies in 1984 when AT&T was divided into seven Regional Bell Operation Companies (RBOC). The change has shaped the fabrics of society. Paper mail to electronic mail, shopping mode to online shopping, and libraries to digital libraries have changed day-to-day life. All the changes are the results of advanced communication and networking technologies. In this section, we explain the concepts and terms that will help the reader understand different communication and networking technologies described in the later subsections.

Transmission, Data, and Signals

Transmission is the communication of data that is propagated and processed from one point in a network to another point in the network by means of electrical, electromagnetic, or optical signals. The terms **digital** and **analog** are often used in the contexts of transmission, data, and signals. This leads to the following six definitions:

- **Digital data** is a source entity of which meaning has discrete values, such as the text or integers.
- **Analog data** is a source entity of which meaning has continuous values, such as voice or video.
- **Digital signal** is a sequence of voltage pulses that can only be transmitted over certain wired media.
- **Analog signal** is continuously varying electromagnetic wave that could be transmitted over a variety of wired and wireless media.
- **Digital transmission** is a transmission system that is concerned about the content of data it transmits or receives. It can carry either digital data or analog data with either a digital signal or an analog signal.
- **Analog transmission** is a transmission system that is not concerned about the content of data it transmits or receives. It can carry either analog data or digital data with an analog signal only.

Combinations of Data and Signals

There are four possible combinations of data and signals. For each combination, this discussion provides examples of applications and techniques to encode data. Note that a digital transmission system can use each of the four combinations for its applications, but an analog transmission system can only use an analog signal to carry either digital data or analog data.

Digital Data and Analog Signal

One example of this combination is that a PC or terminal connects to a remote host via a telephone network with a voice-grade modem. The modem modulates digital data to the characteristics of an analog carrier signal. These characteristics are the frequency, amplitude, and phase of the carrier signal. The modem uses one or any combination of the following three basic techniques as shown in Figure 5.1.

- **Amplitude-shift keying (ASK):** In this technique, binary values are represented by two different amplitudes of a carrier signal. The binary value 0 could be represented by a lower amplitude, while a binary 1 could be represented by a higher amplitude.
- **Frequency shift keying (FSK):** In this technique, the two binary values are represented by two different frequencies. In the full duplex FSK, one set of frequencies could be designated for one direction of the communication, while the other set could be designated for an opposite direction of the communication. For example, a set of frequencies (1070 Hz, 1270 Hz) is used for one direction of the communication, and the other set (2070 Hz, 2075 Hz) is used for other direction of the communication.
- **Phase-shift keying:** In this technique, the phase of the carrier signal is shifted to a certain degree of phase to represent the binary value. In the simplest case, no shift can represent the binary 0, while a shift of 180° can represent the binary 1. If a carrier signal can produce four phases, such as no shift, shifting to 90°, shifting to 180° and shifting to 270°, then each phase could represent two bits. Therefore, no shifting can represent 00, the shifting to 90° can represent 01, the shifting to 180° can represent 10, and the shifting to 270° can represent 11.

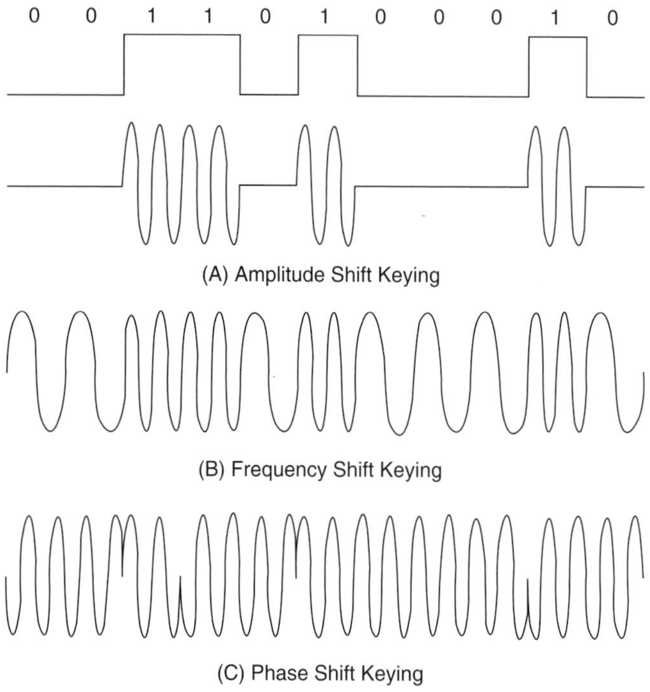

FIGURE 5.1 Three Basic Techniques

Analog Data and Digital Signal

An example of the combination is the *T*1 carrier carrying voice signal from a switching local office to another switching local office via a tandem switch. The voice is converted into a stream of digital bits that are carried by digital signal with some encoding scheme. The receiving system would then convert the stream of digital bits back to the original analog voice. Another example of the combination is compact disc audio.

The basic technique for this combination is called **pulse code modulation** (PCM). PCM takes the samples of analog data at the sampling rate that is twice the highest signal frequency and then assigns each sample to a binary code. The analog data is quantized into 8 steps (0 to 7), an each binary code will have 3 bits assigned to it. According to the Nyquist's sampling theorem, the samples contain all information of the original signal and can be reconstructed from these samples. The generated bit stream will be 011110111111110100011100110.

Digital Data and Digital Signal

One example of the combination is a text file transfer via a baseband ethernet local area network where digital signal is used. There are many different digital encoding techniques for this combination. The techniques are summarized as below:

- Nonreturn to zero level (NRZ-L)
- Nonreturn to zero inverted (NRZI)
- Manchester encoding
- Differential manchester encoding
- Bipolar with 8-zero substitution (B8ZS)
- High-density bipolar 3 zeros (HDB3)

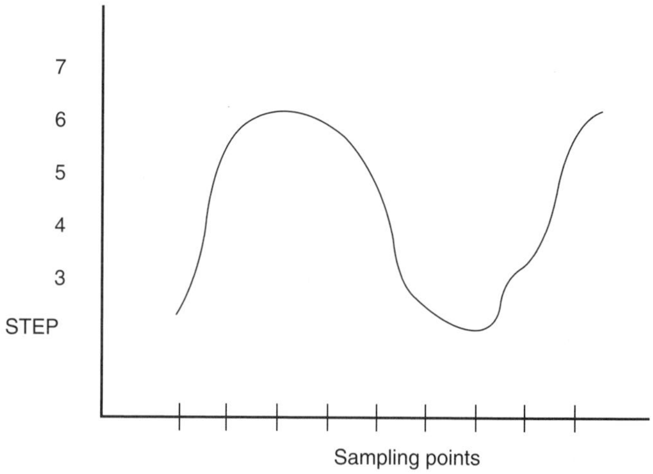

FIGURE 5.2 Pulse Code Modulation

Analog Data and Analog Signal

One example of the combination is the analog CATV or traditional telephone network. The techniques used in this combination are similar **to modulations** of a modem.

The modulation is required to provide a higher frequency for more efficient transmission or allow multiple user transmission using frequency modulation. The three modulation techniques amplitude modulation (AM), phase modulation (PM), and frequency modulation (FM) are used.

5.1.2 Advantages of Digital Transmission

The industry has been gradually converting an analog transmission system into a digital transmission system due to the cost, quality, and capacity of the digital transmission system. The evolution of integrated service digital network (ISDN) and digital subscriber line (DSL) to replace the analog local loop and the conversion of analog CATV into a digital CATV are examples of conversion into a digital transmission system. The following is a list of benefits for a digital transmission system:

- **The digital transmission system is cheaper**. The technique for designing and manufacturing a digital circuitry is more advanced and cheaper. The very large scale integration (VLSI) technique allows for mass production of a circuitry at lower cost.
- **Transmission quality is higher**. With the digital transmission system, transmission noise is not accumulated with the use of a repeater. A repeater recovers an incoming signal based on the content and regenerates a fresh signal; an amplifier used by an analog transmission system amplifies a noise component as well so that noises are accumulated in an analog transmission system.
- **A digital transmission system has higher capacity**. Time division multiplexing (TDM) instead of frequency division multiplexing (FDM) is used in a digital transmission system. TDM shares the entire bandwidth by all users, and the waste of bandwidth can be reduced to a minimum. The granularity for a time slot is also more flexible than a subfrequency of an FDM. Therefore, the bandwidth utilization is much higher for a digital transmission system, allowing digital CTAV to accommodate more than 100 channels.
- **The system enables security and privacy**. The most important tool for transmission security and privacy is the use of encryption. The encryption uses a cipher algorithm to replace information with a ciphertext (output of a cipher algorithm) using some mathematical algorithm. Digital transmission system makes the manipulation of digital information much easier.
- **The system enables easier integration of multiple data types**. The digital transmission system deals with digital information only. This makes for easier control and integration of multimedia applications.

Bit Rate Per Second Versus Baud Rate

Many people use **bit rate per second** (bps) and **baud rate** interchangeably, which is not correct. The bps is defined as the number of binary bits transmitted per second, while Baud rate is defined as the number of signal elements or states transmitted per second. Frequency, amplitude, or phase can be considered as signaling elements. If a modem only uses two frequencies signaling elements in the transmission, the baud rate of the modem is equal to bps, where the higher frequency can represent a binary 1 and a lower frequency can represent a binary 0. If a modem uses four frequencies to represent signaling elements, then each frequency can represent two binary bits. For example, 00 corresponds to the first frequency, 01 corresponds to the second frequency, 10 corresponds to the third frequency, and 11 corresponds to the last frequency. In this case, a baud rate is equal to 2 bps.

Maximum Data Rate

Some factors can limit the maximum transmission rate of a transmission system. Nyquist's theorem specifies the maximum data rate for noiseless condition, whereas the Shannon theorem specifies the maximum data rate under a noise condition.

The **Nyquist theorem** states that a signal with the bandwidth B can be completely reconstructed if 2B samples per second are used. The theorem further states that:

$$R_{\max} = 2B \log_2 M, \tag{5.1}$$

where $R_{\max}$ is the maximum data rate and M is the discrete levels of signal.

For example, if a transmission system like the telephone network has 3000 Hz of bandwidth, then the maximum data rate $= 2 \times 3000 \log_2 2 = 6000$ bits/sec (bps).

The **Shannon theorem** states the maximum data rate as follows:

$$R_{\max} = B \log_2 (1 + S/N), \tag{5.2}$$

where S is the signal power and N is the noise power.

For example, if a system has bandwidth B = 3 kHz with 30-dB quality of transmission line, then the maximum data rate $= 3000 \log_2 (1 + 1000) = 29{,}904$ bps.

5.1.3 Asynchronous and Synchronous Transmission

To determine what bits are constituted as an octet (character), two transmission techniques are used.

- **Asynchronous transmission**. In this technique, each character is enclosed with a start bit, 7 or 8 data bits, an optional parity bit, and stop bits. The gap between character transmission is not necessarily fixed in length of time. It is cheaper but is less efficient. For example, assume that there are 1 stop bit and 1 start bit in the asynchronous transmission with 8 bits of character; the efficiency is only 80% (fixed).

- **Synchronous transmission.** In this technique, all characters are blocked together and transmitted without a gap between two characters being transmitted. It requires more complicated hardware to handle buffering and blocking, but this hardware more efficient. For example, assume that there are 240 characters and 3 SYN characters in one block. Then the number of data bits $= 240 \times 8 = 1920$ bits, the number of SYN character bits $= 3 \times 8 = 24$ bits, and efficiency $= 1920/1944 \sim 99\%$ variable efficiency; the larger the block is, the higher the efficiency.

Transmission Impairments

The signal received may differ from the signal transmitted. The effect will degrade the signal quality for analog signals and introduce bit errors for digital signals. There are three types of **transmission impairments**: attenuation, delay distortion, and noise.

(1) **Attenuation:** The impairment is caused by the strength of signals that degrades with distance over a transmission link. Three factors are related to the attenuation:
   - The received signal should have sufficient strength to be intelligently interpreted by a receiver. An amplifier or a repeater is needed to boost the strength of the signal.
   - A signal should be maintained at a level higher than the noise so that error will not be generated. Again, an amplifier or a repeater can be used.
   - Attenuation is an increasing function of frequency, with more attenuation at higher frequency than at lower frequency. An equalizer can smooth out the effect of attenuation across frequency bands, and an amplifier can amplify high frequencies more than low frequencies.

(2) **Delay distortion:** The velocity of propogation of a signal through a guided medium varies with frequencies; it is fast at the center of the frequency, but it falls off at the two edges of frequencies. Equalization techniques can be used to smooth out the delay distortion. Delay distortion is a major reason for the timing jitter problem, where the receiver clock deviates from the incoming signal in a random fashion so that an incoming signal might arrive earlier or late.

(3) **Noise:** Impairment occurs when an unwanted signal is inserted between transmission and reception. There are four types of noises:
   - **Thermal noise:** This noise is a function of temperature and bandwidth. It cannot be eliminated. The thermal noise is proportional to the temperature and bandwidth as shown in the equation: thermal noise $= K(\text{constant})$ *temperature *bandwith.
   - **Intermodulation noise** This noise is caused by nonlinearity in the transmission system $f1$; $f2$ frequencies could produce a signal at $f1 + f2$ or ABS $(f1 - f2)$ and affect the frequencies at $f1 + f2$ or ABS $(f1 - f2)$.
   - **Cross talk:** This type of noise is caused by electrical coupling in the near by twisted pair or by unwanted signal picked by microwave antennas. For example, sometimes when you are on the telephone, you might hear someone else's conversation due to the cross talk problem.
   - **Impulse noise:** Irregular pulses and short duration of relative high amplitude cause impulse noise. This noise is also caused by lightning and faults in the communication system. It is not an annoyance for analog data, but it is an annoyance for digital data. For example, 0.01 sec at 4800 bps causes 50 bits of distortion.

Error Detection and Recovery

Due to the transmission impairments, a message arriving at the receiving station might be in error. The receiving station will detect the error and might ask the sending station for a retransmission. It is very rare that the receiving station would correct errors. The approach for requesting a retransmission when an error is detected is called **forward correction**. The error detection is mostly done at the data link layer that is the second layer of the network architecture although some error detection schemes might be done at the first layer, the physical layer of the network architecture. The error detection schemes that have been used are described below. The cyclic redundancy check (CRC) is a most commonly used error detection scheme at the data link layer.

- **Single parity checking:** This scheme appends a parity bit to the character to be transmitted to make the total number of binary ones in one character be either odd or even. An odd parity checking or an even parity checking has to be agreed between a transmitter and receiver. The problem of this approach is that the even number of bits changed will make the number of ones for being odd or even, and the parity bit error cannot be detected.
- **Two coordinated parity checking:** The scheme checks for the parities for a block of characters and not just for a single character. The parities are added horizontally and vertically within a block of characters to produce a block check character (BCC). BCC is appended to the end of a block to be transmitted. Table 5.1 illustrates the error detection scheme. Assume that odd parity checking is used.

TABLE 5.1 Error Detection Scheme

Data bits	Parity
0010010	1
1001001	0
0100101	0
0000001	0

In Table 5.1 the last row with numbers 00000010 (including the parity bit) is the BCC. The scheme will not work if there are an even number of bits changed horizontally and/or vertically. Even the BCC will be different, but all is correct for the odd parity checking.

Cyclic Redundancy Check

For a **cyclic redundancy check**, the scheme selects a standard defined 8 bits, 10 bits, 12 bits, 16 bits, or 32 bits of a constant check data. A constant check data is normally represented as a polynomial constant. For example, the polynomial $x^3 + x^2 + 1$ is the check data bit of 1101. A data message that could be thousands of bytes is divided by the polynomial constant with exclusive or bit-by-bit operation. The remainder of the division is the CRC that is appended at the end of a frame or block. With the above example, the CRC will be 3 bits long. The receiving station takes the data it receives and performs the same computation with the same polynomial constant. If the computed CRC is equal to the CRC in the received message, it is fine; otherwise an error has occurred. The CRC error detection scheme cannot detect all errors at all times. In the case of an error pattern being the multiples of the polynomial constant, the error cannot be detected. Therefore, it is very important to select a polynomial constant that has low probability of being generated in a transmission environment.

5.1.4 Sharing of Network Nodes and Transmission Facility Resources

It is very economically infeasible to have one dedicated physical link between any two computers on a network. Networking nodes and the transmission facility must be shared. **Multiplexing** allows for a physical link to be shared by multiple users to fully use the link and reduce the number of input/output (I/O) ports required for a computer. Switching techniques avoid a creation of the mesh or complete topology where two computers are directly connected in a network. Direct connection increase the cost of I/O ports on each computer and transmission links. Multiplexing is also very inflexible without the use of switching if the direct link between two computers is broken.

Multiplexing

As the list below indicates, there are three multiplexing techniques.

- Time division multiplexing (TDM)—synchronous TDM: Multiple digital signals or analog signals carrying digital data can be carried on a single transmission path by interleaving portions of each signal in time. The interleaving interval can be one bit, one octet, or one block of a fixed size of octets. Each signal or connection path takes a fixed time slot but use the whole bandwidth of the link. "Dummy" information will be sent on the slot even if there is no data for the connection. This wastes the capacity of the transmission bandwidth.
- Frequency division multiplexing (FDM): A number of signals can be carried simultaneously on the same transmission link with each signal being modulated into a separate frequency. The bandwidth of each signal is reduced due to the division of multiple channels. Digital and analog data can both be supported. CATV is an example of the use of FDM.
- Statistical TDM (asynchronous TDM or intelligent TDM): The time slots are not preassigned for a signal or connection path. They are allocated on a demand basis. There is no waste of bandwidth. A terminal number is included in the message to identify where the message came from. The output rate is designed to be less than the sum of data rates of all inputs. At the peak load, total inputs might exceed output capacity, then backlog will be built. The trade-off is between the size of buffers and the rates of total inputs to be supported.

5.1.5 Switching

Switching techniques avoid direct connections between any two computers on a network. These techniques not only save the cost of transmission links but also provide the flexible and reliable connections between any two computers on the network. There are two major switching techniques: one is mainly used in the telephone network, and the other is used in data and other telecommunication networks. The message switching is rarely used, but it is described here for completeness.

Circuit Switching

A connection between the calling station and the called station is established on demand for exclusive use of the physical connection. Three phases of a circuit switching connection exist; (1) connection establishment phase requires a separate setup signal to send and the acknowledgement signal to receive; (2) The data transfer phase has no interruption to the data transfer; and (3) the termination phase allows the release of the resource so that other connection, can use the resource. The switching technique does not require buffers or queues and is free of congestion. Long setup time may cause undesirable delay, which is rather inefficient if no data are being transferred. Appropriate applications may have relatively continuous flow, such as voice, sensor, and telemetry inputs.

Message Switching

An entire message, like an e-mail or a file, is transmitted to an intermediate point, stored for a period of time, and an transmitted again toward its destination. The address of the destination is included in the message.

No dedicated path is established between two stations. The delay at each intermediate point is the time to wait for all messages to arrive; plus, there is a queuing delay in waiting for the opportunity to retransmit to next node.

The line efficiency for message switching is greater than for circuit switching because the transmission line is shared over time but not at the same time for multiple messages. Simultaneous availability of sender and receiver is not required, and the message priority can be established in the message switching. Because a connection is not needed, the line has a flexible routing capability and the capability for sending the same message to multiple destinations. With buffering capability, congestion problems will be reduced. Any error detected requires retransmission of the entire message. If the message is a megabytes file, the whole file will need to be retransmitted. No overlap of transmission is possible. Message switching is generally not suitable for the interactive traffic due to a long delay.

Packet Switching

A message is divided into a number of smaller packets, and one packet at a time is sent through the network. There are two approaches in packet switching: (1) **datagram** will not need to establish a connection path, and each packet is treated independently; and (2) the **virtual circuit** approach requires that a connection be established, and all packets will need to go over that connection path. There is less overhead on the retransmission as the packet is smaller; overlap of the transmission is possible. Virtual circuit is more suitable for interactive traffic because it has smaller delay.

6

Communication Network Architecture

Vijay K. Garg
Department of Electrical and Computer Engineering, University of Illinois at Chicago, Chicago, Illinois, USA

Yih-Chen Wang
Lucent Technologies, Naperville, Illinois, USA

6.1 Computer Network Architecture

Computer network architecture refers to a set of rules that allow for connectivity among a large number of computers. This set of rules is also called **communication protocols**. To simplify the complexity of network design, the communication functions are divided into several levels of abstractions. Each level or layer of the protocol is designed in such a way that the change to one layer normally does not affect adjacent layers. The services of higher layers are implemented to use the services provided at lower layers. There are two interfaces at each layer. One is the peer-to-peer protocol between two computers. The other is the service interface to its adjacent layers on the same computer. Peer-to-peer protocol between two computers mostly regards indirect communication, and the direct communication only occurs at the lowest layer or hardware level. Each higher layer of protocol adds its own header information to the data message it receives from its higher layer of protocol before it passes the data message to its lower layer. This process is called **encapsulation**. The receiving system reverses the process, called **decapsulation**, by removing the header at each layer before passing the data message to its upper layer. Two prevalent network architectures are described in this section.

6.1.1 Open System Interconnection

Open System Interconnection (OSI) is an International Standard Organization (ISO) standard that defines computer communication network architecture. It is a well-defined network architecture, but the implementation of its network protocols is very rare. When its draft standard came out in 1985, many predicated that the implementation of network protocols would predominate in the industry. The prediction was incorrect due to the wide use of Internet protocols in 1985. However, OSI represents a very powerful network reference model to which all communication technologies refer their architecture (see Figure 6.1). This also includes the most popular TCP/IP architecture.

OSI divides the communication functions into seven layers of which functions are described in the following list.

- **Physical layer**: This layer is responsible for activating and deactivating physical connections upon request from the data link layer and transmitting bits over a physical connection in a synchronous or asynchronous mode. It also handles very limited error control, like single-character parity checking.
- **Data link layer:** This layer is responsible for establishing and releasing data link connections for use by the network layer.

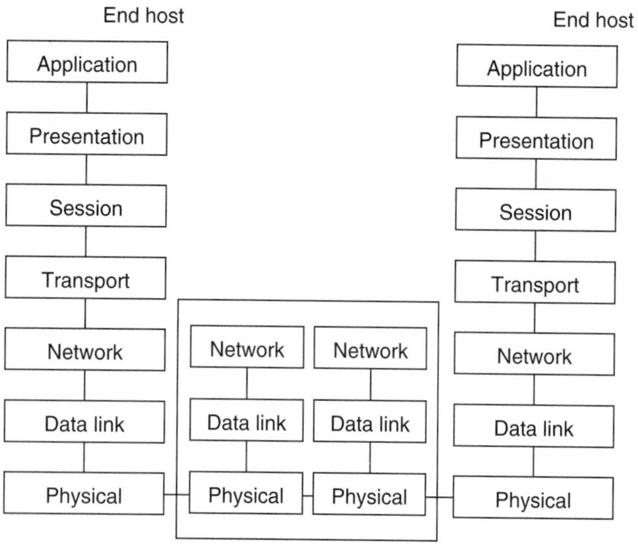

FIGURE 6.1 OSI Network Reference Model

- Network layer: Responsibilities of this layer include providing the data integrity transmission for a point-to-point connection so that data will not be lost or duplicated. The layer accomplishes this task by maintaining a sequential order of frames that are transmitted over a data link connection and detecting and correcting transmission errors with retransmission of the frames, if necessary. The other important function in the data link layer is to provide flow control, which is a way to allow the receiving station to inform the sending station of stopping transmission for a moment so that the receiver will not overload its buffers.
- Network (packet) layer: Two major functions in the network layer are **routing control** and **congestion control**. Routing control is the process for maintaining a routing table and determining optimum routing over a network connection. Congestion control is needed when there are too many packets queued for a system and there is no space to store them. This is normally happens in a datagram-type of connection, where the network resources for a connection are not preallocated. The network layer is also responsible for multiplexing multiple network connections over a data link connection to maximize its use. Flow control is provided at the network layer as well.
- Transport layer: This layer provides error-free end user transmission. To improve the utilization, it multiplexes multiple transport connections over a network connection. It controls data flow to prevent from overloading network resources just like the flow control function provided at the data link and network layers. This layer and layers above it are end-to-end, peer-to-peer protocols, for which their protocol data units (PDU) are processed between two end systems. The layers below the transport layers are the point-to-point, peer-to-peer protocol where the PDUs are processed only between two computer systems connecting together.
- Session layer: Providing management activities for transaction based applications, this layer ties these application streams, together to form an integrated application. For example, a multimedia application may consist of the transport of data, fax, and video streams that are all managed at the session layer as a single application.
- Presentation layer: This layer is responsible for performing any required text formatting or text compression. It negotiates the choice of syntax to be used for data transfer.
- Application layer: To provide an entry point for using OSI protocols is one task of this layer. This task can be accomplished by providing either the Application Programming Interface (API) or standard UNIX I/O functions, like open(0), close(0), read(), and write() functions.

The layer also performs common application functions, such as connection management, and provides specific application functions, like file transfer using File Transfer and Access Management (FTAM), Electronic Mail (X.400), and Virtual Terminal Protocol (VTP).

6.1.2 TCP/IP Network Architecture

The **TCP/IP network architecture** also refers to the Internet architecture. The Transmission Control Protocol (TCP) is a transport layer protocol, and the Internet Protocol (IP) is a network layer protocol. Both protocols were evolved from a earlier packet switching network called ARPANET that was funded by the Department of Defense. The TCP/IP network has been the center of many networking technologies and applications. Many network protocols and applications are running at the top of the TCP/IP protocol. For example, the Voice over IP (VOIP) and the Video Conference application using MBONE are the applications running over the TCP/IP network. The TCP/IP network has the corresponding five layers in the OSI reference model. Figure 6.2 shows the

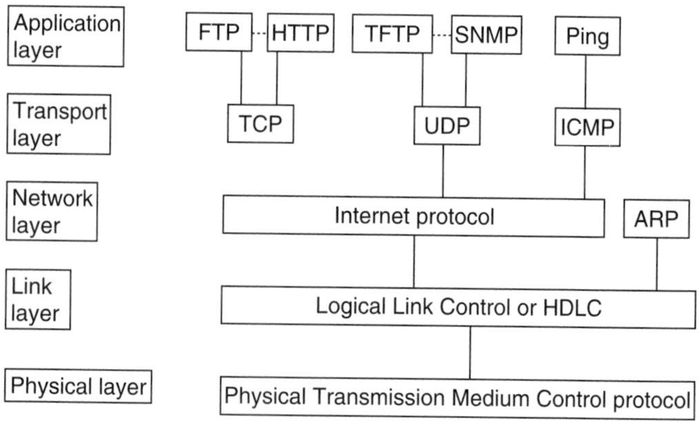

FIGURE 6.2 TCP/IP Network Architecture and Applications

layers of the TCP/IP network and some applications that might exist on the TCP/IP networks. The standard organization for the TCP/IP-related standard is the Internet Engineering Task Force (IETF), which issues Request-for-Comment (RFC) documents. Normally, IETF requires that a prototype implementation be completed before an RFC can be submitted for comments.

6.2 Local Networking Technologies

This section describes the **local area network** (LAN) and **metropolitan area network** (MAN) technologies. The dramatic and continuing decrease in computer hardware cost and an increase in computer capability have increased the usage of single-function systems or workstations. One has the desire to interconnect these single function systems and workstations together for variety reasons. The major reasons are to exchange data between systems, to share expensive resources, to improve the performance for real-time applications, and to improve the productivity of the employees.

LAN is a communication network that supports variety of devices within a small area. The small area could be a single building or a campus. MAN is the communication network that spans a larger geographic area than LAN and could be an area consisting of a few blocks of buildings or even a large metropolitan area. The characteristics of LAN are high data rate, short distance, low error rate, shared medium, and single organization ownership (typically).

This section describes the general technologies used by a LAN, the internetworking with bridges and routers, and the performance of local network. General technologies include the transmission media, topologies, and Medium Access Control (MAC) protocol.

6.2.1 Technologies of Local Networks

The nature of a local network is very closely tied to the technologies it uses. The key technologies are the topology, transmission medium, and MAC protocol. Unlike a wide area network (WAN), LAN is a shared medium network, and the choice of a transmission medium impacts the topology it can use. These key technologies are closely related to each other. For example, a twisted pair medium is more suitable for the star topology. Each type of technology is described in separate subsection in this chapter.

Local Network Topologies

Topology is the manner in which network interface devices and transmission links are interconnected. There are two reasons to study the topologies for LAN as well as for WAN. One is that the mesh (complete) topology, which allows for every two devices to be directly connected via a dedicate link, is economically infeasible. The other reason is that different topologies would provide flexibility for applications and the reliability for the connections.

For each topology, one should consider the **characteristics**, **reliability**, **expandability**, and **performance**. The characteristics of a topology should consider the connectivity of devices, the components required (e.g., tap or splitter, etc.), and the transmission. The reliability addresses how a node failure or cable break would affect the operation of the network. The expandability addresses the ease of adding a new station to the network, and the performance should consider the delay characteristics of the topology. The topologies used in the local networks are star, ring, and bus/tree. Table 6.1 summaries these topologies.

Transmission Media

The **transmission medium** is a physical transmission path between two devices. A transmission medium could be a guided hard-wired medium (e.g., the twisted pair cable used in the telephone network), a coaxial cable used by CATV, or optical fiber. It could be an unguided medium through atmosphere, such as microwave, laser, or infrared. Most LANs use the guided transmission medium in a building. The unguided medium would be used when there is a need to connect two LANs in different buildings and when the digging of the ground between two buildings is very difficult. Until recently, wireless LAN using microwave radio at the same building or nearby buildings has become a very popular way to connect portable devices together in the network with the data rate up to 10 Mbps.

Of the three hard-wired media (e.g., the twisted pair, coaxial cable, and optical fiber), the use of coaxial cable in local networks has started to phase out due to its heavier weight and inflexibility to bend. The twisted pair is popular for lower speeds (100 Mbps or less) of a local network, whereas the optical fiber is more suitable for higher speeds (hundreds of megabits per second or higher) of a local network. Of two unguided media used in local networks, the microwave is used to connect two LANs at different buildings, but both microwave and infrared can be used for wireless LANs.

In choosing a transmission medium for a local network, one should consider the following factors:

- Cost of installation of transmission medium
- Connectivity
- Transmission characteristics
- Topology constraints
- Distance coverage
- Environmental constraints

Medium Access Control Protocol for Local Networks

In the local network environment, multiple stations on the network share a transmission medium. If more than one station is transmitting at the same time, the transmission will be corrupted. **Medium Access Control** (MAC) protocol is

TABLE 6.1 Various Topologies for LAN

Topology	Star	Ring	Bus/Tree
Characteristics	Stations are connected to a central hub or star coupler, which serves as a relay point.	Each station is connected to a repeater, and all repeaters are connected to form a closed loop or ring. Signal is traveling unidirectional along the ring. There is no buffering at each repeater.	More than one station is connected to a shared medium. When a signal is transmitted on the medium, it will travel in both directions; then the signal is absorbed at both ends by the terminators.
Reliability	Station failure would not cause a failure of the network, but a failure on the hub or coupler would disable the network operation. A break of the cable would isolate only one station from the network.	Station failure would not affect the network, but the repeater failure may disable the network unless repeater bypass logic is implemented. A cable break would disable the network transmission unless a self-configuration dual ring network is installed.	Station failure would not affect the network operation, but the cable break would disconnect the network and produce the reflection that could interfere with the transmission.
Expandability	Adding a new station to the network involves only wiring from the station to the hob or the coupler.	If a multiple access unit (MAU) is equipped, the adding of a new station just involves the wiring between the station and the MAU.	The addition of a new station is easier and would not need to bring down the network for addition of a new station.
Performance	The number of ports on hubs or couplers limits the capacity of the network. The delay on a hub or coupler is very small.	There is a least 1-bit delay at each repeater. The round-trip delay would include the propagation delay, transmission delay, and repeater delay. The number of stations on a ring network is limited for the performance concern.	The performance of LAN using bus/tree topology depends on the bandwidth of the network, the number of active stations on the network, and the kind of medium access protocol being used.

needed to regulate an access to the common medium. There are two approaches for control of the multiple transmissions. In the **centralized approach**, a station on the network is designated as a primary station, and the rest of the stations are the secondary stations. All communications have to be polled and selected by the primary station. The primary station polls for a secondary station on the network for transmission and then selects a specific secondary station for the reception of a message. There is no direct transmission between any two secondary stations. In the **distributed approach**, each station has a MAC to decide when it can transmit, and the direct transmission occurs between any two stations. The distributed approach is currently used by most local networks. The advantages and disadvantages of the centralized approach are in the following list. Note that the opposite of the advantages and disadvantages on the list would become the disadvantages and advantages of the distributed approach, respectively.

The advantages of **centralized approach** are as follows:

- Greater control over access for providing priorities, overrides, and bandwidth allocation
- Simple logic at each station
- Avoidance of the coordination problem

The disadvantages of centralized approach are as follows:

- Disabling of entire network due to single point of failure
- Reduction of the efficiency of network operation caused by overhead imposed on the primary station

The MAC protocols for local networks are specified in a series of IEEE802 standards. The fiber data distribution interface (FDDI) for MAN is specified in the ANSI ASC X3T9.5 committee. The popular local network standards are listed as follows:

- IEEE 802.2—Logical Link Control
- IEEE 802.3—Carrier Sense Multiple Access/Collision Detection (CSMA/CD)
- IEEE 802.4—Token Passing Bus
- IEEE 802.5—Token Passing Ring
- IEEE 802.1—Wireless LAN
- ANSI X3T9.5—Fiber Data Distribution Interface

Before we describe each local network standard, we describe the local network architecture. The following subsections address the architecture and each standard.

6.2.2 Local Network Reference Model

Open System Interconnection (OSI) is an International Standard Organization (ISO) standard that defines computer communication network architecture. As mentioned before, all networking technologies describe their network architecture within the OSI reference model. How does a local network fit into the OSI's reference model? We need to determine what layers are required for a local network. The MAC protocol exists for each local network type and is considered part of the data link layer. The MAC protocol has the following deficiencies:

- The protocol only guarantees the transmission of frame onto the medium.
- There is no frame sequencing, and most MAC protocols do not acknowledge the receiving of a frame.
- There is no flow control between two stations.
- There is no retransmission for discard packets.
- It is a data gram-type of protocol, and frame can be lost.

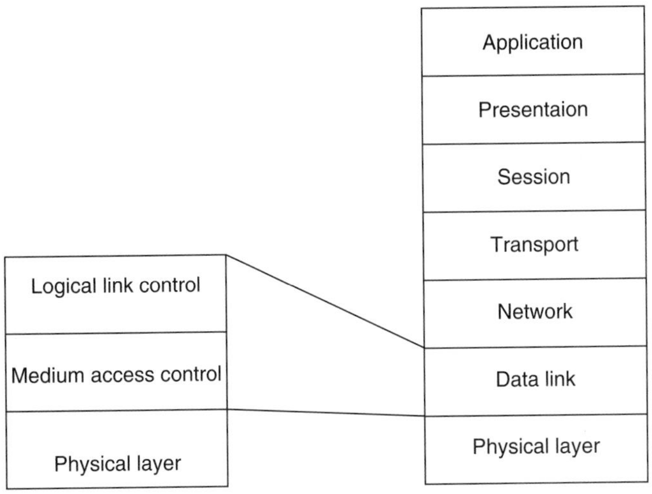

FIGURE 6.3 Local Network Reference Model

With these deficiencies, we need a data link layer above the MAC protocol to make up these deficiencies. Each frame carries the source and destination address in a local network, and all stations on the network will receive the frame through the shared transmission medium. Therefore, a routing function of the network layer within a local network is not required. Figure 6.3 depicts the local network architecture.

6.2.3 IEEE 802.2—Logical Link Control

Logical Link Control (LLC) provides conventional data link protocol functions, such as error control and flow control. LLC is very similar to several famous data link protocols, like Synchronous Data Link Control (SDLC) or High Level Data Link Control (HDLC) protocols. The only difference is that LCC includes the service access points (SAP) information in the frame to allow for multiple applications (programs) on one station to communicate simultaneously with other applications on other stations in the network.

LLC provides the following three services for a local network application:

(1) **Unacknowledged connectionless service**: This is a data gram-type of service, so there is no overhead to establish a connection. It provides no acknowledgment to ensure the delivery of a frame.
(2) **Acknowledged connectionless service**: This service provides acknowledgment to a frame received to relieve the burden in the higher layer.
(3) **Connection-oriented service**: This service provides flow control, sequencing, and error recovery between SAPs. It also allows for multiplexing logical endpoints over a single physical link.

6.2.4 IEEE 802.3—Carrier Sense Multiple Access Collision Detection

Ethernet LAN, which was initially developed by Xerox, Digital, and Intel, used Carrier Sense Multiple Access/Collision Detection (CSMA/CD) technology many years ago. There has been a long evolution path for the CSMA/CD Medium Access Protocol. The following describes the precursors of CSMA/CD technology.

Pure ALOHA

Pure ALOHA technology was developed at the University of Hawaii to interconnect a packet radio network across multiple islands in Hawaii. The protocol has the following procedure:

(1) Frames are transmitted at will.
(2) The sending station listens for the time equal to the maximum possible round-trip delay.
(3) The station that receives the frame has to send an acknowledgment. If the acknowledgment is not received before the timer expires, the frame is retransmitted.
(4) The error is checked at the receiving station, which ignores all of erroneous frames.

The successful transmission for Pure ALOHA is only 18%.

Slotted ALOHA

Improvement of Pure ALOHA led to the creation of Slotted ALOHA. The procedures of the algorithm are as follows:

(1) Time on the channel is organized into a uniform slot, and each slot size is equal to a frame transmission time.
(2) A frame is transmitted only at the beginning of the time slot
(3) If the frame does not go through, the frame is sent again after a random amount of time.

The performance of the Slotted ALOHA is 37% empty slots, 37% success, and 26% collisions. This yields the successful transmission of 37%.

There are three problems with the ALOHA schemes. The first problem is the station not listening to the transmission medium before sending a packet. The second problem regards the station not listening to the transmission medium during the sending of the packet. The last problem involves the station not taking advantage of a much shorter propagation delay than the frame transmission time. The shorter propagation delay provides a fast feedback for the state of the current transmission. Various versions of CSMA have been proposed to solve only the first problem. The variants of CSMA schemes are dealing with what to do if the transmission medium is sensed to be busy. These CSMA schemes are described next.

Nonpersistent CSMA

In **nonpersistent CSMA**, if a frame is waiting for transmission, a station checks if the medium is idle and transmits the frame right away if the medium is idle. If the medium is busy, the station waits for a random amount of time and then repeats the same procedure to check if the medium is idle. The algorithm can reduce the possibility collisions, but there is waste due to the idle time.

1-Persistent CSMA

For **1-persistent CSMA**, the algorithm is used as part of the CSMA/CD. The procedure is as follows:

(1) A station checks if the transmission medium is idle, and then it transmits the frame if it is idle.
(2) If the medium is busy, the station continues to listen until the channel is idle, and then it transmits right away.
(3) If no acknowledgment is received, the station waits for a random amount of time and repeats the same procedure to check if the transmission medium is idle.

The scheme would guarantee a collision if more than one station were waiting for transmitting frames.

P-Persistent CSMA

The **p-persistent CSMA** algorithm takes a moderate approach between nonpersistent and 1-persistent CSMA. It specifies a value; the probability of transmission after detecting the medium is idle. The station first checks if the medium is idle, transmits a frame with the probability *P* if it is idle, and delays one time unit of maximum propagation delay with 1-*P*. If the medium is busy, the station continues to listen until the channel is idle and repeats the same procedure when the medium is idle. In general, at the heavier load, decreasing *P* would reduce the number of collisions. At the lighter load, increasing *P* would avoid the delay and improve the utilization. The value of *P* can be dynamically adjusted based on the traffic load of the network.

CSMA/CD is the result of the evolution of these earlier protocols and the additions of two capabilities to CSMA protocols. The first capability is the listening during the transmission; the second one is the transmission of the minimum frame size to ensure that the transmission time is longer than the propagation delay so that the state of the transmission can be determined. CSMA/CD detects a collision and avoids the unusable transmission of damaged frames. The following describes the procedures of CSMA/CD:

(1) If the medium is idle, the frame is transmitted.
(2) The medium is listened to during the transmission; if collision is detected, a special jamming signal is sent to inform all of stations of the collisions.
(3) After a random amount of time (back-off), there is an attempt to transmit with 1-persistent CSMA.

The **back-off algorithm** uses the delay of 0 to 2 time units for the first 11 attempts and 0 to 1023 time units for 12 to 16 attempts. The transmitting station gives up when it reaches the 16th attempt. This is the **last-in first-out unfair algorithm** and requires imposing the minimum frame size for the purpose of collision detection. In principle, the minimum frame size is based on the signal propagation delay on the network and is different between baseband and broadband networks. The baseband network uses digital signaling, and there is only one channel used for the transmission, while the broadband network uses analog signaling, and it can have more than one channel. One channel is used for transmitting, and another channel can be used for receiving. The baseband network has two times the propagation delay between the farthest stations in the network, and the broadband network has four times the propagation delay from the station to the headend, with two stations close to each other and as far as possible from the "headend." The delay is the minimum transmission time and can be converted into the minimum frame size.

The comparison of baseband and broadband in CSMA/CD schemes is as follows:

- **Different carrier sense (CS)**: Baseband detects the presence of transition between binary 1 and binary 0 on the channels, but broadband performs the actual carrier sense, just like the technique used in the telephone network.
- **Different collision detection (CD) techniques**: Baseband compares the received signal with a collision detection (CD) threshold. If the received signal exceeds the threshold, it claims that the collision is detected. It may fail to detect a collision due to signal attenuation. Broadband performs a bit-by-bit comparison or lets the headend perform collision detection by checking whether higher signal strength is received at the headend. If the headend detects a collision, it sends a jamming signal to the outbound channel.

High-Speed Ethernet-Like LAN

When an ethernet LAN has the speed of 100 Mbps or higher, it is classified as a high-speed ethernet LAN. The **high-speed ethernet-like LAN** is a natural evolution of low-speed traditional 10-Mbps LAN. IEEE specified two standards: 100BASE-T and 100VG-AnyLAN. Table 6.2 provides a comparison between 100BASE-T and 100VG-AnyLAN

TABLE 6.2 Comparison Between 100BASE-T and 100VG-ANYLN

	100BASE-T	100VG-AnyLAN
Standard	Part of IEEE802.3	IEEE802.12
Access protocol	CSMA/CD	Round Robin
Frame format	CSMA/CD 10BASE	CDMA/CD 10BASE or Token Ring
Maximum number of stations	1024	Unspecified
Frame size	1500 octets	1500 or 4500 octets
Distance coverage	210 m	2.5 Km
Topology	Star Wired	Hierarchical Star/Tree

The 100BASE-T network has several options of cabling schemes that are listed as follows:

- **100BASE-TX**: Each station uses two category 5 Unshielded Twisted Pair (UTP) or Shielded Twisted Pair (STP). One pair is for transmitting, and the other pair is for receiving.
- **100BASE-FX**: Each station uses two optical fibers. One is for transmitting, and the other is for receiving.
- **100BASE-T4**: Each station uses four pairs of category 3 UTP, three pairs for transmission, and one pair for collision detection.

Each station is connected to a multipoint repeater, and each input to the repeater is repeated on every output link. If two inputs overlap, a jam signal is transmitted on all output links. The class 1 repeater can support mixtures of transmission media types; a class 2 repeater can support only the same transmission media type, but it can interconnect with another class 2 repeater. Each network can be configured as either a **full duplex** or **half duplex**. The full duplex operation requires a switch hub and not a multipoint repeater. In the full duplex operation, each station would own the entire bandwidth of the network. In the half duplex operation, each station would share the bandwidth of the transmission medium. To support a mixture of speeds on a network, each device can send link integrity pulses to indicate the speed it supports. The process is called **autonegotiation**. The priority order in the technology ability field used in the 100BASE-T autonegotiation scheme is based on the highest common denominator regarding abilities. The 100BASE-T4 medium has the higher priority than the 100BASE-FX, and the 100BASE-T4 provides for the operation over a wider range of twisted pair cables and, therefore, is more flexible. The 10BASE-T is the lowest common denominator and has the lowest priority. In general, a full-duplex capability can also operate in half duplex mode and, hence, has a higher priority. The priority is assigned in the following order from highest to lowest:

- 100BASE-TX full duplex
- 100BASE-T4
- 100BASE-TX
- 10BASE-T full duplex
- 10BASE-T

Gigabit Ethernet

Due to the popularity of ethernet, the IEEE802 committee has also defined the standard for **gigabit ethernet**. A 10-gigabit ethernet has been made commercially available. Some of the requirements for this gigabit ethernet are as follows. This ethernet:

(1) Needs high speed for backbone network;
(2) Should provide a smooth migration from a 100-Mbps ethernet to gigabit ethernet;
(3) Is an alternative to ATM and FDDI for backbone connection; and
(4) Should work for copper wires and optical fiber.

A gigabit ethernet uses the same CSMA/CD protocol in a same-frame format. It can provide for a 100-Mbps traffic load as a backbone network. There are some differences however, with a previous version of ethernet, which are listed as follows:

1. A separate Gigabit Medium Independent Interface (GMII) defines an independent 8-bit parallel transmission and receives synchronous data interface. GMII is optional for all of transmission media except UTP. It is a chip-to-chip interface between MAC and physical hardware.
2. Two encoding schemes are 8B/10B for fiber and shielded copper and a 4-Dimensional 5-Level Pulse Amplitude Modulation are (4D-PAM5) for UTP.
3. A Carrier Extension Enhancement appends a set of symbols to a short MAC frame to a resulting block of at least 4096 bits, so transmission time is longer than the propagation time.

6.2.5 IEEE 802.4—Token Passing Bus Protocol

With the **IEEE 802.4 Token Passing Bus Protocol**, each station is assigned a logical position in an ordered sequence in a physical bus topology and a logical ring. Each station knows its logical preceding and successor addresses. A token frame regulates the access of medium. When finishing a transmission or when time elapses, a token is passed to the next station in the logical ordering. Each station can only transmit a frame when it holds a token *except* when a non token-holding station responds to a poll or a request for acknowledgment. There are several token maintenance events described in the following paragraphs.

Ring Initialization

This event occurs when a logical ring is broken or a network has just powered up and the system needs to determine which station can hold the token first. The event will trigger the following steps at each station:

(1) Arbitrary length (multiples of time slots) of a claim token is issued based on the first two bits of its address field. For example, 00 will be mapped to one time slot, 01 will be mapped to two time slots, 10 will be mapped to three time slots, and 11 will be mapped to four time slots.
(2) A station drops its claim after transmitting its claim token and then hears anything transmitted on the bus; otherwise, it tries to claim a token at the next iteration with the next two address bits. A station uses all of its address bits; if the station succeeds on the last iteration, it is considered a token holder.

Addition to Ring

The occurrence of an event is used for the expansion of a network. This occurrence is used to grant opportunity for a nonparticipating station to join in the network. The event follows these steps:

1. Each station periodically issues a solicit-successor frame to invite any nonparticipating station to join in the network.
2. If there is no response, the station transfers the token to its successor. If there is only one response, the station sets its successor to the requesting station and passes the token to the requesting node. It also indicates who is the requesting station's successor.
3. The station sends a contention resolution frame for the multiple responses and waits for any valid response. Each station can only respond in one of four time slot windows based on its two address bits and refrains from its claim if it hears anything before its window slot arrives. If the token holder station receives a valid set-successor frame, it passes the token to the demander. Otherwise, it tries for the next iteration. Only the stations that responded the previous time are allowed to respond.
4. If an invalid response is received, the station goes into an idle or listening state to avoid the conflict situation in which another station thinks it holds a token.

Deletion of a Station

The occurrence of event is when a station wants to separate itself from the network. The event triggers the following procedures:

1. A station waits for a token to arrive and then sends a set-successor frame to its preceding station to splice its successor.
2. The token holder station passes the token to its current successor as usual.
3. The station will then be deleted from the network in the next go-around.

A duplicated address or a broken logical ring would trigger the error recovery for the token bus protocol. Table 6.3 illustraties the error recovery procedures:

TABLE 6.3 Token Bus Error Recovery Procedures

Condition	Action
Multiple tokens	Drops tokens to 1 or 0
Unaccepted token (no valid frame is received)	Retries the sending token two times
Failed station after two tries of sending token	Sends the message of who follows the failed station frame
Failed receiver after another two tries of sending the message of who follows the frame	Enters listen state
No token activity	Becomes inactive with time out and initializes ring

Comparison of CSMA/CD and Token Bus

An ethernet using CSMA/CD is much more popular than a Token Bus network. The main reasons for this popularity are the cheaper cost for installation and ease of administration. This section compares the pros and cons of the protocols and not the products.

Advantages of CSMA/CD

- Simple algorithm
- Widely used to provide favorable cost and reliability
- Fair access
- Good performance at low to medium load

Disadvantages of CSMA/CD

- Collision detection issues/requirements
- Limit on minimum packet size (72 bytes minimum)
- Nondeterministic delay

Advantages of Token Bus

- Excellent throughput performance
- No special requirements for signal strength
- Deterministic delay
- Class of services with priority schemes
- Stability at higher loads

Disadvantages of Token Bus

- Complicated algorithms
- Overhead for passing token even with *light* traffic

6.2.6 IEEE 802.5 Token Passing Ring Protocol

The **IEEE 802.5 Token Passing Ring** protocol uses the ring topology where repeaters are connected via a transmission medium to form a closed path. Data are transmitted serially bit by bit through the transmission media, and they regenerate the bits at each repeater before they are sent to the next repeater on the ring. Data are transmitted in packets. A repeater should perform the packet reception, transmission, and removal functions. A repeater can be one of three states: **listen**, **transmit**, and **bypass** states. A user station is connected to the repeater and contains the MAPs, which is summarized below:

(1) A station can only transmit data when it holds a free token.
(2) A free token turns into a busy token followed by a packet to be transmitted.
(3) The originating station purges the packet and releases a free token.
(4) To determine when to release the free token, there are three approaches. The **Single Token** protocol does not

release the free token unless the leading edge of the busy token is returned. The **Multiple Token** protocol releases the free token as soon as it has completed transmitting release occurs even if the busy token has not returned). The **Single Frame** protocol waits until all of data has been purged and then releases a free token. In general, if the frame length is smaller than the ring length measured in bit time, the multiple protocols have better throughput. Regardless of which protocol is used, the length of a ring should be long enough to hold a free token.

The token Passing Ring protocol uses two subfields in the frame to recognize whether the frame is copied successfully by the destination and whether a duplicated address is detected. When a frame is returned to the sender, the subfield *C* indicates that the frame is copied to the destination buffer, and the subfield *A* indicates if the duplicated address condition has been detected. A destination station detects the duplicated address condition when it finds out that some other station already marks the *C* bit. Unlike CSMA/CD, Token Passing Ring protocol provides the priority and reservation algorithms in the protocol. The following steps outline the algorithm with notations **Pm**, which is the priority of the message to transmit; **Pr**, which is the priority of the token protocol; and **Rr**, which is the receive reservation priority.

1. A station waits for a free token with a *Pr* less than or equal to *Pm* and then seizes it. If the free token has higher priority (i.e., Pm < Pr), the station can set the Rr field to Pm *only if* Rr is less than Pm and Pm is less than Pr.
2. A station reserves the priority at a busy token by setting the Pm to the Rr field *if* the Rr is less than the Pm.
3. After seizing a token, the token indicator bit is set to 1, the Rr field is set to 0, Pr is unchanged.
4. When releasing a free token, the Pr field is set to the max(Pr, Rr, Pm), and the Rr field is set to max(Rr, Pm).
5. Each station downgrades the priority of a free token to a former level stored in a stack.

Just like the Token Passing Bus protocol, the Token Passing Ring protocol has a token maintenance problem when a station detects no token activity or detects a persistently circulating busy token by issuing a new token. When a station detects no token activity for a period of time, the network enters the initialization procedure to claim a token. A monitor station on the network sets the monitor bit in the frame. It will then remove the frame and issue a new token if the monitor station has seen the monitor bit set. This ring protocol enables a fair access with a "round robin" scheme, but it is inefficient under a lightly loaded system because the station that has a frame to be transmitted has to wait for the token to arrive. Unlike CSMA/CD, a station can transmit a frame right away as long as no other station is transmitting. A ring network, however, has many advantages over a bus network. Some major advantages are less noise and larger distance coverage resulting from the use of repeaters and a high-speed link for the use of optical fiber. The major problem with a ring network is the accumulation of a timing jitter along the ring and across multiple repeaters. This effect significantly limits the size of a ring network.

6.2.7 IEEE 802.11—Wireless LANs

The increased demands for mobility and flexibility in daily life are demands that lead the development from wired LANs to wireless LANs (WLANs). Today a WLAN can offer users high bit rates to meet the requirements of bandwidth consuming services, such as video conferences and streaming videos. With this in mind, a user of a WLAN will have high demands on the system and will not accept too much degradation in performance to achieve mobility and flexibility. This will in turn put high demands on the design of WLANs of the future. In this section we first discuss the WLAN architecture, deployment and security issues. Finally, future trends in wireless technology are also discussed.

Architecture of WLAN

WLAN consists of access points and terminals that have a WLAN connectivity. Finding the optimal locations for access points is important and can be achieved by measuring the relative signal strength of the access points. Placing the access points in a corporation network opens an access way to the resources in an intranet. With wired LANs, an intruder must first gain physical access to the building before he or she can plug a computer to the network and eavesdrop on the traffic. The intranet is typically considered secure even though employees can cause security breaches and data are transmitted unencrypted. If the information transmitted in the corporation network is extremely valuable to the corporation, the WLAN interface should be protected from unauthorized users and eavesdropping. The obvious way to extend the intranet with a wireless LAN is to connect the access points directly to the intranet as illustrated in Figure 6.4.

The WLAN standard is defined in the **IEEE 802.11 WLAN protocol**, an extension of the IEEE 802.3 standard for wired LAN. The term LAN applies to any type of WLAN. It can be based on 802.11a, 802.11b, and 802.11g. A new standard called IEEE 802.11n is evolving to provide hundreds of megabits per second data rate. Another name for WLAN is wireless fidelity (Wi-Fi). Hotspots are another big development in the Wi-Fi field. Hot spots are usually public spaces such as libraries, malls, parks, beaches, airports, hotels, coffee shops, and restaurants where Wi-Fi Internet access is offered for a price or, even in some cases, offered for free. It is predicted that the number of hot spots could peak at 150,000 by 2005 before eventually declining because of unprofitable hot spots being deactivated. Of course, sometimes these hot spots can be

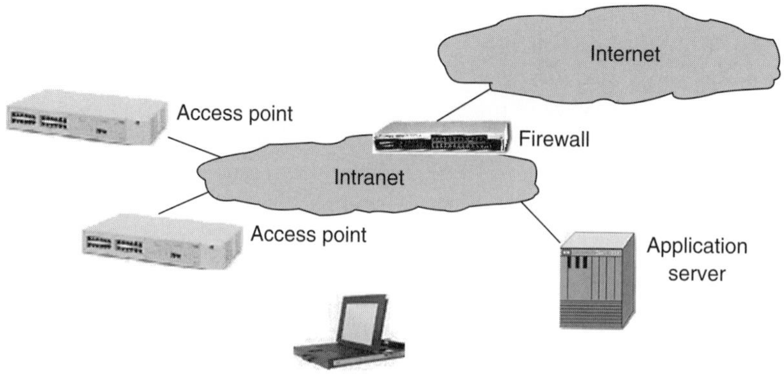

FIGURE 6.4 Access Points of WLAN

somewhat confining because of one needs to make payment arrangements and also obtain a password for each use. Different hot spots are also often run by different Internet service providers, so paying just one provider will not always guarantee service at all hot spot zones. In addition, most hot spots only offer minimal if any security. Most clerks at these places have no idea of how to turn on the security features of their deployed WLAN. IEEE 802.11 network architecture can be configured in two different ways. One way is the **independent basic service set** (IBSS), also often referred to as an ad hoc network or even a peer-to-peer mode (see Figure 6.5). In an ad hoc network, 802.11 wireless stations in the same cell or in the transmission range of each other communicate directly with one another without using any connection to a wired network. This kind of network is usually only formed on a temporary basis and put up and taken down rather quickly. These networks usually are of small size and have a focused purpose. A business meeting in a conference room where employees bring laptops to share information is a good example of this type of network.

The second way and the most frequent way an IEEE 802.11 network architecture can be configured is by the **basic service set** (BSS), also often referred to as an infrastructure network (see Figure 6.6). In an infrastructure network, one station, called an access point (AP), acts as a central station for each cell, and many overlapping cells can make up a network. If there is more than one cell, the network is actually called an extended service set (ESS), and each cell is actually considered a basic service set (BSS) (see Figure 6.7). In an ESS, each BSS is connected together by means of a distribution system (DS). So these types of networks can have one or more access points. These access points serve as an ethernet bridge between the wireless LAN and a wired network infrastructure. These access points are not mobile and are considered part of the wired network infrastructure. The wireless stations can be mobile and can roam from one cell to the next, allowing for seamless coverage throughout the whole service area of the ESS.

WLAN Security and Deployment Issues

WLAN technology provides tremendous convenience to many mobile users, but the initial deployment rate was not as expected. Security has represented the major obstacle to the

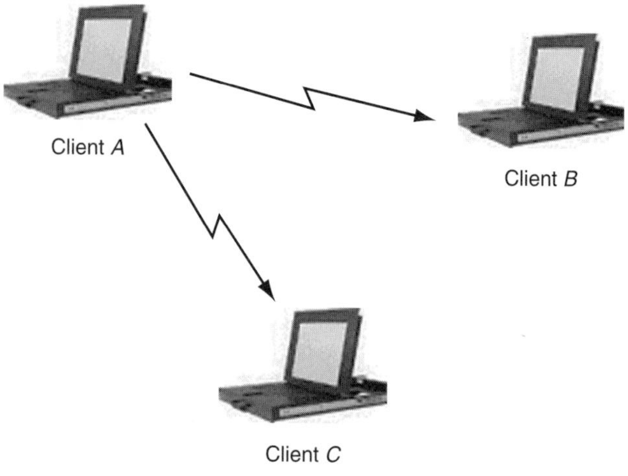

FIGURE 6.5 Example Ad Hoc Network

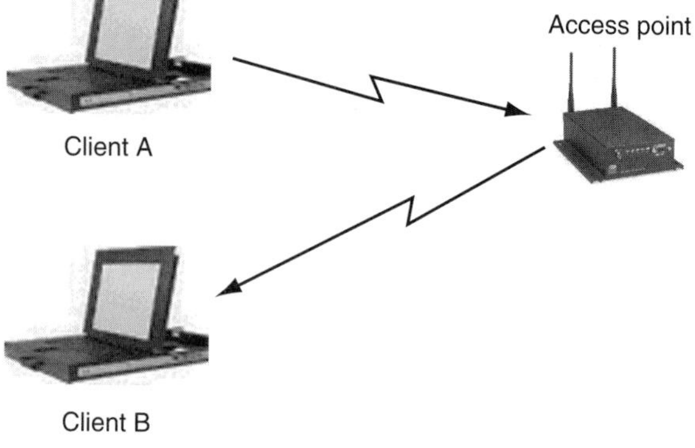

FIGURE 6.6 Basic Service Set: Example Infrastructure Network

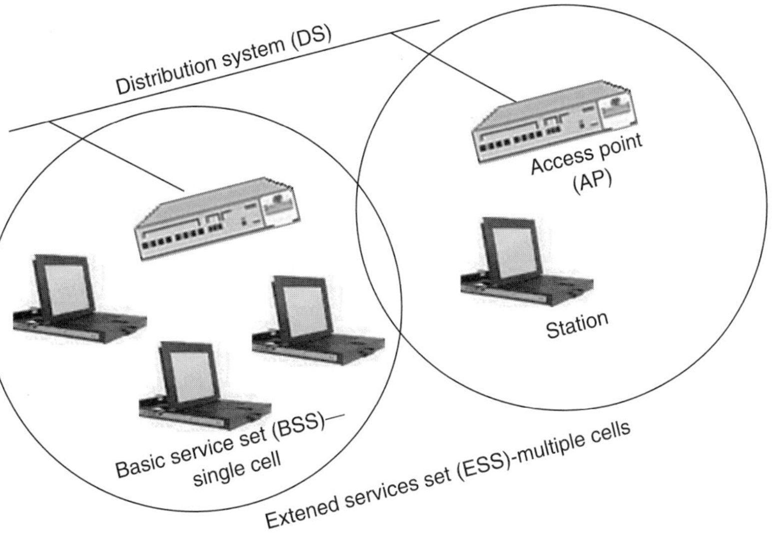

FIGURE 6.7 Extended Service Set—Infrastructure Mode

widespread adoption of WLAN. About 50% of IT managers still consider WLAN security being a major concern to deploy the technology throughout the company. Two popular security standards have been implemented in WLAN networks. They are WEP (Wireless Equivalent Protocol) and WPA (Wi-Fi Protected Access).

WEP is implanted in 802.11b protocol and works at Layer 2 to encrypt all over the air transmission using 40/64 bit keys. Unfortunately, these keys are not strong and are easier to break. Furthermore, it only authenticates a device, not a user. **WPA** is a security enhancement that tries to solve two problems of WEP—weak data encryption and lack of user authentication capability. In order to provide a stronger encryption algorithm, WPA employs Temporal key Integrity Protocol (TKIP), which provides per-packet key to make the hacker's job more difficult. WPA also implements the Extensible Authentication Protocol (ESP), which works with a RADIUS authentication server to authenticate and authorize a user using WLAN. Virtual Private Network (VPN) can also be used to regulate access to a company's corporate network from WLAN, either in a private network or public access network. VPN is placed behind the wireless access points to authenticate and authorize a user.

WLAN deployment can be either centralized or distributed. In the central approach, the access points are mainly dumb devices with the intelligence existing in a central switch. The access points are just radio stations. In the distributed approach, each access point has intelligence and is configured with access-control information, user authentication, and policy administration. The centralized approach may save operational cost but may not be as efficient as the distributed approach.

Future WLAN Technological Developments

Although it is very risky to speculate on future developments in a field that is changing as fast as wireless communications, there does seem to be a certain amount of consensus about broad trends.

Communicating Anytime, Anywhere, and in Any Mode

There seems to be a wide acceptance of the notion of being able to communicate anywhere, anytime, and in any mode. Software programmable radios allow a subscriber unit to adapt its modulation, multiple-access method, and other characteristics to be able to communicate with a wide range of different systems and more efficiently support diverse voice, data, image, and video requirements. Likewise, a single base station unit will be able to adapt its characteristics to support older generations of subscriber equipment while retaining the ability to support future developments through software changes.

Extending Multimedia and Broadband Services to Mobile Users

Providing for the electronic transport of multimedia services is a challenging one because of the inherent differences in voice, data, image, and video traffic as explained earlier. Extending multimedia and bandwidth on-demand capabilities into the wireless environment presents a number of challenges. In the mobile, wireless world, available bandwidths (and, hence, transmission rates) are typically limited by spectrum scarcity, and the transmission links between the mobile and base stations are often characterized by high error rates and rapidly changing performance. In addition, a mobile terminal that is connected to one port on the network at one moment may be connected to another one the next moment. Despite these

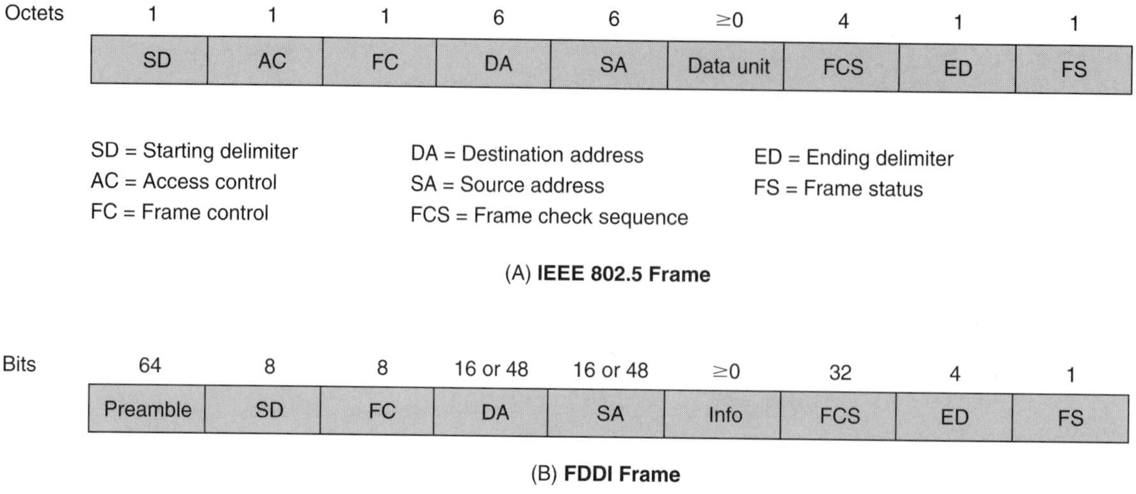

FIGURE 6.8 Frame Formats of IEEE 802.5 and FDDI.

challenges, there is a considerable amount of research interest in wireless ATM, and an entire issue of *IEEE Personal Communications* was devoted to the topic.

Embedded Radios

There is also a long-term trend toward, for want of a better term, **embedded radios**. This trend is comparable to the trend in microelectronics that is resulting in computer chips being installed in such ordinary items as automobiles, washing machines, sewing machines, and other consumer appliances and office equipment. Radios are even being installed in humans in the form of pacemakers and heart monitors. These devices are less visible than personal computers, computer workstations, and mainframe computers, but their impact on everyday life is significant nevertheless.

6.2.8 ANSI X3T9.5—Fiber Data Distribution Interface

The other type of LAN is called **metropolitan area network** (MAN). MAN has characteristics of high capacity of supporting at least 100-Mbps speed and more than 500 stations on the network. It has a larger geographic scope than LAN provides support for integrated data types, and has the provision of dual cables to improve throughput and reliability

Fiber data distribution interface (FDDI) is an example of MAN that is an **ANSI X3T9.5 standard**, proposed by the ANSI XT9.5 study group. Each field in the MAC frame can either be represented as a symbol (4 bits, for nondata) or in bits format (data or address). The IEEE 802.5 token ring frame format is similar to the FDDI frame format. The FDDI frame includes a preamble to help in clocking because this is desirable for high-speed communication. FDDI does not include an access control (AC) field that is used for priority reservation scheme. FDDI uses a capacity allocation scheme instead of priority reservation. Figure 6.8 shows the frame formats of IEEE 802.5 and FDDI.

FDDI MAC protocol is very similar to IEEE 802.5–Token Passing Ring protocol, but there are differences described as follows:

- After absorbing the free token, an FDDI station starts to transmit data frames. In IEEE 802.5—Token Passing Ring MAC, the token type bit is changed to a busy token type, and the data is appended to the token.
- An FDDI station frees the token right after transmission of a data frame, and it will not wait for the data frame return. In IEEE 802.5, a station will not free a token until the leading edge of the data frame returns.
- FDDI MAC uses the Time-Token Protocol (TTP) for both synchronous and asynchronous services to all stations. If a token arrives earlier based on the rotation timer, a station can optionally send the asynchronous data. If the token is late, only the synchronous data can be sent. In IEEE 802.5, MAC protocol is based on explicit priority reservation. Both protocols allow the network to respond to changes in traffic load, but FDDI supports more steady load because lower-priority traffic may have more opportunity to send when a token arrives early.
- The use of a restricted token will allow for two stations to have multiframe dialog capability to interchange long sequences of data frames and acknowledgments. This would improve the performance of the application that uses the capability.

The data-encoding scheme used by FDDI is called 4B/5B, which encodes 4-bit data within a 5-bit cell. There is no more than three consecutive zero bits in a cell, and at least two transitions occur in a 5-bit cell. The binary bit values are represented with non return to zero inverted (NRZI), where the transition at the beginning of the bit time denotes a binary 1 for that bit time, and no transition indicates a binary 0. Only 16 out of 32 code patterns are used for data, and other patterns are used for control symbols. Timing jitter is one of transmission impairments in the data communication. The deviation of clock recovery occurs when a receiver attempts to recover clocking as well as data. Due to the high speed of transmission, the deviation of the clock is more severe for FDDI than for IEEE 802.5. The centralized clocking used by the IEEE 802.5 network is inappropriate for 100 Mbps, and it requires a complicated and expensive phase lock loop circuitry. Distributed clocking is therefore used by FDDI. With distributed clocking, each station recovers a clock from its incoming signal and transmits out at station clock speed. Each station also maintains its own elastic buffer, unlike the IEEE 802.5 network that only designates one station to have the elastic buffer.

FDDI specifies three reliability requirements of providing an automatic bypass for a bad or power-off station, using dual rings for easy reconfiguration when one ring is broken, and installing a wiring concentrator. Table 6.4 summarizes the comparison between IEEE 802.5 Token Passing Ring and ANSI X3T9.5 FDDI.

FDDI is more reliable than other local network systems, but it is discouraged due to its high cost. There is a trend to migrate FDDI into switched backbone fast ethernet for the following additional reasons:

- More bandwidth to the desktops would require increasing the backbone capacity
- Because of continued evolution of ethernet architecture and its speed, it would make business and technical sense to migrate FDDI backbone into switched fast ethernet. Table 6.5 shows the comparison of FDDI and fast ethernet.

TABLE 6.4 Comparing IEEE 802.5 and X3T9.5 FDDI

ANSI X3T9.5 FDDI	IEEE 802.5 Token Passing Ring
Fiber or twisted pair as transmission medium	Twisted pair or optical fiber as transmission medium
100 Mbps	4, 16, or 100 Mbps
Reliability specification	No reliability specification
Maximum of 1000 stations	Maximum of 250 stations
4B/5B encoding	Differential Manchester
Distributed clocking	Centralized clocking
Maximum of 4500 octets of frame sizes	Maximum of 4550 octets for a 4-Mbps network and 18200 octets for 16-Mbps and 100-Mbps networks
Time token rotation	Priority reservation
Token release right after the transmission	Token release after busy token comes back

TABLE 6.5 Comparing FDDI and Fast Ethernet

	FDDI	Fast ethernet
Reliability	Self-healing dual rings	Can provide one backup connection
Maximum frame size	4500 octets	1518 octets
Performance	Sustained performance with increasing number of stations	Whole network owned by each user with the switched Ethernet
Distance	Up to 32 Km with fiber	Up to 32 Km with fiber
Price	Close to $1000/port	$100–$150/port

6.3 Local Network Internetworking Using Bridges or Routers

A bridge is a device that operates at the MAC level to connect with a similar type of local network, and a router is a device that operates at the third layer of protocol to connect dissimilar types of local networks, including a WAN. Internetworking using **bridges** or **routers** are required for Communication with multiple LANs in the organization, Communication with a LAN in a different location, and communication with the Internet.

A simple bridge contains just the forwarding function, but a more sophisticated bridge would offer additional functions, like dynamic routing, priority, and congestion control. There are several reasons to use a bridge for connecting multiple LANs:

- **Reliability**: To avoid a single point of failure, a network can be divided into several self-contained units.
- **Performance**: Placing users into several smaller LANs rather than in a large LAN would reduce the contention time and improve the performance.
- **Security**: It is more desirable to keep different types of traffic separate to improve security (e.g., work group communication can be on the same LAN).
- **Location**: Different floors or buildings requires a bridge to connect these locations.

The communication path between two LANs/MANs can be connected either by a single bridge or multiple bridges. In the case of connecting with multiple bridges, a point-to-point protocol, like High-Level Data Link Control (HDLC) or an X.25 connection, can be used between two bridges. Bridges do not modify the content of the frames they receive but must contain addressing and routing intelligence. The bridge can also act as a multiple port switch connecting more than two LANs.

The routing approaches of bridges are as complicated as routers. The basic routing operation is to avoid forwarding frames in a closed loop. Each bridge should make a decision about whether the frame should be forwarded and what LAN should forward the frame next. The following approaches are used for routing with bridges:

- **Fixed routing**: A central routing directory is created at the network control computer. The directory contains the source-destination pairs of LANs and the bridge ID for each pair of LANs. Each bridge on the network has a table that contains the source-destination pairs of LAN for each of LANs attached to the bridge.
- **Transparent bridge**: The algorithm was developed by IEEE 802.1 and intended to interconnect similar LAN or dissimilar LAN (IEEE 802.3, IEEE 802.4, and IEEE 802.5) and learn in which side of LAN a MAC station would reside by examining the source and destination addresses. The main idea is for the bridges to select the ports over which they will forward frames. The algorithm uses graph theory to construct a spanning tree to avoid a closed loop to forward the frame. The algorithm selects one root bridge for the network, one designated port for each bridge, and one designated bridge for each LAN. All frames will be forwarded toward the root bridge via the designated bridge and the designated port. The root bridge then forwards the frames via all of its ports. The root bridge has the smallest ID, and each bridge computes the shortest path to the root and notes which port is on the path. The designated port is the preferred path to the root. Each LAN selects a designated bridge that will be responsible for forwarding frames toward the root bridge. The winner is the bridge closer to the root and has the smaller ID.
- **Source routing**: The routing algorithm was developed by IEEE 802.5. Each source station determines the route that the frame will take and includes a sequence of LAN and bridge identifiers in the route information field (RIF). The route information is obtained by discovery of broadcast frames sent by a source station. The approach requires the changes (additional bits) to the MAC frame format
- **Source routing transparent** (SRT): This approach allows for the interconnection of LANs by a mixture of source routing and transparent bridging. The route information indicator (RII) bit in MAC source address determines which algorithm will be used. If the RII is equal to 0, then the frame is handled by transparent bridge logic; if it equals to 1, it is handled by the source routing logic.

Routers in a LAN are used to connect LANs and the Internet. The addressing scope of the router is at the network layer (e.g., IP). A router makes a decision about how to route the packets. The routing protocol can be either internal within a domain network or external between two domain networks.

6.3.1 Performance of Local Networks

The objectives to study the performance of a LAN is to understand the factors that affect the performance of a local network, to understand the relative performance of various local network protocols, and to apply the knowledge in the design and configuration of a local network. The common characteristics between LAN and MAN are that a shared medium is used, the packet switching technique is used, and the MAP is used. The common measurable parameters for LAN and MAN are described as follows:

- **Delay** (D): The time period between when the time a frame is ready for transmission from a station and the completion of the successful transmission
- **Throughput** (S): Total rate of data being successfully transmitted between stations
- **Utilization** (U): The fraction of total network capacity being used
- **Offered load** (G): The total rate of data presented to network for transmission
- **Input load** (I): The rate of data generated by the stations attached to a local network

How are these parameters are related? The S and U are proportional to G, but they are flat when network capacity is exceeded. The D increases when the number of active users or I is increased.

A local network distinguishes itself as short distance; therefore, the smaller propagation delay would provide instant feedback about the state of a transmission. The frame size could determine the transmission time and also plays an important role in the designing network and affects the performance of a local network. The parameter a is obtained by dividing the propagation time by the transmission time and is the most important single parameter that affects the performance of the local network. The smaller a is, the better performance the network has. If transmission time is normalized to 1, a should indicate the propagation delay of a network. The parameter a can also be used as a good indication of effects on utilization and throughput. In general, the utilization varies inversely with a ($U = 1/(1 + a)$) regardless of which MAC protocol is used. Throughput is affected by the parameter a regardless of the offered load (G). The larger the a is, the lower throughput the network has.

Different MAC protocols produce different types of overhead. For the contention protocols, like ALOHA, CSMA, and CSMA/CD, the overheads are collisions, acknowledgment, and waste of slot time. For the token bus protocol, the overheads are the passing token (even when the network is idle), token transmission, and acknowledgment. For the token ring protocol, the overhead is just waiting for the token when the other station has no data to send.

The factors that affect the performance of LAN/MAN are listed as follows:

- **Capacity**: Affects the parameter a
- **Propagation delay**: Affects the parameter a
- **Frame size**: Affects the parameter a
- **MAC protocols**: Affects throughput and delay

- **Offered load**: Affects throughput and utilization
- **Number of stations**: Affects delay, input load, offered load, and throughput
- **Error rate**: Not quite an issue for LAN/MAN performance

The number of active users in a local network is an important parameter that affects the bound of the performance. Understanding the bounds of the performance would help to determine the size of a LAN. The stability of a LAN can be at the state of either low delay, where the capacity is greater than the offered load; of high delay, where more time is spent on controlling access to network, and little time is spent on actual data transmission; or the unbound delay, where the offered load is greater than the capacity. The state of the unbound delay must be avoided in the design. In general, delay is bounded by the number of users (N) multiplied by the transmission time of **Tmsg**. The throughput is bounded by $N/(\text{Tidle} + N^*\text{Tmsg})$, where the **Tidle** is the mean idle time. The number of active stations can be determined by the following equation:

$$N = \text{Tidle}/\text{Tmsg}.$$

Generally, the study of LAN performance made by the IEEE 802 Committee suggests that smaller frame sizes have greater differences in the throughput between token passing and CSMA/CD. CSMA/CD strongly depends on the value of a. Token ring is the least sensitive to the workload. CSMA/CD has the shortest delay under the light load and is most sensitive to the heavy load.

6.4 Conclusion

The International Standard Organization's (ISO) Open System Interconnection (OSI) continues to be the de facto standard for network reference model. All of existing networks and future evolving networks continue to model their network based on OSI's reference model. This chapter introduces a general network architecture using OSI and provides a more detailed description of Local Network and Wireless Local Area Network. WLAN has been extensively used in home/office network and there are increasing numbers of public hotspots worldwide. The seamless interoperability between WLAN and mobile cellular networks has been on the way. It is expected that utilizing IP network to deliver various types of applications becomes a norm. IP Multimedia Subsystem (IMS) is intended to deliver ALL IP network architecture for all types of applications.

Glossary

4B/5B	Encoding scheme to encode 4 bits of data into a 5-bit cell
ANSI	American National Standard Institute
ATM	Asynchronous Transfer Mode
CSMA	Carrier Sense Multiple Access
CSMA/CD	Carrier Sense Multiple Access with Collision Detection
FDDI	Fiber Data Distribution Interface
IEEE 802	An IEEE committee that specifies LAN standards
IMS	IP Multimedia System
LAN	Local Area Network
LLC	Logical Link Control
MAC	Medium Access Control
MAN	Metropolitan Area Network
NRZI	Non-Return Zero Inverted
TTP	Time-Token Protocol
WEP	Wireless Equivalent Protocol
WPA	Wi-Fi Protected Access

References

de Buda, R. (1972). Coherent demodulation of frequency shift keying with low deviation ratio. *IEEE Transactions on Communications COM-20*, 466–470.

Garg, V.K., and Wilkes, J.E. (1996). Wireless and personal communications systems. Engelwood Cliffs, NJ: Prentice Hall.

Murota, K., and Hirade, K. (1981). GMSK modulation for digital mobile radio technology *IEEE Transactions on Communications COM-29* (7), pages. 000–000.

Pasupathy, S. (1979). Minimal shift keying: A spectrally efficient modulation. *IEEE Communications Magazine*, Vol., pages. 00, 000–000.

Proakis, J.G. (1989). *Digital communication.* New York: McGraw-Hill. 00, 000–000.

Schwartz, M., Bennett, W., and Stein, S. (1966). *Communications systems and techniques.* New York: McGraw-Hill, 00, 000–000.

Sklar, B. (1988). *Digital communications: Fundamental and applications.* Englewood Cliff, NJ: Prentice Hall.

(1989). Special issue on bandwidth and power efficient coded modulation. *IEEE Journal of Selected Area in Communications 7.* 000–000.

(1991). Special issue on bandwidth and power efficient modulation. *IEEE Communications Magazine 29.* 000–000.

Stalling, W. (1988). *Data and computer communications.* New York: MacMillan Publishing. 00, 000–000.

Wozencraft, J.M., and Jacobs, I.M. (1965). *Principals of communications engineering.* New York: John Wiley Sons. 00, 000–000.

Ziemer, R.E., and Peterson, R.L. (1992). *Introduction to digital communication.* New York: Macmillan Publishing. 00, 000–000.

Ziemer, R.E., and Tranter, W.H. (1990). *Principle of communications.* Boston: Houghton-Mifflin. 00, 000–000.

7

Wireless Network Access Technologies

Vijay K. Garg
Department of Electrical and Computer Engineering, University of Illinois at Chicago, Chicago, Illinois, USA

Yih-Chen Wang
Lucent Technologies, Naperville, Illinois, USA

7.1 Access Technologies

The problem of a cellular system boils down to the choice of the way to share a common pool of radio channels between users. Users can gain access to any channel (each user is not always assigned to the same channel). A channel can be thought of as merely a portion of the limited radio resource that is temporarily allocated for a specific purpose, such as someone's phone call. A multiple access method is a definition of how the radio spectrum is divided into channels and how channels are allocated to the many users of the systems. The sharing of spectrum is required to achieve high capacity by simultaneously allocating the available bandwidth to multiple users. There are three possible multiple access methods: **frequency-division multiple access** (FDMA), **time-division multiple access** (TDMA), and **code-division multiple access** (CDMA).

7.1.1 FDMA

FDMA assigns individual channels to individual users. It can be seen from Figure 7.1 that each user is allocated a unique frequency channel. These channels are assigned on demand to subscribers who request service. Guard bands are maintained between adjacent signal spectra to minimize cross talk between channels. During the period of the call, no other user can share the same frequency band. In frequency division duplex (FDD) systems, the users are assigned a channel as a pair of frequencies; one frequency is used for the upward channel, while the other frequency is used for the downward channel.

FDMA is used by analog systems, such as AMPS, NMT, or Radiocom 2000. The **advantages of FDMA** are the following:

- The complexity of FDMA systems is lower when compared to TDMA and CDMA systems, though this is changing as digital signal processing methods improve for TDMA and CDMA.
- FDMA it is technically simple to implement.
- A capacity increase can be obtained by reducing the information bit rate and using efficient digital codes.
- Since FDMA is a continuous transmission scheme, fewer bits are needed for overhead purposes as compared to TDMA.

The **disadvantages of FDMA** include the following:

- Only modest capacity improvements could be expected from a given spectrum allocation.
- FDMA wastes bandwidth. If an FDMA channel is not in use, then it sits idle and cannot be used by other users to increase or share capacity.
- FDMA systems have higher cell site system costs compared to TDMA systems because of the need to use costly bandpass filters to eliminate spurious radiation at the base station.
- The FDMA mobile unit uses duplexers since both the transmitter and receiver operate at the same time. A duplexer adds weight, size, and cost to a radio transmitter and can limit the minimum size of a subscriber unit.
- FDMA requires tight RF filtering to minimize adjacent channel interference.

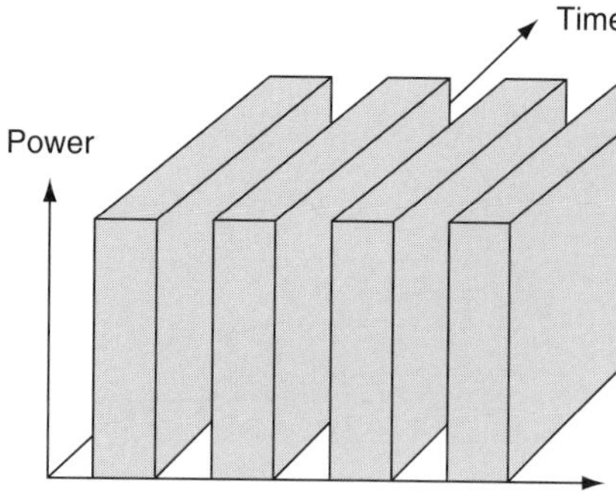

FIGURE 7.1 Frequency-Division Multiple Access (FDMA)

- The maximum bit rate per channel is fixed and small, inhibiting the flexibility in bit-rate capability that may be a requirement for 3G applications in the future.

7.1.2 TDMA

TDMA relies on the fact that the audio signal has been digitized (i.e., divided into a number of millisecond-long packets). It allocates a single frequency channel for a short period of time and then moves to another channel. The digital samples from a single transmitter occupy different time slots in several bands at the same time, as shown in Figure 7.2.

In a TDMA system, the radio spectrum is divided into time slots, and in each slot, only one user is allowed to either transmit or receive. It can be seen from Figure 7.2 that each user occupies a cyclically repeating time slot, so a channel may be thought of as a particular time slot that reoccurs every frame, where several time slots make up a frame. Since the transmission for any user is noncontinuous, digital modulation must be used with TDMA. The transmission from various users is interlaced into a repeating frame structure as shown in Figure 7.3. Each frame is made up of a preamble, information bits addressed to various stations, and trail bits. The function of the **preamble** is to provide identification and incidental information and to allow synchronization of the slot at the intended receiver. Guard times are used between each user's transmission to minimize cross talk between channels. In a TDMA/TDD system, half of the time slots in the frame information message would be used for the forward link channels and half would be used for reverse link channels. In TDMA/FDD systems, an identical or similar frame structure would be used for either forward or reverse transmission, but the carrier frequencies would be different for the forward and reverse links.

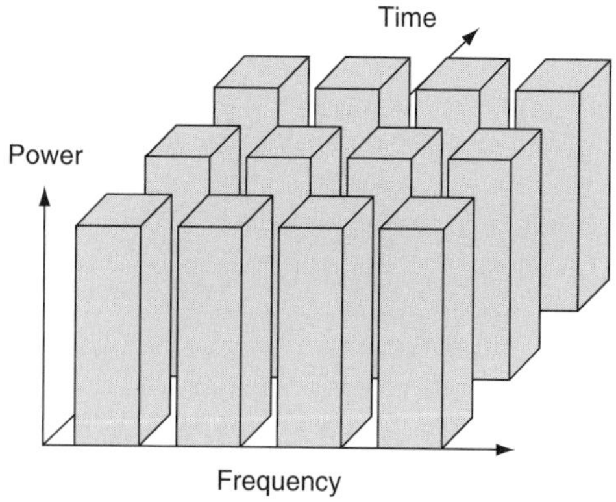

FIGURE 7.2 Time-Division Multiple Access (TDMA)

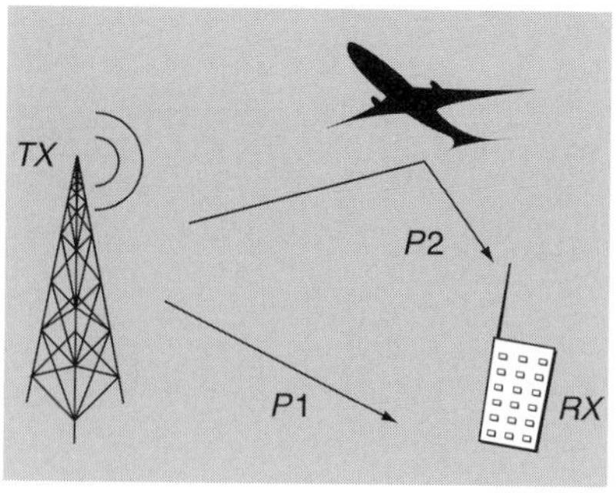

FIGURE 7.3 Multipath Interference

GSM uses a TDMA technique, where the carrier is 200 kHz wide and supports eight full rate channels. A channel (roughly) consists of the recurrence every 4.6 ms of a time slot of 0.58 ms.

The **advantages of TDMA** are the following:

- TDMA permits a flexible bit rate, not only for multiples of a basic single channel rate but also submultiples for low bit-rate multicast traffic.
- The handoff process in TDMA is much simpler for a subscriber unit because it is able to listen for other base stations during idle time slots.
- TDMA potentially integrates into VLSI without narrow band filters, giving a low-cost floor in volume production.
- TDMA uses different time slots for transmission and reception; thus, duplexers are not required.
- TDMA has the advantage in that it enables allocation of different numbers of time slots per frame to different users. Thus, bandwidth can be supplied on demand to different users by concatenating or reassigning time slots based on priority.
- TDMA can be easily adapted to the transmission of data as well as voice communication.
- TDMA offers the ability to carry data rates of 64 kbps to 120 Mbps (expandable in multiples of 64 kbps). This enables operators to offer personal communication-like services including fax, voice-band data, and short message services (SMSs) as well as bandwidth-intensive applications, such as multimedia and videoconferencing.
- Unlike spread-spectrum techniques that can suffer from interference among users, all of whom are on the same frequency band and transmitting at the same time, TDMA's technology, which separates users in time,

ensures that they will not experience interference from other simultaneous transmissions.

- TDMA also provides the user with extended battery life and talk time because the mobile is only transmitting a portion of the time (From one-third to one-tenth) during conversations.
- TDMA installations offer substantial savings in base-station equipment, space, and maintenance, an important factor as cell sizes grow ever smaller.
- TDMA is the most cost-effective technology for upgrading a current analog system to digital.
- TDMA is the only technology that offers an efficient use of hierarchical cell structures (HCSs) offering pico, micro, and macrocells. HCSs allow coverage for the system to be tailored to support specific traffic and service needs. By using this approach, system capacities of more than 40-times AMPS can be achieved in a cost-efficient way.
- Because of its inherent compatibility with FDMA analog systems, TDMA allows service compatibility with the use of dual-mode handsets.

The **disadvantages of TDMA** are listed here:

- TDMA requires a substantial amount of signal processing for matched filtering and correlation detection for synchronizing with a time slot.
- each user has a predefined time slot. However, users roaming from one cell to another are not allotted a time slot. Thus, if all the time slots in the next cell are already occupied, a call might well be disconnected. Likewise, if all the time slots in the cell in which a user happens to be are already occupied, a user will not receive a dial tone.
- TDMA is subjected to multipath distortion. A signal coming from a tower to a handset might come from any one of several directions. It might have bounced off several different buildings before arriving (see Figure 7.3), which can cause interference.

One way of getting around this interference is to put a time limit on the system. The system will be designed to receive, treat, and process a signal within a certain time limit. After the time limit has expired, the system ignores signals. The sensitivity of the system depends on how far it processes the multipath frequencies. Even at thousandths of seconds, these multipath signals cause problems.

All cellular architectures, whether microcell- or macrocell-based, have a unique set of propagation problems. Macrocells are particularly affected by multipath signal loss—a phenomenon usually occurring at the cell fringes where reflection and refraction may weaken or cancel a signal.

Frequency and time division multiplexing can be combined (i.e., a channel can use a certain frequency band for a certain amount of time). This scheme is more robust against frequency selective interference (i.e., interference in a certain small frequency band). In addition, this scheme provides some (weak) protection against tapping. GSM uses this combination of frequency and time division multiplexing for transmission between a mobile phone and a base station.

7.1.3 CDMA

In CDMA, each user is assigned a unique code sequence for encoding an information-bearing signal. The receiver, knowing the code sequences of the user, decodes a received signal after reception and recovers the original data. This is possible since the cross correlations between the code of the desired user and the codes of the other users are small. Because the bandwidth of the code signal is chosen to be much larger than the bandwidth of the information-bearing signal, the encoding process enlarges (spreads) the spectrum of the signal and is therefore also known as **spread-spectrum modulation**. The resulting signal is also called a spread-spectrum signal.

CDMA is commonly explained by the "cocktail party image," where groups of people of different languages can communicate simultaneously, despite the surrounding noise. For a group of people speaking the same language, the rest of the people in the room are received as noise. Knowing the language they are talking allows them to filter out this noise and understand each other. If someone records the "noise," and knows different languages, he or she will be able, playing the tape several times, to extract the various conversations taking place in the various languages. With sufficient processing capacity, all conversations can be extracted simultaneously.

CDMA receivers separate communication channels by means of a pesudo-random modulation that is applied and removed in the digital domain (with these famous codes), not on the basis of frequency. Multiple users occupy the same frequency band.

The use of CDMA permits average interference among all users, thus avoiding the dimension of a network for the worst case. It thus permits to optimize the use of spectrum efficiency. It also efficiently supports variable bit-rate services.

This multiple access method reduces the peak transmitter power, thus reducing the power amplifier consumption and then increasing the battery efficiency.

CDMA reduces the average transmitted power, and increases the battery efficiency. It allows a reuse factor of 1 (i.e., the same frequency is used in adjacent cells), thus avoiding the need for frequency planning. On the other hand, code planning is required, but this is less difficult than frequency planning, as the code reuse pattern is much larger than the frequency reuse pattern commonly encountered in FDMA systems.

However, CDMA requires in particular a complex and very accurate power control, which is a key factor for the system capacity and its proper operation.

In direct sequence CDMA (DS-CDMA), the modulated data signal is directly modulated by a digital, discrete-time, discrete-valued code signal. The data signal can be either analog or digital; in most cases, it is digital. In the case of a digital signal, the data modulation is often omitted, and the data signal is directly multiplied by the code signal; the resulting signal modulates the wideband carrier.

Basic DS-CMA elements include the RAKE receiver, power control, soft handover, interfrequency handover, and multiuser detection. Key CDMA system attributes include increased frequency reuse, efficient variable rate speech compression, enhanced RF power control, lower average transmit power, ability to simultaneously receive and combine several signals to increase service reliability, seamless handoff, extended battery life, and advanced features.

IS-95 CDMA

The IS-95 CDMA air interface standard, after the first revision in 1995, was termed **IS-95A**; it specifies the air interface for cellular, 800-MHz frequency band. ANSI J-STD-008 specifies the air interface for 1900 MHz in PCS. It differs from IS-95A primarily in the frequency plan and in call processing related to subscriber station identity, such as paging and call origination. TSB74 specifies the Rate Set 2 standard. IS-95B merges the IS-95A, ANSI J-STD-008, and TSB74 standards. In addition, it specifies high-speed data operation using up to eight parallel codes, resulting in a maximum bit rate of 115.2 Kb/s. **Table 7.1** lists the main parameters of the IS-95 CDMA air interface.

Wideband CDMA

The 3G air interface standardization for the schemes based on CMA focuses on two main types of wideband CDMA: **network asynchronous** and **network synchronous**. In network asynchronous schemes, the base stations are not synchronized; in network synchronous schemes, the base stations are synchronized to each other within a few microseconds.

A network asynchronous CDMA proposal is WCDMA in European Telecommunication Standard Institute (ETSI) and Association of Radio Industries and Businesses (ARIB). A network synchronous wideband CDMA scheme has been proposed by CDMA2000. The main technical approaches of WCDMA and CDMA2000 systems are chip rate, downlink channel structure, and network synchronization. CDMA2000 uses a chip rate of 3.6864 Mc/s for the 5-MHz band allocation with the direct spread downlink and a 1.2288 Mc/s chip rate for the multicarrier downlink. WCDMA uses direct spread with a chip rate of 4.096 Mc/s.

WCDMA

The WCDMA scheme was developed as a joint effort between ETSI and ARIB in 1997. The ETSI WCDMA scheme was developed from the FMA2 scheme in Europe and the ARIB WCDMA from the Core-A scheme in Japan. The uplink of the WCDMA scheme is based mainly on the FMA2 scheme and the downlink on the Core-A scheme. **Table 7.2** lists the main parameters of WCDMA.

CDMA2000

Within standardization committee TIA TR45.4, the subcommittee TR45.5.4 was responsible for the selection of the basic CDMA2000 concept. For all the other wideband CMA schemes, the goal has been to provide data rates for the IMT-2000 performance requirements of at least 144 Kb/s in a vehicular environments, 384 Kb/s in a pedestrian environment, and 2.048 Mb/s in an indoor office environment. The main

TABLE 7.1 IS-95 CDMA Air Interface Parameters

Bandwidth	1.25 MHz
Chip rate	1.2288 Mc/s
Frequency band uplink	• 869–894 MHz • 1930–1980 MHz
Frequency band downlink	• 824–849 MHz • 1850–1910 MHz
Frame length	20 ms
Bit rates	• Rate set 1: 9.6 kb/s • Rate set 2: 14.4 kb/s • IS-95B: 115.2 kb/s
Speech code	• QCELP 8 kb/s • ACELP 13 kb/s
Soft handover	Yes
Power control	• Uplink: Open loop + fast closed loop • Downlink: Slow quality loop
Number of RAKE fingers	Four
Spreading codes	Walsh + long *M*-sequence

TABLE 7.2 Parameters of WCDMA

Channel bandwidth	1.25, 5, 10, 20 MHz
Downlink RF channel structure	Direct spread
Chip rate	1.024/4.096/8.192/16.384 Mc/s
Roll-off factor for chip shaping	0.22
Frame length	10 ms/20 ms (optional)
Spreading modulation	• Balanced QPSK (downlink) • Dual channel (uplink) • Complex spreading circuit
Coherent detection	User-dedicated time multiplexed pilot (downlink and uplink); no common pilot in downlink
Channel multiplexing in uplink	Control and pilot channel time multiplexed *I* and *Q* multiplexing for data and control channel
Multirate	Variable spreading and multicode
Spreading factors	4–256
Power control	Open- and fast-closed loop (1.6 KHz)
Spreading (downlink)	Variable length orthogonal sequences for channel separation Gold sequence 2^{18} for cell and user separation (truncated cycle 10 ms)
Spreading (uplink)	Variable-length orthogonal sequences for channel separation Gold sequence 2^{41} for user separation (different time shifts in *I* and *Q* channel, truncated cycle 10 ms)
Handover	• Soft handover • Interfrequency handover

TABLE 7.3 CDMA2000 Parameter Summary

Channel bandwidth	1.25, 5, 10, 20 MHz
Downlink RF channel structure	Direct spread or multicarrier
Chip rate	• 1.2288/3.6864/7.3728/11.0593/14.7456 Mc/s for direct spread • $n\times$ 1.2288 Mc/s ($n = 1, 3, 6, 9, 12$) for multicarrier
Roll-off factor	Similar to IS-95
Frame length	• 20 ms for data and control • 5 ms for control information on the fundamental and dedicated control channel
Spreading modulation	• Balanced QPSK (downlink) • Dual channel QPSK (uplink) • Complex spreading circuit
Coherent detection	• Pilot time multiplexed with PC and EIB (uplink) • Common continuous pilot channel and auxiliary pilot (downlink)
Channel multiplexing in uplink	• Control, pilot, fundamental, and supplemental code multiplexed • *I* and *O* multiplexing for data and control channel
Multirate	Variable spreading and multicode
Spreading factors	4–256
Power control	Open- and fast-closed loop (800 Hz, higher rates under study)
Spreading (downlink)	Variable-length Walsh sequences for channel separation, *M*-sequence 2^{15} (same sequence with time shift used in different cells, different sequence in *I* and *Q* channel)
Spreading (uplink)	• Variable-length orthogonal sequences for channel separation *M*-sequence 2^{15} (same sequence for all users, different sequence in *I* and *Q* channels) *M*-sequence $2^{41} - 1$ for user separation (different time shifts for different users)
Handover	• Soft handover • Interfrequency handover

TABLE 7.4 Comparison of FDMA, TDMA, and CDMA

Approach	FDMA	TDMA	CDMA
Idea	Segment the frequency band into disjoint sub-bands	Segment sending time into disjoint time slots, demanding driven or fixed patterns	Spread of the spectrum using orthogonal codes
Terminals	Every terminal with own frequency, uninterrupted	All terminals active for short periods of time on the same frequency	All terminals can be active at the same place at the same time, uninterrupted
Signal separation	Filtering in the frequency domain	Synchronization in the time domain	Code plus RAKE receivers
Advantages	Simple, established, robust	Established, fully digital, flexible	Flexible, less planning needed, soft handover
Disadvantages	Inflexible, wasteful of spectrum	Guard space needed, synchronization difficult	Complex receivers, needs more complicated power control for senders

focus of standardization has been providing 144 Kb/s and 384 Kb/s with approximately 5-MHz bandwidth. The main parameters of CDMA2000 are listed in Table 7.3

7.2 Comparisons of FDMA, TDMA, and CDMA

To conclude this section, we make a comparison of the three basic access technologies discussed in Table 7.4. The table shows the MAC schemes without combination with other schemes. However, in real systems, the MAC schemes always occur in combinations such as those used in GSM and IS-95.

The primary advantage of DS-CDMA is its ability to tolerate a fair amount of interfering signals compared to FDMA and TDMA, which typically cannot tolerate any such interference. With DS-CMA, adjacent microcells share the same frequencies, whereas with FDMA and TDMA, it is not feasible for adjacent microcells to share the same frequencies because of interference.

8

Convergence of Networking Technologies

Vijay K. Garg
Department of Electrical and Computer Engineering, University of Illinois at Chicago, Chicago, Illinois, USA

Yih-Chen Wang
Lucent Technologies, Naperville, Illinois, USA

8.1 Convergence

The next generation of wireless networks will be flexible, open, and standards-based. These networks will facilitate the convergence of wireless networks, the PSTN, and the Internet and will provide voice, data, and video services to platforms such as cell phones, PDAs, laptops, PCs, digital cameras, and a plethora of new portable devices.

Advances in optical networking will increase the capacity of a core network by stuffing vastly more information into each fiber. Next-generation media interface protocols, such as SIP and MEGACO, have already proven more effective than SS7 and H.323 in a convergence network of wireless systems, wireline systems, and the Internet.

The key to wireless convergence is IP and softswitch. This software-on-a-switch is the heart of the next-generation wireless networks that provides integration of the wireless networks, the Internet, and the PSTN. This is the elusive Holy Grail of communications—any-to-any end point interoperability, enhanced services, and flexible billing are all on one network.

8.1.1 Session Initiation Protocol

Session initiation protocol (SIP) is an **application-layer signaling protocol** for creating, modifying, and terminating sessions with one or more participants. SIP **sessions** include Internet multimedia conferences, Internet telephone calls, and multimedia distribution. SIP invitations create sessions carrying **session descriptions**, which allow participants to agree on a set of compatible media types. SIP supports **user mobility** by proxying and redirecting requests to the user's current location. Users can **register** their current locations. SIP makes minimal assumptions about the underlying transport protocol. It can be extended easily with additional capabilities and is an enabling technology for providing innovative new services that integrate multimedia with Internet services such as the World Wide Web, e-mail, instant messaging, and presence.

The core SIP protocol, specified in Internet Engineering Task Force (IETF) Request for Comments (RFC) 2543, is built on foundation IP protocols, such as HTTP and Simple Mail Transfer Protocol (SMTP). As such, it supports a request/response transaction model that is text-based, similar to e-mail, and self-describing; in addition, SIP inherited many of the features from standard Internet protocols. For example, it uses many of HTTPv1.1 header fields, supports Uniform Resource Identifier/Universal Resource Locator (URI/URL) for addressing, and employs the Multimedia Internet Mail Encapsulation (MIME) protocol for message payload description. It is lightweight as the baseline SIP consists only of six methods. Extensions are being proposed by adding new methods, new headers, and new message body types to support additional applications such as interworking with legacy systems, cable, or wireless applications. An SIP initiation scenario is shown in Figure 8.1.

The standard SIP architectural components are the following:

- **An SIP client** is an end system with the SIP **User Agent (UA)** residing in it. The user agent consists of two

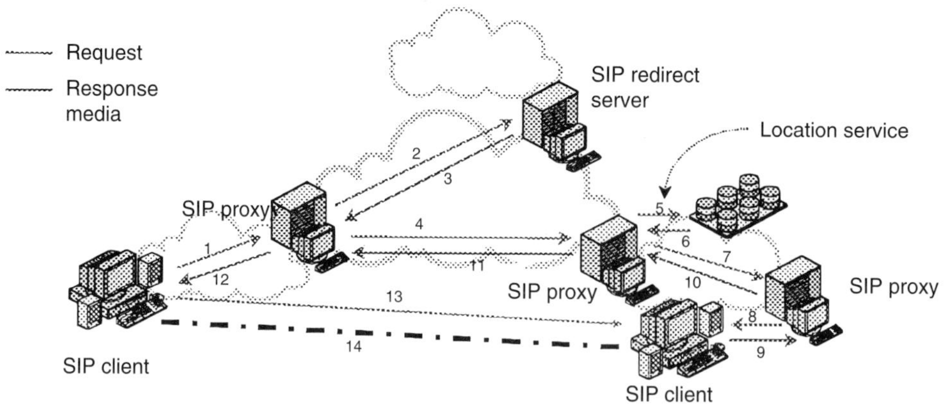

FIGURE 8.1 Session Initiation in SIP

components: the **User Agent Client** (**UAC**) is responsible for sending SIP requests, and the **User Agent Server** (**UAS**) listens for incoming requests, and prompts a user or executes a program to determine responses.

- **A proxy server** is responsible for routing and delivering messages to the called party. It receives requests and forwards them to another server (called a **next-hop server**), which may be another proxy server, a UAS, or a redirect server. It can *fork* a request, sending copies to multiple next-hop servers at once. This allows a call setup request to try many different locations at once. It can also forward the invitation to a multicast group. A proxy server can be call-stateful, stateful (transaction-stateful), or stateless.
- **A redirect server** also receives requests and determines net-hop server(s). Instead of forwarding the request there, however, it returns the address(es) of the next-hop server(s) to the sender.
- **A locator service** is used by an SIP server to obtain information about a callee's possible locations. It is outside the scope of SIP. It can be anything, such as LDAP, whois, whois++, POST corporate database, local file, or result of program execution (IN).
- **A registration server** receives updates on the current locations of users. It is typically colocated with a proxy redirect server. The server may make its information available through the location server.

Often hailed as more flexible than H.323, SIP is an application-layer control protocol that can establish, modify, and terminate sessions or calls. SIP is text-based and light-

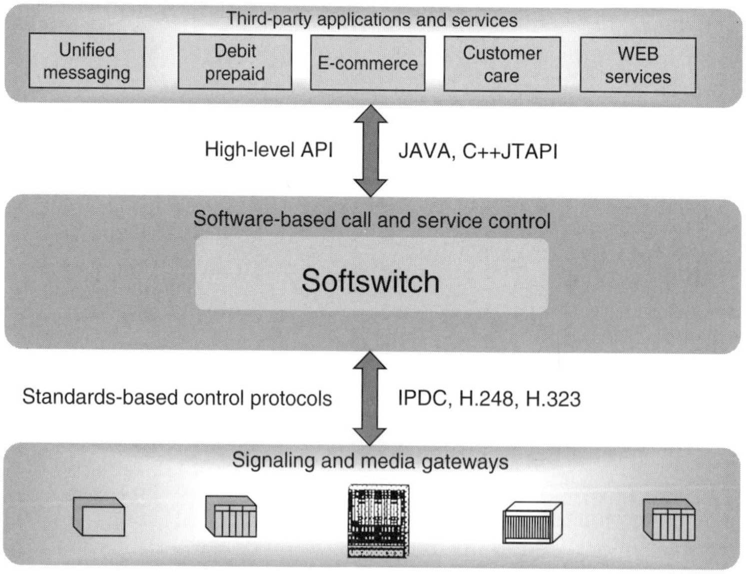

FIGURE 8.2 The Softswitch Architecture

weight, and it uses a simple invitation–acceptance message structure. Other benefits of SIP over H.323 include scalability, service richness, lower latency, faster speed, and ability to distribute for carrier-grade reliability.

SIP is a new protocol currently under development. The core SIP application specified in IETF RFC 2543 is being implemented and tested in SIP bake-offs. Many extensions of SIP are being proposed and are under discussion. Holes and gaps exist for system-level deployment. For example, QoS, security, services, and operations issues are still being debated. Interworking mappings of SIP with H.323 and ISUP/BICC are not yet finalized (Internet draft status).

8.1.2 Softswitch

Softswitch is an all-encompassing term for the next-generation communications systems that employ open standards to create integrated networks with a decoupled service intelligence capable of carrying voice, video, and data traffic more efficiently and with far greater value-added service potential than is possible with existing circuit-switched networks. Softswitch-based networks will enable service providers to support new, enhanced voice service features as well as offer new types of multimedia applications in addition to integrating existing wireline and wireless voice services with advanced data and video services.

The separation of call control and services (Figure 8.2) from the underlying transport network is a key enabling feature of softswitch-based networks. Indeed, the very approach of building switched networks with extracted service intelligence represents a major break-through when compared to the circuit-switched approach of combining the transport hardware, call control, and service logic together into a single, proprietary piece of equipment.

Softswitches can be used in 3G and 4G core networks to provide call control, mobility management, and an open service creation environment for carriers. A softswitch can deliver foundation mobility functions of roaming, location updates, subscriber profile management, intersystem handover, and RAN interworking to wireless service providers.

8.1.3 All IP Architecture

The Internet is unprecedented in its impact on the world community of industries, institutions, and individuals. In some way, the Internet has touched most of our lives in terms of how we communicate, how we promote our products, how we teach our children, and how we invest our time. No media adoption curve has been faster than the Internet's. In the United States alone, it took almost 40 years for 50 million people to use radio and 15 years for 50 million people to use Television and cellular communications. Internet users reached the 50-million mark in just 5 years.

During that time, the world became increasingly mobile, defined by the "take-it-with-you philosophy" we have developed regarding information and our access to it. For the wireless cellular industry, that shift in attitude has created the opportunity to add mobility to Internet accessibility—effectively allowing subscribers to carry the power of the Internet with them anywhere at any time.

The convergence of wireless and Internet usage is already underway. Globally, Internet users as a whole are projected to increase from about 200 million at present to almost 1 billion by the year 2005. During the same period of time, global wireless subscribers are expected to increase from 300 million to over a billion.

With these market dynamics in mind, several industry-leading businesses have agreed that next-generation wireless

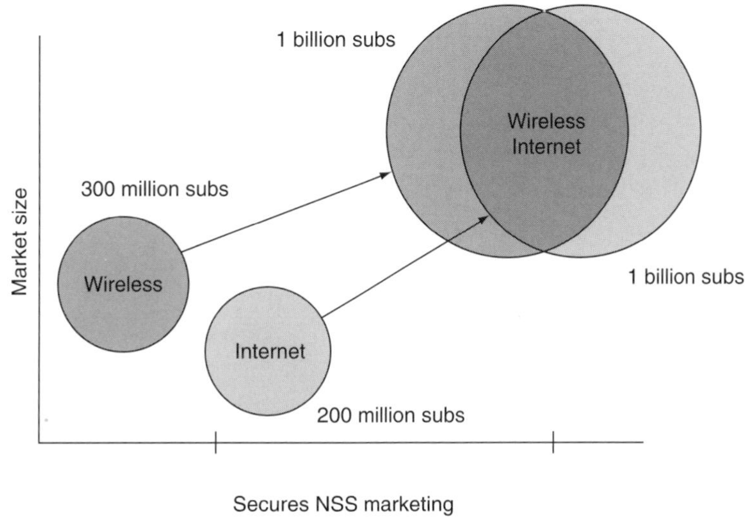

FIGURE 8.3 Projected Wireless Internet Convergence

networks will leverage the packet-based technology of IP. This strategy provides operators with the unique opportunity to deliver a multitude of new services to mobile cellular subscribers in a manner more customizable than previously possible (see Figure 8.3).

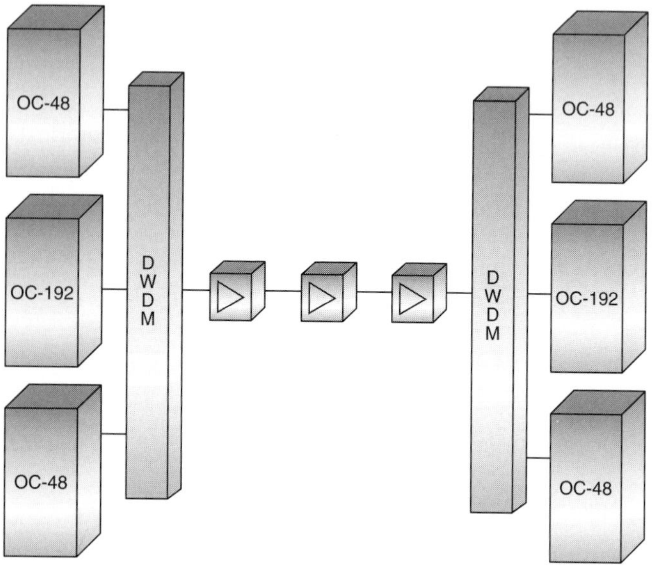

FIGURE 8.4 DWDM Systems and Optical Amplifiers

As the industry continues to invest heavily in advancing IP technology for supporting real-time applications such as voice with reliable service and toll quality, it is expected to further accelerate the introduction of new network capabilities that are defined within IP standards for network implementations.

8.2 Optical Networking

As networks face increasing bandwidth demand and diminishing fiber availability, network providers are moving toward a crucial milestone in network evolution: the **optical network**. Optical networks, based on the emergence of the optical layer in transport networks, provide higher capacity and reduced costs for new applications such as the Internet, video and multimedia interaction, and advanced digital services.

Optical networks began with wavelength division multiplexing (WDM), which arose to provide additional capacity on existing fibers. The components of the optical network are defined according to how the wavelengths are transmitted, groomed, or implemented in the network. Viewing the network from a layered approach, the optical network requires the addition of an optical layer. To help define network functionality, networks are divided into several different physical or virtual layers. The first layer, the **services layer,** is where the services—such as data traffic—enter the telecommunications network. The next layer, **SONET,** provides restoration, performance monitoring, and provisioning that is transparent to the first layer.

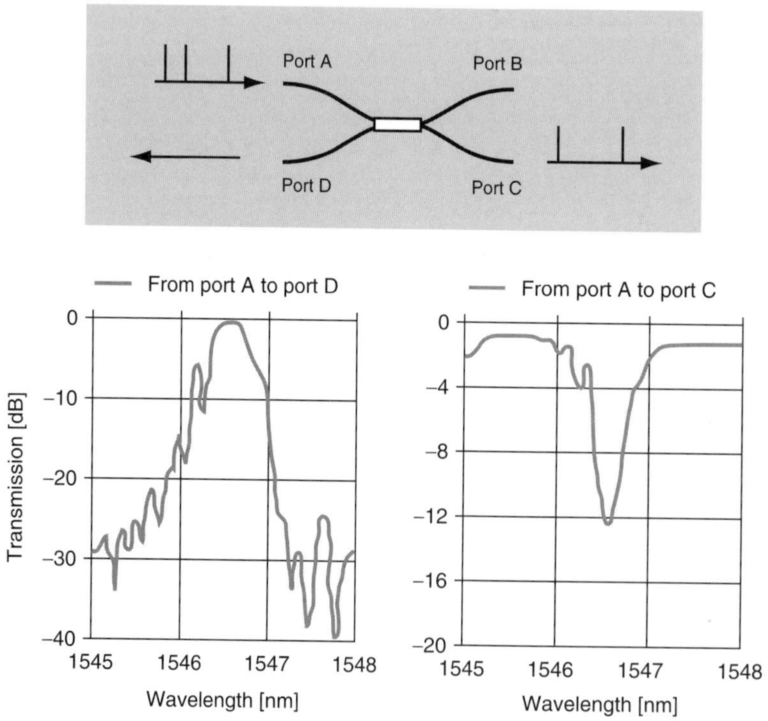

FIGURE 8.5 In Fiber Bragg Grating Technology: Optical A/D Multiplexer

8.2.1 Dense Wavelength Division Multiplexing

As optical filters and laser technology improved, the ability to combine more than two signal wavelengths on a fiber became a reality. **Dense wavelength division multiplexing** (DWDM) combines multiple signals on the same fiber, ranging up to 40 or 80 channels. By implementing DWDM systems and optical amplifiers, networks can provide a variety of bit rates (i.e., OC–48 or OC–192) and a multitude of channels over a single fiber (see Figure 8.4). The wavelengths used are all in the range in which optical amplifiers perform optimally, typically from about 1530 nm to 1565 nm.

Optical Amplifiers

The performance of **optical amplifiers** has improved significantly—with current amplifiers providing significantly lower noise and flatter gain—which is essential to DWDM systems. The total power of amplifiers also has steadily increased, with amplifiers approaching +20-dBm outputs, which is many orders of magnitude more powerful than the first amplifiers.

Narrow Band Lasers

Without a narrow, stable, and coherent light source, none of the optical components would be of any value in the optical network. Advanced **lasers with narrow bandwidths** provide the narrow wavelength source that is the individual channel in optical networks. Typically, long-haul applications use externally modulated lasers, while shorter applications can use integrated laser technologies.

Fiber Bragg Gratings

Commercially available **fiber Bragg gratings** have been important components for enabling WDM and optical networks. A fiber Bragg grating is a small section of fiber that has been modified to create periodic changes in the index of refraction. Depending on the space between the changes, a certain frequency of light—the Bragg resonance wavelength—is reflected back, while all other wavelengths pass through (see Figure 8.5). The wavelength-specific properties of the grating make fiber Bragg gratings useful in implementing optical add/drop multiplexers. Bragg gratings also are being developed to aid in dispersion compensation and signal filtering as well.

Thin-Film Substrates

Another essential technology for optical networks is the **thin-film substrate**. By coating a thin glass or polymer substrate with a thin interference film of dielectric material, the substrate can be made to pass through only a specific wavelength and reflect all others. By integrating several of these components, many optical network devices are created, including multiplexers, demultiplexers, and add/drop devices.

IX

CONTROLS AND SYSTEMS

Michael Sain
*Department of Electrical Engineering,
University of Notre Dame,
Notre Dame, Indiana, USA*

The question of feedback in circuits and systems can be viewed in a denumerable number of different ways. It is, therefore, no longer possible to give a coherent, yet brief, overview of the subject. I am reminded of the story of the hotel with an infinite number of rooms. In such a hotel, there is always room for another guest. To accommodate the next guest, even if all the rooms are occupied, one has only to ask each room occupant to move to the room with the next higher number. When this is done, all the current occupants will again be in a room, and room one will be available for the new guest!

Nonetheless, one can set up some equivalence classes on feedback in circuits and systems. In one of these constructions, it is possible to focus upon whether the feedback was introduced deliberately by the designer, or whether it just happened to be present in a circuit or system which was not designed with any particular notion of feedback in mind. A great many readers will likely identify with the first of these two classes, in which the feedback has been incorporated by specific choice, with sensors and actuators and processors which transform measurements into commands to achieve a given set of goals. Indeed, this way of looking at things was apparent early in the modern literature. Circuit persons will also recognize the second class of possibilities, in as much as feedback in electronic amplifiers was being discussed almost at the same time that the systems viewpoint was being put into place.

By now we realize that feedback is a question of relationships. When we have, for instance, two equations in two unknowns, the age-old process of eliminating one of the two unknowns can be interpreted as a feedback construction. The more interrelationships which exist, the more "feedback" that exists. Feedback, then, is present whether we explicitly plan for it or not. It would seem provident, therefore, to become better acquainted with some of the features and uses of feedback, so that we can understand more clearly what it is doing in our circuit or system, even if we did not have feedback in mind.

To obtain this type of insight, we shall focus primarily upon the first class above, in which the feedback is deliberately put into play. In such situations, one big variable is the nature of the process which is being controlled by the use of feedback. One may be controlling speed, or pressure, or temperature, or voltage, or current, or just about any variable that one can imagine. To capture at least a bit of this flavor of the subject, the chapter addresses a wide range of different processes which are being placed under control. A second big variable is the design purpose for which the feedback is applied. Perhaps it is for stabilization, tracking, or amplification. Perhaps it is for coping with process model errors which arise because the process is too complicated to model exactly—assuming that we have the knowledge to do so. Perhaps it is for coping with disturbances, such as noise interfering with the transfer of data from one portal of a circuit to another. Or perhaps it is because we cannot get good estimates of key circuit and system parameters, or we cannot obtain a big enough budget to make all the measurements we want, or to install all the actuators we wish.

In the face of all these sorts of issues, we approach the task here by creating a diverse sample of the sorts of issues which can arise. One can create entire handbooks on controls; and that has been done. In fact, one can create encyclopedias on controls; and that has been done. Here we have only a section, and we approach the question by taking samples. This is entirely in the spirit of the subject; and the task of the reader is to try to master the behavior of the whole body of knowledge from these samples! Welcome to the experience of feedback circuits and systems!!!

1

Algebraic Topics in Control

Cheryl B. Schrader
College of Engineering, Boise State University, Boise, Idaho, USA

1.1 Introduction

Control engineers often are referred to as mathematicians in disguise. Indeed, a firm foundation in mathematics is essential for success in the control arena. Many practicing control engineers delve into the intricacies of one or more particular fields of mathematics. Consequently, many of the sections in this chapter on controls rely heavily on algebraic considerations or have algebraic roots. This chapter purports to provide a lively introduction for such investigations by discussing common algebraic topics that occur in control. Along the way, examples of engineering applications are pointed out to the interested reader.

From its very first introduction in elementary school, the term **algebra** conveys a certain level of abstraction. Most people associate algebra with a collection of arithmetic operations combined with representative symbols such as x and y. As undergraduate students, engineers become comfortable with a discussion of fields and vector spaces along with their axiomatic relationships, and this chapter begins here. In the true sense, vector spaces do not form an algebra equation without the additional concept of vector multiplication (tensor product), as mentioned later in Section 1.3. However, square matrices of dimension $n \times n$ do form an algebra equation although it is not commutative. How the associated operations are defined in this algebra along with interesting and important characteristics, are illustrated in some detail. This discussion provides a firm foundation for an understanding of algebra for the engineer. Concluding this chapter are examples of other types of algebras and applications of interest to the control engineer.

Control theory and the tools used in control theory are particularly useful in the study of biomedical engineering and medicine, and the applications extend from the "head to the toes." Implantable organs, such as the heart, ear, kidney, and retina, all require exacting control systems that are designed with a significant number of parameters and need to work reliably 100% of the time. Other capabilities under development, such as remote surgery, also require extensive control systems because a minor slip could be life-threatening. The controller needs to be able to differentiate between intentional hand movement and a tremor on the part of the surgeon.

Standard algebraic notation is employed and should not be unfamiliar to the informed reader. MATLAB, the tool many engineers use for testing and development (and that is used here for numerical examples) was first developed in the late

1970s and written in Fortran. The name originated from the term *matrix laboratory*. The examples are generated using Version 5.3 (release 11), and the same vectors and matrices are carried throughout this chapter. MATLAB is a powerful problem-solving environment with many built-in debugging tools. It can generate world-class graphics that can be imported to almost any file format for inclusion in reports and presentations. As the reader will no doubt observe, MATLAB is quickly assimilable; with a little knowledge of history and intuition, actual MATLAB commands are easy to identify. For a detailed introduction on MATLAB, see Higham and Higham (2000).

1.2 Vector Spaces Over Fields and Modules Over Rings

It is assumed that the reader already has some knowledge and experience with vector spaces and fields. What may not be as readily apparent is that the reader, subsequently, also has experience with modules and rings. This section briefly describes the essence of these ideas in a mathematical sense.

Real numbers, complex numbers, and binary numbers all are **fields**. Specifically a field F is a nonempty set F and two binary operations, addition ($+$) and multiplication, that together satisfy the following properties for all $a, b, c \in F$:

1. **Associativity**: $(a+b)+c = a+(b+c); (ab)c = a(bc)$.
2. **Commutativity**: $a+b = b+a$; $ab = ba$.
3. **Distributivity**: $a(b+c) = (ab)+(ac)$.
4. **Additive identity**: $\exists 0 \in F \ni a+0 = a$.
5. **Multiplicative identity**: $\exists 1 \in F \ni a1 = a$.
6. **Additive inverse**: For every $a \in F, \exists b \in F \ni a+b = 0$. [Notation note: $b = -a$].
7. **Multiplicative inverse**: For every nonzero, $a \in F$, $\exists b \in F \ni ab = 1$. [Notation note: $b = a^{-1}$].

It is commonly known with real numbers that multiplication distributes over addition, and both additive and multiplicative inverses exist. In the case where a multiplicative inverse does not exist, but properties 1 through 6 hold (such as with integers), then the set does not form a field but is categorized as a **commutative ring**. If property 2 also does not hold, then the correct terminology is a **ring**.

To speak of an **additive group**, a single operation is used (addition) along with a nonempty set G, satisfying additive properties 1, 4, and 6. If in addition the operation is commutative (as described in property 2), then the additive group is **abelian**.

To discuss an F-**vector space** V, one simply requires a nonempty set V and a field F that together with binary operations $+: V \times V \rightarrow V$ and $*: F \times V \rightarrow V$ satisfy the following axioms for all elements $v, w \in V$ and $a, b \in F$:

1. V and $+$ form an additive abelian group.
2. $a*(v+w) = (a*v)+(a*w)$.
3. $(a+b)*v = (a*v)+(b*v)$.
4. $(ab)*v = a*(b*v)$.
5. $1*v = v$.

Vectors are elements of V, and scalars are elements of F. Often, one uses the terminology **vector space** V **over the field** F. What form vectors may take will be examined more closely in Section 1.3. For the purposes of this treatise, and for the following MATLAB examples, the field of real numbers, $\Re$, will be used most often. The reader is urged to remember that any choice of field is allowed.

As a direct generalization of a vector space, a **module** M replaces the underlying field by a ring R. Technically speaking, M is a **left module** because the scalar appears left of the module element. In an analogous fashion, a **module** M **over the ring** R is an R-**module** M. From this discussion, it is apparent that if R is also a field, R-modules are merely vector spaces. In working with modules, it is important to remember to avoid any vector space results relying on division by a nonzero scalar. It is precisely this notion that leads to the extremely powerful application of modules in system analysis, controllability, and observability.

1.3 Matrices and Matrix Algebra

1.3.1 Standard Matrix Notions and Examples

By the time an engineer completes his or her undergraduate studies, familiarity exists with matrices and matrix operations through multivariable calculus or linear algebra. Moreover, a graduate should have a firm understanding of the importance in regarding matrices as representations of linear operators. For a detailed discussion of linear operations and their related matrix representations, see Schrader and Sain (2000). Of course, standard matrix notions can be generalized and expressed using rings and fields, which is the approach used here.

Consider the set of **arrays** (or **matrices**):

$$A = \begin{bmatrix} a_{11} & a_{12} & \cdots & a_{1n} \\ a_{21} & a_{22} & \cdots & a_{2n} \\ \vdots & \vdots & \ddots & \vdots \\ a_{m1} & a_{m2} & \cdots & a_{mn} \end{bmatrix}, \tag{1.1}$$

with m rows, n columns, and **elements** a_{ij} from a field F (or alternatively, $A \in F^{m \times n}$). Define **matrix addition** $(A+B=C)$ element-wise $(a_{ij}+b_{ij}=c_{ij})$ for $A, B,$ and $C \in F^{m \times n}$; $i = 1, 2, \ldots, m$; and $j = 1, 2, \ldots, n$. An additive identity is the zero matrix, an $m \times n$ matrix whose every element is the additive identity in F. For example, if F is the ring of integers or the field of real or complex numbers, then every element is simply 0. An additive inverse of a matrix A with elements $a_{ij} \in F$ is found element by element by solving for $b_{ij} \in F$ in $a_{ij} + b_{ij} = 0$. Thus, $b_{ij} = -a_{ij}$ and the notation generalizes to the matrix level; that is, $B = -A$.

Such a discussion leads naturally to the concept of **scalar multiplication**. Consider a field element $r \in F$ and a matrix over the same field $A \in F^{m\times n}$. The product $P = rA \in F^{m\times n}$ is calculated element-wise:

$$p_{ij} = ra_{ij};\ i = 1, 2, \ldots, m,\ j = 1, 2, \ldots, n. \tag{1.2}$$

It is simple to see that the additive inverse $B = -A$ is merely the scalar multiplication by -1 of the matrix A.

Example 1

```
>> a = [1 2 3]; b = [4; 5; 6]; c = [7 8 9];
>> a + b
??? Error using ==> +
Matrix dimensions must agree.
>> a + c
ans =
    8    10    12
>> -b
ans =
    -4
    -5
    -6
```

$$a = [1\ \ 2\ \ 3].\qquad b = \begin{bmatrix} 4 \\ 5 \\ 6 \end{bmatrix}.\qquad c = [7\ \ 8\ \ 9].$$

To discuss the multiplication of two matrices A and B over the same field F represented by AB, one is restricted to choose matrices where the number of columns of A equal the number of rows of B. Such matrices are said to be **conformable**. The resultant matrix, $C = AB$, has its number of rows equal to the number of rows of A and its number of columns equal to the number of columns of B. Hence:

$$A \in F^{m\times n} \text{ and } B \in F^{n\times p} \Rightarrow C \in F^{m\times p}. \tag{1.3}$$

Each element in C is found using one row of A and one column of B at a time:

$$c_{ij} = \sum_{k=1}^{n} a_{ik} b_{kj}. \tag{1.4}$$

It is important to note that **matrix multiplication** is not commutative, in general, although it is associative and distributive with respect to matrix addition.

Example 2

```
Consider the same a, b, and c from Example 1.
>> a * b
ans =
    32
>> b * a
ans =
    4    8   12
    5   10   12
    6   12   18
>> a * c
??? Error using ==> *
Inner matrix dimensions must agree.
```

$$a = [1\ \ 2\ \ 3].\qquad b = \begin{bmatrix} 4 \\ 5 \\ 6 \end{bmatrix}.\qquad c = [7\ \ 8\ \ 9].$$

For the special case where A has the same number of rows as columns (A is **square**) and:

$$A = \begin{bmatrix} 1 & 0 & \ldots & 0 \\ 0 & 1 & \ldots & 0 \\ \vdots & \vdots & \ddots & \vdots \\ 0 & 0 & \ldots & 1 \end{bmatrix}, \tag{1.5}$$

$AB = B$ regardless of choice of B. In such case, A is said to be the **identity matrix** and is most often represented by the symbol I. The identity matrix is an example of a **diagonal matrix**, where all $a_{ij} = 0$ when $i \neq j$, written as $A = \text{diag}\{1, 1, \ldots, 1\}$. In MATLAB, $I^{n\times n}$ is generated with the command **eye** (n). Taken all together, the set of square matrices over F, matrix addition, matrix multiplication, and the zero and identity matrices satisfy the six properties of a ring. A discussion of a multiplicative inverse in the context of matrices is left for Section 1.4.

Example 3

```
>> d = [1 1 1; 1 2 3; 1 3 6]; e = [1 1 1; 4 2 1; 9 3 1];
>> d * e
ans =
14    6    3
36   14    6
67   25   10
>> e * d
ans =
 3    6   10
 7   11   16
13   18   24
>> d * eye (3)
ans =
1  1  1
1  2  3
1  3  6
```

$$d = \begin{bmatrix} 1 & 1 & 1 \\ 1 & 2 & 3 \\ 1 & 3 & 6 \end{bmatrix}.\qquad e = \begin{bmatrix} 1 & 1 & 1 \\ 4 & 2 & 1 \\ 9 & 3 & 1 \end{bmatrix}.$$

A matrix with one column or one row is commonly referred to as a **column vector** or **row vector**, respectively. It can be shown easily that the set of all n tuples for $n > 0$ satisfies the axioms of a vector space over the field of real numbers.

Although many engineers tend to think of this *n*tuple example as synonymous with the definition of a vector space, this interpretation is limiting. For example, the set of polynomials of degree less than $m > 0$ with real coefficients satisfies all axioms of a vector space taking into account standard polynomial addition and multiplication. Section 1.5 will examine polynomials in greater detail.

The most interesting observation when considering *n*tuple vectors is that while one can multiply scalars together or multiply vectors by scalars, one simply cannot multiply two vectors together. Example 2 illustrates this dilemma. The concept of vector multiplication together with a vector space extends to the concept of an **algebra**. This is the additional mathematical understanding one must master to incorporate the common manifestation of vector multiplication found in the **tensor product**.

1.4 Square Matrix Functions: Determinants and Inverses

Square matrices ($n \times n$) have many special properties and functions associated with them. In this section, determinants will be examined first because they are important in the study of matrices and their control applications. Following a discussion of determinants, this section will examine more fully the concept of a matrix multiplicative inverse.

1.4.1 Determinants

Determinants can be viewed quite naturally using a detailed description of vector multiplication and the tensor product. For details regarding this approach, the reader is referred to Sain and Schrader (2000). In the current examination, determinants will be approached from the matrix algebra perspective.

Originally attributed to Vandermonde, a determinant is a scalar value or single algebraic expression resulting from a square matrix. It is, therefore, appropriate to begin with square matrices that are also scalars, $A = F^{1\times 1}$. In such case, the **determinant** of A, more commonly written as det (A) or $|A|$, is the scalar itself:

$$\det(A) = \det([a_{11}]) = a_{11}. \tag{1.6}$$

Now consider $A \in F^{2\times 2}$ and introduce the concept of the **minor** of a_{ij}, which is defined as the determinant of the submatrix that results when row i and column j are removed. For a general 2×2 matrix:

$$A = \begin{bmatrix} a_{11} & a_{12} \\ a_{21} & a_{22} \end{bmatrix}, \tag{1.7}$$

and it is obvious that:

- minor of $a_{11} = \det(a_{22}) = a_{22}$.
- minor of $a_{12} = \det(a_{21}) = a_{21}$.
- minor of $a_{21} = \det(a_{12}) = a_{12}$.
- minor of $a_{22} = \det(a_{11}) = a_{11}$.

The actual calculation of the determinant of equation 1.7 relies on minors according to the algorithm following:

Step 1. Choose any row i (or column j).
Step 2. Multiply each element a_{ik} (or a_{kj}) in that row (or column) by its minor and by $(-1)^{i+k}$ (or $(-1)^{k+j}$).
Step 3. Add the results from step 2.

The multiplication of a minor by ± 1, appropriately, is termed a **cofactor**. Following the previous algorithm, the choice of column 2 ($j = 2$) yields:

$$a_{12}a_{21}(-1) + a_{22}a_{11}(+1). \tag{1.8}$$

The value of the determinant is independent of the choice of row or column; it makes no difference whatsoever. Determinants of 2×2 matrices are fairly straightforward in that the two elements along the main diagonal are multiplied together and then the product of the remaining two elements is subtracted.

An explanation of determinants for matrices larger than 2×2 is best understood first with 3×3 matrices to help grasp the notation, which can then be expanded to $n \times n$ matrices in general. Consider first:

$$\det(A) = \begin{vmatrix} a_{11} & a_{12} & a_{13} \\ a_{21} & a_{22} & a_{23} \\ a_{31} & a_{32} & a_{33} \end{vmatrix}, \tag{1.9}$$

and expand about row 1 according to the three-step algorithm. Then:

$$\det(A) = a_{11}\begin{vmatrix} a_{22} & a_{23} \\ a_{32} & a_{33} \end{vmatrix} - a_{12}\begin{vmatrix} a_{21} & a_{23} \\ a_{31} & a_{33} \end{vmatrix} + a_{13}\begin{vmatrix} a_{21} & a_{22} \\ a_{31} & a_{32} \end{vmatrix} \tag{1.10}$$

$$= a_{11}(a_{22}a_{33} - a_{23}a_{32}) - a_{12}(a_{21}a_{33} - a_{23}a_{31}) + a_{13}(a_{21}a_{32} - a_{22}a_{31}) \tag{1.11}$$

$$= a_{11}a_{22}a_{33} + a_{12}a_{23}a_{31} + a_{13}a_{21}a_{32} - a_{12}a_{21}a_{33} - a_{11}a_{23}a_{32} - a_{13}a_{22}a_{31}. \tag{1.12}$$

Similar to the 2×2 case, there exists a relatively simple way to find the determinant of equation 1.9. Write the first two columns next to the original columns as follows:

$$\begin{matrix} a_{11} & a_{12} & a_{13} & a_{11} & a_{12} \\ a_{21} & a_{22} & a_{23} & a_{21} & a_{22} \\ a_{31} & a_{32} & a_{33} & a_{31} & a_{32} \end{matrix}. \tag{1.13}$$

Form the first three products of equation 1.12 by starting with a_{11} in equation 1.13 and drawing a diagonal downward to the right. Repeat with a_{12} and with a_{13}, forming two more diagonals. The last three products of equation 1.12 are formed by subtracting left diagonals. Begin with a_{12}, draw a diagonal down to the left, and then repeat with a_{11} and a_{13}. For $n \times n$ matrices where $n > 3$, such simple characterizations do not exist. Thus, one is forced to define determinants in general using symbols to describe the expansions illustrated in equations 1.10 through 1.12.

In general, determinants of $n \times n$ matrices can be computed using cofactors, which in turn require finding determinants of $(n-1) \times (n-1)$ matrices and so on. Such is the case for $n = 2$ and $n = 3$. Define A_{ij} as the cofactor formed by (1) deleting the *i*th row and the *j*th column of A, (2) calculating the determinant of the $(n-1) \times (n-1)$ submatrix, and (3) multiplying by $(-1)^{i+j}$. Expanding about any row i produces:

$$\det(A) = \sum_{k=1}^{n} a_{ik} A_{ik}, \tag{1.14}$$

or about any column j:

$$\det(A) = \sum_{k=1}^{n} a_{kj} A_{kj}. \tag{1.15}$$

1.4.2 Inverses

Any square matrix $A \in F^{n \times n}$ with a nonzero determinant has a unique inverse $A^{-1} \in F^{n \times n}$, which satisfies:

$$AA^{-1} = A^{-1}A = I. \tag{1.16}$$

Matrix inverses can be determined by forming the **cofactor matrix**:

$$\widetilde{A} = \begin{bmatrix} A_{11} & A_{12} & \dots & A_{1n} \\ A_{21} & A_{22} & \dots & A_{2n} \\ \vdots & \vdots & \ddots & \vdots \\ A_{n1} & A_{n2} & \dots & A_{nn} \end{bmatrix}, \tag{1.17}$$

taking its transpose, $\widetilde{A}^T$ (by interchanging the rows and columns of $\widetilde{A}$), and scaling:

$$A^{-1} = \frac{\widetilde{A}^T}{\det(A)}. \tag{1.18}$$

The matrix $\widetilde{A}^T$ is termed the **adjoint** of A and is calculated in MATLAB using **adj (*A*)**.

Matrices with nonzero determinants are said to be **nonsingular** or of **full rank**. For nonsquare matrices $A \in F^{m \times n}$, the rank is the number of linearly independent columns of A over the field F and is $\leq \min(m, n)$. Related commands in MATLAB are **rank** and **null**. Simultaneous linear equations may have a different number of solutions, and how many solutions exist is related to the rank of A. To solve simultaneous equations, MATLAB provides the backslash command (\) to determine a vector $x \in F^n$, satisfying $Ax = y$ for $A \in F^{m \times n}$ and $y \in F^m$. Given $A \in F^{m \times n}$ and $Y \in F^{m \times p}$, the backslash operator can be used to solve for $X \in F^{n \times p}$ in $AX = Y$. Directly solving for the matrix inverse, assuming one exists, can be avoided by solving for X in $AX = I$. This is usually faster and more accurate because Cholesky or LU factorizations are employed. More advanced factorizations contribute to solving the sparse system problem (A is large and has few nonzero elements) that occurs, for example, in electrical power systems.

All is not lost in terms of inverses in the case of nonsquare matrices $A \in F^{m \times n}$ or rank-deficient square matrices. Define the pseudoinverse $A^+ \in F^{n \times m}$ by:

$$A^+ = (A^T A)^{-1} A^T. \tag{1.19}$$

Note that $A^+A = I$ but AA^+ may not. Hence, A^+ is also termed a **left inverse**. There exists, as one immediately assumes, an analogous derivation for a **right inverse.** An interesting recent study uses pseudoinverses in crptanalysis to prove insecurity.

Example 4

Use the same *d* and *e* as in example 3.

$$d = \begin{bmatrix} 1 & 1 & 1 \\ 1 & 2 & 3 \\ 1 & 3 & 6 \end{bmatrix}. \quad e = \begin{bmatrix} 1 & 1 & 1 \\ 4 & 2 & 1 \\ 9 & 3 & 1 \end{bmatrix}. \quad f = \begin{bmatrix} 1 & 1 & 1 \\ 1 & 2 & 3 \\ 2 & 4 & 6 \end{bmatrix}.$$

```
>> det(d)
ans =
    1
>> det(e)
ans =
   -2
>> inv(d)
>> ans =
     3  -3   1
    -3   5  -2
     1  -2   1
>> inv(d) * d
>> ans =
    1  0  0
    0  1  0
    0  0  1
>> pinv(e)
ans =
     0.5000  -1.0000   0.5000
    -2.5000   4.0000  -1.5000
     3.0000  -3.0000   1.0000
>> f = [1 1 1; 1 2 3; 2 4 6]
Warning: Matrix is singular to working precision.
```

```
ans =

     Inf  Inf  Inf
     Inf  Inf  Inf
     Inf  Inf  Inf

>> det(f)

ans =

     0.
```

1.5 The Algebra of Polynomials

Polynomials in the indeterminate λ with coefficients from a field F, usually represented by $F[\lambda]$, form a commutative ring. The set of polynomials of degree less than some integer $m > 0$ with real coefficients along with the standard addition and multiplication of polynomials form a vector space. Such polynomials can be described by:

$$p(\lambda) = a_m\lambda^m + a_{m-1}\lambda^{m-1} + \cdots + a_1\lambda + a_0 = \sum_{k=0}^{m} a_k\lambda^k, \quad (1.20)$$

for $a_i \in \Re$ and $a_m \neq 0$. The **degree** (or **order**) of the polynomial in equation 1.20 is m. Further, if $a_m = 1$, the polynomial is called **monic**.

One of the most important results of classical algebra is that polynomials with real coefficients can be factored into real linear factors and real quadratic factors. Any polynomial $\Re[\lambda]$ has at most m **roots** (or **zeros**). If the same matrix is considered over the field of complex numbers $\mathbb{C}$, then it has exactly m roots. Nonreal complex roots must occur in complex conjugate pairs. Note that there are essentially three problems related to polynomials: **evaluation, root finding**, and **curve fitting**—and MATLAB functions **polyval, roots**, and **polyfit** address all three.

Now this section presents at a concrete discussion of **algebras**. An (associative) **algebra** over a field F is a set S, which is both a vector space and a ring over F such that the additive group structures are the same, and for all $s, t \in S$, and $a \in F$, the following axiom holds:

$$a(st) = (as)t = s(at). \quad (1.21)$$

The relationship of equation 1.21 combines the concepts of a ring with those of a vector space. Consider now the ring $F[\lambda]$, which has also a vector space structure. Addition and the additive identity, 0, are the same for both structures. Moreover, for two polynomials $p(\lambda)$, $q(\lambda) \in F[\lambda]$ and $a \in F$, equation 1.21 is satisfied. Thus, polynomials form an associative algebra. The question may arise as to why the terminology **matrix algebra** was used previously. It is easy to show that the set of $n \times n$ matrices over F is an associative algebra by recognizing that multiplication of a matrix A by a scalar r can be written $rA = (rI)A$.

Finally, a ratio of two polynomials is a rational function:

$$f(\lambda) = \frac{n(\lambda)}{d(\lambda)}. \quad (1.22)$$

Rational functions occur most often in control as transfer functions for single-input and single-output systems. Systems with more than one input and more than one output can be described by transfer function matrices. These matrices can be realized as matrix polynomial fractions or as a polynomial matrix "divided" by a polynomial. Much research has been accomplished on this topic, particularly in its relationship to multivariable system analysis and control. Inequalities involving polynomial matrices, "inverses," and optimization problems are also becoming increasingly important.

1.6 Characteristic and Singular Values

1.6.1 Characteristic Values

Characteristic values—also called eigenvalues, characteristic or latent roots, proper or spectral values—describe a square matrix representation of a linear operator that is independent of the basis chosen. The **characteristic polynomial** of $A \in \mathbb{C}^{n\times n}$ is $p(\lambda) = \det(\lambda I - A)$, whose roots are the **characteristic values** of A. Here, matrices are considered over the complex field to admit the possibility of complex roots. The **characteristic equation**, $p(\lambda) = 0$, is of degree n and has n roots. Characteristic values depend on special matrix properties of A. Another important result known as the **Cayley-Hamilton theorem** is that a matrix A satisfies its own characteristic equation; that is, $p(A) = 0$.

The idea behind characteristic values is to describe the action of a matrix on a vector by a single scalar value:

$$Ax = \lambda x, \quad (1.23)$$

where $x \in \mathbb{C}^n$ and $\lambda \in \mathbb{C}$. Equation 1.23 can be written as:

$$(\lambda I - A)x = 0, \quad (1.24)$$

which has nonzero solutions x (**characteristic vectors** or **eigenvectors**) only if λ is a characteristic value of A. Characteristic values are often used in control to determine system stability.

Following what is known about polynomials, complex characteristic values occur in complex conjugate pairs, as do their associated characteristic vectors. Characteristic values are unique although characteristic vectors are not. This makes their actual computation (e.g., using MATLAB) somewhat arbitrary. The function **poly(A)** returns the characteristic polynomial for a matrix A. Then the **roots** command can be utilized to find the characteristic values although this is not a numerically reliable process. Alternatively, one can call the function **eig**, which can also return characteristic vectors as well.

The same function can be used to solve the generalized eigenvector problem in the case where not all eigenvalues are distinct. Eigenvectors associated with distinct eigenvalues are linearly independent; however, eigenvectors associated with repeated eigenvalues may be linearly dependent. The generalized eigenvector problem produces linearly independent eigenvectors by recognizing one can be rather clever in eigenvector choice. The result is that the non-uniqueness of eigenvectors can be used as an advantage. Complete eigenstructure assignment can be approached through the feedback control problem, which has been applied, for example, in active control of structures.

What makes the linear independence of eigenvectors important is that a matrix of such vectors can be used as a **similarity transformation** Q to relate a matrix A to its **normal** or **Jordan canonical form**:

$$\hat{A} = \operatorname{diag}\{J_1, J_2, \ldots, J_k\}, \tag{1.25}$$

where each J_i is a Jordan block of the form:

$$\begin{bmatrix} \lambda & 1 & 0 & \ldots & 0 \\ 0 & \lambda & 1 & \ldots & 0 \\ \vdots & \vdots & \ddots & \ddots & \vdots \\ 0 & 0 & \ldots & \lambda & 1 \\ 0 & 0 & \ldots & 0 & \lambda \end{bmatrix}. \tag{1.26}$$

For the distinct eigenvalue case, $\hat{A}$ is purely diagonal. For the repeated eigenvalue case, $\hat{A}$ is as close to diagonal as possible. The matrices A and $\hat{A}$ are said to be *similar* and are related by:

$$\hat{A} = Q^{-1}AQ. \tag{1.27}$$

Additional details on eigenvectors and linear transformations in general can be found in Schrader and Sain (2000). Canonical forms such as the companion, controllable, and modal or observable forms are commonly used for control purposes. The key is in always remembering, through the appropriate linear transformation, how a canonical form is related to the original matrix. This is much like dropping breadcrumbs behind you as you move from one place to another so that you can always find your way home.

1.6.2 Singular Values

Unique scalars are associated with nonsquare matrices $A \in F^{m\times n}$, and these are termed **singular values**. In general, a matrix A of rank r can be described by its **singular value decomposition** (SVD):

$$A = U\Sigma V^*, \tag{1.28}$$

where $\Sigma \in F^{m\times n}$ is $\operatorname{diag}\{\sigma_1, \sigma_2, \ldots, \sigma_r, 0, \ldots, 0\}$ and where and $U \in F^{m\times m}$ and $V \in F^{n\times n}$ are **unitary** ($UU^* = U^*U = I$ and $V^*V = VV^* = I$). Here the * notation is used to represent the **complex conjugate** (or **Hermitian**) **transpose**. The elements σ_i are related by:

$$\sigma_1 \geq \sigma_2 \geq \cdots \geq \sigma_r > 0, \tag{1.29}$$

which are the singular values of A; the columns of U are the **left singular vectors**, and the columns of V the **right singular vectors**. Singular values are unique although singular vectors are not. The SVD is quite useful and reliable in determining the rank of a constant matrix or its pseudoinverse and in the realization of transfer function matrices. By construction, singular values are the positive square root of the eigenvalues of A^*A.

Example 5

Use the same e as before.

```
>> [V, D] = eigs(e)
V =                                   Eigenvectors in columns

   0.2738    0.3487    0.2014
   0.5006   -0.1162   -0.7710
   0.8213   -0.9300    0.6042

D =                                   Eigenvalues along diagonal

   5.8284         0         0
        0   -2.0000         0
        0         0    0.1716

>> [U, S, V] = svd(e)
U =                                   Left singular vectors

   0.1324    0.8014    0.5833
   0.4264    0.4852   -0.7634
   0.8948   -0.3498    0.2775

S =                                   Singular values along diagonal

  10.6496         0         0
        0    1.2507         0
        0         0    0.1502

V =                                   Right singular vectors

   0.9288   -0.3244    0.1793
   0.3446    0.5777   -0.7400
   0.1365    0.7490    0.6483

>> [V, J] = jordan(e)
V =                                   Linear transformation

   0.5294    0.3497    0.1209
  -0.1765    0.6394   -0.4629
  -1.4118    1.0490    0.3627

J =                                   Jordan form

  -2.0000    0.0000    0.0000
  -0.0000    5.8284    0.0000
        0    0.0000    0.1716
```

```
>> V\e*V
ans =
   -2.0000   0.0000   0.0000
   -0.0000   5.8284   0.0000
         0   0.0000   0.1716
```

1.7 Nonassociative Algebras

Thus far, this chapter has considered only associative algebras, so now it turns its attention to other algebras for the sake of completeness. Other types of algebras may evolve by generalizing the concept of associative algebras (e.g., alternative algebras) or by redefining the associative product (e.g., Lie algebras). Consider an associative algebra A over a field F and replace the standard associative product of two elements $x, y \in A$, xy, with the *commutator* product $xy - yx$. Note that associativity holds only in specific cases. By redefining the associative product, we obtain a nonassociative algebra called the **Lie algebra**. Similarly, if one chooses the **anticommutator** product, $1/2(xy + yx)$, we obtain the **Jordan algebra**, also nonassociative. Nonassociative algebras are becoming increasingly important in control and quantum computers and in autonomous vehicle motion planning.

The reader is most likely familiar with any number of algebras or classes of algebras such as exterior, multilinear, Lie, Jordan, composition, alternative, differential, nonassociative, Boolean, commutative, or division. Some of these listed algebras may be misnomers according to the strict definition of an **algebra**. Of course, the point is that algebra is much more than operations. It is an axiomatic framework governed by a global view of the whole construct. It is a way of describing how a basic element behaves. And that, after all, is control in its most rudimentary sense.

1.8 Biosystems Applications

Biosystems present many potential applications for the use of algebra, including the areas of biomechanics, tissue engineering, image processing (e.g., CAT and MRI), and biosignal processing (e.g., EMG and EEG). The determination of which variables are dependent and which are independent is important in the design of gait analysis and prostheses. For example, in gait analysis, is the angle between the hip and femur on the opposite side of the body important for proper ankle movement? Is the angle on the same side important? These questions can be answered by taking multiple measurements and solving for dependent and independent variables. Tissue engineering can involve the interaction of materials (e.g., implants, lasers, etc.) with tissues of the body, such as implants using feedback to identify the amount of sugar in the blood for diabetics. The application of control theory to quantum systems, including molecular control, lasing without inversion, NMR spectroscopy, and quantum information systems, has become of great interest in recent years. Medical image analysis can be highly dependent on fundamental algebraic concepts and system theoretic techniques, and bilinear models are widely used for nonlinear biomedical signal and image processing. In medical technology development, the microelectromechanical and nanoelectromechanical phenomena and processes produce new challenges for the control engineer. For noninvasive data systems, such as multisensorial electrophysiology and biomagnetism, sensor array analysis depends on advanced algebraic techniques.

The interested reader should look in the more recent journals and conference proceedings to see other exciting topics that address algebraic notions. For biomedical engineering, the prominent resources are the *IEEE Transactions on Biomedical Engineering*, the *IEEE Transactions on Neural and Rehabilitation*, and the annual *International Conference of the IEEE Engineering in Medicine and Biology Society*. The American Society of Mechanical Engineers (ASME) has a bioengineering division with an active publications and conference schedule.

References

Higham, D.J., and Higham, N.J. (2000). *MATLAB guide*. SIAM: Philadelphia.

Sain, M.K., and Schrader, C.B. (2000). *Bilinear operators and matrices*. In Wai-Kai Chen (Ed.), *Mathematics for circuits and filters*. New York: CRC Press.

Schrader, C.B., and Sain, M.K. (2000). *Linear operators and matrices*. In Wai-Kai Chen (Ed.), *Mathematics for circuits and filters*. New York: CRC Press.

2

Stability

Derong Liu
Department of Electrical and Computer Engineering, University of Illinois at Chicago, Chicago, Illinois, USA

2.1 Introduction

For a given control system, stability is usually the most fundamental question to ask. If the system is linear and time-invariant, many stability criteria are available. Among them are the **Routh's stability criterion** and the **Nyquist stability criterion**. If the system is nonlinear or linear but time-varying, however, then such stability criteria do not apply.

Stability is mostly concerned with dynamical systems. It is a fundamental requirement in most engineering applications that the systems designed are stable. In control engineering applications, the design of a control system must first guarantee the stability of the overall system. That is to say, one of the design objectives that must be guaranteed in a controller design for a given system is stability. Stability is always a requirement independent of the technique used in the controller design, be it by **linear control, adaptive control, optimal control, intelligent control**, and the like.

2.2 Stability Concepts

Let $r(t)$, $y(t)$, and $h(t)$ be the input, output, and the impulse responses of a linear time-invariant system, respectively. For an nth order system, its initial conditions are given by:

$$y^{(k)}(0) = \left.\frac{d^k y(t)}{dt^k}\right|_{t=0} \qquad k = 0, 1, \cdots, n-1.$$

Zero initial conditions of an nth order system means that:

$$y^{(k)}(0) = 0, \; k = 0, 1, \cdots, n-1.$$

Finite initial conditions of an nth order system imply that:

$$\left|y^{(k)}(0)\right| \leq \eta, \; k = 0, 1, \cdots, n-1,$$

for some positive number $\eta < \infty$.

2.2.1 Bounded-Input Bounded-Output Stability

With zero initial conditions, a linear-time invariant system is said to be **bounded-input bounded-output** (BIBO) stable, or simply stable, if its output $y(t)$ is bounded in response to a bounded input $r(t)$. A system is said to be unstable if it is not BIBO stable.

The stability of a linear system can be determined from the location of the poles of the closed-loop transfer function in the s plane. If any of these poles lie in the right half of the s plane, with increasing time they give rise to the dominant mode, and the transient response increases monotonically or oscillates with increasing amplitude. This represents an unstable system. For such a system, as soon as the power is turned on, the output may increase with time. If no saturation takes place in the system and no mechanical stop is provided, then the system may eventually be subjected to damage and fail since the response of a real physical system cannot increase indefinitely. Hence, closed-loop poles in the right half of the s plane are not permissible in the usual linear control systems. If all closed-loop poles lie to the left of the $j\omega$ axis, any transient response eventually reaches equilibrium. This represents a stable system. Therefore, for BIBO stability, the roots of the characteristic equation, or the poles of the transfer function $H(s)$, must all lie in the left half of the s plane. Note that when a system has roots on the $j\omega$ axis, it is unstable by this definition.

2.2.2 Zero-Input Stability

If the zero-input response $y(t)$, subject to the finite initial conditions, reaches zero as time t approaches infinity, the system is said to be zero-input stable, or stable; otherwise, the system is unstable.

The **zero-input stability** defined here can also be equivalently stated as follows. A linear time-invariant system is zero-input stable if for any set of finite initial conditions, there exists a positive number M that depends on the initial condition, such that:

(1) $|y(t)| \leq M < \infty$ for all $t \geq 0$; and
(2) $\lim_{t \to \infty} |y(t)| = 0$.

Because the second condition requires that the magnitude of $y(t)$ reaches zero as time approaches infinity, the zero-input stability is also known as the **asymptotic stability**.

For linear time-invariant systems, BIBO stability and zero-input stability all have the same requirement: the roots of the characteristic equations must be located in the left half of the s plane. Thus, if a system is BIBO stable, it must also be zero-input or asymptotically stable. For this reason, the present discussion will simply refer to the stability of linear time-invariant systems without mentioning BIBO or zero-input. This chapter also often refers to the situation when the characteristic equation has simple roots on the $j\omega$ axis and none in the right half of the s plane as **marginally stable** or **marginally unstable**.

Example 1

A system with (closed-loop) transfer function is given by:

$$H(s) = \frac{1}{(s+1)(s+2)(s+3)},$$

which will be stable. A system with transfer function is given by:

$$H(s) = \frac{s+1}{(s-1)(s^2+2s+3)},$$

which will be unstable. A system with transfer function is given by:

$$H(s) = \frac{3(s+3)}{(s+2)(s^2+5)},$$

which will be unstable or marginally stable.

2.3 Stability Criteria

Whether a linear system is stable or unstable is a property of the system itself and does not depend on the input or driving function of the system. The poles of the input or driving function do not affect the property of stability of the system, but they contribute only to steady-state response terms in the solution. Thus, the problem of closed-loop stability can be solved readily by choosing no closed-loop poles in the right half of the s plane, including the $j\omega$ axis.

The most common representation of a linear system is given by its closed-loop transfer function in the form of:

$$\frac{B(s)}{A(s)} = \frac{b_0 s^m + b_1 s^{m-1} + \cdots + b_{m-1}s + b_m}{a_0 s^n + a_1 s^{n-1} + \cdots + a_{n-1}s + a_n},$$

where the as and bs are constants and $m \leq n$. A simple criterion, known as Routh's stability criterion, enables the determination of the number of closed-loop poles that lie in the right half of the s plane without having to factor the polynomial $A(s)$.

2.3.1 Routh's Stability Criterion

1) Write the polynomial in s in the following form:

$$a_0 s^n + a_1 s^{n-1} + \cdots + a_{n-1}s + a_n = 0, \qquad (2.1)$$

where the coefficients are real quantities. Assume that $a_n \neq 0$; that is, any zero root has been removed.

2) If any of the coefficients are zero or negative in the presence of at least one positive coefficient, there is a root or are roots that are imaginary or have positive real parts. In such a case, the system is not stable.

3) If all coefficients are positive, arrange the coefficients of the polynomial in rows and columns according to the following pattern:

s^n	a_0	a_2	a_4	a_6	$\cdots$
s^{n-1}	a_1	a_3	a_5	a_7	$\cdots$
s^{n-2}	b_1	b_2	b_3	b_4	$\cdots$
s^{n-3}	c_1	c_2	c_3	c_4	$\cdots$
s^{n-4}	d_1	d_2	d_3	d_4	$\cdots$
$\vdots$	$\vdots$	$\vdots$			
s^2	e_1	e_2			
s^1	f_1				
s^0	g_1				

The coefficients b_1, b_2, b_3, and so on, are evaluated as follows:

$$b_1 = \frac{a_1 a_2 - a_0 a_3}{a_1}.$$

$$b_2 = \frac{a_1 a_4 - a_0 a_5}{a_1}.$$

$$b_3 = \frac{a_1 a_6 - a_0 a_7}{a_1}.$$

$$\vdots$$

The evaluation of the *bs* is continued until the remaining ones are all zero. The same pattern of cross-multiplying the coefficients of the two previous rows is followed in evaluating the *cs*, *ds*, and so on. That is:

$$c_1 = \frac{b_1 a_3 - a_1 b_2}{b_1}.$$

$$c_2 = \frac{b_1 a_5 - a_1 b_3}{b_1}.$$

$$c_3 = \frac{b_1 a_7 - a_1 b_4}{b_1}.$$

$$\vdots$$

$$d_1 = \frac{c_1 b_2 - b_1 c_2}{c_1}.$$

$$d_2 = \frac{c_1 b_3 - b_1 c_3}{c_1}.$$

$$\vdots$$

This process is continued until the *n*th row has been completed. The complete array of coefficients is triangular. Note that in developing the array, an entire row may be divided or multiplied by a positive number to simplify the subsequent numerical calculation without altering the stability conclusion.

4) The number of roots in equation 2.1 with positive real parts is equal to the number of changes in the sign of the coefficients of the first column of the array.

It is noted that the exact values of the terms in the first column need not be known; instead, only the signs are needed. The necessary and sufficient conditions that all roots of equations 2.1 lie in the left half of the *s* plane are that all the coefficients of equation 2.1 be positive and all terms in the first column of the array have positive sign.

Example 2

Consider the following polynomial:

$$s^4 + 2s^3 + 3s^2 + 4s + 5 = 0.$$

According to the procedure, the following array can be formed.

s^4	1	3	5
s^3	2	4	0
s^2	1	5	
s^1	-6		
s^0	5		

In this example, the number of changes in the sign of the coefficients in the first column is two. This means that there are two roots with positive real parts. Note that the result is unchanged when the coefficients of any row are multiplied or divided by a positive number to simplify the computation. If the polynomial in the present example is the denominator of a closed-loop transfer function, the corresponding system will be unstable.

Routh's stability criterion is also known as the **Routh-Hurwitz criterion.**

2.3.2 Nyquist Stability Criterion

For a system shown in Figure 2.1, the closed-loop transfer function is given by:

$$\frac{Y(s)}{R(s)} = \frac{G(s)}{1 + G(s)H(s)}.$$

For stability, all roots of the following characteristic equation must lie in the left half of the *s* plane:

$$1 + G(s)H(s) = 0.$$

The Nyquist stability criterion applies to cases when $G(s)H(s)$ has neither poles nor zeros on the $j\omega$ axis. In the system shown in Figure 2.1, if the open-loop transfer function $G(s)H(s)$ has P poles in the right half of the *s* plane and $\lim_{s \to \infty} G(s)H(s) =$ constant, then for stability, the $G(j\omega)H(j\omega)$ locus, as ω varies from $-\infty$ to ∞, must circle the $-1 + j0$ point P times in the counterclockwise direction.

This criterion can be expressed as:

$$Z = N + P,$$

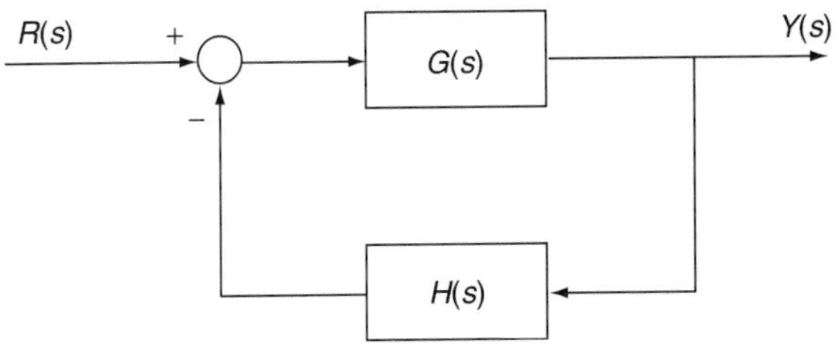

FIGURE 2.1 A Closed-Loop System

where Z = number of zeros of $1 + G(s)H(s)$ in the right half of the s plane, N = number of clockwise encirclements of the $-1 + j0$ point, and P = number of poles of $G(s)H(s)$ in the right half of the s plane. If P is not zero for a stable control system, $Z = 0$ or $N = -P$ must be true, which means that there must be P counterclockwise encirclements of the $-1 + j0$ point.

Example 3

Consider a closed-loop system whose open-loop transfer function is given by:

$$G(s)H(s) = \frac{K}{(T_1 s + 1)(T_2 s + 1)},$$

where $T_1 > 0$ and $T_2 > 0$.

A plot of $G(j\omega)H(j\omega)$ is shown in Figure 2.2. Because $G(s)H(s)$ does not have any poles in the right half of the s plane and the $-1 + j0$ point is not encircled by the $G(j\omega)H(j\omega)$ locus, this system is stable for any positive values of K, T_1, and T_2.

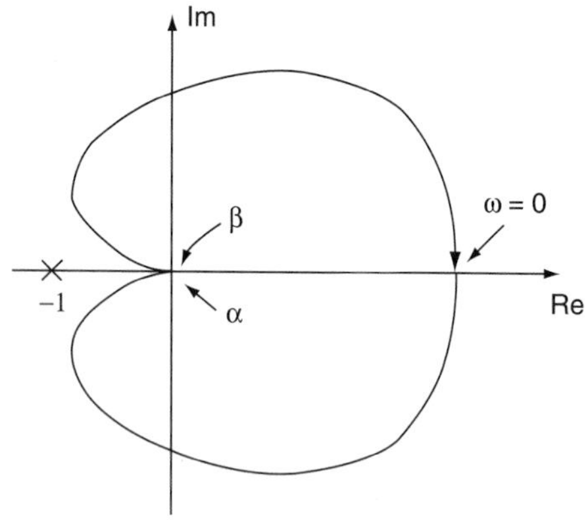

FIGURE 2.2 Polar Plot of $G(j\omega)H(j\omega)$. The α corresponds to $\omega = +\infty$, and β corresponds to $\omega = -\infty$.

Note that in the preceding criterion, it is assumed that $G(s)H(s)$ has neither poles nor zeros on the $j\omega$ axis. In this case, the criterion is derived by considering the Nyquist contour in Figure 2.3(A), which encloses the entire right half of the s plane. When $G(s)H(s)$ has poles and/or zeros on the $j\omega$ axis, the contour must be modified as in Figure 2.3(B).

The Nyquist stability criterion also applies to cases when $G(s)H(s)$ has poles and/or zeros on the $j\omega$ axis. In the system shown in Figure 2.1, if the open-loop transfer function $G(s)H(s)$ has P poles in the right half of the s plane, then for stability, the $G(s)H(s)$ locus (as a representative point s traces on the modified Nyquist path in the clockwise direction) must circle the $-1 + j0$ point P times in the counterclockwise direction.

Example 4

A closed-loop system has the following open-loop transfer function:

$$G(s)H(s) = \frac{K(T_2 s + 1)}{s^2(T_1 s + 1)},$$

whose stability depends on the relative magnitude of T_1 and T_2. Plots of the locus $G(s)H(s)$ for three cases, $T_1 < T_2$, $T_1 = T_2$, and $T_1 > T_2$, are shown in Figure 2.4. For $T_1 < T_2$, the locus of $G(s)H(s)$ does not encircle the $-1 + j0$ point, and the closed-loop system is stable. For $T_1 = T_2$, the locus of $G(s)H(s)$ passes through the $-1 + j0$ point, which indicates that the closed-loop poles are located on the $j\omega$ axis. For $T_1 > T_2$, the locus of $G(s)H(s)$ encircles the $-1 + j0$ point twice in the clockwise direction. Thus, the closed-loop system has two closed-loop poles in the right half of the s plane, and the system is unstable.

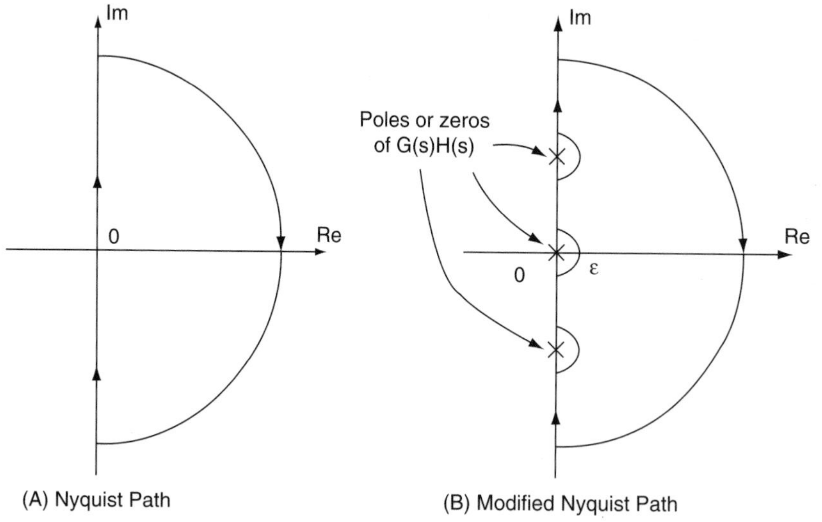

FIGURE 2.3 The Nyquist Contour

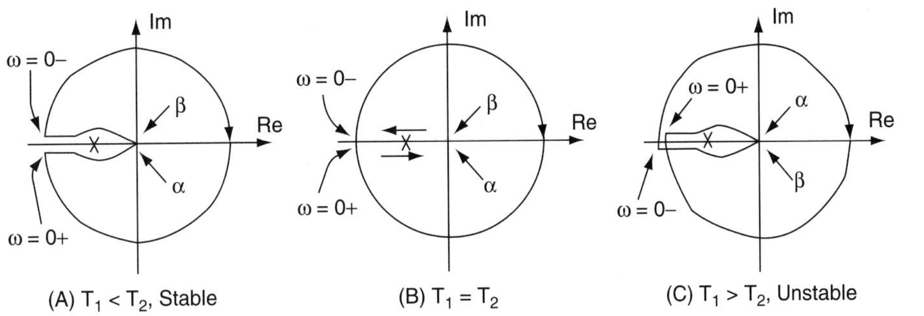

FIGURE 2.4 Polar Plots in the *GH* Plane. For Figure 2.4(B), $G(j\omega)H(j\omega)$ passes through the $-1 + j0$ point. In all of the figures, α corresponds to $\omega = +\infty$ and β corresponds to $\omega = -\infty$.

2.3.3 Phase Margin

The **phase margin** is the amount of additional phase lag at the gain crossover frequency required to bring the system to the verge of instability. The crossover frequency is the frequency at which $|G(j\omega)|$, the magnitude of the open-loop transfer function, is unity. The phase margin γ is 180° plus the phase angle ϕ of the open-loop transfer function at the gain crossover frequency, or:

$$\gamma = 180^\circ + \phi.$$

2.3.4 Gain Margin

The **gain margin** is defined as the reciprocal of the magnitude $|G(j\omega)|$ at the frequency at which the phase angle is -180°. Defining the phase crossover frequency ω_1 to be the frequency at which the phase angle of the open-loop transfer function equals -180° gives the gain margin K_g:

$$K_g = \frac{1}{|G(j\omega_1)|}.$$

In terms of decibels:

$$K_g \mathrm{dB} = 20 \log K_g = -20 \log |G(j\omega_1)|.$$

Phase margin and gain margin are illustrated in Figures 2.5 and 2.6.

For a stable minimum-phase system, the gain margin indicates how much the gain can be increased before the system becomes unstable. For an unstable system, the gain margin indicates how much the gain must be decreased to make the system stable.

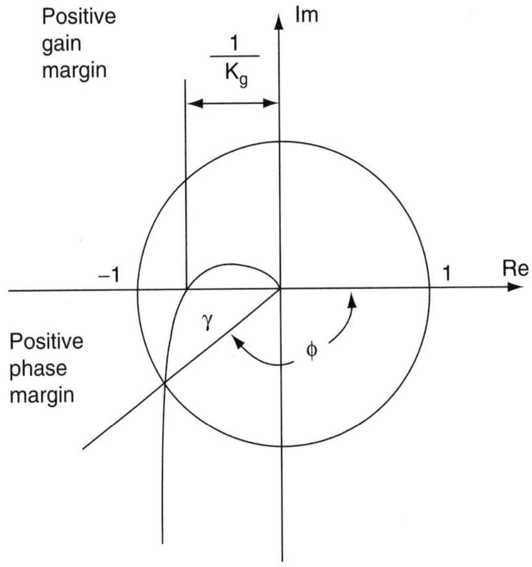

FIGURE 2.5 Phase and Gain Margin of a Stable System

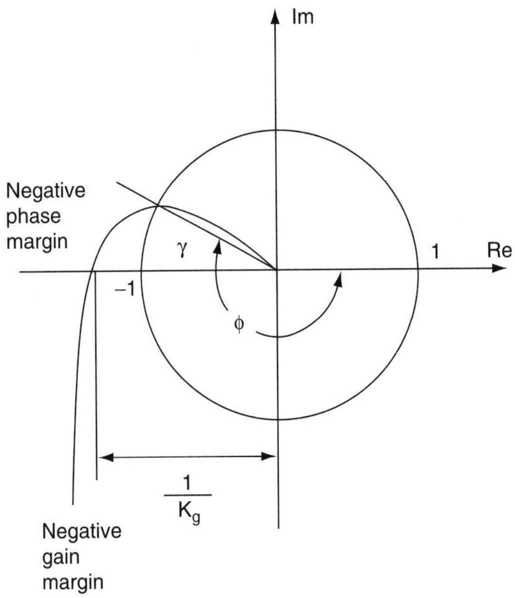

FIGURE 2.6 Phase and Gain Margin of an Unstable System

2.4 Lyapunov Stability Concepts

When the system in question is nonlinear or linear but time varying, the preceding stability criteria do not apply any more.

The alternative, **Lyapunov stability theory**, will be introduced as follows.

Use $\|x\|$ to denote the norm of $x \in R^n$. One example of vector norms is given by:

$$\|x\| = \sqrt{\sum_{i=1}^{n} x_i^2},$$

which is usually called the Euclidean norm or the l_2-norm. A neighborhood of the origin specified by $h > 0$ will be denoted by $B(h) = \{x \in R^n: \|x\| < h\}$. Consider continuous-time dynamical systems described by differential equations of the form:

$$\dot{x} = f(x),\ t \geq 0, \tag{2.2}$$

where $x \in R^n$ and $\dot{x}$ represent the derivative of x with respect to time t. It is assumed that $f: R^n \to R^n$ or $f: B(h) \to R^n$ for some $h > 0$. It is also assumed that $f(x)$ is continuous in x.

2.4.1 Equilibrium Point

The **equilibrium point** is in the state space at which the dynamical system will stay if it starts from that point. For systems described by equation 2.2, an equilibrium point x_e satisfies the condition that $f(x_e) = 0$.

Other terms for equilibrium point include **stationary point, singular point, critical point**, and **rest position.**

Example 5

Consider the simple pendulum system described by the equation:

$$\begin{cases} \dot{x}_1 & = x_2, \\ \dot{x}_2 & = -k \sin x_1, \end{cases} \tag{2.3}$$

where $k > 0$.

Equation 2.3 describes a dynamical system in R^2. The x_1 denotes the angle of the pendulum from the vertical position, and x_2 denotes the angular velocity. This system has two equilibrium points. One of them is located as shown in Figure 2.7(left), and the other is located as shown in Figure 2.7(right). Mathematically, these two equilibrium points can be represented by $(2k\pi, 0)$ and $(2k\pi + \pi, 0)$, respectively, for $k = 0, \pm 1, \pm 2, \ldots, N$.

2.4.2 Isolated Equilibrium Point

An equilibrium point x_e is called isolated if there is an $h > 0$ such that $B(x_e, h) \triangleq \{x \in R^n: \|x - x_e\| < h\}$ contains no equilibrium point other than x_e itself, where $\|\cdot\|$ represents any equivalent vector norm on R^n.

Both equilibrium points in example 5 are isolated equilibrium points in R^2.

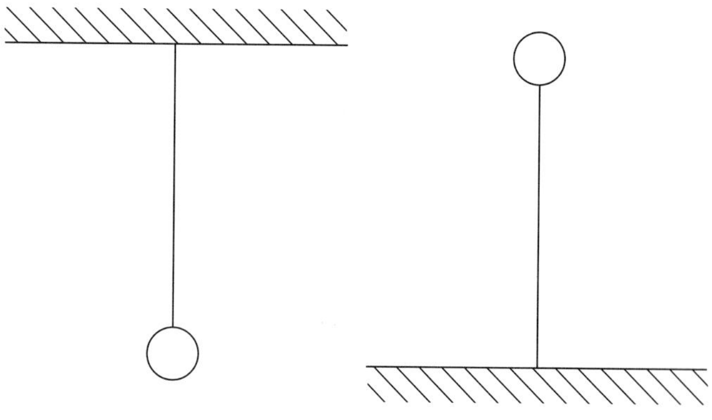

FIGURE 2.7 The Two Equilibrium Points of a Pendulum

Example 6

Consider a system described by equations:

$$\begin{aligned} \dot{x}_1 &= -ax_1 + bx_1x_2. \\ \dot{x}_2 &= -bx_1x_2. \end{aligned} \tag{2.4}$$

The $a > 0$ and $b > 0$ are constants. This example does not have any isolated equilibrium because every point on the x_2 axis (i.e., $x_1 = 0$) is an equilibrium point for equation 2.4.

Example 7

The linear system is described by:

$$\dot{x} = A(t)x,$$

which has a unique equilibrium that is at the origin if $A(t)$ is nonsingular for all $t \geq 0$.

Example 8

Assume that in equation 2.2, f is continuously differentiable with respect to all of its arguments, and let:

$$J(x_e) = \left.\frac{\partial f(x)}{\partial x}\right|_{x=x_e},$$

where $\partial f/\partial x$ is the $n \times n$ **Jacobian matrix** defined by

$$\frac{\partial f}{\partial x} = \left[\frac{\partial f_i}{\partial x_j}\right].$$

If $f(x_e) = 0$, and $J(x_e)$ is nonsingular, then x_e is an isolated equilibrium of system 2.2.

In stability theory, it is usually assumed that a given equilibrium point is an isolated equilibrium.

Stability concepts will be introduced for equilibrium at the origin (i.e., for $x_e = 0$). If the equilibrium x_e of equation 2.2 is not at the origin, one can always transform equation 2.2 by letting $z = x - x_e$. After such a transformation, the new system will have an equilibrium at the origin. Equation 2.2 will then become:

$$\frac{dz}{dt} = F(z),$$

where $F(z) = f(z + x_e)$. Therefore, without loss of generality, one can assume that system 2.2 has an isolated equilibrium point at the origin.

2.4.3 Stability

The equilibrium $x_e = 0$ of equation 2.2 is **stable** if for every $\varepsilon > 0$ there exists a $\delta(\varepsilon) > 0$ such that:

$$\|x(t)\| < \varepsilon \text{ for all } t \geq 0,$$

whenever $\|x(0)\| < \delta(\varepsilon)$.

The behavior of a stable equilibrium point can be depicted in Figure 2.8 for the case $x \in R^2$. By choosing the initial points in a sufficiently small neighborhood of the origin specified by δ, the trajectory of the system can be forced to lie entirely inside a neighborhood of the origin specified by ε.

The equilibrium $x_e = 0$ of equation 2.2 is **asymptotically stable** if the following statements are true:

1) The equilibrium is stable.
2) If there exists an $\eta > 0$ such that $\lim_{t \to \infty} x(t) = 0$ whenever $\|x(0)\| < \eta$.

The set of all $x(0) \in R^n$ such that $x(t) \to 0$ as $t \to \infty$ is called the **domain of attraction** of the equilibrium $x_e = 0$ of equation 2.2.

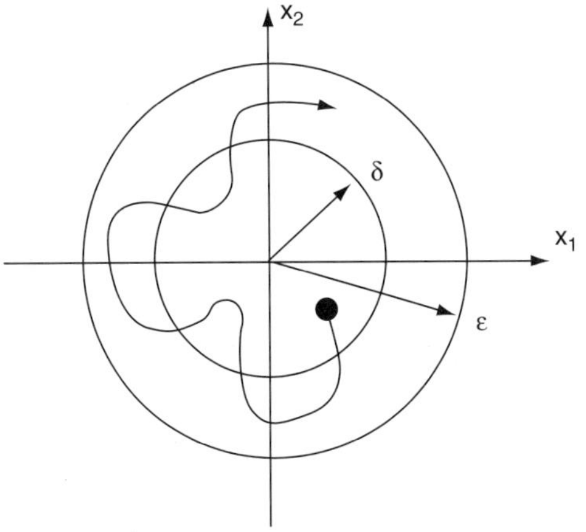

FIGURE 2.8 Typical Behavior of the Trajectory in the Vicinity of a Stable Equilibrium. This is true whenever $x(0) \leq \delta(\varepsilon)$.

The equilibrium $x_e = 0$ of equation 2.2 is **exponentially stable** if there exists an $\alpha > 0$, and for every $\varepsilon > 0$, there exists a $\delta(\varepsilon) > 0$ such that:

$$\|x(t)\| \leq \varepsilon e^{-\alpha t} \text{ for all } t \geq 0,$$

whenever $\|x(0)\| \leq \delta(\varepsilon)$.

The equilibrium $x_e = 0$ of equation 2.2 is **unstable** if the equilibrium is not stable.

The preceeding concepts pertain to local properties of an equilibrium. The following concepts characterize some global properties of an equilibrium.

2.4.4 Global Properties of an Equilibrium

The equilibrium $x_e = 0$ of equation 2.2 is **asymptotically stable in the large** if it is stable and if every solution of the equation tends to zero as $t \to \infty$.

The other term used often for asymptotic stability in the large is **global stability**.

The equilibrium $x_e = 0$ of equation 2.2 is **exponentially stable in the large** (or **globally exponentially stable**) if there exists an $\alpha > 0$ and for any $\beta > 0$, there exists a $k(\beta) > 0$ such that:

$$\|x(t)\| < k(\beta)\|x(0)\|e^{-\alpha t} \text{ for all } t \geq 0,$$

whenever $\|x(0)\| < \beta$.

The preceding concepts are referred to as stability (and instability) **in the sense of Lyapunov**.

2.5 Lyapunov Stability of Linear Time-Invariant Systems

Consider linear systems described by:

$$\dot{x} = Ax, \; t \geq 0 \quad \text{if A is nonsingular.} \tag{2.5}$$

System 2.5 has a unique equilibrium $x_e = 0$.

The equilibrium $x_e = 0$ of equation 2.5 is stable if all eigenvalues of A have nonpositive real parts and if every eigenvalue of A that has a zero real part is a simple zero of the characteristic polynomial of A.

The equilibrium $x_e = 0$ of equation 2.5 is asymptotically stable if and only if all eigenvalues of A have negative real parts.

A real $n \times n$ matrix A is called **stable** or a **Hurwitz matrix** if all of its eigenvalues have negative real parts. If at least one of the eigenvalues has a positive real part, then A is called unstable.

Thus, the equilibrium $x_e = 0$ of equation 2.5 is asymptotically stable if and only if A is stable. If A is unstable, then $x_e = 0$ is unstable.

2.6 Lyapunov Stability Results

The most important fact about the Lyapunov stability theory is to determine the stability properties of an equilibrium of a system 2.2 without having to solve equation 2.2.

2.6.1 Definiteness of a Function

A continuous function $v: R^n \to R$ [resp., $v: B(h) \to R$] is said to be **positive definite** if:

1) $v(0) = 0$, and
2) $v(x) > 0$ for all $x \neq 0$ [resp., $0 < \|x\| \leq r$ for some $r > 0$].

A continuous function v is said to be **negative definite** if $-v$ is a positive definite function.

Positive semidefinite: A continuous function $v: R^n \to R$ [resp., $v: B(h) \to R$] is said to be **positive semidefinite** if:

1) $v(0) = 0$, and
2) $v(x) \geq 0$ for all $x \in B(r)$ for some $r > 0$.

A continuous function v is said to be **negative semidefinite** if $-v$ is a positive semidefinite function.

A continuous function $v: R^n \to R$ is said to be **radially unbounded** if:

1) $v(0) = 0$,
2) $v(x) > 0$ for all $x \neq 0$, and
3) $v(x) \to \infty$ as $\|x\| \to \infty$.

Example 9

The function $v: R^3 \to R$ is given by:

$$v(x) = x_1^2 + x_2^2 + x_3^2,$$

which is positive definite and radially unbounded.

Example 10

The function $v: R^3 \to R$ is given by:

$$v(x) = x_1^2 + (x_2 - x_3)^2,$$

which is positive semidefinite. It is not positive definite because it is zero for all $x \in R^3$ such that $x_1 = 0$ and $x_2 = x_3$.

Example 11

The function $v: R^3 \to R$ is given by:

$$v(x) = x_1^2 + x_2^2 + x_3^2 - \left(x_1^2 + x_2^2 + x_3^2\right)^3,$$

which is positive definite in the interior of the ball given by $x_1^2 + x_2^2 + x_3^2 < 1$. It is not radially unbounded since $v(x) < 0$ when $x_1^2 + x_2^2 + x_3^2 > 1$.

Example 12

The function $v: R^3 \to R$ is given by:

$$v(x) = \frac{x_1^4}{1 + x_1^4} + x_2^4 + x_3^4,$$

which is positive definite but not radially unbounded.

2.6.2 Principle Lyapunov Stability Theorems

- If there exists a continuously differentiable and positive definite function v and its derivative (with respect to t) along the solutions of equation 2.2 given by:

$$\dot{v}_{(2.2)} = \sum_{i=1}^{n} \frac{\partial v}{\partial x_i} f_i(x) = \nabla v(x)^T f(x),$$

is negative semidefinite (or identically zero), then the equilibrium $x_e = 0$ of equation 2.2 is stable.
- If there exists a continuously differentiable and positive definite function v with a negative definite derivative $\dot{v}_{(2.2)}$, then the equilibrium $x_e = 0$ of equation 2.2 is asymptotically stable.
- If there exists a continuously differentiable, positive definite, and radially unbounded function v with a negative definite derivative $\dot{v}_{(2.2)}$, then the equilibrium $x_e = 0$ of equation 2.2 is asymptotically stable in the large.

Example 13

Consider the simple pendulum in example 5. If the following is chosen:

$$v(x) = k(1 - \cos x_1) + \frac{1}{2} x_2^2,$$

then this function is continuously differentiable, $v(0) = 0$, and v is positive definite. Along the solutions of equation 2.3, the following is true:

$$\dot{v}_{(2.3)} = (k \sin x_1)\dot{x}_1 + x_2 \dot{x}_2 = (k \sin x_1)x_2 + x_2(-k \sin x_1) = 0.$$

Therefore, the equilibrium $x_e = 0$ of the simple pendulum system is stable.

Example 14

Consider the system:

$$\begin{aligned} \dot{x}_1 &= (x_1 - c_2 x_2)(x_1^2 + x_2^2 - 1). \\ \dot{x}_2 &= (c_1 x_1 + x_2)(x_1^2 + x_2 - 1). \end{aligned} \tag{2.6}$$

In these equations, $c_1 > 1$ and $c_2 > 0$. If the following is chosen:

$$v(x) = c_1 x_1^2 + c_2 x_2^2,$$

it can be verified that:

$$\dot{v}_{(2.6)} = 2(c_1x_1^2 + c_2x_2^2)(x_1^2 + x_2^2 - 1).$$

Because v is positive definite and $\dot{v}_{(2.6)}$ is negative definite for $x_1^2 + x_2^2 < 1$, the equilibrium $x_e = 0$ of equation 2.6 is asymptotically stable.

Example 15
Consider the system:

$$\begin{aligned}\dot{x}_1 &= x_2 - c_1x_1(x_1^2 + x_2^2).\\ \dot{x}_2 &= -x_1 - cx_2(x_1^2 + x_2).\end{aligned} \tag{2.7}$$

In these equations, $c > 0$. If the following is chosen:

$$v(x) = x_1^2 + x_2^2,$$

it can be verified that:

$$\dot{v}_{(2.7)} = -2c(x_1^2 + x_2^2)^2.$$

Because v is positive definite and radially unbounded and $\dot{v}_{(2.7)}$ is negative definite, the equilibrium $x_e = 0$ of equation 2.7 is asymptotically stable in the large.

Lyapunov instability theorem: The equilibrium $x_e = 0$ of equation 2.2 is unstable if there exists a continuously differentiable function v such that $\dot{v}_{(2.2)}$ is negative definite (positive definite) and if in every neighborhood of the origin there are points x such that $v(x) < 0[v(x) > 0]$.

Example 16
Consider the system:

$$\begin{aligned}\dot{x}_1 &= c_1x_1 + x_1x_2.\\ \dot{x}_2 &= -c_2x_2 + x_1^2.\end{aligned} \tag{2.8}$$

The $c_1 > 0$ and $c_2 > 0$ are constants. If the following is chosen:

$$v(x) = x_2^2 - x_1^2,$$

then the following is true:

$$\dot{v}_{(2.8)} = -2(c_1x_1^2 + c_2x_2^2),$$

Because $\dot{v}_{(2.8)}$ is negative definite and in every neighborhood of the origin $v(x) < 0$ when $|x_1| < |x_2|$, the equilibrium $x_e = 0$ of (2.8) is unstable.

References

Hahn, W. (1967). *Stability of motion.* Berlin: Springer-Verlag.
Kuo, B.C. (2002). *Automatic control systems.* (8th Ed.). New York: John Wiley & Sons.
Miller, R.K., and Michel, A.N. (1982). *Ordinary differential equations.* New York: Academic Press.
Ogata, K. (2001). *Modern control engineering.* (3d Ed.). Upper Saddle River, NJ: Prentice Hall.

3

Robust Multivariable Control

Oscar R. González
Department of Electrical and Computer Engineering, Old Dominion University, Norfolk, Virginia, USA

Atul G. Kelkar
Department of Mechanical Engineering, Iowa State University, Ames, Iowa, USA

3.1 Introduction

A mathematical model of the physical process that needs to be controlled is needed for control system design and analysis. This model could be derived from first principles, obtained via system identification, or created by one engineer and then given to another. Regardless of the origin of the model, the control engineer needs to completely understand it and its limitations. In some processes, it is possible to consider models with a **single input and output** (SISO) for which a vast amount of literature is available. In many other cases, it is necessary to consider models with **multiple inputs and/or outputs** (MIMO). This chapter presents a selection of modeling, analysis, and design topics for multivariable, finite-dimensional, causal, linear, time-invariant systems that generalize the results for SISO systems. The reader is encouraged to consult the references at the end of this chapter for additional topics in multivariable control systems and for the proofs to the lemmas and theorems presented.

3.2 Modeling

The derivation of a mathematical model for a linear, time-invariant system typically starts by writing differential equations relating the inputs to a standard set of variables, such as loop currents and the configuration variables (e.g., displacements and velocities). In the time domain, the differential equations can be written as follows:

$$P(\mathcal{D})\xi(t) = Q(\mathcal{D})u(t), \tag{3.1}$$

where $\mathcal{D} \overset{\Delta}{=} d/dt$ is a differential operator, $u(t) \in \mathcal{R}^m$ is a vector of inputs, $\xi(t) \in \mathcal{R}^l$ is a vector of the standard variables called the partial state variables, and $P(\mathcal{D})$ and $Q(\mathcal{D})$ are differential operator matrices of compatible dimensions. The vector of output variables is, in general, represented by a linear, differential combination of the partial state variables and the inputs:

$$y(t) = R(\mathcal{D})\xi(t) + W(\mathcal{D})u(t), \tag{3.2}$$

where $y(t) \in \mathcal{R}^p$ is the vector of outputs and where $R(\mathcal{D})$ and $W(\mathcal{D})$ are differential operator matrices of compatible dimensions. The system representation introduced in equations 3.1 and 3.2 is commonly referred to as the **polynomial matrix description** (PMD) and is the most natural representation for many engineering processes. For analysis and design, a state-space representation that is an equivalent representation of proper[1] PMDs is more appropriate. The state-space representation of a system with input $u(t)$ and output $y(t)$ is as follows:

$$\dot{x}(t) = Ax(t) + Bu(t),\ y(t) = Cx(t) + Du(t), \tag{3.3}$$

where $x(t) \in \mathcal{R}^n$ is a vector of state variables, any minimal set of variables at time t_0 that together with the input $u(t)$, $t > t_0$ completely characterizes the response of the system for $t > t_0$. The state-space representation as defined in equation 3.3 will be represented by the four-tuple (A, B, C, D). The study of the fundamental properties of the state-space representation is covered, for example, in a first graduate course in linear systems (Antsaklis and Michel, 1997.) Because any state-space and PMD representation of a given system are equivalent, they have similar properties. For example, the eigenvalues of A are the roots of $|P(\lambda)|$, where $P(\lambda)$ is the polynomial matrix in λ found by replacing $\mathcal{D}$ with λ. Furthermore, the state-space representation is state controllable (observable) if and only if the equivalent PMD is controllable (observable).

The generalization of an SISO transfer function is a transfer function matrix, which is found by taking the Laplace transform of equation 3.3 with zero initial conditions. This operation yields:

$$G(s) = C(sI - A)^{-1}B + D, \tag{3.4}$$

where $G(s) \in \mathcal{R}(s)^{p\times m}$ is a proper rational transfer function matrix consisting of $p \times m$ SISO transfer function entries. The transfer function matrix can also be written in terms of an equivalent PMD. Because a transfer function matrix as in the SISO case represents only the controllable and observable dynamics of a system, consider the following controllable and observable PMD:

$$D_L(s)\xi(s) = N_L(s)u(s),\ y(s) = \xi(s), \tag{3.5}$$

where $D_L(s) \in \mathcal{R}[s]^{p\times p}$ and $N_L(s) \in \mathcal{R}[s]^{p\times m}$ are left coprime polynomial matrices in s. The PMD in equation 3.5 is controllable since $D_L(s)$ and $N_L(s)$ and $N_L(s)$ are left coprime; otherwise, it would only be observable. The PMD is a realization of the transfer function matrix in equation 3.4 since $G(s) = D_L(s)^{-1}N_L(s)$. The poles of a transfer function matrix $G(s)$ are the roots of the pole polynomial of $G(s)$ where the pole polynomial is given by the least common denominator of all nonzero minors of $G(s)$. So, every pole appears as the pole of at least one SISO transfer function entry of $G(s)$. The MIMO poles can also be found as the roots of $|D_L(s)|$. As in SISO systems, the set of poles is a subset of the set of eigenvalues of A *if* A and $G(s)$ satisfy equation 3.4. There are several definitions of multivariable zeros (Schrader and Sain, 1989). The transmission zero definition states that $s = s_z$ is a zero if the rank of $N_L(s_z)$ is less than the rank of $N_L(s)$. An important difference between SISO and transmission zeros is that a transmission zero does not have to be a zero of any SISO transfer function entry of $G(s)$. Another difference is that it is possible for $G(s)$ to have a pole equal to a zero. To understand other properties of poles and transmission zeros, such as multiplicities, the Smith-McMillan form of $G(s)$ can be used.

To understand the effect of the poles of $G(s)$ on a particular output, consider $y_i(s)$, the ith response given by:

$$y_i(s) = \sum_{j=1}^{m} G_{ij}(s)u_j(s), \tag{3.6}$$

where $G_{ij}(s)$ is the ijth entry of $G(s)$ and where $u_j(s)$ is the Laplace transform of the jth input. Equation 3.6 shows that in MIMO systems, the outputs are linear combinations of the inputs where the weights are rational transfer functions. Only the poles of the entries in the ith row of $G(s)$ can affect $y_i(s)$, depending on which inputs are nonzero. More insight can be gained by solving for $y_i(s)$ in a state-space representation of the system. Because the poles correspond only to the controllable and observable eigenvalues of A, let equation 3.3 be a minimal, that is, a controllable and observable realization of $G(s)$ with n states. The Laplace transform of the ith output is given by:

$$y_i(s) = (C_i(sI - A)^{-1}B + D_i)u(s), \tag{3.7}$$

where C_i and D_i are the ith rows of C and D, respectively. If A has distinct eigenvalues, the partial fraction expansion of the response is given by:

$$y_i(s) = \left(\sum_{\ell=1}^{n} \frac{1}{s - \lambda_\ell} C_i v_\ell w_\ell^H B + D_i\right) u(s), \tag{3.8}$$

where v_ℓ, $w_\ell \in C^n$ are right and left eigenvectors of A, respectively, and ^{H} denotes complex-conjugate transpose. The effect of the ℓth pole on the ith output is determined by the $1 \times n$ residue matrix $C_i v_\ell w_\ell^H B$ and the product of this residue times the vector of inputs $u(s)$. If each entry in the residue matrix is small, then this pole will have little effect on $y_i(t)$. It is also possible for some entries of the residue matrix not to be small but the product of a residue with the vector of inputs to still be small. This indicates that for some directions of the input vector, a pole may have a more significant effect than in other directions. The concept of input directions is unique to MIMO systems, and it will be discussed again in the next subsection.

Stability is an important system property. This section is mostly interested in two types: **asymptotic** and **bounded-input and bounded-output** (BIBO) stability. Asymptotic stability is a property of an internal representation, such as a state-space. The representation in 3.3 is said to be asymptotically stable if the solutions of $\dot{x}(t) = Ax(t)$ approach the origin

[1] A PMD is proper if $\lim_{\lambda\to\infty} (R(\lambda)P^{-1}(\lambda)Q(\lambda) + W(\lambda))$ exists and is finite. In fact, if the PMD in equations 3.1 and 3.2 is equivalent to the state-space in equation 3.3, then $\lim_{\lambda\to\infty} (R(\lambda)P^{-1}(\lambda)Q(\lambda) + W(\lambda)) = D$.

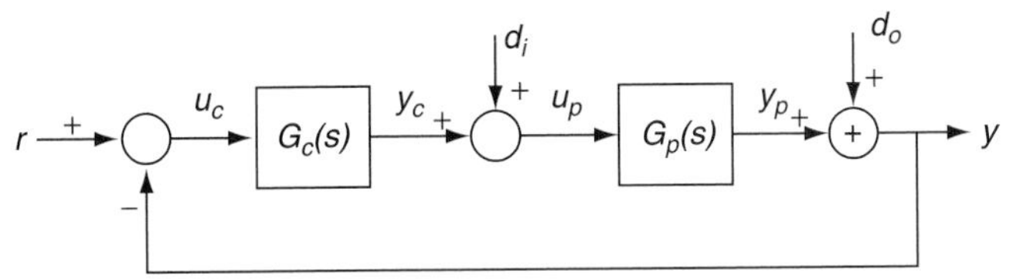

FIGURE 3.1 Classical Unity Feedback Configuration

for all initial conditions $x(0)$. The test for asymptotic stability is that the eigenvalues of A have negative real parts. Asymptotic stability implies BIBO stability, which is a property of an external representation. For BIBO stability, only the poles of a transfer function matrix need to have negative real parts.

Consider now the classical unity feedback configuration in Figure 3.1, where $r(t) \in \mathcal{R}^p$ denotes a vector of reference inputs and where $d_i(t) \in \mathcal{R}^m$ and $d_o(t) \in \mathcal{R}^p$ denote vectors of disturbance inputs at the input and output of the plant, respectively. Assume that the state-space representations (transfer function matrices) of the plant and controller are given by $(A_p, B_p, C_p, D_p)(G_p(s))$ and $(A_c, B_c, C_c, D_c)(G_c(s))$, respectively. The representation of the closed system will be well-formed if the dimensions ($G_p(s) \in \mathcal{R}_p(s)^{p\times m}$ and $G_c(s) \in \mathcal{R}_p(s)^{m\times p}$) are compatible and if $|I + D_pD_c| \neq 0$. The state representation of the closed-loop system is of the form:

$$\dot{x}_{cl}(t) = A_{cl}x_{cl}(t) + B_{cl}w(t), \quad y(t) = C_{cl}x_{cl}(t) + D_{cl}w(t), \quad (3.9)$$

where $x_{cl}(t) = [x_p(t), x_c(t)]^T$ and $w(t) = [r(t), d_o(t), d_i(t)]^T$. The properties of the closed-loop system are determined by analyzing $(A_{cl}(t), B_{cl}(t), C_{cl}(t), D_{cl}(t))$. For example, the closed-loop system is asymptotically stable if the eigenvalues of $A_{cl}(t)$ have negative real parts. The closed-loop system is then said to be internally stable.

The analysis and design of complex feedback configurations is simplified by using the general closed-loop block diagram in Figure 3.2, where $w(t)$ is the vector of all exogenous inputs, $y_K(t)$ is the vector of controller outputs, $z(t)$ is the vector of performance variables, and $u_K(t)$ is the vector of inputs to the controller. The top block is called the two-input and two-output plant, $\boldsymbol{P}$, and the bottom one corresponds to the matrix of controllers denoted by $\boldsymbol{K}$ that is formed after all the controllers have been pulled out of the closed-loop system. As a

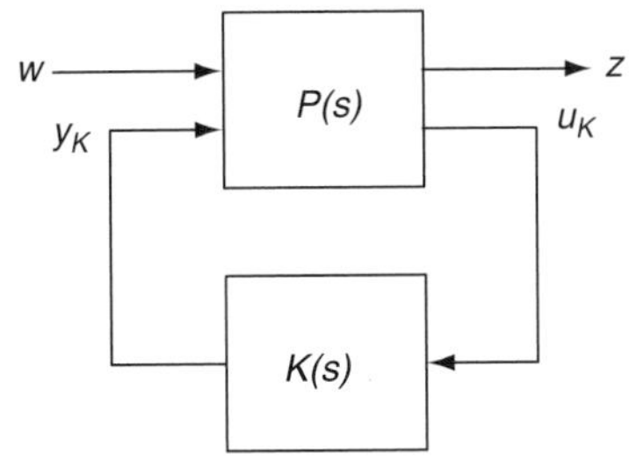

FIGURE 3.2 General Closed-Loop Block Diagram

trivial example, the classical unity feedback is represented in the general block diagram with the following transfer function matrices:

$$\begin{bmatrix} z(s) \\ u_K(s) \end{bmatrix} = \boldsymbol{P}(s)\begin{bmatrix} w(s) \\ y_K(s) \end{bmatrix}, y_K(s) = \boldsymbol{K}(s)u_K(s),$$

where:

$$\boldsymbol{P}(s) = \begin{bmatrix} P_{11}(s) & P_{12}(s) \\ P_{21}(s) & P_{22}(s) \end{bmatrix} = \begin{bmatrix} I & -I & -G_p(s) & -G_p(s) \\ 0 & 0 & 0 & I \\ I & -I & -G_p(s) & -G_p(s) \end{bmatrix}.$$

In this example, $\boldsymbol{K}(s) = G_c(s)$, the exogenous inputs vector is $w(t) = [r(t), d_o(t), d_i(t)]^T$, the vector of performance variables has been taken to be $z(t) = [r(t) - y(t), y_K(t)]^T$, and the vector of controller's inputs and outputs is simply given by $u_K(t) = u_c(t)$ and $y_K(t) = y_c(t)$, respectively. The first performance variable is the tracking error, $e(s) \triangleq r(s) - y(s)$. A state-space representation of the two-input and two-output plant is the following:

$$\dot{x}_P(t) = A_p x_P(t) + \begin{bmatrix} 0 & 0 & B_p & B_p \end{bmatrix}\begin{bmatrix} w(t) \\ y_K(t) \end{bmatrix}. \quad (3.10)$$

$$\begin{bmatrix} z(t) \\ u_K(t) \end{bmatrix} = \begin{bmatrix} C_p \\ 0 \\ -C_p \end{bmatrix} x_P(t) + \begin{bmatrix} I & 0 & -D_p & -D_p \\ 0 & 0 & 0 & I \\ I & 0 & -D_p & -D_p \end{bmatrix}\begin{bmatrix} w(t) \\ y_K(t) \end{bmatrix}. \quad (3.11)$$

The closed-loop transfer function matrix from the exogenous inputs to the performance variables is as follows:

$$z(s) = T_{zw}(s)w(s).$$

The closed-loop transfer function matrix is in fact a lower linear fractional transformation (LFT) of $\boldsymbol{P}(s)$ with respect to $\boldsymbol{K}(s)$ as given by:

$$T_{zw}(s) = \mathcal{F}_\ell(\boldsymbol{P}(s), \boldsymbol{K}(s)) = P_{11}(s) + P_{12}(s)\boldsymbol{K}(s)(I - P_{22}(s)\boldsymbol{K}(s))^{-1}P_{21}(s). \quad (3.12)$$

Based on this LFT, the general closed-loop system is well-formed if $|I - P_{22}(\infty)\boldsymbol{K}(\infty)| \neq 0$. In the trivial example of a classical unity feedback system, this reduces to $|I - P_{22}(\infty)\boldsymbol{K}(\infty)| = |I + D_pD_c| \neq 0$.

3.3 Performance Analysis

The purpose of the controller is to make the closed-loop system meet the desired specifications regardless of the uncertainty present. Examples of specifications are that the closed-loop system should be asymptotically stable, the steady-state error for each output channel should be small, and these stability and performance specifications should be maintained not only for the nominal plant model but also for all models in a specified uncertainty set. In this case, the closed-loop system will be said to have the properties of robust stability and robust performance.

Consider the classical unity feedback configuration in Figure 3.1 with an additional sensor noise vector $\eta(s)$ so that $u_c(s) = r(s) - (\eta(s) + y(s))$. The performance analysis is dependent on the following three types of transfer function matrices:

- **Return ratio:** $L_o(s) = G_p(s)G_c(s)$.
- **Sensitivity:** $S_o(s) = (I + L_o(s))^{-1}$.
- **Complementary sensitivity:** $T_o(s) = L_o(s)(I + L_o(s))^{-1}$.

These transfer function matrices are defined when the loop is broken at the output to the plant. Similar definitions follow when the loop is broken at the plant's input. The input/output and input/error relations can be written as shown next:

$$y(s) = S_o(s)d_o(s) + S_o(s)G_p(s)d_i(s) + T_o(s)r(s) - T_o(s)\eta(s). \quad (3.13)$$

$$e(s) = -S_o(s)d_o(s) - S_o(s)G_p(s)d_i(s) + (I - T_o(s))r(s) + T_o(s)\eta(s). \quad (3.14)$$

Notice that the mapping from the reference inputs, $r(s)$, to the errors contributed by them, $e_r(s)$, is given by $I - T_o(s)$. In this case, if the closed-loop system is internally stable, the steady-state tracking error contributed by a vector of step reference inputs $r(s) = R_0 \frac{1}{s}$, $R_0 \in \mathcal{R}^p$ is found using Laplace's final value theorem to be $e_r(\infty) = (I - T_o(0))R_0$. Thus, the error contributions to each output channel will be zero if in addition to internal stability, $T_o(0) = I$. Because the sensitivity and complementary sensitivity transfer function matrices satisfy:

$$S_o(s) + T_o(s) = I, \quad (3.15)$$

then $S_o(0) = 0_{p \times p}$ results in $e_r(\infty) = 0$. A sufficient condition for the dc gain of the sensitivity matrix to vanish is that every entry of $L_o(s)$ must have at least one pole at the origin. This is the generalization of system type to MIMO systems. Furthermore, if $S_o(0) = 0_{p \times p}$, the steady-state error contributed by a vector of step functions at the output disturbances, $d_o(s)$, will also be zero. An additional condition is needed for zero steady-state error contributed by a vector of step functions at the input disturbances, $d_i(s)$. A sufficient condition is that the plant has no entries with poles at the origin.

Making $T(0) = I$ results in the desired zero steady-state errors to vectors of step functions at the reference inputs and at the input and output disturbances. This choice, however, has the undesirable effect of making the error contributed by the dc component of the sensor noise not to be attenuated. This is a common trade-off in control systems, which does not affect the desired performance as long as the signal-to-noise ratio for low frequencies is made sufficiently high.

Zero steady-state error is possible for step inputs by appropriately including an exogenous model of the exosystem in the feedback loop (Gonzalez and Antsaklis, 1991). If zero steady-state error is not needed, then it will be necessary to make the mappings from the four exogenous inputs to the tracking error in equation 3.14 *small.* Three ways to determine the size of a transfer function matrix are presented in the following subsection.

To simplify the presentation, assume from now on that the plant is square with $p = m$. In addition, since the physical units used for input and output signals may lead to errors of different orders of magnitude, it is useful to normalize or scale the magnitudes of the plant's inputs and outputs. Procedures to perform scaling of MIMO systems are presented in, for example, Skogestad and Postlethwaite (1996). One approach is to normalize the plant's inputs and outputs so that the magnitude of each error is less than one. An alternative and common choice is to include the normalization in the frequency-dependent weights to be introduced for control system design.

3.3.1 MIMO Frequency Response and System Gains

To determine the **frequency response** of a BIBO stable transfer function matrix, $G(s)$, let its input be the vector of complex exponentials $u(t) = \underline{u}e^{j\omega t}$, $\underline{u} \in C^p$; then, the steady-state response is a complex exponential vector of the same frequency with amplitudes and phases changed by $G(s)_{|s=j\omega}$. Let the steady-state response be given by $y_{ss}(t) = \underline{y}e^{j\omega t}$, $\underline{y} \in C^p$, then the complex vectors $\underline{u}$ and $\underline{y}$ are related by:

$$\underline{y} = G(j\omega)\underline{u}. \quad (3.16)$$

The complex matrix $G(j\omega) \in C^{p\times p}$ is called the frequency response matrix, and it can be used to determine the size of $G(s)$ at a particular frequency ω.

In general, the size of $G(s)$ is defined as the gain from an input to its corresponding output. If the input is $u(t) = \underline{u}e^{j\omega t}$, the gain of $G(s)$ at ω can be defined at steady-state to be the ratio of the Euclidean vector norms $\|\underline{y}\|/\|\underline{u}\|$. This concept of gain is bounded as follows:

$$\underline{\sigma}(\omega) = \min_{\|\underline{u}\|\neq 0} \frac{\|G(j\omega)\underline{u}\|}{\|\underline{u}\|} \leqslant \frac{\|G(j\omega)\underline{u}\|}{\|\underline{u}\|} \leqslant \max_{\|\underline{u}\|\neq 0} \frac{\|G(j\omega)\underline{u}\|}{\|\underline{u}\|} = \bar{\sigma}(\omega), \quad (3.17)$$

where $\bar{\sigma}(\omega)$ and $\underline{\sigma}(\omega)$ are the largest and smallest singular values of $G(j\omega)$. These bounds are used to define the size of a TFM as follows:

- **Large** $G(s)$ is said to be *large* at ω if $\underline{\sigma}(G(j\omega))$ is large.
- **Small** $G(s)$ is said to be *small* at ω if $\bar{\sigma}(G(j\omega))$ is small.

The application and the scaling of the model determine what is meant by large or small singular values. In general, $\underline{\sigma}(G(j\omega)) \gg 1$ indicates that $G(s)$ is large at ω. Similarly, $\bar{\sigma}(G(j\omega)) \ll 1$ indicates that $G(s)$ is small at ω.

A graphical representation of the frequency response consists of plotting the maximum and minimum singular values of $G(j\omega)$ versus $\log(\omega)$. This is the generalization of Bode's magnitude plot to MIMO systems.

To get a better understanding of the role of singular values, substitute the singular value decomposition (SVD) of $G(j\omega)$ in equation 3.16:

$$\underline{y} = Y(\omega)\Sigma(\omega)U(\omega)^H\underline{u}, \quad (3.18)$$

where $\Sigma(\omega) \in \mathcal{C}^{p\times p}$ is diag $\{\bar{\sigma}(\omega) = \sigma_1(\omega), \ \sigma_2(\omega), \ldots, \sigma_p(\omega) = \underline{\sigma}(\omega)\}$ and where $Y(\omega)$ and $U(\omega) \in \mathcal{C}^{p\times p}$ are unitary matrices (see Section 9, Chapter 1). The diagonal entries of $\Sigma(\omega)$ are called the singular values of $G(s)$, and they depend on frequency. They are ordered from the largest to the smallest. Now, rewrite (equation 3.18 in terms of the columns of $Y(\omega)$ and $U(\omega)$, and let $Y(\omega) = [y_1(\omega)\cdots y_p(\omega)]$ and $U(\omega) = [u_1(\omega)\cdots u_p(\omega)]$. Since the columns of $U(\omega)$ form an orthonormal basis of C^p, let $\underline{\alpha}$ be the representation of $\underline{u}$ with respect to $\{u_1(\omega), \ldots, u_p(\omega)\}$. Substituting $\underline{u} = U(\omega)\underline{\alpha}$ in equation 3.18 gives the following:

$$\underline{y} = Y(\omega)\Sigma(\omega)\underline{\alpha} = \sum_{i=1}^{p}\sigma_i(\omega)\alpha_i(\omega)y_i(\omega), \quad (3.19)$$

where $\underline{\alpha} = [\alpha_1(\omega)\cdots\alpha_p(\omega)]^T$. This equation shows that the representation of $\underline{y}$ in terms of the columns of $Y(\omega)$ (also an orthonormal basis) is as written here:

$$\Sigma(\omega)\underline{\alpha} = [\bar{\sigma}(\omega)\alpha_1(\omega), \quad \sigma_2(\omega)\alpha_2(\omega)\cdots\underline{\sigma}(\omega)(\omega)\alpha_p(\omega)]^T.$$

The columns of $U(\omega)$ can be called the **principal input directions**, and the columns of $Y(\omega)$ are the **principal output directions**. If the input is parallel to a principal input direction, then the steady-state response will be along the corresponding principal output direction scaled by the corresponding singular value. In this sense, the singular values quantify the size of the effect of a transfer function matrix on particular input directions.

This characterization of size of a TFM will be used to develop quantitative measures of performance when the inputs are sinusoids in a specified frequency band. Two other measures will also be useful, and they can be defined as induced gains when the inputs and outputs belong to specified signal spaces.

Consider first the set of energy signals that are vectors of functions u that map $\mathcal{R}$ into $\mathcal{R}^p$. The signal u is said to have finite energy if:

$$\|u\|_2 = \left(\int_0^\infty \|u(\tau)\|^2 d\tau\right)^{1/2} < \infty, \quad (3.20)$$

where $\|u\|_2$ is the norm of the vector signal and $\|u(\tau)\|$ is the Euclidean norm of a vector in $\mathcal{R}^p$. To pose and solve optimal control problems, it is useful to consider the most general set of energy signals. The desired set is the Lebesgue space of all square integrable functions. This set is denoted by $\mathcal{L}_{2+} \stackrel{\Delta}{=} \mathcal{L}_2[0, \infty)$. For convenience, the spatial dimension of vectors with entries in $\mathcal{L}_{2+}$ will not be included. A useful measure of the size of the transfer function matrix is the induced system gain. If the system is represented by the mapping $G: \mathcal{L}_2 \rightarrow \mathcal{L}_2$, then the induced system gain is as follows:

$$\sup_{u\neq 0}\frac{\|Gu\|_2}{\|u\|_2} = \sup_{\|u\|_2=1}\|Gu\|_2 = \|G\|_\infty, \quad (3.21)$$

where $\|G\|_\infty$ is the induced system norm. This is the $\mathcal{H}_\infty$ norm of the system that gives the maximum gain of the system to a vector of sinusoids in the worst possible direction and worst possible frequency. If $G(s)$ is proper and stable, then $\|G\|_\infty = \sup_\omega \bar{\sigma}\{G(j\omega)\}$ is finite. Another popular measure of a system is the $\mathcal{H}_2$ norm. One interpretation of the $\mathcal{H}_2$ norm is as the gain from a white noise input with unit variance to the power of the output. The power is defined as:

$$\|y\|_{pow} = \left(\lim_{T\to\infty}\frac{1}{2T}\int_{-T}^{T}\|y(\tau)\|^2 d\tau\right)^{1/2},$$

where $\|y\|_{pow}$ is only a seminorm.

3.3.2 Performance Measures

Consider again the classical unity feedback configuration in Figure 3.1 with an additional sensor noise vector $\eta(s)$ so that $u_c(s) = r(s) - (\eta(s) + y(s))$. If $r(t)$, $d_i(t)$, and $d_o(t)$ are vectors of sinusoids up to a frequency ω_{low}, then adequate steady-state performance to these inputs requires that $\bar{\sigma}(S_o(j\omega)) \ll 1$ and $\bar{\sigma}(S_o(j\omega)G_p(j\omega)) \ll 1$ for $\omega < \omega_{low}$. The former requirement is met if and only if $\bar{\sigma}(L_o(j\omega)) \gg 1$ for $\omega < \omega_{low}$. If $G_c(s)$ is invertible, the second requirement is met if $\underline{\sigma}(G_c(j\omega)) \gg 1$ for $\omega < \omega_{low}$. If the signal-to-noise ratio becomes poor for $\omega > \omega_{high}$, then acceptable performance at steady-state requires attenuating the high-frequency components of the noise by making $\bar{\sigma}(T_o(j\omega)) \ll 1$ for $\omega > \omega_{high}$. This is accomplished by making $\bar{\sigma}(L_o(j\omega)) \ll 1$ for $\omega > \omega_{high}$. These are just some of the design guidelines that will lead to acceptable designs. There are additional trade-offs and performance limitations that need to be taken into account, including guidelines for roll-offs during the mid-frequencies.

A convenient way to combine the low-, mid-, and high-frequency requirements is to introduce frequency-dependent weights. These are included in $\boldsymbol{P}(s)$ when the general block diagram in Figure 3.2 is formed. The weights are typically added on the exogenous input channels and the performance output channels. The former weights serve to include in the model spectral information about the inputs. The latter weights are important in design to emphasize the frequency bands where the performance outputs need to be minimized. A possible design problem that results is to find a proper compensator $G_c(s)$ so that $\|T_{z\omega}\|_\infty < 1$. This and other control problems will be discussed in the following sections.

3.4 Stability Theorems

In this section, a brief introduction will be given to various stability criteria that can be used to determine the stability of a closed-loop control system.

3.4.1 Nyquist Criteria

Consider a negative feedback interconnection of a plant $G_p(s)$ and controller $G_c(s)$ in Figure 3.1. Let n_{Lo+} denote the number of unstable poles of the return ratio $L_o(s)$. Then the Nyquist stability criteria is given by the following theorem:

Stability Theorem: The closed-loop system consisting of the negative feedback interconnection of $G_p(s)$ and $G_c(s)$ is internally stable if and only if $n_{Lo+} = n_{Gp+} + n_{Gc+}$ and the Nyquist plot of $|I + L_o(s)|$ encircles the origin n_{Lo+} times in the anticlockwise direction and does not pass through the origin.

The first condition guarantees that the closed-loop system has no unstable hidden modes. The second condition gives a MIMO generalization of the Nyquist plot in terms of the determinant of $I + L_o(s)$. Because of the determinant, the Nyquist plot of a scalar times $I + L_o(s)$ is not simply a scaled version of the Nyquist plot of $|I + L_o(s)|$.

3.4.2 Small Gain Criteria

Another criteria that is often used to determine internal stability of the feedback interconnection is the **small gain theorem**, which is based on limiting the loop gain of the system. This theorem is central to the analysis of robust stability. The small gain theorem states that if the feedback interconnection of two proper and stable systems has a loop-gain product less than unity, then the closed-loop system is internally stable. There exist several versions of this theorem. One version is given next.

Theorem: Consider the feedback system in Figure 3.1, where systems G_p and G_c are proper and stable. Then, the feedback system is internally stable if:

$$\|G_p\|_\infty \|G_c\|_\infty < 1.$$

The stability theorems already given can be used in the analysis and synthesis of control systems. These theorems are also useful to determine conditions for robust stability and performance of the closed-loop system with the real plant as discussed in the following section.

3.5 Robust Stability

Controller design uses a nominal plant model. The error between the nominal model and the real plant arises primarily from two sources: **unmodeled dynamics** and **parametric uncertainties**. If the controller design does not take these errors into account, it cannot guarantee the performance of the closed-loop system with the real plant nor guarantee that the closed-loop system will be stable. Therefore, it is important to design controllers that will maintain closed-loop stability in spite of erroneous design models and uncertainties in parameter values. Controllers so designed are said to impart stability robustness to the closed-loop system.

To analyze stability robustness, consider the unity feedback system in Figure 3.1. A basic plant uncertainty representation is $G_p(s) = G_{po}(s) + \Delta_a(s)$, where $G_{po}(s)$ is the nominal plant model and where $\Delta_a(s)$ is the additive uncertainty. Other uncertainty representations include the output multiplicative uncertainty, $G_p(s) = (I + \Delta_o(s))G_{po}(s)$, and the input multiplicative one, $G_p(s) = G_{po}(s)(I + \Delta_i(s))$. For design purposes, it helps to normalize the uncertainty representations. For example, consider that the real plant is represented with an output multiplicative uncertainty, and let $\Delta_o(s) = W_o(s)\tilde{\Delta}_o(s)$, where $\tilde{\Delta}_o(s)$ and $W_o(s)$ are proper and stable with $\|\tilde{\Delta}_o(s)\|_\infty \leq 1$ and where $W_o(s)$ is a frequency-dependent scaling matrix. In this case, the unity feedback system in Figure 3.1 will be robustly stable if and only if $\|T_o W_o\|_\infty \leq 1$.

A more general result is possible that is independent of the particular types of uncertainties needed to represent the real plant. Consider the general block diagram in Figure 3.2: if all the normalized uncertainty blocks are pulled out, it results in a new general block diagram shown in Figure 3.3 that is

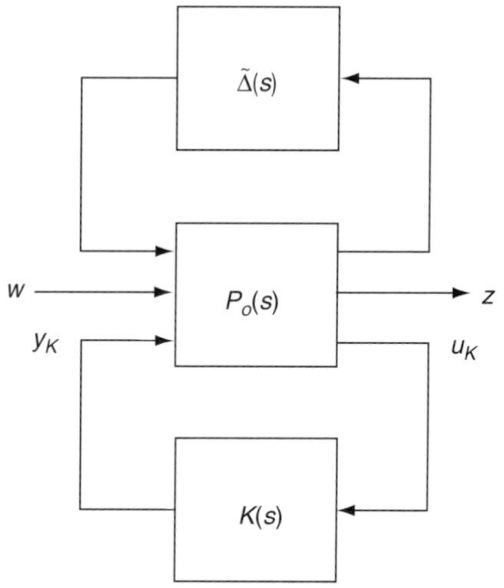

FIGURE 3.3 Closed-Loop System with Uncertainties Pulled Out

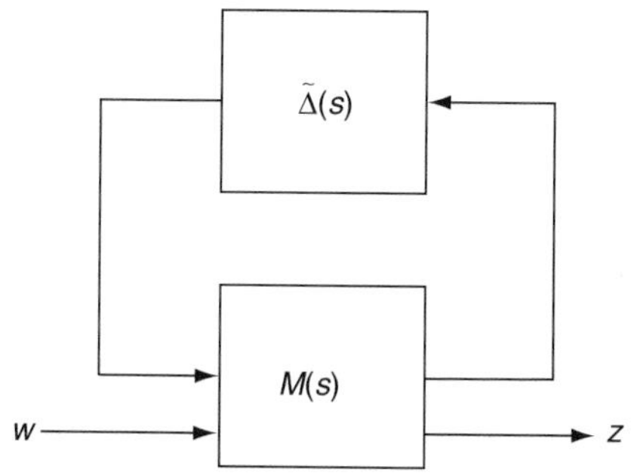

FIGURE 3.4 Simplified Closed-Loop System with Uncertainties Pulled Out

useful for robustness analysis. In Figure 3.3, $\tilde{\Delta}(s)$ is the block diagonal matrix of all the normalized uncertainty blocks ($\|\tilde{\Delta}(s)\|_\infty \leq 1$), and $\boldsymbol{P}_o(s)$ is the augmented nominal plant, including the uncertainty frequency weights. $\boldsymbol{P}_0(s)$ is assumed to be stabilized by $\boldsymbol{K}(s)$. Using a lower LFT, the bottom loop in Figure 3.3 can be closed resulting in Figure 3.4, where $\boldsymbol{M}(s)$ is proper and stable. If $\boldsymbol{M}(s)$ is partitioned corresponding to the two vector inputs and outputs and if $\|M_{11}\|_\infty < 1$, then the closed-loop system is robustly stable. These results that make use of the unstructured uncertainties lead to very conservative results. One way to reduce the conservativeness is to take advantage of the structure in the block diagonal matrix $\tilde{\Delta}(s)$ as done with the structured singular values.

3.6 Linear Quadratic Regulator and Gaussian Control Problems

3.6.1 Linear Quadratic Regulator Formulation

The linear quadratic regulator (LQR) is a classical optimal control problem used by many control engineers. Solutions to LQR are easy to compute and can typically be used to compute a baseline design useful for comparison. A formulation of the LQR problem considers the state equation of the plant:

$$\dot{x}(t) = Ax(t) + Bu(t). \tag{3.22}$$

The formulation also considers the following quadratic cost function of the states and control input:

$$J = \frac{1}{2}x^T(t_f)Sx(t_f) + \frac{1}{2}\int_0^{t_f} x^T(t)Qx(t) + u^T(t)Ru(t)dt, \tag{3.23}$$

where $S = S^T \geq 0$, $Q = Q^T \geq 0$ and $R = R^T > 0$.

To minimize the cost function, consider the Hamiltonian system with state and costate ($p(t)$) dynamics given by:

$$\begin{bmatrix} \dot{x}(t) \\ \dot{p}(t) \end{bmatrix} = \begin{bmatrix} A & -BR^{-1}B^T \\ -Q & -A^T \end{bmatrix} \begin{bmatrix} x(t) \\ p(t) \end{bmatrix}.$$

$$\text{with initial conditions:} \quad x(0) = x_0; \quad p(t_f) = Sx(t_f). \tag{3.24}$$

The optimal controller uses full-state feedback and is given by:

$$u(t) = -R^{-1}B^TP(t)x(t) = -K(t)x(t),$$

where $P(t)$ is the solution of the matrix Riccati equation:

$$\dot{P}(t) = -P(t)A - A^TP(t) + P(t)BR^{-1}B^TP(t) - Q,$$

which is solved backward in time starting at $P(t_f) = S$. The optimal feedback gain matrix $K(t)$ is given by:

$$K = R^{-1}B^TP(t),$$

and the optimal cost is as follows:

$$J_{\text{opt}} = \frac{1}{2}x^T(0)P(0)x(0).$$

3.6.2 Linear Quadratic Gaussian Formulation

The linear quadratic Gaussian (LQG) control problem is an optimal control problem where a quadratic cost function is minimized when the plant has random initial conditions, white noise disturbance input, and white measurement noise. The typical implementation of the LQR solution requires that the plant states be estimated, which can be posed as an LQG problem. The plant is described by the following state and output equations:

$$\begin{aligned} \dot{x}(t) &= Ax(t) + B_uu(t) + B_ww(t). \\ y_m(t) &= C_mx(t) + v(t) \text{ (measurement output)}. \\ y_p(t) &= C_px(t) \text{ (performance output)}. \end{aligned} \tag{3.25}$$

The $v(t)$ and $w(t)$ are uncorrelated zero-mean Gaussian noise processes; that is, $w(t)$ and $\upsilon(t)$ are white noise processes with covariances satisfying:

$$E\left\{\begin{bmatrix} w(t) \\ v(t) \end{bmatrix}[w^T(t+\tau), v^T(t+\tau)]\right\} = \begin{bmatrix} W & 0 \\ 0 & V \end{bmatrix}\delta(\tau).$$

The quadratic cost function that is to be minimized is given by:

$$J = E\left[\frac{1}{2}x^T(t_f)Sx(t_f) + \frac{1}{2}\int_0^{t_f} x^T(t)Qx(t) + u^T(t)Ru(t)dt\right], \tag{3.26}$$

where $S = S^T \geq 0$, $Q = Q^T \geq 0$ and $R = R^T > 0$. The optimal controller is a full-state feedback controller and is given by:

$$u(t) = -K(t)\hat{x}(t),$$

where $\hat{x}(t)$ is the Kalman state estimate. The closed-loop state equation can then be given by:

$$\begin{bmatrix} \dot{x}(t) \\ \dot{e}(t) \end{bmatrix} = A_{cl}(t)\begin{bmatrix} x(t) \\ e(t) \end{bmatrix} + B_{cl}(t)\begin{bmatrix} w(t) \\ v(t) \end{bmatrix},$$

where:

$$A_{cl} = \begin{bmatrix} A - B_uK(t) & B_uK(t) \\ 0 & A - G(t)C_m \end{bmatrix}, \; B_{cl} = \begin{bmatrix} B_w & 0 \\ B_w & -G(t) \end{bmatrix}.$$

The closed-loop state covariance matrix is as follows:

$$\dot{\sum}_{\tilde{x}}(t) = A_{cl}(t)\sum_{\tilde{x}}(t) + \sum_{\tilde{x}}(t)A_{cl}^T(t) + B_{cl}(t)\begin{bmatrix} W & 0 \\ 0 & V \end{bmatrix}B_{cl}^T(t),$$

where $\tilde{x}(t) = [x(t),\, e(t)]^T$, and:

$$\sum_{\tilde{x}}(0) = E\begin{bmatrix} x(0)x^T(0) & x(0)e^T(0) \\ e(0)x^T(0) & e(0)e^T(0) \end{bmatrix}.$$

The performance output covariance is given by:

$$\sum_{y_p}(t) = [\,C_p \quad 0\,]\sum_{\tilde{x}}(t)\begin{bmatrix} C_p^T \\ 0 \end{bmatrix} = C_p\sum_{\tilde{x}}C_p^T,$$

and the input covariance is given by:

$$\sum_u(t) = [\,-K(t)\;K(t)\,]\sum_{\tilde{x}}(t)\begin{bmatrix} -K^T(t) \\ K^T(t) \end{bmatrix}.$$

The cost is as follows:

$$J = \frac{1}{2}\mathrm{Tr}\left\{\begin{bmatrix} S & 0 \\ 0 & 0 \end{bmatrix}\sum_{\tilde{x}}(t_f) + \int_0^{t_f}\begin{bmatrix} Q + K^T(t)RK(t) & -K^T(t)RK(t) \\ -K^T(t)RK(t) & K^T(t)RK(t) \end{bmatrix}\sum_{\tilde{x}}(t)dt\right\}.$$

Remarks

For convenience of implementation, the steady-state solution obtained in the limit as $t_f \to \infty$ is used. The LQG control design is optimized to reject white noise disturbances; however, it can be modified to handle constant disturbances via feed-forward and integral control. The selection of gains in the feed-forward case can be done in a similar way as in the case of tracking system design. Prior to implementation, the robustness of LQG designs needs to be evaluated since there is no guarantee that any useful robustness will be obtained. For constant disturbance rejection via integral control, one needs to use integral Kalman filter and also integral state feedback. The integral LQG can be used both for disturbance rejection and tracking. Similar modifications are possible to handle tracking of time-varying reference inputs.

3.7 H_∞ Control

The LQR and LQG can also be posed as two-norm optimization problems (referred to as H_2 control problems). If the optimization problem is posed using the H_∞ norm as the cost function, the H_∞ formulation results.

The **H_∞ control problem** can be defined in terms of the general closed-loop block diagram in Figure 3.2 where the exogenous signals are included in the vector $\omega(s)$ and an appropriate choice of performance variables is given by the vector $z(s)$. The TFM $\boldsymbol{P}(s)$ includes frequency-dependent weights and appropriate normalization as described previously. Consider the following realization of $\boldsymbol{P}(s)$:

$$\dot{x}_P(t) = Ax_P(t) + \begin{bmatrix} B_1 \vdots B_2 \end{bmatrix}\begin{bmatrix} w(t) \\ \cdots \\ y_K(t) \end{bmatrix} \tag{3.27}$$

$$\begin{bmatrix} z(t) \\ \cdots \\ u_K(t) \end{bmatrix} = \begin{bmatrix} C_1 \\ \cdots \\ C_2 \end{bmatrix}x_P(t) + \begin{bmatrix} 0 & \vdots & D_{12} \\ \cdots & & \cdots \\ D_{21} & \vdots & 0 \end{bmatrix}\begin{bmatrix} w(t) \\ \cdots \\ y_K(t) \end{bmatrix}, \tag{3.28}$$

where:

$$D_{21}B_1^T = 0. \tag{3.29}$$

$$D_{21}D_{21}^T = I. \tag{3.30}$$

$$D_{12}^TC_1 = 0. \tag{3.31}$$

$$D_{12}^TD_{12}^T = I. \tag{3.32}$$

The control objective is to design a feedback controller that internally stabilizes the closed-loop system such that the ∞-norm of the mapping T_{zw} is bounded:

$$\|T_{zw}\|_\infty = \sup_{\|w(t)\|_2 \neq 0}\frac{\|z(t)\|_2}{\|w(t)\|_2} < \gamma. \tag{3.33}$$

A suboptimal controller satisfying the above mentioned objective exists if positive semidefinite solutions to the following two Riccati equations are possible:

$$\dot{P}(t) = P(t)A + A^TP(t) - P(t)(B_2B_2^T - \gamma^{-2}B_1B_1^T)P(t) + C_1^TC_1. \tag{3.34}$$

$$\dot{Q}(t) = AQ(t) + Q(t)A^T - Q(t)(C_2^TC_2 - \gamma^{-2}C_1^TC_1)Q(t) + B_1B_1^T. \tag{3.35}$$

Moreover, the solutions $P(t)$ and $Q(t)$ satisfy the following:

$$\rho(P(t)Q(t)) < \gamma^2. \tag{3.36}$$

The Hamiltonian systems corresponding to the two Riccati equations can be obtained similarly to how they were obtained for the LQG case. The H_∞ controller, which satisfies the mentioned bound, is given by:

$$\begin{aligned}\dot{x}_c &= A_c(t)x_c(t) + B_c(t)u_K(t).\\ y_K(t) &= C_c(t)x_c(t).\end{aligned} \tag{3.37}$$

The matrices $A_c(t)$, $B_c(t)$, and $C_c(t)$ are given as follows:

$$\begin{aligned}A_c(t) &= A + \gamma^{-2}B_1B_1^TP(t) - B_2B_2^TP(t)\\ &\quad - [I - \gamma^{-2}Q(t)P(t)]^{-1}Q(t)C_2^TC_2.\\ B_c(t) &= [I - \gamma^{-2}Q(t)P(t)]^{-1}Q(t)C_2^T.\\ C_c(t) &= -B_1^TP(t).\end{aligned} \tag{3.38}$$

In practice, the steady-state solution of the H_∞ control problem is often desired. The steady-state solution not only simplifies the controller implementation but also renders a closed-loop system time-invariant, simplifying the robustness and performance analysis. For the steady-state H_∞ control problem, a suboptimal solution exists *if* and only if the following conditions are satisfied. The first condition is that the algebraic Riccati equation:

$$0 = PA + A^TP - P(B_2B_2^T - \gamma^{-2}B_1B_1^T)P + C_1^TC_1,$$

has a positive semidefinite solution P such that $[A - (B_2B_2^T - \gamma^{-2}B_1B_1^T)P]$ is stable. The second condition is that the algebraic Riccati equation:

$$0 = AQ + QA^T - Q(C_2^TC_2 - \gamma^{-2}C_1^TC_1)Q + B_1B_1^T,$$

has a positive semidefinite solution Q such that $[A - Q(C_2^TC_2 - \gamma^{-2}C_1^TC_1)]$ is stable. The third condition is $\rho(PQ) < \gamma^2$.

3.8 Passivity-Based Control

Passivity is an important property of dynamic systems. A large class of physical systems, such as flexible space structures with collocated and compatible actuators and sensors, can be classified as being naturally passive. A passive system can be robustly stabilized by any strictly passive controller despite unmodeled dynamics and parametric uncertainties. This important stability characteristic has attracted much attention of researchers in the control of passive systems. This section presents selected definitions and stability theorems for passive linear systems.

3.8.1 Passivity of Linear Systems

For finite-dimensional linear, time-invariant (LTI) systems, passivity is equivalent to **positive realness** of the transfer function (Newcomb, 1966; Desoer and Vidyasagar, 1975). The concept of strict positive realness has also been defined in the literature and is closely related to strict passivity.

Let $G(s)$ denote a $p \times p$ matrix whose elements are proper rational functions of the complex variable s. The $G(s)$ is said to be stable if all its elements are analytic in $Re(s) \geq 0$. Let the conjugate-transpose of a complex matrix H be denoted by H^H.

Definition 1

A $p \times p$ rational matrix $G(s)$ is said to be positive real (PR) if:

- All elements of $G(s)$ are analytic in $Re(s) > 0$
- $G(s) + G^H(s) \geq 0$ in $Re(s) > 0$, or equivalently:
  - Poles on the imaginary axis are simple and have non-negative-definite residues
  - $G(j\omega) + G^H(j\omega) \geq 0$ for $\omega \in (-\infty, \infty)$

Various definitions of **strictly positive real** (SPR) systems are found in the literature (Kelkar and Joshi, 1996). Given below is the definition of a class of SPR systems: marginally, strictly, and positive-real (MSPR) systems.

Definition 2

A $p \times p$ rational matrix $G(s)$ is said to be **marginally strictly positive real** (MSPR) if it is positive real and the following is true:

$$G(j\omega) + G^H(j\omega) > 0 \; for \; \omega \in (-\infty, \infty).$$

Definition 2 (Joshi and Gupta, 1996) gives the least restrictive class of SPR systems. If $G(s)$ is MSPR, it can be expressed as $G(s) = G_1(s) + G_2(s)$, where $G_2(s)$ is weak SPR (Kelkar and Joshi, 1996) and where all the poles of $G_1(s)$ are purely imaginary (Joshi and Gupta, 1996).

3.8.2 State-Space Characterization of PR Systems

For LTI systems, the state-space characterization of positive real (PR) conditions results in the Kalman-Yakubovich-Popov (KYP) lemma. In Lozano-Leal and Joshi (1990), the KYP lemma was extended to WSPR systems, in Joshi and Gupta (1996), it was extended to MSPR systems. These extensions are given next.

Let (A, B, C, D) denote an nth-order minimal realization of the $p \times p$ transfer function matrix $G(s)$. The following lemma then gives the state-space characterization of WSPR system.

The Lozano-Leal and Joshi (1990) Lemma: The $G(s)$ is WSPR if and only if there exist real matrices: $P = P^T > 0$, $P \in R^{n\times n}$, $L \in R^{p\times n}$, and $W \in R^{p\times p}$, such that:

$$\begin{aligned}A^TP + PA &= -L^TL.\\ C &= B^TP + W^TL.\\ W^TW &= D + D^T.\end{aligned} \tag{3.39}$$

In these equations, (A, B, L, W) is controllable and observable or minimal, and $F(s) = W + L(sI - A)^{-1}B$ is minimum phase. If $G(s)$ is MSPR, it can be expressed as $G(s) = G_1(s) + G_2(s)$, where $G_2(s)$ is WSPR and all the poles of $G_1(s)$ are purely imaginary (Josh and Gupta, 1996). Let (A_2, B_2, C_2, D) denote an n_2^{th}-order minimal realization of $G_2(s)$, the stable part of $G(s)$. The following lemma is an extension of the KYP lemma to the MSPR case.

The Josh and Gupta (1996) Lemma: If $G(s)$ is MSPR, there exist real matrices: $P = P^T > 0$, $P \in R^{n \times n}$, $\mathfrak{L} \in R^{p \times n_2}$ and $W \in R^{p \times p}$, such that equation 3.39 holds with:

$$L = [0_{p \times n_1}, \mathfrak{L}_{p \times n2}], \quad (3.40)$$

where $(A_2, B_2, \mathfrak{L}, W)$ is minimal and $F(s) = W + L(sI - A)^{-1}B = W + \mathfrak{L}(sI - A_2)^{-1}B_2$ is minimum phase.

3.8.3 Stability of PR Systems

The stability theorem for a feedback interconnection of a PR and a MSPR system is given next.

LMI form of PR Lemma: An alternate form of KYP Lemma can be given in terms of the following Linear Matrix Inequality (LMI). A system (A, B, C, D) is said to be PR if it satisfies:

$$\begin{bmatrix} A^TP + PA & PB \\ B^TP & 0 \end{bmatrix} + \begin{bmatrix} C & D \\ 0 & I \end{bmatrix}^T \begin{bmatrix} U & W \\ W^T & V \end{bmatrix} \begin{bmatrix} C & D \\ 0 & I \end{bmatrix} < 0 \quad (3.41)$$

$$P = P_T > 0, \quad (3.42)$$

where $U = 0$, $V = 0$, and $W = -I$. This LMI condition is convenient to use in the case of checking PR-ness of MIMO systems. This LMI is a special case of dissipativity LMI (Kelkar and Joshi, 1996).

Stability Theorem

The closed-loop system consisting of negative feedback interconnection of $G_p(s)$ and $G_c(s)$ (Figure 3.1) is globally asymptotically stable if $G_p(s)$ is PR, $G_c(s)$ is MSPR, and none of the purely imaginary poles of $G_c(s)$ is a transmission zero of $G_p(s)$ (Joshi and Gupta, 1996).

Note that in the theorem systems $G_p(s)$ and $G_c(s)$ can be interchanged. Some nonlinear extensions of these results are also obtained in Isidori *et al.* (1999). Passivity-based controllers based on these fundamental stability results have proven to be highly effective in robustly controlling inherently passive linear and nonlinear systems.

Most physical systems, however, are not inherently passive, and passivity-based control methods cannot extend directly to such systems. For example, unstable systems and acoustic systems are not passive. One possible method of making these nonpassive systems amenable to passivity-based control is to *passify* them using suitable compensation. If the compensated system is ensured to be *robustly passive* despite plant uncertainties, it can be robustly stabilized by any MSPR controller. In Kelkar and Joshi (1997), various passification tech-

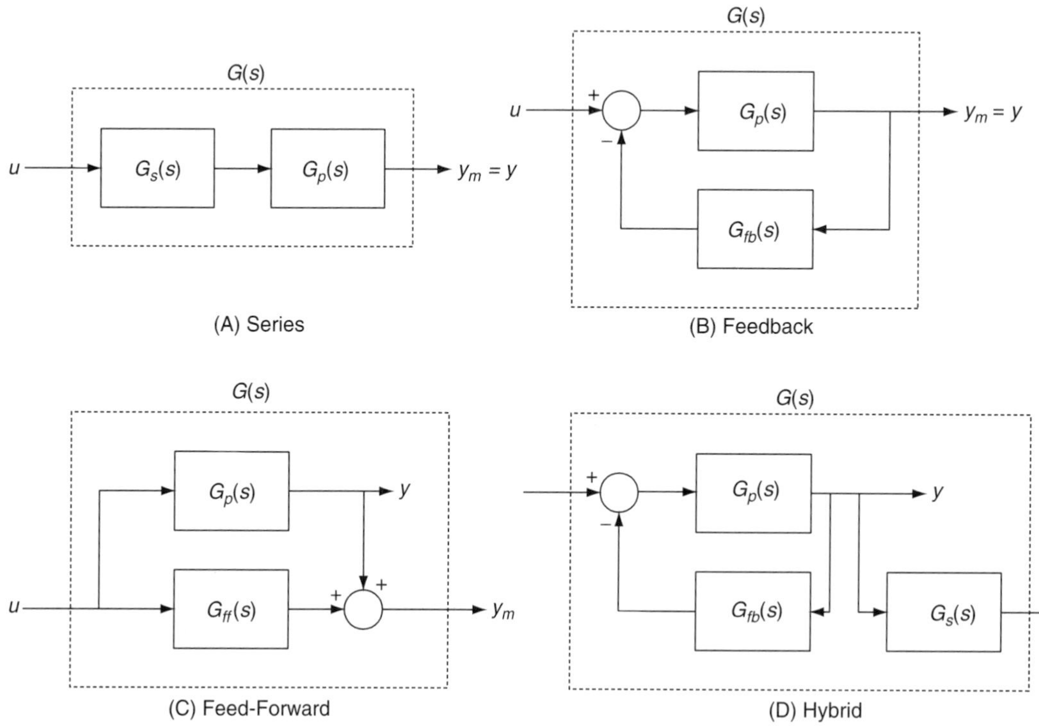

FIGURE 3.5 Methods of Passification

niques are presented, and some numerical examples are given, demonstrating the use of such techniques. A brief review of these methods is given next.

3.8.4 Passification Methods

The four passification methods in Figure 3.5 **series**, **feedback**, **feed-forward**, and **hybrid passification** are given in Kelkar and Joshi (1997) for finite-dimensional linear time-invariant nonpassive systems as shown. Once passified, the system can be controlled by any MSPR or weakly SPR (WSPR) controller (Isidori *et al.*, 1999). In Figure 3.5, the system with input $u(t)$ and output $y_m(t)$ $(G(s))$ represents the passified system. The type of passification to be used depends on the dynamic characteristics of the unpassified plant. For example, the system having unstable poles will require feedback passification, whereas the system having nonminimum phase zeros (i.e., having unstable zero dynamics) will require feed-forward passification. Some systems may require a combination of the basic passification methods. For SISO systems, the passification process is easier than for MIMO systems. The reason is that for SISO systems, only the phase plot needs to be checked to determine passivity, whereas in the case of MIMO systems, the KYP lemma conditions have to be checked. One numerical technique that can be used to check the KYP lemma is linear matrix inequality (LMI)-based PR conditions. The solution of the LMI can be done using the LMI tool box in MATLAB or another semidefinite programming package.

One important thing to be noted here is that, in the case of inherently passive systems, the use of an MSPR controller guarantees stability robustness to unmodeled dynamics and parametric uncertainties; however, in the case of nonpassive systems that are rendered passive using passifying compensation, stability robustness depends on the robustness of the passification. That is, the problem of **robust stability** is transformed into the problem of **robust passification**. In Kelkar and Joshi (1998) a number of sufficient conditions are derived to check the robustness of the passification.

3.9 Conclusion

This chapter has presented some of the fundamental tools in the analysis and design of linear, time-invariant, continuous-time, robust, multivariable control systems. The chapter starts with an introduction to modeling. The analysis tools include basic measures of performance, frequency response, and stability theorems. Linear quadratic, H_∞ and passivity-based control synthesis techniques were also introduced.

References

Antsaklis, P.J., and Michel, A.N. (1997). *Linear systems.* New York: McGraw-Hill.

Desoer, C.A., and Vidyasagar, M. (1975). *Feedback systems: Input output properties.* New York: Academic Press.

Gonzalez, O.R., and Antsaklis, P.J. (1991). Internal models in regulation, stabilization, and tracking. *International Journal of Control 53*(2), 411–430.

Green, M., and Limebeer, D.J. N. (1995). *Linear robust control.* Englewood Cliffs, NJ: Prentice Hall.

Isidori, A., Joshi, S.M., and Kelkar, A.G. (1999). Asymptotic stability of interconnected passive nonlinear systems. *International Journal of Robust and Nonlinear Control 9*, 261–273.

Joshi, S.M., and Gupta, S. (1996). On a class of marginally stable positive-real systems. *IEEE Transactions on Automatic Control 41*(1), 152–155.

Kelkar, A.G., and Joshi, S.M. (1996). *Control of nonlinear multibody flexible space structures,* Lecture notes in control and information sciences, Vol. 221. New York: Springer-Verlag.

Kelkar, A.G., and Joshi, S.M. (1997). Robust control of nonpassive systems via passification. *Proceedings of the American Control Conference 5*, 2657–2661.

Kelkar, A.G., and Joshi, S.M. (1998). Robust passification and control of nonpassive systems. *Proceedings of the American Control Conference*, 3133–3137.

Lozano-Leal, R., and Joshi, S.M. (1990). Strictly positive real functions revisited. *IEEE Transactions on Automatic Control 35*(11), 1243–1245.

Lozano-Leal, R., and Joshi, S.M. (1990). On the design of dissipative LQG-type controllers. In P. Dorato and R.K. Yedavalli (Eds.) *Recent advances in robust control.* New York: IEEE Press.

Maciejowski, J.M. (1989). *Multivariable feedback design.* Reading, MA: Addison-Wesley.

Newcomb, R.W. (1966). *Linear multiport synthesis.* New York: McGraw-Hill.

Schrader, C.B., and Sain, M.K. (1966). Research on system zeros: A survey. *International Journal of Control 50*(4), 1407–1433.

Skogestad, S., and Postlethwaite, I. (1996). *Multivariable feedback control.* West Sussex, England: John Wiley Sons.

4

State Estimation

Jay Farrell
Department of Electrical Engineering, University of California, Riverside, California, USA

4.1 Introduction

The **state of a system** is the *minimal* set of information required to *completely* summarize the status of the system at an initial time t_0. Because the state is a complete summary of the status of the system, it is immaterial how the system managed to get to its state at t_0. In addition, knowledge of the state at t_0, of the inputs to the system for $t > t_0$, and of the dynamics of the system allows the state to be determined for all $t > t_0$. Therefore, the concept of **state** allows time to be divided into past ($t < t_0$), present ($t = t_0$), and future ($t > t_0$), with the state at t_0 summarizing the information from the past that is necessary (together with knowledge of the system inputs) to predict the future evolution of the system.

The goal of the control system designer is to specify the inputs to a dynamic system to force the system to perform some useful purpose. This task can be interpreted as forcing the system from its state at t_0 to a desired state at a future time $t_1 > t_0$. Obtaining accurate knowledge of the state at t_0 and for all t between t_0 and t_1 is often critical to the completion of this control objective. This process is referred to as state estimation.

4.2 State-Space Representations

For dynamic systems described by a finite number of ordinary differential equations, the state of the dynamic system can be represented as a vector $\boldsymbol{x}(t)$ in **continuous-time**. For dynamic systems described by a finite number of ordinary difference equations, the state of the dynamic system can be represented as a vector $\boldsymbol{x}(k)$ in **discrete-time**. In either case, $\boldsymbol{x} \in \Re^n$. The vector representation of the state allows numerous tools from linear algebra to be applied to the analysis and design of state estimation and control systems. The state-space model in continuous-time representation is the following:

$$\begin{aligned} \dot{\boldsymbol{x}}(t) &= \boldsymbol{F}(t)\boldsymbol{x}(t) + \boldsymbol{G}_u\boldsymbol{u}(t) + \boldsymbol{G}_\omega\omega(t). \\ \boldsymbol{y}(t) &= \boldsymbol{H}(t)\boldsymbol{x}(t) + \nu(t). \end{aligned} \quad (4.1)$$

The $\boldsymbol{u} \in \Re^m$ is the deterministic control input, $\omega \in \Re^q$ is a stochastic (process noise) vector, $\boldsymbol{y} \in \Re^p$ is the measured output, and $\nu \in \Re^p$ is a stochastic (measurement noise) vector. The process and measurement noise vectors are each assumed to be Gaussian white noise processes:

$$E\langle \omega(t)\rangle = 0 \quad E\langle \omega(t_1), \omega(t_2)\rangle = Q(t_1)\delta(t_1 - t_2). \tag{4.2}$$
$$E\langle \nu(t)\rangle = 0 \quad E\langle \nu(t_1), \nu(t_2)\rangle = R(t_1)\delta(t_1 - t_2).$$

The $\delta(t)$ is the Dirac delta function. The power spectral density matrices $Q(t)$ and $R(t)$ are symmetric positive definite for all t.

State estimation may be desired for two reasons. First, when the measurement matrix $H \neq I$, then the state is not directly measured; however, knowledge of the full state may be desirable (e.g., for control). Second, even if $H = I$, it may be beneficial to filter the measurements to decrease the effects of the measurement noise. Filtering can be considered as a state estimation process.

State estimation and control processes are often implemented via a computer. Therefore, for implementation purposes, it is usually convenient to work with the discrete-time state-space representation:

$$\begin{aligned} x(k+1) &= \Phi_x(k)x(k) + \Gamma_u u(k) + \Gamma_\omega \omega_d(k) \\ y(k) &= H(k)x(k) + \nu(k). \end{aligned} \tag{4.3}$$

In equation 4.3, $x(k) = x(kT)$, k is an integer, and T is a fixed sample period. If F, G_u, and G_ω are time invariant over the interval $(kT, (k+1)T]$, then $\Phi_x(k)$, Γ_u, Γ_ω, and the statistics of the discrete-time Gaussian white noise process $\omega_d(k)$ can be determined so that the solutions to equation 4.1 at the sample times $t = kT$ have the same first- and second-order statistics as the solution of equation 4.3. Both $\nu(k)$ and $\omega_d(k)$ are discrete-time Gaussian white noise sequences with covariance matrices defined by:

$$E\langle \nu(k)\nu(j)^T\rangle = \begin{cases} R(k) & \text{if } k = j, \\ 0 & \text{otherwise;} \end{cases} \quad \text{and,}$$
$$E\langle \omega_d(k)\omega_d(j)^T\rangle = \begin{cases} Q_d(k) & \text{if } k = j, \\ 0 & \text{otherwise.} \end{cases}$$

The fact that linear operations on Gaussian random variables yields Gaussian random variables and the assumptions that $u(k)$ is deterministic and $\omega_d(k)$ and $\nu(k)$ are Gaussian random variables allow $x(k)$ and $y(k)$ to be modeled as Gaussian random variables. Therefore, the distributions of $x(k)$ and $y(k)$ are completely characterized by their first- and second-order statistics, which may be time varying.

The mean and variance of the state [i.e., $\bar{x}(k) = E\langle x(k)\rangle$ and $P(k) = E\langle (x(k) - \bar{x}(k))(x(k) - \bar{x}(k))^T\rangle$] can be propagated through time using the model of equation 4.3 as:

$$\bar{x}(k+1) = \Phi(k)\bar{x}(k) + \Gamma_u u(k). \tag{4.4}$$

$$P(k+1) = \Phi(k)P(k)\Phi^T(k) + \Gamma_\omega Q_d(k)\Gamma_\omega^T. \tag{4.5}$$

The superscript T denotes a transpose. Note that when initial conditions $\bar{x}(k_o)$, $P(k_o)$; signal $u(k)$; noise statistics $Q_d(k)$; and the state-space model $\Phi(k)$, Γ_u, Γ_ω are known for $k \in [k_o, k_f]$, these equations allow propagation of the mean and variance of the state for all $k \in [k_o, k_f]$. At any time instant of interest, the mean and variance of the output (i.e., $\bar{y}$ and P_y), can be computed as:

$$\bar{y}(k) = H(k)\bar{x}(k) \tag{4.6}$$

$$P_y(k) = H(k)P(k)H^T(k) + R(k). \tag{4.7}$$

4.3 Recursive State Estimation

Equations 4.4 through 4.5 provide the mean and the covariance of the state through time based only on the initial mean state vector and its error covariance matrix. When measurements $y(k)$ are available, the measurements can be used to improve the accuracy of an estimate of the state vector at time k. The symbols $\hat{x}^-(k)$ and $\hat{x}^+(k)$ are used to denote the estimate of $x(k)$ before and after incorporating the measurement, respectively.

Similarly, the symbols $P_{\hat{x}}^-(k)$ and $P_{\hat{x}}^+(k)$ are used to denote the error covariance matrices corresponding to $\hat{x}^-(k)$ and $\hat{x}^+(k)$, respectively. This section presents the time propagation and measurement update equations for both the state estimate and its error covariance. The equations are presented in a form that is valid for any linear unbiased measurement correction. These equations contain a gain matrix K that determines the estimator performance. The choice of K to minimize the error covariance $P_{\hat{x}}^+(k)$ will be of interest.

For an unbiased linear measurement, the update will have the form:

$$\hat{x}^+(k) = \hat{x}^-(k) + K(k)(y(k) - \hat{y}^-(k)), \tag{4.8}$$

where $\hat{y}^-(k) = H(k)\hat{x}^-(k)$. The error covariance of $\hat{x}^+(k)$ is the following:

$$\begin{aligned} P_{\hat{x}}^+(k) &= (I - K(k)H(k))P_{\hat{x}}^-(k)(I - K(k)H(k))^T \\ &\quad + K(k)R(k)K^T(k). \end{aligned} \tag{4.9}$$

$K(k)$ is a possibly time-varying state estimation gain vector to be designed. If no measurement is available at time k, then $K(k) = 0$, which yields $\hat{x}^+(k) = \hat{x}^-(k)$ and $P_{\hat{x}}^+(k) = P_{\hat{x}}^-(k)$. If a measurement is available, and the state estimator is designed well, then $P_{\hat{x}}^+(k) \leq P_{\hat{x}}^-(k)$. In either case, the time propagation of the state estimate and its error covariance matrix is achieved by:

$$\hat{x}^-(k+1) = \Phi(k)\hat{x}^+(k) + \hat{\Gamma}_u u(k) \tag{4.10}$$

$$P_{\hat{x}}^-(k+1) = \Phi(k)P_{\hat{x}}^+(k)\Phi^T(k) + \Gamma_\omega Q_d(k)\Gamma_\omega^T. \tag{4.11}$$

At least two issues are of interest relative to the state estimation problem. First, does there exist a state estimation gain vector $\boldsymbol{K}(k)$ such that $\hat{\boldsymbol{x}}$ is guaranteed to converge to $\boldsymbol{x}$ regardless of initial condition and the sequence $\boldsymbol{u}(k)$? Second, how should the designer select the gain vector $\boldsymbol{K}(k)$?

The first issue raises the question of observability. A linear time-invariant system is observable if the following matrix has rank n:

$$\left[\boldsymbol{H}^T, \boldsymbol{\Phi}^T\boldsymbol{H}^T, \ldots, (\boldsymbol{\Phi}^T)^n\boldsymbol{H}^T\right].$$

When a system is observable, then it is guaranteed that a stabilizing gain vector $\boldsymbol{K}$ exists. Assuming that the system of interest is observable, the remainder of this chapter discusses the design and analysis of state estimators.

Figure 4.1 portrays the state estimator in conjunction with the system of interest. The system of interest is depicted in the upper left. The state estimator is superimposed on a gray background in the lower right.

This interconnected system will be referred to as the **state estimation system**. The figure motivates several important comments. First, although the state estimator has only n states, the state estimation system has $2n$ states. Second, the inputs to the state estimation system are the deterministic input u and the stochastic inputs ω and ν. Third, the inputs to the state estimator are the deterministic input u and the measured plant output y. The state space model for the state estimation system is the following:

$$\begin{bmatrix} \boldsymbol{x}(k+1) \\ \hat{\boldsymbol{x}}^-(k+1) \end{bmatrix} = \begin{bmatrix} \boldsymbol{\Phi}_x & 0 \\ \boldsymbol{L}\boldsymbol{H}_x & \boldsymbol{\Phi} - \boldsymbol{L}\boldsymbol{H} \end{bmatrix} \begin{bmatrix} \boldsymbol{x}(k) \\ \hat{\boldsymbol{x}}^-(k) \end{bmatrix} + \begin{bmatrix} \boldsymbol{\Gamma}_u & \boldsymbol{\Gamma}_\omega & \boldsymbol{0} \\ \hat{\boldsymbol{\Gamma}}_u & \boldsymbol{0} & \boldsymbol{L} \end{bmatrix} \begin{bmatrix} u(k) \\ \omega(k) \\ \nu(k) \end{bmatrix}, \tag{4.12}$$

where $\boldsymbol{L} = \boldsymbol{\Phi}\boldsymbol{K}$.

Based on this state-space model, with the assumption that the system is time invariant, the transfer function from ν to $\hat{y}$ is as written here:

$$G_\nu(z) = \boldsymbol{H}[z\boldsymbol{I} - (\boldsymbol{\Phi} - \boldsymbol{L}\boldsymbol{H})]^{-1}\boldsymbol{L}, \tag{4.13}$$

where z is the discrete-time unit advance operator. Assuming that $\boldsymbol{H} = \boldsymbol{H}_x$, the transfer function from u to r is as follows:

$$G_u(z) = \boldsymbol{H}[z\boldsymbol{I} - (\boldsymbol{\Phi} - \boldsymbol{L}\boldsymbol{H})]^{-1} \left[(z\boldsymbol{I} - \boldsymbol{\Phi})(z\boldsymbol{I} - \boldsymbol{\Phi}_x)^{-1}\boldsymbol{\Gamma}_u - (z\boldsymbol{I} - \boldsymbol{\Phi}_x)(z\boldsymbol{I} - \boldsymbol{\Phi}_x)^{-1}\hat{\boldsymbol{\Gamma}}_u\right]. \tag{4.14}$$

Therefore, if $\boldsymbol{\Gamma}_u = \hat{\boldsymbol{\Gamma}}_u$ and $\boldsymbol{\Phi}_x = \boldsymbol{\Phi}$, then this transfer function is identically zero. Assuming again that $\boldsymbol{H} = \boldsymbol{H}_x$, the transfer function from ω to r is the following:

$$G_\omega(z) = \boldsymbol{H}[z\boldsymbol{I} - (\boldsymbol{\Phi} - \boldsymbol{L}\boldsymbol{H})]^{-1}[z\boldsymbol{I} - \boldsymbol{\Phi}][z\boldsymbol{I} - \boldsymbol{\Phi}_x]^{-1}\boldsymbol{\Gamma}_\omega. \tag{4.15}$$

In the special case where, in addition, $\boldsymbol{\Phi}_x = \boldsymbol{\Phi}$, the transfer function $\boldsymbol{G}_\omega(z)$ has n identical poles and zeros. This transfer functions is often stated as:

$$G_\omega(z) = \boldsymbol{H}[z\boldsymbol{I} - (\boldsymbol{\Phi} - \boldsymbol{L}\boldsymbol{H})]^{-1}\boldsymbol{\Gamma}_\omega, \tag{4.16}$$

where n pole and zero cancellations have occurred. These pole-zero cancellations and therefore the validity of equation 4.16 are dependent on the exact modeling assumption and the stability of the canceled poles.

4.4 State Estimator Design Approaches

Two categories of design approaches are commonly discussed: **Luenberger observers** and **Kalman filters**.

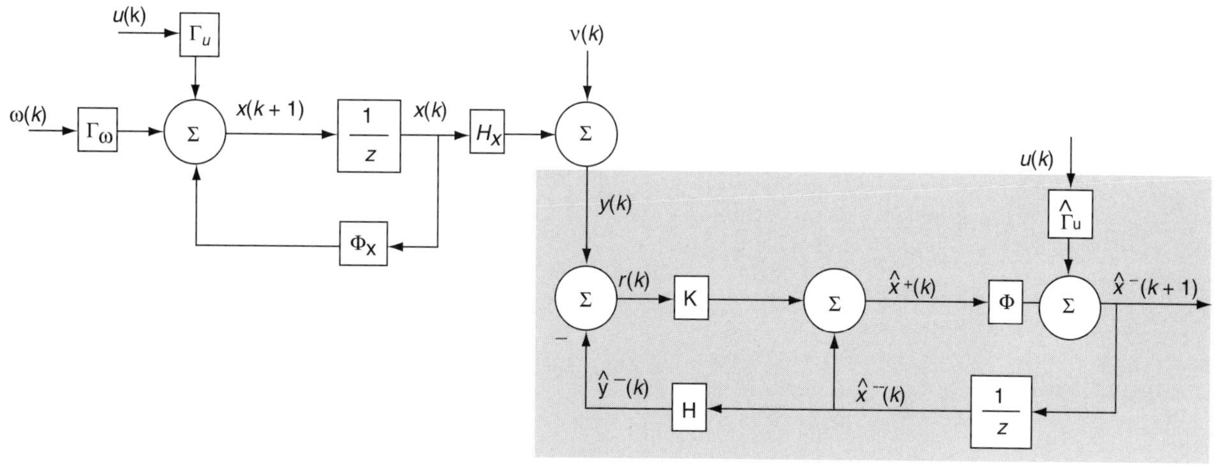

FIGURE 4.1 State Estimation System Block Diagram

4.4.1 Luenberger Observer

The time propagation and measurement update equations 4.8 through 4.11 can be combined. The combined equation for the estimate and its error covariance *prior* to the measurement correction at time $k+1$ are as follows:

$$\hat{x}^-(k+1) = \Phi(I - KH)\hat{x}^-(k) + \Gamma_u u(k) + \Phi K y(k). \quad (4.17)$$

$$P_{\hat{x}}^-(k+1) = \Phi\left[(I - KH)P_{\hat{x}}^-(k)(I - KH)^T + KRK^T\right]\Phi^T + \Gamma_\omega Q_d \Gamma_\omega^T. \quad (4.18)$$

The time indices have been dropped for convenience of notation. The combined equation for the estimate and its error covariance *posterior* to the measurement update are the following:

$$\hat{x}^+(k+1) = (I - KH)\Phi\hat{\mathbf{x}}^+(k) + (I - KH)\Gamma_u \mathbf{u}(k) + Ky(k+1). \quad (4.19)$$

$$P_{\hat{x}}^+(k+1) = (I - KH)[\Phi P_{\hat{x}}^+(k)\Phi^T + \Gamma_\omega Q_d \Gamma_\omega^T](I - KH)^T + KRK^T. \quad (4.20)$$

Note that even if the system is time invariant (i.e., Φ, K, H, Γ_u, Γ_ω all constant), both the state estimate and its error covariance matrix still change with time.

Assuming perfect modeling (i.e., $\Phi = \Phi_x$, etc.) and a time-invariant system, the discrete-time dynamics of the state estimation error $\tilde{x}^- = x - \hat{x}^-$ can be computed using equations 4.3 and 4.17:

$$\tilde{x}^-(k+1) = \Phi\tilde{x}^-(k) + \Gamma_\omega \omega(k) + \Phi K(y(k) - \hat{y}(k)). \quad (4.21)$$

$$= \Phi(I - KH)\tilde{x}^-(k) + \Gamma_\omega \omega(k) - \Phi K v(k). \quad (4.22)$$

Equation 4.22 shows that stability of the state estimate requires that the eigenvalues of $(\Phi - LH)$ have a magnitude less than one, where $L = \Phi K$.

The **Luenberger observer** uses a time-invariant gain vector L to place the eigenvalues of the observer state transition matrix $(\Phi - LH)$ at specified locations in the unit circle $|z| < 1$. When the output y is a scalar, there exists a unique L that will place the n eigenvalues at their specified locations. When y is a vector of measurements, then portions of the eigenvectors of $(\Phi - LH)$ can also be specified by the designer. The observer pole placement problem is dual to the controller pole placement problem.

4.4.2 Kalman Filter

Equations 4.8 and 4.9 illustrate the trade-offs that are implicit in the selection of the gain matrix K. The state estimation error after the measurement has two contributions. The first is the state estimation error propagated from the previous time instant:

$$(I - K(k)H(k))P_{\hat{x}}^-(k)(I - K(k)H(k))^T.$$

The second is the state estimation error due to measurement noise at the current time:

$$K(k)RK^T(k).$$

The **Kalman filter algorithm** produces the time varying gain sequence $K(k)$ that minimizes[1] *trace* $(P_{\hat{x}}^+(k))$ at every measurement instant k.

The Kalman filter gain sequence is computed as:

$$K(k) = P_{\hat{x}}^-(k)H(k)^T[H(k)P_{\hat{x}}^-(k)H(k)^T + R(k)]^{-1}. \quad (4.23)$$

For a time-varying linear system driven by Gaussian white noise, the Kalman filter gain sequence can be derived as the optimal estimator from the least-squares, maximum-likelihood, and mean-squares perspectives. The estimator discussion of this chapter was initiated with the assumed "linear update" of equation 4.8. This assumption is not restrictive in the sense that it can be shown for linear systems with additive noise Gaussian white measurement and process noise that a linear measurement update is optimal (i.e., a nonlinear estimator will not achieve better performance). The Kalman filter also has the property that the measurement residual $r(k)$ is a white noise sequence.

For the Kalman filter, even if the system is time invariant (i.e., Φ, $\Gamma_{\mathbf{u}}$, Γ_ω, H, R, and Q constant), the Kalman filter may be time varying. When a steady-state Kalman filter exists, the steady-state Kalman filter gain is:

$$K = P_{ss}^- H^T[HP_{ss}^- H^T + R]^{-1}, \quad (4.24)$$

where P_{ss}^- is the steady-state solution of equation 4.18, which is the discrete-time algebraic Riccati equation:

$$P_{ss}^- = \Phi\left[P_{ss}^- - P_{ss}^- H^T[HP_{ss}^- H^T + R]^{-1}HP_{ss}^-\right]\Phi^T + \Gamma_\omega Q_d \Gamma_\omega^T. \quad (4.25)$$

This form of the equation is particular to the Kalman filter and does not hold for general estimators. In steady-state, $P_{\hat{x}}^-$ is constant with value P_{ss}^-. In steady-state, $P_{\hat{x}}^+$ is also constant with value $(I - KH)P_{ss}^-$. Note that $P_{\hat{x}}^-$ and $P_{\hat{x}}^+$ are *not* equal in steady-state.

Let $\hat{z}(k) = C\hat{x}(k)$ be a linear combination of the states that is of interest. When the steady-state Kalman filter exists, the transfer function from y to $\hat{z}$ is the following:

$$C[zI - \Phi(I - KH)]^{-1}K. \quad (4.26)$$

[1] If $D = [d_{ij}]$ is an $n \times n$ matrix, then *trace* $(D) = \sum_{i=1}^n d_{ii}$.

Equation 4.26 is the infinite impulse response Weiner filter for estimating z from y. Finite impulse response Weiner filters could be constructed from the Markov parameters of this steady-state Kalman filter.

4.5 Performance Analysis

This chapter has presented two methods for the design of state estimators; however, the task is not complete. For example, the Kalman filter methodology would require the designer to develop and use a full stochastic model of the system, process noise, actuators, and sensors. Because the Kalman filter approach assumes a linear model driven by white noise, any colored noise process will necessitate the augmentation of additional states to the model so that the colored noise process is represented as a linear system with white noise inputs.

The construction of such models is not straightforward. Often due to data and experimental limitations, the modeling process does not yield a single exact model but a range of possible models. The full model may also be too complex to allow a full Kalman filter implementation with online covariance propagation. Even when the full implementation is possible, additional analysis may show that nearly equivalent and possibly more robust performance can be achieved with fewer computations for an implementation using a reduced order model.

The goal of this section is to present methods for the analysis of suboptimal or reduced order filters. The idea of the analysis approach is not difficult, but the implementation of computationally efficient algorithms can be time consuming. The interested reader should consult the references at the end of this chapter.

4.5.1 Error Budgeting

Error budgeting is a methodology for determining how much each error source contributes to the overall estimation error. The methodology uses covariance propagation equations 4.9 and 4.11 that are repeated below for convenience:

$$\boldsymbol{P}_{\tilde{x}}^{+}(k) = (\boldsymbol{I} - \boldsymbol{K}(k)\boldsymbol{H}(k))\boldsymbol{P}_{\tilde{x}}^{-}(k)(\boldsymbol{I} - \boldsymbol{K}(k)\boldsymbol{H}(k))^T + \boldsymbol{K}(k)\boldsymbol{R}\boldsymbol{K}^T(k). \quad (4.27)$$

$$\boldsymbol{P}_{\tilde{x}}^{-}(k+1) = \boldsymbol{\Phi}(k)\boldsymbol{P}_{\tilde{x}}^{+}(k)\boldsymbol{\Phi}^T(k) + \boldsymbol{\Gamma}_{\omega}\boldsymbol{Q}_d(k)\boldsymbol{\Gamma}_{\omega}^T. \quad (4.28)$$

Note that if the gain sequence $\boldsymbol{K}(k)$ is known, then equations 4.27 and 4.28 are linear in P, $\boldsymbol{Q}_d$, and $\boldsymbol{R}$. Therefore, the superposition principle can be applied. The error budgeting procedure can therefore be divided into three steps, where the second step has several subparts. The error budgeting approach is presented here assuming that the Kalman filtering approach is used and that the system model is known and correct. Suboptimal filtering is discussed in the next subsection.

Assuming that $\boldsymbol{\Phi}$, $\boldsymbol{P}_{\tilde{x}}(0)$, $\boldsymbol{\Gamma}_{\omega}$, $\boldsymbol{Q}_d$, $\boldsymbol{H}$ and $\boldsymbol{R}$ are known for each k, the procedure is as follows.

- **Step 1**: Iterate equations 4.27 and 4.28 for the time duration of interest. Compute $\boldsymbol{K}(k)$ over the time interval using equation 4.23. This gain sequence is stored and used in each repetition of step 2. It is critical that the same gain sequence be used throughout step 2.
- **Step 2**: Divide the error sources into q mutually exclusive subgroups. For simplicity of this discussion, let $q = 3$ and the subgroups be (1) $\boldsymbol{P}_{\tilde{x}}(0)$, (2) $\boldsymbol{Q}_d$, and (3) $\boldsymbol{R}$. In general, each component of $\boldsymbol{P}_{\tilde{x}}(0)$, $\boldsymbol{Q}_d$, and $\boldsymbol{R}$ could be in its own subgroup. The main constraint is that each error component must be in exactly one subgroup.

  Equations 4.27 and 4.28 are now iterated over the time duration of interest, for q repetitions. Each repetition uses the estimator gain sequence determined in step 1. For the ith repetition $(q-1)$, error sources are set to zero, with only the ith subgroup of error sources being nonzero. Denote the sequence of covariance matrices that results from the ith repetition as $\boldsymbol{P}_{\tilde{x}}^{-}(k, i)$ and $\boldsymbol{P}_{\tilde{x}}^{+}(k, i)$.
- **Step 3**: The covariance matrices $\boldsymbol{P}_{\tilde{x}}^{-}(k, i)$ and $\boldsymbol{P}_{\tilde{x}}^{+}(k, i)$ (usually just the diagonal elements) are compared to identify which of the q error sources is dominant at any time k. Identification of the dominant sources allows the sensor or model development efforts to focus on the most effective directions. Identification of insignificant error sources motivates areas of possible cost savings through either the elimination of sensors or the purchase of lower cost sensors.

In step 1, it is useful as a check to also save the covariance matrices. The check is that the sum of the covariance matrices from each iteration of step 2 should equal the corresponding covariance matrix from step 1:

$$\boldsymbol{P}_{\tilde{x}}^{-}(k) = \sum_{i=1}^{q} \boldsymbol{P}_{\tilde{x}}^{-}(k, i) \text{ and } \boldsymbol{P}_{\tilde{x}}^{+}(k) = \sum_{i=1}^{q} \boldsymbol{P}_{\tilde{x}}^{+}(k, i), \quad (4.29)$$

for all k in the time span of interest.

4.5.2 Covariance Analysis

Covariance analysis is concerned with the iteration of equations 4.27 and 4.28 to characterize the expected filter performance. These equations are applicable for any estimator gain sequence. The error budgeting analysis of the previous section is a form of covariance analysis.

A second form of covariance analysis is the iteration of the same equations for suboptimal gain sequences. **Suboptimal gain sequences** are often of interest to avoid the need for online computation of the error covariance equations, which are typically the most computationally expensive portion of the Kalman algorithm. Examples of suboptimal gain sequences are the Luenberger gain, the steady-state Kalman filter gain, or a curve fit to the time-varying Kalman gain sequence. Iteration of equations 4.27 and 4.28 once for the optimal gain sequence and once for a suboptimal gain sequence allows analysis of the amount of performance degradation expected from the suboptimal approach. Note that this paragraph has only considered the case where the design model matches the actual system, but a suboptimal gain sequence is used. The case where the filter model is distinct from the system model is discussed in the remainder of this section.

Consider a system described by:

$$\left.\begin{aligned} x(k+1) &= \Phi_x \mathbf{x}(k) + \Gamma_\omega \omega_d(k) \\ y(k) &= H_x(k)\mathbf{x}(k) + \nu(k) \end{aligned}\right\}, \quad (4.30)$$

and an estimator design model described by:

$$\left.\begin{aligned} \hat{x}(k+1) &= \Phi \hat{x}(k) + \Gamma \hat{\omega}_d(k) \\ \hat{y}(k) &= H(k)\hat{x}(k) + \hat{\nu}(k). \end{aligned}\right\}. \quad (4.31)$$

The time propagation and measurement update of the state estimate are as follows:

$$\left.\begin{aligned} \hat{x}^-(k+1) &= \Phi \hat{x}^+(k) \\ \hat{x}^+(k) &= \hat{x}^-(k) + \hat{K}(\mathbf{y}(k) - \hat{y}^-(k)) \end{aligned}\right\}, \quad (4.32)$$

where $P(0)$ and the properties of $\hat{\nu}$ and $\hat{\omega}_d$ are accounted for in the design of the estimator gain sequence $\hat{K}(k)$.

The remainder of this section is concerned with calculation of the error variance of $z = (Vx - \hat{x})$ in the actual implemented system where the design model of the estimator does not match the "truth model" for the actual system. In this model structure, $dim(x)$ is usually larger than $dim(\hat{x})$. The matrix V is defined to select the appropriate linear combination of the actual system states that correspond to the estimator states.

To analyze the performance of the coupled system, the joint state space representation for the system 4.30 and the implemented estimator 4.32 will be required:

$$\begin{bmatrix} x^-(k+1) \\ \hat{x}^-(k+1) \end{bmatrix} = \begin{bmatrix} \Phi_x(k) & 0 \\ 0 & \Phi(k) \end{bmatrix} \begin{bmatrix} x^+(k) \\ \hat{x}^+(k) \end{bmatrix} + \begin{bmatrix} \Gamma_\omega(k) \\ 0 \end{bmatrix} \omega_d(k) \quad (4.33)$$

$$z(k) = [V - \mathbf{I}] \begin{bmatrix} x(k) \\ \hat{x}(k) \end{bmatrix} \quad (4.34)$$

$$y^-(k) = [H_x(k), 0] \begin{bmatrix} x^-(k) \\ \hat{x}^-(k) \end{bmatrix} + \nu(k) \quad (4.35)$$

$$\hat{y}^-(k) = [0, \mathbf{H}(k)] \begin{bmatrix} x^-(k) \\ \hat{x}^-(k) \end{bmatrix} \quad (4.36)$$

$$\mathbf{K}(k) = \begin{bmatrix} 0 \\ \hat{K}(k) \end{bmatrix} \quad (4.37)$$

$$\begin{bmatrix} \mathbf{x}^+(k) \\ \hat{\mathbf{x}}^+(k) \end{bmatrix} = \begin{bmatrix} x^-(k) \\ \hat{x}^-(k) \end{bmatrix} + K(\mathrm{y}^-(k) - \hat{y}^-(k)) \quad (4.38)$$

This expression represents the actual time propagation of the real system and the estimate. Since the estimate is calculated exactly,[2] no process driving noise is represented in the corresponding rows of equation 4.33. In addition, the zeros concatenated into the estimation gain vector account for the fact that the estimator corrections do not affect the state of the actual system. For this coupled system, the covariance propagates between sampling times according to:

$$P_{11}^-(k+1) = \Phi_x P_{11}^+ \Phi_x^T + \Gamma_\omega Q_d \Gamma_\omega^T. \quad (4.39)$$

$$P_{12}^-(k+1) = \Phi_x P_{12}^+ \Phi^T. \quad (4.40)$$

$$P_{22}^-(k+1) = \Phi P_{22}^+ \Phi^T. \quad (4.41)$$

The system is altered by measurement updates according to:

$$P_{11}^+(k) = P_{11}^-. \quad (4.42)$$

$$P_{12}^+(k) = P_{12}^-\left(I - H^T\hat{K}^T\right) + P_{11}^- H_x^T \hat{K}^T. \quad (4.43)$$

$$\begin{aligned} P_{22}^+(k) = {} & \left(I - \hat{K}H\right)P_{22}^-\left(I - H^T\hat{K}^T\right) + \hat{K}H_x P_{11}^- H_x^T \hat{K}^T \\ & + \hat{K}R\hat{K}^T + \hat{K}H_x P_{12}^-\left(I - H^T\hat{K}^T\right) \\ & + \left(I - \hat{K}H\right)P_{21}^- H_x^T \hat{K}^T. \end{aligned} \quad (4.44)$$

In equations 4.39 through 4.44, $P_{11} = cov(x, x)$, $P_{21}^T = P_{12} = cov(x, \hat{x})$, $P_{22} = cov(\hat{x}, \hat{x})$. The error variance of the variable of interest, z, is defined at any instant by:

$$cov(z) = P_z = VP_{11}V^T + P_{22} - VP_{12} - P_{21}V^T. \quad (4.45)$$

Because the matrix P_z depends on K, the covariance analysis can be repeated for different gain sequences to compare the predicted performance of the alternative implementations. For given gain sequences, the covariance analysis can be repeated for different values of Φ_x, Q_d, and R to determine the sensitivity of the performance to the design assumptions.

4.6 Implementation Issues

4.6.1 Minimizing Latency Due to Computation

Because the Kalman filter covariance equations and gain computation do not depend on the data, the equations can be reorganized to minimize the latency between the time that

[2] Quantization error can be modeled by a second additive noise source affecting the estimate but is dropped in this analysis for convenience. In floating point processors, the effect is small.

the measurements become available and the time that the measurement-corrected state estimates are generated.

Assume that $\mathbf{\Phi}(k), \boldsymbol{H}(k+1), \boldsymbol{Q}_d(k)$, and $\boldsymbol{R}(k+1)$ are known at time k. Also, assume that $\hat{\boldsymbol{x}}^+(k)$ has just been computed and posted at time k:

1. Compute the error covariance corresponding to $\hat{\boldsymbol{x}}^+(k)$ using equation 4.9:

$$\boldsymbol{P}^+(k) = [\boldsymbol{I} - \boldsymbol{K}(k)\boldsymbol{H}(k)]\boldsymbol{P}^-(k)[\boldsymbol{I} - \boldsymbol{K}(k)\boldsymbol{H}(k)]^T + \boldsymbol{K}(k)\boldsymbol{R}(k)\boldsymbol{K}(k)^T.$$

2. Propagate the state estimate covariance to the next time instant using equation 4.11:

$$\boldsymbol{P}^-(k+1) = \mathbf{\Phi}(k)\boldsymbol{P}^+(k)\mathbf{\Phi}(k)^T + \mathbf{\Gamma}_\omega \boldsymbol{Q}_d(k)\boldsymbol{\Gamma}_\omega^T.$$

3. Precompute the Kalman gain vector for the upcoming measurement using equation 4.23:

$$\boldsymbol{K}(k+1) = \boldsymbol{P}^-(k+1)\boldsymbol{H}(k+1)^T (\boldsymbol{H}(k+1)\boldsymbol{P}^-(k+1)\boldsymbol{H}(k+1)^T + \boldsymbol{R}(k+1))^{-1}.$$

4. As soon as $\boldsymbol{u}(k)$ is available, propagate the state estimate to the next time instant using equation 4.10:

$$\hat{\boldsymbol{x}}^-(k+1) = \mathbf{\Phi}(k)\hat{\boldsymbol{x}}^+(k) + \mathbf{\Gamma}_u \boldsymbol{u}(k).$$

5. When $\boldsymbol{y}(k+1)$ becomes available, the only computation required for the corrected state estimate is as follows:

$$\hat{\boldsymbol{x}}^+(k+1) = \hat{\boldsymbol{x}}^-(k+1) + \boldsymbol{K}(k+1)[\boldsymbol{y}(k+1) - \boldsymbol{H}(k+1)\hat{\boldsymbol{x}}^-(k+1)].$$

Step 5 is completed with high priority. Immediately after completion of step 5, $\hat{\boldsymbol{x}}^+(k+1)$ is posted for use by other systems. Then, k is incremented by 1 and as CPU time becomes available, steps 1 through 4 are completed as low priority processes as long as they are complete before $\boldsymbol{y}(k+1)$ becomes available (again). In step 1, equation 4.9 is used because the equation is applicable to any estimation gain vector, while the alternative form $\boldsymbol{P}^+(k) = [\boldsymbol{I} - \boldsymbol{K}(k)\mathbf{H}(k)]\mathbf{P}^-(k)$ is applicable only for the Kalman gain vector. Also, equation 4.9 has better numeric properties. Note that the portions of step 5 involving $\hat{\boldsymbol{x}}^-(k+1)$ could also be precomputed to allow additional latency reduction. If $\mathbf{\Phi}(k)$, $\boldsymbol{H}(k)$, $\boldsymbol{Q}_d(k)$, and $\boldsymbol{R}(k)$ can be determined off-line, then steps 1 through 3 can be computed off-line, so that only steps 4 and 5 occur online. If a Luenberger estimator is used, steps 1 and 2 are eliminated, and $\boldsymbol{K} = \boldsymbol{L}$ is fixed; therefore, only the time propagation and measurement updates of steps 4 and 5 are required online.

4.6.2 Scalar Processing

The Kalman filter algorithms have been presented, for notational convenience, in a vector form with m measurements. The portion of the measurement update that requires the most computing operations (i.e., flops) is the covariance update and gain vector calculation. For example, the following standard algorithms can be programmed to require $(\frac{3}{2}n^2m + \frac{3}{2}nm^2 + nm + m^3 + \frac{1}{2}m^2 + \frac{1}{2}m)$ flops:

$$\boldsymbol{K} = \boldsymbol{P}^-\boldsymbol{H}^T(\boldsymbol{R} + \boldsymbol{H}\boldsymbol{P}^-\boldsymbol{H}^T). \tag{4.46}$$

$$\mathbf{P}^+ = (\boldsymbol{I} - \boldsymbol{K}\boldsymbol{H})\boldsymbol{P}^-. \tag{4.47}$$

Alternatively, when the measurements are independent, the measurements can be equivalently treated as m sequential measurements with a zero width time interval between measurements.

Define the following as:

$$\boldsymbol{P}_1 = \boldsymbol{P}^-(k) \quad \hat{\boldsymbol{x}}_1 = \hat{\boldsymbol{x}}^-(k)$$
$$\boldsymbol{H} = \begin{bmatrix} \boldsymbol{H}_1 \\ \vdots \\ \boldsymbol{H}_m \end{bmatrix} \quad \boldsymbol{R} = \begin{bmatrix} R_1 & \cdots & 0 \\ \vdots & & \vdots \\ 0 & \cdots & R_m \end{bmatrix}. \tag{4.48}$$

Then, the equivalent scalar measurement processing algorithm is for $i = 1$ to m:

$$\boldsymbol{K}_i = \frac{\boldsymbol{P}_i\boldsymbol{H}_i^T}{R_i + \boldsymbol{H}_i\boldsymbol{P}_i\boldsymbol{H}_i^T}. \tag{4.49}$$

$$\hat{\boldsymbol{x}}_{i+1} = \hat{\boldsymbol{x}}_i + \boldsymbol{K}_i(\bar{y}_i - \boldsymbol{H}_i\hat{\boldsymbol{x}}_i). \tag{4.50}$$

$$\boldsymbol{P}_{i+1} = (\boldsymbol{I} - \boldsymbol{K}_i\boldsymbol{H}_i)\boldsymbol{P}_i. \tag{4.51}$$

In addition, the state and error covariance matrix posterior to the set of m measurements are defined by:

$$\hat{\boldsymbol{x}}^+(k) = \hat{\boldsymbol{x}}_{m+1} \tag{4.52}$$

$$\boldsymbol{P}^+(k) = \boldsymbol{P}_{m+1}. \tag{4.53}$$

The total number of computations for the m scalar measurement updates is $m(\frac{3}{2}n^2 + \frac{5}{2}n)$ plus m scalar divisions. Thus, it can be seen that m scalar updates are computationally cheaper for any m.

At the completion of the m scalar measurements, $\hat{\boldsymbol{x}}^+(k)$ and $\boldsymbol{P}^+(k)$ will be identical to the values that would have been computed by the corresponding vector measurement update. The state feedback measurement vectors $\boldsymbol{K}_i$ corresponding to the scalar updates are *not* equal to the columns of the state feedback gain matrix that would result from the corresponding vector update. This is due to the different ordering of the updates affecting the error covariance matrix $\boldsymbol{P}_i$ at the intermediate steps during the scalar updates.

4.6.3 Complementary Filtering

In applications, it often happens that multiple noisy measurements of a signal are available, and it is of interest to

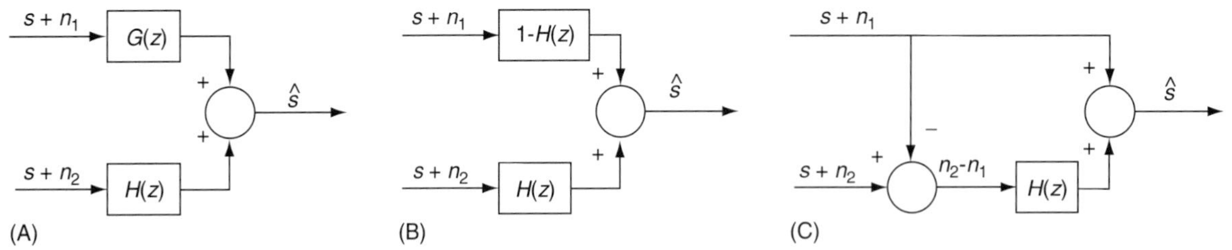

FIGURE 4.2 Block diagrams for the Estimation of a Signal s from Noisy Measurements. (A) Generic Filter Structure. (B) Complementary Filter Constrained to Not Distort s. (C) Feed-forward Complementary Filter.

combine the measurements to obtain an improved or even optimal estimate of the signal. Such a situation is illustrated in Figure 4.2(A), where two measurements of the signal s are available:

$$s_1 = s + n_1 \text{ and } s_2 = s + n_2 \tag{4.54}$$

The measurement noise processes n_1 and n_2 are random independent processes with known spectral densities, and s can be either random or deterministic; however, the signal s is not known, so it cannot be interpreted as the u term in equation 4.54. The objective is to design filters $G(z)$ and $H(z)$ such that $\hat{s} = G(z)s_1 + H(z)s_2$. If the estimator is designed with the constraint that the signal s cannot be distorted by the estimation process, then $G(z) = 1 - H(z)$. This constrained estimator, referred to as a complementary filter, is shown in Figure 4.2(B). Restructuring the complementary filter block diagram as shown in Figure 4.2(C) has two advantages. First, only a single filter is required. Second, the input to $H(z)$ is a random signal with known spectral density. Therefore, the filter design problem is properly structured for the Kalman or Weiner filter design techniques. The objective of the filter is to accurately estimate $-\hat{n}_1$. Additional computational advantages result that will be stated in the following example.

4.7 Example: Inertial Navigation System Error Estimation

As an illustrative example of the estimation concepts presented in this chapter, consider the following example of the calibration of a simplified inertial navigation system (INS) using an independent measurement of position:

$$p_m = p + \nu,$$

where ν is assumed to be Gaussian white noise with $cov(\nu) = \sigma_\nu^2 = (0.01\ m)^2$ and where p is the true position.

The inertial navigation system computations, which are depicted by the left block diagram in Figure 4.3, are the following:

$$\dot{p}_c = v_c, \quad \dot{v}_c = a_m, \tag{4.55}$$

where a_m is the measured acceleration that is corrupted by a bias plus Gaussian white noise, v_c is the computed velocity, and p_c is the computed position. The bias is assumed to be a random walk:

$$\dot{b} = \omega,$$

where ω is a Gaussian white noise process. Note that the INS computation block diagram can be decomposed into signal

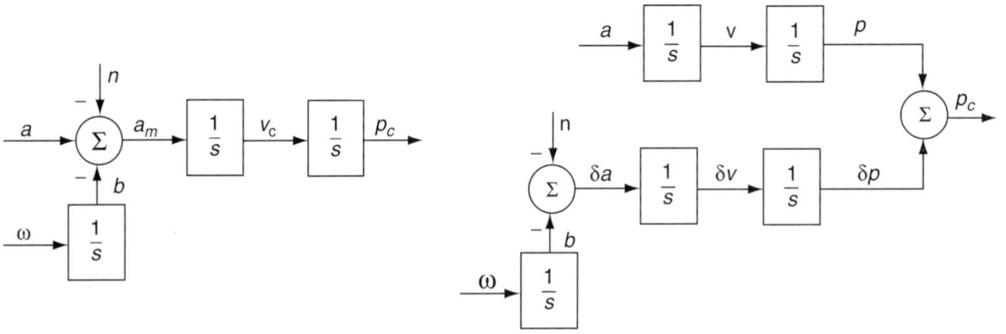

FIGURE 4.3 Inertial Navigation System Block Diagrams. (A) Block Diagram of the Actual INS Computations (B) Block Diagram Decomposing the INS into Signal and Error Channels

and noise channels as illustrated in Figure 4.3(B). This block diagram illustrates that the position p_c computed by the INS can be considered as the sum of the true position p and a colored noise process δp. The state-space model for δp is as seen here:

$$\begin{bmatrix} \delta\dot{p} \\ \delta\dot{v} \\ \dot{b} \end{bmatrix} = \begin{bmatrix} 0 & 1 & 0 \\ 0 & 0 & 1 \\ 0 & 0 & 0 \end{bmatrix} \begin{bmatrix} \delta p \\ \delta v \\ b \end{bmatrix} + \begin{bmatrix} 0 & 0 \\ 0 & 1 \\ 1 & 0 \end{bmatrix} \begin{bmatrix} \omega \\ n \end{bmatrix}. \quad (4.56)$$

$$\delta p = [1 \quad 0 \quad 0] \begin{bmatrix} \delta p \\ \delta v \\ b \end{bmatrix}. \quad (4.57)$$

The model specifies the $\boldsymbol{F}$, $\boldsymbol{H}$, and Γ_w matrices. Note that $\Gamma_u = 0$. The spectral density of w and n are denoted $\sigma_w^2 = 4 \times 10^{-10} \frac{m^2}{s^5}$ and $\sigma_n^2 = 10^{-4} \frac{m^2}{s^3}$, which correspond to an inexpensive solid-state device.

The feed-forward complementary filter to combine p_c and p_m to estimate p is shown in Figure 4.4. In this figure, all signals are illustrated in continuous time to clearly show the integrals involved. Both the INS and the filter would typically be implemented via computer in discrete-time representation. The analysis that follows will also be performed in discrete time. This feed-forward complementary filter implementation has several useful features in addition to those previously noted. First, the inputs to the filter $H(s)$ are stationary stochastic quantities unaffected by p, v, or a. In typical INS applications, this independence is not true. Second, in the discrete-time implementation, the INS update rate can be much higher that the rate at which the position measurements are made. For Kalman filter implementations, this fact significantly reduces the computational load because the time propagation and measurement updates of the estimation error covariance matrices, which are the most computationally expensive portion of the algorithm, occur at the lower rate of the position measurement.

Consider the design of the complementary filter as a Kalman filter, where $\boldsymbol{H} = [1, 0, 0]$, $\boldsymbol{R} = \sigma_v^2$, and:

$$\Phi = \begin{bmatrix} 1 & T & \frac{1}{2}T^2 \\ 0 & 1 & T \\ 0 & 0 & 1 \end{bmatrix} \text{ and}$$

$$\boldsymbol{Q}_d = \begin{bmatrix} \frac{\sigma_n^2 T^3}{3} + \frac{\sigma_\omega^2 T^5}{20} & \frac{\sigma_n^2 T^2}{2} + \frac{\sigma_\omega^2 T^4}{8} & \frac{\sigma_\omega^2 T^3}{6} \\ \frac{\sigma_n^2 T^2}{2} + \frac{\sigma_\omega^2 T^4}{8} & \sigma_n^2 T + \frac{\sigma_\omega^2 T^3}{3} & \frac{\sigma_\omega^2 T^2}{2} \\ \frac{\sigma_\omega^2 T^3}{6} & \frac{\sigma_\omega^2 T^2}{2} & \sigma_\omega^2 T \end{bmatrix}. \quad (4.58)$$

The left column of Figure 4.5 shows the covariance analysis using the method of Section 4.5.2 for a scenario where the position aiding is available at 1 Hz for the first 60 sec but not available for the subsequent 15 sec. In the first 60 sec the position and velocity error standard deviations (STDs) have effectively reached their steady-state accuracy. These plots clearly show that error accumulates during each 1 sec of INS integration and is reduced by each 1 Hz position update. The accelerometer bias error is still being slowly corrected at the end of the 60 sec of position corrections.

Two natural questions relevant to the system design are what are the dominant error sources causing the growth of the INS errors between the 1-Hz corrections during the first 60 sec, and what are the dominant error sources that cause the INS error growth during the period when the position aiding is no longer available. Both of these question are answerable by the error budgeting analysis described in Section 4.5.1. The error budget analysis plot for the position estimate is shown in Figure 4.5(D). This figure shows the error contributions from ω, n, v, and the total error from all sources. The plot for the error contribution from $P(0)$ has not been included because it is insignificant for the time period shown. Note that the figure plots the standard deviations (STDs). To perform the superposition check that the errors from the individual sources add up to the total error at every time instant, the plotted

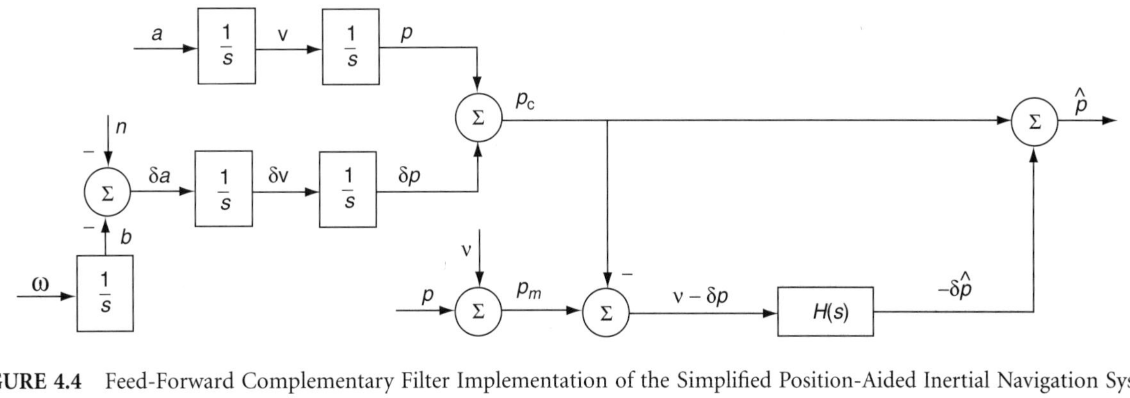

FIGURE 4.4 Feed-Forward Complementary Filter Implementation of the Simplified Position-Aided Inertial Navigation System

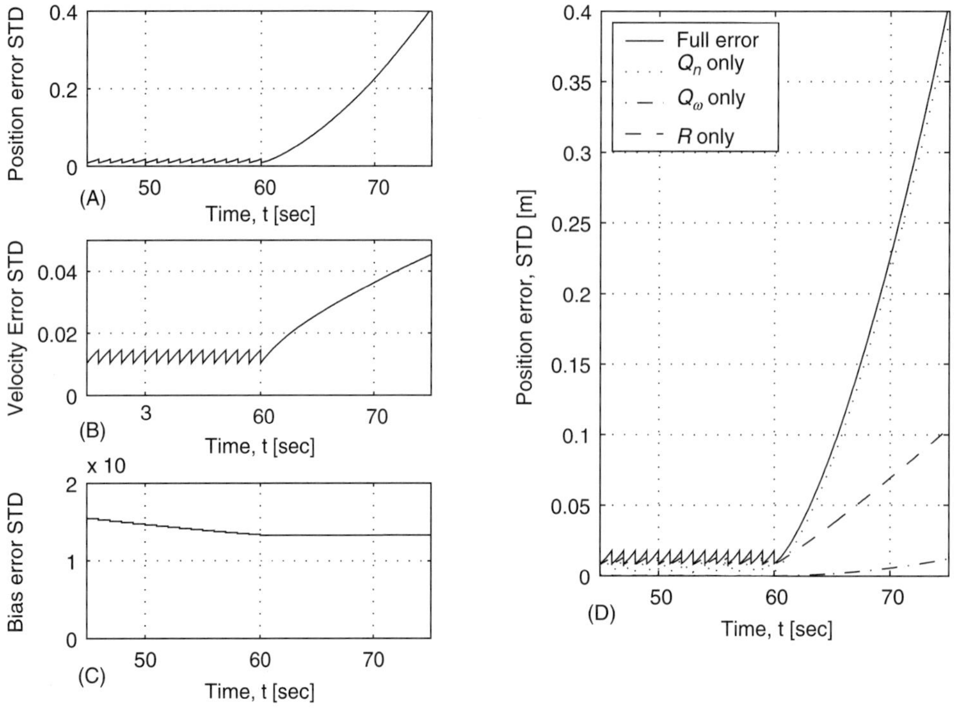

FIGURE 4.5 Covariance and Error Budget Analysis for the Position-Aided INS Example (A) Position Error in m/s Versus Time (B) Velocity Error in m/s Versus Time. (C) Accelerometer Bias Error in m/s Versus Time (D) Position Error Budget Analysis

quantities must be squared prior to the addition. The figure clearly shows that by 62 sec, 2 sec after the position aiding has been removed, the acceleration measurement noise n has become the dominant error source. The accelerometer bias drift term is rather insignificant. Therefore, if the design objective were to maintain position accuracy less than 0.3 m for a longer period of time, then either an independent velocity aiding measurement or an accelerometer with less measurement noise would be required.

This example has been designed as a simplified analysis of a carrier phase differential global positioning system-aided INS. A full error analysis is significantly more complicated for a few reasons. First, the INS equations are nonlinear. The implementation therefore requires a more advanced estimator implementation, such as an extended Kalman filter. Second, the INS error equations are functions of the acceleration and angular rate of the inertial measurement unit. In addition, the linearized measurement matrix is a function of the satellite positions and is therefore time varying. Therefore, the state-space model is time varying, and the full error analysis would be dependent on assumed vehicle maneuvers and satellite configurations.

4.8 Further Reading

Various aspects of linear algebra and the state-space methodology are discussed in Brogan (1991). The Luenberger observer is presented in Luenberger (1966). The Kalman filter is presented in Kalman (1960, 1961). Estimation theory and estimator design are discussed in several texts (Brown and Hwang, 1992; Gelb *et al.*, 1974; Maybeck, 1979). Practical aspects of estimation theory including state augmentation, suboptimal filter analysis, and applications are discussed in Farrell and Barth (1999), Gelb et al. (1974), and Maybeck (year). Markov parameters and FIR approximations to IIR filters are discussed in Mendel (1995) and Moore (1981). Efficient computation of the discrete-time state transition matrix and process noise covariance matrix are discussed in Van Loan (1978). The full implementation of the GPS/INS system corresponding to the example is described in Farrell (2000).

References

Brogan, W.L. (1991). *Modern control theory.* Englewood Cliffs, NJ: Prentice Hall.

Brown, R.G., and Hwang, Y.C. (1992). *Introduction to random signals and applied Kalman filtering.* (2d ed.). New York: John Wiley & Sons.

Farrell, J.A., Givargis, T., and Barth, M., (2000). Real-time differential carrier phase GPS-aided INS. *IEEE Transactions on Control Systems Technology 8*(4), 709–721.

Farrell, J.A., and Barth, M. (1999). *The Global Positioning System and inertial navigation: Theory and practice.* McGraw-Hill.

Gelb, A., Kasper, J.F. Jr., Nash, R.A., Jr., Price, C.F., and Sutherland, A.A. (1974). *Applied optimal estimation.* Cambridge, MA: MIT Press.

Kalman, R.E. (1960). A new approach to linear filtering and prediction problems. *ASME Journal of Basic Engineering, Ser. D 82*, 34–45.

Kalman, R.E., and Bucy, R.S. (1961). New results in linear filtering and prediction theory. *ASME Journal of Nasic Engineering, Ser. D 83*, 95–108.

Luenberger, D. (1966). Observers for multivarible systems. *IEEE Transactions on Automated Control 11*, 190–197.

Maybeck, P.S. (1979). *Stochastic models, estimation, and control.* Vol. 1–3. San Diego, CA: Academic Press.

Mendel, J.M. (1995). *Lessons in estimation theory for signal processing, communications, and control.* Englewood Cliffs, NJ: Prentice Hall.

Moore, B. (1981). Principal component analysis in linear systems: Control, ability, observability, and model reduction. *IEEE Transactions on Automatic Control AC-26*, 17–32.

Van Loan, C. (1978). Computing integrals involving the matrix exponential. *IEEE Transactions on Automatic Control AC-23*, 395–404.

5

Cost-Cumulants and Risk-Sensitive Control

Chang-Hee Won
Department of Electrical Engineering, University of North Dakota, Grand Forks, North Dakota, USA

5.1 Introduction

Cost-cumulant control, also known as statistical control, is an optimal control method that minimizes a linear combination of quadratic cost cumulants. **Risk-sensitive control** is an optimal control method that minimizes the exponential of the quadratic cost criterion. This is equivalent to optimizing a denumerable sum of all the cost cumulants.

Optimal control theory deals with the optimization, either minimization or maximization, of a given cost criterion. **Linear-quadratic-Gaussian control**, **minimum cost-variance control**, and **risk-sensitive control** are discussed in terms of cost cumulants. Figure 5.1 presents an overview of the optimal control and the relationships among different optimal control methods.

5.2 Linear-Quadratic-Gaussian Control

The **linear quadratic Gaussian (LQG) control** method optimizes the mean, which is the first cumulant, of a quadratic cost criterion (Anderson and Moore, 1989; Davis, 1977; Kwarkernaak and Sivan, 1972).

Typical system dynamics for LQG control are given by the stochastic equation:

$$\begin{aligned} dx(t) &= Ax(t)dt + Bk(t, x)dt + E(t)dw(t). \\ y(t)dt &= Cx(t)dt + dv(t). \end{aligned} \tag{5.1}$$

Here $w(t)$ and $v(t)$ are vector Brownian motions. The meaning of a **Brownian motion**, such as $w(t)$, can be given directly or in terms of its differential $dw(t)$. In the latter case, the $dw(t)$ is a Gaussian random process with zero mean, covariance matrix $W\,dt$, and independent increments. A similar description applies to $dv(t)$, with covariance $V\,dt$. It is assumed that $dw(t)$ and $dv(t)$ are independent. The matrices A, B, C, and E are of compatible size. It should be remarked that the formalism of equation 5.1 is that of a stochastic differential equation. Intuitively, one thinks of dividing both sides of this equation by dt to obtain the more colloquial form. But the formal derivative of a Brownian motion—which is known as white noise—is not a well-defined random process, and this motivates an alternate way of thinking.

The quadratic cost criterion is given by:

$$J(k) = \int_0^{t_F} (x'Qx + k'Rk)dt. \tag{5.2}$$

The weighting matrix Q is a symmetric and positive semidefinite matrix, and R is a symmetric and positive definite matrix.

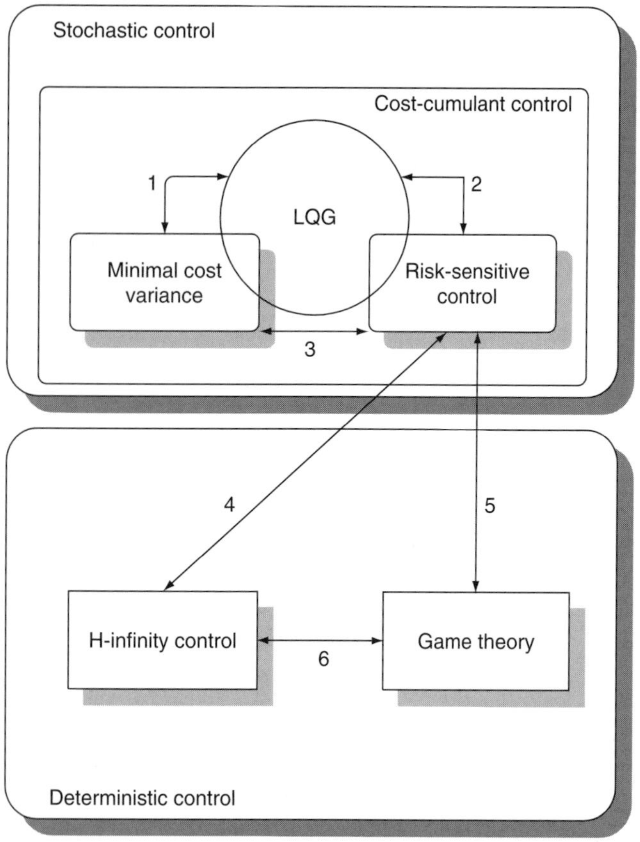

FIGURE 5.1 Relationship Between Various Optimal and Robust Control Methods. (Sain *et al.*, 2000; Jacobson, 1973; Won and Sain, 1995; Glover and Doyle, 1988; Whittle, 1990; Rhee and Speyer, 1992; Jacobson, 1973; Runolfsson, 1994; Uchida, 1989; Basar and Bernhard, 1991.)

The LQG control problem then becomes a minimization of the mean quadratic cost over feedback controller k:

$$J^* = \min_k E\{J(k)\}. \qquad (5.3)$$

The full-state feedback control problem is to choose the control k as a function of the state x so that the cost criterion of equation 5.3 is minimized. The general partial observation or output feedback control problem is to choose the control k as a function of the observation y so that the cost of equation 5.3 is minimized.

Assume now that the problem has a solution of the quadratic form $\frac{1}{2}x'\Pi x$. The matrix Π can be found from the **Riccati equation:**

$$0 = \dot{\Pi}(t) + Q + A'\Pi(t) + \Pi(t)A - \Pi(t)BR^{-1}B'\Pi(t), \qquad (5.4)$$

where $\Pi(t_F) = 0$.

Then the **full-state feedback optimal controller** is given by (Davis, 1977):

$$k(t, x) = -R^{-1}B'\Pi(t)x(t). \qquad (5.5)$$

The solution of the output feedback LQG problem is found using the **certainty equivalence principle**. The optimal control is found using a Kalman filter, where an optimal estimate $\hat{x}$ is obtained such that $E\{(x - \hat{x})'(x - \hat{x})\}$ is minimum. Then this estimate is used as if it were an exact measurement of the state to solve the deterministic LQG control.

For the output feedback case, the estimated states are given by:

$$\frac{d\hat{x}}{dt} = Ax + Bk + PCV^{-1}(y - C\hat{x}), \qquad (5.6)$$

where P satisfies the **forward Riccati equation:**

$$\dot{P}(t) = W + AP(t) + P(t)A' - P(t)C'V^{-1}CP(t). \qquad (5.7)$$

In equation 5.7, the initial condition is $P(0) = \text{cov}(x_0)$. Finally the **optimal output feedback controller** is given as:

$$k(t, x) = -R^{-1}B'\Pi(t)\hat{x}(t). \qquad (5.8)$$

5.3 Cost-Cumulant Control

5.3.1 Minimal Cost Variance Control

Minimum cost variance (MCV) control is a special case of cost-cumulant or statistical control where the second cumulant, variance, is minimized, whereas the first cumulant, mean, is kept at a prespecified level.

Here, open-loop MCV and full-state feedback MCV control laws are discussed. An **open-loop control law** is a function u: $[0, t_F] \rightarrow u$ where u is some specified allowable set of control values. A **closed-loop or feedback control law** is a function that depends on time and the past evolution of the process [i.e., $u(t, x(s); 0 \leq s \leq t)$].

Open-Loop MCV Control

Consider a linear system (Sain and Liberty, 1971):

$$dx(t) = A(t)x(t)dt + B(t)u(t)dt + E(t)dw(t), \qquad (5.9)$$

and the performance measure:

$$J = \int_0^{t_F} [x'(t)Qx(t) + u'(t)Ru(t)]dt + x'(t_F)Q_F x(t_F), \qquad (5.10)$$

where $w(t)$ is zero mean with white characteristics relative to the system, t_F is the fixed final time, $x(t) \in \mathbb{R}^n$ is the state of the system, and $u(t) \in \mathbb{R}^m$ is the control action. Note that:

$$E\{dw(t)dw'(t)\} = W\,dt. \qquad (5.11)$$

The fundamental idea behind minimal cost-variance control is to minimize the variance of the cost criterion J:

$$J_{MV} = VAR_k\{J\}, \qquad (5.12)$$

while satisfying a constraint:

$$E_k\{J\} = M, \tag{5.13}$$

where J is the cost criterion and where the subscript k on E denotes the expectation based on a control law k generating the control action $u(t)$ from the state $x(t)$ or from a measurement history arising from that state. By means of a Lagrange multiplier μ corresponding to the constraint of equation 5.13, one can form the function:

$$J_{MV} = \mu(E_k\{J\} - M) + VAR_k\{J\}, \tag{5.14}$$

which is equivalent to minimizing:

$$\hat{J}_{MV} = \mu E_k\{J\} + VAR_k\{J\}. \tag{5.15}$$

A Riccati solution to $\hat{J}_{MV}$ minimization is developed for the open-loop case:

$$u(t) = k(t, x(0)). \tag{5.16}$$

The solution is based on the differential equations:

$$\dot{z}(t) = A(t)z(t) - \frac{1}{2}B(t)R^{-1}B'(t)\hat{\rho}(t). \tag{5.17}$$

$$\dot{\hat{\rho}}(t) = -A'(t)\hat{\rho}(t) - 2Qz(t) - 8\mu Q\nu(t). \tag{5.18}$$

$$\dot{\nu}(t) = A(t)\nu(t) + E(t)WE'(t)y(t). \tag{5.19}$$

$$\dot{y}(t) = -A'(t)y(t) - Qz(t). \tag{5.20}$$

These equations have the boundary conditions:

$$z(0) = x(0). \tag{5.21}$$

$$\hat{\rho}(t_F) = 2Q_F\, z(t_F) + 8\mu Q_F\nu(t_F). \tag{5.22}$$

$$\nu(0) = 0. \tag{5.23}$$

$$y(t_F) = Q_F z(t_F). \tag{5.24}$$

The equations also have the control action relationship:

$$\nu(t) = -\frac{1}{2}R^{-1}B(t)\rho(t). \tag{5.25}$$

The variable $z(t)$ is the mathematical expectation of $x(t)$. The variable $\rho(t)$ corresponds to the costate variable of optimal control theory because it is the variable that enforces the differential equation constraint between $z(t)$ and $\nu(t)$. The variable $\nu(t)$ and $y(t)$ are introduced to reduce the integro-differential equation.

Full-State Feedback Minimal Cost-Variance Control

Consider the Ito sense stochastic differential equation (SDE) with control (Sain *et al.*, 2000):

$$dx(t) = [A(t)x(t) + B(t)k(t, x)]\ dt + E(t)\ dw(t).$$

The equation also has the cost criterion:

$$J(t, x(t), k) = \int_t^{t_F} [x(t)'Qx(t) + k'(t, x)R(t)k(t, x)]ds + x'(t_F)Q_F x(t_F). \tag{5.26}$$

In MCV control, we define a class of admissible controllers, and then the cost variance is minimized within that class of controllers. Define $V_1(t, x; k) = E\{J(t, x(t), k)|x(t) = x\}$ and $V_2(t, x; k) = E\{J^2(t, x(t), k)|x(t) = x\}$. A function M is an admissible mean cost criterion if there exists an admissible control law k such that:

$$V_1(t, x; k) = M(t, x), \tag{5.27}$$

for all $t \in [0, t_F]$ and $x \in \mathbb{R}^n$.

A minimal mean cost-control law k_M^* satisfies $V_1(t, x; k_M^*) = V_1^*(t, x) \leq V_1(t, x; k)$ for $t \in T$, $x \in \mathbb{R}^n$ and for k, an admissible control law. An MCV control law $k_{V|M}^*$ satisfies $V_2(t, x; k_{V|M}^*) = V_2^*(t, x) \leq V_2(t, x; k)$ for $t \in T$, $x \in \mathbb{R}^n$ whenever k is admissible. The corresponding minimal cost variance is given by $V^*(t, x) = V_2^*(t, x) - M^2(t, x)$ for $t \in T$, $x \in \mathbb{R}^n$. Here the full-state feedback solution of the MCV control problem is presented for a linear system and a quadratic cost criterion.

Then the linear optimal MCV controller is given by (Sain *et al.*, 2000):

$$K_{V|M}^*(t, x) = -R^{-1}(t)B'(t)[\mathcal{M}(t) + \gamma(t)V(t)]x,$$

where $\mathcal{M}$ and $\mathcal{V}$ are the solutions of the **coupled Riccati-type equations** (suppressing the time argument):

$$0 = \dot{\mathcal{M}} + A'\mathcal{M} + \mathcal{M}A + Q - \mathcal{M}BR^{-1}B'\mathcal{M} + \gamma^2\mathcal{M}BR^{-1}B'\mathcal{V}. \tag{5.28}$$

$$0 = \dot{\mathcal{V}} + 4\mathcal{M}EWE'\mathcal{M} + A'\mathcal{V} + \mathcal{V}A - \mathcal{M}BR^{-1}B'\mathcal{V} - \mathcal{V}BR^{-1}B'\mathcal{M} - 2\gamma\mathcal{V}BR^{-1}B'\mathcal{V}, \tag{5.29}$$

with boundary conditions $\mathcal{M}(t_F) = Q_F$ and $\mathcal{V}(t_F) = 0$. Once again, if γ approaches zero, classic LQG results are obtained.

This MCV idea can be generalized to minimize any cost cumulants. Viewing the cost function as a random variable and optimizing any cost cumulant is called cost-cumulant or statistical control.

5.4 Risk-Sensitive Control

A large class of control systems can be described in state-variable form by the stochastic equations (Anderson and Moore, 1989; Whittle, 1996):

$$dx(t) = Ax(t)dt + Bk(t, x)dt + dw(t).$$
$$y(t)dt = Cx(t)dt + dv(t). \quad (5.30)$$

Here, $x(t)$ is a $2n$-dimensional state vector, $k(t,x)$ is an m-dimensional input vector, $w(t)$ is a q-dimensional disturbance vector of Brownian motions, $y(t)$ is a p-dimensional vector of output measurements, and $v(t)$ is an r-dimensional output noise vector of Brownian motions that affect the measurements being taken.

The risk-sensitive cost criterion is given by:

$$J_{RS}(\theta) = -\theta^{-1} \log E_k\{e^{-\theta J}\}, \quad (5.31)$$

where J is the classical quadratic cost criterion:

$$J = \int_0^{t_F} (x'Qx + k'Rk)dt. \quad (5.32)$$

The RS control problem then becomes a minimization of the cost $J_{RS}(\theta)$ over feedback controller k:

$$J^*_{RS}(\theta) = \min_u J_{RS}(\theta). \quad (5.33)$$

Assume a solution of the quadratic form $\frac{1}{2}x'\Pi x - \sigma' x+$ (terms independent of x). The matrix Π can be found from the Riccati-type equation:

$$0 = \dot{\Pi}(t) + Q + A'\Pi(t) + \Pi(t)A - \Pi(t)\left(BR^{-1}B' + \theta W\right)\Pi(t), \quad (5.34)$$

where $\Pi(t_F) = 0$.

Then, the full-state feedback optimal controller is given by (Whittle, 1996):

$$k(t, x) = -R^{-1}B'\Pi(t)x(t) + R^{-1}B'\sigma(t), \quad (5.35)$$

where $\dot{\sigma}(t) + (A - \Pi(B'R^{-1}B' + \theta W))'\ \sigma(t) = 0$ is a backward linear equation. The matrix P satisfies the forward Riccati-type equation:

$$\dot{P}(t) = W + AP(t) + P(t)A' - P(t)\left(C'V^{-1}C + \theta Q\right)P(t), \quad (5.36)$$

where $P(0) = (x_0)$. The updating equation for the risk-sensitive Kalman filter is given by:

$$\frac{d\mathring{x}}{dt} = Ax + Bk + PCV^{-1}(y - C\mathring{x}) - \theta PQ\mathring{x}, \quad (5.37)$$

where $\mathring{x}(0) = 0$. $\mathring{x}$ denotes the mean of x conditional on the initial information, current observation history, and previous control history. Finally, the optimal output feedback controller is given as (Whittle, 1996):

$$k(t, x) = -R^{-1}B'\Pi(t)\hat{x}(t) + R^{-1}B'\sigma(t), \quad (5.38)$$

where $\hat{x}$ is the minimal-stress estimate of x, given by:

$$\hat{x}(t) = (I + \theta P(t)\Pi(t))^{-1}(\mathring{x}(t) + \theta P(t)\sigma(t)). \quad (5.39)$$

As θ approaches zero, the cost criterion of equation 5.31 becomes $E_k\{J\}$, and the matrices Π and P are obtained from the Riccati equations:

$$O = \dot{\Pi}(t) + Q + A'\Pi(t) + \Pi(t)A - \Pi(t)BR^{-1}B'\Pi(t). \quad (5.40)$$

$$\dot{P}(t) = W + AP(t) + P(t)A' - P(t)C'V^{-1}CP(t). \quad (5.41)$$

Thus, the classic LQG result is obtained as θ approaches zero:

5.5 Relationship Between Risk-Sensitive and Cost-Cumulant Control

To see the relationship between RS and cost-cumulant control, consider a cost criterion:

$$J = \int_0^{t_F} [x(t)'Qx(t) + k'(t,x)R(t)k(t,x)]ds + x'(t_F)Q_F x(t_F). \quad (5.42)$$

Classical LQG control minimizes the first cumulant or the mean of the cost criterion of equation 5.42. In MCV control, the second cumulant of equation 5.42 is minimized while the mean is kept at a prespecified level. Furthermore, RS control minimizes an infinite linear combination of the cost cumulants. To see this, consider an RS cost criterion:

$$J_{RS} = -\theta^{-1} \log (E\{\exp(-\theta J)\}), \quad (5.43)$$

where θ is a real parameter and E denotes expectation. Then, the moment-generating function or the first characteristic function is given by:

$$\phi(s) = E \exp(-sJ). \quad (5.44)$$

The cumulant generating function $\psi(s)$ is defined by:

$$\psi(s) = \log \phi(s) = \sum_{i=1}^{\infty} \frac{(-1)^i}{i!} \beta_i s^i, \quad (5.45)$$

in which the $\{\beta_i\}$ are known as the **cumulants** or sometimes as the **semi-invariants** of J. Now by comparing equations 5.43, 5.44, and 5.45, it is noted that:

$$J_{RS} = \left(-\theta^{-1}\right)\left\{\sum_{i=1}^{\infty} \frac{(-1)^i}{i!} \beta_i(J)(\theta)^i\right\}, \quad (5.46)$$

where $\beta_i(J)$ denotes the ith cumulant of J with respect to the control law k. Thus, it is important to note that the RS cost criterion is an infinite linear combination of the cost cumulants. Moreover, approximating to the second order:

$$\begin{aligned} J_{RS} &= \beta_1(J) - \frac{\theta}{2}\beta_2(J) + O(\theta^2) \\ &= E\{J\} - \frac{\theta}{2} VAR\{J\} + O(\theta^2). \end{aligned} \tag{5.47}$$

Therefore, the minimal cost mean and minimal cost variance problems can be viewed as first- and second-order approximations of the RS control problem respectively. Minimizing the $VAR\{J\}$ under the restriction that the first cumulant $E\{J\}$ exists is called the **minimal cost variance** (MCV) **problem**. Moreover, minimizing any linear combination of cost cumulants under certain restrictions would be called cost-cumulant or statistical control. Thus, classical LQG control (optimization of the first cumulant), MCV control (optimization of the second cumulant), and RS control (optimization of the infinite number of cumulants) are all special cases of the cost-cumulant control problem.

5.6 Applications

An application of risk-sensitive control to satellite attitude maneuver is given in this section. An application of minimal cost variance control to an earthquake structure control is also given here. For linear quadratic Gaussian applications, see Anderson and Moore (1989), Fleming and Rishel (1975) and Kwarkernaak and Sivan (1972). For more risk-sensitive control examples, refer to Bensoussan (1992) and Whittle (1996).

5.6.1 Risk-Sensitive Control Applied to Satellite Attitude Maneuver

This subsection shows the simulation results associated with the model of a geostationary satellite equipped with a bias momentum wheel on the third axis of body frame. This model assumes that the disturbance torque is Gaussian white noise. A stochastic RS controller is then applied. For this model, small attitude angle and roll/yaw dynamics are assumed to be decoupled from the pitch dynamics.

A roll/yaw attitude model of the geostationary satellite is simplified as the following linear differential equation of $h_w \gg \max\{I_i, \omega_c\}$:

$$\begin{aligned} dx(t) = &\begin{bmatrix} 0 & 0 & 1 & 0 \\ 0 & 0 & 0 & 1 \\ -\frac{h_w\omega_c}{I_{11}} & 0 & 0 & -\frac{h_w}{I_{11}} \\ 0 & -\frac{h_w\omega_c}{I_{22}} & \frac{h_w}{I_{22}} & 0 \end{bmatrix} x(t)dt \\ &+ \begin{bmatrix} 0 \\ 0 \\ \frac{B_e}{I_{11}}\cos(\alpha) \\ \frac{B_e}{I_{22}}\sin(\alpha) \end{bmatrix} m(t)dt + \begin{bmatrix} 0 \\ 0 \\ \frac{1}{I_{11}} \\ \frac{1}{I_{22}} \end{bmatrix} dw(t) \end{aligned} \tag{5.48}$$

$$dy(t) = I_{4\times 4}x(t)dt + dv(t) \tag{5.49}$$

The dw/dt is Gaussian white noise representing the disturbance torque, dv/dt is Gaussian white noise representing the measurement noise, h_w is the wheel momentum, α is the angle that the positive roll axis makes with the magnetic torquer, ω_c is the orbital rate, I_{ii} is the moment of inertia of the *i*th axis, $x = [\gamma, r, \dot{\gamma}, \dot{r}]$ is the state with yaw (γ), roll (r), m is a dipole moment of the magnetic torquer (control), $B_e = 1.07 \times 10^{-7}$ telsa is the nominal magnetic field strength, and $I_{4\times 4}$ is an identity matrix with dimension four. The expected value of dw/dt is zero with $E\{dw/dt \times dw/dt\} = 0.7B_e$, and the expected value of dv/dt is zero with $E\{dv/dt \times dv/dt'\} = 1 \times 10^{-7}$. Here $\theta = 5 \times 10^{-2}$ was chosen for the demonstration purpose, but this risk-sensitivity parameter, θ should be viewed as another design parameter just like the weighting matrices Q and R. By varying this θ, different performance and stability results can be obtained. Theoretically, all θ that give a solution to the Riccati equation 5.36 are possible. The next example shows how to choose this risk-sensitivity parameter to obtain larger stability margin. The constants for the operational mode are given as $I_{11} = 1988\,\text{kg}\cdot\text{m}^2$, $I_{22} = 1876\,\text{kg}\cdot\text{m}^2$, $I_{12} = I_{21} = 0$, $h_w = 55\ \text{kg}\cdot\text{m}^2/s$, $\omega_c = 0.00418\ \text{deg}/s$, and $\theta = 60$ deg. These values are actual parameters of the geostationary satellite. The initial condition is [0.5 deg, 0, 0, 0.007 deg/sec]. Finally, the weighting matrices are chosen to be $Q = I_{4\times 4}$ and $R = 1 \times 10^{-10}$.

In this model, the states are measured with the sensor noise, dv/dt. A Kalman filter is then used to estimate the states. The following simulations are performed using MATLAB, a software package. The RS controller is found using equation 5.35. Note that both yaw and roll angles reduce to a value close to the origin. Figure 5.2 shows the roll and yaw angles with respect to time variation. After about 3 hours, both roll and yaw angles stay below 0.1 degree. Initially, large control action is needed, but after 3 hours or so, less than 300 Atm^2 magnetic

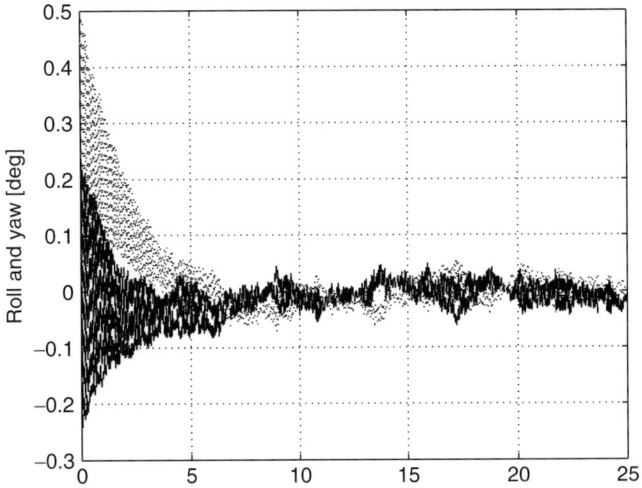

FIGURE 5.2 Roll (Dark) and Yaw (Light) Versus Time, RS Control

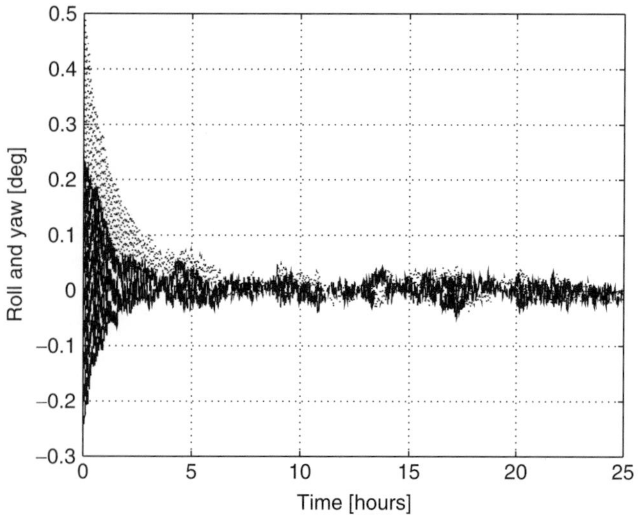

FIGURE 5.3 Roll (Dark) and Yaw (Light) Versus Time, LQG Control

torque is required. It is important to note that despite the external disturbances, RS control law produces good performance.

To compare the results with the well-known LQG controller, the system was simulated with an LQG controller. The θ approached infinity in equation 5.36. Note that equation 5.36 becomes a classical Riccati equation as θ goes to infinity. This is shown in Figure 5.3. Note that in LQG case, it takes longer for yaw and roll angles to fall below 0.1 degree, and the variation in the angles are larger than the RS case. Thus, in this sense, RS controller outperforms the LQG controller.

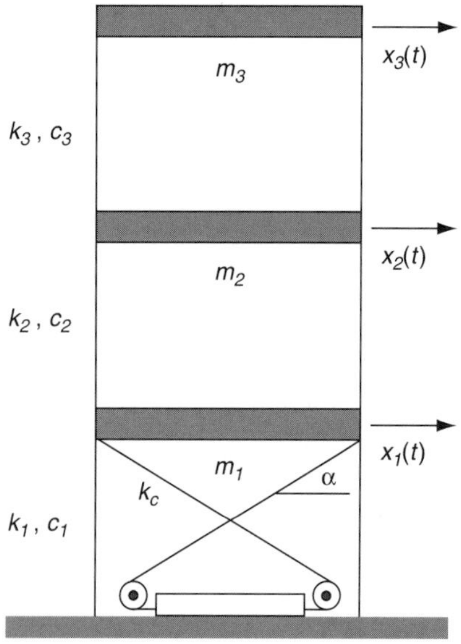

FIGURE 5.4 Schematic Diagram for Three Degree-of-Freedom Structure

5.6.2 MCV Control Applied to Seismic Protection of Structures

A 3DOF, single-bay structure with an active tendon controller as shown in Figure 5.4 is considered here. The structure is subject to a one-dimensional earthquake excitation. If a simple shear frame model for the structure is assumed, the governing equations of motion in state space form can be written as:

$$dx(t) = \begin{bmatrix} 0 & I \\ -M_s^{-1}K_s & -M_s^{-1}C_s \end{bmatrix} x(t)dt + \begin{bmatrix} 0 \\ M_s^{-1}B_s \end{bmatrix} u(t)dt + \begin{bmatrix} 0 \\ -\Gamma_s \end{bmatrix} dw(t),$$

where the following apply:

$$M_s = \begin{bmatrix} m_1 & 0 & 0 \\ 0 & m_2 & 0 \\ 0 & 0 & m_3 \end{bmatrix}, \ B_s = \begin{bmatrix} -4k_c \cos\alpha \\ 0 \\ 0 \end{bmatrix}.$$

$$C_s = \begin{bmatrix} c_1 + c_2 & -c_2 & 0 \\ -c_2 & c_2 + c_3 & -c_3 \\ 0 & -c_3 & c_3 \end{bmatrix}, \ \Gamma_s = \begin{bmatrix} 1 \\ 1 \\ 1 \end{bmatrix}.$$

$$K_s = \begin{bmatrix} k_1 + k_2 & -k_2 & 0 \\ -k_2 & k_2 + k_3 & -k_3 \\ 0 & -k_3 & k_3 \end{bmatrix}.$$

The m_i, c_i, k_i are the mass, damping, and stiffness, respectively, associated with the ith floor of the building. The k_c is the stiffness of the tendon. The Brownian motion $w(t)$ with $E\{dw(t)\} = 0$ and $E\{dw(t)dw'(t)\} = Wdt$; in this example, $W = 1.00 \times 2\pi \ \text{in}^2/\text{sec}^3$. The parameters were chosen to

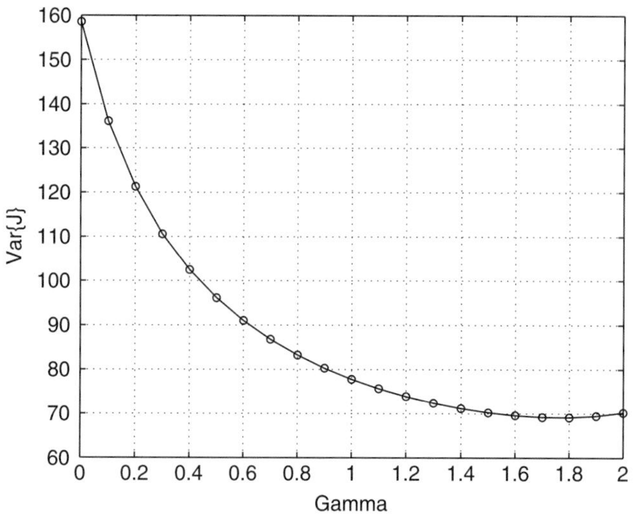

FIGURE 5.5 Optimal Variance: Full-State Feedback, MCV, 3DOF

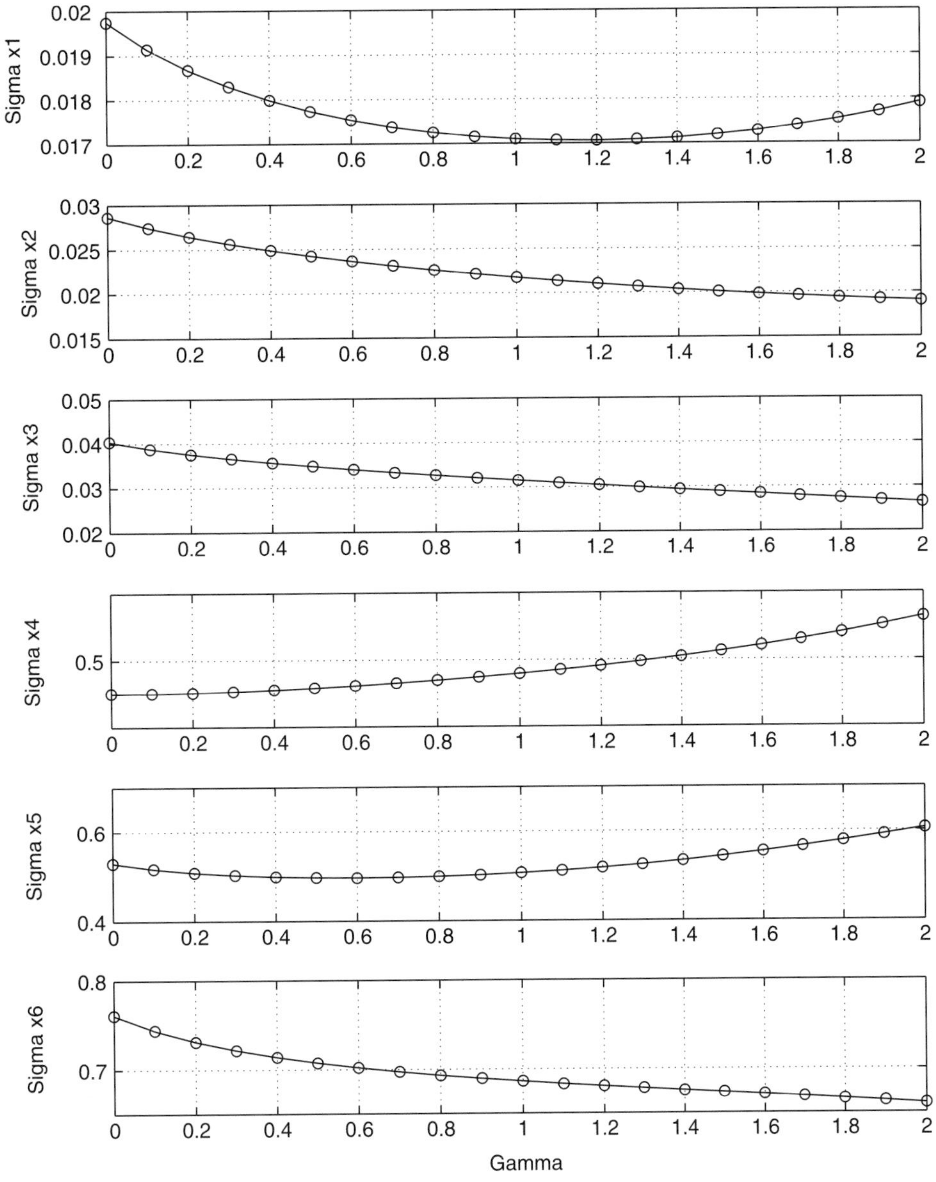

FIGURE 5.6 Displacements and Velocities. Full-State Feedback, MCV, 3DOF

match modal frequencies and dampings of an experimental structure. The cost criterion is given by:

$$J = \int_0^{t_F} \left(z'(t) K_s z(t) + k_c u^2(t) \right) dt,$$

together with $R = k_c$, where z is a vector of floor displacements and $x = [z \; \dot{z}]'$.

Figure 5.5 shows that the variance of the cost criterion decreases as γ increases. Note that the $\gamma = 0$ point corresponds to the classical LQG case.

Figure 5.6 shows the RMS displacement responses of first (σ_{x_1}), second (σ_{x_2}), and third (σ_{x_3}) floor and the RMS velocity responses of first (σ_{x_4}), second (σ_{x_5}), and third (σ_{x_6}) floor, respectively, versus the MCV parameter γ. It is important to note that both third floor RMS displacement and velocity responses can be decreased by choosing large γ.

5.7 Conclusions

This chapter describes linear-quadratic-Gaussian (LQG), minimal cost variance (MCV), and risk-sensitive (RS) controls in terms of the cost cumulants. Cost cumulant control, which is also called statistical control, views the optimization criterion as a random variable and minimizes any cumulant of the optimization criterion. Then LQG, MCV, and RS are all special cases of cost cumulant control where in LQG the mean, in MCV the variance, and in RS all cumulants of the cost function are optimized. This chapter provides the optimal controllers for the LQG, MCV, and RS methods. Finally, satellite attitude control application using RS controller and building control application using MCV controller are described.

References

Anderson, B.D.O., and Moore, J.B. (1989). *Optimal control, linear quadratic methods.* Englewood Cliffs, NJ: Prentice Hall.

Basar, T., and Bernhard, P. (1991). H_∞-*optimal control and related minimax design problems.* Boston: Birkhauser.

Bensoussan, A. (1992). *Stochastic control of partially observable systems.* Cambridge: Cambridge University Press.

Davis, M.H.A. (1977). *Linear estimation and stochastic control.* London: Halsted Press.

Fleming, W.H., and Rishel, R.W. (1975). *Deterministic and stochastic optimal control.* New York: Springer-Verlag.

Glover, K., and Doyle, J.C. (1988). State-space formulae for all stabilizing controllers that satisfy H_∞-norm bound and relations to risk sensitivity. *Systems and Control Letters 11*, 167–172.

Jacobson, D.H. (1973). Optimal stochastic linear systems with exponential performance criteria and their relationship to deterministic differential games, *IEEE Transactions on Automatic Control*, AC-18, 124–131.

Kwakernaak, H., and Sivan, R. (1972). *Linear optimal control systems.* New York: John Wiley & Sons.

Rhee, I., and Speyer, J. (1992). Application of a game theoretic controller to a benchmark problem. *Journal of Guidance, Control, and Dynamics 15*(5), 1076–1081.

Runolfsson, T. (1994). The equivalence between infinite-horizon optimal control of stochastic systems with exponential-of-integral performance index and stochastic differential games. *IEEE Transactions on Automatic Control 39*(8), 1551–1563.

Sain, M.K., and Liberty, S.R. (1971). Performance measure densities for a class of LQG control systems. *IEEE Transactions on Automatic Control AC-16* (5), 431–439.

Sain, M.K., Won, C.H., and Spencer, Jr., B.F. (1992). Cumulant minimization and robust control. In Duncan, T.E., and Pasik-Duncan, B. (Eds.) *Stochastic Theory and Adaptive Control Lecture Notes in Control and Information Services 184* Berlin: Springer-Verlag, pp. 411–425.

Sain, M.K., Won, C.H., Spencer, Jr., B.F., and Liberty, S.R. (2000). Cumulants and risk-sensitive control: A cost mean and variance theory with application to seismic protection of structures. In J.A. Filar, V. Gaitsgory, and K. Mizukami (Eds.), *Advances in Dynamic Games and Applications, Annals of the International Society of Dynamic Games*, Vol. 5. Boston: Birkhauser.

Uchida, K., and Fujita, M. (1989). On the central controller: Characterizations via differential games and LEQG control problems. *Systems and Control Letters 15*(1), 9–13.

Whittle, P. (1996). *Optimal control, basics and beyond.* New York: John Wiley & Sons.

Won, C.H. (1995). *Cost cumulants in risk-sensitive and minimal cost variance control*, Ph.D. Dissertation. University of Notre Dame.

6

Frequency Domain System Identification

Gang Jin
Ford Motor Company,
Dearborn, Michigan, USA

6.1 Introduction

A general procedure for the frequency domain identification of multiple inputs/multiple outputs (MIMO) linear time invariant systems is illustrated in Figure 6.1. Typically, one starts with the experimental frequency response function (FRF) of the test system. These FRF data may either be computed from the saved input/output measurement data or measured directly online by a spectrum analyzer. Based on these data, the matrix fraction (MF) or the polynomial matrix (PM) curve-fitting technique is applied to find a transfer function matrix (TFM) that closely fits into the FRF data. Detailed algorithms for the curve-fitting are introduced in Section 6.2. Frequently, for the purposes of simulation and control, one needs a state-space realization of the system. This may be achieved by various linear system realization algorithms. In particular, the eigensystem realization algorithm (ERA) is presented in Section 6.3 for this purpose, thanks to its many successes in previous application studies. The Markov parameters, based on the parameters from which the state-space model will be derived, can be easily generated from the identified transfer function matrix. Finally, as a measure of performance, the model FRF is computed and is compared to the experimental FRF. This is illustrated in Section 6.4 by means of two experimental application examples.

6.2 Frequency Domain Curve-Fitting

Frequency domain curve-fitting is a technique to fit a TFM closely into the observed FRF data. Like other system identification techniques, this is a two-step procedure: **model structure selection** and **model parameter optimization**. In this context, the first step is to parameterize the TFM in some special forms. Two such forms are introduced in the following: the matrix fraction (MF) parameterization and the polynomial matrix (PM) parameterization. This is always a critical step in the identification because it will generally lead to quite different parameter optimization algorithms and resulting model properties. In particular, this section shows that for the MF form, the parameters can be optimized by means of linear least squares (LLS) solutions. As for the PM parameterization, one has to resort to some nonlinear techniques; specifically, this section introduces the celebrated Gauss-Newton (GN) method. On the other hand, the PM parameterization offers more flexibility in the sense that it allows the designer to specify certain properties of the identified model (e.g., fixed zeros in any input/output channels). This feature may be quite desirable as shown by the application studies in Section 6.4.

Before starting the discussion, it is important to make clear the notations that will be used throughout this section. Assume the test system has r input excitation channels and

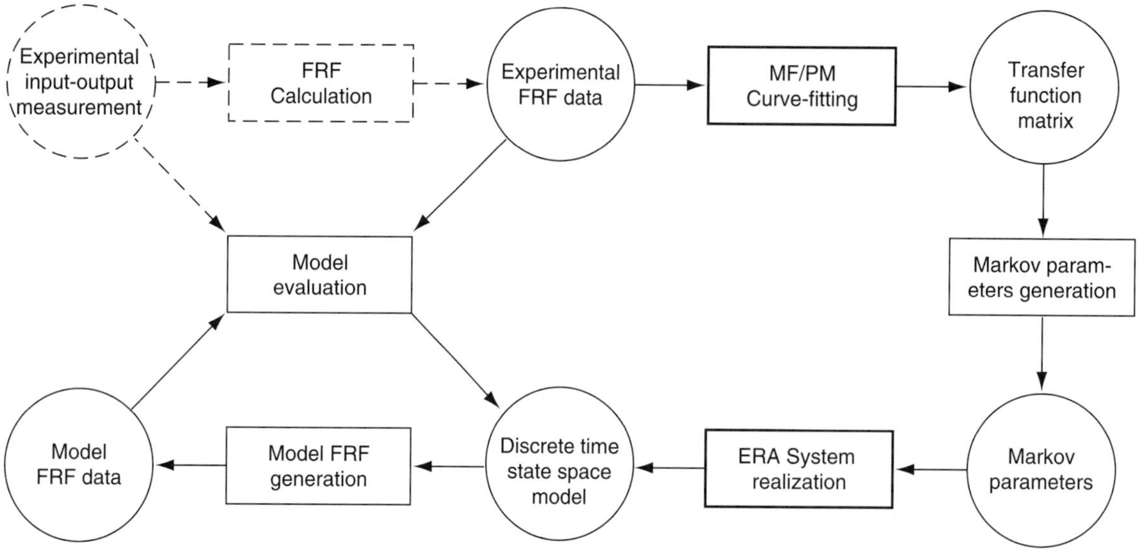

FIGURE 6.1 General Procedure of Frequency Domain System Identification. Bolded components imply critical steps; dashed components imply steps may be excluded.

m output measurement channels. Use $\{\mathcal{G}(\omega_i)\}_{i=1,\ldots l}$ to denote the observed FRF data based on which the TFM $G(z^{-1})$ will be estimated. To evaluate $G(z^{-1})$ at discrete frequencies ω_i, use the map $z(\omega_i) = e^{j2\pi\omega_i/w_s}$, with w_s being the sampling frequency. The curve-fitting error is measured by Frobenius norm $\|\cdot\|_F$ for matrices and by 2-norm $\|\cdot\|_2$ for vectors. Use I_n to denote an identity matrix of dimensions $n \times n$.

6.2.1 Matrix Fraction Parameterization

The **matrix fraction** (MF) **parameterization** of a TFM takes the following form:

$$G(z^{-1}) = Q^{-1}(z^{-1})R(z^{-1}), \qquad (6.1)$$

where:

$$Q(z^{-1}) = I_m + Q_1 z^{-1} + Q_2 z^{-2} + \cdots + Q_q z^{-q}, \qquad (6.2)$$

$$R(z^{-1}) = R_0 + R_1 z^{-1} + R_2 z^{-2} + \cdots + R_p z^{-p}. \qquad (6.3)$$

The constant matrices $Q_1, \ldots, Q_q$ and $R_0, R_1, \ldots, R_p$ are referred to as **observer Markov parameters** in Juang (1994). From the same reference, the reader may find more detailed material for the MF curve-fitting discussed here and the ERA realization algorithm discussed in a later section. To simplify the notation, without loss of generality, assume $p = q$.

To fit the TFM $G(z^{-1})$ as in equation 6.1 into the observed FRF data $\{\mathcal{G}(\omega_i)\}_{i=1,\ldots,l}$, one may solve the following parameter optimization problem:

$$G^*(z^{-1}) = \arg \min_{G=Q^{-1}R} \sum_{i=1}^{l} \|Q(z^{-1}(\omega_i))\mathcal{G}(\omega_i) - R(z^{-1}(\omega_i))\|_F^2. \quad (6.4)$$

Substituting the $Q(z^{-1})$ and $R(z^{-1})$ polynomials in equations 6.2 and 6.3 and vectorizing the summation, equation 6.4 is changed into the form:

$$G^*(z^{-1}) = \arg \min_{G=Q^{-1}R} \|\Phi\Theta + \Psi\|_F, \qquad (6.5)$$

where:

$$\Phi = \begin{bmatrix} z^{-1}(\omega_1)\mathcal{G}^T(\omega_1) & \ldots & z^{-p}(\omega_1)\mathcal{G}^T(\omega_1) & -I_r & -z^{-1}(\omega_1)I_r & \ldots & -z^{-p}(\omega_1)I_r \\ \vdots & \ddots & \vdots & \vdots & \vdots & \ddots & \vdots \\ z^{-1}(\omega_l)\mathcal{G}^T(\omega_l) & \ldots & z^{-p}(\omega_l)\mathcal{G}^T(\omega_l) & -I_r & -z^{-1}(\omega_l)I_r & \ldots & -z^{-p}(\omega_l)I_r \end{bmatrix}. \qquad (6.6)$$

$$\Psi^T = [\mathcal{G}(\omega_1) \cdots \mathcal{G}(\omega_l)]. \qquad (6.7)$$

$$\Theta^T = [Q_1 \cdots Q_p \; R_0 \; R_1 \cdots R_p]. \qquad (6.8)$$

Thus, the MF curve-fitting has been reduced to a standard LLS problem, which can be solved by various efficient algorithms (e.g., the QR factorization approach quoted in algorithm of equations 6.29 through 6.31).

6.2.2 Polynomial Matrix Parameterization

The **polynomial matrix** (PM) **parameterization** of a TFM has the form:

$$G(z^{-1}) = \frac{B(z^{-1})}{\alpha(z^{-1})}, \qquad (6.9)$$

where:

$$B(z^{-1}) = B_0 + B_1 z^{-1} + B_2 z^{-2} + \cdots + B_p z^{-p}. \qquad (6.10)$$

$$\alpha(z^{-1}) = 1 + a_1 z^{-1} + a_2 z^{-2} + \cdots + a_q z^{-q}. \qquad (6.11)$$

These equations are the numerator polynomial matrix and the minimal polynomial of the TFM, respectively. Assume that the orders of the numerator and denominator are equal (i.e., $p = q$). The goal of parameter optimization is to find $G^*(z^{-1})$ with a prespecified order p such that the estimation error is minimized:

$$G^*(z^{-1}) = \arg\min_G \sum_{i=1}^{l} w^2(\omega_i)\|\mathcal{G}(\omega_i) - G(z^{-1}(\omega_i))\|_F^2 \quad (6.12)$$

Here, an additional term $w(\cdot)$ is included to allow desirable frequency weighting on the estimation error. In the following, $G(z^{-1})$ in equation 6.12 will be parametized in a way that allows the inclusion of the fixed zeros. This is done first for the single input/single output (SISO) case; then, it is generalized to the MIMO case.

Let the numerator polynomial of the SISO system be the following:

$$B(z^{-1}) = \bar{B}(z^{-1}) \cdot \tilde{B}(z^{-1}) \quad (6.13)$$

$$= \bar{\psi}^T\bar{b} \cdot \tilde{\psi}^T\tilde{b}, \quad (6.14)$$

where $\bar{b} = [\bar{b}_0, \bar{b}_1, \ldots, \bar{b}_{p_s}]^T$ is the numerator parameter vector corresponding to the fixed zeros; $\tilde{b} = [\tilde{b}_0, \tilde{b}_1, \ldots, \tilde{b}_{p-p_s}]^T$ is the to-be-estimated numerator parameter vector; and $\bar{\psi} = [1, z^{-1}, \ldots, z^{-p_s}]^T$ and $\tilde{\psi} = [1, z^{-1}, \ldots, z^{-(p-p_s)}]^T$ are the corresponding z-vectors. Similarly, the denominator polynomial is as follows:

$$\alpha(z^{-1}) = 1 + \phi^T a, \quad (6.15)$$

where $a = [a_1, \ldots, a_p]^T$ and $\phi = [z^{-1}, \ldots, z^{-p}]^T$. The estimation error to be minimized is then:

$$F = \sum_{i=1}^{l}\left| w(\omega_i)\left(\mathcal{G}(\omega_i) - \frac{\bar{\psi}^T(\omega_i)\bar{b} \cdot \tilde{\psi}^T(\omega_i)\tilde{b}}{1 + \phi^T(\omega_i)a}\right)\right|^2 \quad (6.16)$$

$$= \sum_{i=1}^{l}\left|\frac{w(\omega_i)}{1+\phi^T(\omega_i)a}\left(\mathcal{G}(\omega_i) - [\bar{\psi}^T(\omega_i)\bar{b}\cdot\tilde{\psi}^T(\omega_i) - \mathcal{G}(\omega_i)\cdot\phi^T(\omega_i)]\begin{bmatrix}\tilde{b}\\ a\end{bmatrix}\right)\right|^2, \quad (6.17)$$

where for simplicity $\bar{\psi}$ and $\tilde{\psi}$ are written as functions of ω_i.

For the MIMO case, equation 6.17 is generalized to (recall that m and r denote the numbers of output and input channels respectively):

$$F = \sum_{j=1}^{m}\sum_{k=1}^{r}\sum_{i=1}^{l}\left|\frac{w(\omega_i)}{1+\phi^T(\omega_i)a}\left(\mathcal{G}_{jk}(\omega_i) - [\bar{\psi}_{jk}^T(\omega_i)\bar{b}_{jk}\cdot\tilde{\psi}_{jk}^T(\omega_i) - \mathcal{G}_{jk}(\omega_i)\cdot\phi^T(\omega_i)]\begin{bmatrix}\tilde{b}_{jk}\\ a\end{bmatrix}\right)\right|^2. \quad (6.18)$$

Finally, the right-hand side of equation 6.18 is vectorized for standard optimizations:

$$F(\theta) = \|W(a)(y - H\theta)\|_2^2, \quad (6.19)$$

where $\theta = [\tilde{b}_{11}^T, \tilde{b}_{12}^T, \ldots, \tilde{b}_{mr}^T, a^T]^T$, y is a vector containing the measured FRF data, and $W(a)$ is a weighting function with variate a.

Readers should have no difficulties to derive the detailed expressions of $W(a)$ and H. For the special case when there is no fixed zeros (i.e., $\bar{\psi}^T\bar{b} = 1$ in equation 6.14), the results are given in Bayard (1992). It is important to point out that $W(a)$ and H have the following structure:

$$W(a) = \begin{bmatrix} W(a) & 0 & \cdots & 0 \\ 0 & W(a) & \ddots & \vdots \\ \vdots & \ddots & \ddots & 0 \\ 0 & \cdots & 0 & W(a) \end{bmatrix}. \quad (6.20)$$

$$H = \begin{bmatrix} \Psi_{11} & 0 & \cdots & 0 & \Phi_{11} \\ 0 & \Psi_{12} & \ddots & \vdots & \Phi_{12} \\ \vdots & \ddots & \ddots & 0 & \vdots \\ 0 & \cdots & 0 & \Psi_{mr} & \Phi_{mr} \end{bmatrix}. \quad (6.21)$$

These structures enable the design of an efficient optimization algorithm. This will be discussed in the next section.

6.2.3 Least Squares Optimization Algorithms

This section serves two purposes. First, it gives a brief (but general) account on a few of the most important parameter optimization algorithms, namely the **linear least squares** (LLS) method, the Newton's method, and the Gauss-Newton's (GN) method. Second, the discussion applies some of the methods to the parameter optimization problems arising from the curve-fitting process. In particular, this section presents a fast algorithm based on the GN method to the minimization of equation 6.19. Excellent textbooks in this field are abundant, and this discussion only refers the reader to a few of them: Gill *et al.*, (1981), Dennis and Schnabel (1996), and Stewart (1973).

General Algorithm Development

Let $F(\theta)$ be the scalar-valued multivariate objective function to be minimized. If the first and second derivatives of F are available, a local quadratic model of the objective function may be obtained by taking the first three terms of the Taylor-series expansion about a point θ_k in the parameter vector space:

$$F(\theta_k + p) \approx F_k + g_k^T p + \frac{1}{2} p^T G_k p, \tag{6.22}$$

where p denotes a step in the parameter space and where F_k, g_k, and G_k denote the value, gradient, and Hessian of the objective function at θ_k. Equation 6.22 indicates that to find a local minimum of the objective function, an iterative searching procedure is required. The celebrated **Newton's method** is defined by choosing the step $p = p_k$ so that $\theta_k + p_k$ is a stationary point of the quadratic model in 6.22. This amounts to solving the following linear equation:

$$G_k p_k = -g_k. \tag{6.23}$$

In the system identification case, the objective function $F(\theta)$ is often in the **sums of squares** form, as in equation 6.19:

$$F(\theta) = \frac{1}{2}\sum_{i=1}^{n} f_i(\theta)^2 = \frac{1}{2}\|f(\theta)\|_2^2, \tag{6.24}$$

where f_i is the ith component of the vector f. To implement the Newton's method, the gradient and Hessian of F are calculated as:

$$g(\theta) = J(\theta)^T f(\theta) \tag{6.25}$$

$$G(\theta) = J(\theta)^T J(\theta) + Q(\theta), \tag{6.26}$$

where $J(\theta)$ is the Jacobian matrix of f and where $Q(\theta) = \sum_{i=1}^{n} f_i(\theta) G_i(\theta)$, with $G_i(\theta)$ being the Hessian of $f_i(\theta)$. The Newton's equation 6.23 thus becomes as follows:

$$(J(\theta_k))^T J(\theta_k) + Q(\theta_k))p_k = -J(\theta_k)^T f(\theta_k). \tag{6.27}$$

When $\|f(\theta_k)\|$ is small, $\|Q(\theta_k)\|$ is usually small and is often omitted from equation 6.27. If this is the case, then solving p_k from 6.27 is equivalent to solving the following linear least squares (LLS) problem:

$$p_k = \arg\min_p \frac{1}{2}\|J(\theta_k)p + f(\theta_k)\|_2^2. \tag{6.28}$$

Equation 6.28 gives the **Gauss-Newton method**. The LLS problem is often solved by the QR factorization method given in the following algorithm.

Algorithm 1 (QR Factorization and LLS Solution)
Let $A \in R^{m \times n}$ have full column rank. Then A can be uniquely factorized into the form:

$$A = QR, \tag{6.29}$$

where Q has orthonormal columns and where R is upper triangular with positive diagonal elements. The unique solution of the LLS problem:

$$\min_{x \in R^n} \|Ax - y\|_2^2, \tag{6.30}$$

is given by:

$$x^* = R^{-1} Q^T y. \tag{6.31}$$

Application to the Curve-Fitting Problems
This discussion now returns to the PM curve-fitting problem in equation 6.19. By letting $f = \boldsymbol{W}(a)(\boldsymbol{y} - \boldsymbol{H}\theta)$, the Gauss-Newton method in equation 6.28 may be applied. It turns

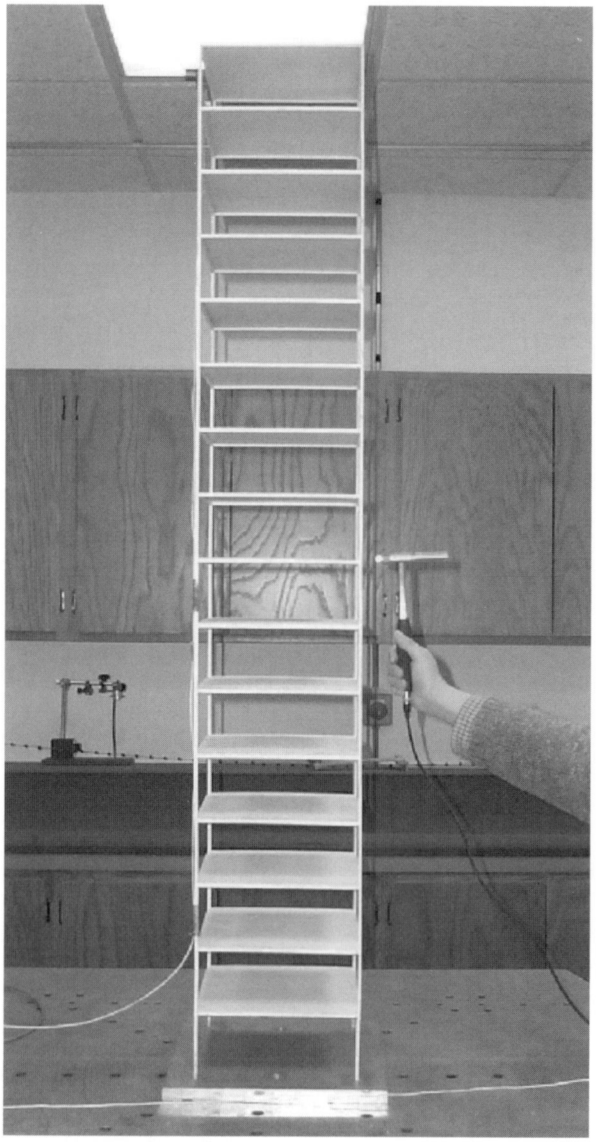

FIGURE 6.2 Picture of the 16-Story Structure

TABLE 6.1 Key Identification Parameters

	Curve-fitting			System realization		
Test structure	Type	p	q	α	β	n
16-Story	MF	16	16	32	32	10
Benchmark	PM	8	8	12	30	8

out that the Jacobian matrix $J(\theta_k)$ of f has an identical structure as H in equation 6.21. Thus, instead of solving equation 6.28 directly, the following decomposition of LLS problems may be applied.

Algorithm 2 (Decomposition of LLS Problem)

Let $A \in R^{m\times n}$ have full column rank. Decompose A into $A = [A_1, A_2]$, with $A_1 \in R^{m\times p}$ and $A_2 \in R^{m\times q}$. Let the unique QR factorization of A_1 be $A_1 = Q_1R_1$. Then the unique solution x^* of the LLS problem of equation 6.30 has the form $x^* = [x_1^{*T}, x_2^{*T}]^T$, with x_2^* and x_1^* being the unique solutions of the following LLS problems:

$$\min_{x_2\in R^q} \quad \|\tilde{A}_2x_2 - y\|_2^2. \tag{6.32}$$

$$\min_{x_1\in R^p} \quad \|A_1x_1 - (y - A_2x_2^*)\|_2^2. \tag{6.33}$$

In these equations, $\tilde{A}_2 = (I - Q_1Q_1^T)A_2$.

In practice, $J(\theta_k)$ is divided into $J(\theta_k) = [J_1(\theta_k), J_2(\theta_k)]$, with $J_1(\theta_k)$ corresponding to the block-diagonal terms and $J_2(\theta_k)$ the last block matrix column in $J(\theta_k)$. Solving the LLS problems with $J_2(\theta_k)$ and $J_1(\theta_k)$ thus corresponds to updating the denominator a and the numerators b_{jk} estimations. Moreover, due to its block-diagonal structure, the LLS problem with $J_1(\theta_k)$ should be further decomposed (i.e., the LLS solutions of b_{jk} are independently solved for each j and k). Thus, the computational cost of the optimization algorithm is significantly reduced.

To complete the discussion of the algorithm, note that the initial values for the Gauss-Newton iteration may be generated by the classical Sanathanan-Koerner (SK) iteration composed of a sequence of reweighted LLS problems:

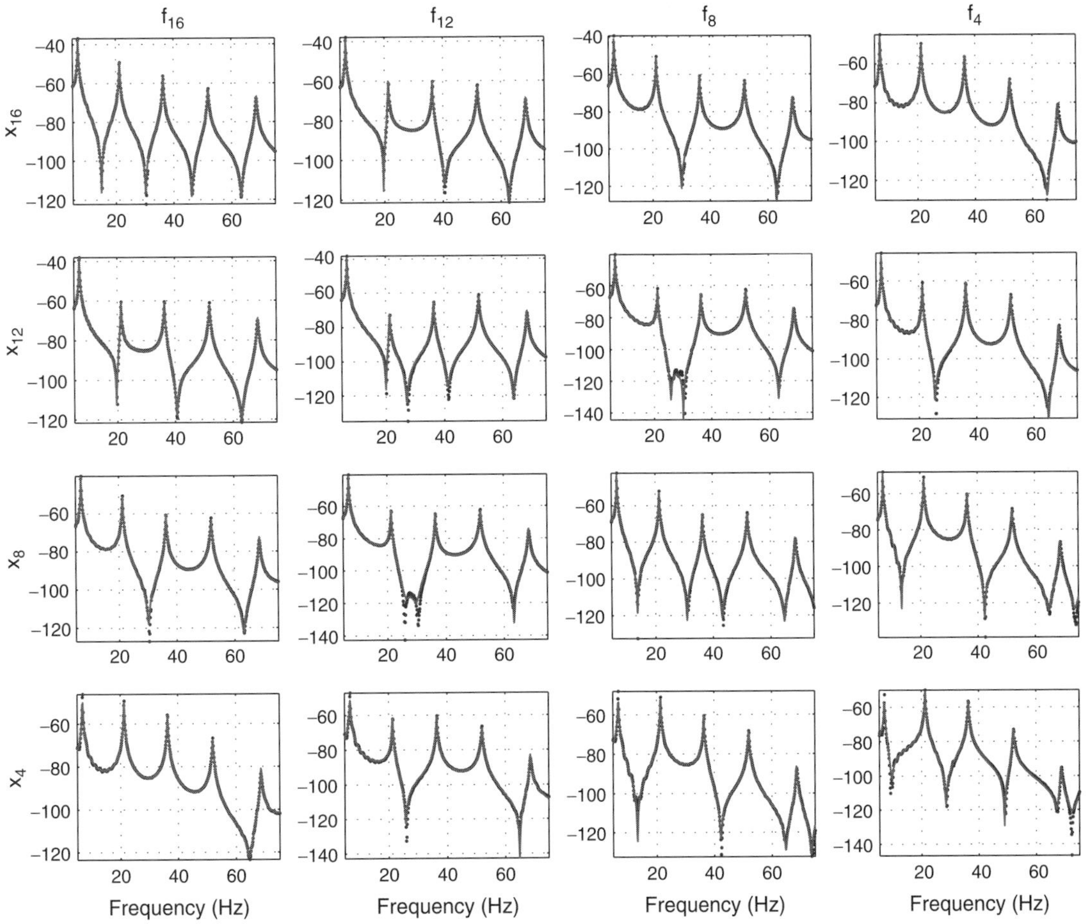

FIGURE 6.3 Comparison of Experimental and Model FRF for the 16-Story Structure: Magnitude Plot. f_j denotes the input force on the jth floor; x_j denotes the output displacement of the jth floor; dotted lines are for measurement data; solid lines are for model output.

$$\theta_{k+1} = \arg \min_{\theta} \| \boldsymbol{W}(a_k)(\boldsymbol{y} - \boldsymbol{H}\theta)\|_2^2, \quad (6.34)$$

with initial condition $\theta_0 = 0$.

6.3 State-Space System Realization

System realization is a technique to determine an internal state-space description for a system given with an external description, typically its TFM or impulse response. The name reflects the fact that if a state-space description is available, an electronic circuit can be built in a straightforward manner to realize the system response. There is a great amount of literature on this subject both from a system theoretical point of view (Antsaklis and Michel, 1997) and from a practical system identification point of view (Juang, 1994). In the following, a well-developed method in the second category, the eigensystem realization algorithm (ERA), is selected to construct a model in the state space form. First presented are the formulas to generate the Markov parameters from the TFM, which are the starting point for the ERA method.

6.3.1 Markov Parameters Generation

To calculate the **Markov parameters** $Y_0, Y_1, Y_2, \ldots$ from the system TFM, first note that:

$$G(z^{-1}) = \sum_{i=0}^{\infty} Y_i z^{-i}. \quad (6.35)$$

For the case when $G(z^{-1})$ is parameterized in the MF form (i.e., $G(z^{-1}) = Q^{-1}(z^{-1})R(z^{-1})$), the system Markov parameters can be determined from:

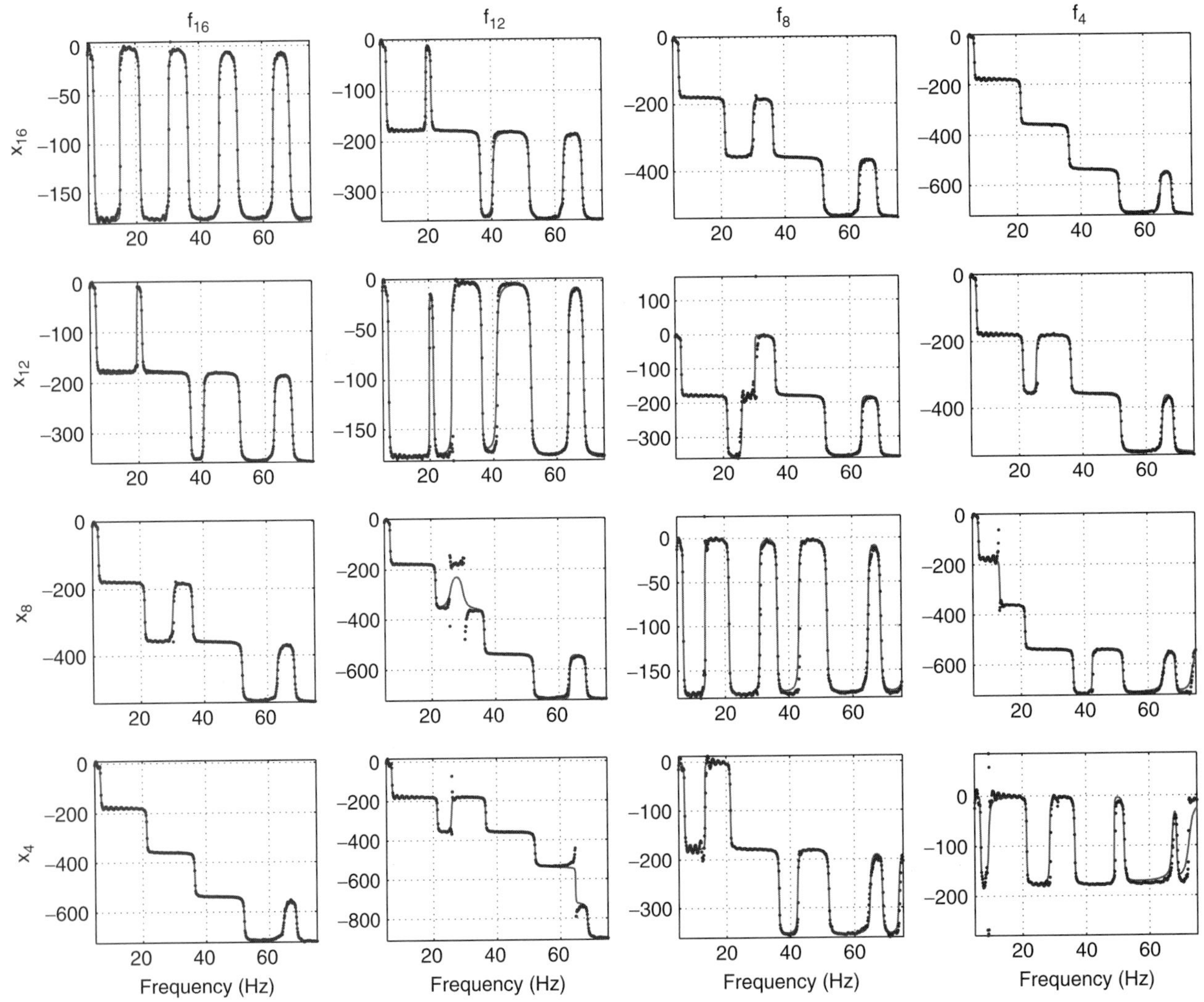

FIGURE 6.4 Comparison of Experimental and Model FRF for the 16-Story Structure: Phase Plot. Same notations are used as in Figure 6.3.

FIGURE 6.5 Picture of the Seismic-AMD Benchmark Structure

$$\left(\sum_{i=0}^{p} Q_i z^{-i}\right)\left(\sum_{i=0}^{\infty} Y_i z^{-i}\right) = \sum_{i=0}^{p} R_i z^{-i}, \tag{6.36}$$

by the following iterative calculations starting from $Y_0 = R_0$:

$$Y_k = \begin{cases} R_k - \sum_{i=1}^{k} Q_i Y_{k-i}, & \text{for } k = 1, \ldots, p. \\ -\sum_{i=1}^{p} Q_i Y_{k-i}, & \text{for } k = p+1, \ldots, \infty. \end{cases} \tag{6.37}$$

If the TFM is parameterized in the PM form, the derivation of the system Markov parameters is almost the same: one starts with $Y_0 = B_0$, and continues with the following iterative procedure:

$$Y_k = \begin{cases} B_k - \sum_{i=1}^{k} a_i Y_{k-i}, & \text{for } k = 1, \ldots, p. \\ -\sum_{i=1}^{p} a_i Y_{k-i}, & \text{for } k = p+1, \ldots, \infty. \end{cases} \tag{6.38}$$

TABLE 6.2 PM Iteration Record

Iteration type	Iteration index	FRF residue	α Step norm	β Step norm
	1	651.47	100%	100%
S	2	159.46	1.91%	14.8%
K	3	156.75	0.16%	1.36%
	4	156.74	0.24%	0.84%
	1	137.03	10.9%	21.8%
	2	133.47	4.23%	8.31%
G	3	128.03	0.56%	2.15%
N	4	123.81	0.21%	0.50%
	5	123.34	0.40%	1.07%
	6	123.48	0.15%	0.37%

6.3.2 The ERA Method

To solve for a state-space model (A, B, C, D) using the **ERA method**, first form the generalized Hankel matrices:

$$H(k-1) = \begin{bmatrix} Y_k & Y_{k+1} & \cdots & Y_{k+\beta-1} \\ Y_{k+1} & Y_{k+2} & \cdots & Y_{k+\beta} \\ \vdots & \vdots & \ddots & \vdots \\ Y_{k+\alpha-1} & Y_{k+\alpha} & \cdots & Y_{k+\alpha+\beta-2} \end{bmatrix}. \tag{6.39}$$

Note that in general, α and β are chosen to be the smallest numbers such that $H(k)$ has as large row and column ranks are possible. Additional suggestions to determine their optimal values are given in Juang (1994). Let the singular value decomposition of $H(0)$ be $H(0) = U\Sigma V^T$, and let n denote the index where the singular values have the largest drop in magnitude. Then, $H(0)$ can be approximated by:

$$H(0) \approx U_n \Sigma_n V_n^T, \tag{6.40}$$

where U_n and V_n are the first n columns of U and V, respectively, and Σ_n is the diagonal matrix containing the largest n singular values of $H(0)$. Finally, an nth order state-space realization (A, B, C, D) can be calculated by:

$$\begin{aligned} A &= \Sigma_n^{-1/2} U_n^T H(1) V_n \Sigma_n^{-1/2}, \quad B = \Sigma_n^{1/2} V_n^T E_r, \\ C &= E_m^T U_n \Sigma_n^{1/2}, \quad D = Y_0, \end{aligned} \tag{6.41}$$

where E_r and E_m^T are the elementary matrices that pick out the first r (the number of system inputs) columns and first m (the number of system outputs) rows of their multiplicands, respectively.

6.4 Application Studies

This section presents two experimental level application studies conducted in the Structural Dynamics and Control/

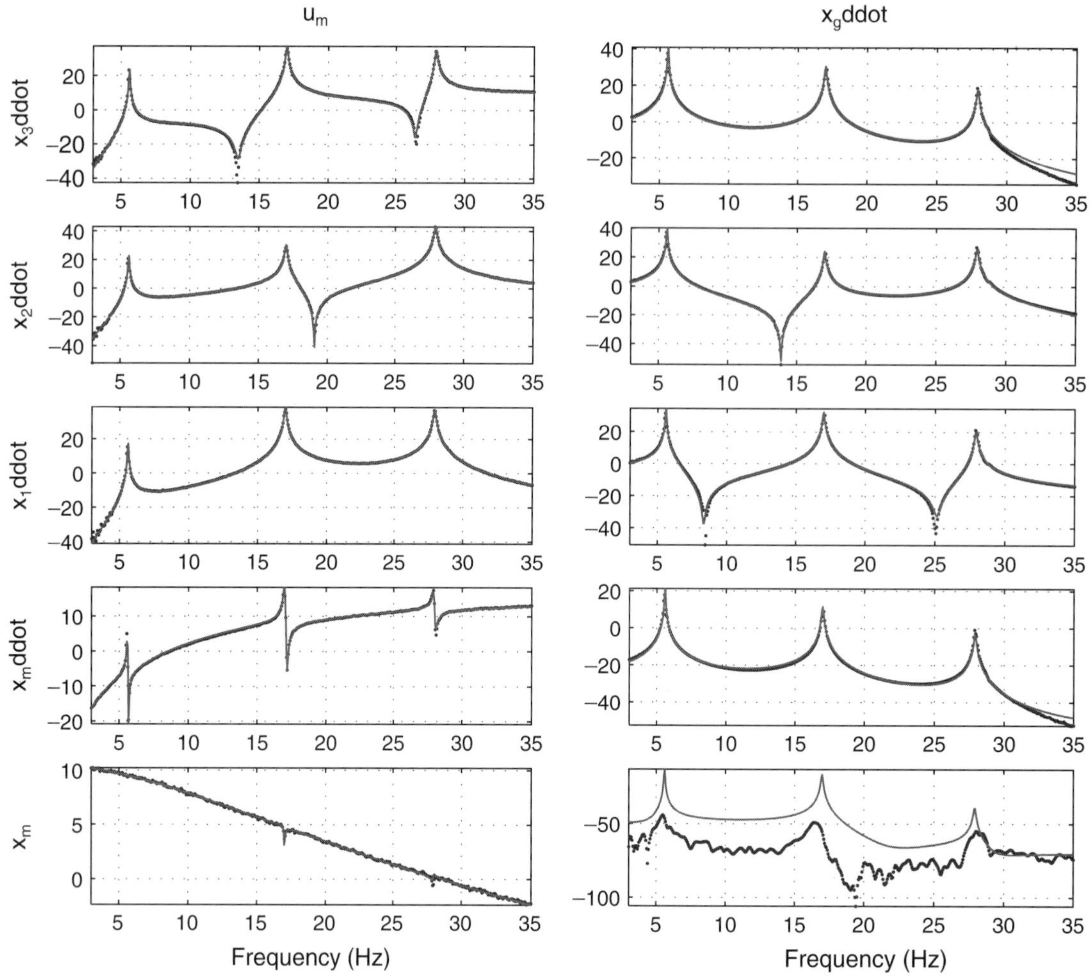

FIGURE 6.6 Comparison of Experimental and Model FRF for the Seismic-AMD Benchmark Structure: Magnitude Plot. u_m denotes the input command to the AMD; $x_g ddot$ denotes the input ground acceleration to the structure; $x_j ddot$ denotes the output acceleration of the *j*th floor; $x_m ddot$ denotes the output acceleration of the AMD; x_m denotes the output displacement of the AMD; dotted lines are for measurement data; solid lines are for model output.

Earthquake Engineering Laboratory (SDC/EEL) at University of Notre Dame. This section only presents the results pertinent to identification studies discussed so far in this chapter. For detailed information about these experiments, including experimental setups and/or control developments, the reader may refer to Jin *et al.* (2000), Jin (2002), and Dyke *et al.* (1994), respectively.

6.4.1 Identification of a 16-Story Structure

The first identification target is a 16-story steel structure model shown in Figure 6.2. The system is excited by impulse force produced by a PCB hammer and applied individually at the 16th, 12th, 8th, and 4th floors. The accelerations of these floors are selected as the system measurement outputs and are sensed by PCB accelerometers. The goal of the identification is to capture accurately the first five pairs of the complex poles of the structure. For this purpose, a DSPT Siglab spectrum analyzer is used to measure the FRF data. The sampling rate is set at 256 Hz, and the frequency resolution is set at 0.125 Hz. The experimental FRF is preconditioned to eliminate the second order direct current (dc) zeros from acceleration measurement. The MF parameterization is chosen for the curve-fitting, which is complemented by the ERA method for state-space realization. The key identification parameters are given in Table 6.1. The final discrete-time state-space realization has 10 states. The magnitude and phase plots of its transfer functions are compared to the experimental FRF data in Figures 6.3 and 6.4. Excellent agreements are found in all but the very high frequency range. The mismatch there is primarily due to the unmodeled high-frequency dynamics.

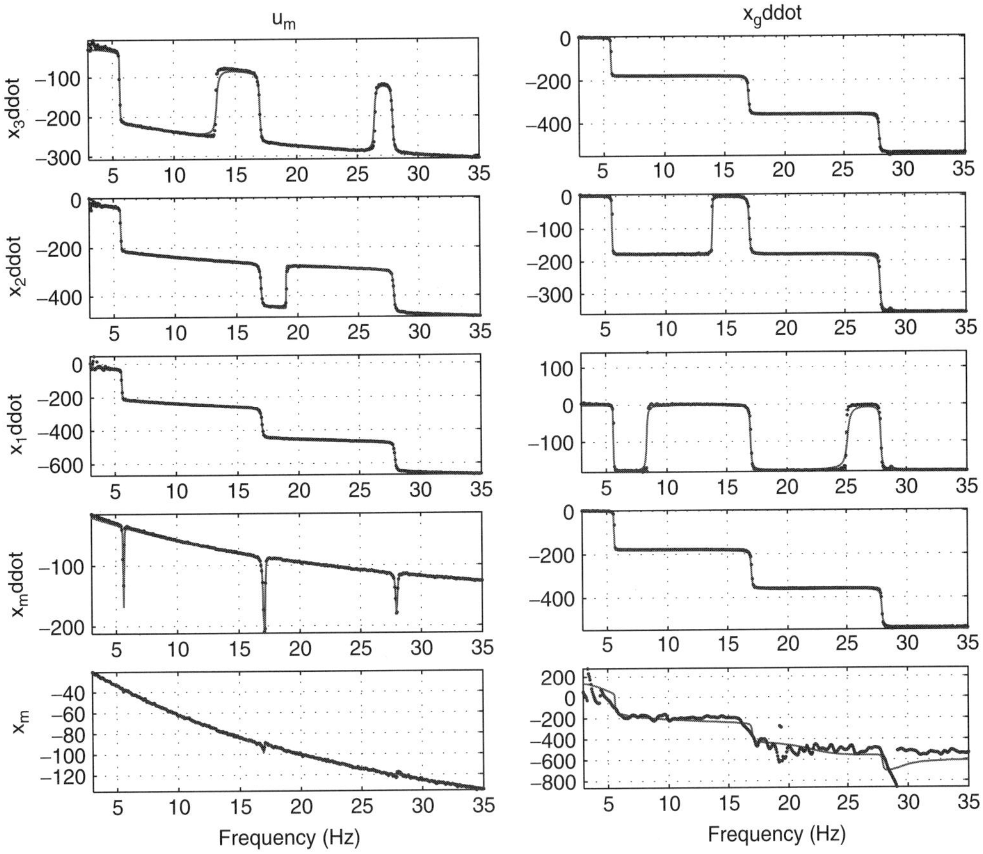

FIGURE 6.7 Comparison of Experimental and Model FRF for the Seismic-AMD Benchmark Structure: Phase Plot. Same notations are used as in Figure 6.6.

6.4.2 Identification of the Seismic-Active Mass Driver Benchmark Structure

The second identification target of this discussion is a three-story steel structure model, with an active mass driver (AMD) installed on the third floor to reduce the vibration of the structure due to simulated earthquakes. A picture of the system is given in Figure 6.5. This system has been used recently as the ASCE first-generation seismic-AMD benchmark study. The benchmark structure has two input excitations: the voltage command sent to the AMD by the control computer and the ground acceleration generated by a seismic shaker. The system responses are measured by four accelerometers for the three floors and the AMD and one linear variable differential transformer (LVDT) for the displacement of the AMD. The sampling rate is 256 Hz, and the frequency resolution is 0.0625 Hz. Due to noise and nonlinearity, only the frequency range of 3 to 35 Hz of the FRF data is considered to be accurate and, thus, this range is used for the identification.

A preliminary curve-fitting is carried out using the MF parameterization. The identified model matches the experimental data accurately in all but the low-frequency range of channels corresponding to the AMD command input and the acceleration outputs. A detailed analytical modeling of the system reveals that there are four (respectively two) fixed dc zeros from AMD command input to the structure (respectively AMD) acceleration outputs:

$$\lim_{s\to 0} \frac{G_{\ddot{x}_i u_m}(s)}{s^4} = k_i, \quad i = 1, 2, 3. \tag{6.42}$$

$$\lim_{s\to 0} \frac{G_{\ddot{x}_m u_m}(s)}{s^2} = k_m. \tag{6.43}$$

The $G_{\ddot{x}_i u_m}$ and $G_{\ddot{x}_m u_m}$ are the transfer functions from AMD command input u_m to structure and AMD acceleration outputs, respectively. The k_i and k_m are the static gains of these transfer functions with the fixed dc zeroes removed. These fixed zeros dictate the use of the PM curve-fitting technique to explicitly include such a priori information.

Again, key identification parameters are presented in Table 6.1. The outputs of the parameter optimization iterations are documented in Table 6.2. The final discrete-time state-space realization has eight states as predicted by the analytical modeling. The magnitude and phase plots of its transfer functions are compared to the experimental FRF data in Figures 6.6 and 6.7. All the input output channels are identified accurately except for the (5,2) element, which corresponds to the ground acceleration input and the displacement output of the AMD. The poor fitting there is caused by the extremely low signal-to-noise ratio.

6.5 Conclusion

This chapter discusses the identification of linear dynamic systems using frequency domain measurement data. After outlining a general modeling procedure, the two major computation steps, frequency domain curve-fitting and state-space system realization, are illustration with detailed numerical routines. The algorithms employ TFM models in the form of matrix fraction or polynomial matrix and require respectively linear or nonlinear parameter optimizations. Finally, the proposed identification schemes are validated through the modeling of two experimental test structures.

References

Antsaklis, P., and Michel, A. (1997). *Linear systems.* New York: McGraw-Hill.

Bayard, D. (1992). Multivariable frequency domain identification via two-norm minimization. *Proceedings of the American Control Conference,* 1253–1257.

Dennis, Jr., J.E., and Schnabel, R.B. (1996). *Numerical methods for unconstrained optimization and nonlinear equations.* Philadelphia: SIAM Press.

Dyke, S., Spencer, Jr., B., Belknap, A., Ferrell, K., Quast, P., and Sain, M. (1994). Absolute acceleration feedback control strategies for the active mass driver. *Proceedings of the World Conference on Structural Control 2,* TP1:51–TP1:60.

Gill, P.E., Murray, W., and Wright, M.H. (1981). *Practical optimization.* New York: Academic Press.

Jin, G., Sain, M.K., and Spencer, Jr., B.F. (2000). Frequency domain identification with fixed zeros: First generation seismic-AMD benchmark. *Proceedings of the American Control Conference,* 981–985.

Jin, G. (2002). *System identification for controlled structures in civil engineering application: Algorithm development and experimental verification.* Ph.D. Dissertation, University of Notre Dame.

Juang, J.N. (1994). *Applied system identification.* Englewood Cliffs, NJ: Prentice Hall.

Stewart, G.W. (1973). *Introduction to matrix computations.* New York: Academic Press.

7

Modeling Interconnected Systems: A Functional Perspective

Stanley R. Liberty
Academic Affairs, Bradley University, Peoria, Illinois, USA

7.1 Introduction

This chapter, adapted from Liberty and Saeks (1973, 1974), contains a particular viewpoint on the mathematical modeling of systems that focuses on an explicit algebraic description of system connection information. The resulting system representation is called the **component connection model**. The component connection approach to system modeling provides useful ways to tackle a variety of system problems and has significant conceptual value. The general applicability of this viewpoint to some of these system problems is presented here at a conceptual level. Readers interested in specific examples illustrating the component connection model should first refer to Decarlo and Saeks (1981). Additional references are provided on specific applications of the component connection philosophy to system problems. Readers with a clear understanding of the general concept of function should have no difficulty comprehending the material in this section.

7.2 The Component Connection Model

A system, in simplistic mathematical terms, is a mapping from a set of inputs (signals) to a set of outputs (signals) (i.e., an input/output relation with a unique output corresponding to each input). (In using this external system description, the context of an internal system state has been deliberately suppressed for simplicity of presentation in this chapter.) This may be abstractly notated by:

$$y = Su, \tag{7.1}$$

where S is the symbol for the input/output relation (the system). On the other hand, one may think of the physical system symbolized by S as an interconnection of components. Engineering experience explains that a particular S is determined by two factors: the types of components in the system and the way in which the components are interconnected. This latter observation on the physical structure of a system gives rise to the **component connection viewpoint**.

Again, thinking of a physical system, the system's connection structure can be held fixed while the types of system components or their values change. Experience explains that for each distinct set of components or component values, a unique input/output relation is determined. From a mathematical perspective, a given connection description determines a mapping of the internal component parameters to the input/output relation.

Now, one can generalize this slightly by thinking of the physical system as an interconnection of subsystems, each with its own input/output relation. Note that a subsystem may be either a discrete component, an interconnection of components, or an interconnection of "smaller" subsystems. This hierarchical structure is germane to the component connection philosophy and is precisely the fundamental concept of large-scale system theory.

Thinking in this way, one notes that a given connection description determines a system-valued function of a system-valued variable. Because this function is entirely determined by the connections, it is called the connection function:

$$S = f(Z), \tag{7.2}$$

where Z symbolizes the internal subsystem descriptions or component parameters and where f symbolizes the connection function. In the special case where the subsystems represented by Z are linear and time-invariant, Z may be interpreted as a matrix of component transfer functions, gains, or immitances (the precise physical interpretation is unimportant), which is also the case for S. One should not, however, restrict the thinking process to linear time-invariant systems. In addition, it is important to point out that, even when all of the subsystems are linear, the connection function f is nonlinear.

What may be surprising is that for most commonly encountered systems, the nonlinear function f can be completely described by four matrices that permit the connection function to be manipulated quite readily. Indeed, feasible analytical and computational techniques based on this point of view have been developed and applied to problems in system analysis (Singh and Trauboth, 1973; Prasad and Reiss, 1970; Trauboth and Prasad, 1970), sensitivity studies (Ransom, 1973, 1972; Ransom and Saeks, 1975; DeCarlo, 1983), synthesis (Ransom, 1973), stability analysis (Lu and Liu, 1972), optimization (Richardson, 1969; Richardson *et al.*, 1969), well-posedness (Ransom, 1972; Singh, 1972), and fault analysis (Ransom, 1973, 1972; Saeks, 1970; Saeks *et al.*, 1972; Ransom and Saeks, 1973a, 1973b; DeCarlo and Rapisarda, 1988; Reisig and DeCarlo, 1987; Rapisarda and DeCarlo, 1983; Garzia, 1971). The reason for this core of linearity in the f description is that connection information is generally contained in "conservation" laws that, upon the proper choice of system variables, yield linear "connection" equations.

To obtain an intuitive feel for what types of manipulations one performs on f to solve the systems problems previously mentioned, consider, for example, a classical passive network synthesis problem. In this problem, one is given the input/output relation S, and one knows what type of network structure is desired. For example, a ladder synthesis problem statement consists of a transfer function specification and the "connection information" of ladder structure. The S and f are specified, and it is desired to determine Z (the components). Clearly, to find Z, one must invert f. In this case, one seeks a right inverse of f because the uniqueness of Z is not a concern and f is viewed as "onto" the class of transfer functions of interest.

If, on the other hand, one is attempting to identify the value of a component in an interconnected system of components given external measurements and knowledge of the connections, then one seeks a left inverse of f. Here, the concern may be with uniqueness, and f is viewed as one-to-one and into the set of input/output relations.

Another classical system problem encountered in design analysis is that of determining the sensitivity of the input/output relation S to variations in certain component values. In this case, differentiation of the connection function f with respect to the component parameters of interest is the major constituent of the sensitivity analysis. Specific applications of the component connection model to sensitivity analysis are contained in Ransom (1973, 1972), Ransom and Saeks (1975), and DeCarlo (1983).

The conceptual description just described should provide the reader with a "feel" for the component connection philosophy. This chapter now provides more detailed insight.

7.2.1 Detailed Insight to Component Connection Philosophy

Although an abstract functional interpretation of the connections in a system is conceptually valid, it is of no practical value unless one has a specific and computationally viable representation of the function. Fortunately, there is such a representation, and the conceptual problem of inverting or differentiating the connection function may actually be carried out.

Classically, in circuit and system theory, connections are represented by some type of graph (e.g., linear graph, bond graph, signal flow graph) or block diagram. This graphical depiction of connection information can readily be converted into a set of linear algebraic constraints on the circuit or system variables for subsequent analysis. As such, it is natural to adopt an algebraic model of the connections from the start.

The precise form of the component connection model with its algebraic description of a system's connection information may be seen by examining Figure 7.1. Figure 7.1(A) shows a box depiction of a system with inputs u and outputs y. The inner box labeled Z represents the system components, and the outer donut shaped area represents the system connections (e.g., scalers, adders, Kirchhoff law constraints, etc.). Now, if Z represents a matrix of component input/output relations, the following may be abstractly written:

$$b = Za, \tag{7.3}$$

where b denotes the vector of component output variables and a the vector of component input variables. The component connection model can be obtained by redrawing the system of Figure 7.1(A) as in Figure 7.1(B), where the components and connections have been separated. Thus, the overall system is composed of two interconnected boxes. One box contains the components as described by equation 7.3, and the second donut-shaped box contains the connections. The donut-shaped box has inputs u and b and outputs a and y. Finally,

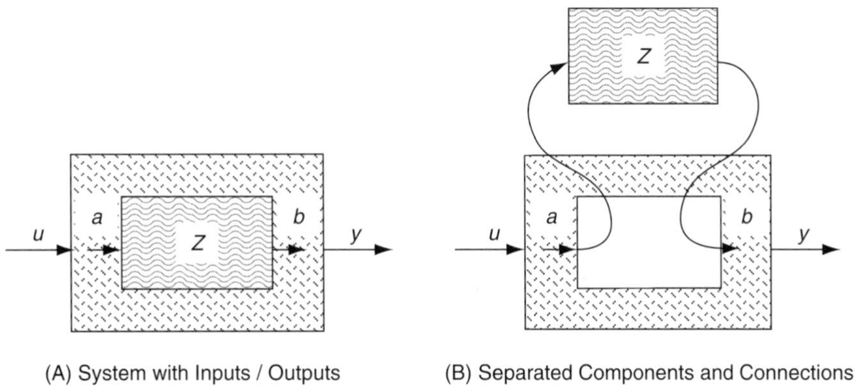

FIGURE 7.1 A System as an Interconnection of Components

since the connections are characterized entirely by linear algebraic constraints, the donut-shaped box may be mathematically modeled (i.e., the connections) by the matrix equation:

$$\begin{bmatrix} a \\ y \end{bmatrix} = \begin{bmatrix} L_{11} & L_{12} \\ L_{21} & L_{22} \end{bmatrix} \begin{bmatrix} b \\ u \end{bmatrix}. \tag{7.4}$$

The component connection model is thus a pair of equations, 7.3 and 7.4, with the matrix Z representing the components, and the set of four L matrices representing the connections.

An interesting observation can be made here. In the very special case where each component is an integrator, the following is true:

$$\dot{b} = a. \tag{7.5}$$

Substituting equation 7.5 into 7.4 yields:

$$\begin{aligned} \dot{b} &= L_{11} b + L_{12} u. \\ y &= L_{21} b + L_{22} u. \end{aligned} \tag{7.6}$$

The reader will recognize these equations as the familiar "state model" of linear dynamical system theory. Thus, intuitively, the component connection model may be viewed as a generalization of the linear state model. In fact, the L matrices describe how to interconnect subsystems to form a large-scale system S, just as integrators used to be patched together on an analog computer to simulate a system. If one views the L matrices as a generalization of the state model, where the integrators have been replaced by general components, a beneficial cross fertilization between component connection theory and classical system theory results. It is possible and indeed useful to talk about controllable or observable connections (Singh, 1972).

The component connection model was first used intuitively by Prasad and Trauboth (Singh and Trauboth, 1973; Prasad and Reiss, 1970; Trauboth and Prasad, 1970) and formalized by Saeks (Saeks, 1970; Saeks *et al.*, 1972) for application in the fault isolation problem. Existence conditions for the L matrices were first studied by Prasad (Prasad and Reiss, 1970) and later by Singh (1972) who have shown that the existence of L matrices is a reasonable assumption. This is essentially the same type of assumption that one makes when assuming that the state equations of a system exist. As seen above, if the components in the system are integrators, then the L matrices are precisely the state matrices.

The advantage of the component connection model over the classical graphical and diagrammatic connection models is that, being inherently algebraic, the component connection model is readily manipulated and amenable to numerical techniques. Moreover, this model unifies the various graphical and diagrammatic models. In fact, electronic circuits, which are commonly described by hybrid block diagram-linear graph models, are handled as readily as passive circuits or analog computer diagrams [see DeCarlo and Saeks (1981) for some simple examples].

Using the component connection model, the desired representation for the connection function is obtained. This is illustrated by a linear example since the concept can be more intuitively understood in the linear case. It is important to point out again, however, that more generality is actually present. Viewing Z as a matrix of transfer functions, the overall system transfer function is desired. Simultaneous solution of equations 7.1, 7.3, and 7.4 yields:

$$S = L_{22} + L_{21}(1 - ZL_{11})^{-1} ZL_{12} = f(Z), \tag{7.7}$$

and we have a representation of the connection function in terms of the four L matrices even though the connection function is nonlinear in Z. This matrix representation facilitates manipulation of the connection function as shown later (Ransom, 1973, 1972; Saeks, 1970; Ransom and Saeks, 1973b, 1973a). The first observation that the connection function could be so represented was made by Saeks (1970), while the first exploitation of the concept is attributable to Ransom (1973, 1972) and Saeks (1973b, 1973a).

7.3 System Identification

In a **system identification problem**, one is normally asked to identify the mathematical input/output relation S from input/output data. In theory, this is straightforward if the system is truly linear (Chang, 1973), but in practice this may be difficult due to noisy data and finite precision arithmetic. The nonlinear case is difficult in general because series approximations must be made that may not converge rapidly enough for large signal modeling (Bello *et al.*, 1973). In both linear and nonlinear identification techniques, random inputs are commonly used as probes (Bello *et al.*, 1973; Cooper and McGillem, 1971; Lee and Schetzen, 1961). The identification of S is then carried out by mathematical operations on the system input and output autocorrelations and the cross-autocorrelations between input and output.

This section does not contain a discussion of such techniques, but some observations are presented. It should be noted that if only input/output information is assumed known (as is the case in standard identification techniques), then system identification does not supply any information on the system structure or the types or values of components. In such cases, the component connection model cannot be used. In many applications, however, connection information and component type information are available. In such cases, the component connection model may be applicable. Indeed, for identification schemes using random inputs, it is easily shown that the component correlation functions and the overall system correlation functions are related by:

$$R_{yy} = L_{21}R_{bb}L_{21}^T + R_{yu}L_{22}^T + L_{22}R_{uy} - L_{22}R_{uu}L_{22}^T. \tag{7.8}$$

$$R_{aa} = L_{11}R_{bb}L_{11}^T + L_{11}R_{bu}L_{12}^T + L_{12}R_{ub}L_{11}^T + L_{12}R_{uu}L_{12}^T. \tag{7.9}$$

$$L_{21}R_{bu} = [R_{yu} - L_{22}R_{uu}]. \tag{7.10}$$

$$R_{ab} = L_{11}R_{bb} + L_{12}R_{ub}. \tag{7.11}$$

The R_{jk} is the crosscorrelation of signal j and signal k. If the left inverse of L_{21} exists, then the subsystem correlation functions can be determined and used to identify the subsystems via standard identification techniques. However, in general, this left inverse does not exist and either the type information or the psuedo-inverse of L_{21} must be used to yield subsystem models.

7.4 Simulation

There are numerical challenges in computer simulation of large-scale systems. These difficulties are due to the simultaneous solution of large numbers of differential equations. The efficiency in solving a single differential equation is entirely determined by the numerical techniques employed. However, in analyzing a large-scale system from the component connection viewpoint, the starting point is a specified collection of decoupled component differential equations and a set of coupled algebraic connection equations. Thus, it is possible to take advantage of the special form of the equations characterizing a large-scale dynamical system and to formulate analysis procedures that are more efficient than those which might result from a purely numerical study of the coupled differential equation characterizing the entire system.

Such a procedure, namely a **relaxation scheme**, wherein one integrates each of the component differential equations by separately iterating through the connection equations to obtain the solution of the overall system, was implemented over 30 years ago at the Marshall Space Flight Center (Prasad and Reiss, 1970; Trauboth and Prasad, 1970; Saeks, 1969). The feasibility of the scheme was verified, and significant computer memory savings resulted. The key to this approach of the numerical analysis of a large-scale system is that numerical routines are applied to the decoupled composite component differential equations rather than an overall equation for the entire system. Thus, the complexity of the numerical computations is determined by the complexity of the largest component in the system. In contrast, if the numerical procedures are applied directly to a set of differential equations characterizing the entire system, the complexity of the numerical computations is determined by the complexity of the entire system (the sum of the complexities of all the components).

Since 1970, computer simulation software has evolved substantially, and packages that deal with large-scale systems either explicitly or implicitly exploit the component connection philosophy. One of the more recent software developments for modeling and simulation of large-scale heterogeneous systems is an object-oriented language called Modelica; see Mattson *et al.* (1998) for additional information.

7.5 Fault Analysis

Fault analysis is similar to system identification in that the system is reidentified to determine if the component input/output relations have changed. It should be clear that connection information is essential to detection of a faulty component. As before, note that fault analysis is not restricted to linear systems. To illustrate some of the technique, the discussion will focus on a linear time-invariant situation.

To attack the fault analysis problem, assume that there is a given set of external system parameters measured at a finite set of frequencies $S(\omega_i)$, $i = 1, 2, \ldots, n$ and that there are the connection matrices, L_{11}, L_{12}, L_{21}, and L_{22}. Compute or approximate $Z(\omega)$. For simplicity of presentation, assume that $L_{22} = 0$. There is no loss of generality with this assumption since one can always replace $S(\omega_i)$ by $\tilde{S}(\omega_i) = S(\omega_i) - L_{22}$ and then work with the system whose measured external parameters are given by $\tilde{S}$ and whose connection matrices are $\tilde{L}_{11} = L_{11}$, $\tilde{L}_{12} = L_{12}$, $\tilde{L}_{21} = L_{21}$, and $\tilde{L}_{22} = 0$.

In the most elementary form of the fault analysis problem, assume that $S(\omega)$ is given at only a single frequency and compute $Z(\omega)$ exactly. This form of the problem was first studied by Saeks (1970) and then extended by Saeks and colleagues (Ransom and Saeks, 1975; Saeks *et al.*, 1972). Its solution is based on the observation that the connection function f can be decomposed into two functions (under the assumption that $L_{22} = 0$) as:

$$f = hog. \tag{7.12}$$

Here, g is a nonlinear function that maps the component parameter matrix Z into an intermediary matrix R via:

$$R = g(Z) = (1 - ZL_{11})^{-1}Z, \tag{7.13}$$

and h is a linear function that maps R into the external parameter matrix S via:

$$S = h(R) = L_{21}RL_{12}. \tag{7.14}$$

The left inverse of f is given in terms of g and h via:

$$f^{-L} = g^{-L}oh^{-L}. \tag{7.15}$$

A little algebra will reveal that g^{-L} always exists and is given by the formula:

$$Z = g^{-L}(R) = (1 + RL_{11})^{-L}R. \tag{7.16}$$

Thus, the fault isolation problem is reduced to that of finding the left inverse of the linear function h. This is most easily done by working with the matrix representation of h as:

$$\underline{h} = [L_{12}^T \otimes L_{21}], \tag{7.17}$$

where $\otimes$ is the matrix tensor (Kronecker) product (Ransom, 1972; Saeks *et al.*, 1972; Rao and Mitra, 1971). Standard matrix algebraic techniques can be used to compute h^{-L}. Indeed, the rows of $\underline{h}$ can each be identified with external input/output "gains." Thus, the problem of choosing a minimal set of external parameters for fault analysis is reduced to the problem of choosing a minimal set of rows in $\underline{h}$ that render its columns linearly independent (Singh, 1972). There are difficulties with this technique in certain cases, but fortunately, a number of approaches have been developed for alleviating these (Ransom, 1973, 1972; Ransom and Saeks, 1973b).

7.6 Concluding Remarks

Although this chapter has not provided a complete or detailed explanation of component connection philosophy, the reader should now have a fundamental understanding of the power and utility of the philosophy. Even though the component connection approach to modeling large-scale systems originated over three decades ago, it is likely that its full potential has not been realized. There are several areas of research in which the component connection approach might lead to new results. One of these is the area of optimal decentralized control of large-scale systems.

References

Bello, P.A. *et al.* (1973). Nonlinear system modeling and analysis. *Report RADC-TR-73-178*. Rome Air Development Center, AFSC, Griffiss AFB, New York.

Chang, R.R. (1973). System identification from input/output data. *M.S. Thesis*, Department of Electrical Engineering, University of Notre Dame.

Cooper, G.R., and McGillem, C.D. (1971). Probabilistic methods of signal and system analysis. New York: Holt, Rinehart, and Winston.

DeCarlo, R.A. (1983). Sensitivity calculations using the component connection model. *International Journal of Circuit Theory and Applications 12*(3), 288–291.

DeCarlo, R.A., and Rapisarda, L. (1988). Fault diagnosis under A limited-fault assumption and limited test-point availability. *IEEE Transactions on Circuits Systems and Signal Processing 4*(4), 481–509.

DeCarlo, R.A., and Saeks, R.E. (1981). *Interconnected dynamical systems*. New York: Marcel Dekker.

Garzia, R.F. (1971). Fault isolation computer methods. *NASA Technical Report CR-1758*. Marshall Space Flight Center, Huntsville.

Lee, Y.W., and Schetzen, M. (1961). Quarterly progress report 60. *Research Laboratory of Electronics*. Cambridge: MIT.

Liberty, S.R., and Saeks, R.E. (1973). The component connection model in system identification, analysis, and design. *Proceedings of the Joint EMP Technical Meeting*. 61–76.

Liberty, S.R., and Saeks, R.E. (1974). The component connection model in circuit and system theory. *Proceedings of the European Conference on Circuit Theory and Design*. 141–146.

Lu, F., and Liu, R.W. (1972). Stability of large-scale dynamical systems. *Technical Memorandum EE-7201*, University of Notre Dame.

Mattson, S.E., Elmqvist, H., and Otter, M. (1998). Physical system modeling with modelica. *Journal of Control Engineering Practice 6*, 501–510.

Prasad, N.S., and Reiss, J. (1970). The digital simulation of interconnected systems. *Proceedings of the Conference of the International Association of Cybernetics*.

Ransom, M.N. (1972). A functional approach to large-scale dynamical systems. *Proceedings of the 10th Allerton Conference on Circuits and Systems*. 48–55.

Ransom, M.N. (1972). On-state equations of RLC large-scale dynamical systems. *Technical Memorandum EE-7214*, University of Notre Dame.

Ransom, M.N. (1973). A functional approach to the connection of a large-scale dynamical systems. Ph.D. Thesis, University of Notre Dame.

Ransom, M.N., and Saeks, R.E. (1973a). Fault isolation with insufficient measurements. *IEEE Transactions on Circuit Theory CT-20*(4), 416–417.

Ransom, M.N., and Saeks, R.E. (1973b). Fault isolation via term expansion. *Proceedings of the Third Pittsburgh Symposium on Modeling and Simulation.* 224–228.

Ransom, M.N., and Saeks, R.E. (1975). The connection function–theory and application. *IEEE Transactions on Circuit Theory and Applications 3,* 5–21.

Rao, C.R., and Mitra, S.K. (1971). Generalized inverse matrices and its applications. New York: John Wiley & Sons.

Rapisarda, L., and DeCarlo, R.A. (1983). Analog mulitifrequency fault diagnosis. *IEEE Transactions on Circuits and Systems* CAS-30(4), 223–234.

Reisig, D., and DeCarlo, R.A. (1987). A method of analog-digital multiple fault diagnosis. *IEEE Transactions on Circuit Theory and Applications 15,* 1–22.

Richardson, M.H. (1969). Optimization of large-scale discrete-time systems by a component problem method. Ph.D. Thesis, University of Notre Dame.

Richardson, M.H., Leake, R.J., and Saeks, R.E. (1969). A component connection formulation for large-scale discrete-time system optimization. *Proceedings of the Third Asilomar Conference on Circuit and Systems.* 665–670.

Saeks, R.E. (1969). Studies in system simulation. *NASA Technical Memorandum 53868.* George Marshall Space Flight Center, Huntsville.

Saeks, R.E. (1970). Fault isolation, component decoupling, and the connection groupoid. *NASA-ASEE Summer Faculty Fellowship Program Research Reports.* 505–534. Auburn University.

Saeks, R.E., Singh, S.P., and Liu, R.W. (1972). Fault isolation via components simulation. *IEEE Transactions on Circuit Theory CT-19,* 634–660.

Singh, S.P. (1972). Structural properties of large-scale dynamical systems. Ph.D. Thesis, University of Notre Dame.

Singh, S.P., and Trauboth, H. (1973). MARSYAS. *IEEE Circuits and Systems Society Newsletter 7.*

Trauboth, H., and Prasad, N.S. (1970). MARSYAS—A software system for the digital simulation of physical systems. *Proceedings of the Spring Joint Computing Conference.* 223–235.

8

Fault-Tolerant Control

Gary G. Yen
Intelligent Systems and Control Laboratory, School of Electrical and Computer Engineering, Oklahoma State University, Stillwater, Oklahoma, USA

8.1 Introduction

While most research attention has been focused on fault detection and diagnosis, much less research effort has been dedicated to "general" failure accommodation mainly because of the lack of well-developed control theory and techniques for general nonlinear systems. Because of the inherent complexity of nonlinear systems, most of model-based analytical redundancy fault diagnosis and accommodation studies deal with the linear system that is subject to simple additive or multiplicative faults. This assumption has limited the system's effectiveness and usefulness in practical applications. In this research work, the online fault accommodation control problems under catastrophic system failures are investigated. The main interest is dealing with the unanticipated system component failures in the most general formulation. Through discrete-time Lyapunov stability theory, the necessary and sufficient conditions to guarantee the system online stability and performance under failures are derived, and a systematic procedure and technique for proper fault accommodation under the unanticipated failures are developed. The approach is to combine the control technique derived from discrete-time Lyapunov theory with the modern intelligent technique that is capable of **self-optimization** and **online adaptation** for real-time failure estimation. A complete architecture of fault diagnosis and accommodation has also been presented by incorporating the developed intelligent fault-tolerant control (FTC) scheme with a cost-effective fault-detection scheme and a multiple model-based failure diagnosis process to efficiently handle the false alarms and the accommodation of both the anticipated and unanticipated failures in online situations.

8.2 Overview of Fault Diagnosis and Accommodation

Because of the increasing demands of system safety and reliability in modern engineering design, research issues dealing with failure diagnosis and accommodation have attracted significant attention in the control society, as the first page of references at the end of this chapter indicates. System failures caused by unexpected interference or aging of system components will possibly result in changes or changing of system dynamics. Thus, the original design under the fault-free condition is no longer reliable and possibly leads to instability. More serious problems, such as the survivability of the system, may arise when the instability of the system may cause the loss of human life. The research work dealing with the underlying problems is usually referred to as **fault diagnosis and accommodation** (FDA). The major objectives of FDA or FTC is to detect and isolate the encountered failures and to take the necessary actions to prevent the system from getting and unstable and maintain the successful control mission.

Traditional FDA approaches are based on so-called hardware redundancy where extra components are used as backup in case of failures. Because of the additional cost, space, weight, and complexity of incorporating redundant hardware, model-based methods (in the spirit of analytical redundancy) with inexpensive and high performance microprocessors have dominated the FDA research activities. The major reason for the prevalence of this approach is information processing techniques using powerful computing devices and memory systems can be used to establish the necessary redundancy without the need of hardware instrumentation in the system (Polycarpou and Helmicki, 1995). Under the model-based analytical redundancy, system behaviors are compared with the analytically obtained values through a mathematical model. The resulting differences are so-called residuals (Gertler, 1988). In the ideal situation, the residuals will be zeros in the fault-free system, and any deviation will be interpreted as an indication of faults. This is rarely true, however, in practice with the presence of measurement noise disturbances and modeling errors. The deviation can be the combinational results of noises, disturbances, modeling errors, and faults. Naturally, with the presence of significant noises, statistical analysis of the residuals becomes a reasonable procedure to generate a logical pattern called the **signature of the failure** for the proper fault detection and isolation. Many research activities have also been dedicated to investigate a proper residual generation to facilitate the fault isolation process (Garcia and Frank, 1999; Garcia *et al.*, 1998; Gertler and Hu, 1999; Gertler *et al.*, 1995; Gertler and Kunwer, 1995). Generally speaking, residual generation, statistical testing, and logical analysis are usually combined as the three stages of the fault detection and isolation (Gertler, 1988). Due to the inherent complexity of nonlinear systems, most model-based analytical redundancy fault diagnosis studies deal with the linear system that is subject to simple additive or multiplicative faults. This assumption has limited its effectiveness and usefulness in practical applications (Polycarpou and Helmicki, 1995; Polycarpou and Vemuri, 1995). A series of research works that is devoted to more general failure cases is reported in Polycarpou and Helmicki (1995), Polycarpou and Vemuri (1995), Zhang *et al.*, (1999, 2000), Vemuri and Polycarpou (1997) and Trunov and Polycarpou (1999). Although significant progress has been made in the theoretical analysis for fault detection and sensitivity conditions, more practical and complicated fault detection and diagnosis (FDD) problems, such as detection and diagnosis of possible multiple failures, still remain to be solved. The representative fault diagnosis methods are shown in Figure 8.1.

Similar to fault detection and diagnosis research work, most of the fault accommodation schemes are mainly designed based on the powerful and well-developed linear design methodology to obtain the desired objectives. Typical approaches for failure accommodation techniques include the pseudo-

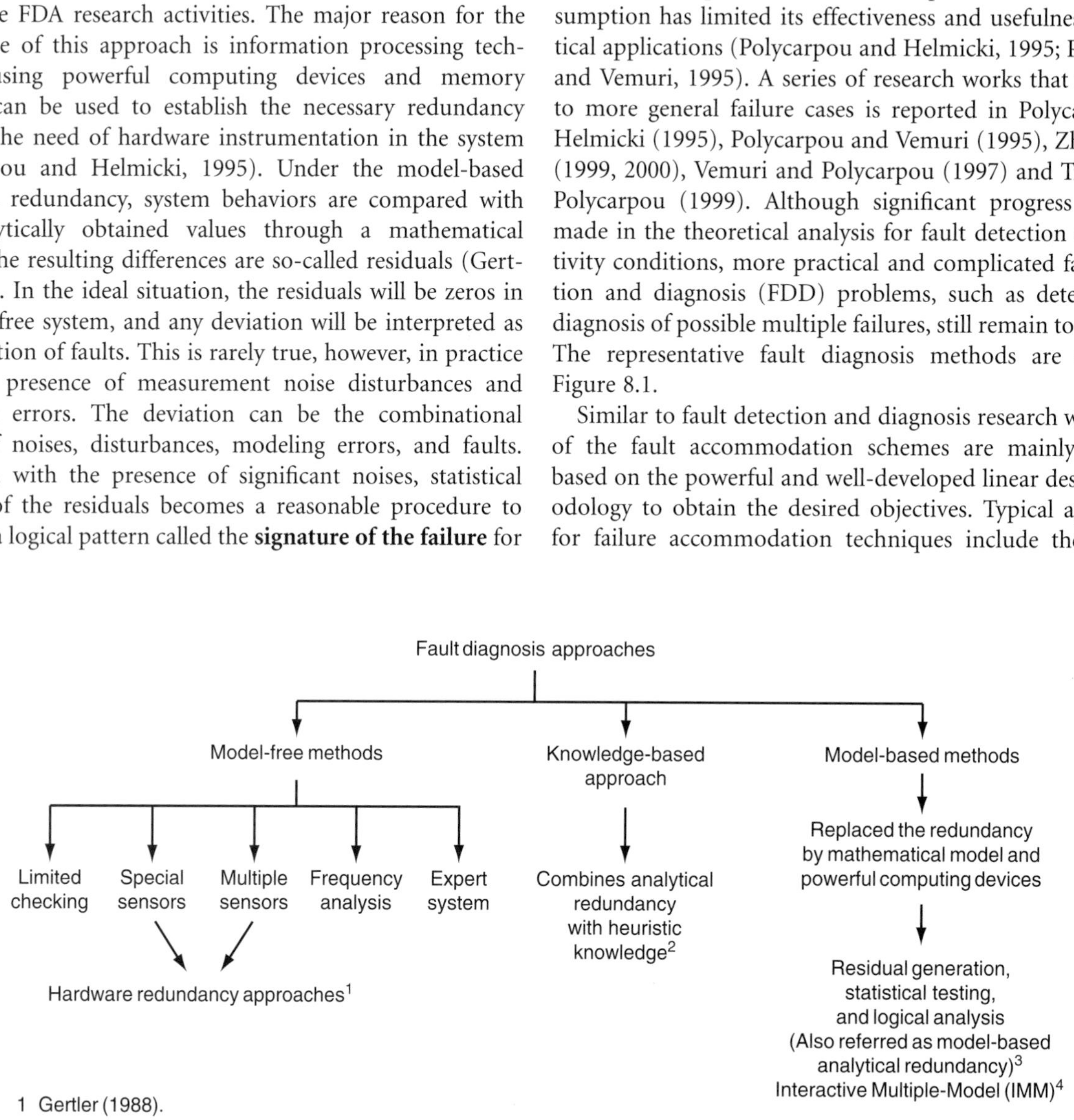

1 Gertler (1988).
2 Frank (1990).
3 Chow and Willsky (1984); Emani-Naeini (1988); Ge and Feng (1988); Lou et al (1986); Frank (1990); Gertler (1988); Polycorpou and Helmicki (1995); Polycarpou and Vemuri (1995); Jiang (1994); Garcia and Frank (1999); Garcia et al. (1998); Gertler and Hu (1999); Gertler et al. (1995); Gertler and Kunwer (1995); Zitzler and Thiele (1999); Maybeck and Stevens (1991); Zhang and Li (1998); Zhang and Jiang (1999); Laparo et al. (1991).
4 Zhang and Li (1998); Zhang and Jiang (1999); Laparo et al. (1991).

FIGURE 8.1 Typical FD Methods

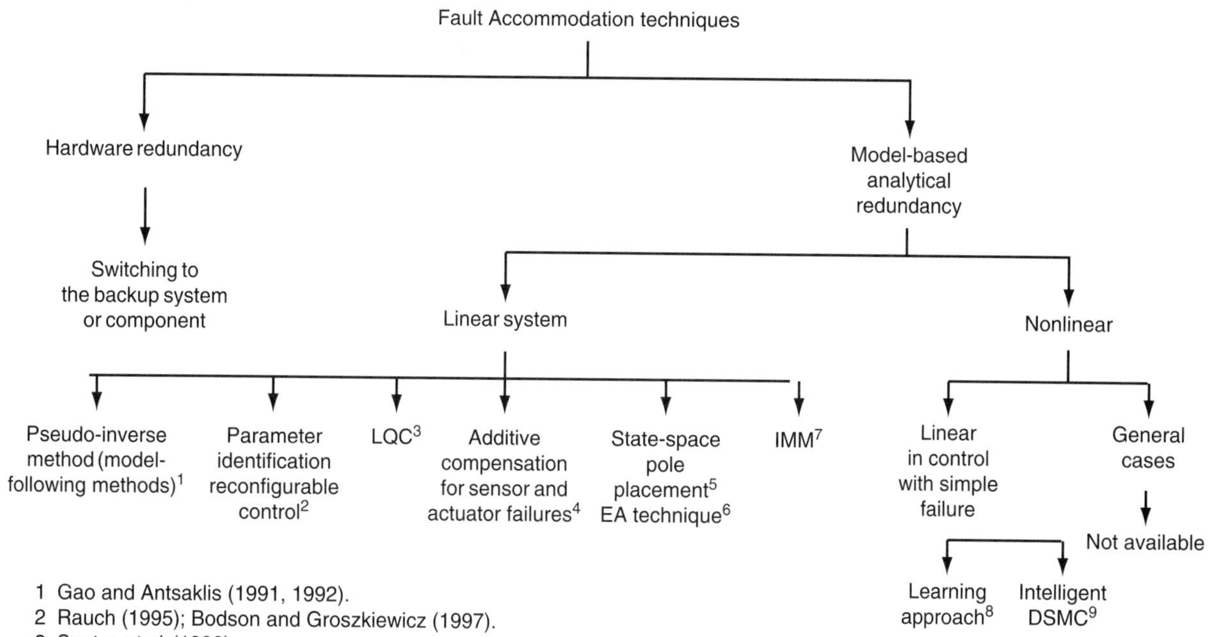

1 Gao and Antsaklis (1991, 1992).
2 Rauch (1995); Bodson and Groszkiewicz (1997).
3 Sauter *et al.* (1998).
4 Noura *et al.* (2000); Theilliol *et al.* (1998).
5 Jiang and Zhao (1998, 1999).
6 Jiang (1994).
7 Maybeck and Stevens (1991); Polycarpal and Vemuri (1995).
8 Polycarpou and Helmicki (1995); Polycarpou and Vemuri (1995).
9 Zhang *et al.* (1999).

FIGURE 8.2 Typical FA Approaches

inverse method or model-following method (Gao and Antsaklis, 1991, 1992), eigenstructure assignment (Jiang, 1994), LQC (Sauter *et al.*, 1998), additive compensation for sensor and actuator failures (Noura *et al.*, 2000; Theilliol *et al.*, 1998), reconfiguration control with parameter identification (Rauch, 1995; Bodson and Grosz-Kiewicz, 1997), state-space pole placement (Jiang and Zhao, 1998, 1999); and (interactive) multiple model method (IMM) (Maybeck and Stevens, 1991; Zhang and Li, 1998; Zhang and Jiang, 1999) as shown in Figure 8.2. However, this is rarely the case in practice since all the systems are inherently nonlinear, and the system dynamics under failure situations are more likely to be nonlinear and time varying. The failure situations can be further categorized into anticipated and unanticipated faults where the anticipated ones are referred to as the known faults based on the prior knowledge of the system or possibly the history of system behavior, and the unanticipated ones are the unexpected failure situations, which have to be identified online. In general, the recognition and accommodation of the anticipated failures are considered relatively easier to solve because of the availability of the prior information and sufficient time for the development of solutions (i.e., off-line). The details of the systematic procedure of the proper failure isolation and accommodation for anticipated faults, such as the generation of residuals, fault signature identification, and the selection logic of the proper control actions, however, are still left unanswered. For unanticipated failures under general format, the online identification and accommodation are even more difficult and rarely touched. When a dynamic system encounters failures possibly caused by unexpected interferences, it is not a reasonable approach to assume certain types of dynamic change caused by those unexpected failures. Of course, the faulty system is possibly uncontrollable if the failure is very serious and fatal. It is absolutely crucial to take early actions to properly control the system behavior in time to prevent the failure from causing more serious loss if the system under failures is controllable at that time.

Because the successful fault accommodation relies on precise failure diagnosis and both processes have to be successfully achieved in the *online and real-time* situation for the unanticipated failures, it has been almost impossible to accomplish and guarantee the system safety based solely on the contemporary control technique with insufficient and/or imprecise information of the failure. Nevertheless, it is believed that technology breakthrough will not happen overnight. Instead, it occurs with slow progress, one step at a time. An early work (Yen and Ho, 2000), where two online control laws were developed and reported for a special nonlinear system with unanticipated component failures, focuses on dealing with the **general unanticipated system component failures** for the **general nonlinear** system. Through discrete-time Lyapunov stability theory, the necessary and sufficient conditions to guarantee the system online stability and performance under failures were derived, and a systematic procedure and technique for proper fault accommodation under unanticipated failures were developed. The approach is to combine the control technique derived from discrete-time Lyapunov stability theory with the modern intelligent technique that is capable of **self-optimization** and **online adaptation** for real-time failure estimation. A more complete architecture of fault diagnosis and accommodation has also been presented by incorporating the developed intelligent fault-tolerant control technique with a cost-effective fault detection scheme and a multiple model-based failure diagnosis process to efficiently handle the false alarms, to accommodate the anticipated failures, and to reduce the unnecessary control effort and computational complexity in online situations.

This chapter is organized as follows. In Section 8.3, the online fault accommodation control problems of interest are defined. In Section 8.4, through theoretical analysis, the necessary and sufficient conditions to maintain the system's online stability are derived together with the proposed intelligent fault accommodation technique. The complete multiple model-based FDA architecture is presented in Section 8.5 with the suggested cost-effective fault detection and diagnosis scheme. An online simulation study is provided in Section 8.6 to demonstrate the effectiveness of the proposed online accommodation technique in various failure scenarios. The conclusion is included in Section 8.7 with the discussion of the current and future research directions.

8.3 Problem Statement

A general n-input and m-output dynamic system can be described by equation 8.1.

$$\begin{aligned} &y_l(k+d) = f_l(\bar{y}_1, \bar{y}_2, \ldots \bar{y}_m, \bar{u}_1, \bar{u}_2, \ldots \bar{u}_n).\\ &\bar{y}_i = \{y_i(k+d-1), y_i(k+d-2), \ldots, y_i(k+d-p_i)\}.\\ &\bar{u}_j = \{u_j(k), u_j(k-1), \ldots u_j(k-q_j)\}.\\ &p_i, q_j \in \Re^+, i = 1, 2, \ldots, m., j = 1, 2, \ldots, n., \text{ and } l = 1, 2, \ldots, m. \end{aligned} \quad (8.1)$$

In equation 8.1, $f_l: \Re^P \times \Re^Q \mapsto \Re$, with $P = \sum_{i=1}^m p_i$ and $Q = \sum_{j=1}^n q_j$ as the mathematical realization of the system dynamics for the lth output. The $y_l, y_i, u_j \in \Re$ are the lth and ith system outputs and jth input, respectively. The d is the relative degree of the system (the smallest delay from the input signal to the system output). In general, f_l may not be readily available in mathematical format all the time due to the difficulty of modeling a complex dynamic system. It is possible, however, to develop a realization to describe the system behavior off-line with a known bounded uncertainty in the desired working region of the system using all the existing modeling techniques; technique include the modern intelligent technology, such as artificial neural networks or fuzzy logics provided there is enough computing resource and there is sufficient time for the development of the realization (Narendra and Parthasathy, 1990; Narendra and Mukhopaelhyay, 1997; Nie and Linkens, 1993) as shown in equation 8.2:

$$y_l(k+d) = \hat{f}_l(\bar{y}_1, \bar{y}_2, \ldots, \bar{y}_m, \bar{u}_1, \bar{u}_2, \ldots \bar{u}_n) + \eta_l(y, u), \quad (8.2)$$

where $\|\eta_1(y, u)\|_2 \leq \delta_0$, $\forall (y, u) \subseteq (\mathbf{Y}, \mathbf{U})$, and $(\mathbf{Y}, \mathbf{U})$ represent the desired working region, and $\delta_0 \in \Re$ is a known constant. Thus, $\hat{f}_l$, the realization of the real system with a known bounded uncertainty in the desired working region of the system, will be either a mathematical, numerical or a combined realization; it is assumed that this realization is developed off-line and available. Equation 8.1 denotes a healthy system under the fault-free situation and equation 8.2 is the corresponding nominal model. Under different component failures, the system dynamics are represented by the following equation:

$$\begin{aligned} y_l(k+d) &= f_l(\bar{y}_1, \bar{y}_2, \ldots \bar{y}_m, \bar{u}_1, \bar{u}_2, \ldots \bar{u}_n)\\ &+ \sum_{v=1}^{r} \beta_v^l(k - T_v^l) F_v^l(\bar{y}_1, \bar{y}_2, \ldots \bar{y}_m, \bar{u}_1, \bar{u}_2, \ldots \bar{u}_n, k), \end{aligned} \quad (8.3)$$

where $F_v^l(\cdot): \Re^P \times \Re^Q \times \Re^+ \mapsto \Re$ with $P = \sum_{i=1}^m p_i$ and $Q = \sum_{j=1}^n q_j$ representing the dynamic change (a general **time-varying** function depends on past system outputs, past control inputs, and the current control input) caused by the unknown and possibly unanticipated failure mode ^{v} for the lth output. The $F_v^l(\cdot)$, $\beta_v^l(\cdot)$, and T_v^l are assumed unknown due to the possible occurrence of unanticipated failures. The r is the number of system failures. All the cases in which $r > 1$ are referred to as multiple-failure cases. Two typical faults, incipient faults and abrupt faults, are considered to be involved online. Their characteristics can be described by the time-varying constant gain, $\beta_v^l(\cdot)$, shown in Figure 8.3.

In Figure 8.3, $\alpha_v^l \in \Re^+$ is an unknown constant that defines the time profile of the incipient failure mode v, and $U(k)$ denotes the unit step function. Abrupt failures are used to represent the sudden change of the system dynamics, and the incipient failures are used to describe the slow-varying aging effect of the system component. The control objective is to generate appropriate control signals to stabilize the system and, possibly, drive the system outputs back to the desired trajectories, $y_1(k+d) \in \Re$, $\forall l = 1, 2, \ldots, m$, in online situations with the presence of the abrupt and/or incipient faults. Under the presence of component failures, the system dynamics may change dramatically or keep changing with time, and the change cannot be detected until it actually happens. Thus,

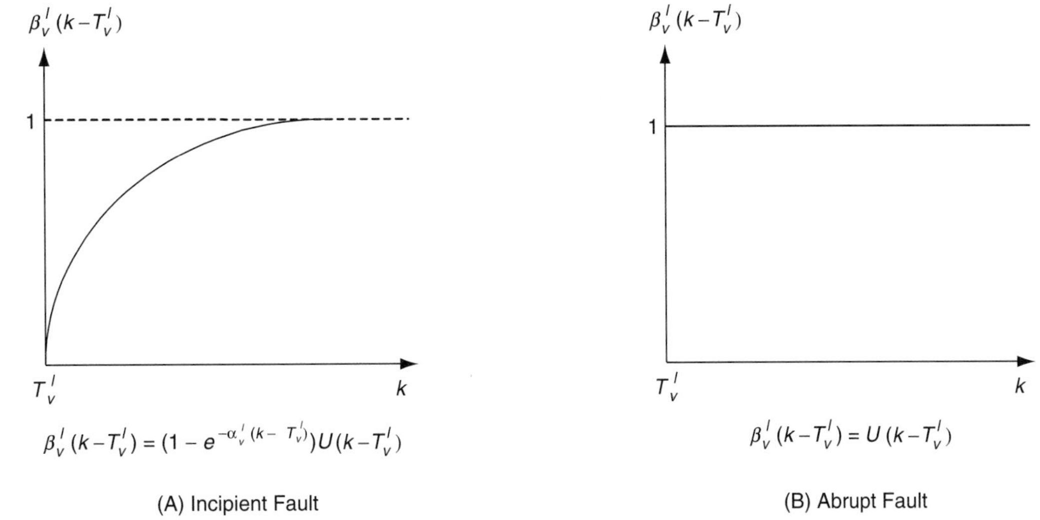

FIGURE 8.3 Time-Varying Constant Gains

the fault accommodation problem has to be solved in the online real-time fashion, which significantly increases the complexity and technical difficulty. To avoid instability problems caused by failures, the online, real-time adjustment of the control law or the reconfiguration of the control actions becomes essential.

8.4 Online Fault Accommodation Control

In this section, the general online fault accommodation problems will be analyzed from both theoretical and realistic points of view. The major focus of interest is to develop an online failure accommodation technique for the general nonlinear dynamic system under general unanticipated catastrophic system failures.

Without loss of generality, let $d = 1$ and consider the SISO system to simplify the analysis. Define a sliding surface function representing the desired dynamics as shown in equation 8.4:

$$S(k) = \frac{y_d(k) - y_d(k-1)}{\Delta t} - \frac{y(k) - y(k-1)}{\Delta t} + a(y_d(k) - y(k)), \tag{8.4}$$

where $y_d(k)$ and $y(k)$ represent the desired system output and the actual system output at time step k, respectively. The Δt represents the sampling period, and $a \in \Re^+$ defines how fast the system output will converge to the desired output. If Lyapunov function is defined as $V(y(k)) = S^2(y(k))$, make $V(y(k+1)) \leq V(y(k))$ and $\forall k \geq 0$, and the sliding surface will be an equation of the tracking error. This implies that the desired dynamics represented by $S = 0$ are attractive, and the desired trajectory can be obtained.

Apparently, the controller's design objective becomes seeking the control input that will satisfy $S^2(k+1) < S^2(k)$, which is equivalent to $[S(k+1) + S(k)][S(k+1) - S(k)] < 0$. This is the same as satisfying the following inequalities:

$$\begin{aligned} -S(k) < S(k+1) < S(k) \quad &\text{when } S(k) > 0. \\ S(k) < S(k+1) < -S(k) \quad &\text{when } S(k) < 0. \end{aligned} \tag{8.5}$$

For $S(k) > 0$, plugging in equation (8.4) yields:

$$\begin{aligned} -S(k) < \frac{y_d(k+1) - y_d(k)}{\Delta t} - \frac{y(k+1) - y(k)}{\Delta t} \\ + a(y_d(k+1) - y(k+1)) < S(k). \end{aligned} \tag{8.6}$$

Reorganizing the inequality results in:

$$-S(k) - \bar{Y}(k) < (-a - \frac{1}{\Delta t})y(k+1) < S(k) - \bar{Y}(k), \tag{8.7}$$

where $\bar{Y}(k) = \dfrac{y_d(k+1) - y_d(k) + y(k)}{\Delta t} + ay_d(k+1)$. This can be further simplified as:

$$(\bar{Y}(k) + S(k))\left(a + \frac{1}{\Delta t}\right)^{-1} > y(k+1) > (\bar{Y}(k) - S(k))\left(a + \frac{1}{\Delta t}\right)^{-1}. \tag{8.8}$$

For $S(k) < 0$, the following equation applies:

$$(\bar{Y}(k) - S(k))\left(a + \frac{1}{\Delta t}\right)^{-1} > y(k+1) > (\bar{Y}(k) + S(k))\left(a + \frac{1}{\Delta t}\right)^{-1}. \tag{8.9}$$

Notice that the left-hand side and the right-hand side of the inequalities in equations 8.8 and 8.9 are known and can be computed at each time-step. Thus, the online fault-tolerant control problems become finding the effective control signal that satisfies the inequality of equation 8.8 when $S(k) > 0$ or equation 8.9 when $S(k) < 0$ at every time-step.

Let $y(k+1) = \bar{\Theta}[\bar{y}, \bar{u}]$, which represents the system dynamics under failures, where $\bar{y}$ and $\bar{u}$ represent the regression vectors of system outputs and inputs, respectively. Based on the implicit function theorem (Apostol, 1974), the control law can be written as $u(k) = \bar{G}[\bar{y}, \bar{Y}(a + 1/\Delta t)^{-1}, \bar{u}\backslash\{u(k)\}]$ provided $\bar{G}$ exists [i.e., $\bar{u}\backslash\{u(k)\}$ denotes the set containing the regression vector of input excluding the current item, $u(k)$. Because the nonexistence of $\bar{G}$ corresponds to the cases where the system becomes uncontrollable under failure situations, the existence problem becomes trivial. Unfortunately, the realization of $\bar{G}$ cannot be provided without knowing the true structures of the system dynamics and the failure dynamics. Thus, the control law cannot be implemented in reality. Through the modern computational intelligence approaches, the effective control input satisfying inequalities of equation 8.8 or 8.9 can be found using optimization algorithms without the complete realization of $\bar{G}$. The systematic procedure is described as follows.

The unknown failure dynamics can be realized through an online estimator. The true system output, $y(k+1)$, can be approximated by the sum of the outputs from the nominal model, $N\hat{y}(k+1)$, and the online estimator, $nf\hat{y}(k+1)$, as follows:

$$\begin{aligned} y(k+1) &= Ny(k+1) + fy(k+1) \\ &= N\hat{y}(k+1) + N\tilde{y}(k+1) + nfy(k+1) + nf\tilde{y}(k+1). \end{aligned} \tag{8.10}$$

In equation 8.10, the following is true:

- $Ny(k+1)$ is the output of the actual system.
- $fy(k+1)$ is the output of the failure dynamics.
- $N\hat{y}(k+1)$ is the output of the nominal model.
- $N\tilde{y}(k+1)$ is the remaining uncertainty between the nominal system and the nominal model.
- $nfy(k+1)$ is the output of the online estimator.
- $nf\tilde{y}(k+1)$ is the remaining uncertainty between the estimator and the failure dynamics.

In addition, the following applies:

$$\begin{aligned} Ny(k+1) &= N\hat{y}(k+1) + N\tilde{y}(k+1);\ fy(k+1) \\ &= nfy(k+1) + nf\tilde{y}(k+1). \end{aligned}$$

The using equation 8.10, the inequalities become for $s(k) > 0$:

$$\begin{aligned} &(S(k)+\bar{Y}(k))\left(a+\frac{1}{\Delta t}\right)^{-1} > N\hat{y}(k+1) + N\tilde{y}(k+1) + nfy(k+1) \\ &+nf\tilde{y}(k+1) > (\bar{Y}(k) - S(k))\left(a+\frac{1}{\Delta t}\right)^{-1}. \end{aligned} \tag{8.11}$$

For $S(k) < 0$, the inequalities become the following:

$$\begin{aligned} &(\bar{Y}(k) - S(k))\left(a+\frac{1}{\Delta t}\right)^{-1} > N\hat{y}(k+1) + N\tilde{y}(k+1) + nfy(k+1) \\ &+ nf\tilde{y}(k+1) > (\bar{Y}(k) + S(k))\left(a+\frac{1}{\Delta t}\right)^{-1}. \end{aligned} \tag{8.12}$$

Modern intelligent optimization techniques, such as genetic algorithms, immune algorithms, simulated annealing, and reinforcement learning, have been exploited in a variety of areas and applications (Chun *et al.*, 1997; Chun *et al.*, 1998; Marchesi *et al.*, 1994; Tanaka and Yoshida, 1999; Marchesi, 1998; Juang *et al.*, 2000; Offner, 2000; Ho and Huang, 2000). Although the effectiveness in achieving successful optimization objectives has been demonstrated, most of them are still applied in off-line situations due to the time-consuming iterative process. From the computational complexity point of view, and the well-known and efficient gradient descent algorithm will be considered and used in the rest of this chapter because of its popularity and effectiveness in online applications. The optimization procedure is shown as follows.

The desired point at every time-step is written as:

$$\begin{aligned} \text{Desire}(k) &= \left[(\bar{Y}(k) + S(k))\left(a+\frac{1}{\Delta t}\right)^{-1}\right. \\ &\left. +(\bar{Y}(k) - S(k))\left(a+\frac{1}{\Delta t}\right)^{-1}\right]/2 \\ &= \bar{Y}(k)\left(a+\frac{1}{\Delta t}\right)^{-1}. \end{aligned} \tag{8.13}$$

Define the error as:

$$\begin{aligned} \text{Error}(k) &= \text{Desire}(k) - N\hat{y}(k+1) - N\tilde{y}(k+1) \\ &\quad - nfy(k+1) - nf\tilde{y}(k+1). \end{aligned} \tag{8.14}$$

The effective control input can be searched based on the gradient descent algorithm for square error:

$$\begin{aligned} \frac{\partial \text{Error}(k)^2}{\partial u(k)} &= 2\ \text{Error}(k)\frac{\partial\ \text{Error}(k)}{\partial u(k)} \\ &= -2\ \text{Error}(k)\left[\frac{\partial N\hat{y}(k+1)}{\partial u(k)} + \frac{\partial N\tilde{y}(k+1)}{\partial u(k)}\right. \\ &\left. + \frac{\partial nfy(k+1)}{\partial u(k)} + \frac{\partial nf\tilde{y}(k+1)}{\partial u(k)}\right]. \end{aligned} \tag{8.15}$$

The resulting control input will be updated by:

$$u(k)_{\text{new}} = u(k)_{\text{old}} - \alpha\frac{\partial\ \text{Error}(k)^2}{\partial u(k)}, \tag{8.16}$$

where α is the learning rate parameter. The searching procedure is repeated until inequalities of equations 8.11 and 8.12 hold, the control input converges, or the maximum number of iterations is reached. Of course, the term $nf\tilde{y}(k+1)$, the remaining uncertainty of the failure dynamics, and the remaining uncertainty of the nominal system $N\tilde{y}(k+1)$ are unknown; the terms $\partial nf\tilde{y}(k+1)/\partial u(k)$ and $\partial N\tilde{y}(k+1)/\partial u(k)$ cannot be computed either. So, the actual searching procedure is based on the approximated values:

$$\hat{\text{Error}}(k) = \text{Desire}(k) - N\hat{y}(k+1) - nf\hat{y}(k+1). \tag{8.17}$$

$$\frac{\partial \hat{\text{Error}}(k)^2}{\partial u(k)} = -2\hat{\text{Error}}(k)\left[\frac{\partial N\hat{y}(k+1)}{\partial u(k)} + \frac{\partial nf\hat{y}(k+1)}{\partial u(k)}\right]. \tag{8.18}$$

At every time step, the desired point of $\bar{Y}(k)(a+1/\Delta t)^{-1}$ is computed, and the effective control signal is searched to ensure the actual result is as close to the desired point as possible through the realizations of the nominal system dynamics and failure dynamics. Observing inequalities of equations 8.11 and 8.12 closely, if $nf\tilde{y}(k+1)$ and $N\tilde{y}(k+1)$, the remaining uncertainty of the failure dynamics and the nominal system are bounded, combining these results with equations 8.13 and 8.14 determines that the desired dynamics, represented by S function, are also bounded. This can be proven and summarized in theorem 8.1:

8.4.1 Theorem 1

If the nominal system model is accurate and precise enough such that $N\tilde{y}(k+1)$, the remaining uncertainty of the nominal system, is bounded by $\sup_{\forall k>T_f}\{|N\tilde{y}(k+1)|\}$ and the online estimator is accurate enough such that the remaining uncertainty of the failure dynamics, $nf\tilde{y}(k+1)$, is bounded by the least upper bound, $\sup_{\forall k>T_f}\{|nf\tilde{y}(k+1)|\}$, and $\Delta Error(k)$, then the error after the searching effort of the optimization algorithm is finite (i.e., bounded by $\sup_{\forall k>T_f}\{|\Delta Error(k)|\}$). In addition, the system stability after time-step T_f under arbitrary unanticipated system failures is guaranteed in an online situation, and the sliding surface function, S, defined by the system performance error is also bounded as follows:

$$\boldsymbol{\Sigma \le S(k+1) \le \Xi},$$

where the following two equations apply:

$$\Sigma = -\left[\sup_{\forall k>T_f}\{|N\tilde{y}(k+1)|\} + \sup_{\forall k>T_f}\{|nf\tilde{y}(k+1)|\} + \sup_{\forall k>T_f}\{|\Delta\text{Error}(k)|\}\right]\left(a+\frac{1}{\Delta t}\right).$$

$$\Xi = \left[\sup_{\forall k>T_f}\{|N\tilde{y}(k+1)|\} + \sup_{\forall k>T_f}\{|nf\tilde{y}(k+1)|\} + \sup_{\forall k>T_f}\{|\Delta\text{Error}(k)|\}\right]\left(\frac{a+1}{\Delta t}\right).$$

The sliding surface function, S, is defined as:

$$S(k) = \frac{y_d(k)-y_d(k-1)}{\Delta t} - \frac{y(k)-y(k-1)}{\Delta t} + a(y_d(k)-y(k));\ a>0,$$

where $y_d(k)$ is the desired trajectory at time step k.

Proof

Let $\Delta\text{Error}(k)$ represent the error after the searching effort of the optimization algorithm. Then:

$$\Delta\text{Error}(k) = \text{Desire}(k) - N\hat{y}(k+1) - nf\hat{y}(k+1).$$

By equation 8.13:

$$\bar{Y}(k)\left(a+\frac{1}{\Delta t}\right)^{-1} - \Delta\text{Error}(k) = N\hat{y}(k+1) + nf\hat{y}(k+1). \tag{8.19}$$

For $S(k) > 0$ plugging in equation 8.19 into inequality 8.11, the result is as follows:

$$(\bar{Y}(k)+S(k))\left(a+\frac{1}{\Delta t}\right)^{-1} > \bar{Y}(k)\left(a+\frac{1}{\Delta t}\right)^{-1} - \Delta Error(k) + N\tilde{y}(k+1) + nf\tilde{y}(k+1) > (\bar{Y}(k)-S(k))\left(a+\frac{1}{\Delta t}\right)^{-1}.$$

Simplifing the inequality yields these two functions:

$$S(k) > [N\tilde{y}(k+1) + nf\tilde{y}(k+1) - \Delta\text{Error}(k)]\left(a+\frac{1}{\Delta t}\right).$$
$$-S(k) < -[N\tilde{y}(k+1) + nf\tilde{y}(k+1) - \Delta\text{Error}(k)]\left(a+\frac{1}{\Delta t}\right). \tag{8.20}$$

Since $S(k) > 0$, $-S(k) < [\sup_{\forall k>T_f}\{|N\tilde{y}(k+1)|\} + \sup_{\forall k>T_f}\{|nf\tilde{y}(k+1)|\} + \sup_{\forall k>T_f}\{|\Delta\text{Error}(k)|\}](a+1/\Delta t$ is always true.

By assumptions 1, 2, and 3, the following inequalities will hold for the worst condition:

$$S(k) > [\sup_{\forall k>T_f}\{|N\tilde{y}(k+1)|\} + \sup_{\forall k>T_f}\{|nf\tilde{y}(k+1)|\} + \sup_{\forall k>T_f}\{|\Delta\text{Error}(k)|\}]\left(a+\frac{1}{\Delta t}\right). \tag{8.21}$$

Apparently, $[\sup_{\forall k>T_f}\{|N\tilde{y}(k+1)|\} + \sup_{\forall k>T_f}\{|nf\tilde{y}(k+1)|\} + \sup_{\forall k>T_f}\{|\Delta\text{Error}(k)|\}](a+1/\Delta t) = \inf_{\forall k>T_f}\{S(k)\}$, which is the greatest lower bound of $S(k)$. In addition, $-[\sup_{\forall k>T_f}\{|N\tilde{y}(k+1)|\} + \sup_{\forall k>T_f}\{|nf\tilde{y}(k+1)|\} + \sup_{\forall k>T_f}\{|\Delta\text{Error}(k)|\}](a+1/\Delta t) = \sup_{\forall k>T_f}\{-S(k)\}$, which is the least upper bound of $-S(k)$. Since $S(k) > S(k+1)$ and $-S(k) < -S(k+1)$, the following inequalities will always hold:

$$[\sup_{\forall k>T_f}\{|N\tilde{y}(k+1)|\} + \sup_{\forall k>T_f}\{|nf\tilde{y}(k+1)|\} + \sup_{\forall k>T_f}\{|\Delta\text{Error}(k)|\}]\left(a+\frac{1}{\Delta t}\right) \geq S(k+1).$$
$$-[\sup_{\forall k>T_f}\{|N\tilde{y}(k+1)|\} + \sup_{\forall k>T_f}\{|nf\tilde{y}(k+1)|\} + \sup_{\forall k>T_f}\{|\Delta\text{Error}(k)|\}](a+\frac{1}{\Delta t}) \leq -S(k+1). \quad (8.22)$$

These hold true for both situations, $S(k+1) > 0$ and $S(k+1) < 0$, which implies:

$$\Sigma \leq S(k+1) \leq \Xi, \quad (8.23)$$

where:

$$\Sigma = -[\sup_{\forall k>T_f}\{|N\tilde{y}(k+1)|\} + \sup_{\forall k>T_f}\{|nf\tilde{y}(k+1)|\} + \sup_{\forall k>T_f}\{|\Delta\text{Error}(k)|\}]\left(a+\frac{1}{\Delta t}\right).$$
$$\Xi = [\sup_{\forall k>T_f}\{|N\tilde{y}(k+1)|\} + \sup_{\forall k>T_f}\{|nf\tilde{y}(k+1)|\} + \sup_{\forall k>T_f}\{|\Delta\text{Error}(k)|\}]\left(a+\frac{1}{\Delta t}\right).$$

For $S(k) < 0$, plugging in equation 8.19 into inequality 8.12, the result is the following:

$$(\bar{Y}(k)-S(k))\left(a+\frac{1}{\Delta t}\right)^{-1} > \bar{Y}(k)\left(a+\frac{1}{\Delta t}\right)^{-1} - \Delta\text{Error}(k)$$
$$+N\tilde{y}(k+1)+nf\tilde{y}(k+1) > (\bar{Y}(k)+S(k))\left(a+\frac{1}{\Delta t}\right)^{-1}.$$

Simplifing the inequality gets:

$$-S(k) > [N\tilde{y}(k+1)+nf\tilde{y}(k+1)-\Delta\text{Error}(k)]\left(a+\frac{1}{\Delta t}\right).$$
$$S(k) < [N\tilde{y}(k+1)+nf\tilde{y}(k+1)-\Delta\text{Error}(k)]\left(a+\frac{1}{\Delta t}\right). \quad (8.24)$$

Because $S(k) < 0$, $S(k) < [\sup_{\forall k>T_f}\{|N\tilde{y}(k+1)|\} + \sup_{\forall k>T_f}\{|nf\tilde{y}(k+1)|\} + \sup_{\forall k>T_f}\{|\Delta\text{Error}(k)|\}](a+1/\Delta t)$ is always true. By assumptions 1, 2, and 3, the following inequalities will hold for the worst condition:

$$-S(k) > [\sup_{\forall k>T_f}\{|N\tilde{y}(k+1)|\} + \sup_{\forall k>T_f}\{|nf\tilde{y}(k+1)|\} + \sup_{\forall k>T_f}\{|\Delta\text{Error}(k)|\}]\left(a+\frac{1}{\Delta t}\right). \quad (8.25)$$

Apparently, $[\sup_{\forall k>T_f}\{|N\tilde{y}(k+1)|\} + \sup_{\forall k>T_f}\{|nf\tilde{y}(k+1)|\} + \sup_{\forall k>T_f}\{|\Delta\text{Error}(k)|\}]\ (a+1/\Delta t) = \inf_{\forall k>T_f}\{-S(k)\}$, which is the greatest lower bound of -$S(k)$.

In addition $-[\sup_{\forall k>T_f}\{|N\tilde{y}(k+1)|\} + \sup_{\forall k>T_f}\{|nf\tilde{y}(k+1)|\} + \sup_{\forall k>T_f}\{|\Delta\text{Error}(k)|\}](a+1/\Delta t) = \sup_{\forall k>T_f} S(k)\}$, which is the least upper bound of $S(k)$. For both of these equations, since $-S(k) > S(k+1)$ and $S(k) < -S(k+1)$, the following inequalities will always hold:

$$[\sup_{\forall k>T_f}\{|N\tilde{y}(k+1)|\} + \sup_{\forall k>T_f}\{|nf\tilde{y}(k+1)|\} + \sup_{\forall k>T_f}\{|\Delta\text{Error}(k)|\}]\left(a+\frac{1}{\Delta t}\right) \geq S(k+1).$$
$$-[\sup_{\forall k>T_f}\{|N\tilde{y}(k+1)|\} + \sup_{\forall k>T_f}\{|nf\tilde{y}(k+1)|\} + \sup_{\forall k>T_f}\{|\Delta\text{Error}(k)|\}]\left(a+\frac{1}{\Delta t}\right) \leq -S(k+1). \quad (8.26)$$

This is true for both situations, $S(k+1) > 0$ and $S(k+1) < 0$, which implies:

$$\Sigma \leq S(k+1) \leq \Xi, \quad (8.27)$$

where:

$$\Sigma = -[\sup_{\forall k>T_f}\{|N\tilde{y}(k+1)|\} + \sup_{\forall k>T_f}\{|nf\tilde{y}(k+1)|\} + \sup_{\forall k>T_f}\{|\Delta\text{Error}(k)|\}]\left(a+\frac{1}{\Delta t}\right).$$
$$\Xi = [\sup_{\forall k>T_f}\{|N\tilde{y}(k+1)|\} + \sup_{\forall k>T_f}\{|nf\tilde{y}(k+1)|\} + \sup_{\forall k>T_f}\{|\Delta\text{Error}(k)|\}]\left(a+\frac{1}{\Delta t}\right).$$

The result is exactly the same result as for equation 8.23. Thus, the sliding surface function, S, is bounded by the value defined by the least upper bounds of the remaining

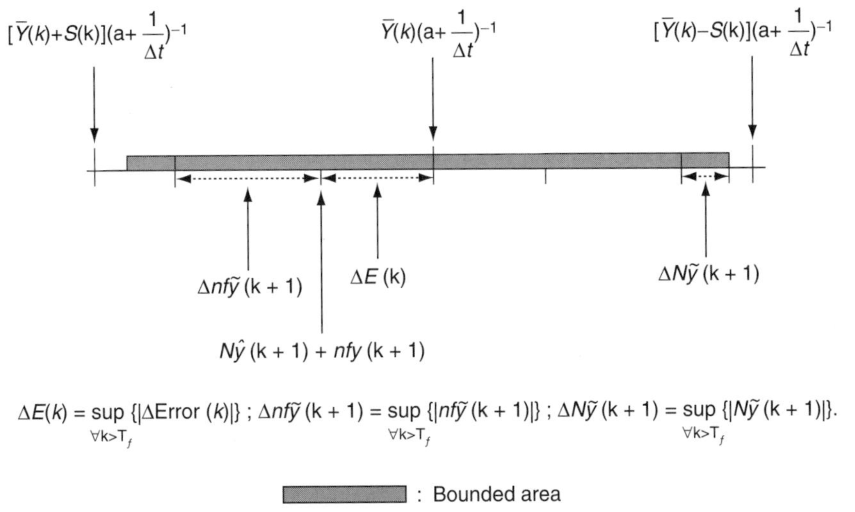

FIGURE 8.4 The Bound of the Sliding Surface Function

uncertainties of the nominal system, the failure dynamics, and the error by the optimization algorithm.

The discrete-time Lyapunov stability theory indicates that the control problem can be solved as long as **the numerical value of the failure dynamics** is realized at each **time-step**, which is a measure of how far the failure drives the system dynamics away from the desired dynamics. Based on the above theoretical analysis, the system under unexpected catastrophic failures can be stabilized online, and the performance can be recovered provided an effective online estimator for the unknown failure dynamics such that the necessary and sufficient condition in theorem 1 is satisfied. Moreover, since the online estimator is used to provide the approximated numerical value of the failure dynamics at each time-step based on the most recent measurements (i.e., the failure may be time varying), no specific structure or dynamics is required for the estimator. In other words, only a static function approximator that approximates the most recent behavior of the failure is needed for the control purpose. Figure 8.4 indicates how the S function is bounded by the upper bounds of the nominal model uncertainty, $\Delta N\tilde{y}(k+1)$, optimization error, $\Delta E(k)$, and the prediction error of the failure dynamics, $\Delta nf\tilde{y}(k+1)$.

8.4.2 Online Learning of the Failure Dynamics

With the universal approximation capability for any piecewise continuous function to any degree of accuracy (Hordik *et al.*, 1989), **artificial neural networks** become one of the most promising candidates for the online control problems of our interest. In this research work, neural network is exploited and used as the online estimator for the unknown failure dynamics. Some important features of the online learning using neural networks should first be addressed here. The structure of the online estimator needs to be decided (i.e., in neural networks, the number of layers, number of neurons in each layer, and neuron transfer functions have to be specified). It is known that neural networks are sensitive to the number of neurons in the hidden layers. Too few neurons can result in underfitting problems (poor approximation), and too many neurons may contribute to an overfitting problem, where all the training patterns are well fit, but the fitting curve may take wild oscillations between the training data points (Demuth and Beale, 1998). The criterion for stopping of the training process is another important issue in the real applications. If the mean square error of the estimator is forced to reach a very small value, the estimator may perform poorly for the new input data slightly away from the training patterns. This is the well-known generalization problem. Besides, in the real applications, the training patterns may contain some noise since they are the measurements from real sensors. The estimator may adjust itself to fit the noise instead of the real failure dynamics. Some methods proposed to improve these problems, such as early stopping criterion and generalization network training algorithms, may be useful to remedy these situations (Demoth and Beale, 1998; Mackey, 1992)

In the online situation, the number of input output data for the training process becomes a very important design parameter. The system dynamics may keep changing because of different fault situations (i.e., the incipient fault, abrupt fault, and multiple faults). Apparently, using all input/output measurements to train the online estimator does not make too much sense since it is possible to use invalid training patterns to mislead the estimator, and it is also unrealistic for online applications. In other words, only finite and limited number of data sets should be used as training patterns to adjust the parameters of the estimator. A reasonable way is to use the most recent input/output measurements. A set, B, that contains the most recent measurements in a fixed length of a time-shifting data window is used to collect the training patterns:

$$B = \{(p(m), t(m) | p \in \Re^S;\ t \in \Re^T;\ k - j + 1 \leq m \leq k\}, \qquad (8.28)$$

where $p(m)$ and $t(m)$ are the network input vector and desired output vector at time-step m, respectively. The k is the current time-step, and j represents the length of the time-shifting data window, which is a design parameter. This parameter has to be decided based on the system computational capability, sampling rate, and the performance requirement. In addition, the maximum number of the effective control signal searching iterations is another important design parameter in real-time applications. The number has to be within an allowable range according to the system computational capacity in the online situation. For gradient descent-type of optimization algorithms, a time-varying learning rate can be used to possibly reduce the searching time.

8.5 Architecture of Multiple Model-Based Fault Diagnosis and Accommodation

Although the online control law in closed form is not available for general online failure accommodation due to the unknown dynamic changes caused by the failures, based on the above analysis and theorem, the successful control mission can still be achieved provided sufficient online computational power and the satisfaction of the necessary and sufficient conditions. In other words, a substantial amount of computational cost may be paid for maintaining the control mission. In many real systems, the failures could be well-known or anticipated according to the history of system behavior and/or the aging degree of the system components. For those known and/or expected faults, it is apparently not necessary to waste the computational source for failure accommodation. Those fault patterns or signatures and the corresponding control actions can be developed off-line and prestored in a database for an online control purpose. The appropriate online failure accommodation actions are then suggested by a fault detection and diagnosis scheme that detects and identifies the failure patterns online. This idea has been adopted in many (interactive) multiple model-based FTC research works (Maybeck and Stevens, 1991; Zhang and Li, 1998; Zhang and Jiang, 1999; Laparo *et al.*, 1991).

Although there is no solid theoretical result to guarantee the stability of multiple model switching, this idea has attracted substantial attention and been widely used in many areas (Murray-Smith and Johansen, 1997). Inspired by the same spirit, a basic architecture of an intelligent fault diagnosis and accommodation framework is suggested as shown in Figure 8.5. The developed intelligent online fault-tolerant

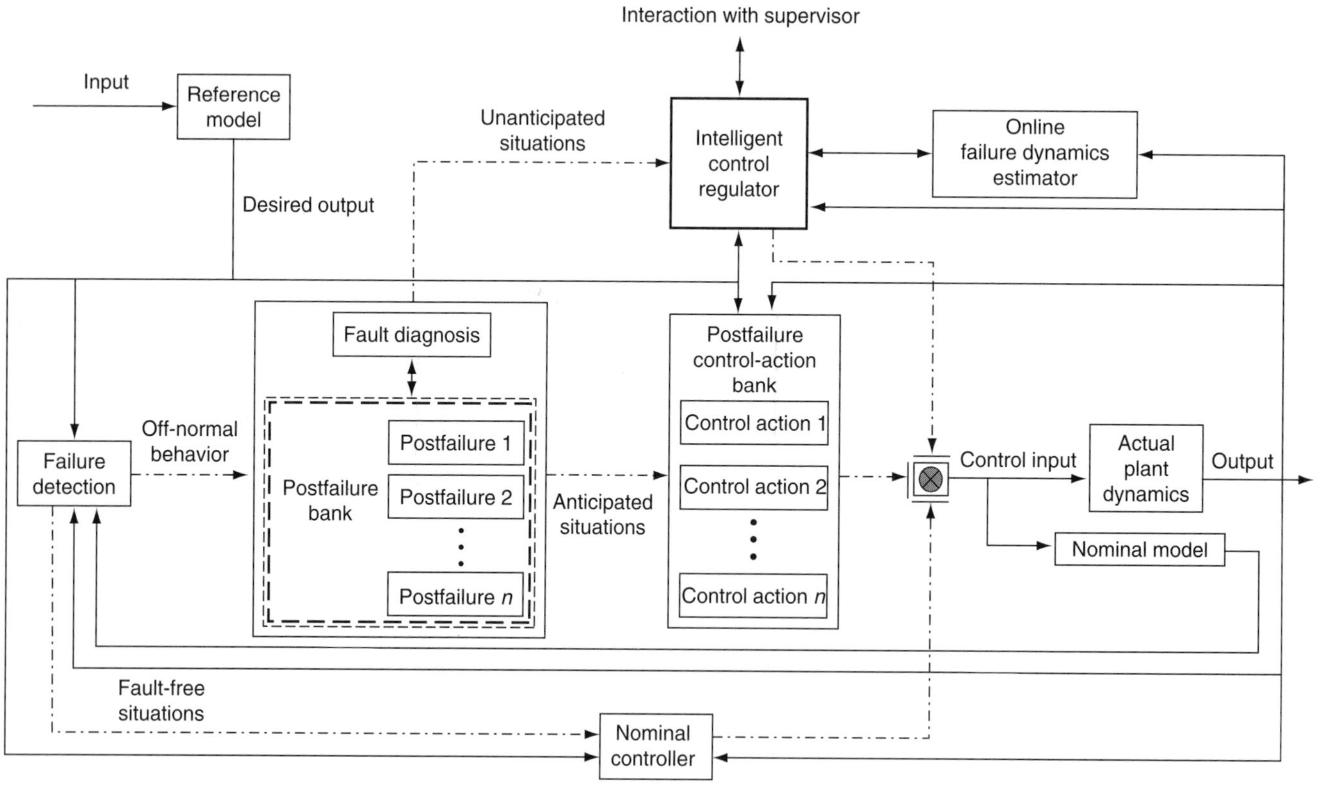

FIGURE 8.5 A Multiple Model-Based Fault Diagnosis and Accommodation Architecture

control technique incorporates a separate fault detection scheme, a failure diagnosis mechanism, and postfailure control actions to form a more sophisticated and complete FDA methodology. An intelligent control regulator is parallel with the postfailure control actions and the nominal controller to emphasize that, in online situations, the effective control actions that may come from one of the three sources [the **nominal controller** (corresponding to fault-free situations), the **post-failure control actions** (corresponding to expected failure modes), and the **intelligent control regulator** (corresponding to the unanticipated failure scenarios)] are decided based on system behavior or health. The intelligent fault accommodation technique will be applied only when it is necessary for control purposes. Under this framework, the unnecessary computational waste for anticipated failures and false alarms could be avoided. A clearer picture of how this idea works is depicted in a flow chart shown in Figure 8.6. The system "health" is continuously monitored by the fault-detection scheme with the knowledge of the nominal system behavior every certain period of time. Any abnormal behavior will trigger the failure diagnosis mechanism to analyze the situation and further decide which control actions should be taken. If the failure is recognized as an expected fault by comparing the failure signatures with those of the existing post failures, the corresponding postfailure control actions will be selected as the current effective control commands. Otherwise, the developed intelligent FTC technique is initialized. Notice that the dashed line shown in Figure 8.5 indicates that only one action will be taken at every time instant.

8.5.1 The Dilemma of Online Fault Detection and Diagnosis

With the presence of measurement noises, disturbances, and modeling errors, the problems of fault detection and diagnosis, such as sensitivity, robustness, false alarm, missed detection, and failure isolability, are substantially difficult issues to solve. In spite of many research efforts dedicated to addressing the FDD problems (Zhang *et al.*, 1999, 2000; Vemuri and Polycarpou, 1997; Trunov and Polycarpou, 1999; Zhang and Li,

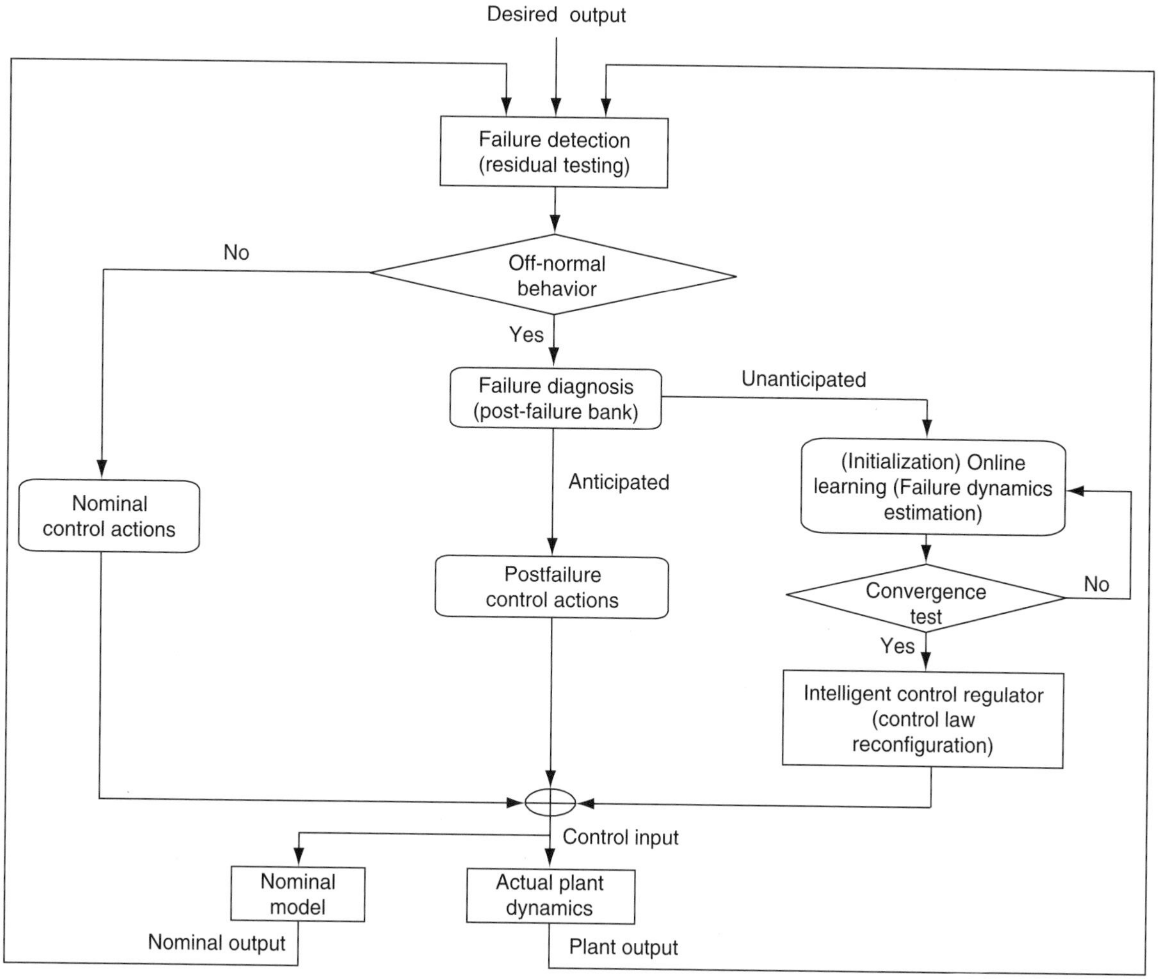

FIGURE 8.6 The Flow Chart of the Multiple Model-Based FDA Framework

1998; Zhang and Jiang, 1999; Laparo *et al.*, 1991), complete online fault detection and diagnosis are still impossible at present due to the inherent complexity of the problems and the time constraint in online operations. Until the invention of breakthrough technology for FDD, from the realistic point of view, finding a better trade-off solution for real implementation based on contemporary technology is the best an engineer can do. Moreover, from the system online safety point of view, false alarms are always preferable over missed detection. Thus, a conservative fault-detection scheme with least computational cost is a good choice to guarantee avoidance of missed detection in online situations. The problem of failure isolation is even more difficult to solve when there is more than one fault involved. Similarly, the conservative diagnostic attitude should be preferable since the price of the misdiagnosis and mistreatment could be instability and unaffordable loss. A possible cost-effective and conservative fault-detection and diagnosis schemes used in the later online simulation section is shown in equations 8.29 and 8.30, respectively:

$$\psi_f = \frac{1}{\omega_f} \sum_{k=k_0-\omega_f+1}^{k_0} (ny(k) - y(k))^2. \quad (8.29)$$

$$\psi_{f\mathrm{diag}_i} = \sum_{k=k_0-\omega_{f\mathrm{diag}+1}}^{k_0} ([y(k) - ny(k)] - pf_i(k))^2. \quad (8.30)$$

In equations 8.29 and 8.30, k_0 is the current time-step, pf_i, and ny represent the time domain signatures of the postfailure model i and the nominal model, respectively. Equation 8.29 will examine the system's state every ω_f time-steps. In cases of failure alarms caused by unexpected disturbances or measurement noises, the detection scheme will eventually recognize the false alarm situations by comparing the signatures of the actual system behavior with the nominal system using a prespecified threshold value and recommend nominal control actions to avoid the unnecessary control effort after the effect resulting from unexpected disturbances or noises decay. Sharing the spirit of the multiple model approach, the effective control actions to achieve successful failure accommodations for anticipated faults are selected based on the matching conditions of the signatures between the actual failures and the multiple model-based failures. The fault isolation process is to compare the most recent time domain signatures between the actual failure and the postfailures. The differences between the actual measurements and the outputs of the nominal model are considered as the outputs from the failure dynamics, and they are compared with the "signatures," outputs, from the postfailure models within a certain length of time-shifting window, $\omega_{f\mathrm{diag}}$. A prespecified threshold value is used to compare with $\psi_{f\mathrm{diag}_i}$ for the proper selection of the anticipated failure condition. If none of the "signatures" of the anticipated failures meets the criterion, the system status will be switched to the unanticipated failure situation, and the intelligent online FTC approach will be initialized. The ω_f, $\omega_{f\,diag}$, and the threshold values of fault detection and diagnosis are the design parameters that should be selected under considerations of computational capacity, modeling uncertainty, expected noises, and the accuracies of the postfailure model.

8.6 Simulation Study and Discussions

8.6.1 Online Fault Accommodation Technique for Unanticipated System Failures

Consider a multiple inputs and/or outputs (MIMO) system under different failures as shown in equation 8.31:

$$\begin{aligned} y_1(k+1) &= \frac{y_1(k)}{1+y_2(k)^2} + u_1(k)^2 + u_2(k)^2 - 16u_1(k) - 20u_2(k) + \Delta f_1(k). \\ y_2(k+1) &= \frac{y_1(k)y_2(k)}{1+y_2(k)^2} + u_1(k)u_2(k) + 20u_1(k) - 5u_2(k) + \Delta f_2(k). \\ \Delta f_1(k) &= \beta_{10}(k-T_{10}) \times 0.1 \times \frac{k-25}{20} y_1(k)\cos(u_1(k)) + \beta_{11}(k-T_{11}) \\ &\quad \times 0.6 y_1(k) y_2(k). \\ \Delta f_2(k) &= \beta_{20}(k-T_{20}) \times 0.1 \times y_1(k)y_2(k). \end{aligned} \quad (8.31)$$

In equation 8.31, $\beta_{10}(k - T_{10}) = U(k - T_{10})$, $\beta_{11}(k - T_{11}) = U(k - T_{11})$, $\beta_{20}(k - T_{20}) = U(k - T_{20})$, $T_{10} = 25$, $T_{20} = 15$, and $T_{11} = 123$. No information about the failures is assumed known. The nominal system is first realized through a multilayer perceptron (MLP) network with 4 input neurons, 75 hidden neurons, and 2 output neurons (4–75–2). The 2000 input/output training patterns are collected by supplying uniformly distributed random inputs varying from −1.5 to 1.5 to the nominal system in equation (8.31). A 4–40–2 MLP network is used as a nominal controller trained off-line using the similar idea discussed in Section 8.4 [i.e., use equations 8.13 through 8.18 with the pretrained nominal model (4–75–2 MLP network) to collect training patterns]. A 3–4–2 MLP network is chosen as the online estimator to approximate the most recent behavior of the unanticipated multiple failure dynamics online (i.e., Δf_1, Δf_2), and the Levenberg-Marquardt with Bayesian regularization algorithm (Demouth and Beale, 1998; Mackey, 1992) is used in all the training processes (i.e., off-line training for the nominal model and nominal controller; online training for the failure estimator). Figures 8.7 and 8.8 show the system responses for the first output and the second output together with the desired outputs, respectively. The failures have been properly accommodated by online adjustment of the control signals, while the nominal controller alone fails to maintain the system stability under multiple failures. Figures 8.9 and 8.10 show the online estimations together with the numerical values of the actual failures at each time-step.

The actual control inputs are shown in Figures 8.11 and 8.12. In this example, the length of the time-shifting data

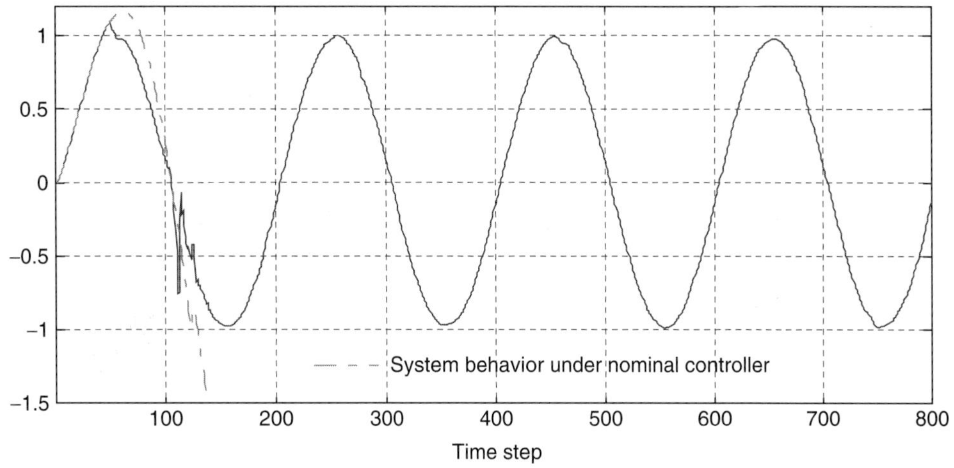

FIGURE 8.7 System Response, $y1$, Versus Desired Output, $y1d$ (MIMO System; 25-data Window). Solid line: System output $y1$/Dashed line: Desired output $y1d$.

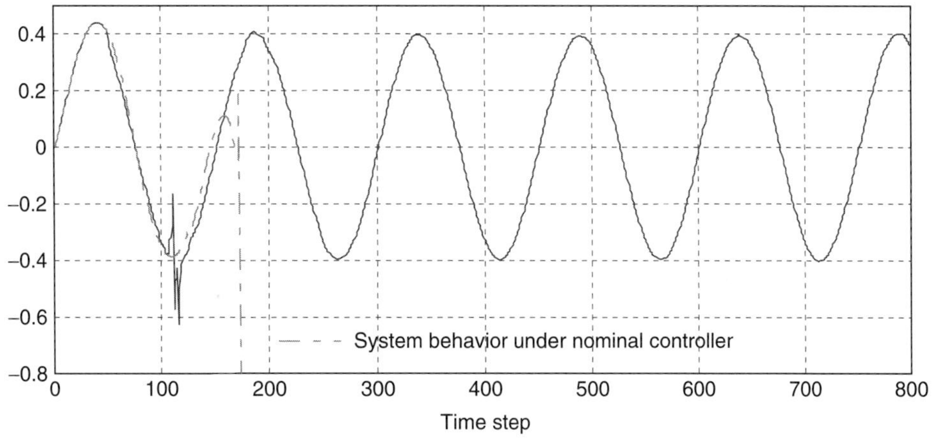

FIGURE 8.8 System Response, $y2$, Versus Desired Output, $y2d$ (MIMO System; 25-Data Window). Solid line: System output $y2$/Dashed line: Desired output $y2d$.

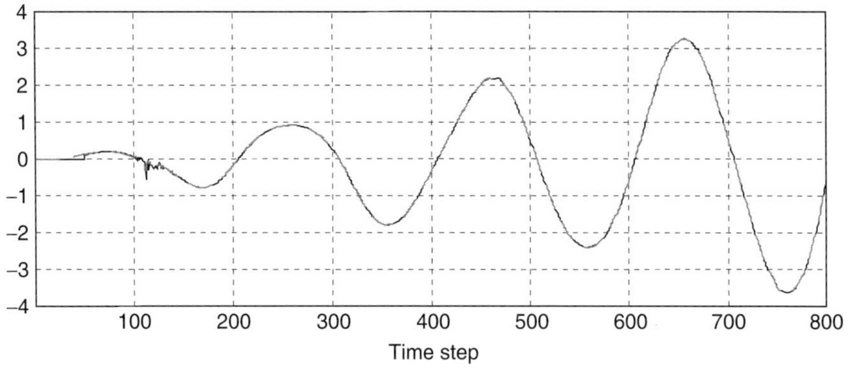

FIGURE 8.9 Output Prediction, $nfy1$, from the Online Estimator Versus Actual Failure Dynamics Output, $fy1$ (MIMO System; 25-Data Window). Solid line: Output ($nfy1$) of online estimator/Dashed line: Actual failure output ($fy1$).

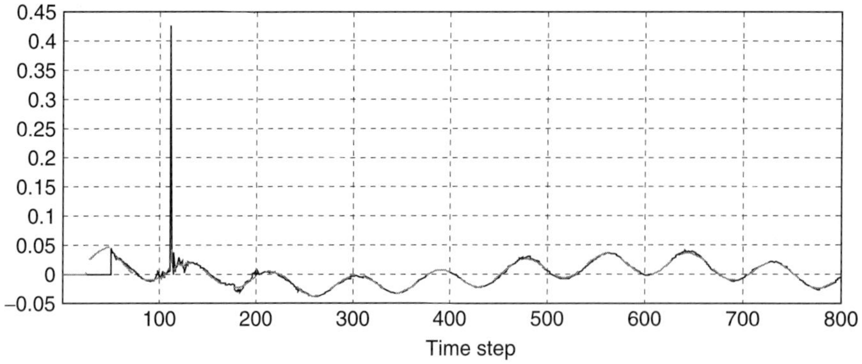

FIGURE 8.10 Output Prediction, *nfy*2, from the Online Estimator Versus Actual Failure Dynamics Output, *fy*2 (MIMO System; 25-Data Window). Solid line: Output (*nfy*2) of online estimator/Dashed line: Actual failure output (*fy*2).

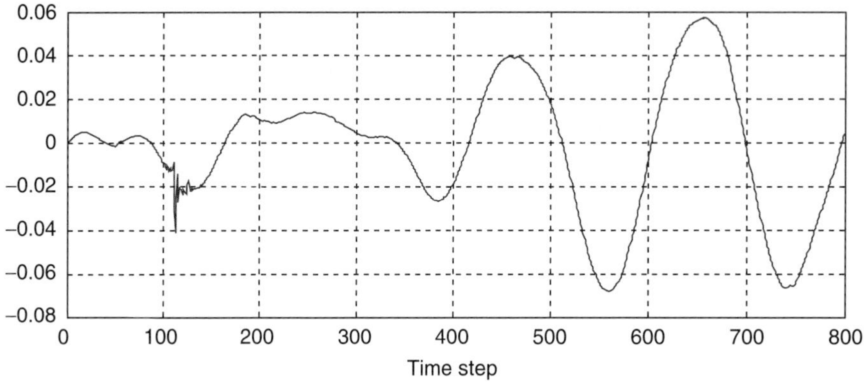

FIGURE 8.11 Control Input *u*1 (MIMO System; 25-Data Window)

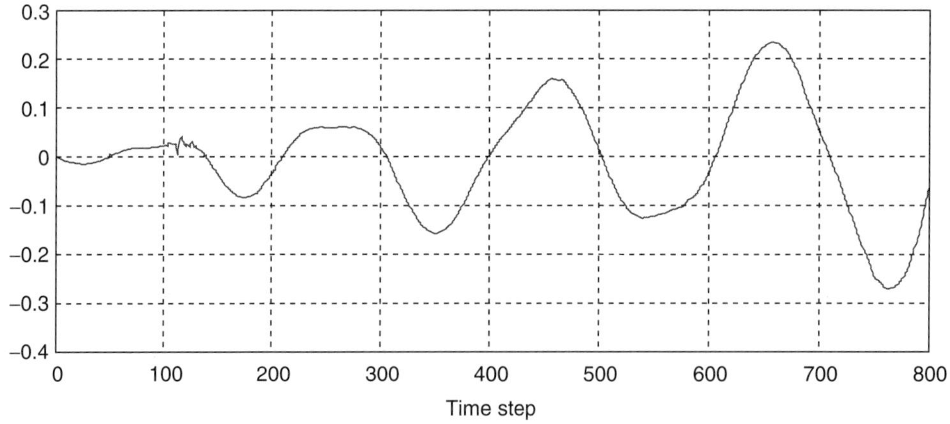

FIGURE 8.12 Control Input *u*2 (MIMO System; 25-Data Window)

window is selected as 25. Simulation tests show that this is a good trade-off number under the constraints of the system performance and the computational complexity. Two *S* functions are defined separately for system output 1 and 2 in the form of equation (8.4) with $a = 10$. A simple mean value of the two estimated gradient directions realized through the *NN* online estimator is used for searching the effective control inputs. This approach is based on the assumption that the searching directions of the effective control signals to accommodate failure dynamics do not conflict, which may not always be true under unanticipated catastrophic system failures. In general, this becomes a multiobjective optimization problem that remains to be an open issue (Zitzler and Thiele, 1999).

8.6.2 Intelligent FDA framework

To obtain a deeper insight into the online FDA problems in the real applications, a separate simulation study has been performed to test the proposed FDA framework. The design parameters, including the threshold value of failure alarms, threshold value in the failure diagnosis process, lengths of the evaluating windows for fault detection and for failure diagnosis, are preselected as $7.0 \times e^{-5}$, $5.0 \times e^{-6}$, 5, and 10, respectively. Consider a nominal system described in equation 8.32 where the system dynamics is not in linear-in-control format, and the nominal model is first realized by a 3–30–1 MLP network. The nominal controller is designed using the same technique described in Subsection 8.6.1:

$$y(k+1) = 0.3y(k) + 0.6y(k-1) + u(k)^2 - 5u(k). \quad (8.32)$$

Three postfailure modes are shown in equation 8.33:

- **Postfailure 1:** $pf_1(k+1) = -(0.3y(k) + 0.6y(k-1) + u(k)^2 - 5u(k)) + 0.06y(k)^2 + 4u(k).$ (8.33a)
- **Postfailure 2:** $pf_2(k+1) = 0.05 \times y(k) \times \cos(u(k)).$ (8.33b)
- **Postfailure 3:** $pf_3(k+1) = -0.8 \times u(k)^2 + 3.5u(k).$ (8.33c)

To simulate the failures in a real dynamic system, the specific mathematical formats of these failure dynamics are assumed unknown and have to be realized by separate network models. The idea is described in the following steps:

1. Place the system under failure situations (i.e., using the nominal mathematical model with the failure dynamics in simulation stage).
2. Feed in 3000 uniformly distributed random input signals varying from −1.5 to 1.5 with selected initial conditions.
3. Collect the training patterns. The desired outputs are computed by using the differences between the system outputs and the outputs from the nominal model.
4. Train the model for the failure dynamics.

Three separate MLP networks with different structures, 3–30–1, 2–20–1, 1–20–1, are used to realize the three post failures offline, respectively, by following the above steps. The postfailure control actions are also realized by using *NN* controllers (i.e., 3–20–1, 3–20–1, and 3–10–1 MLP networks), and they are obtained using the same technique as described in Section 5.1. The preselected value for the diagnosis process is $1.0 \times e^{-4}$ and the rest of the design parameters are the same as those in Subsection 8.6.1.

Scenario 1

Consider the failure situation containing an **abrupt anticipated failure mode 3** starting at time step 70 and an **incipient unanticipated time-varying failure** with the corresponding time profile shown in equation 8.34:

$$f_2(\cdot) = \frac{ky(k)}{320 \times (1 + y(k-1)^2)}; \qquad (8.34)$$
$$\beta_2(k - T_2) = \left(1 - e^{-0.2\times(k-T_2)}\right) U(k - T_2);\ T_2 = 314.$$

Figures 8.13 through 8.15 show the online simulation results under the intelligent FDA framework. The abrupt anticipated failure mode 3 that occurred at time-step 70 is quickly recognized by the diagnosis process, and the effective control commands are then switched to the corresponding postfailure control actions. This is clearly seen in the online system status plot shown in Figure 8.14 and the system online response shown in Figure 8.13. Observing Figure 8.15 closely, one can also find that the time domain signatures of actual failure dynamics and the postfailure mode 3 are almost undistinguishable from time-step 80 to 320. After the second change of the system dynamics caused by another time-varying incipient failure, the discrepancy between the signatures are getting large, which indicates that an unanticipated situation has occurred.

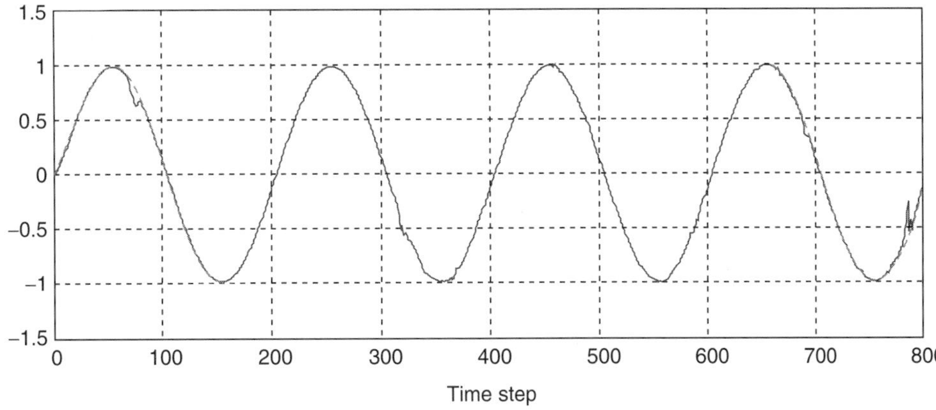

FIGURE 8.13 System Response Versus Desired Output (Scenario 1; General Case; FDA Simulation). Solid line: Acutal output/Dashed line: Desired output.

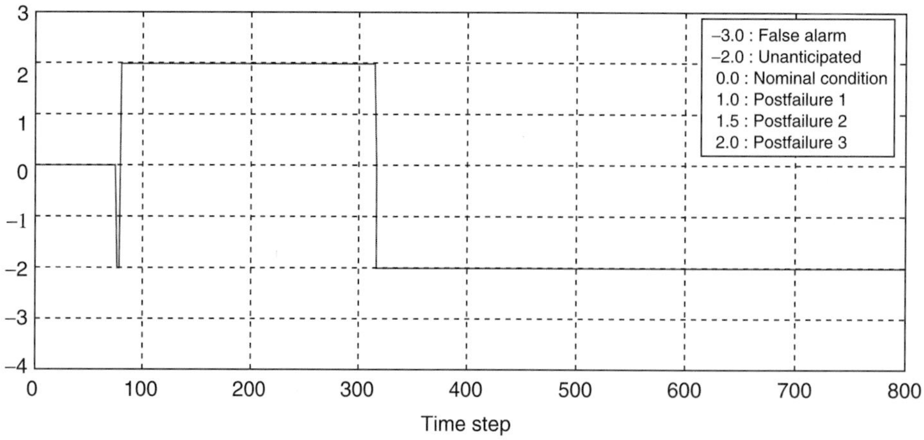

FIGURE 8.14 System Status (Scenario 1; General Case; FDA Simulation)

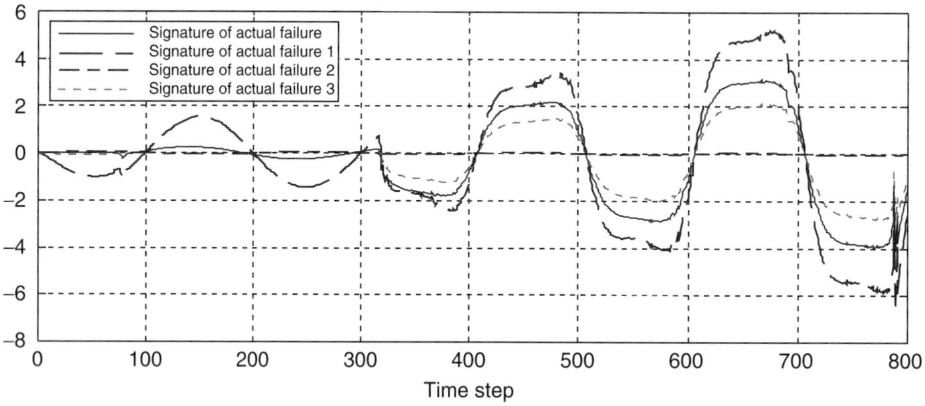

FIGURE 8.15 Signatures of Postfailures Versus Actual Failure Dynamics (Scenario 1; General Case; FDA Simulation)

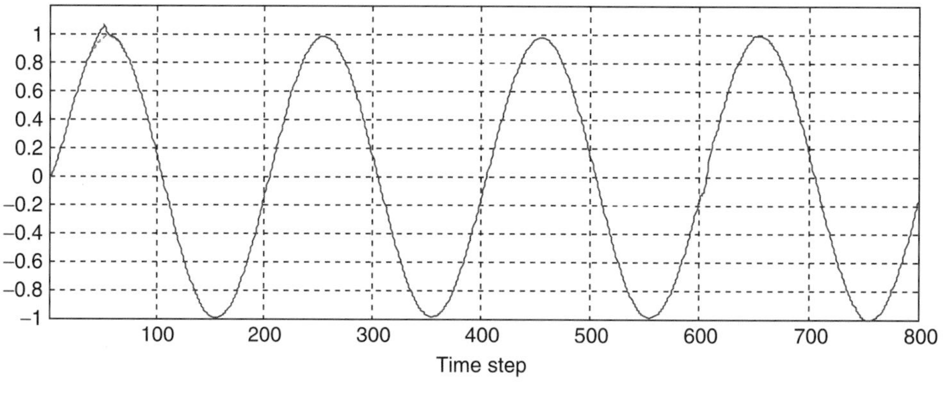

FIGURE 8.16 System Response Versus desired Output (Scenario 2; General Case; FDA Simulation). Solid line: Acutal output/Dashed line: Desired output.

Scenario 2

Consider the failure situation containing an **incipient anticipated failure mode 2** with the time profile shown in equation (8.35) and an **abrupt anticipated failure mode 3** starting at time-step 601:

$$\beta_1(k - T_1) = (1 - e^{-0.009\times(k-T_1)})U(k - T_1);\ T_1 = 25. \quad (8.35)$$

Figures 8.16 through 8.18 are the plots of the test results. Due to the slow variation of the failure dynamics (with $\alpha_1 = 0.009$), the incipient fault is not identified as quickly as an abrupt

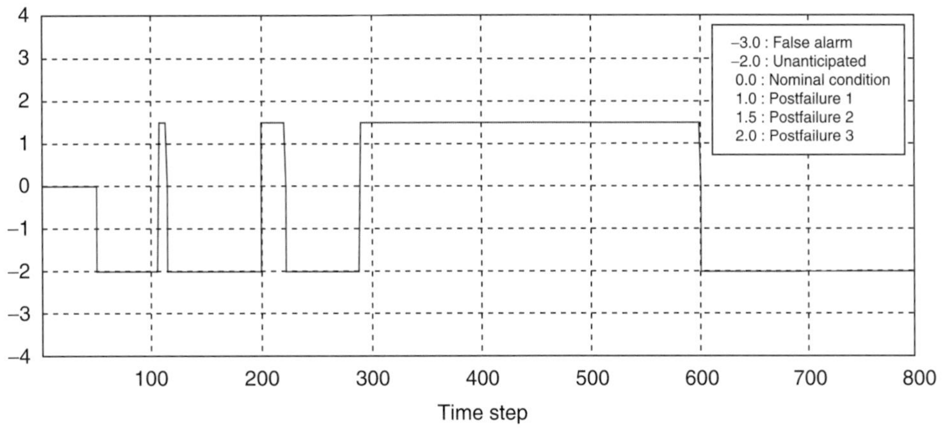

FIGURE 8.17 System Status (Scenario 2; General Case; FDA Simulation)

failure since this incipient time profile will take almost 300 time-steps to converge (i.e., 300 time-steps are required for the time profile to reach 0.9328). Figure 8.17 shows this expected result. The failure diagnosis scheme cannot be sure of the actual failure situation until the time-step almost reaches 300. This is also a correct decision since both the failure dynamics and the corresponding control actions are realized by *NN* under the abrupt failure situation. However, diagnostic oscillation is not observed before time-step 290. It appears that the diagnostic threshold value, $1.0 \times e^{-4}$, may be too big for the accuracy of the *NN* postfailure 2. Observing Figure 8.18 closely, it is easy to tell there are signature differences between the actual failure and the postfailure 2 from time-step 100 to 220. In other words, the *NN* postfailure model 2 in this case may be accurate enough to use a smaller threshold value to have a better diagnostic report. This result suggests that, for better failure diagnosis, the selection of the diagnostic threshold value should be failure-dependent and also should be chosen based on the accuracy of the *NN* postfailure model.

8.6.3 False Alarm Situations

The nominal system and the nominal controller in the general case are selected to test the false alarm situation under the intelligent FDA architecture. To simulate the false failure alarm possibly caused by *unexpected* measurement noises, uniformly distributed random white noises are generated and added to the measurements, which are used for fault detection defined in equation. The unknown and unexpected white noises are generated during three time periods: time-steps 78 to 81, 212 to 215, and 465 to 466. The added noises are varying from −0.3 to 0.3, -0.23 to 0.23, and -0.85 to 0.85, respectively. The simulation results are plotted and shown in Figures 8.19 through 8.21.

Figure 8.20 indicates how the proposed FDA framework reacts to the false alarm situation caused by unknown noises. Once the contaminated measurements trigger the alarm, the system behavior is immediately examined by failure diagnosis, and, since the measurements are contaminated by noises, it will not be recognized as any one of the anticipated failures.

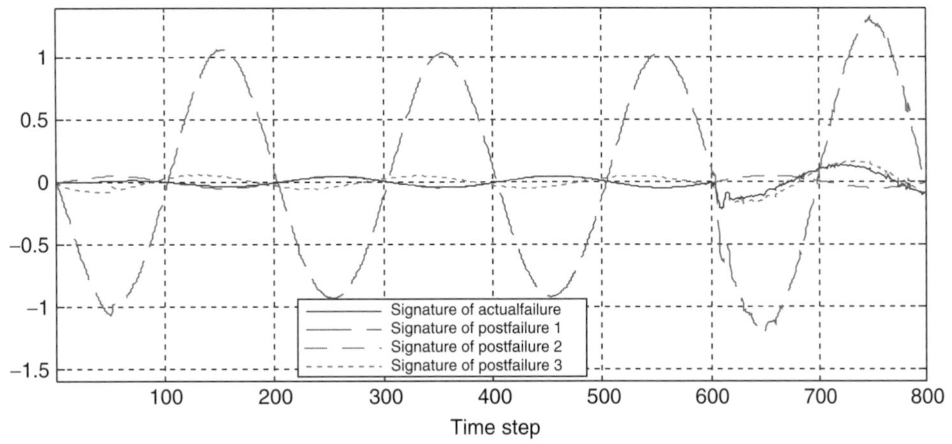

FIGURE 8.18 Signatures of Postfailures Versus Actual Failure Dynamics (Scenario 2; General Case; FDA Simulation)

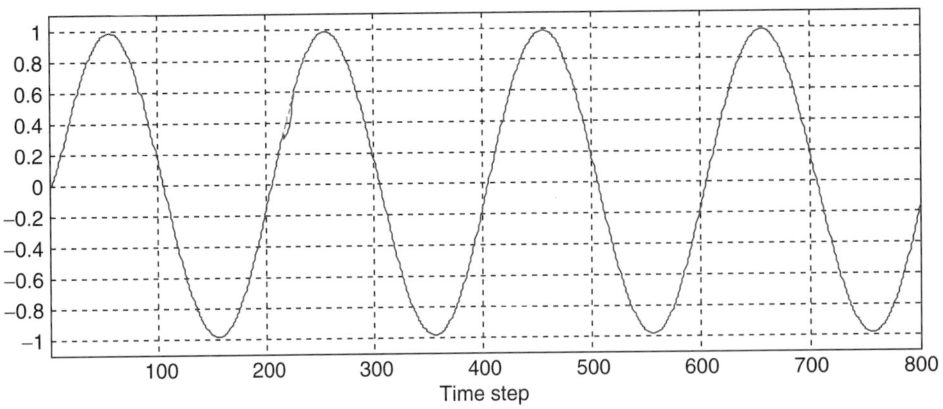

FIGURE 8.19 System Response Versus Desired Output (False Alarm Situations; FDA Simulation). Solid line: Acutal output/Dashed line: Desired output.

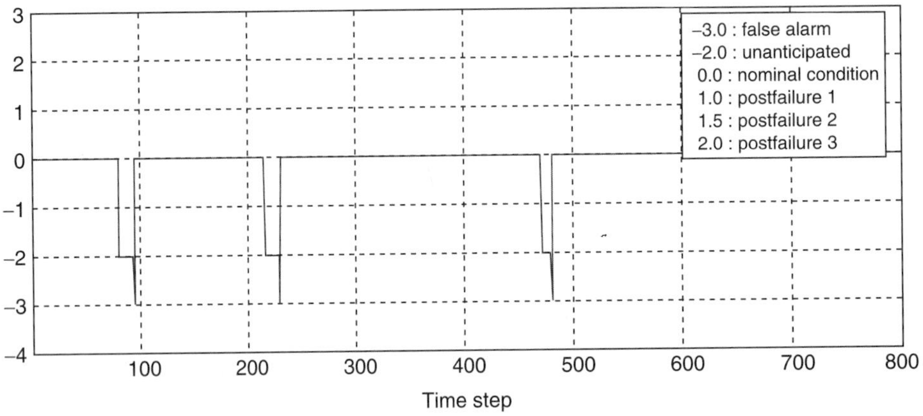

FIGURE 8.20 System Status (False Alarm Situations; FDA Simulation)

Thus, the diagnostic result will suggest an unanticipated situation. However, after the effect resulting from unexpected noises decays, the fault-detection scheme discovers that the system behavior is as normal as it is under the nominal situation. After double checking that this situation, the fault-detection scheme flags a false alarm signal, and the system status is switched back to the nominal condition. Figure 8.19 shows some slight deviations of the system output away from the desired trajectory during the periods of noisy data, which is apparently caused by the fact that the contaminated measurements are used directly in the fault-detection scheme and in the unanticipated failure conditions as the learning targets for failure accommodation.

8.6.4 Comments and Discussions

Many simulation tests have been omitted in this chapter due to space limitation. The following summaries, however, can be drawn:

1. The design parameters of the fault-detection scheme, such as the length of the time-shifting evaluating window and threshold values, have direct effects on the system performance. Short-evaluating window in the fault-detection scheme may result in a sensitive failure detector, while a long one may appear to be too slow to take proper accommodation.
2. Similar conditions also exist for failure diagnosis. The smaller threshold value gives more restriction in failure recognition and thus provides a more conservative diagnostic result. It also implies that more computational cost may possibly be spent due to the conservative attitude. On the other hand, if the value appears to be too big, large uncertainty may exist in the diagnostic result. Thus, the control actions may switch among nominal controller, postfailure control actions, and intelligent control regulator. The best design of these parameters should be on system-dependent bases.

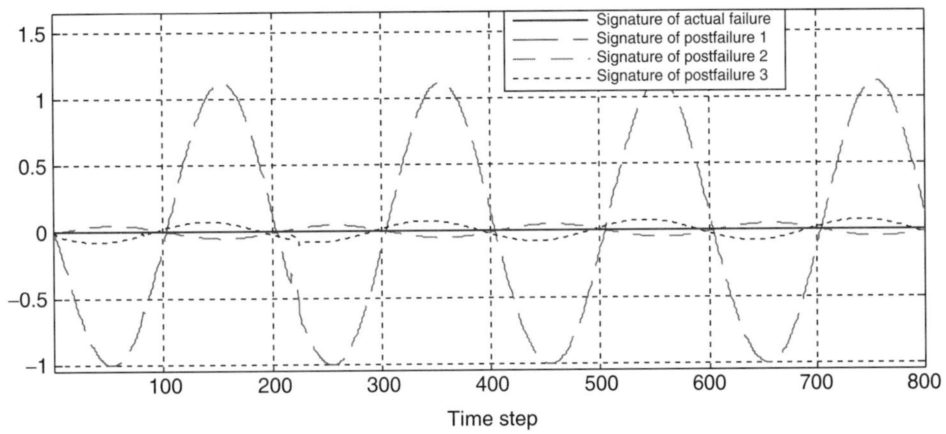

FIGURE 8.21 Signatures of Postfailures Versus Actual Failure Dynamics (False Alarm Situations; FDA Simulation)

8.7 Conclusion

In research work presented in this chapter, the online fault accommodation control problems under various system failures were investigated. The major interest was focused on dealing with the *unanticipated* system failures in the general formulation. Through discrete-time Lyapunov stability theory, the necessary and sufficient conditions to guarantee the system online stability and performance under failures were derived together with an online fault accommodation control technique. In general, the effectiveness of the developed online fault accommodation technique for unanticipated system failures was demonstrated through the simulation study. Based on modern intelligent techniques, the unexpected failures could be identified and properly accommodated online without the complete realization of the failure dynamics. The price paid for this achievement of the successful control mission relies on a certain degree of computational complexity. Simulation results indicate that, under the Levenberg-Marquardt training algorithm with Bayesian regularization (Demuth and Beale, 1998; Mackey, 1992), the online learning of the unanticipated failure dynamics usually converges within 10 iterations, and the online simulation speed can reach two to three time-steps per second under the Intel Pentium II 450 dual processors. Although this may not be fast enough in many real-time control systems that require a higher sampling rate, it is believed that with the continuous performance improvement of microprocessors and semiconductor technology, the proposed online fault accommodation technique will be implemented online in most of the real-time control systems in near future.

A complete architecture for intelligent fault diagnosis and accommodation has also been presented by incorporating the proposed intelligent fault-tolerant control technique with a cost-effective fault-detection scheme and a multiple model-based failure diagnosis process to efficiently handle the false alarms and the accommodation of the anticipated failure modes. Simulation results indicate that unnecessary control effort and computational complexity have been significantly reduced in online situations when the failures are anticipated. Under the multiple model-based failure diagnosis process together with the postfailure control actions, a successful fault isolation mission is quickly reached through the multiple model failure recognition. System performance recovery can be obtained through the multiple model switching in the postfailure control actions.

To obtain a sophisticated understanding for the speed limit of the developed online fault accommodation technique under the realtime control system with currently available microprocessors, an experimental online real-time fault-tolerant control test bed is currently under construction for examining the proposed control framework in real hardware. The major attention of the future research work is also focused on testing the online fault accommodation control framework in general MIMO cases where the control problems are both theoretically and technically complicated since the searching of effective control signals to accommodate the multiple failures becomes a multiobjective optimization problem, which is still one of the remaining open research issues. Thus, how to systematically reorganize the priorities of the control missions related to both system stability and performance for the general MIMO system under multiple failures in the online situation becomes challenging work.

References

Apostol, T. (1974). *Mathematical analysis.* Menlo Park, CA: Addison-Wesley.

Bodson, M., and Groszkiewicz, J. (1997). Multivariable adaptive algorithms for reconfigurable flight control. *IEEE Transactions on Control Systems Technology* 5(2), 217–229.

Chow, E., and Willsky, A. (1984). Analytical redundancy and the design of robust failure detection systems. *IEEE Transactions on Automatic Control* 29(7), 603–614.

Chun, J., Jung, H., and Hahn, S. (1998). A study on comparison of optimization performances between immune algorithm and other heuristic algorithms. *IEEE Transactions on Magnetics 34*(5), 2972–2975.

Chun, J., Kim, M., Jung, H., and Hong, S. (1997). Shape optimization of electromagnetic devices using immune algorithm. *IEEE Transactions on Magnetics 33*(2), 1876–1879.

Demuth, H., and Beale, M. (1998). *Neural Network Tool Box User's Guide*, Version 3. Natick, MA: The Math Works.

Emani-Naeini, A., Akhter, M., and Rock, S. (1988). Effect of model uncertainty on failure detection: The threshold selector. *IEEE Transactions on Automatic Control 33*(2), 1106–1115.

Frank, P. (1990). Fault diagnosis in dynamic systems using analytical and knowledge-based redundancy—A survey and some new results. *Automatica 26*(3), 459–474.

Gao, Z., and Antsaklis, P. (1991). Stability of the pseudo-inverse method for reconfigurable control systems. *International Journal of Control 53*(3), 717–729.

Gao, Z., and Antsaklis, P. (1992). Reconfigurable control system and design via perfect model following. *International Journal of Control 56*(4), 783–798.

Garcia, E., and Frank, P. (1999). Multiple fault isolation in linear systems. *Proceedings of the 38th IEEE Conference on Decision and Control* 3114–3115.

Garcia, E., Seliger, R., and Frank, P. (1998). Nonlinear decoupling approach to fault isolation in linear systems. *Proceedings of the American Control Conference* 2867–2871.

Ge, W., and Feng, C. (1988). Detection of faulty components via robust observation. *International Journal of Control 47*(2), 581–599.

Gertler, J. (1988). Survey of model-based failure detection and isolation in complex plants. *IEEE Control Systems Magazine 8*(6), 3–11.

Gertler, J., Costin, M., Fang, X., Kowalczuk, Z., Kumwer, M., and Monajemy, R. (1995). Model-based diagnosis for automotive engines. *IEEE Transactions on Control Systems Technology 3*(10), 61–69.

Gertler, J., and Hu, Y. (1999). Direct identification of optimal nonlinear parity models. *Proceedings of the 1999 IEEE International Symposium on Computer-Aided Control System Design* 164–169.

Gertler, J., and Kunwer, M. (1995). Optimal residual decoupling for structured diagnosis and disturbance insensitivity. *International Journal of Control 61*, 395–421.

Ho, S., and Huang, M. (2000). An efficient quadratic curve approximation using an intelligent genetic algorithm. *Proceedings of the Genetic and Evolutionary Computation Conference* 766.

Hornik, K., Stinchombe, M., and White, H. (1989). Multilayer feedforward networks are universal approximators. *Neural Networks 2*(5), 359–366.

Jiang, J. (1994). Design of reconfigurable control systems using eigenstructure assignment. *International Journal of Control 59*(2), 395–410.

Jiang, J., and Zhao, Q. (1998). Fault-tolerant control system synthesis using imprecise fault identification and reconfigurable control. *Proceedings of the IEEE International Symposium on Intelligent Control* 169–174.

Jiang, J., and Zhao, Q. (1999). Reconfigurable control based on imprecise fault identification. *Proceedings of the American Control Conference* 114–118.

Juang, C., Lin, J., and Lin, C. (2000). Genetic reinforcement learning through symbiotic evolution for fuzzy controller design. *IEEE Transactions on Systems, Man, and Cybernetics 30*(2), 290–302.

Laparo, K., Buchner, M., and Vasudeva, K. (1991). Leak detection in an experimental heat exchanger process: A multiple model approach. *IEEE Transactions on Automatic Control 36*(2), 167–177.

Lou, X., Willsky, A., and Verghese, G. (1986). Optimally robust redundancy relations for failure detection in uncertain systems. *Automatica 22*(3), 333–344.

Mackey, D. (1992). Bayesian interpolation. *Neural Computation 4*(3), 415–447.

Marchesi, M. (1998). A new class of optimization algorithm for circuit design and modeling. *IEEE International Symposium on Circuits and Systems 2*, 1691–1695.

Marchesi, M., Molinari, G., and Repetto, M. (1994). A parallel-simulated annealing algorithm for the design of magnetic structures. *IEEE Transactions on Magnetics 30*(5), 3439–3442.

Maybeck, P., and Stevens, R. (1991). Reconfigurable flight control via multiple model adaptive control methods. *IEEE Transactions on Aerospace and Electronic Systems 27*(3), 470–480.

Murray-Smith, R., and Johansen, J. (1997). *Multiple model approaches to modeling and control.* Bristol, PA: Taylor & Francis.

Narendra, K., and Mukhopadhyay, S. (1997). Adaptive control using neural networks and approximate models. *IEEE Transactions on Neural Networks 8*(3), 475–485.

Narendra, K., and Parthasathy, K. (1990). Identification and control of dynamical systems using neural networks. *IEEE Transactions on Neural Networks 1*(1), 4–27.

Nie, J., and Linkens, D. (1993). Learning control using fuzzied self-organizing radial basis function network. *IEEE Transactions on Fuzzy Systems 1*(4), 280–287.

Noura, H., Sauter, D., and Theilliol, D. (2000). Fault-tolerant control in dynamics systems: Application of a winding machine. *IEEE Control Systems Magazine 20*(1), 33–49.

Ottner, S. (2000). Optimizing television commercial air-time by means of a genetic algorithm. *Proceedings of the Genetic and Evolutionary Computation Conference* 761.

Polycarpou, M., and Helmicki, A. (1995). Automated fault detection and accommodation: A learning systems approach. *IEEE Transactions on Systems, Man, and Cybernetics 25*(11), 1447–1458.

Polycarpou, M., and Vemuri, A. (1995). Learning methodology for failure detection and accommodation. *IEEE Control Systems Magazine 15*(3), 16–24.

Rauch, H. (1995). Autonomous control reconfiguration. *IEEE Control Systems Magazine 15*(6), 37–48.

Sauter, D., Hamelin, F., and Noura, H. (1998). Fault-tolerant control in dynamic systems using convex optimization. *Proceedings of the IEEE International Symposium on Intelligent Control* 187–192.

Tanaka, Y., and Yoshida, T. (1999). An applications of reinforcement learning to manufacturing scheduling problems. *IEEE International Conference on Systems, Man, and Cybernetics 4*, 534–539.

Theilliol, D., Noura, H., and Sauter, D. (1998). Fault-tolerant control method for actuator and component faults. *Proceedings of the 37th IEEE Conference on Decision and Control* 604–609.

Trunov, A., and Polycarpou, M. (1999). Automated fault diagnosis in nonlinear multivariable systems using a learning methodology. *IEEE Transactions on Neural Networks 11*(1), 1–11.

Vemuri, A., and Polycarpou, M. (1997). Robust nonlinear fault diagnosis in input/output systems. *International Journal of Control 68*(2), 343–360.

Yen, G. (1994). Reconfigurable learning control in large space structures. *IEEE Transactions on Control Systems Technology 2*(4), 362–370.

Yen, G., and Ho, L. (2000). Fault-tolerant control: An intelligent sliding model control strategy. *Proceedings of the American Control Conference* 4204–4208.

Zhang, Y., and Jiang, J. (1999). An interacting multiple model-based fault detection, diagnosis, and fault-tolerant control approach. *Proceedings of the 38th Conference on Decision and Control* 3593–3598.

Zhang, Y., and Li, X. (1998). Detection and diagnosis of sensor and actuator failures using IMM estimator. *IEEE Transactions on Aerospace and Electronic Systems 34*(4), 1293–1313.

Zhang, X., Parisini, T., and Polycarpou, M. (1999). Robust parametric fault detection and isolation for nonlinear systems. *Proceedings of the 38th IEEE Conference on Decision and Control* 3102–3107.

Zhang, X., Polycarpou, M., and Parisini, T. (2000). Abrupt and incipient fault isolation of nonlinear uncertainty systems. *Proceedings of the American Control Conference* 3713–3717.

Zitzler, E., and Thiele, L. (1999). Multiobjective evolutionary algorithms: A comparative case study and the strength Pareto approach. *IEEE Transactions on Evolutionary Computation 3*(4), 257–271.

9

Gain-Scheduled Controllers

Christopher J. Bett
Raytheon Integrated Defense Systems, Tewksbury, Massachusetts, USA

9.1 Introduction

Gain scheduling is a practical and powerful method for the control of nonlinear systems. A gain-scheduled controller is formed by interpolating between a set of linear controllers derived for a corresponding set of plant linearizations associated with several operating points. The interpolation is based on exogenous scheduling parameters that vary slowly and capture the plant nonlinearities. The resulting controller is a linear system whose parameters (gains) are adjusted (scheduled) as a function of the exogenous scheduling variables.

Gain scheduling is attractive because it exploits linear control design methods that are theoretically mature, are well-understood, and involve computationally efficient synthesis procedures. Because the control design is based on linear approximations to the plant, the designer can guarantee that at each operating point, the feedback system has the desired stability and performance properties. In addition, because the controller gains are adjusted to reflect the plant operating conditions, the gain-scheduled control system has the potential of responding rapidly to changing operating conditions. The gain-scheduling method has been proven sound by successful application throughout industry.

Although the gain-scheduling approach is a simple extension of linear design methods, the technique can guarantee closed-loop stability and performance only in a local sense. Stability and performance properties of the gain-scheduled designs must be inferred from extensive simulation studies. The operating region where stability and performance is ensured is difficult to quantify. Furthermore, the designer must typically resort to heuristic methods to derive the scheduling law.

This chapter describes the historical approach to gain scheduling, discussing options for developing the scheduling law and noting some of the gain-scheduling lessons that have received theoretical justification in recent years. More recent gain scheduling results are presented for the class of systems modeled as linear parameter varying (LPV) systems. For LPV systems, some of the difficulties associated with gain scheduling designs are alleviated. Specifically, recent theoretical developments have made it possible to simultaneously synthesize the controller and scheduling law to guarantee stability and performance over a known set of parameter variations.

9.2 Gain-Scheduling Design Through Linearization

In the classical approach to gain scheduling, the nonlinear plant is linearized about a set of operating conditions or equilibrium points. For each linearized plant, a linear controller is derived to satisfy applicable performance requirements, such as disturbance rejection or transient response specifications. The gain-scheduled controller is formed by interpolating, or **scheduling**, among the set of linear control laws.

Gain scheduling is typically performed on an exogenous scheduling parameter that characterizes the operating environment. In a missile guidance application, for example, autopilot gains might be scheduled on temperature, air pressure, Mach number, or a combination of these. The scheduling parameter will be denoted by an s-dimensional vector, θ, that takes values in a compact subset, $\Theta \subset \Re^s$, called the **parameter set.** Associated with Θ is a signal space known as the **parameter variation set,** $\mathcal{F}_\Theta$, that is the set of all continuous functions mapping $\Re^+$ into Θ.

The notation $\theta \in \mathcal{F}_\Theta$ denotes a function in the parameter variation set; $\theta \in \Theta$ denotes a vector in a compact subset of $\Re^s$. Note that both $\mathcal{F}_\Theta$ and, for instance, $\mathcal{L}^n_\infty$ represent signal spaces. Technically, $\mathcal{F}_\Theta \subset \mathcal{L}^s_\infty$ since $\mathcal{F}_\Theta$ consists of supremum bounded s-dimensional vectors that vary continuously in time. In this discussion, $\mathcal{F}_\Theta$ will always refer to parameter signals or parameter variations; $\mathcal{L}^n_\infty$ will refer to signals in the plant input, output, or state-space.

9.2.1 Linearization

Consider a nonlinear plant with dynamics represented by:

$$\begin{aligned} \dot{\bar{x}} &= f(\bar{x}, w, \bar{u}, \theta). \\ \bar{z} &= g(\bar{x}, w, \bar{u}, \theta). \\ \bar{y} &= h(\bar{x}, w, \bar{u}, \theta). \end{aligned} \tag{9.1}$$

The $\bar{x}$ is the $n \times 1$ state vector, w is the $n_w \times 1$ vector of exogenous disturbances, u is the $n_u \times 1$ control input vector, z is the $n_z \times 1$ error vector, and y is the $n_y \times 1$ measurement vector. Time dependency has been suppressed for notational convenience. For fixed θ, the nonlinear plant can be linearized in the usual fashion to yield:

$$\begin{bmatrix} \dot{x} \\ \hline z \\ y \end{bmatrix} = \left[\begin{array}{c|cc} A(\theta) & B_w(\theta) & B_u(\theta) \\ \hline C_z(\theta) & D_{zw}(\theta) & D_{zu}(\theta) \\ C_y(\theta) & D_{yw}(\theta) & D_{yu}(\theta) \end{array}\right] \begin{bmatrix} x \\ \hline w \\ u \end{bmatrix} =: P(\theta) \begin{bmatrix} x \\ \hline w \\ u \end{bmatrix}, \tag{9.2}$$

where

$$P(\theta) = \left.\left[\begin{array}{c|cc} \frac{\partial}{\partial x} f(x, w, u, \theta) & \frac{\partial}{\partial w} f(x, w, u, \theta) & \frac{\partial}{\partial u} f(x, w, u, \theta) \\ \hline \frac{\partial}{\partial x} g(x, w, u, \theta) & \frac{\partial}{\partial w} g(x, w, u, \theta) & \frac{\partial}{\partial u} g(x, w, u, \theta) \\ \frac{\partial}{\partial x} h(x, w, u, \theta) & \frac{\partial}{\partial w} h(x, w, u, \theta) & \frac{\partial}{\partial u} h(x, w, u, \theta) \end{array}\right]\right|_{(x, w, u)=(x_{eq}(\theta), 0, u_{eq}(\theta))},$$

and the pair $(x_{eq}(\theta), u_{eq}(\theta))$ represents the equilibrium point of the nonlinear system for a fixed value of θ; that is, the pair satisfies:

$$\begin{aligned} 0 &= f(x_{eq}(\theta), 0, u_{eq}(\theta), \theta). \\ z_{eq} &= g(x_{eq}(\theta), 0, u_{eq}(\theta), \theta). \\ y_{eq} &= h(x_{eq}(\theta), 0, u_{eq}(\theta), \theta). \end{aligned} \tag{9.3}$$

The input, output, and state vectors for the linearized system in equation 9.2 are the deviations from the operating point for a fixed value of θ:

$$\begin{aligned} x &:= \bar{x} - x_{eq}(\theta). \\ u &:= \bar{u} - u_{eq}(\theta). \\ z &:= \bar{z} - z_{eq}(\theta). \\ y &:= \bar{y} - y_{eq}(\theta). \end{aligned}$$

9.2.2 Scheduling Law

For linearizations associated with a fixed set of scheduling parameters, $\Theta_0 := \{\theta_i | i = 1, \ldots, N\} \subset \Theta$, there are the linear controllers:

$$\begin{bmatrix} \dot{v} \\ \hline u \end{bmatrix} = \left[\begin{array}{c|c} F_i & G_i \\ \hline H_i & J_i \end{array}\right] \begin{bmatrix} v \\ \hline y \end{bmatrix} =: K_i \begin{bmatrix} v \\ \hline y \end{bmatrix}, \tag{9.4}$$

which are designed for each $\theta_i \in \Theta_0$. The linear controllers can be designed by any applicable design technique. Here, v is the $m \times 1$ controller state vector, and the controller matrices are of the proper dimension.

The scheduling law is chosen, typically, in ad hoc fashion as either a continuous or discrete scheduling law. In continuous scheduling, parameters of the gain scheduled controller vary continuously with the scheduling parameter and are commonly formed through interpolation of the fixed-point controller parameters. The interpolated controller parameters may be feedback gains, elements of the controller matrices, or controller poles/zeros. While the fixed-point controllers are linear, the parameter interpolation method need obey no such restriction.

As an example, suppose that θ is a scalar and that the fixed operating points are ordered so that $\theta_1 < \theta_2 < \cdots < \theta_N$. A simple, continuous scheduling law is then given by:

$$K = \begin{cases} K_1 & \theta < \theta_1 \\ \left(\dfrac{\theta - \theta_i}{\theta_{i+1} - \theta_i}\right) K_i + \left(\dfrac{\theta_{i+1} - \theta}{\theta_{i+1} - \theta_i}\right) K_{i+1} & \theta_i \leq \theta \leq \theta_{i+1}, i = 1, \ldots, N-1 \\ K_N & \theta \geq \theta_N \end{cases} \tag{9.5}$$

Note that no extrapolation is performed for $\theta < \theta_1$ or $\theta \geq \theta_N$. If the parameter is known to vary significantly beyond (θ_1, θ_N), then additional design points should be chosen.

In discrete scheduling, parameters of the gain-scheduled controller are switched when the parameter trajectory evolves out of a predefined subset of the parameter set, $\Theta_i \subset \Theta$. This is true, for instance, when the distance between the current operating condition and the design point associated with the current controller parameters exceeds a design threshold. Expanding on the preceding example, a simple discrete scheduling law is given by:

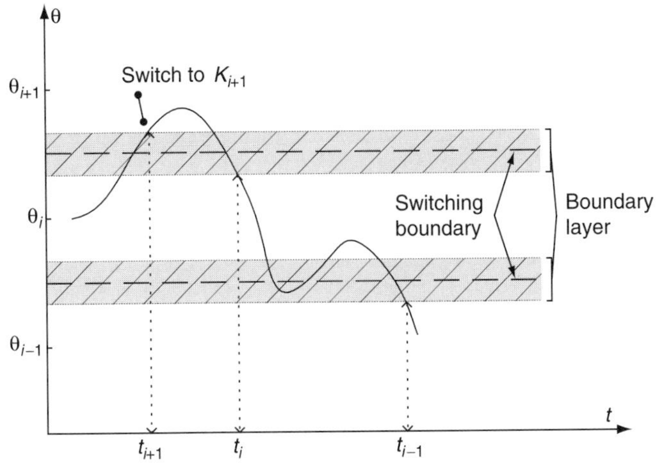

FIGURE 9.1 Hysteresis in Discrete Scheduling

$$K = \begin{cases} K_1 & \theta \leq \frac{1}{2}(\theta_1 + \theta_2). \\ K_i & \frac{1}{2}(\theta_{i-1} + \theta_i) < \theta \leq \frac{1}{2}(\theta_i + \theta_{i+1}), i = 2, \ldots, N-1. \\ K_N & \theta > \frac{1}{2}(\theta_{N-1} + \theta_N). \end{cases} \tag{9.6}$$

Often, the switching law will employ a hysteresis to prevent chattering (fast switching) if the scheduling parameter evolves near a switching boundary. The hysteresis simply delays the switch to new controller parameters until the parameter has moved beyond a boundary layer encompassing the switching boundary, as illustrated in Figure 9.1.

9.2.3 Discussion

It is important to remember that the gain scheduling methods just described are, in general, justified through repeated use and do not have rigorous theoretical justification. The methods are generally implemented in an ad hoc manner appropriate to the problem at hand; achievable performance may be sensitive to the form of scheduling law and the form of the controller matrices.

In some cases, scheduling parameters are chosen to be functions of the plant variables to make the problem more tractable from a design standpoint. Because the scheduling parameter is no longer independent of the plant variables (e.g., $\partial\theta/\partial x \neq 0$), the closed-loop system formed by the linearized plant and linear controllers does not match the linearized closed-loop plant. The differences in the closed-loop realizations are due to hidden coupling modes whose presence may degrade or limit achievable performance. Although there is no comprehensive theory for removing these modes from the gain-scheduled design, examples are present in the literature (Nichols *et al.*, 1993). As a rule, the designer must exhibit additional caution in ensuring that the scheduling parameter does not vary too rapidly.

For discrete scheduling designs, abrupt changes in the controller output can occur at switching instants. These changes may cause excitation of unmodeled plant dynamics, degrading performance and possibly leading to catastrophic failure. To mitigate these risks, one must resort to techniques such as bumpless transfer (Hanus *et al.*, 1987) or ensure that the design can tolerate the switching through other methods (Bett and Lemon, 1999).

In both the continuous, and discrete gain-scheduling designs, stability and performance of the gain-scheduled system can, in general, only be guaranteed in a neighborhood of the current operating condition. Specifically, the gain-scheduled design is not guaranteed stable or guaranteed to exhibit desired performance properties for all $\theta \in \Theta$, or for all $\theta \in \mathcal{F}_\Theta$. It is the latter of these two conditions that is the most difficult to guarantee. Further assurances of stability and performance must be derived from extensive simulation and testing. Successful designs typically involve scheduling on a slowly changing variable that captures the nonlinear behavior of the plant.

The difficulties associated with applying the gain-scheduling techniques just described stem from the local nature of the results and the lack of formal methods for deriving the scheduling law. These difficulties manifest themselves in increased simulation, test, and redesign. For the so-called linear parameter varying plants, these difficulties can be alleviated. For linear parameter varying plants, modern linear robust control techniques can be applied to the design of a gain-scheduled controller without the need for *ad hoc* interpolation methods.

9.3 Gain Scheduling for Linear Parameter Varying Systems

9.3.1 LPV Systems

A linear parameter varying system may be viewed as a linear system whose parameters depend on an exogenous variable, as stated in the following definition.

Definition 1 (LPV System)

Given a compact set $\Theta \subset \Re^s$ and continuous functions A: $\Re^s \to \Re^{n\times n}$, B: $\Re^s \to \Re^{n\times n_w}$, C: $\Re^s \to \Re^{n_z\times n}$, and D: $\Re^s \to \Re^{n_z\times n_w}$, an nth order linear parameter varying (LPV) system is a dynamical system whose dynamics evolve as:

$$\begin{bmatrix} \dot{x}(t) \\ z(t) \end{bmatrix} = \begin{bmatrix} A(\theta(t)) & B(\theta(t)) \\ C(\theta(t)) & D(\theta(t)) \end{bmatrix} \begin{bmatrix} x(t) \\ w(t) \end{bmatrix}, \tag{9.7}$$

where $\theta \in \mathcal{F}_\Theta$.

Although the trajectory of $\theta(t)$ is not known a priori, it is assumed to be measurable in real-time, thus providing real-time information on the behavior of the system. LPV systems can occur naturally (an LTI plant subject to a time-varying parameter uncertainty) or can be formed through linearization of a nonlinear plant. Given the latter interpretation, a gain-scheduled controller can be designed for the LPV system using the approach described in Section 9.2. The inherent structure of the LPV system, however, lends itself to the application of linear robust control design methods, and recent theoretical results have led to gain-scheduled controller synthesis approaches for LPV systems that form the controller and scheduling law directly.

The remainder of this section is concerned with determining an LPV controller, $K(\theta)$, with state-space realization:

$$\begin{aligned} \dot{v} &= A_k(\theta)v + B_k(\theta)y. \\ u &= C_k(\theta)v + D_k(\theta)y. \end{aligned} \tag{9.8}$$

Equation 9.8 applies to the LPV plant, $P(\theta)$, with LPV state-space realization:

$$\begin{aligned} \dot{x} &= A(\theta)x + B_w(\theta)w + B_u(\theta)u. \\ z &= C_z(\theta)x + D_{zu}(\theta)u. \\ y &= C_y(\theta)x + D_{yw}(\theta)w. \end{aligned} \tag{9.9}$$

The closed-loop LPV system shown in Figure 9.2, denoted $\mathcal{F}_\ell(P(\theta), K(\theta))$, is asymptotically stable and posesses the appropriate performance attributes. Synthesis conditions for induced-$\mathcal{L}_2$ and induced-$\mathcal{L}_\infty$ are presented. In both cases, gain-scheduled controller synthesis is achieved by solving a set of linear matrix inequalities.

9.3.2 Controller Synthesis for Induced-$\mathcal{L}_2$ Performance

The induced-$\mathcal{L}_2$ norm is commonly interpreted as the gain applied to the energy of an input signal; designing a controller to limit the induced-$\mathcal{L}_2$ norm of various system loops ensures that internal or output signal energy does not exceed prescribed thresholds. This has obvious implications for overall system performance. The design objective is commonly applied in loop shaping and disturbance rejection problems when the energy spectral density of the input disturbance is known (or bounded). Induced-$\mathcal{L}_2$ controller synthesis is also known as $\mathcal{H}_\infty$ synthesis.

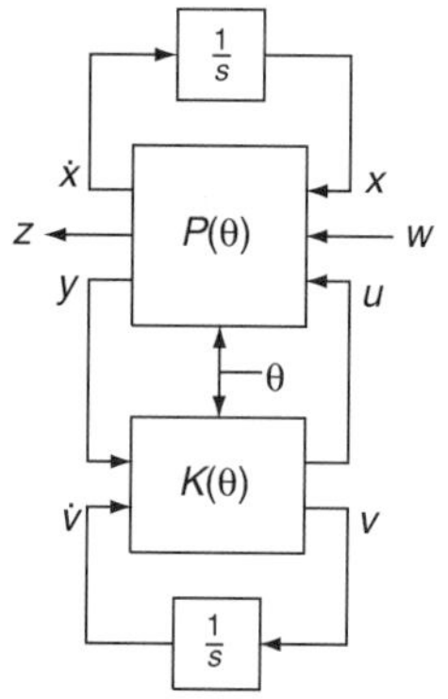

FIGURE 9.2 LPV Control System

The induced-$\mathcal{L}_2$ gain is defined in the usual manner from the associated signal norm.

Definition 2 (Induced-$\mathcal{L}_2$ Gain)
The two-norm of a signal $f: \Re^+ \to \Re^n$ is defined as:

$$\|f\|_2 := \left(\int_0^\infty f'(t)f(t)dt \right)^{1/2}.$$

The $\mathcal{L}_2^n$ denotes the normed n-dimensional signal space in the usual fashion under the above signal norm definition. Consider the LPV system of definition 9.1. For any $\theta \in \mathcal{F}_\Theta$ with $x(t_0) = 0$, the causal linear input/output mapping, $G_\theta: \mathcal{L}_2^{n_w} \to \mathcal{L}_2^{n_z}$, of the linear time-varying system described in equation 9.7 is defined as:

$$G_\theta w(t) = \int_{t_0}^{t} C(\theta(\tau))\Phi_\theta(t, \tau)B(\theta(\tau))w(\tau)d\tau + D(\theta(t))w(t),$$

where $\Phi_\theta(t, t_0)$ is the associated state transition matrix. The induced-$\mathcal{L}_2$ norm of G_θ is given by:

$$\|G_\theta\|_{i2} := \sup_{w \in \mathcal{L}_2^{nw},\, \|w\|_2 < 1} \|G_\theta w\|_2. \tag{9.10}$$

The induced-$\mathcal{L}_2$ synthesis conditions presented here are valid under the assumptions that, for all $\theta \in \Theta$:

- $(A(\theta), B_u(\theta), C_y(\theta))$ are controllable and detectable;
- $D_{zu}(\theta)$ has full column rank; and
- $D_{yw}(\theta)$ has full row rank.

Fix a performance level $\gamma > 0$ and consider the LPV plant in equation 9.9 under the preceding assumptions. A sufficient condition for the existence of a finite-dimensional LPV controller $K(\theta)$ that internally stabilizes the system and renders:

$$\|\mathcal{F}_\ell(P(\theta), K(\theta))\|_{i,2} < \gamma,$$

for all $\theta \in \mathcal{F}_\Theta$ is the existence of matrices $\hat{A} \in \Re^{n \times n}$, $\hat{B} \in \Re^{n \times n_y}$, $\hat{C} \in \Re^{n_u \times n}$, and $\hat{D} \in \Re^{n_u \times n_y}$, along with $n \times n$ positive definite symmetric matrices X and Y, so that for all $\theta \in \Theta$:

$$R(\theta) < 0 \quad \text{and} \quad \begin{bmatrix} X & I \\ I & Y \end{bmatrix} > 0, \tag{9.11}$$

where the (i, j)th blocks of the symmetric matrix $R(\theta)$ are given by:

$$[R(\theta)]_{1,1} = XA(\theta) + A(\theta)'X + \hat{B}C_y(\theta) + C_y(\theta)'\hat{B}'.$$
$$[R(\theta)]_{1,3} = XB_w(\theta) + \hat{B}D_{yw}(\theta).$$
$$[R(\theta)]_{2,1} = \hat{A}' + A(\theta) + B_u(\theta)D_kC_y(\theta).$$
$$[R(\theta)]_{2,2} = A(\theta)Y + YA(\theta)' + B_u(\theta)\hat{C} + \hat{C}'B_u(\theta)'.$$
$$[R(\theta)]_{2,3} = B_w(\theta) + B_u(\theta)D_kD_{yw}(\theta). \quad (9.12)$$
$$[R(\theta)]_{3,3} = -\gamma I_{n_w}.$$
$$[R(\theta)]_{4,1} = C_z(\theta) + D_{zu}(\theta)D_kC_y(\theta).$$
$$[R(\theta)]_{4,2} = C_z(\theta)Y + D_{zu}(\theta)\hat{C}.$$
$$[R(\theta)]_{4,3} = D_{zu}(\theta)D_kD_{yw}(\theta).$$
$$[R(\theta)]_{4,4} = -\gamma I_{n_z}.$$

If the matrix inequalities admit a solution, then a parameter-dependent controller is found by solving the factorization problem for matrices M and N such that:

$$I - XY = NM'. \quad (9.13)$$

The controller is also found by forming the parameter-dependent controller matrices from:

$$A_k(\theta) = N^{-1}(\hat{A} - X(A(\theta) - B_u(\theta)D_kC_y(\theta))Y - \hat{B}C_y(\theta)Y - XB_u(\theta)\hat{C})(M^{-1})'. \quad (9.14)$$
$$B_k(\theta) = N^{-1}(\hat{B} - xB_u(\theta)D_k). \quad (9.15)$$
$$C_k(\theta) = (\hat{C} - D_kC_yY)(M^{-1})'. \quad (9.16)$$

The matrix inequality conditions of equation 9.11 represent a set of constraints that must be satisfied for each $\theta \in \Theta$; hence, they represent an infinite number of constraints. There are a variety of methods to make the problem tractable. The most attractive require a simple parameter-set geometry and linear fractional parameter dependence. When Θ is a polytope and the parameter dependence is multiaffine, then it is sufficient to evaluate the matrix inequalities simultaneously at the vertices of the polytope. If the dimension of Θ is small enough, the resulting set of linear matrix inequalities (LMIs) can typically be solved quickly using commercially available software on desktop computers. If the parameter dependence in the system matrix coefficients is linear fractional, then a technique known as the S-procedure (Boyd *et al.*, 1994) may be applied, though this technique can yield conservative results (Apkarian and Adams, 1998).

When the parameter set is not polytopic, one must typically resort to gridding the parameter set and simultaneously solving a set of LMIs defined at each grid point. The grid is refined until it is apparent (through solution convergence) that the solution will be valid for points not on the grid. Figure 9.3 illustrates the gridding approach for a parameter set defined by $\Theta := \{(\theta_1, \theta_2): \theta_2 = \theta_1^2, \theta_1 \in [-1, 1]\}$. Note that as the grid is refined, the number of LMI constraints multiplies; LPV controller synthesis may become impractical from a computational standpoint.

Another approach is to attempt to satisfy the conditions for a polytopic set that contains Θ. In the preceding example, the synthesis problem could be approached by solving a set of LMIs defined at the vertices of the polytope $\{(\theta_1, \theta_2): \theta_1 \in [-1, 1], \theta_2 \in [0, 1]\}$, as illustrated in Figure 9.3. The results of this alternative are potentially conservative since they must account for parameter variations that will not occur; worse, controllability or observability may be lost at some points in the larger parameter set so that the conditions of equation 9.11 do not admit a solution for any $\gamma > 0$.

9.3.3 Controller Synthesis for Induced-$\mathcal{L}_\infty$ Performance

A measure of system performance that frequently arises in control problems is the **peak value** of an appropriately selected

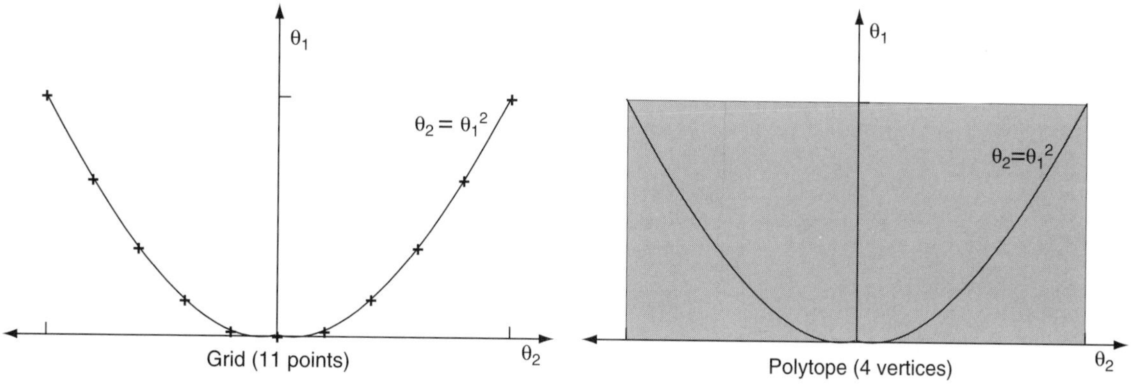

FIGURE 9.3 Alternatives for Nonpolytopic Parameter Sets

plant signal. Control problems in which this performance measure arise are prevalent in the literature: motor control problems with electrical (voltage or current) or mechanical (motion) restrictions and process control problems with chemical concentration restrictions are two examples. In each of these examples, violation of the amplitude restrictions can lead to performance degradation and possibly catastrophic system failure. In these types of performance problems, energy-based design techniques, such as the technique described in Subsection 9.3.1 are often inadequate. The problem considered in this section is that of synthesizing LPV controllers so that the controlled system satisfies bounded amplitude performance constraints; the approach presented here is based on results reported in Bett and Lemmon (1997).

The induced system gains are defined in the usual manner from the associated signal norms.

Definition 3 (Induced-$\mathcal{L}_\infty$ Gain)
The infinity norm of a signal $f\colon \Re^+ \rightarrow \Re^n$ is defined as:

$$\|f\|_\infty := \operatorname*{ess\,sup}_{t\geq 0} \|f(t)\|,$$

where $\|\cdot\|$ denotes the Euclidean vector norm and where $\mathcal{L}_\infty^n$ denotes the normed n-dimensional signal space in the usual fashion under the above signal norm definition. Consider the LPV system of definition 1. For any $\theta \in \mathcal{F}_\Theta$ with $x(t_0) = 0$, the causal linear input/output mapping, $H_\theta\colon \mathcal{L}_\infty^{n_w} \rightarrow \mathcal{L}_\infty^{n_z}$, of the linear time-varying system described in equation 9.7 is defined as:

$$H_\theta w(t) = \int_{t_0}^{t} C(\theta(\tau))\Phi_\theta(t, \tau)B(\theta(\tau))w(\tau)d\tau + D(\theta(t))w(t),$$

where $\Phi_\theta(t, t_0)$ is the associated state transition matrix. The induced-$\mathcal{L}_\infty$ norm of H_θ is given by:

$$\|H_\theta\|_{i\infty} := \sup_{w\in\mathcal{L}_\infty^{n_w},\ \|w\|_\infty<1} \|H_\theta w\|_\infty. \tag{9.17}$$

The induced-$\mathcal{L}_\infty$ synthesis conditions presented here are valid under the assumptions that, for all $\theta \in \Theta$:

- $(A(\theta),\ B_u(\theta),\ C_y(\theta))$ are controllable and detectable;
- $D_{yw}(\theta)$ has full row rank; and
- $B_w(\theta)D'_{yw}(\theta) = 0$.

Fix any number $\alpha > 0$ and performance level γ, and consider the LPV plant in equation 9.9 under the preceding assumptions. A sufficient condition for the existence of a strictly proper and finite-dimensional LPV controller $K(\theta)$ that internally stabilizes the system and renders:

$$\|\mathcal{F}_\ell(P(\theta),\ K(\theta))\|_{i,\infty} < \gamma,$$

for all $\theta \in \mathcal{F}_\Theta$, is the existence of matrices $Z = Z' > 0$ and V such that:

$$\begin{bmatrix} A(\theta)(Z - Y_\alpha) + (Z - Y_\alpha)A'(\theta) + B_u(\theta)V & Y_\alpha C'_y(\theta) \\ +V'B'_u(\theta) + \alpha(Z - Y_\alpha) & \\ C_y(\theta)Y_\alpha & -\frac{1}{\alpha}D_{yw}(\theta)D'_{yw}(\theta) \end{bmatrix} \leq 0 \tag{9.18}$$

$$\begin{bmatrix} (Z - Y_\alpha) & (Z - Y_\alpha)C'_z(\theta) + V'D'_{zu}(\theta) \\ C_z(\theta)(Z - Y_\alpha) + D_{zu}(\theta)V & \gamma^2 I - C_z(\theta)Y_\alpha C'_z(\theta) \end{bmatrix} > 0 \tag{9.19}$$

These equations apply for all $\theta \in \Theta$, where Y_α is the minimal stabilizing solution of the Riccati inequality:

$$\begin{aligned} &\left(A(\theta) + \frac{1}{2}\alpha I\right)Y_\alpha + Y_\alpha\left(A(\theta) + \frac{1}{2}\alpha I\right)' \\ &-\alpha Y_\alpha C'_2(\theta)D_{21}(\theta)(D'_{21}(\theta))^{-1}C_2(\theta)Y_\alpha + \frac{1}{\alpha}B_1(\theta)B'_1(\theta) \leq 0 \end{aligned} \tag{9.20}$$

(Y_α is the minimal stabilizing solution of equation 9.20 if for any other positive definite solution, Y, to equation 9.20 $Y_\alpha - Y$ is negative definite.) In addition, if the above matrix inequalities admit a solution, then one controller that renders

$$\|\mathcal{F}_\ell(P(\theta),\ K(\theta))\|_{i,\infty} < \gamma$$

for all $\theta \in \mathcal{F}_\Theta$ is given by:

$$K = \left[\begin{array}{c|c} A(\theta) + B_u(\theta)\bar{J} - \alpha Y_\alpha C'_y(\theta) & \alpha Y_\alpha C_y{}'(\theta)(D_{yw}(\theta)D_{yw}(\theta)')^{-1} \\ (D_{yw}(\theta)D_{yw}(\theta)')^{-1}C_y(\theta) & \\ \hline \bar{J} & 0 \end{array}\right], \tag{9.21}$$

where $\bar{J} := V(Z - Y_\alpha)^{-1}$.

If Θ is a singleton, then the above conditions are strengthened to necessary and sufficient. It should also be noted that the restriction that Y_α in equation 9.20 be a minimal stabilizing solution is overly restrictive for the existence of a self-scheduled controller. If this restriction, however, is removed, the results do not reduce to necessary and sufficient conditions in the LTI case.

The preceding result represents the key analytical device for the synthesis of LPV controllers satisfying prespecified amplitude performance constraints. Gain-scheduled controller synthesis is accomplished in an iterative fashion: perform a line search for values of $\alpha > 0$, solving (at each α) the synthesis conditions of equations 9.18 through 9.20 for the required matrices and/or the required performance level, γ. When the required performance has been achieved, the LPV controller is formed from the realization in equation 9.21.

The bounded amplitude performance results are intuitively satisfying because they are similar in flavor to the $\mathcal{H}_\infty$ results of Becker and Packard (1994); parameter-dependent controller synthesis requires the feasibility of a "controller" Riccati inequality (equation 9.18), an "observer" Riccati inequality (equation 9.20), and a "spectral radius" coupling inequality (equation 9.19). Thus, the approach to bounded amplitude control described here requires the same type of numerical solutions as, for example, induced-$\mathcal{L}_2$ control, though the bounded amplitude problem may be more computationally intensive due to the line search over α.

As in the induced-$\mathcal{L}_2$ problem, the synthesis conditions are represented by a set of constraints that must be satisfied for each $\theta \in \Theta$ and therefore represent an infinite number of constraints. The techniques discussed for the induced-$\mathcal{L}_2$ problem are applicable here as well.

9.4 Conclusions

Historically, gain scheduled controllers have been comprised of a set of linear controllers whose gains are adjusted according to a scheduling law derived via heuristic methods, such as interpolating on gains, poles and zeros. Gain scheduling via the linearization techniques described in Section 9.2 has traditionally been justified through successful application. For additional theoretical justification, see Rugh (1991) and Shamma and Athans (1990, 1991). Related theoretical work can be found in the area of extended linearization Baumann and Rugh (1986), which can be viewed as gain scheduling on input, output, or state variables rather than exogenous scheduling variables. This approach involves nonlinear control design in the spirit of the gain-scheduling approach described in Section 9.2.

Gain-scheduling methods for LPV systems have emerged only recently. The relationship between gain scheduling via linearization and parameter-dependent systems is explained in Shamma and Athans (1990, 1991). Early work on parameter-dependent controllers is covered in Kamen and Khargonekar (1984); more recently, parameter-dependent synthesis approaches based on Lyapunov and small-gain techniques have appeared (Becker and Packard, 1994; Packard, 1994; Apkarian and Gahinet, 1995) and frame the synthesis problem as a convex optimization problem. These approaches generalize $\mathcal{H}_\infty$ techniques for LTI systems to LPV systems. Other methods are described in Apkarian and Adams (1998) and Apkarian *et al.* (1995).

The primary advantage of dealing with LPV systems is that difficulties associated with the derivation of the scheduling law and verifying performance of the controlled system are alleviated. The methods applied to LPV systems inherently account for the time-varying nature of the plant and handle the entire parameter set with a single design. The need for extensive simulations can be significantly reduced.

The LPV synthesis conditions described here make use of a fixed Lyapunov function (as opposed to one that depends on the scheduling variables) to characterize stability and performance. Such approaches are potentially conservative because they allow for arbitrary rates of variation in the scheduling variables, a generalization that may render the synthesis conditions infeasible. A significant improvement over such techniques can be obtained by exploiting the concept of parameter-dependent Lyapunov functions. Parameter-dependent Lyapunov functions allow incorporation of knowledge on the parameter variation rate into the analysis or synthesis technique and can lead to less conservative answers. Implementation of solutions based on parameter-dependent Lyapunov functions require not only the real-time measurement of the parameter but also its time derivative. This is generally prohibitive because such values are either unavailable or difficult to estimate in real-time. Generally speaking, there is no systematic rule for selecting the functional dependence of the Laypunov function on the parameter. The key idea is to mimic the parameter dependence of the plant in the Lyapunov function variables.

References

Apkarian, P., and Adams, R.J. (1998). Advanced gain-scheduling techniques for uncertain systems. *IEEE Transactions on Control Systems Technology* 6, 21–32.

Apkarian, P., and Gahinet, P. (1995). A convex characterization of gain-scheduled H_∞ controllers. *IEEE Transactions on Automatic Control* 40, 853–864.

Apkarian, P., Gahinet, P., and Becker, G. (1995). Self-scheduled H_∞ control of linear parameter-varying systems: A design example. *Automatica* 31, 1251–1261.

Baumann, W.T., and Rugh, W.J. (1986). Feedback control of nonlinear systems by extended linearization. *IEEE Transactions on Automatic Control* AC-31, 40–46.

Becker, G., and Packard, A. (1994). Robust performance of linear parametrically varying systems using parametrically dependent linear feedback. *Systems and Control Letters* 23, 205–215.

Bett, C.J., and Lemmon, M.D. (1997). Sufficient conditions for self-scheduled bounded amplitude control. *Technical Report ISIS-97-009*, Department of Electrical Engineering, University of Notre Dame.

Bett, C.J., and Lemmon, M.D. (1999). Bounded amplitude performance of switched LPV systems with applications to hybrid systems. *Automatica* 35, 491–503.

Boyd, S., El Gaoui, L., Feron, E., and Balakrishnan, V. (1994). Linear matrix inequalities in system and control theory. Philadelphia: Society for Industrial and Applied Mathematics.

Hanus, R., Kinnaert, M., and Henrotte, J.L. (1987). Conditioning technique, a general antiwindup and bumbless transfer method. *Automatica* 23, 729–739.

Kamen, E.W., and Khargonekar, P.P. (1984). On the control of linear systems whose coefficients are functions of parameters. *IEEE Transactions on Automatic Control* AC-29, 25–33.

Nichols, R.A., Reichert, R.T., and Rugh, W.J. (1993). Gain scheduling for *H*-infinity controllers: A flight control example. *IEEE Transactions on Control Systems Technology* 1, 69–79.

Packard, A. (1994). Gain scheduling via linear fractional transformations. *Systems and Control Letters* 22, 79–92.

Rugh, W.J. (1991). Analytical framework for gain scheduling. *IEEE Control Systems Magazine*, 11, 79–84.

Shamma, J.S., and Athans, M. (1990). Analysis of gain-scheduled control for nonlinear plants. *IEEE Transactions on Automatic Control*, 35, 898–907.

Shamma, J.S., and Athans, M. (1991). Guaranteed properties of gain-scheduled control for linear parameter-varying plants. *Automatica* 27, 559–564.

10

Sliding-Mode Control Methodologies for Regulating Idle Speed in Internal Combustion Engines

Stephen Yurkovich
Center for Automotive Research, The Ohio State University, Columbus, Ohio, USA

Xiaoqiu Li
Cummins Engine, Columbus, Indiana, USA

10.1 Introduction

Generally speaking, sliding-mode control (SMC) is a robust nonlinear control algorithm that employs discontinuous control to force system state trajectories to lie on some prescribed sliding surface (Utkin, 1992; DeCarlo *et al.*, 2000, 1996, 1988). Used widely for its robustness to model parameter uncertainties and external disturbances, SMC methodology has had great success in a wide variety of practical applications due to its ease of implementation and its favorable robustness properties. It can be argued, in fact, that this tendency toward successful application (a tendency enjoyed by only a few nonlinear design methodologies) has given rise to several new areas of active research in the general field of SMC. For example, until recent years, the bulk of research in SMC has been on systems without time delay and on continuous-time systems; thus, two other areas of research are addressed in this chapter, namely **SMC for systems with inherent delay** in the state and input and SMC for **discrete-time, sampled-data systems.**

An important issue in practical applications is how to address transport delay inherent in many physical systems. Although relatively little attention has been given to SMC for this class of systems in the open literature, several good works have recently appeared that address some of the associated difficulties. For example, necessary and sufficient conditions for the existence of sliding modes for a multidimensional system are given in Jafarov (1990), where the asymptotic stability of the system on the sliding surface using a Lyapunov-Krasovsky functional is proved. Drakunov and Utkin (1992) give the two-stage design procedure for dynamic systems that can be written in differential-difference equations in the block form. A linear transformation for distributed time-delay systems is proposed in Zheng *et al.* (1995), and the developed SMC algorithm is applied to liquid propellent rocket motors. In Hu *et al.* (1998), a transformation for systems with delay only in the control is used to design a sliding-mode controller for the transformed systems; the advantages over state feedback control are given. Sufficient conditions for sliding-mode control of systems with uncertain, continuously differentiable, and bounded time delay in the states are given in Sinha *et al.* (1997), using an invariant cone of positive initial functions. Finally, we note that El-Khazali (1998) proposes an output feedback sliding-mode control design method for similar uncertain time-delay systems.

With respect to the second area, discrete-time systems, it is well-known that "chattering" or oscillation problems occur at the boundary layer when the controls are implemented by digital computers at a given sampling frequency (Furuta, 1990). For this reason, researchers have been investigating the development of discrete sliding-mode control (DSMC) for more than 10 years; see, for example, the works of Furuta (1990), Sira-Ramirez (1991), Spurgeon (1991), and Kaynak and Denker (1993). One particular direction of research along these lines has been toward the coupling of adaptation with DSMC, primarily because the discrete sliding-mode control might not exist in the presence of system parameter uncertainties or external disturbances; this, in fact, is one of the main differences between DSMC and continuous sliding-mode control. There have appeared several good works on adaptive DSMC for linear time-invariant systems. For example, Chan (1997) proposes a nonswitching type of adaptive DSMC design method for a minimum-phase plant in the input/output form with bounded disturbances, whereas Furuta suggests using a switching type control law (Furuta, 1993). Both of these works prove that the state trajectories will converge to a small neighborhood of the defined sliding surface in finite time. Bartolini *et al.* (1995) combined adaptive control with sliding-mode control for discrete linear systems in state-space forms.

This chapter describes two recently developed theoretical results, extensions to the existing theory highlighted in the previous paragraphs for SMC based on Li and Yurkovich (2000, 2001). In the first technique (Li and Yurkovich, 2001), where SMC is extended for a class of point-delayed systems, a linear transformation is applied to convert the delayed system to a delay-free system whose spectra embed all the unstable poles of the original system with a given stability margin. It can be proven that the delayed system is asymptotically stable with SMC, based on the delay-free system. In the second approach, an adaptive SMC design method for discrete nonlinear systems, where explicit knowledge of the system dynamics is not available, is developed. A three layer feed-forward neural network is used as a function approximator for the unknown dynamics. The control law is designed based on the outputs of the approximators, and the sliding surface is defined in terms of a stable polynomial of the system outputs.

The focus for applications of these new techniques is on a representative problem from automotive engineering, that of idle speed control for the typical spark-ignited, internal combustion (IC) engine. Idle is one of the most frequently encountered operating conditions in city driving. Thus, there is motivation for adjustment to the lowest possible idle speed (i.e., while keeping smooth engine performance) as a way of reducing fuel consumption. Factors that most affect the idle speed are the intake-airflow and the ignition timing. The idle speed can therefore be controlled by changing the air quantity (charge adjustment) and/or the ignition timing. In this chapter, we will be concerned with disturbance torques acting on the plant model and being able to maintain nearly constant idle speed. These disturbance torques may arise from several sources, such as the engagement of the automatic transmission, the turning-on (cycling) of the air conditioner unit, or the effects of loading with the power steering system.

Characteristics inherent to the IC engine idle speed control problem—an induction-to-power delay and finite sensor/actuator response time—can lead to design challenges with respect to disturbance rejection and/or lead to problems of chattering with traditional SMC designs. For this reason, the idle speed control problem is chosen as a focus for application here, and both techniques for SMC described in this chapter (for point-delay systems and adaptive DSMC) are applied to the problem, where simulation and experimental results are provided.

10.2 SMC for Systems with Delay

In this section, we describe an extension of an idea that first appeared in Fiagbedzi and Pearson (1986, 1987) for transforming delayed systems. Dividing the unstable poles of the delayed system with a given stability margin into N sets, where each set contains n unstable poles, N characteristic matrices can be found that inherit the nN unstable poles. A linear transformation can then be applied to convert the original delayed system to a delay-free system with the state matrix being the direct sum of the N characteristic matrices. The novelty of the approach described here is in showing that SMC can then be designed based on the delay-free system, and **asymptotic stability** can be shown for *both* **delay-free** and **delayed systems** on certain sliding surfaces. The reaching condition is proved in the case of bounded and matched external disturbance, and the sliding surface is chosen by the means of a Lyapunov function. It should be noted that the sliding surface is a function of previous system states and controls and has intuitive appeal for time-delay systems.

10.2.1 Problem Formulation

We consider a system with point delay in both states and control described by:

$$\Sigma: \dot{x}(t) = A_0x(t) + A_1x(t-r) + B_0u(t) + B_1u(t-h) + Df(t), \quad (10.1)$$

where r and h are the delays with the system states and control respectively, $x \in \mathcal{R}^n$ represents the system states, $u \in \mathcal{R}^m$ represents the control, $f \in \mathcal{R}^k$ is the (bounded) disturbance, A_0 and A_1 are $n \times n$ matrices, B_0 and B_1 are $n \times m$ matrices, and D is an $n \times k$ matrix.

Denote $x_t(\theta) = x(t+\theta)$ for $\theta \in [-r, 0]$, and $u_t(\tau) = u(t+\tau)$ for $\tau \in [-h, 0]$. The initial conditions are given as $x(\theta) = x_0(\theta)$, $\theta \in [-r, 0]$ and $u(\tau) = u_0(\tau)$, $\tau \in [-h, 0]$, where $x_0 \in \mathcal{C}([-r, 0]; \mathcal{R}^n)$ and $u_0 \in \mathcal{C}([-h, 0]; \mathcal{R}^m)$; that is, x_0 is a real continuous function of $\theta \in [-r, 0]$, and u_0 is a real continuous function of $\tau \in [-h, 0]$.

10.2.2 Controller Design

Denote a non-negative real number ν_0 as the required "stability margin," and define the set of "unstable" poles (Fiagbedzi and Pearson, 1986) of Σ (given by equation 10.1) as:

$$\sigma_u(\Sigma) = \{s \in \mathcal{C} : \det(sI - A_0 - A_1e^{-rs}) = 0,\ Re(s) \geq -\nu_0\}. \quad (10.2)$$

Because the number of the "unstable" poles is finite, one can generally find ways to divide $\sigma_u(\Sigma)$ as:

$$\sigma_u(\Sigma) = \Lambda_c = \Lambda_1 \cup \Lambda_2 \cup \ldots \cup \Lambda_N, \quad (10.3)$$

where Λ_i, $i = 1, 2 \ldots N$ includes n "unstable" poles. Assuming, for each Λ_i, there exists $A^{(i)}$ satisfying the characteristic matrix equation, then:

$$A^{(i)} = A_0 + e^{-rA^{(i)}}A_1, \quad (10.4)$$

and $\sigma(A^{(i)}) = \Lambda_i$; hence, one can obtain:

$$A_c = \bigoplus_{i=1}^{N} A^{(i)} \text{ and } \sigma(A_c) = \Lambda_c, \quad (10.5)$$

where $\oplus$ denotes the direct sum.

Defining a linear transformation as:

$$\begin{aligned} z_c(t) =& (\mathcal{T}_{AC}(x, u))(t) = \sum_{i=1}^{N}(e_i \otimes I_n)x(t) \\ &+ \int_{-r}^{0} e^{A_c\theta}\sum_{i=1}^{N}(e_i \otimes I_n)A_1x(t-r-\theta)d\theta \\ &+ \int_{-h}^{0} e^{A_c\tau}\sum_{i=1}^{N}(e_i \otimes I_n)B_1u(t-h-\tau)d\tau, \end{aligned} \quad (10.6)$$

where e_i is an N-dimensional unit column vector whose entries are 0 except for the ith entry, which is 1, $e_i \otimes I_n = [0_{n\times n}, \ldots, 0_{n\times n}, I_{n\times n}, 0_{n\times n}, \ldots, 0_{n\times n}]^T$. This is true with $I_{n\times n}$ as the ith entry, and one can obtain a delay-free system:

$$\Sigma': \dot{z}(t) = A_cz(t) + B_cu(t) + D_cf(t), \quad (10.7)$$

where $B_c = \Sigma_{i=1}^{N} e_i \otimes (B_0 + e^{-hA^{(i)}}B_1)$, A_c is defined in equation 10.5, and $D_c = \Sigma_{i=1}^{N} e_i \otimes D$.

- **Assumption 1**: The characteristic matrices $A^{(i)}$, $i = 1, 2, \ldots, N$, exist [i.e., the n left eigenvectors of Σ corresponding to Λ_i are linearly independent (Fiagbedzi and Pearson, 1986)].
- **Assumption 2:** Σ is spectrally controllable (Olbrot, 1978):
$$\begin{aligned} &\operatorname{rank}\left[sI - A_0 - e^{-rs}A_1 | B_0 + e^{-hs}B_1\right] = n, \\ &\forall s \in \{s \in \mathcal{C},\ Re(s) \geq -\nu_0\}. \end{aligned} \quad (10.8)$$
- **Assumption 3**: The disturbance $D_cf(t)$ satisfies the matching conditions (i.e., D_c can be written in the form $D_c = B_cE$, where E is an $m \times k$ matrix). That is, $D = (B_o + \exp^{-hA^{(i)}}B_1)E$ for $i = 1, 2, \ldots, N$.

With this problem definition and these assumptions, we will need the following lemma. **Lemma 1**, (A_c, B_c) is controllable if **assumption 2** holds and the spectra of $A^{(i)}$, $i = 1, 2, \ldots N$ do not overlap (Fiagbedzi and Pearson, 1986).

We are now in a position to describe the controller design for systems of equations 10.1 and 10.7. Generally speaking, sliding-mode controller design consists of two steps, described next.

Sliding Surface Construction

An appropriate **sliding surface** on which the system has desired behavior must be constructed. Because (A_c, B_c) is controllable, one can find a feedback gain matrix $K \in \mathcal{R}^{m\times nN}$ such that $A_c + B_cK$ is asymptotically stable. Denote $A_s = A_c + B_cK$. Then there exists a positive definite matrix P satisfying the Ricatti inequality:

$$PA_s + A_s^TP < 0. \quad (10.9)$$

This leads to the following important lemma needed for a subsequent stability proof.

Lemma 2 The system equation 10.7 is asymptotically stable on the sliding surface:

$$S(t) = B_c^TPz(t) = 0. \quad (10.10)$$

Because the disturbance is matched, it will not affect the system behavior on the sliding surface. Write the sliding surface as:

$$S = B_c^TPz = Cz, \quad (10.11)$$

where $C = B_c^T P \in \mathcal{R}^{m\times nN}$. Substituting equation 10.6 into equation 10.11, we obtain the switching functional:

$$S(t) = C\left[\sum_{i=1}^{N}(e_i \otimes I_n)x(t) + \int_{-r}^{0} e^{A_c\theta}\sum_{i=1}^{N}(e_i \otimes I_n)A_1 x(t-r-\theta)d\theta + \int_{-h}^{0} e^{A_c\tau}\sum_{i=1}^{N}(e_i \otimes I_n)B_1 u(t-h-\tau)d\tau\right]. \quad (10.12)$$

Partitioning C as $C = [C^1, C^2, \ldots, C^N]$, where $C^i \in \mathcal{R}^{m\times n}$ and $i = 1, 2, \ldots, N$, rewrite S as:

$$S(t) = \left(\sum_{i=1}^{N} C^i\right)x(t) + \int_{-r}^{0}\left(\sum_{i=1}^{N} C^i e^{A^{(i)}\theta}\right)A_1 x(t-r-\theta)d\theta + \int_{-h}^{0}\left(\sum_{i=1}^{N} C^i e^{A^{(i)}\tau}\right)B_1 u(t-h-\tau)d\tau. \quad (10.13)$$

The system is initially on the switching surface at t_r, $S(t) = 0$ and $\dot{S}(t) = 0$ for $t \geq t_r$. Furthermore:

$$\dot{S}(t) = C\dot{z}(t) = CA_c z(t) + CB_c u(t) + CD_c f(t)$$
$$= CA_c\left\{\sum_{i=1}^{N}(e_i \otimes I_n)x(t) + \int_{-r}^{0} e^{A_c\theta}\sum_{i=1}^{N}(e_i \otimes I_n)A_1 x(t-r-\theta)d\theta + \int_{-h}^{0} e^{A_c\tau}\sum_{i=1}^{N}(e_i \otimes I_n)B_1 u(t-h-\tau)d\tau\right\} + CB_c u(t) + CD_c f(t). \quad (10.14)$$

Assuming $(CB_c)^{-1}$ exists, one can usually obtain the equivalent control by letting $\dot{S}(t) = 0$. However, the disturbance term $CD_c f(t)$ is unknown. Hence, we will instead calculate an **equivalent control for the nominal system** given by:

$$u_{eq,\,nom}(t) = -(CB_c)^{-1}CA_c\left\{\sum_{i=1}^{N}(e_i \otimes I_n)x(t) + \int_{-r}^{0} e^{A_c\theta}\sum_{i=1}^{N}(e_i \otimes I_n)A_1 x(t-r-\theta)d\theta + \int_{-h}^{0} e^{A_c\tau}\sum_{i=1}^{N}(e_i \otimes I_n)B_1 u(t-h-\tau)d\tau\right\}. \quad (10.15)$$

Again, partitioning $CA_c \in \mathcal{R}^{m\times nN}$ as $CA_c = [\bar{C}^1, \bar{C}^2, \ldots, \bar{C}^N]$, where $\bar{C}^i \in \mathcal{R}^{m\times n}$, we have:

$$u_{eq,\,nom}(t) = -(CB_c)^{-1}\left\{\left(\sum_{i=1}^{N}\bar{C}^i\right)x(t) + \int_{-r}^{0}\left(\sum_{i=1}^{N}\bar{C}^i e^{A^{(i)}\theta}\right)A_1 x(t-r-\theta)d\theta + \int_{-h}^{0}\left(\sum_{i=1}^{N}\bar{C}^i e^{A^{(i)}\tau}\right)B_1 u(t-h-\tau)d\tau\right\}. \quad (10.16)$$

Note that $u_{eq,\,nom}$ is not the equivalent control for the system of equation 10.7. Rather, it is calculated solely for the purpose of control design.

Control Law Design

A **suitable control law** should be chosen to satisfy the reaching condition (i.e., global stability of the switching surface, $S = 0$). Specifically, for a Lyapunov function, $V = S^T S/2$, satisfaction of the inequality $\dot{V} = S^T\dot{S} < 0$ should be guaranteed.

With the choice of:

$$\dot{S} = -(CB_c)^{-1}D_1 sgn(S), \quad (10.17)$$

where D_1 is a diagonal matrix with elements $d_i > 0$, $i = 1, 2, \ldots, m$ and $\text{sgn}(S) = [\text{sgn}(S_1), \ldots, \text{sgn}(S_m)]^T$; one can obtain the control with:

$$u(t) = -(CB_c)^{-1}\left\{\left(\sum_{i=1}^{N}\bar{C}^i\right)x(t) + \int_{-r}^{0}\left(\sum_{i=1}^{N}\bar{C}^i e^{A^{(i)}\theta}\right)A_1 x(t-r-\theta)d\theta + \int_{-h}^{0}\left(\sum_{i=1}^{N}\bar{C}^i e^{A^{(i)}\tau}\right)B_1 u(t-h-\tau)d\tau\right\} - (CB_c)^{-1}D_1\text{sgn}(S). \quad (10.18)$$

Stability Analysis

We now state, without proof [see Li and Yurkovich (2001)] the following **stability results**:

- **Theorem 1**: Suppose the disturbance is bounded and $\|f(x, t)\|_2 \leq \beta$. The system given by equation 10.7 with control given in equation 10.18 is asymptotically stable if $d_0 > \|CD\|_2\beta$, where $d_0 = \min(d_1, d_2, \ldots, d_m)$.
- **Theorem 2**: The system given by equation 10.1 with control given in equation 10.18 is asymptotically stable with a stability margin of ν_0 if $\sigma_u(\Sigma) \subseteq \bigcup_{i=1}^{N}\sigma(A^{(i)})$ and $\sigma(A^{(i)})\bigcap\sigma(A^j) = \varnothing$ for $i \neq j$, where $\varnothing$ denotes the void set.

10.3 Discrete Adaptive Sliding-Mode Control

Let us now move to the second method described in this chapter and give the theoretical development prior to presenting application results for both methods. As an extension to the works described in the previous sections, in this section we describe an **adaptive DSMC design method** for **nonlinear systems** having an **input/output form**. Explicit knowledge of the system dynamics is not required; instead, feed-forward neural networks are applied to approximate the unknown part of the system dynamics based on past system outputs and control inputs. The sliding surface, defined in terms of a stable polynomial of the system tracking error, is used to tune the neural network structure online using gradient methods. The **sliding-mode control law** is given based on the approximated system dynamics. It can be proved that the state trajectories will converge to a small sliding sector whose size could be as small as desired provided that the structure of the neural networks is large enough.

10.3.1 Problem Formulation

Consider a discrete nonlinear single-input, single-output feedback linearizing system defined by:

$$y(k+d) = \Phi[\lambda(k)] + \Gamma[\lambda(k)]u(k), \tag{10.19}$$

where $\Phi[\bullet]$ and $\Gamma[\bullet]$ are smooth nonlinear functions that might be unknown and where $\lambda(k)$ is a vector consisting of previous control inputs and system outputs given by:

$$\begin{aligned}\lambda(k) = [&y(k), y(k-1), \ldots, y(k-n), u(k-1),\\ &u(k-2), \ldots, u(k-m)]^T \in S_\lambda \subset \mathcal{R}^{n+m+1}.\end{aligned} \tag{10.20}$$

In addition, the $y(k)$ is the system output, the $u(k)$ is the control input satisfying $u(k) \in S_u \subset \mathcal{R}$, the S_λ and S_u denote complex sets, the n and m are constants that are not necessarily known, and the d is the (known) relative degree of the system.

The control objective is to have the system output $y(k)$ track a known reference input $r(k)$. Specifically, the objective of sliding-mode control is to steer the system state trajectories in a sliding surface that is defined in terms of a stable polynomial of the tracking error and to maintain the subsequent motion of the state trajectories on this surface.

10.3.2 Controller Design

Discrete Sliding-Mode Control

As an initial step in designing the control law, let us assume $\Phi[\bullet]$ and $\Gamma[\bullet]$ are known functions. Define the discrete sliding surface as:

$$s(k+d) = C(q^{-1})e(k+d), \tag{10.21}$$

where $C(q^{-1}) = 1 + c_0 q^{-1} + \ldots + c_l q^{-1}$ is a Hurwitz polynomial, q^{-1} is a unit-delay operator, and $e(k+d) = y(k+d) - r(k+d)$ is the tracking error.

The equivalent control method (Drakunov and Utkin, 1992) can be applied to ensure $s(k+d) = 0$. Obviously if $s(k) = 0$ for all k, the error dynamics will be determined by $C(q^{-1})$ only. For a stable polynomial $C(q^{-1})$, $\lim_{k\to\infty} e(k) = 0$ if $\lim_{k\to\infty} s(k) = 0$. Hence, instead of trying to make the tracking error, $e(k)$, equal to zero directly, the control goal is changed to enforce the state trajectories on the manifold $s(k) = 0$ at each sampling instant. Note that the state trajectories can deviate from the sliding surface between the sampling points, which is different from continuous time sliding-mode control in which the state trajectories are kept on the sliding manifold at all times.

Upon substitution for the error in equation 10.21, we have the following:

$$\begin{aligned}s(k+d) &= C(q^{-1})y(k+d) - C(q^{-1})r(k+d)\\ &= C(q^{-1})\Phi[\lambda(k)] + C(q^{-1})\Gamma[\lambda(k)]u(k)\\ &\quad - C(q^{-1})r(k+d)\\ &= F[\lambda'(k)] + G[\lambda'(k)]u(k) - C(q^{-1})r(k+d).\end{aligned} \tag{10.22}$$

In equation 10.22, $F[\lambda'(k)] = C(q^{-1})\Phi[\lambda(k)]$ and $G[\lambda'(k)] = C(q^{-1})\Gamma[\lambda(k)]$, assumed to be bounded away from zero. Moreover, $\lambda'(k) = [y(k), y(k-1), \ldots, y(k-n-l), u(k-1), u(k-2), \ldots, u(k-m-l)]$.

Using the equivalent control method, we obtain the control $u_{eq}(k)$ by letting $s(k+d) = 0$, resulting in:

$$u_{eq}(k) = -G[\lambda'(k)]^{-1}\{F[\lambda'(k)] - C(q^{-1})r(k+d)\}. \tag{10.23}$$

After reaching the sliding surface, the system dynamics are described by these two equations:

$$s(k+d) = F[\lambda'(k)] + G[\lambda'(k)]u(k) - C(q^{-1})r(k+d) = 0. \tag{10.24}$$

$$\begin{aligned}y(k+d) =& \Phi[\lambda(k)] - \Gamma[\lambda(k)]G[\lambda'(k)]^{-1}\{F[\lambda'(k)]\\ &- C(q^{-1})r(k+d)\}.\end{aligned} \tag{10.25}$$

Adaptive Sliding-Mode Control

In most applications, exact knowledge of $F[\bullet]$ and $G[\bullet]$ is not possible. As a solution, some methodology should be used to approximate these nonlinear functions. Three-layer feed-forward neural networks are applied in this work, with output in the form:

$$y = \sum_{i=1}^{p} \omega_i \mathcal{H}\left(\sum_{j=1}^{r} \omega_{ij} x(j) + \hat{\omega}_j\right) + \hat{\omega}, \tag{10.26}$$

where p is the number of the neural nodes in the hidden layer; r is the number of the inputs; ω_i, ω_{ij}, and $(i = 1, 2, \ldots, p,$

$j = 1, 2, \ldots, r$) are the weights; $\hat{\omega}_j$ and $\hat{\omega}$ are the bias terms; and $\mathcal{H}$ is the squashing function that should be continuous, nonconstant, bounded, monotonically increasing (Funahashi, 1989), and differentiable (Chin and Coats, 1986). A hyperbolic tangent function is used for $\mathcal{H}$ in this work.

Denote the network output by $y = \mathcal{F}(x, A)$, where A represents the network structure and contains all the network weights and bias terms (i.e., $A = [\hat{\omega}, \omega_1, \ldots, \omega_p, \hat{\omega}_1, \ldots, \hat{\omega}_r, \omega_{11}, \ldots, \omega_{1r}, \omega_{21}, \ldots, \omega_{2r}, \ldots, \omega_{p1}, \ldots, \omega_{pr}]^T$). Define the change of the network output with respect to A as:

$$\zeta(x, A) = \frac{\partial \mathcal{F}(x, A)}{\partial A}. \tag{10.27}$$

It has been shown (Hornik *et al.*, 1989) that for any nonlinear function $f(x)$, there exists a neural network with the structure A^* that satisfies $|\mathcal{F}(x, A^*) - f(x)| \leq \varepsilon$ given any small positive number ε (characteristics of universal approximators). Note that ε could be as small as desired, but decreasing ε generally increases the size of the neural networks.

Denoting $\mathcal{F}(x, A)$ as the actual output of the neural networks, the representation error could be written in the form:

$$\begin{aligned} f(x) - \mathcal{F}(x, A) &= f(x) - \mathcal{F}(x, A^*) + \mathcal{F}(x, A^*) - \mathcal{F}(x, A) \\ &= \mathcal{O}(\varepsilon) + \mathcal{F}(x, A^*) - \mathcal{F}(x, A) \\ &= -\phi^T\zeta + \mathcal{O}(|\phi|^2) + \mathcal{O}(\varepsilon) \\ &= -\phi^T\zeta + \alpha. \end{aligned} \tag{10.28}$$

In equation 10.28, $\phi = A - A^*$, $\mathcal{O}(\varepsilon)$ is the ideal approximation error, $\mathcal{O}(|\phi|^2)$ represents residual terms after linearization, and $\alpha = \mathcal{O}(|\phi|^2) + \mathcal{O}(\varepsilon) \leq \bar{\alpha}$.

Rewrite the control law of equation 10.23 as:

$$u(k) = -\hat{G}[\lambda'(k)]^{-1}\{\hat{F}[\lambda'(k)] - C(q^{-1})r(k+d)\}, \tag{10.29}$$

where $\hat{F}[\bullet]$ and $\hat{G}[\bullet]$ represent the approximation of the nonlinear functions F and G, respectively.

The sliding dynamics become the following:

$$\begin{aligned} s(k+d) &= F[\lambda'(k)] + G[\lambda'(k)]u(k) - C(q^{-1})r(k+d) \\ &= F[\lambda'(k)] - \hat{F}[\lambda'(k)] + (G[\lambda'(k)] - \hat{G}[\lambda'(k)])u(k) \\ &= F[\lambda'(k)] - \mathcal{F}[\lambda'(k), A_f(k+d-1)] \\ &\quad + (G[\lambda'(k)] - \mathcal{F}[\lambda'(k), A_g(k+d-1)])u(k) \\ &= -\phi_f^T(k+d-1)\zeta_f(k) + \alpha_f(k) - \phi_g^T(k+d-1)\zeta_g(k)u(k) \\ &\quad + \alpha_g(k)u(k) \\ &= -\phi^T(k+d-1)\zeta(k) + \alpha(k), \end{aligned} \tag{10.30}$$

where the following equations hold true:

$$\zeta_f(k) = \frac{\partial \mathcal{F}[\lambda'(k), A_f(k+d-1)]}{\partial A_f(k+d-1)}. \tag{10.31}$$

$$\zeta_g(k) = \frac{\partial \mathcal{F}[\lambda'(k), A_g(k+d-1)]}{\partial A_g(k+d-1)}. \tag{10.32}$$

$$\phi(k+d-1) = [\phi_f^T(k+d-1), \phi_g^T(k+d-1)]^T. \tag{10.33}$$

$$\zeta(k) = [\zeta_f^T(k), \zeta_g^T(k)u(k)]^T. \tag{10.34}$$

$$\alpha(k) = \alpha_f(k) + \alpha_g(k)u(k) \leq \bar{\alpha}. \tag{10.35}$$

Defining a sliding sector $S = \{s(k) \| s(k)| \leq \gamma\bar{\alpha}\}$, where $1 < \gamma < \infty$, we have the following theorem; for proof, see Hornik (1989).

- **Theorem 3:** With the adaptation law:

$$\phi(k) = \phi(k-1) + \frac{\kappa(k)\eta\zeta(k-d)}{1 + |\zeta(k-d)|^2}s(k), \tag{10.36}$$

where $\kappa(k) = \begin{cases} 0 & s(k) \in S \\ 1 & \text{otherwise,} \end{cases}$ and $0 < \eta < 2(1 - \frac{1}{\gamma})$.

1) $|\phi(k+1) - \phi(k)|$ will converge to zero;
2) $\lim_{k \to \infty} s(k) = 0$.

Calculation of $\zeta(k-d)$ in the statement of the theorem 3 is carried out using equation 10.27, with equation 10.31 and equation 10.32 as appropriate. We also note that a projection algorithm could be used to ensure $\hat{G}[\bullet]$ is bounded away from zero. Finally, in the case that we do not know the number of past outputs and control inputs that characterize the system (i.e., precise values of n and m), trial and error procedures can be used.

10.4 Application: IC Engine Idle Speed Control

The importance of idle speed control is due in part to vehicles (in a typical drive cycle) spending a large percentage of fuel in the idle condition. A satisfactory idle speed control algorithm should regulate the engine operating at a low speed while effectively rejecting typical torque disturbances due to accessory loads, such as the power steering pump, the air conditioning compressor, the various electrical loads, the engagement of the automatic transmission, and so on. A production idle speed controller generally consists of an anticipatory term using the information from the accessories, a proportional-integral feedback control term operating on the air mass signal, and a proportional feedback term for the spark timing. A survey of idle speed models and control methodologies is given by Hrovat *et al.* (1996).

To minimize fuel consumption, the engine should operate at a low speed; however, low idle speed could induce large engine speed variations or even cause the engine to stall in the presence of torque disturbances. Hence, a suitable control law should keep the idle speed as low as possible while minimizing speed variations and maintaining the capability of torque disturbance rejection. One of the main difficulties of idle speed control lies with the induction-to-torque delay; as an additional challenge in this application, the desired idle speed is set to 611 RPM instead of the normal (production) idle speed of 740 RPM.

10.4.1 Engine Model for Idle Speed Control

A highly simplified two-input (idle bypass valve opening and spark advance), two-output (engine speed and intake manifold pressure) **idle speed control model** for IC engines was developed by Yurkovich and Simpson (1997) and used in this work. The model includes intake manifold dynamics, induction-to-power delay, and engine rotational dynamics encompassed in the equations:

$$\frac{dp_m(\theta)}{d\theta} + \frac{\eta_v V_d}{4\pi V_m} p_m(\theta) = K_1 \frac{\alpha(\theta)}{\omega(\theta)}. \quad (10.37)$$

$$\frac{d\omega(\theta)}{d\theta} = -\frac{B}{J} + \frac{\tau_e(\theta)}{J\omega(\theta)}. \quad (10.38)$$

$$\tau_e(\theta) = K_\tau p_m(\theta - \theta_d) + K_\delta \Delta\delta(\theta) + \tau_f(\theta). \quad (10.39)$$

The parameters for a Ford V-8 engine are shown in Table 10.1, and all of the variables used in these dynamical equations are defined in the Appendix.

Although simple in nature, we emphasize that this model encompasses the essential dynamics needed for control design. Note that the model is constructed in the crank angle domain instead of the time domain. Because the engine inherently divides its continuous physical processes into four distinct events (intake, compression, power, and exhaust), representation of the engine dynamics in the crank angle domain (as opposed to the time domain) is intuitively appealing and has certain advantages for control purposes, particularly for the idle speed control problem (Yurkovich and Simpson, 1997; Chin and Coats, 1986).

Letting $K_2 = \eta_v V_d / 4\pi V_m$ and linearizing equations 10.37, 10.38, and 10.39 about the nominal operating point (ω_0, p_{m0}) (using the notation Δ to denote increments), the state variable form of this model, with $x(\theta) = [\Delta\omega(\theta), \Delta p_m(\theta)]^T$ and control input $u(\theta) = [\Delta\alpha(\theta), \Delta\delta(\theta)]^T$ as well as $f(\theta) = \tau_f(\theta)$, is as follows:

$$\dot{x}(\theta) = A_0 x(\theta) + A_1 x(\theta - \theta_d) + B_0 u(\theta) + Df, \quad (10.40)$$

where:

$$A_0 = \begin{pmatrix} \frac{-\tau_{e_o}}{J\omega_o^2} & 0 \\ \frac{-K_1\alpha_o}{\omega_o^2} & -K_2 \end{pmatrix}, \quad A_1 = \begin{pmatrix} 0 & \frac{K_\tau}{J\omega_o} \\ 0 & 0 \end{pmatrix},$$

$$B_0 = \begin{pmatrix} 0 & \frac{K_\delta}{J\omega_o} \\ \frac{K_1}{\omega_o} & 0 \end{pmatrix}, \text{ and } D = \begin{pmatrix} \frac{1}{J\omega_o} \\ 0 \end{pmatrix}.$$

TABLE 10.1 The V-8 Engine Parameters

Parameter	Value	Units
η_v	0.55	Dimensionless
V_d	0.0046	m^3
V_m	0.0029	m^3
J	0.0843	$Nm - sec^2/rad$
K_τ	5.7143	Nm/kPa
K_1	110	(kPa/sec)/deg
B	0.592	Nm − sec
θ_d	315	deg
ω_0	740	RPM

10.4.2 Engine Test Cell

A 4.6 L Ford V-8 fuel injected engine is used in this work. The engine has been removed from the vehicle and mounted on an engine test stand. The data is processed using a TMS320C30 digital signal processing (DSP) system, and two sensors are available for engine speed measurements. An optical encoder provides an analog pulse every two crankshaft degrees, while a flywheel sensor outputs a sinusoidal signal with a peak every crankshaft degree. The output of the optical encoder is used as a "clock" for the data sampling, and the output of the flywheel sensor provides the value of the engine speed.

For repeatability in the experiments, an electric load is added to the engine crankshaft through the alternator for the torque disturbances. This is accomplished via 12 vehicle headlights, which is equivalent to the disturbance encountered when the power steering pump or the air conditioning compressor is activated. When the nominal idle speed is 611 RPM, the load is about 12 Nm.

The mass airflow entering the engine is controlled by a pulse width-modulated signal to the idle bypass (actuator) solenoid, and the spark advance control is accomplished via a special circuit designed for these applications.

10.5 Application of SMC for Point-Delayed Systems

We now discuss results of the application of the **continuous SMC design for point-delayed systems** on the idle speed control problem described in the preceding section.

An overview of the control scheme and plant dynamics is shown in Figure 10.1. The characteristic polynomial of the system given by equation 10.40 is $(s + \tau_{e_o}/J\omega_o^2)(s + K_2) + K_1\alpha_o K_\tau e^{-\theta_d s}/J\omega_o^3$. Choosing the stability margin $\nu_0 = 1$, one can find four "unstable" poles, $\{-0.8656 \pm j0.8684,$

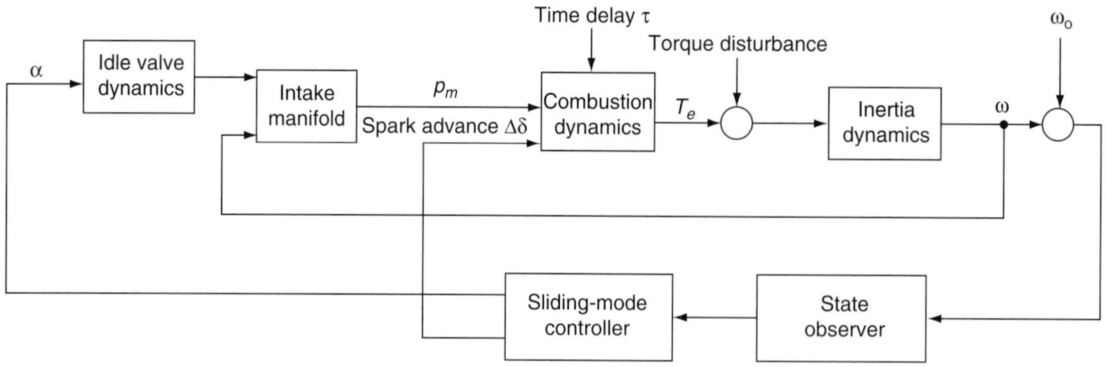

FIGURE 10.1 Point-Delayed SMC Structure

$-0.0680 \pm j0.1123\} = \Lambda_1 \cup \Lambda_2\}$. Hence, $N = 2$. Furthermore, ranks $[sI - A_0 - A_1 e^{-\theta_d s} | B_0] = 2$ for all $s \in \{s \in C,\ Re(s) \geq -\nu_0\}$, and the system is spectrally controllable.

- For $s = -0.8656 + j0.8684 \in \Lambda_1$:

$$Q = \begin{pmatrix} -0.8656 & 1 \\ 0.8684 & 0 \end{pmatrix},\ J = \begin{pmatrix} -0.8656 & -0.8684 \\ 0.8684 & -0.8656 \end{pmatrix}, \text{ and}$$

$$A^{(1)} = Q^{-1} J Q = \begin{pmatrix} -1.7312 & 1 \\ -1.5034 & 0 \end{pmatrix}.$$

- For $s = -0.0680 + j0.1123 \in \Lambda_2$:

$$Q = \begin{pmatrix} -0.0680 & 1 \\ 0.1123 & 0 \end{pmatrix},\ J = \begin{pmatrix} -0.0680 & -0.1123 \\ 0.1123 & -0.0680 \end{pmatrix}, \text{ and}$$

$$A^{(2)} = Q^{-1} J Q = \begin{pmatrix} -0.0604 & 1.0046 \\ -0.0126 & -0.0756 \end{pmatrix}.$$

- The control matrix is the direct sum of $A^{(1)}$ and $A^{(2)}$; that is:

$$A_c = \bigoplus_{i=1}^{2} A^{(i)} = \begin{pmatrix} -1.7312 & 1 & 0 & 0 \\ -1.5034 & 0 & 0 & 0 \\ 0 & 0 & -0.0604 & 1.0046 \\ 0 & 0 & -0.0126 & -0.0756 \end{pmatrix}, \quad (10.41)$$

where $B_c = \sum_{i=1}^{2} e_i \otimes B_0$, and $D_c = \sum_{i=1}^{2} e_i \otimes D$.

Using a linear transformation:

$$z(\theta) = \sum_{i=1}^{2} (e_i \otimes I_n) x(\theta) + \int_{-\theta_d}^{0} e^{A_c \tau} \sum_{i=1}^{2} (e_i \otimes I_n) A_1 x(\theta - \theta_d - \tau) d\tau, \quad (10.42)$$

we obtain the delay-free system:

$$\dot{z}(\theta) = A_c z(\theta) + B_c u(\theta) + D_c f(\theta). \quad (10.43)$$

To reduce the steady-state error, choose the sliding surface as:

$$\begin{aligned} S &= C_1 z(\theta) + C_2 \int_0^{\theta} z(\theta') d\theta' \\ &= C_1 \left[\sum_{i=1}^{2} (e_i \otimes I_n) x(\theta) + \int_{-\theta_d}^{0} e^{A_c \tau} \sum_{i=1}^{2} (e_i \otimes I_n) A_1 x(\theta - \theta_d - \tau) d\tau \right] \\ &\quad + C_2 \int_0^{\theta} \left[\sum_{i=1}^{2} (e_i \otimes I_n) x(\theta') \right. \\ &\quad \left. + \int_{-\theta_d}^{0} e^{A_c \tau} \sum_{i=1}^{2} (e_i \otimes I_n) A_1 x(\theta' - \theta_d - \tau) d\tau \right] d\theta', \end{aligned} \quad (10.44)$$

where, for simplicity, $C_1 = \begin{pmatrix} 0 & 0 & 1 & 0 \\ 0 & 0 & 0 & 1 \end{pmatrix}$, and $C_2 = \begin{pmatrix} 0 & 0 & c_1 & 0 \\ 0 & 0 & 0 & c_2 \end{pmatrix}$, $c_1,\ c_2 > 0$. $C_1 B_c = B_0$.

To increase the reaching speed, we choose:

$$\dot{S}(\theta) = -D_1 sgn(S(\theta)) - D_2 S(\theta), \quad (10.45)$$

where D_1 and D_2 are diagonal matrices with positive diagonal elements. Hence, the control law is as written here:

$$\begin{aligned} u(\theta) &= -B_0^{-1}(C_1 A_c + C_2) \left[\sum_{i=1}^{2} (e_i \otimes I_n) x(\theta) \right. \\ &\quad \left. + \int_{-\theta_d}^{0} e^{A_c \tau} \sum_{i=1}^{2} (e_i \otimes I_n) A_1 x(\theta - \theta_d - \tau) d\tau \right] \\ &\quad - B_0^{-1} D_1 sgn(S(\theta)) - B_0^{-1} D_2 S(\theta). \end{aligned} \quad (10.46)$$

Because the manifold pressure sensor is located upstream from the idle valve actuator (after the throttle plate), it cannot measure the pressure changes caused by the idle valve actuator. Thus, a state observer is necessary. Moreover, since the torque disturbance appears in the system state equations, it must also be estimated to estimate the intake manifold pressure accurately. Standard observer design methods are employed in the following simulation results (Li and Yurkovich, 2000).

10.5.1 Simulation Results

Figure 10.2 gives the engine speed variations and control inputs. A step disturbance of 15 Nm is loaded to the system after 20 engine cycles and unloaded after 150 engine cycles. The integral term in the control law is discretized using the trapezoidal rule with step size $\Delta\tau = 4$ crankshaft degrees. The plant is integrated using Euler's rule with sampling time $T = 32$ crankshaft degrees. It is evident that the controller maintains the idle speed at 611 RPM, and there is no steady-state error. After the torque disturbance is added to the system, the engine speed drops to 582 RPM, with a deviation of 29 RPM from the desired idle speed of 611 RPM. After the disturbance is removed, the engine speed increases to 642 RPM, with a deviation of 41 RPM. The disturbance is completely rejected in a few engine cycles for both loading and unloading cases (with residual oscillations due to the chattering in the spark advance). The controller thus has the capability of disturbance rejection for a typical "bad" disturbance. Note that the engine speed does not oscillate significantly although both of the control inputs have chattering characteristics (physical actuators would, of course, not be able to respond in this manner).

Figures 10.3 and 10.4 show the robustness of the controller to system parameter uncertainties. Variations in K_δ and K_I are considered because they are crucial to controller design. Figure 10.3 gives the engine speed variations to a 15 Nm disturbance after 20 engine cycles for K_δ varying from 0.25 to 8. One can see that the controller can stabilize the system and reject the disturbance for all the cases. As K_δ decreases, however, the engine speed deviation from the nominal idle speed increases, whereas the chattering becomes significant when K_δ increases. Further increasing K_δ would lead to instability. Figure 10.4 gives the engine speed for $K_I = 11$ and 220. Increasing K_I would give higher chattering and even instability.

10.5.2 Experimental Results

The experimental results (engine speed variations and control inputs) are plotted in Figure 10.5 The integral term in the control law is discretized using the trapezoidal rule with $\Delta\tau = 32$ crankshaft degrees, while the control law is calculated every 120 crankshaft degrees. The electrical load is applied manually after about 64 engine cycles and removed after about 183 engine cycles.

From these results, one can see that the controller has the capability to reject the disturbance after about 10 engine cycles. After the torque disturbance is loaded, the engine speed drops to 545 RPM, which has a difference of 66 RPM from the nominal

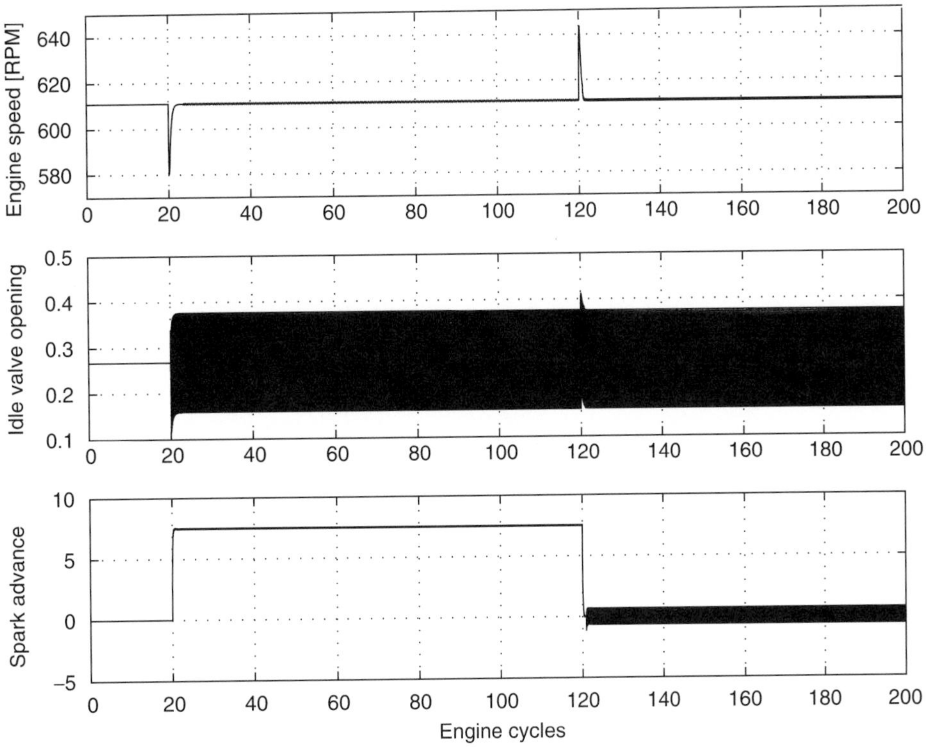

FIGURE 10.2 Engine Speed and Control Inputs (Simulation)

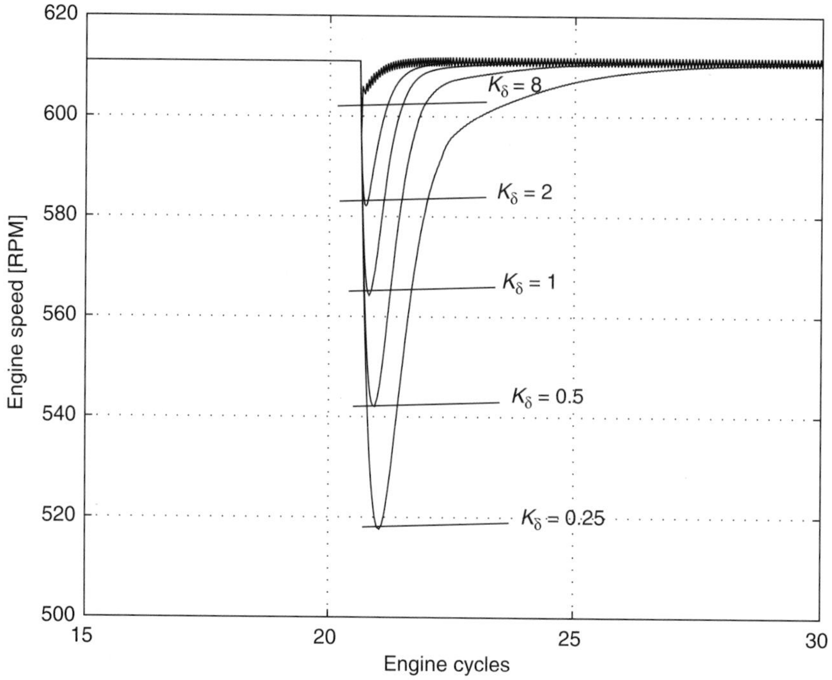

FIGURE 10.3 Engine Speed Variations with Varying K_δ

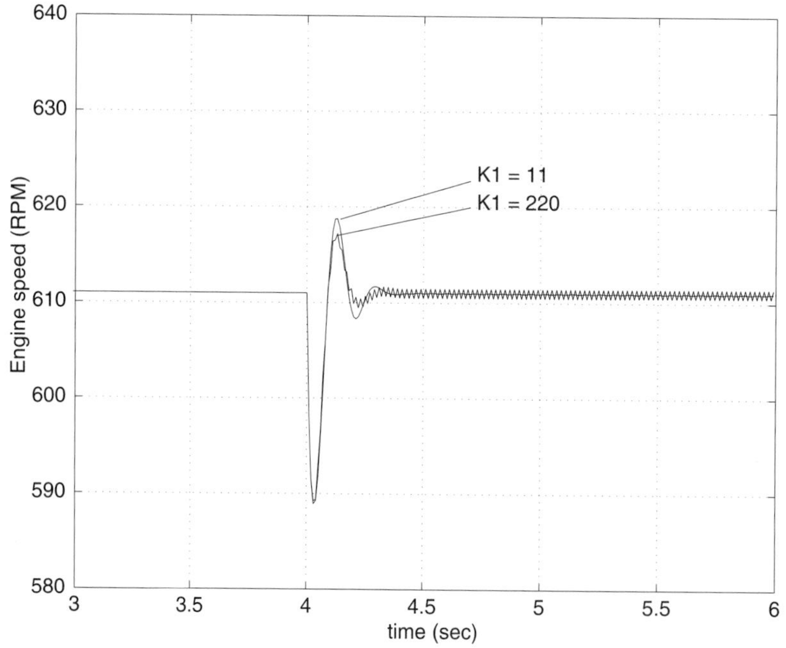

FIGURE 10.4 Engine Speed Variations with Varying K_1

idle speed. The maximum engine speed is 652 RPM when the load is removed, with a difference of 41 RPM. The engine speed variations for the production electrical engine control (EEC) IV controller, which also uses both mass airflow and spark advance as control inputs, are plotted in Figure 10.6 for comparison. Note that the production controller uses a set point (nominal) of 740 RPM. The minimum engine speed is 601 RPM, with a difference of 139 RPM from the nominal idle speed of 740 RPM. The maximum engine speed is 837 RPM, with a difference of 126 RPM. The 5% rising time is about 40 engine cycles for the loading case and even longer for the unloading case. Hence, one can conclude that the proposed sliding-mode controller gives

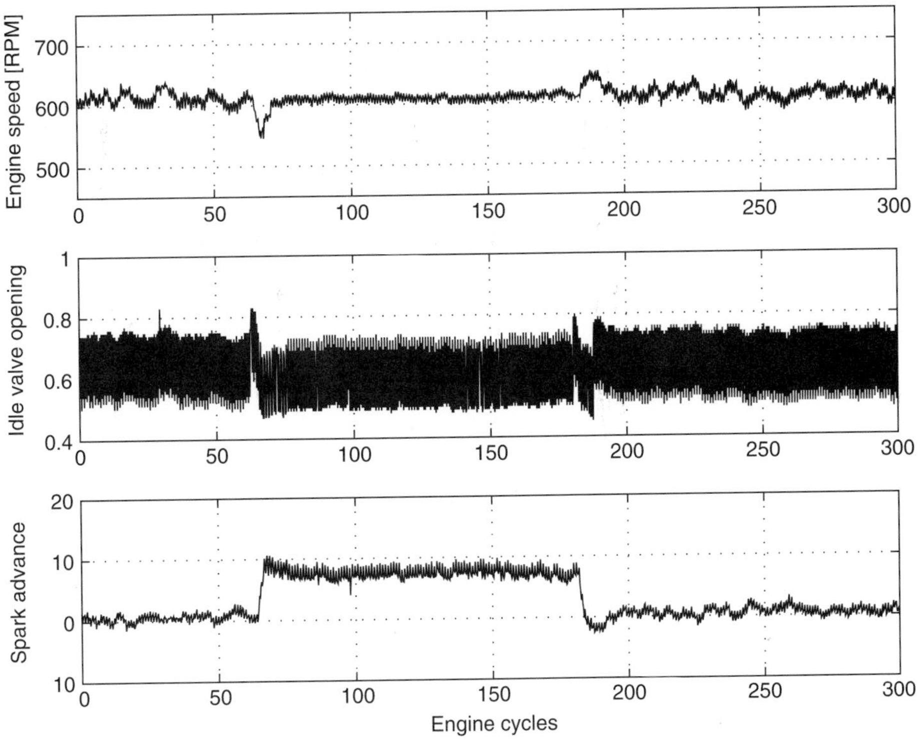

FIGURE 10.5 Point-Delayed SMC Experimental Results

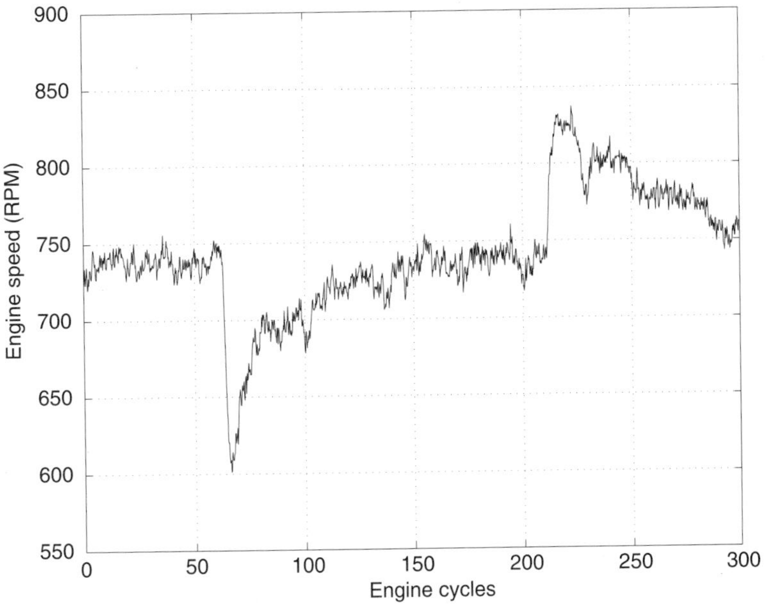

FIGURE 10.6 Engine Speed Using Production Controller (for Comparison)

much better control performance than the production EEC IV controller in this "out-of-vehicle" test. It is important to note that the control input signals shown in Figure 10.5 (idle valve opening and spark advance) are the **commanded inputs**. The actuators themselves, particularly in the case of the idle speed control valve, do not necessarily have the bandwidth capabilities to respond at such frequencies. With the pulse-width modulated controller implementation, in fact, much of the high-frequency chattering effect is essentially filtered out. This, of course, is typical in sliding-mode control applications.

10.6 Application of Adaptive DSMC

Referring now to the development and notation from Section 10.3, we next discuss the results of the **adaptive DSMC** as applied to **the idle speed control problem**. An overview of the control scheme and plant dynamics is shown in Figure 10.7.

10.6.1 Simulations

To convert equations 10.37 through 10.39 and equation 10.19, we differentiate the system output variable, $\omega(k)$, until the system input variable, $\alpha(k)$, appears on the right-hand side of the equation. Denoting $K_2 = \eta_\upsilon V_d/4\pi V_m$ and differentiating results in:

$$J\omega(\theta)\ddot{\omega}(\theta) + J(\dot{\omega}(\theta))^2 + B\dot{\omega}(\theta) = K_\tau K_1 \frac{\alpha(\theta-\theta_d)}{\omega(\theta-\theta_d)} - K_2 J\omega(\theta)\dot{\omega}(\theta) - K_2 B\omega(\theta). \quad (10.47)$$

Discretizing the above equation using Euler's rule, it is easy to show that:

$$\omega(k+1) = f_1[\omega(k), \omega(k-1)] + g_1[\omega(k), \omega(k-1), \omega(k-4)]\alpha(k-4), \quad (10.48)$$

where $f_1[\bullet]$ and $g_1[\bullet]$ are nonlinear functions and continuous in their arguments. In these derivations, it is assumed that $\theta_d = 3T$ with $T = 120^\circ$ crank angle.

To get the form of equation 10.19, we need to advance the time index forward:

$$\begin{aligned}\omega(k+2) &= f_1[\omega(k+1), \omega(k)] + g_1[\omega(k+1), \omega(k), \omega(k-3)]\alpha(k-3)\\ &= f_1\{f_1[\omega(k), \omega(k-1)] + g_1[\omega(k), \omega(k-1), \omega(k-4)]\alpha(k-4), \omega(k)\}\\ &\quad + g_1\{f_1[\omega(k), \omega(k-1)] + g_1[\omega(k), \omega(k-1), \omega(k-4)]\alpha(k-4), \omega(k-1), \omega(k-3)\}\alpha(k-3)\\ &= f_2[\omega(k), \omega(k-1), \omega(k-4), \alpha(k-4)]\\ &\quad + g_2[\omega(k), \omega(k-1), \omega(k-3), \omega(k-4), \alpha(k-4)], \alpha(k-3).\end{aligned}$$

...

$$\begin{aligned}\omega(k+5) = {} & f_5[\omega(k), \omega(k-1), \ldots, \omega(k-4), \alpha(k-1), \ldots, \alpha(k-4)]\\ & + g_5[\omega(k), \omega(k-1), \ldots, \omega(k-4), \alpha(k-1), \ldots, \alpha(k-4)]\alpha(k).\end{aligned} \quad (10.49)$$

Comparing with equation 10.19, we have $n = 4$, $m = 4$, and the relative degree of the system is $d = 5$. Note that we could also have determined the values of m and n by trial and error, as mentioned earlier.

Choosing $C(q^{-1}) = 1 + c_0 q^{-1}$ with $c_0 = 0.3$, one obtains:

$$\begin{aligned}s(k+5) &= C(q^{-1})f_5[\omega(k), \omega(k-1), \ldots, \omega(k-4), \alpha(k-1), \ldots, \alpha(k-4)]\\ &\quad + C(q^{-1})g_5[\omega(k), \omega(k-1), \ldots, \omega(k-4), \alpha(k-1), \ldots, \alpha(k-4)]\alpha(k) - C(q^{-1})r\\ &= F[\lambda(k)] + G[\lambda(k)]\alpha(k) - C(q^{-1})r,\end{aligned} \quad (10.50)$$

where $\lambda(k) = [\omega(k-1), \omega(k-2), \ldots, \omega(k-5), \alpha(k-1), \ldots, \alpha(k-5)]$ is the regression vector, and $r = 611$ RPM is the desired engine speed.

The control law is as written here:

$$u(k) = -\hat{G}[\lambda(k)]^{-1}\{\hat{F}[\lambda(k)] + C(q^{-1})r\}. \quad (10.51)$$

A three-layer feed-forward neural network, with 11 inputs and 5 hidden nodes, is used to approximate both $F[\bullet]$ and $G[\bullet]$. A projection algorithm is applied to ensure $0 < g_{\min} < G[\bullet]$, while $u(k)$ is bounded by $0.05 \le u(k) \le 0.95$. Simulation results are shown in Figure 10.8. As before, a step disturbance of 15 Nm is applied to the system as a load torque after 40 engine cycles and then removed (unloaded) after 120 engine cycles. Elements of the neural network structure vector A are initially chosen as random numbers between 0 and 1, and the data shown in the figures are collected after training the neural networks for 500 engine cycles. Engine speed variations are evident, but it can be seen that the controller is capable of rejecting typical bad disturbances. We note the presence of a small steady-state error (about 0.5 RPM). This is because the controller can only bring the system state trajectories into a

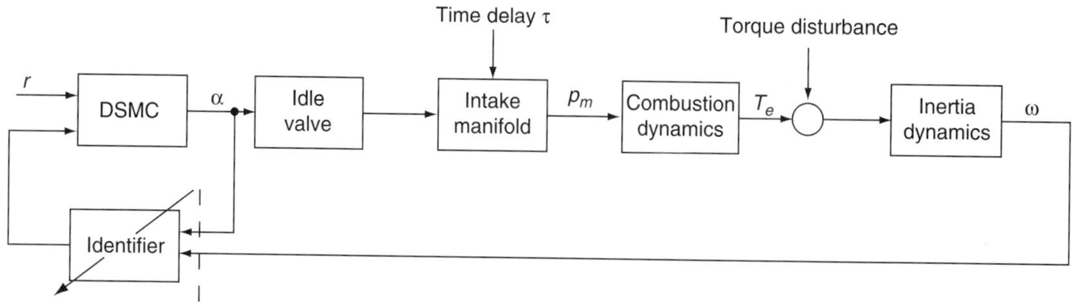

FIGURE 10.7 Adaptive DSMC Structure

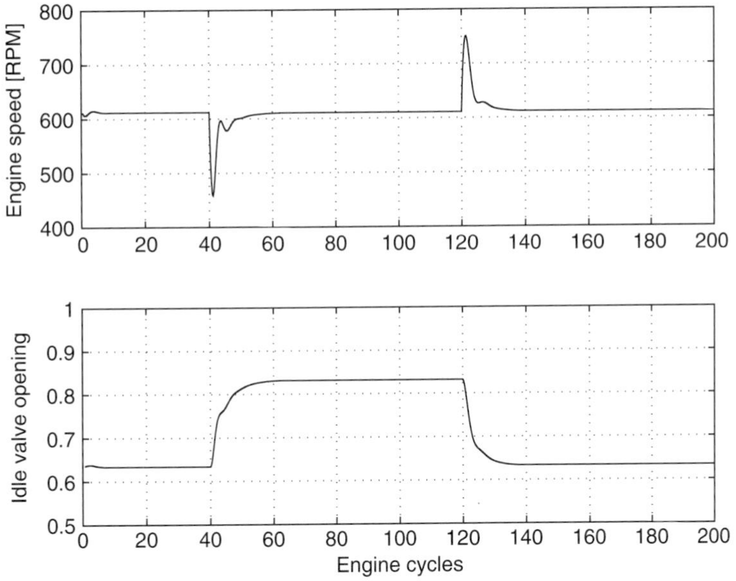

FIGURE 10.8 Engine Speed Variations and Control Input (Simulation)

small predefined sliding sector, which, in this case, is chosen as 0.2. The simulation input is shown as the idle bypass valve opening.

10.6.2 Experimental Results

The engine speed variations for the adaptive DSMC are plotted in Figure 10.9. To shorten the computation time for a step, we use $\lambda(k) = [\omega(k), \omega(k-1), \alpha(k-1)]$ as the regression vector for the neural network (obtained by trial and error). The disturbance torque (electrical load) is applied after 65 engine cycles and removed after 271 engine cycles. The engine speed drops to 532 RPM after the torque disturbance is loaded, which has a deviation of 79 RPM from the desired idle speed of 611 RPM. The engine speed increases to 681 RPM after the torque load is removed, with a difference of 70 RPM. The disturbance is rejected (5% rise time) after about 22 engine cycles for the loading case and 25 engine cycles for the unloading case. The minimum engine speed is 601 RPM, with a difference of 139 RPM from the nominal idle speed of 740 RPM. Once again, this response may be compared with that of the production controller, which uses both mass airflow and spark advance as control inputs, shown in Figure 10.6. When comparing Figure 10.9 to Figure 10.6, it is important to note that the production controller regulates to a nominal engine speed of 740 RPM (as opposed to 611 for the controller of Figure 10.9). We therefore conclude that the adaptive DSMC as developed, *which uses only the idle valve opening as control input*, exhibits superior performance over the production controller that uses two control inputs.

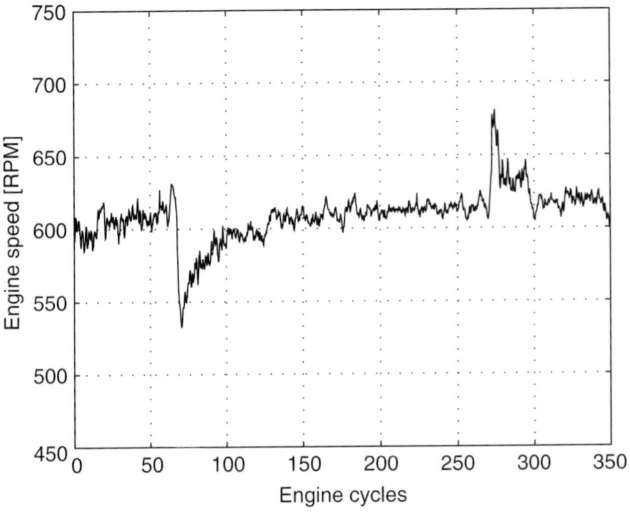

FIGURE 10.9 Adaptive DSMC Experimental Results

10.7 Summary

It is well-known that delays in system states and control input degrade controller performance and worsen the so-called chattering problems for sliding-mode control. Two recent approaches to counteract the effects of system delay and chattering in SMC were presented in this chapter.

For the first method, described fully in Li and Yurkovich (2001), a continuous SMC design method was given for systems with point delay in both states and control. A linear transformation was applied to convert the delay system to a delay-free system whose spectra embed all the unstable poles of the original system within a given stability margin. SMC was then designed based on the delay-free system. Asymptotic stability is ensured for both the delay-free system and the original delayed system in the case of bounded and matched disturbance for the designed control law. The control design method was applied to the IC engine idle speed control problem, not for stabilization (because the engine is stable itself during idling) but for compensation with the effects of the delay.

With the second method, described in more detail in Li and Yurkovich (2000), a discrete adaptive sliding-mode control design algorithm was presented for nonlinear systems. Feedforward neural networks are used to approximate the unknown system dynamics, where system outputs and previous control inputs are used as the inputs for the function approximators, and the defined sliding surface is applied to tune their structure online using gradient methods. There are three main advantages to the adaptive DSMC algorithm described herein:

1. Explicit knowledge of the system dynamics is not necessary for the controller design.
2. The controller is adaptive to the system parameter uncertainties and external disturbances.
3. The time delay in the system can be addressed by increasing the relative order of the system.

Simulation and experimental results for the IC engine idle speed control problem showed that both SMC controllers developed are capable of maintaining a low idle speed (611 RPM) while rejecting torque disturbances.

Acknowledgments

We wish to gratefully acknowledge the many helpful suggestions and numerous discussions on this work with Professor Vadim Utkin at the Ohio State University. For the experimental aspects, we wish to thank Professor Giorgio Rizzoni for use of the laboratory facilities and for giving us expert advice along the way. Finally, we also wish to acknowledge the Center for Automotive Research at Ohio State for fellowship support of second author Xiaoqiu Li.

Appendix: IC Engine Nomenclature

p_m : Intake manifold pressure in kPa
η_v : Volumetric efficiency
V_d : Engine displacement in m^3
V_m : Manifold volume in m^3
ω : Engine speed in rad/sec
K_1 : Coefficient relating throttle opening to delayed manifold pressure in kPa/sec/deg
J : Engine inertia in $Nm - sec^2/rad$
B : Lumped damping coefficient in $Nm - sec/rad$
K_τ : Coefficient relating engine torque to manifold pressure in Nm/kPa
$K\delta$: Coefficient relating engine torque to spark timing in Nm/deg
$\Delta\delta$: Spark advance in deg
θ : Crank angle in rad
θ_d : Induction-to-power delay in rad
τ_e : Net engine torque in Nm
τ_f : Disturbance torque in Nm

References

Bartolini, G., Ferrara, A., and Utkin, V.I. (1995). Adaptive sliding-mode control in discrete-time systems. *Automatica 31*, 769–773.

Chan, C.Y. (1997). Discrete adaptive sliding-mode tracking controller. *Automatica 33*, 999–1002.

Chen, F.C., and Khalil, H.K. (1995). Adaptive controls of a class of nonlinear discrete-time systems using neural networks. *IEEE Transactions Automatic Control AC-40*, 791–801.

Chin, Y.K., and Coats, F.E. (1986). Engine dynamics: time-based versus crank-angle based. *Society of Automotive Engineers 860412.*

DeCarlo, R., Drakunov, S., and Li, X. (1996). A unifying characterization of sliding-mode control: A Lyapunov approach. *ASME Journal of Dynamic Systems, Measurement, and Control, Special Issue on Variable Structure Systems, 122*(4), 708–718.

DeCarlo, R., Zak, S., and Drakunov, S. (1996). Variable structure sliding mode controller design. In William Levine (Ed.), *The control handbook.* Boca Raton, Florida: CRC Press.

DeCarlo, R., Zak, S., and Matthews, G. (1988). Variable structure control of nonlinear dynamic systems: A tutorial. *Proceedings of IEEE 76*(3), 212–232.

Drakunov, S.V., and Utkin, V.I. (1992). Sliding-mode control in dynamic systems. *International Journal of Control 55*, 1029–1037.

El-Khazali, R. (1998). Variable structure robust control of uncertain time-delay systems. *Automatica, 34*, 327–332.

Fiagbedzi, Y.A., and Pearson, A.E. (1986). Feedback stabilization of linear autonomous time lag systems. *IEEE Transactions on Automatic Control AC-31*, 847–855.

Fiagbedzi, Y.A., and Pearson, A.E. (1987). A multistage reduction technique for feedback stabilizing distributed time-lag systems. *Automatica 23*, 311–326.

Funahashi, K. (1989). On the approximate realization of continuous mappings by neural networks. *Neural Networks 2*, 183–192.

Furuta, K. (1990). Sliding-mode control of discrete system. *Systems Control Letters, 14*, 145–152.

Furuta, K. (1997). VSS type self-tuning control. *IEEE Transactions on Industrial Electronics* 37–44.

Hornik, K., Stinchcombe, M., and White, H. (1989). Multilayer feedforward neural networks are universal approximators. *Neural Networks 2*, 359–366.

Hrovat, D., and Sun, J. (1996). Models and control methodologies for IC engine idle speed control design. *Proceedings of the 13th IFAC World Congress* 243–248.

Hu, K.J., Basker, V.R., and Crisalle, O.D. (1998). Sliding-mode control of uncertain input-delay systems. *Proceedings of the American Control Conference*, 564–568.

Jafarov, E.M. (1990). Analysis and synthesis of multidimensional SVS with delays in sliding-modes. *Proceedings of the 11th IFAC World Congress*, 46–49.

Kaynak, O., and Denker, A. (1993). Discrete-time sliding-mode control in the presence of system uncertainty. *International Journal of Control 57*, 1177–1189.

Kotta, U. (1989) Comments on the stability of discrete-time sliding-mode control systems. *IEEE Transactions on Automatic Control*, 1021–1022.

Li, X., and Yurkovich, S. (2000). Neural network-based, discrete adaptive sliding-mode control for idle speed regulation in IC engines. *ASME Journal of Dynamic Systems, Measurement, and Control 122*(2), 269–275.

Li, X., and Yurkovich, S. (2001). Sliding-mode control of delayed systems with application to engine idle speed control *IEEE Transactions on Control System Technology 9* (6), 802–810.

Olbrot, A.W. (1978). Stabilizability, detectability, and spectrum assignment for linear autonomous systems with general time delays. *IEEE Transactions for Automatic Control AC-23*, 887–890.

Sinha, A.S.C., El-Sharkawy, M., and Rizkalla, M. (1997). Sliding-mode control of uncertain delay differential systems. *Nonlinear Analysis, Theory, Methods, & Applications 30*, 1075–1086.

Sira-Ramirez, S. (1991). Nonlinear discrete variable structure systems in quasi-sliding mode. *International Journal of Control 54*, 1171–1187.

Spurgeon, S.K. (1991). Sliding-mode control design for uncertain discrete-time systems. *Proceedings of the 30th IEEE Conference on Decision and Control*, 2136–2141.

Utkin, V.I. (1992). *Sliding modes in control and optimization.* Springer.

Yurkovich, S., and Simpson, M. (1997). Crank-angle domain modeling and control for idle speed. *Society of Automotive Engineers Transactions, Journal of Engines 106, 34–41.*

Zheng, F., Cheng, M., and Gao, W.B. (1995). Variable structure control of time-delay systems with a simulation study on stabilizing combustion in liquid propellent rocket motors. *Automatica 31*, 1030–1037.

11

Nonlinear Input/Output Control: Volterra Synthesis

Patrick M. Sain
Raytheon Company,
El Segundo, California, USA

11.1 Introduction

A power series expansion, provided it exists, often provides a useful representation of a nonlinear plant. The same can be said for the desired closed-loop input/output map of a feedback control system. Given a Volterra series representation for each of these components, then **Volterra feedback synthesis** (VFS) can be used to design and realize a nonlinear controller that uses a finite consecutive number of Volterra kernels of the plant and desired closed-loop input/output map. A striking feature of this method is that it permits the specification of a nonlinear desired closed-loop behavior. The control design takes place in the frequency domain and is realized as an interconnected set of linear systems. The controller possesses an interesting recursive structure that is readily exploited to produce an equivalent reduced-order realization. In fact, this simplified controller implementation can be computed to an arbitrarily high order in an automatic manner.

The VFS approach is reasonably general; the specific development presented herein draws upon the total synthesis problem (TSP) framework, and so this chapter begins with concise descriptions of the TSP paradigm, Volterra plant representation, and controller synthesis. The length of the general formulas at this point is only to be admired; the intrepid reader will note, no doubt with some relief, that the following section describes an equivalent simplified reduced-order implementation of the controllers represented by these formulas, complete with block diagrams. An example application is provided using a based-isolated single degree of freedom structure with nonlinear hysteretic damping.

11.2 Problem Definition Using Total Synthesis

Based on Rugh's results (1981) using Volterra series to represent nonlinear systems, Al-Baiyat and Sain (1986, 1989) applied the use of Volterra operators to nonlinear regulator design in the context of the **total synthesis problem** (TSP) in 1986. Since then, Sain *et al.* (1990, 1991) have used the TSP framework to apply Volterra operators in nonlinear servomechanism design, and in 1995, Doyle *et al.* (1995) cast the method into a model-predictive control design. Al-Baiyat showed that the Volterra operators comprising the controller can be realized as interconnections of linear systems, and Sain (1997) derived an equivalent reduced-order realization. For brevity, a partial linearization regulator design (Sain *et al.*, 1997b) is presented herein. The approach extends to servomechanisms, and the general controller design is given by Sain (1997).

Let R, U, and Y denote real vector spaces of dimensions p, m, and p, respectively representing the spaces of requests, plant inputs, and plant outputs. Let P: $U \rightarrow Y$ denote an input output description of a nonlinear plant. Define the desired closed-loop response to a command by T: $R \rightarrow Y$ and the desired plant input for a command by M: $R \rightarrow U$. The operators P, T, M, and E are assumed to have Volterra series

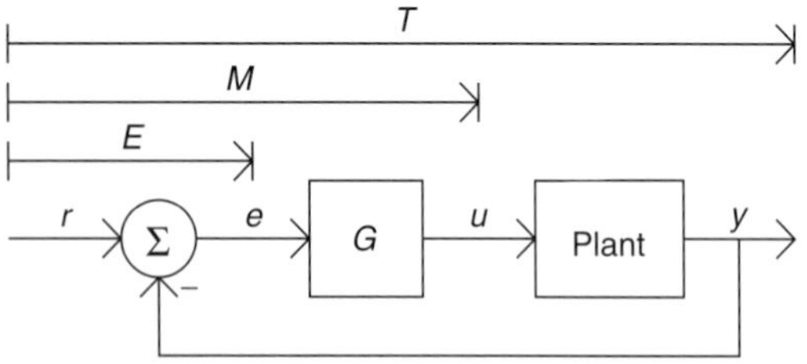

FIGURE 11.1 TSP Regulator Configuration

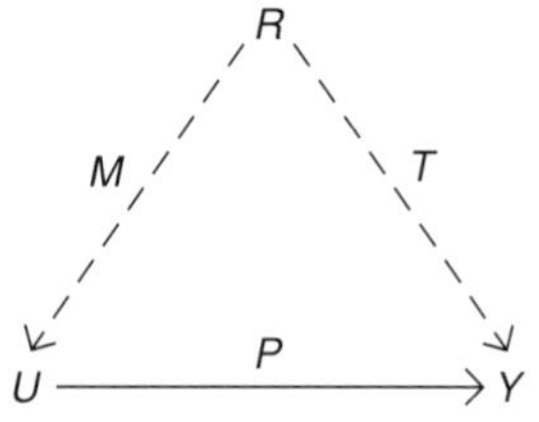

FIGURE 11.2 Commutative Diagram

representations at the point or in the region of operation. An illustration of these operators for a regulator design is given in Figure 11.1 In general, a pair (M, T) is desired such that $T = P \ o \ M$ and the diagram in Figure 11.2 commutes. In the sequel, T is given, and the objective is to find and realize a controller G: $Y \rightarrow U$ such that $M = G \ o \ E$, where E: $R \rightarrow Y$; in this case, Y is the space of output errors.

11.3 Plant Representation

Within the TSP framework, consider linear analytic plants, that is, plants for which the state and output equations are given by:

$$\dot{x} = f(x) + g(x)u, \quad x(0) = x_0. \tag{11.1}$$

$$y = h(x). \tag{11.2}$$

In equations 11.1 and 11.2, $x \in X$; $u \in U$; $y \in Y$; f, g: $X \rightarrow X$; $g(x)$: $U \rightarrow X$; h: $X \rightarrow Y$; and f, g, and h are analytic in x. Such plants can be represented using a so-called bilinear approximation, obtained by denoting the approximation of the state and output equations $f_1(x, u)$ and $f_2(x)$, respectively and by taking a multivariable Taylor series expansion about the operating point (x_0, u_0), truncating all terms of order higher than n, yielding:

$$f_1(x_0, u_0) = \sum_{i=1}^{n} A_{1i}(x_0, u_0)\tilde{x}^{[i]} + \sum_{i=1}^{n-1} D_{1i}(x_0, u_0)\tilde{x}^{[i]} \otimes \tilde{u} + D_{10}\tilde{u}. \tag{11.3}$$

$$f_2(x_0) = \sum_{i=1}^{n} C_{1i}(x_0, u_0)\tilde{x}^{[i]}. \tag{11.4}$$

In these equations, $\tilde{x} = x - x_0$, $\tilde{u} = u - u_0$, and $x^{[i]} = x \otimes \cdots \otimes x$, the i-fold Kronecker tensor product of x with itself. To reduce notational complexity, $\otimes$ is assumed to have precedence over matrix and scalar multiplication. Suppressing the explicit dependence on (x_0, u_0) and if f_1^j denotes $\partial^j f_1/\partial x^j$, then the following results:

$$f_1^j = \sum_{i=j}^{n} A_{ji}\tilde{x}^{[i]} + \sum_{i=j}^{n-1} D_{j(i-1)}\tilde{x}^{[i]} \otimes \tilde{u}. \tag{11.5}$$

$$\frac{d}{dt}\begin{bmatrix} \tilde{x} \\ \tilde{x}^{[2]} \\ \vdots \\ \tilde{x}^{[n]} \end{bmatrix} = \begin{bmatrix} A_{11} & A_{12} & \cdots & A_{1n} \\ 0 & A_{22} & \cdots & A_{2n} \\ \vdots & \vdots & \ddots & \vdots \\ 0 & 0 & \cdots & A_{nn} \end{bmatrix}\begin{bmatrix} \tilde{x} \\ \tilde{x}^{[2]} \\ \vdots \\ \tilde{x}^{[n]} \end{bmatrix} + \begin{bmatrix} D_{11} & D_{12} & \cdots & D_{1(n-1)} \\ D_{21} & D_{22} & \cdots & D_{2(n-1)} \\ \vdots & \vdots & \ddots & \vdots \\ 0 & 0 & \cdots & D_{n(n-1)} \end{bmatrix}\begin{bmatrix} \tilde{x} \\ \tilde{x}^{[2]} \\ \vdots \\ \tilde{x}^{[n]} \end{bmatrix} \otimes \tilde{u} + \begin{bmatrix} D_{10} \\ 0 \\ \vdots \\ 0 \end{bmatrix}\tilde{u}. \tag{11.6}$$

Equations 11.5 and 11.6 apply for $2 \leq j \leq n$, or, more compactly, $\dot{\tilde{\tilde{x}}} = A\tilde{\tilde{x}} + D\tilde{\tilde{x}} \otimes \tilde{u} + B\tilde{u}, y = C\tilde{\tilde{x}}, \tilde{\tilde{x}}(0) = \tilde{\tilde{x}}_0$. In the sequel, the notation A_k denotes the matrix partition composed of the leftmost k partitions of the top k rows of A; C_k denotes the leftmost k partitions of C; and D_k denotes the leftmost $k - 1$ partitions in the top k rows of D, with $D_1 = D_{10}$.

For a multiple-input, multiple-output, finite-dimensional, causal, time-invariant plant, define the homogenous multilinear Volterra operator as:

$$P_i[u(t)] = \int_0^t \cdots \int_0^{\tau_{i-1}} p_i(\tau_1, \ldots, \tau_i)u(t - \tau_1) \otimes \cdots \otimes u(t - \tau_i)d\tau_i \cdots d\tau_1, \tag{11.7}$$

where $t \geq \tau_1 \geq \cdots \geq \tau_i \geq 0$ and where p_i is its ith Volterra kernel. A Volterra representation then has the form (Al-Baiyat and Sain, 1986, 1989):

$$y(t) = \sum_{i=1}^{\infty} P_i[u(t)]. \tag{11.8}$$

The convolutional nature of this representation makes it expedient to work in the transform domain. Define the multidimensional Laplace transform of the ith Volterra kernel as (Bussgang *et al.*, 1974):

$$P_i(s_1, \ldots, s_i) = \int_0^\infty \cdots \int_0^\infty p_i(\tau_1, \ldots, \tau_i) e^{-(s_1\tau_1+\cdots+s_i\tau_i)} d\tau_1 \ldots d\tau_i. \quad (11.9)$$

Al-Baiyat and Sain (1986) showed that for a stationary time invariant bilinear system of the form given above, where $\overline{P}_i^j = ((s_1 + \cdots + s_j)\ I - A_i)^{-1} D_i$, the transforms of the first three kernels are as follows:

$$P_1(s) = C_1 \overline{P}_1^1. \quad (11.10)$$

$$P_2(s_1, s_2) = C_2 \overline{P}_2^2 \{\overline{P}_1^1 \otimes I_m\}. \quad (11.11)$$

$$P_3(s_1, s_2, s_3) = C_3 \overline{P}_3^3 \{[\overline{P}_2^2 \{\overline{P}_1^1 \otimes I_m\}] \otimes I_m\}. \quad (11.12)$$

The general form is the following:

$$P_j(s_1, \ldots, s_j) = C_j \overline{P}_j^j \{[\overline{P}_{j-1}^{j-1} \{[\cdots \overline{P}_2^2 \{\overline{P}_1^1 \otimes I_m\} \cdots] \otimes I_m\}] \otimes I_m\}. \quad (11.13)$$

11.4 Controller Design

Now, the goal is to formulate an expression for the Volterra kernels of the controller G in the frequency domain in terms of the kernels of the plant, given above, and those of the desired closed-loop map, T. The development follows that given in Al-Baiyat Sain (1986). Similarly to the Volterra operator $P_i[u(t)]$, define Volterra operators associated with the maps T and M such that:

$$y(t) = \sum_{j=1}^{\infty} T_j[r(t)]. \quad (11.14)$$

$$u(t) = \sum_{k=1}^{\infty} M_k[r(t)]. \quad (11.15)$$

To obtain a relation between the operators P_i, T_j, and M_k, replace the request signal $r(t)$ by $cr(t)$, where c is an arbitrary constant; because Volterra operators are multilinear, equating expressions for y yields:

$$\sum_{j=1}^{\infty} c^j T_j[r(t)] = \sum_{i=1}^{\infty} P_i \left(\sum_{j=1}^{\infty} c^j M_j[r(t)] \right). \quad (11.16)$$

Let r_i denote $r(t - \tau_i)$ and define:

$$\sum_{i=1}^{\infty} P_i \left(\sum_{j=1}^{\infty} c^j M_j[r(t)] \right) = \sum_{j_1=1}^{\infty} \cdots \sum_{j_i=1}^{\infty} c^{j_1+\cdots+j_i} P_i \big(M_{j_1}[r(t)], \ldots, M_{j_i}[r(t)] \big), \quad (11.17)$$

where the following is true:

$$P_i\big(M_{j_1}[r(t)], \ldots, M_{j_i}[r(t)]\big) = \int_0^t \int_0^{\tau_1} \cdots \int_0^{\tau_{i-1}} p_i(\tau_1, \ldots, \tau_i) M_{j_1}(r_1) \otimes \cdots \otimes M_{j_i}(r_i) d\tau_i \ldots d\tau_1. \quad (11.18)$$

Equating output equation expressions and suppressing the argument $[r(t)]$ yields:

$$\sum_{j=1}^{\infty} c^j T_j = \sum_{i=1}^{\infty} \left(\sum_{j_1=1}^{\infty} \cdots \sum_{j_i=1}^{\infty} c^{j_1+\cdots+j_i} P_i(M_{j_1}, \ldots, M_{j_i}) \right). \quad (11.19)$$

Equating powers of c on both sides of equation 11.19 gives:

$$T_1 = P_1 M_1. \quad (11.20)$$

$$T_2 = P_1 M_2 + P_2(M_1, M_1). \quad (11.21)$$

$$T_3 = P_1 M_3 + P_2(M_1, M_2) + P_2(M_2, M_1) + P_3(M_1, M_1, M_1). \quad (11.22)$$

For $\kappa_i = \sum_{j=1}^{i} k_j$, $\kappa_0 = 0$:

$$T_i = P_1 M_i + \sum_{j=2}^{i} \left(\sum_{k_1=1}^{i-j+1} \sum_{k_2=1}^{i-j-k_1+2} \cdots \sum_{k_{j-1}=1}^{i-\kappa_{j-2}-1} P_j(M_{k_1}, M_{k_2}, \ldots, M_{k_{j-1}}, M_{i-\kappa_j-1}) \right). \quad (11.23)$$

Applying the multidimensional Laplace transform yields:

$$T_1(s) = P_1(s) M_1(s). \quad (11.24)$$

$$T_2(s_1, s_2) = P_1(s_1 + s_2) M_2(s_1, s_2) + P_2(s_1, s_2)[M_1(s_1) \otimes M(s_2)]. \quad (11.25)$$

$$\begin{aligned} T_3(s_1, s_2, s_3) = {} & P_1(s_1 + s_2 + s_3) M_3(s_1, s_2, s_3) \quad (11.26) \\ & + P_2(s_1, s_2 + s_3)[M_1(s_1) \otimes M_2(s_2, s_3)] \\ & + P_2(s_1 + s_2, s_3)[M_2(s_1, s_2) \otimes M_1(s_3)] \\ & + P_3(s_1, s_2, s_3)[M_1(s_1) \otimes M_1(s_2) \otimes M_1(s_3)]. \quad (11.27) \end{aligned}$$

$$\begin{aligned} T_i(s_1, \ldots, s_i) = {} & P_1(s_1 + \cdots + s_i) M_i(s_1, \ldots, s_i) \\ & + \sum_{j=2}^{i} \left\{ \sum_{k_1=1}^{i-j+1} \sum_{k_2=1}^{i-j-k_1+2} \cdots \sum_{k_{j-1}=1}^{i-\kappa_{j-2}-1} \right. \\ & P_j(s_1 + \cdots + s_{\kappa_1}, s_{\kappa_1+1} + \cdots + s_{\kappa_2}, \ldots, s_{\kappa_{j-1}+1} + \cdots + s_i) \\ & [M_{k_1}(s_1, \ldots, s_{\kappa_1}) \otimes M_{k_2}(s_{\kappa_1+1}, \ldots, s_{\kappa_2}) \otimes \cdots \otimes \\ & \left. M_{i-\kappa_{j-1}}(s_{\kappa_{j-1}+1}, \ldots, s_i)] \right\}. \quad (11.28) \end{aligned}$$

Observe that if the pair $(M_1(s), T_1(s))$ is chosen for $P_1(s)$, then one can proceed to the design equation involving $P_2(s_1, s_2)$, where the pair $(M_2(s_1, s_2)$ and $T_2(s_1, s_2))$ is chosen and so on. To complete the design, consider the synthesis of the controller

G that realizes the desired mappings (M, T) shown in Figure 11.1 Note that the similarity of the relation $M = G \; o \; E$ permits expressions for the operators M_j to be obtained in a manner similar to the one for the operators T_i. Also from Figure 11.1, note that $e = r - y$, which can be written as:

$$\sum_{i=1}^{\infty} E_i[r(t)] = I[r(t)] - \sum_{i=1}^{\infty} T_i[r(t)], \qquad (11.29)$$

where I is the identity operator, and, thus, $E_1(s) = I - T_1(s)$. In addition, for $i > 1$, $E_i(s_1, \ldots, s_i) = -T_i(s_1, \ldots, s_i)$. A direct relation between the Volterra kernels of the operators T, M, and G can be found if the design is well-posed on the first order or linear level, implying that $E_1(s)$ has an inverse. As T is usually open to choice, the latter requirement is reasonable.

11.5 Simplified Partial Linearization Controller Design

Consider designing a feedback system for a given nonlinear system so that $T_i = 0$, $1 < i < n$, where T_i is the ith Volterra kernel of the closed loop. Such a design is called **partial linearization**, and it has the desirable attribute of greatly lessening the complexity of the controller design equations. Both the general controller design and the partial linearization controller design, as derived using the equations above, yield realizations with repeated structures. Collecting terms with like coefficients yields the simplified controller design. The result is presented here for the partial linearization controller, with the general result given by Sain (1997).

Suppose for T that the second and higher order kernels are zero. Then, if $P_1(s)$ inverts, the following results:

$$M_1(s) = P_1^{-1}(s) T_1(s). \qquad (11.30)$$

$$M_2(s_1, s_2) = -P_1^{-1}(s_1 + s_2) P_2(s_1, s_2)[M_1(s_1) \otimes M_1(s_2)]. \qquad (11.31)$$

$$\begin{aligned} M_3(s_1, s_2, s_3) = {} & -P_1^{-1}(s_1 + s_2 + s_3)\{P_2(s_1, s_2 + s_3)[M_1(s_1) \\ & \otimes M_2(s_2, s_3)] + P_2(s_1 + s_2, s_3)[M_2(s_1, s_2) \otimes M_1(s_3)] \\ & + P_3(s_1, s_2, s_3)[M_1(s_1) \otimes M_1(s_2) \otimes M_1(s_3)]\}. \qquad (11.32) \end{aligned}$$

In general, the following is true:

$$\begin{aligned} & M_i(s_1, \ldots, s_i) = -P_1^{-1}(s_1 + \ldots + s_i) \\ & \sum_{j=2}^{i} \left\{ \sum_{k_1=1}^{i-j+1} \sum_{k_2=1}^{i-j-k_1+2} \cdots \sum_{k_{j-1}=1}^{i-\kappa_{j-2}-1} \right. \\ & P_j(s_1 + \ldots + s_{\kappa_1}, s_{\kappa_1+1} + \ldots + s_{\kappa_2}, \ldots, s_{\kappa_{j-1}+1} + \ldots + s_i) \\ & \big[M_{k_1}(s_1, \ldots, s_{\kappa_1}) \otimes M_{k_2}(s_{\kappa_1+1}, \ldots, s_{\kappa_2}) \\ & \left. \otimes \ldots \otimes M_{i-\kappa_{j-1}(s_{\kappa_{j-1}+1}, \ldots, s_i)}\big] \right\}. \qquad (11.33) \end{aligned}$$

A nice feature of these equations is their recursive nature. Once $M_1(s)$ has been found, it can be used to obtain an expression for $M_2(s_1, s_2)$ and so forth. Under partial linearization, the operators are represented by $E_k = 0$, $k > 1$. Assuming $E_1(s)$ and $P_1(s)$ to be invertible, then the kernels of the controller G are given by:

$$G_1(s) = P_1^{-1}(s) T_1(s) E_1^{-1}(s). \qquad (11.34)$$

$$G_2(s_1, s_2) = -P_1^{-1}(s_1 + s_2) P_2(s_1, s_2)[G_1(s_1) \otimes G_1(s_2)]. \qquad (11.35)$$

$$\begin{aligned} G_3(s_1, s_2, s_3) = {} & -P_1^{-1}(s_1 + s_2 + s_3)\{P_2(s_1, s_2 + s_3)[G_1(s_1) \\ & \otimes G_2(s_2, s_3)] + P_2(s_1 + s_2, s_3)[G_2(s_1, s_2) \otimes G_1(s_3)] \\ & + P_3(s_1, s_2, s_3)[G_1(s_1) \otimes G_1(s_2) \otimes G_1(s_3)]\}. \qquad (11.36) \end{aligned}$$

In general:

$$\begin{aligned} & G_i(s_1, \ldots, s_i) = -P_1^{-1}(s_1 + \ldots + s_i) \\ & \sum_{j=2}^{i} \left\{ \sum_{k_1=1}^{i-j+1} \sum_{k_2=1}^{i-j-k_1+2} \cdots \sum_{k_{j-1}=1}^{i-\kappa_{j-2}-1} \right. \\ & P_j(s_1 + \ldots + s_{\kappa_1}, s_{\kappa_1+1} + \ldots + s_{\kappa_2}, \ldots, s_{\kappa_{j-1}+1} + \ldots + s_i) \\ & \big[G_{k_1}(s_1, \ldots, s_{\kappa_1}) \otimes G_{k_2}(s_{\kappa_1+1}, \ldots, s_{\kappa_2}) \\ & \left. \otimes \ldots \otimes G_{i-\kappa_{j-1}}(s_{\kappa_{j-1}-1}, \ldots, s_i)\big] \right\}. \qquad (11.37) \end{aligned}$$

Again, note the recursive nature of the controller design equations. Once $G_1(s)$ has been computed, it can be used to find $G_2(s_1, s_2)$ and so on.

The next step in the realization process is to substitute the expression for the plant kernels P_i in the frequency domain. Given $\bar{G}_i^j = G_i(s_j, s_{j+1}, \ldots, s_{j+i-1})$, then:

$$\bar{G}_2^1 = -(C_1 \bar{P}_1^2)^{-1} C_2 \bar{P}_2^2 \{[\bar{P}_1^1 \bar{G}_1^1] \otimes \bar{G}_1^2\}. \qquad (11.38)$$

$$\begin{aligned} \bar{G}_3^1 = {} & -(C_1 \bar{P}_1^3)^{-1} \{C_2 \bar{P}_2^3 \{[\bar{P}_1^1 \bar{G}_1^1] \otimes \bar{G}_2^2\} + C_2 \bar{P}_2^3 \{[\bar{P}_1^2 \bar{G}_2^1] \otimes \bar{G}_1^3\} \\ & + C_3 \bar{P}_3^3 \{[\bar{P}_2^2 \{[\bar{P}_1^1 \bar{G}_1^1] \otimes \bar{G}_1^2\}] \otimes \bar{G}_1^3\}\}. \qquad (11.39) \end{aligned}$$

$$\begin{aligned} \bar{G}_i^1 = {} & -(C_1 \bar{P}_1^i)^{-1} \sum_{j=2}^{i} \left\{ \sum_{k_1=1}^{i-j+1} \sum_{k_2=1}^{i-j-\kappa_1+2} \sum_{k_3=1}^{i-j-\kappa_2+3} \cdots \sum_{k_{j-1}=1}^{i-\kappa_{j-2}-1} \right. \\ & C_j \bar{P}_j^i \Big\{ \Big[\bar{P}_{j-1}^{\kappa_{j-1}} \Big\{ \Big[\cdots \Big[\bar{P}_2^{\kappa_2} \Big\{ \Big[\bar{P}_1^{\kappa_1} \bar{G}_{k_1}^1 \Big] \otimes \bar{G}_{k_2}^{\kappa_1+1} \Big\} \Big] \cdots \Big] \\ & \left. \otimes \bar{G}_{k_{j-1}}^{\kappa_{j-2}+1} \Big\} \Big] \otimes \bar{G}_{i-\kappa_{j-1}}^{\kappa_{j-1}+1} \Big\} \right\}. \qquad (11.40) \end{aligned}$$

Each of equations 11.38 through 11.40 can be realized as a set of interconnected linear subsystems. These realizations possess a number of interesting features that are nicely illustrated using block diagrams. In doing so, certain notational conveniences are employed.

First, the superscripts on the quantities $\bar{P}_i^j$ and $\bar{G}_i^j$ in equations 11.38 through 11.40 are redundant and can be neglected without loss (Sain, 1997). The proof lies in observing that the ordering of terms in a controller design equation, together with the established use of parentheses and brackets, are sufficient to determine what the superscripts should be.

Second, if $e \in Y$ represents the output error of the closed loop, and δ_{11} is its Laplace transform, then let $\delta_{11}^{(i)}$ denote the i-fold tensor product of δ_{11} with itself, then define:

$$\beta_{i1} = \sum_{k=1}^{i} \bar{G}_k^1\left(\delta_{11}^{(k)}\right), \tag{11.41}$$

making β_{i1} the Laplace transform of the ith order controller output.

Finally, the layout of the block diagrams follows certain conventions. If the controller design equations are fully expanded and each written on one line, then their addends, as read from left to right, appear in the block diagrams proceeding from top to bottom.

The fundamental observation behind the simplification process is that quantities present in the realization for $\bar{G}_i^1$ reappear in the realizations for $\bar{G}_j^1$, $j > i$. The motivation behind the simplification process is that in computing the output of a given controller component $\bar{G}_i^1$, the most efficient algorithm is to compute these repeated quantities only once, store them in memory, and then reuse them as needed. The reduction process is thus a problem of identifying a set Γ of elements common to a sequence of controller components $\bar{G}_i^1$, $i = 1, 2, 3, \ldots, n$, and determining which ones to save for later reuse.

In computing the n components $\bar{G}_i^1(\delta_{11}^{(i)})$, $1 \le i \le n$, of the output of an nth order controller, many quantities are computed repeatedly. Therefore, a system of notation is now introduced with the primary goal of eliminating redundant calculations by identifying members of the common set Γ and the secondary goal of further simplifying the controller output calculation. The identification of Γ presented here is not unique. Therefore, the notation Γ_0 will be used to denote the particular identification described herein and to distinguish it from the general notion of the common set represented by Γ. Here and in the sequel, superscripts are suppressed.

The first members of Γ_0 have already been introduced, namely β_{i1}. Given $\alpha_{ij} \in \Gamma_0$ and $2 \le j \le i$ defined by $\alpha_{ij} = \bar{P}_{j-1}\beta_{(i-1)(j-1)}$, then:

$$\beta_{ij} = \sum_{k=1}^{i-j+1} \alpha_{(i-k+1)j} \otimes \beta_{k1}. \tag{11.42}$$

The simplified partial linearization nth order controller output can now be calculated as follows. First, compute the output of the first order controller, $\beta_{11} = P_1^{-1}T_1(I - T_1)^{-1}\delta_{11}$, and then the output of the nth order controller for $n \ge 2$ is computed using:

$$\beta_{n1} = \beta_{11} - \sum_{i=2}^{n} \sum_{j=2}^{i} P_1^{-1} C_j \bar{P}_j \beta_{ij}. \tag{11.43}$$

The members of the common set Γ_0, (namely α_{ij} and β_{ij}), are computed and stored the first time they are encountered and then recalled as needed. In the sequel, the equation $g_{ij} = -P_1^{-1}C_j\bar{P}_j\beta_{ij}$ and the following identifications are convenient:

$$g_i = \sum_{j=2}^{i} -P_1^{-1} C_j \bar{P}_j \beta_{ij} \tag{11.44}$$

$$= \sum_{j=2}^{i} g_{ij}. \tag{11.45}$$

The above algorithm for calculating the output of the simplified partial linearization controller takes full advantage of the recursive nature of the design equations, reducing the number of states and floating-point computations. Furthermore, the simplified form is amenable to the construction of general software routines, capable of computing the controller output for an arbitrary value of the controller order n.

The simplified block diagrams for the second third-order controller components are shown in Figures 11.3 and 11.4. A general form for the ith order component is shown in Figure 11.5.

11.6 SDOF Base-Isolated Structure Example

Consider a single degree of freedom (SDOF) structure sitting on a base isolation system consisting of hysteretically damped bearings, and let the structure be subject to ground acceleration $\ddot{x}_g(t)$ and a control force $f(t)$ supplied by a hydraulic actuator. Such a system is shown in Figure 11.6 The forces exerted by the individual bearings are modeled collectively. The equation of motion for the horizontal displacement of the mass m of the structure is as follows:

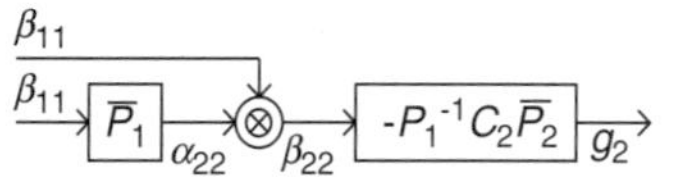

FIGURE 11.3 Simplified G_2 Realization

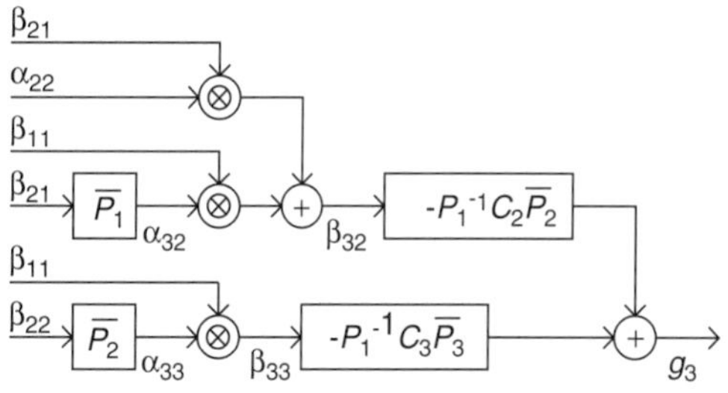

FIGURE 11.4 Simplified G_3 Realization

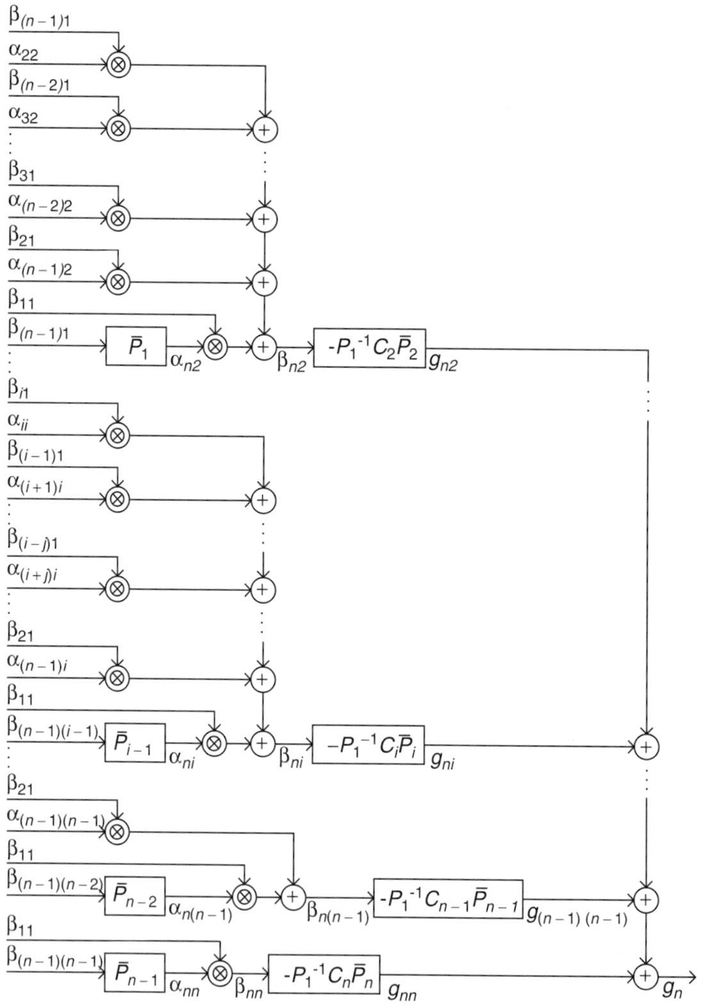

FIGURE 11.5 General Form for Simplified Realization

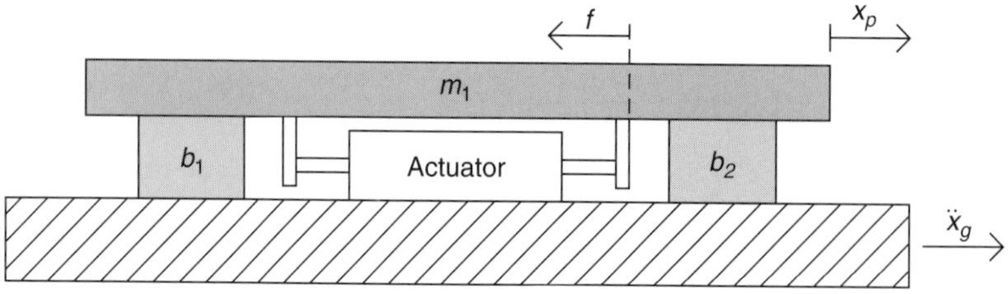

FIGURE 11.6 A Base-Isolated SDOF Structure of Mass m_1. The structure is supported by hysteretically damped bearings b_1 and b_2 that are subject to ground acceleration $\ddot{x}_g$ and control force f.

$$f(t) = m[\ddot{x}_p(t) + \ddot{x}_g(t)] + c\dot{x}_p(t) + kx_p(t) + Q[x(t), \dot{x}(t)]. \tag{11.46}$$

Descriptions of the variables and parameter values are given in Table 11.1 Both the bearings and the actuator affect the stiffness and damping coefficients c and k, and it should be understood that the values identified for these parameters provided in Table 11.1 are meant to represent the behavior of the system with the actuator and bearings in place. The hysteretic restoring force Q generated by the base isolating bearings is given by:

$$Q[x_p(t), \dot{x}_p(t)] = \alpha \frac{F_y}{Y} x_p(t) + (1 - \alpha)F_y z(t), \tag{11.47}$$

TABLE 11.1 Parameters for the SDOF System

Variable	Description	Value
c	Damping coefficient	600π kg/s
f	Force applied by actuator	(N)
k	Stiffness coefficient	$3 \times 10^6 \pi^2$ kg/s^2
m	Mass of structure	3000 kg
Q	Hysteretic restoring force of base-isolation bearing	N
$\ddot{x}_g$	Horizontal ground acceleration	m/s^2
x_p	Horizontal displacement of the mass m	m
f_p	Natural undamped frequency	5 Hz
ξ_p	Damping ratio	1%

TABLE 11.2 Parameters for the Hysteretically Damped Base-Isolation System

Variable	Description	Value
A	Hysteresis loop-shaping parameter	1
F_y	Force of equivalent hysteretic damper	40 k
n	Parameter controlling the smoothness of the transition from elastic to plastic response	2
x_p	Horizontal displacement	m
Y	Yield displacement of equivalent hysteretic damper	1 mm
z	Hysteretic displacement	
α	Post yielding to preyielding stiffness ratio	0.2
β	Hysteresis loop-shaping parameter	-0.25
γ	Hysteresis loop-shaping parameter	0.75

where $z(t)$ is a dimensionless quantity representing hysteretic displacement (Fan *et al.*, 1988; Sain *et al.*, 1997) satisfying the first-order differential equation:

$$Y\dot{z}(t) = \gamma|\dot{x}_p(t)||z(t)|^{n-1}z(t) - \beta\dot{x}_p(t)|z(t)|^n + A\dot{x}_p(t). \quad (11.48)$$

Table 11.2 lists descriptions of the variables and parameter values. The actuator force f applied to the SDOF structure is governed by the equation (Dyke *et al.*, 1995; DeSilva, 1989):

$$\dot{f}(t) = \frac{2BA_f k_q \gamma_f}{V}[u(t) - x_p(t)] - \frac{2Bk_c}{V}f(t) - \frac{2BA_f^2}{V}\dot{x}_p(t). \quad (11.49)$$

The $u(t)$ is the control signal generated by the controller. Descriptions of the quantities in equation 11.49 are provided in Table 11.3.

Defining $A_z = \frac{A}{Y}$, $\beta_z = \frac{\beta}{Y}$, and $\gamma_z = \frac{\gamma}{Y}$ and using the fact that $|z| = z \text{ sgn } z$, then:

$$\dot{z}(t) = [A_z - (\beta_z + \gamma_z \text{ sgn } \dot{x}_p \text{ sgn } z)z^n(t)]\dot{x}_p(t) \quad (11.50)$$

$$= [A_z + \sigma(t)z^n(t)]\dot{x}_p(t), \quad (11.51)$$

where $\sigma(t) = -[\beta_z + \gamma_z \text{ sgn } \dot{x}_p(t) \text{ sgn } z(t)]$. Substitution of the expression for Q into the equation of motion of the

TABLE 11.3 Parameters for the Actuator

Variable	Description	Value
A_f	Cross-sectional area of actuator	m^2
B	Bulk modulus of hydraulic fluid	N/m^2
f	Force applied to SDOF structure by actuator	N
g	Proportional feedback gain	2.5
k_c	Flow–pressure coefficient	m$^4 \cdot$ s/kg
k_q	Flow gain	m^2/s
x_p	Horizontal displacement	m
V	Characteristic hydraulic fluid volume	m^3
$2BA_f k_q G/V$	Displacement coefficient	3.6484×10^8 N/m/s
$2Bk_c/V$	Force feedback coefficient	66.67 1/s
$2BA^2/V$	Velocity feedback coefficient	2.1891×10^7 N/m

structure yields the following system of equations. For convenience, the explicit dependence on time is suppressed:

$$\ddot{x}_p = -\left(\frac{k}{m} + \alpha\frac{F_y}{Ym}\right)x_p - \frac{c}{m}\dot{x}_p + \frac{1}{m}f(t) - (1-\alpha)\frac{Fy}{m}z - \ddot{x}_g. \quad (11.52)$$

$$\dot{f} = -\frac{2BA_f k_q \gamma_f}{V}x_p - \frac{2BA_f^2}{V}\dot{x}_p - \frac{2Bk_c}{V}f(t) + \frac{2BA_f k_q \gamma_f}{V}u. \quad (11.53)$$

$$\dot{z} = (A_z + \sigma z^n)\dot{x}_p. \quad (11.54)$$

Given z with $n = 2$, with $z_0 = z(t_\sigma)$ and $x_{p0} = x_p(t_\sigma)$ (where t_σ represents the most recent point in time where $\sigma(t)$ switched value yields (for $\sigma \geq 0$ and $\sigma < 0$) closed-form expressions for z are readily obtained (Sain, 1997).

An example application of a third-order VFS partial linearization regulator design for the SDOF structure described previously is simulated in equations 11.55 to 11.57, using the simplified controller design derived above. Choosing the state vector $[x_1\ x_2\ x_3]' = [x_p\ \dot{x}_p\ f]'$, where z can be expressed as previously noted, depending on the sign of σ, the resulting state-space description has the form $y = x_1$, and the following results:

$$\dot{x}_1 = x_2. \quad (11.55)$$

$$\dot{x}_2 = -\left(\frac{k}{m} + \alpha\frac{F_y}{Ym}\right)x_1 - \frac{c}{m}x_2 + \frac{1}{m}x_3 - (1-\alpha)\frac{F_y}{m}z. \quad (11.56)$$

$$\dot{x}_3 = -\frac{2BA_f k_q \gamma_f}{V}x_1 - \frac{2BA_f^2}{V}x_2 - \frac{2Bk_c}{V}x_3 + \frac{2BA_f k_q \gamma_f}{V}u. \quad (11.57)$$

This example represents a preliminary test of the control system to demonstrate that the actuator can be commanded to move the first floor of the structure to a specific location. Under the hypothesis that the VFS control designs work to make small errors smaller, a command request $r = \bar{x}_1 = 0.004$ m was specified, and a constant offset of $y_0 = 0.001$ m, representing a biased position sensor, was added to the measured output of the plant. Regulation about a nonzero set point

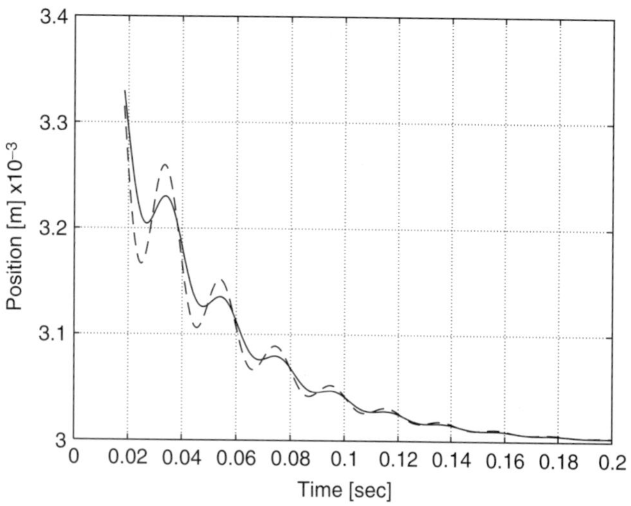

FIGURE 11.7 Position of First Floor Following Initial Transient for Linear Controller Design (Dashed) and Third-Order Controller Design (Solid)

was chosen because the higher order Taylor series coefficients of the control model are zero if $\bar{x}_1 = 0$. Primarily for computational convenience, the desired closed-loop map $T_1(s)$ was given four poles at $s = -1000$ in this example.

A result is shown in Figure 11.7 The initial transient response was omitted so as not to obscure the effect of the higher order controller. The response of the closed-loop system, meaning the position of the first floor, x_1, is shown for the linear controller design and the third-order controller design. Note that the magnitude of the oscillations in the response as the output tends toward the commanded position are much smaller for the third-order design when compared to those produced by the linear controller design.

11.7 Conclusion

Volterra feedback synthesis is shown to provide a capability for specifying nonlinear input output behavior for a closed-loop feedback system. The procedure is developed for the class of linear analytic plants that admit Volterra series representation, and a simplified means of systematically realizing the Volterra kernels and computing the output of the resulting controller is provided, up to an arbitrary order. The simplified form is recursive in nature and lends itself to an automated design and simulation procedure. The procedure is applied to a regulator design for a base-isolated SDOF structure with hysteretic damping.

References

Al-Baiyat, S.A., and Sain, M.K. (1989). A Volterra method for nonlinear control design. *Preprints IFAC Symposium on Nonlinear Control System Design*, 76–81.

Al-Baiyat, S.A., and Sain, M.K. (1986). Control design with transfer functions associated to higher order Volterra kernels. *Proceedings of the 25th IEEE Conference on Decision and Control*, 1306–1311.

Bussgang, J.J., Ehrman, L., and Graham, J. (1974). Analysis of nonlinear systems with multiple inputs. *Proceedings of the IEEE* 62, 1088–1119.

DeSilva, C.W. (1989). *Control sensors and actuators.* Englewood Cliffs, NJ: Prentice Hall.

Doyle, III, F.J., Ogunnaike, B.A., and Pearson, R.K. (1995). Nonlinear model-based control using second-order Volterra models. *Automatica* 31, 697–714.

Dyke, S.J., Spencer, Jr., B.F., Quast, P., and Sain, M.K. (1995). Role of control-structure interaction in protective system design. *ASCE Journal of Engineering Mechanisms* 121, 322–338.

Fan, F.G., Ahmadi, G., and Tadjbakhsh, I.G. (1988). Base isolation of a multi-story building under a harmonic ground motion—A comparison of performances of various systems. *Technical Report.* NCEER-88-0010. National Center for Earthquake Engineering Research, State University of New York at Buffalo.

Rugh, W.J. (1981). *Nonlinear system theory: The Volterra/Wiener approach.* Baltimore, MD: Johns Hopkins University Press.

Sain, P.M. (1997). *Volterra control synthesis, hysteresis models, and magnetorheological structure protection.* Ph.D. dissertation. Department of Electrical Engineering., University of Notre Dame, Notre Dame, Indiana.

Sain, P.M., Sain, M.K., and Spencer, Jr., B.F. (1997). Models for hysteresis and application to structural control. *Proceedings of the American Control Conference*, 16–20.

Sain, P.M., Sain, M.K., and Spencer, Jr., B.F. (1997). Volterra feedback synthesis: A systematic algorithm for simplified Volterra controller design and realization. *Proceedings of Thirty-Fifth Annual Allerton Conference on Communication, Control, and Computing*, 1053–1062.

Sain, P.M., Sain, M.K., and Michel, A.N. (1991). Nonlinear model-matching design of servomechanisms. *Proceedings of the First IFAC Symposium on Design Methods of Control Systems*, 594–599.

Sain, P.M., Sain, M.K., and Michel, A.N. (1990). On coordinated feed-forward excitation of nonlinear servomechanisms. *Proceedings of the American Control Conference*, 1695–1700.

12

Intelligent Control of Nonlinear Systems with a Time-Varying Structure

Raúl Ordóñez
Department of Electrical and Computer Engineering, University of Dayton, Dayton, Ohio, USA

Kevin M. Passino
Department of Electrical and Computer Engineering, The Ohio State University, Columbus, Ohio, USA

12.1 Introduction

The field of nonlinear adaptive control developed rapidly in the 1990s. Work by Polycarpou and Ioannu (1991) and others gave birth to an important branch of adaptive control theory: the nonlinear online function approximation-based control, which includes neural (Polycarpou, 1996) and fuzzy (Su and Stepanenko, 1994) approaches (note that several other relevant works are focused on neural and fuzzy control, many of them cited in the references within the above publications). The neural and fuzzy approaches are most of the time equivalent, differing between each other only for the structure of the approximator chosen (Spooner and Passino, 1996). Most of these publications deal with indirect adaptive control, trying first to identify the dynamics of the systems and eventually generating a control input according to the certainty equivalence principle (with some modification to add robustness to the control law), whereas very few authors (Spooner and Passino, 1996; Rovithakis and Christodoulou, 1995) use the direct approach in which the controller *directly* generates the control input to guarantee stability.

Plants whose dynamics can be expressed in the so-called **strict feedback form** have been considered, and techniques like **back-stepping** and **adaptive back-stepping** (Krstic *et al.*, 1995) have emerged for their control. Publications of Polycarpou (1996) and Polycarpou and Mears (1998) present an extension of the tuning functions approach in which the nonlinearities of the strict feedback system are not assumed to be parametric uncertainties but are rather completely unknown nonlinearities to be approximated online with nonlinearly parameterized function approximators. Both the adaptive methods in Krstic *et al.* (1995) and in Polycarpou (1996) and Polycarpou and Mears (1998) attempt to approximate the dynamics of the plant online, so they may be classified as indirect adaptive schemes.

In this chapter, we combine an extension of the class of strict feedback systems considered in Polycarpou (1996) and Polycarpou and Mears (1998) with the concept of a dynamic structure that depends on time; this combination allows us to propose a class of nonlinear systems with a time-varying structure, for which we develop a *direct* adaptive control approach. This class of systems is a generalization of the class of strict feedback systems traditionally considered in the literature. Moreover, the direct adaptive control developed here is, to our knowledge, the first of its kind in this context, and it presents several advantages with respect to indirect adaptive methods, including the fact that it needs less plant information to be implemented.

12.2 Direct Adaptive Control

Consider the class of continuous time nonlinear systems given by:

$$\dot{x}_i = \sum_{j=1}^{R} \rho_j(\upsilon)\left(\phi_i^j(X_i) + \psi_i^j(X_i)x_{i+1}\right).$$
$$\dot{x}_n = \sum_{j=1}^{R} \rho_j(\upsilon)\left(\phi_n^j(X_n) + \psi_n^j(X_n)u\right). \quad (12.1)$$

The $i = 1, 2, \ldots, n-1$, $X_i = [x_1, \ldots, x_i]^\top$, $X_n \in \mathbb{R}^n$ is the state vector, which we assume measurable, and $u \in \mathbb{R}$ is the control input. The variable $\upsilon \in \mathbb{R}^q$ may be an additional input or a possibly exogenous **scheduling variable**. We assume that υ and its derivatives up to and including the $(n-1)^{th}$ one are bounded and available for measurement, which may imply that υ is given by an external dynamical system. The functions ρ_j, $j = 1, \ldots, R$ may be considered to be **interpolating functions** that produce the time-varying structural nature of system (Polycarpou and Ioannou, 1991) because they combine R systems in strict feedback form (given by the ϕ_i^j and ψ_i^j functions, $i = 1, \ldots, n$, $j = 1 \ldots, R$), and the combination depends on time through the variable υ (thereby, the dynamics of the plant may be different at each time point depending on the scheduling variable). Here, we assume that the functions ρ_j are n times continuously differentiable and that they satisfy, for all $\upsilon \in \mathbb{R}^q$, $\sum_{j=1}^{R} \rho_j(\upsilon) < \infty$ and $\partial^i \rho_j(\upsilon)/\partial\upsilon^i < \infty$. Denote for convenience $\phi_i^c(X_i, \upsilon) = \sum_{j=1}^{R} \rho_j(\upsilon)\phi_i^j(X_i)$ and $\psi_i^c(X_i, \upsilon) = \sum_{j=1}^{R} \rho_j(\upsilon)\psi_i^j(X_i)$. We will assume that ϕ_i^c and ψ_i^c are sufficiently smooth in their arguments and that they satisfy (for all $X_i \in \mathbb{R}^i$ and $\upsilon \in \mathbb{R}^q$) $i = 1, \ldots, n$, $\phi_i^c(0, \upsilon) = 0$ and $\psi_i^c(X_i, \upsilon) \neq 0$.

Here, we will develop a direct adaptive control method for the class of systems in equation 12.1. We assume that the interpolation functions ρ_j are known, but the functions ϕ_i^j and ψ_i^j (which constitute the underlying time-varying dynamics of the system) are unknown. In an indirect adaptive methodology, one would attempt to identify the unknown functions and then construct a stabilizing control law based on the approximations to the plant dynamics. Here, however, we will postulate the existence of an ideal control law (based on the assumption that the plant belongs to the class of systems of equation 12.1) that possesses some desired stabilizing properties, and then we devise adaptation laws that attempt to approximate the ideal control equation. This approximation will be performed within a compact set $\mathcal{S}_{x_n} \subset \mathbb{R}^n$ of arbitrary size that contains the origin. In this manner, the results obtained are semiglobal, in the sense that they are valid as long as the state remains within $\mathcal{S}_{x_n}$; this set can be made as large as desired by the designer. In particular, with enough plant information, the set can be made large enough so that the state never exits it because, as will be shown, a bound can be placed on the state transient. Furthermore, as will be indicated, the stability can be made global by using bounding control terms.

For each vector X_i, we will assume the existence of a compact set $\mathcal{S}_{x_i} \subset \mathbb{R}^i$ specified by the designer. We will consider trajectories in the compact sets $\mathcal{S}_{x_i}$, $i = 1, \ldots, n$, where the sets are constructed such that $\mathcal{S}_{x_i} \subset \mathcal{S}_{x_{i+1}}$, for $i = 1, \ldots, n-1$. We assume the existence of bounds $\underline{\psi}_i^c$, $\bar{\psi}_i^c \in \mathbb{R}$, and $\psi_{i_d}^c \in \mathbb{R}$, $i = 1, \ldots, n$ (*not necessarily known*), such that for all $\upsilon \in \mathbb{R}^q$ and $X_i \in \mathcal{S}_{x_i}$, $i = 1, \ldots, n$:

$$0 < \underline{\psi}_i^c \leq \psi_i^c(X_i, \upsilon) \leq \bar{\psi}_i^c < \infty.$$
$$|\dot{\psi}_i^c| = \left|\sum_{j=1}^{R}\left(\frac{\partial\rho_j(\upsilon)}{\partial\upsilon}\dot{\upsilon}\psi_i^j(X_i) + \rho_j(\upsilon)\frac{\partial\psi_i^j(X_i)}{\partial X_i}\dot{X}_i\right)\right| \leq \psi_{i_d}^c. \quad (12.2)$$

This assumption implies that the affine terms in the plant dynamics have a bounded gain and a bounded rate of change. Since the functions ψ_i^c are assumed continuous, they are therefore bounded in $\mathcal{S}_x$. Similarly, note that even though the term $|\dot{X}_i|$ may not necessarily be globally bounded, it will have a constant bound in $\mathcal{S}_x$ due to the continuity assumptions we make. Therefore, the assumption of equation 12.2 will always be satisfied in $\mathcal{S}_{x_n}$. Moreover, in the simplest of cases, the first part of assumption of equation 12.2 is satisfied globally when the functions ψ_i^j are constant or sector bounded for all $X_i \in \mathbb{R}^i$.

The class of plants of equation 12.1 is, to our knowledge, the most general class of systems considered so far within the context of adaptive control based on back-stepping. In particular, in Krstić *et al.* (1995), which are indirect adaptive approaches, the input functions ψ_i^j are assumed to be constant for $i = 1, \ldots, n$. This assumption allows the authors of those works to perform a simpler stability analysis, which becomes more complex in the general case (Ordoñez and Passino, 2001). The addition of the interpolation functions ρ_j, $j = 1, \ldots, R$ also extends the class of strict feedback systems to one including systems with a time-varying structure (Ordoñez and Passino, 2000a) as well as systems falling in the domain of gain scheduling (where the plant dynamics are identified at different operating points and then interpolated between using a scheduling variable). Note that if we let $R = 1$ and $\rho_1(\upsilon) = 1$ for all υ, together with $\psi_i^c = 1$, $i = 1, \ldots, n$, we have the particular case considered in Polycarpou (1996) and Polycarpou and Mears (1998).

The direct approach presented here has several advantages with respect to indirect approaches such as in Krstić *et al.* (1995), Polycarpou (1996), and Polycarpou and Mears (1998). In particular, bounds on the input functions ψ_i^j are only assumed to exist and need *neither to be known nor to be estimated.* This is because the ideal law is formulated so that there is not an explicit need to include information about the bounds in the actual control law. Moreover, although assumption of equation 12.2 appears to be more restrictive than what is needed in the indirect adaptive case, it is in fact not so because the stability results are semiglobal [i.e., since we are operating in the compact sets $\mathcal{S}_{x_n}$, continuity of the affine terms automatically implies the satisfaction of the second part of assumption 12.2].

12.2.1 Direct Adaptive Control Theorem

Next, we state our main result and then show its proof. For convenience, we use the notation $\nu_i = [\upsilon, \dot{\upsilon}, \ldots, \upsilon^{(i-1)}] \in \mathbb{R}^{q\times i}$, $i = 1, \ldots, n$.

Theorem

Consider system of equation 12.1 with the state vector X_n measurable and the scheduling matrix ν_{n-1} measurable and bounded, together with the above stated assumptions on ϕ_i^c, ψ_i^c, ρ_j, and equation 12.2. Assume also that $\nu_i(0) \in \mathcal{S}_{\upsilon_i} \subset \mathbb{R}^{q\times i}$, $X_n(0) \in \mathcal{S}_{x_i} \subset \mathbb{R}^i$, $i = 1, \ldots, n$ where $\mathcal{S}_{\upsilon_i}$ and $\mathcal{S}_{x_i}$ are compact sets specified by the designer and large enough that ν_i and X_i do not exit them. Consider the diffeomorphism $z_1 = x_1$, $z_i = x_i - \hat{\alpha}_{i-1} - \alpha_{i-1}^s$, $i = 2, \ldots, n$, with $\hat{\alpha}_i(X_i, \nu_i) = \sum_{j=1}^R \rho_j(\upsilon)\hat{\theta}^\top_{\alpha_i^j}\zeta_{\alpha_i^j}(X_i, \nu_i)$, and $\alpha_i^s(z_i, z_{i-1}) = -k_i z_i - z_{i-1}$, with $k_i > 0$ and $z_0 = 0$. Assume the functions $\zeta_{\alpha_i^j}(X_i, \nu_i)$ to be at least $n - i$ times continuously differentiable; to satisfy, for $i = 1, \ldots, n$, $j = 1, \ldots, R$:

$$\left|\frac{\partial^{n-i}\zeta_{\alpha_i^j}}{\partial[X_i, \nu_i]^{n-i}}\right| < \infty. \tag{12.3}$$

Consider the adaptation laws for the parameter vectors $\hat{\theta}_{\alpha_i^j} \in \mathbb{R}^{N_{\alpha_i^j}}$, $N_{\alpha_i^j} \in \mathbb{N}$, $\dot{\hat{\theta}}_{\alpha_i^j} = -\rho_j \gamma_{\alpha_i^j}\zeta_{\alpha_i^j} z_i - \sigma_{\alpha_i^j}\hat{\theta}_{\alpha_i^j}$, where $\gamma_{\alpha_i^j} > 0$, $\sigma_{\alpha_i^j} > 0$, $i = 1, \ldots, n$, $j = 1, \ldots, R$ are design parameters. Then, the control law $u = \hat{\alpha}_n + \alpha_n^s$ guarantees boundedness of all signals and convergence of the states to the residual set:

$$\mathcal{D}_d = \left\{X_n \in \mathcal{R}^n \colon \sum_{i=1}^n z_i^2 \le \frac{2\underline{\psi}_m W_d}{\beta_d}\right\}, \tag{12.4}$$

where $\underline{\psi}_m = \min_{1\le i\le n}\bar{\psi}_i^c$, β_d is a constant and where W_d measures approximation errors and ideal parameter sizes, and its magnitude can be reduced through the choice of the design constants k_i, $\gamma_{\alpha_i^j}$, and $\sigma_{\alpha_i^j}$.

Proof

The proof requires n steps and is performed inductively. First, let $z_1 = x_1$ and $z_2 = x_2 - \hat{\alpha}_1 - \alpha_1^s$, where $\hat{\alpha}_1$ is the approximation to an ideal signal α_1^* ("ideal" in the sense that if we had $\hat{\alpha}_1 = \alpha_1^*$, we would have a globally asymptotically stable closed loop without need for the stabilizing term α_1^s), and α_1^s will be given below. Let $c_1 > 0$ be a constant such that $c_1 > \psi_{1d}^c/2\psi_1^c$ and $\alpha_1^*(x_1, \nu) = (1/\psi_1^c)(-\phi_1^c - c_1 z_1)$. Because the ideal control α_1^* is smooth, it may be approximated with arbitrary accuracy for ν and x_1 within the compact sets $\mathcal{S}_{\nu_1} \subset \mathbb{R}^q$ and $\mathcal{S}_{x_1} \subset \mathbb{R}$, respectively, as long as the size of the approximator can be made arbitrarily large.

For approximators of finite size, let $\alpha_1^*(x_1, \nu) = \sum_{j=1}^R \rho_j(\nu)\theta^{*\top}_{\alpha_1^j}\zeta_{\alpha_1^j}(\nu, x_1) + \delta_{\alpha_1}(\nu, x_1)$, where the parameter vectors $\theta^*_{\alpha_1^j} \in \mathbb{R}^{N_{\alpha_1^j}}$ and $N_{\alpha_1^j} \in \mathbb{N}$ are optimum in the sense that they minimize the representation error δ_{α_1} over the set $\mathcal{S}_{x_1} \times \mathcal{S}_{\nu_1}$ and suitable compact parameter spaces $\Omega_{\alpha_1^j}$, in addition, $\zeta_{\alpha_1^j}(x_1, \nu)$ are defined via the choice of the approximator structure [see Ordoñez and Passino (2000b) for an example of a choice for $\zeta_{\alpha_1^j}$]. The parameter sets $\Omega_{\alpha_1^j}$ are simply mathematical artifacts. As a result of the stability proof, the approximator parameters are bounded using the adaptation laws in theorem 1, so $\Omega_{\alpha_1^j}$ does not need to be defined explicitly, and no parameter projection (or any other "artificial" means of keeping the parameters bounded) is required. The representation error δ_{α_1} arises because the sizes $N_{\alpha_i^j}$ are finite, but it may be made arbitrarily small within $\mathcal{S}_{x_1} \times \mathcal{S}_{\nu_1}$ by increasing $N_{\alpha_i^j}$ (i.e., we assume the chosen approximator structures possess the **universal approximation property**). In this way, there exists a constant bound $d_{\alpha_1} > 0$ such that $|\delta_{\alpha_1}| \le d_{\alpha_1} < \infty$. To make the proof logically consistent, however, we need to assume that some knowledge about this bound and a bound on $\theta^*_{\alpha_1^j}$ is available (since, in this case, it becomes possible to guarantee a priori that $\mathcal{S}_{x_1} \times \mathcal{S}_{\nu_1}$ is large enough). However, in practice, some amount of redesign may be required because these bounds are typically guessed by the designer.

Let $\Phi_{\alpha_1^j} = \hat{\theta}_{\alpha_1^j} - \theta^*_{\alpha_1^j}$ denote the parameter error and approximate α_1^* with $\hat{\alpha}_1(x_1, \nu, \hat{\theta}_{\alpha_1^j}; j = 1, \ldots, R) = \sum_{j=1}^R \rho_j(\nu)\hat{\theta}^\top_{\alpha_1^j}\zeta_{\alpha_1^j}(x_1, \nu)$. Hence, we have a linear in the parameters approximator with parameter vectors $\hat{\theta}_{\alpha_1^j}$. Note that the structural dependence on time of system of equation 12.1 is reflected in the controller because $\hat{\alpha}_1$ can be viewed as using the functions $\rho_j(\nu)$ to interpolate between "local" controllers of the form $\hat{\theta}^\top_{\alpha_1^j}\zeta_{\alpha_1^j}(x_1, \nu)$, respectively. Notice that since the functions ρ_j are assumed continuous and ν bounded, the signal $\hat{\alpha}_1$ is well defined for all $\nu \in \mathcal{S}_{\nu_1}$.

Consider the dynamics of the transformed state $\dot{z}_1 = \phi_1^c + \psi_1^c(z_2 + \hat{\alpha}_1 + \alpha_1^s) + \psi_1^c(\alpha_1^* - \alpha_1^*) = -c_1 z_1 + \psi_1^c z_2 + \psi_1^c(\hat{\alpha}_1 - \alpha_1^*) + \psi_1^c\alpha_1^s = -c_1 z_1 + \psi_1^c z_2 + \psi_1^c\left(\sum_{j=1}^R \rho_j \Phi^\top_{\alpha_1^j}\zeta_{\alpha_1^j} - \delta_{\alpha_1^j}\right) + \psi_1^c\alpha_1^s$. Let $V_1 = \frac{1}{2\psi_1^c}z_1^2 + \frac{1}{2}\sum_{j=1}^R \Phi^\top_{\alpha_1^j}\Phi_{\alpha_1^j}/\gamma_{\alpha_1^j}$, and examine its derivative $\dot{V}_1 = (2\psi_1^c(2z_1\dot{z}_1) - 2z_1^2\dot{\psi}_1^c)/(4\psi_1^{c^2}) + \sum_{j=1}^R \Phi^\top_{\alpha_1^j}\dot{\Phi}_{\alpha_1^j}/\gamma_{\alpha_1^j}$. Using the expression for $\dot{z}_1$, $\dot{V}_1 = -c_1 z_1^2/\psi_1^c + z_1 z_2 + z_1\sum_{j=1}^R \rho_j\Phi^\top_{\alpha_1^j}\zeta_{\alpha_1^j} - z_1\delta_{\alpha_1^j} + z_1\alpha_1^s - \frac{1}{2}z_1^2\dot{\psi}_1^c/\psi_1^{c^2} + \sum_{j=1}^R \Phi^\top_{\alpha_1^j}\dot{\Phi}_{\alpha_1^j}/\gamma_{\alpha_1^j}$. Choose the adaptation law $\dot{\hat{\theta}}_{\alpha_1^j} = \dot{\Phi}_{\alpha_1^j} = -\rho_j\gamma_{\alpha_1^j}\zeta_{\alpha_1^j}z_1 - \sigma_{\alpha_1^j}\hat{\theta}_{\alpha_1^j}$ with design constants $\gamma_{\alpha_1^j} > 0$, $\sigma_{\alpha_1^j} > 0$, $j = 1, \ldots, R$ (we think of $\sigma_{\alpha_1^j}\hat{\theta}_{\alpha_1^j}$ as a "leakage term"). Also, note that for any constant $k_1 > 0$, $-z_1\delta_{\alpha_1^j} \le |z_1| d_{\alpha_1} \le k_1 z_1^2 + d_{\alpha_1}^2/(4k_1)$. We pick $\alpha_1^s = -k_1 z_1$.

Notice also that in completing squares, $-\Phi_{\alpha_1^j}^{\top}\hat{\theta}_{\alpha_1^j} = -\Phi_{\alpha_1^j}^{\top}(\Phi_{\alpha_1^j} + \theta^*_{\alpha_1^j}) \leq -|\Phi_{\alpha_1^j}|^2/2 + |\theta^*_{\alpha_1^j}|^2/2$. Finally, observe that $-z_1^2 \psi_1^c(c_1 + \dot{\psi}_1^c/2\psi_1^c) \leq -z_1^2/\psi_1^c(c_1 - \psi_{1_d}^c/2\underline{\psi}_1^c) \leq -\bar{c}_1 z_1^2/\psi_1^c$ with $\bar{c}_1 = c_1 - \psi_{1_d}^c/2\underline{\psi}_1^c > 0$. Then, we obtain $\dot{V}_1 \leq -\bar{c}_1 z_1^2/\psi_1^c - \frac{1}{2}\sum_{j=1}^{R}\sigma_{\alpha_1^j}|\Phi_{\alpha_1^j}|^2/\gamma_{\alpha_1^j} + z_1 z_2 + d_{\alpha_1}^2/4k_1 + \frac{1}{2}\sum_{j=1}^{R}\sigma_{\alpha_1^j}|\Phi_{\alpha_1^j}|^2/\gamma_{\alpha_1^j} + z_1 z_2 + d_{\alpha_1}^2/4k_1 + \frac{1}{2}\sum_{j=1}^{R}\sigma_{\alpha_1^j}\theta^*_{\alpha_1^j}/\gamma_{\alpha_1^j}$. This completes the first step of the proof.

We may continue in this manner up to the nth step (we omit intermediate steps for brevity), where we have $z_n = x_n - \hat{\alpha}_{n-1} - \alpha^s_{n-1}$, with $\hat{\alpha}_{n-1}$ and α^s_{n-1} defined as in theorem 1. Consider the ideal signal $\alpha_n^*(X_n, \nu_n) = (1/\psi_n^c)(\phi_n^c - c_n z_n + \dot{\hat{\alpha}}_{n-1} + \dot{\alpha}^s_{n-1})$ with $c_n > \psi^c_{n_d}/(2\underline{\psi}_n^c)$. Notice that, even though the terms $\hat{\theta}_{\alpha_{n-1}^j}$ appear in α_n^* through the partial derivatives in $\dot{\hat{\alpha}}_{n-1}$, $\hat{\theta}_{\alpha_{n-1}^j}$ does not need to be an input to α_n^*, since the resulting product of the partial derivatives and $\dot{\hat{\theta}}_{\alpha_{n-1}^j}$ can be expressed in terms of $z_1, \ldots, z_{n-1}$, ν and $\sigma_{\alpha_{n-1}^j}$ $\hat{\alpha}_{n-1}$. To simplify the notation, however, we will omit the dependencies on inputs other than X_i and ν_i, but bearing in mind that, when implementing this method, more inputs may be required to satisfy the proof. Note also that by assumption in equation 12.3, $|\alpha_n^*| < \infty$ for bounded arguments. Therefore, we may represent α_n^* with $\alpha_n^*(X_n, \nu_n) = \sum_{j=1}^{R}\rho_j(\nu)\theta^{*\top}_{\alpha_n^j}\zeta_{\alpha_n^j}(X_n, \nu_n) + \delta_{\alpha_n}(X_n, \nu_n)$ for $X_n \in \mathcal{S}_{X_n} \subset \mathbb{R}^n$ and $\nu_n \in \mathcal{S}_{\nu_n} \subset \mathbb{R}^{q\times n}$. The parameter vector $\theta^*_{\alpha_n^j} \in \mathbb{R}^{N\alpha_n^j}$, $N_{\alpha_n^j} \in \mathbb{N}$ is an optimum within a compact parameter set Ω_{α_n}, in a sense similar to $\theta^*_{\alpha_1^j}$, so that for $(X_n, \nu_n) \in \mathcal{S}_{X_n} \times \mathcal{S}_{\nu_n}, |\delta_{\alpha_n}| \leq d_{\alpha_n} < \infty$ for some bound $d_{\alpha_n} > 0$. Let $\Phi_{\alpha_n^j} = \hat{\theta}_{\alpha_n^j} - \theta^*_{\alpha_n^j}$, and consider the approximation $\hat{\alpha}_n$ as given in theorem 1. The control law $u = \hat{\alpha}_n + \alpha_n^s$ yields $\dot{z}_n = \phi_n^c + \psi_n^c(\hat{\alpha}_n + \alpha_n^s) - \dot{\hat{\alpha}}_{n-1} - \dot{\alpha}^s_{n-1} + \psi_n^c(\alpha_n^* - \alpha_n^*) = -c_n z_n + \psi_n^c\left(\sum_{j=1}^{R}\rho_j(\nu)\Phi^{\top}_{\alpha_n^j}\zeta_{\alpha_n^j} - \delta_{\alpha_n}\right) + \psi_n^c\alpha_n^s$. Choose the Lyapunov function candidate $V = V_{n-1} + \frac{1}{2\psi_n^c}z_n^2 + \frac{1}{2}\sum_{j=1}^{R}\Phi^{\top}_{\alpha_n^j}\Phi_{\alpha_n^j}/\gamma_{\alpha_n^j}$, and examine its derivative $\dot{V} = \dot{V}_{n-1} - c_n z_n^2/\psi_n^c + z_n\sum_{j=1}^{R}\rho_j(\nu)\Phi^{\top}_{\alpha_n^j}\zeta_{\alpha_n^j} - z_n\delta_{\alpha_n} + z_n\alpha_n^s - \frac{1}{2}z_n^2\dot{\psi}_n^c/\psi_n^{c2} + \sum_{j=1}^{R}\Phi^{\top}_{\alpha_n^j}\dot{\Phi}_{\alpha_n^j}/\gamma_{\alpha_n^j}$. One can show inductively that $\dot{V}_{n-1} \leq -\sum_{i=1}^{n-1}\bar{c}_i z_i^2/\psi_i^c - \frac{1}{2}\sum_{i=1}^{n-1}\sum_{j=1}^{R}\sigma_{\alpha_i^j}|\Phi_{\alpha_i^j}|^2\gamma_{\alpha_i^j} + z_{n-1}z_n + \sum_{i=1}^{n-1}d_{\alpha_i^2}/(4k_i) + \frac{1}{2}\sum_{i=1}^{n-1}\sum_{j=1}^{R}\sigma_{\alpha_i^j}|\theta^*_{\alpha_i^j}|^2/\gamma_{\alpha_i^j}$ with constants $\bar{c}_i = c_i - \psi_{i_d}/(2\underline{\psi}_i^c) > 0$, $i = 1, \ldots, n$. The choice of adaptation laws for $\theta_{\alpha_n^j}$ and of α_n^s in theorem 1, together with the observations that $-(\sigma_{\alpha_n^j}/\gamma_{\alpha_n^j})\Phi^{\top}_{\alpha_n^j}\hat{\theta}_{\alpha_n^j} \leq -(\sigma_{\alpha_n^j}/\gamma_{\alpha_n^j})(|\Phi_{\alpha_n^j}|^2/2) + \sigma_{\alpha_n^j}/\gamma_{\alpha_n^j}|\theta^*_{\alpha_n^j}|^2/2$, $z_n\delta_{\alpha_n^j} \leq k_n z_n^2 + d_{\alpha_n}/4k_n$, with $k_n > 0$ and $(-z_n^2/\psi_n^c)(c_n + \dot{\psi}_n^c/2\psi_n^c) \leq -\bar{c}_n z_n^2/\psi_n^c$ imply the following:

$$\dot{V} \leq -\sum_{i=1}^{n}\frac{\bar{c}_i z_i^2}{\bar{\psi}_i^c} - \frac{1}{2}\sum_{i=1}^{n}\sum_{j=1}^{R}\sigma_{\alpha_i^j}\frac{|\Phi_{\alpha_i^j}|^2}{\gamma_{\alpha_i^j}} + W_d, \tag{12.5}$$

where W_d contains the combined effects of representation errors and ideal parameter sizes and is given by $W_d = \sum_{i=1}^{n} d^2_{\alpha_i}/4k_i + 1/2\sum_{i=1}^{n}\sum_{j=1}^{R}\sigma_{\alpha_i^j}|\theta_{\alpha_i^j}|^2\ \gamma_{\alpha_i^j}$. Note that if $\sum_{i=1}^{n}\bar{c}_i z_i^2/\psi_i^c \geq W_d$ or $1/2\sum_{i=1}^{n}\sum_{j=1}^{R}\sigma_{\alpha_i^j}|\Phi_{\alpha_i^j}|^2/\gamma_{\alpha_i^j} \geq W_d$, then we have $\dot{V} \leq 0$. Furthermore, letting $\underline{\psi}_m = \min_{1\leq i\leq n}(\underline{\psi}_i^c)$ and $\bar{\psi}_m = \max_{1\leq i\leq n}(\bar{\psi}_i^c)$ and defining $\bar{c}_0 = \min_{1\leq i\leq n}(\bar{c}_i)$, $\psi_m = \underline{\psi}_m/\bar{\psi}_m$, and $\sigma_0 = \min_{1\leq i\leq n,\, 1\leq j\leq R}(\sigma_{\alpha_i^j})$, we have $-\sum_{i=1}^{n}\bar{c}_i z_i^2/\bar{\psi}_i^c \leq -\bar{c}_0\sum_{i=1}^{n} z_i^2/\bar{\psi}_i^c = -\bar{c}_0\sum_{i=1}^{n}(z_i^2/\psi_i^c)(\psi_i^c/\bar{\psi}_i^c) \leq -\bar{c}_0\sum_{i=1}^{n}(z_i^2/\psi_i^c)(\underline{\psi}_i^c/\bar{\psi}_i^c) \leq -\bar{c}_0\psi_m\sum_{i=1}^{n} z_i^2/\psi_i^c$ and $-\frac{1}{2}\sum_{i=1}^{n}\sum_{j=1}^{R}\sigma_{\alpha_i^j}|\Phi_{\alpha_i^j}|^2/\gamma_{\alpha_i^j} \leq -\sigma_0\frac{1}{2}\sum_{i=1}^{n}\sum_{j=1}^{R}|\Phi_{\alpha_i^j}|^2/\gamma_{\alpha_i^j}$. Then, letting $\beta_d = \min(2\bar{c}_0\psi_m, \sigma_0)$, we have that if:

$$V = \frac{1}{2}\sum_{i=1}^{n}\frac{z_i^2}{\psi_i^c} + \frac{1}{2}\sum_{i=1}^{n}\sum_{j=1}^{R}\frac{|\Phi_{\alpha_i^j}|^2}{\gamma_{\alpha_i^j}} \geq V_0, \tag{12.6}$$

with $V_0 = W_d/\beta_d$, then $\dot{V} \leq 0$ and all signals in the closed loop are bounded. Furthermore, we have $\dot{V} \leq -\beta_d V + W_d$, which implies that $0 \leq V(t) \leq W_d/\beta_d + (V(0) - W_d/\beta_d)\, e^{-\beta_d t}$, so both the transformed states and the parameter error vectors converge to a bounded set. Finally, we conclude from the upper bound on $V(t)$ that the state vector X_n converges to the residual set of equation 12.4

Remark 1

The representation error bounds and the size of the ideal parameter vectors are assumed known because they affect the size of the residual set to which the states converge. It is possible to augment the direct adaptive algorithm with "auto-tuning" capabilities similar to Polycarpou and Mears (1998), which would relax the need for these bounds.

Furthermore, note that the stability result of theorem 1 is semiglobal in the sense that it is valid in the compact sets $\mathcal{S}_{\nu_i}$ and $\mathcal{S}_{x_i}$, $i = 1, \ldots, n$, which can be made arbitrarily large. The stability result may be made global by adding a high-gain bounding control term to the control law. Such a term may be particularly useful when, due to a complete lack of a priori knowledge, the control designer is unable to guarantee that the compact sets $\mathcal{S}_{x_i}$, $i = 1, \ldots, n$, are large enough so that the state will not exit them before the controller has time to bring the state inside D_d; moreover, it may also happen that due to a poor design and poor system knowledge, D_d is not contained in $\mathcal{S}_{x_n}$. In this case, too, bounding control terms may be helpful until the design is refined and improved. Using bounding control, however, requires explicit knowledge of functional upper bounds of $|\psi_i^c(\nu, X_i)|$ and also of the lower bounds $\underline{\psi}_i^c$, $i = 1, \ldots, n$, whose knowledge we do not mandate in theorem 1. Bounding terms may be added to the diffeomorphism in theorem 1, but we do not present the analysis since it is similar to the one we present here and it is algebraically tedious; we simply note, though, that the bounding terms have to be smooth (because they need to be differentiable), so they need to be defined in terms of smooth approximations to the sign, saturation, and absolute value functions that are typically used in this approach.

Remark 2

If the bounds $\underline{\psi}_i^c$, $\bar{\psi}_i^c$, and $\psi_{i_d}^c$ are known, it becomes possible for the designer to directly set the constants c_i in the control law. Notice that with knowledge of these bounds, the term $\underline{\psi}_m$ is also known, and we can pick constants c_i such that $c_i > \psi_{i_d}^c / 2\underline{\psi}_i^c$. Define the auxiliary functions $\eta_i = c_i z_i$. We may explicitly set the constant c_i in α_i^* if we let η_i be an input to the ith approximator structure (i.e., if we let $\alpha_i^*(X_i, \nu_i, \dot{X}_{ri}, \eta_i) = \sum_{j=1}^R \rho_j(\upsilon)\theta_{\alpha_i^j}^{*\top} \zeta_{\alpha_i^j}(X_i, \nu_i, \dot{X}_{r_i}, \eta_i) + \delta_\alpha$).

Then, the approximators used in the control procedure are given by $\hat{\alpha}_i(X_i, \nu_i, \dot{X}_{r_i}, \eta_i) = \sum_{j=1}^R \rho_j(\nu)\hat{\theta}_{\alpha_i^j}^\top \zeta_{\alpha_i^j}(X_i, \nu_i, \dot{X}_{ri}, \eta_i)$, and the stability analysis can be carried out as expected.

12.2.2 Performance Analysis: $\mathcal{L}_2$ Bounds and Transient Design

The stability result of theorem 1 is useful because it indicates conditions to obtain a stable closed-loop behavior for a plant belonging to the class given by equation 12.1. However, it is not immediately clear how to choose the several design constants to improve the control performance. Here, we concentrate on the tracking problem and present design guidelines with respect to an $\mathcal{L}_2$ bound on the tracking error. We are interested in having x_1 track the reference model state x_{r_1} of the reference model $\dot{x}_{r_i} = x_{r_{i+1}}$, $i = 1, 2, \ldots, n-1$, $\dot{x}_{r_n} = f_r(X_{r_n}, r)$ with bounded reference input $r(t) \in \mathbb{R}$. Now, we need to use the diffeomorphism $z_1 = x_1 - x_{r_1}$, $z_i = x_i - \hat{\alpha}_{i-1} - \alpha_{i-1}^s$, $i = 2, \ldots, n$, with $\alpha_1^*(x_1, \upsilon, \dot{x}_{r_1}) = 1/\psi_i^c(-\phi_1^c - c_1 z_1 + x_{r_2})$ and $\alpha_i^*(X_i, \nu_i, \dot{X}_{r_i}) = 1/\psi_i^c(-\phi_i^c - c_i z_i + \dot{\hat{\alpha}}_i + \dot{\alpha}_i^s)$ for $i = 2, \ldots, n$. The stability proof needs to be modified accordingly, and it can be shown that the tracking error $|x_1 - x_{r_1}|$ converges to a neighborhood of size $\sqrt{2\underline{\psi}_m W_d/\beta_d}$.

From the upper bound on $V(t)$, we can write $V(t) \leq W_d/\beta_d + V(0)e^{-\beta_d t}$. From here, it follows that $\frac{1}{2}\sum_{i=1}^n z_i^2(t)/\psi_i^c(t) \leq W_d/\beta_d + (\frac{1}{2}\sum_{i=1}^n z_i^2(0)/\psi_i^c(0) + \frac{1}{2}\sum_{i=1}^n \sum_{j=1}^R = |\Phi_{\alpha_i^j}(0)|^2/\gamma_{\alpha_i^j})e^{-\beta_d t}$. The terms $z_i(0)$ depend on the design constants in a complex manner. For this reason, rather than trying to take them into account in the design procedure, we follow the trajectory initialization approach taken in Krstić *et al.* (1995), which allows the designer to set $z_i(0) = 0$, $i = 1, \ldots, n$ by an appropriate choice of the reference model's initial conditions. In our case, in addition to the assumption that it is possible to set the initial conditions of the reference model, we will have to assume certain invertibility conditions on the approximators. In particular, because $z_1(0) = x_1(0) - x_{r_1}(0)$, we need to set $x_{r_1}(0) = x_1(0)$ for $z_1(0) = 0$.

For the ith transformed state z_i, $i = 2, \ldots, n$, $z_i(0) = x_i(0) - \hat{\alpha}_{i-1}(0) - \alpha_{i-1}^s(0)$. Notice that $\alpha_{i-1}^s(0) = \alpha_{i-1}^s(z_{i-1}(0), z_{i-2}(0))$, so that if $z_{i-1}(0) = 0$ and $z_{i-2}(0) = 0$, we have $\alpha_{i-1}^s(0) = 0$. In particular, notice that this holds for $i = 2$. In this case, to set $z_2(0) = 0$, we need to have $\hat{\alpha}_1(x_1(0), \upsilon(0), x_{r_2}(0)) = x_2(0)$. This equation can be solved analytically (or numerically) for $x_{r_2}(0)$, provided $\partial\hat{\alpha}_1/\partial x_{r_2}|_{t=0} \neq 0$. This is not an unreasonable condition because it depends on the choice of approximator structure the designer makes. The structure can be chosen so that it satisfies this condition. Granted this is the case, it clearly holds that $\alpha_2^s(0) = 0$, and the same procedure can be inductively carried out for $i = 3, \ldots, n$, with the choices $\hat{\alpha}_{i-1}(X_{i-1}(0), \nu_{i-1}(0), x_{r_i}(0)) = x_i(0)$.

This procedure yields the simpler bound $\sum_{i=1}^n z_i^2(t) \leq 2\underline{\psi}_m W_d/\beta_d + \underline{\psi}_m(\sum_{i=1}^n \sum_{j=1}^R |\Phi_{\alpha_i^j}(0)|^2/\gamma_{\alpha_i^j})e^{-\beta_d t}$. We would like to make this bound small so that the transient excursion of the tracking error is small. Notice that we do not have direct control on the size of β_d because this term depends on the unknown constants c_i, which appear in the ideal signals α_i^*. Even though it is not necessary to be able to set β_d to reduce the size of the bound, it is possible to do so if the bounds $\underline{\psi}_i^c$, $\bar{\psi}_i^c$, and $\psi_{i_d}^c$ are known.

At this point, it becomes more clear how to choose the constants to achieve a smaller bound. Recalling the expression of W_d, note that one may first want to have $\beta_d > 1$, so W_d is not made larger when divided by β_d and so the convergence is faster. This may be achieved by setting c_i such that $2\bar{c}_i\psi_m > 1$ (if enough knowledge is available to do so) and $\sigma_{\alpha_i^j} > 1$. However, having a large $\sigma_{\alpha_i^j}$ makes W_d larger; this can be offset, however, by also choosing the ratio $\sigma_{\alpha_i^j}/\gamma_{\alpha_i^j} < 1$ or smaller. Finally, it is clear that making k_i larger reduces the effects of the representation errors and, therefore, makes W_d smaller. Observe that there is enough design freedom to make W_d small and β_d large independently of each other.

These simple guidelines may become very useful when performing a real control design. Moreover, notice that the bound on $\sum_{i=1}^n z_i^2(t)$ makes it possible to specify the compact sets of the approximators so that, even throughout the transient, it can be guaranteed that the states will remain within the compact sets without the need for a global bounding control term. This has been a recurrent shortcoming of many online function approximation-based methods, and the explicit bound on the transient makes it possible to overcome it.

These simple guidelines, which are obvious under careful observation, may become very useful when performing a real control design. The next section illustrates these guidelines and shows their practical effect in a real application.

12.3 Application: Direct Adaptive Wing Rock Regulation with Varying Angle of Attack

Subsonic wing rock is a nonlinear phenomenon experienced by an aircraft with slender delta wings, in which limit cycle roll and roll rate oscillations or unstable behavior are experienced by the aircraft with pointed forebodies at high angles of attack. Wing rock may diminish flight effectiveness or even present serious danger due to potential instability of the aircraft. Here, we will apply the method of local control laws of theorem 1 to the problem of **wing rock regulation**.

Other approaches to this problem can be found in Joshi *et al.* (1998), Singh *et al.* (1995), Luo and Lan (1993), and Krstić *et al.* (1995), among others. In Singh *et al.* (1995), the authors present conventional adaptive and neural adaptive control methods for wing rock control. In Luo and Lan (1993), an optimal feedback control using Beecham-Titchener's averaging technique is applied. They present a single-neuron controller trained with backpropagation to regulate wing rock, and this controller is tested in a wind tunnel. In Krstić *et al.* (1995) the authors use the tuning functions method of adaptive back-stepping to develop a wing rock regulator.

It is interesting to note that all these methods are developed at a *fixed* angle of attack and, in some cases, are tested at another angle close to the design point, which serves to help the researchers claim robustness of the designs. Here, the problem is considered in a more general setting, where the angle of attack is allowed to vary with time according to the evolution of an external dynamical system (which may represent the commands of the pilot together with the aircraft dynamics). As will be noted subsequently, the dynamics of the wing rock phenomenon change nonlinearly with the angle of attack, which makes the problem of developing controllers that are robust against angle of attack a challenging one. However, this problem fits the class of time-varying systems (Polycarpou and Ioannu, 1991) considered in this chapter, so development of a controller that can operate at all angles of attack is greatly simplified by following theorem 1.

There exist several analytical nonlinear models that characterize the phenomenon of wing rock (Hsu and Lan, 1985; Nayfeh *et al.*, 1989; Elzebda *et al.*, 1989). The model we use here is the one presented in Nayfeh *et al.* (1989) and Elzebda *et al.* (1989), which has the advantage over the model in Hsu and Lan (1985) of being differentiable and, according to the authors, slightly more accurate. This model is given by:

$$\ddot{\phi} = -\omega_j^2\phi + \mu_1^j\dot{\phi} + b_1^j\dot{\phi}^3 + \mu_2^j\phi^2\dot{\phi} + b_2^j\phi\dot{\phi}^2 + g\delta_a, \quad (7)$$

where ϕ is the roll angle, δ_a is the output of an actuator with first-order dynamics, $g = 1.5$ is an input gain, and $\omega_j^2 = -c_1 a_1^j$, $\mu_1^j = c_1 a_2^j - c_2$, $b_1^j = c_1 a_3^j$, $\mu_2^j = c_1 a_4^j$, and $b_2^j = c_1 a_5^j$ are system coefficients that depend on the parameters a_i^j, which in turn are functions of the angle of attack, denoted here by υ (aircraft notation conventions dictate the use of α as the angle of attack; however, to avoid confusions with the notation here, we will use υ instead). From Nayfeh *et al.* (1989), we let $c_1 = 0.354$ and $c_2 = 0.001$, constants given by the physical parameters of a delta wing used in wind tunnel experiments in Levin and Katz (1984) to develop the analytical model of equation 12.7. In Nayfeh *et al.* (1989), four angles of attack are considered, at which the coefficients a_i^j are given. We added three points to the table in Nayfeh *et al.* (1989) by assuming that the functions passing through the points a_i^j are approximately piecewise linear (a reasonable assumption, considering the plots presented in this publication). Thus, the points used are given in Table 12.1, where the points at $v = 17, 19$ and 23.75 have been added to the table in question.

TABLE 12.1 Parameters for the Coefficients in the Wing Rock Model

v	a_1^j	a_2^j	a_3^j	a_4^j	a_5^j
15	−0.01026	−0.02117	−0.14181	0.99735	−0.83478
17	−0.02007	−0.0102	−0.0837	0.63333	−0.5034
19	−0.0298	0.000818	−0.0255	0.2692	−0.1719
21.5	−0.04207	0.01456	0.04714	−0.18583	0.24234
22.5	−0.04681	0.01966	0.05671	−0.22691	0.59065
23.75	−0.0518	0.0261	0.065	−0.2933	1.0294
25	−0.05686	0.03254	0.07334	−0.3597	1.4681

To build a smooth, time-varying model of the wing rock that depends on the angle of attack v, we will consider the interpolation functions:

$$\rho_j(v) = \frac{e^{-\left((v-v_j)/s_j\right)^2}}{\sum_{l=1}^{7} e^{-((-v_l)/s_l)^2}}, \quad (12.8)$$

where the centers v_j and spreads s_j, $j = 1, \ldots, 7$, are given in Table 12.2. Notice that the interpolation functions of equation 12.8 satisfy the assumptions stated in theorem 1.

To test the accuracy of the interpolations, let:

$$a_i(v) = \sum_{j=1}^{7} \rho_j(v) a_i^j, \quad (12.9)$$

for $i = 1, \ldots, 5$. Figure 12.1 contains the plots of the interpolated coefficients $a_i(v)$ (solid lines) as well as the data points in Table 12.1 marked by circles. We see that the interpolations are generally close to the data points, so we may consider the resulting time-varying model accurate enough.

We will assume the control input u affects the wing through an actuator with linear, first-order dynamics. To express the model in the form of equation 12.1, we let $x_1 = \phi$, $x_2 = \dot{\phi}$, and $x_3 = \delta_a$. Then, the time-varying wing rock model is given by:

TABLE 12.2 Centers and Spreads for Wing Rock Interpolation Functions

j	1	2	3	4	5	6	7
v_j	15	17	19	21.5	22.5	23.75	25
s_j	1.5	1.5	1.5	2.0	1	1	1

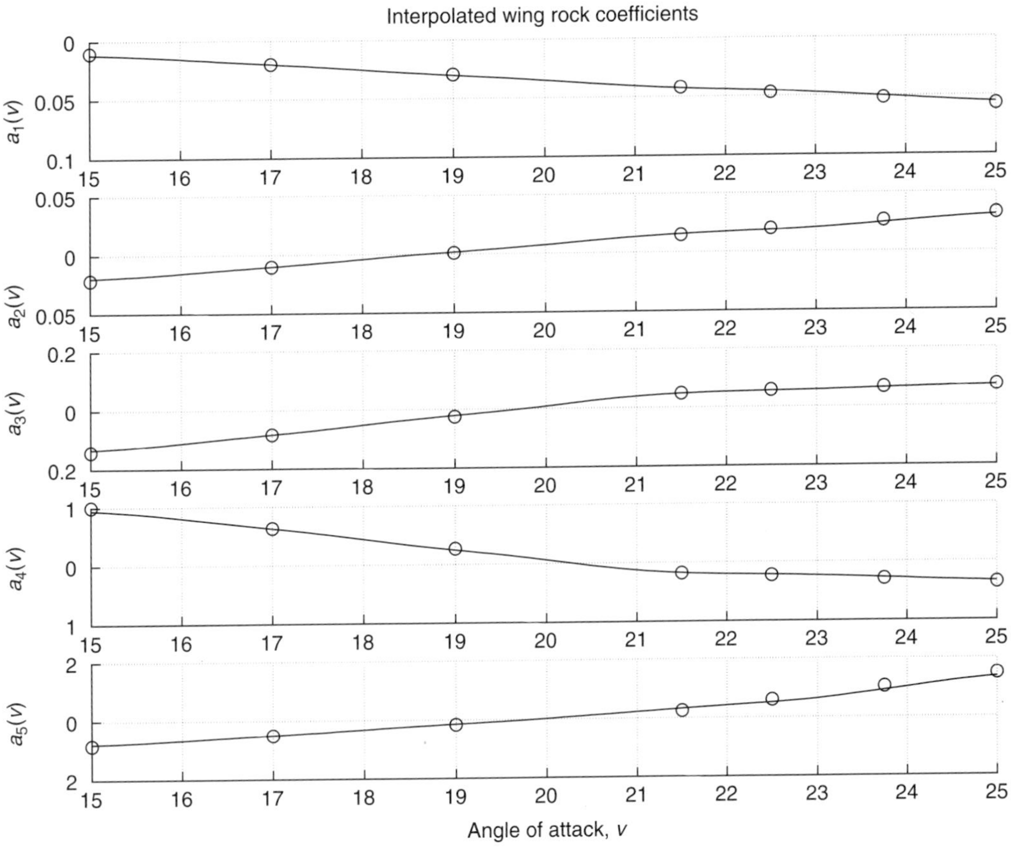

FIGURE 12.1 Interpolated Coefficients for the Time-Varying Wing Rock Model

$$\dot{x}_1 = x_2.$$
$$\dot{x}_2 = \sum_{j=1}^{7} \rho_j(v)\left(-w_j^2\phi + \mu_1^j\dot{\phi} + b_1^j\dot{\phi}^3 + \mu_2^j\phi^2\dot{\phi} + b_2^j\phi\dot{\phi}^2\right) + gx_3.$$
$$\dot{x}_3 = -\frac{1}{\tau}x_3 + \frac{1}{\tau}u. \qquad (12.10)$$

The actuator time constant is $\tau = 1/15$. We will assume that the angle of attack v varies according to an exogenous dynamical system:

$$\begin{bmatrix} \dot{v}_1 \\ \dot{v}_2 \end{bmatrix} = \begin{bmatrix} 0 & 25 \\ -25 & -10 \end{bmatrix} \begin{bmatrix} v_1 \\ v_2 \end{bmatrix} + \begin{bmatrix} 0 \\ 500 \end{bmatrix} + \begin{bmatrix} 0 \\ 62.5 \end{bmatrix} r, \qquad (12.11)$$

where $v_1 = v$, $v_2 = \dot{v}$, and r is a command input that can take values between minus one and one. System 12.11 has its poles at $-5 \pm 24.5i$ ($i = \sqrt{-1}$), and its equilibrium is at $v_1 = 20$ and $v_2 = 0$.

According to the analysis performed in Nayfeh *et al.* (1989), the wing rock system has a stable focus at the origin for angles of attack v less than approximately 19.5. For larger angles, the origin becomes an unstable equilibrium, and a limit cycle appears around it. In both cases, however, the system is unstable and may diverge to infinity if the initial conditions are large enough (since we are dealing with angles, such a divergence means that the wings rotate faster and faster). The problem we consider here has the angle of attack varying within the range between 15 and 25, so the qualitative behavior of equation 12.10 changes periodically, as v becomes respectively smaller or larger than 19.5. To gain a better insight into how the dynamic behavior of the wing rock phenomenon changes qualitatively with v, consider Figure 12.2, where we let the system start at the initial condition $X_3(0) = [-4, 0, 0]$ and $v_2(0) = [20, 0]$. Initially, we set $r = 1$, so the angle of attack stabilizes at 22.5, and we let the system run in open loop for 200 sec. We observe that x_1 and x_2 are approaching a limit cycle, which would be reached if the system were allowed to run for a longer time: however, at $t = 200$, we let $r = -1$ (this is marked by an arrow in Figure 12.2), so the angle of attack changes and after a short transient stabilizes at 17.5. Not being close enough to the origin to be attracted by the local stable focus, the system starts to diverge.

We will consider two designs with the purpose of illustrating how different design constants may in fact reduce or enlarge the $\mathcal{L}_2$ bound stated in Subsection 12.2.2. In order to do so, we will let the controller track the state x_{r_1} of a reference trajectory given by:

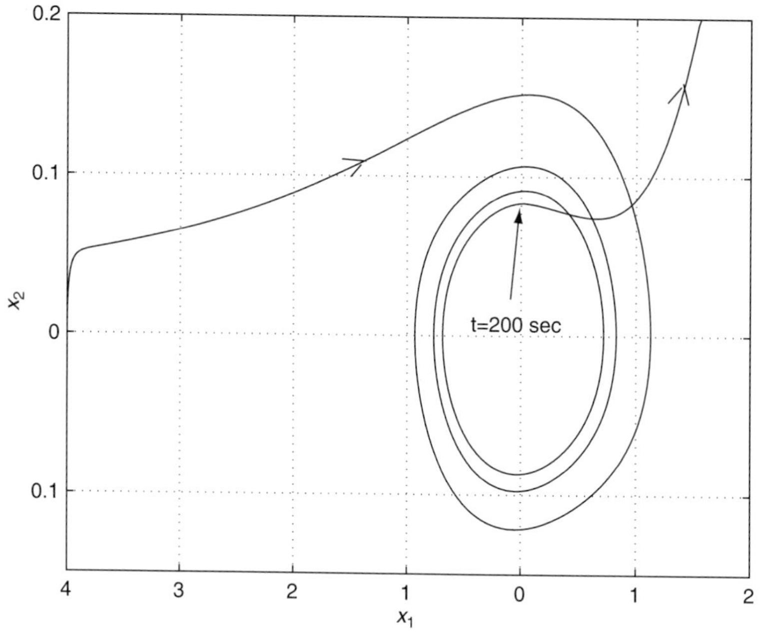

FIGURE 12.2 Qualitative Change in Wing Rock Dynamics with Varying Angle of Attack

$$\begin{aligned}\dot{x}_{r_1} &= x_{r_2}.\\ \dot{x}_{r_2} &= x_{r_3}.\\ \dot{x}_{r_3} &= -45x_{r_1} - 39x_{r_2} - 11x_{r_3}.\end{aligned} \tag{12.12}$$

That is, these equations represent a stable linear system with poles at -5, -3 and -3.

We use radial basis function neural networks (Moody and Darken, 1989) as the approximators. Since v has its equilibrium at 20, we use $\bar{v} = v - 20$ instead of v as input to the approximators. For $\hat{\alpha}_1(x_1, \bar{v}, x_{r_2})$, we choose for $j = 1, \ldots, 7$:

$$\zeta_{\alpha_{1j}} = \left[1, \frac{\exp\left((x_1 - c_{x_k})^2/s_{x_1}^2)\right) \exp\left((\bar{v} - c_{v_l})^2/s_{v_1}^2\right) \exp\left((x_{r_2} - c_{r_m})^2/s_{r_2}^2\right)}{\sum_{d=1}^{2}\sum_{p=1}^{2}\sum_{q=1}^{2} \exp\left((x_1 - c_{x_d})^2/s_{x_1}^2\right) \exp\left((\bar{v} - c_{v_p})^2/s_{v_1}^2\right) \exp\left((x_{r_2} - c_{r_q})^2/s_{r_2}^2\right)} \right]^{\top}, \tag{12.13}$$

for $k = 1, 2$, $l = 1, 2$, and $m = 1, 2$, with the centers c_{x_k}, c_{v_l}, and c_{r_m} evenly spaced along the intervals $[-15, 15], [-5, 5]$, and $[-5, 5]$, respectively, and the spreads $s_{x_1} = 30$, $s_{v_1} = 10$, and $s_{r_1} = 10$ (i.e., $\mathcal{S}_{x_1} = \{x_1 \in \mathbb{R}: -15 \leq x_1 \leq 15\}$ and $\mathcal{S}_{v_1} = \{v \in \mathbb{R}: 15 \leq v \leq 25\}$; the other compact sets are defined in a similar manner).

For $\hat{\alpha}_2(X_2, v_2, x_{r_3})$, we space three centers evenly on the interval $[-15, 15]$ for x_1 and x_2, two centers evenly spaced on the interval $[-5, 5]$ for v_1 and v_2, and two centers evenly spaced on the interval $[-10, 10]$ for x_{r_3}. The spreads are chosen as $s_{x_1} = s_{x_2} = 15$, $s_{v_1} = s_{v_2} = 10$, and $s_{r_2} = 20$. The functions $\zeta_{\alpha_2^j}$ are all chosen the same for $j = 1, \ldots, 7$ and with a similar structure to equation 12.13. The same is done for $\hat{\alpha}_3(X_3, v_3, \dot{x}_{r_3})$, where we evenly space the centers for x_i and v_i, $i = 1, 2, 3 (v_3 = \dot{v}_2)$, along the intervals $[-20, 20]$ and $[-5, 5]$, respectively, and along $[-100, 100]$ for $\dot{x}_{r_3}$, where we use three centers for each x_i, two for each v_i, and two for $\dot{x}_{r_3}$. The spreads are $s_{x_i} = 20$, $s_{v_i} = 10$, $i = 1, 2, 3$, and $s_{r_3} = 200$. Again, for all $\zeta_{\alpha_2^j}$ and $\zeta_{\alpha_3^j,\, j=1,\ldots,7}$, we replace v by $\bar{v}$ in v_i, $i = 2, 3$.

All the coefficient vectors are initialized with zeros. Since the approximator's parameters are guaranteed to be bounded, we do not need to set explicitly the parameter spaces $\Omega_{\alpha_i^j}$, and no parameter projection is used. We will use the tracking diffeomorphism given in Subsection 12.2.2, and we will also perform trajectory initialization to be able to better apply the analysis there. Because of the choice we make for the system's initial conditions ($X_3(0) = [-4, 0, 0]$) and because all parameter coefficients are picked initially equal to zero, the choice $X_{r_3}(0) = X_3(0) = [-4, 0, 0]$ guarantees $z_i(0) = 0$ and $i = 1, 2, 3$.

We will let the reference to the angle of attack system alternate between -1 and 1 every 0.5 sec, and we choose $v_2(0) = [20, 0]$ as the initial condition for the angle of attack system. We will consider two designs. In the first one, we let $k_1 = 0.3$, $k_2 = 0.6$, and $k_3 = 0.4$. For the adaptation laws, we pick $\gamma_{\alpha_1^j} = 0.1$, $\sigma_{\alpha_1^j} = 0.2$, $\gamma_{\alpha_s^j} = 0.5$, and $\sigma_{\alpha_1^j} = 0.1$ as well as $\gamma_{\alpha_3^j} = 0.2$, $\sigma_{\alpha_1^j} = 0.3$, and $j = 1, \ldots, 7$. Figures 12.3 and 12.4 show the control results. In Figure 12.3 we first observe the plant's behavior in open loop, plotted with a dashed line. The

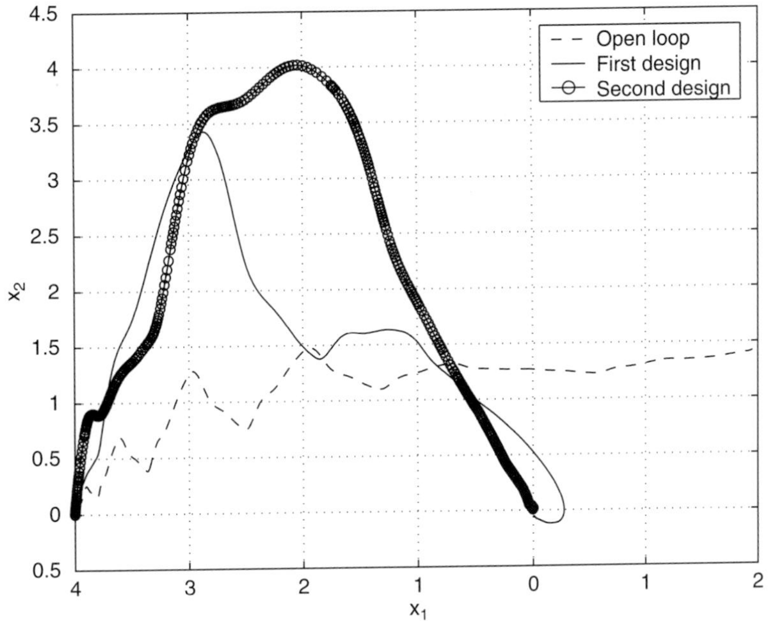

FIGURE 12.3 Direct Adaptive Wing Rock Regulation: Roll and Roll Rate

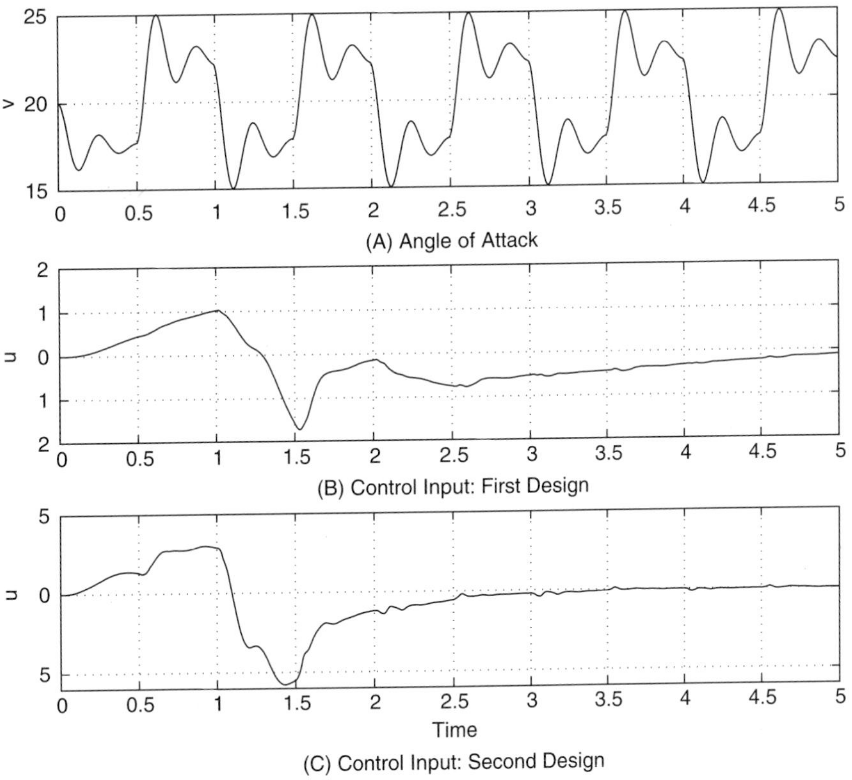

FIGURE 12.4 Angle of Attack and Control Input

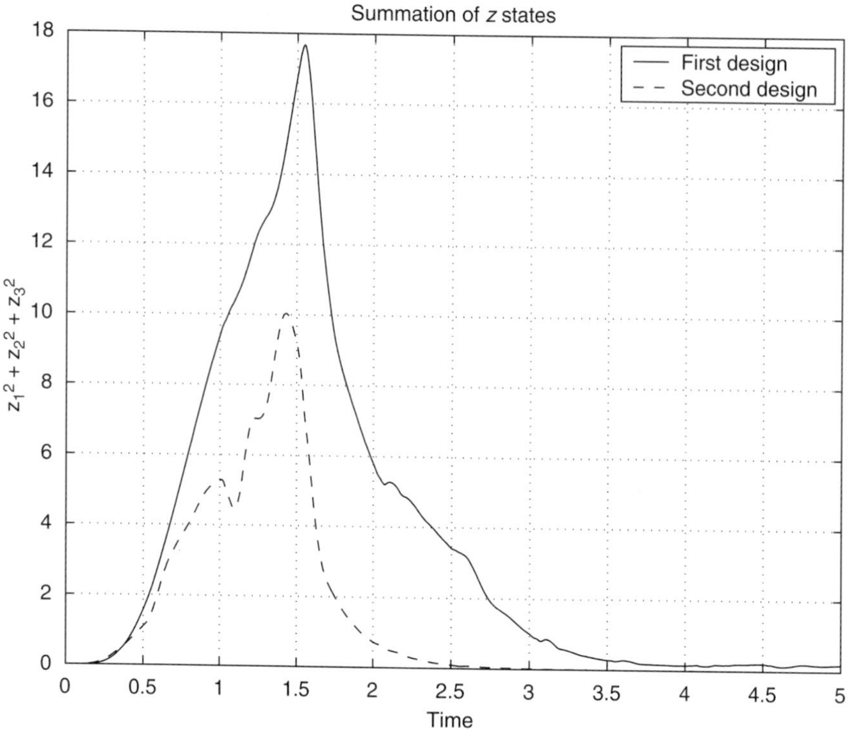

FIGURE 12.5 Different Plots of $\sum_{i=1}^{3} z_i^2(t)$ for the Two Designs

oscillations are due to the changing angle of attack; eventually, if allowed to run uncontrolled, the system diverges. The solid line in Figure 12.3 corresponds to the first design just described. Notice that x_1 has an overshoot beyond zero. In Figure 12.4(A), we can see the evolution of the angle of attack. In Figure 12.4(B), we observe the control input generated by this first design.

Next, with the purpose of improving the controller's performance, we follow the guidelines given in subsection 12.2.2 and change the design constants in the following manner: we let $k_1 = 1$, $k_2 = 3$, $k_3 = 2.5$, $\gamma_{\alpha_1^j} = 0.1$, $\sigma_{\alpha_1^j} = 0.1$, $\gamma_{\alpha_2^j} = 1.5$, $\sigma_{\alpha_2^j} = 0.5$, $\gamma_{\alpha_3^j} = 1.2$, and $\sigma_{\alpha_2^j} = 0.4$, $j = 1, \ldots, 7$. Notice that we have chosen larger constants k_i as well as larger $\gamma_{\alpha_i^j}$, which we expect should help reduce the size of W_d. At the same time, we have set the ratios between $\sigma_{\alpha_i^j}$ and their respective $\gamma_{\alpha_i^j}$ to be less than or equal to one, which was not the case in the first design.

Figure 12.3 shows the resulting closed-loop behavior, plotted in a solid line with circles, and the bottom plot in Figure 12.4 corresponds to the control input generated by the second design. Although it cannot be seen from Figure 12.3, the convergence to zero is about twice as fast with the second design than with the first one, taking less than 4 secs for x_1 and x_2 to become small enough, versus about 8 sec with the first design. Furthermore, there is now no overshoot for x_1. Note that in the second design, x_2 reaches a slightly larger value than in the first design; this results from the fact that the $\mathcal{L}_2$ bound in subsection 12.2.2 is a measure of $\sum_{i=1}^{3} z_i^2$ and not of $\sum_{i=1}^{3} x_i^2$. This means that the only conclusion one can reach for sure is that x_1 will have a smaller bound in the second design. However, observe in Figure 12.5 the plots of $\sum_{i=1}^{3} z_i^2(t)$ for both designs. The solid line, corresponding to the first design, is always above the dashed line, which corresponds to the second design, as expected.

12.4 Conclusion

In this chapter, we have developed a direct adaptive control method for a class of uncertain nonlinear systems with a time-varying structure using a Lyapunov approach to construct the stability proofs. The systems we considered are composed of a finite number of "pieces," or dynamic subsystems, that are interpolated by functions depending on a possibly exogenous scheduling variable. We assumed that each piece is in strict feedback form and showed that the methods yield stability of all signals in the closed loop as well as show convergence of the state vector to a residual set around the equilibrium, whose size can be set by the choice of several design parameters

We argued that the direct adaptive method presents several advantages over indirect methods in general, including the need for a smaller amount of information about the plant and a simpler design. We also provided design guidelines based on $\mathcal{L}_2$ bounds on the transient, and we argued that this bound makes it possible to precisely determine how large the compact sets for the function approximators should be so that the states do not exit them. Finally, we applied the direct adaptive method to the problem of wing rock regulation, where the wing rock dynamics are uncertain and where the angle of attack is allowed to vary with time.

References

Elzebda, J.M., Nayfeh, A.H., and Mook, D.T. (1989). Development of an analytical model of wing rock for slender delta wings. *AIAA Journal of Aircraft 26*, 737–743.

Hsu, C.H., and Lan, E. (1985). Theory of wing rock. *AIAA Journal of Aircraft 22*, 920–924.

Joshi, S.V., Sreenatha, A.G., and Chandrasekhar, J. (1998). Suppresion of wing rock of slender delta wings using a single neuron controller. *IEEE Transactions on Control Systems Technology 6*, 671–677.

Krstić, M., Kanellakopoulos, I., and Kokotović, P. (1995). *Nonlinear and adaptive control design.* New York: John Wiley & Sons.

Levin, D., and Katz, J. (1984). Dynamic load measurements with delta wings undergoing self-induced roll oscillations. *AIAA Journal of Aircraft 21*, 30–36.

Luo, J., and Lan, C.E. (1993). Control of wing rock motion of slender delta wings. *Journal of Guidance Control Dynamics 16*, 225–231.

Moody, J., and Darken, C. (1989). Fast-learning in networks of locally tuned processing units. *Neural Computation 1*, 281–294.

Nayfeh, A.H., Elzebda, J.M., and Mook, D.T. (1989). Analytical study of the subsonic wing rock phenomenon for slender delta wings. *AIAA Journal of Aircraft 26*(9), 805–809.

Ordóñez, R., and Passino, K.M. (2001). Indirect adaptive control for a class of nonlinear systems with a time-varying structure. *International Journal of Control* 74, 701–717.

Ordóñez, R., and Passino, K.M. (2000a). Adaptive control for a class of nonlinear systems with a time-varying structure. *IEEE Transactions on Automatic Control* 46, 152–155.

Ordóñez, R., and Passino, K.M. (2000b). Wing rock regulation with a time-varying angle of attack. *Proceedings of the International Symposium Intelligent Control/IEEE Mediterranean Conference on Automation and Control* 49–54.

Polycarpou, M.M., (1996). Stable adaptive neural control scheme for nonlinear systems. *IEEE Transactions on Automatic Control 41* 447–451.

Polycarpou, M.M., and Ioannou, P.A. (1991). Identification and control of nonlinear systems using neural network models: Design and stability analysis. *Electrical Engineering: Systems Report 91-09-01*, University of Southern California.

Polycarpou, M.M., and Mears, M.J. (1998). Stable adaptive tracking of uncertain systems using nonlinearly parametrized online approximators. *International Journal of Control 70*, 363–384.

Rovithakis, G.A., and Christodoulou, M.A. (1995). Direct adaptive regulation of unknown nonlinear dynamical systems via dynamic neural networks. *IEEE Transactions on Systems, Man, and Cybernetics 25*, 1578–1995.

Singh, S.N., Yim, W., and Wells, W.R. (1995). Direct adaptive and neural control of the wing rock motion of slender delta wings. *Journal of Guidance Control Dynamics 18*, 25–30.

Spooner, J.T., and Passino, K.M. (1996). Stable adaptive control using fuzzy systems and neural networks. *IEEE Transactions in Fuzzy Systems 4*, 339–359.

Su, C.Y., and Stepanenko, Y. (1994). Adaptive control of a class of nonlinear systems with fuzzy logic. *IEEE Transactions on Fuzzy Systems 2*, 285–294.

13

Direct Learning by Reinforcement*

Jennie Si[†]
Department of Electrical Engineering, Arizona State University, Tempe, Arizona, USA

13.1 Introduction

This chapter focuses on a systematic treatment for developing a generic online learning control system based on the fundamental principle of **reinforcement learning** or, more specifically, **neural dynamic programming**. This online learning system improves its performance over time in two aspects. First, it learns from its own mistakes through the reinforcement signal from the external environment and tries to reinforce its action to improve future performance. Second, system states associated with the positive reinforcement are memorized through a network learning process where in the future, similar states will be more positively associated with a control action leading to a positive reinforcement. This discussion also introduces a successful candidate of online learning control design. Real-time learning algorithms are derived for individual components in the learning system. Some analytical insight are provided to give guidelines on the learning process that takes place in each module of the online learning control system. The performance of the online learning controller is measured by its learning speed, success rate of learning, and the degree to meet the learning control objective. The overall learning control system performance is tested on a single cart–pole balancing problem and a pendulum swing-up and balancing task.

13.2 A General Framework for Direct Learning Through Association and Reinforcement

Consider a class of learning decision and control problems in terms of optimizing a performance measure over time with the following constraints. First, a model of the environment or the system that interacts with the learner is not available a priori. The environment/system can be stochastic, nonlinear, and subject to change. Second, learning takes place "on-the-fly" while interacting with the environment. Third, even though measurements from the environment are available from one decision and control step to the next, a final outcome of the learning process from a generated sequence of decisions and controls comes as a delayed signal in an indicative "win or lose" format. Figure 13.1 is a schematic diagram of an online learning control scheme. The binary reinforcement signal $r(t)$ is provided from the external environment and is either a 0 or a -1 corresponding to success or failure, respectively.

* Portions reprinted, with permission, from the 2001 *IEEE Transactions on Neural Networks* 2(2), 264–276.

[†] Supported by NSF under grant ECS-0002098.

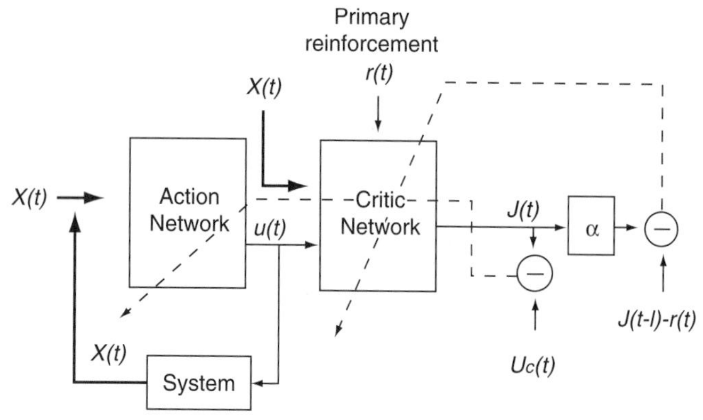

FIGURE 13.1 Schematic Diagram for Implementations of Neural Dynamic Programming as a Direct Learning Mechanism. The solid lines represent signal flow, and the dashed lines are the paths for parameter tuning.

In our direct online learning control design, the controller is "naive" when it just starts to control; the action network and the critic network are both randomly initialized in their weights and parameters. Once a system state is observed, an action will be subsequently produced based on the parameters in the action network. A "better" control value under the specific system state will lead to a more balanced equation of the principle of optimality. This set of system operations will be reinforced through memory or association between states and control output in the action network. Otherwise, the control value will be adjusted through tuning the weights in the action network to make the equation of the principle of optimality more balanced.

To be more quantitative, consider the critic network as depicted in Figure 13.1. The output of the critic element, the J function, approximates the discounted total **reward-to-go**. Specifically, it approximates $R(t)$ at time t given by:

$$R(t) = r(t+1) + \alpha r(t+2) + \cdots, \tag{13.1}$$

where $R(t)$ is the future accumulative reward-to-go value at time t and where α is a discount factor for the infinite-horizon problem ($0 < \alpha < 1$). We have used $\alpha = 0.95$ in our implementations; $r(t+1)$ is the external reinforcement value at time $t+1$.

13.2.1 The Critic Network

The **critic network** is used to provide $J(t)$ as an approximate of $R(t)$ in equation 13.1. The prediction error for the critic element is defined as:

$$e_c(t) = \alpha J(t) - [J(t-1) - r(t)], \tag{13.2}$$

and the objective function to be minimized in the critic network is as follows:

$$E_c(t) = \frac{1}{2} e_c^2(t). \tag{13.3}$$

The weight update rule for the critic network is a gradient-based adaptation given by:

$$w_c(t+1) = w_c(t) + \Delta w_c(t). \tag{13.4}$$

$$\Delta w_c(t) = l_c(t)\left[-\frac{\partial E_c(t)}{\partial w_c(t)}\right]. \tag{13.5}$$

$$\frac{\partial E_c(t)}{\partial w_c(t)} = \left[-\frac{\partial E_c(t)\partial J(t)}{\partial J(t)\partial w_c(t)}\right]. \tag{13.6}$$

In equations 13.4 to 13.6, $l_c(t) > 0$ is the learning rate of the critic network at time t, which usually decreases with time to a small value, and w_c is the weight vector in the critic network.

13.2.2 The Action Network

The principle in adapting the **action network** is to indirectly back-propagate the error between the desired ultimate objective, denoted by U_c, and the approximate J function from the critic network. Since 0 is defined as the reinforcement signal for success, U_c is set to 0 in the discussion is design paradigm and in the following case studies. In the action network, the state measurements are used as inputs to create a control as the output of the network. In turn, the action network can be implemented by either a linear or a nonlinear network, depending on the complexity of the problem. The weight updating in the action network can be formulated as follows. Let the following be true:

$$e_a(t) = J(t) - U_c(t). \tag{13.7}$$

The weights in the action network are updated to minimize the following performance error measure:

$$E_a(t) = \frac{1}{2} e_a^2(t). \tag{13.8}$$

The update algorithm is then similar to the one in the critic network. By a gradient descent rule:

$$w_a(t+1) = w_a(t) + \Delta w_a(t). \tag{13.9}$$

$$\Delta w_a(t) = l_a(t)\left[-\frac{\partial E_a(t)}{\partial w_a(t)}\right]. \tag{13.10}$$

$$\frac{\partial E_a(t)}{\partial w_a(t)} = \frac{\partial E_a(t)}{\partial J(t)}\frac{\partial J(t)}{\partial u(t)}\frac{\partial u(t)}{\partial w_a(t)}. \tag{13.11}$$

In equations 13.9 to 13.11, $l_a(t) > 0$ is the learning rate of the action network at time t, which usually decreases with time to a small value, and w_a is the weight vector in the action network.

13.2.3 Online Learning Algorithms

This chapter is online learning configuration introduced in the previous subsections involves two major components in the learning system: the **action network** and the **critic network.** The following devises learning algorithms and elaborates how learning takes place in each of the two modules. In the discussion is NDP design, both the action network and the critic network are nonlinear multilayer feed-forward networks. In these designs, one hidden layer is used in each network. The neural network structure for the nonlinear, multilayer critic network is shown in Figure 13.2. In the critic network, the output $J(t)$ will be of the form:

$$J(t) = \sum_{i=1}^{N_h} w_{c_i}^{(2)}(t) p_i(t). \tag{13.12}$$

$$p_i(t) = \frac{1 - \exp^{-q_i(t)}}{1 + \exp^{-q_i(t)}}, \; i = 1, \ldots, N_h. \tag{13.13}$$

$$q_i(t) = \sum_{j=1}^{n+1} w_{c_{ij}}^{(1)}(t) x_j(t), \; i = 1, \ldots, N_h. \tag{13.14}$$

The q_i is the ith hidden node input of the critic network, and p_i is the corresponding output of the hidden node. The N_h is the total number of hidden nodes in the critic network, and $n+1$ is the total number of inputs into the critic network including the analog action value $u(t)$ from the action network. By applying the chain rule, the adaptation of the critic network is summarized below.

(1) For $\Delta w_c^{(2)}$ (hidden to output layer):

$$\Delta w_{ci}^{(2)}(t) = l_c(t)\left[-\frac{\partial E_c(t)}{\partial w_{c_i}^{(2)}(t)}\right], \tag{13.15}$$

$$\frac{\partial E_c(t)}{\partial w_{c_i}^{(2)}(t)} = \frac{\partial E_c(t)}{\partial J(t)} \frac{\partial J(t)}{\partial w_{c_i}^{(2)}(t)} = \alpha e_c(t) p_i(t). \tag{13.16}$$

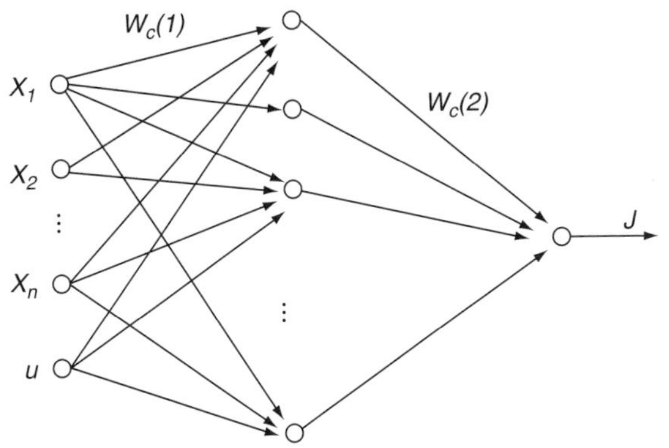

FIGURE 13.2 Schematic Diagram for the Implementation of a Nonlinear Critic Network Using a Feed-Forward Network with one Hidden Layer

(2) For $\Delta w_c^{(1)}$ (input to hidden layer):

$$\Delta w_{c_{ij}}^{(1)}(t) = l_c(t)\left[-\frac{\partial E_c(t)}{\partial w_{c_{ij}}^{(1)}(t)}\right], \tag{13.17}$$

$$\frac{\partial E_c(t)}{\partial w_{c_{ij}}^{(1)}(t)} = \frac{\partial E_c(t)}{\partial J(t)} \frac{\partial J(t)}{\partial p_i(t)} \frac{\partial p_i(t)}{\partial q_i(t)} \frac{\partial q_i(t)}{\partial w_{c_{ij}}^{(1)}(t)} \tag{13.18}$$

$$= \alpha e_c(t) w_{c_i}^{(2)}(t) \left[\frac{1}{2}\left(1 - p_i^2(t)\right)\right] x_j(t). \tag{13.19}$$

Now, investigate the adaptation in the action network, which is implemented by a feed-forward network similar to the one in Figure 13.2, except that the inputs are the n measured states and the output is the action $u(t)$. The following are the associated equations for the action network:

$$u(t) = \frac{1 - \exp^{-u(t)}}{1 + \exp^{-u(t)}}. \tag{13.20}$$

$$v(t) = \sum_{i=1}^{N_h} w_{a_i}^{(2)}(t) g_i(t). \tag{13.21}$$

$$g_i(t) = \frac{1 - \exp^{-h_i(t)}}{1 + \exp^{-h_i(t)}}, \; i = 1, \ldots, N_h. \tag{13.22}$$

$$h_i(t) = \sum_{j=1}^{n} w_{a_{ij}}^{(1)}(t) x_j(t), \; i = 1, \ldots, N_h. \tag{13.23}$$

The v is the input to the action node, and g_i and h_i are the output and the input of the hidden nodes of the action network, respectively. Since the action network inputs the state measurements only, there is no $(n+1)th$ term in equation 13.23 as in the critic network (see equation 13.14 for comparison). The update rule for the nonlinear multilayer action network also contains two sets of equations:

(1) For $\Delta w_a^{(2)}$ (hidden to output layer):

$$\Delta w_{a_i}^{(2)}(t) = l_a(t)\left[-\frac{\partial E_a(t)}{\partial w_{a_i}^{(2)}(t)}\right], \tag{13.24}$$

$$\frac{\partial E_a(t)}{\partial w_{a_i}^{(2)}(t)} = \frac{\partial E_a(t)}{\partial J(t)} \frac{\partial J(t)}{\partial u(t)} \frac{\partial u(t)}{\partial v(t)} \frac{\partial v(t)}{\partial w_{a_i}^{(2)}(t)}, \tag{13.25}$$

$$= e_a(t)\left[\frac{1}{2}\left(1 - u^2(t)\right)\right] g_i(t) \sum_{i=1}^{N_h} \left[w_{c_i}^{(2)}(t) \frac{1}{2}\left(1 - p_i^2(t)\right) w_{c_{i,n+1}}^{(1)}(t)\right]. \tag{13.26}$$

In equations 13.24 to 13.26, $\partial J(t)/\partial u(t)$ is obtained by changing variables and by a chain rule. The result is the summation term, and $w_{c_{i,n+1}}^{(1)}$ is the weight associated with the input element from the action network.

(2) For $\Delta w_a^{(1)}$ (input to hidden layer):

$$\Delta w_{a_{ij}}^{(1)}(t) = l_a(t)\left[-\frac{\partial E_a(t)}{\partial w_{a_{ij}}^{(1)}(t)}\right], \tag{13.27}$$

$$\frac{\partial E_a(t)}{\partial w_{a_{ij}}^{(1)}(t)} = \frac{\partial E_a(t)}{\partial J(t)}\frac{\partial J(t)}{\partial u(t)}\frac{\partial u(t)}{\partial v(t)}\frac{\partial v(t)}{\partial g_i(t)}\frac{\partial g_i(t)}{\partial h_i(t)}\frac{\partial h_i(t)}{\partial w_{a_{ij}}^{(1)}(t)} \tag{13.28}$$

$$= e_a(t)\left[\frac{1}{2}\left(1-u^2(t)\right)\right]w_{a_i}^{(2)}(t)\left[\frac{1}{2}\left(1-g_i^2(t)\right)\right]x_j(t).$$

$$\sum_{i=1}^{N_h}\left[w_{c_i}^{(2)}(t)\frac{1}{2}\left(1-p_i^2(t)\right)w_{c_i,n+1}^{(1)}(t)\right]. \tag{13.29}$$

In implementation, equations 13.16 and 13.19 are used to update the weights in the critic network and equations 13.26 and 13.29 are used to update the weights in the action network.

13.3 Analytical Characteristics of an Online NDP Learning Process

This section is dedicated to expositions of analytical properties of the online learning algorithms in the context of **neural dynamic programming** (NDP). It is important to note that in contrast to usual neural network applications, there is no readily available training sets of input/output pairs to be used for approximating J^* in the sense of least-squares-fit in NDP applications. Both the control action u and the approximated J function are updated according to an error function that changes from one time-step to the next. Therefore, the convergence argument for the steepest descent algorithm does not hold valid for any of the two networks, action or critic. This results in a simulation approach to evaluate the cost-to-go function J for a given control action u. The online learning takes place, aiming at iteratively improving the control policies based on simulation outcomes. This creates analytical and computational difficulties that do not arise in a more typical neural network training context. The closest analytical results in terms of approximating J function was obtained by Tsitsiklis (1997) where a linear in parameter function approximator was used to approximate the J function. The limit of convergence was characterized as the solution to a set of interpretable linear equations, and a bound was placed on the resulting approximation error.

It is worth pointing out that the existing implementations of NDP are usually computationally very intensive (Bertsekas and Tsitsiklis, 1996) and often require a considerable amount of trial and error. Most of the computations and experimentations with different approaches were conducted offline. The following paragraphs provide some analytical insight on the online learning process for proposed NDP designs in this chapter. Specifically, the stochastic approximation argument is used to reveal the asymptotic performance of this discussion's online NDP learning algorithms in an averaged sense for the action and the critic networks under certain conditions.

13.3.1 Stochastic Approximation Algorithms

The original work in **recursive stochastic approximation algorithms** was introduced by Robbins and Monro (1951), who developed and analyzed a recursive procedure for finding the root of a real-valued function $g(w)$ of a real variable w. The function is not known, but noise-corrupted observations could be taken at values of w selected by the experimenter.

A function $g(w)$ with the form $g(w) = Ex[f(w)]$ ($Ex[]$ is the expectation operator) is called a regression function of $f(w)$ and, conversely, $f(w)$ is called a sample function of $g(w)$. The following conditions are needed to obtain the Robbins-Monro algorithm (Robbins and Monro, 1951):

- **C1**: The $g(w)$ has a single root w^*, $g(w^*) = 0$, and:

$$g(w) < 0 \text{ if } w < w^*.$$

$$g(w) > 0 \text{ if } w > w^*.$$

  This first condition is assumed with little loss of generality since most functions of a single root not satisfying this condition can be made to do so by multiplying the function by -1.
- **C2**: The variance of $f(w)$ from $g(w)$ is finite:

$$\sigma^2(w) = Ex[g(w) - f(w)]^2 < \infty. \tag{13.30}$$

- **C3**:

$$|g(w)| < B_1|w - w^*| + B_0 < \infty. \tag{13.31}$$

  This third condition is a very mild condition. The values of B_1 and B_0 need not be known to prove the validity of the algorithm. As long as the root lies in some finite interval, the existence of B_1 and B_0 can always be assumed.

If the conditions C1 through C3 are satisfied, the algorithm from Robbins and Monro (1951) can be used to iteratively seek the root w^* of the function $g(w)$:

$$w(t+1) = w(t) - l(t)f[w(t)], \tag{13.32}$$

where $l(t)$ is a sequence of positive numbers that satisfy the following conditions:

$$\begin{aligned} &1)\ \lim_{t\to\infty} l(t) = 0 \\ &2)\ \sum_{t=0}^{\infty} l(t) = \infty \\ &3)\ \sum_{t=0}^{\infty} l^2(t) < \infty. \end{aligned} \tag{13.33}$$

Furthermore, $w(t)$ will converge toward w^* in the mean square error sense and with probability of value 1:

$$\lim_{t\to\infty} Ex[\|w(t) - w^*\|^2] = 0. \quad (13.34)$$

$$\text{Prob}\left\{\lim_{t\to\infty} w(t) = w^*\right\} = 1. \quad (13.35)$$

The convergence with probability 1 in equation 13.35 is also called **convergence almost truly.** In this chapter, the Robbins-Monro algorithm is applied to optimization problems (Kusher and Yin, 1997). In that setting, $g(w) = \partial E/\partial W$, where E is an objective function to be optimized. If E has a local optimum at w^*, $g(w)$ will satisfy the condition C1 locally at w^*. If E has a quadratic form, $g(w)$ will satisfy the condition C1 globally.

13.3.2 Convergence in Statistical Average for Action and Critic Networks

Neural dynamic programming is still in its early stage of development. The problem is not trivial due to several consecutive learning segments being updated simultaneously. A practically effective on-line learning mechanism and a step-by-step analytical guide for the learning process do not coexist at this time. This chapter is dedicated to reliable implementations of NDP algorithms for solving a general class of on-line learning control problems. As demonstrated in previous sections, experimental results in this direction are very encouraging. This section provides some asymptotic convergence results for each component of the NDP system. The Robbins-Monro algorithm provided in the previous section is the main tool to obtain results. Throughout this chapter, it is implied that the state measurements are samples of a continuous state-space. Specifically, the discussion assumes without loss of generality that the input $X_j \in \chi \subset \mathcal{R}^n$ has discrete probability density $p(X) = \sum_{j=1}^{N} p_j \delta(X - X_j)$, where $\delta()$ is the delta function.

The following paragraphs analyze one component of the NDP system at a time. When one component (e.g., the action network) is under consideration, the other component (e.g., the critic network) is considered to have completed learning; their weights do not change anymore.

To examine the learning process taking place in the action network, the following objective function for the action network is defined as:

$$\begin{aligned}\tilde{E}_a &= \frac{1}{2}\sum_i p_i[J(X_i) - U_c]^2 \\ &= \frac{1}{2}Ex[(J - U_c)^2].\end{aligned} \quad (13.36)$$

It can be seen that equation 13.36 is an "averaged" error square between the estimated J and a final desired value U_c. To contrast this notion, equation 13.8 is an "instantaneous" error square between the two. To obtain a (local) minimum for the averaged error measure in equation 13.38, the Robbins-Monro algorithm can be applied by first taking a derivative of this error with respect to the parameters, which are the weights in the action network in this case. Let:

$$\tilde{e}_a = J - U_c. \quad (13.37)$$

Since J is smooth in $\tilde{w}_a$, and $\tilde{w}_a$ belongs to a bounded set, the derivative of $\tilde{E}_a$ with respect to the weights of the action network is then of the form:

$$\frac{\partial \tilde{E}_a}{\partial \tilde{w}_a} = Ex\left[\tilde{e}_a \frac{\partial \tilde{e}_a}{\partial \tilde{w}_a}\right]. \quad (13.38)$$

According to the Robbins-Monro algorithm, the root (can be a local root) of $\partial\tilde{E}_a/\partial\tilde{w}_a$ as a function of $\tilde{w}_a$ can be obtained by the following recursive procedure *if* the root exists and *if* the step size $l_a(t)$ meets all the requirements described in equation 13.35:

$$\tilde{w}_a(t+1) = \tilde{w}_a(t) - l_a(t)\left[\tilde{e}_a \frac{\partial \tilde{e}_a}{\partial \tilde{w}_a}\right]. \quad (13.39)$$

Equation 13.37 may be considered as an instantaneous error between a sample of the J function and the desired value U_c. Therefore, equation 13.39 is equivalent to the update equation for the action network given in equations 13.9 to 13.11. From this viewpoint, the online action network updating rule of these equations 13.9 to 13.11 is actually converging to a (local) minimum of the error square between the J function and the desired value U_c in a statistical average sense. Or in other words, even if these equations represent a reduction in instantaneous error square at each iterative time-step, the action network updating rule asymptotically reaches a (local) minimum of the statistical average of $(J - U_c)^2$.

By the same token, a similar framework can be constructed to describe the convergence of the critic network. Recall that the residual of the principle of optimality equation to be balanced by the critic network is of the following form:

$$e_c(t) = \alpha J(t) - J(t-1) + r(t). \quad (13.40)$$

The instantaneous error square of this residual is given as:

$$E_c(t) = \frac{1}{2}e_c^2(t). \quad (13.41)$$

Instead of the instantaneous error square, let:

$$\tilde{E}_c = Ex[E_c], \quad (13.42)$$

and assume that the expectation is well-defined over the discrete state measurements. The derivative of $\tilde{E}_c$ with respect to the weights of the critic network is then of the form:

$$\frac{\partial \tilde{E}_c}{\partial \tilde{w}_c} = Ex\left[e_c \frac{\partial e_c}{\partial \tilde{w}_c}\right]. \quad (13.43)$$

According to the Robbins-Monro algorithm, the root (can be a local root) of $\partial \tilde{E}_c / \partial \tilde{w}_c$ as a function of $\tilde{w}_c$ can be obtained by the following recursive procedure *if* the root exists and *if* the step size $l_c(t)$ meets all the requirements described in equation 13.35:

$$\tilde{w}_c(t+1) = \tilde{w}_c(t) - l_c(t)\left[e_c \frac{\partial e_c}{\partial \tilde{w}_c}\right]. \quad (13.44)$$

Therefore, equation 13.44 is equivalent to the update rule for the critic network given in equations 13.4 to 13.6. From this viewpoint, the online critic network update rule of equations 13.4 to 13.6 is actually converging to a (local) minimum of the residual square of the equation of the principle of optimality in a statistical average sense.

13.4 Example 1

The proposed NDP design has been implemented on a **single cart–pole balancing problem**. To begin with, the self-learning controller has no prior knowledge about the plant but only of online measurements. The objective is to balance a single pole mounted on a cart, which can move either to the right or to the left on a bounded, horizontal track. The goal for the learning controller is to provide a force (applied to the cart) of a fixed magnitude in either the right or the left direction so that the pole stands balanced and avoids hitting the track boundaries. The controller receives reinforcement only after the pole has fallen.

To provide the learning controller measured states as inputs to the action and the critic networks, the cart–pole system was simulated on a digital computer using a detailed model that includes all of the nonlinearities and reactive forces of the physical system such as frictions. Note that these simulated states would be the measured ones in real-time applications.

13.4.1 The Cart–Pole Balancing Problem

The cart–pole system used in the current study is the same as the one in Barto *et al.* (1983):

$$\frac{d^2\theta}{dt^2} = \frac{g \sin\theta + \cos\theta[-F - ml\dot{\theta}^2 \sin\theta + \mu_c \mathrm{sgn}(\dot{x})] - \frac{\mu_p \dot{\theta}}{ml}}{l\left(\frac{4}{3} - \frac{m \cos^2\theta}{m_c + m}\right)}. \quad (13.45)$$

$$\frac{d^2x}{dt^2} = \frac{F + ml[\dot{\theta}^2 \sin\theta - \ddot{\theta}\cos\theta] - \mu_c \mathrm{sgn}(\dot{x})}{m_c + m}. \quad (13.46)$$

In equations 13.45 and 13.46, these variables and values apply:

- $g = 9.8\,\mathrm{m/s^2}$, acceleration due to gravity
- $m_c = 1.0\,\mathrm{kg}$, mass of cart
- $m = 0.1\,\mathrm{kg}$, mass of pole
- $l = 0.5\,\mathrm{m}$, half-pole length
- $\mu_c = 0.0005$, coefficient of friction of cart on track
- $\mu_p = 0.000002$, coefficient of friction of pole on cart
- $F = \pm 10$ Newtons, force applied to cart's center of mass
- $\mathrm{sgn}(x) = \begin{cases} 1, & \text{if } x > 0. \\ 0, & \text{if } x = 0. \\ -1, & \text{if } x < 0. \end{cases}$

The nonlinear differential equations of 13.45 and 13.48 are numerically solved by a fourth-order Runge-Kutta method. This model provides four state variables: (1) $x(t)$, position of the cart on the track; (2) $\theta(t)$, angle of the pole with respect to the vertical position; (3) $\dot{x}(t)$, cart velocity; and (4) $\theta(t)$, angular velocity.

In the current study, a run consists of a maximum of 1,000 consecutive trials. It is considered successful if the last trial (trial number less than 1,000) of the run has lasted 600,000 time-steps. Otherwise, if the controller is unable to learn to balance the cart–pole within 1,000 trials (i.e., none of the 1,000 trials has lasted over 600,000 time-steps), then the run is considered unsuccessful. This chapter's simulations have used 0.02 sec for each time-step, and a trial is a complete process from start to fall. A pole is considered fallen when the pole is outside the range of $[-12°, 12°]$ and/or the cart is beyond the range of $[-2.4, 2.4]$ meters in reference to the central position on the track. Note that although the force F applied to the cart is binary, the control $u(t)$ fed into the critic network as shown in Figure 13.1 is continuous.

13.4.2 Simulation Results

Several experiments were conducted to evaluate the effectiveness of this chapter's learning control designs. The parameters used in the simulations are summarized in Table 13.1 with the proper notations defined in the following:

- $l_c(0)$: Initial learning rate of the critic network
- $l_a(0)$: Initial learning rate of the action network
- $l_c(t)$: Learning rate of the critic network at time t, which is decreased by 0.05 every 5 time-steps until it reaches 0.005 and stays at $l_c(f) = 0.005$ thereafter
- $l_a(t)$: Learning rate of the action network at time t, which is decreased by 0.05 every 5 time-steps until it reaches 0.005 and stays at $l_a(f) = 0.005$ thereafter
- N_c: Internal cycle of the critic network
- N_a: Internal cycle of the action network
- T_c: Internal training error threshold for the critic network
- T_a: Internal training error threshold for the action network
- N_h: Number of hidden nodes

TABLE 13.1 Summary of Parameters Used in Obtaining the Results Given in Table 13.2

Parameter	$l_c(0)$	$l_a(0)$	$l_c(f)$	$l_a(f)$	*
Value	0.3	0.3	0.005	0.005	*
Parameter	N_c	N_a	T_c	T_a	N_h
Value	50	100	0.05	0.005	6

Note that the weights in the action and the critic networks were trained using their internal cycles, N_a and N_c, respectively. That is, within each time-step, the weights of the two networks were updated for at most N_a and N_c times, respectively, or stopped once the internal training error threshold T_a and T_c has been met.

To be more realistic, both a sensor noise and an actuator noise have been added to the state measurements and the action network output. Specifically, the actuator noise has been implemented through $u(t) = u(t) + \rho$, where ρ is a uniformly distributed random variable. For the sensor noise, both uniform and Gaussian random variables were added to the angle measurements θ. The uniform state sensor noise was implemented through $\theta = (1 + \text{noise percentage}) \times \theta$. Gaussian sensor noise was zero mean with specified variance.

The proposed configuration of neural dynamic programming has been evaluated, and the results are summarized in Table 13.2. The simulation results summarized in Table 13.2 were obtained through averaged runs. Specifically, 100 runs were performed to obtain the results reported here. Each run was initialized to random conditions in terms of network weights. If a run is successful, the number of trials it took to balance the cart–pole is then recorded. The number of trials listed in the table corresponds to the one averaged over all of the successful runs. Therefore, there is a need to record the percentage of successful runs out of 100. This number is also recorded in the table. A good configuration is the one with a high percentage of successful runs as well as a low average number of trials needed to learn to perform the balancing task.

TABLE 13.2 Performance Evaluation of NDP Learning Controller when Balancing a Cart-Pole System

Noise type	Success rate	Number of trials
Noise-free	100%	6
Uniform 5% actuator	100%	8
Uniform 10% actuator	100%	14
Uniform 5% sensor	100%	32
Uniform 10% sensor	100%	54
Gaussian $\sigma^2 = 0.1$ sensor	100%	164
Gaussian $\sigma^2 = 0.2$ sensor	100%	193

Note: The second column represents the percentage of successful runs out of 100. The third column depicts the average number of trials it took to learn to balance the cart–pole. The average is taken over the successful runs.

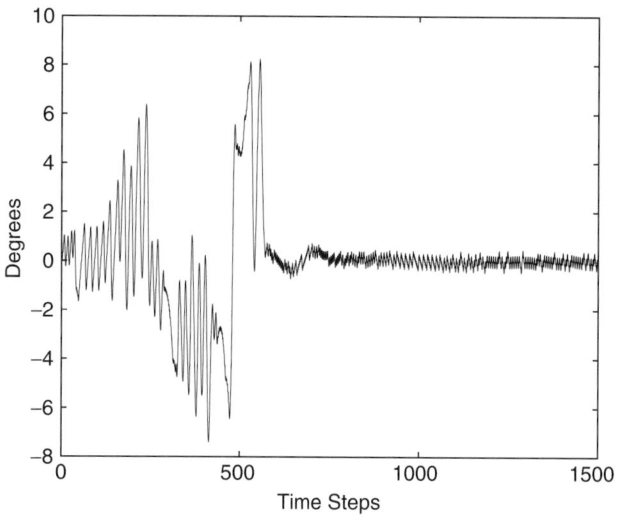

FIGURE 13.3 A Typical Angle Trajectory During a Successful Learning Trial for the NDP Controller when the System is Free of Noise

Figure 13.3 shows a typical movement or trajectory of the pendulum angle under an NDP controller for a successful learning trial. The system under consideration is not subject to any noise. Figure 13.4 represents a summary of typical statistics of the learning process in histograms. It contains vertical angle histograms when the system learns to balance the cart–pole using ideal state measurements without noise corruption.

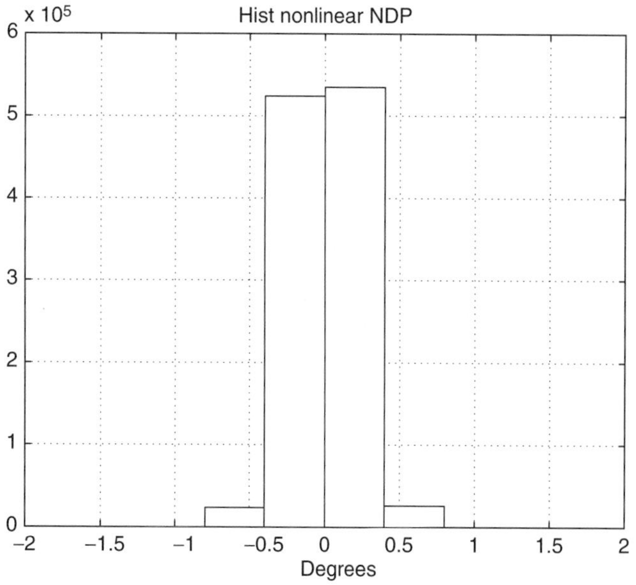

FIGURE 13.4 Histogram of Angle Variations Under the Control of NDP On-Linear Learning Mechanism in the Single Cart–Pole Problem. The system is free of noise in this case.

13.5 Example 2

This section examines the performance of the proposed NDP design in a **pendulum swing-up and balancing task**. The case under study is identical to the one in Santamaria *et al.* (1996).

The pendulum is held by one end and can swing in a vertical plane. The pendulum is actuated by a motor that applied a torque at the hanging point. The dynamics of the pendulum are as follows:

$$\frac{d\omega}{dt} = \frac{3}{4\,ml^2}(F + mlgsin(\theta)). \tag{13.47}$$

$$\frac{d\theta}{dt} = \omega. \tag{13.48}$$

In equations 13.47 and 13.48, $m = 1/3$ and $l = 3/2$ are the mass and length of the pendulum bar, respectively, and $g = 9.8$ is the gravity. The action is the angular acceleration F, and it is bounded between -3 and 3, namely $F_{\min} = -3$ and $F_{\max} = 3$. A control action is applied every four time-steps. The system states are the current angle θ and the angular velocity ω. This task requires the controller to not only swing up the bar but also to balance it at the top position. The pendulum initially sits still at $\theta = \pi$. This task is considered difficult in the sense that (1) there is no closed-form analytical solution for the optimal solution, and complex numerical methods are required to compute it, and (2) the maximum and minimum angular acceleration values are not strong enough to move the pendulum straight up from the starting state without first creating angular momentum (Santamaria *et al.*, 1996).

In this study, a run consists of a maximum of 100 consecutive trials. It is considered successful if the last trial (trial number less than 100) of the run has lasted 800 time-steps (with a step size of 0.05 sec). Otherwise, if the NDP controller is unable to swing up and keep the pendulum balanced at the top within 100 trials (i.e., none of the 100 trails has lasted over 800 time-steps), then the run is considered unsuccessful. In this discussion's simulations, a trial is either terminated at the end of the 800 times-steps or when the angular velocity of the pendulum is greater than 2π (i.e., $\omega > 2\pi$).

The following paragraphs explain two implementation scenarios with different settings in reinforcement signal r. In setting 1, $r = 0$ when the angle displacement is within 90° from the position of $\theta = 0$; $r = -0.4$ when the angle is in the lower half of the plane: and $r = -1$ when the angular velocity $\omega > 2\pi$. In setting 2, $r = 0$ when the angle displacement is within 10° from the position of $\theta = 0$; $r = -0.4$ when the angle is in the remaining area of the plane; and $r = -1$ when the angular velocity $\omega > 2\pi$.

This chapter is proposed NDP configuration is then used to perform the described task. The same configuration and the same learning parameters as those in the first case study are used. NDP controller performance is summarized in Table 13.3. The simulation results summarized in the table were obtained through averaged runs. Specifically, 60 runs were performed to obtain the results reported here. Note that more runs have been used than those in Santamaria *et al.*

TABLE 13.3 Performance Evaluation of NDP Learning Controller to Swing Up and Then Balance a Pendulum

Reinforcement implementation	Success rate	Number of trials
Setting 1	100%	4.2
Setting 2	96%	3.5

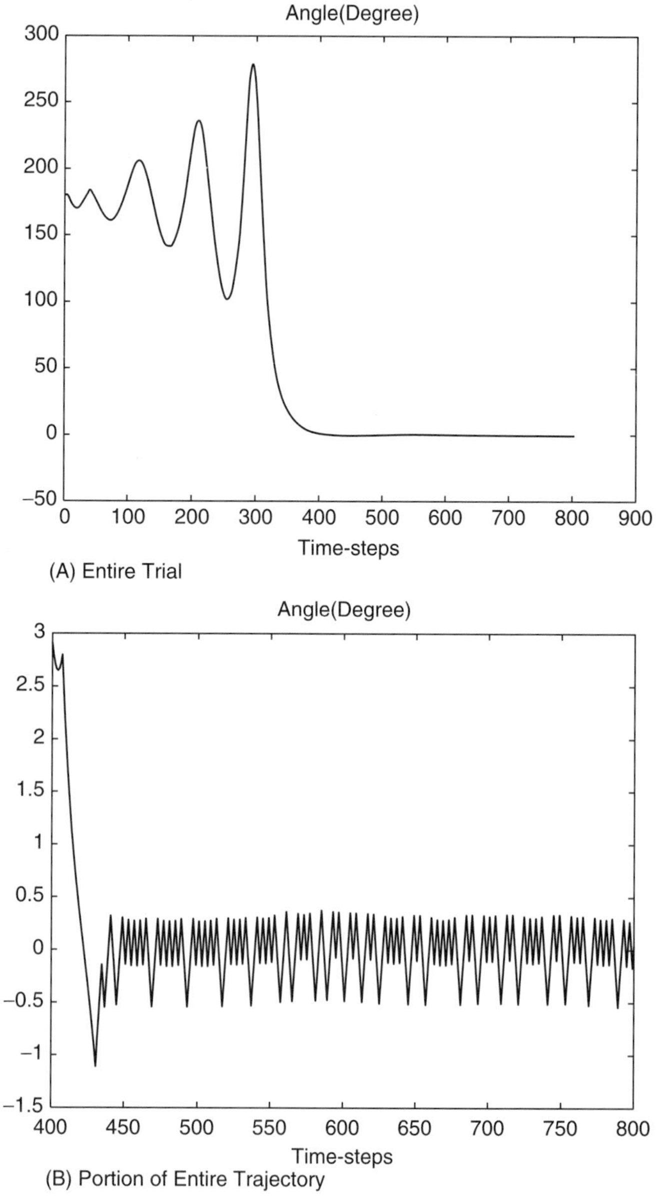

FIGURE 13.5 A Typical Angle Trajectory During a Successful Learning Trial for the NDP Controller in the Pendulum Swing Up and Balancing Task

(1996) (which was 36) to generate the final result statistics. Every other simulation condition has been kept the same as that in Santamaria *et al.* (1996). Each run was initialized to $\theta = \pi$ and $\omega = 0$. The number of trials listed in the table corresponds to the one averaged over all of the successful runs. The percentage of successful runs out of 60 was also recorded in the table. Figure 13.5 shows a typical trajectory of the pendulum angle under an NDP controller for a successful learning trial. This trajectory is characteristic for both setting 1 and setting 2.

The second column represents the percentage of successful runs out of 60. The third column depicts the average number of trials it took to learn to successfully perform the task. The average is taken over the successful runs.

13.6 Conclusion

This chapter is an introduction to a learning control scheme that can be implemented in real time. It may be viewed as a model independent approach to the adaptive critic designs. The chapter demonstrates the implementation details and learning results using two illustrative examples. The chapter also provides a view on the convergence property of the learning process under the assumption that the two network models, the critic and the action, can be learned separately. The analysis of the complete learning process remains to be an open issue.

References

Barto, A.G., Sutton, R.S., and Anderson, C.W. (1983). Neuron-like adaptive elements that can solve difficult learning control problems. *IEEE Transactions on Systems, Man, and Cybernetics 13,* 834–847.

Bertsekas, D.P., and Tsitsiklis, J.N. (1996). *Neuro-dynamic programming.* Belmont, MA: Athena Scientific.

Kushner, H.J., and Yin, G.G. (1997). *Stochastic approximation algorithms and applications.* New York: Springer.

Robbins, H., and Monro, S. (1951). A stochastic approximation method. *Annals of Mathematics and Statistics 22,* 400–407.

Santamaria, J.C., Sutton, R.S., and Ram, A. (1996). Experiments with reinforcement learning in problems with continuous state and action spaces. *COINS Technical Report 96–88,* University of Massachussetts, Amherst.

Tsitsiklis, J.N., and Van Roy, B. (1997). An analysis of temporal-difference learning with function approximation. *IEEE Transactions on Automatic Control* 42(5), 674–690.

Werbos, P.J. (1990). A menu of design for reinforcement learning over time. *Neural Networks for Control,* W.T. Miller III, R.S. Sutton, and P.J. Werbos (eds.), MIT Press, Cambridge, MA.

Werbos, P.J. (1992). Approximate dynamic programming for real-time control and neural modeling. Handbook of Intelligent Control. D. White and D. Sofge (eds.), Van Nostrand Reinhold, New York.

14

Software Technologies for Complex Control Systems

Bonnie S. Heck
School of Electrical and Computer Engineering, Georgia Institute of Technology, Atlanta, Georgia, USA

14.1 Introduction

Control systems and computers enjoy a very successful partnership that began modestly in the 1940s when the theoretical framework for sampled data control systems was first established. The early implementation of the control was done with minicomputers that were rather bulky and sensitive to environmental conditions (e.g., heat, humidity, etc.) and, hence, not well suited for many applications. The popularity of digital control grew dramatically shortly after microprocessors were introduced in the early 1970s. The main considerations of these early digital controllers were speed of the processor, effect of quantization, and effect of sampling rates. The software considerations revolved around the size of the executable program (in terms of memory requirements) and the speed at which it executed (to increase the sampling rate). Many early control system programs were written in assembly code because compiled C programs and Pascal programs were often too large and too slow. As computer technology advanced, the emphasis in digital control research was placed on analysis and on the development of sophisticated algorithms.

New developments in software engineering promise to make revolutionary advances in the way control systems are implemented. Moreover, advances in the implementation may inspire a new generation of digital control algorithms that take advantage of new software technologies. The challenge is that control systems are unique among computer applications since they rely on hard real-time computations. Too slow of an update rate (or an erratic update rate) means that the closed-loop system may experience stability problems. In other computer applications, such as telecommunications, slow update rates may simply degrade performance and irritate customers rather than destabilize the system.

The need for speed of operation historically resulted in tight coupling of software. **Tight coupling** means that different software modules are intertwined so that they share data or control flow information. Modification to one software module may necessitate modifications to all modules that are coupled with it. Although this is advantageous for the speed of real-time applications, it makes system integration a nightmare for large projects. Consider the flight control for an aircraft: there are many different control algorithms, sensors, and actuators that need to be integrated. Multiple processors are used generally on board. Integrating all the software for the different components through a common communications network is a huge task and often is the bottleneck in any updates to the flight controls, especially when the software is tightly coupled. Tight coupling also severely complicates the flight certification of software, which is a verification and validation procedure used to prove that the software is stable and will not have run-time errors.

As computers have become faster and as communication networks have become more efficient in the flow of information, the practice of providing speed through tight coupling of control system software modules has become less necessary. Replacing the need for speed is the need for ease of system integration and system evolution. A very desirable feature is **plug and play**, which refers to ready interchangeability of modules and system reconfiguration with minimal software or protocol changes. For example, as new sensors become available, it would be nice to pull out the old sensors and put in the new sensors with drivers that are "plug and play" with the rest of the system. Also, for the sake of adaptability, it would be helpful to have **run-time reconfiguration** of the software (such as updating or changing algorithms online in real-time). Another desirable feature would be **interoperability** among different processor platforms and different languages. For example, suppose a processor embedded in an application is running one type of operating system and a desktop computer attached to the system is running on another operating system. Data transfer between these two systems is not trivial; for example, the memory byte ordering is different between Windows-based platforms and Unix platforms, and word-size differences on 32-bit versus 64-bit processor architectures cause incompatibilities.

This chapter gives a brief tutorial on some software technologies that are important for control applications. While software is employed both in design and in implementation, this chapter concentrates on implementation technologies. This is a hot topic of current research, so readers should consult the latest information when using these technologies.

14.2 Objects and Components: Software Technologies

One of the major advances in computer science over the last few decades has been **object-oriented programming** (OOP). People who learned programming before 1990 would find OOP somewhat elusive to define and even more elusive to understand. The OOP, however, has taken hold in the constructs of the newer languages, such as Java and C++ (the object-oriented version of C). The motivation for using OOP is for the ease of evolution of software and the reuse of code for different applications. OOP builds programs in a modular fashion, where the software modules are called **objects**. Objects have well-defined interfaces that are used to interact with other objects. Each interface specifies a set of object attributes that are visible to other objects and a set of functions (or "methods") that represent the behavior or activities the object can perform. The term **class** refers to the object type and is analogous to a generic plan for building objects (like the software version of the term blueprint), whereas the objects themselves are the specific cases. Classes tend to be organized in a hierarchical relationship: the most common features of different classes are grouped together in one high-level class called a **superclass**, with specialization in **subclasses**. The class structure makes it easier to reuse modular portions of the software for multiple applications because changes can be localized in object classes.

There are three main mechanisms used in object-oriented programs:

- **Encapsulation** refers to objects being self-contained modules with well-defined interfaces for interaction with other objects. The variables used in an object are generally limited in scope so that they are *not* accessible outside of the object (unless they are specifically declared as "public" variables). An object's interface that allows it to interact with the outside world is kept constant even as the internal workings may be modified and updated, so modification to one object does not require corresponding modifications to other objects. This encapsulation property enables software updates to be accomplished more efficiently and with less risk of error.
- **Inheritance** is a property relating to the hierarchy of objects. A superclass of objects has certain properties common to all subclasses that inherit from it. The subclasses can reuse predefined methods (or function calls) of the superclass without having to redefine the methods themselves. Thus, inheritance promotes software reuse. When applying the software to a new application, only certain methods need to be specialized; the rest can remain unchanged.
- **Polymorphism** is a property relating to the ability of a subclass of a superclass to tailor itself for a specific application or override the methods defined in the superclass. Thus, one method in a superclass might mean different things in different subclasses. The primary advantage is that one set of operations can be applied to objects, and the appropriate implementation of each operation is chosen dependent on the class of object to which it is applied.

These concepts can be made more concrete by applying the OOP paradigm to sensors used in flight controls, such as the Global Positioning System (GPS), radar altimeter, and Inertial Measurement Unit (IMU). These **sensors**, typically purchased commercial-off-the-shelf (COTS), include the actual sensor hardware, a digital output, and a software driver (usually in C) for a processor to read the output. Each of these sensors constitutes an object. Moreover, all sensors perform a common function and have certain common tasks: initializing hardware, opening the data port, closing the data port, and the periodic reading of data during operation. Therefore, you can create a generic class of objects called "Sensor" that encapsulates common attributes and procedures. The hierarchy of this structure is illustrated in Figure 14.1. The Sensor class has methods (or function calls) defined for the common sensor tasks, and the three different sensor subclasses inherit

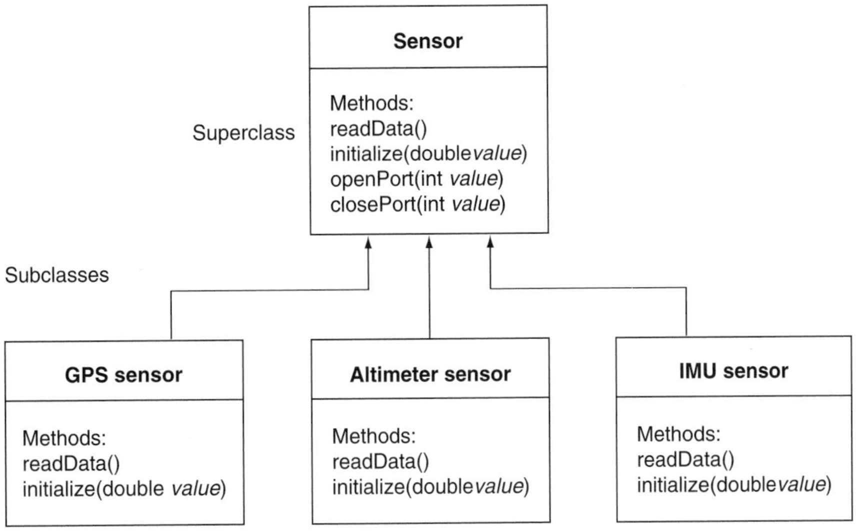

FIGURE 14.1 An OOP Class Diagram of Sensor Objects

these methods. These methods form the interface that all other objects can use to interact with these sensors (such as a control algorithm reading the current data). A Java example of a method call made from a control algorithm is $y =$ gps.readData(), where gps is an object of the GPS class and readData() is the method within the object that can be accessed publicly by other objects.

Because, as stated previously, encapsulation refers to very modular objects with constant interfaces (even as the object's internal code changes), a GPS from one vendor may be replaced by one from another vendor. As long as the method call is not changed, no other object in the system will have to be changed. Inheritance is reflected by the methods that are present in the Sensor class and used by the subclasses. In this example, you could define the methods openPort(int *value*) and closePort(int *value*) in the superclass where the argument is the number of the port to be used. Even though these methods are not explicitly defined in the subclasses, the method call gps.openPort(int *value*) has meaning because the GPS class inherits the method from the superclass.

Polymorphism can be seen in the initialize and the readData methods, which can apply to any sensor object but require different implementation depending on the type of sensor. For this purpose, the subclasses all have these methods defined, which override the method defined in the superclass. Why then would you bother to define these methods in the superclass when they are defined more specifically in the subclasses? The answer lies in the way the methods can be called from the control algorithm object. As defined, the control algorithm can group all the sensor objects into one vector; the method call would only need to be made once (on the vector object), and each sensor's appropriate method would be used automatically. This demonstrates polymorphism, where the same method call means something different to the different subclasses. Polymorphism removes the need for loops or complex case statements to determine which initialization routine to run.

The desire for more extensive reusability of software than is provided by OOP gave rise to a new concept in the late 1990s termed **software components**, which are software modules that are designed to be reused in different software applications (Szyperski, 1998). The term arises from the engineering term **component**, referring to the pieces of equipment (preferably COTS equipment) that an engineer integrates together to form a larger application. The idea is that software for large applications could also be composed of individual software modules that may be available as COTS. A software component can be an object, a group of objects, or a non-object-oriented software module. In this chapter, the term **component** will be used to represent a generic, reusable module in a control system that may be composed entirely of software, hardware, or a combination of both. The term **distributed components** is often used to describe components that are linked via a communication network.

As an example of these concepts, consider an object-oriented model of a typical flight control system. The flight control consists of a low-level stability augmentation control as well as a high-level control, such as guidance, and each one may have a dedicated processor. There are many sensors on board such as a GPS, altimeter, and an IMU. The actuators are the servomechanisms that drive the control surfaces (the ailerons, the rudder, and the elevator). The object-oriented modeling method subdivides the flight controls into components as shown in Figure 14.2. Each bubble represents an object (or more generally, a component). Some of the components consist of only software (such as the controllers), one component consists of physical features (the vehicle dynam-

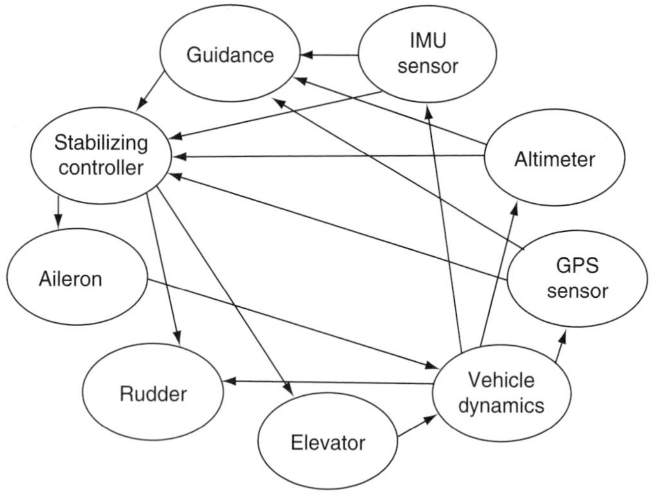

FIGURE 14.2 Objects (or Components) of a Typical Flight Control System. The links represent communication pathways.

ics), and the rest consist of both hardware and software (specifically, the software drivers for the actuators and sensors). Each of these components can be represented entirely by a software model; for example, the vehicle dynamics could be replaced by a simulation for modeling purposes. The resulting modeled software components could be applied in numerous aircraft flight control systems, demonstrating the reusability of software components. Finally, some of the links shown in the figure may be made along a communication network, so these components may be described as distributed components.

While the use of OOP and software components have become well established in computer science arenas, such as desktop computing and network software for managing local and wide area networks, the use of these concepts in the field of control systems is newer and challenges the real-time aspects of the software.

14.3 Layered Architectures

In a layered architecture, objects are designed using a building block mentality. The bottom layer is composed of objects that perform low-level, often tedious functions. The next layer has somewhat higher functionality and makes calls to the objects in the lower layer. Each successive layer upward is more high-level in its functionality. A common way of explaining this layering is that the details are "abstracted away," meaning that some of the tedious details needed to perform the function are hidden from the higher level objects simply by delegating them to the lower levels. In a layered architecture, the object calls are all downward. This architecture is the motivation for the libraries of common function calls that are available with most high-level languages, including the **application programming interfaces** (APIs) that are available for object-oriented languages such as Java.

A simple analogy in hardware to layered architecture is the design of digital circuits. At the bottom layer of the design are **transistors**. Transistors are composed together to form Boolean logic gates at the next layer, such as NAND and OR. At the next layer are devices such as **decoders** and **multiplexers** that are composed of logic gates. These devices then become the building blocks for higher layers. A complex digital circuit could be designed at the transistor level, or it could be designed at the multiplexer/decoder level. The transistor-level design would most likely yield a more efficient implementation but would be much more difficult and time-consuming to the person designing the circuit. In a similar way, this concept of abstracting away the details is the motivation for programming using higher level languages rather than programming at the assembly code level. Programming at the assembly code level produces a more efficient code but is much more tedious and harder to troubleshoot. The terms **level** and **layer** will be used throughout the rest of this chapter to reflect layers of abstraction in a layered architecture.

14.4 Networked Communications

Complex control systems often have multiple processors that must be interconnected through a communication network. All communication networks have protocols that are the specifications (or rules) by which the data is transferred. Most common model for data networks is the **open systems interconnection** (OSI) **model** (Tanenbaum, 1996), which has seven layers of protocols: the **physical layer** (the hardware that actually transmits and receives the data) at the bottom and the **application layer** (the software level that accesses mail and does FTP functions) at the top. In between are the layers that check for errors, perform routing (such as IP) tasks, create rules for providing a reliable connection including breaking information into packets (such as TCP), and handle collisions of packets (such as CSMA/CD). Typical data networks, such as ethernet, send packets of information in a nondeterministic or "bursty" manner. The desire to have a more deterministic, steady stream of information in applications such as voice and video transmission prompted the design of the **asynchronous transfer mode** (ATM) model, which is an alternative to the OSI model and has the potential for good applicability to networked control systems.

A layer of abstraction can be added on top of the protocol layers to build communication models. The three common types of models are the following:

- ***Point–to–point:*** In this model, all communication is one-to-one between pairs of entities. A common example for this type of communication is a telephone call between two parties.

- ***Client–server:*** This model is a many-to-one communication model where many clients can be connected simultaneously to one server. The server acts as a centralized source of data.
- ***Publish–subscribe:*** In this model, many components interact with one another in a many-to-many architecture. To achieve this, this publish–subscribe model introduces an intermediate level between the components that acts to mediate the communications. Components can be designated as **publishers**, which produce data, and **subscribers**, which require data as input. An example of a publisher is a sensor that produces measurements. An example of a subscriber is a control algorithm that requires the measurements from the sensor. In a complex control system, there may be many publishers and many subscribers. For example, a process control system may have distributed components using many processors, or at least many separate algorithms all using similar data, and numerous sensors producing data.

14.5 Middleware

Integrating different software applications either in a single processor or across a network requires detailed knowledge of the operating systems on the connected machines. Moreover, the use of networks requires knowledge of communication protocols that dictate how packets of data are sent and verified. Programming at the operating system level or at the network level adds a layer of complexity to the process that most control engineers would prefer to avoid. As an alternative, a new layer, called **middleware**, is inserted between the operating system and the application software as shown in Figure 14.3. The middleware sets up the communication protocols and interfaces with the operating systems for the machines on the network to allow different software applications on the different machines (or on the same machines) to communicate. As such, middleware insulates application users from the lower level operations. You can see how useful middleware is by examining the interaction of different application software programs running on one machine. For example, as long as software vendors build to an interface standard, users are able to cut and paste from one window to another application running in a different window.

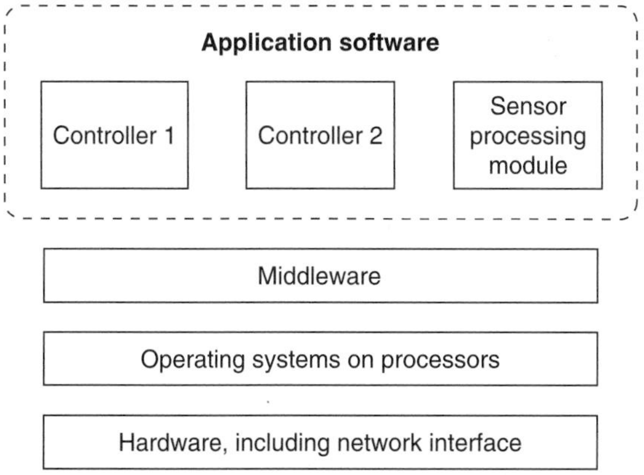

FIGURE 14.3 Architecture Showing how Middleware Acts to Insulate the Application Software from the Low-Level Computer Functions

When middleware is used as the basis for integrating applications, it can be seen as a **software substrate**. To be effective and easy to use as a substrate, middleware should have certain properties. First, it must have a standard interface so that all application software programs that use the standard can be integrated easily. Second, it must have **heterogeneous interoperability**, that is, be able to integrate application software that is written in different languages and/or running on different processor platforms. The ultimate goal for control systems would be to have middleware that has plug-and-play capability. Also, for adaptability sake, it would be helpful to have run-time reconfiguration of the software components (such as updating or changing algorithms online in real-time). The control engineering field is not yet to this point in its use of middleware. More discussion of the available products is given in the last section of this chapter; first, more fundamental concepts on software engineering are introduced.

Consider the use of **object-oriented middleware** to provide the distributed communication channel for the publish/subscribe method of communications. Object-oriented middleware treats the software and hardware components connected to it as objects. Ordinarily, when one object on a distributed communication network calls a method of another object, it would have to know which object had that method and be able to locate it on the network. The middleware makes this method request transparent to the object that makes the call by using **remote procedure calls** (RPCs). Because all objects have a standard interface and all coupling between objects is mediated through the communication channel (that is, the objects are not required to be coupled together directly), adding or removing objects from the overall system can be done in a much simpler manner than is possible if the components are all tightly coupled. Consider three different alternatives for object-oriented middleware: **DCOM**, **CORBA**, and **Java-based technologies** (including Jini technology) (Szyperski, 1998).

14.5.1 Distributed Component Object Model

Component Object Model (COM) was developed by Microsoft and is used to integrate different software applications on platforms running Microsoft Windows. Distributed Component Object Model (DCOM) extends this standard to software applications that may be running on different Windows-based platforms and are connected via a network. Applicability to other platforms is under development.

14.5.2 Common Object Request Broker Architecture

Common Object Request Broker Architecture (CORBA) is a software standard developed by a consortium called the Object Management Group (OMG). OMG has approximately 800 member companies who develop and adopt products that abide by the standards set forth by OMG. A basic feature of CORBA is the **Object Request Broker** (ORB) that handles remote procedure calls. When an object calls a method of another object distributed elsewhere on the network, the ORB intercepts the call and directs it. In this manner, the calling object does not need to know the location of the remote object. The middleware itself is available in different languages that all abide by the CORBA standard; for example, the Java 2 Platform Standard Edition, v1.3, includes a CORBA ORB. The use of CORBA in control engineering is rather new; see Wills *et al.* (2000) for details on the development of a product that is based on a CORBA substrate. Some high-level features of CORBA, however, make it very attractive for future control engineering products used to integrate systems. Its main benefits are that it has a well-accepted standard for the interface, it allows for interoperability of software components written in different languages (including Java and C++) and/or running on different platforms, it handles all network communication protocols, and it provides for run-time reconfiguration of software components. Components can also be integrated that are not object-oriented (such as legacy code) by wrapping the code in a layer that provides an object-oriented interface. Services are continually being added to the ORB structure that extend the capabilities of CORBA. For example, while CORBA is implemented as a client/server communication model, a layer (called an event service) has been added that emulates the publish/subscribe behavior.

14.5.3 Java-Based Technologies

The third common way of integrating components is with Java-based technologies. Java is an object-oriented language that is platform independent and is open source (meaning that the source code is not proprietary), so it has good promise for use in middleware development. Java is based on a layered architecture with new APIs being developed on a continuing basis. Some of the APIs that enable Java to be used for distributed applications are the Remote Method Invocation (RMI) API, the RMI over Internet Inter-ORB Protocol (IIOP) API, and the Java ORB API. The RMI API provides the means for distributed objects to interface through remote method calls (similar to remote procedure calls but specialized to Java objects). The RMI IIOP API provides the remote method call capability using a CORBA ORB. The Java ORB API provides a Java-based CORBA ORB. It should be noted that Java components can interact with non-Java components (such as those written in C) through the use of the Java Native Interface (JNI). JNI works by providing a standard for writing Java code to wrap around non-Java code; the combined code then can be treated like a Java object.

Although many of these Java APIs provide services useful in middleware, a complete package is available in Java: the Jini network technology. Jini is a software architecture for integrating hardware over a network. The lower layers of the architecture handle network protocol decisions, so these decisions are transparent to the user who interfaces at the higher layers. The user also has the option of performing remote method calls using the Java RMI technology, CORBA services, or XML. Thus, Jini can be used as a complementary technology to CORBA. There is a strong continuing effort in the Java community to build new APIs that are increasingly aimed at making Java the basis for integrating multiple components across a network with real-time specifications. Because this field changes rapidly, the reader is encouraged to visit the Sun Microsystems Java Web site for updates.

The following example illustrates how middleware can be used to facilitate control system integration and evolution of the system as new hardware components or new software algorithms become available. Consider the flight control example introduced in Figure 14.2. This control system is shown in Figure 14.4 integrated into a publish/subscribe model. The publishing components (those that produce data) are shown with arrows pointing away from them, and the subscribing components (those that require data) are shown with arrows pointing toward them. Note that some components are both publishers and subscribers.

To illustrate the communication procedure, consider the guidance module that requires altitude information. It subscribes to this data from the software substrate. When subscribing, it declares the data that it needs and the rate at which it must receive the data. The GPS and the altimeter both publish this data, so the middleware must have a strategy for deciding which of these sensors to use in retrieving data. One strategy is to give both sensors weight values (Pardo-Castellote *et al.*, 1999). The middleware sends data to the subscriber from the publisher that has a higher weight. These weights can be changed in real-time by the middleware so that alternate sensors can be used based on current flight conditions. The subscribers themselves are not changed in this process. In a similar manner, an additional sensor might be added to the system with a prescribed weight. If that weight is higher than the existing publishers of that data, the subscribers will receive the data from the new component. Note that this is a simplified example; in reality, there is usually a filtering module between the sensors and controllers that may be used to fuse sensor data. This module may be integrated as a separate component that both publishes and subscribes to data.

As mentioned previously, remote procedure calls may be used to access the methods of one object that is remote from the calling object. This is inefficient for real-time applications, such as a stabilizing controller needing sensor measurements at

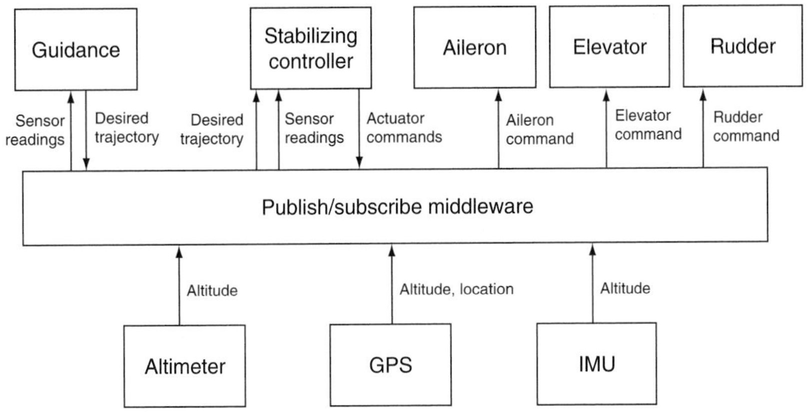

FIGURE 14.4 Flight Control Example Implemented Using Middleware That Implements a Publish/subscribe Communication Model

a high rate. One way that the controller can get its data quickly is for the middleware to duplicate the data and the sensor methods in the memory that is local to the controller program. The middleware can update the data at the specified rate by putting this information into a memory location that is accessed by the subscribing object. Hence, the stabilizing controller can call a method of a sensor object, and it is actually calling the duplicate (also called the replicate or the cache copy). Such a local method call is much faster than a remote method call across the communication channel.

14.6 Real-Time Applications

As mentioned previously, digital control systems are much more sensitive to time delays than are other software applications. There are three types of timed events that occur in typical control system: **periodic events** (such as updating control signals at a given rate), **asynchronous events** (such as set point adjustments from a supervisory control), and **sporadic events** (such as a fault occurrence). A long time delay that is incurred when processing an event could cause catastrophic behavior, such as instability or failure to recover from a fault. Making the problem more difficult to manage and to analyze is the fact that operations in typical networked computer systems occur in a nondeterministic manner, giving rise to varying time delays. The concern over how this nondeterministic behavior affects different software applications has prompted a great deal of research in the area of **real-time computing**.

Consider the real-time operation of a complex control system where different components are connected via a network. There are three places where the time delays occur: the **data transfer over the network**, the **software applications running on the processors**, and the **middleware that integrates the components**.

Consider first the network communications. Performance of a communication network is commonly measured in terms of time delays (e.g., due to access delay, message delay, transmission delays), reliability or accuracy of data, and throughput (e.g., max data rate divided by the data size). For control systems, the time delay is the most critical measure due to stability and performance considerations. To address the special concerns of control systems, some specialized protocols have been written that provide for a dedicated control area network (such as DeviceNet). The alternative is to use a standard data network such as ethernet. Data networks typically use protocols (such as TCP/IP) that result in a nondeterministic transfer of data, as opposed to the smaller but more frequent data packet transfers conducted by control area networks. Results examining the use of ethernet for real-time applications are given in Schneider *et al.* (2000), while a comparison of ethernet (using CSMA/CD) to a token ring bus and to control area networks is given in Fend-Li Lian *et al.* (2001). Networks using ATM technology have good performance with other real-time applications, such as video and voice, and have good potential for future use in control applications.

Next, consider real-time behavior of software applications. While high-level languages do have function calls that seem to imply real-time, the timings are not exact. For example, a sleep(20) command implemented in a Java thread would seem to make the thread pause for 20 msec. However, the thread may actually sleep for 19 msec or 21 msec. This can be explained by examining how an operating system on a processor handles several tasks that are running concurrently (known as **multitasking** or **multithreading**: the system must schedule a dedicated processor time to each of these different tasks. As a result, tasks are typically scheduled in a nondeterministic manner, which gives rise to the resulting **soft real-time** behavior.

While the processor itself does have an absolute clock, a real-time operating system (e.g., LinuxOS, VxWorks, Sun/Chorus

ClassiX, and QNX) must be used to get hard real-time performance out of the applications. There are products (rather kernels) that can be installed on a system with a non-real-time operating system, such as the Real-Time Linux modification of the popular Linux Operating System. Here, hard real-time applications coexist with the normal linux kernel and hard real-time tasks are always given priority for execution. The normal linux kernel as a whole is only executed when slack time is available. Most computers use hardware components to perform tasks that need real-time performance, such as video cards on desktop machines or DSP boards in signal processing applications. These tasks can be performed in parallel without the need for the nondeterministic uncertainty introduced by scheduling of processor time. Another alternative is to use separate dedicated processors to perform each concurrent task, such as having one processor perform low-level control loops and a separate processor performing high-level loops such as fault detection and supervisory control. Moreover, the software application must be able to take advantage of real-time operations. While C has been used successfully in control system implementation, Java lacks good real-time applicability. This may change in the near future because there is a large effort to develop real-time Java capability as evidenced by the release of Real-Time Java Specification (Bollella *et al.*, 2000).

Finally, the middleware must run in real-time. For example, the original version of CORBA did not support hard real-time operations because of overhead in the client/server implementation due to the ORB intercepting the remote method calls, redirecting them, and marshalling or demarshalling data to translate data values from their local representations to a common network protocol format. CORBA also lacked real-time scheduling capabilities. Extensions have been made to build a real-time CORBA (also known as TAO) (O'Ryan *et al.*, 2000). As described previously, using local replicas helps speed up the real-time behavior of the middleware. Further, real-time applications require the middleware to schedule events based on **quality of service** (QoS) measures. Typical QoS parameters include desired rates for periodic events (e.g., updates from sensors), deadlines for when a task must be completed (e.g., when the new command must be sent to the actuator), and priority values indicating the relative importance of different procedures. A real-time middleware must also time stamp transactions and prescribe efficient and fast memory usage.

14.7 Software Tools for Control Applications

Software tools for control systems are aimed at either design and analysis or at implementation. The most common design and analysis tool is MATLAB from Mathworks. MATLAB is often used in conjunction with the Simulink tool box to provide a graphical user interface along with some expanded simulation capabilities. To ease system implementation, the Real-Time Workshop tool box can be used to generate C code from a Simulink block diagram. Mathworks has a continuing effort to increase the efficiency of code generation to improve the speed of processing and to reduce the size of the compiled program (an important feature for embedded processors).

In this chapter, emphasis has been placed on using middleware for integrating complex control systems. As mentioned previously, middleware is beneficial for its ability to hide the network communications decisions from the user, its reuse in integrating multiple control system applications, and its ease of system evolution and reconfiguration. A further benefit from some middleware is its ability to integrate software written in different languages and running on different platforms. The development of middleware for control system applications is rather new. One commercial product that uses a real-time publish and subscribe communication model is NDDS made by Real-Time Innovations (RTI). NDDS has good performance for relatively fast periodic events, which is the case for the low-level stability augmentation controller used in flight controls. RTI also markets ControlShell, a graphics tool that allows integration of control systems in an object-oriented fashion. Consult the RTI Web site for more information on the publish and subscribe communications model for use in controls. While no commercial products currently use CORBA for control systems, a CORBA-based software substrate for controls is under development (Wills *et al.*, 2001). Extensions to CORBA and links to CORBA-based products for other applications can be found on the OMG Web site.

Acknowledgments

The author would like to thank Raymond Garcia, Suresh Kannan, and Dr. Linda Wills for their valuable comments. The author would also like to acknowledge the support of DARPA under the Software Enabled Control Program headed by Dr. Helen Gill.

References

Bollella, G., Dibble, P., Furr, S., Gosling, J., Hardin, D., Turnbull, M. (2000). *The real-time specification for Java.* New York: Addison-Wesley.

Lian, F.L., Moyne, F.L.J., and Tilbury, D. (2001). Performance evaluation of control networks: Ethernet, ControlNet, and DeviceNet. *IEEE Control Systems Magazine 21*(1), 66–83.

O'Ryan, C., Schmidt, D., Kuhns, F., Spivak, M., Parsons, J., Pyarali, I., and Levine, D. (2000). Evaluating policies and mechanisms to support distributed real-time applications with CORBA 3.0. *IEEE Real-Time Technology and Applications Symposium.* 188–197.

Pardo-Castellote, G., Schneider, S., and Hamilton, M. (1999). *NDDS: The real-time publish-subscribe middleware.* Retrieved January 11, 2001: www.rti.com/products/ndds/literature.html

Schneider, S. (2000). Making ethernet work in real-time. *Sensors Online.* Retrieved August 17, 2004: www.sensormag.com/articles/article_index. Also available as: Can ethernet be real-time? Retrieved January 11, 2001: www.rti.com/products/ndds/literature.html

Szyperski, C. (1998). *Component software: Beyond object-oriented programming.* New York: Addison-Wesley.

Tanenbaum, A. (1996). *Distributed operating systems.* Englewood Cliffs, NJ: Prentice Hall.

Wills, L., Kannan, S., Sanders, S., Guler, M., Heck, B., Prasad, J.V.R., Vachtsevanos, G., Schrage, D. (2001). An open platform for reconfigurable control. *IEE Control Systems Magazine 21*(3), 49–64.

Index

B

D

Q